# Climate Change 2022
# Mitigation of Climate Change

## Working Group III Contribution to the Sixth Assessment Report of the Intergovernmental Panel on Climate Change

## Volume 2


**Edited by**

**Priyadarshi R. Shukla**
Co-Chair, Working Group III

**Jim Skea**
Co-Chair, Working Group III

**Raphael Slade**
Head of TSU (Science)

**Roger Fradera**
Head of TSU (Operations)

**Minal Pathak**
Senior Scientist

**Alaa Al Khourdajie**
Senior Scientist

**Malek Belkacemi**
IT/Web Manager

**Renée van Diemen**
Senior Scientist

**Apoorva Hasija**
Publication Manager

**Géninha Lisboa**
TSU Administrator

**Sigourney Luz**
Communications Manager

**Juliette Malley**
Senior Administrator

**David McCollum**
Senior Scientist

**Shreya Some**
Scientist

**Purvi Vyas**
Science Officer


T0377621

CAMBRIDGE
UNIVERSITY PRESS

Shaftesbury Road, Cambridge CB2 8EA, United Kingdom
One Liberty Plaza, 20th Floor, New York, NY 10006, USA
477 Williamstown Road, Port Melbourne, VIC 3207, Australia
314–321, 3rd Floor, Plot 3, Splendor Forum, Jasola District Centre, New Delhi – 110025, India
103 Penang Road, #05-06/07, Visioncrest Commercial, Singapore 238467

Cambridge University Press is part of Cambridge University Press & Assessment, a department of the University of Cambridge. We share the University's mission to contribute to society through the pursuit of education, learning and research at the highest international levels of excellence.

www.cambridge.org
Information on this title: www.cambridge.org/9781009157933
DOI: 10.1017/9781009157926

© Intergovernmental Panel on Climate Change 2022

This work is in copyright. It is subject to statutory exceptions and to the provisions of relevant licensing agreements; with the exception of the Creative Commons version the link for which is provided below, no reproduction of any part of this work may take place without the written permission of Cambridge University Press.

An online version of this work is published at doi.org/10.1017/9781009157926 under a Creative Commons Open Access license CC-BY-NC-ND 4.0 which permits re-use, distribution and reproduction in any medium for non-commercial purposes providing appropriate credit to the original work is given. You may not distribute derivative works without permission. To view a copy of this license, visit https://creativecommons.org/licenses/by-nc-nd/4.0

All versions of this work may contain content reproduced under license from third parties. Permission to reproduce this third-party content must be obtained from these third-parties directly. When citing this work, please include a reference to the DOI 10.1017/9781009157926

First published 2022

Printed in the United Kingdom by TJ Books Limited, Padstow Cornwall

*A catalogue record for this publication is available from the British Library.*

ISBN – 2 Volume Set: 978-1-009-15793-3 Paperback
ISBN – Volume 1: 978-1-009-42390-8 Paperback
ISBN – Volume 2: 978-1-009-42391-5 Paperback

Cambridge University Press & Assessment has no responsibility for the persistence or accuracy of URLs for external or third-party internet websites referred to in this publication and does not guarantee that any content on such websites is, or will remain, accurate or appropriate.

Electronic copies of this report are available from the IPCC website www.ipcc.ch.

Front cover photograph: Matt Bridgestock, Director and Architect at John Gilbert Architects

All International Energy Agency (IEA) Data, IEA Further Data and Derived Data has been sourced from https://www.iea.org/data-and-statistics.

# 10 Transport

**Coordinating Lead Authors:**
Paulina Jaramillo (the United States of America), Suzana Kahn Ribeiro (Brazil), Peter Newman (Australia)

**Lead Authors:**
Subash Dhar (India/Denmark), Ogheneruona E. Diemuodeke (Nigeria), Tsutomu Kajino (Japan), David Simon Lee (United Kingdom), Sudarmanto Budi Nugroho (Indonesia), Xunmin Ou (China), Anders Hammer Strømman (Norway), Jake Whitehead (Australia)

**Contributing Authors:**
Otavio Cavalett (Brazil), Michael Craig (the United States of America), Felix Creutzig (Germany), Nick Cumpsty (United Kingdom), Maria Josefina Figueroa Meza (Venezuela/Denmark), Maximilian Held (Switzerland), Christine Hung (Norway), Alan Jenn (the United States of America), Paul Kishimoto (Canada), Takashi Kuzuya (Japan), William F. Lamb (Germany/United Kingdom), Shivika Mittal (India), Helene Muri (Norway), Minal Pathak (India), Daniel Posen (Canada), Simon Robertson (Australia), Andrea Santos (Brazil), Karen C. Seto (the United States of America), Rohit Sharma (India), Benjamin K. Sovacool (Denmark/United Kingdom), Linda Steg (the Netherlands), Lorenzo Usai (Italy), Aranya Venkatesh (the United States of America/India), Sonia Yeh (Sweden)

**Review Editors:**
Lynette Cheah (Singapore), Ralph E.H. Sims (New Zealand)

**Chapter Scientist:**
Yuan Gao (China)

**This chapter should be cited as:**
Jaramillo, P., S. Kahn Ribeiro, P. Newman, S. Dhar, O.E. Diemuodeke, T. Kajino, D.S. Lee, S.B. Nugroho, X. Ou, A. Hammer Strømman, J. Whitehead, 2022: Transport. In IPCC, 2022: *Climate Change 2022: Mitigation of Climate Change. Contribution of Working Group III to the Sixth Assessment Report of the Intergovernmental Panel on Climate Change* [P.R. Shukla, J. Skea, R. Slade, A. Al Khourdajie, R. van Diemen, D. McCollum, M. Pathak, S. Some, P. Vyas, R. Fradera, M. Belkacemi, A. Hasija, G. Lisboa, S. Luz, J. Malley, (eds.)]. Cambridge University Press, Cambridge, UK and New York, NY, USA. doi: 10.1017/9781009157926.012

# Table of Contents

## Executive Summary

**Meeting climate mitigation goals would require transformative changes in the transport sector (*high confidence*).** In 2019, direct greenhouse gas (GHG) emissions from the transport sector were 8.7 $GtCO_2$-eq (up from 5.0 $GtCO_2$-eq in 1990) and accounted for 23% of global energy-related $CO_2$ emissions. 70% of direct transport emissions came from road vehicles, while 1%, 11%, and 12% came from rail, shipping, and aviation, respectively. Emissions from shipping and aviation continue to grow rapidly. Transport-related emissions in developing regions of the world have increased more rapidly than in Europe or North America, a trend that is likely to continue in coming decades (*high confidence*). {10.1, 10.5, 10.6}

**Since the IPCC's Fifth Assessment Report (AR5) there has been a growing awareness of the need for demand management solutions combined with new technologies, such as the rapidly growing use of electromobility for land transport and the emerging options in advanced biofuels and hydrogen-based fuels for shipping and aviation.** There is a growing need for systemic infrastructure changes that enable behavioural modifications and reductions in demand for transport services that can in turn reduce energy demand. The response to the COVID-19 pandemic has also shown that behavioural interventions can reduce transport-related GHG emissions. For example, COVID-19-based lockdowns have confirmed the transformative value of telecommuting replacing significant numbers of work and personal journeys as well as promoting local active transport. There are growing opportunities to implement strategies that drive behavioural change and support the adoption of new transport technology options. {Chapter 5, 10.2, 10.3, 10.4, 10.8}

**Changes in urban form, behaviour programmes, the circular economy, the shared economy, and digitalisation trends can support systemic changes that lead to reductions in demand for transport services or expand the use of more efficient transport modes (*high confidence*).** Cities can reduce their transport-related fuel consumption by around 25% through combinations of more compact land use and the provision of less car-dependent transport infrastructure. Appropriate infrastructure, including protected pedestrian and bike pathways, can also support much greater localised active travel.[1] Transport demand management incentives are expected to be necessary to support these systemic changes (*high confidence*). There is mixed evidence of the effect of circular economy initiatives, shared economy initiatives, and digitalisation on demand for transport services. For example, while dematerialisation can reduce the amount of material that needs to be transported to manufacturing facilities, an increase in online shopping with priority delivery can increase demand for freight transport. Similarly, while teleworking could reduce travel demand, increased ridesharing could increase vehicle-km travelled. {Chapter 1, Chapter 5, 10.2, 10.8}

**Battery electric vehicles (BEVs) have lower lifecycle greenhouse gas emissions than internal combustion engine vehicles (ICEVs) when BEVs are charged with low-carbon electricity (*high confidence*).** Electromobility is being rapidly implemented in micromobility (e-autorickshaws, e-scooters, e-bikes), in transit systems, especially buses, and, to a lesser degree, in the electrification of personal vehicles. BEVs could also have the added benefit of supporting grid operations. The commercial availability of mature lithium-ion batteries (LIBs) has underpinned this growth in electromobility.

As global battery production increases, unit costs are declining. Further efforts to reduce the GHG footprint of battery production, however, are essential for maximising the mitigation potential of BEVs. The continued growth of electromobility for land transport would require investments in electric charging and related grid infrastructure (*high confidence*). Electromobility powered by low-carbon electricity has the potential to rapidly reduce transport GHG and can be applied with multiple co-benefits in the developing world's growing cities (*high confidence*). {10.3, 10.4, 10.8}

**Land-based, long-range, heavy-duty trucks can be decarbonised through battery electric haulage (including the use of electric road systems), complemented by hydrogen- and biofuel-based fuels in some contexts (*medium confidence*). These same technologies and expanded use of available electric rail systems can support rail decarbonisation (*medium confidence*).** Initial deployments of battery electric, hydrogen- and bio-based haulage are underway, and commercial operations of some of these technologies are considered feasible by 2030 (*medium confidence*). These technologies nevertheless face challenges regarding driving range, capital and operating costs, and infrastructure availability. In particular, fuel cell durability, high energy consumption, and costs continue to challenge the commercialisation of hydrogen-based fuel cell vehicles. Increased capacity for low-carbon hydrogen production would also be essential for hydrogen-based fuels to serve as an emissions reduction strategy (*high confidence*). {10.3, 10.4, 10.8}

**Decarbonisation options for shipping and aviation still require R&D, though advanced biofuels, ammonia, and synthetic fuels are emerging as viable options (*medium confidence*).** Increased efficiency has been insufficient to limit the emissions from shipping and aviation, and natural gas-based fuels are likely inadequate to meet stringent decarbonisation goals for these segments (*high confidence*). High energy density, low-carbon fuels are required, but they have not yet reached commercial scale. Advanced biofuels could provide low-carbon jet fuel (*medium confidence*). The production of synthetic fuels using low-carbon hydrogen with $CO_2$ captured through direct air capture (DAC) or bioenergy with carbon capture and storage (BECCS) could provide jet and marine fuels but these options still require demonstration at scale (*low confidence*). Ammonia produced with low-carbon hydrogen could also serve as a marine fuel (*medium confidence*). Deployment of these fuels requires reductions in production costs. {10.2, 10.3, 10.4, 10.5, 10.6, 10.8}

---

1 Active travel is travel that requires physical effort, for example journeys made by walking or cycling.

**Scenarios from bottom-up and top-down models indicate that without intervention, $CO_2$ emissions from transport could grow in the range of 16% and 50% by 2050 (*medium confidence*).** The scenarios literature projects continued growth in demand for freight and passenger services, particularly in developing countries in Africa and Asia (*high confidence*). This growth is projected to take place across all transport modes. Increases in demand notwithstanding, scenarios that limit warming to 1.5°C with no or limited overshoot suggest that a 59% reduction (42–68% interquartile range) in transport-related $CO_2$ emissions by 2050, compared to modelled 2020 levels is required. While many global scenarios place greater reliance on emissions reduction in sectors other than transport, a quarter of the 1.5°C degree scenarios describe transport-related $CO_2$ emissions reductions in excess of 68% (relative to modelled 2020 levels) (*medium confidence*). Illustrative mitigation pathways 1.5 renewables (REN) and 1.5 low demand (LD) describe emission reductions of 80% and 90% in the transport sector, respectively, by 2050. Transport-related emission reductions, however, may not happen uniformly across regions. For example, transport emissions from the Developed Countries and Eastern European and West-Central Asian countries decrease from 2020 levels by 2050 across all scenarios compatible with a 1.5°C goal (C1–C2 group), but could increase in Africa, Asia and Pacific, Latin America and Caribbean, and the Middle East in some of these scenarios.[2] {10.7}

The scenarios literature indicates that fuel and technology shifts are crucial to reducing carbon emissions to meet temperature goals. In general terms, electrification tends to play the key role in land-based transport, but biofuels and hydrogen (and derivatives) could play a role in decarbonisation of freight in some contexts (*high confidence*). Biofuels and hydrogen (and derivatives) are likely more prominent in shipping and aviation (*high confidence*). The shifts towards these alternative fuels must occur alongside shifts towards clean technologies in other sectors (*high confidence*). {10.7}

**There is a growing awareness of the need to plan for the significant expansion of low-carbon energy infrastructure, including low-carbon power generation and hydrogen production, to support emissions reductions in the transport sector (*high confidence*).** Integrated energy planning and operations that take into account energy demand and system constraints across all sectors (transport, buildings, and industry) offer the opportunity to leverage sectoral synergies and avoid inefficient allocation of energy resources. Integrated planning of transport and power infrastructure would be particularly useful in developing countries where 'greenfield' development doesn't suffer from constraints imposed by legacy systems. {10.3, 10.4, 10.8}

**The deployment of low-carbon aviation and shipping fuels that support decarbonisation of the transport sector could require changes to national and international governance structures (*medium confidence*).** Currently, the Paris Agreement does not specifically cover emissions from international shipping and aviation. Instead, accounting for emissions from international transport in the Nationally Determined Contributions is at the discretion of each country. While the International Civil Aviation Organization (ICAO) and International Maritime Organization (IMO) have established emissions reductions targets, only strategies to improve fuel efficiency and reduce demand have been pursued, and there has been minimal commitment to new technologies. Some authors in the literature have argued that including international shipping and aviation under the Paris Agreement could spur stronger decarbonisation efforts in these segments. {10.5, 10.6, 10.7}

**There are growing concerns about resource availability, labour rights, non-climate environmental impacts, and costs of critical minerals needed for LIBs (*medium confidence*).** Emerging national strategies on critical minerals and the requirements from major vehicle manufacturers are leading to new, more geographically diverse mines. The standardisation of battery modules and packaging within and across vehicle platforms, as well as increased focus on design for recyclability, are important. Given the high degree of potential recyclability of LIBs, a nearly closed-loop system in the future could mitigate concerns about critical mineral issues (*medium confidence*). {10.3, 10.8}

**Legislated climate strategies are emerging at all levels of government and, together with pledges for personal choices, could spur the deployment of demand- and supply-side transport mitigation strategies (*medium confidence*).** At the local level, legislation can support local transport plans that include commitments or pledges from local institutions to encourage behaviour change by adopting an organisational culture that motivates sustainable behaviour, with inputs from the creative arts. Such institution-led mechanisms could include bike-to-work campaigns, free transport passes, parking charges, or eliminating car benefits. Community-based solutions like solar sharing, community charging, and mobility as a service can generate new opportunities to facilitate low-carbon transport futures. At the regional and national levels, legislation can include vehicle and fuel efficiency standards, R&D support, and large-scale investments in low-carbon transport infrastructure. {10.8, Chapter 15}



2 See Annex II Table 1 for details of regional groupings used in this report.

## 10.1 Introduction and Overview

This chapter examines the transport sector's role in climate change mitigation. It appraises the transport system's interactions beyond the technology of vehicles and fuels to include the full lifecycle analysis of mitigation options, a review of enabling conditions, and metrics that can facilitate advancing transport decarbonisation goals. The chapter assesses developments in the systems of land-based transport and introduces, as a new feature since AR5, two separate sections focusing on the trends and challenges in aviation and shipping. The chapter assesses the future trajectories emerging from global, energy, and national scenarios and concludes with a discussion on enabling conditions for transformative change in the sector.

This section (Section 10.1) discusses how transport relates to virtually all the Sustainable Development Goals (SDGs), the trends and drivers making transport a big contributor to greenhouse gas (GHG) emissions, the impacts climate change is having on transport that can be addressed as part of mitigation, and the overview of emerging transport disruptions with potential to shape a low-carbon transport pathway.

**Table 10.1 | Transport and the Sustainable Development Goals: Synergies and trade-offs.**

| | Sustainable Development Goals: Synergies and trade-offs | | | | |
|---|---|---|---|---|---|
| | Basic human needs | Earth preconditions | Sustainable resource use | Social and economic development | Universal values |
| | 1 No poverty; 2 Zero hunger; 3 Good health and well-being | 13 Climate action; 14 Life below water; 15 Life on land | 6 Clean water and sanitation; 7 Affordable and clean energy; 12 Responsible consumption and production | 8 Decent work and economic growth; 9 Industry, innovation and infrastructure; 11 Sustainable cities and communities | 4 Quality education; 5 Gender equality; 10 Reduced inequalities; 17 Partnerships for the goals |
| Transport-related topics (low-carbon transport; active transport; electric vehicles) Advances in vehicle technology; Improved public transport system | – Lower air pollution contributes to positive health outcomes.<br>– Energy access can contribute to poverty alleviation.<br>– Transport planning is a major player in reducing poverty in cities.<br>– Access to healthcare.<br>– Diseases from air pollution.<br>– Injuries and deaths from traffic accidents.<br>– Reduced driving-induced stress.<br>– Links between active transport and good health with positive effects of walking and cycling.<br>– Improving road accessibility to disabled users.<br>– Reduce time spent on transport/mobility. | – Reduction of GHG emissions along the entire value chain, e.g., Well-to-Wheel.<br>– Further development addressing minor GHG emissions and pollutants.<br>– Transport oriented to sustainable development.<br>– Circular economy principle applied to transport. | – Share of renewable energy use.<br>– Energy efficiency of vehicles.<br>– Clean and affordable energy off-grid.<br>– Reduce material consumption during production, lifecycle analysis of vehicles and their operations including entire value chains.<br>– Closed loop carbon and nutrient cycle linked to circular economy. | – Role of transport for economic and human development.<br>– Decarbonised public transport rather than private vehicle use.<br>– Transport oriented to sustainable development.<br>– Sustainable transport infrastructure and systems for cities and rural areas.<br>– Affordability of mobility services, this can also be covered under 'universal access' to public transport.<br>– Accessibility vs mobility: mobility to opportunities; transport equity; development as freedom.<br>– Positive economic growth (employment) outcomes due to resource efficiency and lower productive energy cost.<br>– Role of transport provision in accessing work, reconfiguration of social norms, as working from home.<br>– Transport manufacturers as key employers changing role of transport-related labour due to platform economy, and innovations in autonomous vehicles. | – Gender equality in transport.<br>– Reduced inequalities.<br>– Enables access to quality education.<br>– Partnership for the goals. |
| References | Grant et al. 2016; Haines et al. 2017; Cheng et al. 2018; Nieuwenhuijsen 2018; Smith et al. 2018; Sofiev et al. 2018; Peden and Puvanachandra 2019; King and Krizek 2020; Macmillan et al. 2020 | Farzaneh et al. 2019; see particularly following chapters. | SLoCaT 2019; see particularly following chapters. | Bruun and Givoni 2015; Pojani and Stead 2015; Hensher 2017; ATAG 2018; Grzelakowski 2018; Weiss et al. 2018; Brussel et al. 2019; Gota et al. 2019; Mohammadi et al. 2019; Peden and Puvanachandra 2019; SLoCaT 2019; Xu et al. 2019 | Hernandez 2018; Prati 2018; Levin and Faith-Ell 2019; Vecchio et al. 2020 |

### 10.1.1 Transport and the Sustainable Development Goals

The adoption of the 2030 Agenda for Sustainable Development by the United Nations (UN) has renewed international efforts to pursue and accurately measure global actions towards sustainable development (United Nations 2015). The 17 SDGs set out the overall goals that are further specified by 169 targets and 232 SDG indicators, many of which relate to transport (United Nations 2017; Lisowski et al. 2020). A sustainable transport system provides safe, inclusive, affordable, and clean passenger and freight mobility for current and future generations (Williams 2017; Litman 2021) so transport is particularly linked to SDGs 3, 7, 8, 9, 11, 12, and 13 (Move Humanity 2018; IRP 2019; WBA 2019; SLoCaT 2019; Yin 2019). Table 10.1 summarises transport-related topics for these SDGs and corresponding research. Section 17.3.3.7 also provides a cross-sectoral overview of synergies and trade-offs between climate change mitigation and the SDGs.

### 10.1.2 Trends, Drivers and the Critical Role of Transport in GHG Growth

The transport sector directly emitted around 8.9 Gtonnes (Gt) of carbon dioxide equivalent ($CO_2$-eq) in 2019, up from 5.1 $GtCO_2$-eq in 1990 (Figure 10.1). Global transport was the fourth largest source of GHG emissions in 2019 following the power, industry, and the agriculture,

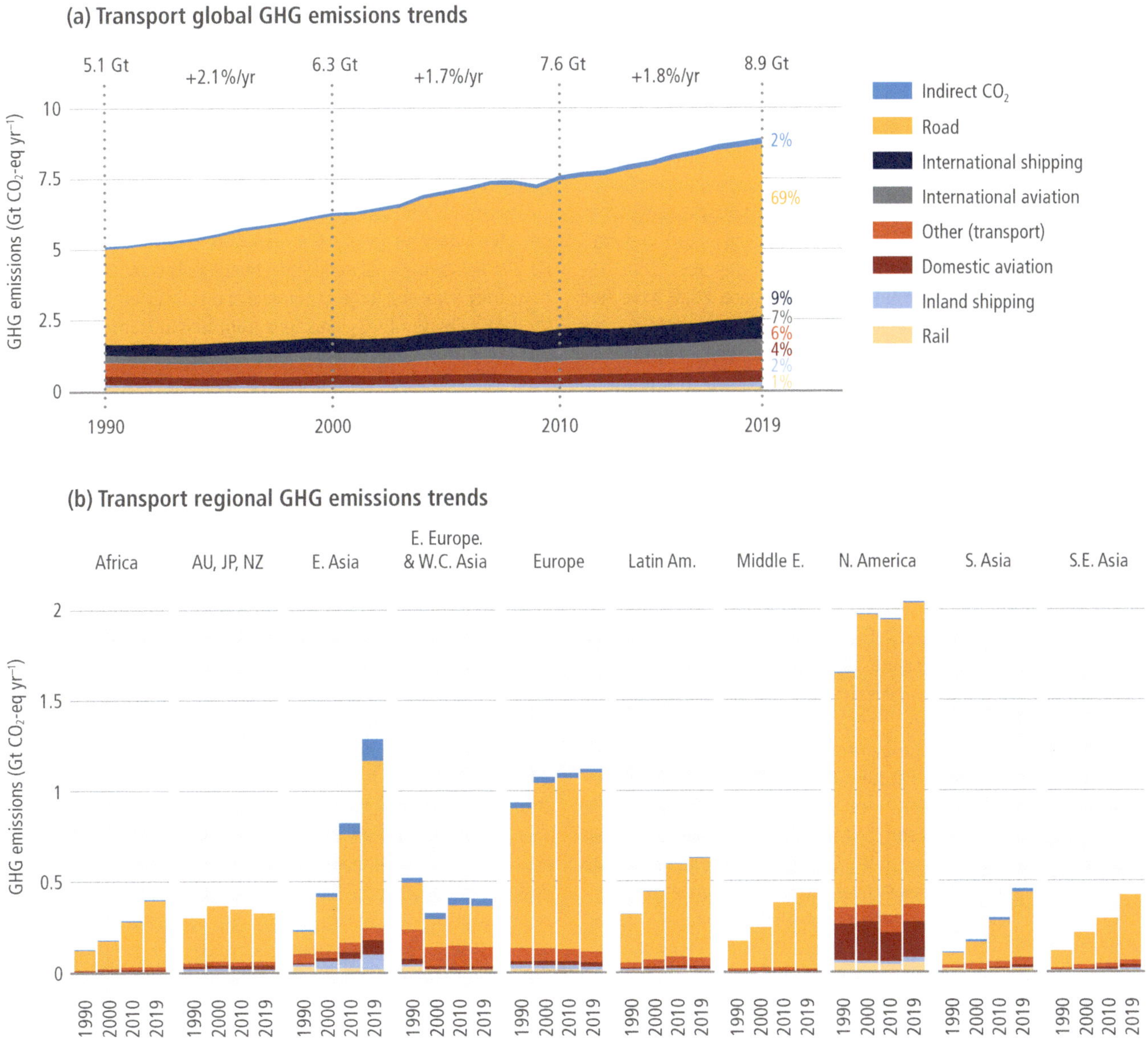


**Figure 10.1 | Global and regional transport greenhouse gas emissions trends.** Indirect emissions from electricity and heat consumed in transport are shown in panel **(a)** and are primarily linked to the electrification of rail systems. These indirect emissions do not include the full lifecycle emissions of transportation systems (e.g., vehicle manufacturing and infrastructure), which are assessed in Section 10.4. International aviation and shipping are included in panel (a) but excluded from panel (b). Indirect emissions from fuel production, vehicle manufacturing and infrastructure construction are not included in the sector total. Source: adapted from Lamb et al. (2021) using data from Minx et al. (2021).

forestry and land use (AFOLU) sectors. In absolute terms, the transport sector accounts for roughly 15% of total GHG emissions and about 23% of global energy-related $CO_2$ emissions (IEA 2020a). Transport-related GHG emissions have increased fast over the last two decades, and since 2010, the sector's emissions have increased faster than for any other end-use sector, averaging +1.8% annual growth (Section 10.7). Addressing emissions from transport is crucial for GHG mitigation strategies across many countries, as the sector represents the largest energy consuming sector in 40% of countries worldwide. In most remaining countries, transport is the second largest energy-consuming sector, reflecting different levels of urbanisation and land use patterns, speed of demographic changes and socio-economic development (IEA 2012; Gota et al. 2019; Hasan et al. 2019; Xie et al. 2019).

As of 2019, the largest source of transport emissions is the movement of passengers and freight in road transport (6.1 $GtCO_2$-eq, 69% of the sector's total). International shipping is the second largest emission source, contributing 0.8 $GtCO_2$-eq (9% of the sector's total), and international aviation is third with 0.6 $GtCO_2$-eq (7% of the sector's total). All other transport emissions sources, including rail, have been relatively trivial in comparison, totalling 1.4 $GtCO_2$-eq in 2019. Between 2010 and 2019, international aviation had among the fastest growing GHG emissions among all segments (+3.4% per year), while road transport remained one of the fastest growing (+1.7% per year) among all global energy-using sectors. Note that the COVID-19-induced economic lockdowns implemented since 2020 have had a very substantial impact on transport emissions – higher than any other sector (Chapter 2). Preliminary estimates from Crippa et al. (2021) suggest that global transport $CO_2$ emissions declined to 7.6 $GtCO_2$ in 2020, a reduction of 11.6% compared to 2019 (Crippa et al. 2021; Minx et al. 2021). These lockdowns affected all transport segments, and particularly international aviation (estimated 45% reduction in 2020 global $CO_2$ emissions), road transport (–10%), and domestic aviation (–9.3%). By comparison, aggregate $CO_2$ emissions across all sectors are estimated to have declined by 5.1% as a result of the COVID-19 pandemic (Section 2.2.2).

Growth in transport-related GHG emissions has taken place across most world regions (see Figure 10.1b). Between 1990 and 2019, growth in emissions was relatively slow in Europe, Australia, Japan and New Zealand, Eurasia, and North America while it was unprecedently fast in other regions. Driven by economic and population growth, the annual growth rates in Eastern Asia, Southern Asia, South-East Asia and Pacific, and Africa were 6.1%, 5.2%, 4.7%, and 4.1%, respectively. Latin America and the Middle East have seen somewhat slower growth in transport-related GHG emissions (annual growth rates of 2.4% and 3.3%, respectively) (ITF 2019; Minx et al. 2021). Section 10.7 provides a more detailed comparison of global transport emissions trends with those from regional and sub-sectoral studies.

The rapid growth in global transport emissions is primarily a result of the fast growth in global transport activity levels, which grew by 73% between 2000 and 2018. Passenger and freight activity growth have outpaced energy efficiency and fuel economy improvements in this period (ITF 2019). The global increase in passenger travel activities has taken place almost entirely in non-OECD countries, often starting from low motorisation rates (SLoCaT 2018a). Passenger cars, two- and three-wheelers, and mini buses contribute about 75% of passenger transport-related $CO_2$ emissions, while collective transport services (bus and railways) generate about 7% of the passenger transport-related $CO_2$ emissions despite covering a fifth of passenger transport globally (Rodrigue 2017; Halim et al. 2018; Sheng et al. 2018; SLoCaT 2018a; Gota et al. 2019). While alternative lighter powertrains have great potential for mitigating GHG emissions from cars, the trend has been towards increasing vehicle size and engine power within all vehicle size classes, driven by consumer preferences towards larger sport utility vehicles (SUVs) (IEA 2020a). On a global scale, SUV sales have been constantly growing in the last decade, with 40% of the vehicles sold in 2019 being SUVs (IEA 2020a) (Section 10.4, Box 10.3).

Indirect emissions from electricity and heat shown in Figure 10.1 account for only a small fraction of current emissions from the transport sector (2%) and are associated with electrification of certain modes like rail or bus transport (Lamb et al. 2021). Increasing transport electrification will affect indirect emissions, especially where carbon-intense electricity grids operate.

Global freight transport, measured in tonne-kilometres (tkm), grew by 68% between 2000 and 2015 and is projected to grow 3.3 times by 2050 (ITF 2019). If unchecked, this growth will make decarbonisation of freight transport very difficult (McKinnon 2018; ITF 2019). International trade and global supply chains from industries frequently involving large geographical distances are responsible for the fast increase of $CO_2$ emissions from freight transport (Yeh et al. 2017; McKinnon 2018), which are growing faster than emissions from passenger transport (Lamb et al. 2021). Heavy-duty vehicles (HDVs) make a disproportionate contribution to air pollution, relative to their global numbers, because of their substantial emissions of particulate matter and of black carbon with high short-term warming potentials (Anenberg et al. 2019).

On-road passenger and freight vehicles dominate global transport-related $CO_2$ emissions and offer the largest mitigation potential (Taptich et al. 2016; Halim et al. 2018). This chapter examines a wide range of possible transport emission reduction strategies. These strategies can be categorised under the 'Avoid-Shift-Improve' (ASI) framework described in Chapter 5 (Taptich et al. 2016). 'Avoid' strategies reduce total vehicle travel. They include compact communities and other policies that minimise travel distances and promote efficient transport through pricing and demand management programmes. 'Shift' strategies shift travel from higher-emitting to lower-emitting modes. These strategies include more multimodal planning that improves active and collective transport modes, complete streets roadway design, high occupant vehicle priority strategies that favour shared modes, Mobility as a Service (MaaS), and multimodal navigation and payment apps. 'Improve' strategies reduce per-kilometre emission rates. These strategies include hybrid and electric vehicle incentives, lower-carbon and cleaner fuels, high-emitting vehicle scrappage programmes, and efficient driving and anti-idling campaigns (Lutsey and Sperling 2012; Gota et al. 2015). These topics are assessed within the rest of this chapter, including how combinations of ASI with new technologies can potentially lead from incremental interventions into low-carbon transformative transport improvements that include social and equity benefits (Section 10.8).

### 10.1.3 Climate Adaptation on the Transport Sector

Climate change impacts such as extremely high temperatures, intense rainfall leading to flooding, more intense winds and/or storms, and sea level rise can seriously impact transport infrastructure, operations, and mobility for road, rail, shipping, and aviation. Studies since AR5 confirm that serious challenges to all transport infrastructures are increasing, with consequent delays or derailing (Miao et al. 2018; Moretti and Loprencipe 2018; Pérez-Morales et al. 2019; Palin et al. 2021). These impacts have been increasingly documented but, according to Forzieri et al. (2018), little is known about the risks of multiple climate extremes on critical infrastructures at local to continental scales. All roads, bridges, rail systems, and ports are likely to be affected to some extent. Flexible pavements are particularly vulnerable to extreme high temperatures that can cause permanent deformation and crumbling of asphalt (Underwood et al. 2017; Qiao et al. 2019). Rail systems are also vulnerable, with a variety of hazards, both meteorological and non-meteorological, affecting railway asset lifetimes. Severe impacts on railway infrastructure and operations can arise from the occurrence of temperatures below freezing, excess precipitation, storms and wildfires (Thaduri et al. 2020; Palin et al. 2021) as can impacts on underground transport systems (Forero-Ortiz et al. 2020).

Most countries are examining opportunities for combined mitigation-adaptation efforts, using the need to mitigate climate change through transport-related GHG emissions reductions and reduction of pollutants as the basis for adaptation action (Thornbush et al. 2013; Wang et al. 2020). For example, urban sprawl indirectly affects climate processes, increasing emissions and vulnerability, which worsens the potential to adapt (Congedo and Munafò 2014; Macchi and Tiepolo 2014). Hence, using a range of forms of rapid transit as structuring elements for urban growth can mitigate climate change-related risks as well as emissions, reducing impacts on new infrastructure, often in more vulnerable areas (Newman et al. 2017). Such changes are increasingly seen as having economic benefit (Ha et al. 2017), especially in developing nations (Chang 2016; Monioudi et al. 2018).

Since AR5 there has been a growing awareness of the potential and actual impacts from global sea level rise due to climate change on transport systems (Dawson et al. 2016; Rasmussen et al. 2018; IPCC 2019; Noland et al. 2019), particularly on port facilities (Stephenson et al. 2018; Yang et al. 2018b; Pérez-Morales et al. 2019). Similarly, recent studies suggest changes in global jet streams could affect the aviation sector (Staples et al. 2018; Becken and Shuker 2019), and extreme weather conditions can affect runways (heat buckling) and aircraft lift. Combined, climate impacts on aviation could result in payload restrictions and disruptions (Coffel et al. 2017; Monioudi et al. 2018). According to Williams (2017), studies have indicated that the amount of moderate-or-greater clear-air turbulence on transatlantic flight routes in winter will increase significantly in the future as the climate changes. More research is needed to fully understand climate-induced risks to transportation systems.

### 10.1.4 Transport Disruption and Transformation

Available evidence suggests that transport-related $CO_2$ emissions would need to be restricted to about 2 to 3 Gt in 2050 (1.5°C scenario-1.5DS, B2DS), or about 70 to 80% below 2015 levels, to meet the goals set in the Paris Agreement. It also indicates that a balanced and inter-modal application of Avoid, Shift, and Improve measures is capable of yielding an estimated reduction in transport emissions of 2.39 $GtCO_2$-equivalent by 2030 and 5.74 $GtCO_2$-equivalent by 2050 (IPCC 2018; Gota et al. 2019). Such a transformative decarbonisation of the global transport system requires, in addition to technological changes, a paradigm shift that ensures prioritisation of high-accessibility transport solutions that minimise the amount of mobility required to meet people's needs, and favours transit and active transport modes (Lee and Handy 2018; SLoCaT 2021). These changes are sometimes called disruptive as they are frequently surprising in how they accelerate through a technological system.

The assessment of transport innovations and their mitigation potentials is at the core of how this chapter examines the possibilities for changing transport-related GHG trajectories. The transport technology innovation literature analysed in this chapter emphasises how a mixture of mitigation technology options and social changes are now converging and how, in combination, they may have potential to accelerate trends toward a low-carbon transport transition. Such changes are considered disruptive or transformative (Sprei 2018). Of the current transport trends covered in the literature, this chapter focuses on three key technology and policy areas: electro-mobility in land-based transport vehicles, new fuels for ships and planes, and overall demand reductions and efficiency. These strategies are seen as being necessary to integrate at all levels of governance and, in combination with the creation of fast, extensive, and affordable multimodal public transport networks, can help achieve multiple advantages in accordance with SDGs

Electrification of passenger transport in light-duty vehicles (LDVs) is well underway as a commercial process with socio-technical transformative potential and will be examined in detail in Sections 10.3 and 10.4. But the rapid mainstreaming of electric vehicles (EVs) will still need enabling conditions for land transport to achieve the shift away from petroleum fuels, as outlined in Chapter 3 and detailed in Section 10.8. The other mitigation options reviewed in this chapter are so far only incremental and are less commercial, especially shipping and aviation fuels, so stronger enabling conditions are likely, as detailed further in Sections 10.5 to 10.8. The enabling conditions that would be needed for the development of an emerging technological solution for such fuels are likely to be very different from those for electromobility, but nevertheless they both will need demand and efficiency changes to ensure they are equitable and inclusive.

Section 10.2 sets out the transformation of transport through examining systemic changes that affect demand for transport services and the efficiency of the system. Section 10.3 looks at the most promising technological innovations in vehicles and fuels. The next three sections (10.4, 10.5, and 10.6) examine mitigation options for land transport, aviation, and shipping. Section 10.7 describes the space of solutions assessed in a range of integrated modelling and sectoral transport scenarios. Finally, Section 10.8 sets out what would be needed for the most transformative scenario that can manage to achieve the broad goals set out in Chapter 3 and the transport goals set out in Section 10.7.

## 10.2 Systemic Changes in the Transport Sector

Systemic change is the emergence of new organisational patterns that affect the structure of a system. While much attention has been given to engine and fuel technologies to mitigate GHG emissions from the transport sector, population dynamics, finance and economic systems, urban form, culture, and policy also drive emissions from the sector. Thus, systemic change requires innovations in these components. These systemic changes offer the opportunity to decouple transport emissions from economic growth. In turn, such decoupling allows environmental improvements like reduced GHG emissions without loss of economic activity (UNEP 2011; UNEP 2013; Newman et al. 2017; IPCC 2018).

There is evidence that suggests decoupling of transport emissions and economic growth is already happening in developed and developing countries. Europe and China have shown the most dramatic changes (Huizenga et al. 2015; Gao and Newman 2018; SLoCaT 2018b) and many cities are demonstrating decoupling of transport-related emissions through new net zero urban economic activity (Loo and Banister 2016; SLoCaT 2018a). A continued and accelerated decoupling of the growth of transport-related GHG emissions from economic growth is crucial for meeting the SDGs, as outlined in Section 10.1. This section focuses on several overlapping components of systemic change in the transport sector that affect the drivers of GHG emissions: urban form, physical geography, and infrastructure; behaviour and mode choice; and new demand concepts. Table 10.3 at the end of the section provides a high-level summary of the effect of these systemic changes on emissions from the transport sector.

### 10.2.1 Urban Form, Physical Geography, and Transport Infrastructure

The physical characteristics that make up built areas define the urban form. These physical characteristics include the shape, size, density, and configuration of the human settlements. Urban form is intrinsically coupled with the infrastructure that allows human settlements to operate. In the context of the transport sector, urban form and urban infrastructure influence the time and cost of travel, which, in turn, drive travel demand and modal choice (Marchetti and Ausubel 2001; Newman and Kenworthy 2015).

Throughout history, three main urban fabrics have developed, each with different effects on transport patterns based on a fixed travel time budget of around one hour (Newman et al. 2016). The high-density urban fabric developed over the past several millennia favoured walking and active transport for only a few kilometres (km). In the mid-19th century, urban settlements developed a medium-density fabric that favoured trains and trams traveling over 10 to 30 km corridors. Finally, since the mid-20th century, urban form has favoured automobile travel, enabling mass movement between 50 and 60 km. Table 10.2 describes the effect of these urban fabrics on GHG emissions and other well-being indicators.

Since AR5, urban design has increasingly been seen as a major way to influence the GHG emissions from urban transport systems. Indeed, research suggests that implementing urban form changes could reduce GHG emissions from urban transport by 25% in 2050, compared with a business-as-usual scenario (Creutzig et al. 2015b; Creutzig 2016). Researchers have identified a variety of variables to study the relationship between urban form and transport-related GHG emissions. Three notable aspects summarise these relationships: urban space utilisation, urban spatial form, and urban transportation infrastructure (Tian et al. 2020). Urban density (population or employment density) and land-use mix define the urban space utilisation. Increases in urban density and mixed function can effectively reduce per capita car use by reducing the number of trips and shortening travel distances. Similarly, the continuity of urban space and the dispersion of centres reduces travel distances (Tian et al. 2020), though such changes are rarely achieved without shifting transport infrastructure investments away from road capacity increases (Newman and Kenworthy 2015; McIntosh et al. 2017). For example, increased investment in public transport coverage, optimal transfer plans, shorter transit travel time, and improved transit travel efficiency make public transit more attractive (Heinen et al. 2017; Nugroho et al. 2018a; Nugroho et al. 2018b) and hence increase density and land values (Sharma and Newman 2020). Similarly, forgoing the development of major roads for the development of pedestrian and bike pathways enhances the attractiveness of active transport modes (Zahabi et al. 2016; Keall et al. 2018; Tian et al. 2020).

**Table 10.2 | The systemic effect of city form and transport emissions.**

| Annual transport emissions and co-benefits | Walking urban fabric | Transit urban fabric | Automobile urban fabric |
|---|---|---|---|
| Transport GHG | 4 tonnes per person | 6 tonnes per person | 8 tonnes per person |
| Health benefits from walkability | High | Medium | Low |
| Equity of locational accessibility | High | Medium | Low |
| Construction and household waste | 0.87 tonnes per person | 1.13 tonnes per person | 1.59 tonnes per person |
| Water consumption | 35 kilolitre per person | 42 kilolitre per person | 70 kilolitre per person |
| Land | 133 square metres per person | 214 square metres per person | 547 square metres per person |
| Economics of infrastructure and transport operations | High | Medium | Low |

Source: Newman et al. (2016); Thomson and Newman (2018); Seto et al. (2021).

Ultimately, infrastructure investments influence the structural dependence on cars, which in turn influence the lock-in or path dependency of transport options with their greenhouse emissions (Newman et al. 2015; Grieco and Urry 2016). The 21st century saw a new trend to reach peak car use in some countries as a result of a revival in walking and transit use (Grieco and Urry 2016; Newman et al. 2017; Gota et al. 2019). While some cities continue on a trend towards reaching peak car use on a per-capita basis, for example Shanghai and Beijing (Gao and Newman 2020), there is a need for increased investments in urban form strategies that can continue to reduce car dependency around the world.

### Cross-Chapter Box 7 | Urban Form: Simultaneously Reducing Urban Transport Emissions, Avoiding Infrastructure Lock-in, and Providing Accessible Services

**Authors:** Felix Creutzig (Germany), Karen C. Seto (the United States of America), Peter Newman (Australia)

Urban transport is responsible for about 8% of global $CO_2$ emissions or 3 $GtCO_2$ per year (Chapters 5 and 8). In contrast to energy supply technologies, urban transport directly interacts with mobility lifestyles (Section 5.4). Similarly, non-GHG emission externalities, such as congestion, air pollution, noise, and safety, directly affect urban quality of life, and result in considerable welfare losses. Low-carbon, highly accessible urban design is not only a major mitigation option, it also provides for more inclusive city services related to well-being (Sections 5.1 and 5.2). Urban planning and design of cities for people are central to realise emission reductions without relying simply on technologies, though the modes of transport favoured will influence the ability to overcome the lock-in around automobile use (Gehl 2010; Creutzig et al. 2015b).

Where lock-in has occurred, other strategies may alleviate the GHG emissions burden. Urban planning still plays a key role in recreating local hubs. Available land can be used to build rail-based transit, made financially viable by profiting from land value captured around stations (Ratner and Goetz 2013). Shared or pooled mobility can offer flexible on-demand mobility solutions that are efficient also in suburbs and for integrating with longer commuting trips (ITF 2017).

Global emissions trajectories of urban transport will be decided in rapidly urbanising Asia and Africa. Urban transport-related GHG emissions are driven by incomes and car ownership but there is considerable variation among cities with similar income and car ownership levels (Newman and Kenworthy 2015). While electrification is a key strategy to decarbonise urban transport, urban infrastructures can make a difference of up to a factor of 10 in energy use and induced GHG emissions (Erdogan 2020). Ongoing urbanisation patterns risk future lock-in of induced demand on GHG emissions, constraining lifestyles to energy-intensive and high $CO_2$-related technologies (Erickson and Tempest 2015; Seto et al. 2016) (Sections 5.4, 8.2.3 and 10.2.1). Instead, climate solutions can be locked into urban policies and infrastructures (Ürge-Vorsatz et al. 2018) especially through the enhancement of the walking and transit urban fabric. Avoiding urban sprawl, associated with several externalities (Dieleman and Wegener 2004), is a necessary decarbonisation condition, and can be guided macro-economically by increasing fuel prices and marginal costs of motorised transport (Creutzig 2014). Resulting urban forms not only reduce GHG emission from transport but also from buildings, as greater compactness results in reduced thermal loss (Borck and Brueckner 2018). Health benefits from reduced car dependence are an increasing element driving this policy agenda (Speck 2018) (Section 10.8).

Low-carbon highly accessible urban design is not only a major mitigation option, it also provides for more inclusive city services related to well-being (Sections 5.1 and 5.2). Solutions involve planning cities around walkable sub-centres, where multiple destinations, such as shopping, jobs, leisure activities, and others, can be accessed within a 10 minute walk or bicycle ride (Newman and Kenworthy 2006). Overall, the mitigation potential of urban planning is about 25% in 2050 compared with a business-as-usual scenario (Creutzig et al. 2015a; Creutzig et al. 2015b). Much higher levels of decarbonisation can be achieved if cities take on a regenerative development approach and act as geo-engineering systems on the atmosphere (Thomson and Newman 2016).

### 10.2.2 Behaviour and Mode Choice

Behaviour continues to be a major source of interest in the decarbonisation of transport as it directly addresses demand. Behaviour is about people's actions based on their preferences. Chapter 5 described an 'Avoid, Shift, Improve' process for demand-side changes that affect sectoral emissions. This section discusses some of the drivers of behaviour related to the transport sector and how they link to this 'Avoid, Shift, Improve' process.

**Avoid: the effect of prices and income on demand.** Research has shown that household income and price have a strong influence on people's preferences for transport services (Bakhat et al. 2017; Palmer et al. 2018). The relationship between income and demand is defined by the income elasticity of demand. For example, research suggests that in China, older and wealthier populations continued to show a preference for car travel (Yang et al. 2019) while younger and low-income travellers sought variety in transport modes (Song et al. 2018). Similarly, Bergantino et al. (2018b) evaluated the income elasticity of transport by mode in the UK. They found that the income elasticity for private cars is 0.714, while the income elasticities of rail and bus use are 3.253 (the greater elasticity, the more the demand will grow or decline, depending on income). Research has also shown a positive relationship between income and demand for air travel, with income elasticities of air travel demand being positive and as large as 2 (Gallet and Doucouliagos 2014; Valdes 2015; Hakim and Merkert 2016; Hakim and Merkert 2019; Hanson et al. 2022). A survey in 98 Indian cities also showed income as the main factor influencing travel demand (Ahmad and de Oliveira 2016). Thus, as incomes and wealth across the globe rise, demand for travel is likely to increase as well.

The price elasticity of demand measures changes in demand as a result of changes in the prices of the services. In a meta-analysis of the price elasticity of energy demand, Labandeira et al. (2017) report the average long-term price elasticity of demand for gasoline and diesel to be –0.773 and –0.443, respectively. That is, demand will decline with increasing prices. A similar analysis of long-term data in the United States (US), the United Kingdom (UK), Sweden, Australia, and Germany reports the gasoline price elasticity of demand for car travel (as measured through vehicle-kilometre – vkm – per capita) ranges between –0.1 and –0.4 (Bastian et al. 2016). For rail travel, the price elasticity of demand has been found to range between –1.05 and –1.1 (Zeng et al. 2021). Similarly, price elasticities for air travel range from –0.53 to –1.91 depending on various factors such as purpose of travel (business or leisure), season, and month and day of departure (Morlotti et al. 2017). The price elasticities of demand suggest that car use is inelastic to prices, while train use is relatively inelastic to the cost of using rail. Conversely, consumers seem to be more responsive to the cost of flying, so that strategies that increase the cost of flying are likely to contribute to some avoidance of aviation-related GHG emissions.

While the literature continues to show that time, cost, and income dominate people's travel choices (Ahmad and de Oliveira 2016; Capurso et al. 2019; He et al. 2020), there is also evidence of a role for personal values, and environmental values in particular, shaping choices within these structural limitations (Bouman and Steg 2019). For example, individuals are more likely to drive less when they care about the environment (De Groot et al. 2008; Abrahamse et al. 2009; Jakovcevic and Steg 2013; Hiratsuka et al. 2018; Ünal et al. 2019). Moreover, emotional and symbolic factors affect the level of car use (Steg 2005). Differences in behaviour may also result due to differences in gender, age, norms, values, and social status. For example, women have been shown to be more sensitive to parking pricing than men (Simićević et al. 2020).

Finally, structural shocks, such as a financial crisis, a pandemic, or the impacts of climate change could affect the price and income elasticities of demand for transport services (van Ruijven et al. 2019). COVID-19 lockdowns reduced travel demand by 19% (aviation by 32%) and some of the patterns that have emerged from the lockdowns could permanently change the elasticity of demand for transport (Tirachini and Cats 2020; Hendrickson and Rilett 2020; Newman 2020a; SLoCaT 2021; Hanson et al. 2022). In particular, the COVID-19 lockdowns have spurred two major trends: electronic communications replacing many work and personal travel requirements; and revitalised local active transport and e-micromobility (Newman 2020a; SLoCaT 2021). The permanence of these changes post-COVID-19 is uncertain but possible (Earley and Newman 2021) (Cross-Chapter Box 1 in Chapter 1). However, these changes will require growth of infrastructure for better ICT bandwidths in developing countries, and better provision for micromobility in all cities.

**Shift: mode choice for urban and intercity transport.** Shifting demand patterns (as opposed to avoiding demand) can be particularly important in decarbonising the transport sector. As a result, the cross-elasticity of demand across transport modes is of particular interest for understanding the opportunities for modal shift. The cross-elasticity represents the demand effect on mode i (e.g., bus) when an attribute of mode j (e.g., rail) changes marginally. Studies on the cross-elasticities of mode choice for urban travel suggest that the cross-elasticity for car demand is low, but the cross-elasticities of walking, bus, and rail with respect to cars are relatively large (Fearnley et al. 2017; Wardman et al. 2018). In practice, these cross-elasticities suggest that car drivers are not very responsive to increased prices for public transit, but transit users are responsive to reductions in the cost of driving. When looking at the cross-elasticities of public transit options (bus vs metro vs rail), research suggests that consumers are particularly sensitive to in-vehicle and waiting time when choosing public transit modes (Fearnley et al. 2018). These general results provide additional evidence that increasing the use of active and public transport requires interventions that make car use more expensive while making public transit more convenient (e.g., smart apps that tell the user the exact time for transit arrival (Box 10.1)).

The literature on mode competition for intercity travel reveals that while cost of travel is a significant factor (Zhang et al. 2017), sensitivity decreases with increasing income as well as when the cost of the trip was paid by someone else (Capurso et al. 2019). Some research suggests little competition between bus and air travel but the cross-elasticity between air and rail suggest strong interactions (Wardman et al. 2018). Price reduction strategies such as discounted rail fares could enhance the switch from air travel to high-speed rail. Both air fares and flight frequency impact high-speed rail (HSR) usage (Zhang et al. 2019b). Airline companies reduce fares on routes that are directly competing with HSR (Bergantino et al. 2018a) and charge high fares on non-HSR routes (Xia and Zhang 2016). On the Rome to Milan route, better frequency and connections, and low costs of HSR resulting from competition between HSR companies have significantly reduced air travel and shares of buses and cars (Desmaris and Croccolo 2018).

Finally, and as noted in Chapter 5, recent research shows that individual, social, and infrastructure factors also affect people's mode choices. For example, perceptions about common travel behaviour (what people perceive to be 'normal' behaviour) influence their travel mode choice. The research suggests that well-informed individuals whose personal norms match low-carbon objectives, and who believe they have control over their decisions, are most motivated to shift mode. Nonetheless, such individual and social norms can only marginally influence mode choice unless infrastructure factors can enable reasonable time and cost savings (Convery and Williams 2019; Javaid et al. 2020; Feng et al. 2020; Wang et al. 2021).

**Improve: consumer preferences for improved and alternative vehicles.** While reductions in demand for travel and changes in mode choice can contribute to reducing GHG emissions from the transport sector, cars are likely to continue to play a prominent role. As a result, improving the performance of cars will be crucial for the decarbonisation of the transport sector. Sections 10.3 and 10.4 describe the technological options available for reduced $CO_2$ emissions from vehicles. The effectiveness in deploying such technologies will partly depend on consumer preferences and their effect on adoption rates. Given the expanded availability of electric vehicles, there is also a growing body of work on the drivers of vehicle choice. A survey in Nanjing found women had more diverse travel purposes than men, resulting in a greater acceptance of electric bikes (Lin et al. 2017). Individuals are more likely to adopt an electric vehicle (EV) when they think this adoption benefits the environment or implies a positive personal attribute (Noppers et al. 2014; Noppers et al. 2015; Haustein and Jensen 2018). Other work suggests that people's preference for EVs depends upon vehicle attributes, infrastructure availability, and policies that promote EV adoption, specifically, purchasing and operating costs, driving range, charging duration, vehicle performance, and brand diversity (Liao et al. 2016). Behaviour change to enable transport transformations will need to make the most of these factors while also working on the more structural issues of time, space, and cost.

### 10.2.3 New Demand Concepts

Structural and behavioural choices that drive transport-related GHG emissions, such as time and cost based on geography of freight and urban fabric, are likely to continue to be major factors. But there is also a variation within each structural choice that is based around personal demand factors related to values that indirectly change choices in transport. Chapter 5 identified three megatrends that affect demand for services, including circular economy, the shared economy, and digitalisation. These three megatrends can have specific effect on transport emissions, as described below.

**Circular economy.** The problem of resources and their environmental impacts is driving the move to a circular economy (Bleischwitz et al. 2017). Circular economy principles include increased material efficiency, reusing or extending product lifetimes, recycling, and green logistics. Dematerialisation, the reduction in the quantity of the materials used in the production of one unit of output, is a circular economy principle that can affect the operations and emissions of the transport sector, as reductions in the quantities of materials used reduce transport needs, while reductions in the weight of products improve the efficiency of transporting them. Dematerialisation can occur through more efficient production processes but also when a new product is developed to provide the same functionality as multiple products. The best example of this trend is a smart phone, which provides the service of at least 22 other former devices (Rifkin 2019). A move to declutter lifestyles can also drive dematerialisation (Whitmarsh et al. 2017). Some potential for dematerialisation has been suggested due to 3-D printing, which would also reduce transport emissions through localised production of product components (d'Aveni 2015; UNCTAD 2018). There is evidence to suggest, however, that reductions in material use resulting from more efficient product design or manufacturing are offset by increased consumer demand (Kasulaitis et al. 2019). Whether or not dematerialisation can lead to reduction of emissions from the transport sector is still an open question that requires evaluating the entire product ecosystem (Van Loon et al. 2014; Coroama et al. 2015; Kasulaitis et al. 2019).

**Shared economy.** Shared mobility is arguably the most rapidly growing and evolving sector of the sharing economy and includes bike sharing, e-scooter sharing, car sharing, and on-demand mobility (Greenblatt and Shaheen 2015). The values of creating a more shared economy are related to both reduced demand and greater efficiency, as well as the notion of community well-being associated with the act of sharing instead of simply owning for oneself (Maginn et al. 2018; Sharp 2018). The literature on shared mobility is expanding, but there is much uncertainty about the effect shared mobility will have on transport demand and associated emissions (Nijland and Jordy 2017; ITF 2018a; Tikoudis et al. 2021).

Asia represents the largest car-sharing region with 58% of worldwide membership and 43% of global fleets deployed (Dhar et al. 2020). Europe accounts for 29% of worldwide members and 37% of shared vehicle fleets (Shaheen et al. 2018). Ride-sourcing and carpooling systems are among the many new entrants in the short-term shared mobility options. On-demand transport options complemented with technology have enhanced the possibility of upscaling (Alonso-González et al. 2018). Car sharing could provide the same level of service as taxis, but taxis could be three times more expensive (Cuevas et al. 2016). The sharing economy, as an emerging economic-technological phenomenon (Kaplan and Haenlein 2010), is likely to be a key driver of demand for transport of goods although data shows increasing container movement due to online shopping (Suel and Polak 2018).

There is growing evidence that this more structured form of behavioural change through shared economy practices, supported by a larger group than a single family, has a much greater potential to save transport emissions, especially when complemented with decarbonised grid electricity (Greenblatt and Shaheen 2015; Sharp 2018). Carpooling, for example, could result in an 11% reduction in vehicle-kilometres and a 12% reduction in emissions, as carpooling requires less empty or non-productive passenger-kilometres (pkm) (ITF 2020a; ITF 2020b). However, the use of local shared mobility systems such as on-demand transport may create more transport emissions if there is an overall modal shift out of transit (ITF 2018a; Schaller 2018). Similarly, some work suggests that commercial shared vehicle services such as Uber and Lyft are leading to increased vehicle km travelled (and associated GHG emissions) in part due to deadheading (Schaller 2018; Tirachini and Gomez-Lobo 2020; Ward et al. 2021). Successful providers compete by optimising personal comfort and convenience rather than enabling a sharing culture (Eckhardt and Bardhi 2015), and concerns have been raised regarding the wider societal impacts of these systems and for specific user groups such as older people (Fitt 2018; Marsden 2018). Concerns have also been expressed over the financial viability of demand-responsive transport systems (Ryley et al. 2014; Marsden 2018), how the mainstreaming of shared mobility systems can be institutionalised equitably, and the operation and governance of existing systems that are only mode- and operator-focused (Akyelken et al. 2018; Jittrapirom et al. 2018; Pangbourne et al. 2020; Marsden 2018).

**Digitalisation.** In the context of the transport sector, digitalisation has enabled teleworking, which in turn reduces travel demand. On the other hand, the prevalence of online shopping, enabled by the digital economy, could have mixed effects on transport emissions (Le et al. 2021). For example, online shopping could reduce vehicle-kilometres travelled but the move to expedited or rush delivery could mitigate some benefits as it prevents consolidation of freight (Jaller and Pahwa 2020).

Digitalisation could also lead to systemic changes by enabling smart mobility. The smart mobility paradigm refers to the process and practices of assimilation of ICTs and other sophisticated high-technology innovations into transport (Noy and Givoni 2018). Smart mobility can be used to influence transport demand and efficiency (Benevolo et al. 2016). The synergies of emerging technologies (ICT, internet of things, big data) and shared economy could overcome some of the challenges facing the adoption of emerging technologies (Marletto 2014; Chen et al. 2016; Weiss et al. 2018; Taiebat and Xu 2019) and enable the expected large growth in emerging cities to be more sustainable (Docherty et al. 2018). However, ICT, in particular the internet of things (IoT), could also cause more global energy demand (Hittinger and Jaramillo 2019). Box 10.1 summarises the main smart technologies being adopted rapidly by cities across the world and their use in transport. There is a growing body of literature about the effect of smart technology (including sensors guiding vehicles) on the demand for transport services. Smart technologies can improve competitiveness of transit and active transport over personal vehicle use by combining the introduction of new electro-mobility that improves time and cost along with behaviour change factors (Pålsson et al. 2017; SLoCaT 2018a; SLoCaT 2018b; SLoCaT2021). However, it is unclear what the net effect of smart technology on GHG emissions from the transport sector will be (Debnath et al. 2014; Lenz and Heinrichs 2017).



## Box 10.1 | Smart City Technologies and Transport

**Information and communication technology (ICT).** ICT is at the core of smart mobility and will provide the avenue for data to be collected and shared across the mobility system. The use of ICT can help cities by providing real-time information on mobility options that can inform those using private vehicles, along with transit users or those using bikes or walking. ICT can help with ticketing and payment for transit or for road user charges (Tafidis et al. 2017; Gössling 2018) when combined with other technologies such as Blockchain (Hargroves et al. 2020).

**Internet of Things sensors.** Sensors can be used to collect data to improve road safety, improve fuel efficiency of vehicles, and reduce $CO_2$ emissions (Kubba and Jiang 2014; Kavitha et al. 2018). Sensors can also provide data to digitally simulate transport planning options, inform the greater utilisation of existing infrastructure and modal interconnections, and significantly improve disaster and emergency responses (Hargroves et al. 2017). In particular, IoT sensors can be used to inform the operation of fast-moving trackless trams and their associated last-mile connectivity shuttles as part of a transit activated corridor (Newman et al. 2019, 2021).

**Mobility as a Service.** New, app-based mobility platforms will allow for the integration of different transport modes (such as last-mile travel, shared transit, and even micro-transit such as scooters or bikes) into easy-to-use platforms. By integrating these modes, users will be able to navigate from A to B to C based on which modes are most efficient, with the necessary bookings and payments being made through one service. With smart city planning, these platforms can steer users towards shared and rapid transit (which should be the centrepiece of these systems), rather than encourage more people to opt for the perceived convenience of booking a single-passenger ride (Becker et al. 2020). In low-density car-dependent cities, however, MaaS services such as the use of electric scooters/bikes are less effective as the distances are too long and they do not enable the easy sharing that can happen in dense station precincts (Jittrapirom et al. 2017).

**Artificial intelligence (AI) and big data analytics.** The rapidly growing level of technology enablement of vehicles and urban infrastructure, combined with the growing ability to analyse larger and larger data sets, presents a significant opportunity for transport planning, design, and operation in the future. These technologies are used together to enable decisions about what kind of transport planning is used down particular corridors. Options such as predictive congestion management of roads and freeways, simulating planning options, and advanced shared transit scheduling can provide value to new and existing transit systems (Toole et al. 2015; Anda et al. 2017; Hargroves et al. 2017).

**Blockchain or distributed ledger technology.** Blockchain technology provides a non-hackable database that can be programmed to enable shared services like a local, solar microgrid where both solar and shared electric vehicles can be managed (Green and Newman 2017). Blockchain can be used for many transport-related applications including being the basis of MaaS or any local shared mobility service as it facilitates shared activity without intermediary controls. Other applications include verified vehicle ownership documentation, establishing identification, real-time road user pricing, congestion zone charging, vehicle-generated collision information, collection of tolls and charges, enhanced freight tracking and authenticity, and automated car parking and payments (Hargroves et al. 2020). This type of functionality will be particularly valuable for urban regeneration along a transit activated corridor, where it can be used for managing shared solar in and around station precincts as well as managing shared vehicles linked to the whole transport system (Newman et al. 2021). This technology can also be used for road user charging along any corridor and by businesses accessing any services and in managing freight (Carter and Koh 2018; Nguyen et al. 2019; Hargroves et al. 2020; Sedlmeir et al. 2020).

Autonomous vehicles are the other emerging transport technology that have the potential to significantly improve ride quality and safety. Planes and high-speed trains are already largely autonomous as they are guided in all their movements, especially coming into stations and airports, although that does not necessarily mean they are driverless. Automation is also being used in new on-road transit systems like trackless trams (Ndlovu and Newman 2020)). Private vehicles are being fitted with more and more levels of autonomy and many are being trialled as 'driverless' in cities (Aria et al. 2016; Skeete 2018). If autonomous systems can be used to help on-road transit become more time- and cost-competitive with cars, then the kind of transformative and disruptive changes needed to assist decarbonisation of transport become more feasible (Bösch et al. 2018; Kassens-Noor et al. 2020; Abe 2021). Similarly, vehicle automation could improve vehicle efficiency and reduce congestion, which would in turn reduce emissions (Vahidi and Sciarretta 2018; Massar et al. 2021). On the other hand, if autonomous cars make driving more convenient, they could reduce demand for transit (Auld et al. 2017; Sonnleitner et al. 2021). Paradoxically, autonomous cars could provide access to marginal groups such as the elderly, people with disabilities, and those who cannot drive, which could in turn increase travel demand (as measured by pkm) (Harper et al. 2016).

Heavy haulage trucks in the mining industry are already autonomous (Gaber et al. 2021) and automation of long-haul trucks may happen sooner than automation of LDVs (Hancock et al. 2019). Autonomous trucks may facilitate route and speed optimisation, and reduce fuel use, which can in turn reduce emissions (Nasri et al. 2018; Paddeu and Denby 2021). There is growing interest in using drones for package delivery. Drones could have lower impacts than ground-based delivery and, if deployed carefully, drones could reduce energy use and GHG emissions from freight transport (Stolaroff et al. 2018). Overall, some commentators are optimistic that smart and autonomous technologies can transform the GHG emissions from the transport sector (Seba 2014; Rifkin 2019; Sedlmeir et al. 2020). Others are more sanguine unless policy interventions can enable the technologies to be used for purposes that include zero carbon and the SDGs (Faisal et al. 2019; Hancock et al. 2019).

### 10.2.4 Overall Perspectives on Systemic Change

The interactions between systemic factors set out here and technology factors discussed in much more detail in the next sections show that there is always going to be a need to integrate both approaches.



**Table 10.3 | Components of systemic change and their impacts on the transport sector.**

| Systemic change | Mechanisms through which it affects emissions in transport sector and is likely to affect emissions |
|---|---|
| Changes in urban form | Denser, more compact polycentric cities with mixed land use patterns can reduce the distance between where people live, work, and pursue leisure activities, which can reduce travel demand. Case studies suggest that these changes in urban form could reduce transport-related GHG emissions between 4 to 25%, depending on the setting (Creutzig et al. 2015a; Creutzig et al.2015b; Pan et al. 2020). |
| Investments in transit and active transport infrastructure | Improving public transit systems and building infrastructure to support active transport modes (walking and biking) could reduce car travel. Case studies suggest that active mobility could reduce emissions from urban transport by 2% to 10% depending on the setting (Creutzig et al. 2016; Zahabi et al. 2016; Keall et al. 2018; Gilby et al. 2019; Neves and Brand 2019; Bagheri et al. 2020; Ivanova et al. 2020; Brand et al. 2021). A shift to public transit modes can likely offer significant emissions reductions, but estimates are uncertain. |
| Changes in economic structures | Higher demand as a result of higher incomes could increase emissions, particularly from aviation and shipping. Higher prices could have the opposite effect and reduce emissions. Structural changes associated with financial crises, pandemics, or the impacts of climate change could affect the elasticity of demand in uncertain ways. Thus, the effect of changes in economic structures on the GHG emissions from the transport sectors is uncertain. |
| Teleworking | A move towards a digital economy that allows workers to work and access information remotely could reduce travel demand. Case studies suggest that teleworking could reduce transport emissions by 20% in some instances, but likely by 1%, at most, across the entire transport system (Roth et al. 2008; O'Keefe et al. 2016; Shabanpour et al. 2018; O'Brien and Aliabadi 2020). |
| Dematerialisation of the economy | A reduction in goods needed due to combining multiple functions into one device would reduce the need for transport. Reduced weights associated with dematerialisation would improve the efficiency of freight transport. However, emissions reductions from these efforts are likely dwarfed by increased consumption of goods. |
| Supply chain management | Supply chains could be optimised to reduce the movement or travel distance of product components. Logistics planning could optimise the use of transport infrastructure to increase utilisation rates and decrease travel. The effect of these strategies on the GHG emissions from the transport sector is uncertain. |
| e-commerce | The effect of e-commerce on transport emissions is uncertain. Increased e-commerce would reduce demand for trips to stores but could increase demand for freight transport (particularly last-mile delivery) (Jaller and Pahwa 2020; Le et al. 2021). |
| Smart mobility | ICT and smart city technologies can be used to improve the efficiency of operating the transport system. Furthermore, smart technologies can improve competitiveness of transit and active transport over personal vehicle use by streamlining mobility options to compete with private cars. The effect of smart mobility on the GHG emissions from the transport sector is uncertain (Creutzig 2021). |
| Shared mobility | Shared mobility could increase utilisation rates of LDVs, thus improving the efficiency of the system. However, shared mobility could also divert users from transit systems or active transport modes. Studies on ride-sourcing have reported both potential for reductions and increases in transport-related emissions (Schaller 2018; Ward et al. 2021). Other case studies suggests that carpooling to replace 20% of private car trips could result in a 12% reduction in GHG emissions (ITF 2020a; ITF 2020b). Thus, the effect of shared mobility on transport-related GHG emissions is highly uncertain. |
| Vehicle automation | Vehicle automation could have positive or negative effects on emissions. Improved transit operations, more efficient traffic management, and better routing for light- and heavy-duty transport could reduce emissions (Nasri et al. 2018; Vahidi and Sciarretta 2018; Massar et al. 2021; Paddeu and Denby 2021). However, autonomous cars could make car travel more convenient, removing users from transit systems and increasing access to marginalised groups, which would in turn increase vehicle-kilometre travelled (Harper et al. 2016; Auld et al. 2017; Sonnleitner et al. 2021). Drones could reduce energy use and GHG emissions from freight transport (Stolaroff et al. 2018). |

Good technology that has the potential to transform transport will not be used unless it fulfils broad mobility and accessibility objectives related to time, cost, and well-being. Chapter 5 has set out three transport transformations based on demand-side factors with highly transformative potential. Table 10.3 provides a summary of these systemic changes and their likely impact on GHG emissions. Note that the quantitative estimates provided in the table may not be additive and the combined effect of these strategies on GHG emissions from the transport sector require additional analysis.

## 10.3 Transport Technology Innovations for Decarbonisation

This section focuses on vehicle technology and low-carbon fuel innovations to support decarbonisation of the transport sector. Figure 10.2 summarises the major pathways reviewed in this section.

The advancements in energy carriers described in Figure 10.2 are discussed in greater detail in Chapter 6 (Energy) and Chapter 11 (Industry) but the review presented in this chapter highlights their application in the transport sector. This section pays attention to the advancements in alternative fuels, electric, and fuel cell technologies since AR5.

### 10.3.1 Alternative Fuels – An Option for Decarbonising Internal Combustion Engines

The average fuel consumption of new internal combustion engine (ICE) vehicles has improved significantly in recent years due to more stringent emissions regulations. However, improvements are now slowing down. The average fuel consumption of LDVs decreased by only 0.7 % between 2016 and 2017, reaching 7.2 litres of gasoline-equivalent (Lg-eq) per 100 km in 2017, much slower than the

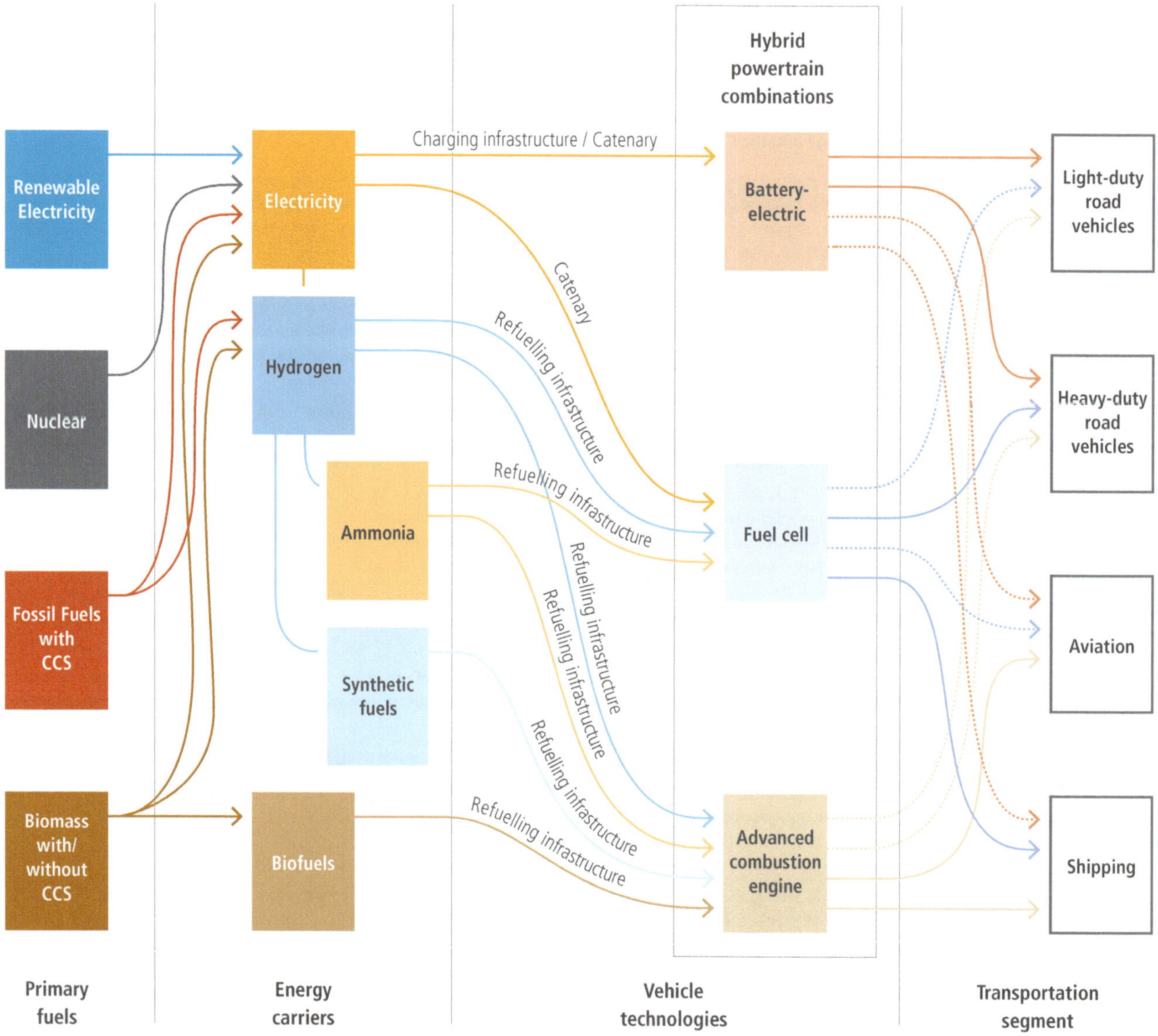


**Figure 10.2 | Energy pathways for low-carbon transport technologies.** Primary energy sources are shown in the far left, while the segments of the transport system are in the far right. Energy carriers and vehicle technologies are represented in the middle. Primary pathways are shown with solid lines, while dotted lines represent secondary pathways.

**Table 10.4 | Engine technologies to reduce emissions from light-duty ICE vehicles and their implementation stage.** Table nomenclature: GDI = Gasoline direct injection, VVT = Variable valve technology, CDA = Cylinder deactivation, CR = compression ratio, GDCI = Gasoline direct injection compression ignition, EGR = exhaust gas recirculation, RCCI = Reactivity controlled compression ignition, GCI = Gasoline compression ignition. Source: Joshi (2020).

| Implementation stage | Engine technology | $CO_2$ reduction (%) |
|---|---|---|
| Implemented | Baseline: GDI, turbo, stoichiometry | 0 |
| Development | Atkinson cycle (+ VVT) | 3–5 |
| | Dynamic CDA + Mild hybrid or Miller | 10–15 |
| | Lean-burn GDI | 10–20 |
| | Variable CR | 10 |
| | Spark assisted GCI | 10 |
| | GDCI | 15–25 |
| | Water injection | 5–10 |
| | Pre-chamber concepts | 15–20 |
| | Homogeneous lean | 15–20 |
| | Dedicated EGR | 15–20 |
| | 2-stroke opposed-piston diesel | 25–35 |
| | RCCI | 20–30 |

1.8 % improvement per year between 2005 and 2016 (GFEI 2020). Table 10.4 summarises recent and forthcoming improvements to ICE technologies and their effect on emissions from these vehicles. However, these improvements are not sufficient to meet deep decarbonisation levels in the transport sector. While there is significant and growing interest in electric and fuel-cell vehicles, future scenarios indicate that a large number of LDV may continue to be operated by ICE in conventional, hybrid, and plug-in hybrid configurations over the next 30 years (IEA 2019a), unless they are regulated away through ICE vehicle sales bans (as some nations have announced) (IEA 2021a). Moreover, ICE technologies are likely to remain the prevalent options for shipping and aviation. Thus, reducing $CO_2$ and other emissions from ICEs through the use of low-carbon or zero-carbon fuels is essential to a balanced strategy for limiting atmospheric pollutant levels. Such alternative fuels for ICE vehicles include natural gas-based fuels, biofuels, ammonia, and other synthetic fuels.

**Natural Gas.** Natural gas could be used as an alternative fuel to replace gasoline and diesel. Natural gas in vehicles can be used as compressed natural gas (CNG) and liquefied natural gas (LNG). CNG is gaseous at relatively high pressure (10 to 25 megapascal (MPa)) and temperature (–40 to 30°C). In contrast, LNG is used in liquid form at relatively low pressure (0.1 MPa) and temperature (–160°C). Therefore, CNG is particularly suitable for commercial vehicles and light- to medium-duty vehicles, whereas LNG is better suited to replace diesel in HDVs (Dubov et al. 2020; Dziewiatkowski et al. 2020; Yaïci and Ribberink 2021). CNG vehicles have been widely deployed in some regions, particularly in Asian-Pacific countries. For example, there are about 6 million CNG vehicles in China, the most of any country (Qin et al. 2020). However, only 20% of vehicles that operate using CNG were originally designed as CNG vehicles, with the rest being gasoline-fuelled vehicles that have been converted to operate with CNG (Chala et al. 2018).

Natural gas-based vehicles have certain advantages over conventional fuel-powered ICE vehicles, including lower emissions of criteria air pollutants, no soot or particulate, low carbon to Hydrogen ratio, moderate noise, a wide range of flammability limits, and high octane numbers (Kim 2019; Bayat and Ghazikhani 2020). Furthermore, the technology readiness level (TRL) of natural gas vehicles is very high (TRL 8–9), with direct modification of existing gasoline and diesel vehicles possible (Transport and Environment 2018; Peters et al. 2021; Sahoo and Srivastava 2021). On the other hand, methane emissions from the natural gas supply chain and tailpipe $CO_2$ emissions remain a significant concern (Trivedi et al. 2020). As a result, natural gas as a transition transportation fuel may be limited due to better alternative options being available and due to regulatory pressure to decarbonise the transport sector rapidly. For example, the International Maritime Office (IMO) has set a target of 40% less carbon intensity in shipping by 2030, which cannot be obtained by simply switching to natural gas.

**Biofuels.** Since AR5, the faster than anticipated adoption of electromobility, primarily for LDVs, has partially shifted the debate around the primary use of biofuels from land transport to the shipping and aviation sectors (IEA 2017a; Davis et al. 2018). At the same time, other studies highlight that biofuels may have to complement electromobility in road transport, particularly in developing countries, offering relevant mitigation opportunities in the short- and mid-term (up to 2050) (IEA 2021b). An important advantage of biofuels is that they can be converted into energy carriers compatible with existing technologies, including current powertrains and fuel infrastructure. Also, biofuels can diversify the supply of transport fuel, raise energy self-sufficiency in many countries, and be used as a strategy to diversify and strengthen the agro-industrial sector (Puricelli et al. 2021). The use of biofuels as a mitigation strategy is driven by a combination of factors, including not only the costs and technology readiness levels of the different biofuel conversion technologies, but also the availability and costs of both biomass feedstocks and alternative mitigation options, and the relative speed and scale of the energy transition in energy and transport sectors (Box 10.2).

Many studies have addressed the lifecycle emissions of biofuel conversion pathways for land transport, aviation, and marine applications (Koeble et al. 2017; Staples et al. 2018; Tanzer et al. 2019). Bioenergy technologies generally struggle to compete with existing fossil fuel-based ones because of the higher costs involved. However, the extent of the cost gap depends critically on the availability and costs of biomass feedstock (IEA 2021b). Ethanol from corn and sugarcane is commercially available in countries such as Brazil and the US. Biodiesel from oil crops and hydro-processed esters and fatty acids are available in various countries, notably in Europe and parts of Southeast Asia. On the infrastructure side, biomethane blending is being implemented in some regions of the US and Europe, particularly in Germany, with the help of policy measures (IEA 2021b). While many of these biofuel conversion technologies could also be implemented using seaweed feedstock options, these value chains are not yet mature (Jiang et al. 2016).

## Box 10.2 | Bridging Land Use and Feedstock Conversion Footprints for Biofuels

Under specific conditions, biofuels may represent an important climate mitigation strategy for the transport sector (Daioglou et al. 2020; Muratori et al. 2020). Both the IPCC Special Report on Global Warming of 1.5°C and the IPCC Special Report on Climate Change and Land highlighted that biofuels could be associated with climate mitigation co-benefits and adverse side effects to many SDGs. These side effects depend on context-specific conditions, including deployment scale, associated land-use changes and agricultural management practices (Section 7.4.4 and Box 7.10). There is broad agreement in the literature that the most important factors in determining the climate footprint of biofuels are the land use and land-use change characteristics associated with biofuel deployment scenarios (Elshout et al. 2015; Daioglou et al. 2020). This issue is covered in more detail in Box 7.1. While the mitigation literature primarily focuses on the GHG-related climate forcings, note that land is an integral part of the climate system through multiple geophysical and geochemical mechanisms (albedo, evaporation, etc.). For example, Sections 2.2.7 and 7.3.4 in the AR6 WGI report indicate that geophysical aspects of historical land-use change outweigh the geochemical effects, leading to a net cooling effect. The land-related carbon footprints of biofuels presented in Sections 10.4–10.6 are adopted from Chapter 7 (Section 7.4.4, Box 7, and Figure 7.1). The results show how the land-related footprint increases due to an increased outtake of biomass, as estimated with different models that rely on global supply scenarios of biomass for energy and fuel of 100 exajoules (EJ). The integrated assessment models and scenarios used include the EMF 33 scenarios (IAM-EMF33), from partial models with constant land cover (PM-CLC), and from partial models with natural regrowth (PM-NGR). These results are combined with both biomass cultivation emission ranges for advanced biofuels aligned with Koeble et al. (2017), El Akkari et al. (2018), Jeswani et al. (2020), and Puricelli et al. (2021) and conversion efficiencies and conversion phase emissions as described in Table 10.5. The modelled footprints resulting from land-use changes related to delivering 100 EJ of biomass at global level are in the range of 3–77 g$CO_2$-eq per MJ of advanced biofuel (median 38 g$CO_2$-eq $MJ^{-1}$) at an aggregate level for Integrated Assessment Models (IAMs) and partial models with constant land cover (Daioglou et al. 2020; Rose et al. 2020). The results for partial models with natural regrowth are much higher (91–246 $CO_2$-eq $MJ^{-1}$ advanced biofuel). The latter ranges may appear in contrast with the results from the scenario literature in Section 10.7, where biofuels play a role in many scenarios compatible with low warming levels. This contrast is a result of different underlying modelling practices. The general modelling approach used for the scenarios in the AR6 database accounts for the land-use change and all other GHG emissions along a given transformation trajectory, enabling assessments of the warming level incurred. The results labelled 'EMF33' and 'partial models with constant land cover' are obtained with this modelling approach. The results in the category 'partial models with natural regrowth' attribute additional $CO_2$ emissions to the bioenergy system, corresponding to estimated uptake of $CO_2$ in a counterfactual scenario where land is not used for bioenergy, but instead subject to natural vegetation regrowth. While the partial analysis provides insights into the implications of alternative land-use strategies, such analysis does not identify the actual emissions of bioenergy production. As a result, the partial analysis is not compatible with the identification of warming levels incurred by an individual transformation trajectory, and therefore not aligned with the general approach applied for the scenarios in the AR6 database.

More details on land-use change impacts and the potential to deliver the projected demands of biofuels at the global level are further addressed in Chapter 7. While, in general, the above results cover most of the variety of GHG range intensities of biofuel options presented in the literature, the more specific life cycle assessment (LCA) literature should be consulted when considering specific combinations of biomass feedstock and conversion technologies in specific regions.

Technologies to produce advanced biofuels from lignocellulosic feedstocks have suffered from slow technology development and are still struggling to achieve full commercial scale. Their uptake is likely to require carbon pricing and/or other regulatory measures, such as clean fuel standards in the transport sector or blending mandates. Several commercial-scale advanced biofuels projects are in development in many parts of the world, encompassing a wide selection of technologies and feedstock choices, including carbon capture and sequestration (CCS) that supports carbon dioxide removal. The success of these projects is vital to moving forward the development of advanced biofuels and bringing many of the advanced biofuels value chains closer to the market (IEA 2021b). Finally, biofuel production and distribution supply chains involve notable transport and logistical challenges that need to be overcome (Mawhood et al. 2016; Skeer et al. 2016; IEA 2017a; Puricelli et al. 2021).

Table 10.5 summarises performance data for different biofuel technologies, while Figure 10.3 shows the technology readiness levels.

Within the aviation sector, jet fuels produced from biomass resources (so-called sustainable aviation fuels, or SAF) could offer significant climate mitigation opportunities under the right policy circumstances. Despite the growing interest in aviation biofuels, demand and production volumes remain negligible compared to conventional fossil aviation fuels. Nearly all flights powered by biofuels have used fuels derived from vegetable oils and fats, and the blending level of biofuels into conventional aviation fuels for testing is up to 50% today (Mawhood et al. 2016). To date, only one facility in the US is regularly producing sustainable aviation fuels based on waste oil feedstocks. The potential to scale up bio-based SAF volumes is severely restricted by the lack of low-cost and sustainable feedstock options (Chapter 7). Lignocellulosic feedstocks are considered to have great potential for

**Table 10.5 | Ranges of efficiency, GHG emissions, and relative costs of selected biofuel conversion technologies for road, marine, and aviation biofuels.**

| Main application | Conversion technology | Energy efficiency of conversion[a] | GHG emissions of conversion process ($gCO_2$-eq per MJ of fuel)[b] | Relative cost of conversion process |
|---|---|---|---|---|
| Road | Lignocellulosic ethanol | 35%[c] | 5[d] | Medium |
| Road/aviation | Gasification and Fischer-Tropsch synthesis | 57%[e] | <1[d] | High |
| Road | Ethanol from sugar and starch | 60–70%[f] | 1–31[d] | Low |
| Road | Biodiesel from oil crops | 95%[g] | 12–30[d] | Low |
| Marine | Upgraded pyrolysis oil | 30–61%[h] | 1–4[h] | Medium |
| Aviation/marine | Hydro-processed esters and fatty acids | 80%[i] | 3[i] | Medium |
| Aviation | Alcohol to jet | 90%[j] | <1[k] | High |
| Road/marine | Biomethane from residues | 60%[l] | n/a | Low |
| Marine/aviation | Hydrothermal liquefaction | 35–69%[h] | <1[h] | High |
| Aviation | Sugars to hydrocarbons | 65%[m] | 15[m] | High |
| Road | Gasification and syngas fermentation | 40%[n] | 30–40[n] | High |

Notes: [a] Calculated as liquid fuels output divided by energy in feedstock entering the conversion plant; [b] GHG emissions here refers only to the conversion process. Impacts form the different biomass options are not included here as they are addressed in Chapter 7; [c] Olofsson et al. (2017); [d] Koeble et al. (2017); [e] Simell et al. (2014); [f] de Souza Dias et al. (2015); [g] Castanheira et al. (2015); [h] Tanzer et al. (2019); [i] Klein et al. (2018); [j] Narula et al. (2017); [k] de Jong et al. (2017); [l] Salman et al. (2017); [m] Moreira et al. (2014); Roy et al. (2015); Handler et al. (2016); [n] Salman et al. (2017); Moreira et al. (2014); Roy et al. (2015); Handler et al. (2016).



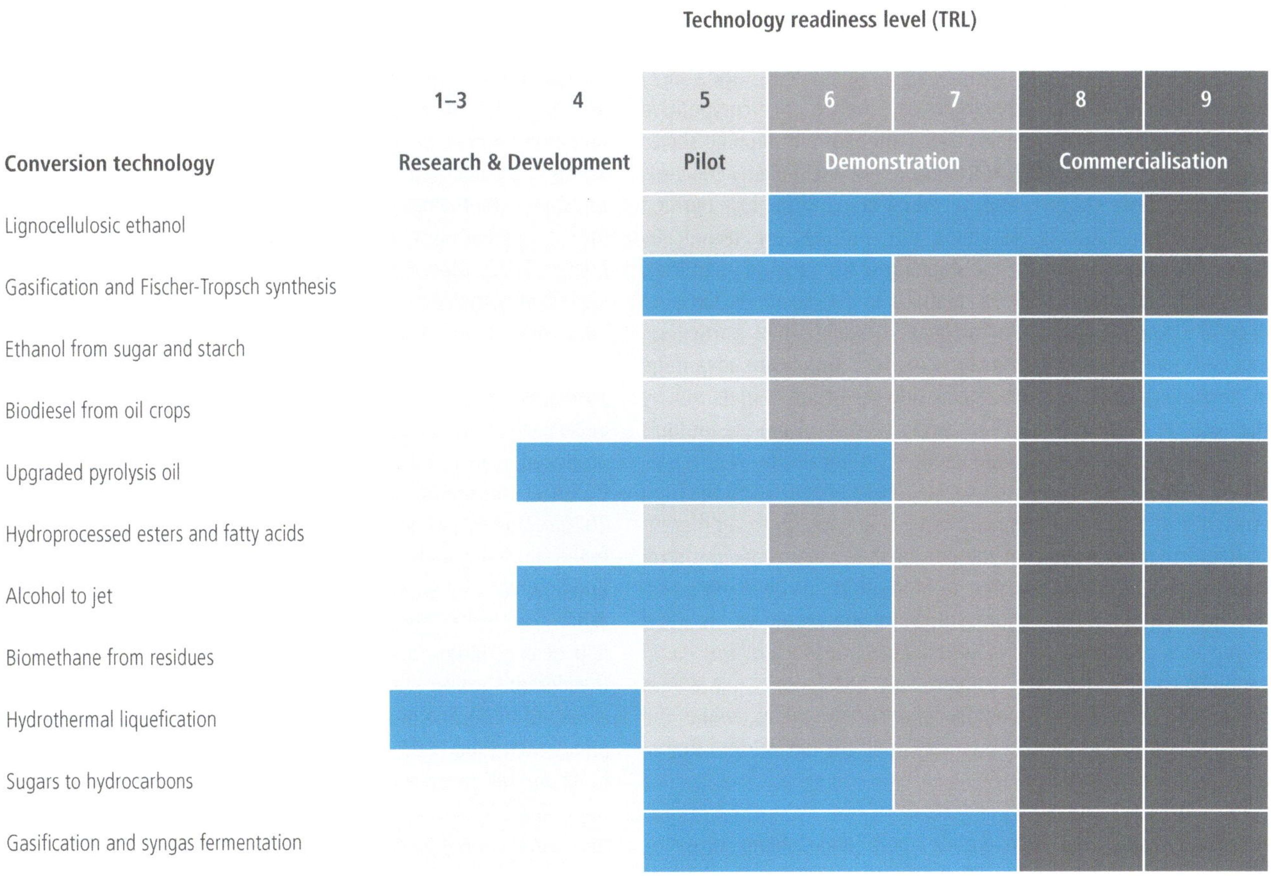


**Figure 10.3 | Commercialisation status of selected biofuels conversion technologies.** The blue boxes represent the current technology readiness level of each conversion technology. Source: based on Mawhood et al. (2016), Skeer et al. (2016), IEA (2017a), and Puricelli et al. (2021).

the production of financially competitive bio-based SAF in many regions. However, production facilities involve significant capital investment and estimated levelised costs are typically more than twice the selling price of conventional jet fuel. In some cases (notably for vegetable oils), the feedstock price is already higher than that of fossil jet fuel (Mawhood et al. 2016). Some promising technological routes for producing SAF from lignocellulosic feedstocks are below technology readiness level (TRL) 6 (pilot scale), with just a few players involved in the development of these technologies. Although it would be physically possible to address the mid-century projections for substantial use of biofuels in the aviation sector (according to the International Energy Agency (IEA) and other sectoral organisations (ICAO 2017)), this fuel deployment scale could only be achieved with very large capital investments in bio-based SAF production infrastructure, and substantial policy support.

In comparison to the aviation sector, the prospects for technology deployment are better in the shipping sector. The advantage of shipping fuels is that marine engines have a much higher operational flexibility on a mix of fuels, and shipping fuels do not need to undergo as extensive refining processes as road and aviation fuels to be considered drop-in. However, biofuels in marine engines have only been tested at an experimental or demonstration stage, leaving open the question about the scalability of the operations, including logistics issues. Similar to the aviation sector, securing a reliable, sustainable biomass feedstock supply and mature processing technologies to produce price-competitive biofuels at a large scale remains a challenge for the shipping sector (Hsieh and Felby 2017). Other drawbacks include industry concerns about oxidation, storage, and microbial stability for less purified or more crude biofuels. Assuming that biofuels are technically developed and available for the shipping sector in large quantities, a wider initial introduction of biofuels in the sector is likely to depend upon increased environmental regulation of particulate and GHG emissions. Biofuels may also offer a significant advantage in meeting ambitious sulphur emission reduction targets set by the sectoral organisations. More extensive use of marine biofuels will most likely be first implemented in inner-city waterways, inland river freight routes, and coastal green zones. Given the high efficiency of the diesel engine, a large-scale switch to a different standard marine propulsion method in the near to medium-term future seems unlikely. Thus, much of the effort has been placed on developing biofuels compatible with diesel engines. So far, biodiesel blends look promising, as it is used in land transport. Hydrotreated vegetable oil (HVO) is also a technically good alternative and is compatible with current engines and supply chains, while the introduction of multifuel engines may open the market for ethanol fuels (Hsieh and Felby 2017).

**Ammonia.** At room temperature and atmospheric pressure, ammonia is a colourless gas with a distinct odour. Due to relatively mild conditions for liquefaction, ammonia is transferred and stored as a liquefied or compressed gas and has been used as an essential industrial chemical resource for many products. In addition, since ammonia does not contain carbon, it has attracted attention as a carbon-neutral fuel that can also improve combustion efficiency (Gill et al. 2012). Furthermore, ammonia could also serve as a hydrogen carrier and be used in fuel cells. These characteristics have driven increased interest in the low-carbon production of ammonia, which would have to be coupled to low-carbon hydrogen production (with low-carbon electricity providing the needed energy or with CCS).

For conventional internal combustion engines, the use of ammonia remains challenging due to the relatively low burning velocity and high ignition temperature. Therefore, Frigo and Gentili (2014) have suggested a dual-fuelled spark ignition engine operated by liquid ammonia and hydrogen, where hydrogen is generated from ammonia using the thermal energy of exhaust gas. On the other hand, the high-octane number of ammonia means good knocking resistance of spark ignition engines and is promising for improving thermal efficiency. For compression ignition engines, the high-ignition temperature of ammonia requires a high compression ratio, causing an increase in mechanical friction. Since Gray et al. (1966), many studies have shown that the compression ratio can be reduced by mixing ammonia with secondary fuels such as diesel and hydrogen with low self-ignition temperatures, as summarised by Dimitriou and Javaid (2020). Using a secondary fuel with a high cetane number and the adoption of a suitable fuel injection timing has enabled highly efficient combustion of compression ignition engines in the dual fuel mode with ammonia ratios up to 95% (Dimitriou and Javaid 2020). One major challenge for realising an ammonia-fuelled engine is the reduction of unburned ammonia, as described in Section 6.4.5 (Reiter and Kong 2011). Processes being examined include the use of exhaust gas recirculation (EGR) (Pochet et al. 2017) and after treatment systems. However, these processes require space, which is a constraint for LDVs and air transport but more practical for ships. Shipbuilders are developing an ammonia engine based on the existing diesel dual-fuel engine to launch a service in 2025 (Brown 2019; MAN-ES 2019). Ammonia could therefore contribute significantly to decarbonisation in the shipping sector (Section 10.6), with potential niche applications elsewhere.

**Synthetic fuels.** Synthetic fuels can contribute to transport decarbonisation through synthesis from electrolytic hydrogen produced with low-carbon electricity or hydrogen produced with CCS, and captured $CO_2$ using the Fischer-Tropsch process (Liu et al. 2020a). Due to similar properties of synthetic fuels to those of fossil fuels, synthetic fuels can reduce GHG emissions in both existing and new vehicles without significant changes to the engine design. While the Fischer-Tropsch process is a well-established technology (Liu et al. 2020a), low-carbon synthetic fuel production is still at the demonstration stage. Even though their production costs are expected to decline in the future due to lower renewable electricity prices, increased scale of production, and learning effects, synthetic fuels are still up to three times more expensive than conventional fossil fuels (Section 6.6.2.4). Furthermore, since the production of synthetic fuels involves thermodynamic conversion loss, there is a concern that the total energy efficiency is lower than that of electric vehicles (Yugo and Soler 2019). Given these high costs and limited scales, the adoption of synthetic fuels will likely focus on the aviation, shipping, and long-distance road transport segments, where decarbonisation by electrification is more challenging. In particular, synthetic fuels are considered promising as an aviation fuel (Section 10.5).

### 10.3.2 Electric Technologies

Widespread electrification of the transport sector is likely crucial for reducing transport emissions and depends on appropriate electrical energy storage systems (EES). However, large-scale diffusion of EES depends on improvements in energy density (energy stored per unit volume), specific energy (energy stored per unit weight), and costs (Cano et al. 2018). Recent trends suggest EES-enabled vehicles are on a path to becoming the leading technology for LDVs, but their contribution to heavy-duty freight is more uncertain.

**Electrochemical storage of light and medium-duty vehicles.** Electrochemical storage, i.e., batteries, are one of the most promising forms of energy storage for the transport sector and have dramatically improved in their commerciality since AR5. Rechargeable batteries are of primary interest for applications within the transport sector, with a range of mature and emerging chemistries able to support the electrification of vehicles. The most significant change since AR5 and SPR1.5 is the dramatic rise in lithium-ion batteries (LIB), which has enabled electromobility to become a major feature of decarbonisation.

Before the recent growth in market share of LIBs, lead-acid batteries, nickel batteries, high-temperature sodium batteries, and redox flow batteries were of particular interest for the transport sector (Placke et al. 2017). Due to their low costs, lead-acid batteries have been used in smaller automotive vehicles, e.g., e-scooters and e-rickshaws (Dhar et al. 2017). However, their application in electric vehicles will be limited due to their low specific energy (Andwari et al. 2017). Nickel-metal hydride (NiMH) batteries have a better energy density than lead-acid batteries and have been well optimised for regenerative braking (Cano et al. 2018). As a result, NiMH batteries were the battery of choice for hybrid electric vehicles (HEVs). Ni-Cadmium (NiCd) batteries have energy densities lower than NiMH batteries and cost around ten times more than lead-acid batteries (Table 6.5). For this reason, NiCd batteries do not have major prospects within automotive applications. There are also no examples of high-temperature sodium or redox flow batteries being used within automotive applications.

Commercial application of LIBs in automotive applications started around 2000 when the price of LIBs was more than USD1000 per kWh (Schmidt et al. 2017). By 2020, the battery manufacturing capacity for automotive applications was around 300 GWh per year (IEA 2021a). Furthermore, by 2020, the average battery pack cost had come down to USD137 per kWh, a reduction of 89% in real terms since 2010 (Henze 2020). Further improvements in specific energy, energy density (Nykvist et al. 2015; Placke et al. 2017) and battery service life (Liu et al. 2017) of LIBs are expected through additional design optimisation (Table 6.5). These advances are expected to lead to EVs with even longer driving ranges, further supporting the uptake of LIBs for transport applications (Cano et al. 2018). However, the performance of LIBs under freezing and high temperatures is a concern (Liu et al. 2017) for reliability. Auto manufacturers have some pre-heating systems for batteries to see that they perform well in very cold conditions (Wu et al. 2020).

For EVs sold in 2018, the material demand was about 11 kilotonnes (kt) of optimised lithium, 15 kt of cobalt, 11 kt of manganese, and 34 kt of nickel (IEA 2019a; IEA 2021a). IEA projections for 2030 in the EV 30@30 scenario show that the demand for these materials would increase by 30 times for lithium and around 25 times for cobalt. While there are efforts to move away from expensive materials such as cobalt (IEA 2019a; IEA 2021a), dependence on lithium will remain, which may be a cause of concern (Olivetti et al. 2017; You and Manthiram 2018). A more detailed discussion on resource constraints for lithium is provided in Box 10.6.

Externalities from resource extraction are another concern, though current volumes of lithium are much smaller than other metals (steel, aluminium). As a result, lithium was not even mentioned in UNEP's global resource outlook (IRP 2019). Nonetheless, it is essential to manage demand and limit externalities since the demand for lithium is going to increase many times in the future. Reuse of LIBs used in EVs for stationary energy applications can help in reducing the demand for LIBs. However, the main challenges are the difficulty in accessing the information on the health of batteries to be recycled and technical problems in remanufacturing the batteries for their second life (Ahmadi et al. 2017). Recycling lithium from used batteries could be another possible supply source (Winslow et al. 2018). While further R&D is required for commercialisation (Ling et al., 2018), recent efforts at recycling LIBs are very encouraging (Ma et al. 2021). The standardisation of battery modules and packaging within and across vehicle platforms, increased focus on design for recyclability, and supportive regulation are important to enable higher recycling rates for LIBs (Harper et al. 2019).

Several next-generation battery chemistries are often referred to as post-LIBs (Placke et al. 2017). These chemistries include metal-sulphur, metal-air, metal-ion (besides Li), and all-solid-state batteries. The long development cycles of the automotive industry (Cano et al. 2018) and the advantages of LIBs in terms of energy density and cycle life (Table 6.5) mean that it is unlikely that post-LIB technologies will replace LIBs in the next decade. However, lithium-sulphur, lithium-air, and zinc-air have emerged as potential alternatives for LIBs. These emerging chemistries may also be used to supplement LIBs in dual-battery configurations, to extend the driving range at lower costs or with higher energy density (Cano et al. 2018). Lithium-sulphur (Li-S) batteries have a lithium metal anode with a higher theoretical capacity than lithium-ion anodes and much lower-cost sulphur cathodes relative to typical Li-ion insertion cathodes (Manthiram et al. 2014). As a result, Li-S batteries are much cheaper than LIB to manufacture and have a higher energy density (Table 6.5). Conversely, these batteries face challenges from sulphur cathodes, such as low conductivity of the sulphur and lithium sulphide phases, and the relatively high solubility of sulphur species in common lithium battery electrolytes, leading to low cycle life (Cano et al. 2018). Lithium-air batteries offer a further improvement in specific energy and energy density above Li–S batteries owing to their use of atmospheric oxygen as a cathode in place of sulphur. However, their demonstrated cycle life is much lower (Table 6.5). Lithium-air batteries also have low specific power. Therefore, lithium-air require an extra battery for practical applications (Cano et al. 2018). Finally, zinc–air batteries could more likely be used in future EVs because of their more advanced technology status and higher practically achievable energy density (Fu et al. 2017). Like Li-air batteries, their poor specific power and energy efficiency will probably prevent zinc-air batteries from being used as a primary energy source for EVs. Still, they could be promising when used in a dual-battery configuration (Cano et al. 2018).

The technological readiness of batteries is a crucial parameter in the advancement of EVs (Manzetti and Mariasiu 2015). Energy density, power density, cycle life, calendar life, and the cost per kWh are the pertinent parameters for comparing the technological readiness of various battery technologies (Manzetti and Mariasiu 2015; Andwari et al. 2017; Lajunen et al. 2018). Table 6.5 provides a summary of the values of these parameters for alternative battery technologies. LIBs comprehensively dominate the other battery types and are at a readiness level where they can be applied for land transport applications (cars, scooters, electrically-assisted cycles) and at battery pack costs below USD150 per kWh, making EVs cost-competitive with conventional vehicles (Nykvist et al. 2019). In 2020 the stock of battery electric LDVs had crossed the 10 million mark (IEA 2021a). Schmidt et al. (2017) project that the cost of a battery pack for LIBs will reach USD100 per kWh by 2030, but more recent trends show this could happen much earlier. For example, according to IEA, battery pack costs could be as low as USD80 per kWh by 2030 (IEA 2019a). In addition, there are clear trends that now vehicle manufacturers are offering vehicles with bigger batteries, greater driving ranges, higher top speeds, faster acceleration, and all size categories (Nykvist et al. 2019). In 2020 there were over 600,000 battery electric buses and over 31,000 battery electric trucks operating globally (IEA 2021a).

LIBs are not currently envisaged to be suitable for long-haul transport. However, several battery technologies are under development (Table 6.5), which could further enhance the competitiveness of EVs and expand their applicability to very short-haul aviation and ships, especially smaller vehicles. Li-S, Li-air, and Zn-air hold the highest potential for these segments (Cano et al. 2018). All three of these technologies rely on making use of relatively inexpensive elements, which can help bring down battery costs (Cano et al. 2018). The main challenge these technologies face is in terms of the cycle life. Out of the three, Li-S has already been used for applications in unmanned aerial vehicles (Fotouhi et al., 2017) due to relatively high specific energy (almost double the state of the art LIBs). However, even with low cycle life, Li-air and Zn-air hold good prospects for commercialisation as range extender batteries for long-range road transport and with vehicles that are typically used for city driving (Cano et al. 2018).

**Alternative electricity storage technologies for heavy-duty transport.** While LIBs described in the previous section are driving the electrification of LDVs, their application to railways, aviation, ships, and large vehicles faces challenges due to the higher power requirements of these applications. The use of a capacitor with a higher power density than LIBs could be suitable for the electrification of such vehicles. It is one of the solutions for regenerating large and instantaneous energy from regenerative brakes. Classical capacitors generally show more attractive characteristics in power density (8000–10,000 watts per kilogram (W/kg)) than batteries. However, the energy density is poor (1–4 watt-hours per kilogram (Wh/kg)) compared to batteries, and there is an issue of self-discharge (González et al. 2016; Poonam et al. 2019). To improve the energy density, electrochemical double layer capacitors (EDLCs; supercapacitor) and hybrid capacitors (10–24 Wh/kg, 900–9000 W/kg at the product level) such as Li-ion capacitors have been developed. The highest energy density of the LIC system (100–140 Wh/kg in the research stage) are approaching that of the Li-ion battery systems (80–240 Wh/kg in the product stage) (Naoi et al. 2012; Panja et al. 2020). Examples of effective use of capacitors include a 12-tonne truck with a capacitor-based kinetic energy recovery system that has been reported to save up to 32% of the fuel use of a standard truck (Kamdar 2017). Similarly, an EDLC bank applied to electric railway systems has been shown to result in a 10% reduction in power consumption per day (Takahashi et al. 2017). Finally, systems in which capacitors are mounted on an electric bus for charging at a stop have been put into practical use, for example by a trackless tram (Newman et al. 2019). At the bus stop, the capacitor is charged at 600 kW for 10 about 40 seconds, which provides enough power for about 5 to 10 km (Newman et al. 2019). In addition, more durable capacitors can achieve a longer life than LIB systems (ADB 2018).

Hybrid energy storage (HES) systems, which combine a capacitor and a battery, achieve both high power and high energy, solving problems such as capacity loss of the battery and self-discharge of the capacitor. In these systems, the capacitor absorbs the steeper power, while the LIB handles the steady power, thereby reducing the power loss of the EV to half. Furthermore, since the in-rush current of the battery is suppressed, there is an improvement in the reliability of the LIB (Noumi et al. 2014). In a hybrid diesel train, 8.2% of the regenerative energy is lost due to batteries' limited charge-discharge performance; however, using an EDLC with batteries can save this energy (Takahashi et al. 2017; Mayrink et al. 2020).

The development of power storage devices and advanced integrated system approaches, including power electronics circuits such as HES and their control technologies, are important for the electrification of mobility. These technologies are solutions that could promote the electrification of systems, reduce costs, and contribute to the social environment through multiple outcomes in the decarbonisation agenda.

### 10.3.3 Fuel Cell Technologies

In harder-to-electrify transport segments, such as heavy-duty vehicles, shipping, and aviation, hydrogen holds significant promise for delivering emissions reductions if it is produced using low-carbon energy sources. In particular, hydrogen fuel cells are seen as an emerging option to power larger vehicles for land-based transport (Tokimatsu et al. 2016; IPCC 2018; IEA 2019b). Despite this potential, further advancements in technological and economic maturity will be required in order for hydrogen fuel cells to play a greater role. While this section focuses primarily on hydrogen fuel cells, ammonia and methanol fuel cells may also emerge as options for low power applications.

During the last decade, hydrogen fuel cell vehicles (HFCVs) have attracted growing attention, with fuel cell technology improving through research and development. Fuel cell systems cost 80% to 95% less than they did in the early 2000s, at approximately USD50 per kW for light-duty (80 kW) and $100 per kW for medium-heavy-duty (160 kW). These costs are approaching the US Department of Energy's (US DOE) goal of USD40 per kW in 2025 at a production target of 500,000 systems per year (IEA 2019c). In addition to cost reductions, the power density of fuel cell stacks has now reached around 3.0 kilowatt per litre (kW/l) and average durability has improved to approximately 2000 to 3000 hours (Jouin et al. 2016; Kurtz et al. 2019). Despite these improvements, fuel cell systems are not yet mature for many commercial applications. For example, the US DOE has outlined that for hydrogen fuel cell articulated trucks (semi-trailers) to compete with diesel vehicles, fuel cell durability will need to reach 30,000 hours (US DOE 2019). While some fuel cell buses have demonstrated durability close to these targets (Eudy and Post 2018a), another review of light fuel cell vehicles found maximum durability of 4000 hours (Kurtz et al. 2019). As more fuel cell vehicles are trialled, it is expected that further real-world data will become available to track ongoing fuel cell durability improvements.

Ammonia and methanol fuel cells are considered to be less mature than hydrogen fuel cells. However, they offer the benefit of using a more easily transported fuel that can be directly used without converting to hydrogen (Zhao et al. 2019). Conversely, both methanol and ammonia are toxic, and in the case of methanol fuel cells, carbon dioxide is released as a by-product of generating electricity with the fuel cell (Zhao et al. 2019). Due to the lower power output, methanol and ammonia fuel cells are also not well suited to heavy-duty vehicles (Jeerh et al. 2021). They are therefore unlikely to compete with hydrogen fuel cells. However, ammonia and methanol could be converted to hydrogen at refuelling stations as an alternative to being directly used in fuel cells (Zhao et al. 2019).

Several FCV-related technologies are fully ready for demonstration and early market deployment, however, further research and development will be required to achieve full-scale commercialisation, likely from 2030 onwards (Staffell et al. 2019; Energy Transitions Commission 2020; IEA 2021b). Some reports argue that it may be possible to achieve serial production of fuel cell heavy-duty trucks in the late 2020s, with comparable costs to diesel vehicles achieved after 2030 (Jordbakker et al. 2018). Over the next decade or so, hydrogen FCVs could become cost-competitive for various transport applications, potentially including long-haul trucks, marine ships, and aviation (Hydrogen Council 2017; FCHEA 2019; FCHJU 2019; BloombergNEF 2020; Hydrogen Council 2020). The speed of fuel cell system cost reduction is a key factor for achieving widespread uptake. Yet, experts disagree on the relationship between the scale of fuel cell demand, cost, and performance improvements (Cano et al. 2018). Costs of light-, medium-, and heavy-duty fuel cell powertrains have decreased by orders of magnitude with further reductions of a factor of two expected with continued technological progress (Whiston et al. 2019). For example, the costs of platinum for fuel cell stacks have decreased by an order of magnitude (Staffell et al. 2019); current generation FCVs use approximately 0.25 g/kW platinum and a further reduction of 50–80% is expected by 2030 (Hao et al. 2019).

Hydrogen is likely to take diverse roles in the future energy system: as a fuel in industry and buildings, as well as transport, and as energy storage for variable renewable electricity. Further research is required to understand better how a hydrogen transport fuel supply system fits within the larger hydrogen energy system, especially in terms of integration within existing infrastructure, such as the electricity grid and the natural gas pipeline system (IEA 2015).

Strong and durable policies would be needed to enable widespread use of hydrogen as a transport fuel and to sustain momentum during a multi-decade transition period for hydrogen FCVs to become cost-competitive with electric vehicles (Hydrogen Council 2017; FCHEA 2019; FCHJU 2019; IEA 2019c; BNEF 2020; Hydrogen Council 2020). The analysis suggests that hydrogen is likely to have strategic and niche roles in transport, particularly in long-haul shipping and aviation. With continuing improvements, hydrogen and electrification will likely play a role in decarbonising heavy-duty road and rail vehicles.

### 10.3.4 Refuelling and Charging Infrastructure

The transport sector relies on liquid gasoline, and diesel for land-based transport, jet fuel for aviation, and heavy fuel oil for shipping. Extensive infrastructure for refuelling liquid fossil fuels already exists. Ammonia, synthetic fuels, and biofuels have emerged as alternative fuels for powering combustion engines and turbines used in land, shipping, and aviation (Figure 10.2). Synthetic fuels such as e-methanol and Fischer-Tropsch liquids have similar physical properties and could be used with existing fossil fuel infrastructure (Yugo and Soler, 2019). Similarly, biofuels have been used in several countries together with fossil fuels (Panoutsou et al. 2021). Ammonia is a liquid, but only under pressure, and therefore will not be compatible with liquid fossil fuel refuelling infrastructure. Ammonia is, however, widely used as a fertiliser and chemical raw material and 10% of annual ammonia production is transported via sea (Gallucci 2021). As such, a number of port facilities include ammonia storage and transport infrastructure and the shipping industry has experience in handling ammonia (Gallucci 2021). This infrastructure would likely need to be extended in order to support the use of ammonia as a fuel for shipping and therefore ports are likely to be the primary sites for these new refuelling facilities.

EVs and HFCV require separate infrastructure than liquid fuels. The successful diffusion of new vehicle technologies is dependent on the preceding deployment of infrastructure (Leibowicz 2018), so that the deployment of new charging and refuelling infrastructure will be critical for supporting the uptake of emerging transport technologies like EVs and HFCVs, where it makes sense for each to be deployed. As a result, there is likely a need for simultaneous investment in both infrastructure and vehicle technologies to accelerate decarbonisation of the transport sector.

**Charging infrastructure.** Charging infrastructure is important for a number of key reasons. From a consumer perspective, robust and reliable charging infrastructure networks are required to build confidence in the technology and overcome the often-cited barrier of 'range anxiety' (She et al. 2017). Range anxiety is where consumers do not have confidence that an EV will meet their driving range requirements. For LDVs, the majority of charging (75–90%) has been reported to take place at or near homes (Figenbaum 2017; Webb et al. 2019; Wenig et al. 2019). Charging at home is a particularly significant factor in the adoption of EVs as consumers are less willing to purchase an EV without home charging (Berkeley et al. 2017; Funke and Plötz 2017; Nicholas et al. 2017). However, home charging may not be an option for all consumers. For example, apartment dwellers may face specific challenges in installing charging infrastructure (Hall and Lutsey 2020). Thus, the provision of public charging infrastructure is another avenue for alleviating range anxiety, facilitating longer distance travel in EVs, and in turn, encouraging adoption (Hall and Lutsey 2017; Melliger et al. 2018; Narassimhan and Johnson 2018; Melton et al. 2020). Currently, approximately 10% of charging occurs at public locations, roughly split equally between alternating current (AC) (slower) and direct current (DC) (fast) charging (Figenbaum 2017; Webb et al. 2019; Wenig et al. 2019). Deploying charging infrastructure at workplaces and commuter car parks is also important, particularly as vehicles are parked at these locations for many hours. Indeed, around 15–30% of EV charging currently occurs at these locations (Figenbaum 2017; Webb et al. 2019; Wenig et al. 2019). It has been suggested that automakers and utilities could provide support for the installation of home charging infrastructure (Hardman et al. 2018), while policymakers can provide support for public charging. Such support could come via supportive planning policy, building regulations, and financial support. Policy support could also incentivise the deployment of charging stations at workplaces and commuter car parks. Charging at these locations would have the added benefit of using excess solar energy generated during the day (Hardman et al. 2018; Webb et al. 2019).



While charging infrastructure is of high importance for the electrification of light-duty vehicles, arguably it is even more important for heavy-duty vehicles, given the costs of high-power charging infrastructure. It is estimated that the installed cost of fast-charging hardware can vary between approximately USD45,000 to USD200,000 per charger, depending on the charging rate, the number of chargers per site, and other site conditions (Hall and Lutsey 2019; Nelder and Rogers 2019; Nicholas 2019). Deployment of shared charging infrastructure at key transport hubs, such as bus and truck depots, freight distribution centres, marine shipping ports and airports, can encourage a transition to electric vehicles across the heavy transport segments. Furthermore, if charging infrastructure sites are designed to cater for both light- and heavy-duty vehicles, infrastructure costs could decrease by increasing utilisation across multiple applications and/or fleets (Nelder and Rogers 2019).

There are two types of charging infrastructure for electric vehicles: conductive charging involving a physical connection and wireless/induction charging. The majority of charging infrastructure deployed today for light- and heavy-duty vehicles is conductive. However, wireless charging technologies are beginning to emerge – particularly for applications like bus rapid transit – with vehicles able to charge autonomously while parked and/or in motion (IRENA 2019). For road vehicles, electric road systems, or road electrification, is also emerging as an alternative form of conductive charging infrastructure that replaces a physical plug (Ainalis et al. 2020; Hill et al. 2020). This type of charging infrastructure is particularly relevant for road freight where load demand is higher. Road electrification can take the form of a charging rail built into the road pavement, run along the side of the road, through overhead catenary power lines – similar to electrical infrastructure used for rail – or at recharging facilities at stations along the route. This infrastructure can also be used to directly power other electrified powertrains, such as hybrid and HFCV (Hardman et al. 2018; Hill et al. 2020).

Charging infrastructure also varies in terms of the level of charging power. For light vehicles, charging infrastructure is generally up to 350 kW, which provides approximately 350 kilometres for every 10 minutes of charging. For larger vehicles, like buses and trucks, charging infrastructure is generally up to 600 kW, providing around 50–100 km for every 10 minutes of charging (depending on the size of the vehicle). Finally, even higher-power charging infrastructure is currently being developed at rates greater than 1 MW, particularly for long-haul trucks and for short-haul marine shipping and aviation. For example, one of the largest electric ferries in the world, currently operating in Denmark, uses a 4.4 MW charger (Heinemann et al. 2020).

Finally, there are several different charging standards, varying across transport segments and across geographical locations. Like electrical appliances, different EV charging connectors and sockets have emerged in different regions, such as CCS2 in Europe (ECA 2021), GB/T in China (Hove and Sandalow 2019). Achieving interoperability between charging stations is seen as another important issue for policymakers to address to provide transparent data to the market on where EV chargers are located and a consistent approach to paying for charging sessions (van der Kam and Bekkers 2020). Interoperability could also play an important role in enabling smart charging infrastructure (Neaimeh and Andersen 2020).

**Smart charging: electric vehicle-grid integration strategies.** EVs provide several opportunities for supporting electricity grids if appropriately integrated. Conversely, a lack of integration could negatively affect the grid, particularly if several vehicles are charged in parallel at higher charging rates during peak demand periods (Webb et al. 2019; Jochem et al. 2021). There are three primary approaches to EV charging. In unmanaged charging, EVs are charged ad hoc, whenever connected, regardless of conditions on the broader electricity grid (Webb et al. 2019; Jochem et al. 2021). Second, in managed charging, EVs are charged during periods beneficial to the grid, e.g., at periods of high renewable generation and/or low demand. Managed charging also allows utilities to regulate the rate of charge and can thus provide frequency and regulation services to the grid (Weis et al. 2014). Finally, in bidirectional charging or vehicle-to-grid (V2G), EVs are generally subject to managed charging, but an extension provides the ability to export electricity from the vehicle's battery back to the building and/or wider electricity grid (Ercan et al. 2016; Noel et al. 2019; Jochem et al. 2021). The term 'smart charging' has become an umbrella term to encompass both managed charging (often referred to as V1G) and V2G. For electric utilities, smart charging strategies can provide back-up power, support load balancing, reduce peak loads (Zhuk et al. 2016; Noel et al. 2019;

Jochem et al. 2021), reduce the uncertainty in forecasts of daily and hourly electrical loads (Peng et al. 2012), and allow greater utilisation of generation capacity (Hajimiragha et al. 2010; Madzharov et al. 2014).

Smart charging strategies can also enhance the climate benefits of EVs (Yuan et al. 2021). Controlled charging can help avoid high-carbon electricity sources, decarbonisation of the ancillary service markets, or peak shaving of high-carbon electricity sources (Jochem et al. 2021). V2G-capable EVs can result in even lower total emissions, particularly when compared to other alternatives (Reddy et al. 2016). Noel et al. (2019) analysed V2G pathways in Denmark and noted that at a penetration rate of 75% by 2030, USD34 billion in social benefits could be accrued (through things like displaced pollution). These social benefits translate to USD1,200 per vehicle. V2G-capable EVs were found to have the potential to reduce carbon emissions compared to a conventional gasoline vehicle by up to 59%, assuming optimised charging schedules (Hoehne and Chester 2016).

Projections of energy storage suggest smart charging strategies will come to play a significant role in future energy systems. Assessment of different energy storage technologies for Europe showed that V2G offered the most storage potential compared to other options and could account for 200 GW of installed capacity by 2060, whereas utility-scale batteries and pumped hydro storage could provide 160 GW of storage capacity (Després et al. 2017). Another study found that EVs with controlled charging could provide similar services to stationary storage but at a far lower cost (Coignard et al. 2018). While most deployments of smart charging strategies are still at the pilot stage, the number of projects continues to expand, with the V2G Hub documenting at least 90 V2G projects across 22 countries in 2021 (Vehicle to Grid 2021). Policymakers have an important role in facilitating collaboration between vehicle manufacturers, electricity utilities, infrastructure providers, and consumers to enable smart charging strategies and ensure EVs can support grid stability and the uptake of renewable energy. This is a critical part of decarbonising transport.

**Hydrogen infrastructure.** HFCVs are reliant on the development of widespread and convenient hydrogen refuelling stations (FCHEA 2019; IEA 2019c; BNEF 2020). Globally, there are around 540 hydrogen refuelling stations, with the majority located in North America, Europe, Japan, and China (IEA 2021a). Approximately 70% of these refuelling stations are open to the public (Coignard et al. 2018). Typical refuelling stations currently have a refuelling capacity of 100 to 350 kg/day (CARB 2019; CARB 2020; H2 Tools 2020; AFDC 2021). At most, current hydrogen refuelling stations have daily capacities under 500 kg a day (Liu et al. 2020b).

The design of hydrogen refuelling stations depends on the choice of methods for hydrogen supply and delivery, compression and storage, and the dispensing strategy. Hydrogen supply could happen via on-site production or via transport and delivery of hydrogen produced off-site. At the compression stage, hydrogen is compressed to achieve the pressure needed for economic stationary and vehicle storage. This pressure depends on the storage strategy. Hydrogen can be stored as a liquid or a gas. Hydrogen can also be dispensed to vehicles as a gas or a liquid, depending on the design of the vehicles (though it tests the extremes of temperature range and storage capacity for an industrial product). The technological and economic development of each of these components continues to be researched.

If hydrogen is produced off site in a large centralised plant, it must be stored and delivered to refuelling stations. The cost of hydrogen delivery depends on the amount of hydrogen delivered, the delivery distance, the storage method (compressed gas or cryogenic liquid), and the delivery mode (truck or pipeline). Table 10.6 describes the three primary options for hydrogen delivery. Most hydrogen refuelling stations today are supplied by trucks and, very occasionally, hydrogen pipelines. Gaseous tube trailers could also be used to deliver hydrogen in the near term, or over shorter distances, due to the low fixed cost (although the variable cost is high). Both liquefied truck trailers and pipelines are recognised as options in the medium to long term as they have higher capacities and lower costs over longer distances (FCHJU 2019; Li et al. 2020; EU 2021). Alternatively, hydrogen can be produced on site using a small-scale on-site electrolyser or steam methane reforming unit combined with CCS. Hydrogen is generally dispensed to vehicles as a compressed gas at pressures 350 or 700 bar, or as liquified hydrogen at –253°C (Hydrogen Council 2020).

The costs for hydrogen refuelling stations vary widely and remain uncertain for the future (IEA 2019c). The IEA reports that the investment cost for one hydrogen refuelling station ranges between USD0.6 million and USD2 million for hydrogen at a pressure of 700 bar and a delivery capacity of 1300 kg per day. The investment cost of hydrogen refuelling stations with lower refuelling capacities (~50 kg $H_2$ per day) delivered at lower pressure (350 bar) range between USD0.15–1.6 million. A separate estimate by the International Council for Clean Transport suggests that at a capacity of 600 kg of hydrogen per day, the capital cost of a single refuelling station would be approximately USD1.8 million (ICCT 2017). Given the high investment costs for hydrogen refuelling stations, low utilisation can translate into a high price for delivered hydrogen. In Europe, most pumps operate at less than 10% capacity. For small refuelling stations with a capacity of 50 kg $H_2$ per day, this utilisation rate translates to a high price of around USD15–25 per kg $H_2$ – in line with current retail prices (IEA 2019c). The dispensed cost of hydrogen is also highly correlated with the cost of electricity, when $H_2$ is produced using electrolysis, which is required to produce low-carbon hydrogen.

**Table 10.6 | Overview of three transport technologies for hydrogen delivery in the transport sector showing relative differences.** Source: IEA (2019c).

| | Capacity | Delivery distance | Energy loss | Fixed costs | Variable costs | Deployment phase |
|---|---|---|---|---|---|---|
| Gaseous tube trailers | Low | Low | Low | Low | High | Near term |
| Liquefied truck trailers | Medium | High | High | Medium | Medium | Medium to long term |
| Hydrogen pipelines | High | High | Low | High | Low | Medium to long term |

## 10.4 Decarbonisation of Land-based Transport

### 10.4.1 Light-duty Vehicles for Passenger Transport

LDVs represent the main mode of transport for private citizens (ITF 2019) and currently represent the largest share of transport emissions globally (IEA 2019d). Currently, powertrains depending on gasoline and diesel fuels remain the dominant technology in the LDV segment (IEA 2019d). HEVs, and fully battery electric vehicles (BEVs), however, have become increasingly popular in recent years (IEA 2021a). Correspondingly, the number of lifecycle assessment (LCA) studies investigating HEVs, BEVs, and fuel cell vehicles have increased. While historically the focus has been on the tailpipe emissions of LDVs, LCA studies demonstrate the importance of including emissions from the entire vehicle value chain, particularly for alternative powertrain technologies.

Figure 10.4 presents the cumulative lifecycle emissions for selected powertrain technologies and fuel chain combinations for compact and mid-sized LDVs. This figure summarises the harmonised findings from the academic literature reviewed and the data submitted through an IPCC data collection effort, as described in Appendix 10.1 (Hawkins et al. 2013; Messagie et al. 2014; Bauer et al. 2015; Tong et al. 2015b; Ellingsen et al. 2016; Gao et al. 2016; Kim and Wallington 2016; Cai et al. 2017; Evangelisti et al. 2017; Ke et al. 2017; Lombardi et al. 2017; Miotti et al. 2017; Valente et al. 2017; Cox et al. 2018; de Souza et al. 2018; Elgowainy et al. 2018; Luk et al. 2018; Bekel and Pauliuk 2019; Cusenza et al. 2019; Hoque et al. 2019; IEA 2019a; Rosenfeld et al. 2019; Shen et al. 2019; Wang et al. 2019; Wu et al. 2019; Ambrose et al. 2020; Benajes et al. 2020; Hill et al. 2020; Knobloch et al. 2020; Prussi et al. 2020; Qiao et al. 2020; Wolfram et al. 2020; Zheng et al. 2020; Sacchi 2021; Valente et al. 2021). The values in the figure (and the remaining figures in this section) depend on the 100-year global warming potential (GWP) used in each study, which may differ from the recent GWP updates from WGI. However, it is unlikely that the qualitative insights gained from the figures in this section would change using the update 100-year GWP values.

Furthermore, note that the carbon footprint of biofuels used in Figure 10.4 are aggregate numbers not specific to any individual value chain or fuel type. They are derived by combining land use-related carbon emissions from Chapter 7 with conversion efficiencies and emissions as described in Section 10.3. Specifically, land-use footprints derived from the three modelling approaches employed here are: i) Integrated Assessment Models – Energy Modelling Forum 33 (IAM EMF33); ii) Partial models assuming constant land cover (CLC), and, iii) Partial models using natural regrowth (NRG). The emissions factors used here correspond to scenarios where global production of biomass for energy purposes are 100 EJ/year, with lower emissions factors expected at lower levels of consumption and vice versa. Further details are available in Box 10.2 and Chapter 7.

The tailpipe emissions and fuel consumption reported in the literature generally do not use empirical emissions data. Rather, they tend to report fuel efficiency using driving cycles such as New European Driving Cycle or the US Environmental Protection Agency Federal Test Procedure. As a result, depending on the driving cycle used, operating emissions reported in literature are possibly underestimated by as much as 15–38%, in comparison to real driving emissions (Fontaras et al. 2017; Tsiakmakis et al. 2017; Triantafyllopoulos et al. 2019). The extent of these underestimations, however, varies between powertrain types, engine sizes, driving behaviour and environment.

Current average lifecycle impacts of mid-size ICEVs span from approximately 65 $gCO_2$-eq $pkm^{-1}$ to 210 $gCO_2$-eq $pkm^{-1}$, with both values stemming from ICEVs running on biofuels. Between this range of values, the current reference technologies are found, with diesel-powered ICEVs having total median lifecycle impacts of 130 $gCO_2$-eq $pkm^{-1}$ and gasoline-fuelled vehicle 160 $gCO_2$-eq $pkm^{-1}$. Fuel consumption dominates the lifecycle emissions of ICEVs, with approximately 75% of these emissions arising from the tailpipe and fuel chain.

HEVs and plug-in HEVs (PHEVs) vary in terms of degree of powertrain electrification. HEVs mainly rely on regenerative braking for charging the battery. PHEVs combine regenerative braking with external power sources for charging the battery. Operating emissions intensity is highly dependent on the degree to which electrified driving is performed, which in turn is user- and route-dependent. For PHEVs, emissions intensity is also dependent on the source of the electricity for charging. HEV and PHEV production impacts are comparable to the emissions generated for producing ICEVs as the batteries are generally small compared to those of BEVs. Current HEVs may reduce emissions compared to ICEVs by up to 30%, depending on the fuel, yielding median lifecycle intensities varying between 60 $gCO_2$-eq $pkm^{-1}$ (biofuels, EMF33) and 165–170 $gCO_2$-eq $pkm^{-1}$ (biofuels, partial models NRG). Within this wide range, all the combinations of electric and fossil-fuelled driving can be found, as well as the lifecycle intensity for driving 100% on fossil fuel. Because HEVs rely on combustion as the main energy conversion process, they offer limited mitigation opportunities. However, HEVs represent a suitable temporary solution, yielding a moderate mitigation potential, in areas where the electricity mix is currently so carbon intensive that the use of PHEVs and BEVs is not an effective mitigation solution (Wolfram and Wiedmann 2017; Wu et al. 2019).

In contrast to HEVs, PHEVs may provide greater opportunities for use-phase emissions reductions for LDVs. These increased potential benefits are due to the ability to charge the battery with low-carbon electricity and the longer full-electric range in comparison to HEVs (Laberteaux et al. 2019). Consumer behaviour (e.g., utility factor (UF) and charging patterns), manufacturer settings, and access to renewable electricity for charging strongly influence the total operational impacts (Wu et al. 2019). The UF is a weighting of the percentage of distance covered using the electric charge (charge depleting (CD) stage) versus the distance covered using the internal combustion engine (charge sustaining (CS) stage) (Paffumi et al. 2018). When the PHEV operates in CS mode, the internal combustion engine is used for propulsion and to maintain the state of charge of the battery within a certain range, together with regenerative braking (Plötz et al. 2018; Raghavan and Tal 2020). When running in CS mode, PHEVs have a reduced mitigation potential and have impacts comparable to those of HEVs. On the other hand, when the PHEV operates in CD mode, the battery alone provides the required

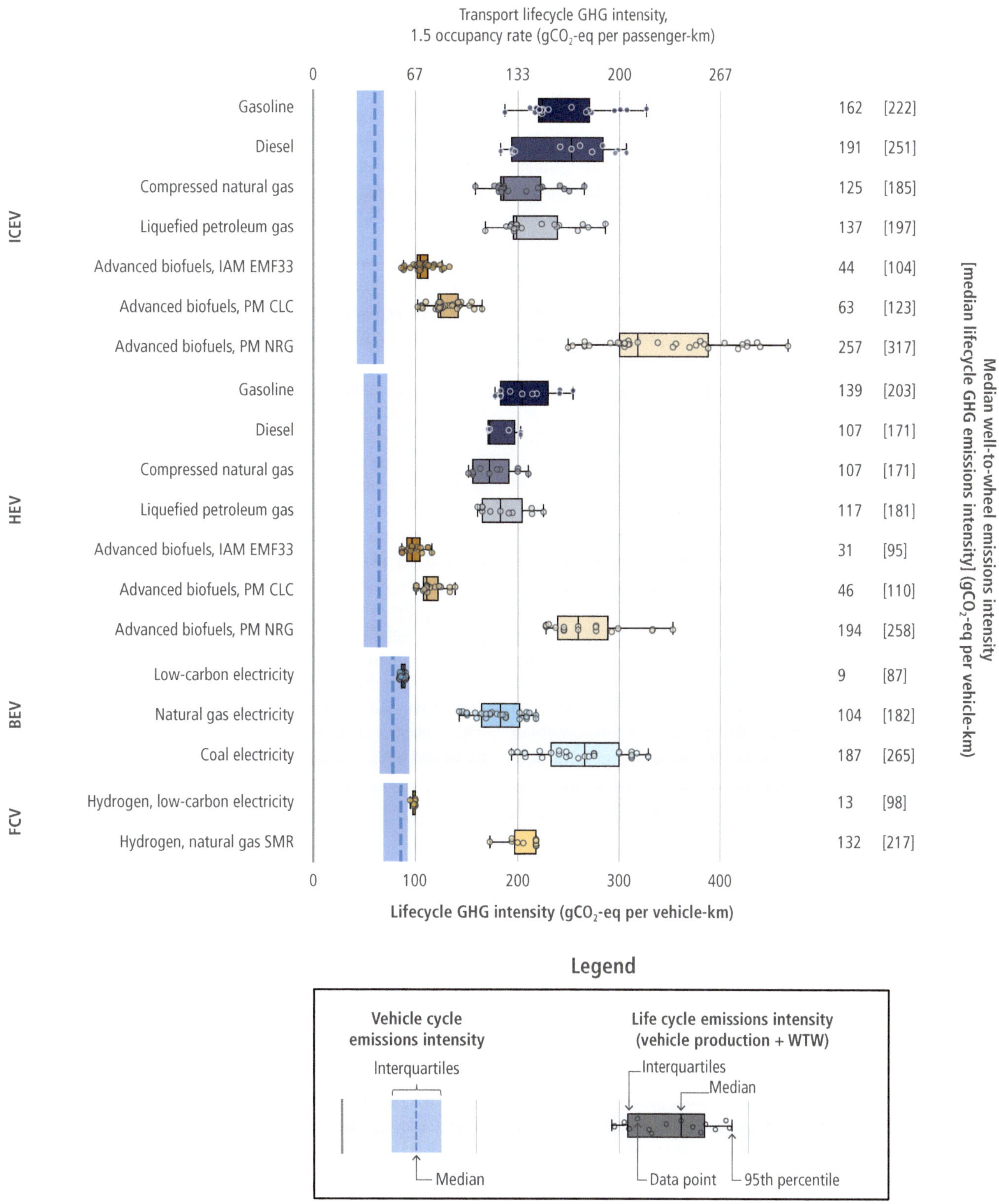


**Figure 10.4 | Life cycle greenhouse gas emissions intensities for mid-sized light-duty vehicle and fuel technologies from the literature.** The primary x-axis reports units in $gCO_2$-eq $vkm^{-1}$, assuming a vehicle life of 180,000 km. The secondary x-axis uses units of $gCO_2$-eq $pkm^{-1}$, assuming a 1.5 occupancy rate. The values in the figure rely on the 100-year GWP value embedded in the source data, which may differ slightly from the updated 100-year GWP values from WGI. The shaded area represents the interquartile range for combined vehicle manufacturing and end-of-life phases. The length of the box and whiskers represent the interquartile range of the operation phase for different fuel chains, while their placement on the x-axis represents the absolute lifecycle climate intensity, that is, includes manufacturing and end-of-life phases. Each individual marker indicates a data point. 'Advanced biofuels' refers to the use of second-generation biofuels and their respective conversion and cultivation emission factors. 'IAM EMF33' refers to emissions factors for advanced biofuels derived from simulation results from the integrated assessment models EMF33 scenarios. 'PM' refers to partial models, where 'CLC' is with constant land cover and 'NRG' is with natural regrowth. 'Hydrogen, low-carbon electricity' is produced via electrolysis using low-carbon electricity. 'Hydrogen, natural gas SMR' refers to fuels produced via steam methane reforming of natural gas.

propulsion energy (Plötz et al. 2018; Raghavan and Tal 2020). Thus, in CD mode, PHEVs hold potential for higher mitigation potential, due to the possibility of charging the battery with low-carbon electricity sources. Consequently, the UF greatly influences the lifecycle emissions of PHEVs. The current peer-reviewed literature presents a wide range of UFs mainly due to varying testing protocols applied for estimating the fuel efficiency and user behaviour (Pavlovic et al. 2017; Paffumi et al. 2018; Plötz et al. 2018; Plötz et al. 2020; Raghavan and Tal 2020; Hao et al. 2021). These factors make it difficult to harmonise and compare impacts across PHEV studies. Due to the low number of appropriate PHEV studies relative to the other LDV technologies and the complications in harmonising available PHEV results, this technology is omitted from Figure 10.4. However, due to the dual operating nature of PHEV vehicles, one can expect that the lifecycle GHG emissions intensities for these vehicles will lie between those of their ICEV and BEV counterparts of similar size and performance.

Currently, BEVs have higher manufacturing emissions than equivalently-sized ICEVs, with median emissions of 14 $tCO_2$-eq per vehicle against approximately 10 $tCO_2$-eq per vehicle of their mid-sized fossil-fuelled counterparts. These higher production emissions of BEVs are largely attributed to the battery pack manufacturing and to the additional power electronics required. As manufacturing technology and capacity utilisation improve and globalise to regions with low-carbon electricity, battery manufacturing emissions will likely decrease. Due to the higher energy efficiency of the electric powertrain, BEVs may compensate for these higher production emissions in the driving phase. However, the mitigation ability of this technology relative to ICEVs is highly dependent on the electricity mix used to charge the vehicle. As a consequence of the variety of energy sources available today, current BEVs have a wide range of potential average lifecycle impacts, ranging between 60 and 180 $gCO_2$-eq $pkm^{-1}$ with electricity generated from wind and coal, respectively. The ability to achieve large carbon reductions via vehicle electrification is thus highly dependent on the generation of low-carbon electricity, with the greatest mitigation effects achieved when charging the battery with low-carbon electricity. The literature suggests that current BEVs, if manufactured on low-carbon electricity as well as operated on low-carbon electricity would have footprints as low 22 $gCO_2$-eq $pkm^{-1}$ for a compact-sized car (Ellingsen et al. 2014; Ellingsen et al. 2016). This value suggests a reduction potential of around 85% compared to similarly-sized fossil fuel vehicles (median values). Furthermore, BEVs have a co-benefit of reducing local air pollutants that are responsible for human health complications, particularly in densely-populated areas (Hawkins et al. 2013; Ke et al. 2017).

As with BEVs, current HFCVs have higher production emissions than similarly-sized ICEVs and BEVs, generating on average approximately 15 $tCO_2$-eq per vehicle. As with BEVs, the lifecycle impacts of FCVs are highly dependent on the fuel chain. To date, the most common method of hydrogen production is steam methane reforming of natural gas (Khojasteh Salkuyeh et al. 2017), which is relatively carbon intensive, resulting in lifecycle emissions of approximately 88 $gCO_2$-eq $pkm^{-1}$. Current literature covering lifecycle impacts of FCVs shows that vehicles fuelled with hydrogen produced from steam methane reforming of natural gas offer little or no mitigation potential over ICEVs. Other available hydrogen fuel chains vary widely in carbon intensity, depending on the synthesis method and the energy source used (electrolysis or steam methane reforming; fossil fuels or renewables). The least carbon-intensive hydrogen pathways rely on electrolysis powered by low-carbon electricity. Compared to ICEVs and BEVs, FCVs for LDVs are at a lower technology readiness level, as discussed in section 10.3.

## Box 10.3 | Vehicle Size Trends and Implications on the Fuel Efficiency of LDVs

**Vehicle size trends.** On a global scale, SUV sales have been constantly growing in the last decade, with 39% of the vehicles sold in 2018 being SUVs (IEA 2019d). If the trend towards increasing vehicle size and engine power continues, it may result in higher overall emissions from the LDV fleet (relative to smaller vehicles with the same powertrain technology). The magnitude of the influence vehicle mass has on fuel efficiency varies with the powertrain, which have different efficiencies. Box 10.3 Figure 1 highlights this relationship using data from the same literature used to create Figure 10.4. Higher powertrain efficiency results in lower energy losses in operation, and thus requires less energy input to move a given mass than a powertrain of lower efficiency. This pattern is illustrated by the more gradual slope of BEVs in Box 10.3 Figure 1. The trend towards bigger and heavier vehicles, with consequently higher use phase emissions, can be somewhat offset by improvements in powertrain design, fuel efficiency, lightweighting, and aerodynamics (Gargoloff et al. 2018; Wolfram et al. 2020). The potential improvements provided by these strategies are case specific and not thoroughly evaluated in the literature, either individually or as a combination of multiple strategies.

**Lightweighting.** There is an increasing use of advanced materials such as high-strength steel, aluminium, carbon fibre, and polymer composites for vehicle lightweighting (Hottle et al. 2017). These materials reduce the mass of the vehicle and thereby also reduce the fuel or energy required to drive. Lightweighted components often have higher production emissions than the components they replace due to the advanced materials used (Kim and Wallington 2016). Despite these higher production emissions, some studies suggest that the reduced fuel consumption over the lifetime of the lightweighted vehicle may provide a net mitigation effect in comparison to a non-lightweighted vehicle (Kim and Wallington 2013; Hottle et al. 2017; Milovanoff et al. 2019; Upadhyayula et al. 2019; Wolfram et al. 2020). However, multiple recent publications have found that in some cases, depending on, for example, vehicle size and carbon intensity of the lightweighting materials employed, the GHG emissions avoided due to improved fuel efficiency do not

*Box 10.3 (continued)*

offset the higher manufacturing emissions of the vehicle (Luk et al. 2018; Wu et al. 2019). In addition, these advanced materials may be challenging to recycle in a way that retains their high technical performance (Meng et al. 2017).

**Co-effects on particulate matter.** Lightweighting may also alleviate the particulate matter (PM) emissions arising from road and brake wear. BEVs are generally heavier than their ICEV counterparts, which may potentially cause higher stress on road surfaces and tyres, with consequently higher PM emissions per kilometre driven (Timmers and Achten 2016). Regenerative braking in HEVs, BEVs and FCVs, however, reduces the mechanical braking required, and therefore may compensate for the higher brake wear emissions from these heavier vehicle types. In addition, BEVs have no tailpipe emissions, which further offsets the increased PM emissions from road and tyre wear. Therefore, lightweighting strategies may offer a carbon and particulates mitigation effect; however, in some cases, other technological options may reduce $CO_2$ emissions even further.

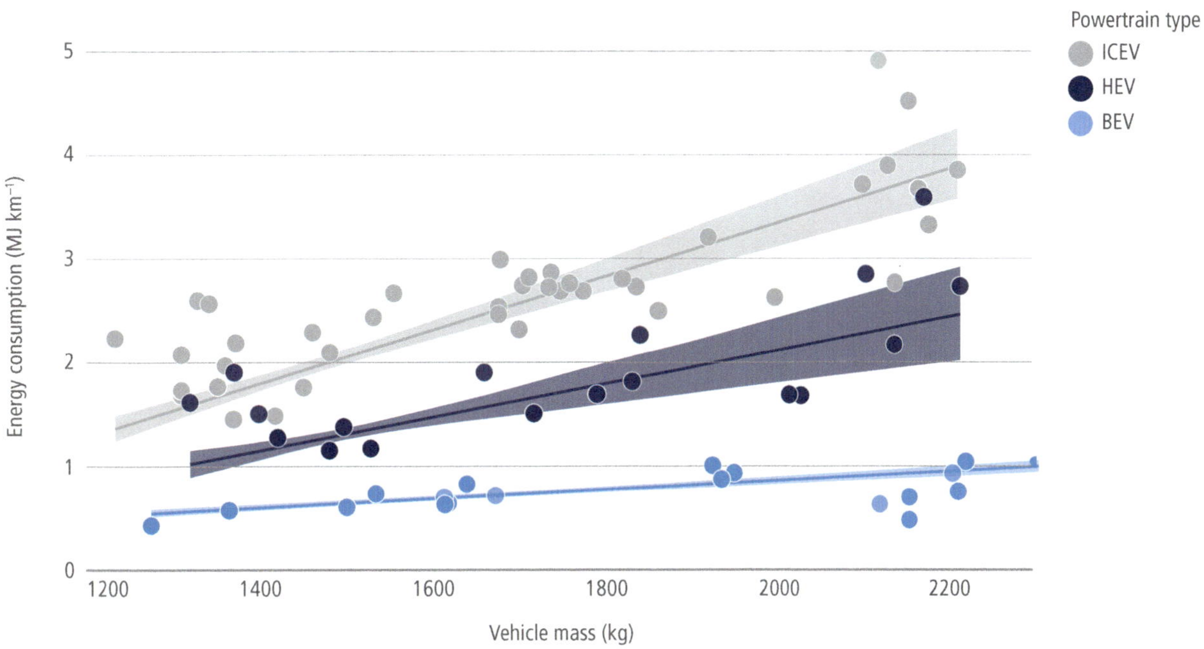


**Box 10.3, Figure 1 | Illustration of energy consumption as a function of vehicle size (using mass as a proxy) and powertrain technology.** FCVs omitted due to lacking data.

Two-wheelers, consisting mainly of lower-powered mopeds and higher-powered motorcycles, are popular for personal transport in densely populated cities, especially in developing countries. LCA studies for this class of vehicle are relatively uncommon compared to four-wheeled LDVs. In the available results, however, two-wheelers exhibit similar trends for the different powertrain technologies as the LDVs, with electric powertrains having higher production emissions, but usually lower operating emissions. The lifecycle emissions intensity for two-wheelers is also generally lower than four-wheeled LDVs on a vehicle-kilometre basis. However, two-wheelers generally cannot carry as many passengers as four-wheeled LDVs. Thus, on a passenger-kilometre basis, a fully occupied passenger vehicle may still have lower emissions than a fully occupied two-wheeler. However, today, most passenger vehicles have relatively low occupancy and thus have a correspondingly high emissions intensity on a pkm basis. This points to the importance of utilisation of passenger vehicles at higher occupancies to reduce the lifecycle intensity of LDVs on a pkm basis. For example, the median emissions intensity of a gasoline passenger vehicle is 222 $gCO_2$-eq $vkm^{-1}$, and 160 $gCO_2$-eq $vkm^{-1}$ for a gasoline two-wheeler (Cox and Mutel 2018). At a maximum occupancy factor of four and two passengers, respectively, the transport emissions intensity for these vehicles is 55 and 80 $gCO_2$-eq $pkm^{-1}$. Under the same occupancy rates assumption, BEV two-wheelers recharged on the average European electricity mix, achieve lower lifecycle GHG intensities than BEV four-wheeled LDVs. On the other hand, FCV two-wheelers with hydrogen produced via steam methane reforming present higher GHG intensity than their four-wheeled counterparts, when compared on a pkm basis at high occupancy rates.

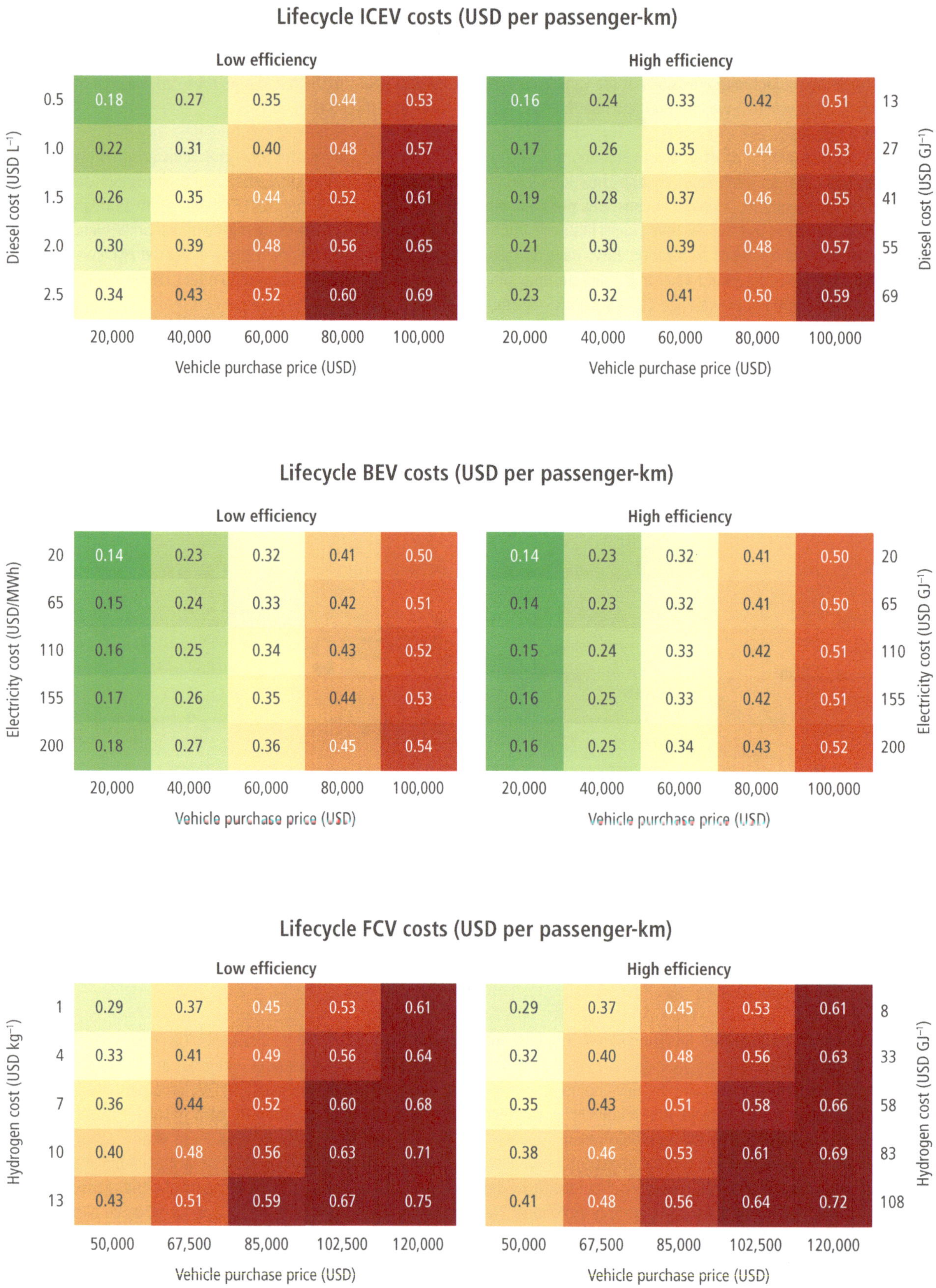


**Figure 10.5 | LCC for light-duty internal combustion engine vehicles, battery electric vehicles, and hydrogen fuel cell vehicles.** The results for ICEVs represent the LCC of a vehicle running on gasoline. However, these values are also representative for ICEVs running on diesel as the costs ranges in the literature for these two solutions are similar. The secondary y-axis depicts the cost of the different energy carriers normalised in USD per gigajoule for easier cross-comparability.

ICEV, HEV, and PHEV technologies, which are powered using combustion engines, have limited potential for deep reduction of GHG emissions. Biofuels offer good mitigation potential if low land-use change emissions are incurred (e.g., the IAM EMF33 and partial models, CLC biofuels pathways shown in Figure 10.4). The literature shows large variability, depending on the method of calculating associated land-use changes. Resolving these apparent methodological differences is important to consolidating the role biofuels may play in mitigation, as well as the issues raised in Chapter 7 about the conflicts over land use. The mitigation potential of battery and fuel cell vehicles is strongly dependent on the carbon intensity of their production and the energy carriers used in operation. However, these technologies likely offer the highest potential for reducing emissions from LDVs. Prior work on the diffusion dynamics of transport technologies suggests that 'the diffusion of infrastructure precedes the adoption of vehicles, which precedes the expansion of travel' (Leibowicz 2018). These dynamics reinforce the argument for strong investments in both the energy infrastructure and the vehicle technologies.

To successfully transition towards LDVs utilising low-carbon fuels or energy sources, the technologies need to be accessible to as many people as possible, which requires competitive costs compared to conventional diesel and gasoline vehicles. The lifecycle costs (LCCs) of LDVs depend on the purchasing costs of the vehicles, their efficiency, the fuel costs, and the discount rate. Figure 10.5 shows the results of a parametric analysis of LCC for diesel LDVs, BEVs, and FCVs. The range of vehicle efficiencies captured in Figure 10.5 is the same as the range used for Figure 10.4, while the ranges for fuel costs and vehicle purchase prices come from the literature. The assumed discount rate for this parametric analysis is 3%. Appendix 10.2 includes the details about the method and underlying data used to create this figure.

Figure 10.5 shows the range of LCC, in USD per passenger-kilometre, for different powertrain technologies, and the influence of vehicle efficiency (low or high), vehicle purchase price, and fuel/electricity cost on the overall LCC. For consistency with Figure 10.4, an occupancy rate of 1.5 is assumed. Mid-sized ICEVs have a purchase price of USD20,000–40,000, and average fuel costs are in the range of USD1–1.5 per litre. With these conditions, the LCC of fossil-fuelled LDVs span between USD0.22–0.35 $pkm^{-1}$ or between USD0.17–0.28 $pkm^{-1}$, for low- and high-efficiency ICEVs respectively (Figure 10.5).

BEVs have higher purchase prices than ICEVs, though a sharp decline has been observed since AR5. Due to the rapid development of the lithium-ion battery technology over the years (Schmidt et al. 2017) and the introduction of subsidies in several countries, BEVs are quickly reaching cost parity with ICEVs. Mid-sized BEVs' average purchase prices are in the range of USD30,000–50,000 but the levelised cost of electricity shows a larger spread (USD65–200/MWh) depending on the geographical location and the technology (Chapter 6). Therefore, assuming purchase price parity between ICEVs and BEVs, BEVs show lower LCC (Figure 10.5) due to higher efficiency and the lower cost of electricity compared to fossil fuels on a per-gigajoule (GJ) basis (secondary y-axis on Figure 10.5).

FCVs represent the most expensive solution for LDV, mainly due to the currently higher purchase price of the vehicle itself. However, given the lower technology readiness level of FCVs and the current efforts in the research and development of this technology, FCVs could become a viable technology for LDVs in the coming years. The issues regarding the extra energy involved in creating the hydrogen and its delivery to refuelling sites remain, however. The levelised cost of hydrogen on a per GJ basis is lower than conventional fossil fuels but higher than electricity. In addition, within the levelised cost of hydrogen, there are significant cost differences between the hydrogen-producing technologies. Conventional technologies such as coal gasification and steam methane reforming of natural gas, both with and without carbon capture and storage, represent the cheapest options (Bekel and Pauliuk 2019; Parkinson et al. 2019; Khzouz et al. 2020; Al-Qahtani et al. 2021). Hydrogen produced via electrolysis is currently the most expensive technology, but with significant potential cost reductions due to the current technology readiness level.

### 10.4.2 Transit Technologies for Passenger Transport

Buses provide urban and peri-urban transport services to millions of people around the world and a growing number of transport agencies are exploring alternative-fuelled buses. Alternative technologies to conventional diesel-powered buses include buses powered with CNG, LNG, synthetic fuels, and biofuels (e.g., biodiesel, renewable diesel, dimethyl ether); diesel hybrid-electric buses; battery electric buses; electric catenary buses; and hydrogen fuel cell buses. Rail is an alternative mode of transit that could support decarbonisation of land-based passenger mobility. Electric rail systems can provide urban services (light rail and metro systems), as well as longer-distance transport. Indeed, many cities of the world already have extensive metro systems, and regions like China, Japan and Europe have a robust high-speed intercity railway network. Intercity rail transport can be powered with electricity, however, fossil fuels are still prevalent for long-distance rail passenger transport in some regions. Battery electric long-distance trains may be a future option for these areas.

Figure 10.6 shows the lifecycle GHG emissions from different powertrain and fuel technologies for buses and passenger rail. The data in each panel came from a number of relevant scientific studies (Cai et al. 2015; Tong et al. 2015a; Dimoula et al. 2016; de Bortoli et al. 2017; Valente et al. 2017; Meynerts et al. 2018; IEA 2019e; de Bortoli and Christoforou 2020; Hill et al. 2020; Liu et al. 2020a; Valente et al. 2021). The width of the bar represents the variability in available estimates, which is primarily driven by variability in reported vehicle efficiency, size, or drive cycle. While some bars overlap, the Figure may not fully capture correlations between results. For example, low efficiency associated with aggressive drive cycles may drive the upper end of the emission ranges for multiple technologies; thus, an overlap does not necessarily suggest uncertainty regarding which vehicle type would have lower emissions for a comparable trip. Additionally, reported lifecycle emissions do not include embodied GHG emissions associated with infrastructure construction and maintenance. These embodied emissions are potentially a larger fraction of lifecycle emissions for rail than for other transport modes (Chester and Horvath 2012; Chester et al. 2013). One study reported values ranging from 10–25 $gCO_2$ per passenger-kilometre (International Union of Railways 2016), although embodied emissions from rail are known to vary widely across case studies (Olugbenga et al. 2019). These caveats are also applicable to the other figures in this section.

Figure 10.6 highlights that BEV and FCV buses and passenger rail powered with low-carbon electricity or low-carbon hydrogen, could offer reductions in GHG emissions compared to diesel-powered buses or diesel-powered passenger rail. However, and not surprisingly, these technologies would offer only little emissions reductions if power generation and hydrogen production rely on fossil fuels. While buses powered with CNG and LNG could offer some reductions compared to diesel-powered buses, these reductions are unlikely to be sufficient to contribute to deep decarbonisation of the transport sector and they may slow down conversion to low- or zero-carbon options already commercially available. Biodiesel and renewable diesel fuels (from sources with low upstream emissions and low risk of induced land-use change) could offer important near-term reductions for buses and passenger rail, as these fuels can often be used with existing vehicle infrastructure. They could also be used for long haul trucks and trains, shipping and aviation as discussed below and in later sections.

There has been growing interest in the production of synthetic fuels from $CO_2$ produced by direct air capture (DAC) processes. Figure 10.6 includes the lifecycle GHG emissions from buses and passenger rail powered with synthetic diesel produced through a DAC system paired with a Fischer-Tropsch (FT) process, based on Liu et al. (2020a). This process requires the use of hydrogen (as shown in Figure 10.2), so the emissions factors of the resulting fuel depend on the emissions intensity of hydrogen production. An electricity emissions factor less than 140 $gCO_2$-eq $kWh^{-1}$ would be required for this pathway to achieve lower emissions than petroleum diesel (Liu et al. 2020a); for example, this would be equivalent to a 75% wind and 25% natural gas electricity mix (Appendix 10.1). If the process relied on steam methane reforming for hydrogen production or fossil-based power generation, synthetic diesel from the DAC-FT process would not provide GHG emissions reductions compared to conventional diesel. DAC-FT from low-carbon energy sources appears to be promising from an emissions standpoint and could warrant the R&D and demonstration attention outlined in the rest of the chapter, but it cannot be contemplated as a decarbonisation strategy without the availability of low-carbon hydrogen.

At high occupancy, both bus and rail transport offer substantial GHG reduction potential per pkm, even compared with the lowest-emitting private vehicle options. Even at 20% occupancy, bus and rail may still offer emission reductions compared to passenger cars, especially notable when comparing BEVs with low-carbon electricity (the lowest-emission option for all technologies) across the three modes. Only when comparing a fossil fuel-powered bus at low occupancy with a low-carbon powered car at high occupancy is this conclusion reversed. Use of public transit systems, especially those that rely on buses and passenger rail fuelled with the low-carbon fuels previously described, would thus support efforts to decarbonise the transport sector. Use of these public transit systems will depend on urban design and consumer preferences (Section 10.2, Chapters 5 and 8), which in turn depend on time, costs, and behavioural choices.

Figure 10.7 shows the results of a parametric analysis of the LCCs of transit technologies with the highest potential for GHG emissions reductions. As with Figure 10.5, the vehicle efficiency ranges are the same as those from the LCA estimates (80% occupancy). Vehicle, fuel, and maintenance costs represent ranges in the literature (Eudy and Post 2018b; IEA 2019e; Argonne National Laboratory 2020; BNEF 2020; Eudy and Post 2020; Hydrogen Council 2020; IEA 2020b; IEA 2020c; IRENA 2020; Johnson et al. 2020; Burnham et al. 2021; IEA 2021c; IEA 2021d; US Energy Information Administration 2021), and the discount rate is 3% where applicable. Appendix 10.2 provides the details behind these estimates. The panels for the ICEV can represent buses and passenger trains powered with any form of diesel, whether derived from petroleum, synthetic hydrocarbons, or biofuels. For reference, global average automotive diesel prices from 2015–2020 fluctuated around USD1 per litre, and the 2019 world average industrial electricity price was approximately USD100 per MWh (IEA 2021d). Retail hydrogen prices in excess of USD13 per kilogram have been observed (Eudy and Post 2018a; Argonne National Laboratory 2020; Burnham et al. 2021) though current production cost estimates for hydrogen produced from electrolysis are far lower (IRENA 2020) (and as reported in Chapter 6), at around USD5–7 per kg with future forecasts as low as USD1 per kg (BNEF 2020; Hydrogen Council 2020; IRENA 2020) (and as reported in Chapter 6).

Under most parameter combinations, rail is the most cost-effective option, followed by buses, both of which are an order of magnitude cheaper than passenger vehicles. Note that costs per pkm are strongly influenced by occupancy assumptions; at low occupancy (e.g., <20% for buses and <10% for rail), the cost of transit approaches the LCC for passenger cars. For diesel rail and buses, cost ranges are driven by fuel costs, whereas vehicles are both important drivers for electric or hydrogen modes due to high costs (but also large projected improvements) associated with batteries and fuel cell stacks. Whereas the current state of ICEV technologies is best represented by cheap vehicles and low fuel costs for diesel (top left of each panel), these costs are likely to rise in future due to stronger emission/efficiency regulations and rising crude oil prices. On the contrary, the current status of alternative fuels is better represented by high capital costs and mid-to-high fuel costs (right side of each panel; mid-to-bottom rows), but technology costs are anticipated to fall with increasing experience, research, and development. Thus, while electric rail is already competitive with diesel rail, and electric buses are competitive with diesel buses in the low efficiency case, improvements are still required in battery costs to compete against modern diesel buses on high efficiency routes, at current diesel costs. Similarly, improvements to both vehicle cost and fuel costs are required for hydrogen vehicles to become cost effective compared to their diesel or electric counterparts. At either the upper end of the diesel cost range (bottom row of ICEV panels), or within the 2030–2050 projections for battery costs, fuel cell costs and hydrogen costs (top left of BEV and FCV panels), both battery- and hydrogen-powered vehicles become financially attractive.

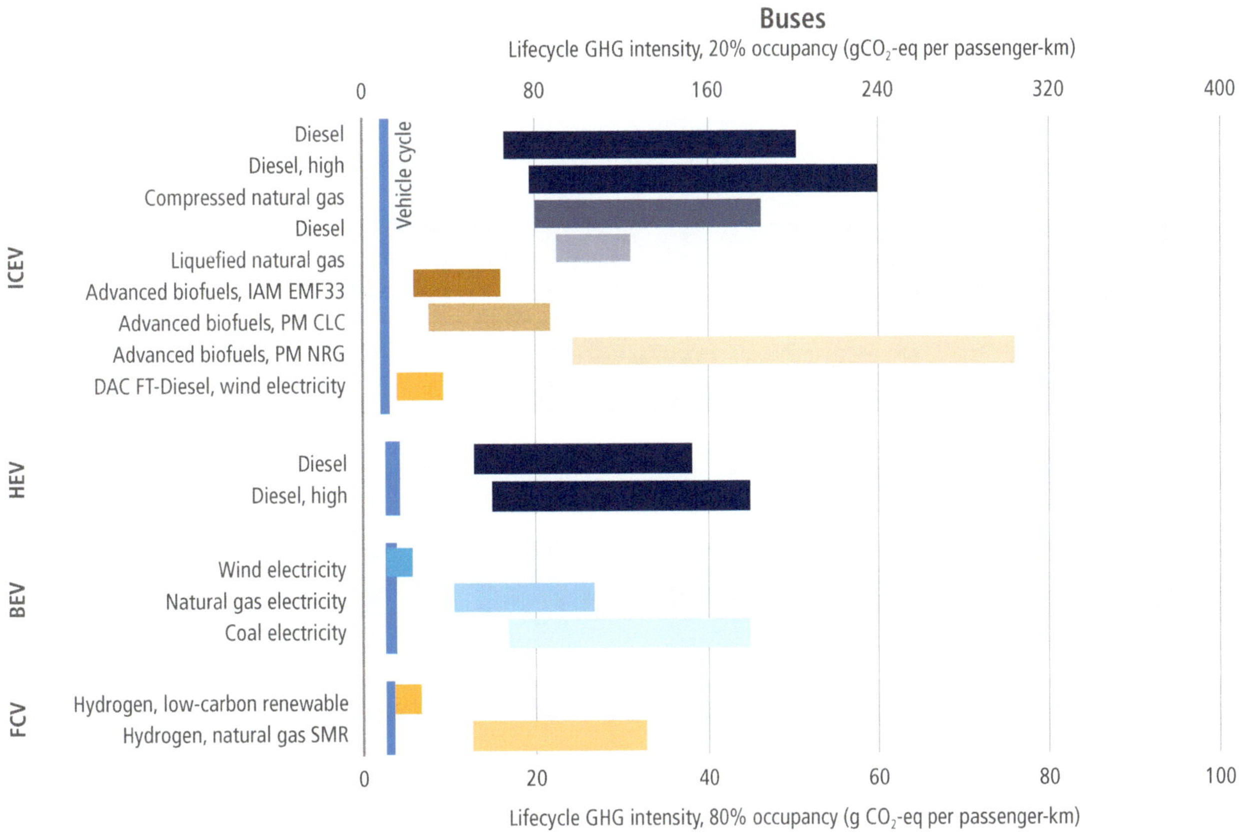


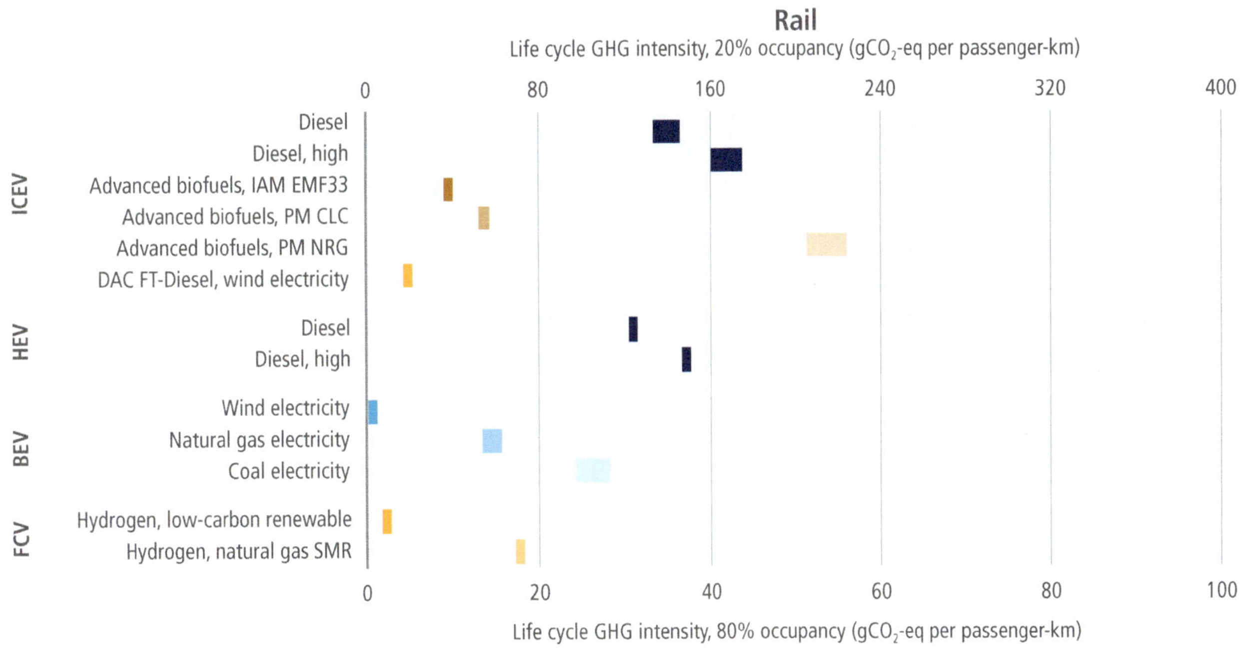


**Figure 10.6 | Lifecycle greenhouse gas intensity of land-based bus and rail technologies.** Each bar represents the range of the lifecycle estimates, bounded by minimum and maximum energy use per passenger-kilometre, as reported for each fuel/powertrain combination. The ranges are driven by differences in vehicle characteristics and operating efficiency. For energy sources with highly variable upstream emissions, low, medium and/or high representative values are shown as separate rows. The primary x-axis shows lifecycle GHG emissions, in $gCO_2$-eq $pkm^{-1}$, assuming 80% occupancy; the secondary x-axis assumes 20% occupancy. The values in the figure rely on the 100-year GWP value embedded in the source data, which may differ slightly from the updated 100-year GWP values from WGI. For buses, the main bars show full lifecycle, with vertical bars disaggregating the vehicle cycle. 'Diesel, high' references emissions factors for diesel from oil sands. 'advanced biofuels', refers to the use of second-generation biofuels and their respective conversion and cultivation emissions factors. 'IAM EMF33' refers to emissions factors for advanced biofuels derived from simulation results from the integrated assessment models EMF33 scenarios. 'PM' refers to partial models, where 'CLC' is with constant land cover and 'NRG' is with natural regrowth. 'DAC FT-Diesel, wind electricity' refers to Fischer-Tropsch diesel produced via a $CO_2$ direct air capture process that uses wind electricity. 'Hydrogen, low-carbon renewable' refers to fuels produced via electrolysis using low-carbon electricity. 'Hydrogen, natural gas SMR' refers to fuels produced via steam methane reforming of natural gas. Results for ICEVs with 'high emissions DAC FT-Diesel from natural gas' are not included here since the lifecycle emissions are estimated to be substantially higher than petroleum diesel ICEVs.

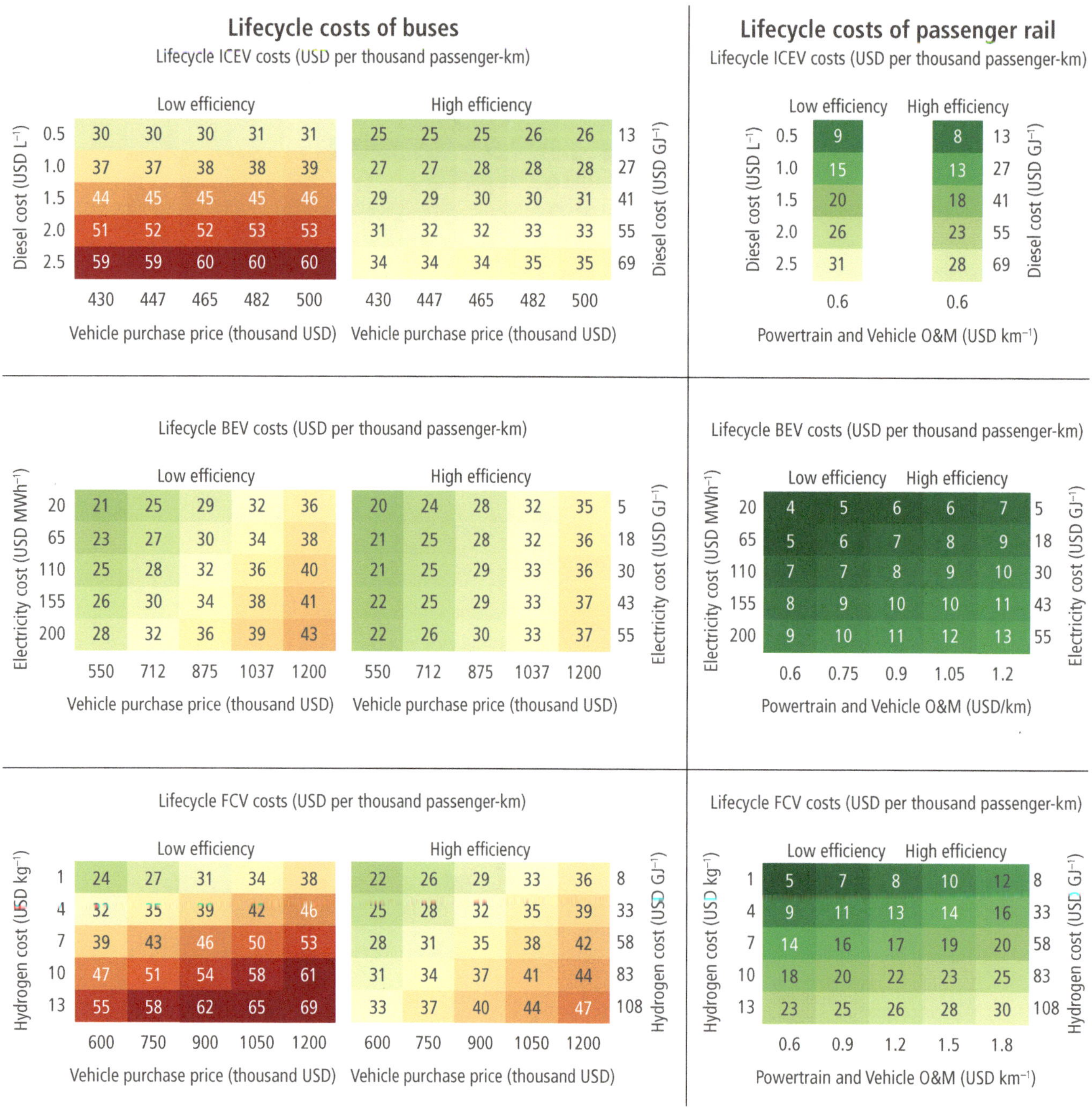


**Figure 10.7 | Lifecycle costs for internal combustion engine vehicles, battery electric vehicles, and hydrogen fuel cell vehicles for buses and passenger rail.** The range of efficiencies for each vehicle type are consistent with the range of efficiencies in Figure 10.6 (80% occupancy). The results for the ICEV can be used to evaluate the lifecycle costs of ICE buses and passenger rail operated with any form of diesel, whether from petroleum, synthetic hydrocarbons, or biofuel, as the range of efficiencies of vehicles operating with all these fuels is similar. The secondary y-axis depicts the cost of the different energy carriers normalised in USD/GJ for easier cross-comparability.

### 10.4.3 Land-based Freight Transport

As is the case with passenger transport, there is growing interest in alternative fuels that could reduce GHG emissions from freight transport. Natural gas-based fuels (e.g., CNG, LNG) are an example, however these may not lead to drastic reductions in GHG emissions compared to diesel. Natural gas-powered vehicles have been discussed as a means to mitigate air quality impacts (Khan et al. 2015; Cai et al. 2017; Pan et al. 2020), but those impacts are not the focus of this review. Decarbonisation of medium- and heavy-duty trucks would likely require the use of low-carbon electricity in battery electric trucks, low-carbon hydrogen or ammonia in fuel-cell trucks, or bio-based fuels (from sources with low upstream emissions and low risk of induced land-use change) used in ICE trucks.

Freight rail is also a major mode for the inland movement of goods. Trains are more energy efficient (per tkm) than trucks, so expanded use of rail systems (particularly in developing countries where

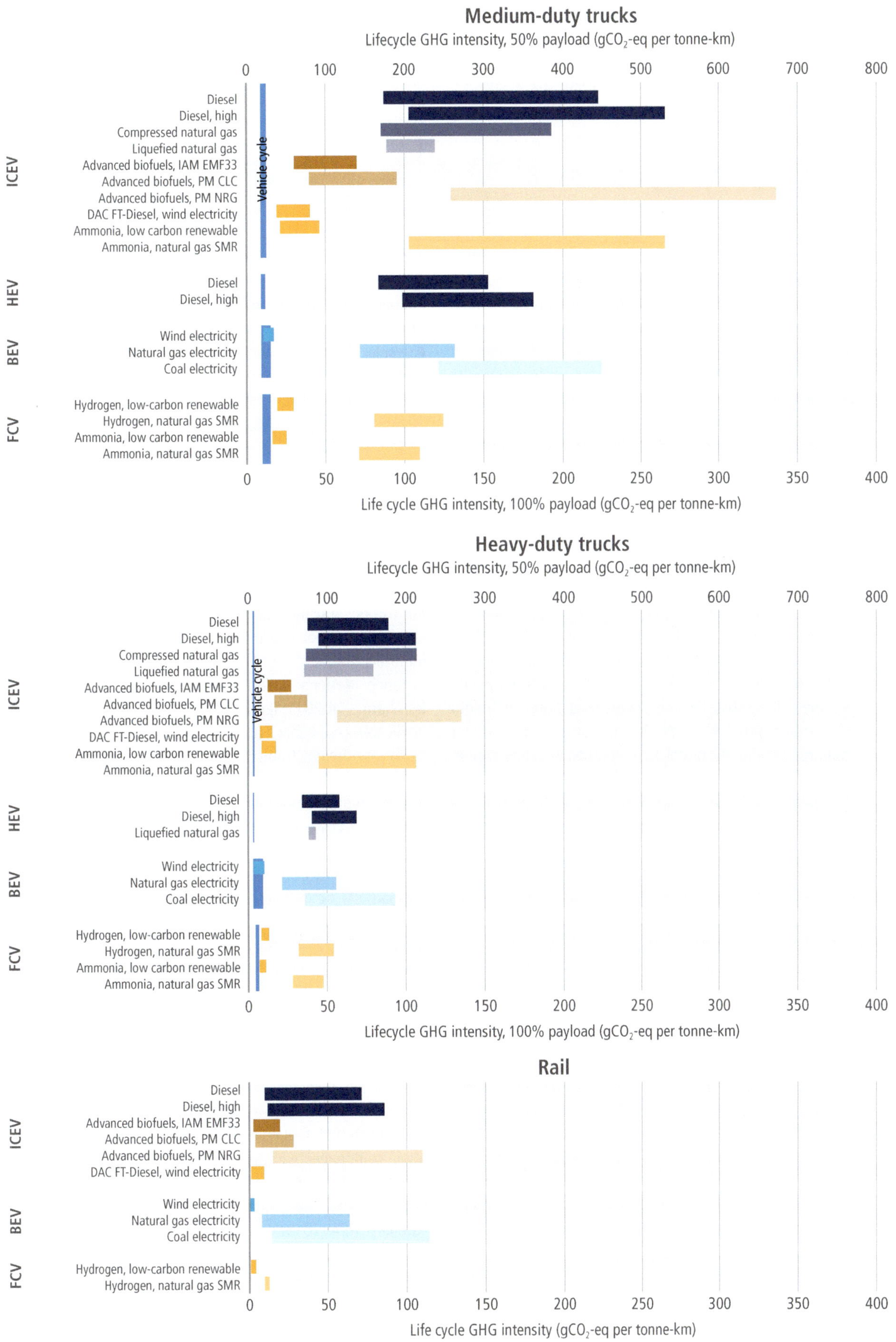


**Figure 10.8 | Lifecycle greenhouse gas intensity of land-based freight technologies and fuel types.**

**Figure 10.8 (continued): Lifecycle greenhouse gas intensity of land-based freight technologies and fuel types.** Each bar represents the range of the lifecycle estimates, bounded by minimum and maximum energy use per tkm, as reported for each fuel/powertrain combination. The ranges are driven by differences in vehicle characteristics and operating efficiency. For energy sources with highly variable upstream emissions, low, medium and/or high representative values are shown as separate rows. For trucks, the primary x-axis shows lifecycle GHG emissions, in $gCO_2$-eq $tkm^{-1}$, assuming 100% payload; the secondary x-axis assumes 50% payload. The values in the figure rely on the 100-year GWP value embedded in the source data, which may differ slightly from the updated 100-year GWP values from WGI. For rail, values represent average payloads. For trucks, main bars show full lifecycle, with vertical bars disaggregating the vehicle cycle. 'Diesel, high' references emissions factors for diesel from oil sands. 'Advanced biofuels' refers to the use of second-generation biofuels and their respective conversion and cultivation emission factors. 'IAM EMF33' refers to emissions factors for advanced biofuels derived from simulation results from the EMF33 scenarios. 'PM' refers to partial models, where 'CLC' is with constant land cover and 'NRG' is with natural regrowth. DAC FT-Diesel, wind electricity refers to Fischer-Tropsch diesel produced via a $CO_2$ direct air capture process that uses wind electricity. 'Ammonia and Hydrogen, low-carbon renewable' refers to fuels produced via electrolysis using low-carbon electricity. 'Ammonia and Hydrogen, natural gas SMR' refers to fuels produced via steam methane reforming of natural gas.

demand for goods could grow exponentially) could provide carbon abatement opportunities. While diesel-based locomotives are still a major mode of propulsion used in freight rail, interest in low-carbon propulsion technologies is growing. Electricity already powers freight rail in many European countries using overhead catenaries. Other low-carbon technologies for rail may include advanced storage technologies, biofuels, synthetic fuels, ammonia, or hydrogen.

Figure 10.8 presents a review of lifecycle GHG emissions from land-based freight technologies (heavy- and medium-duty trucks, and rail). Each panel within the figure represents data in GHG emissions per tonne-kilometre of freight transported by different technology and/or fuel types, as indicated by the labels to the left. The data in each panel came from a number of relevant scientific studies (Tong et al. 2015a; Frattini et al. 2016; Nahlik et al. 2016; Zhao et al. 2016; CE Delft 2017; Isaac and Fulton 2017; Song et al. 2017; Valente et al. 2017; Cooper and Balcombe 2019; Lajevardi et al. 2019; Hill et al. 2020; Liu et al. 2020a; Merchan et al. 2020; Prussi et al. 2020; Gray et al. 2021; Valente et al. 2021). Similar to the results for buses, technologies that offer substantial emissions reductions for freight include: ICEV trucks powered with the low-carbon variants for biofuels, ammonia or synthetic diesel; BEVs charged with low-carbon electricity; and FCVs powered with renewable-based electrolytic hydrogen, or ammonia. Since ammonia and Fischer-Tropsch diesel are produced from hydrogen, their emissions are higher than the source hydrogen, but their logistical advantages over hydrogen are also a consideration (Section 10.3).

Trucks exhibit economies of scale in fuel consumption, with heavy-duty trucks generally showing lower emissions per tkm than medium-duty trucks. Comparing the lifecycle GHG emissions from trucks and rail, it is clear that rail using internal combustion engines is more carbon efficient than using internal combustion trucks. Note that the rail emissions are reported for an average representative payload, while the trucks are presented at 50% and 100% payload, based on available data. The comparison between trucks and rail powered with electricity or hydrogen is less clear – especially considering that these values omit embodied GHG from infrastructure construction. One study reported embodied rail infrastructure emissions of 15 $gCO_2$ per tonne-kilometre for rail (International Union of Railways 2016), although such embodied emissions from rail are known to vary widely across case studies (Olugbenga et al. 2019). Regardless, trucks and rail with low-carbon electricity or low-carbon hydrogen have substantially lower emissions than incumbent technologies.

For trucks, Figure 10.8 includes two x-axes representing two different assumptions about their payload, which substantially influence emissions per tonne-kilometre. These results highlight the importance of truckload planning as an emissions reduction mechanism, for example, as also shown in Kaack et al. (2018). Several studies also point to improvements in vehicle efficiency as an important mechanism to reduce emissions from freight transport (Taptich et al. 2016; Kaack et al. 2018). However, projections for diesel vehicles using such efficiencies beyond 2030 are promising, but still far higher emitting than vehicles powered with low-carbon sources.

Figure 10.9 shows the results of a parametric analysis of the LCC of trucks and freight rail technologies with the highest potential for deep GHG reductions. As with Figure 10.8, the vehicle efficiency ranges are the same as those from the LCA estimates (80% payload for trucks; effective payload as reported by original studies for rail). Vehicle, fuel and maintenance costs represent ranges in the literature (Moultak et al. 2017; Eudy and Post 2018b; IEA 2019e; Argonne National Laboratory 2020; BNEF 2020; IRENA 2020; Burnham et al. 2021; IEA 2021c), and the discount rate is 3% where applicable (Appendix 10.2). The panels for the ICEV can represent trucks and freight trains powered with any form of diesel, whether derived from petroleum, synthetic hydrocarbons, or biofuels. See discussion preceding Figure 10.7 for additional details about current global fuel costs. Under most parameter combinations, rail is the more cost-effective option, but the high efficiency case for trucks (representing fuel-efficient vehicles, favourable drive cycles and high payload) can be more cost-effective than the low efficiency case for rail (representing systems with higher fuel consumption and lower payload). For BEV trucks, cost ranges are driven by vehicle purchase price due to the large batteries required and the associated wide range between their current high costs and anticipated future cost reductions. For all other truck and rail technologies, fuel cost ranges play a larger role. Similar to transit technologies, the current state of freight ICEV technologies is best represented by cheap vehicles and low fuel costs for diesel (top left of each panel), and the current status of alternative fuels is better represented by high capital costs and mid-to-high fuel costs (right side of each panel; mid-to-bottom rows), with expected future increases in ICEV LCC and decreases in alternative fuel vehicle LCC. Electric and hydrogen freight rail are potentially already competitive with diesel rail (especially electric catenary (IEA 2019e)), but low data availability (especially for hydrogen efficiency ranges) and wide ranges for reported diesel rail efficiency (likely encompassing low capacity utilisation) makes this comparison challenging. Alternative fuel trucks are currently more expensive than diesel trucks, but future increases in diesel costs or a respective decrease in hydrogen costs or in BEV capital

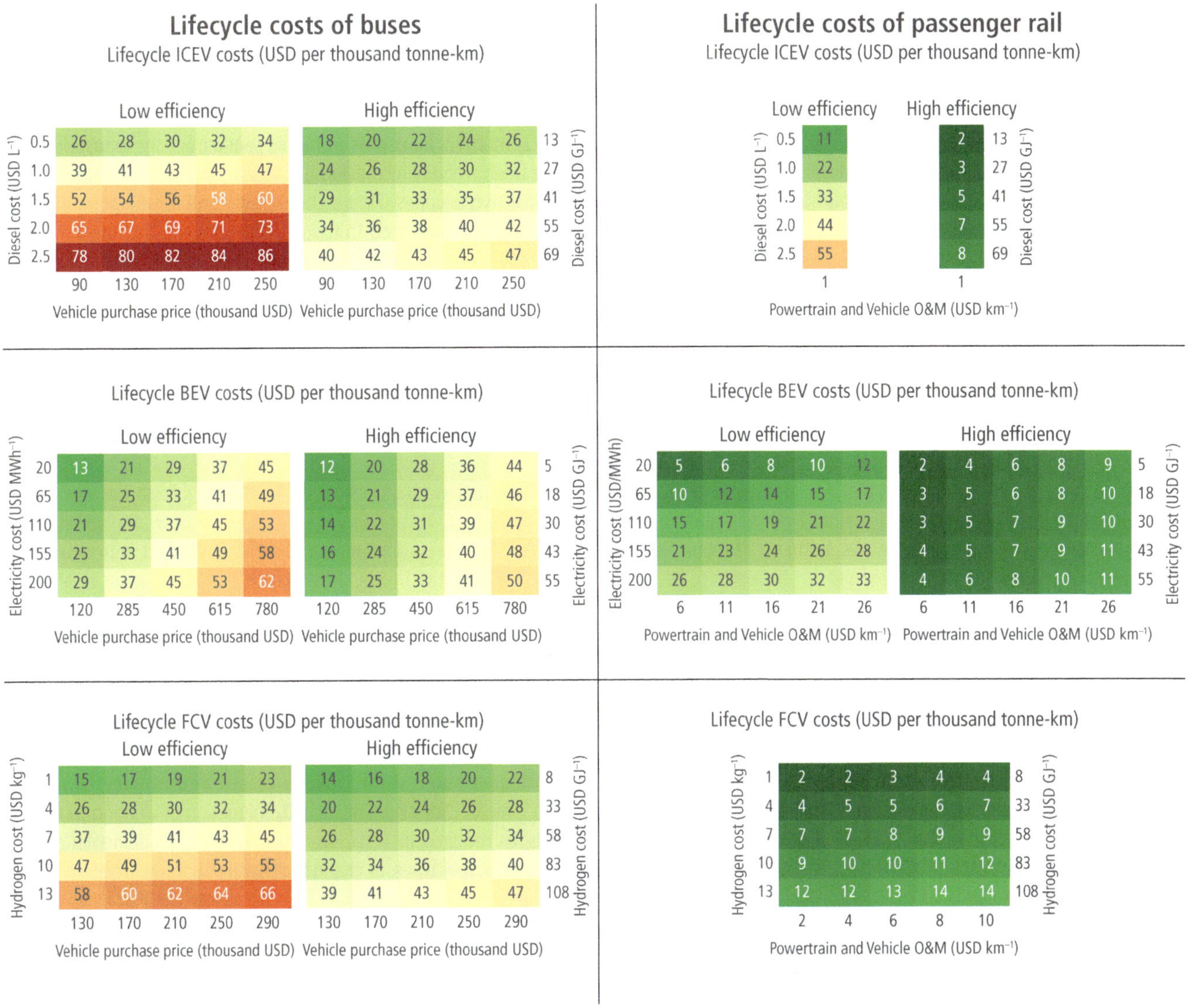


**Figure 10.9 | Life cycle costs for internal combustion engine vehicles, battery electric vehicles, and hydrogen fuel cell vehicles for heavy-duty trucks and freight rail.** The range of efficiencies for each vehicle type are consistent with the range of efficiencies in Figure 10.8. The results for ICEV can be used to evaluate the lifecycle costs of ICE trucks and freight rail operated with any form of diesel, whether from petroleum, synthetic hydrocarbons, or biofuels, as the range of efficiencies of vehicles operating with all these fuels is similar. The secondary y-axis depicts the cost of the different energy carriers normalised in USD per GJ for easier cross-comparability.

costs (especially the battery) would enable either alternative fuel technology to become financially attractive. These results are largely consistent with raw results reported in existing literature, which suggest ambiguity over whether BEV trucks are already competitive, but more consistency that hydrogen is not yet competitive, but could be in future (Zhao et al. 2016; Moultak et al. 2017; Sen et al. 2017; White and Sintov 2017; Zhou et al. 2017; Mareev et al. 2018; Yang et al. 2018a; El Hannach et al. 2019; Lajevardi et al. 2019; Tanco et al. 2019; Burke and Sinha 2020; Jones et al. 2020). There is limited data available on the LCC for freight rail, but at least one study IEA (2019g) suggests that electric catenary rail is likely to have similar costs to diesel rail, while battery electric trains remain more expensive and hydrogen rail could become cheaper under forward-looking cost reduction scenarios.

### 10.4.4 Abatement Costs

Taken together, the results in this section suggest a range of cost-effective opportunities to reduce GHG emissions from land-based transport. Mode shift from cars to passenger transit (bus or rail) can reduce GHG emissions while also reducing LCCs, resulting in a negative abatement cost. Likewise, increasing the utilisation of vehicles (i.e., % occupancy for passenger vehicles or % payload for freight vehicles) simultaneously decreases emissions and costs per pkm or per tkm, respectively. Within a given mode, alternative fuel sources also show strong potential to reduce emissions at minimal added costs. For LDVs, BEVs can offer emissions reductions with LCCs that are already approaching that for conventional ICEVs. For transit and freight, near-term abatement costs for the low-carbon BEV and FCV options relative to their diesel counterparts range from near USD0/tonne $CO_2$-eq (e.g., BEV buses and BEV passenger rail) into the hundreds or even low thousands of dollars per tonne $CO_2$-eq (e.g., for heavy-duty BEV and FCV trucks at current vehicle and fuel costs). With projected future declines in storage, fuel cell, and low-carbon hydrogen fuel costs, however, both BEV and FCV technologies can likewise offer GHG reductions at negative abatement costs across all land-transport modes in 2030 and beyond. Further information about costs and potentials is available in Chapter 12.

## 10.5 Decarbonisation of Aviation

This section addresses the potential for reducing GHG emissions from aviation. The overriding constraint on developments in technology and energy efficiency for this sector is safety. Governance is complex in that international aviation comes under the International Civil Aviation Organization (ICAO), a specialised UN agency. The measures to reduce GHG emissions that are considered include both in-sector (technology, operations, fuels) and out of sector (market-based measures, high-speed rail modal shift/substitution). Demand management is not explicitly considered in this section, as it was discussed in 10.2. A limited range of scenarios to 2050 and beyond are available and assessed at the end of the section.

10

### 10.5.1 Historical and Current Emissions from Aviation

Aviation is widely recognised as a 'hard-to-decarbonise' sector (Gota et al. 2019) having a strong dependency on liquid fossil fuels and an infrastructure that has long 'lock-in' timescales, resulting in slow fleet turnover times. The principal GHG emitted is $CO_2$ from the combustion of fossil fuel aviation kerosene ('Jet-A'), although its non-$CO_2$ emissions can also affect climate (Section 10.5.2). International emissions of $CO_2$ are about 65% of the total emissions from aviation (Fleming and de Lépinay 2019), which totalled approximately 1 Gt of $CO_2$ in 2018. Emissions from this segment of the transport sector have been steadily increasing at rates of around 2.5% per year over the last two decades (Figure 10.10), although for the period 2010 to 2018 the rate increased to roughly 4% per year. The latest available data (2018) indicate that aviation is responsible for approximately 2.4% of total anthropogenic emissions of $CO_2$ (including land-use change) on an annual basis (using IEA data, IATA data and global emissions data of Le Quéré et al. (2018b)).

### 10.5.2 Short-lived Climate Forcers and Aviation

Aviation's net warming effect results from its historical and current emissions of $CO_2$, and non-$CO_2$ emissions of water vapour, soot, sulphur dioxide (from sulphur in the fuel), and nitrogen oxides ($NO_x$, = NO + $NO_2$) (IPCC 1999; Lee et al. 2021; Szopa et al. 2021). Although the effective radiative forcing (ERF) of $CO_2$ from historic aviation emissions is not currently the largest forcing term, it is difficult to address because of the sector's current dependency on fossil-based hydrocarbon fuels and the longevity of $CO_2$. A residual of emissions of $CO_2$ today will still have a warming effect in many thousands of years (Archer et al. 2009; Canadell et al. 2021) whereas water vapour, soot, and $NO_x$ emissions will have long ceased to contribute to warming after some decades. As a result, $CO_2$ mitigation of aviation to net zero levels, as required in 1.5°C scenarios, requires fundamental shifts in technology, fuel types, or changes of behaviour or demand.

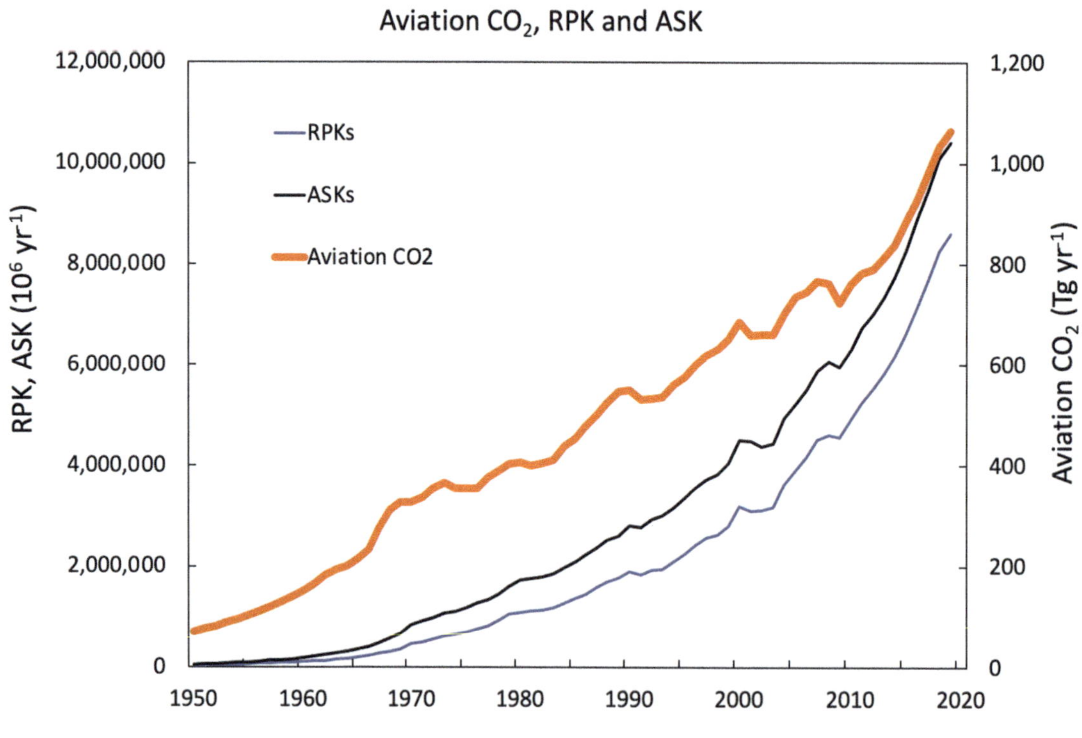


**Figure 10.10 | Historical global emissions of $CO_2$ from aviation, along with capacity and transport work (given in available seat kilometres, ASK; revenue passenger-kilometres, RPK).** Source: adapted from Lee et al. (2021) using IEA and other data.

The non-$CO_2$ effects of aviation on climate fall into the category of short-lived climate forcers (SLCFs). Emissions of $NO_x$ currently result in net positive warming from the formation of short-term ozone (warming) and the destruction of ambient methane (cooling). If the conditions are suitable, emissions of soot and water vapour can trigger the formation of contrails (Kärcher 2018), which can spread to form extensive contrail-cirrus cloud coverage. Such cloud coverage is estimated to have a combined ERF that is about 57% of the current net ERF of global aviation (Lee et al. 2021), although a comparison of cirrus cloud observations under pre- and post-COVID-19 pandemic conditions suggest that this forcing could be smaller (Digby et al. 2021). Additional effects from aviation from aerosol-cloud interactions on high-level ice clouds through soot (Chen and Gettelman 2013; Zhou and Penner 2014; Penner et al. 2018), and lower-level warm clouds through sulphur (Righi et al. 2013; Kapadia et al. 2016) are highly uncertain, with no best estimates available (Lee et al. 2021). In total, the net ERF from aviation's non-$CO_2$ SLCFs is estimated to be approximately 66% of aviation's current total forcing. It is important to note that the fraction of non-$CO_2$ forcing to total forcing is not a fixed quantity and is dependent on the recent history of growth (or otherwise) of $CO_2$ emissions (Klöwer et al. 2021). The non-$CO_2$ effects from aviation are the subject of discussion for mitigation options (Arrowsmith et al. 2020). However, the issues are complex, potentially involving technological and operational trade-offs with $CO_2$.

### 10.5.3 Mitigation Potential of Fuels, Operations, Energy Efficiency, and Market-based Measures

**Technology options for engine and airframe.** For every kilogram of jet fuel combusted, 3.16 kg $CO_2$ is emitted. Engine and airframe manufacturers' primary objective, after safety issues, is to reduce direct operating costs, which are highly dependent on fuel burn. Large investments have gone into engine technology and aircraft aerodynamics to improve fuel burn per kilometre (Cumpsty et al. 2019). There have been major step changes in engine technology over time, from early turbojet engines to larger turbofan engines. However, the basic configuration of an aircraft has remained more or less the same for decades and will likely remain at least to 2037 (Cumpsty et al. 2019). Airframes performance has improved over the years with better wing design, but large incremental gains have become much harder as the technology has matured. For twin-aisle aircraft, generally used for long ranges, fuel-burn is a pressing concern and there have been several all-new aircraft designs with improvements in their lift-to-drag ratio (Cumpsty et al. 2019). The principal opportunities for fuel reduction come from improvements in aerodynamic efficiency, aircraft mass reduction, and propulsion system improvements. In the future, Cumpsty et al. (2019) suggest that the highest rate of fuel burn reduction achievable for new aircraft is likely to be no more than about 1.3% per year, which is well short of ICAO's aspirational goal of 2% global annual average fuel efficiency improvement. Radically different aircraft shapes, like the blended wing body (where the wings are not distinct from the fuselage), are likely to use about 10% less fuel than future advanced aircraft of conventional form (Cumpsty et al. 2019). Such improvements would be 'one-off' gains, do not compensate for growth in emissions of $CO_2$ expected to be in excess of 2% per annum, and would take a decade or more to penetrate the fleet completely. Thus, the literature does not support the idea that there are large improvements to be made in the energy efficiency of aviation that keep pace with the projected growth in air transport.

**Operational improvements for navigation.** From a global perspective, aircraft navigation is relatively efficient, with many long-haul routes travelling close to great circle trajectories, and avoiding headwinds that increase fuel consumption. The ICAO estimates that flight inefficiencies on a global basis are currently of the order 2% to 6% (ICAO 2019), while Fleming and de Lépinay (2019) project operational improvements (air traffic management) of up to 13% on a regional basis by 2050. 'Intermediate stop operations' have been suggested, whereby longer-distance travel is broken into flight legs, obviating the need to carry fuel for the whole mission. Linke et al. (2017) modelled this operational behaviour on a global basis and calculated a fuel saving of 4.8% over a base case in which normal fuel loads were carried. However, this approach increases the number of landing/take-off cycles at airports. 'Formation flying', which has the potential to reduce fuel burn on feasible routes, has also been proposed (Xu et al. 2014; Marks et al. 2021).

**Alternative biofuels, synthetic fuels, and liquid hydrogen.** As noted above, the scope for reducing $CO_2$ emissions from aviation through improved airplane technology or operations is limited and unable to keep up with the projected growth, let al.ne reduce beyond the present emission rate at projected levels of demand (assuming post-pandemic recovery of traffic). Thus, the literature outlined here suggests that the only way for demand for aviation to continue to grow without increasing $CO_2$ emissions is to employ alternative lower-carbon bio- or synthetic aviation fuels (Klöwer et al. 2021). For shorter ranges, flights of light planes carrying up to 50 passengers may be able to use electric power (Sahoo et al. 2020) but these planes are a small proportion of the global aviation fleet (Epstein and O'Flarity 2019; Langford and Hall 2020) and account for less than 12% of current aviation $CO_2$ emissions. Alternative lower-carbon footprint fuels have been certified for use over recent years, principally from bio-feedstocks, but are not yet widely available at economic prices (Kandaramath Hari et al. 2015; Capaz et al. 2021a). In addition, alternative fuels from bio-feedstocks have variable carbon footprints because of different lifecycle emissions associated with various production methods and associated land-use change (de Jong et al. 2017; Staples et al. 2018; Capaz et al. 2021b; Zhao et al. 2021).

The development of 'sustainable aviation fuels' (referred to as 'SAFs') that can reduce aviation's carbon footprint is a growing area of interest and research. Alternative aviation fuels to replace fossil-based kerosene have to be certified to an equivalent standard as Jet-A for a variety of parameters associated with safety issues. Currently, the organisation responsible for aviation fuel standards, ASTM International, has certified seven different types of sustainable aviation fuels with maximum blends ranging from 10% to 50% (Chiaramonti 2019). Effectively, these blend requirements limit the amount of non-hydrocarbon fuel (e.g., methanol) that can be added at present. While there currently is a minimum level of aromatic hydrocarbon contained in jet fuel to prevent 'O-ring' shrinkage in the fuel seals (Khandelwal et al. 2018), this minimum level can likely be lower in the medium to long term, with the added benefits of reduced soot formation and reduced contrail cirrus formation (Bier et al. 2017; Bier and Burkhardt 2019).

Bio-based fuels can be produced using a variety of feedstocks including cultivated feedstock crops, crop residues, municipal solid waste, waste fats, oils and greases, wood products and forestry residues (Staples et al. 2018). Each of these different sources can have different associated lifecycle emissions, such that they are not net zero $CO_2$ emissions but have associated emissions of $CO_2$ or other GHGs from their production and distribution (Section 10.3, Box 10.2). In addition, associated land-use change emissions of $CO_2$ represent a constraint in climate change mitigation potential with biofuel (Staples et al. 2017) and have inherent large uncertainties (Plevin et al. 2010). Other sustainability issues include food vs fuel arguments, water resource use, and impacts on biodiversity. Cost-effective production, feedstock availability, and certification costs are also relevant (Kandaramath Hari et al. 2015). Nonetheless, bio-based SAFs have been estimated to achieve lifecycle emissions reductions ranging between approximately 2% and 70% under a wide range of scenarios (Staples et al. 2018). For a set of European aviation demand scenarios, Kousoulidou and Lonza (2016) estimated that the fuel demand in 2030 would be about 100 million tonnes of oil equivalent and biokerosene (HEFA/HVO) penetration would provide around 2% of the total fuel demand at that date. Several issues limit the expansion of biokerosene for aviation, the primary one being the current cost of fossil fuel compared to the costs of SAF production (Capaz et al. 2021a). Other hybrid pathways, for example the hydrogenation of biofuels (the hydrogen assumed to be generated with low-carbon energy), could increase the output and improve the economic feasibility of bio-based SAF (Hannula 2016; Albrecht et al. 2017).

Costs remain a major barrier for bio-SAF, which cost around three times the price of kerosene (Kandaramath Hari et al. 2015). Clearly, for SAFs to be economically competitive, large adjustments in prices of fossil fuels or the introduction of policies is required. Staples et al. (2018) estimated that in order to introduce bio-SAFs that reduce lifecycle GHG emissions by at least 50% by 2050, prices and policies were necessary for incentivisation. They estimate the need for 268 new biorefineries per year and capital investments of approximately 22 to 88 billion USD2015 per year between 2020 and 2050. Wise et al. (2017) suggest that carbon prices would help leverage production and availability.

Various pathways have been discussed for the production of non-bio SAFs such as power-to-liquid pathways (Schmidt et al. 2018), sometimes termed 'electro-fuels' (Goldmann et al. 2018), or more generalised 'Power-to-X' pathways (Kober and Bauer 2019). This process would involve the use of low-carbon electricity, $CO_2$, and water to synthesise jet fuel through the Fischer-Tropsch process or methanol synthesis. Hydrogen would be produced via an electrochemical process, powered by low-carbon energy and combined with $CO_2$ captured directly from the atmosphere or through BECCS. The energy requirement from photovoltaics has been estimated to be of the order 14 to 20 EJ to phase out aviation fossil fuel by 2050 (Gössling et al. 2021a). These synthetic fuels have potential for large lifecycle emissions reductions (Schmidt et al. 2016). In comparison to bio-SAF production, the implementation of the processes is in its infancy. However, assuming availability of low-carbon energy electricity, these fuels have much smaller land and water requirements than bio-SAF. Low carbon-energy supply, scalable technology, and therefore costs, represent barriers. Scheelhaase et al. (2019) review current estimates of costs, which are estimated to be approximately four to six times the price of fossil kerosene.

Liquid hydrogen ($LH_2$) as a fuel has been discussed for aeronautical applications since the 1950s (Brewer 1991) and a few experimental aircraft have flown using such a fuel. Experimental, small aircraft have also flown using hydrogen fuel cells. Although the fuel has an energy density per unit mass about three times greater than kerosene, it has a much lower energy density per unit volume (approximately factor 4 (McKinsey 2020)). The increased volume requirement makes the fuel less attractive for aviation since it would require the wings to be thickened or fuel to take up space in the fuselage. Bicer and Dincer (2017) found that $LH_2$-powered aircraft compared favourably to conventional kerosene-powered aircraft on a lifecycle basis, providing that the $LH_2$ was generated from low-carbon energy sources (0.014 $kgCO_2$ per tkm compared with 1.03 $kgCO_2$ per tkm for an unspecified passenger aircraft). However, Ramos Pereira et al. (2014) also made a lifecycle comparison and found much smaller benefits of $LH_2$-powered aircraft (manufactured from low-carbon energy) compared with conventional fossil kerosene. The two studies expose the sensitivities of boundaries and assumptions in the analyses. Shreyas Harsha (2014) and Rondinelli et al. (2017) conclude that there are many infrastructural barriers but that the environmental benefits of low-carbon-based $LH_2$ could be considerable. Khandelwal et al. (2013) take a more optimistic view of the prospect of $LH_2$-powered aircraft but envisage them within a hydrogen-oriented energy economy. A recently commissioned study by the European Union (EU)'s Clean Sky undertaking, (McKinsey 2020) addresses many of the aspects of the opportunities and obstacles in developing $LH_2$-powered aircraft. The report provides an optimistic view of the feasibility of developing such aircraft for short to medium haul but makes clear that new aircraft designs (such as blended-wing body aircraft) would be needed for longer distances.

The non-$CO_2$ impacts of $LH_2$-powered aircrafts remain poorly understood. The emission index of water vapour would be much larger (estimated to be 2.6 times greater by Ström and Gierens (2002)) than for conventional fuels), and the occurrence of contrails may increase but have lower ERF because of the lower optical depth (Marquart et al. 2005). Moreover, contrails primarily form on soot particles from kerosene-powered aircraft, which would be absent from $LH_2$ exhaust (Kärcher 2018). The overall effect is currently unknown as there are no measurements. Potentially, $NO_x$ emissions could be lower with combustor redesign (Khandelwal et al. 2013).

In conclusion, there are favourable arguments for $LH_2$-powered aircraft, both on an efficiency basis (Verstraete 2013) and an overall reduction in GHG emissions, even on an lifecycle basis. However, $LH_2$ requires redesign of the aircraft, particularly for long-haul operations. Similarly, there would be a need for expanded infrastructure for fuel manufacture, storage, and distribution at airports, which is likely to be more easily overcome if there is a more general move towards a hydrogen-based energy economy.

**Technological and operational trade-offs between $CO_2$ and non-$CO_2$ effects.** Since aviation has additional non-$CO_2$ warming effects, there has been some discussion as to whether these can be addressed by either technological or operational means. For example, improved fuel efficiency has resulted from high overall pressure ratio engines with large bypass ratios. This improvement has increased pressure and temperature at the combustor inlet, with a resultant tendency to increase thermal $NO_x$ formation in the combustor. Combustor technology aims to reduce this increase, but it represents a potential technology trade-off whereby $NO_x$ control may be at the expense of extra fuel efficiency. Estimating the benefits or disbenefits of $CO_2$ (proportional to fuel burned) vs $NO_x$ in terms of climate is complex (Freeman et al. 2018).

Any global warming potential/temperature change potential type emissions equivalency calculation always involves the user selection of a time horizon over which the calculation is made, which is a *subjective* choice (Fuglestvedt et al. 2010). In general, the longer the time horizon, the more important $CO_2$ becomes in comparison with a short-lived climate forcing agent. So, for example, a net (overall) aviation GWP for a 20-year time horizon is 4.0 times that of $CO_2$ alone, but only 1.7 over a 100-year time horizon. Correspondingly, a GTP for a 20-year time horizon is 1.3, but it is 1.1 for 100 years (Lee et al. 2021).

A widely discussed opportunity for mitigation of non-$CO_2$ emissions from aviation is the avoidance of persistent contrails that can form contrail cirrus. Contrails only form in ice-supersaturated air below a critical temperature threshold (Kärcher 2018). It is therefore feasible to alter flight trajectories to avoid such areas conducive to contrail formation, since ice-supersaturated areas tend to be tens to hundreds of kilometres in the horizontal and only a few 100 metres in the vertical extent (Gierens et al. 1997). Theoretical approaches show that avoidance is possible on a flight-by-flight basis (Matthes et al. 2017; Teoh et al. 2020). Case studies have shown that flight planning according to trajectories with minimal climate impact can substantially (up to 50%) reduce the aircraft's net climate impacts despite small additional $CO_2$ emissions (Niklaß et al. 2019). However, any estimate of the net benefit or disbenefit depends firstly on the assumed magnitude of the contrail cirrus ERF effect (itself rather uncertain, assessed with a low confidence level) and upon the choice of metric and time horizon applied. While this is a potentially feasible mitigation option, notwithstanding the $CO_2$ per contrail trade-off question, meteorological models cannot currently predict the formation of persistent contrails with sufficient accuracy in time and space (Gierens et al. 2020); this mitigation option is speculated to take of the order of up to a decade to mature (Arrowsmith et al. 2020).

**Market-based offsetting measures.** The EU introduced aviation into its $CO_2$ emissions trading scheme (ETS) in 2012. Currently, the EU-ETS for aviation includes all flights within the EU as well as to and from Eastern European and West-Central Asian states. Globally, ICAO agreed in 2016 to commence, in 2020, the 'Carbon Offsetting and Reduction Scheme for International Aviation' (CORSIA). The pandemic subsequently resulted in the baseline being changed to 2019.

CORSIA has a phased implementation, with an initial pilot phase (2021–2023) and a first phase (2024–2026) in which states will participate voluntarily. The second phase will then start in 2026–2035, and all states will participate unless exempted. States may be exempted if they have lower aviation activity levels or based on their UN development status. As of September 2021, 109 ICAO Member States will voluntarily be participating in CORSIA starting in 2022. In terms of routes, only those where both States connecting the route are participating are included. There will be a special review of CORSIA by the end of 2032 to determine the termination of the scheme, its extension, or any other changes to the scheme beyond 2035.

By its nature, CORSIA does not lead to a reduction in in-sector emissions from aviation since the programme deals mostly in approved offsets. At its best, CORSIA is a transition arrangement to allow aviation to reduce its impact in a more meaningful way later. From 2021 onwards, operators can reduce their CORSIA offsetting requirements by claiming emissions reductions from 'CORSIA Eligible Fuels' that have demonstrably reduced lifecycle emissions. These fuels are currently available at greater costs than the offsets (Capaz et al. 2021a). As a result, most currently approved CORSIA offsets are avoided emissions, which raises the issue of additionality (Warnecke et al. 2019). The nature of avoided emissions is to prevent an emission that was otherwise considered to be going to occur, for example, prevented deforestation. Avoided emissions are 'reductions' (over a counterfactual) and purchased from other sectors that withhold from an intended emission (Becken and Mackey 2017), such that if additionality were established, a maximum of 50% of the intended emissions are avoided. Some researchers suggest that avoided deforestation offsets are not a meaningful reduction, since deforestation continues to be a net source of $CO_2$ emissions (Mackey et al. 2013; Friedlingstein et al. 2020).

**Modal shift to high-speed rail.** Due to the limitations of the current suite of aviation mitigation strategies, the potential for high-speed rail (HSR) is of increasing interest (Givoni and Banister 2006; Chen 2017; Bi et al. 2019). The IEA's *Net Zero by 2050* roadmap suggests significant behavioural change, with more regional flights shifting to HSR in the Net Zero Emissions by 2050 scenario pathway (IEA 2021e). For HSR services to be highly competitive with air travel, the optimal distance between the departure and arrival points has been found to be in the approximate range of 400 to 800 km (Bows et al. 2008; Rothengatter 2010), although in the case of China's HSR operations, this range can be extended out to 1000 km, with corresponding air services having experienced significant demand reduction upon HSR service commencement (Lawrence et al. 2019). In some instances, negative effects on air traffic, air fare, and flight frequency have occurred at medium-haul distances such as HSR services in China on the Wuhan–Guangzhou route (1069 km) and the Beijing–Shanghai route (1318 km) (Fu et al. 2015; Zhang and Zhang 2016; Chen 2017; Li et al. 2019; Ma et al. 2019). This competition at medium-haul distances is contrary to that which has been experienced in European and other markets and may be attributable to China having developed a comprehensive network with hub stations, higher average speeds, and an integrated domestic market with strong patronage (Zhang et al. 2019a).

The LCA literature suggests that the GHG emissions associated with HSR vary depending on spatial, temporal, and operational specifics (Åkerman 2011; Baron et al. 2011; Chester and Horvath 2012; Yue et al. 2015; Hoyos et al. 2016; Jones et al. 2017; Robertson 2016; Robertson 2018; Lin et al. 2019). These studies found a wide range of approximately 10 to 110 $gCO_2 pkm^{-1}$ for HSR. This range is principally attributable to the sensitivity of operational parameters such as the HSR passenger seating capacity, load factor, composition of renewable and non-renewable energy sources in electricity production, rolling stock energy efficiency and patronage (i.e., ridership both actual and forecast), and line-haul infrastructure specifics (e.g., tunnelling and aerial structure requirements for a particular corridor) (Åkerman 2011; Chester and Horvath 2012; Yue et al. 2015; Newman et al. 2018; Robertson 2018). The prospect for HSR services providing freight carriage (especially online purchases) is also growing rapidly (Strale 2016; Bi et al. 2019; Liang and Tan 2019) with a demonstrated emissions reduction potential from such operations (Hoffrichter et al. 2012). However, additional supportive policies will most likely be required (Strale 2016; Watson et al. 2019). Limiting emissions avoidance assessments for HSR modal substitution to account only for $CO_2$ emissions ignores aviation's non-$CO_2$ effects (Section 10.5.2), and likely results in an under-representation of the climate benefits of HSR replacing flights.

HSR modal substitution can generate a contra-effect if the air traffic departure and arrival slots that become available as the result of the modal shift are simply reallocated to additional air services (Givoni and Banister 2006; Givoni and Dobruszkes 2013; Jiang and Zhang 2016; Cornet et al. 2018; Zhang et al. 2019a). Furthermore, HSR services have the potential to increase air traffic at a hub airport through improved networks but this effect can vary based on the distance of the HSR stations from airports (Jiang and Zhang 2014; Xia and Zhang 2016; Zhang et al. 2019b; Liu et al. 2019). Such rebound effects could be managed through policy interventions. For example, in 2021 the French government regulated that all airlines operating in France suspend domestic airline flights on routes if a direct rail alternative with a travel time of less than 2.5 hours is available. Other air travel demand reduction measures that have been proposed include regulations to ban frequent flyer reward schemes, mandates that all marketing of air travel declare flight emissions information to the prospective consumer (i.e., the carbon footprint of the nominated flight), the introduction of a progressive 'Air Miles Levy' as well as the inclusion of all taxes and duties that are presently exempt from air ticketing (Carmichael 2019). Moreover, China has the highest use of HSR in the world in part due to its network and competitive speeds and in part due to heavy regulation of the airline industry, in particular restrictions imposed on low-cost air carrier entry and subsidisation of HSR (Li et al. 2019). These air travel demand reduction strategies may induce shifts to other alternative modes in addition to stimulating HSR ridership.

Despite the risk of a rebound effect, and due to the probable reality of an incremental adoption of sustainable aviation fuel technology in the coming decades, the commencement of appropriate HSR services has the potential to provide, particularly in the short- to medium-term, additional means of aviation emissions mitigation.

### 10.5.4 Assessment of Aviation-specific Projections and Scenarios

The most recent projection from ICAO (prior to the COVID-19 pandemic) for international traffic (mid-range growth) is shown in Figure 10.11. This projection shows the different contributions of mitigation measures from two levels of improved technology, as well as improvements in air traffic management and infrastructure use. The projections indicate an increase in $CO_2$ emissions by a factor of 2.2 in 2050 over 2020 levels for the most optimistic set of mitigation assumptions. The high/low traffic growth assumptions would indicate increases by factors of 2.8 and 1.1, respectively in 2050, over 2020 levels (again, for the most optimistic mitigation assumptions).

The International Energy Agency has published several long-term aviation scenarios since AR5 within a broader scope of energy projections. Their first set of aviation scenarios include a 'reference technology scenario', a '2°C Scenario' and a 'Beyond 2°C Scenario'. The scenarios are simplified in assuming a range of growth rates and technological/operational improvements (IEA 2017b). Mitigation measures brought about by policy and regulation are treated in a broad-brush manner, noting possible uses of taxes, carbon pricing, price and regulatory signals to promote innovation.

The IEA has more recently presented aviation scenarios to 2070 in their 'Sustainable Development Scenario' that assume some limited reduction in demand post-COVID-19, and potential technology improvements in addition to direct reductions in fossil kerosene usage from substitution of biofuels and synthetic fuels (IEA 2021b). There is much uncertainty in how aviation will recover from the COVID-19 pandemic but, in this scenario, air travel returns to 2019 levels in three years, and then continues to expand, driven by income. Government policies could dampen demand (12% lower by 2040 than the IEA 'Stated Policies Scenario', which envisages growth at 3.4% per year, which in turn is lower than ICAO at 4.3%). Mitigation takes place largely by fuel substitution with lower-carbon biofuels and synthetic fuels, with a smaller contribution from technology. Approximately 85% of the actual cumulative $CO_2$ emissions (to 2070) are attributed to use of fuel at their lowest technology readiness level of 'Prototype', which is largely made up of biofuels and synthetic fuels, as shown in Figure 10.12. Details of the technological scenarios and the fuel availability/uptake assumptions are given in IEA (2021b), which also makes clear that the relevant policies are not currently in place to make any such scenario happen.

Within the Coupled Model Intercomparison Project Phase 6 emissions database, a range of aviation emissions scenarios for a range of Shared Socio-economic Pathway (SSP) scenarios are available (Figure 10.13). This Figure suggests that by 2050, direct emissions from aviation could be 1.5 to 6.5 (5–95th percentile) times higher than in the 2020 model year under the scenarios that

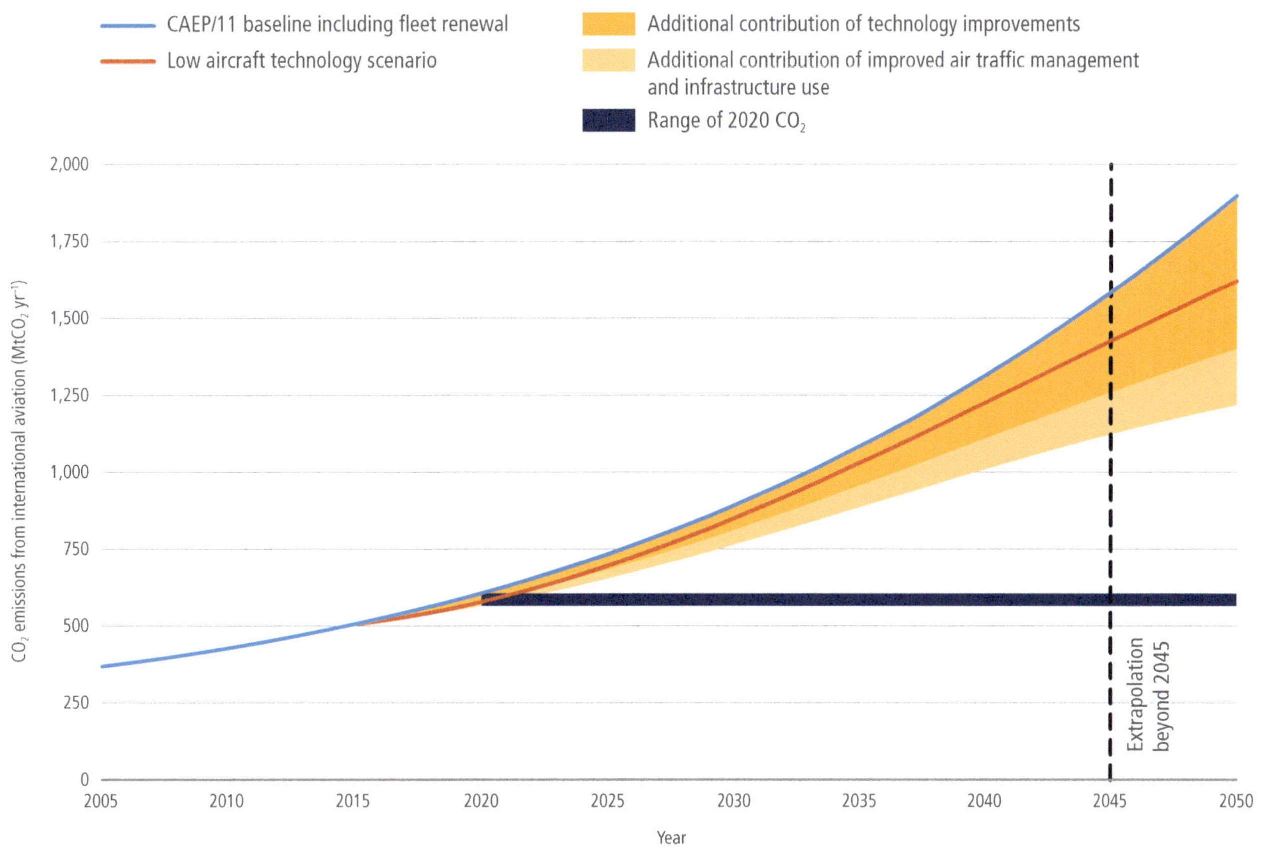


**Figure 10.11 | Projections of international aviation emissions of $CO_2$. Data in Mt $yr^{-1}$, to 2050, showing contributions of improved technology and air traffic management and infrastructure use to emissions reductions to 2050.** Data from Fleming and de Lépinay (2019); projections made pre-COVID-19 global pandemic.

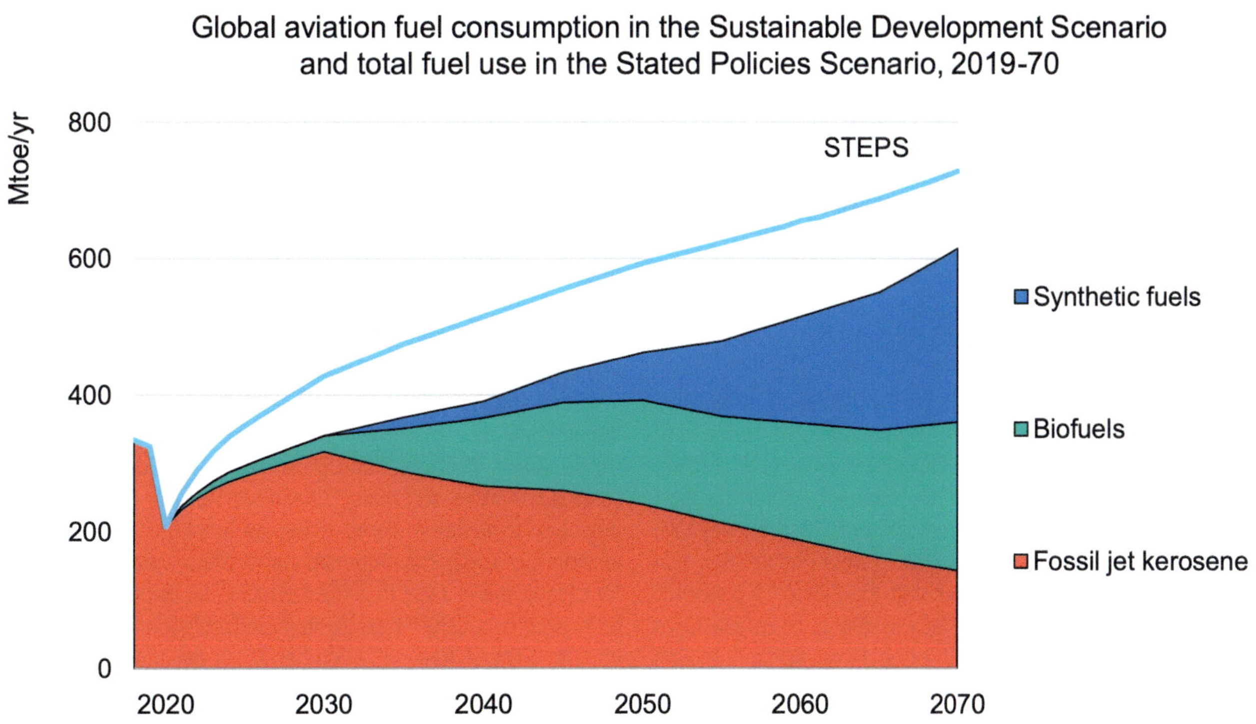


**Figure 10.12 | The International Energy Agency's scenario of future aviation fuel consumption for the States Policies Scenario ('STEPS') and composition of aviation fuel use in the Sustainable Development Scenario.** Source: adapted from IEA (2021b).

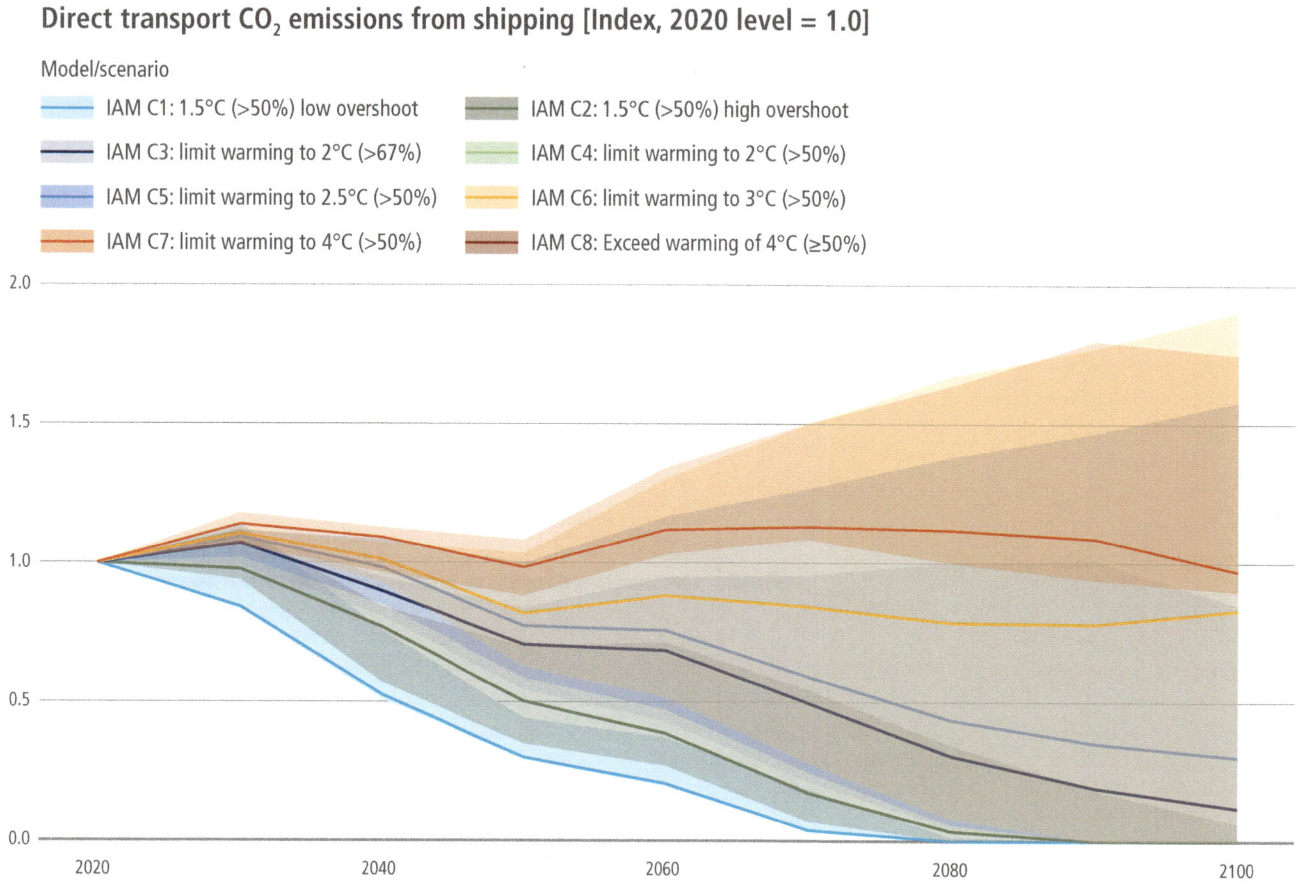


**Figure 10.13 | $CO_2$ emissions from AR6 aviation scenarios indexed to 2020 modelled year.** Data from the AR6 scenario database.

exceed warming of 4°C during the 21st century with a likelihood of 50% or greater (C8). In the C1 (which limit warming to 1.5°C (>50%) during the 2st century with no or limited overshoot) and C2 (which return warming to 1.5°C (>50%) during the 2st century after a high overshoot) scenarios, aviation emissions could still be up to 2.5 times higher in 2050 than in the 2020 model year (95th percentile) but may need to decrease by 10% by 2050 (5th percentile).

The COVID-19 pandemic of 2020 has changed many activities and, consequentially, associated emissions quite dramatically (Le Quéré et al. 2018b; Friedlingstein et al. 2020; Liu et al. 2020c; UNEP 2020). Aviation was particularly affected, with a reduction in commercial flights in April 2020 of about 74% over 2019 levels, with some recovery over the following months, remaining at 42% lower as of October 2020 (Petchenik 2021). The industry is considering a range of potential recovery scenarios, with the International Air Transport Association (IATA) speculating that recovery to 2019 levels may take up until 2024 (Earley and Newman 2021) (Cross-Chapter Box 1 in Chapter 1). Others suggest, however, that the COVID-19 pandemic and increased costs as a result of feed-in quotas or carbon taxes could slow down the rate of growth of air travel demand, though global demand in 2050 would still grow 57%–187% between 2018 and 2050 (instead of 250% in a baseline recovery scenario) (Gössling et al. 2021a).

### 10.5.5 Accountability and Governance Options

Under Article 2.2 of the Kyoto Protocol, Annex I countries were called to *'…pursue limitation or reduction of emissions of GHGs not controlled by the Montreal Protocol from aviation and marine bunker fuels, working through the International Civil Aviation Organization and the International Maritime Organization, respectively.'* The Paris Agreement is different, in that ICAO (and the IMO) are not named. As a result, the Paris Agreement, through the NDCs, seemingly covers $CO_2$ emissions from domestic aviation (currently 35% of the global total from aviation) but does not cover emissions from international flights. A number of states and regions, including the UK, France, Sweden, and Norway, have declared their intentions to include international aviation in their net zero commitments, while the EU, New Zealand, California, and Denmark are considering doing the same (Committee on Climate Change 2019). The Paris Agreement describes temperature-based goals, such that it is unclear how emissions of GHGs from international aviation would be accounted for. Clearly, this is a less than ideal situation for clarity of governance of international GHG emissions from both aviation and shipping. At its 40th General Assembly (October 2019) the ICAO requested its Council to *'…continue to explore the feasibility of a long-term global aspirational goal for international aviation, through conducting detailed studies assessing the attainability and impacts of any goals proposed, including the impact on growth as well as costs in all countries, especially developing countries, for the progress of the work to be presented to the 41st Session of the ICAO Assembly'.* What form this goal will take is unclear until work is presented to the 41st Assembly (Autumn, 2022). It is likely, however, that new accountability and governance structures will be needed to support decarbonisation of the aviation sector.

## 10.6 Decarbonisation of Shipping

Maritime transport is considered one of the key cornerstones enabling globalisation (Kumar and Hoffmann 2002). But as for aviation, shipping has its challenges in decarbonisation, with a strong dependency on fossil fuels without major changes since AR5. At the same time, the sector has a range of opportunities that could help reduce emissions through not only changing fuels, but also by increasing energy efficiency, optimising operations and ship design, reducing demand, improving regulations, as well as other options that will be reviewed in this section.

### 10.6.1 Historical and Current Emissions from Shipping

Maritime transport volume has increased by 250% over the past 40 years, reaching an all-time high of 11 billion tonnes of transported goods in 2018 (UNCTAD 2019). This growth in transport volumes has resulted in continued growth in GHG emissions from the shipping sector, despite an improvement in the carbon intensity of ship operations, especially since 2014. The estimated total emissions from maritime transport can vary depending on data set and calculation method, but range over 600–1100 $MtCO_2\ yr^{-1}$ over the past decade (Figure 10.14), corresponding to 2–3% of total anthropogenic emissions. The legend in Figure 10.14 refers to the following data sources: Endresen et al. (2003), Eyring et al. (2005), Dalsøren et al. (2009), DNV GL (DNV GL 2019), CAMS-GLOB-SHIP (Jalkanen et al. 2014; Granier et al. 2019), EDGAR (Crippa et al. 2019), Hoesly et al. (2018), Johansson et al. (2017), ICCT (Olmer et al. 2017), the IMO GHG Studies; IMO 2nd (Buhaug et al. 2009), IMO 3rd (Smith et al. 2014), IMO 4th-vessel and IMO 4th-voyage (Faber et al. 2020), and Kramel et al. (2021).

### 10.6.2 Short-lived Climate Forcers and Shipping

Like aviation, shipping is also a source of emissions of SLCFs as described in Section 10.5, including nitrogen oxides ($NO_x$), sulphur oxides ($SO_x$, such as $SO_2$ and $SO_4$), carbon monoxide (CO), black carbon, and non-methane volatile organic compounds (NMVOCs) (Szopa et al. 2021). Though SLCF have a shorter lifetime than the associated $CO_2$ emissions, these short-lived forcers can have both a cooling effect (e.g., $SO_x$) or a warming effect (e.g., ozone from $NO_x$). The cooling from the SLCF from a pulse emission will decay rapidly and diminish after a couple of decades, while the warming from the long-lived substances lasts for centuries (Szopa et al. 2021).

Emissions of SLCF from shipping not only affect the climate, but also the environment, air quality, and human health. Maritime transport has been shown to be a major contributor to coastal air quality degradation (Zhao et al. 2013; Jalkanen et al. 2014; Viana et al. 2014; Goldsworthy and Goldsworthy 2015; Goldsworthy 2017). Sulphur emissions may contribute towards acidification of the ocean (Hassellöv et al. 2013). Furthermore, increases in sulphur deposition on the oceans have also been shown to increase the flux of $CO_2$ from the oceans to the atmosphere (Hassellöv et al. 2013). To address the risks of $SO_x$ emissions from shipping, there is now a cap on the on the sulphur content permissible in marine fuels (IMO 2013). There is also significant uncertainty about the impacts of pollutants emitted from ships on the marine environment (Blasco et al. 2014).



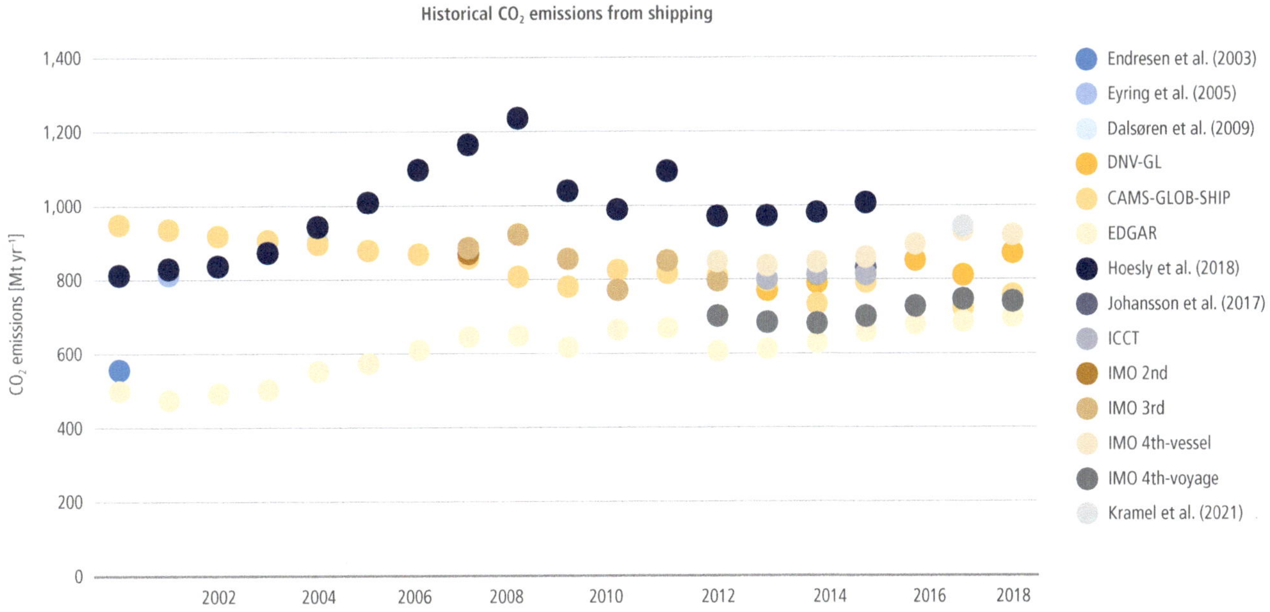


**Figure 10.14 | $CO_2$ emissions (Mt $yr^{-1}$) from shipping 2000–2018.** Data from various inventories as shown in the label.

Pollution control is implemented to varying degrees in the modelling of the SSP scenarios (Rao et al. 2017); for example, SSPs 1 and 5 assume that increasing concern for health and the environment result in more stringent air pollution policies than today (Szopa et al. 2021). There is a downward trend in $SO_x$ and $NO_x$ emissions from shipping in all the SSPs, in compliance with regulations. The SLCF emissions reduction efforts, within the maritime sector, are also contributing towards achieving the UN SDGs. In essence, while long-lived GHGs are important for long-term mitigation targets, accounting for short-lived climate forcers is important both for current and near-term forcing levels as well as broader air pollution and SDG implications.

### 10.6.3 Shipping in the Arctic

Shipping in the Arctic is a topic of increasing interest. The reduction of Arctic summer sea ice increases the access to the northern sea routes (Smith and Stephenson 2013; Melia et al. 2016; Aksenov et al. 2017; Fox-Kemper et al. 2021). Literature and public discourse has sometimes portrayed this trend as positive (Zhang et al. 2016b), as it allows for shorter shipping routes, for example between Asia and Europe, with estimated travel time savings of 25–40% (Aksenov et al. 2017). However, the acceleration of Arctic cryosphere melt and reduced sea ice that enable Arctic shipping reduce surface albedo and amplify climate warming (Eyring et al. 2021). Furthermore, local air pollutants can play different roles in the Arctic. For example, black carbon emissions reduce albedo and absorb heat in air, on snow and ice (Browse et al. 2013; Kang et al. 2020; Messner 2020; Eyring et al. 2021). Finally, changing routing from Suez to the northern sea routes may reduce total emissions for a voyage, but also shifts emissions from low to high latitudes. Changing the location of the emissions adds complexity to the assessment of the climatic impacts of Arctic shipping, as the local conditions are different and the SLCF may have a different impact on clouds, precipitation, albedo and local environment (Dalsøren et al. 2013; Fuglestvedt et al. 2014; Marelle et al. 2016). Observations have shown that 5–25% of air pollution in the Arctic stems from shipping activity within the Arctic itself (Aliabadi et al. 2015). Emissions outside the Arctic can affect Arctic climate, and changes within the Arctic may have global climate impacts. Both modelling and observations have shown that aerosol emissions from shipping can have a significant effect on air pollution and shortwave radiative forcing (Ødemark et al. 2012; Peters et al. 2012; Dalsøren et al. 2013; Roiger et al. 2014; Righi et al. 2015; Marelle et al. 2016).

Increased Arctic shipping activity may also pose increased risks to local marine ecosystems and coastal communities from invasive species, underwater noise, and pollution (Halliday et al. 2017; IPCC 2019). Greater levels of Arctic maritime transport and tourism have political, as well as socio-economic, implications for trade, and nations and economies reliant on the traditional shipping corridors. There has been an increase in activity from cargo, tankers, supply, and fishing vessels in particular (Winther et al. 2014; Zhao et al. 2015). Projections indicate more navigable Arctic waters in the coming decades (Smith and Stephenson 2013; Melia et al. 2016) and continued increases in transport volumes through the northern sea routes (Corbett et al. 2010; Lasserre and Pelletier 2011; Winther et al. 2014). Emission patterns and quantities, however, are also likely to change with future regulations from IMO, and depend on technology developments, and activity levels which may depend upon geopolitics, commodity pricing, trade, natural resource extraction, insurance costs, taxes, and tourism demand (Johnston et al. 2017). The need to include indigenous peoples' voices when shaping policies and governance of shipping activities in the high north is increasing (Dawson et al. 2020).

The Arctic climate and environment pose unique hazards and challenges with regard to safe and efficient shipping operations: low temperature challenges, implications for vessel design, evacuation and rescue systems, communications, oil spills, variable sea ice, and meteorological conditions (Buixadé Farré et al. 2014). To understand the total implications of shipping in the Arctic, including its climate impacts, a holistic view of synergies, trade-offs, and co-benefits is needed, with assessments of impacts on not only the physical climate, but also the local environment and ecosystems. To further ensure safe operations in the Arctic waters, close monitoring of activities may be valuable.

### 10.6.4 Mitigation Potential of Fuels, Operations and Energy Efficiency

A range of vessel mitigation options for the international fleet exist and are presented in this section. A variety of feedstocks and energy carriers can be considered for shipping. As feedstocks, fuels from biomass (advanced biofuels), fuels produced from renewable electricity and $CO_2$ capture from flue gas or the air (electro-, e-, or power-fuels), and fuels produced via thermochemical processes (solar fuels) can be considered. As energy carriers, synthetic fuels and the direct use of electricity (stored in batteries) are of relevance. The most prominent synthetic fuels discussed in the literature are hydrogen, ammonia, methane, methanol, and synthetic hydrocarbon diesel. Figure 10.15 shows the emissions reductions potential for alternative energy carriers that have been identified as having the highest potential to mitigate operational emissions from the sector (Chatzinikolaou and Ventikos 2014; Brynolf et al. 2014; Teeter and Cleary 2014; Traut et al. 2014; Lindstad et al. 2015; Psaraftis 2015; Seddiek 2015; Tillig et al. 2015; Winkel et al. 2016; DNV GL 2017; Bicer and Dincer 2018a; Biernacki et al. 2018; Bongartz et al. 2018; Gilbert et al. 2018; Hua et al. 2018; ITF 2018b; Singh et al. 2018; Balcombe et al. 2019; Hansson et al. 2019; Sharafian et al. 2019; Winebrake et al. 2019; Czermański et al. 2020; Faber et al. 2020; Hansson et al. 2020; Kim et al. 2020; Liu et al. 2020a; Nguyen et al. 2020; Perčić et al. 2020; Sadeghi et al. 2020; Seithe et al. 2020; Xing et al. 2020; Valente et al. 2021; Stolz et al. 2021).

Low-carbon hydrogen and ammonia are seen to have positive potential as a decarbonised shipping fuel. Hydrogen and ammonia, when produced from renewables or coupled to CCS as opposed to mainly by fossil fuels with high lifecycle emissions (Bhandari et al. 2014), may contribute to significant $CO_2$-eq reductions of up to 70–80% compared to low-sulphur heavy fuel oil (Bicer and Dincer 2018b; Gilbert et al. 2018). These fuels have their own unique transport and storage challenges as ammonia requires a pilot fuel due to difficulty in combustion, and ammonia combustion could lead to elevated levels of $NO_x$, $N_2O$, or $NH_3$ emissions depending on engine technology used (DNV GL 2020). There is a need for the further development of technology and procedures for safe storage and handling of fuels such as hydrogen and ammonia, both onboard and onshore, for faster uptake (Hoegh-Guldberg et al. 2019), but they remain an encouraging decarbonisation option for shipping in the next decade.

While methanol produced from fossil sources induces an emissions increase of +7.5% (+44%), e-methanol (via hydrogen from electrolysis based on renewable energy and carbon from direct air capture) reduces emissions by 80% (82%). In general, several synthetic fuels, such as synthetic diesel, methane, methanol, ethanol, and dimethyl ether could in principle be used for shipping (Horvath et al. 2018). The mitigation potential of these is fully dependent on the sourcing of the hydrogen and carbon required for their synthesis.

As noted in Section 10.3, LNG has been found to have a relatively limited mitigation potential and may not be viewed as a low-carbon alternative, but has a higher availability than other fuel options (Gilbert et al. 2018). Emissions reductions across the full fuel lifecycle are found in the order of 10%, with ranges reported from –30% (reduction) to +8% (increase), if switching from heavy fuel oil to LNG, as indicated in Figure 10.15 (Bengtsson et al. 2011). Regardless of the production pathway, the literature points to the risk of methane slip (emissions of unburnt methane especially at low engine loads and from transport to ports) from LNG-fuelled vessels, with no current regulation on emissions caps (Anderson et al. 2015; Ushakov et al. 2019; Peng et al. 2020). Leakage rates are a critical point for the total climate impact of LNG as a fuel, where high pressure engines remedy this more than low pressure ones. As discussed in Section 10.3, some consider LNG as a transition fuel, while some literature points to the risk of stranded assets due to the increasing decarbonisation regulation from IMO and the challenge of meeting IMO's 2030 emissions reductions targets using this fuel.

In addition to fossil and e-fuels, advanced biofuels might play a role to provide the energy demand for future shipping. Biomass is presently used to produce alcohol fuels (such as ethanol and methanol), liquid biogas, or biodiesel that can be used for shipping and could reduce $CO_2$ emissions from this segment. As explained in Box 10.2 and Chapter 7, the GHG footprint associated with biofuels is strongly dependent on the incurred land use and land-use change emissions. Advanced biofuels from processing cellulose rather than sugar are likely to be more attractive in terms of the quantities required but are not commercially available (Section 10.3). The estimates of emissions reductions from biofuels shown in Figure 10.15 rely on data from the Integrated Assessment Models – Energy Modelling Forum 33 (IAM EMF33), partial models assuming constant land cover (CLC), and partial models using natural growth (NRG). Box 10.2 and Section 10.4 include a more detailed description of the assumptions underlying these models and their estimates. The results based on IAM EMF33 and CLC suggests median mitigation potential of around 73% for advanced biofuels in shipping, while the NRG-based results suggest increased emissions from biofuels. The EMF33 and CLC results rely on modelling approaches compatible with the scenarios in the AR6 database (Chapter 6 and Box 7.7).

In addition to fuels, there are other measures that may aid the transition to low-carbon shipping. The amounts and speed of uptake of alternative low- or zero-carbon fuels in ports depend upon investments in infrastructure – including bunkering infrastructure, refinery readiness, reliable supply of the fuels, as well as sustainable production. The ship lifetime and age also play a role; retrofitting ships to accommodate engines and fuel systems for new fuel types may not be an option for older vessels. As such, operational efficiency becomes more important (Bullock et al. 2020). There is some potential to continue to improve the energy efficiency of vessels through operational changes (Traut et al. 2018), reducing the speed or 'slow steaming' (Bullock et al. 2020), and improved efficiency in port operations (Viktorelius and Lundh 2019; Poulsen and Sampson 2020). There is also a growing interest in onboard technologies for capturing carbon, with prototype ships underway showing 65–90% potential reduction in $CO_2$ emissions (Luo and Wang 2017; Awoyomi et al. 2020; Japan Ship Technology Research Association et al. 2020). Challenges identified include $CO_2$ capture efficiency (Zhou and Wang 2014), increased operating costs, and limited onboard power supply (Fang et al. 2019). Furthermore, designing $CO_2$ storage tanks for transport to shore may pose a challenge, as the volume and weight of captured $CO_2$ could be up to four times more than standard oil (Decarre et al. 2010).



Changes in design and engineering provide potential for reducing emissions from shipping through a range of measures, for example by optimising hull design and vessel shape, power and propulsion systems that include wind- or solar-assisted propulsion, and through improved operations of vessels and ports. Figure 10.15 shows that such measures may decrease emissions by 5–40%, though with a broad range in potential (Bouman et al. 2017). Nuclear propulsion could decrease emissions from individual vessels by 98%. Battery- or hybrid-electric ships have been identified as a means to reduce emissions in short-sea shipping such as ferries and inland waterways (Gagatsi et al. 2016), which may also importantly reduce near-shore SLCF pollution (Nguyen et al. 2020). Figure 10.15 shows that the median emissions from electric ships can be about 40% lower than equivalent fossil-based vessels but can vary widely. The wide reduction potential of battery electric propulsion is due to different assumptions about the $CO_2$ intensity of the electricity used and the levels of $CO_2$ footprints associated with battery production.

Although projections indicate continued increase in freight demand in the future, demand-side reductions could contribute to mitigation. The development of autonomous systems may play a role (Colling and Hekkenberg 2020; Liu et al. 2021) while 3-D printing can reduce all forms of freight as parts and products can be printed instead of shipped (UNCTAD 2018). As more than 40% of transported freight is fossil fuels, a lessened demand for such products in low-emissions scenarios should contribute to reducing the overall maritime transport needs and hence emissions in the future (Sharmina et al. 2017). An increase in alternative fuels, on the other hand, may increase freight demand (Mander et al. 2012). Potentials for demand-side reduction in shipping emissions may arise from improving processes around logistics and packaging, and further taxes and charges could serve as leverage for reducing demand and emissions.

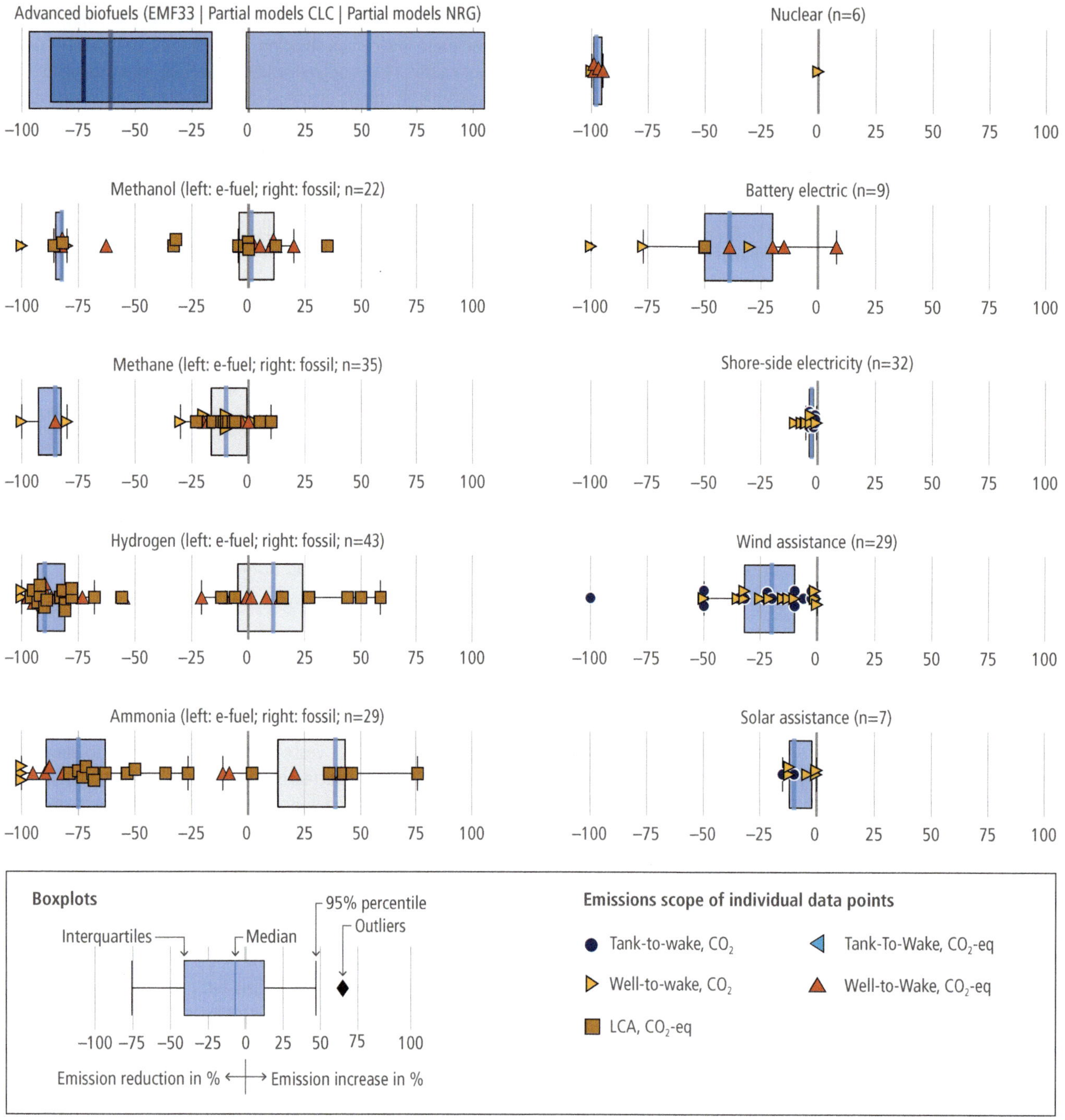


**Figure 10.15 | Emissions reductions potential of alternative fuels compared to conventional fuels in the shipping sector.** The x-axis is reported in %. Each individual marker represents a data point from the literature, where the brown square indicates a full LCA $CO_2$-eq value; light blue triangles tank-to-wake $CO_2$-eq; red triangles well-to-wake $CO_2$-eq; yellow triangles well-to-wake $CO_2$; and dark blue circles tank-to-wake $CO_2$ emissions reduction potentials. The values in the Figure rely on the 100-year GWP value embedded in the source data, which may differ slightly with the updated 100-year GWP values from WGI. 'n' indicates the number of data points per sub-panel. Grey shaded boxes represent data where the energy comes from fossil resources, and blue from low-carbon renewable energy sources. 'Advanced biofuels EMF33' refers to emissions factors derived from simulation results from the integrated assessment models EMF33 scenarios (darkest coloured box in top left panel). Biofuels partial models CLC refers to partial models with constant land cover. Biofuels partial models NRG refers to partial models with natural regrowth. For ammonia and hydrogen, low-carbon fuel is produced via electrolysis using low-carbon electricity, and 'fossil' refers to fuels produced via steam methane reforming of natural gas.

The coming decade is projected to be costly for the shipping sector, as it is preparing to meet the 2030 and 2050 emissions reduction targets set by the IMO (UNCTAD 2018). With enough investments, incentives, and regulation, substantial reductions of $CO_2$ emissions from shipping could be achieved through alternative energy carriers. The literature suggests that their cost could be manyfold higher than for conventional fuels, which in itself could reduce demand for shipping, and hence its emissions, but could make the transition difficult. R&D may help reduce these costs. The literature points to the need for developing technology roadmaps for enabling the maritime transport sector to get on to pathways for decarbonisation early enough to reach global goals (Kuramochi et al. 2018). Accounting for the full lifecycle emissions of the vessels and the fuels is required to meet the overall long-term objectives of cutting GHG and SLCF emissions. The urgency of implementing measures for reducing emissions is considered to be high, considering the lifetime of vessels is typically 20 years, if not more.

### 10.6.5 Accountability and Governance Options

Regulatory frameworks for the shipping sector have been developed over time and will continue to be through bodies such as the IMO, which was established by the UN to manage international shipping. The IMO strategy involves a 50% reduction in GHG emissions from international shipping by 2050 compared to 2008 (IMO 2018). The strategy includes a reduction in carbon intensity of international shipping by at least 40% by 2030, and 70% by 2050, compared to 2008. IMO furthermore aims for the sectoral phase-out of GHG emissions as soon as possible this century.

In 2020, the IMO approved the short-term goal-based measure to reduce the carbon intensity of existing international vessels. This measure addresses both technical and operational strategies. The operational element is represented by a Carbon Intensity Indicator (CII), and the technical element is represented by the Energy Efficiency Existing Ship Index (EEXI), which will apply to ships from 2023. The EEXI builds upon the Energy Efficiency Design Index (EEDI), which is a legally-binding mitigation regulation for newbuild ships, established as a series of baselines for the amount of fuel ships may burn for a particular cargo-carrying capacity. The EEDI differs per ship segment. For example, ships built in 2022 and beyond should be 50% more energy efficient than those built in 2013. This legislation aims to reduce GHG emissions in particular. Energy efficiency may be improved by several of the mitigation options outlined above. The Ship Energy Efficiency Management Plan (SEEMP) is seen as the international governance instrument to improve energy efficiency and hence emissions from ships. SEEMP is a measure to enable changes to operational measures and retrofits (see Johnson et al. 2013). The combination of EEXI, EEDI, and SEEMP may reduce emissions by 23% by 2030 compared to a 'no policy' scenario (Sims et al. 2014). With regards to accountability, it is mandatory for ships greater than or equal to 5000 gross tonnage to collect fuel consumption data, as well as specified data. Such as for transport work. Similarly, the EU Monitoring, Reporting and Verification Regulation requires mandatory reporting of a vessel's fuel consumption when operating in European waters.

Policy choices may enable or hinder changes, and gaps in governance structures may, to some degree, hinder the objectives of mechanisms like SEEMP to improve energy efficiency and emissions. Policies may be developed to incentivise investments in necessary changes to the global fleet and related infrastructures. The literature argues that regulations and incentives that motivate mitigation through speed optimisation, ship efficiency improvements, and retrofits with lower-carbon technologies at a sub-global scale may contribute to immediate reductions in $CO_2$ emissions from the sector (Bows-Larkin 2015). The role of the financial sector, through initiatives such as the Poseidon Principle, which limit lending to companies that fail to uphold environmental standards, could also become increasingly important (Sumaila et al. 2021).

It has been proposed to make shipping corporations accountable for their emissions by making it mandatory to disclose their vessels' emissions reductions (Rahim et al. 2016). Market-based mechanisms may increasingly encourage ship operators to comply with IMO GHG regulations. Development of policies such as carbon pricing or taxation to enable a business case for adopting low-carbon fuels could be a near-term priority for acceleration of transformation of the sector (Hoegh-Guldberg et al. 2019). The EU is considering including shipping in its carbon trading system, with the details still to be agreed upon but expected to come into force in 2023, along with the CII. The proposition is that shipowners who conduct voyages within Europe, or start or end at an EU port, will have to pay for carbon permits to cover the $CO_2$ emitted by their vessel.

Regulations exist also to limit emissions of air pollution from shipping with the aim to improve environment and health impacts from shipping in ports and coastal communities. In sulphur emission control areas (SECAs), the maximum permissible sulphur content in marine fuels is 0.10% mass/mass. These are further tightened by the IMO legislation on reducing marine fuel sulphur content to a maximum of 0.5% in 2020 outside SECAs, compared to 3.5% permissible since 2012 (MARPOL Convention). The MARPOL Annex VI also limits the emissions of ozone-depleting substances and ozone precursors, $NO_x$, and volatile organic compounds from tankers (Mertens et al. 2018). The implementation of the emission control areas have been shown to reduce the impacts on health and the environment (Viana et al. 2015).

While there are many governance and regulatory initiatives that help reduce emissions from the shipping sector, few are transformative on their own, unless zero-carbon fuels can become available at a reasonable cost as suggested in Sections 10.3 and below.

### 10.6.6 Transformation Trajectories for the Maritime Sector

Figure 10.16 shows $CO_2$ emissions from shipping in scenarios from the AR6 database and the Fourth GHG study by the IMO (Faber et al. 2020). Panel (a) shows that $CO_2$ emissions from shipping go down by 33–70% (5–95th% percentile) by 2050 in the C1 and C2 scenarios, which limit warming to 1.5°C (>50%) during the 21st century with no or limited overshoot or return warming to 1.5°C (>50%) during

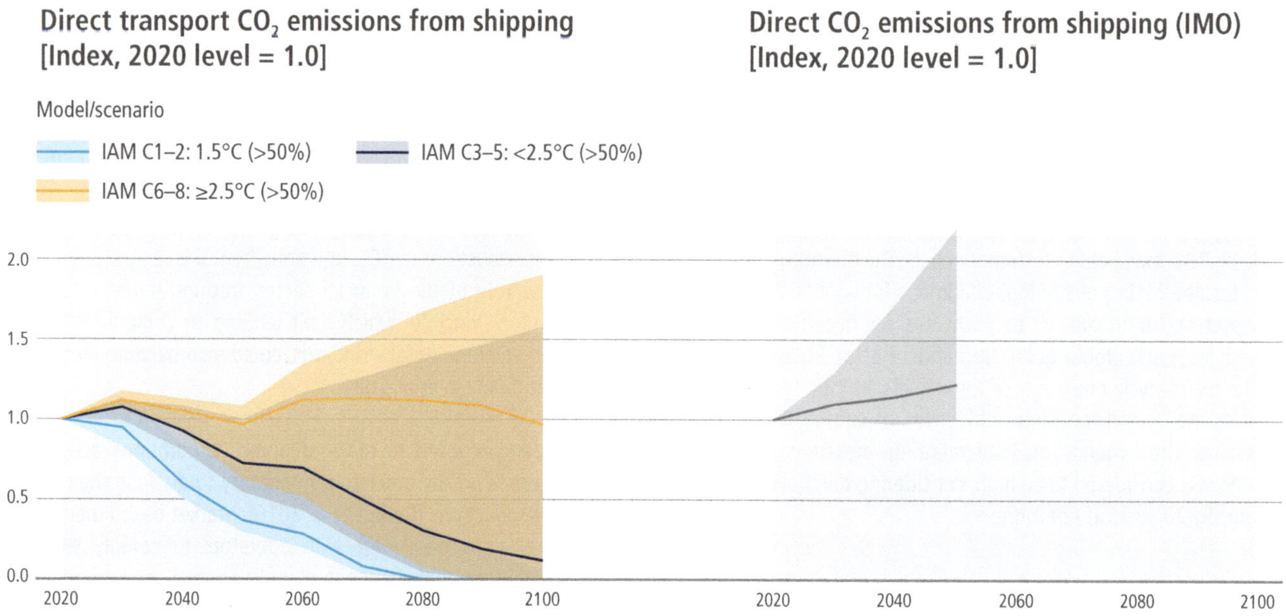


**Figure 10.16 | $CO_2$ emissions from shipping scenarios indexed to 2020 modelled year.** Panel **(a)** scenarios from the AR6 database. Panel **(b)** scenarios from the Fourth IMO GHG Study (Faber et al., 2020). Figures show median, 5th and 95th percentile (shaded area) for each scenario group.



the 21st century after a high overshoot. By 2080, median values for the same set of scenarios reach net zero $CO_2$ emissions. IAMs often do not report emissions pathways for shipping transport and the sector is underrepresented in most IAMs (Esmeijer et al. 2020). Hence pathways established outside IAMs can be different for the sector. Indeed, the IMO projections for growth in transport demand (Faber et al. 2020) indicate increases of 40–100% by 2050 for the global fleet. Faber and et al. (2020), at the same time predict reductions in trade for fossil fuels dependent on decarbonisation trajectories. The energy efficiency improvements of the vessels in these scenarios are typically of 20–30%. This offsets some of the increases from higher demand in the future scenarios. Fuels assessed by the Fourth IMO GHG study were limited to heavy fuel oil, marine gasoil, LNG, and methanol, with a fuels mix ranging from 91–98% conventional fuel use and a small remainder of alternative fuels (primarily LNG and some methanol). Panel (b) shows average fleetwide emissions of $CO_2$ based on these aggregate growth and emissions trajectories from the IMO scenarios. In these scenarios, $CO_2$ emissions from shipping remain stable or grow compared to 2020 modelled levels. These results contrast with the low emissions trajectories in the C1–C2 bin in panel (a). It seems evident that the scenarios in the AR6 database explore a broader solutions space for the sector than the Fourth GHG study by the IMO. However, the 1.5°C–2°C warming goal has led to an IMO 2050 target of 40% reduction in carbon intensity by 2030, which would require emissions reduction efforts to begin immediately. Results from global models suggest the solutions space for deep emissions reductions in shipping is available.

Combinations of measures are likely to be needed for transformative transitioning of the shipping sector to a low-carbon future, particularly if an expected increase in demand for shipping services is realised (Smith et al. 2014; Faber et al. 2020). Both GHG and SLCF emissions decrease significantly in SSP1-1.9, where mitigation is achieved in the most sustainable way (Rao et al. 2017). Conversely, there are no emissions reductions in the scenarios presented by the IMO Fourth GHG study, even though these scenarios incorporate some efficiency improvements and a slight increase in the use of LNG.

Options outlined in this chapter suggest a combination of policies to reduce demand, increase investments by private actors and governments, and develop the technology readiness level of alternative fuels and related infrastructure (especially synthetic fuels). Some literature suggests that battery electric-powered short-distance sea shipping could yield emissions reductions given access to low-carbon electricity. For deep sea shipping, advanced biofuels, hydrogen, ammonia, and synthetic fuels hold potential for significant emissions reductions, depending on GHG characteristics of the fuel chain and resource base. Other options, such as optimisation of speed and hull design and wind-assisted ships, could also combine to make significant contributions by 2050 to further bring emissions down. In total a suite of mitigation options exists or is on the horizon for the maritime sector.

## 10.7 Scenarios from Integrated, Sectoral, and Regional Models

### 10.7.1 Transport Scenario Modelling

This section reviews the results of three types of models that systemically combine options to assess different approaches to generating decarbonisation pathways for the transport system: (i) integrated assessment models (IAM); (ii) global transport energy models (GTEM); and (iii) national transport-energy models (NTEM) (Edelenbosch et al. 2017; Yeh et al. 2017). Common assumptions across the three model types include trajectories of socioeconomic development, technological development, resource availability, policy, and behavioural change. The key differences underlying these models are their depth of technological and behavioural detail versus scope in terms of sectoral and regional coverage. In very general terms, the narrower the scope in terms of sectors and regions, the more depth on spatial, technological, and behavioural detail. A large set of scenarios from these models were collected in a joint effort led by Chapter 3 and supported by Chapter 10 and others. The outcomes from over 100 models have been analysed for this chapter with the methodologies set out in Annex III for the whole report.

GHG emissions from transport are a function of travel demand, travel mode, transport technology, GHG intensity of fuels, and energy efficiency. These drivers can be organised around a group of levers that can advance the decarbonisation of the transport system. The levers thus include reducing travel activity, increasing use of lower-carbon modes, and reducing modal energy intensity and fuel carbon content. This section explores each lever's contributions to the decarbonisation of the transport sector by reviewing the results from the three model types IAM, GTEM, and NTEM.

IAMs integrate factors from other sectors that interact with the transport system endogenously, such as fuel availability and costs. IAMs minimise mitigation costs to achieve a temperature goal *across all sectors of the economy* over a long time horizon (typically to 2100). IAMs typically capture mitigation options for energy and carbon intensity changes with greater technology/fuel details and endogeneity linked to the other sectors. In the scenarios with very large-scale electrification of the transport sector, the coupling with the other sectors in fuel production, storage, and utilisation becomes more important. G-/NTEMs and related regional transport sectoral models have more details on transport demand, technology, behaviours, and policies than IAMs, but treat the interactions with the other sectors exogenously, potentially missing some critical interactions, such as the fuel prices and carbon intensity of electricity. National models have detailed representation of national policies related to transport and energy, sometimes with greater spatial resolution. Compared with IAMs, G-/NTEMs typically have greater detailed representation to explore mitigation options along the activity and mode dimensions where spatial, cultural, and behavioural details can be more explicitly represented. Section 5 in Annex III provides more details about these types of models. Scenarios for shipping and aviation are handled in more detail in sections 10.5 and 10.6, respectively.

This section applies the following categorisation of scenarios (see Table 3.1 for more details):

- C1 (scenarios that limit warming to 1.5°C (>50%) during the 21st century with no or limited overshoot)
- C2 (scenarios that return warming to 1.5°C (>50%) during the 21st century after a high overshoot)
- C3 (scenarios that limit warming to 2°C (>67%) throughout the 21st century)
- C4 (scenarios that limit warming to 2°C (>50%) throughout the 21st century)
- C5 (scenarios that limit warming to 2.5°C (>50%) throughout the 21st century)
- C6 (scenarios that limit warming to 3°C (>50%) throughout the 21st century)
- C7 (scenarios that limit warming to 4°C (>50%) throughout the 21st century)
- C8 (scenarios that exceed warming of 4°C (≥50%) during the 21st century)

A large share of the scenarios was developed prior to 2020. Results from such scenarios are indexed to a modelled (non-COVID) year 2020, referred to as 2020Mod.

### 10.7.2 Global Emissions Trajectories

In 2018, transport emitted 8.5 $GtCO_2$-eq, reaching a near doubling from 1990 levels after two decades of 2% per year emissions growth (Section 10.1). Assessing future trajectories, Figure 10.17 provides an overview of direct $CO_2$ emissions estimates from the transport sector across IAMs (colour bars) and selected global transport models (grey bars). The results from the IAMs are grouped in bins by temperature goal. Global transport energy models are grouped into reference and policy bins, since the transport sector cannot by itself achieve fixed global temperature goals. The policy scenarios in GTEMs and NTEMs cover a wide range of 'non-reference' scenarios, which include, for example, assumptions based on the 'fair share action' principles. In these scenarios, transport emissions reach reductions consistent with the overall emissions trajectories aligning with warming levels of 2°C. These scenarios may also consider strengthening existing transport policies, such as increasing fuel economy standards or large-scale deployments of electric vehicles. In most cases, these Policy scenarios are not necessarily in line with the temperature goals explored by the IAMs.

According to the collection of simulations from the IAM and GTEM models shown in Figure 10.17, global transport emissions could grow up to 2–47% (5–95th percentile) by 2030 and –6–130% by 2050 under the C7 scenarios that limit warming to 4°C (>50%) throughout the 21st century and C8 scenarios that exceed 4°C (≥50%) during the 21st century. Population and GDP growth and the secondary effects, including higher travel service demand per capita and increased freight activities per GDP, drive the growth in emissions in these scenarios (Section 10.7.3). Though transport efficiencies (energy use per pkm travelled and per tkm of goods delivered) are expected to continue to improve in line with the historical trends (Section 10.7.4), total transport emissions would grow due to roughly constant carbon intensity (Section 10.7.5) under the C7 and C8 scenarios that limit warming to 4°C (>50%) throughout the 21st century or exceed 4°C (≥50%) during the 21st century. In these scenarios, Significant increases in emissions (>150% for the medium values by 2050) would come from Asia and Pacific, the Middle East, and Africa. Compared to estimated 2020 levels, in 2050 Developed Countries would have median 25% decrease in transport emissions in the C7 scenarios that limit warming to 4°C (>50%) throughout the 21st century or median 15% increase in transport emissions in the C8 scenarios that exceed warming of 4°C (≥50%) during the 21st century.

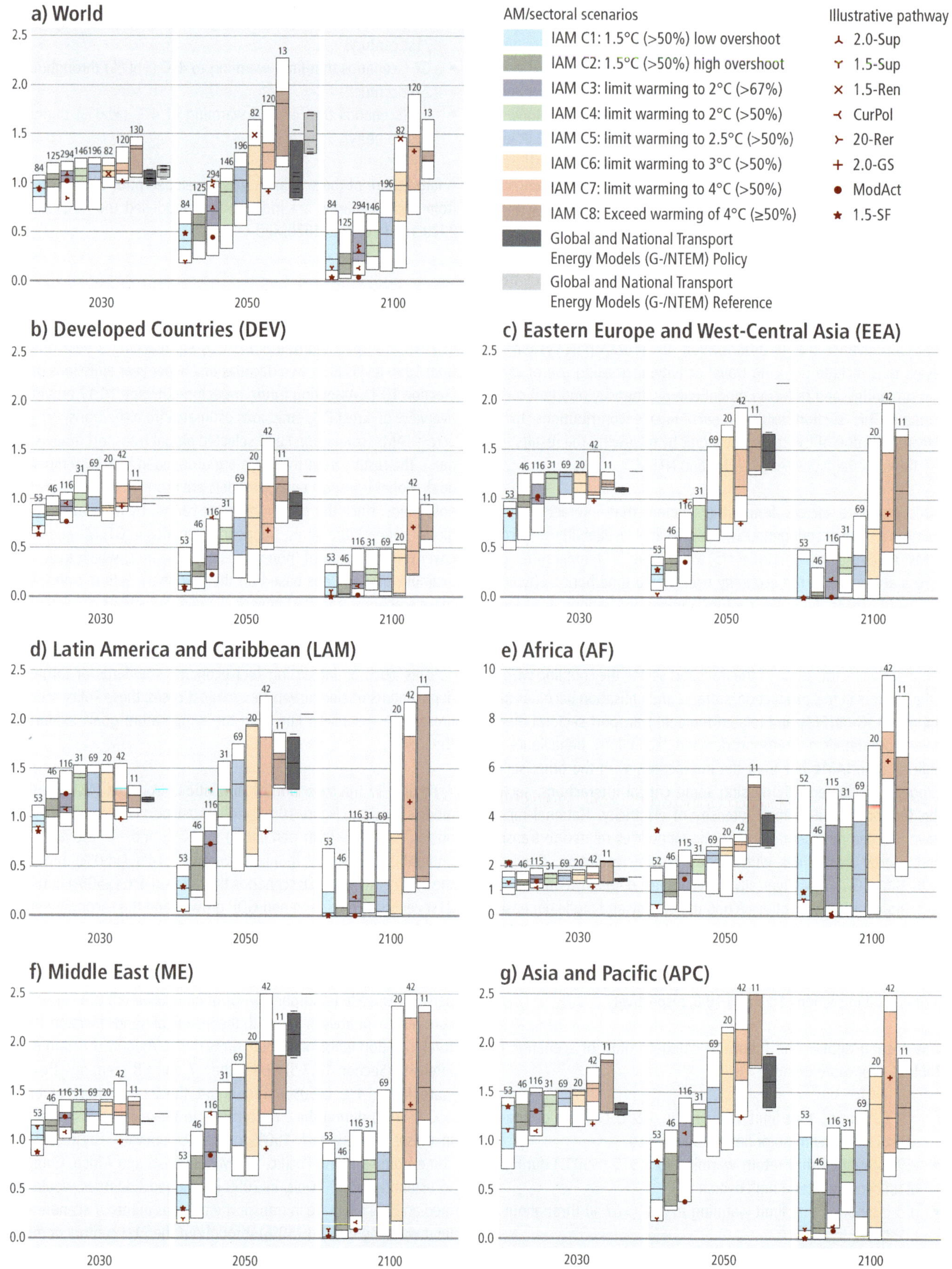


**Figure 10.17 | Direct $CO_2$ emissions from transport in 2030, 2050, and 2100 indexed to 2020 modelled year across R6 Regions and World.** IAM results are grouped by temperature targets. Sectoral studies are grouped by reference and policy categories. Plots show 5–95th percentile, 25–75th percentile, and median. Numbers above the bars indicate the number of scenarios. Data from the AR6 scenario database.

To meet temperature goals, by 2050 global transport emissions would need to decrease by 17% (+67% to –23% for the 5–95th percentile) below 2020Mod levels in the scenarios that limit warming to 2°C (>67%), 2°C (>50%) and 2.5 °C (>50%) throughout the 21st century (C3-C5 scenarios – orange bars), and 47% (14–80% for the 5–95th percentile) in the scenarios that limit warming to 1.5°C (>50%) during the 21st century with no or limited overshoot or return to 1.5°C (>50%) during the 21st century after high overshoot during the 21st century (C1–C2 scenarios – green bars). However, transport-related emission reductions may not happen uniformly across regions. For example, transport emissions from the Developed Countries and Eastern Europe and West Central Asia would decrease from 2020 levels by 2050 across all C1–C2 scenarios, but could increase in Africa, Asia and Pacific, Latin America and Caribbean, and the Middle East, in some of these scenarios. In particular, the median transport emissions in India and Africa could increase by 2050 in C1–C2 scenarios, while the 95th percentile emissions in Asia and Pacific, Latin America and Caribbean, and the Middle East, could be higher in 2050 than in 2020.

The Reference scenario emission pathways from GTEMs described in Figure 10.17 have similar ranges to C7–C8 scenario groups in 2050. The Policy scenarios are roughly in line with C6–C7 scenarios for the world region. The results suggest that the majority of the Policy scenarios examined by the GTEMS reviewed here are in the range of the C3–C6 scenarios examined by the IAMs (Gota et al. 2016; IEA 2017b; Yeh et al. 2017; Fisch-Romito and Guivarch 2019). The NDCs in the transport sector include a mix of measures targeting efficiency improvements of vehicles and trucks; improving public transit services; decarbonising fuels with alternative fuels and technologies including biofuels, fossil- or bio-based natural gas, and electrification; intelligent transport systems; and vehicle restrictions (Gota et al. 2016). Because of the long lag-time for technology turnover, these measures are not expected to change 2030 emissions significantly. However, they could have greater impacts on 2050 emissions.

Several GTEMs not included in AR6 scenario database have examined ambitious $CO_2$ mitigation scenarios. For example, a meta-analysis of scenarios suggests that global transport emissions consistent with warming levels of 2°C, would peak in 2020 at around 7–8 $GtCO_2$ and decrease to 2.5–9.2 Gt for 2°C, with an average of 5.4 Gt by 2050 (Gota et al. 2019). For comparison, the IEA's Sustainable Development Scenario suggests global transport emissions decrease to 3.3 Gt (or 55% reduction from 2020 level) by 2050 (IEA 2021f). The latest IEA *Net Zero by 2050* report proposes transport emissions to be close to zero by 2050 (IEA 2021e). The latter is lower than the interquartile ranges of the C1 group of scenarios from the AR6 database analysed here.

Low-carbon scenarios are also available from national models (Latin America, Brazil, Canada, China, France, Germany, Indonesia, India, Italy, Japan, Mexico, South Africa, UK, US) with a good representation of the transport sector. The low-carbon scenarios are either defined with respect to a global climate stabilisation level of, for example, 2°C/1.5°C Scenario (Dhar et al. 2018), or a $CO_2$ target that is more stringent than what has been considered in the NDCs, such as the net-zero emissions pathways (Bataille et al. 2020; IEA 2021e). These studies have generally used bottom-up models (see Annex III) for the analysis, but in some cases, they are run by national teams using global models (e.g., the Global Change Assessment Model (GCAM) for China and India). National studies show that transport $CO_2$ emissions could decline significantly in low-carbon scenarios in all the developed countries reviewed (Bataille et al. 2015; Kainuma et al. 2015; Hillebrandt et al. 2015; Mathy et al. 2015; Pye et al. 2015; Virdis et al. 2015; Williams et al. 2015; Zhang et al. 2016a) in 2050 from the emissions in 2010 and reductions could vary from 65% to 95%. However, in developing countries reviewed (Di Sbroiavacca et al. 2014; Altieri et al. 2015; Buira and Tovilla 2015; Rovere et al. 2015; Shukla et al. 2015; Siagian et al. 2015; Teng et al. 2015; Dhar et al. 2018), emissions could increase in 2050 in the range of 35% to 83% relative to 2010 levels. Transport $CO_2$ emissions per capita in the developing countries were much lower in 2010 (varying from 0.15 to 1.39 $tCO_2$ per capita) relative to developed countries (varying from 1.76 to 5.95 $tCO_2$ per capita). However, results from national modelling efforts suggest that, by 2050, the $CO_2$ emissions per capita in developed countries (varying from 0.19 to 1.04 $tCO_2$ per capita) could be much lower than in developing countries (varying from 0.21 to 1.7 $tCO_2$ per capita).

The transport scenario literature's mean outcomes suggest that the transport sector may take a less steep emissions reduction trajectory than the cross-sectoral average and still be consistent with the 2°C goal. For example, most of the scenarios that limit or return warming to 1.5°C (>50%) during the 21st century (C1–C2) reach zero emissions by 2060, whereas transport sector emissions are estimated in the range of 20% of the 2020Mod level (4–65% for the 10th to 90th percentiles) by 2100. This finding is in line with perspectives in the literature suggesting that transport is one of the most difficult sectors to decarbonise (Davis et al. 2018). There is, however, quite a spread in the results for 2050. Since temperature warming levels relate to global emissions from all sectors, modelling results from IAMs tend to suggest that in the short and medium term, there might be lower cost mitigation options outside the transport sector. On the other hand, compared with GTEMs/NTEMs, some IAMs may have limited mitigation options available, including technology, behavioural changes, and policy tools especially for aviation and shipping. The models therefore rely on other sectors and/or negative emissions elsewhere to achieve the overall desired warming levels. This potential shortcoming should be kept in mind when interpreting the sectoral results from IAMs.

### 10.7.3 Transport Activity Trajectories

Growth in passenger and freight travel demand is strongly dependent on population growth and GDP. In 2015, transport activities were estimated at around 35–50 trillion pkm, or 5,000–7,000 pkm per person per year, with significant variations among studies (IEA 2017b; ITF 2019). The number of passenger cars in use has grown 45% globally between 2005–2015, with the most significant growth occurring in the developing countries of Asia and the Middle East (119%), Africa (79%), and South and Central America (80%), while growth in Europe and North America is the slowest (21% and 4% respectively) (IOMVM 2021). On the other hand, car ownership

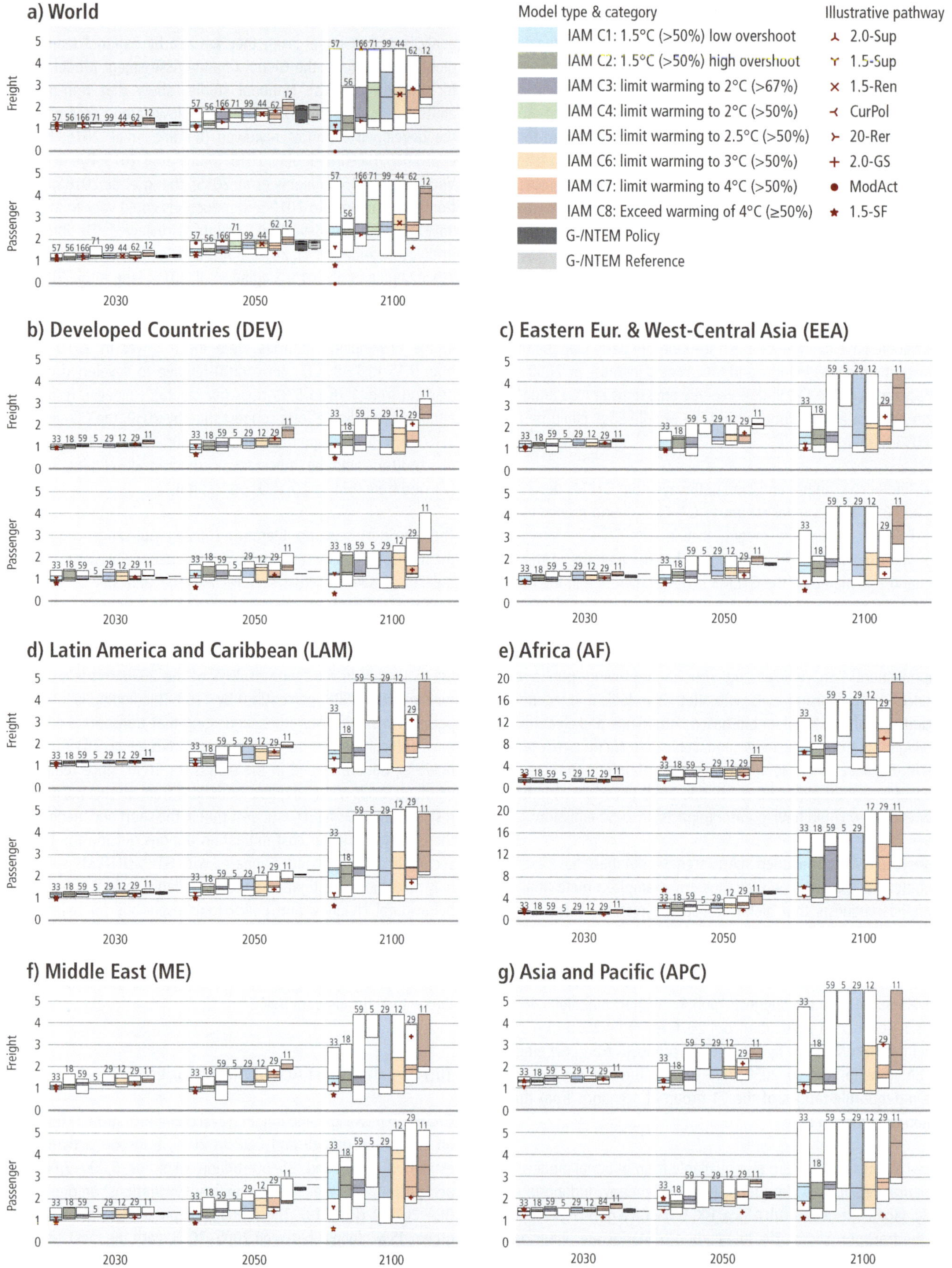


**Figure 10.18 | Transport activity trajectories for passenger (bottom panel) and freight (top panel) in 2030, 2050, and 2100 indexed to 2020 modelled year across R6 Regions and World.** Plots show 5–95th percentile, 25–75th percentile, and median. Numbers above the bars indicate the number of scenarios. Data from the AR6 scenario database.

levels in terms of vehicles per 1000 people in 2015 were low in developing countries of Asia and the Middle East (141), Africa (42), South and Central America (176), while in Europe and North America they are relatively high (581 and 670 respectively) (IOMVM 2021). The growth rate in commercial vehicles (freight and passenger) was 41% between 2005 and 2015, with a somewhat more even growth across developed and developing countries (IOMVM 2021).

Figure 10.18 shows activity trajectories for both freight and passenger transport based on the AR6 database for IAMs. According to demand projections from the IAMs, global passenger and freight transport demand could increase relative to a modelled year 2020 across temperature goals. The median transport demand from IAMs for all the scenarios in line with warming levels below 2.5°C (C1–C5) suggests that global passenger transport demand could grow by 1.14–1.3 times in 2030 and by 1.5–1.8 times in 2050 (1.27–2.33 for the 5–95th percentile across C1–C5 scenarios) relative to modelled 2020 level. Developed regions including North America and Europe exhibit lower growth in passenger demand in 2050 compared to developing countries across all the scenarios. In 2030, most of the global passenger demand growth happens in Africa (44% growth relative to 2020), and Asia and Pacific (57% growth in China and 59% growth in India relative to 2020) in the scenarios that limit warming below 2.5°C (>50%) throughout the 21st century (C5). These regions start from a low level of per capita demand. For example, in India, demand may grow by 84%. However, the per capita demand in 2010 was under 7000 km per person per year (Dhar and Shukla 2015). Similarly, in China, demand may grow by 52%, starting from per capita demand of 8000 km per person per year in 2010 (Pan et al. 2018). The per capita passenger demand in these regions was lower than in developed countries in 2010, but it converges towards the per capita passenger transport demand of advanced economies in less stringent climate scenarios (C6–C7). Demand for passenger travel would grow at a slower rate in the stricter temperature stabilisation scenarios (<2.5°C and 1.5°C scenarios, C1–C5) compared to the scenarios with higher warming levels (C7–C8). The median global passenger demand in the scenarios that limit or return warming to 1.5°C during the 21st century (C1–C2) is 27% lower in 2050 relative to C8.

Due to limited data availability, globally consistent freight data is difficult to obtain. In 2015, global freight demand was estimated to be 108 trillion tkm, most of which was transported by sea (ITF 2019). The growth rates of freight service demand vary dramatically among different regions: over the 1975–2015 period, road freight activity in India increased more than 9-fold, 30-fold in China, and 2.5-fold in the US (Mulholland et al. 2018). Global freight demand continues to grow but at a slower rate compared to passenger demand across all the scenarios in 2050 compared to modelled 2020 values. Global median freight demand could increase by 1.17–1.28 times in 2030 and 1.18–1.7 times in 2050 in all the scenarios with warming below 2.5°C (C1–C5). Like passenger transport, the models suggest that a large share of growth occurs in Africa and Asian regions (59% growth in India and 50% growth in China in 2030 relative to a modelled year 2020) in the C5 scenarios that limit warming below 2.5°C (>50%) throughout the 21st century. Global median freight demand grows more slowly in the stringent temperature stabilisation scenarios, and is 40% and 22% lower in 2050 in the scenarios that limit or return warming to 1.5°C (>50%) during the 21st century (C1–C2) and below 2.5°C scenarios (C3–C4), respectively, compared to scenarios with warming levels of above 4°C (C8).

GTEMs show broad ranges for future travel demand, particularly for the freight sector. These results show more dependency on models than on baseline or policy scenarios. According to ITF Transport Outlook (ITF 2019), global passenger transport and freight demand could more than double by 2050 in a business-as-usual scenario. Mulholland et al. (2018) suggest the freight sector could grow 2.4-fold over 2015–2050 in the reference scenario, with the majority of growth attributable to developing countries. The IEA suggests a more modest increase in passenger transport, from 51 trillion pkm in 2014 to 110 trillion pkm in 2060, in a reference scenario without climate policies and a climate scenario that would limit emissions below 2°C. The demand for land-based freight transport in 2060 is, however, slightly lower in the climate scenario (116 trillion tkm) compared to the reference scenario (130 trillion tkm) (IEA 2017b). The ITF, however, suggests that ambitious decarbonisation policies could reduce global demand for passenger transport by 13–20% in 2050, compared to the business-as-usual scenario (ITF 2019; ITF 2021). The reduction in vehicle travel through shared mobility could reduce emissions from urban passenger transport by 30% compared to the business-as-usual scenario. Others suggest that reductions larger than 25%, on average, for both passenger and freight in 2030 and 2050 may be needed to achieve very low carbon emissions pathways (Fisch-Romito and Guivarch 2019). In the absence of large-scale carbon dioxide removal, few global studies highlight the need for significant demand reduction in critical sectors (aviation, shipping and road freight) in well below 2°C scenarios (van Vuuren et al. 2018; Grant et al. 2021; Sharmina et al. 2021).

Many models find small differences in passenger transport demand across temperature goals because IAM models rely on historical relationships between population, GDP, and demand for services to estimate future demand. This assumption poses a limitation to the modelling efforts, as mitigation efforts would likely increase travel costs that could result in lower transport demand (Zhang et al. 2018). In most models, demand is typically an exogenous input. These models often assume mode shifts of activities from the most carbon-intensive modes (driving and flying for passenger travel and trucking for freight) to less carbon-intensive modes (public transit and passenger rail, and freight rail) to reduce emissions.

Traditionally there is a disconnection between IAM models and bottom-up sectoral or city-based models due to the different scale (both spatial and temporal) and focus (climate mitigation vs urban pollution, safety (Creutzig 2016)). The proliferation of shared and on-demand mobility solutions is leading to rebound effects for travel demand (Chen and Kockelman 2016; Coulombel et al. 2019) and this is a new challenge for modelling. Some IAM studies have recently begun to explore demand-side solutions for reducing transport demand to achieve very low-carbon scenarios through a combination of culture and low-carbon lifestyle (Creutzig et al. 2018; van Vuuren et al. 2018); urban development (Creutzig et al. 2015a); increased vehicle occupancy (Grubler et al. 2018); improved logistics and streamlined supply chains for the freight sector (Mulholland et al. 2018); and disruptive low-carbon innovation, described as technological and business model innovations offering 'novel value propositions to consumers and which can reduce GHG emissions if adopted at scale' (Wilson et al. 2019). In the literature from national models, demand has been differentiated between conventional and sustainable development scenarios through narratives built around policies, projects, and programmes envisaged at the national level (Dhar and Shukla 2015; Shukla et al. 2015) and price elasticities of travel demand (Dhar et al. 2018). However, a greater understanding of the mechanisms underlying energy-relevant decisions and behaviours (Brosch et al. 2016), and the motivations for sustainable behaviour (Steg et al. 2015), are critically needed to realise these solutions.

Overall, passenger and freight activity are likely to continue to grow rapidly under the C7 (>3.0°C) scenarios, but most growth would occur in developing countries. Most models treat travel demand exogenously following the growth of population and GDP, but they have limited representation of responses to price changes, policy incentives, behavioural shifts, nor innovative mobility solutions that can be expected to occur in more stringent mitigation scenarios. Chapter 5 provides a more detailed discussion of the opportunities for demand changes that may result from social and behavioural interventions.



### 10.7.4 Transport Modes Trajectories

Globally over the last century, shares of faster transport modes have generally increased with increasing passenger travel demand (Schäfer 2017; Schafer and Victor 2000). For short- to medium-distance travel, private cars have displaced public transit, particularly in OECD countries, due to a variety of factors, including faster travel times in many circumstances (Liao et al. 2020); consumers increasingly valuing time and convenience with GDP growth; and broader transport policies, such as provision of road versus public transit infrastructure (Mattioli et al. 2020). For long-distance travel, travel via aviation for leisure and business has increased (Lee et al. 2021). These trends do not hold in all countries and cities, as many now have rail transit that is faster than driving (Newman et al. 2015). For instance, public transport demand rose from 1990 through to 2016 in France, Denmark, and Finland (eurostat 2019). In general, smaller and denser countries and cities with higher or increasing urbanisation rates tend to have greater success in increasing public transport share. However, other factors, like privatisation of public transit (Bayliss and Mattioli 2018) and urban form (ITF 2021), also play a role. Different transport modes can provide passenger and freight services, affecting the emissions trajectories for the sector.

Figure 10.19 shows activity trajectories for freight and passenger transport through 2100 relative to a modelled year 2020 across different modes, based on the AR6 database for IAMs and global transport models. Globally, climate scenarios from IAMs, and policy and reference scenarios from global transport models, indicate increasing demand for freight and passenger transport via most modes through 2100 (Yeh et al. 2017; Mulholland et al. 2018; Zhang et al. 2018; Khalili et al. 2019). Road passenger transport exhibits a similar increase (roughly tripling) through 2100 across scenarios. For road passenger transport, scenarios that limit or return warming to 1.5°C during the 21st century (C1–C2) have a smaller increase from modelled 2020 levels (median increase of 2.4 times modelled 2020 levels) than do scenarios with higher warming levels (C3–C8) (median increase of 2.7–2.8 times modelled 2020 levels). There are similar patterns for passenger road transport via light-duty vehicle, for which median increases from modelled 2020 levels are smaller for C1–C2 (3 times larger) than for C3–C5 (3.1 times larger) or C6–C7 (3.2 times larger). Passenger transport via aviation exhibits a 2.2 times median increase relative to modelled 2020 levels under C1–C2 and C3–C5 scenarios but exhibits a 6.2 times increase under C6–C8. The only passenger travel mode that exhibits a decline in its median value through 2100 according to IAMs is walking/bicycling, in C3–C5 and C6–C8 scenarios. However, in C1–C2 scenarios, walking/bicycling increases by 1.4 times relative to modelled 2020 levels. At the 5th percentile of IAM solutions (lower edge of bands in Figure 10.19), buses and walking/bicycling for passenger travel both exhibit significant declines.

For freight, Figure 10.19 shows that the largest growth occurs in transport via road (Mulholland et al. 2018). By 2100, global transport models suggest a roughly four-fold increase in median-heavy-duty trucking levels relative to modelled 2020 levels, while IAMs suggest a two- to four-fold increase in freight transport by road by 2100. Notably, the 95th percentile of IAM solutions see road transport by up to 4.7 times through 2100 relative to modelled 2020 levels, regardless of warming level. Other freight transport modes – aviation, international shipping, navigation, and railways – exhibit less growth than road transport. In scenarios that limit or return warming to 1.5°C (>50%) during the 21st century (C1–C2), navigation and rail transport remain largely unchanged and international shipping roughly doubles by 2100. Scenarios with higher warming (i.e., moving from C1–C2 to C6–C8) generally lead to more freight by rail and less freight by international shipping.

Relative to global trajectories, upper-income regions – including North America, Europe, and the Pacific OECD – generally see less growth in passenger road via light-duty vehicle and passenger aviation, given more saturated demand for both. Other regions like China exhibit similar modal trends as the global average, whereas regions such as the African continent and Indian subcontinent exhibit significantly larger shifts, proportionally, in modal transport than the globe. In particular, the African continent represents the starkest departure from global results. Freight and passenger transport modes exhibit significantly greater growth across Africa than globally in all available scenarios. Across Africa, median freight and passenger transport via road from IAMs increases by 5 to 16 times and 4 to 28 times, respectively, across warming levels by 2100 relative to modelled 2020 levels. Even C1 has considerable growth in Africa via both modes (3 to 16 times increase for freight and 4 to 29 times increase for passenger travel at 5th and 95th percentiles of IAM solutions by 2100).

As noted in Section 10.2, commonly explored mitigation options related to mode change include a shift to public transit, shared mobility, and demand reductions through various means, including improved urban form, teleconferences that replace passenger

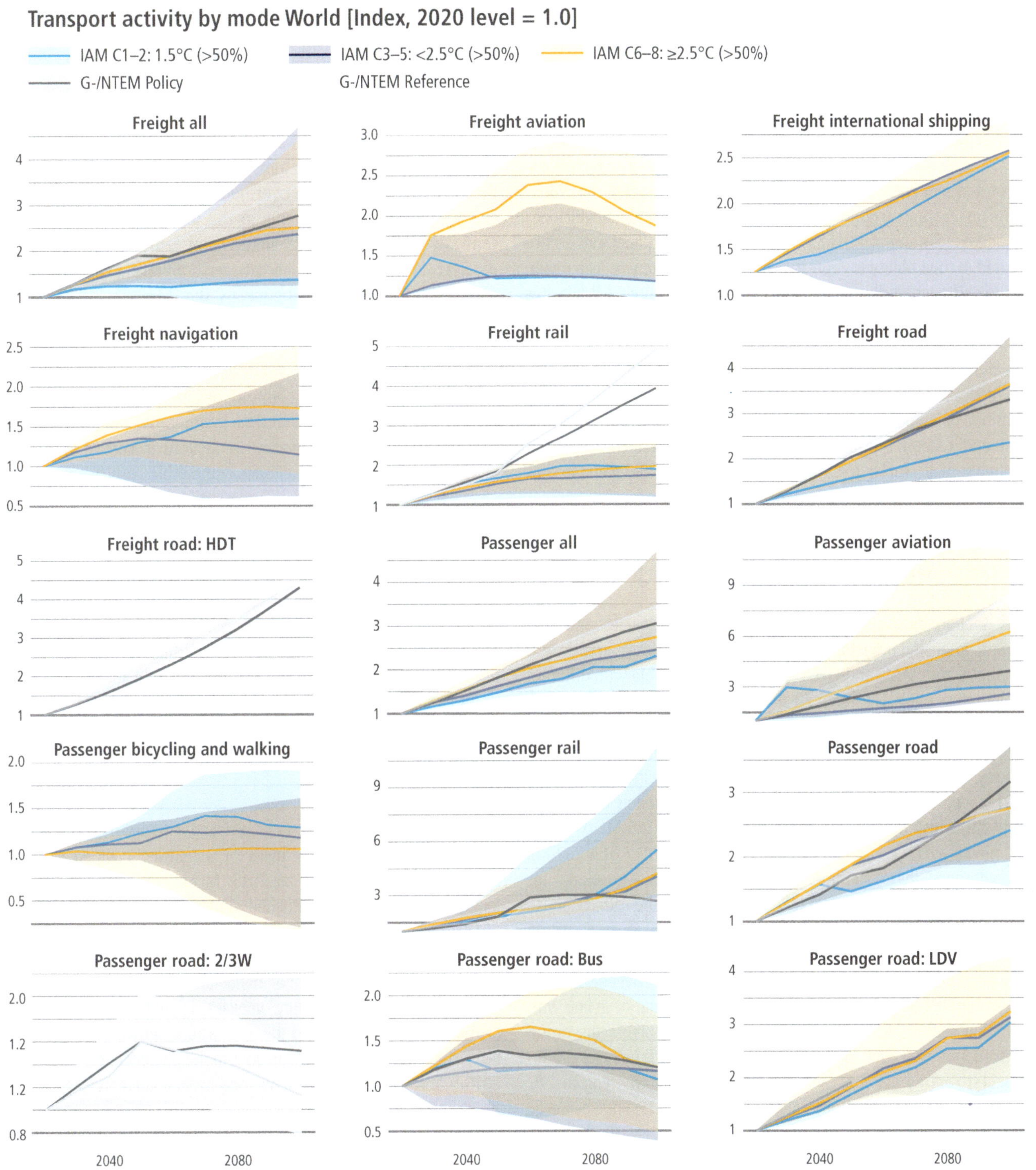


**Figure 10.19 | Transport activity trajectories for passenger and freight across different modes.** Global passenger (billion pkm per year) and freight (billion tkm per year) demand projections relative to a modelled year 2020 index. Results for IAM are for selected stabilisation temperatures by 2100. Also included are global transport models Reference and Policy scenarios. Data from the AR6 scenario database. Trajectories span the 5th to 95th percentiles across models, with a solid line indicating the median value across models.

travel (Creutzig et al. 2018; Grubler et al. 2018; Wilson et al. 2019), improved logistics efficiency, green logistics, and streamlined supply chains for the freight sector (Mulholland et al. 2018). NDCs often prioritise options like bus improvements and enhanced mobility that yield pollution, congestion, and urban development co-benefits, especially in medium- and lower-income countries (Fulton et al. 2017). Conversely, high-income countries, most of which have saturated and entrenched private vehicle ownership, typically focus more on technology options, such as electrification and fuel efficiency standards (Gota et al. 2016). Available IAM and regional models are limited in their ability to represent modal shift strategies. As a result, mode shifts alone do not differentiate climate scenarios. While this lack of representation is a limitation of the models, it is unlikely that such interventions would completely negate the increases in demand the models suggest. Therefore, transport via light-duty vehicle and aviation, freight transport via road, and other modes will likely continue to increase through to the end of the century. Consequently, fuel and carbon efficiency and fuel energy and technology will probably play crucial roles in differentiating climate scenarios, as discussed in the following sub-sections.



### 10.7.5 Energy and Carbon Efficiency Trajectories

This section explores what vehicle energy efficiencies and fuel carbon intensity trajectories, from the data available in the AR6 database from IAMs and GTEMs, could be compatible with different temperature goals. Figure 10.20 shows passenger and freight energy intensity, and fuel carbon intensity, indexed relative to 2020Mod values. The top panel shows passenger energy intensity across all modes. LDVs constitute a major share of this segment. Yeh et al. (2017) report 2.5–2.75 MJ vkm$^{-1}$ in 2020 across models for the LDV segment, which is very close to the IEA estimate of 2.5 MJ vkm$^{-1}$ for the global average fuel consumption for LDVs in 2017 (IEA 2020d). For reference, these numbers correspond to 1.6–1.7 MJ pkm$^{-1}$ for an occupancy rate of 1.5. The following results of the AR6 database are conditional on the corresponding reductions in fuel carbon intensity. Figure 10.20 shows that the scenarios suggest that passenger transport's energy intensity drops to between 10–23% (interquartile ranges across C1–C4) in 2030 for scenarios in line with warming levels below 2°C. In 2050, the medians across the group of scenarios that limit or return warming to 1.5°C (>50%) during the 21st century (C1–C2), and scenarios that limit warming to 2°C (>67% or >50%) throughout the 2st century (C3–C4) suggest energy intensity reductions of 51% and 45–46% respectively. These values correspond to annual average energy efficiency improvement rates of 2.3–2.4% and 2.0–2.1%, respectively, from 2020 to 2050. For reference, the IEA reports an annual energy efficiency improvement rate of 1.85% per year in 2005–16 (IEA 2020d). In contrast, the results from GTEMs suggest lower energy efficiency improvement, with median values for policy scenarios of 39% reduction in 2050, corresponding to annual energy efficiency improvement rates close to 1.6%. The IAM scenarios suggest median energy intensity reductions of passenger transport of 57–61% by the end of the century would align with warming levels of both 1.5°C and 2°C (C1–C4) given the corresponding decarbonisation of the fuels.

The scenarios in line with warming levels of 1.5°C or 2°C goals (C1 to C4) show different trends for freight's energy intensity. The amount of overshoot and differences in demand for freight services and, to some extent, fuel carbon intensities contribute to these differences. For the two scenarios aligning with the warming levels of 1.5°C, the trajectories in 2030 and 2050 are quite different. The median C2 scenario that returns warming to 1.5°C (>50%) during the 21st century after high overshoot takes a trajectory with lower energy intensity improvements in the first half of the century. In contrast, the C1 scenario that limits warming to 1.5°C (>50%) during the 21st century with no or limited overshoot take on a more steadily declining trajectory across the means. The IAMs provide a less clear picture of required energy intensity improvements for freight than for passenger transport associated with different temperature targets. As for the carbon intensity of direct energy used across both passenger and freight, the modelling scenarios suggest very moderate reductions by 2030. The interquartile ranges for the C1 scenarios suggest global average reductions in carbon intensity of 5–10%. Across the other scenarios compatible with warming levels of 1.5°C or 2°C (C2–C4), the interquartile ranges span from 1–6% reductions in carbon intensity of direct energy used for transport. For 2050, the scenarios suggest that dependence on fuel decarbonisation increases with more stringent temperature targets. For the scenarios that limits warming to 1.5°C (>50%) during the 21st century with no or limited overshoot (C1), global carbon intensity of energy used for transport decreases by 37–60% (interquartile range) by 2050 with a mean of 50% reduction. The IAM scenarios in the AR6 database do not suggest full decarbonisation of transport fuels by 2100. The interquartile ranges across the C1–C4 set of scenarios, compatible with warming levels of 2°C and less, span from 61–91% reduction from 2020Mod levels.

Increasing the occupancy rate of passenger transport (Grubler et al. 2018) and reducing empty miles or increasing payload in freight deliveries (Gucwa and Schäfer 2013; McKinnon 2018) via improved logistics efficiency or streamlined supply chains (Mulholland et al. 2018), can present significant opportunities to effectively improve energy efficiency and decrease GHG emissions in transport. However, the recent trends of consumer behaviours have shown a declining occupancy rate of light-duty vehicles in industrialised countries (Schäfer and Yeh 2020), and the accelerating growing preference for SUVs challenges emissions reductions in the passenger car market (IEA 2019d). These trends motivate a strong focus on demand-side options.

Based on the scenario literature, a 51% reduction in median energy intensity of passenger transport and a corresponding 38–50% reduction in median carbon intensity by 2050 would be aligned with transition trajectories yielding warming levels below 1.5°C by the end of the century. For comparison, the LCA literature suggests a switch from current ICEs to current BEVs would yield a reduction in energy intensity well beyond 45% and up to 70%, for a mid-sized vehicle (Section 10.4). Correspondingly, a switch from diesel or gasoline to low-carbon electricity or low-carbon hydrogen would yield carbon intensity reduction beyond the median scenario value. Thus, the LCA literature suggests technologies exist today that would already match and exceed the median energy and carbon intensities values that might be needed by 2050 for low warming levels.

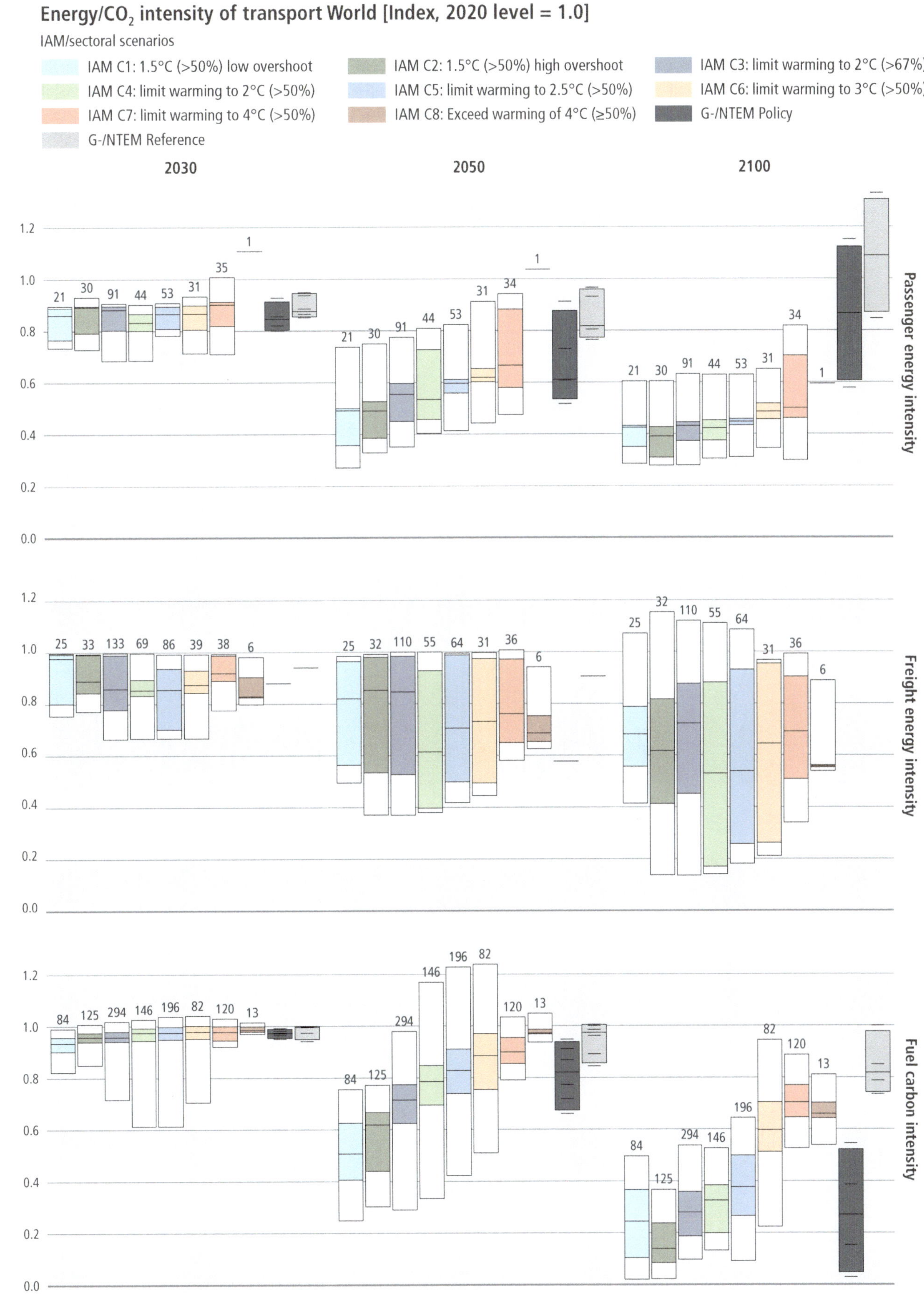


Figure 10.20 | Energy efficiency and carbon intensity of transport in 2030, 2050, and 2100 indexed to 2020 modelled year across scenarios. Plots show 5th/95th percentile, 25th/75th percentile, and median. Numbers above the bars indicate the number of scenarios. Data from the AR6 scenario database.

### 10.7.6 Fuel Energy and Technology Trajectories

Two mechanisms for reducing carbon emissions from the transport sector are fuel switching for current vehicle technologies and transitioning to low-carbon vehicle technologies. Figure 10.21 combines data from IAMs and GTEMs on shares of transport final energy by fuel. These shares account for fuel uses across modes – road, aviation, rail, and shipping – and both passenger and freight transport. Since the technologies have different conversion efficiencies, these shares of final energy by fuel are necessarily different from the shares by service (passenger-km or tonne-km) by fuel and shares of vehicle stock by fuel. For example, a current battery electric LDV powertrain is roughly three times more energy-efficient than a comparable ICE powertrain (Section 10.3, Table 10.9 in Appendix 10.1); thus, fuel

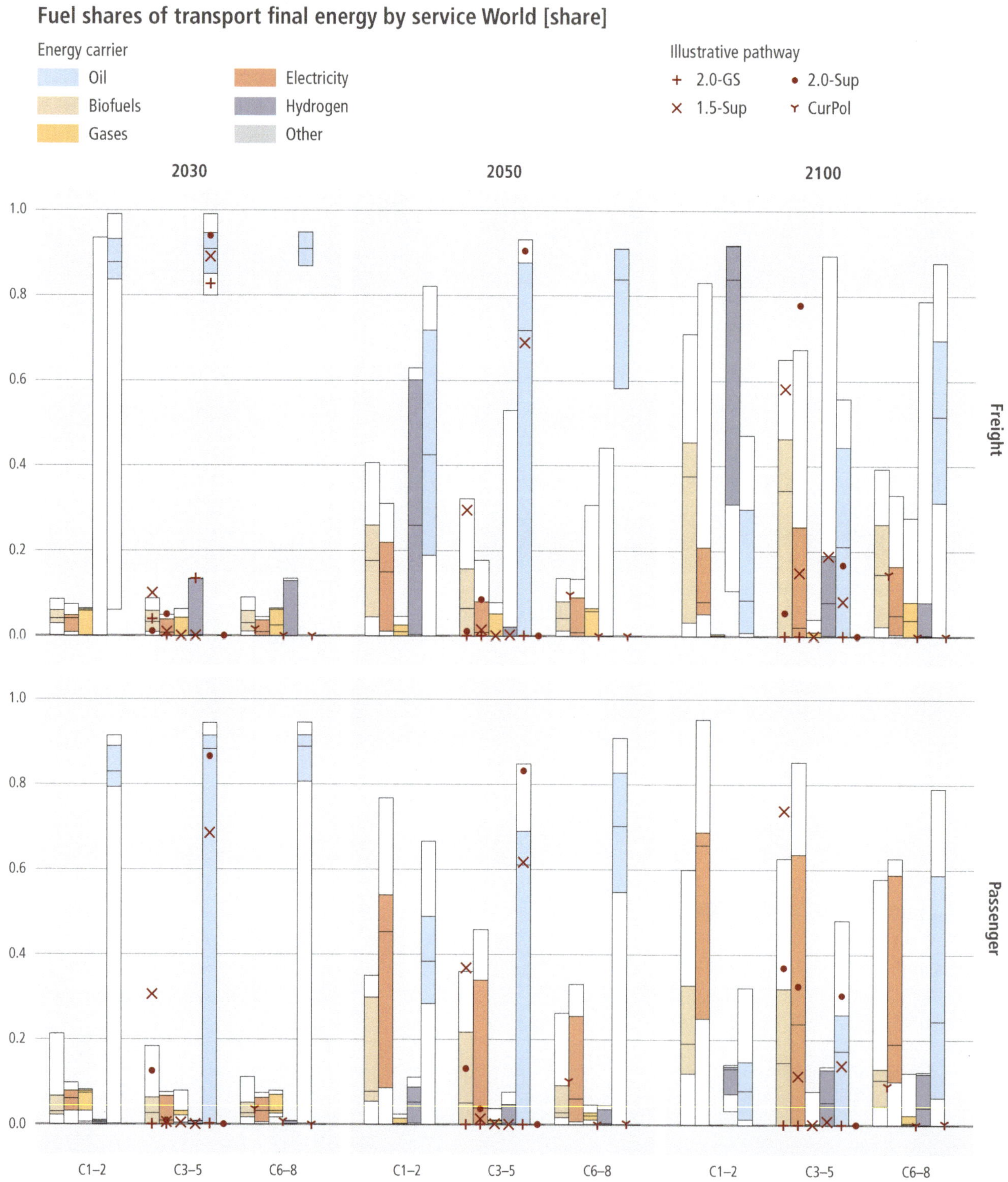


**Figure 10.21 | Global shares of final fuel energy in the transport sector in 2030, 2050, and 2100 for freight and passenger vehicles.** Plots show 10th/90th percentile, 25th–75th percentile, and median. Data from the AR6 scenario database.

shares of 0.25 for electricity and 0.75 for oil could correspond to vehicle stock shares of 0.5 and 0.5, respectively. In general, while models may project that EVs constitute a greater share of road vehicle stock, and provide a greater share of road passenger-kilometres, their share of transport final energy (Figure 10.21) can still remain lower than the final energy share of fuels used in less-efficient (e.g., ICE) vehicles. Thus, the shares of transport final energy by fuel presented in Figure 10.21 should be interpreted with care.

IAM and GTEM scenarios indicate that fuel and technology shifts are crucial to reduce carbon emissions to achieve lower levels of warming (Edelenbosch et al. 2017; IEA 2017b). Across the transport sector, a technology shift towards advanced fuel vehicles is the dominant driver of decarbonisation in model projections. This trend is consistent across climate scenarios, with larger decreases in the final energy share of oil in scenarios that achieve progressively lower levels of warming. Due to efficiency improvements, the higher efficiency of advanced fuel vehicles, and slower progress in the freight sector, the final energy share of oil decreases more rapidly after 2030. By 2050, the final energy shares of electricity, biofuels, and alternative gaseous fuels increase, with shares from electricity generally about twice as high (median values from 10–30% across warming levels) as the shares from biofuels and gases (median values from 5–10%). While IAMs suggest that the final energy share of hydrogen will remain low in 2050, by 2100 the median projections include 5–10% hydrogen in transport final energy.

While few IAMs report final energy shares by transport mode or passenger/freight, several relevant studies provide insights into fuel share trends in passenger LDVs and freight vehicles. The IEA suggests that full LDV electrification would be the most promising low-carbon pathway to meet a 1.75°C goal (IEA 2017b). The MIT Economic Projection and Policy Analysis model focuses on the future deployment of gasoline versus EV technologies in the global LDV stock (Ghandi and Paltsev 2019). These authors estimate that the global stock of vehicles could increase from 1.1 billion vehicles in 2015 up to 1.8 billion by 2050, with a growth in EVs from about 1 million vehicles in 2015 up to 500 million in 2050. These changes are driven primarily by cost projections (mostly battery cost reductions). Similarly, the International Council on Clean Transport (ICCT) indicates that EV technology adoption in the light-duty sector can lead to considerable climate benefits. Their scenarios reach nearly 100% electrification of LDVs globally, leading to global GHG emissions from LDVs ranging from 0% to 50% of 2010 levels in 2050 (Lutsey 2015). Khalili et al.(2019) estimate transport stocks through 2050 under aggressive climate mitigation scenarios that nearly eliminate road transport emissions. They find the demand for passenger transport could triple through 2050, but emissions targets could be met through widespread adoption of BEVs (80% of LDVs) and, to a lesser extent, fuel cell and plug-in hybrid electric vehicles. Contrary to these estimates, the US Energy Information Administration finds small adoption of electrification for LDVs and instead identifies diffusion of natural gas-fuelled LDVs in OECD and, to a greater extent, non-OECD countries through 2040. This trend occurs in a reference and a 'low liquids' case, which lowers LDV ownership growth rates and increases preferences for alternative fuel vehicles. A comprehensive overview of regional technology adoption models across many methodological approaches can be found in Jochem et al. (2018).

In freight transport, studies indicate a shift toward alternative fuels would need to be supplemented by efficiency improvements. The IEA suggests efficiency improvements would be essential for decarbonisation of trucks, aviation, and shipping in the short-to-medium term. At the same time, the IEA suggests that fuel switching to advanced biofuels would be needed to decarbonise freight in the long term (IEA 2019d). Mulholland et al. (2018) investigated the impacts of decarbonising road freight in two scenarios: countries complying with COP21 pledges and a second more ambitious reduction scenario in line with limiting global temperature rise to 1.75°C. Despite the deployment of logistics improvements, high-efficiency technologies, and low-carbon fuels, activity growth leads to a 47% increase in energy demand for road freight while overall GHG emissions from freight increase by 55% (4.8 $GtCO_2$-eq) in 2050 (relative to 2015) in the COP21 scenario. In the 1.75°C scenario, decarbonisation happens primarily through a switch to alternative fuels (hybrid electric and full battery electric trucks), which leads to a 60% reduction in GHG emissions from freight in 2050 relative to 2015. Khalili et al. (2019) also find substantial shifts to alternative fuels in HDVs under aggressive climate mitigation scenarios. Battery electric, hydrogen fuel cell, and plug-in hybrid electric vehicles constitute 50%, 30%, and 15% of heavy-duty vehicles respectively in 2050. They also find 90% of buses would be electrified by 2050.

## Box 10.4 | Three Illustrative Mitigation Pathways

Section 10.7 presents the full set of scenarios in the AR6 database and highlights the broader trends of how the transport sector may transform in order to be compliant with different warming levels. This box elaborates on three illustrative mitigation pathways (IMPs) to exemplify a few different ways the sector may transform. Seven illustrative pathways are introduced in Section 3.2.5. In this box we focus on three of the IMPs: (i) focus on deep renewable energy penetration and electrification (IMP-Ren), (ii) low demand (IMP-LD), and (iii) pathways that align with both Sustainable Development Goals and climate policies (IMP-SP). In particular, the variants of these three scenarios limit warming to 1.5°C with no or limited overshoot (C1).

All of the three selected pathways reach global net zero $CO_2$ emissions across all sectors between 2060 and 2070, but not all reach net zero GHG emissions (Figure 3.4). Panel (a) in Box 10.4, Figure 1 below shows the $CO_2$ trajectories for the transport sector for the selected IMPs. Please note that the year 2020 is modelled in these scenarios, therefore, the scenarios do not reflect the effects of

*Box 10.4 (continued)*

the COVID-19 pandemic. For the low demand scenarios IMP-LD and renewables pathway IMP-Ren, $CO_2$ emissions from the transport sector decreases to 10% and 20% of modelled 2020 levels by 2050 respectively. In contrast, the IMP-SP has a steady decline of transport sector $CO_2$ emissions over the century. By 2050, this scenario has a 50% reduction in emissions compared to modelled 2020 levels. Panels (b), (c) and (d) show energy by different fuels for the three selected IMPs. The IMP-SP yields a drop in energy for transport of about 40% by the end of the century. $CO_2$ emissions reductions are obtained through a phase-out of fossil fuels with electricity and biofuels, complemented by a minor share of hydrogen, by the end of the century. In IMP-Ren, the fuel energy demand at the end of the century is on a par with the 2020 levels, but the fuel mix has shifted towards a larger share of electricity complimented by biofuels and a minor share of hydrogen. For the IMP-LD scenario, the overall fuel demand decreases by 45% compared to 2020 levels by the end of the century. Oil is largely phased out by mid century, with electricity and hydrogen becoming the major fuels in the second half of the century. Across the three IMPs, electricity plays a major role, in combination with biofuels, hydrogen, or both.

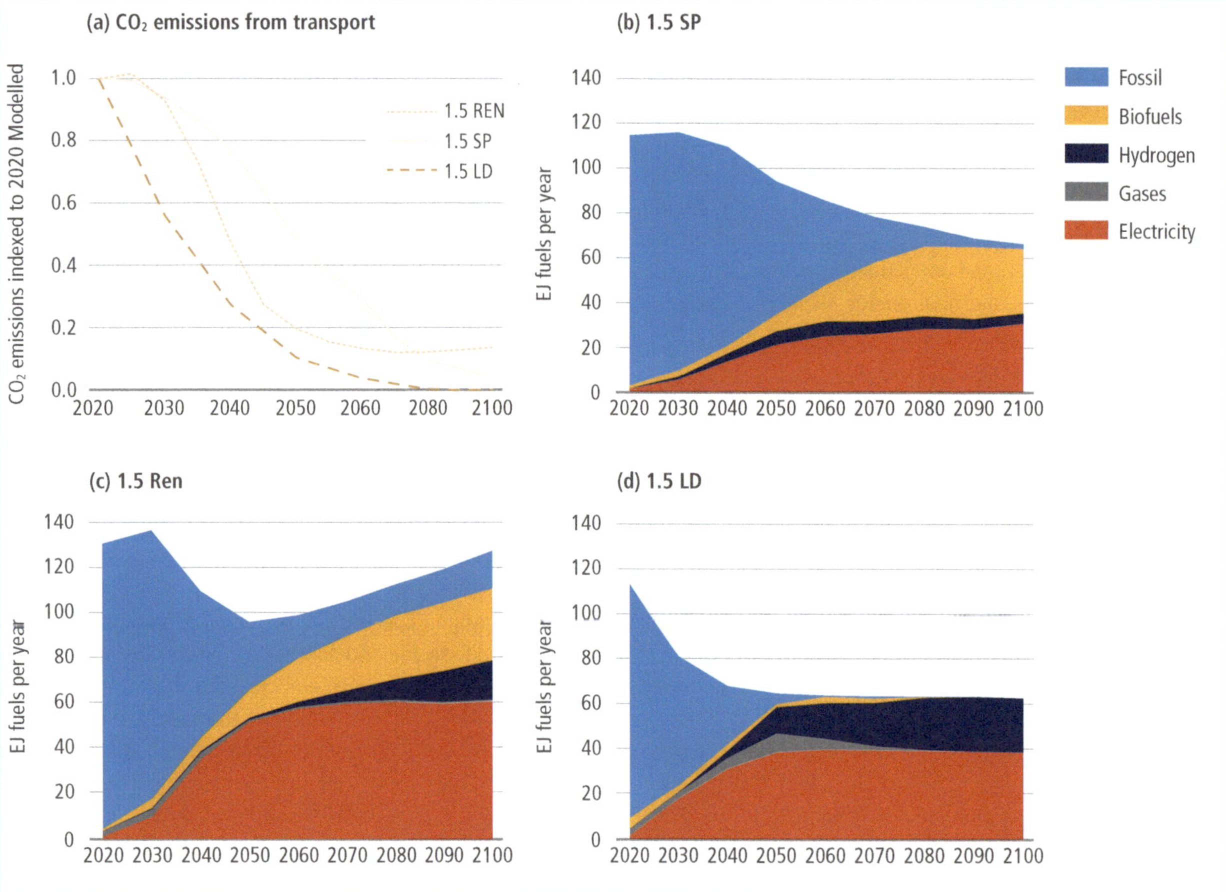


**Box 10.4, Figure 1 | Three Illustrative mitigation pathways for the Transport sector.** Panel **(a)** shows $CO_2$ emissions from the transport sector indexed to simulated non-COVID-2019 2020 levels. Panels **(b)**, **(c)**, and **(d)** show fuels mix to achieve 1.5°C warming through three illustrative mitigation pathways: IMP-SP, 1.5 IMP-Ren and IMP-LD, respectively. All data from IPCC AR6 scenario database.

### 10.7.7 Insights from the Modelling Literature

This section provides an updated, detailed assessment of future transport scenarios from IAM, GTEMs, and NTEMs given a wide range of assumptions and under a set of policy targets and conditions. The scenario modelling tools are necessary to aggregate individual options and understand how they fit into mitigation pathways from a systems perspective. The scenarios suggest that 43% (30–63% for the interquartile ranges) reductions in $CO_2$ emissions from transport (below modelled 2020 levels) by 2050 would be compatible with warming levels of 1.5°C (C1–C2 group). While the global scenarios suggest emissions reductions in energy supply sectors at large precede those in the demand sectors (Section 3.4.1), a subset of the scenarios also demonstrate that more stringent emission reductions in the transport sector are feasible. For example, the illustrative mitigation pathways IMP-REN and IMP-LD suggest emissions reductions of 80%

and 90% respectively are feasible by 2050 *en route* to warming levels of 1.5°C with low or no overshoot by the end of the century.

The scenarios from the different models project continued growth in demand for freight and passenger services, particularly in developing countries. The potential for demand reductions is evident, but the specifics of demand-reduction measures remain less explored by the scenario literature. This limitation notwithstanding, the IAM and GTEMs suggest that interventions that reduce the energy and fuel carbon intensity of transport are likely crucial to successful mitigation strategies.

The scenario literature suggests that serious attempts at carbon mitigation in the transport sector must examine the uptake of alternative fuels. The scenarios described in the IAMs and GTEMs literature decarbonise through a combination of fuels. Across the scenarios, electrification plays a key role, complemented by biofuels and hydrogen. In general terms, electrification tends to play the key role in passenger transport while biofuels and hydrogen are more prominent in the freight segment. The three illustrative mitigation pathways described in Box 10.4 exemplify different ways these technologies may be combined and still be compatible with warming levels of 1.5°C with low or no overshoot. Shifts towards alternative fuels must occur alongside shifts towards clean technologies in other sectors, as all alternative fuels have upstream impacts. Without considering other sectors, fuel shifts would not yield their full mitigation potentials. These collective efforts are particularly important for the electrification of transport, as the transformative mitigation potential is strongly dependent on the decarbonisation of the power sector. In this regard, the scenario literature is well aligned with the LCA literature reviewed in Section 10.4.

The models reviewed in this section would all generally be considered to have a good representation of fuels, technologies, and costs, but they often better represent land transport modes than shipping and aviation. While these models have their strengths in some areas, they have some limitations in other areas, like behavioural aspects. These models are also limited in their ability to account for unexpected technological innovation, such as a breakthrough in heavy vehicle fuels, artificial intelligence, autonomy and big data, even the extent of digital communications replacing travel (Section 10.2). As a result of these limitations, the models cannot yet provide an exhaustive set of options for decarbonising the transport sectors. These limitations notwithstanding, the models can find solutions encompassing the transport sector and its interactions with other sectors that are compatible with stringent emissions mitigation efforts. The solutions space of transportation technology trajectories is therefore wider than explored by the models, so there is still a need to better understand how all options in combination may support the transformative mitigation targets.

## 10.8 Enabling Conditions

### 10.8.1 Conclusions Across the Chapter

This final section draws some conclusions from the chapter and provides an overview-based feasibility assessment of the major transport mitigation options, as well as a description of emerging issues. The section ends by outlining an integrated framework for enabling the transformative changes that are emerging and required to meet the potential transformative scenarios from Section 10.7.

Transport is becoming a major focus for mitigation as its GHG emissions are large and growing faster than those of other sectors, especially in aviation and shipping. The scenarios literature suggests that without mitigation actions, transport emissions could grow by up to 65% by 2050. Alternatively, successful deployment of mitigation strategies could reduce sectoral emissions by 68%, which would be consistent with the goal of limiting temperature change to 1.5°C above pre-industrial levels. This chapter has reviewed the literature on all aspects of transport and has featured three special points of focus: (i) a survey of lifecycle analysis from the academic and industry community that uses these tools; (ii) surveying the modelling community for top-down and bottom-up approaches to identify decarbonisation pathways for the transport sector, and (iii) for the first time in the IPCC, separate sections on shipping and aviation. The analysis of the literature suggests three crucial components for the decarbonisation of the transport sector: demand and efficiency strategies, electromobility, and alternative fuels for shipping and aviation.

The challenge of decarbonisation requires a transition of the socio-technical system, which depends on the combination of technological innovation and societal change (Geels et al. 2017). A socio-technical system includes technology, regulation, user practices and markets, cultural meaning, infrastructure, maintenance networks, and supply networks (Geels 2005) (Cross Chapter Box 12 in Chapter 16). The multi-level perspective (MLP) is a framework that provides insights to assist policymakers when devising transformative transition policies (Rip and Kemp 1998; Geels 2002). Under the MLP framework, strategies are grouped into three different categories. The Micro level (niche) category includes strategies where innovation differs radically to that of the incumbent socio-technical system. The niche provides technological innovations a protected space during development and usually requires considerable R&D and demonstrations. In the Meso level (regime) state, demonstrations begin to emerge as options that can be adopted by leading groups who begin to overcome lock-in barriers from previous technological dependence. Finally, in the Macro level (landscape) stage, mainstreaming happens, and the socio-technical system enables innovations to break through. Figure 10.22 maps the MLP stages for the major mitigation strategies identified in this chapter.

**Demand and behaviour.** While technology options receive substantial attention in this chapter, there are many social and equity issues that cannot be neglected in any transformative change to mitigate climate change. Transport systems are socio-economic systems that include systemic factors that are developing into potentially transformative drivers of emissions from the sector. These systemic drivers include, for example, changes in urban form that minimise automobile dependence and reduce stranded assets; behaviour change programmes that emphasise shared values and economies; smart technologies that enable better and more equitable options for transit and active transport as well as integrated approaches to using autonomous vehicles; new ways of

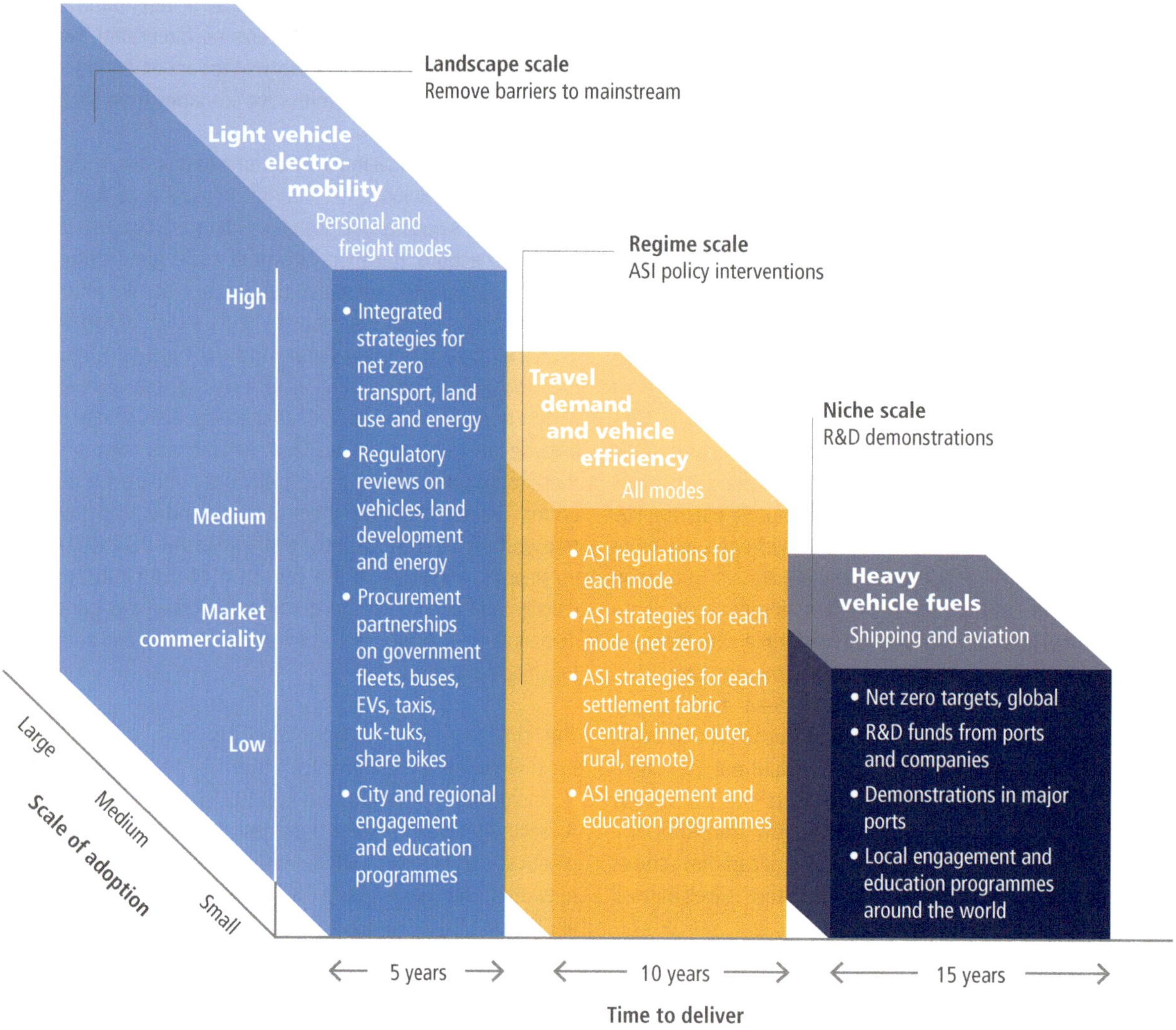


**Figure 10.22 | Mitigation options and enabling conditions for transport.** Niche scale includes strategies that still require innovation.

enabling electric charging systems to fit into electricity grids, creating synergistic benefits to grids, improving the value of electric transit, and reducing range anxiety for EV users; and new concepts for the future economy such as circular economy, dematerialisation, and shared economy that have the potential to affect the structure of the transport sector. The efficacy of demand reduction and efficiency opportunities depends on the degree of prioritisation and focus by government policy. Figure 10.22 suggests that innovative demand and efficiency strategies are at the regime scales. While these strategies are moving beyond R&D, they are not mainstreamed yet and have been shown to work much more effectively if combined with technology changes, as has been outlined in the transformative scenarios from Section 10.7 and in Chapter 5.

**Electromobility in land-based transport.** Since AR5, there has been a significant breakthrough in the opportunities to reduce transport GHG emissions in an economically efficient way due to electrification of land-based vehicle systems, which are now commercially available. EV technologies are particularly well established for light-duty passenger vehicles, including micromobility. Furthermore, there are positive developments to enable EV technologies for buses, light- and medium-duty trucks, and some rail applications (though advanced biofuels and hydrogen may also contribute to the decarbonisation of these vehicles in some contexts). In developing countries, where micromobility and public transit account for a large share of travel, EVs are ideal to support mitigation of emissions. Finally, demand for critical materials needed for batteries has become a focus of attention, as described in Box 10.6.

Electromobility options are moving from regime to landscape levels. This transition is evident in the trend of incumbent automobile manufacturers producing an increasing range of EVs in response to demand, policy, and regulatory signals. EVs for light-duty passenger travel are largely commercial and likely to become competitive with ICE vehicles in the early 2020s (Dia 2019; Bond et al. 2020; Koasidis et al. 2020). As these adopted technologies increase throughout cities and regions, governments and energy suppliers will have to deploy new infrastructure to support them, including reliable low-carbon grids and charging stations (Sierzchula et al. 2014). In addition, regulatory reviews will be necessary to ensure equitable transition and achievement of SDGs, addressing the multitude of possible barriers that may be present due to the incumbency of traditional automotive manufacturers and associated supporting elements of the socio-technical system (Newman 2020b) (Chapter 6). Similarly, new partnerships between government, industry, and communities will be needed to support the transition to electromobility. These partnerships could be particularly effective at supporting engagement and education programmes (Newman 2020b) (Chapter 8).

Deployment of electromobility is not limited to developed countries. The transportation sector in low- and middle-income countries includes millions of gas-powered motorcycles within cities across Africa, South-East Asia, and South America (Posada et al. 2011; Ehebrecht et al. 2018). Many of these motorcycles function as taxis. In Kampala, Uganda, estimates place the number of motorcycle taxis, known locally as *boda bodas*, at around 40,000 (Ehebrecht et al. 2018). The popularity of the motorcycle for personal and taxi use is due to many factors including lower upfront costs, lack of regulation, and mobility in highly congested urban contexts (Posada et al. 2011; UNECE 2018). While motorcycles are often seen as a more fuel-efficient alternative, emissions can be worse from two-wheelers than cars, particularly nitrogen oxides (NOx), carbon monoxide (CO), and hydrocarbon emissions (Vasic and Weilenmann 2006; Ehebrecht et al. 2018). These two-wheeler emissions contribute to dangerous levels of air pollution across many cities in low- and middle-income countries. In Kampala, for example, air pollution levels frequently exceed levels deemed safe for humans by the World Health Organization (Kampala Capital City Authority 2018; World Health Organization 2018; Airqo 2020). To mitigate local and environmental impacts, electric *boda boda* providers are emerging in many cities, including Zembo in Kampala and Ampersand in Kigali, Rwanda.

Bulawayo, the second-largest city in Zimbabwe, is also looking at opportunities for deploying electromobility solutions. The city is now growing again after a difficult recent history, and there is a new emphasis on achieving the Sustainable Development Goals (City of Bulawayo 2020a; City of Bulawayo 2020b). With these goals in mind, Bulawayo is seeking opportunities for investment that can enable leapfrogging in private, fossil fuel vehicle ownership. In particular, trackless trams, paired with solar energy, have emerged as a potential pathway forward (Kazunga 2019). Trackless trams are a new battery-based mid-tier transit system that could enable urban development around stations and that use solar energy for powering both transit and the surrounding buildings (Newman et al. 2019). The new trams are rail-like in their capacities and speed, providing a vastly better mobility system that is decarbonised and enables low transport costs (Ndlovu and Newman 2020). While this concept is only under consideration in Bulawayo, climate funding could enable the wider deployment of such projects in developing countries.

**Fuels for aviation and shipping.** Despite technology improvements for land-based transport, equivalent technologies for long distance aviation and shipping remain elusive. Alternative fuels for use in long-range aviation and shipping are restricted to the niche level. The aviation sector is increasingly looking towards synthetic fuels using low-carbon combined with $CO_2$ from direct air capture, while shipping is moving towards ammonia produced using low-carbon hydrogen. Biofuels are also of interest for these segments. To move out of the niche level, there is a need to set deployment targets to support breakthroughs in these fuels. Similarly, there is a need for regulatory changes to remove barriers in new procurement systems that accommodate uncertainty and risks inherent in the early adoption of new technologies and infrastructure (Borén 2019; Sclar et al. 2019; Marinaro et al. 2020). R&D programmes and demonstration trials are the best focus for achieving fuels for such systems. Finally, there is a need for regulatory changes. Such regulatory changes need to be coordinated through ICAO and IMO as well as with national implementation tools related to the Paris Agreement (see Box 10.5). Long-term visions, including creative exercises for cities and regions, will be required, providing a protected space for the purpose of trialling new technologies (Borén 2019; Geels 2019).

### 10.8.2 Feasibility Assessment

Figure 10.23 sets out the feasibility of the core mitigation options using the six criteria created for the cross-sectoral analysis. This feasibility assessment outlines how the conclusions outlined in Section 10.8.1 fit into the broader criteria created for feasibility in the whole AR6 report and that emphasise the SDGs. Figure 10.23 highlights that there is *high confidence* that demand reductions and mode shift can be feasible as the basis of a GHG emissions mitigation strategy for the transport sector. However, demand-side interventions work best when integrated with technology changes. The technologies that can support such changes have a range of potential limitations as well as opportunities. EVs have a reliance on renewable resources (wind, solar, and hydro) for power generation, which could pose constraints on geophysical resources, land use, and water use. Furthermore, expanding the deployment of EVs requires a rapid deployment of new power generation capacity and charging infrastructure. The overall feasibility of electric vehicles for land transport is likely high and their adoption is accelerating. HFCVs for land transport would also have constraints related to geophysical resource needs, land use, and water use. These constraints are likely higher than for EVs, since producing hydrogen with electricity reduces the overall efficiency of meeting travel demand. Furthermore, the infrastructure needed to produce, transport, and deliver hydrogen is under-developed and would require significant R&D and a rapid scale-up. Thus, the feasibility of HFCV is likely lower than for EVs. Biofuels could be used in all segments of the transport sector, but there may be some concerns about their feasibility. Specifically, there are concerns about land use, water use, impacts on water quality and eutrophication, and biodiversity impacts. Advanced biofuels could mitigate some concerns and the feasibility of using these fuels likely varies by world region. The feasibility assessment for alternative fuels for shipping and aviation suggests that hydrogen-based fuels like ammonia and synthetic fuels have the lowest technology readiness of all mitigation options considered in this chapter. Reliance on electrolytic hydrogen for the production of these fuels poses concerns about land and water use. Using ammonia for shipping could pose risks for air quality and toxic discharges to the environment. The DAC/BECCS infrastructure that would be needed to produce synthetic fuel does not yet exist. Thus, the feasibility suggests that the technologies for producing and using these hydrogen-based fuels for transport are in their infancy.

**Figure 10.23 | Summary of the extent to which different factors would enable or inhibit the deployment of mitigation options in transport.** Blue bars indicate the extent to which the indicator enables the implementation of the option (E) and orange bars indicate the extent to which an indicator is a barrier (B) to the deployment of the option, relative to the maximum possible barriers and enablers assessed. An 'X' signifies the indicator is not applicable or does not affect the feasibility of the option, while a forward slash indicates that there is no or limited evidence whether the indicator affects the feasibility of the option. The shading indicates the level of confidence, with darker shading signifying higher levels of confidence. Appendix 10.3 provides an overview of the extent to which the feasibility of options may differ across context (e.g., region), time (e.g., 2030 versus 2050), and scale (e.g., small versus large), and includes a line of sight on which the assessment is based. The assessment method is explained in Annex II.11.

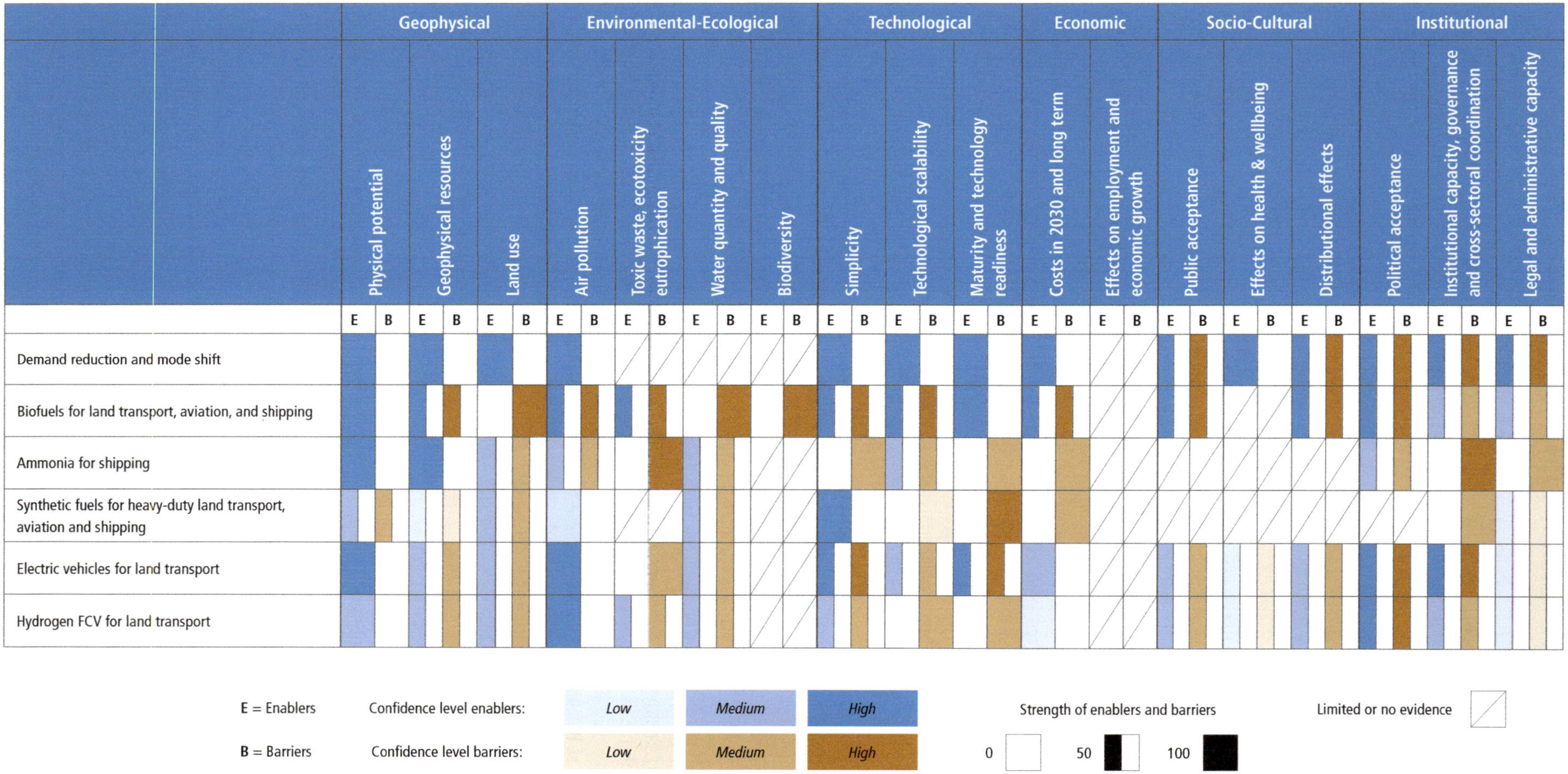

### 10.8.3 Emerging Transport Issues

**Planning for integration with the power sector:** Decarbonising the transport sector will require significant growth in low-carbon electricity to power EVs, and more so for producing energy-intensive fuels, such as hydrogen, ammonia and synthetic fuels. Higher electricity demand will necessitate greater expansion of the power sector and increase land use. The strategic use of energy-intensive fuels, focused on harder-to-decarbonise transport segments, can minimise the increase in electricity demand. Additionally, integrated planning of transport and power infrastructure could enable sectoral synergies and reduce the environmental, social, and economic impacts of decarbonising transport and energy. For example, smart charging of EVs could support more efficient grid operations. Hydrogen production, which is likely crucial for the decarbonisation of shipping and aviation, could also serve as storage for electricity produced during low-demand periods. Integrated planning of transport and power infrastructure would be particularly useful in developing countries where 'greenfield' development doesn't suffer from constraints imposed by legacy systems.

**Shipping and aviation governance:** Strategies to deliver fuels in sufficient quantity for aviation and shipping to achieve transformative targets are growing in intensity and often feature the need to review international and national governance. Some authors in the literature have argued that the governance of the international transport systems could be included in the Paris Agreement process (Gençsü and Hino 2015; Lee 2018; Traut et al. 2018). Box 10.6 sets out these issues.

## Box 10.5 | Governance Options for Shipping and Aviation

Whenever borders are crossed, the aviation and shipping sector creates international emissions that are not assigned to states' Nationally Declared Contributions under the Paris Agreement. Emissions from these segments are rapidly growing (apart from COVID-19 affecting aviation) and are projected to grow between 60% to 220% by 2050 (IPCC 2018; UNEP 2020). Currently, the International Civil Aviation Organization (ICAO) and the International Marine Organization (IMO), specialised UN Agencies, are responsible for accounting and suggesting options for managing these emissions.

### Transformational goals?

ICAO has two global aspirational goals for the international aviation sector: 2% annual fuel efficiency improvement through 2050; and carbon neutral growth from 2020 onwards. To achieve these goals, ICAO has established CORSIA – Carbon Offsetting and Reduction Scheme for International Aviation, a market-based programme.

In 2018, IMO adopted an Initial Strategy on the reduction of GHG emissions from ships. This strategy calls for a reduction of the carbon intensity of new ships through implementation of further phases of the energy efficiency design index (EEDI). The IMO calls for a 40% reduction of the carbon intensity of international shipping by 2030, and is striving for a 70% reduction by 2050. Such reductions in carbon intensity would result in an overall decline in emissions of 50% in 2050 (relative to 2008).

These goals are likely insufficiently transformative for the decarbonisation of aviation or shipping, though they are moving towards a start of decarbonisation at a period in history where the options are still not clear, as set out in Sections 10.5 and 10.6.

### Regulations?

The ICAO is not a regulatory agency, but rather produces standards and recommended practices that are adopted in national and international legislation. IMO does publish 'regulations' but does not have powers of enforcement. Non-compliance can be regulated by nation states if they so desire, as a ship's MARPOL certificate, issued by the flag state of the ship, means there is some responsibility for states with global shipping fleets.

### Paris?

Some authors in the literature have argued that emissions from international aviation and shipping should be part of the Paris Agreement (Gençsü and Hino 2015; Lee 2018; Traut et al. 2018; Rayner 2021), arguing that the shipping and aviation industries would prefer emissions to be treated under an international regime rather than a national-oriented regime. If international aviation and shipping emissions were a part of the Paris Agreement, it may remove something of the present ambiguity about responsibilities. However, inclusion in the Paris Agreement is unlikely to fundamentally change emissions trends unless targets and enforcement mechanisms are developed, by ICAO and IMO or by nation states through global processes.

*Box 10.5 (continued)*

**Individual nations?**

If international regulations are not made, then the transformation of aviation and shipping will be left to individual nations. In 2020, Switzerland approved a new $CO_2$ tax on flights (The Swiss Parliament 2020), with part of its revenues earmarked for the development of synthetic aviation fuels, to cover up to 80% of their additional costs compared to fossil jet fuel (Energieradar 2020). Appropriate financing frameworks will be a key to the large-scale market adoption of these fuels. Egli et al. (2019) suggest that the successful design of investment policies for solar and wind power over the past 20 years could serve as a model for future synthetic aviation fuels production projects 'attracting a broad spectrum of investors in order to create competition that drives down financing cost', and with state investment banks building 'investor confidence in new technologies.' These national investment policies would provide the key enablers for successful deployments.

**Managing critical minerals:** Critical minerals are required to manufacture lithium-ion batteries (LIB) and other renewable power technologies. There has been growing awareness that critical minerals may face challenges related to resource availability, labour rights, and costs. Box 10.6 sets out the issues, showing how emerging national strategies on critical minerals, along with requirements from major vehicle manufacturers, are addressing the need for rapid development of new mines with a more balanced geography, less use of cobalt through continuing LIB innovations, and a focus on recycling batteries. The standardisation of battery modules and packaging within and across vehicle platforms, as well as increased focus on design for recyclability, are important. Given the high degree of potential recyclability of LIBs, a near closed-loop system in the future would be a feasible opportunity to minimise critical mineral issues.



## Box 10.6 | Critical Minerals and The Future of Electromobility and Renewables

The global transition towards renewable energy technologies and battery systems necessarily involves materials, markets, and supply chains on a hitherto unknown scale and scope. This has raised concerns regarding mineral requirements central to the feasibility of the energy transition. Constituent materials required for the development of these low-carbon technologies are regarded as 'critical' materials (US Geological Survey 2018; Commonwealth of Australia 2019; Lee et al. 2020; Marinaro et al. 2020; Sovacool et al. 2020). 'Critical materials' are critical because of their economic or national security importance, or high risk of supply disruption. Many of these materials and rare earth elements (REEs) as 'technologically critical', not only due to their strategic or economic importance but the risk of short supply or price volatility (Marinaro et al. 2020). In addition to these indicators, production growth and market dynamics are also incorporated into screening tools to assess emerging trends in material commodities that are deemed as fundamental to the well-being of the nation (NSTC 2018).

The critical materials identified by most nations are: REEs neodymium and dysprosium for permanent magnets in wind turbines and electric motors; lithium and cobalt, primarily for batteries though many other metals are involved; and, cadmium, tellurium, selenium, gallium and indium for solar PV manufacture (Valero et al. 2018; Giurco et al. 2019). Predictions are that the transition to a clean energy world will be significantly energy intensive (World Bank Group 2017; Sovacool et al. 2020), putting pressure on the supply chain for many of the metals and materials required.

Governance of the sustainability of mining and processing of many of these materials, in areas generally known for their variable environmental stewardship, remains inadequate and often a source for conflict. Sovacool et al. (2020) propose four holistic recommendations for improvement to make these industries more efficient and resilient: diversification of mining enterprises for local ownership and livelihood benefit; improved traceability of material sources and transparency of mining enterprises; exploration of alternative resources; and the incorporation of minerals into climate and energy planning by connecting to the NDCs under the Paris Agreement.

**Resource constraints?**

Valero et al. (2018) highlight that the demand for many of the REEs and other critical minerals will, at the current rate of renewable energy infrastructure growth, increase by 3000 times or more by 2050. Some believe this growth may reach constraints in supply (Giurco et al. 2019). Others suggest that the minerals involved are not likely to physically run out (Sovacool et al. 2020) if well managed, especially as markets are found in other parts of the world (for example the transition away from lithium from brine lakes to hard rock sources). Lithium hydroxide, more suitable for batteries, now competes well, in terms of cost, when extracted from rock sources (Azevedo et al. 2018) due to the ability to more easily create high quality lithium hydroxide from rock sources, even though brines provide a cheaper

*Box 10.6 (continued)*

source of lithium (Kavanagh et al. 2018). Australia has proven resources of all the Li-ion battery minerals and has a strategy for their ethical and transparent production (Commonwealth of Australia 2019). Changes in the technology have also been used to reduce need for certain critical minerals (Månberger and Stenqvist 2018). Recycling of all the minerals is not yet well developed but is likely to be increasingly important (Habib and Wenzel 2014; World Bank Group 2017; Giurco et al. 2019; Golroudbary et al. 2019).

### International collaboration

There have been many instances since the 1950s when the supply of essential minerals has been restricted by nations in times of conflict and world tensions, but international trade has continued under the framework of the World Trade Organization. Keeping access open to critical minerals needed for the low-carbon transition will be an essential role of the international community as the need for local manufacture of such renewable and electromobility technologies will be necessary for local economies. Nassar et al. (2020) report that over the past 30 years the US has become increasingly reliant in imports to meet domestic demand for minerals, including REEs. In terms of heavy REEs, essential for permanent magnets for wind turbines, China has a near-monopoly on REE processing, though other mines and manufacturing facilities are now responding to these constrained markets (Stegen 2015; Gulley et al. 2018; Gulley et al. 2019; Yan et al. 2020). China, on the other hand, is reliant on other nations for the supply of other critical metals, particularly cobalt and lithium for batteries.

A number of critical materials strategies have now been developed by nations developing the manufacturing base of new power and transport technologies. Some of these strategies pay particular attention to the supply of lithium (Martin et al. 2017; Hache et al. 2019). For example, Horizon 2020, a substantial EU Research and Innovation programme, couples research and innovation in science, industry, and society to foster a circular economy in Europe, thus reducing bottlenecks in the EU nations. Similarly CREEN (Canada Rare Earth Elements Network) is supporting the US–EU–Japan resource partnership with Australia (Klossek et al. 2016).

As renewables and electromobility-based development leapfrog into the developing world it will be important to ensure the critical minerals issues are managed for local security of supply as well as participation in the mining and processing of such minerals to enable countries to develop their own employment around renewables and electromobility (Sovacool et al. 2020).



**Enabling creative foresight:** Human culture has always had a creative instinct that enables the future to be better dealt with through imagination (Montgomery 2017). Science and engineering have often been preceded by artistic expressions; for example Jules Verne first dreamed of the hydrogen future in 1874 in his novel *The Mysterious Island*. Autonomous vehicles have regularly occupied the minds of science fiction authors and filmmakers (Braun 2019). Such narratives, scenario building, and foresighting are increasingly seen as a part of the climate change mitigation process (Lennon et al. 2015; Muiderman et al. 2020) and can 'liberate oppressed imaginaries' (Luque-Ayala 2018). Barber (2021) emphasised the important role of positive images about the future instead of dystopian visions and the impossibility of business-as-usual futures.

Transport visions can be a part of this cultural change as well as the more frequently presented visions of renewable energy (Wentland 2016; Breyer et al. 2017). There are some emerging technologies, like Maglev, Hyperloop, and drones that are likely to continue the electrification of transport even further (Daim 2021) and which are only recently at the imagination stage. Decarbonised visions for heavy vehicle systems appear to be a core need from the assessment of technologies in this chapter. Such visioning or foresighting requires deliberative processes and the literature contains a growing list of transport success stories based on such processes (Weymouth and Hartz-Karp 2015). Ultimately, reducing GHG emissions from the transport sector would benefit from creative visions that integrate a broad set of ideas about technologies, urban and infrastructure planning (including transport, electricity, and telecommunications infrastructure), and human behaviour and at the same time can create opportunities to achieve the SDGs.

**Enabling transport climate emergency plans, local pledges and net zero strategies:** National, regional and local governments are now producing transport plans with a climate emergency focus (Jaeger et al. 2015; Pollard 2019). Such plans are often grounded in the goals of the Paris Agreement, based around local low-carbon transport roadmaps that contain targets for and involve commitments or pledges from local stakeholders, such as workplaces, local community groups, and civil society organisations. Pledges often include phasing out fossil fuel-based cars, buses, and trucks (Plötz et al. 2020), strategies to meet the targets through infrastructure, urban regeneration and incentives, and detailed programmes to help citizens adopt change. These institution-led mechanisms could include bike-to-work campaigns, free transport passes, parking charges, or eliminating car benefits. Community-based solutions like solar sharing, community charging, and mobility as a service can generate new opportunities to facilitate low-carbon transport futures. Cities in India and China have established these transport roadmaps, which are also supported by the United Nations Centre for Regional Development's Environmentally Sustainable Transport programme (Baeumler et al. 2012; Pathak and Shukla 2016; UNCRD 2020). There have been concerns raised that these pledges may be used to delay climate action in some cases (Lamb et al. 2020) but such pledges can be calculated at a personal level and applied through every level of activity from individual, household, neighbourhood, business, city, nation or groups of nations (Meyer and Newman 2020) and are increasingly being demonstrated through shared communities and local activism (Bloomberg and Pope 2017; Sharp 2018; Figueres and Rivett-Carnac 2020). Finally, the world's major financing institutions are also engaging in decarbonisation efforts by requiring their recipients to commit to Net Zero Strategies before they can receive their funding (Robins 2018; Newman 2020a) (Chapter 15, Cross-Chapter Box 1 in Chapter 1). As a result, transparent methods are emerging for calculating what these financing requirements mean for transport by companies, cities, regions, and infrastructure projects (Chapters 8 and 15). The continued engagement of financial institutions may, like in other sectors, become a major factor in enabling transformative futures for transport as long as governance and communities continue to express the need for such change.

### 10.8.4 Tools and Strategies to Enable Decarbonisation of the Transport Sector

Using the right tools and strategies is crucial for the successful deployment of mitigation options. Table 10.7 summarises the tools and strategies required to enable electromobility, new fuels for aviation and shipping, and the more social aspects of demand efficiency.

**Table 10.7 | Tools and strategies for enabling mitigation options to achieve transformative scenarios.**

| Tools and strategies | Travel demand reduction (TDR) and fuel/vehicle efficiency | Light vehicle electromobility systems | Alternative fuel systems for Shipping and Aviation |
|---|---|---|---|
| Education and R&D | TDR can be assisted with digitalisation, connected autonomous vehicle, EVs and mobility as a Service (Marsden et al. 2018; Shaheen et al. 2018).<br>Knowledge gaps on TDR exist for longer distance travel (intercity); non-mandatory trips (leisure; social trips), and travel by older people. Travel demand foresighting tools can be open source (Marsden 2018). | Behaviour change programmes help EVs become more mainstream. R&D will help on the socio-economic structures that impede adoption of EVs, the urban structures that enable reduced car dependence, and how EVs can assist grids (Newman 2010; Taiebat and Xu 2019; Seto et al. 2021). | R&D is critical for new fuels and to test the full lifecycle costs of various heavy vehicle options (Marinaro et al. 2020). |
| Access and equity | TDR programmes in cities can be inequitable. To avoid such inequities, there is a need for better links to spatial and economic development (Marsden et al.2018), mindful of diverse local priorities, personal freedom and personal data (Box 10.1). | Significant equity issues with EVs in the transition period can be overcome with programmes that enable affordable electric mobility, especially public transit (IRENA 2016). | Shipping is mostly freight and is less of a problem but aviation has big equity issues (Bows-Larkin 2015). |
| Financing economic incentives and partnerships | Carbon budget implications of different demand futures should be published and used to help incentivise net zero projects (Marsden 2018). Business and community pledges for net zero can be set up in partnership agreements (Section 10.8.3). | Multiple opportunities for financing, economic incentives, and partnerships with clear economic benefits can be assured, especially using the role of value capture in enabling such benefits. The nexus between EVs and the electricity grid needs opportunities to demonstrate positive partnership projects (Zhang et al. 2014; Mahmud et al. 2018; Newman et al. 2018; Sovacool et al. 2018; Sharma and Newman 2020). | Taking R&D into demonstration projects is the main stage for heavy vehicle options and these are best done as partnerships. Government assistance will greatly assist in such projects as well as an R&D levy. Abolishing fossil fuel subsidies and imposing carbon taxes is likely to help in the early stages of heavy vehicle transitions (Sclar et al. 2019). |
| Co-benefits and overcoming fragmentation | Programmes that focus on people-centred solutions for future mobility, with more pluralistic and feasible sets of outcomes for all people, can be successful. They need to focus on more than simple benefit-cost ratios and include well-being and livelihoods, considering transport as a system rather than loosely connected modes, as well as behaviour change programmes (Barter and Raad 2000; Newman 2010; Martens 2020). | The SDG benefits of zero-carbon light vehicle transport systems are being demonstrated and can now be quantified as nations mainstream this transition. Projects with transit and sustainable housing are more able to show such benefits. New benefit-cost ratio methods that focus on health benefits in productivity are now favouring transit and active transport (Buonocore et al. 2019; UK DoT 2019; Hamilton et al. 2021). | Heavy vehicle systems can also demonstrate SDG co-benefits if formulated with these in mind. Demonstrations of how innovations can also help SDGs will attract more funding. Such projects need cross-government consideration (Pradhan et al. 2017). |
| Regulation and assessment | Implementing a flexible regulatory framework is needed for most TDR (Li and Pye 2018). Regulatory assessment can help with potential additional (cyber) security risks due to digitalisation, autonomous vehicles, the internet of things, and big data (Shaheen and Cohen 2019). Assessment tools and methods need to take account of greater diversity of population, regions, blurring of modes, and distinct spatial characteristics (Newman and Kenworthy 2015). | With zero-carbon light vehicle systems rapidly growing, the need for a regulated target and assessment of regulatory barriers can assist each city and region to transition more effectively. Regulating EVs for government fleets and recharge infrastructure can establish incentives (Bocken et al. 2016). | Zero-carbon heavy vehicle systems need to have regulatory barrier assessments as they are being evaluated in R&D demonstrations (Sclar et al. 2019). |

| Tools and strategies | Travel demand reduction (TDR) and fuel/vehicle efficiency | Light vehicle electromobility systems | Alternative fuel systems for Shipping and Aviation |
|---|---|---|---|
| Governance and institutional capacity | TDR works better if adaptive decision-making approaches focus on more inclusive and whole-of-system benefit-cost ratios (Marsden 2018; Yang et al. 2020). | Governance and institutional capacity can now provide international exchanges and education programmes based on successful cities and nations, enabling light vehicle decarbonisation to create more efficient and effective policy mechanisms towards self-sustaining markets (Greene et al. 2014; Skjølsvold and Ryghaug 2019). | Governance and institutional capacity can help make significant progress if targets are backed with levies for not complying. Carbon taxes would also affect these segments. A review of international transport governance is likely (Makan and Heyns 2018). |
| Enabling infrastructure | Ensuring space for active transport and urban activities is taken from road space will be necessary in some places (Gössling et al. 2021b).<br>Increasing the proportion of infrastructure that supports walking in urban areas will structurally enable reductions in car use (Newman and Kenworthy 2015) (Section 10.2).<br>Creating transit activated corridors of transit-oriented development-based rail or mid-tier transit using value capture for financing will create inherently less car dependence (McIntosh et al. 2017; Newman et al. 2019). | Large-scale electrification of LDVs requires expansion of low-carbon power systems, while charging or battery swapping infrastructure is needed for some segments (Gnann et al. 2018; Ahmad et al. 2020). | In addition to increasing the capabilities to produce low- or zero-carbon fuels for shipping and aviation, there is a need to invest in supporting infrastructure including low-carbon power generation. New hydrogen delivery and refuelling infrastructure may be needed (Maggio et al. 2019). For zero-carbon synthetic fuels, infrastructure is needed to support carbon capture and $CO_2$ transport to fuel production facilities (Edwards and Celia 2018). |

Frequently Asked Questions (FAQs)

## FAQ 10.1 | How important is electromobility in decarbonising transport and are there major constraints in battery minerals?

Electromobility is the biggest change in transport since AR5. When powered with low-carbon electricity, electric vehicles (EVs) provide a mechanism for major GHG emissions reductions from the largest sources in the transport sectors, including cars, motorbikes, autorickshaws, buses and trucks. The mitigation potential of EVs depends on the decarbonisation of the power system. EVs can be charged by home or business renewable power before or in parallel to the transition to grid-based low-carbon power.

Electromobility is happening rapidly in micromobility (e-autorickshaws, e-scooters, e-bikes) and in transit systems, especially buses. EV adoption is also accelerating for personal cars. EVs can be used in grid stabilisation through smart charging applications.

The state-of-the-art lithium-Ion batteries (LIBs) available in 2020 are superior to alternative cell technologies in terms of battery life, energy density, specific energy, and cost. The expected further improvements in LIBs suggest these chemistries will remain superior to alternative battery technologies in the medium term, and therefore LIBs will continue to dominate the electric vehicle market.

Dependence on LIB metals will remain, which may be a concern from the perspective of resource availability and costs. However, the demand for such metals is much lower than the reserves available, with many new mines starting up in response to the new market, particularly in a diversity of places.

Recycling batteries will significantly reduce long-term resource requirements. The standardisation of battery modules and packaging within and across vehicle platforms, as well as increased focus on design for recyclability, are important. Many mobility manufacturers and governments are considering battery recycling issues to ensure the process is mainstreamed.



The most significant enabling condition in electromobility is to provide electric recharging opportunities and an integration strategy so that vehicles support the grid.

## FAQ 10.2 | How hard is it to decarbonise heavy vehicles in transport like long-haul trucks, ships and planes?

There are few obvious solutions to decarbonising heavy vehicles like international ships and planes. The main focus has been increased efficiency, which so far has not prevented these large vehicles from becoming the fastest-growing source of GHG globally. These vehicles likely need alternative fuels that can be fitted to the present propulsion systems. Emerging demonstrations suggest that ammonia, advanced biofuels, or synthetic fuels could become commercial.

Electric propulsion using hydrogen fuel cells or Li-ion batteries could work with short-haul aviation and shipping, but the large long-lived vessels and aircraft likely need alternative liquid fuels for most major long-distance functions.

Advanced biofuels, if sourced from resources with low GHG footprints, offer decarbonisation opportunities. As shown in Chapters 2, 6, and 12, there are multiple issues constraining traditional biofuels. Sustainable land management and feedstocks, as well as R&D efforts to improve lignocellulosic conversion routes, are key to maximising the mitigation potential from advanced biofuels.

Synthetic jet and marine fuels can be made using $CO_2$ captured with DAC/BECCS and low-carbon hydrogen. These fuels may also have less contrails-based climate impacts and lower emissions of local air pollutants. However, these fuels still require significant R&D and demonstration.

The deployment of low-carbon aviation and shipping fuels that support decarbonisation of the transport sector will likely require changes to national and international governance structures.

Frequently Asked Questions (FAQs)

## FAQ 10.3 | How can governments, communities and individuals reduce demand and be more efficient in consuming transport energy?

Cities can reduce their transport-related fuel consumption by around 25% through combinations of more compact land use and less car-dependent transport infrastructure.

More traditional programmes for reducing unnecessary high-energy travel through behaviour change programmes (e.g., taxes on fuel, parking, and vehicles, or subsidies for alternative low-carbon modes) continue to be evaluated, with mixed results due to the dominance of time savings in an individual's decision-making.

The circular economy, the shared economy, and digitalisation trends can support systemic changes that lead to reductions in demand for transport services or expand the use of more efficient transport modes.

COVID-19 lockdowns have confirmed the transformative value of telecommuting, replacing significant numbers of work and personal journeys, as well as promoting local active transport. These changes may not last and impacts on productivity and health are still to be fully evaluated.

Solutions for individual households and businesses involving pledges and shared communities that set new cultural means of reducing fossil fuel consumption, especially in transport, are setting out new approaches for how climate change mitigation can be achieved.

## References


Abe, R., 2021: Preferences of urban rail users for first- and last-mile autonomous vehicles: Price and service elasticities of demand in a multimodal environment. *Transp. Res. Part C Emerg. Technol.*, **126(103105)**, 103105, doi:10.1016/j.trc.2021.103105.

Abrahamse, W., L. Steg, R. Gifford, and C. Vlek, 2009: Factors influencing car use for commuting and the intention to reduce it: A question of self-interest or morality? *Transp. Res. Part F Traffic Psychol. Behav.*, **12(4)**, 317–324, doi:10.1016/j.trf.2009.04.004.

ADB, 2018: *Sustainable Transport Solutions. Low-Carbon Buses in the People's Republic of China 12*. Asian Development Bank, Metro Manila, Philippines, 72 pp., https://www.adb.org/publications/sustainable-transport-solutions-peoples-republic-china.

AFDC, 2021: Hydrogen Fueling Station Locations. The US Department of Energy, https://afdc.energy.gov/fuels/hydrogen_stations.html.

Ahmad, F., M. Saad Alam, I. Saad Alsaidan, and S.M. Shariff, 2020: Battery swapping station for electric vehicles: opportunities and challenges. *IET Smart Grid*, **3(3)**, 280–286, doi:10.1049/iet-stg.2019.0059.

Ahmad, S. and J.A. Puppim de Oliveira, 2016: Determinants of urban mobility in India: Lessons for promoting sustainable and inclusive urban transportation in developing countries. *Transp. Policy*, **50**, 106–114, doi:10.1016/j.tranpol.2016.04.014.

Ahmadi, L., S.B. Young, M. Fowler, R.A. Fraser, and M.A. Achachlouei, 2017: A cascaded life cycle: reuse of electric vehicle lithium-ion battery packs in energy storage systems. *Int. J. Life Cycle Assess.*, **22(1)**, 111–124, doi:10.1007/s11367-015-0959-7.

Ainalis, D., C. Thorne, and D. Cebon, 2020: *Decarbonising the UK's Long-Haul Road Freight at Minimum Economic Cost*. Centre for Sustainable Road Freight, Cambridge, UK, 34 pp.

Airqo, 2020: Know your air. Makere Unviersity in Uganda. https://airqo.net/.

Åkerman, J., 2011: The role of high-speed rail in mitigating climate change – The Swedish case Europabanan from a life cycle perspective. *Transp. Res. Part D Transp. Environ.*, **16(3)**, 208–217, doi:10.1016/j.trd.2010.12.004.

Aksenov, Y. et al., 2017: On the future navigability of Arctic sea routes: High-resolution projections of the Arctic Ocean and sea ice. *Mar. Policy*, **75**, 300–317, doi:10.1016/j.marpol.2015.12.027.

Akyelken, N., D. Banister, and M. Givoni, 2018: The Sustainability of Shared Mobility in London: The Dilemma for Governance. *Sustainability*, **10(2)**, 420, doi:10.3390/su10020420.

Al-Qahtani, A., B. Parkinson, K. Hellgardt, N. Shah, and G. Guillen-Gosalbez, 2021: Uncovering the true cost of hydrogen production routes using life cycle monetisation. *Appl. Energy*, **281**, 115958, doi:10.1016/j.apenergy.2020.115958.

Albrecht, F.G., D.H. König, N. Baucks, and R.-U. Dietrich, 2017: A standardized methodology for the techno-economic evaluation of alternative fuels – A case study. *Fuel*, **194**, 511–526, doi:10.1016/j.fuel.2016.12.003.

Aliabadi, A.A., R.M. Staebler, and S. Sharma, 2015: Air quality monitoring in communities of the Canadian Arctic during the high shipping season with a focus on local and marine pollution. *Atmos. Chem. Phys.*, **15(5)**, 2651–2673, doi:10.5194/acp-15-2651-2015.

Alonso-González, M.J., T. Liu, O. Cats, N. Van Oort, and S. Hoogendoorn, 2018: The Potential of Demand-Responsive Transport as a Complement to Public Transport: An Assessment Framework and an Empirical Evaluation. *Transp. Res. Rec. J. Transp. Res. Board*, **2672(8)**, 879–889, doi:10.1177/0361198118790842.

Altieri, K. et al., 2015: *Pathways to deep decarbonization in South Africa*. Sustainable Development Solutions Network (SDSN) and Institute for Sustainable Development and International Relations (IDDRI), 218 pp., https://www.africaportal.org/publications/pathways-to-deep-decarbonization-in-south-africa/.

Ambrose, H., A. Kendall, M. Lozano, S. Wachche, and L. Fulton, 2020: Trends in life cycle greenhouse gas emissions of future light duty electric vehicles. *Transportation Research Part D: Transport and Environment*, **81**, 102287, doi:10.1016/j.trd.2020.102287.

Ampersand, 2020: Ampersand. https://www.ampersand.solar/.

Anda, C., A. Erath, and P.J. Fourie, 2017: Transport modelling in the age of big data. *Int. J. Urban Sci.*, **21**(sup1), 19–42, doi:10.1080/12265934.2017.1281150.

Anderson, M., K. Salo, and E. Fridell, 2015: Particle- and Gaseous Emissions from an LNG Powered Ship. *Environ. Sci. Technol.*, **49(20)**, 12568–12575, doi:10.1021/acs.est.5b02678.

Andwari, A.M., A. Pesiridis, S. Rajoo, R. Martinez-Botas, and V. Esfahanian, 2017: A review of Battery Electric Vehicle technology and readiness levels. *Renew. Sustain. Energy Rev.*, **78**(February), 414–430, doi:10.1016/j.rser.2017.03.138.

Anenberg, S., J. Miller, D. Henze, and R. Minjares, 2019: *A global snapshot of the air pollution-related health impacts of transportation sector emissions in 2010 and 2015*. International Council on Clean Transportation, Washington, DC, USA, 55 pp.

Archer, D. et al., 2009: Atmospheric Lifetime of Fossil Fuel Carbon Dioxide. *Annu. Rev. Earth Planet. Sci.*, **37(1)**, 117–134, doi:10.1146/annurev.earth.031208.100206.

Argonne National Laboratory, 2020: AFLEET Tool. https://greet.es.anl.gov/afleet_tool.

Aria, E., J. Olstam, and C. Schwietering, 2016: Investigation of Automated Vehicle Effects on Driver's Behavior and Traffic Performance. *Transp. Res. Procedia*, **15**, 761–770, doi:10.1016/j.trpro.2016.06.063.

Arrowsmith, S. et al., 2020: *Updated analysis of the non-$CO_2$ climate impacts of aviation and potential policy measures pursuant to the EU Emissions Trading System Directive Article 30(4)*. European Union Aviation Safety Agency, 193 pp.

ATAG, 2018: *Aviation: Benefits Beyond Borders*. Air Transport Action Group, Geneva, Switzerland.

Auld, J., V. Sokolov, and T.S. Stephens, 2017: Analysis of the effects of connected-automated vehicle technologies on travel demand. *Transp. Res. Rec. J. Transp. Res. Board*, **2625**, 1–8, doi:10.3141/2625-01.

Awoyomi, A., K. Patchigolla, and E.J. Anthony, 2020: Process and Economic Evaluation of an Onboard Capture System for LNG-Fueled $CO_2$ Carriers. *Ind. Eng. Chem. Res.*, **59(15)**, 6951–6960, doi:10.1021/acs.iecr.9b04659.

Azevedo, M. et al., 2018: *Lithium and cobalt: A tale of two commodities*. McKinsey & Co., https://www.mckinsey.com/~/media/mckinsey/industries/metals%20and%20mining/our%20insights/lithium%20and%20cobalt%20a%20tale%20of%20two%20commodities/lithium-and-cobalt-a-tale-of-two-commodities.pdf.

Baeumler, A., E. Ijjasz-Vasquez, and S. Mehndiratta, 2012: *Sustainable low-carbon city development in China*. World Bank Publications, Washington, DC, USA.

Bagheri, M., M.N. Mladenović, I. Kosonen, and J.K. Nurminen, 2020: Analysis of Potential Shift to Low-Carbon Urban Travel Modes: A Computational Framework Based on High-Resolution Smartphone Data. *Sustainability*, **12(15)**, 5901, doi:10.3390/su12155901.

Bakhat, M., X. Labandeira, J.M. Labeaga, and X. López-Otero, 2017: Elasticities of transport fuels at times of economic crisis: An empirical analysis for Spain. *Energy Econ.*, **68(1)**, 66–80, doi:10.1016/j.eneco.2017.10.019.

Balcombe, P. et al., 2019: How to decarbonise international shipping: Options for fuels, technologies and policies. *Energy Convers. Manag.*, **182**(December 2018), 72–88, doi:10.1016/j.enconman.2018.12.080.

Barber, J., 2021: The Challenge of Imagining Sustainable Futures: Climate Fiction in Sustainability Communication. In: *The Sustainability Communication Reader* [Weder, F., L. Krainer, and M. Karmasin, (eds.)], Springer Fachmedien Wiesbaden, Wiesbaden, Germany, pp. 143–160.

Baron, T., G. Martinetti, and D. Pepion, 2011: *Carbon footprint of high speed rail*. International Union of Railways, Paris, France, 58 pp.

Barter, P.A. and T. Raad, 2000: *Taking Steps – A Community Action Guide to People-Centred, Equitable and Sustainable Urban Transport*. Sustainable Transport Action Network for Asia and the Pacific, Kuala Lumpur, Malaysia.

Bastian, A., M. Börjesson, and J. Eliasson, 2016: Explaining "peak car" with economic variables. *Transp. Res. Part A Policy Pract.*, **88**, 236–250, doi:10.1016/J.TRA.2016.04.005.

Bataille, C., D. Sawyer, and N. Melton, 2015: *Pathways to deep decarbonization in Canada*. Sustainable Development Solutions Network, 56 pp., https://cmcghg.com/wp-content/uploads/2015/07/Final-Canada-DDPP-Country-Report-July-14.pdf.

Bataille, C. et al., 2020: Net-zero deep decarbonization pathways in Latin America: Challenges and opportunities. *Energy Strateg. Rev.*, **30**, 100510, doi:10.1016/j.esr.2020.100510.

Bauer, C. et al., 2015: The environmental performance of current and future passenger vehicles: Life cycle assessment based on a novel scenario analysis framework. *Applied Energy*, **157**, 871–883, doi:10.1016/j.apenergy.2015.01.019.

Bayat, Y. and M. Ghazikhani, 2020: Experimental investigation of compressed natural gas using in an indirect injection diesel engine at different conditions. *J. Clean. Prod.*, **271**, 122450, doi:10.1016/j.jclepro.2020.122450.

Bayliss, K. and G. Mattioli, 2018: *Privatisation, inequality and poverty in the UK: Briefing prepared for UN Rapporteur on extreme poverty and human rights*, Sustainability Research Institute (SRI), School of Earth and Environment, The University of Leeds, Leeds, UK.

Becken, S. and B. Mackey, 2017: What role for offsetting aviation greenhouse gas emissions in a deep-cut carbon world? *J. Air Transp. Manag.*, **63**, 71–83, doi:10.1016/j.jairtraman.2017.05.009.

Becken, S. and J. Shuker, 2019: A framework to help destinations manage carbon risk from aviation emissions. *Tour. Manag.*, **71**, 294–304, doi:10.1016/j.tourman.2018.10.023.

Becker, H., M. Balac, F. Ciari, and K.W. Axhausen, 2020: Assessing the welfare impacts of Shared Mobility and Mobility as a Service (MaaS). *Transp. Res. Part A Policy Pract.*, **131**, 228–243, doi:10.1016/J.TRA.2019.09.027.

Bekel, K. and S. Pauliuk, 2019: Prospective cost and environmental impact assessment of battery and fuel cell electric vehicles in Germany. *Int. J. Life Cycle Assess.*, **24(12)**, 2220–2237, doi:10.1007/s11367-019-01640-8.

Benajes, J., G. Antonio, J. Monsalve-Serrano, and S. Martinez-Boggio, 2020: Emissions Reduction from Passenger Cars with RCCI Plug-in Hybrid Electric Vehicle Technology. *Applied Thermal Engineering*, **164**, 114430, doi:10.1016/j.applthermaleng.2019.114430.

Benevolo, C., R.P. Dameri, and B. D'Auria, 2016: Smart Mobility in Smart City. In: *Lecture Notes in Information Systems and Organisation* [Torre, T., A.M. Braccini, and R. Spinelli, (eds.)], Vol. 11, Springer Heidelberg, Cham, Switzerland, pp. 13–28.

Bengtsson, S., K. Andersson, and E. Fridell, 2011: A comparative life cycle assessment of marine fuels: Liquefied natural gas and three other fossil fuels. *Proc. Inst. Mech. Eng. Part M J. Eng. Marit. Environ.*, **225(2)**, 97–110, doi:10.1177/1475090211402136.

Bergantino, A.S., M. Capurso and J. Toner., 2018a: *Households' expenditure elasticities for transport products and services: a geo-demographic approach*. Working Papers 18_3, SIET Società Italiana di Economia dei Trasporti e della Logistica.

Bergantino, A.S., C. Capozza, and M. Capurso, 2018b: Pricing strategies: who leads and who follows in the air and rail passenger markets in Italy. *Appl. Econ.*, **50(46)**, 4937–4953, doi:10.1080/00036846.2018.1459039.

Berkeley, N., D. Bailey, A. Jones, and D. Jarvis, 2017: Assessing the transition towards Battery Electric Vehicles: A Multi-Level Perspective on drivers of, and barriers to, take up. *Transportation Research Part A: Policy and Practice*, **106**, 320–332, doi:10.1016/j.tra.2017.10.004.

Bhandari, R., C.A. Trudewind, and P. Zapp, 2014: Life cycle assessment of hydrogen production via electrolysis – a review. *J. Clean. Prod.*, **85**, 151–163, doi:10.1016/j.jclepro.2013.07.048.

Bi, M., S. He, and W. (Ato) Xu, 2019: Express delivery with high-speed railway: Definitely feasible or just a publicity stunt. *Transp. Res. Part A Policy Pract.*, **120**, 165–187, doi:10.1016/j.tra.2018.12.011.

Bicer, Y. and I. Dincer, 2017: Life cycle evaluation of hydrogen and other potential fuels for aircrafts. *Int. J. Hydrogen Energy*, **42(16)**, 10722–10738, doi:10.1016/j.ijhydene.2016.12.119.

Bicer, Y. and I. Dincer, 2018a: Environmental impact categories of hydrogen and ammonia driven transoceanic maritime vehicles: A comparative evaluation. *Int. J. Hydrogen Energy*, **43(9)**, 4583–4596, doi:10.1016/j.ijhydene.2017.07.110.

Bicer, Y. and I. Dincer, 2018b: Clean fuel options with hydrogen for sea transportation: A life cycle approach. *Int. J. Hydrogen Energy*, **43(2)**, 1179–1193, doi:10.1016/j.ijhydene.2017.10.157.

Bier, A. and U. Burkhardt, 2019: Variability in Contrail Ice Nucleation and Its Dependence on Soot Number Emissions. *J. Geophys. Res. Atmos.*, **124(6)**, 3384–3400, doi:10.1029/2018JD029155.

Bier, A., U. Burkhardt, and L. Bock, 2017: Synoptic Control of Contrail Cirrus Life Cycles and Their Modification Due to Reduced Soot Number Emissions. *J. Geophys. Res. Atmos.*, **122(21)**, 11, 584–11, 603, doi:10.1002/2017JD027011.

Biernacki, P., T. Röther, W. Paul, P. Werner, and S. Steinigeweg, 2018: Environmental impact of the excess electricity conversion into methanol. *J. Clean. Prod.*, **191**, 87–98, doi:10.1016/j.jclepro.2018.04.232.

Blasco, J., V. Durán-Grados, M. Hampel, and J. Moreno-Gutiérrez, 2014: Towards an integrated environmental risk assessment of emissions from ships' propulsion systems. *Environ. Int.*, **66**, 44–47, doi:10.1016/J.ENVINT.2014.01.014.

Bleischwitz, R. et al., 2017: *Routledge handbook of the resource nexus*. Routledge, New York and London.

Bloomberg, M. and C. Pope, 2017: *Climate of hope: How cities, businesses, and citizens can save the planet*. St. Martin's Press, New York and London.

BNEF, 2020: *Hydrogen Economy Outlook Key messages*. Bloomberg Finance L. P., https://data.bloomberglp.com/professional/sites/24/BNEF-Hydrogen-Economy-Outlook-Key-Messages-30-Mar-2020.pdf.

Bocken, N.M.P., I. de Pauw, C. Bakker, and B. van der Grinten, 2016: Product design and business model strategies for a circular economy. *J. Ind. Prod. Eng.*, **33(5)**, 308–320, doi:10.1080/21681015.2016.1172124.

Bond, K., E. Vaughan, and H. Benham, 2020: *Decline and Fall: The size and vulnerability of the fossil fuel system*. Carbon Tracker.

Bongartz, D. et al., 2018: Comparison of light-duty transportation fuels produced from renewable hydrogen and green carbon dioxide. *Appl. Energy*, **231**, 757–767, doi:10.1016/j.apenergy.2018.09.106.

Borck, R. and J.K. Brueckner, 2018: Optimal energy taxation in cities. *J. Assoc. Environ. Resour. Econ.*, **5(2)**, 481–516, doi:10.1086/695614.

Borén, S., 2019: Electric buses' sustainability effects, noise, energy use, and costs. *Int. J. Sustain. Transp.*, **14(12)**, 956–971, doi:10.1080/15568318.2019.1666324.

Bösch, P.M., F. Ciari, and K.W. Axhausen, 2018: Transport Policy Optimization with Autonomous Vehicles. *Transp. Res. Rec. J. Transp. Res. Board*, **2672(8)**, 698–707, doi:10.1177/0361198118791391.

Bouman, E.A., E. Lindstad, A.I. Rialland, and A.H. Strømman, 2017: State-of-the-art technologies, measures, and potential for reducing GHG emissions from shipping – A review. *Transp. Res. Part D Transp. Environ.*, **52**, 408–421, doi:10.1016/j.trd.2017.03.022.

Bouman, T. and L. Steg, 2019: Motivating Society-wide Pro-environmental Change. *One Earth*, **1(1)**, 27–30, doi:10.1016/j.oneear.2019.08.002.

Bows-Larkin, A., 2015: All adrift: aviation, shipping, and climate change policy. *Clim. Policy*, **15(6)**, 681–702, doi:10.1080/14693062.2014.965125.

Bows, A., K. Anderson, and P. Upham, 2008: *Aviation and climate change: Lessons for European policy*. Routledge, New York and London.

Brand, C. et al., 2021: The climate change mitigation effects of daily active travel in cities. *Transp. Res. Part D Transp. Environ.*, **93**, 102764, doi:10.1016/j.trd.2021.102764.

Braun, R., 2019: Autonomous vehicles: from science fiction to sustainable future. In: *Mobilities, Literature, Culture* [Aguiar, M., C. Mathieson, and L. Pearce, (eds.)], Palgrave Macmillan, Cham, Switzerland, pp. 259–280.

Brewer, G.D., 1991: *Hydrogen aircraft technology*. Routledge, New York and London.

Breyer, C., S. Heinonen, and J. Ruotsalainen, 2017: New consciousness: A societal and energetic vision for rebalancing humankind within the limits of planet Earth. *Technol. Forecast. Soc. Change*, **114**, 7–15, doi:10.1016/j.techfore.2016.06.029.

Brosch, T., D. Sander, and M.K. Patel, 2016: Editorial: Behavioral Insights for a Sustainable Energy Transition. *Front. Energy Res.*, **4**, 15, doi:10.3389/fenrg.2016.00015.

Brown, T., 2019: *MAN Energy Solutions: An Ammonia Engine for the Maritime Sector; Ammonia Energy Association*. https://www.ammoniaenergy.org/articles/man-energy-solutions-an-ammonia-engine-for-the-maritime-sector/.

Browse, J., K.S. Carslaw, A. Schmidt, and J.J. Corbett, 2013: Impact of future Arctic shipping on high-latitude black carbon deposition. *Geophys. Res. Lett.*, **40(16)**, 4459–4463, doi:10.1002/grl.50876.

10

Brussel, M., M. Zuidgeest, K. Pfeffer, and M. van Maarseveen, 2019: Access or Accessibility? A Critique of the Urban Transport SDG Indicator. *ISPRS Int. J. Geo-Information*, **8(2)**, 67, doi:10.3390/ijgi8020067.

Bruun, E. and M. Givoni, 2015: Sustainable mobility: Six research routes to steer transport policy. *Nature*, **523(7558)**, 29–31, doi:10.1038/523029a.

Brynolf, S., E. Fridell, and K. Andersson, 2014: Environmental assessment of marine fuels: Liquefied natural gas, liquefied biogas, methanol and bio-methanol. *J. Clean. Prod.*, **74** (July), 86–95, doi:10.1016/j.jclepro.2014.03.052.

Buhaug, Ø. et al., 2009: *2nd IMO GHG Study 2009*. International Maritime Organization, London, UK, 220 pp.

Buira, D. and J. Tovilla, 2015: *Pathways to deep decarbonization in Mexico*. Sustainable Development Solutions Network (SDSN) and Institute for Sustainable Development and International Relations (IDDRI), 48 pp.

Buixadé Farré, A. et al., 2014: Commercial Arctic shipping through the Northeast Passage: routes, resources, governance, technology, and infrastructure. *Polar Geogr.*, **37(4)**, 298–324, doi:10.1080/1088937X.2014.965769.

Bullock, S., J. Mason, J. Broderick, and A. Larkin, 2020: Shipping and the Paris climate agreement: a focus on committed emissions. *BMC Energy*, **2(1)**, 5, doi:10.1186/s42500-020-00015-2.

Buonocore, J.J. et al., 2019: Metrics for the sustainable development goals: renewable energy and transportation. *Palgrave Commun.*, **5(1)**, 136, doi:10.1057/s41599-019-0336-4.

Burke, A. and A.K. Sinha, 2020: *Technology, Sustainability, and Marketing of Battery Electric and Hydrogen Fuel Cell Medium-Duty and Heavy-Duty Trucks and Buses in 2020–2040*. California Department of Transportation, Sacramento, CA, USA.

Burnham, A. et al., 2021: *Comprehensive total cost of ownership quantification for vehicles with different size classes and powertrains*. Argonne National Laboratory, US Department of Energy, Lemont, IL, USA.

C E Delft, 2017: *STREAM Freight Transport 2016: Emissions of freight transport modes – Version 2*. CE Delft, Delft, Netherlands, 86 pp.

Cai, H., A. Burnham, M. Wang, W. Hang, and A. Vyas, 2015: *The GREET Model Expansion for Well-to-Wheels Analysis of Heavy-Duty Vehicles*. Argonne National Laboratory, US Department of Energy, Lemont, IL, USA.

Cai, H., A. Burnham, R. Chen, and M. Wang, 2017: Wells to wheels: Environmental implications of natural gas as a transportation fuel. *Energy Policy*, **109**(January), 565–578, doi:10.1016/j.enpol.2017.07.041.

Canadell, J.G., P.M.S. Monteiro, M.H. Costa, L. Cotrim da Cunha, P.M. Cox, A.V. Eliseev, S. Henson, M. Ishii, S. Jaccard, C. Koven, A. Lohila, P.K. Patra, S. Piao, J. Rogelj, S. Syampungani, S. Zaehle, and K. Zickfeld, 2021: Global Carbon and other Biogeochemical Cycles and Feedbacks. In *Climate Change 2021: The Physical Science Basis. Contribution of Working Group I to the Sixth Assessment Report of the Intergovernmental Panel on Climate Change* [Masson-Delmotte, V., P. Zhai, A. Pirani, S.L. Connors, C. Péan, S. Berger, N. Caud, Y. Chen, L. Goldfarb, M.I. Gomis, M. Huang, K. Leitzell, E. Lonnoy, J.B.R. Matthews, T.K. Maycock, T. Waterfield, O. Yelekçi, R. Yu, and B. Zhou (eds.)]. Cambridge University Press Cambridge, UK and New York, NY, USA. In Press.

Cano, Z. et al., 2018: Batteries and fuel cells for emerging electric vehicle markets. *Nat. Energy*, **3**, 279–289, doi:10.1038/s41560-018-0108-1.

Capaz, R.S., E. Guida, J.E.A. Seabra, P. Osseweijer, and J.A. Posada, 2021a: Mitigating carbon emissions through sustainable aviation fuels: costs and potential. *Biofuels, Bioprod. Biorefining*, **15(2)**, 502–524, doi:10.1002/bbb.2168.

Capaz, R.S., J.A. Posada, P. Osseweijer, and J.E.A. Seabra, 2021b: The carbon footprint of alternative jet fuels produced in Brazil: exploring different approaches. *Resour. Conserv. Recycl.*, **166**, 105260, doi:10.1016/j.resconrec.2020.105260.

Capurso, M., S. Hess, and T. Dekker, 2019: Modelling the role of consideration of alternatives in mode choice: An application on the Rome-Milan corridor. *Transp. Res. Part A Policy Pract.*, **129**, 170–184, doi:10.1016/j.tra.2019.07.011.

CARB, 2019: *Annual Evaluation of Fuel Cell Electric Vehicle Deployment & Hydrogen Fuel Station Network Development*. California Air Resources Board, Sacramento, CA, USA.

CARB, 2020: *California Hydrogen Infrastructure Tool (CHIT):* California Air Resources Board, Sacramento, CA, USA.

Carmichael, R., 2019: *Behaviour change, public engagement and Net Zero, a report for the Committee on Climate Change*. Committee on Climate Change, London, UK, 81 pp.

Carter, C. and L. Koh, 2018: *Blockchain disruption in transport: are you decentralised yet?* Transport Systems Catapult, Milton Keynes, UK, 47 pp.

Castanheira, É.G., R. Grisoli, S. Coelho, G. Anderi Da Silva, and F. Freire, 2015: Life-cycle assessment of soybean-based biodiesel in Europe: Comparing grain, oil and biodiesel import from Brazil. *J. Clean. Prod.*, **102**, 188–201, doi:10.1016/j.jclepro.2015.04.036.

Chala, G.T., A.R.A. Aziz, and F.Y. Hagos, 2018: Natural Gas Engine Technologies: Challenges and Energy Sustainability Issue. *Energies*, **11(11)**, doi:10.3390/en11112934.

Chang, S.E., 2016: Socioeconomic Impacts of Infrastructure Disruptions. In: *Oxford Research Encyclopedia of Natural Hazard Science*, Oxford University Press, Oxford, UK.

Chatzinikolaou, S.D. and N.P. Ventikos, 2014: Assessment of ship emissions in a life cycle perspective. 3rd International Energy, Life Cycle Assessment, and Sustainability Workshop & Symposium (ELCAS3), 7–9 July 2014, Nisyros, Greece.

Chen, C.C. and A. Gettelman, 2013: Simulated radiative forcing from contrails and contrail cirrus. *Atmos. Chem. Phys.*, **13(24)**, 12525–12536, doi:10.5194/ACP-13-12525-2013.

Chen, T.D. and K.M. Kockelman, 2016: Carsharing's life-cycle impacts on energy use and greenhouse gas emissions. *Transp. Res. Part D Transp. Environ.*, **47**, 276–284, doi:10.1016/j.trd.2016.05.012.

Chen, T.D., K.M. Kockelman, and J.P. Hanna, 2016: Operations of a shared, autonomous, electric vehicle fleet: Implications of vehicle & charging infrastructure decisions. *Transp. Res. Part A Policy Pract.*, **94**, 243–254, doi:10.1016/j.tra.2016.08.020.

Chen, Z., 2017: Impacts of high-speed rail on domestic air transportation in China. *J. Transp. Geogr.*, **62**, 184–196, doi:10.1016/j.jtrangeo.2017.04.002.

Cheng, X. et al., 2018: Comparing road safety performance across countries: Do data source and type of mortality indicator matter? *Accid. Anal. Prev.*, **121**, 129–133, doi:10.1016/j.aap.2018.09.012.

Chester, M. and A. Horvath, 2012: High-speed rail with emerging automobiles and aircraft can reduce environmental impacts in California's future. *Environ. Res. Lett.*, **7(3)**, 30412, doi:10.1088/1748-9326/7/3/034012.

Chester, M., S. Pincetl, Z. Elizabeth, W. Eisenstein, and J. Matute, 2013: Infrastructure and automobile shifts: Positioning transit to reduce life-cycle environmental impacts for urban sustainability goals. *Environ. Res. Lett.*, **8(1)**, 15041, doi:10.1088/1748-9326/8/1/015041.

Chiaramonti, D., 2019: Sustainable Aviation Fuels: the challenge of decarbonization. *Energy Procedia*, **158**, 1202–1207, doi:10.1016/j.egypro.2019.01.308.

City of Bulawayo, 2020a: *Report of Study – Master Plan 2019–2034,* City of Bulawayo, Bulawayo, Zimbabwe.

City of Bulawayo, 2020b: *Corporate Strategy 2020–2024,* City of Bulawayo, Bulawayo, Zimbabwe.

Coffel, E.D., T.R. Thompson, and R.M. Horton, 2017: The impacts of rising temperatures on aircraft takeoff performance. *Clim. Change*, **144(2)**, 381–388, doi:10.1007/s10584-017-2018-9.

Coignard, J., S. Saxena, J. Greenblatt, and D. Wang, 2018: Clean vehicles as an enabler for a clean electricity grid. *Environ. Res. Lett.*, **13(5)**, 054031, doi:10.1088/1748-9326/aabe97.

Colling, A. and R. Hekkenberg, 2020: Waterborne platooning in the short sea shipping sector. *Transp. Res. Part C Emerg. Technol.*, **120**, 102778, doi:10.1016/J.TRC.2020.102778.

Committee on Climate Change, 2019: *Net Zero: The UK's contribution to stopping global warming*. Committee on Climate Change, London, UK, 275 pp.

Commonwealth of Australia, 2019: *Australia's Critical Minerals Strategy 2019*. Commonwealth of Australia, 22 pp.

Congedo, L. and M. Munafò, 2014: Urban Sprawl as a Factor of Vulnerability to Climate Change: Monitoring Land Cover Change in Dar es Salaam. In: *Climate Change Vulnerability in Southern African Cities* [Macchi, S. and M. Tiepolo, (eds.)], Springer, Cham, Switzerland, pp. 73–88.

Convery, S. and B. Williams, 2019: Determinants of Transport Mode Choice for Non-Commuting Trips: The Roles of Transport, Land Use and Socio-Demographic Characteristics. *Urban Sci.*, **3(3)**, 82, doi:10.3390/urbansci3030082.

Cooper, J. and P. Balcombe, 2019: Life cycle environmental impacts of natural gas drivetrains used in road freighting. *Procedia CIRP*, **80**, 334–339, doi:10.1016/j.procir.2019.01.070.

Corbett, J.J. et al., 2010: Arctic shipping emissions inventories and future scenarios. *Atmos. Chem. Phys.*, **10(19)**, 9689–9704, doi:10.5194/acp-10-9689-2010.

Cornet, Y., G. Dudley, and D. Banister, 2018: High Speed Rail: Implications for carbon emissions and biodiversity. *Case Stud. Transp. Policy*, **6(3)**, 376–390, doi:10.1016/j.cstp.2017.08.007.

Coroama, V.C., Å. Moberg, and L.M. Hilty, 2015: Dematerialization Through Electronic Media? In: *ICT innovations for sustainability. Advances in Intelligent Systems and Computing* [Hilty, L.M. and B. Aebischer, (eds.)], Springer, Cham, Switzerland, pp. 405–421.

Coulombel, N., V. Boutueil, L. Liu, V. Viguié, and B. Yin, 2019: Substantial rebound effects in urban ridesharing: Simulating travel decisions in Paris, France. *Transp. Res. Part D Transp. Environ.*, **71**, 110–126, doi:10.1016/j.trd.2018.12.006.

Cox, B.L. and C.L. Mutel, 2018: The environmental and cost performance of current and future motorcycles. *Appl. Energy*, **212**, 1013–1024, doi:10.1016/j.apenergy.2017.12.100.

Cox, B., et al., 2018: Uncertain Environmental Footprint of Current and Future Battery Electric Vehicles. *Environ. Sci. Technol.*, **52(8)**, 4989–4995, doi:10.1021/acs.est.8b00261.

Creutzig, F., 2014: How fuel prices determine public transport infrastructure, modal shares and urban form. *Urban Clim.*, **10**(P1), 63–76, doi:10.1016/j.uclim.2014.09.003.

Creutzig, F., 2016: Evolving Narratives of Low-Carbon Futures in Transportation. *Transp. Rev.*, **36(3)**, 341–360, doi:10.1080/01441647.2015.1079277.

Creutzig, F., 2021: *Making Smart Mobility Sustainable: How to Leverage the Potential of Smart and Shared Mobility to Mitigate Climate Change*. Israel Public Policy Institute and Heinrich Böll Foundation Tel Aviv, Tel Aviv, Israel, 18 pp.

Creutzig, F. et al., 2015a: Transport: A roadblock to climate change mitigation? *Science*, **350(6263)**, 911–912, doi:10.1126/science.aac8033.

Creutzig, F., G. Baiocchi, R. Bierkandt, P.-P. Pichler, and K.C. Seto, 2015b: Global typology of urban energy use and potentials for an urbanization mitigation wedge. *Proc. Natl. Acad. Sci.*, **112(20)**, 6283–6288, doi:10.1073/pnas.1315545112.

Creutzig, F. et al., 2016: Beyond Technology: Demand-Side Solutions for Climate Change Mitigation. *Annu. Rev. Environ. Resour.*, **41(1)**, 173–198, doi:10.1146/annurev-environ-110615-085428.

Creutzig, F. et al., 2018: Towards demand-side solutions for mitigating climate change. *Nat. Clim. Change*, **8(4)**, 260–263, doi:10.1038/s41558-018-0121-1.

Crippa, M. et al., 2019: *Fossil $CO_2$ and GHG emissions of all world countries – 2019 Report, EUR 29849 EN*. Publications Office of the European Union, Luxembourg.

Crippa, M. et al., 2021: Global anthropogenic emissions in urban areas: patterns, trends, and challenges. *Environ. Res. Lett.*, **16(7)**, 074033, doi:10.1088/1748-9326/ac00e2.

Cuevas, V., M. Estrada, and J.M. Salanova, 2016: Management of On-demand Transport Services in Urban Contexts. Barcelona Case Study. *Transp. Res. Procedia*, **13**, 155–165, doi:10.1016/j.trpro.2016.05.016.

Cumpsty, N. et al., 2019: *Independent expert integrated technology goals assessment and review for engines and aircraft*. International Civil Aviation Organization, Montréal, Canada, 225 pp.

Cusenza, M.A. et al., 2019: Energy and environmental assessment of a traction lithium-ion battery pack for plug-in hybrid electric vehicles. *Journal of Cleaner Production*, **215**, 634–649, doi:10.1016/j.jclepro.2019.01.056.

Czermański, E., B. Pawłowska, A. Oniszczuk-Jastrząbek, and G.T. Cirella, 2020: Decarbonization of Maritime Transport: Analysis of External Costs. *Front. Energy Res.*, **8**, 28, doi:10.3389/fenrg.2020.00028.

d'Aveni, R., 2015: The 3-D printing revolution. *Harvard Business Review*, May, Harvard Business School Publishing, Cambridge, MA, USA.

Daim, T.U., 2021: *Roadmapping Future Technologies, Products and Services*. Springer, London, UK.

Daioglou, V. et al., 2020: Progress and barriers in understanding and preventing indirect land-use change. *Biofuels, Bioprod. Biorefining*, **14(5)**, 924–934, doi:10.1002/bbb.2124.

Dalsøren, S.B. et al., 2009: Update on emissions and environmental impacts from the international fleet of ships: the contribution from major ship types and ports. *Atmos. Chem. Phys.*, **9(6)**, 2171–2194, doi:10.5194/acp-9-2171-2009.

Dalsøren, S.B. et al., 2013: Environmental impacts of shipping in 2030 with a particular focus on the Arctic region. *Atmos. Chem. Phys.*, **13(4)**, 1941–1955, doi:10.5194/acp-13-1941-2013.

Davis, S.J. et al., 2018: Net-zero emissions energy systems. *Science*, **360(6396)**, eaas9793, doi:10.1126/science.aas9793.

Dawson, D., J. Shaw, and W. Roland Gehrels, 2016: Sea-level rise impacts on transport infrastructure: The notorious case of the coastal railway line at Dawlish, England. *J. Transp. Geogr.*, **51**, 97–109, doi:10.1016/j.jtrangeo.2015.11.009.

Dawson, J. et al., 2020: Infusing Inuit and local knowledge into the low impact shipping corridors: An adaptation to increased shipping activity and climate change in Arctic Canada. *Environ. Sci. Policy*, **105**(November 2019), 19–36, doi:10.1016/j.envsci.2019.11.013.

de Bortoli, A. and Z. Christoforou, 2020: Consequential LCA for territorial and multimodal transportation policies: method and application to the free-floating e-scooter disruption in Paris. *J. Clean. Prod.*, **273**, 122898, doi:10.1016/j.jclepro.2020.122898.

de Bortoli, A., A. Féraille, and F. Leurent, 2017: Life Cycle Assessment to Support Decision-Making in Transportation Planning: A Case of French Bus Rapid Transit. *Transportation Research Board 96th Annual Meeting*, Washington, DC, USA.

De Groot, J.I., L. Steg, and M. Dicke, 2008: Transportation Trends from a Moral Perspective: Value Orientations. In: *New Transportation Research Progress* [Gustavsson, F.N., (ed.)], Nova Science Publishers, New York, pp. 67–91.

de Jong, S. et al., 2017: Life-cycle analysis of greenhouse gas emissions from renewable jet fuel production. *Biotechnol. Biofuels*, **10(1)**, 64, doi:10.1186/s13068-017-0739-7.

de Souza, L.L.P., et al., 2018: Comparative environmental life cycle assessment of conventional vehicles with different fuel options, plug-in hybrid and electric vehicles for a sustainable transportation system in Brazil. *Journal of Cleaner Production*, **203**, 444–468, doi:10.1016/j.jclepro.2018.08.236.

de Souza Dias, M.O. et al., 2015: Sugarcane processing for ethanol and sugar in Brazil. *Environ. Dev.*, **15**, doi:10.1016/j.envdev.2015.03.004.

Debnath, A.K., H.C. Chin, M.M. Haque, and B. Yuen, 2014: A methodological framework for benchmarking smart transport cities. *Cities*, **37**, 47–56, doi:10.1016/j.cities.2013.11.004.

Decarre, S., J. Berthiaud, N. Butin, and J.L. Guillaume-Combecave, 2010: $CO_2$ maritime transportation. *Int. J. Greenh. Gas Control*, **4(5)**, 857–864, doi:10.1016/J.IJGGC.2010.05.005.

Desmaris, C. and F. Croccolo, 2018: The HSR competition in Italy: How are the regulatory design and practices concerned? *Res. Transp. Econ.*, **69**, 290–299, doi:10.1016/j.retrec.2018.05.004.

Després, J. et al., 2017: Storage as a flexibility option in power systems with high shares of variable renewable energy sources: a POLES-based analysis. *Energy Econ.*, **64**, 638–650, doi:10.1016/j.eneco.2016.03.006.

Dhar, S. and P.R. Shukla, 2015: Low carbon scenarios for transport in India: Co-benefits analysis. *Energy Policy*, **81**, 186–198, doi:10.1016/j.enpol.2014.11.026.

Dhar, S., M. Pathak, and P.R. Shukla, 2017: Electric vehicles and India's low carbon passenger transport: a long-term co-benefits assessment. *J. Clean. Prod.*, **146**, 139–148, doi:10.1016/j.jclepro.2016.05.111.

Dhar, S., M. Pathak, and P.R. Shukla, 2018: Transformation of India's transport sector under global warming of 2°C and 1.5°C scenario. *J. Clean. Prod.*, **172**, 417–427, doi:10.1016/j.jclepro.2017.10.076.

Dhar, S., M. Pathak, and P.R. Shukla, 2020: Transformation of India's steel and cement industry in a sustainable 1.5°C world. *Energy Policy*, **137**, 111104, doi:10.1016/j.enpol.2019.111104.

Di Sbroiavacca, N., G. Nadal, F. Lallana, J. Falzon, and K. Calvin, 2014: Emissions reduction scenarios in the Argentinean Energy Sector. *Energy Econ.*, **56**, 552–563, doi:10.1016/j.eneco.2015.03.021.

Dia, H., 2019: Rethinking Urban Mobility: Unlocking the Benefits of Vehicle Electrification. In: *Decarbonising the Built Environment: Charting the Transition* [Newton, P., D. Prasad, A. Sproul, and S. White, (eds.)], Palgrave Macmillan, Singapore, pp. 83–98.

Dieleman, F. and M. Wegener, 2004: Compact city and urban sprawl. *Built Environ.*, **30(4)**, 308–323, doi:10.2148/benv.30.4.308.57151.

Digby, R.A.R., N.P. Gillett, A.H. Monahan, and J.N.S. Cole, 2021: An Observational Constraint on Aviation-Induced Cirrus From the COVID-19-Induced Flight Disruption. *Geophys. Res. Lett.*, **48(20)**, e2021GL095882-e2021GL095882, doi:10.1029/2021GL095882.

Dimitriou, P. and R. Javaid, 2020: A review of ammonia as a compression ignition engine fuel. *Int. J. Hydrogen Energy*, **45(11)**, 7098–7118, doi:10.1016/J.IJHYDENE.2019.12.209.

Dimoula, V., F. Kehagia, and A. Tsakalidis, 2016: A Holistic Approach for Estimating Carbon Emissions of Road and Rail Transport Systems. *Aerosol Air Qual. Res.*, **16**, 61–68, doi:10.4209/aaqr.2015.05.0313.

Docherty, I., G. Marsden, and J. Anable, 2018: The governance of smart mobility. *Transp. Res. Part A Policy Pract.*, **115**(October 2017), 114–125, doi:10.1016/j.tra.2017.09.012.

DOE-US, 2019: *Hydrogen Class 8 Long Haul Truck Targets*. US Department of Energy, Lemont, IL, USA, 31 pp.

DNV GL, 2017: *Low Carbon Shipping Towards 2050*. D N V GL, Høvik, Norway.

DNV GL, 2019: *Maritime Forecast To 2050 Energy Transition Outlook 2019*. D N V GL, Høvik, Norway, 118 pp.

DNV GL, 2020: *Ammonia as a marine fuel*. D N V GL, Høvik, Norway, 48 pp.

Dubov, G.M., D.S. Trukhmanov, and S.A. Nokhrin, 2020: The Use of Alternative Fuel for Heavy-Duty Dump Trucks as a Way to Reduce the Anthropogenic Impact on the Environment. *IOP Conf. Ser. Earth Environ. Sci.*, **459**, 42059, doi:10.1088/1755-1315/459/4/042059.

Dziewiatkowski, M., D. Szpica, and A. Borawski, 2020: Evaluation of impact of combustion engine controller adaptation process on level of exhaust gas emissions in gasoline and compressed natural gas supply process. *Eng. Rural Dev.*, **19**, 541–548, doi:10.22616/ERDev.2020.19.TF122.

Earley, R. and P. Newman, 2021: Transport in the Aftermath of COVID-19: Lessons Learned and Future Directions. *J. Transp. Technol.*, **11(02)**, 109–127, doi:10.4236/jtts.2021.112007.

ECA, 2021: Infrastructure for charging electric vehicles: more charging stations but uneven deployment makes travel across the EU complicated. European Court of Auditors, Luxembourg, https://www.eca.europa.eu/en/Pages/DocItem.aspx?did=58260.

Eckhardt, G.M. and F. Bardhi, 2015: The Sharing Economy isn't About Sharing At All. *Harv. Bus. Rev.*, https://hbr.org/2015/01/the-sharing-economy-isnt-about-sharing-at-all.

Edelenbosch, O.Y. et al., 2017: Decomposing passenger transport futures: Comparing results of global integrated assessment models. *Transp. Res. Part D Transp. Environ.*, **55**, 281–293, doi:10.1016/j.trd.2016.07.003.

Edwards, R.W.J. and M.A. Celia, 2018: Infrastructure to enable deployment of carbon capture, utilization, and storage in the United States. *Proc. Natl. Acad. Sci.*, **115(38)**, E8815–E8824, doi:10.1073/pnas.1806504115.

Egli, F., B. Steffen, and T.S. Schmidt, 2019: Bias in energy system models with uniform cost of capital assumption. *Nat. Commun.*, **10(1)**, 4588, doi:10.1038/s41467-019-12468-z.

Ehebrecht, D., D. Heinrichs, and B. Lenz, 2018: Motorcycle taxis in sub-Saharan Africa: Current knowledge, implications for the debate on "informal" transport and research needs. *J. Transp. Geogr.*, **69**, 242–256, doi:10.1016/j.jtrangeo.2018.05.006.

El Akkari, M., O. Réchauchère, A. Bispo, B. Gabrielle, and D. Makowski, 2018: A meta-analysis of the greenhouse gas abatement of bioenergy factoring in land use changes. *Sci. Rep.*, **8(1)**, 8563, doi:10.1038/s41598-018-26712-x.

El Hannach, M., P. Ahmadi, L. Guzman, S. Pickup, and E. Kjeang, 2019: Life cycle assessment of hydrogen and diesel dual-fuel class 8 heavy duty trucks. *Int. J. Hydrogen Energy*, **44(16)**, 8575–8584, doi:10.1016/J.IJHYDENE.2019.02.027.

Elgowainy, A., et al., 2018: Current and Future United States Light-Duty Vehicle Pathways: Cradle-to-Grave Lifecycle Greenhouse Gas Emissions and Economic Assessment. *Environmental Science & Technology*, **52**, 2392–2399, doi:10.1021/acs.est.7b06006.

Ellingsen, L.A.-W., B. Singh, and A.H. Strømman, 2016: The size and range effect: lifecycle greenhouse gas emissions of electric vehicles. *Environ. Res. Lett.*, **11(5)**, 54010, doi:10.1088/1748-9326/11/5/054010.

Ellingsen, L.A.-W. et al., 2014: Life Cycle Assessment of a Lithium-Ion Battery Vehicle Pack. *J. Ind. Ecol.*, **18(1)**, 113–124, doi:10.1111/jiec.12072.

Elshout, P.M.F. et al., 2015: Greenhouse-gas payback times for crop-based biofuels. *Nat. Clim. Change*, **5(6)**, 604–610, doi:10.1038/nclimate2642.

Endresen, Ø. et al., 2003: Emission from international sea transportation and environmental impact. *J. Geophys. Res. Atmos.*, **108(17)**, doi:10.1029/2002jd002898.

Energieradar, 2020: $CO_2$-Gesetz

Energy Transitions Commission, 2020: *Making Mission Possible: Delivering a Net-Zero Economy*. Energy Transitions Commission, London, UK, 89 pp.

Epstein, A.H. and S.M. O'Flarity, 2019: Considerations for Reducing Aviation's $CO_2$ with Aircraft Electric Propulsion. *J. Propuls. Power*, **35(3)**, 572–582, doi:10.2514/1.B37015.

Ercan, T., M. Noori, Y. Zhao, and O. Tatari, 2016: On the Front Lines of a Sustainable Transportation Fleet: Applications of Vehicle-to-Grid Technology for Transit and School Buses. *Energies*, **9(4)**, 230, doi:10.3390/en9040230.

Erdogan, S., 2020: Analyzing the environmental Kuznets curve hypothesis: The role of disaggregated transport infrastructure investments. *Sustain. Cities Soc.*, **61**, 102338, doi:10.1016/j.scs.2020.102338.

Erickson, P. and K. Tempest, 2015: *Keeping cities green: Avoiding carbon lock-in due to urban development*. Stockholm Environment Institute, Seattle, WA, USA, 28 pp.

Esmeijer, K., M. Den Elzen, and H. Van Soest, 2020: *Analysing International Shipping and Aviation Emission Projections of Integrated Assessment Models*. PBL Netherlands Environmental Assessment Agency, The Hague, Nertherlands, 39 pp.

EU, 2021: *Assessment of hydrogen delivery options*. The European Commission's Joint Research Centre, Brussels, Belgium, 4 pp.

Eudy, L. and M. Post, 2018a: *Zero-Emission Bus Evaluation Results: Orange County Transportation Authority Fuel Cell Electric Bus*. US Department of Transportation, Washington, DC, USA, 48 pp.

Eudy, L. and M. Post, 2018b: *Fuel Cell Buses in U.S. Transit Fleets: Current Status 2018*. National Renewable Energy Laboratory, Golden, CO, USA, 50 pp.

Eudy, L. and M. Post, 2020: *Fuel Cell Buses in U.S. Transit Fleets: Current Status 2020*. National Renewable Energy Laboratory, Golden, CO, USA, 57 pp.

eurostat, 2019: Modal split of passenger transport. https://ec.europa.eu/eurostat/web/products-datasets/-/t2020_rk310.

Evangelisti, S., C. Tagliaferri, D.J.L. Brett, and P. Lettieri, 2017: Life cycle assessment of a polymer electrolyte membrane fuel cell system for passenger vehicles. *Journal of Cleaner Production*, **142**, 4339–4355, doi:10.1016/j.jclepro.2016.11.159.

Eyring, V., H.W. Köhler, J. Van Aardenne, and A. Lauer, 2005: Emissions from international shipping: 1. The last 50 years. *J. Geophys. Res. D Atmos.*, **110(17)**, 171–182, doi:10.1029/2004JD005619.

Eyring, V., N.P. Gillett, K.M. Achuta Rao, R. Barimalala, M. Barreiro Parrillo, N. Bellouin, C. Cassou, P.J. Durack, Y. Kosaka, S. McGregor, S. Min, O. Morgenstern, and Y. Sun, 2021: Human Influence on the Climate System. In *Climate Change 2021: The Physical Science Basis. Contribution of Working Group I to the Sixth Assessment Report of the Intergovernmental Panel on Climate Change* [Masson-Delmotte, V., P. Zhai, A. Pirani, S.L. Connors, C. Péan, S. Berger, N. Caud, Y. Chen, L. Goldfarb, M.I. Gomis, M. Huang, K. Leitzell, E. Lonnoy, J.B.R. Matthews, T.K. Maycock, T. Waterfield, O. Yelekçi, R. Yu, and B. Zhou (eds.)]. Cambridge University Press, Cambridge, UK and New York, NY, USA.

Faber, J. et al., 2020: *Fourth IMO GHG Study 2020 – Final report*. International Maritime Organization, London, UK, 524 pp.

Faisal, A., T. Yigitcanlar, M. Kamruzzaman, and G. Currie, 2019: Understanding autonomous vehicles: A systematic literature review on capability, impact, planning and policy. *J. Transp. Land Use*, **12(1)**, 45–72, doi:10.5198/jtlu.2019.1405.

Fang, S. et al., 2019: Optimal Sizing of Shipboard Carbon Capture System for Maritime Greenhouse Emission Control. *IEEE Trans. Ind. Appl.*, **55(6)**, 5543–5553, doi:10.1109/TIA.2019.2934088.

Farzaneh, H., J.A.P. de Oliveira, B. McLellan, and H. Ohgaki, 2019: Towards a Low Emission Transport System: Evaluating the Public Health and Environmental Benefits. *Energies*, **12(19)**, 3747, doi:10.3390/en12193747.

FCHEA, 2019: *Roadmap to a US Hydrogen Economy: Reudcing emissions and driving growth across the nation*. The Fuel Cell and Hydrogen Energy Association, Washington, DC, USA, 96 pp.

FCHJU, 2019: *Hydrogen roadmap Europe: A sustainable pathway for the European energy transtition*. Fuel Cells and Hydrogen 2 Joint Undertaking, Brussels, Belgium, 70 pp.

Fearnley, N. et al., 2017: Triggers of Urban Passenger Mode Shift – State of the Art and Model Evidence. *Transp. Res. Procedia*, **26**, 62–80, doi:10.1016/J.TRPRO.2017.07.009.

Fearnley, N. et al., 2018: Competition and substitution between public transport modes. *Res. Transp. Econ.*, **69**, 51–58, doi:10.1016/j.retrec.2018.05.005.

Feng, C., Y.-S. Xia, and L.-X. Sun, 2020: Structural and social-economic determinants of China's transport low-carbon development under the background of aging and industrial migration. *Environ. Res.*, **188**, 109701, doi:10.1016/j.envres.2020.109701.

Figenbaum, E., 2017: Perspectives on Norway's supercharged electric vehicle policy. *Environ. Innov. Soc. Transitions*, **25**, 14–34, doi:10.1016/j.eist.2016.11.002.

Figueres, C. and T. Rivett-Carnac, 2020: *The Future we choose: Surviving the climate crisis*. Vintage, New York.

Fisch-Romito, V. and C. Guivarch, 2019: Transportation infrastructures in a low carbon world: An evaluation of investment needs and their determinants. *Transp. Res. Part D Transp. Environ.*, **72**(May), 203–219, doi:10.1016/j.trd.2019.04.014.

Fitt, H., 2018: Exploring How Older People Might Experience Future Transport Systems. In: *Geographies of Transport and Ageing* [Curl, A. and C. Musselwhite, (eds.)], Springer International Publishing, Cham, Switzerland, pp. 199–225.

Fleming, G.G. and I. de Lépinay, 2019: *Environmental Trends in Aviation to 2050*. International Civil Aviation Organization, Montréal, Canada, 7 pp.

Fontaras, G. et al., 2017: The difference between reported and real-world $CO_2$ emissions: How much improvement can be expected by WLTP introduction? *Transp. Res. Procedia*, **25**, 3933–3943, doi:10.1016/j.trpro.2017.05.333.

Forero-Ortiz, E., E. Martínez-Gomariz M. Cañas Porcuna, L. Locatelli, and B. Russo, 2020: Flood Risk Assessment in an Underground Railway System under the Impact of Climate Change – A Case Study of the Barcelona Metro. *Sustainability*, **12(13)**, 5291, doi:10.3390/su12135291.

Forzieri, G. et al., 2018: Escalating impacts of climate extremes on critical infrastructures in Europe. *Glob. Environ. Change*, **48**, 97–107, doi:10.1016/j.gloenvcha.2017.11.007.

Fotouhi, A., D.J. Auger, K. Propp, and S. Longo, 2017: Electric vehicle battery parameter identification and SOC observability analysis: NiMH and Li-S case studies. *IET Power Electron.*, **10(11)**, 1289–1297, doi:10.1049/iet-pel.2016.0777.

Fox-Kemper, B., H.T. Hewitt, C. Xiao, G. Aðalgeirsdóttir, S.S. Drijfhout, T.L. Edwards, N.R. Golledge, M. Hemer, R.E. Kopp, G. Krinner, A. Mix, D. Notz, S. Nowicki, I.S. Nurhati, L. Ruiz, J.-B. Sallée, A.B.A. Slangen, and Y. Yu, 2021: Ocean, Cryosphere and Sea Level Change. In *Climate Change 2021: The Physical Science Basis. Contribution of Working Group I to the Sixth Assessment Report of the Intergovernmental Panel on Climate Change* [MassonDelmotte, V., P. Zhai, A. Pirani, S.L. Connors, C. Péan, S. Berger, N. Caud, Y. Chen, L. Goldfarb, M.I. Gomis, M. Huang, K. Leitzell, E. Lonnoy, J.B.R. Matthews, T.K. Maycock, T. Waterfield, O. Yelekçi, R. Yu, and B. Zhou (eds.)]. Cambridge University Press, Cambridge, UK and New York, NY, USA. In Press.

Frattini, D. et al., 2016: A system approach in energy evaluation of different renewable energies sources integration in ammonia production plants. *Renew. Energy*, **99**, 472–482, doi:10.1016/J.RENENE.2016.07.040.

Freeman, S., D.S. Lee, L.L. Lim, A. Skowron, and R.R. De León, 2018: Trading off Aircraft Fuel Burn and $NO_x$ Emissions for Optimal Climate Policy. *Environ. Sci. Technol.*, **52(5)**, 2498–2505, doi:10.1021/acs.est.7b05719.

Friedlingstein, P. et al., 2020: Global Carbon Budget 2020. *Earth Syst. Sci. Data*, **12(4)**, 3269–3340, doi:10.5194/essd-12-3269-2020.

Frigo, S. and R. Gentili, 2014: Further evolution of an ammonia fuelled range extender for hybrid vehicles. *Proceedings of the NH3 Fuel Association*, Des Moines, IA, USA, September 21–24.

Fu, H., L. Nie, L. Meng, B.R. Sperry, and Z. He, 2015: A hierarchical line planning approach for a large-scale high speed rail network: The China case. *Transp. Res. Part A Policy Pract.*, **75**, 61–83, doi:10.1016/j.tra.2015.03.013.

Fu, J. et al., 2017: Electrically rechargeable zinc–air batteries: progress, challenges, and perspectives. *Adv. Mater.*, **29(7)**, 1604685, doi: 10.1002/adma.201604685.

Fuglestvedt, J.S. et al., 2010: Transport impacts on atmosphere and climate: Metrics. *Atmos. Environ.*, **44(37)**, 4648–4677, doi:10.1016/j.atmosenv.2009.04.044.

Fuglestvedt, J.S. et al., 2014: Climate penalty for shifting shipping to the Arctic. *Environ. Sci. Technol.*, **48(22)**, 13273–13279, doi:10.1021/es502379d.

Fulton, L., A. Mejia, M. Arioli, K. Dematera, and O. Lah, 2017: Climate change mitigation pathways for Southeast Asia: $CO_2$ emissions reduction policies for the energy and transport sectors. *Sustainability*, **9(7)**, doi:10.3390/su9071160.

Funke, S.A. and P. Plötz, 2017: *A techno-economic analysis of fast charging needs in Germany for different ranges of battery electric vehicles*. European Battery, Hybrid and Fuel Cell Electric Vehicle Congress, Geneva, Switzerland, 14–16 March.

Gaber, T., Y. El Jazouli, E. Eldesouky, and A. Ali, 2021: Autonomous Haulage Systems in the Mining Industry: Cybersecurity, Communication and Safety Issues and Challenges. *Electronics*, **10(11)**, 1357, doi:10.3390/electronics10111357.

Gagatsi, E., T. Estrup, and A. Halatsis, 2016: Exploring the Potentials of Electrical Waterborne Transport in Europe: The E-ferry Concept. *Transp. Res. Procedia*, **14**, 1571–1580, doi:10.1016/j.trpro.2016.05.122.

Gallet, C.A. and H. Doucouliagos, 2014: The income elasticity of air travel: A meta-analysis. *Ann. Tour. Res.*, **49**, 141–155, doi:10.1016/J.ANNALS.2014.09.006.

Gallucci, M., 2021: The Ammonia Solution: Ammonia engines and fuel cells in cargo ships could slash their carbon emissions. *IEEE Spectr.*, **58(3)**, 44–50, doi:10.1109/MSPEC.2021.9370109.

Gao, T. et al., 2016: A Rechargeable Al/S Battery with an Ionic-Liquid Electrolyte. *Angewandte Chemie International Edition*, **55**, 9898–9901, doi:10.1002/anie.201603531.

Gao, Y. and P. Newman, 2018: Beijing's Peak Car Transition: Hope for Emerging Cities in the 1.5°C Agenda. *Urban Plan.*, **3(2)**, 82–93, doi:10.17645/up.v3i2.1246.

Gao, Y. and P. Newman, 2020: Are Beijing and Shanghai automobile dependent cities? In: *Handbook on Transport and Urban Transformation in China* [Chen, C.-L., H. Pan, Q. Shen, and J.J. Wang, (eds.)], Edward Elgar Publishing, Cheltenham, UK, pp. 229–245.

Gargoloff, J. et al., 2018: Robust Optimization for Real World $CO_2$ Reduction. SAE Technical Paper 2018-37-0015, 2018, https://doi.org/10.4271/2018-37-0015.

Geels, F.W., 2002: Technological transitions as evolutionary reconfiguration processes: A multi-level perspective and a case-study. *Res. Policy*, **31(8–9)**, 1257–1274, doi:10.1016/S0048-7333(02)00062-8.

Geels, F.W., 2005: *Technological transitions and system innovations: a co-evolutionary and socio-technical analysis*. Edward Elgar Publishing, Cheltenham, UK.

Geels, F.W., 2019: Socio-technical transitions to sustainability: a review of criticisms and elaborations of the Multi-Level Perspective. *Curr. Opin. Environ. Sustain.*, **39**, 187–201, doi:10.1016/j.cosust.2019.06.009.

Geels, F.W., B.K. Sovacool, T. Schwanen, and S. Sorrell, 2017: Sociotechnical transitions for deep decarbonization. *Science*, **357(6357)**, 1242–1244, doi:10.1126/science.aao3760.

Gehl, J., 2010: *Cities for People*. Island Press, Washington, DC, USA, 288 pp.

Gençsü, I. and M. Hino, 2015: Raising Ambition to Reduce International Aviation and Maritime Emissions. Contributing paper for *Seizing the Global Opportunity: Partnerships for Better Growth and a Better Climate*. [Davis, M. and S. Chatwin, (eds.)]. London, UK, and Washington, DC, USA, 24 pp.

GFEI, 2020: *Vehicle Efficiency and Electrification: A Global Status Report*. Global Fuel Economy Initiative, London, UK.

Ghandi, A. and S. Paltsev, 2019: Projecting a Deployment of Light-Duty Internal Combustion and Electric Vehicles in Economy-Wide Models. **4307**(Ldv), MIT Joint Program on the Science and Policy of Global Change, Technical Note 17, Cambridge, MA, USA, 42 pp.

Gierens, K., S. Matthes, and S. Rohs, 2020: How Well Can Persistent Contrails Be Predicted? *Aerospace*, **7(12)**, 169, doi:10.3390/aerospace7120169.

Gierens, K.M., U. Schumann, H.G.J. Smit, M. Helten, and G. Zängl, 1997: Determination of humidity and temperature fluctuations based on MOZAIC data and parametrisation of persistent contrail coverage for general circulation models. *Ann. Geophys.*, **15(8)**, 1057–1066, doi:10.1007/s00585-997-1057-3.

Gilbert, P. et al., 2018: Assessment of full life-cycle air emissions of alternative shipping fuels. *J. Clean. Prod.*, **172(2018)**, 855–866, doi:10.1016/j.jclepro.2017.10.165.

Gilby, S. et al., 2019: *Sustainable Lifestyles Policy and Practice: Challenges and Way Forward – Annex: Case Studies*. Institute for Global Environmental Strategies, Hayama, Japan, 144 pp.

Gill, S.S., G.S. Chatha, A. Tsolakis, S.E. Golunski, and A.P.E. York, 2012: Assessing the effects of partially decarbonising a diesel engine by co-fuelling with dissociated ammonia. *Int. J. Hydrogen Energy*, **37(7)**, 6074–6083, doi:10.1016/J.IJHYDENE.2011.12.137.

Giurco, D., E. Dominish, N. Florin, T. Watari, and B. McLellan, 2019: Requirements for Minerals and Metals for 100% Renewable Scenarios. In: *Achieving the Paris Climate Agreement Goals* [Teske, S., (ed.)], Springer International Publishing, Cham, Switzerland, pp. 437–457.

Givoni, M. and D. Banister, 2006: Airline and railway integration. *Transp. Policy*, **13(5)**, 386–397.

Givoni, M. and F. Dobruszkes, 2013: A Review of Ex-Post Evidence for Mode Substitution and Induced Demand Following the Introduction of High-Speed Rail. *Transp. Rev.*, **33(6)**, 720–742, doi:10.1080/01441647.2013.853707.

Gnann, T. et al., 2018: Fast charging infrastructure for electric vehicles: Today's situation and future needs. *Transp. Res. Part D Transp. Environ.*, **62**, 314–329, doi:10.1016/j.trd.2018.03.004.

Goldmann, A. et al., 2018: A Study on Electrofuels in Aviation. *Energies*, **11(2)**, 392, doi:10.3390/en11020392.

Goldsworthy, B., 2017: Spatial and temporal allocation of ship exhaust emissions in Australian coastal waters using AIS data: Analysis and treatment of data gaps. *Atmos. Environ.*, **163**, 77–86, doi:10.1016/J.ATMOSENV.2017.05.028.

Goldsworthy, L. and B. Goldsworthy, 2015: Modelling of ship engine exhaust emissions in ports and extensive coastal waters based on terrestrial AIS data – An Australian case study. *Environ. Model. Softw.*, **63**, 45–60, doi:10.1016/J.ENVSOFT.2014.09.009.

Golroudbary, S.R., D. Calisaya-Azpilcueta, and A. Kraslawski, 2019: The Life Cycle of Energy Consumption and Greenhouse Gas Emissions from Critical Minerals Recycling: Case of Lithium-ion Batteries. *Procedia CIRP*, **80**, 316–321, doi:10.1016/j.procir.2019.01.003.

González, A., E. Goikolea, J.A. Barrena, and R. Mysyk, 2016: Review on supercapacitors: Technologies and materials. *Renew. Sustain. Energy Rev.*, **58(115–119)**, 1189–1206, doi:10.1016/j.rser.2015.12.249.

Gössling, S., 2018: ICT and transport behavior: A conceptual review. *Int. J. Sustain. Transp.*, **12(3)**, 153–164, doi:10.1080/15568318.2017.1338318.

Gössling, S., A. Humpe, F. Fichert, and F. Creutzig, 2021a: COVID-19 and pathways to low-carbon air transport until 2050. *Environ. Res. Lett.*, **16(3)**, 034063, doi:10.1088/1748-9326/abe90b.

Gössling, S., J. Nicolosi, and T. Litman, 2021b: The Health Cost of Transport in Cities. *Curr. Environ. Heal. Reports*, **8(2)**, 196–201, doi:10.1007/s40572-021-00308-6.

Gota, S., C. Huizenaga, K. Peet, and G. Kaar, 2015: *Emission Reduction Poetential in the Transport Sector by 2030*. Paris Process on Mobility and Climate, 54 pp., https://www.gcca.eu/sites/default/files/2019-12/Emission-Reduction-Potential-in-the-Transport-Sector-by-2030.pdf.

Gota, S., C. Huizenaga, K. Peet, and G. Kaar, 2016: *Nationally Determined Contributions (NDCs) Offer Opportunities for Ambitious Action on Transport and Climate Change*. Paris Process on Mobility and Climate, 58 pp., https://slocat.net/wp-content/uploads/2020/02/PPMC_2016_NDCs-Offer-Opportunities-for-Ambitious-Action.pdf.

Gota, S., C. Huizenga, K. Peet, N. Medimorec, and S. Bakker, 2019: Decarbonising transport to achieve Paris Agreement targets. *Energy Effic.*, **12(2)**, 363–386, doi:10.1007/s12053-018-9671-3.

Granier, C. et al., 2019: *The Copernicus Atmosphere Monitoring Service global and regional emissions*. Copernicus Atmosphere Monitoring Service (CAMS) report, 54 pp., doi:10.24380/d0bn-kx16.

Grant, N., A. Hawkes, S. Mittal, and A. Gambhir, 2021: Confronting mitigation deterrence in low-carbon scenarios. *Environ. Res. Lett.*, **16(6)**, 64099, doi:10.1088/1748-9326/ac0749.

Grant, R., G. Goldsmith, D. Gracy, and D. Johnson, 2016: Better Transportation to Health Care Will Improve Child Health and Lower Costs. *Adv. Pediatr.*, **63(1)**, 389–401, doi:10.1016/j.yapd.2016.04.003.

Gray, J.T., E. Dimitroff, N.T. Meckel, and R.D. Quillian, 1966: Ammonia Fuel – Engine Compatibility and Combustion. SAE Technical Paper 660156, 1966, https://doi.org/10.4271/660156.

Gray, N., S. McDonagh, R. O'Shea, B. Smyth, and J.D. Murphy, 2021: Decarbonising ships, planes and trucks: An analysis of suitable low-carbon fuels for the maritime, aviation and haulage sectors. *Adv. Appl. Energy*, **1**, 100008, doi:10.1016/j.adapen.2021.100008.

Green, J. and P. Newman, 2017: Citizen utilities: The emerging power paradigm. *Energy Policy*, **105**, 283–293, doi:10.1016/j.enpol.2017.02.004.

Greenblatt, J.B. and S. Shaheen, 2015: Automated Vehicles, On-Demand Mobility, and Environmental Impacts. *Curr. Sustain. Energy Reports*, **2(3)**, 74–81, doi:10.1007/s40518-015-0038-5.

Greene, D.L., S. Park, and C. Liu, 2014: Public policy and the transition to electric drive vehicles in the U.S.: The role of the zero emission vehicles mandates. *Energy Strateg. Rev.*, **5**, 66–77, doi:10.1016/j.esr.2014.10.005.

Grieco, M. and J. Urry, 2016: *Mobilities: New perspectives on transport and society*. Routledge, New York and London, 384 pp.

Grubler, A. et al. 2018: A low energy demand scenario for meeting the 1.5°C target and sustainable development goals without negative emission technologies. *Nat. Energy*, **3**, 515–527, doi:10.1038/s41560-018-0172-6.

Grzelakowski, A.S., 2018: Transport conditions of the global economy. *Transp. Econ. Logist.*, **80**, 75–84, doi:10.26881/etil.2018.80.08.

Gucwa, M. and A. Schäfer, 2013: The impact of scale on energy intensity in freight transportation. *Transp. Res. Part D Transp. Environ.*, **23**, 41–49, doi:10.1016/j.trd.2013.03.008.

Gulley, A.L., N.T. Nassar, and S. Xun, 2018: China, the United States, and competition for resources that enable emerging technologies. *Proc. Natl. Acad. Sci.*, **115(16)**, 4111–4115, doi:10.1073/pnas.1717152115.

Gulley, A.L., E.A. McCullough, and K.B. Shedd, 2019: China's domestic and foreign influence in the global cobalt supply chain. *Resour. Policy*, **62**, 317–323, doi:10.1016/j.resourpol.2019.03.015.

Guo, J. et al., 2020: Statistically enhanced model of oil sands operations: Well-to-wheel comparison of in situ oil sands pathways. *Energy*, **208**, 118250, doi:10.1016/j.energy.2020.118250.

H2 Tools, 2020: *International Hydrogen Fueling Stations*. Pacific Northwest National Laboratory, Washington, DC, USA, https://h2tools.org/hyarc/hydrogen-data/international-hydrogen-fueling-stations.

Ha, S., K. Kim, K. Kim, H. Jeong, and H. Kim, 2017: Reliability Approach in Economic Assessment of Adapting Infrastructure to Climate Change. *J. Manag. Eng.*, **33(5)**, 4017022, doi:10.1061/(asce)me.1943-5479.0000530.

Habib, K. and H. Wenzel, 2014: Exploring rare earths supply constraints for the emerging clean energy technologies and the role of recycling. *J. Clean. Prod.*, **84(1)**, 348–359, doi:10.1016/j.jclepro.2014.04.035.

Hache, E., G.S. Seck, M. Simoen, C. Bonnet, and S. Carcanague, 2019: Critical raw materials and transportation sector electrification: A detailed bottom-up analysis in world transport. *Appl. Energy*, **240**(April), 6–25, doi:10.1016/j.apenergy.2019.02.057.

Haines, A. et al., 2017: Short-lived climate pollutant mitigation and the Sustainable Development Goals. *Nat. Clim. Change*, **7(12)**, 863–869, doi:10.1038/s41558-017-0012-x.

Hajimiragha, A., C.A. Canizares, M.W. Fowler, and A. Elkamel, 2010: Optimal Transition to Plug-In Hybrid Electric Vehicles in Ontario, Canada, Considering the Electricity-Grid Limitations. *IEEE Trans. Ind. Electron.*, **57(2)**, 690–701, doi:10.1109/TIE.2009.2025711.

Hakim, M.M. and R. Merkert, 2016: The causal relationship between air transport and economic growth: Empirical evidence from South Asia. *J. Transp. Geogr.*, **56**, 120–127, doi:10.1016/j.jtrangeo.2016.09.006.

Hakim, M.M. and R. Merkert, 2019: Econometric evidence on the determinants of air transport in South Asian countries. *Transp. Policy*, **83**, 120–126, doi:10.1016/j.tranpol.2017.12.003.

Halim, R., L. Kirstein, O. Merk, and L. Martinez, 2018: Decarbonization Pathways for International Maritime Transport: A Model-Based Policy Impact Assessment. *Sustainability*, **10(7)**, 2243, doi:10.3390/su10072243.

Hall, D. and N. Lutsey, 2017: *Literature review on power utility best practices regarding electric vehicles.* International Council on Clean Transportation, Washington, DC, USA, 52 pp.

Hall, D. and N. Lutsey, 2019: *Estimating the infrastructure needs and costs for the launch of zero-emission trucks*. International Council on Clean Transportation, Washington, DC, USA, 37 pp. https://theicct.org/sites/default/files/publications/ICCT_EV_HDVs_Infrastructure_20190809.pdf.

Hall, D. and N. Lutsey, 2020: *Electric vehicle charging guide for cities*. International Council on Clean Transportation, Washington, DC, USA, 24 pp.

Halliday, W.D., S.J. Insley, R.C. Hilliard, T. de Jong, and M.K. Pine, 2017: Potential impacts of shipping noise on marine mammals in the western Canadian Arctic. *Mar. Pollut. Bull.*, **123(1–2)**, 73–82, doi:10.1016/J.MARPOLBUL.2017.09.027.

Hamilton, I. et al., 2021: The public health implications of the Paris Agreement: a modelling study. *Lancet Planet. Heal.*, **5(2)**, e74–e83, doi:10.1016/S2542-5196(**20**)30249-7.

Hancock, P.A., I. Nourbakhsh, and J. Stewart, 2019: On the future of transportation in an era of automated and autonomous vehicles. *Proc. Natl. Acad. Sci.*, **116(16)**, 7684–7691, doi:10.1073/pnas.1805770115.

Handler, R.M., D.R. Shonnard, E.M. Griffing, A. Lai, and I. Palou-Rivera, 2016: Life Cycle Assessments of Ethanol Production via Gas Fermentation: Anticipated Greenhouse Gas Emissions for Cellulosic and Waste Gas Feedstocks. *Ind. Eng. Chem. Res.*, **55(12)**, doi:10.1021/acs.iecr.5b03215.

Hannula, I., 2016: Hydrogen enhancement potential of synthetic biofuels manufacture in the European context: A techno-economic assessment. *Energy*, **104**, 199–212, doi:10.1016/j.energy.2016.03.119.

Hanson, D., T. Toru Delibasi, M. Gatti, and S. Cohen, 2022: How do changes in economic activity affect air passenger traffic? The use of state-dependent income elasticities to improve aviation forecasts. *J. Air Transp. Manag.*, **98**, 102147, doi:10.1016/J.JAIRTRAMAN.2021.102147.

Hansson, J., S. Månsson, S. Brynolf, and M. Grahn, 2019: Alternative marine fuels: Prospects based on multi-criteria decision analysis involving Swedish stakeholders. *Biomass and Bioenergy*, **126**, 159–173, doi:10.1016/j.biombioe.2019.05.008.

Hansson, J., S. Brynolf, E. Fridell, and M. Lehtveer, 2020: The potential role of ammonia as marine fuel – based on energy systems modeling and multi-criteria decision analysis. *Sustainability*, **12(8)**, 10–14, doi:10.3390/SU12083265.

Hao, H. et al., 2019: Securing Platinum-Group Metals for Transport Low-Carbon Transition. *One Earth*, **1(1)**, 117–125, doi:10.1016/j.oneear.2019.08.012.

Hao, X., Y. Yuan, H. Wang, and Y. Sun, 2021: Actual Electricity Utility Factor of Plug-In Hybrid Electric Vehicles in Typical Chinese Cities Considering Charging Pattern Heterogeneity. *World Electr. Veh. J.*, **12(4)**, 169, doi:10.3390/wevj12040169.

Hardman, S. et al., 2018: A review of consumer preferences of and interactions with electric vehicle charging infrastructure. *Transp. Res. Part D Transp. Environ.*, **62**, 508–523, doi:10.1016/j.trd.2018.04.002.

Hargroves, K., B. Stantic, D. Conley, D. Ho, and G. Grant, 2017: *Big Data, Technology and Transport – The State of Play*. Sustainable Built Environment National Research Centre, Bentley, Australia, 25 pp.

Hargroves, K., B. Stantic, and D. Allen, 2020: *Exploring the potential for artificial intelligence and blockchain to enhance transport*. Sustainable Built Environment National Research Centre, Nathan, Australia, 32 pp.

Harper, C.D., C.T. Hendrickson, S. Mangones, and C. Samaras, 2016: Estimating potential increases in travel with autonomous vehicles for the non-driving, elderly and people with travel-restrictive medical conditions. *Transp. Res. Part C Emerg. Technol.*, **72**, 1–9, doi:10.1016/J.TRC.2016.09.003.

Harper, G. et al., 2019: Recycling lithium-ion batteries from electric vehicles. *Nature*, **575(7781)**, 75–86, doi:10.1038/s41586-019-1682-5.

Harris, A., D. Soban, B.M. Smyth, and R. Best, 2020: A probabilistic fleet analysis for energy consumption, life cycle cost and greenhouse gas emissions modelling of bus technologies. *Appl. Energy*, **261**, 114422, doi:10.1016/J.APENERGY.2019.114422.

Harsha, S., 2014: Liquid hydrogen as aviation fuel and its relative performance with commercial aircraft fuel. *Int. J. Mech. Eng. Robot. Res.*, **1**, 73–77.

Hasan, M.A., D.J. Frame, R. Chapman, and K.M. Archie, 2019: Emissions from the road transport sector of New Zealand: key drivers and challenges. *Environ. Sci. Pollut. Res.*, **26(23)**, 23937–23957, doi:10.1007/s11356-019-05734-6.

Hassellöv, I.M., D.R. Turner, A. Lauer, and J.J. Corbett, 2013: Shipping contributes to ocean acidification. *Geophys. Res. Lett.*, **40(11)**, 2731–2736, doi:10.1002/grl.50521.

Haustein, S. and A.F. Jensen, 2018: Factors of electric vehicle adoption: A comparison of conventional and electric car users based on an extended theory of planned behavior. *Int. J. Sustain. Transp.*, **12(7)**, 484–496, doi:10.1080/15568318.2017.1398790.

Hawkins, T.R., B. Singh, G. Majeau-Bettez, and A.H. Strømman, 2013: Comparative Environmental Life Cycle Assessment of Conventional and Electric Vehicles. *J. Ind. Ecol.*, **17(1)**, doi:10.1111/j.1530-9290.2012.00532.x.

He, B.-J., D. Zhao, and Z. Gou, 2020: Integration of Low-Carbon Eco-City, Green Campus and Green Building in China. In: *Green Guilding in developing countries* [Gou, Z., (ed.)], Springer, Cham, Switzerland, pp. 49–78.

Heinemann, T., H. Mikkelsen, A. Kortsari, and L. Mitropoulos, 2020: *D7.5 Final validation and evaluation report, E-Ferry Project.* European Commission, 173 pp.

Heinen, E., A. Harshfield, J. Panter, R. Mackett, and D. Ogilvie, 2017: Does exposure to new transport infrastructure result in modal shifts? Patterns of change in commute mode choices in a four-year quasi-experimental cohort study. *J. Transp. Heal.*, **6**, 396–410, doi:10.1016/j.jth.2017.07.009.

Hendrickson, C. and L.R. Rilett, 2020: The COVID-19 Pandemic and Transportation Engineering. *J. Transp. Eng. Part A Syst.*, **146(7)**, 1820001, doi:10.1061/JTEPBS.0000418.

Hensher, D.A., 2017: Future bus transport contracts under a mobility as a service (MaaS) regime in the digital age: Are they likely to change? *Transp. Res. Part A Policy Pract.*, **98**, 86–96, doi:10.1016/J.TRA.2017.02.006.

Henze, V., 2020: *Battery Pack Prices Cited Below $100/kWh for the First Time in 2020, While Market Average Sits at $137/kWh*. BloombergNEF. https://about.bnef.com/blog/battery-pack-prices-cited-below-100-kwh-for-the-first-time-in-2020-while-market-average-sits-at-137-kwh/.

Hernandez, D., 2018: Uneven mobilities, uneven opportunities: Social distribution of public transport accessibility to jobs and education in Montevideo. *J. Transp. Geogr.*, **67**, 119–125, doi:10.1016/j.jtrangeo.2017.08.017.

Hill, N. et al., 2020: *Determining the environmental impacts of conventional and alternatively fuelled vehicles through LCA*. European Commission Publications Office, Luxembourg.

Hiratsuka, J., G. Perlaviciute, and L. Steg, 2018: Testing VBN theory in Japan: Relationships between values, beliefs, norms, and acceptability and expected effects of a car pricing policy. *Transp. Res. Part F Traffic Psychol. Behav.*, **53**, 74–83, doi:10.1016/j.trf.2017.12.015.

Hittinger, E. and P. Jaramillo, 2019: Internet of Things: Energy boon or bane? *Science*, **364(6438)**, 326–328, doi:10.1126/science.aau8825.

Hoegh-Guldberg, O. et al., 2019: *The Ocean as a Solution to Climate Change: Five Opportunities for Action*. World Resources Institute, Washington, DC, USA, 116 pp.

Hoehne, C.G. and M.V. Chester, 2016: Optimizing plug-in electric vehicle and vehicle-to-grid charge scheduling to minimize carbon emissions. *Energy*, **115**, 646–657, doi:10.1016/j.energy.2016.09.057.

Hoesly, R.M. et al., 2018: Historical (1750–2014) anthropogenic emissions of reactive gases and aerosols from the Community Emissions Data System (CEDS). *Geosci. Model Dev.*, **11(1)**, 369–408, doi:10.5194/gmd-11-369-2018.

Hoffrichter, A., J. Silmon, S. Iwnicki, S. Hillmansen, and C. Roberts, 2012: Rail freight in 2035 – traction energy analysis for high-performance freight trains. *Proc. Inst. Mech. Eng. Part F J. Rail Rapid Transit*, **226(6)**, 568–574, doi:10.1177/0954409712441753.

Hoque, N., W. Biswas, I. Mazhar, and I. Howard, 2019: Environmental life cycle assessment of alternative fuels for Western Australia's transport sector. *Atmosphere*, **10**, 1–27, doi:10.3390/atmos10070398.

Horvath, S., M. Fasihi, and C. Breyer, 2018: Techno-economic analysis of a decarbonized shipping sector: Technology suggestions for a fleet in 2030 and 2040. *Energy Convers. Manag.*, **164**, 230–241, doi:10.1016/J.ENCONMAN.2018.02.098.

Hottle, T., C. Caffrey, J. McDonald, and R. Dodder, 2017: Critical factors affecting life cycle assessments of material choice for vehicle mass reduction. *Transp. Res. Part D Transp. Environ.*, **56**(October), 241–257, doi:10.1016/j.trd.2017.08.010.

Hove, A. and D. Sandalow, 2019: *Electric vehicle charging in China and the United States*. Columbia, Center on Global Energy Policy, New York, NY, USA, 88 pp.

Hoyos, D., G. Bueno, and I. Capellán-Pérez, 2016: Environmental assessment of high-speed rail. In: *Evaluating High-Speed Rail* [Albalate, D. and G. Bel, (eds.)], Routledge, New York and London, pp. 135–155.

Hsieh, C. and C. Felby, 2017: *Biofuels for the marine shipping sector*. IEA Bioenergy, 86 pp., https://www.ieabioenergy.com/wp-content/uploads/2018/02/Marine-biofuel-report-final-Oct-2017.pdf.

Hua, J., C.-W. Cheng, and D.-S. Hwang, 2018: Total life cycle emissions of post-Panamax containerships powered by conventional fuel or natural gas. https://doi.org/10.1080/10962247.2018.1505675, **69(2)**, 131–144, doi:10.1080/10962247.2018.1505675.

Huizenga, C., K. Peet, and S. Gota, 2015: *Analysis on national transport sector emissions: 1990–2012*. Paris Process on Mobility and Climate, Beijing, China, 14 pp.

Hydrogen Council, 2017: *How Hydrogen Empowers the Energy Transition*. Hydrogen Council, Brussels, Belgium, 28 pp.

Hydrogen Council, 2020: *Path to hydrogen competitiveness A cost perspective*. Hydrogen Council, Brussels, Belgium, 88 pp.

ICAO, 2017: *Sustainable Aviation Fuels Guide*. International Civil Aviation Organization, Montréal, Canada, 65 pp.

ICAO, 2019: *2019 Environmental Report, Aviation and Environment: Destination Green The Next Chapter*. International Civil Aviation Organization, Montréal, Canada 376 pp.

ICCT, 2017: *Developing hydrogen fueling infrastructure for fuel cell vehicles: A status update*. International Council on Clean Transportation, Washington, DC, USA, 22 pp.

IEA, 2012: *Technology Roadmap – Fuel Economy of Road Vehicles*. International Energy Agency, Paris, France, 50 pp.

IEA, 2015: *Technology roadmap – Hydrogen and fuel cells*. International Energy Agency, Paris, France, 80 pp.

IEA, 2017a: *Technology Roadmap: Delivering Sustainable Bioenergy*. International Energy Agency, Paris, France, 94 pp.

IEA, 2017b: *Energy Technology Perspectives 2017: Catalysing Energy Technology Transformations*. International Energy Agency, Paris, France, 441 pp.

IEA, 2019a: *Global EV Outlook 2019*. International Energy Agency, Paris, France, 232 pp.

IEA, 2019b: *$CO_2$ emissions from fuel combustion*. International Energy Agency, Paris, France, 165 pp.

IEA, 2019c: *The Future of Hydrogen. Seizing today's opportunities*. International Energy Agency (IEA), Paris, France, 203 pp.

IEA, 2019d: *World Energy Outlook 2019*. International Energy Agency, Paris, France, 810 pp.

IEA, 2019e: *The Future of Rail*. International Energy Agency, Paris, France, 175 pp.

IEA, 2020a: *Tracking Transport 2020*. International Energy Agency, Paris, France.

IEA, 2020b: *World Energy Outlook*. International Energy Agency, Paris, France.

IEA, 2020c: *IEA Bioenergy Summary Series: Advanced Biofuels – Potential for Cost Reduction*. International Energy Agency, International Energy Agency, Paris, France. 3 pp.

IEA, 2020d: *Fuel Consumption of Cars and Vans*. International Energy Agency, Paris, France.

IEA, 2021a: *Global EV Outlook 2021*. International Energy Agency, Paris, France, 101 pp.

IEA, 2021b: *Energy Technology Perspectives 2020*. International Energy Agency, Paris, France, 400 pp.

IEA, 2021c: *Energy Prices and Taxes Statistics*. International Energy Agency, Paris, France.

IEA, 2021d: *Energy Prices: Overview*. International Energy Agency, Paris, France.

IEA, 2021e: *Net Zero by 2050*. International Energy Agency, Paris, France, 224 pp.

IEA, 2021f: *World Energy Model Documentation: 2020 Version*. International Energy Agency, Paris, France, 112 pp.

IMO, 2013: *MARPOL: Annex VI and NTC 2008 with guidelines for implementation*. International Maritime Organization, London, UK.

IMO, 2018: *Note by the International Maritime Organization to the UNFCCC Talanoa Dialogue*. International Maritime Organization, 27 pp.

International Union of Railways, 2016: *Carbon Footprint of Railway Infrastructure: Comparing existing methodologies on typical corridors, Recommendations for harmonized approach*. International Union of Railways, Paris, France, 54 pp.

IOMVM, 2021: *World Vehicles in Use*. International Organization of Motor Vehicle Manufacturers, https://www.oica.net/category/vehicles-in-use/.

IPCC, 1999: *Aviation and the Global Atmosphere: A Special Report of the Intergovernmental Panel on Climate Change*. [Penner, J.E., D.H. Lister, D.J. Griggs, D.J. Dokken, and M. McFarland (eds.)]. Cambridge University Press, Cambridge, UK and New York, NY, USA, 373 pp.

IPCC, 2018: Global Warming of 1.5°C. *An IPCC Special Report on the impacts of global warming of 1.5°C above pre-industrial levels and related global greenhouse gas emission pathways, in the context of strengthening the global response to the threat of climate change, sustainable development, and efforts to eradicate poverty* [Masson-Delmotte, V., P. Zhai, H.-O. Pörtner, D. Roberts, J. Skea, P.R. Shukla, A. Pirani, W. Moufouma-Okia, C. Péan, R. Pidcock, S. Connors, J.B.R. Matthews, Y. Chen, X. Zhou, M.I. Gomis, E. Lonnoy, T. Maycock, M. Tignor, and T. Waterfield (eds.)]. Cambridge University Press, Cambridge, UK, and New York, NY, USA.

IPCC, 2019: *IPCC Special Report on the Ocean and Cryosphere in a Changing Climate* [H.-O. Pörtner, D.C. Roberts, V. Masson-Delmotte, P. Zhai, M. Tignor, E. Poloczanska, K. Mintenbeck, A. Alegría, M. Nicolai, A. Okem, J. Petzold, B. Rama, N.M. Weyer (eds.)]. Cambridge University Press, Cambridge, UK, and New York, NY, USA.

IRENA, 2016: *The renewable energy route to sustainable transport*. The International Renewable Energy Agency, Abu Dhabi, 72 pp.

IRENA, 2019: *Navigating the way to a renewable future: Solutions to decarbonise shipping, Preliminary findings*. International Renewable Energy Agency, Abu Dhabi, 36 pp.

IRENA, 2020: *Green Hydrogen Cost Reduction: Scaling up Electrolysers to Meet the 1.5°C Climate Goal*. International Renewable Energy Agency, Abu Dhabi, 105 pp.

IRP, 2019: *Global Resources Outlook 2019: Natural Resources for the Future We Want*. United Nations Environment Programme, Nairobi, Kenya, 169 pp.

Isaac, R. and L. Fulton, 2017: Propulsion Systems for 21st Century Rail. *World Conference on Transport Research*, Transportation Research Procedia, 10–15 July 2016, Shanghai, China, https://pdfs.semanticscholar.org/52ba/7add103ba4564d379a04d4ca7837bdca7763.pdf.

ITF, 2017: *Shared Mobility Simulations for Helsinki Case-Specific Policy Analysis*. OECD Publishing, Paris, France, 97 pp.

ITF, 2018a: *Transition to Shared Mobility: How large cities can deliver inclusive transport services*. OECD Publishing, Paris, France, 55 pp.

ITF, 2018b: *Decarbonising Maritime Transport: Pathways to zero-carbon shipping by 2035*. OECD Publishing, Paris, France, 86 pp.

ITF, 2019: *ITF Transport Outlook 2019*. OECD Publishing, Paris, France, 200 pp.

ITF, 2020a: *Shared Mobility Simulations for Lyon*. OECD Publishing, Paris, France, 91 pp.

ITF, 2020b: *Good to Go? Assessing the Environmental Performance of New Mobility*. OECD Publishing, Paris, France, 87 pp.

ITF, 2021: *ITF Transport Outlook 2021*. OECD, Paris, France, 249 pp.

Ivanova, D. et al., 2020: Quantifying the potential for climate change mitigation of consumption options. *Environ. Res. Lett.*, **15(9)**, 093001, doi:10.1088/1748-9326/ab8589.

Jaeger, A., S.B. Nugroho, E. Zusman, R. Nakano, and R. Daggy, 2015: Governing sustainable low-carbon transport in Indonesia: An assessment of provincial transport plans. *Nat. Resour. Forum*, **39(1)**, 27–40, doi:10.1111/1477-8947.12066.

Jakovcevic, A. and L. Steg, 2013: Sustainable transportation in Argentina: Values, beliefs, norms and car use reduction. *Transp. Res. Part F Traffic Psychol. Behav.*, **20**, 70–79, doi:10.1016/j.trf.2013.05.005.

Jalkanen, J.-P., L. Johansson, and J. Kukkonen, 2014: A Comprehensive Inventory of the Ship Traffic Exhaust Emissions in the Baltic Sea from 2006 to 2009. *Ambio*, **43(3)**, 311–324, doi:10.1007/s13280-013-0389-3.

Jaller, M. and A. Pahwa, 2020: Evaluating the environmental impacts of online shopping: A behavioral and transportation approach. *Transp. Res. Part D Transp. Environ.*, **80**, 102223, doi:10.1016/J.TRD.2020.102223.

Japan Ship Technology Research Association, The Nippon Foundation, and Ministry of Land Infrastructure Transport and Tourism, 2020: *Roadmap to Zero Emission from International Shipping*. Japan Ship Technology Research Association (JSTRA) and Ministry of Land, Infrastructure, Transport and Tourism (MLIT), The Nippon Foundation, Japan, 136 pp.

Javaid, A., F. Creutzig, and S. Bamberg, 2020: Determinants of low-carbon transport mode adoption: systematic review of reviews. *Environ. Res. Lett.*, **15(10)**, 103002, doi:10.1088/1748-9326/ABA032.

Jeerh, G., M. Zhang, and S. Tao, 2021: Recent progress in ammonia fuel cells and their potential applications. *J. Mater. Chem. A*, **9(2)**, 727–752, doi:10.1039/D0TA08810B.

Jeswani, H.K., A. Chilvers, and A. Azapagic, 2020: Environmental sustainability of biofuels: a review. *Proc. R. Soc. A Math. Phys. Eng. Sci.*, **476(2243)**, 20200351, doi:10.1098/rspa.2020.0351.

Jiang, C. and A. Zhang, 2014: Effects of high-speed rail and airline cooperation under hub airport capacity constraint. *Transp. Res. Part B Methodol.*, **60**, 33–49, doi:10.1016/j.trb.2013.12.002.

Jiang, C. and A. Zhang, 2016: Airline network choice and market coverage under high-speed rail competition. *Transp. Res. Part A Policy Pract.*, **92**, 248–260, doi:10.1016/j.tra.2016.06.008.

Jiang, R., K.N. Ingle, and A. Golberg, 2016: Macroalgae (seaweed) for liquid transportation biofuel production: what is next? *Algal Res.*, **14**, 48–57, doi:10.1016/J.ALGAL.2016.01.001.

Jittrapirom, P. et al., 2017: Mobility as a service: A critical review of definitions, assessments of schemes, and key challenges. *Urban Plan.*, **2(2)**, 13–25, doi:10.17645/up.v2i2.931.

Jittrapirom, P., V. Marchau, R. van der Heijden, and H. Meurs, 2018: Dynamic adaptive policymaking for implementing Mobility-as-a Service (MaaS). *Res. Transp. Bus. Manag.*, **27**, 46–55, doi:10.1016/j.rtbm.2018.07.001.

Jochem, P., J.J. Gómez Vilchez, A. Ensslen, J. Schäuble, and W. Fichtner, 2018: Methods for forecasting the market penetration of electric drivetrains in the passenger car market. *Transp. Rev.*, **38(3)**, 322–348, doi:10.1080/01441647.2017.1326538.

Jochem, P., J. Whitehead, and E. Dütschke, 2021: The Impact of Electric Vehicles on Energy Systems. *Int. Encycl. Transp.*, **1**, 560–565, doi:10.1016/B978-0-08-102671-7.10515-9.

Johansson, L., J.P. Jalkanen, and J. Kukkonen, 2017: Global assessment of shipping emissions in 2015 on a high spatial and temporal resolution. *Atmos. Environ.*, **167**, 403–415, doi:10.1016/j.atmosenv.2017.08.042.

10 Johnson, C., E. Nobler, L. Eudy, and M. Jeffers, 2020: *Financial Analysis of Battery Electric Transit Buses*. National Renewable Energy Laboratory, Golden, CO, USA, 45 pp.

Johnson, H., M. Johansson, K. Andersson, and B. Södahl, 2013: Will the ship energy efficiency management plan reduce $CO_2$ emissions? A comparison with ISO 50001 and the ISM code. *Marit. Policy Manag.*, **40(2)**, 177–190, doi:10.1080/03088839.2012.757373.

Johnston, M., J. Dawson, E. De Souza, and E.J. Stewart, 2017: Management challenges for the fastest growing marine shipping sector in Arctic Canada: Pleasure crafts. *Polar Rec. (Gr. Brit).*, **53(1)**, 67–78, doi:10.1017/S0032247416000565.

Jones, H., F. Moura, and T. Domingos, 2017: Life cycle assessment of high-speed rail: a case study in Portugal. *Int. J. Life Cycle Assess.*, **22(3)**, 410–422, doi:10.1007/s11367-016-1177-7.

Jones, J., A. Genovese, and A. Tob-Ogu, 2020: Hydrogen vehicles in urban logistics: A total cost of ownership analysis and some policy implications. *Renew. Sustain. Energy Rev.*, **119**, 109595, doi:10.1016/J.RSER.2019.109595.

Jordbakker, G.N., A. Amundsen, I. Sundvor, E. Figenbaum, and I.B. Hovi, 2018: *Technological maturity level and market introduction timeline of zero-emission heavy-duty vehicles*. Institute of Transport Economics, Oslo, Norway, 82 pp.

Joshi, A., 2020: Review of vehicle engine efficiency and emissions. *SAE Int. J. Adv. & Curr. Prac. in Mobility*, **2(5)**, 2479–2507, doi:10.4271/2020-01-0352.

Jouin, M. et al., 2016: Estimating the end-of-life of PEM fuel cells: Guidelines and metrics. *Appl. Energy*, **177**, 87–97, doi:10.1016/j.apenergy.2016.05.076.

Kaack, L.H., P. Vaishnav, M.G. Morgan, I.L. Azevedo, and S. Rai, 2018: Decarbonizing intraregional freight systems with a focus on modal shift. *Environ. Res. Lett.*, **13(8)**, 083001, doi:10.1088/1748-9326/aad56c.

Kainuma, M., T. Masui, K. Oshiro, and G. Hibino, 2015: *Pathways to deep decarbonization in Japan*. Sustainable Development Solutions Network (SDSN) and Institute for Sustainable Development and International Relations (IDDR), 41 pp.

Kamdar, A., 2017: Ultracapacitor kinetic energy recovery systems in road transport vehicles: is it a viable retrofit option for reducing fuel consumption and $CO_2$ emissions? *Southern African Transport Conference*, Johannesburg, South Africa.

Kampala City Capital Authority, 2018: Kampala's Air Quality Is Six Times Worse Than Global Standards. Kampala City Capital Authority, Kampala, Uganda, https://www.kcca.go.ug/news/316/#.Yc2H3GhBzIU.

Kandaramath Hari, T., Z. Yaakob, and N.N. Binitha, 2015: Aviation biofuel from renewable resources: Routes, opportunities and challenges. *Renew. Sustain. Energy Rev.*, **42**, 1234–1244, doi:10.1016/j.rser.2014.10.095.

Kang, S., Y. Zhang, Y. Qian, and H. Wang, 2020: A review of black carbon in snow and ice and its impact on the cryosphere. *Earth-Science Rev.*, **210**, 103346, doi:10.1016/J.EARSCIREV.2020.103346.

Kapadia, Z.Z. et al., 2016: Impacts of aviation fuel sulfur content on climate and human health. *Atmos. Chem. Phys.*, **16(16)**, 10521–10541, doi:10.5194/acp-16-10521-2016.

Kaplan, A.M., and M. Haenlein, 2010: Users of the world, unite! The challenges and opportunities of Social Media. *Bus. Horiz.*, **53(1)**, 59–68, doi:10.1016/j.bushor.2009.09.003.

Kärcher, B., 2018: Formation and radiative forcing of contrail cirrus. *Nat. Commun. 2018 91*, **9(1)**, 1–17, doi:10.1038/s41467-018-04068-0.

Kassens-Noor, E. et al., 2020: Sociomobility of the 21st century: Autonomous vehicles, planning, and the future city. *Transp. Policy*, **99**, 329–335, doi:10.1016/j.tranpol.2020.08.022.

Kasulaitis, B.V, C.W. Babbitt, and A.K. Krock, 2019: Dematerialization and the Circular Economy: Comparing Strategies to Reduce Material Impacts of the Consumer Electronic Product Ecosystem. *J. Ind. Ecol.*, **23(1)**, 119–132, doi:10.1111/jiec.12756.

Kavanagh, L., J. Keohane, G. Garcia Cabellos, A. Lloyd, and J. Cleary, 2018: Global Lithium Sources – Industrial Use and Future in the Electric Vehicle Industry: A Review. *Resources*, **7(3)**, 57, doi:10.3390/resources7030057.

Kavitha, N. et al., 2018: A Smart Monitoring System in Vehicles. *J. Pure Appl. Math.*, **118(20)**, 1485–1490.

Kazunga, O., 2019: Bulawayo dreams of 'Trackless Tram'. *Chronicle*, https://www.chronicle.co.zw/bulawayo-dreams-of-trackless-tram/.

Ke, W., S. Zhang, X. He, Y. Wu, and J. Hao, 2017: Well-to-wheels energy consumption and emissions of electric vehicles: Mid-term implications from real-world features and air pollution control progress. *Appl. Energy*, **188**, 367–377, doi:10.1016/J.APENERGY.2016.12.011.

Keall, M.D., C. Shaw, R. Chapman, and P. Howden-Chapman, 2018: Reductions in carbon dioxide emissions from an intervention to promote cycling and walking: A case study from New Zealand. *Transp. Res. Part D Transp. Environ.*, **65**, 687–696, doi:10.1016/J.TRD.2018.10.004.

Kemfert, C., P. Opitz, T. Traber, and L. Handrich, 2015: *Pathways to deep decarbonization in Germany*. Sustainable Development Solutions Network (SDSN) and Institute for Sustainable Development and International Relations (IDDRI), 84 pp.

Khalili, S., E. Rantanen, D. Bogdanov, and C. Breyer, 2019: Global Transportation Demand Development with Impacts on the Energy Demand and Greenhouse Gas Emissions in a Climate-Constrained World. *Energies*, **12(20)**, 3870, doi:10.3390/en12203870.

Khan, M.I., T. Yasmin, and A. Shakoor, 2015: Technical overview of compressed natural gas (CNG) as a transportation fuel. *Renew. Sustain. Energy Rev.*, **51**, 785–797, doi:10.1016/j.rser.2015.06.053.

Khandelwal, B., A. Karakurt, P.R. Sekaran, V. Sethi, and R. Singh, 2013: Hydrogen powered aircraft: The future of air transport. *Prog. Aerosp. Sci.*, **60**, 45–59, doi:10.1016/j.paerosci.2012.12.002.

Khandelwal, B., C.J. Wijesinghe, and S. Sriraman, 2018: Effect of Alternative Fuels on Emissions and Engine Compatibility. In: *Energy for Propulsion* [Runchal, A.K., A.K. Gupta, A. Kushari, A. De, and S.K. Aggarwal, (eds.)], Springer, Singapore, pp. 27–50.

Khojasteh Salkuyeh, Y., B.A. Saville, and H.L. MacLean, 2017: Techno-economic analysis and life cycle assessment of hydrogen production from natural gas using current and emerging technologies. *Int. J. Hydrogen Energy*, **42(30)**, 18894–18909, doi:10.1016/j.ijhydene.2017.05.219.

Khzouz, M. et al., 2020: Life Cycle Costing Analysis: Tools and Applications for Determining Hydrogen Production Cost for Fuel Cell Vehicle Technology. *Energies*, **13(15)**, doi:10.3390/en13153783.

Kim, E.S., 2019: Structural integrity evaluation of CNG pressure vessel with defects caused by heat treatment using numerical analysis. *J. Mech. Sci. Technol.*, **33(11)**, 5297–5302, doi:10.1007/s12206-019-1021-7.

Kim, H.C. and T.J. Wallington, 2013: Life-cycle energy and greenhouse gas emission benefits of lightweighting in automobiles: Review and harmonization. *Environ. Sci. Technol.*, **47(12)**, 6089–6097, doi:10.1021/es3042115.

Kim, H.C. and T.J. Wallington, 2016: Life Cycle Assessment of Vehicle Lightweighting: A Physics-Based Model To Estimate Use-Phase Fuel Consumption of Electrified Vehicles. *Environ. Sci. Technol.*, **50(20)**, 11226–11233, doi:10.1021/acs.est.6b02059.

Kim, K., G. Roh, W. Kim, and K. Chun, 2020: A Preliminary Study on an Alternative Ship Propulsion System Fueled by Ammonia: Environmental and Economic Assessments. *J. Mar. Sci. Eng.*, **8(3)**, 183, doi:10.3390/jmse8030183.

King, D.A. and K.J. Krizek, 2020: The power of reforming streets to boost access for human-scaled vehicles. *Transp. Res. Part D Transp. Environ.*, **83**, 102336, doi:10.1016/j.trd.2020.102336.

Klein, B.C. et al., 2018: Techno-economic and environmental assessment of renewable jet fuel production in integrated Brazilian sugarcane biorefineries. *Appl. Energy*, **209**, doi:10.1016/j.apenergy.2017.10.079.

Klossek, P., J. Kullik, and K.G. van den Boogaart, 2016: A systemic approach to the problems of the rare earth market. *Resour. Policy*, **50**, 131–140, doi:10.1016/j.resourpol.2016.09.005.

Klöwer, M. et al., 2021: Quantifying aviation's contribution to global warming. *Environ. Res. Lett.*, **16(10)**, 104027, doi:10.1088/1748-9326/ac286e.

Knobloch, F., et al., 2020: Net emission reductions from electric cars and heat pumps in 59 world regions over time. *Nature Sustainability*, **3**, 437–447, doi:10.1038/s41893-020-0488-7.

Koasidis, K. et al., 2020: Many Miles to Paris: A Sectoral Innovation System Analysis of the Transport Sector in Norway and Canada in Light of the Paris Agreement. *Sustainability*, **12(14)**, 5832, doi:10.3390/su12145832.

Kober, T. and C. Bauer, 2019: *Perspectives of Power-to-X technologies in Switzerland*. 40 pp.

Koeble, R. et al. 2017: *Definition of input data to assess GHG default emissions from biofuels in EU legislation*. Version 1c –July, European Commission, Joint Research Centre, Publications Office, doi:10.2790/658143.

Kousoulidou, M. and L. Lonza, 2016: Biofuels in aviation: Fuel demand and $CO_2$ emissions evolution in Europe toward 2030. *Transp. Res. Part D Transp. Environ.*, **46**, 166–181, doi:10.1016/j.trd.2016.03.018.

Kramel, D. et al., 2021: Global Shipping Emissions from a Well-to-Wake Perspective: The MariTEAM Model. *Environ. Sci. Technol.*, **55(22)**, 15040–15050, doi:10.1021/acs.est.1c03937.

Kubba, A. and K. Jiang, 2014: A Comprehensive Study on Technologies of Tyre Monitoring Systems and Possible Energy Solutions. *Sensors*, **14(6)**, 10306–10345, doi:10.3390/s140610306.

Kumar, S. and F. Hoffmann, 2002: *Globalization: The maritime nexus. Handbook of maritime economics*. Maritime Press, London, UK.

Kuramochi, T. et al., 2018: Ten key short-term sectoral benchmarks to limit warming to 1.5°C. *Clim. Policy*, **18(3)**, 287–305, doi:10.1080/14693062.2017.1397495.

Kurtz, J.M., S. Sprik, G. Saur, and S. Onorato, 2019: *Fuel cell electric vehicle durability and fuel cell performance*. National Renewable Energy Laboratory, Golden, CO, USA, 20 pp.

Labandeira, X., J.M. Labeaga, and X. López-Otero, 2017: A meta-analysis on the price elasticity of energy demand. *Energy Policy*, **102**, 549–568, doi:10.1016/j.enpol.2017.01.002.

Laberteaux, K.P., K. Hamza, and J. Willard, 2019: Optimizing the electric range of plug-in vehicles via fuel economy simulations of real-world driving in California. *Transp. Res. Part D Transp. Environ.*, **73**(June), 15–33, doi:10.1016/j.trd.2019.05.013.

Lajevardi, S.M., J. Axsen, and C. Crawford, 2019: Comparing alternative heavy-duty drivetrains based on GHG emissions, ownership and abatement costs: Simulations of freight routes in British Columbia. *Transp. Res. Part D Transp. Environ.*, **76**(September), 19–55, doi:10.1016/j.trd.2019.08.031.

Lajunen, A., P. Sainio, L. Laurila, J. Pippuri-mäkeläinen, and K. Tammi, 2018: Overview of Powertrain Electrification and Future Scenarios for Non-Road Mobile Machinery. *Energies*, **11(5)**, 1184, doi:10.3390/en11051184.

Lamb, W.F. et al., 2020: Discourses of climate delay. *Glob. Sustain.*, **3**, e17, doi:10.1017/sus.2020.13.

Lamb, W.F. et al., 2021: A review of trends and drivers of greenhouse gas emissions by sector from 1990 to 2018. *Environ. Res. Lett.*, **16(7)**, 073005, doi:10.1088/1748-9326/abee4e.

Langford, J.S. and D.K. Hall, 2020: Electrified Aircraft Propulsion. *Bridge*, **50(2)**, 21–27.

Lasserre, F. and S. Pelletier, 2011: Polar super seaways? Maritime transport in the Arctic: an analysis of shipowners' intentions. *J. Transp. Geogr.*, **19(6)**, 1465–1473, doi:10.1016/j.jtrangeo.2011.08.006.

Lawrence, M., R. Bullock, and Z. Liu, 2019: *China's high-speed rail development*. World Bank Publications, Washington, DC, USA.

Le, H.T.K., A.L. Carrel, and H. Shah, 2021: Impacts of online shopping on travel demand: a systematic review. *Transp. Rev.*, 42:3, 273–295. doi:10.1080/01441647.2021.1961917.

Le Quéré, C. et al., 2018a: Global Carbon Budget 2017. *Earth Syst. Sci. Data*, **10(1)**, 405–448, doi:10.5194/essd-10-405-2018.

Le Quéré, C. et al., 2018b: Global Carbon Budget 2018. *Earth Syst. Sci. Data*, **10(4)**, 2141–2194, doi:10.5194/essd-10-2141-2018.

Lee, A.E. and S.L. Handy, 2018: Leaving level-of-service behind: The implications of a shift to VMT impact metrics. *Res. Transp. Bus. Manag.*, **29**, 14–25, doi:10.1016/j.rtbm.2018.02.003.

Lee, D.S., 2018: *International aviation and the Paris Agreement temperature goals*. Department of Transport, UK, 15 pp.

Lee, D.S. et al., 2021: The contribution of global aviation to anthropogenic climate forcing for 2000 to 2018. *Atmos. Environ.*, **244**, doi:10.1016/j.atmosenv.2020.117834.

Lee, J. et al., 2020: Reviewing the material and metal security of low-carbon energy transitions. *Renew. Sustain. Energy Rev.*, **124**, 109789, doi:10.1016/j.rser.2020.109789.

Leibowicz, B.D., 2018: Policy recommendations for a transition to sustainable mobility based on historical diffusion dynamics of transport systems. *Energy Policy*, **119**(April 2017), 357–366, doi:10.1016/j.enpol.2018.04.066.

Lennon, M., H.A. Baer, and M. Singer, 2015: The Anthropology of Climate Change: An Integrated Critical Perspective. *Hum. Ecol.*, **43(6)**, 871–873, doi:10.1007/s10745-015-9794-5.

Lenz, B. and D. Heinrichs, 2017: What Can We Learn from Smart Urban Mobility Technologies? *IEEE Pervasive Comput.*, **16(2)**, 84–86, doi:10.1109/MPRV.2017.27.

Levin, L. and C. Faith-Ell, 2019: How to Apply Gender Equality Goals in Transport and Infrastructure Planning. In: *Integrating Gender into Transport Planning* [Scholten, C.L. and T. Joelsson, (eds.)], Palgrave Macmillan, Cham, Switzerland, pp. 89–118.

Li, H., J. Strauss, and L. Lu, 2019: The impact of high-speed rail on civil aviation in China. *Transp. Policy*, **74**, 187–200, doi:10.1016/j.tranpol.2018.11.015.

Li, P.-H. and S. Pye, 2018: Assessing the benefits of demand-side flexibility in residential and transport sectors from an integrated energy systems perspective. *Appl. Energy*, **228**, 965–979, doi:10.1016/j.apenergy.2018.06.153.

Li, X.J., J.D. Allen, J.A. Stager, and A.Y. Ku, 2020: Paths to low-cost hydrogen energy at a scale for transportation applications in the USA and China via liquid-hydrogen distribution networks. *Clean Energy*, **4(1)**, doi:10.1093/ce/zkz033.

Liang, X.-H. and K.-H. Tan, 2019: Market potential and approaches of parcels and mail by high speed rail in China. *Case Stud. Transp. Policy*, **7(3)**, 583–597, doi:10.1016/j.cstp.2019.07.008.

Liao, F., E. Molin, and B. van Wee, 2016: Consumer preferences for electric vehicles: a literature review. *Transp. Rev.*, **37(3)**, 252–275, doi:10.1080/01441647.2016.1230794.

Liao, Y., J. Gil, R.H.M. Pereira, S. Yeh, and V. Verendel, 2020: Disparities in travel times between car and transit: Spatiotemporal patterns in cities. *Sci. Rep.*, **10(1)**, 1–12, doi:10.1038/s41598-020-61077-0.

Lin, J., H. Li, W. Huang, W. Xu, and S. Cheng, 2019: A Carbon Footprint of High-Speed Railways in China: A Case Study of the Beijing-Shanghai Line. *J. Ind. Ecol.*, **23(4)**, 869–878, doi:10.1111/jiec.12824.

Lin, X., P. Wells, and B.K. Sovacool, 2017: Benign mobility? Electric bicycles, sustainable transport consumption behaviour and socio-technical transitions in Nanjing, China. *Transp. Res. Part A Policy Pract.*, **103**, 223–234, doi:10.1016/j.tra.2017.06.014.

Lindstad, H. et al., 2015: *GHG emission reduction potential of EU-related maritime transport and on its impacts*. TNO, Delft, Netherlands, 130 pp.

Linke, F., V. Grewe, and V. Gollnick, 2017: The Implications of Intermediate Stop Operations on Aviation Emissions and Climate. *Meteorol. Zeitschrift*, **26(6)**, 697–709, doi:10.1127/metz/2017/0763.

Lisowski, S. et al., 2020: Criteria-based approach to select relevant environmental SDG indicators for the automobile industry. *Sustainability*, **12(21)**, 1–22, doi:10.3390/su12218811.

Litman, T., 2021: *Well Measured: Developing Indicators for Sustainable and Livable Transport Planning*. Victoria Transport Policy Institute, Victoria, Canada, 117 pp. https://www.vtpi.org/wellmeas.pdf.

Littlefield, J. et al., 2019: *Life Cycle Analysis of Natural Gas Extraction and Power Generation*. National Energy Technology Laboratory, Pittsburgh, PA, USA, 374 pp. https://www.netl.doe.gov/energy-analysis/details?id=7C7809C2-49AC-4CE0-AC72-3C8F8A4D87AD.

Littlefield, J.A., J. Marriott, G.A. Schivley, and T.J. Skone, 2017: Synthesis of recent ground-level methane emission measurements from the U.S. natural gas supply chain. *J. Clean. Prod.*, **148**, 118–126, doi:10.1016/J.JCLEPRO.2017.01.101.

Liu, C.M., N.K. Sandhu, S.T. McCoy, and J.A. Bergerson, 2020a: A life cycle assessment of greenhouse gas emissions from direct air capture and Fischer-Tropsch fuel production. *Sustain. Energy Fuels*, **4(6)**, 3129–3142, doi:10.1039/C9SE00479C.

Liu, J., A.W.K. Law, and O. Duru, 2021: Abatement of atmospheric pollutant emissions with autonomous shipping in maritime transportation using Bayesian probabilistic forecasting. *Atmos. Environ.*, **261**(July), 118593, doi:10.1016/j.atmosenv.2021.118593.

Liu, N., F. Xie, Z. Lin, and M. Jin, 2020b: Evaluating national hydrogen refueling infrastructure requirement and economic competitiveness of fuel cell electric long-haul trucks. *Mitig. Adapt. Strateg. Glob. Change*, **25(3)**, 477–493, doi:10.1007/s11027-019-09896-z.

Liu, S., Y. Wan, H.-K. Ha, Y. Yoshida, and A. Zhang, 2019: Impact of high-speed rail network development on airport traffic and traffic distribution: Evidence from China and Japan. *Transp. Res. Part A Policy Pract.*, **127**, 115–135, doi:10.1016/j.tra.2019.07.015.

Liu, Y., B. Yang, X. Dong, Y. Wang, and Y. Xia, 2017: A Simple Prelithiation Strategy To Build a High-Rate and Long-Life Lithium-Ion Battery with Improved Low-Temperature Performance. *Angew. Chemie Int. Ed.*, **56(52)**, doi:10.1002/anie.201710555.

Liu, Z. et al., 2020c: Near-real-time monitoring of global $CO_2$ emissions reveals the effects of the COVID-19 pandemic. *Nat. Commun.*, **11(1)**, 5172, doi:10.1038/s41467-020-18922-7.

Lombardi, L., L. Tribioli, R. Cozzolino, and G. Bella, 2017: Comparative environmental assessment of conventional, electric, hybrid, and fuel cell powertrains based on LCA. *The International Journal of Life Cycle Assessment*, **22**, 1989–2006, doi:10.1007/s11367-017-1294-y.

Loo, B.P.Y. and D. Banister, 2016: Decoupling transport from economic growth: Extending the debate to include environmental and social externalities. *J. Transp. Geogr.*, **57**, 134–144, doi:10.1016/j.jtrangeo.2016.10.006.

Luk, J.M., H.C. Kim, R.D. De Kleine, T.J. Wallington, and H.L. MacLean, 2018: Greenhouse gas emission benefits of vehicle lightweighting: Monte Carlo probabalistic analysis of the multi material lightweight vehicle glider. *Transp. Res. Part D Transp. Environ.*, **62**(February), 1–10, doi:10.1016/j.trd.2018.02.006.

Luo, X. and M. Wang, 2017: Study of solvent-based carbon capture for cargo ships through process modelling and simulation. *Appl. Energy*, **195**, 402–413, doi:10.1016/J.APENERGY.2017.03.027.

Luque-Ayala, A., 2018: Post-development carbon. In: *Rethinking Urban Transitions*, Routledge, New York and London, pp. 224–241.

Lutsey, N., 2015: *Global Climate Change Mitigation Potential From a Transition To Electric Vehicles*. International Council on Clean Transportation, Washington, DC, USA, 17 pp.

Lutsey, N. and D. Sperling, 2012: Regulatory adaptation: Accommodating electric vehicles in a petroleum world. *Energy Policy*, **45**, 308–316, doi:10.1016/j.enpol.2012.02.038.

Ma, W., Q. Wang, H. Yang, A. Zhang, and Y. Zhang, 2019: Effects of Beijing-Shanghai high-speed rail on air travel: Passenger types, airline groups and tacit collusion. *Res. Transp. Econ.*, **74**, 64–76, doi:10.1016/j.retrec.2018.12.002.

Ma, X. et al., 2021: Recycled cathode materials enabled superior performance for lithium-ion batteries. *Joule*, 5**(11)**, 2955–2970, doi:10.1016/j.joule.2021.09.005.

Macchi, S. and M. Tiepolo (eds.), 2014: *Climate Change Vulnerability in Southern African Cities*. Springer International Publishing, Cham, Switzerland, 258 pp.

Mackey, B. et al., 2013: Untangling the confusion around land carbon science and climate change mitigation policy. *Nat. Clim. Change*, **3(6)**, 552–557, doi:10.1038/nclimate1804.

Macmillan, A. et al., 2020: Suburb-level changes for active transport to meet the SDGs: Causal theory and a New Zealand case study. *Sci. Total Environ.*, **714**, doi:10.1016/j.scitotenv.2020.136678.

Madzharov, D., E. Delarue, and W. D'haeseleer, 2014: Integrating electric vehicles as flexible load in unit commitment modeling. *Energy*, **65**, 285–294, doi:10.1016/j.energy.2013.12.009.

Maggio, G., A. Nicita, and G. Squadrito, 2019: How the hydrogen production from RES could change energy and fuel markets: A review of recent literature. *Int. J. Hydrogen Energy*, **44(23)**, 11371–11384, doi:10.1016/j.ijhydene.2019.03.121.

Maginn, P.J., P. Burton, and C. Legacy, 2018: Disruptive Urbanism? Implications of the 'Sharing Economy' for Cities, Regions, and Urban Policy. *Urban Policy Res.*, **36(4)**, 393–398, doi:10.1080/08111146.2018.1555909.

Mahmud, M.A.P., N. Huda, S.H. Farjana, and C. Lang, 2018: Environmental sustainability assessment of hydropower plant in Europe using life cycle assessment. *IOP Conf. Ser. Mater. Sci. Eng.*, **351(1)**, 012006, doi:10.1088/1757-899X/351/1/012006.

Makan, H. and G.J. Heyns, 2018: Sustainable supply chain initiatives in reducing greenhouse gas emission within the road freight industry. *J. Transp. Supply Chain Manag.*, **12**, doi:10.4102/jtscm.v12i0.365.

MAN-ES, 2019: *Technical Paper: Engineering the future two-stroke green-ammonia engine*. MAN Energy Solutions, Copenhagen, Denmark, 20 pp.

Månberger, A. and B. Stenqvist, 2018: Global metal flows in the renewable energy transition: Exploring the effects of substitutes, technological mix and development. *Energy Policy*, **119**, 226–241, doi:10.1016/j.enpol.2018.04.056.

Mander, S., C. Walsh, P. Gilbert, M. Traut, and A. Bows, 2012: Decarbonizing the UK energy system and the implications for UK shipping. *Carbon Manag.*, **3(6)**, 601–614, doi:10.4155/cmt.12.67.

Manthiram, A., Y. Fu, S.-H. Chung, C. Zu, and Y.-S. Su, 2014: Rechargeable Lithium–Sulfur Batteries. *Chem. Rev.*, **114(23)**, 11751–11787, doi:10.1021/cr500062v.

Manzetti, S. and F. Mariasiu, 2015: Electric vehicle battery technologies: From present state to future systems. *Renew. Sustain. Energy Rev.*, **51**, 1004–1012, doi:10.1016/j.rser.2015.07.010.

Marchetti, C. and J. Ausubel, 2001: The Evolution of Transport. *The Industrial Physicist*, 7 **(2)**, 20–24.

Mareev, I., J. Becker, and D. Sauer, 2018: Battery Dimensioning and Life Cycle Costs Analysis for a Heavy-Duty Truck Considering the Requirements of Long-Haul Transportation. *Energies*, **11(1)**, 55, doi:10.3390/en11010055.

Marelle, L. et al., 2016: Air quality and radiative impacts of Arctic shipping emissions in the summertime in northern Norway: From the local to the regional scale. *Atmos. Chem. Phys.*, **16(4)**, 2359–2379, doi:10.5194/acp-16-2359-2016.

Marinaro, M. et al., 2020: Bringing forward the development of battery cells for automotive applications: Perspective of R&D activities in China, Japan, the EU and the USA. *J. Power Sources*, **459**, 228073, doi:10.1016/j.jpowsour.2020.228073.

Marks, T. et al., 2021: Climate Impact Mitigation Potential of Formation Flight. *Aerospace*, **8(1)**, 14, doi:10.3390/aerospace8010014.

Marletto, G., 2014: Car and the city: Socio-technical transition pathways to 2030. *Technol. Forecast. Soc. Change*, **87**, 164–178, doi:10.1016/j.techfore.2013.12.013.

Marquart, S., M. Ponater, L. Ström, and K. Gierens, 2005: An upgraded estimate of the radiative forcing of cryoplane contrails. *Meteorol. Zeitschrift*, **14**, 573–582, doi:10.1127/0941-2948/2005/0057.

Marsden, G., 2018: Planning for autonomous vehicles? Questions of purpose, place and pace. *Plann Theory Pract.*, **19(5)**.

Marsden, G., J. Dales, P. Jones, E. Seagriff, and N.J. Spurling, 2018: *All Change?: The future of travel demand and its implications for policy and planning*. First Report of the Commission on Travel Demand, http://www.demand.ac.uk/wp-content/uploads/2018/04/FutureTravel_report_final.pdf.

Martens, K., 2020: *A people-centred approach to accessibility*. International Trasnport Forum, Paris, France, 29 pp.

Martin, G., L. Rentsch, M. Höck, and M. Bertau, 2017: Lithium market research – global supply, future demand and price development. *Energy Storage Mater.*, **6**, 171–179, doi: 10.1016/j.ensm.2016.11.004.

Massar, M. et al., 2021: Impacts of autonomous vehicles on greenhouse gas emissions – positive or negative? *Int. J. Environ. Res. Public Health*, **18(11)**, 5567, doi:10.3390/ijerph18115567.

Mathy, S., P. Criqui, and J.-C. Hourcade, 2015: *Pathways to deep decarbonization in France*. Sustainable Development Solutions Network (SDSN) and Institute for Sustainable Development and International Relations (IDDRI), 60 pp., https://irp.cdn-website.com/6f2c9f57/files/uploaded/DDPP_FRA.pdf.

Matthes, S. et al., 2017: A Concept for Multi-Criteria Environmental Assessment of Aircraft Trajectories. *Aerospace*, **4(3)**, 42, doi:10.3390/aerospace4030042.

Mattioli, G., C. Roberts, J.K. Steinberger, and A. Brown, 2020: The political economy of car dependence: A systems of provision approach. *Energy Res. Soc. Sci.*, **66**, 101486, doi:10.1016/j.erss.2020.101486.

Mawhood, R., E. Gazis, S. de Jong, R. Hoefnagels, and R. Slade, 2016: Production pathways for renewable jet fuel: a review of commercialization status and future prospects. *Biofuels, Bioprod. Biorefining*, **10**, 462–484, doi:10.1002/bbb.1644.

Mayrink, S. et al., 2020: Regenerative Braking for Energy Recovering in Diesel-Electric Freight Trains: A Technical and Economic Evaluation. *Energies*, **13(4)**, doi:10.3390/en13040963.

McIntosh, J.R., P. Newman, R. Trubka, and J. Kenworthy, 2017: Framework for land value capture from investments in transit in car-dependent cities. *J. Transp. Land Use*, **10(1)**, 155–185, doi:10.5198/JTLU.2015.531.

McKinnon, A., 2018: *Decarbonizing Logistics: Distributing Goods in a Low Carbon World*. Kogan Page, London, UK.

McKinsey, 2020: *Hydrogen-powered aviation: A fact-based study of hydrogen technology, economics, and climate impact by 2050*. Clean Sky 2 and Fuel Cells and Hydrogen 2 Joint Undertakings, Brussels, Belgium, 96 pp.

Melia, N., K. Haines, and E. Hawkins, 2016: Sea ice decline and 21st century trans-Arctic shipping routes. *Geophys. Res. Lett.*, **43(18)**, 9720–9728, doi:10.1002/2016GL069315.

Melliger, M.A., O.P.R. van Vliet, and H. Liimatainen, 2018: Anxiety vs reality – Sufficiency of battery electric vehicle range in Switzerland and Finland. *Transp. Res. Part D Transp. Environ.*, **65**, 101–115, doi:10.1016/j.trd.2018.08.011.

Melton, N., J. Axsen, and B. Moawad, 2020: Which plug-in electric vehicle policies are best? A multi-criteria evaluation framework applied to Canada. *Energy Res. Soc. Sci.*, **64**, 101411, doi:10.1016/J.ERSS.2019.101411.

Meng, F., J. Mckechnie, T. Turner, K.H. Wong, and S.J. Pickering, 2017: Environmental Aspects of Use of Recycled Carbon Fiber Composites in Automotive Applications. *Environ. Sci. Technol.*, **51(21)**, 12727–12736, doi:10.1021/acs.est.7b04069.

Merchan, A.L., S. Belboom, and A. Léonard, 2020: Life cycle assessment of rail freight transport in Belgium. *Clean Technol. Environ. Policy*, **22(5)**, 1109–1131, doi:10.1007/s10098-020-01853-8.

Mertens, M., V. Grewe, V.S. Rieger, and P. Jöckel, 2018: Revisiting the contribution of land transport and shipping emissions to tropospheric ozone. *Atmos. Chem. Phys.*, **18(8)**, 5567–5588, doi:10.5194/ACP-18-5567-2018.

Messagie, M., et al., 2014: A Range-Based Vehicle Life Cycle Assessment Incorporating Variability in the Environmental Assessment of Different Vehicle Technologies and Fuels. *Energies*, **7**, 1467–1482, doi:10.3390/en7031467.

Messner, S., 2020: Future Arctic shipping, black carbon emissions, and climate change. In: *Maritime Transport and Regional Sustainability* [Ng, A.K.Y., J. Monios, and C. Jiang, (eds.)], Elsevier, pp. 195–208, https://doi.org/10.1016/B978-0-12-819134-7.00012-5.

Meyer, K. and P. Newman, 2020: *Planetary Accounting: Quantifying How to Live Within Planetary Limits at Different Scales of Human Activity*. Springer Nature, Singapore.

Meynerts, L. et al., 2018: Life Cycle Assessment of a Hybrid Train – Comparison of Different Propulsion Systems. *Procedia CIRP*, **69**, 511–516, doi:10.1016/j.procir.2017.11.035.

Miao, Q., M.K. Feeney, F. Zhang, E.W. Welch, and P.S. Sriraj, 2018: Through the storm: Transit agency management in response to climate change. *Transp. Res. Part D Transp. Environ.*, **63**, 421–432, doi:10.1016/j.trd.2018.06.005.

Milovanoff, A. et al., 2019: A Dynamic Fleet Model of U.S Light-Duty Vehicle Lightweighting and Associated Greenhouse Gas Emissions from 2016 to 2050. *Environ. Sci. Technol.*, **53(4)**, 2199–2208, doi:10.1021/acs.est.8b04249.

Minx, J.C. et al., 2021: A comprehensive and synthetic dataset for global, regional, and national greenhouse gas emissions by sector 1970–2018 with an extension to 2019. *Earth Syst. Sci. Data*, **13(11)**, 5213–5252, doi:10.5194/essd-13-5213-2021.

Miotti, M., J. Hofer, and C. Bauer, 2017: Integrated environmental and economic assessment of current and future fuel cell vehicles. *International Journal of Life Cycle Assessment*, **22**, 94–110, doi:10.1007/s11367-015-0986-4.

Mohammadi, A., F. Elsaid, and L. Amador-Jiminez, 2019: Optimizing transit maintenance and rehabilitation to support human development and sustainability: A case study of Costa Rica's railroad network. *Int. J. Sustain. Transp.*, **13(7)**, 497–510, doi:10.1080/15568318.2018.1486488.

Monioudi, I. N. et al., 2018: Climate change impacts on critical international transportation assets of Caribbean Small Island Developing States (SIDS): the case of Jamaica and Saint Lucia. *Reg. Environ. Change*, **18(8)**, 2211–2225, doi:10.1007/s10113-018-1360-4.

Montgomery, J., 2017: *The new wealth of cities: city dynamics and the fifth wave*. Routledge, New York and London.

Moreira, M., A.C. Gurgel, and J.E.A. Seabra, 2014: Life Cycle Greenhouse Gas Emissions of Sugar Cane Renewable Jet Fuel. *Environ. Sci. Technol.*, **48(24)**, 14756–14763, doi:10.1021/es503217g.

Moretti, L. and G. Loprencipe, 2018: Climate Change and Transport Infrastructures: State of the Art. *Sustainability*, **10(11)**, 4098, doi:10.3390/su10114098.

Morlotti, C., M. Cattaneo, P. Malighetti, and R. Redondi, 2017: Multi-dimensional price elasticity for leisure and business destinations in the low-cost air transport market: Evidence from easyJet. *Tour. Manag.*, **61**, 23–34, doi:10.1016/j.tourman.2017.01.009.

Moultak, M., N. Lutsey, and D. Hall, 2017: *Transitioning to Zero-Emission Heavy-Duty Freight Vehicles*. The International Council on Clean Transportation, Washington, DC, USA, 53 pp.

Move Humanity, 2018: *Closing the SDG Budget Gap*. Pica Publishing Ltd, 49 pp., https://movehumanity.org/wp-content/uploads/2018/12/MOVE-HUMANITY-REPORT-WEB-V6-201218.pdf.

Muiderman, K., A. Gupta, J. Vervoort, and F. Biermann, 2020: Four approaches to anticipatory climate governance: Different conceptions of the future and implications for the present. *WIREs Clim. Change*, **11(6)**, e673–e673, doi:10.1002/wcc.673.

Mulholland, E., J. Teter, P. Cazzola, Z. McDonald, and B.P. Ó Gallachóir, 2018: The long haul towards decarbonising road freight – A global assessment to 2050. *Appl. Energy*, **216**, 678–693, doi:10.1016/J.APENERGY.2018.01.058.

Muratori, M. et al., 2020: EMF-33 insights on bioenergy with carbon capture and storage (BECCS). *Clim. Change*, **163(3)**, 1621–1637, doi:10.1007/s10584-020-02784-5.

Nahlik, M.J., A.T. Kaehr, M.V. Chester, A. Horvath, and M.N. Taptich, 2016: Goods Movement Life Cycle Assessment for Greenhouse Gas Reduction Goals. *J. Ind. Ecol.*, **20(2)**, 317–328, doi:10.1111/jiec.12277.

Naoi, K., S. Ishimoto, J. Miyamoto, and W. Naoi, 2012: Second generation 'nanohybrid supercapacitor': evolution of capacitive energy storage devices. *Energy Environ. Sci.*, **5(11)**, 9363–9373, doi:10.1039/C2EE21675B.

Narassimhan, E. and C. Johnson, 2018: The role of demand-side incentives and charging infrastructure on plug-in electric vehicle adoption: analysis of US States. *Environ. Res. Lett.*, **13(7)**, 074032, doi:10.1088/1748-9326/aad0f8.

Narula, C. K., B.H. Davison, and Z. Li, 2020. Zeolitic catalytic conversion of alcohols to hydrocarbon fractions with reduced gaseous hydrocarbon content. United States. https://www.osti.gov/servlets/purl/1651075.

Nassar, N.T., Alonso, E., and Brainard, J.L., 2020, Investigation of U.S. Foreign Reliance on Critical Minerals—U.S. Geological Survey Technical Input Document in Response to Executive Order No. 13953 Signed September 30, 2020 (Ver. 1.1, December 7, 2020): U.S. Geological Survey Open-File Report 2020–1127, 37 pp., https://doi.org/10.3133/ofr20201127.

Nasri, M.I., T. Bektaş, and G. Laporte, 2018: Route and speed optimization for autonomous trucks. *Comput. Oper. Res.*, **100**, 89–101, doi:10.1016/J.COR.2018.07.015.

Ndlovu, V. and P. Newman, 2020: Leapfrog Technology and How It Applies to Trackless Tram. *J. Transp. Technol.*, **10(03)**, 198–213, doi:10.4236/jtts.2020.103013.

Neaimeh, M. and P.B. Andersen, 2020: Mind the gap – open communication protocols for vehicle grid integration. *Energy Informatics*, **3(1)**, 1–17, doi:10.1186/s42162-020-0103-1.

Nelder, C. and E. Rogers, 2019: *Reducing EV charging infrastructure costs*. Rocky Mountain Institute, 49 pp., https://rmi.org/ev-charging-costs.

Neves, A. and C. Brand, 2019: Assessing the potential for carbon emissions savings from replacing short car trips with walking and cycling using a mixed GPS-travel diary approach. *Transp. Res. Part A Policy Pract.*, **123**, 130–146, doi:10.1016/J.TRA.2018.08.022.

Newman, P., 2010: Sustainable cities of the future: The behavior change driver. *Sustain. Dev. L. Pol'y*, **11(1)**, 7–10.

Newman, P., 2020a: COVID, CITIES and CLIMATE: Historical Precedents and Potential Transitions for the New Economy. *Urban Sci.*, **4(3)**, 32, doi:10.3390/urbansci4030032.

Newman, P., 2020b: Cool planning: How urban planning can mainstream responses to climate change. *Cities*, **103**, doi:10.1016/j.cities.2020.102651.

Newman, P. and J. Kenworthy, 2006: Urban Design to Reduce Automobile Dependence. *Opolis*, **2(1)**, 35–52.

Newman, P. and J. Kenworthy, 2015: *The End of Automobile Dependence: How Cities are Moving Beyond Car-Based Planning*. Island Press/Center for Resource Economics, Washington, DC, USA, 320 pp.

Newman, P., A. Matan, and J. McIntosh, 2015: Urban transport and sustainable development. In: *Routledge International Handbook of Sustainable Development*, Routledge, New York and London, pp. 337–350.

Newman, P., L. Kosonen, and J. Kenworthy, 2016: Theory of urban fabrics: planning the walking, transit/public transport and automobile/motor car cities for reduced car dependency. *Town Plan. Rev.*, **87(4)**, 429–458, doi:10.3828/tpr.2016.28.

Newman, P., T. Beatley, and H. Boyer, 2017: *Resilient cities: Overcoming fossil fuel dependence*. Island Press/Center for Resource Economics, Washington, DC, USA, 253 pp.

Newman, P., S. Davies-Slate, and E. Jones, 2018: The Entrepreneur Rail Model: Funding urban rail through majority private investment in urban regeneration. *Res. Transp. Econ.*, **67**, 19–28, doi:10.1016/j.retrec.2017.04.005.

Newman, P. et al., 2019: The Trackless Tram: Is It the Transit and City Shaping Catalyst We Have Been Waiting for? *J. Transp. Technol.*, **09(01)**, 31–55, doi:10.4236/jtts.2019.91003.

Newman, P., S. Davies-Slate, D. Conley, K. Hargroves, and M. Mouritz, 2021: From TOD to TAC: Why and How Transport and Urban Policy Needs to Shift to Regenerating Main Road Corridors with New Transit Systems. *Urban Sci.*, **5(3)**, 52, doi:10.3390/urbansci5030052.

Nguyen, H.P. et al., 2020: The electric propulsion system as a green solution for management strategy of $CO_2$ emission in ocean shipping: A comprehensive review. *Int. Trans. Electr. Energy Syst.*, 2021; 31: e12580, doi:10.1002/2050-7038.12580.

Nguyen, T.H., J. Partala, and S. Pirttikangas, 2019: Blockchain-Based Mobility-as-a-Service. 28th International Conference on Computer Communication and Networks, Valencia, Spain.

Nicholas, M., 2019: *Estimating electric vehicle charging infrastructure costs across major U.S. metropolitan areas*. The International Council on Clean Transportation, Washington, DC, USA, 11 pp.

Nicholas, M., G. Tal, and T. Turrentine, 2017: *Advanced plug-in electric vehicle travel and charging behavior interim report*. Institute of Transportation Studies, University of California, Davis, CA, USA, 38 pp.

Nieuwenhuijsen, M.J., 2018: Influence of urban and transport planning and the city environment on cardiovascular disease. *Nat. Rev. Cardiol.*, **15(7)**, 432–438, doi:10.1038/s41569-018-0003-2.

Nijland, H. and J. van Meerkerk, 2017: Mobility and environmental impacts of car sharing in the Netherlands. *Environ. Innov. Soc. Transitions*, **23**, 84–91, doi:10.1016/j.eist.2017.02.001.

Niklaß, M. et al., 2019: Potential to reduce the climate impact of aviation by climate restricted airspaces. *Transp. Policy*, **83**, 102–110, doi:10.1016/j.tranpol.2016.12.010.

Noel, L., J. Kester, G.Z. de Rubens, and B.K. Sovacool, 2019: *Vehicle-to-Grid: A Sociotechnical Transition Beyond Electric Mobility*. Palgrave Macmillan, Cham, Switzerland.

Noland, R.B., S. Wang, S. Kulp, and B.H. Strauss, 2019: Employment accessibility and rising seas. *Transp. Res. Part D Transp. Environ.*, **77**, 560–572, doi:10.1016/j.trd.2019.09.017.

Noppers, E.H., K. Keizer, J.W. Bolderdijk, and L. Steg, 2014: The adoption of sustainable innovations: Driven by symbolic and environmental motives. *Glob. Environ. Change*, **25(1)**, 52–62, doi:10.1016/j.gloenvcha.2014.01.012.

Noppers, E.H., K. Keizer, M. Bockarjova, and L. Steg, 2015: The adoption of sustainable innovations: The role of instrumental, environmental, and symbolic attributes for earlier and later adopters. *J. Environ. Psychol.*, **44**, 74–84, doi:10.1016/j.jenvp.2015.09.002.

Noy, K. and M. Givoni, 2018: Is 'Smart Mobility' Sustainable? Examining the Views and Beliefs of Transport's Technological Entrepreneurs. *Sustainability*, **10(2)**, 422, doi:10.3390/su10020422.


10

NSTC, 2018: *Assessment of Critical Minerals: Updated Application of Screening Methodology*. National Science and Technology Council, USA, 10 pp., https://trumpwhitehouse.archives.gov/wp-content/uploads/2018/02/Assessment-of-Critical-Minerals-Update-2018.pdf.

Nugroho, S.B., E. Zusman, and R. Nakano, 2018a: Pedestrianisation programs and its impacts on the willingness to increase walking distance in Indonesian cities: The case of Bandung City and Bogor City, Indonesia. *Transportation Systems in the Connected Era – Proceedings of the 23rd International Conference of Hong Kong Society for Transportation Studies,* Hong Kong Society for Transportation Studies, Hong Kong, https://www.iges.or.jp/en/pub/pedestrianisation-programs-and-its-impacts/en.

Nugroho, S.B. et al., 2018b: Does the Improvement of Pedestrian Facilities Increase Willingness to Walk? The Case of Bandung, Indonesia. *Transportation Research Board 97th Annual Meeting*, Washington, DC, USA.

Nykvist, B. and M. Nilsson, 2015: Rapidly falling costs of battery packs for electric vehicles. *Nat. Clim. Change*, **5(4)**, 329–332, doi:10.1038/nclimate2564.

Nykvist, B., F. Sprei, and M. Nilsson, 2019: Assessing the progress toward lower priced long range battery electric vehicles. *Energy Policy*, **124**, 144–155, doi:10.1016/j.enpol.2018.09.035.

O'Brien, W. and F. Yazdani Aliabadi, 2020: Does telecommuting save energy? A critical review of quantitative studies and their research methods. *Energy Build.*, **225**, 110298, doi:10.1016/j.enbuild.2020.110298.

O'Keefe, P., B. Caulfield, W. Brazil, and P. White, 2016: The impacts of telecommuting in Dublin. *Res. Transp. Econ.*, **57**, 13–20, doi:10.1016/J.RETREC.2016.06.010.

Ødemark, K. et al., 2012: Short-lived climate forcers from current shipping and petroleum activities in the Arctic. *Atmos. Chem. Phys.*, **12(4)**, 1979–1993, doi:10.5194/acp-12-1979-2012.

Olivetti, E.A., G. Ceder, G.G. Gaustad, and X. Fu, 2017: Lithium-Ion Battery Supply Chain Considerations: Analysis of Potential Bottlenecks in Critical Metals. *Joule*, **1(2)**, 229–243, doi:10.1016/j.joule.2017.08.019.

Olmer, N., B. Comer, B. Roy, X. Mao, and D. Rutherford, 2017: *Greenhouse Gas Emissions from Global Shipping, 2013–2015*. International Council on Clean Transportation, Washington, DC, USA, 38 pp.

Olofsson, J., Z. Barta, P. Börjesson, and O. Wallberg, 2017: Integrating enzyme fermentation in lignocellulosic ethanol production: Life-cycle assessment and techno-economic analysis. *Biotechnol. Biofuels*, **10**, 51–64, doi:10.1186/s13068-017-0733-0.

Olugbenga, O., N. Kalyviotis, and S. Saxe, 2019: Embodied emissions in rail infrastructure: A critical literature review. *Environ. Res. Lett.*, **14(12)**, 123002, doi:10.1088/1748-9326/ab442f.

Paddeu, D. and J. Denby, 2021: Decarbonising road freight: Is truck automation and platooning an opportunity? *Clean Technol. Environ. Policy*, **1**, doi:10.1007/s10098-020-02020-9.

Paffumi, E., M. De Gennaro, and G. Martini, 2018: Alternative utility factor versus the SAE J2841 standard method for PHEV and BEV applications. *Transp. Policy*, **68**(April), 80–97, doi:10.1016/j.tranpol.2018.02.014.

Palin, E.J., I.S. Oslakovic, K. Gavin, and A. Quinn, 2021: Implications of climate change for railway infrastructure. *Wiley Interdiscip. Rev. Clim. Change*, **12(5)**, e728–e728, doi:10.1002/WCC.728.

Palmer, K., J.E. Tate, Z. Wadud, and J. Nellthorp, 2018: Total cost of ownership and market share for hybrid and electric vehicles in the UK, US and Japan. *Appl. Energy*, **209**(November 2017), 108–119, doi:10.1016/j.apenergy.2017.10.089.

Pålsson, H., F. Pettersson, and L. Winslott Hiselius, 2017: Energy consumption in e-commerce versus conventional trade channels – Insights into packaging, the last mile, unsold products and product returns. *J. Clean. Prod.*, **164**, 765–778, doi:10.1016/j.jclepro.2017.06.242.

Pan, D. et al., 2020: Methane emissions from natural gas vehicles in China. *Nat. Commun.*, **11(1)**, 1–10, doi:10.1038/s41467-020-18141-0.

Pan, X., H. Wang, L. Wang, and W. Chen, 2018: Decarbonization of China's transportation sector: In light of national mitigation toward the Paris Agreement goals. *Energy*, **155**, 853–864, doi:10.1016/j.energy.2018.04.144.

Pangbourne, K., M.N. Mladenović, D. Stead, and D. Milakis, 2020: Questioning mobility as a service: Unanticipated implications for society and governance. *Transp. Res. Part A Policy Pract.*, **131**, 35–49, doi:10.1016/j.tra.2019.09.033.

Panja, T. et al., 2020: Fabrication of high-performance dual carbon Li-ion hybrid capacitor: mass balancing approach to improve the energy-power density and cycle life. *Sci. Rep.*, **10(1)**, doi:10.1038/s41598-020-67216-x.

Panoutsou, C. et al., 2021: Advanced biofuels to decarbonise European transport by 2030: Markets, challenges, and policies that impact their successful market uptake. *Energy Strateg. Rev.*, **34**, 100633, doi:10.1016/j.esr.2021.100633.

Parkinson, B., P. Balcombe, J.F. Speirs, A.D. Hawkes, and K. Hellgardt, 2019: Levelized cost of $CO_2$ mitigation from hydrogen production routes. *Energy Environ. Sci.*, **12(1)**, 19–40, doi:10.1039/c8ee02079e.

Pathak, M. and P.R. Shukla, 2016: Co-benefits of low carbon passenger transport actions in Indian cities: Case study of Ahmedabad. *Transp. Res. Part D Transp. Environ.*, **44**, 303–316, doi:10.1016/j.trd.2015.07.013.

Pavlovic, J. et al., 2017: The Impact of WLTP on the Official Fuel Consumption and Electric Range of Plug-in Hybrid Electric Vehicles in Europe. *SAE Tech. Pap.*, **2017** (September), doi:10.4271/2017-24-0133.

Peden, M.M. and P. Puvanachandra, 2019: Looking back on 10 years of global road safety. *Int. Health*, **11(5)**, 327–330, doi:10.1093/inthealth/ihz042.

Peng, M., L. Liu, and C. Jiang, 2012: A review on the economic dispatch and risk management of the large-scale plug-in electric vehicles (PHEVs)-penetrated power systems. *Renew. Sustain. Energy Rev.*, **16(3)**, 1508–1515, doi:10.1016/j.rser.2011.12.009.

Peng, W. et al., 2020: Comprehensive analysis of the air quality impacts of switching a marine vessel from diesel fuel to natural gas. *Environ. Pollut.*, **266**, doi:10.1016/j.envpol.2020.115404.

Penner, J.E., C. Zhou, A. Garnier, and D.L. Mitchell, 2018: Anthropogenic aerosol indirect effects in cirrus clouds. *J. Geophys. Res. Atmos.*, **123(20)**, 11–652, doi:10.1029/2018JD029204.

Perčić, M., N. Vladimir, and A. Fan, 2020: Life-cycle cost assessment of alternative marine fuels to reduce the carbon footprint in short-sea shipping: A case study of Croatia. *Appl. Energy*, **279**, 115848, doi:10.1016/J.APENERGY.2020.115848.

Pereira, S.R., T. Fontes, and M.C. Coelho, 2014: Can hydrogen or natural gas be alternatives for aviation? – A life cycle assessment. *Int. J. Hydrogen Energy*, **39(25)**, 13266–13275, doi:10.1016/j.ijhydene.2014.06.146.

Pérez-Morales, A., F. Gomariz-Castillo, and P. Pardo-Zaragoza, 2019: Vulnerability of Transport Networks to Multi-Scenario Flooding and Optimum Location of Emergency Management Centers. *Water*, **11(6)**, 1197, doi:10.3390/w11061197.

Petchenik, I., 2021: Commercial flights down 42% in 2020. *Flightradar24 Blog*. https://www.flightradar24.com/blog/commercial-flights-down-42-in-2020/.

Peters, K., P. Stier, J. Quaas, and H. Graßl, 2012: Aerosol indirect effects from shipping emissions: Sensitivity studies with the global aerosol-climate model ECHAM-HAM. *Atmos. Chem. Phys.*, **12(13)**, 5985–6007, doi:10.5194/ACP-12-5985-2012.

Peters, R. et al., 2021: Future power train solutions for long-haul trucks. *Sustainability*, **13(4)**, 1–59, doi:10.3390/su13042225.

Placke, T., R. Kloepsch, S. Dühnen, and M. Winter, 2017: Lithium ion, lithium metal, and alternative rechargeable battery technologies: the odyssey for high energy density. *J. Solid State Electrochem.*, **21(7)**, 1939–1964, doi:10.1007/s10008-017-3610-7.

Plevin, R.J., A.D. Jones, M.S. Torn, and H.K. Gibbs, 2010: Greenhouse Gas Emissions from Biofuels' Indirect Land Use Change Are Uncertain but May Be Much Greater than Previously Estimated. *Environ. Sci. Technol.*, **44(21)**, 8015–8021, doi:10.1021/es101946t.

Plötz, P., S.Á. Funke, and P. Jochem, 2018: Empirical fuel consumption and $CO_2$ emissions of plug-in hybrid electric vehicles. *J. Ind. Ecol.*, **22(4)**, 773–784, doi:10.1111/jiec.12623.

Plötz, P., C. Moll, G. Bieker, P. Mock, and Y. Li, 2020: Real-world usage of plug-in hybrid electric vehicles. International Council on Clean Transportation, Berlin, Germany Pochet, M., I. Truedsson, F. Foucher, H. Jeanmart, and F. Contino, 2017: Ammonia-Hydrogen Blends in Homogeneous-Charge Compression-Ignition Engine. *13th International Conference on Engines & Vehicles*, 10.

Pojani, D. and D. Stead, 2015: Sustainable Urban Transport in the Developing World: Beyond Megacities. *Sustainability*, **7(6)**, 7784–7805, doi:10.3390/SU7067784.

Pollard, S.G., 2019: Imagining the net zero emissions city: Urban climate governance in the City of Melbourne, Australia. In: *The role of non-state actors in the green transition: Building a sustainable future* [Hoff, J., Q. Gausset, and S. Lex, (eds.)], Routledge, New York and London, pp. 211–229.

Poonam, K. Sharma, A. Arora, and S.K. Tripathi, 2019: Review of supercapacitors: Materials and devices. *J. Energy Storage*, **21**, 801–825, doi:10.1016/J.EST.2019.01.010.

Posada, F., F. Kamakate, and A. Bandivadekar, 2011: *Sustainable Management of Two- and Three-Wheelers in Asia*. International Council on Clean Transportation, 15 pp. https://theicct.org/sites/default/files/publications/ICCT_Asia23wheelers_2011.pdf.

Poulsen, R.T. and H. Sampson, 2020: A swift turnaround? Abating shipping greenhouse gas emissions via port call optimization. *Transp. Res. Part D Transp. Environ.*, **86**, 102460, doi:10.1016/J.TRD.2020.102460.

Pradhan, P., L. Costa, D. Rybski, W. Lucht, and J.P. Kropp, 2017: A Systematic Study of Sustainable Development Goal (SDG) Interactions. *Earth's Futur.*, **5(11)**, 1169–1179, doi:10.1002/2017EF000632.

Prati, G., 2018: Gender equality and women's participation in transport cycling. *J. Transp. Geogr.*, **66**, 369–375, doi:10.1016/j.jtrangeo.2017.11.003.

Prussi, M., M. Yugo, L. De Prada, M. Padella, and M. Edwards, 2020: *JEC Well-To-Wheels report v5*. Publications Office of the European Union, Luxembourg, 135 pp, doi:10.2760/100379.

Psaraftis, H.N., 2016: Green Maritime Transportation: Market Based Measures. In: *Green Transportation Logistics: The Quest for Win-Win Solutions* [Psaraftis, H.N., (ed.)], Springer International Publishing, Cham, Switzerland, pp. 267–297.

Puricelli, S. et al., 2021: A review on biofuels for light-duty vehicles in Europe. *Renew. Sustain. Energy Rev.*, **137**, 110398, doi:10.1016/J.RSER.2020.110398.

Pye, S., G. Anandarajah, B. Fais, C. McGlade, and N. Strachan, 2015: *Pathways to deep decarbonization in the United Kingdom*. Sustainable Development Solutions Network (SDSN) and Institute for Sustainable Development and International Relations (IDDRI).

Qiao, Q., et al., 2020: Life cycle cost and GHG emission benefits of electric vehicles in China. *Transp. Res. Part D: Transport and Environment*, **86**, 102418, doi:10.1016/j.trd.2020.102418.

Qiao, Y., J. Santos, A.M.K. Stoner, and G. Flinstch, 2019: Climate change impacts on asphalt road pavement construction and maintenance: An economic life cycle assessment of adaptation measures in the State of Virginia, United States. *J. Ind. Ecol.*, 24, 342–355.

Qin, X., L. Liang, X. Zhang, G. Deng, and H. Liang, 2020: Structural Design and Stress Analysis of a Fully-Wrapped Composite CNG Gas Cylinder With Nominal Working Pressure of 30 MPa. *Proceedings of the Volume 3: Design and Analysis*, American Society of Mechanical Engineers, V003T03A008. ASME, https://doi.org/10.1115/PVP2020-21521.

Raghavan, S.S. and G. Tal, 2020: Plug-in hybrid electric vehicle observed utility factor: Why the observed electrification performance differ from expectations. *Journal of Sustainable Transportation*, 16:2, 105–136, doi:10.1080/15568318.2020.1849469.

Rahim, M.M., M.T. Islam, and S. Kuruppu, 2016: Regulating global shipping corporations' accountability for reducing greenhouse gas emissions in the seas. *Mar. Policy*, **69**, 159–170, doi:10.1016/j.marpol.2016.04.018.

Rao, S. et al., 2017: Future air pollution in the Shared Socio-economic Pathways. **42**, 346–358, doi:10.1016/j.gloenvcha.2016.05.012.

Rasmussen, D.J. et al., 2018: Extreme sea level implications of 1.5°C, 2.0°C, and 2.5°C temperature stabilization targets in the 21st and 22nd centuries. *Environ. Res. Lett.*, **13(3)**, doi:10.1088/1748-9326/aaac87.

Ratner, K.A., and A.R. Goetz, 2013: The reshaping of land use and urban form in Denver through transit-oriented development. *Cities*, **30(1)**, 31–46, doi:10.1016/J.CITIES.2012.08.007.

Rayner, T., 2021: Taking the slow route to decarbonisation? Developing climate governance for international transport. *Earth Syst. Gov.*, **8**, 100100, doi:10.1016/j.esg.2021.100100.

Reddy, K.S., L.K. Panwar, R. Kumar, and B.K. Panigrahi, 2016: Distributed resource scheduling in smart grid with electric vehicle deployment using fireworks algorithm. *J. Mod. Power Syst. Clean Energy*, **4(2)**, 188–199, doi:10.1007/s40565-016-0195-6.

Reiter, A.J. and S.C. Kong, 2011: Combustion and emissions characteristics of compression-ignition engine using dual ammonia-diesel fuel. *Fuel*, **90(1)**, 87–97, doi:10.1016/J.FUEL.2010.07.055.

Rifkin, J., 2019: *The Green New Deal: Why the Fossil Fuel Civilization Will Collapse by 2028, and the Bold Economic Plan to Save Life on Earth.*, St. Martin's Press.

Righi, M., J. Hendricks, and R. Sausen, 2013: The global impact of the transport sectors on atmospheric aerosol: simulations for year 2000 emissions. *Atmos. Chem. Phys.*, **13(19)**, 9939–9970, doi:10.5194/acp-13-9939-2013.

Righi, M., J. Hendricks, and R. Sausen, 2015: The global impact of the transport sectors on atmospheric aerosol in 2030 – Part 1: Land transport and shipping. *Atmos. Chem. Phys.*, **15(2)**, 633–651, doi:10.5194/ACP-15-633-2015.

Rip, A. and R. Kemp, 1998: Technological change. In: *Human choice and climate change: Vol. II, Resources and Technology,* Battelle Press, Colombus [Rayner, S. and Malone E.L. (eds.)], Colombus, OH, USA, pp. 327–399.

Robertson, S., 2016: The potential mitigation of $CO_2$ emissions via modal substitution of high-speed rail for short-haul air travel from a life cycle perspective – An Australian case study. *Transp. Res. Part D Transp. Environ.*, **46**, 365–380, doi:10.1016/j.trd.2016.04.015.

Robertson, S., 2018: A carbon footprint analysis of renewable energy technology adoption in the modal substitution of high-speed rail for short-haul air travel in Australia. *Int. J. Sustain. Transp.*, **12(4)**, 299–312, doi:10.1080/15568318.2017.1363331.

Robins, N., 2018: *The Road to Net-Zero Finance: A report prepared by the Advisory Group on Finance for the UK's Climate Change Committee*, Climate Change Committee, London, UK, 42 pp.

Rodrigue, J.-P., C. Comtois, and B. Slack, 2017: *The Geography of Transport Systems*. Fourth Edition. Routledge, New York, NY, USA, 440 pp.

Roiger, A. et al., 2014: Quantifying Emerging Local Anthropogenic Emissions in the Arctic Region: The ACCESS Aircraft Campaign Experiment. *Bull. Am. Meteorol. Soc.*, **96(3)**, 441–460, doi:10.1175/BAMS-D-13-00169.1.

Rondinelli, S., A. Gardi, R. Kapoor, and R. Sabatini, 2017: Benefits and challenges of liquid hydrogen fuels in commercial aviation. *Int. J. Sustain. Aviat.*, **3(3)**, 200, doi:10.1504/IJSA.2017.086845.

Rose, S.K. et al., 2020: An overview of the Energy Modeling Forum 33rd study: assessing large-scale global bioenergy deployment for managing climate change. *Clim. Change*, **163(3)**, 1539–1551, doi:10.1007/s10584-020-02945-6.

Rosenfeld, D.C., J. Lindorfer, and K. Fazeni-Fraisl, 2019: Comparison of advanced fuels—Which technology can win from the life cycle perspective? *Journal of Cleaner Production*, **238**, 117879, doi:10.1016/j.jclepro.2019.117879.

Roth, K.W., T. Rhodes, and R. Ponoum, 2008: The energy and greenhouse gas emission impacts of telecommuting in the U.S. 2008 IEEE International Symposium on Electronics and the Environment, IEEE, pp. 1–6.

Rothengatter, W., 2010: Competition between airlines and high-speed rail. In: *Critical issues in air transport economics and business* [Macário, R. and E. Van de Voorde, (eds.)], Routledge, pp. 329–352.

Rovere, E.L. La, C. Gesteira, C. Grottera, and W. Wills, 2015: *Pathways to deep decarbonization in Brazil*. Sustainable Development Solutions Network (SDSN) and Institute for Sustainable Development and International Relations (IDDRI), 44 pp.

Roy, P., A. Dutta, and B. Deen, 2015: Greenhouse gas emissions and production cost of ethanol produced from biosyngas fermentation process. *Bioresour. Technol.*, **192**, 185–191, doi:10.1016/J.BIORTECH.2015.05.056.

Ryley, T.J., P.A. Stanley, M.P. Enoch, A.M. Zanni, and M.A. Quddus, 2014: Investigating the contribution of Demand Responsive Transport to a sustainable local public transport system. *Res. Transp. Econ.*, **48**, 364–372, doi:10.1016/j.retrec.2014.09.064.

Sacchi, R., 2021: Carculator: Prospective environmental and economic life cycle assessment of vehicles made blazing fast. Available from: https://github.com/romainsacchi/carculator/blob/master/docs/introduction.rst.

Sadeghi, S., S. Ghandehariun, and M.A. Rosen, 2020: Comparative economic and life cycle assessment of solar-based hydrogen production for oil and gas industries. *Energy*, **208**, 118347, doi:10.1016/j.energy.2020.118347.

Sahoo, S. and D.K. Srivastava, 2021: Effect of compression ratio on engine knock, performance, combustion and emission characteristics of a bi-fuel CNG engine. *Energy*, **233**, 121144, doi:10.1016/J.ENERGY.2021.121144.

Sahoo, S., X. Zhao, and K. Kyprianidis, 2020: A Review of Concepts, Benefits, and Challenges for Future Electrical Propulsion-Based Aircraft. *Aerosp. 2020*, **7(4)**, 44, doi:10.3390/AEROSPACE7040044.

Salman, C.A., S. Schwede, E. Thorin, and J. Yan, 2017: Enhancing biomethane production by integrating pyrolysis and anaerobic digestion processes. *Appl. Energy*, **204**, 1074–1083, doi:10.1016/j.apenergy.2017.05.006.

Schafer, A. and D.G. Victor, 2000: The future mobility of the world population. *Transp. Res. Part A Policy Pract.*, **34(3)**, 171–205, doi:10.1016/S0965-8564(98)00071-8.

Schäfer, A.W., 2017: Long-term trends in domestic US passenger travel: the past 110 years and the next 90. *Transportation (Amst).*, **44(2)**, 293–310, doi:10.1007/s11116-015-9638-6.

Schäfer, A.W. and S. Yeh, 2020: A holistic analysis of passenger travel energy and greenhouse gas intensities. *Nat. Sustain.*, **3(6)**, doi:10.1038/s41893-020-0514-9.

Schaller, B., 2018: *The New Automobility: Lyft, Uber and the Future of American Cities*. Schaller Consulting, Brooklyn, NY, USA, 41 pp. http://www.schallerconsult.com/rideservices/automobility.pdf.

Scheelhaase, J., S. Maertens, and W. Grimme, 2019: Synthetic fuels in aviation – Current barriers and potential political measures. *Transp. Res. Procedia*, **43**, 21–30.

Schmidt, O., A. Hawkes, A. Gambhir, and I. Staffell, 2017: The future cost of electrical energy storage based on experience rates. *Nat. Energy*, **2(8)**, 17110, doi:10.1038/nenergy.2017.110.

Schmidt, P., W. Weindorf, A. Roth, V. Batteiger, and F. Riegel, 2016: *Power-to-Liquids: Potentials and Perspectives for the Future Supply of Renewable Aviation Fuel*. Umweltbundesamt, Dessau-Roßlau, Germany, 36 pp.

Schmidt, P., V. Batteiger, A. Roth, W. Weindorf, and T. Raksha, 2018: Power-to-Liquids as Renewable Fuel Option for Aviation: A Review. *Chemie Ing. Tech.*, **90(1–2)**, 127–140, doi:10.1002/cite.201700129.

Sclar, R., C. Gorguinpour, S. Castellanos, and X. Li, 2019: *Barriers to Adopting Electric Buses*. World Resources Institute, Washington, DC, USA, 60 pp.

Seba, T., 2014: *Clean Disruption of Energy and Transportation: How Silicon Valley Will Make Oil, Nuclear, Natural Gas, Coal, Electric Utilities and Conventional Cars Obsolete by 2030*. Clean Planet Ventures, Silicon Valley, CA, USA.

Seddiek, I.S., 2015: An Overview: Environmental and Economic Strategies for Improving Quality of Ships Exhaust Gases. *Int. J. Marit. Eng.*, **157**(A1), 53–64, doi:10.3940/rina.ijme.2015.a1.311.

Sedlmeir, J., H.U. Buhl, G. Fridgen, and R. Keller, 2020: The Energy Consumption of Blockchain Technology: Beyond Myth. *Bus. Inf. Syst. Eng.*, **62(6)**, 599–608, doi:10.1007/s12599-020-00656-x.

Seithe, G.J., A. Bonou, D. Giannopoulos, C.A. Georgopoulou, and M. Founti, 2020: Maritime Transport in a Life Cycle Perspective: How Fuels, Vessel Types, and Operational Profiles Influence Energy Demand and Greenhouse Gas Emissions. *Energies*, **13(11)**, 2739, doi:10.3390/en13112739.

Sen, B., T. Ercan, and O. Tatari, 2017: Does a battery-electric truck make a difference? Life cycle emissions, costs, and externality analysis of alternative fuel-powered Class 8 heavy-duty trucks in the United States. *J. Clean. Prod.*, **141**, 110–121, doi:10.1016/J.JCLEPRO.2016.09.046.

Seto, K.C. et al., 2016: Carbon Lock-In: Types, Causes, and Policy Implications. *Annu. Rev. Environ. Resour.*, **41(1)**, 425–452, doi:10.1146/annurev-environ-110615-085934.

Seto, K.C. et al., 2021: From Low- to Net-Zero Carbon Cities: The Next Global Agenda. *Annu. Rev. Environ. Resour.*, **46**, 23.1–23.9, doi:10.1146/annurev-environ-050120.

Shabanpour, R., N. Golshani, M. Tayarani, J. Auld, and A. (Kouros) Mohammadian, 2018: Analysis of telecommuting behavior and impacts on travel demand and the environment. *Transp. Res. Part D Transp. Environ.*, **62**, 563–576, doi:10.1016/j.trd.2018.04.003.

Shaheen, S. and A. Cohen, 2019: Shared ride services in North America: definitions, impacts, and the future of pooling. *Transp. Rev.*, **39(4)**, 427–442, doi:10.1080/01441647.2018.1497728.

Shaheen, S., A. Cohen, and M. Jaffee, 2018: *Innovative Mobility: Carsharing Outlook*. University of California, Berkeley, CA, USA, 8 pp.

Sharafian, A., P. Blomerus, and W. Mérida, 2019: Natural gas as a ship fuel: Assessment of greenhouse gas and air pollutant reduction potential. *Energy Policy*, **131**, 332–346, doi:10.1016/j.enpol.2019.05.015.

Sharma, R. and P. Newman, 2020: Land Value Capture Tools: Integrating Transit and Land Use through Finance to Enable Economic Value Creation. *Morden Econ.*, **11(4)**, 938–964, doi:10.4236/me.2020.114070.

Sharmina, M., C. McGlade, P. Gilbert, and A. Larkin, 2017: Global energy scenarios and their implications for future shipped trade. *Mar. Policy*, **84**, 12–21, doi:10.1016/j.marpol.2017.06.025.

Sharmina, M. et al., 2021: Decarbonising the critical sectors of aviation, shipping, road freight and industry to limit warming to 1.5–2°C. *Clim. Policy*, **21(4)**, 455–474, doi:10.1080/14693062.2020.1831430.

Sharp, D., 2018: Sharing Cities for Urban Transformation: Narrative, Policy and Practice. *Urban Policy Res.*, **36(4)**, 513–526, doi:10.1080/08111146.2017.1421533.

She, Z.-Y., Qing Sun, J.-J. Ma, and B.-C. Xie, 2017: What are the barriers to widespread adoption of battery electric vehicles? A survey of public perception in Tianjin, China. *Transp. Policy*, **56**, 29–40, doi:10.1016/J.TRANPOL.2017.03.001.

Shen, W., W. Han, T.J. Wallington, and S.L. Winkler, 2019: China Electricity Generation Greenhouse Gas Emission Intensity in 2030: Implications for Electric Vehicles. *Environ. Sci. Technol.*, **53**, 6063–6072, doi:10.1021/acs.est.8b05264.

Sheng, Y., X. Shi, and B. Su, 2018: Re-analyzing the economic impact of a global bunker emissions charge. *Energy Econ.*, **74**, 107–119, doi:10.1016/j.eneco.2018.05.035.

Shukla, P., S. Dhar, M. Pathak, D. Mahadevia, and A. Garg, 2015: *Pathways to Deep Decarbonization in India.* Sustainable Development Solutions Network (SDSN) and Institute for Sustainable Development and International Relations (IDDRI), 63 pp.

Siagian, U.W.R. et al., 2015: *Pathways to deep decarbonization in Indonesia*. Sustainable Development Solutions Network (SDSN) and Institute for Sustainable Development and International Relations (IDDRI), 48 pp.

Sierzchula, W., S. Bakker, K. Maat, and B. van Wee, 2014: The influence of financial incentives and other socio-economic factors on electric vehicle adoption. *Energy Policy*, **68(0)**, 183–194, doi:10.1016/j.enpol.2014.01.043.

Simell, P. et al., 2014: Clean syngas from biomass – process development and concept assessment. *Biomass Convers. Biorefinery*, **4**, 357–370, doi:10.1007/s13399-014-0121-y.

Simićević, J., V. Molan, and N. Milosavljević, 2020: Informal park-and-ride behaviour. *Put i saobraćaj*, **66(1)**, 9–13, doi:10.31075/PIS.66.01.02.

Sims R., R. Schaeffer, F. Creutzig, X. Cruz-Núñez, M. D'Agosto, D. Dimitriu, M.J. Figueroa Meza, L. Fulton, S. Kobayashi, O. Lah, A. McKinnon, P. Newman, M. Ouyang, J.J. Schauer, D. Sperling, and G. Tiwari, 2014: Transport. In: *Climate Change 2014: Mitigation of Climate Change. Contribution of Working Group III to the Fifth Assessment Report of the Intergovernmental Panel on Climate Change* [Edenhofer, O., R. Pichs-Madruga, Y. Sokona, E. Farahani, S. Kadner, K. Seyboth, A. Adler, I. Baum, S. Brunner, P. Eickemeier, B. Kriemann, J. Savolainen, S. Schlömer, C. von Stechow, T. Zwickel and J.C. Minx (eds.)]. Cambridge University Press, Cambridge, UK and New York, NY, USA.

Singh, V., I. Dincer, and M.A. Rosen, 2018: Life Cycle Assessment of Ammonia Production Methods. In: *Exergetic, Energetic and Environmental Dimensions* [Dincer, I., C.O. Colpan, and O.B.T.-E. Kizilkan (eds.)], Elsevier, pp. 935–959, doi:10.1016/B978-0-12-813734-5.00053-6.

Skeer, J., F. Boshell, and M. Ayuso, 2016: Technology Innovation Outlook for Advanced Liquid Biofuels in Transport. *ACS Energy Lett.*, **1**, 724–725, doi:10.1021/acsenergylett.6b00290.

Skeete, J.-P., 2018: Level 5 autonomy: The new face of disruption in road transport. *Technol. Forecast. Soc. Change*, **134**, 22–34, doi:10.1016/j.techfore.2018.05.003.

Skjølsvold, T.M. and M. Ryghaug, 2020: Temporal echoes and cross-geography policy effects: Multiple levels of transition governance and the electric vehicle breakthrough. *Environ. Innov. Soc. Transitions*, **35**, 232–240, doi:10.1016/j.eist.2019.06.004.

SLoCaT, 2018a: *Transport and Climate Change: Global Status Report 2018*. SLoCaT, 184 pp., http://slocat.net/tcc-gsr.

SLoCaT, 2018b: *Sustainable Development Goals & Transport*. SLoCaT, http://www.slocat.net/sdgs-transport.

SLoCaT, 2019: *Sustainable Transport: A Critical Driver to Achieve the Sustainable Development Goals An analysis of 2016–2019 Voluntary National Reviews*. SLoCaT, 44 pp., www.slocat.net/vnr.

SLoCaT, 2021: *SLOCAT Transport and Climate Change Global Status Report Tracking Trends in a Time of Change: The Need for Radical Action Towards Sustainable Transport Decarbonisation*. Transport and Climate Change Global Status Report, 2nd Edition, SLoCaT, 365 pp., www.tcc-gsr.com.

Smith, L.C. and S.R. Stephenson, 2013: New Trans-Arctic shipping routes navigable by midcentury. *Proc. Natl. Acad. Sci.*, **110(13)**, E1191–5, doi:10.1073/pnas.1214212110.

Smith, M.S. et al., 2018: Advancing sustainability science for the SDGs. *Sustain. Sci.*, **13(6)**, 1483–1487, doi:10.1007/s11625-018-0645-3.

Smith Stegen, K., 2015: Heavy rare earths, permanent magnets, and renewable energies: An imminent crisis. *Energy Policy*, **79**, 1–8, doi:10.1016/j.enpol.2014.12.015.

Smith, T.W.P. et al., 2014: *Third International Maritime Organization Green House Gas study*. International Maritime Organization, London, UK.

Sofiev, M. et al., 2018: Cleaner fuels for ships provide public health benefits with climate tradeoffs. *Nat. Commun.*, **9(1)**, 406, doi:10.1038/s41467-017-02774-9.

Song, F., S. Hess, and T. Dekker, 2018: Accounting for the impact of variety-seeking: Theory and application to HSR-air intermodality in China. *J. Air Transp. Manag.*, **69**, 99–111, doi:10.1016/j.jairtraman.2018.02.008.

Song, H., X. Ou, J. Yuan, M. Yu, and C. Wang, 2017: Energy consumption and greenhouse gas emissions of diesel/LNG heavy-duty vehicle fleets in China based on a bottom-up model analysis. *Energy*, **140**, 966–978, doi:10.1016/j.energy.2017.09.011.

Sonnleitner, J., M. Friedrich, and E. Richter, 2021: Impacts of highly automated vehicles on travel demand: macroscopic modeling methods and some results. *Transportation* (2021), 1–24, doi:10.1007/s11116-021-10199-z.

Sovacool, B.K., L. Noel, J. Axsen, and W. Kempton, 2018: The neglected social dimensions to a vehicle-to-grid (V2G) transition: a critical and systematic review. *Environ. Res. Lett.*, **13(1)**, 13001, doi:10.1088/1748-9326/aa9c6d.

Sovacool, B.K. et al., 2020: Sustainable minerals and metals for a low-carbon future. *Science*, **367(6473)**, 30–33, doi:10.1126/science.aaz6003.

Speck, J., 2018: *Walkable City Rules*. Island Press/Center for Resource Economics, Washington, DC, USA.

Sprei, F., 2018: Disrupting mobility. *Energy Res. Soc. Sci.*, **37**, 238–242, doi:10.1016/j.erss.2017.10.029.

Szopa, S., V. Naik, B. Adhikary, P. Artaxo, T. Berntsen, W.D. Collins, S. Fuzzi, L. Gallardo, A. Kiendler Scharr, Z. Klimont, H. Liao, N. Unger, and P. Zanis, 2021: Short-Lived Climate Forcers. In *Climate Change 2021: The Physical Science Basis. Contribution of Working Group I to the Sixth Assessment Report of the Intergovernmental Panel on Climate Change* [Masson-Delmotte, V., P. Zhai, A. Pirani, S.L. Connors, C. Péan, S. Berger, N. Caud, Y. Chen, L. Goldfarb, M.I. Gomis, M. Huang, K. Leitzell, E. Lonnoy, J.B.R. Matthews, T.K. Maycock, T. Waterfield, O. Yelekçi, R. Yu, and B. Zhou (eds.)]. Cambridge University Press, Cambridge, UK and New York, NY, USA. In Press.

Staffell, I. et al., 2019: The role of hydrogen and fuel cells in the global energy system. *Energy Environ. Sci.*, **12(2)**, 463–491, doi:10.1039/c8ee01157e.

Staples, M.D., R. Malina, and S.R.H. Barrett, 2017: The limits of bioenergy for mitigating global life-cycle greenhouse gas emissions from fossil fuels. *Nat. Energy*, **2(2)**, 16202, doi:10.1038/nenergy.2016.202.

Staples, M.D., R. Malina, P. Suresh, J.I. Hileman, and S.R.H. Barrett, 2018: Aviation $CO_2$ emissions reductions from the use of alternative jet fuels. *Energy Policy*, **114**, 342–354, doi:10.1016/j.enpol.2017.12.007.

Steg, L., 2005: Car use: Lust and must. Instrumental, symbolic and affective motives for car use. *Transp. Res. Part A Policy Pract.*, **39(2–3)**, special issue, 147–162, doi:10.1016/j.tra.2004.07.001.

Steg, L., G. Perlaviciute, and E. van der Werff, 2015: Understanding the human dimensions of a sustainable energy transition. *Front. Psychol.*, **6**(June), 1–17, doi:10.3389/fpsyg.2015.00805.

Stephenson, S.R. et al., 2018: Climatic Responses to Future Trans-Arctic Shipping. *Geophys. Res. Lett.*, **45(18)**, 9898–9908, doi:10.1029/2018GL078969.

Stolaroff, J.K. et al., 2018: Energy use and life cycle greenhouse gas emissions of drones for commercial package delivery. *Nat. Commun.*, **9(1)**, 409, doi:10.1038/s41467-017-02411-5.

Stolz, B., M. Held, G. Georges, and K. Boulouchos, 2021: The $CO_2$ reduction potential of shore-side electricity in Europe. *Appl. Energy*, **285**, 116425, doi:10.1016/j.apenergy.2020.116425.

Strale, M., 2016: High-speed rail for freight: Potential developments and impacts on urban dynamics. *Open Transp. J.*, **10(1)**, doi:10.2174/1874447801610010057.

Ström, L. and K. Gierens, 2002: First simulations of cryoplane contrails. *J. Geophys. Res. Atmos.*, **107**(D18), AAC 2-1, doi:10.1029/2001JD000838.

Suel, E. and J.W. Polak, 2018: Incorporating online shopping into travel demand modelling: challenges, progress, and opportunities. *Transp. Rev.*, **38(5)**, 576–601, doi:10.1080/01441647.2017.1381864.

Sumaila, U.R. et al., 2021: Financing a sustainable ocean economy. *Nat. Commun. 2021 121*, **12(1)**, 1–11, doi:10.1038/s41467-021-23168-y.

Tafidis, P. et al., 2017: Exploring the impact of ICT on urban mobility in heterogenic regions. *Transp. Res. Procedia*, **27**, 309–316, doi:10.1016/j.trpro.2017.12.030.

Taiebat, M. and M. Xu, 2019: Synergies of four emerging technologies for accelerated adoption of electric vehicles: Shared mobility, wireless charging, vehicle-to-grid, and vehicle automation. *J. Clean. Prod.*, **230**, 794–797, doi:10.1016/j.jclepro.2019.05.142.

Takahashi, S., T. Uemura, S. Kon, and H. Nebashi, 2017: Effect evaluation of electric double-layer capacitor (EDLC) for regenerative braking energy utilization in sinjuku line, Seibu railway. IECON 2017 – 43rd Annual Conference of the IEEE Industrial Electronics Society, IEEE, 3881–3883.

Tanco, M., L. Cat, and S. Garat, 2019: A break-even analysis for battery electric trucks in Latin America. *J. Clean. Prod.*, **228**, 1354–1367, doi:10.1016/J.JCLEPRO.2019.04.168.

Tanzer, S.E., J. Posada, S. Geraedts, and A. Ramírez, 2019: Lignocellulosic marine biofuel: Technoeconomic and environmental assessment for production in Brazil and Sweden. *J. Clean. Prod.*, **239**, 117845, doi:10.1016/j.jclepro.2019.117845.

Taptich, M.N., A. Horvath, and M.V. Chester, 2016: Worldwide Greenhouse Gas Reduction Potentials in Transportation by 2050. *J. Ind. Ecol.*, **20(2)**, 329–340, doi:10.1111/jiec.12391.

Teeter, J.L. and S.A. Cleary, 2014: Decentralized oceans: Sail-solar shipping for sustainable development in SIDS. *Nat. Resour. Forum*, **38(3)**, 182–192, doi:10.1111/1477-8947.12048.

Teng, F. et al., 2015: *Pathways to deep decarbonization in China*. Sustainable Development Solutions Network (SDSN) and Institute for Sustainable Development and International Relations (IDDRI), 44 pp.

Teoh, R., U. Schumann, A. Majumdar, and M.E.J. Stettler, 2020: Mitigating the Climate Forcing of Aircraft Contrails by Small-Scale Diversions and Technology Adoption. *Environ. Sci. Technol.*, **54(5)**, 2941–2950, doi:10.1021/ACS.EST.9B05608.

Thaduri, A., D. Galar, and U. Kumar, 2020: Space weather climate impacts on railway infrastructure. *Int. J. Syst. Assur. Eng. Manag.*, **11**, 267–281, doi:10.1007/S13198-020-01003-9.

The Swiss Parliament, 2020: *Un oui franc à la taxe sur les billets d'avion*. (Press release, 15 January 2020).

Thomson, G. and P. Newman, 2016: Geoengineering in the Anthropocene through Regenerative Urbanism. *Geosciences*, **6(4)**, 46, doi:10.3390/geosciences6040046.

Thomson, G. and P. Newman, 2018: Urban fabrics and urban metabolism – from sustainable to regenerative cities. *Resour. Conserv. Recycl.*, **132**, 218–229, doi:10.1016/j.resconrec.2017.01.010.

Thornbush, M., O. Golubchikov, and S. Bouzarovski, 2013: Sustainable cities targeted by combined mitigation–adaptation efforts for future-proofing. *Sustain. Cities Soc.*, **9**, 1–9, doi:10.1016/j.scs.2013.01.003.

Tian, X.L., C.J. An, and Z.K. Chen, 2020: Assessing the Impact of Urban Form on the Greenhouse Gas Emissions from Household Vehicles: A Review. *J. Environ. Informatics Lett.*, doi:10.3808/jeil.202000029.

Tikoudis, I. et al., 2021: Ridesharing services and urban transport $CO_2$ emissions: Simulation-based evidence from 247 cities. *Transp. Res. Part D Transp. Environ.*, **97**, 102923, doi:10.1016/j.trd.2021.102923.

Tillig, F., W. Mao, and J.W. Ringsberg, 2015: *Systems modelling for energy-efficient shipping*. Chalmers University of Technology, Gothenburg, Sweden, 40 pp.

Timmers, V.R.J.H. and P.A.J. Achten, 2016: Non-exhaust PM emissions from electric vehicles. *Atmos. Environ.*, **134**, 10–17, doi:10.1016/j.atmosenv.2016.03.017.

Tirachini, A. and O. Cats, 2020: COVID-19 and Public Transportation: Current Assessment, Prospects, and Research Needs. *J. Public Transp.*, **22(1)**, 1, doi:10.5038/2375-0901.22.1.1.

Tirachini, A. and A. Gomez-Lobo, 2020: Does ride-hailing increase or decrease vehicle kilometers traveled (VKT)? A simulation approach for Santiago de Chile. *Int. J. Sustain. Transp.*, **14(3)**, 187–204, doi:10.1080/15568318.2018.1539146.

Tokimatsu, K. et al., 2016: Role of innovative technologies under the global zero emissions scenarios. *Appl. Energy*, **162**, 1483–1493, doi:10.1016/j.apenergy.2015.02.051.

Tong, F., P. Jaramillo, and I.M.L. Azevedo, 2015a: Comparison of Life Cycle Greenhouse Gases from Natural Gas Pathways for Medium and Heavy-Duty Vehicles. *Environ. Sci. Technol.*, **49(12)**, 7123–7133, doi:10.1021/es5052759.

Tong, F., P. Jaramillo, and I.M.L. Azevedo, 2015b: Comparison of Life Cycle Greenhouse Gases from Natural Gas Pathways for Light-Duty Vehicles. *Energy and Fuels*, **29**, 6008–6018, doi:10.1021/acs.energyfuels.5b01063.

Toole, J.L. et al., 2015: The path most traveled: Travel demand estimation using big data resources. *Transp. Res. Part C Emerg. Technol.*, **58**, 162–177, doi:10.1016/j.trc.2015.04.022.

Transport and Environment, 2018: *CNG and LNG for vehicles and ships – the facts*. Transport & Environment, Brussels, Belgium, 72 pp.

Traut, M. et al., 2014: Propulsive power contribution of a kite and a Flettner rotor on selected shipping routes. *Appl. Energy*, **113**, 362–372, doi:10.1016/j.apenergy.2013.07.026.

Traut, M. et al., 2018: $CO_2$ abatement goals for international shipping. *Clim. Policy*, **18(8)**, 1066–1075, doi:10.1080/14693062.2018.1461059.

Triantafyllopoulos, G. et al., 2019: A study on the $CO_2$ and $NO_x$ emissions performance of Euro 6 diesel vehicles under various chassis dynamometer and on-road conditions including latest regulatory provisions. *Sci. Total Environ.*, **666**(x), 337–346, doi:10.1016/j.scitotenv.2019.02.144.

Trivedi, S., R. Prasad, A. Mishra, A. Kalam, and P. Yadav, 2020: Current scenario of CNG vehicular pollution and their possible abatement technologies: an overview. *Environ. Sci. Pollut. Res.*, **27(32)**, 39977–40000, doi:10.1007/s11356-020-10361-7.

Tsiakmakis, S. et al., 2017: *From NEDC to WLTP: effect on the type-approval $CO_2$ emissions of light-duty vehicles*. Publications Office of the European Union, Luxembourg, 50 pp.

UK DoT, 2019: Transport, health, and wellbeing.

Ünal, A.B., L. Steg, and J. Granskaya, 2019: "To support or not to support, that is the question". Testing the VBN theory in predicting support for car use reduction policies in Russia. *Transp. Res. Part A Policy Pract.*, **119**(October 2018), 73–81, doi:10.1016/j.tra.2018.10.042.

UNCRD, 2020: *Environmentally Sustainable Transport (EST) Initiative and Programs in Asia*. United Nations Centre for Regional Development, Nagoya, Japan, 27 pp., https://www.unescap.org/sites/default/files/3b.3_AsianEnvironSustainTransportInitiative_CRC%20Mohanty.pdf.

UNCTAD, 2018: *50 Years of Review of Maritime Transport, 1968–2018: Reflecting on the past, exploring the future*. United Nations Conference on Trade and Development, Geneva, Switzerland, 97 pp.

UNCTAD, 2019: *Review of maritime transport 2019*. United Nations Conference on Trade and Development, New York, 132 pp., https://unctad.org/system/files/official-document/rmt2019_en.pdf.

Underwood, B.S., Z. Guido, P. Gudipudi, and Y. Feinberg, 2017: Increased costs to US pavement infrastructure from future temperature rise. *Nat. Clim. Change*, **7(10)**, 704–707, doi:10.1038/nclimate3390.

UNECE, 2018: *Road Safety Performance Review Uganda*. United Nations Economic Commission for Africa and United Nations Economic Commission for Europe Information Service, Geneva, Switzerland.

UNEP, 2011: *Decoupling Natural Resource Use and Environmental Impacts from Economic Growth*. United Nations Environment Programme, Nairobi, Kenya, 58 pp.

UNEP, 2013: *City-level decoupling: urban resource flows and the governance of infrastructure transitions.* United Nations Environment Programme, Nairobi, Kenya.

UNEP, 2020: *Emissions Gap Report 2020*. United Nations Environment Programme, Nairobi, Kenya, 128 pp.

United Nations, 2015: *Transforming our world: the 2030 Agenda for Sustainable Development*. United Nations, New York, 35 pp., https://www.un.org/ga/search/view_doc.asp?symbol=A/RES/70/1&Lang=E.

United Nations, 2017: *Work of the Statistical Commission pertaining to the 2030 Agenda for Sustainable Development*. United Nations, New York, 25 pp., https://ggim.un.org/documents/a_res_71_313.pdf.

Upadhyayula, V.K.K., A.G. Parvatker, A. Baroth, and K. Shanmugam, 2019: Lightweighting and electrification strategies for improving environmental performance of passenger cars in India by 2030: A critical perspective based on life cycle assessment. *J. Clean. Prod.*, **209**, 1604–1613, doi:10.1016/j.jclepro.2018.11.153.

Ürge-Vorsatz, D. et al., 2018: Locking in positive climate responses in cities. *Nat. Clim. Change*, **8(3)**, 174–177, doi:10.1038/s41558-018-0100-6.


10

US Energy Information Administration, *2021: Annual Energy Outlook*. US Energy Information Administration, Washington, DC, USA, https://www.eia.gov/outlooks/aeo/.

US Geological Survey, 2018: USGS Critical Minerals Annual Review. *Mining Engineering*, **71(5)**, 35–47.

Ushakov, S., D. Stenersen, and P.M. Einang, 2019: Methane slip from gas fuelled ships: a comprehensive summary based on measurement data. *J. Mar. Sci. Technol.*, **24(4)**, 1308–1325, doi:10.1007/s00773-018-00622-z.

Vahidi, A. and A. Sciarretta, 2018: Energy saving potentials of connected and automated vehicles. *Transp. Res. Part C Emerg. Technol.*, **95**, 822–843, doi:10.1016/j.trc.2018.09.001.

Valdes, V., 2015: Determinants of air travel demand in Middle Income Countries. *J. Air Transp. Manag.*, **42**, 75–84, doi:10.1016/J.JAIRTRAMAN.2014.09.002.

Valente, A., D. Iribarren, and J. Dufour, 2017: Harmonised life-cycle global warming impact of renewable hydrogen. *J. Clean. Prod.*, **149(15)**, 762–772, doi:10.1016/j.jclepro.2017.02.163.

Valente, A., D. Iribarren, and J. Dufour, 2021: Harmonised carbon and energy footprints of fossil hydrogen. *Int. J. Hydrogen Energy*, **46(33)**, 17587–17594, doi:10.1016/j.ijhydene.2020.03.074.

Valero, A., A. Valero, G. Calvo, and A. Ortego, 2018: Material bottlenecks in the future development of green technologies. *Renew. Sustain. Energy Rev.*, **93**, 178–200, doi:10.1016/j.rser.2018.05.041.

van der Kam, M. and R. Bekkers, 2020: *Comparative analysis of standardized protocols for EV roaming*. Netherlands Knowledge Platform for Public Charging Infrastructure, Utrecht, Netherlands, 42 pp.

van Loon, P., A.C. McKinnon, L. Deketele, and J. Dewaele, 2014: The growth of online retailing: a review of its carbon impacts. *Carbon Manag.*, **5(3)**, 285–292, doi:10.1080/17583004.2014.982395.

van Ruijven, B.J., E. De Cian, and I. Sue Wing, 2019: Amplification of future energy demand growth due to climate change. *Nat. Commun.*, **10(1)**, 1–12, doi:10.1038/s41467-019-10399-3.

van Vuuren, D.P. et al., 2018: Alternative pathways to the 1.5°C target reduce the need for negative emission technologies. *Nat. Clim. Change*, **8(5)**, 391–397, doi:10.1038/s41558-018-0119-8.

Vasic, A.-M. and M. Weilenmann, 2006: Comparison of Real-World Emissions from Two-Wheelers and Passenger Cars. *Environ. Sci. Technol.*, **40(1)**, doi:10.1021/es0481023.

Vecchio, G., I. Tiznado-Aitken, and R. Hurtubia, 2020: Transport and equity in Latin America: a critical review of socially oriented accessibility assessments. *Transp. Rev.*, **40(3)**, 354–381, doi:10.1080/01441647.2020.1711828.

Verstraete, D., 2013: Long range transport aircraft using hydrogen fuel. *Int. J. Hydrogen Energy*, **38(34)**, 14824–14831, doi:10.1016/j.ijhydene.2013.09.021.

Viana, M. et al., 2014: Impact of maritime transport emissions on coastal air quality in Europe. *Atmos. Environ.*, **90**, 96–105, doi:10.1016/j.atmosenv.2014.03.046.

Viana, M. et al., 2015: Environmental and health benefits from designating the Marmara Sea and the Turkish Straits as an Emission Control Area (ECA). *Environ. Sci. Technol.*, **49(6)**, 3304–3313, doi:10.1021/es5049946.

Viktorelius, M. and M. Lundh, 2019: Energy efficiency at sea: An activity theoretical perspective on operational energy efficiency in maritime transport. *Energy Res. Soc. Sci.*, **52**, 1–9, doi:10.1016/j.erss.2019.01.021.

Virdis, M.R. et al., 2015: *Pathways to deep decarbonization in Italy*. Sustainable Development Solutions Network (SDSN) and Institute for Sustainable Development and International Relations (IDDRI), 71 pp.

Wang, M. et al., 2019: Summary of Expansions and Updates in GREET 2019. doi:10.2172/1569562.

Wang, T. et al., 2020: Impact analysis of climate change on rail systems for adaptation planning: A UK case. *Transp. Res. Part D Transp. Environ.*, **83**, 102324, doi:10.1016/J.TRD.2020.102324.

Wang, T., B. Shen, C. Han Springer, and J. Hou, 2021: What prevents us from taking low-carbon actions? A comprehensive review of influencing factors affecting low-carbon behaviors. *Energy Res. Soc. Sci.*, **71**, 101844, doi:10.1016/j.erss.2020.101844.

Ward, J.W., J.J. Michalek, and C. Samaras, 2021: Air Pollution, Greenhouse Gas, and Traffic Externality Benefits and Costs of Shifting Private Vehicle Travel to Ridesourcing Services. *Environ. Sci. Technol.*, **55**, 13174–13185, doi:10.1021/ACS.EST.1C01641.

Wardman, M., J. Toner, N. Fearnley, S. Flügel, and M. Killi, 2018: Review and meta-analysis of inter-modal cross-elasticity evidence. *Transp. Res. Part A Policy Pract.*, **118**, 662–681, doi:10.1016/j.tra.2018.10.002.

Warnecke, C., L. Schneider, T. Day, S. La Hoz Theuer, and H. Fearnehough, 2019: Robust eligibility criteria essential for new global scheme to offset aviation emissions. *Nat. Clim. Change 2019 93*, **9(3)**, 218–221, doi:10.1038/s41558-019-0415-y.

Watson, I., A. Ali, and A. Bayyati, 2019: Freight transport using high-speed railways. *Int. J. Transp. Dev. Integr.*, **3(2)**, 103–116, doi:10.2495/TDI-V3-N2-103-116.

WBA, 2019: *Climate and Energy Benchmark*. World Benchmarking Alliance, Amsterdam, https://www.worldbenchmarkingalliance.org/climate-and-energy-benchmark/.

Webb, J., J. Whitehead, and C. Wilson, 2019: Who Will Fuel Your Electric Vehicle in the Future? You or Your Utility? In: *Consumer, Prosumer, Prosumager* [Sioshansi, F., (ed.)], Elsevier, pp. 407–429, doi: 10.1016/B978-0-12-816835-6.00018-8.

Weis, A., P. Jaramillo, and J. Michalek, 2014: Estimating the potential of controlled plug-in hybrid electric vehicle charging to reduce operational and capacity expansion costs for electric power systems with high wind penetration. *Appl. Energy*, **115**, 190–204, doi:10.1016/j.apenergy.2013.10.017.

Weiss, D.J. et al., 2018: A global map of travel time to cities to assess inequalities in accessibility in 2015. *Nature*, **553(7688)**, 333–336, doi:10.1038/nature25181.

Wenig, J., M. Sodenkamp, and T. Staake, 2019: Battery versus infrastructure: Tradeoffs between battery capacity and charging infrastructure for plug-in hybrid electric vehicles. *Appl. Energy*, **255**, 113787, doi:10.1016/J.APENERGY.2019.113787.

Wentland, A., 2016: Imagining and enacting the future of the German energy transition: electric vehicles as grid infrastructure. *Innov. Eur. J. Soc. Sci. Res.*, **29(3)**, 285–302, doi:10.1080/13511610.2016.1159946.

Weymouth, R. and J. Hartz-Karp, 2015: Deliberative collaborative governance as a democratic reform to resolve wicked problems and improve trust. *J. Econ. Soc. Policy*, **17(1)**, 62–95.

Whiston, M.M. et al., 2019: Expert assessments of the cost and expected future performance of proton exchange membrane fuel cells for vehicles. *Proc. Natl. Acad. Sci.*, **116(11)**, 4899–4904, doi:10.1073/pnas.1804221116.

White, L.V. and N.D. Sintov, 2017: You are what you drive: Environmentalist and social innovator symbolism drives electric vehicle adoption intentions. *Transp. Res. Part A Policy Pract.*, **99**, 94–113, doi:10.1016/j.tra.2017.03.008.

Whitmarsh, L., S. Capstick, and N. Nash, 2017: Who is reducing their material consumption and why? A cross-cultural analysis of dematerialization behaviours. *Philos. Trans. R. Soc. A Math. Phys. Eng. Sci.*, **375(2095)**, doi:10.1098/rsta.2016.0376.

Williams, J.H., B. Haley, F. Kahrl, and J. Moore, 2015: *Pathways to Deep Decarbonization in the United States*. Sustainable Development Solutions Network (SDSN) and Institute for Sustainable Development and International Relations (IDDRI), 118 pp.

Williams, K. (ed.), 2017: *Spatial Planning, Urban Form and Sustainable Transport*. Routledge, London, UK.

Wilson, C., H. Pettifor, E. Cassar, L. Kerr, and M. Wilson, 2019: The potential contribution of disruptive low-carbon innovations to 1.5°C climate mitigation. *Energy Effic.*, **12(2)**, 423–440, doi:10.1007/s12053-018-9679-8.

Winebrake, J.J., J.J. Corbett, F. Umar, and D. Yuska, 2019: Pollution Tradeoffs for Conventional and Natural Gas-Based Marine Fuels. *Sustainability*, **11(8)**, 2235, doi:10.3390/su11082235.

Winkel, R., U. Weddige, D. Johnsen, V. Hoen, and S. Papaefthimiou, 2016: Shore Side Electricity in Europe: Potential and environmental benefits. *Energy Policy*, **88**, 584–593, doi:10.1016/j.enpol.2015.07.013.

Winslow, K.M.K. M., S.J.S.J. Laux and T.G.T.G. Townsend, 2018: A review on the growing concern and potential management strategies of waste lithium-ion batteries. *Resour. Conserv. Recycl.*, **129**, 263–277, doi:10.1016/j.resconrec.2017.11.001.

Winther, M. et al., 2014: Emission inventories for ships in the arctic based on satellite sampled AIS data. *Atmos. Environ.*, **91**, 1–14, doi:10.1016/j.atmosenv.2014.03.006.

Wise, M., M. Muratori, and P. Kyle, 2017: Biojet fuels and emissions mitigation in aviation: An integrated assessment modeling analysis. *Transp. Res. Part D Transp. Environ.*, **52**, 244–253, doi:10.1016/j.trd.2017.03.006.

Wolfram, P. and T. Wiedmann, 2017: Electrifying Australian transport: Hybrid life cycle analysis of a transition to electric light-duty vehicles and renewable electricity. *Appl. Energy*, **206** (August), 531–540, doi:10.1016/j.apenergy.2017.08.219.

Wolfram, P., Q. Tu, N. Heeren, S. Pauliuk, and E.G. Hertwich, 2021: Material efficiency and climate change mitigation of passenger vehicles. *J. Ind. Ecol.*, **25(2)**, 494–510, doi:10.1111/jiec.13067.

World Bank, 2017: *The Growing Role of Minerals and Metals for a Low Carbon Future*. World Bank, Washington, DC, USA, 112 pp.

World Health Organization, 2018: *Ambient (outdoor) air pollution*. World Health Organization,Geneva, Switzerland https://www.who.int/news-room/fact-sheets/detail/ambient-(outdoor)-air-quality-and-health.

Wu, D. et al., 2019: Regional Heterogeneity in the Emissions Benefits of Electrified and Lightweighted Light-Duty Vehicles. *Environ. Sci. Technol.*, **53(18)**, 10560–10570, doi:10.1021/acs.est.9b00648.

Wu, S., R. Xiong, H. Li, V. Nian, and S. Ma, 2020: The state of the art on preheating lithium-ion batteries in cold weather. *J. Energy Storage*, **27**, doi:10.1016/j.est.2019.101059.

Xia, W. and A. Zhang, 2016: High-speed rail and air transport competition and cooperation: A vertical differentiation approach. *Transp. Res. Part B Methodol.*, **94**, 456–481, doi:10.1016/j.trb.2016.10.006.

Xie, R., L. Huang, B. Tian, and J. Fang, 2019: Differences in Changes in Carbon Dioxide Emissions among China's Transportation Subsectors: A Structural Decomposition Analysis. *Emerg. Mark. Financ. Trade*, **55(6)**, 1294–1311, doi:10.1080/1540496X.2018.1526076.

Xing, H., S. Spence, and H. Chen, 2020: A comprehensive review on countermeasures for $CO_2$ emissions from ships. *Renew. Sustain. Energy Rev.*, **134**, 110222, doi:10.1016/j.rser.2020.110222.

Xu, J., S.A. Ning, G. Bower, and I. Kroo, 2014: Aircraft Route Optimization for Formation Flight. *J. Aircr.*, **51(2)**, 490–501, doi:10.2514/1.C032154.

Xu, J., J. Bai, and J. Chen, 2019: An Improved Indicator System for Evaluating the Progress of Sustainable Development Goals (SDGs) Sub-Target 9.1 in County Level. *Sustainability*, **11(17)**, 4783, doi:10.3390/su11174783.

Yaïci, W. and H. Ribberink, 2021: Feasibility Study of Medium- and Heavy-Duty Compressed Renewable/Natural Gas Vehicles in Canada. *J. Energy Resour. Technol.*, **143(9)**, doi:10.1115/ES2020-1617.

Yan, W. et al., 2020: Rethinking Chinese supply resilience of critical metals in lithium-ion batteries. *J. Clean. Prod.*, **256**, 120719, doi:10.1016/j.jclepro.2020.120719.

Yang, H., M. Dijst, J. Feng, and D. Ettema, 2019: Mode choice in access and egress stages of high-speed railway travelers in China. *J. Transp. Land Use*, **12(1)**, 701–721, doi:10.5198/jtlu.2019.1420.

Yang, J., A.-O. Purevjav, and S. Li, 2020: The Marginal Cost of Traffic Congestion and Road Pricing: Evidence from a Natural Experiment in Beijing. *Am. Econ. J. Econ. Policy*, **12(1)**, 418–453, doi:10.1257/pol.20170195.

Yang, L., C. Hao, and Y. Chai, 2018a: Life Cycle Assessment of Commercial Delivery Trucks: Diesel, Plug-In Electric, and Battery-Swap Electric. *Sustainability.* **10(12)**, 4547, doi:10.3390/SU10124547.

Yang, Z. et al., 2018b: Risk and cost evaluation of port adaptation measures to climate change impacts. *Transp. Res. Part D Transp. Environ.*, **61**, 444–458, doi:10.1016/j.trd.2017.03.004.

Yeh, S. et al., 2017: Detailed assessment of global transport-energy models' structures and projections. *Transp. Res. Part D Transp. Environ.*, **55**, 294–309, doi:10.1016/j.trd.2016.11.001.

Yin, W., 2019: Integrating Sustainable Development Goals into the Belt and Road Initiative: Would It Be a New Model for Green and Sustainable Investment? *Sustainability*, **11(24)**, 6991, doi:10.3390/su11246991.

You, Y. and A. Manthiram, 2018: Progress in High-Voltage Cathode Materials for Rechargeable Sodium-Ion Batteries. *Adv. Energy Mater.*, **8(2)**, 1701785, doi:10.1002/aenm.201701785.

Yuan, M., J.Z. Thellufsen, H. Lund, and Y. Liang, 2021: The electrification of transportation in energy transition. *Energy*, **236**, doi:10.1016/j.energy.2021.121564.

Yue, Y. et al., 2015: Life cycle assessment of High Speed Rail in China. *Transp. Res. Part D Transp. Environ.*, **41**, 367–376, doi:10.1016/J.TRD.2015.10.005.

Yugo, M. and A. Soler, 2019: A look into the role of e-fuels in the transport system in Europe (2030–2050) (literature review). *Concawe Rev.*, **28(1)**, 4–22.

Zahabi, S.A.H., A. Chang, L.F. Miranda-Moreno, and Z. Patterson, 2016: Exploring the link between the neighborhood typologies, bicycle infrastructure and commuting cycling over time and the potential impact on commuter GHG emissions. *Transp. Res. Part D Transp. Environ.*, **47**, 89–103, doi:10.1016/J.TRD.2016.05.008.

Zeng, Y., B. Ran, N. Zhang, and X. Yang, 2021: Estimating the Price Elasticity of Train Travel Demand and Its Variation Rules and Application in Energy Used and $CO_2$ Emissions. *Sustainability. 2021, Vol. 13, Page 475*, **13(2)**, 475, doi:10.3390/SU13020475.

Zhang, A., Y. Wan, and H. Yang, 2019a: Impacts of high-speed rail on airlines, airports and regional economies: A survey of recent research. *Transp. Policy*, **81**, A1–A19, doi:10.1016/J.TRANPOL.2019.06.010.

Zhang, H., W. Chen, and W. Huang, 2016a: TIMES modelling of transport sector in China and USA: Comparisons from a decarbonization perspective. *Appl. Energy*, **162**, 1505–1514, doi:10.1016/j.apenergy.2015.08.124.

Zhang, R., S. Fujimori, H. Dai, and T. Hanaoka, 2018: Contribution of the transport sector to climate change mitigation: Insights from a global passenger transport model coupled with a computable general equilibrium model. *Appl. Energy*, **211**(November 2017), 76–88, doi:10.1016/j.apenergy.2017.10.103.

Zhang, R., D. Johnson, W. Zhao, and C. Nash, 2019b: Competition of airline and high-speed rail in terms of price and frequency: Empirical study from China. *Transp. Policy*, **78**, 8–18, doi:10.1016/j.tranpol.2019.03.008.

Zhang, X., J. Xie, R. Rao, and Y. Liang, 2014: Policy Incentives for the Adoption of Electric Vehicles across Countries. *Sustainability*, **6(11)**, 8056–8078, doi:10.3390/su6118056.

Zhang, Y.-S., E.-J. Yao, and X. Sun, 2017: Impact analysis of HSR fare discount strategy on HSR share in Beijing–Shanghai transportation corridor. *Transp. Lett.*, **9(4)**, 215–227, doi:10.1080/19427867.2016.1253635.

Zhang, Y. and A. Zhang, 2016: Determinants of air passenger flows in China and gravity model: Deregulation, LCCs, and high-speed rail. *J. Transp. Econ. Policy*, **50(3)**, 287–303, doi:10.2139/ssrn.2775501.

Zhang, Y., Q. Meng, and L. Zhang, 2016b: Is the Northern Sea Route attractive to shipping companies? Some insights from recent ship traffic data. *Mar. Policy*, **73**, 53–60, doi:10.1016/j.marpol.2016.07.030.

Zhao, L. et al., 2015: Techno-economic analysis of bioethanol production from lignocellulosic biomass in China: Dilute-acid pretreatment and enzymatic hydrolysis of corn stover. *Energies*, **8(5)**, doi:10.3390/en8054096.

Zhao, M. et al., 2013: Characteristics and ship traffic source identification of air pollutants in China's largest port. *Atmos. Environ.*, **64**, 277–286, doi:10.1016/J.ATMOSENV.2012.10.007.

Zhao, X., F. Taheripour, R. Malina, M.D. Staples, and W.E. Tyner, 2021: Estimating induced land use change emissions for sustainable aviation biofuel pathways. *Sci. Total Environ.*, **779**, 146238, doi:10.1016/j.scitotenv.2021.146238.

Zhao, Y., N.C. Onat, M. Kucukvar, and O. Tatari, 2016: Carbon and energy footprints of electric delivery trucks: A hybrid multi-regional input-output life cycle assessment. *Transp. Res. Part D Transp. Environ.*, **47**, 195–207, doi:10.1016/j.trd.2016.05.014.

Zhao, Y. et al., 2019: An Efficient Direct Ammonia Fuel Cell for Affordable Carbon-Neutral Transportation. *Joule*, **3(10)**, 2472–2484, doi:10.1016/j.joule.2019.07.005.

Zheng, Y., et al., 2020: Well-to-wheels greenhouse gas and air pollutant emissions from battery electric vehicles in China. *Mitigation and Adaptation Strategies for Global Change*, **25**, 355–370, doi:10.1007/s11027-019-09890-5.

Zhou, C. and J.E. Penner, 2014: Aircraft soot indirect effect on large-scale cirrus clouds: Is the indirect forcing by aircraft soot positive or negative? *J. Geophys. Res. Atmos.*, **119(19)**, 11,303–311,320, doi:10.1002/2014JD021914.

Zhou, P. and H. Wang, 2014: Carbon capture and storage – Solidification and storage of carbon dioxide captured on ships. *Ocean Eng.*, **91**, 172–180, doi:10.1016/j.oceaneng.2014.09.006.

Zhou, T., M.J. Roorda, H.L. MacLean, and J. Luk, 2017: Life cycle GHG emissions and lifetime costs of medium-duty diesel and battery electric trucks in Toronto, Canada. *Transp. Res. Part D Transp. Environ.*, **55**, 91–98, doi:10.1016/j.trd.2017.06.019.

Zhuk, A., Y. Zeigarnik, E. Buzoverov, and A. Sheindlin, 2016: Managing peak loads in energy grids: Comparative economic analysis. *Energy Policy*, **88**, 39–44, doi:10.1016/j.enpol.2015.10.006.

## Appendix 10.1: Data and Methods for Life Cycle Assessment

### IPCC Lifecycle Assessment Data Collection Effort

In mid-2020, the IPCC, in collaboration with the Norwegian University of Science and Technology, released a request for data from the lifecycle assessment (LCA) community, to estimate the lifecycle greenhouse (GHG) emissions of various passenger and freight transport pathways. The data requested included information about vehicle and fuel types, vintages, vehicle efficiency, payload, emissions from vehicle and battery manufacturing, and fuel cycle emissions factors, among others.

Data submissions were received from approximately 20 research groups, referencing around 30 unique publications. These submissions were supplemented by an additional 20 studies from the literature. While much of this literature was focused on LDVs and trucks, relatively few studies referenced bus and rail pathways.

### Harmonisation method

First, the datapoints were separated into categories based on the approximate classification (e.g., heavy-duty vs medium-duty trucks), powertrain (i.e., internal combustion engines (ICEV), hybrid electric vehicles (HEV), battery electric vehicles (BEV), fuel cell vehicles (FCV)), and fuel combination. For each category of vehicle/powertrain/fuel, a simplified LCA that harmonises values from across the reviewed studies was constructed, using the following basic equation:

$$Lifecycle\ GHG\ intensity = \frac{FC}{P} * EF + \frac{VC}{P * LVKT}$$

Where:

- Lifecycle GHG intensity represents the normalised lifecycle GHG emissions associated with each transportation mode, measured in $gCO_2$-eq per passenger-kilometre (pkm)or $gCO_2$-eq per tonne-kilometre (tkm).
- FC is the fuel consumption of the vehicle in megajoules (MJ) or kilowatt hours (kWh) per km.
- P represents the payload (measured in tonnes of cargo) or number of passengers, at a specified utilisation capacity (e.g., 50% payload or 80% occupancy).
- EF is an emissions factor representing the lifecycle GHG intensity of the fuel used, measured in $gCO_2$-eq $MJ^{-1}$ or $gCO_2$-eq $kWh^{-1}$. A single representative EF value is selected for each fuel type. When a given fuel type can be generated in different ways with substantially different upstream emissions factors (e.g., hydrogen from methane steam reforming vs hydrogen from water electrolysis), these are treated as two different fuel categories. The fuel emissions factors that were used are presented in Table 10.8.
- VC are the vehicle cycle emissions of the vehicle, measured in $gCO_2$-eq per vehicle. This may include vehicle manufacturing, maintenance and end of life, or just manufacturing.
- LVKT is the lifetime vehicle kilometres travelled.

Note: for plug-in hybrid electric vehicles (PHEV), the value of FC/P*EF is a weighted sum of this aggregate term for each of battery and diesel/gasoline operation.

Fuel emissions factors used are presented in Table 10.8. Note that the fuel emissions factors were compiled from several studies that used different global warming potential (GWP) values in their underlying assumptions, and therefore the numbers reported here

**Table 10.8 | Fuel emissions factors used to estimate lifecycle greenhouse gas (GHG) emissions of passenger and freight transport pathways.**

| Fuel | Emissions factor | Units | Source |
|---|---|---|---|
| Gasoline | 92 | $gCO_2$-eq $MJ^{-1}$ | Submissions to IPCC data call (median) |
| Diesel | 92 | $gCO_2$-eq $MJ^{-1}$ | Submissions to IPCC data call (median) |
| Diesel, high | 110 | $gCO_2$-eq $MJ^{-1}$ | Diesel from oil sands: average of in-situ pathways (Guo et al. 2020) |
| Biofuels, IAM EMF33 | 25 | $gCO_2$-eq $MJ^{-1}$ | From Chapter 7 |
| Biofuels, partial models CLC | 36 | $gCO_2$-eq $MJ^{-1}$ | From Chapter 7 |
| Biofuels, partial models NG | 141 | $gCO_2$-eq $MJ^{-1}$ | From Chapter 7 |
| Compressed natural gas | 71 | $gCO_2$-eq $MJ^{-1}$ | Submissions to IPCC data call (median) |
| Liquefied natural gas | 76 | $gCO_2$-eq $MJ^{-1}$ | Submissions to IPCC data call (median) |
| Liquefied petroleum gas | 78 | $gCO_2$-eq $MJ^{-1}$ | Submissions to IPCC data call (median) |
| DAC FT-Diesel, wind electricity | 12 | $gCO_2$-eq $MJ^{-1}$ | From electrolytic hydrogen produced using low-carbon electricity (Liu et al. 2020a) |
| DAC FT-Diesel, natural gas electricity | 370 | $gCO_2$-eq $MJ^{-1}$ | From electrolytic hydrogen produced using natural gas electricity; extrapolated from Liu et al. (2020a) |
| Ammonia, low carbon renewable | 3.2 | $gCO_2$-eq $MJ^{-1}$ | From electrolytic hydrogen produced using low-carbon electricity via Haber-Bosch (Gray et al. 2021) |
| Ammonia, natural gas SMR | 110 | $gCO_2$-eq $MJ^{-1}$ | From $H_2$ derived from natural gas steam methane reforming; via Haber-Bosch (Frattini et al. 2016) |
| Hydrogen, low carbon renewable | 10 | $gCO_2$-eq $MJ^{-1}$ | From electrolysis with low-carbon electricity (Valente et al. 2021) |
| Hydrogen, natural gas SMR | 95 | $gCO_2$-eq $MJ^{-1}$ | From steam-methane reforming of fossil fuels |
| Wind electricity | 9.3 | $gCO_2$-eq $kWh^{-1}$ | Submissions to IPCC data call (median) |
| Natural gas electricity | 537 | $gCO_2$-eq $kWh^{-1}$ | Submissions to IPCC data call (median) |
| Coal electricity | 965 | $gCO_2$-eq $kWh^{-1}$ | Submissions to IPCC data call (median) |

may be slightly different if the 100-year global warming potential (GWP100) from the AR6 had been used. This difference would be small given the small contribution from non-$CO_2$ gases to the total lifecycle emissions. For example, methane ($CH_4$) emissions exist in the lifecycle of natural gas supply chains or natural gas-dependent supply chains such as hydrogen from steam methane reforming (SMR). Recent data from the US suggests emissions of approximately 0.2–0.3 $gCH_4$ per MJ natural gas (Littlefield et al. 2017, 2019), which would range by no more than 1–2 $gCO_2$-eq per MJ natural gas (<3% of natural gas lifecycle emissions) when converting from a GWP100 of 25 (AR4) or 36 (AR5) to the current (AR6) GWP100 of 29.8.

For LDVs, the entire distribution of estimated lifecycle emissions is presented for each vehicle/powertrain/fuel category (as a boxplot) in Figure 10.4. For trucks, rail and buses, only the low and high estimates are presented (as solid bars) in Figures 10.6 and 10.8, since the number of datapoints were not sufficient to present as a distribution. Table 10.9 presents the low and high estimates of fuel efficiency for each category. The references used are reported in the main text.

For transit and freight, the lifecycle harmonisation exercise allows two aggregate parameters to vary from the low to high among submitted values within each category: FC/P and VC/P. Aggregate parameters are used to capture internal correlations (e.g., fuel consumption and payload; both depend heavily on vehicle size) and are presented in Tables 10.10 to 10.14. The references used are reported in the main text.

**Table 10.9 | Range of fuel efficiencies for light-duty vehicles by fuel and powertrain category, per vehicle kilometre.**

| Fuel | Powertrain | Fuel efficiency (MJ per vehicle-km) | | Electric efficiency (kWh per vehicle-km) | |
|---|---|---|---|---|---|
| | | Low | High | Low | High |
| Compression ignition | ICEV | 1.34 | 2.6 | | |
| Spark ignition | ICEV | 1.37 | 2.88 | | |
| Spark ignition | HEV | 1.22 | 2.05 | | |
| Compression ignition | HEV | 1.15 | 1.51 | | |
| Electricity | BEV | | | 0.12 | 0.242 |
| Hydrogen | FCV | 1.14 | 1.39 | | |

**Table 10.10 | Range of fuel efficiencies for buses by fuel and powertrain category, at 80% occupancy.**

| Fuel | Powertrain | Fuel efficiency (MJ per passenger-km) | | Electric efficiency (kWh per passenger-km) | |
|---|---|---|---|---|---|
| | | Low | High | Low | High |
| Diesel | ICEV | 0.16 | 0.52 | | |
| CNG | ICEV | 0.25 | 0.61 | | |
| LNG | ICEV | 0.27 | 0.37 | | |
| Biodiesel | ICEV | 0.16 | 0.52 | | |
| DAC FT-Diesel | ICEV | 0.16 | 0.52 | | |
| Diesel | HEV | 0.11 | 0.37 | | |
| Electricity | BEV | | | 0.01 | 0.04 |
| Hydrogen | FCV | 0.11 | 0.31 | | |

**Table 10.11 | Range of fuel efficiencies for passenger rail by fuel and powertrain category, at 80% occupancy.**

| Fuel | Powertrain | Fuel efficiency (MJ per passenger-km) | | Electric efficiency (kWh per passenger-km) | |
|---|---|---|---|---|---|
| | | Low | High | Low | High |
| Diesel | ICEV | 0.36 | 0.40 | | |
| Biofuels | ICEV | 0.36 | 0.40 | | |
| DAC FT-Diesel | ICEV | 0.36 | 0.40 | | |
| Diesel | HEV | 0.33 | 0.33 | | |
| Electricity | BEV | | | 0.03 | 0.03 |
| Hydrogen[a] | FCV | 0.18 | 0.18 | | |

[a] Occupancy corresponds to average European occupancy rates (IEA 2019e).

Table 10.12 | Range of fuel efficiencies for heavy-duty truck by fuel and powertrain category, at 100% payload.

| Fuel | Powertrain | Fuel efficiency (MJ per tonne-km) | | Electric efficiency (kWh per tonne-km) | |
|---|---|---|---|---|---|
| | | Low | High | Low | High |
| Diesel | ICEV | 0.38 | 0.93 | | |
| CNG | ICEV | 0.48 | 1.45 | | |
| LNG | ICEV | 0.43 | 1.00 | | |
| Biofuels | ICEV | 0.38 | 0.93 | | |
| Ammonia[a] | ICEV | 0.38 | 0.93 | | |
| DAC FT-Diesel | ICEV | 0.38 | 0.93 | | |
| Diesel | HEV | 0.34 | 0.59 | | |
| LNG | HEV | 0.46 | 0.51 | | |
| Electricity | BEV | | | 0.03 | 0.09 |
| Hydrogen | FCV | 0.25 | 0.43 | | |
| Ammonia[b] | FCV | 0.25 | 0.43 | | |

[a] Ammonia ICEV trucks are assumed to have the same fuel economy as diesel ICEVs due to lack of data.
[b] Ammonia FCV trucks are assumed to have the same fuel economy as hydrogen FCVs due to lack of data.

Table 10.13 | Range of fuel efficiencies for medium-duty truck by fuel and powertrain category, at 100% payload.

| Fuel | Powertrain | Fuel efficiency (MJ per tonne-km) | | Electric efficiency (kWh per tonne-km) | |
|---|---|---|---|---|---|
| | | Low | High | Low | High |
| Diesel | ICEV | 0.85 | 2.30 | | |
| CNG | ICEV | 1.08 | 2.54 | | |
| LNG | ICEV | 1.05 | 1.41 | | |
| Biofuels | ICEV | 0.85 | 2.30 | | |
| Ammonia[a] | ICEV | 0.85 | 2.30 | | |
| DAC FT-Diesel | ICEV | 0.85 | 2.30 | | |
| Diesel | HEV | 0.81 | 1.54 | | |
| Electricity | BEV | | | 0.12 | 0.22 |
| Hydrogen | FCV | 0.65 | 0.99 | | |
| Ammonia[b] | FCV | 0.65 | 0.99 | | |

[a] Ammonia ICEV trucks are assumed to have the same fuel economy as diesel ICEVs due to lack of data.
[b] Ammonia FCV trucks are assumed to have the same fuel economy as Hydrogen FCVs due to lack of data.

Table 10.14 | Range of fuel efficiencies for freight rail by fuel and powertrain category, at an average payload.

| Fuel | Powertrain | Fuel efficiency (M per /tonne-km) | | Electric efficiency (kWh per tonne-km) | |
|---|---|---|---|---|---|
| | | Low | High | Low | High |
| Diesel | ICEV | 0.11 | 0.78 | | |
| Biodiesel | ICEV | 0.11 | 0.78 | | |
| DAC FT-Diesel | ICEV | 0.11 | 0.78 | | |
| Electricity | BEV | | | 0.01 | 0.12 |
| Hydrogen | FCV | 0.10 | 0.10 | | |

## Appendix 10.2: Data and Assumptions for Lifecycle Cost Analysis

### Fuel cost ranges

For diesel, a range of USD0.5–2.5 per litre is used based on historic diesel costs across all OECD countries reported in the IEA Energy Prices and Taxes Statistics database (IEA 2021c) since 2010. The lower end of this range is consistent with the minimum projected value from the 2021 US Annual Energy Outlook (low oil price scenario, USD0.55 $l^{-1}$) (US Energy Information Administration 2021). The upper end of the range encompasses both the maximum diesel price observed in the 2021 US Annual Energy Outlook projections (high oil price scenario, USD1.5 $l^{-1}$) (US Energy Information Administration 2021), and the diesel price that would correspond to the 2020 IEA World Energy Outlook crude oil price projections (Stated Policies scenario) (IEA 2020b), assuming the historical price relationship between crude oil and diesel is maintained (USD1.5 $l^{-1}$). For reference, the IEA reports current world-average automotive diesel costs to be around 1 USD $l^{-1}$ (IEA 2021d). The selected range also captures the current range of production costs for values for bio-based and synthetic diesels (EUR51–144 $MWh^{-1}$, corresponding to USD0.6–1.70 $l^{-1}$), which are generally still higher than wholesale petroleum diesel costs (EUR30–50 $MWh^{-1}$, corresponding to USD0.35–0.6 $l^{-1}$), as reported by IEA (IEA 2020c). This range also encompasses costs for synthesised electrofuels from electrolytic hydrogen, as reported in Chapter 6 (USD1.6 $l^{-1}$).

10

The range of electricity costs used here is consistent with the range of levelised cost of electricity estimates presented in Chapter 6 (USD20–200 $MWh^{-1}$).

For hydrogen, a range of USD1 to USD13 per kilogram is used. The upper end of this range corresponds approximately to reported retail costs in the US (Eudy and Post 2018b; Argonne National Laboratory 2020; Burnham et al. 2021). Despite the high upper bound, lower costs (USD6–7 $kg^{-1}$) are already consistent with recent cost estimates of hydrogen produced via electrolysis (Chapter 6) and current production cost estimates from IRENA (IRENA 2020). The lower end of the range (USD1 $kg^{-1}$) corresponds to projected future price decreases for electrolytic hydrogen (BNEF 2020; Hydrogen Council 2020; IRENA 2020), and is consistent with projections from Chapter 6 for the low end of long-term future prices for fossil hydrogen with CCS.

### Vehicle efficiencies

The vehicle efficiencies used in developing the lifecycle cost estimates were derived from the harmonised ranges used to develop lifecycle GHG estimates and are presented in Tables 10.9 to 10.14.

### Other inputs to bus cost model

For buses, a 40-foot North American transit bus with a passenger capacity of 50, lifetime of 15 years, and an annual distance travelled of 72,400 km based on data in the ANL AFLEET model (Argonne National Laboratory 2020) is assumed. Maintenance costs were assumed to be USD0.63 per km for ICEV buses and USD0.38 per km for BEV and ICEV buses, also based on data from the AFLEET model (Argonne National Laboratory 2020). For ICEV and BEV purchase costs, data from the National Renewable Energy Laboratory (Johnson et al. 2020) is used for bounding ranges (USD430,000 to 500,000 for ICEV and USD579,000 to 1,200,000 for BEV), which encompass the default values from AFLEET model (Argonne National Laboratory 2020). Note that wider ranges are available in the literature (e.g., as low as USD120,000 per bus in Burnham et al. (2021) and Harris et al. (2020)); but these are not included in the sensitivity analysis to avoid conflating disparate vehicles. For FCV buses, the upper bound of the purchase price range (USD1,200,000) represents current costs in the US (Argonne National Laboratory 2020; Eudy and Post 2020), and the lower bound represents the target future value from the US Department of Energy (Eudy and Post 2020).

### Other inputs to rail cost model

For freight and passenger rail, powertrain and vehicle operation and maintenance costs in USD per km from the IEA Future of Rail report (IEA 2019e) (IEA Figure 2.14 for passenger rail and IEA Figure 2.15 for freight rail) are used as a proxy for non-fuel costs. The ranges span conservative and forward-looking cases. In addition, the range for BEV rail ranges encompass short- and long-distance trains – corresponding to 100–200 km for passenger rail, and 400–750 km for freight rail. Note that all values exclude the base vehicle costs, but they are expected not to be significant as they are amortised over the lifetime distance travelled. For freight rail, a network that is representative of North America is assumed, with a payload of 2800 tonnes per train (IEA Figure 1.17), assumed to be utilised at 100%, with a lifetime of 10 years, and an average distance travelled of 120,000 km $yr^{-1}$. For BEV freight rail, the range in powertrain costs is driven by battery costs of USD250–600 $kWh^{-1}$, while for FCV freight rail, the range in powertrain costs is driven by fuel cell stack costs of USD50–1000 $kW^{-1}$. For passenger rail, a network that is representative of Europe is assumed, with an average occupancy of 180 passengers per train (IEA Figure 1.14), with a lifetime of 10 years, and an average distance travelled of 115,000 km per year.

### Other inputs to truck cost model

Capital cost ranges vary widely in the literature depending on the exact truck model, size and other assumptions. For ICEVs in this analysis, the lower bound (USD90,000) corresponds to the 2020 estimate for China from Moultak et al. (2017), and the upper bound (USD250,000) corresponds to the 2030 projection for the US from the same study. These values encompass the full range reported by Argonne (Burnham et al. 2021). The lower bound BEV cost (USD120,000) is taken from 2030 projections for China (Moultak et al. 2017) and the upper bound (USD780,000) is taken from 2020 cost estimates in the US (class 8 sleeper cab tractor) (Burnham et al. 2021). The lower bound for FCV trucks (USD130,000) corresponds to the 2050 estimate for class 8 sleeper cab tractors from Argonne National Laboratory and the upper bound (USD290,000) corresponds to the 2020 estimate from the same study (Burnham et al. 2021). These values span the full range reported by Moultak et al. (2017) for the US, Europe and China from 2020–2030.

The analysis uses a truck lifetime of 10 years and annual distance travelled of 140,000 km based on Burnham et al. (2021). An effective payload of 17 tonnes (80% of maximum payload of 21 tonnes) is assumed based on reported average effective payload submitted by Argonne National Laboratory in response to the IPCC LCA data collection call. A discount rate of 3% is used, based on Burnham et al. (2021) and consistent with the social discount rate from Chapter 3. Maintenance costs are assumed to be USD0.15 $km^{-1}$ for ICEV trucks and USD0.09 $km^{-1}$ for BEV and FCV trucks, as reported in Burnham et al. (2021).

## Appendix 10.3: Line of Sight for Feasibility Assessment

| | Geophysical | | |
|---|---|---|---|
| | Physical potential | Geophysical recourses | Land use |
| **Demand reduction and mode shift** | + | + | + |
| *Role of contexts* | Adoption of Avoid Shift Improve approach along with improving fuel efficiency will have negligible physical constraints; they can be implemented across the countries. | Reduction in demand, fuel efficiency and demand management measures such as Clean Air Zones and parking policies will reduce negative impact on land use and resource consumption – without any constraints in terms of available resources. | Reduction in demand, increase in fuel efficiency and demand management measures will have a positive impact on land use as compared to 'without' them – no likely adverse constraints in terms of limited land use (such as decline in biofuel). |
| *Line of sight* | Holguín-Veras, J. and I. Sánchez-Díaz, 2016: Freight Demand Management and the Potential of Receiver-Led Consolidation programs. *Transp. Res. Part A Policy Pract.*, **84**, 109–130, doi:10.1016/j.tra.2015.06.013.<br>Creutzig, F. et al., 2018: Towards demand-side solutions for mitigating climate change. *Nat. Clim. Change*, **8(4)**, 260–263, doi:10.1038/s41558-018-0121-1.<br>Rajé, F., 2017: *Transport, Demand Management and Social Inclusion: The need for ethnic perspectives*. Routledge, London, UK, 184 pp.<br>Dumortier, J., M. Carriquiry, and A. Elobeid, 2021: Where does all the biofuel go? Fuel efficiency gains and its effects on global agricultural production. *Energy Policy*, **148**, 111909, doi:10.1016/j.enpol.2020.111909. | | |
| **Biofuels for land transport, aviation, and shipping** | + | ± | – |
| *Role of contexts* | Climate conditions are an important factor for bioenergy viability. Land availability constraints might be expected for bioenergy deployment. | Land and synthetic fertilisers are examples of limited resources to deploy large-scale biofuels, however the extent of these restrictions will depend on local and context specific conditions. | Implementing biofuels may require additional land use. However, it will depend on context and local specific conditions. |
| *Line of sight* | Daioglou, V., J.C. Doelman, B. Wicke, A. Faaij, and D.P. van Vuuren, 2019: Integrated assessment of biomass supply and demand in climate change mitigation scenarios. *Glob. Environ. Change*, **54**, 88–101, doi:10.1016/j.gloenvcha.2018.11.012.<br>Roe, S. et al., 2021: Land-based measures to mitigate climate change: Potential and feasibility by country. *Glob. Change Biol.*, **27(23)**, 6025–6058, doi:10.1111/gcb.15873. | | |
| **Ammonia for shipping** | + | + | ± |
| *Role of contexts* | A global ammonia supply chain is already established; the primary requirement for delivering greater carbon emissions reductions will be through the production of ammonia using green hydrogen or CCS. | The use of ammonia would reduce reliance on fossil fuels for shipping and is expected to reduce reliance on natural resources when produced using green hydrogen. The primary resource requirements will be the supply of renewable electricity and clean water to produce green hydrogen, from which ammonia can be produced. | No major changes in land use for the vehicle. Increases may occur if the hydrogen is produced through electrolysis and renewable energy sources or hydrogen production with CCS. |
| *Line of sight* | Bicer, Y. and I. Dincer, 2018: Clean fuel options with hydrogen for sea transportation: A life cycle approach. *Int. J. Hydrogen Energy*, **43(2)**, 1179–1193, doi:10.1016/j.ijhydene.2017.10.157.<br>Gilbert, P. et al., 2018: Assessment of full life-cycle air emissions of alternative shipping fuels. *J. Clean. Prod.*, **172**, 855–866, doi:10.1016/j.jclepro.2017.10.165. | | |
| **Synthetic fuels for heavy-duty land transport, aviation, and shipping (e.g., DAC-FT)** | ± | ± | ± |
| *Role of contexts* | Fischer Tropsch chemistry is well established; pilot scale direct air capture (DAC) plants are already in operation;<br>– Does not qualify as a mitigation option except in regions with very low-carbon electricity. | + Gasification can use a wide range of feedstocks; DAC can be applied in a wide range of locations<br>– Limited information available on potential limits related to large input energy requirements, or water use and required sorbents for DAC. | No major changes in land use for the vehicle. Potential increases in land use for electricity generation (especially solar, wind or hydropower) for $CO_2$ capture and fuel production; likely lower land use than crop-based biofuels. |

| | Geophysical | | |
|---|---|---|---|
| | Physical potential | Geophysical recourses | Land use |
| *Line of sight* | Realmonte, G. et al., 2019: An inter-model assessment of the role of direct air capture in deep mitigation pathways. *Nat. Commun.*, **10(1)**, 3277, doi:10.1038/s41467-019-10842-5.<br>Liu, C.M., N.K. Sandhu, S.T. McCoy, and J.A. Bergerson, 2020: A life cycle assessment of greenhouse gas emissions from direct air capture and Fischer-Tropsch fuel production. *Sustain. Energy Fuels*, **4(6)**, 3129–3142, doi:10.1039/C9SE00479C.<br>Ueckerdt, F. et al., 2021: Potential and risks of hydrogen-based e-fuels in climate change mitigation. *Nat. Clim. Change*, **11(5)**, 384–393, doi:10.1038/s41558-021-01032-7. | Realmonte, G. et al., 2019: An inter-model assessment of the role of direct air capture in deep mitigation pathways. *Nat. Commun.*, **10(1)**, 3277, doi:10.1038/s41467-019-10842-5. | |
| **Electric vehicles for land transport** | + | ± | ± |
| *Role of contexts* | Electromobility is being adopted across a range of land transport options including light-duty vehicles, trains and some heavy-duty vehicles, suggesting no physical constraints. | Current dominant battery chemistry relies on minerals that may face supply constraints, including lithium, cobalt, and nickel. Regional supply/availability varies. Alternative chemistries exist; recycling may likewise alleviate critical material concerns. Similar supply constraints may exist for some renewable electricity sources (e.g., solar) required to support EVs. May reduce critical materials required for catalytic converters in ICEVs (e.g., platinum, palladium, rhodium). | No major changes in land use for the vehicle. Potential increases in land use for electricity generation (especially solar, wind or hydropower) and mineral extraction, but may be partially offset by a decrease in land use for fossil fuel production; likely lower land use than crop-based biofuels, or technologies with higher electricity use (e.g., those based on electrolytic hydrogen). |
| *Line of sight* | IEA, 2021: *Global EV Outlook 2021*. International Energy Agency, Paris, France, 101 pp. | Jones, B., R.J.R. Elliott, and V. Nguyen-Tien, 2020: The EV revolution: The road ahead for critical raw materials demand. *Appl. Energy*, **280**, 115072, doi:10.1016/J.APENERGY.2020.115072.<br>Xu, C. et al., 2020: Future material demand for automotive lithium-based batteries. *Commun. Mater. 2020 11*, **1(1)**, 1–10, doi:10.1038/s43246-020-00095-x.<br>IEA, 2021: *The Role of Critical Minerals in Clean Energy Transitions*. International Energy Agency, Paris, France, 287 pp.<br>Zhang, J. et al., 2016: Assessing Economic Modulation of Future Critical Materials Use: The Case of Automotive-Related Platinum Group Metals. *Environ. Sci. Technol.*, **50(14)**, 7687–7695, doi:10.1021/ACS.EST.5B04654.<br>Milovanoff, A., I.D. Posen, and H.L. MacLean, 2020: Electrification of light-duty vehicle fleet al.ne will not meet mitigation targets. *Nat. Clim. Change*, **10(12)**, 1102–1107, doi:10.1038/s41558-020-00921-7. | Arent, D. et al., 2014: Implications of high renewable electricity penetration in the U.S. for water use, greenhouse gas emissions, land-use, and materials supply. *Appl. Energy*, **123**, 368–377, doi:10.1016/j.apenergy.2013.12.022.<br>Orsi, F., 2021: On the sustainability of electric vehicles: What about their impacts on land use? *Sustain. Cities Soc.*, **66**, 102680, doi:10.1016/J.SCS.2020.102680. |
| **Hydrogen FCV for land transport** | + | ± | ± |
| *Role of contexts* | The use of fuel cells in the transport sector is growing, and will potentially be important in heavy-duty land transport applications. | FCVs are reliant on critical minerals for manufacturing fuel cells, electric motors and supporting batteries. Platinum is the primary potential resource constraint for fuel cells; however, its use may decrease as the technology develops, and platinum is highly recyclable. | |
| *Line of sight* | IEA, 2020: *Global EV Outlook 2020*. Paris, France, 276 pp. | Hao, H. et al., 2019: Securing Platinum-Group Metals for Transport Low-Carbon Transition. *One Earth*, **1(1)**, 117–125, doi:10.1016/j.oneear.2019.08.012.<br>Rasmussen, K.D., H. Wenzel, C. Bangs, E. Petavratzi, and G. Liu, 2019: Platinum Demand and Potential Bottlenecks in the Global Green Transition: A Dynamic Material Flow Analysis. *Environ. Sci. Technol.*, **53(19)**, 11541–11551, doi:10.1021/acs.est.9b01912. | Orsi, F., 2021: On the sustainability of electric vehicles: What about their impacts on land use? *Sustain. Cities Soc.*, **66**, 102680, doi:10.1016/J.SCS.2020.102680. |

| | Environmental-ecological | | | |
|---|---|---|---|---|
| | Air pollution | Toxic waste, ecotoxicity eutrophication | Water quantity and quality | Biodiversity |
| **Demand reduction and mode shift** | + | 0 | 0 | 0 |
| *Role of contexts* | Reduction in demand, increase in fuel efficiency and demand management measures will improve air quality. | | | Reduction in demand, fuel efficiency and demand management measures such as Clean Air Zones and parking Policies will reduce need for roads and protect biodiversity. |
| *Line of sight* | Creutzig, F. et al., 2018: Towards demand-side solutions for mitigating climate change. *Nat. Clim. Change*, **8(4)**, 260–263, doi:10.1038/s41558-018-0121-1.<br>Dumortier, J., M. Carriquiry, and A. Elobeid, 2021: Where does all the biofuel go? Fuel efficiency gains and its effects on global agricultural production. *Energy Policy*, **148**, 111909, doi:10.1016/j.enpol.2020.111909.<br>Ambarwati, L., R. Verhaeghe, B. van Arem, and A.J. Pel, 2016: The influence of integrated space–transport development strategies on air pollution in urban areas. *Transp. Res. Part D Transp. Environ.*, **44**, 134–146, doi:10.1016/j.trd.2016.02.015.<br>DEFRA and DoT, 2020: *Clean Air Zone Framework: Principles for setting up Clean Air Zones in England.*, Department of Environment Food & Rural Affairs/Department of Transport, Government of UK, London, UK, 35 pp. | | | |
| **Biofuels for land transport, aviation, and shipping** | ± | ± | – | – |
| *Role of contexts* | Biofuels may improve air quality due to reduction in the emission of some pollutants, such as SOx and particulate matter, in relation to fossil fuels. Evidence is mixed for other pollutants such as NOx. The biofuels supply chain (e.g., due to increased fertiliser use) may negatively impact air quality. | Increased use of fertilisers and agrochemicals due to biofuel production may increase impacts in ecotoxicity and eutrophication; some biofuels may be less toxic than fossil fuel counterparts. | Increasing production of biofuels may increase pressure on water resources due to the need for irrigation. However, some biofuel options may also improve these aspects in respect to conventional agriculture. These impacts will depend on specific local conditions. | Additional land use for biofuels may increase pressure on biodiversity. However, biofuel can also increase biodiversity depending on the previous land use. These impacts will depend on specific local conditions and previous land uses. |
| *Line of sight* | Robertson, G.P. et al., 2017: Cellulosic biofuel contributions to a sustainable energy future: Choices and outcomes. *Science*, **356(6345)**, doi:10.1126/science.aal2324.<br>Humpenöder, F. et al., 2018: Large-scale bioenergy production: how to resolve sustainability trade-offs? *Environ. Res. Lett.*, **13(2)**, 024011, doi:10.1088/1748-9326/aa9e3b.<br>Ai, Z., N. Hanasaki, V. Heck, T. Hasegawa, and S. Fujimori, 2021: Global bioenergy with carbon capture and storage potential is largely constrained by sustainable irrigation. *Nat. Sustain.*, **4(10)**, 884–891, doi:10.1038/s41893-021-00740-4. | | | |
| **Ammonia for shipping** | ± | – | ± | **Limited Evidence (LE)** |
| *Role of contexts* | If produced from green hydrogen or coupled with CCS, ammonia could reduce short-lived climate forcers and particulate matter precursors including black carbon and $SO_2$. However, the combustion of ammonia could lead to elevated levels of nitrogen oxides and ammonia emissions. | Ammonia is highly toxic, and therefore requires special handling procedures to avoid potentially catastrophic leaks into the environment. That said, large volumes of ammonia are already safely transported internationally due to a high level of understanding of safe handling procedures. Additionally, the use of ammonia in shipping presents a risk of eutrophication and ecotoxicity from the release of ammonia into the water system – either via a fuel leak or via unburnt ammonia emissions. | May increase or decrease water footprint depending on the upstream energy source. | Lack of studies assessing the potential impacts of the technology on biodiversity. |
| *Line of sight* | Bicer, Y. and I. Dincer, 2018: Clean fuel options with hydrogen for sea transportation: A life cycle approach. *Int. J. Hydrogen Energy*, **43**(2), 1179–1193, doi:10.1016/j.ijhydene.2017.10.157.<br>Gilbert, P. et al., 2018: Assessment of full life-cycle air emissions of alternative shipping fuels. *J. Clean. Prod.*, **172**(2018), 855–866, doi:10.1016/j.jclepro.2017.10.165.<br>ABS, 2020: *Ammonia as a Marine Fuel*. American Bureau of Shipping, Spring, 28 pp. | | | |

| | Environmental-ecological | | | |
|---|---|---|---|---|
| | **Air pollution** | **Toxic waste, ecotoxicity eutrophication** | **Water quantity and quality** | **Biodiversity** |
| **Synthetic fuels for heavy-duty land transport, aviation, and shipping (e.g., DAC-FT)** | + | NE | ± | LE |
| *Role of contexts* | Potential reductions in air pollutants related to reduced presence of sulphur, metals, and other contaminants; improvements likely smaller than for electric vehicles or hydrogen fuel cell vehicles. | | DAC requires significant amounts of water, which may be a limitation in water stressed areas; typically uses less water than crop-based biofuels. | Potential biodiversity issues related to electricity generation; however fossil fuel supply chains also adversely impact biodiversity; net effect is unknown. |
| *Line of sight* | Beyersdorf, A.J. et al., 2014: Reductions in aircraft particulate emissions due to the use of Fischer–Tropsch fuels. *Atmos. Chem. Phys.*, **14(1)**, 11–23, doi:10.5194/acp-14-11-2014.<br>Lobo, P., D.E. Hagen, and P.D. Whitefield, 2011: Comparison of PM Emissions from a Commercial Jet Engine Burning Conventional, Biomass, and Fischer–Tropsch Fuels. *Environ. Sci. Technol.*, **45(24)**, 10744–10749, doi:10.1021/es201902e.<br>Gill, S.S., A. Tsolakis, K.D. Dearn, and J. Rodríguez-Fernández, 2011: Combustion characteristics and emissions of Fischer–Tropsch diesel fuels in IC engines. *Prog. Energy Combust. Sci.*, **37(4)**, 503–523, doi:10.1016/j.pecs.2010.09.001. | | Realmonte, G. et al., 2019: An inter-model assessment of the role of direct air capture in deep mitigation pathways. *Nat. Commun.*, **10(1)**, 3277, doi:10.1038/s41467-019-10842-5.<br>Byers, E.A., J.W. Hall, J.M. Amezaga, G.M. O'Donnell, and A. Leathard, 2016: Water and climate risks to power generation with carbon capture and storage. *Environ. Res. Lett.*, **11(2)**, 024011, doi:10.1088/1748-9326/11/2/024011. | |
| **Electric vehicles for land transport** | + | – | ± | LE |
| *Role of contexts* | Elimination of tailpipe emissions. If powered by nuclear or renewables, large overall improvements in air pollution. Even if powered partially by fossil fuel electricity, tailpipe emissions tend to occur closer to population and thus typically have larger impact on human health than powerplant emissions; negative air quality impacts may occur, but only in fossil fuel-heavy grids. | Some toxic waste associated with mining and processing of metals for batteries and some renewable electricity supply chains (production and disposal). | May increase or decrease water footprint depending on the upstream electricity source. | Potential biodiversity issues related to electricity generation; however fossil fuel supply chains also adversely impact biodiversity; net effect is unknown. |
| *Line of sight* | Requia, W.J., M. Mohamed, C.D. Higgins, A. Arain, and M. Ferguson, 2018: How clean are electric vehicles? Evidence-based review of the effects of electric mobility on air pollutants, greenhouse gas emissions and human health. *Atmos. Environ.*, **185**, 64–77, doi:10.1016/J.ATMOSENV.2018.04.040.<br>Horton, D.E. et al., 2021: Effect of adoption of electric vehicles on public health and air pollution in China: a modelling study. *Lancet Planet. Heal.*, doi:10.1016/s2542-5196(21)00092-9.<br>Gai, Y. et al., 2020: Health and climate benefits of Electric Vehicle Deployment in the Greater Toronto and Hamilton Area. *Environ. Pollut.*, **265**, 114983, doi:10.1016/j.envpol.2020.114983.<br>Choma, E.F., J.S. Evans, J.K. Hammitt, J.A. Gómez-Ibáñez, and J.D. Spengler, 2020: Assessing the health impacts of electric vehicles through air pollution in the United States. *Environ. Int.*, **144**, 106015, doi:10.1016/j.envint.2020.106015.<br>Schnell, J.L. et al., 2019: Air quality impacts from the electrification of light-duty passenger vehicles in the United States. *Atmos. Environ.*, **208**, 95–102, doi:10.1016/j.atmosenv.2019.04.003.<br>Tessum, C.W., J.D. Hill, and J.D. Marshall, 2014: Life cycle air quality impacts of conventional and alternative light-duty transportation in the United States. *Proc. Natl. Acad. Sci.*, **111(52)**, 18490–18495, doi:10.1073/pnas.1406853111. | Lattanzio, R.K. and C.E. Clark, 2020: *Environmental Effects of Battery Electric and Internal Combustion Engine Vehicles.*, Congressional Research Service, Washington, DC, USA, 41 pp.<br>Puig-Samper Naranjo, G., D. Bolonio, M.F. Ortega, and M.-J. García-Martínez, 2021: Comparative life cycle assessment of conventional, electric and hybrid passenger vehicles in Spain. *J. Clean. Prod.*, **291**, 125883, doi:10.1016/j.jclepro.2021.125883.<br>Bicer, Y. and I. Dincer, 2017: Comparative life cycle assessment of hydrogen, methanol and electric vehicles from well to wheel. *Int. J. Hydrogen Energy*, **42(6)**, 3767–3777, doi:10.1016/j.ijhydene.2016.07.252.<br>Hawkins, T.R., B. Singh, G. Majeau-Bettez, and A.H. Strømman, 2013: Comparative Environmental Life Cycle Assessment of Conventional and Electric Vehicles. *J. Ind. Ecol.*, **17(1)**, doi:10.1111/j.1530-9290.2012.00532.x. | Onat, N.C., M. Kucukvar, and O. Tatari, 2018: Well-to-wheel water footprints of conventional versus electric vehicles in the United States: A state-based comparative analysis. *J. Clean. Prod.*, **204**, 788–802, doi:10.1016/j.jclepro.2018.09.010.<br>Kim, H.C. et al., 2016: Life Cycle Water Use of Ford Focus Gasoline and Ford Focus Electric Vehicles. *J. Ind. Ecol.*, **20(5)**, 1122–1133, doi:10.1111/jiec.12329.<br>Wang, L. et al., 2020: Life cycle water use of gasoline and electric light-duty vehicles in China. *Resour. Conserv. Recycl.*, **154**, 104628, doi:10.1016/j.resconrec.2019.104628. | |

| | Environmental-ecological | | | |
|---|---|---|---|---|
| | Air pollution | Toxic waste, ecotoxicity eutrophication | Water quantity and quality | Biodiversity |
| **Hydrogen FCV for land transport** | + | ± | ± | LE |
| *Role of contexts* | Fuel cells' only tailpipe emission is water vapour. However, blue hydrogen production pathways may generate air pollutants near the production sites. Overall, FCV would reduce emissions of criteria air pollutants. | Mining of platinum group metals may generate additional stress on the environment, compared to conventional technologies. Furthermore, the recycling of fuel cell stacks can generate additional impacts. | May increase or decrease water footprint depending on the upstream energy source. | Lack of studies assessing the potential impacts of the technology on biodiversity. |
| *Line of sight* | Wang, Q., M. Xue, B. Le Lin, Z. Lei, and Z. Zhang, 2020: Well-to-wheel analysis of energy consumption, greenhouse gas and air pollutants emissions of hydrogen fuel cell vehicle in China. *J. Clean. Prod.*, **275**, doi:10.1016/j.jclepro.2020.123061. | Velandia Vargas, J.E. and J.E.A. Seabra, 2021: Fuel-cell technologies for private vehicles in Brazil: Environmental mirage or prospective romance? A comparative life cycle assessment of PEMFC and SOFC light-duty vehicles. *Sci. Total Environ.*, **798**, 149265, doi:10.1016/j.scitotenv.2021.149265.<br>Bohnes, F.A., J.S. Gregg, and A. Laurent, 2017: Environmental Impacts of Future Urban Deployment of Electric Vehicles: Assessment Framework and Case Study of Copenhagen for 2016–2030. *Environ. Sci. Technol.*, **51(23)**, 13995–14005, doi:10.1021/acs.est.7b01780. | | |

| | Technological | | |
|---|---|---|---|
| | Simplicity | Technological scalability | Maturity and technology readiness |
| **Demand reduction and mode shift** | + | + | + |
| *Role of contexts* | Application of demand reduction and fuel efficiency measures can be scaled and developing countries can leapfrog to most advanced technology. India skipped Euro V, and implemented Euro VI from IV, but this shift will require investment in the short term. | Technology to deliver demand reduction and fuel efficiency is readily available. | Significant economic benefit in short and long term. |
| *Line of sight* | Vashist, D., N. Kumar, and M. Bindra, 2017: Technical Challenges in Shifting from BS IV to BS-VI Automotive Emissions Norms by 2020 in India: A Review. *Arch. Curr. Res. Int.*, **8(1)**, 1–8, doi:10.9734/ACRI/2017/33781.<br>DEFRA and DoT, 2020: *Clean Air Zone Framework: Principles for setting up Clean Air Zones in England.*, Department of Environment Food & Rural Affairs/Department of Transport, Government of UK, London, UK, 35 pp. | | |
| **Biofuels for land transport, aviation, and shipping** | ± | ± | + |
| *Role of contexts* | Typically based on internal combustion engines, similar to fossil fuels, however, may require engine recalibration. | Biofuels are scalable and may benefit from economies of scale; potential for scale up of sustainable crop production may be limited. | There are many biofuels technologies that are already at commercial scale, while some technologies for advanced biofuels are still under development. |
| *Line of sight* | Mawhood, R., E. Gazis, S. de Jong, R. Hoefnagels, and R. Slade, 2016: Production pathways for renewable jet fuel: a review of commercialization status and future prospects. *Biofuels, Bioprod. Biorefining*, **10**, 462–484, doi:10.1002/bbb.1644.<br>Puricelli, S. et al., 2021: A review on biofuels for light-duty vehicles in Europe. *Renew. Sustain. Energy Rev.*, **137**, 110398, doi:10.1016/J.RSER.2020.110398. | | |
| **Ammonia for shipping** | – | ± | ± |
| *Role of contexts* | Requires either new engines or retrofits for existing engines. It is likely some ammonia will need to be mixed with a secondary fuel due its relatively low burning velocity and high ignition temperature. This would likely require existing powertrains to be modified to accept dual fuel mixes, including ammonia. Exhaust treatment systems are also required to deal with the release of unburnt ammonia emissions. | Ammonia supply chains are well established; transport and storage more feasible than hydrogen; scalability of electrolytic production routes remains a challenge for producing low-GHG ammonia. | The production, transport and storage of ammonia is mature based on existing international supply chains. The use of ammonia in ships is still at the early stages of research and development. Further research and development will be required for ammonia to be widely used in shipping, including improving the efficiency of combustion, and treatment of exhaust emissions. Ammonia could also potentially be used in fuel cell powertrains in the future, but the development of this technology is even less mature at present. |

| | Technological | | |
|---|---|---|---|
| | Simplicity | Technological scalability | Maturity and technology readiness |
| *Line of sight* | Frigo, S., R. Gentili, and F. De Angelis, 2014: Further Insight into the Possibility to Fuel a SI Engine with Ammonia plus Hydrogen. SAE Technical Paper 2014-32-008, doi:10.4271/2014-32-0082.<br>Dimitriou, P. and R. Javaid, 2020: A review of ammonia as a compression ignition engine fuel. *Int. J. Hydrogen Energy*, **45(11)**, 7098–7118, doi:10.1016/J.IJHYDENE.2019.12.209.<br>MAN Energy Solutions, 2019: *Engineering the future two-stroke green-ammonia engine*. MAN Energy Solutions, Copenhagen, Denmark, 20 pp. | | |
| **Synthetic fuels for heavy-duty land transport, aviation, and shipping (e.g., DAC-FT)** | + | – | – |
| *Role of contexts* | Can produce drop-in fuels, which use existing engine technologies. | Rate at which DAC or other carbon capture can be scaled up is likely a limiting factor; large energy inputs (requiring substantial new low-carbon energy resources), and sorbent requirements likely to be a challenge. | Some processes (e.g., Fischer Tropsch) are well established, but DAC and BECCS are still at demonstration stage. |
| *Line of sight* | Sutter, D., M. van der Spek, and M. Mazzotti, 2019: 110th Anniversary: Evaluation of $CO_2$-Based and $CO_2$-Free Synthetic Fuel Systems Using a Net-Zero-$CO_2$-Emission Framework. *Ind. Eng. Chem. Res.*, **58(43)**, 19958–19972, doi:10.1021/acs.iecr.9b00880.<br>The Royal Society, 2019: *Sustainable synthetic carbon based fuels for transport: Policy Brief*. The Royal Society, London, UK, 46 pp. | The Royal Society, 2019: *Sustainable synthetic carbon based fuels for transport: Policy Brief*. The Royal Society, London, UK, 46 pp.<br>Realmonte, G. et al., 2019: An inter-model assessment of the role of direct air capture in deep mitigation pathways. *Nat. Commun.*, **10(1)**, 3277, doi:10.1038/s41467-019-10842-5. | Liu, C.M., N.K. Sandhu, S.T. McCoy, and J.A. Bergerson, 2020: A life cycle assessment of greenhouse gas emissions from direct air capture and Fischer-Tropsch fuel production. *Sustain. Energy Fuels*, **4(6)**, 3129–3142, doi:10.1039/C9SE00479C. |
| **Electric vehicles for land transport** | ± | ± | ± |
| *Role of contexts* | Fewer engine components; lower maintenance requirements than conventional vehicles; potential concerns surrounding battery size/weight, charging time, and battery life. | Widespread application already feasible; some limits to adoption in remote communities or long-haul freight; at large scale, may positively or negatively impact electric grid functioning depending on charging behaviour and grid integration strategy. | + Technology is mature for light-duty vehicles;<br>– Improvements in battery capacity and density as well as charging speed required for heavy-duty applications. |
| *Line of sight* | Burnham, A. et al., 2021: *Comprehensive total cost of ownership quantification for vehicles with different size classes and powertrains*. Argonne National Laboratory, US Department of Energy, Lemont, IL, USA, 227 pp. | IEA, 2021: *Global EV Outlook 2021*. International Energy Agency, Paris, France,101 pp.<br>Milovanoff, A., I.D. Posen, and H.L. MacLean, 2020: Electrification of light-duty vehicle fleet al.ne will not meet mitigation targets. *Nat. Clim. Change*, **10(12)**, 1102–1107, doi:10.1038/s41558-020-00921-7.<br>Crozier, C., T. Morstyn, and M. McCulloch, 2020: The opportunity for smart charging to mitigate the impact of electric vehicles on transmission and distribution systems. *Appl. Energy*, **268**, 114973, doi:10.1016/j.apenergy.2020.114973.<br>Kapustin, N O. and D.A. Grushevenko, 2020: Long-term electric vehicles outlook and their potential impact on electric grid. *Energy Policy*, **137**, 111103, doi:10.1016/j.enpol.2019.111103.<br>Das, H.S., M.M. Rahman, S. Li, and C.W. Tan, 2020: Electric vehicles standards, charging infrastructure, and impact on grid integration: A technological review. *Renew. Sustain. Energy Rev.*, **120**, 109618, doi:10.1016/j.rser.2019.109618.<br>Liimatainen, H., O. van Vliet, and D. Aplyn, 2019: The potential of electric trucks – An international commodity-level analysis. *Appl. Energy*, **236**, 804–814, doi:10.1016/j.apenergy.2018.12.017.<br>Forrest, K., M. Mac Kinnon, B. Tarroja, and S. Samuelsen, 2020: Estimating the technical feasibility of fuel cell and battery electric vehicles for the medium and heavy duty sectors in California. *Appl. Energy*, **276**, 115439, doi:10.1016/j.apenergy.2020.115439. | IEA, 2021: *Global EV Outlook 2021*. International Energy Agency, Paris, France, 101 pp.<br>Smith, D. et al., 2020: *Medium- and Heavy-Duty Vehicle Electrification: An Assessment of Technology and Knowledge Gaps*. Oak Ridge National Laboratory, Oak Ridge, TN, USA, 85 pp.<br>Forrest, K., M. Mac Kinnon, B. Tarroja, and S. Samuelsen, 2020: Estimating the technical feasibility of fuel cell and battery electric vehicles for the medium and heavy duty sectors in California. *Appl. Energy*, **276**, 115439, doi:10.1016/j.apenergy.2020.115439. |

| | Technological | | |
|---|---|---|---|
| | Simplicity | Technological scalability | Maturity and technology readiness |
| **Hydrogen FCV for land transport** | ± | – | – |
| *Role of contexts* | Lower maintenance requirements compared to conventional technologies; potential issues with on-vehicle hydrogen storage, fuel cell degradation and lifetime; fewer weight and refuelling time barriers compared to electric vehicles. | Currently the refuelling infrastructure is limited, but it is growing at the pace of the technology deployment. Challenges exist with transport and distribution of hydrogen. Electrolytic hydrogen not currently produced at scale. | The technology is already available to users for light-duty vehicle applications and buses, but further improvements in fuel cell technology are needed. Use in heavy-duty applications is currently constrained. Maturity and technology readiness level can vary for different parts of the supply chain, and is lower than for EVs. |
| *Line of sight* | Trencher, G., A. Taeihagh, and M. Yarime, 2020: Overcoming barriers to developing and diffusing fuel-cell vehicles: Governance strategies and experiences in Japan. *Energy Policy*, **142**, 111533, doi:10.1016/j.enpol.2020.111533. | Pollet, B.G., S.S. Kocha, and I. Staffell, 2019: Current status of automotive fuel cells for sustainable transport. *Curr. Opin. Electrochem.*, **16** (May), 1–6, doi:10.1016/j.coelec.2019.04.021. | Wang, J., H. Wang, and Y. Fan, 2018: Techno-Economic Challenges of Fuel Cell Commercialization. *Engineering*, **4(3)**, 352–360, doi:10.1016/j.eng.2018.05.007.<br>Kampker, A. et al., 2020: Challenges towards large-scale fuel cell production: Results of an expert assessment study. *Int. J. Hydrogen Energy*, **45(53)**, 29288–29296, doi:10.1016/J.IJHYDENE.2020.07.180. |

| | 4. Economic | |
|---|---|---|
| | Costs in 2030 and long term | Employment effects and economic growth |
| **Demand reduction and mode shift** | + | LE |
| *Role of contexts* | Significant economic benefit in short and long term. | |
| *Line of sight* | Creutzig, F. et al., 2018: Towards demand-side solutions for mitigating climate change. *Nat. Clim. Change*, **8(4)**, 260–263, doi:10.1038/s41558-018-0121-1.<br>The UK, 2020: *The Green Book*. HM Treasury, London, UK, https://www.gov.uk/government/publications/the-green-book-appraisal-and-evaluation-in-central-governent/the-green-book-2020. | |
| **Biofuels for land transport, aviation, and shipping** | ± | LE |
| *Role of contexts* | Some biofuels are already cost competitive with fossil fuels. In the future, reduction of costs for advanced biofuels may be a challenge. | Biofuels are expected to increase job creation in comparison to fossil fuel alternatives. This is still to be further demonstrated. |
| *Line of sight* | Daioglou, V. et al., 2020: Bioenergy technologies in long-run climate change mitigation: results from the EMF-33 study. *Clim. Change*, **163(3)**, 1603–1620, doi:10.1007/s10584-020-02799-y.<br>Brown, A., et al., 2020. *Advanced Biofuels – Potential for Cost Reduction*. IEA Bioenergy, Paris, France, 88. | |
| **Ammonia for shipping** | – | NE |
| *Role of contexts* | Green ammonia is likely to be significantly more expensive than conventional fuels for the coming decades. | |
| *Line of sight* | Energy Transitions Commission, 2021. *Making the hydrogen economy possible*. Energy Transitions Commission, 92 pp. https://energy-transitions.org/wp-content/uploads/2021/04/ETC-Global-Hydrogen-Report.pdf.<br>Energy Transitions Commission, 2020. *The First Wave: A blueprint for commercial-scale zero-emission shipping pilots*. Energy Transitions Commission, 102 pp. https://www.energy-transitions.org/wp-content/uploads/2020/11/The-first-wave.pdf. | |
| **Synthetic fuels for heavy-duty land transport, aviation, and shipping (e.g., DAC-FT)** | – | NE |
| *Role of contexts* | Large uncertainty on future costs but expected to remain higher than conventional fuels for the coming decades. | |

| | 4. Economic | |
|---|---|---|
| | Costs in 2030 and long term | Employment effects and economic growth |
| *Line of sight* | Ueckerdt, F. et al., 2021: Potential and risks of hydrogen-based e-fuels in climate change mitigation. *Nat. Clim. Change*, **11(5)**, 384–393, doi:10.1038/s41558-021-01032-7.<br>Zang, G. et al., 2021: Synthetic Methanol/Fischer–Tropsch Fuel Production Capacity, Cost, and Carbon Intensity Utilizing $CO_2$ from Industrial and Power Plants in the United States. *Environ. Sci. Technol.*, **55(11)**, 7595–7604, doi:10.1021/acs.est.0c08674.<br>Scheelhaase, J., S. Maertens, and W. Grimme, 2019: Synthetic fuels in aviation – Current barriers and potential political measures. *Transp. Res. Procedia*, **43**, 21–30, doi:10.1016/j.trpro.2019.12.015. | |
| **Electric vehicles for land transport** | + | LE |
| *Role of contexts* | Lifecycle costs for electric vehicles are anticipated to be lower than for conventional vehicles by 2030; *high confidence* for light-duty vehicles; *lower confidence* for heavy-duty applications. | Some grey studies exist on employment effects of electric vehicles; however, the peer-reviewed literature is not well developed. |
| *Line of sight* | IEA, 2021a: *Global EV Outlook 2021*. International Energy Agency, Paris, France, 101 pp.<br>Liimatainen, H., O. van Vliet, and D. Aplyn, 2019: The potential of electric trucks – An international commodity-level analysis. *Appl. Energy*, **236**, 804–814, doi:10.1016/j.apenergy.2018.12.017.<br>Kapustin, N.O. and D.A. Grushevenko, 2020: Long-term electric vehicles outlook and their potential impact on electric grid. *Energy Policy*, **137**, 111103, doi:10.1016/j.enpol.2019.111103.<br>Forrest, K., M. Mac Kinnon, B. Tarroja, and S. Samuelsen, 2020: Estimating the technical feasibility of fuel cell and battery electric vehicles for the medium and heavy duty sectors in California. *Appl. Energy*, **276**, 115439, doi:10.1016/j.apenergy.2020.115439. | |
| **Hydrogen FCV for land transport** | + | LE |
| *Role of contexts* | Lifecycle costs for hydrogen fuel cell vehicles projected to be competitive with conventional vehicles in future, however high uncertainty remains. | Some studies exist on employment effects of hydrogen economy; however, the literature is not well developed and does not apply directly to FCVs. |
| *Line of sight* | Miotti, M., J. Hofer, and C. Bauer, 2017: Integrated environmental and economic assessment of current and future fuel cell vehicles. *Int. J. Life Cycle Assess.*, **22(1)**, 94–110, doi:10.1007/s11367-015-0986-4.<br>Ruffini, E. and M. Wei, 2018: Future costs of fuel cell electric vehicles in California using a learning rate approach. *Energy*, **150**, 329–341, doi:10.1016/j.energy.2018.02.071.<br>Olabi, A.G., T. Wilberforce, and M.A. Abdelkareem, 2021: Fuel cell application in the automotive industry and future perspective. *Energy*, **214**, 118955, doi:10.1016/j.energy.2020.118955. | |

| | Socio-cultural | | |
|---|---|---|---|
| | Public acceptance | Effects on health & well-being | Distributional effects |
| **Demand reduction and mode shift** | ± | + | ± |
| *Role of contexts* | Public support for some measures, such as emissions charging schemes, can be mixed initially, they are likely to gain acceptance as benefits are realised and/or focused. Such as recent COVID-19 road network changes in London. | Significant economic health and well-being benefits. | Some measures, such as travel restrictions, emission charging schemes and others, can have mixed distributional effects initially (e.g., on accessibility). |
| *Line of sight* | Winter, A.K. and H. Le, 2020: Mediating an invisible policy problem: Nottingham's rejection of congestion charging. *Local Environ.*, **25(6)**, 463–471, doi:10.1080/13549839.2020.1753668.<br>Creutzig, F. et al., 2018: Towards demand-side solutions for mitigating climate change. *Nat. Clim. Change*, **8(4)**, 260–263, doi:10.1038/s41558-018-0121-1.<br>DEFRA and DoT, 2020: *Clean Air Zone Framework: Principles for setting up Clean Air Zones in England*., Department of Environment Food & Rural Affairs/Department of Transport, Government of UK, London, UK, 35 pp.<br>Adhikari, M., L.P. Ghimire, Y. Kim, P. Aryal, and S.B. Khadka, 2020: Identification and Analysis of Barriers against Electric Vehicle Use. *Sustainability*, **12(12)**, 4850, doi:10.3390/su12124850.<br>TfL (2020) London Streetspace changes. https://www.pgweb.uk/planning-all-subjects/quieter-neighbourhoods/2847-120-doctors-and-nurses-urge-continuation-of-low-traffic-neighbourhoods-and-cycle-lanes-schemes. | | |

| | Socio-cultural | | |
|---|---|---|---|
| | Public acceptance | Effects on health & well-being | Distributional effects |
| **Biofuels for land transport, aviation, and shipping** | ± | LE | ± |
| *Role of contexts* | Varied public acceptance of biofuel options is observed in different regions of the world. | No known impacts. | Food security but agricultural economies. |
| *Line of sight* | Løkke, S., E. Aramendia, and J. Malskær, 2021: A review of public opinion on liquid biofuels in the EU: Current knowledge and future challenges. *Biomass and Bioenergy*, **150**, 106094, doi:10.1016/j.biombioe.2021.106094.<br>Taufik, D. and H. Dagevos, 2021: Driving public acceptance (instead of skepticism) of technologies enabling bioenergy production: A corporate social responsibility perspective. *J. Clean. Prod.*, **324**, 129273, doi:10.1016/j.jclepro.2021.129273. | | |
| **Ammonia for shipping** | LE | LE | LE |
| *Role of contexts* | Some concerns in industry regarding handling of hazardous fuel; limited evidence overall. | | |
| *Line of sight* | N/A | | |
| **Synthetic fuels for heavy-duty land transport, aviation, and shipping (e.g., DAC-FT)** | LE | LE | NE |
| *Role of contexts* | Currently low public awareness of the technology and little evidence regarding associated perceptions. | No known impacts. | |
| *Line of sight* | N/A | | |
| **Electric vehicles for land transport** | ± | ± | ± |
| *Role of contexts* | Growing public acceptance, especially in some jurisdictions (e.g., majority of light-duty vehicle sales in Norway are electric), but wide differences across regions; range anxiety remains a barrier among some groups. | No major impacts; some potential for reduced noise, which can improve well-being of city residents but may adversely affect pedestrian safety. | Higher vehicle purchase price and access to off-road parking limits access for some disadvantaged groups; potentially insufficient infrastructure for adoption in rural communities (initially); air quality improvements may disproportionately benefit disadvantaged groups, but may also shift some impacts onto communities in close proximity to electricity generators. |
| *Line of sight* | Coffman, M., P. Bernstein, and S. Wee, 2017: Electric vehicles revisited: a review of factors that affect adoption. *Transp. Rev.*, **37(1)**, 79–93, doi:10.1080/01441647.2016.1217282.<br>Burkert, A., H. Fechtner, and B. Schmuelling, 2021: Interdisciplinary Analysis of Social Acceptance Regarding Electric Vehicles with a Focus on Charging Infrastructure and Driving Range in Germany. *World Electr. Veh. J.*, **12(1)**, 25, doi:10.3390/wevj12010025.<br>Wang, N., L. Tang, and H. Pan, 2018b: Analysis of public acceptance of electric vehicles: An empirical study in Shanghai. *Technol. Forecast. Soc. Change*, **126**, 284–291, doi:10.1016/j.techfore.2017.09.011. | Campello-Vicente, H., R. Peral-Orts, N. Campillo-Davo, and E. Velasco-Sanchez, 2017: The effect of electric vehicles on urban noise maps. *Appl. Acoust.*, **116**, 59–64, doi:10.1016/j.apacoust.2016.09.018. | Canepa, K., S. Hardman, and G. Tal, 2019: An early look at plug-in electric vehicle adoption in disadvantaged communities in California. *Transp. Policy*, **78**, 19–30, doi:10.1016/j.tranpol.2019.03.009.<br>Brown, M.A., A. Soni, M.V Lapsa, K. Southworth, and M. Cox, 2020: High energy burden and low-income energy affordability: conclusions from a literature review. *Prog. Energy*, **2(4)**, 42003, doi:10.1088/2516-1083/abb954. |
| **Hydrogen FCV for land transport** | ± | ± | ± |
| *Role of contexts* | Public acceptance is growing in countries where the technology is being promoted and subsidised. However, sparse infrastructure, high costs and perceived safety concerns are currently barriers to a widespread deployment of the technology. | No major impacts: some potential for reduced noise, which can improve well-being of city residents but may adversely affect pedestrian safety. | Higher vehicle purchase price limits access for some disadvantaged groups; potentially insufficient infrastructure for adoption in rural communities (initially); air quality improvements may disproportionately benefit disadvantaged groups. |

| | Socio-cultural | | |
|---|---|---|---|
| | Public acceptance | Effects on health & well-being | Distributional effects |
| *Line of sight* | Itaoka, K., A. Saito, and K. Sasaki, 2017: Public perception on hydrogen infrastructure in Japan: Influence of rollout of commercial fuel cell vehicles. *Int. J. Hydrogen Energy*, **42(11)**, 7290–7296, doi:10.1016/j.ijhydene.2016.10.123.<br>Canepa, K., S. Hardman, and G. Tal, 2019: An early look at plug-in electric vehicle adoption in disadvantaged communities in California. *Transp. Policy*, **78**, 19–30, doi:10.1016/j.tranpol.2019.03.009.<br>Brown, M.A., A. Soni, M. V Lapsa, K. Southworth, and M. Cox, 2020: High energy burden and low-income energy affordability: conclusions from a literature review. *Prog. Energy*, **2(4)**, 42003, doi:10.1088/2516-1083/abb954.<br>Trencher, G., 2020: Strategies to accelerate the production and diffusion of fuel cell electric vehicles: Experiences from California. *Energy Reports*, doi:10.1016/j.egyr.2020.09.008. | | |

| | Institutional | | |
|---|---|---|---|
| | Political acceptance | Institutional capacity and governance, cross-sectoral coordination | Legal and administrative feasibility |
| **Demand reduction and mode shift** | ± | ± | ± |
| *Role of contexts* | Public support for some measures, such as emissions charging schemes, can be mixed initially, it is likely to gain acceptance as benefits are realised and/or focused. Such as recent COVID-19 road network changes in London. | Some local authorities have limited capacity to deliver demand management measures as compared to other developed authorities. However, this can be mitigated to optioneering processes to select the preferred measures in the local context. | Legal air quality limits are forcing cities and countries to implement travel demand reduction and fuel efficiency measures, such as in the UK and Europe. However, there may be legal and administrative changes in delivery of measures. |
| *Line of sight* | Winter, A.K. and H. Le, 2020: Mediating an invisible policy problem: Nottingham's rejection of congestion charging. *Local Environ.*, **25(6)**, 463–471, doi:10.1080/13549839.2020.1753668.<br>Creutzig, F. et al., 2018: Towards demand-side solutions for mitigating climate change. *Nat. Clim. Change*, **8(4)**, 260–263, doi:10.1038/s41558-018-0121-1.<br>DEFRA and DoT, 2020: *Clean Air Zone Framework: Principles for setting up Clean Air Zones in England.*, Department of Environment Food & Rural Affairs/Department of Transport, Government of UK, London, U35 pp.<br>TfL (2020) London Streetspace changes. https://www.pgweb.uk/planning-all-subjects/quieter-neighbourhoods/2847-120-doctors-and-nurses-urge-continuation-of-low-traffic-neighbourhoods-and-cycle-lanes-schemes. | | |
| **Biofuels for land transport, aviation, and shipping** | ± | ± | ± |
| *Role of contexts* | Varied political support for biofuels deployment in different regions of the world. | There is varied institutional capacity to coordinate biofuels deployment in different regions of the world. | There are different legal contexts and barriers for biofuels implementation on different regions of the world. |
| *Line of sight* | Lynd, L.R., 2017: The grand challenge of cellulosic biofuels. *Nat. Biotechnol.*, **35(10)**, 912–915, doi:10.1038/nbt.3976.<br>Markel, E., C. Sims, and B.C. English, 2018: Policy uncertainty and the optimal investment decisions of second-generation biofuel producers. *Energy Econ.*, **76**, 89–100, doi:10.1016/j.eneco.2018.09.017. | | |
| **Ammonia for shipping** | ± | - | - |
| *Role of contexts* | Varied political support for deployment in different regions of the world. | The major contributor to marine emissions is international shipping, which falls under the jurisdiction of the International Maritime Organization. Coordination with international governments will be required. | Potential challenges related to emissions regulations. |
| *Line of sight* | Hoegh-Guldberg, O. et al., 2019: *The Ocean as a Solution to Climate Change: Five Opportunities for Action*. World Resources Institute, Washington D. C., 116 pp.<br>Energy Transitions Commission, 2021. *Making the hydrogen economy possible*. Energy Transitions Commission, https://energy-transitions.org/wp-content/uploads/2021/04/ETC-Global-Hydrogen-Report.pdf.<br>Energy Transitions Commission, 2020. *The First Wave: A blueprint for commercial-scale zero-emission shipping pilots*. Energy Transitions Commission, https://www.energy-transitions.org/wp-content/uploads/2020/11/The-first-wave.pdf. | | |

| | Institutional | | |
|---|---|---|---|
| | Political acceptance | Institutional capacity and governance, cross-sectoral coordination | Legal and administrative feasibility |
| **Synthetic fuels for heavy-duty land transport, aviation, and shipping (e.g., DAC-FT)** | LE | – | ± |
| *Role of contexts* | Plans for adoption of technology remain at early stage; political acceptance not known. | Synthetic fuel use in aviation and marine shipping requires international coordination; challenges exist related to carbon accounting frameworks for utilisation of $CO_2$; likely fewer barriers for use of fuel in land transport applications. | Legal barriers exist for synthetic fuel use in aviation; need for development of $CO_2$ capture markets; drop-in fuels are compatible with existing fuel standards in many jurisdictions. |
| *Line of sight* | Scheelhaase, J., S. Maertens, and W. Grimme, 2019: Synthetic fuels in aviation – Current barriers and potential political measures. *Transp. Res. Procedia*, **43**, 21–30, doi:10.1016/j.trpro.2019.12.015. | | |
| **Electric vehicles for land transport** | ± | ± | ± |
| *Role of contexts* | Varied political support for deployment in different regions of the world. | Coordination needed between transport sector (including vehicle manufacturers; charging infrastructure) and power sector (including increased generation and transmission; capacity to handle demand peaks). Institutional capacity is variable. | Compatible with urban low emission zones; grid integration may require market and regulatory changes. |
| *Line of sight* | Milovanoff, A., I.D. Posen, and H.L. MacLean, 2020: Electrification of light-duty vehicle fleet al.ne will not meet mitigation targets. *Nat. Clim. Change*, **10(12)**, 1102–1107, doi:10.1038/s41558-020-00921-7.<br>IEA, 2021: *Global EV Outlook 2021*. International Energy Agency, Paris, France, 101 pp. | | |
| **Hydrogen FCV for land transport** | ± | ± | ± |
| *Role of contexts* | Varied political support for deployment in different regions of the world. | Coordination needed across sector (including vehicle manufacturers, hydrogen producers and refuelling infrastructure). Institutional capacity is variable. | Compatible with urban low emission zones; fuel distribution network may require market and regulatory changes. |
| *Line of sight* | Itaoka, K., A. Saito, and K. Sasaki, 2017: Public perception on hydrogen infrastructure in Japan: Influence of rollout of commercial fuel cell vehicles. *Int. J. Hydrogen Energy*, **42(11)**, 7290–7296, doi:10.1016/j.ijhydene.2016.10.123. | | |

# 11

# Industry


**Coordinating Lead Authors:**
Igor A. Bashmakov (the Russian Federation), Lars J. Nilsson (Sweden)

**Lead Authors:**
Adolf Acquaye (Ghana/United Kingdom), Christopher Bataille (Canada), Jonathan M. Cullen (New Zealand/United Kingdom), Stéphane de la Rue du Can (the United States of America), Manfred Fischedick (Germany), Yong Geng (China), Kanako Tanaka (Japan)

**Contributing Authors:**
Fredric Bauer (Sweden), Ali Hasanbeigi (the United States of America), Peter Levi (United Kingdom), Anna Myshak (the Russian Federation), Daniel Perczyk (Argentina), Cedric Philibert (France), Sascha Samadi (Germany)

**Review Editors:**
Nick Campbell (France/United Kingdom), Ramón Pichs-Madruga (Cuba)

**Chapter Scientist:**
Siyue Guo (China)



**This chapter should be cited as:**
Bashmakov, I.A., L.J. Nilsson, A. Acquaye, C. Bataille, J.M. Cullen, S. de la Rue du Can, M. Fischedick, Y. Geng, K. Tanaka, 2022: Industry. In IPCC, 2022: *Climate Change 2022: Mitigation of Climate Change. Contribution of Working Group III to the Sixth Assessment Report of the Intergovernmental Panel on Climate Change* [P.R. Shukla, J. Skea, R. Slade, A. Al Khourdajie, R. van Diemen, D. McCollum, M. Pathak, S. Some, P. Vyas, R. Fradera, M. Belkacemi, A. Hasija, G. Lisboa, S. Luz, J. Malley, (eds.)]. Cambridge University Press, Cambridge, UK and New York, NY, USA. doi: 10.1017/9781009157926.013

# Table of Contents

## Executive Summary

**The Paris Agreement, the Sustainable Development Goals (SDGs) and the COVID-19 pandemic provide a new context for the evolution of industry and the mitigation of industry greenhouse gas (GHG) emissions (*high confidence*).** This chapter is focused on what is new since AR5. It emphasises the energy and emissions intensive basic materials industries and key strategies for reaching net zero emissions. {11.1.1}

**Net zero $CO_2$ emissions from the industrial sector are possible but challenging (*high confidence*).** Energy efficiency will continue to be important. Reduced materials demand, material efficiency, and circular economy solutions can reduce the need for primary production. Primary production options include switching to new processes that use low to zero GHG energy carriers and feedstocks (e.g., electricity, hydrogen, biofuels, and carbon capture and utilisation (CCU) for carbon feedstock), and carbon capture and storage (CCS) for remaining $CO_2$. These options require substantial scaling up of electricity, hydrogen, recycling, $CO_2$, and other infrastructure, as well as phase-out or conversion of existing industrial plants. While improvements in the GHG intensities of major basic materials have nearly stagnated over the last 30 years, analysis of historical technology shifts and newly available technologies indicate these intensities can be reduced to net zero emissions by mid-century. {11.2, 11.3, 11.4}

**Whatever metric is used, industrial emissions have been growing faster since 2000 than emissions in any other sector, driven by increased basic materials extraction and production (*high confidence*).** GHG emissions attributed to the industrial sector originate from fuel combustion, process emissions, product use and waste, which jointly accounted for 14.1 $GtCO_2$-eq or 24% of all direct anthropogenic emissions in 2019, second behind the energy transformation sector. Industry is a leading GHG emitter – 20 $GtCO_2$-eq or 34% of global emissions in 2019 – if indirect emissions from power and heat generation are included. The share of emissions originating from direct fuel combustion is decreasing and was 7 $GtCO_2$-eq, 50% of direct industrial emissions in 2019. {11.2.2}

**Global material intensity (in-use stock of manufactured capital, in tonnes per unit of GDP is increasing (*high confidence*).** In-use stock of manufactured capital per capita has been growing faster than GDP per capita since 2000. Total global in-use stock of manufactured capital grew by 3.4% $yr^{-1}$ in 2000–2019. At the same time, per capita material stocks in several developed countries have stopped growing, showing a decoupling from GDP per capita. {11.2.1, 11.3.1}

**Plastic is the material for which demand has been growing the strongest since 1970 (*high confidence*).** The current >99% reliance on fossil feedstock, very low recycling, and high emissions from petrochemical processes is a challenge for reaching net zero emissions. At the same time, plastics are important for reducing emissions elsewhere, for example, light-weighting vehicles. There are as yet no shared visions for fossil-free plastics, but several possibilities. {11.4.1.3}

**Scenario analyses show that significant cuts in global GHG emissions and even close to net zero emissions from GHG intensive industry (e.g., steel, plastics, ammonia, and cement) can be achieved by 2050 by deploying multiple available and emerging options (*medium confidence*).** Cutting industry emissions significantly requires a reorientation from the historic focus on important but incremental improvements (e.g., energy efficiency) to transformational changes in energy and feedstock sourcing, materials efficiency, and more circular material flows. {11.3, 11.4}

**Key climate mitigation options such as materials efficiency, circular material flows and emerging primary processes, are not well represented in climate change scenario modelling and integrated assessment models, albeit with some progress in recent years (*high confidence*).** The character of these interventions (e.g., appearing in many forms across complex value chains, making cost estimates difficult) combined with the limited data on new fossil-free primary processes help explain why they are less represented in models than, for example, CCS. As a result, overall mitigation costs and the need for CCS may be overestimated. {11.4.2.1}

**Electrification is emerging as a key mitigation option for industry (*high confidence*).** Electricity is a versatile energy carrier, potentially produced from abundant renewable energy sources or other low carbon options; regional resources and preferences will vary. Using electricity directly, or indirectly via hydrogen from electrolysis for high temperature and chemical feedstock requirements, offers many options to reduce emissions. It also can provide substantial grid balancing services, for example through electrolysis and storage of hydrogen for chemical process use or demand response. {11.3.5}

**Carbon is a key building block in organic chemicals, fuels and materials, and will remain important (*high confidence*).** In order to reach net zero $CO_2$ emissions for the carbon needed in society (e.g., plastics, wood, aviation fuels, solvents, etc.), it is important to close the use loops for carbon and carbon dioxide through increased circularity with mechanical and chemical recycling, more efficient use of biomass feedstock with the addition of low GHG hydrogen to increase product yields (e.g., for biomethane and methanol), and potentially direct air capture of $CO_2$ as a new carbon source. {11.3, 11.4.1}

**Production costs for very low to zero emissions basic materials may be high but the cost for final consumers and the general economy will be low (*medium confidence*).** Costs and emissions reductions potential in industry, and especially heavy industry, are highly contingent on innovation, commercialisation, and market uptake policy. Technologies exist to take all industry sectors to very low or zero emissions but require 5 to 15 years of intensive innovation, commercialisation, and policy to ensure uptake. Mitigation costs are in the rough range of USD50–150 $tCO_2$-$eq^{-1}$, with wide variation within and outside this band. This affects competitiveness and requires supporting policy. Although production cost increases can be significant, they translate to very small increases in the costs for final products, typically less than a few percent depending on product, assumptions, and system boundaries. {11.4.1.5}

**There are several technological options for very low to zero emissions steel, but their uptake will require integrated material efficiency, recycling, and production decarbonisation policies (*high confidence*).** Material efficiency can potentially reduce steel demand by up to 40% based on design for less steel use, long life, reuse, constructability, and low contamination recycling. Secondary production through high quality recycling must be maximised. Production decarbonisation will also be required, starting with the retrofitting of existing facilities for partial fuel switching (e.g., to biomass or hydrogen), CCU and CCS, followed by very low and zero emissions production based on high-capture CCS or direct hydrogen, or electrolytic iron ore reduction followed by an electric arc furnace. {11.3.2, 11.4.1.1}

**There are several current and near-horizon options to greatly reduce cement and concrete emissions. Producer, user, and regulator education, as well as innovation and commercialisation policy are needed (*medium confidence*).** Cement and concrete are currently overused because they are inexpensive, durable, and ubiquitous, and consumption decisions typically do not give weight to their production emissions. Basic material efficiency efforts to use only well-made concrete thoughtfully and only where needed (e.g., using right-sized, prefabricated components) could reduce emissions by 24–50% through lower demand for clinker. Cementitious material substitution with various materials (e.g., ground limestone and calcined clays) can reduce process calcination emissions by up to 50% and occasionally much more. Until a very low GHG emissions alternative binder to Portland cement is commercialised, which does not look promising in the near to medium term, CCS will be essential for eliminating the limestone calcination process emissions for making clinker, which currently represent 60% of GHG emissions in best available technology plants. {11.3.2, 11.3.6, 11.4.1.2}

**While several technological options exist for decarbonising the main industrial feedstock chemicals and their derivatives, the costs vary widely (*high confidence*).** Fossil fuel-based feedstocks are inexpensive and still without carbon pricing, and their biomass- and electricity-based replacements will likely be more expensive. The chemical industry consumes large amounts of hydrogen, ammonia, methanol, carbon monoxide, ethylene, propylene, benzene, toluene, and mixed xylenes and aromatics from fossil feedstock, and from these basic chemicals produces tens of thousands of derivative end-use chemicals. Hydrogen, biogenic or air-capture carbon, and collected plastic waste for the primary feedstocks can greatly reduce total emissions. Biogenic carbon feedstock is likely to be limited due to competing land uses. {11.4.1.3}

**Light industry and manufacturing can be largely decarbonised through switching to low GHG fuels (e.g., biofuels and hydrogen) and electricity (e.g., for electrothermal heating and heat pumps) (*high confidence*).** Most of these technologies are already mature, for example, for low temperature heat, but a major challenge is the current low cost of fossil methane and coal relative to low and zero GHG electricity, hydrogen, and biofuels. {11.4.1.4}

**The pulp and paper industry has significant biogenic carbon emissions but relatively small fossil carbon emissions. Pulp mills have access to biomass residues and by-products and in paper mills the use of process heat at low to medium temperatures allows for electrification (*high confidence*).** Competition for feedstock will increase if wood substitutes for building materials and petrochemicals feedstock. The pulp and paper industry can also be a source of biogenic carbon dioxide and carbon for organic chemicals feedstock and carbon dioxide removal (CDR) using CCS. {11.4.1.4}

**The geographical distribution of renewable resources has implications for industry (*medium confidence*).** The potential for zero emission electricity and low-cost hydrogen from electrolysis powered by solar and wind, or hydrogen from other very low emission sources, may reshape where currently energy and emissions intensive basic materials production is located, how value chains are organised, trade patterns, and what gets transported in international shipping. Regions with bountiful solar and wind resources, or low fugitive methane co-located with CCS geology, may become exporters of hydrogen or hydrogen carriers such as methanol and ammonia, or home to the production of iron and steel, organic platform chemicals, and other energy-intensive basic materials. {11.2, 11.4 and Box 11.1}

**The level of policy maturity and experience varies widely across the mitigation options (*high confidence*).** Energy efficiency is a well-established policy field with decades of experience from voluntary and negotiated agreements, regulations, energy auditing and demand side-management (DSM) programmes (see AR5). In contrast, materials demand management and efficiency are not well understood and addressed from a policy perspective. Barriers to recycling that policy could address are often specific to the different material loops (e.g., copper contamination for steel and lack of technologies or poor economics for plastics) or waste management systems. For electrification and fuel switching the focus has so far been mainly on innovation and developing technical supply-side solutions rather than creating market demand. {11.5.2, 11.6}

**Industry has so far largely been sheltered from the impacts of climate policy and carbon pricing due to concerns for competitiveness and carbon leakage (*high confidence*).** New industrial development policy approaches needed for realising a transition to net zero GHG emissions are emerging. The transition requires a clear direction towards net zero, technology development, market demand for low-carbon materials and products, governance capacity and learning, socially inclusive phase-out plans, as well as international coordination of climate and trade policies. It requires comprehensive and sequential industrial policy strategies leading to immediate action as well as preparedness for future decarbonisation, governance at different levels (from international to local), and integration with other policy domains. {11.6}

## 11.1 Introduction and New Developments

### 11.1.1 About This Chapter

The AR5 was published in 2014. The Paris Agreement and the 17 Sustainable Development Goals (SDGs) were adopted in 2015. An increasing number of countries have since announced ambitions to be carbon neutral by 2045–2060. The COVID-19 pandemic shocked the global economy in 2020 and motivated economic stimulus with demands for green recovery and concerns for economic security. All this has created a new context and a growing recognition that all industry, including the energy and emissions intensive industries, need to reach net zero GHG emissions. There is an ongoing mind shift around the opportunities to do so, with electrification and hydrogen emerging among key mitigation options as a result of renewable electricity costs falling rapidly. On the demand side there has been renewed attention to end-use demand, material efficiency, and more and better-quality recycling measures. This chapter takes its starting point in this new context and emphasises the need for deploying innovative processes and practices in order to limit the global warming to 1.5°C or 2°C (IPCC 2018a).

The industrial sector includes ores and minerals mining, manufacturing, construction and waste management. It is the largest source of global GHG and $CO_2$ emissions, which include direct and indirect fuel-combustion-related emissions, emissions from industrial processes and products use, as well as from waste. This chapter is focused on heavy industry – the high temperature heat and process emissions intensive basic materials industries that account for 65% of industrial GHG and over 70% of industrial $CO_2$ emissions (waste excluded), where deployment of near-zero emissions technologies can be more challenging due to capital intensity and equipment lifetimes compared with other manufacturing industries. The transition of heavy industries to zero emissions requires supplementing the traditional toolkit of energy and process efficiency, fuel switching, electrification, and decarbonisation of power with material end-use demand management and efficiency, circular economy, fossil-free feedstocks, carbon capture and utilisation (CCU), and carbon capture and storage (CCS). Energy efficiency was extensively treated in AR5 and remains a key mitigation option. This chapter is focused mainly on new options and developments since AR5, highlighting measures along the whole value chains that are required to approach zero emissions in primary materials production.

### 11.1.2 Approach to Understanding Industrial Emissions

The Kaya identity offers a useful tool of decomposing emission sources and their drivers, as well as of weighing the mitigation options. The one presented below (Equation 11.1) builds on the previous assessments (IPCC 2014, 2018b; Hoegh-Guldberg et al. 2018), and reflect a material stock-driven services-oriented vision to better highlight the growing importance of industrial processes (dominated in emissions increments in 2010–2019), product use and waste in driving emissions. Services delivery (nutrition, shelter, mobility, education, etc.; see Chapter 5 for more detail) not only requires energy and material flows (fuels, food, feed, fertilisers, packaging, etc.), but also material stocks (buildings, roads, vehicles, machinery, etc.), the mass of which has already exceeded 1000 Gt (Krausmann et al. 2018). As material efficiency appears to be an important mitigation option, material intensity or productivity (material extraction or consumption versus GDP (Oberle et al. 2019; Hertwich et al. 2020)) is reflected in the identity with two dimensions: as material stock intensity of GDP (tonnes per dollar) and material intensity of building and operating accumulated in-use stock.[1] For sub-global analysis the ratio of domestically used materials to total material production becomes important to reflect outsourced materials production and distinguish between territorial and consumption-based emissions. The identity for industry differs significantly from that for sectors with where combustion emissions dominate (Lamb et al. 2021).



1 Accumulated material stock initially was introduced in the analysis of past trends (Krausmann et al. 2018; Wiedenhofer et al. 2019), but recently it was incorporated in different forms in the long-term projections for the whole economy (Krausmann et al. 2020) and for some sectors (buildings and cars in Hertwich et al. (2020)) with a steadily improving regional resolution (Krausmann et al. 2020).

Recent progress in data availability that allows the integration of major emission sources along with socio-economic metabolism, material flows and stock analysis enriches the identity for industry from a perspective of possible policy interventions (Bashmakov 2021):

$$GHG = POP \cdot \frac{GDP}{POP} \cdot \frac{MStock}{GDP} \cdot \left[\frac{MPR + MSE}{MStock} \cdot Dm \cdot \left(\frac{E}{(MPR + MSE)} \cdot \frac{(GHGed + GHGeind)}{E} + \frac{GHGoth}{MPR + MSE}\right)\right]$$

Equation 11.1

**Equation 11.1 Table 1 | Variables, Factors, Policies and Drivers**

| Variables | Factors | Policies and drivers | |
|---|---|---|---|
| $POP$ | Population | Demographic policies | Demand decarbonisation |
| $\frac{GDP}{POP}$ | Services (expressed via *GDP* – final consumption and investments needed to maintain and expand stock) per capita | Sufficiency and demand management (reduction) | |
| $\frac{MStock}{GDP}$ | Material stock (*MStock* – accumulated in-use stocks of materials embodied in manufactured fixed capital) intensity of *GDP* | Material stock efficiency improvement | |
| $\frac{MPR + MSE}{MStock}$ | Material inputs (both virgin (primary materials extraction, *MPR*) and recycled (secondary materials use, *MSE*)) per unit of in-use material stock | Material efficiency, substitution and circular economy | |
| $Dm$ | Share of allocated emissions – consumption vs production emissions accounting (valid only for sub-global levels)* | Trade policies including carbon leakage issues (localisation versus globalisation) | CBAM |
| $\frac{E}{(MPR + MSE)}$ | Sum of energy use for basic material production (*Em*), processing and other operational industrial energy use (*Eoind*) per unit of material inputs | Energy efficiency of basic materials production and other industrial processes | Production decarbonisation |
| $\frac{(GHGed + GHGeind)}{E}$ | Direct (*GHGed*) and indirect (*GHGeind*) combustion-related industrial emissions per unit of energy | Electrification, fuel switching, and energy decarbonisation (hydrogen, CCUS-fuels) | |
| $\frac{GHGoth}{MPR + MSE}$ | Emissions from industrial processes and product use, waste, F-gases, indirect nitrogen emissions per unit of produced materials | Feedstock decarbonisation (hydrogen), CCUS-industrial processes, waste and F-gases management | |

*$Dm$=1, when territorial emission is considered, and $Dm$ equals the ratio of domestically used materials to total material production for the consumption-based emission accounting). CBAM – carbon border adjustment mechanism.

Factors in Equation 11.1 are interconnected by either positive or negative feedbacks: scrap-based production or light-weighing improves operational energy efficiency, while growing application of carbon capture, use and storage (CCUS) brings it down and increases material demands (Hertwich et al. 2019; IEA 2020a, 2021a). There are different ways to disaggregate Equation 11.1: by industrial subsectors (Bashmakov 2021); by reservoirs of material stock (buildings, infrastructure, vehicles, machinery and appliances, packaging, etc.); by regions and countries (where carbon leakage becomes relevant); by products and production chains (material extraction, production of basic materials, basic materials processing, production of final industrial products); by traditional and low carbon technologies used; and by stages of products' lives including recycling.

An industrial transition to net zero emissions is possible when the three last multipliers in Equation 11.1 (in square parentheses) are approaching zero. Contributions from different drivers (energy efficiency, low carbon electricity and heat, material efficiency, switching to low carbon feedstock and CCUS) to this evolution vary with time. Energy efficiency dominates in the short- and medium term and potentially long term (in the range of 10–40% by 2050) (IPCC 2018a; Crijns-Graus et al. 2020; IEA 2020a), but for deep decarbonisation trajectories, contributions from the other drivers steadily grow, as the share of non-energy sources in industrial emissions rises and new technologies to address mitigation from these sources mature (Material Economics 2019; CEMBUREAU 2020; BP 2020; Hertwich et al. 2020, 2019; IEA 2021a, 2020a; Saygin and Gielen 2021) (Figure 11.1).

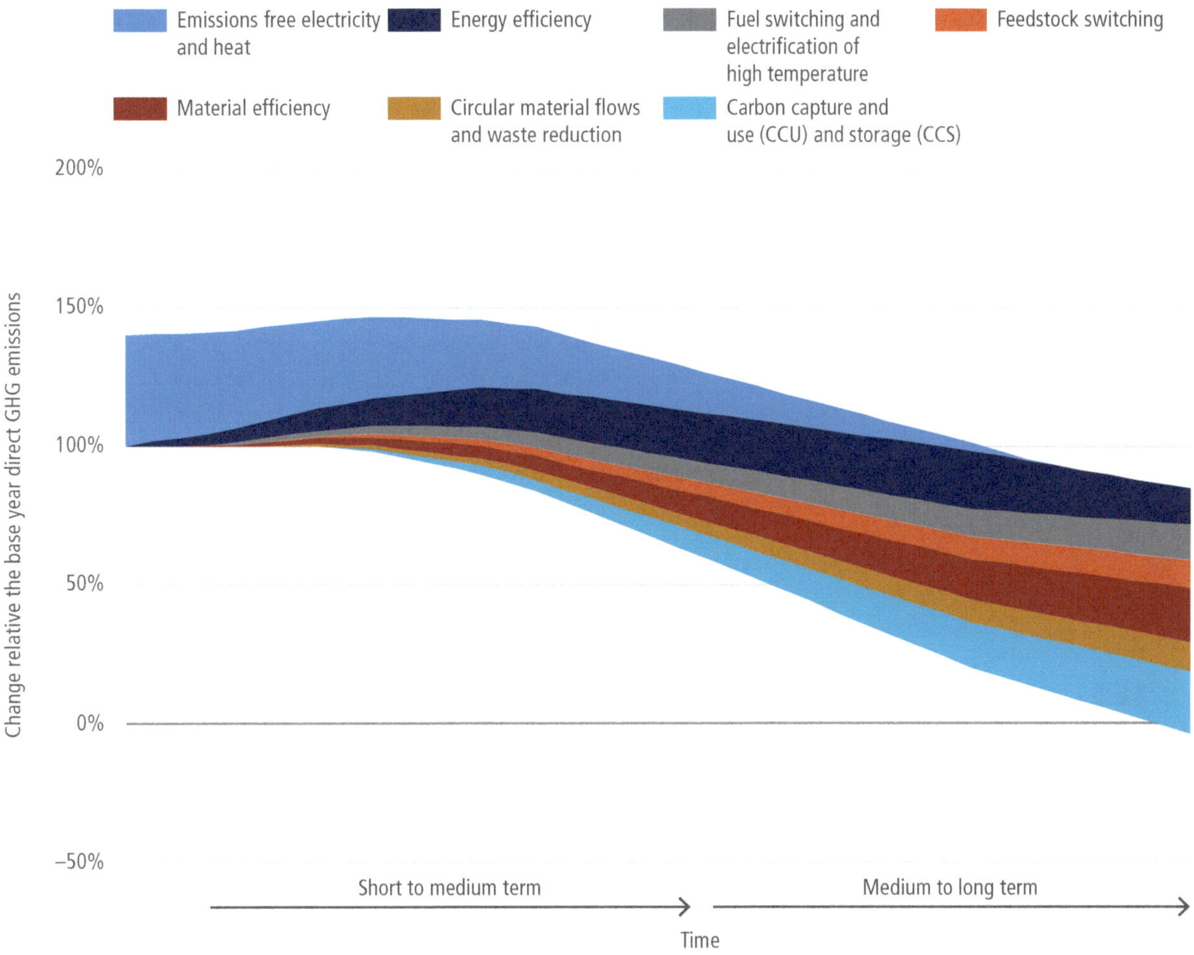


**Mitigation options**

| | Short to medium term | Medium to long term |
|---|---|---|
| Decarbonising production | – Reduction of indirect emissions via lower-carbon electricity and heat supply | – Provision of emissions-free elctricity and high temperature heat |
| Decarbonising production | – Energy efficiency improvements to best available technologies | – Energy efficiency approaching thermodynamic minimums |
| Decarbonising production | – Fuel switching, biomass and electricity use for high temperature process heat | – Deep low-carbon electrification, green hydrogen use |
| Decarbonising production | – Partial substitution of high-carbon feedstock | – Zero emissions feedstock (green hydrogen, biomass) for basic materials production |
| Decarbonising production | – Small scale and sectorally narrow concentrated $CO_2$ flow CCUS | – Broad-scale, large-scale concentrated $CO_2$ flow and possibly post-combustion CCUS |
| Decarbonising demand | – Material efficiency and substitution | – Ecodesign, material efficiency, demand reduction |
| Decarbonising demand | – Increasing recycling rates | – Circular material flows and effective industrial waste management |

Base year and contributions from the drivers are only illustrative. Drivers' contribution varies across industries. Indirect emissions reduction is considered as outcome of mitigation activites in the energy sector; see Chapter 6.

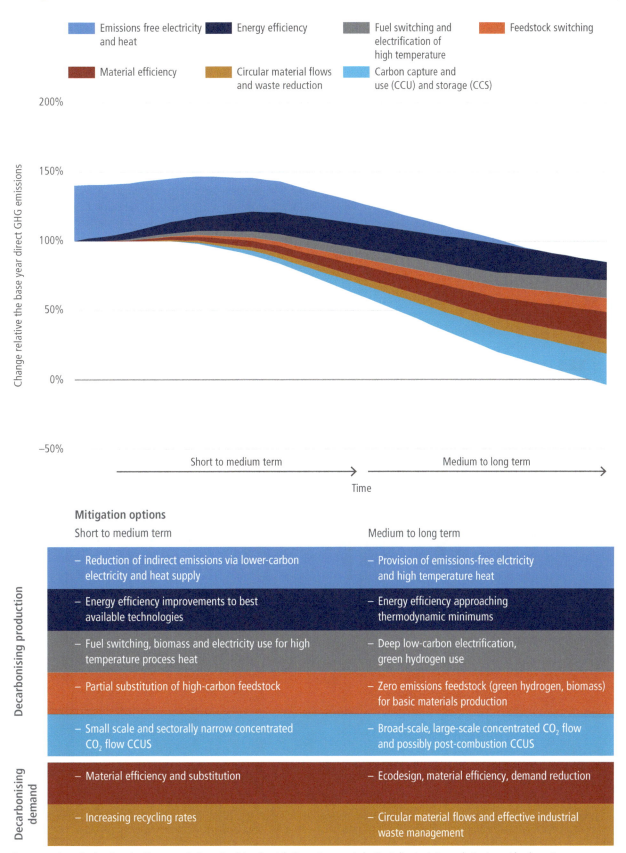

**Figure 11.1 | Stylised composition and contributions from different drivers to the transition of industry to net zero emissions.**

## 11.2 New Trends in Emissions and Industrial Development

### 11.2.1 Major Drivers

The use of materials is deeply coupled with economic development and growth. For centuries, humanity has been producing and using hundreds of materials (Ashby 2012), the diversity of which skyrocketed in the recent half-century to achieve the desired performance and functionality of multiple products (density; hardness; compressive strength; melting point, resistance to mechanical and thermal shocks and to corrosion; transparency; heat- or electricity conductivity; chemical neutrality or activity, to name a few). New functions drive the growth of material complexity of products; for example, a modern computer chip embodies over 60 different elements (Graedel et al. 2015).

Key factors driving up industrial GHG emissions since 1900 include population and per capita GDP,[2] while energy efficiency and non-combustion GHG emissions intensity (from industrial processes and waste) has been pushing it down. Material efficiency factors – material stock intensity of GDP and ratio of extraction, processing and recycling of materials per unit of built capital along with combustion-related emissions intensity factors and electrification – were cyclically switching their contributions with relatively limited overall impact. Growing recycling allowed for replacement of some energy-intensive virgin materials and thus contributed to mitigation. In 2014–2019, a combination of these drivers allowed for a slowdown in the growth of industrial GHG emissions to below 1% (Figure 11.2 and Table 11.1), while to match a net zero emissions trajectory it should decline by 2% $yr^{-1}$ in 2020–2030 and by 8.9% $yr^{-1}$ in 2030–2050 (IEA 2021a).

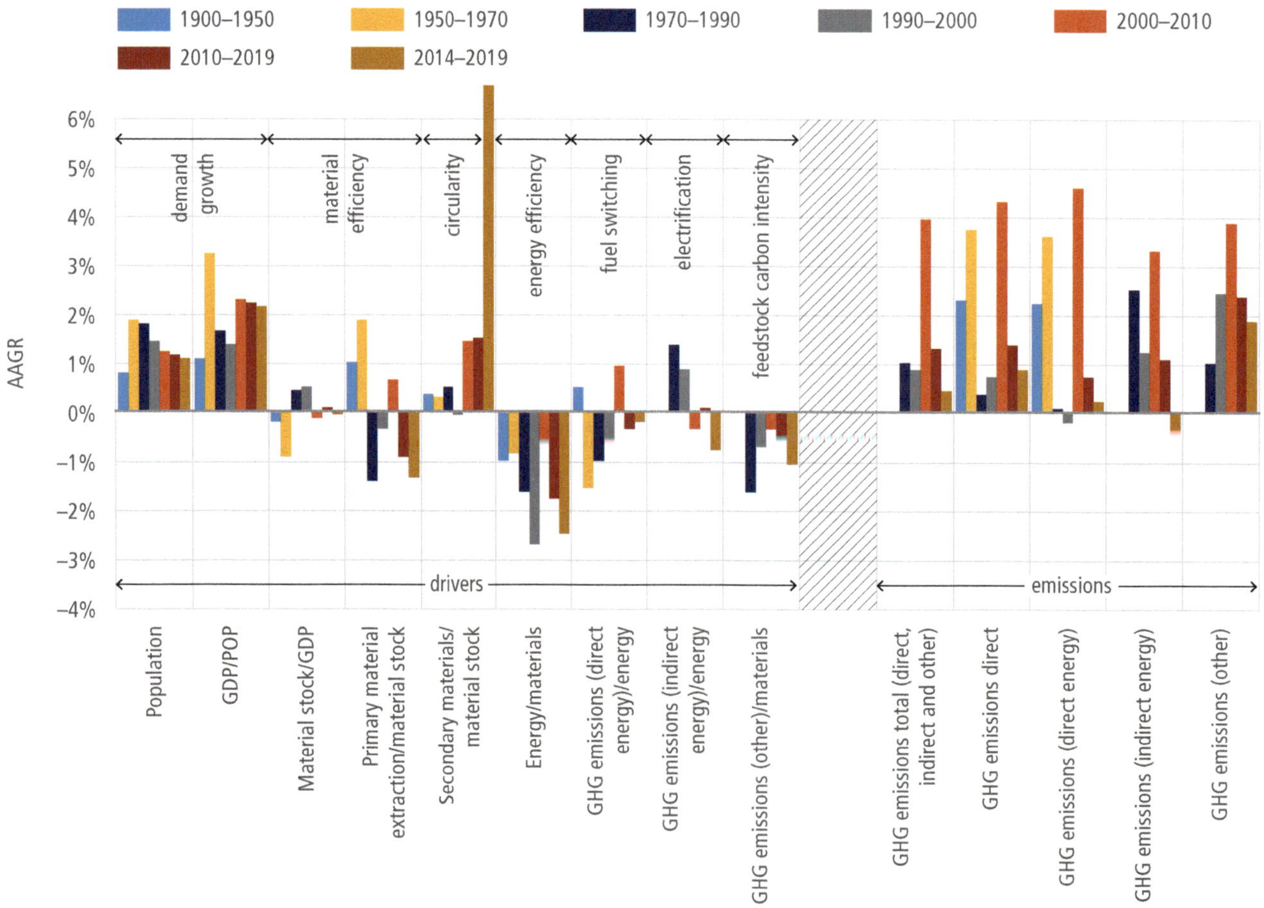


**Figure 11.2 | Average annual growth rates of industrial sector GHG emissions and drivers (1900–2019). Before 1970, GHG emission (other) is limited to that from cement production. Waste emission is excluded. Primary material extraction excludes fuels and biomass. Presented factors correspond directly to Equation 11.1.** Sources: population before 1950 and GDP before 1960: Maddison Project (2018); population from 1950 to 1970: UN (2015); population and GDP for 1960–2020: World Bank (2021); data on material stock, extraction, and use of secondary materials: Wiedenhofer et al. (2019); data on material extraction: UNEP and IRP (2020); industrial energy use for 1900–1970: IIASA (2018), for 1971–2019: IEA (2021b); data on industrial GHG emissions for 1900–1970: CDIAC (2017), for 1970–2019: data from Crippa et al. (2021) and Minx et al. (2021).

[2] In 2020 this factor played on the reduction side as the COVID-19 crisis led to a global decline in demand for basic materials, respective energy use and emissions by 3–5 % (IEA 2020a).

There are two major concepts of **material efficiency** (*ME*). The broader one highlights demand reduction via policies promoting more intensive use, assuming sufficient (excluding luxury) living space or car ownership providing appropriate service levels – housing days or miles driven and life-time extension (Hertwich et al. 2019, 2020). This approach focuses on dematerialisation of society (Lechtenböhmer and Fischedick 2020), where a 'dematerialisation multiplier' (Pauliuk et al. 2021) limits both material stock and GDP growth, as progressively fewer materials are required to build and operate the physical in-use stock to deliver sufficient services. According to the IRP (2020), reducing floor space demand by 20% via shared and smaller housing compared to the reference scenario would decrease Group of Seven (G7) countries' GHG emissions from the material-cycle of residential construction up to 70% in 2050. The narrower concept ignores demand and sufficiency aspects and focuses on supply chains considering *ME* as less basic materials use to produce a certain final product, for example, a car or a metre squared of living space (OECD 2019a; IEA 2020a). No matter if the broader or the narrower concept of *ME* is applied, in 1970–2019 it did not contribute much to the decoupling of industrial emissions from GDP. This is expected to change in the future (Figure 11.2).

Material efficiency analysis mostly uses material intensity or productivity indicators, which compare material extraction or consumption with GDP (Oberle et al. 2019; Hertwich et al. 2020). Those indicators are functions of **material stock intensity of GDP** (tonnes per dollar) and material intensity of building and operating accumulated in-use stock. Coupling services or GDP with the built stock allows for a better evaluation of demand for primary basic materials (Müller et al. 2011; Liu et al. 2013; Liu and Müller 2013; Pauliuk et al. 2013a; Cao et al. 2017; Wiedenhofer et al. 2019; Hertwich et al. 2020; Krausmann et al. 2020). Since 1970 material stock growth driven by industrialisation and urbanisation slightly exceeded that of GDP and there was no decoupling,[3] so in Kaya-like identities material stock may effectively replace GDP. There are different methods to estimate the former (see reviews in Pauliuk et al. (2015, 2019) and Wiedenhofer et al. (2019), the results of which are presented for major basic materials with some geographical resolution (Liu and Müller 2013; Pauliuk et al. 2013a) or globally (Graedel et al. 2011; Geyer et al. 2017; Krausmann et al. 2018; Pauliuk et al. 2019; Wiedenhofer et al. 2019; International Aluminium Institute 2021a).

For a subset of materials, such as solid wood, paper, plastics, iron/steel, aluminium, copper, other metals/minerals, concrete, asphalt, bricks, aggregate, and glass, total in-use stock escalated from 36 Gt back in 1900 to 186 Gt in 1970, 572 Gt in 2000, and 960 Gt in 2015, and by 2020 it exceeded 1,100 Gt, or 145 tonnes per capita (Krausmann et al. 2018, 2020; Wiedenhofer et al. 2019). In 1900–2019, the stock grew 31-fold, which is strongly coupled with GDP growth (36-fold). As the UK experience shows, material stock intensity of GDP may ultimately decline after services fully dominate GDP, and this allows for material productivity improvements to achieve absolute reduction in material use, as stock expansion slows down (Streeck et al. 2020). While the composition of basic materials within the stock of manufactured capital was evolving significantly, overall stock use associated with a unit of GDP has been evolving over the last half-century in a quite narrow range of 7.7–8.6 t per USD1000 (2017 purchasing power parity (PPP)) showing neither signs of decoupling from GDP, nor saturation as of yet. Mineral building materials (concrete, asphalt, bricks, aggregate, and glass) dominate the stock volume by mass (94.6% of the whole stock, with the share of concrete alone standing at 43.5%), followed by metals (3.5%) and solid wood (1.4%). The largest part of in-use stock of our 'cementing societies' (Cao et al. 2017) is constituted by concrete: about 417 Gt in 2015; Krausmann et al. (2018) extrapolated this to 478 Gt (65 tonnes per capita) in 2018, which contains about 88 Gt of cement.[4] The iron and steel stock is assessed at 25–35 Gt (Wiedenhofer et al. 2019; Gielen et al. 2020; Wang et al. 2021), while the plastics stock reached 2.5–3.2 Gt (Geyer et al. 2017; Wiedenhofer et al. 2019; Saygin and Gielen 2021) and the aluminium stock approached 1.1 Gt (International Aluminium Institute 2021a), or just 0.1% of the total. In sharp contrast to global energy intensity, which has more than halved since 1900 (Bashmakov 2019), in 2019 material stock intensity (in-use stock of manufactured capital per GDP) was only 14% below the 1900 level, but 15% above the 1970 level. In-use stock per capita has been growing faster than GDP per capita since 2000 (Figure 11.3). The growth rate of total in-use stock of manufactured capital was 3.8% in 1971–2000 and 3.5% in 2000–2019, or 32–35 Gt $yr^{-1}$, to which concrete and aggregates contributed 88%. Recent demand for stockbuilding materials was 51–54 Gt $yr^{-1}$, to which recycled materials recently contributed only about 10% of material input. About 46–49 Gt $yr^{-1}$ was virgin inputs, which after accounting for processing waste and short-lived products (over 8 Gt $yr^{-1}$) scale up to 54–58 Gt $yr^{-1}$ of primary extraction (Krausmann et al. 2017, 2018; UNEP and IRP 2020). The above indicates that we have only begun to exploit the potential for recycling and circularity more broadly.

Total **extraction of all basic materials** (including biomass and fuels) in 2017 reached 92 Gt $yr^{-1}$, which is 13 times above the 1900 level (Figure 11.3).[5] When recycled resources are added, total material inputs exceed 100 Gt (Circle Economy 2020). In Equation 11.1 *MPR* represents only material inputs to the stock, excluding dissipative use – biomass (food and feed) and combusted fuels. Total extraction of stock building materials (metal ores and non-metallic minerals) in 2017 reached 55 Gt $yr^{-1}$.[6] In 1970–2018, it grew 4.3-fold and the ratio of *MPR* to accumulated in-use capital has nearly been constant since 1990 along with ratio to GDP (Figure 11.3).

End-of-life waste from accumulated stocks along with (re)-manufacturing and construction waste is assessed at 16 Gt $yr^{-1}$ in 2014 and can be extrapolated in 2018 to 19 Gt $yr^{-1}$ (Krausmann et al. 2018; Wiedenhofer et al. 2019), or 1.8% from stock of manufactured

3 This conclusion is also valid separately for developed countries and rest of the world (Krausmann et al. 2020).

4 Cement stock for 2014 was estimated at 75 Gt (Cao et al. 2020).

5 IRP (2020) estimate 2017 material extraction at 94 Gt $yr^{-1}$.

6 It approaches 60 Gt $yr^{-1}$ after construction and furniture wood and feedstock fuels are added (Krausmann et al. 2018; Wiedenhofer et al. 2019; UNEP and IRP 2020).

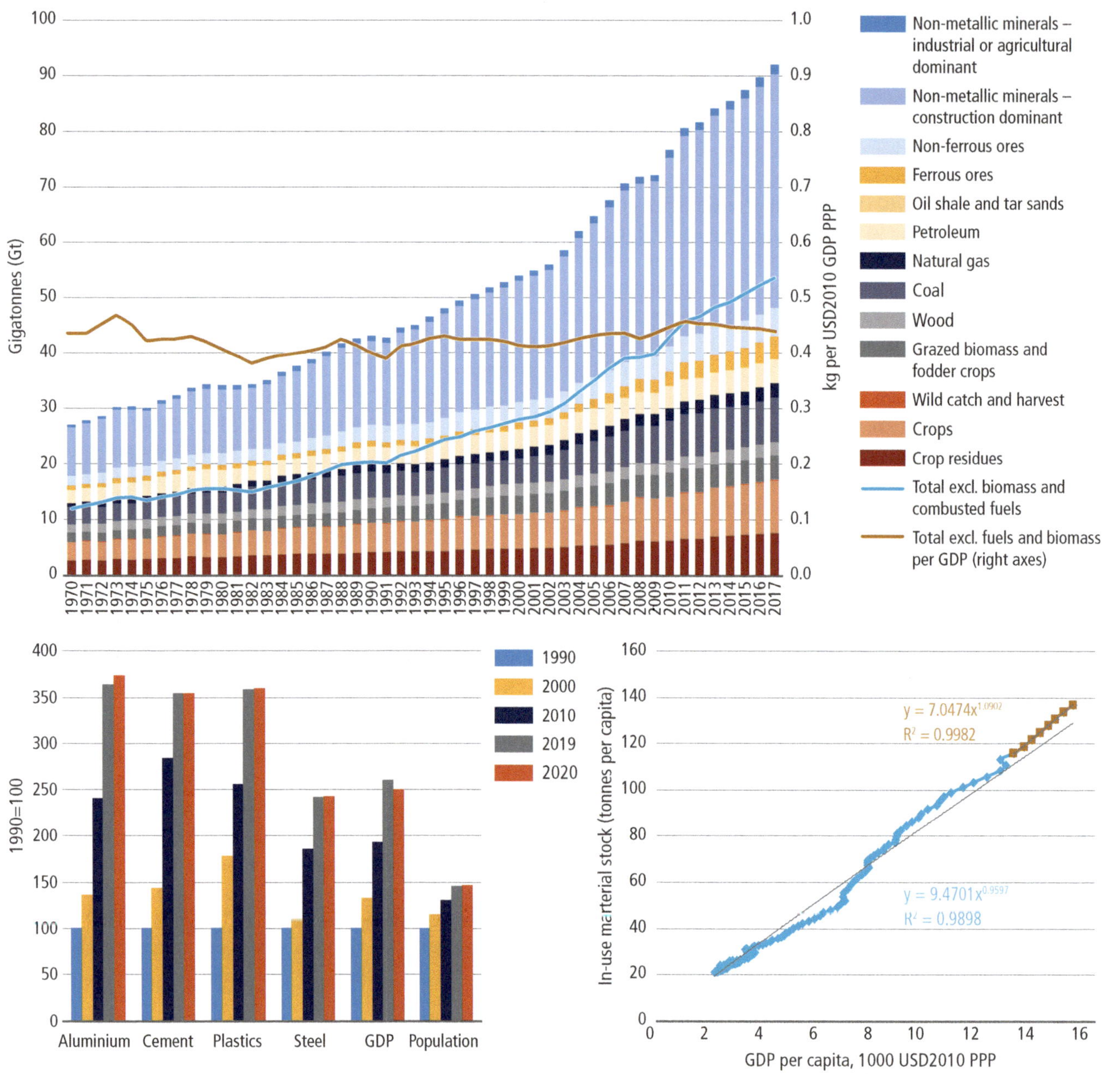


**Figure 11.3 | Raw natural materials extraction since 1970.** In windows: left – growth of population, GDP and basic materials production (1990 = 100) in 1990–2020; right – in-use stock per capita vs income level (1900–2018; brown dots are for 2000–2018). The regressions provided show that for more recent years elasticity of material stock to GDP was greater than unity, comparing with the lower unity in preceding years. Source: developed based on Maddison Project (2018); Wiedenhofer et al. (2019); IEA (2020b); UNEP and IRP (2020); International Aluminium Institute (2021a); Statista (2021a,b); U.S. Geological Survey (2021); World Bank (2021); World Steel Association (2021).

capital. Less than 6 Gt $yr^{-1}$ was recycled and used to build the stock (about 10% of inputs).[7] While the circularity gap is still large, and limited circularity was engineered into accumulated stocks,[8] **material recycling** mitigated some GHG emissions by replacing energy-intensive virgin materials.[9] When the stock saturates, in closed material loops the end-of-life materials waste has to be equal to material input, and primary production therefore has to be equal to end-of-life waste multiplied by unity minus recycling rate. When the latter grows, as the linear metabolism is replaced with the circular one, the share of primary materials production in total material input declines.

7 Mayer et al. (2019) found that in 2010–2014 the secondary-to-primary materials ratio for the EU-28 was slightly below 9%.

8 According to Circle Economy (2020) 8.6 Gt $yr^{-1}$ or 8.6% of total inputs for all resources.

9 Environmental impacts of secondary materials are much (up to an order of magnitude) lower compared to primary materials (OECD 2019a; IEA 2021a; Wang et al. 2021), but to enable and mobilise circularity benefits it requires social system and industrial designing transformation (Oberle et al. 2019).

Recycling rates for metals are higher than for other materials: the end-of-life scrap input ratio for 13 metals is over 50%, and stays in the range of 25–50% for another ten, but even for metals recycling flows fail to match the required inputs (Graedel et al. 2011). Globally, despite overall recycling rates being at 85%, the all-scrap ratio for steel production in recent years stays close to 35–38% (Gielen et al. 2020; IEA 2021b) ranging from 22% in China (only 10% in 2015) to 69% in the US and to 83% in Turkey (BIR 2020). For end-of-life scrap this ratio declined from 30% in 1995–2010 to 21–25% after 2010 (Gielen et al. 2020; Wang et al. 2021).

For aluminium, the share of scrap-based production grew from 17% in 1962 to 34% in 2010 and stabilised at this level until 2019, while the share of end-of-life scrap grew from 1.5% in 1962 to nearly 20% in 2019 (International Aluminium Institute 2021a). The global recycling (mostly mechanical) rate for plastics is only 9–10%[10] (Geyer et al. 2017; Saygin and Gielen 2021), and that for paper progressed from 34% in 1990 to 44% in 2000 and to over 50% in 2014–2018 (IEA 2020b).

The limited impacts of material efficiency factors on industrial GHG emissions trends reflect the lack of integration of material efficiency in energy and climate policies which partly results from the inadequacy of monitored indicators to inform policy debates and set targets;[11] lack of high-level political focus and industrial lobbying; uncoordinated policy across institutions and sequential nature of decision-making along supply chains; carbon pricing policy lock-in with upstream sectors failing to pass carbon costs on to downstream sectors (due to compensation mechanisms to reduce carbon leakage) and so have no incentives to exploit such options as light-weighting, reusing, remanufacturing, recycling, diverting scrap, extending product lives, using products more intensely, improving process yields, and substituting materials (Skelton and Allwood 2017; Gonzalez Hernandez et al. 2018b; Hilton et al. 2018). Poor progress with material efficiency is part of the reason why industrial GHG emissions are perceived as 'hard to abate', and many industrial low-carbon trajectories to 2050 leave up to 40% of emissions in place (Material Economics 2019; IEA 2021a). The importance of this factor activation rises as in-use material stock is expected to scale up by a factor of 2.2–2.7 to reach 2215–2720 Gt by 2050 (Krausmann et al. 2020). Material extraction in turn is expected to rise to 140–200 Gt yr$^{-1}$ by 2060 (OECD 2019a; Hertwich et al. 2020) providing unsustainable pressure on climate and environment and calling for fundamental improvements in material productivity.

In 2014–2019, the average annual growth rate (AAGR) of global **industrial energy use** was 0.4% compared to 3.2% in 2000–2014, following new policies and trends, particularly demonstrated by China[12] (IEA 2020b,d). Whatever metric is applied, industry (coal transformation, mining, quarrying, manufacturing and construction) driven mostly by material production, dominates global energy consumption. About two fifths of energy produced globally goes to industry, directly or indirectly. Direct energy use (including energy used in coal transformation) accounts for nearly 30% of total final energy consumption. When supplemented by non-energy use, the share for the post-AR5 period (2015–2019) stands on average close to 40% of final energy consumption, and at 28.5% of primary energy use.[13] With an account of indirect energy use for the generation of power and centralised heat to be consumed in industry, the latter scales up to 37%. Industrial energy use may be split by: material production and extraction (including coal transformation): 51% on average for 2015–2019; non-energy use (mostly chemical feedstock): 22%[14]; and other energy use (equipment, machinery, food and tobacco, textiles, leather, etc.): 27%. Energy use for material production and feedstock[15] makes about three quarters (73%) of industrial energy consumption and is responsible for 77% of its increment in 2015–2019 (based on IEA 2021a).

For over a century, **industrial energy efficiency** improvements have partially offset growth in GHG emissions. Industrial energy use per tonne of extracted materials (ores and building materials as a proxy for materials going through the whole production chain to final products) fell by 20% in 2000–2019 and by 15% in 2010–2019, accelerated driven by high energy prices to 2.4% yr$^{-1}$ in 2014–2019, matching the values observed back in 1990–2000 (Figure 11.2). Assessed per value added using market exchange rates, industrial energy intensity globally dropped by 12% in 2010–2018, after its 4% decline in 2000–2010, resulting in 2000–2018 decline by 15% (IEA 2020b,a). The 2020 COVID crisis slowed down energy intensity improvements by shifting industrial output towards more energy-intensive basic materials (IEA 2020e). Specific energy consumption per tonne of iron and steel, chemicals and cement production in 2019 was about 20% below the 2000 level (IEA 2020b,a). This progress is driven by moving towards best available technologies (BATs) for each product through new and highly efficient production facilities in China, India and elsewhere, and by the contribution from recycled scrap metals, paper and cardboard.

Physical energy intensity for the production of materials typically declines and then stabilises at the BAT level once the market is saturated, unless a transformative new technology enters the market (Gutowski et al. 2013; Crijns-Graus et al. 2020; IEA 2021a). Thus, the energy saving effect of switching to secondary used material comes to the forefront, as energy consumption per tonne for many basic primary materials approach the BATs. This highlights the need to push towards circular economy, materials efficiency, reduced demand, and

11

[10] IEA (2021a) assesses the global plastics collection rate at 17% for 2020.

[11] Significant progress with data and indicators was reached in recent years with the development of several global coverage material flows datasets (Oberle et al. 2019).

[12] China contributed three quarters of global industrial energy use increment in 2000–2014. Since 2014 China's share in global industrial energy use has slowly declined, reaching about a third in 2018 (IEA 2020d).

[13] This is close to 28.8% average 1900–2018 share of industrial energy use in global primary energy consumption. This share shows a slow decline trend (0.01% yr$^{-1}$) in response to the growing share of services in global GDP, with about 60-year-long cycles.

[14] Industry also produces goods traditionally used as feedstock – hydrogen and ammonia – which in the future may be widely used as energy carriers.

[15] Mapping global flows of fuel feedstock allows for better tailoring of downstream mitigation options for chemical products (Levi and Cullen 2018).

fundamental process changes (e.g., towards electricity and hydrogen-based steel making). Improved recycling rates allow for a substantial reduction in energy use along the whole production chain – material extraction, production, and assembling – which is in great excess of energy used for collection, separation, treatment, and scrap recycling minus energy used for scrap landfilling. The International Energy Agency (IEA 2019b) estimates that by increasing the recycling content of fabricated metals, average specific energy consumption (SEC) for steel and aluminium may be halved by 2060. Focusing on whole systems 'integrative design' expands efficiency resource much beyond the sum of potentials for individual technologies. Material efficiency coupled with energy efficiency can deliver much greater savings than energy efficiency alone. Gonzalez Hernandez et al. (2018b) stress that presently about half of steel or aluminium are scrapped in production or oversized for targeted services. They show that resource efficiency expressed in exergy as a single metric for both material and energy efficiency for the global iron and steel sector is only 33%, while secondary steel-making is about twice as efficient (66%) as ore-based production (29%). While shifting globally in ore-based production from the average to the best available level can save 6.4 EJ $yr^{-1}$, the saving potential of shifting to secondary steel-making is 8 EJ $yr^{-1}$, and is limited mostly by scrap availability and steel quality requirements.

### 11.2.2 New Trends in Emissions

GHG emissions attributable to the industrial sector (see Chapter 2) in 2019 originate from industrial fuel combustion (7.1 $GtCO_2$-eq directly and about 5.9 Gt indirectly from electricity and heat generation[16]); industrial processes (4.5 $GtCO_2$-eq) and products use (0.2 Gt), as well as from waste (2.3 Gt) (Figure 11.4a,b). Overall industrial direct GHG emissions amount to 14.1 $GtCO_2$-eq (Figure 11.4c and Table 11.1), and scales up to 20 $GtCO_2$-eq after indirect emissions are added,[17] putting industry (24%, direct emissions) second after the energy sector in total GHG emissions and lifting it to the leading position after indirect emissions are allocated (34% in 2019).[18] The corresponding shares for 1990–2000 were 21% for direct emissions and 30% for both direct and indirect (Crippa et al. 2021; Lamb et al. 2021; Minx et al. 2021). As the industrial sector is expected to decarbonise slower than other sectors it will keep this leading position for the coming decades (IEA 2021a). In 2000–2010, total industrial emissions grew faster (3.8% $yr^{-1}$) than in any other sector (see Chapter 2), mostly due to the dynamics shown by basic materials extraction and production. Industry contributed nearly half (45%) of overall incremental global GHG emissions in the 21st century.

Industrial sector GHG emissions accounting is complicated by carbon storage in products (Levi and Cullen 2018). About 35% of chemicals' mass is $CO_2$, which is emitted at use stage – decomposition of fertilisers, or plastic waste incineration (Saygin and Gielen 2021), and sinks. Recarbonation and mineralisation of alkaline industrial materials and wastes (also known as the 'sponge effect') provide 0.6–1 $GtCO_2$ $yr^{-1}$ uptake by cement-containing products[19] (Cao et al. 2020; Guo et al. 2021); see Section 11.3.6 for further discussion in decarbonisation context.

In 1970–1990, industrial direct combustion-related emissions were growing modestly, and in 1990–2000 even switched to a slowly declining trend, steadily losing their share in overall industrial emissions. Electrification was the major driver behind both indirect and total industrial emissions in those years. This quiet evolution was interrupted in the beginning of the 21st century, when total emissions increased by 60–68% depending on the metric applied (the fastest growth ever seen). In 2000–2019 iron, steel and cement absolute GHGs increased more than any other period in history (Bashmakov 2021). Emissions froze temporarily in 2014–2016, partly in the wake of the financial crisis, but returned to their growth trajectory in 2017–2019 (Figure 11.4a).

The largest incremental contributors to industrial emissions in 2010–2019 were industrial processes at 40%, then indirect emissions (25%), and only then direct combustion (21%), followed by waste (14%; Figure 11.4). Therefore, to stop emission growth and to switch to a zero-carbon pathway more mitigation efforts should be focused on industrial processes, product use and waste decarbonisation, along with the transition to low-carbon electrification (Hertwich et al. 2020).

Basic materials production dominates both direct industrial GHG emissions (about 62%, waste excluded)[20] as well as direct industrial $CO_2$ emissions (70%), led by iron and steel, cement, chemicals, and non-ferrous metals (Figure 11.4e). Basic materials also contribute 60% to indirect emissions. In a zero-carbon power world, with industry lagging behind in the decarbonisation of high-temperature processes and feedstock, it may replace the energy sector as the largest generator of indirect emissions embodied in capital stock.[21] According to Circle Economy (2020) and Hertwich et al. (2020), GHG

16 Indirect emissions are assessed based on the EDGAR database (Crippa et al. 2021). The IEA database reports 6 Gt of $CO_2$ for 2019 (IEA 2020f).

17 Based on Crippa et al. (2021) and Minx et al. (2021). In 2019, industrial $CO_2$-only emissions were 10.4 $GtCO_2$, which due to wider industrial processes and product use (IPPU) coverage exceeds the $CO_2$ emission assessed by the IEA (2021a) at 8.9 Gt for 2019 and at 8.4–8.5 Gt for 2020.

18 According to the IEA (2020f), industry fuel combustion $CO_2$-only emissions contributed 24% to total combustion emissions, but combined with indirect emission it accounted for 43% in 2018.

19 There are suggestions to incorporate carbon uptake by cement-containing products in IPCC methodology for national GHG inventories (Stripple et al. 2018).

20 Crippa et al. (2021) and the IEA (2020a) assess materials-related scope 1 + 2 (direct and indirect emissions) correspondingly at 10.3 for 2019 and at 10.7 for 2018. Hertwich (2021) updated estimates for the global cradle-to-gate material-production-related GHG emissions for 2018 at 11.8 Gt (5.1 Gt for metals, 3.7 Gt for non-metallic minerals, 1.8 Gt for plastics and rubber, 1 Gt for wood) – which is about 69% of direct and indirect industrial emissions (waste excluded). These assessments are consistent as transportation of basic materials contributes around 1 $GtCO_2$-eq. to GHG emissions.

21 According to Hertwich et al. (2020), of the 11.5 $GtCO_2$-eq 2015 global materials GHG footprint about 5 Gt were embodied in buildings and infrastructure, and nearly 3 Gt in machinery, vehicles, and electronics.

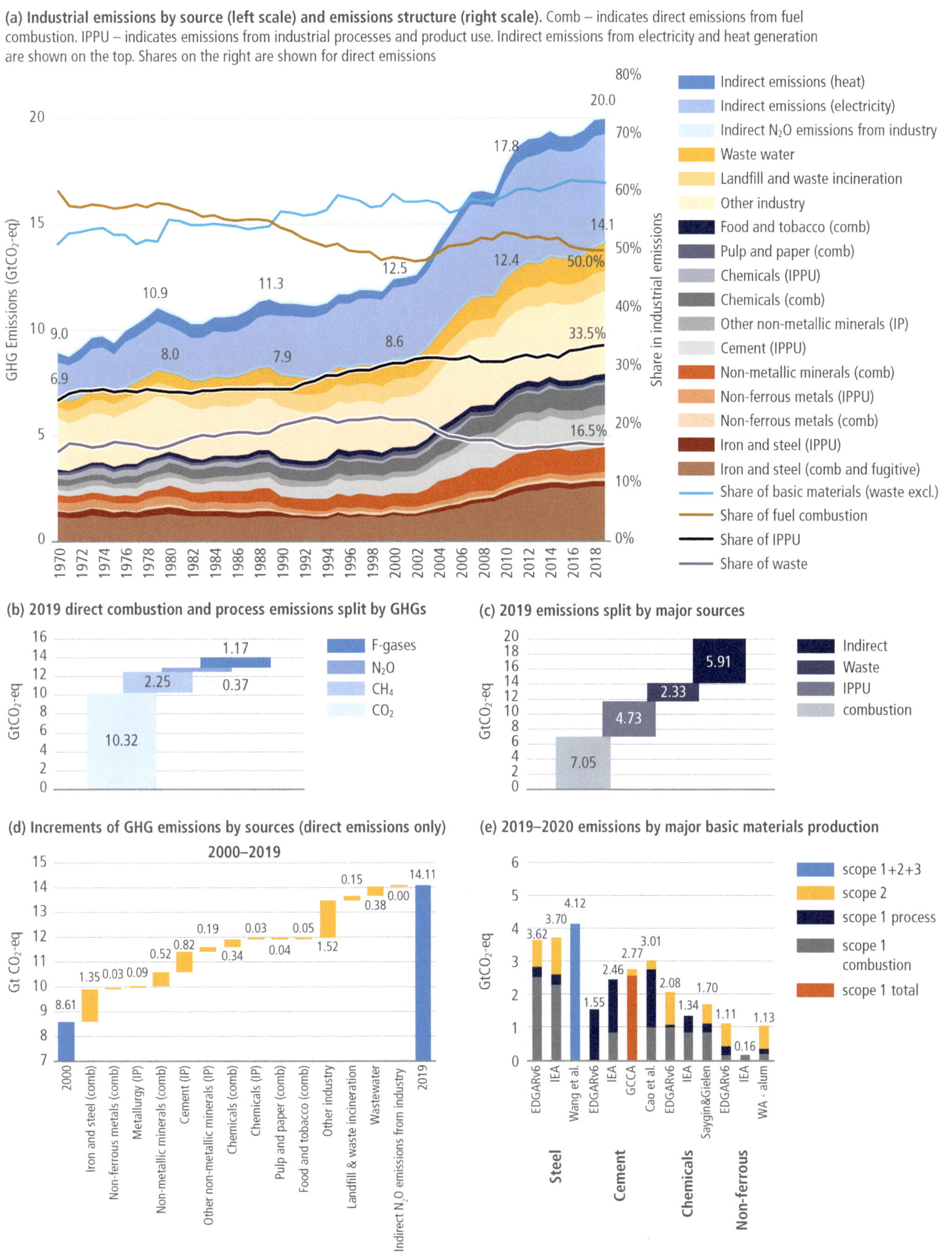


**Figure 11.4 | Industrial sector direct global greenhouse gas (GHG) emissions.** Source: calculated based on emissions data from Crippa et al. (2021) and Minx et al. (2021). Indirect emissions were assessed using IEA (2021b). For (e): Cao et al. (2020); IEA (2020b, 2021a); GCCA (2021a); International Aluminium Institute (2021a); and Wang et al. (2021).

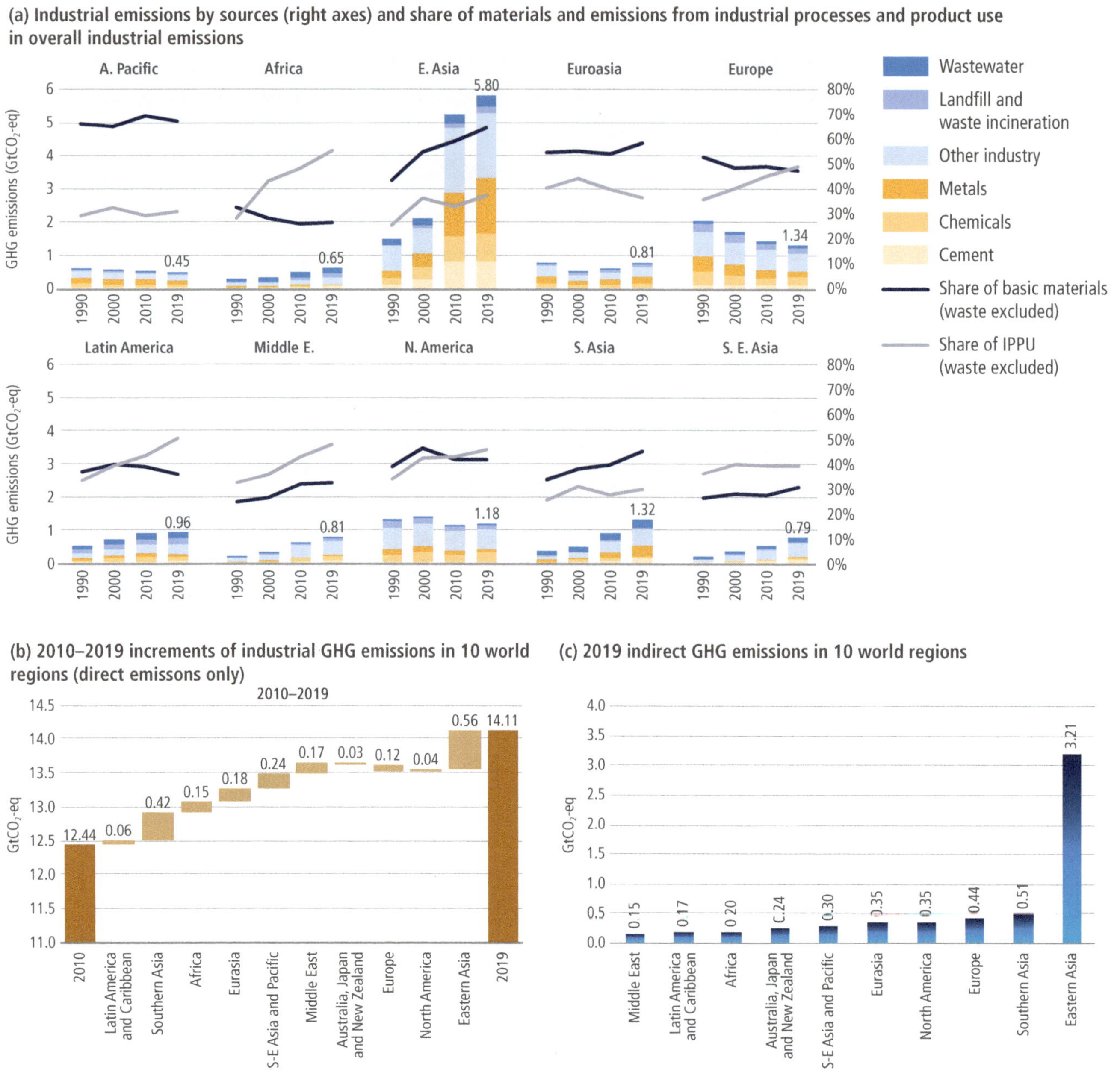


**Figure 11.5 | Industrial sector greenhouse gas (GHG) emissions in 10 world regions (1990–2019).** Source: calculated based on emissions data from Crippa et al. (2021). Indirect emissions were assessed using IEA (2021b).

emissions embodied in buildings and infrastructure, machinery and transport equipment exceed 50% of their present carbon footprint.

In 1970–2000, direct GHG emissions per unit of energy showed a steady decline interrupted by noticeable growth in 2001–2018 driven by the fast expansion of steel and cement production (Figure 11.5; IEA 2021a). Non-energy-related GHG emissions per unit of extracted materials decline continuously, as the share of not carbon intensive building materials (aggregates and sand) grows.

|Iron and steel carbon intensity stagnated in 1995–2015 due to rapid growth in carbon-intensive production in some countries (Wang et al. 2021). For aluminium carbon intensity declined in 2010–2019 by only 2% (International Aluminium Institute 2021a). The carbon intensity of cement-making since 2010 is down by only 4%. In 1990–2019 it fell by 19.5%, mostly due to energy efficiency improvements (by 18.5%) as the carbon intensity of the fuel mix declined only by 3% (GCCA 2021b). Historical analysis shows the carbon intensity of steel production has declined with 'stop and go' patterns in 50–60-year cycles, reflective of the major jumps in best available technology (BAT). From 1900 to 1935 and from 1960 to 1990 specific scope 1 + 2 + 3 emissions fell by 1.5–2.5 $tCO_2$ per tonne, or as much as needed now to achieve net zero. While historical declines were mostly due to commissioning large capacities with new technologies, with total emissions growing, by 2050 and beyond the decline will likely materialise via new ultra-low emission capacity replacements pushing absolute emissions to net zero (Bataille et al. 2021b).

Table 11.1 | Dynamics and structure of industrial greenhouse gas (GHG) emissions.

| | | Average annual growth rates | | | | Share in total industrial sector emissions | | | | | 2019 emissions $MtCO_2$-eq |
|---|---|---|---|---|---|---|---|---|---|---|---|
| | | 1971–1990 | 1991–2000 | 2000–2010 | 2011–2019 | 1970 | 1990 | 2000 | 2010 | 2019 | |
| Direct $CO_2$ emissions from fuel combustion | Mining (excl. fuels), manufacturing industries and construction | 0.13% | –0.18% | 4.62% | 0.77% | 45.8% | 37.3% | 33.2% | 36.6% | 34.9% | 6981 |
| | Iron and steel | 0.20% | 0.13% | 5.62% | 2.28% | 12.4% | 10.2% | 9.4% | 11.4% | 12.4% | 2481 |
| | Chemical and petrochemical | 3.66% | 1.54% | 3.16% | 1.19% | 3.0% | 4.9% | 5.2% | 4.9% | 4.9% | 977 |
| | Non-ferrous metals | 2.12% | 3.20% | 1.12% | 1.36% | 0.7% | 0.8% | 1.0% | 0.8% | 0.8% | 163 |
| | Non-metallic minerals | 2.91% | 1.88% | 6.24% | –0.04% | 3.3% | 4.6% | 5.0% | 6.5% | 5.7% | 1148 |
| | Paper, pulp and printing | 0.78% | 2.79% | 0.09% | –2.69% | 1.4% | 1.3% | 1.5% | 1.1% | 0.7% | 150 |
| | Food and tobacco | 2.55% | 1.50% | 3.03% | –1.04% | 1.3% | 1.6% | 1.7% | 1.6% | 1.3% | 265 |
| | Other | –1.55% | –2.89% | 4.61% | –0.22% | 23.8% | 13.8% | 9.4% | 10.3% | 9.0% | 1797 |
| Indirect emissions – electricity | | 2.87% | 2.06% | 3.00% | –0.87% | 17.6% | 24.6% | 27.3% | 25.8% | 21.2% | 4236 |
| Indirect emissions – heat | | 2.08% | –3.09% | 2.53% | 9.83% | 5.6% | 6.7% | 4.5% | 4.0% | 8.3% | 1663 |
| Industrial processes $CO_2$ | Total | 1.45% | 2.16% | 5.00% | 1.93% | 11.0% | 11.6% | 13.0% | 14.9% | 15.7% | 3144 |
| | Non-metallic minerals | 2.22% | 2.36% | 5.66% | 1.67% | 5.7% | 7.0% | 8.0% | 9.7% | 10.0% | 2008 |
| | Chemical and petrochemical | 4.51% | 2.52% | 3.50% | 2.01% | 1.5% | 2.9% | 3.4% | 3.4% | 3.6% | 720 |
| | Metallurgy | –3.11% | 0.37% | 5.16% | 3.10% | 3.6% | 1.5% | 1.4% | 1.7% | 2.0% | 391 |
| | Other | 1.55% | 2.30% | –1.21% | 2.89% | 0.1% | 0.2% | 0.2% | 0.1% | 0.1% | 25 |
| Industrial product use GHG | | –0.22% | –0.49% | –1.02% | 0.41% | 2.7% | 2.0% | 1.7% | 1.1% | 1.0% | 204 |
| Other non-$CO_2$ GHG | | –0.60% | 5.20% | 4.29% | 3.20% | 5.5% | 3.9% | 5.8% | 6.2% | 7.3% | 1470 |
| Waste GHG | | 1.94% | 1.35% | 1.22% | 1.57% | 11.9% | 13.8% | 14.4% | 11.4% | 11.6% | 2327 |
| Total GHG | | 1.16% | 0.98% | 3.61% | 1.32% | 100.0% | 100.0% | 100.0% | 100.0% | 100.0% | 20,025 |

Source: calculated based on Crippa et al. (2021); IEA (2021b); and Minx et al. (2021).

### 11.2.3 Industrial Development Patterns and Supply Chains (Regional)

The dramatic increase in industrial emissions after 2000 is clearly associated with economic growth in Asia, which dominated both absolute and incremental emissions (Figure 11.5a,b).

More recent 2010 to 2019 trends show that regional contributions to additional emissions are distributed more evenly, while a large part still comes from Asian countries, where both rates of economic growth and the share of industrial emissions much exceed the global average. All other regions also contributed to total industrial GHG emissions. Structural shifts towards emissions from industrial processes and products use are common for many regions (Figure 11.5a).

**Economic development.** Regional differences in emission trends are determined by the differences observed in economic development, trade and supply chain patterns. The major source of industrial emissions is production of energy-intensive materials, such as iron and steel, chemicals and petrochemicals, non-ferrous metals and non-metallic products. Steel and cement are key inputs to urbanisation and infrastructure development (buildings and infrastructure are responsible for about three fourths of the steel stock). Application of a 'services-stock-flow-emissions' perspective (Wiedenhofer et al. 2019; Bashmakov 2021; Haberl et al. 2021) shows that relationship patterns between stages of economic development, per capita stocks and flows of materials are not trivial with some clear transition points. Cao et al. (2017) mapped countries by four progressive stages in cement stock per capita S-shape evolution as a function of income and urbanisation: initial stage for developing countries with a low level and slow linear growth; take-off stage with accelerated growth; slowdown stage; and finally a shrinking stage (represented by just a few countries with very high incomes exceeding 40,000 USD2010 per capita) and urbanisation levels above 80%. Bleischwitz et al. (2018) use a similar approach with five stages to study material saturation effects for apparent consumption and stocks per capita for steel, cement, aluminium, and copper. This logic may be generalised to other materials from which in-use stock is built. While globally cement in-use stock is about 12 tonnes per capita, in developed countries it is 15–30 tonnes per capita, but the order of magnitude is lower in developing states with high per capita escalation rates (Cao et al. 2017). When stocks for some materials saturate – per capita stock peaks – the 'scrap age' is coming (Pauliuk et al. 2013a). Steel in-use stock has already saturated in advanced economies at 14 ± 2 tonnes per capita due to largely completed urbanisation and infrastructure developments, and a switch towards services-dominated economy. This saturation level is three to four times that of the present global average, which is below 4 tonnes per capita (Pauliuk et al. 2013a; Graedel et al. 2011; Wiedenhofer et al. 2019). China is entering the maturing stage of steel and cement consumption, resulting in a moderate projection of additional demand followed by expected industrial emissions peaking in the next 10 to 15 years (Zhou et al. 2013; Bleischwitz et al. 2018; OECD 2019a; Wu et al. 2019; Zhou et al. 2020). But many developing countries are still urbanising, and the growing need for infrastructure services results in additional demand for steel and cement. Materials intensity of the global economy is projected by OECD (2019a) to decline at 1.3% $yr^{-1}$ until 2060, driven by improving resource efficiency and the switch to circular economy, but with a projected tripling of global GDP it means a doubling of projected materials use (OECD 2019a). Under the business-as-usual scenario, India's demand for steel may more than quadruple over the next 30 years (de la Rue du Can et al. 2019; Dhar et al. 2020). In the IEA (2021a) net-zero-energy scenario, the saturation effect along with material efficiency counterbalances activity effects and keeps demand growth for basic materials modest while escalate demand for critical materials (copper, lithium, nickel, graphite, cobalt and others).

**International trade and supply chain.** In Equation 11.1 the share of allocated emissions (*Dm*) equals unity when territorial emission is considered, and to the ratio of domestically used materials to total material production for consumption-based emission accounting. Tracking consumption-based emissions provides additional insights in the global effectiveness of national climate policies. Carbon emissions embodied in international trade are estimated to account for 20–30% of global carbon emissions (Meng et al. 2018; OECD.Stat 2019) and are the reason for different emissions patterns of OECD versus non-OECD countries (Chapter 2).

Based on OECD.Stat (2019) datasets, 2015 $CO_2$ emissions embodied in internationally traded industrial products (manufacturing and mining, excluding fuels) by all countries are assessed at 3 $GtCO_2$, or 30% of direct $CO_2$ emissions in the industrial sector as reported by Crippa et al. (2021). OECD countries collectively have reduced territorial emissions (shares of basic materials in direct emissions in those regions decline (Figure 11.5b), but demonstrated no progress in reducing outsourced emissions embedded in imported industrial products (Arto and Dietzenbacher 2014; OECD.Stat 2019). Accounting for net carbon emissions embodied in international trade of only industrial products (1283 million $tCO_2$ in 2015) escalates direct OECD industrial $CO_2$ emissions (1333 million $tCO_2$ of energy-related and 502 million $tCO_2$ of industrial processes) 1.7 fold, 2.3-fold for the US, 1.5-fold for the EU, and more than triples it for the UK, while cutting (*Dm*) by a third for China and Russia (OECD.Stat 2019; IEA 2020f). In most OECD economies, the amount of $CO_2$ embodied in net import from non-OECD countries is equal to, or even greater than, the size of their Paris 2030 emissions reduction commitments. In the UK, the Parliament Committee on Energy and Climate Change requested that a consumption-based inventory be complementarily used to assess the effectiveness of domestic climate policy in delivering absolute global emissions reductions (Barrett et al. 2013; UKCCC 2019a). It should be noted that the other side of the coin is that exports from countries with lower production carbon intensities can lead to overall less emissions than if production took place in countries with high carbon intensities, which may become critical in the global evolution toward lower emissions. The evolution of *Dm* to the date was driven mostly by factors other than carbon regulation often equipped with carbon leakage prevention tools. Empirical tests have failed to date to detect meaningful 'carbon leakage' and impacts of carbon prices on net import, direct foreign investments, volumes of production, value added, employment, profits, and innovation in industry (Sartor 2013; Branger et al. 2016; Saussay and Sato 2018; Ellis et al. 2019; Naegele and Zaklan 2019; Acworth et al. 2020; Carratù et al. 2020; Pyrka et al. 2020; Zachmann and McWilliams 2020). In the coming years, availability of large low-cost renewable electricity potential and cheap hydrogen may become a new driver for relocation of such carbon intensive industries as steel production (Bataille 2020a; Gielen et al. 2020; Bataille et al. 2021a; Saygin and Gielen 2021).

## 11.3 Technological Developments and Options

The following overview of technical developments and mitigation options which relate to the industrial sector is organised in six equally important strategies: (i) demand for materials, (ii) materials efficiency, (iii) circular economy and industrial waste, (iv) energy efficiency, (v) electrification and fuel switching, and (vi) CCUS, feedstock and biogenic carbon. Each strategy is described in detail, followed by a discussion of possible overlaps and interactions between strategies and how conflicts and synergies can be addressed through integration of the approaches.

### 11.3.1 Demand for Materials

Demand for materials is a key driver of energy consumption and $CO_2$ emissions in the industrial sector. Rapid growth in material demand over the last quarter century has seen demand for key energy-intensive materials increase 2.5- to 3.5-fold (Figure 11.6), with growth linked to, and often exceeding, population growth and economic development. The International Energy Agency (IEA) explains, 'as economies develop, urbanise, consume more goods and build up their infrastructure, material demand per capita tends to increase considerably. Once industrialised, an economy's material demand may level off and perhaps even begin to decline' (IEA 2019b).

The Kaya-like identity presented earlier in the chapter (Equation 11.1) suggests that material demand can be decoupled from population and economic development by two means: (i) reducing the accumulated material stock (*MStock*) used to deliver material

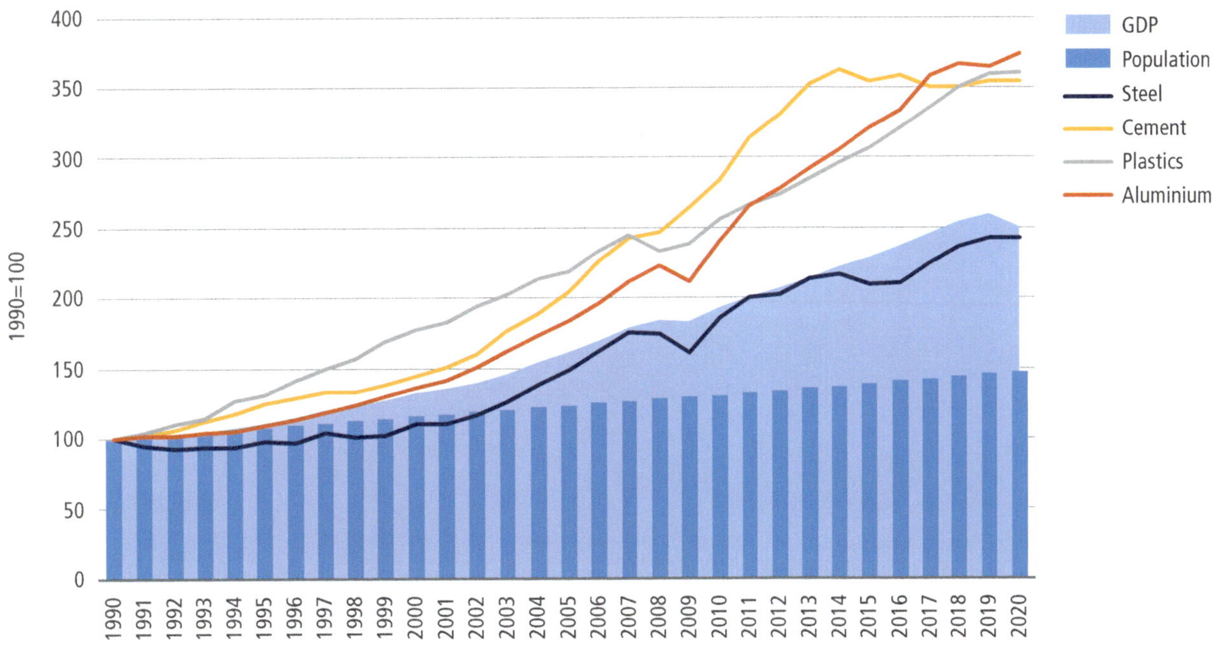


**Figure 11.6 | Growth in global demand for selected key materials and global population, 1990–2019.** Notes: based on global values, shown indexed to 1990 levels (=100). Steel refers to crude steel production. Aluminium refers to primary aluminium production. Plastic refers to the production of a subset of key thermoplastic resins. Cement and concrete follow similar demand patterns. Sources: 1990–2018: IEA (2020b). 2019–2020: GCCA (2021a); International Aluminium Institute (2021a); Statista (2021b); U.S. Geological Survey (2021); World Bank (2021); World Steel Association (2021).

services; and (ii) reducing the material (*MPR* + *MSE*) required to maintain material stocks (*MStock*). Such material demand reduction strategies are linked upstream to material efficiency strategies (the delivery of goods and services with less material demand, and thus energy and emissions) and to demand reduction behaviours, through concepts such as sufficiency, sustainable consumption and social practice theory (Spangenberg and Lorek 2019). Materials demand can also be influenced through urban planning, building codes and related socio-cultural norms that shape the overall demand for square metres per capita of floor space, mobility and transport infrastructures (Chapter 5).

Modelling suggests that per capita material stocks saturate (level off) in developed countries and decouple from GDP. Pauliuk et al. (2013b) demonstrated this saturation effect in an analysis of in-use steel stocks in 200 countries, showing that per capita steel in stocks in countries with a long industrial history (e.g., USA, UK, Germany) had saturation levels between 11 and 16 tonnes. More recently, Bleischwitz et al. (2018) confirmed the occurrence of a saturation effect for four materials (steel, cement, aluminium and copper) in four industrialised countries (Germany, Japan, UK and USA) together with China. These findings have led to the revision of some material demand forecasts, which previously had been based solely on population and economic trends.

The saturation effect for material stocks is critical for managing material demand in **developed countries**. Materials are required to meet demand for the creation of new stocks and the maintenance of existing stocks (Gutowski et al. 2017). Once saturation is attained the need for new stocks is minimised, and materials are only required for replacing old stocks and maintenance. Saturation allows material efficiency strategies (such as light-weight design, longer lifetimes, and more intense use) to reduce the required per capita level of material stocks, and material circularity strategies (closing material loops through remanufacture, reuse and recycling) to lessen the energy and carbon impacts required to maintain the material stock. However, it should be noted that some materials still show little evidence of saturation (i.e., plastics, see Box 11.2). Furthermore, meeting climate change targets in developed countries will require the construction of new low-carbon infrastructures (i.e., renewable energy generation, new energy distribution and storage systems, electric vehicles and building heating systems) which may increase demand for emissions intensive materials (i.e., steel, concrete and glass).

For **developing countries**, who are still far from saturation levels, strong growth for new products and the creation of new infrastructure capacity may still drive global material demand. However, there is an expectation that economic development can be achieved at lower per capita material stock levels, based on the careful deployment of material efficiency and circularity by design (Grubler et al. 2018).

### 11.3.2 Material Efficiency

Material efficiency (*ME*) – the delivery of goods and services with less material – is increasingly seen as an important strategy for reducing GHG emissions in industry (IEA 2017, 2019b). Options to improve *ME* exist at every stage in the lifecycle of materials and products, as shown in Figure 11.7. This includes: designing products which are lighter, optimising to maintain the end-use service while

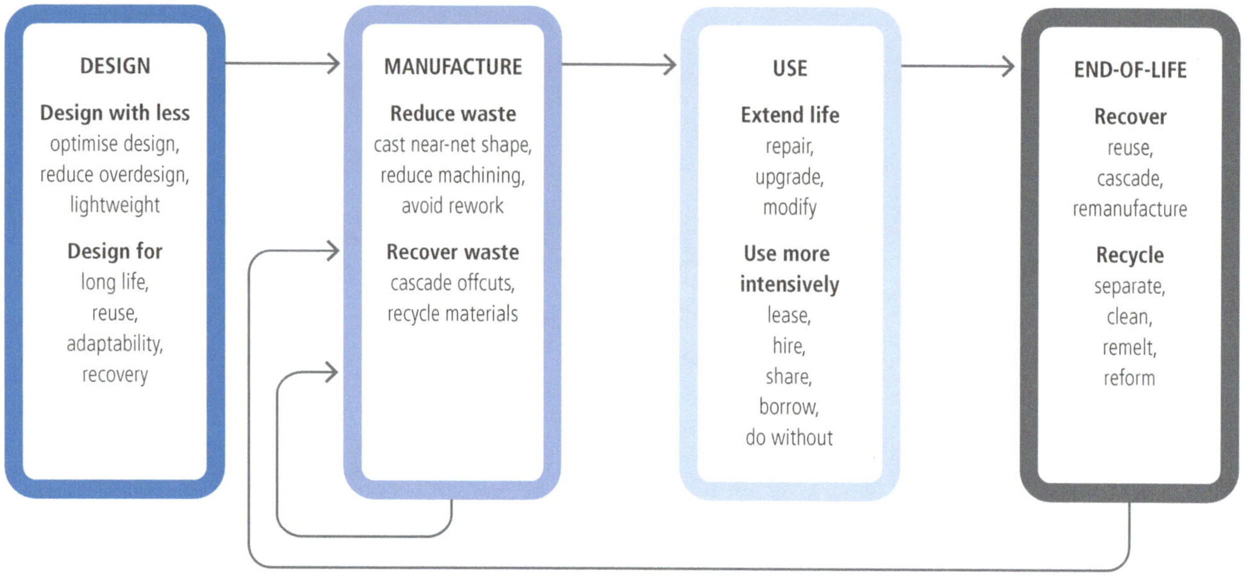


**Figure 11.7 | Material efficiency (*ME*) strategies across the value chain.** Source: derived from strategies in Allwood et al. (2012).

minimising material use, designing for circular principles (i.e., longer life, reusability, repairability, and ease of high-quality recycling); pushing manufacturing and fabrication process to use materials and energy more efficiently and recover material wastes; increasing the capacity, intensity of use, and lifetimes of product in use; improving the recovery of materials at the end of life, through improved remanufacturing, reuse and recycling processes. For more specific examples see Allwood et al. (2012); Lovins (2018); Hertwich et al. (2019); Scott et al. (2019); and Rissman et al. (2020).

*ME* provides plentiful options to reduce emissions, yet because interventions are dispersed across supply chains and span many different stakeholders, this makes assessing mitigation potentials and costs more challenging. For this reason, *ME* interventions have traditionally been under-represented in climate change scenario modelling and integrated assessment models (IAMs) (Grubler et al. 2018; Allwood 2018). However, two advances in the modelling of materials flows have underpinned the recent emergence of *ME* options being included in climate scenario modelling.

Firstly, over many years, the academic community has built up detailed global material-flow maps of the processing steps involved in making energy-intensive materials. Some prominent recent examples include: steel (Gonzalez Hernandez et al. 2018b), pulp and paper (Van Ewijk et al. 2018), petrochemicals (Levi and Cullen 2018). In addition, material-flow maps at the regional and sectoral levels have flourished, for example: steel (Serrenho et al. 2016) and cement (Shanks et al. 2019) in the UK; automotive sheet-metal (Horton et al. 2019); and steel-powder applications (Azevedo et al. 2018). The detailed and transparent physical mapping of material supply chains in this manner enables *ME* interventions to be traced back to where emissions are released, and allows these options to be compared against decarbonisation and traditional energy efficiency measures (Levi and Cullen 2018). For example, a recent analysis by Hertwich et al. (2019) makes the link between *ME* strategies and reducing GHG emissions in buildings, vehicles and electronics, while Gonzalez Hernandez et al. (2018a) examines leveraging *ME* as a climate strategy in European Union (EU) policy. Research to explore the combined analysis of materials and energy, using exergy analysis (for steel: Gonzalez Hernandez et al. 2018b) allows promising comparisons across industrial sectors.

Secondly, many *ME* interventions result in immediate GHG emissions savings (short-term), for example, light-weighting products, reusing today's product components, and improving manufacturing yields. Yet, for other *ME* actions emissions savings are delayed temporally (long-term). For example, designing a product for future reuse, or with a longer life, only reaps emissions savings at the end of the product life, when emissions for a replacement product are avoided. Many durable products have long lifetimes (cars >10 years, buildings >40 years) which requires dynamic modelling of material stocks, over time, to enable these actions to be included in scenario modelling activities. Consequently, much effort has been invested recently to model material stocks in use, to estimate their lifetimes, and anticipate the future waste and replenishment materials to maintain existing stocks and grow the material stock base. Dynamic material models have been applied to material and product sectors, at the country and global level. These include, for example: vehicles stocks in the UK (Serrenho et al. 2017; Craglia and Cullen 2020) and in China (Liu et al. 2020); buildings stocks in the UK (Cabrera Serrenho et al. 2019), China (Hong et al. 2016; Cao et al. 2018, 2019) and the European Union (Sandberg et al. 2016); electronic equipment in Switzerland (Thiébaud et al. 2017); specific material stocks, such as cement (Cao et al. 2020, 2017), construction materials (Sverdrup et al. 2017; Habert et al. 2020), plastics (Geyer et al. 2017), copper (Daehn et al. 2017), and all metals (Elshkaki et al. 2018); all materials in China (Jiang et al. 2019), Switzerland (Heeren and Hellweg 2019) and the world (Krausmann et al. 2017).

These two advances in the knowledge base have allowed the initial inclusion of some *ME* strategies in energy and climate change scenario models. The International Energy Agency (IEA) first created a *ME* scenario (MES) in 2015, with an estimated 17% reduction in industrial energy demand in 2040 (IEA 2015). The World Energy Outlook report includes a dedicated sub-chapter with calculations explicitly on industrial material efficiency (IEA 2019c). They also include *ME* options in their modelling frameworks and reporting, for example for petrochemicals (IEA 2018a), and in the Material Efficiency in Clean Energy Transitions report (IEA 2019b). In Grubler et al. (2018) 1.5°C Low Energy Demand (LED) scenario, global material output decreases by 20% from today, by 2050, with one-third due to dematerialisation, and two-thirds due to *ME*, resulting in significant emissions savings. Material Economics' analysis of Industrial Transformation 2050 (Material Economics 2019), found that resource efficiency and circular economy measures (i.e., *ME*) could almost halve the 530 $MtCO_2\ yr^{-1}$ emitted by the basic materials sectors in the EU by 2050. Finally, the Emissions Gap Report, UNEP (2019) includes an assessment of potential material efficiency savings in residential buildings and cars.

Clearly, more work is required to fully integrate *ME* strategies into mainstream climate change models and future scenarios. Efforts are focused on endogenising *ME* strategies within climate change modelling, assessing the synergies and trade-offs which exist between energy efficiency and *ME* interventions, and building up data for the assessment of emissions saved and the cost of mitigation from real *ME* actions. This requires analysts to work in cross-disciplinary teams and to engage with stakeholders from across the full breadth of material supply chains. Efforts should be prioritised to foster engagement between the IAM community and emerging *ME* models based in the Life Cycle Assessment, Resource Efficiency, and Industrial Ecology communities (see also Sharmina et al. 2021).

### 11.3.3 Circular Economy and Industrial Waste

Circular economy (CE) is another effective approach to mitigate industrial GHG emissions and has been widely promoted worldwide since the fourth IPCC assessment report (AR4). From an industrial point of view, CE focuses on closing the loop for materials and energy flows by incorporating policies and strategies for more efficient energy, materials and water consumption, while emitting minimal waste to the environment (Geng et al. 2013). Moving away from a linear mode of production (sometimes referred to as an 'extract-produce-use-discard' model), CE promotes the design of durable goods that can be easily repaired, with components that can be reused, remanufactured, and recycled (Wiebe et al. 2019). In particular, since CE promotes reduction, reuse and recycling, a large amount of energy and GHG-intense virgin material processing can be reduced, leading to significant carbon emission reductions. For example, in the case of aluminium, the energy efficiency of primary production is relatively close to best available technology (Figure 11.8), while switching to production using recycled materials requires only about 5% as much energy (Section 11.4.1.4). However, careful evaluation is needed from a lifecycle perspective since some recycling activities may be energy- and emission-intensive, for example, the chemical recycling of plastics (Section 11.4.1.3).

As one systemic approach, CE can be seen as conducted at different levels, namely, at the micro level (within a single company, such as process integration and cleaner production), meso level (between three or more companies, such as industrial symbiosis or eco-industrial parks) and macro level (cross-sectoral cooperation, such as urban symbiosis or a regional eco-industrial network). Each level requires different tools and policies, such as CE-oriented incentive and tax policies (macro level), and eco-design regulations (micro level). This section is focused on industry and a broader discussion of the CE concept is found in Box 12.2 and Section 5.3.4.2.

**Micro level:** More firms have begun to implement the concept of CE, particularly multi-national companies, since they believe that multiple benefits can be obtained from CE efforts, and it has become common across sectors (D'Amato et al. 2019). Typical CE tools and policies at this level include cleaner production, eco-design, environmental labelling, process synthesis, and green procurement. For instance, leading chemical companies are incorporating CE into their industrial practices, for example, through the design of more recyclable plastics, a differentiated and market-driven portfolio of resins, films and adhesives that deliver a total package that is more sustainable, cost-efficient and capable of meeting new packaging and plastics preferences. Problematically, at the same time the plastics industry is improving recyclability, it has, for example, been expanding into markets without recycling capacity (Mah 2021). Similarly, automakers are pursuing strategies to increase the portion of new vehicles that are fully recyclable when they reach the end of life, with increasing ambitions for using recycled material, largely motivated by end-of-life vehicle regulations. This will require networks that are available to collect and sort all the materials in vehicles, and policy incentives to do it (Wiebe et al. 2019; Soo et al. 2021).

**Meso level:** Industrial parks first appeared in Manchester, UK, at the end of the 19th century and they have been implemented in industrialised countries for maximising energy and material efficiency, which also has merit for $CO_2$-emissions reduction, as stated in AR5. Industrial parks reduce the cost of infrastructure and utilities by concentrating industrial activities in planned areas, and are typically founded around large, long-term anchor companies. Complementary industries and services provided by industrial parks can entail diversified effects on the surrounding region and stimulate regional development (Huang et al. 2019a). This is crucial for small and medium enterprises (SMEs) because they often lack access to information and funds for sophisticated technologies.

Typical CE tools and policies at this level include sustainable supply chains and industrial symbiosis. A common platform for sharing information and enhancing communication among industrial stakeholders through the application of information and telecommunication technologies is helpful for facilitating the creation of industrial symbiosis. The main benefit of industrial symbiosis is the overall reduction of both virgin materials and final wastes, as well as reduced/avoided transportation costs from by-product exchanges among tenant companies, which can specifically help small- and medium-sized enterprises to improve their growth and competitiveness. From a climate perspective, this indicates significant industrial emission mitigation since the extraction, processing of virgin materials and the final disposal of industrial wastes are more energy intensive. Also, careful site selection of such parks can facilitate the use of renewable energy. Due to these advantages, eco-industrial parks have been actively promoted, especially in East Asian countries, such as China, Japan and the Republic of Korea (South Korea), where national indicators and governance exist (Geng et al. 2019). For instance, the successful implementation of industrial symbiosis at Dalian Economic and Technological Development Zone has achieved significant co-benefits, including GHG-emission reduction, economic and social benefits, and improved ecosystem functions (Liu et al. 2018). Another case at Ulsan industrial park, South Korea, estimated that 60,522 tonnes of $CO_2$ were avoided annually through industrial symbiosis between two companies (Kim et al. 2018b). The case of China shows the great potential of implementing these measures, estimating 111 million tonnes of $CO_2$ equivalent will be reduced in 213 national-level industrial parks in 2030 compared with 2015 (Guo et al. 2018). As such, South Korea's national eco-industrial park project has reduced over 4.7 million tonnes of $CO_2$ equivalent through their industrial symbiosis efforts (Park et al. 2019). Meso-level CE solutions have been identified as essential for industrial decarbonisation (Section 11.4.3). Moreover, waste prevention as the top of the so-called 'waste hierarchy' can be promoted on the meso level for specific materials or product systems. For instance, the European Environment Agency published a report on plastic waste prevention approaches in all 28 EU-member states (Wilts and Bakas 2019). However, challenges exist for industrial symbiosis activities, such as inter-firm contractual uncertainties, the lack of synergy infrastructure, and the regulations that hamper reuse and recycling. Therefore, necessary legal reforms are needed to address these implementation barriers.

**Macro level:** The macro level uses both micro- and meso-level tools within a broader policy strategy, addressing the specific challenge of CE as a cross-cutting policy (Wilts et al. 2016). More synergy opportunities exist beyond the boundary of one industrial park. This indicates the necessity of scaling up industrial symbiosis to urban symbiosis. Urban symbiosis is defined as the use of by-products (waste) from cities as alternative raw materials for energy sources for industrial operations (Sun et al. 2017). It is based on synergistic opportunity arising from geographic proximity through the transfer of physical sources (waste materials) for environmental and economic benefits. Japan is the first country to promote urban symbiosis. For instance, the Kawasaki urban symbiosis efforts can save over 114,000 tonnes of $CO_2$ emissions annually (Ohnishi et al. 2017). Another simulation study indicates that Shanghai (the largest Chinese city) has the potential to save up to 16.8 $MtCO_2$ through recycling all the available wastes (Dong et al. 2018). As such, the simulation of urban-energy-symbiosis networks in Ulsan, South Korea, indicates that 243,396 $tCO_2^{-1}$ $yr^{-1}$ emission and USD48 million $yr^{-1}$ fuel cost can be saved (Kim et al. 2018a). Moreover, Wiebe et al. (2019) estimate that the adoption of the CE can lead to a significantly lower global material extraction compared to a baseline. Their global results range from a decrease of about 27% in metal extraction to 8% in fossil fuel extraction and use, 8% in forestry products, and about 7% in non-metallic minerals, indicating significant climate change benefits. A macro-perspective calculation on the circulation of iron in Japan's future society shows that $CO_2$ emissions from the steel sector can be reduced by 56% as per the following assumptions: the amount recovered from social stock is the same as the amount of inflow, and all scrap was used domestically, and the export of steel products is halved (LCS 2018). A key challenge is to go beyond ensuring proper waste management to setting metrics, targets and incentives to preserve the incorporated value in specific waste streams. Estimations for Germany have shown that despite recycling rates of 64% for all solid-waste streams, these activities only lead to a resource-use reduction of only 18% (Steger et al. 2019). In general, the identification of the most appropriate CE method for different countries requires understanding and information exchange on background conditions, local policies and myriad other factors influencing material flows from the local up to the global level (Tapia Carlos et al. 2019). Also, an information platform should be created at the national level so that all the stakeholders can share their CE technologies and expertise, information (such as materials/energy/water consumption data), and identify the potential synergy opportunities.

### 11.3.4 Energy Efficiency

Energy efficiency in industry is an important mitigation option and central in keeping 1.5°C within reach (IPCC SR1.5). It has long been recognised as the first mitigation option in industry (Yeen Chan and Kantamaneni 2016; Nadel and Ungar 2019; IEA 2021a). It allows reduction of the necessary scale of deployment for low-carbon energy supplies and associated mitigation costs (Energy Transitions Commission 2018). The efficiency potentials are greatest in the non-energy-intensive industries and are often relatively limited in energy-intensive ones, such as steel (Pardo and Moya 2013; Kuramochi 2016; Arens et al. 2017). Deep decarbonisation in these subsectors requires fundamental process changes but energy efficiency remains important to reduce costs and the need for low-carbon energy supplies.

Below, we focus mainly on the technical progress and on new options that are reflected in the literature since AR5 and refer the reader there for a broader and deeper treatment of energy efficiency. Digitalisation and the development of industrial high-temperature heat pumps are two notable technology developments that can facilitate energy efficiency improvements.

Industrial energy efficiency can be improved through multiple technologies and practices (Tanaka 2011; Fawkes et al. 2016; Lovins 2018; Crijns-Graus et al. 2020; IEA 2020a). There are two parallel processes in improvement of specific energy consumption (SEC): progress in energy-efficient BAT and moving the SEC of industrial plants towards BAT. Both slow down as theoretical thermodynamic minimums are approached (Gutowski et al. 2013). For the last several decades the focus has been on effective spreading of BAT technologies through application of policies for worldwide diffusion of energy-saving technologies (Section 11.6). As a result the SEC for

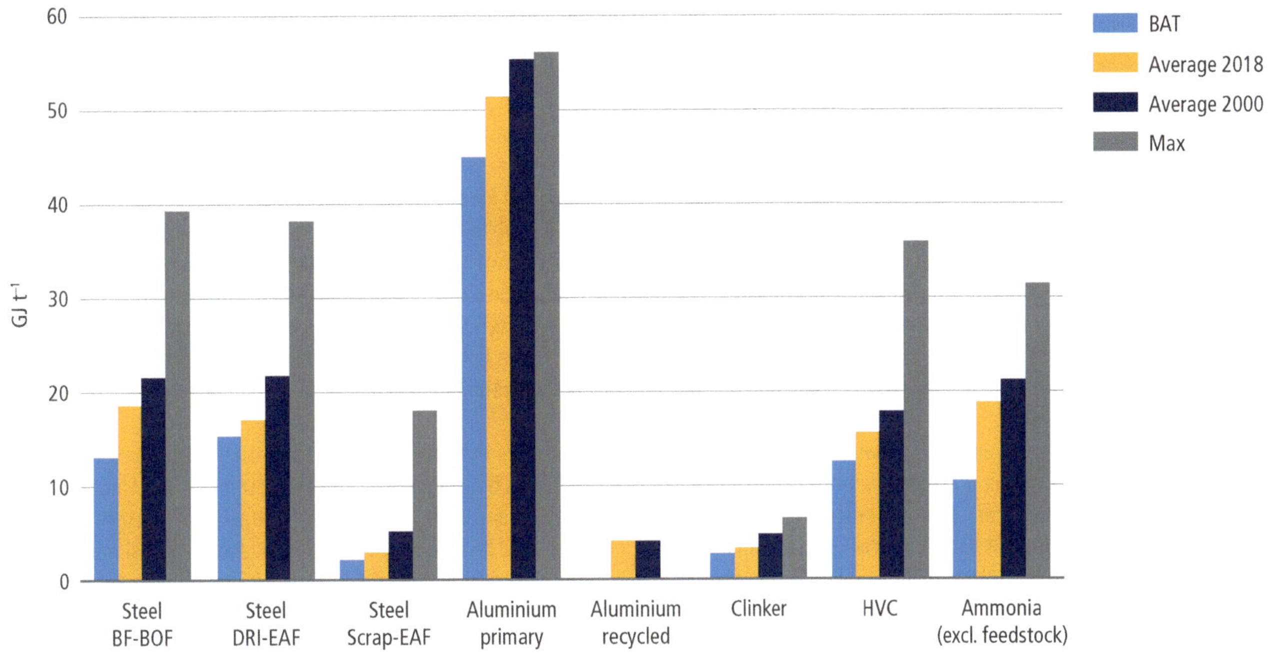


**Figure 11.8 | Energy efficiency indicators for basic material production.** Energy accounting is based on final energy use. Sectoral boundaries for steel are as defined in IEA (2020c).Sources: calculated based on UNIDO (2010); Saygin et al. (2011); Hasanbeigi et al. (2012); Moya and Pardo (2013); Napp et al. (2014); WBCSD (2016); IEA (2017, 2018b); IEA and WBCSD (2018); IEA (2019b, 2020c); Crijns-Graus et al. (2020); IEA (2020b); International Aluminium Institute (2020).

many basic primary materials is approaching BAT and there are signs that energy efficiency improvements have been slowing down over recent decades (IEA 2019d, 2020a, 2021a) (Figure 11.8).

#### 11.3.4.1 Heat-use Energy Efficiency Improvement

While about 10% of global GHG emissions originate from combustion to produce high-temperature heat for basic material production processes (Sandalow et al. 2019), limited efforts have been made to decarbonise heat production. There is still a large potential for using various grades of waste heat and the development of high-temperature heat pumps facilitates its use. NEDO (2019) applies a 'Reduce, Reuse, and Recycle' concept for improved energy efficiency, and we use this frame our discussion of heat efficiency.

*Reduce* refers to reducing heat needs via improved thermal insulation, for example, where porous type insulators have been developed with thermal conductivity half of what is traditionally achieved by heat-resistant bricks under conditions of high compressive strength (Fukushima and Yoshizawa 2016). *Reuse* refers to waste heat recovery. A study for the EU identified a waste heat potential of about 300 TWh $yr^{-1}$, corresponding to about 10% of total energy use in industry. About 50% of this was below 200°C, about 25% at temperatures 200°C–500°C, and 25% at temperatures of 500°C and above (Papapetrou et al. 2018). A survey conducted in Japan showed that 9% of the input energy is lost as waste heat, of which heat below 199°C accounts for 68% and that below 149°C was 29% (NEDO 2019). McBrien et al. (2016) identified that in the steel sector process heat recovery presently saves 1.8 GJ per tonne of hot rolled steel, while integrated across all production processes heat recovery with conventional heat exchange could save 2.5 GJ $t^{-1}$, and it scales up to 3.0 GJ $t^{-1}$ using an alternative heat exchange that recovers energy from hot steel. High-temperature industrial heat pumps represent a new and important development for upgrading waste heat and at the same time they facilitate electrification. One recent example is a high-temperature heat pump that can raise temperatures up to 165°C at a coefficient of performance (COP) of 3.5 by recovering heat from unused hot water (35°C–65°C) (Arpagaus et al. 2018). Commercially available heat pumps can deliver 100°C–150°C but at least up to 280°C is feasible (Zühlsdorf et al. 2019). Mechanical vapour recompression avoids the loss of latent heat by condensation, then it acts as a highly efficient heat pump with a 5–10 COP (Philibert 2017a).

Waste heat to power (WHP), or *Recycle* in NEDO's terms, is also an under-utilised option. For example, a study for the cement, glass and iron industries in China showed that current technology enables only 7–13% of waste heat to be used for power generation. With improved technologies, potentially 40–57% of waste heat with temperatures above 150°C could be used for power generation via heat recovery. Thermal power fluctuations can be a challenge and negatively affect the operation and economic feasibility of heat recovery power systems such as steam and/or organic Rankine cycle. In such cases, latent heat storage technology and intermediate storage units may be applied (Jiménez-Arreola et al. 2018). The development of thermoelectric conversion materials that produce power from unused heat and energy harvested from a higher temperature environment is also progressing, with several possible applications in industrial processes (Gayner and Kar 2016; Jood et al. 2018; Lv et al. 2018; Ohta et al. 2018). A potential early application in industry is to power wireless sensors, a niche that uses microwatts or milliwatts, and avoid power cables (Champier 2017).

#### 11.3.4.2 Smart Energy Management

Energy management systems to reduce energy costs in an integrated and systematic manner were first developed in the 1970s, mainly in low-energy-resource countries, for example, by establishing energy managers and institutionalising management targets (Tanaka 2011). Strategic energy management has since then evolved and been promoted through the establishment of dedicated organisational infrastructures for energy-use optimisation, such as ISO-50001 which specifies the requirements for establishing, implementing, maintaining, and improving an energy management system (Biel and Glock 2016; Tunnessen and Macri 2017). Digitalisation, sometimes referred to as Industry 4.0, facilitates further improvements in process control and optimisation through technology development involving sensors, communications, analytics, digital twins, machine learning, virtual reality, and other simulation and computing technologies (Rogers 2018), all of which can improve energy efficiency. One example is combustion control systems, where big data analysis of factors affecting boiler efficiency, operation optimisation and load forecasting have shown that it can lead to energy savings of 9% (Wang et al. 2017).

Smart energy systems with real-time monitoring allow for optimisation of innovative technologies, energy demand response, balancing of energy supply and demand including that on real-time pricing, and product quality management, and prediction and reduction of idle time for workers and robots (ERIA 2016; Pusnik et al. 2016; ISO 2018; Legorburu and Smith 2018; Ferrero et al. 2020; Nimbalkar et al. 2020). The IEA estimated that smart manufacturing could deliver 15 EJ in energy savings between 2014 and 2030 (IEA 2019d). Smart manufacturing systems that integrate manufacturing intelligence in real time through the entire production operation have not been yet widely spread in the industry. Examples have been demonstrated and integrated in real operation in the electrical appliance assembly industry (Yoshimoto 2016). Combining process controls and automation allows cost optimisation and improved productivity (Edgar and Pistikopoulos 2018).

### 11.3.5 Electrification and Fuel Switching

The principle of electrification and fuel switching as a GHG mitigation strategy is that industries, to the extent possible, switch their end uses of energy from a high GHG intensity energy carrier to a lower or zero intensity one, including both its direct and indirect production and end-use GHG emissions. In general, and non-exclusively, this implies a transition from coal (about 0.09 $tCO_2$ $GJ^{-1}$ on combustion), refined petroleum products (about 0.07 $tCO_2$ $GJ^{-1}$), and natural gas (about 0.05 $tCO_2$ $GJ^{-1}$) to biofuels, direct solar heating, electricity, hydrogen, ammonia, or net zero synthetic hydrocarbon fuels. Switching to these energy carriers is not necessarily lower emitting, however; how they are made matters.

Fuel switching has already been observed to reduce direct combustion $CO_2$ emissions in many jurisdictions. There are significant debates about the net effect of upstream fossil fuel production and fugitive emissions, but observers have noted that in the case of US power generation it would take a leakage rate of about 2.7% from natural gas production to undo the direct fuel switching from coal mitigation effect, and the value is likely higher in most cases (Alvarez et al. 2012; Hausfather 2015). Coal mine methane emissions are also estimated to be substantially higher than previously assessed (Kholod et al. 2020). Alvarez et al. (2018) estimated US fugitive emissions (not including the Permian) at 2.3% of supply, 60% more than previously estimated, while recent Canadian papers indicate fugitive emissions are at least 50% more than reported (Chan et al. 2020; MacKay et al. 2021). However, given the potential for energy supply infrastructure lock-in effects (Tong et al. 2019), purely fossil fuel to fossil fuel switching is a limited and potentially dangerous strategy unless it is used very carefully and in a limited way.

Biofuels come in many forms, including ones that are nearly identical to fossil fuels but sourced from biogenic sources. Solid biomass, either direct from wood chips, lignin or processed pellets, is the most commonly used renewable fuel in industry today and is occasionally used in cement kilns and boilers. Biomethane, biomethanol, and bioethanol are all commercially made today using fermentation and anaerobic digestion techniques and are mostly 'drop-in' compatible with fossil fuel equivalents. In principle they cycle carbon in and out of the atmosphere, but their lifecycle GHG intensities are typically not GHG neutral due to land-use changes, soil carbon depletion, fertiliser use, and other dynamics (Hepburn et al. 2019), and are highly case specific. Most commercial biofuel feedstocks come from agricultural (e.g., corn) and food waste sources, and the feedstock is limited; to meet higher levels of biomass use a transition to using higher cellulose feedstocks like straw, switchgrass and wood waste, available in much larger quantities, must be fully commercialised and deployed. Significant efforts have been made to make ethanol from cellulosic biomass, which promises much higher quantities, lower costs, and lower intensities, but commercialisation efforts, with a few exceptions, have largely not succeeded (Padella et al. 2019). The IEA estimates, however, that up to 20% of today's fossil methane use, including by industry, could be met with biomethane (IEA 2020g) by 2040, using a mixture of feedstocks and production techniques. Biofuel use may also be critical for producing negative emissions when combined with carbon capture and storage (i.e., bioenergy with carbon capture and storage – BECCS). Most production routes for biofuels, biochemicals and biogas generate large side streams of concentrated $CO_2$ which is easily captured, and which could become a source of negative emissions (Sanchez et al. 2018) (Section 11.3.6). Finally, it should be noted that biofuel combustion can, if inadequately controlled, have substantial negative local air quality effects, with implications for SDGs 3, 7 and 11.

There is a large identified potential for direct solar heating in industry, especially in regions with strong solar insolation and sectors with lower heat needs (<180°C), for example, food and beverage processing, textiles, and pulp and paper (Schoeneberger et al. 2020). The key challenges to adoption are site and use specificity, capital intensity, and a lack of standardised mass manufacturing for equipment and a supply chain to provide them.

Switching to electricity for end uses, or 'direct electrification', is a highly discussed strategy for net zero industrial decarbonisation

(Lechtenböhmer et al. 2016; Palm et al. 2016; Åhman et al. 2017; Axelson et al. 2018; Bataille et al. 2018a; Davis et al. 2018; UKCCC 2019b; Material Economics 2019). Electricity is a flexible energy carrier that can be made from many forms of primary energy, with high potential process improvements in terms of end-use efficiency (Eyre 2021), quality and process controllability, digitisability, and no direct local air pollutants (McMillan et al. 2016; Jadun et al. 2017; Deason et al. 2018; Mai et al. 2018). The net-GHG effect of electrification is contingent on how the electricity is made, and because total output increases can be expected, for full effect it should be made with a very low GHG intensity primary source (i.e., <50 g $CO_2$ $kWh^{-1}$: e.g., hydroelectricity, nuclear energy, wind, solar photovoltaics, or fossil fuels with 95+% carbon capture and storage (IPCC 2014)). This has strong implications for the electricity sector and its generation mix when the goal is a net-zero-emissions electricity system. Despite their falling costs, progressively higher mixes of variable wind and solar on a given grid will require support from grid flexibility sources, including demand response, more transmission, storage on multiple time scales, or firm low-to-negative emissions generation sources (e.g., nuclear energy, hydrogen fuel cells or turbines, biofuels, fossil or biofuels with CCS, and geothermal) to moderate costs (Jenkins et al. 2018; Sepulveda et al. 2018; Williams et al. 2021). Regions that may be slower to reduce the GHG intensity of their electricity production will likely need to consider more aggressive use of other measures, like energy and material efficiency or bioenergy.

The long-term potential for full-process electrification is a very sector-by-sector and process-by-process phenomenon, with differing energy and capacity needs, load profiles, stock turnover, capacity for demand response, and characteristics of decision-makers. Industrial electrification is most viable in the near term in cases with: minimal retrofitting and rebuild in processes; with relatively low local electricity costs; where the degree of process complexity and process integration is more limited and extensive process re-engineering would not be required; where combined heat and power is not used; where induction heating technologies are viable; and where process heating temperatures are lower (Deason et al. 2018).

For these reasons, lighter, manufacturing-orientated industries are more readily electrifiable than heavier industry like steel, cement, chemicals and other sectors with high heat and feedstock needs. Steam boilers, curing, drying and small-scale process heating, with typically lower maximum heat temperature needs (<200°C–250°C) are readily electrifiable with appropriate fossil-fuel-to-electricity price ratios (accounting for capital costs and efficiencies), and direct induction and infrared heating are available for higher temperature needs. These practices are uncommon outside regions with ample hydroelectric power due to the currently relatively low cost of coal, natural gas and heating oil, and especially when there is no carbon combustion cost. Madeddu et al. (2020) argue up to 78% of Europe's industrial energy requirements are electrifiable through existing commercial technologies. In contrast, Mai et al. (2018) saw only a moderate industrial heat supply electrification in their high-electrification scenario for the US. Electrification has also been explored in: raw and recycled steel (Fischedick et al. 2014b; Vogl et al. 2018); ammonia (Bazzanella and Ausfelder 2017; Philibert 2017a); and chemicals (Palm et al. 2016; Bazzanella and Ausfelder 2017). While most chemical production of feedstock chemicals (e.g., $H_2$, $NH_3$, CO, $CH_3OH$, $C_2H_4$, $C_2H_6$ and $C_2H_5OH$) is done thermo-catalytically today, it is feasible to use direct electrocatalytic production, by itself or in combination with utilisation of previously captured carbon sources if a fossil fuel feedstock is used, or well-known bio-catalytic (e.g., fermentation) and thermo-catalytic processes (Bazzanella and Ausfelder 2017; De Luna et al. 2019; Kätelhön et al. 2019). It may even be commercially possible to electrify cement sintering and calcination through plasma or microwave options (Material Economics 2019).

Increased electrification of industry will result in increased overall demand for electricity. For example, 75 TWh of electricity was used by steel in the EU in 2015 (out of the 1000 TWh total used by industry), Material Economics (2019), varying between their new process, circularity and CCUS scenarios, projects increased demand to 355 (+373%), 214 (+185%) and 238 (+217%) TWh. These values are consistent with Vogl et al. (2018), which projects a tripling of electricity demand in the German or Swedish steel industries if hydrogen-direct reduced iron and electric arc furnace steel-making (DRI EAFs) replaces BF-BOFs. Material Economics (2019) was conservative with its use of electricity in chemical production, making preferential use of biofeedstocks and some CCUS, and electricity demand still rose from 118 TWh to 510, 395 and 413 TWh in their three scenarios. Bazzanella and Ausfelder (2017), exploring deeper reductions from the chemical sector using more electrochemistry, projected scenarios with higher electricity demands of 960–4900 TWh (140% of the projected available clean electricity at the time) with maximum electricity use. In counterpoint, however, with revised wind capabilities and costs, the IEA (2019e) Offshore Wind Outlook indicates that ten times the current EU electricity use could be produced if necessary. Greater use of electro-catalytic versus thermo-catalytic chemistry, as projected by De Luna et al. (2019), could greatly reduce these electricity needs, but the technology readiness levels are currently low. Finally, the UKCCC (2019b), which focused primarily on CCS for industry in its 'Further Ambition' scenario (the UK currently consumes about 300 TWh), in its supplementary 'Further Electrification' scenario projects an additional 300 TWh for general electrolysis needs and another 200 TWh for synthetic fuel production.

While it has been demonstrated that almost any heating end use can be directly electrified, this would imply very high instantaneous thermal loads for blast furnace-basic oxygen furnace (BF-BOF) steel production, limestone calcination for cement and lime production, and other end uses where flame-front (1000°C–1700°C) temperatures are currently needed. This indicates a possible need for another energy carrier to minimise instantaneous generation and transmission needs. These needs can be met at varying current and potential future costs using: bioliquids or gases hydrogen, ammonia, or net zero synthetic hydrocarbons or alcohols.

Broadly speaking, **hydrogen** can contribute to a cleaner energy system in two ways: (i) existing applications of hydrogen (e.g., nitrogen fertiliser production, refinery upgrading) can use hydrogen produced using alternative, cleaner production methods; (ii) new applications can use low-GHG hydrogen as an alternative to current fuels and inputs, or as a complement to the greater use of electricity in these applications. In these cases – for example, in transport, heating, industry (e.g., hydrogen-direct reduced iron and steel production) and electricity – hydrogen can be used in its pure form, or be converted to hydrogen-based fuels, including ammonia, or synthetic net zero hydrocarbons and alcohols such as methane or methanol (IEA 2019f). The IEA states that hydrogen could be used to help integrate more renewables, including by enhancing storage options and 'exporting sunshine and wind' from places with abundant resources; decarbonise steel, chemicals, trucks, ships and planes; and boost energy security by diversifying the fuel mix and providing flexibility to balance grids (IEA 2019f).

Around 70 Mt $yr^{-1}$ of pure hydrogen is produced today: 76% from natural gas and 23% from coal, resulting in emissions of roughly 830 $MtCO_2$ $yr^{-1}$ in 2016/17 (IEA 2019f), or 4.7% of global industrial direct and indirect emissions (waste excluded; Table 11.1). Fuels refining (about 410 $MtCO_2$ $yr^{-1}$) and production of ammonia (420 $MtCO_2$ $yr^{-1}$) largely dominate its uses. Another 45 Mt hydrogen is being produced along with other gases, on purpose or as by-products, and used as fuel, to make methanol or as a chemical reactant (IEA 2019f). Very low and potentially zero GHG (depending on the energy source) hydrogen can be made via: electrolysis separation of water into hydrogen and oxygen (Glenk and Reichelstein 2019), also known as 'green $H_2$'; electrothermal separation of water, as done in some nuclear plants (Bicer and Dincer 2017); partial oxidation of coal or naphtha or steam/auto methane reforming (SMR/ATR) combined with CCS (Leeson et al. 2017), or 'blue $H_2$'; methane pyrolysis, where the hydrogen and carbon are separated thermally and the carbon is left as a solid (Abbas and Wan Daud 2010; Ashik et al. 2015), or via biomass gasification (Ericsson 2017), which could be negative emissions if the $CO_2$ from the gasification process is sequestered. All these processes would in turn need to be run using very low or zero GHG energy carriers for the resulting hydrogen to also be low in GHG emissions.

**Ammonia production**, made from hydrogen and nitrogen using the Haber-Bosch process, is the most voluminous chemical produced from fossil fuels, being used as feedstock for nitrogen fertilisers and explosives, as well as a cleanser, a refrigerant, and for other uses. Most ammonia is made today using methane as the hydrogen feedstock and heat source but has been made using electrolysis-based hydrogen in the past, and there are several announced investments to resume doing so. If ammonia is used as a combustion fuel, care must be taken to avoid $N_2O$ as a GHG and $NO_x$ in general as a local air pollutant.

Hydrogen can also be combined with low-to-zero net GHG carbon (Section 11.3.6) and oxygen and made into **methane, methanol** and other potential net zero **synthetic hydrocarbons and alcohol** energy carriers using methanation, steam reforming and Fischer-Tropsch processes, all of which can provide higher degrees of storable and shippable high-temperature energy using known industrial processes in novel combinations (Bataille et al. 2018a; Davis et al. 2018). If the hydrogen and oxygen is accessed via electrolysis, the terms 'power-to-fuel' or 'e-fuels' are often used (Ueckerdt et al. 2021). Given their carbon content, if used as fuels, their carbon will eventually be oxidised and emitted as $CO_2$ to the atmosphere. This makes their net-GHG intensity dependent on the carbon source (Hepburn et al. 2019), with recycled fossil fuels, biocarbon and direct air capture carbon all having very different net-$CO_2$ impacts – see section 11.3.6 on CCS and CCU for elaboration.



## Box 11.1 | Hydrogen in Industry

The 'hydrogen economy' is a long-touted vision for the energy and transport sectors, and one that has gone through hype-cycles since the energy crises in the 1970s (Melton et al. 2016). The widely varying visions of hydrogen futures have mainly been associated with fuel cells in vehicles, small-scale decentralised cogeneration of heat and electricity, and to a certain extent energy storage for electricity (Eames et al. 2006; Syniak and Petrov 2008). However, nearly all hydrogen currently produced is used in industry, mainly for hydrotreating in oil refineries, to produce ammonia, and in other chemical processes, and it is mostly made using fossil fuels.

In the context of net zero emissions, new visions are emerging in which hydrogen has a central role to play in decarbonising industry. Near-term industrial applications for hydrogen include feeding it into ammonia production for fertilisers, while a more novel application would be as a replacement for coal as the reductant in steel-making, being piloted by the HYBRIT project in Sweden 2020–2021, and many companies have initiated hydrogen steel-making projects. As shown in Sections 11.3.5 and 11.3.6, there are many other potential applications of hydrogen, some of which are still relatively unexplored. Hydrogen can also be used to produce various lower-GHG hydrocarbons and alcohols for fuels and chemical feedstocks using carbon from biogenic sources or direct air capture of $CO_2$ (Ericsson 2017; Huang et al. 2020).

The geographical distribution of the potential for hydrogen from electrolysis powered by renewables like solar and wind, nuclear electrothermally produced hydrogen, and hydrogen from fossil gas with CCS may reshape where heavy industry is located, how value chains are organised, and what gets transported in international shipping (Bataille 2020a; Gielen et al. 2020; Bataille et al. 2021a; Saygin and Gielen 2021). Regions with bountiful renewables resources, nuclear, or methane co-located with CCS geology may become exporters of hydrogen or hydrogen carriers such as methanol and ammonia, or home to the production of iron and steel, organic platform chemicals, and other energy-intensive basic materials. This in turn may generate new trade patterns and needs for bulk transport.

### 11.3.6 CCS, CCU, Carbon Sources, Feedstocks, and Fuels

Carbon is an important and highly flexible building block for a wide range of fuels, organic chemicals and materials including methanol, ethanol, olefins, plastics, textiles, and wood and paper products. In this chapter we define CCS as requiring return of $CO_2$ from combustion or process gases or ambient air to the geosphere for geological time periods (i.e., thousands of years) (IPCC 2005; IEA 2009; Bruhn et al. 2016; IEA 2019g). CCU is defined as being where carbon (as CO or $CO_2$) is captured from one process and reused for another, reducing emissions from the initial process, but is then potentially but not necessarily released to the atmosphere in following processes (Bruhn et al. 2016; Detz and van der Zwaan 2019; Tanzer and Ramírez 2019). In both cases the net effect on atmospheric emissions depends on the initial source of the carbon, be it from a fossil fuel, from biomass, or from direct air capture (Cuéllar-Franca and Azapagic 2015; Hepburn et al. 2019) and the duration of storage or use, which can vary from days to millennia.

While CCS and CCU share common capture technologies, what happens to the $CO_2$ and therefore the strategies that will employ them can be very different. CCS can help maintain near-$CO_2$ neutrality for fossil $CO_2$ that passes through the process, with highly varying partially negative emissions if the source is biogenic (Hepburn et al. 2019), and fully negative emissions if the source is air capture, all not considering the energy used to drive the above processes. CCS has been covered in other IPCC publications at length, for example, IPCC (2005), and in most mitigation-oriented assessments since, for example, the IEA's Energy Technology Perspectives (ETP) 2020 and Net Zero scenario reports (IEA 2021a, 2020a). The potentials and costs for CCS in industry vary considerably due to the diversity of industrial processes (Leeson et al. 2017), as well as the volume and purity of different flows of $CO_2$ (Naims 2016); Kearns et al. (2021) provide a recent review. As a general rule it is not possible to capture all the $CO_2$ emissions from an industrial plant. To achieve zero or negative emissions, CCS would need to be combined with some use of sustainably sourced biofuel or feedstock, or the remaining emissions would need to be offset by carbon dioxide removal (CDR) elsewhere.

For concentrated $CO_2$ sources (e.g., cleaning of wellhead formation gas to make it suitable for the pipeline network, hydrogen production using steam methane reforming, ethanol fermentation, or from combustion of fossil fuels with oxygen in a nitrogen-free environment, i.e., 'oxycombustion') CCS is already amenable to commercial oil and gas reinjection techniques used to eliminate hydrogen sulphide gas and brines at prices of USD10–40 $tCO_2$-eq$^{-1}$ sequestered (Wilson et al. 2003; Leeson et al. 2017). Most currently operating CCS facilities take advantage of concentrated $CO_2$ flows, for example, from formation gas cleaning on the Snoevit and Sleipner platforms in Norway, from syngas production for the Al Reyadah DRI steel plant in Abu Dhabi, and from SMR hydrogen production on the Quest upgrader in Alberta. Since concentrated process $CO_2$ emissions are often exempted from existing cap and trade systems, these opportunities for CCS have largely gone unexploited. Many existing projects partially owe their existence to the utilisation of the captured $CO_2$ for enhanced oil recovery, which in many cases counts as both CCS and CCU because of the permanent nature of the $CO_2$ disposal upon injection if sealed properly (Mac Dowell et al. 2017). There are several industrial CCS strategies and pilot projects working to take advantage of the relative ease of concentrated $CO_2$ disposal (e.g., LEILAC for limestone calcination process emissions from cement production, HISARNA direct oxycombustion smelting for steel) (Bataille 2020a). An emerging option for storing carbon is methane pyrolysis by which methane is split into hydrogen and solid carbon that may subsequently be stored (Schneider et al. 2020).

There are several post-combustion CCS projects underway globally (IEA 2019g), generally focused on energy production and processing rather than industry. Their costs are higher but evolving downward – Giannaris et al. (2020) suggest USD47 $tCO_2^{-1}$ for a follow-up 90% capture power generation plant based on learnings from the Saskpower Boundary Dam pilot – but crucially these costs are higher than implicit and explicit carbon prices almost everywhere, resulting in limited investment and learning in these technologies. A key challenge with all CCS strategies, however, is building a gathering and transport network for $CO_2$, especially from dispersed existing sites; hence most pilot projects are built near EOR/geological storage sites, and the movement towards industrial clustering in the EU and UK (UKCCC 2019b), and as suggested in IEA (2019f).

In the case of CCU, CO and $CO_2$ are captured and subsequently converted into valuable products (e.g., building materials, chemicals and synthetic fuels) (Styring et al. 2011; Bruhn et al. 2016; Artz et al. 2018; Brynolf et al. 2018; Daggash et al. 2018; Breyer et al. 2019; Kätelhön et al. 2019; Vreys et al. 2019). CCU has been envisioned as part of the 'circular economy' but conflicting expectations on CCU and its association or not with CCS leads to different and contested framings (Palm and Nikoleris 2021). The duration of the $CO_2$ storage in these products varies from days to millennia according to the application, potentially but not necessarily replacing new fossil, biomass or direct air capture feedstocks, before meeting one of several possible fates: permanent burial, decomposition, recycling or combustion, all with differing GHG implications. While the environmental assessment of CCS projects is relatively straightforward, however, this is not the case for CCU technologies. The net-GHG mitigation impact of CCU depends on several factors (e.g., the capture rate, the energy requirements, the lifetime of utilisation products, the production route that is substituted, and associated room for improvement along the traditional route) and has to be determined by lifecycle $CO_2$ or GHG analysis (e.g., Nocito and Dibenedetto 2020; and Bruhn et al. 2016). For example, steel-mill gases containing carbon monoxide and carbon dioxide can be used as feedstock together with hydrogen for producing chemicals. In this way, the carbon originally contained in the coke used in the blast furnace is used again, or cascaded, and emissions are reduced but not brought to zero. If fossil-sourced $CO_2$ is only reused once and then emitted, the maximum reduction is 50% (Tanzer and Ramírez 2019). The logic of using steel-mill CO and $CO_2$ could equally be applied to gasified biomass, however, with a far lower net-GHG footprint, likely negative, which CCU fed by fossil fuels cannot be if end-use combustion is involved.

Partly because of the complexity of the lifecycle analysis accounting, the literature on CCU is not always consistent in terms of the net-GHG impacts of strategies. For example, Artz et al. (2018), focused not

11

just on GHG mitigation but multi-attribute improvements to chemical processes from reutilisation of $CO_2$, suggests the largest reduction in the absolute amount of GHGs from $CO_2$ reutilisation could be achieved by the coupling of highly concentrated $CO_2$ sources with carbon-free hydrogen or electrons from low GHG power in so called 'power-to-fuel' scenarios. From the point of view of maximising GHG mitigation using surplus 'curtailed' renewable power, however, Daggash et al. (2018) instead indicates the best use would be for direct air capture and CCS. These results depend on what system is being measured, and what the objective is.

There are several potential crucial transitional roles for synthetic hydrocarbons and alcohols (e.g., methane, methanol, ethanol, ethylene, diesel and jet fuel) constructed using fossil, biomass or direct carbon capture (DAC) and CCU (Breyer et al. 2015; Dimitriou et al. 2015; Sternberg and Bardow 2015; Fasihi et al. 2017; Bataille et al. 2018a; Bataille 2020a). They can allow reductions in the GHG intensity of high-value legacy transport, industry and real estate that currently runs on fossil fuels but cannot be easily or readily retrofitted. They can be used by existing long-lived energy and feedstock infrastructure, transport and storage, which can compensate for seasonal supply fluctuations and contribute to enhancing energy security (Ampelli et al. 2015). Finally, they can reduce the GHG intensity of end uses that are very difficult to run on electricity, hydrogen or ammonia (e.g., long-haul aviation). However, their equivalent mitigation cost today would be very high (USD960–1440 $tCO_2$-$eq^{-1}$), with the potential to fall to USD24–324 $tCO_2$-$eq^{-1}$) with commercial economies of scale, with very high uncertainty (Hepburn et al. 2019; IEA 2020a; Ueckerdt et al. 2021).

A very large and important uncertainty is the long-term demand for hydrocarbon and alcohol fuels (whether fossil-, biomass- or DAC-based), chemical feedstocks (e.g., methanol and ethylene) and materials, and competition for biomass feedstock with other priorities, including agriculture, biodiversity and other proximate land-use needs, as well as need for negative emissions through BECCS. The current global plastics production of around 350 Mt $yr^{-1}$ is almost entirely based on petroleum feedstock and recycling rates are very low. If this or future demand were to be 100% biomass-based it would require tens of exajoules of biomass feedstock (Meys et al. 2021). If demand can be lowered and recycling increased (mechanical as well as chemical) the demand for biomass feedstock can be much lower (Material Economics 2019). Promising routes in the short-term would be to utilise $CO_2$ from anaerobic digestion for biogas and fermentation for ethanol in the production of methane or methanol (Ericsson 2017); methanol can be converted into ethylene and propylene in a methanol-to-olefins process and used in the production of plastics (Box 11.2). New process configurations where hydrogen is integrated into biomass conversion routes to increase yields and utilise all carbon in the feedstock are relatively unexplored (Ericsson 2017; De Luna et al. 2019).

There are widely varying estimates of the capacity of CCU to reduce GHG emissions and meet the net zero objective. According to Hepburn et al. (2019), the estimated potential for the scale of $CO_2$ utilisation in fuels varies widely, from 1 to 4.2 $GtCO_2$ $yr^{-1}$, reflecting uncertainties in potential market penetration, requiring carbon prices of around USD40 to 80 $tCO_2^{-1}$, increasing over time. The high end represents a future in which synthetic fuels have sizeable market shares, due to cost reductions and policy drivers. The low end – which is itself considerable – represents very modest penetration into the methane and fuels markets, but it could also be an overestimate if $CO_2$-derived products do not become cost competitive with alternative clean energy vectors such as hydrogen or ammonia, or with direct sequestration. Brynolf et al. (2018) indicates that a key cost variable will be the cost of electrolysers for producing hydrogen. Kätelhön et al. (2019) estimate that up to 3.5 GtC $yr^{-1}$ could be displaced from chemical production by 2030 using CCU, but this would require clean electricity equivalent to 55% of estimated global power production, at the same time other sectors' demand would also be rising. Mac Dowell et al. (2017) suggest that while CCU, and specifically $CO_2$-based enhanced oil recovery, may be an important economic incentive for early CCS projects (up to 4–8% of required mitigation by 2050), it is unlikely the chemical conversion of $CO_2$ for CCU will account for more than 1% of overall mitigation.

Finally, there is another class of CCU activities associated with carbonation of alkaline industrial wastes (including iron and steel slags, coal fly ash, mining and mineral processing wastes, incinerator residues, cement and concrete wastes, and pulp and paper mill wastes) using waste or atmospheric $CO_2$. Given the large volume of alkaline wastes produced by industry, capture estimates are as high as 4 $GtCO_2$ $yr^{-1}$ (Cuéllar-Franca and Azapagic 2015; Ebrahimi et al. 2017; Kaliyavaradhan and Ling 2017; Pasquier et al. 2018; Huang et al. 2019c; Pan et al. 2020; Zhang et al. 2020) However, as some alkaline wastes are already used directly as supplementary cementitious materials to reduce clinker-to-cement ratios, and their abundant availability in the future is questionable (e.g., steel blast furnace slag and coal fly ash), there will be a strong competition between mitigation uses (Section 11.4.2), and the potential for direct removal by carbonation is estimated at about 1 $GtCO_2$ $yr^{-1}$ (Renforth 2019).

The above CCU literature has identified that there may be a highly unpredictable competition between fossil, biogenic and direct air capture carbon to provide highly uncertain chemical feedstock, material and fuel needs. Fossil waste carbon will likely initially be plentiful but will add to net atmospheric $CO_2$ when released. Biogenic carbon is variably, partially net-negative, but the available stock will be finite and compete with biodiversity and agriculture needs for land. Direct air capture carbon will require significant amounts of low-GHG electricity or methane with high-capture rate CCS (Keith et al. 2018). There are clearly strong interactive effects between low-carbon electrification, switching to biomass, hydrogen, ammonia, synthetic hydrocarbons via CCU, and CCS.

### 11.3.7 Strategy Interactions and Integration

In this section we conceptually address interactions between service demand, service product intensity, product material efficiency, energy efficiency, electrication and fuel switching, CCU and CCS, and what conflicts and synergies may exist. Post AR5 a substantial literature has emerged, see Rissman et al. (2020), that addresses integrated

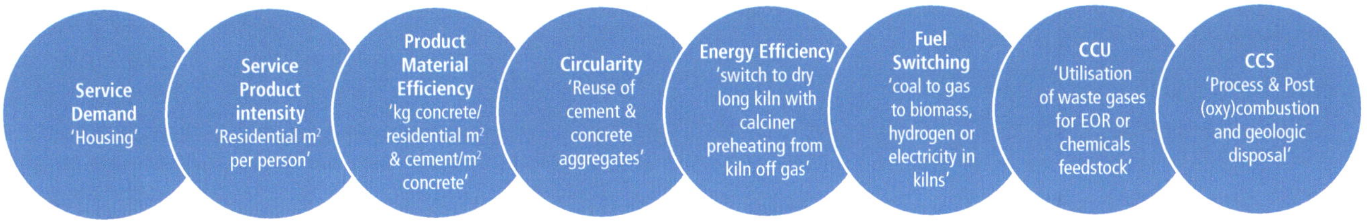


**Figure 11.9 | Fully interactive, non-sequential strategies for decarbonising industry.**

and interactive technical deep decarbonisation pathways for GHG-intense industrial sectors, and how they interact with the rest of the economy (Denis-Ryan et al. 2016; Åhman et al. 2017; Wesseling et al. 2017; Axelson et al. 2018; Davis et al. 2018; Bataille et al. 2018a; Bataille 2020a). It is a common finding across this literature and a related scenario literature (Energy Transitions Commission 2018; Material Economics 2019; UKCCC 2019a,b; IEA 2019b, 2020a; CAT 2020; IEA 2021a) that deep decarbonisation of industry requires integrating all available options. There is no 'silver bullet' and so all behavioural and technological options have to be mobilised, with more emphasis required on the policy mechanisms necessary to engage a challenging transition in the coming decades in highly competitive, currently GHG-intense, price-sensitive sectors with long-lived capital stock (Wesseling et al. 2017; Bataille et al. 2018a; Bataille 2020a), discussed in the final section of this chapter.

While the strategies are not sequential and interact strongly, we discuss them in the order given. Reduced demand through reduced service demand and product intensity per service unit (Grubler et al. 2018; van Vuuren et al. 2018) reduces the need for the next six strategies. Greater material efficiency (see earlier sections) reduces the need for the next five, and so on – see Figure 11.9 above.

Circular economy introduces itself throughout, but mainly at the front end when designing materials and processes to be more materially efficient, efficient in use, and easy to recycle, and at the back end, when a material or product's services life has come to end, and it is time for recycling or sustainable disposal (Murray et al. 2017; Korhonen et al. 2018). The entire chain's potential will be maximised when these strategies are designed in ahead of time instead of considered on assembly, or as a retrofit (Allwood et al. 2012; Gonzalez Hernandez et al. 2018a; IEA 2019b; Material Economics 2019; Bataille 2020a). For example, when designing a building: (i) Is the building shell, interior mass and ducting orientated for passive heating and cooling, and can the shell and roof have building-integrated solar PV or added easily, with hard-to-retrofit wiring already incorporated? (ii) Are steel and high-quality concrete only used where really needed (i.e., for shear, tension and compression strength), can sections be prefabricated off-site, can other materials be substituted, such as wood? (iii) Can the interior fittings be built with easy-to-recycle plastics or other sustainably disposable materials (e.g., wood)? (iv) Can this building potentially serve multiple purposes through its anticipated lifetime, are service conduits oversized and easy to access for retrofitting? (v) When it is time to be taken apart, can pieces be reused, and all componnents recycled at high purity levels, for example, can all the copper wiring be easily be found and removed, are the steel beams clearly tagged with their content? The answers to these questions will be very regionally and site specific, and require revision of educational curricula for the entire supply chain, as well as revision of building codes.

Energy efficiency is a critical strategy for net zero transitions and enabling clean electrification (IEA 2021a). Improving the efficiency of energy services provision reduces the need for material intensive energy supply, energy storage, CCU and CCS infrastructure, and limits generation and transmission expansion to reduce an ever-higherdemand, with associated generation, transmission, and distribution losses. Using electricity efficiently can help reduces peak demand and the need for peaking plants (currently often powered by fossil fuels), and energy storage systems.

Electrification and final energy efficiency are deeply entangled, because switching to electricity from fossil fuels in most cases improves GJ for GJ end-use energy efficiency: resistance heaters are almost 100% efficient, heat pumps can be 300–400% efficient, induction melting can improve mixing and temperature control, and electric vehicle motors typically translate 90–95% of input electricity to motor drive in contrast to 35–45% for a large, modern internal combustion engine. Overall, the combined effect could be 40% lower global final energy demand assuming renewable electricity is used (Eyre 2021).

There are potentially complicated physical and market fuel switching relationships between low-GHG electricity, bioliquids and gases, hydrogen, ammonia, and synthetic hydrocarbons constructed using CCU, with remaining $CO_2$ potentially being disposed of using CCS. Whether or not they compete for a wide range of end uses and primary demand needs will be regional and whether or not infrastructure is available to supply them. Regions with less than optimal renewable energy resources, or not sufficient to meet growing needs, could potentially indirectly import them as liquid or compressed hydrogen, ammonia or synthetic hydrocarbon feedstocks made in regions with abundant resources (Armijo and Philibert 2020; Bataille 2020a). Large-scale CCU and CCS applications need additional basic materials to build corresponding infrastructure and energy to operate it, thus reducing overall material and energy efficiencies.

There are different roles for different actors in relation to the different mitigation strategies (exemplified in Table 11.2), with institutions and supply chains developed to widely varying levels, for example, while energy efficiency is a relatively mature strategy with an established supply chain, material efficiency is not.

**Table 11.2 | Examples of the potential roles of different actors in relation to different mitigation strategies indicating the importance of engaging a wide set of actors across all mitigation strategies.**

| Sectors | Demand control measures (DM) | Materials efficiency (ME) | Circular economy | Energy efficiency | Electrification, hydrogen and fuel switching | CCU | CCS |
|---|---|---|---|---|---|---|---|
| Architectural and engineering firms | Build awareness on the material demand implications of e.g., building codes, urban planning and infrastructure. | Education of designers, architects and engineers, etc. Develop design tools. Map material flows. | Design and build for e.g., repurpose, reuse and recycle. Improve transparency on volumes and flows. | Maintain high expertise, knowledge sharing, transparency, and benchmarking. | Support innovation. Share best practice. Design for dynamic demand response for grid balancing. | Develop allocation rules, monitoring and transparency. Coordination and collaboration across sectors. | Transparency, monitoring and labelling. Coordination and collaboration for transport and disposal infrastructure. |
| Industry and service sector | Digital solutions to reduce office space and travel. Service-oriented business models for lower product demand. | Design for durability and light weight. Minimise industry scrap. | Design for reuse and recycling. Use recycled feedstock and develop industrial symbiosis. | Maintain energy management systems. | Develop and deploy new technologies in production, engage with lead markets. | Develop new technologies. Engage in new value chains and collaborations for sourcing carbon. | Plan for CCS where possible and phase-out of non-retrofittable plants where necessary. |
| International bodies | Best practice sharing. Knowledge building on demand options. | Progressivity in international standards (e.g., ISO). | Transparency and regulation around products, waste handling, trade, and recycling. | Maintain efforts for sharing good practice and knowledge. | Coordinate innovation efforts, technology transfer, lead markets, and trade policies. | Coordinate and develop accounting and standards. Ensure transparency. | Align regulation to facilitate export, transport, and storage. |
| Regional and national government, and cities | Reconsider spatial planning and regulation that has demand implications. | Procurement guidelines and better indicators. Standards and building codes. | Regulation on product design (e.g., Ecodesign Directive). Collect material-flow data. | Continue energy efficiency policies such as incentives, standards, labels, and disclosure requirements. | R&D and electricity infrastructure. Policy strategies for making investment viable (including carbon pricing instruments). | Align regulation to facilitate implementation and ensure accountability for emissions. | Develop regulation and make investment viable. Resolve long-term liabilities. |
| Civil society and consumer organisations | Information and advocacy related to social norms. | Strengthen lobby efforts and awareness around e.g., planned obsolescence. | Engage in standards, monitoring and transparency. | Monitor progress. | Information on embodied emissions. Assess renewable electricity and grid expansion. | Develop standards and accounting rules. | Ensure transparency and accountability. |

## 11.4 Sector Mitigation Pathways and Cross-sector Implications

This section continues the discussion of the various mitigation options and strategy elements introduced in Section 11.3 and makes them explicit for the most relevant industry sectors. For the various sectors, Section 11.4.1 concludes with a tabular overview of key technologies and processes, their technology readiness level (TRL), potential timing of market penetration, mitigation potential and assessment of associated mitigation costs.

An integrated sequencing of mature short-term actions and less mature longer-term actions is crucial to avoid lock-in effects. Temporal implementation and discussion of the general quantitative role of the different options to achieve net zero emissions in the industrial sectors is core to the second part of the section (Section 11.4.2), where industry-wide mitigation pathways are analysed. This comprises the collection and discussion of mitigation scenarios available in the literature with a high technological resolution for the industry sector in addition to a set of illustrative global and national GHG mitigation scenarios selected from chapters 3 and 4, representing different GHG mitigation ambitions and different pathways to achieve certain mitigation targets. Comparing technology-focused sector-based scenarios with more top-down-oriented scenario approaches allows for a reciprocal assessment of both perspectives and helps to identify robust elements for the transformation of the sector. Comparison of real-world conditions within the sector (e.g., industry structure and logics, investment cycles, market behaviour, power, and institutional capacity) and the transformative pathways described in the scenarios helps researchers, analysts, governments, and all stakeholders understand the need not only for technological change, but for structural (e.g., new value chains, markets, infrastructures, and sectoral couplings) and behavioural (e.g., design practices and business models) change at multiple levels.

When undergoing a transformative process, it is obvious that interactions occur within the sector but also on a cross-sectoral basis. Relevant interactions are identified and discussed in the third and fourth part of the subsection. Changes are induced along the whole value chain, i.e., switching to an alternative (climate-friendly, e.g., low-GHG hydrogen-based) steel-making process has substantial impacts on the value chain, associated sub-suppliers, and electricity and coal outputs. In addition, cross-sectoral interactions are discussed. This includes feedback loops with other end-use chapters, for example, higher material demand through market penetration of some GHG mitigation technologies or measures (e.g., insulation materials for buildings, steel for windmills) and lower demand through others (e.g., less steel for fossil fuel extraction, transport and processing), or substantial additional demand of critical materials (e.g., the widely varying demands for copper, lithium, nickel, cobalt and rare earths for producing windmills, solar panels, and batteries). Generally, if consumption- (or behaviour-) driven additional material demand creates scarcity it becomes important to increase efforts on material efficiency, substitution, recycling/reuse, and sustainable consumption patterns.

### 11.4.1 Sector-specific Mitigation Potential and Costs

Based on the general discussion of strategies across industry in Section 11.3, this subsection focuses on the sector perspective and provides insights into the sector-specific mitigation technologies and potentials. As industry is comprised of many different subsectors, the discussion here has its focus on the most important sources of GHG emissions, that is, steel, cement and concrete, as well as chemicals, before other sectors are discussed.

#### 11.4.1.1 Steel

For the period leading up to 2020, in terms of end-use allocation globally, approximately 40% of steel is used for structures, 20% for industrial equipment, 18% for consumer products, 13% for infrastructure, and 10% for vehicles (Bataille 2020b). The global production of crude steel increased by 41% between 2008 and 2020 (World Steel Association 2021) and its GHG emissions, depending on the scope covered, is 3.7–4.1 $GtCO_2$-eq. It represented 20% of total global direct industrial emissions in 2019 accounting for coke oven and blast furnace gases use (Crippa et al. 2021; Lamb et al. 2021; Minx et al. 2021; Olivier and Peters 2018; World Steel Association 2021; IEA 2020a) (Figure 11.4 and Table 11.1). Steel production can be divided into primary production based on iron ore and secondary production based on steel scrap. The blast furnace-basic oxygen furnace route (BF-BOF) is the main primary steel route globally, while the electric arc furnace (EAF) is the preferred process for the less energy and emissions-intensive melting and alloying of recycled steel scrap. The direct reduced iron (DRI) route is a lesser-used route that replaces BFs for reducing iron ore, usually followed by an EAF. In 2019, 73% of global crude steel production was produced in BF-BOFs, while 26% was produced in EAFs, a nominal 5.6% of which is DRI (World Steel Association 2021).

An estimated 15% energy efficiency improvement is possible within the BF-BOF process (Figure 11.8). Several options exist for deep-GHG emissions reductions in steel-production processes (Fischedick et al. 2014b; Leeson et al. 2017; Axelson et al. 2018; Vogl et al. 2018; Bataille 2020a; Holappa 2020; Rissman et al. 2020; Fan and Friedmann 2021; Wang et al. 2021).Each could reduce specific $CO_2$ emissions of primary steel production by 80% or more relative to today's dominant BF-BOF route if input streams are based on carbon-free energy and feedstock sources or if they deploy high-capture CCS:

- **Increasing the share of the secondary route** can bring down emissions quickly and potential emissions savings are significant, from a global average 2.3 $tCO_2^{-1}$ per tonne steel in BF-BOFs down to 0.3 (or less) $tCO_2^{-1}$ per tonne steel in EAFs (Pauliuk et al. 2013a; Zhou et al. 2019), the latter depending on scrap preheating and electricity GHG intensity. However, realising this potential is dependent on the availability of regional and global scrap supplies and requires careful sorting and scrap management, especially to eliminate copper contamination (Daehn et al. 2017). There is significant uncertainty about how much new scrap will be available and usable (Xylia et al. 2018; IEA 2019b; Wang et al. 2021). Most steel is recycled already; the gains are mainly to be made in quality (i.e., separation from contaminants like copper). End-of-life scrap availability and its contribution to steel production will increase as in use stock saturates in many countries (Xylia et al. 2016).
- **BF-BOFs with CCU or CCS.** Abdul Quader et al. (2016) and Fan and Friedmann (2021) indicate that it would be difficult to retrofit BF-BOFs beyond 50% capture, which is insufficient for long-term emission targets but may be useful in some cases for avoiding cumulative emissions where other options are not available. However, BF-BOFs need their furnaces relined every 15–25 years (IEA 2021a; Vogl et al. 2021b), at a cost of 80–100% of a new build, and this would be an opportunity to build a new facility designed for 90%+ capture (e.g., fewer $CO_2$ outlets). This would depend upon access to transport to geology appropriate for CCS.
- **Methane-based syngas (hydrogen and carbon monoxide) direct reduced iron (DRI) with CCS.** Most DRI facilities currently use a methane-based syngas of $H_2$ and CO as both reductant and fuel (some use coal). A syngas DRI-EAF steel-making facility has been operating in Abu Dhabi since 2016 that captures carbon emitted from the DRI furnace (where it is a co-reductant with hydrogen) and sends it to a nearby oil field for enhanced oil recovery.
- **Hydrogen-based direct reduced iron (H-DRI)** is based on the already commercialised DRI technology but using only hydrogen as the reductant; pure hydrogen has already been used commercially by Circored in Trinidad 1999–2008. The reduction process of iron ore is typically followed by an EAF for smelting. During a transitional period, DRI could start with methane or a mixture of methane and hydrogen as some of the methane ($\leq$30% hydrogen can be substituted with green or blue hydrogen without the need to change the process). If the hydrogen is produced based on carbon-free sources, this steel-production process can be nearly $CO_2$ neutral (Vogl et al. 2018).
- **In the aqueous electrolysis route** (small-scale piloted as Siderwin during the EU ULCOS programme), the iron ore is bathed in an electrolyte solution and an electric current is used to remove the oxygen, followed by an electric arc furnace for melting and alloying.
- In the *molten* **oxide electrolysis** route, an electric current is used to directly reduce and melt the iron ore using electrolysis in one step, followed by alloying. These processes both promise

a significant increase in energy efficiency compared with the direct reduced iron (DRI) and blast furnace routes (Cavaliere 2019). If the electricity used is based on carbon-free sources, this steel-production process can be nearly $CO_2$ neutral. Both processes would require supplemental carbon, but this is typically only up to 0.05% per tonne steel, with a maximum of 2.1%. Aqueous electrolysis is possible with today's electrode technologies, while molten oxide electrolysis would require advances in high-temperature electrodes.

- **The HIsarna® process** is a new type of coal-based smelting reduction process, which allows certain agglomeration stages (coking plant, sintering/pelletising) to be dispensed with. The iron ore, with a certain amount of steel scrap, is directly reduced to pig iron in a single reactor. This process is suitable to be combined with CCS technology because of its relatively easy to capture and pure $CO_2$ exhaust gas flow. $CO_2$ emission reductions of 80% are believed to be realisable relative to the conventional blast furnace route (Abdul Quader et al. 2016). The total GHG balance also depends on further processing in a basic oxygen furnace or in an EAF. The HIsarna process was small-scale piloted under the EU ULCOS program.
- **Hydrogen co-firing in BF-BOFs** can potentially reduce emission by 30–40%, referring to experimental work by the Course50 projects and Thyssen Krupp, but coke is required to maintain stack integrity beyond that.

Reflecting the different conditions at existing and potential future plant sites, when choosing one of the above options a combination of different measures and structural changes (including electricity, hydrogen and CCU or CCS infrastructure needs) will likely be necessary in the future to achieve deep reductions in $CO_2$ emissions of steel production.

In addition, increases in material efficiency (e.g., more targeted steel use per vehicle, building or piece of infrastructure) and increases in the intensity of product use (e.g., sharing cars instead of owning them) can contribute significantly to reduce emissions by reducing the need for steel production. The IEA (2019b) suggested that up to 24% of cement and 40% of steel demand could be plausibly reduced through strong material efficiency efforts by 2060. Potential material efficiency contribution for the EU is estimated to be much higher – 48% (Material Economics 2019). Recycling would cut the average $CO_2$ emissions per tonne of steel produced by 60% (Material Economics 2019), but globally by 2050 secondary steel production is limited to 40–56% in various scenarios (IEA 2019b), with 46% in the IEA (2021a) and up to 56% in 2050 in Xylia et al. (2016). It may scale up to 68% by 2070 (Xylia et al. 2016). CCU and more directly CCS are other options to reduce GHG emissions but depend on the full lifecycle net GHGs that can be allocated to the process (Section 11.3.6). Bio-based fuels can also substitute for some of the coal input, but due to other demands for biomass this strategy is likely to be limited to specific cases.

Abatement costs for these strategies vary considerably from case to case and for each a plausible cost range is difficult to establish; compare this with **Table 11.3** (Fischedick et al. 2014b; Leeson et al. 2017; Axelson et al. 2018; Vogl et al. 2018; Fan and Friedmann 2021; Wang et al. 2021). A key point is that while cost of production increases are significant, the effect on final end uses is typically very small (Rootzén and Johnsson 2016), with significant policy consequences (see Section 11.6 on public and private lead markets for cleaner materials).

#### 11.4.1.2 Cement and Concrete

The cement sector is regarded as a sector where mitigation options are especially narrow (Energy Transitions Commission 2018; Habert et al. 2020). Cement is used as the glue to hold together sand, gravel and stone aggregates to make concrete, the most consumed manufactured substance globally. The production of cement has been increasing faster than the global population since the middle of the last century (Scrivener et al. 2018). Despite significant improvements in energy efficiency over the last couple of decades (e.g., a systematic move from wet to dry kilns with calciner preheaters feeding off the kilns), the direct emissions of cement production (the sum of energy and process emissions) are estimated to be 2.1–2.5 $GtCO_2$-eq in 2019 or 14–17% of total global direct industrial GHG emissions (Lehne and Preston 2018; Bataille 2020a; Sanjuán et al. 2020; Crippa et al. 2021; Hertwich 2021; Lamb et al. 2021) (Figure 11.4). Typically, about 40% of these direct emissions originate from process heating (e.g., for calcium carbonate (limestone) decomposition into calcium oxide at 850°C or higher, directly followed by combination with cementitious materials at about 1450°C to make clinker), while 60% are process $CO_2$ emissions from the calcium carbonate decomposition (Kajaste and Hurme 2016; IEA and WBCSD 2018; Andrew 2019). Some of the $CO_2$ is reabsorbed into concrete products and can be seen as avoided during the decades-long life of the products; estimates of this flux vary between 15 and 30% of the direct emissions (Stripple et al. 2018; Andersson et al. 2019; Schneider 2019; Cao et al. 2020; GCCA 2021a). Some companies are mixing $CO_2$ into hardening concrete, both to dispose of the $CO_2$ and more importantly reduce the need for binder (Lim et al. 2019).

One of the simplest and most effective ways to reduce cement and concrete emissions is to make stronger concrete through better mixing and aggregate sizing and dispersal; poorly and well-made concrete can vary in strength by a factor of four for a given volume (Fechner and Kray 2012; Habert et al. 2020). This argues for a refocus of the market away from 'one size fits all', often bagged cements to professionally mixed clinker, cementitious material and filler mixtures appropriate to the needs of the end use.

Architects, engineers and contractors also tend to overbuild with cement because it is cheap as well as corrosion- and water-resistant. Buildings and infrastructure can be purposefully designed to minimise cement use to its essential uses (e.g., compression strength and corrosion-resistance), and replace its use with other materials (e.g., wood, stone and other fibres) for non-essential uses. This could reduce cement use by 20–30% (Imbabi et al. 2012; Brinkerhoff and GLDNV 2015; D'Alessandro et al. 2016; Lehne and Preston 2018; IEA 2019b; Shanks et al. 2019; Habert et al. 2020).

Because so much of the emissions from concrete come from the limestone calcination to make clinker, anything that reduces use of clinker for a given amount of concrete reduces its GHG intensity.

While 95% Portland cement is common in some markets, it is typically not necessary for all end-use applications, and many markets will add blast furnace slag, coal fly ash, or natural pozzolanic materials to replace cement as supplementary cementitious materials; 71% was the global average clinker content of cement in 2019 (IEA 2020a). All these materials are limited in volume, but a combination of roughly two to three parts ground limestone and one part specially selected, calcined clays can also be used to replace clinker (Fechner and Kray 2012; Lehne and Preston 2018; Habert et al. 2020). Local building codes determine what mixes of cementitious materials are allowed for given uses and would need to be modified to allow these alternative mixtures where appropriate.

Ordinary Portland cement process $CO_2$ emissions cannot be avoided or reduced through the use of non-fossil energy sources. For this reason, CCS technology, which could capture just the process emissions (e.g., the EU LEILAC project, which concentrates the process emissions from the limestone calciner, see following paragraph) or both the energy and process-related $CO_2$ emissions, is often mentioned as a potentially important element of an ambitious mitigation strategy in the cement sector. Different types of CCS processes can be deployed, including post-combustion technologies such as amine scrubbing and membrane-assisted $CO_2$-liquefation, oxycombustion in a low-to-zero nitrogen environment (full or partial) to produce a concentrated $CO_2$ stream for capture and disposal, or calcium-looping (Dean et al. 2011). The IEA puts cement CCS technologies at the technology readiness level (TRL) 6–8 (IEA 2020h). These approaches have different strengths and weaknesses concerning emission abatement potential, primary energy consumption, costs and retrofittability (Hills et al. 2016; Gardarsdottir et al. 2019; Voldsund et al. 2019). Use of biomass energy combined with CCS has the possibility of generating partial negative emissions, with the caveats introduced in Section 11.3.6 (Hepburn et al. 2019).

The energy-related emissions of cement production can also be reduced by using bioenergy solids, liquids or gases (TRL 9) (IEA and WBCSD 2018), hydrogen or electricity (TRL 4 according to IEA (2020h)) for generating the high-temperature heat at the calciner – hydrogen and bioenergy co-burning could be complementary due to their respective fast-vs-slow combustion characteristics. In an approach pursued by the LEILAC research project, the calcination process step is carried out in a steel vessel that is heated indirectly using natural gas (Hills et al. 2017). The LEILAC approach makes it possible to capture the process-related emissions in a comparatively pure $CO_2$ stream, which reduces the energy required for $CO_2$ capture and purification. This technology (LEILAC in combination with CCS) could reduce total furnace emissions by up to 85% compared with an unabated, fossil fuelled cement plant, depending on the type of energy sources used for heating (Hills et al. 2017). In principle, the LEILAC approach allows the eventual potential electrification of the calciner by electrically heating the steel enclosure instead of using fossil burners.

In the long run, if some combination of material efficiency, better mixing and aggregate sizing, cementitious material substitution and 90%+ capture CCS with supplemental bioenergy are not feasible in some regions or at all to achieve near-zero emissions, alternatives to limestone-based ordinary Portland cement may be needed. There are several highly regional alternative chemistries in use that provide partial reductions (Fechner and Kray 2012; Lehne and Preston 2018; Habert et al. 2020), for example, carbonatable calcium silicate clinkers, and there have been pilot projects with magnesium-oxide-based cements, which could be negative emissions. Lower carbon cement chemistries are not nearly as widely available as limestone deposits (Material Economics 2019), and would require new materials testing protocols, codes, pilots and demonstrations.

Any substantial changes in cement and concrete material efficiency or production decarbonisation, however, will require comprehensive education and continuing re-education for cement producers, architects, engineers, contractors and small, non-professional users of cements. It will also require changes to building codes, standards, certification, labeling, procurement, incentives, and a range of polices to help create the market will be needed, as well as those for information disclosure, and certification for quality. Even an end-of-pipe solution like CCS will require infrastructure for transport and disposal. Abatement costs for these strategies vary considerably from case to case and for each a plausible cost range is difficult to establish, but they are summarised in Table 11.3 from the following literature and other sources (Wilson et al. 2003; Fechner and Kray 2012; Leeson et al. 2017; Moore 2017; Lehne and Preston 2018; IEA 2019f; Habert et al. 2020).

#### 11.4.1.3 Chemicals

The chemical industry produces a broad range of products that are used in a wide variety of applications. The products range from plastics and rubbers to fertilisers, solvents, and specialty chemicals such as food additives and pharmaceuticals. The industry is the largest industrial energy user and its direct emissions were about 1.1–1.7 $GtCO_2$-eq or about 10% of total global direct industrial emissions in 2019 (Olivier and Peters 2018; IEA 2019f; Crippa et al. 2021; Lamb et al. 2021; Minx et al. 2021) (Figure 11.4 and Table 11.1). With regard to energy requirements and $CO_2$ emissions, ammonia, methanol, olefins, and chlorine production are of great importance (Boulamanti and Moya Rivera 2017). Ammonia is primarily used for nitrogen fertilisers, methanol for adhesives, resins, and fuels, whereas olefins and chlorine are mainly used for the production of polymers, which are the main components of plastics.

Technologies and process changes that enable the decarbonisation of chemicals production are specific to individual processes. Although energy efficiency in the sector has steadily improved over the past decades (Boulamanti and Moya Rivera 2017; IEA 2018a) (Figure 11.8), a significant share of the emissions is caused by the need for heat and steam in the production of primary chemicals (Bazzanella and Ausfelder 2017) (Box 11.2). This energy is currently supplied almost exclusively through fossil fuels which could be substituted with bioenergy, hydrogen, or low or zero carbon electricity, for example, using electric boilers or high-temperature heat pumps (Bazzanella and Ausfelder 2017; Thunman et al. 2019; Saygin and Gielen 2021). The chemical industry has among the largest potentials for industrial energy demand to be electrified with existing technologies, indicating the possibility for a rapid reduction of energy-related emissions (Madeddu et al. 2020).

The production of ammonia causes most $CO_2$ emissions in the chemical industry, about 30% according to the IEA (2018a) and nearly one third according to Crippa et al. (2021), Lamb et al. (2021) and Minx et al. (2021). Ammonia is produced in a catalytic reaction between nitrogen and hydrogen – the latter most often produced through natural gas reforming (Stork et al. 2018; Material Economics 2019) and in some regions through coal gasification, which has several times higher associated $CO_2$ emissions. Future low-carbon options include hydrogen from electrolysis using low- or zero-carbon energy sources (Philibert 2017a), natural gas reforming with CCS, or methane pyrolysis, a process in which methane is transformed into hydrogen and solid carbon (Bazzanella and Ausfelder 2017; Material Economics 2019; (Section 11.3.5 and Box 11.1). Electrifying ammonia production would lead to a decrease in total primary energy demand compared to conventional production, but a significant efficiency improvement potential remains in novel synthesis processes (Wang et al. 2018; Faria 2021). Combining renewable energy sources and flexibility measures in the production process could allow for low-carbon ammonia production on all continents (Fasihi et al. 2021). Steam cracking of naphtha and natural gas liquids for the production of olefins (i.e., ethylene, propylene and butylene), and other high-value chemicals is the second most $CO_2$-emitting process in the chemical industry, accounting for another almost 20% of the emissions from the subsector (IEA 2018a). Future lower-carbon options include electrifying the heat supply in the steam cracker as described above, although this will not remove the associated process emissions from the cracking reaction itself or from the combustion of the by-products. Further in the future, electrocatalysis of carbon monoxide, methanol, ethanol, ethylene and formic acid could allow direct electric recombination of waste chemical products into new intermediate products (De Luna et al. 2019).

A ranking of key emerging technologies with likely deployment dates from the present to 2025 relevant for the chemical industry identified different carbon capture processes together with electrolytic hydrogen production as being of very high importance to reach net zero emissions (IEA 2020a). Methane pyrolysis, electrified steam cracking, and the biomass-based routes for ethanol-to-ethylene and lignin-to-BTX were ranked as being of medium importance. While macro-level analyses show that large-scale use of carbon circulation through CCU is possible in the chemical industry as primary strategy, it would be very energy intensive and the climate impact depends significantly on the source of and process for capturing the $CO_2$ (Artz et al. 2018; Kätelhön et al. 2019; Müller et al. 2020). Significant synergies can be found when combining circular CCU approaches with virgin carbon feedstocks from biomass (Bachmann et al. 2021; Meys et al. 2021).

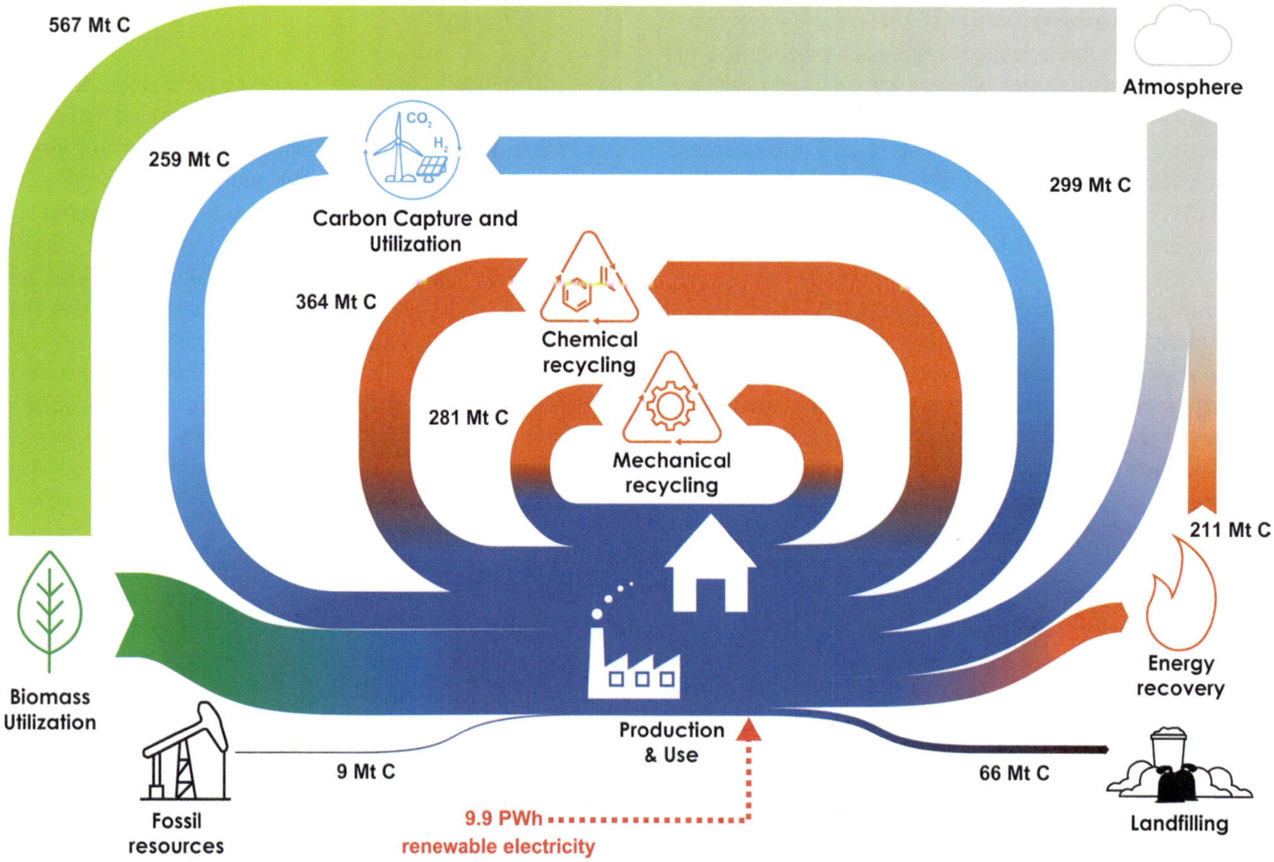


**Figure 11.10 Feedstock supply and waste treatment in a scenario with a combination of mitigation measures in a pathway for low-carbon plastics.** Source: From Meys et al., "Achieving net-zero greenhouse gas emission plastics by a circular carbon economy". *Science*, 374(6563), 71–76, DOI: 10.1126/science.abg9853. Reprinted with permission from AAAS.

In a net zero world carbon will still be needed for many chemical products, but the sector must also address the lifecycle emissions of its products which arise in the use phase, for example, $CO_2$ released from urea fertilisers, or at the end of life, for example, the incineration of waste plastics which was estimated to emit 100 Mt globally in 2015 (Zheng and Suh 2019). Reducing lifecycle emissions can partly be achieved by closing the material cycles starting with material and product design planning for reuse, remanufacturing, and recycling of products – ending up with chemical recycling which yields recycled feedstock that substitutes virgin feedstocks for various chemical processes (Rahimi and García 2017; Smet and Linder 2019).However, the chemical recycling processes which are most well-studied are pyrolytic processes which are energy intensive and have significant losses of carbon to off-gases and solid residues (Dogu et al. 2021; Davidson et al. 2021). They are thus associated with significant $CO_2$ emissions, which can even be larger in systems with chemical recycling than energy recovery (Meys et al. 2020). Further, the products from many pyrolytic chemical recycling processes are primarily fuels, which then in their subsequent use will emit all contained carbon as $CO_2$ (Vollmer et al. 2020). Achieving carbon neutrality would thus require this $CO_2$ either to be recirculated through energy-consuming synthesis routes or to be captured and stored (Geyer et al. 2017; Lopez et al. 2018; Material Economics 2019; Thunman et al. 2019). As all chemical products are unlikely to fit into chemical recycling systems, CCS can be used to capture and store a large share of their end-of-life emissions when combined with waste combustion plants or heat-demanding facilities like cement kilns (Leeson et al. 2017; Tang and You 2018).

Reducing emissions involves demand-side measures, for example, efficient end use, materials efficiency and slowing demand growth, as well as recycling where possible to reduce the need for primary production. The following strategies for primary production of organic chemicals which will continue to need a carbon source are key in avoiding the GHG emissions of chemical products throughout their lifecycles:

**Recycled feedstocks**: *Chemical recycling* of plastics unsuitable for mechanical recycling was already mentioned. Through *pyrolysis* of old plastics, both gas and a naphtha-like pyrolysis oil can be generated, a share of which could replace fossil naphtha as a feedstock in the steam cracker (Honus et al. 2018a,b). Alternatively, waste plastics could be *gasified* and combined with low-carbon hydrogen to a syngas, for example, the production and methanol and derivatives (Lopez et al. 2018; Stork et al. 2018). Other chemical recycling options include polymer selective chemolysis, catalytic cracking, and hydrocracking (Ragaert et al. 2017). Carbon losses and process emissions must be minimised and it may thus be necessary to combine chemical recycling with CCS to reach near-zero emissions (Thunman et al. 2019; Smet and Linder 2019; Meys et al. 2021).

**Biomass feedstocks:** Substituting fossil carbon at the inception of a product lifecycle for carbon from renewable sources processed in designated biotechnological processes (Lee et al. 2019; Hatti-Kaul et al. 2020) using specific biomass resources (Isikgor and Becer 2015) or residual streams already available (Abdelaziz et al. 2016). Routes with thermochemical and catalytic processes, such as pyrolysis and subsequent catalytic upgrading, are also available (Jing et al. 2019).

**Synthetic feedstocks:** Carbon captured with direct air capture or from point sources (bioenergy, chemical recycling, or during a transition period from industrial-processes-emitting fossil $CO_2$) can be combined with low-GHG hydrogen into a syngas for further valorisation (Kätelhön et al. 2019). Thus, low-carbon methanol can be produced and used in methanol-to-olefins/aromatics (MTO/MTA) processes, substituting the steam cracker (Gogate 2019) or Fischer-Tropsch processes could produce synthetic hydrocarbons.

Reflecting the diversity of the sector, the listed options can only be illustrative. The above-listed strategies all rely on low-carbon energy to reach near-zero emissions. In considering mitigation strategies for the sector it will be key to focus on those for which there is a clear path towards (close to) zero emissions, with high (carbon) yields over the full product value chain and minimal fossil resource use for both energy and feedstocks (Saygin and Gielen 2021), with CCU and CCS employed for all remnant carbon flows. The necessity of combining mitigation approaches in the chemicals industry with low-carbon energy was recently highlighted in an analysis (Figure 11.10) which showed how the combined use of different recycling options, carbon capture, and biomass feedstocks was most effective at reducing global lifecycle emissions from plastics (Meys et al. 2021). While most of the chemical processes for doing all the above are well known and have been used commercially at least partly, they have not been used at large scale and in an integrated way. In the past, external conditions (e.g., availability and price of fossil feedstocks) have not set the necessary incentives to implement alternative routes and to avoid emitting combustion- and process-related $CO_2$ emissions to the atmosphere. Most of these processes will very likely be more costly than using fossil fuels and full-scale commercialisation would require significant policy support and the implementation of dedicated lead markets (Wesseling et al. 2017; Bataille et al. 2018a; Material Economics 2019; Wyns et al. 2019). As in other subsectors, abatement costs for the various strategies vary considerably across regions and products, making it difficult to establish a plausible cost range for each (Bazzanella and Ausfelder 2017; Philibert 2017a; Philibert 2017b; Axelson et al. 2018; IEA 2018a; De Luna et al. 2019; Saygin and Gielen 2021).

## Box 11.2 | Plastics and Climate Change

The global production of plastics has increased rapidly over the past 70 years, with a compound annual growth rate (CAGR) of 8.4%, about 2.5 times the growth rate for global GDP (Geyer et al. 2017) and higher than other materials since 1970 (IEA 2019b). Global production of plastics is now more than 400 million tonnes, including synthetic fibres (IEA 2019b) The per capita use of plastics is still up to 20 times higher in developed countries than in developing countries with low signs of saturation and the potential for an increased use is thus still very large (IEA 2018a). Plastics is the largest output category from the petrochemical industry, which as a whole currently uses about 14% of petroleum and 8% of natural gas (IEA 2018a). Forecasts for plastic production assuming continued growth at recent rates of about 3.5% point towards a doubled production by 2035, following record-breaking investments in new and increased production capacity based on petroleum and gas in recent years (CIEL 2017; Bauer and Fontenit 2021). IEA forecasts show that even in a world where transport demand for oil falls considerably by 2050 from the current about 100 mbpd, feedstock demand for chemicals will rise from about 12 mbpd to 15–18 mbpd (IEA 2019b). Projections for increasing plastic production as well as petroleum use, together with the lack of investments in breakthrough low-emission technologies, do not align with necessary emission reductions.

About half of the petroleum that goes into the chemical industry is used for producing plastics, and a significant share of this is combusted or lost in the energy-intensive production processes, primarily the steam cracker. GHG emissions from plastic production depend on the feedstock used (ethane-based production is associated with lower emissions than naphtha-based), the type of plastic produced (production of simple polyolefins is associated with lower emissions than more complex plastics such as polystyrene), and the contextual energy system (e.g., the GHG intensity of the electricity used) but weighted averages have been estimated to be 1.8 $tCO_2$-eq $t^{-1}$ for North American production (Daniel Posen et al. 2017) and 2.3 $tCO_2$-eq $t^{-1}$ for European production (Material Economics 2019). In regions more dependent on coal electricity production the numbers are likely to be higher, and several times higher for chemical production using coal as a feedstock – coal-based MTO has seven times higher emissions than olefins from steam cracking (Xiang et al. 2014). Coal-based plastic and chemicals production has over the past decade been developed and deployed primarily in China (Yang et al. 2019). The production of plastics was thus conservatively estimated to emit 1085 $MtCO_2$-eq $yr^{-1}$ in 2015 (Zheng and Suh 2019). Downstream compounding and conversion of plastics was estimated to emit another 535 $MtCO_2$-eq $yr^{-1}$, while end-of-life treatment added 161 $MtCO_2$-eq $yr^{-1}$. While incineration of plastic waste was the cause of only 5% of global plastic lifecycle emissions, in regions with waste-to-energy infrastructures this share is significantly larger, for example, 13% of lifecycle emissions in Europe (Ive Vanderreydt et al. 2021). The effective recycling rate of plastics remains low relating to a wide range of issues such as insufficient collection systems, sorting capacity, contaminants and quality deficiencies in recycled plastics, design of plastics integrated in complex products such as electronics and vehicles, heterogenous plastics used in packaging, and illegal international trade.



#### 11.4.1.4 Other Industry Sectors

The other big sources of direct global industrial combustion and process $CO_2$ emissions are light manufacturing and industry (9.7% in 2016), non-ferrous metals like aluminium (3.1%), pulp and paper (1.1%), and food and tobacco (1.9%) (Bataille 2020a; Crippa et al. 2021; Lamb et al. 2021).

*Light manufacturing and industry*

Light manufacturing and industry represent a very diverse sector in terms of energy service needs (e.g., motive power, ventilation, drying, heating, compressed air, etc.) and it comprises both small and large plants in different geographical contexts. Most of the direct fossil fuel use is for heating and drying, and it can be replaced with low-GHG electricity through direct resistance, high-temperature heat pumps and mechanical vapour recompression, induction, infrared, or other electrothermal processes (Lechtenböhmer et al. 2016; Bamigbetan et al. 2017). Madeddu et al. (2020) argue up to 78% of Europe's industrial energy requirements are electrifiable through existing commercial technologies and 99% with the addition of new technologies currently under development. Direct solar heating is possible for low temperature needs (<100°C) and concentrating solar for higher temperatures. Commercially available heat pumps can deliver 100°C–150°C but at least up to 280°C is feasible (Zühlsdorf et al. 2019). Plasma torches using electricity can be used where high temperatures (>1000°C) are required, but hydrogen, biogenic or synthetic combustible hydrocarbons (methane, methanol, ethanol, LPG, etc.) can also be used (Bataille et al. 2018a).

There is also a large potential for energy savings through cascading in industrial clusters similar to the one at Kalundborg, Denmark. Waste heat can be passed at lower and lower temperatures from facility to facility or circulated as low-grade steam or hot water, and boosted as necessary using heat pumps and direct heating. Such geographic clusters would also enable lower-cost infrastructure for hydrogen production and storage as well as $CO_2$ gathering, transport and disposal (IEA 2019f).

*Aluminium and other non-ferrous metals*

Demand for aluminium comes from a variety of end uses where a reasonable cost, light-weight metal is desirable. It has historically been used in aircraft, window frames, strollers, and beverage containers. As fuel economy has become more desirable and design improvements have allowed crush bodies made of aluminium instead of steel, aluminium has become progressively more attractive for cars. Primary aluminium demand is total demand (100 Mt $yr^{-1}$ in 2020) net of manufacturing waste reuse (14% of virgin and recycled input) and end-of-life recycling (about 20% of what reaches market). Primary aluminium consumption rose from under 20 Mt $yr^{-1}$ in 1995 to over 66 Mt primary ingot production in 2020 (International Aluminium Institute change to 2021c). The International Aluminium Institute (2021a) expects total aluminium consumption to reach 150–290 Mt $yr^{-1}$ by 2050 with primary aluminium contributing 69–170 Mt and secondary recycled 91–120 Mt (as in-use stock triples or quadruples). The OECD forecasts increases in demand by 2060 for primary aluminium to 139 Mt $yr^{-1}$ and for secondary aluminium to 71 Mt $yr^{-1}$ (OECD 2019a). Primary (as opposed to recycled) aluminium is generally made in a two-stage process, often geographically separated. In the first stage aluminium oxide is extracted from bauxite ore (often with other trace elements) using the Bayer hydrometallurgical process, which requires up to 200°C heat when sodium hydroxide is used to leach the aluminium oxide, and up to 1000°C for kilning. This is followed by electrolytic separation of the oxygen from the elemental aluminium using the Hall-Héroult process, by far the most energy-intense part of making aluminium. This process has large potential emissions from the electricity used (12.5 MWh per tonne aluminium BAT, 14–15 MWh per tonne average). From bauxite mine to aluminium ingot, reported total global average emissions are between 12 and 17.6 $tCO_2$-eq per tonne of aluminium, depending on estimates and assumptions made[22] (Saevarsdottir et al. 2020). About 10% of this, 1.5 tonnes of direct $CO_2$ per tonne of aluminium are currently emitted as the graphite electrodes are depleted and combine with oxygen, and if less than optimal conditions are maintained, perfluorocarbons can be emitted with widely varying GHG intensity, up to the equivalent of 2 $tCO_2$-eq per tonne of aluminium. PFC emissions, however, have been greatly reduced globally and almost eliminated in well-run facilities. Aluminium, if it is not contaminated, is highly recyclable and requires 1/20 of the energy required to produce virgin aluminium; increasing aluminium recycling rates from the 20–25% global average is a key emissions reduction strategy (Haraldsson and Johansson 2018).

The use of low- and zero-GHG electricity (e.g., historically from hydropower) can reduce the indirect emissions associated with making aluminium. A public-private partnership with financial support from the province of Québec and the Canadian federal government has recently announced a fundamental modification to the Hall-Héroult process by which the graphite electrode process emissions can be eliminated by substitution of inert electrodes. This technology is slated to be available in 2024 and is potentially retrofittable to existing facilities (Saevarsdottir et al. 2020).

Smelting and otherwise processing of other non-ferrous metals like nickel, zinc, copper, magnesium and titanium with less overall emissions have relatively similar emissions reduction strategies (Bataille and Stiebert 2018): (i) Increase material efficiency; (ii) Increase recycling of existing stock; (iii) Pursue ore-extraction processes (e.g., hydro- and electro-metallurgy) that allow more use of low-carbon electricity as opposed to pyrometallurgy, which uses heat to melt and separate the ore after it has been crushed. These processes have been used occasionally in the past but have generally not been used due to the relatively inexpensive nature of fossil fuels.

*Pulp and paper*

The pulp and paper industry (PPI) is a small net-emitter of $CO_2$, assuming the feedstock is sustainably sourced (Chapter 7), but it has large emissions of biogenic $CO_2$ from feedstock (700–800 Mt $yr^{-1}$) (Tanzer et al. 2021). It includes pulp mills, integrated pulp and paper mills, and paper mills using virgin pulpwood and other fibre sources, residues and co-products from wood products manufacturing, and recycled paper as feedstock. Pulp mills typically have access to bioenergy in the chemical pulping processes to cover most or all of heat and electricity needs, for example, through chemicals recovery boilers and steam turbines in the kraft process. Mechanical pulping mainly uses electricity for energy; decarbonisation thus depends on grid emission factors. With the exception of the lime kiln in kraft pulp mills, process temperature needs are typically less than or equal to 150°C to 200°C, mainly steam for heating and drying. This means that this sector can be relatively easily decarbonised through continued energy efficiency, fuel switching and electrification, including use of high-temperature heat pumps (Ericsson and Nilsson 2018). Electrification of pulp mills could, in the longer term, make bio-residues currently used internally for energy, available as a carbon source for chemicals (Meys et al. 2021). The PPI also has the capabilities, resources and knowledge, to implement these changes. Inertia is mainly caused by equipment turnover rates, relative fuel and electricity prices, and the profitability of investments.

A larger and more challenging issue is how the forestry industry can contribute to the decarbonisation of other sectors and how biogenic carbon will be used in a fossil-free society, for example, through developing the forest-based bioeconomy (Pülzl et al. 2014; Bauer 2018). In recent years the concept of biorefineries has gained increasing traction. Most examples involve innovations for taking by-products or diverting small streams to produce fuels, chemicals and bio-composites that can replace fossil-based products, but there is little common vision on what really constitutes a biorefinery (Bauer et al. 2017). Some of these options have limited scalability and the cellulose fibre remains the core product even in the relatively large shift from paper production to textiles fibre production.

Pulp mills have been identified as promising candidates for post-combustion capture and CCS (Onarheim et al. 2017), which could allow some degree of net-negative emissions. For deep decarbonisation across all sectors, notably switching to biomass feedstock for fuels, organic chemicals and plastics, the availability of biogenic carbon (in biomass or as biogenic $CO_2$; Chapter 7) becomes an issue. A scenario where biogenic carbon is CCU as feedstock implies large demands



[22] According to the International Aluminium Institute (2021b), scope 3 (cradle to gate) emissions from the aluminium industry in 2018 reached 1.127 $GtCO_2$-eq or 17.6 $tCO_2$-eq per tonne of primary aluminium. In the Beyond 2°C Scenario (B2DS) it is expected to be reduced to 2.5 $tCO_2$-eq per tonne.

for hydrogen, completely new value chains and more closed carbon loops, all areas which are as yet largely unexplored (Ericsson 2017; Meys et al. 2021).

#### 11.4.1.5 Overview of Estimates of Specific Mitigation Potential and Abatement Costs of Key Technologies and Processes for Main Industry Sectors

Climate-policy-related literature focusing on deep industrial emission reductions has expanded rapidly since AR5. An increasing body of research proposes deep decarbonisation pathways for energy-intensive industries (Figure 11.13). Bataille et al. (2018a) address the question of whether it is possible to reduce GHG emissions to very low, zero, or negative levels, and identifies preliminary technological and policy elements that may allow the transition, including the use of policy to drive technological innovation and uptake. Material Economics (2019), the IEA (2019b), the Energy Transitions Commission (2018) and Climate Action Tracker (CAT; 2020) take steps to identify pathways integrating energy efficiency, material efficiency, circular economy and innovative technologies options to cut GHG emissions across basic materials and value chains. The key conclusion is that net zero $CO_2$ emissions from the largest sources (steel, plastics, ammonia, and cement) could be achieved by 2050 by deploying already available multiple options packaged in different ways (Davis et al. 2018; Material Economics 2019; UKCCC 2019b). The studies assume that for those technologies that have a kind of breakthrough technology status further technological development and significant cost reduction can be expected.

Table 11.3, modified from Bataille (2020a) and built from McMillan et al. (2016); Bazzanella and Ausfelder (2017); Philibert (2017a); Wesseling et al. (2017); Axelson et al. (2018); Bataille et al. (2018a) Davis et al. (2018); Energy Transitions Commission (2018); IEA (2019f, 2020c); Material Economics (2019); and UKCCC (2019b), presents carbon intensities that could be achieved by implementing mitigation options in major basic material industries, mitigation potential, estimates for mitigation costs, TRL and potential year of market introduction (Figure 11.13).

Table 11.3 acknowledges that for many carbon-intensive products a large variety of novel processes, inputs and practices capable of providing very deep emission reductions are already available and emerging. However, their application is subject to different economic and structural limitations, therefore in the scenarios assuming deep decarbonisation by 2050–2060 different technological mixes can be observed (Section 11.4.2).

While deep GHG emissions reduction potential is assessed for various regions, assessment of associated costs is limited to only a few regions; nevertheless those analyses may be illustrative at the global scale. UKCCC (2019b) provides costs assessments for different industrial subsectors (Table 11.3) for the UK. They provide three ranges: core, more ambitious, and when energy and material efficiency are limited. The core options range from 2–85 GBP2019 $tCO_2\text{-eq}^{-1}$ (e.g., reduction in GHG emissions by about 50% by 2050 applying energy efficiency (EE), *ME*, CCS, biomass and electrification). The more ambitious options are estimated at 32–119 GBP2019 $tCO_2\text{-eq}^{-1}$ (e.g., 90% emissions reduction via widespread deployment of hydrogen, electrification or bioenergy for stationary industrial heat/combustion). Finally, costs range from 33–299 GBP $tCO_2\text{-eq}^{-1}$ when energy and material efficiency are limited.

In Material Economics (2019), costs are provided for separate technologies and subsectors, and also by pathways, each including new industrial processes, circular economy and CCS components in different proportions, allowing for the transition to net zero industrial emission in the EU by 2050. That means that the study provides information about the three main mid- to long-term options which could enable a wide abatement of GHG emissions. Given different electricity-price scenarios, average abatement costs associated with the circular economy-dominated pathway are: 12–75 EUR2019 $tCO_2\text{-eq}^{-1}$; for the carbon capture-dominated pathway 79 EUR2019 $tCO_2\text{-eq}^{-1}$; and for the new processes-dominated scenario 91 EUR2019 $tCO_2\text{-eq}^{-1}$. Consequently, net-zero-emission pathways are about 3–25% costlier compared to the baseline (Material Economics 2019). According to the Energy Transitions Commission (2018), cement decarbonisation would cost on average USD110–130 $tCO_2^{-1}$ depending on the cost scenario. Rootzén and Johnsson (2016) state that $CO_2$ avoidance costs for the cement industry vary from 25 to 110 EUR $tCO_2^{-1}$, depending on the capture option considered and on the assumptions made with respect to the different cost items involved. According to the Energy Transitions Commission (2018), steel can be decarbonised on average at USD60 $tCO_2^{-1}$, with highly varying costs depending on low-carbon electricity prices.

For customers of final products, information on the potential impact of supply-side decarbonisation on final prices may be more useful than that of $CO_2$ abatement costs. A different approach has been developed to assess the costs of mitigation by estimating the potential impacts of supply-side decarbonisation on final product prices. Material Economics (2019) shows that with deep decarbonisation, depending on the pathway, steel costs grow by 20–30%; plastics by 20–45%; ammonia by 15–60%; and cement (not concrete) by 70–115%. While these are large and problematic cost increases for material producers working with low margins in a competitive market, final end-use product price increases are far less, for example, a car becomes 0.5% more expensive, supported by both Rootzén and Johnsson (2016) and the Energy Transitions Commission (2018). For comparison, Rootzén and Johnsson (2017) found that decarbonising cement-making, while doubling the cost of cement, would add <1% to the costs of a residential building; the Energy Transitions Commission (2018) found concrete would be 10–30% more expensive, adding USD15,000 or 3% to the price of a house including land value. Finally, the IEA (2020a) estimated the impact on end-use prices are rather small, even in a net zero scenario; they find price increases of 0.2% for a car and 0.6% for a house, based on higher costs for steel and cement respectively.

Thus, the price impact scales down going across the value chain and might be acceptable for a significant share of customers. However, it has to be reflected that the cumulative price increase could be more significant if several different zero-carbon materials (e.g., steel, plastics and aluminium) in the production process of a certain product

**Table 11.3 | Technological potentials and costs for deep decarbonisation of basic industries.** Percentages of maximum reduction are multiplicative, not additive.

| Sector | Current intensity ($tCO_2$-eq $t^{-1}$) | Potential GHG reduction | NASA TRL | Cost per tonne $CO_2$-eq (USD2019 $tCO_2$-eq$^{-1}$ for percentage of emissions) ? = unknown | Year available, assuming policy drivers |
|---|---|---|---|---|---|
| **Iron and steel** | | | | | |
| Current intensity – all steel (worldsteel) | 1.83 | | | | |
| Current intensity – ~BF-BOF/Best BF-BOF and NG-DRI (with near-zero GHG electricity) | 2.3/1.8 and 0.7 | | | | |
| Current intensity – EAF (depends on electricity intensity & pre-heating fuel) | ≥0 | Up to 99% | | | |
| Material efficiency (IEA 2019 'Material Efficiency…') | | Up to 40% | 9 | Subject to supply chain building codes and education | Today |
| More recycling; depends on available stock, recycling network, quality of scrap, availability of DRI for dilution | | Highly regional, growing with time | 9 | Subject to logistical, transport, sorting and recycling equipment costs | Today |
| BF-BOF with top gas recirculation and CCU/S[a] | | 60% | 6–7 | USD70–130 $t^{-1}$ | 2025–2030 |
| Syngas ($H_2$ & CO) DRI EAF with concentrated flow CCU/S | | ≥ 90% | 9 | ≥USD40 $t^{-1}$ | Today |
| Hisarna with concentrated $CO_2$ capture[b] | | 80–90% | 7 | USD40–70 $t^{-1}$ | 2025 |
| Hydrogen DRI EAF[c] – fossil hydrogen with CCS is in operation, electrolysis-based hydrogen scheduled for 2026 | | Up to 99% | 7 | USD39–79 $t^{-1}$ and USD46 $MWh^{-1}$ [d] | 2025 |
| Aqueous (e.g., SIDERWIN) or Molten Oxide (e.g., Boston Metals) Electrolysis (MOE)[e] | | Up to 99% | 3–5 | ? | 2035–2040 |
| **Cement and concrete** | | | | | |
| Current intensity, about 60% is limestone calcination | 0.55 | | | | |
| Building design to minimise concrete (IEA 2019b, 2020a) | | Up to 24% | 9 | Low, education, design and logistics related | 2025 |
| Alternative lower-GHG fuels, e.g., waste (biofuels and hydrogen, see above) | | 40% | 9 | Cost of alt. fuels | Today |
| CCUS for process heating & $CaCO_3$ calcination $CO_2$ (e.g., LEILAC, possible retrofit)[f] | | 99% calc., ≤90% heat | 5–7 | ≤USD40$t^{-1}$ calc. ≤USD120$t^{-1}$ heat | 2025 |
| Clinker substitution (e.g., limestone + calcined clays)[g] | | 40–50% | 9 | Near zero, education, logistics, building code revisions | Today |
| Use of multi-sized and well-dispersed aggregates[d] | | Up to 75% | 9 | Near zero | Today |
| Magnesium or ultramafic cements[d] | | Negative? | 1–4 | ? | 2040 |
| **Aluminium and other non-ferrous metals** | | | | | |
| Current Al intensity, from hydro- to coal-based electricity production. 1.5 $tCO_2$ are produced by graphite electrode decay | 1.5 $t^{-1}$ + electricity required (i.e., 10 $t^{-1}$ (NG) to 18 $t^{-1}$ (coal)) | | | | |
| Inert electrodes and green electricity[h] | | 100% | 6–7 | Relatively low | 2024 |
| Hydro/electrolytic smelting (with $CO_2$ CCUS if necessary) | | Up to 99% | 3–9 | Ore-specific | <2030 |
| **Chemicals (see also cross cutting feedstocks above)i** | | | | | |
| Catalysis of ammonia from low-/zero-GHG hydrogen $H_2$ | 1.6 (NG), 2.5 (naptha), 3.8 (coal) | ≤99% | 9 | Cost of $H_2$ | Today |
| Electrocatalysis: $CH_4$, $CH_3OH$, $C_2H_5OH$, CO, olefins[j] | | Up to 99% | 3 | Cost: elec., $H_2$, $CO_x$ | 2030 |
| Catalysis of olefins from: (m)ethanol, $H_2$ and $CO_x$ directly | | 9% | 9, 3 | Cost: $H_2$ and $CO_x$ | <2030 |
| End-use plastics, mainly CCUS and recycling | 1.3–4.2, about 2.4 | 94% | 5–6 | USD150–240 $t^{-1}$ | 2030? |
| **Pulp and paper** | | | | | |
| Full biomass firing, including lime kilns | | 60–75% | 9 | About USD50 $t^{-1}$ | Today |
| **Other manufacturing** | | | | | |
| Electrification using current tech (boilers, 90°C–140°C heat pumps | | 99% | 9 | Cost: elec. vs NG | 2025 |
| Using new tech (induction, plasma heating) | | 99% | 3–6 | | 2025 |

| Sector | Current intensity ($tCO_2$-eq $t^{-1}$) | Potential GHG reduction | NASA TRL | Cost per tonne $CO_2$-eq (USD2019 $tCO_2$-eq$^{-1}$ for percentage of emissions) ? = unknown | Year available, assuming policy drivers |
|---|---|---|---|---|---|
| **Cross-cutting (CCUS, $H_2$, net zero $C_oO_xH_y$ fuels/feedstocks)** | | | | | |
| CCUS of post-combustion $CO_2$ diluted in nitrogen[e] | | Up to 90% | 6–7 | ≤USD120 $t^{-1}$ | 2025 |
| CCUS of concentrated $CO_2$[e] | | 99% | 9 | ≤USD40 $t^{-1}$ | Today |
| $H_2$ production: steam or auto-thermal $CH_4$ reforming with CCS[e] | | SMR ≤90%<br>ATR >90% | 6*,<br>9** | 56% @≤USD40 $t^{-1}$ chem**,<br>≤USD120 heat*,+20%/kg | ≤2025 |
| $H_2$ production: coal with CCUS[e] | | ≤90% | 6 | 25–50% per $H_2$ $kg^{-1}$ | ≤2025 |
| $H_2$ production: alkaline or PEM electrolysis[k] | | 99% | 9 | About USD50 $t^{-1}$ or<br><USD20–30 $MWh^{-1}$ | Today |
| $H_2$ production: reversible solid oxide fuel electrolysis[j] | | 99% | 6–8 | About 40USD $t^{-1}$<br>or <USD40 $MWh^{-1}$ | 2025 |
| $H_2$ production: $CH_4$ pyrolysis or catalytic cracking[l] | | 99% | 5 | ? | 2030? |
| Hydrogen as $CH_4$ replacement | | ≤10% | 9 | See above | Today |
| Biogas or liquid replacement hydrocarbons | | 60–90% | 9 | Biomass USD per $GJ^{-1}$;<br>≥USD50 $t^{-1}$, uncertain | Today |
| Anaerobic digestion/fermentation: $CH_4$, $CH_3OH$ and $C_2H_5OH$[m] | | Up to –99% | 9 | Biomass cost | Today |
| Methane or methanol from $H_2$ and $CO_x$ (CCUS for excess). Maximum –50% reduction if C source is FF | | 50–99% | 6–9 | Cost: $H_2$ and $CO_x$ | Today |
| 850°C woody biomass gasification with CCS for excess carbon: CO, $CO_2$, $H_2$, $H_2O$, $CH_4$, $C_2H_4$ and $C_6H_6$[n] | | Could be negative | 7–8 | About USD50–75 $t^{-1}$, uncertain | Today |
| Direct air capture for short- and long-chain $C_oO_xH_y$[o] | | Up to 99% | 3 | Cost: E, $H_2$, $CO_x$<br>about USD94–232 $t^{-1}$ | ≤2030 |

[a] Data for CCS costs for steel-making: Birat (2012); Leeson et al. (2017); and Axelson et al. (2018).

[b] Data for Hisarna: Axelson et al. (2018).

[c] Data for hydrogen DRI electric arc furnaces: Fischedick et al. (2014b) and Vogl et al. (2018).

[d] Converted from EUR2018 34–68 $t^{-1}$ and EUR2018 40 $MWh^{-1}$.

[e] Data for Molten Oxide Electrolysis (also known as SIDERWIN): Fischedick et al. 2014b and Axelson et al. 2018. The TRLs differ by source, the value provided is from Axelson et al. (2018), based on UCLOS SIDERWIN.

[f] Data for making hydrogen from SMR and ATR with CCUS: Leeson et al. (2017); Moore (2017); and IEA (2019f). The cost of CCS disposal of concentrated sources of $CO_2$ at USD15–40 $tCO_2$-eq$^{-1}$ is well established as commercial for direct or EOR purposes and is based on the long-standing practice of disposing of hydrogen sulphide and oil brines underground. Wilson et al. (2003) and Leeson et al. (2017). There is a wide variance, however, in estimated $tCO_2$-eq$^{-1}$ break-even prices for industrial post-combustion capture of $CO_2$ from sources highly diluted in nitrogen (e.g., Leeson et al. (2017) at USD60–170 $tCO_2$-eq$^{-1}$), but most fall under USD120 $tCO_2$-eq$^{-1}$.

[g] Data for clinker substitution and use of well-mixed and multi-sized aggregates: Fechner and Kray 2012; Lehne and Preston 2018; and Habert et al. 2020).

[h] Rio Tinto, Alcoa and Apple have partnered with the governments of Québec and Canada to form a coalition to commercialise inert as opposed to sacrificial graphite electrodes by 2024, thereby making the standard Hall-Héroult process very low emissions if low-carbon electricity is used.

[i] Data and other information: Bazzanella and Ausfelder (2017); Axelson et al. (2018); IEA (2018a); De Luna et al. (2019); and Philibert (2017b,a).

[j] See De Luna et al. (2019) for a state-of-the-art review of electrocatalysis, or direct recombination of organic molecules using electricity and catalysts.

[k] Data for hydrogen production from electrolysis: Bazzanella and Ausfelder (2017); Philibert (2017a); Philibert (2017b); IEA (2019f); and Armijo and Philibert (2020).

[l] Data for methane pyrolysis to make hydrogen: Abbas and Wan Daud (2010). Data for hydrogen production from methane catalytic cracking: Amin et al. (2011) and Ashik et al. (2015).

[m] Data for anaerobic digestion or fermentation for the production of methane, methanol and ethanol: De Luna et al. (2019).

[n] Data for woody biomass gasification: Li et al. (2019) and van der Meijden et al. (2011).

[o] Data on direct air capture of $CO_2$: Keith et al. (2018) and Fasihi et al. (2019).

have to be combined, indicating the importance of material efficiency being applied along with production decarbonisation.

### 11.4.2 Transformation Pathways

To discuss the general role and temporal implementation of the different options for achieving a net zero GHG emissions industry, mitigation pathways will be analysed. This starts with showing the results of IAM-based scenarios followed by specific studies which provide much higher technological resolution and allow a much deeper look into the interplay of different mitigation strategies. The comparison of more technology-focused sector-based scenarios with top-down-oriented scenarios provides the opportunity for a reciprocal assessment across different modelling philosophies and helps to identify robust elements for the transformation of the sector. Only some of the scenarios available in the literature allow for at least rough estimates of the necessary investments and give direction about relevant investment cycles and potential risks of stranded or depreciated assets. In some specific cases cost comparisons can be translated into expected difference costs not only for the overall sector, but also for relevant materials or even consumer products.

#### 11.4.2.1 Central Results From (Top-down) Scenarios Analysis and Illustrative Mitigation Pathways Discussion

Chapter 3 conducted a comprehensive analysis of scenarios based on IAMs. The resulting database comprises more than 1000 model-based scenarios published in the literature. The scenarios span a broad range along temperature categories from rather baseline-like scenarios to the description of pathways that are compatible with the 1.5°C target. Comparative discussion of scenarios allows some insights with regard to the relevance of mitigation strategies for the industry sector (Figure 11.11).

The main results from the Chapter 3 analysis from an industry perspective are:

- While all scenarios show a decline in energy and carbon intensity over time, final energy demand and associated industry-related $CO_2$ emissions increase in many scenarios. Only ambitious scenarios (category C1) show significant reduction in final energy demand in 2030, more or less constant demand in 2050, but increasing demand in 2100, driven by growing material use throughout the 21st century. While carbon intensity shrinks over time, energy related $CO_2$-emissions decline after 2030 even in less ambitious scenarios, but particularly in those pursuing a temperature increase below 2°C. Reduction of $CO_2$ emissions in the sector are achieved through a combination of technologies which includes nearly all options that have been discussed in this chapter (Sections 11.3 and 11.4.1). However, there are big differences with regard to the intensity by which the various options are implemented in the scenarios. This is particularly true for CCS for industrial applications and material efficiency and material demand management (i.e., service demand, service product intensity). The latter options are still under-represented in many global IAMs.
- There are only a few scenarios which allow net-negative $CO_2$ emissions for the industry for the second half of the century, while most scenarios assessed (including the majority of 1.5°C scenarios) end up with still significant positive $CO_2$ emissions. In comparison to the whole system most scenarios expect a slower decrease of industry-related emissions.
- There is a great – up to a factor of two – difference in assumptions about the GHG mitigation potential associated with different carbon cost levels between IAMs and sector-specific industry models. Consequently, IAMs pick up mitigation options slower or later (or not at all) than models which are more technologically

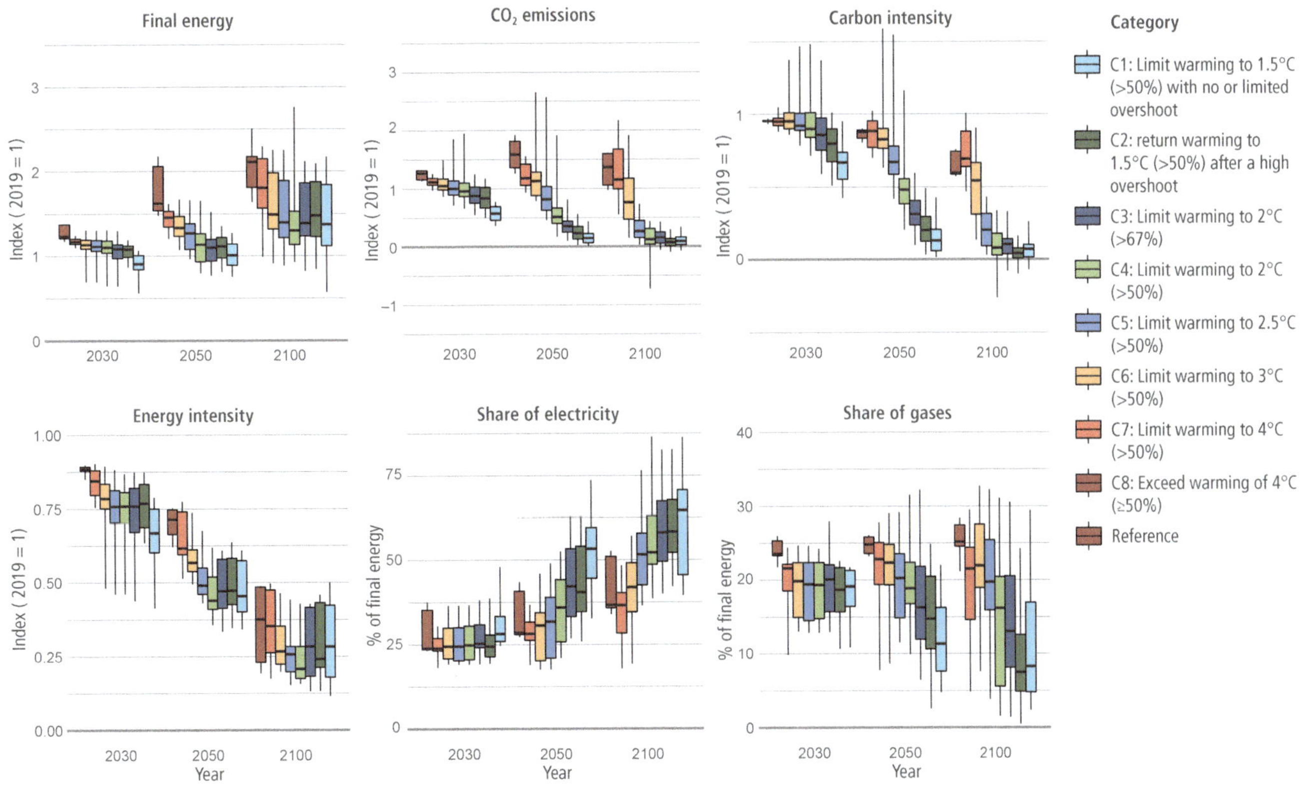


**Figure 11.11 Industrial final energy (top left), $CO_2$ emissions (top middle), energy intensity (bottom left), carbon intensity (top right), share of electricity (bottom middle), and share of gases (bottom right).** Energy intensity is final energy per unit of GDP. Carbon intensity is $CO_2$ emissions per EJ of final energy. The first four indicators are indexed to 2019, where values less than 1 indicate a reduction. Industrial-sector $CO_2$ emissions include fuel-combustion emissions only. Boxes indicate the interquartile range, the median is shown with a horizontal black line, while vertical lines show the 5 to 95% interval. Source: data are from the AR6 database; only scenarios that pass the vetting criteria are included (Section 3.2).

detailed. Due to their top-down perspective IAMs to date have not been able to represent the high complexity of industries in terms of the broad variety of technologies and processes (particularly circularity aspects) and to fully reflect the dynamics of the sector. In addition, as energy and carbon price elasticities are still not completely understood, primarily cost-driven models have their limitations. However, there are several ongoing activities to bring more engineering knowledge and technological details into the IAM models (Kermeli et al. 2021).

In addition to the more aggregated discussion, the IAMs illustrative mitigation pathways (IMPs) allow a deeper look into the transformation pathways related to the scenarios. For the illustrative mitigation pathways (IMPs) approach, sets of scenarios have been selected which represent different levels of GHG mitigation ambitions, scenarios which rely on different key strategies or even exclude some mitigation options, represent delayed actions or SDG-oriented pathways. For more detailed information about the selection see Section 3.3.2. Figure 11.12 compares for a selected number of key variables the results of IMPs and puts them in the context of the whole sample of IAMs scenario results for three temperature categories.

With growing mitigation ambition final energy demand is significantly lower in comparison of a current policy pathway (CurPol) and a scenario that explores the impact of further moderate actions (ModAct). Based on the underlying assumptions, scenarios IMP-SP and IMP-LD are characterised by the lowest final energy demand, triggered by high energy efficiency improvement rates as well as additional demand side measures, while a scenario with extensive use of CDR in the industry and the energy sectors to achieve net-negative emissions (IMP-Neg) leads to a significant increase in final energy demand. Scenario IMP-GS represents a pathway where mitigation action is gradually strengthened by 2030 compared to pre-COP 26 Nationally Determined Contributions (NDCs) shows the lowest final energy demand. All ambitious IMPs show substantially increasing contributions from electricity, with electricity's end-use share more than doubling for some of them by 2050 and more than tripling by 2100. The share of hydrogen shows a flatter curve for many scenarios, reaching 5% (IMP-Ren) in 2050 and up to 20% in 2100 for some scenarios (Ren, LD). Those scenarios that have a strong focus on renewable energy electrification show high shares of hydrogen in the sector. In comparison to sector-specific and national studies which show typically a range between 5 and 15% by 2050, many IAM IMPs expect hydrogen to play a less important role. Results for industrial CCS

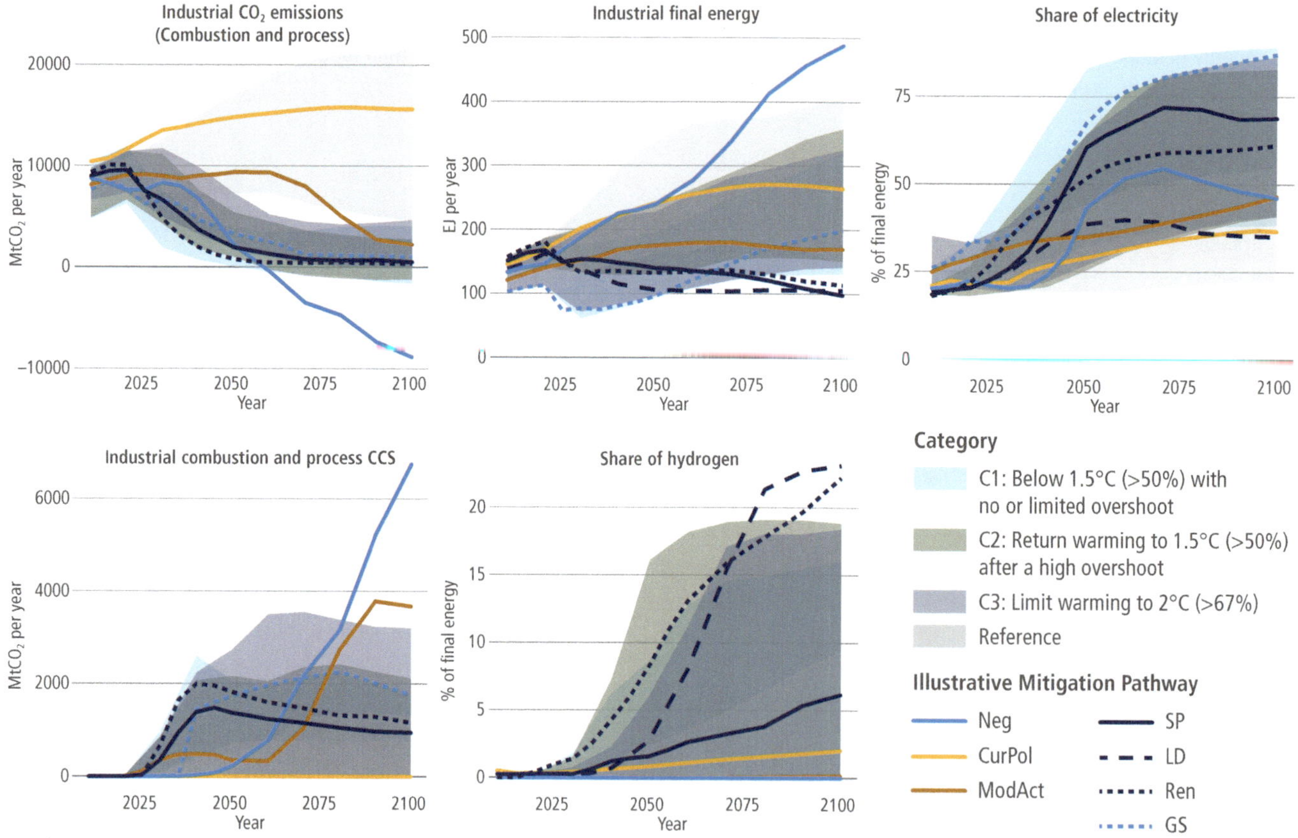


**Figure 11.12 | Comparison of industry-sector-related $CO_2$ emissions (including process emissions), final energy demand, share of electricity and hydrogen in the final energy mix, and industrial carbon capture and storage (CCS) for different mitigation scenarios representing illustrative mitigation pathways and the full sample of integrated assessment models (IAM) scenario results for three temperature categories (figure based on scenario database).** Indicators in the Illustrative Mitigation Pathways (lines) and the 5–95% range of reference, 1.5°C and 2°C scenarios (shaded areas). The selected IMPs reflect the following characteristics: opportunities for reducing demand (IMP-LD; low demand), the role of deep renewable energy penetration and electrification (IMP-Ren; renewables), extensive use of carbon dioxide removal (CDR) in the industry and the energy sectors to achieve net-negative emissions (IMP-Neg), insights into how shifting development can lead to deep emission reductions and achieve sustainable development goals (IMP-SP; shifting pathways), and insights into how slower short-term emissions reductions can be compensated by very fast emission reductions later on (IMP-GS; gradual strengthening). Furthermore, two scenarios were selected to illustrate the consequences of current policies and pledges; these are CurPol (Current Policies) and ModAct (Moderate Action), and are referred to as Pathways Illustrative of Higher Emissions. Source: data are from the AR6 database; only scenarios that pass the vetting criteria are included (Section 3.2).

show a broad variety of contributions, with the GS scenario (where hydrogen is not relevant as a mitigation option) representing the upper bound to 2050, with almost 2 $GtCO_2\ yr^{-1}$ captured and stored by 2050. Beyond 2050 the upper bound is associated with scenario IMP-Neg associated with extensive use of CDR in the industry and energy sectors to achieve net-negative emissions in the second half of the century – more than 6 $GtCO_2\ yr^{-1}$ is captured and stored in 2100 (this represents roughly 60% of 2018 direct $CO_2$ emissions of the sector).

#### 11.4.2.2 In-depth Discussion and 'Reality' Check of Pathways From Specific Sector Scenarios

Since AR5 a number of studies providing a high technological level of detail for the industry sector have been released which describe how the industry sector can significantly reduce its GHG emissions until the middle of the century. Many of these studies try to specifically reflect the particular industry sector characteristics and barriers that hinder industry to follow an optimal transformation pathway. They vary in respect to different characteristics. In respect to their geographical scope, some studies analyse the prospects for industry sector decarbonisation on a global level (IEA 2017a; Energy Transitions Commission 2018; Grubler et al. 2018; IEA 2020a, 2019b, 2020c; Tchung-Ming et al. 2018); regional level, for example, European Commission (2018) and Material Economics (2019); or country level – studies for China, from where most industry-related emissions come (e.g., Zhou et al. 2019).[23] In regard to sectoral scope, some studies include the entire industry sector, while others focus on selected GHG emission intensive sectors, such as steel, chemicals and/or concrete. Most of the scenarios focus solely on $CO_2$ emissions, that is non-$CO_2$ emissions of the industrial sector are neglected.[24]

Industry sector mitigation studies also differ in regard to whether they develop coherent scenarios or whether they focus on discussing and analysing selected key mitigation strategies, without deriving full energy and emission scenarios. Coherent scenarios are developed in IEA (2017); Energy Transitions Commission (2018); Grubler et al. (2018); Tchung-Ming et al. (2018); IEA (2019b, 2020a,c); IEA (2021a); and IRENA (2021) on the global level, and in Climact (2018); European Commission (2018); and Material Economics (2019) on the European level. Recent literature analysing selected key mitigation strategies, for example IEA (2019b) and Material Economics (2019) has focused either exclusively or to a large extent on analysing the potential of materials efficiency and circular economy measures to reduce the need for primary raw materials relative to a business-as-usual development. The IEA (2021a, 2020a) also provides deep insights in to single mitigation strategies for the industry sector, particularly the role of CCS. The following discussion mainly concentrates on scenarios from the IEA. It has to be acknowledged that they only represent a small segment of the huge scenario family (see the scenario database in Chapter 3), but this approach enables to show the chronological evolution of scenarios coming from the same institution, using the same modelling approach (which allows a technology-rich analytical backcasting approach), but reflect additional requests that emerge over time (Table 11.5). In the 2DS scenario from the 'Energy Technology Perspectives (ETP)' study (IEA 2017), which intends to describe in great technological detail how the global energy system could transform by 2060 so as to be in line with limiting global warming to below 2°C, total $CO_2$ emissions are 74% lower in 2060 than in 2014, while only 39% lower in the industry sector. The Beyond 2°C Scenario (B2DS) of the same study intends to show how far known clean energy technologies (including those that lead to negative emissions) could go if pushed to their practical limits, allowing the future temperature increase to be limited to 'well below' 2°C and lowering total $CO_2$ emissions by 100% by 2060 and by 75% relative to 2014 in the industry sector.

Technologies penetration assumed in the CTS scenario by 2060 allows for an industrial emission cut of 45% from 2017 levels and a 50% cut against projected 2060 emissions in the Reference Technology Scenario (RTS) from the same study (IEA 2019b), similar to IEA's 2DS scenario. Energy efficiency improvements and deployment of BATs contribute 46% to cumulative emission reduction in 2018–2060, while fuel switching (15%), material efficiency (19%) and deployment of innovative processes (20%) provide the rest. IEA (2020a,c) which continues the Energy Technology Perspectives series include the new Sustainable Development Scenario (SDS) to describe a trajectory for emissions consistent with reaching global 'net zero' $CO_2$ emissions by around 2070.[25] In 2070 the net zero balance is reached through a compensation of the remaining $CO_2$ emissions (fossil fuel combustion and industrial processes still lead to around 3 $GtCO_2$) by a combination of BECCS and to a lesser degree direct air capture and storage. In IEA (2020c) the Faster Innovation Case (FIC) shows a possibility to reach a net zero emissions level globally already in 2050, assuming that technology development and market penetration can be significantly accelerated. Innovation plays a major role in this scenario as almost half of all the additional emissions reductions in 2050 relative to the reference case would be from technologies that are in an early stage of development and have not yet reached the market today (IEA 2020c). The most ambitious IEA scenario NZE2050 (IEA 2021a) describes a pathway reaching net zero emissions at system level by 2050. With 0.52 $GtCO_2$ industry-related $CO_2$ emissions (including process emissions) it ends up 94% below 2018 levels in 2050. Remaining emissions in the industry sector have to be compensated by negative emissions (e.g., via DAC).

11

[23] In addition, there are many other studies available which have developed country-specific, technologically detailed scenarios for industry decarbonisation (e.g., Gerbert et al. 2018) and a few which have investigated the decarbonisation prospects of individual industrial clusters (Schneider 2019), but these types of studies are not discussed here.

[24] Most of the global mitigation scenarios solely focus on $CO_2$ emissions. Non-$CO_2$ emissions make up only a small share of the industry sector's current $CO_2$-eq. emissions and include $N_2O$ emissions (e.g., from nitric and adipic acid production), $CH_4$ emissions (e.g., from chemical production and iron and steel production) and various F-gases (such as perfluorocarbons from primary aluminium production and semiconductor manufacturing) (USEPA and ICF 2012; Gambhir et al. 2017). Mitigation options for these non-$CO_2$ emissions are discussed in Gambhir et al. (2017).

[25] Following the description of IEA SDS 2020 would limit the global temperature rise to below 1.8°C with a 66% probability if $CO_2$ emissions remain at net zero after 2070. If $CO_2$ emissions were to fall below net zero after 2070, then this would increase the possibility of reaching 1.5°C by the end of the century (IEA 2020c).

Table 11.4 | Perspectives on industrial sector mitigation potential (comparison of different IEA scenarios).

| Reduction of direct $CO_2$ emissions | Scenario assumptions[a] | IEA (2017, 2020c,i, 2021a) | | IEA (2019b) | IEA (2020a,c) | |
|---|---|---|---|---|---|---|
| | | 2030 | 2050 | 2060 | 2050 | 2070 |
| **Baseline direct emissions from industrial sector** | | | | | | |
| Reference Technology Scenario (RTS) | Industry sector improvements in energy consumption and $CO_2$ emissions are incremental, in line with currently implemented and announced policies and targets. | 9.8 $GtCO_2$ | 10.4 $GtCO_2$ | 9.7 $GtCO_2$ | | |
| **Emissions reduction potential** | | | | | | |
| 2°C Scenario (2DS) | Assumes the decoupling of production in industry from $CO_2$-emissions growth across the sector that would be compatible with limiting the rise in global mean temperature to 2°C by 2100. | –7% vs 2014[a]<br>–20% vs RTS[b] | –39% vs 2014[b]<br>–50% vs RTS[b] | | | |
| Beyond 2°C Scenario (B2DS) | Pushes the available $CO_2$ abatement options in industry to their feasible limits in order to aim for the 'well below 2°C' target. | –28% vs 2014<br>–38% vs RTS | –75% vs 2014<br>–80% vs RTS | | | |
| Clean Technology Scenario (CTS) | Strong focus on clean technologies. Energy efficiency and deployment of BATs contribute 46% to cumulative emission reduction in 2018–2060; fuel switch –15%; material efficiency – 19%; deployment of innovative processes – 20%. | | | 5 Gt $CO_2$ or –45% vs 2017 level and –50% from 2060 RTS level | | |
| Sustainable Development Scenario 2020 (SDS 2020) | Leads to net zero emissions globally by 2070. Remaining emissions in some sectors (including industry) in 2070 will be compensated by negative emissions in other areas (e.g., through BECCS and DAC). | | | | ~ 4.0 $GtCO_2$ | ~ 0.6 $GtCO_2$ |
| Net zero emissions (NZE, 2021) | Net zero emissions across all sectors are reached already by 2050. | –23%<br>(i.e., 2.1 $GtCO_2$)<br>vs 2018. | –94%<br>(i.e., 8.4 $GtCO_2$)<br>vs 2018 | | | |
| Faster Innovation Case (FIC) | Achieves net-zero emissions status already by 2050 based on accelerated development and market penetration of technologies which have currently not yet reached the market. | | | | 0.8 Gt $CO_2$<br>(mainly steel and chemical industry) | |

[a] Based on bottom-up technology modelling of five energy-intensive industry subsectors (cement, iron and steel, chemicals and petrochemicals, aluminium, and pulp and paper).
[b] Industrial direct $CO_2$ emissions reached 8.3 $GtCO_2$ in 2014, 24% of global $CO_2$ emissions.
Source: IEA (2017, 2019b, 2020a, 2020c,i, 2021a).

Two studies complement the discussion of the IEA scenarios and are related to the IEA database.[26] The ETC Supply Side scenario builds on the ETP 2017 study, investigating additional emission reduction potentials in the emissions-intensive sectors such as heavy industry and heavy-duty transport so as to be able to reach net zero emissions by the middle of the century. The LED scenario (Grubler et al. 2018) also builds on the ETP 2017 study, but focuses on the possible potential of very far-reaching efforts to reduce future material demand.

A comparison of the different mitigation scenarios shows that they depend on how individual mitigation strategies in the industry sector (Figure 11.13) are assessed. The use of CCS, for example, is in many scenarios assessed as very important, while other scenarios indicate that ambitious mitigation levels can be achieved without CCS in the industry sector. CCS plays a major role in the B2DS scenario (3.2 $GtCO_2$ in 2050), the ETC Supply Side scenario (5.4 $GtCO_2$ in 2050) and the IEA (2020a, 2021a) scenarios (e.g., 2.8 Gt $CO_2$ in NZE2050 in 2050, roughly one half of the captured $CO_2$ is related to cement production),

[26] Other global mitigation scenarios (e.g., from Tchung-Ming et al. (2018) and Shell Sky Scenario from Shell (2018)) are not included in the following scenario comparison as these studies' energy and emission base year data on the industry sector deviates considerably from the other three studies included in the comparison, which all use IEA data. Furthermore, unlike the other studies, Tchung-Ming et al. (2018) do not provide detailed information on the steel, chemicals and concrete subsectors. Not included here but worth mentioning are many other sector-specific studies, for example Napp et al. (2019, 2014), which consider more technologically advanced decarbonisation routes for the sector.

while it is explicitly excluded in the LED scenario. In the latter scenario, on the other hand, considerable emission reductions are assumed to be achieved by far-reaching reductions in material demand relative to a baseline development. In other words, the analysed scenarios also suggest that to reach very strong emission reductions from the industry sector either CCS needs to be deployed to a great extent or considerable material demand reductions will need to be realised. Such demand reductions only play a minor role in the 2DS scenario and no role in the ETC Supply Side scenario. The SDS described in IEA (2020a) provides a pathway where both CCS and material efficiency contribute significantly. In SDS material efficiency is a relevant factor in several parts of industry, explicitly steel, cement, and chemicals. Combining the different material efficiency options including a substantial part lifetime extension (particularly of buildings) leads to 29% less steel production by 2070, 26% less cement production, and 25% less chemicals production respectively in comparison to the reference line used in the study (Stated Policy Scenario: STEPS). Sector- or subsector-specific analysis supports the growing role of material efficiency. For the global chemical and petrochemical sector, Saygin and Gielen (2021) point out that circular economy (including recycling) has to cover 16% of the necessary reduction that is needed for the implementation of a 1.5°C scenario.

In all scenarios, the relevance of biomass and electricity in industrial final energy demand increases, especially in the more ambitious scenarios NZE2050, SDS, ETC Supply Side and LED. While in all scenarios, electrification becomes more and more important, hydrogen or hydrogen-derived fuels, on the other hand, do not contribute to industrial final energy demand by the middle of the century in 2DS and B2DS, while LED (1% final energy share in 2050) and particularly ETC Supply Side (25% final energy share in 2050) consider hydrogen or hydrogen-derived fuels as a significant option. In the updated IEA scenarios hydrogen and hydrogen-based fuels already play a more important role. In the SDS share in industry, final energy is around 10% (IEA 2020a) and in the Faster Innovation Case around 12% (IEA 2020c) in 2050. In the latter case this is based on the assumption that by 2050 on average each year 22 hydrogen-based steel plants come into operation (IEA 2020c). In SDS around 60% of the hydrogen is produced on-site via water electrolysis while the remaining 40% is generated in fossil fuel plants (methane reforming) coupled with CCS facilities. In the NZE2050 scenario biomass/biomethane (13%/3%), hydrogen (3%), natural gas with CCUS (4%), and coal with CCUS (4%) are responsible for 27% of the final energy demand of the sector. This is much more than in 2018, starting here from roughly 6% (only biomass). Direct use of electricity still plays a bigger role in the analysis, as share of electricity increases in NZE2050 from 22% in 2018 to 28% in 2030 and 46% in 2050 (with 15% a part of the electricity is used to produce hydrogen). This is reflecting the effect that since the publication of older IEA reports more direct electric applications for the sector become available. In NZE2050 approximately 25% of total heat used in the sector is electrified directly with heat pumps or indirectly with synthetic fuels already by 2030.

For B2DS it is assumed that most of the available abatement options in the industry sector are pushed to their feasible limits. That leads to cumulative direct $CO_2$ emissions reductions compared to 2DS which come from: energy efficiency improvements and BAT deployment (42%), innovative processes and CCS (37%), switching to lower carbon fuels and feedstocks (13%), and material efficiency strategies in manufacturing processes (8%). Energy efficiency improvements are particularly important in the first time period.

The IEA World Energy Outlook indicates energy efficiency improvement in the 2020 to 2030 period as a major basis to switch from STEPS (stated policies) to the SDS (net zero emissions by 2070) pathway (IEA 2020i, 2021c). For many energy-intensive industries annual efficiency gains have to be almost doubled (e.g., from 0.6% $yr^{-1}$ to 1.0% $yr^{-1}$ for cement production) to contribute sufficiently to the overall goal. If net zero $CO_2$ emissions should be achieved already by 2050 as pursued in the NZE2050 scenario (IEA 2020i, 2021c) further accelerating energy efficiency improvements are necessary (e.g., for cement, annual efficiency gains of 1.75%), leading to the effect that in 2030 many processes are implemented closely to their technological limits. In total, sector final energy demand can be held nearly constant at 2018 levels until 2050 and decoupled from product demand growth.

The comparative analysis leads to the point that the relevance of individual mitigation strategies in different scenarios depends not only on a scenario's level of ambition. Instead, implicit or explicit assumptions about: (i) the costs associated with each strategy, (ii) future technological progress and availability of individual technologies, and (iii) the future public or political acceptance of individual strategies are likely to be main reasons for the observed differences between the analysed scenarios. For many energy-intensive products, technologies capable of deep emission cuts are already available. Their application is subject to different economic and resources constraints (incremental investment needs, product prices escalation, requirements for escalation of new low-carbon power generation). To fully exploit potential availability of carbon-free energy sources (e.g., electricity or hydrogen and related derivates) is a fundamental prerequisite and marks the strong interdependencies between the industry and the energy sector.

Assessment of the scenario literature allows to conclude that under specific conditions strong $CO_2$-emission reductions in the industry sector by 2050–2070 and even net-zero-emission pathways are possible. However, there is no consensus on the most plausible or most desirable mix of key mitigation strategies to be pursued. In addition it has to be stressed that suitable pathways are very country-specific and depend on the economic structure, resource potentials, technological competences, and political preferences and processes of the country or region in question (Bataille 2020a).

There is a consensus among the scenarios that a significant shift is needed from a transition process in the past mainly based on marginal (incremental) changes (with a strong focus on energy efficiency efforts) to one based on transformational change. To limit the barriers that are associated with transformational change, besides overcoming the valley of death for technologies or processes with breakthrough character, it is required to carefully identify structural change processes which are connected with substantial changes of the existing system (including the whole process chain). This has to be done at an early stage and has to be linked with considerations about preparatory measures which are able to flank the changes and to foster the establishment of new structures (Section 11.6). The right sequencing of the various mitigation options and building appropriate bridges between the different strategies are important. Rissman et al. (2020) proposes three phases of technologies deployment for the industry sector: (i) energy/material efficiency improvement (mainly incremental) and electrification in combination with demonstration projects for new technologies potentially important in subsequent phases (2020–2035), (ii) structural shifts based on technologies which reach maturity in phase (i) such as CCS and alternative materials (2035–2050), (iii) widespread deployment for technologies that are nascent today like molten oxide electrolysis-based steel-making. There are no strong boundaries between the different phases and all phases have to be accompanied by effective policies like R&D programmes and market pull incentives.

Taking the steel sector as an illustrative example, sector-specific scenarios examining the possibility to reach GHG reduction beyond 80% (CAT 2020; Bataille et al. 2021b; IEA 2021a; Vogl et al. 2021b) indicate that robust measures comprise direct reduction of iron (DRI) with hydrogen in combination with efforts to further close the loops and increase availability of scrap metal (reducing the demand for primary steel). As hydrogen-based DRI might not be a fully mature technology before 2030 (depending on further developments of the policy framework and technological progress), risk of path dependencies has to be taken into consideration when reinvestments in existing production capacities will be required in the coming years. For existing plants, implementation of energy efficiency measures (e.g., utilisation of waste heat, improvement of high-temperature pumps) could build a bridge for further mitigation measures but have only limited unexhausted potential. As many GHG mitigation measures are associated with high investment costs and missing operating experience, a step-by-step implementing process might be an appropriate strategy to avoid investment leakage (given the mostly long operation times, investment cycles have to be used so as not to miss opportunities) and to gain experience. In the case of steel, companies can start with the integration of a natural gas-based direct reduced iron furnace feeding the reduced iron to an existing blast furnace, blending and later replacing the natural gas by hydrogen in a second stage, and later transitioning to a full hydrogen DRI EAF or molten oxide electrolysis EAF, all without disturbing the local upstream and downstream supply chains.

It is worth mentioning the flexibility of implementing transformational changes not the least depends on the age profile and projected longevity of existing capital stock, especially the willingness to accept the intentional or market-based stranding of high GHG intensity investments. This is a relevant aspect in all producing countries, but particularly in those countries with a rather young industry structure (i.e., comparative low age of existing facilities on average). Tong et al. (2019) suggest that in China, using the survival rate as a proxy, less than 10% of existing cement or steel production facilities will reach their end of operation time by 2050. Vogl et al. (2021b) argue that the mean blast furnace campaign is considerably shorter than used in Tong et al.(2019), at only 17 years between furnace relining, which suggests there is more room for retrofitting with clean steel major process technologies than generally assumed. Bataille et al. (2021b) found if very low carbon intensity processes were mandatory starting in 2025, given the lifetimes of existing facilities, major steel process lifetimes of up to 27 years would still make a full retrofit cycle with low-carbon processes possible.

In general, early adoption of new technologies plays a major role. Considering the long operation time (lifetime) of industrial facilities (e.g., steel mills and cement kilns) early adoption of new technologies is needed to avoid lock-in. For the SDS 2020 scenario, the IEA (2020h) calculated the potential cumulative reduction of $CO_2$ emissions from the steel, cement and chemicals sector to be around 57 $GtCO_2$ if

**Table 11.5 | Contribution to emission reduction of different mitigation strategies for net zero emissions pathways (range represents three different pathways for the industry sector in Europe; each related scenario focuses on different key strategies).**[27]

| | Steel | Plastics | Ammonia | Cement |
|---|---|---|---|---|
| | Contribution to emission reduction (%) (range represents the three different pathways of the study) | | | |
| Circularity | 5–27 | 15-28 | 13–22 | 10–44 |
| Energy efficiency | 5–23 | 2–9 | | 1–5 |
| Fossil fuels and waste fuels | 9–41 | 0–27 | | 0–51 |
| Decarbonised electricity | 36–59 | 16–22 | 25–84 | 29–71 |
| Biomass for fuel or feedstock | 5–9 | 18–22 | | 0–9 |
| End-of-life plastic | | 16–35 | | |
| CCS | 5–34 | 0–31 | 0–57 | 29–79 |
| | Required electrification level | | | |
| Growth of electricity demand (times compared with 2015) | 3–5 | 3–4 | | 2–5 |
| | Investments and production costs escalation | | | |
| Investment needs growth (% versus BAU) | 25–65 | 122–199 | 6–26 | 22–49 |
| Cost of production (% versus BAU) | +2–20 | +20–43 | +15–111 | +70–115 |

Source: Material Economics (2019).

[27] Note: In the described scenarios CCS was not taken into consideration as a mitigation option by the authors.

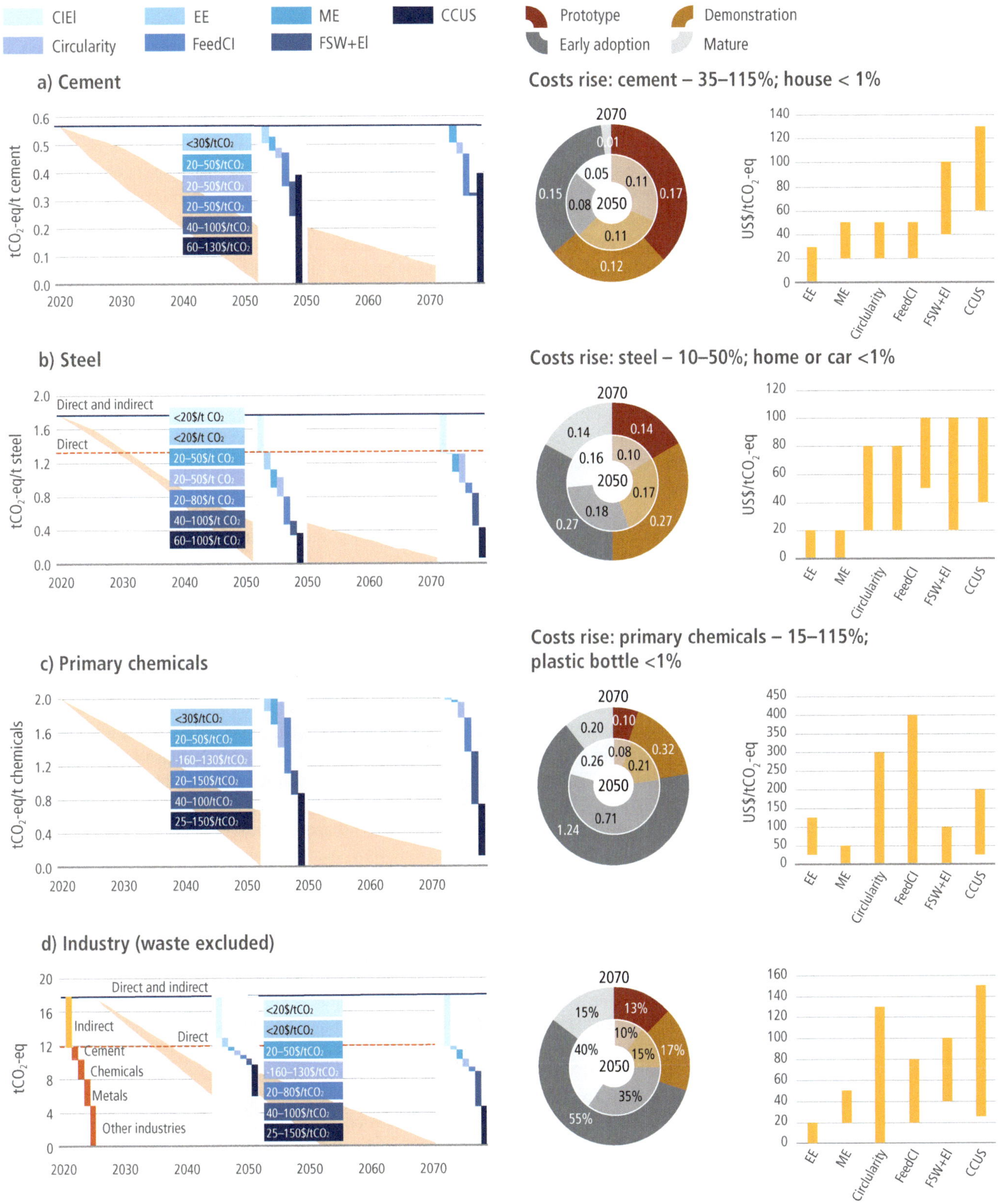


**Figure 11.13 | Potentials and costs for zero-carbon mitigation options for industry and basic materials:** ***CIEL – carbon intensity of electricity for indirect emissions; EE – energy efficiency; ME – material efficiency; Circularity – material flows (clinker substituted by coal fly ash, blast furnace slag or other by-products and waste, steel scrap, plastic recycling, etc.); FeedCI – feedstock carbon intensity (hydrogen, biomass, novel cement, natural clinker substitutes); FSW+El – fuel switch and processes electrification with low-carbon electricity.*** Ranges for mitigation options are shown based on bottom-up studies for grouped technologies packages, not for single technologies. In circles, contribution to mitigation from technologies based on their readiness are shown for 2050 (2040) and 2070. Direct emissions include fuel combustion and process emissions. Indirect emissions include emissions attributed to consumed electricity and purchased heat. For basic chemicals only methanol, ammonia and high-value chemicals are considered. The total for industry doesn't include emissions from waste. Base values for 2020 for direct and indirect emissions were calculated using 2019 GHG emission data (Crippa et al. 2021) and data for materials production from World Steel Association (2020a) and IEA (2021d). Negative mitigation costs for some options like Circularity are not reflected. Data from sources: Pauliuk et al. (2013a); Fawkes et al. (2016); WBCSD (2016); Bazzanella and Ausfelder (2017); IEA (2018a, 2019b,g,h, 2020a,c, 2021a); Lehne and Preston (2018); Scrivener et al. (2018); EUROFER (2019); Friedmann et al. (2019); Material Economics (2019); Sandalow et al. (2019); CAT (2020); CEMBUREAU (2020); Gielen et al. (2020); Habert et al. (2020); World Steel Association (2020b); Bataille (2020a); GCCA (2021a); and Saygin and Gielen (2021).

production technology is changed at its first mandatory retrofit, typically 25 years, rather than at 40 years (typical retrofitted lifetime) (Figure 11.14). Net zero pathways require that the new facilities are based on zero- or near-zero emissions technologies from 2030 onwards (IEA 2021c).

Another important finding is that material efficiency and demand management are still not well represented in the scenario literature. Besides IEA (2020a) two of the few exceptions are Material Economics (2019) for the EU and Zhou et al. (2019) for China. Zhou et al. (2019) describe a consistent mitigation pathway (Reinventing Fire scenario) for China where in 2050 $CO_2$ emissions are at a level 42% below 2010 emissions. Around 13% of the reduction is related to less material demand, mainly based on extension of building and infrastructure lifetime, as well as reduction of material losses in the production process and application of higher quality materials particularly high-quality cement (Zhou et al. 2019). For buildings and cars, Pauliuk et al. (2021) analysed the potential role of material efficiency and demand management strategies on material demand to be covered by the industry sector.

For the four subsectors in industry with high emissions, Table 11.5 shows results from Material Economics (2019) for the EU. The combination of circularity, material and energy efficiency, fossil and waste fuels mix, electrification, hydrogen, CCS and biomass use varies from scenario to scenario with none of these options ignored, but trade-offs are required.

The analysis of net zero emission pathways requires significantly higher investments compared to business as usual (BAU): 25–65% for steel, 6–26% for ammonia, 22–49% for cement, and with 122–199% the highest number for plastics (Material Economics 2019).

While sector-specific cost analyses are rare in general, there are scenarios indicating that pathways to net zero $CO_2$ emissions in the emissions-intensive sectors can be realised with limited additional costs. According to the Energy Transitions Commission (2018), deep decarbonisation from four major industry subsectors (plastics, steel, aluminium and cement) is achievable on a global level with cumulative incremental capital investments (2015–2050) limited to about 0.1% of aggregate GDP over that period. UKCCC (2019a) assesses that total incremental costs (compared to a theoretical scenario with no climate change policy action at all) for cutting industrial emissions by 90% by 2050 is 0.2% of expected 2050 UK GDP (UKCCC 2019a). The additional investment is 0.2% of gross fixed capital formation (Material Economics 2019). The IEA (2020a) indicates the required annual incremental global investment in heavy industry is approximately 40 billion 2019USD $yr^{-1}$ moving from STEPS to the SDS scenario (2020–2040), rising to USD55 billion $yr^{-1}$ (2040–2070), effectively 0.05–0.07% of global annual GDP today.

Finally, a new literature is emerging, based on the new sectoral electrification, hydrogen- and CCS- based technologies listed in previous sections, considering the possibility of rearranging standard supply and process chains using regional and international trade in intermediate materials like primary iron, clinker and chemical feedstocks, to reduce global emissions by moving production of these materials to regions with large and inexpensive renewable energy potential or CCS geology (Bataille 2020a; Gielen et al. 2020; Bataille et al. 2021a; Saygin and Gielen 2021).

In a sequence of sectoral- and industry-wide figures above (Figure 11.13), it is shown – starting in the present on the left and moving through 2050 to 2070 on the right, how much separate mitigation strategies can contribute and how they are integrated in the literature to reach near-zero emissions. For cement, steel and primary chemicals GHG intensities are presented, and for all industry absolute GHG emissions are displayed. Effects of the following mitigation strategies are reflected: energy efficiency, material efficiency, circularity/recycling, feedstock carbon intensity, fuel switching, CCU and CCS. Contributions of technologies split by their readiness for 2050 and 2070 are provided along with ranges of mitigation costs for achieving near-zero emissions for each strategy, accompanied by ranges of associated basic materials cost escalations and driven by these final products' prices increments.

### 11.4.3 Cross-sectoral Interactions and Societal Pressure on Industry

Mitigation involves greater integration and coupling between sectors. This is widely recognised, for example, in the case of electrification of transport (Sections 6.6.2 and 10.3.1), but it has been less explored for industrial decarbonisation. Industry is a complex web of subsectors and intersectoral interaction and dependence, with associated mitigation opportunities and co-benefits and costs (OECD 2019b; Mendez-Alva et al. 2021). Implementation of the mitigation options assessed in Section 11.3 will result in new sectoral couplings, value chains, and business models but also in the phasing out of old ones. Notably, electrification in industry, hydrogen and sourcing of non-fossil carbon involves profound changes to how industry interacts with electricity systems and how industrial subsectors interact. For example, the chemicals and forestry industries will become much more coupled if various forms of biogenic carbon become an important feedstock for plastics (Figure 11.10). Clinker substitution with blast furnace slag in the cement industry is a well-established way of reducing $CO_2$ emissions (Fechner and Kray 2012), but this slag will no longer be available if blast furnaces are phased out. Furthermore, additional material demand resulting from mitigation in other sectors, as well as adaptation and the importance of material efficiency improvements, are issues that have attracted increasing attention since AR5 (IEA 2019b; Bleischwitz 2020; Hertwich et al. 2020). How future material will be affected under different climate scenarios is underexplored and typically not accounted for in modelling (Bataille et al. 2021a).

Using industrial waste heat for space heating, via district heating, is an established practice that still has a large potential with large quantities of low-grade heat being wasted (Fang et al. 2015). For Denmark it is estimated that 5.1% of district heating demand could be met with waste heat (Bühler et al. 2017) and for four towns studied in Austria 3–35% of total heat demand could be met (Karner et al. 2016). A European study shows that temporal heat demand flexibility could allow for up to 100% utilisation of excess heat from industry (Karner et al. 2018). A study of a Swedish chemicals complex estimated that 30–50% of excess heat generated on-site could be recovered with payback periods below three years (Eriksson et al. 2018).

A European study found that most of the industrial symbiosis or clustering synergies today are in the chemicals sector with shared streams of energy, water, and carbon dioxide (Mendez-Alva et al. 2021). For future mitigation, the UKCCC (2019b) finds that industrial clustering may be essential for achieving the necessary efficiencies of scale and to build the infrastructure needed for industrial electrification; carbon capture, transport and disposal; hydrogen production and storage; heat cascading between industries and to other potential heat users (e.g., residential and commercial buildings).

With increasing shares of renewable electricity production there is a growing interest in industrial demand response, storage and hybrid solutions with on-site PV and combined heat and power (CHP) (Shoreh et al. 2016; Scheubel et al. 2017; Schriever and Halstrup 2018). With future industrial electrification, and in particular with hydrogen used as reduction agent in iron-making or as feedstock in the chemicals industry, the level of interaction between industry and power systems becomes very high. Large amounts of coking coal, or oil and gas as petrochemical energy and feedstock, are then replaced by electricity. For example, Meys et al. (2021) estimates a staggering future electricity demand of 10,000 TWh in a scenario for a net zero emissions plastics production of 1100 Mt in 2050 (see Section 11.3.5 for other estimates of electricity demand). Much of this electricity is used to produce hydrogen to allow for CCU and this provides a very large potential flexible demand if electrolysers are combined with hydrogen storage. Vogl et al. (2018) describe how hydrogen DRI and EAF steel plants can be highly flexible in their electricity demand by storing hydrogen or hot-briquetted iron and increasing the share of scrap in EAF. The IEA (2019f) Future of Hydrogen report suggests that hydrogen production and storage networks could be in locations with already existing hydrogen production and storage, for example, chemical industries, and that these could be ideal for system load balancing and demand response, and in the case of district heating systems – for heat cascading.

The climate awareness that investors, shareholders, and customers demand from companies has been increasing steadily. It is reflected in the growing number of environmental management, carbon footprint accounting, benchmarking and reporting schemes (e.g., the Carbon Disclosure Project, Task Force on Climate-Related Financial Disclosures, Environmental Product Declarations, and others, e.g., Qian et al. 2018) requiring companies to disclose both direct and indirect GHG emissions, and creating explicit (for regulatory schemes) as well as implicit GHG liabilities. This requires harmonised and widely accepted methods for environmental and carbon footprint accounting (Bashmakov et al. 2021b). From an investor perspective there are both physical risks (e.g., potential damages from climate change to business) and transition risks (e.g., premature devaluation of assets driven by new policies and technologies deployment and changes in public and private consumer preferences (NGFS 2019a)). Accompanied by reputational risks this leads to increased attention to Sustainable and Responsible Investment (SRI) principles and increased demands from investors, consumers and governments on climate and sustainability reporting and disclosure (NGFS 2019b). For example, Japan's Keidanren promotes a scheme by different industries to reduce GHG through the global value chain, including material procurement, product-use stages, and disposal, regardless of geographical origin, with provided quantitative visualisation (Keidanren (Japan Business Federation) 2018). The EU adopted a non-financial disclosure directive in 2014 (Kinderman 2020) and a Taxonomy for Sustainable Finance in 2019 (Section 15.6.1).

### 11.4.4 Links to Climate Change and Adaptation

Sectors that are particularly vulnerable to climate change include agriculture, forestry, fisheries and aquaculture, and their downstream processing industries (Bezner et al. 2021). Many of the energy-intensive industries are located based on access to fresh water (e.g., pulp and paper) or sea transport (e.g., petrochemicals). Risks of major concern for industry include disrupted supply chains and energy supplies due to extreme weather events, as well as risks associated with droughts, floods with dirty water, sea level rise and storm surges (Dodman et al. 2021). Adaptation measures may in turn affect the demand for basic materials (e.g., steel and cement), for example, increased demand to build sea walls and protect infrastructure, but we have not found any estimates of the potential demand. Increased heat stress is unsafe for outdoor labourers and can reduce worker productivity, for example, in outdoor construction, resource extraction and waste handling (Ranasinghe et al. 2021).

## 11.5 Industrial Infrastructure, Policy, and Sustainable Development Goal Contexts

### 11.5.1 Existing Industry Infrastructures

Countries are at different stages of different economic development paths. Some are already industrialised, while developing and emerging economies are on earlier take-off stages or accelerated growth stages and have yet to build the basic infrastructure needed to allow for basic mobility, housing, sanitation, and other services (Section 11.2.3). The available in-use stock of material per capita and in each country therefore differs significantly, and transition pathways will require a different mix of strategies, depending on each country's material demand to build, maintain, and operate stock of long-lived assets. Industrialised economies have much greater opportunities for reusing and recycling materials, while emerging economies have greater opportunities to avoid carbon lock-in. The IEA projected that more than 90% of the additional 2050 production of key materials will originate in non-OECD countries (IEA 2017). As incomes rise in emerging economies, the industry sector will grow in tandem to meet the increased demand for the manufactured goods and raw materials essential for infrastructure development. The energy and feedstocks needed to support this growth are likely to constitute a large portion of the increase in the emerging economies' GHG emissions in the future unless new low-carbon pathways are identified and promoted.

Emissions are typically categorised by the territory, subsector or group of technologies from which they emanate. An alternative subdivision is that between existing sources that will continue to generate emissions in the future, and those that are yet to be built (Erickson et al. 2015). The rate of emissions from existing assets will eventually tend to zero, but in a timeframe that is relevant to existing climate and energy goals, the cumulative contribution to emissions from existing infrastructure and equipment is likely to be substantial. Aside from the magnitude of the contribution, the distinction between emissions from existing and forthcoming assets is instructive because of the difference in approach to mitigation that may be necessary or desirable in each instance to avoid getting locked into decades of highly carbon-intensive operations (Lecocq and Shalizi 2014).

Details of the methodologies to assess 'carbon lock-in' or 'committed emissions' differ across studies but the core components of the approaches adopted are common to each: an account of the existing level of emissions for the scope being assessed is established; this level is projected forward with a stylised decay function that is informed by assessments of the current age and typical lifetimes of the underlying assets. From this, a cumulative emissions estimate is calculated. The future emissions intensity of the operated assets is usually assumed to remain constant, implying that nothing is done to retrofit with mitigating technologies (e.g., carbon capture) or alter the way in which the plant is operated (e.g., switching to an alternative fuel or feedstock). While the quantities of emissions derived are often referred to as 'committed' or 'locked-in', their occurrence is of course dependent on a suite of economic, technology and policy developments that are highly uncertain.

Data on the current age profile and typical lifetimes of emissions-intensive industrial equipment are difficult to procure and verify and most of the studies conducted in this area contain little detail on the global industrial sector. Two recent studies are exceptions, both of which cover the global energy system, but contain detailed and novel analysis on the industrial sector (Tong et al. 2019; IEA 2020a). Tong et al. (2019) use unit-level data from China's Ministry of Ecology and Environment to obtain a more robust estimate of the age profile of existing capacity in the cement and iron and steel sectors in the country. The IEA (2020a) uses proprietary global capacity datasets for the iron and steel, cement and chemicals sectors, and historic energy consumption data for the remaining industry sectors as a proxy for the rate of historic capacity build-up.

Both studies come to similar estimates on the average age of cement plants and blast furnaces in China of around 10–12 years old, which are the figures for which they have overlapping coverage. Both studies also use the same assumption of the typical lifetime of assets in these sectors of 40 years, whereas the IEA (2020a) study uses 30 years for chemical sector assets and 25 years for other industrial sectors. The studies come to differing estimates of cumulative emissions by 2050 from the industry sector; 196 $GtCO_2$ in the IEA (2020a) study, and 162 $GtCO_2$ in the Tong et al. (2019) study. This difference is attributable to a differing scope of emissions, with the IEA (2020a) study including industrial process emissions (which for the cement sector in particular are substantial) in addition to the energy-related emissions quantities accounted for in the Tong et al. (2019) study. After correcting for this difference in scope, the emissions estimates compare favourably.

The IEA (2020a) study provides supplementary analysis for the industry sector, examining the impact of considering investment cycles alongside the typical lifetimes assumed in its core analysis of emissions from existing industrial assets. For three heavy industry sectors – iron and steel, cement, and chemicals – the decay function applied to emissions from existing assets is re-simulated using a 25-year investment cycle assumption (Figure 11.14). This is 15 years shorter than the typical lifetimes assumed for assets in the iron and steel and cement sectors, and five years shorter than that considered for the chemical sector. The shorter timeframe for the investment cycle is a simplified way of representing the intermediate investments that are made to extend the life of a plant, such as the re-lining of a blast furnace, which can occur multiple times during the lifetime of an installation. These investments can often be similar in magnitude to that of replacing the installation, and they represent key points for intervention to reduce emissions. The findings of this supplementary analysis are that around 40%, or 60 $GtCO_2$, could be avoided by 2050 if near-zero emissions options are available to replace this capacity, or units are retired, retrofitted or refurbished in a way that significantly mitigates emissions (e.g., retrofitting carbon capture, or fuel or process switching to utilise bioenergy or low-carbon hydrogen).

As this review was being finalised several papers were released that somewhat contradict the Tong et al. (2010) results (Bataille et al. 2021b; Vogl et al. 2021b). Broadly speaking, these papers argue that while

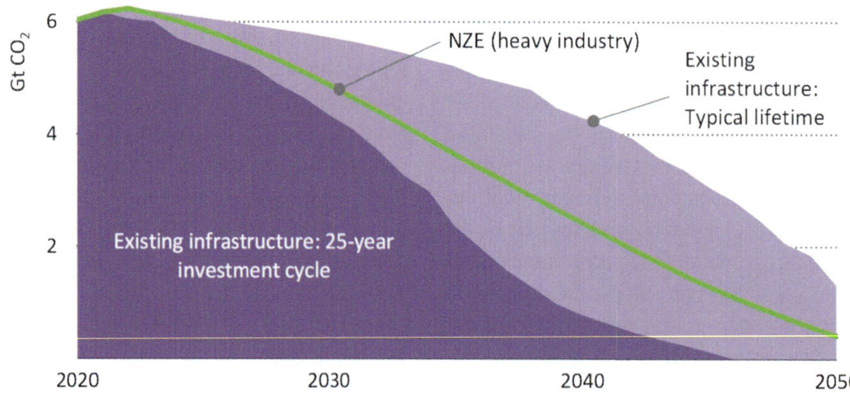


**Figure 11.14 | $CO_2$ emissions from existing heavy industrial assets in the NZE.** Source: International Energy Agency (2021), Net Zero by 2050, IEA, Paris.

high-emitting facilities may last for a long time, be difficult to shut down early, and are inherent to local boarder supply chains, individual major processes that are currently highly GHG intense, such as blast furnaces and basic oxygen smelters, could be retired and replaced during major retrofits on much shorter time cycles of 15 to 25 years.

The cost of retrofitting or retiring a plant before the end of its lifetime depends on plant-specific conditions as well as a range of economic, technology and policy developments. For industrial decarbonisation it may be a greater challenge to accelerate the development and deployment of zero-emission technologies and systems than to handle the economic costs of retiring existing assets before end of life. The 'lock-in' also goes beyond the lifetime of key process units, such as blast furnaces and crackers, since they are typically part of large integrated plants or clusters with industrial symbiosis, as well as infrastructures with feedstock storage, ports, and pipelines. Individual industrial plants are often just a small part of a complex network of many facilities in an industrial supply chain. In that sense, current assessments of 'carbon lock-in' rely on simplifications due to the high the complexity of industry.

Conditions are also subsector and context specific in terms of mitigation options, industry structures, markets, value chains and geographical location. For example, the hydrogen steel-making joint venture in Sweden involves three different companies headquartered in Sweden (in mining, electricity and steel-making, respectively), two of which are state-owned, with a shared vision and access to iron ore, fossil-free electricity and high-end steel markets (Kushnir et al. 2020). In contrast, chemical clusters may consist of several organisations that are subsidiaries to large multinational corporations with headquarters across the world, that also compete in different markets. Even in the presence of a local vision for sustainability this makes it difficult to engage in formalised collaboration or get support from headquarters (Bauer and Fuenfschilling 2019).

Furthermore, it is relevant to consider also institutional and behavioural lock-in (Seto et al. 2016). On one side, existing high-emitting practices may be favoured through formal and informal institutions (e.g., regulations and social norms or expectations, respectively), for example, around building construction and food packaging. On the other side, mitigation options may face corresponding institutional barriers. Examples include how cars are conventionally scrapped (i.e., crushed, leading to copper contamination of steel) rather than being dismantled, or slow permitting procedures for new infrastructure and industrial installations for reducing emissions.

### 11.5.2 Current Industrial and Broader Policy Context

The basic motivation for industrial policy historically has been economic development and wealth creation. Industrial policy can be progressive and promote new developments or be protective to help infant or declining industries. It may also involve the phase-out of industries, including efforts to retrain workers and create new jobs. Industrial policy is not one policy intervention but rather the combined effects of many policy instruments that are coordinated towards an industrial goal. Industrial policies can be classified as being either vertical or horizontal depending on whether singular sectors or technologies are targeted (e.g., through R&D, tariffs and subsidies) or the whole economy (e.g., education, infrastructure, and general tax policies). The horizontal policies are not always thought of as industrial policy, although taking a broad view, including policy coordination and institution building, is important for industrial policy to be effective (see e.g., Andreoni and Chang 2019).

In the past ten years there has been increasing interest and attention to industrial policy. One driver is the desire to retain industry or re-industrialise in regions within Europe and North America where industry has a long record of declining shares of GDP. The need for economic growth and poverty eradication is a key driver in developing countries. An important aspect is the need to meet the 'dual challenge of creating wealth for a growing population while staying within planetary boundaries' (Altenburg and Assman 2017). The need for industrial policy that supports environmental goals and green growth has been analysed by Rodrik (2014); Aiginger (2014); Warwick (2013); and Busch et al. (2018). Similar ideas are taken up in OECD reports on green growth (OECD 2011) and system innovation (OECD 2015). However, these approaches to green industrial policy and innovation tend to focus on opportunities for manufacturing industries to develop through new markets for cleaner technologies. They rarely include explicit attention to the necessity of zero emissions and the profound changes in production, use and recycling of basic materials that this entails. This may also involve the phase-out or repurposing of industries that currently rely on fossil fuels and feedstock.

The policy implications of zero emissions for heavy industries are relatively unexplored, although some analyses in this direction are available (e.g., Åhman et al. 2017; Philibert 2017a; Wesseling et al. 2017; Bataille et al. 2018a; Wyns et al. 2019; Bataille 2020a; Fan and Friedmann 2021). For industry, there has been a long time focus on energy efficiency policies through voluntary and negotiated agreements, energy management and audit schemes, and various programmes targeting industry (Fischedick et al. 2014a). Since AR5, interest in circular economy policies has increased and they have become more prevalent across regions and countries, including the EU, China, USA., Japan and Brazil (e.g., McDowall et al. 2017; Ranta et al. 2018; Geng et al. 2019). For electrification and CCUS, efforts are nascent and mainly focused on technology development and demonstrations. Policies for demand reduction and materials efficiency are still relatively unexplored (e.g., Pollitt et al. 2020 and IEA 2019b). Since zero emissions in industry is a new governance challenge it will be important to build awareness and institutional capacity in industrialised as well as developing countries.

In the context of climate change policy, it is fair to say that industry has so far been sheltered from the increasing costs that decarbonisation may entail. This is particularly true for the energy- and emissions-intensive industries where cost increases and lost competitiveness may lead to carbon leakage (i.e., that industry relocates to regions with less stringent climate policies). Heavy industries typically pay no or very low energy taxes and where carbon pricing exists (e.g., in the European Trading Scheme) they are sheltered through free allocation of emission permits and potentially compensated for resulting electricity price increases. For example, Okereke and McDaniels

(2012) show how the European steel industry was successful in avoiding cost increases and how information asymmetry in the policy process was important for that purpose.

### 11.5.3 Co-benefits of Mitigation Strategies and Sustainable Development Goals

The deployment of climate change mitigation strategies is primarily influenced by its costs and potential, but also by other broader sustainable development factors such as the Sustainable Development Goals (SDGs). Mitigation actions therefore are to be considered through the prism of impacts on achieving other economic, social and environmental goals. Those impacts are classified as co-benefits when they are positive or as risk when they are negative. Co-benefits can serve as additional drivers, while risks can inhibit the deployment of available mitigation options. Actions taken to mitigate climate change have direct and indirect interactions with SDGs, both positive (synergies) or negative (trade-offs) (Fuso Nerini et al. 2019).

Given the wide range of stakeholders involved in climate actions and their (often contradictory) interests and priorities, the nature of co-benefits and risk can affect decision-making processes and the behaviour of stakeholders (Labella et al. 2020). Co-benefits form an important driver supporting the adoption of mitigation strategies, yet are commonly overlooked in policymaking. Karlsson et al. (2020), based on a review of 239 peer-reviewed articles concluded that diverse co-benefit categories, including air, soil and water quality, diet, physical activity, biodiversity, economic performance, and energy security, are prevalent in the literature.

#### 11.5.3.1 Sustainable Development Goals Co-benefits Through Material Efficiency and Demand Reduction

Material efficiency, an important mitigation option (SDG 13, climate action) for heavy industries, is yet to be fully acknowledged and leveraged (Gonzalez Hernandez et al. 2018a; Sudmant et al. 2018; Dawkins et al. 2019). Material efficiency directly addresses SDG 12 (responsible production and consumption) but also provides opportunities to reduce the pressures and impacts on environmental systems (SDG 6, clean water and sanitation) (Olivetti and Cullen 2018). Exploiting material efficiency usually requires new business models and provides potential co-benefits of increased employment and economic opportunities (SDG 8, decent work and economic growth).

Material efficiency also provides co-benefits through infrastructural development (SDG 9, industry, innovation and infrastructure) (Mathews et al. 2018) to support the wide range of potential material efficiency strategies including light-weighting, reusing, remanufacturing, recycling, diverting scrap, extending product lives, using products more intensely, improving process yields, and substituting materials (Allwood et al. 2011). Worrell et al. (2016) also emphasises how material efficiency improvements, in addition to limiting the impacts of climate change help deliver sustainable production and consumption co-benefits through environmental stewardship. Binder and Blankenberg (2017) and Dhandra (2019) show that sustainable consumption is positively related to life satisfaction and subjective well-being (SDG 3), and Guillen-Royo (2019) adds positive associations with happiness and life satisfaction.

The reduction in excessive consumption and demand for products and services generates a reduction in post-consumption waste and so enhances clear water and sanitation (SDG 6) (Govindan 2018; Minelgaitė and Liobikienė 2019), and reduces waste along product supply chains and lifecycles (SDG 12) (Genovese et al. 2017; UNSD 2020). At the risk side there are possible reductions of employment, incomes, sales taxes from the material extraction and processing activities, considered as excessive for sustainable consumption (Thomas 2003).

#### 11.5.3.2 Sustainable Development Goals Co-benefits From Circular Economy and Industrial Waste

While the circular economy concept first emerged in the context of waste avoidance, resource depletion, closed-loop recycling, etc., it has now evolved as a tool for a broader systemic national policy due to its potential wider benefits (Geng et al. 2013). It represents new circular business models that encourage design for reuse and to improve material recovery and recycling, and so represents a departure from the traditional linear production and consumption systems (with landfilling at the end), with a wide range of potential co-benefits to a wide range of SDGs (Guo et al. 2016; Genovese et al. 2017; Schroeder et al. 2019; UNSD 2020).

Genovese et al. (2017) articulates the advantages from an environmental and responsible consumption and production point of view (SDG 12). Many studies have outlined new business models based on the circular economy that foster sustainable economic growth and the generation of new jobs (SDG 8) (Antikainen and Valkokari 2016), as well as global competitiveness and innovation in business and the industrial sector (Pieroni et al. 2019), such as its potential synergies with industry 4.0 (Garcia-Muiña et al. 2018).

Following a review of the literature, Schroeder et al. (2019) identified linkages between circular economy practices and SDGs based on a relationship scoring system, and highlighted that such SDGs as SDG 6 (clean water and sanitation), SDG 7 (affordable and clean energy), SDG 8 (decent work and economic growth), SDG 12 (responsible consumption and production), and SDG 15 (life on land) all strongly benefit from circular economy practices. With the potential to impact on all stages of the value chain (micro, meso and macro level of the economy), circular economy has also been identified as a key industrial strategy to managing waste across sectors.

Chatziaras et al. (2016) highlights the co-benefit to SDG 7 (affordable and clean energy) resulting from waste-derived fuel for the cement industry. Through the management of industrial waste using circular economy practices, studies such as Geng et al. (2012) and Bonato and Orsini (2017) have pointed out co-benefits to SDGs beyond clear environmental and economic benefits, highlighting how it also benefits SDG 3 and 11 through improved social relations between industrial sectors and local societies, and improved public environmental awareness and public health levels.

#### 11.5.3.3 Sustainable Development Goals Co-benefits From Energy Efficiency

Beyond the very direct links between energy and climate change, reliable, clean, and affordable energy (SDG 7) presents a cross-cutting issue, central to all SDGs and fundamental to development, and energy efficiency enables its provision by reducing the direct supply and necessary infrastructure required. Energy efficiency improvements can be delivered through multiple technical options and tested policies, delivering energy and resource savings simultaneously with other socio-economic and environmental co-benefits. At the macro level, this includes enhancement of energy security (SDG 16, peace, justice and strong institutions) delivered through clean low-carbon energy systems (Fankhauser and Jotzo 2018). Much of the literature, including Sari and Akkaya (2016), Allan et al. (2017) and Garrett-Peltier (2017), points out that energy efficiency improvements deliver superior employment opportunities (SDG 8 – decent work and economic growth), while a limited number of studies have reported that it can negatively impact employment in fuel supply sectors (Costantini et al. 2018).

Many studies report that energy efficiency improvements are essential for supporting overall economic growth, contributing to positive changes in multi-factor productivity (SDGs 8 and 9 – decent work and economic growth and industry, innovation, and infrastructure) (Lambert et al. 2014; Bataille and Melton 2017; Rajbhandari and Zhang 2018; Bashmakov 2019; Stern 2019) through industrial innovation (SDG 9) (Kang and Lee 2016), with some dissent (e.g., Mahmood and Ahmad 2018). Improved energy efficiency against a background of growing energy prices helps industrial plants stay competitive (Bashmakov and Myshak 2018). Energy efficiency allows continued economic growth under strong environmental regulation. Given that energy efficiency measures reduce the combustion of fossil fuels it leads to reduced air pollution at industrial sites (Williams et al. 2012) and better indoor comfort at working places.

Since less energy supply infrastructure is needed in cities and less energy is needed to produce materials such as cement and concrete, and metals, energy efficiency indirectly supports 'sustainable cities and communities' (SDG 11) (Di Foggia 2018). In addition, energy efficiency in industry reflects achievements in meeting SDG 12 (responsible consumption and production).

#### 11.5.3.4 Sustainable Development Goals Co-benefits From Electrification and Fuel Switching

A key, generally underappreciated SDG benefit of electrification is improved urban and indoor air quality (at working places as well) and associated health benefits (SDG 3) from clean electrification (SDG 7) of industrial facilities (IEA 2016). With energy being such an important cross-cutting issue to sustainable development, some SDGs, such as SDGs 1, 3, 4 and 5 (Harmelink et al. 2018) are co-beneficiaries to using electrification and fuel switching as a climate action mitigation option.

#### 11.5.3.5 Sustainable Development Goals Co-benefits from Carbon Capture and Utilisation, and Carbon Capture and Storage

CCU and CCS have been identified as playing key roles in the transition of industry to net zero. Advancements in the development and deployment of both CCS and CCU foster climate action (SDG 13). Other co-benefits for CCS include control of non-$CO_2$ pollutants (SDG 3), direct foreign investment and know-how (SDG 9), enhanced oil recovery from existing resources, and diversified employment prospects and skills (SDG 8) (Bonner 2017). For CCU, the main co-benefit related contributions are expected within the context of energy transition processes, and in societal advancements that are linked to technological progress (Olfe-Kräutlein 2020). Therefore, the expectations are that the deployment of CCU technologies would have least potential for meeting the SDG targets relating to society/people, compared with the anticipated contributions to the pillars of ecology and economy.

These mitigation options carry a large number of risks as well. The high cost of the capture and storage process not only limit the technology penetration, but also make energy and products more expensive (risk to SDG 7), potential leaks from undersea or underground $CO_2$ storages carries risks for achieving SDGs 6, 14 and 15. While there are economic costs involved with the deployment of CCS and CCU (Bataille et al. 2018a), there are also significant economic and developmental costs associated with taking no action, because of the potential negative impact of climate change. CCS and CCU have been argued as providing public good (Bergstrom and Ty 2017) and co-benefits to key SDGs (Schipper et al. 2011). On the other hand, Fan et al. (2018) among others have noted the potential lock-in of existing energy structures due to CCS. Refer to Table 17.1 for CCS and CCU co-benefits with respect to other sector chapters.

## 11.6 Policy Approaches and Strategies

Industrial decarbonisation is technically possible on the mid-century horizon, but requires scale up of technology development and deployment, multi-institutional coordination, and sectoral and national industrial policies with detailed subsectoral and regional mitigation pathways and transparent monitoring and evaluation processes (Åhman et al. 2017; Wesseling et al. 2017; Bataille et al. 2018a; Rissman et al. 2020; Nilsson et al. 2021). Transitions of industrial systems entail innovations, plant and technology phase-outs, changes across and within existing value chains, new sectoral couplings, and large investments in enabling electricity, hydrogen, and other infrastructures. Low-carbon transitions are likely to be contested, non-linear and require a multi-level perspective policy approach that addresses a large spectrum of social, political, cultural and technical changes as well as accompanying phase-out policies, and involve a wide range of actors, including civil society groups, local authorities, labour unions and industry associations e(Geels et al. 2017; Rogge and Johnstone 2017; Yamada and Tanaka 2019; Koasidis et al. 2020). See also Cross-Chapter Box 12.

Deployment of the mitigation options presented in this chapter (Sections 11.3 and 11.4) needs support from a mix of policy instruments including: GHG pricing coupled with border adjustments or other economic signals for trade-exposed industries; robust government support for research, development, and deployment; energy, material and emissions standards; recycling policies; sectoral technology roadmaps; market pull policies; and support for new infrastructure (Figure 11.15) ( Flanagan et al. 2011; Rogge et al. 2017; Bataille et al. 2018a; Tvinnereim and Mehling 2018; Creutzig 2019; Bataille 2020a; Rissman et al. 2020). The combination of the above will depend on specific sectoral market barriers, technology maturity, and local political and social acceptance (Hoppmann et al. 2013; Rogge and Reichardt 2016). Industrial decarbonisation policies need to be innovative and definitive about net zero $CO_2$ emissions to trigger the level of investment needed for the profound changes in production, use and recycling of basic materials needed (Nilsson et al. 2021). Inclusive and transparent governance that assesses industry decarbonisation progress, monitors innovation and accountability, and provides regular recommendations for policy adjustments is also important for progressing (Mathy et al. 2016; Bataille 2020a).

The level of policy experience and institutional capacity needed varies widely across the mitigation options. In many countries,

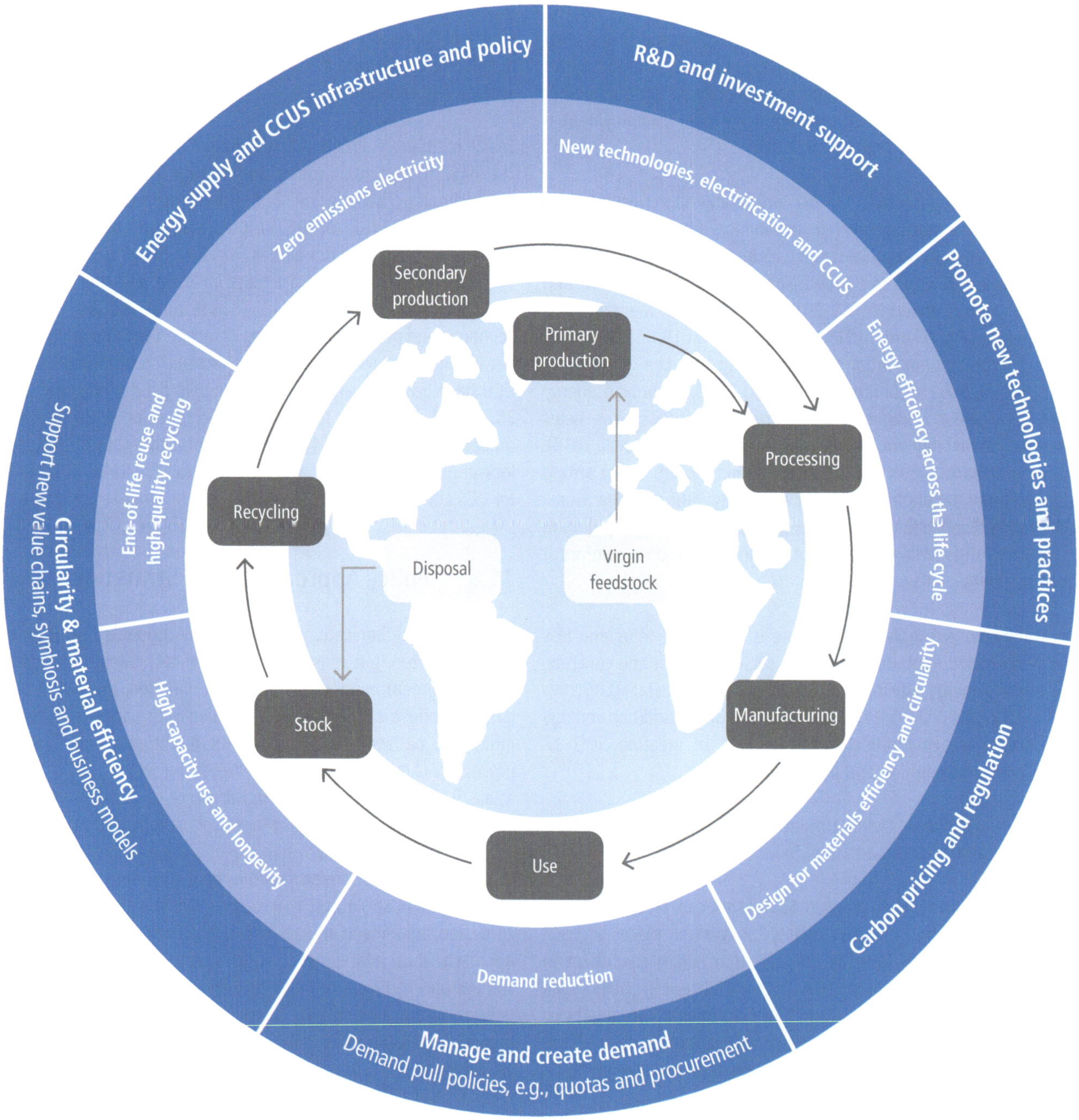


Figure 11.15 | Schematic figure showing the lifecycle of materials (green), mitigation options (light blue) and policy approaches (dark blue).

energy efficiency is a well-established policy field with decades of experience from voluntary and negotiated agreements, regulations, standards, energy audits, and demand-side management (DSM) programmes (see AR5), but there are also many countries where the application of energy efficiency policy is absent or nascent (see AR5) (Tanaka 2011; Fischedick et al. 2014a; García-Quevedo and Jové-Llopis 2021; Saunders et al. 2021). The application of DSM and load flexibility will also need to grow with electrification and renewable energy integration.

Materials efficiency and circular economy are not well understood from a policy perspective and were for a long time neglected in low-GHG industry roadmaps although they may represent significant potential (Allwood et al. 2011; Gonzalez Hernandez et al. 2018b; IEA 2019b, 2020a; Calisto Friant et al. 2021; Polverini 2021). Material efficiency is also neglected in products design, architectural and civil engineering education, infrastructure and building codes, and urban planning (Section 5.6) (Braun et al. 2018; Orr et al. 2019). For example, the overuse of steel and concrete in construction is well documented but policies or strategies (e.g., design guidelines or regulation) for improving the situation are lacking (Dunant et al. 2018; Shanks et al. 2019). Various circular economy solutions are gaining interest from policymakers with examples such as regulations and economic incentives for repair and reuse, initiatives to reduce planned obsolescence, and setting targets for recycling. Barriers that policies need to address are often specific to the different material loops (e.g., copper contamination for steel and lack of technologies or poor economics for plastics).

There is also a growing interest from policymakers in electrification and fuel switching but the focus has been mainly on innovation and on developing technical production-side solutions rather than on creating markets for enabling demand for low-carbon products, although the concept of green public procurement is gaining traction. The situation is similar for CCU and CCS. Low-carbon technologies adoption represents an additional cost to producers, and this must be handled through fiscal incentives like tax benefits, GHG pricing, green subsidies, regulation and permit procedures. For example, the 45Q tax credit provides some incentives to reduce investor risk for CCS and attract private investment in the USA (Ochu and Friedmann 2021).

Since industrial decarbonisation is only recently emerging as a policy field there is little international collaboration on facilitation (Oberthür et al. 2021). Given that most key materials markets are global and competitive, unless there is much greater global governance to contribute to the decarbonisation of GHG-intensive industry through intergovernmental and transnational institutions it is questionable that the world will achieve industry decarbonisation by 2050.

As GHG pricing, through GHG taxes or cap and trade schemes, has remained a central avenue for climate policy, this section begins with a review of how the industrial sector has been concerned with these instruments. The rest of the section is then structured into five key topics, following insights on key failures that policy must address to enable and support large-scale transformations as well as the need for complementary mixes of policies to achieve this goal (Weber and Rohracher 2012; Rogge and Reichardt 2016; Grillitsch et al. 2019). The section describes how the need to focus on long-term transitions rather than incremental changes can be managed through the planning and strategising of transition pathways; discusses the role of research, development, and innovation policy; highlights the need for enabling low-carbon demand and market creation; reflects on the necessity of establishing and maintaining a level of knowledge and capacity in the policy domain about the industrial transition challenge; and points to the critical importance of coherence across geographical and policy contexts. The section concludes with a reflection on how different groups of actors needs to take up different parts of the responsibility for mitigating climate change in the industrial sector.

### 11.6.1 GHG Prices and GHG Markets

Internalising the cost of GHG emissions in consumer choices and producer investment decisions has been a major strategy promoted by economists and considered by policymakers to mitigate emissions cost-effectively and to incentivise low-GHG innovations in a purportedly technology neutral way (Stiglitz et al. 2017; Boyce 2018). In the absence of a coordinated effort, individual countries, regions and cities have implemented carbon-pricing schemes. As of 23 August 2021, 64 carbon schemes have been implemented or are scheduled by law for implementation, covering 22.5% of global GHG emissions (World Bank 2020), 35 of which are carbon taxes, primarily implemented on a national level and 29 of which are emissions trading schemes, spread across national and sub-national jurisdictions.

Assessments of pricing mechanisms show generally that they lead to reduced emissions, even in sectors that receive free allocation such as industry (Martin et al. 2016; Haites et al. 2018; Narassimhan et al. 2018; Metcalf 2019; Bayer and Aklin 2020). However, questions remain as to whether these schemes can bring emissions down fast enough to reach the Paris Agreement goals (Boyce 2018; Tvinnereim and Mehling 2018; World Bank Group 2019). Most carbon prices are well below the levels needed to motivate investments in high-cost options that are needed to reach net zero emissions (Section 11.4.1.5). Among the 64 carbon-price schemes implemented worldwide today, only nine have carbon prices above USD40 (World Bank 2020). These are all based in Europe and include EU Emissions Trading System (ETS) (above USD40 since March 2021), Switzerland ETS, and seven countries with carbon taxes. Furthermore, emissions-intensive and trade-exposed (EITE) industries are typically allowed exemptions and receive provisions that shelter them from any significant cost increase in virtually all pricing schemes (Haites 2018). These provisions have been allocated due to concerns about loss of competitiveness and carbon leakage which result from relocation and increased imports from jurisdictions with no, or weak, GHG emission regulations (Branger and Quirion 2014a; Branger and Quirion 2014b; Jakob 2021a). Embodied emissions in international trade accounts for one quarter of global $CO_2$ emissions in 2015 (Moran et al. 2018) and has increased significantly over the past few decades, representing a significant challenge to competitiveness related to climate policy. CBAM, or CBA are trade-based mechanisms designed to 'equalise' the carbon costs for domestic and foreign producers. They are increasingly being considered by policymakers to address carbon leakage and create a level playing field for products produced in jurisdiction with no, or lower, carbon price (Mehling et al. 2019; Markkanen et al. 2021). On 14 July 2021, the European Commission adopted a proposal for a CBAM that requires importers of aluminium, cement, iron and steel, electricity and fertiliser to buy certificates at the ETS price for the emissions embedded in the imported products (European Commission 2021; Mörsdorf 2021). CBAMs should be crafted very carefully, to meet technical and legal challenges (Jakob et al. 2014; Sakai and Barrett 2016; Rocchi et al. 2018; Cosbey et al. 2019; Joltreau and Sommerfeld 2019; Pyrka et al. 2020). Technical challenges arise because estimating the price adjustment requires reliable data on the GHG content of products imported as well as a clear understanding of the climate policy implications from the countries of imports. Application of pricing tools in industry requires standardisation (benchmarking) of carbon-intensity assessments at products, installations, enterprises, countries, regions, and the global level. The limited number of existing benchmarking systems are not yet harmonised and thus not able to fulfill this function effectively. This limits the scope of products that can potentially be covered by CBAM-type policies (Bashmakov et al. 2021a).

Legal challenges arise because CBAM can be perceived as a protectionist measure violating the principle of non-discrimination under the regulations of the World Trade Organization (WTO). However the absence of GHG prices can also been perceived as a subsidy for fossil fuel-based production (Stiglitz 2006; Al Khourdajie and Finus 2020; Kuusi et al. 2020). Another argument supporting CBAM implementation is the possibility to induce low-GHG investment in non-regulated regions (Cosbey et al. 2019).

Thus far, California is the only jurisdiction that has implemented CBA tariffs applied on electricity imports from neighbouring states and provides insights on how a CBA can work in practice by using 'default' GHG emissions intensity benchmarks (Fowlie et al. 2021). CBAM is an approach likely to be applied first to a few selected energy-intensive industries that are at risk of carbon leakage, as the EU is considering. The implementation of CBA needs to balance applicability versus fairness of treatment. An option recently proposed is an individual adjustment mechanism to give companies exporting to the EU the option to demonstrate their actual carbon intensity (Mehling and Ritz 2020). Any CBAMs will have to comply with multilaterally agreed rules under the WTO Agreements to be implemented.

The adoption of CBAM by different countries may evolve into the formation of a climate club where countries would align on specific elements of climate regulation (e.g., primary iron or clinker intensity) to facilitate implementation and incentivise countries to join (Nordhaus 2015; Hagen and Schneider 2021; Tagliapietra and Wolff 2021a,b). However, not all countries have the same abilities to report, adapt and transition to low-carbon production. The implications of CBAMs on trade relationships should be considered to avoid country divide and separation from a common goal of global decarbonisation (Michaelowa et al. 2019; Kuusi et al. 2020; Banerjee 2021; Eicke et al. 2021; Bashmakov 2021). The globalisation of markets and the fragmentation of supply chains complicates the assignment of responsibility for GHG emissions mitigations related to trade (Jakob et al. 2021). Production-based carbon-price schemes minimise the incentives for downstream carbon abatement due to the imperfect pass through of carbon costs and therefore overlook demand-side solutions such as material efficiency (Skelton and Allwood 2017; Baker 2018). An alternative approach is to set the carbon pricing downstream on the consumption of carbon-intensive materials, whether they are imported or produced locally (Neuhoff et al. 2015, 2019; Munnings et al. 2019). However, implementation of consumption-based GHG pricing is also challenged by the need of product GHG traceability and enforcement transaction costs (Jakob et al. 2014; Munnings et al. 2019). Hybrid approaches are also considered (Neuhoff et al. 2015; Bataille et al. 2018a; Jakob et al. 2021). The efficacy of GHG prices to achieve major industry decarbonisation has been challenged by additional real world implementation problems, such as highly regionally fragmented GHG markets (Boyce 2018; Tvinnereim and Mehling 2018) and the difficult social acceptance of price increases (Bailey et al. 2012; Raymond 2019). The higher GHG prices likely needed to incentivise industry to adopt low-GHG solutions pose social equity issues and resistance (Grainger and Kolstad 2010; Bataille et al. 2018b; Hourcade et al. 2018; Huang et al. 2019b; Wang et al. 2019). GHG pricing is also associated with promoting mainly incremental low-cost options and not investments in radical technical change or the transformation of socio-technical systems (Grubb et al. 2014; Vogt-Schilb et al. 2018; Stiglitz 2019; Rosenbloom et al. 2020). Transparent and strategic management of cap-and-trade proceeds toward inclusive decarbonisation transition that support high abatement cost options can contribute toward easing these shortcomings (Carl and Fedor 2016; Raymond 2019). In California, Senate Bill 535 (De León, Statutes of 2012) require that at least a quarter of the proceeds go to projects that provide a benefit to disadvantaged communities (California Climate Investments 2020).

Clear and firm emission reduction caps towards 2050 are essential for sending strong signals to businesses. However, many researchers recognise that complementary policies must be developed to set current production and consumption patterns toward a path consistent with achieving the Paris Agreement goals as cap-and-trade or carbon taxes are not enough (Schmalensee and Stavins 2017; Vogt-Schilb and Hallegatte 2017; Bataille et al. 2018b; Kirchner et al. 2019). In this broader policy context, proceeds from pricing schemes can be used to support the deployment of options with near-term abatement costs that are too high to be incentivised by the prevailing carbon price, but which show substantial cost-reduction potential with scale and learning, and to ensure a just transition (Wang and Lo 2021).

### 11.6.2 Transition Pathways Planning and Strategies

Decarbonising the industry sector requires transitioning how material and products are produced and used today to development pathways that include the strategies outlined in Sections 11.3 and 11.4 and Figure 11.15. Such broad approaches require the development of transition planning that assesses the impacts of the different strategies and considers local conditions and social challenges that may result from conflicts with established practices and interests, with planning and strategies directly linked to these challenges.

Governments have traditionally used voluntary agreements or mandatory energy or emission reduction targets to achieve emission reduction for specific emission-intensive sectors (e.g., UK Climate Change Agreements; India Performance, Achieve and Trade scheme). Sector visions, roadmaps and pathways combined with a larger context of socio-economic goals, with clear objectives and policy direction, are needed for every industrial sector to achieve decarbonisation and at the time of writing they are emerging for some sectors. Grillitsch et al. (2019b) working from the socio-technical transitions literature, focuses on the need for maintaining 'directionality' for innovation (e.g., towards net zero transformation), the capacity for iterative technological and policy 'experimentation' and learning, 'demand articulation' (e.g., engagement of material efficiency and high value circularity), and 'policy coordination' as four main framing challenges. Wesseling et al. (2017b) bridges from the socio-technical transitions literature to a world more recognisable by executives and engineers, composed of structural components that include actors (e.g., firms, trade associations, government, research organisations, consumers, etc.), institutions (e.g., legal structures, norms, values and formal policies or regulations), technologies (e.g., facilities, infrastructure) and system interactions.

Several studies (Åhman et al. 2017; Bataille et al. 2018a; Material Economics 2019; Wyns et al. 2019) offer detailed transition plans using roughly the same five overarching strategies: (i) policies to encourage material efficiency and high quality circularity; (ii) 'supply push' R&D and early commercialisation as well as 'demand pull' to develop niche markets and help emerging technologies cross 'the valley of death'; (iii) GHG pricing or regulations with competitiveness provisions to trigger innovation and systemic GHG reduction; (iv) long-run, low-cost finance mechanisms to enable investment and reduce risk; (v) infrastructure planning and construction (e.g. $CO_2$ transport and disposal, electricity and hydrogen transmission and storage), and institutional support (e.g., labour market training and transition support; electricity market reform). Wesseling et al. (2017b) and (Bataille et al. 2018a) further add a step to conduct ongoing stakeholder engagements, including stakeholders with effective 'veto' power (i.e., firms, unions, government, communities, indigenous groups), to share and gather information, educate, debate, and build consensus for a robust, politically resilient policy package. This engagement of stakeholders can also bring on new supply chain collaborations and bridge the cost pass-through challenge (e.g., the Swedish HYBRIT steel project, or the ELYSIS consortium, with plans to bring fully commercialised inert electrodes for bauxite electrolysis to market by 2024).

Detailed sectoral roadmaps that assess the technical, economic, social and political opportunities and provide a clear path to low-GHG development are needed to guide policy designs. For example, the German state of North Rhine Westphalia passed a Climate Process Law that resulted in the adoption of a Climate Protection Plan that set subsector targets through a transparent stakeholder engagement process based on scenario development and identification of low-GHG options (Lechtenböhmer et al. 2015), see Box 11.3. Another example is the UK set of Industrial Decarbonisation and Energy Efficiency Roadmaps to 2050 as well as the UK Strategic Growth Plan, which are accompanied by Action Plans for each energy-intensive subsector.



## Box 11.3 | IN4Climate NRW – Initiative for a Climate-friendly Industry in North Rhine-Westphalia (NRW)

IN4Climate NRW (www.in4climate.nrw) was launched in September 2019 by the state government of North Rhine-Westphalia (IN4climate.NRW 2019) as a platform for collaboration between representatives from industry, science and politics. IN4climate.NRW offers a common space to develop innovative strategies for a carbon-neutral industrial sector, bringing together different perspectives and competencies.

North Rhine-Westphalia is Germany's industrial heartland. Around 19% of North Rhine-Westphalia's GHGs have their origin in the industry sector. Consequently, the sector bears a particular responsibility when it comes to climate protection, but the state is also a source of high-quality jobs and export value. The NRW government understands that the state's current competitive advantage can only be maintained if the regional industry positions itself as a front runner for becoming GHG-neutral.

In working together across different branches (more than 30 companies representing mainly steel, cement, chemical, aluminium industry, refineries and energy utilities) and enabling a direct interaction between industry and government officials, IN4Climate provides a benefit to the participating companies. People from the different areas are working together in so-called innovation teams and underlying working groups with a self-organised process of setting their milestones and working schedule while reflecting long-term needs as well as short-term requirements based on political or societal discussions.

The innovation teams aim to identify and set concrete impulses for development and implementation of breakthrough technologies, specify necessary infrastructures (e.g., for hydrogen production, storage and transport) and appropriate policy settings (i.e., integrated state, national and European policy mix). They also include an attempt to create a discourse between the public and the industry sectors as a kind of sounding board for the early detection of barriers and obstacles.

*Box 11.3 (continued)*

The initiative has been successful so far, for example, having developed a clear vision for a hydrogen strategy and an associated policy framework as well as a broader decarbonisation strategy for the whole sector. It is present at the national level as well as at the European level. Being successful and unique, IN4Climate is useful as a blueprint for other regions and is often visited by companies and administration staff from other German states.

It is particularly the so far missing intensive and dedicated cooperation across industrial subsectors that can be seen as a success factor. Facing substantial transformation needs associated with structural changes and infrastructure challenges, very often solutions can't be provided and realised by a single sector but need cooperation and coordination. Even more, chicken-and-egg problems like the construction of new infrastructures (e.g., for hydrogen and $CO_2$ disposal) require cooperation and new modes of collaboration. IN4Climate provides the necessary link for this.

### 11.6.3 Technological Research, Development, and Innovation

Policies for research, development, and innovation (RDI) for industry are present in most countries but it is only recently, and mainly in developed countries, that decarbonisation of emissions-intensive industries has been prioritised (Åhman et al. 2017; Nilsson et al. 2021). Emission-intensive industries are characterised by large dominant actors and mature process technologies with high fixed cost, long payback times and low profit margins on the primary production side of the value chain. Investments in RDI are commonly low and aimed at incremental improvements to processes and products (Wesseling et al. 2017).

#### 11.6.3.1 Applied Research

Investing in RDI for low-GHG process emissions is risky and uncompetitive in the absence of convincing climate policy. Research investment should be guided by assessing options, technology readiness levels, and roadmaps towards technology demonstration and commercialisation. The potential GHG and environmental implications need to be assessed early on to assess the sustainability implications and to direct research needs (Yao and Masanet 2018; Zimmerman et al. 2020). Strategic areas for RDI can be focused on a set of possible process options for producing basic materials using fossil-free energy and feedstock, or CCU and CCS (Sections 11.3.5 and 11.3.6). Policies to enhance RDI include public funding for applied research, technological and business model experimentation, pilot and demonstration projects, as well as support for education and training – which further have the positive side effect of leading to spill-overs and network effects through labour market mobility and collaboration (Nemet et al. 2018). Innovative business models will not emerge if the transition is not considered along the full value chain with a focus on materials efficiency, circularity, and new roles for industry in a transitioning energy system, including possibly providing demand response for electricity through designed-in flexibility, for example, by combining electrolysis hydrogen production with substantial storage (Vogl et al. 2018).

Fostering collaborative innovation across sectors through the support of knowledge sharing and capabilities building is important as mitigation options involve new or stronger sectoral couplings (Tönjes et al. 2020). One example is linking chemicals to forestry in the upscaling of forest bio-refineries, although it has proven to be difficult to engage a diverse group of actors in such collaborations (Karltorp and Sandén 2012; Bauer et al. 2018). Heterogeneous collaboration and knowledge exchange can be encouraged through conscious design of RDI programs and by supporting network initiatives involving diverse actor groups (Van Rijnsoever et al. 2015; Söderholm et al. 2019).

#### 11.6.3.2 Policy Support From Demonstration to Market

Applied research is relatively inexpensive compared to piloting, demonstrations, and early commercialisation, and arguably a lot of it has already been done for the key technologies that need to climb the technology readiness ladder (see Table 11.3). This includes electricity and hydrogen-based processes, electro-thermal technologies, high-temperature heat pumps, catalysis, lightweight building construction, low embodied carbon construction materials, etc. Demonstration to market strategies can be particularly successful when the complete supply chain is considered. A prominent example of such an integrated supply chain approach is the UK Offshore Wind Accelerator Project. Coordinated by the UK Carbon Trust and working with wind turbine manufacturers, the project looked across the potential supply chain for floating offshore wind and identified what components manufacturers could innovate and produce by themselves, and where there were gaps beyond the capability of any one firm. This process led to several key areas of work where the government and firms could work together; once the concepts were piloted and proven, the firms went back into a competitive mode. The project illustrates the potential importance of third parties, including government, in creating platforms and opportunities for cross-industry exchange and collaboration (Tönjes et al. 2020).

11

Pilot and demonstration projects funded through public-private partnerships contributes to risk mitigation for industries and helps inform on the feasibility, performance, costs and environmental impacts of decarbonisation technologies. Most countries already maintain government research and deployment programs. For example, Horizon Europe has a total budget of 95.5 billion EUR (USD117 billion) for 2021–2027, of which 30% will be directed to green technology research. The EU has conducted several demonstration projects for emission-intensive industries, such as the Ultra-Low Carbon Steel (ULCOS) project (Abdul Quader et al. 2016), which led to several small-scale pilots that are now going to larger-scale firm pilots (e.g., HISARNA, HYBRIT and SIDERWIN). Supported by the EU, several cement firms are working together on the cement LEILAC project, where a new form of limestone calciner is being developed to concentrate the process $CO_2$ emerging from quicklime production (about 60% of cement emissions) for eventual utilisation or geological storage (as one of many options for cement, see for example, Plaza et al. 2020). If LEILAC works, it is conceivable that existing cement plants globally that are located near CCS opportunities could have their emissions reduced by 60% with one major retrofit of the kiln.

Once a technology has been demonstrated with scale-up potential, the next stage is commercialisation. This is a very expensive stage, where costs are not yet compensated by revenue (see, e.g., Åhman et al. 2018 and Nemet et al. 2018). The H-DRI, SIDERWIN and LEILAC examples are all at the stage of scaling up. Given the resource requirement, a diversified portfolio of investors and support is required to share the risk. LEILAC includes several firms, as did the UK Offshore Wind Accelerator. Government funds are also required and could be refunded in the future through an equity position, royalty or tax. Fast-growing economies, which are adding new industrial capacity, can provide opportunities to pilot, demonstrate and scale up new technologies, as shown by the rapid expansion of electric vehicle and solar panel production in China, which contributed to driving down costs (Nemet 2019; Hsieh et al. 2020; Jackson et al. 2021).

Finally, large capital flows towards deployment of low-GHG solutions will not materialise without a growing demand for low-carbon materials and products that allows business opportunities. Policy will thus be needed to support the first niche markets which are essential for refining new decarbonised technologies, troubleshooting, and for building manufacturing economies of scale. Market creation does however go beyond the nurturing, shielding, and empowerment of early niches (Smith and Raven 2012; Raven et al. 2016) and must also consider how to significantly reshape existing markets to create space for decarbonised solutions and crowd out fossil-based ones (Mazzucato 2016).

### 11.6.4 Market Pull

The perception of an increasing durable demand for low-GHG products induces manufacturers to invest in decarbonisation strategies (Olatunji et al. 2019). Policies can support and accelerate this process by creating niche markets, stimulating demand for low-carbon products through procurement and financing and by addressing informational and other market barriers.

#### 11.6.4.1 Public Procurement

Governments spend a large portion of their budget on the provision of products and material through infrastructure development, general equipment, and miscellaneous goods. The OECD estimates that an average of 30% of general government expenditure goes to public procurements in OECD countries, representing 12.6% of GDP, which makes government a powerful market actor (OECD 2021). Public procurement can therefore create a significant market pull and be used to pursue strategic environmental goals (Ghisetti 2017). Local, regional and national authorities can use their purchasing power to create niche markets and to guarantee demand for low-GHG products and material (Wesseling and Edquist 2018; Muslemani et al. 2021). In some cases, governments will have to adapt government procurement policies that are not well suited for the procurement of products and services that focus on the decarbonisation benefits and longer-term procurement commitments of emissions-reducing technologies and projects (Ghisetti 2017). Implementation can be challenged by the complexity of criteria, the lack of credible information to check GHG intensities and the added time needed for selection (Geng and Doberstein 2008; Testa et al. 2012; Bratt et al. 2013; Zhu et al. 2013; Cheng et al. 2018; Liu et al. 2019b). To ease these hurdles, the EU commission has developed environmental criteria that can be directly inserted in tender documents (Igarashi et al. 2015; European Commission 2016). These criteria are voluntary, and the extent of their application varies across public authorities (Michelsen and de Boer 2009; Bratt et al. 2013; Testa et al. 2016). In the Netherlands, companies achieving a desirable certification level under the national $CO_2$ Performance Ladder obtain a competitive advantage in public procurement (Rietbergen and Blok 2013; Rietbergen et al. 2015). Globally, many countries have implemented green product procurement or sustainable procurement following Sustainable Development Goal (SDG) 12 – 'Responsible consumption and production' (UNEP 2017). Public procurement is also developing at sub-national levels. For example, the state of California in the United States of America passed the Buy Clean California Act (AB 262) that establishes maximum acceptable global warming potentials for eligible steel and glass construction materials for public procurement (USGBC-LA 2018) (Box 11.4).

## Box 11.4 | Buy Clean California Act

In October 2017, California passed Assembly Bill (AB) 262, the Buy Clean California Act, a new law requiring state-funded building projects to consider the global warming potential (GWP) of certain construction materials during procurement. The goal of AB 262 is to use California's substantial purchasing power to buy low-carbon products. Such low-carbon public procurement will directly reduce emissions by using lower-carbon products, and indirectly by sending a market signal to manufacturers to reduce their emissions in order to stay competitive in California.

The bill requirements are two-pronged: as of January 2020, manufacturers of eligible materials must submit a facility-specific environmental product declaration (EPD), and the eligible materials must demonstrate (through submitted EPDs) GWP below the product-specific compliance limits defined by the state Department of General Services (DGS), which will regulate policy implementation. The eligible materials include structural steel, carbon steel rebar, flat glass, and mineral wool insulation. In January 2021, the DGS published maximum acceptable GWP limits for each product category set at the industry average of facility-specific GWP for each material. Beginning 1 July 2021, awarding authorities were required to verify GWP compliance for all eligible materials (USGBC-LA 2018; DGS 2020).

Prior to adoption of the Buy Clean California Act, the California Department of Transportation (Caltrans) had been evaluating the use of lifecycle assessment and EPDs in evaluating materials. In addition to the materials specified in Buy Clean California Act (noted above), the Caltrans project includes materials used extensively in transportation (concrete, asphalt, and aggregate). Also, the California High-Speed Rail project had begun using EPDs as part of its procurement process. The High-Speed Rail Sustainability Report states that the construction projects will: (i) require EPDs for construction materials including steel products and concrete mix designs, and (ii) require 'optimized lifecycle scores for major materials' and include additional strategies to reduce impacts across the life cycle of the project (Simonen et al. 2019).

Several other states such as Washington, Minnesota, Oregon, Colorado, New York and New Jersey are developing similar types of Buy Clean regulations (Simonen et al. 2019; BGA 2020).



#### 11.6.4.2 Private Procurement

The number of companies producing sustainability reports has increased rapidly over the last decade (Jackson and Belkhir 2018) and so has the number of pledges to carbon neutrality announced. This trend has mainly been driven by consumer concerns, investor requests, and as a business strategy to gain a competitive advantage (Higgins and Coffey 2016; Ibáñez-Forés et al. 2016; Koberg and Longoni 2019). For example, Apple and the governments of Québec and Canada are the financier and lead market maker in the Elysis consortium to bring inert electrodes to market for bauxite smelting to make zero-GHG aluminium. Aluminium is a very small fraction of the cost of a laptop or smartphone, so even expensive low-emissions aluminium adds to Apple's brand at very little cost per unit sold. Some countries are also requiring corporate to report their emissions. For example, the French government requires companies with 500 or more employees and financial institutions to report Corporate Social Responsibility (CSR) and disclose publicly Scope 1 (direct emissions), Scope 2 (indirect emissions from purchased electricity) and Scope 3 (emissions from supply chain impacts and consumer usage and end-of-life recycling practices) emissions (Mason et al. 2016).

The most common climate mitigation strategies used by corporates are to set emissions reduction targets in line with the Paris Agreement goals through science-based targets (SBTs) and to develop internal carbon pricing (Kuo and Chang 2021). The SBT initiative records that 338 SBT companies reduced their emissions by 302 $MtCO_2$-eq between 2015 and 2019 (SBTi 2021). As of August 2021, 858 companies had set SBT and over 2000 companies across the world currently use internal carbon pricing with a median internal carbon price of USD25 per metric tonne of $CO_2$-eq (Bartlett et al. 2021). The most determined companies have developed internal GHG abatement strategies that incorporate their supply chains' emissions (Martí et al. 2015; Gillingham et al. 2017; Tost et al. 2020) and design procurement contracts that encourage or require their suppliers to also improve their product GHG footprint (Liu et al. 2019a). For many corporations, the emissions impact within their supply chain far exceeds their operations direct emissions (CDP 2019). Therefore, the opportunities to reduce emissions through purchasing goods and services from the supply chain (Scope 3) have much greater potentials than from direct emissions.

However, these trends have to be approached with caution as some of the emissions reductions are not direct emissions reductions from companies' operations, instead often from offset projects of varying quality (Chrobak 2021). There is a lack of consistency and comparability in the way firms are reporting emissions, which limits the possibilities to assess companies' actual ambition and progress (Sullivan and Gouldson 2012; Burritt and Schaltegger 2014; Liu et al. 2015; Rietbergen et al. 2015; Blanco et al. 2016). More research is needed to assess the current impacts of corporate voluntary climate actions and if these efforts meet the Paris Agreement's goals (Rietbergen et al. 2015; Wang and Sueyoshi 2018). It will be critically important that the international corporate accounting frameworks, standards, and related guidance (e.g., GHG Protocol) be maintained and improved to reflect evolving needs in the global market and to allow for comparison of objectives and progress.

#### 11.6.4.3 GHG Content Certifications

The development of GHG labels corresponds to a growing demand from consumers desiring information about the climate impacts of their consumption (Darnall et al. 2012; Tan et al. 2014; Feucht and Zander 2018). GHG labels fill this information gap by empowering consumers' purchasing decisions and creating higher value for low-GHG products and materials (Vanclay et al. 2011; Cohen and Vandenbergh 2012). The willingness to pay for lower-GHG products has been found to be positive but to depend on socio-economic consumer characteristics, cultural preferences and the product considered (Shuai et al. 2014; de-Magistris and Gracia 2016; Tait et al. 2016; Li et al. 2017; Feucht and Zander 2018). Companies and governments that favour low-GHG products and who are seeking to achieve environmental, social, and governance (ESG) goals also need readily available and reliable information about the GHG content of products and materials they purchase and produce (Long and Young 2016; Munasinghe et al. 2016).

Numerous methodologies have been developed by public and private organisations to meet the needs for credible and comparable environmental metrics at the product and organisation levels. Most follow lifecycle assessment standards as described in ISO 14040 and ISO 14044, ISO 14067 for climate change footprint only and ISO 14025 (2006) for environmental product declarations (EPD), but the way system boundaries are applied in practice varies (Wu et al. 2014; Liu et al. 2016). Adoption has been challenged by the complexity and the profusion of applications which contribute to confuse stakeholders (Gadema and Oglethorpe 2011; Guenther et al. 2012; Brécard 2014). The options of applying different system boundaries and allocation principles involve value judgements that in turn influence the results (Tanaka 2008; Finnveden et al. 2009; McManus et al. 2015; Overland 2019). A more systematic and coordinated international approach based on transparent and reliable data and methodologies is needed to induce global low-GHG market development (Pandey et al. 2011; Darnall et al. 2012; Tan et al. 2014).

Within the context of GHG content certifications and EPD development, more transparency is needed to increase international comparability and to validate claims to meet consumers demand for low-GHG material and products (Rangelov et al. 2021). Greater automation, publicly available reference databases, benchmarking systems and increased stakeholder collaboration can also support the important role of conveying credible emissions information between producers, traders and consumers.

#### 11.6.4.4 Performance Standards and Codes

Policymakers can set minimum performance standards or maximum emission content specifications through legislation to increase the use of low-GHG materials and products by mandating the adoption of low-GHG production and construction processes while requiring material and resource efficiency aspects.

Construction of buildings represented 11% of energy and process-related $CO_2$ emissions globally in 2018 (IEA and UNEP 2019). The share of embodied emissions in construction is increasing as building energy efficiency is improving and energy supply is decarbonised (Chastas et al. 2016). As a result, jurisdictions are increasingly considering new requirements in building codes to reduce embodied emissions. This is the case of France's new building code which is shifting from a thermal regulation (RT 2012) to an environmental regulation (RE 2020) to include embodied GHG LCA metrics for encouraging use of low-GHG building materials (Ministère de la Transition écologique et solidaire 2018; Schwarz et al. 2020). The 2018 International Green Construction Code (IGCC) provides technical requirements that can be adopted by jurisdictions for encouraging low-GHG building construction, which also covers minimum longevity and durability of structural, building envelope, and hardscape materials (Art. 1001.3.2.3) (Celadyn 2014). Low-GHG building rating systems, such as LEEDs, are voluntary standards which include specific requirements on material resources in their rating scale. Trade-offs between energy performance achievement and material used in building construction needs to be further assessed and considered as low-GHG building code requirements develop. Local governments can also lead the way by adopting standards for construction. This is the case of the county of Marin in California which specifies maximum embodied carbon in $kgCO_2$-eq $m^{-3}$ and maximum ordinary Portland cement content in lbs/yd$^3$ for different levels of concrete compressive strength (Marin County 2021).

Governments are also turning their attention to developing standards to increase the durability of products and materials by requiring options for maintenance, reparability, reusability, upgradability, recyclability and waste handling. For example, the EU Ecodesign Directive includes new requirements for manufacturers to make available for a minimum of seven to 10 years spare parts to repair household equipment (Talens Peiró et al. 2020; Calisto Friant et al. 2021; Nikolaou and Tsagarakis 2021). The European Commission plans to widen the resource efficiency requirements beyond energy-related products to cover products such as textiles and furniture as well as high-impact intermediary products such as steel, cement and chemicals in a new sustainable product policy legislative initiative. (Domenech and Bahn-Walkowiak 2019; Llorente-González and Vence 2019; European Commission 2020; Polverini 2021).

Further research is needed to understand how different international and national frameworks, codes, and standards that focus on emissions can work in unison to amplify their mutually desired outcomes. Building performance and market instrument trading frameworks recognised globally do not always incentivise the same outcomes due to the differences in market approach. LCA metrics are a useful tool to help assess optimal options for ultimate emission reduction objectives (Röck et al. 2020; Shadram et al. 2020).

#### 11.6.4.5 Financial Incentives

Fossil-free basic materials production will often lead to higher costs of production, for example, 20–40% more for steel, 70–115% more for cement, and potentially 15–60% for chemicals (Material Economics 2019). There is a nascent literature on what are effectively material 'feed-in-tariffs' to bridge the commercialisation 'valley of death' (Wilson and Grubler 2011) of early development of low-GHG materials (Bataille et al. 2018a; Neuhoff et al. 2018; Sartor and Bataille 2019; Wyns et al. 2019). Renewable electricity support schemes have typically been price-based (e.g., production subsidies and feed-in-tariffs) or volume-based (e.g., quota obligations and certificate schemes) and both principles can be applied when thinking about low-GHG materials. Auction schemes are typically used for larger-scale projects, for example, offshore wind parks.

Based on how feed-in-tariffs worked, a contract for difference (CfD) could guarantee a minimum and higher-than-market price for a given volume of early low-GHG materials. CfDs could be based on a minimum effective GHG price reflecting parity with the costs of current higher-emitting technologies, or directly on the higher base capital and operating costs for a lower-GHG material (Richstein 2017; Chiappinelli et al. 2019; Sartor and Bataille 2019; Vogl et al. 2021a). CfDs can also be offered through low-GHG material procurement where an agreed price offsets the incremental cost of buying low-GHG content product or material. Private firms, by themselves or collectively, can also guarantee a higher than market price for low-GHG materials from their supplier for marketing purposes (Bataille et al. 2018a; Bataille 2020a). Reverse auctions (by which the lowest bidder gets the production subsidy) for low-GHG materials is also an option but it remains to be analysed and explored. While these financial incentive schemes have been implemented for renewable energy, their application to incentivise and support low-GHG material production have yet to be developed and implemented. The German government is currently developing a draft law which will allow companies that commit to cut GHG emissions by more than half using innovative technologies to bid for 10-year CfDs with a guaranteed price for low-carbon steel, chemical and cement products (Agora Energiewende and Wuppertal Institut 2019; BMU 2021).

New and innovative financial market contracts for basic materials that represent low-carbon varieties of conventional materials are emerging. This is the case of aluminium for which quantity of low-GHG production already exist in countries where hydroelectric power is a common power source. Market developments will allow for low-GHG aluminium to trade at a premium rate as demand develops. For example, Harbor Aluminium has launched a green aluminium spot premium at the end of October 2019 and the London Metal Exchange has introduced a 'green aluminium' spot exchange contract. (LME 2020; Das 2021).

#### 11.6.4.6 Extended Producer Responsibility

Extended producer responsibility (EPR) systems are increasingly used by policymakers to require producers to take responsibility for the end life of their outputs and to cover the cost of recycling of materials or otherwise responsibly managing problematic wastes (Kaza et al. 2018). According to the OECD, there are about 400 EPR systems in operation worldwide, three quarters of which have been established over the last two decades. One third of EPR systems cover small consumer electronic equipment, followed by packaging and tyres (each 17%), vehicles, lead-acid batteries and a range of other products (OECD 2016).

While the economic value of some discarded materials such as steel, paper and aluminium is generally high enough to justify the cost and efforts of recycling, at current rates of 85%, above 60%, and 43%, respectively (Graedel et al. 2011; Cullen and Allwood 2013), others like plastic or concrete have a much lower re-circularity value (Graedel et al. 2011). Most plastic waste ends up in landfills or dumped in the environment, with 9% recycled and 12% incinerated globally (Geyer et al. 2017; UNEP 2018). Collected waste plastics from OECD countries were largely exported to China until a ban in 2018 required OECD countries to review their practices (Qu et al. 2019). EPR schemes may thus need to be strengthened to actually achieve a reduced use of virgin GHG-intensive materials. The potential for re-circularity of unreacted cement and aggregates in concrete is increasing as new standards and requirement develops. For example, concrete fines are now standardised as a new cement constituent in the European standardisation CEN/TC 51 – 'cements and construction limes'.

### Box 11.5 | Circular Economy Policy

The implementation of a circular economy relies on the operationalisation of the R-imperatives or strategies which extend from the original 3Rs: Reduce, Reuse and Recycle, with the addition of Refuse, Reduce, Resell/Reuse, Repair, Refurbish, Remanufacture, Repurpose, Recycle, Recover (energy), Re-mine and more (Reike et al. 2018). The R implementation strategies are diverse across countries (Ghisellini et al. 2016; Kalmykova et al. 2018) but, in practice, the lower forms of retention of materials, such as recycling and recover (energy), often dominate. The lack of policies for higher retention of material use such as Reduce, Reuse, Repair and Remanufacture is due to institutional failures, lack of coordination and lack of strong advocates (Gonzalez Hernandez et al. 2018a).

Policies addressing market barriers to circular business development need to demonstrate that circular products meet quality performance standards, ensure that the full environmental costs are reflected in market prices and foster market opportunities for circular products exchange, notably through industrial symbiosis clusters and trading platforms (Kirchherr et al. 2018; OECD 2019a; Hartley et al. 2020; Hertwich 2020). Policy levels span from micro (such as consumer or company) to meso (eco-industrial parks) and macro (provinces, regions and cities) (Geng et al. 2019). The creation of eco-industry parks ('industrial clusters') has been encouraged by governments to facilitate waste exchanges between facilities, where by-products from one industry are used as a feedstock to

*Box 11.5 (continued)*

another (Ding and Hua 2012; Jiao and Boons 2014; Shi and Yu 2014; Tian et al. 2014; Winans et al. 2017). Systematic assessment of wastes and resources is carried out to assess possible exchange between different supply chains and identify synergies of waste streams that include metal scraps, waste plastics, water heat, bagasse, paper, wood scraps, ash, sludge and others (Ding and Hua 2012; Shi and Yu 2014).

The development of data collection and indicators is nascent and need to ramp up to quantify the impacts and provide evidence to improve circular economy and materials efficiency policies. Policymakers need to leverage the potential socio-economic opportunities of transitioning to circular economies (Llorente-González and Vence 2020), which shows positive GDP growth and job creation by shifting to more labour-intensive recycling plants and repair services than resource-extraction activities (WRAP and Alliance Green 2015; Cambridge Econometrics et al. 2018). The International Labour Organization estimates that worldwide employment would grow by 0.1% by 2030 under a circular economy scenario (ILO 2018). However questions remain if the type of jobs created are concentrated in low-wage labour-intensive circular activities which may need targeted policy instruments to improve working conditions (Llorente-González and Vence 2020).

### 11.6.5 Knowledge and Capacity

It is important that government bodies, academia and other actors strengthen their knowledge and capacities for the broad transformational changes envisioned for industry. In Japan, industry has been voluntarily working on GHG reduction, under the Framework of Keidanren's Commitment to a Low-carbon Society since 2009. Government and scientific experts regularly review their commitments and discuss results, monitoring methods, and reconsidering goals. Industry federations/associations can obtain advice in the follow-up meetings from other industries and academics. The energy and transport sectors have decades of building institutions and expertise, whereas industrial decarbonisation is largely a new policy domain. Most countries have experience in energy efficiency policies, some areas of research and innovation, waste management, regulations for operational permits and pollution control, worker safety and perhaps fuel switching. There is less experience with market demand pull policies although low-GHG public procurement is increasingly being tested. Circular economy policies are evolving but potential policies for managing material demand growth are less understood. Material efficiency policies through, for example, product standards or regulation against planned obsolescence are nascent but relatively unexplored (Gonzalez Hernandez et al. 2018a).

All this argues for active co-oversight, management and assessment by government, firms, sector associations and other actors, in effect the formation of an active industrial policy that includes decarbonisation in its broader mandate of economic and social development (OECD 2019b; Bataille 2020a). This could draw from the quadruple helix innovation model, which considers the role of government, universities, the private sector, the natural environment and social systems to foster collaboration in innovation (Carayannis and Campbell 2019; Durán-Romero et al. 2020). Important aspects of governance include mechanisms for monitoring, transparency, and accountability. It may involve the development of new evaluation approaches, including a greater focus on *ex ante* evaluations and assessment of, for example, readiness and capacities, rather than *ex post* evaluations of outcomes. Such organisational routines for learning have been identified as a key aspect of policy capacity to govern evolutionary processes (Karo and Kattel 2018; Kattel and Mazzucato 2018). Although many governments have adopted ideas of focusing resources on the mission or challenge of climate change mitigation, comparisons between Western and East Asian contexts show significant differences in the implementation of governance structures (Karo 2018; Mazzucato et al. 2020; Wanzenböck et al. 2020). Overall, improved knowledge and stronger expertise is important also to handle information asymmetries and the risk of regulatory capture.



### 11.6.6 Policy Coherence and integration

Industrial net zero transitions, while technically feasible, involve not just a shift in production technology but major shifts in demand, material efficiency, circularity, supply chain structure and geographic location, labour training and adaptation, finance, and industrial policy. This transition must also link decarbonisation to larger environmental and social goals (e.g,. air and water quality, low-GHG growth, poverty alleviation, sustainable development goals) (OECD 2019b).

Although there is little evidence of carbon leakage so far it will be ever more important to strive for coherence in climate and trade policies as some countries take the lead in decarbonising internationally traded basic materials (Jakob 2021b). At the time of writing the previously academic debate on this issue is shifting to real policymaking through debates and negotiations around carbon border adjustment (Section 11.6.1) and sectoral agreements or climate clubs (Nordhaus 2015; Åhman et al. 2017; Jakob 2021a; Nilsson et al. 2021). The climate and trade policy integration should also consider what is sometimes called positive leakage, that is that heavy industry production moves to where it is easier to reach zero emissions. As a result, policy should go beyond border measures to include, for example, international technology cooperation and transfer and development of shared lead markets.

Energy-intensive production steps may move where clean resources are most abundant and relatively inexpensive (Gielen et al. 2020; Bataille et al. 2021a). For example, steel-making has historically located itself near iron ore and coal resources whereas in the future it may be located near iron ore and zero-GHG electricity or close to carbon storage sites (Fischedick et al. 2014b; Vogl et al. 2018; Bataille 2020a). This indicates large changes in industrial and supply chain structure, with directly associated needs for employment and skills. Some sectors will grow, and some will shrink, with differing skill needs. Each new workforce cohort needs the general specific skill to provide the employment that is needed at each stage in the transition, implicating a need for coordination with policies for education and retraining.

Depending on what mixes of deep decarbonisation strategies are followed in a given region (e.g., material efficiency, electrification, hydrogen, biomass, CCU and CCS), infrastructure will need to be planned, financed and constructed. The UKCCC Net Zero Technical Report describes the infrastructure needs for achieving net zero GHG in the UK by 2050 for every sector of the economy (UKCCC 2019b). Transportation would be facilitated with pipelines or ships to allow transfer of captured $CO_2$ for utilisation and disposal, and associated institutional frameworks (IEAGHG 2021). Electrification will require market design and transmission to support increased generation, transmission, and flexible demand. Hydrogen, CCU, and CCS will require significant new or adapted infrastructure. Hydrogen and $CO_2$ pipelines, and expanded electricity transmission, have natural monopoly characteristics which are normally governed and planned by national and regional grid operators and their regulators. Industrial clustering (also known as eco-parks), such as those planned in Rotterdam (Netherlands) and Teeside (UK), would allow more physical and cost-effective sharing of electricity, CCU, CCS, and hydrogen infrastructure but is dependent on physical planning, permitting, and infrastructure policies.

Costing analysis (Chapter 15) indicates an increased upfront need for financial capital which requires policies to encourage long-term, patient capital that reflects society's preferences for investment in industrial decarbonisation and the minimum 10 or more years horizon before there are significant new commercially available processes.

All the above indicate the need for general industrial policy as part of a coherent general economic, taxation, investment, employment and social policy for climate change mitigation (Wesseling et al. 2017; Bataille et al. 2018a; Wyns et al. 2019; Nilsson et al. 2021).

### 11.6.7 Roles and Responsibilities

While all climate policy requires topic-specific adaptive governance for long-term effectiveness (Mathy et al. 2016), deep decarbonisation of heavy industry has special governance challenges, different from those for the electricity, transport or buildings sectors (Åhman et al. 2017; Wesseling et al. 2017; Bataille et al. 2018a). Competition is strong, investments are rare, capital intensive and very 'lumpy'. In an atmosphere where transformative innovation is required the process is very capital-focused with non-diversifiable risks unless several companies are involved. There are significant infrastructure needs for electricity, hydrogen, and CCS and CCU. Given there is no 'natural' market for low-emissions materials, there is a need to manage both the supply and demand sides of the market, especially in early phase through lead supplier and markets. Finally, there is a very high probability of surprises and substantial learning, which could affect policy choice, direction, and stringency.

Different types of actors thus have to play different but coordinated roles and responsibilities in developing, supporting, and implementing policies for an industrial transition. Table 11.6 below shows how the different core parts of integrated policymaking for an industrial transition may depend on efforts from different actors groups and highlights the responsibility of these actor groups in developing a progressive and enabling policy context for the transition. This includes policymakers at local, national, and international arenas as well as civil society organisations, industry firms, and interest organisations.

**Table 11.6 | Examples of the potential roles of different actors in key policy and governance areas for a low-GHG transition to indicate the importance of agency and wide stakeholder engagement in the governance of industrial decarbonisation.**

| Actors | Direction: planning and strategising pathways to net zero | Innovation: RD&D for new technologies and other solutions | Market creation: create and shape demand-pull for various solutions | Knowledge and capacity: build institutional capacity across various actors | Coherence: establish international and national policy coherence |
|---|---|---|---|---|---|
| International bodies and multilateral collaboration | More attention to industry in NDCs. Monitor progress and identify gaps. Develop international roadmaps. | Include heavy industry decarbonisation in technology cooperation (e.g., Mission Innovation). | International standards, benchmarking systems, and GHG labels. Allow for creation and protection of lead markets. | Support knowledge building and sharing on industrial decarbonisation. | Align other conventions and arenas (e.g., WTO) with climate targets and include heavy industry transitions in negotiations. |
| Regional and national government and cities | Require net zero strategies in permitting. Set targets and facilitate roadmaps at various levels. Sunset clauses and phase-out agreements for polluting plants. | Experimentation for recycling, materials efficiency, and demand management. Hydrogen, electrification, and other infrastructure. | Public procurement for innovation and lead markets. Green infrastructure investments. | Develop policy expertise for industrial transformation. Support and facilitate material efficiency and circular solutions through design standards, building codes, recycling, and waste policy. | Support vertical policy coherence (i.e., international, national, city level). |

| Actors | Direction: planning and strategising pathways to net zero | Innovation: RD&D for new technologies and other solutions | Market creation: create and shape demand-pull for various solutions | Knowledge and capacity: build institutional capacity across various actors | Coherence: establish international and national policy coherence |
|---|---|---|---|---|---|
| Civil society | Monitor and evaluate leaders and laggards. Support transparency. | Engage in responsible innovation programs, experimentation, and social innovation. | Progressive labelling, standards and criteria for low emissions materials and products (e.g., LCA-based), including updating. | Engage in policy processes and build capacity on industrial decarbonisation. Support consumer information and knowledge. | Monitor and support policy coherence and coordination across policy domains (trade, climate, waste, etc.). |
| Industrial sectors and associations | Adopt net zero emissions targets, roadmaps, and policy strategies for reaching them. Assess whole value chains, scope 3 emissions and new business models. | Share best practice. Coordination and collaboration. Efficient markets for new technology (e.g., licensing). | Work across (new) value chains to establish lead markets for low emissions materials as well as for materials efficiency and circularity. | Education and retraining for designers, engineers, architects, etc. Information sharing and transparency to reduce information asymmetry. | Coordination across policy domains (trade, climate, waste, etc.). Explore sectoral couplings, new value chains and location of heavy industry. |
| Corporations and companies | Set zero emissions targets and develop corporate- and plant-level roadmaps for reaching targets. | Lead and participate in R&D, pilots, and demonstrations. Increase and direct R&D efforts at reaching net zero. | Marketing and procurement of low-emissions materials and products. Include Scope 3 emissions to assess impact and mitigation strategies. | Engage in value chains for increased recycling and materials efficiency. Build knowledge and capacity for reorientation and transformation. | MNCs avoid race to the bottom, and strategically account for high carbon price as part of transition strategy. |

## 11.7 Knowledge Gaps

An increasing body of research proposes deep decarbonisation pathways for energy-intensive industries including mitigation options such as materials efficiency, circular economy and new primary processes. These options are under-represented in climate change scenario modelling and integrated assessment models, some of which do not even reflect evolution of demand for basic materials, which is a key driver behind energy consumption and GHG emissions in the industrial sector. As a result, no agreement is reached so far between bottom-up and top-down studies on the effectiveness and costs for many promising mitigation options, their respective roles, sequencing and packaging within various mitigation pathways.

A significant shift is needed from the transition process of the past mainly based on marginal and incremental changes, with a strong focus on energy efficiency efforts, to one grounded in transformational change where there is limited knowledge of how to implement such change effectively.

There is a knowledge gap on comparable, comprehensive, and detailed quantitative information on costs and potentials associated with the mitigation options for deep decarbonisation in industry, as cost estimates are not often comparable due to the regional or country focus, differences in costs metrics, currencies, discount rates, and energy prices across studies and regions.

A very large and important uncertainty is the availability of biomass for deep decarbonisation pathways due to competition for biomass feedstock with other priorities and the extent to which electrification can reduce the demand for bioenergy in the industry, transport and energy sectors.

CCS and CCU are important mitigation options in industry, for which the potentials and costs vary considerably depending on the diversity of industrial processes, the volume and purity of carbon dioxide flows, the energy requirements, the lifetime of utilisation products and the production route.

The effectiveness of mitigation policies in industry is poorly known, as so far the sector has largely been sheltered from the impacts of climate policy due to the concerns of competitiveness and carbon leakage. There is a lack of integration of material efficiency and circularity with energy and climate policies which partly results from the inadequacy of monitored indicators to inform policy debates and set targets, a lack of high-level political focus, a history of strong industrial lobbying, uncoordinated policy across subsectors and institutions, and the sequential nature of decision-making along supply chains.

Industry as a whole is a very complex web of sectors, subsectors and inter-sectoral interactions and dependence, with diverse associated mitigation opportunities and co-benefits and costs. Additional knowledge is needed to understand sectoral interactions in the transformation processes.

Industrial climate mitigation policy is supplemental to many other policy instruments developed to reach multiple industrial goals, for the range of stakeholders with their interest and priorities reflecting the assessment of co-benefits and risk and affecting decision-making processes and behaviour of stakeholders. Better knowledge is needed to identify the co-benefits for the adoption of climate change mitigation strategies.

Frequently Asked Questions (FAQs)

## FAQ 11.1 | What are the key options to reduce industrial emissions?

Industry has a diverse set of greenhouse gas (GHG) emission sources across subsectors. To decarbonise industry requires that we pursue several options simultaneously. These include energy efficiency, materials demand management, improving materials efficiency, more circular material flows, electrification, as well as carbon capture and utilisation (CCU) and carbon capture and storage (CCS). Improved materials efficiency and recycling reduces the need for primary resource extraction and the energy-intensive primary processing steps. Future recycling may include chemical recycling of plastics if quality requirements make mechanical recycling difficult. One approach, albeit energy intensive, is to break down waste plastics to produce new monomer building blocks, potentially based on biogenic carbon and hydrogen instead of fossil feedstock. Hydrogen can also be used as a reduction agent instead of coke and coal in ironmaking. Process emissions from cement production can be captured and stored or used as feedstock for chemicals and materials. Electricity and hydrogen needs can be very large but the potential for renewable electricity, possibly in combination with other low carbon options, is not a limiting factor.

## FAQ 11.2 | How costly is industrial decarbonisation and will there be synergies or conflicts with sustainable development?

In most cases and in early stages of deployment, decarbonisation through electrification or CCS will make the primary production of basic materials such as cement, steel, or polyethylene more expensive. However, demand management, energy and materials efficiency, and more circular material flows can dampen the effect of such cost increases. In addition, the cost of energy-intensive materials is typically a very small part of the total price of products, such as an appliance, a bottle of soda or a building, so the effect on consumers is very small. Getting actors to pay more for zero-emission materials is a challenge in supply chains with a strong focus on competitiveness and cutting costs, but it is not a significant problem for the broader economy. Reduced demand for services such as square metres of living space or kilometres of car travel is an option where material living standards are already high. If material living standards are very low, increased material use is often needed for more sustainable development. The options of materials and energy efficiency, and more circular material flows, generally have synergies with sustainable development. Increased use of electricity, hydrogen, CCU and CCS may have both positive and negative implications for sustainable development and thus require careful assessment and implementation for different contexts.

## FAQ 11.3 | What needs to happen for a low-carbon industry transition?

Broad and sequential policy strategies for industrial development and decarbonisation that pursue several mitigation options at the same time are more likely to result in resource-efficient and cost-effective emission reductions. Industrial decarbonisation is a relatively new field and thus building capacity for industrial transition governance is motivated. For example, policy to support materials efficiency or fundamental technology shifts in primary processes is less developed than energy efficiency policy and carbon pricing. Based on shared visions or pathways for a zero-emission industry, industrial policy needs to support development of new technologies and solutions as well as market creation for low- and zero-emission materials and products. This implies coordination across several policy domains including research and innovation, waste and recycling, product standards, digitalisation, taxes, regional development, infrastructure, public procurement, permit procedures and more to make the transition to a carbon neutral industry. International competition means that trade rules must be evolved to not conflict with industrial decarbonisation. Some local and regional economies may be disadvantaged from the transition which can motivate re-education and other support.



11

## References


Abbas, H.F. and Wan Daud, W.M.A., 2010: Hydrogen production by methane decomposition: A review. *Int. J. Hydrogen Energy*, **35(3)**, 1160–1190, doi:10.1016/j.ijhydene.2009.11.036.

Abdelaziz, O.Y. et al., 2016: Biological valorization of low molecular weight lignin. *Biotechnol. Adv.*, **34(8)**, 1318–1346, doi:10.1016/j.biotechadv.2016.10.001.

Abdul Quader, M.,S. Ahmed, S.Z. Dawal, and Y. Nukman, 2016: Present needs, recent progress and future trends of energy-efficient Ultra-Low Carbon Dioxide ($CO_2$) Steelmaking (ULCOS) program. *Renew. Sustain. Energy Rev.*, **55**, 537–549, doi:10.1016/j.rser.2015.10.101.

Acworth, W., C. Kardish, and K. Kellner, 2020: *Carbon Leakage and Deep Decarbonization: Future-proofing. Carbon Leakage Protection*. ICAP, Berlin, Germany, 76 pp.

Agora Energiewende, and Wuppertal Institut, 2019: *Climate-Neutral Industry. Key technologies and policy options for steel, chemicals and cement*. Executive Summary.

Åhman, M., L.J. Nilsson, and B. Johansson, 2017: Global climate policy and deep decarbonization of energy-intensive industries. *Clim. Policy*, **17(5)**, 634–649, doi:10.1080/14693062.2016.1167009.

Åhman, M., J.B. Skjaerseth, and P.O. Eikeland, 2018: Demonstrating climate mitigation technologies: An early assessment of the NER 300 programme. *Energy Policy*, **117**, 100–107, doi:10.1016/j.enpol.2018.02.032.

Aiginger, K., 2014: *Industrial Policy for a sustainable growth path*. Vienna, Austria, 33 pp. https://www.oecd.org/economy/Industrial-Policy-for-a-sustainable-growth-path.pdf.

Al Khourdajie, A. and M. Finus, 2020: Measures to enhance the effectiveness of international climate agreements: The case of border carbon adjustments. *Eur. Econ. Rev.*, **124**, 103405, doi:10.1016/j.euroecorev.2020.103405.

Allan, G., P. McGregor, and K. Swales, 2017: Greening regional development: employment in low-carbon and renewable energy activities. *Reg. Stud.*, **51(8)**, 1270–1280, doi:10.1080/00343404.2016.1205184.

Allwood, J.M., 2018: Unrealistic techno-optimism is holding back progress on resource efficiency. *Nat. Mater.*, **17(12)**, 1050–1051, doi:10.1038/s41563-018-0229-8.

Allwood, J.M., M.F. Ashby, T.G. Gutowski, and E. Worrell, 2011: Material efficiency: A white paper. *Resour. Conserv. Recycl.*, **55(3)**, 362–381, doi:10.1016/j.resconrec.2010.11.002.

Allwood, J.M. et al., 2012: *Sustainable materials: With both eyes open*. 1st ed. UIT Cambridge, UK, 410 pp.

Altenburg, T. and C. Assman, 2017: *Green Industrial Policy: Concepts, Policies, Country Experiences*. Geneva, Switerland, Bonn, Germany, 221 pp. https://www.unido.org/sites/default/files/files/2017-12/green_industrial_policy_book.pdf.

Alvarez, R.A., S.W. Pacala, J.J. Winebrake, W.L. Chameides, and S.P. Hamburg, 2012: Greater focus needed on methane leakage from natural gas infrastructure. *Proc. Natl. Acad. Sci.*, **109(17)**, 6435–6440, doi:10.1073/pnas.1202407109.

Alvarez, R.A. et al., 2018: Assessment of methane emissions from the U.S. oil and gas supply chain. *Science*, **361(6398)**, pp 186–188, doi:10.1126/science.aar7204.

Amin, A.M., E. Croiset, and W. Epling, 2011: Review of methane catalytic cracking for hydrogen production. *Int. J. Hydrogen Energy*, **36(4)**, 2904–2935, doi:10.1016/j.ijhydene.2010.11.035.

Ampelli, C., S. Perathoner, and G. Centi, 2015: $CO_2$ utilization: an enabling element to move to a resource- and energy-efficient chemical and fuel production. *Philos. Trans. R. Soc. A Math. Phys. Eng. Sci.*, **373(2037)**, doi:10.1098/rsta.2014.0177.

Andersson, R., H. Stripple, T. Gustafsson, and C. Ljungkrantz, 2019: Carbonation as a method to improve climate performance for cement based material. *Cem. Concr. Res.*, **124**(July), 105819, doi:10.1016/j.cemconres.2019.105819.

Andreoni, A. and H.J. Chang, 2019: The political economy of industrial policy: Structural interdependencies, policy alignment and conflict management. *Struct. Change Econ. Dyn.*, **48**, 136–150, doi:10.1016/j.strueco.2018.10.007.

Andrew, R.M., 2019: Global $CO_2$ emissions from cement production, 1928–2018. *Earth Syst. Sci. Data*, **11(4)**, 1675–1710, doi:10.5194/essd-11-1675-2019.

Antikainen, M. and K. Valkokari, 2016: A Framework for Sustainable Circular Business Model Innovation. *Technol. Innov. Manag. Rev.*, **6(7)**, 5–12, doi:10.22215/timreview/1000.

Arens, M., E. Worrell, W. Eichhammer, A. Hasanbeigi, and Q. Zhang, 2017: Pathways to a low-carbon iron and steel industry in the medium-term – the case of Germany. *J. Clean. Prod.*, **163**, 84–98, doi:10.1016/j.jclepro.2015.12.097.

Armijo, J. and C. Philibert, 2020: Flexible production of green hydrogen and ammonia from variable solar and wind energy: Case study of Chile and Argentina. *Int. J. Hydrogen Energy*, **45(3)**, 1541–1558, doi:10.1016/j.ijhydene.2019.11.028.

Arpagaus, C., F. Bless, M. Uhlmann, J. Schiffmann, and S.S. Bertsch, 2018: High temperature heat pumps: Market overview, state of the art, research status, refrigerants, and application potentials. *Energy*, **152**, 985–1010, doi:10.1016/j.energy.2018.03.166.

Arto, I. and E. Dietzenbacher, 2014: Drivers of the growth in global greenhouse gas emissions. *Environ. Sci. Technol.*, **48(10)**, 5388–5394, doi:10.1021/es5005347.

Artz, J. et al., 2018: Sustainable Conversion of Carbon Dioxide: An Integrated Review of Catalysis and Life Cycle Assessment. *Chem. Rev.*, **118(2)**, 434–504, doi:10.1021/acs.chemrev.7b00435.

Ashby, M., 2012: *Materials and the Environment: Eco-informed Material Choice:* 2nd ed. Elsevier/Butterworth-Heinemann, Amsterdam, the Netherlands, Boston, USA 616 pp.

Ashik, U.P.M., W.M.A. Wan Daud, and H.F. Abbas, 2015: Production of greenhouse gas free hydrogen by thermocatalytic decomposition of methane – A review. *Renew. Sustain. Energy Rev.*, **44**, 221–256, doi:10.1016/j.rser.2014.12.025.

Axelson, M., I. Robson, G. Khandekar, and T. Wyns, 2018: *Breaking through: Industrial Low-CO2 Technologies on the Horizon*. Brussels, Belgium, 92 pp. www.ies.be/Breaking-Through_Report_13072018.

Azevedo, J.M.C., A. CabreraSerrenho, and J.M. Allwood, 2018: Energy and material efficiency of steel powder metallurgy. *Powder Technol.*, **328**, 329–336, doi:10.1016/j.powtec.2018.01.009.

Bachmann, M. et al., 2021: Renewable carbon feedstock for polymers: Environmental benefits from synergistic use of biomass and CO2. *Faraday Discuss.*, **230(0)**, 227–246, doi:10.1039/d0fd00134a.

Bailey, I., I. MacGill, R. Passey, and H. Compston, 2012: The fall (and rise) of carbon pricing in Australia: A political strategy analysis of the carbon pollution reduction scheme. *Env. Polit.*, 21(5), 691–711, doi:10.1080/09644016.2012.705066.

Baker, L., 2018: Of embodied emissions and inequality: Rethinking energy consumption. *Energy Res. Soc. Sci.*, **36**, 52–60, doi:10.1016/j.erss.2017.09.027.

Bamigbetan, O., T.M. Eikevik, P. Nekså, and M. Bantle, 2017: Review of vapour compression heat pumps for high temperature heating using natural working fluids. *Int. J. Refrig.*, **80**, 197–211, doi:10.1016/j.ijrefrig.2017.04.021.

Banerjee, S., 2021: Conjugation of border and domestic carbon adjustment and implications under production and consumption-based accounting of India's National Emission Inventory: A recursive dynamic CGE analysis. *Struct. Change Econ. Dyn.*, **57**, 68–86, doi:10.1016/j.strueco.2021.01.007.

Barrett, J. et al., 2013: Consumption-based GHG emission accounting: a UK case study. *Clim. Policy*, **13(4)**, 451–470, doi:10.1080/14693062.2013.788858.

Bartlett, N., T. Coleman, and S. Schmidt, 2021: *Putting a Price on Carbon. The state of internal carbon pricing by corporates globally*. 23 pp. CDP, London, UK.

Bashmakov, I. and A. Myshak, 2018: "Minus 1" and Energy Costs Constants: Sectorial Implications. *J. Energy*, **2018**, 1–24, doi:10.1155/2018/8962437.

Bashmakov, I. et al., 2021a: *CBAM: Implications for the Russian economy Center for Energy Efficiency - XXI Moscow*. Moscow, Russia, 22 pp.

Bashmakov, I.A., 2019: Energy efficiency and economic growth. *Vopr. Ekon.*, **2019(10)**, 32–63, doi:10.32609/0042-8736-2019-10-32-63.

Bashmakov, I.A., 2021: Greenhouse gas emissions caused by global steel industry: the past, the present and the future. *Ferr. Metall. Bull. Sci., Tech. Econ. Inf.*, **77(8)**, 882–901, doi:10.32339/0135-5910-2021-9-1071-1086.

Bashmakov, I.A., D.O. Skobelev, K.B. Borisov, and T.V. Guseva, 2021b: Benchmarking systems for greenhouse gases specific emissions in steel industry. *Ferr. Metall. Bull. Sci., Tech. Econ. Inf.*, **77**(9), 1071–1086, doi:10.32339/0135-5910-2021-9-1071-1086.

Bataille, C., 2020a: Physical and policy pathways to net-zero emissions industry. *WIRES Wiley Interdiscip. Rev.*, **e633(2)**, doi:10.1002/wcc.633.

Bataille, C., 2020b: *Low and zero emissions in the steel and cement industries: Barriers, technologies and policies*. 42 pp. OECD Green Growth Papers, No. 2020/02, OECD Publishing, Paris.

Bataille, C. and N. Melton, 2017: Energy efficiency and economic growth: A retrospective CGE analysis for Canada from 2002 to 2012. *Energy Econ.*, **64**, 118–130, doi:10.1016/j.eneco.2017.03.008.

Bataille, C., and S. Stiebert, 2018: *The transition toward very low carbon heavy industry in the Canadian context: Detailed technical and policy analysis and recommendations for the iron & steel, chemicals, forestry products & packaging, and base metal mining & processing sectors. Phase II of the Canadian Heavy Industry Deep Decarbonization Project. The Industrial Gas Users Association (IGUA)*. 84 pp.

Bataille, C. et al., 2018a: A review of technology and policy deep decarbonization pathway options for making energy intensive industry production consistent with the Paris Agreement. *J. Clean. Prod.*, **187**, 960–973, doi:10.1016/j.jclepro.2018.03.107.

Bataille, C., C. Guivarch, S. Hallegatte, J. Rogelj, and H. Waisman, 2018b: Carbon prices across countries. *Nat. Clim. Change*, **8(8)**, 648–650, doi:10.1038/s41558-018-0239-1.

Bataille, C., L. Nilsson, and F. Jotzo, 2021a: Industry in a net-zero emissions world – uprooting of supply chains, broader policy thinking, and how to model it all. *Energy Strateg. Rev.* (in press)

Bataille, C., S. Steibert, and F.G.N. Li, 2021b: *Global facility level net-zero steel pathways: technical report on the first scenarios of the Net-zero Steel Project*, Paris, France, 34 pp.

Bauer, F., 2018: Narratives of biorefinery innovation for the bioeconomy: Conflict, consensus or confusion? *Environ. Innov. Soc. Transitions*, **28**, 96–107, doi:10.1016/j.eist.2018.01.005.

Bauer, F. and L. Fuenfschilling, 2019: Local initiatives and global regimes – Multi-scalar transition dynamics in the chemical industry. *J. Clean. Prod.*, **216**, 172–183, doi:10.1016/j.jclepro.2019.01.140.

Bauer, F. and G. Fontenit, 2021: Plastic dinosaurs – Digging deep into the accelerating carbon lock-in of plastics. *Energy Policy*, **156**, 112418, doi:10.1016/j.enpol.2021.112418.

Bauer, F., L. Coenen, T. Hansen, K. McCormick, and Y.V. Palgan, 2017: Technological innovation systems for biorefineries: a review of the literature. *Biofuels, Bioprod. Biorefining*, **11**(3), 534–548, doi:10.1002/bbb.1767.

Bauer, F., T. Hansen and H. Hellsmark, 2018: Innovation in the bioeconomy– dynamics of biorefinery innovation networks. *Technol. Anal. Strateg. Manag.*, **30(8)**, 935–947, doi:10.1080/09537325.2018.1425386.

Bayer, P. and M. Aklin, 2020: The European Union Emissions Trading System reduced $CO_2$ emissions despite low prices. *Proc. Natl. Acad. Sci. U. S. A.*, **117(16)**, 8804–8812, doi:10.1073/pnas.1918128117.

Bazzanella, A. M. and F. Ausfelder, 2017: *Technology study: Low carbon energy and feedstock for the European Chemical Industry*. 166 pp. DECHEMA Gesellschaft für Chemische Technik und Biotechnologie e.V. https://dechema.de/dechema_media/Technology_study_Low_carbon_energy_and_feedstock_for_the_European_chemical_industry-p-20002750.pdf.

Bergstrom, J.C. and D. Ty, 2017: Economics of Carbon Capture and Storage. In: *Recent Advances in Carbon Capture and Storage*, InTech. Recent Advances in Carbon Capture and Storage, DOI10.5772/62966, 266 pp.

Bezner Kerr, R., T. Hasegawa, R. Lasco, I. Bhatt, D. Deryng, A. Farrell, H. Gurney-Smith, H. Ju, S. Lluch-Cota, F. Meza, G. Nelson, H. Neufeldt, and P. Thornton, 2021: Food, fibre, and other ecosystem products. In: *Climate Change 2022: Impacts, Adaptation and Vulnerability of Working Group II to the Sixth Assessment Report of the Intergovernmental Panel on Climate Change* [H.-O. Pörtner, D.C. Roberts, M. Tignor, E.S. Poloczanska, K. Mintenbeck, A. Alegría, M. Craig, S. Langsdorf, S. Löschke, V. Möller, A. Okem, B. Rama (eds.)]. Cambridge University Press, Cambridge, UK and New York, NY, USA. In press.

BGA, 2020: BlueGreen Alliance | Buy Clean. https://www.bluegreenalliance.org/work-issue/buy-clean/ (Accessed November 29, 2020).

Bicer, Y. and I. Dincer, 2017: Life cycle assessment of nuclear-based hydrogen and ammonia production options:A comparative evaluation. *Int. J. Hydrogen Energy*, **42(33)**, 21559–21570, doi:10.1016/j.ijhydene.2017.02.002.

Biel, K. and C.H. Glock, 2016: Systematic literature review of decision support models for energy-efficient production planning. *Comput. Ind. Eng.*, **101**, 243–259, doi:10.1016/j.cie.2016.08.021.

Binder, M., and A.K. Blankenberg, 2017: Green lifestyles and subjective well-being: More about self-image than actual behavior? *J. Econ. Behav. Organ.*, **137**, 304–323, doi:10.1016/j.jebo.2017.03.009.

BIR, 2020: *World Steel Recycling in Figures 2015–2019.*, Brussels, Belgium, 40 pp. Bureau of International Recycling, Brussels, Belgium.

Birat, J.P., 2012: Sustainability footprint of steelmaking byproducts. *Ironmak. Steelmak.*, **39(4)**, 270–277, doi:10.1179/1743281211Y.0000000054.

Blanco, C., F. Caro and C.J. Corbett, 2016: The state of supply chain carbon footprinting: analysis of CDP disclosures by US firms. *J. Clean. Prod.*, **135**, 1189–1197, doi:10.1016/J.JCLEPRO.2016.06.132.

Bleischwitz, R., 2020: Mineral resources in the age of climate adaptation and resilience. *J. Ind. Ecol.*, **24**(2), 291–299, doi:10.1111/jiec.12951.

Bleischwitz, R., V. Nechifor, M. Winning, B. Huang, and Y. Geng, 2018: Extrapolation or saturation – Revisiting growth patterns, development stages and decoupling. *Glob. Environ. Change*, **48**, 86–96, doi:10.1016/j.gloenvcha.2017.11.008.

BMU, 2021: *Eckpunkte für eine Förderrichtlinie Klimaschutzverträge zur Umsetzung des Pilotprogramms "Carbon Contracts for Difference."* 5 pp. Bundesministerium für Umwelt, Naturschutz und nukleare Sicherheit (BMU) https://www.bmu.de/fileadmin/Daten_BMU/Download_PDF/Klimaschutz/eckpunktepapier_klimaschutzvertraege_ccfd_bf.pdf.

Bonato, D. and R. Orsini, 2017: Urban Circular Economy: The New Frontier for European Cities' Sustainable Development. In: *Sustainable Cities and Communities Design Handbook: Green Engineering, Architecture, and Technology*, Elsevier, Oxford, UK, Cambridge, USA. pp. 235–245.

Bonner, M., 2017: *An exploration of the opportunities to promote carbon capture and storage (CCS) in the United Nations framework convention on Climate Change (UNFCCC)*. 19 pp. Docklands Australia. https://www.globalccsinstitute.com/archive/hub/publications/201643/1706-ccs-opportunities-unfccc-comms-fd.1-fina.pdf.

Boulamanti A., and J.A. Moya Rivera, 2017: *Energy efficiency and GHG emissions: Prospective scenarios for the Chemical and Petrochemical Industry*. 229 pp. European Commission, Joint Research Centre, Petten, The Netherlands.

11

Boyce, J.K., 2018: Carbon Pricing: Effectiveness and Equity. *Ecol. Econ.*, **150**, 52–61, doi:10.1016/J.ECOLECON.2018.03.030.

BP, 2020: *Statistical Review of World Energy*. 65 pp. BP, London. https://www.bp.com/en/global/corporate/energy-economics/statistical-review-of-world-energy.html.

Branger, F. and P. Quirion, 2014a: Climate policy and the "carbon haven" effect. *Wiley Interdiscip. Rev. Clim. Change*, **5(1)**, 53–71, doi:10.1002/wcc.245.

Branger, F., and P. Quirion, 2014b: Would border carbon adjustments prevent carbon leakage and heavy industry competitiveness losses? Insights from a meta-analysis of recent economic studies. *Ecol. Econ.*, **99**, 29–39, doi:10.1016/j.ecolecon.2013.12.010.

Branger, F., P. Quirion, and J. Chevallier, 2016: Carbon leakage and competitiveness of cement and steel industries under the EU ETS: Much ado about nothing. *Energy J.*, **37(3)**, 109–135, doi:10.5547/01956574.37.3.fbra.

Bratt, C., S. Hallstedt, K.-H. Robèrt, G. Broman, and J. Oldmark, 2013: Assessment of criteria development for public procurement from a strategic sustainability perspective. *J. Clean. Prod.*, **52**, 309–316, doi:10.1016/J.JCLEPRO.2013.02.007.

Braun, A., P. Kleine-Moellhoff, V. Reichenberger, and S. Seiter, 2018: Case Study Analysing Potentials to Improve Material Efficiency in Manufacturing Supply Chains, Considering Circular Economy Aspects. *Sustainability*, **10(3)**, 880, doi:10.3390/su10030880.

Brécard, D., 2014: Consumer confusion over the profusion of eco-labels: Lessons from a double differentiation model. *Resour. Energy Econ.*, **37**, 64–84, doi:10.1016/j.reseneeco.2013.10.002.

Breyer, C., E. Tsupari, V. Tikka, and P. Vainikka, 2015: Power-to-Gas as an Emerging Profitable Business Through Creating an Integrated Value Chain. *Energy Procedia*, **73**, 182–189, doi:10.1016/j.egypro.2015.07.668.

Breyer, C., M. Fasihi, C. Bajamundi, and F. Creutzig, 2019: Direct Air Capture of $CO_2$: A Key Technology for Ambitious Climate Change Mitigation. *Joule*, **3(9)**, 2053–2057, doi:10.1016/j.joule.2019.08.010.

Brinkerhoff, P. and GLDNV, 2015: *Industrial Decarbonisation & Energy Efficiency Roadmaps to 2050: Cement*. 94 pp. Report prepared for UK Department of Energy and Climate Change and Department for Business, Innovation and Skills. WSP and Parsons Brinckerhoff and DNV GL, New York, USA and Bærum, Norway. https://www.gov.uk/government/uploads/system/uploads/attachment_data/file/416674/Cement_Report.pdf.

Bruhn, T., H. Naims, and B. Olfe-Kräutlein, 2016: Separating the debate on $CO_2$ utilisation from carbon capture and storage. *Environ. Sci. Policy*, **60**, 38–43, doi:10.1016/j.envsci.2016.03.001.

Brynolf, S., M. Taljegard, M. Grahn, and J. Hansson, 2018: Electrofuels for the transport sector: A review of production costs. *Renew. Sustain. Energy Rev.*, **81**, 1887–1905, doi:10.1016/j.rser.2017.05.288.

Bühler, F., S. Petrović, K. Karlsson, and B. Elmegaard, 2017: Industrial excess heat for district heating in Denmark. *Appl. Energy*, **205** (August), 991–1001, doi:10.1016/j.apenergy.2017.08.032.

Burritt, R. and S. Schaltegger, 2014: Accounting towards sustainability in production and supply chains. *Br. Account. Rev.*, **46(4)**, 327–343, doi:10.1016/j.bar.2014.10.001.

Busch, J., T.J. Foxon, and P.G. Taylor, 2018: Designing industrial strategy for a low carbon transformation. *Environ. Innov. Soc. Transitions*, **29**, 114–125, doi:10.1016/j.eist.2018.07.005.

Cabrera Serrenho, A., M. Drewniok, C. Dunant, and J.M. Allwood, 2019: Testing the greenhouse gas emissions reduction potential of alternative strategies for the English housing stock. *Resour. Conserv. Recycl.*, **144**, 267–275, doi:10.1016/j.resconrec.2019.02.001.

California Climate Investments, 2020: *Annual Report to the Legislature on California Climate Investments Using Cap-and-Trade Auction Proceeds*, Sacramento, CA, 126 pp. https://ww2.arb.ca.gov/sites/default/files/auction-proceeds/2020_cci_annual_report.pdf.

Calisto Friant, M., W.J.V. Vermeulen, and R. Salomone, 2021: Analysing European Union circular economy policies: words versus actions. *Sustain. Prod. Consum.*, **27**, 337–353, doi:10.1016/j.spc.2020.11.001.

Cambridge Econometrics, Trinomics and ICF, 2018: *Impacts of circular economy policies on the labour market*. 78 pp. https://www.camecon.com/wp-content/uploads/2019/01/Circular-Economy-DG-Env-final-report.pdf.

Cao, Z., L. Shen, A.N. Løvik, D.B. Müller, and G. Liu, 2017: Elaborating the History of Our Cementing Societies: An in-Use Stock Perspective. *Environ. Sci. Technol.*, **51(19)**, 11468–11475, doi:10.1021/acs.est.7b03077.

Cao, Z. et al., 2018: A Probabilistic Dynamic Material Flow Analysis Model for Chinese Urban Housing Stock. *J. Ind. Ecol.*, **22(2)**, 377–391, doi:10.1111/jiec.12579.

Cao, Z., G. Liu, S. Zhong, H. Dai, and S. Pauliuk, 2019: Integrating Dynamic Material Flow Analysis and Computable General Equilibrium Models for Both Mass and Monetary Balances in Prospective Modeling: A Case for the Chinese Building Sector. *Environ. Sci. Technol.*, **53**(1), 224–233, doi:10.1021/acs.est.8b03633.

Cao, Z. et al., 2020: The sponge effect and carbon emission mitigation potentials of the global cement cycle. *Nat. Commun.*, **11(1)**, 3777, doi:10.1038/s41467-020-17583-w.

Carayannis, E.G. and D.F.J. Campbell, 2019: *Smart Quintuple Helix Innovation Systems*. Springer International Publishing, Cham, Switzerland, 51–54 pp.

Carl, J. and D. Fedor, 2016: Tracking global carbon revenues: A survey of carbon taxes versus cap-and-trade in the real world. *Energy Policy*, **96**, 50–77, doi:10.1016/j.enpol.2016.05.023.

Carratù, M., B. Chiarini, and P. Piselli, 2020: Effects of European emission unit allowance auctions on corporate profitability. *Energy Policy*, **144**, 111584, doi:10.1016/j.enpol.2020.111584.

CAT, 2020: *Paris Agreement Compatible Sectoral Benchmark*. 67 pp. https://climateactiontracker.org/documents/753/CAT_2020-07-10_ParisAgreementBenchmarks_FullReport.pdf.

Cavaliere, P., 2019: Electrolysis of Iron Ores: Most Efficient Technologies for Greenhouse Emissions Abatement. In: *Clean Ironmaking and Steelmaking Processes*, Springer International Publishing, Cham, Switzerland, pp. 555–576.

CDIAC, 2017: *CDIAC Fossil-Fuel $CO_2$ Emissions*. PA, USA. https://cdiac.ess-dive.lbl.gov/trends/emis/tre_glob_2014.html (Accessed December 20, 2019).

CDP, 2019: *Cascading commitments: driving ambitious action through supply chain engagement*. New York, USA, 39 pp. https://cdn.cdp.net/cdp-production/cms/reports/documents/000/004/072/original/CDP_Supply_Chain_Report_2019.pdf?1550490556. (Accessed December 16, 2019).

Celadyn, W., 2014: *Durability of buildings and sustainable architecture*. Working paper from Technical Transactions Architecture, 7-A, 17–26 pp. https://www.semanticscholar.org/paper/Durability-of-buildings-and-sustainable-Celadyn/b0ddee2a11b8fd6e573f804924d02f89ffbc7b7b

CEMBUREAU, 2020: *Cementing the European Green Deal*. Brussels, Belgium, 7 pp. https://cembureau.eu/media/kuxd32gi/cembureau-2050-roadmap_final-version_web.pdf.

Champier, D., 2017: Thermoelectric generators: A review of applications. *Energy Convers. Manag.*, **140**, 167–181, doi:10.1016/j.enconman.2017.02.070.

Chan, E. et al., 2020: Eight-Year Estimates of Methane Emissions from Oil and Gas Operations in Western Canada Are Nearly Twice Those Reported in Inventories. *Environ. Sci. Technol.*, **54(23)**, 14899–14909, doi:10.1021/acs.est.0c04117.

Chan, Y. and R. Kantamaneni, 2016: *Study of Energy Efficiency and Energy Saving Potential in Industry and on Possible Policy Mechanisms*. London, UK, 456 pp. https://energy.ec.europa.eu/system/files/2016-01/151201%2520DG%2520ENER%2520Industrial%2520EE%2520study%2520-%2520final%2520report_clean_stc_0.pdf. (Accessed August 28, 2021).

Chastas, P., T. Theodosiou, and D. Bikas, 2016: Embodied energy in residential buildings-towards the nearly zero energy building: A literature review. *Build. Environ.*, **105**, 267–282, doi:10.1016/j.buildenv.2016.05.040.

Chatziaras, N., C.S. Psomopoulos, and N.J. Themelis, 2016: Use of waste derived fuels in cement industry: a review. *Manag. Environ. Qual. An Int. J.*, **27**(2), 178–193, doi:10.1108/MEQ-01-2015-0012.

Cheng, W., A. Appolloni, A. D'Amato, and Q. Zhu, 2018: Green Public Procurement, missing concepts and future trends – A critical review. *J. Clean. Prod.*, **176**, 770–784, doi:10.1016/J.JCLEPRO.2017.12.027.

Chiappinelli, O. et al., 2019: *Inclusive transformation of the European materials sector*. Climate Strategies, London, UK. 29 pp. https://climatestrategies.org/publication/inclusive-transformation-materials/.

Chrobak, U., 2021: Corporate Climate Pledges Pile Up—Will It Matter? *Engineering*, **7**(8), 1044–1046, doi:10.1016/j.eng.2021.06.011.

CIEL, 2017: *How Fracked Gas, Cheap Oil, and Unburnable Coal are Driving the Plastics Boom*., Washington, DC., 9 pp. https://www.ciel.org/wp-content/uploads/2017/09/Fueling-Plastics-How-Fracked-Gas-Cheap-Oil-and-Unburnable-Coal-are-Driving-the-Plastics-Boom.pdf.

Circle Economy, 2020: *Circularity gap report 2020*. The Hague, Netherlands, 69 pp. https://assets.website-files.com/5e185aa4d27bcf348400ed82/5e26ead616b6d1d157ff4293_20200120%20-%20CGR%20Global%20-%20Report%20web%20single%20page%20-%20210x297mm%20-%20compressed.pdf. (Accessed December 18, 2020).

Climact, 2018: *Net zero by 2050: from whether to how – zero emissions pathways to the Europe we want*. The Hague, The Netherlands, 68 pp. https://europeanclimate.org/wp-content/uploads/2019/11/09-18-net-zero-by-2050-from-whether-to-how.pdf. (Accessed March 27, 2021).

Cohen, M.A., and M.P. Vandenbergh, 2012: The potential role of carbon labeling in a green economy. *Energy Econ.*, **34**, S53–S63, doi:10.1016/J.ENECO.2012.08.032.

Cosbey, A.,S. Droege, C. Fischer, and C. Munnings, 2019: Developing Guidance for Implementing Border Carbon Adjustments: Lessons, Cautions, and Research Needs from the Literature. *Rev. Environ. Econ. Policy*, **13**(1), 3–22, doi:10.1093/reep/rey020.

Costantini, V., F. Crespi, and E. Paglialunga, 2018: The employment impact of private and public actions for energy efficiency: Evidence from European industries. *Energy Policy*, **119**, 250–267, doi:10.1016/j.enpol.2018.04.035.

Craglia, M. and J. Cullen, 2020: Modelling transport emissions in an uncertain future: What actions make a difference? *Transp. Res. Part D Transp. Environ.*, **89**, 102614, doi:10.1016/j.trd.2020.102614.

Creutzig, F., 2019: The Mitigation Trinity: Coordinating Policies to Escalate Climate Mitigation. *One Earth*, **1(1)**, 76–85, doi:10.1016/j.oneear.2019.08.007.

Crijns-Graus, W., H. Yue, S. Zhang, K. Kermeli, and E. Worrell, 2020: Energy Efficiency Improvement Opportunities in the Global Industrial Sector. In: *Encyclopedia of Renewable and Sustainable Materials*, Elsevier, Oxford, UK and Cambridge, USA, pp. 377–388.

Crippa, M. et al., 2021: EDGAR v6.0 Greenhouse Gas Emissions. *Eur. Comm. Jt. Res. Cent. [Dataset]*, doi:http://data.europa.eu/89h/97a67d67-c62e-4826-b873-9d972c4f670b.

Cuéllar-Franca, R.M. and A. Azapagic, 2015: Carbon capture, storage and utilisation technologies: A critical analysis and comparison of their life cycle environmental impacts. *J. $CO_2$ Util.*, **9**, 82–102, doi:10.1016/j.jcou.2014.12.001.

Cullen, J. M. and J. M. Allwood, 2013: Mapping the global flow of aluminum: From liquid aluminum to end-use goods. *Environ. Sci. Technol.*, **47(7)**, 3057–3064, doi:10.1021/es304256s.

D'Alessandro, A. et al., 2016: Innovative concretes for low-carbon constructions: a review. *Int. J. Low-Carbon Technol.*, 12(3), 289–309, doi:10.1093/ijlct/ctw013.

D'Amato, D., J. Korhonen, and A. Toppinen, 2019: Circular, Green, and Bio Economy: How Do Companies in Land-Use Intensive Sectors Align with Sustainability Concepts? *Ecol. Econ.*, **158** (January), 116–133, doi:10.1016/j.ecolecon.2018.12.026.

Daehn, K.E., A. Cabrera Serrenho, and J.M. Allwood, 2017: How Will Copper Contamination Constrain Future Global Steel Recycling? *Environ. Sci. Technol.*, **51(11)**, 6599–6606, doi:10.1021/acs.est.7b00997.

Daggash, H.A. et al., 2018: Closing the carbon cycle to maximise climate change mitigation: power-to-methanol vs. power-to-direct air capture. *Sustain. Energy Fuels*, **2(6)**, 1153–1169, doi:10.1039/C8SE00061A.

Daniel Posen, I., P. Jaramillo, A.E. Landis, and W. Michael Griffin, 2017: Greenhouse gas mitigation for U.S. plastics production: Energy first, feedstocks later. *Environ. Res. Lett.*, **12(3)**, 034024, doi:10.1088/1748-9326/aa60a7.

Darnall, N., C. Ponting, and D.A. Vazquez-Brust, 2012: Why Consumers Buy Green. In: *Green Growth: Managing the Transition to a Sustainable Economy*, Springer Netherlands, Dordrecht, pp. 287–308.

Das, S., 2021: The quest for low carbon aluminum: Developing a sustainability index. *Light Met. Age*, **79(1)**, 34–43.

Davidson, M.G., R.A. Furlong, and M.C. McManus, 2021: Developments in the life cycle assessment of chemical recycling of plastic waste – A review. *J. Clean. Prod.*, **293**, 126163, doi:10.1016/j.jclepro.2021.126163.

Davis, S.J. et al., 2018: Net-zero emissions energy systems. *Science.*, **360**(6396), doi:10.1126/science.aas9793.

Dawkins, E. et al., 2019: Advancing sustainable consumption at the local government level: A literature review. *J. Clean. Prod.*, **231**, 1450–1462, doi:10.1016/j.jclepro.2019.05.176.

de la Rue du Can, S. et al., 2019: Modeling India's energy future using a bottom-up approach. *Appl. Energy*, **238**, 1108–1125, doi:10.1016/J.APENERGY.2019.01.065.

De Luna, P. et al., 2019: What would it take for renewably powered electrosynthesis to displace petrochemical processes? *Science*, **364(6438)**, eaav3506, doi:10.1126/science.aav3506.

de-Magistris, T. and A. Gracia, 2016: Consumers' willingness-to-pay for sustainable food products: the case of organically and locally grown almonds in Spain. *J. Clean. Prod.*, **118**, 97–104, doi:10.1016/J.JCLEPRO.2016.01.050.

Dean, C.C., J. Blamey, N.H. Florin, M.J. Al-Jeboori, and P.S. Fennell, 2011: The calcium looping cycle for $CO_2$ capture from power generation, cement manufacture and hydrogen production. *Chem. Eng. Res. Des.*, **89(6)**, 836–855, doi:10.1016/j.cherd.2010.10.013.

Deason, J., M. Wei, G. Leventis, S. Smith, and L. Schwartz, 2018: *Electrification of buildings and industry in the United States*. Berkeley, USA, 65 pp. http://eta-publications.lbl.gov/sites/default/files/electrification_of_buildings_and_industry_final_0.pdf. (Accessed March 27, 2021).

Denis-Ryan, A., C. Bataille, and F. Jotzo, 2016: Managing carbon-intensive materials in a decarbonizing world without a global price on carbon. *Clim. Policy*, **16(sup1)**, S110–S128, doi:10.1080/14693062.2016.1176008.

De Smet, M. and M. Linder, 2019: *A circular economy for plastics – Insights from research and innovation to inform policy and funding decisions*. Brussels, Belgium, 239 pp. https://www.hbm4eu.eu/wp-content/uploads/2019/03/2019_RI_Report_A-circular-economy-for-plastics.pdf. (Accessed December 15, 2019).

Detz, R.J. and B. van der Zwaan, 2019: Transitioning towards negative $CO_2$ emissions. *Energy Policy*, **133**(January), 110938, doi:10.1016/j.enpol.2019.110938.

DGS, 2020: Buy Clean California Act – Department of General Services (DGS), CA, USA. https://www.dgs.ca.gov/PD/Resources/Page-Content/Procurement-Division-Resources-List-Folder/Buy-Clean-California-Act. (Accessed October 29, 2021).

Dhandra, T.K., 2019: Achieving triple dividend through mindfulness: More sustainable consumption, less unsustainable consumption and more life satisfaction. *Ecol. Econ.*, **161**, 83–90, doi:10.1016/j.ecolecon.2019.03.021.

Dhar, S., M. Pathak, and P.R. Shukla, 2020: Transformation of India's steel and cement industry in a sustainable 1.5°C world. *Energy Policy*, **137**, 111104, doi:10.1016/j.enpol.2019.111104.

Di Foggia, G., 2018: Energy efficiency measures in buildings for achieving sustainable development goals. *Heliyon*, **4(11)**, e00953, doi:10.1016/j.heliyon.2018.e00953.

Dimitriou, I. et al., 2015: Carbon dioxide utilisation for production of transport fuels: process and economic analysis. *Energy Environ. Sci.*, **8(6)**, 1775–1789, doi:10.1039/C4EE04117H.

Ding, J. and W. Hua, 2012: Featured chemical industrial parks in China: History, current status and outlook. *Resour. Conserv. Recycl.*, **63**, 43–53, doi:10.1016/J.RESCONREC.2012.03.001.

Dodman, D., B. Hayward, M. Pelling, V. Castan Broto, W. Chow, E. Chu, R. Dawson, L. Khirfan, T. McPhearson, A. Prakash, Y. Zheng, and G. Ziervogel, 2022: Cities, Settlements and Key Infrastructure. In: *Climate Change 2022: Impacts, Adaptation, and Vulnerability. Contribution of Working Group II to the Sixth Assessment Report of the Intergovernmental Panel on Climate Change* [H.-O. Pörtner, D.C. Roberts, M. Tignor, E.S. Poloczanska, K. Mintenbeck, A. Alegría, M. Craig, S. Langsdorf, S. Löschke, V. Möller, A. Okem, B. Rama (eds.)]. Cambridge University Press, Cambridge, UK and New York, NY, USA. In press.

Dogu, O. et al., 2021: The chemistry of chemical recycling of solid plastic waste via pyrolysis and gasification: State-of-the-art, challenges, and future directions. *Progress in Energy and Combustion Science*, Vol. 84, 100901. doi.org/10.1016/j.pecs.2020.100901

Domenech, T. and B. Bahn-Walkowiak, 2019: Transition Towards a Resource Efficient Circular Economy in Europe: Policy Lessons From the EU and the Member States. *Ecol. Econ.*, **155**, 7–19, doi:10.1016/j.ecolecon.2017.11.001.

Dong, H., Y. Geng, X. Yu, and J. Li, 2018: Uncovering energy saving and carbon reduction potential from recycling wastes: A case of Shanghai in China. *J. Clean. Prod.*, **205**, 27–35, doi:10.1016/j.jclepro.2018.08.343.

Dunant, C.F., M.P. Drewniok, S. Eleftheriadis, J.M. Cullen, and J.M. Allwood, 2018: Regularity and optimisation practice in steel structural frames in real design cases. *Resour. Conserv. Recycl.*, **134**, 294–302, doi:10.1016/j.resconrec.2018.01.009.

Durán-Romero, G. et al., 2020: Bridging the gap between circular economy and climate change mitigation policies through eco-innovations and Quintuple Helix Model. *Technol. Forecast. Soc. Change*, **160**, 120246, doi:10.1016/j.techfore.2020.120246.

Eames, M., W. McDowall, M. Hodson, and S. Marvin, 2006: Negotiating contested visions and place-specific expectations of the hydrogen economy. *Technol. Anal. Strateg. Manag.*, **18(3–4)**, 361–374, doi:10.1080/09537320600777127.

Ebrahimi, A. et al., 2017: Sustainable transformation of fly ash industrial waste into a construction cement blend via CO2 carbonation. *J. Clean. Prod.*, **156**, 660–669, doi:10.1016/j.jclepro.2017.04.037.

Edgar, T.F. and E.N. Pistikopoulos, 2018: Smart manufacturing and energy systems. *Comput. Chem. Eng.*, **114**, 130–144, doi:10.1016/j.compchemeng.2017.10.027.

Eicke, L., S. Weko, M. Apergi, and A. Marian, 2021: Pulling up the carbon ladder? Decarbonization, dependence, and third-country risks from the European carbon border adjustment mechanism. *Energy Res. Soc. Sci.*, **80**, 102240, doi:10.1016/j.erss.2021.102240.

Ellis, J., D. Nachtigall, and F. Venmans, 2019: *Carbon pricing and competitiveness: Are they at odds?* OECD Publishing, Paris, France, 25 pp. (Accessed August 27, 2021).

Elshkaki, A., T.E. Graedel, L. Ciacci, and B.K. Reck, 2018: Resource Demand Scenarios for the Major Metals. *Environ. Sci. Technol.*, **52(5)**, 2491–2497, doi:10.1021/acs.est.7b05154.

Energy Transitions Commission, 2018: *Mission Possible: Reaching net-zero carbon emissions from harder-to-abate sectors by mid-century*. London, UK, 171 pp. http://www.energy-transitions.org/sites/default/files/ETC_Mission Possible_FullReport.pdf. (Accessed March 27, 2021).

ERIA, 2016: Exploring Energy-Saving Potential for Industrial Sector Using Factory, Energy Management System. In: *Study on the Advancement of the Energy Management System in the East Asia Summit Region ERIA Research Project Report 2015 17* [Iida, Y., S. Inoue, and Y. Li, (eds.)], pp. 19–52.

Erickson, P., S. Kartha, M. Lazarus, and K. Tempest, 2015: Assessing carbon lock-in. *Environ. Res. Lett.*, **10(8)**, 084023, doi:10.1088/1748-9326/10/8/084023.

Ericsson, K., 2017: *Biogenic carbon dioxide as feedstock for production of chemicals and fuels: A techno-economic assessment with a European perspective*. Environmental and Energy System Studies, Lund, Sweden, 47 pp. https://portal.research.lu.se/portal/en/publications/biogenic-carbon-dioxide-as-feedstock-for-production-of-chemicals-and-fuels(67d3a737-cf7c-4109-bc4f-a6346956d6a2).html. (Accessed March 27, 2021).

Ericsson, K. and L.J. Nilsson, 2018: *Climate innovations in the paper industry: Prospects for decarbonisation*. REINVENT project, Lund, Sweden, 33 pp. https://static1.squarespace.com/static/59f0cb986957da5faf64971e/t/5dc1acfb29bc520c15c858fc/1572973823092/%28updated%29D2.4+Climate+innovations+in+the+paper+industry.pdf. (Accessed March 27, 2021).

Eriksson, L., M. Morandin, and S. Harvey, 2018: A feasibility study of improved heat recovery and excess heat export at a Swedish chemical complex site. *Int. J. Energy Res.*, **42(4)**, 1580–1593, doi:10.1002/er.3950.

EUROFER, 2019: *Low Carbon Roadmap. Pathways to a CO2-neutral European Steel Industry*., Brussels, Belgium, 18 pp. https://www.eurofer.eu/assets/Uploads/EUROFER-Low-Carbon-Roadmap-Pathways-to-a-CO2-neutral-European-Steel-Industry.pdf.

European Commission, 2016: *Buying Green! A handbook on green public procurement. [Comprando verde! Un manual de Compras Públicas Verdes]*. 3rd ed., Brussels, Belgium.

European Commission, 2018: *A Clean Planet for all – A European long-term strategic vision for a prosperous, modern, competitive and climate neutral economy*. Brussles, Belgium, 25 pp. https://eur-lex.europa.eu/legal-content/EN/TXT/PDF/?uri=CELEX:52018DC0773&from=EN. (Accessed December 15, 2019).

European Commission, 2020: *A new Circular Economy Action Plan*., Brussels, Belgium., 20 pp. https://eur-lex.europa.eu/legal-content/EN/TXT/?qid=1583933814386&uri=COM:2020:98:FIN.

European Commission, 2021: *Proposal for a REGULATION OF THE EUROPEAN PARLIAMENT AND OF THE COUNCIL establishing a carbon border adjustment mechanism*. Brussels, Belgium.

Eyre, N., 2021: From using heat to using work: reconceptualising the zero carbon energy transition. *Energy Effic.*, **14(7)**, 77, doi:10.1007/s12053-021-09982-9.

Fan, J.-L., M. Xu, F. Li, L. Yang, and X. Zhang, 2018: Carbon capture and storage (CCS) retrofit potential of coal-fired power plants in China: The technology lock-in and cost optimization perspective. *Appl. Energy*, **229**, 326–334, doi:10.1016/j.apenergy.2018.07.117.

Fan, Z. and S.J. Friedmann, 2021: Low-carbon production of iron and steel: Technology options, economic assessment, and policy. *Joule*, **5**(4), 829–862, doi:10.1016/j.joule.2021.02.018.

Fang, H., J. Xia, and Y. Jiang, 2015: Key issues and solutions in a district heating system using low-grade industrial waste heat. *Energy*, **86**, 589–602, doi:10.1016/j.energy.2015.04.052.

Fankhauser, S. and F. Jotzo, 2018: Economic growth and development with low-carbon energy. *Wiley Interdiscip. Rev. Clim. Change*, **9(1)**, e495, doi:10.1002/wcc.495.

Faria, J.A., 2021: Renaissance of ammonia synthesis for sustainable production of energy and fertilizers. *Current Opinion in Green and Sustainable Chemistry*, 29, 100466. 10.1016/j.cogsc.2021.100466.

Fasihi, M., D. Bogdanov, and C. Breyer, 2017: Long-Term Hydrocarbon Trade Options for the Maghreb Region and Europe—Renewable Energy Based Synthetic Fuels for a Net Zero Emissions World. *Sustainability*, **9**(2), 306, doi:10.3390/su9020306.

Fasihi, M., O. Efimova, and C. Breyer, 2019: Techno-economic assessment of $CO_2$ direct air capture plants. *J. Clean. Prod.*, **224**, 957–980, doi:10.1016/j.jclepro.2019.03.086.

Fasihi, M., R. Weiss, J. Savolainen, and C. Breyer, 2021: Global potential of green ammonia based on hybrid PV-wind power plants. *Appl. Energy*, **294**, 116170, doi:10.1016/j.apenergy.2020.116170.

Fawkes, S., K. Oung, and D. Thorpe, 2016: *Best Practices and Case Studies for Industrial Energy Efficiency Improvement – An Introduction for Policy Makers*. Copenhagen, Denmark, 171 pp. https://europa.eu/capacity4dev/file/31409/download?token=kOYq7O5T.

Fechner, T. and C. Kray, 2012: Attacking location privacy. Proceedings of the 2012 ACM Conference on Ubiquitous Computing - UbiComp '12, New York, NY, USA, ACM Press, 95–98.

Ferrero, R., M. Collotta, M.V. Bueno-Delgado, and H.C. Chen, 2020: Smart management energy systems in industry 4.0. *Energies*, **13**(2), 382, doi:10.3390/en13020382.

Feucht, Y. and K. Zander, 2018: Consumers' preferences for carbon labels and the underlying reasoning. A mixed methods approach in 6 European countries. *J. Clean. Prod.*, **178**, 740–748, doi:10.1016/J.JCLEPRO.2017.12.236.

Finnveden, G. et al., 2009: Recent developments in Life Cycle Assessment. *J. Environ. Manage.*, **91(1)**, 1–21, doi:10.1016/j.jenvman.2009.06.018.

Fischedick, M., J. Roy, A. Abdel-Aziz, A. Acquaye, J.M. Allwood, J.-P. Ceron, Y. Geng, H. Kheshgi, A. Lanza, D. Perczyk, L. Price, E. Santalla, C. Sheinbaum, and K. Tanaka, 2014: Industry. In: *Climate Change 2014: Mitigation of Climate Change. Contribution of Working Group III to the Fifth Assessment Report of the Intergovernmental Panel on Climate Change* [Edenhofer, O., R. Pichs-Madruga, Y. Sokona, E. Farahani, S. Kadner, K. Seyboth, A. Adler, I. Baum, S. Brunner, P. Eickemeier, B. Kriemann, J. Savolainen, S. Schlömer, C. von Stechow, T. Zwickel and J.C. Minx (eds.)]. Cambridge University Press, Cambridge, UK and New York, NY, USA, pp. 739–810.

Fischedick, M., J. Marzinkowski, P. Winzer, and M. Weigel, 2014b: Techno-economic evaluation of innovative steel production technologies. *J. Clean. Prod.*, **84(1)**, 563–580, doi:10.1016/j.jclepro.2014.05.063.

Flanagan, K., E. Uyarra, and M. Laranja, 2011: Reconceptualising the "policy mix" for innovation. *Res. Policy*, **40(5)**, 702–713, doi:10.1016/j.respol.2011.02.005.

Fowlie, M., C. Petersen, and M. Reguant, 2021: Border Carbon Adjustments When Carbon Intensity Varies across Producers: Evidence from California. *AEA Pap. Proc.*, **111**, 401–405, doi:10.1257/pandp.20211073.

Friedmann, J., Fan, Z and Tang, K. (2019): *Low-Carbon Heat Solutions for Heavy Industry: Sources, Options, and Costs Today (Centre on Global Energy Policy)*. New York, NY, USA 98 pp.

Fukushima, M. and Y. I. Yoshizawa, 2016: Fabrication of highly porous honeycomb-shaped mullite-zirconia insulators by gelation freezing. *Adv. Powder Technol.*, **27**(3), 908–913, doi:10.1016/j.apt.2016.02.015.

Fuso Nerini, F. et al., 2019: Connecting climate action with other Sustainable Development Goals. *Nat. Sustain*, **2(8)**, 674–680, doi:10.1038/s41893-019-0334-y.

Gadema, Z. and D. Oglethorpe, 2011: The use and usefulness of carbon labelling food: A policy perspective from a survey of UK supermarket shoppers. *Food Policy*, **36(6)**, 815–822, doi:10.1016/J.FOODPOL.2011.08.001.

Gambhir, A. et al., 2017: The contribution of non-$CO_2$ greenhouse gas mitigation to achieving long-term temperature goals. *Energies*, **10(5)**, 602, doi:10.3390/en10050602.

Garcia-Muiña, F.E., R. González-Sánchez, A.M. Ferrari, and D. Settembre-Blundo, 2018: The paradigms of Industry 4.0 and circular economy as enabling drivers for the competitiveness of businesses and territories: The case of an Italian ceramic tiles manufacturing company. *Soc. Sci.*, **7(12)**, 255, doi:10.3390/socsci7120255.

García-Quevedo, J. and E. Jové-Llopis, 2021: Environmental policies and energy efficiency investments. An industry-level analysis. *Energy Policy*, **156**, 112461, doi:10.1016/j.enpol.2021.112461.

Gardarsdottir, S. et al., 2019: Comparison of Technologies for $CO_2$ Capture from Cement Production—Part 2: Cost Analysis. *Energies*, **12(3)**, 542, doi:10.3390/en12030542.

Garrett-Peltier, H., 2017: Green versus brown: Comparing the employment impacts of energy efficiency, renewable energy, and fossil fuels using an input-output model. *Econ. Model.*, **61**, 439–447, doi:10.1016/j.econmod.2016.11.012.

Gayner, C. and K. K. Kar, 2016: Recent advances in thermoelectric materials. *Prog. Mater. Sci.*, **83**, 330–382, doi:10.1016/j.pmatsci.2016.07.002.

GCCA, 2021a: *The GCCA 2050 Cement and Concrete Industry Roadmap for Net Zero Concrete*. London, UK, 46 pp. https://gccassociation.org/concretefuture/wp-content/uploads/2021/10/GCCA-Concrete-Future-Roadmap-Document-AW.pdf. Accessed 1 November 2021.

GCCA, 2021b: *GNR–GCCA in Numbers*. https://gccassociation.org/sustainability-innovation/gnr-gcca-in-numbers/ (Accessed August 27, 2021).

Geels, F.W., B.K. Sovacool, T. Schwanen, and S. Sorrell, 2017: The Socio-Technical Dynamics of Low-Carbon Transitions. *Joule*, **1(3)**, 463–479, doi:10.1016/j.joule.2017.09.018.

Geng, Y., and B. Doberstein, 2008: Greening government procurement in developing countries: Building capacity in China. *J. Environ. Manage.*, **88(4)**, 932–938, doi:10.1016/J.JENVMAN.2007.04.016.

Geng, Y., J. Fu, J. Sarkis, and B. Xue, 2012: Towards a national circular economy indicator system in China: An evaluation and critical analysis. *J. Clean. Prod.*, **23(1)**, 216–224, doi:10.1016/j.jclepro.2011.07.005.

Geng, Y., J. Sarkis, S. Ulgiati, and P. Zhang, 2013: Measuring China's circular economy. *Science*, **340(6127)**, 1526–1527, doi:10.1126/science.1227059.

Geng, Y., J. Sarkis, and R. Bleischwitz, 2019: How to globalize the circular economy. *Nature*, **565**(7738), 153–155, doi:10.1038/d41586-019-00017-z.

Genovese, A., A.A. Acquaye, A. Figueroa, and S.C.L. Koh, 2017: Sustainable supply chain management and the transition towards a circular economy: Evidence and some applications. *Omega (UK)*, **66**, 344–357, doi:10.1016/j.omega.2015.05.015.

Gerbert, P. et al., 2018: *Klimapfade für Deutschland (Climate paths for Germany)*. BCG and Prognos. Berlin, Germany, 286 pp. https://www.zvei.org/fileadmin/user_upload/Presse_und_MedienPublikationen/2018/Januar/Klimapfade_fuer_Deutschland_BDI-Studie_/Klimapfade-fuer-Deutschland-BDI-Studie-12-01-2018.pdf. (Accessed December 18, 2019).

Geyer, R., J.R. Jambeck, and K.L. Law, 2017: Production, use, and fate of all plastics ever made. *Sci. Adv.*, **3(7)**, doi:10.1126/sciadv.1700782.

Ghisellini, P., C. Cialani, and S. Ulgiati, 2016: A review on circular economy: the expected transition to a balanced interplay of environmental and economic systems. *J. Clean. Prod.*, **114**, 11–32, doi:10.1016/J.JCLEPRO.2015.09.007.

Ghisetti, C., 2017: Demand-pull and environmental innovations: Estimating the effects of innovative public procurement. *Technol. Forecast. Soc. Change*, **125**, 178–187, doi:10.1016/j.techfore.2017.07.020.

Giannaris, S., C. Bruce, B. Jacobs, W. Srisang, and D. Janowczyk, 2020: Implementing a second generation CCS facility on a coal fired power station – results of a feasibility study to retrofit SaskPower's Shand power station with CCS. *Greenh. Gases Sci. Technol.*, **10(3)**, 506–518, doi:10.1002/ghg.1989.

Gielen, D., D. Saygin, E. Taibi, and J. Birat, 2020: Renewables-based decarbonization and relocation of iron and steel making: A case study. *J. Ind. Ecol.*, **24(5)**, 1113–1125, doi:10.1111/jiec.12997.

Gillingham, K., S. Carattini, and D. Esty, 2017: Lessons from first campus carbon-pricing scheme. *Nature*, **551(7678)**, 27–29, doi:10.1038/551027a.

Glenk, G. and S. Reichelstein, 2019: Economics of converting renewable power to hydrogen. *Nat. Energy*, **4(3)**, 216–222, doi:10.1038/s41560-019-0326-1.

Gogate, M. R., 2019: Methanol-to-olefins process technology: current status and future prospects. *Pet. Sci. Technol.*, **37(5)**, 559–565, doi:10.1080/10916466.2018.1555589.

Gonzalez Hernandez, A., S. Cooper-Searle, A.C.H. Skelton, and J.M. Cullen, 2018a: Leveraging material efficiency as an energy and climate instrument for heavy industries in the EU. *Energy Policy*, **120**, 533–549, doi:10.1016/j.enpol.2018.05.055.

Gonzalez Hernandez, A., L. Paoli, and J.M. Cullen, 2018b: How resource-efficient is the global steel industry? *Resour. Conserv. Recycl.*, **133**, 132–145, doi:10.1016/j.resconrec.2018.02.008.

Govindan, K., 2018: Sustainable consumption and production in the food supply chain: A conceptual framework. *Int. J. Prod. Econ.*, **195**, 419–431, doi:10.1016/j.ijpe.2017.03.003.

Graedel, T.E. et al., 2011: *Recycling rates of metals: a status report*. Paris, France, 44 pp.

Graedel, T.E., E.M. Harper, N.T. Nassar, and B.K. Reck, 2015: On the materials basis of modern society. *Proc. Natl. Acad. Sci.*, **112(20)**, 6295–6300, doi:10.1073/pnas.1312752110.

Grainger, C.A. and C.D. Kolstad, 2010: Who Pays a Price on Carbon? *Environ. Resour. Econ.*, **46(3)**, 359–376, doi:10.1007/s10640-010-9345-x.

Grillitsch, M., T. Hansen, L. Coenen, J. Miörner, and J. Moodysson, 2019: Innovation policy for system-wide transformation: The case of strategic innovation programmes (SIPs) in Sweden. *Res. Policy*, **48(4)**, 1048–1061, doi:10.1016/j.respol.2018.10.004.

Grubb, M., Hourcade J. and Neuhoff K., 2014: *Planetary Economics*. Routledge, Oxfordshire, UK, 548 pp.

Grubler, A. et al., 2018: A low energy demand scenario for meeting the 1.5°C target and sustainable development goals without negative emission technologies. *Nat. Energy*, **3(6)**, 515–527, doi:10.1038/s41560-018-0172-6.

Guenther, M., C.M. Saunders, and P.R. Tait, 2012: Carbon labeling and consumer attitudes. *Carbon Manag.*, **3(5)**, 445–455, doi:10.4155/cmt.12.50.

Guillen-Royo, M., 2019: Sustainable consumption and wellbeing: Does online shopping matter? *J. Clean. Prod.*, **229**, 1112–1124, doi:10.1016/j.jclepro.2019.05.061.

Guo, B., Y. Geng, T. Sterr, L. Dong, and Y. Liu, 2016: Evaluation of promoting industrial symbiosis in a chemical industrial park: A case of Midong. *J. Clean. Prod.*, **135**, 995–1008, doi:10.1016/j.jclepro.2016.07.006.

Guo, R. et al., 2021: Global $CO_2$ uptake by cement from 1930 to 2019. *Earth Syst. Sci. Data*, **13(4)**, 1791–1805, doi:10.5194/essd-13-1791-2021.

Guo, Y., J. Tian, N. Zang, Y. Gao, and L. Chen, 2018: The Role of Industrial Parks in Mitigating Greenhouse Gas Emissions from China. *Environ. Sci. Technol.*, **52(14)**, 7754–7762, doi:10.1021/acs.est.8b00537.

Gutowski, T., D. Cooper, and S. Sahni, 2017: Why we use more materials. *Philos. Trans. R. Soc. A Math. Phys. Eng. Sci.*, **375(2095)**, 20160368, doi:10.1098/rsta.2016.0368.

Gutowski, T.G., S. Sahni, J.M. Allwood, M.F. Ashby, and E. Worrell, 2013: The energy required to produce materials: Constraints on energy-intensity improvements, parameters of demand. *Philos. Trans. R. Soc. A Math. Phys. Eng. Sci.*, **371(1986)**, 20120003, doi:10.1098/rsta.2012.0003.

Haberl, H. et al., 2021: Stocks, flows, services and practices: Nexus approaches to sustainable social metabolism. *Ecol. Econ.*, **182**, 106949, doi:10.1016/j.ecolecon.2021.106949.

Habert, G. et al., 2020: Environmental impacts and decarbonization strategies in the cement and concrete industries. *Nat. Rev. Earth Environ.*, **1(11)**, 559–573, doi:10.1038/s43017-020-0093-3.

Hagen, A. and J. Schneider, 2021: Trade sanctions and the stability of climate coalitions. *J. Environ. Econ. Manage.*, **109**, 102504, doi:10.1016/j.jeem.2021.102504.

Haites, E., 2018: Carbon taxes and greenhouse gas emissions trading systems: what have we learned? *Clim. Policy*, **18(8)**, 955–966, doi:10.1080/14693062.2018.1492897.

Haites, E. et al., 2018: *Experience With Carbon Taxes And Greenhouse Gas Emissions Trading Systems*. *Duke Environmental Law & Policy Forum*, Vol 29, pp 109–182.

Haraldsson, J. and M. T. Johansson, 2018: Review of measures for improved energy efficiency in production-related processes in the aluminium industry – From electrolysis to recycling. *Renew. Sustainability. Energy Rev.*, **93**(May), 525–548, doi:10.1016/j.rser.2018.05.043.

Harmelink, M., M. Beerepoot, and A. Puranasamriddhi, 2018: *Energy Access projects and assessment of their Contribution to the Sustainable Development Goals: SDG1, SDG3, SDG4 and SDG5*. Bangkok, Thailand, 56 pp. https://www.asia-pacific.undp.org/content/rbap/en/home/library/climate-and-disaster-resilience/energy-access-projects-and-SDG-benefits.html. (Accessed March 27, 2021).

Hartley, K., R. van Santen, and J. Kirchherr, 2020: Policies for transitioning towards a circular economy: Expectations from the European Union (EU). *Resour. Conserv. Recycl.*, **155**, 104634, doi:10.1016/j.resconrec.2019.104634.

Hasanbeigi, A., L. Price, and E. Lin, 2012: Emerging energy-efficiency and CO2 emission-reduction technologies for cement and concrete production: A technical review. *Renew. Sustainability. Energy Rev.*, **16(8)**, 6220–6238, doi:10.1016/j.rser.2012.07.019.

Hatti-Kaul, R., L.J. Nilsson, B. Zhang, N. Rehnberg, and S. Lundmark, 2020: Designing Biobased Recyclable Polymers for Plastics. *Trends Biotechnol.*, **38(1)**, 50–67, doi:10.1016/j.tibtech.2019.04.011.

Hausfather, Z., 2015: Bounding the climate viability of natural gas as a bridge fuel to displace coal. *Energy Policy*, **86**, 286–294, doi:10.1016/j.enpol.2015.07.012.

Heeren, N. and S. Hellweg, 2019: Tracking Construction Material over Space and Time: Prospective and Geo-referenced Modeling of Building Stocks and Construction Material Flows. *J. Ind. Ecol.*, **23(1)**, 253–267, doi:10.1111/jiec.12739.

Hepburn, C. et al., 2019: The technological and economic prospects for $CO_2$ utilization and removal. *Nature*, **575**(7781), 87–97, doi:10.1038/s41586-019-1681-6.

Hertwich, E., R. Lifset, S. Pauliuk, and N. Heeren, 2020: *Resource Efficiency and Climate Change: Material Efficiency Strategies for a low-carbon Future*. Paris, France, 155 pp.

Hertwich, E.G., 2020: Carbon fueling complex global value chains tripled in the period 1995–2012. *Energy Econ.*, **86**, 104651, doi:10.1016/j.eneco.2019.104651.

Hertwich, E.G., 2021: Increased carbon footprint of materials production driven by rise in investments. *Nat. Geosciences*, **14(3)**, 151–155, doi:10.1038/s41561-021-00690-8.

Hertwich, E.G. et al., 2019: Material efficiency strategies to reducing greenhouse gas emissions associated with buildings, vehicles, and electronics – A review. *Environ. Res. Lett.*, **14(4)**, 043004, doi:10.1088/1748-9326/ab0fe3.

Higgins, C. and B. Coffey, 2016: Improving how sustainability reports drive change: a critical discourse analysis. *J. Clean. Prod.*, **136**, 18–29, doi:10.1016/J.JCLEPRO.2016.01.101.

Hills, T., D. Leeson, N. Florin, and P. Fennell, 2016: Carbon Capture in the Cement Industry: Technologies, Progress, and Retrofitting. *Environ. Sci. Technol.*, **50(1)**, 368–377, doi:10.1021/acs.est.5b03508.

Hills, T.P., M. Sceats, D. Rennie, and P. Fennell, 2017: LEILAC: Low Cost $CO_2$ Capture for the Cement and Lime Industries. *Energy Procedia*, **114** (July), 6166–6170.

Hilton, B. et al., 2018: *Re-defining Value – The Manufacturing. Revolution. Remanufacturing, Refurbishment, Repair and Direct Reuse in the Circular Economy*. Nairobi, Kenya, 267 pp.

Hoegh-Guldberg, O., D. Jacob, M. Taylor, M. Bindi, S. Brown, I. Camilloni, A. Diedhiou, R. Djalante, K.L. Ebi, F. Engelbrecht, J. Guiot, Y. Hijioka, S. Mehrotra, A. Payne, S.I. Seneviratne, A. Thomas, R. Warren, and G. Zhou, 2018: Impacts of 1.5°C Global Warming on Natural and Human Systems. In: *Global Warming of 1.5°C. An IPCC Special Report on the impacts of global warming of 1.5°C above pre-industrial levels and related global greenhouse gas emission pathways, in the context of strengthening the global response to the threat of climate change, sustainable development, and efforts to eradicate poverty* [Masson-Delmotte, V., P. Zhai, H.-O. Pörtner, D. Roberts, J. Skea, P.R. Shukla, A. Pirani, W. Moufouma-Okia, C. Péan, R. Pidcock, S. Connors, J.B.R. Matthews, Y. Chen, X. Zhou, M.I. Gomis, E. Lonnoy, T. Maycock, M. Tignor, and T. Waterfield eds.)]. Cambridge University Press, Cambridge, UK, and New York, NY, USA.

Holappa, L., 2020: A general vision for reduction of energy consumption and $CO_2$ emissions from the steel industry. *Metals (Basel).*, **10(9)**, 1–20, doi:10.3390/met10091117.

Hong, L. et al., 2016: Building stock dynamics and its impacts on materials and energy demand in China. *Energy Policy*, **94**, 47–55, doi:10.1016/j.enpol.2016.03.024.

Honus, S., S. Kumagai, G. Fedorko, V. Molnár, and T. Yoshioka, 2018a: Pyrolysis gases produced from individual and mixed PE, PP, PS, PVC, and PET—Part I: Production and physical properties. *Fuel*, **221**, 346–360, doi:10.1016/j.fuel.2018.02.074.

Honus, S., S. Kumagai, V. Molnár, G. Fedorko, and T. Yoshioka, 2018b: Pyrolysis gases produced from individual and mixed PE, PP, PS, PVC, and PET—Part II: Fuel characteristics. *Fuel*, **221**, 361–373, doi:10.1016/j.fuel.2018.02.075.

Hoppmann, J., M. Peters, M. Schneider, and V.H. Hoffmann, 2013: The two faces of market support – How deployment policies affect technological exploration and exploitation in the solar photovoltaic industry. *Res. Policy*, **42(4)**, 989–1003, doi:10.1016/j.respol.2013.01.002.

Horton, P.M., J.M. Allwood, and C. Cleaver, 2019: Implementing material efficiency in practice: A case study to improve the material utilisation of automotive sheet metal components. *Resour. Conserv. Recycl.*, **145**, 49–66, doi:10.1016/j.resconrec.2019.02.012.

Hourcade, J.C., A. Pottier, and E. Espagne, 2018: Social value of mitigation activities and forms of carbon pricing. *Int. Econ.*, **155**, 8–18, doi:10.1016/j.inteco.2018.06.001.

Hsieh, I.Y.L., M.S. Pan, and W.H. Green, 2020: Transition to electric vehicles in China: Implications for private motorization rate and battery market. *Energy Policy*, **144**, 111654, doi:10.1016/j.enpol.2020.111654.

Huang, B. et al., 2019a: Review of the development of China's Eco-industrial Park standard system. *Resour. Conserv. Recycl.*, **140**, 137–144, doi:10.1016/j.resconrec.2018.09.013.

Huang, H. et al., 2019b: Emissions trading systems and social equity: A CGE assessment for China. *Appl. Energy*, **235**, 1254–1265, doi:10.1016/J.APENERGY.2018.11.056.

Huang, H. et al., 2019c: Life-cycle assessment of emerging $CO_2$ mineral carbonation-cured concrete blocks: Comparative analysis of $CO_2$ reduction potential and optimization of environmental impacts. *J. Clean. Prod.*, **241**, 118359, doi:10.1016/j.jclepro.2019.118359.

Huang, Z., G. Grlm, J. Schaidle, and L. Tao, 2020: Using waste $CO_2$ to increase ethanol production from corn ethanol biorefineries: Techno-economic analysis. *Appl. Energy*, **280**, 115964, doi:10.1016/j.apenergy.2020.115964.

Ibáñez-Forés, V., B. Pacheco-Blanco, S.F. Capuz-Rizo, and M.D. Bovea, 2016: Environmental Product Declarations: Exploring their evolution and the factors affecting their demand in Europe. *J. Clean. Prod.*, **116**, 157–169, doi:10.1016/j.jclepro.2015.12.078.

IEA, 2009: *Technology roadmap: carbon capture and storage*. International Energy Agency (IEA), Paris, France, 52 pp. https://www.iea.org/reports/technology-roadmap-carbon-capture-and-storage-2009. (Accessed March 27, 2021).

IEA, 2015: *World Energy Outlook 2015*. Paris, France, 700 pp. https://www.iea.org/reports/world-energy-outlook-2015. (Accessed March 27, 2021).

IEA, 2016: *Energy and Air Pollution (Special Report)*. Paris, France, 266 pp. https://www.iea.org/reports/energy-and-air-pollution. (Accessed October 30, 2021).

IEA, 2017: *Energy Technology Perspectives 2017*. Paris, France, 438 pp. https://www.iea.org/reports/energy-technology-perspectives-2017. (Accessed December 15, 2019).

IEA, 2018a: *The future of petrochemicals: towards more sustainable plastics and fertilisers*. Paris, France, 66 pp. https://www.iea.org/reports/the-future-of-petrochemicals. (Accessed December 15, 2019).

IEA, 2018b: *Energy Efficiency 2018. Analysis and outlooks to 2040*. Paris, France, 170 pp. https://www.iea.org/reports/energy-efficiency-2018. (Accessed December 14, 2019).

IEA, 2019a: $CO_2$ Emissions from Fuel Combustion online data service. data.iea.org/payment/products/115-co2-emissions-from-fuel-combustion-2018-edition.aspx. (Accessed December 20, 2020).

IEA, 2019b: *Material efficiency in clean energy transitions*. OECD, Paris, France, 158 pp. https://www.iea.org/reports/material-efficiency-in-clean-energy-transitions. (Accessed December 15, 2019).

IEA, 2019c: *World Energy Outlook 2019*. Paris, France, 807 pp. https://www.iea.org/reports/world-energy-outlook-2019. (Accessed December 16, 2019).

IEA, 2019d: *Energy Efficiency 2019*. Paris, France, 107 pp. https://www.iea.org/reports/energy-efficiency-2019. (Accessed December 19, 2020).

IEA, 2019e: *Offshore Wind Outlook 2019*. Paris, France, 96 pp. https://www.iea.org/reports/offshore-wind-outlook-2019. (Accessed December 23, 2019).

IEA, 2019f: *The Future of Hydrogen*. Paris, France, 199 pp. https://www.iea.org/reports/the-future-of-hydrogen. (Accessed December 21, 2019).

IEA, 2019g: *Transforming industry through CCUS*. Paris, France. 58 pp. https://www.iea.org/publications/reports/TransformingIndustrythroughCCUS/. (Accessed December 16, 2019).

IEA, 2019h: *Putting $CO_2$ to use*. Paris, France, 83 pp. https://www.iea.org/reports/putting-co2-to-use (Accessed December 23, 2020).

IEA, 2020a: *Energy Technology Perspective 2020*. Paris, France 397 pp. https://www.iea.org/reports/energy-technology-perspectives-2020 (Accessed September 23, 2020).

IEA, 2020b: Tracking industry 2020. https://www.iea.org/reports/tracking-industry-2020 (Accessed December 20, 2020).

IEA, 2020c: *Iron and Steel Technology Roadmap Towards more sustainable steelmaking – Part of the Energy Technology Perspectives series*. Paris, France, 187 pp. https://www.iea.org/reports/iron-and-steel-technology-roadmap. (Accessed October 29 2021).

IEA, 2020d: World energy balances and statistics. https://www.iea.org/data-and-statistics/data-product/world-energy-balances (Accessed October 29, 2021).

IEA, 2020e: *Energy Efficiency 2020*. Paris, France, 102 pp. https://www.iea.org/reports/energy-efficiency-2020 (Accessed December, 19, 2020).

IEA, 2020f: $CO_2$ emissions from fuel combustion database. https://www.iea.org/reports/co2-emissions-from-fuel-combustion-overview (Accessed December 20, 2020).

IEA, 2020g: *Outlook for biogas and biomethane*. Paris, France, 91 pp. https://www.iea.org/reports/outlook-for-biogas-and-biomethane-prospects-for-organic-growth. (Accessed December, 19, 2020).

IEA, 2020h: *Energy technology perspective special report – clean energy innovation*. Paris, France, 182 pp. https://webstore.iea.org/download/direct/4022?fileName=Energy_Technology_Perspectives_2020_-_Special_Report_on_Clean_Energy_Innovation.pdf. (Accessed October, 23, 2020).

IEA, 2020i: *World Energy Outlook 2020*. Paris, France. 461 pp. https://www.iea.org/events/world-energy-outlook-2020. (Accessed March, 27, 2021).

IEA, 2021a: *Net Zero by 2050: A Roadmap for the Global Energy Sector*. Paris, France, 222 pp. https://www.iea.org/reports/net-zero-by-2050 (Accessed July, 19, 2021).

IEA, 2021b: $CO_2$ Emissions from Fuel Combustion online data service. data.iea.org/payment/products/115-co2-emissions-from-fuel-combustion-2021-edition.aspx. (Accessed August 27, 2021).

IEA, 2021c: *World Energy Outlook 2021*. Paris, France, 383 pp. https://www.iea.org/reports/world-energy-outlook-2021. (Accessed November, 3, 2021).

IEA, 2021d: World energy balances. https://www.iea.org/data-and-statistics/data-product/world-energy-balances (Accessed October 29, 2020).

IEA and WBCSD, 2018: *Technology roadmap – low-carbon transition in the cement industry*. Paris, France, 61 pp. https://webstore.iea.org/download/direct/1008?fileName=TechnologyRoadmapLowCarbonTransitionintheCementIndustry.pdf. (Accessed December, 15, 2019).

IEA and UNEP, 2019: *Global Status Report for Buildings and Construction*. Paris, France, 41 pp. https://www.iea.org/reports/global-status-report-for-buildings-and-construction-2019 (Accessed January, 6, 2020).

IEAGHG, 2021: *Exporting $CO_2$ for Offshore Storage – The London Protocol's Export Amendment and Associated Guidelines and Guidance*. Paris, France. 13 pp. https://www.club-co2.fr/files/2021/04/IEAGHG-2021-TR02-Exporting-CO2-for-Offshore-Storage-The-London-Protocol-s-Export-Amendment-and-Associated-Guidelines-and-Guidance.pdf. (Accessed Novemberr, 13, 2021).


11

Igarashi, M., L. de Boer, and O. Michelsen, 2015: Investigating the anatomy of supplier selection in green public procurement. *J. Clean. Prod.*, **108**, 442–450, doi:10.1016/J.JCLEPRO.2015.08.010.

IIASA, 2018: PFU Database. https://www.iiasa.ac.at/web/home/research/researchPrograms/TransitionstoNewTechnologies/PFUDB.en.html (Accessed December 21, 2019).

ILO, 2018: *World Employment and Social Outlook: Trends 2018*. International Labour Organization (ILO), Geneva, Switzerland, 81 pp. https://www.ilo.org/wcmsp5/groups/public/---dgreports/---dcomm/---publ/documents/publication/wcms_615594.pdf. (Accessed December 20, 2020).

Imbabi, M.S., C. Carrigan, and S. McKenna, 2012: Trends and developments in green cement and concrete technology. *Int. J. Sustain. Built Environ.*, **1(2)**, 194–216, doi:10.1016/j.ijsbe.2013.05.001.

IN4climate.NRW, 2019: IN4climate.NRW. https://www.in4climate.nrw/en/index/ (Accessed December 20, 2020).

International Aluminium Institute, 2020: World Aluminium Institute Database. https://international-aluminium.org/statistics/primary-aluminium-smelting-energy-intensity/ (Accessed December 20, 2020).

International Aluminium Institute, 2021a: Aluminium Cycle 2019. https://alucycle.international-aluminium.org/public-access/ (Accessed August 28, 2021).

International Aluminium Institute, 2021b: *Aluminium Sector Greenhouse Gas Pathways to 2050*. London, UK, 20 pp.

International Aluminium Institute, 2021c: International aluminum institute statistics. https://alucycle.international-aluminium.org/public-access/ (Accessed December 21, 2020).

IPCC, 2005: *IPCC Special Report on Carbon Dioxide Capture and Storage*. Prepared by Working Group III of the Intergovernmental Panel on Climate Change. [Metz, B., O. Davidson, H.C. De Coninck, M. Loos, and L.A. Meyer, (eds.)]. Cambridge University Press, Cambridge, UK and New York, NY, USA, 442 pp.

IPCC, 2014: *Climate Change 2014: Mitigation of Climate Change. Contribution of Working Group III to the Fifth Assessment Report of the Intergovernmental Panel on Climate Change* [Edenhofer, O., R. Pichs-Madruga, Y. Sokona, E. Farahani, S. Kadner, K. Seyboth, A. Adler, I. Baum, S. Brunner, P. Eickemeier, B. Kriemann, J. Savolainen, S. Schlömer, C. von Stechow, T. Zwickel and J.C. Minx (eds.)]. Cambridge University Press, Cambridge, UK and New York, NY, USA.

IPCC, 2018a: *Global Warming of 1.5°C. An IPCC Special Report on the impacts of global warming of 1.5°C above pre-industrial levels and related global greenhouse gas emission pathways, in the context of strengthening the global response to the threat of climate change, sustainable development, and efforts to eradicate poverty* [Masson-Delmotte, V., P. Zhai, H.-O. Pörtner, D. Roberts, J. Skea, P.R. Shukla, A. Pirani, W. Moufouma-Okia, C. Péan, R. Pidcock, S. Connors, J.B.R. Matthews, Y. Chen, X. Zhou, M.I. Gomis, E. Lonnoy, T. Maycock, M. Tignor, and T. Waterfield eds.)]. Cambridge University Press, Cambridge, UK, and New York, NY, USA.

IPCC, 2018b: Summary for Policymakers. In: *Global Warming of 1.5°C. An IPCC Special Report on the impacts of global warming of 1.5°C above pre-industrial levels and related global greenhouse gas emission pathways, in the context of strengthening the global response to the threat of climate change, sustainable development, and efforts to eradicate poverty* [Masson-Delmotte, V., P. Zhai, H.-O. Pörtner, D. Roberts, J. Skea, P.R. Shukla, A. Pirani, W. Moufouma-Okia, C. Péan, R. Pidcock, S. Connors, J.B.R. Matthews, Y. Chen, X. Zhou, M.I. Gomis, E. Lonnoy, T. Maycock, M. Tignor, and T. Waterfield eds.)]. Cambridge University Press, Cambridge, UK, and New York, NY, USA.

IRENA, 2021: *World Energy Transitions Outlook: 1.5°C Pathway*. Abu Dhabi, United Arab Emirates, 53 pp.

Isikgor, F.H. and C.R. Becer, 2015: Lignocellulosic biomass: a sustainable platform for the production of bio-based chemicals and polymers. *Polym. Chem.*, **6(25)**, 4497–4559, doi:10.1039/c5py00263j.

ISO, 2018: ISO 50001 ENERGY MANAGEMENT. https://www.iso.org/iso-50001-energy-management.html (Accessed November 30, 2020).

Vanderreydt, I., Rommens, T., Tenhunen, A., Fogh Mortensen, L., and Tange, I., 2021: *Greenhouse gas emissions and natural capital implications of plastics (including biobased plastics) — Eionet Portal*. Flanders, Belgium, 59 pp. https://www.eionet.europa.eu/etcs/etc-wmge/products/greenhouse-gas-emissions-and-natural-capital-implications-of-plastics-including-biobased-plastics (Accessed August 28, 2021).

Jackson, J. and L. Belkhir, 2018: Assigning firm-level GHGE reductions based on national goals – Mathematical model and empirical evidence. *J. Clean. Prod.*, **170**, 76–84, doi:10.1016/j.jclepro.2017.09.075.

Jackson, M.M., J.I. Lewis, and X. Zhang, 2021: A green expansion: China's role in the global deployment and transfer of solar photovoltaic technology. *Energy Sustain. Dev.*, **60**, 90–101, doi:10.1016/j.esd.2020.12.006.

Jadun, P. et al., 2017: *Electrification futures study: end-use electric technology cost and performance projections through 2050*. Golden, CO: National Renewable Energy Laboratory. NREL/TP-6A20-70485. 94 pp. https://www.nrel.gov/docs/fy18osti/70485.pdf. (Accessed December 22, 2019).

Jakob, M., 2021a: Why carbon leakage matters and what can be done against it. *One Earth*, **4(5)**, 609–614, doi:10.1016/j.oneear.2021.04.010.

Jakob, M., 2021b: Climate policy and international trade – A critical appraisal of the literature. *Energy Policy*, **156** (July), 112399, doi:10.1016/j.enpol.2021.112399.

Jakob, M., J.C. Steckel, and O. Edenhofer, 2014: Consumption-versus production-based emission policies. *Annu. Rev. Resour. Econ.*, **6(1)**, 297–318, doi:10.1146/annurev-resource-100913-012342.

Jakob, M., H. Ward, and J.C. Steckel, 2021: Sharing responsibility for trade-related emissions based on economic benefits. *Glob. Environ. Change*, **66**, 102207, doi:10.1016/j.gloenvcha.2020.102207.

Jenkins, J.D., M. Luke, and S. Thernstrom, 2018: Getting to Zero Carbon Emissions in the Electric Power Sector. *Joule*, **2(12)**, 2498–2510, doi:10.1016/j.joule.2018.11.013.

Jiang, M. et al., 2019: Provincial and sector-level material footprints in China. *Proc. Natl. Acad. Sci. U. S. A.*, **116(52)**, 26484–26490, doi:10.1073/pnas.1903028116.

Jiao, W. and F. Boons, 2014: Toward a research agenda for policy intervention and facilitation to enhance industrial symbiosis based on a comprehensive literature review. *J. Clean. Prod.*, **67**, 14–25, doi:10.1016/J.JCLEPRO.2013.12.050.

Jiménez-Arreola, M. et al., 2018: Thermal power fluctuations in waste heat to power systems: An overview on the challenges and current solutions. *Appl. Therm. Eng.*, **134**, 576–584, doi:10.1016/j.applthermaleng.2018.02.033.

Jing, Y., Y. Guo, Q. Xia, X. Liu, and Y. Wang, 2019: Catalytic Production of Value-Added Chemicals and Liquid Fuels from Lignocellulosic Biomass. *Chem*, **5(10)**, 2520–2546, doi:10.1016/j.chempr.2019.05.022.

Joltreau, E. and K. Sommerfeld, 2019: Why does emissions trading under the EU Emissions Trading System (ETS) not affect firms' competitiveness? Empirical findings from the literature. *Clim. Policy*, **19(4)**, 453–471, doi:10.1080/14693062.2018.1502145.

Jood, P., M. Ohta, A. Yamamoto, and M.G. Kanatzidis, 2018: Excessively Doped PbTe with Ge-Induced Nanostructures Enables High-Efficiency Thermoelectric Modules. *Joule*, **2(7)**, 1339–1355, doi:10.1016/j.joule.2018.04.025.

Kajaste, R. and M. Hurme, 2016: Cement industry greenhouse gas emissions – management options and abatement cost. *J. Clean. Prod.*, **112**, 4041–4052, doi:10.1016/j.jclepro.2015.07.055.

Kaliyavaradhan, S.K. and T.-C. Ling, 2017: Potential of $CO_2$ sequestration through construction and demolition (C&D) waste – An overview. *J. $CO_2$ Util.*, **20** (June), 234–242, doi:10.1016/j.jcou.2017.05.014.

Kalmykova, Y., M. Sadagopan, and L. Rosado, 2018: Circular economy – From review of theories and practices to development of implementation tools. *Resour. Conserv. Recycl.*, **135**, 190–201, doi:10.1016/J.RESCONREC.2017.10.034.

11

Kang, D. and D.H. Lee, 2016: Energy and environment efficiency of industry and its productivity effect. *J. Clean. Prod.*, **135**, 184–193, doi:10.1016/j.jclepro.2016.06.042.

Karlsson, M., E. Alfredsson, and N. Westling, 2020: Climate policy co-benefits: a review. *Clim. Policy*, **20(3)**, 292–316, doi:10.1080/14693062.2020.1724070.

Karltorp, K. and B.A. Sandén, 2012: Explaining regime destabilisation in the pulp and paper industry. *Environ. Innov. Soc. Transitions*, **2**, 66–81, doi:10.1016/j.eist.2011.12.001.

Karner, K., M. Theissing, and T. Kienberger, 2016: Energy efficiency for industries through synergies with urban areas. *J. Clean. Prod.*, **119**(2016), 167–177, doi:10.1016/j.jclepro.2016.02.010.

Karner, K., R. McKenna, M. Klobasa, and T. Kienberger, 2018: Industrial excess heat recovery in industry-city networks: a technical, environmental and economic assessment of heat flexibility. *J. Clean. Prod.*, **193**, 771–783, doi:10.1016/j.jclepro.2018.05.045.

Karo, E., 2018: Mission-oriented innovation policies and bureaucracies in East Asia. *Ind. Corp. Change*, **27(5)**, 867–881, doi:10.1093/icc/dty031.

Karo, E. and R. Kattel, 2018: Innovation and the State: Towards an Evolutionary Theory of Policy Capacity. In: Wu, X., Howlett, M., Ramesh, M. (eds) *Policy Capacity and Governance*, Studies in the Political Economy of Public Policy. Palgrave Macmillan, Cham, Switzerland, pp. 123–150. https://doi.org/10.1007/978-3-319-54675-9_6.

Kätelhön, A., R. Meys, S. Deutz, S. Suh, and A. Bardow, 2019: Climate change mitigation potential of carbon capture and utilization in the chemical industry. *Proc. Natl. Acad. Sci.*, **166(23)**, 11187–11194, doi:10.1073/pnas.1821029116.

Kattel, R. and M. Mazzucato, 2018: Mission-oriented innovation policy and dynamic capabilities in the public sector. *Ind. Corp. Change*, **27(5)**, 787–801, doi:10.1093/icc/dty032.

Kaza, S., L. Yao, P. Bhada-Tata, and F. Van Woerden, 2018: *What a Waste 2.0: A Global Snapshot of Solid Waste Management to 2050*. World Bank, Washington D.C., USA, 272 pp. https://openknowledge.worldbank.org/handle/10986/30317. (Accessed March 27, 2021).

Kearns, D., H. Liu, and C. Consoli, 2021: *TECHNOLOGY READINESS AND COSTS OF CCS*. GCCSI. Washington D.C., USA. www.globalccsinstitute.com/resources/multimedia-library/technology-readiness-and-costs-of-ccs/. Acccessed 28 August 2021.

Keidanren (Japan Business Federation), 2018: *Contributing to Avoided Emissions the the Global Value Chain – A new approach to climate change measures by private actors*. Tokyo, Japan, 71 pp. http://www.keidanren.or.jp/en/policy/vape/gvc2018.pdf. (Accessed March 27, 2021).

Keith, D.W., G. Holmes, D. St. Angelo, and K. Heidel, 2018: A Process for Capturing $CO_2$ from the Atmosphere. *Joule*, **2(8)**, 1573–1594, doi:10.1016/j.joule.2018.05.006.

Kermeli, K. et al., 2021: Improving material projections in Integrated Assessment Models: The use of a stock-based versus a flow-based approach for the iron and steel industry. *Energy*, Vol. 239 122434, doi:10.1016/j.energy.2021.122434.

Kholod, N. et al., 2020: Global methane emissions from coal mining to continue growing even with declining coal production. *J. Clean. Prod.*, **256**, 120489, doi:10.1016/j.jclepro.2020.120489.

Kim, H.W. et al., 2018a: Co-benefit potential of industrial and urban symbiosis using waste heat from industrial park in Ulsan, Korea. *Resour. Conserv. Recycl.*, **135**, 225–234, doi:10.1016/j.resconrec.2017.09.027.

Kim, H.W., S. Ohnishi, M. Fujii, T. Fujita, and H. S. Park, 2018b: Evaluation and Allocation of Greenhouse Gas Reductions in Industrial Symbiosis. *J. Ind. Ecol.*, **22(2)**, 275–287, doi:10.1111/jiec.12539.

Kinderman, D., 2020: The challenges of upward regulatory harmonization: The case of sustainability reporting in the European Union. *Regul. Gov.*, **14(4)**, 674–697, doi:10.1111/rego.12240.

Kirchherr, J. et al., 2018: Barriers to the Circular Economy: Evidence From the European Union (EU). *Ecol. Econ.*, **150**, 264–272, doi:10.1016/j.ecolecon.2018.04.028.

Kirchner, M., J. Schmidt, and S. Wehrle, 2019: Exploiting Synergy of Carbon Pricing and Other Policy Instruments for Deep Decarbonization. *Joule*, **3(4)**, 891–893, doi:10.1016/j.joule.2019.03.006.

Koasidis, K. et al., 2020: The UK and German low-carbon industry transitions from a sectoral innovation and system failures perspective. *Energies*, **13(18)**, 4994, doi:10.3390/en13194994.

Koberg, E. and A. Longoni, 2019: A systematic review of sustainable supply chain management in global supply chains. *J. Clean. Prod.*, **207**, 1084–1098, doi:10.1016/J.JCLEPRO.2018.10.033.

Korhonen, J., A. Honkasalo, and J. Seppälä, 2018: Circular Economy: The Concept and its Limitations. *Ecol. Econ.*, **143**, 37–46, doi:10.1016/j.ecolecon.2017.06.041.

Krausmann, F. et al., 2017: Global socioeconomic material stocks rise 23-fold over the 20th century and require half of annual resource use. *Proc. Natl. Acad. Sci. U. S. A.*, **114(8)**, 1880–1885, doi:10.1073/pnas.1613773114.

Krausmann, F., C. Lauk, W. Haas, and D. Wiedenhofer, 2018: From resource extraction to outflows of wastes and emissions: The socioeconomic metabolism of the global economy, 1900–2015. *Glob. Environ. Change*, **52**, 131–140, doi:10.1016/j.gloenvcha.2018.07.003.

Krausmann, F., D. Wiedenhofer, and H. Haberl, 2020: Growing stocks of buildings, infrastructures and machinery as key challenge for compliance with climate targets. *Glob. Environ. Change*, **61**, 102034, doi:10.1016/j.gloenvcha.2020.102034.

Kuo, L. and B.-G. Chang, 2021: Ambitious corporate climate action: Impacts of science-based target and internal carbon pricing on carbon management reputation – Evidence from Japan. *Sustain. Prod. Consum.*, **27**, 1830–1840, doi:10.1016/j.spc.2021.04.025.

Kuramochi, T., 2016: Assessment of midterm $CO_2$ emissions reduction potential in the iron and steel industry: a case of Japan. *J. Clean. Prod.*, **132(132)**, 81–97, doi:10.1016/j.jclepro.2015.02.055.

Kushnir, D., T. Hansen, V. Vogl, and M. Åhman, 2020: Adopting hydrogen direct reduction for the Swedish steel industry: A technological innovation system (TIS) study. *J. Clean. Prod.*, **242**, 118185, doi:10.1016/j.jclepro.2019.118185.

Kuusi, T. et al., 2020: *Carbon Border Adjustment Mechanisms and Their Economic Impact on Finland and the EU*. Helsinki, Finland, 152 pp. http://urn.fi/URN:ISBN:978-952-287-922-6. (Accessed October 21, 2019).

Labella, Á., K. Koasidis, A. Nikas, A. Arsenopoulos, and H. Doukas, 2020: APOLLO: A Fuzzy Multi-criteria Group Decision-Making Tool in Support of Climate Policy. *Int. J. Comput. Intell. Syst.*, **13(1)**, 1539, doi:10.2991/ijcis.d.200924.002.

Lamb, W.F. et al., 2021: A review of trends and drivers of greenhouse gas emissions by sector from 1990 to 2018. *Environ. Res. Lett.*, **16(7)**, 073005, doi:10.1088/1748-9326/abee4e.

Lambert, J.G., C.A.S. Hall, S. Balogh, A. Gupta, and M. Arnold, 2014: Energy, EROI and quality of life. *Energy Policy*, **64**, 153–167, doi:10.1016/j.enpol.2013.07.001.

LCS, 2018: *Toward Future Low-Carbon Society using Scrap Iron Recycling (Vol. 2)*. Center for Low Carbin Society Strategy, Tokyo, Japan, 17 pp. https://www.jst.go.jp/lcs/pdf/fy2018-pp-18.pdf. (Accessed December, 15, 2019)

Lechtenböhmer, S. and M. Fischedick, 2020: *An Integrated Climate-Industrial Policy as the Core of the European Green Deal in Brief*. Wuppertal Institute, Wuppertal, Germany, 8 pp. https://epub.wupperinst.org/frontdoor/deliver/index/docId/7483/file/7483_Climate-Industrial-Policy.pdf. (Accessed March 27, 2021).

Lechtenböhmer, S., C. Schneider, M.Y. Roche, and S. Höller, 2015: Re-industrialisation and low-carbon economy-can they go together? Results from stakeholder-based scenarios for energy-intensive industries in the German state of North Rhine Westphalia. *Energies*, **8(10)**, 11404–11429, doi:10.3390/en81011404.

Lechtenböhmer, S., L.J. Nilsson, M. Åhman, and C. Schneider, 2016: Decarbonising the energy intensive basic materials industry through electrification – Implications for future EU electricity demand. *Energy*, **115**, 1623–1631, doi:10.1016/j.energy.2016.07.110.

Lecocq, F. and Z. Shalizi, 2014: The economics of targeted mitigation in infrastructure. *Clim. Policy*, **14(2)**, 187–208, doi:10.1080/14693062.2014.861657.

Lee, S.Y. et al., 2019: A comprehensive metabolic map for production of bio-based chemicals. *Nat. Catal.*, **2(1)**, 18–33, doi:10.1038/s41929-018-0212-4.

Leeson, D., P. Fennell, N. Shah, C. Petit, and N. Mac Dowell, 2017: A Techno-economic analysis and systematic review of carbon capture and storage (CCS) applied to the iron and steel, cement, oil refining and pulp and paper industries. *Int. J. Greenh. Gas Control*, **61** (June), 71–84, doi:10.1016/j.ijggc.2017.03.020.

Legorburu, G. and A.D. Smith, 2018: Energy modeling framework for optimizing heat recovery in a seasonal food processing facility. *Appl. Energy*, **229**, 151–162, doi:10.1016/j.apenergy.2018.07.097.

Lehne, J. and F. Preston, 2018: *Making Concrete Change Innovation in Low-carbon Cement and Concrete*. Chatham House, London, UK. 122 pp.

Levi, P.G. and J.M. Cullen, 2018: Mapping Global Flows of Chemicals: From Fossil Fuel Feedstocks to Chemical Products. *Environ. Sci. Technol.*, **52(4)**, 1725–1734, doi:10.1021/acs.est.7b04573.

Li, Q., R. Long and H. Chen, 2017: Empirical study of the willingness of consumers to purchase low-carbon products by considering carbon labels: A case study. *J. Clean. Prod.*, **161**, 1237–1250, doi:10.1016/J.JCLEPRO.2017.04.154.

Li, S. et al., 2019: Recent advances in hydrogen production by thermo-catalytic conversion of biomass. *Int. J. Hydrogen Energy*, **44(28)**, 14266–14278, doi:10.1016/j.ijhydene.2019.03.018.

Lim, T., B.R. Ellis, and S.J. Skerlos, 2019: Mitigating $CO_2$ emissions of concrete manufacturing through $CO_2$-enabled binder reduction. *Environ. Res. Lett.*, **14(11)**, 114014, doi:10.1088/1748-9326/ab466e.

Liu, C., W. Chen, and J. Mu, 2019a: Retailer's multi-tier green procurement contract in the presence of suppliers' reference point effect. *Comput. Ind. Eng.*, **131**, 242–258, doi:10.1016/j.cie.2019.03.013.

Liu, G., C.E. Bangs, and D. B. Müller, 2013: Stock dynamics and emission pathways of the global aluminium cycle. *Nat. Clim. Change*, **3(4)**, 338–342, doi:10.1038/nclimate1698.

Liu, G. and D. B. Müller, 2013: Centennial evolution of aluminum in-use stocks on our aluminized planet. *Environ. Sci. Technol.*, **47(9)**, 4882–4888, doi:10.1021/es305108p.

Liu, J., J. Xue, L. Yang, and B. Shi, 2019b: Enhancing green public procurement practices in local governments: Chinese evidence based on a new research framework. *J. Clean. Prod.*, **211**, 842–854, doi:10.1016/J.JCLEPRO.2018.11.151.

Liu, M. et al., 2020: End-of-life passenger vehicles recycling decision system in China based on dynamic material flow analysis and life cycle assessment. *Waste Manag.*, **117**, 81–92, doi:10.1016/j.wasman.2020.08.002.

Liu, T., Q. Wang, and B. Su, 2016: A review of carbon labeling: Standards, implementation, and impact. *Renew. Sustain. Energy Rev.*, **53**, 68–79, doi:10.1016/J.RSER.2015.08.050.

Liu, Z. et al., 2015: Reduced carbon emission estimates from fossil fuel combustion and cement production in China. *Nature*, **524(7565)**, 335–338, doi:10.1038/nature14677.

Liu, Z. et al., 2018: Co-benefits accounting for the implementation of eco-industrial development strategies in the scale of industrial park based on emergy analysis. *Renew. Sustain. Energy Rev.*, **81**, 1522–1529, doi:10.1016/j.rser.2017.05.226.

Llorente-González, L.J. and X. Vence, 2019: Decoupling or "decaffing"? The underlying conceptualization of circular economy in the European union monitoring framework. *Sustainability*, **11(18)**, 4898, doi:10.3390/su11184898.

Llorente-González, L.J., and X. Vence, 2020: How labour-intensive is the circular economy? A policy-orientated structural analysis of the repair, reuse and recycling activities in the European Union. *Resour. Conserv. Recycl.*, **162**, 105033, doi:10.1016/j.resconrec.2020.105033.

LME, 2020: *London Metal Exchange - Sustainability Discussion Paper*. London, UK, 20 pp.

Long, T.B. and W. Young, 2016: An exploration of intervention options to enhance the management of supply chain greenhouse gas emissions in the UK. *J. Clean. Prod.*, **112**, 1834–1848, doi:10.1016/J.JCLEPRO.2015.02.074.

Lopez, G. et al., 2018: Recent advances in the gasification of waste plastics. A critical overview. *Renew. Sustain. Energy Rev.*, **82**, 576–596, doi:10.1016/j.rser.2017.09.032.

Lovins, A.B., 2018: How big is the energy efficiency resource? *Environ. Res. Lett.*, **13(9)**, 090401, doi:10.1088/1748-9326/aad965.

Lundberg, S., P.O. Marklund, E. Strömbäck, and D. Sundström, 2015: Using public procurement to implement environmental policy: an empirical analysis. *Environ. Econ. Policy Stud.*, **17(4)**, 487–520, doi:10.1007/s10018-015-0102-9.

Lv, S. et al., 2018: Study of different heat exchange technologies influence on the performance of thermoelectric generators. *Energy Convers. Manag.*, **156**, 167–177, doi:10.1016/j.enconman.2017.11.011.

Mac Dowell, N., P.S. Fennell, N. Shah, and G.C. Maitland, 2017: The role of $CO_2$ capture and utilization in mitigating climate change. *Nat. Clim. Change*, **7(4)**, 243–249, doi:10.1038/nclimate3231.

MacKay, K. et al., 2021: Methane emissions from upstream oil and gas production in Canada are underestimated. *Sci. Rep.*, **11(1)**, 8041, doi:10.1038/s41598-021-87610-3.

Maddison Project, 2018: Maddison Project Database. http://www.ggdc.net/maddison; https://datacatalog.worldbank.org/dataset/global-economic-monitor (Accessed December 20, 2019).

Madeddu, S. et al., 2020: The CO2 reduction potential for the European industry via direct electrification of heat supply (power-to-heat). *Environ. Res. Lett.*, **15**(12), 124004, doi:10.1088/1748-9326/abbd02.

Mah, A., 2021: Future-Proofing Capitalism: The Paradox of the Circular Economy for Plastics. *Glob. Environ. Polit.*, **21(2)**, 121–142, doi:10.1162/glep_a_00594.

Mahmood, T. and E. Ahmad, 2018: The relationship of energy intensity with economic growth: Evidence for European economies. *Energy Strateg. Rev.*, **20**, 90–98, doi:10.1016/j.esr.2018.02.002.

Mai, T.T. et al., 2018: *Electrification Futures Study: Scenarios of Electric Technology Adoption and Power Consumption for the United States*. Golden, CO, USA, 129 pp.

Marin County, 2021: *Title 19 – MARIN COUNTY BUILDING CODE*. A Codification of the General Ordinances of the County of Marin, California, Marin County, USA.

Markkanen, S. et al., 2021: *On the Borderline: The EU CBAM and its place in the world of trade*. Cambridge, UK, 73 pp.

Martí, J.M.C., J.S. Tancrez, and R.W. Seifert, 2015: Carbon footprint and responsiveness trade-offs in supply chain network design. *Int. J. Prod. Econ.*, 166, 129–142. doi:10.1016/j.ijpe.2015.04.016.

Martin, R., M. Muûls, and U.J. Wagner, 2016: The impact of the European Union emissions trading scheme on regulated firms: What is the evidence after ten years? *Rev. Environ. Econ. Policy*, **10(1)**, 129–148, doi:10.1093/reep/rev016.

Mason, A., W. Martindale, A. Heath, and S. Chatterjee, 2016: *French Energy Transition Law – Global Investor Briefing*. PRI, London, UK. 15 pp. https://www.iigcc.org/resource/french-energy-transition-law-global-investor-briefing/. (Accessed October 2, 2021).

Material Economics, 2019: *Industrial Transformation 2050: Pathways to net-zero emisisons from EU Heavy Industry*. Cambridge, UK. 207 pp. https://materialeconomics.com/publications/industrial-transformation-2050. (Accessed December 14, 2019).

11

Mathews, J.A., H. Tan, and M. Hu, 2018: Moving to a Circular Economy in China: Transforming Industrial Parks into Eco-industrial Parks. *Calif. Manage. Rev.*, **60(3)**, 157–181, doi:10.1177/0008125617752692.

Mathy, S., P. Criqui, K. Knoop, M. Fischedick, and S. Samadi, 2016: Uncertainty management and the dynamic adjustment of deep decarbonization pathways. *Clim. Policy*, **16(sup1)**, S47–S62, doi:10.1080/14693062.2016.1179618.

Mayer, A. et al., 2019: Measuring Progress towards a Circular Economy: A Monitoring Framework for Economy-wide Material Loop Closing in the EU28. *J. Ind. Ecol.*, **23(1)**, 62–76, doi:10.1111/jiec.12809.

Mazzucato, M., 2016: From market fixing to market-creating: a new framework for innovation policy. *Ind. Innov.*, **23(2)**, 140–156, doi:10.1080/13662716.2016.1146124.

Mazzucato, M., R. Kattel, and J. Ryan-Collins, 2020: Challenge-Driven Innovation Policy: Towards a New Policy Toolkit. *J. Ind. Compet. Trade*, **20(2)**, 421–437, doi:10.1007/s10842-019-00329-w.

McBrien, M., A.C. Serrenho, and J.M. Allwood, 2016: Potential for energy savings by heat recovery in an integrated steel supply chain. *Appl. Therm. Eng.*, **103**, 592–606, doi:10.1016/J.APPLTHERMALENG.2016.04.099.

McDowall, W. et al., 2017: Circular Economy Policies in China and Europe. *J. Ind. Ecol.*, **21(3)**, 651–661, doi:10.1111/jiec.12597.

McManus, M. C. et al., 2015: Challenge clusters facing LCA in environmental decision-making – what we can learn from biofuels. *Int. J. Life Cycle Assess.*, **20(10)**, 1399–1414, doi:10.1007/s11367-015-0930-7.

McMillan, C. et al., 2016: *Generation and Use of Thermal Energy in the U.S. Industrial Sector and Opportunities to Reduce its Carbon Emissions*. Denver, USA, 172 pp. http://www.nrel.gov/docs/fy17osti/66763.pdf. (Accessed March 27, 2021).

Mehling, M.A. and R.A. Ritz, 2020: *Going beyond default intensities in an EU carbon border adjustment mechanism*. Climate Strategies, London, UK. 23 pp.

Mehling, M.A., H. Van Asselt, K. Das, S. Droege, and C. Verkuijl, 2019: Designing Border Carbon Adjustments for Enhanced Climate Action. *Am. J. Int. Law*, **113(3)**, 433–481, doi:10.1017/ajil.2019.22.

Melton, N., J. Axsen, and D. Sperling, 2016: Moving beyond alternative fuel hype to decarbonize transportation. *Nat. Energy*, **1(3)**, 1–10, doi:10.1038/nenergy.2016.13.

Mendez-Alva, F., H. Cervo, G. Krese, and G. Van Eetvelde, 2021: Industrial symbiosis profiles in energy-intensive industries: Sectoral insights from open databases. *J. Clean. Prod.*, **314** (June), 128031, doi:10.1016/j.jclepro.2021.128031.

Meng, B., G.P. Peters, Z. Wang, and M. Li, 2018: Tracing $CO_2$ emissions in global value chains. *Energy Econ.*, **73**, 24–42, doi:10.1016/j.eneco.2018.05.013.

Metcalf, G.E., 2019: *On the Economics of a Carbon Tax for the United States*. 405–484 pp.

Meys, R. et al., 2020: Towards a circular economy for plastic packaging wastes – the environmental potential of chemical recycling. *Resour. Conserv. Recycl.*, **162**, 105010, doi:10.1016/j.resconrec.2020.105010.

Meys, R. et al., 2021: Achieving net-zero greenhouse gas emission plastics by a circular carbon economy. *Science*, **374(6563)**, 71–76, doi:10.1126/science.abg9853.

Michaelowa, A., I. Shishlov, and D. Brescia, 2019: Evolution of international carbon markets: lessons for the Paris Agreement. *Wiley Interdiscip. Rev. Clim. Change*, **10(6)**, 1–24, doi:10.1002/wcc.613.

Michelsen, O. and L. de Boer, 2009: Green procurement in Norway; a survey of practices at the municipal and county level. *J. Environ. Manage.*, **91(1)**, 160–167, doi:10.1016/J.JENVMAN.2009.08.001.

Minelgaitė, A. and G. Liobikienė, 2019: Waste problem in European Union and its influence on waste management behaviours. *Sci. Total Environ.*, **667**, 86–93, doi:10.1016/j.scitotenv.2019.02.313.

Ministère de la Transition écologique et solidaire, 2018: *Projet de Stratégie Nationale Bas-Carbone – La transition écologique et solidaire vers la neutralité carbone*. Copenhagen, Denmark, 151 pp.

Minx, J.C. et al., 2021: A comprehensive and synthetic dataset for global, regional, and national greenhouse gas emissions by sector 1970–2018 with an extension to 2019. *Earth Syst. Sci. Data*,13, 5213–5252,. https://doi.org/10.5194/essd-13-5213-2021.

Moore, J., 2017: Thermal Hydrogen: An emissions free hydrocarbon economy. *Int. J. Hydrogen Energy*, **42(17)**, 12047–12063, doi:10.1016/j.ijhydene.2017.03.182.

Moran, D., A. Hasanbeigi, and C. Springer, 2018: *The Carbon Loophole in Climate Policy*. 65 pp. KGM & Associates Pty Ltd. and Global Efficiency Intelligence, LLC. Sydney, Australia and FL, USA.

Mörsdorf, G., 2021: A simple fix for carbon leakage? Assessing the environmental effectiveness of the EU carbon border adjustment. *Energy Policy*, 161112596, doi:10.1016/j.enpol.2021.112596.

Moya, J.A., and N. Pardo, 2013: The potential for improvements in energy efficiency and $CO_2$ emissions in the EU27 iron and steel industry under different payback periods. *J. Clean. Prod.*, **52**, 71–83, doi:10.1016/j.jclepro.2013.02.028.

Müller, D.B., T. Wang, and B. Duval, 2011: Patterns of iron use in societal evolution. *Environ. Sci. Technol.*, **45(1)**, 182–188, doi:10.1021/es102273t.

Müller, L.J. et al., 2020: The carbon footprint of the carbon feedstock $CO_2$. *Energy Environ. Sci.*, **13(9)**, 2979–2992, doi:10.1039/d0ee01530j.

Munasinghe, M., P. Jayasinghe, V. Ralapanawe, and A. Gajanayake, 2016: Supply/value chain analysis of carbon and energy footprint of garment manufacturing in Sri Lanka. *Sustain. Prod. Consum.*, **5**, 51–64, doi:10.1016/J.SPC.2015.12.001.

Munnings, C., W. Acworth, O. Sartor, Y.G. Kim, and K. Neuhoff, 2019: Pricing carbon consumption: synthesizing an emerging trend. *Clim. Policy*, **19(1)**, 92–107, doi:10.1080/14693062.2018.1457508.

Murray, A., K. Skene, and K. Haynes, 2017: The Circular Economy: An Interdisciplinary Exploration of the Concept and Application in a Global Context. *J. Bus. Ethics*, **140(3)**, 369–380, doi:10.1007/s10551-015-2693-2.

Muslemani, H., X. Liang, K. Kaesehage, F. Ascui, and J. Wilson, 2021: Opportunities and challenges for decarbonizing steel production by creating markets for 'green steel' products. *J. Clean. Prod.*, **315**, 128127, doi:10.1016/J.JCLEPRO.2021.128127.

Nadel, S. and L. Ungar, 2019: *Halfway There: Energy Efficiency Can Cut Energy Use and Greenhouse Gas Emissions in Half by 2050*. ACEEE, Washington D.C., USA, 63 pp. https://www.aceee.org/sites/default/files/publications/researchreports/u1907.pdf. (Accessed August 28, 2021).

Naegele, H. and A. Zaklan, 2019: Does the EU ETS cause carbon leakage in European manufacturing? *J. Environ. Econ. Manage.*, **93**, 125–147, doi:10.1016/j.jeem.2018.11.004.

Naims, H., 2016: Economics of carbon dioxide capture and utilization—a supply and demand perspective. *Environ. Sci. Pollut. Res.*, **23(22)**, 22226–22241, doi:10.1007/s11356-016-6810-2.

Napp, T.A., A. Gambhir, T.P. Hills, N. Florin, and P. Fennell, 2014: A review of the technologies, economics and policy instruments for decarbonising energy-intensive manufacturing industries. *Renew. Sustain. Energy Rev.*, **30**, 616–640, doi:10.1016/j.rser.2013.10.036.

Napp, T. A. et al., 2019: The role of advanced demand-sector technologies and energy demand reduction in achieving ambitious carbon budgets. *Appl. Energy*, **238**(April 2018), 351–367, doi:10.1016/j.apenergy.2019.01.033.

Narassimhan, E., K.S. Gallagher, S. Koester, and J.R. Alejo, 2018: Carbon pricing in practice: a review of existing emissions trading systems. *Clim. Policy*, **18(8)**, 967–991, doi:10.1080/14693062.2018.1467827.

NEDO, 2019: *Survey of waste heat in the industrial field*. Kanagawa, Japan. 239 pp. http://www.thermat.jp/HainetsuChousa/HainetsuReport.pdf. (Accessed December 15, 2019).

Nemet, G.F., 2019: *How solar energy became cheap: A model for low-carbon innovation*. Routledge, Abingdon, Oxon, UK and New York, NY, USA, 238 pp.

Nemet, G.F., V. Zipperer, and M. Kraus, 2018: The valley of death, the technology pork barrel, and public support for large demonstration projects. *Energy Policy*, **119**, 154–167, doi:10.1016/j.enpol.2018.04.008.

Neuhoff, K. et al., 2015: *Inclusion of Consumption of Carbon Intensive Commodities in Carbon Pricing Mechanisms*. 6 pp. Climate Strategies., London, UK. https://www.diw.de/documents/dokumentenarchiv/17/diw_01.c.523297.de/policy-brief-ioc.pdf (Accessed December, 23, 2019).

Neuhoff, K. et al., 2018: *Filling gaps in the policy package to decarbonise production and use of materials*. Berlin, Germany, 41 pp. https://climatestrategies.org/wp-content/uploads/2018/06/CS-DIW_report-designed-2.pdf. (Accessed December 16, 2019).

Neuhoff, K. et al., 2019: *Building blocks for a climate-neutral European industrial sector*. Climate Strategies. London, UK. 33 pp. https://climatestrategies.org/wp-content/uploads/2019/10/Building-Blocks-for-a-Climate-Neutral-European-Industrial-Sector.pdf. (Accessed October, 26, 2021).

NGFS, 2019a: *A call for action: Climate change as a source of financial risk*. NGFS. Paris, France. 39 pp. https://www.ngfs.net/sites/default/files/medias/documents/ngfs_first_comprehensive_report_-_17042019_0.pdf. (Accessed December, 15, 2019).

NGFS, 2019b: *A sustainable and responsible investment guide for central banks' portfolio management*. NGFS. Paris, France. 39 pp. https://www.ngfs.net/sites/default/files/medias/documents/ngfs-a-sustainable-and-responsible-investment-guide.pdf. (Accessed December, 15, 2019).

Nikolaou, I.E. and K.P. Tsagarakis, 2021: An introduction to circular economy and sustainability: Some existing lessons and future directions. *Sustain. Prod. Consum.*, **28**, 600–609, doi:10.1016/j.spc.2021.06.017.

Nilsson, L.J. et al., 2021: An industrial policy framework for transforming energy and emissions intensive industries towards zero emissions. *Clim. Policy*, **21(8)**, 1053–1065, doi:10.1080/14693062.2021.1957665.

Nimbalkar, S. et al., 2020: Enhancing operational performance and productivity benefits in breweries through smart manufacturing technologies. *J. Adv. Manuf. Process.*, **2(4)**, doi:10.1002/amp2.10064.

Nocito, F. and A. Dibenedetto, 2020: Atmospheric $CO_2$ mitigation technologies: carbon capture utilization and storage. *Curr. Opin. Green Sustain. Chem.*, **21**, 34–43, doi:10.1016/j.cogsc.2019.10.002.

Nordhaus, W., 2015: Climate clubs: Overcoming free-riding in international climate policy. *Am. Econ. Rev.*, **105(4)**, 1339–1370, doi:10.1257/aer.15000001.

Oberle, B. et al., 2019: *Global Resources Outlook 2019: Natural Resources for the Future We Want*. Paris, France, 168 pp. https://www.resourcepanel.org/reports/global-resources-outlook. (Accessed March 27, 2021).

Oberthür, S.,G. Khandekar and T. Wyns, 2021: Global governance for the decarbonization of energy-intensive industries: Great potential underexploited. *Earth Syst. Gov.*, **8**, 100072, doi:10.1016/j.esg.2020.100072.

Ochu, E.R., and S.J. Friedmann, 2021: CCUS in a net-zero U.S. power sector: Policy design, rates, and project finance. *Electr. J.*, **34(7)**, 107000, doi:10.1016/j.tej.2021.107000.

OECD.Stat, 2019: OECD.Stat 2019. https://stats.oecd.org/Index.aspx?DataSetCode=IO_GHG_2019 (Accessed December 20, 2020).

OECD, 2011: *Towards Green Growth*. Paris, France, 24 pp. https://www.oecd.org/greengrowth/48012345.pdf. (Accessed December 15, 2019).

OECD, 2015: *System Innovation: Synthesis Report*. Paris, France, 101 pp. http://www.pte.pl/pliki/2/1/OECD System.pdf. (Accessed March 27, 2021).

OECD, 2016: *Extended Producer Responsibility*. OECD. Paris, France. 6 pp.

OECD, 2019a: *Global Material Resources Outlook to 2060*. OECD, Paris, France.210 pp.

OECD, 2019b: Moving to sustainable industrial production. In: *Accelerating Climate Action: Refocusing Policies through a Well-being Lens*, pp. 65–88.

OECD, 2021: *Government at a Glance 2021*. OECD, Paris, France, 281 pp.

Ohnishi, S., H. Dong, Y. Geng, M. Fujii, and T. Fujita, 2017: A comprehensive evaluation on industrial & urban symbiosis by combining MFA, carbon footprint and emergy methods – Case of Kawasaki, Japan. *Ecol. Indic.*, **73**, 513–524, doi:10.1016/j.ecolind.2016.10.016.

Ohta, H., S.W. Kim, S. Kaneki, A. Yamamoto, and T. Hashizume, 2018: High Thermoelectric Power Factor of High-Mobility 2D Electron Gas. *Adv. Sci.*, **5(1)**, 1700696, doi:10.1002/advs.201700696.

Okereke, C. and D. McDaniels, 2012: To what extent are EU steel companies susceptible to competitive loss due to climate policy? *Energy Policy*, **46**, 203–215, doi:10.1016/j.enpol.2012.03.052.

Olatunji, O.O. et al., 2019: Competitive advantage of carbon efficient supply chain in manufacturing industry. *J. Clean. Prod.*, **238**, 117937, doi:10.1016/j.jclepro.2019.117937.

Olfe-Kräutlein, B., 2020: Advancing CCU Technologies Pursuant to the SDGs: A Challenge for Policy Making. *Front. Energy Res.*, **8(198)**, doi:10.3389/fenrg.2020.00198.

Olivetti, E.A., and J.M. Cullen, 2018: Toward a sustainable materials system. *Science*, **360(6396)**, 1396–1398, doi:10.1126/science.aat6821.

Olivier, J.G J., and J.A.H.W., Peters, 2018: *Trends in global $CO_2$ and total greenhouse gas emissions – 2018 report*. The Hague, the Netherlands, 53 pp. https://www.pbl.nl/sites/default/files/downloads/pbl-2018-trends-in-global-co2-and-total-greenhouse-gas-emissons-2018-report_3125_0.pdf.

Onarheim, K., S. Santos, P. Kangas, and V. Hankalin, 2017: Performance and costs of CCS in the pulp and paper industry part 1: Performance of amine-based post-combustion $CO_2$ capture. *Int. J. Greenh. Gas Control*, **59**, 58–73, doi:10.1016/j.ijggc.2017.02.008.

Orr, J. et al., 2019: Minimising energy in construction: Practitioners' views on material efficiency. *Resour. Conserv. Recycl.*, **140**, 125–136, doi:10.1016/j.resconrec.2018.09.015.

Overland, I., 2019: The geopolitics of renewable energy: Debunking four emerging myths. *Energy Res. Soc. Sci.*, **49**, 36–40, doi:10.1016/j.erss.2018.10.018.

Padella, M., A. O'Connell, and M. Prussi, 2019: What is still Limiting the Deployment of Cellulosic Ethanol? Analysis of the Current Status of the Sector. *Appl. Sci.*, **9(21)**, 4523, doi:10.3390/app9214523.

Palm, E. and A. Nikoleris, 2021: Conflicting expectations on carbon dioxide utilisation. *Technol. Anal. Strateg. Manag.*, **33(2)**, 217–228, doi:10.1080/09537325.2020.1810225.

Palm, E., L.J. Nilsson, and M. Åhman, 2016: Electricity-based plastics and their potential demand for electricity and carbon dioxide. *J. Clean. Prod.*, **129**, 548–555, doi:10.1016/j.jclepro.2016.03.158.

Pan, S.Y. et al., 2020: $CO_2$ mineralization and utilization by alkaline solid wastes for potential carbon reduction. *Nat. Sustain.*, **3(5)**, 399–405, doi:10.1038/s41893-020-0486-9.

Pandey, D., M. Agrawal, and J.S. Pandey, 2011: Carbon footprint: current methods of estimation. *Environ. Monit. Assess.*, **178(1–4)**, 135–160, doi:10.1007/s10661-010-1678-y.

Papapetrou, M., G. Kosmadakis, A. Cipollina, U. La Commare, and G. Micale, 2018: Industrial waste heat: Estimation of the technically available resource in the EU per industrial sector, temperature level and country. *Appl. Therm. Eng.*, **138**, 207–216, doi:10.1016/j.applthermaleng.2018.04.043.

Pardo, N. and J.A. Moya, 2013: Prospective scenarios on energy efficiency and $CO_2$ emissions in the European Iron and Steel industry. *Energy*, **54**, 113–128, doi:10.1016/j.energy.2013.03.015.

Park, J., J.M. Park, and H.S. Park, 2019: Scaling-Up of Industrial Symbiosis in the Korean National Eco-Industrial Park Program: Examining Its Evolution over the 10 Years between 2005–2014. *J. Ind. Ecol.*, **23**(1), 197–207, doi:10.1111/jiec.12749.

Pasquier, L.-C., N. Kemache, J. Mocellin, J.-F. Blais, and G. Mercier, 2018: Waste Concrete Valorization; Aggregates and Mineral Carbonation Feedstock Production. *Geosciences*, **8(9)**, 342, doi:10.3390/geosciences8090342.

Pauliuk, S., R.L. Milford, D.B. Müller, and J.M. Allwood, 2013a: The steel scrap age. *Environ. Sci. Technol.*, **47(7)**, 3448–3454, doi:10.1021/es303149z.

Pauliuk, S., T. Wang, and D.B. Müller, 2013b: Steel all over the world: Estimating in-use stocks of iron for 200 countries. *Resour. Conserv. Recycl.*, **71**, 22–30, doi:10.1016/j.resconrec.2012.11.008.

Pauliuk, S., G. Majeau-Bettez, and D.B. Müller, 2015: A general system structure and accounting framework for socioeconomic metabolism. *J. Ind. Ecol.*, **19(5)**, 728–741, doi:10.1111/jiec.12306.

Pauliuk, S., N. Heeren, M.M. Hasan, and D.B. Müller, 2019: A general data model for socioeconomic metabolism and its implementation in an industrial ecology data commons prototype. *J. Ind. Ecol.*, **23(5)**, 1016–1027, doi:10.1111/jiec.12890.

Pauliuk, S. et al., 2021: Global scenarios of resource and emission savings from material efficiency in residential buildings and cars. *Nat. Commun.*, **12(1)**, 5097, doi:10.1038/s41467-021-25300-4.

Philibert, C., 2017a: *Renewable Energy for Industry: From Green Energy to Green Materials and Fuels*. IEA, Paris, France, 67 pp. https://www.iea.org/publications/insights/insightpublications/Renewable_Energy_for_Industry.pdf.

Philibert, C., 2017b: *Producing ammonia and fertilizers: new opportunities from renewables.* Renewable Energy Division, IEA. Paris, France, 6 pp. http://www.iea.org/media/news/2017/FertilizermanufacturingRenewables_1605.pdf.

Pieroni, M.P.P., T.C. McAloone, and D.C.A. Pigosso, 2019: Business model innovation for circular economy and sustainability: A review of approaches. *J. Clean. Prod.*, **215**, 198–216, doi:10.1016/j.jclepro.2019.01.036.

Plaza, M.G., S. Martínez, and F. Rubiera, 2020: $CO_2$ capture, use, and storage in the cement industry: State of the art and expectations. *Energies*, **13(21)**, 5692, doi:10.3390/en13215692.

Pollitt, H., K. Neuhoff, and X. Lin, 2020: The impact of implementing a consumption charge on carbon-intensive materials in Europe. *Clim. Policy*, **20(sup1)**, S74–S89, doi:10.1080/14693062.2019.1605969.

Polverini, D., 2021: Regulating the circular economy within the ecodesign directive: Progress so far, methodological challenges and outlook. *Sustain. Prod. Consum.*, **27**, 1113–1123, doi:10.1016/j.spc.2021.02.023.

Pülzl, H., D. Kleinschmit, and B. Arts, 2014: Bioeconomy – an emerging meta-discourse affecting forest discourses? *Scand. J. For. Res.*, **29(4)**, 386–393, doi:10.1080/02827581.2014.920044.

Pusnik, M., F. Al-Mansour, B. Sucic, and A.F. Gubina, 2016: Gap analysis of industrial energy management systems in Slovenia. *Energy*, **108**, 41–49, doi:10.1016/j.energy.2015.10.141.

Pyrka, M., J. Boratyński, I. Tobiasz, R. Jeszke, and M. Sekuła, 2020: *The Effects of the Implementation of the Border Tax Adjustment in the Context of More Stringent EU Climate Policy until 2030*. Institute of Environmental Protection - National Research InstituteWarsaw, Poland, 48 pp.

Qian, W., J. Hörisch, and S. Schaltegger, 2018: Environmental management accounting and its effects on carbon management and disclosure quality. *J. Clean. Prod.*, **174**, 1608–1619, doi:10.1016/j.jclepro.2017.11.092.

Qu, S. et al., 2019: Implications of China's foreign waste ban on the global circular economy. *Resour. Conserv. Recycl.*, **144**, 252–255, doi:10.1016/j.resconrec.2019.01.004.

Ragaert, K., L. Delva, and K. Van Geem, 2017: Mechanical and chemical recycling of solid plastic waste. *Waste Manag.*, **69**, 24–58, doi:10.1016/j.wasman.2017.07.044.

Rahimi, A. and J. M. García, 2017: Chemical recycling of waste plastics for new materials production. *Nat. Rev. Chem.*, **1(6)**, 0046, doi:10.1038/s41570-017-0046.

Rajbhandari, A. and F. Zhang, 2018: Does energy efficiency promote economic growth? Evidence from a multicountry and multisectoral panel dataset. *Energy Econ.*, **69**, 128–139, doi:10.1016/j.eneco.2017.11.007.

Ranasinghe, R., A.C. Ruane, R. Vautard, N. Arnell, E. Coppola, F.A. Cruz, S. Dessai, A.S. Islam, M. Rahimi, D. Ruiz Carrascal, J. Sillmann, M.B. Sylla, C. Tebaldi, W. Wang, R. Zaaboul, 2021, 2021: Climate Change Information for Regional Impact and for Risk Assessment. In: *Climate Change 2021: The Physical Science Basis. Contribution of Working Group I to the Sixth Assessment Report of the Intergovernmental Panel on Climate Change* [Masson-Delmotte, V., P. Zhai, A. Pirani, S. L. Connors, C. Péan, S. Berger, N. Caud, Y. Chen, L. Goldfarb, M. I. Gomis, M. Huang, K. Leitzell, E. Lonnoy, J. B. R. Matthews, T. K. Maycock, T. Waterfield, O. Yelekçi, R. Yu and B. Zhou (eds.)]. Cambridge University Press, Cambridge, UK and New York, NY, USA.

Rangelov, M., H. Dylla, A. Mukherjee, and N. Sivaneswaran, 2021: Use of environmental product declarations (EPDs) of pavement materials in the United States of America (USA) to ensure environmental impact reductions. *J. Clean. Prod.*, **283**, 124619, doi:10.1016/j.jclepro.2020.124619.

Ranta, V., L. Aarikka-Stenroos, P. Ritala, and S. J. Mäkinen, 2018: Exploring institutional drivers and barriers of the circular economy: A cross-regional comparison of China, the US, and Europe. *Resour. Conserv. Recycl.*, **135**, 70–82, doi:10.1016/j.resconrec.2017.08.017.

Raven, R., F. Kern, B. Verhees, and A. Smith, 2016: Niche construction and empowerment through socio-political work. A meta-analysis of six low-carbon technology cases. *Environ. Innov. Soc. Transitions*, **18**, 164–180, doi:10.1016/j.eist.2015.02.002.

Raymond, L., 2019: Policy perspective: Building political support for carbon pricing – Lessons from cap-and-trade policies. *Energy Policy*, **134**, 110986, doi:10.1016/j.enpol.2019.110986.

Reike, D., W.J.V. Vermeulen, and S. Witjes, 2018: The circular economy: New or Refurbished as CE 3.0? — Exploring Controversies in the Conceptualization of the Circular Economy through a Focus on History and Resource Value Retention Options. *Resour. Conserv. Recycl.*, **135**, 246–264, doi:10.1016/j.resconrec.2017.08.027.

Renforth, P., 2019: The negative emission potential of alkaline materials. *Nat. Commun.*, **10(1)**, 1401, doi:10.1038/s41467-019-09475-5.

Richstein, J.C., 2017: *Project-Based Carbon Contracts: A Way to Finance Innovative Low-Carbon Investments*. DIW Berlin Discussion Paper No. 1714, 18 pp. doi.org/10.2139/ssrn.3109302

Rietbergen, M.G and K. Blok, 2013: Assessing the potential impact of the $CO_2$ Performance Ladder on the reduction of carbon dioxide emissions in the Netherlands. *J. Clean. Prod.*, **52**, 33–45, doi:10.1016/j.jclepro.2013.03.027.

Rietbergen, M.G., A. Van Rheede, and K. Blok, 2015: The target-setting process in the $CO_2$ Performance Ladder: Does it lead to ambitious goals for carbon dioxide emission reduction? *J. Clean. Prod.*, **103**, 549–561, doi:10.1016/j.jclepro.2014.09.046.

Rissman, J. et al., 2020: Technologies and policies to decarbonize global industry: Review and assessment of mitigation drivers through 2070. *Appl. Energy*, **266** (November 2019), 114848, doi:10.1016/j.apenergy.2020.114848.

Rocchi, P., M. Serrano, J. Roca, and I. Arto, 2018: Border Carbon Adjustments Based on Avoided Emissions: Addressing the Challenge of Its Design. *Ecol. Econ.*, **145**, 126–136, doi:10.1016/j.ecolecon.2017.08.003.

Röck, M. et al., 2020: Embodied GHG emissions of buildings – The hidden challenge for effective climate change mitigation. *Appl. Energy*, **258**, 114107, doi:10.1016/j.apenergy.2019.114107.

Rodrik, D., 2014: Green industrial policy. *Oxford Rev. Econ. Policy*, **30(3)**, 469–491, doi:10.1093/oxrep/gru025.

Rogers, E., 2018: Integrating smart manufacturing and strategic energy management programs. In: *ECEEE 2018 industry proceedings*, Berlin, Germany, pp. 23–31.

Rogge, K.S. and K. Reichardt, 2016: Policy mixes for sustainability transitions: An extended concept and framework for analysis. *Res. Policy*, **45(8)**, 1620–1635, doi:10.1016/j.respol.2016.04.004.

Rogge, K.S. and P. Johnstone, 2017: Exploring the role of phase-out policies for low-carbon energy transitions: The case of the German Energiewende. *Energy Res. Soc. Sci.*, **33**, 128–137, doi:10.1016/j.erss.2017.10.004.

Rogge, K.S., F. Kern, and M. Howlett, 2017: Conceptual and empirical advances in analysing policy mixes for energy transitions. *Energy Res. Soc. Sci.*, **33** (October), 1–10, doi:10.1016/j.erss.2017.09.025.

Rootzén, J. and F. Johnsson, 2016: Paying the full price of steel – Perspectives on the cost of reducing carbon dioxide emissions from the steel industry. *Energy Policy*, **98**, 459–469, doi:10.1016/j.enpol.2016.09.021.

Rootzén, J. and F. Johnsson, 2017: Managing the costs of CO2 abatement in the cement industry. *Clim. Policy*, **17(6)**, 781–800, doi:10.1080/14693062.2016.1191007.

Rosenbloom, D., J. Markard, F.W. Geels, and L. Fuenfschilling, 2020: Why carbon pricing is not sufficient to mitigate climate change—and how "sustainability transition policy" can help. *Proc. Natl. Acad. Sci.*, **117(16)**, 8664–8668.

Saevarsdottir, G., H. Kvande, and B.J. Welch, 2020: Aluminum Production in the Times of Climate Change: The Global Challenge to Reduce the Carbon Footprint and Prevent Carbon Leakage. *JOM*, **72(1)**, 296–308, doi:10.1007/s11837-019-03918-6.

Sakai, M. and J. Barrett, 2016: Border carbon adjustments: Addressing emissions embodied in trade. *Energy Policy*, **92**, 102–110, doi:10.1016/j.enpol.2016.01.038.

Samadi, S. and C. Barthel, 2020: Meta-Analysis of industry sector transformation strategies in German, European, and global deep decarbonization scenarios. *ECEEE Industrial Summer Study Proceedings.*, 445–455. https://epub.wupperinst.org/frontdoor/deliver/index/docId/7576/file/7576_Samadi.pdf

Sanchez, D.L., N. Johnson, S.T. McCoy, P.A. Turner, and K.J. Mach, 2018: Near-term deployment of carbon capture and sequestration from biorefineries in the United States. *Proc. Natl. Acad. Sci.*, **115(19)**, 4875–4880, doi:10.1073/pnas.1719695115.

Sandalow, D. et al., 2019: *Industrial heat decarbonization roadmap*. ICCEF. Bonn, Germany, 75 pp. https://www.icef.go.jp/pdf/summary/roadmap/icef2019_roadmap.pdf. (Accessed March 27, 2021).

Sandberg, N.H. et al., 2016: Dynamic building stock modelling: Application to 11 European countries to support the energy efficiency and retrofit ambitions of the EU. *Energy Build.*, **132**, 26–38, doi:10.1016/j.enbuild.2016.05.100.

Sanjuán, M.Á., C. Andrade, P. Mora, and A. Zaragoza, 2020: Carbon Dioxide Uptake by Cement-Based Materials: A Spanish Case Study. *Appl. Sci.*, **10(1)**, 339, doi:10.3390/app10010339.

Sari, A. and M. Akkaya, 2016: Contribution of Renewable Energy Potential to Sustainable Employment. *Procedia - Soc. Behav. Sci.*, **229**, 316–325, doi:10.1016/j.sbspro.2016.07.142.

Sartor, O., 2013: *Carbon Leakage in the Primary Aluminium Sector: What Evidence after 6.5 Years of the EU ETS?* Elsevier BV. Amsterdam, Netherlands 20 pp.

Sartor, O., and C. Bataille, 2019: *IDDRI Policy Brief: Decarbonising basic materials in Europe: How Carbon Contracts-for-Difference could help bring breakthrough technologies to market*. IDDRI. Paris, France. 16 pp. https://www.iddri.org/en/publications-and-events/study/decarbonising-basic-materials-europe. (Accessed March, 27, 2021).

Saunders, H.D. et al., 2021: Energy Efficiency: What Has Research Delivered in the Last 40 Years? *Annu. Rev. Environ. Resour.*, **46(1)**, 135–165, doi:10.1146/annurev-environ-012320-084937.

Saussay, A. and M. Sato., 2018: *The impacts of energy prices on industrial foreign investment location: evidence from global firm level data*. Grantham Research on Climate Change and the Environment. London, UK. 63 pp.

Saygin, D. and D. Gielen, 2021: Zero-emission pathway for the global chemical and petrochemical sector. *Energies*, **14(13)**, 3772, doi:10.3390/en14133772.

Saygin, D., E. Worrell, M.K. Patel, and D.J. Gielen, 2011: Benchmarking the energy use of energy-intensive industries in industrialized and in developing countries. *Energy*, **36(11)**, 6661–6673, doi:10.1016/j.energy.2011.08.025.

SBTi, 2021: *From Ambition To Impact: How Companies Are Reducing Emissions At Scale With Science-Based Targets. Science Based Targets Initiative (SBTi) Annual Progress Report, 2020*. 76 pp. https://sciencebasedtargets.org/resources/files/SBTiProgressReport2020.pdf. (Accessed August, 21, 2021).

Scheubel, C., T. Zipperle, and P. Tzscheutschler, 2017: Modeling of industrial-scale hybrid renewable energy systems (HRES) – The profitability of decentralized supply for industry. *Renew. Energy*, **108**, 52–63, doi:10.1016/j.renene.2017.02.038.

Schipper, L., E. Deakin, and C. McAndrews, 2011: Carbon Dioxide Emissions from Urban Road Transport in Latin America: $CO_2$ Reduction as a Co-Benefit of Transport Strategies. In: Rothengatter, W., Hayashi, Y., Schade, W. (eds) Transport Moving to Climate Intelligence. Transportation Research, Economics and Policy. Springer, New York, NY. pp. 111–127. https://doi.org/10.1007/978-1-4419-7643-7_8

Schmalensee, R. and R.N. Stavins, 2017: Lessons learned from three decades of experience with cap and trade. *Rev. Environ. Econ. Policy*, **11(1)**, 59–79, doi:10.1093/reep/rew017.

Schneider, M., 2019: The cement industry on the way to a low-carbon future. *Cem. Concr. Res.*, **124**, 105792, doi:10.1016/j.cemconres.2019.105792.

Schneider, S., S. Bajohr, F. Graf, and T. Kolb, 2020: State of the Art of Hydrogen Production via Pyrolysis of Natural Gas. *ChemBioEng Rev.*, **7(5)**, 150–158, doi:10.1002/cben.202000014.

Schoeneberger, C.A. et al., 2020: Solar for industrial process heat: A review of technologies, analysis approaches, and potential applications in the United States. *Energy*, **206**, 118083, doi:10.1016/j.energy.2020.118083.

Schriever, M. and D. Halstrup, 2018: Exploring the adoption in transitioning markets: Empirical findings and implications on energy storage solutions-acceptance in the German manufacturing industry. *Energy Policy*, **120**, 460–468, doi:10.1016/j.enpol.2018.03.029.

Schroeder, P., K. Anggraeni, and U. Weber, 2019: The Relevance of Circular Economy Practices to the Sustainable Development Goals. *J. Ind. Ecol.*, **23(1)**, 77–95, doi:10.1111/jiec.12732.

Schwarz, M., C. Nakhle, and C. Knoeri, 2020: Innovative designs of building energy codes for building decarbonization and their implementation challenges. *J. Clean. Prod.*, **248**, 119260, doi:10.1016/j.jclepro.2019.119260.

Scott, K., J. Giesekam, J. Barrett, and A. Owen, 2019: Bridging the climate mitigation gap with economy-wide material productivity. *J. Ind. Ecol.*, **23(4)**, 918–931, doi:10.1111/jiec.12831.

Scrivener, K.L., V.M. John, and E.M. Gartner, 2018: Eco-efficient cements: Potential economically viable solutions for a low-CO2 cement-based materials industry. *Cem. Concr. Res.*, **114**(December), 2–26, doi:10.1016/j.cemconres.2018.03.015.

Sepulveda, N. A., J. D. Jenkins, F. J. de Sisternes, and R. K. Lester, 2018: The Role of Firm Low-Carbon Electricity Resources in Deep Decarbonization of Power Generation. *Joule*, **2(11)**, 2403–2420, doi:10.1016/j.joule.2018.08.006.

Serrenho, A.C., Z.S. Mourão, J. Norman, J. M. Cullen, and J.M. Allwood, 2016: The influence of UK emissions reduction targets on the emissions of the global steel industry. *Resour. Conserv. Recycl.*, **107**, 174–184, doi:10.1016/j.resconrec.2016.01.001.

Serrenho, A.C., J.B. Norman, and J.M. Allwood, 2017: The impact of reducing car weight on global emissions: The future fleet in Great Britain. *Philos. Trans. R. Soc. A Math. Phys. Eng. Sci.*, **375(2095)**, 20160364, doi:10.1098/rsta.2016.0364.

Seto, K. C. et al., 2016: Carbon Lock-In: Types, Causes, and Policy Implications. *Annu. Rev. Environ. Resour.*, **41**(1), 425–452, doi:10.1146/annurev-environ-110615-085934.

Shadram, F., S. Bhattacharjee, S. Lidelöw, J. Mukkavaara, and T. Olofsson, 2020: Exploring the trade-off in life cycle energy of building retrofit through optimization. *Appl. Energy*, **269**, 115083, doi:10.1016/j.apenergy.2020.115083.

Shanks, W. et al., 2019: How much cement can we do without? Lessons from cement material flows in the UK. *Resour. Conserv. Recycl.*, **141**, 441–454, doi:10.1016/j.resconrec.2018.11.002.

Sharmina, M. et al., 2021: Decarbonising the critical sectors of aviation, shipping, road freight and industry to limit warming to 1.5–2°C. *Clim. Policy*, **21(4)**, 455–474, doi:10.1080/14693062.2020.1831430.

Shell, 2018: *Shell Scenarios Sky: Meeting the Goals of Paris Agreement*. The Hague, the Netherlands, 69 pp. https://www.shell.com/energy-and-innovation/the-energy-future/scenarios/shell-scenario-sky.html. (Accessed December 23, 2019).


11

Shi, L. and B. Yu, 2014: Eco-industrial parks from strategic niches to development mainstream: The cases of China. *Sustainability.*, **6(9)**, 6325–6331, doi:10.3390/su6096325.

Shoreh, M.H., P. Siano, M. Shafie-khah, V. Loia, and J.P.S. Catalão, 2016: A survey of industrial applications of Demand Response. *Electr. Power Syst. Res.*, **141**, 31–49, doi:10.1016/j.epsr.2016.07.008.

Shuai, C.-M., L.-P. Ding, Y.-K. Zhang, Q. Guo, and J. Shuai, 2014: How consumers are willing to pay for low-carbon products? – Results from a carbon-labeling scenario experiment in China. *J. Clean. Prod.*, **83**, 366–373, doi:10.1016/J.JCLEPRO.2014.07.008.

Simonen, K. et al., 2019: *Buy Clean Washington Study*. UW, WSU and CWU. Washington D.C., USA. 173 pp.

Skelton, A.C.H. and J.M. Allwood, 2017: The carbon price: a toothless tool for material efficiency? *Philos. Trans. R. Soc. A Math. Phys. Eng. Sci.*, **375(2095)**, 20160374, doi:10.1098/rsta.2016.0374.

Smith, A. and R. Raven, 2012: What is protective space? Reconsidering niches in transitions to sustainability. *Res. Policy*, **41(6)**, 1025–1036, doi:10.1016/j.respol.2011.12.012.

Söderholm, P. et al., 2019: Technological development for sustainability: The role of network management in the innovation policy mix. *Technol. Forecast. Soc. Change*, **138**, 309–323, doi:10.1016/j.techfore.2018.10.010.

Soo, V.K. et al., 2021: The influence of end-of-life regulation on vehicle material circularity: A comparison of Europe, Japan, Australia and the US. *Resour. Conserv. Recycl.*, **168** (November 2020), 105294, doi:10.1016/j.resconrec.2020.105294.

Spangenberg, J.H. and S. Lorek, 2019: Sufficiency and consumer behaviour: From theory to policy. *Energy Policy*, **129**, 1070–1079, doi:10.1016/j.enpol.2019.03.013.

Statista, 2021a: World steel production and scrap consumption 2020. https://www.statista.com/statistics/270835/world-steel-production-and-scrap-consumption/ (Accessed August 28, 2021).

Statista, 2021b: Global plastic production 1950–2020. https://www.statista.com/statistics/282732/global-production-of-plastics-since-1950/ (Accessed August 28, 2021).

Steger, S. et al., 2019: *Stoffstromorientierte Ermittlung des Beitrags der Sekundärrohstoffwirtschaft zur Schonung von Primärrohstoffen und Steigerung der Ressourcenproduktivität*. Dessau, Germany, 391 pp. https://www.umweltbundesamt.de/publikationen/stoffstromorientierte-ermittlung-des-beitrags-der. (Accessed December 15, 2019).

Stern, D., 2019: *Routledge Handbook of Energy Economics*. [Soytaş, U. and R. Sarı, (eds.)]. Routledge, London, UK.

Sternberg, A. and A. Bardow, 2015: Power-to-What? – Environmental assessment of energy storage systems. *Energy Environ. Sci.*, **8(2)**, 389–400, doi:10.1039/C4EE03051F.

Stiglitz, J., 2006: A New Agenda for Global Warming. *Econ. Voice*, **3(7)**, doi:10.2202/1553-3832.1210.

Stiglitz, J.E., 2019: Addressing climate change through price and non-price interventions. *Eur. Econ. Rev.*, **119**, 594–612, doi:10.1016/j.euroecorev.2019.05.007.

Stiglitz, J.E. et al., 2017: *Report of the High-Level Commission on Carbon Prices.* World Bank. Washington D.C., USA. 61 pp, doi:10.7916/D8-W2NC-4103.

Stork, M., J. de Beer, N. Lintmeijer, and B. den Ouden, 2018: *Chemistry for Climate: Acting on the need for speed. Roadmap for the Dutch Chemical Industry towards 2050*. KL Utrecht, the Netherlands, 101 pp. https://www.vnci.nl/Content/Files/file/Downloads/VNCI_Routekaart-2050.pdf. (Accessed March 27, 2021).

Streeck, J., D. Wiedenhofer, F. Krausmann, and H. Haberl, 2020: Stock-flow relations in the socio-economic metabolism of the United Kingdom 1800–2017. *Resour. Conserv. Recycl.*, **161**, 104960, doi:10.1016/j.resconrec.2020.104960.

Stripple, H., C. Ljungkrantz, T. Gustafsson, and R. Andersson, 2018: *$CO_2$ uptake in cement-containing products. Background and calculation models for IPCC implementation*. IVL. Stockholm, Sweden. 66 pp.

Styring, P., D. Jansen, H. Coninick, ·De, H. Reith, and K. Armstrong, 2011: *Carbon Capture and Utilisation in the green economy*. New York, USA, 60 pp. http://co2chem.co.uk/wp-content/uploads/2012/06/CCU in the green economy report.pdf. (Accessed March 27, 2021).

Sudmant, A., A. Gouldson, J. Millward-Hopkins, K. Scott, and J. Barrett, 2018: Producer cities and consumer cities: Using production- and consumption-based carbon accounts to guide climate action in China, the UK, and the US. *J. Clean. Prod.*, **176**, 654–662, doi:10.1016/j.jclepro.2017.12.139.

Sullivan, R. and A. Gouldson, 2012: Does voluntary carbon reporting meet investors' needs? *J. Clean. Prod.*, **36**, 60–67, doi:10.1016/J.JCLEPRO.2012.02.020.

Sun, L. et al., 2017: Eco-benefits assessment on urban industrial symbiosis based on material flows analysis and emergy evaluation approach: A case of Liuzhou city, China. *Resour. Conserv. Recycl.*, **119**, 78–88, doi:10.1016/j.resconrec.2016.06.007.

Sverdrup, H.U., D. Koca, and P. Schlyter, 2017: A Simple System Dynamics Model for the Global Production Rate of Sand, Gravel, Crushed Rock and Stone, Market Prices and Long-Term Supply Embedded into the WORLD6 Model. *Biophys. Econ. Resour. Qual.*, **2(2)**, 8, doi:10.1007/s41247-017-0023-2.

Tagliapietra, S. and G.B. Wolff, 2021a: Form a climate club: United States, European Union and China. *Nature*, **591(7851)**, 526–528, doi:10.1038/d41586-021-00736-2.

Tagliapietra, S. and G.B. Wolff, 2021b: Conditions are ideal for a new climate club. *Energy Policy*, **158**, 112527, doi:10.1016/j.enpol.2021.112527.

Tait, P., C. Saunders, M. Guenther, and P. Rutherford, 2016: Emerging versus developed economy consumer willingness to pay for environmentally sustainable food production: a choice experiment approach comparing Indian, Chinese and United Kingdom lamb consumers. *J. Clean. Prod.*, **124**, 65–72, doi:10.1016/J.JCLEPRO.2016.02.088.

Talens Peiró, L., D. Polverini, F. Ardente, and F. Mathieux, 2020: Advances towards circular economy policies in the EU: The new Ecodesign regulation of enterprise servers. *Resour. Conserv. Recycl.*, **154**, 104426, doi:10.1016/j.resconrec.2019.104426.

Tan, M.Q.B., R.B.H. Tan, and H.H. Khoo, 2014: Prospects of carbon labelling – a life cycle point of view. *J. Clean. Prod.*, **72**, 76–88, doi:10.1016/J.JCLEPRO.2012.09.035.

Tanaka, K., 2008: Assessment of energy efficiency performance measures in industry and their application for policy. *Energy Policy*, **36(8)**, 2887–2902, doi:10.1016/j.enpol.2008.03.032.

Tanaka, K., 2011: Review of policies and measures for energy efficiency in industry sector. *Energy Policy*, **39(10)**, 6532–6550, doi:10.1016/j.enpol.2011.07.058.

Tang, Y. and F. You, 2018: Multicriteria Environmental and Economic Analysis of Municipal Solid Waste Incineration Power Plant with Carbon Capture and Separation from the Life-Cycle Perspective. *ACS Sustain. Chem. Eng.*, **6(1)**, 937–956, doi:10.1021/acssuschemeng.7b03283.

Tanzer, S.E. and A. Ramírez, 2019: When are negative emissions negative emissions? *Energy Environ. Sci.*, **12(4)**, 1210–1218, doi:10.1039/c8ee03338b.

Tanzer, S.E., K. Blok, and A. Ramírez, 2021: Decarbonising Industry via BECCS: Promising Sectors, Challenges, and Techno-economic Limits of Negative Emissions. *Curr. Sustain. Energy Reports*, 8(4), 253–262, doi:10.1007/s40518-021-00195-3.

Tapia Carlos et al., 2019: *Circular Economy and Territorial Consequences*. Luxembourg, 84 pp. https://www.espon.eu/circular-economy.

Tchung-Ming, S., A. Diaz Vazquez, and K. Keramidas, 2018: *Global Energy and Climate Outlook 2018: Greenhouse gas emissions and energy balances*. JRC. Sevilla, Spain. 244 pp.

Testa, F., F. Iraldo, M. Frey, and T. Daddi, 2012: What factors influence the uptake of GPP (green public procurement) practices? New evidence from an Italian survey. *Ecol. Econ.*, **82**, 88–96, doi:10.1016/J.ECOLECON.2012.07.011.

Testa, F., E. Annunziata, F. Iraldo, and M. Frey, 2016: Drawbacks and opportunities of green public procurement: An effective tool for sustainable production. *J. Clean. Prod.*, **112**, 1893–1900, doi:10.1016/j.jclepro.2014.09.092.

Thiébaud, E., L.M. Hilty, M. Schluep, and M. Faulstich, 2017: Use, Storage, and Disposal of Electronic Equipment in Switzerland. *Environ. Sci. Technol.*, **51(8)**, 4494–4502, doi:10.1021/acs.est.6b06336.

Thomas, V. M., 2003: Demand and Dematerialization Impacts of Second-Hand Markets: Reuse or More Use? *J. Ind. Ecol.*, **7(2)**, 65–78, doi:10.1162/108819803322564352.

Thunman, H. et al., 2019: Circular use of plastics-transformation of existing petrochemical clusters into thermochemical recycling plants with 100% plastics recovery. *Sustain. Mater. Technol.*, **22**, e00124, doi:10.1016/j.susmat.2019.e00124.

Tian, J., W. Liu, B. Lai, X. Li, and L. Chen, 2014: Study of the performance of eco-industrial park development in China. *J. Clean. Prod.*, **64**, 486–494, doi:10.1016/J.JCLEPRO.2013.08.005.

Tong, D. et al., 2019: Committed emissions from existing energy infrastructure jeopardize 1.5°C climate target. *Nature*, **572(7769)**, 373–377, doi:10.1038/s41586-019-1364-3.

Tönjes, A., K. Knoop, T. Lechtenböhmer, H. Mölter, and K. Witte, 2020: Dynamics of cross-industry low-carbon innovation in energy intensive industries. *ECEEE Industrial Summer Study Proceedings.*, 467–476. https://www.eceee.org/library/conference_proceedings/eceee_Industrial_Summer_Study/2020/6-deep-decarbonisation-of-industry/dynamics-of-cross-industry-low-carbon-innovation-in-energy-intensive-industries/. (Accessed March 27, 2021).

Tost, M., M. Hitch, S. Lutter, S. Feiel, and P. Moser, 2020: Carbon prices for meeting the Paris agreement and their impact on key metals. *Extr. Ind. Soc.*, **7(2)**, 593–599, doi:10.1016/j.exis.2020.01.012.

Tunnessen, W. and D. Macri, 2017: Plant-level Goal and Recognition Programs as a Strategic Energy Management Tool. *2017 ACEEE Summer Study on Energy Efficiency in Industry*, Denver, CO, USA, 176–187. https://www.aceee.org/files/proceedings/2017/data/polopoly_fs/1.3687835.1501159018!/fileserver/file/790245/filename/0036_0053_000055.pdf. (Accessed August 28, 2021).

Tvinnereim, E. and M. Mehling, 2018: Carbon pricing and deep decarbonisation. *Energy Policy*, **121**, 185–189, doi:10.1016/j.enpol.2018.06.020.

U.S. Geological Survey, 2021: *Mineral Commodity Summaries*. Reston, VA, USA. 200 pp.

Ueckerdt, F. et al., 2021: Potential and risks of hydrogen-based e-fuels in climate change mitigation. *Nat. Clim. Change*, **11(5)**, 384–393, doi:10.1038/s41558-021-01032-7.

UKCCC, 2019a: *Net Zero: The UK's contribution to stopping global warming*. London, UK, 275 pp. https://www.theccc.org.uk/wp-content/uploads/2019/05/Net-Zero-The-UKs-contribution-to-stopping-global-warming.pdf.

UKCCC, 2019b: *Net Zero Technical Report*. London, UK, 302 pp. https://www.theccc.org.uk/publication/net-zero-technical-report/.

UN, 2015: World Population Prospects 2015. https://population.un.org/wpp/ ((Accessed December 20, 2020)).

UNEP, 2017: *2017 Global Review of Sustainable Public Procurement*. Paris, France, 124 pp.

UNEP, 2018: *Single-use plastics: A roadmap for sustainability*. Nairobi, Kenya, 104 pp.

UNEP, 2019: *Emissions Gap Report 2019*. Paris, France. 81 pp. https://www.unenvironment.org/resources/emissions-gap-report-2019. (Accessed December 19, 2020).

UNEP, and IRP, 2020: Global Material Flows Database. https://www.resourcepanel.org/global-material-flows-database (Accessed December 20, 2020).

UNIDO, 2010: *Global Industrial Energy Efficiency Benchmarking. An Energy Policy Tool*. Vienna, Austria. 58 pp.

UNSD, 2020: *Global indicator framework for the Sustainable Development Goals and targets of the 2030 Agenda for Sustainable Development*. 21 pp. https://unstats.un.org/sdgs/indicators/Global Indicator Framework after 2020 review_Eng.pdf. (Accessed October 19, 2021).

USEPA, and ICF, 2012: *Global Anthropogenic Non-$CO_2$ Greenhouse Gas Emissions: 1990–2030*. Washington D.C., USA, 176 pp. https://www.epa.gov/sites/production/files/2016-08/documents/epa_global_nonco2_projections_dec2012.pdf. (Accessed December 15, 2019).

USGBC-LA, 2018: Buy Clean California – USGBC LA. https://usgbc-la.org/programs/buy-clean-california/ (Accessed October 29, 2021).

van der Meijden, C.M., L.P.L.M. Rabou, B.J. Vreugdenhil, and R. Smit, 2011: Large Scale Production of Bio Methane from Wood. The International Gas Union Research Conference IGRC, Seoul, South Korea, 16 pp ftp://ftp.ecn.nl/pub/www/library/report/2011/m11098.pdf. (Accessed March, 27, 2021)

Van Ewijk, S., J.A. Stegemann, and P. Ekins, 2018: Global life cycle paper flows, recycling metrics, and material efficiency. *J. Ind. Ecol.*, **22(4)**, 686–693, doi:10.1111/jiec.12613.

Van Rijnsoever, F.J., J. Van Den Berg, J. Koch, and M.P. Hekkert, 2015: Smart innovation policy: How network position and project composition affect the diversity of an emerging technology. *Res. Policy*, **44(5)**, 1094–1107, doi:10.1016/j.respol.2014.12.004.

van Vuuren, D.P. et al., 2018: Alternative pathways to the 1.5°C target reduce the need for negative emission technologies. *Nat. Clim. Change*, **8(5)**, 391–397, doi:10.1038/s41558-018-0119-8.

Vanclay, J.K. et al., 2011: Customer Response to Carbon Labelling of Groceries. *J. Consum. Policy*, **34(1)**, 153–160, doi:10.1007/s10603-010-9140-7.

Vogl, V., M. Åhman, and L.J. Nilsson, 2018: Assessment of hydrogen direct reduction for fossil-free steelmaking. *J. Clean. Prod.*, **203**, 736–745, doi:10.1016/j.jclepro.2018.08.279.

Vogl, V., M. Åhman, and L.J. Nilsson, 2021a: The making of green steel in the EU: a policy evaluation for the early commercialization phase. *Clim. Policy*, **21(1)**, 78–92, doi:10.1080/14693062.2020.1803040.

Vogl, V., O. Olsson, and B. Nykvist, 2021b: Phasing out the blast furnace to meet global climate targets. *Joule*, **5(10)**, 2646–2662, doi:10.1016/j.joule.2021.09.007.

Vogt-Schilb, A. and S. Hallegatte, 2017: Climate policies and nationally determined contributions: reconciling the needed ambition with the political economy. *Wiley Interdiscip. Rev. Energy Environ.*, **6(6)**, e256, doi:10.1002/wene.256.

Vogt-Schilb, A., G. Meunier, and S. Hallegatte, 2018: When starting with the most expensive option makes sense: Optimal timing, cost and sectoral allocation of abatement investment. *J. Environ. Econ. Manage.*, **88**, 210–233, doi:10.1016/j.jeem.2017.12.001.

Voldsund, M. et al., 2019: Comparison of technologies for CO2 capture from cement production—Part 1: Technical evaluation. *Energies*, **12(3)**, 559, doi:10.3390/en12030559.

Vollmer, I. et al., 2020: Beyond Mechanical Recycling: Giving New Life to Plastic Waste. *Angew. Chemie Int. Ed.*, **59(36)**, 15402–15423, doi:10.1002/anie.201915651.

Vreys, K., S. Lizin, M. Van Dael, J. Tharakan, and R. Malina, 2019: Exploring the future of carbon capture and utilisation by combining an international Delphi study with local scenario development. *Resour. Conserv. Recycl.*, **146**, 484–501, doi:10.1016/j.resconrec.2019.01.027.

Wang, D.D., and T. Sueyoshi, 2018: Climate change mitigation targets set by global firms: Overview and implications for renewable energy. *Renew. Sustain. Energy Rev.*, **94**, 386–398, doi:10.1016/j.rser.2018.06.024.

Wang, N., J.C.K. Lee, J. Zhang, H. Chen, and H. Li, 2018: Evaluation of Urban circular economy development: An empirical research of 40 cities in China. *J. Clean. Prod.*, **180**, 876–887, doi:10.1016/j.jclepro.2018.01.089.

Wang, P. et al., 2021: Efficiency stagnation in global steel production urges joint supply- and demand-side mitigation efforts. *Nat. Commun.*, **12(1)**, 2066, doi:10.1038/s41467-021-22245-6.

Wang, Q. et al., 2019: Distributional impact of carbon pricing in Chinese provinces. *Energy Econ.*, **81**, 327–340, doi:10.1016/J.ENECO.2019.04.003.
Wang, W., C. Li, and S. Wang, 2017: Big data used in energy efficiency and load forecasting of heating boilers. *2017 36th Chinese Control Conference (CCC)*, IEEE, 5638–5641.
Wang, X. and K. Lo, 2021: Just transition: A conceptual review. *Energy Res. Soc. Sci.*, **82**, 102291, doi:10.1016/j.erss.2021.102291.
Wanzenböck, I., J.H. Wesseling, K. Frenken, M.P. Hekkert, and K.M. Weber, 2020: A framework for mission-oriented innovation policy: Alternative pathways through the problem–solution space. *Sci. Public Policy*, 47(4), 474–489doi:10.1093/scipol/scaa027.
Warwick, K., 2013: *Beyond Industrial Policy: Emerging Issues and New Trends*. OECD Paris, France.
WBCSD, 2016: *Cement Industry Energy and $CO_2$ Performance. Getting the Numbers Right*. Geneva, Switzerland. 20 pp. https://docs.wbcsd.org/2016/12/GNR.pdf. (Accessed December, 19, 2020).
Weber, K.M. and H. Rohracher, 2012: Legitimizing research, technology and innovation policies for transformative change: Combining insights from innovation systems and multi-level perspective in a comprehensive "failures" framework. *Res. Policy*, **41(6)**, 1037–1047, doi:10.1016/j.respol.2011.10.015.
Wesseling, J.H., and C. Edquist, 2018: Public procurement for innovation to help meet societal challenges: A review and case study. *Sci. Public Policy*, **45(4)**, 493–502, doi:10.1093/SCIPOL/SCY013.
Wesseling, J.H. et al., 2017: The transition of energy intensive processing industries towards deep decarbonization: Characteristics and implications for future research. *Renew. Sustain. Energy Rev.*, **79**(January), 1303–1313, doi:10.1016/j.rser.2017.05.156.
Wiebe, K.S., M. Harsdorff, G. Montt, M.S. Simas, and R. Wood, 2019: Global Circular Economy Scenario in a Multiregional Input-Output Framework. *Environ. Sci. Technol.*, **53(11)**, 6362–6373, doi:10.1021/acs.est.9b01208.
Wiedenhofer, D., T. Fishman, C. Lauk, W. Haas, and F. Krausmann, 2019: Integrating Material Stock Dynamics Into Economy-Wide Material Flow Accounting: Concepts, Modelling, and Global Application for 1900–2050. *Ecol. Econ.*, **156**, 121–133, doi:10.1016/j.ecolecon.2018.09.010.
Williams, C., A. Hasanbeigi, G. Wu, and L. Price, 2012: *International Experiences with Quantifying the Co-Benefits of Energy-Efficiency and Greenhouse-Gas Mitigation Programs and Policies*. Berkeley, CA, USA, 102 pp.
Williams, J.H. et al., 2021: Carbon-Neutral Pathways for the United States. *AGU Adv.*, **2(1)**, doi:10.1029/2020AV000284.
Wilson, C. and A. Grubler, 2011: Lessons from the history of technological change for clean energy scenarios and policies. *Nat. Resour. Forum*, **35(3)**, 165–184, doi:10.1111/j.1477-8947.2011.01386.x.
Wilson, E.J., T.L. Johnson, and D.W. Keith, 2003: Regulating the Ultimate Sink: Managing the Risks of Geologic $CO_2$ Storage. *Environ. Sci. Technol.*, **37(16)**, 3476–3483, doi:https://doi.org/10.1021/es021038+.
Wilts, H. and I. Bakas, 2019: *Preventing plastic waste in Europe*. Copenhagen, Denmark, 57 pp. https://www.eea.europa.eu/publications/preventing-plastic-waste-in-europe.
Wilts, H., N. von Gries, and B. Bahn-Walkowiak, 2016: From waste management to resource efficiency-the need for policy mixes. *Sustainability.*, **8(7)**, 1–16, doi:10.3390/su8070622.
Winans, K., A. Kendall, and H. Deng, 2017: The history and current applications of the circular economy concept. *Renew. Sustain. Energy Rev.*, **68**, 825–833, doi:10.1016/j.rser.2016.09.123.
World Bank, 2020: Carbon Pricing Dashboard | Up-to-date overview of carbon pricing initiatives. *World Bank*. https://carbonpricingdashboard.worldbank.org/ (Accessed October 26, 2021).
World Bank, 2021: World Bank Development Indicators. https://datatopics.worldbank.org/world-development-indicators/ (Accessed August 28, 2021).
World Bank Group, 2019: *State and Trends of Carbon Pricing 2019*. Washington, DC, USA, 94 pp.
World Steel Association, 2020a: *Statistical yearbook 2010–2020*. Brussels, Belgium. 42 pp. https://www.worldsteel.org/steel-by-topic/statistics/steel-statistical-yearbook.html. (Accessed December, 19, 2020).
World Steel Association, 2020b: *Steel's contribution to a low carbon future and climate resilient societies*. Brussels, Belgium. 6 pp. https://www.worldsteel.org/en/dam/jcr:7ec64bc1-c51c-439b-84b8-94496686b8c6/Position_paper_climate_2020_vfinal.pdf. (Accessed March, 27, 2021).
World Steel Association, 2021: *World Steel in Figures 2021*. Brussels, Belgium. 32 pp. https://www.worldsteel.org/en/dam/jcr:976723ed-74b3-47b4-92f6-81b6a452b86e/World%2520Steel%2520in%2520Figures%25202021.pdf. (Accessed August, 28, 2021).
Worrell, E., J. Allwood, and T. Gutowski, 2016: The Role of Material Efficiency in Environmental Stewardship. *Annu. Rev. Environ. Resour.*, **41(1)**, 575–598, doi:10.1146/annurev-environ-110615-085737.
WRAP and Green Alliance, 2015: *Employment and the circular economy: Job creation in a more resource efficient Britain*. Banbury, UK, 24 pp. https://wrap.org.uk/sites/files/wrap/Employment and the circular economy summary.pdf. (Accessed March 27, 2021).
Wu, P., S.P. Low, B. Xia, and J. Zuo, 2014: Achieving transparency in carbon labelling for construction materials – Lessons from current assessment standards and carbon labels. *Environ. Sci. Policy*, **44**, 11–25, doi:10.1016/J.ENVSCI.2014.07.009.
Wu, R., Y. Geng, X. Cui, Z. Gao, and Z. Liu, 2019: Reasons for recent stagnancy of carbon emissions in China's industrial sectors. *Energy*, **172**, 457–466, doi:10.1016/j.energy.2019.01.156.
Wyns, T., G. Khandekar, M. Axelson, O. Sartor, and K. Neuhoff, 2019: *Industrial Transformation 2050 – Towards an Industrial strategy for a Climate Neutral Europe*. Brussels, Belgium, 79 pp. https://www.ies.be/files/Industrial_Transformation_2050_0.pdf.
Xiang, D., Y. Qian, Y. Man, and S. Yang, 2014: Techno-economic analysis of the coal-to-olefins process in comparison with the oil-to-olefins process. *Appl. Energy*, **113(113)**, 639–647, doi:10.1016/j.apenergy.2013.08.013.
Xylia, M., S. Silveira, J. Duerinck, and F. Meinke-Huben, 2016: Worldwide resource efficient steel production. In: *Eceee Industrial Summer Study Proceedings*, Stockholm, Sweden. pp. 321–333.
Xylia, M., S. Silveira, J. Duerinck, and F. Meinke-Hubeny, 2018: Weighing regional scrap availability in global pathways for steel production processes. *Energy Effic.*, **11(5)**, 1135–1159, doi:10.1007/s12053-017-9583-7.
Yamada, K. and K. Tanaka, 2019: *Promotion of Constructing Zero Carbon Society: Effectiveness of Quantitative Evaluation of Technology and System for Sustainable Economic Developmente*. T20 Japan, Tokyo, Japan. 19 pp. https://t20japan.org/policy-brief-promotion-constructing-zero-carbon-society/. (Accessed March 27, 2021).
Yang, Y. et al., 2019: Progress in coal chemical technologies of China. *Rev. Chem. Eng.*, **36(1)**, 21–66, doi:10.1515/revce-2017-0026.
Yao, Y. and E. Masanet, 2018: Life-cycle modeling framework for generating energy and greenhouse gas emissions inventory of emerging technologies in the chemical industry. *J. Clean. Prod.*, **172**, 768–777, doi:10.1016/j.jclepro.2017.10.125.
Yoshimoto, Y., 2016: Realization of productivity improvement and energy saving by data utilization to a factory. *Spring meeting, the Japan Society for Precision Engineering*, pp. 443–444. https://www.jstage.jst.go.jp/article/pscjspe/2016S/0/2016S_443/_pdf/-char/ja.
Zachmann, G. and B. McWilliams, 2020: *A European carbon border tax: much pain, little gain*. Bruegel, Brussels, Belgium, 19 pp.
Zhang, Z. et al., 2020: Recent advances in carbon dioxide utilization. *Renew. Sustain. Energy Rev.*, **125**(March), 109799, doi:10.1016/j.rser.2020.109799.
Zheng, J. and S. Suh, 2019: Strategies to reduce the global carbon footprint of plastics. *Nat. Clim. Change*, **9(5)**, 374–378, doi:10.1038/s41558-019-0459-z.
Zhou, N. et al., 2013: China's energy and emissions outlook to 2050: Perspectives from bottom-up energy end-use model. *Energy Policy*, **53**, 51–62, doi:10.1016/j.enpol.2012.09.065.

Zhou, N. et al., 2019: A roadmap for China to peak carbon dioxide emissions and achieve a 20% share of non-fossil fuels in primary energy by 2030. *Appl. Energy*, **239**(February), 793–819, doi:10.1016/j.apenergy.2019.01.154.

Zhou, N. et al., 2020: *China Energy Outlook: Understanding China's Energy and Emissions Trends*. Berkeley, USA, 135 pp.

Zhu, Q., Y. Geng, and J. Sarkis, 2013: Motivating green public procurement in China: An individual level perspective. *J. Environ. Manage.*, **126**, 85–95, doi:10.1016/J.JENVMAN.2013.04.009.

Zimmerman, J.B., P.T. Anastas, H.C. Erythropel, and W. Leitner, 2020: Designing for a green chemistry future., **367(6476)**, 397–400, doi:10.1126/science.aay3060.

Zühlsdorf, B., F. Bühler, M. Bantle, and B. Elmegaard, 2019: Analysis of technologies and potentials for heat pump-based process heat supply above 150°C. *Energy Convers. Manag. X*, **2**, 100011, doi:10.1016/j.ecmx.2019.100011.

12

# Cross-sectoral Perspectives


**Coordinating Lead Authors:**
Mustafa Babiker (Sudan/Saudi Arabia), Göran Berndes (Sweden)

**Lead Authors:**
Kornelis Blok (the Netherlands), Brett Cohen (South Africa), Annette Cowie (Australia), Oliver Geden (Germany), Veronika Ginzburg (the Russian Federation), Adrian Leip (Italy/Germany), Pete Smith (United Kingdom), Masahiro Sugiyama (Japan), Francis Yamba (Zambia)

**Contributing Authors:**
Alaa Al Khourdajie (United Kingdom/Syria), Almut Arneth (Germany), Inês Margarida Lima de Azevedo (Portugal/the United States of America), Christopher Bataille (Canada), David Beerling (United Kingdom), Rachel Bezner Kerr (the United States of America/Canada), Jessie Bradley (the Netherlands), Holly Jean Buck (the United States of America), Luisa F. Cabeza (Spain), Katherine Calvin (the United States of America), Donovan Campbell (Jamaica), Jofre Carnicer Cols (Spain), Vassilis Daioglou (Greece), Mathijs Harmsen (the Netherlands), Lena Höglund-Isaksson (Sweden), Joanna I. House (United Kingdom), David Keller (Germany/the United States of America), Kiane de Kleijne (the Netherlands), Susanna Kugelberg (Sweden), Igor Makarov (the Russian Federation), Francisco Meza (Chile), Jan Christoph Minx (Germany), Michael Morecroft (United Kingdom), Gert-Jan Nabuurs (the Netherlands), Henry Neufeldt (Denmark/Germany), Aleksandra Novikova (Germany/the Russian Federation), Sudarmanto Budi Nugroho (Indonesia), Andreas Oschlies (Germany), Camille Parmesan (United Kingdom/the United States of America), Glen P. Peters (Norway/Australia), Joseph Poore (United Kingdom), Joana Portugal-Pereira (Brazil), Julio C. Postigo (the United States of America/Peru), Prajal Pradhan (Germany/Nepal), Phil Renforth (United Kingdom), Marta G. Rivera-Ferre (Spain), Stephanie Roe (the Philippines/the United States of America), Pramod K. Singh (India), Raphael Slade (United Kingdom), Stephen M. Smith (United Kingdom), Maria Cristina Tirado von der Pahlen (the United States of America/Spain), Daniela Toribio Ramirez (Mexico)

**Review Editors:**
Gilberto Jannuzzi (Brazil), Andy Reisinger (New Zealand)

**Chapter Scientists:**
Kiane de Kleijne (the Netherlands), Eveline María Vásquez-Arroyo (Peru/Brazil)



**This chapter should be cited as:**
Babiker, M., G. Berndes, K. Blok, B. Cohen, A. Cowie, O. Geden, V. Ginzburg, A. Leip, P. Smith, M. Sugiyama, F. Yamba, 2022: Cross-sectoral perspectives. In IPCC, 2022: *Climate Change 2022: Mitigation of Climate Change. Contribution of Working Group III to the Sixth Assessment Report of the Intergovernmental Panel on Climate Change* [P.R. Shukla, J. Skea, R. Slade, A. Al Khourdajie, R. van Diemen, D. McCollum, M. Pathak, S. Some, P. Vyas, R. Fradera, M. Belkacemi, A. Hasija, G. Lisboa, S. Luz, J. Malley, (eds.)]. Cambridge University Press, Cambridge, UK and New York, NY, USA. doi: 10.1017/9781009157926.014

# Table of Contents

## Executive Summary

**The total emission mitigation potential achievable by the year 2030, calculated based on sectoral assessments, is sufficient to reduce global greenhouse gas emissions to half of the current (2019) level or less (*robust evidence, high agreement*).** This potential (32–44 $GtCO_2$-eq) requires implementation of a wide range of mitigation options. Options with mitigation costs lower than USD20 $tCO_2^{-1}$ make up more than half of this potential and are available for all sectors. {12.2, Table 12.3}

**Carbon dioxide removal (CDR) is a necessary element to achieve net zero $CO_2$ and greenhouse gas (GHG) emissions both globally and nationally, counterbalancing residual emissions from hard-to-transition sectors. It is a key element in scenarios that limit warming to 2°C (>67%) or lower by 2100 (*robust evidence, high agreement*).** Implementation strategies need to reflect that CDR methods differ in terms of removal process, timescale of carbon storage, technological maturity, mitigation potential, cost, co-benefits, adverse side effects, and governance requirements. All Illustrative Mitigation Pathways (IMPs) use land-based biological CDR (primarily afforestation/reforestation (A/R)) and/or bioenergy with carbon capture and storage (BECCS) and some include direct air carbon capture and storage (DACCS). As a median value (5–95% range) across the scenarios that limit warming to 2°C (>67%) or lower, cumulative volumes of BECCS, $CO_2$ removal from AFOLU (mainly A/R), and DACCS reach 328 (168–763) gigatonnes of $CO_2$ equivalent ($GtCO_2$), 252 (20–418) $GtCO_2$, and 29 (0–339) $GtCO_2$ for the 2020–2100 period, with annual volumes at 2.75 (0.52–9.45) $GtCO_2$ $yr^{-1}$ for BECCS, 2.98 (0.23–6.38) $GtCO_2$ $yr^{-1}$ for the $CO_2$ removal from AFOLU (mainly A/R), and 0.02 (0–1.74) $GtCO_2$ $yr^{-1}$ for DACCS, in 2050. {12.3, Cross-Chapter Box 8 in this chapter}

**Despite limited current deployment, moderate to large future mitigation potentials are estimated for direct air carbon capture and sequestration (DACCS), enhanced weathering (EW) and ocean-based CDR methods (including ocean alkalinity enhancement and ocean fertilisation) (*medium evidence, medium agreement*).** The potential for DACCS (5–40 $GtCO_2$ $yr^{-1}$) is limited mainly by requirements for low-carbon energy and by cost (USD100–300 (full range: USD84–386) $tCO_2^{-1}$). DACCS is currently at a medium technology readiness level. EW has the potential to remove 2–4 (full range: <1 to about 100) $GtCO_2$ $yr^{-1}$, at costs ranging from USD50 to 200 (full range: USD24–578) $tCO_2^{-1}$. Ocean-based methods have a combined potential to remove 1–100 $GtCO_2$ $yr^{-1}$ at costs of USD40–500 $tCO_2^{-1}$, but their feasibility is uncertain due to possible side effects on the marine environment. EW and ocean-based methods are currently at a low technology readiness level. {12.3}

**Realising the full mitigation potential from the food system requires change at all stages from producer to consumer and waste management, which can be facilitated through integrated policy packages (*robust evidence, high agreement*).** Some 23–42% of global GHG emissions are associated with food systems, while there is still widespread food insecurity and malnutrition. Absolute GHG emissions from food systems increased from 14 to 17 $GtCO_2$-eq $yr^{-1}$ in the period 1990–2018. Both supply and demand-side measures are important to reduce the GHG intensity of food systems. Integrated food policy packages based on a combination of market-based, administrative, informative, and behavioural policies can reduce cost compared to uncoordinated interventions, address multiple sustainability goals, and increase acceptance across stakeholders and civil society (*limited evidence, medium agreement*). {7.2, 7.4, 12.4}

**Diets high in plant protein and low in meat and dairy are associated with lower GHG emissions (*robust evidence, high agreement*).** Ruminant meat shows the highest GHG intensity. Beef from dairy systems has lower emissions intensity than beef from beef herds (8–23 and 17–94 $kgCO_2$-eq per 100 g protein, respectively) when a share of emissions is allocated to dairy products. The wide variation in emissions reflects differences in production systems, which range from intensive feedlots with stock raised largely on grains through to rangeland and transhumance production systems. Where appropriate, a shift to diets with a higher share of plant protein, moderate intake of animal-source foods and reduced intake of added sugars, salt and saturated fats could lead to substantial decreases in GHG emissions. Benefits would also include reduced land occupation and nutrient losses to the surrounding environment, while at the same time providing health benefits and reducing mortality from diet-related non-communicable diseases. {7.4.5, 12.4}

**Emerging food technologies such as cellular fermentation, cultured meat, plant-based alternatives to animal-based food products, and controlled-environment agriculture, can bring substantial reductions in direct GHG emissions from food production (*limited evidence, high agreement*).** These technologies have lower land, water, and nutrient footprints, and address concerns over animal welfare. Access to low-carbon energy is needed to realise the full mitigation potential, as some emerging technologies are relatively more energy intensive. This also holds for deployment of cold chain and packaging technologies, which can help reduce food loss and waste, but increase energy and materials use in the food system. (*limited evidence, high agreement*). {11.4.1.3, 12.4}

**Scenarios that limit warming to 2°C (>67%) or lower by 2100 commonly involve extensive mitigation in the agriculture, forestry and other land use (AFOLU) sector that at the same time provides biomass for mitigation in other sectors. Bioenergy is the most land intensive renewable energy option, but the total land occupation of other renewable energy options can become significant in high deployment scenarios (*robust evidence, high agreement*).** Growing demands for food, feed, biomaterials, and non-fossil fuels increase the competition for land and biomass while climate change creates additional stresses on land, exacerbating existing risks to livelihoods, biodiversity, human and ecosystem health, infrastructure, and food systems. Appropriate integration of bioenergy and other bio-based systems, and of other mitigation options, with existing land and biomass uses can improve resource use efficiency, mitigate pressures on natural ecosystems and support adaptation through measures to combat land degradation, enhance food security, and improve resilience through maintenance of the productivity of the land resource base (*medium evidence, high agreement*). {3.2.5, 3.4.6, 12.5}

**Bio-based products as part of a circular bioeconomy have potential to support adaptation and mitigation. Key to maximising benefits and managing trade-offs are sectoral integration, transparent governance, and stakeholder involvement (*high confidence*).** A sustainable bioeconomy relying on biomass resources will need to be supported by technology innovation and international cooperation and governance of global trade to disincentivise environmental and social externalities (*medium confidence*). {12.5, Cross-Working Group Box 3 in this chapter}

**Coordinated, cross-sectoral approaches to climate change mitigation should be adopted to target synergies and minimise trade-offs between sectors and with respect to sustainable development (*robust evidence, high agreement*).** This requires integrated planning using multiple-objective-multiple-impact policy frameworks. Strong interdependencies and cross-sectoral linkages create both opportunities for synergies and the need to address trade-offs related to mitigation options and technologies. This can only be done if coordinated sectoral approaches to climate change mitigation policies that mainstream these interactions are adopted. Integrated planning and cross-sectoral alignment of climate change policies are particularly evident in developing countries' Nationally Determined Contributions (NDCs) pledged under the Paris Agreement, where key priority sectors such as agriculture and energy are closely aligned between the proposed mitigation and adaptation actions in the context of sustainable development and the Sustainable Development Goals (SDGs). {12.6.2}

**Carbon leakage is a critical cross-sectoral and cross-country consequence of differentiated climate policy (*robust evidence, medium agreement*).** Carbon leakage occurs when mitigation measures implemented in one country/sector lead to increased emissions in other countries/sectors. Global commodity value chains and associated international transport are important mechanisms of carbon leakage. Reducing emissions from the value chain and transportation can offer opportunities to mitigate three elements of cross-sectoral spillovers and related leakage: (i) domestic cross-sectoral spillovers within the same country; (ii) international spillovers within a single sector resulting from substitution of domestic production of carbon-intensive goods with their imports from abroad; and (iii) international cross-sectoral spillovers among sectors in different countries. {12.6.3}

**Cross-sectoral considerations in mitigation finance are critical for the effectiveness of mitigation action as well as for balancing the often conflicting social, developmental, and environmental policy goals at the sectoral level (*medium evidence, medium agreement*).** True resource mobilisation plans that properly address mitigation costs and benefits at sectoral level cannot be developed in isolation from their cross-sectoral implications. There is an urgent need for multilateral financing institutions to align their frameworks and delivery mechanisms including the use of blended financing to facilitate cross-sectoral solutions as opposed to causing competition for resources among sectors. {12.6.4}

**Understanding the co-benefits and trade-offs associated with mitigation is key to supporting societies to prioritise among the various sectoral policy options (*medium evidence, medium agreement*).** For example, CDR options can have positive impacts on ecosystem services and the SDGs, but also potential adverse side effects; transforming food systems has potential co-benefits for several SDGs, but also trade-offs; and land-based mitigation measures may have multiple co-benefits but may also be associated with trade-offs among environmental, social, and economic objectives. Therefore, the possible implementation of the different sectoral mitigation options would depend on how societies prioritise mitigation versus other products and services, including food, material well-being, nature conservation and biodiversity protection, as well as on other considerations such as society's future dependence on CDR and on carbon-based energy and materials. {12.3, 12.4, 12.5, 12.6.1}

**Governance of CDR, food systems and land-based mitigation can support effective and equitable policy implementation (*medium evidence, high agreement*).** Effectively responding to climate change while advancing sustainable development will require coordinated efforts among a diverse set of state- and non-state-actors on global, multinational, national, and sub-national levels. Governance arrangements in public policy domains that cut through traditional sectors are confronted with specific challenges, such as establishing reliable systems for monitoring, reporting and verification (MRV) that allow evaluation of mitigation outcomes and co-benefits. Effectively integrating CDR into mitigation portfolios can build on already existing rules, procedures and instruments for emissions abatement. Additionally, to accelerate research, development, and demonstration, and to incentivise CDR deployment, a political commitment to formal integration into existing climate policy frameworks is required, including reliable MRV of carbon flows. Food systems governance may be pioneered through local food policy initiatives complemented by national and international initiatives, but governance on the national level tends to be fragmented, and thus have limited capacity to address structural issues like inequities in access. The governance of land-based mitigation, including land-based CDR, can draw on lessons from previous experience with regulating biofuels and forest carbon; however, integrating these insights requires governance that goes beyond project-level approaches and emphasises integrated land use-planning and management within the frame of the SDGs. {7.4, Box 7.2, 7.6, 12.3.3, 12.4, 12.5}

## 12.1 Introduction

### 12.1.1 Chapter Overview

The scope of this chapter was motivated by the need for a succinct bottom-up cross-sectoral view of greenhouse gas (GHG) emissions mitigation coupled with the desire to provide systemic perspectives on critical mitigation potentials and options that go beyond individual sectors and cover cross-sectoral topics such as food systems, land systems, and carbon dioxide removal (CDR) methods. Driven by this motivation, Chapter 12 provides a focused thematic assessment of CDR methods and food systems, followed by consideration of land-related impacts of mitigation options (land-based CDR and other mitigation options that occupy land) and other cross-sectoral impacts of mitigation, with emphasis on synergies and trade-offs between mitigation options, and between mitigation and other environmental and socio-economic objectives. The systems focus is unique to the Sixth Assessment Report (AR6) of the IPCC and is of critical policy relevance as it informs coordinated approaches to planning interventions that deliver multiple benefits and minimise trade-offs, and coordinated policy approaches to support such planning, to tap relatively under-explored areas for the strengthening and acceleration of mitigation efforts in the short to medium term, and for dealing with residual emissions in hard-to-transition sectors in the medium to long term.

Table 12.1 presents an overview of the cross-sectoral perspectives addressed in Chapter 12, mapping the chapter's main themes to the sectoral and global chapters in this report. These mappings reflect the cross-sectoral aspects of mitigation options in the context of sustainable development, sectoral policy interactions, governance, implications in terms of international trade, spillover effects, and competitiveness, and cross-sectoral financing options for mitigation. While some cross-sector technologies are covered in more detail in sectoral chapters, this chapter covers important cross-sectoral linkages and provides synthesis concerning costs and potentials of mitigation options, and co-benefits and trade-offs that can be associated with deployment of mitigation options. Additionally, Chapter 12 covers CDR methods and specific considerations related to land use and food systems, complementing Chapter 7. The literature assessed in the chapter includes both peer-reviewed and grey literature since the Fifth Assessment Report (AR5) of the IPCC, including the IPCC Special Report on Global Warming of 1.5°C (SR1.5), the IPCC Special Report on Climate Change and Land (SRCCL) and the IPCC Special Report on the Ocean and Cryosphere in a Changing Climate (SROCC). Knowledge gaps are identified and reflected where encountered, as well as in a separate section. Finally, a strong link is maintained with sectoral chapters and the relevant global chapters of this report to ensure consistency.

### 12.1.2 Chapter Content

Chapters 5 to 11 assess outcomes from mitigation measures that are applicable in individual sectors, and potential co-benefits and adverse side effects of these individual measures. Chapter 12 brings together the cross-sectoral aspects of these assessments including synergies and trade-offs as well as the implications of measures that have application in more than one sector and measures whose implementation in one sector impacts implementation in other sectors.

Taking stock of the sectoral mitigation assessments, Chapter 12 provides a summary synthesis of sectoral mitigation costs and potentials in the short and long term along with comparison to the top-down integrated assessment model (IAM) assessment literature of Chapter 3 and the national/regional assessment literature of Chapter 4.

In the context of cross-sectoral synergies and trade-offs, the chapter identifies a number of mitigation measures that have application in more than one sector. Examples include measures involving product and material circularity, which contribute to mitigation of GHG emissions in a number of ways, such as treatment of organic waste to reduce methane emissions, avoid emissions through generation of renewable energy, and reduce emissions through substitution of synthetic fertilisers. Low-carbon energy technologies such as solar and wind may be used for grid electricity supply, as embedded generation in the buildings sector (e.g., rooftop solar) and for energy supply in the agriculture sector. Nuclear and bio-based thermal electric generation can provide multiple synergies including base load to augment solar and wind, district heating, and seawater desalination. Grid-integrated hydrogen systems can buffer variability of solar and wind power and are being explored as a mitigation option in the transport and industry sectors. Carbon capture and storage (CCS) has potential application in a number of industrial processes (cement, iron and steel, petroleum refining and pulp and paper) and the fossil fuel electricity sector. When coupled with energy recovery from biomass (BECCS), CCS can help to provide $CO_2$ removal from the atmosphere. On the demand side, electric vehicles are also considered an option for balancing variable power, energy efficiency options find application across the sectors, as does reducing demand for goods and services, and improving material use efficiency. Focused inquiry into these areas of cross-sectoral perspectives is provided for CDR, food systems, and land-based mitigation options.

A range of examples of where mitigation measures result in cross-sectoral interactions and integration is identified. The mitigation potential of electric vehicles, including plug-in hybrids, is linked to the extent of decarbonisation of the electricity grid, as well as to the liquid fuel supply emissions profile. Making buildings energy positive, where excess energy is used to charge vehicles, can increase the potential of electric and hybrid vehicles. Advanced process control and process optimisation in industry can reduce energy demand and material inputs, which in turn can reduce emissions linked to resource extraction and manufacturing. Trees and green roofs planted to counter urban heat islands reduce the demand for energy for air conditioning and simultaneously sequester carbon. Material and product circularity contributes to mitigation, such as treatment of organic waste to reduce methane emissions, generate renewable energy, and to substitute for synthetic fertilisers.

The chapter also discusses cross-sectoral mitigation potential related to diffusion of general-purpose technologies (GPT), such as electrification, digitalisation, and hydrogen. Examples include the use of hydrogen as an energy carrier, which, when coupled with low-carbon energy, has potential for driving mitigation in energy, industry, transport, and buildings (Box 12.5), and digitalisation has the potential for reducing GHG emissions through energy savings across multiple sectors.

The efficient realisation of the above examples of cross-sectoral mitigation would require careful design of government interventions across planning, policy, finance, governance, and capacity building fronts. In this respect, Chapter 12 assesses literature on cross-sectoral integrated policies, cross-sectoral financing solutions, cross-sectoral spillovers and competitiveness effects, and on cross-sectoral governance for climate change mitigation.

Finally, in the context of cross-sectoral synergies and trade-offs, the chapter assesses the non-climate mitigation co-benefits and adverse effects in relation to SDGs, building on the fast-growing literature on the non-climate impacts of mitigation.

### 12.1.3 Chapter Layout

The chapter is mapped into seven sections. Cost and potentials of mitigation technologies are discussed in Section 12.2, where a comparative assessment and a summary of sectoral mitigation cost and potentials is provided in coordination with the sectoral Chapters 5 to 11, along with a comparison to aggregate cost and potentials based on IAM outputs presented in Chapter 3.

Section 12.3 provides a synthesis of the state and potential contribution of CDR methods for addressing climate change. CDR options associated with the agriculture, forestry and other land use (AFOLU) and energy sectors are dealt with in Chapters 6 and 7 and synthesised in Section 12.3. Other methods, not dealt with elsewhere, are covered in more detail. A comparative assessment is provided for the different CDR options in terms of costs, potentials, governance, impacts and risks, and synergies and trade-offs.

Section 12.4 assesses the literature on food systems and GHG emissions. The term 'food system' refers to a composite of elements (environment, people, inputs, processes, infrastructures, institutions, etc.) and activities that relate to the production, processing, distribution, preparation and consumption of food, and the outputs of these activities, including socio-economic and environmental outcomes. Climate change mitigation opportunities and related implications for sustainable development and adaptation are assessed, including those arising from food production, landscape impacts, supply chain and distribution, and diet shifts.

Section 12.5 provides a cross-sectoral perspective on land occupation and related impacts, risks and opportunities associated with land-based mitigation options as well as mitigation options that are not designated land based, yet occupy land. It builds on SRCCL and Chapter 7 in this report, which covers mitigation in AFOLU, including biomass production for mitigation in other sectors. In addition to an assessment of biophysical and socio-economic risks, impacts and opportunities, this section includes a Cross-Working Group Box (WGII and WGIII) on Mitigation and Adaptation via the Bioeconomy, and a Box on Land Degradation Neutrality as a framework to manage trade-offs in land-based mitigation.

Section 12.6 provides a cross-sectoral perspective on mitigation, co-benefits, and trade-offs, including those related to sustainable development and adaptation. The synthesised sectoral mitigation

**Table 12.1 | An overview of cross-sector perspectives addressed in Chapter 12.**

| | Sectoral chapters | | | | | | | Global chapters | | | | |
|---|---|---|---|---|---|---|---|---|---|---|---|---|
| Chapter 12 themes | Chapter 5 | Chapter 6 | Chapter 7 | Chapter 8 | Chapter 9 | Chapter 10 | Chapter 11 | Chapter 13 | Chapter 14 | Chapter 15 | Chapter 16 | Chapter 17 |
| **Costs & potentials** | Change in demand | Renewables<br>CCU<br>CCS<br>Nuclear | Land-use change | Urban planning<br>Cities<br>Demographics | Standards<br>Electrification | Hybridisation<br>Electric vehicles<br>Fuel economy<br>Decoupling | Technology<br>Biomass<br>CCU<br>CCS | Enabling of mitigation | | Finance of mitigation | | Synergies and trade-offs with SDGs |
| CDR | | BECCS | Land-based CDR | | Carbon storage in buildings | | | | International governance | | | |
| **Food systems** | Food demand<br>Well-being | Energy demand of some emerging mitigation options | Agricultural production<br>Demand-side measures | Urban food systems; controlled-environment agriculture | | Food transport | Food processing and packaging | Food system transformation | Governance | | | Food system and SDGs |
| **Mitigation & land use** | | Land use/ occupation: bioenergy hydro solar windnuclear | A/R Biomass production<br>Bioenergy<br>Biochar | | Land use and biomass supply | Land use and biomass supply | Land use and biomass supply | | Governance | | | Co-benefits and adverse side effects |
| **Cross-sectoral perspectives** | Electrification, Hydrogen, Digitalisation, Circularity, Synergies, Trade-offs, Spillovers | | | | | | | Policy interactions<br>Policy packages<br>Case studies<br>Value chain and carbon leakage | Governance<br>Leakage | Blended financing | General-urpose technologies<br>Electrification<br>Hydrogen | SDGs<br>co-benefits<br>Trade-offs<br>Adaptation |

synergies and trade-offs are mapped into options/technologies, policies, international trade, and finance domains. Cross-sectoral mitigation technologies fall into three categories in which the implementation of the technology: (i) occurs in parallel in more than one sector; (ii) could involve interaction between sectors, and/or (iii) could create resource competition among sectors. Policies that have direct sectoral effects include specific policies for reducing GHG emissions and non-climate policies that yield GHG emissions reductions as co-benefits. Policies may also have indirect cross-sectoral effects, including synergies and trade-offs that may, in addition, spill over to other countries.

Section 12.7 provides an overview of knowledge gaps, which could be used to inform further research.

## 12.2 Aggregation of Sectoral Costs and Potentials

The aim of this section is to provide a consolidated overview of the net emissions reduction potentials and costs for mitigation options available in the various sectors dealt with in the sectoral Chapters 6, 7, 9, 10 and 11 of this assessment report. This overview provides policymakers with an understanding of which options are more or less important in terms of mitigating emissions in the short term (here interpreted as 2030), and which ones are more or less costly. The intention is not to provide a high level of accuracy for each technology cost or potential, but rather to indicate relative importance on a global scale and whether costs are low, intermediate or high. The section starts with an introduction (Section 12.2.1), providing definitions and the background. Next, ranges of net emission reduction potentials and the associated costs for the year 2030 are presented (Section 12.2.2) and compared to earlier estimates and with the outputs of IAMs (Section 12.2.3). Finally, an outlook to the year 2050 is provided (Section 12.2.4).

### 12.2.1 Introduction

The term 'mitigation potential' is used here to report the quantity of net greenhouse gas emissions reductions that can be achieved by a given mitigation option relative to a specified reference scenario. The net greenhouse gas emission reduction is the sum of reduced emissions and enhanced sinks. Several types of potential can be distinguished. The technical potential is the mitigation potential constrained by theoretical limits in addition to the availability of technology and practices. Quantification of technical potentials primarily takes into account technical considerations, but social, economic and/or environmental considerations are sometimes also considered, if these represent strong barriers to the deployment of an option. The economic potential, being the potential reported in this section, is the proportion of the technical potential for which the social benefits exceed the social costs, taking into account a social discount rate and the value of externalities (see Annex I: Glossary). In this section, only externalities related to greenhouse gas emissions are taken into account. They are represented by using different cost cut-off levels of options in terms of USD per tonne of avoided $CO_2$-eq emissions. Other potentials, such as market potentials, could also be considered, but they are not included in this section.

The analysis presented here is based, as far as possible, on information contained in Chapters 6, 7, 9, 10 and 11, where costs and potentials, referred to here as 'sectoral mitigation potentials' have been discussed for each individual sector. In the past, these were designated as bottom-up potentials, in contrast to the top-down potentials that are obtained from integrated energy-economic models and IAMs. However, IAMs increasingly include 'bottom-up' elements, which makes the distinction less clear. Still, sectoral studies often have more technical and economic detail than IAMs. They may also provide more up-to-date information on technology options and associated costs. However, aggregation of results from sectoral studies is more complex, and although interactions and overlap are corrected for as far as possible in this analysis, it is recognised that such systemic effects are much more rigorously taken into account in IAMs. A comparison is made between the sectoral results and the outcomes of the IAMs in Section 12.2.3.

Costs of mitigation options will change over time. For many technologies, costs will reduce as a result of technological learning. An attempt has been made to take into account the average, implementation-weighted costs until 2030. However, the underlying literature did not always allow such costs to be presented. For the year 2030, the results are presented similarly to AR4, with a breakdown of the potential in 'cost bins'. For the year 2050, a more qualitative approach is provided. The origins of the cost data in this section are mostly based on studies carried out in the period 2015–2020. Given the wide range of the cost bins that are used in this section it is not meaningful (and often not possible) to convert to USD values for one specific year. This may lead to some extra uncertainty, but this is expected to be relatively small.

As indicated previously, net emission reduction potentials are presented based on comparison with a reference scenario. Unfortunately, not all costs and potentials found in the literature are determined against the same reference scenarios. In this assessment, reference scenarios are based on what were assumed current-policy scenarios in the period 2015–2019. Typical reference scenarios are the Shared Socio-economic Pathway (SSP2) scenarios (Fricko et al. 2017) and the Current Policies scenario from the World Energy Outlook (WEO) 2019 (IEA 2019). They can both be considered scenarios with middle-of-the-road expectations on population growth and economic development, but there are still some differences between the two (Table 12.2). The net emissions reduction potentials reported here were generally based on analyses carried out before 2020, so the impact of the COVID-19 pandemic was not taken into account. For comparison, the Stated Policies scenario of the World Energy Outlook 2020 (IEA 2020a) is also shown, one of the scenarios in which the impact of COVID-19 was considered. Variations of up to 10% between the different reference scenarios exist with respect to macro-variables such as total primary energy use and total GHG emissions. The potential estimates presented below should be interpreted against this background. The total emissions under the reference scenarios in 2030 are expected to be in the range of 54 to 68 $GtCO_2$-eq $yr^{-1}$ with a median of 60 $GtCO_2$-eq $yr^{-1}$ (Table 4.1).

**Table 12.2 | Key characteristics of the scenarios used as a reference for determining costs and potentials.** The values are for the year 2030.

| | SSP2 reference (MESSAGE-GLOBIOM) (Fricko et al. 2017) | All reference scenarios median (25th–75th percentiles in parenthesis) (AR6 scenarios database, IIASA, 2021) | WEO-2019 (Current Policies) (IEA 2019) | WEO-2020 (Stated Policies) (IEA 2020a) | AR6 WG III Chapter 4 (Chapter 4, Table 4.1) |
|---|---|---|---|---|---|
| Real GDP (purchasing power parity, PPP) ($10^{12}$ USD) | 158 (USD2010) | 159 (154–171) | 3.6% p.a.↑ (2018 to 2030) | 2.9% p.a.↑ (2019 to 2030) | |
| Population (billion) | 8.30 | 8.30 (8.20–8.34) | 8.60 | | |
| Total primary energy use (EJ) | 627 | 670 (635–718) | 710 | 660 | |
| Total final energy use (EJ) | 499 | 480 (457–508) | 502 | 472 | |
| Energy-related $CO_2$ emissions (Gt) | 33.0 | 37.9 (34.7–41.4) | 37.4 | 33.2[a] | 37 (35–45) |
| $CO_2$ emissions energy and industry (Gt) | 37.9 | 42.3 (39.0–45.8) | | 36.0 | |
| Total $CO_2$ emissions (Gt) | 40.6 | 45.7 (41.8–49.4) | | | 43 (38–51) |
| Total greenhouse gas emissions ($GtCO_2$-eq) | 52.7 | 59.7 (55.0–65.8) | | | 60 (54–68) |

[a] The difference between WEO-2020 and WEO-2019 is partly explained by the fact that WEO-2019 had two different reference scenarios: Current Policies and Stated Policies. WEO-2020 has only one reference: the Stated Policies Scenario, which 'is based on today's policy settings'. The Stated Policies Scenario in WEO-2019 had energy-related emissions of 34.9 $GtCO_2$. EJ, exajoules (1 x 1018 joules); p.a., per annum.

For the energy sector the potentials are determined using the World Energy Outlook 2019 Current Policies Scenario as a reference (IEA 2019). However, for the economic assessment, more recent Levelised Costs of Electricity (LCOEs) for different electricity generating technologies were used (IEA 2020a). For the AFOLU sector, the potentials were derived from a variety of studies. It may be expected that the best estimates, as averages, match with the reference in a middle-of-the-road scenario. For the buildings sector, the Current Policies scenario of World Energy Outlook 2019 (IEA 2019) was used as a reference. For the transport sector, the references of the underlying sources were used. For the industry sector, the scenarios used have emissions that are slightly higher than in the Current Policies scenario from the World Energy Outlook 2019 (IEA 2019).



### 12.2.2 Costs and Potentials of Options for 2030

In this section, we present an overview of mitigation options per sector. An overview of net emissions reduction potentials for different mitigation options is presented in Table 12.3.

Firstly, a brief overview of the process of data collection is presented, with a more detailed overview being found in Supplementary Material 12.SM.1.2. For the energy sector, the starting point for the determination of the emissions reduction potentials was the Emissions Gap Report (UNEP 2017), but new literature was also assessed, and a few studies that provide updated estimates of the mitigation potentials were included. It was found that higher mitigation potentials than in the UNEP report are now reported for solar and wind energy, but at the same time electricity production by solar and wind energy in the reference scenario has increased, compared to earlier versions of the World Energy Outlook. The net effect is a modest increase in the average value of the potential, and a wider uncertainty range. Costs of electricity-generating technologies are discussed in Section 6.4.7, with a summary of LCOEs from the literature being presented in Section 6.4.7. Mitigation costs of electricity production technology depend on local conditions and on the baseline technology being displaced, and it is difficult to determine the distribution over the cost ranges used in this assessment. However, it is possible to indicate a broad cost range for these technologies. These cost ranges are presented in Table 12.3. For onshore wind and utility-scale solar energy, there is strong evidence that despite regional differences in resource potential and cost, a large part of the mitigation potential can be found in the negative cost category or at cost parity with fossil fuel-based options. This is also the case for nuclear energy in some regions. Other technologies show mostly positive mitigation costs, the highest mitigation costs are for CCS and bioelectricity with CCS, for details see Supplementary Material 12.SM.1.2.

For the AFOLU sector, assessments of global net emissions reduction studies were provided in Table 7.3. The number of studies depends on the type of mitigation action, but ranges from five to nine. Each of these studies relies on a much larger number of underlying data sources. From these studies, emissions reduction ranges and best estimates were derived. The studies presented refer to different years in the period 2020 to 2050, and the mitigation potential presented for AFOLU primarily refers to the average over the period 2020 to 2050. However, because most of the activities involve storage of carbon in stocks that accumulate carbon, or conversely decay over time (e.g., forests, mangroves, peatland soils, agricultural soils, wood products), the 2020 to 2050 average provides a good approximation of the amount of permanent atmospheric $CO_2$ mitigation that could be available at a given price in 2030. The exception is BECCS, which is in an early upscaling phase, so the potential estimated by Chapter 7 as an average for the 2020 to 2050 period is not included in Table 12.3. Note that for the energy sector a mitigation potential for BECCS is provided in Table 12.3.

The emissions reduction potentials for the buildings sector were based on the analysis by Chapter 9 authors of a large number of sectoral studies for individual countries or regions. In total, the chapter analysed the results of 67 studies that assess the potential of technological energy efficiency and onsite renewable energy production and use, and the results of 11 studies that assess the potential of sufficiency measures helping avoid demand for energy and materials. The sufficiency measures were included in models by reorganisation of human activities; efficient design, planning, and use of building space; higher density of building and settlement inhabitancy; redefining and downsizing goods and equipment, limiting their use to health, living, and working standards, and their sharing. Most of these studies targeted 2050 for the decarbonisation of buildings; the potentials in 2030 reported here rely on the estimates for 2030 provided by these studies or on the interpolated estimates targeting these 2050 figures. Based on these individual country studies, regional aggregate emissions reduction percentages were found. The potential estimates were assembled in the order sufficiency, efficiency, renewable options, correcting the amount of the potential at each step for the interaction with preceding measures. Note that the option 'Enhanced use of wood products' was analysed by Chapter 7, but is listed under the buildings sector in Table 12.3, as such enhanced use of wood takes place predominantly in the construction sector.

For the transport sector, Chapter 10 provided data on the emissions reduction potential for shipping. For the other transportation modes, additional sources were used to achieve a complete overview of emissions reduction potentials (for further details, see Supplementary Material 12.SM.1.2). A limited number of estimates for global emissions reduction potential is available: the total number of sources is about 10, and some estimates rely on just one source. The data have been coordinated with Chapter 10 authors.

For the industrial sector, global emissions reduction potentials per technology class per sector were derived by Chapter 11 authors, using primarily sectoral or technology-oriented literature. The analysis is based on about 75 studies, including sectoral assessments (Sections 11.4.1 and 11.4.2 and Figure 11.13).

For methane emissions reduction from oil and gas operations, coal mining, waste treatment and wastewater, an analysis was done, based on three major data sources in this area (Harmsen et al. 2019; US EPA 2019; Höglund-Isaksson et al. 2020); for oil and gas operations this was complemented by IEA (2021a). A similar analysis for reductions of emissions of fluorinated gases was carried out based on analysis by the same institutes (Purohit and Höglund-Isaksson 2017; Harmsen et al. 2019; US EPA 2019). Data for CDR options not discussed previously (such as DACCS and enhanced weathering) were taken from Section 12.3. For more details about data sources and data processing, see Supplementary Material 12.SM.1.2.

In Table 12.4 mitigation potentials for all gases are presented in $GtCO_2$-eq. For most sectors the mitigation potentials (notably for methane emissions reductions from coal, oil and gas, waste and wastewater) have been converted to $CO_2$-eq using global warming potential (GWP) values as presented in AR6 WGIII (Cross-Chapter Box 2 in Chapter 2). However, the underlying literature did not always accommodate this, in which cases older GWP values apply. Given the uncertainty ranges in the mitigation potentials in Table 12.3, the impact on the results of using different GWP values is considered to be very small.

For all options, uncertainty ranges of the mitigation potentials are given in Table 12.3. As far as possible, the ranges represent the variation in assessments found in the literature. This is the case for wind and solar energy, for the AFOLU options, for the methane mitigation options (coal, oil and gas, waste and wastewater) and for fluorinated gas mitigation. For the latter options, some variability exists for each cost bin, but aggregated over cost ranges the variation is much smaller, typically ±50%. For the buildings sector and the industrial sector options, the uncertainty in the mitigation potential is estimated by the lead authors of those chapters. For options for which only limited sources were available, an uncertainty range of ±50% was used. Overall, the uncertainty range per option is typically in the range of ±20% to ±60%.

Despite these uncertainties, clearly a number of options with high potentials can be identified, including solar energy, wind energy, reducing conversion of forests and other natural ecosystems, and restoration of forests and other natural ecosystems. As mid-range values, they each represent 4 to 7% of total reference emissions for 2030. Soil carbon sequestration in agriculture and fuel switching in industry can also be considered as options with high potential, although it should be noted that these options consist of a number of discernible sub-options, see Table 12.3. It can be observed that for each sector, a variety of options is available. Many of the smaller options each make up 1 to 2% of the reference emissions for 2030. Within this group of smaller options there are some categories that, summed together, stand out as substantial: the energy efficiency options and the methane mitigations options.

Costs are highly variable across the options. All sectors have several options for which at least part of the potential has mitigation costs below USD20 $tCO_2^{-1}$. The only exception is the industrial sector, in which only energy efficiency is available below this cost level. At the same time, a substantial part of the emissions reduction potential comes at higher cost, much being in the USD20 to 100 $tCO_2^{-1}$ cost ranges. All sectors have substantial additional potential in these cost ranges; only for transportation is this limited. Aggregation of the potentials per cost bin shows that the potential in these cost bins is marginally smaller than in the two cheapest cost bins. For some options, potential was identified in the 100 to 200 $tCO_2^{-1}$ cost bin. The mitigation potentials identified in this cost range make up only a small part of the total mitigation potential.

**Table 12.3 | Detailed overview of global net GHG emissions reduction potentials ($GtCO_2$-eq) in the various cost categories for the year 2030.** Note that potentials within and across sectors cannot be summed, as the adoption of some options may affect the mitigation potentials of other options. Only monetary costs and benefits of options are taken into account. Negative costs occur when the benefits are higher than the costs. For wind energy, for example, this is the case if production costs are lower than those of the fossil alternatives. Ranges are indicated for each option separately, or indicated for the sector as a whole (see Notes column); they reflect full ranges. Cost ranges are not cumulative, e.g., to obtain the full potential below USD50 $tCO_2$-eq$^{-1}$, the potentials in the cost bins <USD0, USD0–20 and USD20–50 $tCO_2$-eq$^{-1}$ need to be summed together.

| Emissions reduction options (including carbon sequestration options) | Cost categories (USD $tCO_2$-eq$^{-1}$) | | | | | Notes |
|---|---|---|---|---|---|---|
| | <0 | 0–20 | 20–50 | 50–100 | 100–200 | |
| **Energy sector** | | | | | | **Cost ranges are derived as ranges of LCOEs for different electricity generating technologies and the potentials are updated from UNEP (2017).** |
| Wind energy | 2.1–5.6 (majority in <0 range) | | | | | Costs for system integration of intermittent renewables are not included, but these are expected to have limited impact until 2030 and will depend on market design and cross-sectoral integration. |
| Solar energy | 2.0–7.0 (majority in <0 range) | | | | | |
| Nuclear energy | 0.88 ± 50% | | | | | |
| Bioelectricity | | | | 0.86 ± 50% | | Biomass use for indoor heating and industrial heat is not included here. Currently, about 90% of renewable industrial heat consumption is bio-based, mainly in industries that can use their own biomass waste and residues (IEA, 2020). |
| Hydropower | | 0.32 ± 50% | | | | Mitigation costs show large variation and may end up beyond these ranges. |
| Geothermal energy | | 0.74 ± 50% | | | | Mitigation costs show large variation and may end up beyond these ranges. |
| Carbon capture and storage (CCS) | | | | 0.54 ± 50% | | |
| Bioelectricity with CCS | | | | 0.30 ± 50% | | |
| $CH_4$ emissions reduction from coal mining | 0.04 (0.01–0.06) | 0.41 (0.15–0.64) | 0.03 (0.02–0.05) | 0.02 (0.01–0.03) | | |
| $CH_4$ emissions reduction from oil and gas operations | 0.31 (0.12–0.56) | 0.61 (0.23–1.30) | 0.07 (0.03–0.20) | 0.06 (0.00–0.29) | 0.10 (0–0.29) | |
| **Land-based mitigation options (including agriculture and forestry)** | | | | | | **Potentials for AFOLU are averages for the period 2020–2050 and represent a proxy for mitigation in 2030.**<br>**Technical potentials listed below include the potentials already listed in the previous columns.**<br>**Note that in Table 7.3 the same potentials are listed, but they are cumulative over the cost bins.** |
| Carbon sequestration in agriculture (soil carbon sequestration, agroforestry and biochar application) | | 0.50 (0.38–0.60) | 0.73 (0.5–1.0) | 2.21 (0.6–3.9) | | Technical potential: 9.5 (range 1.1–25.3). |
| $CH_4$ and $N_2O$ emissions reduction in agriculture (reduced enteric fermentation, improved manure management, nutrient management, rice cultivation) | | 0.35 (0.11–0.84) | – | 0.28 (0.19–0.46) | | Technical potential: 1.7 (range 0.5–3.2).<br>GWPs used from AR4 and AR5. |
| Protection of natural ecosystems (avoid deforestation, loss and degradation of peatlands, coastal wetlands and grasslands) | | 2.28 (1.7–2.9) | 0.12 (0.06–0.18) | 1.63 (1.3–4.2) | 0.22 (0.09–0.45) | Technical potential 6.2 (range 2.8–14.4). |
| Restoration (afforestation, reforestation, peatland restoration, coastal wetland restoration) | | 0.15 | 0.57 (0.2–1.5) | 1.46 (0.6–2.3) | 0.66 (0.4–1.1) | Technical potential 5.0 (range 1.1–12.3). |
| Improved forest management, fire management | | 0.38 (0.32–0.44) | – | 0.78 (0.32–1.44) | | Technical potential 1.8 (range 1.1–2.8). |
| Reduction of food loss and food waste | | | | | | Feasible potential 0.5 (0.1–0.9).<br>Technical potential 0.7 (0.1–1.6).<br>Estimates reflect direct mitigation from diverted agricultural production only, not including land use effects. |

<table>
<tr><th rowspan="2">Emissions reduction options (including carbon sequestration options)</th><th colspan="5">Cost categories (USD $tCO_2$-eq$^{-1}$)</th><th rowspan="2">Notes</th></tr>
<tr><th><0</th><th>0–20</th><th>20–50</th><th>50–100</th><th>100–200</th></tr>
<tr><td>Shift to sustainable healthy diets</td><td></td><td></td><td></td><td></td><td></td><td>Feasible potential 1.7 (1.0–2.7).<br>Technical potential 3.5 (2.1–5.5).<br>Estimates reflect direct mitigation from diverted agricultural production only, not including land-use effects.</td></tr>
<tr><td>Buildings</td><td></td><td></td><td></td><td></td><td></td><td>To avoid double-counting, the numbers were corrected for the potential overlap between options in the order sufficiency, efficiency, renewable measures and they could be therefore added up. In 2050, much larger and cheaper potential is available (see Section 9.6); the potential in 2030 is lower and more expensive, mostly due to various feasibility constraints.</td></tr>
<tr><td>Sufficiency to avoid demand for energy services (e.g., efficient building use and increased inhabitancy and density)</td><td>0.56<br>(0.28–0.84)</td><td></td><td></td><td></td><td></td><td></td></tr>
<tr><td>Efficient lighting, appliances and equipment, including information and communications technologies, water heating and cooking technologies</td><td>0.73<br>(0.54–0.91)</td><td></td><td></td><td></td><td></td><td></td></tr>
<tr><td>New buildings with very high energy performance (change in construction methods, management and operation of buildings, efficient heating, ventilation and air conditioning)</td><td></td><td colspan="3">0.35<br>(0.26–0.53)</td><td>0.83<br>(0.62–1.24)</td><td></td></tr>
<tr><td>Onsite renewable production and use (often backed-up with demand-side flexibility and digitalisation measures, typically installed in very new high energy performance buildings)</td><td></td><td colspan="3">0.20<br>(0.15–0.30)</td><td>0.27<br>(0.20–0.40)</td><td></td></tr>
<tr><td>Improvement of existing building stock (thermal efficiency of building envelopes, management and operation of buildings, and efficient heating, ventilation and air conditioning leading to 'deep' energy savings)</td><td></td><td colspan="4">0.27<br>(0.20–0.34)</td><td>Additionally, there is 0.50 (range 0.37–0.62) $GtCO_2$-eq of potential above a price of USD200 $tCO_2$-eq$^{-1}$.</td></tr>
<tr><td>Enhanced use of wood products</td><td></td><td></td><td></td><td></td><td></td><td>Technical potential 1.0 (range 0.04–3.7).<br>Economic potential 0.38 (range 0.3–0.5) (varying carbon prices). Potential is mainly in the construction sector.</td></tr>
<tr><td>Transport</td><td></td><td></td><td></td><td></td><td></td><td>Options for the transportation sector have an uncertainty of ±50%.</td></tr>
<tr><td>Light duty vehicles – fuel efficiency</td><td>0.6</td><td></td><td></td><td></td><td></td><td></td></tr>
<tr><td>Light duty vehicles – electric vehicles</td><td></td><td></td><td></td><td></td><td></td><td>Estimated potential is 0.5-0.7 $GtCO_2$-eq, depending on the carbon intensity of the electricity supplied to the vehicles. Mitigation costs are variable.</td></tr>
<tr><td>Light duty vehicles – shift to public transport</td><td>0.5</td><td></td><td></td><td></td><td></td><td></td></tr>
<tr><td>Light duty vehicles – shift to bikes and e-bikes</td><td>0.2</td><td></td><td></td><td></td><td></td><td></td></tr>
<tr><td>Heavy duty vehicles – fuel efficiency</td><td>0.4</td><td></td><td></td><td></td><td></td><td></td></tr>
<tr><td>Heavy duty vehicles – electric vehicles</td><td></td><td></td><td></td><td></td><td></td><td>Estimated potential is 0.2 $GtCO_2$-eq.<br>Mitigation costs are variable.</td></tr>
<tr><td>Heavy duty vehicles – shift to rail</td><td></td><td></td><td></td><td></td><td></td><td>No data available.</td></tr>
<tr><td>Shipping – efficiency, optimisation, biofuels</td><td>0.5<br>(0.4–0.7)</td><td></td><td></td><td></td><td></td><td></td></tr>
<tr><td>Aviation – energy efficiency</td><td>0.12–0.32</td><td></td><td></td><td></td><td></td><td>Limited evidence.</td></tr>
<tr><td>Biofuels</td><td></td><td colspan="3">0.6–0.8</td><td></td><td></td></tr>
</table>

12

| Emissions reduction options (including carbon sequestration options) | Cost categories (USD $tCO_2$-eq$^{-1}$) | | | | | Notes |
|---|---|---|---|---|---|---|
| | <0 | 0–20 | 20–50 | 50–100 | 100–200 | |
| **Industry** | | | | | | **The numbers for the industry sector typically have an uncertainty of ±25%, unless indicated differently.** **The numbers are corrected for overlap between the options, except for the 0.15 $GtCO_2$ potential in the highest cost bin. For the rest they can be aggregated to provide full potentials.** |
| Energy efficiency | | 1.14 | | | | This only applies to more efficient use of fuels. More efficient use of electricity is not included. |
| Material efficiency | | | 0.93 | | | |
| Circularity (enhanced recycling) | | | 0.48 | | | |
| Fuel switching | | | 1.28 | 0.67 | 0.15 | |
| Feedstock decarbonisation, process change | | | | 0.38 | | |
| Carbon capture, utilisation and storage (CCU and CCS) | | | | | 0.15 (0.08–0.36) | |
| Cementitious material substitution | | | 0.28 | | | |
| Reduction of non-$CO_2$ emissions | | 0.2 | | | | |
| **Cross-sectorial** | | | | | | |
| Emission reduction of fluorinated gases | 0.26 (0.01–0.50) | 0.68 (0.55–0.90) | 0.18 (0.01–0.42) | 0.09 (0–0.20) | 0.03 (0–0.05) | GWPs not updated. |
| Reduction of $CH_4$ emissions from solid waste | 0.33 (0.24–0.43) | 0.11 (0.03–0.15) | 0.06 (0.03–0.08) | 0.04 (0.01–0.10) | 0.08 (0.02–0.12) | |
| Reduction of $CH_4$ emissions from wastewater | 0.02 (0–0.05) | 0.03 (0.01–0.05) | 0.04 (0.01–0.07) | 0.03 (0.02–0.04) | 0.07 (0.01–0.16) | |
| Direct air carbon capture and storage (DACCS) | | | | | very small | There is potential in these categories, but given the current technology readiness levels, for 2030 the potential is limited. Also, it is not certain whether the costs will have dropped below 200 USD $tCO_2^{-1}$ before 2030. In the longer term, much larger potentials are projected, see Section 12.3.1. |
| Enhanced weathering | | | | | very small | |

It could be that there is limited potential in this range; however, a more plausible explanation, supported by several authors of sectoral chapters, is that this cost range is relatively unexplored.

In this assessment, the emphasis is on the specific mitigation costs of the various options, and these are often considered as an indicator to prioritise options. However, in such a prioritisation, other elements will also play a role, like the development of technology for the longer term (Section 12.2.4) and the need to optimise investments over longer time periods, see for example Vogt-Schilb et al. (2018) who argue that sometimes it makes sense to start with implementing the most expensive option.

In this section, an overview of emissions mitigation options for the year 2030 was presented. The overview of the mitigation potential is based on a variety of approaches, relying on a large number of sources, and the number of sources varied strongly from sector to sector. The main conclusions from this section are: (i) there is a variety of options per sector, (ii) per sector the options combined show significant mitigation potential, (iii) there are a few major options and a lot of smaller ones, and (iv) more than half of the potential comes at costs below USD20 $tCO_2^{-1}$ (between sectors: *medium* to *robust evidence*, *high agreement*).

### 12.2.3 Aggregation of Sectoral Results and Comparison with Earlier Analyses and Integrated Assessment Models

In this section, the mitigation potentials are aggregated per sector, and then to the global economy. These potentials, which are based on sectoral analysis, are then compared to the results from earlier assessments and the results from IAMs. Given the incompleteness of data on the mitigation potential at mitigation costs larger than USD100 $tCO_2^{-1}$, the focus will be on options with mitigation costs below USD100 $tCO_2^{-1}$.

As suggested previously, the overview presented in Table 12.3 should be interpreted with care, as the implementation of one option may affect the mitigation potential of another option. Most sectoral chapters have supplied mitigation potentials that were already adjusted for overlap and mutual influences (industry, buildings, AFOLU). For the energy sector, interactions between the options will occur, but parallel implementation of all the options seems to be possible; if all options at costs levels below USD100 $tCO_2^{-1}$ were implemented, this would lead to an additional power generation with no direct $CO_2$ emissions of 41% of the total projected generation in 2030. This seems to be possible, but as higher penetrations are relatively unexplored, we apply a smaller uncertainty range at the high end. For the calculation of the aggregate potentials in the energy sector, error propagation rules were applied. For the transport sector, there will be interaction between the technical measures on the one hand and the modal shift measures on the other hand. Given the small mitigation contribution of the modal shift options, these interactions will be negligible. The resulting aggregate mitigation potentials and their uncertainty ranges per (sub)sector are given in Table 12.4 (columns indicated 'AR6'). This overview confirms the large potentials per sector, even when taking the uncertainty ranges into account.

Calculating aggregated mitigation potentials for the global economy requires that interactions between sectors also need to be taken into account (Section 12.6). First of all, there may be overlap between the electricity supply sector and the electricity demand sectors: if the electricity sector is extensively decarbonised, the avoided emissions due to electricity efficiency measures and local electricity production will be significantly reduced. Therefore, this demand-side mitigation potential is only taken into account for 25% (reflecting the degree of further decarbonisation of the power sector) in the cross-sectoral aggregation. For the other demand sectors, this problem does not arise. The industry sector did not provide estimates for electricity efficiency improvement and in the transport sector the utilisation of electricity to date is very low. Electrification options may occur in all sectors, but this enhances the mitigation potential in combination with a decreased carbon intensity of the power sector. For other energy sector options, such as methane emissions reduction from coal, oil and natural gas operations, the situation is more complex. The total emissions reduction potential for fossil fuels in the other sectors is high. Should this potential be realised, this would lead to a reduction of the potential reported here. However, reducing fossil fuel use also leads to a reduction in the upstream $CH_4$ (methane) emissions, so in the case of reducing fossil fuel use, these upstream emissions will also be avoided, so no overestimate of the aggregate emissions reduction potential occurs.

The total potential, given these corrections for overlap, leads to a mid-range value for the total mitigation potential at costs below USD100 $tCO_2$-$eq^{-1}$ of 38 $GtCO_2$-eq. Given the fact that it is not to be expected that mitigation potentials of the various sectors are mutually correlated, that is, it is not to be expected that mitigation potentials are all on the high side or all on the low side, the ranges are aggregated using error propagation rules, which leads to a range for the mitigation potential of 32 to 44 $GtCO_2$-eq.

Mitigation costs and potentials for 2030 have been presented previously, notably in AR4 Chapter 11 on Mitigation from a Cross-sectoral Perspective (Barker et al. 2007) and the Emissions Gap Report (UNEP 2017). Note that AR5 did not provide emissions reduction potentials in this form. The aggregated potentials reported

**Table 12.4 | Overview of aggregate sectoral net GHG emissions reduction potentials ($GtCO_2$-eq) for the year 2030 at costs below USD100 $tCO_2$-$eq^{-1}$.** Comparisons with earlier assessments are also provided. Note that sectors are not entirely comparable across the three different estimates.

| Sector | Mitigation potentials at costs less than USD100 $tCO_2$-$eq^{-1}$ | | | | |
|---|---|---|---|---|---|
| | AR6 best estimate | AR6 range | AR4 (Barker et al. 2007) | UNEP2017 best estimate (UNEP 2017) | UNEP 2017 range (UNEP 2017) |
| Electricity sector | 11.0 | 7.9–12.5 | 6.2–9.3 | 10.3 | 9.5–11.0 |
| Other energy sector (methane) | 1.6 | 1.1–2.1 | | 2.2 | 1.7–2.6 |
| Agriculture | 4.1 | 1.7–6.7 | 2.3–6.4 | 4.8 | 3.6–6.0 |
| Forestry and other land use-related options | 7.3 | 3.9–13.1 | 1.3–4.2 | 5.3 | 4.1–6.5 |
| AFOLU demand-side options (estimates reflect direct mitigation from **diverted agricultural production only, not including land-use effects)** | 2.2 | 1.1–3.6 | | | 1.3–3.4 |
| Buildings (potentials up to USD200 **$tCO_2$-$eq^{-1}$ in parentheses)** | Dir 0.7<br>(1.1)<br>Ind 1.3<br>(2.1)<br>Tot 2.0<br>(3.2) | 0.5–1.0<br>(0.7–1.5)<br>0.9–1.8<br>(1.5–3.1)<br>1.4–2.9<br>(2.3–4.6) | Dir 2.3–2.9<br>Ind 3.0–3.8<br>Tot 5.4–6.7 | Dir 1.9<br>Ind 4.0<br>Tot 5.9 | Dir 1.6–2.1 |
| Transport | 3.8 | 1.9–5.7 | 1.6–2.5 | 4.7 | 4.1–5.3 |
| Industry | Dir 5.4 | 4.0–6.7 | Dir 2.3–4.9<br>Ind 0.83<br>Tot 3.1–5.7 | Dir 3.9<br>Ind 1.9<br>Tot 5.8 | Dir 3.0–4.8 |
| Fluorinated gases (all sectors) | 1.2 | 0.7–1.5 | NE | 1.5 | 1.2–1.8 |
| Waste and wastewater | 0.7 | 0.6–0.8 | 0.4–1.0 | 0.4 | 0.3–0.5 |
| Enhanced weathering | – | – | – | 1.0 | 0.7–1.2 |
| **Total of all sectors** | **38** | **32–44** | **15.8–31.1** | **38** | **35–41** |

Note: Dir = reduction of direct emissions, Ind = reduction of indirect emissions (related to electricity production), Tot = reduction of total emissions, NE = not estimated, AR4: Table 11.3, UNEP-2017: Chapter 4.

here are higher than those estimated in AR4. Note, however, that AR4 suggested the potentials were underestimated by 10 to 15%, but a higher potential still remains in the current assessment. In a sector-by-sector comparison, higher potentials than in AR4 can be observed especially for the energy sector and the forestry sector, and to a more limited extent for the industry sector and the transport sector. For the energy sector, the change can largely be explained by the higher estimates for wind and solar energy and the improved understanding of how to integrate high shares of intermittent renewable energy sources into power systems. For industry and transport, the higher potentials can be partly explained by the inclusion of more options, like recycling and material efficiency (for industry) and electric transportation and modal shifts for transport. For buildings, a lower potential can be observed compared to AR4, one reason is that the 2030 reference direct and indirect emissions were estimated as 45% and 11% higher in AR4 than they were in AR6 (signalling a much quicker actual switch to electricity than was thought 15 to 20 years ago, among other reasons). The other reason for a difference is that the scenarios considered in AR4 had 25 to 30 years between their start year until the target year of 2030 and the scenarios reviewed in AR6 have only 10 to 15 years before 2030. The current retrofitting rates of existing buildings and penetration rates of nearly zero-energy buildings do not allow for decarbonisation of the sector over 10 to 15 years, but they do over a longer time period. A much larger potential than reported here for 2030 can still be realised in the timeframe up to 2050 (Section 9.6.2).

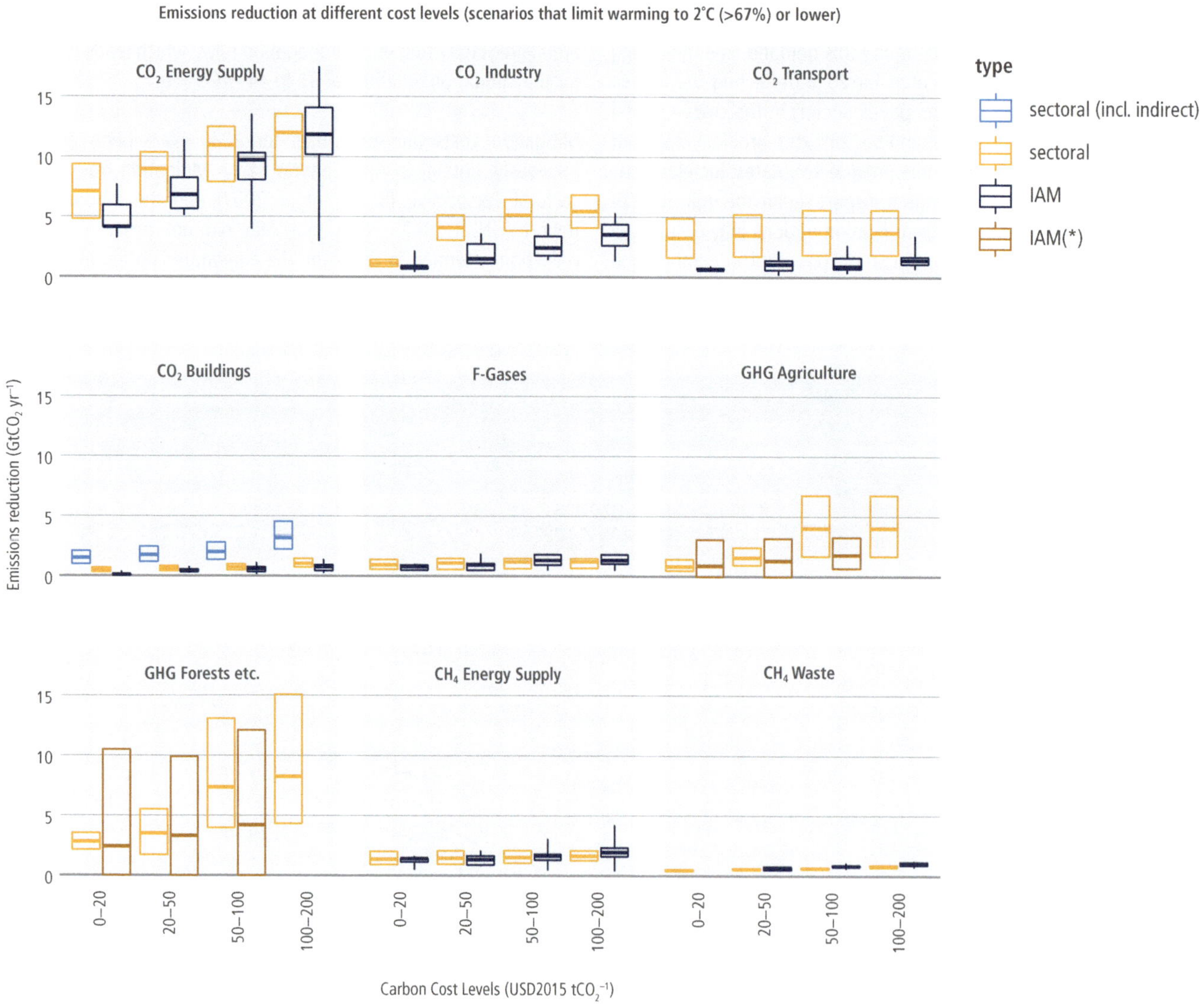


**Figure 12.1 | Comparison of sectoral estimates for emissions reduction potential with the emissions reductions calculated using IAMs.** Emission reductions calculated using IAMs are given as box plots of global emissions reductions for each sector (dark blue and brown) at different global carbon cost levels (horizontal axis) for 2030, based on all scenarios that limit warming to 2°C (>67%) or lower (see Chapter 3) in the AR6 scenarios database (IIASA 2021). For IAMs, the cost levels correspond to the levels of the carbon price. Hinges in the dark blue box plots represent the interquartile ranges and whiskers extend to 5th and 95th percentiles while the hinges in the brown box plots describe the full range, and the middle point indicates the mean, not the median. In yellow, the estimates from the sectoral analysis are given. In all cases, only direct emissions reductions are presented, except for the light-blue boxes (for buildings), which include indirect emissions reductions. The light-blue boxes are only given for reasons of completeness. For buildings the dark-blue boxes should be compared with the yellow boxes. Light-blue and yellow boxes represent the full ranges of estimates. For IAMs, global carbon prices are applied, which are subject to significant uncertainty.

Another global analysis was done by McKinsey (2009), which presents a marginal abatement cost curve for 2030, suggesting a total potential of 38 $GtCO_2$-eq (note that the reference for that study is 70 $GtCO_2$-eq, which is at the high end of the reference range used in this assessment).

The potentials reported here are comparable with UNEP (2017). Note that material for the energy sector from the UNEP report was partly reused in this analysis. Furthermore, some options for the transport sector (aviation and biofuels) were identical to the estimates in the UNEP report. The remaining mitigation potentials are all based on new – and much more extended – assessment. There are some notable changes. The AR6 mitigation potential for forestry is substantially larger. For buildings the potential is smaller, mainly related to the smaller mitigation potential for electric appliances than in the UNEP report. But overall, the estimates of the total mitigation potential are well aligned, which confirms there is substantial consistency across various emissions reduction estimates.

The results of the sectoral mitigation potentials are also compared with mitigation impacts as calculated by IAMs. To this end, cumulative sectoral potentials over cost ranges were determined, based on the information in Table 12.3. For options that are in various cost ranges, we assumed that they are evenly distributed over these cost ranges. The only exception is wind and solar energy, for which it is indicated that the majority of the mitigation potential is in the negative cost range. It was assumed that the fraction in the negative cost range was 60%; the remainder is evenly distributed over the other cost ranges. These cumulative potentials were compared with emissions reductions realised in IAMs at certain price levels for $CO_2$. Note that these price levels selected in IAMs are average price levels – not all IAMs use globally uniform carbon prices, so underlying these cost levels, there may be regional differentiation. Data were taken from the AR6 scenarios database. Note that, strictly speaking, not all models in the database are IAMs; in this analysis all models in the database were used, but the term IAMs is used as shorthand in the text that follows. All scenarios that limit warming to 2°C (>67%) or lower are included for the comparison (i.e., the categories of scenarios C1 to C3 in Chapter 3). A comparison per sector is provided in Figure 12.1. It is important to note that two different things are compared in this figure: on the one hand emissions reduction potentials and on the other hand realisations of (part of) the potential within the context of a certain scenario. Having said that, a number of lessons can be learned from the comparison of both.

For the energy supply sector, the emissions reductions projected by the IAMs are for the higher cost levels comparable with the potentials found in the sectoral analysis. But at lower cost levels, the emissions reductions as projected by IAMs are smaller than for the sectoral analysis. This is likely due to the fact that high costs for solar energy and wind energy are assumed in IAM models (Krey et al. 2019; Shiraki and Sugiyama 2020). This is not surprising, as the scenario database comprises studies dating back to 2015. A more detailed comparison for the power sector is given in Figure 12.2. Both the sectoral analysis and the IAMs find that both solar and wind energy in particular show strong growth potential, although there is a continuing role for other low-carbon technologies, like nuclear energy and hydropower.

For the AFOLU sector, the sectoral studies provide net emissions reduction potentials comparable with projections from the IAMs at costs levels up to USD50 $tCO_2$-$eq^{-1}$. However, beyond that level the mitigation potential found in the sectoral analysis is larger than in the IAMs. For agriculture, it can be explained by the fact that carbon sequestration options, like soil carbon, biochar and agroforestry, have little to no representation in IAMs. Similarly, for forestry and other land use-related options, the protection and restoration of other ecosystems than forests (peatland, coastal wetlands and savannas) are not represented in IAMs. Also note that some IAM baselines already have small carbon prices, which induce land-based mitigation, while in others, mitigation, particularly from reduced deforestation, is part of the storyline even without an implemented

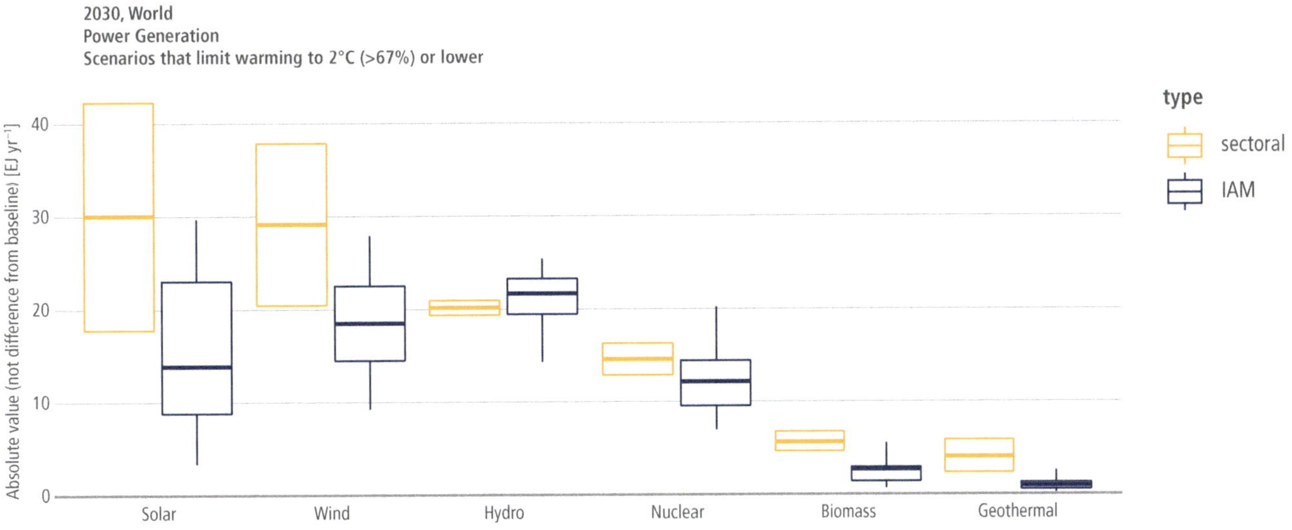


**Figure 12.2 | Electricity production in 2030 as calculated by IAMs (dark blue), compared with electricity production potentials found in the sectoral analysis (yellow).** Cost cut-offs at USD100 $tCO_2^{-1}$ are applied to both electricity production in 2030 as calculated by IAMs and electricity production potentials found in the sectoral analyses. Hinges in the dark-blue box plots represent the interquartile ranges and whiskers extend to the 5th and 95th percentiles, while the hinges in the yellow box plots describe the full range.

carbon price. Both of these effects dampen the mitigation potential available in the USD100 $tCO_2$-$eq^{-1}$ carbon price scenario from IAMs. Furthermore, estimates of mitigation through forestry and other land use-related options from the AR6 IAM scenario database represent the net emissions from A/R and deforestation, thus are likely to be lower than the sectoral estimates of A/R potential expressed as gross removals.

For the buildings and transport sectors, the sectoral mitigation potentials are higher than those projected by the IAMs. The difference in the transport sector is particularly significant. One possible explanation is that options with negative costs are already included in the reference. In addition, some options, like avoiding demand for energy services in the building sector and model shift in transportation, are less well represented in IAMs.

For the industry sector, the sectoral emissions reduction potentials are somewhat higher than those reported on average by IAMs. The difference can well be explained by the fact that most IAMs do not include circularity options like material efficiency and recycling; these options together account for 1.5 $GtCO_2$-eq at costs levels from USD20 $tCO_2$-$eq^{-1}$ onwards.

For mitigation of emissions of methane and fluorinated gases, the comparability between the sectoral results and IAMs is good.

Overall, it is concluded that there are differences between the sectoral analyses and the IAM outcomes, but most of the differences can be explained by the exclusion of specific options in most IAMs. This comparability confirms the reliability of the sectoral analysis of emissions reduction potential. It also demonstrates the added value of sectoral analyses of mitigation potentials: they can more rapidly adapt to changes in price levels of technologies and adopt new options for emissions mitigation.

In this section, the information on individual options reported in Section 12.2.2 to sectoral and economy-wide totals has been aggregated. It is concluded that, based on the sectoral analysis, the global mitigation potential is in the range of 32 to 44 $GtCO_2$-eq. This mitigation potential is substantially higher than that reported in AR4, but it is comparable to the more recent estimate by UNEP (2017). Differences exist with the results of IAMs, but most of these can be well explained. The conclusion that the global potential is in this range can be drawn with *high agreement* and *robust evidence*.

Given the median projection of the reference emissions of 60 $GtCO_2$-eq in 2030, the range of mitigation potentials presented here is sufficient to bring down global emissions in the year 2030 to a level of 16 to 28 $GtCO_2$-eq. Taking into account that there is a range in reference projections for 2030 of 54 to 68 $GtCO_2$-eq, the resulting emissions level shows a wider range: 12 to 31 $GtCO_2$-eq. This is about, or below half, the most recent (2019) emissions value of 59 ± 6.6 $GtCO_2$-eq (*high confidence*).

### 12.2.4 Sectoral Findings on Emission Pathways until 2050

As noted previously, a more qualitative approach is followed and less quantitative information is presented for 2050. The sectoral results are summarised in Table 12.5. In addition to the many technologies that already play a role by 2030 (Table 12.3) additional technologies may be needed for deep decarbonisation, for example for managing power systems with high shares of intermittent renewable sources and for providing new fuels and associated infrastructure for sectors that are hard to decarbonise. New processes also play an important role, notably for industrial processes. In general, stronger sector coupling is needed, particularly increased integration of energy end use and supply sectors.



**Table 12.5 | Mitigation options and their characteristics for 2050.**

| Sector | Major options | Degree to which net zero-GHG is possible |
|---|---|---|
| Energy sector. | Range of supply-side options possible (see 2030 overview).<br>Increased share of electricity in final energy use.<br>Potentially important role for hydrogen, ammonia, etc. | Zero $CO_2$ energy system is possible. |
| Agriculture, forestry and other land use (AFOLU). | Options comparable to those in 2030. Permanence is important. | Some hard-to-abate activities will still have positive emissions, but for the sector as a whole, net negative emissions are possible through carbon sequestration in agriculture and forestry. |
| Buildings. | Sufficiency, high performance new and existing buildings with efficient heating, ventilation, and air conditioning, especially heat pumps, building management and operation, efficient appliances, and onsite renewables backed up with demand flexibility and digitalisation measures. | At least 8.2 $GtCO_2$ or 61% reduction, as compared to the baseline is possible with options on the demand side. This is a low estimate, because in some developing regions literature is not sufficient to derive a comprehensive estimate. Nearly net zero $CO_2$ emissions is possible if grid electricity will also be decarbonised. Carbon storage in buildings provides CDR. |
| Transport. | Electrification can become a major option for many transport modes. For long-haul trucking, ships and aviation, in addition biofuels, hydrogen and potentially synthetic fuels can be applied. | To a large extent if the electricity sector is fully decarbonised and the deployment of alternative fuels for long-haul trucking, aviation and shipping is successful. |
| Industry. | Stronger role for material efficiency and recycling.<br>Full decarbonisation through new processes; CCS, CCU and hydrogen can become dominant. | Approx. 85% reduction is possible. Net zero $CO_2$ emissions is possible with retrofitting and early retirement. |
| Cross-sectoral. | Direct air carbon capture and storage.<br>Enhanced weathering.<br>Ocean-based methods. | Contributes CDR to support net zero GHG by counterbalancing sectoral emissions. |

## 12.3 Carbon Dioxide Removal

Carbon dioxide removal (CDR) refers to a cluster of technologies, practices, and approaches that remove and sequester carbon dioxide from the atmosphere and durably store the carbon in geological, terrestrial, or ocean reservoirs, or in products. Despite the common feature of removing carbon dioxide, CDR methods can be very different (Smith et al. 2017). There are proposed methods for removal of non-$CO_2$ greenhouse gases such as methane (Jackson et al. 2019; Jackson et al. 2021) but scarcity of literature on these methods prevents assessment here.

A number of CDR methods (e.g., afforestation/reforestation (A/R), bioenergy with carbon capture and storage (BECCS), soil carbon sequestration (SCS), biochar, wetland/peatland restoration and coastal restoration) are dealt with elsewhere in this report (Chapters 6 and 7). These methods are synthesised in Section 12.3.2. Others, not dealt with elsewhere, – direct air carbon capture and storage (DACCS), enhanced weathering (EW) of minerals and ocean-based approaches including ocean fertilisation (OF) and ocean alkalinity enhancement (OAE) – are discussed in Sections 12.3.1.1 to 12.3.1.3 below (see also IPCC 2019b and AR6 WGI, Section 5.6). Some methods, such as BECCS and DACCS, involve carbon storage in geological formations, which is discussed in Chapter 6. The climate system and the carbon cycle responses to CDR deployment and each method's physical and biogeochemical characteristics such as storage form and duration are assessed in Chapters 4 and 5 of the AR6 WGI report.

### Cross-Chapter Box 8 | Carbon Dioxide Removal: Key Characteristics and Multiple Roles in Mitigation Strategies

**Authors:** Oliver Geden (Germany), Alaa Al Khourdajie (United Kingdom/Syria), Christopher Bataille (Canada), Göran Berndes (Sweden), Holly Jean Buck (the United States of America), Katherine Calvin (the United States of America), Annette Cowie (Australia), Kiane de Kleijne (the Netherlands), Jan Christoph Minx (Germany), Gert-Jan Nabuurs (the Netherlands), Glen P. Peters (Norway/Australia), Andy Reisinger (New Zealand), Pete Smith (United Kingdom), Masahiro Sugiyama (Japan)

Carbon dioxide removal (CDR) is a necessary element of mitigation portfolios to achieve net zero $CO_2$ and GHG emissions both globally and nationally, counterbalancing residual emissions from hard-to-transition sectors such as industry, transport and agriculture. CDR is a key element in scenarios that limit warming to 2°C (>67%) or lower, regardless of whether global emissions reach near-zero, net zero or net-negative levels (Sections 3.3, 3.4, 3.5 and 12.3). While national mitigation portfolios aiming at net zero or net-negative emissions will need to include some level of CDR, the choice of methods and the scale and timing of their deployment will depend on the ambition for gross emissions reductions, how sustainability and feasibility constraints are managed, and how political preferences and social acceptability evolve (Section 12.3.3). This box gives an overview of CDR methods, presents a categorisation based on the key characteristics of removal processes and storage timescales, and clarifies the multiple roles of CDR in mitigation strategies. The term 'negative emissions' is used in this report only when referring to the net emissions outcome at a systems level (e.g., 'net negative emissions' at global, national, sectoral or supply chain levels).

#### Categorisation of the main CDR methods

CDR refers to anthropogenic activities that remove $CO_2$ from the atmosphere and store it durably in geological, terrestrial, or ocean reservoirs, or in products. It includes anthropogenic enhancement of biological, geochemical or chemical $CO_2$ sinks, but excludes natural $CO_2$ uptake not directly caused by human activities. Increases in land carbon sink strength due to $CO_2$ fertilisation or other indirect effects of human activities are not considered CDR (see Glossary). Carbon capture and storage (CCS) and carbon capture and utilisation (CCU) applied to $CO_2$ from fossil fuel use are not CDR methods as they do not remove $CO_2$ from the atmosphere. CCS and CCU can, however, be part of CDR methods if the $CO_2$ has been captured from the atmosphere, either indirectly in the form of biomass or directly from ambient air, and stored durably in geological reservoirs or products (Sections 11.3.6 and 12.3).

There are many different CDR methods and associated implementation options (Cross-Chapter Box 8, Figure 1). Some of these methods (including afforestation and improved forest management, wetland restoration and soil carbon sequestration (SCS)) have been practised for decades to millennia, although not necessarily with the intention of removing carbon from the atmosphere. Conversely, methods such as direct air carbon capture and storage (DACCS), bioenergy with carbon capture and storage (BECCS) and enhanced weathering are novel, and while experience is growing, their demonstration and deployment are limited in scale. CDR methods have been categorised in different ways in the literature, highlighting different characteristics. In this report, as in AR6 WGI, the categorisation is based on the role of CDR methods in the carbon cycle, that is, on the removal process (*land-based biological*; *ocean-based biological*; *geochemical*; *chemical*) and on the timescale of storage (*decades to centuries*; *centuries to millennia*; *ten thousand years or longer*). The time scale of storage is closely linked to the storage medium: carbon stored in ocean reservoirs (through enhanced weathering, ocean alkalinity enhancement or ocean fertilisation) and in geological formations (through BECCS or DACCS) generally has longer storage times and is less vulnerable to reversal through human actions or disturbances such as drought and wildfire than carbon stored in terrestrial reservoirs (vegetation, soil). Furthermore, carbon stored in vegetation or through SCS has

12

*Cross-Chapter Box 8 (continued)*

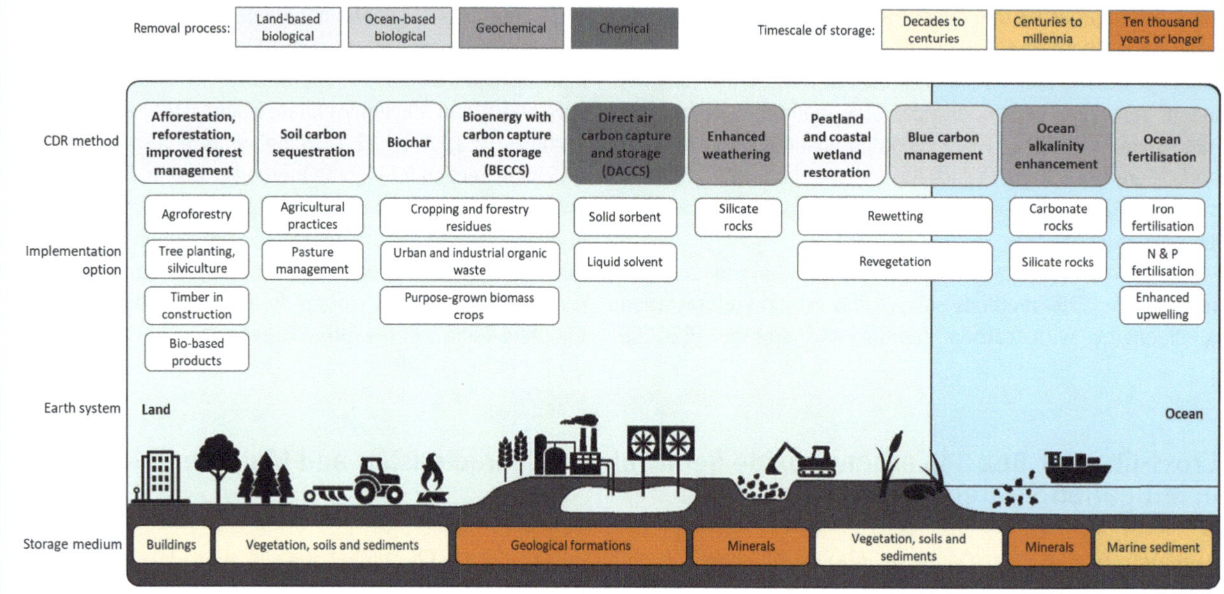


**Cross-Chapter Box 8, Figure 1 | Carbon dioxide removal taxonomy. Methods are categorised based on removal process (grey shades) and storage medium (for which timescales of storage are given, yellow/brown shades).** Main implementation options are included for each CDR method. Note that specific land-based implementation options can be associated with several CDR methods, for example, agroforestry can support soil carbon sequestration and provide biomass for biochar or BECCS. Source: adapted from Minx et al. (2018).

shorter storage times and is more vulnerable than carbon stored in buildings as wood products; as biochar in soils, cement and other materials; or in chemical products made from biomass or potentially through direct air (Fuss et al. 2018; Minx et al. 2018; NASEM 2019) capture (Section 11.3.6; AR6 WGI, Figure 5.36). Within the same category (e.g., land-based biological CDR) options often differ with respect to other dynamic or context-specific dimensions, such as mitigation potential, cost, potential for co-benefits and adverse side effects, and technology readiness level (Table 12.6).

### Roles of CDR in mitigation strategies

Within ambitious mitigation strategies at global or national levels, CDR cannot serve as a substitute for deep emissions reductions but can fulfil multiple complementary roles: it can (i) further reduce net $CO_2$ or GHG emission levels in the near-term; (ii) counterbalance residual emissions from hard-to-transition sectors, such as $CO_2$ from industrial activities and long-distance transport (e.g., aviation, shipping), or methane and nitrous oxide from agriculture, in order to help reach net zero $CO_2$ or GHG emissions in the mid-term; (iii) achieve and sustain net-negative $CO_2$ or GHG emissions in the long-term, by deploying CDR at levels exceeding annual residual gross $CO_2$ or GHG emissions (Sections 2.7.3 and 3.5).

In general, these roles of CDR are not mutually exclusive and can exist in parallel. For example, achieving net zero $CO_2$ or GHG emissions globally might involve some countries already reaching net-negative levels at the time of global net zero, allowing other countries more time to achieve this. Equally, achieving net-negative $CO_2$ emissions globally, which could address a potential temperature overshoot by lowering atmospheric $CO_2$ concentrations, does not necessarily involve all countries reaching net-negative levels (Rajamani et al. 2021; Rogelj et al. 2021) (Cross-Chapter Box 3 in Chapter 3).

Cross-Chapter Box 8, Figure 2 shows these multiple roles of CDR in a stylised ambitious mitigation pathway that can be applied to global and national levels. While such mitigation pathways will differ in their shape and exact composition, they include the same basic components: $CO_2$ emissions from fossil sources, $CO_2$ emissions from managed land, non-$CO_2$ emissions, and various forms of CDR. Cross-Chapter Box 8, Figure 2 also illustrates the importance of distinguishing between gross $CO_2$ removals from the atmosphere through deployment of CDR methods and the net emissions outcome (i.e., gross emissions minus gross removals).

CDR methods currently deployed on managed land, such as afforestation or reforestation and improved forest management, lead to $CO_2$ removals already today, even when net emissions from land use are still positive, for example, when gross emissions from deforestation and draining peatlands exceed gross removals from afforestation or reforestation and ecosystem conservation (Sections 2.2 and 7.2;

*Cross-Chapter Box 8 (continued)*

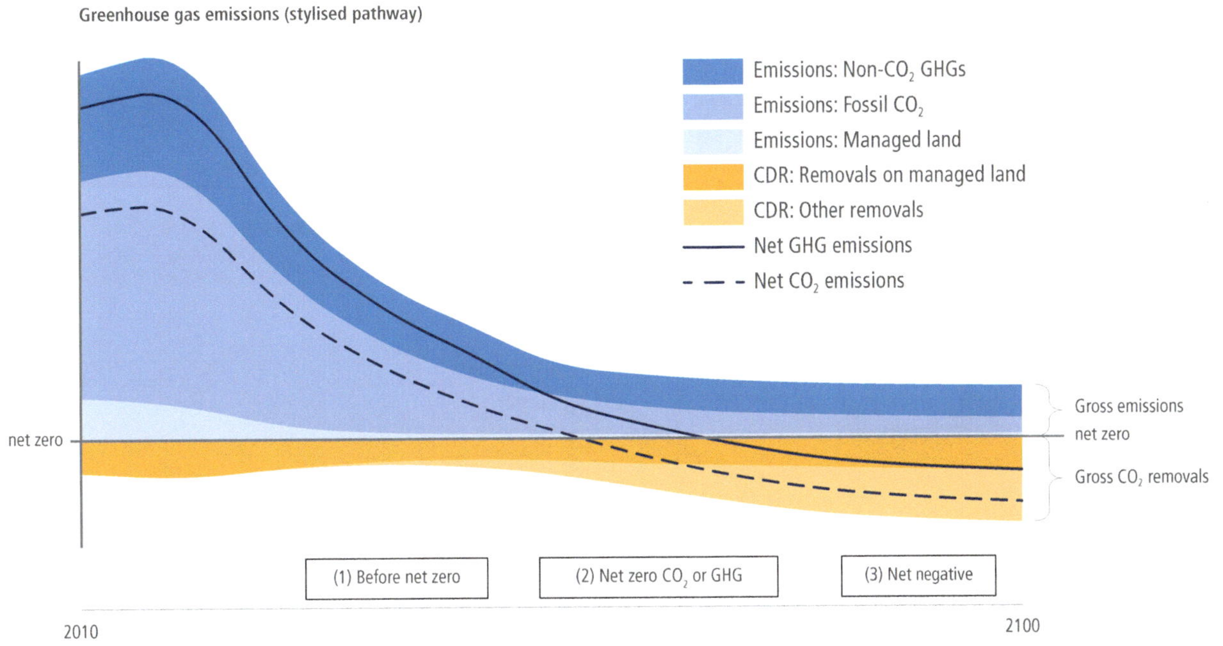


**Cross-Chapter Box 8, Figure 2 | Roles of CDR in global or national mitigation strategies.** Stylised pathway showing multiple functions of CDR in different phases of ambitious mitigation: (1) further reducing net $CO_2$ or GHG emissions levels in near-term; (2) counterbalancing residual emissions to help reach net zero $CO_2$ or GHG emissions in the mid-term; (3) achieving and sustaining net-negative $CO_2$ or GHG emissions in the long-term.

Cross-Chapter Box 6 in Chapter 7). As there are currently no removal methods for non-$CO_2$ gases that have progressed beyond conceptual discussions (Jackson et al. 2021), achieving net zero GHG implies gross $CO_2$ removals to counterbalance residual emissions of both $CO_2$ and non-$CO_2$ gases, applying 100-year global warming potential (GWP100) as the metric for reporting $CO_2$-equivalent emissions, as required for emissions reporting under the Rulebook of the Paris Agreement (Cross-Chapter Box 2 in Chapter 2).

Net zero $CO_2$ emissions will be achieved earlier than net zero GHG emissions. As volumes of residual non-$CO_2$ emissions are expected to be significant, this time-lag could reach one to several decades, depending on the respective size and composition of residual GHG emissions at the time of net zero $CO_2$ emissions. Furthermore, counterbalancing residual non-$CO_2$ emissions by $CO_2$ removals will lead to net-negative $CO_2$ emissions at the time of net zero GHG emissions (Cross-Chapter Box 3 in Chapter 3).



While many governments have included A/R and other forestry measures in their NDCs under the Paris Agreement (Moe and Røttereng 2018; Fyson and Jeffery 2019; Mace et al. 2021), and a few countries also mention BECCS, DACCS and enhanced weathering in their mid-century low emission development strategies (Buylova et al. 2021), very few are pursuing the integration of a broad range of CDR methods into national mitigation portfolios so far (Schenuit et al. 2021) (Box 12.1). There are concerns that the prospect of large-scale CDR could, depending on the design of mitigation strategies, obstruct near-term emissions reduction efforts (Lenzi et al. 2018; Markusson et al. 2018), mask insufficient policy interventions (Geden 2016; Carton 2019), might lead to an overreliance on technologies that are still in their infancy (Anderson and Peters 2016; Larkin et al. 2018; Grant et al. 2021), could overburden future generations (Lenzi 2018; Shue 2018; Bednar et al. 2019) might evoke new conflicts over equitable burden-sharing (Pozo et al. 2020; Lee et al. 2021; Mohan et al. 2021), could impact food security, biodiversity or land rights (Buck 2016; Boysen et al. 2017; Dooley and Kartha 2018; Hurlbert et al. 2019; Dooley et al. 2021), or might be perceived negatively by stakeholders and broader public audiences (Royal Society and Royal Academy of Engineering 2018; Colvin et al. 2020). Conversely, without considering different timescales of carbon storage (Fuss et al. 2018; Hepburn et al. 2019) and implementation of reliable measurement, reporting and verification of carbon flows (Mace et al. 2021), CDR deployment might not deliver the intended benefit of removing $CO_2$ durably from the atmosphere. Furthermore, without appropriate incentive schemes and market designs (Honegger et al. 2021b), CDR implementation options could see under-investment. The many challenges in research, development and demonstration of novel approaches, to advance innovation according to broader societal objectives and to bring down costs, could delay their scaling up and deployment (Nemet et al. 2018). Depending on the scale

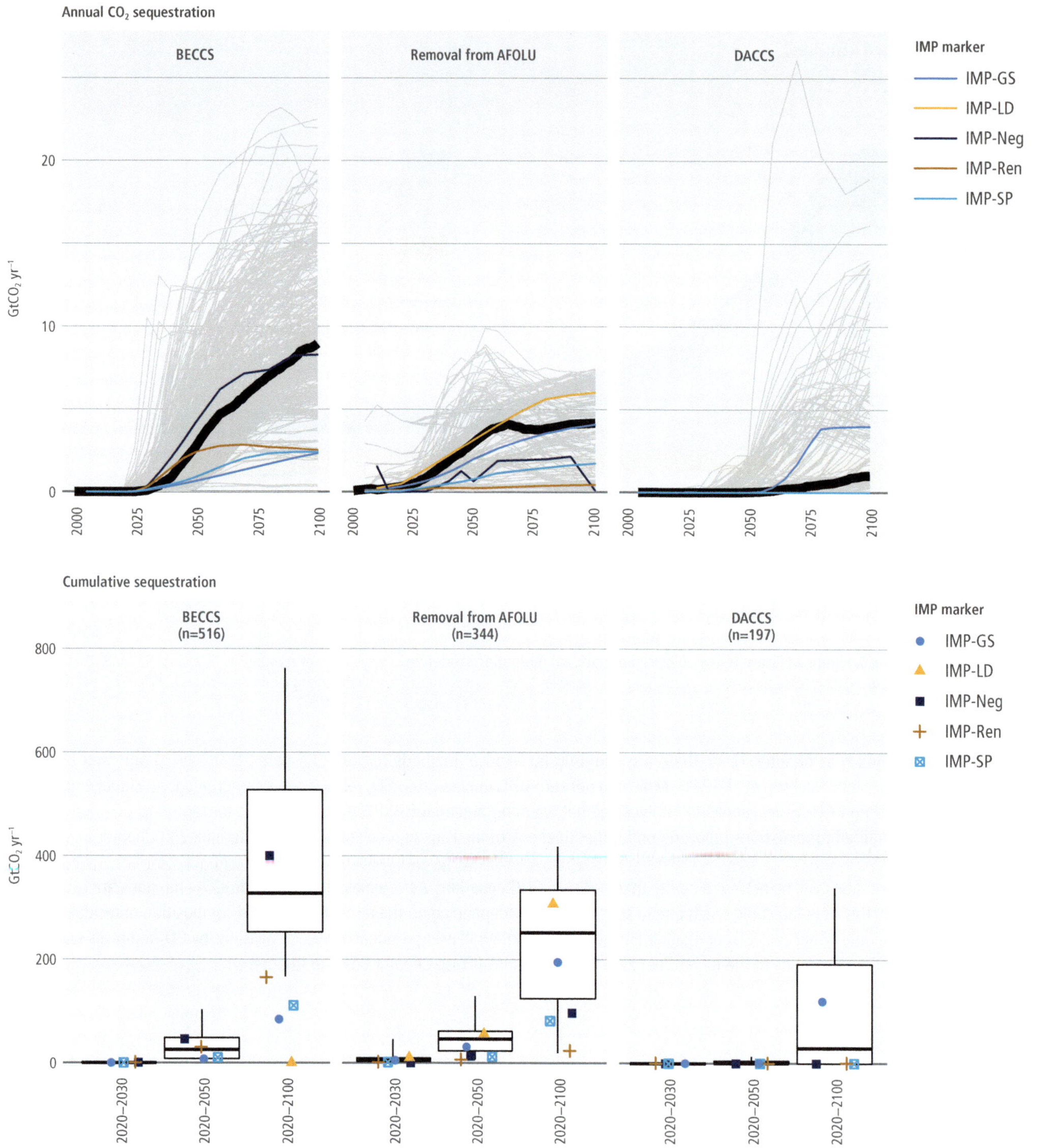


**Figure 12.3 | Sequestration through three predominant CDR methods: BECCS, $CO_2$ removal from AFOLU (mainly A/R), and DACCS (upper panels) annual sequestration and (lower panels) cumulative sequestration.** The IAM scenarios described in the figure correspond to those that limit warming to 2°C (>67%) or lower. The black line in each of the upper panels indicates the median of all the scenarios in categories C1 to C3. Hinges in the lower panels represent the interquartile ranges while whiskers extend to 5th and 95th percentiles. The IMPs are highlighted with colours, as shown in the key. The number of scenarios is indicated in the header of each panel. The number of scenarios with a non-zero DACCS value is 146.

and deployment scenario, CDR methods could bring about various co-benefits and adverse side effects (see below). All this highlights the need for appropriate CDR governance and policies (Section 12.3.3).

The volumes of future global CDR deployment assumed in IAM-based mitigation scenarios are large compared to current volumes of deployment, which presents a challenge since rapid and sustained upscaling from a small base is particularly difficult (de Coninck et al. 2018; Nemet et al. 2018; Hanna et al. 2021). All Illustrative Mitigation Pathways (IMPs) that limit warming to 2°C (>67%) or lower use some form of CDR. Across the full range of similarly ambitious IAM scenarios (scenario categories C1 to C3; see Section 3.3), the reported annual $CO_2$ removal from AFOLU (mainly A/R) reaches 0.86 [0.01–4.11] $GtCO_2\ yr^{-1}$ by 2030, 2.98 [0.23–6.38] $GtCO_2\ yr^{-1}$ by 2050, and 4.19 [0.1–6.91] $GtCO_2\ yr^{-1}$ by 2100 (values are the medians and bracketed values denote the 5–95th percentile range[1]). The annual BECCS deployment is 0.08 [0–1.09] $GtCO_2\ yr^{-1}$, 2.75 [0.52–9.45] $GtCO_2\ yr^{-1}$, and 8.96 [2.63–16.15] $GtCO_2\ yr^{-1}$ for these years, respectively. The annual DACCS deployment eaches 0 [0–0.02] $GtCO_2\ yr^{-1}$ by 2030, 0.02 [0–1.74] $GtCO_2\ yr^{-1}$ by 2050, and 1.02 [0–12.6] $GtCO_2\ yr^{-1}$ by 2100 (Figure 12.3).[2] Reported cumulative volumes of BECCS, $CO_2$ removal from AFOLU, and DACCS reach 328 [168–763] $GtCO_2$, 252 [20–418] $GtCO_2$, and 29 [0–339] $GtCO_2$ for the 2020–2100 period, respectively. Reaching the higher end of CDR volumes is subject to issues regarding their feasibility (see below), especially if achieved with only a limited number of CDR methods. Recent studies have identified some drivers for large-scale CDR deployment in IAM scenarios, including insufficient representation of variable renewables, a high discount rate that tends to increase initial carbon budget overshoot and therefore inflates usage of CDR to achieve net-negative emissions at later times, omission of CDR methods aside from BECCS and A/R (Emmerling et al. 2019; Hilaire et al. 2019; Köberle 2019), and limited deployment of demand-side options (Grubler et al. 2018; van Vuuren et al. 2018; Daioglou et al. 2019). The levels of CDR in IAMs in modelled pathways would change depending on the allowable overshoot of policy targets such as temperature or radiative forcing and the costs of non-CDR mitigation options (Johansson et al. 2020; van der Wijst et al. 2021) (Section 3.2.2).

While many CDR methods are gradually being explored, IAM scenarios have focused mostly on BECCS and A/R (Tavoni and Socolow 2013; Fuhrman et al. 2019; Rickels et al. 2019; Calvin et al. 2021; Diniz Oliveira et al. 2021). Although some IAM studies have also included other methods such as DACCS (Chen and Tavoni 2013; Marcucci et al. 2017; Realmonte et al. 2019; Fuhrman et al. 2020; Akimoto et al. 2021; Fuhrman et al. 2021a), enhanced weathering (Strefler et al. 2021), SCS and biochar (Holz et al. 2018) there is much less literature compared to studies on BECCS (Hilaire et al. 2019). A large-scale coordinated IAM study on BECCS ('EMF-33') has been conducted (Muratori et al. 2020; Rose et al. 2020) but none exists for other CDR methods. A recent review proposes a combination of various CDR methods (Fuss et al. 2018) but more in-depth literature on such a portfolio approach is limited (Strefler et al. 2021). A multi-criteria analysis has identified pathways with CDR portfolios different from least-cost pathways often dominated by BECCS and A/R (Rueda et al. 2021).

At the national and regional levels, the role of land-based biological CDR methods has long been analysed, but there is little detailed techno-economic assessment of the role of other CDR methods. There is a small but emerging literature providing such assessments for developed countries (Kraxner et al. 2014; Baik et al. 2018; Daggash et al. 2018; Patrizio et al. 2018; Sanchez et al. 2018; Breyer et al. 2019; Kato and Kurosawa 2019; Larsen et al. 2019; McQueen et al. 2020; Bistline and Blanford 2021; García-Freites et al. 2021; Jackson et al. 2021; Kato and Kurosawa 2021; Negri et al. 2021) while the literature outside developed countries is limited (Alatiq et al. 2021; Fuhrman et al. 2021b; Weng et al. 2021).

In IAMs, CDR is contributed mainly by the energy sector (through BECCS) and AFOLU (through A/R) (Figure 12.3). IAMs are starting to include other CDR methods, such as DACCS and enhanced weathering (Section 12.3.1), which are yet to be attributed to specific sectors in IAMs. Following IPCC guidance for UNFCCC inventories, A/R and SCS are reported in land use, land-use change and forestry (LULUCF), while BECCS would be reported in the sector where the carbon capture occurs, that is, the energy sector in the case of electricity and heat production, and the industry sector for BECCS linked to manufacturing (e.g., steel or hydrogen) (Tanzer et al. 2020; Bui et al. 2021; Tanzer et al. 2021).

### 12.3.1 CDR Methods Not Assessed Elsewhere in This Report: DACCS, Enhanced Weathering and Ocean-based Approaches

This section assesses the CDR methods that are not carried out solely within conventional sectors and so are not covered in other parts of the report: direct air carbon capture and storage, enhanced weathering, and ocean-based approaches. It provides an overview of each CDR method: their costs, potentials, risks and impacts, co-benefits, and their role in mitigation pathways. Since these processes, approaches and technologies have medium to low technology readiness levels, they are subject to significant uncertainty.

#### 12.3.1.1 Direct Air Carbon Capture and Storage (DACCS)

Direct air capture (DAC) is a chemical process to capture ambient $CO_2$ from the atmosphere. Captured $CO_2$ can be stored underground (direct air carbon capture and storage, DACCS) or utilised in products (direct air carbon capture and utilisation, DACCU). DACCS shares with conventional CCS the transport and storage components but is distinct in its capture part. Because $CO_2$ is a well-mixed GHG, DACCS can be sited relatively flexibly, though its locational flexibility is constrained by the availability of low-carbon energy and storage sites. Capturing the $CO_2$ involves three basic steps: (i) contacting the air, (ii) capturing on a liquid or solid sorbent or a liquid solvent, and (iii) regeneration of the solvent or the sorbent (with heat, moisture and/or pressure). After capture, the $CO_2$ stream can be stored underground or utilised. The duration of storage is an important consideration; geological reservoirs or mineralisation result in removal for more than 1000 years. The duration of the removal through DACCU (Breyer et al. 2019) varies with the lifetime of respective products (Wilcox et al. 2017; Bui et al. 2018; Fuss et al. 2018; Gunnarsson et al. 2018; Royal Society and Royal Academy of Engineering 2018; Creutzig et al. 2019), ranging from weeks to months for synthetic fuels to centuries or more for building materials (e.g., concrete cured

[1] Cumulative levels of CDR from AFOLU cannot be quantified precisely given that: (i) some pathways assess CDR deployment relative to a baseline; and (ii) different models use different reporting methodologies that in some cases combine gross emissions and removals in AFOLU. Total CDR from AFOLU equals or exceeds the net negative emissions mentioned.

[2] We use representative options for labels of each variable reported in the AR6 scenarios database.

using mineral carbonation) (Hepburn et al. 2019). The efficiency and environmental impacts of DACCS and DACCU options depend on the carbon intensity of the energy input (electricity and heat) and other lifecycle assessment (LCA) considerations (Zimmerman 2018; Jacobson 2019). See Chapters 6 and 11 for further details regarding carbon capture and utilisation. Another key consideration is the net carbon $CO_2$ removal of DACCS over its lifecycle (Madhu et al. 2021). Deutz and Bardow (2021) and Terlouw et al. (2021) demonstrated that the life-cycle net emissions of DACCS systems can be negative, even for existing supply chains and some current energy mixes. They found that the GHG intensity of energy sources is a key factor.

DAC options can be differentiated by the specific chemical processes used to capture ambient $CO_2$ from the air and recover it from the sorbent (Fasihi et al. 2019). The main categories are (i) liquid solvents with high-temperature regeneration, (ii) solid sorbents with low-temperature regeneration and (iii) regenerating by moisturising of solid sorbents. Other approaches such as electro-swing (Voskian and Hatton 2019) have been proposed but are less developed. Compared to other CDR methods, the primary barrier to upscaling DAC is its high cost and large energy requirement (*high confidence*) (Nemet et al. 2018), which can be reduced through innovation. It has therefore attracted entrepreneurs and private investments (IEA 2020b).

**Status:** There are some demonstration projects by start-up companies and academic researchers, who are developing various types of DAC, including aqueous potassium solvent with calcium carbonation and solid sorbents with heat regeneration (NASEM 2019). These projects are supported mostly by private investments and grants or sometimes serve utilisation niche markets (e.g., $CO_2$ for beverages, greenhouses, enhanced oil recovery). As of 2021, there are more than ten plants worldwide, with a scale of $ktCO_2$ $yr^{-1}$ or smaller (Larsen et al. 2019; NASEM 2019; IEA 2020b). Because of the fundamental difference in the $CO_2$ concentration at the capture stage, DACCS does not benefit directly from research, development and demonstration (RD&D) of conventional CCS. Public RD&D programmes dedicated to DAC have therefore been proposed (Larsen et al. 2019; NASEM 2019). Possible research topics include development of new liquid solvents, novel solid sorbents, and novel equipment or system designs, and the need for third-party evaluation of techno-economic aspects has also been emphasised (NASEM 2019). However, since basic research does not appear to be a primary barrier, both NASEM (2019) and Larsen et al. (2019) argue for a stronger focus on demonstration in the US context. Though the US and UK governments have begun funding DACCS research (IEA 2020b), the scale of R&D activities is limited.

**Costs:** As the process captures dilute $CO_2$ (~0.04%) from the ambient air, it is less efficient and more costly than conventional carbon capture applied to power plants and industrial installations (with a $CO_2$ concentration of ~10%) (*high confidence*). The cost of a liquid solvent system is dominated by the energy cost (because of the much higher energy demand for $CO_2$ regeneration, which reduces the efficiency) while capital costs account for a significant share of the cost of solid sorbent systems (Fasihi et al. 2019). The range of the DAC cost estimates found in the literature is wide (USD60–1000 $tCO_2^{-1}$) (Fuss et al. 2018) partly because different studies assume different use cases, differing phases (first plant vs *n*th plant) (Lackner et al. 2012), different configurations, and disparate system boundaries. Estimates of industrial origin are often on the lower side (Ishimoto et al. 2017). Fuss et al. (2018) suggest a cost range of USD600–1000 $tCO_2^{-1}$ for first-of-a-kind plants, and USD100–300 $tCO_2^{-1}$ as experience accumulates. An expert elicitation study found a similar cost level for 2050 with a median of around USD200 $tCO_2^{-1}$ (Shayegh et al. 2021) (*medium evidence, medium agreement*). NASEM (2019) systematically evaluated the costs of different designs and found a range of 84–386 USD2015 $tCO_2^{-1}$ for the designs currently considered by active technology developers. This cost range excludes the site-specific costs of transportation or storage.

**Potentials:** There is no specific study on the potential of DACCS but the literature has assumed that the technical potential is virtually unlimited provided that high energy requirements could be met (*medium evidence, high agreement*) (Marcucci et al. 2017; Fuss et al. 2018; Lawrence et al. 2018) since DACCS encounters fewer non-cost constraints than any other CDR method. Focusing only on the Maghreb region, Breyer et al. (2020) reported an optimistic potential 150 $GtCO_2$ at less than USD61 $tCO_2^{-1}$ for 2050. Fuss et al. (2018) suggest a potential of 0.5–5 $GtCO_2$ $yr^{-1}$ by 2050 because of environmental side effects and limits to underground storage. In addition to the ultimate potentials, Realmonte et al. (2019) noted the rate of scale-up as a strong constraint on deployment. Meckling and Biber (2021) discuss a policy roadmap to address the political economy for upscaling. More systematic analysis on potentials is necessary; first and foremost on national and regional levels, including the requirements for low-carbon heat and power, water and material demand, availability of geological storage and the need for land in case of low-density energy sources such as solar or wind power.

**Risks and impacts:** DACCS requires a considerable amount of energy (*high confidence*), depending on the type of technology, water, and make-up sorbents, while its land footprint is small compared to other CDR methods (Smith et al. 2016). Yet, depending on the source of energy for DACCS (e.g., renewables vs nuclear), DACCS could require a significant land footprint (NASEM 2019; Sekera and Lichtenberger 2020). The theoretical minimum energy requirement for separating $CO_2$ gas from the air is about 0.5 GJ $tCO_2^{-1}$ (Socolow et al. 2011). Fasihi et al. (2019) reviewed the published estimates of energy requirements and found that for the current technologies, the total energy requirement is about 4–10 GJ $tCO_2^{-1}$, with heat accounting for about 80% and electricity about 20% (McQueen et al. 2021). At a 10 $GtCO_2$ $yr^{-1}$ sequestration scale, this would translate into 40–100 exajoules (EJ) $yr^{-1}$ of energy consumption (32–80 EJ $yr^{-1}$ for heat and 8–20 EJ $yr^{-1}$ electricity), which can be contrasted with the current primary energy supply of about 600 EJ $yr^{-1}$ and electricity generation of about 100 EJ $yr^{-1}$. For the solid sorbent technology, low-temperature heat could be sourced from heat pumps powered by low-carbon sources such as renewables (Breyer et al. 2020), waste heat (Beuttler et al. 2019), and nuclear energy (Sandalow et al. 2018). Unless sourced from a clean source, this amount of energy could cause environmental damage (Jacobson 2019). Because DACCS is an open system, water lost from evaporation must be replenished. Water loss varies, depending on technology (including adjustable factors such as the concentration of the liquid solvent) as well as environmental conditions (e.g., temperate vs tropical climates). For a liquid solvent system, it can be 0–50 $tH_2O\ tCO_2^{-1}$ (Fasihi et al. 2019). A water loss rate of about 1–10 $tH_2O\ tCO_2^{-1}$ (Socolow et al. 2011) would translate into about 10–100 $GtH_2O$ (10–100 $km^3$) to capture 10 $GtCO_2$ from the atmosphere. Some solid sorbent technologies actually produce water as a by-product, for example 0.8–2 $tH_2O\ tCO_2^{-1}$ for a solid-sorbent technology with heat regeneration (Beuttler et al. 2019; Fasihi et al. 2019). Large-scale deployment of DACCS would also require a significant quantity of materials, and energy to produce them (Chatterjee and Huang 2020). Hydroxide solutions are currently being produced as a by-product of chlorine but replacement (make-up) requirement of such materials at scale exceeds the current market supply (Realmonte et al. 2019). The land requirements for DAC units are not large enough to be of concern (Madhu et al. 2021). Furthermore, these can be placed on unproductive lands, in contrast to biological CDR. Nevertheless, to ensure that $CO_2$-depleted air does not enter the air contactor of an adjacent DAC system, there must be enough space between DAC units, similar to wind power turbines. Considering this, Socolow et al. (2011) estimated a land footprint of 1.5 $km^2\ MtCO_2^{-1}$. In contrast, large energy requirements can lead to significant footprints if low-density energy sources (e.g., solar PV) are used (Smith et al. 2016). For the issues associated with $CO_2$ utilisation and storage, see Chapter 6.

**Co-benefits:** While Wohland et al. (2018) proposed solid sorbent-based DAC plants as a Power-to-X technology that could use excess renewable power (at times of low or even negative prices), such operation would add additional costs. Installations would need to be designed for intermittent operations (i.e., at low load factors) which would negatively affect capital and operation costs (Daggash et al. 2018; Sandalow et al. 2018) as a high time-resolution model suggests a high utilisation rate (Breyer et al. 2020). Solid sorbent DAC designs can potentially remove more water from the ambient air than needed for regeneration, thereby delivering surplus water that would contribute to SDG 6 (clean water and sanitation) in arid regions (Sandalow et al. 2018; Fasihi et al. 2019).

**Trade-offs and spillover effects:** Liquid solvent DACCS systems need substantial amounts of water (Fasihi et al. 2019), although much less than BECCS systems (Smith et al. 2016), which could negatively affect SDG 6 (clean water and sanitation). Although the high energy demand of DACCS could affect SDG 7 (affordable and clean energy) negatively through potential competition or positively through learning effects (Beuttler et al. 2019), its impact has not been thoroughly assessed yet.

**Role in mitigation pathways:** There are a few IAM studies that have explicitly incorporated DACCS. Stringent emissions constraints in these studies lead to high carbon prices, allowing DACCS to play an important role in mitigation. Chen and Tavoni (2013) examined the role of DACCS in an IAM (WITCH) and found that incorporating DACCS reduces the overall cost of mitigation and tends to postpone the timing of mitigation. The scale of capture goes up to 37 $GtCO_2\ yr^{-1}$ in 2100. Akimoto et al. (2021) introduced DACCS in the IAM DNE21+, and also found the long-term marginal cost of abatement is significantly reduced by DACCS. Marcucci et al. (2017) ran MERGE-ETL, an integrated model with endogenous learning, and showed that DACCS allows for a model solution for the 1.5°C target, and that DACCS substitutes for BECCS under stringent targets. In their analysis, DACCS captures up to 38.3 $GtCO_2\ yr^{-1}$ in 2100. Realmonte et al. (2019) modelled two types of DACCS (based on liquid and solid sorbents) with two IAMs (TIAM-Grantham and WITCH), and showed that in deep mitigation scenarios, DACCS complements, rather than substitutes, other CDR methods such as BECCS, and that DACCS is effective at containing mitigation costs. At the national scale, Larsen et al. (2019) utilised the Regional Investment and Operations (RIO) Platform coupled with the Energy PATHWAYS model, and explicitly represented DAC in US energy systems scenarios. They found that in a scenario that reaches net zero emissions by 2045, about 0.6 $GtCO_2$ or 1.8 $GtCO_2$ of DACCS would be deployed, depending on the availability of biological carbon sinks and bioenergy. The modelling supporting the European Commission's initial proposal for net zero GHG emissions by 2050 incorporated DAC, with the captured $CO_2$ used for both synthetic fuel production (DACCU) and storage (DACCS) (Capros et al. 2019). Fuhrman et al. (2021a) evaluated the role of DACCS across five shared socio-economic pathways with the GCAM modelling framework and identified a substantial role for DACCS in mitigation and a decreased pressure on land and water resources from BECCS, even under the assumption of limited energy efficiency improvement and conservative cost declines of DACCS technologies. The newest iteration of the World Economic Outlook by IEA (2021b) deploys CDR on a limited scale, and DACCS removes 0.6 $GtCO_2$ in 2050 for its Net Zero $CO_2$ Emissions scenario.

Status, costs, potentials, risk and impacts, co-benefits, trade-offs and spillover effects and the role in mitigation pathways of DACCS are summarised in Table 12.6.

#### 12.3.1.2 Enhanced Weathering

Enhanced weathering involves (i) the mining of rocks containing minerals that naturally absorb $CO_2$ from the atmosphere over geological timescales (as they become exposed to the atmosphere through geological weathering), (ii) the comminution of these rocks to increase the surface area, and (iii) the spreading of these crushed rocks on soils (or in the ocean/coastal environments; Section 12.3.1.3) so that they react with atmospheric $CO_2$ (Schuiling and Krijgsman 2006; Hartmann et al. 2013; Beerling et al. 2018; Goll et al. 2021). Construction waste and waste materials from mining can also be used as a source material for enhanced weathering. Silicate rocks such as basalt, containing minerals rich in calcium and magnesium and lacking metal ions such as nickel and chromium, are most suitable for enhanced weathering (Beerling et al. 2018); they reduce soil solution acidity during dissolution, and promote the chemical transformation of $CO_2$ to bicarbonate ions. The bicarbonate ions can precipitate in soils and drainage waters as a solid carbonate mineral (Manning 2008), or remain dissolved and increase alkalinity levels in the ocean when the water reaches the sea (Renforth and Henderson 2017). The modelling study by Cipolla et al. (2021) found that rate of weathering is greater in high rainfall environments, and was increased by organic matter amendment.

**Status:** Enhanced weathering has been demonstrated in the laboratory and in small-scale field trials (TRL 3–4) but has yet to be demonstrated at scale (Beerling et al. 2018; Amann et al. 2020).

12

The chemical reactions are well understood (Manning 2008; Gillman 1980; Gillman et al. 2001), but the behaviour of the crushed rocks in the field and potential co-benefits and adverse side effects of enhanced weathering require further research (Beerling et al. 2018). Small-scale laboratory experiments have calculated weathering rates that are orders of magnitude slower than the theoretical limit for mass transfer-controlled forsterite (Renforth et al. 2015; Amann et al. 2020) and basalt dissolution (Kelland et al. 2020). Uncertainty surrounding silicate mineral dissolution rates in soils, the fate of the released products, the extent of legacy reserves of mining by-products that might be exploited, location and availability of rock extraction sites, and the impact on ecosystems remain poorly quantified and require further research to better understand feasibility (Renforth 2012; Moosdorf et al. 2014; Beerling et al. 2018). Closely monitored, large-scale demonstration projects would allow these aspects to be studied (Smith et al. 2019a; Beerling et al. 2020).

**Costs:** Fuss et al. (2018), in a systematic review of the costs and potentials of CDR methods including enhanced weathering, note that costs are closely related to the source of the rock and the technology used for rock grinding and material transport (Renforth 2012; Hartmann et al. 2013; Strefler et al. 2018). Due to differences in the methods and assumptions between studies, literature ranges are highly uncertain and range from USD15–40 $tCO_2^{-1}$ to USD3460 $tCO_2^{-1}$ (Köhler et al. 2010; Taylor et al. 2016). Renforth (2012) reported operational costs in the UK of applying mafic rocks (rocks with high magnesium and iron silicate mineral concentrations) of USD70–578 $tCO_2^{-1}$, and for ultramafic rocks (rocks rich in magnesium and iron silicate minerals but with very low silica content – the low silica content enhances weathering rates) of USD24–123 $tCO_2^{-1}$. Beerling et al. (2020) combined a spatially resolved weathering model with a techno-economic assessment to suggest costs of between USD54–220 $tCO_2^{-1}$ (with a weighted mean of USD118–128 $tCO_2^{-1}$). Fuss et al. (2018) suggested an author judgement cost range of USD50–200 $tCO_2^{-1}$ for a potential of 2–4 $GtCO_2\ yr^{-1}$ from 2050, excluding biological storage.

**Potentials:** In a systematic review of the costs and potentials of enhanced weathering, Fuss et al. (2018) report a wide range of potentials (*limited evidence, low agreement*). The highest reported regional sequestration potential, 88.1 $GtCO_2\ yr^{-1}$, is reported for the spreading of pulverised rock over a very large land area in the tropics, a region considered promising given the higher temperatures and greater rainfall (Taylor et al. 2016). Considering cropland areas only, the potential carbon removal was estimated by Strefler et al. (2018) to be 95 $GtCO_2\ yr^{-1}$ for dunite and 4.9 $GtCO_2\ yr^{-1}$ for basalt. Slightly lower potentials were estimated by Lenton (2014) where the potential of carbon removal by enhanced weathering (including adding carbonate and olivine to both oceans and soils) was estimated to be 3.7 $GtCO_2\ yr^{-1}$ by 2100, but with mean annual removal an order of magnitude less at 0.2 GtC-eq $yr^{-1}$ (Lenton 2014). The estimates reported in Smith et al. (2016) are based on the potential estimates of Lenton (2014). Beerling et al. (2020) estimate that up to 2 $GtCO_2\ yr^{-1}$ could be removed by 2050 by spreading basalt onto 35–59% (weighted mean 53%) of agricultural land of 12 countries. Fuss et al. (2018) provide an author judgement range for potential of 2–4 $GtCO_2\ yr^{-1}$ for 2050.

**Risks and impacts:** Mining of rocks for enhanced weathering will have local impacts and carries risks similar to those associated with the mining of mineral construction aggregates, with the possible additional risk of greater dust generation from fine comminution and land application. In addition to direct habitat destruction and increased traffic to access mining sites, there could be adverse impacts on local water quality (Younger and Wolkersdorfer 2004).

**Co-benefits:** Enhanced weathering can improve plant growth by pH modification and increased mineral supply (Kantola et al. 2017; Beerling et al. 2018), can enhance SCS in some soils (Beerling et al. 2018) thereby protecting against soil erosion (Wright and Upadhyaya 1998), and increasing the cation exchange capacity, resulting in increased nutrient retention and availability (Gillman 1980; Baldock and Skjemstad 2000; Gillman et al. 2001; Manning 2010; Guntzer et al. 2012; Tubana et al. 2016; Yu et al. 2017; Haque et al. 2019; Smith et al. 2019a). Through these actions, it can contribute to SDG 2 (zero hunger), SDG 15 (life on land) (by reducing land demand for croplands), SDG 13 (climate action) (through CDR), SDG 14 (life below water) (by ameliorating ocean acidification) and SDG 6 (clean water and sanitation) (Smith et al. 2019a). To more directly ameliorate ocean acidification while increasing CDR and reducing impacts on land ecosystems, alkaline minerals could instead be directly added to the ocean (Section 12.3.1.3). There are potential benefits in poverty reduction through employment of local workers in mining (Pegg 2006).

**Trade-offs and spillover effects:** Air quality could be adversely affected by the spreading of rock dust (Edwards et al. 2017), though this can partly be ameliorated by water-spraying (Grundnig et al. 2006). As noted above, any significant expansion of the mining industry would require careful assessment to avoid possible detrimental effects on biodiversity (Amundson et al. 2015). The processing of an additional 10 billion tonnes of rock would require up to 3000 Terawatt-hours of energy, which could represent approximately 0.1–6 % of global electricity use in 2100. The emissions associated with this additional energy generation may reduce the net carbon dioxide removal by up to 30% with present-day grid average emissions, but this efficiency loss would decrease with low-carbon power (Beerling et al. 2020).

**Role in mitigation pathways:** Only one study to date has included enhanced weathering in an integrated assessment model to explore mitigation pathways (Strefler et al. 2021).

Status, costs, potentials, risk and impacts, co-benefits, trade-offs and spillover effects and the role in mitigation pathways of enhanced weathering are summarised in Table 12.6.

#### 12.3.1.3 Ocean-based Methods

The ocean, which covers over 70% of the Earth's surface, contains about 38,000 gigatonnes of carbon, some 45 times more than the present atmosphere, and oceanic uptake has already consumed close to 30–40% of anthropogenic carbon emissions (Sabine et al. 2004; Gruber et al. 2019). The ocean is characterised by diverse biogeochemical cycles involving carbon, and ocean circulation has much longer timescales than the atmosphere, meaning that additional anthropogenic carbon could potentially be stored in the ocean for centuries to millennia for methods that increase deep ocean-dissolved carbon concentrations or temporarily bury the carbon; or essentially permanently (over ten thousand years) for methods that store the carbon in mineral forms or as ions by increasing alkalinity (Siegel et al., 2021) (Cross-Chapter Box 8, Figure 1). A wide range of methods and implementation options for marine CDR have been proposed (Gattuso et al. 2018; Hoegh-Guldberg et al. 2018; GESAMP 2019). The most studied ocean-based CDR methods are ocean fertilisation, alkalinity enhancement (including electrochemical methods) and intensification of biologically-driven carbon fluxes and storage in marine ecosystems, referred to as 'blue carbon'. The mitigation potentials, costs, co-benefits and trade-offs of these three options are discussed below. Less well studied are methods including artificial upwelling, terrestrial biomass dumping into oceans, direct $CO_2$ removal from seawater (with CCS), and sinking marine biomass into the deep ocean or harvesting it for bioenergy (with CCS) or biochar (GESAMP 2019). These methods are summarised briefly below. Potential climate response and influence on the carbon budget of ocean-based CDR methods are discussed in WGI AR6, Chapter 5.

*Ocean fertilisation (OF)*

One natural mechanism of carbon transfer from the atmosphere to the deep ocean is the ocean biological pump, which is driven by the sinking of organic particles from the upper ocean. These particles derive ultimately from primary production by phytoplankton and most of them are remineralised within the upper ocean with only a small fraction reaching the deep ocean where the carbon can be sequestered on centennial and longer timescales. Increasing nutrient availability would stimulate uptake of $CO_2$ through phytoplankton photosynthesis producing organic matter, some of which would be exported into the deep ocean, sequestering carbon. In areas of the ocean where macronutrients (nitrogen, phosphorus) are available in sufficient quantities (about 25% of the total area), the growth of phytoplankton is limited by the lack of trace elements such as iron. Thus, OF CDR can be based on two implementation options to increase the productivity of phytoplankton (Minx et al. 2018): macronutrient enrichment and micronutrient enrichment. A third option, highlighted in GESAMP (2019), is based on fertilisation for fish stock enhancement, for instance, as naturally occurs in eastern boundary current systems. Iron fertilisation is the best-studied OF option to date, but knowledge so far is still inadequate to predict global ecological and biogeochemical consequences.

**Status:** OF has a natural analogue: periods of glaciation in the geological past are associated with changes in deposition of dust containing iron into the ocean. Increased formation of phytoplankton has also been observed during seasonal deposition of dust from the Arabian Peninsula and ash deposition on the ocean surface after volcanic eruptions (Achterberg et al. 2013; Jaccard et al., 2013; Olgun et al. 2013; Martínez-García et al. 2014). OF options may appear technologically feasible, and enhancement of photosynthesis and $CO_2$ uptake from surface waters is confirmed by a number of field experiments conducted in different areas of the ocean, but there is scientific uncertainty about the proportion of newly-formed organic carbon that is transferred to deep ocean, and the longevity of storage (Blain et al. 2008; Williamson et al. 2012; Trull et al. 2015). The efficiency of OF also depends on the region and experimental conditions, especially in relation to the availability of other nutrients, light and temperature (Aumont and Bopp 2006). In the case of macronutrients, very large quantities are needed and the proposed scaling of this technique has been viewed as unrealistic (Williamson and Bodle 2016).

**Costs:** Ocean fertilisation costs depend on nutrient production and its delivery to the application area (Jones 2014). The costs range from USD2 $tCO_2^{-1}$ for fertilisation with iron (Boyd 2008) to USD457 $tCO_2^{-1}$ for nitrate (Harrison 2013). Reported costs for macronutrient application at USD20 $tCO_2^{-1}$ (Jones 2014) contrast with higher estimates by (Harrison 2013) reporting that low costs are due to overestimation of sequestration capacity and underestimation of logistical costs. The median of OF cost estimates, USD230 $tCO_2^{-1}$ (Gattuso et al., 2021) indicates low cost-effectiveness, albeit uncertainties are large.

**Potentials:** Theoretical calculations indicate that organic carbon export increases 2–20 kg per gram of iron added, but experiments indicate much lower efficiency: a significant part of the $CO_2$ can be emitted back the atmosphere because much of the organic carbon produced is remineralised in the upper ocean. Efficiency also varies with location (Bopp et al. 2013). Between studies, there are substantial differences in the ratio of iron added to carbon fixed photosynthetically, and in the ratio of iron added to carbon eventually sequestered (Trull et al. 2015), which has implications both for the success of this strategy and its cost. Estimates indicate potentially achievable net sequestration rates of 1–3 $GtCO_2\ yr^{-1}$ for iron fertilisation, translating into cumulative CDR of 100–300 $GtCO_2$ by 2100 (Ryaboshapko and Revokatova 2015; Minx et al. 2018), whereas OF with macronutrients has a higher theoretical potential of 5.5 $GtCO_2\ yr^{-1}$ (Harrison 2017; Gattuso et al. 2021). Modelling studies show a maximum effect on atmospheric $CO_2$ of 15–45 parts per million volume in 2100 (Zeebe and Archer 2005; Aumont and Bopp 2006; Keller et al. 2014; Gattuso et al. 2021).

**Risks and impacts:** Several of the mesoscale iron enrichment experiments have seen the emergence of potentially toxic species of diatoms (Silver et al. 2010; Trick et al. 2010). There is also (limited) evidence of increased concentrations of other GHGs such as methane and nitrous oxide during the subsurface decomposition of the sinking particles from iron-stimulated blooms (Law 2008). Impacts on marine biology and food web structure are not well known, however OF at large scale could cause changes in nutrient distributions or anoxia in subsurface water (Fuhrman and Capone 1991; DFO 2010). Other potential risks are perturbation to marine ecosystems via reorganisation of community structure, enhanced deep ocean acidification (Oschlies et al. 2010) and effects on human food supply.

**Co-benefits:** Co-benefits of OF include a potential increase in fish biomass through enhanced biological production (Minx et al. 2018) and reduced ocean acidification in the short term in the upper ocean (by $CO_2$ removal), though it could be enhanced in the long term in the ocean interior (by $CO_2$ release) (Oschlies et al., 2010; Gattuso et al. 2018).

**Trade-offs and spillover effects:** Potential drawbacks include subsurface ocean acidification and deoxygenation (Cao and Caldeira 2010; Oschlies et al., 2010; Williamson et al. 2012); altered regional meridional nutrient supply and fundamental alteration of food webs (GESAMP 2019); and increased production of $N_2O$ and $CH_4$ (Jin and Gruber 2003; Lampitt et al. 2008). Ocean fertilisation is considered to have negative consequences for eight SDGs, and a combination of both positive and negative consequences for seven SDGs (Honegger et al. 2020).

*Ocean Aakalinity enhancement (OAE)*

CDR through 'ocean alkalinity enhancement' or 'artificial ocean alkalinisation' (Renforth and Henderson 2017) can be based on: (i) the dissolution of natural alkaline minerals that are added directly to the ocean or coastal environments; (ii) the dissolution of such minerals upstream from the ocean (e.g., enhanced weathering, Section 12.3.1.2); (iii) the addition of synthetic alkaline materials directly to the ocean or upstream; and (iv) electrochemical processing of seawater. In the case of (ii), minerals are dissolved on land and the dissolution products are conveyed to the ocean through runoff and river flow. These processes result in chemical transformation of $CO_2$ and sequestration as bicarbonate and carbonate ions ($HCO_3^-$, $CO_3^{2-}$) in the ocean. Imbalances between the input and removal fluxes of alkalinity can result in changes in global oceanic alkalinity and therefore the capacity of the ocean to store carbon. Such alkalinity-induced changes in partitioning of carbon between atmosphere and ocean are thought to play an important role in controlling climate change on timescales of 1000 years and longer (e.g., Zeebe 2012). The residence time of dissolved inorganic carbon in the deep ocean is around 100,000 years. However, residence time may decrease if alkalinity is reduced by a net increase in carbonate minerals by either increased formation (precipitation) or reduced dissolution of carbonate (Renforth and Henderson 2017). The alkalinity of seawater could potentially also be increased by electrochemical methods, either directly by reactions at the cathode that increase the alkalinity of the surrounding solution that can be discharged into the ocean, or by forcing the precipitation of solid alkaline materials (e.g., hydroxide minerals) that can then be added to the ocean (e.g., Rau et al. 2013; La Plante et al. 2021).

**Status:** OAE has been demonstrated by a small number of laboratory experiments (in addition to enhanced weathering, Section 12.3.1.2). The use of enhanced ocean alkalinity for carbon storage was first proposed by Kheshgi (1995) who considered the creation of highly reactive lime that would readily dissolve in the surface ocean and sequester $CO_2$. An alternative method proposed the dissolution of carbonate minerals (e.g., calcium carbonate) in the presence of waste flue gas $CO_2$ and seawater as a means capturing $CO_2$ and converting it to bicarbonate ions (Rau and Caldeira 1999; Rau 2011). House et al. (2007) proposed the creation of alkalinity in the ocean through electrolysis. The fate of the stored carbon is the same for these proposals (i.e., $HCO_3^-$ and $CO_3^{2-}$ ions), but the reaction pathway is different. Enhanced weathering of silicate minerals such as olivine could add alkalinity to the ocean, for example, by placing olivine sand in coastal areas (Meysman and Montserrat 2017; Montserrat et al. 2017). Some authors suggest use of maritime transport to discharge calcium hydroxide (slaked lime) (Caserini et al. 2021).

**Costs:** Techno-economic assessments of OAE largely focus on quantifying overall energy and carbon balances. Cost ranges are USD40–260 $tCO_2^{-1}$ (Fuss et al. 2018). Considering life-cycle carbon and energy balances for various OAE options, adding lime (or other reactive calcium or magnesium oxide/hydroxides) to the ocean would cost USD64–260 $tCO_2^{-1}$ (Renforth et al. 2013; Renforth & Kruger 2013; Caserini et al. 2019). Rau (2008) and Rau et al. (2018) estimate that electrochemical processes for increasing ocean alkalinity may have a net cost of USD3–160 $tCO_2^{-1}$, largely depending on energy cost and co-product ($H_2$) market value. In the case of direct addition of alkaline minerals to the ocean (i.e., without calcination), the cost is estimated to be USD20–50 $tCO_2^{-1}$ (Harvey 2008; Köhler et al. 2013; Renforth and Henderson 2017).

**Potentials:** For OAE, the ocean theoretically has the capacity to store thousands of $GtCO_2$ (cumulatively) without exceeding pre-industrial levels of carbonate saturation (Renforth and Henderson 2017) if the impacts were distributed evenly across the surface ocean. The potential of increasing ocean alkalinity may be constrained by the capability to extract, process, and react minerals (Section 12.3.1.2); the demand for co-benefits (see below), or to minimise impacts around points of addition. Important challenges with respect to the detailed quantification of the $CO_2$ sequestration efficiency include nonstoichiometric dissolution, reversed weathering and potential pore water saturation in the case of adding minerals to shallow coastal environments (Meysman and Montserrat 2017). Fuss et al. (2018) suggest storage potentials of 1–100 $GtCO_2$ $yr^{-1}$. (González and Ilyina 2016) suggested that addition of 114 picomoles of alkalinity to the surface ocean could remove 3400 $GtCO_2$ from the atmosphere.

**Risks and impacts:** For OAE, the local impact of increasing alkalinity on ocean chemistry can depend on the speed at which the impacted seawater is diluted/circulated and the exchange of $CO_2$ from the atmosphere (Bach et al. 2019). Also, more extreme carbonate chemistry perturbations due to non-equilibrated alkalinity could affect local marine biota (Bach et al. 2019), although biological impacts are largely unknown. Air-equilibrated seawater has a much lower potential to perturb seawater carbonate chemistry. However, seawater with slow air-sea gas exchange, in which alkalinity increases, consumes $CO_2$ from the surrounding water without immediate replenishment from the atmosphere, which would increase seawater pH and saturation states and may impact marine biota (Meysman and Montserrat 2017; Montserrat et al. 2017). It may be possible to use this effect to ameliorate ocean acidification. Like enhanced weathering, some proposals may result in the dissolution products of silicate minerals (e.g., silicon, iron, potassium, nickel) being supplied to ocean ecosystems (Meysman and Montserrat 2017; Montserrat et al. 2017). Ecological and biogeochemical consequences of OAE largely depend on the minerals used. When natural minerals such as olivine are used, the release of additional Si and Fe could have fertilising effects (Bach et al. 2019). In addition to perturbations to marine ecosystems via reorganisation of community structure, potentially adverse effects of OAE that should be studied include the release of toxic trace metals from some deposited minerals (Hartmann et al. 2013).

**Co-benefits:** Intentional addition of alkalinity to the oceans through OAE would decrease the risk to ocean ecosystems caused by the $CO_2$-induced impact of ocean acidification on marine biota and the global carbon cycle (Doney et al. 2009; Köhler et al. 2010; Rau et al. 2012; Williamson and Turley 2012; Albright et al. 2016; Bach et al. 2019). OAE could be jointly implemented with enhanced weathering (Section 12.3.1.2), spreading the finely crushed rock in the ocean rather than on land. Regional alkalinisation could be effective in protecting coral reefs against acidification (Feng et al. 2016; Mongin et al., 2021) and coastal OAE could be part of a broader strategy for geochemical management of the coastal zone, safeguarding specific coastal ecosystems, such as important shellfisheries, from the adverse impact of ocean acidification (Meysman and Montserrat 2017).

**Trade-offs and spillover effects:** There is a paucity of research on biological effects of alkalinity addition. The very few studies that have explored the impact of elevated alkalinity on ocean ecosystems have largely been limited to single species experiments (Cripps et al. 2013; Gore et al. 2019) and a constrained field study quantifying the net calcification response of a coral reef flat to alkalinity enhancement (Albright et al. 2016). The addition rate would have to be great enough to overcome mixing of the local seawater with the ambient environment, but not sufficient to detrimentally impact ecosystems. More research is required to assess locations in which this may be feasible, and how such a scheme may operate (Renforth and Henderson 2017). The environmental impact of large-scale release of natural dissolution products into the coastal environment will strongly depend on the scale of olivine application, the characteristics of the coastal water body (e.g., residence time) and the particular biota present (e.g., coral reefs will react differently compared with seagrasses) (Meysman and Montserrat 2017). Model simulations (González et al. 2018) suggest that termination of OAE implemented on a massive scale under a high $CO_2$ emission scenario (Representative Concentration Pathway 8.5) might pose high risks to biological systems sensitive to rapid environmental changes because it would cause a sharp increase in ocean acidification. For example, OAE termination would lead to a decrease in surface pH in warm shallow regions where vulnerable coral reefs are located, and a drop in the carbonate saturation state. However, other studies with lower levels of OAE have shown no termination effect (Keller et al., 2014).

*Blue carbon management*

The term 'blue carbon' was used originally to refer to biological carbon sequestration in all marine ecosystems, but it is increasingly applied to CDR associated with rooted vegetation in the coastal zone, such as tidal marshes, mangroves and seagrasses. Potential for carbon sequestration in other coastal and non-coastal ecosystems, such as macroalgae (e.g., kelp), is debated (Krause-Jensen and Duarte, 2016; Krause-Jensen et al., 2018). In this report, blue carbon refers to CDR through coastal blue carbon management.

**Status:** In recent years, there has been increasing research on the potential, effectiveness, risks, and possibility of enhancing $CO_2$ sequestration in shallow coastal ecosystems (Duarte, 2017). About 20% of the countries that are signatories to the Paris Agreement refer to blue carbon approaches for climate change mitigation in their NDCs and are moving toward measuring blue carbon in inventories. About 40% of those same countries have pledged to manage shallow coastal ecosystems for climate change adaptation (Kuwae and Hori 2019).

**Costs:** There are large differences in the cost of CDR applying blue carbon management methods between different ecosystems (and at the local level). Median values are estimated as USD240, 30,000, and 7800 $tCO_2^{-1}$, respectively for mangroves, salt marsh and seagrass habitats (Gattuso et al. 2021). Currently estimated cost effectiveness (for climate change mitigation) is very low (Siikamäki et al. 2012; Bayraktarov et al. 2016; Narayan et al. 2016).

**Potentials:** Globally, the total potential carbon sequestration rate through blue carbon CDR is estimated in the range 0.02–0.08 $GtCO_2\ yr^{-1}$ (Wilcox et al. 2017; National Academies of Sciences 2019). Gattuso et al. (2021) estimate the theoretical cumulative potential of coastal blue carbon management by 2100 to be 95 $GtCO_2$, taking into account the maximum area that can be occupied by these habitats and historic losses of mangroves, seagrass and salt marsh ecosystems.

**Risks and impacts:** For blue carbon management, potential risks relate to the high sensitivity of coastal ecosystems to external impacts associated with both degradation and attempts to increase carbon sequestration. Under expected future warming, sea level rise and changes in coastal management, blue carbon ecosystems are at risk, and their stored carbon is at risk of being lost (Bindoff et al. 2019).

**Co-benefits:** Blue carbon management provides many non-climatic benefits and can contribute to ecosystem-based adaptation, also reducing emissions associated with habitat degradation and loss (Howard et al. 2017; Hamilton and Friess 2018). Shallow coastal ecosystems have been severely affected by human activity; significant areas have already been deforested or degraded and continue to be denuded. These processes are accompanied by carbon emissions. The conservation and restoration of coastal ecosystems, which will lead to increased carbon sequestration, is also essential for the preservation of basic ecosystem services, and healthy ecosystems tend to be more resilient to the effects of climate change.



**Trade-offs and spillover effects:** Blue carbon management schemes should consist of a mix of restoration, conservation and areal increase, including complex engineering interventions that enhance natural capital, safeguard their resilience and the ecosystem services they provide, and decrease the sensitivity of such ecosystems to further disturbances.

*Overview of other ocean-based CDR approaches*

**Artificial upwelling:** This concept uses pipes or other methods to pump nutrient-rich deep ocean water to the surface where it has a fertilising effect (see OF section). To achieve $CO_2$ removal at a Gt magnitude, modelling studies have shown that artificial upwelling

**Figure 12.4 | Summary of the extent to which different factors would enable or inhibit the deployment of the carbon dioxide removal methods DACCS, EW, ocean fertilisation and blue carbon management.** Blue bars indicate the extent to which the indicator enables the implementation of the CDR method (E) and orange bars indicate the extent to which an indicator is a barrier (B) to the deployment of the method, relative to the maximum possible barriers and enablers assessed. An 'X' signifies the indicator is not applicable or does not affect the feasibility of the method, while a forward slash indicates that there is no or limited evidence whether the indicator affects the feasibility of the method. The shading indicates the level of confidence, with darker shading signifying higher levels of confidence. Supplementary Material 12.SM.B provides an overview of the factors affecting the feasibility of CDR methods and how they differ across contexts (e.g., region), time (e.g., 2030 versus 2050), and scale (e.g., small versus large), and includes a line of sight on which the assessment is based. The assessment methodology is explained in Annex II, Part IV, Section 11.

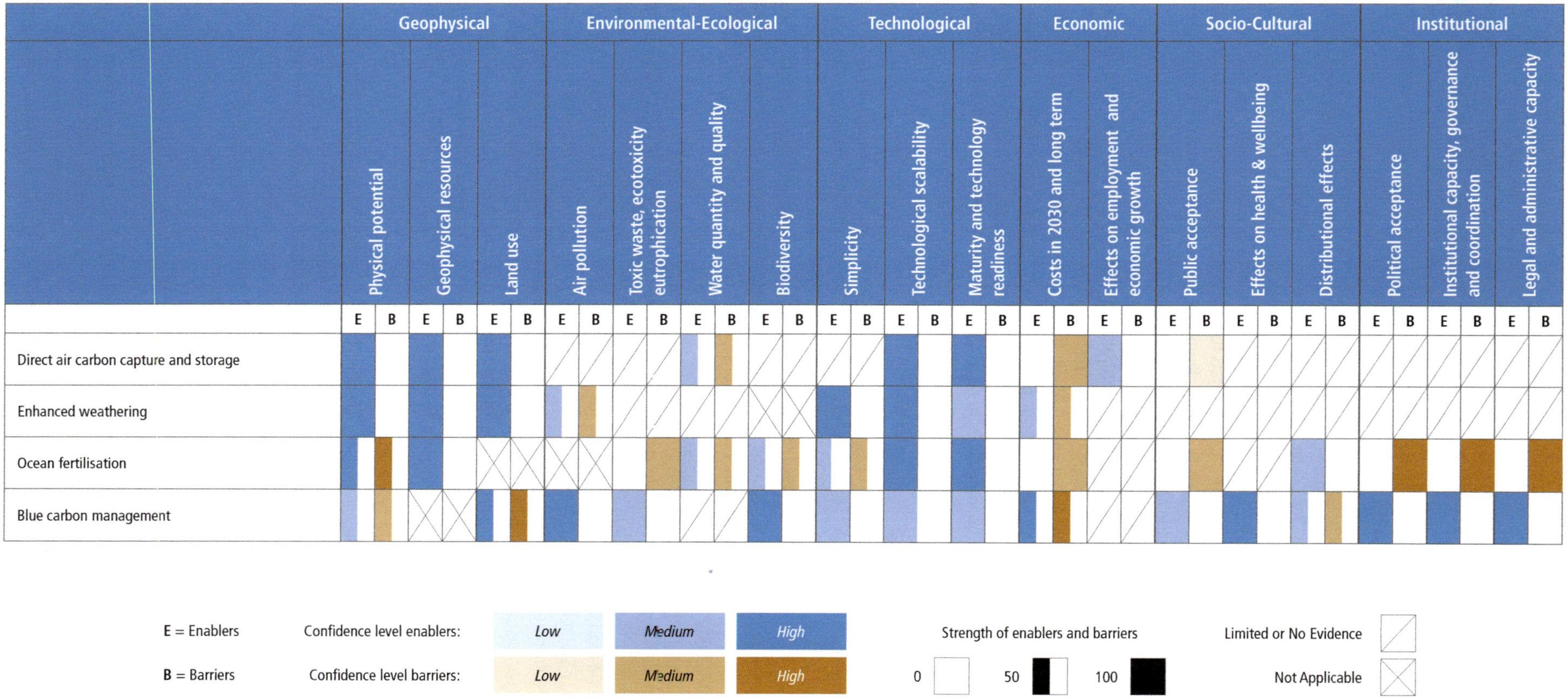

would have to be implemented on a massive scale (over 50% of the ocean to deliver maximum rate of 10$GtCO_2$ $yr^{-1}$ under RCP8.5) (Oschlies et al., 2010, Keller et al. 2014). Because the deep water is much colder than surface water, at massive scale this could cool the Earth's surface by several degrees, but the cooling effect would cease as the deeper ocean warms, and would reverse, leading to rapid warming, if the pumping ceased (Oschlies et al., 2010, Keller et al. 2014).

Furthermore, the cooling would also severely alter atmospheric circulation and precipitation patterns (Kwiatkowski et al. 2015). Several upwelling approaches have been developed and tested (Pan et al., 2016) and more R&D is underway.

**Terrestrial biomass dumping:** There are proposals to sink terrestrial biomass (crop residues or logs) into the deep ocean as a means of sequestering carbon (Strand and Benford 2009). Sinking biochar has also been proposed (Miller and Orton, 2021). Decomposition would be inhibited by the cold and sometimes hypoxic/anoxic environment on the ocean floor, and absence of bacteria that decompose terrestrial lignocellulosic biomass, so storage timescale is estimated at hundreds to thousands of years (Strand and Benford 2009) (Burdige 2005). Potential side effects on marine ecosystems, chemistry, or circulation have not been thoroughly assessed. Neither have these concepts been evaluated with respect to the impacts on land from enhanced transfer of nutrients and organic matter to the ocean, nor the relative merits of alternative applications of residues and biochar as an energy source or soil amendment (Chapter 7).

**Marine biomass CDR options:** Proposals have been made to grow macroalgae (Duarte et al., 2017) for BECCS (N'Yeurt et al. 2012; Duarte et al. 2013; Chen et al., 2015), to sink cultured macroalgae into the deep sea, or to use marine algae for biochar (Roberts et al., 2015). Naturally-growing sargassum has also been considered for these purposes (Bach et al., 2021). Froehlich et al. (2019) found a substantial area of the ocean (about 48 million $km^2$) suitable for farming seaweed. N'Yeurt et al. (2012) suggested that converting 9% of the oceans to macroalgal aquaculture could take up 19 $GtCO_2$ in biomass, generate 12 Gt per annum of biogas, and the $CO_2$ produced by burning the biogas could be captured and sequestered. Productivity of farmed macroalgae in the open ocean could potentially be enhanced through fertilising via artificial upwelling (Fan et al., 2020) or through cultivation platforms that dive at night to access nutrient-rich waters below the, often nutrient-limited, surface ocean. If the biomass were sunk, it is unknown how long the carbon would remain in the deep ocean and what the additional impacts would be. Research and development on macroalgae cultivation and use is currently underway in multiple parts of the world, though not necessarily directly focused on CDR.

**Extraction of $CO_2$ from seawater (with storage):** $CO_2$ can be extracted by applying a vacuum, or by purging with a gas low in $CO_2$ (Koweek et al., 2016). $CO_2$ stripping can also be accomplished by acidifying seawater with a mineral acid, or through electrodialysis and electrolysis, to convert bicarbonate ions ($HCO_3^-$) to $CO_2$ (Willauer et al., 2017; Eisaman et al., 2018; Digdaya et al., 2020; Eisaman 2020; Sharifian et al., 2021). The removal of $CO_2$ from the ocean surface leads to undersaturation in the water, thus forcing $CO_2$ to move from the atmosphere into the ocean to restore equilibrium. Electrochemical seawater $CO_2$ extraction has been modelled, prototyped, and analysed from a techno-economic perspective (Eisaman et al., 2012; Willauer et al., 2017; de Lannoy et al., 2018; Eisaman et al., 2018a; Eisaman et al., 2018b).

Status, costs, potentials, risk and impacts, co-benefits, trade-offs and spillover effects and the role in mitigation pathways of ocean-based approaches are summarised in Table 12.6.

#### 12.3.1.4 Feasibility Assessment

Following the framework presented in Section 6.4 and Annex II, Part IV, Section 11, a multi-dimensional feasibility assessment of the CDR methods covered here is provided in Figure 12.4, taking into account the assessment presented in this section. Both DACCS and EW perform positively on the geophysical and technological dimensions while for ocean-based approaches performance is mixed. There is limited evidence to assess social-cultural, environmental/ecological, and institutional dimensions as the literature is still nascent for DACCS and EW, while these aspects are positive for blue carbon and mixed or negative for ocean fertilisation. On the economic dimension, the cost is assessed negatively for all CDR methods.

### 12.3.2 Consideration of Methods Assessed in Sectoral Chapters: A/R, Biochar, BECCS, Soil Carbon Sequestration

**Status:** BECCS, afforestation/reforestation (A/R), soil carbon sequestration (SCS) and biochar are land-based biological CDR methods (Smith et al. 2016). BECCS combines biomass use for energy with CCS to capture and store the biogenic carbon geologically (Section 6.4.2.6); A/R and SCS involve fixing atmospheric carbon in biomass and soils, and biochar involves converting biomass to biochar and using it as a soil amendment. These CDR methods can be associated with both co-benefits and adverse side effects (Smith et al. 2016; Hurlbert et al. 2019; Mbow et al. 2019; Olsson et al. 2019; Schleicher et al. 2019; Smith et al. 2019b; Babin et al. 2021; Dooley et al. 2021) (Sections 7.4 and 12.5).

Among CDR methods, BECCS and A/R are most commonly selected by IAMs to meet the requirements of scenarios that limit warming to 2°C (>67%) or lower. This is partially because of the long lead time required to refine IAMs to include additional methods and update techno-economic parameters. Currently, few IAMs represent SCS or biochar (Frank et al. 2017). Given the removal potential of SCS and biochar and some potential co-benefits, more efforts should be made to include these methods within IAMs, so that their mitigation potential can be compared to other CDR methods, along with possible co-benefits and adverse side effects (Smith et al. 2016; Rogelj et al. 2018) (Section 12.5).

**Potential:** The technical potential for BECCS by 2050 is estimated at 0.5–11.3 $GtCO_2$-eq $yr^{-1}$ (Table 7.3). These potentials do not include avoided emissions resulting from the use of heat, electricity and/or fuels provided by the BECCS system, which depend on substitution patterns, conversion efficiencies, and supply chain emissions for the

BECCS and substituted energy systems (Box 7.7). The mitigation effect of BECCS also depends on how deployment affects land carbon stocks and sink strength (Section 7.4.4).

As detailed in Chapter 7, the technical potential for gross removals realised through A/R in 2050 is 0.5–10.1 $GtCO_2$-eq $yr^{-1}$, and for improved forest management the potential is 1–2.1 $GtCO_2$-eq $yr^{-1}$ (including both CDR and emissions reduction). Technical potential for SCS in 2050 is estimated to be 0.6–9.4 $GtCO_2$-eq $yr^{-1}$, for agroforestry it is 0.3–9.4 $GtCO_2$-eq $yr^{-1}$, and for biochar it is 0.2–6.6 $GtCO_2$-eq $yr^{-1}$. Peatland and coastal wetland restoration have a technical potential of 0.5–2.1 $GtCO_2$-eq $yr^{-1}$ in 2050, with an estimated 80% of the potential being CDR. Note that these potentials reflect only biophysical and technological conditions and become reduced when factoring in economic, environmental, socio-cultural and institutional constraints (Table 12.6).

**Costs:** Costs across technologies vary substantially (Smith et al. 2016) and were estimated to be USD15–400 $tCO_2^{-1}$ for BECSS, USD0–240 $tCO_2^{-1}$ for A/R, –USD45 to +USD100 $tCO_2^{-1}$ for SCS and USD10–345 $tCO_2^{-1}$ for biochar. Fuss et al. (2018) estimated abatement cost ranges for BECCS, A/R, SCS and biochar to be 100–200, 5–50, 0–100, and 30–120 $tCO_2$-$eq^{-1}$ respectively, corresponding to 2100 potentials. Ranges for economic potential (<USD100 $tCO_2^{-1}$) reported in Chapter 7 are 0.5–3.0 $GtCO_2$ $yr^{-1}$ (A/R); 0.6–1.9 $GtCO_2$ $yr^{-1}$ (improved forest management); 0.7–2.5 $GtCO_2$ $yr^{-1}$ (SCS); 0.4–1.1 $GtCO_2$ $yr^{-1}$ (agroforestry); 0.3–1.8 $GtCO_2$ $yr^{-1}$ (biochar); and 0.2–0.8 $GtCO_2$ $yr^{-1}$ (peatland and coastal wetland restoration).

**Risks, impacts, and co-benefits:** a brief summary of risks, impacts and co-benefits is provided here and more detail is provided in Chapter 7 and Section 12.5. A/R and biomass production for BECCS and biochar potentially compete for land, water and other resources, implying possible adverse outcomes for ecosystem health, biodiversity, livelihoods and food security (*medium evidence, high agreement*) (Smith et al. 2016; Heck et al. 2018; Hurlbert et al. 2019; Mbow et al. 2019) (Chapter 7). SCS requires the addition of nitrogen and phosphorus to maintain stoichiometry of soil organic matter, leading to a potential risk of eutrophication (Fuss et al. 2018). Apart from possible negative effects associated with biomass supply, adverse side effects from biochar are relatively low if the biomass is uncontaminated (Tisserant and Cherubini 2019).

Possible climate risks relate to direct and/or indirect land carbon losses (A/R, BECCS, biochar), increased $N_2O$ emissions (BECCS, SCS), saturation and non-permanence of carbon storage (A/R, SCS) (Jia et al. 2019; Smith et al. 2019b) (Chapter 7), and potential $CO_2$ leakage from deep geological reservoirs (BECCS) (Chapter 6). Land cover change associated with A/R and biomass supply for BECCS and biochar may cause albedo changes that reduce mitigation effectiveness (Fuss et al. 2018; Jia et al. 2019). Potentially unfavourable albedo change resulting from biochar use can be minimised by incorporating biochar into the soil (Fuss et al. 2018) (Chapter 7).

Concerning co-benefits, A/R and biomass production for BECCS or biochar could improve soil carbon, nutrient and water cycling (*robust evidence, high agreement*), and contribute to market opportunities, employment and local livelihoods, economic diversification, energy security, and technology development and transfer (*medium evidence, high agreement*) (Fuss et al. 2018) (Chapter 7). It may contribute to reduction of other air pollutants, health benefits, and reduced dependency on imported fossil fuels. A/R can improve biodiversity if native and diverse species are used (Fuss et al. 2018). For biochar, additional co-benefits include increased crop yields, reduced drought impacts, and reduced $CH_4$ and $N_2O$ emissions from soils (Joseph et al., 2021) (Section 7.4.5.2). SCS can improve soil quality and resilience and improve agricultural productivity and food security (Frank et al. 2017; Smith et al. 2019b).

**Role in mitigation pathways:** Biomass use for BECCS in 2050 is 61 EJ $yr^{-1}$ (13–208 EJ $yr^{-1}$, 5–95th percentile range) in scenarios limiting warming to 1.5°C (>50%) with no or limited overshoot (C1, excluding traditional energy). This corresponds to 5.3 $GtCO_2$ $yr^{-1}$ (1.1–18 $GtCO_2$ $yr^{-1}$) CDR, if assuming 28 kg C $GJ^{-1}$ biomass carbon content and 85% capture rate in BECCS systems. In scenarios that limit warming to 2°C (>67%) (C3), biomass use for BECCS in 2050 is 28 EJ $yr^{-1}$ (0–96 EJ $yr^{-1}$, 5–95th percentile range), corresponding to 2.4 $GtCO_2$ $yr^{-1}$ (0–8.3 $GtCO_2$ $yr^{-1}$) CDR. Cumulative $CO_2$ removal from AFOLU (mainly through A/R), as reported from models, in the period 2020 to 2100 is 262 $GtCO_2$ (17–397 $GtCO_2$) and 209 $GtCO_2$ (20–415 $GtCO_2$) in C1 and C3 scenarios, respectively (5–95th percentile range).

Uncertainties remain in two main areas: the availability of land and biomass, which is affected by many factors (Anandarajah et al. 2018) (Chapter 7), and the role of other mitigation measures including CDR methods other than A/R and BECCS. Strong near-term climate change mitigation to limit overshoot, and deployment of CDR methods other than A/R and BECCS, may significantly reduce the contribution of these CDR methods in scenarios limiting warming to 1.5°C or 2°C (Köberle 2019; Hasegawa et al. 2021).

**Trade-offs and spillovers:** Some land-based biological CDR methods, such as BECCS and A/R, demand land. Combining mitigation strategies has the potential to increase overall carbon sequestration rates (Humpenöder et al. 2014). However, these CDR methods may also compete for resources (Frank et al. 2017). Land-based mitigation approaches currently propose the use of forests (i) as a source of woody biomass for bioenergy and various biomaterials and (ii) for carbon sequestration in vegetation, soils, and forest products. Forests are therefore required to provide both provisioning (biomass feedstock) and regulating (carbon sequestration) ecosystem services. This multifaceted strategy has the potential to result in trade-offs (Makkonen et al. 2015). Some land-based mitigation options could conflict with biodiversity goals, e.g., A/R using monoculture plantations can reduce species richness when introduced into (semi-)natural grasslands (Smith et al. 2019a; Dooley et al. 2021). When trade-offs exist between biodiversity protection and mitigation objectives, biodiversity is typically given a lower priority, especially if the mitigation option is considered risk-free and economically feasible (Pörtner et al. 2021). Approaches that promote synergies, such as sustainable forest management, reducing deforestation rates, cultivation of perennial crops for bioenergy in sustainable farming practices, and mixed-species forests in A/R, can

**Table 12.6 | Summary of status, costs, potentials, risk and impacts, co-benefits, trade-offs and spillover effects and the role in mitigation pathways for CDR methods.** Technology readiness level (TRL) is a measure of maturity of the CDR method. Scores range from 1 (basic principles defined) to 9 (proven in operational environment). Author judgement ranges (assessed by authors in the literature) are shown, with full literature ranges shown in brackets.

| CDR method | Status (TRL) | Cost (USD $tCO_2^{-1}$) | Mitigation Potential ($GtCO_2\ yr^{-1}$) | Risk and impacts | Co-benefits | Trade-offs and spillover effects | Role in modelled mitigation pathways | Section |
|---|---|---|---|---|---|---|---|---|
| DACCS | 6 | 100–300 (84–386) | 5–40 | Increased energy and water use | Water produced (solid sorbent DAC designs only) | Potentially increased emissions from water supply and energy generation | In a few IAMs; DACCS complements other CDR methods | 12.3.1.1 |
| Enhanced weathering | 3–4 | 50–200 (24–578) | 2–4 (<1–95) | Mining impacts; air quality impacts of rock dust when spreading on soil | Enhanced plant growth, reduced erosion, enhanced soil carbon, reduced soil acidity, enhanced soil water retention | Potentially increased emissions from water supply and energy generation | In a few IAMs; EW complements other CDR methods | 12.3.1.2 |
| Ocean alkalinity enhancement | 1–2 | 40–260 | 1–100 | Increased seawater pH and saturation states may impact marine biota. Possible release of nutritive or toxic elements and compounds. Mining impacts | Limiting ocean acidification | Potentially increased emissions of $CO_2$ and dust from mining, transport and deployment operations | No data | 12.3.1.3 |
| Ocean fertilisation | 1–2 | 50–500 | 1–3 | Nutrient redistribution, restructuring of the ecosystem, enhanced oxygen consumption and acidification in deeper waters, potential for decadal-to-millennial-scale return to the atmosphere of nearly all the extra carbon removed, risks of unintended side effects | Increased productivity and fisheries, reduced upper ocean acidification | Subsurface ocean acidification, deoxygenation; altered meridional supply of macro-nutrients as they are utilised in the iron-fertilised region and become unavailable for transport to, and utilisation in, other regions, fundamental alteration of food webs, biodiversity | No data | 12.3.1.3 |
| Blue carbon management in coastal ecosystems | 2–3 | Insufficient data, estimates range from ~100 to ~10,000 | <1 | If degraded or lost, coastal blue carbon ecosystems are likely to release most of their carbon back to the atmosphere; potential for sediment contaminants, toxicity, bioaccumulation and biomagnification in organisms; issues related to altering degradability of coastal plants; use of subtidal areas for tidal wetland carbon removal; effect of shoreline modifications on sediment redeposition and natural marsh accretion; abusive use of coastal blue carbon as means to reclaim land for purposes that degrade capacity for carbon removal | Potential for many non-climatic benefits and can contribute to ecosystem-based adaptation, coastal protection, increased biodiversity, reduced upper ocean acidification; could potentially benefit human nutrition or produce fertiliser for terrestrial agriculture, anti-methanogenic feed additive, or as an industrial or materials feedstock | If degraded or lost, coastal blue carbon ecosystems are likely to release most of their carbon back to the atmosphere. The full delivery of the benefits at their maximum global capacity will require years to decades to be achieved | Not incorporated in IAMs, but in some bottom-up studies: small contribution | 12.3.1.3, 7.4 |
| BECCS | 5–6 | 15–400 | 0.5–11 | Competition for land and water resources, to grow biomass feedstock. Biodiversity and carbon stock loss if from unsustainable biomass harvest | Reduction of air pollutants; fuel security, optimal use of residues, additional income, health benefits and if implemented well can enhance biodiversity, soil health and land carbon | Competition for land with biodiversity conservation and food production | Substantial contribution in IAMs and bottom-up sectoral studies | 7.4 |

| CDR method | Status (TRL) | Cost (USD $tCO_2^{-1}$) | Mitigation Potential ($GtCO_2\ yr^{-1}$) | Risk and impacts | Co-benefits | Trade-offs and spillover effects | Role in modelled mitigation pathways | Section |
|---|---|---|---|---|---|---|---|---|
| Afforestation/ reforestation | 8–9 | 0–240 | 0.5–10 | Reversal of carbon removal through wildfire, disease, pests may occur. Reduced catchment water yield and lower groundwater level if species and biome are inappropriate | Enhanced employment and local livelihoods, improved biodiversity, improved renewable wood products provision, soil carbon and nutrient cycling. Possibly less pressure on primary forest | Inappropriate deployment at large scale can lead to competition for land with biodiversity conservation and food production | Substantial contribution in IAMs and also in bottom-up sectoral studies | 7.4 |
| Biochar | 6–7 | 10–345 | 0.3–6.6 | Particulate and GHG emissions from production; biodiversity and carbon stock loss from unsustainable biomass harvest | Increased crop yields and reduced non-$CO_2$ emissions from soil; resilience to drought | Environmental impacts associated with particulate matter; competition for biomass resource | In development – not yet in global mitigation pathways simulated by IAMs | 7.4 |
| Soil carbon sequestration in croplands and grasslands | 8–9 | -45–100 | 0.6–9.3 | Risk of increased nitrous oxide emissions due to higher levels of organic nitrogen in the soil; risk of reversal of carbon sequestration | Improved soil quality, resilience and agricultural productivity | Attempts to increase carbon sequestration potential at the expense of production. Net addition per hectare is very small; hard to monitor | In development – not yet in global mitigation pathways simulated by IAMs; in bottom-up studies: with medium contribution | 7.4 |
| Peatland and coastal wetland restoration | 8–9 | Insufficient data | 0.5–2.1 | Reversal of carbon removal in drought or future disturbance. Risk of increased methane emissions | Enhanced employment and local livelihoods, increased productivity of fisheries, improved biodiversity, soil carbon and nutrient cycling | Competition for land for food production on some peatlands used for food production | Not in IAMs but some bottom-up studies with medium contribution | 7.4 |
| Agroforestry | 8–9 | Insufficient data | 0.3–9.4 | Risk that some land area lost from food production; requires high skills | Enhanced employment and local livelihoods, variety of products, improved soil quality, more resilient systems | Some trade-off with agricultural crop production, but enhanced biodiversity, and resilience of system | No data from IAMs, but in bottom-up sectoral studies. with medium contribution | 7.4 |
| Improved forest management | 8–9 | Insufficient data | 0.1–2.1 | If improved management is understood as merely intensification involving increased fertiliser use and introduced species, then it could reduce biodiversity and increase eutrophication | In case of sustainable forest management, it leads to enhanced employment and local livelihoods, enhanced biodiversity, improved productivity | If it involves increased fertiliser use and introduced species, it could reduce biodiversity and increase eutrophication and upstream GHG emissions | No data from IAMs, but in bottom-up sectoral studies with medium contribution | 7.4 |

mitigate biodiversity impacts and even improve ecosystem capacity to support biodiversity while mitigating climate change (Pörtner et al. 2021) (Section 12.5). Systematic land-use planning could help to deliver land-based mitigation options that also limit trade-offs with biodiversity (Longva et al. 2017) (Cross-Working Group Box 3: Mitigation and Adaptation via the Bioeconomy, in this chapter).

Status, costs, potentials, risk and impacts, co-benefits, trade-offs and spillover effects and the role in mitigation pathways of A/R, biochar, SCS, peatland and coastal wetland restoration, agroforestry and forest management are summarised in Table 12.6. See also Section 12.5.

### 12.3.3 CDR Governance and Policies

As shown in Cross-Chapter Box 8 in this chapter, CDR fulfils multiple functions in different phases of ambitious mitigation: (i) further reducing net $CO_2$ or GHG emission levels in the near term; (ii) counterbalancing residual emissions (from hard-to-transition sectors like transport, industry, or agriculture) to help reach net zero $CO_2$ or GHG emissions in the mid term; (iii) achieving and sustaining net-negative $CO_2$ or GHG emissions in the long term. While inclusion of emissions and removals on managed land (LULUCF) is mandatory for developed countries under UNFCCC inventory rules (Grassi et al. 2021), not all Annex I countries have included land-based biological removals when setting domestic mitigation targets in the past, but updated NDCs for 2030 indicate a shift, most notably in the European Union (Gheuens and Oberthür 2021; Schenuit et al. 2021). The early literature on CDR governance and policy has been mainly conceptual rather than empirical, focusing on high-level principles (see the concerns listed in the introduction to Section 12.3) and the representation of CDR in global mitigation scenarios (Section 3.2.2). However, with the widespread adoption of net zero targets and the recognition that CDR is a necessary element of mitigation portfolios to achieve net zero $CO_2$ or GHG emissions, countries with national net-zero emissions targets have begun to integrate CDR into modelled national mitigation pathways, increase research, development and demonstration (RD&D) efforts on CDR methods, and consider CDR-specific incentives and policies (Honegger et al. 2021b; Schenuit et al. 2021) (Box 12.1). Nevertheless, this increasing consideration of CDR has not yet extended to net-negative targets and policies to achieve these. While the use of CDR at levels that would lead to net negative $CO_2$ or GHG emissions in the long term has been assumed in most global mitigation scenarios that limit warming to 1.5°C, net-negative emissions trajectories and BECCS as the main CDR method modelled to achieve these have not been mirrored by corresponding UNFCCC decisions so far (Fridahl 2017; Mohan et al. 2021). Likewise, only a few national long-term mitigation plans or legal acts entail a vision for net-negative GHG emissions (Buylova et al. 2021), for example Finland, Sweden, Germany and Fiji.

For countries with emissions targets aiming for net zero or lower, the core governance question is not whether CDR should be mobilised or not, but which CDR methods governments want to see deployed by whom, by when, at which volumes and in which ways (Minx et al. 2018; Bellamy and Geden 2019). The choice of CDR methods and the scale and timing of their deployment will depend on the respective ambitions for gross emissions reductions, how sustainability and feasibility constraints are managed, and how political preferences and social acceptability evolve (Bellamy 2018; Forster et al. 2020; Fuss et al. 2020; Waller et al. 2020; Clery et al. 2021; Iyer et al. 2021; Rogelj et al. 2021). As examples of emerging CDR policymaking at (sub-)national levels show, policymakers are beginning to incorporate CDR methods beyond those currently dominating global mitigation scenarios, that is, BECCS and afforestation/reforestation (Bellamy and Geden 2019; Buylova et al. 2021; Schenuit et al. 2021; Uden et al. 2021) (Box 12.1). CDR policymaking is faced with the need to consider method-specific timescales of $CO_2$ storage, as well as challenges in MRV and accounting, potential co-benefits, adverse side effects, interactions with adaptation and trade-offs with SDGs (Dooley and Kartha 2018; McLaren et al. 2019; Buck et al. 2020; Honegger et al. 2020; Brander et al. 2021; Dooley et al. 2021; Mace et al. 2021) (Table 12.6). Therefore, CDR governance and policymaking are expected to focus on responsibly incentivising RD&D and targeted deployment, building on both technical and governance experience with already widely practised CDR methods like afforestation/reforestation (Lomax et al. 2015; Field and Mach 2017; Bellamy 2018; Carton et al. 2020; VonHedemann et al. 2020), as well as learning from two decades of slow-moving CCS deployment (Buck 2021; Martin-Roberts et al. 2021; Wang et al. 2021). For some less well-understood methods and implementation options, such as ocean alkalinisation or enhanced weathering, investment in RD&D can help in understanding the risks, rewards, and uncertainties of deployment (Nemet et al. 2018; Fajardy et al. 2019; Burns and Corbett 2020; Goll et al. 2021).



## Box 12.1 | Case Study: Emerging CDR Policy, Research and Development in the United Kingdom

Climate change mitigation policies in the UK have been motivated since 2008 by a domestic, legally-binding framework. This framework includes a 2050 target for net zero greenhouse gas emissions, interim targets and an independent advisory body called the Climate Change Committee (Muinzer 2019). It has led successive UK governments to publish mitigation plans to 2050, causing policy to be more forward looking (Averchenkova et al. 2021).

The UK's targets include emissions and removals from LULUCF. In 2008 the target for 2050 was an economy-wide net emissions reduction of at least 80% below 1990 levels. Even the first government plans to achieve this target proposed deployment of removal methods, specifically afforestation and wood in construction, increased soil carbon and BECCS (HM Government 2011).

*Box 12.1 (continued)*

Adoption of the Paris Agreement in 2015 caused the government to change the legislated 2050 target to a reduction of at least 100% (i.e., net zero). Since then, removal of $CO_2$ and other greenhouse gases has received greater prominence as a distinct topic. The most recent national plan (published October 2021) proposes deployment not only of the methods mentioned above, but also DACCS, biochar and enhanced weathering. The government has committed to amend accounting of UK targets to include a wider range of removal methods beyond LULUCF, and set a target of 5 $MtCO_2$ $yr^{-1}$ from methods such as BECCS, DACCS and enhanced weathering by 2030. It is consulting on markets and incentives for deployment, and exploring new requirements for MRV (HM Government 2021).

In parallel to these policy developments, the UK funds research into technical, environmental and social aspects of removal (Lezaun et al. 2021). Research on some elements (e.g., forestry, CCS, soils, bioenergy) have been funded for well over a decade, but the first programme dedicated to greenhouse gas removal ran during 2017–2021. This has been followed by two new programmes with greater focus on demonstration, totalling GBP100 million over four years (HM Government 2021). A wide variety of methods is supported in these programmes, covering approaches such as $CO_2$ capture from seawater and capture of methane from cattle, in addition to those included already in national mitigation scenarios.

Deployment of removal methods has lagged behind expectations, as national targets for tree planting are not being met and infrastructure for $CO_2$ transport and storage is not yet in place (Climate Change Committee 2021). While public awareness around carbon removal is low, studies indicate support in general, provided it is perceived as enhancing rather than impeding action to reduce emissions (Cox et al. 2020a).

Since the enhancement of carbon sinks is a form of climate change mitigation (Honegger et al. 2021a), CDR governance challenges will in many respects be similar to those around emissions reduction measures, as will policy instruments like RD&D funding, carbon pricing, tax or investment credits, certification schemes, and public procurement (Sections 13.4, 13.6, 14.4 and 14.5). Effectively integrating CDR into mitigation portfolios can build on already existing rules, procedures and instruments for emissions abatement (Torvanger 2019; Fridahl et al. 2020; Zakkour et al. 2020; Honegger et al. 2021b; Mace et al. 2021; Rickels et al. 2021). Additionally, to accelerate RD&D and to incentivise CDR deployment, a political commitment to formal integration into existing climate policy frameworks is required (*robust evidence, high agreement*) (Lomax et al. 2015; Geden et al. 2018; Honegger and Reiner 2018; VonHedemann et al. 2020; Schenuit et al. 2021). To avoid CDR being misperceived as a substitute for deep emissions reductions, the prioritisation of emissions cuts can be signalled and achieved with differentiated target setting for reductions and removals (Geden et al. 2019; McLaren et al. 2019). Similarly, sub-targets are conceivable for different types of CDR, to prioritise preferred methods according to characteristics such as removal processes or timescales of storage (Smith 2021).

IPCC guidance on quantifying removals is available for land-based biological CDR methods (IPCC 2006, 2019), but has yet to be developed for other CDR methods (Royal Society and Royal Academy of Engineering 2018). Challenges with development of estimation algorithms, data collection, and attribution between sectors and countries will need to be overcome (Luisetti et al. 2020; Wedding et al. 2021). Trusted methodologies for MRV, required to enable private sector participation. will need to address the permanence, leakage, and saturation challenges with land- and ocean-based biological methods (Mace et al. 2021). Protocols that also capture social and ecological co-benefits could encourage the adoption of biological CDR methods such as SCS, biochar, A/R and blue carbon management (*robust evidence, high agreement*) (VonHedemann et al. 2020; Macreadie et al. 2021).

Private capital and companies, impact investors, and philanthropy will play a role in technical demonstrations and bringing down costs, as well as creating demand for carbon removal products on voluntary markets, which companies may purchase to fulfil corporate social responsibility-driven targets (Friedmann 2019; Fuss et al. 2020; Joppa et al. 2021). Niche markets can provide entry points for limited deployment of novel CDR methods (Cox and Edwards 2019), but targeting currently existing revenue streams by using $CO_2$ captured from the atmosphere in Enhanced Oil Recovery and other utilisation routes (Mackler et al. 2021; Meckling and Biber 2021) is contested, and highlights the importance of choosing appropriate system boundaries when assessing supply chains (Tanzer and Ramírez 2019; Brander et al. 2021). While the private sector will play a distinct role in scaling CDR, governments will need to commit to developing infrastructure for the transport and storage of $CO_2$, including financing, permitting, and regulating liabilities (Sanchez et al. 2018; Mace et al. 2021; Mackler et al. 2021).

International governance considerations include global technology transfer around CDR implementation options (Batres et al. 2021); land use change that could affect food production and land condition and cause conflict around land tenure and access (Dooley and Kartha 2018; Hurlbert et al. 2019; Milne et al. 2019); and efforts to create sustainable and just supply chains for CDR (Fajardy and Mac Dowell 2020; Tan et al. 2021), such as resources used for BECCS, enhanced weathering, or ocean alkalinisation. International governance would be particularly important for methods posing transboundary risks, especially for ocean-based methods. Specific regulations have so far only been developed in the context of the London Protocol, an international treaty that explicitly regulates ocean fertilisation and allows parties to govern other marine CDR methods like ocean alkalinity enhancement (GESAMP 2019; Burns and Corbett 2020; Boettcher et al. 2021) (Section 14.4.5).

Engagement of civil society organisations and publics will be important for shaping CDR policy and deployment (*medium evidence, high agreement*). Public awareness of CDR and its role in national net zero emissions strategies is generally very low (Cox et al. 2020a), and perceptions differ across countries and between methods (Bertram and Merk 2020; Spence et al. 2021; Sweet et al. 2021; Wenger et al. 2021). When awareness increases, social processes will shape political attitudes on CDR (Shrum et al. 2020), as will efforts to frame particular CDR methods as 'natural' or 'technological' (Osaka et al. 2021), and the policy instruments chosen to support CDR (Bellamy et al. 2019). Lack of confidence in CDR implementation options from both publics and investors, and lack of trust in project developers (Cox et al. 2020b) have hampered support for CCS (Thomas et al. 2018) and are expected to affect deployment of CDR methods with geological storage (Gough and Mander 2019). On local and regional scales, CDR projects will need to consider air and water quality, impacts to human health, energy needs, land use and ecological integrity, and local community engagement and procedural justice. Bottom-up and community-driven strategies are important for deploying equitable carbon removal projects (Batres et al. 2021; Hansson et al. 2021).

## 12.4 Food systems

### 12.4.1 Introduction

This section complements Chapter 7 by reviewing recent estimates of food system emissions and assessing options beyond the agriculture, forestry and land use sectors to mitigate food systems GHG emissions. A food system approach enables identification of cross-sectoral mitigation opportunities including both technological and behavioural options. Further, a system approach permits evaluation of policies that do not necessarily directly target primary producers or consumers, but other food system actors, with possibly higher mitigation efficiency. A food system approach was introduced in the IPCC Special Report on Climate Change and Land (SRCCL) (Mbow et al. 2019). Besides major knowledge gaps in the quantification of food system GHG emissions (Section 12.4.2), the SRCCL authors identified as major knowledge gaps the understanding of the dynamics of dietary change (including behavioural patterns, the adoption of plant-based dietary patterns, and interaction with human health and nutrition of sustainable healthy diets and associated feedbacks); and instruments and mechanisms to accelerate transitions towards sustainable and healthy food systems.

Sufficient food and adequate nutrition are fundamental human needs (HLPE 2020; Ingram 2020). Food needs to be grown and processed, transported and distributed, and finally prepared and consumed. Food systems range from traditional, involving only few people and short supply chains, to modern food systems, comprising complex webs involving large numbers of stakeholders and processes that grow and transform food commodities into food products and distribute them globally (Gómez and Ricketts 2013; HLPE 2017). A 'food system' includes all food chain activities (production, processing, distribution, preparation, consumption of food) and the management of food loss and wastes. It also includes institutions and infrastructures influencing any of these activities, as well as people and systems impacted (HLPE 2017; FAO 2018a). Food choices are determined by the food environment, consisting of the 'physical, economic, political and socio-cultural context in which consumers engage with the food system to acquire, prepare and consume food' (HLPE 2017). Food system outcomes encompass food and nutrition, productivity, profit and livelihood of food producers and other actors in food value chains, but also social outcomes and the impact on the environment (Zurek et al. 2018). 'Sustainable healthy diets' have been defined by FAO and WHO (FAO and WHO 2019) as 'dietary patterns that promote all dimensions of individuals' health and wellbeing; have low environmental pressure and impact; are accessible, affordable, safe and equitable; and are culturally acceptable'.

The SRCCL estimated overall global anthropogenic emissions from food systems to range between 10.8 and 19.1 $GtCO_2$-eq $yr^{-1}$, equivalent to 21–37% of total anthropogenic emissions (Mbow et al. 2019; Rosenzweig et al. 2020a). The authors identified major knowledge gaps for the GHG emissions inventories of food systems, particularly in providing disaggregated emissions from the food industry and transportation. The food system approach taken in the SRCCL (Mbow et al. 2019) evaluates the synergies and trade-offs of food system response options and their implications for food security, climate change adaptation and mitigation. This integrated framework allows the identification of fundamental attributes of responses to maximise co-benefits, while avoiding maladaptation measures and adverse side effects. A food system approach supports the design of interconnected climate policy responses to tackle climate change, incorporating perspectives of producers and consumers. The SRCCL (Mbow et al. 2019) found that the technical mitigation potential by 2050 of demand-side responses at 0.7–8.0 $GtCO_2$-eq $yr^{-1}$ is comparable to supply-side options at 2.3–9.6 $GtCO_2$-eq $yr^{-1}$. This shows that mitigation actions need to go beyond food producers and suppliers to incorporate dietary changes and consumers' behavioural patterns and reveals that producers and consumers need to work together to reduce GHG emissions.

Though total production of calories is sufficient for the world population (Wood et al. 2018; Benton et al. 2019), availability and access to food is unequally distributed, and there is a lack of nutrient-dense foods, fruit and vegetables (Berners-Lee et al. 2018; KC et al. 2018). In 2019, close to 750 million people were food insecure. An estimated 2 billion people lacked adequate access to safe and nutritious food in both quality and quantity (FAO et al. 2020). Two billion adults are overweight or obese through inadequate nutrition, with an upward trend globally (FAO et al. 2019). Low intake of fruit and vegetables is further aggravated by high intake rates of refined grains, sugar and sodium, together leading to a high risk of non-communicable diseases such as cardiovascular disease and type 2 diabetes (Springmann et al. 2016; Clark et al. 2018; Clark et al. 2019; GBD 2017 Diet Collaborators et al. 2019; Willett et al. 2019) (*robust evidence, high agreement*). At least 340 million children under five years of age experience lack of vitamins or other essential bio-available nutrients, including almost 200 million suffering from stunting, wasting or overweight (UNICEF 2019).

Bodirsky et al. (2020) find that the global prevalence of overweight will increase to 39–52% of world population in 2050 (from 29% in 2010; range across the Shared Socio-economic Pathways studied), and the prevalence of obesity to 13–20% (9% in 2010). The prevalence of underweight people was predicted to approximately halve, with absolute numbers stagnating at 0.4–0.7 billion. Although many studies represent future pathways of diets and food systems, there are few holistic and consistent narratives and quantification of the future pathways of diets and food systems (Mitter et al. 2020; Mora et al. 2020). Alternative pathways for improved diets and food systems have been developed, emphasising climate, environmental and health co-benefits (Bajželj et al. 2014; Hedenus et al. 2014; Damerau et al. 2016; Weindl et al. 2017a; Weindl et al. 2017b; Springmann et al. 2018a; Bodirsky et al. 2020; Prudhomme et al. 2020; Hamilton et al. 2021), reduced food waste and closing yield gaps (Bajželj et al. 2014; Pradhan et al. 2014), nitrogen management (Bodirsky et al. 2014), urban and peri-urban agriculture (Kriewald et al. 2019) and different sustainability targets (Henry et al. 2018b). The UN Food and Agriculture Organization (FAO) has examined three alternative food system scenarios: 'business as usual', 'towards sustainability', and 'stratified societies' (FAO 2018b). Others have identified research priorities or changes in legislation needed to support adoption of improved food systems (Mylona et al. 2018).

Malnutrition aggravates susceptibility of children to various infectious diseases (França et al. 2009; Farhadi and Ovchinnikov 2018), and infectious diseases can also decrease nutrient uptake, thereby promoting malnutrition (Farhadi and Ovchinnikov 2018). Contamination of food with bacteria, viruses, parasites and microbial toxins can cause foodborne illnesses (Ricci et al. 2017; Abebe et al. 2020; Gallo et al. 2020), foodborne substances such as food additives and specific proteins can cause adverse reactions, and contamination with toxic chemical substances used in agriculture and food processing can lead to poisoning or chronic diseases (Gallo et al. 2020). Further, health risks from food systems may originate from the use of antibiotics in livestock production and the occurrence of anti-microbial resistance in pathogens (ECDC et al. 2015; Bennani et al. 2020), or zoonotic diseases such as COVID-19 (Gan et al. 2020; Patterson et al. 2020; Vågsholm et al. 2020).

Modern food systems are highly consolidated, through vertical and horizontal integration (Swinnen and Maertens 2007; Folke et al. 2019). This consolidation has led to uneven distribution of power across the food value chain, with influence concentrated among a few actors in the post-farmgate food supply chain (e.g., large food processors and retailers), and has contributed to a loss of indigenous agriculture and food systems, for example on Pacific Islands (Vogliano et al. 2020). While agricultural producers contribute a higher proportion of GHG emissions compared with other actors in the supply chain, they have relatively little power to change the system (Clapp 2019; Group of Chief Scientific Advisors 2020; Leip et al. 2021).

In 2016, the agriculture, fisheries, and forestry sectors employed 29% of working people; employment within these sectors was 4% in developed countries, down from 9% in 1995, and 57% in least developed countries, down from 71% in 1995 (World Bank 2021). Employment in other (non-agriculture) food system sectors, such as the food processing industry and service sectors, differs between food systems. The share of total non-farm food system employment ranges from 10% in traditional food systems (e.g., sub-Saharan Africa), to over 50% in food systems in transition (e.g., Brazil), to high shares (80%) in modern food systems (e.g., US) (Townsend et al. 2017). The share of the food expenditures that farmers receive is decreasing; at the global level, this share has been estimated at 27% in 2015 (Yi et al. 2021).

### 12.4.2 GHG Emissions from Food Systems

#### 12.4.2.1 Sectoral Contribution of GHG Emissions from Food Systems

New calculations using the EDGAR v6.0 (Crippa et al. 2021a) and FAOSTAT (FAO 2021) databases provide territorial-based food system GHG emissions by country globally for the period 1990 to 2018 (Crippa et al. 2021b). The data are calculated based on a combination of country-specific data and aggregated information as described by Crippa et al. (2021b) and Tubiello et al. (2021). The data show that, in 2018, 17 $GtCO_2$-eq $yr^{-1}$ (95% confidence range 13–23 $GtCO_2$-eq $yr^{-1}$, calculated according to Solazzo et al. (2020)) were associated with the production, processing, distribution, consumption of food and management of food system residues. This corresponded to 31% (range 23–42%) of total anthropogenic GHG emissions of 54 $GtCO_2$-eq $yr^{-1}$. Based on the IPCC sectoral classification (Table 12.7 and Figure 12.5), the largest contribution of food systems GHG emissions in 2018 was from agriculture, that is, livestock and crop production systems (6.3 $GtCO_2$-eq $yr^{-1}$, range 2.6–11.9) and land use, land use change and forestry (LULUCF) (4.0 $GtCO_2$-eq $yr^{-1}$, range 2.1–5.9) (Figure 12.5). Emissions from energy use were 3.9 $GtCO_2$-eq $yr^{-1}$ (3.6–4.4), waste management 1.7 $GtCO_2$-eq $yr^{-1}$ (0.9–2.6), and industrial processes and product use 0.9 $GtCO_2$-eq $yr^{-1}$ (0.6–1.1). The share of GHG emissions from food systems generated outside the AFOLU (agriculture and LULUCF) sectors has increased over recent decades, from 28% in 1990 to 39% in 2018.

**Energy:** Emissions from energy use occur throughout the food supply chain. In 2018, the main contributions came from energy industries supplying electricity and heat (970 $MtCO_2$-eq $yr^{-1}$), manufacturing and construction (920 $MtCO_2$-eq $yr^{-1}$, of which 29% was attributable to the food, beverage, and tobacco industry), and transport (760 $MtCO_2$-eq $yr^{-1}$). These emissions were almost entirely as $CO_2$. Energy emissions from forestry and fisheries amounted to 480 $MtCO_2$-eq $yr^{-1}$, with 91% of emissions as $CO_2$. Emissions from residential and commercial fuel combustion contributed 250 $MtCO_2$-eq $yr^{-1}$ (79% of emissions as $CO_2$, and with emissions of 1.7 $MtCH_4$ $yr^{-1}$) and 130 $MtCO_2$-eq $yr^{-1}$ (with 98% of emissions as $CO_2$), respectively.

Refrigeration uses an estimated 43% of energy in the retail sector (Behfar et al. 2018) and significantly increases fuel consumption during distribution. Besides being energy intensive, supermarket refrigeration also contributes to GHG emissions through leakage of refrigerants (fluorinated gases, or F-gases), although their contribution to food system GHG emissions is estimated to be minor (Crippa et al. 2021b). The cold chain accounts for approximately 1% of global GHG emissions, but as the volume of refrigerators per capita in developing countries is reported to be one order of magnitude lower than in developed countries (19 $m^3$ versus 200 $m^3$ refrigerated storage capacity per 1000 inhabitants), the importance of refrigeration to total GHG emissions is expected to increase (James and James 2010). Although refrigeration gives rise to GHG emissions, both household refrigeration and effective cold chains could contribute to a substantial reduction in losses of perishable food and thus in emissions associated with food provision (University of Birmingham 2018; James and James 2010). A trade-off exists between reducing food waste and increased refrigeration emissions, with the benefits depending on type of produce, location and technologies used (Sustainable Cooling for All 2018; Wu et al. 2019).

Transport has overall a minor importance for food system GHG emissions, with a share of 5% to 6% (Poore and Nemecek 2018; Crippa et al. 2021b). The largest contributor to food system transport GHG emissions was road transport (92%), followed by marine shipping (4%), rail (3%), and aviation (1%). Only looking at energy needs, air or road transport consumes one order of magnitude higher energy (road: 70–80 MJ $t^{-1}$ $km^{-1}$; aviation: 100–200 MJ $t^{-1}$ $km^{-1}$) than marine shipping (10–20 MJ $t^{-1}$ $km^{-1}$) or rail (8–10 MJ $t^{-1}$ $km^{-1}$) (FAO 2011). For specific food products with high water content, relatively low agricultural emissions and high average transport distances, the share of transport in total GHG emissions can be over 40% (e.g., bananas, with total global average GHG emissions of 0.7 $kgCO_2$-eq $kg^{-1}$) (Poore and Nemecek 2018), but transport is a minor source of GHG emissions for most food products (Poore and Nemecek 2018).

**Industry:** Direct industrial emissions associated with food systems are generated by the refrigerants industry (580 $MtCO_2$-eq $yr^{-1}$ as F-gases) and the fertiliser industry for ammonia production (280 $MtCO_2$-eq $yr^{-1}$ as $CO_2$) and nitric acid (60 $MtCO_2$-eq $yr^{-1}$ as $N_2O$). The industry sector data account for $CO_2$ stored in urea (–50 $MtCO_2$-eq $yr^{-1}$). Packaging contributed about 6% of total food system emissions (0.98 $GtCO_2$-eq $yr^{-1}$, 91% as $CO_2$, with $CH_4$ emissions of 2.8 Mt $CH_4$ $yr^{-1}$). Major emissions sources are pulp and paper (60 $MtCO_2$-eq $yr^{-1}$) and aluminium (30 $MtCO_2$-eq $yr^{-1}$), with ferrous metals, glass, and plastics making a smaller contribution. High shares of emissions from packaging are found for beverages and some fruit and vegetables (Poore and Nemecek 2018).

**Waste:** Management of waste generated in the food system (including food waste, wastewater, packaging waste, etc.) leads to biogenic GHG emissions, and contributed 1.7 $GtCO_2$-eq $yr^{-1}$ to food systems' GHG emissions in 2018. Of these emissions, 55% were from domestic and commercial wastewater (30 $MtCH_4$ $yr^{-1}$ and 310 $ktN_2O$ $yr^{-1}$), 36% from solid waste management (20 $MtCH_4$ $yr^{-1}$ and 310 $ktN_2O$ $yr^{-1}$), and 8% from industrial wastewater (4 $MtCH_4$ $yr^{-1}$ and 80 $ktN_2O$ $yr^{-1}$). Emissions from waste incineration and other waste management systems contributed 1%.

**Table 12.7 | GHG emissions from food systems by sector according to IPCC classification in Mt gas $yr^{-1}$ and food systems' share of total anthropogenic GHG emissions in 1990 and 2015.**

| Sector | $CO_2$ | $CH_4$ | $N_2O$ | F-gases | GHG | $CO_2$ | $CH_4$ | $N_2O$ | F-gases | GHG |
|---|---|---|---|---|---|---|---|---|---|---|
| | Emissions (Mt gas $yr^{-1}$) | | | | | Share of total sectoral emissions (%) | | | | |
| | 1990 | | | | | | | | | |
| 1 Energy | 2212 | 10 | 0 | – | 2583 | 10.5 | 10.2 | 26.7 | – | 10.7 |
| 2 Industrial processes | 190 | 0 | 0 | 0 | 263 | 14.5 | 0 | 38 | 4.8 | 16.2 |
| 3 Solvent and Other Product Use | 0 | – | – | – | 0 | 0.2 | – | – | – | 0.2 |
| 4 Agriculture | 102 | 142 | 5 | – | 5370 | 100 | 100 | 99.2 | – | 99.8 |
| 5 LULUCF | 4946 | – | 0 | – | 5080 | 181 | – | 194 | – | 182 |
| 6 Waste | 3 | 40 | 0 | – | 1155 | 29 | 72.4 | 99.1 | – | 73.2 |
| **Total** | **7453** | **192** | **6** | **0** | **14452** | **29.3** | **65.2** | **84.5** | **4.8** | **40.3** |
| Total ($MtCO_2$-eq $yr^{-1}$) | 7453 | 5243 | 1755 | 0 | 14452 | 29.3 | 63.9 | 84.5 | 0.3 | 40.3 |
| | 2015 | | | | | | | | | |
| 1 Energy | 3449 | 13 | 0 | – | 3927 | 10.1 | 9.5 | 24.1 | – | 10.2 |
| 2 Industrial processes | 242 | 0 | 0 | 0 | 881 | 7.9 | 0 | 28.6 | 58 | 20.1 |
| 3 Solvent and Other Product Use | 7 | – | – | – | 7 | 4.1 | – | – | – | 3.6 |
| 4 Agriculture | 140 | 161 | 7 | – | 6326 | 100 | 100 | 99.1 | – | 99.7 |
| 5 LULUCF | 3823 | – | 1 | – | 3982 | 190 | – | 229 | – | 191 |
| 6 Waste | 5 | 58 | 0 | – | 1699 | 30.6 | 71.8 | 99.1 | – | 72.9 |
| **Total** | **7666** | **231** | **8** | **0** | **16821** | **19.3** | **61.6** | **83.7** | **58** | **31.1** |
| Total ($MtCO_2$-eq $yr_{-1}$) | 7666 | 6317 | 2256 | 581 | 16821 | 19.3 | 60.2 | 83.7 | 53.6 | 31.1 |

Notes: Agricultural emissions include the emissions from the whole sector; biomass production for non-food use currently not differentiated. Non-food system AFOLU emissions are negative (that is, a net carbon sink), therefore the share of AFOLU food system emissions is >100. Source: EDGARv6 (Crippa et al. 2019; Crippa et al. 2021b), and FAOSTAT (FAO 2021). LULUCF: land use, land-use change and forestry.

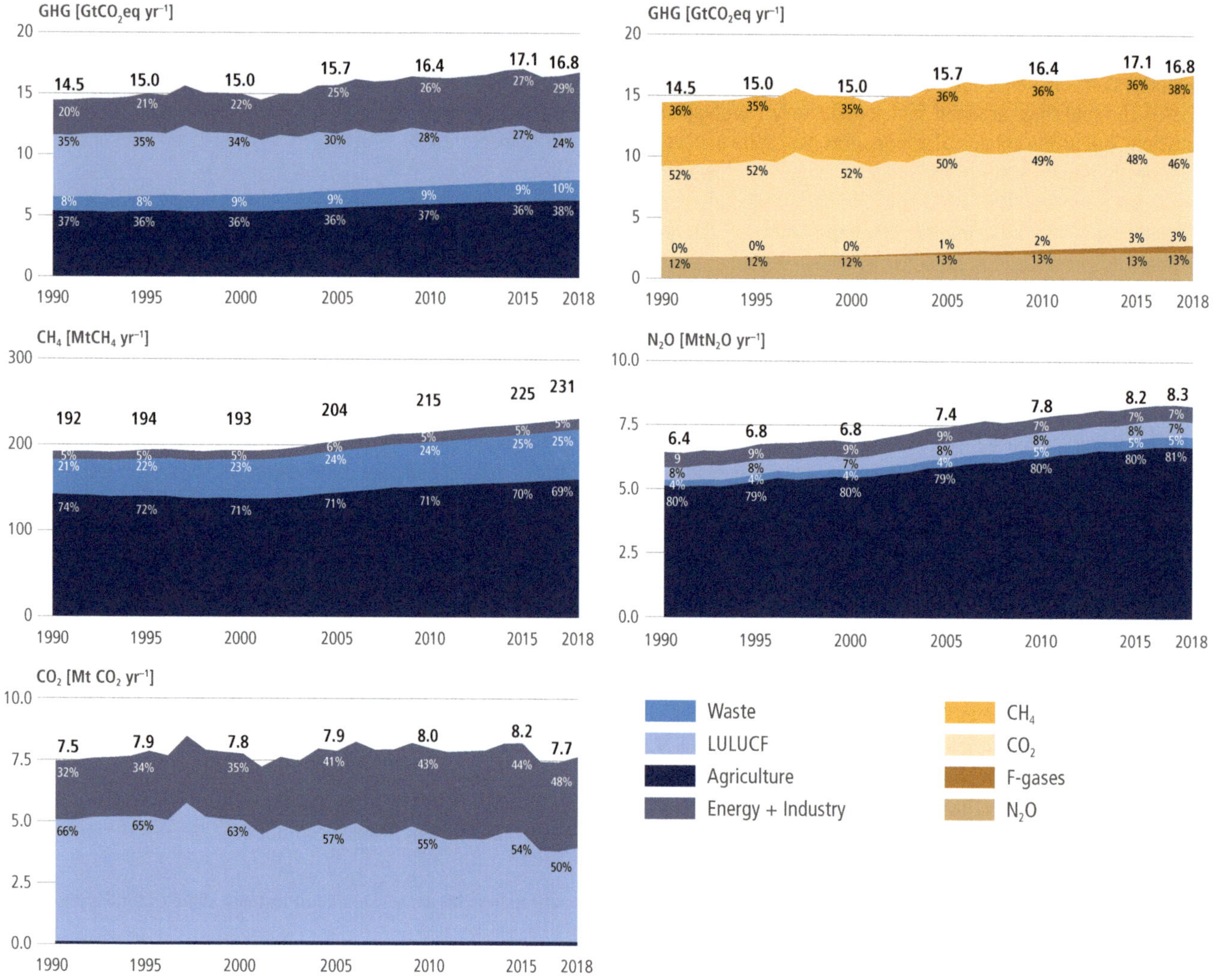


**Figure 12.5 | Food system GHG emissions from the agriculture, LULUCF, waste, and energy & industry sectors.** Source: Crippa et al. (2021b).



### 12.4.2.2 GHG Intensities of Food Commodities

There is high variability in the GHG emissions of different food products and production systems (Figure 12.6). GHG emissions intensities – measured using attributional lifecycle assessment, considering the full supply chain, expressed as $CO_2$-eq per kg of product or per kg of protein – are generally highest for ruminant meat, cheese, and certain crustacean species (e.g., farmed shrimp and prawns, trawled lobster) (Nijdam et al. 2012; Clark and Tilman 2017; Clune et al. 2017; Hilborn et al. 2018; Poore and Nemecek 2018) (*robust evidence, high agreement*). Generally, beef from dairy systems has a lower footprint (8–23 kg$CO_2$-eq per 100 g protein than beef from beef herds (17–94 kg$CO_2$-eq per 100 g protein (Figure 12.6, re-calculated from Poore and Nemecek (2018) using AR6 GWPs based on a 100year horizon) (*medium evidence*, *high agreement*). The wide variation in emissions from beef reflects differences in production systems, which range from intensive feedlots with stock raised largely on grains through to rangeland and transhumance production systems. Dairy systems are generally more intensive production systems, with higher digestibility feed than beef systems. Further, emissions from dairy systems are shared between milk and meat, which brings GHG footprints of beef from dairy herds closer to those of meat from monogastric animals, with emissions intensities of pork (4.4–13 kg$CO_2$-eq per 100 g protein) and poultry meat (2.3–11 kg$CO_2$-eq per 100 g protein) (Poore and Nemecek 2018).

Emissions intensities for farmed fish ranged from 2.4–11 kg$CO_2$-eq per 100 g protein (Poore and Nemecek 2018). For Norwegian seafood, large differences have been found ranging from 1.1 kg$CO_2$-eq kg$^{-1}$ edible product for herring to more than 8 kg$CO_2$-eq kg$^{-1}$ edible product for salmon shipped by road and ferry from Oslo to Paris (Winther et al. 2020). For capture fish, large differences in emissions have been found, ranging from 0.2–7.9 kg$CO_2$-eq kg$^{-1}$ landed fish (Parker et al. 2018), although an environmental comparison of capture fish to farmed foods should include other indicators such as overfishing. Plant-based foods generally have lower GHG emissions (–2.2 to +4.5 kg$CO_2$-eq per 100 g protein) than farmed animal-based foods (Nijdam et al. 2012; Clark and Tilman 2017; Clune et al. 2017; Hilborn et al. 2018; Poore and Nemecek 2018) (*robust evidence, high agreement*). Several plant-based foods are associated with emissions from land use change, for example, palm oil, soy and coffee (Poore and Nemecek 2018), although emissions intensities are context

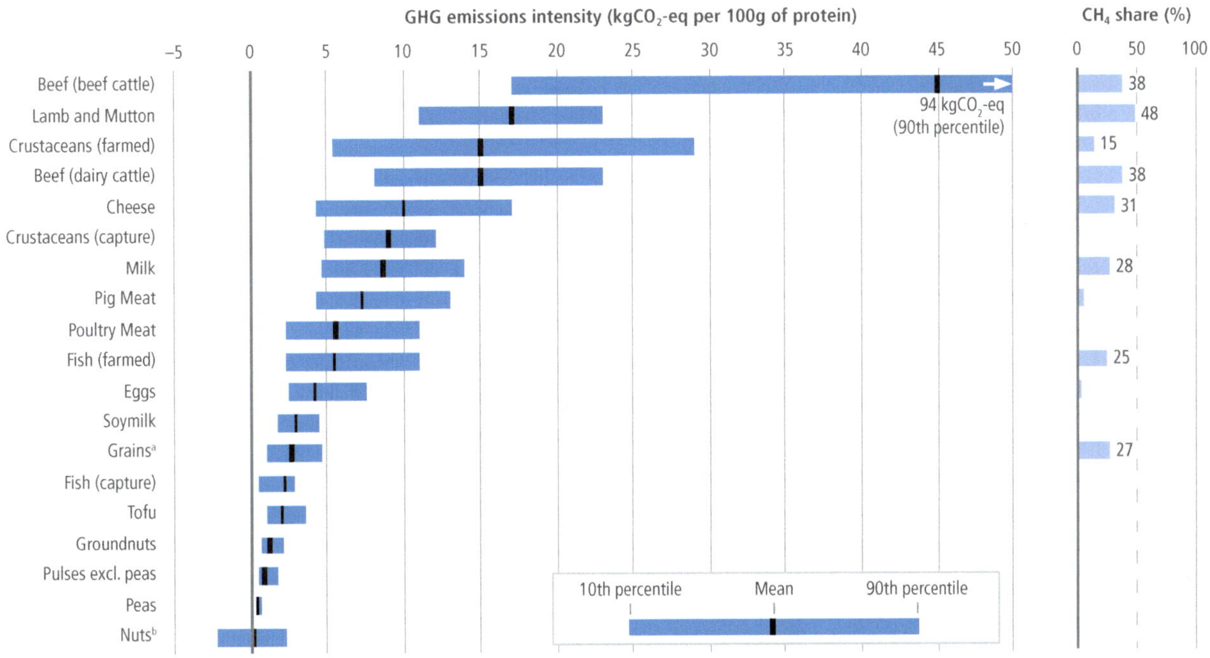


**Figure 12.6 | Ranges of GHG intensities [kg$CO_2$-eq per 100 g protein, 10–90th percentile] in protein-rich foods, quantified via a meta-analysis of attributional lifecycle assessment studies using economic allocation.** Aggregation of $CO_2$, $CH_4$, and $N_2O$ emissions in Poore and Nemecek (2018) updated to use AR6 100-year GWP. Data for capture fish, crustaceans, and cephalopods from Parker et al. (2018), with post-farm data from Poore and Nemecek (2018), where the ranges represent differences across species groups. $CH_4$ emissions include emissions from manure management, enteric fermentation, and flooded rice only. [a] Grains are not generally classed as protein-rich, but they provide about 41% of global protein intake. Here grains are a weighted average of wheat, maize, oats, and rice by global protein intake. [b] Conversion of annual to perennial crops can lead to carbon sequestration in woody biomass and soil, shown as negative emissions intensity. Source: data from Poore and Nemecek (2018); Parker et al. (2018).

specific (Meijaard et al. 2020) and for plant-based proteins, GHG footprints per serving remain lower than those of animal source proteins (Kim et al. 2019).

In traditional production systems, especially in developing countries, livestock serve multiple functions, providing draught power, fertiliser, investment and social status, besides constituting an important source of nutrients (Weiler et al. 2014). In landscapes dominated by forests or cropland, semi-natural pastures grazed by ruminants provide heterogeneity that supports biodiversity (Röös et al. 2016). Grazing on marginal land and the use of crop residues and food waste can provide human-edible food with lower demands for cropland (Röös et al. 2016; Van Zanten et al. 2018; Van Hal et al. 2019). Animal protein requires more land than vegetable protein, so switching consumption from animal to vegetable proteins could reduce the pressure on land resources and potentially enable additional mitigation through expansion of natural ecosystems, storing carbon while supporting biodiversity, or reforestation to sequester carbon and enhance wood supply capacity for the production of bio-based products substituting fossil fuels, plastics, cement, etc. (Schmidinger and Stehfest 2012; Searchinger et al. 2018b; Hayek et al. 2021). At the same time, alternatives to animal-based meat and other livestock products are being developed (Figure 12.6). Their increasing visibility in supermarkets and catering services, as well as falling production prices, could make meat substitutes competitive in one to two decades (Gerhardt et al. 2019). However, uncertainty around their uptake creates uncertainty around their effect on future GHG emissions.

#### 12.4.2.3 Territorial National Per Capita GHG Emissions from Food Systems

Food systems are connected to other societal systems, such as the energy system, financial system, and transport system (Leip et al. 2021). Also, food systems are dynamic and continuously changing and adapting to existing and anticipated future conditions. Food production systems are very diverse and vary by farm size, intensity level, farm specialisation, technological level, production methods (e.g., organic, conventional, etc.), with differing environmental and social consequences (Václavík et al. 2013; Fanzo 2017; Herrero et al. 2017; Herrero et al. 2021).

Various frameworks have been proposed to assess sustainability of food systems, including metrics and indicators on environmental, health, economic and equity issues, pointing to the importance of recognising the multi-dimensionality of food system outcomes (Gustafson et al. 2016; Chaudhary et al. 2018; Hallström et al. 2018; Zurek et al. 2018; Eme et al. 2019; Béné et al. 2020; Hebinck et al. 2021). Data platforms are being developed, but so far comprehensive data for evidence-based food system policy are lacking (Fanzo et al. 2020).

To visualise several food systems dimensions in a GHG context, Figure 12.7 shows GHG emissions per capita and year for regional country aggregates (Crippa et al. 2021a; Crippa et al. 2021b), indicated by the size of the bubbles. The GHG emissions presented here are based on territorial accounting similar to the UNFCCC GHG

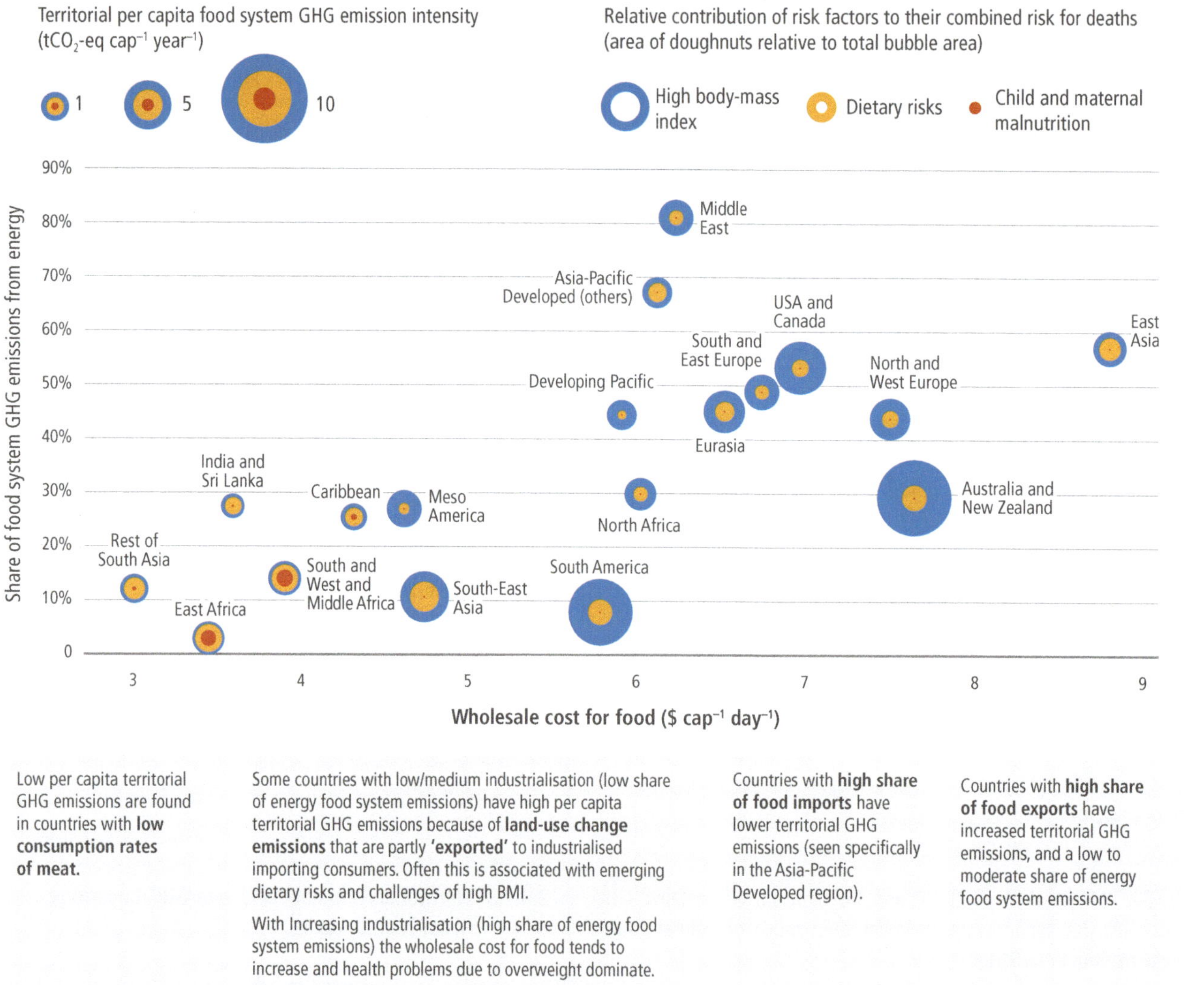


**Figure 12.7 | Regional differences in health outcomes, territorial per capita GHG emissions from national food systems, and share of food system GHG emissions from energy use.** GHG emissions are calculated according to the IPCC Tier 1 approach and are assigned to the country where they occur, not necessarily where the food is consumed. Health outcome is expressed as relative contribution of each of the following risk factors to their combined risk for deaths: child and maternal malnutrition (red), dietary risks (yellow) or high body mass index (blue). Sources: wholesale cost of food per capita: Springmann et al. 2021); territorial food system GHG emissions: EDGAR v.6, Crippa et al. (2021a), recalculated according to Crippa et al. (2021b) using AR6 GWPs; deaths attributed to dietary factors: IHME (2018); GBD 2017 Diet Collaborators et al. (2019).



inventories: emissions are assigned to the country where they occur, not where food is consumed (Crippa et al. 2021a; Crippa et al. 2021b) (Section 12.4.2.1). The colours of the bubbles indicate the relative contribution of the following risk factors to deaths, according to the classification used in the Global Burden of Disease Study: child and maternal malnutrition (red, deficiencies of iron, zinc or Vitamin A, or low birth weight or child growth failure), dietary risks (yellow, for example diets low in vegetables, legumes, whole grains or diets high in red and processed meat and sugar-sweetened beverages) or high body mass index (blue). The combined contribution of these three risk factors to total deaths varies strongly and is between 28% and 88% of total deaths. Figure 12.7 shows that dietary risk factors are prevalent throughout all regions. Though not a complete measure of the health impact of food, these were selected as a proxy for nutritional adequacy and balance of diets, avoidance of food insecurity, over- or mal-nutrition and associated non-communicable diseases (GBD 2017 Diet Collaborators 2018; GBD 2017 Diet Collaborators et al. 2019).

The share of GHG emissions from energy use is taken as a proxy for the structure of food supply in a region (Section 12.4.1), and the cost for food as a proxy for the structure of the demand side and the access to (healthy) food (Chen et al. 2016; Finaret and Masters 2019; Hirvonen et al. 2019; HLPE 2020; Springmann et al. 2021), though acknowledging the limitations of such a simplification.

While total food system emissions in 2018 range between 0.9 and 8.5 $tCO_2$-eq per capita per year between regions, the share of energy emissions relative to energy and land-based (agriculture and food system land-use change) emissions ranges between 3% and 78%. Regional expenditures for food range from USD3.0–8.8 per capita per day (Figure 12.7), though there is high variability within countries and the costs of nutrient-adequate diets often exceeds those of diets delivering adequate energy (Hirvonen et al. 2019; Bai et al. 2020; FAO et al. 2020). Thus, low-income households in industrialised countries can also be affected by food insecurity (Penne and Goedemé 2020).

### 12.4.3 Mitigation Opportunities

GHG emissions from food systems can be reduced by targeting direct or indirect GHG emissions in the supply chain including enhanced carbon sequestration, by introducing sustainable production methods such as agroecological approaches which can reduce system-level GHG emissions of conventional food production and also enhance resilience (HLPE 2019), by substituting food products with high GHG intensities with others of lower GHG intensities, by reducing food over-consumption, and/or by reducing food loss and waste. The substitution of food products with others that are more sustainable and/or healthier is often called 'dietary shift'.

Clark et al. (2020) showed that even if fossil fuel emissions were eliminated immediately, food system emissions alone would jeopardise the achievement of the 1.5°C target and threaten the 2°C target. They concluded that both demand-side and supply-side strategies are needed, including a shift to a diet with lower GHG intensity and rich in plant-based 'conventional' foods (e.g., pulses, nuts), or new food products that could support dietary shift. Such dietary shift needs to overcome socio-cultural, knowledge, and economic barriers to significantly achieve GHG mitigation (Section 12.4.5).

Food losses occur at the farm, post-harvest and during the food processing/wholesale stages of a food supply chain, while in the final retail and consumption stages the term food waste is used (HLPE 2014). Typically, food losses are linked to technical issues such as lack of infrastructure and storage, while food waste is often caused by socio-economic and behavioural factors. Mitigation opportunities through reducing food waste and loss exist in all food supply chain stages and are described in the sub-sections below.

Food system mitigation opportunities are divided into five categories as given in Table 12.8:

- Food production from agriculture, aquaculture, and fisheries (Chapter 7.4 and Section 12.4.3.1)
- Controlled-environment agriculture (Section 12.4.3.2)
- Emerging food production technologies (Section 12.4.3.3)
- Food processing industries (Section 12.4.3.4)
- Storage and distribution (Section 12.4.3.5)

Food system mitigation opportunities can be either incremental or transformative (Kugelberg et al. 2021). Incremental options are based on mature technologies, for which processes and causalities are understood, and their implementation is generally accepted by society. They do not require a substantial change in the way food is produced, processed, or consumed and might lead to a (slight) shift in production systems or preferences. Transformative mitigation opportunities have wider food system implications and usually coincide with a significant change in food choices. They are based on technologies that are not yet mature and are expected to require further innovation (Klerkx and Rose 2020), and/or mature technologies that might already be part of some food systems but are not yet widely accepted and have transformative potential if applied at large scale, for example consumption of insects (Raheem et al. 2019a). Many emerging technologies might be seen as a further step in agronomic development where land-intensive production methods relying on the availability of naturally-available nutrients and water are successively replaced with crop variants and cultivation practices reducing these dependencies at the cost of larger energy input (Winiwarter et al. 2014). Others suggest a shift to agroecological approaches combining new scientific insights with local knowledge and cultural values (HLPE 2019). Food system transformation can lead to regime shifts or (fast) disruptions (Pereira et al. 2020) if driven by events that are out of control of private or public measures and have a 'crisis' character (e.g., BSE) (Skuce et al. 2013).

Table 12.8 summarises the main characteristics of food system mitigation opportunities, their effect on GHG emissions, and associated co-benefits and adverse effects.

Agricultural food production systems range from smallholder subsistence farms to large animal production factories, in open spaces, greenhouses, rural areas or urban settings.

**Dietary shift:** Studies demonstrate that a shift to diets rich in plant-based foods, particularly pulses, nuts, fruits and vegetables, such as vegetarian, pescatarian or vegan diets, could lead to substantial reduction of greenhouse gas emissions as compared to current dietary patterns in most industrialised countries, while also providing health benefits and reducing mortality from diet-related non-communicable diseases (Springmann et al. 2018a; Chen et al. 2019; Willett et al. 2019; Bodirsky et al. 2020; Costa Leite et al. 2020; Ernstoff et al. 2020; Jarmul et al. 2020; Semba et al. 2020; Theurl et al. 2020; Hamilton et al. 2021).

Pulses such as beans, chickpeas, or lentils, have a protein composition complementary to cereals, providing together all essential amino acids (Foyer et al. 2016; McDermott and Wyatt 2017). Bio-availability of proteins in foods is influenced by several factors, including amino acid composition, presence of anti-nutritional factors, and preparation method (Hertzler et al. 2020; Weindl et al. 2020; Semba et al. 2021). Soy beans, in particular, have a well-balanced amino acid profile with high bio-availability (Leinonen et al. 2019). Pulses are part of most traditional diets (Semba et al. 2021) and supply up to 10–35% of protein in low-income countries, but consumption decreases with increasing income and they are globally only a minor share of the diet (McDermott and Wyatt 2017). Pulses play a key role in crop rotations, fixing nitrogen and breaking disease cycles, but yields of pulses are relatively low and have seen small yield increases relative to those of cereals (Foyer et al. 2016; McDermott and Wyatt 2017; Barbieri et al. 2021; Semba et al. 2021).

**Technological innovations:** have made food production more efficient since the onset of agriculture (Winiwarter et al. 2014; Herrero et al. 2020). Emerging technologies include digital agriculture

Table 12.8 | Food system mitigation opportunities.

| Food system mitigation options<br>I: incremental; T: transformative | | | Direct and indirect effect on GHG mitigation<br>D: direct emissions except emissions from energy use; E: energy demand; M: material demand; FL: food losses; FW: food waste<br>Direction of effect on GHG mitigation: + increased mitigation; 0 neutral; – decreased mitigation | Co-benefits/adverse effects<br>H: health aspects; A: animal welfare; R: resource use; L: land demand; E: ecosystem services; 0: neutral<br>+ co-benefits; – adverse effects | Source |
|---|---|---|---|---|---|
| Food from agriculture, aquaculture and fisheries | (I) | Dietary shift, in particular increased share of plant-based protein sources | D+ ↓ GHG footprint | A+ Animal welfare<br>L+ Land sparing<br>H+ Good nutritional properties, potentially ↓ risk from zoonotic diseases, pesticides and antibiotics | 1–5 |
| | (I/T) | Digital agriculture | D+ ↑ Logistics | L+ Land sparing<br>R+ ↑ Resource use efficiencies | 6–7 |
| | (T) | Gene technology | D+ ↑ Productivity or efficiency | H+ ↑ Nutritional quality<br>E0 ↓ Use of agrochemicals; ↑ probability of off-target impacts | 7–11 |
| | (I) | Sustainable intensification, Land-use optimisation | D+ ↓ GHG footprint<br>E0 Mixed effects | L+ Land sparing<br>R– Might ↑ pollution/biodiversity loss | 7, 12 |
| | (I) | Agroecology | D+ ↓ GHG/area, positive micro-climatic effects<br>E+ ↓ Energy, possibly ↓ transport<br>FL+ Circular approaches | E+ Focus on co-benefits/ecosystem services<br>R+ Circular, ↑ nutrient and water use efficiencies | 13–17 |
| Controlled-environment agriculture | (T) | Soilless agriculture | D+ ↑ productivity, weather independent<br>FL+ harvest on demand<br>E- Currently ↑ energy demand, but ↓ transport, building spaces can be used for renewable energy | R+ Controlled loops ↑ nutrient and water use efficiency<br>L+ Land sparing<br>H+ Crop breeding can be optimised for taste and/or nutritional quality | 18–24 |
| Emerging food production technologies | (T) | Insects | D0 Good feed conversion efficiency<br>FW+ Can be fed on food waste | H0 Good nutritional qualities but attention to allergies and food safety issues required | 25–28 |
| | (I/T) | Algae and bivalves | D+ ↓ GHG footprints | A+ Animal welfare<br>L+ Land sparing<br>H+ Good nutritional qualities; risk of heavy metal and pathogen contamination<br>R+ Biofiltration of nutrient-polluted waters | 29–32 |
| | (I/T) | Plant-based alternatives to animal-based food products | D+ No emissions from animals, ↓ inputs for feed | A+ Animal welfare<br>L+ Land sparing<br>H+ Potentially ↓ risk from zoonotic diseases, pesticides and antibiotics; but ↑ processing demand | 31–33 |
| | (T) | Cellular agriculture (including cultured meat, microbial protein) | D+ No emissions from animals, high protein conversion efficiency<br>E– ↑ Energy need<br>FLW+ ↓ Food loss and waste | A+ Animal welfare<br>R+ ↓ Emissions of reactive nitrogen or other pollutants<br>H0 Potentially ↓ risk from zoonotic diseases, pesticides and antibiotics; ↑ research on safety aspects needed | 3, 24<br>34–42 |

| Food system mitigation options<br>I: incremental; T: transformative | | | Direct and indirect effect on GHG mitigation<br>D: direct emissions except emissions from energy use; E: energy demand; M: material demand; FL: food losses; FW: food waste<br>Direction of effect on GHG mitigation: + increased mitigation; 0 neutral; – decreased mitigation | Co-benefits/adverse effects<br>H: health aspects; A: animal welfare; R: resource use; L: land demand; E: ecosystem services; 0: neutral<br>+ co-benefits; – adverse effects | Source |
|---|---|---|---|---|---|
| Food processing and packaging | (I) | Valorisation of by-products, food loss and waste logistics and management | M+ Substitution of bio-based materials<br>FL+ ↓ of food losses | | 43–44 |
| | (I) | Food conservation | FW+ ↓ Food waste<br>EC ↑ energy demand but also energy savings possible (e.g., refrigeration, transport) | | 45–46 |
| | (I) | Smart packaging and other technologies | FW+ ↓ Food waste<br>M0 ↑ Material demand and ↑ material-efficiency<br>E0 ↑ Energy demand; energy savings possible | H+ Possibly ↑ freshness/reduced food safety risks | 46–49 |
| | (I) | Energy efficiency | E+ ↓ Energy | | 50 |
| Storage and distribution | (I) | Improved logistics | D+ ↓ Transport emissions<br>FL+ ↓ Losses in transport<br>FW– Easier access to food could ↑ food waste | | 46–47<br>51–53 |
| | (I) | Specific measures to reduce food waste in retail and food catering | FW+ ↓ Food waste<br>E+ ↓ Downstream energy demand<br>M+ ↓ Downstream material demand | | 54–56 |
| | (I) | Alternative fuels/ transport modes | D+ ↓ Emissions from transport | | |
| | (I) | Energy efficiency | E+ ↓ Energy in refrigeration, lightening, climatisation | | 57–58 |
| | (I) | Replacing refrigerants | D+ ↓ Emissions from the cold chain | | 50<br>59–60 |

Sources: [1] McDermott and Wyatt (2017); [2] Foyer et al. (2016); [3] Semba et al. (2021); [4] Weindl et al. (2020); [5] Hertzler et al. (2020); [6] Finger et al. (2019); [7] Herrero et al. (2020); [8] Steinwand and Ronald (2020); [9] Zhang et al. (2020a); [10] Ansari et al. (2020); [11] Eckerstorfer et al. (2021); [12] Folberth et al. (2020); [13] HLPE (2019); [14] Wezel et al. (2009); [15] Van Zanten et al. (2018); [16] Van Zanten et al. (2019); [17] van Hal et al. (2019); [18] Beacham et al. (2019); [19] Benke and Tomkins (2017); [20] Gómez and Gennaro Izzo (2018); [21] Maucieri et al. (2018); [22] Rufí-Salís et al. (2020); [23] Shamshiri et al. (2018); [24] Graamans et al. (2018); [25] Fasolin et al. (2019); [26] Garofalo et al. (2019); [27] Parodi et al. (2018); [28] Varelas (2019); [29] Gentry et al. (2020); [30] Peñalver et al. (2020); [31] Torres-Tiji et al. (2020); [32] Willer and Aldridge (2020); [33] Fresán et al. (2019); [34] Mejia et al. (2019); [35] Tuomisto (2019); [36] Thorrez and Vandenburgh (2019); [37] Tuomisto and Teixeira de Mattos (2011); [38] Mattick et al. (2015); [39] Mattick (2018); [40] Souza Filho et al. (2019); [41] Chriki and Hocquette (2020); [42] Hadi and Brightwell (2021); [43] Göbel et al. (2015); [44] Caldeira et al. (2020); [45] Silva and Sanjuán (2019); [46] FAO (2019a); [47] Molina-Besch et al. (2019); [48] Poyatos-Racionero et al. (2018); [49] Müller and Schmid (2019); [50] Niles et al. (2018); [51] Lindh et al. (2016); [52] Wohner et al. (2019); [53] Bajželj et al. (2020); [54] Buisman et al. (2019); [55] Albizzati et al. (2019); [56] Liu et al. (2016); [57] Chaomuang et al. (2017); [58] Lemma et al. (2014); [59] McLinden et al. (2017); [60] Gullo et al. (2017). Food from Agriculture, Aquaculture, and Fisheries.

(using advanced sensors, big data), gene technology (crop bio-fortification, genome editing, crop innovations), sustainable intensification (automation of processes, improved inputs, precision agriculture) (Herrero et al. 2020), or multi-trophic aquaculture approaches (Knowler et al. 2020; Sanz-Lazaro and Sanchez-Jerez 2020), though literature on aquaculture and fisheries in the context of GHG mitigation is limited.

Such technologies may contribute to a reduction of GHG emissions at the food system level, enhanced provision of food, better consideration of ecosystem services, and/or contribute to nutrition-sensitive agriculture, for example, by increasing the nutritional quality of staple crops, increasing the palatability of leguminous crops such as lupines, or increasing the agronomic efficiency or resilience of crops with good nutritional characteristics.

For details on agricultural mitigation opportunities refer to Section 7.4.

#### 12.4.3.1 Controlled-environment Agriculture

Controlled-environment agriculture is mainly based on hydroponic or aquaponic cultivation systems that do not require soil. Aquaponics combine hydroponics with a re-circulating aquaculture compartment for integrated production of plants and fish (Junge et al. 2017; Maucieri et al. 2018), while aeroponics is a further development of hydroponics that replaces water as a growing medium with a mist of nutrient solution (Al-Kodmany 2018). Aquaponics could potentially produce proteins in urban farms, but the technology is not yet mature and its economic and environmental performance is unclear (Love et al. 2015; O'Sullivan et al. 2019).

Controlled-environment agriculture is often undertaken in urban environments to take advantage of short supply chains (O'Sullivan et al. 2019), and might use abandoned buildings or be integrated in supermarkets, producing for example herbs 'on demand'.

Optimising growing conditions, hydroponic systems achieve higher yields than un-conditioned agriculture (O'Sullivan et al. 2019); and yields can be further enhanced in $CO_2$-enriched atmospheres (Shamshiri et al. 2018; Armanda et al. 2019). By using existing spaces or modular systems that can be vertically stacked, this technology minimises land demand, however it is energy intensive and requires large financial investments. So far, only a few crops are commercially produced in vertical farms, including lettuce and other leafy greens, herbs and some vegetables, due to their short growth period and high value (Benke and Tomkins 2017; Armanda et al. 2019; Beacham et al. 2019; O'Sullivan et al. 2019). Through breeding, other crops could reach commercial feasibility, or crops with improved taste or nutritional characteristics can be grown (O'Sullivan et al. 2019).

In controlled-environment agriculture, photosynthesis is fuelled by artificial light through LEDs or a combination of natural light with LEDs. Control of the wave band and light cycle of the LEDs and micro-climate can be used to optimise photosynthetic activity, yield and crop quality (Gómez and Gennaro Izzo 2018; Shamshiri et al. 2018).

Co-benefits of controlled-environment agriculture include minimising water and nutrient losses as well as agro-chemical use (Al-Kodmany 2018; Shamshiri et al. 2018; Armanda et al. 2019; Farfan et al. 2019; O'Sullivan et al. 2019; Rufí-Salís et al. 2020) (*robust evidence, high agreement*). Water is recycled in a closed system and additionally some plants generate fresh water by evaporation from grey or black water, and high nutrient use efficiencies are possible. Food production from controlled-environment agriculture is independent of weather conditions and able to satisfy some consumer demand for locally-produced fresh and diverse produce throughout the year (Benke and Tomkins 2017; Al-Kodmany 2018; O'Sullivan et al. 2019).

Controlled-environment agriculture is a very energy intensive technology (mainly for cooling) and its GHG intensity depends therefore crucially on the source of the energy. Options for reducing GHG intensity include reducing energy use through improved lighting and cooling efficiency or by employing low-carbon energy sources, potentially integrated into the building structure (Benke and Tomkins 2017).

Comprehensive studies assessing the GHG balance of controlled-environment agriculture are lacking. The overall GHG emissions from controlled-environment agriculture is therefore uncertain and depends on the balance of reduced GHG emissions from production and distribution and reduced land requirements, versus increased external energy needs.

#### 12.4.3.2 Emerging Foods and Production Technologies

A diverse range of novel food products and production systems are emerging, that are proposed to reduce GHG emissions from food production, mainly by replacing conventional animal-source food with alternative protein sources. Assessments of the potential of dietary changes are given in Sections 5.3 and 7.4. Here, we assess the GHG intensities of emerging food production technologies. This includes products such as insects, algae, mussels and products from bio-refineries, some of which have been consumed in certain societies and/or in smaller quantities (Pikaar et al. 2018; Jönsson et al. 2019; Govorushko 2019; Raheem et al. 2019a; Souza Filho et al. 2019). The novel aspect considered here is the scale at which they are proposed to replace conventional food with the aim to reduce both negative health and environmental impacts. To fully realise the health benefits, dietary shifts should also encompass a reduction in consumption of added sugars, salt, saturated fats, and potentially harmful additives (Curtain and Grafenauer 2019; Fardet and Rock 2019; Petersen et al. 2021).

Meat analogues have attracted substantial venture capital, and production costs have dropped considerably in the last decade, with some reaching market maturity (Mouat and Prince 2018; Santo et al. 2020), but there is uncertainty whether they will 'disrupt' the food market or remain niche products. According to Kumar et al. (2017), the demand for plant-based meat analogues is expected to increase as their production is relatively cheap and they satisfy consumer demands with regard to health and environmental concerns as well as ethical and religious requirements. Consumer acceptance is still low for some options, especially insects (Aiking and de Boer 2019) and cultured meat (Chriki and Hocquette 2020; Siegrist and Hartmann 2020).

**Insects:** Farmed edible insects have a higher feed conversion ratio than other animals farmed for food, and have short reproduction periods with high biomass production rates (Halloran et al. 2016). Insects have good nutritional qualities (Parodi et al. 2018). They are suited as a protein source for both humans and livestock, with high protein content and favourable fatty acid composition (Fasolin et al. 2019; Raheem et al. 2019b). If used as feed, they can grow on food waste and manure; if used as food, food safety concerns and regulations can restrict the use of manure (Raheem et al. 2019b) or food waste (Varelas 2019) as growing substrates, and the dangers of pathogenic or toxigenic microorganisms and incidences of anti-microbial resistance need to be managed (Garofalo et al. 2019).

**Algae and bivalves** have a high protein content and a favourable nutrient profile and can play a role in providing sustainable food. Bivalves are high in omega-3 fatty acids and vitamin B12 and therefore well suited as replacement of conventional meats, and have a lower GHG footprint (Parodi et al. 2018; Willer and Aldridge 2020). Micro- and macro algae are rich in omega-3 and omega-6 fatty acids, anti-oxidants and vitamins (Parodi et al. 2018; Peñalver et al. 2020; Torres-Tiji et al. 2020). Kim et al. (2019) show that diets with modest amounts of animals low on the food chain such as forage fish, bivalves, or insects have similar GHG intensities to vegan diets. Algae and bi-valves can be used to filter nutrients from waters, though care is required to avoid accumulation of hazardous substances (Gentry et al. 2020; Willer and Aldridge 2020).

**Plant-based meat, milk and egg analogues:** Demand for plant-based proteins is increasing and incentivising the development of protein crop varieties with improved agronomic performance and/or nutritional quality (Santo et al. 2020). There is also an emerging market for meat replacements based on plant proteins, such as pulses, cereals, soya, algae and other ingredients mainly used to imitate the taste, texture and nutritional profiles of animal-source food (Kumar et al. 2017; Boukid 2021). Currently, the majority of plant-based meat analogues is based on soy (Semba et al. 2021). While other products still serve a niche market, their share is growing rapidly and some studies project a sizeable share within a decade (Kumar et al. 2017; Jönsson et al. 2019). In particular, plant-based milk alternatives have seen large increases in market share (Jönsson et al. 2019). A LCA of 56 plant-based meat analogues showed mean GHG intensities (farm to factory) of 0.21–0.23 $kgCO_2$-eq per 100 g of product or 20 g of protein for all assessed protein sources (Fresán et al. 2019). Higher footprints were found in the meta-review by Santo et al. (2020). Including preparation, Meija et al. (2019) found higher emissions for burgers and sausages as compared to minced products.

**Cellular agriculture:** The use of fungi, algae and bacteria is an old process (beer, bread, yoghurt) and serves, among others, for the preservation of products. The concept of cellular agriculture (Mattick 2018) covers bio-technological processes that use micro-organisms to produce acellular (fermentation-based cellular agriculture) or cellular products. Yeasts, fungi or bacteria can synthesise acellular products such as haem, milk and egg proteins, or protein-rich animal feed, other food ingredients, and pharmaceutical and material products (Rischer et al. 2020; Mendly-Zambo et al. 2021). Cellular products include cell tissues such as muscle cells to grow cultured meat, fish or other cells (Post 2012; Rischer et al. 2020) and products where the micro-organisms will be eaten themselves (Pikaar et al. 2018; Sillman et al. 2019; Schade et al. 2020). Single cell proteins, combined with photovoltaic electricity generation and direct air capture of carbon dioxide, are proposed as highly land- and energy-efficient alternatives to plant-based protein (Leger et al. 2021). Some microbial proteins are produced in a 'bioreactor' and use Haber-Bosch nitrogen and vegetable sugars or atmospheric $CO_2$ as source of nitrogen and carbon (Pikaar et al. 2018; Simsa et al. 2019). Cultured meat is currently at the research stage and some challenges remain, such as the need for animal-based ingredients to ensure fast and effective growth of muscle cells; tissue engineering to create different meat products; production at scale and at competitive costs; and regulatory barriers (Post 2012; Stephens et al. 2018; Rubio et al. 2019; Tuomisto 2019; Post et al. 2020). Only a few studies to date have quantified the GHG emissions of microbial proteins or cultured meat, suggesting GHG emissions at the level of poultry meat (Tuomisto and Teixeira de Mattos 2011; Mattick et al. 2015; Souza Filho et al. 2019; Tuomisto 2019).

A review of LCA studies on different plant-based, animal source and nine 'future food' protein sources (Parodi et al. 2018) concluded that insects, macro-algae, mussels, mycoproteins and cultured meat show similar GHG intensities per unit of protein (mean values ranging 0.3–3.1 $kgCO_2$-eq per 100 g protein), comparable to milk, eggs, and tuna (mean values ranging 1.2–5.4 $kgCO_2$-eq per 100 g protein); while *chlorella* and *spirulina* consume more energy per unit of protein and were associated with higher GHG emissions (mean values ranging 11–13 $kgCO_2$-eq per 100 g protein). As the main source of GHG emissions from insects and cellular agriculture foods is energy consumption, their GHG intensity improves with increased use of low-carbon energy (Smetana et al. 2015; Parodi et al. 2018; Pikaar et al. 2018).

Future foods offer other benefits such as lower land requirements, controlled systems with reduced losses of water and nutrients, increased resilience, and possibly reduced hazards from pesticide and antibiotics use and zoonotic diseases, although more research is needed including on allergenic and other safety aspects, and possibly reduced protein bioavailability (Alexander et al. 2017; Parodi et al. 2018; Stephens et al. 2018; Fasolin et al. 2019; Chriki and Hocquette 2020; Santo et al. 2020; Hadi and Brightwell 2021; Tzachor et al. 2021) (*medium evidence, high agreement*). Research is needed also on the effect of processing (Wickramasinghe et al. 2021), though a randomised crossover trial comparing appetising plant foods with meat alternatives found several beneficial and no adverse effects from the consumption of the plant-based meats (Crimarco et al. 2020).

#### 12.4.3.3 Food Processing and Packaging

Food processing includes preparation and preservation of fresh commodities (fruit and vegetables, meat, seafood and dairy products), grain milling, production of baked goods, and manufacture of pre-prepared foods and meals. Food processors range from small local operations to large multinational food producers, producing food for local to global markets. The importance of food processing and preservation is particularly evident in developing countries which lack cold chains for the preservation and distribution of fresh perishable products such as fresh fish (Adeyeye and Oyewole 2016; Adeyeye 2017).

Mitigation in food processing largely focuses on reducing food waste and fossil energy usage during the processing itself, as well as in the transport, packaging and storage of food products for distribution and sale (Silva and Sanjuán 2019). Reducing food waste provides emissions savings by reducing wastage of primary inputs required for food production. Another mitigation route, contributing to the circular bioeconomy (Section 12.6.1.2 and Cross-Working Group Box 3 in this chapter), is by valorisation of food processing by-products through recovery of nutrients and/or energy. No global analyses of the emissions savings potential from the processing step in the value chain could be found.

Reduced food waste during food processing can be achieved by seeking alternative processing routes (Atuonwu et al. 2018), improved communication along the food value chain (Göbel et al. 2015), optimisation of food processing facilities, reducing contamination, and limiting damages and spillage (HLPE 2014). Optimisation of food packaging also plays an important role in reducing food waste, in that it can extend product shelf life; protect against damage during transport and handling; prevent spoilage; facilitate easy opening and emptying; and communicate storage and preparation information to consumers (Molina-Besch et al. 2019).

Developments in smart packaging are increasingly contributing to reducing food waste along the food value chain. Strategies for reducing the environmental impact of packaging include using less, and more sustainable, materials and a shift to reusable packaging (Coelho et al. 2020). Active packaging increases shelf life through regulating the environment inside the packaging, including levels of oxygen, moisture and chemicals released as the food ages (Emanuel and Sandhu 2019). Intelligent packaging communicates information on the freshness of the food through indicator labels (Poyatos-Racionero et al. 2018), and data carriers can store information on conditions such as temperature along the entire food chain (Müller and Schmid 2019).

LCA can be used to evaluate the benefits and trade-offs associated with different processing or packaging types (Silva and Sanjuán 2019). Some options, such as aluminium, steel and glass, require high energy investment in manufacture when produced from primary materials, with significant savings in energy through recycling being possible (Camaratta et al. 2020). However, these materials are inert in landfill. Other packaging options, such as paper and biodegradable packaging, may require a lower energy investment during manufacture, but may require larger land area and can release methane when consigned to anaerobic landfill where there is no methane recovery. Nevertheless, packaging accounts for only 1–12% (typically around 5%) of the GHG emissions in the lifecycle of a food system (Wohner et al. 2019; Crippa et al. 2021b), suggesting that its benefits can often outweigh the emissions associated with the packaging itself.

The second component of mitigation in food processing relates to reduction in fossil energy use. Opportunities include energy efficiency in processes (also discussed in Section 11.3), the use of heat and electricity from low-carbon energy sources in processing (Chapter 6), through off-grid thermal processing (sun drying, food smoking) and improving logistics efficiencies. Energy-intensive processes with energy-saving potential include milling and refining (oil seeds, corn, sugar), drying, and food safety practices such as sterilisation and pasteurisation (Niles et al. 2018). Packaging also plays a role: reduced transport energy can be achieved through reducing the mass of goods transported and improving packing densities in transport vehicles (Lindh et al. 2016; Molina-Besch et al. 2019; Wohner et al. 2019). Choice of packaging also influences refrigeration energy requirements during transport and storage.

#### 12.4.3.4 Storage and Distribution

Transport mitigation options along the supply chain include improved logistics, the use of alternative fuels and transport modes, and reduced transport distances. Logistics and alternative fuels and transport modes are discussed in Chapter 10. Transport emissions might increase with increasing demand for a diversity of foods as developing countries become more affluent. New technologies that enable food on demand or online food shopping systems might further increase emissions from food transport; however, the consequences are uncertain and might also entail a shift from individual traffic to bulk transport. The impact on food waste is also uncertain as more targeted delivery options could reduce food waste, but easier access to a wider range of food could also foster over-supply and increase food waste. Mitigation opportunities in food transport are inherently linked to decarbonisation of the transport sector (Chapter 10).

Retail and the food service industry are the main factors shaping the external food environment or 'food entry points'; they are the 'physical spaces where food is obtained; the built environment that allows consumers to access these spaces' (HLPE 2017). These industries have significant influence on consumers' choices and can play a role in reducing GHG emissions from food systems. Opportunities are available for optimisation of inventories in response to consumer demands through advanced IT systems (Niles et al. 2018), and for discounting foods close to sell-by dates, which can serve to reduce both food spoilage and wastage (Buisman et al. 2019).

As one of the highest contributors to energy demand at this stage in the food value chain, refrigeration has received a strong focus in mitigation. Efficient refrigeration options include advanced refrigeration temperature control systems, and installation of more efficient refrigerators, air curtains and closed display fridges (Chaomuang et al. 2017). Also related to reducing emissions from cooling and refrigeration is the replacement of hydrofluorocarbons which have very high GWPs with lower GWP alternatives (Niles et al. 2018). The use of propane, isobutane, ammonia, hydrofluoroolefins and $CO_2$ (refrigerant R744) are among those that are being explored, with varying success (McLinden et al. 2017). In recent years, due to restrictions on high GWP-refrigerants, a considerable growth in the market availability of appliances and systems with non-fluorinated refrigerants has been seen (Eckert et al. 2021).

Energy efficiency alternatives generic to buildings more broadly are also relevant here, including efficient lighting, heating, ventilation, and air conditioning systems and building management, with ventilation being a particularly high energy user in retail, that warrants attention (Kolokotroni et al. 2015).

In developing countries particularly, better infrastructure for transportation and expansion of processing and manufacturing industries can significantly reduce food losses, particularly of highly perishable food (Niles et al. 2018; FAO 2019a).

### 12.4.4 Enabling Food System Transformation

Food system mitigation potentials in AFOLU are assessed in Section 7.4, and food system mitigation potentials linked to demand-side measures are assessed in Chapter 5. Studies suggest that implementing supply- and demand-side policies in combination makes ambitious mitigation targets easier to achieve (Clark et al. 2020; Global Panel on Agriculture and Food Systems for Nutrition 2020; Temme et al. 2020; Latka et al. 2021a) (*high agreement*, *limited evidence*).

**Table 12.9 | Assessment of food system policies targeting (post-farm gate) food chain actors and consumers.**

| | Level<br>G: global/multinational;<br>N: national; L: local | Transformative potential | Environmental effectiveness | Feasibility | Distributional effects | Cost | Co-benefits[a] and adverse side effect | Implications for coordination, coherence and consistency in policy package[b] |
|---|---|---|---|---|---|---|---|---|
| **Integrated food policy packages** | **NL** | | | | can be controlled | cost efficient | + balanced, addresses multiple sustainability goals | Reduces cost of uncoordinated interventions; increases acceptance across stakeholders and civil society (*robust evidence*, *high agreement*) |
| Taxes on food products | GN | | | | regressive | low[#1] | – unintended substitution effects | High enforcing effect on other food policies; higher acceptance if compensation or hypothecated taxes (*medium evidence*, *high agreement*) |
| GHG taxes on food | GN | | | | regressive | low[#2] | – unintended substitution effects<br>+ high spillover effect | Supportive, enabling effect on other food policies, agricultural/fishery policies; requires changes in power distribution and trade agreements (*medium evidence*, *medium agreement*) |
| Trade policies | G | | | | impacts global distribution | complex effects | + counters leakage effects<br>+/– effects on market structure and jobs | Requires changes in existing trade agreements (*medium evidence*, *high agreement*) |
| Investment into research and innovation | GN | | | | none | medium | + high spillover effect<br>+ converging with digital society | Can fill targeted gaps for coordinated policy packages (e.g., monitoring methods) (*robust evidence*, *high agreement*) |
| Food and marketing regulations | N | | | | | low | | Can be supportive; might be supportive to realise innovation; voluntary standards might be less effective (*medium evidence*, *medium agreement*) |
| Organisational-level procurement policies | NL | | | | | low | + can address multiple sustainability goals | Enabling effect on other food policies; reaches large share of population (*medium evidence*, *high agreement*) |
| Sustainable food-based dietary guidelines | GNL | | | | none | low | + can address multiple sustainability goals | Little attention so far on environmental aspects; can serve as benchmark for other policies (labels, food formulation standards, etc.) (*medium evidence*, *medium agreement*) |
| Food labels/ information | GNL | | | | education level relevant | low | + empowers citizens<br>+ increases awareness<br>+ multiple objectives | Effective mainly as part of a policy package; incorporation of other objectives (e.g., animal welfare, fair trade); higher effect if mandatory (*medium evidence*, *medium agreement*) |
| Nudges | NL | | | | none | low | + possibly counteracting information deficits in population subgroups | High enabling effect on other food policies (*medium evidence*, *high agreement*) |

Effect of measures: negative; none/unclear; slightly positive; positive

Notes: [#1] Minimum level to be effective 20% price increase; [#2] Minimum level to be effective USD50–80 $tCO_2$-eq. [a] In addition, all interventions are assumed to address health and climate change mitigation. [b] Requires coordination between policy areas, participation of stakeholders, transparent methods and indicators to manage trade-offs and prioritisation between possibly conflicting objectives; and suitable indicators for monitoring and evaluation against objectives.

The trends in the global and national food systems towards a globalisation of food supply chains and increasing dominance of supermarkets and large corporate food processors (Dries et al. 2004; Neven and Reardon 2004; Baker and Friel 2016; Andam et al. 2018; Popkin and Reardon 2018; Reardon et al. 2019; Pereira et al. 2020) have led to environmental, food insecurity and malnutrition problems. Studies therefore call for a transformation of current global and national food systems to solve these problems (Schösler and Boer 2018; McBey et al. 2019; Kugelberg et al. 2021). This has not yet been successful, including due to insufficient coordination between relevant food system policies (Weber et al. 2020) (*medium evidence, high agreement*).

Different elements of food systems are currently governed by separate policy areas that in most countries scarcely interact or cooperate (Termeer et al. 2018; iPES Food 2019). This compartmentalisation makes the identification of synergetic and antagonistic effects difficult and faces the possibility of failure due to unintended and unanticipated negative impacts on other policy areas and consequently lack of agreement and social acceptance (Mylona et al. 2018; Brouwer et al. 2020; Mausch et al. 2020; Hebinck et al. 2021) (Section 12.4.5). This could be overcome through cooperation across several policy areas (Sections 12.6.2 and 13.7), in particular agriculture, nutrition, health, trade, climate and environment, and an inclusive and transparent governance structure (Termeer et al. 2018; Bhunnoo 2019; Diercks et al. 2019; Herrero et al. 2021; iPES Food 2019; Mausch et al. 2020; Kugelberg et al. 2021), making use of potential spillover effects (Kanter et al. 2020; OECD 2021).

Transformation of food systems may come from technological, social or institutional innovations that start as niches but can potentially lead to rapid changes, including changes in social conventions (Centola et al. 2018; Benton et al. 2019).



Where calories and ruminant animal-source food are consumed in excess of health guidelines, reduction of excess meat (and dairy) consumption is among the most effective measures to mitigate GHG emissions, with a high potential for environment, health, food security, biodiversity, and animal welfare co-benefits (Hedenus et al. 2014; Springmann et al. 2018a; Chai et al. 2019; Chen et al. 2019; Kim et al. 2019; Willett et al. 2019; Semba et al. 2020; Theurl et al. 2020; Hamilton et al. 2021; Stylianou et al. 2021) (*robust evidence, high agreement*). Dietary changes are relevant for several SDGs, in addition to SDG 13 (climate action), including SDG 2 (zero hunger), SDG 3 (good health and well-being), SDG 6 (clean water and sanitation), SDG 12 (responsible consumption and production), SDG 14 (life below water) and SDG 15 (life on land) (Bruce M et al. 2018; Mbow et al. 2019; Vanham et al. 2019; Herrero et al. 2021) (Section 12.6.1). However, behavioural change towards diets of lower environmental impact and higher nutritional qualities faces barriers both from agricultural producers and consumers (Apostolidis and McLeay 2016; Aiking and de Boer 2018; de Boer et al. 2018; Milford et al. 2019), and requires policy packages that combine informative instruments with behavioural, administrative and/or market-based instruments, and are attentive to the needs of, and engage, all food system stakeholders including civil society networks, and change the food environment (Cornelsen et al. 2015; Kraak et al. 2017; Stoll-Kleemann and Schmidt 2017; El Bilali 2019; iPES Food 2019; Milford et al. 2019; Temme et al. 2020) (Section 12.4.1) (*robust evidence, high agreement*).

Table 12.9 summarises the implications of a range of policy instruments discussed in more detail in the following sub-sections and highlights the benefits of integrated policy packages. Furthermore, Table 12.9 assesses transformative potential, environmental effectiveness, feasibility, distributional effect, cost, and cost-benefits and trade-offs of individual policy instruments, as well as their potential role as part of coherent policy packages. Table 12.9 shows that information and behavioural policy instruments can have significant but small effects in changing diets (*robust evidence, medium agreement*), but are mutually enforcing and might be essential to lower barriers and increase acceptance of market-based and administrative instruments (*medium evidence, high agreement*).

The policy instruments are assessed in relation to shifting food consumption and production towards increased sustainability and health. This includes lowering GHG emissions, although not in all cases is this the primary focus of the instrument, and in some cases lowering GHG emissions may not even be explicitly mentioned.

#### 12.4.4.1 Market-based Instruments

**Taxes and subsidies:** Food-based taxes have largely been implemented to reduce non-communicable diseases and sugar intake, particularly those targeting sugar-sweetened beverages (WHO 2019). Many health-related organisations recommend the introduction of such taxes to improve the nutritional quality of marketed products and consumers' diets (Wright et al. 2017; Park and Yu 2019; WHO 2019), even though the impacts of food taxes are complex due to cross-price and substitution effects and supplier reactions (Cornelsen et al. 2015; Green et al. 2019; Blakely et al. 2020) and can have a regressive effect (WHO 2019). Subsidies and taxes are found to be effective in changing dietary behaviour at levels above 20% price increase (Cornelsen et al. 2015; Niebylski et al. 2015; Nakhimovsky et al. 2016; Hagenaars et al. 2017; Mozaffarian et al. 2018), even though longer-term effects are scarcely studied (Cornelsen et al. 2015) and effects of sugar tax with tax rates lower than 20% have been observed for low-income groups (Temme et al. 2020).

Modelling results show only small consumption shifts with moderate meat price increases; and high price increases are required to reach mitigation targets, even though model predictions become highly uncertain due to lack of observational data (Mazzocchi 2017; Bonnet et al. 2018; Fellmann et al. 2018; Zech and Schneider 2019; Latka et al. 2021b). Taxes applied at the consumer level are found to be more effective than levying the taxes on the production side (Springmann et al. 2017).

Unilateral taxes on food with high GHG intensities have been shown to induce increases in net export flows, which could reduce global prices and increase global demand. Indirect effects on GHG mitigation therefore could be reduced by up to 70–90% of national results (Fellmann et al. 2018; Zech and Schneider 2019) (*limited evidence, high agreement*). The global mitigation potential for GHG taxation of food products at USD52 $kgCO_2$-$eq^{-1}$ has been estimated at 1 $GtCO_2$-eq $yr^{-1}$ (Springmann et al. 2017).

Studies have shown that taxes can improve the nutritional quality of diets and reduce GHG emissions from the food system, particularly if accompanied by other policies that increase acceptance and elasticity, and reduce regressive and distributional problems (Niebylski et al. 2015; Hagenaars et al. 2017; Mazzocchi 2017; Springmann et al. 2017; Wright et al. 2017; Henderson et al. 2018; Säll 2018; FAO et al. 2020; Penne and Goedemé 2020) (*robust evidence, high agreement*).

**Trade:** Since the middle of the last century, global trade in agricultural products has contributed to boosting productivity and reducing commodity prices, while also incentivising national subsidies for farmers to remain competitive in the global market (Benton et al. 2019). Trade liberalisation has been coined as an essential element of sustainable food systems, and as one element required to achieve sustainable development, that can shift pressure to regions where the resources are less scarce (Wood et al. 2018; Traverso and Schiavo 2020). However, Clapp (2017) argues that the main economic benefit of trade liberalisation flows to large transnational firms. Benton and Bailey (2019) argue that low food prices in the second half of last century contributed to both yield and food waste increases, and to a focus on staple crops to the disadvantage of nutrient-dense foods. However, global trade can also contribute to economic benefits such as jobs and income, reduce food insecurity and facilitate access to nutrients (Wood et al. 2018; Hoff et al. 2019; Traverso and Schiavo 2020; Geyik et al. 2021) and has contributed to increased food supply diversity (Kummu et al. 2020). The relevance of trade for food security, and adaptation and mitigation of agricultural production, has also been discussed in Mbow et al. (2019).

Trade policies can be used to protect national food system measures, by requiring front-of-package labels, or to impose border taxes on unhealthy products (Thow and Nisbett 2019). For example, in the frame of the Pacific Obesity Prevention in Communities project, the Fijian government implemented three measures (out of seven proposed) that eliminated import duties on fruits and vegetables, and imposed 15% import duties on unhealthy oils (Latu et al. 2018). Trade agreements, however, have the potential to undermine national efforts to improve public health (Unar-Munguía et al. 2019). GHG mitigation efforts in food supply chains can be counteracted by GHG leakage, with a general increase of environmental and social impact in developing countries exporting food products, and a decrease in the developed countries importing food products (Fellmann et al. 2018; Sandström et al. 2018; Wiedmann and Lenzen 2018). The demand for agricultural commodities has also been associated with tropical deforestation, though a robust estimate on the extent of embodied deforestation in food commodities is not available (Pendrill et al. 2019).

**Investment into research and innovation:** El Bilali (2019) assessed research gaps in the food system transition literature and found a need to develop comparative studies that enable the assessment of spatial variability and scalability of food system transitions. The author found also that the role of private industry and corporate business is scarcely researched, although they could play a major role in food system transitions.

The InterAcademy Partnership assessed how research can contribute to providing the required evidence and opportunities for food system transitions, with a focus on climate change impacts and mitigation (IAP 2018). The project builds on four regional assessments of opportunities and challenges on food and nutrition security in Africa (NASAC 2018), the Americas (IANAS 2018), Asia (AASSA 2018), and Europe (EASAC 2017). The Partnership concludes with a set of research questions around food systems, that need to be better understood: (i) how are sustainable food systems constituted in different contexts and at different scales? (ii) how can transition towards sustainable food systems be achieved? and (iii) how can success and failure be measured along sustainability dimensions including climate mitigation?

#### 12.4.4.2 Regulatory and Administrative Instruments

**Marketing regulations:** Currently, 16 countries regulate marketing of unhealthy food to children, mainly on television and in schools (Taillie et al. 2019), and many other efforts are ongoing across the globe (European Commission 2019). The aim to counter the increase in obesity in children and target products high in saturated fats, trans-fatty acids, free sugars and/or salt (WHO 2010) was endorsed by 192 countries (Kovic et al. 2018). Nutrition and health claims for products are used by industry to increase sales, for example in the sport sector or for breakfast cereals. They can be informative, but can also be misleading if misused for promoting unhealthy food (Whalen et al. 2018; Ghosh and Sen 2019; Sussman et al. 2019).

Strong statutory marketing regulations can significantly reduce the exposure of children to, and sales of, unhealthy food compared with voluntary restrictions (Kovic et al. 2018; Temme et al. 2020). Data on effectiveness of marketing regulations with a broader food sustainability scope are not available. On the other hand, regulations that mobilise private investment into emerging food production technologies can be instrumental in curbing the cost and making them competitive (Bianchi et al. 2018a).

**Voluntary sustainability standards:** Voluntary sustainability standards are developed either by a public entity or by private organisations to respond to consumers' demands for social and environmental standards (Fiorini et al. 2019). For example, the Dutch Green Protein Alliance, an alliance of government, industry, NGOs and academia, formulated a goal to shift the ratio of protein consumption from 60% animal source proteins currently to 40% by 2050 (Aiking and de Boer 2020), and Cool Food Pledge signatories (organisations that serve food, such as restaurants, hospitals and universities) committed to a 25% reduction in GHG emissions by 2030, compared with 2015 (Cool Food 2020). For firms, obtaining certification under such schemes can be costly, and costs are generally borne by the producers and/or supply chain stakeholders (Fiorini et al. 2019). The effectiveness of private voluntary sustainability standards is uncertain. Cazzolla Gatti et al. (2019) have investigated the effectiveness of the Roundtable on Sustainable Palm Oil on halting forest loss and habitat degradation in Southeast Asia and concluded that production of certified palm oil continued to lead to deforestation.

**Organisational procurement:** Green public procurement is a policy that aims to create additional demand for sustainable products (Bergmann Madsen 2018; Mazzocchi and Marino 2019) or decrease demand for less sustainable products (e.g., the introduction of 'Meatless Monday' by the Norwegian Armed Forces) (Cheng et al. 2018; Gava et al. 2018; Milford and Kildal 2019; Wilts et al. 2019). To improve dietary choices, organisations can increase the price of unsustainable options while decreasing the price of sustainable ones, or employ information or choice architecture measures (Goggins and Rau 2016; Goggins 2018). Procurement guidelines exist at global, national, organisational or local levels (Noonan et al. 2013; Neto and Gama Caldas 2018). Procurement rules in schools or public canteens increase the accessibility of healthy food and can improve dietary behaviour and decrease purchases of unhealthy food (Cheng et al. 2018; Temme et al. 2020).

**Food regulations:** Novel foods based on insects, microbial proteins or cellular agriculture must go through authorisation processes to ensure compliance with food safety standards before they can be sold to consumers. Several countries have 'novel food' regulations governing the approval of foods for human consumption. For example, the European Commission, in its update of the Novel Food Regulation in 2015, expanded its definition of novel food to include food from cell cultures, or that produced from animals by non-traditional breeding techniques (EU 2015).

For animal product analogues, regulatory pathways and procedures (Stephens et al. 2018) and terminology issues (defining equivalence questions) (Carrenõ and Dolle 2018; Pisanello and Ferraris 2018) need clarification, as does their relation to religious rules (Chriki and Hocquette 2020).

Examples of legislation targeting food waste include the French ban on wasting food approaching best-before dates, requiring its donation to charity organisations (Global Alliance for the Future of Food 2020). In Japan, the Food Waste Recycling Law set targets for food waste recycling for industries in the food sector for 2020, ranging between 50% for restaurants and 95% for food manufacturers (Liu et al. 2016).

12.4.4.3 Informative Instruments.

**Sustainable food-based dietary guidelines:** National food-based dietary guidelines (FBDGs) provide science-based recommendations on food group consumption quantities. They are available for 94, mostly upper- and middle-income, countries globally (Wijesinha-Bettoni et al. 2021), are adapted to national cultural and socio-economic context, and can be used as a benchmark for food formulation standards for public and private food procurement, or to inform citizens (Bechthold et al. 2018; Temme et al. 2020). Most FBDGs are based on health considerations and only a few mention environmental sustainability aspects (Bechthold et al. 2018; Ritchie et al. 2018; Ahmed et al. 2019; Springmann et al. 2020). Implementation of FBDGs so far focuses largely in the education and health sectors, with few countries also using their potential for guiding food system policies in other sectors (Wijesinha-Bettoni et al. 2021).

Despite the fact that 1.5 billion people follow a vegetarian diet from choice or necessity, and that the position statements of various nutrition societies point out that vegetarian diets are adequate if well planned, few FBDGs give recommendations for vegetarian diets (Costa Leite et al. 2020). An increase in consumption of plant-based food is a recurring recommendation in FBDGs, though an explicit reduction or limit of animal-source proteins is not often included, with the exception of red or processed meat (Temme et al. 2020). To account for changing dietary trends, however, FBDGs need to incorporate sustainability aspects (Herforth et al. 2019). A healthy diet respecting planetary boundaries has been proposed by Willett et al. (2019), though some authors have questioned the validity of the nutritional (Zagmutt et al. 2019) or environmental implications, such as water use (Vanham et al. 2020). In October 2019, 14 global cities pledged to adhere to this 'planetary health diet' (C40 Cities 2019).

**Education on food/nutrition and environment:** Some consumers are reluctant to adopt sustainable healthy dietary patterns because of a lack of awareness of the environmental and health consequences of what they eat, but also out of suspicion towards alternatives that are perceived as not 'natural' and that seem to be difficult to integrate into their daily dietary habits (Hartmann and Siegrist 2017; Stephens et al. 2018; McBey et al. 2019; Siegrist and Hartmann 2020) or simply lack of knowledge on how to prepare or eat unfamiliar foods (El Bilali 2019; Aiking and de Boer 2020; Temme et al. 2020). Misconceptions may contribute, for example, to the belief that packaging or 'food miles' dominate the climate impact of food (Macdiarmid et al. 2016). However, spillover effects can induce sustainable behaviour from 'entry points' such as concerns about food waste (El Bilali 2019). Early-life experiences are crucial determinants for adopting healthy and sustainable lifestyles (Bascopé et al. 2019; McBey et al. 2019), so improved understanding of sustainability aspects in the education of public health practitioners and in university education is proposed (Wegener et al. 2018). Investment in education, particularly of women (Vermeulen et al. 2020), might lower the barrier for stronger policies to be accepted and effective (McBey et al. 2019; Temme et al. 2020) (*medium evidence, high agreement*).

**Food labels:** Instruments to improve transparency and information on food sustainability aspects are based on the assumption of the 'rational' consumer. Information gives the necessary freedom of choice, but also the responsibility to make the 'right choice' (Kersh 2015; Bucher et al. 2016). Studies find a lack of consumer awareness about the link between own food choices and environmental effect (Grebitus et al. 2016; Leach et al. 2016; Hartmann and Siegrist 2017; de Boer et al. 2018) and so effective messaging is required to raise awareness and acceptance of potentially stricter food system policies.

Back-of-package labels usually provide detailed nutritional information (Temple 2019). Front-of-package labels simplify and interpret the information: for example, the traffic light system or the Nutri-Score label used in France (Kanter et al. 2018b) and the health star rating used in Australia and New Zealand (Shahid et al. 2020) provide an aggregate rating based on product attributes such as energy, sugar, saturated fat and fibre content; other labels warn against frequent consumption (e.g., in the 1990s Finland introduced a mandatory warning for products high in salt; the keyhole label was introduced in Sweden in 1989 (Storcksdieck genannt Bonsmann et al. 2020); and 'high in' (energy/saturated fat/sugar) labels were introduced in Chile in 2016 to reduce obesity (Corvalán et al. 2019)). Front-of-package labels serve also as an incentive to industry to produce healthier or more sustainable products, or can serve as a marketing strategy (Van Loo et al. 2014; Apostolidis and McLeay 2016; Kanter et al. 2018b). Carbon footprint labels can be difficult for consumers to understand (Hyland et al. 2017), and simple, interpretative summary indicators used on front-of-package labels (e.g., traffic lights) are more effective than more complex ones (Bauer and Reisch 2019; Ikonen et al. 2019; Temple 2019; Tørris and Mobekk 2019) (*robust evidence, high agreement*). Reviews find mixed results but overall a positive effect of food labels in improving direct purchasing decisions (Hieke and Harris 2016; Sarink et al. 2016; Anastasiou et al. 2019; Shangguan et al. 2019; Temple 2019), and in raising levels of awareness, thus possibly increasing success of other policy instruments (Apostolidis and McLeay 2016; Samant and Seo 2016; Al-Khudairy et al. 2019; Miller et al. 2019; Temple 2019) (*medium evidence, high agreement*).

#### 12.4.4.4 Behavioural Instruments

***Choice architecture:*** Information is more effective if accompanied by reinforcement through structural changes or by changing the food environment, such as through product placement in supermarkets, to overcome the intention–behaviour gap (Bucher et al. 2016; Broers et al. 2017; Tørris and Mobekk 2019). Behavioural change strategies have also been shown to improve efficiencies of school food programmes (Marcano-Olivier et al. 2020).

Environmental considerations rank behind financial, health, or sensory factors for determining citizens' food choices (Leach et al. 2016; Hartmann and Siegrist 2017; Neff et al. 2018; Rose 2018; Gustafson et al. 2019). There is evidence that choice architecture ('nudging') can be effective in influencing purchase decisions, but regulators do not normally explore this option (Broers et al. 2017). Examples of green nudging include making the sustainable option the default option, enhancing visibility, accessibility of, or exposure to, sustainable products and reducing visibility and accessibility of unsustainable products, or increasing the salience of healthy sustainable choices through social norms or food labels (Bucher et al. 2016; Wilson et al. 2016; Broers et al. 2017; Al-Khudairy et al. 2019; Bauer and Reisch 2019; Ferrari et al. 2019; Weinrich and Elshiewy 2019; Cialdini and Jacobson 2021). Available evidence suggests that choice architecture measures are relatively inexpensive and easy to implement (Ferrari et al. 2019; Tørris and Mobekk 2019), they are a preferred solution if a restriction of choices is to be avoided (Wilson et al. 2016; Kraak et al. 2017; Vecchio and Cavallo 2019), and can be effective (Arno and Thomas 2016; Bucher et al. 2016; Bianchi et al. 2018b; Cadario and Chandon 2018) if embedded in policy packages (Wilson et al. 2016; Tørris and Mobekk 2019) (*medium evidence, high agreement*).

Choice architecture measures are also facilitated by growing market shares of animal-free protein sources taken up by discount chains and fast food companies, that enhance visibility of new products and ease integration into daily life for consumers, particularly if sustainable products are similar to the products they substitute (Slade 2018). This effect can be further increased by media and role models (Elgaaied-Gambier et al. 2018).

### 12.4.5 Food Systems Governance

To support the policies outlined in Section 12.4.4, food system governance depends on the cooperation of actors across traditional sectors in several policy areas, in particular agriculture, nutrition, health, trade, climate, and environment (Termeer et al. 2018; Bhunnoo 2019; Diercks et al. 2019; iPES Food 2019; Rosenzweig et al. 2020b). Top-down integration, mandatory mainstreaming, or boundary-spanning structures like public-private partnerships may be introduced to promote coordination (Termeer et al. 2018). 'Flow-centric' rather than territory-centric governance combined with private governance mechanisms has enabled codes of conduct and certification schemes (Eakin et al. 2017), for example the Roundtable on Sustainable Palm Oil (RSPO), as well as commodity chain transparency initiatives and platforms like Trase (Meijaard et al. 2020; Pirard et al. 2020). Trade agreements are an emerging arena of governance in which improving GHG performance may be an objective, and trade agreements can involve sustainability assessments.

Research on food system governance is mostly non-empirical or case study based, which means that there is limited understanding of which governance arrangements work in specific social and ecological contexts to produce particular food system outcomes (Delaney et al. 2018). Research has identified a number of desirable attributes in food systems governance, including adaptive governance (Termeer et al. 2018), a systems perspective (Whitfield et al. 2018), governance that considers food system resilience (Ericksen 2008; Moragues-Faus et al. 2017; Meyer 2020), transparency, participation of civil society (Candel 2014; Duncan 2015;), and cross-scale governance (Moragues-Faus et al. 2017).

Food systems governance has multiple targets and objectives, not least contributing to the achievement of the SDGs. GHG emissions from food systems can be impacted by both interventions targeted at different parts of the food system and interventions in other systems, such as reducing deforestation or promoting reforestation (Lee et al. 2019). For example, policies targeting health can contribute to diet shifts away from red meat, while also influencing GHG emissions (Springmann et al. 2018b; Semba et al. 2020); national and local food self-sufficiency policies may also have GHG impacts (Kriewald et al. 2019; Loon et al. 2019). Cross-sectoral governance could enhance synergies between reduced GHG emissions from food systems and other goals; however, integrative paradigms for cross-sectoral governance between food and other sectors have faced implementation challenges (Delaney et al. 2018). For example, in the late 2000s, the water-energy-food nexus emerged as a framework for cross-sectoral governance, but has not been well integrated into policy (Urbinatti et al. 2020), perhaps because of perceptions that it is an academic concept, or that it takes a technical-administrative view of governance; simply adopting the paradigm is not sufficient to develop effective nexus governance (Cairns and Krzywoszynska 2016; Weitz et al. 2017; Pahl-Wostl et al. 2018). Other policy paradigms and theoretical frameworks that aim to integrate food systems governance include system transition, agroecology, multifunctionality in agriculture (Andrée et al. 2018), climate-smart agriculture (Taylor 2018) and the circular economy (Box 12.4). Cross-sectoral coordination on food systems and climate governance could be aided by internal recognition and ownership by agencies, dedicated budgets for cross-sectoral projects, and consistency in budgets (Pardoe et al. 2018) (Boxes 12.1 and 12.2).

Food systems governance is still fragmented at national levels, which means that there may be a proliferation of efforts that cannot be scaled and are ineffective (Candel 2014). National policies can be complemented or possibly pioneered by initiatives at the local level (de Boer et al. 2018; Rose 2018). The city-region has been proposed as a useful focus for food system governance (Vermeulen et al. 2020); for example, the Milan Urban Food Policy Pact involves 180 global cities committed to integrative food system strategies (Candel 2019; Moragues-Faus 2021). Local food policy groups and councils that assemble stakeholders from government, civil society, and the private sector have formed trans-local networks of place-based local food policy groups, with over two hundred food policy councils worldwide (Andrée et al. 2018). However, the fluidity and lack of clear agendas and membership structures may hinder their ability to confront fundamental structural issues like unsustainable diets or inequities in food access (Santo and Moragues-Faus 2019).

Early characterisations of food systems governance featured a binary distinction between global and local scales, but this has been replaced by a relational approach where the local governance is seen as a process that relies on the interconnections between scales (Lever et al. 2019). Cross-scalar governance is not simply an aggregation of local groups, but involves the telecoupling of distant systems; for example, transnational NGO networks have been able to link coffee retailers in the global North with producers in the global South via international NGOs concerned about deforestation and social justice (Eakin et al. 2017). Global governance institutions like the Committee on World Food Security can promote policy coherence globally and reinforce accountability at all levels (McKeon 2015), as can norm-setting efforts like the Voluntary Guidelines for the Responsible Governance of Tenure of Land, Fisheries and Forests (FAO 2012). Global multi-stakeholder processes like the UN Food Systems Summit can foster the development of principles for guiding further actions based on sound scientific evidence. The European Commission's Farm to Fork strategy aims to promote policy coherence in food policy at EU and national levels, and could be the exemplar of a genuinely integrated food policy (Schebesta and Candel 2020).

## Box 12.2 | Case Study: The Finnish Food2030 Strategy

Until 2016, the strategic goals of Finnish food policy were split between different programmes and ministries, resulting in fragmented national oversight of the Finnish food system. To enable policy coordination, a national food strategy was adopted in 2017 called Food2030 (Government of Finland 2017). Food2030 embodies a holistic food system approach and addresses multiple outcomes of the food system, including the competitiveness of the food supply chain and the development of local, organic and climate-friendly food production, as well as responsible and sustainable consumption.

The specific policy mix covers a range of policy instruments to enable changes in agro-food supply, processing and societal norms (Kugelberg et al. 2021). The government provides targeted funding and knowledge support to drive technological innovations on climate solutions to reduce emissions from food and in the agriculture, forestry and land use sectors. In addition, the Finnish government applies administrative means, such as legislation, advice, guidance on public procurement and support schemes to diversify and increase organic food production to 20% of arable land, which in turn improve the opportunities for small-scale food production and steer public bodies to purchase local and organic food. The Finnish government applies educational and informative instruments to enable a shift to healthy and sustainable dietary behaviours. The policy objective is to reduce consumption of meat and replace it with other sources of protein, aligned with nutrition recommendations and avoiding food waste. The Ministry of Agriculture and Forestry, in collaboration with the Finnish Farmer's unions and the Union of Swedish-speaking Farmers and Forest Owners in Finland, ran a two-year multi-media campaign in 2018 with key messages on the sustainability, traceability and safety of locally-produced food (Ministry of Agriculture and Forestry 2021). A 'Food Facts' website project (Luke 2021), funded by the Ministry of Agriculture and Forestry in collaboration with the Natural Resources Institute Finland and the Finnish Food Safety Authority, helps to raise knowledge about food, which could shape responsible individual food behaviour, for example choosing local and sustainable foods and reducing food waste.

A critical enabler for developing a shared food system strategy across sectors and political party boundaries was the implementation of a one-year inclusive, deliberative and consensual stakeholder engagement process. A wide range of stakeholders could exert real influence during the vision-building process, resulting in strong agreement on key policy objectives, and subsequently an important leverage point to policy change (Kugelberg et al. 2021). Moreover, cross-sectoral coordination of Food2030 and the government's wider climate action programmes are enabled by a number of institutional mechanisms and collaborative structures, for example the advisory board for the food chain, formally established during the agenda-setting stage of Food2030, inter-ministerial committees to guide and assess policy implementation, and Our Common Dining Table, a multi-stakeholder partnership that assembles 18 food system actors to engage in reflexive discussions about the Finnish food system.

*Box 12.2 (continued)*

Critical barriers to strategy and policy formulation include a lack of attention to integrated impact assessments (Kugelberg et al. 2021), which blurs a transparent overview of potential trade-offs and hidden conflicts. There were few policy evaluations from independent organisations to inform policymaking, reducing the opportunities for more progressive policy approaches. Monitoring and food policy evaluation is very close to the ministry in charge, which hampers critical thinking about policy measures (Hildén et al. 2014). In addition, there is a lack of standardised indicators covering the whole food system, which hinders comprehensive oversight of progress towards a sustainable food system (Kanter et al. 2018a). Some of the problems related to monitoring, reporting and verification (MRV) are typical for countries in the EU. To improve, MRV will probably require structural changes, such as efforts to build up institutional capacity and application of new technology, development of standardised indicators covering the whole food system, regulations on transparency and verification, and mechanisms to enable reflexive discussions between business, farmers, public, NGOs and the government (Meadowcroft and Steurer 2018; Kanter et al. 2020).

## 12.5 Land-related Impacts, Risks and Opportunities Associated with Mitigation Options

### 12.5.1 Introduction

This section provides a cross-sectoral perspective on land occupation and related impacts, risks and opportunities associated with land-based mitigation options, as well as mitigation options that are not designated land-based, yet occupy land. It builds on Chapter 7, which covers mitigation in agriculture, forestry and other land use (AFOLU, including future availability of biomass resources for mitigation in other sectors. It complements Section 12.4, which covers mitigation inherent in the food system, as well as Chapters 6, 9, 10 and 11, which cover mitigation in the energy, transport, building and industry sectors, and Chapters 3 and 4 which cover land and biomass use, primarily in energy applications, in mitigation and development pathways in the near- to mid-term (Chapter 4) and in pathways compatible with long-term goals (Chapter 3).

The deployment of climate change mitigation options often affects land and water conditions, and ecosystem capacity to support biodiversity and a range of ecosystem services (IPCC 2019a; IPBES 2019) (*robust evidence*, *high agreement*). It can increase or decrease terrestrial carbon stocks and sink strength, hence impacting the mitigation effect positively or negatively. As for any other land uses, impacts, risks and opportunities associated with mitigation options that occupy land depend on deployment strategy and on contextual factors that vary geographically and over time (Doelman et al. 2018; Hurlbert et al. 2019; Smith et al. 2019a; Wu et al. 2020) (*robust evidence*, *high agreement*).

The IPCC Special Report on Global Warming of 1.5°C (SR1.5) found that large areas may be utilised for A/R and energy crops in modelled pathways limiting warming to 1.5°C (Rogelj et al. 2018). The SRCCL investigated the implications of land-based mitigation measures for land degradation, food security and climate change adaptation. It focused on identification of synergies and trade-offs associated with individual land-based mitigation measures (Smith et al. 2019b). In this section we expand beyond the scope of the Special Report on Climate Change and Land (SRCCL) assessment to include also mitigation measures that occupy land while not being considered land-based measures, we discuss ways to minimise potential adverse effects, and we consider the potential for synergies through integrating mitigation measures with other land uses, by applying a systems perspective that seeks to meet multiple objectives from multi-functional landscapes. Mitigation measures with zero land occupation, e.g., offshore wind and kelp farming, are not considered.

### 12.5.2 Land Occupation Associated with Different Mitigation Options

As reported in Chapter 3, in scenarios limiting warming to 1.5°C (>50%) with no or limited overshoot, median area dedicated for energy crops in 2050 is 1.99 (0.56 to 4.82) million square kilometres ($Mkm^2$) and median forest area increased 3.22 (–0.67 to 8.90) $Mkm^2$ in the period 2019 to 2050 (5–95th percentile range, scenario category C1). For comparison, the total global areas of forests, cropland and pasture (in 2015) are in the SRCCL estimated at about 40 $Mkm^2$, 15.6 $Mkm^2$, and 27.3 $Mkm^2$, respectively (additionally, 21 $Mkm^2$ of savannahs and shrublands are also used for grazing) (IPCC 2019a). The SRCCL concluded that conversion of land for A/R and bioenergy crops at the scale commonly found in pathways limiting warming to 1.5°C or 2°C is associated with multiple feasibility and sustainability constraints, including land carbon losses (*high confidence*). Pathways in which warming exceeds 1.5°C require less land-based mitigation, but the impacts of higher temperatures on regional climate and land, including land degradation, desertification, and food insecurity, become more severe (Smith et al. 2019b).

Depending on emissions-reduction targets, the portfolio of mitigation options chosen, and the policies developed to support their implementation, different land-use pathways can arise with large differences in resulting agricultural and forest area. Some response options can be more effective when applied together (Smith et al. 2019b); for example, dietary change, efficiency increases, and reduced wastage can reduce emissions as well as the pressure on land resources, potentially enabling additional land-based mitigation such as A/R and cultivation of biomass crops for biochar, bioenergy and other bio-based products. The SRCCL (Smith et al. 2019b) report that dietary change combined with reduction in food loss and waste can reduce the land requirement for food production by up to 5.8 $Mkm^2$ (0.8–2.4 $Mkm^2$ for dietary change; about 2 $Mkm^2$ for reduced post-harvest losses, and 1.4 $Mkm^2$ for reduced food waste) (Parodi et al. 2018; Springmann et al. 2018; Clark et al. 2020; Rosenzweig et al. 2020b) (Sections 7.4 and 12.4). Stronger mitigation action in the near term targeting non-$CO_2$ emissions reduction and deployment of other CDR options (DACCS, enhanced weathering, ocean-based approaches; see Section 12.3) can reduce the land requirement for land-based mitigation (Obersteiner et al. 2018; van Vuuren et al. 2018).

Global integrated assessment models (IAMs) provide insights into the roles of land-based mitigation in pathways limiting warming to 1.5°C or 2°C; interaction between land-based and other mitigation options such as wind and solar power; influence of land-based mitigation on food markets, land use and land carbon; and the role of BECCS vis-à-vis other CDR options (Chapter 3). However, IAMs do not capture more subtle changes in land management and in the associated industrial/energy systems due to relatively coarse temporal and spatial resolution, and limited representation of land quality and feedstocks/management practices, interactions between biomass production and conversion systems, and local context, for example, governance of land use (Daioglou et al. 2019; Rose et al. 2020; Welfle et al. 2020; Calvin et al. 2021). A/R have generally been modelled as forests managed for carbon sequestration alone, rather than forestry providing both carbon sequestration and biomass supply (Calvin et al. 2021). Because IAMs do not include options to integrate new biomass production with existing agricultural and forestry systems (Paré et al. 2016; Mansuy et al. 2018; Cossel et al. 2019; Braghiroli and Passarini 2020; Djomo et al. 2020; Moreira et al. 2020; Strapasson et al. 2020; Rinke Dias de Souza et al. 2021), they may over-estimate the total additional land area required for biomass production. On the other hand, some integrated biomass production systems may prove less attractive to landholders than growing biomass crops in large blocks, from logistic, economic, or other points of view (Ssegane et al. 2016; Busch 2017; Ferrarini et al. 2017).

Land occupation associated with mitigation options other than A/R and bioenergy is rarely quantified in global scenarios. Stressing large uncertainties (e.g., type of biomass used and share of solar PV integrated in buildings), Luderer et al. (2019) modelled land occupation and land transformation associated with a range of alternative power system decarbonisation pathways in the context of a global 2°C climate stabilisation effort. On a per-megawatt hour (MWh) basis, bioelectricity with CCS was most land intensive, followed by hydropower, coal with CCS, and concentrated solar power (CSP), which in turn were around five times as land-intensive as wind and solar photovoltaics (PV). A review of studies of power densities (electricity generation per unit land area) confirmed the relatively larger land occupation associated with biopower, although hydropower overlaps with biopower (van Zalk and Behrens 2018). This study also quantifies the low land occupation of nuclear energy, similar to fossil energy sources.

The land occupation of PV depends on the share of ground-mounted versus buildings-integrated PV, the latter assumed to reach 75% share by 2050 (Luderer et al. 2019). van de Ven et al. (2021) assumed a 3% share of urbanised land in 2050 available for rooftop PV; Capellán-Pérez et al. (2017) and Dupont et al. (2020) report 2–3% availability of urbanised surface area, when considering factors such as roof slopes and shadows between buildings, and threshold relating to energy return on investment. Land occupation of solar technologies is considered to be underestimated in studies assuming ideal conditions, with real occupation being five to ten times higher (De Castro et al. 2013; MacKay 2013; Ong et al. 2013; Smil 2015; Capellán-Pérez et al. 2017).

Production of hydrogen and synthetic hydrocarbon fuels via electrolysis and hydrocarbon synthesis is subject to conversion losses that vary depending on technology, system integration and source of carbon (Wulf et al. 2020; Ince et al. 2021) (Sections 6.4.4.1 and 6.4.5.1). Indicative electricity-to-hydrocarbon fuel efficiency loss is estimated at about 60% (Ueckerdt et al. 2021). The advantage of smaller land occupation for solar, wind, hydro and nuclear, compared with biomass-based options, is therefore smaller for hydrocarbon fuels than for electricity. Furthermore, biofuels are often co-produced with other bio-based products, which further reduces their land occupation, although comparisons are complicated by inconsistent approaches to allocating land occupation between co-products (Ahlgren et al. 2015; Czyrnek-Delêtre et al. 2017).

Note that comparisons on a per-MWh basis do not reflect the GHG emissions associated with the power options, or that the different options serve different functions in power systems. Reservoir hydropower and biomass-based dispatchable power can complement other balancing options (e.g., battery storage, grid extensions and demand-side management (Göransson and Johnsson 2018) (Chapter 6) to provide power stability and quality needed in power systems with large amounts of variable electricity generation from wind and solar power plants. Furthermore, the requirements of transport in grids, pipelines and so on differ. For example, electricity from buildings-integrated PV can be used in the same location as it is generated.

The character of land occupation, and, consequently, the associated impacts (Section 12.5.3), vary considerably among mitigation options and also for the same option depending on geographic location, scale, system design and deployment strategy (Olsson et al. 2019; Ioannidis and Koutsoyiannis 2020; van de Ven et al. 2021). Land occupation associated with different mitigation options can be large uniform areas (e.g., large solar farms, reservoir hydropower dams, or tree plantations), or more distributed, such as wind turbines, solar PV, and patches of biomass cultivation integrated with other land uses in heterogeneous landscapes (Cacho et al. 2018; Jager and Kreig 2018; Correa et al. 2019; Englund et al. 2020a). Studies with broader scope, covering total land use requirement induced by plant infrastructure, provide a more complete picture of land footprints. For example, Wu et al. (2021) quantified a land footprint for the infrastructure of a pilot solar plant being three times the onsite land area. Sonter et al. (2020b) found significant overlap of mining areas (82% targeting materials needed for renewable energy production) and biodiversity conservation sites and priorities, suggesting that strategic planning is critical to address mining threats to biodiversity (Section 12.5.4) along with recycling and exploration of alternative technologies that use that use abundant minerals (Box 10.6).

There are also situations where expanding mitigation is more or less decoupled from additional land use. The use of organic consumer waste, harvest residues and processing side-streams in the agriculture and forestry sectors can support significant volumes of bio-based products with relatively lower land-use change risks than dedicated biomass production systems (Hanssen et al. 2019; Spinelli et al. 2019; Mouratiadou et al. 2020). Such uses can provide waste management solutions while increasing the mitigation achieved from the land that is already used for agricultural and forest production. Bioenergy accounts for about 90% of renewable heat used in industrial applications, mainly in industries that can use their own biomass waste and residues, such as the pulp and paper industry, food industry, and ethanol production plants (IEA 2020c) (Chapters 6 and 11). Heat and electricity produced on-site from side-streams but not needed for the industrial processes can be sold to other users, such as district heating systems. Surplus waste and residues can also be used to produce solid and liquid biofuels, or be used as feedstock in other industries such as the petrochemical industry (IRENA 2018; Lock and Whittle 2018; Thunman et al. 2018; IRENA 2019; Haus et al. 2020) (Chapters 6 and 11). Electrification and improved process efficiencies can reduce GHG emissions and increase the share of harvested biomass that is used for production of bio-based products (Johnsson et al. 2019; Madeddu et al. 2020; Lipiäinen and Vakkilainen 2021; Rahnama Mobarakeh et al. 2021; Silva et al. 2021) (Chapter 11). Besides integrating solar thermal panels and solar PV into buildings and other infrastructure, floating solar PV panels in, for example, hydropower dams (Ranjbaran et al. 2019; Cagle et al. 2020; Haas et al. 2020; Lee et al. 2020; Gonzalez Sanchez et al. 2021), and over canals (Lee et al. 2020; McKuin et al. 2021) could decouple renewable energy generation from land use while simultaneously reducing evaporation losses and potentially mitigating aquatic weed growth and climate change impacts on water body temperature and stratification (Cagle et al. 2020; Exley et al. 2021; Gadzanku et al. 2021; Solomin et al. 2021).

### 12.5.3 Consequences of Land Occupation: Biophysical and Socio-economic Risks, Impacts and Opportunities

Land occupation associated with mitigation options can present challenges related to impacts and trade-offs, but can also provide opportunities and in different ways support the achievement of additional societal objectives, including adaptation to climate change. This section focuses on mitigation options that have significant risks, impacts and/or co-benefits with respect to land resources, food security and the environment. Bioenergy (with or without CCS), biochar and bio-based products require biomass feedstocks that can be obtained from purpose-grown crops, residues from conventional agriculture and forestry systems, or from biomass wastes, each with different implications for the land. Here we consider separately (i) 'biomass-based systems', including dedicated biomass crops (e.g., perennial grasses, short rotation woody crops) and biomass produced as a co-product of conventional agricultural production (e.g., maize stover), and (ii) 'afforestation/reforestation', including forests established for ecological restoration and plantations grown for forest products and agroforestry, where biomass may also be a co-product. We then discuss impacts and opportunities common to both systems, before considering impacts and opportunities associated with non-land-based mitigation options that nevertheless occupy land.

*Biomass-based systems*

Mitigation options that are based on the use of biomass, that is, bioenergy/BECCS, biochar, wood buildings, and other bio-based products, can have different positive and negative effects depending on the character of the mitigation option, the land use, the biomass conversion process, how the bio-based products are used and what other product they substitute (Leskinen et al. 2018; Howard et al. 2021; Myllyviita et al. 2021). The impacts of the same mitigation option can therefore vary significantly and the outcome in addition depends on previous land/biomass use (Cowie et al. 2021). As biomass-based systems commonly produce multiple food, material and energy products, it is difficult to disentangle impacts associated with individual bio-based products (Ahlgren et al. 2015; Djomo et al. 2017; Obydenkova et al. 2021). As for other mitigation options, governance has a critical influence on outcome, but larger scale and higher expansion rate generally translates into higher risk for negative outcomes such as competition for scarce land, freshwater and phosphorous resources, displacement of natural ecosystems, and diminishing capacity of agroecosystems to support biodiversity and essential ecosystem services, especially if produced without sustainable land management and in inappropriate contexts (Popp et al. 2017; Dooley and Kartha 2018; Hasegawa et al. 2018; Heck et al. 2018; Humpenöder et al. 2018; Fujimori et al. 2019; Hurlbert et al. 2019; IPBES 2019; Smith et al. 2019b; Drews et al. 2020; Hasegawa et al. 2020; Schulze et al. 2020; Stenzel et al. 2021) (*medium evidence*, *high agreement*).

Removal of crop and forestry residues can cause land degradation through soil erosion and decline in nutrients and soil organic matter (Cherubin et al. 2018) (*robust evidence*, *high agreement*). These risks can be reduced by retaining a proportion of the residues to protect the soil surface from erosion and moisture loss and maintain or increase soil organic matter (Section 7.4.3.6); incorporating a perennial groundcover into annual cropping systems (Moore et al. 2019); and by replacing nutrients removed, such as by applying ash from bioenergy combustion plants (Kludze et al. 2013; Harris et al. 2015; Warren Raffa et al. 2015; de Jong et al. 2017) while safeguarding against contamination risks (Pettersson et al. 2020) (*medium evidence*, *high agreement*). Besides topography, soil, and climate conditions, sustainable residue removal rates also depend on the fate of extracted biomass. For example, to maintain the same level of soil organic carbon, the harvest of straw, if used for combustion (which would return no carbon to fields), was estimated to be only 26% of the rate that could be extracted if used for anaerobic digestion involving return of recalcitrant carbon to fields (Hansen et al. 2020). Similarly, biomass pyrolysis produces biochar which can be returned to soils to counteract carbon losses associated with biomass extraction (Joseph et al. 2021; Lehmann et al. 2021).

Expansion of biomass crops, especially monocultures of exotic species, can pose risks to natural ecosystems and biodiversity through introduction of invasive species and land use change, also impacting

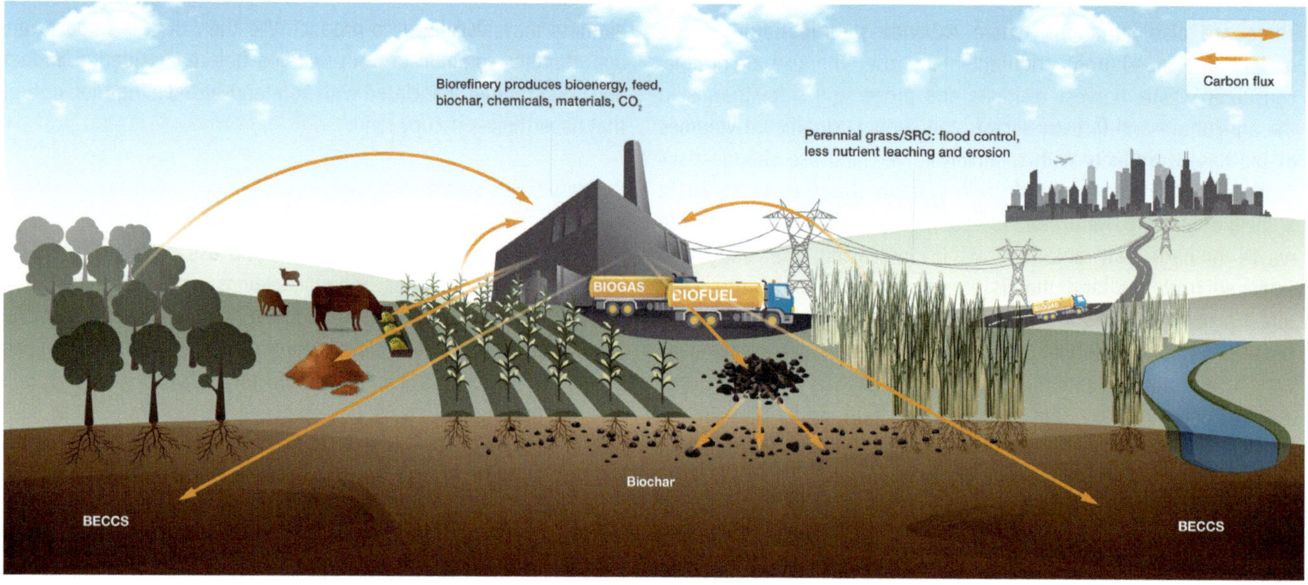


**Figure 12.8 | Overview of opportunities related to selected land-based climate change mitigation options.**

the mitigation value (*robust evidence, high agreement*) (Liu et al. 2014; El Akkari et al. 2018). Cultivation of conventional oil, sugar, and starch crops tends to have larger negative impact than lignocellulosic crops (Núñez-Regueiro et al. 2020). Social and environmental outcomes can be enhanced through integration of suitable plants (such as perennial grasses and short rotation woody crops) into agricultural landscapes (within crop rotations or through strategic localisation, for example as contour belts, along fencelines and riparian buffers). Such integrated systems can provide shelter for livestock, retention of nutrients and sediment, erosion control, pollination, pest and disease control, and flood regulation (*robust evidence, high agreement*) (Berndes et al. 2008; Christen and Dalgaard 2013; Asbjornsen et al. 2014; Holland et al. 2015; Ssegane et al. 2015; Dauber and Miyake 2016; Milner et al. 2016; Ssegane and Negri 2016; Styles et al. 2016; Zheng et al. 2016; Ferrarini et al. 2017; Crews et al. 2018; Henry et al. 2018a; Zalesny et al. 2019; Osorio et al. 2019; Englund et al. 2020b; Englund et al. 2021) (Figure 12.8, Box 12.3, and Cross-Working Group Box 3 in this chapter). Many of the land use practices described above align with agroecology principles (AR6 WGII Section 5.14, AR6 WGII Box 5.11 and AR6 WGII Cross-Chapter Box NATURAL) and can simultaneously contribute to climate change mitigation, climate change adaptation and reduced risk of land degradation (IPCC 2019a) (*robust evidence, high agreement*).

12

### *Afforestation/reforestation (A/R)*

When A/R activities comprise the establishment of natural forests, the risk to land is primarily associated with potential displacement of previous land use to new locations, which could indirectly cause land-use change including deforestation (Sections 7.4.2 and 7.6.2.4). A/R (including agroforestry) aimed at providing timber, fibre, biomass, non-timber resources and other ecosystem services can provide renewable resources to society and long-term livelihoods for communities. Forest management and harvesting regimes around the world will adjust in different ways as society seeks to meet climate goals. The outcome depends on forest type, climate, forest ownership and the character and product portfolio of the associated forest industry (Lauri et al. 2019; Favero et al. 2020). How forest carbon stocks, biodiversity, hydrology, and so on are affected by changes in forest management and harvesting in turn depends on both management practices and the characteristics of the forest ecosystems (Eales et al. 2018; Griscom et al. 2018; Kondo et al. 2018; Nieminen et al. 2018; Thom et al. 2018; Runting et al. 2019; Tharammal et al. 2019) (*robust evidence, medium agreement*). As described above, the GHG savings achieved from producing and using bio-based products will in addition depend on the character of existing societal systems, including technical infrastructure and markets, as this determines the product substitution patterns.

Environmental and socio-economic co-benefits are enhanced when ecological restoration principles are applied (Gann et al. 2019) along with effective planning at landscape level and strong governance (Morgan et al., 2020). For example, restoration of natural vegetation and establishing plantations on degraded land enable organic matter to accumulate in the soil and have potential to deliver significant co-benefits for biodiversity, land resource condition and livelihoods (Box 12.3 and Cross-Working Group Box 3 in this chapter). Tree planting and agroforestry on cleared land can deliver biodiversity benefits (Seddon et al. 2009; Kavanagh and Stanton 2012; Law et al. 2014), with biodiversity outcomes influenced by block size, configuration and species mix (Cunningham et al. 2015; Paul et al. 2016) (*robust evidence, high agreement*).

### *Risks and opportunities common to biomass production and A/R mitigation options*

Biomass-based systems and A/R can contribute to addressing land degradation through land rehabilitation or restoration (Box 12.3). Land-based mitigation options that produce biomass for bioenergy/BECCS or biochar through land *rehabilitation* rather than land *restoration* imply a trade-off between production / carbon sequestration and biodiversity outcomes (Hua et al. 2016; Cowie et al. 2018). Restoration, seeking to establish native vegetation with the aim to maximise ecosystem integrity, landscape connectivity, and conservation of on-ground carbon stock, will have higher biodiversity benefits than rehabilitation measures (Lin et al. 2013). However, sequestration rate declines as forests mature, and the sequestered carbon is vulnerable to loss through disturbance such as wildfire, so there is a higher risk of reversal of the mitigation benefit compared with use of biomass for substitution of fossil fuels and GHG-intensive building materials (Russell and Kumar 2017; Dugan et al. 2018; Anderegg et al. 2020). Trade-offs between different ecosystem services, and between societal objectives including climate change mitigation and adaptation, can be managed through integrated landscape approaches that aim to create a mosaic of land uses, including conservation, agriculture, forestry and settlements (Freeman et al. 2015; Nielsen 2016; Reed et al. 2016; Sayer et al. 2017) where each is sited with consideration of land potential and socio-economic objectives and context (Cowie et al. 2018) (*limited evidence*, *high agreement*).

Impacts of biomass production and A/R on the hydrological cycle and water availability and quality depend on scale, location, previous land use/cover and type of biomass production system. For example, extraction of logging residues in forests managed for timber production has little effect on hydrological flows, while land-use change to establish dedicated biomass production can have a significant effect (Teter et al. 2018; Drews et al. 2020). Deployment of A/R can affect temperature, albedo and precipitation locally and regionally, and can mitigate or enhance the effects of climate change in the affected areas (Stenzel et al. 2021b) (Section 7.2.4). A/R activities can increase evapotranspiration, impacting groundwater and downstream water availability, but can also result in increased infiltration to groundwater and improved water quality (Farley et al. 2005; Zhang et al. 2016; Zhang et al. 2017; Lu et al. 2018) and can be beneficial where historical clearing has caused soil salinisation and stream salinity (Farrington and Salama 1996; Marcar 2016). There is *limited evidence* that very large-scale land-use or vegetation cover changes can alter regional climate and precipitation patterns, for example downwind precipitation depends on upwind evapotranspiration from forests and other vegetation (Keys et al. 2016; Ellison et al. 2017; van der Ent and Tuinenburg 2017).

Another example of beneficial effects includes perennial grasses and woody crops planted to intercept runoff and subsurface lateral flow, reducing nitrate entering groundwater and surface waterbodies (Femeena et al. 2018; Woodbury et al. 2018; Griffiths et al. 2019). In India, Garg et al. (2011) found desirable effects as a result of planting Jatropha on wastelands previously used for grazing (which could continue in the Jatropha plantations): soil evaporation was reduced, as a larger share of the rainfall was channelled to plant transpiration and groundwater recharge, and less runoff resulted in reduced soil erosion and improved downstream water conditions. Thus, adverse effects can be reduced and synergies achieved when plantings are sited carefully, with consideration of potential hydrological impacts (Davis et al. 2013).

Several biomass conversion technologies can generate co-benefits for land and water. Anaerobic digestion of organic wastes (e.g., food waste, manure) produces a nutrient-rich digestate and biogas that can be utilised for heating and cooking or upgraded for use in electricity generation, industrial processes, or as transportation fuel (Chapter 6) (Parsaee et al. 2019; Hamelin et al. 2021). The digestate is a rich source of nitrogen, phosphorus and other plant nutrients, and its application to farmland returns exported nutrients as well as carbon (Cowie 2020b). Studies have identified potential risks, including manganese toxicity, copper and zinc contamination, and ammonia emissions, compared with application of undigested animal manure (Nkoa 2014). Although the anaerobic digestion process reduces pathogen risk compared with undigested manure feedstocks, it does not destroy all pathogens (Nag et al. 2019). Leakage of methane is a significant risk that needs to be managed, to ensure mitigation potential is achieved (Bruun et al. 2014). Anaerobic digestion of wastewater, such as sugarcane vinasse, reduces methane emissions and pollution loading as well as producing biogas (Parsaee et al. 2019).

Biorefineries can convert biomass to food, feed and biomaterials along with bioenergy (Aristizábal-Marulanda and Cardona Alzate 2019; Schmidt et al. 2019). Biorefinery plants are commonly characterised by high process integration to achieve high resource use efficiency, minimise waste production and energy requirements, and maintain flexibility towards changing markets for raw materials and products (Schmidt et al. 2019). Emerging technologies can convert biomass that is indigestible for monogastric animals or humans (e.g., algae, grass, clover or alfalfa) into food and feed products. For example, lactic acid bacteria can facilitate the use of green plant biomass such as grasses and clover to produce a protein concentrate suitable for animal feed and other products for material or energy use (Lübeck and Lübeck 2019). Selection of crops suitable for co-production of protein feed along with biofuels and other bio-based products can significantly reduce the land conversion pressure by reducing the need to cultivate other crops (e.g., soybean) for animal feed (Bentsen and Møller 2017; Solati et al. 2018). Thus, such solutions, using alternatives to high-input, high-emissions grain-based feed, can enable sustainable intensification of agricultural systems with reduced environmental impacts (Jørgensen and Lærke 2016). The use of seaweed and algae as biorefinery feedstock can facilitate recirculation of nutrients from waters to agricultural land, thus reducing eutrophication while substituting purpose-grown feed (Thomas et al. 2021).



Pyrolysis can convert organic wastes, including agricultural and forestry residues, food waste, manure, poultry litter and sewage sludge, into combustible gas and biochar, which can be used as a soil amendment (Joseph et al. 2021; Schmidt et al. 2021) (Chapter 7). Pyrolysis facilitates nutrient recovery from biomass residues, enabling return to farmland as biochar, noting, however, that a large fraction of nitrogen is lost during pyrolysis (Joseph et al. 2021). Conversion to biochar aids the logistics of transport and land application of materials such as sewage sludge, by reducing mass and volume, improving flow properties, stability and uniformity, and decreasing odour. Pyrolysis is well suited for materials that may be contaminated with pathogens, microplastics, and per- and polyfluoroalkyl substances, such as abattoir and sewage wastes, removing these risks, and reduces availability of heavy metals in feedstock (Joseph et al. 2021). Applying biochar to soil sequesters biochar-carbon for hundreds to thousands of years and can further increase soil carbon by reducing mineralisation of soil organic matter and newly added plant carbon (Singh et al. 2012; Wang et al. 2016a; Weng et al. 2017; Lehmann et al. 2021). Biochars can improve a range of soil properties, but effects vary depending on biochar properties, which are determined by feedstock and production conditions (Singh et al. 2012; Wang et al. 2016a), and on the soil properties where biochar is applied (Razzaghi et al. 2020). Biochars can increase nutrient availability, reduce leaching losses (Singh et al. 2010; Haider et al. 2017) and enhance crop yields, particularly in infertile acidic soils (Jeffery et al. 2017), thus supporting food security under changing climate. Biochars can enhance infiltration and soil water-holding capacity, reducing runoff and leaching, increasing water retention in the landscape and improving drought tolerance and resilience to climate change (Quin et al. 2014; Omondi et al. 2016). (See Chapter 7 for a review of biochar's potential contribution to climate change mitigation.)

Both A/R and dedicated biomass production could have adverse impacts on food security and cause indirect land-use change if deployed in locations used for food production (IPCC 2019a). But the degree of impact associated with a certain mitigation option also depends on how deployment takes place and the rate and total scale of deployment. The highest increases in food insecurity due to deployment of land-based mitigation are expected to occur in sub-Saharan Africa and Asia (Hasegawa et al. 2018). The land area that could be used for bioenergy or other land-based mitigation options with low to moderate risks to food security depends on patterns of socio-economic development, reaching limits between 1 and 4 million $km^2$ (Hurlbert et al. 2019; IPCC 2019a; Smith et al. 2019c).

The use of less productive, degraded/marginal lands has received attention as an option for biomass production and other land-based mitigation that can improve the productive and adaptive capacity of the lands (Liu et al. 2017; Qin et al. 2018; Dias et al. 2021; Kreig et al. 2021) (Section 7.4.4 and Cross-Working Group Box 3 in this chapter). The potential is however uncertain as biomass growth rates may be low, a variety of assessment approaches have been used, and the identification of degraded/marginal land as 'available' has been contested, as much low productivity land is used informally by impoverished communities, particularly for grazing, or may be economically infeasible or environmentally undesirable for development of energy crops (*medium evidence*, *low agreement*) (Baka 2013; Fritz et al. 2013; Haberl et al. 2013; Baka 2014).

As many of the SDGs are closely linked to land use, the identification and promotion of mitigation options that rely on land uses described above can support a growing use of bio-based products while advancing several SDGs, such as SDG 2 (zero hunger), SDG 6 (clean water and sanitation), SDG 7 (affordable and clean energy) and SDG 15 (life on land) (Fritsche et al. 2017; IRP 2019; Blair et al. 2021). Policies supporting the target of Land Degradation Neutrality (LDN) (SDG 15.3) encourage planning of measures to counteract loss of productive land due to unsustainable agricultural practices and land conversion, through sustainable land management and strategic restoration and rehabilitation of degraded land (Cowie et al. 2018). LDN can thus be an incentive for land-based mitigation measures that build carbon in vegetation and soil, and can provide impetus for land-use planning to achieve multifunctional landscapes that integrate land-based mitigation with other land uses (Box 12.3). The application of sustainable land management practices that build soil carbon will enhance the productivity and resilience of crop and forestry systems, thereby enhancing biomass production (Henry et al. 2018a). Non-bio-based mitigation options can enhance land-based mitigation: (i) enhanced weathering, that is, adding ground silicate rock to soil to take up atmospheric $CO_2$ through chemical weathering (Section 12.3), could supply nutrients and alleviate soil acidity, thereby boosting productivity of biomass crops and A/R, particularly when combined with biochar application (Haque et al. 2019; De Oliveira Garcia et al. 2020; Buss et al. 2021); and (ii) land rehabilitation and enhanced landscape diversity through production of biomass crops could simultaneously contribute to climate change mitigation, climate change adaptation, addressing land degradation, increasing biodiversity and improving food security in the longer term (Mackey et al. 2020) (Chapter 7).

*Wind power*

The land requirement and impacts (including visual and noise impacts) of onshore wind turbines depend on the size and type of installation, and location (Ioannidis and Koutsoyiannis 2020). Wind power and agriculture can coexist in beneficial ways and wind power production on agriculture land is well established (Fritsche et al. 2017; Miller and Keith 2018a). Spatial planning and local stakeholder engagement can reduce opposition due to visual landscape impacts and noise (Frolova et al. 2019; Hevia-Koch and Ladenburg 2019). Repowering, that is, replacing with higher capacity wind turbines, can mitigate additional land requirement associated with deployment towards higher share of wind in power systems (Pryor et al. 2020).

Mortality and disturbance risks to birds, bats and insects are major ecological concerns associated with wind farms (Thaxter et al. 2017; Cook et al. 2018; Heuck et al. 2019; Coppes et al. 2020; Choi et al. 2020; Fernández-Bellon 2020; Marques et al. 2020; Voigt 2021). Careful siting is critical (May et al. 2021), while painting blades to increase the visibility can also reduce mortality due to collision (May et al. 2020). Theoretical studies have suggested that wind turbines could lead to warmer night temperatures due to atmospheric mixing (Keith et al. 2004), later confirmed through observation (Zhou et al. 2013), although Vautard et al. (2014) found limited impact at scales consistent with climate policies. More recent studies report mixed results: indications that the warming effect could be substantial with widespread deployment (Miller and Keith 2018b) and conversely limited impacts on regional climate at 20% of US electricity from wind. (Pryor et al. 2020).

*Solar power*

As for wind power, land impacts of solar power depend on the location, size and type of installation (Ioannidis and Koutsoyiannis 2020). Establishment of large-scale solar farms could have positive or negative environmental effects at the site of deployment, depending on the location. Solar PV and CSP power installations can lock away land areas, displacing other uses (Mohan 2017). Solar

PV can be deployed in ways that enhance agriculture: for example, Hassanpour Adeh et al. (2018) found that biomass production and water use efficiency of pasture increased under elevated solar panels. PV systems under development may achieve significant power generation without diminishing agricultural output (Miskin et al. 2019). Global mapping of solar panel efficiency showed that croplands, grasslands and wetlands are located in regions with the greatest solar PV potential (Adeh et al. 2019). Dual-use agrivoltaic systems are being developed that overcome previously recognised negative impact on crop growth, mainly due to shadows (Marrou et al. 2013a; Marrou et al. 2013b; Armstrong et al. 2016), thus facilitating synergistic co-location of solar photovoltaic power and cropping (Adeh et al. 2019; Miskin et al. 2019). Assessment of the potential for optimising deployment of solar PV and energy crops on abandoned cropland areas produced an estimate of the technical potential for optimal combination at 125 EJ per year (Leirpoll et al. 2021).

Deserts can be well suited for solar PV and CSP farms, especially at low latitudes where global horizontal irradiance is high, as there is lower competition for land and land carbon loss is minimal, although remote locations may pose challenges for power distribution (Xu et al. 2016). Solar arrays can reduce the albedo, particularly in desert landscapes, which can lead to local temperature increases and regional impacts on wind patterns (Millstein and Menon 2011). Modelling studies suggest that large-scale wind and solar farms, for example in the Sahara (Li et al. 2018), could increase rainfall through reduced albedo and increased surface roughness, stimulating vegetation growth and further increasing regional rainfall (Li et al. 2018) (*limited evidence*). Besides impacts at the site of deployment, wind and solar power affect land through mining of critical minerals required by these technologies (Viebahn et al. 2015; McLellan et al. 2016; Carrara et al. 2020).

*Nuclear power*

Nuclear power has land impacts and risks associated with mining operations (Falck 2015; Winde et al. 2017; Srivastava et al. 2020) and disposal of spent fuel (IAEA 2006a; Ewing et al. 2016; Bruno et al. 2020), but the land occupation is small compared to many other mitigation options. Substantial volumes of water are required for cooling (Liao et al. 2016), as for all thermal power plants, but most of this water is returned to rivers and other water bodies after use (Sesma Martín and Rubio-Varas 2017). Negative impacts on aquatic systems can occur due to chemical and thermal pollution loading (Fricko et al. 2016; Raptis et al. 2016; Bonansea et al. 2020). The major risk to land from nuclear power is that a nuclear accident leads to radioactive contamination. An extreme example, the 1986 Chernobyl accident in Ukraine, resulted in radioactive contamination across Europe. Most of the fallout concentrated in Belarus, Ukraine and Russia, where some 125,000 $km^2$ of land (more than a third of which was in agricultural use) was contaminated. About 350,000 people were relocated away from these areas (IAEA 2006b; Sovacool 2008). About 116,000 people were permanently evacuated from the 4200 km Chernobyl exclusion zone (IAEA 2006a). New reactor designs with passive and enhanced safety systems reduce the risk of such accidents significantly (Section 6.4.2.4). An example of alternatives to land reclamation for productive purposes, a national biosphere reserve has been established around Chernobyl to conserve, enhance and manage carbon stocks and biodiversity (Deryabina et al. 2015; Ewing et al. 2016), although invertebrate and plant populations are affected (Mousseau and Møller 2014; Mousseau and Møller 2020).

*Hydropower*

Reservoir hydropower projects submerge areas as dams are established for water storage. Hydropower can be associated with significant and highly varying land occupation and carbon footprint (Poff and Schmidt 2016; Scherer and Pfister 2016a; dos Santos et al. 2017; Ocko and Hamburg 2019). The flooding of land causes $CH_4$ emissions due to the anaerobic decomposition of submerged vegetation and there is also a loss of carbon sequestration due to mortality of submerged vegetation. The size of GHG emissions depends on the amount of vegetation submerged. The carbon in accumulated sediments in reservoirs may be released to the atmosphere as $CO_2$ and $CH_4$ upon decommissioning of dams, and while uncertain, estimates indicate that these emissions can make up a significant part of the cumulative GHG emissions of hydroelectric power plants (Moran et al. 2018; Almeida et al. 2019; Ocko and Hamburg 2019). Positive radiative forcing due to lower albedo of hydropower reservoirs compared to surrounding landscapes can reduce mitigation contribution significantly (Wohlfahrt et al. 2021).

Hydropower can have high water usage due to evaporation from dams (Scherer and Pfister 2016b). Hydropower projects may impact aquatic ecology and biodiversity, necessitate the relocation of local communities living within or near the reservoir or construction sites, and affect downstream communities (in positive or negative ways) (Moran et al. 2018; Barbarossa et al. 2020). Displacement as well as resettlement schemes can have both socio-economic and environmental consequences including those associated with establishment of new agricultural land (Ahsan and Ahmad 2016; Nguyen et al. 2017). Dam construction may also stimulate migration into the affected region, which can lead to deforestation and other negative impacts (Chen et al. 2015). Impacts can be mitigated through basin-scale dam planning that considers GHG emissions along with social and ecological effects (Almeida et al. 2019). Land occupation is minimal for run-of-river hydropower installations, but without storage they have no resilience to drought and installations inhibit dispersal and migration of organisms (Lange et al. 2018). Reservoir hydropower schemes can regulate water flows and reduce flood damage to agricultural production (Amjath-Babu et al. 2019). On the other hand, severe flooding due to failure of hydropower dams has caused fatalities, damage to infrastructure and loss of productive land (Farrington and Salama 1996; Farley et al. 2005; Zhang et al. 2016; Marcar 2016; Zhang et al. 2017; Kalinina et al. 2018; Lu et al. 2018).

### 12.5.4 Governance of Land-related Impacts of Mitigation Options

The land sector (Chapter 7) contributes to mitigation via emissions reduction and enhancement of land carbon sinks, and by providing biomass for mitigation in other sectors. Key challenges for governance

12

of land-based mitigation include social and environmental safeguards (Duchelle et al. 2017; Sills et al. 2017; Larson et al. 2018); insufficient financing (Turnhout et al. 2017); capturing co-benefits; ensuring additionality; addressing non-permanence of carbon sequestration; monitoring, reporting, and verification (MRV) of emissions reduction and carbon dioxide removals; and avoiding leakage or spillover effects. Governance approaches to addressing these challenges are discussed in Section 7.6, and include MRV systems and integrity criteria for project-level emissions trading; payments for ecosystem services; land-use planning and land zoning; certification schemes, standards and codes of practice.

With respect to renewable energy options that occupy land, the focus of governance has been directed to technological adoption and public acceptance (Sequeira and Santos 2018), rather than land use. Recent work has found that spatial processes shape the emerging energy transition, creating zones of friction between global investors, national and local governments, and civil society (Jepson and Caldas 2017; McEwan 2017). For example, Yenneti et al. (2016) have argued that hydropower and ground-based solar parks in India, which have involved enclosure of lands designated as degraded, displacing pastoral use by vulnerable communities, have constituted forms of spatial injustice. Hydropower leads to dam-induced displacement, and though this can be addressed through compensation mechanisms, governance is complicated by a lack of transparency in resettlement data (Kirchherr et al. 2016; Kirchherr et al. 2019). Renewable energy production is resulting in new land conflict frontiers where degraded land is framed as having mitigation value such as for palm oil production and wind power in Mexico (Backhouse and Lehmann 2020); land use conflict as well as impacts on wildlife from large-scale solar installations have also emerged in the southwestern United States (Mulvaney 2017). The renewable energy transition also involves the extraction of critical minerals used in renewable energy technologies, such as lithium and cobalt. Governance challenges include the lack of transparent greenhouse gas accounting for mining activities (Lee et al. 2020a), and threats to biodiversity from land disturbance, which require strategic planning to address (Sonter et al. 2020a). Strategic spatial planning is needed more generally to address trade-offs between using land for renewable energy and food: for example, agriculture can be co-located with solar photovoltaics (Barron-Gafford et al. 2019) or wind power (Miller and Keith 2018a). Integrative spatial planning can integrate renewable energy with not just agriculture, but mobility and housing (Hurlbert et al. 2019). Integrated planning is needed to avoid scalar pitfalls, and local and regional contextualised governance solutions need to be sited within a planetary frame of reference (Biermann et al. 2016). Greater planning and coordination are also needed to ensure co-benefits from land-based mitigation (Box 12.3) as well as from CDR and efforts to reduce food systems emissions.

In emerging domains for governance such as land-based mitigation, global institutions, private sector networks and civil society organisations are playing key roles in terms of norm-setting. The shared languages and theoretical frameworks, or cognitive linkages (Pattberg et al. 2018), that arise with polycentric governance can not only be helpful in creating expectations and establishing benchmarks for (in)appropriate practices where enforceable 'hard law' is missing (Karlsson-Vinkhuyzen et al. 2018; Gajevic Sayegh 2020), they can also form the basis of voluntary guidelines or niche markets (Box 12.3) However, the ability to apply participatory processes for developing voluntary guidelines and other participatory norm-setting endeavours varies from place to place. Social and cultural norms shape the ability of women, youth, and different ethnic groups to participate in governance fora, such as those around agroecological transformation (Anderson et al. 2019). Furthermore, establishing new norms alone does not solve structural challenges such as lack of access to food, nor does it confront power imbalances, or provide mechanisms to deal with uncooperative actors (Morrison et al. 2019).



## Box 12.3 | Land Degradation Neutrality as a Framework to Manage Trade-offs in Land-based Mitigation

The United Nations Convention to Combat Desertification (UNCCD) introduced the concept of Land Degradation Neutrality (LDN), defined as 'a state whereby the amount and quality of land resources necessary to support ecosystem functions and services and enhance food security remain stable or increase within specified temporal and spatial scales and ecosystems' (UNCCD 2015), and it has been adopted as a target of SDG 15 (life on land). At December 2020, 124 (mostly developing) countries had committed to pursue voluntary LDN targets.

The goal of LDN is to maintain or enhance land-based natural capital, and its associated ecosystem services, such as provision of food and regulation of water and climate, while enhancing the resilience of the communities that depend on the land. LDN encourages a dual-pronged approach promoting sustainable land management (SLM) to avoid or reduce land degradation, combined with strategic effort in land restoration and rehabilitation to reverse degradation on degraded lands and thereby deliver the target of 'no net loss' of productive land (Orr et al. 2017).

In the context of LDN, land restoration refers to actions undertaken with the aim of reinstating ecosystem functionality, whereas land rehabilitation refers to actions undertaken with a goal of provision of goods and services (Cowie et al. 2018). Restoration interventions can include destocking to encourage regeneration of native vegetation; shelter belts of local species established from seed or seedlings, strategically located to provide wildlife corridors and link habitat; and rewetting drained peatland. 'Farmer-managed natural regeneration' is a low-cost restoration approach in which regeneration of tree stumps and roots is encouraged, stabilising soil and

*Box 12.3 (continued)*

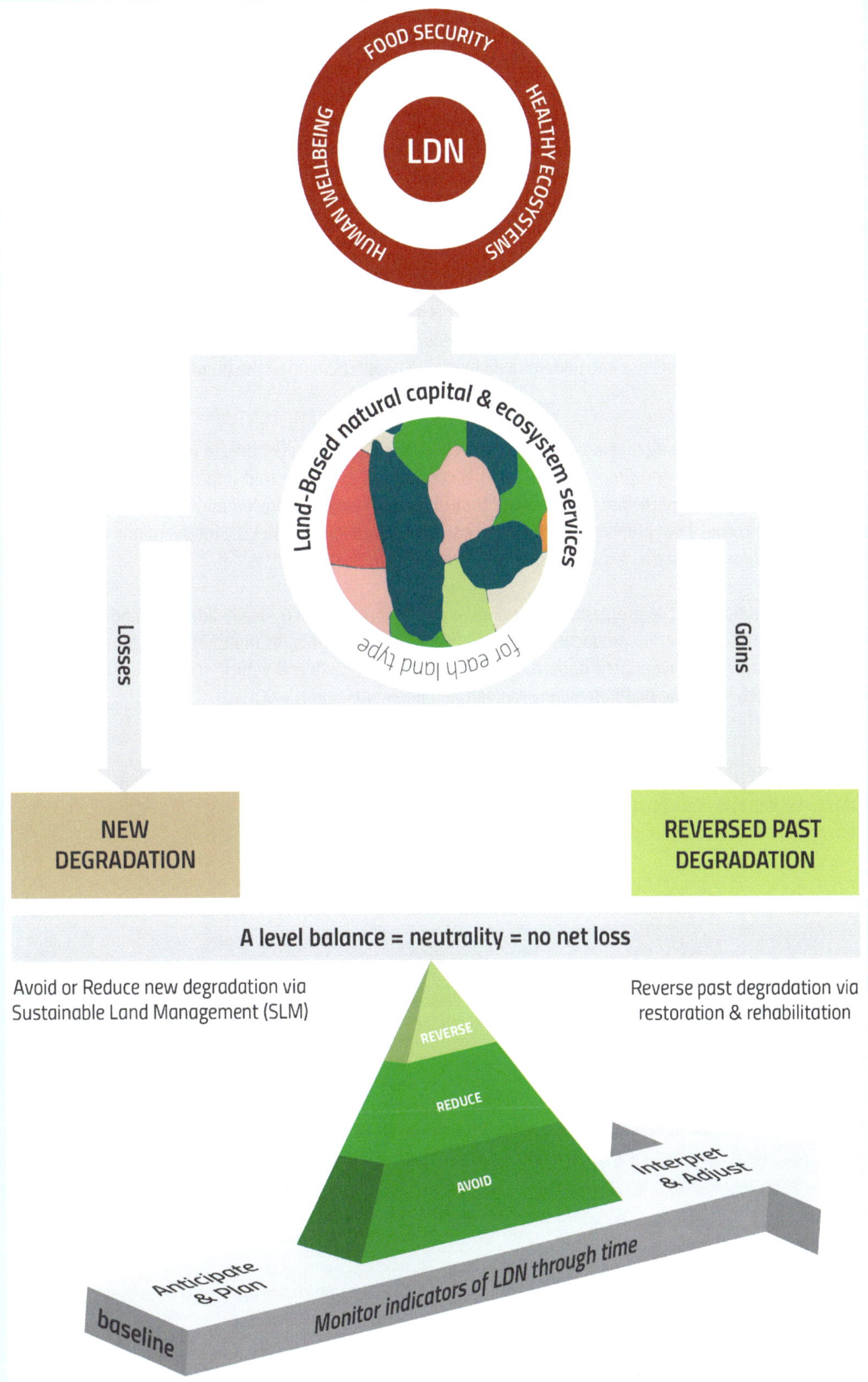


**Box 12.3, Figure 1 | Schematic illustrating the elements of the Land Degradation Neutrality conceptual framework.** Source: Cowie et al. (2018). Used with permission.

12

*Box 12.3 (continued)*

enhancing soil nutrients and organic matter levels (Chomba et al. 2020; Lohbeck et al. 2020). Rehabilitation actions include establishment of energy crops, or afforestation with fast-growing exotic trees to sequester carbon or produce timber. Application of biochar can facilitate rehabilitation by enhancing nutrient retention and water-holding capacity, and stimulating microbial activity (Cowie 2020a).

SLM, rehabilitation and restoration activities undertaken towards national LDN targets have potential to deliver substantial CDR through carbon sequestration in vegetation and soil. In addition, biomass production, for bioenergy or biochar, could be an economically viable land use option for reversing degradation, through rehabilitation. Alternatively, a focus on ecological restoration (Gann et al. 2019) as the strategy for reversing degradation will deliver greater biodiversity benefits.

Achieving neutrality requires estimating the likely impacts of land-use and land-management decisions, to determine the area of land, of each land type, that is likely to be degraded (Orr et al. 2017). This information is used to plan interventions to reverse degradation on an equal area of the same land type. Therefore, pursuit of LDN requires concerted and coordinated efforts to integrate LDN objectives into land-use planning and land management, underpinned by sound understanding of the human–environment system and effective governance mechanisms.

Countries are advised to apply a landscape-scale approach for planning LDN interventions, in which land uses are matched to land potential, and resilience of current and proposed land uses is considered, to ensure that improvement in land condition is likely to be maintained (Cowie 2020a). A participatory approach that enables effective representation of all stakeholders is encouraged, to facilitate equitable outcomes from planning decisions, recognising that decisions on LDN interventions are likely to involve trade-offs between various environmental and socio-economic objectives (Schulze et al. 2021).

Planning and implementation of LDN programmes provides a framework in which locally-adapted land-based mitigation options can be integrated with use of land for production, conservation and settlements, in multifunctional landscapes where trade-offs are recognised and managed, and synergistic opportunities are sought. LDN is thus a vehicle to focus collaboration in pursuit of the multiple land-based objectives of the multilateral environmental agreements and the SDGs.

Table 12.10 collates risks, impacts and opportunities associated with different mitigation options that occupy land.

**Table 12.10 | Summary of impacts, risks and co-benefits associated with land occupation by mitigation options considered in Section 12.5.**

| Mitigation option | Impacts and risks | Opportunities for co-benefits |
|---|---|---|
| **Non-bio-based options that may displace food production** | | |
| Solar farms | Land use competition; loss of soil carbon; heat island effect (scale dependent) (Sections 12.5.3 and 12.5.4) | Target areas unsuitable for agriculture such as deserts (Section 12.5.3) |
| Hydropower (dams) | Land use competition; displacement of natural ecosystems; $CO_2$ and $CH_4$ emissions (Sections 12.5.3 and 12.5.4) | Water storage (including for irrigation) and regulation of water flows; pumped storage can store excess energy from other renewable generation sources (Section 12.5.3) |
| **Non-bio-based options that can (to a varying degree) be integrated with food production** | | |
| Wind turbines | May affect local/regional weather and climate (scale dependent); impacts on wildlife; visual impacts (Section 12.5.3) | Design and siting informed by visual landscape impacts, relevant habitats, and flight trajectories of migratory birds (Section 12.5.3) |
| Solar panels | Land use competition (Section 12.5.3) | Integration with buildings and other infrastructure; integration with food production is being explored (Section 12.5.2) |
| Enhanced weathering (EW) | Disturbance at sites of extraction; ineffective in low rainfall regions (Section 12.3.1.2) | Increased crop yields and biomass production through nutrient supply and increasing pH of acid soils; synergies with biochar (Section 12.5.3) |
| **Bio-based options that may displace existing food production** | | |
| Afforestation/ reforestation (A/R) | Land use competition, potentially leading to indirect land use change; reduced water availability; loss of biodiversity (Section 12.5.3) | Strategic siting to minimise adverse impacts on hydrology, land use, biodiversity (Section 12.5.3) |
| Biomass crops | Land use competition, potentially leading to indirect land-use change; reduced water availability; reduced soil fertility; loss of biodiversity (Section 12.5.3) | Strategic siting to minimise adverse impacts/enhance beneficial effects on land use, landscape variability, biodiversity, soil organic matter, hydrology and water quality (Section 12.5.3) |

| Mitigation option | Impacts and risks | Opportunities for co-benefits |
|---|---|---|
| **Bio-based options that can (to a varying degree) be combined with food production** | | |
| Agroforestry | Competition with adjacent crops and pastures reduces yields (Section 7.4.3.3) | Shelter for stock and crops, diversification, biomass production, increases soil organic matter and soil fertility; increased biodiversity and perennial vegetation enhance beneficial organisms; can reduce need for pesticides (Sections 7.4.3.3 and 12.5.3) |
| Soil carbon management in croplands and grasslands | Increase in nitrous oxide emissions if fertiliser used to enhance crop production; reduced cereal production through increased crop legumes and pasture phases could lead to indirect land use change (Sections 7.4.3.1 and 7.4.3.6) | Increasing soil organic matter improves soil health, increases crop and pasture yields and resilience to drought, can reduce fertiliser requirement, nutrient leaching and need for land use change (Section 7.4.3.1) |
| Biochar addition to soil | Land use competition if biochar is produced from purpose-grown biomass. Loss of forest carbon stock and impacts on biodiversity if biomass is harvested unsustainably. (Section 12.5.3) | Facilitate beneficial use of organic residues, to return nutrients to farmland. Increased land productivity; increased carbon sequestration in vegetation and soil; increased nutrient-use efficiency, and reduced requirement for chemical fertiliser (Sections 7.4.3.2 and 12.5.3) |
| Harvest residue extraction and use for bioenergy, biochar and other bio-products | Decline in soil organic matter and soil fertility (Section 12.5.3) | Nutrients returned to soil e.g., as ash; reduced fuel load and wildfire risk (Sections 7.4.3.2 and 12.5.3) |
| Manure management (i.e., for biogas) | Risk of fugitive emissions<br>Can contain pathogens<br>(Sections 7.4.3.7 and 12.5.3) | Biogas as renewable energy source, digestate as soil amendment (Section 12.5.3) |
| **Options that do not occupy land used for food production** | | |
| Management of organic waste (food waste, biosolids, organic component of municipal solid waste) | Can contain contaminants (heavy metals, persistent organic pollutants, pathogens) (Section 12.5.3) | Processing using anaerobic digestion or pyrolysis produces renewable gas and soil amendment, enabling return of nutrients to farmland. (Note that some feedstock nitrogen is lost in pyrolysis) (Section 12.5.3) |
| A/R and biomass production on degraded non-forested land (e.g., abandoned agricultural land) | High labour and material inputs can be needed; abandoned land can support informal grazing and have significant biodiversity value. Reduced water availability (Section 12.5.3) | Application of biochar can re-establish nutrient cycling; bioenergy crops can add organic matter, restoring soil fertility, and can remove heavy metals, enabling food production (Sections 7.4.3.2 and 12.5.3) |

12

## Cross-Working Group Box 3 | Mitigation and Adaptation via the Bioeconomy

**Authors:** Henry Neufeldt (Denmark/Germany), Göran Berndes (Sweden), Almut Arneth (Germany), Rachel Bezner Kerr (the United States of America/Canada), Luisa F Cabeza (Spain), Donovan Campbell (Jamaica), Jofre Carnicer Cols (Spain), Annette Cowie (Australia), Vassilis Daioglou (Greece), Joanna House (United Kingdom), Adrian Leip (Italy/Germany), Francisco Meza (Chile), Michael Morecroft (United Kingdom), Gert-Jan Nabuurs (Netherlands), Camille Parmesan (United Kingdom/the United States of America), Julio C. Postigo (the United States of America/Peru), Marta G. Rivera-Ferre (Spain), Raphael Slade (United Kingdom), Maria Cristina Tirado von der Pahlen (the United States of America/Spain), Pramod K. Singh (India), Pete Smith (United Kingdom)

### Summary statement

The growing demand for biomass offers both opportunities and challenges to mitigate and adapt to climate change and natural resource constraints (*high confidence*). Increased technology innovation, stakeholder integration and transparent governance structures and procedures at local to global scales are key to successful bioeconomy deployment maximising benefits and managing trade-offs (*high confidence*).

Limited global land and biomass resources accompanied by growing demands for food, feed, fibre, and fuels, together with prospects for a paradigm shift towards phasing out fossil fuels, set the frame for potentially fierce competition for land[3] and biomass to meet burgeoning demands, even as climate change increasingly limits natural resource potentials (*high confidence*).

3 For lack of space, the focus is on land only, although the bioeconomy also includes sea-related bioresources.

*Cross-Working Group Box 3 (continued)*

Sustainable agriculture and forestry, technology innovation in bio-based production within a circular economy, and international cooperation and governance of global trade in products to reflect and disincentivise their environmental and social externalities, can provide mitigation and adaptation via bioeconomy development that responds to the needs and perspectives of multiple stakeholders to achieve outcomes that maximise synergies while limiting trade-offs (*high confidence*).

**Background**

There is *high confidence* that climate change, population growth and changes in per capita consumption will increase pressures on managed as well as natural and semi-natural ecosystems, exacerbating existing risks to livelihoods, biodiversity, human and ecosystem health, infrastructure, and food systems (Conijn et al. 2018; IPCC 2018; IPCC 2019a; Lade et al. 2020). At the same time, many global mitigation scenarios presented in IPCC assessment reports rely on large GHG emissions reduction in the AFOLU sector and concurrent deployment of reforestation/afforestation and biomass use in a multitude of applications (Rogelj et al. 2018; Hanssen et al. 2020) (AR6 WGI Chapters 4 and 5, AR6 WGIII Chapters 3 and 7).

Given the finite availability of natural resources, there are invariably trade-offs that complicate land-based mitigation unless land productivity can be enhanced without undermining ecosystem services (Obersteiner et al. 2016; Campbell et al. 2017; Caron et al. 2018; Conijn et al. 2018; Heck et al. 2018; Searchinger 2018a; Smith et al. 2019). Management intensities can often be adapted to local conditions with consideration of other functions and ecosystem services, but at a global scale the challenge remains to avoid further deforestation and degradation of intact ecosystems, in particular biodiversity-rich systems (AR6 WGII Cross-Chapter Box NATURAL), while meeting the growing demands. Further, increased land-use competition can affect food prices and impact food security and livelihoods (To and Grafton 2015; Chakravorty et al. 2017), with possible knock-on effects related to civil unrest (Abbott et al. 2017; D'Odorico et al. 2018).

**Developing new bio-based solutions while mitigating overall biomass demand growth**

Many existing bio-based products have significant mitigation potential. Increased use of wood in buildings can reduce GHG emissions from cement and steel production while providing carbon storage (Churkina et al. 2020). Substitution of fossil fuels with biomass in manufacture of cement and steel can reduce GHG emissions where these materials are difficult to replace. Dispatchable power based on biomass can provide power stability and quality as the contribution from solar and wind power increases (AR6 WGIII Chapter 6), and biofuels can contribute to reducing fossil fuel emissions in the transport and industry sectors (AR6 WGIII Chapters 10 and 11). The use of bio-based plastics, chemicals and packaging could be increased, and biorefineries can achieve high resource-use efficiency in converting biomass into food, feed, fuels and other bio-based products (Aristizábal-Marulanda and Cardona Alzate 2019; Schmidt et al. 2019). There is also scope for substituting existing bio-based products with more benign products. For example, cellulose-based textiles can replace cotton, which requires large amounts of water, chemical fertilisers and pesticides to ensure high yields.

While increasing and diversified use of biomass can reduce the need for fossil fuels and other GHG-intensive products, unfavourable GHG balances may limit the mitigation value. Growth in biomass use may in the longer term also be constrained by the need to protect biodiversity and ecosystems' capacity to support essential ecosystem services. Biomass use may also be constrained by water scarcity and other resource scarcities, and/or challenges related to public perception and acceptance due to impacts caused by biomass production and use. Energy conservation and efficiency measures and deployment of technologies and systems that do not rely on carbon, such as carbon-free electricity supporting, *inter alia*, electrification of transport as well as industry processes and residential heating (IPCC 2018; UNEP 2019), can constrain the growth in biomass demand when countries seek to phase out fossil fuels and other GHG-intensive products while providing an acceptable standard of living. Nevertheless, demand for bio-based products may become high where full decoupling from carbon is difficult to achieve (e.g., aviation, bio-based plastics and chemicals) or where carbon storage is an associated benefit (e.g., wood buildings, BECCS, biochar for soil amendments), leading to challenging trade-offs (e.g., food security, biodiversity) that need to be managed in environmentally sustainable and socially just ways.

Changes on the demand side as well as improvements in resource-use efficiencies within the global food and other bio-based systems can also reduce pressures on the remaining land resources. For example, dietary changes toward more plant-based food (where appropriate) and reduced food waste can provide climate change mitigation along with health benefits (Willett et al. 2019) (AR6 WGIII Sections 7.4 and 12.4) and other co-benefits with regard to food security, adaptation and land use (Mbow et al. 2019; Smith et al. 2019a) (AR6 WGII Chapter 5). Advancements in the provision of novel food and feed sources (e.g., cultured meat, insects, grass-based protein feed and cellular agriculture) can also limit the pressures on finite natural resources (Parodi et al. 2018; Zabaniotou 2018) (AR6 WGIII Section 12.4).

*Cross-Working Group Box 3 (continued)*

## Circular bioeconomy

Circular economy approaches (AR6 WGIII Section 12.6) are commonly depicted by two cycles, where the biological cycle focuses on regeneration in the biosphere and the technical cycle focuses on reuse, refurbishment and recycling to maintain value and maximise material recovery (Mayer et al. 2019a). Biogenic carbon flows and resources are part of the biological carbon cycle, but carbon-based products can be included in, and affect, both the biological and the technical carbon cycles (Kirchherr et al. 2017; Winans et al. 2017; Velenturf et al. 2019). The integration of circular economy and bioeconomy principles has been discussed in relation to organic waste management (Teigiserova et al. 2020), societal transition and policy development (European Commission 2018; Bugge et al. 2019) as well as COVID-19 recovery strategies (Palahí et al. 2020). To maintain the natural resource base, circular bioeconomy emphasises sustainable land use and the return of biomass and nutrients to the biosphere when it leaves the technical cycle.

Scarcity is an argument for adopting circular economy principles for the management of biomass, as for non-renewable resources. Waste avoidance, product reuse and material recycling keep down resource use while maintaining product and material value. However, reuse and recycling are not always feasible, for example when biofuels are used for transport and bio-based biodegradable chemicals are used to reduce ecological impacts, where losses to the environment are unavoidable. A balanced approach to management of biomass resources could start from the perspective of value preservation within the carbon cycle, with possible routes for biomass use based on the carbon budget defined by the Paris Agreement, principles for sustainable land use and natural ecosystem protection.

## Land-use opportunities and challenges in the bioeconomy

Analyses of synergies and trade-offs between adaptation and mitigation in the agriculture and forestry sectors show that outcomes depend on context, design and implementation, so actions have to be tailored to the specific conditions to minimise adverse effects (Kongsager 2018). This is supported in literature analysing the nexus between land, water, energy and food in the context of climate change, which consistently concludes that addressing these different domains together rather than in isolation would enhance synergies and reduce trade-offs (Obersteiner et al. 2016; D'Odorico et al. 2018; Soto Golcher and Visseren-Hamakers 2018; Froese et al. 2019; Momblanch et al. 2019).

Nature-based solutions addressing climate change can provide opportunities for sustainable livelihoods as well as multiple ecosystem services, such as flood risk management through floodplain restoration, saltmarshes, mangroves or peat renaturation (UNEP 2021; AR6 WGII Cross-Chapter Box NATURAL). Climate-smart agriculture can increase productivity while enhancing resilience and reducing GHG emissions inherent to production (Lipper et al. 2014; Bell et al. 2018; FAO 2019b; Singh and Chudasama 2021). Similarly, climate-smart forestry considers the whole value chain and integrates climate objectives into forest sector management through multiple measures (from strict reserves to more intensively managed forests) providing mitigation and adaptation benefits (Nabuurs et al. 2018; Verkerk et al. 2020) (AR6 WGIII Section 7.3).

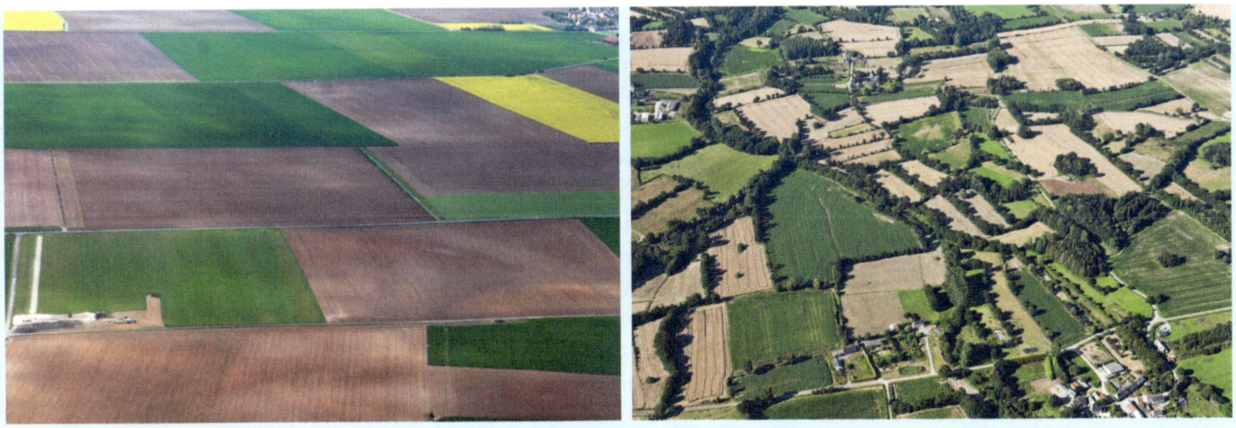

**Cross-Working Group Box 3, Figure 1 | Left:** High-input intensive agriculture, aiming for high yields of a few crop species, with large fields and no semi-natural habitats. **Right:** Agroecological agriculture, supplying a range of ecosystem services, relying on biodiversity and crop and animal diversity instead of external inputs, and integrating plant and animal production, with smaller fields and presence of semi-natural habitats. Source: Reprinted by permission from Springer Nature Customer Service Centre GmbH: Springer Nature, *Nature Sustainability*, Towards better representation of organic agriculture in life cycle assessment, Hayo M. G. van der Werf et al. © 2020.

*Cross-Working Group Box 3 (continued)*

Agroecological approaches can be integrated into a wide range of land management practices to support a sustainable bioeconomy and address equity considerations (HLPE 2019). Relevant land-use practices, such as agroforestry, intercropping, organic amendments, cover crops and rotational grazing, can provide mitigation and support adaption to climate change via food security, livelihoods, biodiversity and health co-benefits (Ponisio et al. 2015; Garibaldi et al. 2016; D'Annolfo et al. 2017; Bezner Kerr et al. 2019; Clark et al. 2019b; Córdova et al. 2019; HLPE 2019; Mbow et al. 2019; Renard and Tilman 2019; Sinclair et al. 2019; Bharucha et al. 2020; Bezner Kerr et al. 2021) (AR6 WGII Cross-Chapter Box NATURAL). Strategic integration of appropriate biomass production systems into agricultural landscapes can provide biomass for bioenergy and other bio-based products while providing co-benefits such as enhanced landscape diversity, habitat quality, retention of nutrients and sediment, erosion control, climate regulation, flood regulation, pollination and biological pest and disease control (Christen and Dalgaard 2013; Asbjornsen et al. 2014; Holland et al. 2015; Ssegane et al. 2015; Dauber and Miyake 2016; Milner et al. 2016; Ssegane and Negri 2016; Styles et al. 2016; Zumpf et al. 2017; Cacho et al. 2018; Alam and Dwivedi 2019; Cubins et al. 2019; HLPE 2019; Olsson et al. 2019; Zalesny et al. 2019; Englund et al. 2020) (AR6 WGIII Box 12.3). Such approaches can help limit environmental impacts from intensive agriculture while maintaining or increasing land productivity and biomass output.

Transitions from conventional to new biomass production and conversion systems include challenges related to cross-sector integration and limited experience with new crops and land use practices, including needs for specialised equipment (Thornton and Herrero 2015; HLPE 2019) (AR6 WGII Section 5.10). Introduction of agroecological approaches and integrated biomass/food crop production can result in lower food crop yields per hectare, particularly during transition phases, potentially causing indirect landuse change, but can also support higher and more stable yields, reduce costs, and increase profitability under climate change (Muller et al. 2017; Seufert and Ramakutty 2017; Barbieri et al. 2019; HLPE 2019; Sinclair et al. 2019; Smith et al. 2019a; Smith et al. 2020). Crop diversification, organic amendments, and biological pest control (HLPE 2019) can reduce input costs and risks of occupational pesticide exposure and food and water contamination (González-Alzaga et al. 2014; EFSA 2017; Mie et al. 2017), reduce farmers' vulnerability to climate change (e.g., droughts and spread of pests and diseases affecting plant and animal health) (Delcour et al. 2015; FAO 2020) and enhance provisioning and sustaining ecosystem services, such as pollination (D'Annolfo et al. 2017; Sinclair et al. 2019).

Barriers toward wider implementation include absence of policies that compensate land owners for providing enhanced ecosystem services and other environmental benefits, which can help overcome short-term losses during the transition from conventional practices before longer-term benefits can accrue. Other barriers include limited access to markets, knowledge gaps, financial, technological or labour constraints, lack of extension support and insecure land tenure (Jacobi et al. 2017; Kongsager 2017; Hernández-Morcillo et al. 2018; Iiyama et al. 2018; HLPE 2019). Regional-level agroecology transitions may be facilitated by co-learning platforms, farmer networks, private sector, civil society groups, regional and local administration and other incentive structures (e.g., price premiums, access to credit, regulation) (Coe et al. 2014; Pérez-Marin et al. 2017; Mier y Terán Giménez Cacho et al. 2018; HLPE 2019; Valencia et al. 2019; SAEPEA 2020). With the right incentives, improvements can be made with regard to profitability, making alternatives more attractive to land owners.



### Governing the solution space

Literature analysing the synergies and trade-offs between competing demands for land suggest that solutions are highly contextualised in terms of their environmental, socio-economic and governance-related characteristics, making it difficult to devise generic solutions (Haasnoot et al. 2020). Aspects of spatial and temporal scale can further enhance the complexity, for instance where transboundary effects across jurisdictions or upstream-downstream characteristics need to be considered, or where climate change trajectories might alter relevant biogeophysical dynamics (Postigo and Young 2021). Nonetheless, there is broad agreement that taking the needs and perspectives of multiple stakeholders into account in a transparent process during negotiations improves the chances of achieving outcomes that maximise synergies while limiting trade-offs (Ariti et al. 2018; Metternicht 2018; Favretto et al. 2020; Kopáček 2021; Muscat et al. 2021). Yet differences in agency and power between stakeholders or anticipated changes in access to or control of resources can undermine negotiation results even if there is a common understanding of the overarching benefits of more integrated environmental agreements and the need for greater coordination and cooperation to avoid longer-term losses to all (Aarts and Leeuwis 2010; Weitz et al. 2017). There is also the risk that strong local participatory processes can become disconnected from broader national plans, and thus fail to support the achievement of national targets. Thus, connection between levels is needed to ensure that ambition for transformative change is not derailed at local level (Aarts and Leeuwis 2010; Postigo and Young 2021).

Decisions on land uses between biomass production for food, feed, fibre or fuel, as well as nature conservation or restoration and other uses (e.g., mining, urban infrastructure), depend on differences in perspectives and values. Because the availability of land for diverse biomass uses is invariably limited, setting priorities for land-use allocations therefore first depends on making the perspectives underlying what is considered as 'high-value' explicit (Fischer et al. 2007; Garnett et al. 2015; De Boer and Van Ittersum 2018;

*Cross-Working Group Box 3 (continued)*

Muscat et al. 2020). Decisions can then be made transparently based on societal norms, needs and the available resource base. Prioritisation of land use for the common good therefore requires societal consensus building embedded in the socio-economic and cultural fabric of regions, societies and communities. Integration of local decision-making with national planning ensures local actions complement national development objectives.

International trade in the global economy today provides important opportunities to connect producers and consumers, effectively buffering price volatilities and potentially offering producer countries access to global markets, which can be seen as an effective adaptation measure (Baldos and Hertel 2015; Costinot et al. 2016; Hertel and Baldos 2016; Gouel and Laborde 2021) (AR6 WGII Section 5.11). But there is also clear evidence that international trade and the global economy can enhance price volatility, lead to food price spikes and affect food security due to climate and other shocks, as seen recently due to the COVID-19 pandemic (Cottrell et al. 2019; WFP-FSIN 2020; Verschuur et al. 2021) (AR6 WGII Section 5.12). The continued strong demand for food and other bio-based products, mainly from high- and middle-income countries, therefore requires better cooperation between nations and global governance of trade to more accurately reflect and disincentivise their environmental and social externalities. Trade in agricultural and extractive products driving land-use change in tropical forest and savanna biomes is of major concern because of the biodiversity impacts and GHG emissions incurred in their provision (Hosonuma et al. 2012; Forest Trends 2014; Smith et al. 2014; Henders et al. 2015; Curtis et al. 2018; Pendrill et al. 2019; Seymour and Harris 2019; Kissinger et al. 2021) (AR6 WGII Tropical Forests Cross-Chapter Paper).

In summary, there is significant scope for optimising use of land resources to produce more biomass while reducing adverse effects (*high confidence*). Context-specific prioritisation, technology innovation in bio-based production, integrative policies, coordinated institutions and improved governance mechanisms to enhance synergies and minimise trade-offs can mitigate the pressure on managed as well as natural and semi-natural ecosystems (*medium confidence*). Yet, energy conservation and efficiency measures, and deployment of technologies and systems that do not rely on carbon-based energy and materials, are essential for mitigating biomass demand growth as countries pursue ambitious climate goals (*high confidence*).

## 12.6 Other Cross-sectoral Implications of Mitigation

This section presents further cross-sectoral considerations related to GHG mitigation. Firstly, various cross-sectoral perspectives on mitigation actions are presented. Then, sectoral policy interactions are presented. Finally, implications in terms of international trade spillover effects and competitiveness, and finance flows and related spillover effects at the sectoral level, are addressed.

### 12.6.1 Cross-sectoral Perspectives on Mitigation Action

Chapters 5 to 11 present mitigation measures applicable in individual sectors, and potential co-benefits and adverse side effects[4] of these individual measures. This section builds on the sectoral analysis of mitigation action from a cross-sectoral perspective. Firstly, Section 12.6.1.1 brings together some of the observations presented in the sectoral chapters to show how different mitigation actions in different sectors can contribute to the same co-benefits and result in the same adverse side effects, thereby demonstrating the potential synergistic effects. The links between these co-benefits and adverse side effects and the SDGs is also demonstrated. In Section 12.6.1.2, the focus turns from sector-specific mitigation measures to mitigation measures which have cross-sectoral implications, including measures that have application in more than one sector and measures where implementation in one sector impacts on implementation in another. Finally, Section 12.6.1.3 notes the cross-sectoral relevance of a selection of general-purpose technologies, a topic that is covered further in Chapter 16.

#### 12.6.1.1 A Cross-sectoral Perspective on Co-benefits and Adverse Side Effects of Mitigation Measures, and Links with the SDGs

A body of literature has been developed which addresses the co-benefits of climate mitigation action (Karlsson et al. 2020). Adverse side effects of mitigation are also well documented. Co-benefits and adverse side effects in individual sectors and associated with individual mitigation measures are discussed in the individual sector chapters (Sections 5.2, 6.7.7, 7.4, 7.6, 8.2, 8.4, 9.8, 10.1.1 and 11.5.3), as well as in previous IPCC General and Special Assessment reports. The term 'co-impacts' has been proposed to capture both the co-benefits and adverse side effects of mitigation. An alternative framing is one of multiple objectives, where climate change mitigation is placed alongside other objectives when assessing policy decisions

4 Here, the term co-benefits is used to refer to the additional benefits to society and the environment that are realised in parallel with emissions reductions, while an understanding of adverse side effects highlights where policy- and decision makers are required to make trade-offs between mitigation benefits and other impacts. The choice of language differs to some degree in other chapters.

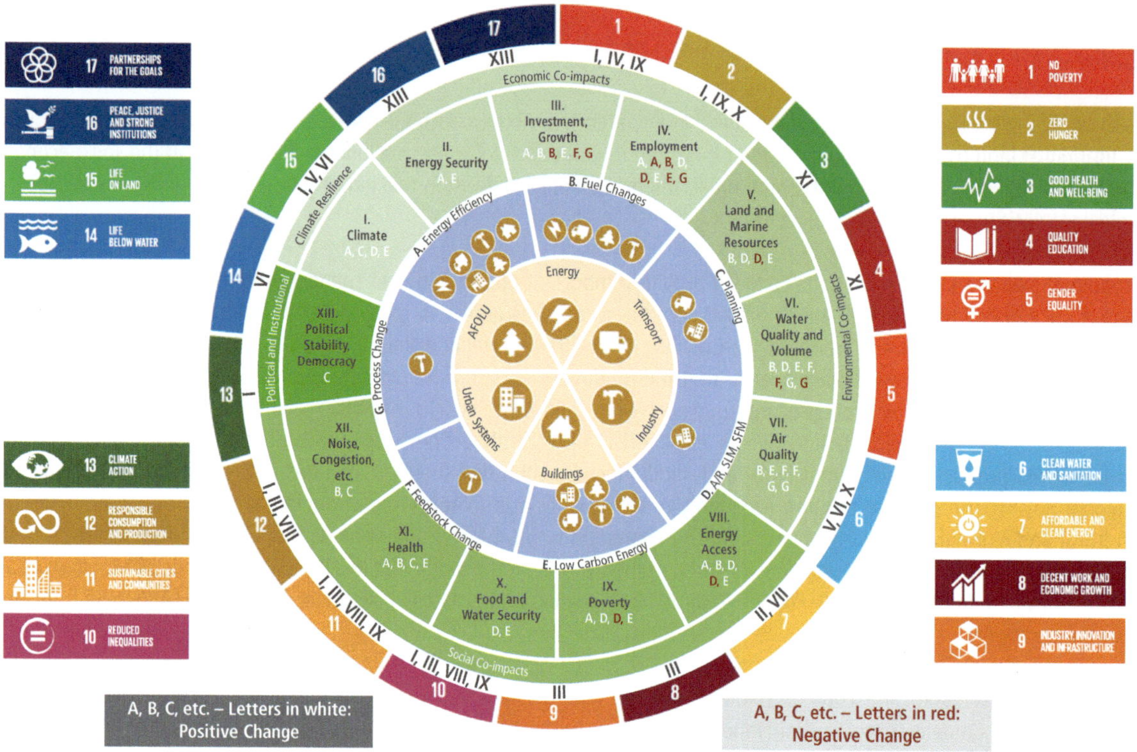


**Figure 12.9 | Co-benefits and adverse side effects of mitigation actions with links to the SDGs.** The inner circle represents the sectors in which mitigation occurs. The second circle shows different generic types of mitigation actions (A to G), with the symbols showing which sectors they are applicable to. The third circle indicates different types of climate related co-benefits (green letters) and adverse side effects (red letters) that may be observed as a result of implementing each of the mitigation actions. Here I relates to climate resilience, II-IV economic co-impacts, V-VII environmental, VIII-XII social, and XIII political and institutional. The final circle maps co-benefits and adverse side effects relevant to the SDGs. Source: re-used with permission from Cohen et al. (2021).

(Ürge-Vorsatz et al. 2014; Mayrhofer and Gupta 2016; Cohen et al. 2017; Bhardwaj et al. 2019).

The identification and assessment of co-benefits has been argued to serve a number of functions (Section 1.4) including using them as leverage for securing financial support for implementation, providing justification of actions which provide a balance of both short- and long-term benefits and obtaining stakeholder buy-in (*robust evidence*, *low agreement*) (Karlsson et al. 2020). Assessment of adverse side effects has been suggested to be useful in avoiding unforeseen negative impacts of mitigation and providing policy- and decision-makers with the information required to make informed trade-offs between climate and other benefits of actions (Ürge-Vorsatz et al. 2014; Bhardwaj et al. 2019; Cohen et al. 2019) (*high evidence*, *low agreement*).

Various approaches to identifying and organising co-impacts in specific contexts and across sectors have been proposed towards providing more comparable and standardised analyses. However, consistent quantification of co-impacts, including cost-benefit analysis, and the utilisation of the resulting information, remain a challenge (Ürge-Vorsatz et al. 2014; Floater et al. 2016; Mayrhofer and Gupta 2016; Cohen et al. 2019; Karlsson et al. 2020). This challenge is further exacerbated when considering that co-impacts of a mitigation measure in one sector can either enhance or reduce the co-impacts associated with mitigation in another, or the achievement of co-benefits in one geographic location can lead to adverse side effects in another. For example, the production of lithium for batteries for energy storage has the potential to contribute to protecting water resources and reducing wastes associated with coal-fired power in many parts of the world, but mining of lithium has the potential for creating water and waste challenges if not managed properly (Agusdinata et al. 2018; Kaunda 2020).

While earlier literature has suggested that co-impacts assessments can support adoption of climate mitigation action, a more recent body of literature has suggested limitations in such framing (Ryan 2015; Bernauer and McGrath 2016; Walker et al. 2018). Presenting general information on co-impacts as a component of a mitigation analysis does not always lead to increased support for climate mitigation action. Rather, the most effective framing is determined by factors relating to local context, type of mitigation action under consideration and target stakeholder group. More work has been identified to be required to bring context into planning co-impacts assessments and communication thereof (Ryan 2015; Bernauer and McGrath 2016; Walker et al. 2018) (*low evidence*, *low agreement*).

An area where the strong link between the cross-sectoral co-impacts of mitigation action and global government policies is being clearly considered is in the achievement of the SDGs (Obergassel et al. 2017; Doukas et al. 2018; Markkanen and Anger-Kraavi 2019; Smith et al. 2019; van Soest et al. 2019) (Chapters 1 and 17, individual sectoral chapters). Figure 12.9 demonstrates these relationships from a cross-sectoral perspective. It shows the links between sectors which give rise to emissions, the mitigation measures that can find application in the sector, and co-benefits and adverse side effects of mitigation measures and the SDGs (noting that the figure is not intended to be comprehensive). Such a framing of co-impacts from a cross-sectoral perspective in the context of the SDGs could help to further support climate mitigation action, particularly within the context of the Paris Agreement (Gomez-Echeverri 2018) (*medium evidence*, *medium agreement*). Literature sources utilised in the compilation of this diagram are presented in Supplementary Material 12.SM.3.

#### 12.6.1.2 Mitigation Measures from a Cross-sectoral Perspective

Three aspects of mitigation from a cross-sectoral perspective are considered, following Barker et al. (2007):

- mitigation measures used in more than one sector;
- implications of mitigation measures for interaction and integration between sectors; and
- competition among sectors for scarce resources.

A number of mitigation measures find application in more than one sector. Renewable energy technologies such as solar and wind may be used for grid electricity supply, as embedded generation in the buildings sector and for energy supply in the agriculture sector (Shahsavari and Akbari 2018) (Chapters 6, 7 and 8). Hydrogen and fuel cells, coupled with low-carbon energy technologies for producing the hydrogen, are being explored in transport, urban heat, industry and for balancing electricity supply (Dodds et al. 2015; Staffell et al. 2019) (Chapters 6, 8 and 11). Electric vehicles are considered an option for balancing variable power (Kempton and Tomić 2005; Liu and Zhong 2019). Carbon capture and storage (CCS) and carbon capture and utilisation (CCU) have potential application in a number of industrial processes (cement, iron and steel, petroleum refining and pulp and paper) (Leeson et al. 2017; Garcia and Berghout 2019) (Chapters 6 and 11) and the fossil fuel electricity sector (Chapter 6). When coupled with energy recovery from biomass, CCS can provide a carbon sink (BECCS) (Section 12.5). On the demand side, energy efficiency options find application across the sectors (Chapters 6, 8, 9, 10, and 11), as do reducing demand for goods and services (Chapter 5) and improving material efficiency (Section 11.3.2).

A range of examples where mitigation measures result in cross-sectoral interactions and integration is identified. The mitigation potential of electric vehicles, including plug-in hybrids, is linked to the extent of decarbonisation of the electricity grid, as well as to the liquid fuel supply emissions profile (Lutsey 2015). Making buildings energy positive, where excess energy is used to charge vehicles, can increase the potential of electric and hybrid vehicles (Zhou et al. 2019). Advanced process control and process optimisation in industry can reduce energy demand and material inputs (Section 11.3), which in turn can reduce emissions linked to resource extraction and manufacturing. Reductions in coal-fired power generation through replacement with renewables or nuclear power result in a reduction in coal mining and its associated emissions. Increased recycling results in a reduction in emissions from primary resource extraction. CCU can contribute to the transition to more renewable energy systems via power-to-X technologies, which enables the production of $CO_2$-based fuels/e-fuels and chemicals using carbon dioxide and hydrogen (Breyer et al. 2015; Anwar et al. 2020). Certain emissions reductions in the AFOLU sector are contingent on energy sector decarbonisation. Trees and green roofs planted to counter urban heat islands reduce the demand for energy for air conditioning and simultaneously sequester carbon (Kim and Coseo 2018; Kuronuma et al. 2018). Recycling of organic waste avoids methane generation if the waste would have been disposed of in landfill sites, can generate renewable energy if treated through anaerobic digestion, and can reduce requirements for synthetic fertiliser production if the nutrient value is recovered (Creutzig et al. 2015). Liquid transport biofuels link to the land, energy and transport sectors (Section 12.5.2.2).

Demand-side mitigation measures, discussed in Chapter 5, also have cross-sectoral implications which need to be taken into account when calculating mitigation potentials. Residential electrification has the potential to reduce emissions associated with lighting and heating, particularly in developing countries where these are currently met by fossil fuels and using inefficient technologies, but will increase demand for electricity (Chapters 5 and 8 and Sections 6.6.2.3 and 8.4.3.1). Many industrial processes can also be electrified in the move away from fossil reductants and direct energy carriers (Chapter 11). The impact of electrification on electricity sector emissions will depend on whether electricity generation is based on fossil fuels in the absence of CCS or low-carbon energy sources (Chapter 5).

At the same time, saving electricity in all sectors reduces the demand for electricity, thereby reducing mitigation potential of renewables and CCS. Demand-side flexibility measures and electrification of vehicle fleets are supportive of more intermittent renewable energy supply options (Sections 6.3.7, 6.4.3.1 and 10.3.4). Production of maize, wheat, rice and fresh produce requires lower energy inputs on a lifecycle basis than poultry, pork and ruminant-based meats (Clark and Tilman 2017) (Section 12.4). It also requires less land area per kilocalorie or protein output (Clark and Tilman 2017; Poore and Nemecek 2018), so replacing meat with these products makes land available for sequestration, biodiversity or other societal needs. However, production of co-products of the meat industry, such as leather and wool, is reduced, resulting in a need for substitutes. Further discussion and examples of cross-sectoral implications of mitigation, with respect to cost and potentials, are presented in Section 12.2. One final example on this topic included here is that of circular economy (Box 12.4).



Finally, in terms of competition among sectors for scarce resources, this issue is often considered in the assessments of mitigation potentials linked to bioenergy and diets (vegetable vs animal food products), land use and water (*robust evidence*, *high agreement*) (Section 12.5 and Cross-Working Group Box 3 in this Chapter). It is, however, also relevant elsewhere. Constraints have been identified in the supply of indium, tellurium, silver, lithium, nickel and platinum that are required for implementation of some specific renewable energy technologies (Watari et al. 2018; Moreau et al. 2019). Other studies have shown constraints in supply of cobalt, one of the key elements used in production of lithium-ion batteries, which has been assessed for mitigation potential in energy, transport and buildings sectors (*medium evidence*, *high agreement*) (Jaffe 2017; Olivetti et al. 2017), although alternatives to cobalt are being developed (Olivetti et al. 2017; Watari et al. 2018).

## Box 12.4 | Circular Economy from a Cross-Sectoral Perspective

Circular economy approaches consider the entire lifecycle of goods and services, and seek to design out waste and pollution, keep products and materials in use, and regenerate natural systems (The Ellen MacArthur Foundation 2013; CIRAIG 2015). The use of circular economy for rethinking how society's needs for goods and services is delivered in such a way as to minimise resource use and environmental impact and maximise societal benefit has been discussed elsewhere in this assessment report (Chapter 5 and Section 5.3.4). A wide range of potential application areas is identified, from food systems to bio-based products to plastics to metals and minerals to manufactured goods. Circular economy approaches are implicitly cross-sectoral, impacting the energy, industrial, AFOLU, waste and other sectors. They will have climate and non-climate co-benefits and trade-offs. The scientific literature mainly investigates incremental measures claiming but not demonstrating mitigation; highest mitigation potential is found in the industry, energy, and transport sectors; mid-range potential in the waste and building sectors; and lowest mitigation gains in agriculture (Cantzler et al. 2020). Circular economy thinking has been identified to support increased resilience to the physical effects of climate change and contribute to meeting other SDGs, notably SDG 12 (responsible consumption and production) (The Ellen MacArthur Foundation 2019).

Circular economy approaches to deployment of low-carbon infrastructure have been suggested to be important to optimise resource use and mitigate environmental and societal impacts caused by extraction and manufacturing of composite and critical materials as well as infrastructure decommissioning (Jensen and Skelton 2018; Sica et al. 2018; Salim et al. 2019; Watari et al. 2019; Jensen et al. 2020; Mignacca et al. 2020). The circular carbon economy is an approach inspired by the circular economy principles that rely on a combination of technologies, including CCU, CCS and CDR, to enable transition pathways especially relevant in economies dependent on fossil fuel exports (Lee et al. 2017; Alshammari 2020; Morrow and Thompson 2020; Zakkour et al. 2020). The integration of circular economy and bioeconomy principles (Cross-Working Group Box 3 in this chapter) is conceptualised in relation to policy development (European Commission 2018) as well as COVID-19 recovery strategies (Palahí et al. 2020), emphasising the use of renewable energy sources and sustainable management of ecosystems with transformation of biological resources into food, feed, energy and biomaterials.

At this stage, however, there is no single global agreement on how circular economy principles are best implemented, and differential government support for circular economy interventions is observed in different jurisdictions.

#### 12.6.1.3 Cross-sectoral Considerations Relating to Emerging General-purpose Technologies

General-purpose technologies (GPTs) include, but are not limited to, additive manufacturing, artificial intelligence, biotechnology, hydrogen, digitalisation, electrification, nanotechnology and robots (de Coninck et al. 2018). Many of the individual sectoral chapters have identified the roles that such technologies can have in supporting mitigation of GHG emissions. Section 16.2.2.3 presents an overview of the individual technologies and specific applications thereof.

In this chapter, which focuses on cross-sectoral implications of mitigation, it is highlighted that certain of these GPTs will find application across the sectors, and there will be synergies and trade-offs when utilising these technologies in more than sector. One example here is the use of hydrogen as an energy carrier, which, when coupled with low-carbon energy, has potential for driving mitigation in energy, industry, transport, and buildings. The increased uptake of hydrogen across the economy requires establishment of hydrogen production, transport and storage infrastructure which could simultaneously support multiple sectors, although there is the potential to utilise existing infrastructure in some parts of the world (Alanne and Cao 2017).

Box 12.5 provides further details on hydrogen in the context of cross-sectoral mitigation specifically, while further details on the role of hydrogen in individual sectors are provided in Chapters 6, 8. 9, 10 and 11. In contrast, the benefits of digitalisation, which could potentially give rise to substantial energy savings across multiple sectors, need to be traded off against demand for electricity to operate consumer devices, data centres, and data networks. Measures are required to increase energy efficiency of these technologies (IEA 2017). Section 5.3.4.1 of this report provides further information on energy and emissions benefits and costs of digitalisation.

With respect to co-impacts of GPTs, the other focus of this chapter, it is highlighted that assessment of the environmental, social and economic implications of such technologies is challenging and context specific, with multiple potential cross-sectoral linkages (de Coninck et al. 2018). Each GPT would need to be explored in context of what it is being used for, and potentially in the geographical context, in order to understand the co-impacts of its use.

## Box 12.5 | Hydrogen in the Context of Cross-sectoral Mitigation Options

Interest in hydrogen as an intermediary energy carrier has grown rapidly in the years since the 5th Assessment Report of WGIII (AR5) was published. This is reflected in this WGIII assessment report, where the term 'hydrogen' is used more than five times more often than in AR5. In Chapter 6 of this report, it is shown that hydrogen can be produced with low carbon impact from fossil fuels (Section 6.4.2.6), renewable electricity and nuclear energy (Section 6.4.5.1), or biomass (Section 6.4.2.5). In the energy sector, hydrogen is one of the options for storage of energy in low-carbon electricity systems (Sections 6.4.4.1 and 6.6.2.2). But, also importantly, hydrogen can be produced to be used as a fuel for sectors that are hard to decarbonise; this is possible directly in the form of hydrogen, but also in the form of ammonia or other energy carriers (Section 6.4.5.1). In the transport sector, fuel cell engines (Section 10.3.3) running on hydrogen can become important, especially for heavy duty vehicles (Section 10.4.3). In the industry sector hydrogen already plays an important role in the chemical sector (for ammonia and methanol production) (Box 11.1 in Chapter 11) and in the fuel sector (in oil refinery processes and for biofuel production) (IEA 2019b). Beyond the production of ammonia and methanol for both established and novel applications, the largest potential industrial application for low-carbon hydrogen is seen in steel-making (Section 11.4.1.1). Hydrogen and hydrogen derivatives can play a further role as substitute energy carriers (Section 11.3.5) and for the production of intermediate chemical products such as methanol, ethanol and ethylene when combined with CCU (Section 11.3.6). For the building sector, the exploration of the usefulness of hydrogen is at an early stage (Box 9.4).

An overview report (IEA 2019b) already sees opportunities in 2030 for buildings, road freight and passenger vehicles. This report also suggests a high potential application in iron and steel production, aviation and maritime transport, and for electricity storage. Several industry roadmaps have been published that map out a possible role for hydrogen until 2050. The most well known and ambitious is the roadmap by the Hydrogen Council (2017), which sketches a global scenario leading to 78 EJ hydrogen use in 2050, mainly for transport, industrial feedstock, industrial energy and to a lesser extent for buildings and power generation. Hydrogen makes up 18% of total final energy use in this vision. An analysis by IRENA on hydrogen from renewable sources comes to a substantially lower number: 8 EJ (excluding hydrogen use in power production and feedstock uses). On a regional level, most roadmaps and scenarios have been published for the European Union, for example by the Fuel Cell and Hydrogen Joint Undertaking (Blanco et al. 2018; EC 2018; FCH 2019; Navigant 2019). All these reports have scenario variants with hydrogen share in final energy use of 10% to over 20% by 2050.

When it comes to the production of low-carbon hydrogen, the focus of the attention is on production using electricity from renewable sources via electrolysis, so-called 'green hydrogen'. However, 'blue hydrogen', produced out of natural gas with CCS, is also often considered. Since a significantly increasing role for hydrogen would require considerable infrastructure investments and would affect existing trade flows in raw materials, governments have started to set up national hydrogen strategies, both potential exporting (e.g., Australia) and importing (e.g., Japan) countries (METI 2017; COAG Energy Council 2019).

As already reported in Chapter 6 (Section 6.2.4.1), production costs of green hydrogen are expected to come down from the current levels of above USD100 $MWh^{-1}$. Price expectations are: EUR40–60 $MWh^{-1}$ for both green and blue hydrogen production in the EU by 2050 (Navigant 2019) with production costs already being lower in North Africa; 42–87 USD $MWh^{-1}$ for green hydrogen in 2030 and 20–41 USD $MWh^{-1}$ in 2050 (BNEF 2020); EUR75 $MWh^{-1}$ in 2030 (Glenk and Reichelstein 2019). For fossil-based technologies combined with CCS, prices may range from USD33–80 $MWh^{-1}$ (Table 6.8). Such prices can make hydrogen competitive for industrial feedstock applications, and probably for several transportation modes in combination with fuel cells, but without further incentives, not necessarily for stationary applications in the coming decades: wholesale natural gas prices are expected to range from USD7–31 $MWh^{-1}$ across regions and scenarios, according to the World Energy Outlook (IEA 2020a); coal prices mostly are even lower than natural gas prices (all fossil fuel prices refer to unabated technology and untaxed fuels). The evaluation of macro-economic impacts is relatively rare. A study by Mayer et al. (2019b) indicated that a shift to hydrogen in iron and steel production would lead to regional GDP losses in the range of 0.4–2.7% in 2050 across EU+3, with some regions making gains under a low-cost electricity scenario.

The IAM scenarios imply a modest role played by hydrogen, with some scenarios featuring higher levels of penetration. The consumption of hydrogen is projected to increase by 2050 and onwards in scenarios likely limiting global warming to 2°C or below, and the median share of hydrogen in total final energy consumption is 2.1% in 2050 and 5.1% in 2100 (Box 12.4, Figure 1) (Numbers are based on the AR6 scenarios database). There is large variety in hydrogen shares, but the values of 10% and more of final energy use that occur in many roadmaps are only rarely reached in the scenarios. Hydrogen is predominantly used in the industry and transportation sectors. In the scenarios, hydrogen is produced mostly by electrolysis and by biomass energy conversion with CCS (Box 12.5, Figure 1). Natural gas with CCS is expected to play only a modest role; here a distinct difference between the roadmaps quoted before and the IAM results is observed.

*Box 12.5 (continued)*

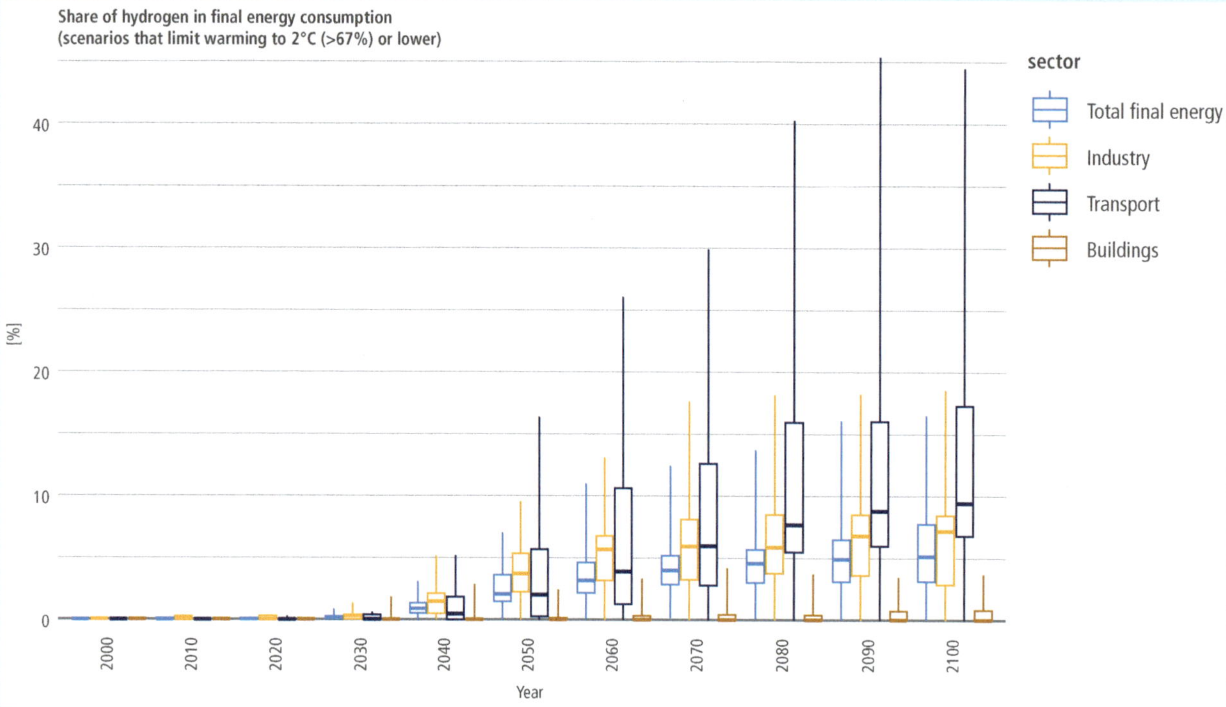


**Box 12.5, Figure 1 | Fraction of hydrogen (light blue) in total final energy consumption, and for each sector.** Hinges represent the interquartile ranges and whiskers extend to 5th and 95th percentiles. Source: AR6 scenarios database.

It is concluded that there is increasing confidence that hydrogen can play a significant role, especially in the transport sector and the industrial sector. However, there is much less agreement on timing and volumes, and there is also a range of perspectives on the role of the various production methods of hydrogen.

12

### 12.6.2 Sectoral Policy Interactions (Synergies and Trade-offs)

A taxonomy of policy types and attributes is provided by Section 13.6. In addition, the sectoral chapters provide an in-depth discussion of important mitigation policy issues such as policy overlaps, policy mixes, and policy interaction as well as policy design considerations and governance. The point of departure for the assessment in this chapter is a focus on cross-sectoral perspectives aiming at maximising policy synergies and minimising policy trade-offs.

**Synergies and trade-offs resulting from mitigation policies are not clearly discernible from either sector-level studies or global and regional top-down studies. Rather, they would require a cross-sectoral integrated policy framework** (von Stechow et al. 2015; Monier et al. 2018; Pardoe et al. 2018; Singh et al. 2019) or multiple-objective-multiple-impact policy assessment framework identifying key co-impacts and avoiding trade-offs (*robust evidence, high agreement*) (Ürge-Vorsatz et al. 2014).

Sectoral studies typically cover differentiated response measures while the IAM literature mostly uses uniform efficient market-based measures. This has important implications for understanding the differences in magnitude and distribution of mitigation costs and potentials of Section 12.2 (Karplus et al. 2013; Rausch and Karplus 2014). There is a comprehensive literature on the efficiency of uniform carbon pricing compared to sector-specific mitigation approaches, but relatively less literature on the distributional impacts of carbon taxes and measures to mitigate potential adverse distributional impacts (Rausch and Karplus 2014; Rausch and Reilly 2015; Wang et al. 2016b; Åhman et al. 2017; Mu et al. 2018). For example, in terms of cross-sectoral distributional implications, studies find negative competitiveness impacts for the energy-intensive industries (*robust evidence, medium agreement*) (Rausch and Karplus 2014; Wang et al. 2016b; Åhman et al. 2017).

Strong interdependencies and cross-sectoral linkages create both opportunities for synergies and the need to address trade-offs. This calls for coordinated sectoral approaches to climate change mitigation policies that mainstream these interactions (Pardoe et al. 2018). Such an approach is also called for in the context of cross-sectoral interactions of adaptation and mitigation measures, examples are in the agriculture, biodiversity, forests, urban, and water sectors (Arent et al. 2014; Berry et al. 2015; Di Gregorio et al. 2017). Integrated planning and cross-sectoral alignment of climate change policies are particularly evident in developing countries' NDCs pledged under the Paris Agreement, where key priority sectors such as agriculture and energy are closely aligned between the proposed mitigation and adaptation actions in the context of sustainable development and the SDGs. An example is the integration between climate-smart agriculture and low-carbon energy (*robust evidence*, *high agreement*) (Antwi-Agyei et al. 2018; England et al. 2018). Yet, there appear to be significant challenges relating to institutional capacity and resources to coordinate and implement such cross-sectoral policy alignment, particularly in developing country contexts (Antwi-Agyei et al. 2018).

Another dimension of climate change policy interactions in the literature is related to trade-offs and synergies between climate change mitigation and other societal objectives. For example, in mitigation policies related to energy, trade-offs and synergies between universal electricity access and climate change mitigation would call for complementary policies such as pro-poor tariffs, fuel subsidies, and broadly integrated policy packages (Dagnachew et al. 2018). In agriculture and forestry, research suggests that integrated policy programmes enhance mitigation potentials across the land-use-agriculture-forestry nexus and lead to synergies and positive spillovers (Galik et al. 2019). To maximise synergies and deal with trade-offs in such a cross-sectoral context, evidence-based/informed and holistic policy analysis approaches like nexus approaches and multi-target back-casting approaches that take into account unanticipated outcomes and indirect consequences would be needed (*robust evidence*, *high agreement*) (Klausbruckner et al. 2016; Hoff et al. 2019; van der Voorn et al. 2020) (Box 12.6).

The consequences of large-scale land-based mitigation for food security, biodiversity (Dasgupta 2021), the state of soil, water resources, and so on can be significant, depending on many factors, such as economic development (including distributional aspects), international trade patterns, agronomic development, diets, land-use governance and policy design, and not least climate change itself (Winchester and Reilly 2015; Fujimori et al. 2018; Hasegawa et al. 2018; Van Meijl et al. 2018). Policies and regulations that address other aspects apart from climate change can indirectly influence the attractiveness of land-based mitigation options. For example, farmers may find it attractive to shift from annual food/feed crops to perennial grasses and short rotation woody crops (suitable for bioenergy) if the previous land uses become increasingly restricted due to impacts on groundwater quality and eutrophication of water bodies (*robust evidence*, *medium agreement*) (Sections 12.4 and 12.5).

Finally, there are knowledge gaps in the literature particularly in relation to policy scalability and the extent and magnitude of policy interactions when scaling the policy to a level consistent with low GHG emissions pathways such as 2°C and 1.5°C.

## Box 12.6 | Case Study: Sahara Forest Project in Aqaba, Jordan

### Nexus framing

Shifting to renewable (in particular solar) energy reduces dependency on fossil fuel imports and greenhouse gas emissions, which is crucial for mitigating climate change. Employing renewable energy for desalination of seawater and for cooling of greenhouses in integrated production systems can enhance water availability, increase crop productivity and generate co-products and co-benefits (e.g., algae, fish, dryland restoration, greening of the desert).

### Nexus opportunities

The Sahara Forest project integrated production system uses amply available natural resources, namely solar energy and seawater, for improving water availability and agricultural/biomass production, while simultaneously providing new employment opportunities. Using hydroponic systems and humidity in the air, water needs for food production are 50% lower compared to other greenhouses.

### Technical and economic nexus solutions

Several major technologies are combined in the Sahara Forest Project, namely electricity production through the use of solar power (PV or CSP), freshwater production through seawater desalination using renewable energy, seawater-cooled greenhouses for food production, and outdoor revegetation using run-off from the greenhouses.

### Stakeholders involved

The key stakeholders which benefit from such an integrated production system are from the water sector, which urgently requires an augmentation of irrigation (and other) water, and the agricultural sector, which relies on the additional desalinated water to maintain and increase agricultural production. The project also involves public and private sector partners from Jordan and abroad, with little engagement of civil society so far.

*Box 12.6 (continued)*

**Framework conditions**

The Sahara Forest Project has been implemented at pilot scale so far, including the first pilot with one hectare and one greenhouse pilot in Qatar and a larger 'launch station' with three hectares and two greenhouses in Jordan. These pilots have been funded by international organisations such as the Norwegian Ministry of Climate and Environment, Norwegian Ministry of Foreign Affairs and the European Union. Alignment with national policies, institutions and funding, as well as upscaling of the project, is underway or planned.

**Monitoring and evaluation and next steps**

The multi-sectoral planning and investments that are needed to upscale the project require cooperation among the water, agriculture, and energy sectors and an active involvement of local actors, private companies, and investors. These cooperation and involvement mechanisms are currently being established in Jordan. Given the emphasis on the economic value of the project, public-private partnerships are considered as the appropriate business and governance model, when the project is upscaled. Scenarios for upscaling (seawater use primarily in low-lying areas close to the sea, to avoid energy-intensive pumping) include 50 MW of CSP, 50 hectares of greenhouses, which would produce 34,000 tonnes of vegetables annually, provide employment for over 800 people, and sequester more than 8000 tonnes of $CO_2$-eq annually.

Source: SFP Foundation; Hoff et al. (2019).

### 12.6.3 International Trade Spillover Effects and Competitiveness

International spillovers of mitigation policies are effects that carbon-abatement measures implemented in one country have on sectors in other countries. These effects include (i) carbon leakage in manufacture; (ii) the effects on energy trade flows and incomes related to fossil fuel exports from major exporters; (iii) technology and knowledge spillovers; and (iv) transfer of norms and preferences via various approaches to establish sustainability requirements on traded goods, such as EU-RED and environmental labelling systems to guide consumer choices (*robust evidence, medium agreement*). This section focuses on cross-sectoral aspects of international spillovers related to the first two effects.

#### 12.6.3.1 Cross-sectoral Aspects of Carbon Leakage

Carbon leakage occurs when mitigation measures implemented in one country or sector lead to a rise in emissions in other countries or sectors. Three types of spillovers are possible: (i) domestic cross-sectoral spillovers when mitigation policy in one sector leads to the re-allocation of labour and capital towards the other sectors of the same country; (ii) international spillovers within a single sector when mitigation policy leads to substitution of domestic production of carbon-intensive goods with their imports from abroad; and (iii) international cross-sectoral spillovers when mitigation policy in one sector in one country leads to the rise in emissions in other sectors in other countries. While the first two are described in Section 13.6, this section focuses on the third. Though some papers address this type of leakage, there is still a significant lack of knowledge on this topic.

One possible channel of cross-sectoral international carbon leakage is through global value chains. Mitigation policy in one country not only leads to shifts in competitiveness across industries producing final goods but also across those producing raw materials and intermediary goods all over the world.

This type of leakage is especially important because the countries that provide basic materials are usually emerging or developing economies, many of which have no or limited regulation of GHG emissions. For this reason, foreign direct investment in developing economies usually leads to an increase in emissions (Kivyiro and Arminen 2014; Shahbaz et al. 2015; Bakhsh et al. 2017): in the case of basic materials the effect of expansion of economic activity on emissions exceeds the effect of technological spillovers, while for developed countries the effect is opposite (Shahbaz et al. 2015; Pazienza 2019). Meng et al. (2018) calculated that environmental cost for generating one unit of GDP through international trade was 1.4 times higher than that through domestic production in 1995. By 2009, this difference increased to 1.8 times. Carbon leakage due to the differences in environmental regulation was the main driver of this increase.

In order to address emissions leakage through global value chains, Liu and Fan (2017) propose the value-added-based emissions accounting principle, which makes it possible to account for GHG emissions within the context of the economic benefit principle. Davis et al. (2011) notice that the analysis of value chains gives an opportunity to find the point where regulation would be the most efficient and the least vulnerable to leakage. For instance, transaction costs of global climate policy and the risks of leakage may be reduced if emissions are regulated at the extraction stage as there are far fewer agents involved in this process than in burning of fossil fuels or consumption of energy-intensive goods. Li et al. (2020) calls for coordinated efforts to reduce emissions embodied in trade flows in pairs of the economies with the highest leakage, such as China and the United States, China and Germany, China and Japan, Russia and Germany.

Unfortunately, these proposals either face difficulties in collection and verification of data on emissions along value chains or require a high level of international cooperation, which is hardly achievable at the moment. Neuhoff et al. (2016) and Pollitt et al. (2020) focus on the regulation of emissions embodied in global value chains through national policy instruments. They propose implementation of a charge on consumption of imported basic materials into the European emissions trading system. Such a charge, equivalent to around EUR80 $tCO_2^{-1}$, could reduce the EU's total $CO_2$ emissions by up to 10% by 2050 (Pollitt et al. 2020) without significant effects on competitiveness. This proposal is very close to the carbon border adjustment introduced in the EU and described in more detail in Sections 13.2 and 13.6.

Cross-sectoral effects of carbon leakage also occur through the multiplier effect, when the mitigation policy in any sector in country A leads to the increase of relative competitiveness and therefore production of the same sector in country B, which automatically leads to the expansion of economic activity in other sectors of country B. This expansion may in turn lead to the rise of production and emissions in country A as a result of feedback effects. These spillovers should be taken into consideration while designing climate policy, along with potential synergies that may appear due to joint efforts. However, the scale of these effects with regards to leakage should not be overestimated. Even for intrasectoral leakage, many *ex ante* modelling studies generally suggest limited carbon leakage rates (Chapter 13). Intersectoral leakage should be even less significant. Interregional spillover and feedback effects are well studied in China (Zhang 2017; Ning et al. 2019). Even within a single country, interregional spillover effects are much lower than intraregional effects, and feedback effects are even less intense. Cross-sectoral spillovers across national borders as a result of mitigation policy should be even smaller, although these are less well studied. In future, if the differences in carbon price between regions increase, leakage through cross-sectoral multipliers may play a more important role.

Another important cross-sectoral aspect of carbon leakage concerns the transport sector. If mitigation policy leads to the substitution of domestic carbon-intensive production with imports, one of the side effects of this substitution is the rise of emissions from transportation of imported goods. International transport is responsible for about a third of worldwide trade-related emissions, and over 75% of emissions for major manufacturing categories (Cristea et al. 2013). Carbon leakage would potentially increase the emissions from transportation significantly as the trade of major consuming economies of the EU and US would shift towards distant trading partners in East and South Asia. Meng et al. (2018) consider more distant transportation as one of the major contributors to the rise in emissions embodied in international trade from 1995 to 2009.

Emissions leakage due to international trade, investment and value chains is a significant obstacle to more ambitious climate policies in many regions. However, it does not mean that disruption of trade would reduce global emissions. Zhang et al. (2020) show that deglobalisation and the drop in international trade may result in emissions reductions in the short term, but in the longer term it will make each country build more complete industrial systems to satisfy their final demand, although they have comparative disadvantages in some production stages. As a result, emissions would increase. According to Zhang et al. (2020), for China, the decrease of the degree of global value chain participation (which ranges from 0 to 1) by 0.1 would lead to an increase in gross carbon intensity of China's exports of 11.7%. On distributional implications, Parrado and De Cian (2014) report that trade-driven spillover effects transmitted through imports of materials and equipment result in significant inter-sectoral distributional effects, with some sectors witnessing substantial expansion in activity and emissions and others witnessing a decline in activities and emissions.

It should also be mentioned that international trade leads to important knowledge and technology spillovers (Sections 16.3 and 16.5) and is critically important for achieving other Sustainable Development Goals (Section 12.6.1). Any policies imposing additional barriers to international trade should therefore be implemented with great caution and require comprehensive evaluation of various economic, social and environmental effects.

#### 12.6.3.2 The Spillover Effects on the Energy Sector

Cross-sectoral trade-related spillovers of mitigation policies include their effect on energy prices. Other things being equal, regulation of emissions of industrial producers decreases the demand for fossil fuels that would reduce prices and encourage the rise of fossil fuel consumption in regions with no or weaker climate policies (*robust evidence, medium agreement*).

Arroyo-Currás et al. (2015) studied the energy channel of carbon leakage with the REMIND IAM of the global economy. They came to the conclusion that the leakage rate through the energy channel is less than 16% of the emissions reductions of regions who introduce climate policies first. This result did not differ much for different sizes and compositions of the early mover coalition.

Bauer et al. (2015) built a multi-model scenario ensemble for the analysis of energy-related spillovers of mitigation policies and revealed huge uncertainty: energy-related carbon leakage rates varied from negative values to 50%, primarily depending on the trends in inter-fuel substitution.

Another kind of spillover in the energy sector concerns the 'green paradox': announcement of future climate policies causes an increase in production and trade in fossil fuels in the short term (Jensen et al. 2015; Kotlikoff et al. 2016). The delayed carbon tax should therefore be higher than an immediately implemented carbon tax in order to achieve the same temperature target (van der Ploeg 2016). Studies also make a distinction between a 'weak' and 'strong' green paradox (Gerlagh 2011). The former refers to a short-term rise in emissions in response to climate policy, while the latter refers to rising cumulative damage.

The green paradox may work in different ways for different kinds of fossil fuels. For instance, Coulomb and Henriet (2018) show that climate policies in the transport and power-generation sectors increase the discounted profits of the owners of conventional oil and gas, compared to the no-regulation baseline, but will decrease these profits for coal and unconventional oil and gas producers.

Many studies also distinguish different policy measures by the scale of green paradox they provide. The immediate carbon tax is the first-best instrument from the perspective of global welfare. Delayed carbon tax leads to some green paradox but less than in the case of support for renewables (Michielsen 2014; van der Ploeg and Rezai 2019). With respect to the latter, support for renewable electricity has a lower green paradox than support for biofuels (Michielsen 2014; Gronwald et al. 2017). The existence of the green paradox is an additional argument in favour of more decisive climate policy now: any postponements will lead to additional consumption of fossil fuels and consequently the need for more ambitious and costly efforts in future.

The effect of fossil fuel production expansion as a result of anticipated climate policy may be compensated by the effect of divestment. Delayed climate policy creates incentives for investors to divest from fossil fuels. Bauer et al. (2018) show that this divestment effect is stronger and thus announcing of climate policies leads to the reduction of energy-related emissions.

The implication of the effects of mitigation policies through the energy-related spillovers channel is of particular significance to oil-exporting countries (*medium evidence*, *medium agreement*). Emissions-reduction measures lead to decreasing demand for fossil fuels and consequently to the decrease in exports from major oil- and gas-exporting countries. The case of Russia is one of the most illustrative. Makarov et al. (2020) show that the fulfilment by Paris Agreement Parties of their NDCs would lead to 25% reduction of Russia's energy exports by 2030 with significant reduction of its economic growth rates. At the same time, the domestic consumption of fossil fuels is anticipated to increase in response to the drop in external demand that would provoke carbon leakage (Orlov and Aaheim 2017). Such spillovers demonstrate the need for dialogue between exporters and importers of fossil fuels while implementing the mitigation policies.



### 12.6.4 Implications of Finance for Cross-sectoral Mitigation Synergies and Trade-offs

Finance is a principal enabler of GHG mitigation and an essential component of countries' NDC packages submitted under the Paris Agreement (UNFCCC 2016). The assessment of investment requirements for mitigation along with their financing at sectoral levels are addressed in detail by sectoral chapters while the assessment of financial sources, instruments, and the overall mitigation financing gap is addressed by Chapter 15 (Sections 15.3, 15.4, and 15.5). The focus in this chapter with respect to finance is on the scope and potential for financing integrated solutions that create synergies between and among sectors.

Cross-sectoral considerations in mitigation finance are critical for the effectiveness of mitigation action as well as for balancing the often conflicting social, developmental and environmental policy goals at the sectoral level. True measures of mitigation policy impacts and hence plans for resource mobilisation that properly address costs and benefits cannot be developed in isolation from their cross-sectoral implications. Unaddressed cross-sectoral coordination and interdependency issues are identified as major constraints in raising the necessary financial resources for mitigation in a number of countries (Bazilian et al. 2011; Welsch et al. 2014; Hoff et al. 2019a).

Integrated financial solutions to leverage synergies between sectors, as opposed to purely sector-based financing, at international, national, and local levels are needed to scale up GHG mitigation potentials. At the international level, finance from multilateral development banks (MDBs) is a major source of GHG mitigation finance in developing countries (*medium evidence, medium agreement*) (World Bank Group 2015; Ha et al. 2016; Bhattacharya et al. 2016; Bhattacharya et al. 2018). In 2018, MDBs reported a total of USD30.165 billion in financial commitments to climate change mitigation, with 71% of total mitigation finance being committed through investment loans and the rest in the form of equity, guarantees, and other instruments. GHG reduction activities eligible for MDB finance are limited to those compatible with low-emission pathways recognising the importance of long-term structural changes, such as the shift in energy production to low-carbon energy technologies and the modal shift to low-carbon modes of transport leveraging both greenfield and energy efficiency projects. Sector-wise, the MDBs' mitigation finance for 2018 is allocated to renewable energy (29%), transport (18%), energy efficiency (18%), lower-carbon and efficient energy generation (7%), agriculture, forestry and land use (8%), waste and wastewater (8%), and other sectors (12%) (MDB 2019). Unfortunately, due to institutional and incentives issues, MDB finance has mostly focused on sectoral solutions and has not been able to properly leverage cross-sectoral synergies. At the national level, applied research has shown that integrated modelling of land, energy and water resources not only has the potential to identify superior solutions, but also reveals important differences in terms of investment requirements and required financing arrangements compared to the traditional sectoral financing toolkits (Welsch et al. 2014). Agriculture, forestry, nature-based solutions and other forms of land use are promising sectors for leveraging financing solutions to scale up GHG mitigation efforts (Section 15.4). Moving to more productive and resilient forms of land use is a complex task, given the cross-cutting nature of land use, which necessarily results in apparent trade-offs between mitigation, adaptation, and development objectives. Finance is one area to manage these trade-offs where there may be opportunities to redirect the hundreds of billions spent annually on land use around the world towards green activities, without sacrificing either productivity or economic development (Falconer et al. 2015). Nonetheless, that would require active public support in design of land-use mitigation and adaptation strategies, coordination between public and private instruments across land use sectors, and leveraging of policy and financial instruments to redirect finance toward greener land-use practices (*limited evidence, medium agreement*). For example, the Welsch et al. (2014) study on Mauritius shows that the promotion of a local biofuel industry from sugar cane could be economically favourable in the absence of water constraints, leading to a reduction in petroleum imports and GHG emissions while enhancing energy security. Yet, under a water-constrained scenario as a result of climate change, the need for additional energy to expand irrigation to previously rain-fed sugar plantations and to power desalination plants yields the opposite result in terms of GHG emissions and energy costs, making biofuels a sub-optimal option, and negatively affects their economics and the prospects for financing.

At the local level, integrated planning and financing are needed to achieve more sustainable outcomes. For example, at a city level, integration is needed across sectors such as transport, energy systems, buildings, sewage and solid waste to optimise emissions footprints. How a city is designed will affect transportation demands, which makes it either more or less difficult to implement efficient public transportation, leading in turn to more or fewer emissions. Under such cases, solutions in terms of public and private investment paths and financing policies based on purely internal sector considerations are bound to cause adverse impacts on other sectors and poor overall outcomes (Gouldson et al. 2016).

Availability and access to finance are among the major barriers to GHG emissions mitigation across various sectors and technology options (*robust evidence, high agreement*). Resource maturity mismatches and risk exposure are two main factors limiting ability of commercial banks and other private lenders to contribute to green finance (Mazzucato and Semieniuk 2018). At all levels, mobilising the necessary resources to leverage cross-sectoral mitigation synergies would require the combination of public and private financial sources (Jensen and Dowlatabadi 2018). Traditional public financing would be required to synergise mitigation across sectors where the risk-return and time profiles of investment are not sufficiently attractive for the business sector. Over the years, private development financing through public-private partnerships and other related variants has been a growing source of finance to leverage cross-sectoral synergies and manage trade-offs (Anbumozhi and Timilsina 2018; Attridge and Engen 2019; Ishiwatari et al. 2019). Promoting such blended approaches to finance along with result-based financing architectures to strengthen delivery institutions are advocated as effective means to mainstream cross-sectoral mitigation finance (*limited evidence, high agreement*) (Attridge and Engen 2019; Ishiwatari et al. 2019). The World Bank group and the International Financial Corporation have used the blended finance results-based approach to climate financing that addresses institutional, infrastructure, and service needs across sectors targeting developing countries and marginalised communities (GPRBA 2019; IDA 2019).

## 12.7 Knowledge Gaps

Finally, the literature review and analysis in Chapter 12 has taken account of the post-AR5 literature available and accessible to the chapter authors. Nonetheless, the assessment of the chapter is incomplete without mentioning knowledge gaps encountered during the assessment. These knowledge gaps include:

1. Interactions (synergies and trade-offs) between different CDR methods when deployed together are under-researched:
   - co-benefits and trade-offs with biodiversity and ecosystem services associated with the implementation of CDR methods.
   - constraining technical costs and potentials for CDR methods to define realistically achievable costs and potentials. Such research is useful for improving the representation of CDR methods in IAMs and country-level mitigation pathway modelling.
2. More work is required on how framing and communication of mitigation actions in terms of mitigation versus co-benefits potential affects public support in different contexts.
3. Additional research work is required to determine the cross-sectoral mitigation potential of emerging general-purpose technologies.
4. There is a lack of literature on mitigation finance frameworks promoting cross-sectoral mitigation linkages.
5. Additional research is needed to better quantify the net GHG emissions and co-benefits and adverse effects of emerging food technologies.
   - Research in social and behavioural sciences should invest in assessing effectiveness of instruments aiming at shifting food choices in different national contexts.
   - A better evidence basis is required to understand synergistic effects of policies in food system policy packages.
6. There is a lack of literature on regional and global mitigation potential of biomass production systems that are strategically deployed in agriculture and forestry landscapes, to achieve specific co-benefits.
7. There is a lack of knowledge on land occupation and associated co-benefits and adverse side effects from large-scale deployment of non-AFOLU mitigation options, and how such options can be integrated with agriculture and forestry to maximise synergies and minimise trade-offs.

Frequently Asked Questions (FAQs)

## FAQ 12.1 | How could new technologies to remove carbon dioxide from the atmosphere contribute to climate change mitigation?

Limiting the increase in warming to well below 2°C, and achieving net zero $CO_2$ or GHG emissions, will require anthropogenic $CO_2$ removal from the atmosphere.

The carbon dioxide removal (CDR) methods studied so far have different removal potentials, costs, co-benefits and side effects. Some biological methods for achieving CDR, like afforestation/reforestation or wetland restoration, have long been practised. If implemented well, these practices can provide a range of co-benefits, but they can also have adverse side effects such as biodiversity loss or food price increases. Other chemical and geochemical approaches to CDR include direct air carbon capture and storage (DACCS), enhanced weathering or ocean alkalinity enhancement. They are generally less vulnerable to reversal than biological methods.

DACCS uses chemicals that bind to $CO_2$ directly from the air; the $CO_2$ is then removed from the sorbent and stored underground or mineralised. Enhanced weathering involves the mining of rocks containing minerals that naturally absorb $CO_2$ from the atmosphere over geological timescales, which are crushed to increase the surface area and spread on soils (or elsewhere) where they absorb atmospheric $CO_2$. Ocean alkalinity enhancement involves the extraction, processing, and dissolution of minerals and addition to the ocean where they enhance sequestration of $CO_2$ as bicarbonate and carbonate ions in the ocean.

## FAQ 12.2 | Why is it important to assess mitigation measures from a systemic perspective, rather than only looking at their potential to reduce greenhouse gas (GHG) emissions?

Mitigation measures do not only reduce GHGs, but have wider impacts. They can result in decreases or increases in GHG emissions in another sector or part of the value chain from where they are applied. They can have wider environmental (e.g., air and water pollution, biodiversity), social (e.g., employment creation, health) and economic (e.g., growth, investment) co-benefits or adverse side effects. Mitigation and adaptation can also be linked. Taking these considerations into account can help to enhance the benefits of mitigation action, and avoid unintended consequences, as well as provide a stronger case for achieving political and societal support and raising the finances required for implementation.

## FAQ 12.3 | Why do we need a food systems approach for assessing GHG emissions and mitigation opportunities from food systems?

Activities associated with the food system caused about one-third of total anthropogenic GHG emissions in 2015, distributed across all sectors. Agriculture and fisheries produce crops and animal-source food, which are partly processed in the food industry, packed, distributed, retailed, cooked, and finally eaten. Each step is associated with resource use, waste generation, and GHG emissions.

A food systems approach helps identify critical areas as well as novel and alternative approaches to mitigation on both the supply side and the demand side of the food system. But complex co-impacts need to be considered and mitigation measures tailored to the specific context. International cooperation and governance of global food trade can support both mitigation and adaptation.

There is large scope for emissions reduction in both cropland and grazing production, and also in food processing, storage and distribution. Emerging options such as plant-based alternatives to animal food products and food from cellular agriculture are receiving increasing attention, but their mitigation potential is still uncertain and depends on the GHG intensity of associated energy systems due to relatively high energy needs. Diet changes can reduce GHG emissions and also improve health in groups with excess consumption of calories and animal food products, which is mainly prevalent in developed countries. Reductions in food loss and waste can help reduce GHG emissions further.

Recommendations to buy local food and avoid packaging can contribute to reducing GHG emissions but should not be generalised, as trade-offs exist with food waste, GHG footprint at farm gate, and accessibility to diverse healthy diets.

## References


Aarts, N. and C. Leeuwis, 2010: Participation and Power: Reflections on the Role of Government in Land Use Planning and Rural Development. *J. Agric. Educ. Ext.*, **16(2)**, 131–145, doi:10.1080/13892241003651381.

AASSA, 2018: *Opportunities and challenges for research on food and nutrition security and agriculture in Asia*. Association of Academies and Societies of Sciences in Asia, Gyeonggi-do, Republic of Korea, 70 pp.

Abbott, M., M. Bazilian, D. Egel, and H.H. Willis, 2017: Examining the food–energy–water and conflict nexus. *Curr. Opin. Chem. Eng.*, **18**, 55–60, doi:10.1016/j.coche.2017.10.002.

Abebe, E., G. Gugsa, and M. Ahmed, 2020: Review on Major Food-Borne Zoonotic Bacterial Pathogens. *J. Trop. Med.*, **2020**, doi:10.1155/2020/4674235.

Achterberg, E.P. et al., 2013: Natural iron fertilization by the Eyjafjallajökull volcanic eruption. *Geophys. Res. Lett.*, **40(5)**, 921–926, doi:10.1002/grl.50221.

Adeh, E.H., S.P. Good, M. Calaf, and C.W. Higgins, 2019: Solar PV Power Potential is Greatest Over Croplands. *Sci. Rep.*, **9(1)**, 11442, doi:10.1038/s41598-019-47803-3.

Adeyeye, S.A.O., 2017: The role of food processing and appropriate storage technologies in ensuring food security and food availability in Africa. *Nutr. Food Sci.*, **47(1)**, 122–139, doi:10.1108/NFS-03-2016-0037.

Adeyeye, S.A.O. and O.B. Oyewole, 2016: An Overview of Traditional Fish Smoking In Africa. *J. Culin. Sci. Technol.*, **14(3)**, 198–215, doi:10.1080/15428052.2015.1102785.

Agusdinata, D.B., W. Liu, H. Eakin, and H. Romero, 2018: Socio-environmental impacts of lithium mineral extraction: Towards a research agenda. *Environ. Res. Lett.*, **13(12)**, 123001, doi:10.1088/1748-9326/aae9b1.

Ahlgren, S. et al., 2015: Review of methodological choices in LCA of biorefinery systems - key issues and recommendations. *Biofuels, Bioprod. Biorefining*, **9(5)**, 606–619, doi:10.1002/BBB.1563.

Åhman, M., L.J. Nilsson, and B. Johansson, 2017: Global climate policy and deep decarbonization of energy-intensive industries. *Clim. Policy*, **17(5)**, 634–649, doi:10.1080/14693062.2016.1167009.

Ahmed, S., S. Downs, and J. Fanzo, 2019: Advancing an Integrative Framework to Evaluate Sustainability in National Dietary Guidelines. *Front. Sustain. Food Syst.*, **3**(September), 1–20, doi:10.3389/fsufs.2019.00076.

Ahsan, R. and M. Ahmad, 2016: Development, displacement and resettlement a challenge for social sustainability: A study on mega development project (Bakun Dam) in Sarawak. *Int'l J. Adv. Agric. Environ. Engg.*, **3(1)**, 47–51.

Aiking, H. and J. de Boer, 2019: Protein and sustainability – the potential of insects. *J. Insects as Food Feed*, **5(1)**, 3–7, doi:10.3920/JIFF2018.0011.

Aiking, H. and J. de Boer, 2020: The next protein transition. *Trends Food Sci. Technol.*, **105**(May), 515–522, doi:10.1016/j.tifs.2018.07.008.

Akimoto, K., F. Sano, J. Oda, H. Kanaboshi, and Y. Nakano, 2021: Climate change mitigation measures for global net-zero emissions and the roles of CO2 capture and utilization and direct air capture. *Energy Clim. Change*, **2**, 100057, doi:10.1016/j.egycc.2021.100057.

Al-Khudairy, L., O.A. Uthman, R. Walmsley, S. Johnson, and O. Oyebode, 2019: Choice architecture interventions to improve diet and/or dietary behaviour by healthcare staff in high-income countries: A systematic review. *BMJ Open*, **9(1)**, 1–16, doi:10.1136/bmjopen-2018-023687.

Al-Kodmany, K., 2018: The vertical farm: A review of developments and implications for the vertical city. *Buildings*, **8(2)**, 24, doi:10.3390/buildings8020024.

Alam, A. and P. Dwivedi, 2019: Modeling site suitability and production potential of carinata-based sustainable jet fuel in the southeastern United States. *J. Clean. Prod.*, **239**, 117817, doi:10.1016/J.JCLEPRO.2019.117817.

Alanne, K. and S. Cao, 2017: Zero-energy hydrogen economy (ZEH2E) for buildings and communities including personal mobility. *Renew. Sustain. Energy Rev.*, **71**(October 2016), 697–711, doi:10.1016/j.rser.2016.12.098.

Alatiq, A. et al., 2021: Assessment of the carbon abatement and removal opportunities of the Arabian Gulf Countries. *Clean Energy*, **5(2)**, 340–353, doi:10.1093/ce/zkab015.

Albizzati, P.F., D. Tonini, C.B. Chammard, and T.F. Astrup, 2019: Valorisation of surplus food in the French retail sector: Environmental and economic impacts. *Waste Manag.*, **90**, 141–151, doi:10.1016/j.wasman.2019.04.034.

Albright, R. et al., 2016: Reversal of ocean acidification enhances net coral reef calcification. *Nature*, **531(7594)**, 362–365, doi:10.1038/nature17155.

Alexander, P. et al., 2017: Could consumption of insects, cultured meat or imitation meat reduce global agricultural land use? *Glob. Food Sec.*, **15**, 22–32, doi:10.1016/j.gfs.2017.04.001.

Almeida, R.M. et al., 2019: Reducing greenhouse gas emissions of Amazon hydropower with strategic dam planning. *Nat. Commun.*, **10(1)**, 4281, doi:10.1038/s41467-019-12179-5.

Alshammari, Y.M., 2020: Achieving Climate Targets via the Circular Carbon Economy: The Case of Saudi Arabia. *C—Journal Carbon Res.*, **6(3)**, 54, doi:10.3390/c6030054.

Amann, T. et al., 2020: Enhanced Weathering and related element fluxes – a cropland mesocosm approach. *Biogeosciences*, **17(1)**, 103–119, doi:10.5194/bg-17-103-2020.

Amjath-Babu, T.S. et al., 2019: Integrated modelling of the impacts of hydropower projects on the water-food-energy nexus in a transboundary Himalayan river basin. *Appl. Energy*, **239** (April 2019), 494–503, doi:10.1016/j.apenergy.2019.01.147.

Amundson, R. et al., 2015: Soil and human security in the 21st century. *Science*, **348(6235)**, doi:10.1126/science.1261071.

Anandarajah, G., O. Dessens, and W. McDowall, 2018: The Future for Bioenergy Systems: The Role of BECCS? In: *Biomass Energy with Carbon Capture and Storage (BECCS), Wiley Online Books*, pp. 205–226.

Anastasiou, K., M. Miller, and K. Dickinson, 2019: The relationship between food label use and dietary intake in adults: A systematic review. *Appetite*, **138**(April), 280–291, doi:10.1016/j.appet.2019.03.025.

Anbumozhi, V., and P. Timilsina, 2018: Leveraging Private Finance Through Public Finance: Role of International Financial Institutions. In: *Financing for Low-carbon Energy Transition: Unlocking the Potential of Private Capital* [Anbumozhi, V., K. Kalirajan, and F. Kimura, (eds.)], Springer Singapore, Singapore, pp. 317–334.

Andam, K.S., D. Tschirley, S.B. Asante, R.M. Al-Hassan, and X. Diao, 2018: The transformation of urban food systems in Ghana: Findings from inventories of processed products. *Outlook Agric.*, **47(3)**, 233–243, doi:10.1177/0030727018785918.

Anderegg, W.R.L. et al., 2020: Climate-driven risks to the climate mitigation potential of forests. *Science*, **368(6497)**, doi:10.1126/SCIENCE.AAZ7005.

Anderson, C., J. Bruil, M.J. Chappell, C. Kiss, and M. Pimbert, 2019: From Transition to Domains of Transformation: Getting to Sustainable and Just Food Systems through Agroecology. *Sustainability*, **11**, 5272, doi:10.3390/su11195272.

Anderson, K. and G. Peters, 2016: The trouble with negative emissions. *Science*, **354(6309)**, 182–183, doi:10.1126/science.aah4567.

Andrée, P., M. Coulas, and P. Ballamingie, 2018: Governance recommendations from forty years of national food strategy development in Canada and beyond. *Can. Food Stud. / La Rev. Can. des études sur l'alimentation*, **5(3)**, 6–27, doi:10.15353/cfs-rcea.v5i3.283.

Ansari, W.A. et al., 2020: *Genome editing in cereals: Approaches, applications and challenges*. *Int. J. Mol. Sci.* **21(11)**, doi.org/10.3390/ijms21114040.

Antwi-Agyei, P., A.J. Dougill, T.P. Agyekum, and L.C. Stringer, 2018: Alignment between nationally determined contributions and the sustainable development goals for West Africa. *Clim. Policy*, **18(10)**, 1296–1312, doi:10.1080/14693062.2018.1431199.

Anwar, M.N. et al., 2020: CO2 utilization: Turning greenhouse gas into fuels and valuable products. *J. Environ. Manage.*, **260**, doi:10.1016/j.jenvman.2019.110059.

Apostolidis, C. and F. McLeay, 2016: Should we stop meating like this? Reducing meat consumption through substitution. *Food Policy*, **65**, 74–89, doi:10.1016/j.foodpol.2016.11.002.

Arent, D. et al., 2014: Implications of high renewable electricity penetration in the U.S. for water use, greenhouse gas emissions, land-use, and materials supply. *Appl. Energy*, **123**, 368–377, doi:10.1016/j.apenergy.2013.12.022.

Aristizábal-Marulanda, V. and C.A. Cardona Alzate, 2019: Methods for designing and assessing biorefineries: Review. *Biofuels, Bioprod. Biorefining*, **13(3)**, 789–808, doi:10.1002/bbb.1961.

Ariti, A.T., J. van Vliet, and P.H. Verburg, 2018: Farmers' participation in the development of land use policies for the Central Rift Valley of Ethiopia. *Land use policy*, **71**, 129–137, doi:10.1016/J.LANDUSEPOL.2017.11.051.

Armanda, D.T., J.B. Guinée, and A. Tukker, 2019: The second green revolution: Innovative urban agriculture's contribution to food security and sustainability – A review. *Glob. Food Sec.*, **22**(August), 13–24, doi:10.1016/j.gfs.2019.08.002.

Armstrong, A., N.J. Ostle, and J. Whitaker, 2016: Solar park microclimate and vegetation management effects on grassland carbon cycling. *Environ. Res. Lett.*, **11(7)**, 74016, doi:10.1088/1748-9326/11/7/074016.

Arno, A. and S. Thomas, 2016: The efficacy of nudge theory strategies in influencing adult dietary behaviour: A systematic review and meta-analysis. *BMC Public Health*, **16(1)**, 1–11, doi:10.1186/s12889-016-3272-x.

Arroyo-Currás, T. et al., 2015: Carbon leakage in a fragmented climate regime: The dynamic response of global energy markets. *Technol. Forecast. Soc. Change*, **90**(PA), 192–203, doi:10.1016/j.techfore.2013.10.002.

Asbjornsen, H. et al., 2014: Targeting perennial vegetation in agricultural landscapes for enhancing ecosystem services. *Renew. Agric. Food Syst.*, **29(2)**, 101–125, doi:10.1017/S1742170512000385.

Attridge, S. and L. Engen, 2019: *Blended finance in the poorest countries: The need for a better approach*. ODI, London, 75 pp.

Atuonwu, J.C. et al., 2018: Comparative assessment of innovative and conventional food preservation technologies: Process energy performance and greenhouse gas emissions. *Innov. Food Sci. Emerg. Technol.*, **50**, 174–187, doi:10.1016/j.ifset.2018.09.008.

Aumont, O. and L. Bopp, 2006: Globalizing results from ocean in situ iron fertilization studies. *Global Biogeochem. Cycles*, **20(2)**, doi:10.1029/2005GB002591.

Averchenkova, A., S. Fankhauser, and J.J. Finnegan, 2021: The impact of strategic climate legislation: evidence from expert interviews on the UK Climate Change Act. *Clim. Policy*, **21(2)**, 251–263, doi:10.1080/14693062.2020.1819190.

Babin, A., C. Vaneeckhaute, and M.C. Iliuta, 2021: Potential and challenges of bioenergy with carbon capture and storage as a carbon-negative energy source: A review. *Biomass and Bioenergy*, **146**, 105968, doi:10.1016/j.biombioe.2021.105968.

Bach, L.T., S.J. Gill, R.E.M. Rickaby, S. Gore, and P. Renforth, 2019: CO2 Removal With Enhanced Weathering and Ocean Alkalinity Enhancement: Potential Risks and Co-benefits for Marine Pelagic Ecosystems. *Front. Clim.*, **1**, 7, doi:10.3389/fclim.2019.00007.

Bach, L.T. et al., 2021: Testing the climate intervention potential of ocean afforestation using the Great Atlantic Sargassum Belt. *Nat. Commun.*, **12(1)**, 2556, doi:10.1038/s41467-021-22837-2.

Backhouse, M. and R. Lehmann, 2020: New 'renewable' frontiers: contested palm oil plantations and wind energy projects in Brazil and Mexico. *J. Land Use Sci.*, **15(2–3)**, 373–388, doi:10.1080/1747423X.2019.1648577.

Bai, Y., R. Alemu, S.A. Block, D. Headey, and W.A. Masters, 2021: Cost and affordability of nutritious diets at retail prices: Evidence from 177 countries. *Food Policy*, **99**, 101983, doi:10.1016/j.foodpol.2020.101983.

Baik, E. et al., 2018: Geospatial analysis of near-term potential for carbon-negative bioenergy in the United States. *Proc. Natl. Acad. Sci.*, **115(13)**, 3290–3295, doi:10.1073/pnas.1720338115.

Bajželj, B. et al., 2014: Importance of food-demand management for climate mitigation. *Nat. Clim. Change*, **4(10)**, 924–929, doi:10.1038/nclimate2353.

Bajželj, B., T.E. Quested, E. Röös, and R.P.J. Swannell, 2020: The role of reducing food waste for resilient food systems. *Ecosyst. Serv.*, **45**(June), 101140, doi:10.1016/j.ecoser.2020.101140.

Baka, J., 2013: The Political Construction of Wasteland: Governmentality, Land Acquisition and Social Inequality in South India. *Dev. Change*, **44(2)**, 409–428, doi:10.1111/dech.12018.

Baka, J., 2014: What wastelands? A critique of biofuel policy discourse in South India. *Geoforum*, **54**, 315–323, doi:10.1016/j.geoforum.2013.08.007.

Baker, P. and S. Friel, 2016: Food systems transformations, ultra-processed food markets and the nutrition transition in Asia. *Global. Health*, **12(1)**, 80, doi:10.1186/s12992-016-0223-3.

Bakhsh, K., S. Rose, M.F. Ali, N. Ahmad, and M. Shahbaz, 2017: Economic growth, $CO_2$ emissions, renewable waste and FDI relation in Pakistan: New evidences from 3SLS. *J. Environ. Manage.*, **196**, 627–632, doi:10.1016/j.jenvman.2017.03.029.

Baldock, J.A. and J.O. Skjemstad, 2000: Role of the soil matrix and minerals in protecting natural organic materials against biological attack. *Org. Geochem.*, **31(7–8)**, 697–710, doi:10.1016/S0146-6380(00)00049-8.

Baldos, U.L.C. and T.W. Hertel, 2015: The role of international trade in managing food security risks from climate change. *Food Secur.*, **7(2)**, doi:10.1007/s12571-015-0435-z.

Barbarossa, V. et al., 2020: Impacts of current and future large dams on the geographic range connectivity of freshwater fish worldwide. *Proc. Natl. Acad. Sci.*, **117(7)**, 3648 LP – 3655, doi:10.1073/pnas.1912776117.

Barbieri, P., S. Pellerin, V. Seufert, and T. Nesme, 2019: Changes in crop rotations would impact food production in an organically farmed world. *Nat. Sustain.*, **2(5)**, 378–385, doi:10.1038/s41893-019-0259-5.

Barbieri, P. et al., 2021: Global option space for organic agriculture is delimited by nitrogen availability. *Nat. Food*, 2, 363–372, doi:10.1038/s43016-021-00276-y.

Barker, T. et al., 2007: Mitigation from a cross-sectoral perspective. In: *Climate Change 2007: Mitigation. Contribution of Working Group III to the Fourth Assessment Report of the Intergovernmental Panel on Climate Change* [Metz, B., O.R. Davidson, P.R. Bosch, R. Dave, and L.A. Meyer, (eds.)]. Cambridge University Press, Cambridge, UK and New York, NY, USA, pp. 621–687.

Barron-Gafford, G.A. et al., 2019: Agrivoltaics provide mutual benefits across the food–energy–water nexus in drylands. *Nat. Sustain.*, **2(9)**, 848–855, doi:10.1038/s41893-019-0364-5.

Bascopé, M., P. Perasso, and K. Reiss, 2019: Systematic review of education for sustainable development at an early stage: Cornerstones and pedagogical approaches for teacher professional development. *Sustainability*, **11(3)**, doi:10.3390/su11030719.

Batres, M. et al., 2021: Environmental and climate justice and technological carbon removal. *Electr. J.*, **34(7)**, 107002, doi:10.1016/j.tej.2021.107002.

Bauer, J.M. and L.A. Reisch, 2019: Behavioural Insights and (Un)healthy Dietary Choices: a Review of Current Evidence. *J. Consum. Policy*, **42(1)**, 3–45, doi:10.1007/s10603-018-9387-y.

Bauer, N. et al., 2015: CO2 emission mitigation and fossil fuel markets: Dynamic and international aspects of climate policies. *Technol. Forecast. Soc. Change*, **90**(PA), 243–256, doi:10.1016/j.techfore.2013.09.009.

Bauer, N. et al., 2018: Global energy sector emission reductions and bioenergy use: overview of the bioenergy demand phase of the EMF-33 model comparison. *Clim. Change*, **163**, 1553–1568, doi:10.1007/s10584-018-2226-y.

Bayraktarov, E. et al., 2016: The cost and feasibility of marine coastal restoration. *Ecol. Appl.*, **26(4)**, 1055–1074, doi:10.1890/15-1077.

12

Bazilian, M. et al., 2011: Considering the energy, water and food nexus: Towards an integrated modelling approach. *Energy Policy*, **39(12)**, 7896–7906, doi:10.1016/j.enpol.2011.09.039.

Beacham, A.M., L.H. Vickers, and J.M. Monaghan, 2019: Vertical farming: a summary of approaches to growing skywards. *J. Hortic. Sci. Biotechnol.*, **94(3)**, 277–283, doi:10.1080/14620316.2019.1574214.

Bechthold, A., H. Boeing, I. Tetens, L. Schwingshackl, and U. Nöthlings, 2018: Perspective: Food-Based Dietary Guidelines in Europe—Scientific Concepts, Current Status, and Perspectives. *Adv. Nutr.*, **9(5)**, 544–560, doi:10.1093/advances/nmy033.

Bednar, J., M. Obersteiner, and F. Wagner, 2019: On the financial viability of negative emissions. *Nat. Commun.*, **10(1)**, 1783, doi:10.1038/s41467-019-09782-x.

Beerling, D.J. et al., 2018: Farming with crops and rocks to address global climate, food and soil security. *Nat. Plants*, **4**, 138–147, doi:10.1038/s41477-018-0108-y.

Beerling, D.J. et al., 2020: Potential for large-scale CO2 removal via enhanced rock weathering with croplands. *Nature*, **583**, 242–248, doi:10.1038/s41586-020-2448-9.

Behfar, A., D. Yuill, and Y. Yu, 2018: Supermarket system characteristics and operating faults (RP-1615). *Sci. Technol. Built Environ.*, **24(10)**, 1104–1113, doi:10.1080/23744731.2018.1479614.

Bell, P. et al., 2018: *A Practical Guide to Climate-Smart Agriculture Technologies in Africa*. CCAFS Working Paper no. 224. CGIAR Research Program on Climate Change, Agriculture and Food Security, Wageningen, Netherlands, http://www.ccafs.cgiar.org.

Bellamy, R., 2018: Incentivize negative emissions responsibly. *Nat. Energy*, **3(7)**, 532–534, doi:10.1038/s41560-018-0156-6.

Bellamy, R. and O. Geden, 2019: Govern $CO_2$ removal from the ground up. *Nat. Geosci.*, **12(11)**, 874–876, doi:10.1038/s41561-019-0475-7.

Bellamy, R., J. Lezaun, and J. Palmer, 2019: Perceptions of bioenergy with carbon capture and storage in different policy scenarios. *Nat. Commun.*, **10(1)**, 743, doi:10.1038/s41467-019-08592-5.

Béné, C. et al., 2020: Global drivers of food system (un)sustainability: A multi-country correlation analysis. *PLoS One*, **15(4)**, e0231071, doi:10.1371/journal.pone.0231071.

Benke, K. and B. Tomkins, 2017: Future food-production systems: vertical farming and controlled-environment agriculture. *Sustain. Sci. Pract. Policy*, **13(1)**, 13–26, doi:10.1080/15487733.2017.1394054.

Bennani, H. et al., 2020: Overview of evidence of antimicrobial use and antimicrobial resistance in the food chain. *Antibiotics*, **9(2)**, 1–18, doi:10.3390/antibiotics9020049.

Benton, T.G., T. Bailey, and R. Bailey, 2019: The paradox of productivity: agricultural productivity promotes food system inefficiency. *Glob. Sustain.*, **2**, 1–8, doi:10.1017/sus.2019.3.

Bentsen, N.S. and I.M. Møller, 2017: Solar energy conserved in biomass: Sustainable bioenergy use and reduction of land use change. *Renew. Sustain. Energy Rev.*, **71**, 954–958, doi.org/10.1016/j.rser.2016.12.124.

Bergmann Madsen, B., 2018: Copenhagen: Organic Conversion in Public Kitchens., https://webgate.ec.europa.eu/dyna/bp-portal/getfile.cfm?fileid=332.

Bernauer, T. and L.F. McGrath, 2016: Simple reframing unlikely to boost public support for climate policy. *Nat. Clim. Change*, **6(7)**, 680–683, doi:10.1038/nclimate2948.

Berndes, G., P. Börjesson, M. Ostwald, and M. Palm, 2008: Multifunctional biomass production systems –an overview with presentation of specific applications in India and Sweden. *Biofuels, Bioprod. Biorefining*, **2(1)**, 16–25, doi:10.1002/bbb.52.

Berners-Lee, M., C. Kennelly, R. Watson, and C.N. Hewitt, 2018: Current global food production is sufficient to meet human nutritional needs in 2050 provided there is radical societal adaptation. *Elem Sci Anth*, **6(1)**, 52, doi:10.1525/elementa.310.

Berry, P.M. et al., 2015: Cross-sectoral interactions of adaptation and mitigation measures. *Clim. Change*, **128(3–4)**, 381–393, doi:10.1007/s10584-014-1214-0.

Bertram, C. and C. Merk, 2020: Public Perceptions of Ocean-Based Carbon Dioxide Removal: The Nature-Engineering Divide? *Front. Clim.*, **2**, 31, doi:10.3389/fclim.2020.594194.

Beuttler, C., L. Charles, and J. Wurzbacher, 2019: The Role of Direct Air Capture in Mitigation of Anthropogenic Greenhouse Gas Emissions. *Front. Clim.*, **1**, 10, doi:10.3389/fclim.2019.00010.

Bezner Kerr, R. et al., 2019: Participatory agroecological research on climate change adaptation improves smallholder farmer household food security and dietary diversity in Malawi. *Agric. Ecosyst. Environ.*, **279**, 109–121, doi:10.1016/j.agee.2019.04.004.

Bezner Kerr, R. et al., 2021: Can agroecology improve food security and nutrition? A review. *Glob. Food Sec.*, **29**, 100540, doi:10.1016/J.GFS.2021.100540.

Bhardwaj, A., M. Joshi, R. Khosla, and N.K. Dubash, 2019: More priorities, more problems? Decision-making with multiple energy, development and climate objectives. *Energy Res. Soc. Sci.*, **49**, 143–157, doi:10.1016/J.ERSS.2018.11.003.

Bharucha, Z.P., S.B. Mitjans, and J. Pretty, 2020: Towards redesign at scale through zero budget natural farming in Andhra Pradesh, India. *Int. J. Agric. Sustain.*, **18(1)**, doi:10.1080/14735903.2019.1694465.

Bhattacharya, A., J.P. Meltzer, J. Oppenheim, Z. Qureshi, and L.N. Stern, 2016: *Delivering on sustainable infrastructure for better development and better climate*. Brookings Institution, Washington, DC, USA, 160 pp. https://www.brookings.edu/research/delivering-on-sustainable-infrastructure-for-better-development-and-better-climate/ (Accessed December 3, 2019).

Bhattacharya, A., H. Kharas, M. Plant, and A. Prizzon, 2018: *The new global agenda and the future of the multilateral development bank system*. Brookings Institution, Washington, DC, USA, 24 pp. https://www.brookings.edu/research/the-new-global-agenda-and-the-future-of-the-multilateral-development-bank-system/ (Accessed December 3, 2019).

Bhunnoo, R., 2019: The need for a food-systems approach to policy making. *Lancet*, **393(10176)**, 1097–1098, doi:10.1016/S0140-6736(18)32754-5.

Bianchi, E., C. Bowyer, J.A. Morrison, R. Vos, and L. Wellesley, 2018a: *Redirecting investment for a global food system that is sustainable and promotes healthy diets*. Economics Discussion Papers 2018-69, Kiel Institute for the World Economy, Kiel, Germany, http://hdl.handle.net/10419/182388.

Bianchi, F., E. Garnett, C. Dorsel, P. Aveyard, and S.A. Jebb, 2018b: Restructuring physical micro-environments to reduce the demand for meat: a systematic review and qualitative comparative analysis. *Lancet Planet. Heal.*, **2(9)**, e384–e397, doi:10.1016/S2542-5196(18)30188-8.

Biermann, F. et al., 2016: Down to Earth: Contextualizing the Anthropocene. *Glob. Environ. Change*, **39**, 341–350, doi.org/10.1016/j.gloenvcha.2015.11.004.

Bindoff, N.L., W.W.L. Cheung, J.G. Kairo, J. Arístegui, V.A. Guinder, R. Hallberg, N. Hilmi, N. Jiao, M.S. Karim, L. Levin, S. O'Donoghue, S.R. Purca Cuicapusa, B. Rinkevich, T. Suga, A. Tagliabue, and P. Williamson, 2019: Changing Ocean, Marine Ecosystems, and Dependent Communities. In: *IPCC Special Report on the Ocean and Cryosphere in a Changing Climate* [Pörtner, H.-O., D.C. Roberts, V. Masson-Delmotte, P. Zhai, M. Tignor, E. Poloczanska, K. Mintenbeck, A. Alegría, M. Nicolai, A. Okem, J. Petzold, B. Rama and N.M. Weyer (eds.)]. Cambridge University Press, Cambridge, UK, and New York, NY, USA, pp. 447–587.

Bistline, J.E.T. and G.J. Blanford, 2021: Impact of carbon dioxide removal technologies on deep decarbonization of the electric power sector. *Nat. Commun.*, **12(1)**, 3732, doi:10.1038/s41467-021-23554-6.

Blain, S., B. Quéguiner, and T. Trull, 2008: The natural iron fertilization experiment KEOPS (KErguelen Ocean and Plateau compared Study): An overview. *Deep Sea Res. Part II Top. Stud. Oceanogr.*, **55(5–7)**, 559–565, doi:10.1016/J.DSR2.2008.01.002.

Blair, M.J., B. Gagnon, A. Klain, and B. Kulišić, 2021: Contribution of Biomass Supply Chains for Bioenergy to Sustainable Development Goals. *L. 2021*, **10(2)**, 181, doi:10.3390/LAND10020181.

Blakely, T. et al., 2020: The effect of food taxes and subsidies on population health and health costs: a modelling study. *Lancet Public Heal.*, **5(7)**, e404–e413, doi:10.1016/S2468-2667(20)30116-X. Blanco, H., W. Nijs, J. Ruf, and A. Faaij, 2018: Potential for hydrogen and Power-to-Liquid in a low-carbon EU energy system using cost optimization. *Appl. Energy*, **232**, 617–639, doi:10.1016/j.apenergy.2018.09.216.

BNEF, 2020: *Electric Vehicle Outlook 2020*. BloombergNEF, London, UK, https://about.bnef.com/electric-vehicle-outlook/.

Bodirsky, B.L. et al., 2014: Reactive nitrogen requirements to feed the world in 2050 and potential to mitigate nitrogen pollution. *Nat. Commun.*, **5(1)**, 3858, doi:10.1038/ncomms4858.

Bodirsky, B.L. et al., 2020: The Ongoing Nutrition Transition Thwarts Long-Term Targets for Food Security, Public Health and Environmental Protection. *Sci. Rep.*, **10(1)**, 1–14, doi:10.1038/s41598-020-75213-3.

Boettcher, M. et al., 2021: Navigating Potential Hype and Opportunity in Governing Marine Carbon Removal. *Front. Clim.*, **3**, 47, doi:10.3389/fclim.2021.664456.

Bonansea, M. et al., 2020: Assessing water surface temperature from Landsat imagery and its relationship with a nuclear power plant. doi.org/10.1080/02626667.2020.1845342, **66(1)**, 50–58, doi:10.1080/02626667.2020.1845342.

Bonnet, C., Z. Bouamra-Mechemache, and T. Corre, 2018: An Environmental Tax Towards More Sustainable Food: Empirical Evidence of the Consumption of Animal Products in France. *Ecol. Econ.*, **147**(January), 48–61, doi:10.1016/j.ecolecon.2017.12.032.

Bopp, L. et al., 2013: Multiple stressors of ocean ecosystems in the 21st century: Projections with CMIP5 models. *Biogeosciences*, **10(10)**, 6225–6245, doi:10.5194/BG-10-6225-2013.

Boukid, F., 2021: Plant-based meat analogues: from niche to mainstream. *Eur. Food Res. Technol.*, **247(2)**, 297–308, doi:10.1007/s00217-020-03630-9.

Boyd, P.W., 2008: Introduction and synthesis. *Mar. Ecol. Prog. Ser.*, **364**, 213–218, doi:10.3354/meps07541.

Boysen, L.R. et al., 2017: The limits to global-warming mitigation by terrestrial carbon removal. *Earth's Futur.*, **5(5)**, 463–474, doi:10.1002/2016EF000469.

Braghiroli, F.L., and L. Passarini, 2020: Valorization of Biomass Residues from Forest Operations and Wood Manufacturing Presents a Wide Range of Sustainable and Innovative Possibilities. *Curr. For. Reports 2020 62*, **6(2)**, 172–183, doi:10.1007/S40725-020-00112-9.

Brander, M., F. Ascui, V. Scott, and S. Tett, 2021: Carbon accounting for negative emissions technologies. *Clim. Policy*, **21(5)**, 699–717, doi:10.1080/14693062.2021.1878009.

Breyer, C., E. Tsupari, V. Tikka, and P. Vainikka, 2015: Power-to-gas as an emerging profitable business through creating an integrated value chain. *Energy Procedia*, **73**, 182–189, 10.1016/J.EGYPRO.2015.07.668

Breyer, C., M. Fasihi, C. Bajamundi, and F. Creutzig, 2019: Direct Air Capture of $CO_2$: A Key Technology for Ambitious Climate Change Mitigation. *Joule*, **3(9)**, 2053–2057, doi:10.1016/j.joule.2019.08.010.

Breyer, C., M. Fasihi, and A. Aghahosseini, 2020: Carbon dioxide direct air capture for effective climate change mitigation based on renewable electricity: a new type of energy system sector coupling. *Mitig. Adapt. Strateg. Glob. Change*, **25(1)**, 43–65, doi:10.1007/s11027-019-9847-y.

Broers, V.J.V., C. De Breucker, S. Van Den Broucke, and O. Luminet, 2017: A systematic review and meta-analysis of the effectiveness of nudging to increase fruit and vegetable choice. *Eur. J. Public Health*, **27(5)**, 912–920, doi:10.1093/eurpub/ckx085.

Brouwer, I.D., J. McDermott, and R. Ruben, 2020: Food systems everywhere: Improving relevance in practice. *Glob. Food Sec.*, **26**(June), 100398, doi:10.1016/j.gfs.2020.100398.

Bruno, J., L. Duro, and F. Diaz-Maurin, 2020: Spent nuclear fuel and disposal. In: Advances in Nuclear Fuel Chemistry [Piro, M.H.H (ed.)]., Woodhead Publishing, Cambridge, UK, pp 527–553, doi:10.1016/B978-0-08-102571-0.00014-8.

Bruun, S., L.S. Jensen, V.T. Khanh Vu, and S. Sommer, 2014: Small-scale household biogas digesters: An option for global warming mitigation or a potential climate bomb? *Renew. Sustain. Energy Rev.*, **33**, 736–741, doi:10.1016/j.rser.2014.02.033.

Bucher, T. et al., 2016: Nudging consumers towards healthier choices: a systematic review of positional influences on food choice. *Br. J. Nutr.*, **115(12)**, 2252–2263, doi:10.1017/S0007114516001653.

Buck, H.J., 2016: Rapid scale-up of negative emissions technologies: social barriers and social implications. *Clim. Change*, **139(2)**, 155–167, doi:10.1007/s10584-016-1770-6.

Buck, H.J., 2021: Social science for the next decade of carbon capture and storage. *Electr. J.*, **34(7)**, 107003, doi:10.1016/j.tej.2021.107003.

Buck, H.J., J. Furhman, D.R. Morrow, D.L. Sanchez, and F.M. Wang, 2020: Adaptation and Carbon Removal. *One Earth*, **3(4)**, 425–435, doi:10.1016/j.oneear.2020.09.008.

Bugge, M.M., S. Bolwig, T. Hansen, and A.N. Tanner, 2019: Theoretical perspectives on innovation for waste valorisation in the bioeconomy. In: *From Waste to Value* [Klitkou, A., A.M. Fevolden, and M. Capasso, (eds.)], Routledge, London, UK, pp. 51–70.

Bui, M. et al., 2018: Carbon capture and storage (CCS): the way forward. *Energy Environ. Sci.*, **11(5)**, 1062–1176, doi:10.1039/C7EE02342A. Bui, M., D. Zhang, M. Fajardy, and N. Mac Dowell, 2021: Delivering carbon negative electricity, heat and hydrogen with BECCS – Comparing the options. *Int. J. Hydrogen Energy*, **46(29)**, 15298–15321, doi:10.1016/j.ijhydene.2021.02.042.

Buisman, M.E., R. Haijema, and J.M. Bloemhof-Ruwaard, 2019: Discounting and dynamic shelf life to reduce fresh food waste at retailers. *Int. J. Prod. Econ.*, **209**, 274–284, doi:10.1016/j.ijpe.2017.07.016.

Burdige, D.J., 2005: Burial of terrestrial organic matter in marine sediments: A re-assessment. *Global Biogeochem. Cycles*, **19(4)**, n/a-n/a, doi:10.1029/2004GB002368.

Burns, W. and C.R. Corbett, 2020: Antacids for the Sea? Artificial Ocean Alkalinization and Climate Change. *One Earth*, **3(2)**, 154–156, doi:10.1016/j.oneear.2020.07.016.

Busch, G., 2017: A spatial explicit scenario method to support participative regional land-use decisions regarding economic and ecological options of short rotation coppice (SRC) for renewable energy production on arable land: case study application for the Göttingen district, Germany. *Energy, Sustain. Soc. 2017 71*, **7(1)**, 1–23, doi:10.1186/S13705-017-0105-4.

Buss, W., K. Yeates, E.J. Rohling, and J. Borevitz, 2021: Enhancing natural cycles in agro-ecosystems to boost plant carbon capture and soil storage. *Oxford Open Clim. Change*, **1(1)**, doi:10.1093/OXFCLM/KGAB006.

Buylova, A., M. Fridahl, N. Nasiritousi, and G. Reischl, 2021: Cancel (Out) Emissions? The Envisaged Role of Carbon Dioxide Removal Technologies in Long-Term National Climate Strategies. *Front. Clim.*, **3**, 63, doi:10.3389/fclim.2021.675499.

C40 Cities, 2019: *Good Food Cities: Achieving a Planetary Health Diet for All*. 2 pp. https://c40-production-images.s3.amazonaws.com/press_releases/images/415_C40_Good_Food_Cities_Declaration_EN_Final_-_CLEAN_3_.original.pdf?1570699994.

Cacho, J.F., M.C. Negri, C.R. Zumpf, and P. Campbell, 2018: Introducing perennial biomass crops into agricultural landscapes to address water quality challenges and provide other environmental services. *Wiley Interdiscip. Rev. Energy Environ.*, **7(2)**, e275, doi:10.1002/wene.275.

Cadario, R. and P. Chandon, 2018: Which Healthy Eating Nudges Work Best? A Meta-Analysis of Field Experiments. *Marketing Science 39(3):465-486*, https://doi.org/10.1287/mksc.2018.1128.

Cagle, A.E. et al., 2020: The Land Sparing, Water Surface Use Efficiency, and Water Surface Transformation of Floating Photovoltaic Solar Energy Installations. *Sustainability*, **12(19)**, 8154, doi:10.3390/su12198154.

12

Cairns, R. and A. Krzywoszynska, 2016: Anatomy of a buzzword: The emergence of 'the water-energy-food nexus' in UK natural resource debates. *Environ. Sci. Policy*, **64**, 164–170, doi:10.1016/j.envsci.2016.07.007.

Caldeira, C. et al., 2020: Sustainability of food waste biorefinery: A review on valorisation pathways, techno-economic constraints, and environmental assessment. *Bioresour. Technol.*, **312**(March), 123575, doi:10.1016/j.biortech.2020.123575.

Calvin, K. et al., 2021: Bioenergy for climate change mitigation: scale and sustainability. *GCB Bioenergy*, **13(9)**, 1346–1371, doi:10.1111/gcbb.12863.

Camaratta, R., T.M. Volkmer, and A.G. Osorio, 2020: Embodied energy in beverage packaging. *J. Environ. Manage.*, **260**, 110172, doi.org/10.1016/j.jenvman.2020.110172.

Campbell, B.M. et al., 2017: Agriculture production as a major driver of the earth system exceeding planetary boundaries. *Ecol. Soc.*, **22(4)**, doi:10.5751/ES-09595-220408.

Campbell, B.M. et al., 2018: Urgent action to combat climate change and its impacts (SDG 13): transforming agriculture and food systems. *Curr. Opin. Environ. Sustain.*, **34**(Sdg 13), 13–20, doi:10.1016/j.cosust.2018.06.005.

Candel, J.J.L., 2014: Food security governance: a systematic literature review. *Food Secur.*, **6(4)**, 585–601, doi:10.1007/s12571-014-0364-2.

Candel, J.J.L., 2019: What's on the menu? A global assessment of MUFPP signatory cities' food strategies. *Agroecol. Sustain. Food Syst.*, **44(7)**, 1–28, doi:10.1080/21683565.2019.1648357.

Cantzler, J. et al., 2020: Saving resources and the climate? A systematic review of the circular economy and its mitigation potential. *Environ. Res. Lett.*, **15(12)**, 123001, doi:10.1088/1748-9326/abbeb7.

Cao, L. and K. Caldeira, 2010: Can ocean iron fertilization mitigate ocean acidification? *Clim. Chang. 2010 991*, **99(1)**, 303–311, doi:10.1007/S10584-010-9799-4.

Capellán-Pérez, I., C. de Castro, and I. Arto, 2017: Assessing vulnerabilities and limits in the transition to renewable energies: Land requirements under 100% solar energy scenarios. *Renew. Sustain. Energy Rev.*, **77**, 760–782, doi:10.1016/J.RSER.2017.03.137.

Capros, P. et al., 2019: Energy-system modelling of the EU strategy towards climate-neutrality. *Energy Policy*, **134**, 110960, doi:10.1016/j.enpol.2019.110960.

Caron, P. et al., 2018: Food systems for sustainable development: proposals for a profound four-part transformation. *Agron. Sustain. Dev.*, **38**, doi:10.1007/s13593-018-0519-1.

Carrara, S., P. Alves Dias, B. Plazzotta, C. Pavel, and European Commission. Joint Research Centre., 2020: Raw materials demand for wind and solar PV technologies in the transition towards a decarbonised energy system. *JCR*, EUR 30095 EN, Publications Office of the European Union, Luxembourg, 2020, doi:10.2760/160859.

Carrenõ, I. and T. Dolle, 2018: Tofu steaks? Developments on the naming and marketing of plant-based foods in the aftermath of the TofuTown judgement. *Eur. J. Risk Regul.*, **9(3)**, 575–584, doi:10.1017/err.2018.43.

Carton, W., 2019: "Fixing" Climate Change by Mortgaging the Future: Negative Emissions, Spatiotemporal Fixes, and the Political Economy of Delay. *Antipode*, **51(3)**, 750–769, doi:10.1111/anti.12532.

Carton, W., A. Asiyanbi, S. Beck, H.J. Buck, and J.F. Lund, 2020: Negative emissions and the long history of carbon removal. *WIREs Clim. Change*, **11(6)**, e671, doi:10.1002/wcc.671.

Caserini, S. et al., 2019: Affordable CO2 negative emission through hydrogen from biomass, ocean liming, and CO2 storage. *Mitig. Adapt. Strateg. Glob. Change*, **24(7)**, 1231–1248, doi:10.1007/s11027-018-9835-7.

Caserini, S. et al., 2021: Potential of Maritime Transport for Ocean Liming and Atmospheric CO2 Removal. *Front. Clim.*, **3**, 22, doi:10.3389/fclim.2021.575900.

Cazzolla Gatti, R., J. Liang, A. Velichevskaya, and M. Zhou, 2019: Sustainable palm oil may not be so sustainable. *Sci. Total Environ.*, **652**, 48–51, doi:10.1016/j.scitotenv.2018.10.222.

Centola, D., J. Becker, D. Brackbill, and A. Baronchelli, 2018: Experimental evidence for tipping points in social convention. *Science*, **360(6393)**, 1116–1119, doi:10.1126/science.aas8827.

Chai, B.C. et al., 2019: Which Diet Has the Least Environmental Impact on Our Planet? A Systematic Review of Vegan, Vegetarian and Omnivorous Diets. *Sustainability*, **11(15)**, 4110, doi:10.3390/su11154110.

Chakravorty, U., M.H. Hubert, M. Moreaux, and L. Nøstbakken, 2017: Long-Run Impact of Biofuels on Food Prices. *Scand. J. Econ.*, **119(3)**, doi:10.1111/sjoe.12177.

Chaomuang, N., D. Flick, and O. Laguerre, 2017: Experimental and numerical investigation of the performance of retail refrigerated display cabinets. *Trends Food Sci. Technol.*, **70**, 95–104, doi:10.1016/j.tifs.2017.10.007.

Chatterjee, S. and K.-W. Huang, 2020: Unrealistic energy and materials requirement for direct air capture in deep mitigation pathways. *Nat. Commun.*, **11(1)**, 3287, doi:10.1038/s41467-020-17203-7.

Chaudhary, A., D.I. Gustafson, and A. Mathys, 2018: Multi-indicator sustainability assessment of global food systems. *Nat. Commun.*, **9(1)**, 848, doi:10.1038/s41467-018-03308-7.

Chen, C. and M. Tavoni, 2013: Direct air capture of CO2 and climate stabilization: A model based assessment. *Clim. Change*, **118(1)**, 59–72, doi:10.1007/s10584-013-0714-7.

Chen, C., A. Chaudhary, and A. Mathys, 2019: Dietary Change Scenarios and Implications for Environmental, Nutrition, Human Health and Economic Dimensions of Food Sustainability. *Nutrients*, **11(4)**, 856, doi:10.3390/nu11040856.

Chen, D., E.C. Jaenicke, and R.J. Volpe, 2016: Food Environments and Obesity: Household Diet Expenditure Versus Food Deserts. *Am. J. Public Health*, **106(5)**, 881–888, doi:10.2105/AJPH.2016.303048.

Chen, G., R.P. Powers, L.M.T. de Carvalho, and B. Mora, 2015: Spatiotemporal patterns of tropical deforestation and forest degradation in response to the operation of the Tucuruí hydroelectric dam in the Amazon basin. *Appl. Geogr.*, **63**, 1–8, doi.org/10.1016/j.apgeog.2015.06.001.

Chen, H., D. Zhou, G. Luo, S. Zhang, and J. Chen, 2015: Macroalgae for biofuels production: Progress and perspectives. *Renew. Sustain. Energy Rev.*, **47**, 427–437, doi:10.1016/j.rser.2015.03.086.

Cheng, W., A. Appolloni, A. D'Amato, and Q. Zhu, 2018: Green Public Procurement, missing concepts and future trends – A critical review. *J. Clean. Prod.*, **176**, 770–784, doi:10.1016/j.jclepro.2017.12.027.

Cherubin, M.R. et al., 2018: Crop residue harvest for bioenergy production and its implications on soil functioning and plant growth: A review. *Sci. Agric.*, **75**, 255–272, https://doi.org/10.1590/1678-992X-2016-0459

Choi, D.Y., T.W. Wittig, and B.M. Kluever, 2020: An evaluation of bird and bat mortality at wind turbines in the Northeastern United States. *PLoS One*, **15(8)**, e0238034.

Chomba, S., F. Sinclair, P. Savadogo, M. Bourne, and M. Lohbeck, 2020: Opportunities and Constraints for Using Farmer Managed Natural Regeneration for Land Restoration in Sub-Saharan Africa. *Front. For. Glob. Change*, **0**, 122, doi:10.3389/FFGC.2020.571679.

Chriki, S., and J.F. Hocquette, 2020: The Myth of Cultured Meat: A Review. *Front. Nutr.*, **7**(February), 1–9, doi:10.3389/fnut.2020.00007.

Christen, B. and T. Dalgaard, 2013: Buffers for biomass production in temperate European agriculture: A review and synthesis on function, ecosystem services and implementation. *Biomass and Bioenergy*, **55**, 53–67, doi:10.1016/j.biombioe.2012.09.053.

Churkina, G. et al., 2020: Buildings as a global carbon sink. *Nat. Sustain.*, **3**, 269–276, doi:10.1038/s41893-019-0462-4.

Cialdini, R.B. and R.P. Jacobson, 2021: Influences of social norms on climate change-related behaviors. *Curr. Opin. Behav. Sci.*, **42**, 1–8, doi:10.1016/j.cobeha.2021.01.005.

Cipolla, G., S. Calabrese, L.V. Noto, and A. Porporato, 2021: The role of hydrology on enhanced weathering for carbon sequestration II. From hydroclimatic scenarios to carbon-sequestration efficiencies. *Adv. Water Resour.*, **154**, 103949, doi:10.1016/j.advwatres.2021.103949.

CIRAIG, 2015: *Circular Economy: a critical literature review of concepts*. International Reference Centre for the Life Cycle of Products, Processes and Services (CIRAIG), Montréal, QC, Canada, 53 pp.

Clapp, J., 2017: The trade-ification of the food sustainability agenda. *J. Peasant Stud.*, **44(2)**, 335–353, doi:10.1080/03066150.2016.1250077.

Clapp, J., 2019: The rise of financial investment and common ownership in global agrifood firms. *Rev. Int. Polit. Econ.*, **26(4)**, 604–629, doi:10.1080/09692290.2019.1597755.

Clark, M. and D. Tilman, 2017: Comparative analysis of environmental impacts of agricultural production systems, agricultural input efficiency, and food choice. *Environ. Res. Lett.*, **12(6)**, 064016, doi:10.1088/1748-9326/aa6cd5.

Clark, M., J. Hill, and D. Tilman, 2018: The Diet, Health, and Environment Trilemma. *Annu. Rev. Environ. Resour.*, **43(1)**, 109–134, doi:10.1146/annurev-environ-102017-025957.

Clark, M.A., M. Springmann, J. Hill, and D. Tilman, 2019: Multiple health and environmental impacts of foods. *Proc. Natl. Acad. Sci.*, **116(46)**, 23357–23362, doi:10.1073/pnas.1906908116.

Clark, M.A. et al., 2020: Global food system emissions could preclude achieving the 1.5° and 2°C climate change targets. *Science*, **370(6517)**, 705–708, doi:10.1126/science.aba7357.

Clery, D.S. et al., 2021: Bringing greenhouse gas removal down to earth: Stakeholder supply chain appraisals reveal complex challenges. *Glob. Environ. Change*, **71**, 102369, doi:10.1016/j.gloenvcha.2021.102369.

Climate Change Committee, 2021: *Progress in reducing emissions: 2021 Report to Parliament*. Climate Change Committee, London, UK, 223 pp. www.theccc.org.uk/publication/2021-progress-report-to-parliament/.

Clune, S., E. Crossin, and K. Verghese, 2017: Systematic review of greenhouse gas emissions for different fresh food categories. *J. Clean. Prod.*, **140**, 766–783, doi:10.1016/j.jclepro.2016.04.082.

COAG Energy Council, 2019: *Austrialia's National Hydrogen Strategy*. Commonwealth of Australia, 136 pp.

Coe, R., F. Sinclair, and E. Barrios, 2014: Scaling up agroforestry requires research 'in' rather than 'for' development. *Curr. Opin. Environ. Sustain.*, **6(1)**, 73–77, doi:10.1016/j.cosust.2013.10.013.

Coelho, P.M., B. Corona, R. ten Klooster, and E. Worrell, 2020: Sustainability of reusable packaging–Current situation and trends. *Resour. Conserv. Recycl. X*, **6** (March), 100037, doi:10.1016/j.rcrx.2020.100037.

Cohen, B., E. Tyler, and M. Torres Gunfaus, 2017: Lessons from co-impacts assessment under the Mitigation Action Plans and Scenarios (MAPS) Programme. *Clim. Policy*, **17(8)**, 1065–1075, doi:10.1080/14693062.2016.1222258.

Cohen, B. et al., 2019: Multi-criteria decision analysis in policy-making for climate mitigation and development. *Clim. Dev.*, **11(3)**, 212–222, doi:10.1080/17565529.2018.1445612.

Cohen, B., A. Cowie, M. Babiker, A. Leip, and P. Smith, 2021: Co-benefits and trade-offs of climate change mitigation actions and the Sustainable Development Goals. *Sustain. Prod. Consum.*, **26**, 805–813, doi:10.1016/j.spc.2020.12.034.

Colvin, R.M. et al., 2020: Learning from the Climate Change Debate to Avoid Polarisation on Negative Emissions. *Environ. Commun.*, **14(1)**, 23–35, doi:10.1080/17524032.2019.1630463.

Conijn, J.G., P.S. Bindraban, J.J. Schröder, and R.E.E. Jongschaap, 2018: Can our global food system meet food demand within planetary boundaries? *Agric. Ecosyst. Environ.*, **251**, 244–256, doi:10.1016/j.agee.2017.06.001.

Cook, A.S.C. P., E.M. Humphreys, F. Bennet, E.A. Masden, and N.H.K. Burton, 2018: Quantifying avian avoidance of offshore wind turbines: Current evidence and key knowledge gaps. *Mar. Environ. Res.*, **140**, 278–288, doi.org/10.1016/j.marenvres.2018.06.017.

Cool Food, 2020: The Cool Food Pledge. https://coolfood.org/pledge/ (Accessed December 1, 2020).

Coppes, J. et al., 2020: The impact of wind energy facilities on grouse: a systematic review. *J. Ornithol.*, **161(1)**, 1–15, doi:10.1007/s10336-019-01696-1.

Córdova, R., N.J. Hogarth, and M. Kanninen, 2019: Mountain Farming Systems' Exposure and Sensitivity to Climate Change and Variability: Agroforestry and Conventional Agriculture Systems Compared in Ecuador's Indigenous Territory of Kayambi People. *Sustain.*, **11(9)**, doi:10.3390/su11092623.

Cornelsen, L., R. Green, A. Dangour, and R. Smith, 2015: Why fat taxes won't make us thin. *J. Public Heal. (United Kingdom)*, **37(1)**, 18–23, doi:10.1093/pubmed/fdu032.

Correa, D.F. et al., 2019: Towards the implementation of sustainable biofuel production systems. *Renew. Sustain. Energy Rev.*, **107**, 250–263, doi:10.1016/j.rser.2019.03.005.

Corvalán, C., M. Reyes, M.L. Garmendia, and R. Uauy, 2019: Structural responses to the obesity and non-communicable diseases epidemic: Update on the Chilean law of food labelling and advertising. *Obes. Rev.*, **20(3)**, 367–374, doi:10.1111/obr.12802.

Cossel, M. Von et al., 2019: Prospects of Bioenergy Cropping Systems for A More Social-Ecologically Sound Bioeconomy. *Agron. 2019, Vol. 9, Page 605*, **9(10)**, 605, doi:10.3390/AGRONOMY9100605.

Costa Leite, J., S. Caldeira, B. Watzl, and J. Wollgast, 2020: Healthy low nitrogen footprint diets. *Glob. Food Sec.*, **24**, doi:10.1016/j.gfs.2019.100342.

Costinot, A., D. Donaldson, and C. Smith, 2016: Evolving Comparative Advantage and the Impact of Climate Change in Agricultural Markets: Evidence from 1.7 Million Fields around the World. *J. Polit. Econ.*, **124(1)**, 205–248, doi:10.1086/684719.

Cottrell, R.S. et al., 2019: Food production shocks across land and sea. *Nat. Sustain.*, **2(2)**, doi:10.1038/s41893-018-0210-1.

Coulomb, R. and F. Henriet, 2018: The Grey Paradox: How fossil-fuel owners can benefit from carbon taxation. *J. Environ. Econ. Manage.*, **87**, 206–223, doi:10.1016/j.jeem.2017.07.001.

Cowie, A., 2020a: *Guidelines for Land Degradation Neutrality: A report prepared for stapgethe Scientific and Technical Advisory Panel of the Global Environment Facility*. Washington DC, USA, 60 pp. https://catalogue.unccd.int/1474_LDN_Technical_Report_web_version.pdf.

Cowie, A.L., 2020b: Bioenergy in the circular economy. In: *Handbook of the Circular Economy*, [Brandão, M., Lazarevic, D., and Finnveden, G. (eds.)] Edward Elgar Publishing, Cheltenham, UK, pp. 382–395.

Cowie, A.L. et al., 2018: Land in balance: The scientific conceptual framework for Land Degradation Neutrality. *Environ. Sci. Policy*, **79**, 25–35, doi:10.1016/j.envsci.2017.10.011.

Cowie, A.L. et al., 2021: Applying a science-based systems perspective to dispel misconceptions about climate effects of forest bioenergy. *GCB Bioenergy*, **13(8)**, 1210–1231, doi:10.1111/GCBB.12844.

Cox, E. and N.R. Edwards, 2019: Beyond carbon pricing: policy levers for negative emissions technologies. *Clim. Policy*, **19(9)**, 1144–1156, doi:10.1080/14693062.2019.1634509.

Cox, E., E. Spence, and N. Pidgeon, 2020a: Public perceptions of carbon dioxide removal in the United States and the United Kingdom. *Nat. Clim. Change*, **10(8)**, 744–749, doi:10.1038/s41558-020-0823-z.

Cox, E., E. Spence, and N. Pidgeon, 2020b: Incumbency, Trust and the Monsanto Effect: Stakeholder Discourses on Greenhouse Gas Removal. *Environ. Values*, **29(2)**, 197–220, doi:10.3197/096327119X15678473650947.

Creutzig, F. et al., 2015: Bioenergy and climate change mitigation: An assessment. *GCB Bioenergy*, **7(5)**, 916–944, doi:10.1111/gcbb.12205.

Creutzig, F. et al., 2019: The mutual dependence of negative emission technologies and energy systems. *Energy Environ. Sci.*, **12(6)**, 1805–1817, doi:10.1039/C8EE03682A. Crimarco, A. et al., 2020: A randomized crossover trial on the effect of plant-based compared with animal-based meat on trimethylamine-N-oxide and cardiovascular disease risk factors in generally healthy adults: Study with Appetizing Plantfood-Meat Eating Alternative Trial (SWAP-MEAT). *Am. J. Clin. Nutr.*, **112(5)**, 1188–1199, doi:10.1093/ajcn/nqaa203.

Crippa, M. et al., 2019: *Fossil $CO_2$ and GHG emissions of all world countries - 2019 Report*. Publications Office of the European Union, Luxembourg, 251 pp.

Crippa, M. et al., 2021a: EDGAR v6.0 Greenhouse Gas Emissions.

Crippa, M., E. Solazzo, D. Guizzardi, F. Monforti-Ferrario, and A. Leip, 2021b: Food systems are responsible for a third of global anthropogenic GHG emissions. *Nat. Food*,. **2**, 198–209, https://www.nature.com/articles/s43016-021-00225-9.

Cripps, G., S. Widdicombe, J. Spicer, and H. Findlay, 2013: Biological impacts of enhanced alkalinity in Carcinus maenas. *Mar. Pollut. Bull.*, **71(1–2)**, 190–198, doi:10.1016/j.marpolbul.2013.03.015.

Cristea, A., D. Hummels, L. Puzzello, and M. Avetisyan, 2013: Trade and the greenhouse gas emissions from international freight transport. *J. Environ. Econ. Manage.*, **65(1)**, 153–173, doi:10.1016/j.jeem.2012.06.002.

Cubins, J.A. et al., 2019: Management of pennycress as a winter annual cash cover crop. A review. *Agron. Sustain. Dev.*, **39(5)**, 1–11, doi:10.1007/S13593-019-0592-0.

Cunningham, S.C. et al., 2015: Balancing the environmental benefits of reforestation in agricultural regions. *Perspect. Plant Ecol. Evol. Syst.*, **17(4)**, 301–317, doi:10.1016/j.ppees.2015.06.001.

Curtain, F. and S. Grafenauer, 2019: Plant-Based Meat Substitutes in the Flexitarian Age: An Audit of Products on Supermarket Shelves. *Nutrients*, **11(11)**, 2603, doi:10.3390/nu11112603.

Curtis, P.G., C.M. Slay, N.L. Harris, A. Tyukavina, and M.C. Hansen, 2018: Classifying drivers of global forest loss. *Science*, **361(6407)**, 1108–1111, doi:10.1126/science.aau3445.

Czyrnek-Delêtre, M.M., B.M. Smyth, and J.D. Murphy, 2017: Beyond carbon and energy: The challenge in setting guidelines for life cycle assessment of biofuel systems. *Renew. Energy*, **105**, 436–448, doi:10.1016/J.RENENE.2016.11.043.

D'Annolfo, R., B. Gemmill-Herren, B. Graeub, and L.A. Garibaldi, 2017: A review of social and economic performance of agroecology. *Int. J. Agric. Sustain.*, **15(6)**, doi:10.1080/14735903.2017.1398123.

D'Odorico, P. et al., 2018: The Global Food-Energy-Water Nexus. *Rev. Geophys.*, **56(3)**, 456–531, doi:10.1029/2017RG000591.

Daggash, H.A. et al., 2018: Closing the carbon cycle to maximise climate change mitigation: power-to-methanol vs. power-to-direct air capture. *Sustain. Energy Fuels*, **2(6)**, 1153–1169, doi:10.1039/C8SE00061A. Dagnachew, A.G., P.L. Lucas, A.F. Hof, and D.P. van Vuuren, 2018: Trade-offs and synergies between universal electricity access and climate change mitigation in Sub-Saharan Africa. *Energy Policy*, **114**, 355–366, doi:10.1016/j.enpol.2017.12.023.

Daioglou, V., J.C. Doelman, B. Wicke, A. Faaij, and D.P. van Vuuren, 2019: Integrated assessment of biomass supply and demand in climate change mitigation scenarios. *Glob. Environ. Change*, **54**, 88–101, doi:10.1016/j.gloenvcha.2018.11.012.

Damerau, K., A.G. Patt, and O.P.R. van Vliet, 2016: Water saving potentials and possible trade-offs for future food and energy supply. *Glob. Environ. Change*, **39**, 15–25, doi:10.1016/j.gloenvcha.2016.03.014.

Dasgupta, P., 2021: *The Economics of Biodiversity: The Dasgupta Review*. HM Treasury, London, UK, 610 pp. https://www.gov.uk/government/publications/final-report-the-economics-of-biodiversity-the-dasgupta-review (Accessed August 24, 2021).

Dauber, J. and S. Miyake, 2016: To integrate or to segregate food crop and energy crop cultivation at the landscape scale? Perspectives on biodiversity conservation in agriculture in Europe. *Energy. Sustain. Soc.*, **6**(1), 25, doi:10.1186/s13705-016-0089-5.

Davis, S.C. et al., 2013: Management swing potential for bioenergy crops. *GCB Bioenergy*, **5**(6), 623–638, doi:10.1111/gcbb.12042.

Davis, S.J., G.P. Peters, and K. Caldeira, 2011: The supply chain of CO2 emissions. *Proc. Natl. Acad. Sci.*, **108(45)**, 18554–18559, doi:10.1073/pnas.1107409108.

De Boer, I.J.M. and M.K. Van Ittersum, 2018: Circularity in agricultural production. Wageningen University & Research, Wageningen, Netherlands, https://edepot.wur.nl/470625.

de Boer, J., H. Aiking, J. De Boer, and H. Aiking, 2018: Prospects for pro-environmental protein consumption in Europe: Cultural, culinary, economic and psychological factors. *Appetite*, **121**, 29–40, doi:10.1016/j.appet.2017.10.042.

De Castro, C., M. Mediavilla, L.J. Miguel, and F. Frechoso, 2013: Global solar electric potential: A review of their technical and sustainable limits. *Renew. Sustain. Energy Rev.*, **28**, 824–835, doi:10.1016/J.RSER.2013.08.040.

de Coninck, H., A. Revi, M. Babiker, P. Bertoldi, M. Buckeridge, A. Cartwright, W. Dong, J. Ford, S. Fuss, J.-C. Hourcade, D. Ley, R. Mechler, P. Newman, A. Revokatova, S. Schultz, L. Steg, and T. Sugiyama, 2018: Strengthening and implementing the global response. In: *Global warming of 1.5°C. An IPCC Special Report on the impacts of global warming of 1.5°C above pre-industrial levels and related global greenhouse gas emission pathways, in the context of strengthening the global response to the threat of climate change* [Masson-Delmotte, V., P. Zhai, H.-O. Pörtner, D. Roberts, J. Skea, P.R. Shukla, A. Pirani, W. Moufouma-Okia, C. Péan, R. Pidcock, S. Connors, J.B.R. Matthews, Y. Chen, X. Zhou, M.I. Gomis, E. Lonnoy, T. Maycock, M. Tignor, and T. Waterfield (eds.)], Cambridge University Press, Cambridge, UK, and New York, NY, USA, pp. 313–443.

de Jong, J., C. Akselsson, G. Egnell, S. Löfgren, and B.A. Olsson, 2017: Realizing the energy potential of forest biomass in Sweden – How much is environmentally sustainable? *For. Ecol. Manage.*, **383**, 3–16, doi:10.1016/J.FORECO.2016.06.028.

de Lannoy, C.F. et al., 2018: Indirect ocean capture of atmospheric CO2: Part I. Prototype of a negative emissions technology. *Int. J. Greenh. Gas Control*, **70**, 243–253, doi:10.1016/J.IJGGC.2017.10.007.

De Oliveira Garcia, W. et al., 2020: Impacts of enhanced weathering on biomass production for negative emission technologies and soil hydrology. *Biogeosciences*, **17(7)**, 2107–2133, doi:10.5194/BG-17-2107-2020.

Delaney, A. et al., 2018: Governance of food systems across scales in times of social-ecological change: a review of indicators. *Food Secur.*, **10(2)**, 287–310, doi:10.1007/s12571-018-0770-y.

Delcour, I., P. Spanoghe, and M. Uyttendaele, 2015: Literature review: Impact of climate change on pesticide use. *Food Res. Int.*, **68**, 7–15, doi.org/10.1016/j.foodres.2014.09.030.

Deryabina, T.G. et al., 2015: Long-term census data reveal abundant wildlife populations at Chernobyl. *Curr. Biol.*, **25(19)**, R824–R826, doi.org/10.1016/j.cub.2015.08.017.

Deutz, S. and A. Bardow, 2021: Life-cycle assessment of an industrial direct air capture process based on temperature–vacuum swing adsorption. *Nat. Energy*, **6(2)**, 203–213, doi:10.1038/s41560-020-00771-9.

DFO, 2010: *Ocean Fertilization: Mitigating environmental impacts of future scientific research*. DFO Canadian Science Advisory Secretarat, Ottawa, Canada, 14 pp.

Di Gregorio, M. et al., 2017: Climate policy integration in the land use sector: Mitigation, adaptation and sustainable development linkages. *Environ. Sci. Policy*, **67**, 35–43, doi:10.1016/j.envsci.2016.11.004.

Dias, T.A. da C., E.E.S. Lora, D.M.Y. Maya, and O.A. del Olmo, 2021: Global potential assessment of available land for bioenergy projects in 2050 within food security limits. *Land use policy*, **105**, 105346, doi:10.1016/J.LANDUSEPOL.2021.105346.

Diercks, G., H. Larsen, and F. Steward, 2019: Transformative innovation policy: Addressing variety in an emerging policy paradigm. *Res. Policy*, **48(4)**, 880–894, doi:10.1016/j.respol.2018.10.028.

Digdaya, I.A. et al., 2020: A direct coupled electrochemical system for capture and conversion of CO2 from oceanwater. *Nat. Commun.*, **11(1)**, 4412, doi:10.1038/s41467-020-18232-y.

Diniz Oliveira, T. et al., 2021: A mixed-effect model approach for assessing land-based mitigation in integrated assessment models: A regional perspective. *Glob. Chang. Biol.*, **27(19)**, 4671–4685, doi:10.1111/gcb.15738.

Djomo, S.N. et al., 2017: Solving the multifunctionality dilemma in biorefineries with a novel hybrid mass–energy allocation method. *GCB Bioenergy*, **9(11)**, 1674–1686, doi:10.1111/GCBB.12461.

Djomo, S.N. et al., 2020: Green proteins: An energy-efficient solution for increased self-sufficiency in protein in Europe. *Biofuels, Bioprod. Biorefining*, **14(3)**, 605–619, doi:10.1002/BBB.2098.

Dodds, P.E. et al., 2015: Hydrogen and fuel cell technologies for heating: A review. *Int. J. Hydrogen Energy*, **40(5)**, 2065–2083, doi:10.1016/j.ijhydene.2014.11.059.

Doelman, J.C. et al., 2018: Exploring SSP land-use dynamics using the IMAGE model: Regional and gridded scenarios of land-use change and land-based climate change mitigation. *Glob. Environ. Change*, **48**, 119–135, doi:10.1016/j.gloenvcha.2017.11.014.

Doney, S.C., V.J. Fabry, R.A. Feely, and J.A. Kleypas, 2009: Ocean Acidification: The Other $CO_2$ Problem. *Ann. Rev. Mar. Sci.*, **1(1)**, 169–192, doi:10.1146/annurev.marine.010908.163834.

Dooley, K. and S. Kartha, 2018: Land-based negative emissions: risks for climate mitigation and impacts on sustainable development. *Int. Environ. Agreements Polit. Law Econ.*, **18(1)**, 79–98, doi:10.1007/s10784-017-9382-9.

Dooley, K., E. Harrould-Kolieb, and A. Talberg, 2021: Carbon-dioxide Removal and Biodiversity: A Threat Identification Framework. *Glob. Policy*, **12**(S1), 34–44, doi:10.1111/1758-5899.12828.

dos Santos, M.A. et al., 2017: Estimates of GHG emissions by hydroelectric reservoirs: The Brazilian case. *Energy*, **133**, 99–107, doi:10.1016/j.energy.2017.05.082.

Doukas, H., A. Nikas, M. González-Eguino, I. Arto, and A. Anger-Kraavi, 2018: From Integrated to Integrative: Delivering on the Paris Agreement. *Sustainability*, **10(7)**, 2299, doi:10.3390/su10072299.

Drews, M., M.A.D. Larsen, and J.G. Peña Balderrama, 2020: Projected water usage and land-use-change emissions from biomass production (2015–2050). *Energy Strateg. Rev.*, **29**, 100487, doi:10.1016/J.ESR.2020.100487.

Dries, L., T. Reardon, and J.F.M. Swinnen, 2004: The rapid rise of supermarkets in Central and Eastern Europe: Implications for the agrifood sector and rural development. *Dev. Policy Rev.*, **22(5)**, 525–556, doi:10.1111/j.1467-7679.2004.00264.x.

Duarte, C.M., I.J. Losada, I.E. Hendriks, I. Mazarrasa, and N. Marbà, 2013: The role of coastal plant communities for climate change mitigation and adaptation. *Nat. Clim. Change*, **3(11)**, 961–968, doi:10.1038/nclimate1970.

Duarte, C.M., J. Wu, X. Xiao, A. Bruhn, and D. Krause-Jensen, 2017: Can Seaweed Farming Play a Role in Climate Change Mitigation and Adaptation? *Front. Mar. Sci.*, **4**, 100, doi:10.3389/fmars.2017.00100.

Duchelle, A.E. et al., 2017: Balancing carrots and sticks in REDD+: implications for social safeguards. *Ecol. Soc.*, **22(3)**, art2, doi:10.5751/ES-09334-220302.

Dugan, A.J. et al., 2018: A systems approach to assess climate change mitigation options in landscapes of the United States forest sector. *Carbon Balance Manag.*, **13(1)**, 13, doi:10.1186/s13021-018-0100-x.

Duncan, J., 2015: "Greening" global food governance. *Can. Food Stud. / La Rev. Can. des études sur l'alimentation*, **2(2)**, 335, doi:10.15353/cfs-rcea.v2i2.104.

Dupont, E., R. Koppelaar, and H. Jeanmart, 2020: Global available solar energy under physical and energy return on investment constraints. *Appl. Energy*, **257**, 113968, doi:10.1016/J.APENERGY.2019.113968.

Eakin, H., X. Rueda, and A. Mahanti, 2017: Transforming governance in telecoupled food systems. *Ecol. Soc.*, **22(4)**, art32, doi:10.5751/ES-09831-220432.

Eales, J. et al., 2018: What is the effect of prescribed burning in temperate and boreal forest on biodiversity, beyond pyrophilous and saproxylic species? A systematic review. *Environ. Evid.*, **7(1)**, 19, doi:10.1186/s13750-018-0131-5.

EASAC, 2017: *Opportunities and challenges for research on food and nutrition security and agriculture in Europe. EASAC policy report 34.* German National Academy of Sciences, Halle (Saale), Germany.

EC, 2018: In-Depth Analysis in Support of the Commission Communication COM (2018) 773 "A Clean Planet for All.", 393 pp.

ECDC, EFSA, and EMA, 2015: ECDC/EFSA/EMA first joint report on the integrated analysis of the consumption of antimicrobial agents and occurrence of antimicrobial resistance in bacteria from humans and food-producing animals. *EFSA J.*, **13(1)**, 4006, doi:10.2903/j.efsa.2015.4006.

Eckerstorfer, M.F. et al., 2021: Biosafety of genome editing applications in plant breeding: Considerations for a focused case-specific risk assessment in the eu. *BioTech*, **10(3)**, 1–14, doi:10.3390/biotech10030010.

Eckert, M., M. Kauffeld, and V. Siegismund, eds., 2021: *Natural Refrigerants: Applications and Practical Guidelines*. Vde Verlag GmbH, Berlin, Germany.

Edwards, D.P. et al., 2017: Climate change mitigation: potential benefits and pitfalls of enhanced rock weathering in tropical agriculture. *Biol. Lett.*, **13(4)**, 20160715, doi:10.1098/rsbl.2016.0715.

EFSA, 2017: Scientific Opinion of the PPR Panel on the follow-up of the findings of the External Scientific Report 'Literature review of epidemiological studies linking exposure to pesticides and health effects.' *EFSA J.*, **15(10)**, doi:10.2903/j.efsa.2017.5007.

Eisaman, M.D., 2020: Negative Emissions Technologies: The Tradeoffs of Air-Capture Economics. *Joule*, **4(3)**, 516–520, doi:10.1016/J.JOULE.2020.02.007.

Eisaman, M.D. et al., 2012: CO2 extraction from seawater using bipolar membrane electrodialysis. *Energy Environ. Sci.*, **5(6)**, 7346–7352, doi:10.1039/C2EE03393C. Eisaman, M.D. et al., 2018: Indirect ocean capture of atmospheric CO2: Part II. Understanding the cost of negative emissions. *Int. J. Greenh. Gas Control*, **70**, 254–261, doi:10.1016/J.IJGGC.2018.02.020.

El Akkari, M., O. Réchauchère, A. Bispo, B. Gabrielle, and D. Makowski, 2018: A meta-analysis of the greenhouse gas abatement of bioenergy factoring in land use changes. *Sci. Rep.*, **8**, doi:10.1038/s41598-018-26712-x.

El Bilali, H., 2019: Research on agro-food sustainability transitions: A systematic review of research themes and an analysis of research gaps. *J. Clean. Prod.*, **221**, 353–364, doi:10.1016/j.jclepro.2019.02.232.

Elgaaied-Gambier, L., E. Monnot, and F. Reniou, 2018: Using descriptive norm appeals effectively to promote green behavior. *J. Bus. Res.*, **82**, 179–191, doi:10.1016/j.jbusres.2017.09.032.

Ellison, D. et al., 2017: Trees, forests and water: Cool insights for a hot world. *Glob. Environ. Change*, **43**, 51–61, doi:10.1016/j.gloenvcha.2017.01.002.

Emanuel, N. and H.K. Sandhu, 2019: Food Packaging Development: Recent Perspective. *Journal of Thin Films, Coating Science Technology and Application*, **6(3)**, ISSN: 2455-3344

Eme, P., J. Douwes, N. Kim, S. Foliaki, and B. Burlingame, 2019: Review of Methodologies for Assessing Sustainable Diets and Potential for Development of Harmonised Indicators. *Int. J. Environ. Res. Public Health*, **16(7)**, 1184, doi:10.3390/ijerph16071184.

Emmerling, J. et al., 2019: The role of the discount rate for emission pathways and negative emissions. *Environ. Res. Lett.*, **14(10)**, doi:10.1088/1748-9326/ab3cc9.

England, M.I., L.C. Stringer, A.J. Dougill, and S. Afionis, 2018: How do sectoral policies support climate compatible development? An empirical analysis focusing on southern Africa. *Environ. Sci. Policy*, **79**, 9–15, doi:10.1016/j.envsci.2017.10.009.

Englund, O. et al., 2020a: Beneficial land use change: Strategic expansion of new biomass plantations can reduce environmental impacts from EU agriculture. *Glob. Environ. Change*, **60**, 101990, doi:10.1016/j.gloenvcha.2019.101990.

Englund, O. et al., 2020b: Multifunctional perennial production systems for bioenergy: performance and progress. *Wiley Interdiscip. Rev. Energy Environ.*, **9(5)**, e375, doi:10.1002/WENE.375.

Englund, O. et al., 2021: Strategic deployment of riparian buffers and windbreaks in Europe can co-deliver biomass and environmental benefits. *Commun. Earth Environ. 2021 21*, **2(1)**, 1–18, doi:10.1038/s43247-021-00247-y.

Erb, K.-H. et al., 2018: Unexpectedly large impact of forest management and grazing on global vegetation biomass. *Nature*, **553(7686)**, 73–76, doi:10.1038/nature25138.

Ericksen, P.J., 2008: What Is the Vulnerability of a Food System to Global Environmental Change? *Ecol. Soc.*, **13(2)**, art14, doi:10.5751/ES-02475-130214.

Ernstoff, A. et al., 2020: Towards win–win policies for healthy and sustainable diets in switzerland. *Nutrients*, **12(9)**, 1–24, doi:10.3390/nu12092745.

EU, 2015: Regulation (EU) 2015/2283 of the European Parliament and of the Council of 25 November 2015 on novel foods, amending Regulation (EU) No 1169/2011 of the European Parliament and of the Council and repealing Regulation (EC) No 258/97 of the European Parliam. *Off. J. Eur. Union*, **327(258)**, 1–22.

European Commission, 2018: *Sustainable & circular bioeconomy, the European way Outcome report*. Brussels, Belgium.

European Commission, J.R.C., 2019: Restrictions on marketing of food, non-alcoholic and alcoholic beverages to protect health. [Dataset] PID: doi:http://data.europa.eu/89h/a5798df4-da80-4576-9502-218d6c2fff19.

Ewing, R.C., R.A. Whittleston, and B.W.D. Yardley, 2016: Geological Disposal of Nuclear Waste: a Primer. *Elements*, **12(4)**, 233–237, doi:10.2113/GSELEMENTS.12.4.233.

Exley, G., A. Armstrong, T. Page, and I.D. Jones, 2021: Floating photovoltaics could mitigate climate change impacts on water body temperature and stratification. *Sol. Energy*, **219**, 24–33, doi:10.1016/J.SOLENER.2021.01.076.

Fajardy, M. and N. Mac Dowell, 2020: Recognizing the Value of Collaboration in Delivering Carbon Dioxide Removal. *One Earth*, **3(2)**, 214–225, doi:10.1016/j.oneear.2020.07.014.

Fajardy, M., P. Patrizio, H.A. Daggash, and N. Mac Dowell, 2019: Negative Emissions: Priorities for Research and Policy Design. *Front. Clim.*, **1**, 6, doi:10.3389/fclim.2019.00006.

Falck, W.E., 2015: Radioactive and other environmental contamination from uranium mining and milling. In: *Environmental Remediation and Restoration of Contaminated Nuclear and Norm Sites* [van Velzen, L. ed.)], pp 3–34, doi:10.1016/B978-1-78242-231-0.00001-6.

Falconer, A., C. Parker, P. Keenlyside, A. Dontenville, and J. Wilkinson, 2015: *Three tools to unlock finance for land-use mitigation and adaptation*. Climate Focus and Climate Policy Initiative, 41 pp. https://climatepolicyinitiative.org/publication/three-tools-to-unlock-finance-for-land-use-mitigation-and-adaptation/ (Accessed December 3, 2019).

Fan, W. et al., 2020: A sea trial of enhancing carbon removal from Chinese coastal waters by stimulating seaweed cultivation through artificial upwelling. *Appl. Ocean Res.*, **101**, 102260, doi:10.1016/J.APOR.2020.102260.

Fanzo, J., 2017: From big to small: the significance of smallholder farms in the global food system. *Lancet Planet. Heal.*, **1(1)**, e15--e16, doi:10.1016/S2542-5196(17)30011-6.

Fanzo, J. et al., 2020: The Food Systems Dashboard is a new tool to inform better food policy. *Nat. Food*, **1(5)**, 243–246, doi:10.1038/s43016-020-0077-y.

FAO, 2011: *"Energy-Smart" Food for People Climate- Issue Paper*. Food and Agriculture Organization of the United Nations, Rome, Italy, 78 pp.

FAO, 2012: *Voluntary Guidelines on the Responsible Governance of Tenure of Land, Fisheries and Forests in the Context of National Food Security*. Food and Agriculture Organization of the United Nations, Rome, Italy, 77 pp.

FAO, 2018a: *Sustainable food systems: Concept and framework*. Food and Agriculture Organization of the United Nations, Rome, Italy, http://www.fao.org/3/ca2079en/CA2079EN.pdf.

FAO, 2018b: *The future of food and agriculture: Alternative pathways to 2050*. Food and Agriculture Organization of the United Nations, Rome, Italy, 228 pp.

FAO, 2019a: *The State of Food and Agriculture 2019. Moving forward on food loss and waste reduction*. Food and Agriculture Organization of the United Nations, Rome, Italy.

FAO, 2019b: *Climate-smart agriculture and the Sustainable Development Goals: Mapping interlinkages, synergies and trade-offs and guidelines for integrated implementation*. Food and Agriculture Organization of the United Nations, Rome, Italy, 144 pp. http://www.fao.org/climate-smart-agriculture/resources/publications/en/.

FAO, 2020: *Climate change: Unpacking the burden on food safety*. Food and Agriculture Organization of the United Nations,, Rome, Italy, 1–176 pp.

FAO, 2021: *The share of food systems in total greenhouse gas emissions. Global, regional and country trends*. FAOSTAT Analytical Brief. Food and Agriculture Organization of the United Nations, Rome, Italy,

FAO, and WHO, 2019: *Sustainable healthy diets - Guiding principles*. Food and Agriculture Organization of the United Nations, Rome, Italy.

FAO, IFAD, UNICEF, WFP, and WHO, 2019: *The state of food security and nutrition in the world. Safeguarding against economic slowdwns*. Food and Agriculture Organization of the United Nations, Rome, Italy, 202 pp.

FAO, IFAD, UNICEF, WFP, and WHO, 2020: *The State of Food Security and Nutrition in the World 2020. Transforming food systems for affordable healthy diets*. Food and Agriculture Organization of the United Nations, Rome, Italy.

Fardet, A. and E. Rock, 2019: Ultra-processed foods: A new holistic paradigm? *Trends Food Sci. Technol.*, **93**(February), 174–184, doi:10.1016/j.tifs.2019.09.016.

Farfan, J., A. Lohrmann, and C. Breyer, 2019: Integration of greenhouse agriculture to the energy infrastructure as an alimentary solution. *Renew. Sustain. Energy Rev.*, **110**(May), 368–377, doi:10.1016/j.rser.2019.04.084.

Farhadi, S. and R. Ovchinnikov, 2018: The relationship between nutrition and infectious diseases: A review. *Biomed. Biotechnol. Res. J.*, **2(3)**, 168, doi:10.4103/bbrj.bbrj_69_18.

Farley, K.A., E.G. Jobbagy, and R.B. Jackson, 2005: Effects of afforestation on water yield: a global synthesis with implications for policy. *Glob. Chang. Biol.*, **11(10)**, 1565–1576, doi:10.1111/j.1365-2486.2005.01011.x.

Farrington, P. and R.B. Salama, 1996: Controlling dryland salinity by planting trees in the best hydrogeological setting. *L. Degrad. Dev.*, **7(3)**, 183–204, doi:10.1002/(SICI)1099-145X(199609)7:3<183::AID-LDR221>3.0.CO;2-Y.

Fasihi, M., O. Efimova, and C. Breyer, 2019: Techno-economic assessment of CO2 direct air capture plants. *J. Clean. Prod.*, **224**, 957–980, doi:10.1016/j.jclepro.2019.03.086.

Fasolin, L.H. et al., 2019: Emergent food proteins – Towards sustainability, health and innovation. *Food Res. Int.*, **125**(July), 108586, doi:10.1016/j.foodres.2019.108586.

Favero, A., A. Daigneault, and B. Sohngen, 2020: Forests: Carbon sequestration, biomass energy, or both? *Sci. Adv.*, **6(13)**, doi:10.1126/sciadv.aay6792.

Favretto, N. et al., 2020: Delivering Climate-Development Co-Benefits through Multi-Stakeholder Forestry Projects in Madagascar: Opportunities and Challenges. *Land*, **9(5)**, 157, doi:10.3390/LAND9050157.

FCH, 2019: *Hydrogen Roadmap Europe*. Fuel Cells and Hydrogen 2 Joint Undertaking, Luxembourg, 70 pp.

Fellmann, T. et al., 2018: Major challenges of integrating agriculture into climate change mitigation policy frameworks. *Mitig. Adapt. Strateg. Glob. Change*, **23(3)**, 451–468, doi:10.1007/s11027-017-9743-2.

Femeena, P.V, K.P. Sudheer, R. Cibin, and I. Chaubey, 2018: Spatial optimization of cropping pattern for sustainable food and biofuel production with minimal downstream pollution. *J. Environ. Manage.*, **212**, 198–209, doi:10.1016/j.jenvman.2018.01.060.

Feng, E.Y., D.P. Keller, W. Koeve, and A. Oschlies, 2016: Could artificial ocean alkalinization protect tropical coral ecosystems from ocean acidification? *Environ. Res. Lett.*, **11(7)**, 074008, doi:10.1088/1748-9326/11/7/074008.

Fernández-Bellon, D., 2020: Limited accessibility and bias in wildlife-wind energy knowledge: A bilingual systematic review of a globally distributed bird group. *Sci. Total Environ.*, **737**, 140238, doi.org/10.1016/j.scitotenv.2020.140238.

Ferrari, L., A. Cavaliere, E. De Marchi, and A. Banterle, 2019: Can nudging improve the environmental impact of food supply chain? A systematic review. *Trends Food Sci. Technol.*, **91**(April), 184–192, doi:10.1016/j.tifs.2019.07.004.

Ferrarini, A., P. Serra, M. Almagro, M. Trevisan, and S. Amaducci, 2017: Multiple ecosystem services provision and biomass logistics management in bioenergy buffers: A state-of-the-art review. *Renew. Sustain. Energy Rev.*, **73**, 277–290, doi.org/10.1016/j.rser.2017.01.052.

Field, C.B., and K.J. Mach, 2017: Rightsizing carbon dioxide removal. *Science*, **356(6339)**, 706–707, doi:10.1126/science.aam9726.

Finaret, A.B. and W.A. Masters, 2019: Beyond Calories: The New Economics of Nutrition. *Annu. Rev. Resour. Econ.*, **11(1)**, 237–259, doi:10.1146/annurev-resource-100518-094053.

Finger, R., S.M. Swinton, N. El Benni, and A. Walter, 2019: Precision Farming at the Nexus of Agricultural Production and the Environment. *Annu. Rev. Resour. Econ.*, **11(1)**, 313–335, doi:10.1146/annurev-resource-100518-093929.

Fiorini, M. et al., 2019: Institutional design of voluntary sustainability standards systems: Evidence from a new database. *Dev. Policy Rev.*, **37(S2)**, O193–O212, doi:10.1111/dpr.12379.

Fischer, J. et al., 2007: Mind the sustainability gap. *Trends Ecol. Evol.*, **22(12)**, 621–4, doi:10.1016/j.tree.2007.08.016.

Floater, G. et al., 2016: *Co-benefits of urban climate action: A framework for cities*. Economics of Green Cities Programme, LSE Cities, London School of Economics and Political Science, London, UK, 86 pp. http://eprints.lse.ac.uk/id/eprint/68876.

Folberth, C. et al., 2020: The global cropland-sparing potential of high-yield farming. *Nat. Sustain.*, **3(4)**, 281–289, doi:10.1038/s41893-020-0505-x.

Folke, C. et al., 2019: Transnational corporations and the challenge of biosphere stewardship. *Nat. Ecol. Evol.*, **3(10)**, 1396–1403, doi:10.1038/s41559-019-0978-z.

Forest Trends, 2014: *Consumer goods and deforestation: An analysis of the extent and nature of illegality in forest conversation for agriculture and timber plantations*, Washington, DC, USA.

Forster, J., N.E. Vaughan, C. Gough, I. Lorenzoni, and J. Chilvers, 2020: Mapping feasibilities of greenhouse gas removal: Key issues, gaps and opening up assessments. *Glob. Environ. Change*, **63**, 102073, doi:10.1016/j.gloenvcha.2020.102073.

Foyer, C.H. et al., 2016: Neglecting legumes has compromised human health and sustainable food production. *Nat. Plants*, **2(8)**, 1–10, doi:10.1038/NPLANTS.2016.112.

França, T. et al., 2009: Impact of malnutrition on immunity and infection. *J. Venom. Anim. Toxins Incl. Trop. Dis.*, **15(3)**, 374–390, doi:10.1590/S1678-91992009000300003.

Frank, S. et al., 2017: Reducing greenhouse gas emissions in agriculture without compromising food security? *Environ. Res. Lett.*, **12(10)**, 105004, doi:10.1088/1748-9326/aa8c83.

Freeman, O.E., L.A. Duguma, and P.A. Minang, 2015: Operationalizing the integrated landscape approach in practice. *Ecol. Soc.*, **20(1)**, doi:10.5751/ES-07175-200124.

Fresán, U., M.A. Mejia, W.J. Craig, K. Jaceldo-Siegl, and J. Sabaté, 2019: Meat Analogs from Different Protein Sources: A Comparison of Their Sustainability and Nutritional Content. *Sustainability*, **11(12)**, 3231, doi:10.3390/su11123231.

Fricko, O. et al., 2016: Energy sector water use implications of a 2°C climate policy. *Environ. Res. Lett.*, **11(3)**, 034011, doi:10.1088/1748-9326/11/3/034011.

Fricko, O. et al., 2017: The marker quantification of the Shared Socioeconomic Pathway 2: A middle-of-the-road scenario for the 21st century. *Glob. Environ. Change,* **42**, 251–267, doi:10.1016/j.gloenvcha.2016.06.004.

Fridahl, M., 2017: Socio-political prioritization of bioenergy with carbon capture and storage. *Energy Policy*, **104**, 89–99, doi:10.1016/J.ENPOL.2017.01.050.

Fridahl, M., R. Bellamy, A. Hansson, and S. Haikola, 2020: Mapping Multi-Level Policy Incentives for Bioenergy With Carbon Capture and Storage in Sweden. *Front. Clim.*, **2**, 25, doi:10.3389/fclim.2020.604787.

Friedmann, S.J., 2019: Engineered CO2 Removal, Climate Restoration, and Humility. *Front. Clim.*, **1**, 3, doi:10.3389/fclim.2019.00003.

Fritsche, U.R. et al., 2017: *Global Land Outlook Working Paper: Energy and Land Use*. UNCCD and IRENA, Darmstadt, Germany, 60 pp.

Fritz, S. et al., 2013: Downgrading Recent Estimates of Land Available for Biofuel Production. *Environ. Sci. Technol.*, **47(3)**, 1688–1694, 130128103203003, doi:10.1021/es303141h.

Froehlich, H.E., J.C. Afflerbach, M. Frazier, and B.S. Halpern, 2019: Blue Growth Potential to Mitigate Climate Change through Seaweed Offsetting. *Curr. Biol.*, **29(18)**, 3087-3093.e3, doi:10.1016/j.cub.2019.07.041.

Froese, R., J. Schilling, Froehse, and Schilling, 2019: The Nexus of Climate Change, Land Use, and Conflicts. *Curr. Clim. Chang. Reports*, **5**, 24–35, doi:10.1007/s40641-019-00122-1.

Frolova, M. et al., 2019: Effects of renewable energy on landscape in Europe: Comparison of hydro, wind, solar, bio-, geothermal and infrastructure energy landscapes. *Hungarian Geogr. Bull.*, **68(4)**, 317–339, doi:10.15201/hungeobull.68.4.1.

Fuhrman, J., H. McJeon, S.C. Doney, W. Shobe, and A.F. Clarens, 2019: From Zero to Hero?: Why Integrated Assessment Modeling of Negative Emissions Technologies Is Hard and How We Can Do Better. *Front. Clim.*, **1**, 11, doi:10.3389/fclim.2019.00011.

Fuhrman, J. et al., 2020: Food–energy–water implications of negative emissions technologies in a +1.5°C future. *Nat. Clim. Change,* **10(10)**, 920–927, doi:10.1038/s41558-020-0876-z.

Fuhrman, J. et al., 2021a: The Role of Direct Air Capture and Negative Emissions Technologies in the Shared Socioeconomic Pathways towards +1.5°C and +2°C Futures. *Environ. Res. Lett.*, **16(11)**, 114012, doi:10.1088/1748-9326/ac2db0.

Fuhrman, J. et al., 2021b: The role of negative emissions in meeting China's 2060 carbon neutrality goal. *Oxford Open Clim. Change,* **1(1)**, doi:10.1093/oxfclm/kgab004.

Fuhrman, J.A. and D.G. Capone, 1991: Possible biogeochemical consequences of ocean fertilization. *Limnol. Oceanogr.*, **36(8)**, 1951–1959, doi:10.4319/lo.1991.36.8.1951.

Fujimori, S. et al., 2018: Inclusive climate change mitigation and food security policy under 1.5°C climate goal. *Environ. Res. Lett.*, **13(7)**, doi:10.1088/1748-9326/aad0f7.

Fujimori, S. et al., 2019: A multi-model assessment of food security implications of climate change mitigation. *Nat. Sustain. 2019 25*, **2(5)**, 386–396, doi:10.1038/s41893-019-0286-2.

Fuss, S. et al., 2018: Negative emissions - Part 2: Costs, potentials and side effects. *Environ. Res. Lett.*, **13(6)**, 063002, doi:10.1088/1748-9326/aabf9f.

Fuss, S. et al., 2020: Moving toward Net-Zero Emissions Requires New Alliances for Carbon Dioxide Removal. *One Earth*, **3**, 145–149, doi:10.1016/j.oneear.2020.08.002.

Fyson, C.L. and M.L. Jeffery, 2019: Ambiguity in the Land Use Component of Mitigation Contributions Toward the Paris Agreement Goals. *Earth's Futur.*, **7(8)**, 873–891, doi:10.1029/2019EF001190.

Gadzanku, S., H. Mirletz, N. Lee, J. Daw, and A. Warren, 2021: Benefits and Critical Knowledge Gaps in Determining the Role of Floating Photovoltaics in the Energy-Water-Food Nexus. *Sustain. 2021,* **13(8)**, 4317, doi:10.3390/SU13084317.

Gajevic Sayegh, A., 2020: Moral duties, compliance and polycentric climate governance. *Int. Environ. Agreements Polit. Law Econ.*, **20(3)**, 483–506, doi:10.1007/s10784-020-09494-4.

Galik, C.S., G.S. Latta, and C. Gambino, 2019: Piecemeal or combined? Assessing greenhouse gas mitigation spillovers in US forest and agriculture policy portfolios. *Clim. Policy*, **19(10)**, 1270–1283, doi:10.1080/14693062.2019.1663719.

Gallo, M., L. Ferrara, A. Calogero, D. Montesano, and D. Naviglio, 2020: Relationships between food and diseases: What to know to ensure food safety. *Food Res. Int.*, **137**(January), 109414, doi:10.1016/j.foodres.2020.109414.

Gan, Y. et al., 2020: Research progress on coronavirus prevention and control in animal-source foods. *J. Multidiscip. Healthc.*, **13**, 743–751, doi:10.2147/JMDH.S265059.

Gann, G.D. et al., 2019: International principles and standards for the practice of ecological restoration. Second edition. *Restor. Ecol.*, **27**(S1), S1–S46, doi:10.1111/rec.13035.

García-Freites, S., C. Gough, and M. Röder, 2021: The greenhouse gas removal potential of bioenergy with carbon capture and storage (BECCS) to support the UK's net-zero emission target. *Biomass and Bioenergy*, **151**, 106164, doi:10.1016/j.biombioe.2021.106164.

Garcia, M. and N. Berghout, 2019: Toward a common method of cost-review for carbon capture technologies in the industrial sector: cement and iron and steel plants. *Int. J. Greenh. Gas Control*, **87**, 142–158, doi:10.1016/j.ijggc.2019.05.005.

Garg, K.K., L. Karlberg, S.P. Wani, and G. Berndes, 2011: Jatropha production on wastelands in India: opportunities and trade-offs for soil and water management at the watershed scale. *Biofuels, Bioprod. Biorefining*, **5(4)**, 410–430, doi:10.1002/bbb.312.

Garibaldi, L.A. et al., 2016: Mutually beneficial pollinator diversity and crop yield outcomes in small and large farms. *Science*, **351(6271)**, 388–391, doi:10.1126/science.aac7287.

Garnett, T., E. Röös, and D. Little, 2015: *Lean, green, mean, obscene...? What is efficiency? And is it sustainable? Animal production and consumption reconsidered*. Food Climate Research Network, University of Oxford, Oxford, UK, 48 pp.

Garofalo, C. et al., 2019: Current knowledge on the microbiota of edible insects intended for human consumption: A state-of-the-art review. *Food Res. Int.*, **125**(June), 108527, doi:10.1016/j.foodres.2019.108527.

Gattuso, J.-P. et al., 2018: Ocean solutions to address climate change and its effects on marine ecosystems. *Front. Mar. Sci.*, **5**, 337, doi:10.3389/fmars.2018.00337.

Gattuso, J.-P., P. Williamson, C.M. Duarte, and A.K. Magnan, 2021: The Potential for Ocean-Based Climate Action: Negative Emissions Technologies and Beyond. *Front. Clim.*, **2**, 37, doi:10.3389/fclim.2020.575716.

Gava, O. et al., 2018: A Reflection of the Use of the Life Cycle Assessment Tool for Agri-Food Sustainability. *Sustainability*, **11(1)**, 71, doi:10.3390/su11010071.

GBD 2017 Diet Collaborators, 2018: Global, regional, and national comparative risk assessment of 84 behavioural, environmental and occupational, and metabolic risks or clusters of risks for 195 countries and territories, 1990–2017: a systematic analysis for the Global Burden of Disease Stu. *Lancet*, **392(10159)**, 1923–1994, doi:10.1016/S0140-6736(18)32225-6.

GBD 2017 Diet Collaborators et al., 2019: Health effects of dietary risks in 195 countries, 1990–2017: a systematic analysis for the Global Burden of Disease Study 2017. *Lancet*, **393(10184)**, 1958–1972, doi:10.1016/S0140-6736(19)30041-8.

Geden, O., 2016: The Paris Agreement and the inherent inconsistency of climate policymaking. *Wiley Interdiscip. Rev. Clim. Change*, **7(6)**, 790–797, doi:10.1002/wcc.427.

Geden, O., G.P. Peters, and V. Scott, 2019: Targeting carbon dioxide removal in the European Union. *Clim. Policy*, **19(4)**, 487–494, doi:10.1080/14693062.2018.1536600.

Geden, O., V. Scott, and J. Palmer, 2018: Integrating carbon dioxide removal into EU climate policy: Prospects for a paradigm shift. *Wiley Interdiscip. Rev. Clim. Change*, **9**, 1–10, doi:10.1002/wcc.521.

Gentry, R.R. et al., 2020: Exploring the potential for marine aquaculture to contribute to ecosystem services. *Rev. Aquac.*, **12(2)**, 499–512, doi:10.1111/raq.12328.

Gerhardt, C. et al., 2019: *How Will Cultured Meat and Meat Alternatives Disrupt the Agricultural and Food Industry?* A. T. Kearney, 20 pp.

Gerlagh, R., 2011: Too much oil. *CESifo Econ. Stud.*, **57(1)**, 79–102, doi:10.1093/cesifo/ifq004.

GESAMP, 2019: *High level review of a wide range of proposed marine geoengineering techniques*. International Maritime Organization, London, UK, 144 pp.

Geyik, O., M. Hadjikakou, B. Karapinar, and B.A. Bryan, 2021: Does global food trade close the dietary nutrient gap for the world's poorest nations? *Glob. Food Sec.*, **28** (December 2020), 100490, doi:10.1016/j.gfs.2021.100490.

Gheuens, J. and S. Oberthür, 2021: Eu climate and energy policy: How myopic is it? *Polit. Gov.*, **9(3)**, 337–347, doi:10.17645/pag.v9i3.4320.

Ghosh, N. and C.K. Sen, 2019: The Promise of Dietary Supplements: Research Rigor and Marketing Claims. In: *Nutrition and Enhanced Sports Performance*, Elsevier, pp. 759–766.

Gillman, G.P., 1980: The Effect of Crushed Basalt Scoria on the Cation Exchange Properties of a Highly Weathered Soil. *Soil Sci. Soc. Am. J.*, **44(3)**, 465–468, doi:10.2136/sssaj1980.03615995004400030005x.

Gillman, G.P., D.C. Burkett, and R.J. Coventry, 2001: A laboratory study of application of basalt dust to highly weathered soils: Effect on soil cation chemistry. *Aust. J. Soil Res.*, **39(4)**, 799–811, doi:10.1071/SR00073.

Glenk, G. and S. Reichelstein, 2019: Economics of converting renewable power to hydrogen. *Nat. Energy*, **4(3)**, 216–222, doi:10.1038/s41560-019-0326-1.

Global Alliance for the Future of Food, 2020: *Systemic Solutions for Healthy Food Systems: A Guide to Government Action*. Global Alliance for the Future of Food,.

Zimmermann A.W. et al., 2020: Techno-Economic Assessment Guidelines for CO2 Utilization. Front. Energy Res. 8(5) doi: 10.3389/fenrg.2020.00005

Global Panel on Agriculture and Food Systems for Nutrition, 2020: *Future Food Systems: For people, our planet, and prosperity*. London, UK,.

Göbel, C., N. Langen, A. Blumenthal, P. Teitscheid, and G. Ritter, 2015: Cutting food waste through cooperation along the food supply chain. *Sustainability*, **7(2)**, 1429–1445, doi:10.3390/su7021429.

Goggins, G., 2018: Developing a sustainable food strategy for large organizations: The importance of context in shaping procurement and consumption practices. *Bus. Strateg. Environ.*, **27(7)**, 838–848, doi:10.1002/bse.2035.

Goggins, G. and H. Rau, 2016: Beyond calorie counting: Assessing the sustainability of food provided for public consumption. *J. Clean. Prod.*, **112**, 257–266, doi:10.1016/j.jclepro.2015.06.035.

Goll, D.S. et al., 2021: Potential CO2 removal from enhanced weathering by ecosystem responses to powdered rock. *Nat. Geosci.*, **14(8)**, 545–549, doi:10.1038/s41561-021-00798-x.

Gomez-Echeverri, L., 2018: Climate and development: Enhancing impact through stronger linkages in the implementation of the Paris Agreement and the Sustainable Development Goals (SDGs). *Philos. Trans. R. Soc. A Math. Phys. Eng. Sci.*, **376(2119)**, doi:10.1098/rsta.2016.0444.

Gómez, C. and L. Gennaro Izzo, 2018: Increasing efficiency of crop production with LEDs. *AIMS Agric. Food*, **3(2)**, 135–153, doi:10.3934/agrfood.2018.2.135.

Gómez, M.I. and K.D. Ricketts, 2013: Food value chain transformations in developing countries: Selected hypotheses on nutritional implications. *Food Policy*, **42(13)**, 139–150, doi:10.1016/j.foodpol.2013.06.010.

González-Alzaga, B. et al., 2014: A systematic review of neurodevelopmental effects of prenatal and postnatal organophosphate pesticide exposure. *Toxicol. Lett.*, **230(2)**, 104–21, doi:10.1016/j.toxlet.2013.11.019.

González, M.F. and T. Ilyina, 2016: Impacts of artificial ocean alkalinization on the carbon cycle and climate in Earth system simulations. *Geophys. Res. Lett.*, **43(12)**, 6493–6502, doi:10.1002/2016GL068576.

González, M.F., T. Ilyina, S. Sonntag, and H. Schmidt, 2018: Enhanced Rates of Regional Warming and Ocean Acidification After Termination of Large-Scale Ocean Alkalinization. *Geophys. Res. Lett.*, **45(14)**, 7120–7129, doi:10.1029/2018GL077847.

Gonzalez Sanchez, R., I. Kougias, M. Moner-Girona, F. Fahl, and A. Jäger-Waldau, 2021: Assessment of floating solar photovoltaics potential in existing hydropower reservoirs in Africa. *Renew. Energy*, **169**, 687–699, doi:10.1016/J.RENENE.2021.01.041.

Göransson, L. and F. Johnsson, 2018: A comparison of variation management strategies for wind power integration in different electricity system contexts. *Wind Energy*, **21(10)**, 837–854, doi:10.1002/WE.2198.

Gore, S., P. Renforth, and R. Perkins, 2019: The potential environmental response to increasing ocean alkalinity for negative emissions. *Mitig. Adapt. Strateg. Glob. Change*, **24(7)**, 1191–1211, doi:10.1007/s11027-018-9830-z.

Gouel, C. and D. Laborde, 2021: The crucial role of domestic and international market-mediated adaptation to climate change. *J. Environ. Econ. Manage.*, **106**, 102408, doi:10.1016/j.jeem.2020.102408.

Gough, C. and S. Mander, 2019: Beyond Social Acceptability: Applying Lessons from CCS Social Science to Support Deployment of BECCS. *Curr. Sustain. Energy Reports*, **6(4)**, 116–123, doi:10.1007/s40518-019-00137-0.

Gouldson, A. et al., 2016: Cities and climate change mitigation: Economic opportunities and governance challenges in Asia. *Cities*, **54**, 11–19, doi.org/10.1016/j.cities.2015.10.010.

Government of Finland, 2017: *Government report on food policy: Food 2030 – Finland feeds us and the world*. Helsinki, Finland, 42 pp.

Govorushko, S., 2019: Global status of insects as food and feed source: A review. *Trends Food Sci. Technol.*, **91**(July 2018), 436–445, doi:10.1016/j.tifs.2019.07.032.

GPRBA, 2019: New Perspectives on Results-Based Blended Finance for Cities: Innovative Finance Solutions for Climate-Smart Infrastructure. Global Partnership for Results-Based Approaches, Washington, DC, USA, 82 pp.

Graamans, L., E. Baeza, A. van den Dobbelsteen, I. Tsafaras, and C. Stanghellini, 2018: Plant factories versus greenhouses: Comparison of resource use efficiency. *Agric. Syst.*, **160** (July 2017), 31–43, doi:10.1016/j.agsy.2017.11.003.

Grant, N., A. Hawkes, S. Mittal, and A. Gambhir, 2021: The policy implications of an uncertain carbon dioxide removal potential. *Joule*, **5(10)**, 2593–2605, doi:10.1016/j.joule.2021.09.004.

Grassi, G. et al., 2021: Critical adjustment of land mitigation pathways for assessing countries' climate progress. *Nat. Clim. Change*, **11(5)**, 425–434, doi:10.1038/s41558-021-01033-6.

Grebitus, C., B. Steiner, and M.M. Veeman, 2016: Paying for sustainability: A cross-cultural analysis of consumers' valuations of food and non-food products labeled for carbon and water footprints. *J. Behav. Exp. Econ.* **63**, 50–58, doi:10.1016/j.socec.2016.05.003.

Gren, I.M., E. Moberg, S. Säll, and E. Röös, 2019: Design of a climate tax on food consumption: Examples of tomatoes and beef in Sweden. *J. Clean. Prod.*, **211**, 1576–1585, doi:10.1016/j.jclepro.2018.11.238.

Griffiths, N.A. et al., 2019: Environmental effects of short-rotation woody crops for bioenergy: What is and isn't known. *GCB Bioenergy*, **11(4)**, 554–572, doi:10.1111/GCBB.12536.

Griscom, B.W., R.C. Goodman, Z. Burivalova, and F.E. Putz, 2018: Carbon and Biodiversity Impacts of Intensive Versus Extensive Tropical Forestry. *Conserv. Lett.*, **11(1)**, e12362, doi:10.1111/conl.12362.

Gronwald, M., N. Van Long, and L. Roepke, 2017: Simultaneous Supplies of Dirty Energy and Capacity Constrained Clean Energy: Is There a Green Paradox? *Environ. Resour. Econ.*, **68(1)**, 47–64, doi:10.1007/s10640-017-0151-6.

Group of Chief Scientific Advisors, 2020: *Towards a Sustainable Food System*. European Commission, DG for Research and Innovation, Brussels, Belgium.

Gruber, N. et al., 2019: The oceanic sink for anthropogenic CO2 from 1994 to 2007. *Science*, **363(6432)**, 1193–1199, doi:10.1126/science.aau5153.

Grubler, A. et al., 2018: A low energy demand scenario for meeting the 1.5°C target and sustainable development goals without negative emission technologies. *Nat. Energy*, **3(6)**, 515–527, doi:10.1038/s41560-018-0172-6.

Grundnig, P.W. et al., 2006: Influence of air humidity on the suppression of fugitive dust by using a water-spraying system. *China Particuology*, **4(5)**, 229–233, doi:10.1016/S1672-2515(07)60265-6.

Gullo, P., K. Tsamos, A. Hafner, Y. Ge, and S.A. Tassou, 2017: State-of-the-art technologies for transcritical R744 refrigeration systems – a theoretical assessment of energy advantages for European food retail industry. *Energy Procedia*, **123**, 46–53, doi:10.1016/J.EGYPRO.2017.07.283.

Gunnarsson, I. et al., 2018: The rapid and cost-effective capture and subsurface mineral storage of carbon and sulfur at the CarbFix2 site. *Int. J. Greenh. Gas Control*, **79**, 117–126, doi:10.1016/j.ijggc.2018.08.014.

Guntzer, F., C. Keller, and J.-D. Meunier, 2012: Benefits of plant silicon for crops: a review. *Agron. Sustain. Dev.*, **32(1)**, 201–213, doi:10.1007/s13593-011-0039-8.

Gustafson, D.I. et al., 2016: Seven food system metrics of sustainable nutrition security. *Sustainability*, **8(3)**, 196, doi:10.3390/su8030196.

Gustafson, D.I., M.S. Edge, T.S. Griffin, A.M. Kendall, and S.D. Kass, 2019: Growing Progress in the Evolving Science, Business, and Policy of Sustainable Nutrition. *Curr. Dev. Nutr.*, **3(6)**, 1–5, doi:10.1093/cdn/nzz059.

Ha, S., T. Hale, and P. Ogden, 2016: Climate Finance in and between Developing Countries: An Emerging Opportunity to Build On. *Glob. Policy*, **7(1)**, 102–108, doi:10.1111/1758-5899.12293.

Haas, J. et al., 2020: Floating photovoltaic plants: Ecological impacts versus hydropower operation flexibility. *Energy Convers. Manag.*, **206**, 112414, doi:10.1016/j.enconman.2019.112414.

Haasnoot, M. et al., 2020: Defining the solution space to accelerate climate change adaptation. *Reg. Environ. Change*, **20(2)**, doi:10.1007/s10113-020-01623-8.

Haberl, H. et al., 2013: Bioenergy: How much can we expect for 2050? *Environ. Res. Lett.*, **8(3)**, doi:10.1088/1748-9326/8/3/031004.

Hadi, J. and G. Brightwell, 2021: Safety of alternative proteins: Technological, environmental and regulatory aspects of cultured meat, plant-based meat, insect protein and single-cell protein. *Foods*, **10(6)**, doi:10.3390/foods10061226.

Hagenaars, L.L., P.P.T. Jeurissen, and N.S. Klazinga, 2017: The taxation of unhealthy energy-dense foods (EDFs) and sugar-sweetened beverages (SSBs): An overview of patterns observed in the policy content and policy context of 13 case studies. *Health Policy (New. York)*, **121(8)**, 887–894, doi:10.1016/j.healthpol.2017.06.011.

Haider, G., D. Steffens, G. Moser, C. Müller, and C.I. Kammann, 2017: Biochar reduced nitrate leaching and improved soil moisture content without yield improvements in a four-year field study. *Agric. Ecosyst. Environ.*, **237**, 80–94, doi:10.1016/j.agee.2016.12.019.

Halloran, A., N. Roos, J. Eilenberg, A. Cerutti, and S. Bruun, 2016: Life cycle assessment of edible insects for food protein: a review. *Agron. Sustain. Dev.*, **36(4)**, 57, doi:10.1007/s13593-016-0392-8.

Hallström, E., J. Davis, A. Woodhouse, and U. Sonesson, 2018: Using dietary quality scores to assess sustainability of food products and human diets: A systematic review. *Ecol. Indic.*, **93**(September 2017), 219–230, doi:10.1016/j.ecolind.2018.04.071.

Hamelin, L., H.B. Møller, and U. Jørgensen, 2021: Harnessing the full potential of biomethane towards tomorrow's bioeconomy: A national case study coupling sustainable agricultural intensification, emerging biogas technologies and energy system analysis. *Renew. Sustain. Energy Rev.*, **138**, 110506, doi:10.1016/J.RSER.2020.110506.

Hamilton, I. et al., 2021: The public health implications of the Paris Agreement: a modelling study. *Lancet Planet. Heal.*, **5(2)**, e74–e83, doi:10.1016/S2542-5196(20)30249-7.

Hamilton, S.E., and D.A. Friess, 2018: Global carbon stocks and potential emissions due to mangrove deforestation from 2000 to 2012. *Nat. Clim. Change*, **8(3)**, 240–244, doi:10.1038/s41558-018-0090-4.

Han Weng, Z. et al., 2017: Biochar built soil carbon over a decade by stabilizing rhizodeposits. *Nat. Clim. Change*, **7(5)**, 371–376, doi:10.1038/nclimate3276.

Hanna, R., A. Abdulla, Y. Xu, and D.G. Victor, 2021: Emergency deployment of direct air capture as a response to the climate crisis. *Nat. Commun.*, **12(1)**, 368, doi:10.1038/s41467-020-20437-0.

Hansen, J.H., L. Hamelin, A. Taghizadeh-Toosi, J.E. Olesen, and H. Wenzel, 2020: Agricultural residues bioenergy potential that sustain soil carbon depends on energy conversion pathways. *GCB Bioenergy*, **12(11)**, 1002–1013, doi:10.1111/GCBB.12733.

Hanssen, S.V. et al., 2019: Biomass residues as twenty-first century bioenergy feedstock—a comparison of eight integrated assessment models. *Clim. Chang. 2019 1633*, **163(3)**, 1569–1586, doi:10.1007/S10584-019-02539-X. Hanssen, S. V et al., 2020: The climate change mitigation potential of bioenergy with carbon capture and storage. *Nat. Clim. Change*, **10(11)**, 1023–1029, doi:10.1038/s41558-020-0885-y.

Hansson, A. et al., 2021: Biochar as multi-purpose sustainable technology: experiences from projects in Tanzania. *Environ. Dev. Sustain.*, **23(4)**, 5182–5214, doi:10.1007/s10668-020-00809-8.

Haque, F., R.M. Santos, A. Dutta, M. Thimmanagari, and Y.W. Chiang, 2019: Co-Benefits of Wollastonite Weathering in Agriculture: CO2 Sequestration and Promoted Plant Growth. *ACS Omega*, **4(1)**, 1425–1433, doi:10.1021/acsomega.8b02477.

Harmsen, J.H.M. et al., 2019: Long-term marginal abatement cost curves of non-CO2 greenhouse gases. *Environ. Sci. Policy*, **99**, 136–149, doi:10.1016/j.envsci.2019.05.013.

Harris, Z.M., R. Spake, and G. Taylor, 2015: Land use change to bioenergy: A meta-analysis of soil carbon and GHG emissions. *Biomass and Bioenergy*, **82**, 27–39, doi:10.1016/j.biombioe.2015.05.008.

Harrison, D.P., 2013: A method for estimating the cost to sequester carbon dioxide by delivering iron to the ocean. *Int. J. Glob. Warm.*, **5(3)**, 231–254, doi:10.1504/ijgw.2013.055360.

Harrison, D.P., 2017: Global negative emissions capacity of ocean macronutrient fertilization. *Environ. Res. Lett.*, **12(3)**, 035001, doi:10.1088/1748-9326/AA5EF5.

Hartmann, C. and M. Siegrist, 2017: Consumer perception and behaviour regarding sustainable protein consumption: A systematic review. *Trends Food Sci. Technol.*, **61**, 11–25, doi:10.1016/j.tifs.2016.12.006.

Hartmann, J. et al., 2013: Enhanced chemical weathering as a geoengineering strategy to reduce atmospheric carbon dioxide, supply nutrients, and mitigate ocean acidification. *Rev. Geophys.*, **51(2)**, 113–149, doi:10.1002/rog.20004.

Harvey, L.D.D., 2008: Mitigating the atmospheric CO2 increase and ocean acidification by adding limestone powder to upwelling regions. *J. Geophys. Res.*, **113**(C4), C04028, doi:10.1029/2007JC004373.

Hasegawa, T. et al., 2018: Risk of increased food insecurity under stringent global climate change mitigation policy. *Nat. Clim. Change*, **8(8)**, 699–703, doi:10.1038/s41558-018-0230-x.

Hasegawa, T. et al., 2020: Food security under high bioenergy demand toward long-term climate goals. *Clim. Chang. 2020 1633*, **163(3)**, 1587–1601, doi:10.1007/S10584-020-02838-8.

Hasegawa, T. et al., 2021: Land-based implications of early climate actions without global net-negative emissions. *Nat. Sustain.*, **4**, 1052–1059, doi:10.1038/s41893-021-00772-w.

Hassanpour Adeh, E., J.S. Selker, and C.W. Higgins, 2018: Remarkable agrivoltaic influence on soil moisture, micrometeorology and water-use efficiency. *PLoS One*, **13(11)**, e0203256, doi:10.1371/journal.pone.0203256.

Haus, S., L. Björnsson, and P. Börjesson, 2020: Lignocellulosic Ethanol in a Greenhouse Gas Emission Reduction Obligation System—A Case Study of Swedish Sawdust Based-Ethanol Production. *Energies 2020, Vol. 13, Page 1048*, **13(5)**, 1048, doi:10.3390/EN13051048.

Hayek, M.N., H. Harwatt, W.J. Ripple, and N.D. Mueller, 2021: The carbon opportunity cost of animal-sourced food production on land. *Nat. Sustain.*, **4** (January), 21–24, doi:10.1038/s41893-020-00603-4.

Hebinck, A. et al., 2021: A Sustainability Compass for policy navigation to sustainable food systems. *Glob. Food Sec.*, **9** (June), 100546, doi:10.1016/j.gfs.2021.100546.

Heck, V., D. Gerten, W. Lucht, and A. Popp, 2018: Biomass-based negative emissions difficult to reconcile with planetary boundaries. *Nat. Clim. Change*, **8(2)**, 151–155, doi:10.1038/s41558-017-0064-y.

Hedenus, F., S. Wirsenius, and D.J.A. Johansson, 2014: The importance of reduced meat and dairy consumption for meeting stringent climate change targets. *Clim. Change*, **124**, 79–91, doi:10.1007/s10584-014-1104-5.

Henders, S., U.M. Persson, and T. Kastner, 2015: Trading forests: Land-use change and carbon emissions embodied in production and exports of forest-risk commodities. *Environ. Res. Lett.*, **10(12)**, doi:10.1088/1748-9326/10/12/125012.

Henderson, B. et al., 2018: The power and pain of market-based carbon policies: a global application to greenhouse gases from ruminant livestock production. *Mitig. Adapt. Strateg. Glob. Change*, **23(3)**, 349–369, doi:10.1007/s11027-017-9737-0.

Henry, B., B. Murphy, and A. Cowie, 2018a: *Sustainable Land Management for Environmental Benefits and Food Security A synthesis report for the GEF*. GEF, Washington DC, USA, 127 pp.

Henry, R.C. et al., 2018b: Food supply and bioenergy production within the global cropland planetary boundary. *PLoS One*, **13(3)**, e0194695, doi:10.1371/journal.pone.0194695.

Hepburn, C. et al., 2019: The technological and economic prospects for CO2 utilization and removal. *Nature*, **575(7781)**, 87–97, doi:10.1038/s41586-019-1681-6.

Herforth, A. et al., 2019: A Global Review of Food-Based Dietary Guidelines. *Adv. Nutr.*, **10(4)**, 590–605, doi:10.1093/advances/nmy130.

Hernández-Morcillo, M., P. Burgess, J. Mirck, A. Pantera, and T. Plieninger, 2018: Scanning agroforestry-based solutions for climate change mitigation and adaptation in Europe. *Environ. Sci. Policy*, **80** (February) 44–52, doi:10.1016/j.envsci.2017.11.013.

Herrero, M. et al., 2017: Farming and the geography of nutrient production for human use: a transdisciplinary analysis. *Lancet Planet. Heal.*, **1(1)**, e33--e42, doi:10.1016/S2542-5196(17)30007-4.

Herrero, M. et al., 2020: Innovation can accelerate the transition towards a sustainable food system. *Nat. Food*, **1(5)**, 266–272, doi:10.1038/s43016-020-0074-1.

Herrero, M. et al., 2021: Articulating the effect of food systems innovation on the Sustainable Development Goals. *Lancet Planet. Heal.*, **5(1)**, e50–e62, doi:10.1016/S2542-5196(20)30277-1.

Hertel, T.W. and U.L.C. Baldos, 2016: Attaining food and environmental security in an era of globalization. *Glob. Environ. Change*, **41**, 195–205, doi:10.1016/j.gloenvcha.2016.10.006.

Hertzler, S.R., J.C. Lieblein-Boff, M. Weiler, and C. Allgeier, 2020: Plant proteins: Assessing their nutritional quality and effects on health and physical function. *Nutrients*, **12(12)**, 1–27, doi:10.3390/nu12123704.

Heuck, C. et al., 2019: Wind turbines in high quality habitat cause disproportionate increases in collision mortality of the white-tailed eagle. *Biol. Conserv.*, **236**, 44–51, doi.org/10.1016/j.biocon.2019.05.018.

Hevia-Koch, P. and J. Ladenburg, 2019: Where should wind energy be located? A review of preferences and visualisation approaches for wind turbine locations. *Energy Res. Soc. Sci.*, **53**, 23–33, doi.org/10.1016/j.erss.2019.02.010.

Hieke, S. and J.L. Harris, 2016: Nutrition information and front-of-pack labelling: Issues in effectiveness. *Public Health Nutr.*, **19(12)**, 2103–2105, doi:10.1017/S1368980016001890.

Hilaire, J. et al., 2019: Negative emissions and international climate goals—learning from and about mitigation scenarios. *Clim. Change*, **157(2)**, 189–219, doi:10.1007/s10584-019-02516-4.

Hilborn, R., J. Banobi, S.J. Hall, T. Pucylowski, and T.E. Walsworth, 2018: The environmental cost of animal source foods. *Front. Ecol. Environ.*, **16(6)**, 329–335, doi:10.1002/fee.1822.

Hildén, M., A. Jordan, and T. Rayner, 2014: Climate policy innovation: developing an evaluation perspective. *Env. Polit.*, **23(5)**, 884–905, doi:10.1080/09644016.2014.924205.

Hirvonen, K., Y. Bai, D. Headey, and W.A. Masters, 2019: Affordability of the EAT–Lancet reference diet: a global analysis. *Lancet Glob. Heal.*, **8**, e59–66, doi:10.1016/S2214-109X(19)30447-4.

HLPE, 2014: *Food Losses and Waste in the Context of Sustainable Food Systems. A Report by the High Level Panel of Experts on Food Security and Nutrition of the Committee on World Food Security.* High Level Panel of Experts on Food Security and Nutrition, Rome, Italy, 6 pp.

HLPE, 2017: *Nutrition and food systems. A report by The High Level Panel of Experts on Food Security and Nutrition Nutrition on World Food Security*. High Level Panel of Experts on Food Security and Nutrition, Rome, Italy, 11 pp.

HLPE, 2019: Agroecological and other innovative approaches for sustainable agriculture and food systems that enhance food security and nutrition. A report by the High Level Panel of Experts on Food Security and Nutrition of the Committee on World Food Security., Rome, Italy, 162 pp.

HLPE, 2020: *Food security and nutrition: building a global narrative towards 2030. V0 Draft report*. Food and Agriculture Organization of the United Nations, Rome, Italy.

HM Government, 2011: *The carbon plan: delivering our low carbon future*. HM Government, London, UK, 218 pp.

HM Government, 2021: *Net Zero Strategy: Build Back Greener*. HM Government, London, UK, 367 pp.

Hoegh-Guldberg, O., D. Jacob, M. Taylor, M. Bindi, S. Brown, I. Camilloni, A. Diedhiou, R. Djalante, K.L. Ebi, F. Engelbrecht, J. Guiot, Y. Hijioka, S. Mehrotra, A. Payne, S.I. Seneviratne, A. Thomas, R. Warren, and G. Zhou, 2018: Impacts of 1.5°C Global Warming on Natural and Human Systems. In: *Global Warming of 1.5°C. An IPCC special report on the impacts of global warming of 1.5°C above pre-industrial levels and related global greenhouse gas emission pathways, in the context of strengthening the global response to the threat of climate change, sustainable development, and efforts to eradicate poverty.* [Masson-Delmotte, V., P. Zhai, H.-O. Pörtner, D. Roberts, J. Skea, P.R. Shukla, A. Pirani, W. Moufouma-Okia, C. Péan, R. Pidcock, S. Connors, J.B.R. Matthews, Y. Chen, X. Zhou, M.I. Gomis, E. Lonnoy, T. Maycock, M. Tignor, and T. Waterfield (eds.)]. Cambridge University Press, Cambridge, UK, and New York, NY, USA.

Hoff, H. et al., 2019a: A Nexus Approach for the MENA Region—From Concept to Knowledge to Action. *Front. Environ. Sci.*, **7**(APR), 48, doi:10.3389/fenvs.2019.00048.

Hoff, H. et al., 2019b: *International spillovers in SDG implementation The case of soy from Argentina*. SEI Policy Brief, Stockholm Environment Institute. Stockholm, Sweden, 8 pp.

Höglund-Isaksson, L. et al., 2020: Technical potentials and costs for reducing global anthropogenic methane emissions in the 2050 timeframe – results from the GAINS model. *Environ. Res. Commun.*, **2(2)**, 25004, doi:10.1088/2515-7620/ab7457.

Holland, R.A. et al., 2015: A synthesis of the ecosystem services impact of second generation bioenergy crop production. *Renew. Sustain. Energy Rev.*, **46**, 30–40, doi:10.1016/j.rser.2015.02.003.

Holz, C., L.S. Siegel, E. Johnston, A.P. Jones, and J. Sterman, 2018: Ratcheting ambition to limit warming to 1.5°C–trade-offs between emission reductions and carbon dioxide removal. *Environ. Res. Lett.*, **13(6)**, 64028, doi:10.1088/1748-9326/aac0c1.

Honegger, M. and D. Reiner, 2018: The political economy of negative emissions technologies: consequences for international policy design. *Clim. Policy*, **18(3)**, 306–321, doi:10.1080/14693062.2017.1413322.

Honegger, M., A. Michaelowa, and J. Roy, 2020: Potential implications of carbon dioxide removal for the sustainable development goals. *Clim. Policy*, **21(5)**, 678–698, doi:10.1080/14693062.2020.1843388.

Honegger, M., W. Burns, and D.R. Morrow, 2021a: Is carbon dioxide removal 'mitigation of climate change'? *Rev. Eur. Comp. Int. Environ. Law*, **00**, 1–9, doi:10.1111/reel.12401.

Honegger, M., M. Poralla, A. Michaelowa, and H.-M. Ahonen, 2021b: Who Is Paying for Carbon Dioxide Removal? Designing Policy Instruments for Mobilizing Negative Emissions Technologies. *Front. Clim.*, **3**, 50, doi:10.3389/fclim.2021.672996.

Hosonuma, N. et al., 2012: An assessment of deforestation and forest degradation drivers in developing countries. *Environ. Res. Lett.*, **7(4)**, doi:10.1088/1748-9326/7/4/044009.

House, K.Z., C.H. House, D.P. Schrag, and M.J. Aziz, 2007: Electrochemical Acceleration of Chemical Weathering as an Energetically Feasible Approach to Mitigating Anthropogenic Climate Change. *Environ. Sci. Technol.*, **41(24)**, 8464–8470, doi:10.1021/es0701816.

Howard, C., C.C. Dymond, V.C. Griess, D. Tolkien-Spurr, and G.C. van Kooten, 2021: Wood product carbon substitution benefits: a critical review of assumptions. *Carbon Balanc. Manag. 2021 161*, **16(1)**, 1–11, doi:10.1186/S13021-021-00171-W.

Howard, J. et al., 2017: Clarifying the role of coastal and marine systems in climate mitigation. *Front. Ecol. Environ.*, **15(1)**, 42–50, doi:10.1002/FEE.1451.

Hua, F. et al., 2016: Opportunities for biodiversity gains under the world's largest reforestation programme. *Nat. Commun.*, **7**, doi:10.1038/ncomms12717.

Humpenöder, F. et al., 2014: Investigating afforestation and bioenergy CCS as climate change mitigation strategies. *Environ. Res. Lett.*, **9(6)**, 064029, doi:10.1088/1748-9326/9/6/064029.

Humpenöder, F. et al., 2018: Large-scale bioenergy production: how to resolve sustainability trade-offs? *Environ. Res. Lett.*, **13(2)**, 024011, doi:10.1088/1748-9326/AA9E3B.

Hurlbert, M., J. Krishnaswamy, E. Davin, F.X. Johnson, C.F. Mena, J. Morton, S. Myeong, D. Viner, K. Warner, A. Wreford, S. Zakieldeen, Z. Zommers, 2019: Risk Management and Decision Making in Relation to Sustainable Development. In: *Climate Change and Land: An IPCC Special Report on climate change, desertification, land degradation, sustainable land management, food security, and greenhouse gas fluxes in terrestrial ecosystems* [Shukla, P.R., J. Skea, E. Calvo Buendia, V. Masson-Delmotte, H.-O. Pörtner, D.C. Roberts, P. Zhai, R. Slade, S. Connors, R. van Diemen, M. Ferrat, E. Haughey, S. Luz, S. Neogi, M. Pathak, J. Petzold, J. Portugal Pereira, P. Vyas, E. Huntley, K. Kissick, M. Belkacemi, J. Malley, (eds.)]. Cambridge University Press, Cambridge, UK, and New York, NY, USA.

Hydrogen Council, 2017: *Hydrogen Scaling up. A sustainable pathway for the global energy transition*. Hydrogen Council, 80 pp.

Hyland, J.J., M. Henchion, M. McCarthy, and S.N. McCarthy, 2017: The role of meat in strategies to achieve a sustainable diet lower in greenhouse gas emissions: A review. *Meat Sci.*, **132**, 189–195, doi:10.1016/j.meatsci.2017.04.014.

IAEA, 2006a: *Environmental Consequences of the Chernobyl Accident and their Remediation: Twenty Years of Experience. Report of the Chernobyl Forum Expert Group 'Environment".'* International Atomic Energy Agency, Vienna, Austria, 180 pp.

IAEA, 2006b: *Environmental Consequences of the Chernobyl Accident and their Remediation: Twenty Years of Experience*. International Atomic Energy Agency, Vienna, Austria.

IANAS, 2018: *Opportunities and challenges for research on food and nutrition security and agriculture in the Americas. Regional analysis prepared from country assessments by IANAS*. Inter-American Network of Academies of Sciences, Mexico City, Mexico, 49 pp.

IAP, 2018: *Opportunities for future research and innovation on food and nutrition security and agriculture. The InterAcademy Partnership's global perspective. Synthesis by IAP based on four regional academy network studies*. InterAcademy Partnership, Trieste, Italy, and Washington, DC, USA, 94 pp.

IDA, 2019: International Development Association. http://ida.worldbank.org/ (Accessed December 15, 2019).

IEA, 2017: *Digitalization & Energy*. International Energy Agency, Paris, France, 188 pp.

IEA, 2018: *World Energy Outlook 2018*. International Energy Agency, Paris, France, 661 pp.

IEA, 2019a: *World Energy Outlook 2019*. International Energy Agency, Paris, France, 810 pp.

IEA, 2019b: *The Future of Hydrogen. Seizing today's opportunities*. International Energy Agency, Paris, France, 203 pp.

IEA, 2020a: *World Energy Outlook 2020*. International Energy Agency, Paris, France, 461 pp.

IEA, 2020b: *Energy Technology Perspectives 2020*. International Energy Agency, Paris, France, 397 pp.

IEA, 2020c: *Renewables 2020*. International Energy Agency, Paris, France, https://www.iea.org/reports/renewables-2020 (Accessed December 23, 2020).

IEA, 2021a: Methane Tracker Database. https://www.iea.org/articles/methane-tracker-database (Accessed September 20, 2021).

IEA, 2021b: *World Energy Outlook 2021*. International Energy Agency, Paris, France, 383 pp.

IHME, 2018: *Findings from the Global Burden of Disease Study 2017*. Institute for Health Metrics and Evaluation, Seattle, USA, 27 pp.

IIASA, 2021: AR6 Scenario Explorer (submissions) hosted by IIASA. International Institute for Applied Systems Analysis, https://data.ene.iiasa.ac.at/ar6-scenario-submission/ (Accessed October 14, 2021).

Iiyama, M. et al., 2018: Addressing the paradox – the divergence between smallholders' preference and actual adoption of agricultural innovations. *Int. J. Agric. Sustain.*, **16(6)**, 472–485, doi:10.1080/14735903.2018.1539384.

Ikonen, I., F. Sotgiu, A. Aydinli, and P.W.J. Verlegh, 2019: Consumer effects of front-of-package nutrition labeling: an interdisciplinary meta-analysis. *J. Acad. Mark. Sci.*, **48**, 360–383, doi:10.1007/s11747-019-00663-9.

Ingram, J., 2020: Nutrition security is more than food security. *Nat. Food*, **1(1)**, 2–2, doi:10.1038/s43016-019-0002-4.

Ince, A.C., C.O. Colpan, A. Hagen, and M.F. Serincan, 2021: Modeling and simulation of Power-to-X systems: A review. *Fuel*, **304**, 121354, doi:10.1016/j.fuel.2021.121354.

Ioannidis, R., and D. Koutsoyiannis, 2020: A review of land use, visibility and public perception of renewable energy in the context of landscape impact. *Appl. Energy*, **276**, 115367, doi:10.1016/j.apenergy.2020.115367.

IPBES, 2019: Summary for Policymakers. In: *Global Assessment Report on Biodiversity and Ecosystem Services* [Brondizio, E.S., J. Settele, S. Díaz, and H.T. Ngo, (eds.)], Intergovernmental Science-Policy Platform on Biodiversity and Ecosystem Services, Bonn, Germany.

IPCC, 2006: *2006 IPCC Guidelines for National Greenhouse Gas Inventories — IPCC*. [Eggelston, S., L. Buendia, K. Miwa, T. Ngara, and K. Tanabe, (eds.)]. Institute for Global Environmental Strategies for the IPCC,.

IPCC, 2018: Summary for Policymakers. In: *Global Warming of 1.5°C an IPCC special report on the impacts of global warming of 1.5°C above pre-industrial levels and related global greenhouse gas emission pathways, in the context of strengthening the global response to the threat of climate change, sustainable development, and efforts to eradicate poverty* [Masson-Delmotte, V., P. Zhai, H.-O. Pörtner, D. Roberts, J. Skea, P.R. Shukla, A. Pirani, W. Moufouma-Okia, C. Péan, R. Pidcock, S. Connors, J.B.R. Matthews, Y. Chen, X. Zhou, M.I. Gomis, E. Lonnoy, T. Maycock, M. Tignor, and T. Waterfield (eds.)]. World Meteorological Organization, Geneva, Switzerland, 32 pp.

IPCC, 2019a: *Climate Change and Land. An IPCC Special Report on climate change, desertification, land degradation, sustainable land management, food security, and greenhouse gas fluxes in terrestrial ecosystems* [Shukla, P.R., J. Skea, E. Calvo Buendia, V. Masson-Delmotte, H.-O. Pörtner, D. C. Roberts, P. Zhai, R. Slade, S. Connors, R. van Diemen, M. Ferrat, E. Haughey, S. Luz, S. Neogi, M. Pathak, J. Petzold, J. Portugal Pereira, P. Vyas, E. Huntley, K. Kissick, M. Belkacemi, J. Malley, (eds.)]. In press.

IPCC, 2019b: *IPCC Special Report on the Ocean and Cryosphere in a Changing Climate* [H.-O. Pörtner, D.C. Roberts, V. Masson-Delmotte, P. Zhai, M. Tignor, E. Poloczanska, K. Mintenbeck, A. Alegría, M. Nicolai, A. Okem, J. Petzold, B. Rama, N.M. Weyer (eds.)]. Cambridge University Press, Cambridge, UK and New York, NY, USA, 755 pp. https://doi.org/10.1017/9781009157964.

iPES Food, 2019: *Towards a Common Food Policy for the European Union: The Policy Reform and Realignment that is Required to Build Sustainable Food Systems in Europe*. International Panel of Experts on Sustainable Food Systems, Brussels, Belgium, 112 pp.

IRENA, 2018: *Hydrogen from renewable power: Technology outlook for the energy transition*. International Renewable Energy Agency, Abu Dhabi, 52 pp.

IRENA, 2019: *Bioenergy from boreal forests: Swedish approach to sustainable wood use*. Abu Dhabi, https://www.irena.org/publications/2019/Mar/Bioenergy-from-boreal-forests-Swedish-approach-to-sustainable-wood-use (Accessed November 9, 2021).

IRP, 2019: *Land Restoration for Achieving the Sustainable Development Goals: An International Resource Panel Think Piece*. [Herrick, J.E. et al., (eds.)]. United Nations Environment Programme, Nairobi, Kenya, 135 pp. https://www.resourcepanel.org/reports/land-restoration-achieving-sustainable-development-goals (Accessed December 5, 2019).

Ishimoto, Y. et al., 2017: *Putting Costs of Direct Air Capture in Context*. Forum for Climate Engineering Assessment, Washington, DC, USA, 21 pp.

Ishiwatari, M. et al., 2019: *Climate Fragility Risks (CFR) In Development Sectors: Six Principles for Managing Synergies and Trade-Offs*. The University of Tokyo, United Nations University Institute for the Advanced Study of Sustainability, Keio University, Ritsumeikan Asia Pacific University, Institute for Global Environmental Strategies, Japan, https://collections.unu.edu/view/UNU:7334#.XSYRAOfvY5o.mendeley.

Iyer, G. et al., 2021: The role of carbon dioxide removal in net-zero emissions pledges. *Energy Clim. Change*, **2**, 100043, doi:10.1016/j.egycc.2021.100043.

Jaccard, S.L. et al., 2013: Two Modes of Change in Southern Ocean Productivity Over the Past Million Years. *Science*, **339(6126)**, 1419–1423, doi:10.1126/science.1227545.

Jackson, R.B., E.I. Solomon, J.G. Canadell, M. Cargnello, and C.B. Field, 2019: Methane removal and atmospheric restoration. *Nat. Sustain.*, **2(6)**, 436–438, doi:10.1038/s41893-019-0299-x.

Jackson, R.B. et al., 2021: Atmospheric methane removal: a research agenda. *Philos. Trans. R. Soc. A Math. Phys. Eng. Sci.*, **379(2210)**, 20200454, doi:10.1098/rsta.2020.0454.

Jacobi, J., S. Rist, and M.A. Altieri, 2017: Incentives and disincentives for diversified agroforestry systems from different actors' perspectives in Bolivia. *Int. J. Agric. Sustain.*, **15(4)**, doi:10.1080/14735903.2017.1332140.

Jacobson, M., 2019: The Health and Climate Impacts of Carbon Capture and Direct Air Capture. *Energy Environ. Sci.*, **12**, doi:10.1039/C9EE02709B. Jaffe, S., 2017: Vulnerable Links in the Lithium-Ion Battery Supply Chain. *Joule*, **1(2)**, 225–228, doi:10.1016/J.JOULE.2017.09.021.

Jager, H.I. and J.A.F. Kreig, 2018: Designing landscapes for biomass production and wildlife. *Glob. Ecol. Conserv.*, **16**, doi:10.1016/j.gecco.2018.e00490.

James, S.J., and C. James, 2010: The food cold-chain and climate change. *Food Res. Int.*, **43(7)**, 1944–1956, doi:10.1016/j.foodres.2010.02.001.

Jarmul, S. et al., 2020: Climate change mitigation through dietary change: a systematic review of empirical and modelling studies on the environmental footprints and health effects of 'sustainable diets.' *Environ. Res. Lett.*, **15(12)**, 123014, doi:10.1088/1748-9326/abc2f7.

Jeffery, S. et al., 2017: Biochar boosts tropical but not temperate crop yields. *Environ. Res. Lett.*, **12(5)**, doi:10.1088/1748-9326/aa67bd.

Jensen, J.P., and K. Skelton, 2018: Wind turbine blade recycling: Experiences, challenges and possibilities in a circular economy. *Renew. Sustain. Energy Rev.*, **97**, 165–176, doi:10.1016/j.rser.2018.08.041.

Jensen, P.D., P. Purnell, and A.P.M. Velenturf, 2020: Highlighting the need to embed circular economy in low carbon infrastructure decommissioning: The

12

case of offshore wind. *Sustain. Prod. Consum.*, **24**, 266–280, doi:10.1016/j.spc.2020.07.012.

Jensen, S., K. Mohlin, K. Pittel, and T. Sterner, 2015: An Introduction to the Green Paradox: The Unintended Consequences of Climate Policies. *Rev. Environ. Econ. Policy*, **9(2)**, 246–265, doi:10.1093/reep/rev010.

Jensen, T. and H. Dowlatabadi, 2018: Challenges in financing public sector low-carbon initiatives: lessons from private finance for a school district in British Columbia, Canada. *Clim. Policy*, **18(7)**, 878–888, doi:10.1080/14693062.2017.1387512.

Jepson, W. and M. Caldas, 2017: Changing energy systems and land-use change. *J. Land Use Sci.*, **12(6)**, 405–406, doi:10.1080/1747423X.2017.1408889.

Jia, G., E. Shevliakova, P. Artaxo, N. De Noblet-Ducoudré, R. Houghton, J. House, K. Kitajima, C. Lennard, A. Popp, A. Sirin, R. Sukumar, L. Verchot, 2019: Land–Climate Interactions. In: *Climate Change and Land. An IPCC Special Report on climate change, desertification, land degradation, sustainable land management, food security, and greenhouse gas fluxes in terrestrial ecosystems.* [P.R. Shukla, J. Skea, E. Calvo Buendia, V. Masson-Delmotte, H.-O. Pörtner, D.C. Roberts, P. Zhai, R. Slade, S. Connors, R. van Diemen, M. Ferrat, E. Haughey, S. Luz, S. Neogi, M. Pathak, J. Petzold, J. Portugal Pereira, P. Vyas, E. Huntley, K. Kissick, M. Belkacemi, J. Malley, (eds.)]. Cambridge University Press, Cambridge, UK, and New York, NY, USA.

Jin, X., and N. Gruber, 2003: Offsetting the radiative benefit of ocean iron fertilization by enhancing N2O emissions. *Geophys. Res. Lett.*, **30(24)**, 2249, doi:10.1029/2003GL018458.

Johansson, D.J.A., C. Azar, M. Lehtveer, and G.P. Peters, 2020: The role of negative carbon emissions in reaching the Paris climate targets: The impact of target formulation in integrated assessment models. *Environ. Res. Lett.*, **15(12)**, 124024, doi:10.1088/1748-9326/abc3f0.

Johnsson, S., E. Andersson, P. Thollander, and M. Karlsson, 2019: Energy savings and greenhouse gas mitigation potential in the Swedish wood industry. *Energy*, **187**, 115919, doi:10.1016/J.ENERGY.2019.115919.

Jones, I.S.., 2014: The cost of carbon management using ocean nourishment. *Int. J. Clim. Chang. Strateg. Manag.*, **6(4)**, 391–400, doi:10.1108/IJCCSM-11-2012-0063.

Jönsson, E., T. Linné, and A. McCrow-Young, 2019: Many Meats and Many Milks? The Ontological Politics of a Proposed Post-animal Revolution. *Sci. Cult. (Lond).*, **28(1)**, 70–97, doi:10.1080/09505431.2018.1544232.

Joppa, L. et al., 2021: Microsoft's million-tonne CO2-removal purchase - lessons for net zero. *Nature*, **597(7878)**, 629–632, doi:10.1038/d41586-021-02606-3.

Jørgensen, U. and P.E. Lærke, 2016: Perennial grasses for sustainable European protein production. In: *Perennial Biomass Crops for a Resource-Constrained World.*

Joseph, S. et al., 2021: How biochar works, and when it doesn't: A review of mechanisms controlling soil and plant responses to biochar. *GCB Bioenergy*, **13(11)**, 1731–1764, doi:10.1111/gcbb.12885.

Junge, R., B. König, M. Villarroel, T. Komives, and M.H. Jijakli, 2017: Strategic points in aquaponics. *Water (Switzerland)*, **9(3)**, 1–9, doi:10.3390/w9030182.

Kalinina, A., M. Spada, and P. Burgherr, 2018: Application of a Bayesian hierarchical modeling for risk assessment of accidents at hydropower dams. *Saf. Sci.*, **110**, 164–177, doi:10.1016/J.SSCI.2018.08.006.

Kanter, D.R. et al., 2018a: Evaluating agricultural trade-offs in the age of sustainable development. *Agric. Syst.*, **163**, 73–88, doi:10.1016/j.agsy.2016.09.010.

Kanter, D.R. et al., 2020: Nitrogen pollution policy beyond the farm. *Nat. Food*, **1(1)**, 27–32, doi:10.1038/s43016-019-0001-5.

Kanter, R., L. Vanderlee, and S. Vandevijvere, 2018b: Front-of-package nutrition labelling policy: Global progress and future directions. *Public Health Nutr.*, **21(8)**, 1399–1408, doi:10.1017/S1368980018000010.

Kantola, I.B., M.D. Masters, D.J. Beerling, S.P. Long, and E.H. DeLucia, 2017: Potential of global croplands and bioenergy crops for climate change mitigation through deployment for enhanced weathering. *Biol. Lett.*, **13(4)**, 20160714, doi:10.1098/rsbl.2016.0714.

Karlsson-Vinkhuyzen, S.I. et al., 2018: Entry into force and then? The Paris agreement and state accountability. *Clim. Policy*, **18(5)**, 593–599, doi:10.1080/14693062.2017.1331904.

Karlsson, M., E. Alfredsson, and N. Westling, 2020: Climate policy co-benefits: a review. *Clim. Policy*, **20(3)**, 1–25, doi:10.1080/14693062.2020.1724070.

Karplus, V.J., S. Paltsev, M. Babiker, and J.M. Reilly, 2013: Should a vehicle fuel economy standard be combined with an economy-wide greenhouse gas emissions constraint? Implications for energy and climate policy in the United States. *Energy Econ.*, **36**, 322–333, doi:10.1016/j.eneco.2012.09.001.

Kato, E. and A. Kurosawa, 2019: Evaluation of Japanese energy system toward 2050 with TIMES-Japan - Deep decarbonization pathways. *Energy Procedia*, **158**, 4141–4146, doi:10.1016/j.egypro.2019.01.818.

Kato, E. and A. Kurosawa, 2021: Role of negative emissions technologies (NETs) and innovative technologies in transition of Japan's energy systems toward net-zero CO2 emissions. *Sustain. Sci.*, **16(2)**, 463–475, doi:10.1007/s11625-021-00908-z.

Kaunda, R.B., 2020: Potential environmental impacts of lithium mining. *J. Energy Nat. Resour. Law*, **38(3)**, 237–244, doi:10.1080/02646811.2020.1754596.

Kavanagh, R.P. and M.A. Stanton, 2012: Koalas use young *Eucalyptus* plantations in an agricultural landscape on the Liverpool Plains, New South Wales. *Ecol. Manag. Restor.*, **13(3)**, 297–305, doi:10.1111/emr.12005.

Kc, B. K. et al., 2018: When too much isn't enough: Does current food production meet global nutritional needs? *PLoS One*, **13(10)**, 1–16, doi:10.1371/journal.pone.0205683.

Keith, D.W. et al., 2004: The influence of large-scale wind power on global climate. *Proc. Natl. Acad. Sci.*, **101(46)**, 16115–16120, doi:10.1073/pnas.0406930101.

Kelland, M.E. et al., 2020: Increased yield and CO2 sequestration potential with the C4 cereal Sorghum bicolor cultivated in basaltic rock dust-amended agricultural soil. *Glob. Chang. Biol.*, **26(6)**, 3658–3676, doi:10.1111/GCB.15089.

Keller, D.P., E.Y. Feng, and A. Oschlies, 2014: Potential climate engineering effectiveness and side effects during a high carbon dioxide-emission scenario. *Nat. Commun.*, **5**, 3304, doi:10.1038/ncomms4304.

Kempton, W. and J. Tomić, 2005: Vehicle-to-grid power implementation: From stabilizing the grid to supporting large-scale renewable energy. *J. Power Sources*, **144(1)**, 280–294, doi:10.1016/j.jpowsour.2004.12.022.

Kersh, R., 2015: Of nannies and nudges: The current state of U.S. obesity policymaking. *Public Health*, **129(8)**, 1083–1091, doi:10.1016/j.puhe.2015.05.018.

Keys, P.W., L. Wang-Erlandsson, and L.J. Gordon, 2016: Revealing Invisible Water: Moisture Recycling as an Ecosystem Service. *PLoS One*, **11(3)**, e0151993.

Kheshgi, H.S., 1995: Sequestering atmospheric carbon dioxide by increasing ocean alkalinity. *Energy*, **20(9)**, 915–922, doi:10.1016/0360-5442(95)00035-F.

Kim, B.F. et al., 2019: Country-specific dietary shifts to mitigate climate and water crises. *Glob. Environ. Change*, **62** (June 2018), 101926, doi:10.1016/j.gloenvcha.2019.05.010.

Kim, G. and P. Coseo, 2018: Urban Park Systems to Support Sustainability: The Role of Urban Park Systems in Hot Arid Urban Climates. *Forests*, **9(7)**, 439, doi:10.3390/f9070439.

Kirchherr, J., H. Pohlner, and K.J. Charles, 2016: Cleaning up the big muddy: A meta-synthesis of the research on the social impact of dams. *Environ. Impact Assess. Rev.*, **60**, 115–125, doi:10.1016/j.eiar.2016.02.007.

Kirchherr, J., D. Reike, and M. Hekkert, 2017: Conceptualizing the circular economy: An analysis of 114 definitions. *Resour. Conserv. Recycl.*, **127**, 221–232, doi:10.1016/j.resconrec.2017.09.005.

Kirchherr, J., M.-P. Ahrenshop, and K. Charles, 2019: Resettlement lies: Suggestive evidence from 29 large dam projects. *World Dev.*, **114**, 208–219, doi:10.1016/j.worlddev.2018.10.003.

Kissinger, G., M. Brockhaus, and S.R. Bush, 2021: Policy integration as a means to address policy fragmentation: Assessing the role of Vietnam's national REDD+ action plan in the central highlands. *Environ. Sci. Policy*, **119**, 85–92, doi:10.1016/j.envsci.2021.02.011.

Kivyiro, P. and H. Arminen, 2014: Carbon dioxide emissions, energy consumption, economic growth, and foreign direct investment: Causality analysis for Sub-Saharan Africa. *Energy*, **74**, 595–606, doi.org/10.1016/j.energy.2014.07.025.

Klausbruckner, C., H. Annegarn, L.R.F. Henneman, and P. Rafaj, 2016: A policy review of synergies and trade-offs in South African climate change mitigation and air pollution control strategies. *Environ. Sci. Policy*, **57**, 70–78, doi:10.1016/j.envsci.2015.12.001.

Klerkx, L. and D. Rose, 2020: Dealing with the game-changing technologies of Agriculture 4.0: How do we manage diversity and responsibility in food system transition pathways? *Glob. Food Sec.*, **24**(October 2019), 100347, doi:10.1016/j.gfs.2019.100347.

Kludze, H. et al., 2013: Estimating sustainable crop residue removal rates and costs based on soil organic matter dynamics and rotational complexity. *Biomass and Bioenergy*, **56**, 607–618, doi.org/10.1016/j.biombioe.2013.05.036.

Knowler, D. et al., 2020: The economics of Integrated Multi-Trophic Aquaculture: where are we now and where do we need to go? *Rev. Aquac.*, **12(3)**, 1579–1594, doi:10.1111/raq.12399.

Köberle, A.C., 2019: The Value of BECCS in IAMs: a Review. *Curr. Sustain. Energy Reports*, **6(4)**, 107–115, doi:10.1007/s40518-019-00142-3.

Köhler, P., J. Hartmann, and D.A. Wolf-Gladrow, 2010: Geoengineering potential of artificially enhanced silicate weathering of olivine. *Proc. Natl. Acad. Sci.*, **107(47)**, 20228–20233, doi:10.1073/pnas.1000545107.

Köhler, P., J.F. Abrams, C. Volker, J. Hauck, and D.A. Wolf-Gladrow, 2013: Geoengineering impact of open ocean dissolution of olivine on atmospheric CO2, surface ocean pH and marine biology. *Environ. Res. Lett.*, **8(1)**, 014009, doi:10.1088/1748-9326/8/1/014009.

Kolokotroni, M., S.A. Tassou, and B.L. Gowreesunker, 2015: Energy aspects and ventilation of food retail buildings. *Adv. Build. Energy Res.*, **9(1)**, 1–19, doi:10.1080/17512549.2014.897252.

Kondo, M. et al., 2018: Plant Regrowth as a Driver of Recent Enhancement of Terrestrial CO2 Uptake. *Geophys. Res. Lett.*, **45(10)**, 4820–4830, doi:10.1029/2018GL077633.

Kongsager, R., 2017: Barriers to the adoption of alley cropping as a climate-smart agriculture practice: Lessons from maize cultivation among the Maya in southern Belize. *Forests*, **8(7)**, 260, doi:10.3390/f8070260.

Kongsager, R., 2018: Linking Climate Change Adaptation and Mitigation: A Review with Evidence from the Land-Use Sectors. *Land*, **7**, 158, doi:10.3390/land7040158.

Kopáček, M., 2021: Land-Use Planning and the Public: Is There an Optimal Degree of Civic Participation? *L. 2021*, **10(1)**, 90, doi:10.3390/LAND10010090.

Kotlikoff, L.J., A. Polbin, and A. Zubarev, 2016: *Will the Paris Accord Accelerate Climate Change?* National Bureau of Economic Research, Cambridge, MA, USA, 44 pp. https://www.nber.org/system/files/working_papers/w22731/w22731.pdf.

Kovic, Y., J.K. Noel, J.A. Ungemack, and J.A. Burleson, 2018: The impact of junk food marketing regulations on food sales: an ecological study. *Obes. Rev.*, **19(6)**, 761–769, doi:10.1111/obr.12678.

Koweek, D.A., D.A. Mucciarone, and R.B. Dunbar, 2016: Bubble Stripping as a Tool to Reduce High Dissolved CO2 in Coastal Marine Ecosystems. *Environ. Sci. Technol.*, **50(7)**, 3790–3797, doi:10.1021/ACS.EST.5B04733/SUPPL_FILE/ES5B04733_SI_001.PDF.

Kraak, V.I., T. Englund, S. Misyak, and E.L. Serrano, 2017: A novel marketing mix and choice architecture framework to nudge restaurant customers toward healthy food environments to reduce obesity in the United States. *Obes. Rev.*, **18(8)**, 852–868, doi:10.1111/obr.12553.

Krause-Jensen, D. and C.M. Duarte, 2016: Substantial role of macroalgae in marine carbon sequestration. *Nat. Geosci.*, **9(10)**, 737–742, doi:10.1038/ngeo2790.

Krause-Jensen, D. et al., 2018: Sequestration of macroalgal carbon: the elephant in the Blue Carbon room. *Biol. Lett.*, **14(6)**, 20180236, doi:10.1098/rsbl.2018.0236.

Kraxner, F. et al., 2014: BECCS in South Korea—Analyzing the negative emissions potential of bioenergy as a mitigation tool. *Renew. Energy*, **61**, 102–108, doi:10.1016/J.RENENE.2012.09.064.

Kreig, J.A.F., E. Parish, and H.I. Jager, 2021: Growing grasses in unprofitable areas of US Midwest croplands could increase species richness. *Biol. Conserv.*, **261**, 109289, doi:10.1016/J.BIOCON.2021.109289.

Krey, V. et al., 2019: Looking under the hood: A comparison of techno-economic assumptions across national and global integrated assessment models. *Energy*, **172**, 1254–1267, doi:10.1016/j.energy.2018.12.131.

Kriewald, S., P. Pradhan, L. Costa, A.G.C. Ros, and J.P. Kropp, 2019: Hungry cities: how local food self-sufficiency relates to climate change, diets, and urbanisation. *Environ. Res. Lett.*, **14(9)**, 094007, doi:10.1088/1748-9326/ab2d56.

Kugelberg, S. et al., 2021: Implications of a food system approach for policy agenda-setting design. *Glob. Food Sec.*, **28**(Forthcoming), 100451, doi:10.1016/j.gfs.2020.100451.

Kumar, P. et al., 2017: Meat analogues: Health promising sustainable meat substitutes. *Crit. Rev. Food Sci. Nutr.*, **57(5)**, 923–932, doi:10.1080/10408398.2014.939739.

Kummu, M. et al., 2020: Interplay of trade and food system resilience: Gains on supply diversity over time at the cost of trade independency. *Glob. Food Sec.*, **24** (February), 100360, doi:10.1016/j.gfs.2020.100360.

Kuronuma, T. et al., 2018: CO2 Payoff of extensive green roofs with different vegetation species. *Sustainability*, **10(7)**, 1–12, doi:10.3390/su10072256.

Kuwae, T., and M. Hori, 2019: The Future of Blue Carbon: Addressing Global Environmental Issues. In: *Blue Carbon in Shallow Coastal Ecosystems* [Kuwae, T. and M. Hori, (eds.)], Springer Singapore, Singapore, pp. 347–373.

Kwiatkowski, L., K.L. Ricke, and K. Caldeira, 2015: Atmospheric consequences of disruption of the ocean thermocline. *Environ. Res. Lett.*, **10(3)**, 034016, doi:10.1088/1748-9326/10/3/034016.

La Plante, E.C. et al., 2021: Saline Water-Based Mineralization Pathway for Gigatonne-Scale CO2 Management. *ACS Sustain. Chem. Eng.*, **9(3)**, 1073–1089, doi:10.1021/acssuschemeng.0c08561.

Lackner, K.S. et al., 2012: The urgency of the development of CO 2 capture from ambient air. *Proc. Natl. Acad. Sci.*, **109(33)**, 13156–13162, doi:10.1073/pnas.1108765109.

Lade, S.J. et al., 2020: Human impacts on planetary boundaries amplified by Earth system interactions. *Nat. Sustain.*, **3**, 119–128, doi:10.1038/s41893-019-0454-4.

Lampitt, R. et al., 2008: Ocean fertilization: a potential means of geoengineering? *Philos. Trans. R. Soc. A Math. Phys. Eng. Sci.*, **366(1882)**, 3919–3945, doi:10.1098/RSTA.2008.0139.

Lange, K. et al., 2018: Basin-scale effects of small hydropower on biodiversity dynamics. *Front. Ecol. Environ.*, **16(7)**, 397–404, doi:10.1002/fee.1823.

Larkin, A., J. Kuriakose, M. Sharmina, and K. Anderson, 2018: What if negative emission technologies fail at scale? Implications of the Paris Agreement for big emitting nations. *Clim. Policy*, **18(6)**, 690–714, doi:10.1080/14693062.2017.1346498.

Larsen, J., W. Herndon, M. Grant, and P. Marster, 2019: *Capturing Leadership - Policies for the US to Advance Direct Air Capture Technology*. Rhodium Group, New York, USA, 68 pp.

Larson, A.M. et al., 2018: Gender lessons for climate initiatives: A comparative study of REDD+ impacts on subjective wellbeing. *World Dev.*, **108**, 86–102, doi.org/10.1016/j.worlddev.2018.02.027.

12

Latka, C. et al., 2021: Paying the price for environmentally sustainable and healthy EU diets. *Glob. Food Sec.*, **28**, 100437, doi:10.1016/j.gfs.2020.100437.

Latu, C. et al., 2018: Barriers and Facilitators to Food Policy Development in Fiji. *Food Nutr. Bull.*, **39(4)**, 621–631, doi:10.1177/0379572118797083.

Lauri, P. et al., 2019: Global Woody Biomass Harvest Volumes and Forest Area Use Under Different SSP-RCP Scenarios. *J. For. Econ.*, **34(3–4)**, 285–309, doi:10.1561/112.00000504.

Law, B.S., M. Chidel, T. Brassil, G. Turner, and A. Kathuria, 2014: Trends in bird diversity over 12years in response to large-scale eucalypt plantation establishment: Implications for extensive carbon plantings. *For. Ecol. Manage.*, **322**, 58–68, doi:10.1016/j.foreco.2014.02.032.

Law, C.S., 2008: Predicting and monitoring the effects of large-scale ocean iron fertilization on marine trace gas emissions. *Mar. Ecol. Prog. Ser.*, **364**, 283–288, doi:10.3354/meps07549.

Lawrence, M.G. et al., 2018: Evaluating climate geoengineering proposals in the context of the Paris Agreement temperature goals. *Nat. Commun.*, **9(1)**, 3734, doi:10.1038/s41467-018-05938-3.

Leach, A.M. et al., 2016: Environmental impact food labels combining carbon, nitrogen, and water footprints. *Food Policy*, **61**, 213–223, doi:10.1016/j.foodpol.2016.03.006.

Lee, H. et al., 2019: Implementing land-based mitigation to achieve the Paris Agreement in Europe requires food system transformation. *Environ. Res. Lett.*, **14(10)**, doi:10.1088/1748-9326/ab3744.

Lee, J., M. Bazilian, B. Sovacool, and S. Greene, 2020a: Responsible or reckless? A critical review of the environmental and climate assessments of mineral supply chains. *Environ. Res. Lett.*, **15(10)**, 103009, doi:10.1088/1748-9326/ab9f8c.

Lee, K., C. Fyson, and C.-F. Schleussner, 2021: Fair distributions of carbon dioxide removal obligations and implications for effective national net-zero targets. *Environ. Res. Lett.*, **16(9)**, 094001, doi:10.1088/1748-9326/ac1970.

Lee, N. et al., 2020b: Hybrid floating solar photovoltaics-hydropower systems: Benefits and global assessment of technical potential. *Renew. Energy*, **162**, 1415–1427.

Lee, R.P., F. Keller, and B. Meyer, 2017: A concept to support the transformation from a linear to circular carbon economy: net zero emissions, resource efficiency and conservation through a coupling of the energy, chemical and waste management sectors. *Clean Energy*, **1(1)**, 102–113, doi:10.1093/ce/zkx004.

Leeson, D. et al., 2017: A Techno-economic analysis and systematic review of carbon capture and storage (CCS) applied to the iron and steel, cement, oil refining and pulp and paper industries, as well as other high purity sources. *Int. J. Greenh. Gas Control*, **61**(June), 71–84, doi:10.1016/j.ijggc.2017.03.020.

Leger, D. et al., 2021: Photovoltaic-driven microbial protein production can use land and sunlight more efficiently than conventional crops. *Proc. Natl. Acad. Sci.*, **118(26)**, doi:10.1073/pnas.2015025118.

Lehmann, J. et al., 2021: Biochar in climate change mitigation. *Nat. Geosci.*, **14**, 883–892, doi.org/10.1038/s41561-021-00852-8.

Leinonen, I. et al., 2019: Lysine Supply Is a Critical Factor in Achieving Sustainable Global Protein Economy. *Front. Sustain. Food Syst.*, **3** (April), 1–11, doi:10.3389/fsufs.2019.00027.

Leip, A., B.L. Bodirsky, and S. Kugelberg, 2021: The role of nitrogen in achieving sustainable food systems for healthy diets. *Glob. Food Sec.*, **28**, 100408, doi:10.1016/j.gfs.2020.100408.

Leirpoll, M.E. et al., 2021: Optimal combination of bioenergy and solar photovoltaic for renewable energy production on abandoned cropland. *Renew. Energy*, **168**, 45–56, doi.org/10.1016/j.renene.2020.11.159.

Lemma, Y., D. Kitaw, and G. Gatew, 2014: Loss in Perishable Food Supply Chain: An Optimization Approach Literature Review. *Int. J. Sci. Eng. Res.*, **5(5)**, 302–311.

Lenton, T.M., 2014: The Global Potential for Carbon Dioxide Removal. In: *Geoengineering of the Climate System* [Harrison, R.M. and R.E. Hester, (eds.)], The Royal Society of Chemistry, Cambridge, UK, pp. 52–79.

Lenzi, D., 2018: The ethics of negative emissions. *Glob. Sustain.*, **1**, e7, doi:10.1017/sus.2018.5.

Lenzi, D., W.F. Lamb, J. Hilaire, M. Kowarsch, and J.C. Minx, 2018: Don't deploy negative emissions technologies without ethical analysis. *Nature*, **561(7723)**, 303–305, doi:10.1038/d41586-018-06695-5.

Leskinen, P. et al., 2018: *Substitution effects of wood-based products in climate change mitigation*. European Forest Institute, 28 pp.

Lever, J., R. Sonnino, and F. Cheetham, 2019: Reconfiguring local food governance in an age of austerity: towards a place-based approach? *J. Rural Stud.*, **69**, 97–105, doi.org/10.1016/j.jrurstud.2019.04.009.

Lezaun, J., P. Healey, T. Kruger, and S.M. Smith, 2021: Governing Carbon Dioxide Removal in the UK: Lessons Learned and Challenges Ahead. *Front. Clim.*, **3**, 89, doi:10.3389/fclim.2021.673859.

Li, Y. et al., 2018: Climate model shows large-scale wind and solar farms in the Sahara increase rain and vegetation. *Science*, **361(6406)**, 1019–1022, doi:10.1126/science.aar5629.

Li, Y.L., B. Chen, and G.Q. Chen, 2020: Carbon network embodied in international trade: Global structural evolution and its policy implications. *Energy Policy*, **139**, 111316, doi.org/10.1016/j.enpol.2020.111316.

Liao, X., J.W. Hall, and N. Eyre, 2016: Water use in China's thermoelectric power sector. *Glob. Environ. Change*, **41**, 142–152, doi:10.1016/J.GLOENVCHA.2016.09.007.

Lin, B.B., S. Macfadyen, A.R. Renwick, S.A. Cunningham, and N.A. Schellhorn, 2013: Maximizing the Environmental Benefits of Carbon Farming through Ecosystem Service Delivery. *Bioscience*, **63(10)**, 793–803, doi:10.1525/bio.2013.63.10.6.

Lindh, H., H. Williams, A. Olsson, and F. Wikström, 2016: Elucidating the Indirect Contributions of Packaging to Sustainable Development: A Terminology of Packaging Functions and Features. *Packag. Technol. Sci.*, **29(4–5)**, 225–246, doi:10.1002/pts.2197.

Lipiäinen, S., and E. Vakkilainen, 2021: Role of the Finnish forest industry in mitigating global change: energy use and greenhouse gas emissions towards 2035. *Mitig. Adapt. Strateg. Glob. Change*, **26(2)**, 1–19, doi:10.1007/S11027-021-09946-5/FIGURES/3.

Lipper, L. et al., 2014: Climate-smart agriculture for food security. *Nat. Clim. Change*, **4(12)**, 1068–1072, doi:10.1038/nclimate2437.

Liu, C. et al., 2016: Food waste in Japan: Trends, current practices and key challenges. *J. Clean. Prod.*, **133(2016)**, 557–564, doi:10.1016/j.jclepro.2016.06.026.

Liu, H., and X. Fan, 2017: Value-added-based accounting of CO2 emissions: A multi-regional input-output approach. *Sustainability*, **9(12)**, 2220, doi:10.3390/su9122220.

Liu, J., and C. Zhong, 2019: An economic evaluation of the coordination between electric vehicle storage and distributed renewable energy. *Energy*, **186**, 115821, doi:10.1016/J.ENERGY.2019.07.151.

Liu, T. et al., 2017: Bioenergy production on marginal land in Canada: Potential, economic feasibility, and greenhouse gas emissions impacts. *Appl. Energy*, **205**, 477–485, doi:10.1016/J.APENERGY.2017.07.126.

Liu, Y., Y. Xu, F. Zhang, J. Yun, and Z. Shen, 2014: The impact of biofuel plantation on biodiversity: a review. *Chinese Sci. Bull.*, **59(34)**, 4639–4651, doi:10.1007/s11434-014-0639-1.

Lock, P. and L. Whittle, 2018: *Future opportunities for using forest and sawmill residues in Australia*. Australian Bureau of Agricultural and Resource Economics and Sciences, Canberra, Australia, https://www.awe.gov.au/abares/research-topics/forests/forest-economics/forest-economic-research/forest-sawmill-residues-report (Accessed November 9, 2021).

Lohbeck, M. et al., 2020: Drivers of farmer-managed natural regeneration in the Sahel. Lessons for restoration. *Sci. Rep.*, **10**, 15038, doi:10.1038/s41598-020-70746-z.

Lomax, G., M. Workman, T. Lenton, and N. Shah, 2015: Reframing the policy approach to greenhouse gas removal technologies. *Energy Policy*, **78**, 125–136, doi:10.1016/j.enpol.2014.10.002.

Longva, Y. et al., 2017: *The potential effects of land-based mitigation on the climate system and the wider environment: A synthesis of current knowledge in support of policy*. LUC4C, Edinburgh, UK, 73 pp.

Loon, M.P. et al., 2019: Impacts of intensifying or expanding cereal cropping in sub-Saharan Africa on greenhouse gas emissions and food security. *Glob. Chang. Biol.*, **25(11)**, 3720–3730, doi:10.1111/gcb.14783.

Love, D.C., M.S. Uhl, and L. Genello, 2015: Energy and water use of a small-scale raft aquaponics system in Baltimore, Maryland, United States. *Aquac. Eng.*, **68**, 19–27, doi:10.1016/j.aquaeng.2015.07.003.

Lu, C., T. Zhao, X. Shi, and S. Cao, 2018: Ecological restoration by afforestation may increase groundwater depth and create potentially large ecological and water opportunity costs in arid and semiarid China. *J. Clean. Prod.*, **176**, 1213–1222, doi:10.1016/j.jclepro.2016.03.046.

Lübeck, M. and P.S. Lübeck, 2019: Application of lactic acid bacteria in green biorefineries. *FEMS Microbiol. Lett.*, **366(3)**, 1–8, doi:10.1093/femsle/fnz024.

Luderer, G. et al., 2019: Environmental co-benefits and adverse side-effects of alternative power sector decarbonization strategies. *Nat. Commun.*, **10(1)**, doi:10.1038/s41467-019-13067-8.

Luisetti, T. et al., 2020: Climate action requires new accounting guidance and governance frameworks to manage carbon in shelf seas. *Nat. Commun.*, **11(1)**, 4599, doi:10.1038/s41467-020-18242-w.

Luke, 2021: Food facts: Finnish food production methods in international comparison, https://projects.luke.fi/ruokafakta/en/frontpage/.

Lutsey, N., 2015: *Global climate change mitigation potential from a transition to electric vehicles.* Working paper 2015-5, International Council on Clean Transportation.

Macdiarmid, J.I., F. Douglas, and J. Campbell, 2016: Eating like there's no tomorrow: Public awareness of the environmental impact of food and reluctance to eat less meat as part of a sustainable diet. *Appetite*, **96**, 487–493, doi:10.1016/j.appet.2015.10.011.

Mace, M.J., C.L. Fyson, M. Schaeffer, and W.L. Hare, 2021: Large-Scale Carbon Dioxide Removal to Meet the 1.5°C Limit: Key Governance Gaps, Challenges and Priority Responses. *Glob. Policy*, **12**(S1), 67–81, doi:10.1111/1758-5899.12921.

MacKay, D.J.C., 2013: Solar energy in the context of energy use, energy transportation and energy storage. *Philos. Trans. R. Soc. A Math. Phys. Eng. Sci.*, **371(1996)**, doi:10.1098/RSTA.2011.0431.

Mackey, B. et al., 2020: Understanding the importance of primary tropical forest protection as a mitigation strategy. *Mitig. Adapt. Strateg. Glob. Chang. 2020 255*, **25(5)**, 763–787, doi:10.1007/S11027-019-09891-4.

Mackler, S., X. Fishman, and D. Broberg, 2021: A policy agenda for gigaton-scale carbon management. *Electr. J.*, **34(7)**, 106999, doi:10.1016/j.tej.2021.106999.

Macreadie, P.I. et al., 2021: Blue carbon as a natural climate solution. *Nat. Rev. Earth Environ.*, **2**, 826–839, doi:10.1038/s43017-021-00224-1.

Madeddu, S. et al., 2020: The $CO_2$ reduction potential for the European industry via direct electrification of heat supply (power-to-heat). *Environ. Res. Lett.*, **15(12)**, 124004, doi:10.1088/1748-9326/ABBD02.

Madhu, K., S. Pauliuk, S. Dhathri, and F. Creutzig, 2021: Understanding environmental trade-offs and resource demand of direct air capture technologies through comparative life-cycle assessment. *Nat. Energy*, **6**, 1035–1044, doi:10.1038/s41560-021-00922-6.

Makarov, I., H. Chen, and S. Paltsev, 2020: Impacts of climate change policies worldwide on the Russian economy. *Clim. Policy*, **20(10)**, 1242–1256, doi:10.1080/14693062.2020.1781047.

Makkonen, M., S. Huttunen, E. Primmer, A. Repo, and M. Hildén, 2015: Policy coherence in climate change mitigation: An ecosystem service approach to forests as carbon sinks and bioenergy sources. *For. Policy Econ.*, **50**, 153–162, doi:10.1016/j.forpol.2014.09.003.

Manning, D.A.C., 2008: Phosphate Minerals, Environmental Pollution and Sustainable Agriculture. *Elements*, **4(2)**, 105–108, doi:10.2113/GSELEMENTS.4.2.105.

Manning, D.A.C., 2010: Mineral sources of potassium for plant nutrition. A review. *Agron. Sustain. Dev.*, **30(2)**, 281–294, doi:10.1051/agro/2009023.

Mansuy, N. et al., 2018: Salvage harvesting for bioenergy in Canada: From sustainable and integrated supply chain to climate change mitigation. *Wiley Interdiscip. Rev. Energy Environ.*, **7(5)**, e298, doi:10.1002/WENE.298.

Marcano-Olivier, M.I., P.J. Horne, S. Viktor, and M. Erjavec, 2020: Using Nudges to Promote Healthy Food Choices in the School Dining Room: A Systematic Review of Previous Investigations. *J. Sch. Health*, **90(2)**, 143–157, doi:10.1111/josh.12861.

Marcar, N., 2016: Prospects for Managing Salinity in Southern Australia Using Trees on Farmland. In: *Agroforestry for the Management of Waterlogged Saline Soils and Poor-Quality Waters* [Dagar, J. and P. Minhas, (eds.)], Springer, New Delhi, India, pp. 49–71.

Marcucci, A., S. Kypreos, and E. Panos, 2017: The road to achieving the long-term Paris targets: energy transition and the role of direct air capture. *Clim. Change*, **144(2)**, 181–193, doi:10.1007/s10584-017-2051-8.

Markkanen, S. and A. Anger-Kraavi, 2019: Social impacts of climate change mitigation policies and their implications for inequality. *Clim. Policy*, **19(7)**, 827–844, doi:10.1080/14693062.2019.1596873.

Markusson, N., D. McLaren, and D. Tyfield, 2018: Towards a cultural political economy of mitigation deterrence by negative emissions technologies (NETs). *Glob. Sustain.*, **1**, e10, doi:10.1017/sus.2018.10.

Marques, A.T. et al., 2020: Wind turbines cause functional habitat loss for migratory soaring birds. *J. Anim. Ecol.*, **89(1)**, 93–103, doi.org/10.1111/1365-2656.12961.

Marrou, H., L. Dufour, and J. Wery, 2013a: How does a shelter of solar panels influence water flows in a soil-crop system? *Eur. J. Agron.*, **50**, 38–51, doi:10.1016/j.eja.2013.05.004.

Marrou, H., L. Guilioni, L. Dufour, C. Dupraz, and J. Wery, 2013b: Microclimate under agrivoltaic systems: Is crop growth rate affected in the partial shade of solar panels? *Agric. For. Meteorol.*, **177**, 117–132, doi:10.1016/j.agrformet.2013.04.012.

Martin-Roberts, E. et al., 2021: Carbon capture and storage at the end of a lost decade. *One Earth*, **4**, doi:10.1016/j.oneear.2021.10.002.

Martínez-García, A. et al., 2014: Iron fertilization of the subantarctic ocean during the last ice age. *Science*, **343(6177)**, 1347–1350, doi:10.1126/science.1246848.

Mattick, C.S., 2018: Cellular agriculture: The coming revolution in food production. *Bull. At. Sci.*, **74(1)**, 32–35, doi:10.1080/00963402.2017.1413059.

Mattick, C.S., A.E. Landis, B.R. Allenby, and N.J. Genovese, 2015: Anticipatory Life Cycle Analysis of In Vitro Biomass Cultivation for Cultured Meat Production in the United States. *Environ. Sci. Technol.*, **49(19)**, 11941–11949, doi:10.1021/acs.est.5b01614.

Maucieri, C. et al., 2018: Hydroponic systems and water management in aquaponics: A review. *Ital. J. Agron.*, **13(1)**, 1–11, doi:10.4081/ija.2017.1012.

Mausch, K., A. Hall, and C. Hambloch, 2020: Colliding paradigms and trade-offs: Agri-food systems and value chain interventions. *Glob. Food Sec.*, **26**, 1–20, doi:10.1016/j.gfs.2020.100439.

May, R. et al., 2020: Paint it black: Efficacy of increased wind turbine rotor blade visibility to reduce avian fatalities. *Ecol. Evol.*, **10(16)**, 8927–8935, doi.org/10.1002/ece3.6592.

May, R., C.R. Jackson, H. Middel, B.G. Stokke, and F. Verones, 2021: Life-cycle impacts of wind energy development on bird diversity in Norway. *Environ. Impact Assess. Rev.*, **90**, 106635, doi:10.1016/J.EIAR.2021.106635.

Mayer, A. et al., 2019a: Measuring Progress towards a Circular Economy: A Monitoring Framework for Economy-wide Material Loop Closing in the EU28. *J. Ind. Ecol.*, **23(1)**, 62–76, doi:10.1111/jiec.12809.

Mayer, J., G. Bachner, and K.W. Steininger, 2019b: Macroeconomic implications of switching to process-emission-free iron and steel production in Europe. *J. Clean. Prod.*, **210**, 1517–1533, doi:10.1016/j.jclepro.2018.11.118.

Mayrhofer, J.P., and J. Gupta, 2016: The science and politics of co-benefits in climate policy. *Environ. Sci. Policy*, **57**, 22–30, doi:10.1016/j.envsci.2015.11.005.

Mazzocchi, G., and D. Marino, 2019: Does food public procurement boost food democracy? Theories and evidences from some case studies. *Econ. Agro-Alimentare*, **21(2)**, 379–404, doi:10.3280/ECAG2019-002011.

Mazzocchi, M., 2017: *Ex-post evidence on the effectiveness of policies targeted at promoting healthier diets*. Food and Agriculture Organization of the United Nations, Rome, Italy, 17 pp.

Mazzucato, M., and G. Semieniuk, 2018: Financing renewable energy: Who is financing what and why it matters. *Technol. Forecast. Soc. Change*, **127**, 8–22, doi:10.1016/j.techfore.2017.05.021.

Mbow, C., C. Rosenzweig, L.G. Barioni, T.G. Benton, M. Herrero, M. Krishnapillai, E. Liwenga, P. Pradhan, M.G. Rivera-Ferre, T. Sapkota, F.N. Tubiello, Y. Xu, 2019: Food Security. In: *Climate Change and Land: an IPCC special report on climate change, desertification, land degradation, sustainable land management, food security, and greenhouse gas fluxes in terrestrial ecosystems*. [Shukla, P.R., J. Skea, E. Calvo Buendia, V. Masson-Delmotte, H.-O. Pörtner, D.C. Roberts, P. Zhai, R. Slade, S. Connors, R. van Diemen, M. Ferrat, E. Haughey, S. Luz, S. Neogi, M. Pathak, J. Petzold, J. Portugal Pereira, P. Vyas, E. Huntley, K. Kissick, M. Belkacemi, J. Malley, (eds.)]. Cambridge University Press, Cambridge, UK, and New York, NY, USA.

McBey, D., D. Watts, and A.M. Johnstone, 2019: Nudging, formulating new products, and the lifecourse: A qualitative assessment of the viability of three methods for reducing Scottish meat consumption for health, ethical, and environmental reasons. *Appetite*, **142**(July), 104349, doi:10.1016/j.appet.2019.104349.

McDermott, J. and A.J. Wyatt, 2017: The role of pulses in sustainable and healthy food systems. *Ann. N. Y. Acad. Sci.*, **1392(1)**, 30–42, doi:10.1111/nyas.13319.

McEwan, C., 2017: Spatial processes and politics of renewable energy transition: Land, zones and frictions in South Africa. *Polit. Geogr.*, **56**, 1–12, doi.org/10.1016/j.polgeo.2016.10.001.

McKeon, N., 2015: Global food governance in an era of crisis: Lessons from the United Nations Committee on World Food Security. *Can. Food Stud. / La Rev. Can. des études sur l'alimentation*, **2(2)**, 328–334, doi:10.15353/cfs-rcea.v2i2.134.

McKinsey, 2009: *Pathways to a low-carbon economy: Version 2 of the global greenhouse gas abatement cost curve*. McKinsey & Company, 192 pp.

McKuin, B. et al., 2021: Energy and water co-benefits from covering canals with solar panels. *Nat. Sustain.*, **4(7)**, 609–617, doi:10.1038/s41893-021-00693-8.

McLaren, D., D. Tyfield, R. Willis, B. Szerszynski, and N. Markusson, 2019: Beyond "Net-Zero": A Case for Separate Targets for Emissions Reduction and Negative Emissions. *Front. Clim.*, **1**, 4, doi:10.3389/fclim.2019.00004.

McLellan, B.C. et al., 2016: Critical Minerals and Energy–Impacts and Limitations of Moving to Unconventional Resources. *Resour.*, **5(2)**, 19, doi:10.3390/RESOURCES5020019.

McLinden, M.O., J.S. Brown, R. Brignoli, A.F. Kazakov, and P.A. Domanski, 2017: Limited options for low-global-warming-potential refrigerants. *Nat. Commun.*, **8(1)**, 1–9, doi:10.1038/ncomms14476.

McQueen, N. et al., 2020: Cost Analysis of Direct Air Capture and Sequestration Coupled to Low-Carbon Thermal Energy in the United States ⍰. *Environ. Sci. Technol.*, **54**, 7542–7551, doi:10.1021/acs.est.0c00476.

McQueen, N. et al., 2021: A review of direct air capture (DAC): scaling up commercial technologies and innovating for the future. *Prog. Energy*, **3(3)**, 032001, doi:10.1088/2516-1083/abf1ce.

MDB, 2019: *Joint Report on Multilateral Development Banks' Climate Finance. Group of Multilateral Development Banks (MDBs)*. World Bank Group, Washington, DC, USA, 56 pp. http://documents.worldbank.org/curated/en/247461561449155666/Joint-Report-on-Multilateral-Development-Banks-Climate-Finance-2018.

Meadowcroft, J. and R. Steurer, 2018: Assessment practices in the policy and politics cycles: a contribution to reflexive governance for sustainable development? *J. Environ. Policy Plan.*, **20(6)**, 734–751, doi:10.1080/1523908X.2013.829750.

Meckling, J. and E. Biber, 2021: A policy roadmap for negative emissions using direct air capture. *Nat. Commun.*, **12(1)**, 2051, doi:10.1038/s41467-021-22347-1.

Meijaard, E. et al., 2020: The environmental impacts of palm oil in context. *Nat. Plants*, **6(12)**, 1418–1426, doi:10.1038/s41477-020-00813-w.

Mejia, M.A. et al., 2019: Life Cycle Assessment of the Production of a Large Variety of Meat Analogs by Three Diverse Factories. *J. Hunger Environ. Nutr.*, **0(0)**, 1–13, doi:10.1080/19320248.2019.1595251.

Mendly-Zambo, Z., L.J. Powell, and L.L. Newman, 2021: Dairy 3.0: cellular agriculture and the future of milk. *Food, Cult. Soc.*, **24(5)**, 675–693, doi:10.1080/15528014.2021.1888411.

Meng, B., G.P. Peters, Z. Wang, and M. Li, 2018: Tracing CO2 emissions in global value chains. *Energy Econ.*, **73**, 24–42, doi.org/10.1016/j.eneco.2018.05.013.

METI, 2017: *Basic Hydrogen Strategy*. Ministerial Council on Renewable Energy, Hydrogen and Related Issues of Japan, Tokyo, Japan, 37 pp. https://www.meti.go.jp/english/press/2017/1226_003.html.

Metternicht, G., 2018: *Land Use and Spatial Planning*. Springer International Publishing, Cham, Switzerland.

Meyer, M.A., 2020: The role of resilience in food system studies in low- and middle-income countries. *Glob. Food Sec.*, **24**, 100356, doi:10.1016/j.gfs.2020.100356.

Meysman, F.J.R. and F. Montserrat, 2017: Negative CO2 emissions via enhanced silicate weathering in coastal environments. *Biol. Lett.*, **13(4)**, 20160905, doi:10.1098/rsbl.2016.0905.

Michielsen, T.O., 2014: Brown backstops versus the green paradox. *J. Environ. Econ. Manage.*, **68(1)**, 87–110, doi:10.1016/j.jeem.2014.04.004.

Mie, A. et al., 2017: Human health implications of organic food and organic agriculture: a comprehensive review. *Environ. Heal.*, **16(1)**, 111, doi:10.1186/s12940-017-0315-4.

Mier y Terán Giménez Cacho, M. et al., 2018: Bringing agroecology to scale: key drivers and emblematic cases. *Agroecol. Sustain. Food Syst.*, **42(6)**, doi:10.1080/21683565.2018.1443313.

Mignacca, B., G. Locatelli, and A. Velenturf, 2020: Modularisation as enabler of circular economy in energy infrastructure. *Energy Policy*, **139**, 111371, doi:10.1016/j.enpol.2020.111371.

Milford, A.B. and C. Kildal, 2019: Meat Reduction by Force: The Case of "Meatless Monday" in the Norwegian Armed Forces. *Sustainability*, **11(10)**, 2741, doi:10.3390/su11102741.

Milford, A.B., C. Le Mouël, B.L. Bodirsky, and S. Rolinski, 2019: Drivers of meat consumption. *Appetite*, **141**(June), 104313, doi:10.1016/j.appet.2019.06.005.

Miller, C.L. et al., 2019: Are Australians ready for warning labels, marketing bans and sugary drink taxes? Two cross-sectional surveys measuring support for policy responses to sugar-sweetened beverages. *BMJ Open*, **9(6)**, 1–13, doi:10.1136/bmjopen-2018-027962.

Miller, L.A. and P.M. Orton, 2021: Achieving negative emissions through oceanic sequestration of vegetation carbon as Black Pellets. *Clim. Change*, **167(3–4)**, 29, doi:10.1007/s10584-021-03170-5.

Miller, L.M. and D.W. Keith, 2018a: Observation-based solar and wind power capacity factors and power densities. *Environ. Res. Lett.*, **13(10)**, doi:10.1088/1748-9326/aae102.

Miller, L.M. and D.W. Keith, 2018b: Climatic Impacts of Wind Power. *Joule*, **2(12)**, 2618–2632, doi:10.1016/j.joule.2018.09.009.

Millstein, D. and S. Menon, 2011: Regional climate consequences of large-scale cool roof and photovoltaic array deployment. *Environ. Res. Lett.*, **6**, 034001, doi:10.1088/1748-9326/6/3/034001.

Milne, S. et al., 2019: Learning from "actually existing" REDD+: A synthesis of ethnographic findings. *Conserv. Soc.*, **17(1)**, 84–95, doi:10.4103/cs.cs_18_13.

Milner, S. et al., 2016: Potential impacts on ecosystem services of land use transitions to second-generation bioenergy crops in GB. *GCB Bioenergy*, **8(2)**, 317–333, doi:10.1111/gcbb.12263.

Ministry of Agriculture and Forestry, 2021: Kampanjen om matens ursprung ökade uppskattningen av inhemsk mat. Finland, https://mmm.fi/sv/-/ruuan-alkuperaa-korostanut-kampanja-lisasi-kotimaisen-ruuan-arvostusta.

Minx, J.C. et al., 2018: Negative emissions - Part 1: Research landscape and synthesis. *Environ. Res. Lett.*, **13(6)**, 063001, doi:10.1088/1748-9326/aabf9b.

Miskin, C.K. et al., 2019: Sustainable co-production of food and solar power to relax land-use constraints. *Nat. Sustain.*, **2(10)**, 972–980, doi:10.1038/s41893-019-0388-x.

Mitter, H. et al., 2020: Shared Socio-economic Pathways for European agriculture and food systems: The Eur-Agri-SSPs. *Glob. Environ. Change*, **65** (December 2019), 102159, doi:10.1016/j.gloenvcha.2020.102159.

Moe, E. and J.-K.S. Røttereng, 2018: The post-carbon society: Rethinking the international governance of negative emissions. *Energy Res. Soc. Sci.*, **44**, 199–208, doi:10.1016/j.erss.2018.04.031.

Mohan, A., 2017: Whose land is it anyway? Energy futures & land use in India. *Energy Policy*, **110**, 257–262, doi:10.1016/J.ENPOL.2017.08.025.

Mohan, A., O. Geden, M. Fridahl, H.J. Buck, and G.P. Peters, 2021: UNFCCC must confront the political economy of net-negative emissions. *One Earth*, **4(10)**, 1348–1351, doi:10.1016/j.oneear.2021.10.001.

Molina-Besch, K., F. Wikström, and H. Williams, 2019: The environmental impact of packaging in food supply chains—does life cycle assessment of food provide the full picture? *Int. J. Life Cycle Assess.*, **24(1)**, 37–50, doi:10.1007/s11367-018-1500-6.

Momblanch, A. et al., 2019: Untangling the water-food-energy-environment nexus for global change adaptation in a complex Himalayan water resource system. *Sci. Total Environ.*, **655**, 35–47, doi:10.1016/j.scitotenv.2018.11.045.

Mongin, M., M.E. Baird, A. Lenton, C. Neill, and J. Akl, 2021: Reversing ocean acidification along the Great Barrier Reef using alkalinity injection. *Environ. Res. Lett.*, **16(6)**, 064068, doi:10.1088/1748-9326/ac002d.

Monier, E. et al., 2018: Toward a consistent modeling framework to assess multi-sectoral climate impacts. *Nat. Commun.*, **9(1)**, doi:10.1038/s41467-018-02984-9.

Montserrat, F. et al., 2017: Olivine Dissolution in Seawater: Implications for CO2 Sequestration through Enhanced Weathering in Coastal Environments. *Environ. Sci. Technol.*, **51(7)**, 3960–3972, doi:10.1021/acs.est.6b05942.

Moore, K.J. et al., 2019: Regenerating Agricultural Landscapes with Perennial Groundcover for Intensive Crop Production. *Agron. 2019, Vol. 9, Page 458*, **9(8)**, 458, doi:10.3390/AGRONOMY9080458.

Moosdorf, N., P. Renforth, and J. Hartmann, 2014: Carbon Dioxide Efficiency of Terrestrial Enhanced Weathering. *Environ. Sci. Technol.*, **48(9)**, 4809–4816, doi:10.1021/es4052022.

Mora, O. et al., 2020: Exploring the future of land use and food security: A new set of global scenarios. *PLoS One*, **15** (7 July), 1–29, doi:10.1371/journal.pone.0235597.

Moragues-Faus, A., 2021: The emergence of city food networks: Rescaling the impact of urban food policies. *Food Policy*, **103**, 102107, doi:10.1016/j.foodpol.2021.102107.

Moragues-Faus, A., R. Sonnino, and T. Marsden, 2017: Exploring European food system vulnerabilities: Towards integrated food security governance. *Environ. Sci. Policy*, **75**(C), 184–215, doi:10.1016/j.envsci.2017.05.015.

Moran, E.F., M.C. Lopez, N. Moore, N. Müller, and D.W. Hyndman, 2018: Sustainable hydropower in the 21st century. *Proc. Natl. Acad. Sci.*, **115(47)**, 11891 LP – 11898, doi:10.1073/pnas.1809426115.

Moreau, V., P. Dos Reis, and F. Vuille, 2019: Enough Metals? Resource Constraints to Supply a Fully Renewable Energy System. *Resources*, **8(1)**, 29, doi:10.3390/resources8010029.

Moreira, M.M.R. et al., 2020: Socio-environmental and land-use impacts of double-cropped maize ethanol in Brazil. *Nat. Sustain. 2020 33*, **3(3)**, 209–216, doi:10.1038/s41893-019-0456-2.

Morrison, T.H. et al., 2019: The black box of power in polycentric environmental governance. *Glob. Environ. Change*, **57**, 101934, doi.org/10.1016/j.gloenvcha.2019.101934.

Morrow, D.R. and M.S. Thompson, 2020: *Reduce, Remove, Recycle: Clarifying the Overlap between Carbon Removal and CCUS*. ICRLP Working Paper No. 2, Institute for Carbon Removal Law and Policy, American University, Washington, DC, USA, http://research.american.edu/carbonremoval/wp-content/uploads/sites/3/2020/12/reduce-remove-recycle_final.pdf.

Mouat, M.J. and R. Prince, 2018: Cultured meat and cowless milk: on making markets for animal-free food. *J. Cult. Econ.*, **11(4)**, 315–329, doi:10.1080/17530350.2018.1452277.

Mouratiadou, I. et al., 2020: Sustainable intensification of crop residue exploitation for bioenergy: Opportunities and challenges. *GCB Bioenergy*, **12(1)**, 71–89, doi:10.1111/gcbb.12649.

Mousseau, T.A. and A.P. Møller, 2014: Genetic and Ecological Studies of Animals in Chernobyl and Fukushima. *J. Hered.*, **105(5)**, 704–709, doi:10.1093/JHERED/ESU040.

Mousseau, T.A. and A.P. Møller, 2020: Plants in the Light of Ionizing Radiation: What Have We Learned From Chernobyl, Fukushima, and Other "Hot" Places? *Front. Plant Sci.*, **0**, 552, doi:10.3389/FPLS.2020.00552.

Mozaffarian, D., S.Y. Angell, T. Lang, and J.A. Rivera, 2018: Role of government policy in nutrition—barriers to and opportunities for healthier eating. *BMJ*, **361**, k2426, doi:10.1136/bmj.k2426.

Mu, Y., S. Evans, C. Wang, and W. Cai, 2018: How will sectoral coverage affect the efficiency of an emissions trading system? A CGE-based case study of China. *Appl. Energy*, **227**, 403–414, doi:10.1016/j.apenergy.2017.08.072.

Muinzer, T., 2019: *Climate and Energy Governance for the UK Low Carbon Transition: The Climate Change Act 2008*. 1st ed. Palgrave Macmillan, Cham, Switzerland, 146 pp.

Muller, A. et al., 2017: Strategies for feeding the world more sustainably with organic agriculture. *Nat. Commun.*, **8(1290)**, doi:10.1038/s41467-017-01410-w.

Müller, P. and M. Schmid, 2019: Intelligent packaging in the food sector: A brief overview. *Foods*, **8(1)**, doi:10.3390/foods8010016.

Mulvaney, D., 2017: Identifying the roots of Green Civil War over utility-scale solar energy projects on public lands across the American Southwest. *J. Land Use Sci.*, **12**, 493–515, doi:10.1080/1747423X.2017.1379566.

Muratori, M. et al., 2020: EMF-33 insights on bioenergy with carbon capture and storage (BECCS). *Clim. Change*, **163(3)**, 1621–1637, doi:10.1007/s10584-020-02784-5.

Muscat, A., E.M. de Olde, I.J.M. de Boer, and R. Ripoll-Bosch, 2020: The battle for biomass: A systematic review of food-feed-fuel competition. *Glob. Food Sec.*, **25** (June), doi:10.1016/j.gfs.2019.100330.

Muscat, A., E.M. de Olde, Z. Kovacic, I.J.M. de Boer, and R. Ripoll-Bosch, 2021: Food, energy or biomaterials? Policy coherence across agro-food and bioeconomy policy domains in the EU. *Environ. Sci. Policy*, **123**, 21–30, doi:10.1016/J.ENVSCI.2021.05.001.

Myllyviita, T., S. Soimakallio, J. Judl, and J. Seppälä, 2021: Wood substitution potential in greenhouse gas emission reduction–review on current state and application of displacement factors. *For. Ecosyst. 2021 81*, **8(1)**, 1–18, doi:10.1186/S40663-021-00326-8.

Mylona, K. et al., 2018: Viewpoint: Future of food safety and nutrition - Seeking win-wins, coping with trade-offs. *Food Policy*, **74** (November 2017), 143–146, doi:10.1016/j.foodpol.2017.12.002.

N'Yeurt, A.D.R., D.P. Chynoweth, M.E. Capron, J.R. Stewart, and M.A. Hasan, 2012: Negative carbon via ocean afforestation. *Process Saf. Environ. Prot.*, **90(6)**, 467–474, doi:10.1016/J.PSEP.2012.10.008.

Nabuurs, G.-J., P.J. Verkerk, M.-J. Schelhaas, J.R. González Olabarria, A. Trasobares, and E. Cienciala, 2018: *Climate-Smart Forestry: mitigation impacts in three European regions*. European Forest Institute.


12

Nag, R. et al., 2019: Anaerobic digestion of agricultural manure and biomass – Critical indicators of risk and knowledge gaps. *Sci. Total Environ.*, **690**, 460–479, doi:10.1016/j.scitotenv.2019.06.512.

Nakhimovsky, S.S. et al., 2016: Taxes on sugar-sweetened beverages to reduce overweight and obesity in middle-income countries: A systematic review. *PLoS One*, **11(9)**, 1–22, doi:10.1371/journal.pone.0163358.

Narayan, S. et al., 2016: *Coastal Wetlands and Flood Damage Reduction: Using Risk Industry-based Models to Assess Natural Defenses in the Northeastern USA*. Lloyd's Tercentenary Research Foundation, London, UK, 23 pp.

NASAC, 2018: *Opportunities and challenges for research on food and nutrition security and agriculture in Africa*. Network of African Science Academies, Nairobi, Kenya.

NASEM, 2019: *Negative Emissions Technologies and Reliable Sequestration: A Research Agenda*. National Academy of Sciences Engineering and Medicine, Washington DC, USA, 510 pp.

Navigant, 2019: *Gas for Climate. The optimal role for gas in a net-zero emissions energy system*. Navigant Netherlands B.V., Utrecht, Netherlands, 231 pp.

Neff, R.A. et al., 2018: Reducing meat consumption in the USA: A nationally representative survey of attitudes and behaviours. *Public Health Nutr.*, **21(10)**, 1835–1844, doi:10.1017/S1368980017004190.

Negri, V. et al., 2021: Life cycle optimization of BECCS supply chains in the European Union. *Appl. Energy*, **298**, 117252, doi:10.1016/j.apenergy.2021.117252.

Nemet, G.F. et al., 2018: Negative emissions - Part 3: Innovation and upscaling. *Environ. Res. Lett.*, **13(6)**, 063003, doi:10.1088/1748-9326/aabff4.

Neto, B. and M. Gama Caldas, 2018: The use of green criteria in the public procurement of food products and catering services: a review of EU schemes. *Environ. Dev. Sustain.*, **20(5)**, 1905–1933, doi:10.1007/s10668-017-9992-y.

Neuhoff, K. et al., 2016: *Inclusion of Consumption of carbon intensive materials in emissions trading – An option for carbon pricing post-2020*. Climate Strategies, 17 pp. https://climatestrategies.org/publication/inclusion-of-consumption-of-carbon-intensive-materials-in-emissions-trading-an-option-for-carbon-pricing-post-2020/ (Accessed December 16, 2020).

Neven, D. and T. Reardon, 2004: The rise of Kenyan supermarkets and the evolution of their horticulture product procurement systems. *Dev. Policy Rev.*, **22(6)**, 669–699, doi:10.1111/j.1467-7679.2004.00271.x.

Nguyen, H.T., T.H. Pham, and L.L. de Bruyn, 2017: Impact of Hydroelectric Dam Development and Resettlement on the Natural and Social Capital of Rural Livelihoods in Bo Hon Village in Central Vietnam. *Sustain. 2017, Vol. 9, Page 1422*, **9(8)**, 1422, doi:10.3390/SU9081422.

Niebylski, M.L., K.A. Redburn, T. Duhaney, and N.R. Campbell, 2015: Healthy food subsidies and unhealthy food taxation: A systematic review of the evidence. *Nutrition*, **31(6)**, 787–795, doi:10.1016/j.nut.2014.12.010.

Nielsen, T.D., 2016: From REDD+ forests to green landscapes? Analyzing the emerging integrated landscape approach discourse in the UNFCCC. *For. Policy Econ.*, **73**, 177–184, doi:10.1016/j.forpol.2016.09.006.

Nieminen, M. et al., 2018: A synthesis of the impacts of ditch network maintenance on the quantity and quality of runoff from drained boreal peatland forests. *Ambio*, **47(5)**, 523–534, doi:10.1007/s13280-017-0966-y.

Nijdam, D., T. Rood, and H. Westhoek, 2012: The price of protein: Review of land use and carbon footprints from life cycle assessments of animal food products and their substitutes. *Food Policy*, **37(6)**, 760–770, doi:10.1016/j.foodpol.2012.08.002.

Niles, M.T. et al., 2018: Climate change mitigation beyond agriculture: a review of food system opportunities and implications. *Renew. Agric. Food Syst.*, **33(3)**, 297–308, doi:10.1017/S1742170518000029.

Ning, Y., L. Miao, T. Ding, and B. Zhang, 2019: Carbon emission spillover and feedback effects in China based on a multiregional input-output model. *Resour. Conserv. Recycl.*, **141**, 211–218, doi:10.1016/j.resconrec.2018.10.022.

Nkoa, R., 2014: Agricultural benefits and environmental risks of soil fertilization with anaerobic digestates: a review. *Agron. Sustain. Dev.*, **34(2)**, 473–492, doi:10.1007/s13593-013-0196-z.

Noonan, K., D. Miller, K. Sell, and D. Rubin, 2013: A procurement-based pathway for promoting public health: Innovative purchasing approaches for state and local government agencies. *J. Public Health Policy*, **34(4)**, 528–537, doi:10.1057/jphp.2013.30.

Núñez-Regueiro, M.M., S.F. Siddiqui, and R.J. Fletcher Jr, 2020: Effects of bioenergy on biodiversity arising from land-use change and crop type. *Conserv. Biol.*, **35(1)** 77–87, doi.org/10.1111/cobi.13452.

O'Sullivan, C.A., G.D. Bonnett, C.L. McIntyre, Z. Hochman, and A.P. Wasson, 2019: Strategies to improve the productivity, product diversity and profitability of urban agriculture. *Agric. Syst.*, **174** (December 2018), 133–144, doi:10.1016/j.agsy.2019.05.007.

Obergassel, W., F. Mersmann, and H. Wang-Helmreich, 2017: Two for one: Integrating the sustainable development agenda with international climate policy. *Gaia*, **26(3)**, 249–253, doi:10.14512/gaia.26.3.8.

Obersteiner, M. et al., 2016: Assessing the land resource–food price nexus of the Sustainable Development Goals. *Sci. Adv.*, **2(9)**, doi:10.1126/sciadv.1501499.

Obersteiner, M. et al., 2018: How to spend a dwindling greenhouse gas budget. *Nat. Clim. Change*, **8(1)**, 7–10, doi:10.1038/s41558-017-0045-1.

Obydenkova, S.V, P.D. Kouris, D.M.J. Smeulders, M.D. Boot, and Y. van der Meer, 2021: Modeling life-cycle inventory for multi-product biorefinery: tracking environmental burdens and evaluation of uncertainty caused by allocation procedure. *Biofuels, Bioprod. Biorefining*, **15(5)**, 1281–1300, doi:10.1002/BBB.2214.

Ocko, I.B. and S.P. Hamburg, 2019: Climate Impacts of Hydropower: Enormous Differences among Facilities and over Time. *Environ. Sci. Technol.*, **53(23)**, 14070–14082, doi:10.1021/acs.est.9b05083.

OECD, 2021: *Making Better Policies for Food Systems*. OECD, Paris, France.

Olgun, N. et al., 2013: Geochemical evidence of oceanic iron fertilization by the Kasatochi volcanic eruption in 2008 and the potential impacts on Pacific sockeye salmon. *Mar. Ecol. Prog. Ser.*, **488**, 81–88, doi:10.3354/meps10403.

Olivetti, E.A., G. Ceder, G.G. Gaustad, and X. Fu, 2017: Lithium-Ion Battery Supply Chain Considerations: Analysis of Potential Bottlenecks in Critical Metals. *Joule*, **1(2)**, 229–243, doi:10.1016/J.JOULE.2017.08.019.

Olsson, L., H. Barbosa, S. Bhadwal, A. Cowie, K. Delusca, D. Flores-Renteria, K. Hermans, E. Jobbagy, W. Kurz, D. Li, D.J. Sonwa, L. Stringer, 2019: Land Degradation. In: *Climate Change and Land: An IPCC Special Report on climate change, desertification, land degradation, sustainable land management, food security, and greenhouse gas fluxes in terrestrial ecosystems.* [Shukla, P.R., J. Skea, E. Calvo Buendia, V. Masson-Delmotte, H.-O. Pörtner, D. C. Roberts, P. Zhai, R. Slade, S. Connors, R. van Diemen, M. Ferrat, E. Haughey, S. Luz, S. Neogi, M. Pathak, J. Petzold, J. Portugal Pereira, P. Vyas, E. Huntley, K. Kissick, M. Belkacemi, J. Malley, (eds.)]. Cambridge University Press, Cambridge, UK, and New York, NY, USA.

Omondi, M.O. et al., 2016: Quantification of biochar effects on soil hydrological properties using meta-analysis of literature data. *Geoderma*, **274**, 28–34, doi:10.1016/j.geoderma.2016.03.029.

Ong, S., C. Campbell, P. Denholm, R. Margolis, and G. Heath, 2013: *Land-Use Requirements for Solar Power Plants in the United States*. National Renewable Energy Laboratory, Denver, CO, USA, 47 pp. https://www.nrel.gov/docs/fy13osti/56290.pdf (Accessed November 9, 2021).

Orlov, A. and A. Aaheim, 2017: Economy-wide effects of international and Russia's climate policies. *Energy Econ.*, **68**, 466–477, doi:10.1016/j.eneco.2017.09.019.

Orr, B.J. et al., 2017: *Scientific conceptual framework for land degradation neutrality*. United Nations Convention to Combat Desertification, Bonn, Germany.

Osaka, S., R. Bellamy, and N. Castree, 2021: Framing "nature-based" solutions to climate change. *WIREs Clim. Change*, **12(5)**, e729, doi:10.1002/wcc.729.

Oschlies, A., 2009: Impact of atmospheric and terrestrial CO2 feedbacks on fertilization-induced marine carbon uptake. *Biogeosciences*, **6(8)**, 1603–1613, doi:10.5194/BG-6-1603-2009.

Oschlies, A., W. Koeve, W. Rickels, and K. Rehdanz, 2010: Side effects and accounting aspects of hypothetical large-scale Southern Ocean iron fertilization. *Biogeosciences*, **7**, 4017–4035, doi:10.5194/bg-7-4017-2010.

Oschlies, A., M. Pahlow, A. Yool, and R.J. Matear, 2010: Climate engineering by artificial ocean upwelling: Channelling the sorcerer's apprentice. *Geophys. Res. Lett.*, **37(4)**, doi:10.1029/2009GL041961.

Osorio, R.J., C.J. Barden, and I.A. Ciampitti, 2019: GIS approach to estimate windbreak crop yield effects in Kansas–Nebraska. *Agrofor. Syst.*, **93(4)**, 1567–1576, doi:10.1007/s10457-018-0270-2.

Pahl-Wostl, C., A. Bhaduri, and A. Bruns, 2018: Editorial special issue: The Nexus of water, energy and food – An environmental governance perspective. *Environ. Sci. Policy*, **90**, 161–222, doi:10.1016/j.envsci.2018.06.021.

Palahí, M. et al., 2020: *Investing in Nature as the true engine of our economy: A 10-point Action Plan for a Circular Bioeconomy of Wellbeing. Knowledge to Action 02*. European Forest Institute.

Pan, Y. et al., 2016: Research progress in artificial upwelling and its potential environmental effects. *Sci. China Earth Sci.*, **59(2)**, 236–248, doi:10.1007/s11430-015-5195-2.

Pardoe, J. et al., 2018: Climate change and the water–energy–food nexus: insights from policy and practice in Tanzania. *Clim. Policy*, **18(7)**, 863–877, doi:10.1080/14693062.2017.1386082.

Paré, D., E. Thiffault, G. Cyr, and L. Guindon, 2016: Quantifying Forest Biomass Mobilisation Potential in the Boreal and Temperate Biomes. In: *Mobilisation of Forest Bioenergy in the Boreal and Temperate* Biomes [Thiffault, E., Smith, C.T., Junginger, M., and Berndes, G. (eds.)]., 36–49, doi:10.1016/B978-0-12-804514-5.00003-2.

Park, H. and S. Yu, 2019: Policy review: Implication of tax on sugar-sweetened beverages for reducing obesity and improving heart health. *Heal. Policy Technol.*, **8(1)**, 92–95, doi:10.1016/j.hlpt.2018.12.002.

Parker, R.W.R. et al., 2018: Fuel use and greenhouse gas emissions of world fisheries. *Nat. Clim. Change*, **8(4)**, 333–337, doi:10.1038/s41558-018-0117-x.

Parodi, A. et al., 2018: The potential of future foods for sustainable and healthy diets. *Nat. Sustain.*, **1(12)**, 782–789, doi:10.1038/s41893-018-0189-7.

Parrado, R. and E. De Cian, 2014: Technology spillovers embodied in international trade: Intertemporal, regional and sectoral effects in a global CGE framework. *Energy Econ.*, **41**, 76–89, doi:10.1016/j.eneco.2013.10.016.

Parsaee, M., M. Kiani Deh Kiani, and K. Karimi, 2019: A review of biogas production from sugarcane vinasse. *Biomass and Bioenergy*, **122**, 117–125, doi.org/10.1016/j.biombioe.2019.01.034.

Patrizio, P. et al., 2018: Reducing US Coal Emissions Can Boost Employment. *Joule*, **2(12)**, 2633–2648, doi:10.1016/J.JOULE.2018.10.004.

Pattberg, P., S. Chan, L. Sanderink, and O. Widerberg, 2018: Linkages: Understanding their Role in Polycentric Governance. In: *Governing Climate Change: Polycentricity in Action?* [Jordan, A., D. Huitema, H. Van Asselt, and J. Forster, (eds.)], Cambridge University Press, Cambridge, UK, pp. 169–187.

Patterson, G.T., L.F. Thomas, L.A. Coyne, and J. Rushton, 2020: Moving health to the heart of agri-food policies; mitigating risk from our food systems. *Glob. Food Sec.*, **26** (August), 100424, doi:10.1016/j.gfs.2020.100424.

Paul, K.I. et al., 2016: Managing reforestation to sequester carbon, increase biodiversity potential and minimize loss of agricultural land. *Land use policy*, **51**, 135–149, doi:10.1016/j.landusepol.2015.10.027.

Pazienza, P., 2019: The impact of FDI in the OECD manufacturing sector on CO2 emission: Evidence and policy issues. *Environ. Impact Assess. Rev.*, **77**, 60–68, doi.org/10.1016/j.eiar.2019.04.002.

Pegg, S., 2006: Mining and poverty reduction: Transforming rhetoric into reality. *J. Clean. Prod.*, **14(3–4)**, 376–387, doi:10.1016/j.jclepro.2004.06.006.

Peñalver, R. et al., 2020: Seaweeds as a functional ingredient for a healthy diet. *Mar. Drugs*, **18(6)**, 1–27, doi:10.3390/md18060301.

Pendrill, F. et al., 2019: Agricultural and forestry trade drives large share of tropical deforestation emissions. *Glob. Environ. Change*, **56** (December 2018), 1–10, doi:10.1016/j.gloenvcha.2019.03.002.

Penne, T. and T. Goedemé, 2020: Can low-income households afford a healthy diet? Insufficient income as a driver of food insecurity in Europe. *Food Policy*, **99** (September), 101978, doi:10.1016/j.foodpol.2020.101978.

Pereira, L.M., S. Drimie, K. Maciejewski, P.B. Tonissen, and R. Biggs, 2020: Food system transformation: Integrating a political–economy and social–ecological approach to regime shifts. *Int. J. Environ. Res. Public Health*, **17(4)**, doi:10.3390/ijerph17041313.

Pérez-Marin, A.M. et al., 2017: Agroecological and social transformations for coexistence with semi-aridity in Brazil. *Sustainability*, **9(6)**, doi:10.3390/su9060990.

Petersen, T., M. Hartmann, and S. Hirsch, 2021: Which meat (substitute) to buy? Is front of package information reliable to identify the healthier and more natural choice? *Food Qual. Prefer.*, **94** (November 2020), 104298, doi:10.1016/j.foodqual.2021.104298.

Pettersson, M., L. Björnsson, and P. Börjesson, 2020: Recycling of ash from co-incineration of waste wood and forest fuels: An overlooked challenge in a circular bioenergy system. *Biomass and Bioenergy*, **142**, 105713, doi:10.1016/J.BIOMBIOE.2020.105713.

Pikaar, I. et al., 2018: Carbon emission avoidance and capture by producing in-reactor microbial biomass based food, feed and slow release fertilizer: Potentials and limitations. *Sci. Total Environ.*, **644**, 1525–1530, doi:10.1016/j.scitotenv.2018.07.089.

Pirard, R. et al., 2020: Corporate ownership and dominance of Indonesia's palm oil supply chains. INFOBRIEF 9, Trase, (January), 7 pp.

Pisanello, D. and L. Ferraris, 2018: Ban on Designating Plant Products as Dairy: Between Market Regulation and Over-Protection of the Consumer. *Eur. J. Risk Regul.*, **9(1)**, 170–176, doi:10.1017/err.2018.4.

Poff, N.L.R. and J.C. Schmidt, 2016: How dams can go with the flow. *Science*, **353(6304)**, 1099–1100, doi:10.1126/science.aah4926.

Pollitt, H., K. Neuhoff, and X. Lin, 2020: The impact of implementing a consumption charge on carbon-intensive materials in Europe. *Clim. Policy*, **20**(sup1), S74–S89, doi:10.1080/14693062.2019.1605969.

Ponisio, L.C. et al., 2015: Diversification practices reduce organic to conventional yield gap. *Proc. R. Soc. B Biol. Sci.*, **282**, 20141396, doi:10.1098/rspb.2014.1396.

Poore, J. and T. Nemecek, 2018a: Reducing food's environmental impacts through producers and consumers. *Science*, **360(6392)**, 987–992, doi:10.1126/science.aaq0216.

Popkin, B.M. and T. Reardon, 2018: Obesity and the food system transformation in Latin America. *Obes. Rev.*, **19(8)**, 1028–1064, doi:10.1111/obr.12694.

Popp, A. et al., 2017: Land-use futures in the shared socio-economic pathways. *Glob. Environ. Change*, **42**, 331–345, doi:10.1016/j.gloenvcha.2016.10.002.

Pörtner, H.-O. et al., 2021: *Scientific outcome of the IPBES-IPCC co-sponsored workshop on biodiversity and climate change (Version 5)*. IPBES secretariat, Bonn, Germany.

Post, M.J., 2012: Cultured meat from stem cells: challenges and prospects. *Meat Sci.*, **92(3)**, 297–301, doi:10.1016/j.meatsci.2012.04.008.

Post, M.J. et al., 2020: Scientific, sustainability and regulatory challenges of cultured meat. *Nat. Food*, **1(7)**, 403–415, doi:10.1038/s43016-020-0112-z.

Postigo, J.C. and K.R. Young, 2021: Preparing for a diminished cryosphere. *Sustain. Sci. 2021*, **16(6)**, 1–4, doi:10.1007/S11625-021-01023-9.

Poyatos-Racionero, E., J.V. Ros-Lis, J.L. Vivancos, and R. Martínez-Máñez, 2018: Recent advances on intelligent packaging as tools to reduce food waste. *J. Clean. Prod.*, **172**, 3398–3409, doi:10.1016/j.jclepro.2017.11.075.

Pozo, C., Á. Galán-Martín, D.M. Reiner, N. Mac Dowell, and G. Guillén-Gosálbez, 2020: Equity in allocating carbon dioxide removal quotas. *Nat. Clim. Change*, **10(7)**, 640–646, doi:10.1038/s41558-020-0802-4.

12

Pradhan, P., M.K.B. Lüdeke, D.E. Reusser, and J.P. Kropp, 2014: Food Self-Sufficiency across scales: How local can we go? *Environ. Sci. Technol.*, **48(16)**, 9463–9470, doi:10.1021/es5005939.

Prudhomme, R., T. Brunelle, P. Dumas, A. Le Moing, and X. Zhang, 2020: Assessing the impact of increased legume production in Europe on global agricultural emissions. *Reg. Environ. Change*, **20(3)**, 91, doi:10.1007/s10113-020-01651-4.

Pryor, S.C., R.J. Barthelmie, and T.J. Shepherd, 2020: 20% of US electricity from wind will have limited impacts on system efficiency and regional climate. *Sci. Reports 2020 101*, **10(1)**, 1–14, doi:10.1038/s41598-019-57371-1.

Purohit, P. and L. Höglund-Isaksson, 2017: Global emissions of fluorinated greenhouse gases 2005–2050 with abatement potentials and costs. *Atmos. Chem. Phys.*, **17(4)**, 2795–2816, doi:10.5194/acp-17-2795-2017.

Qin, Z. et al., 2018: Biomass and biofuels in China: Toward bioenergy resource potentials and their impacts on the environment. *Renew. Sustain. Energy Rev.*, **82**, 2387–2400, doi:10.1016/J.RSER.2017.08.073.

Quin, P.R. et al., 2014: Oil mallee biochar improves soil structural properties-A study with x-ray micro-CT. *Agric. Ecosyst. Environ.*, **191**, 142–149, doi:10.1016/j.agee.2014.03.022.

Raheem, D. et al., 2019a: Traditional consumption of and rearing edible insects in Africa, Asia and Europe. *Crit. Rev. Food Sci. Nutr.*, **59(14)**, 2169–2188, doi:10.1080/10408398.2018.1440191.

Raheem, D. et al., 2019b: Entomophagy: Nutritional, ecological, safety and legislation aspects. *Food Res. Int.*, **126** (September), 108672, doi:10.1016/j.foodres.2019.108672.

Rahnama Mobarakeh, M., M. Santos Silva, and T. Kienberger, 2021: Pulp and Paper Industry: Decarbonisation Technology Assessment to Reach CO2 Neutral Emissions—An Austrian Case Study. *Energies*, **14(4)**, 1161, doi:10.3390/EN14041161.

Rajamani, L. et al., 2021: National 'fair shares' in reducing greenhouse gas emissions within the principled framework of international environmental law. *Clim. Policy*, **21(8)**, 983–1004, doi:10.1080/14693062.2021.1970504.

Ranjbaran, P., H. Yousefi, G.B. Gharehpetian, and F.R. Astaraei, 2019: A review on floating photovoltaic (FPV)power generation units. *Renew. Sustain. Energy Rev.*, **110**, 332–347, doi:10.1016/j.rser.2019.05.015.

Raptis, C.E., M.T.H. van Vliet, and S. Pfister, 2016: Global thermal pollution of rivers from thermoelectric power plants. *Environ. Res. Lett.*, **11(10)**, 104011, doi:10.1088/1748-9326/11/10/104011.

Rau, G.H., Willauer, H.D., Ren, Z.J., 2018: The global potential for converting renewable electricity to negative-CO2-emissions hydrogen. *Nat. Clim. Change*, **8**, 621–625.

Rau, G.H., 2008: Electrochemical splitting of calcium carbonate to increase solution alkalinity: Implications for mitigation of carbon dioxide and ocean acidity. *Environ. Sci. Technol.*, **42(23)**, 8935–8940, doi:10.1021/es800366q.

Rau, G.H., 2011: CO2 mitigation via capture and chemical conversion in seawater. *Env. Sci Technol*, **45(3)**, 1088–1092, doi:10.1021/es102671x.

Rau, G.H. and K. Caldeira, 1999: Enhanced carbonate dissolution: a means of sequestering waste CO2 as ocean bicarbonate. *Energy Convers. Manag.*, **40(17)**, 1803–1813, doi:10.1016/S0196-8904(99)00071-0.

Rau, G.H., E.L. McLeod, and O. Hoegh-Guldberg, 2012: The need for new ocean conservation strategies in a high-carbon dioxide world. *Nat. Clim. Change*, **2(10)**, 720–724, doi:10.1038/Nclimate1555.

Rau, G.H. et al., 2013: Direct electrolytic dissolution of silicate minerals for air CO2 mitigation and carbon-negative H2 production. *Proc. Natl. Acad. Sci.*, **110(25)**, 10095–10100, doi:10.1073/pnas.1222358110.

Rausch, S. and V.J. Karplus, 2014: *Markets versus Regulation: The Efficiency and Distributional Impacts of U.S. Climate Policy Proposals*. Joint Program Report Series Report 263, Massachusetts Institute of Technology, Cambridge, MA, USA, 32 pp. https://globalchange.mit.edu/publication/15897 (Accessed December 3, 2019).

Rausch, S. and J. Reilly, 2015: Carbon Taxes, Deficits, and Energy Policy Interactions. *Natl. Tax J.*, **68(1)**, 157–178.

Razzaghi, F., P.B. Obour, and E. Arthur, 2020: Does biochar improve soil water retention? A systematic review and meta-analysis. *Geoderma*, **361**, 114055, doi:10.1016/J.GEODERMA.2019.114055.

Realmonte, G. et al., 2019: An inter-model assessment of the role of direct air capture in deep mitigation pathways. *Nat. Commun.*, **10(1)**, 1–12, doi:10.1038/s41467-019-10842-5.

Reardon, T. et al., 2019: Rapid transformation of food systems in developing regions: Highlighting the role of agricultural research & innovations. *Agric. Syst.*, **172** (December 2017), 47–59, doi:10.1016/j.agsy.2018.01.022.

Reed, J., J. Van Vianen, E.L. Deakin, J. Barlow, and T. Sunderland, 2016: Integrated landscape approaches to managing social and environmental issues in the tropics: learning from the past to guide the future. *Glob. Chang. Biol.*, **22(7)**, 2540–2554, doi:10.1111/gcb.13284.

Renard, D. and D. Tilman, 2019: National food production stabilized by crop diversity. *Nature*, **571**, 257–260, doi:10.1038/s41586-019-1316-y.

Renforth, P., Jenkins, B.G., Kruger, T., 2013: Engineering challenges of ocean liming. *Energy*, **60**, 442–452, doi:10.1016/j.energy.2013.08.006.

Renforth, P. and Kruger, T., 2013: Coupling Mineral Carbonation and Ocean Liming. *Energy Fuels*, **27(8)**, 4199–4207, doi:10.1021/ef302030w.

Renforth, P., 2012: The potential of enhanced weathering in the UK. *Int. J. Greenh. Gas Control*, **10**, 229–243, doi:10.1016/j.ijggc.2012.06.011.

Renforth, P. and G. Henderson, 2017: Assessing ocean alkalinity for carbon sequestration. *Rev. Geophys.*, **55(3)**, 636–674, doi:10.1002/2016RG000533.

Renforth, P., P.A.E. Pogge von Strandmann, and G.M. Henderson, 2015: The dissolution of olivine added to soil: Implications for enhanced weathering. *Appl. Geochemistry*, **61**, 109–118, doi:10.1016/J.APGEOCHEM.2015.05.016.

Ricci, A. et al., 2017: Guidance on the requirements for the development of microbiological criteria. *EFSA J.*, **15(11)**, doi:10.2903/j.efsa.2017.5052.

Rickels, W., C. Merk, F. Reith, D.P. Keller, and A. Oschlies, 2019: (Mis)conceptions about modeling of negative emissions technologies. *Environ. Res. Lett.*, **14(10)**, 104004, doi:10.1088/1748-9326/ab3ab4.

Rickels, W., A. Proelß, O. Geden, J. Burhenne, and M. Fridahl, 2021: Integrating Carbon Dioxide Removal Into European Emissions Trading. *Front. Clim.*, **3**, 62, doi:10.3389/fclim.2021.690023.

Rinke Dias de Souza, N., T. Lopes Junqueira, and O. Cavalett, 2021: Opportunities and challenges for bioenergy-livestock integrated systems in Brazil. *Ind. Crops Prod.*, **173**, 114091, doi:10.1016/J.INDCROP.2021.114091.

Rischer, H., G.R. Szilvay, and K.M. Oksman-Caldentey, 2020: Cellular agriculture — industrial biotechnology for food and materials. *Curr. Opin. Biotechnol.*, **61**, 128–134, doi:10.1016/j.copbio.2019.12.003.

Ritchie, H., D.S. Reay, and P. Higgins, 2018: The impact of global dietary guidelines on climate change. *Glob. Environ. Change*, **49** (January), 46–55, doi:10.1016/j.gloenvcha.2018.02.005.

Roberts, D.A., N.A. Paul, S.A. Dworjanyn, M.I. Bird, and R. de Nys, 2015: Biochar from commercially cultivated seaweed for soil amelioration. *Sci. Rep.*, **5(1)**, 9665, doi:10.1038/srep09665.

Rogelj, J., D. Shindell, K. Jiang, S. Fifita, P. Forster, V. Ginzburg, C. Handa, H. Kheshgi, S. Kobayashi, E. Kriegler, L. Mundaca, R. Séférian, and M.V. Vilariño, 2018: Mitigation pathways compatible with 1.5°C in the context of sustainable development. In: *Global warming of 1.5°C. An IPCC Special Report on the impacts of global warming of 1.5°C above pre-industrial levels and related global greenhouse gas emission pathways, in the context of strengthening the global response to the threat of climate change, sustainable development, and efforts to eradicate poverty* [Masson-Delmotte, V., P. Zhai, H.-O. Pörtner, D. Roberts, J. Skea, P.R. Shukla, A. Pirani, W. Moufouma-Okia, C. Péan, R. Pidcock, S. Connors, J.B.R. Matthews, Y. Chen, X. Zhou, M.I. Gomis, E. Lonnoy, T. Maycock, M. Tignor, and T. Waterfield (eds.)]. Cambridge University Press, Cambridge, UK, and New York, NY, USA.

Rogelj, J., O. Geden, A. Cowie, and A. Reisinger, 2021: Three ways to improve net-zero emissions targets. *Nature*, **591**, 365–368, doi:10.1038/d41586-021-00662-3.


12

Röös, E., M. Patel, J. Spångberg, G. Carlsson, and L. Rydhmer, 2016: Limiting livestock production to pasture and by-products in a search for sustainable diets. *Food Policy*, **58**, 1–13, doi:10.1016/j.foodpol.2015.10.008.

Rose, D., 2018: Environmental nudges to reduce meat demand. *Lancet Planet. Heal.*, **2(9)**, e374–e375, doi:10.1016/S2542-5196(18)30185-2.

Rose, S.K. et al., 2020: An overview of the Energy Modeling Forum 33rd study: assessing large-scale global bioenergy deployment for managing climate change. *Clim. Change*, **163(3)**, 1539–1551, doi:10.1007/s10584-020-02945-6.

Rosenzweig, C. et al., 2020a: Food system approach offers new opportunities for climate change responses. *Nat. Clim. Change*, 1, 94–97.

Rosenzweig, C. et al., 2020b: Climate change responses benefit from a global food system approach. *Nat. Food*, **1(2)**, 94–97, doi:10.1038/s43016-020-0031-z.

Royal Society, and Royal Academy of Engineering, 2018: *Greenhouse Gas Removal*. Royal Society, London, UK, 134 pp.

Rubio, N.R., K.D. Fish, B.A. Trimmer, and D.L. Kaplan, 2019: In Vitro Insect Muscle for Tissue Engineering Applications. *ACS Biomater. Sci. Eng.*, **5(2)**, 1071–1082, doi:10.1021/acsbiomaterials.8b01261.

Rueda, O., J.M. Mogollón, A. Tukker, and L. Scherer, 2021: Negative-emissions technology portfolios to meet the 1.5°C target. *Glob. Environ. Change*, **67**, 102238, doi:10.1016/j.gloenvcha.2021.102238.

Rufí-Salís, M., M.J. Calvo, A. Petit-Boix, G. Villalba, and X. Gabarrell, 2020: Exploring nutrient recovery from hydroponics in urban agriculture: An environmental assessment. *Resour. Conserv. Recycl.*, **155** (November 2019), 104683, doi:10.1016/j.resconrec.2020.104683.

Runting, R.K. et al., 2019: Larger gains from improved management over sparing–sharing for tropical forests. *Nat. Sustain.*, **2(1)**, 53–61, doi:10.1038/s41893-018-0203-0.

Russell, A.E. and B.M. Kumar, 2017: Forestry for a Low-Carbon Future: Integrating Forests and Wood Products Into Climate Change Strategies. *Environ. Sci. Policy Sustain. Dev.*, **59(2)**, 16–23, doi:10.1080/00139157.2017.1274580.

Ryaboshapko, A.G. and A.P. Revokatova, 2015: A potential role of the negative emission of carbon dioxide in solving the climate problem. *Russ. Meteorol. Hydrol.*, **40(7)**, 443–455, doi:10.3103/S106837391507002X.

Ryan, D., 2015: From commitment to action: a literature review on climate policy implementation at city level. *Clim. Change*, **131(4)**, 519–529, doi:10.1007/s10584-015-1402-6.

SAPEA, Science Advice for Policy by European Academies. (2020). A sustainable food system for the European Union. Berlin: SAPEA. https://doi.org/10.26356/sustainablefood

Salim, H.K., R.A. Stewart, O. Sahin, and M. Dudley, 2019: Drivers, barriers and enablers to end-of-life management of solar photovoltaic and battery energy storage systems: A systematic literature review. *J. Clean. Prod.*, **211**, 537–554, doi:10.1016/j.jclepro.2018.11.229.

Säll, S., 2018: Environmental food taxes and inequalities: Simulation of a meat tax in Sweden. *Food Policy*, **74** (June 2017), 147–153, doi:10.1016/j.foodpol.2017.12.007.

Samant, S.S. and H.S. Seo, 2016: Effects of label understanding level on consumers' visual attention toward sustainability and process-related label claims found on chicken meat products. *Food Qual. Prefer.*, **50**, 48–56, doi:10.1016/j.foodqual.2016.01.002.

Sanchez, D.L., N. Johnson, S.T. McCoy, P.A. Turner, and K.J. Mach, 2018: Near-term deployment of carbon capture and sequestration from biorefineries in the United States. *Proc. Natl. Acad. Sci. U. S. A.*, **115(19)**, 4875–4880, doi:10.1073/pnas.1719695115.

Sandalow, D., J. Friedmann, C. McCormick, and S. McCoy, 2018: *Direct Air Capture of Carbon Dioxide: ICEF Roadmap 2018*. Innovation for Cool Earth Forum, 43 pp.

Sandström, V. et al., 2018: The role of trade in the greenhouse gas footprints of EU diets. *Glob. Food Sec.*, **19** (May), 48–55, doi:10.1016/j.gfs.2018.08.007.

Santo, R. and A. Moragues-Faus, 2019: Towards a trans-local food governance: Exploring the transformative capacity of food policy assemblages in the US and UK. *Geoforum*, **98**, 75–87, doi.org/10.1016/j.geoforum.2018.10.002.

Santo, R.E. et al., 2020: Considering Plant-Based Meat Substitutes and Cell-Based Meats: A Public Health and Food Systems Perspective. *Front. Sustain. Food Syst.*, **4**(August), 1–23, doi:10.3389/fsufs.2020.00134.

Sanz-Lazaro, C. and P. Sanchez-Jerez, 2020: Regional Integrated Multi-Trophic Aquaculture (RIMTA): Spatially separated, ecologically linked. *J. Environ. Manage.*, **271**(July 2019), 110921, doi:10.1016/j.jenvman.2020.110921.

Sarink, D. et al., 2016: The impact of menu energy labelling across socioeconomic groups: A systematic review. *Appetite*, **99**, 59–75, doi:10.1016/j.appet.2015.12.022.

Sayer, J.A. et al., 2017: Measuring the effectiveness of landscape approaches to conservation and development. *Sustain. Sci.*, **12(3)**, 465–476, doi:10.1007/s11625-016-0415-z.

Schade, S., G.I. Stangl, and T. Meier, 2020: Distinct microalgae species for food – Part 2: Comparative life cycle assessment of microalgae and fish for eicosapentaenoic acid (EPA), docosahexaenoic acid (DHA), and protein. *J. Appl. Phycol.*, 32, 2997–3013.

Schebesta, H. and J.J.L. Candel, 2020: Game-changing potential of the EU's Farm to Fork Strategy. *Nat. Food*, **1(10)**, 586–588, doi:10.1038/s43016-020-00166-9.

Schenuit, F. et al., 2021: Carbon Dioxide Removal Policy in the Making: Assessing Developments in 9 OECD Cases. *Front. Clim.*, **3**, 638805, doi:10.3389/fclim.2021.638805.

Scherer, L., and S. Pfister, 2016a: Hydropower's Biogenic Carbon Footprint. *PLoS One*, **11(9)**, e0161947, doi:10.1371/journal.pone.0161947.

Scherer, L. and S. Pfister, 2016b: Global water footprint assessment of hydropower. *Renew. Energy*, **99**, 711–720, doi:10.1016/j.renene.2016.07.021.

Schleicher, J. et al., 2019: Protecting half of the planet could directly affect over one billion people. *Nat. Sustain.*, **2(12)**, 1094–1096, doi:10.1038/s41893-019-0423-y.

Schmidinger, K. and E. Stehfest, 2012: Including $CO_2$ implications of land occupation in LCAs—method and example for livestock products. *Int. J. Life Cycle Assess.*, **17**, 962–972, doi:10.1007/s11367-012-0434-7.

Schmidt, H.-P. et al., 2021: Biochar in agriculture – A systematic review of 26 global meta-analyses. *GCB Bioenergy*, **13(11)**, 1708–1730, doi:10.1111/GCBB.12889.

Schmidt, L.M., L.F. Andersen, C. Dieckmann, A. Lamp, and M. Kaltschmitt, 2019: The Biorefinery Approach. In: *Energy from Organic Materials (Biomass)*, Springer New York, New York, NY, USA, pp. 1383–1412.

Schösler, H. and J. De Boer, 2018: Towards more sustainable diets: Insights from the food philosophies of "gourmets" and their relevance for policy strategies. *Appetite*, **127** (April), 59–68, doi:10.1016/j.appet.2018.04.022.

Schuiling, R.D. and P. Krijgsman, 2006: Enhanced Weathering: An Effective and Cheap Tool to Sequester $CO_2$. *Clim. Change*, **74(1)**, 349–354, doi:10.1007/s10584-005-3485-y.

Schulze, K., Ž. Malek, and P.H. Verburg, 2020: The Impact of Accounting for Future Wood Production in Global Vertebrate Biodiversity Assessments. *Environ. Manag. 2020 663*, **66(3)**, 460–475, doi:10.1007/S00267-020-01322-4.

Schulze, K., Ž. Malek, and P.H. Verburg, 2021: How will land degradation neutrality change future land system patterns? A scenario simulation study. *Environ. Sci. Policy*, **124**, 254–266, doi:10.1016/J.ENVSCI.2021.06.024.

Searchinger, T., R. Waite, C. Hanson and J. Ranganathan, 2018a: *Creating a sustainable food future: A menu of solutions to feed nearly 10 billion people by 2050 (Synthesis Report)*. World Resources Institute, Washington, DC, USA, 96 pp.

Searchinger, T.D., S. Wirsenius, T. Beringer, and P. Dumas, 2018b: Assessing the efficiency of changes in land use for mitigating climate change. *Nature*, **564(7735)**, 249–253, doi:10.1038/s41586-018-0757-z.

Seddon, J., S. Doyle, M. Bourne, R. Maccallum, and S. Briggs, 2009: Biodiversity benefits of alley farming with old man saltbush in central western New South Wales. *Anim. Prod. Sci.*, **49(10)**, 860, doi:10.1071/EA08280.

Sekera, J. and A. Lichtenberger, 2020: Assessing Carbon Capture: Public Policy, Science, and Societal Need. *Biophys. Econ. Sustain.*, **5(3)**, 14, doi:10.1007/s41247-020-00080-5.

Semba, R.D. et al., 2020: Adoption of the 'planetary health diet' has different impacts on countries' greenhouse gas emissions. *Nat. Food*, **1(8)**, 481–484, doi:10.1038/s43016-020-0128-4.

Semba, R.D., R. Ramsing, N. Rahman, K. Kraemer, and M.W. Bloem, 2021: Legumes as a sustainable source of protein in human diets. *Glob. Food Sec.*, **28** (June 2020), 100520, doi:10.1016/j.gfs.2021.100520.

Sequeira, T.N. and M.S. Santos, 2018: Renewable energy and politics: A systematic review and new evidence. *J. Clean. Prod.*, **192**, 553–568, doi.org/10.1016/j.jclepro.2018.04.190.

Sesma Martín, D. and M. del M. Rubio-Varas, 2017: Freshwater for Cooling Needs: A Long-Run Approach to the Nuclear Water Footprint in Spain. *Ecol. Econ.*, **140**, 146–156, doi:10.1016/J.ECOLECON.2017.04.032.

Seufert, V. and N. Ramakutty, 2017: Many shades of gray - The context-dependent performance of organic agriculture. *Sci. Adv.*, **3**(e1602638 NV-3).

Seymour, F. and N.L. Harris, 2019: Reducing tropical deforestation. *Science*, **365(6455)**, 756–757, doi:10.1126/science.aax8546.

SFP Foundation, Sahara Forest Project. https://www.saharaforestproject.com/.

Shahbaz, M., S. Nasreen, F. Abbas, and O. Anis, 2015: Does foreign direct investment impede environmental quality in high-, middle-, and low-income countries? *Energy Econ.*, **51**, 275–287, doi.org/10.1016/j.eneco.2015.06.014.

Shahid, M., B. Neal, and A. Jones, 2020: Uptake of Australia's Health Star Rating System 2014–2019. *Nutr. 2020, Vol. 12, Page 1791*, **12(6)**, 1791, doi:10.3390/NU12061791.

Shahsavari, A. and M. Akbari, 2018: Potential of solar energy in developing countries for reducing energy-related emissions. *Renew. Sustain. Energy Rev.*, **90** (June 2017), 275–291, doi:10.1016/j.rser.2018.03.065.

Shamshiri, R.R. et al., 2018: Advances in greenhouse automation and controlled environment agriculture: A transition to plant factories and urban agriculture. *Int. J. Agric. Biol. Eng.*, **11(1)**, 1–22, doi:10.25165/j.ijabe.20181101.3210.

Shangguan, S. et al., 2019: A Meta-Analysis of Food Labeling Effects on Consumer Diet Behaviors and Industry Practices. *Am. J. Prev. Med.*, **56(2)**, 300–314, doi:10.1016/j.amepre.2018.09.024.

Shayegh, S., V. Bosetti, and M. Tavoni, 2021: Future Prospects of Direct Air Capture Technologies: Insights From an Expert Elicitation Survey. *Front. Clim.*, **3**, 630893, doi:10.3389/fclim.2021.630893.

Shiraki, H. and M. Sugiyama, 2020: Back to the basic: toward improvement of technoeconomic representation in integrated assessment models. *Clim. Change*, **162(1)**, 13–24, doi:10.1007/s10584-020-02731-4.

Shrum, T.R. et al., 2020: Behavioural frameworks to understand public perceptions of and risk response to carbon dioxide removal. *Interface Focus*, **10(5)**, 20200002, doi:10.1098/rsfs.2020.0002.

Shue, H., 2018: Mitigation gambles: uncertainty, urgency and the last gamble possible. *Philos. Trans. R. Soc. A Math. Phys. Eng. Sci.*, **376(2119)**, 20170105, doi:10.1098/rsta.2017.0105.

Sica, D., O. Malandrino, S. Supino, M. Testa, and M.C. Lucchetti, 2018: Management of end-of-life photovoltaic panels as a step towards a circular economy. *Renew. Sustain. Energy Rev.*, **82**, 2934–2945, doi:10.1016/j.rser.2017.10.039.

Siegel, D.A., T. Devries, S.C. Doney, and T. Bell, 2021: Assessing the sequestration time scales of some ocean-based carbon dioxide reduction strategies. *Environ. Res. Lett.*, **16(10)**, 104003, doi:10.1088/1748-9326/AC0BE0.

Siegrist, M. and C. Hartmann, 2020: Consumer acceptance of novel food technologies. *Nat. Food*, **1** (June), 343–350, doi:10.1038/s43016-020-0094-x.

Siikamäki, J., J.N. Sanchirico, and S.L. Jardine, 2012: Global economic potential for reducing carbon dioxide emissions from mangrove loss. *Proc. Natl. Acad. Sci.*, **109(36)**, 14369–14374, doi:10.1073/PNAS.1200519109.

Sillman, J. et al., 2019: Bacterial protein for food and feed generated via renewable energy and direct air capture of CO2: Can it reduce land and water use? *Glob. Food Sec.*, **22**(September), 25–32, doi:10.1016/j.gfs.2019.09.007.

Sills, E.O. et al., 2017: Building the evidence base for REDD+: Study design and methods for evaluating the impacts of conservation interventions on local well-being. *Glob. Environ. Change*, **43**, 148–160, doi.org/10.1016/j.gloenvcha.2017.02.002.

Silva, A.F.R., Y.L. Brasil, K. Koch, and M.C.S. Amaral, 2021: Resource recovery from sugarcane vinasse by anaerobic digestion – A review. *J. Environ. Manage.*, **295**, 113137, doi:10.1016/J.JENVMAN.2021.113137.

Silva, V.L. and N. Sanjuán, 2019: Opening up the black box: A systematic literature review of life cycle assessment in alternative food processing technologies. *J. Food Eng.*, **250** (June 2018), 33–45, doi:10.1016/j.jfoodeng.2019.01.010.

Silver, M.W. et al., 2010: Toxic diatoms and domoic acid in natural and iron enriched waters of the oceanic Pacific. *Proc. Natl. Acad. Sci.*, **107(48)**, 20762–20767, doi:10.1073/pnas.1006968107.

Simsa, R. et al., 2019: Extracellular heme proteins influence bovine myosatellite cell proliferation and the color of cell-based meat. *Foods*, **8(10)**, doi:10.3390/foods8100521.

Sinclair, F. et al., 2019: *Background Paper the Contribution of Agroecological Approaches To Realizing Climate-Resilient Agriculture*. Global Commission on Adaptation, Rotterdam, Netherlands, and Washington, DC, USA, 12 pp. www.gca.org/wp-content/uploads/2020/12/TheContributionsOfAgroecologicalApproaches.pdf.

Singh, A., N. Winchester, and V.J. Karplus, 2019: Evaluating India's climate targets: the implications of economy-wide and sector specific policies. *Clim. Chang. Econ.*, **10(03)**, doi:10.1142/S201000781950009X. Singh, B.P., B.J. Hatton, B. Singh, A.L. Cowie, and A. Kathuria, 2010: Influence of Biochars on Nitrous Oxide Emission and Nitrogen Leaching from Two Contrasting Soils. *J. Environ. Qual.*, **39(4)**, 1224, doi:10.2134/jeq2009.0138.

Singh, B.P., A.L. Cowie, and R.J. Smernik, 2012: Biochar Carbon Stability in a Clayey Soil As a Function of Feedstock and Pyrolysis Temperature. *Environ. Sci. Technol.*, **46(21)**, 11770–11778, doi:10.1021/es302545b.

Singh, P.K. and H. Chudasama, 2021: Pathways for climate change adaptations in arid and semi-arid regions. *J. Clean. Prod.*, **284**, 124744, doi:10.1016/J.JCLEPRO.2020.124744.

Skuce, P.J., E.R. Morgan, J. van Dijk, and M. Mitchell, 2013: Animal health aspects of adaptation to climate change: beating the heat and parasites in a warming Europe. *Animal*, **7 Suppl 2**, 333–345, doi:10.1017/S175173111300075X. Slade, P., 2018: If you build it, will they eat it? Consumer preferences for plant-based and cultured meat burgers. *Appetite*, **125**, 428–437, doi:10.1016/j.appet.2018.02.030.

Smetana, S., A. Mathys, A. Knoch, and V. Heinz, 2015: Meat alternatives: life cycle assessment of most known meat substitutes. *Int. J. Life Cycle Assess.*, **20(9)**, 1254–1267, doi:10.1007/s11367-015-0931-6.

Smil, V., 2015: *Power density: a key to understanding energy sources and uses*. MIT Press, Cambridge, MA, USA, 306 pp.

Smith, P., M. Bustamante, H. Ahammad, H. Clark, H. Dong, E.A. Elsiddig, H. Haberl, R. Harper, J. House, M. Jafari, O. Masera, C. Mbow, N.H. Ravindranath, C.W. Rice, C. Robledo Abad, A. Romanovskaya, F. Sperling, and F. Tubiello, 2014: Agriculture, Forestry and Other Land Use (AFOLU). In: *Climate Change 2014: Mitigation of Climate Change. Contribution of Working Group III to the Fifth Assessment Report of the Intergovernmental Panel on Climate Change* [Edenhofer, O., R. Pichs-Madruga, Y. Sokona, E. Farahani, S. Kadner, K. Seyboth, A. Adler, I. Baum, S. Brunner, P. Eickemeier, B. Kriemann, J. Savolainen, S. Schlömer, C. von Stechow, T. Zwickel and J.C. Minx (eds.)]. Cambridge University Press, Cambridge, UK and New York, NY, USA, pp. 811–922.

Smith, P. et al., 2016: Biophysical and economic limits to negative CO2 emissions. *Nat. Clim. Change*, **6(1)**, 42–50, doi:10.1038/nclimate2870.

Smith, P. et al., 2017: Bridging the gap – Carbon dioxide removal. In: *The UNEP Emissions Gap Report*, United Nations Environment Programme, Nairobi, Kenya, pp. 58–66.

Smith, P. et al., 2019a: Impacts of Land-Based Greenhouse Gas Removal Options on Ecosystem Services and the United Nations Sustainable Development Goals. *Annu. Rev. Environ. Resour.*, **44(1)**, 1–32, doi:10.1146/annurev-environ-101718-033129.

Smith, P., J. Nkem, K. Calvin, D. Campbell, F. Cherubini, G. Grassi, V. Korotkov, A.L. Hoang, S. Lwasa, P. McElwee, E. Nkonya, N. Saigusa, J.-F. Soussana and M.A. Taboada., 2019b: Interlinkages between Desertification, Land Degradation, Food Security and GHG fluxes: Synergies, Trade-offs and Integrated Response Options. In: *Climate Change and Land: an IPCC special report on climate change, desertification, land degradation, sustainable land management, food security, and greenhouse gas fluxes in terrestrial ecosystems.* [P.R. Shukla, J. Skea, E. Calvo Buendia, V. Masson-Delmotte, H.- O. Portner, D. C. Roberts, P. Zhai, R. Slade, S. Connors, R. van Diemen, M. Ferrat, E. Haughey, S. Luz, S. Neogi, M. Pathak, J. Petzold, J. Portugal Pereira, P. Vyas, E. Huntley, K. Kissick, M. Belkacemi and J. Malley, (eds.)]. Cambridge University Press, Cambridge, UK, and New York, NY, USA.

Smith, P. et al., 2020: Which practices co-deliver food security, climate change mitigation and adaptation, and combat land degradation and desertification? *Glob. Chang. Biol.*, **26(3)**, 1532–1575, doi:10.1111/gcb.14878.

Smith, S.M., 2021: A case for transparent net-zero carbon targets. *Commun. Earth Environ.*, **2(1)**, 24, doi:10.1038/s43247-021-00095-w.

Socolow, R. et al., 2011: *Direct Air Capture of CO2 with Chemicals: A Technology Assessment for the APS Panel on Public Affairs*. American Physical Society, 100 pp.

Solati, Z. et al., 2018: Crude protein yield and theoretical extractable true protein of potential biorefinery feedstocks. *Ind. Crops Prod.*, **115**, 214–226, doi.org/10.1016/j.indcrop.2018.02.010.

Solazzo, E. et al., 2020: Uncertainties in the EDGAR emission inventory of greenhouse gases. *Atmos. Chem. Phys. Discuss.*, **2020**, 1–46, doi:10.5194/acp-2020-1102.

Solomin, E., E. Sirotkin, E. Cuce, S.P. Selvanathan, and S. Kumarasamy, 2021: Hybrid Floating Solar Plant Designs: A Review. *Energies 2021, Vol. 14, Page 2751*, **14(10)**, 2751, doi:10.3390/EN14102751.

Sonter, L., M. Dade, J. Watson, and R. Valenta, 2020a: Renewable energy production will exacerbate mining threats to biodiversity. *Nat. Commun.*, **11**, 4174, doi:10.1038/s41467-020-17928-5.

Sonter, L.J., M.C. Dade, J.E.M. Watson, and R.K. Valenta, 2020b: Renewable energy production will exacerbate mining threats to biodiversity. *Nat. Commun. 2020 111*, **11(1)**, 1–6, doi:10.1038/s41467-020-17928-5.

Soto Golcher, C. and I.J. Visseren-Hamakers, 2018: Framing and integration in the global forest, agriculture and climate change nexus. *Environ. and Plan. C: Polit. and Sp.*, **36(8)**, doi:10.1177/2399654418788566.

Souza Filho, P.F., D. Andersson, J.A. Ferreira, and M.J. Taherzadeh, 2019: Mycoprotein: environmental impact and health aspects. *World J. Microbiol. Biotechnol.*, **35(10)**, 1–8, doi:10.1007/s11274-019-2723-9.

Sovacool, B.K., 2008: The costs of failure: A preliminary assessment of major energy accidents, 1907-2007. *Energy Policy*, **36(5)**, 1802–1820, doi:10.1016/j.enpol.2008.01.040.

Spence, E., E. Cox, and N. Pidgeon, 2021: Exploring cross-national public support for the use of enhanced weathering as a land-based carbon dioxide removal strategy. *Clim. Change*, **165(1)**, 23, doi:10.1007/s10584-021-03050-y.

Spinelli, R., R. Visser, R. Björheden, and D. Röser, 2019: Recovering Energy Biomass in Conventional Forest Operations: a Review of Integrated Harvesting Systems. *Curr. For. Reports*, **5(2)**, 90–100, doi:10.1007/s40725-019-00089-0.

Springmann, M., H.C.J. Godfray, M. Rayner, and P. Scarborough, 2016: Analysis and valuation of the health and climate change cobenefits of dietary change. *Proc. Natl. Acad. Sci.*, **113(15)**, 4146–4151, doi:10.1073/pnas.1523119113.

Springmann, M. et al., 2017: Mitigation potential and global health impacts from emissions pricing of food commodities. *Nat. Clim. Change*, **7(1)**, 69–74, doi:10.1038/nclimate3155.

Springmann, M. et al., 2018a: Options for keeping the food system within environmental limits. *Nature*, **562(7728)**, 519–525, doi:10.1038/s41586-018-0594-0.

Springmann, M. et al., 2018b: Health and nutritional aspects of sustainable diet strategies and their association with environmental impacts: a global modelling analysis with country-level detail. *Lancet Planet. Heal.*, **2(10)**, e451–e461, doi:10.1016/S2542-5196(18)30206-7.

Springmann, M. et al., 2020: The healthiness and sustainability of national and global food based dietary guidelines: modelling study. *BMJ*, **44(0)**, m2322, doi:10.1136/bmj.m2322.

Springmann, M., P. Webb, M. Rayner, and P. Scarborough, 2021: The global and regional costs of healthy and sustainable dietary patterns. *Lancet Planet. Heal.*, **5(11)**, E797–E807.

Srivastava, R.R., P. Pathak, and M. Perween, 2020: Environmental and Health Impact Due to Uranium Mining. In: *Uranium in Plants and the Environment* [Gupta, D.K. and C. Walther [eds.] Springer International Publishing, 69–89, doi:10.1007/978-3-030-14961-1_3.

Ssegane, H. and M.C. Negri, 2016: An integrated landscape designed for commodity and bioenergy crops for a tile-drained agricultural watershed. *J. Environ. Qual.*, **45(5)**, 1588–1596, doi:10.2134/jeq2015.10.0518.

Ssegane, H., M.C. Negri, J. Quinn, and M. Urgun-Demirtas, 2015: Multifunctional landscapes: Site characterization and field-scale design to incorporate biomass production into an agricultural system. *Biomass and Bioenergy*, **80**, 179–190, doi:10.1016/j.biombioe.2015.04.012.

Ssegane, H. et al., 2016: The economics of growing shrub willow as a bioenergy buffer on agricultural fields: A case study in the Midwest Corn Belt. *Biofuels, Bioprod. Biorefining*, **10(6)**, 776–789, doi:10.1002/BBB.1679.

Staffell, I. et al., 2019: The role of hydrogen and fuel cells in the global energy system. *Energy Environ. Sci.*, **12(2)**, 463–491, doi:10.1039/c8ee01157e.

Steinwand, M.A., and P.C. Ronald, 2020: Crop biotechnology and the future of food. *Nat. Food*, **1(5)**, 273–283, doi:10.1038/s43016-020-0072-3.

Stenzel, F., D. Gerten, and N. Hanasaki, 2021a: Global scenarios of irrigation water abstractions for bioenergy production: A systematic review. *Hydrol. Earth Syst. Sci.*, **25(4)**, 1711–1726, doi:10.5194/HESS-25-1711-2021.

Stenzel, F. et al., 2021b: Irrigation of biomass plantations may globally increase water stress more than climate change. *Nat. Commun. 2021 121*, **12(1)**, 1–9, doi:10.1038/s41467-021-21640-3.

Stephens, N. et al., 2018: Bringing cultured meat to market: Technical, socio-political, and regulatory challenges in cellular agriculture. *Trends Food Sci. Technol.*, **78**(June 2017), 155–166, doi:10.1016/j.tifs.2018.04.010.

Stoll-Kleemann, S. and U.J. Schmidt, 2017: Reducing meat consumption in developed and transition countries to counter climate change and biodiversity loss: a review of influence factors. *Reg. Environ. Change*, **17(5)**, 1261–1277, doi:10.1007/s10113-016-1057-5.

Storcksdieck genannt Bonsmann, S., G. Marandola, E. Ciriolo, R. Van Bavel, and J. Wollgast, 2020: *Front-of-pack nutrition labelling schemes: a comprehensive review.* EUR 29811 EN, Publications Office of the European Union, Luxembourg.

Strand, S.E. and G. Benford, 2009: Ocean Sequestration of Crop Residue Carbon: Recycling Fossil Fuel Carbon Back to Deep Sediments. *Environ. Sci. Technol.*, **43(4)**, 1000–1007, doi:10.1021/es8015556.

Strapasson, A. et al., 2020: EU land use futures: modelling food, bioenergy and carbon dynamics. *Energy Strateg. Rev.*, **31**, 100545, doi:10.1016/J.ESR.2020.100545.

Strefler, J., T. Amann, N. Bauer, E. Kriegler, and J. Hartmann, 2018: Potential and costs of carbon dioxide removal by enhanced weathering of rocks. *Environ. Res. Lett.*, **13(3)**, 034010, doi:10.1088/1748-9326/aaa9c4.

Strefler, J. et al., 2021: Carbon dioxide removal technologies are not born equal. *Environ. Res. Lett.*, **16(7)**, 074021, doi:10.1088/1748-9326/ac0a11.

Styles, D. et al., 2016: Climate regulation, energy provisioning and water purification: Quantifying ecosystem service delivery of bioenergy willow grown on riparian buffer zones using life cycle assessment. *Ambio*, **45(8)**, 872–884, doi:10.1007/s13280-016-0790-9.

Stylianou, K.S., V.L. Fulgoni, and O. Jolliet, 2021: Prioritization of healthy and sustainable foods for small targeted dietary changes can yield substantial gains for human health and the environment. *Nat. Food*, **2**, 616–627

Sussman, R.L., A.T. McMahon, and E.P. Neale, 2019: An Audit of the Nutrition and Health Claims on Breakfast Cereals in Supermarkets in the Illawarra Region of Australia. *Nutrients*, **11(7)**, doi:10.3390/nu11071604.

Sustainable Energy for All, 2018: *Chilling Prospects: Providing Sustainable Cooling for All*. https://www.seforall.org/sites/default/files/SEforALL_CoolingForAll-Report.pdf.

Sweet, S.K., J.P. Schuldt, J. Lehmann, D.A. Bossio, and D. Woolf, 2021: Perceptions of naturalness predict US public support for Soil Carbon Storage as a climate solution. *Clim. Change*, **166(1)**, 22, doi:10.1007/s10584-021-03121-0.

Swinnen, J.F.M. and M. Maertens, 2007: Globalization, privatization, and vertical coordination in food value chains in developing and transition countries. *Agric. Econ.*, **37**(S1), 89–102, doi:10.1111/j.1574-0862.2007.00237.x.

Taillie, L.S., E. Busey, F.M. Stoltze, and F.R. Dillman Carpentier, 2019: Governmental policies to reduce unhealthy food marketing to children. *Nutr. Rev.*, **77(11)**, 787–816, doi:10.1093/nutrit/nuz021.

Tan, R.R., K.B. Aviso, and S. Bandyopadhyay, 2021: Pinch-based planning of terrestrial carbon management networks. *Clean. Eng. Technol.*, **4**, 100141, doi:10.1016/j.clet.2021.100141.

Tanzer, S.E. and A. Ramírez, 2019: When are negative emissions negative emissions? *Energy Environ. Sci.*, **12(4)**, 1210–1218, doi:10.1039/C8EE03338B.

Tanzer, S.E., K. Blok, and A. Ramírez, 2020: Can bioenergy with carbon capture and storage result in carbon negative steel? *Int. J. Greenh. Gas Control*, **100**, 103104, doi:10.1016/j.ijggc.2020.103104.

Tanzer, S.E., K. Blok, and A. Ramírez, 2021: Decarbonising Industry via BECCS: Promising Sectors, Challenges, and Techno-economic Limits of Negative Emissions. *Curr. Sustain. Energy Reports*, **8**, 253–262, doi:10.1007/s40518-021-00195-3.

Tavoni, M. and R. Socolow, 2013: Modeling meets science and technology: an introduction to a special issue on negative emissions. *Clim. Change*, **118(1)**, 1–14, doi:10.1007/s10584-013-0757-9.

Taylor, L.L. et al., 2016: Enhanced weathering strategies for stabilizing climate and averting ocean acidification. *Nat. Clim. Change*, **6(4)**, 402–406, doi:10.1038/nclimate2882.

Taylor, M., 2018: Climate-smart agriculture: what is it good for? *J. Peasant Stud.*, **45(1)**, 89–107, doi:10.1080/03066150.2017.1312355.

Teigiserova, D.A., L. Hamelin, and M. Thomsen, 2020: Towards transparent valorization of food surplus, waste and loss: Clarifying definitions, food waste hierarchy, and role in the circular economy. *Sci. Total Environ.*, **706**, 136033, doi:10.1016/j.scitotenv.2019.136033.

Temme, E.H.M. et al., 2020: Demand-Side Food Policies for Public and Planetary Health. *Sustainability*, **12(15)**, 5924, doi:10.3390/su12155924.

Temple, N.J., 2019: Front-of-package food labels: A narrative review. *Appetite*, **144** (September 2019), 104485, doi:10.1016/j.appet.2019.104485.

Terlouw, T., K. Treyer, C. Bauer, and M. Mazzotti, 2021: Life Cycle Assessment of Direct Air Carbon Capture and Storage with Low-Carbon Energy Sources. *Environ. Sci. Technol.*, **55(16)**, 11397–11411, doi:10.1021/acs.est.1c03263.

Termeer, C.J.A. M., S. Drimie, J. Ingram, L. Pereira, and M.J. Whittingham, 2018: A diagnostic framework for food system governance arrangements: The case of South Africa. *NJAS - Wageningen J. Life Sci.*, **84** (August 2017), 85–93, doi:10.1016/j.njas.2017.08.001.

Teter, J., S. Yeh, M. Khanna, and G. Berndes, 2018: Water impacts of U.S. biofuels: Insights from an assessment combining economic and biophysical models. *PLoS One*, **13(9)**, e0204298, doi:10.1371/JOURNAL.PONE.0204298.

Tharammal, T., G. Bala, N. Devaraju, and R. Nemani, 2019: A review of the major drivers of the terrestrial carbon uptake: model-based assessments, consensus, and uncertainties. *Environ. Res. Lett.*, **14(9)**, 93005, doi:10.1088/1748-9326/ab3012.

Thaxter, C.B. et al., 2017: Bird and bat species' global vulnerability to collision mortality at wind farms revealed through a trait-based assessment. *Proc. R. Soc. B Biol. Sci.*, **284(1862)**, 20170829, doi:10.1098/rspb.2017.0829.

The Ellen MacArthur Foundation, 2013: *Towards a Circular Economy - Economic and Business Rationale for an Accelerated Transition*. Founding Partners of the Ellen MacArthur Foundation, 98 pp.

The Ellen MacArthur Foundation, 2019: *Completing the Picture: How the Circular Economy Tackles Climate Change*. Ellen MacArthur Foundation, 62 pp.

Theurl, M.C. et al., 2020: Food systems in a zero-deforestation world: Dietary change is more important than intensification for climate targets in 2050. *Sci. Total Environ.*, **735**, 139353, doi:10.1016/j.scitotenv.2020.139353.

Thom, D., W. Rammer, R. Garstenauer, and R. Seidl, 2018: Legacies of past land use have a stronger effect on forest carbon exchange than future climate change in a temperate forest landscape. *Biogeosciences*, **15(18)**, 5699–5713, doi:10.5194/bg-15-5699-2018.

Thomas, G., N. Pidgeon, and E. Roberts, 2018: Ambivalence, naturalness and normality in public perceptions of carbon capture and storage in biomass, fossil energy, and industrial applications in the United Kingdom. *Energy Res. Soc. Sci.*, **46**, 1–9, doi:10.1016/j.erss.2018.06.007.

Thomas, J.-B.E. et al., 2021: Marine biomass for a circular blue-green bioeconomy?: A life cycle perspective on closing nitrogen and phosphorus land-marine loops. *J. Ind. Ecol.*, doi:10.1111/JIEC.13177.

Thornton, P.K. and M. Herrero, 2015: Adapting to climate change in the mixed crop and livestock farming systems in sub-Saharan Africa. *Nat. Clim. Change*, **5(9)**, doi:10.1038/nclimate2754.

Thorrez, L. and H. Vandenburgh, 2019: Challenges in the quest for 'clean meat.' *Nat. Biotechnol.*, **37(3)**, 215–216, doi:10.1038/s41587-019-0043-0.

Thow, A.M. and N. Nisbett, 2019: Trade, nutrition, and sustainable food systems. *Lancet*, **394(10200)**, 716–718, doi:10.1016/S0140-6736(19)31292-9.

Thunman, H. et al., 2018: Advanced biofuel production via gasification – lessons learned from 200 man-years of research activity with Chalmers' research gasifier and the GoBiGas demonstration plant. *Energy Sci. Eng.*, **6(1)**, 6–34, doi:10.1002/ESE3.188.

Tisserant, A. and F. Cherubini, 2019: Potentials, Limitations, Co-Benefits, and Trade-Offs of Biochar Applications to Soils for Climate Change Mitigation. *Land*, **8(12)**, 179, doi:10.3390/LAND8120179.

To, H. and R.Q. Grafton, 2015: Oil prices, biofuels production and food security: past trends and future challenges. *Food Secur.*, **7**, 323–336, doi:10.1007/s12571-015-0438-9.

Torres-Tiji, Y., F.J. Fields, and S.P. Mayfield, 2020: Microalgae as a future food source. *Biotechnol. Adv.*, **41** (August 2019), doi:10.1016/j.biotechadv.2020.107536.

Tørris, C. and H. Mobekk, 2019: Improving Cardiovascular Health through Nudging Healthier Food Choices: A Systematic Review. *Nutrients*, **11(10)**, 1–19, doi:10.3390/nu11102520.

Torvanger, A., 2019: Governance of bioenergy with carbon capture and storage (BECCS): accounting, rewarding, and the Paris agreement. *Clim. Policy*, **19(3)**, 329–341, doi:10.1080/14693062.2018.1509044.

Townsend, R., R. Benfica, and A. Prasann, 2017: Future of Food: Shaping the Food System to Deliver Jobs. World Bank, Washington, DC, USA. Traverso, S. and S. Schiavo, 2020: Fair trade or trade fair? International food trade and cross-border macronutrient flows. *World Dev.*, **132**, 104976, doi:10.1016/j.worlddev.2020.104976.

Trick, C.G. et al., 2010: Iron enrichment stimulates toxic diatom production in high-nitrate, low-chlorophyll areas. *Proc. Natl. Acad. Sci.*, **107(13)**, 5887–5892, doi:10.1073/pnas.0910579107.

Trull, T.W. et al., 2015: Chemometric perspectives on plankton community responses to natural iron fertilisation over and downstream of the

Kerguelen Plateau in the Southern Ocean. *Biogeosciences*, **12**, 1029–1056, doi:10.5194/bg-12-1029-2015.
Tubana, B.S., T. Babu, and L.E. Datnoff, 2016: A Review of Silicon in Soils and Plants and Its Role in US Agriculture. *Soil Sci.*, **181(9/10)**, 1, doi:10.1097/SS.0000000000000179.
Tubiello, F.N. et al., 2021: Greenhouse gas emissions from food systems: Building the evidence base. *Environ. Res. Lett.*, **16(6)**, doi:10.1088/1748-9326/ac018e.
Tuomisto, H.L., 2019: The eco-friendly burger. *EMBO Rep.*, **20(1)**, 1–6, doi:10.15252/embr.201847395.
Tuomisto, H.L. and M.J. Teixeira de Mattos, 2011: Environmental Impacts of Cultured Meat Production. *Environ. Sci. Technol.*, **45(14)**, 6117–6123, doi:10.1021/es200130u.
Turnhout, E. et al., 2017: Envisioning REDD+ in a post-Paris era: between evolving expectations and current practice. *Wiley Interdiscip. Rev. Clim. Change*, **8(1)**, e425, doi:10.1002/wcc.425.
Tzachor, A., C.E. Richards, and L. Holt, 2021: Future foods for risk-resilient diets. *Nat. Food*, **2** (May), doi:10.1038/s43016-021-00269-x.
Uden, S., P. Dargusch, and C. Greig, 2021: Cutting through the noise on negative emissions. *Joule*, **5(8)**, 1956–1970, doi:10.1016/j.joule.2021.06.013.
Ueckerdt, F. et al., 2021: Potential and risks of hydrogen-based e-fuels in climate change mitigation. *Nat. Clim. Chang. 2021 115*, **11(5)**, 384–393, doi:10.1038/s41558-021-01032-7.
Unar-Munguía, M., E. Monterubio Flores, and M.A. Colchero, 2019: Apparent consumption of caloric sweeteners increased after the implementation of NAFTA in Mexico. *Food Policy*, **84(655)**, 103–110, doi:10.1016/j.foodpol.2019.03.004.
UNCCD, 2015, Integration of the Sustainable Development Goals and target into the implementation of the United Nations Convention to Combat Desertification and the Intergovernmental Working Group report on land degradation neutrality: Decision 3/COP 12. ICCD.COP(12)/20/Add.1, https://www.unccd.int/official-documents/cop-12-ankara-2015.
UNEP, 2017: *The Emissions Gap Report 2017*. United Nations Environment Programme, Nairobi, Kenya, 112 pp.
UNEP, 2019: *Emissions Gap Report 2019*. United Nations Environment Programme, Nairobi, Kenya, 108 pp.
UNEP, 2021: *The Adaptation Gap Report 2020*. United Nations Environment Programme, Nairobi, Kenya, https://www.unep.org/resources/adaptation-gap-report-2020.
UNFCCC, 2016: Decision 1/CP.21: Adoption of the Paris Agreement. In: *Report of the Conference of the Parties on its twenty-first session, held in Paris from 30 November to 13 December 2015. Addendum: Part two: Action taken by the Conference of the Parties at its twenty-first session, FCCC/CP/2015/10/Add.1*, United Nations Framework Convention on Climate Change (UNFCCC), pp. 1–36.
UNICEF, 2019: *The State of the World's Children 2019. Children, Food and Nutrition: Growing well in a changing world*. UNICEF, New York, USA, 258 pp.
University of Birmingham, 2018: *A Cool World: Defining the Energy Conundrum of Cooling for All*. University of Birmingham, Birmingham, UK, https://www.birmingham.ac.uk/Documents/college-eps/energy/Publications/2018-clean-cold-report.pdf.
Urbinatti, A.M., M. Dalla Fontana, A. Stirling, and L.L. Giatti, 2020: 'Opening up' the governance of water-energy-food nexus: Towards a science-policy-society interface based on hybridity and humility. *Sci. Total Environ.*, **744**, 140945, doi.org/10.1016/j.scitotenv.2020.140945.
Ürge-Vorsatz, D., S.T. Herrero, N.K. Dubash, and F. Lecocq, 2014: Measuring the Co-Benefits of Climate Change Mitigation. *Annu. Rev. Environ. Resour.*, **39(1)**, 549–582, doi:10.1146/annurev-environ-031312-125456.
US EPA, 2019: *Global Non-CO2 Greenhouse Gas Emission Projections & Mitigation 2015-2050*. United States Environmental Protection Agency, Washington, DC, USA, 78 pp.
Václavík, T., S. Lautenbach, T. Kuemmerle, and R. Seppelt, 2013: Mapping global land system archetypes. *Glob. Environ. Change*, **23(6)**, 1637–1647, doi:10.1016/j.gloenvcha.2013.09.004.
Vågsholm, I., N.S. Arzoomand, and S. Boqvist, 2020: Food Security, Safety, and Sustainability—Getting the Trade-Offs Right. *Front. Sustain. Food Syst.*, **4** (February), 1–14, doi:10.3389/fsufs.2020.00016.
Valencia, V., H. Wittman, and J. Blesh, 2019: Structuring Markets for Resilient Farming Systems. *Agron. Sustain. Dev.*, **39(25)**, doi:10.1007/s13593-019-0572-4.
van de Ven, D.-J. et al., 2021: The potential land requirements and related land use change emissions of solar energy. *Sci. Reports 2021 111*, **11(1)**, 1–12, doi:10.1038/s41598-021-82042-5.
van der Ent, R.J. and O.A. Tuinenburg, 2017: The residence time of water in the atmosphere revisited. *Hydrol. Earth Syst. Sci.*, **21(2)**, 779–790, doi:10.5194/hess-21-779-2017.
van der Ploeg, F., 2016: Second-best carbon taxation in the global economy: The Green Paradox and carbon leakage revisited. *J. Environ. Econ. Manage.*, **78**, 85–105, doi:10.1016/j.jeem.2016.02.006.
van der Ploeg, F. and A. Rezai, 2019: The risk of policy tipping and stranded carbon assets. *J. Environ. Econ. Manage.*, **100** (March), doi:10.1016/j.jeem.2019.102258.
van der Voorn, T., Å. Svenfelt, K.E. Björnberg, E. Fauré, and R. Milestad, 2020: Envisioning carbon-free land use futures for Sweden: a scenario study on conflicts and synergies between environmental policy goals. *Reg. Environ. Change*, **20(2)**, 35, doi:10.1007/s10113-020-01618-5.
van der Werf, H.M.G., M.T. Knudsen, and C. Cederberg, 2020: Towards better representation of organic agriculture in life cycle assessment. *Nat. Sustain.*, **3**, 419–425, doi:10.1038/s41893-020-0489-6.
van der Wijst, K.-I., A.F. Hof, and D.P. van Vuuren, 2021: Costs of avoiding net negative emissions under a carbon budget. *Environ. Res. Lett.*, **16(6)**, 64071, doi:10.1088/1748-9326/ac03d9.
van Hal, O. et al., 2019: Upcycling food leftovers and grass resources through livestock: Impact of livestock system and productivity. *J. Clean. Prod.*, **219**, 485–496, doi:10.1016/j.jclepro.2019.01.329.
Van Loo, E.J., V. Caputo, R.M. Nayga, and W. Verbeke, 2014: Consumers' valuation of sustainability labels on meat. *Food Policy*, **49**(P1), 137–150, doi:10.1016/j.foodpol.2014.07.002.
Van Meijl, H. et al., 2018: Comparing impacts of climate change and mitigation on global agriculture by 2050. *Environ. Res. Lett.*, **13(6)**, 064021, doi:10.1088/1748-9326/aabdc4.
van Soest, H.L. et al., 2019: Analysing interactions among Sustainable Development Goals with Integrated Assessment Models. *Glob. Transitions*, **1**, 210–225, doi:10.1016/j.glt.2019.10.004.
van Vuuren, D.P. et al., 2018: Alternative pathways to the 1.5°C target reduce the need for negative emission technologies. *Nat. Clim. Change*, **8(5)**, 391–397, doi:10.1038/s41558-018-0119-8.
van Zalk, J. and P. Behrens, 2018: The spatial extent of renewable and non-renewable power generation: A review and meta-analysis of power densities and their application in the U.S. *Energy Policy*, **123**, 83–91, doi:10.1016/j.enpol.2018.08.023.
Van Zanten, H.H.E., M.K. Van Ittersum, and I.J.M. De Boer, 2019: The role of farm animals in a circular food system. *Glob. Food Sec.*, **21**, 18–22, doi:10.1016/j.gfs.2019.06.003.
Van Zanten, H.H.E. et al., 2018: Defining a land boundary for sustainable livestock consumption. *Glob. Chang. Biol.*, **24(9)**, 4185–4194, doi:10.1111/gcb.14321.
Vanham, D. et al., 2019: Environmental footprint family to address local to planetary sustainability and deliver on the SDGs. *Sci. Total Environ.*, **693**(June), 133642, doi:10.1016/j.scitotenv.2019.133642.
Vanham, D., M.M. Mekonnen, and A.Y. Hoekstra, 2020: Treenuts and groundnuts in the EAT-Lancet reference diet: Concerns regarding sustainable water use. *Glob. Food Sec.*, **24** (December 2019), doi:10.1016/j.gfs.2020.100357.
Varelas, 2019: Food Wastes as a Potential new Source for Edible Insect Mass Production for Food and Feed: A review. *Fermentation*, **5(3)**, 81, doi:10.3390/fermentation5030081.

Vautard, R. et al., 2014: Regional climate model simulations indicate limited climatic impacts by operational and planned European wind farms. *Nat. Commun.*, **5**, doi:10.1038/ncomms4196.

Vecchio, R., and C. Cavallo, 2019: Increasing healthy food choices through nudges: A systematic review. *Food Qual. Prefer.*, **78** (June 2018), 103714, doi:10.1016/j.foodqual.2019.05.014.

Velenturf, A.P.M. et al., 2019: Circular economy and the matter of integrated resources. *Sci. Total Environ.*, **689**, 963–969, doi:10.1016/j.scitotenv.2019.06.449.

Verkerk, P.J. et al., 2020: Climate-Smart Forestry: the missing link. *For. Policy Econ.*, **115** (June), doi:10.1016/j.forpol.2020.102164.

Vermeulen, S.J., T. Park, C.K. Khoury, and C. Béné, 2020: Changing diets and the transformation of the global food system. *Ann. N. Y. Acad. Sci.*, **1478(1)**, 3–17, doi.org/10.1111/nyas.14446.

Verschuur, J., S. Li, P. Wolski, and F.E.L. Otto, 2021: Climate change as a driver of food insecurity in the 2007 Lesotho-South Africa drought. *Sci. Rep.*, **11(1)**, 1–9, doi:10.1038/s41598-021-83375-x.

Viebahn, P. et al., 2015: Assessing the need for critical minerals to shift the German energy system towards a high proportion of renewables. *Renew. Sustain. Energy Rev.*, **49**, 655–671, doi:10.1016/J.RSER.2015.04.070.

Vogliano, C. et al., 2020: Assessing Diet Quality of Indigenous Food Systems in Three Geographically Distinct Solomon Islands Sites (Melanesia, Pacific Islands). *Nutrients*, **13(1)**, 30, doi:10.3390/nu13010030.

Vogt-Schilb, A., G. Meunier, and S. Hallegatte, 2018: When starting with the most expensive option makes sense: Optimal timing, cost and sectoral allocation of abatement investment. *J. Environ. Econ. Manage.*, **88**, 210–233, doi:10.1016/j.jeem.2017.12.001.

Voigt, C.C., 2021: Insect fatalities at wind turbines as biodiversity sinks. *Conserv. Sci. Pract.*, **3(5)**, e366, doi:10.1111/CSP2.366.

von Stechow, C. et al., 2015: Integrating Global Climate Change Mitigation Goals with Other Sustainability Objectives: A Synthesis. *Annu. Rev. Environ. Resour.*, **40(1)**, 363–394, doi:10.1146/annurev-environ-021113-095626.

VonHedemann, N., Z. Wurtzebach, T.J. Timberlake, E. Sinkular, and C.A. Schultz, 2020: Forest policy and management approaches for carbon dioxide removal. *Interface Focus*, **10(5)**, 20200001, doi:10.1098/rsfs.2020.0001.

Voskian, S., and T.A. Hatton, 2019: Faradaic electro-swing reactive adsorption for CO 2 capture. *Energy Environ. Sci.*, **12(12)**, 3530–3547, doi:10.1039/C9EE02412C.

Walker, B.J.A., T. Kurz, and D. Russel, 2018: Towards an understanding of when non-climate frames can generate public support for climate change policy. *Environ. Behav.*, **50(7)**, 781–806, doi:10.1177/0013916517713299.

Waller, L. et al., 2020: Contested framings of greenhouse gas removal and its feasibility: Social and political dimensions. *WIREs Clim. Change*, **11(4)**, e649, doi.org/10.1002/wcc.649.

Wang, J., Z. Xiong, and Y. Kuzyakov, 2016a: Biochar stability in soil: meta-analysis of decomposition and priming effects. *GCB Bioenergy*, **8(3)**, 512–523, doi:10.1111/gcbb.12266.

Wang, N., K. Akimoto, and G.F. Nemet, 2021: What went wrong? Learning from three decades of carbon capture, utilization and sequestration (CCUS) pilot and demonstration projects. *Energy Policy*, **158**, 112546, doi:10.1016/j.enpol.2021.112546.

Wang, Q., K. Hubacek, K. Feng, Y.M. Wei, and Q.M. Liang, 2016b: Distributional effects of carbon taxation. *Appl. Energy*, **184**, 1123–1131, doi:10.1016/j.apenergy.2016.06.083.

Warren Raffa, D., A. Bogdanski, and P. Tittonell, 2015: How does crop residue removal affect soil organic carbon and yield? A hierarchical analysis of management and environmental factors. *Biomass and Bioenergy*, **81**, 345–355, doi.org/10.1016/j.biombioe.2015.07.022.

Watari, T., B.C. McLellan, S. Ogata, and T. Tezuka, 2018: Analysis of potential for critical metal resource constraints in the international energy agency's long-term low-carbon energy scenarios. *Minerals*, **8(4)**, doi:10.3390/min8040156.

Watari, T. et al., 2019: Total material requirement for the global energy transition to 2050: A focus on transport and electricity. *Resour. Conserv. Recycl.*, **148**, 91–103, doi:10.1016/j.resconrec.2019.05.015.

Weber, H. et al., 2020: What are the ingredients for food systems change towards sustainability? - Insights from the literature. *Environ. Res. Lett.*, **15(11)**, doi:10.1088/1748-9326/ab99fd.

Wedding, L.M. et al., 2021: Incorporating blue carbon sequestration benefits into sub-national climate policies. *Glob. Environ. Change*, **69**, 102206, doi:10.1016/j.gloenvcha.2020.102206.

Wegener, J., D. Fong, and C. Rocha, 2018: Education, practical training and professional development for public health practitioners: a scoping review of the literature and insights for sustainable food system capacity-building. *Public Health Nutr.*, **21(9)**, 1771–1780, doi:10.1017/S1368980017004207.

Weiler, V., H.M.J. Udo, T. Viets, T.A. Crane, and I.J.M. De Boer, 2014: Handling multi-functionality of livestock in a life cycle assessment: The case of smallholder dairying in Kenya. *Curr. Opin. Environ. Sustain.*, **8**, 29–38, doi:10.1016/j.cosust.2014.07.009.

Weindl, I. et al., 2017a: Livestock production and the water challenge of future food supply: Implications of agricultural management and dietary choices. *Glob. Environ. Change*, **47** (March), 121–132, doi:10.1016/j.gloenvcha.2017.09.010.

Weindl, I. et al., 2017b: Livestock and human use of land: Productivity trends and dietary choices as drivers of future land and carbon dynamics. *Glob. Planet. Change*, **159**(April), 1–10, doi:10.1016/j.gloplacha.2017.10.002.

Weindl, I. et al., 2020: Sustainable food protein supply reconciling human and ecosystem health: A Leibniz Position. *Glob. Food Sec.*, **25**, 100367, doi:10.1016/j.gfs.2020.100367.

Weinrich, R., and O. Elshiewy, 2019: Preference and willingness to pay for meat substitutes based on micro-algae. *Appetite*, **142** (October 2018), 104353, doi:10.1016/j.appet.2019.104353.

Weitz, N., C. Strambo, E. Kemp-Benedict, and M. Nilsson, 2017: Closing the governance gaps in the water-energy-food nexus: Insights from integrative governance. *Glob. Environ. Change*, **45**, 165–173, doi.org/10.1016/j.gloenvcha.2017.06.006.

Welfle, A., P. Thornley, and M. Röder, 2020: A review of the role of bioenergy modelling in renewable energy research & policy development. *Biomass and Bioenergy*, **136**, 105542, doi:10.1016/J.BIOMBIOE.2020.105542.

Welsch, M. et al., 2014: Adding value with CLEWS – Modelling the energy system and its interdependencies for Mauritius. *Appl. Energy*, **113**, 1434–1445, doi.org/10.1016/j.apenergy.2013.08.083.

Weng, Y., W. Cai, and C. Wang, 2021: Evaluating the use of BECCS and afforestation under China's carbon-neutral target for 2060. *Appl. Energy*, **299**, 117263, doi:10.1016/j.apenergy.2021.117263.

Wenger, A., M. Stauffacher, and I. Dallo, 2021: Public perception and acceptance of negative emission technologies – framing effects in Switzerland. *Clim. Change*, **167(3)**, 53, doi:10.1007/s10584-021-03150-9.

Wezel, A. et al., 2009: Agroecology as a science, a movement and a practice. *Sustain. Agric.*, **2**, 27–43, doi:10.1007/978-94-007-0394-0_3.

WFP-FSIN, 2020: *Global Report on Food Crises: Joint Analysis for Better Decisions*. United Nations World Food Programme, Food Security Information Network,.

Whalen, R., J. Harrold, S. Child, J. Halford, and E. Boyland, 2018: The health halo trend in UK television food advertising viewed by children: The rise of implicit and explicit health messaging in the promotion of unhealthy foods. *Int. J. Environ. Res. Public Health*, **15(3)**, doi:10.3390/ijerph15030560.

Whitfield, S., A.J. Challinor, and R.M. Rees, 2018: Frontiers in Climate Smart Food Systems: Outlining the Research Space. *Front. Sustain. Food Syst.*, **2**, 2, doi:10.3389/fsufs.2018.00002.

WHO, 2010: *Set of recommendations on the marketing of foods and non-alcoholic beverages to children*. World Health Organization, Geneva, Switzerland, 16 pp.

WHO, 2019: *Health taxes: a primer*. World Health Organization, Geneva, Switzerland.

Wickramasinghe, K. et al., 2021: The shift to plant-based diets: are we missing the point? *Glob. Food Sec.*, **29** (January), 100530, doi:10.1016/j.gfs.2021.100530.

Wiedmann, T. and M. Lenzen, 2018: Environmental and social footprints of international trade. *Nat. Geosci.*, **11(5)**, 314–321, doi:10.1038/s41561-018-0113-9.

Wijesinha-Bettoni, R. et al., 2021: A snapshot of food-based dietary guidelines implementation in selected countries. *Glob. Food Sec.*, **29** (January), 100533, doi:10.1016/j.gfs.2021.100533.

Wilcox, J., P.C. Psarras, and S. Liguori, 2017: Assessment of reasonable opportunities for direct air capture. *Environ. Res. Lett.*, **12(6)**, doi:10.1088/1748-9326/aa6de5.

Willauer, H.D., F. DiMascio, D.R. Hardy, and F.W. Williams, 2017: Development of an Electrolytic Cation Exchange Module for the Simultaneous Extraction of Carbon Dioxide and Hydrogen Gas from Natural Seawater. *Energy and Fuels*, **31(2)**, 1723–1730, doi:10.1021/ACS.ENERGYFUELS.6B02586.

Willer, D.F., and D.C. Aldridge, 2020: Sustainable bivalve farming can deliver food security in the tropics. *Nat. Food*, **1**, 384–388, doi:10.1038/s43016-020-0116-8.

Willett, W. et al., 2019: *Food in the Anthropocene: the EAT–Lancet Commission on healthy diets from sustainable food systems*. The Lancet Commissions, **393(10170)**, 447–492, doi: 10.1016/S0140-6736(18)31788-4.

Williamson, P. and C. Turley, 2012: Ocean acidification in a geoengineering context. *Philos. Trans. R. Soc. A Math. Phys. Eng. Sci.*, **370(1974)**, 4317–4342, doi:10.1098/rsta.2012.0167.

Williamson, P. and R. Bodle, 2016: *Update on climate geoengineering in relation to the Convention on Biological Diversity: Potential impacts and regulatory framework*. Technical Series No.84. Secretariat of the Convention on Biological Diversity, Montreal, Canada, 158 pp.

Williamson, P. et al., 2012: Ocean fertilization for geoengineering: A review of effectiveness, environmental impacts and emerging governance. *Process Saf. Environ. Prot.*, **90(6)**, 475–488, doi:10.1016/J.PSEP.2012.10.007.

Wilson, A.L., E. Buckley, J.D. Buckley, and S. Bogomolova, 2016: Nudging healthier food and beverage choices through salience and priming. Evidence from a systematic review. *Food Qual. Prefer.*, **51**, 47–64, doi:10.1016/j.foodqual.2016.02.009.

Wilts, H., M. O'Brien, and M.O. Brien, 2019: A Policy Mix for Resource Efficiency in the EU: Key Instruments, Challenges and Research Needs. *Ecol. Econ.*, **155** (November 2017), 59–69, doi:10.1016/j.ecolecon.2018.05.004.

Winans, K., A. Kendall, and H. Deng, 2017: The history and current applications of the circular economy concept. *Renew. Sustain. Energy Rev.*, **68**, 825–833, doi:10.1016/j.rser.2016.09.123.

Winchester, N. and J.M. Reilly, 2015: The feasibility, costs, and environmental implications of large-scale biomass energy. *Energy Econ.*, **51**, 188–203, doi:10.1016/j.eneco.2015.06.016.

Winde, F., D. Brugge, A. Nidecker, and U. Ruegg, 2017: Uranium from Africa – An overview on past and current mining activities: Re-appraising associated risks and chances in a global context. *J. African Earth Sci.*, **129**, 759–778, doi:10.1016/J.JAFREARSCI.2016.12.004.

Winiwarter, W., A. Leip, H.L. Tuomisto, and P. Haastrup, 2014: A European perspective of innovations towards mitigation of nitrogen-related greenhouse gases. *Curr. Opin. Environ. Sustain.*, **9–10**, 37–45, doi:10.1016/j.cosust.2014.07.006.

Winther, U., E.S. Hognes, S. Jafarzadeh, and F. Ziegler, 2020: *Greenhouse gas emissions of Norwegian seafood products in 2017*. SINTEF, Trondheim, Norway.

Wohland, J., D. Witthaut, and C.F. Schleussner, 2018: Negative Emission Potential of Direct Air Capture Powered by Renewable Excess Electricity in Europe. *Earth's Futur.*, **6(10)**, 1380–1384, doi:10.1029/2018EF000954.

Wohlfahrt, G., E. Tomelleri, and A. Hammerle, 2021: The albedo–climate penalty of hydropower reservoirs. *Nat. Energy 2021 64*, **6(4)**, 372–377, doi:10.1038/s41560-021-00784-y.

Wohner, B., E. Pauer, V. Heinrich, and M. Tacker, 2019: Packaging-related food losses and waste: An overview of drivers and issues. *Sustainability*, **11(1)**, doi:10.3390/su11010264.

Wood, S.A., M.R. Smith, J. Fanzo, R. Remans, and R.S. Defries, 2018: Trade and the equitability of global food nutrient distribution. *Nat. Sustain.*, **1(1)**, 34–37, doi:10.1038/s41893-017-0008-6.

Woodbury, P.B., A.R. Kemanian, M. Jacobson, and M. Langholtz, 2018: Improving water quality in the Chesapeake Bay using payments for ecosystem services for perennial biomass for bioenergy and biofuel production. *Biomass and Bioenergy*, **114**, 132–142, doi:10.1016/j.biombioe.2017.01.024.

World Bank, 2021: World Development Indicators. World Bank, Washington, DC, USA (Accessed March 19, 2021).

World Bank Group, 2015: *World Bank Group Assistance to Low-Income Fragile and Conflict-Affected States: An Independent Evaulation*. World Bank Group, Washington, DC, USA, 227 pp.

Wright, S.F., and A. Upadhyaya, 1998: A survey of soils for aggregate stability and glomalin, a glycoprotein produced by hyphae of arbuscular mycorrhizal fungi. *Plant Soil*, **198**, 97–107, doi: 10.1023/A:1004347701584.

Wright, A., K.E. Smith, and M. Hellowell, 2017: Policy lessons from health taxes: A systematic review of empirical studies. *BMC Public Health*, **17(1)**, 1–14, doi:10.1186/s12889-017-4497-z.

Wu, G.C. et al., 2020: Low-impact land use pathways to deep decarbonization of electricity. *Environ. Res. Lett.*, **15(7)**, 74044, doi:10.1088/1748-9326/ab87d1.

Wu, W., C. Beretta, P. Cronje, S. Hellweg, and T. Defraeye, 2019: Environmental trade-offs in fresh-fruit cold chains by combining virtual cold chains with life cycle assessment. *Appl. Energy*, **254**, 113586, doi:10.1016/j.apenergy.2019.113586.

Wu, X. et al., 2021: Unveiling land footprint of solar power: A pilot solar tower project in China. *J. Environ. Manage.*, **280**, doi:10.1016/J.JENVMAN.2020.111741.

Wulf, C., P. Zapp, and A. Schreiber, 2020: Review of Power-to-X Demonstration Projects in Europe. *Front. Energy Res.*, **8**, 191, doi:10.3389/fenrg.2020.00191.

Xu, X., K. Vignarooban, B. Xu, K. Hsu, and A.M. Kannan, 2016: Prospects and problems of concentrating solar power technologies for power generation in the desert regions. *Renew. Sustain. Energy Rev.*, **53**, 1106–1131, doi:10.1016/J.RSER.2015.09.015.

Yenneti, K., R. Day, and O. Golubchikov, 2016: Spatial justice and the land politics of renewables: Dispossessing vulnerable communities through solar energy mega-projects. *Geoforum*, **76**, 90–99, doi.org/10.1016/j.geoforum.2016.09.004.

Yi, J. et al., 2021: Post-farmgate food value chains make up most of consumer food expenditures globally. *Nat. Food*, **2** (June), 13–15, doi:10.1038/s43016-021-00279-9.

Younger, P.L. and C. Wolkersdorfer, 2004: Mining Impacts on the Fresh Water Environment: Technical and Managerial Guidelines for Catchment Scale Management. *Mine Water Environ.*, **23(S1)**, s2–s80, doi:10.1007/s10230-004-0028-0.

Yu, G. et al., 2017: Mineral Availability as a Key Regulator of Soil Carbon Storage. *Environ. Sci. Technol.*, **51(9)**, 4960–4969, doi:10.1021/acs.est.7b00305.

Zabaniotou, A., 2018: Redesigning a bioenergy sector in EU in the transition to circular waste-based Bioeconomy-A multidisciplinary review. *J. Clean. Prod.*, **177**, 197–206, doi:10.1016/j.jclepro.2017.12.172.

Zagmutt, F.J., J.G. Pouzou, and S. Costard, 2019: The EAT–Lancet Commission: a flawed approach? *Lancet*, **394(10204)**, 1140–1141, doi:10.1016/S0140-6736(19)31903-8.

Zakkour, P.D. et al., 2020: Progressive supply-side policy under the Paris Agreement to enhance geological carbon storage. *Clim. Policy*, **21**, 1–15, doi:10.1080/14693062.2020.1803039.


12

Zalesny, R.S. et al., 2019: Positive water linkages of producing short rotation poplars and willows for bioenergy and phytotechnologies. *Wiley Interdiscip. Rev. Energy Environ.*, **8(5)**, doi:10.1002/wene.345.

Zech, K.M., and U.A. Schneider, 2019: Carbon leakage and limited efficiency of greenhouse gas taxes on food products. *J. Clean. Prod.*, **213**, 99–103, doi:10.1016/j.jclepro.2018.12.139.

Zeebe, R.E., 2012: History of Seawater Carbonate Chemistry, Atmospheric $CO_2$, and Ocean Acidification. *Annu. Rev. Earth Planet. Sci.*, **40(1)**, 141–165, doi:10.1146/annurev-earth-042711-105521.

Zeebe, R.E. and D. Archer, 2005: Feasibility of ocean fertilization and its impact on future atmospheric CO2 levels. *Geophys. Res. Lett.*, **32(9)**, 1–5, doi:10.1029/2005GL022449.

Zhang, M. et al., 2017: A global review on hydrological responses to forest change across multiple spatial scales: Importance of scale, climate, forest type and hydrological regime. *J. Hydrol.*, **546**, 44–59, doi:10.1016/j.jhydrol.2016.12.040.

Zhang, W., G. Hu, Y. Dang, D.C. Weindorf, and J. Sheng, 2016: Afforestation and the impacts on soil and water conservation at decadal and regional scales in Northwest China. *J. Arid Environ.*, **130**, 98–104, doi:10.1016/j.jaridenv.2016.03.003.

Zhang, Y., 2017: Interregional carbon emission spillover–feedback effects in China. *Energy Policy*, **100**, 138–148, doi.org/10.1016/j.enpol.2016.10.012.

Zhang, Y., M. Pribil, M. Palmgren, and C. Gao, 2020a: A CRISPR way for accelerating improvement of food crops. *Nat. Food*, **1(4)**, 200–205, doi:10.1038/s43016-020-0051-8.

Zhang, Z. et al., 2020b: Production Globalization Makes China’s Exports Cleaner. *One Earth*, **2(5)**, 468–478, doi:10.1016/j.oneear.2020.04.014.

Zheng, X., J. Zhu, and Z. Xing, 2016: Assessment of the effects of shelterbelts on crop yields at the regional scale in Northeast China. *Agric. Syst.*, **143**, 49–60, doi:10.1016/j.agsy.2015.12.008.

Zhou, L., Y. Tian, S. Baidya Roy, Y. Dai, and H. Chen, 2013: Diurnal and seasonal variations of wind farm impacts on land surface temperature over western Texas. *Clim. Dyn.*, **41(2)**, 307–326, doi:10.1007/s00382-012-1485-y.

Zhou, Y., S. Cao, J.L.M. Hensen, and P.D. Lund, 2019: Energy integration and interaction between buildings and vehicles: A state-of-the-art review. *Renew. Sustain. Energy Rev.*, **114**, 109337, doi:10.1016/j.rser.2019.109337.

Zumpf, C., H. Ssegane, M.C. Negri, P. Campbell, and J. Cacho, 2017: Yield and Water Quality Impacts of Field-Scale Integration of Willow into a Continuous Corn Rotation System. *J. Environ. Qual.*, **46(4)**, doi:10.2134/jeq2017.02.0082.

Zurek, M. et al., 2018: Assessing Sustainable Food and Nutrition Security of the EU Food System — An Integrated Approach. *Sustainability*, **10(11)**, 4271, doi:10.3390/su10114271.

# 13

# National and Sub-national Policies and Institutions


**Coordinating Lead Authors:**
Navroz K. Dubash (India), Catherine Mitchell (United Kingdom)

**Lead Authors:**
Elin Lerum Boasson (Norway), Mercy J. Borbor-Córdova (Ecuador), Solomone Fifita (Tonga), Erik Haites (Canada), Mark Jaccard (Canada), Frank Jotzo (Australia), Sasha Naidoo (South Africa), Patricia Romero-Lankao (Mexico/the United States of America), Wei Shen (China/United Kingdom), Mykola Shlapak (Ukraine), Libo Wu (China)

**Contributing Authors:**
Marianne Aasen (Norway), Igor Bashmakov (the Russian Federation), Paolo Bertoldi (Italy), Parth Bhatia (India), Maxwell Boykoff (the United States of America), Jessica Britton (United Kingdom), Sarah Burch (Canada), Charlotte Burns (United Kingdom), Mercedes Bustamante (Brazil), Vanesa Castán Broto (United Kingdom/Spain), Basia Cieszewska (Poland/United Kingdom), Michael Craig (the United States of America), Stephane de la Rue du Can (the United States of America), Manfred Fischedick (Germany), Dana R. Fisher (the United States of America), Amit Garg (India), Oliver Geden (Germany), Robert Germeshausen (Germany), Niklas Höhne (Germany), Angel Hsu (the United States of America/Singapore), Gabriela Iacobuta (Germany), Sébastien Jodoin (Canada), Şiir Kılkış (Turkey), Susanna Kugelberg (Sweden), Matthew Lockwood (United Kingdom), Andreas Löschel (Germany), Cheikh Mbow (Senegal), Brendan Moore (the United States of America/United Kingdom), Yacob Mulugetta (Ethiopia/United Kingdom), Gert-Jan Nabuurs (the Netherlands), Vinnet Ndlovu (Zimbabwe/Australia), Gregory F. Nemet (the United States of America/Canada), Peter Newman (Australia), Lars J. Nilsson (Sweden), Karachepone N. Ninan (India), Grzegorz Peszko (Poland/the United States of America), Marina Povitkina (Sweden), Wang Pu (China), Simone Pulver (the United States of America), Karoline Rogge (Germany/United Kingdom), Joana Setzer (Brazil/United Kingdom), Jale Tosun (Germany)

**Review Editors:**
Alex Godoy (Chile), Xavier Labandeira (Spain)

**Chapter Scientists:**
Parth Bhatia (India), Basia Cieszewska (Poland/United Kingdom)



**This chapter should be cited as:**
Dubash, N.K., C. Mitchell, E.L. Boasson, M.J. Borbor-Cordova, S. Fifita, E. Haites, M. Jaccard, F. Jotzo, S. Naidoo, P. Romero-Lankao, M. Shlapak, W. Shen, L. Wu, 2022: National and sub-national policies and institutions. In IPCC, 2022: *Climate Change 2022: Mitigation of Climate Change. Contribution of Working Group III to the Sixth Assessment Report of the Intergovernmental Panel on Climate Change* [P.R. Shukla, J. Skea, R. Slade, A. Al Khourdajie, R. van Diemen, D. McCollum, M. Pathak, S. Some, P. Vyas, R. Fradera, M. Belkacemi, A. Hasija, G. Lisboa, S. Luz, J. Malley, (eds.)]. Cambridge University Press, Cambridge, UK and New York, NY, USA. doi: 10.1017/9781009157926.015

# Table of Contents



13

## Executive Summary

**Long-term deep emission reductions, including the reduction of emissions to net zero, is best achieved through institutions and governance that nurture new mitigation policies, while at the same time reconsidering existing policies that support continued Greenhouse Gas (GHG) emissions (*robust evidence, high agreement*).** To do so effectively, the scope of climate governance should include both direct efforts to target GHG emissions and indirect opportunities to tackle GHG emissions that result from efforts directed towards other policy objectives. {13.2, 13.5, 13.6, 13.7, 13.9}

**Institutions and governance underpin mitigation by providing the legal basis for action. This includes setting up implementing organisations and the frameworks through which diverse actors interact (*medium evidence, high agreement*).** Institutions can create mitigation and sectoral policy instruments; policy packages for low-carbon system transition; and economy-wide measures for systemic restructuring. {13.2, 13.7, 13.9}

**Policies have had a discernible impact on mitigation for specific countries, sectors, and technologies (*robust evidence, high agreement*), avoiding emissions of several GtCO$^{2}$-eq yr$^{-1}$ (*medium evidence, medium agreement*).** Both market-based and regulatory policies have distinct, but complementary roles. The share of global GHG emissions subject to mitigation policy has increased rapidly in recent years, but big gaps remain in policy coverage, and the stringency of many policies falls short of what is needed to achieve strong mitigation outcomes (*robust evidence, high agreement*). {13.6, Cross-Chapter Box 10 in Chapter 14}

**Climate laws enable mitigation action by signalling the direction of travel, setting targets, mainstreaming mitigation into sector policies, enhancing regulatory certainty, creating law-backed agencies, creating focal points for social mobilisation, and attracting international finance** (*medium evidence, high agreement*). By 2020, 'direct' climate laws primarily focused on GHG reductions were present in 56 countries covering 53% of global emissions, while more than 690 laws, including 'indirect' laws, may also have an effect on mitigation. Among direct laws, 'framework' laws set an overarching legal basis for mitigation either by pursuing a target and implementation approach, or by seeking to mainstream climate objectives through sectoral plans and integrative institutions. {13.2}

**Institutions can enable improved governance by coordinating across sectors, scales and actors, building consensus for action, and setting strategies (*medium evidence, high agreement*).** Institutions are more stable and effective when they are congruous with national context, leading to mitigation-focused institutions in some countries and the pursuit of multiple objectives in others. Sub-national institutions play a complementary role to national institutions by developing locally-relevant visions and plans, addressing policy gaps or limits in national institutions, building local administrative structures and convening actors for place-based decarbonisation. {13.2}

**Sub-national actors are important for mitigation because municipalities and regional governments have jurisdiction over climate-relevant sectors such as land-use, waste and urban policy; are able to experiment with climate solutions; and can forge partnerships with the private sector and internationally to leverage enhanced climate action (*robust evidence, high agreement*).** More than 10,500 cities and nearly 250 regions representing more than 2 billion people have pledged largely voluntary action to reduce emissions. Indirect gains include innovation, establishing norms and developing capacity. However, sub-national actors often lack national support, funding, and capacity to mobilise finance and human resources, and create new institutional competences. {13.5}

**Climate governance is constrained and enabled by domestic structural factors, but it is still possible for actors to make substantial changes (*medium evidence, high agreement*).** Key structural factors are domestic material endowments (such as fossil fuels and land-based resources); domestic political systems; and prevalent ideas, values and belief systems. Developing countries face additional material constraints in climate governance due to development challenges and scarce economic or natural resources. a broad group of actors influence how climate governance develop over time, including a range of civic organisations, encompassing both pro-and anti-climate action groups. {13.3, 13.4}

**Mitigation strategies, instruments and policies that fit with dominant ideas, values and belief systems within a country or within a sector are more easily adopted and implemented (*medium evidence, medium agreement*).** Ideas, values and beliefs may change over time. Policies that bring perceived direct benefits, such as subsidies, usually receive greater support. The awareness of co-benefits for the public increases support of climate policies (*robust evidence, high agreement*). {13.2, 13.3, 13.4}

**Climate litigation is growing and can affect the outcome and ambition of climate governance (*medium evidence, high agreement*).** Since 2015, at least 37 systemic cases have been initiated against states that challenge the overall effort of a state to mitigate or adapt to climate change. If successful, such cases can lead to an increase in a country's overall ambition to tackle climate change. Climate litigation has also successfully challenged governments' authorisations of high-emitting projects setting precedents in favour of climate action. Climate litigation against private sector and financial institutions is also on the rise. {13.4}

**The media shapes the public discourse about climate mitigation. This can usefully build public support to accelerate mitigation action, but may also be used to impede decarbonisation (*medium evidence, high agreement*).** Global media coverage (across a study of 59 countries) has been growing, from about 47,000 stories in 2016–2017 to about 87,000 in 2020–2021. Generally, the media representation of climate science has increased and become more accurate over time. On occasion, the propagation of scientifically misleading information by organised counter-movements has fuelled polarisation, with negative implications for climate policy. {13.4}

**Explicit attention to equity and justice is salient to both social acceptance and fair and effective policymaking for mitigation (*robust evidence, high agreement*).** Distributional implications of alternative climate policy choices can be usefully evaluated at city, local and national scales as an input to policymaking. Institutions and governance frameworks that enable consideration of justice and just transitions are likely to build broader support for climate policymaking. {13.2, 13.6, 13.8, 13.9}

**Carbon pricing is effective in promoting implementation of low-cost emissions reductions (*robust evidence, high agreement*).** While the coverage of emissions trading and carbon taxes has risen to over 20% of global $CO_2$ emissions, both coverage and price are lower than is needed for deep reductions. The design of market mechanisms should be effective as well as efficient, balance distributional goals and find social acceptance. Practical experience has driven progress in market mechanism design, especially of emissions trading schemes (*robust evidence, high agreement*). Carbon pricing is limited in its effect on adoption of higher-cost mitigation options, and where decisions are often not sensitive to price incentives such as in energy efficiency, urban planning, and infrastructure (*robust evidence, medium agreement*). Subsidies have been used to improve energy efficiency, encourage the uptake of renewable energy and other sector-specific emissions saving options (*robust evidence, high agreement*). {13.6}

**Regulatory instruments play an important role in achieving specific mitigation outcomes in sectoral applications (*robust evidence, high agreement*).** Regulation is effective in particular applications and often enjoys greater political support, but tends to be more economically costly, than pricing instruments (*robust evidence, medium agreement*). Flexible forms of regulation (for example, performance standards) have achieved aggregate goals for renewable energy generation, vehicle efficiency and fuel standards, and energy efficiency in buildings and industry (*robust evidence, high agreement*). Infrastructure investment decisions are significant for mitigation because they lock-in high- or low- emissions trajectories over long periods. Information and voluntary programmes can contribute to overall mitigation outcomes (*medium evidence, high agreement*). Designing for overlap and interactions among mitigation policies enhances their effectiveness (*robust evidence, high agreement*). {13.6}

**Removing fossil fuel subsidies would reduce emissions, improve public revenue and macroeconomic performance, and yield other environmental and sustainable development benefits;** subsidy removal may have adverse distributional impacts especially on the most economically vulnerable groups which, in some cases can be mitigated by measures such as redistributing revenue saved, all of which depend on national circumstances (*high confidence*); fossil fuel subsidy removal is projected by various studies (using alternative methodologies) to reduce global $CO_2$ emissions by 1–4%, and GHG emissions by up to 10% by 2030, varying across regions (*medium confidence*). {6.3, 13.6}

**National mitigation policies interact internationally with effects that both support and hinder mitigation action (*medium evidence, high agreement*).** Reductions in demand for fossil fuels tend to negatively affect fossil fuel exporting countries (*medium evidence, high agreement*). Creation of markets for emission reduction credits tends to benefit countries able to supply credits. Policies to support technology development and diffusion tend to have positive spillover effects (*medium evidence, high agreement*). There is no consistent evidence of significant emissions leakage or competitiveness effects between countries, including for emissions-intensive trade-exposed industries covered by emission trading systems (*medium evidence, medium agreement*). {13.6}

**Policy packages are better able to support socio-technical transitions and shifts in development pathways toward low-carbon futures than are individual policies (*robust evidence, high agreement*).** For best effect, they need to be harnessed to a clear vision for change and designed with attention to local governance context. Comprehensiveness in coverage, coherence to ensure complementarity, and consistency of policies with the overarching vision and its objectives are important design criteria. Integration across objectives occurs when a policy package is informed by a clear problem framing and identification of the full range relevant policy sub-systems. {13.7}

**The co-benefits and trade-offs of integrating adaptation and mitigation are most usefully identified and assessed prior to policy making rather than being accidentally discovered (*robust evidence, high agreement*).** This requires strengthening relevant national institutions to reduce silos and overlaps, increasing knowledge exchange at the country and regional levels, and supporting engagement with bilateral and multilateral funding partners. Local governments are well placed to develop policies that generate social and environmental co-benefits but to do so require legal backing and adequate capacity and resources. {13.8}

**Climate change mitigation is accelerated when attention is given to integrated policy and economy-wide approaches, and when enabling conditions (governance, institutions, behaviour, innovation, policy, and finance), are present (*robust evidence, medium agreement*).** Accelerating climate mitigation includes simultaneously weakening high carbon systems and encouraging low-carbon systems; ensuring interaction between adjacent systems (e.g. energy and agriculture); overcoming resistance to policies (e.g., from incumbents in high carbon emitting industries), including by providing transitional support to the vulnerable and negatively affected by distributional impacts; inducing changes in consumer practices and routines; providing transition support; and addressing coordination challenges in policy and governance. {13.7, 13.9}

**Economy-wide packages, including economic stimulus packages, can contribute to shifting sustainable development pathways and achieving net zero outcomes while meeting short term economic goals (*medium evidence, high agreement*).** The 2008–2009 Global Recession showed that policies for sustained economic recovery go beyond short-term fiscal stimulus to include long-term commitments of public spending on the low-carbon economy; pricing reform; addressing affordability; and minimising distributional impacts. COVID-19 spurred stimulus packages and multi-objective recovery policies that may have the potential to meet short-term economic goals while enabling longer-term sustainability goals. {13.9}

## 13.1 Introduction

This chapter assesses national and sub-national policies and institutions. Given the scale and scope of the climate challenge, an immediate challenge for this assessment is defining its scope. Because a very wide range of institutions and policies at multiple scales carry implications for climate change, the approach followed here is to embrace a broad approach. Consequently, institutions and policies discussed include dedicated climate laws and organisations (Section 13.2) and direct mitigation policies such as carbon taxes (Section 13.6), but also those, such as sectoral ministries and their policies (Sections 13.6 and 13.7) and sub-national entities such as regional bodies, cities, and their policies (Section 13.5), the implications of which are salient to mitigation outcomes. This approach recognises that there are important linkages with international climate governance (Chapter 14), notably the role of internationally mandated Nationally Determined Contributions' in stimulating domestic policy development (Section 13.2), transnational networks in spurring sub-national action (Section 13.5), and international effects of domestic policies (Section 13.6).

This encompassing approach to climate governance is also built on a recognition that climate policymaking is routinely formulated in the context of multiple policy objectives such as energy security, energy access, urban development, and mitigation-adaptation linkages. This informs policymaking based on an understanding that to fully maximise direct and indirect climate mitigation potential, maximising co-benefits and minimising trade-offs should be explicitly sought rather than accidentally discovered and policies designed accordingly. This understanding also informs the design of institutions (Section 13.2) and policies (Sections 13.6 and 13.7) as well as the linkage between mitigation and adaptation (Section 13.8).

The chapter also engages with several new developments and an expansion of the literature since AR5.

A growing literature assesses how national policymaking on climate mitigation is dependent on national politics around, and building consensus on, climate action. This, in turn, is shaped by both nationally specific structural features (Section 13.3) and the role of different actors in the policymaking process (Section 13.4). Important new avenues through which climate policy making is shaped, such as climate litigation (Section 13.4.2), and channels for public opinion formation, such as the media (Section 13.4.3) are also assessed. The chapter weaves discussions of the role of justice, understood through a discussion of procedural justice (Section 13.2), distributional justice (Section 13.6) and vulnerability (Section 13.8), and its role in creating public support for climate action (Section 13.9).

A significant new theme is the focus on the dynamic elements of policy making, that is, how policy can be designed to accelerate mitigation. This includes through technological transitions, socio-technical transitions, shifts in development pathways and economy-wide measures. This literature emphasises the importance of examining not just individual policies, but packages of policies (Section 13.7) and how these are enabled by the alignment of policy, institutions, finance, behaviour and innovation (Section 13.9). Also new is attention to the opportunities for economy-wide system change presented by consideration of post-COVID recovery packages, and wider efforts at sustainable economic restructuring (Section 13.9). Consistent with the discussion in Chapter 4, these larger approaches offer opportunities to undertake systemic restructuring and shift development pathways.

Finally, the chapter addresses core themes from earlier assessment reports, but seeks to do so in an enhanced manner. The discussion of climate institutions assesses a growing literature on climate law, as well as both purpose-built climate organisations and the layering of climate responsibilities on existing organisations at national and sub-national scales (Section 13.2). The discussion of policies focuses on an *ex post* assessment of policies, as well as the interaction among them, and learnings on how they can be combined in packages (Sections 13.6 and 13.7). It also lays out a framework for their assessment that encompasses environmental effectiveness, economic effectiveness, distributional outcomes, co-benefits, institutional requirements, as well as a new criterion of transformational potential (Section 13.6).

The aim of this chapter is to assess the full range of the multi-stranded and diverse literature on climate institutions and policy, reflecting the richness of real-world climate governance.

## 13.2 National and Sub-national Institutions and Governance

Institutions and governance arrangements can help address 'policy gaps' and 'implementation gaps' (Cross-Chapter Box 4 in Chapter 4) that hinder climate mitigation. While the need for institutions and governance is universal, individual country approaches vary, based on national approaches and circumstances, as discussed in this section.

Since AR5, the understanding of climate governance has become more encompassing and complex, involving multiple actors, decision-making arenas, levels of decision-making and a variety of political goals. Climate governance sometime directly targets GHG emissions; at other times mitigation results from measures that primarily aim to solve other issues, for instance relating to food production, forest management, energy markets, air pollution, transport systems or technology development, but with mitigation or adaptation effects (Karlsson et al. 2020).

Consistent with usage in this assessment, institutions are rules, norms and conventions that guide, constrain or enable behaviours and practices, including the organisations through which they operate, while governance is the structure, processes and actions that public and private actors use to address societal goals (See Glossary for complete definitions). Multiple terms are used in the literature to discuss climate governance, often varying across countries. Climate laws, or legislation, is passed by legislatures, and often sets the overarching governance context, but the term is also used to refer to legislation that is salient to climate outcomes even if not centrally focused on climate change. National strategies, often referred to as plans, most often operate through executive action by government, set guidance for action and often are not legally binding, although strategies may also be enshrined in law. Both laws and strategies may elaborate targets, or goals, for emissions outcomes, although these are not necessary components of laws and strategies. While laws typically operate at the national level (states may also make laws in federal nations), strategies, plans and targets may also operate at the sub-national level.

This section begins with a discussion of national laws for climate action (Section 13.2.1), followed by a discussion of national strategies (Section 13.2.2). The third section examines institutions (Section13.2.3), including organisations that are established to govern climate actions, and the final section explores sub-national institutions and their challenges in influencing climate mitigation (Section 13.2.4).

### 13.2.1 Climate Laws

National laws that govern climate action often set the legal basis for climate action (Averchenkova et al. 2021). This legal basis can serve several functions: establish a platform for transparent target setting and implementation (Bennett 2018); provide a signal to actors by indicating intent to harness state authority behind climate action (Scotford and Minas 2019); promise enhanced regulatory certainty (Scotford et al. 2017); create law-backed agencies for coordination, compliance and accountability (Scotford and Minas 2019); provide a basis for mainstreaming mitigation into sector action, and create focal points for social mobilisation (*medium evidence*, *high agreement*) (Dubash et al. 2013). For lower/middle income countries, in particular, the existence of a law may also attract international finance by serving as a signal of credibility (Fisher et al. 2017). The realisation of these potential governance gains depends on local context, legal design, successful implementation, and complementary action at different scales.

There are both narrow and broad definitions of what counts as 'climate laws'. The literature distinguishes direct climate laws that explicitly considers climate change causes or impacts – for example through mention of greenhouse gas reductions in its objectives or title (Dubash et al. 2013) – from indirect laws that have 'the capacity to affect mitigation or adaptation' through the subjects they regulate, for example, through promotion of co-benefits, or creation of reporting protocols (Scotford and Minas 2019). Closely related is a 'sectoral approach' based on the layering of climate considerations into existing laws in the absence of an overarching framework law (Rumble 2019). Many countries also adopt executive climate strategies (discussed in Section 13.2), which may either coexist with or substitute for climate laws, and that may also be related to a country's NDC process under the Paris Agreement.

The prevalence of both direct and indirect climate laws has increased considerably since 2007, although definitional differences across studies complicate a clear assessment of their relative importance (*medium evidence*, *high agreement*) (Iacobuta et al. 2018; Nachmany and Setzer 2018). Direct climate laws – with greenhouse gas limitation as a direct objective – had been passed in 56 countries (of 194 studied) covering 53% of emissions in 2020, with most of that rise happening between 2010 and 2015 (Figure 13.1). Both direct and indirect laws – those that have an effect on mitigation even if this is not the primary outcome – is most closely captured by the 'Climate Change Laws of the World' database, which illustrates the same trend of growing prevalence, documenting 694 mitigation-related laws by 2020 versus 558 in 2015 and 342 in 2010 (Nachmany and Setzer 2018; LSE Grantham Research Institute on Climate Change and the Environment 2021).[1] Among these, the majority are accounted for by sectoral indirect laws. For example, a study of Commonwealth countries finds that a majority of these countries have not taken the route of a single overarching law, but rather have an array of laws across different areas, for example, Indian laws on energy efficiency and Ghana's laws on renewable energy promotion (Scotford et al. 2017).

Some direct climate laws may serve as 'framework' laws (Averchenkova et al. 2017; Rumble 2019) that set an overarching legal context within which other legislation and policies operate. Framework laws are intended to provide a coherent legal basis for action, to integrate past legislation in related areas, set clear directions for future policy, and create necessary processes and institutions (*medium evidence*, *medium agreement*) (Townshend et al. 2013; Averchenkova et al. 2017; Fankhauser et al. 2018; Rumble 2019; Averchenkova et al. 2021). There are a variety of approaches to framework laws. Reviews of climate legislation, many of which draw particularly from the long-standing UK Climate Change Act, suggest the need for statutory targets with a long-term direction, shorter term instruments such as carbon budgets to induce action toward targets, a clear assignment of duties and responsibilities including identification of policies and responsibility for their implementation, annual reporting to Parliament; an independent body to support evidence-based decision-making and rules to govern information collection and provision (Barton and Campion 2018; Fankhauser et al. 2018; Abraham-Dukuma et al. 2020; Averchenkova et al. 2021).

However, country examples also suggest other, different approaches to framework laws. Korea's Framework Act on Low Carbon, Green Growth seeks to shift business and society toward green growth through a process of strategy setting and action plans (Jang et al. 2010). Kenya's framework Climate Change Act creates an institutional structure to mainstream climate considerations into sectoral decisions, one of several examples across Africa of efforts to create framework legislation to promote mainstreaming (Rumble 2019). Mexico's General Law on Climate Change includes sectoral emission targets, along with the creation of coordinating institutions across ministries and sub-national authorities (Averchenkova and Guzman Luna 2018). Consequently, different countries have placed emphasis on different aspects of framework laws, although the most widely prevalent approach is that exemplified by the UK.

Climate laws spread through multiple mechanisms, including the impetus provided by international negotiation events, diffusion by

1 Data from climate-laws.org, search for mitigation focused legislation for different time frames. Accessed Oct. 31, 2021.

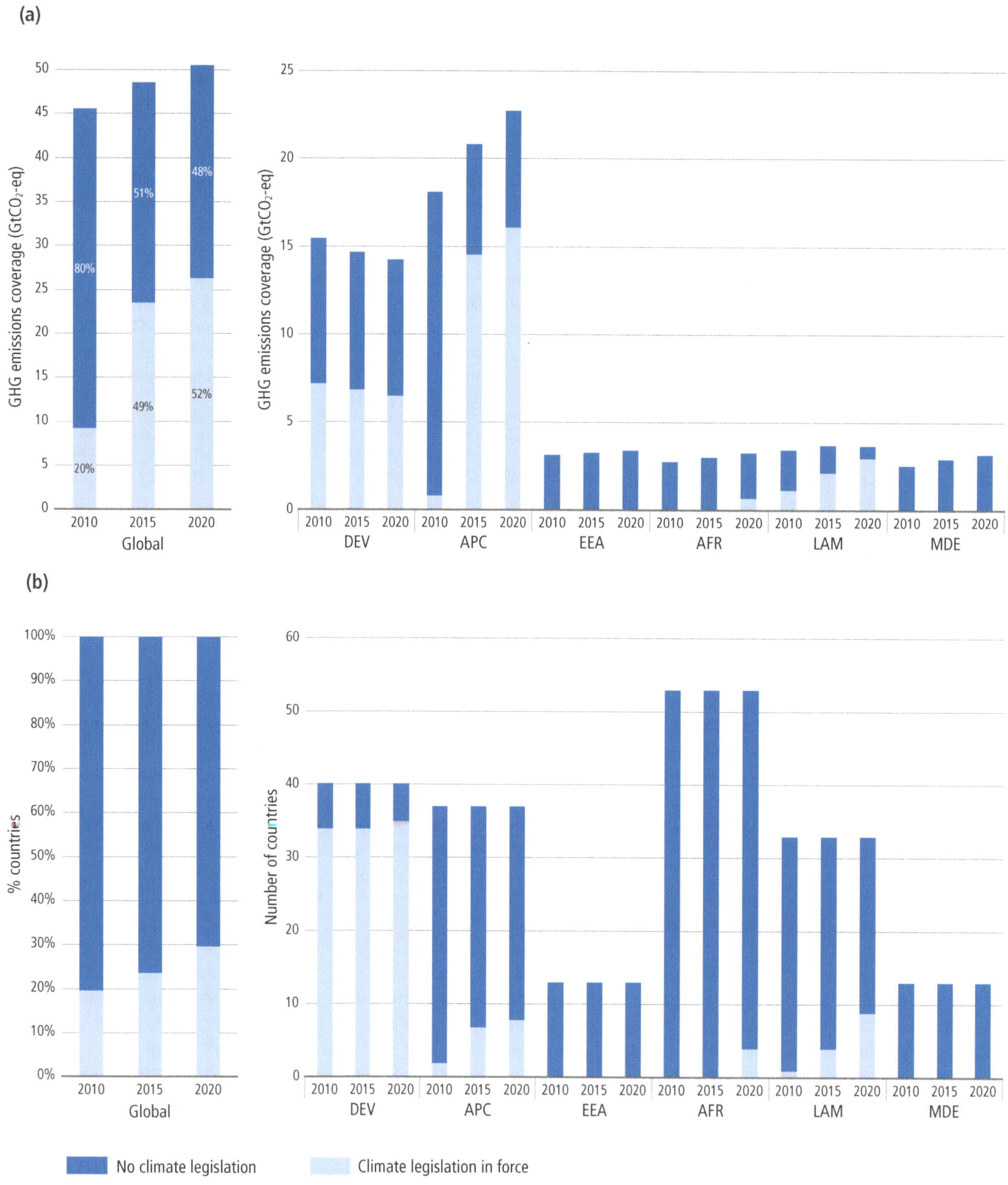


**Figure 13.1 | Prevalence of legislation by emissions and number of countries across regions. Top:** Shares of global GHG emissions under national climate change legislations – in 2010, 2015 and 2020. Emissions data used are for 2019, since emissions shares across regions deviated from past patterns in 2020 due to COVID. **Bottom:** Number of countries with national climate legislation – in 2010, 2015, and 2020. Climate legislation is defined as an act passed by a parliament that includes in its title or objectives reductions in GHGs. AR6 regions: DEV = Developed countries; APC = Asia and Pacific; EEA = Eastern Europe and West-Central Asia; AFR = Africa; LAM = Latin America and the Caribbean; MDE = Middle East. Source: updated and adapted with permission from Iacobuta et al. (2018) to reflect AR6 regional aggregation and recent data.

example across countries, and domestic factors such as business cycles (*medium evidence, medium agreement*). Major landmark events under the UNFCCC have been associated with increases in national legislation (Iacobuta et al. 2018), with a stronger effect in countries where international commitments are binding (Fankhauser et al. 2016). Diffusion through example of legislation from other countries has been documented (Fankhauser et al. 2016; Fleig et al. 2017; Torney 2017; Inderberg 2019; Torney 2019). For example, the UK Climate Change Act was an important influence in pursuing similar acts in Finland and Ireland (Torney 2019) and was also considered in the formulation of Mexico's General Law on Climate Change (Averchenkova and Guzman Luna 2018). The presence of a framework law is positively associated with creation of additional supportive legislation (Fankhauser et al. 2015). Domestic contextual factors can also affect the likelihood of legislation such as a weak business cycle that can impact the political willingness to pass legislation (Fankhauser et al. 2015). In some cases, civil society groups play a role as advocates for legislation, as occurred in the UK (Lockwood 2013; Lorenzoni and Benson 2014; Carter and Childs 2018; Devaney et al. 2020) and in Germany in the build up to passage of their respective Climate Change Act (Flachsland and Levi 2021).

The performance of framework laws suggests a mixed picture. While the structure of the UK Act successfully sets a direction of travel and has resulted in a credible independent body, it performs less well in fostering integration across sectoral areas and providing an enforcement mechanism (Averchenkova et al. 2021). a review of seven European climate change acts concludes that overall targets may not be entirely aligned with planning, reporting and evaluation mechanisms, and that sanction mechanisms are lacking across the board (Nash and Steurer 2019), which limit the scope for legislation to perform its integrative task. These observations suggest the need for careful attention to the design of framework laws.

There is extremely limited evidence on the aggregate effects of climate laws on climate outcomes, although there is a broader literature assessing climate policies (Section 13.6 in this chapter and Cross-Chapter Box 10 in Chapter 14). a single assessment of direct and indirect climate laws as well as relevant executive action across a global database finds a measurable and positive effect: global annual emissions have reduced by about 5.9 $CO_2$ compared to an estimation of what they otherwise would have been (Eskander and Fankhauser 2020). Climate laws require further research, including on the quantification of impact, framework versus sectoral approaches, and the various mechanisms through which laws act – target setting, creating institutional structures, mainstreaming and ensuring compliance.

### 13.2.2 National Strategies and Nationally Determined Contributions

National climate strategies, which are often formulated through executive action, contribute to climate governance in several ways. Strategies enable discussion of low-emissions pathways while accounting for uncertainty, national circumstances and socio-economic objectives (Falduto and Rocha 2020). They frequently set out long term emission goals and possible trajectories over time, with analysis of technological and economic factors (Levin et al. 2018; WRI 2020). This can include quantitative modelling of low-emissions transitions and their economic effects to inform policymakers and stakeholders of potential outcomes (Waisman et al. 2019; Weitzel et al. 2019). Scenario analysis can be used to explore how to make strategies more robust in the face of uncertainty (Sato and Altamirano 2019). Strategies and their regular revision can support long-term structural change by stimulating deliberation and learning (Voß et al. 2009), and to make the link between mitigation and adaptation objectives and actions (Watkiss and Klein 2019; Hans et al. 2020). As part of the Paris Agreement process, several countries have prepared and submitted long-term low-emissions development strategies (Levin et al. 2018), while others have different forms of national climate change strategies independently of the UNFCCC process. Strategies set over time by the European Union are discussed in Box 13.1.

Nationally Determined Contributions (NDCs) prepared under the Paris Agreement may be informed by national strategies (Rocha and Falduto 2019). But the process of preparing NDCs can itself raise political awareness, encourage institutional innovation and coordination, and engage stakeholders (Röser et al. 2020). Nationally Determined Contributions (NDCs) illustrate a diversity of approaches: direct mitigation targets, strategies, plans and actions for low-GHG emission development, or the pursuit of mitigation co-benefits resulting from economic diversification plans and/or adaptation actions (UNFCCC Secretariat 2021). Figure 13.2 shows that the prevalence of emission targets increased across all regions between 2010 and 2020, the period during which the Paris Agreement was reached.

The NDCs vary in their scope, content and time frame, reflecting different national circumstances, and are widely heterogeneous in both stringency and coverage of mitigation efforts (UNFCCC Secretariat 2016, 2021; Pauw et al. 2018; Campagnolo and Davide 2019; Pauw et al. 2019). The mitigation targets in the new or updated NDCs range from economy-wide absolute emission reduction targets to strategies, plans and actions for low-emission development, with specific time frames or implementation periods specified. Less than 10% of parties' NDCs specify when their emissions are expected to peak and some of these parties express their target as a carbon budget (UNFCCC Secretariat 2021). Many long-term strategies submitted by Parties to the UNFCCC refer to net zero emissions or climate neutrality, carbon neutrality, or GHG neutrality with reference to 2050, 2060 or mid-century targets (UNFCCC Secretariat 2021). The growing prevalence and coverage of emission targets is documented in Figure 13.2.

Almost all Parties outlined domestic mitigation measures as key instruments for achieving mitigation targets in specific priority areas such as energy supply (89%), transport (80%), buildings (72%), industry (39%), agriculture (67%), LULUCF (75%) and waste (68%). Renewable energy generation was the most frequently indicated mitigation option (84%), followed by improving energy efficiency of buildings (63%) and multi-sector energy efficiency improvement (48%); afforestation, reforestation and revegetation (48%);

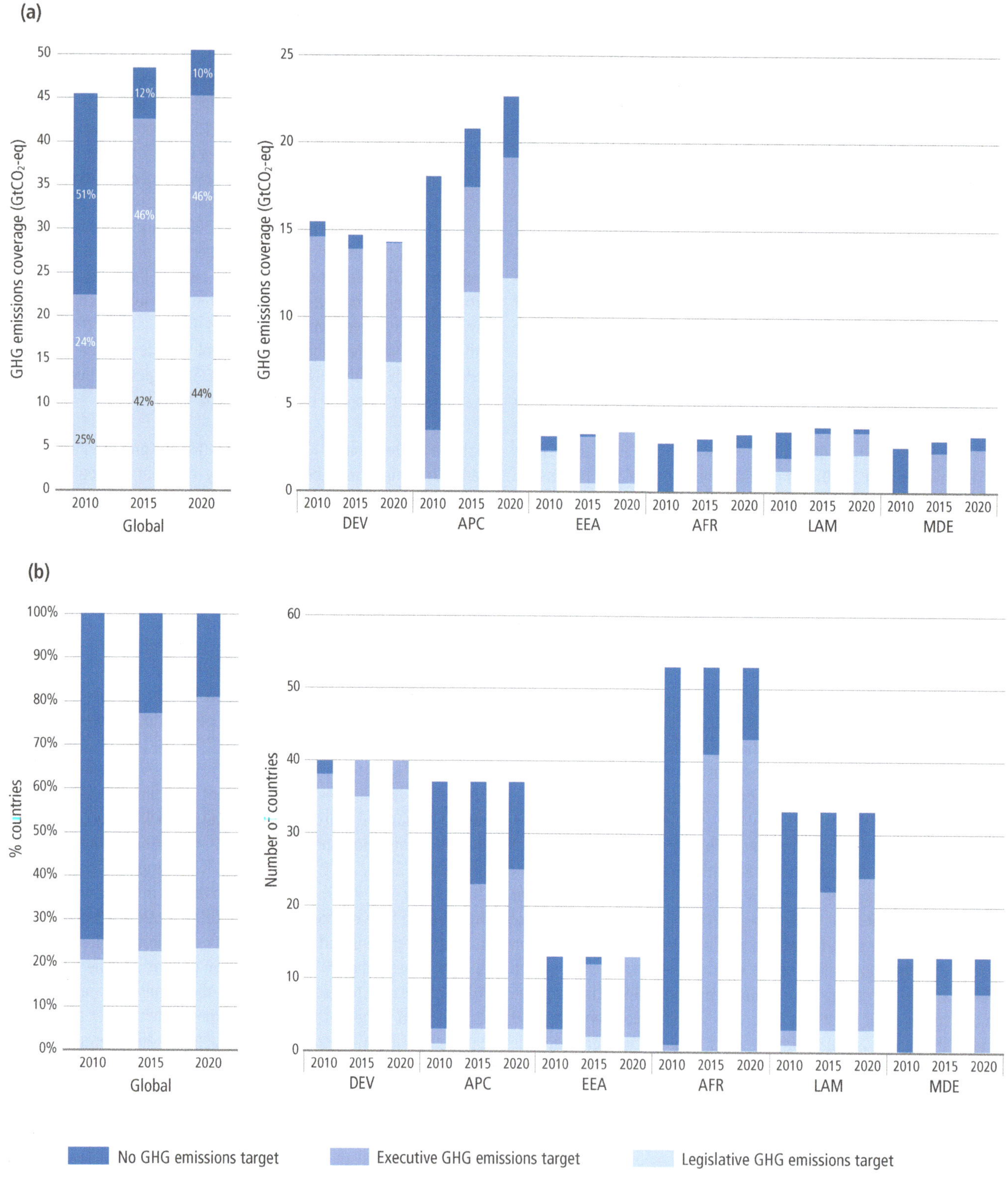


**Figure 13.2 | Prevalence of targets by emissions and number of countries across region. Top:** Shares of global GHG emissions under national climate emission targets – in 2010, 2015 and 2020. Emissions data used are for 2019, since emissions shares across regions deviated from past patterns in 2020 due to COVID. **Bottom:** Number of countries with national climate emission targets – in 2010, 2015, and 2020. Emissions reductions targets were taken into account as a legislative target when they were defined in a law or as part of a country's submission under the Kyoto Protocol, or as an executive target when they were included in a national policy or official submissions under the UNFCCC. Targets were included if they were economy wide or included at least the energy sector. The proportion of national emissions covered are scaled to reflect coverage and whether targets are in GHG or $CO_2$ terms. AR6 regions: DEV = Developed countries; APC = Asia and Pacific; EEA = Eastern Europe and West-Central Asia; AFR = Africa; LAM = Latin America and the Caribbean; MDE = Middle East. Source: updated and adapted with permission from Iacobuta et al. (2018) to reflect AR6 regional aggregation and recent data.

## Box 13.1 | EU Climate Policy Portfolio and the European Green Deal

The European Union (EU)[1] has developed an encompassing climate governance framework (Kulovesi and Oberthür 2020), having ratified the Kyoto Protocol in 2002. In 2003 the EU adopted an Emissions Trading System for sectors with large GHG emitters, which started in 2005. From 2007 to 2009, the EU revised its climate policies, including for vehicle emissions, renewable energy and energy efficiency, and adopted targets for 2020 for GHG emissions reductions, renewable energy shares and energy efficiency improvements. It also adopted in 2009 an Effort Sharing Decision for Member States' emissions reductions for the period 2013–2020 in sectors not covered by the ETS (Boasson and Wettestad 2013; Bertoldi 2018). The ETS has been improved multiple times, including through a 2015 Market Stability Reserve to reduce the surplus of emission allowances (Chaton et al. 2018; Wettestad and Jevnaker 2019). In 2010, the European Commission created a directorate-general (equal to a ministry at the domestic level) for Climate Action. Between 2014 and 2018, the EU agreed on emission reduction targets for 2030 of 30% GHG emission reductions compared to 1990, and again revised its climate policy portfolio including new targets for renewable energies and energy efficiency and a new Effort Sharing Regulation (Fitch-Roy et al. 2019a; Oberthür 2019).

From 2018, climate planning and reporting has been regulated by the EU Governance Regulation (Regulation (EU) 2018/1999), requiring member states to develop detailed and strategic National Energy and Climate Plans (Knodt et al. 2020). In 2019, the European Commission, backed by the European Council (heads of states and government in the EU) and the European Parliament, launched a new broad climate and environment initiative; the 'European Green Deal', implying the revision of many EU polices and introducing the Climate Pact (European Commission 2019a). This roadmap develops a 'new growth strategy for the EU' aimed at reaching climate neutrality by 2050 and spans multiple sectors. In 2020, the European Commission introduced a new climate law establishing the framework for achieving the climate neutrality by 2050 principle, and upgraded its 2030 GHG emission reduction target to at least net 55% reduction, which was adopted in June 2021 (European Commission 2020a). In June 2021, the new policy package 'Fit for 55' was adopted by the Commission; the packages included a proposal for the revision of the ETS, including its extension to shipping and a separate emission trading system for road transport and buildings, a revision of the effort sharing regulation, an amendment of the regulation setting $CO_2$ emission standards for cars and vans, a revision of the energy tax directive, a new carbon border adjustment mechanism, a revision of renewable energy and energy efficiency targets and directives, and a new social fund to make the transition to climate neutrality fair.

and improving energy efficiency of transport (45%) (UNFCCC Secretariat 2021). Parties often communicated mitigation options related to the circular economy, including reducing waste (29%) and recycling waste (30%) and promoting circular economy (25%). Many Parties highlighted policy coherence and synergies between their mitigation measures and development priorities, which included long-term low-emission development strategy(LT-LEDS), the sustainable development goals (SDGs) and, for some, green recovery from the COVID-19 pandemic.

Some countries approach NDCs as an opportunity to integrate mitigation objectives and broader economic shifts or sectoral transformations (*medium evidence*, *medium agreement*). For example, Brazil's 2016 NDC focussed on emissions from land-use change, including agricultural intensification, to align mitigation with a national development strategy of halting deforestation in the Amazon, and increasing livestock production (De Oliveira Silva et al. 2018). While the forest sector accounts for the bulk of Madagascar's mitigation potential, its NDC promotes GHG mitigation in both AFOLU and energy sectors to maximise co-benefits, and achieve a higher number of sustainable development goals (SDGs) (Nogueira et al. 2020).

### 13.2.3 Approaches to National Institutions and Governance

#### 13.2.3.1 The Forms of Climate Institutions

Universal 'best-practice' formulations of organisations may not be applicable across country contexts, but institutions that are suited to national context can be ratcheted up over time in their scope and effectiveness (*medium evidence*, *medium agreement*). National climate institutions take diverse forms because they emerge out of country-specific interactions between national climate politics and existing institutional structures. Certain institutional forms tend to be common across countries, such as expert climate change commissions; a review finds eleven such institutions in existence as of mid-2020. Although this institutional form may be common, these commissions vary in terms of expertise, independence and focus (Abraham-Dukuma et al. 2020), reinforcing the important shaping role of national context.

A review of institutions in eight countries suggests three broad processes through which institutions emerge: 'purpose-built' dedicated institutions focused explicitly on mitigation; 'layering' of mitigation objectives on existing institutions; and 'latent' institutions

[1] The European Union is an international organisation that is discussed here because it plays a large role in shaping climate obligations and policies of its Member States.

## Box 13.2 | Climate Change Institutions in the UK

The central institutional arrangements of climate governance in the UK were established by the 2008 Climate Change Act (CCA): statutory five-year carbon budgets; an independent advisory body, the Committee on Climate Change (CCC); mandatory progress monitoring and reporting to Parliament; and continuous adaptive planning following a five-yearly cycle. The CCC is noteworthy as an innovative institution that has also been emulated by other countries.

The design of the CCC was influenced by the concept of independent central banking (Helm et al. 2003). It has established a reputation for independent high quality analysis and information dissemination, is frequently referred to in Parliament and widely used by other actors in policy debates, all of which suggest a high degree of legitimacy (Averchenkova et al. 2018). However, since the CCC only recommends rather than sets budgets (McGregor et al. 2012), accountability for meeting the carbon budgets works primarily through reputational and political effects rather than legal enforcement.

## Box 13.3 | China's Climate Change Institutions

Climate governance in China features a combination of top-down planning and vertical accountability (Sims Gallagher and Xuan 2019; Teng and Wang 2021). An overarching coordination role is performed by the Leading Group on Carbon Peaking and Carbon Neutrality, appointed by and reporting to the Central Committee of the Chinese Communist Party, and the National Leading Group on Climate Change Response, Energy Conservation, and Emissions Reduction (NLGCCR), headed by the Premier and consisting of more than 30 ministers (Wang et al. 2018a). The Department of Climate Change (DCC) under the Ministry of Ecology and Environment (MEE) is the primary agency in charge of climate issues, with a corresponding local Bureau of Ecology and Environment in each province or city. While MEE is the leading agency for climate policy, the National Development and Reform Commission (NDRC) is the leading agency for setting overall and industry-specific targets in five-year plans, and thus has a key role in coordinating carbon emissions targets with energy and industrial development targets (Wang et al. 2019; Yu 2021). Involvements of ministries related to foreign affairs, public finance, science and technology, as well as sector ministries such as transportation, construction, and manufacturing industries are also needed to push forward sector-specific climate initiatives. At subsidiary levels of government carbon intensity targets are enforced through a 'targets and responsibilities' system that is directly linked to the evaluation of governments' performances (Lin 2012a; Li et al. 2016).

created for other purposes that nonetheless have implications for mitigation outcomes (Dubash 2021). In relatively few countries do new, purpose-built, legally-mandated bodies created specifically for climate mitigation exist although this number is growing; examples include the UK (Averchenkova et al. 2018), China (Teng and Wang 2021), Australia (Keenan et al. 2012) and New Zealand (Timperley 2020). These cases indicate that dedicated and lasting institutions with a strategic long-term focus on mitigation emerge only under conditions of broad national political agreement around climate mitigation as a national priority (Dubash 2021). However, the specific forms of those institutions differ, as illustrated by the case of the UK's Climate Change Committee established as an independent agency (Box 13.2) and China, which is built around a top-down planning structure (Box 13.3).

Where economy-wide institutions do not exist, new institutions may still address sub-sets of the challenge. In Australia, while political conditions resulted in the repeal of an overarching Clean Energy Act in 2014, although a Climate Change Authority continued, other institutions primarily focused on the energy sector such as the Clean Energy Regulator, the Clean Energy Finance Corporation, and the Australia Renewable Agency continued to shape energy outcomes (MacNeil 2021).

Where new dedicated organisations have not emerged, countries may layer climate responsibilities on existing institutions; the addition of mitigation to the responsibilities of the US Environmental Protection Agency is an example (Mildenberger 2021). Layering is also a common approach when climate change is embedded within consideration of multiple objectives of policy. In these cases, climate institutions tend to be layered on sectoral institutions for the pursuit of co-benefits or broader development concerns. Examples include India, where energy security was an important objective of renewable energy promotion policy (Pillai and Dubash 2021), Brazil's mitigation approach focused on sectoral forest policy (Hochstetler 2021) and South Africa's emphasis on job creation as a necessary factor in mitigation policy (Chandrashekeran et al. 2017; Rennkamp 2019). Prior to this process of layering, sectoral institutions, such as in forest and energy sectors, may play an important latent role in shaping climate outcomes, before climate considerations are part of their formal mandate.

New rules and organisations are not only created, they are also dismantled or allowed to wither away. Cases of institutional dismantling or neglect include the Australian Clean Energy Act (Crowley 2017; MacNeil 2021), the Indian Prime Minister's Council on Climate Change, which, while formally functional, effectively does not meet (Pillai and Dubash 2021), and the weakening of climate units inside sectoral ministries in Brazil (Hochstetler 2021). While there is limited literature on the robustness of climate institutions, case studies suggest institutions are more likely to emerge, persist and be effective when institutions map to a framing of climate change that has broad political support (*medium evidence*, *medium agreement*). Thus while mitigation focused framings and institutions may win political support in some countries, in other cases sectorally focused or multiple objectives oriented institutions may be most useful and resilient (Dubash 2021).

#### 13.2.3.2 Addressing Climate Governance Challenges

Climate governance challenges include ensuring coordination, building consensus by mediating conflict, and setting strategy (*medium evidence*, *high agreement*). Coordination is important because climate change is an all-of-economy and society problem that requires cross-sectoral and cross-scale action; building consensus is needed because large-scale transformations can unsettle established interests; and strategy setting is required due to the transformative and time-bound nature of climate mitigation (Dubash et al. 2021). Yet, climate institutions have a mixed record in addressing these challenges.

Institutions that provide coordination, integration across policy areas and mainstreaming are particularly important given the scope and scale of climate change (Candel and Biesbroek 2016; Tosun and Lang 2017) (Section 13.7). Ministries of environment are often appointed as *de facto* agents of coordination, but have been hampered by their limited regulative authority and ability to engage in intra-governmental bargaining with ministries with larger budgets and political heft (Aamodt 2018).

Creation of a high-level coordinating body to coordinate across departments and mainstream climate into sectoral actions is another common approach (Oulu 2015). For example, Kenya has created a National Climate Change Council, which operates through a climate change directorate in the environment ministry to mainstream climate change at the county level (Guey and Bilich 2019). Zhou and Mori (2011) suggest that well-functioning inter-agency coordination mechanisms require support from heads of government, involvement by industry and environment agencies; and engagement by multiple sectoral agencies. However, coordination mechanisms without a clear authority and basis for setting directions run the risk of 'negative coordination', a process through which ministries comment on each other's proposals, removing any ideas that run counter to the interests of their own ministry, leading to even weaker decisions (Flachsland and Levi 2021). Countries with dedicated, new climate institutions tend to have a more explicit and authorised body for climate coordination, such as China's National Leading Group (Box 13.3).

Without explicit coordination with finance ministries, there is a risk of parallel and non-complementary approaches. For example, the South African Treasury pursued a carbon tax without clear indication of how it interfaced with a quantitative sectoral budget approach espoused by the environment ministry (Tyler and Hochstetler 2021). Skovgaard (2012) suggests that there is an important distinction between finance ministries that bring a limiting 'budget frame' to climate action, versus a 'market failure frame' that encourages broader engagement by relevant ministries.

Coordination within federal systems poses additional complexities, such as overlapping authority across jurisdictions, multiple norms in place, and approaches to coordination across scales (Brown 2012). Multi-level governance systems such as the EU can influence the design and functioning of climate policies and institutions in member states, such as Germany (Skjærseth 2017; Jänicke and Wurzel 2019; Flachsland and Levi 2021) and the UK (Lockwood 2021a). In some cases, this can result in distinct European modes of governance as has been suggested occurred in the case of wind energy (Fitch-Roy 2016).

Within countries, institutional platforms allow federal and sub-national governments to negotiate and agree on policy trajectories (Gordon 2015). In Germany, cooperation is channelled through periodic meetings of environment ministers and centre-state working groups (Weidner and Mez 2008; Brown 2012), and in Canada through bilateral negotiations and side-payments between scales of government (Rabe 2007; Gordon 2015). Federal systems might allow for sub-national climate action despite constraints at the federal level, as has occurred in Australia (Gordon 2015; MacNeil 2021) and the United States (Rabe 2011; Jordaan et al. 2019; Bromley-Trujillo and Holman 2020; Thompson et al. 2020). Where agenda-setting rests with the central government, coordination may operate through targets, as with China (Qi and Wu 2013), or frameworks for policy action, as in India (Vihma 2011; Jogesh and Dubash 2015).

Because transition to a low-carbon future is likely to create winners and losers over different time scales; institutions are needed to mediate these interests and build consensus on future pathways (Kuzemko et al. 2016; Lockwood et al. 2017; Finnegan 2019; Mildenberger 2020). Institutions that provide credible knowledge can help support ambition. For example, analysis by the UK Climate Change committee has been harnessed, including by non-state actors, to prevent backsliding on decisions (Lockwood 2021a). Institutions can also help create positive feedback by providing spaces in decision-making for low-carbon interests (Aklin and Urpelainen 2013; Roberts et al. 2018; Lockwood et al. 2017; Finnegan 2019). For example, a renewable energy policy community emerged in China through key agenda setting meetings (Shen 2017), and in India, a National Solar Mission provided a platform for the renewable energy industry (Pillai and Dubash 2021). Conversely, institutions can also exert a drag on change through 'regulatory inertia', as in the case of the UK energy regulator Ofgem, which has exercised veto powers in ways that may limit a low-carbon transition (Lockwood et al. 2017).

Institutions can also create spaces to accommodate concerns of other actors (Upadhyaya et al. 2021). Deliberative bodies, such as Germany's Enquete Commission (Weidner and Mez 2008; Flachsland and Levi 2021) or the Brazilian Forum on Climate Change (Tyler and Hochstetler 2021) provide a space for reconciling competing visions and approaches to climate change. Many countries are creating deliberative bodies to forge 'Just Transition' strategies (Section 13.9).

## Box 13.4 | Procedural Justice

Decision-making consistent with energy and climate justice requires attention to procedural justice (McCauley and Heffron 2018), which includes how decisions are made, and who is involved and has influence on decisions (Sovacool and Dworkin 2015). Procedural justice emphasises the importance of equitable access to decision-making processes and non-discriminatory engagement with all stakeholders (Jenkins et al. 2016), attention to the capability, particularly of marginalised groups, to shape decisions (Holland 2017) and recognition of their specific vulnerabilities in collective political processes (Schlosberg 2012). Consensus-building institutions should avoid reducing normative questions to technical ones, recognising that values, interests and behaviours are all shaped by ongoing climate governance (Ryder 2018; Schwanen 2021). Additionally, communities affected by low-carbon transition may face challenges in articulating their understandings and experiences, which needs to be addressed in the design of climate institutions (Ryder 2018; Schwanen 2021).

Spatially localised alternative discourses of justice are often more recognised socially than national and universal framings of climate justice (Bailey 2017). Participatory forms of governance such as climate assemblies and citizen juries (Ney and Verweij 2015) can help enhance the legitimacy of institutional decisions, even while empirical assessments suggest that these approaches continue to face practical challenges (Devaney et al. 2020; Sandover et al. 2021; Creasy et al. 2021).

## Box 13.5 | South Africa's Monitoring and Evaluation System

South Africa's national monitoring and evaluation system provides high-level guidance on information requirements and assessment methodologies (DEA 2015). The country is developing a comprehensive, integrated National Climate Change Information System, to enable tracking, analysis and enhancement of South Africa's progress towards the country's transition to a low-carbon economy and climate-resilient society (DFFE Republic of South Africa 2021). It includes information on GHG emission reductions achieved, observed and projected climate change, impacts and vulnerabilities, the impact of adaptation and mitigation actions, financial flows and technology transfer activities. South Africa's approach is premised upon continuous learning and improvement through a phased implementation approach (DEA 2019).

a recent innovation is the creation of Citizens' Assemblies that bring together representative samples of citizens to deliberate on policy questions with the intent of informing them (Devaney et al. 2020; Sandover et al. 2021). The ability of institutions to forge agreement also rests on attention to procedural justice (Box 13.4).

Since addressing climate change requires transformative intent and shifting development pathways (Sections 1.6, 3.6, 4.3, 4.4, 13.9, 17.3.2, and Cross-Chapter Box 5 in Chapter 4), institutions that can devise strategies and set trajectories are useful enablers of transformation. Strategy setting often requires an overarching framework such as through framework laws that set targets (Averchenkova et al. 2017), or identify key sectors and opportunities for low-carbon transition (Hochstetler and Kostka 2015) and innovation (UNEP 2018). Few countries have built deliberate and lasting institutions that provide strategic intent, and those that have, have pursued different approaches. The UK's approach rests on five-yearly targets (Box 13.2); Germany requires sectoral budgets enforced through the Bundestag (Flachsland and Levi 2021); and China uses an apex decision-body to set targets (Teng and Wang 2021) (Box 13.3).

Addressing all of these governance concerns – coordination, mediating interests, and strategy setting – require attention to institutional capacity. These include the capacity to address 'upstream' policy issues of agenda setting, framing, analysis and policy design; pursue goals even while mediating interests (Upadhyaya et al. 2021); identify and manage synergies and trade-offs across climate and development objectives (Ürge-Vorsatz et al. 2014; von Stechow et al. 2015; McCollum et al. 2018); identify and choose amongst possible policy options (Howlett and Oliphant 2010); identify areas for transformation and the means to induce innovation (Patt 2017; UNEP 2018); and developing the ability to monitor and evaluate outcomes (Upadhyaya et al. 2021) (Box 13.5). Domorenok et al. (2021) highlight different aspects of the capacity challenge particularly necessary for integrated policy making including: the capacity for horizontal and vertical coordination; implementation capacity including the independence of the state from interests; and administrative capacity required to address compound problems. At a basic level, questions of governmental capacity – the numbers and training of personnel – can shape the choices available for climate institutions and their ability to be strategic (Richerzhagen and Scholz 2008; Harrison and Kostka 2014; Kim 2016). Box 13.5 describes South Africa's approach to building monitoring and evaluation capacity.

The perceived need for attention to institutional capacity is highlighted by the fact that the NDCs of 113 developing countries out of 169 countries studied list capacity building as a condition of NDC implementation (Pauw et al. 2020). While international support for capacity is widely articulated as essential for many countries (Khan et al. 2020), ensuring the form of capacity is appropriate, effective and led domestically remains a challenge (Nago and Krott 2020; Sokona 2021).

### 13.2.4 Institution Building at the Sub-national Level

Jurisdiction over significant mitigation-related arenas like planning, housing and community development reside at the sub-national level. To address linkages between mitigation and local concerns, sub-national actors engage in institution building within a broader socio-economic and political context, with actors and institutions at a multitude of scales shaping the effectiveness of sub-national-scale interventions (Romero-Lankao et al. 2018a). Mitigation policies may demand coordination between sectoral and jurisdictional units that historically have not collaborated; they may require sub-national actors to confront politically sensitive issues such as carbon taxes or increases in utility rates; and they may demand a redistribution of resources to protect endangered ecosystems or vulnerable populations (Hughes and Romero-Lankao 2014).

Sub-national actors have built climate institutions by creating new visions and narratives, by setting new entities or committing existing offices, providing them with funds, staff and legal authority, or by experimenting with innovative solutions that could be transferred to other local governments or scaled nationally (Hoffmann 2011; Hoornweg et al. 2011; Aylett 2015; Hughes and Romero-Lankao 2014; Romero-Lankao et al. 2015; Hughes 2019b). These actors have also created task forces, referendums, coordination of financial and human resources, technical assistance, awareness campaigns and funding (Castán Broto 2017; Romero-Lankao et al. 2018a; Hughes 2019b). National governments can play a key role supporting planning for climate change at the regional and national level, for example, through the articulation of climate change action in national urban politics (Van Den Berg et al. 2018; Cobbinah et al. 2019).

#### 13.2.4.1 Significance of Sub-national Networks

Multi-jurisdictional and multi-sectoral sub-national networks in dozens of countries globally have helped build climate institutions. They have also facilitated social and institutional learning, and addressed gaps in national policy (Holden and Larsen 2015; Jordan et al. 2015; Setzer 2015; Haarstad 2016; Hermwille 2018; Kammerer and Namhata 2018; Rashidi and Patt 2018; Westman and Castan Broto 2018; Lee and Jung 2018; Lee 2019; Schwartz 2019).

Transnational networks have opened opportunities for sub-national actors to play a crucial mitigation role in political stalemates (Jones 2014; Schwartz 2019). The C40, the Global Covenant of Mayors for Climate and Energy, and ICLEI have disseminated information on best practices and promoted knowledge sharing between sub-national governments (Lee 2013; Hakelberg 2014; Heidrich et al. 2016; Kona et al. 2016; Di Gregorio et al. 2020) (Section 14.5.5). Organisations such as the US Carbon Cycle Working Group of the United States Global Change Research Program, the Australian Climate Action Network, and the Mexican Metropolitan Environmental Commission have helped facilitate coordination and learning across multiple jurisdictions and sectors, and connected ambiguous spaces between public, private and civil society actors (Romero-Lankao et al. 2015; Horne and Moloney 2019; Hughes 2019b).

Transnational networks have limited influence on climate policies where national governments exert top-down control (e.g., in the city of Rizhao, China) (Westman et al. 2019); where sub-national actors face political fragmentation, lack regulations, and financial and human resources; or where vertically-integrated governance exists, as in State of São Paulo, Santiago de Chile, and Mexico City (Romero-Lankao et al. 2015; Setzer 2017).

Public support for sub-national climate institutions increases when climate policies are linked to local issues such as travel congestion alleviation or air pollution control (Puppim de Oliveira 2013; Romero-Lankao et al. 2013; Simon Rosenthal et al. 2015; Romero-Lankao et al. 2015; Ryan 2015), or when embedded in development priorities that receive support from the national government or citizens (Jörgensen et al. 2015b; Floater et al. 2016; Dubash et al. 2018). For example, Indian cities have engaged in international climate cooperation seeking innovative solutions to address energy, water and infrastructure problems (Beermann et al. 2016).

#### 13.2.4.2 Factors Influencing Institution Building at the Sub-national Level

Availability of federal funding is a fundamental pillar of city actors' capacity to develop mitigation policies. Administrative structures, such as the presence of a professional city manager and staff assigned specifically to climate efforts (Simon Rosenthal et al. 2015). Cooperation between administrative departments, and the creation of knowledge and data on energy use and emissions are also essential for mitigation planning (Hughes and Romero-Lankao 2014; Ryan 2015). For example, the high technical competency of Tokyo's bureaucracy combined with availability of historical and current data enabled the city's unique cap-and-trade system on large building facilities (Roppongi et al. 2017).

Visions and narratives about the future benefits or risks of climate change are often effectively advanced at the sub-national level, drawing on local governmental abilities to bring together actors involved in place-based decarbonisation across sectors (Hodson and Marvin 2009; Bush et al. 2016; Huang et al. 2018; Prendeville et al. 2018; Levenda et al. 2019). For example, in the plans of 43 C40 Cities, climate action is framed as part of a vision for vibrant, economically prosperous, and socially just cities, that are habitable, secure, resource-efficient, socially and economically inclusive, and competitive internationally (Romero-Lankao and Gnatz 2019).

However, institution building is often constrained by a lack of national support, funding, human resources, coalitions, coordination across old and new organisations, and the ability to create new institutional competences (Valenzuela 2014; Jörgensen et al. 2015a; Ryan 2015; Dubash et al. 2018; Romero-Lankao et al. 2018a; Anderton and Setzer 2018; Cointe 2019; Di Gregorio et al. 2019; Jaccard et al. 2019; Hughes 2019b). Climate mitigation can also be limited by cultural norms and values of policy actors with varying levels of power, and shifting alliances (Lachapelle et al. 2012; Damsø et al. 2016; Giampieri et al. 2019; Romero-Lankao et al. 2018a).

13

**Box 13.6 | Institutionalising Climate Change Within Durban's Local Government**

Durban has effectively linked climate change agendas with ongoing sustainability actions and goals. To do so, adaptation has been broadened to include a just transition to a low-carbon future to address development, energy security and GHG reduction (Roberts et al. 2016).

Durban has mainstreamed climate and justice concerns within local government through strong local leadership by key individuals and departments; included climate concerns within various municipal short-term and long-term planning processes; mobilised civil society; enhanced local and international networking; explored funding opportunities; and restructured institutions (Roberts et al. 2016).

Durban shows that embedding responses to climate change within local government activities requires that climate change is made relevant locally and framed within a broader environmental justice framework (Roberts 2010). Civil society has been key in balancing the influence of the private sector on Durban's dynamic political process (Aylett 2013).

Institution building is constrained by inequities; resources, legal remit, knowledge, and political clout vary widely within and among sub-national governments globally (Jörgensen et al. 2015b; Genus and Theobald 2016; Joffe and Smith 2016; Klinsky 2018; Reckien et al. 2018; Markkanen and Anger-Kraavi 2019). Dominant discourses tend to prioritise scientific and technical expertise and, thus, they focus on infrastructural and economic concerns over the concerns and needs of disadvantaged populations (Heikkinen et al. 2019; Romero-Lankao and Gnatz 2019).

In addition, expert driven, technical solutions such as infrastructural interventions can undermine the knowledge of lower income countries, communities or indigenous knowledge holders, yet are often used by sub-national governments (Ford et al. 2016; Brattland and Mustonen 2018; Nagorny-Koring 2019; Whyte 2017, 2020). Technical solutions, such as electric vehicles or smart grids rarely address the needs and capabilities of disadvantaged communities that may not be able to afford these technologies (Mistry 2014; Romero-Lankao and Nobler 2021). However, mitigation strategies in sectors such as transport and buildings have often focused on technical and market outcomes, the benefits of which are limited to some, while others experience negative externalities or face health risks (Markard 2018; Williams and Doyon 2019; Carley and Konisky 2020). Delivering climate justice requires community-driven approaches to understanding the problem addressing structural inequities and fostering justice, while reducing carbon emissions (Romero-Lankao et al. 2018b; Carley and Konisky 2020; Lewis et al. 2020).

To address this situation requires procedural justice that involves all communities, particularly disadvantaged, in climate mitigation decisions and policies (Box 13.4). Also essential is recognition justice, that addresses past inequities through tools such as subsidies, tariffs, rebates, and other policies (Agyeman 2013; Rydin 2013; UN Habitat 2016). Both tenets are key to ensure the fair distribution of benefits or negative impacts from mitigation policies (distributional justice) (McCauley and Heffron 2018; Lewis et al. 2020). However, the benefits of inclusive approaches are often overlooked in favour of growth oriented mitigation and planning (Rydin 2013; Altenburg 2011; Smith 2019; Lennon 2020). Box 13.6 discusses how the city of Durban has internalised climate change with attention to considerations of justice.

Moreover, deep mitigation requires moving beyond existing technological responses (Mulugetta and Castán Broto 2018) to policies that correspond to the realities of developing countries (Bouteligier 2013). However, best practice approaches tend to be fragmented due to the requirements of different contexts, and often executed as pilot projects that rarely lead to structural change (Nagorny-Koring 2019). Instead, context-specific approaches that include consideration of values, cultures and governance better enable successful translation of best practices (Affolderbach and Schulz 2016; Urpelainen 2018).

## 13.3 Structural Factors that Shape Climate Governance

A growing literature suggests that ambitious climate policy emerges out of strong domestic political support (*medium evidence*, *medium agreement*) (Aklin and Mildenberger 2020; Lamb and Minx 2020; Colgan et al. 2021). Such support is the outcome of political interest constellations and struggles that vary from country to country. Structural factors (such as economic wealth and natural resources, the character of the national political system, and the dominant ideas, values and beliefs) shape how climate change is governed (*medium evidence*, *high agreement*) (Boasson 2015; Hochstetler 2020). This section assesses the ways these structural factors affect political dynamics and decision-making, and ultimately constrain, sustain or enable development of domestic climate governance.

While these structural factors are crucial, they do not determine the outlook of given countries' climate governance, as civic, corporate and/or political groups or individuals can be mobilised and seek to counteract these structural effects, as indicated in the following Section 13.4 that examines the role of various actors and agencies in shaping governance processes. Taken together, Sections 13.3 and 13.4 show that domestic climate governance is not fully constrained by structural factors, but rather that diverse actors can and do achieve substantial changes.

### 13.3.1 Material Endowments

Material endowments are natural and economic resources, such as fossil fuels and renewable energy, forests and land, and economic or financial resources, which tend to shape developments of domestic climate governance (*medium evidence, high agreement*) (Friedrichs and Inderwildi 2013; Lachapelle and Paterson 2013; Bang et al. 2015; Lamb and Minx 2020). Most countries' social and economic systems are largely developed on the basis of their material endowment, and thus they contribute to shape the distribution of political power in that country (Hall and Soskice 2001). Material endowments are by no means the only influencing factor, and actors may succeed to either circumvent or exploit material endowments to impact climate governance (*limited evidence, medium agreement*) (Boasson 2015; Green and Hale 2017; Aklin and Mildenberger 2020).

Since countries are not bound by their material endowment, countries with similar material endowments may differ in climate governance, whereas those with notable differences in material endowments may have similar policies. For instance, countries with rich fossil fuel endowments are found either adopting rather ambitious emission reduction targets and measures, or remaining weak in developing domestic climate policies (Eckersley 2013; Farstad 2019). Further, countries with radically different electricity systems and energy resource potentials are found developing rather similar renewables support schemes such as feed-in-tariff subsidies and competitive tendering programmes (Dobrotkova et al. 2018; Vanegas Cantarero 2020; Boasson et al. 2021). Some policy instruments are widely applied in both developed and developing countries with similar or different material endowment. For example, renewable energy auctions have been experimented by over 100 countries by the end of 2018 (IRENA 2019).

Rich carbon-intensive resources and well developed infrastructure can make low-carbon activities relatively less economically profitable, and negatively influence some perceptions of climate mitigation potential (Bertram et al. 2015a; Erickson et al. 2015). If effective climate policies are introduced despite this, they can alter the importance of country's material endowments in a way that underpin more forceful climate governance over time. For instance, policy interventions to limit fossil fuel exploitation or support renewable energy deployment may change the value of these energy resources over time (Schmitz et al. 2015; Ürge-Vorsatz et al. 2018; Chailleux 2020; Colgan et al. 2021).

Developing countries face additional material constraints in climate governance due to challenges associated with underdevelopment and scarce economic or natural resources (*medium evidence, high agreement*). Hence, many developing countries design domestic climate mitigation policies in combination with policy goals that address various developmental challenges (von Stechow et al. 2016; Deng et al. 2017; Thornton and Comberti 2017; Campagnolo and Davide 2019), such as air quality, urban transportation, energy access, and poverty alleviation (Klausbruckner et al. 2016; Li et al. 2016; Melamed et al. 2016; Slovic et al. 2016; Khreis et al. 2017; Geall et al. 2018; Xie et al. 2018). Combining climate and developmental policies for beneficial synergies should not overlook potential trade-offs and challenges (Dagnachew et al. 2018; Ellis and Tschakert 2019; Peñasco et al. 2021) (Section 13.7.2 for wider discussion).

### 13.3.2 Political Systems

The effectiveness of domestic climate governance will significantly rely on how well it fits with the features of the countries' specific political systems (*limited evidence, high agreement*) (Schmitz 2017; Lamb and Minx 2020). Political systems have developed over generations and constitute a set of formal institutions, such as laws and regulations, bureaucratic structures, political executives, legislative assemblies and political parties (Egeberg 1999; Pierson 2004). Different political systems create differing conditions for climate governance to emerge and evolve, but because political systems are so politically and historically entrenched they are not likely to change quickly even though this could facilitate domestic climate mitigation efforts (*medium evidence, high agreement*) (Duit and Galaz 2008; Boasson et al. 2021). In addition, variations in governance capacities also affect climate policy making and implementation (Meckling and Nahm 2018).

Broader public participation and more open contestation spaces tend to nurture more encompassing climate policies, facilitate stronger commitments to international agreements (Bättig and Bernauer 2009; Böhmelt et al. 2016), achieve more success in decoupling economic growth from $CO_2$ emissions (Lægreid and Povitkina 2018), reduce more $CO_2$ emissions (Clulow 2019; von Stein 2020), and maintain lower deforestation rates (*medium evidence, medium agreement*) (Buitenzorgy and Mol 2011). States with less public participation and contestation space can also develop ambitious climate emission reduction targets and institutions (Zimmer et al. 2015; Eckersley 2016; Han 2017; Engels 2018), but the drivers and effects of climate policies within less open and liberal political contexts has not yet been sufficiently investigated.

Election systems based on proportional representation tend to have lower emissions, higher energy efficiency, higher renewable energy deployment, and more climate friendly investment than systems where leadership candidates have to secure a majority of the votes to be elected (*medium evidence, high agreement*) (Fredriksson and Millimet 2004; Lachapelle and Paterson 2013; Finnegan 2019). Such systems better enable voters supporting ambitious climate positions to influence policymaking (Harrison and Sundstrom 2010; Willis 2018), place less political risks on legislators from additional costs incurred from climate actions on voters (Finnegan 2018, 2019), and strengthen credible commitments to climate policy (Lockwood 2021b). Similarly, rules that govern the relationship between governments and civic societies in decision-making have also been shown to matter in climate governance. Corporatist societies, where economic groups are formally involved in public policy making, have better climate-related outcomes (lower $CO_2$ emissions and higher low-carbon investments) than liberal-pluralist countries, where a larger array of non-governmental organisations compete for informal influence, often through lobbying (*medium evidence, medium agreement*) (Liefferink et al. 2009; Jahn 2016; Finnegan 2018).

Political parties with similar ideological roots in different countries (for instance social democratic or conservative parties) may have different positions on climate governance across countries (Boasson et al. 2021). Nevertheless, on average, a higher share of green parties in a parliament is associated with lower greenhouse gas emissions (Neumayer 2003; Jensen and Spoon 2011; Mourao 2019), and left-wing parties tend to adopt more pro-climate policy positions (*medium evidence*, *high agreement*) (Carter 2013; Tobin 2017; Farstad 2018; Ladrech and Little 2019). There is also evidence, however, that conservative parties in some countries support climate measures (Båtstrand 2015) and consensus can be achieved on climate actions across the political spectrum (Thonig et al. 2021). At the same time, it seems harder to get support for new climate governance initiatives in systems where many political groups can block decision due to many veto points, for instance in systems with bicameralism (the legislature is divided into two separate assemblies) and/or in federalist governments (where regions have national political representation, e.g. USA and Brazil) (*medium evidence*, *high agreement*) (Madden 2014; von Stein 2020) although federal systems hold out the possibility of sub-national action when federal agreement is limited (Section 13.2). There remains a limited literature on the role of green parties and veto points in developing countries (Haynes 1999; Kernecker and Wagner 2019).

In any political system, climate policy adoption and implementation may be obstructed by corrupt practices (Rafaty 2018; Fredriksson and Neumayer 2016) that entail an abuse of entrusted power for private gain (*medium evidence*, *high agreement*) (Treisman 2000). Evidence shows that $CO_2$ emissions levels can be affected by corruption, either through the direct negative effect of corruption on law enforcement, including in the forestry sector (Sundström 2016), or through the negative effect of corruption on countries' income (Welsch 2004). These early findings are reinforced by studies of a global sample of countries (Cole 2007) and from across the developing world (Sahli and Rejeb 2015; Bae et al. 2017; Wang et al. 2018b; Ridzuan et al. 2019; Habib et al. 2020). Corruption also disrupts public support of climate policies by affecting the levels of trust (*medium evidence*, *high agreement*) (Harring 2013; Fairbrother et al. 2019; Davidovic and Harring 2020), which then impact on the compliance of climate policies. More research is required to further understand the causal mechanisms between corrupt practices and emissions.

### 13.3.3 Ideas, Values and Belief Systems

Ideas, values and beliefs affect climate governance by shaping people's perceptions, attitude, and preferences on specific policy and governance issues (*medium evidence*, *high agreement*) (Boasson 2015; McCright et al. 2016b; Schifeling and Hoffman 2019; Leipold et al. 2019; Boasson et al. 2021). While these are often entrenched, they can also change, for instance when facing growing exposures to climate risks, stronger scientific evidence, and dominant public or political discourse (Mayer et al. 2017; Diehl et al. 2021). While change tend to be incremental, the pace of change may vary substantially across countries and specific climate issue areas.

However, new norms sometimes only influence political discussion and not actual governance. For instance, more ambitious climate emission reduction targets may not lead to more effective mitigation actions or policy instruments. Put another way, words do not replace actions (Geden 2016).

Different sets of beliefs can shape climate-related policies, targets, and instruments (Boasson and Wettestad 2013; Boasson 2015; Boasson et al. 2021). First, beliefs link climate governance with social justice concerns; policies, targets and instruments may therefore reflect justice issues (Fuller and McCauley 2016; Reckien et al. 2017; McCauley and Heffron 2018; Routledge et al. 2018; Bäckstrand and Lövbrand 2006, 2019). Second, climate mitigation may be seen as primarily a market correction issue and mitigation compatible with economic growth, as exemplified by ecological modernisation (Mol et al. 2009; Bäckstrand and Lövbrand 2006, 2019), climate capitalism (Newell and Paterson 2010), market logics (Boasson 2015; Boasson et al. 2021) or a global commons approach (Bernstein and Hoffmann 2019). Third, climate governance may be understood relative to policies on technological innovation and progress, often conceptualised as social-technical transformations (Geels et al. 2017a).

Significant variation in ideas, values and beliefs related to climate governance are detected across and within regions, countries, societies, organisations, and individuals (*medium evidence*, *medium agreement*) (Shwom et al. 2015; Boasson et al. 2021; Knox-Hayes 2016; Wettestad and Gulbrandsen 2018). These factors provide the context for climate policymaking and include differences in countries' histories (Aamodt 2018; Aamodt and Boasson 2020); the political culture and regulatory traditions in governing environmental and energy issues (Tosun 2018; Aamodt 2018; Boasson et al. 2021); and even bureaucrats' educational background (Rickards et al. 2014). Structural factors in a country, such as deeply held value systems, are not changed rapidly, just as political systems or natural endowments, are not changed rapidly. Consequently, climate policy and governance is more effective if it takes into account these deep-rooted values and beliefs.

Differences in dominant individual preferences may also be important. The factors that shape individual ideas, values and beliefs about climate governance include trust in politicians, the state and other people in general (Drews and van den Bergh 2016; Harring et al. 2019; Huber et al. 2020), fairness beliefs, variation in political orientation (left leaning more concerned), and class (*medium evidence*, *medium agreement*) (Schmitz et al. 2018; Inglehart and Norris 2017).

Levels of climate change concern on the individual level have increased in most countries (Shwom et al. 2015), and vary with gender (females are more concerned), and place of residence (urban residents are more concerned) (Shwom et al. 2015; McCright et al. 2016a; Ziegler 2017). The higher educated in developing countries tend to be more concerned (Lee et al. 2015) while individuals working in polluting industries tend to oppose forceful climate governance (Bechtel et al. 2019; Mildenberger 2020).

Shifts in mainstream ideas, values and beliefs can underpin changes in climate policy choices and policy outcomes (*limited evidence*, *medium agreement*) (Schleich et al. 2018; Mildenberger and Tingley 2019). For example, emission trading schemes are welcomed as a new regulatory instrument in China in the context of its market-oriented reforms and ideological shift in the past decades (Lo 2013). Based on the study of 167 nation-states and 95 sub-national jurisdictions with carbon pricing, researchers find that that high public belief in climate science underpin adoption of systems that produce a rather high carbon price (Levi et al. 2020). These public opinions need to be identified and leveraged in supporting specific policy choices or changes (Mildenberger and Tingley 2019). Policy support tends to be greater if people believe effective measures are being taken by other actors, including other households (Bostrom et al. 2018; Marlon et al. 2019), and other countries and at the international level (Schleich et al. 2018).

On the other hand, anti-climate ideas or beliefs may arise due to the introduction of more constraining or ambitious climate policies, for example protests in reaction to toll roads in Norway, which increase the cost of driving, or protests in France against increasing carbon taxes (Grossman 2019; Wanvik and Haarstad 2021). The policy implication is that vulnerable or effected groups should be considered when introducing policy change, and that participation, transparency, and good communication all helps to reduce climate-related discontent.

Survey-based studies of public perceptions on hypothetical policy instruments or activities, such as carbon taxes or energy infrastructure, suggest that linking climate policy to other economic and social reforms can increase public support for climate governance (Carattini et al. 2019; Bergquist et al. 2020). People and politicians tend to underestimate other peoples' and politicians' willingness to support mitigation policies (Hurlstone et al. 2014; Mildenberger and Tingley 2019), but if actors are informed about other actors actual perceptions and behaviours this may reduce the tendency to underestimate climate governance support (Mildenberger and Tingley 2019).

## 13.4 Actors Shaping Climate Governance

While Section 13.3 shows that structural factors condition climate governance, their ultimate importance also depends on whether and how various actors are mobilised (Hochstetler 2020; Boasson 2015). a wide range of regional and local governments as well as non-governmental actors have become increasingly engaged in climate governance, for instance through public-private partnerships and transnational networks (Jordan et al. 2015; Dorsch and Flachsland 2017; Jordan et al. 2018) and through the media and litigation, as discussed here.

Climate governance processes result from both slow-moving incremental changes to policy and more rapid bursts of change due to, for example, responses to dramatic weather events, general elections or global climate summits (*medium evidence*, *high agreement*) (Aamodt and Stensdal 2017; Jordan and Moore 2020; Boasson et al. 2021). While Section 13.3 assessed how entrenched structural factors conditions climate governance developments, this section examines how actors are able to alter climate governance by engaging the climate policy process, undertaking litigation and interacting with media.

### 13.4.1 Actors and Agency in the Public Process

A broad array of actors are engaged in shaping mitigation policy processes, including politicians and political parties, corporate actors, citizen groups, indigenous peoples organisations, labour unions and international organisations. Actors aiming to influence the climate-related policymaking process are studied together to understand climate policy dynamics and outcomes (Bulkeley 2000; Fisher 2004; Jost and Jacob 2004; Jasny et al. 2015; Fisher and Leifeld 2019; Jasny and Fisher 2019) and collaboration and influence within climate policy networks (Ingold and Fischer 2014; McAllister et al. 2014; Wagner and Ylä-Anttila 2018; Kammerer et al. 2021). Most research, however, focuses on one particular type of actor.

Political actors are decision-makers, and also influence whether climate governance is perceived as urgent and appropriate (Okereke et al. 2019; Ferrante and Fearnside 2019; Boasson et al. 2021). They include political parties, legislative assemblies and committees, governmental executives and the political leaders of governmental ministries (Boasson 2015). They are more likely to pay attention to climate issues when polling indicates high political salience with the public (Carter 2006, 2014), or when it becomes a contested issue among differing political parties (Boasson et al. 2021). Fluctuations in the public's interest and attention may underpin a disjointed approach in politicians' engagement (Willis 2017, 2018). Policy implementation can be hampered if political actors propose frequent policy changes (Boasson et al. 2021).

Corporate actors often influence policies and their adoption (Pulver and Benney 2013; Mildenberger 2020; Goldberg et al. 2020). Corporate actors acting individually or through industry associations, have worked to sway climate policy in different countries (Falkner 2008; Bernhagen 2008; Newell and Paterson 2010; Meckling 2011; Mildenberger 2020). Their ability varies by country and issue (*medium evidence*, *medium agreement*) (Skjærseth and Skodvin 2010; Boasson and Wettestad 2013; Boasson 2015; Boasson et al. 2021) and depends on material endowments (Moe Singh 2012), access to the political system (Dillon et al. 2018; Mildenberger 2020), and the ability to shape ideas, values and belief systems (Boasson 2015). Corporate actors tend to change their climate policy preferences over time, as indicated by longitudinal studies of some European countries (Boasson and Wettestad 2013; Boasson 2015; Boasson et al. 2021).

Corporate actors are crucial to policy implementation because they are prominent emitters of the greenhouse gases and owners of carbon-intensive technologies and potential providers of solutions as developers, owners and adopters of low emission practices and technologies (Falkner 2008; Perrow and Pulver 2015). Many climate policies and measures rely on businesses' willingness to exploit newly created economic opportunities, such as support schemes for renewable energy and energy efficiency sector or carbon pricing (Olsen 2007; Newell and Paterson 2010; Shen 2015; World Bank 2019). Some corporate actors provide climate solutions, such as renewable energy deployment, and have successfully influenced climate policy development related to feed-in tariffs, taxations, quotas, or emission trading schemes, in the EU (Boasson 2019), Germany (Leiren and Reimer 2018), the USA (Stokes and Breetz 2018), the Nordic countries (Kooij et al. 2018), China (Shen 2017) and Japan (Li et al. 2019).

Fossil fuel industries have been important agenda-setters in many countries, including the USA (Dunlap and McCright 2015; Supran and Oreskes 2017; Downie 2018), the EU (Skjærseth and Skodvin 2010; Boasson and Wettestad 2013), Australia (Ayling 2017), China (Shen and Xie 2018; Tan et al. 2021), India (Schmitz 2017; Blondeel and Van de Graaf 2018), and Mexico (Pulver 2007), with differing positions and impacts across countries (Kim et al. 2016; Nasiritousi 2017). In the US, the oil industry has underpinned emergence of climate scepticism (Dunlap and McCright 2015; Farrell 2016a; Supran and Oreskes 2017), and its spread abroad (Dunlap and Jacques 2013; Engels et al. 2013; Painter and Gavin 2016). Corporate opposition to climate policies is often facilitated by a broad coalition of firms (Cory et al. 2021).

Conservative foundations, sometimes financed by business revenues, have funded a diversity of types of groups, including think-tanks, philanthropic foundations, or activist networks to oppose climate policy (Brulle 2014, 2019). However, there is limited knowledge about the conditions under which actors opposed to climate action succeed in shaping climate governance (Kinniburgh 2019; Martin and Islar 2021).

Some labour unions have developed positions and programmes on climate change (Snell and Fairbrother 2010; Stevins 2013; Räthzel et al. 2018), formed alliances with other actors in the field of climate policy (Stevis 2018) and participated in domestic policy networks on climate change (Just and Jacob 2004), but we know little about their relative importance or success. In countries with significant fossil fuel resources such as Australia, Norway, and the United States, labour unions, particularly industrial unions, tend to contribute to reducing the ambition of domestic climate policies mainly due to the concern of job losses (Mildenberger 2020). Other studies find that the role of labour unions varies across countries (Glynn et al. 2017).

Civil society actors can involve citizens working collectively to change individual behaviours that have climate implications. For example, environmental movements that involve various forms of collective efforts encourage their members to make personal lifestyle changes that reduce their individual carbon footprints (Ergas 2010; Middlemiss 2011; Haenfler et al. 2012; Cronin et al. 2014; Saunders et al. 2014; Büchs et al. 2015; Wynes et al. 2018). These efforts seek to change individual members' consumer behaviours by reducing car-use and flying, shifting to non-fossil fuel sources for individual sources of electricity, and eating less dairy or meat (Cherry 2006; Ergas 2010; Middlemiss 2011; Haenfler et al. 2012; Stuart et al. 2013; Cronin et al. 2014; Saunders et al. 2014; Büchs et al. 2015; Wynes and Nicholas 2017; Wynes et al. 2018; Thøgersen et al. 2021). Consumer/citizen engagement is sometimes encouraged through governmental directives, such as the 'renewable energy communities' granted by the EU renewable energy directive 2018/2001 (The European Parliament and the Council of the European Union 2018). To date, there are only a limited number of case studies that measure the direct effect of participation in these types of movements as it relates to climate outcomes (Saunders et al. 2014; Vestergren et al. 2018, 2019).

Citizens with less access to resources and power also participate by challenging nodes of power – policymakers, regulators, and businesses – to change their behaviours and/or accelerate their efforts. Tactics include lobbying, legal challenges, shareholder activism, coop board stewardship, and voting (Gillan and Starks 2007; Schlozman et al. 2012; Viardot 2013; Bratton and McCahery 2015; Yildiz et al. 2015; Olzak et al. 2016). Citizens provide the labour and political will needed to pressure political and economic actors to enact emission-reducing policies, as well as providing resistance to them (Fox and Brown 1998; Boli and Thomas 1999; Oreskes and Conway 2012; McAdam 2017).

Other citizen engagement involves a range of more confrontational tactics, such as boycotting, striking, protesting, and direct action targeting politicians, policymakers, and businesses (Fisher et al. 2005; Tarrow 2005; Fisher 2010; Saunders et al. 2012; Walgrave et al. 2012; Wahlström et al. 2013; Eilstrup-Sangiovanni and Bondaroff 2014; Hadden 2014, 2015; O'Brien et al. 2018; Chamorel 2019; Cock 2019; 2019b; Hadden and Jasny 2019; Swim et al. 2019). Climate strikes and other more confrontational forms of climate activism have become increasingly common (O'Brien et al. 2018; Evensen 2019; D.A. Fisher 2019; Boulianne et al. 2020; Martiskainen et al. 2020; de Moor et al. 2021; Fisher and Nasrin 2021a). Very few studies look specifically at the effect of these tactics on actual climate-related outcomes and more research is needed to understand the climate effects of citizen engagement and activism (Fisher and Nasrin 2021b).

Citizen engagement has also become common among indigenous groups who tend to have limited structural power but often aim to shape the formation and effects of projects that have implications to climate change. These include opposing extraction and transportation of fossil fuels on their traditional lands (especially in the Americas) (Bebbington and Bury 2013; Hindery 2013; Coryat 2015; Claeys and Delgado Pugley 2017; Wood and Rossiter 2017); large-scale climate mitigation projects that may affect traditional rights (Brannstrom et al. 2017; Moreira et al. 2019; Zárate-Toledo et al. 2019); supporting deployment of small-scale renewable energy initiatives (Thornton and Comberti 2017); seeking to influence the development of REDD+ policies through opposition (Reed 2011); and participation in consultation processes and multi-stakeholder bodies (Bushley 2014; Gebara et al. 2014; Astuti and McGregor 2015; Kashwan 2015; Jodoin 2017). Indigenous groups have been reported to have had some influence on some climate discussions, particularly forest management and siting of renewable energy (Claeys and Delgado Pugley 2017; Jodoin 2017; Thornton and Comberti 2017). Further, more scientific assessments are required on the role of indigenous groups in climate activism and policy (Jodoin 2017; Claeys and Delgado Pugley 2017; Thornton and Comberti 2017).

Activism, including litigation, as well as the tactics of protest and strikes, have played a substantial role in pressuring governments to create environmental laws and environmental agencies tasked

## Box 13.7 | Civic Engagement: The School Strike Movement

On Friday 20 August 2018, Greta Thunberg participated in the first climate school strike. Since then, Fridays for Future – the name of the group coordinating this tactic of skipping school on Fridays to protest inaction on climate change – has spread around the world.

In March 2019, the first *global* climate strike took place, turning out more than one million people around the world (Carrington 2019). Six months later in September 2019, young people and adults responded to a call to participate in climate strikes as part of the 'Global Week for Future' surrounding the UN Climate Action Summit (Thunberg 2019), and the number of participants globally jumped to an estimated six million people (Taylor et al. 2019). Although a handful of studies have reported on who was involved in these strikes, how they were connected, and their messaging (Marris 2019; Wahlström et al. 2019; Evensen 2019; D. Fisher 2019; Boulianne et al. 2020; Bevan et al. 2020; Han and Ahn 2020; Holmberg and Alvinius 2020; Jung et al. 2020; Martiskainen et al. 2020; Thackeray et al. 2020; Trihartono et al. 2020; de Moor et al. 2021; Fisher and Nasrin 2021b), its consequences in terms of political outcomes and emissions reductions have yet to be fully understood (Fisher and Nasrin 2021b).

Although digital activism makes it easier to connect globally, it is unclear how digital technology will affect the youth climate movement, and its effects on carbon emissions. Research suggests that online activism is likely to involve a more limited range of participants and perspectives (Bennett 2013; Elliott and Earl 2018). Digital tactics could also mean that groups are less embedded in communities and less successful at creating durable social ties, factors that have been found to lead to longer term engagement (Tufekci 2017; Rohlinger and Bunnage 2018; Shirky 2010).

with enforcing environmental laws that aimed to maintain clean air and water in countries around the world (*medium evidence*, *high agreement*) (McCloskey 1991; Schreurs 1997; Rucht 1999; Brulle 2000; Steinhardt and Wu 2016; Longhofer et al. 2016; Wong 2018). Several studies find environmental NGOs have a positive effect on reductions in carbon emissions, whether through effects that operate across countries or (Frank et al. 2000; Schofer and Hironaka 2005; Jorgenson et al. 2011; Baxter et al. 2013; Longhofer and Jorgenson 2017; Grant et al. 2018) through impact of NGOs within nations (Shwom 2011; Dietz et al. 2015; Grant and Vasi 2017).

At the same time, other research has documented various forms of backlash against climate policies, both in terms of voting behaviour, as well as other collective efforts (Hill et al. 2010; Williamson et al. 2011; McAdam and Boudet 2012; Wright and Boudet 2012; Walker et al. 2014; Boudet et al. 2016; Fast et al. 2016; Krause et al. 2016; Lyon 2016; Mayer 2016; Stokes 2016; Stokes and Warshaw 2017; Muradian and Pascual 2020; Stokes 2020). In a systematic analysis that includes movements against fossil fuel investments along with those against low-carbon emitting projects around the world, research finds that a quarter of all projects (no matter their targets) were cancelled after facing resistance (Temper et al. 2020).

A range of international organisations can be important, particularly in developing countries, for instance by assisting in framing of national climate governance and supporting the design of climate policies through technical assistance projects (Talaei et al. 2014; Ortega Díaz and Gutiérrez 2018; Kukkonen et al. 2018; Bhamidipati et al. 2019; Charlery and Trærup 2019). Yet for these climate aid initiatives to work effectively requires improved institutional architecture, better appreciation of local contexts, and more inclusive and transparent governance, based on evidence from many multilateral mechanisms like REDD+, CDM, GEF and GCF (Gomez 2013; Arndt and Tarp 2017), and bilateral programmes on energy, agriculture and land-use sectors (Arndt and Tarp 2017; Rogner and Leung 2018; Moss and Bazilian 2018).

### 13.4.2 Shaping Climate Governance Through Litigation

Outside the formal climate policy processes, climate litigation is another important arena for various actors to confront and interact over how climate change should be governed (*robust evidence*, *high agreement*) (Wilensky 2015; Peel and Osofsky 2015, 2018; Bouwer 2018; Setzer and Byrnes 2019; Calzadilla 2019; Setzer and Vanhala 2019; Paiement 2020; Wegener 2020). Climate litigation is an attempt to control, order or influence the behaviour of others in relation to climate governance, and it has been used by a wide variety of litigants (governments, private actors, civil society and individuals) at multiple scales (local, regional, national and international) (Osofsky 2007; Lin 2012b; Keele 2017; McCormick et al. 2018; Peel and Osofsky 2018; Setzer and Vanhala 2019). Climate litigation has become increasingly common (UNEP2020b), but its prevalence varies across countries (*medium evidence*, *high agreement*) (Peel and Osofsky 2015; Wilensky 2015; Bouwer 2018; Lin and Kysar 2020; Setzer and Higham 2021). This is not surprising, given that courts play differing roles across varying political systems and law traditions (La Porta et al. 1998).

This sub-section focuses on relevant climate litigation for policies and institutions. Climate litigation is further discussed in Sections 14.5.1.2 (linkages between mitigation and human rights) and Section 14.5.3 (cross-country implications and international courts/tribunals).

The vast majority of climate cases have emerged in United States, Australia and Europe, and more recently in developing countries (Humby 2018; Kotze and du Plessis 2019; Peel and Lin 2019; Setzer and Benjamin 2019; Zhao et al. 2019; Rodríguez-Garavito 2020).

### Box 13.8 | An Example of Systemic Climate Litigation: Urgenda vs State of the Netherlands

The judgement in *Urgenda vs State of the Netherlands* established the linkage between a state's international duty, domestic actions, and human rights commitments as to the recommendations of IPCC's AR5 (Burgers and Staal 2019; Antonopoulos 2020). It was the first to impose a specific emissions reduction target on a state (de Graaf and Jans 2015; Cox 2016; Loth 2016). The District Court of The Hague ordered the Dutch Government to reduce emissions by at least 25% by the end of 2020. Following the decision of the district court of The Hague in 2015 the Dutch government announced that it would adopt additional measures to achieve the 25% emissions reduction target by 2020 (Mayer 2019). The decision was upheld by the Court of Appeal in 2018 and the Supreme Court in 2019. Since the first judgment in 2015 significant changes in the climate policy environment have been reported, the results of which have included the introduction of a Climate Act and the decision to close all remaining coal fired power plants by 2030 (Verschuuren 2019; Wonneberger and Vliegenthart 2021).

As of 31 May 2021, 1841 cases of climate change litigation from around the world had been identified. Of these, 1387 were filed before courts in the United States, while the remaining 454 were filed in 39 other countries and 13 international or regional courts and tribunals (including the courts of the European Union). Outside the US, Australia (115), the UK (73) and the EU (58) remain the jurisdictions with the highest volume of cases. The majority of cases, 1006, have been filed since 2015 (Setzer and Higham 2021). The number of climate litigation cases in developing countries is also growing. There are at least 58 cases in 18 Global South jurisdictions (*robust evidence*, *high agreement*) (Humby 2018; Kotze and du Plessis 2019; Peel and Lin 2019; Setzer and Benjamin 2019; Zhao et al. 2019; Rodríguez-Garavito 2020; Setzer and Higham 2021).

Overall, courts have also played a more active role for climate governance in democratic political systems (Peel and Osofsky 2015; Eskander et al. 2021). Whether and to what extent differing law traditions and political systems influence the role and importance of climate litigation has, however, not been examined enough scientifically (Setzer and Vanhala 2019; Peel and Osofsky 2020).

The majority of climate change litigation cases are brought against governments, by civic and non-governmental organisations and corporations (Eisenstat 2011; Markell and Ruhl 2012; Wilensky 2015; Fisher et al. 2017; Setzer and Higham 2021). Many, although not all of these cases, seek to ensure that governmental action on climate change is more ambitious, and better aligned with the need to avert or respond to climate impacts identified and predicted by the scientific community (Markell and Ruhl 2012; Setzer and Higham 2021). Climate aligned cases against governments can be divided into two distinct categories: claims challenging the overall effort of a State or its organs to mitigate or adapt to climate change (sometimes referred to as 'systemic climate litigation') (Jackson 2020) and claims regarding authorisation of third-party activity (Bouwer 2018; Gerrard 2021; Ghaleigh 2021).

Systemic climate litigation that seeks an increase in a country's ambition to tackle climate change has been a growing trend since the first court victories in the Urgenda case in the Netherlands (see Box 13.8 below) and the Leghari case in Pakistan in 2015. These cases motivated a wave of similar climate change litigation across the world (Roy and Woerdman 2016; Ferreira 2016; Peeters 2016; Mayer 2019; Paiement 2020; Barritt 2020; Sindico et al. 2021). Between 2015 and 2021, individuals and communities initiated at least 37 cases (including Urgenda and Leghari) against states (Setzer and Higham 2021), challenging the effectiveness of legislation and policy goals (Jackson 2020; Setzer and Higham 2021). Some cases also seek to shape new legal concepts such as 'rights of nature' recognised in the Future Generations case in Colombia (Savaresi and Auz 2019; Rodríguez-Garavito 2020) and 'ecological damage' in the case of Notre Affaire à Tous and others vs France (Torre-Schaub 2021).

Moreover, there are a number of regulatory challenges to state authorisation of high-emitting projects, which differs from systemic cases against states (Bouwer 2018; Hughes 2019a). For instance, the High Court in Pretoria, South Africa, concluded that climate change is a relevant consideration for approving coal-fired power plants (Humby 2018). Similarly, the Federal Court of Australia concluded that the Minister for the Environment owed a duty of care to Australian children in respect to climate impacts when exercising a statutory power to decide whether to authorise a major extension to an existing coal mine (Peel and Markey-Towler 2021).

Climate change litigation has also been brought against corporations by regional or local governments and non-governmental organisations (Wilensky 2015; Ganguly et al. 2018; Foerster 2019). One type of private climate change litigation alleges climate change-related damage and seeks compensation from major carbon polluters (Ganguly et al. 2018; Wewerinke-Singh and Salili 2020). The litigators claim that major oil producers are historically responsible for a significant portion of global greenhouse gas emissions (Heede 2014; Frumhoff et al. 2015; Ekwurzel et al. 2017; Stuart-Smith et al. 2021). These cases rely on advancements in climate science, specifically climate attribution (Marjanac et al. 2017; Marjanac and Patton 2018; McCormick et al. 2018; Minnerop and Otto 2020; Burger et al. 2020b; Stuart-Smith et al. 2021). It is alleged that major carbon emitters had knowledge and awareness of climate change and yet took actions to confound or mislead the public about climate science (Supran and Oreskes 2017). Strategic climate change litigation has also been used to hold corporations to specific human rights responsibilities (Savaresi and Auz 2019; Savaresi and Setzer 2021) (Box 13.8).

In addition to direct cases targeting high emitters, litigation is also now being used to argue against financial investments in the fossil fuel industry (Franta 2017; Colombo 2021). In May 2021, the Hague District Court of the Netherlands issued a ground-breaking judgment holding energy company Royal Dutch Shell (RDS) legally responsible for greenhouse gas emissions from its entire value chain (Macchi and Zeben 2021). Claims have also been brought against banks, pension funds and investment funds for failing to incorporate climate risk into their decision-making, and to disclose climate risk to their beneficiaries (Wasim 2019; Solana 2020; Bowman and Wiseman 2020). These litigation cases also impact on the financial market without directly involving specific financial institutions into the case (Solana 2020) but somehow aim to change their risk perceptions and attitude on high carbon activities (Griffin 2020).

The outcomes of climate litigation can affect the stringency and ambitiousness of climate governance (McCormick et al. 2018; Eskander et al. 2021). In the United States, pro-regulation litigants more commonly win in relation to renewable energy and energy efficiency cases, and more frequently lose in relations to coal-fired power plant cases (McCormick et al. 2018). Outside the US, more than half (58%) of litigation have outcomes that are aligned with climate action (Setzer and Higham 2021). But these cases can also have impacts outside of the legal proceedings before, during and after the case has been brought and decided (Setzer and Vanhala 2019). These impacts include changes in the behaviour of the parties (Peel and Osofsky 2015; Pals 2021), public opinion (Hilson 2019; Burgers 2020), financial and reputational consequences for involved actors (Solana 2020), and impact on further litigation (Barritt 2020). Individual cases have also attracted considerable media attention, which in turn can influence how climate policy is perceived (Nosek 2018; Barritt and Sediti 2019; Hilson 2019; Paiement 2020). While there is evidence to show the influence of some key cases on climate agenda-setting (Wonneberger and Vliegenthart 2021), it is still unclear the extent to which climate litigation actually results in new climate rules and policies (Peel and Osofsky 2018; Setzer and Vanhala 2019; Peel and Osofsky 2020) and to what degree this holds true for all cases (Jodoin et al. 2020). However, there is now increasing academic agreement that climate litigation has become a powerful force in climate governance UNEP 2020b; Burgers 2020). In general, litigations can be applied to constrain both public and private entities, and to shape structural factors mentioned in Section 13.3, such as the beliefs and institutions around climate governance.

### 13.4.3 Media as Communicative Platforms for Shaping Climate Governance

Media is another platform for various actors to present, interpret and shape debates around climate change and its governance (Tindall et al. 2018). The media coverage of climate change has grown steadily since 1980s (O'Neill et al. 2015; Boykoff et al. 2019), but the level and type of coverage differs over time and from country to country (*robust evidence*, *high agreement*) (Boykoff 2011; Schmidt et al. 2013; Schäfer and Schlichting 2014). Media can be a useful conduit to build public support to accelerate mitigation action, but may also be utilised to impede decarbonisation endeavours (Boykoff 2011; O'Neill et al. 2015; Farrell 2016b; Carmichael et al. 2017; Carmichael and Brulle 2018). Different media systems in different regions and countries and with unique cultural and political traditions also affect how climate change is communicated (Eskjær 2013).

A broad variety of media platforms cover climate change issues, including traditional news media, such as newspapers and broadcasting, digital social media (Walter et al. 2018), creative narratives such as climate fiction and films (Svoboda 2016); humour and entertainment media (Brewer and McKnight 2015; Skurka et al. 2018; Boykoff and Osnes 2019); and strategic communications campaigns (Hansen and Machin 2008; Hoewe and Ahern 2017). Media coverage can have far-reaching consequences on policy processes, but we know less about its relative importance compared to other policy shaping factors (*medium evidence*, *medium agreement*) (Liu et al. 2011; Boykoff 2011; Hmielowski et al. 2014).

Popular culture images, science fictions and films of ecological catastrophe can dramatically and emotively convey the dangers of climate change (Bulfin 2017). The overall accuracy of the media coverage on climate change has improved from 2005 to 2019 in the United Kingdom (UK), Australia, New Zealand, Canada, and the USA (McAllister et al. 2021). Moreover, coverage of climate science is increasing. One study (MeCCO) has tracked media coverage of climate change from over 127 sources from 59 countries in North and Latin America, Europe, Middle East, Africa, Asia and Oceania (Boykoff et al. 2021). It shows the number of media science stories in those sources grew steadily from 47,376 per annum to 86,587 per annum between 2017 and 2021 across print, broadcast, digital media and entertainment (Boykoff et al. 2021).

However, increasing media coverage does not always lead to more accurate coverage of climate change mitigation, as it can also spur diffusion of misinformation (Boykoff and Yulsman 2013; van der Linden et al. 2015; Whitmarsh and Corner 2017; Fahy 2018; Painter 2019). In addition, media professionals have at times drawn on the norm of representing both sides of a controversy, bearing the risk of the disproportionate representation of scepticism of anthropogenic climate change despite the convergent agreement in climate science that humans contribute to climate change, (*robust evidence*, *high agreement*) (Freudenburg and Muselli 2010; Boykoff 2013; Painter and Gavin 2016; Tindall et al. 2018; McAllister et al. 2021). This occurs despite increasing consensus among journalists regarding the basic scientific understanding of climate change (Brüggemann and Engesser 2017).

Accurate transference of the climate science has been undermined significantly by climate change counter-movements, particularly in the USA (McCright and Dunlap 2000, 2003; Jacques et al. 2008; Brulle et al. 2012; Boussalis and Coan 2016; Farrell 2016a; Carmichael et al. 2017; Carmichael and Brulle 2018; Boykoff and Farrell 2019; Almiron and Xifra 2019) in both legacy and new/social media environments through misinformation (*robust evidence*, *high agreement*) (van der Linden et al. 2017), including about the causes and consequences of climate change (Brulle 2014; Farrell 2016a; Farrell 2016b; Supran and Oreskes 2017). Misinformation can rapidly spread through social media (Walter et al. 2018). Together with the proliferation of suspicions of 'fake news' and 'post-truth', some traditional and social media contents have fuelled polarisation and partisan divides on climate change in many countries

(Feldman et al. 2017; Hornsey et al. 2018), which can further deter development of new and ambitious climate policy (Tindall et al. 2018). Further, the ideological stance of media also influences the intensity and content of media coverage, in developed and developing countries alike (Dotson et al. 2012; Stoddart and Tindall 2015).

Who dominates the debate on media, and how open the debate can be varies significantly across countries (Takahashi 2011; Poberezhskaya 2015) based on participants' material and technological power. Fossil fuel industries have unique access to mainstream media (Geels 2014) via advertisements, shaping narratives of media reports, and exerting political influence in countries like Australia and the USA (Holmes and Star 2018; Karceski et al. 2020). For social media, novel technical tools, such as automated bots, are emerging to shape climate change discussion on major online platforms such as Twitter (Marlow et al. 2021). Open debates can underpin the adoption of more ambitions climate policy (Lyytimäki 2011). Media coverage on energy saving, patriotism, and social justice in the countries like USA and the UK have helped connect mitigation of climate change with other concerns, thereby raising support to climate action (Leiserowitz 2006; Trope et al. 2007; Doyle 2016; Corner and Clarke 2017; Whitmarsh and Corner 2017; Markowitz and Guckian 2018). Further, media coverage of climate change mitigation has influenced public opinions through discussions on political, economic, scientific and cultural themes about climate change (*medium evidence, high agreement*) (Irwin and Wynne 1996; Smith 2000; Boykoff 2011; O'Neill et al. 2015).

Common challenges in reporting climate change exist around the world (Schmidt et al. 2013; Schäfer and Painter 2021), but particularly so in the developing countries, due to lower capacities, lack of journalists' training in complex climate subjects, and lack of access to clear, timely and understandable climate-related resources and images in newsrooms (*robust evidence, high agreement*) (Harbinson 2006; Shanahan 2009; Broadbent et al. 2016; Lück et al. 2018). Ugandan journalist Patrick Luganda has said, 'Those most at risk from the impacts of climate change typically have had access to the least information about it through mass media.' (Boykoff, 2011), indicating that information availability and capacity is a manifestation of global climate (in)justice.

## 13.5 Sub-national Actors, Networks, and Partnerships

In many countries, sub-national actors and networks are a crucial component of climate mitigation as they have remit over land-use planning, waste management, infrastructure, housing and community development, and their jurisdictions are often where the impacts of climate change are felt (*robust evidence, high agreement*). Depending on the legal framework and other institutional constraints, sub-national actors play crucial roles in developing, delivering and contesting decarbonisation visions and pathways (Schroeder et al. 2013; Ryan 2015; Abbott et al. 2016; Bäckstrand et al. 2017; Amundsen et al. 2018; Fuhr et al. 2018) (Section 13.3.3).

Sub-national actors include organisations, jurisdictions, and networks (e.g., a coalition of cities or state authorities). These are either formal or informal, profit or non-profit and public or private (Avelino and Wittmayer 2016). For example, corporations are formal, private, and for-profit, the state and labour organisations are formal, public, and non-profit, and communities are private, informal, and non-profit. An intermediary sector, crossing the boundaries between private and public, for profit and non-profit, includes energy cooperatives, not-for-profit energy enterprises, and the scientific community (Avelino and Wittmayer 2016).

To address the challenge of climate mitigation, a range of actors across sectors and jurisdictions have created coalitions for climate governance, operating as actor-networks. For example, mitigation policies are particularly effective when they are integrated with co-benefits such as health, biodiversity, and poverty reduction (Romero-Lankao et al. 2018a). Transnational business and public-private partnerships and initiatives, as well as international cooperation at the sub-national and city levels are discussed in Chapter 14.

### 13.5.1 Actor-networks and Policies

The decision adopting the Paris Agreement welcomed contributions of sub-national actors to mobilising and scaling up ambitious climate action (see also Chapter 14). They engage in climate relevant mechanisms, such as the Sustainable Development Goals and the New Urban Agenda. Sub-national actors fill a gap in national policies, participate in transnational and sub-national climate governance networks and facilitate learning and exchange among governmental, community, and private organisations at multiple levels, gathering knowledge and best practices such as emission inventories and risk management tools that can be applied in multiple contexts (Kona et al. 2016; Sharifi and Yamagata 2016; Michaelowa and Michaelowa 2017; Warbroek and Hoppe 2017; Bai et al. 2018; Busch et al. 2018; Hsu et al. 2018; Lee and Jung 2018; Marvin et al. 2018; Romero-Lankao et al. 2018b; Ürge-Vorsatz and Seto 2018; Amundsen et al. 2018; Heikkinen et al. 2019; Hultman et al. 2020).

Sub-national climate change policies exist in more than 142 countries and exemplify the increasing significance of mitigation policy at the sub-national level (Hsu et al. 2018). However, estimations of the number of sub-national actors pledging voluntary climate action are challenging and underreporting is a concern (Hsu et al. 2018; Chan and Morrow 2019). As can be seen in Figure 13.3 more than 10,500 cities and nearly 250 regions representing more than 2 billion people, factoring for overlaps in population between these jurisdictions, have pledged climate action as of December 2020 (Hsu et al. 2020a). More jurisdictions in Europe and North America have pledged action, but in terms of population almost all regions are substantially engaged in sub-national action.

Many of these efforts are organised around transnational or regional networks. For example, a coalition of 130 sub-national (in other words, state, and regional) governments, representing 21% of the global economy and 672 million people, has pledged about 9% emissions reduction compared to a base year (CDP 2020). More than 10,000 cities, representing more than 10% of the global population, participate in the Global Covenant of Mayors, C40 Cities

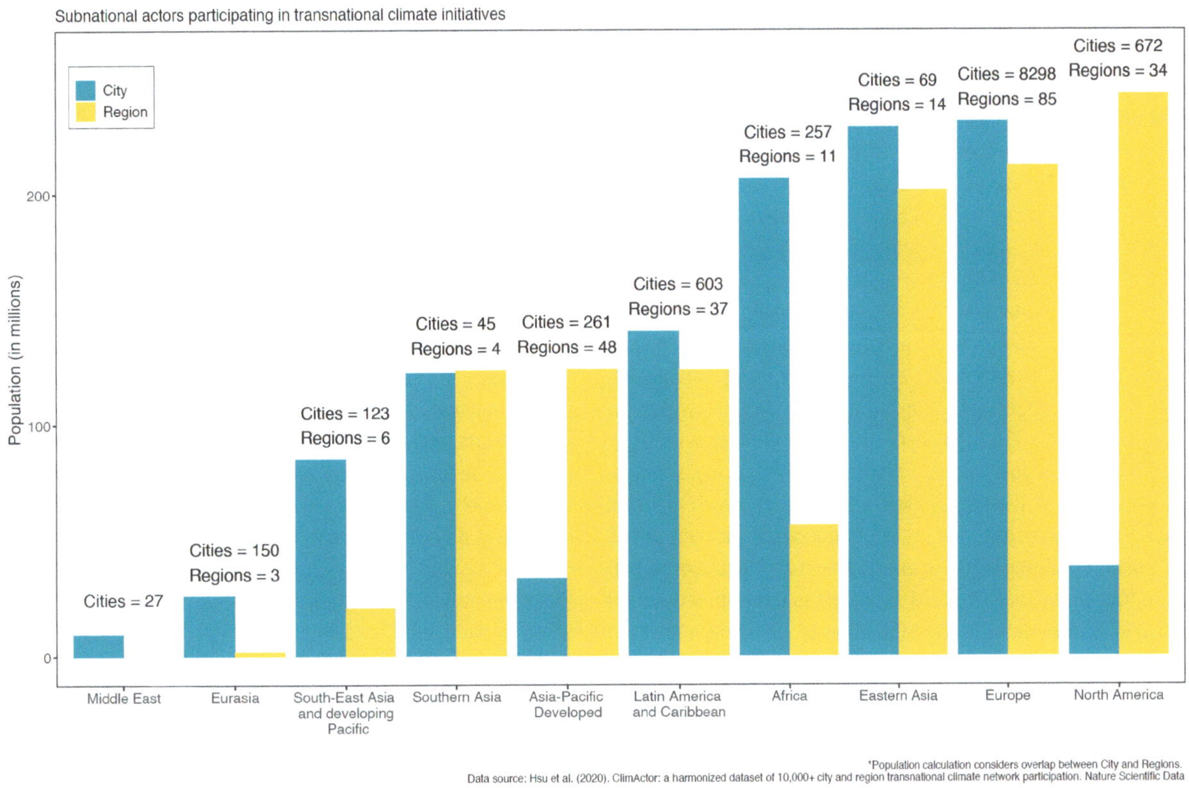


**Figure 13.3 | Sub-national GHG mitigation commitments: Total population by IPCC region.** Population of sub-national actors (cities and regions) recording climate action commitments as captured in the ClimActor dataset. Population calculation considers overlap between City and Regions by only accounting for population once for Cities and Regions that are nested jurisdictions. Source: adapted with permission from Hsu et al. (2020a) to reflect IPCC AR6 aggregation. Compiled in 2020 from multiple sources based on most recent year of data available.

(Global Covenant of Mayors for Climate and Energy 2018), and ICLEI's – Local Governments for Sustainability carbon registry (Hsu et al. 2018). In Europe alone, more than 6000 cities have adopted their own climate action plans (Palermo et al. 2020a) and nearly 300 US sub-national actors – cities and states – were committed to maintaining momentum for climate action as part of the 'We Are Still In' coalition (We Are Still In coalition 2020) in the absence of national US climate legislation. Further, as of October 2020, more than 826 cities and 103 regional governments had made specific pledges to decarbonise, whether in a specific sector (e.g., buildings, electricity, or transport) or through their entire economies, pledging to reduce their overall emissions by at least 80% (NewClimate Institute and Data Driven EnviroLab 2020). Cities such as Barcelona, Spain and Seattle, Washington have adopted net zero goals for 2050 in policy legislation, while many more cities throughout the world, including the Global South such as Addis Ababa in Ethiopia, have net zero targets under consideration (ECIU 2019, 2021).

Sub-national mitigation policies are highlighted below, based on the taxonomy of policies in Section 13.6.1:

a) Economic instruments: as of 2020, there were carbon pricing initiatives (ETS, carbon tax or both) in 24 sub-national jurisdictions (World Bank 2021a). Examples include emission trading systems within North America, such as the Regional Greenhouse Gas Initiative (RGGI) and Western Climate Initiative (which also includes two Canadian provinces); tax rebates for the purchase of EVs; a carbon tax in British Columbia; and a cap-and-trade scheme in Metropolitan Tokyo (Houle et al. 2015; Murray and Rivers 2015; Hibbard et al. 2018; Bernard and Kichian 2019; Raymond 2019; Xiang and Lawley 2019; Chan and Morrow 2019).
b) Regulatory instruments: policies such as land use and transportation planning, performance standards for buildings, utilities, transport electrification, and energy use by public utilities, buildings and fleets are widely prevalent (Bulkeley 2013; Jones 2013; C40 and ARUP 2015; Martinez et al. 2015; Hewitt and Coakley 2019; Palermo et al. 2020b). Policies such as regulatory restrictions, low emission zones, parking controls, delivery planning and freight routes, focus on traffic management and reduction of local air pollution but also have a mitigation impact (Slovic et al. 2016; Khreis et al. 2017; Letnik et al. 2018). For instance, in coordination with national governments, sub-national actors in China, Europe and USA have introduced access to priority lanes, free parking and other strategies fostering the roll-out of EVs (Creutzig 2016; Zhang and Bai 2017; Teske et al. 2018; Zhang and Qin 2018; Romero-Lankao et al. 2021).
c) Land-use planning addresses building form, density, energy, and transport, which are relevant for decarbonisation (Creutzig et al. 2015; Torabi Moghadam et al. 2017; Teske et al. 2018). Its effectiveness is limited by absent or fragmented jurisdiction, financial resources and powers, competition between authorities and policy domains, and national policies that restrict local governments' ability to enact more ambitious policies (Fudge et al. 2016; Gouldson et al. 2016; Petersen 2016). Most rapidly

growing smaller cities in Latin America, Asia and Africa lack capacity for urban planning and enforcement (Romero-Lankao et al. 2015; Creutzig 2016).

d) Other policies: these include information and capacity building, such as carbon labelling aimed at providing carbon footprint information to consumers (Liu et al. 2016); disclosure and benchmarking policies in buildings to increase awareness of energy issues and track mitigation progress (Hsu et al. 2017; Papadopoulos et al. 2018); and procurement guidelines developed by associations (Sustainable Purchasing Leadership Council 2021). For instance, a building retrofit programme was initiated in New York and Melbourne to foster energy efficiency improvements through knowledge provision, training, and consultation (Trencher et al. 2016; Trencher and van der Heijden 2019). Also significant is government provision of public good, services, and infrastructure (Romero Lankao et al. 2019), which includes provision of electric buses or buses on renewable fuels for public transportation (Kamiya and Teter 2019) and zero emission urban freight transport (Quak et al. 2019), sustainable food procurement for public organisations in cities (Smith et al. 2016), decentralised energy resources (Marquardt 2014; Hirt et al. 2021; Kahsar 2021), and green electricity purchase via community choice aggregation programmes and franchise agreements (Armstrong 2019).

### 13.5.2 Partnerships and Experiments

Partnerships, such as those among private and public, or transnational and sub-national entities, have been found to enable better mitigation results in areas outside direct government control such as residential energy use, emissions from local businesses, or private vehicles (Fenwick et al. 2012; Castán Broto and Bulkeley 2013; Aylett 2014; Hamilton et al. 2014; Bulkeley et al. 2016; Wakabayashi and Arimura 2016; Grandin et al. 2018). Partnerships take advantage of investments that match available grants or enable a local energy project, or enhance the scope or impact of mitigation (Burch et al. 2013).

Sub-national actors have also been associated with experiments and laboratories, which promise to achieve the deep change required to address the climate mitigation gap (Smeds and Acuto 2018; Marvin et al. 2018). Experiments span smart technologies, for example, in Malmö, Sweden (Parks 2019), Eco-Art, Transformation-Labs and other approaches that question the cultural basis of current energy regimes and seek reimagined or reinvented futures (Castán Broto and Bulkeley 2013; Guy et al. 2015; Voytenko et al. 2016; Hodson et al. 2018; Peng and Bai 2018; Smeds and Acuto 2018; Culwick et al. 2019; Pereira et al. 2019; Sengers et al. 2019). They may include governance experiments, from formally defined policy experiments to informal initiatives that mobilise new governance concepts (Kivimaa et al. 2017a; Turnheim et al. 2018), and co-design initiatives and grassroots innovations (Martiskainen 2017; Sheikh and Bhaduri 2021). These initiatives often expand the scope for citizen participation. For example, Urban Living Labs foster innovation, coproducing responses to existing problems of energy use, energy poverty and mobility that integrate scientific and expert knowledge with local knowledge and common values (Voytenko et al. 2016; Marvin et al. 2018). The European Network of Living Labs – with a global outreach – has established a model of open and citizen-centric innovation for policy making. The proliferation of Climate Assemblies at the national and sub-national level further emphasises the increasing role that citizens can play in both innovating and planning for carbon mitigation (Sandover et al. 2021).

State and local authorities are often central to initiating and implementing experiments and use an incremental, 'learning by doing' governing approach (Bai et al. 2010; Nevens et al. 2013; Castán Broto and Bulkeley 2013; Mcguirk et al. 2015; Nagorny-Koring and Nochta 2018; Hodson et al. 2018; Peng and Bai 2018; Smeds and Acuto 2018; Culwick et al. 2019; Sengers et al. 2019). Experiments relate to technological learning and changes in policies, practices, services, user behaviour, business models, institutions, and governance (Castán Broto and Bulkeley 2013; Wieczorek et al. 2015; Kivimaa et al. 2017a; Laurent and Pontille 2018; Torrens et al. 2019).

Experimentation has contributed to learning, changes in outcomes when implemented, and shifts in the political landscape (Turnheim et al. 2018). Experiments, however, are often isolated and do not always result in longer-term, more widespread changes. The transformative potential (understood as changes in the fundamental attributes of natural and human systems, see Annex I: Glossary) of experiments is constrained by uncertainty about locally relevant climate change solutions and effects; a lack of comprehensive, and sectorally inclusive national policy frameworks for decarbonisation; budgetary and staffing limitations; and a lack of institutional and political capacity to deliver integrated and planned approaches (Evans and Karvonen 2014; Mcguirk et al. 2015; Bulkeley et al. 2016; Voytenko et al. 2016; Wittmayer et al. 2016; Webb et al. 2017; Grandin et al. 2018; Hölscher et al. 2018; Nagorny-Koring 2019; Sengers et al. 2019).

### 13.5.3 Performance and Global Mitigation Impact

The performance of sub-national actors' mitigation policies have been measured using criteria such as existence of mitigation targets, incentives for mitigation, definition of a baseline, and existence of a monitoring, reporting, and verification procedure (Hsu et al. 2019). Existing evaluations range from small-scale studies assessing the mitigation potential of commitments by sub-national regions, cities and companies in the USA or in 10 high-emitting economies (Roelfsema 2017; Hsu et al. 2019), to larger studies finding that over 9149 cities worldwide could mitigate 1400 $MtCO_2$-eq in 2030 (Global Covenant of Mayors for Climate and Energy 2018; Hsu et al. 2018, 2019). These sub-national mitigation potential estimates vary since a range of approaches exists for accounting for overlaps between sub-national governments and their nested jurisdictions (e.g., states, provinces, and national governments) (Roelfsema et al. 2018; Hsu et al. 2019). One analysis found that the cities of New York, Berlin, London, Greater Toronto, Boston, and Seattle have achieved on average a 0.27 $tCO_2$-eq per capita per year reduction (Kennedy et al. 2012). Hsu et al. (2020c) found that 60% of more than 1000 European cities, representing 6% of the EU's total emissions, are on track to achieving their targets, reducing more than 51 $MtCO_2$-eq. While evidence is

13

limited, there are concerns that implementation challenges persist with city level plans, particularly tied to management of initiatives and engagement of the population (Messori et al. 2020).

Whether participation in transnational climate initiatives impacts sub-national governments' achievement on climate mitigation goals is uncertain. Some find that higher ambition in climate mitigation commitments did not translate into greater mitigation (Kona et al. 2016; Hsu et al. 2019). Other studies associate participation in networks with increased solar photovoltaic systems (PV) investment (Khan and Sovacool 2016; Steffen et al. 2019), and with potential to achieve carbon emissions reductions per capita in line with a global 2°C scenario (Kona et al. 2016).

Reporting networks may attract high-performing actors, suggesting an artificially high level of cities interested in taking climate action or piloting solutions (self-selection bias) that may not be effective elsewhere (van der Heijden 2018). Many studies present a conservative view of potential mitigation impact because they draw upon publicly reported mitigation actions and exclude sub-national actions that are not reported (Kuramochi et al. 2020).

In addition to direct mitigation contributions, climate action partnerships may deliver indirect effects that, while difficult to quantify, ensure long-term change (Chan et al. 2015). Experimentation and policy innovation helps to establish best practices (Hoffmann 2011); set new norms for ambitious climate action that help build coalitions (Chan et al. 2015; Bernstein and Hoffmann 2018); and translate into knowledge sharing or capacity building (Lee and Koski 2012; Hakelberg 2014; Purdon 2015; Acuto and Rayner 2016). Emergent research explores whether, in addition to realising outcomes, mitigation initiatives also provide the resources, skills and networks that governments and other stakeholders currently use to target other development goals (Shaw et al. 2014; Wolfram 2016; Wiedenhofer et al. 2018; Amundsen et al. 2018; Heikkinen et al. 2019).

## 13.6 Policy Instruments and Evaluation

Institutions and governance processes described in previous section result in specific policies, that governments then implement and that shape actions of many stakeholders. This section assesses the empirical experience with the range of policy instruments available to governments with which to shape mitigation outcomes. Section 13.7 that follows deals with how these instruments are combined into packages, and Section 13.9 addresses economy-wide measures and issues.

Many different policy instruments for GHG reduction are in use. They fall into a few major categories that share key characteristics. This section provides one possible taxonomy of these major types of policy instruments, presents a set of criteria for policy evaluation, and synthesises the literature on the most common mitigation policies. The emphasis is on recent empirical evidence on the performance of different policy instruments and lessons that can be drawn from these experiences. This builds on and enhances the AR5 Chapter 15, which provided a more theoretical treatment of policy instruments for mitigation.

### 13.6.1 Taxonomy and Overview of Mitigation Policies

#### 13.6.1.1 Taxonomy of Mitigation Policies

A large number of policies and policy instruments can affect GHG emissions and/or sequestration, whether their primary purpose is climate change mitigation or not. Consequently, consistent with the approach in this chapter, this section adopts a broad interpretation to what is considered mitigation policy. Also, the section recognises the multiplicity of policies that overlap and interact.

Environmental policy instruments, including for climate change mitigation, have long been grouped into three main categories – (i) economic instruments, (ii) regulatory instruments, and (iii) other instruments – although the specific terms differ across disciplines and additional categories are common (Kneese and Schultze 1975; Jaffe and Stavins 1995; Nordhaus 2013; Wurzel et al. 2013). Examples of common policies in each category are shown in Table 13.1, but this is not a comprehensive list. Principles of and empirical experience with the various instruments are synthesised in Sections 13.6.3 to 13.6.5, international interactions are covered in Section 13.6.6.

**Table 13.1 | Classification of mitigation policies.**

| Category | Examples of common types of mitigation policy instruments |
|---|---|
| Economic instruments | Carbon taxes, GHG emissions trading, fossil fuel taxes, tax credits, grants, renewable energy subsidies, fossil fuel subsidy reductions, offsets, R&D subsidies, loan guarantees |
| Regulatory instruments | Energy efficiency standards, renewable portfolio standards, vehicle emission standards, ban on $SF_6$ uses, biofuel content mandates, emission performance standards, methane regulations, land-use controls |
| Other instruments | Information programmes, voluntary agreements, infrastructure, government technology procurement policies, corporate carbon reporting |

#### 13.6.1.2 Coverage of Mitigation Policies

An increasing share of global emissions sources is subject to mitigation policies, though coverage is still incomplete (Eskander and Fankhauser 2020; Nascimento et al. 2021).

While consistent information on global prevalence of policies is not available, in G20 countries the use of various policy instruments has increased steadily over the past two decades (Nascimento et al. 2021). The share of countries that had mitigation policy instruments in place rose across all sectoral categories, albeit to different extents in different sectors and for different policy instruments (Figure 13.4). Among G20 countries the electricity and heat generation has the greatest number of policies in place, and the agriculture and forestry sector the fewest (Nascimento et al. 2021).

The mix of policies has shifted towards more regulatory instruments and carbon pricing relative to information policies and voluntary action (Schmidt and Fleig 2018; Eskander and Fankhauser 2020).

The IEA database, which tracks renewable energy and energy efficiency policies at the national and sub-national levels for about 160 countries, indicates an average of about 225 new renewable energy and energy efficiency policies annually from 2010 through 2019 with a peak in the number of new renewable energy policies in 2011 (IEA 2021).

While an increasing share of $CO_2$ emissions from fossil fuel combustion is subject to mitigation policies, there remain many countries and sectors where no dedicated mitigation policies apply to fuel combustion. Fossil fuel use is subject to energy taxes in the majority but not all jurisdictions, and in some instances, it is subsidised.

The main gaps in current mitigation policy coverage are non-$CO_2$ emissions and $CO_2$ emissions associated with production of industrial materials and chemical feedstocks, which are connected to broader questions of shifting to cleaner production systems (Bataille et al. 2018a; Davis et al. 2018). Sequestration policies focus mainly on forestry and carbon capture and storage (CCS) with limited support for other carbon dioxide removal and use options (Geden et al. 2019; Vonhedemann et al. 2020).

#### 13.6.1.3 Stringency and Overall Effectiveness of Mitigation Policies

The stringency of mitigation policies varies greatly by country, sector and policy (Box 13.9). Stringency can be increased through sequential changes to policies (Pahle et al. 2018).

Estimates of the effective carbon price (as an estimate of overall stringency across policy instruments) differ greatly between countries and sectors (World Bank 2021a). Countries with higher overall effective carbon prices tend to have lower carbon intensity of energy supply and lower emissions intensity of the economy, as shown in an analysis of 42 G20 and OECD countries (OECD 2018). The carbon price that prevails under a carbon tax or ETS is not directly a measure of policy stringency across an economy, as the carbon prices typically only cover a share of total emissions, and rebates or free allowance allocations can limit effectiveness (OECD 2018). At low emissions prices, mitigation incentives are small; as of April 2021, seventeen

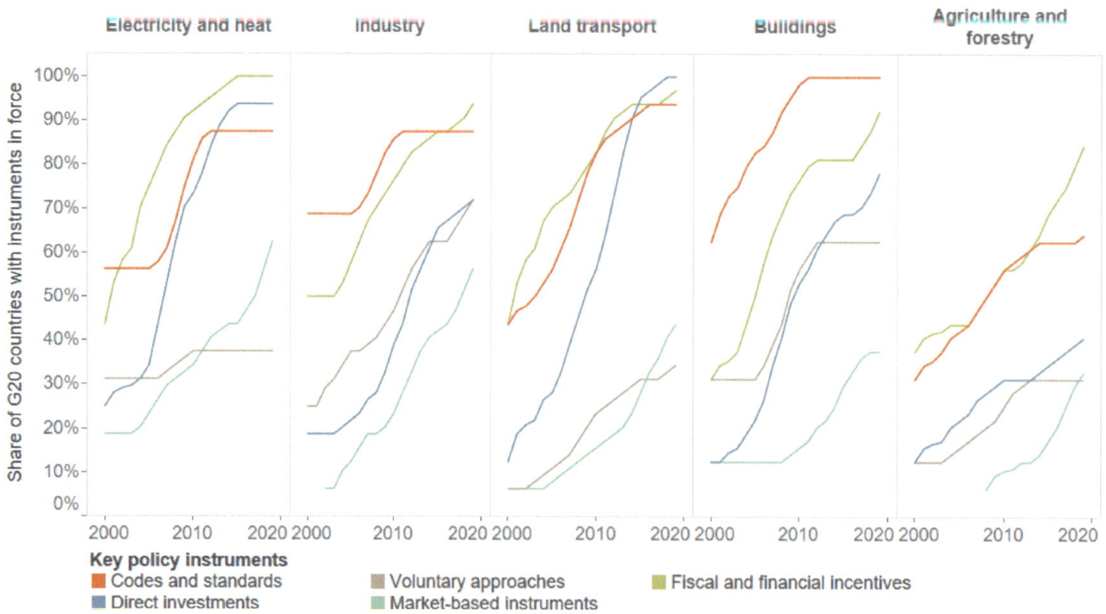


**Figure 13.4 | Share of countries that adopted different policy instruments in different sectors, 2000–2020 (three year moving average).** Source: reproduced with permission from Nascimento et al. (2021).

## Box 13.9 | Comparing the Stringency of Mitigation Policies

Comparing the stringency of policies over time or across jurisdictions is very challenging and there is no single widely accepted metric or methodology (Compston and Bailey 2016; Burck et al. 2019; Tosun and Schnepf 2020; Fekete et al. 2021). Policies are also assessed for their estimated effect on emissions, however, this requires estimation of a counterfactual baseline and isolation of other effects (Cross-Chapter Box 10 in Chapter 14). Economic instruments can be compared on the basis of their price or cost per $tCO_2$-eq. Even that is fraught with complexity in the context of different definitions and estimations for fossil fuel taxes and subsidies. For non-price policies an implicit or equivalent carbon price can be estimated. Factors such as the tax treatment of compliance costs can increase complexity. Accounting for the combined effect of overlapping policies presents additional challenges and such estimates are subject to numerous limitations.

jurisdictions with a carbon pricing policy had a tax rate or allowance price less than USD5 per $tCO_2$ (World Bank 2021a).

Other policies, such as fossil fuel subsidies, may provide incentives to increase emissions thus limiting the effectiveness of the mitigation policy (Section 13.6.3.6). Those effects may be complex and difficult to identify. In most countries trade policy provides an implicit subsidy to $CO_2$ emissions (Shapiro 2020). The analysis of emissions from energy use in buildings in Chapter 9 illustrates the factors that support and counteract mitigation policies.

Furthermore, emissions pricing policies encourage reduction of emissions whose marginal abatement cost is lower than the tax/allowance price, so they have limited impact on emissions with higher abatement costs such as industrial process emissions (Bataille et al. 2018a; Davis et al. 2018). EU ETS emission reductions have been achieved mainly through implementation of low cost measures such as energy efficiency and fuel switching rather than more costly industrial process emissions.

Estimating the overall effectiveness of mitigation policies is difficult because of the need to identify which observed changes in emissions and their drivers are attributable to policy effort and which to other factors. Cross-Chapter Box 10 in Chapter 14 brings together several lines of evidence to indicate that mitigation policies have had a discernible impact on mitigation for specific countries, sectors and technologies and led to avoided global emissions to date by several billion tonnes $CO_2$-eq annually (*medium evidence, medium agreement*).

### 13.6.2 Evaluation Criteria

Policy evaluation is a 'careful, retrospective assessment of merit, worth and value of the administration, output and outcomes of government interventions' (Vedung 2005). The inherent complexity of climate mitigation policies calls for the application of multiple criteria, and reflexiveness of analysis with regard to governments' and societies' objectives for policies (Huitema et al. 2011).

Evaluation of climate mitigation policy tends to focus on the environmental effectiveness and economic efficiency or cost-effectiveness of GHG mitigation policies, with distributional equity sometimes as an additional criterion. In policy design and implementation there is rising interest in co-benefits and side-effects of climate policies, as well as institutional requirements for implementation and the potential of policies to have transformative effect on systems. Table 13.2 elaborates.

**Table 13.2 | Criteria for evaluation and assessment of policy instruments and packages.**

| Criterion | Description |
|---|---|
| Environmental effectiveness | Reducing GHG emissions is the primary goal of mitigation policies and therefore a fundamental criterion in evaluation. Environmental effectiveness has temporal and spatial dimensions. |
| Economic effectiveness | Climate change mitigation policies usually carry economic costs, and/or bring economic benefits other than through avoided future climate change. Economic effectiveness requires minimising costs and maximising benefits. |
| Distributional effects | The costs and benefits of policies are usually distributed unequally among different groups within a society (Zachmann et al. 2018), for example between industry, consumers, taxpayers; poor and rich households; different industries; different regions and countries. Policy design affects distributional effects, and equity can be taken into account in policy design in order to achieve political support for climate policies (Baranzini et al. 2017). |
| Co-benefits, negative side-effects | Climate change mitigation policies can have effects on other objectives, either positive co-benefits (Mayrhofer and Gupta 2016; Karlsson et al. 2020) or negative side-effects. Conversely, impacts on emissions can arise as side-effects of other policies. There can be various interactions between climate change mitigation and the Sustainable Development Goals (Liu et al. 2019). |
| Institutional requirements | Effective implementation of policies requires that specific institutional prerequisites are met. These include effective monitoring of activities or emissions and enforcement, and institutional structures for the design, oversight and revision and updating of policies. Requirements differ between policy instruments. a separate consideration is the overall feasibility of a policy within a jurisdiction, including political feasibility (Jewell and Cherp 2020). |
| Transformative potential | Transformational change is a process that involves profound change resulting in fundamentally different structures (Nalau and Handmer 2015), or a substantial shift in a system's underlying structure (Hermwille et al. 2015). Climate change mitigation policies can be seen has having transformative potential if they can fundamentally change emissions trajectories, or facilitate technologies, practices or products with far lower emissions. |

Not all criteria are applicable to all instruments or in all circumstances and the relative importance of different criteria depend on the objectives in the specific the context. a given policy instrument may score highly on only some assessment criteria. In practice, the empirical evidence seldom exists for assessment of a policy instrument across all criteria.

### 13.6.3 Economic Instruments

Economic instruments, including carbon taxes, emissions trading systems (ETS), purchases of emission reduction credits, subsidies for energy efficiency, renewables and research and development and fossil fuel subsidy removal, provide a financial incentive to reduce emissions. Pricing instruments, especially ETS and carbon taxes, have become more prevalent in recent years (Section 13.6.1). They have proven effective in promoting implementation of the low-cost emissions reductions, and practical experience has driven progress in market mechanism design (*robust evidence*, *high agreement*).

#### 13.6.3.1 Carbon Taxes

A carbon tax is a charge on carbon dioxide or other greenhouse gases imposed on specified emitters or products. In practice features such as exemptions and multiple rates can lead to debate as to whether a specific tax is a carbon tax (Haites 2018). While other taxes can also reduce emissions by increasing the price of GHG emitting products, the result may be inefficient unless the tax rate is proportional to the emissions intensity. a tax on value of fossil fuels, for example, could raise the price on natural gas more than the price of coal, and hence increase emissions if the resulting substitution towards coal were to outweigh reductions in energy use.

As of April 2021, 27 carbon taxes had been implemented by national governments, mostly in Europe (World Bank 2021a). Most of the taxes apply to fossil fuels used for transportation and heating and cover between 3% and 79% of the jurisdiction's emissions. Several countries also tax F-gases. Tax rates vary widely from less than USD1 to over USD137 per $tCO_2$-eq. a few jurisdictions lowered existing fuel taxes when they implemented the carbon tax, thus reducing the effective tax rate (OECD 2021a). How the tax revenue is used varies widely by jurisdiction.

Carbon taxes tend to garner the least public support among possible mitigation policy options (Rhodes et al. 2017; Rabe 2018; Maestre-Andrés et al. 2019; Criqui et al. 2019) although some regulations also meet with opposition (Attari et al. 2009). Policymakers sometimes use the revenue to build support for the tax, allocating some to address regressivity, to address competitiveness claims by industry, to reduce the economic cost by lowering existing taxes, and to fund environmental projects (Gavard et al. 2018; Klenert et al. 2018; Levi et al. 2020).

Carbon tax rates can be adjusted for inflation, increases in income, the effects of technological change, changing policy ambition, or the addition or subtraction of other policies. In practice, numerous jurisdictions have not increased their tax rates annually and some scheduled tax increases have not been implemented (Haites et al. 2018). Predictability of future tax rates helps improve economic performance (Bosetti and Victor 2011; Brunner et al. 2012). Uncertainty about the future existence of a carbon price can hinder investment (Jotzo et al. 2012) and uncertainty about future price levels can increase the resource costs of carbon pricing (Aldy and Armitage 2020).

#### 13.6.3.2 Emission Trading Systems

The most common ETS design – cap-and-trade – sets a limit on aggregate GHG emissions by specified sources, distributes tradable allowances approximately equal to the limit, and requires regulated emitters to submit allowances equal to their verified emissions. The price of allowances is determined by the market, except in cases where government determined price floors or ceilings apply.

ETSs for GHGs were in place in 38 countries as of April 2021 (World Bank 2021a). The EU ETS, which covers 30 countries, was recently displaced by China's national ETS as the largest. ETSs tend to cover emissions by large industrial and electricity generating facilities.[2] Allowance prices as of April 1, 2021 ranged from just over USD1 to USD50, and coverage between 9% and 80% of the jurisdiction's emissions.

Multiple regional pilot ETSs with different designs have been implemented in China since 2013 to provide input to the design of a national system that is to become the world's largest ETS (Jotzo et al. 2018; Qian et al. 2018; Stoerk et al. 2019). Assessments have identified potential improvements to emissions reporting procedures (Zhang et al. 2019) and the pilot ETS designs (Deng et al. 2018). China's national ETS covering over 2200 heat and power plants with annual emissions of about 4 $GtCO_2$took effect in 2021 (World Bank 2021a).

All of the ETSs for which data are available have accumulated surplus allowances which reduces their effectiveness (Haites 2018). Surplus allowances indicate that the caps set earlier were not stringent relative to emissions trends. Most of those ETSs have implemented measures to reduce the surplus including removal/cancellation of allowances and more rapid reduction of the cap. Several ETSs have adopted mechanisms to remove excess allowances from the market when supply is abundant and release additional allowances into the market when the supply is limited, such as the EU 'market stability reserve' (Hepburn et al. 2016; Bruninx et al. 2020). Initial indications are that this mechanism is at least partially successful in stabilising prices in response to short term disruptions such as the COVID-19 economic shock (Gerlagh et al. 2020; Bocklet et al. 2019).

Some ETS also include provisions to limit the range of market prices, making them 'hybrids' (Pizer 2002). a price floor assures a minimum level of policy effect if demand for allowances is low relative to the ETS emissions cap. It is usually implemented through a minimum price at auction, as for example in California's ETS (Borenstein et al. 2019).

[2] The UK was a member of the EU ETS until December 31, 2020. A UK Emissions Trading Scheme (UK ETS) came into effect on 1 January 2021.

a price ceiling allows the government to issue unlimited additional allowances at a pre-determined price to limit the maximum cost of mitigation. Price ceilings have not been activated to date.

#### 13.6.3.3 Evaluation of Carbon Pricing Experience

A carbon tax or GHG ETS increases the prices of emissions intensive goods thus creating incentives to reduce emissions (Stavins 2019) for a comparison of a tax and ETS). The principal advantage of a pricing policy is that it promotes implementation of low-cost reductions; for a carbon tax, reductions whose cost per $tCO_2$-eq reduced is lower than the tax and for an ETS the lowest cost (per $tCO_2$-eq) reductions sufficient to meet the cap. Both a tax and an ETS can be designed to limit adverse economic impacts on regulated sources and emissions leakage.

The corresponding limitations of pricing policies are that they have limited impact on adoption of mitigation measures when decisions are not sensitive to prices and do not encourage adoption of higher cost mitigation measures. Their effectiveness in influencing long-term investments depends on the expectation that the policy will continue and expectations related to future tax rates or allowance prices (Brunner et al. 2012). Other policies can be used in combination with carbon pricing to address these limitations.

The number of pricing policies has increased steadily and covered 21.5% of global GHG emissions in 2020 (World Bank 2021a). Effective coverage is lower because virtually all jurisdictions with a pricing policy have other policies that affect some of the same emissions. For example, a few jurisdictions reduced existing fuel taxes when they introduced their carbon tax thus reducing the effective tax rate, and many jurisdictions have two or more pricing policies

*Environmental effectiveness and co-benefits*

There is abundant evidence that carbon pricing policies reduce emissions. Statistical studies of emissions trends in jurisdictions with and without carbon pricing find a significant impact after controlling for other policies and structural factors (Best et al. 2020; Rafaty et al. 2020). Numerous assessments of specific policies, especially the EU ETS and the British Columbia carbon tax, conclude that most have reduced emissions (*robust evidence*, *high agreement*) (Narassimhan et al. 2018; Haites et al. 2018; Aydin and Esen 2018; Pretis 2019; Andersson 2019; FSR Climate 2019; Metcalf and Stock 2020; Rafaty et al. 2020; Bayer and Aklin 2020; Diaz et al. 2020; Green 2021; Arimura and Abe 2021).

Estimating the emission reductions due to a specific policy is difficult due to the effects of overlapping policies and exogenous factors such as fossil fuel price changes and economic conditions. Studies that attempt to attribute a share of the reductions achieved to the EU ETS place its contribution at 3–25% (FSR Climate 2019; Bayer and Aklin 2020; Chèze et al. 2020). The relationship between a carbon tax and the resulting emission reductions is complex and is influenced by changes in fossil fuel prices, changes in fossil fuel taxes, and other mitigation policies (Aydin and Esen 2018). But the effectiveness of a carbon tax generally is higher in countries where it constitutes a large part of the fossil fuel price (Andersson 2019).

Few of the world's carbon prices are at a level consistent with various estimates of the carbon price needed to meet the Paris Agreement goals. In modelling of mitigation pathways that limit warming to 2°C (>50%)(Section 3.6.1) marginal abatement costs of carbon in 2030 are about 60 to 120 USD2015 per $tCO_2$, and about 170 to 290 USD2015 per $tCO_2$ in pathways that limit warming to 1.5°C (>50%) with no or limited overshoot (Section 3.6). One synthesis study estimates necessary prices at USD40–80 per $tCO_2$ by 2020 (High-Level Commission on Carbon Prices 2017). Only a small minority of carbon pricing schemes in 2021 had prices above USD40 per $tCO_2$, and all of these were in European jurisdictions (World Bank 2021a). Most carbon pricing systems apply only to some share of the total emissions in a jurisdiction, so the headline carbon price is higher than the average carbon price that applies across an economy (World Bank 2021a).

Where ETS or carbon taxes exist, they apply to different proportions of the jurisdiction's greenhouse gas emissions. The share of emissions covered by ETSs in 2020 varied widely, ranged from 9% (Canada) to 80% (California) while the share of emissions covered by carbon taxes ranged from 3% (Latvia and Spain) to 80% (South Africa) (World Bank 2021a).Where carbon pricing policies are effective in reducing GHG emissions, they usually also generate co-benefits including better air quality. For example, a Chinese study of air quality benefits from lower fossil fuel use under carbon pricing suggests that prospective health co-benefits would partially or fully offset the cost of the carbon policy (Li et al. 2018). Depending upon the jurisdiction (for example, if there are fossil fuel subsidies) carbon pricing could also reduce the economic distortions of fossil fuel subsidies, improve energy security through greater reliance on local energy sources and reduce exposure to fossil fuel market volatility. Substantial carbon prices would be in the domestic self-interest of many countries if co-benefits were fully factored in (Parry et al. 2015).

*Economic effectiveness*

Economic theory suggests that carbon pricing policies are on the whole more cost effective than regulations or subsidies at reducing emissions (Gugler et al. 2021). Any mitigation policy imposes costs on the regulated entities. In some cases entities may be able to recover some or all of the costs through higher prices (Neuhoff and Ritz 2019; Cludius et al. 2020). International competition from less stringently regulated firms limits the ability of emissions-intensive, trade-exposed (EITE) firms to raise their prices. Thus, a unilateral mitigation policy creates a risk of adverse economic impacts, including loss of sales, employment, profits, for such firms and associated emissions leakage (Section 13.6.6.1).

Pricing policies can be designed to minimise these risks; free allowances can be issued to EITE participants in an ETS and taxes can provide exemptions or rebates. An extensive *ex post* literature finds no statistically significant adverse impacts on competitiveness or leakage (13.6.6.1).

An *ex post* analysis of European carbon taxes finds no robust evidence of a negative effect on employment or GDP growth (Metcalf and Stock 2020). The British Columbia carbon tax led to a small net increase in employment (Yamazaki 2017) with no significant negative impacts on GDP possibly due to full recycling of the tax revenue (Bernard and Kichian 2021). Few carbon taxes apply to EITE sources (Timilsina 2018), so competitiveness impacts usually are not a particular concern.

Government revenue generated by carbon pricing policies globally was approximately 53 billion USD in 2020 split almost evenly between carbon taxes and ETS allowance sales (World Bank 2021). Revenue raised though carbon pricing is generally considered a relatively efficient form of taxation and a large share of revenue enters general government budgets (Postic and Fetet 2020). Some of the revenue is returned to emitters or earmarked for environmental purposes. Allowance allocation and revenue spending measures have been used to create public support for many carbon pricing policies including at every major reform stage of the EU ETS (Klenert et al. 2018; Dorsch et al. 2020) (Box 5.11).

*Distributional effects*

The most commonly studied distributional impact is the direct impact of a carbon tax on household income. Typically it is regressive; the tax induced increase in energy expenditures represents a larger share of household income for lower income households (Grainger and Kolstad 2010; Timilsina 2018; Dorband et al. 2019; Ohlendorf et al. 2021). Governments can rebate part or all of the revenue to low-income households, or implement other changes to taxation and transfer systems to achieve desired distributional outcomes (Jacobs and van der Ploeg 2019; Saelim 2019; Sallee 2019) (Box 5.11). The full impact of the tax – after any distribution of tax revenue to households and typically adverse effects on investors – generally is less regressive or progressive (Williams III et al. 2015; Goulder et al. 2019). Where the tax revenue is treated as general revenue the government relies on existing income redistribution policies (such as income taxes) and social safety net programmes to address the distributional impacts.

Carbon taxes on fossil fuels have effects similar to the removal of fossil fuel subsidies (Ohlendorf et al. 2021) (Section 13.6.3.6). Even if a carbon tax is progressive it increases prices for fuels, electricity, transport, food and other goods and services that adversely affect the most economically vulnerable. Redistribution of tax revenue is critical to address the adverse impacts on low-income groups (Dorband et al. 2019) (Box 5.11). In countries with a limited capacity to collect taxes and distribute revenues to low-income households, such as some developing countries, carbon taxes may have greater distributional consequences.

Distributional effects have generally not been a significant issue for ETSs. Equity for industrial participants typically is addressed through free allocation of allowances. Impacts on household incomes, with the exception of electricity prices, are too small or indirect to be a concern. Some systems are designed to limit electricity price increases (Petek 2020) or use some revenue for bill assistance to low-income households (RGGI 2019).

*Technological change*

Carbon pricing, especially an ETS that covers industrial sources, stimulates technological change by participants and others (Calel and Dechezleprêtre 2016; FSR Climate 2019; van den Bergh and Savin 2021) (Section 13.6.6.3 and Chapter 16). The purpose of pricing policies is to encourage implementation of the lowest cost mitigation measures. Pricing policies therefore are more likely to stimulate quick, low cost innovation such as fuel switching and energy efficiency, rather than long term, costly technology development such as renewable energy or industrial process technologies (Calel 2020; Lilliestam et al. 2021). To encourage long-term technology development carbon pricing policies need to be complemented by other mitigation and research and development (R&D) policies.

#### 13.6.3.4 Offset Credits

Offset credits are voluntary GHG emission reductions for which tradable credits are issued by a supervisory body (Michaelowa et al. 2019b). a buyer can use purchased credits to offset an equal quantity of its emissions. In a voluntary market governments, firms and individuals purchase credits to offset emissions generated by their actions, such as air travel. a compliance market al.ows specified offset credits to be used for compliance with mitigation policies, especially ETSs, carbon taxes and low-carbon fuel standards. (Newell et al. 2013; Bento et al. 2016; Michaelowa et al. 2019a).

When used for compliance, governments typically specify a maximum quantity of offset credits that can be used, as well as the types of emission reduction actions, the project start dates and the geographic regions eligible credits. Initially, the EU ETS, Swiss ETS and New Zealand ETS accepted credits issued under the Kyoto Protocol (Chapter 14), but they terminated or severely constrained the quantity of international credits allowed for compliance use after 2014 (Shishlov et al. 2016) (Section 13.6.6).

A key question for any offset credit is whether the emission reductions are 'additional': reductions that only happen because of the offset credit payment (Greiner and Michaelowa 2003; Millard-Ball and Ortolano 2010; van Benthem and Kerr 2013; Burke 2016; Bento et al. 2016). To assess additionality and to determine the quantity of credits to be issued, regulators develop methodologies to estimate baseline (business-as-usual) emissions in the absence of offset payments (Newell et al. 2013; Bento et al. 2016). Credits are issued for the difference between the baseline and actual emissions with adjustments for possible emissions increases outside the project boundary (Rosendahl and Strand 2011). Some research suggests that procedural and measurement advances can significantly reduce the risk of severe non-additionality (Mason and Plantinga 2013; Bento et al. 2016; Michaelowa et al. 2019a).

#### 13.6.3.5 Subsidies for Mitigation

Subsidies for mitigation encourage individuals and firms to invest in assets that reduce emissions, changes in processes or innovation. Subsidies have been used to improve energy efficiency, encourage the uptake of renewable energy and other sector-specific emissions saving options (Chapters 6 to 11), and to promote innovation. Targeted subsidies can achieve specific mitigation goals yet have intrinsically narrower coverage than more broad-based pricing instruments. Subsidies are often used not only to achieve emissions reductions but to address market imperfections or to achieve distributional or strategic objectives. Subsidies are often used alongside or in combination with other policy instruments, and are provided at widely differing cost per unit of emissions reduced.

Governments routinely provide direct funding for basic research, subsidies for R&D to private companies, and co-funding of research and deployment with industry (Dzonzi-Undi and Li 2016). Research subsidies have been found to be positively correlated with green product innovation in a study in Germany, Switzerland and Austria (Stucki et al. 2018). Government subsidies for R&D have been found to greatly increase the green innovation performance of energy intensive firms in China (Bai et al. 2019). For more detail see Chapter 16.

Subsidies of different forms are often provided for emissions savings investments to businesses and for the retrofit of buildings for energy efficiency. Emissions reductions from energy efficiencies can often be achieved at low cost, but evidence for some schemes suggests lower effectiveness in emissions reductions than expected *ex ante* (Fowlie et al. 2018; Valentová et al. 2019). Tax credits can be used to encourage firms to produce or invest in low-carbon emission energy and low-emission equipment. Investment subsidies have been found to be more effective in reducing costs and uncertainties in solar energy technologies than production subsidies (Flowers et al. 2016).

Subsidies have been provided extensively and in many countries for the deployment of household rooftop solar systems, and increasingly also for commercial scale renewable energy projects, typically using 'feed-in tariffs' that provide a payment for electricity generated above the market price (Pyrgou et al. 2016). Such schemes have proven effective in deploying renewable energy, but lock-in subsidies for long periods of time. In some cases they provide subsidies at higher levels than would be required to motivate deployment (del Río and Linares 2014). High levels of net subsidies have been shown to diminish incentives for optimal siting of renewable energy installations (Penasco et al. 2019).

A variant of subsidies for deployment of renewable energy are auctioned feed-in tariffs or auctioned contracts-for-difference, where commercial providers bid in a competitive process. Auctions typically lead to lower price premiums (Eberhard and Kåberger 2016; Roberts 2020) but efficient outcomes depend on auction design and market structure (Grashof et al. 2020), although an emergent literature also questions whether spread of auctions is due to performance or the dynamics of the policy formulation process (Fitch-Roy et al. 2019b; Grashof et al. 2020; Grashof 2021). The prequalification requirements or the assessment criteria in the auctions sometimes also include local co-benefits such as local economic diversification (Buckman et al. 2019; White et al. 2021).

Support for rollout clean technologies at high prices can be economically beneficial in the long run if costs are reduced greatly as a function of deployment (Newbery 2018). Deployment support, much of it in the form of feed-in tariffs in Germany, enabled the scaling up of the global solar photovoltaic industry and attendant large reductions in production costs that by 2020 made solar power cost competitive with fossil fuels (Buchholz et al. 2019). There is also evidence for increased innovation activity as a result of solar feed-in tariffs (Böhringer et al. 2017b).

Many governments have also provided subsidies for the purchase of electric vehicles, including with strong effect in China (Ma et al. 2017), Norway (Baldursson et al. 2021) and other countries, and sometimes at relatively high rates (Kong and Hardman 2019).

#### 13.6.3.6 Removal of Fossil Fuel Subsidies

Many governments subsidise fossil fuel consumption and/or production through a variety of mechanisms (Burniaux and Chateau 2014) (Figure 13.5). Different approaches exist to defining the scope and estimating the magnitude of fossil fuel subsidies (Koplow 2018), and all involve estimates, so the magnitudes are uncertain. Rationalising inefficient fossil fuel subsidies is one of the indicators to measure progress toward Sustainable Development Goal 12: Ensure sustainable consumption and production patterns (UNEP 2019a).

Consumption subsidies represent approximately 70% of the total. Most of the subsidies go to petroleum, which accounts for roughly 50% of the consumption subsidies and 75% of the production subsidies (IEA 2020; OECD 2020). Much of the variation in the consumption subsidies is due to fluctuations in the world price of oil which is used as the reference price.

Reducing fossil fuel subsidies would lower $CO_2$ emissions, increase government revenues (Jakob et al. 2015; Dennis 2016; Gass and Echeverria 2017; Rentschler and Bazilian 2017; Monasterolo and Raberto 2019), improve macroeconomic performance (Monasterolo and Raberto 2019), and yield other environmental and sustainable development benefits (*robust evidence*, *medium agreement*) (Jakob et al. 2015; Rentschler and Bazilian 2017; Solarin 2020). The benefits of gasoline subsidies in developing countries accrue mainly to higher income groups, so subsidy reduction usually will reduce inequality (Coady et al. 2015; Dennis 2016; Monasterolo and Raberto 2019; Labeaga et al. 2021). Some subsidies, like tiered electricity rates, benefit low-income groups. Reductions of broad subsidies lead to price increases for fuels, electricity, transport, food and other goods and services that adversely affect the most economically vulnerable (Coady et al. 2015; Zeng and Chen 2016; Rentschler and Bazilian 2017). Distributing some of the revenue saved can mitigate the adverse economic impacts on low-income groups (Dennis 2016; Zeng and Chen 2016; Labeaga et al. 2021; Schaffitzel et al. 2020).

The emissions reduction that could be achieved from fossil fuel subsidy removal depends on the specific context such as magnitude and nature of subsidies, energy prices and demand elasticities, and how the fiscal savings from reduced subsidies are used. Modelling studies of global fossil fuel subsidy removal result in projected emission reductions of between 1% and 10% by 2030 (Delpiazzo et al. 2015; IEA 2015; Jewell et al. 2018; IISD 2019) and between 6.4% and 8.2% by 2050 (Schwanitz et al. 2014; Burniaux and Chateau 2014).

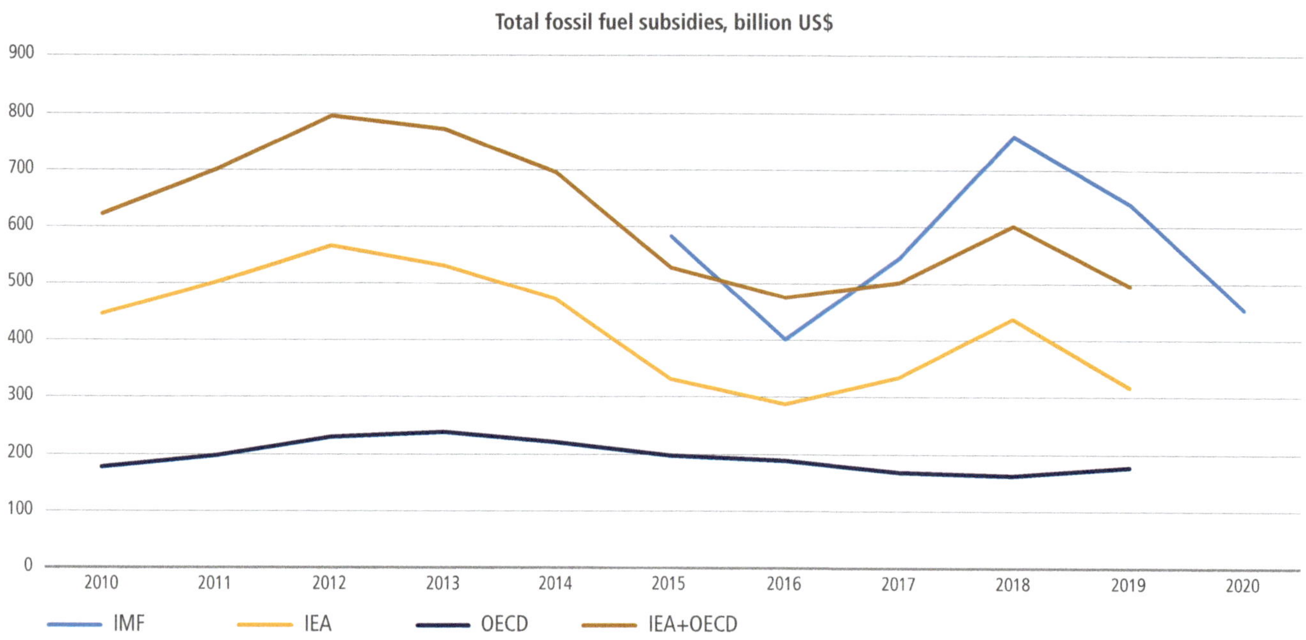


**Figure 13.5 | Total fossil fuel subsidies, 2010–2019, in USD billion (USD2021 for IMF, USD2019 for others).** Source: data from OECD (2020) (43 countries, mainly production subsidies), IEA (2020) (40 countries, mainly consumption subsidies), IMF (Parry et al. 2021; explicit subsidies for all countries).

An extensive literature documents the difficulties of phasing out fossil fuel subsidies (Schmidt et al. 2017; Gass and Echeverria 2017; Skovgaard and van Asselt 2018; Kyle 2018; Perry 2020; Gençsü et al. 2020). Fossil fuel industries lobby to maintain producer subsidies and consumers protest if they are adversely affected by subsidy reductions (Fouquet 2016; Coxhead and Grainger 2018). Yemen (2005 and 2014), Cameroon (2008), Bolivia (2010), Nigeria (2012), Ecuador (2019) all abandoned subsidy reform attempts following public protests (Rentschler and Bazilian 2017; Mahdavi et al. 2020). Indonesia is an example where fossil fuel subsidy removal was successful, helped by social assistance programmes and a communication effort about the benefits of reform (Chelminski 2018; Burke and Kurniawati 2018). To-date instances of fossil fuel subsidy reform or removal have been driven largely by national fiscal and economic considerations (Skovgaard and van Asselt 2019).

### 13.6.4 Regulatory Instruments

Regulatory instruments are applied by governments to cause the adoption of desired processes, technologies, products (including energy products) or outcomes (including emission levels). Failure to comply incurs financial penalties and/or legal sanctions. Regulatory instruments range from performance standards, which prescribe compliance outcomes – and in some cases allow flexibility to achieve compliance, including the trading of credits – to more prescriptive technology-specific standards, also known as command-and-control regulation. Regulatory instruments play an important role to achieve specific mitigation outcomes in sectoral applications (*robust evidence, high agreement*). Mitigation by regulation often enjoys greater political support but tends to be more economically costly than mitigation by pricing instruments (*robust evidence, medium agreement*).

#### 13.6.4.1 Performance Standards, Including Tradable Credits

Performance standards grant regulated entities freedom to choose the technologies and methods to reach a general objective, such as a minimum market share of zero-emission vehicles or of renewable electricity, or a maximum emissions intensity of electricity generated. Tradable performance standards allow regulated entities to trade compliance achievement credits; under-performers can buy surplus credits from over-performers thereby reducing the aggregate cost of compliance (Fischer 2008).

Tradable performance standards have been applied to numerous sectors including electricity generation, personal vehicles, building energy efficiency, appliances, and large industry. An important application is Renewable Portfolio Standards (RPS) for electricity supply, which require that a minimum percentage of electricity is generated from specified renewable sources sometimes including nuclear and fossil fuels with CCS when referred to as a clean electricity standard (Young and Bistline 2018) (Chapter 6). This creates a price incentive to invest in renewable generation capacity. Such incentives can equivalently be created through feed-in tariffs, a form of subsidy (Section 13.6.3) and some jurisdictions have had both instruments (Matsumoto et al. 2017). RPS can differ in features and stringency, and are in operation in many countries and sub-national jurisdictions, including a majority of US states (Carley et al. 2018).

Vehicle emissions standards are a common form of performance standard with flexibility (Chapter 9). a corporate fuel efficiency standard specifies an average energy use and/or GHG emissions per kilometre travelled for vehicles sold by a manufacturer. Another version of this policy, the zero-emission vehicle (ZEV) standard, requires vehicle sellers to achieve minimum requirements for sales of zero-emission vehicles (Bhardwaj et al. 2020). Both instruments allow manufacturers to use tradable credits to achieve compliance.

Low-carbon fuel standards (LCFS), which set an average life-cycle carbon intensity for energy that declines over time, are another example. LCFS are in place in many different jurisdictions (Chapter 9) and have been applied to petroleum products, natural gas, hydrogen and electricity (Yeh et al. 2016). An LCFS allows regulated entities to trade credits creating the potential for high carbon intensity fuel suppliers to cross-subsidise low-carbon intensity transport energy providers including low-carbon biofuels, hydrogen and electricity (Axsen et al. 2020).

Trading and other flexibility mechanisms improve the economic efficiency of standards by harmonising the marginal abatement costs among companies or installations subject to the standard. Nevertheless tradable performance standards are less economically efficient in achieving emissions reductions than carbon pricing, sometimes by a significant amount (Giraudet and Quirion 2008; Chen et al. 2014; Holland et al. 2015; Fox et al. 2017; Zhang et al. 2018).

#### 13.6.4.2 Technology Standards

Technology standards take a more prescriptive approach by requiring a specific technology, process or product. They typically take one of three forms: requirements for specific pollution abatement technologies; requirements for specific production methods; or requirements for specific goods such as energy efficient appliances. They can also take the form of phase-out mandates, as applied for example to planned bans of internal combustion engines for road transport (Bhagavathy and McCulloch 2020), coal use; for example, Germany's decisions to phase out coal (Oei et al. 2020), and some industry processes and products, for example, hydrofluorocarbons (HFCs) and use of sulphur hexafluoride ($SF_6$) in some products (see Box 13.10 on non-$CO_2$ gases). Technology standards are also referred to as command-and-control standards, prescriptive standards, or design standards.

Technology standards are a common climate policy particularly at the sector level (Chapters 6–11). Technology standards tend to score lower in terms of economic efficiency than carbon pricing and performance standards (Besanko 1987). But they may be the best instrument for situations where decisions are not very responsive to price signals such as consumer choices related to energy efficiency and recycling and decisions relating to urban land use and infrastructure choices.

By mandating specific compliance pathways, technology standards risk locking-in a high-cost pathway when lower cost options are available or may emerge through market incentives and innovation (Raff and Walter 2020). Furthermore, standards may require high-cost GHG reductions in one sector while missing low-cost options in another sector. Technology standards can also stifle innovation by blocking alternative technologies from entering the market (Sachs 2012). Benefits of technology standards include their potential to achieve emission reductions in a relatively short time frame and that their effectiveness can be estimated with some confidence (Montgomery et al. 2019).

#### 13.6.4.3 Performance of Regulatory Instruments

Regulatory policy instruments tend to be more economically costly than pricing instruments, as explained above. However, regulatory policies may be preferred for other reasons.

In some cases, regulatory policy can elicit greater political support than pricing policy (Tobler et al. 2012; Lam 2015; Drews and van den Bergh 2016). For example, USA citizens have expressed more support for flexible regulation like the RPS than for carbon taxes (Rabe 2018). And a survey in British Columbia a few years after the simultaneous implementation of a carbon tax and two regulations – the LCFS and a clean electricity standard – found much less strong opposition to the regulations, even after being informed that they were costlier to consumers (Rhodes et al. 2017). The degree of public support for regulations depends, however, on the type of regulation, as outright technology prohibitions can be unpopular (Attari et al. 2009; Cherry et al. 2012).

In comparison to economic instruments, regulatory policies tend to cause greater cost of living increases in percentage terms for lower income consumers – called policy regressivity (Levinson 2019; Davis and Knittel 2019). And unlike carbon taxes, regulations do not generate revenues that can be used to compensate lower income groups.

A renewable energy procurement obligation in South Africa successfully required local hiring with perceived positive results (Walwyn and Brent 2015; Pahle et al. 2016), a clean energy regulation in Korea was perceived to provide greater employment opportunities (Lee 2017), and a UK obligation on energy companies to provide energy retrofits to low-income households improved energy affordability according to participants (Elsharkawy and Rutherford 2018).

From an energy system transformation perspective, technology standards, including phase-out mandates, have particular promise to achieve profound change in specific sectors and technologies (Tvinnereim and Mehling 2018). As such policies change the technologies available in the market, then economic instruments can also have a greater effect (Pahle et al. 2018).

## Box 13.10 | Policies to Limit Emissions of Non-$CO_2$ Gases

Non-$CO_2$ gases weighted by their 100-year GWPs represent approximately 25% of global GHG emissions, of which methane ($CH_4$) accounts for 18%, nitrous oxide ($N_2O$) 4%, and fluorinated gases (HFCs, PFCs, $SF_6$ and $NF_3$) 2% (Minx et al. 2021). Only a small share of these emissions are subject to mitigation policies.

**Methane ($CH_4$).** Anthropogenic sources include agriculture, mainly livestock and rice paddies, fossil fuel extraction and processing, fuel combustion, some industrial processes, landfills, and wastewater treatment (EPA 2019). Atmospheric measurements indicate that methane emissions from fossil fuel production are larger than shown in emissions inventories (Schwietzke et al. 2016). Only a small fraction of global $CH_4$ emissions is regulated. Mitigation policies focus on landfills, coal mines, and oil and gas operations.

Regulations and incentives to capture and utilise methane from coal seams came into effect in China in 2010 (Tan 2018; Tao et al. 2019). Inventory data suggest that emissions peaked and began a slow decline after 2010 (Gao et al. 2020) though satellite data indicate that China's methane emissions, largely attributable to coal mining, continued to rise in line with pre-2010 trends (Miller et al. 2019). Methane emissions from sources including agriculture, waste and industry are included in some offset credit schemes, including the CDM and at national level in Australia's Emissions Reductions Fund (Australian Climate Change Authority 2017) and the Chinese Certified Emission Reduction (CCER) scheme (Lo and Cong 2017).

**Nitrous oxide ($N_2O$).** $N_2O$ emissions are produced by agricultural soil management, livestock waste management, fossil fuel combustion, and adipic acid and nitric acid production (EPA 2019). Most $N_2O$ emissions are not regulated and global emissions have been increasing. $N_2O$ emissions by adipic and nitric acid plants in the EU are covered by the ETS (Winiwarter et al. 2018). $N_2O$ emissions are included in some offset schemes. China, the United States, Singapore, Egypt, and Russia produce 86% of industrial $N_2O$ emissions offering the potential for targeted mitigation action (EPA 2019).

**Hydrofluorocarbons (HFCs).** Most HFCs are used as substitutes for ozone depleting substances. The Kigali Amendment (KA) to the Montreal Protocol will reduce HFC use by 85% by 2047 (UN Environment 2018). To help meet their KA commitments developed country parties have been implementing regulations to limit imports, production and exports of HFCs and to limit specific uses of HFCs.

The EU, for example, issues tradable quota for imports, production and exports of HFCs. Prices of HFCs have increased as expected (Kleinschmidt 2020) which has led to smuggling of HFCs into the EU (European Commission 2019b). HFC use has been slightly (1–6%) below the limit each year from 2015 through 2018 (EEA 2019). China and India released national cooling action plans in 2019, laying out detailed, cross-sectoral plans to provide sustainable, climate friendly, safe and affordable cooling (Dean et al. 2020).

**Perfluorocarbons (PFCs), sulphur hexafluoride ($SF_6$) and nitrogen trifluoride ($NF_3$).** With the exception of $SF_6$, these gases are emitted by industrial activities located in the European Economic Area (EEA) and a limited number (fewer than 30) of other countries. Regulations in Europe, Japan and the USA focus on leak reduction as well as collection and reuse of $SF_6$ from electrical equipment. Other uses of $SF_6$ are banned in Europe (European Union 2014).

PFCs are generated during the aluminium smelting process if the alumina level in the electrolytic bath falls below critical levels (EPA 2019). In Europe these emissions are covered by the EU ETS. The industry is eliminating the emissions through improved process control and a shift to different production processes.

The semiconductor industry uses HFCs, PFCs, $SF_6$ and $NF_3$ for etching and deposition chamber cleaning (EPA 2019) and has a voluntary target of reducing GHG emissions 30% from 2010 by 2020 (World Semiconductor Council 2017). Europe regulates production, import, export, destruction and feedstock use of PFCs and $SF_6$, but not $NF_3$ (EEA 2019). In addition, fluorinated gases are taxed in Denmark, Norway, Slovenia and Spain.

## Box 13.11 | Shadow Cost of Carbon in Regulatory Analysis

In some jurisdictions, public administrations are required to apply a shadow cost of carbon to regulatory analysis.

Traditionally, for example in widespread application in the United States, the shadow cost of carbon is calibrated to an estimate of the social cost of carbon as an approximation of expected future cumulative economic damage from a unit of greenhouse gas emissions (Metcalf and Stock 2017). Social cost of carbon is usually estimated using integrated assessment models and is subject to fundamental uncertainties (Pezzey 2019). An alternative approach, used for example in regulatory analysis in the United Kingdom since 2009, is to define a carbon price that is thought to be consistent with a particular targeted emissions outcome. This approach also requires a number of assumptions, including about future marginal costs of mitigation (Aldy et al. 2021).

In some jurisdictions, the analysis of regulatory instruments is subject to an assessment on the basis of a shadow cost of carbon, which can influence the choice and design of regulations that affect GHG emissions (Box 13.11).

### 13.6.5 Other Policy Instruments

A range of other mitigation policy instruments are in use, often playing a complementary role to pricing and standards.

#### 13.6.5.1 Transition Support Policies

Effective climate change mitigation can cause economic and social disruption where there is transformative change, such as changes in energy systems away from fossil fuels (Section 13.9). Transitional assistance policies can be aimed to ameliorate effects on consumers, workers, communities, corporations or countries (Green and Gambhir 2020) in order to create broad coalitions of supporters or to limit opposition (Vogt-Schilb and Hallegatte 2017).

#### 13.6.5.2 Information Programmes

Information programmes, including energy efficiency labels, energy audits, certification, carbon labelling and information disclosure, are in wide use in particular for energy consumption. They can reduce GHG emissions by promoting voluntary technology choices and behavioural changes by firms and households.

Energy efficiency labelling is in widespread use, including for buildings, and for end users products including cars and appliances. Carbon labelling is used for example for food (Camilleri et al. 2019) and tourism (Gössling and Buckley 2016). Information measures also include specific information systems such as smart electricity meters (Zangheri et al. 2019). Chapters 5 and 9 provide detail.

Information programmes can correct for a range of market failures related to imperfect information and consumer perceptions (Allcott 2016). Alongside mandatory standards (13.6.4), information programmes can nudge firms and consumers to focus on often overlooked operating cost reductions (Carroll et al. 2022). For example, consumers who are shown energy efficiency labels on average buy more energy efficient appliances than those who are not (Stadelmann and Schubert 2018). Information policies can also support the changing of social norms about consumption choices, which have been shown to raise public support for pricing and regulatory policy instruments (Gössling et al. 2020).

Energy audits provide tailored information about potential energy savings and benchmarking of best practices through a network of peers. Typical examples include the United States Better Buildings Challenge that has provided energy audits to support USA commercial and industrial building owners, energy savings have been estimated at 18% to 30% (Asensio and Delmas 2017); and Germany's energy audit scheme for SMEs achieving reductions in energy consumption of 5–70% (Kluczek and Olszewski 2017).

Consumption-oriented policy instruments seek to reduce GHG emissions by changing consumer behaviour directly, via retailers or via the supply chain. Aspects that hold promise are technology lists, supply chain procurement by leading retailers or business associations, a carbon-intensive materials charge and selected infrastructure improvements (Grubb et al. 2020).

The information provided to consumers in labelling programmes is often not detailed enough to yield best possible results (Davis and Metcalf 2016). Providing information about running costs tends to be more effective than providing data on energy use (Damigos et al. 2020). Sound implementation of labelling programmes requires appropriate calculation methodology and tools, training and public awareness (Liang Wong and Krüger 2017). In systems where manufacturers self-report performance of their products, there tends to be misreporting and skewed energy efficiency labelling (Goeschl 2019).

A new form of information programmes are financial accounting standards as frameworks to encourage or require companies to disclose how the transition risks from shifting to a low-carbon economy and physical climate change impacts may affect their business or asset values (Chapter 15). The most prominent such standard was issued in 2017 by the Financial Stability Board's Task Force on Climate-related Financial Disclosures. It has found rapid uptake among regulators and investors (O'Dwyer and Unerman 2020).

Traditionally, corporate reporting has treated climate risks in a highly varied and often minimal way (Foerster et al. 2017). Disclosure of climate-related risks creates incentives for companies to improve their carbon and climate change exposure, and ultimately regulatory standards for climate risk (Eccles and Krzus 2018). Disclosure can also reinforce calls for divestment in fossil fuel assets predominantly promoted by civil society organisations (Ayling and Gunningham 2017), raising moral principles and arguments about the financial risks inherent in fossil fuel investments (Green 2018; Blondeel et al. 2019).

#### 13.6.5.3 Public Procurement and Investment

National, sub-national and local governments determine many aspects of infrastructure planning, fund investment in areas such as energy, transport and the built environment, and purchase goods and services, including for government administration and military provisioning.

Public procurement rules usually mandate cost effectiveness but only in some cases allow or mandate climate change consideration in public purchasing, for example in EU public purchasing guidelines (Martinez Romera and Caranta 2017). Green procurement for buildings has been undertaken in Malaysia (Bohari et al. 2017). a paper cites Taiwan (province of China) green public procurement law, which has contributed to reduced emissions intensity (Tsai 2017). In practice, awareness and knowledge of 'green' public procurement techniques and procedures is decisive for climate-friendly procurement (Testa et al. 2016). Experiences in low-carbon infrastructure procurement point to procedures being tailored to concerns about competition, transaction costs and innovation (Kadefors et al. 2020).

Infrastructure investment decisions lock-in high or low emissions trajectories over long periods. Low-emissions infrastructure can enable or increase productivity of private low-carbon investments (Jaumotte et al. 2021) and is typically only a little more expensive over its lifetime, but faces additional barriers including higher upfront costs, lack of pricing of externalities, or lack of information or aversion to novel products (Granoff et al. 2016). In low-income developing countries, where infrastructure has historically lagged developed countries, some of these hurdles can be exacerbated by overall more difficult conditions for public investment (Gurara et al. 2018).

Governments can also promote low-emissions investments through public-private partnerships and government owned 'green banks' that provide loans on commercial or concessional basis for environmentally friendly private sector investments (David and Venkatachalam 2019; Ziolo et al. 2019). Public funding or financial guarantees such as contracts-for-difference can alleviate financial risk in the early stages of technology deployment, creating pathways to commercial viability (Bataille 2020).

Government provision can also play an important role in economic stimulus programs, including as implemented in response to the pandemic of 2020–2021. Such programmes can support low-emissions infrastructure and equipment, and industrial or business development (Elkerbout et al. 2020; Hainsch et al. 2020; Barbier 2020; Hepburn et al. 2020).

#### 13.6.5.4 Voluntary Agreements

Voluntary Agreements result from negotiations between governments and industrial sectors that commit to achieve agreed goals (Mundaca and Markandya 2016). When used as part of a broader policy framework, they can enhance the cost effectiveness of individual firms in attaining emission reductions while pricing or regulations drive participation in the agreement (Dawson and Segerson 2008).

Public voluntary programmes, where a government regulator develops programs to which industries and firms may choose to participate on a voluntary basis, have been implemented in numerous countries. For example, the United States Environmental Protection Agency introduced numerous voluntary programmes with industry to offer technical support in promoting energy efficiency and emissions reductions, among other initiatives (EPA 2017). a European example is the EU Ecolabel Award programme (European Commission 2020b). Agreements for industrial energy efficiency in Europe (Cornelis 2019) and Japan (Wakabayashi and Arimura 2016) have been particularly effective in addressing information barriers and for smaller companies. The International Civil Aviation Organization's CORSIA scheme (Prussi et al. 2021) is an example of an international industry-based public voluntary programme.

Voluntary agreements are often implemented in conjunction with economic or regulatory instruments, and sometimes are used to gain insights ahead of implementation of regulatory standards, as in the case of energy efficiency PVPs in South Korea (Seok et al. 2021). In some cases, industries use voluntary agreements as partial fulfilment of a regulation (Rezessy and Bertoldi 2011; Langpap 2015). For example, the Netherlands have permitted participating industries to be exempt from certain energy taxes and emissions regulations (Veum 2018).

13

### Box 13.12 | Technology and Research and Development Policy

Private businesses tend to under-invest in research and development because of market failures (Geroski 1995), hence there is a case for governments to support research and technology development. a range of different policy instruments are used, including government funding, preferential tax treatment, intellectual property rules, and policies to support the deployment and diffusion of new technologies. Chapter 16 treats innovation policy in-depth.

### 13.6.6 International Interactions of National Mitigation Policies

One country's mitigation policy can impact other countries in various ways including changes in their GHG emissions (leakage), creation of markets for emission reduction credits, technology development and diffusion (spillovers), and reduction in the value of their fossil fuel resources.

#### 13.6.6.1 Leakage Effects

Compliance with a mitigation policy can affect the emissions of foreign sources via several channels over different time scales (Zhang and Zhang 2017) (Box 13.13 ). The effects may interact and yield a net increase or decrease in emissions. The leakage channel that is of most concern to policymakers is adverse international competitiveness impacts from domestic climate policies.

In principle, implementation of a mitigation policy in one country creates an incentive to shift production of tradable goods whose costs are increased by the policy to other countries with less costly emissions limitation policies (Section 12.6.3). Such 'leakage' could to some extent negate emissions reductions in the first country, depending on the relative emissions intensity of production in both countries.

*Ex ante* modelling studies typically estimate significant leakage for unilateral policies to reduce emissions due to production of emissions intensive products such as steel, aluminium, and cement (Carbone and Rivers 2017). However, the results are highly dependent on assumptions and typically do not reflect policy designs specifically aimed at minimising or preventing leakage (Fowlie and Reguant 2018).

Numerous *ex post* analyses, mainly for the EU ETS, find no evidence of any or significant adverse competitiveness impacts and conclude that there was consequently no or insignificant leakage (*medium evidence*, *medium agreement*) (Branger et al. 2016; Haites et al. 2018; Koch and Basse Mama 2019; FSR Climate 2019; aus dem Moore et al. 2019; Venmans et al. 2020; Kuusi et al. 2020; Verde 2020; Borghesi et al. 2020). This is attributed to large allocations of free allowances to emissions-intensive, trade-exposed sources, relatively low allowance prices, the ability of firms in some sectors to pass costs on to consumers, energy's relatively low share of production costs, and small but statistically significant effects on innovation (Joltreau and Sommerfeld 2019). Few carbon taxes apply to emissions-intensive, trade-exposed sources (Timilsina 2018), so competitiveness impacts usually are not a particular concern.

Policies intended to address leakage include a border carbon adjustment (Ward et al. 2019; Ismer et al. 2020). a border carbon adjustment (BCA) imposes costs – a tax or allowance purchase obligation – on imports of carbon-intensive goods equivalent to those borne by domestic products possibly mirrored by rebates for exports (Böhringer et al. 2012; Fischer and Fox 2012; Zhang 2012; Böhringer et al. 2017c) (Chapter 14). A BCA faces the practical challenge of determining the carbon content of imports (Böhringer et al. 2017a) and the design needs to be consistent with WTO rules and other international agreements (Cosbey et al. 2019;

### Box 13.13 | Possible Sources of Leakage

**Competitiveness:** Mitigation policy raises the costs and product prices of regulated sources which causes production to shift to unregulated sources, increasing their emissions.

**Fossil fuel channel:** Regulated sources reduce their fossil fuel use, which lowers fossil fuel prices and increases consumption and associated emissions by unregulated sources.

**Land-use channel:** Mitigation policies that change land use lead to land use and emissions changes in other jurisdictions (Bastos Lima et al. 2019).

**Terms of trade effect:** Price increases for the products of regulated sources shift consumption to other goods, which raises emissions due to the higher output of those goods.

**Technology channel:** Mitigation policy induces low-carbon innovation, which reduces emissions by sources that adopt the innovations that may include unregulated sources (Gerlagh and Kuik 2007).

**Abatement resource effect:** Regulated sources increase use of clean inputs, which reduces inputs available to unregulated sources and so limits their output and emissions (Baylis et al. 2014).

**Scale channel:** Changes to the output of regulated and unregulated sources affect their emissions intensities so emissions changes are not proportional to output changes (Antweiler et al. 2001).

**Intertemporal channel:** Capital stocks of all sources are fixed initially but change over time affecting the costs, prices, output and emissions of regulated and unregulated products.

Mehling et al. 2019). Model estimates indicate that a BCA reduces but does not eliminate leakage (Branger and Quirion 2014). No BCA has yet been implemented for international trade although such a measure is currently under consideration by some governments.

#### 13.6.6.2 Market for Emission Reduction Credits

A mitigation policy may allow the use of credits issued for emission reductions in other countries for compliance purposes (see also Section 13.6.3.4 on offset credits and Chapter 14 on international credit mechanisms). Creation of international markets for emission reduction credits tends to benefit other countries through financial flows in return for emissions credit sales (*medium evidence, high agreement*).

The EU, New Zealand and Switzerland allowed participants in their emissions trading systems to use credits issued under the Kyoto Protocol mechanisms, including the Clean Development Mechanism (CDM), for compliance. From 2008 through 2014, participants used 3.76 million imported credits for compliance of which 80% were CDM credits (Haites 2016).[3] Use of imported credits has fallen to very low levels since 2014 (World Bank 2014; Shishlov et al. 2016).[4]

The Clean Development Mechanism (CDM) is the world's largest offset programme (Chapter 14). From 2001 to 2019 over 7500 projects with projected emission reductions in excess of 8000 $MtCO_2$-eq were implemented in 114 developing countries using some 140 different emissions reduction methodologies (UNFCCC 2012; UNEP DTU Partnership 2020). Credits reflecting over 2000 $MtCO_2$-eq of emission reductions by 3260 projects have been issued. To address additionality and other concerns the CDM Executive Board frequently updated its approved project methodologies.

#### 13.6.6.3 Technology Spillovers

Mitigation policies stimulate low-carbon R&D by entities subject to those policies and by other domestic and foreign entities (FSR Climate 2019). Policies to support technology development and diffusion tend to have positive spillover effects between countries (*medium evidence, high agreement*) (Section 16.3).

Innovation activity in response to a mitigation policy varies by policy type (Jaffe et al. 2002) and stringency (Johnstone et al. 2012). In addition, many governments have policies to stimulate R&D, further increasing low-carbon R&D activity by domestic researchers. Emitters in other countries may adopt some of the new low-carbon technologies thus reducing emissions elsewhere. Technology development and diffusion is reviewed in Chapter 16.

#### 13.6.6.4 Value of Fossil Fuel Resources

Fossil fuel resources are a significant source of exports, employment and government revenues for many countries. The value of these resources depends on demand for the fuel and competing supplies in the relevant international markets. Discoveries and new production technologies reduce the value of established resources. Mitigation policies that reduce the use of fossil fuels also reduce the value of these resources. A single policy in one country is unlikely to have a noticeable effect on the international price, but similar policies in multiple countries could adversely affect the value of the resources. For fossil fuel exporting countries, mitigation policies consistent with the Paris Agreement goals could result in greater costs from changes in fossil fuel prices due to lower international demand than domestic policy costs (*medium evidence, high agreement*) (Liu et al. 2020).

The impact on the value of established resources will be mitigated, to some extent, by the reduced incentive to explore for and develop new fossil fuel supplies. Nevertheless, efforts to lower global emissions will mean substantially less demand for fossil fuels, with the majority of current coal reserves and large shares of known gas and oil reserves needing to remain unused, with great diversity in impacts between different countries (McGlade and Ekins 2015) (Chapters 3, 6, 15).

Estimates of the potential future loss in value differ greatly. There is uncertainty about remaining future fossil fuel use under different mitigation scenarios, as well as future fossil fuel prices depending on extraction costs, market structures and policies. Estimates of total cumulative fossil fuel revenue lost range between 5–67 trillion USD (Bauer et al. 2015) with an estimate of the net present value of lost profit of around 10 trillion USD (Bauer et al. 2016). Policies that constrain supply of fossil fuels in the context of mitigation objectives could limit financial losses to fossil fuel producers (Chapter 14).

## 13.7 Integrated Policy Packages for Mitigation and Multiple Objectives

Since AR5, the literature on climate policies and policymaking has expanded in two significant directions. First, there is growing recognition that mitigation policy occurs in the context of multiple climate and development objectives (Chapter 4). Different aspects of these linkages are discussed across the AR6 WGIII report, including concepts and framings (Section 1.6.2), shifting sustainable development pathways (Section 4.3 and Cross-Chapter Box 5 in Chapter 4), cross-sectoral interactions (Sections 12.6.1 and 12.6.2), evidence of co-impacts (Section 17.3), links with adaptation (Section 4.4.2) and accelerating the transition (Sections 13.9, 17.1.1, 17.4.5 and 17.4.6). While the concept of development pathways is salient in all countries, it may particularly resonate with policymakers in developing countries focused on providing basic needs and addressing poverty and inequality, including energy poverty (Ahmad 2009; Fuso Nerini et al. 2019; Bel and Teixidó 2020; Caetano et al. 2020; Röser et al. 2020). Consequently, some countries may frame policies predominantly in terms of accelerating mitigation, while in others a multiple objectives approach linked to development pathways may dominate, depending on their specific socio-economic contexts and priorities, governance capacities (McMeekin et al. 2019) and perceptions of historical responsibility (Winkler and Rajamani 2014; Friman and Hjerpe 2015; Winkler et al. 2015; Pan et al. 2017).

[3] 2010 through 2014 for the New Zealand ETS.

[4] All three ETSs were modified after 2012 including provisions that affected compliance use of imported credits.

| | | Framing of outcome | |
|---|---|---|---|
| | | **Enhancing mitigation** | **Addressing multiple objectives of mitigation and development** |
| Approach to policymaking | Shifting incentives | 'Direct mitigation focus' *(Section 13.6; 2.8)*<br>*Objective:* reduce GHG emissions now<br>*Literature:* how to design and implement policy instruments, with attention to distributional and other concerns<br>*Examples:* carbon tax, cap and trade, border carbon adjustment, disclosure policies | 'Co-benefits' *(Sections 17.3; 5.6.2; 12.4.4)*<br>*Objective:* synergies between mitigation and development<br>*Literature:* scope for and policies to realise synergies and avoid trade-offs across climate and development objectives<br>*Examples:* appliance standards, fuel taxes, community forest management, sustainable dietary guidelines, green building codes, packages for air pollution, packages for public transport |
| | Enabling transition | 'Socio-technical transitions' *(Sections 1.7.3; 5.5; 10.8; 6.7; Cross-Chapter Box 12 in Chapter 16)*<br>*Objective:* accelerate low-carbon shifts in socio-technical systems<br>*Literature:* understand socio-technical transition processes, integrated policies for different stages of a technology 'S-curve' and explore structural, social and political elements of transitions<br>*Examples:* packages for renewable energy transition and coal phase-out; diffusion of electric vehicles, process and fuel switching in key industries | 'System transitions to shift development pathways' *(Sections 11.6.6; 7.4.5; 13.9; 17.3.3; Cross-Chapter Box 5 in Chapter 4; Cross-Chapter Box 9 in Chapter 13)*<br>*Objective:* accelerate system transitions and shift development pathways to expand mitigation options and meet other development goals<br>*Literature:* examines how structural development patterns and broad cross-sector and economy-wide measures drive ability to mitigate while achieving development goals through integrated policies and aligning enabling conditions<br>*Examples:* packages for sustainable urbanisation, land-energy-water nexus approaches, green industrial policy, regional just transition plans |

Figure 13.6 | Mapping the landscape of climate policy.

Second, since AR5 there is growing attention to enabling transitions over time. Literature on socio-technical transitions, rooted in innovation studies, highlights the need for different policy focus at different stages of a transition (Geels et al. 2017b,a; Köhler et al. 2019) (Section 1.7.3). Other literature examines how broad patterns of development drive both social and mitigation outcomes through shifts in policies and a re-alignment of enabling conditions (Chapter 4). Explicit efforts to shift development pathways, for example by shifting patterns of energy demand and urbanisation, therefore offer broader mitigation opportunities (Cross-Chapter Box 5 in Chapter 4). Common to both approaches is an emphasis beyond the short term, and attention to enabling longer-term structural shifts in economies and societies.

Taking these trends into account, Figure 13.6 outlines the climate policy landscape, and how it maps to different parts of this Working Group III report. One axis of variation captures alternative framings of desired outcomes in national policymaking – mitigation versus multiple objectives, while the second captures the shift in policymaking from an initial focus on shifting incentives through largely individual policy instruments, to explicit consideration of how policies and economy-wide measures, including those that shift incentives, can combine to enable transitions. As a result, Figure 13.6 represents interconnected policy ideas, but backed by distinct strands of literature. Notably, each of these categories is salient to climate policymaking, although the balance may differ depending on country context.

This section particularly focuses on climate policymaking for transition – both socio-technical transitions and shifts in development pathways, while direct climate policies and co-benefits are addressed in other parts of the report, as indicated in Figure 13.6. This section focuses in particular on lessons for designing policy packages for transitions, and is complemented by discussion in Section 13.8 on integration between adaptation and mitigation, and Section 13.9 on economy-wide measures and the broader enabling conditions necessary to accelerate mitigation.

### 13.7.1 Policy Packages for Low-carbon Sustainable Transitions

Since AR5 an emergent multidisciplinary literature on policy packages, or policy mixes, has emerged that examine how policies may be combined for sustainable low-carbon transitions (Rogge and Reichardt 2016; Kern et al. 2019). This literature covers various sectors including: energy (Rogge et al. 2017); transport (Givoni et al. 2013); industry (Scordato et al. 2018); agri-food (Kalfagianni and Kuik 2017); and forestry (Scullion et al. 2016).

A central theme in the literature is that transitions require policy interventions to address system level changes, thereby going beyond addressing market failures in two ways. First, structural system changes are needed for low-carbon transitions, including building low-carbon infrastructure (or example aligning electricity grids and storage with the requirements of new low-carbon technology), and adjusting existing institutions to low-carbon solutions (for example by reforming electricity market design) (Bak et al. 2017; Patt and Lilliestam 2018). Second, explicit transformational system changes are necessary, including efforts at directing transformations, such as clear direction setting through the elaboration of shared visions, and coordination across diverse actors across different policy fields, such as climate and industrial policy, and across governance levels (Uyarra et al. 2016; Nemet et al. 2017).

There are some specific suggestions for policy packages: Van den Bergh et al. (2021) suggest that innovation support and information provision combined with a carbon tax or market, or adoption subsidy leads to both effective and efficient outcomes. Others question the viability of universally applicable policy packages, and suggest packages need to be tailored to local objectives (del Río 2014) Consequently, much of the literature focuses on broad principles for design of policy packages and mixes, as discussed below.

Comprehensiveness, balance and consistency are important criteria for policy packages or mixes (*robust evidence*, *high agreement*) (Rogge and Reichardt 2016; Scobie 2016; Carter et al. 2018; Santos-lacueva and González 2018). Comprehensiveness assesses the extensiveness of policy packages, including the breadth of system and market failures it addresses (Rogge and Reichardt 2016). For example, instrument mixes that include only moderate carbon pricing, but are complemented by policies supporting new low-carbon technologies and a moratorium on coal-fired power plants may not only be politically more feasible than stringent carbon pricing alone, but may also limit efficiency losses and lower distributional impacts (Bertram et al. 2015b). Balance captures whether policy instruments are deployed in complementary ways given their different purposes, combining for example technology-push approaches such as public R&D with demand-pull approaches such as an energy tax. A combination of technology-push and demand-pull approaches has been shown to support innovation in energy efficient technologies in OECD countries (Costantini et al. 2017). Consistency addresses the alignment of policy instruments among each other and with the policy strategy, which may have multiple and not always consistent objectives (Rogge 2019). Consistency of policy mixes has been identified as an important driver of low-carbon transformation, particularly for renewable energy (Lieu et al. 2018; Rogge and Schleich 2018). Box 13.14 summarises the economics literature on how policies interact, to inform design of packages.

## Box 13.14 | Policy Interactions of Carbon Pricing and Other Instruments

The economics literature provides insights on policy interactions among the multiple overlapping policies that directly or indirectly affect GHG emissions, including when different levels of government are involved. Multiple mitigation policies can be theoretically justified if there are multiple objectives or market failures or to achieve distributional objectives and increase policy effectiveness (Stiglitz 2019). Examples include the coexistence of the EU ETS with vehicle emission standards and energy efficiency standards (Rey et al. 2013), and the fact that 85% of the emissions covered by California's ETS are also subject to other policies (Bang et al. 2017; Mazmanian et al. 2020). Policy interactions are also widespread among energy efficiency policies (Wiese et al. 2018).

Interactive effects can influence the costs of policy outcomes. With multiple overlapping and possibly non-optimal policies, the effect on total cost is not clear. A modelling study of USA mitigation policy finds the costs of using heterogeneous sub-national policies to achieve decarbonisation targets is 10% higher than national uniform policies (Peng et al. 2021). When multiple policy goals are sought, such as mitigation and R&D, a portfolio of optimal policies achieves the goals at significantly lower cost (Fischer and Newell 2008). In some cases, overlapping mitigation policies can raise the cost of mitigation (Böhringer et al. 2016) while lowering the cost of achieving other goals, such as energy efficiency improvements and expansion of renewable energy (Rosenow et al. 2016; Lecuyer and Quirion 2019). It is possible that one or more of the policies is made redundant (Aune and Golombek 2021).

While overlapping policies may raise the cost of mitigation, they increase the likelihood of achieving an emission reduction goal. Policy overlap will lead to different optimal carbon prices across jurisdictions (Bataille et al. 2018b). The existence of overlapping policies will usually increase administrative and compliance costs. However, *ex post* analysis shows that transaction costs of mitigation policies are low and are not a decisive factor in policy choice (Joas and Flachsland 2016).

The effectiveness, as well as economic and distributional effects, of a given mitigation policy will depend on the interactions among all the policies that affect the targeted emissions. Because a market instrument interacts with every other policy that affects the targeted emissions, interactions tend to be more complex for market instruments than for regulations that mandate specific emission reduction actions by targeted sources independent of other policies.

An ETS scheme implemented with existing mitigation policies may be subject to the 'waterbed effect' – emission reductions undertaken by some emitters may be offset by higher emissions by other ETS participants due to overlapping mitigation policies (Schatzki and Stavins 2012). This reduces the impact of the ETS and lowers carbon trading prices (Perino 2018). However *ex post* assessments find net emissions reductions. ETS design features such as a price floor and 'market stability reserve' can limit the waterbed effect (Edenhofer et al. 2017; Kollenberg and Taschini 2019; Narassimhan et al. 2018; FSR Climate 2019).

A carbon tax, unlike the allowance price, does not change in response to the effect of overlapping policies but those policies may reduce emissions by sources subject to the tax and so lower the emission reductions achieved by the tax (Goulder and Stavins 2011).

*Box 13.14 (continued)*

Policy interactions often occur with the introduction of new mitigation policy instruments. For example, in China several sub-national ETSs exist alongside policies to reduce emission intensity, increase energy efficiency and expand renewable energy supplies (Zhang 2015). These quantity-based ETSs interact with many other policies (Duan et al. 2017), for example price-based provincial carbon intensity targets (Qian et al. 2017). They also interact with the level of market regulation; for example, full effectiveness of emissions pricing would require electricity market reform in China (Teng et al. 2017).

Policy packages aimed at low-carbon transitions are more effective when they include elements to enhance the phase out of carbon-intensive technologies and practices – often called exnovation – in addition to supporting low-carbon niches (Kivimaa and Kern 2016; David 2017). Such policies include stringent carbon pricing; changes in regime rules such as design of electricity markets; reduced support for dominant regime technologies such as removing tax deductions for private motor transport based on internal combustion engines; and changes in the balance of representation of incumbents versus new entrants in deliberation and advisory bodies. For example, CGE modelling for China's fossil fuel subsidy reform found that integrating both creation and destabilisation policies is able to reduce rebound effects and make the policy mix more effective (Li et al. 2017). Sweden's pulp and paper industry shows that destabilisation policies including deregulation of the electricity market and a carbon tax were an important complement to support policies (Scordato et al. 2018), and other studies show complementary results for Finland's building sector (Kivimaa et al. 2017b) and Norway's transport and energy sector (Ćetković and Skjærseth 2019).

Policy packages for low-carbon transitions are more successful if they take into account the potential for political contestation and resistance from incumbents who benefit from high-carbon systems (*medium evidence*, *high agreement*) (Geels 2014; Roberts et al. 2018; Kern and Rogge 2018; Rosenbloom 2018). To do so, policies can be sequenced so as to address political obstacles, for example, by initially starting with policies to facilitate the entry of new firms engaged in low-carbon technologies (Pahle et al. 2018). Such policies can generate positive feedbacks by creating constituencies for continuation of those policies, but need to be designed to do so from the outset (Edmondson et al. 2019, 2020). For example, supporting renewable energies through feed-in tariffs can buttress coalitions for more ambitious climate policy, such as through carbon pricing (Meckling et al. 2015). However, negative policy feedback may also arise from ineffective policy instruments that lose public support, or create concentrated losses that arouse oppositional coalitions (Edmondson et al. 2019). Feedback loops can operate through changes in resources available to actors; changes in expectations; and changes in government capacities (Edmondson et al. 2019).

Another promising strategy is to design short-term policies which might help to provide later entry points for more ambitious climate policy (Kriegler et al. 2018) and supportive institutions. The sequencing of policies can build coalitions for climate policy, starting with green industrial policy (e.g. supporting renewable energies through feed-in tariffs) and introducing or making carbon pricing more stringent when supportive coalitions of stringent climate policy have been formed (Meckling et al. 2015). Similarly, investing in supportive institutions, with competencies compatible with low-carbon futures, are a necessary supportive element of transitions (Pahle et al. 2018; Rosenbloom et al. 2019; Domorenok et al. 2021).

### 13.7.2 Policy Integration for Multiple Objectives and Shifting Development Pathways

This sub-section assesses policy integration and packages required to enable shifts in development pathways, with a particular focus on sectoral scale transitions. However, because shifting development pathways requires broad transformative change, it complements discussion on broader shifts in policymaking such as fiscal, educational, and infrastructure policies (Cross-Chapter Box 5 in Chapter 4) and to the alignment of a wide range of enabling conditions required for system transitions (Section 13.9).

In many countries, and particularly when climate policy occurs in the context of sustainable development, policymakers seek to address climate mitigation in the context of multiple economic and social policy objectives (*medium evidence*, *robust agreement*) (Halsnæs et al. 2014; Campagnolo and Davide 2019; Cohen et al. 2019). Studies suggest that co-benefits of climate policies are substantial, especially in relation to air quality, and can yield better mitigation and overall welfare, yet these are commonly overlooked in policymaking (*robust evidence*, *robust agreement*) (Nemet et al. 2010; Ürge-Vorsatz et al. 2014; von Stechow et al. 2015; Mayrhofer and Gupta 2016; Roy et al. 2018; Bhardwaj et al. 2019; Karlsson et al. 2020). Other studies have shown the existence of strong complementarities between the SDGs and realisation of NDC pledges by countries (McCollum et al. 2018). An explicit attention to development pathways can enhance the scope for mitigation, by paying explicit attention to development choices that lock-in or lock-out opportunities for mitigation, such as around land use and infrastructure choices (Cross-Chapter Box 5 in Chapter 4). While the pay-offs are considerable to an approach to mitigation that takes into account linkages to multiple objectives and the opportunity to shift development pathways, there are also associated challenges with implementing this approach to policymaking.

First, spanning policy arenas and addressing multiple objectives places considerable requirements of coordination on the policymaking process (Howlett and del Rio 2015; Obersteiner et al. 2016). Climate policy integration suggests several steps should precede actual policy formulation, beginning with a clear articulation of the policy frame or problem statement (Adelle and Russel 2013; Candel and Biesbroek 2016). For example, a greenhouse gas limitation framework versus a co-benefits framing would likely yield different policy approaches. It is then useful to identify the range of actors and institutions involved in climate governance – the policy subsystem, the goals articulated, the level at which goals are articulated and the links with other related policy goals such as energy security or energy access (Candel and Biesbroek 2016). The adoption of specific packages of policy instruments should ideally follow these prior steps that define the scope of the problem, actors and goals.

In practice, integration has to occur in the context of an already existing policy structure, which suggests the need for finding windows of opportunity to bring about integration, which can be created by international events, alignments with domestic institutional procedures, and openings created by policy entrepreneurs (Garcia Hernandez and Bolwig 2020). Integration also has to occur in the context of existing organisational routines and cultures, which can pose a barrier to integration (Uittenbroek 2016). Experience from the EU suggests that disagreements at the level of policy instruments are amenable to resolution by deliberation, while normative disagreements at the level of objectives require a hierarchical decision structure (Skovgaard 2018). As this discussion suggests, the challenge of integration operates in two dimensions: horizontal – between sectoral authorities such as ministries or policy domains such as forestry – or vertical – either between constitutional levels of power or within the internal mandates and interactions of a sector (Howlett and del Rio 2015; Di Gregorio et al. 2017). There are also important temporal dimensions to policy goals, as policy and benchmarks have to address not just immediate success but also indications of future transformation (Dupont and Oberthür 2012; Dupont 2015).

Second policymaking for shifting development pathways has to account for inherent uncertainties in future development paths (Moallemi and Malekpour 2018; Castrejon-Campos et al. 2020). These uncertainties may be greater in developing countries that are growing rapidly and where structural features of the economy including infrastructure and urbanisation patterns are fluid. For example, reviews of modelling studies of Chinese (Grubb et al. 2015) and Indian emissions futures (Spencer and Dubash 2021) find that differences in projections can substantially be accounted for by alternative assumptions about future economic structural shifts. Consequently, an important design consideration is that policy packages should be robust, that is, perform satisfactorily for all key objectives under a broad range of plausible futures (Kwakkel et al. 2016; Maier et al. 2016; Castrejon-Campos et al. 2020). Such an approach to decision-making can be contrasted with one that tries to design an optimal policy package for the 'best guess' future scenario (Maier et al. 2016). Moreover, policy packages can usefully be adapted dynamically to changing circumstances as part of the policy process (Haasnoot et al. 2013; Hamarat et al. 2014; Maier et al. 2016) including by using exploratory modelling techniques that allow comparison of trade-offs across alternative future scenarios (Hamarat et al. 2014). Another approach is to link quantitative models with a participatory process that enables decision-makers to test the implications of alternative interventions (Moallemi and Malekpour 2018). Rosenbloom et al. (2019) suggest that because policy mixes should adapt to changing circumstances, instead of stability of a particular mix, transitions require embedding policies within a long-term orientation toward a low-carbon economy, including a transition agenda, social legitimacy for this agenda, and an appropriate ecosystem of institutions.

Third, achieving changes in development pathways requires engaging with place-specific context. It requires attention to existing policies, political interests that may gain or lose from a transition, and locally specific governance enablers and disablers. As a result, while there may be approaches that carry over from one context to another, implementation requires careful tailoring of transition approaches to specific policy and governance contexts. Cross-Chapter Box 9 in this chapter summarises case studies of sectoral transitions from other chapters in this report (Chapters 5 to 12) to illustrate this complexity. Broader macroeconomic transformative shifts are discussed in more detail in Section 13.9.

Common to all the sectoral cases in Cross-Chapter Box 9 is a future-oriented vision of sectoral transition often focused on multiple objectives, such as designing tram-based public transport systems in Bulawayo, Zimbabwe to simultaneously stimulate urban centers, create jobs and enable low-carbon transportation. Sectoral transitions are enabled by policy mixes that bring together different combinations of instruments – including regulations, financial incentives, convening, education and outreach, voluntary agreements, procurement and creation of new institutions – to work together in a complementary manner. The effectiveness of a policy mix depends on conditions beyond design considerations and also rests on the larger governance context within which sector transitions occur, which can include enabling and disabling elements. Enabling factors illustrated in Cross-Chapter Box 9 include strong high level political support, for example to address deforestation in Brazil despite powerful logging and farmer interests, or policy design to win over existing private interests, for example, by harnessing distribution networks of kerosene providers to new LPG technology in Indonesia. Disabling conditions include local institutional contexts, such as the lack of tree and land tenure in Ghana, which, along with the monopoly of the state marketing board, posed obstacles to Ghana's low-carbon cocoa transition. These examples emphasise the importance of attention to local context if policy integration and the design of policy mixes are to effectively lead to transitions guided by multiple climate and development objectives.

## Cross-Chapter Box 9 | Case Studies of Integrated Policymaking for Sector Transitions

**Authors:** Parth Bhatia (India), Navroz K. Dubash (India), Igor Bashmakov (the Russian Federation), Paolo Bertoldi (Italy), Mercedes Bustamante (Brazil), Michael Craig (the United States of America), Stephane de la Rue du Can (the United States of America), Manfred Fischedick (Germany) Amit Garg (India), Oliver Geden (Germany), Robert Germeshausen (Germany), Siir Kilkis (Turkey), Susanna Kugelberg (Denmark), Andreas Loeschel (Germany), Cheikh Mbow (Senegal), Yacob Mulugetta (Ethiopia), Gert-Jan Nabuurs (the Netherlands), Vinnet Ndlovu (Zimbabwe/Australia), Peter Newman (Australia), Lars Nilsson (Sweden), Karachepone Ninan (India)

Real world sectoral transitions reinforce critical lessons on policy integration: a high-level strategic goal (Column a in Cross-Chapter Box 9, Table 1), the need for a clear sector outcome framing (column B), a carefully coordinated mix of policy instruments and governance actions (column C), and the importance of context-specific governance factors (column D). Illustrative examples, drawn from sectors, help elucidate the complexity of policymaking in driving sectoral transitions.

**Cross-Chapter Box 9, Table 1 | Case studies of integrated policymaking for sector transitions.**

| A. Illustrative case | B. Objective | C. Policy mix | D. Governance context | |
|---|---|---|---|---|
| | | | Enablers | Barriers |
| Shift in mobility service provision in Kolkata, India *[Box 5.8]* | – Improve system efficiency, sustainability and comfort<br>– Shift public perceptions of public transport | – Strengthen coordination between modes<br>– Formalise and green auto-rickshaws<br>– Procure fuel efficient, comfortable, low-floor AC buses<br>– Ban cycling on busy roads<br>– Deploy policy actors as change-agents, mediating between interest groups | – Cultural norms around informal transport sharing, linked to high levels of social trust<br>– Historically crucial role of buses in transit<br>– App-cab companies shifting norms and formalising mobility sharing<br>– Digitalisation and safety on board | – Complexity: multiple modes with separate networks and meanings<br>– Accommodating and addressing legitimate concerns from social movements about the exclusionary effects of 'premium' fares, cycling bans on busy roads |
| LPG Subsidy ('Zero Kero') Program, Indonesia *[Box 6.3]* | Decrease fiscal expenditures on kerosene subsidies for cooking | – Subsidise provision of Liquefied Petroleum Gas (LPG) cylinders and initial equipment<br>– Convert existing kerosene suppliers to LPG suppliers | – Provincial government and industry support in targeting beneficiaries and implementation<br>– Synergies in kerosene and LPG distribution infrastructures | – Continued user preference for traditional solid fuels<br>– Reduced GHG benefits as subsidy shifted between fossil fuels |
| Action Plan for Prevention and Control of Deforestation in the Legal Amazon, Brazil *[Box 7.9]* | Control deforestation and promote sustainable development | – Expand protected areas; homologation of indigenous lands<br>– Improve inspections, satellite-based monitoring<br>– Restrict public credit for enterprises and municipalities with high deforestation rates<br>– Set up a REDD+ mechanism (Amazon Fund) | – Participatory agenda-setting process<br>– Cross-sectoral consultations on conservation guidelines<br>– Mainstreaming of deforestation in government programmes and projects | – Political polarisation leading to erosion of environmental governance<br>– Reduced representation and independence of civil society in decision-making bodies<br>– Lack of clarity around land ownership |
| Climate Smart Cocoa (CSC) production, Ghana *[Box 7.12]* | – Promote sustainable intensification of cocoa production<br>– Reduce deforestation<br>– Enhance incomes and adaptive capacities | – Distribute shade tree seedlings<br>– Provide access to agronomic information and agrochemical inputs<br>– Design a multi-stakeholder programme including MNCs, farmers and NGOs | – Local resource governance mechanisms ensuring voice for smallholders<br>– Community governance allowed adapting to local context<br>– Private sector role in popularising CSC | – Lack of secure tenure (tree rights)<br>– Bureaucratic and legal hurdles to register trees<br>– State monopoly on cocoa marketing, export |
| Coordination mechanism for joining fragmented urban policymaking in Shanghai, China *[Box 8.3]* | Integrate policymaking across objectives, towards low-carbon urban development | – Combine central targets and evaluation with local flexibility for initiating varied policy experiments<br>– Establish a local leadership team for coordinating cross-sectoral policies involving multiple institutions<br>– Create a direct programme fund for implementation and capacity-building | – Strong vertical linkages between Central and local levels<br>– Mandate for policy learning to inform national policy<br>– Experience with mainstreaming mitigation in related areas (e.g. air pollution) | – Challenging starting point – low share of RE, high dependency on fossil fuels<br>– Continued need for high investments in a developing context |

*Cross-Chapter Box 9 (continued)*

| A. Illustrative case | B. Objective | C. Policy mix | D. Governance context | |
|---|---|---|---|---|
| | | | Enablers | Barriers |
| Policy package for building energy efficiency, EU *[Box 9.SM.1]* | Reduce energy consumption, integrating RE and mitigating GHG emissions from buildings | – Energy performance standards, set at nearly zero energy for new buildings<br>– Energy performance standards for appliances<br>– Energy performance certificates shown during sale<br>– Long-term renovation strategies | – Binding EU-level targets, directives and sectoral effort sharing regulations<br>– Supportive urban policies, coordinated through city partnerships<br>– Funds raised from allowances auctioned under ETS | – Inadequate local technical capacity to implement multiple instruments<br>– Complex governance structure leading to uneven stringency |
| African electromobility – trackless trams with solar in Bulawayo and e-motorbikes in Kampala *[Box 10.4]* | – Leapfrog into a decarbonised transport future<br>– Achieve multiple social benefits beyond mobility provision | – Develop urban centres with solar at station precincts<br>– Public-private partnerships for financing<br>– Sanction demonstration projects for new electric transit and new electric motorbikes (for freight) | – 'Achieving SDGs' was an enabling policy framing<br>– Multi-objective policy process for mobility, mitigation and manufacturing<br>– Potential for funding through climate finance<br>– Co-benefits such as local employment generation | – Economic decline in the first decade of the 21st century<br>– Limited fiscal capacity for public funding of infrastructure<br>– Inadequate charging infrastructure for e-motorbikes |
| Initiative for a climate-friendly industry in North Rhine Westphalia (NRW), Germany *[Box 11.3]* | Collaboratively develop innovative strategies towards a net zero industrial sector, while securing competitiveness | – Build platform to bring together industry, scientists and government in self-organised innovation teams<br>– Intensive cross-branch cooperation to articulate policy/infrastructure needs | – NRW is Germany's industrial heartland, with an export-oriented industrial base<br>– Established government–industry ties<br>– Active discourse between industry and public | Compliance rules preventing in-depth cooperation |
| Food2030 Strategy, Finland *[Box 12.2]* | – Local, organic and climate friendly food production<br>– Responsible and healthy food consumption<br>– A competitive food supply chain | – Target funding and knowledge support for innovations<br>– Apply administrative means (legislation, guidance) to increase organic food production and procurement<br>– Use education and information instruments to shift behaviour (media campaigns, websites) | – Year-long deliberative stakeholder engagement process across sectors<br>– Institutional structures for agenda-setting, guiding policy implementation and reflexive discussions | – Weak role of integrated impact assessments to inform agenda-setting<br>– Monitoring and evaluation close to ministry in charge<br>– Lack of standardised indicators of food system sustainability |

## 13.8 Integrating Adaptation, Mitigation and Sustainable Development

There is growing consensus that integration of adaptation and mitigation will advance progress towards sustainable development, and that ambitious mitigation efforts will reduce the need for adaptation in the long term (*robust evidence*, *high agreement*) (IPCC 2014a). There is no level of mitigation, however, that will completely erase the need for adaptation to climate change (*robust evidence*, *high agreement*) (Mauritsen and Pincus 2017). It is therefore urgent to design and implement a multi-objective policy framework for mitigation, adaptation, and sustainable development that considers issues of equity and long-term developmental pathways across regions (*robust evidence*, *high agreement*) (Jordan et al. 2018; Mills-Novoa and Liverman 2019; Wang and Chen 2019). This section explores the logic behind the integration of adaptation and mitigation in practice (Section 13.8.1), the approaches to this integration including climate-resilient pathways, ecosystem-based solutions, and a nexus approach (Section 13.8.2); examples of the adaption and mitigation relationships and linkages (Section 13.8.3); and enabling and disabling factors for governance of mitigation and adaption.

### 13.8.1 Synergies Between Adaptation and Mitigation

Integrated climate-development actions require a context-specific understanding of synergies and trade-offs with other policy priorities (Figure 13.6) with the aim of implementing mitigation/adaptation policies that reduce GHG emissions while simultaneously strengthening resilience and reducing vulnerability (*robust evidence*, *high agreement*) (Klein et al. 2005; IPCC 2007; Zhao et al. 2018; Mills-Novoa and Liverman 2019; Solecki et al. 2019). Efficient, equitable and inclusive policies which also acknowledge and contribute directly to other pressing priorities such reducing poverty, improving health, providing access to clean water, and fostering sustainable consumption and production practices are helpful for mitigation/adaptation goals (*robust evidence*, *high agreement*) (Landauer et al. 2019; Grafakos et al. 2020).

## Box 13.15 | Adaptation and Mitigation Synergies in Africa

Synergies between mitigation and adaptation actions and sustainable development that can enhance the quality and pace of development in Africa exist at both sectoral and national levels. Available data on NDCs show the top mitigation priorities in African countries include energy, forestry, transport and agriculture and waste, and adaptation priorities focus on agriculture, water, energy and forestry. The energy sector dominates in mitigation actions and the agricultural sector is the main focus of adaptation measures, with the latter sector being a slightly larger source of greenhouse gases than the former (Mbeva et al. 2015; African Development Bank 2019; Nyiwul 2019).

Renewable energy development can support synergies between mitigation and adaptation by stimulating local and national economies through microenterprise development; providing off-grid affordable and accessible solutions; and contributing to poverty reduction through increased locally available resource use and employment and increased technical skills (Nyiwul 2019; Dal Maso et al. 2020). The Paris Agreement's technology transfer and funding mechanisms could reduce renewable energy costs and providing scale economics to local economies.

Barriers to achieving these synergies include the absence of suitable macro-and micro- level policy environments for adaptation and mitigation actions; coherent climate change policy frameworks and governance structures to support adaptation; institutional and capacity deficiencies in climate and policy research such as on data integration and technical analysis; and the high financial needs associated with the cost of mitigation and adaptation (African Development Bank 2019; Nyiwul 2019). Strengthening of national institutions and policies can support maximising synergies and co-benefits between adaptation and mitigation to reduce silos and redundant overlaps, increase knowledge exchange at the country and regional levels, and support engagement with bilateral and multilateral partners and mobilising finance through the mechanisms available (African Development Bank 2019).

Adaptation and mitigation are deeply linked in practice – at the local level, for instance, asset managers address integrated low-carbon resilience to climate change impacts and urban planners do the same (Ürge-Vorsatz et al. 2018; Grafakos et al. 2020) (Table 13.3). Similarly, ecosystem-based (or nature-based) solutions, may generate co-benefits by simultaneously sinking carbon, cooling urban areas through shading, purifying water, improving biodiversity, and offering recreational opportunities that improve public health (Raymond et al. 2017). Accurately identifying and qualitatively or quantitatively assessing these co-benefits (Stadelmann et al. 2014; Leiter and Pringle 2018; Leiter et al. 2019) is central to an integrated adaptation and mitigation policy evaluation.

Some studies press the need to consider the complex ways that power and interests influence how collective decisions are made, and who benefits from and pays for these decisions, of climate policy and to be aware of unintended consequences, especially for vulnerable people living under poor conditions (Mayrhofer and Gupta 2016; De Oliveira Silva et al. 2018). The specific adaptation and mitigation linkages will differ by country and region, as illustrated by Box 13.15.

### 13.8.2 Frameworks That Enable the Integration of Adaption and Mitigation

The IPCC's *Fifth Assessment Report* (AR5) emphasised the importance of climate-resilient pathways – development trajectories that combine adaptation and mitigation through specific actions to achieve the sustainable development goals (Prasad et al. 2009; Lewison et al. 2015; Fankhauser and McDermott 2016; Romero-Lankao et al. 2016; Solecki et al. 2019) – from the household to the state level, since risks and opportunities vary by location and the specific local development context (*robust evidence*, *high agreement*) (IPCC 2014b; Denton et al. 2015).

Synergies between adaptation and mitigation are included in many of the NDCs submitted to the UNFCCC, as part of overall low-emissions climate-resilient development strategies (UNFCCC Secretariat 2016). a majority of developing countries have agreed to develop National Adaptation Plans (NAPs) in which many initiatives contribute simultaneously to the SDGs (Schipper et al. 2020) as well to mitigation efforts (Hönle et al. 2019; Atteridge et al. 2020). For example, developing countries recognise that adaptation actions in sectors such as agriculture, forestry and land-use management can reduce GHGs. Nevertheless, other more complex trade-offs also exist between bioenergy production or reforestation and the land needed for agricultural adaptation and food security (African Development Bank 2019; Hönle et al. 2019; Nyiwul 2019) (Chapter 7). For some of the Small Islands Development States (SIDS), forestry and coastal management, including mangrove planting, saltmarsh and seagrass are sectors that intertwine both mitigation and adaptation (Duarte et al. 2013; Atteridge et al. 2020). Integrated efforts also occur at the city level, such as the Climate Change Action Plan of Wellington City, which includes enhancing forest sinks to increase carbon sequestration while at the same time protecting biodiversity and reducing groundwater runoff as rainfall increases (Grafakos et al. 2019).

To fully maximise their potential co-benefits and trade-offs of integrating adaptation and mitigation, these should be explicitly sought, rather than accidentally discovered (Spencer et al. 2017; Berry et al. 2015), and policies designed to account for both (*robust*

## Box 13.16 | Latin America Region Adaptation Linking Mitigation: REDD+ Lessons

Thirty-three countries in the Latin American region have submitted their NDCs, and 70% of their initiatives have included mitigation and adaptation options focusing on sustainable development (Bárcena et al. 2018; Kissinger et al. 2019). However, most of these policies are disconnected across sectors (Loaiza et al. 2017; Locatelli et al. 2017). National governments have identified their relevant sectors as: energy, agriculture, forestry, land-use change, biodiversity, and water resources (see Figure 1 below). The region houses 57% of the primary forest of the planet. REDD+ aims to reduce GHG while provide ecosystems services to vulnerable communities (Bárcena et al. 2018). Lessons from successful REDD+ programmes include the benefits of a multilevel structure from international to national down to strong community organisation, as well as secure resources funding, with most of the projects relying on external sources of funding (*medium evidence, high agreement*) (Loaiza et al. 2017; Kissinger et al. 2019). However, there is limited evidence of effective adaptation co-benefits, which may be related to the lack of provision of forest standards; a disproportionate focus on mitigation and lack of attention to the well-being of the population in rural and agricultural areas (Kongsager and Corbera 2015).

Conflicts have emerged over political views, government priorities of resources (oil, bioenergy, hydropower), and weak governance among national and local authorities, indigenous groups and other stakeholders such as NGOs which play a critical role in the technological and financial support for the REDD+ initiative (Reed 2011; Kashwan 2015; Gebara et al. 2014; Locatelli et al. 2011, 2017). a more holistic approach which recognises these social, environmental and political drivers would appear to have benefits but assessment is needed to allow evidence-based actionable policy statements.

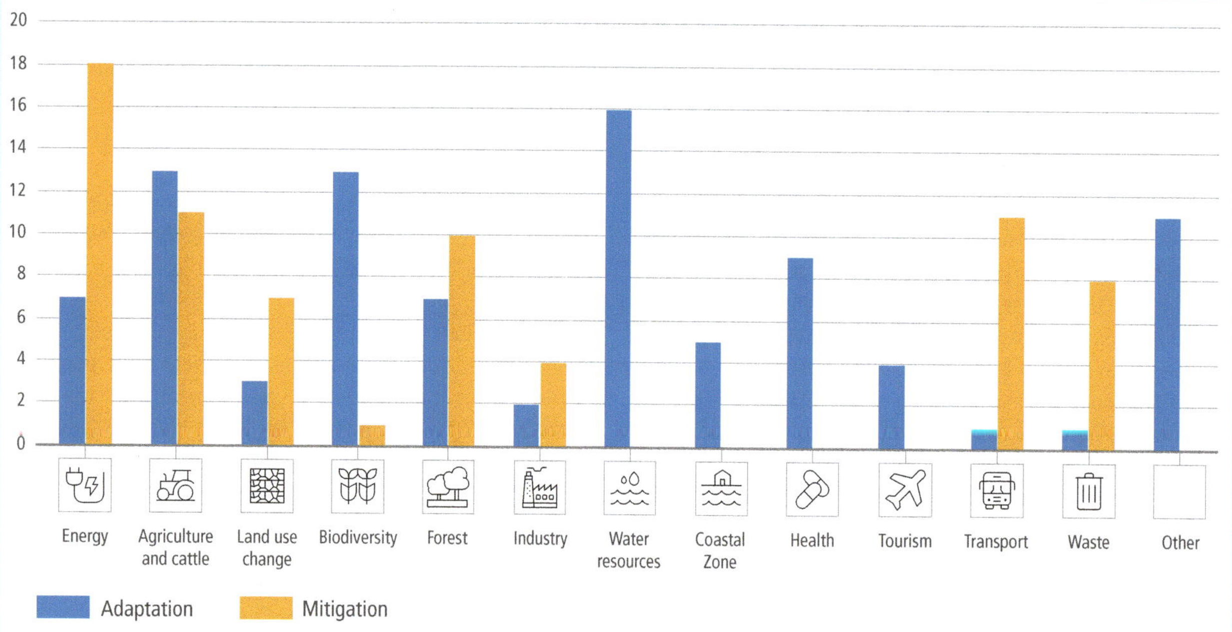


**Box 13.16, Figure 1 | Latin America and Caribbean: high priority sectors for mitigation and adaptation.** Number of countries that name the following sector in their national climate change plans and/or communications. The purple and green bars represent adaptation and mitigation respectively. Source: reproduced with permission from Bárcena et al. (2018).

*evidence, high agreement*) (Caetano et al. 2020). For example, the REDD+ initiative focus on mitigation by carbon sequestration was set up to provide co-benefits such as: nature protection, political inclusion, monetary income, economic opportunities. However, some unintended trade-offs may have occurred such as physical displacement, loss of livelihoods, increased human–wildlife conflicts, property claims, food security concerns, and an unequal distribution of benefits to local population groups (Bushley 2014; Duguma et al. 2014a; Gebara et al. 2014; Kongsager and Corbera 2015; Anderson et al. 2016; Di Gregorio et al. 2016, 2017). Ultimately, ecosystem (or nature-based) strategies, such as the use of wetlands to create accessible recreational areas that improve public health while improving biodiversity, sinking carbon and protecting neighbourhoods from extreme flooding events, may lead to more efficient and cost-effective policies (Klein et al. 2005; Locatelli et al. 2011; Kongsager et al. 2016; Mills-Novoa and Liverman 2019).

The 'nexus' approach is another widely used framework that describes the linkages between water, energy, food, health and other socio-economic factors in some integrated assessment approaches (Rasul and Sharma 2016). The Food-Energy-Water (FEW)nexus, for example, considers how water is required for energy production and supply (and thus tied to mitigation), how energy is needed to treat and transport water, and how both are critical to adaptable and resilient food production systems (Mohtar and Daher 2014; Biggs et al. 2015). Climate change impacts all these dimensions in the form of multi-hazard risk (Froese and Schilling 2019). Although integrative, the FEW nexus faces many challenges including: limited knowledge integration; coordination between different institutions and levels of government; politics and power; cultural values; and ways of managing climate risk (Leck and Roberts 2015; Romero-Lankao et al. 2017; Mercure et al. 2019). More empirical assessment is needed to identify potential overlaps between sectoral portfolios, as this could help to delineate resources allocation for synergies and to avoid trade-offs.

### 13.8.3 Relationships Between Mitigation and Adaptation Measures

There are multiple ways that mitigation and adaptation may be integrated. Table 13.3 sets out those relationships broken down into four areas: adaptation that contributes to mitigation; mitigation that contributes to adaptation; holistic, sustainability first strategies; and trade-offs. The table shows that more holistic and sustainability-oriented policies can open up the possibility for accelerated transitions across multiple priority domains (*robust evidence*, *high agreement*).

Table 13.3 | Relationships between adaptation and mitigation measures.

| Policy/action | Interrelation explained | Reference |
|---|---|---|
| | *Adaptation that contributes to mitigation* | |
| **Coastal adaptation and blue carbon; developing strategies for conservation and restoration of blue carbon ecosystems generating resilient communities and landscapes.**<br>**– Contributes to carbon storage and sequestration.** | Conservation of habitats and ecosystems, protect communities from extreme events, increase food security, and provide ecosystem services. At the same time, restoration of mangroves, tidal marshes, and seagrasses have high rates of carbon sequestration, act as long-term carbon sinks, and are contained within clear national jurisdictions. **Example:** conservation programmes on Brazilian mangroves, Spanish seagrass meadows, the Great Barriers Reef in Australia, and Coastal Management Strategy in New Zealand. | Andresen et al. (2012); Herr and Landis (2016); Duarte (2017); Doll and Oliveira (2017); Howard et al. (2017); Gattuso et al. (2018); Cooley et al. (2019); Karani and Failler (2020); Lovelock and Reef (2020) |
| **Nature-Based Solutions (Nbs); Nature-based solutions are interventions that use the natural functions of healthy ecosystems to protect the environment but also provide numerous economic and social benefits.**<br>**– Contributes to carbon storage and sequestration using individual and clustered trees.** | NbS complement and shares common elements with a wide variety of other approaches to building the resilience of social-ecological systems. Policies at national and sub-national level include community-based adaptation, ecosystem-based disaster risk reduction, climate-smart agriculture, and green infrastructure, and often place emphasis on using participatory and inclusive processes and community/stakeholder engagement. **Examples:** Mexico and the United Kingdom provide support for NbS in their national biodiversity strategies and action plans some related to water management. UK launched the Green Recovery Challenge Fund to create jobs with a focus on tree planting and the rehabilitation of peatlands. | Doswald and Osti (2011); Secretariat of the Convention on Biological Diversity (2019); Ihobe – Environmental Management Agency (2017); Zwierzchowska et al. (2019); Seddon et al. (2020); Choi et al. (2021); OECD (2021b) |
| **Ecosystem-based Adaptation (Eba); use biodiversity and ecosystem services to help people to adapt to the adverse effects of climate change, aiming to maintain and increase the resilience and reduce the vulnerability of ecosystems and people.**<br>**– Contributes to carbon storage and sequestration.** | EbA involves the conservation, sustainable management and restoration of ecosystems, such as forests, grasslands, wetlands, mangroves or coral reefs to reduce the harmful impacts of climate hazards including shifting patterns or levels of rainfall, changes in maximum and minimum temperatures, stronger storms, and increasingly variable climatic conditions. **Examples:** some NDCs include EbA and NbS harmonising national policies (for example: National Adaptation Plan) with other national climate and development policy processes, such as: water resources management plan, disaster risk reduction strategies, land planning codes. | IPBES (2019); Doswald et al. (2014); Secretariat of the Convention on Biological Diversity (2009); McAllister (2007); Colls et al. (2009); Rubio (2017); Raymond et al. (2017); Duarte (2017); Gattuso et al. (2018) |
| **Urban Greening; urban forestry, planting in road reserves and tree planting along main streets.**<br>**– Contributes to carbon storage and sequestration.**<br>**– Energy use reduction.** | Urban afforestation and reforestation produce cooling effect and water retention while helping to reducing carbon dioxide from the atmosphere. Green walls and rooftops increase energy efficiency of buildings and decrease water runoff and provide insulation for the buildings. **Examples:** Wellington City Council and other entities must comply with the New Zealand Emission Trading System regulatory framework that provides guidance and requirements of climate change planning and implementation for both mitigation and adaptation (M&A). | Santamouris (2014); Sharifi and Yamagata (2016); Grafakos et al. (2018); Pasimeni et al. (2019); Anderson et al. (2016) |
| **Climate adaptation plans at city level; sub-national policies that would lead to carbon reduction to support climate mitigation. Contribution to mitigation:**<br>**– Carbon storage and sequestration.**<br>**– Energy use reduction.**<br>**– Renewable energy.** | Cities with Climate Actions Plans include urban spatial planning and capacity-building initiatives. Some cities with adaptation and mitigation combined climate change action plans are: Bangkok, Chicago, Montevideo, Wellington, Durban, Paris, Mexico City, and Melaka. And cities with A&M actions are: Los Angeles, Vancouver, Barcelona, London, Accra, Santiago de Chile, Bogota, Curitiba, and other.<br><br>**Co-benefits generated by climate actions at cities:** heat stress reduction; water scarcity, stormwater and flood management; air quality improvement, human health and well-being, aesthetic/amenity, recreation/tourism, environmental justice, real estate value, food production, green jobs opportunities. | Garcetti (2019); Horne (2020); Barcelona City Council (2018); Greater London Authority (2018); Accra Metropolitan Assembly (2020); Choi et al. (2021); Grafakos et al. (2019); Nakano et al. (2017); Peng and Bai (2018); Zen et al. (2019); Bai et al. (2018) |

| Policy/action | Interrelation explained | Reference |
|---|---|---|
| | **Mitigation that contributes to adaptation** | |
| **Green Infrastructure; policies to support the design and implementation of a hybrid network of natural, semi-natural, and engineered features within, around, and beyond urban areas at all scales, to provide multiple ecosystem services and benefits.**<br>**– Carbon storage and sequestration.**<br>**– Reduced energy consumption.** | Adaptation benefits: flood management, heat stress reduction individually, or jointly, coastal protection, water scarcity management, groundwater resources, ecosystem resilience improvement, air quality, water supply, flood control, water quality improvement, groundwater recharge. Social co-benefits: aesthetic, recreation, environmental education, improved human health/well-being, social cohesion, and poverty reduction. Policy examples: national building code guidelines, flood safety standards, local land-use plans, local building codes, integrated water management for flood control. | Atchison (2019); Conger and Chang (2019); Schoonees et al. (2019); De la Sota et al. (2019); Choi et al. (2021); Zwierzchowska et al. (2019) |
| **REDD+ Strategies; an incentive for developing countries to increase carbon sinks, to protect their forest resources and coastal wetlands. Mostly are national strategies led by the state with contribution of international donors.**<br>**– Contributes to carbon storage and sequestration.**<br>**– Renewable energy.** | REDD+ strategies aim to generate social benefits such as poverty reduction, and ecological services such as water supply, water quality enhancement, conserves soil and water by reducing erosion. For example, indigenous communities of Socio Bosque in Ecuador have sustained livelihoods and maintaining ties to land, place, space, and *cosmovision*. While in Cameroon, upfront contextual inequities with respect to technical capabilities, power, gender, level of education, and wealth have been barriers to individuals' likelihood of participating in and benefiting from the projects. | McBurney (2021); Tegegne et al. (2021); Anderson et al. (2016); Busch et al. (2011); Bushley (2014); Dickson and Kapos (2012); Froese and Schilling (2019); Gebara et al. (2014); Pham et al. (2014); Jodoin (2017) |
| **Household energy-efficiency and renewable energy measures; energy policies may improve socio-economic development.**<br>**– Energy use reduction.** | Energy Efficiency (EE) emerges as a feasible and sustainable solution in Latin America, to minimise energy consumption, increase competitiveness levels and reduce carbon footprint. Achieving high levels of EE in the building sector requires new policies and strengthening their legal framework. Microenterprise development contributes to poverty reductions as renewable energy stimulate local and national economies. | Chan et al. (2017); Silvero et al. (2019); Zabaloy et al. (2019); Alves et al. (2020); Nyiwul (2019); Dal Maso et al. (2020) |
| | **Sustainability first: holistic approaches** | |
| **Integrated community sustainability plans.** | Climate change mitigation and adaptation are embedded in a plan to improve affordability, biodiversity, public health, and other aspects of communities. | Burch et al. (2014); Shaw et al. (2014); Stuart et al. (2016); Dale et al. (2020) |
| **Inclusive future visioning using social-ecological systems or socio-technical systems thinking.** | Participatory processes that highlight the cultural and social dimensions of climate change responses and synergies/trade-offs between priorities rather than an exclusive focus on technical aspects of solutions. | Gillard et al. (2016); Krzywoszynska et al. (2016) |
| **Climate Resilience Cities; integrating New Urban Agenda (NUA), SDGs, climate actions for A&M, and Disaster Risk Reduction (DRR) for local and sub-national governments, and DRR within a multi-hazard approach based on Sendai Framework.** | Resilient cities are including SDGs, targets, A&M options and DRR to build a resilient plan for urban planning, health, life quality and jobs creation.<br>Climate mitigation and sustainable energy actions adopted at the local level are interconnected. For instance, cities with Sustainable Energy and Climate Action Plan, which required the establishment of a baseline emission inventory and the adoption of policy measures, are already showing a tangible achievement regarding sustainable goals. | Barcelona City Council (2018); Garcetti (2019); Accra Metropolitan Assembly (2020); Blok 2016; Giampieri et al. (2019); Gomez Echeverri (2018); Long and Rice (2019); Pasimeni et al. (2019); Romero-Lankao et al. (2016) |
| | **Trade-offs** | |
| **Land-use strategies; for mitigation or adaptation considered in isolation, may cause a conflict in land planning.**<br>**– Carbon storage and sequestration.**<br>**– Energy use reduction.**<br>**– Renewable energy.** | Increasing density of land use, land-use mix and transit connectivity could increase climate stress and reduce green open spaces. It may increase the urban heat island impacting human health, and expose population to coastal inundation. Some of the policies and strategies to minimise this are: land-use planning, zoning, land-use permits, mobilising private finance in the protection of watersheds, integrated coastal zone management, flood safety standards, and other. More assessment is needed prior to new land use to reduce or prevent actions which negatively alter ecosystem services and environmental justice. | O'Donnell (2019); Bush and Doyon (2019); Grafakos et al. (2019); Landauer et al. (2015); Viguié and Hallegatte (2012); Floater et al. (2016); Xu et al. (2019); Landauer et al. (2019) |
| **Low-carbon, net zero and climate change resilient building codes that fail to account for affordability.**<br>**– Energy reduction.**<br>**– Renewable energy.** | Low-carbon or net zero emissions have multi-objective strategies, integrated policies, regulations, and actions at the national and sub-national levels. Trade-offs may be related to policy mechanisms that must be implemented comprehensively, not individually. However, different administrative levels and institutions may create a barrier to inter-sectoral coordination. For example: 'Greening' programmes may produce positive mitigation and adaptation outcomes but may also accelerate displacement and gentrification at city level. | Chaker et al. (2021); del Río and Cerdá (2017); Choi et al. (2021); Papadis and Tsatsaronis (2020); Wolch et al. (2014); Garcia-Lamarca et al. (2021); Haase et al. (2017); Sharifi (2020); Viguié and Hallegatte (2012); del Río (2014) |



### 13.8.3.1 Governing the Linkages Between Mitigation and Adaptation at the Local, Regional, and Global Scales

International policy frameworks, such as the 2015 Paris Agreement, the Sendai Framework for Disaster Disk Reduction, and the New Urban Agenda for sustainable urban systems, provide an integrated approach for both adaptation and mitigation, while promoting sustainable development and climate resilience across scales (from global, regional, to local government actions (*robust evidence*, *high agreement*) (Duguma et al. 2014b; Heidrich et al. 2016; Di Gregorio et al. 2017; Locatelli et al. 2017; Nachmany and Setzer 2018; Mills-Novoa and Liverman 2019). Even so, the specific ways that these linkages are governed vary widely depending on institutional and jurisdictional scale, competing policy priorities, and available capacity (Landauer et al. 2019).

Supranational levels of action such as the EU climate change policy have influenced the development and implementation of Climate Change Action Plans (CCAPs) at the sub-national level (Heidrich et al. 2016; Villarroel Walker et al. 2017; Reckien et al. 2018). While adaptation is gaining prominence and is increasingly included in the NDCs of EU nations, the implementation of adaptation and mitigation by EU states are at different stages (Fleig et al. 2017). Fleig et al. (2017) found that all EU states, with the exception of Hungary, have adopted a framework of laws tackling mitigation and adaptation to climate change. However, an assessment of climate legislation in Europe pointed out that there has been little coordination between mitigation and adaptation, and that implementation varies according to different national conditions (Nachmany et al. 2015). More recently, however, integrated adaptation/mitigation plans have been prepared in Europe under the Covenant of Mayors, in which synergies and trade-offs can be better revealed and assessed (Bertoldi et al. 2020).

Local governments and cities are increasingly emerging as important climate change actors (Gordon and Acuto 2015) (Section 13.5). While cities and local governments are developing Climate Change Action Plans (CCAPs), plans that explicitly integrate the design and implementation of adaptation and mitigation are a minor percentage, with few cities establishing inter-relationships between them (Nordic Council of Ministers 2017; Grafakos et al. 2018). Compared to national climate governance, local governments are more likely to develop and advance climate policies, generating socio-economic or environmental co-benefits, and improve communities' quality of life (Gill et al. 2007; Bowen et al. 2014; Duguma et al. 2014b; Mayrhofer and Gupta 2016; Deng et al. 2017; Hennessey et al. 2017). There may be a disconnect, however, between the responsibility that a particular jurisdiction has over mitigation and adaptation (city officials, for instance) and the scale of resources or capacities that they have available to bring to bear on the problem (regional to national provision of energy and transport) (Di Gregorio et al. 2019; Dale et al. 2020).

### 13.8.4 Integrated Governance Including Equity and Sustainable Development

Climate policy integration carries implications for the pursuit of the SDGs, given that it is nearly impossible to achieve the desired socio-economic gains if fundamental environmental issues, such as climate change, are not addressed (Gomez-Echeverri 2018). Research on climate resilient development pathways (Roy et al. 2018), for instance, argues for long-term policy planning that combines the governance of national climate and SD goals, builds institutional capacity across all sectors, jurisdictions, and actors, and enhances participation and transparency (*robust evidence*, *high agreement*) (Chapter 4 and 17).

In the Global South, climate change policies are often established in the context of sustainable development and of other pressing local priorities (e.g., air pollution, health, and food security). National climate policy in these countries tends to give prominence to adaptation based on country vulnerability, climatic risk, gender-based differences in exposur to that risk, and the importance of local/traditional and indigenous knowledge (Beg et al. 2002; Duguma et al. 2014b). Despite the evidence that integrated mitigation and adaptation policies can be effective and efficient (Klein et al. 2005) and can potentially reduce trade-offs, there is still limited evidence of how such integrated policies would specifically contribute to progress on the SDGs (*robust evidence*, *high agreement*) (Kongsager et al.2016; Di Gregorio et al. 2017; Antwi-Agyei et al. 2018; De Coninck et al. 2018; Campagnolo and Davide 2019).

Where mainstreaming of environmental concerns has been attempted through national plans, they have had success in some cases when backed by strong political commitments that support a vertical coordination structure rather than horizontal structures led by the focus ministry (Nunan et al. 2012). Such political commitments are therefore crucial to success but insufficient in and of themselves (Runhaar et al. 2018; Wamsler et al. 2020). Integration of the budget process is particularly important, as are aligned time frames across different objectives (Saito 2013). Recognition of the functional interactions across policy sectors is improved by a translation of long-term policy objectives into a plan that aligns with integration goals (Corry 2012; Oels 2012; Dupont 2019).

There are important links between inequality, justice and climate change (Ikeme 2003; Bailey 2017). Many of these operate through the benefits, costs and risks of climate action (distributive justice), while others focus on differential participation and recognition of sub-national actors and marginalised groups (procedural justice) (Bulkeley and Castán Broto 2013; Bulkeley et al. 2013; Hughes 2013; Reckien et al. 2018; Romero-Lankao and Gnatz 2019).

Justice principles are rarely incorporated in climate change framing and action (Sovacool and Dworkin 2015; Genus and Theobald 2016; Heikkinen et al. 2019; Romero-Lankao and Gnatz 2019). Yet, equity is salient to mitigation debates, because climate change mitigation policies can have also negative impacts (Brugnach et al. 2017; Ramos-Castillo et al. 2017; Klinsky 2018), exacerbated by poverty, inequality and corruption (Reckien et al. 2018; Markkanen and Anger-Kraavi 2019). The siting of facilities and infrastructure that advance decarbonisation (such as public transit infrastructure, renewable energy facilities and so on) may have implications for environmental justice. Integrated attention to justice in climate, environment and energy, as well as involvement of host communities in siting assessments and decision-making processes, can help to avoid such conflict (McCord et al. 2020; Hughes and Hoffmann 2020). As a result, successful policy integration goes beyond optimising public management routines, and must resolve key trade-offs between actors and objectives (Meadowcroft 2009; Nordbeck and Steurer 2016).

The potential for transformative climate change policy that delivers both adaptation and mitigation is also shaped by a number of enabling and disabling factors tied to governance processes (*robust evidence*, *high agreement*) (Burch et al. 2014) (Section 13.9).

## Box 13.17 | Enabling and Disabling Factors for Integrated Governance of Mitigation and Adaptation

**Ensuring participatory governance and social inclusion.** Interlinkages in the food-energy-water nexus highlight the importance of inclusive processes (Shaw et al. 2014; Nakano et al. 2017; Cook and Chu 2018; Romero-Lankao and Gnatz 2019). The cultivation of urban grassroots innovations and social innovation may accelerate progress (Wolfram and Frantzeskaki 2016), as may the development of carefully-designed climate and energy dialogues that enable learning among multiple stakeholders (Cashore et al. 2019).

**Considering synergies and trade-offs with broader sustainable development priorities.** The explicit consideration of synergies and trade-offs will enable more integrated policy making (Dang et al. 2003; von Stechow et al. 2015). Policy frameworks to do so are just emerging, such as analysis of trade-offs between energy and water policies and agriculture (Huggel et al. 2015; Antwi-Agyei et al. 2018).

**Employing a diverse set of tools to reach targets.** Building codes, land-use plans, public education initiatives, and nature-based solutions such as green ways may impact adaptation and mitigation simultaneously (Burch et al. 2014). Ecological restoration provides another suite of tools, for instance the Brazilian target of restoring and reforesting 0.12 million $km^2$ of forests by 2030, which can enhance biodiversity and ecosystem services while also sinking carbon (Bustamante et al. 2019). Mandatory retrofits to improve indoor air quality can also increase energy efficiency and resilience to climate change impacts (Friel et al. 2011; Houghton 2011).

**Monitoring and evaluating key indicators, beyond only greenhouse gas emissions, such as biodiversity, water quality, and affordability:** An integrated approach requires robust process for collecting data on these indicators. Challenges are related to the limited evidence-base on synergies, co-benefits, and trade-offs across sectors and jurisdictions (Di Gregorio et al. 2016; Kongsager et al. 2016; Locatelli et al. 2017; Zen et al. 2019). Moreover, adaptation policies mostly lack measurable targets or expected outcomes increasing the challenge of designing an integrated framework (OECD 2017).

**Iterative and adaptive management.** Adaptive management helps to address the underlying uncertainty (Kundzewicz et al. 2018) that characterises implementation of integrated approaches to adaptation and mitigation. Policy integration needs to be considered iteratively along the process of development, implementation, and evaluation of climate policies.

**Strategic partnerships that coordinate efforts.** Strategic partnerships among diverse actors, therefore, bring diverse technical skills and capacities to the endeavour (Burch et al. 2016; Islam and Khan 2017). However, realising strategic approaches for joint adaptation and mitigation require adequate financial, technical and human resources.

**Participatory and collaborative planning approaches can help overcome injustices and address power differentials.** Participatory and collaborative planning approaches can provide multiple spaces of deliberation where marginalised voices can be heard (Blue and Medlock 2014; UN Habitat 2016; Castán Broto and Westman 2017; Waisman et al. 2019). These tools organise climate and sustainability action by addressing its democratic deficit and facilitating the recognition of multiple perspectives in environmental planning alongside material limits of development (Agyeman 2013).

# 13.9 Accelerating Mitigation Through Cross-sectoral and Economy-wide System Change

## 13.9.1 Introduction



Section 13.9 assesses literature related to economy wide and cross – sector systemic change as an approach to accelerate climate mitigation.

It focuses specifically on policy and institutions, as two of the six enabling conditions for economy-wide system change and thus provides a third dimension of the role of policy and institutions to climate mitigation. Enabling conditions in general are discussed in Chapter 4 of the SR1.5 (IPCC 2018), as well as Chapter 4 of this report.

This section follows on from Section 13.6 (single policy instruments) and 13.7 (policy packages). Section 13.9 literature follows closely on from Section 13.7 literature on policy packages, which discusses change within one system, although there remains an overlap.

Section 13.9.2 provides a brief introduction to policy and institutions as two of the six dimensions of enabling conditions, and the importance of enabling conditions to systemic change and climate mitigation. Section 13.9.3 briefly introduces actions for transformative justice, which seek to restructure the underlying system framework that produces mitigation inequalities. Section 13.9.4 provides a brief overview of net zero policies and targets (often no more than aspirational), which imply economy-wide measures and system change. Section 13.9.5 assesses the literature arguing for a system restructuring approach to climate mitigation, based on systemic restructuring. Section 13.9.6 assesses the literature on stimulus packages and green new deals which aim for systemic change, sometimes with value for climate mitigation. And finally, Section 13.9.7 assesses emerging literatures which argues that there are existing challenges to accelerating climate mitigation that may be overcome by systemic change and targeted actions.

### 13.9.2 Enabling Acceleration

IPCC AR6 WG3, particularly Chapter 4, following on from the IPCC WG3 SR1.5 (IPCC 2018), has highlighted the importance of enabling conditions for delivering successful climate mitigation actions. The AR6 Glossary term for enabling conditions is: 'enabling conditions include *finance, technological innovation*, strengthening policy instruments, *institutional capacity, multi-level governance*, and changes in *human behaviour* and lifestyles (*medium evidence, high agreement*) (see Glossary). The IPCC SR1.5 report adds to these six dimensions saying enabling conditions also includes 'inclusive processes, attention to power asymmetries and unequal opportunities for development and reconsideration of values' (*medium evidence, high agreement*) (IPCC 2018). Not only is the presence of enabling conditions necessary for delivering the successful implementation of single policy instruments and policy packages, but also for delivering systemic change (*medium evidence, high agreement*) (de Coninck et al. 2018; IPCC 2018; Waisman et al. 2019). The feasibility of 1.5°C compatible pathways is contingent upon enabling conditions for systemic change (*medium evidence, high agreement*) (de Coninck et al. 2018; Waisman et al. 2019).

At the same time, again following on from SR1.5 report, Section 1.8.1 explains that there are six feasibility dimensions of successful delivery of climate goals. These feasibility dimensions include geophysical; environmental and ecological; technological; economic; behaviour and lifestyles and institutional dimensions. The presence or absence of enabling conditions would affect the feasibility of mitigation as well as adaptation pathways and can reduce trade-offs while amplifying synergies between options (Waisman et al. 2019). Policies and institutions, which are two of the six enabling conditions, are therefore central to accelerated mitigation and systemic change. Identifying, and ensuring, the presence of all the enabling conditions for any given goal, including systemic transformation and acceleration of climate mitigation, is an important first step (*medium evidence, medium agreement*) (Roberts et al. 2018; Le Treut et al. 2021; Singh and Chudasama 2021).

### 13.9.3 Transformative Justice Action and Climate Mitigation

Chapter 4 is the lead chapter of this report for justice and climate mitigation issues, and includes an overview of institutions which have been set up to ensure a Just climate transition (Section 4.5). Chapter 13 has sought to integrate justice issues in Section 13.2 in reference to procedural justice and the impact of inequalities on sub-national institutions, Section 13.6 in regard to distribution, and Section 13.8 in relation to integrating mitigation and adaptation policies.

This sub-section introduces the concept of transformative justice as part of measures intending to accelerate mitigation. Fair and effective climate policymaking requires institutional practices to: consider the distributional impacts of climate policy in the design and implementation of every policy (Agyeman 2013; Castán Broto and Westman 2017); align mitigation with other objectives such as inclusion and poverty reduction (Hughes and Hoffmann 2020; Rice et al. 2020; Hess and McKane 2021); represent a variety of voices, especially those of the most vulnerable (Bullard et al. 2008; Temper et al. 2018); and rely on open processes of participation (*robust evidence, high agreement*) (Anguelovski et al. 2016; Bouzarovski et al. 2018; Rice et al. 2020).

Distributive approaches to climate justice address injustices related to access to resources and protection from impacts. There is an important difference between affirmative and transformative justice action (Fraser 1995; Agyeman et al. 2016; Castán Broto and Westman 2019): Affirmative action includes policies and strategies that seek to correct inequitable outcomes without disturbing the underlying political framework while transformative action seeks to correct inequitable outcomes by restructuring the underlying framework that produces inequalities.

Transformative action that responds to distributive justice concerns include economy-wide actions via stimulus packages (such as the European Green Deal and the New Green Deal in the USA) (Section 13.9.5). Other examples are the increasing number of climate litigation suits that are transforming the way distributive dimensions of climate justice are understood (Section 13.4.2).

### 13.9.4 Net Zero Emissions Targets

The last few years have seen a proliferation of net zero emission targets set by national and regional governments, cities as well as companies and institutions (NewClimate Institute and Data Driven EnviroLab 2020; Black et al. 2021; Rogelj et al. 2021) (see also Cross-Chapter Box 3 in Chapter 3). Meeting these targets implies economy-wide systemic change (*medium evidence, high agreement*).

The Energy & Climate Intelligence Unit (ECIU) Net Zero Tracker divides countries into those which have net zero emissions achieved, have it in law, have proposed legislation, have it in policy documents or have emission reduction targets under discussion in some form. a recent study estimated that 131 countries have either adopted, announced or are discussing net zero GHG emissions targets, covering 72% of global emissions (Höhne et al. 2021). Out of those, as of 1 October 2021, the ECIU Net Zero Tracker states that Germany, Sweden, the European Union, Japan, United Kingdom, France, Canada, South Korea, Spain, Denmark, New Zealand, Hungary and Luxembourg have net zero targets set in law (ECIU 2021).

Some have argued that the expansion of these emission reduction targets marks an important increase in climate mitigation momentum since the Paris Agreement of 2015 and the 2018 IPCC Special Report on Global Warming of 1.5°C (Black et al. 2021; Höhne et al. 2021). On the other hand net zero emission targets in their current state vary enormously in scope, quality and transparency – with many countries at the discussion stage – and this makes scrutiny and comparison difficult (NewClimate Institute and Data Driven EnviroLab 2020; Black et al. 2021; Rogelj et al. 2021).

In order to realise the mitigation potential of net zero emission targets some areas within the targets might need to be changed. For example, this includes clearer definitions; well defined time frames and scopes; focusing on direct emission reductions within their own territory; minimal reliance on offsets; scrutiny of use and risks of $CO_2$ removal; attention to equity, near-term action coupled with long-term intent setting; and ongoing monitoring and review (*medium evidence*, *high agreement*) (Levin et al. 2020; NewClimate Institute and Data Driven EnviroLab 2020; Black et al. 2021; Höhne et al. 2021; Rogelj et al. 2021; World Bank 2021b).

### 13.9.5 Systemic Responses for Climate Mitigation

There is now a significant body of work which explicitly states, or implicitly accepts, that systemic change may be necessary to deliver successful climate mitigation, including net zero targets. Newell phrases this as the difference between 'plug and play' mitigation applications where one aspect of a system is changed while everything in the system remains the same compared to systemic change, with change affecting all the system (Newell 2021a,b). This section highlights an emergent, multidisciplinary literature since IPCC AR5, which suggests that acceleration to decarbonised systems via a sustainable development pathway may be better achieved by moving from a single policy instrument or mix of policies approach to a systemic economy-wide approach (Figure 13.6).

The complexity and multi-facetted challenges of rapidly decarbonising our current interconnected systems (such as energy, food, health) in a just way has led Michaelowa et al. (2018) to conclude that implementation of strong mitigation policy packages that are needed requires a systemic change in policymaking.

Multiple modelling assessments of different development and mitigation pathways are available. Most of these analyses which lead to significant climate mitigation assume significant systemic change across social, technological, and economic aspects of a country for example, India (Gupta et al. 2020); Japan (Sugiyama et al. 2021) and the globe (Rogelj et al. 2015; Dejuán et al. 2020).

UNEP (2020) argued that major, long-term sectoral transformation across multiple systems is needed to reach net zero GHG emissions. Bernstein and Hoffmann (2019) and Rockström et al. (2017) argue that the presence of multi-level, multi-sectoral lock-ins of overlapping and interdependent political, economic, technological and cultural forces mean that a new approach of coordinated, cross-economy, systemic climate mitigation is necessary. Creutzig et al. (2018) propose a resetting of the approach to consumption and use of resources to that of demand side solutions, which would have ongoing economy-wide systemic implications.

Others focus more on single system reconfigurations, such as the energy system (Matthes 2017; Tozer 2020); urban systems (Holtz et al. 2018); or the political system (Somerville 2020; Newell and Simms 2020). Becken (2019) argues that only systemic changes at a large scale will be sufficient to break or disrupt existing arrangements and routines in the tourism industry.

Others argue for thinking about mitigation in even wider ways. O'Brien (2018) posits that sector-focused, or a silo approach, to mitigation may need to give way to decisions and policies which reach across sectoral, geographic and political boundaries and involve a broad set of interrelated processes – practical, political and personal. Gillard et al. (Gillard et al. 2016) argue that a response to climate change has to move beyond incremental responses, aiming instead for a society-wide transformation which goes beyond a system perspective to include learning from social theory; while Eyre et al. (2018) argue that moving beyond incremental emissions reductions will require expanding the focus of efforts beyond the technical to include people, and their behaviour and attitudes. Stoddard et al. (2021) argue that 'more sustainable and just futures require a radical reconfiguration of long-run socio-cultural and political economic norms and institutions'. They focus on nine themes: international climate governance, the vested interests of the fossil fuel industry, geopolitics and militarism, economics and financialisation, mitigation modelling, energy supply systems, inequity, high carbon lifestyles and social imaginaries.

### 13.9.6 Economy-wide Measures

Economy-wide stimulus packages which have occurred post COVID-19, and in some cases in response to environmental concerns, have the ability to undermine or aid climate mitigation (*medium evidence*, *high agreement*). Attention in the early efforts of their development and design can contribute to shifting sustainable development pathways and net zero outcomes, while meeting short-term economic goals (*medium evidence*, *high agreement*) (Hepburn et al. 2020; Hanna et al. 2020).

Economy-wide packages, as a way to stimulate and/or restructure domestic economies to deliver particular, desired outcomes is a widely accepted tool of government (for example the Roosevelt's New Deal packages in the USA between 1933 and 1939). a number of country-level stimulus package were put in place after the 2008 Global Recession, and there was support for a Global Green New Deal from UNEP (Steiner 2009; Barbier 2010). Cross-economy structural change packages may provide opportunities for another approach to accelerate climate mitigation.

This approach has already been taken up to some degree by a number of countries/blocs. For example, California as well as Germany, through the German *Energiewende*, are early examples of a USA state and a country which have tried to link their economies to a sustainable future through energy-wide efforts of structural change (Morris and Jungjohann 2016; Burger et al. 2020a).

In addition to these economy-wide measures, there have since been cross-economy Green New Deals implemented such as the European Green Deal (Elkerbout et al. 2020; Hainsch et al. 2020; UNEP 2020a) (Box 13.1) with calls for other New Deals, for example a Blue New Deal (Dundas et al. 2020), or deals to bring together climate and justice goals (Hathaway 2020; MacArthur et al. 2020).

The COVID-19 Pandemic has resulted in global economic recession, which many Governments have responded to with economic stimulus programmes. See also Cross-Chapter Box 1 in Chapter 1 on COVID-19. It has also led to more analysis of the potential of cross-economy stimulus packages to benefit climate goals, including what lessons can be learned from the stimulus packages put in place as a result of the 2008–2009 Global Recession.

The United Nations Environment Programme (UNEP) reviewed the green stimulus plans of the G20 following the 2008–2009 recession to examine what worked; what did not; and the lessons which could be learnt (Barbier 2010). This work was updated (Barbier 2020) and concluded that the constituents of successful green stimulus frameworks were long-term commitments in public spending; pricing reform; ensuring concerns about affordability were overcome; and minimising unwanted distributional impacts. Others argue that post-2008 recession stimulus package outcomes benefited both environmental and industrial objectives and that a long-term policy commitment to the transition to a sustainable, low-carbon economy makes sense from both an environmental and industrial strategy point of view (Fankhauser et al. 2013).

With the outbreak of the COVID-19 Pandemic in 2020, past stimulus packages have been further investigated. One study interviewed 231 central bank officials and identified five key policies for both economic multipliers and climate impacts metrics (Hepburn et al. 2020). These were expenditure on clean physical infrastructure; building energy efficiency retrofits; investment in education and training; natural capital investment; and clean R&D. However, the mix of effective policies may differ in lower and middle income countries: rural support spending was more relevant, while clean R&D was less so. The study illuminated that there were different phases to recovery packages: the initial 'rescue' spending but then a second 'recovery' phase that can be more fairly rated green or not green. Recovery phase policies can deliver both economic and climate goals – co-benefits can be captured (i.e. support for EV infrastructure can also reduce local air pollution etc.) – but package design is important (Hepburn et al. 2020).

Others provide a framework which allows a systematic evaluation of options, given objectives and indicators, for COVID-19 stimulus packages (e.g. Dupont et al. 2020; Jotzo et al. 2020; OECD 2021c). Jotzo et al. (2020) conclude that the programmes that most closely match green stimulus are afforestation and ecosystem restoration programmes, energy efficiency upgrades and RE projects. These type of policies provide short-term goals of COVID-19 while also making progress on longer terms objectives (Jotzo et al. 2020). The IMF concluded that a comprehensive mitigation policy package combining carbon pricing and government green infrastructure spending (that is partly debt financed) can reduce emissions substantially while boosting economic activity, supporting the recovery from the COVID-19 pandemic (Jaumotte et al. 2020).

Conversely, other short-term fiscal or recovery measures in stimulus packages may perpetuate high carbon and environmental damaging systems. These include fossil fuel based infrastructure investment; fiscal incentives for high carbon technologies or projects; waivers or roll-backs of environmental regulation; bailouts of fossil fuel intensive companies without conditions for low-carbon transitions or environmental sustainability (UNEP 2020a; O'Callaghan and Murdock 2021; Vivid Economics 2021).

Of the USD17.2 trillion so far spent on stimulus packages, USD4.8 trillion (28% of the total as of July 2021) is linked to environmental outcomes (Vivid Economics 2021). This study relates to 30 countries: the G20 and 10 others. The packages in EU, Denmark, Canada, France, Spain, the UK, Sweden, Finland and Germany (German Federal Ministry of Finance 2020; Vivid Economics 2021) result in net benefits for the environment. a number of studies provide differing conclusions with respect to net benefits or otherwise for the environment for a number of countries (Climate Action Tracker 2020; UNEP 2020a; Vivid Economics 2021). An OECD database found that, as of mid-July 2021, 21% of economic recovery spending in OECD, EU and Key Partners is allocated to environmentally positive measures (OECD 2021c). O'Callaghan and Murdock (2021) reviewed the 50 countries with the greatest stimulus spend in 2020 and find that 13% of the spend is directed to long-term recovery type measures, of which 18% is spent on green recovery. This is a total of 2.5% of total spend or 368 billion USD on green initiatives.

### 13.9.7 Steps for Acceleration

The multidisciplinary literature exploring how to accelerate climate mitigation and transition to low GHG economies and systems has grown rapidly over the last few years. Acceleration is also confirmed as an important sub-theme of the more specific transition literature (Köhler et al. 2019). While literature focusing on how to accelerate the impact of climate mitigation is derived from empirical evidence, there is very little *ex post* evidence of directed acceleration approaches.

The overlapping discussions of how to accelerate climate mitigation; transition to low-carbon economies; and shift development pathways depends heavily on country-specific dynamics in political coalitions, material endowments, industry strategy, cultural discourses, and civil society pressures (Sections 13.2, 13.3, 13.4, 13.7, and 13.8). Ambition for acceleration at different scales and stringency (whether for cities, country climate policies, country industrial strategies, or national economic restructuring) increase governance challenges, including coordination across stakeholders, institutions, and scales. 'There is therefore no "one-size-fits-all" blueprint for accelerating low-carbon transitions' (*medium evidence, high agreement*) (Geels et al. 2017a; Roberts et al. 2018).

Markard et al. (2020) describe the key challenges to accelerating climate mitigation and sustainability transitions as:

1. The ability for low-carbon innovations to emerge in whole systems. Two critical issues need to occur to overcome this challenge (i) complementary interactions between different elements. For example, in an electricity system, the integration of renewable energy requires complementary storage technologies etc. and (ii) changes in system architecture. Thus, in the accelerating phase, policy has to shift from stimulating singular innovations towards managing wider system transformation.
2. The need for greater interactions between adjacent systems: interactions between multiple systems increases the complexity of the transition. Policies are linked to institutions or government departments, and they are often compartmentalised into different policy areas (e.g. energy policy and transport policy). Increasing and coordinating that interaction adds complexity.
3. The resistance from declining industries; acceleration of sustainability transitions will involve the phase out of unsustainable technologies. As a result, acceleration towards a sustainability transition may be resisted – whether business models, or where jobs are involved. Political struggles and conflicts are an inherent part of accelerating transitions, one strategy to deal with this resistance is to accomplish wide societal support for long-term transition targets and to form broad constituencies of actors in favour of those transitions.
4. The need for changes in consumer practices and routines; this challenge relates to changes in social practices that may be required for mainstreaming of sustainable technologies. For example, electric vehicles require changes in trip planning and refuelling practices. Reducing levels or types of consumption is also desirable.
5. Coordination challenges in policy and governance. There is an increasing complexity of governance which can be overcome by stronger vertical and horizontal policy coordination across systems.

The acceleration literature links two over-arching actions: first, a strategic targeting approach to overcoming the challenges to acceleration by a parallel focus on undermining high carbon systems while simultaneously encouraging low-carbon systems; and second, focusing on a coordinated, cross-economy systemic response, including harnessing enabling conditions (*robust evidence*, *high agreement*) (Rogelj et al. 2015; Geels et al. 2017b; Hvelplund and Djørup 2017; Gomez Echeverri 2018; Markard 2018; Tvinnereim and Mehling 2018; O'Brien 2018; Roberts et al. 2018; Hess 2019; Kotilainen et al. 2019; Victor et al. 2019; European Environment Agency 2019; Rosenbloom and Rinscheid 2020; Newell and Simms 2020; Otto et al. 2020; Strauch 2020; Burger et al. 2020a; Hsu et al. 2020b; Rosenbloom et al. 2020).

Strategic targeting, or the identifying of specific intervention points (Kanger et al. 2020), points of leverage (Abson et al. 2017), or upward cascading tipping points (Sharpe and Lenton 2021), broadly means choosing particular actions which will lead to a greater acceleration of climate mitigation across systems. For example, Dorninger et al.(2020) provide a quantitative systematic review of empirical research addressing sustainability interventions. They take 'leverage points' – places in complex systems where relatively small changes can lead to potentially transformative systemic changes – to classify different interventions according to their potential for system-wide transformative change. They argue that 'deep leverage points' – the goals of a system, its intent, and rules – need to be addressed more directly, and they provide analysis of the food and energy systems.

The strategic choosing of policies and points of intervention is linked to the importance of choosing self-reinforcing actions for acceleration (Rosenbloom et al. 2018; Butler-Sloss et al. 2021; Sharpe and Lenton 2021; Jordan and Moore 2020; Bang 2021). Butler-Sloss et al. (2021) explains the types of self-reinforcing actions (or feedback loops) which can encourage or undermine rapid transformation of energy systems.

An example of this first overarching action, the strategic targeting of the challenges to acceleration, is the focus on undermining carbon-intensive systems, thereby reducing opposition to more generalised acceleration policies, including the encouragement of low-carbon systems (*robust evidence*, *high agreement*) (Hvelplund and Djørup 2017; Rosenbloom 2018; Roberts and Geels 2019; Victor et al. 2019; Rosenbloom et al. 2020; Rosenbloom and Rinscheid 2020). Undermining high carbon systems includes deliberately phasing out unsustainable technologies and systems (Kivimaa and Kern 2016; David 2017; European Environment Agency 2019; Johnsson et al. 2019; UNEP 2019b; Carter and McKenzie 2020; Newell and Simms 2020); confronting the issues of incumbent resistance (Roberts et al. 2018); and avoiding future emissions and energy excess by reducing demand (Rogelj et al. 2015; UNEP 2019b; Victor et al. 2019).

Other strategic goals include tackling the equity and justice issues of 'stranded regions' (Spencer et al. 2018); paying greater attention to system architecture to enable increased acceleration to low-carbon electricity supply, in this case in the wind industry (McMeekin et al. 2019); and the importance of maintaining global ecosystem of low-carbon supply chains (Goldthau and Hughes 2020).

Other strategic goals combine national and global action. For example, global NGO coalitions have formed around strategic policy outcomes such as the 'Keep it in the Ground' movement (Carter and McKenzie 2020), and are supported via coordinated networks, such as the Powering Past Coal Alliance (Jewell et al. 2019), and with knowledge dissemination, for example, the 'Fossil Fuel Cuts Database' (Gaulin and Le Billon 2020).

The second overarching point highlighted by the literature is the benefits of focusing on a coordinated, cross-economy systemic response. Coordination is central to this. For example, coordination of actions and coherent narratives across sectors and cross economy, including within and between all governance levels and scales of actions, is beneficial for acceleration (*robust evidence*, *high agreement*) (Zürn and Faude 2013; Hawkey and Webb 2014; Huttunen et al. 2014; Magro et al. 2014; Warren et al. 2016; Köhler et al. 2019; Kotilainen et al. 2019; McMeekin et al. 2019; Victor et al. 2019; Hsu et al. 2020b). Victor et al. (2019) provide a framework of how to prioritise the most urgent actions for climate mitigation and they give practical case studies of how to improve coordination to accelerate reconfiguration of systems for economy-wide climate mitigation in sectors such as power; cars; shipping; aviation; buildings; cement; and plastics.

However, coordination is a necessary but insufficient condition of acceleration. All enabling conditions are required to deliver systemic transformation (Section 13.9.2).

Other disciplines argue that social transformation is likely to be as important as the technical challenges in a coordinated, cross-economy approach to acceleration. For example, some argue for social tipping interventions (STI) alongside other technical and political interventions so that they can 'activate contagious processes of rapidly spreading technologies, behaviours, social norms, and structural reorganisation' (Otto et al. 2020). They argue that these STIs are *inter alia*: removing fossil fuel subsidies and incentivising decentralised energy generation; building carbon neutral cities; divesting from assets linked to fossil fuels; revealing the moral implications of fossil fuels; strengthening climate education and engagement; and disclosing information of GHG emissions (Otto et al. 2020). Others illuminate the importance of narratives and framings in the take-up (or not) of acceleration actions (Sovacool et al. 2020). Others are optimistic about the possibilities of transformation but also highlight the importance of political economy for rapid and just transitions (Newell and Simms 2020; Newell 2021).

In summary, a synthesis of the multidisciplinary, acceleration literature suggests that climate mitigation is a multifaceted problem which spans cross-economy and society issues, and that solutions to acceleration may lie in coordinated systemic approaches to change and strategic targeting of leverage points. Broadly, this literature agrees on a dual approach of non-incremental systemic change and a targeting of specific acceleration challenges, with tailored actions drawing on enabling conditions. The underlying argument of this is that there is a strategic logic to focusing on actions which undermine high carbon systems at the same time as encouraging low-carbon systems. If high carbon systems are weakened then this may reduce the opposition to policies and actions aimed at accelerating climate mitigation, enabling more support for low-carbon systems. In addition, targeting of actions which may create 'tipping point cascades' which increase the rate of decarbonisation may also be beneficial. Finally, new modes of governance may be better suited to this approach in the context of transformative change.

## 13.10 Further Research

Research has expanded in a number of areas relevant to climate mitigation, yet there is considerable scope to add to knowledge. Key areas for research exist in climate institutions and governance, politics, policies and acceleration of action. In each area there is an overarching need for more *ex post* analysis of impact, more cases from the developing world, and understanding how institutions and policies work in combination with each other.

### 13.10.1 Climate Institutions, Governance and Actors

- The different approaches to framework legislation, how it can be tailored to country context and evolve over time, how it diffuses across countries, and *ex post* analysis of its impact.
- Approaches to mainstreaming climate governance across sectors and at different scales, and developing governmental and non-governmental capacity to bring about long-term low-carbon transformations and associated capacity needs.
- The drivers of sub-national climate action, the scope for coordination or leakage with other scales of action, and the effect, in practice on GHG outcomes.
- Comparative research on how countries develop NDCs, and whether and how that shapes national policy processes.

### 13.10.2 Climate Politics

- The full range of approaches that governments and non-governmental actors may take to overcome lock-in to carbon-intensive activities including through addressing material endowments, cultural values, institutional settings and behaviours.
- The factors that influence emergence of popular movements for and against climate actions, and their direct and indirect impacts.
- The role of civic organisations in climate governance, including religious organisations, consumer groups, indigenous communities, labour unions, and development aid organisations.
- The relationship between climate governance approaches and differing political systems, including the role of corruption on climate governance.
- The impacts of media – traditional and social – on climate mitigation, including the role of disinformation.
- The role of corporate actors in climate governance across a broad range of industries.
- Systematic comparative research on the differing role of climate litigation across various juridical systems.

### 13.10.3 Climate Policies

- Greater *ex post* empirical studies of mitigation policy outcomes, their design features, the impacts of policy instruments under different conditions of implementation, especially in developing countries. Such research needs to assess the effectiveness, economic and distributional effects, co-benefits and side effects, and transformational potential of mitigation policies.
- Understand how packages of policies are designed and implemented, including with attention to local context and trade-offs.
- Policy design and institutional needs for the explicit purpose of net zero transitions.
- Greater understanding of the differences between, and benefits of, policy packages and economy-wide measures for in-system and cross-system structural change.
- Policies and packages for emissions sources that are unregulated or under-regulated, including industrial and non-$CO_2$ emissions.
- The existence and extent of carbon leakage across countries, the relative impact of different channels of leakage, and the implications of policy instruments designed to address leakage.

### 13.10.4 Coordination and Acceleration of Climate Action

- How to ensure a just transition that gains wide popular support through research on actual and perceived distributional effects across countries and contexts.
- How to coordinate and integrate for climate mitigation, between what actors, sectors, governance scale and goals, and how to evaluate.
- Knowledge on the political and policy related links between adaptation and mitigation across sectors and countries.
- Further theoretical and empirical research on the necessary institutional, cultural, social and political conditions to accelerate climate mitigation.
- How to transform developed and developing economies and societies for acceleration, including by shifting development pathways.
- The approaches to, and value of, coordinated, cross economy structural change, including Green New Deal approaches, as a way to accelerate GHG reduction.

Frequently Asked Questions (FAQs)

## FAQ 13.1 | What roles do national play in climate mitigation, and how can they be effective?

Institutions and governance underpin mitigation. Climate laws provide the legal basis for action, organisations through which policies are developed and implemented, and frameworks through which diverse actors interact. Specific organisations, such as expert committees, can inform emission reduction targets, inform the creation of policies and packages, and strengthen accountability. Institutions enable strategic thinking, building consensus among stakeholders and enhanced coordination.

Climate governance is constrained and enabled by countries' political systems, material endowments and their ideas, values and belief systems, which leads to a variety of country-specific approaches to climate mitigation.

Countries follow diverse approaches. Some countries focus on greenhouse gases emissions by adopting comprehensive climate laws and creating dedicated ministries and institutions focused on climate change. Others consider climate change among broader scope of policy objectives, such as poverty alleviation, energy security, economic development and co-benefits of climate actions, with the involvement of existing agencies and ministries. See also FAQ 13.3 on sub-national climate mitigation.

## FAQ 13.2 | What policies and strategies can be applied to combat climate change?

Institutions can enable creation of mitigation and sectoral policy instruments; policy packages for low-carbon system transition, and economy-wide measures for systemic restructuring. Policy instruments to reduce greenhouses gas emissions include economic instruments, regulatory instruments and other approaches.

Economic policy instruments directly influence prices to achieve emission reductions through taxes, permit trading, offset systems, subsidies, and border tax adjustments, and are effective in promoting implementation of low-cost emissions reductions. Regulatory instruments help achieve specific mitigation outcomes particularly in sectoral applications, by establishing technology or performance requirements. Other instruments include information programmes, government provision of goods, services and infrastructure, divestment strategies, and voluntary agreements between governments and private firms.

Climate policy instruments can be sector-specific or economy-wide and could be applied at national, regional, or local levels. Policymakers may directly target GHG emission reduction or seek to achieve multiple objectives, such as urbanisation or energy security, with the effect of reducing emissions. In practice, climate mitigation policy instruments operate in combination with other policy tools, and require attention to the interaction effects between instruments. At all levels of governance, coverage, stringency and design of climate policies define their efficiency in reducing greenhouse gases emissions.

Policy packages, when designed with attention to interactive effects, local governance context, and harnessed to a clear vision for change, are better able to support socio-technical transitions and shifts in development pathways toward low-carbon futures than individual policies. See also Chapter 14 on international climate governance.

## FAQ 13.3 | How can actions at the sub-national level contribute to climate mitigation?

Sub-national actors (for example individuals, organisations, jurisdictions and networks at regional, local and city levels) often have a remit over areas salient to climate mitigation, such as land-use planning, waste management, infrastructure, housing, and community development. Despite constraints on legal authority and dependence on national policy priorities in many countries, sub-national climate change policies exist in more than 120 countries. However, they often lack national support, funding, and capacity, and adequate coordination with other scales. Sub-national climate action in support of specific goals is more likely to succeed when linked to local issues such as travel congestion alleviation, air pollution control.

The main drivers of climate actions at sub-national levels include high levels of citizen concern, jurisdictional authority and funding, institutional capacity, national level support and effective linkage to development objectives. Sub-national governments often initiate and implement policy experiments that could be scaled to other levels of governance.

## References


Aamodt, S., 2018: Environmental Ministries as Climate Policy Drivers: Comparing Brazil and India. *J. Environ. Dev.*, **27(4)**, 355–381, doi:10.1177/1070496518791221.

Aamodt, S. and I. Stensdal, 2017: Seizing policy windows: Policy Influence of climate advocacy coalitions in Brazil, China, and India, 2000–2015. *Glob. Environ. Change*, **46**, 114–125, doi:10.1016/j.gloenvcha.2017.08.006.

Aamodt, S. and E.L. Boasson, 2020: *From Impartial Solutions to Mutual Recognition: Explaining Why the EU Changed its Procedural Climate Justice Preferences*. 1–32 pp, doi:10.2139/ssrn.3541720.

Abbott, K.W., J.F. Green, and R.O. Keohane, 2016: Organizational Ecology and Institutional Change in Global Governance. *Int. Organ.*, **70(2)**, 247–277, doi:10.1017/S0020818315000338.

Abraham-Dukuma, M.C. et al., 2020: Multidisciplinary Composition of Climate Change Commissions: Transnational Trends and Expert Perspectives. *Sustainability*, **12(24)**, 1–23, doi:10.3390/su122410280.

Abson, D.J. et al., 2017: Leverage points for sustainability transformation. *Ambio*, **46(1)**, 30–39, doi:10.1007/s13280-016-0800-y.

Accra Metropolitan Assembly, 2020: *Accra Climate Action Plan: First Five-Year Plan (2020–2025)*. Accra Metropolitan Assembly and C40 Cities, Accra, 70 pp. https://cdn.locomotive.works/sites/5ab410c8a2f42204838f797e/content_entry5ab410faa2f42204838f7990/5ab5605ea2f4220acf45cfa6/files/Accra_Climate_Action_Plan.pdf?1603293785 (Accessed October 17, 2021).

Acuto, M. and S. Rayner, 2016: City networks: breaking gridlocks or forging (new) lock-ins? *Int. Aff.*, **92(5)**, 1147–1166, doi:10.1111/1468-2346.12700.

Adelle, C. and D. Russel, 2013: Climate Policy Integration: a Case of Déjà Vu? *Environ. Policy Gov.*, **23(1)**, 1–12, doi:10.1002/eet.1601.

Affolderbach, J. and C. Schulz, 2016: Mobile transitions: Exploring synergies for urban sustainability research. *Urban Stud.*, **53(9)**, 1942–1957, doi:10.1177/0042098015583784.

African Development Bank, 2019: *Analysis of adaptation components of Africa's Nationally Determined Contributions (NDCs)*. Abidjan, Côte d'Ivoire, 1–46 pp. www.afdb.org/fileadmin/uploads/afdb/Documents/Generic-Documents/Analysis_of_Adaptation_Components_in_African_NDCs_2019.pdf (Accessed March 30, 2021).

Agyeman, J., 2013: *Introducing Just Sustainabilities: Policy, Planning, and Practice*. Bloomsbury Publishing, London, UK, 216 pp.

Agyeman, J., D. Schlosberg, L. Craven, and C. Matthews, 2016: Trends and Directions in Environmental Justice: From Inequity to Everyday Life, Community, and Just Sustainabilities. *Annu. Rev. Environ. Resour.*, **41(1)**, 321–340, doi:10.1146/annurev-environ-110615-090052.

Ahmad, I.H., 2009: *Climate Policy Integration: Towards Operationalization*. United Nations Department of Economic and Social Affairs, New York, 18 pp. https://www.un.org/en/desa/climate-policy-integration-towards-operationalization (Accessed August 14, 2021).

Aklin, M. and J. Urpelainen, 2013: Political competition, path dependence, and the strategy of sustainable energy transitions. *Am. J. Pol. Sci.*, **57(3)**, 643–658, doi:10.1111/ajps.12002.

Aklin, M. and M. Mildenberger, 2020: Prisoners of the Wrong Dilemma: Why distributive conflict, not collective action, characterizes the politics of climate change. *Glob. Environ. Polit.*, **20(4)**, 4–26, doi:10.1162/glep_a_00578.

Aldy, J.E. and S. Armitage, 2020: Cost-Effectiveness Implications of Carbon Price Certainty. *AEA Pap. Proc.*, **110**, 113–118, doi:10.1257/pandp.20201083.

Aldy, J.E., M.J. Kotchen, R.N. Stavins, and J.H. Stock, 2021: Keep climate policy focused on the social cost of carbon. *Science*, **373(6557)**, 850–852, doi:10.1126/science.abi7813.

Allcott, H., 2016: Paternalism and Energy Efficiency: An Overview. *Annu. Rev. Econom.*, **8**, 145–176, doi:10.1146/annurev-economics-080315-015255.

Almiron, N. and J. Xifra, eds., 2019: *Climate Change Denial and Public Relations: Strategic communication and interest groups in climate inaction*. Routledge, London, UK, 268 pp.

Altenburg, T., 2011: Building inclusive innovation systems in developing countries: challenges for IS research. In: *Handbook of Innovation Systems and Developing Countries: Building Domestic Capabilities in a Global Setting* [Lundvall B., K.J. Joseph, C. Chaminade, and J. Vang (eds.)]. Edward Elgar Publishing, London, UK, pp. 33–56.

Alves, F. et al., 2020: Climate change policies and agendas: Facing implementation challenges and guiding responses. *Environ. Sci. Policy*, **104**, 190–198, doi:10.1016/j.envsci.2019.12.001.

Amundsen, H., G.K. Hovelsrud, C. Aall, M. Karlsson, and H. Westskog, 2018: Local governments as drivers for societal transformation: towards the 1.5°C ambition. *Curr. Opin. Environ. Sustain.*, **31**, 23–29, doi:10.1016/j.cosust.2017.12.004.

Anderson, Z.R., K. Kusters, J. McCarthy, and K. Obidzinski, 2016: Green growth rhetoric versus reality: Insights from Indonesia. *Glob. Environ. Change*, **38**, 30–40, doi:10.1016/j.gloenvcha.2016.02.008.

Andersson, J.J., 2019: Carbon Taxes and $CO_2$ Emissions: Sweden as a Case Study. *Am. Econ. J. Econ. Policy*, **11(4)**, 1–30, doi:10.1257/pol.20170144.

Anderton, K. and J. Setzer, 2018: Sub-national climate entrepreneurship: innovative climate action in California and São Paulo. *Reg. Environ. Change*, **18(5)**, 1273–1284, doi:10.1007/s10113-017-1160-2.

Andresen, S., E.L. Boasson, and G. Hønneland, eds., 2012: *International environmental agreements: An introduction*. Routledge, London, UK, 216 pp.

Anguelovski, I. et al., 2016: Equity Impacts of Urban Land-Use Planning for Climate Adaptation. *J. Plan. Educ. Res.*, **36(3)**, 333–348, doi:10.1177/0739456X16645166.

Antonopoulos, I., 2020: Climate Change Effects on Human Rights. In: *Climate Action* [Filho, W.L., A.M. Azul, L. Brandli, P.G. Özuyar, and T. Wall, (eds.)]. Springer, Cham, Switzerland, pp. 159–167.

Antweiler, W., B.R. Copeland, and M.S. Taylor, 2001: Is Free Trade Good for the Environment? *Am. Econ. Rev.*, **91(4)**, 877–908, doi:10.1257/aer.91.4.877.

Antwi-Agyei, P., A.J. Dougill, T.P. Agyekum, and L.C. Stringer, 2018: Alignment between nationally determined contributions and the sustainable development goals for West Africa. *Clim. Policy*, **18(10)**, 1296–1312, doi:10.1080/14693062.2018.1431199.

Arimura, T.H. and T. Abe, 2021: The Impact of the Tokyo Emissions Trading Scheme on Office Buildings: What factor contributed to the emission reduction? *Environ. Econ. Policy Stud.*, **23**, 517–533, doi:10.1007/s10018-020-00271-w.

Armstrong, J.H., 2019: Modeling effective local government climate policies that exceed state targets. *Energy Policy*, **132**, 15–26, doi:10.1016/j.enpol.2019.05.018.

Arndt, C. and F. Tarp, 2017: Aid, Environment and Climate Change. *Rev. Dev. Econ.*, **21(2)**, 285–303, doi:10.1111/rode.12291.

Asensio, O.I. and M.A. Delmas, 2017: The effectiveness of US energy efficiency building labels. *Nat. Energy*, **2(4)**, 17033, doi:10.1038/nenergy.2017.33.

Astuti, R. and A. McGregor, 2015: Governing carbon, transforming forest politics: a case study of Indonesia's REDD+ Task Force. *Asia Pac. Viewp.*, **56(1)**, 21–36, doi:10.1111/apv.12087.

Atchison, J., 2019: Green and Blue Infrastructure in Darwin; Carbon Economies and the Social and Cultural Dimensions of Valuing Urban Mangroves in Australia. *Urban Sci.*, **3(3)**, 86, doi:10.3390/urbansci3030086.

Attari, S.Z. et al., 2009: Preferences for change: Do individuals prefer voluntary actions, soft regulations, or hard regulations to decrease fossil fuel consumption? *Ecol. Econ.*, **68(6)**, 1701–1710, doi:10.1016/j.ecolecon.2008.10.007.

Atteridge, A., C. Verkuijl, and A. Dzebo, 2020: Nationally determined contributions (NDCs) as instruments for promoting national development agendas? An analysis of small island developing states (SIDS). *Clim. Policy*, **20(4)**, 485–498, doi:10.1080/14693062.2019.1605331.

Aune, F.R. and R. Golombek, 2021: Are Carbon Prices Redundant in the 2030 EU Climate and Energy Policy Package? *Energy J.*, **42(3)**, doi:10.5547/01956574.42.3.faun.

aus dem Moore, N., P. Grokurth, and M. Themann, 2019: Multinational corporations and the EU Emissions Trading System: The specter of asset erosion and creeping deindustrialization. *J. Environ. Econ. Manage.*, **94**, doi:10.1016/j.jeem.2018.11.003.

Australian Climate Change Authority, 2017: *Review of the Emissions Reduction Fund*. Commonwealth of Australia, Canberra, 120 pp. https://www.climatechangeauthority.gov.au/sites/default/files/2020-06/CFI%202017%20December/ERF%20Review%20Report.pdf (Accessed March 30, 2021).

Avelino, F. and J.M. Wittmayer, 2016: Shifting Power Relations in Sustainability Transitions: a Multi-actor Perspective. *J. Environ. Policy Plan.*, **18(5)**, 628–649, doi:10.1080/1523908X.2015.1112259.

Averchenkova, A. and S. Guzman Luna, 2018: *Mexico's General Law on Climate Change: Key achievements and challenges ahead*. Grantham Research Institute on Climate Change and the Environment and Centre for Climate Change Economics and Policy, London School of Economics and Political Science, London, 29 pp. https://www.lse.ac.uk/GranthamInstitute/wp-content/uploads/2018/11/Policy_report_Mexico's-General-Law-on-Climate-Change-Key-achievements-and-challenges-ahead-29pp_AverchenkovaGuzman-1.pdf (Accessed March 30, 2021).

Averchenkova, A., S. Fankhauser, and M. Nachmany, 2017: Introduction. In: *Trends in Climate Change Legislation* [Averchenkova, A., S. Fankhauser, and M. Nachmany, (eds.)]. Edward Elgar Publishing, Cheltenham, UK and Northampton, MA, USA, pp. 16.

Averchenkova, A., S. Fankhauser, and J. Finnegan, 2018: *The role of independent bodies in climate governance: the UK's Committee on Climate Change*. Grantham Research Institute on Climate Change and the Environment and Centre for Climate Change Economics and Policy, London School of Economics and Political Science, London, 28 pp. http://www.lse.ac.uk/GranthamInstitute/wp-content/uploads/2018/10/The-role-of-independent-bodies-in-climate-governance-the-UKs-Committee-on-Climate-Change_Averchenkova-et-al.pdf (Accessed March 30, 2021).

Averchenkova, A., S. Fankhauser, and J.J. Finnegan, 2021: The impact of strategic climate legislation: evidence from expert interviews on the UK Climate Change Act. *Clim. Policy*, **21(2)**, 251–263, doi:10.1080/14693062.2020.1819190.

Axsen, J., P. Plötz, and M. Wolinetz, 2020: Crafting strong, integrated policy mixes for deep $CO_2$ mitigation in road transport. *Nat. Clim. Change*, **10(9)**, 809–818, doi:10.1038/s41558-020-0877-y.

Aydin, C. and Ö. Esen, 2018: Reducing $CO_2$ emissions in the EU member states: Do environmental taxes work? *J. Environ. Plan. Manag.*, **61(13)**, 2396–2420, doi:10.1080/09640568.2017.1395731.

Aylett, A., 2013: The Socio-institutional Dynamics of Urban Climate Governance: a Comparative Analysis of Innovation and Change in Durban (KZN, South Africa) and Portland (OR, USA). *Urban Stud.*, **50(7)**, 1386–1402, doi:10.1177/0042098013480968.

Aylett, A., 2014: *Progress and Challenges in the Urban Governance of Climate Change: Results of a Global Survey*. MIT, Cambridge, MA, USA, 68 pp. https://climate-adapt.eea.europa.eu/metadata/publications/progress-and-challenges-in-the-urban-governance-of-climate-change-results-of-a-global-survey/mit_iclei_2014_urbanclimategovernancereport.pdf (Accessed December 12, 2019).

Aylett, A., 2015: Institutionalizing the urban governance of climate change adaptation: Results of an international survey. *Urban Clim.*, **14(1)**, 4–16, doi:10.1016/j.uclim.2015.06.005.

Ayling, J., 2017: a Contest for Legitimacy: The Divestment Movement and the Fossil Fuel Industry. *Law Policy*, **39(4)**, 349–371, doi:10.1111/lapo.12087.

Ayling, J. and N. Gunningham, 2017: Non-state governance and climate policy: the fossil fuel divestment movement. *Clim. Policy*, **17(2)**, 131–149, doi:10.1080/14693062.2015.1094729.

Bäckstrand, K. and E. Lövbrand, 2006: Planting trees to mitigate climate change: Contested discourses of ecological modernization, green governmentality and civic environmentalism. *Glob. Environ. Polit.*, **6(1)**, 50–75, doi:10.1162/glep.2006.6.1.50.

Bäckstrand, K. and E. Lövbrand, 2019: The Road to Paris: Contending Climate Governance Discourses in the Post-Copenhagen Era. *J. Environ. Policy Plan.*, **21(5)**, 519–532, doi:10.1080/1523908X.2016.1150777.

Bäckstrand, K., J.W. Kuyper, B.-O. Linnér, and E. Lövbrand, 2017: Non-state actors in global climate governance: from Copenhagen to Paris and beyond. *Env. Polit.*, **26(4)**, 561–579, doi:10.1080/09644016.2017.1327485.

Bae, J.H., D.D. Li, and M. Rishi, 2017: Determinants of $CO_2$ emission for post-Soviet Union independent countries. *Clim. Policy*, **17(5)**, 591–615, doi:10.1080/14693062.2015.1124751.

Bai, X., B. Roberts, and J. Chen, 2010: Urban sustainability experiments in Asia: Patterns and pathways. *Environ. Sci. Policy*, **13(4)**, 312–325, doi:10.1016/j.envsci.2010.03.011.

Bai, X. et al., 2018: Six research priorities for cities and climate change. *Nature*, **555(7694)**, 23–25, doi:10.1038/d41586-018-02409-z.

Bai, Y., S. Song, J. Jiao, and R. Yang, 2019: The impacts of government R&D subsidies on green innovation: Evidence from Chinese energy-intensive firms. *J. Clean. Prod.*, **233** (October), 819–829, doi:10.1016/j.jclepro.2019.06.107.

Bailey, I., 2017: Spatializing Climate Justice: Justice Claim Making and Carbon Pricing Controversies in Australia. *Ann. Am. Assoc. Geogr.*, **107(5)**, 1128–1143, doi:10.1080/24694452.2017.1293497.

Bak, C., A. Bhattacharya, O. Edenhofer, and B. Knopf, 2017: Towards a comprehensive approach to climate policy, sustainable infrastructure, and finance. *Econ. Open-Access, Open-Assessment E-Journal*, **11(2017–33)**, 1–13, doi:10.5018/economics-ejournal.ja.2017-33.

Baldursson, F.M., N.-H.M. von der Fehr, and E. Lazarczyk, 2021: Electric Vehicles Rollout—Two Case Studies. *Econ. Energy Environ. Policy*, **10(2)**, doi:10.5547/2160-5890.10.2.fbal.

Bang, G., 2021: The United States: conditions for accelerating decarbonisation in a politically divided country. *Int. Environ. Agreements Polit. Law Econ.*, **21(1)**, 43–58, doi:10.1007/s10784-021-09530-x.

Bang, G., A. Underdal, and S. Andresen, eds., 2015: *The Domestic Politics of Global Climate Change: Key Actors in International Climate Cooperation*. Edward Elgar Publishing, Cheltenham, UK and Northampton, MA, USA, 224 pp.

Bang, G., D.G. Victor, and S. Andresen, 2017: California's Cap-and-Trade System: Diffusion and Lessons. *Glob. Environ. Polit.*, **17(3)**, 12–30, doi:10.1162/GLEP_a_00413.

Baranzini, A. et al., 2017: Carbon pricing in climate policy: seven reasons, complementary instruments, and political economy considerations. *Wiley Interdiscip. Rev. Clim. Change*, **8(4)**, e462, doi:10.1002/wcc.462.

Barbier, E.B., 2010: Global Governance: The G20 and a Global Green New Deal. *Econ. Open-Access, Open-Assessment E-Journal*, **4(2010–2)**, 1–35, doi:10.5018/economics-ejournal.ja.2010-2.

Barbier, E.B., 2020: Greening the Post-pandemic Recovery in the G20. *Environ. Resour. Econ.*, **76(4)**, 685–703, doi:10.1007/s10640-020-00437-w.

Barcelona City Council, 2018: *Climate Plan 2018-2030*. Barcelona, Spain, 164 pp. https://www.barcelona.cat/barcelona-pel-clima/sites/default/files/documents/climate_plan_maig.pdf.

Bárcena, A. et al., 2018: *Economics of Climate Change in Latin America and the Caribbean: a graphic view*. Economic Commission for Latin America and the Caribbean (ECLAC), Santiago, Chile, 61 pp. https://www.cepal.org/en/publications/43889-economics-climate-change-latin-america-and-caribbean-graphic-view (Accessed October 25, 2021).

Barritt, E., 2020: Consciously transnational: Urgenda and the shape of climate change litigation: The State of the Netherlands (Ministry of Economic

Affairs and Climate Policy) v Urgenda Foundation. *Environ. Law Rev.*, **22(4)**, 296–305, doi:10.1177/1461452920974493.

Barritt, E. and B. Sediti, 2019: The Symbolic Value of Leghari v Federation of Pakistan: Climate Change Adjudication in the Global South. *King's Law J.*, **30(2)**, doi:10.1080/09615768.2019.1648370.

Barton, B. and J. Campion, 2018: Climate Change Legislation: Law for Sound Policy Making. In: *Innovation in Energy Law and Technology: Dynamic Solutions for Energy Transitions* [Zillman, D., M. Roggenkamp, L. Paddock, and L. Godden, (eds.)]. Vol. 1. Oxford University Press, Oxford, UK, pp. 23–37.

Bastos Lima, M.G., U.M. Persson, and P. Meyfroidt, 2019: Leakage and boosting effects in environmental governance: a framework for analysis. *Environ. Res. Lett.*, **14(10)**, doi:10.1088/1748-9326/ab4551.

Bataille, C. et al., 2018a: a review of technology and policy deep decarbonization pathway options for making energy-intensive industry production consistent with the Paris Agreement. *J. Clean. Prod.*, **187**, 960–973, doi:10.1016/j.jclepro.2018.03.107.

Bataille, C., C. Guivarch, S. Hallegatte, J. Rogelj, and H. Waisman, 2018b: Carbon prices across countries. *Nat. Clim. Change*, **8**, 648–650, doi:10.1038/s41558-018-0239-1.

Bataille, C.G.F., 2020: Physical and policy pathways to net-zero emissions industry. *WIREs Clim. Change*, **11(2)**, e633, doi:10.1002/wcc.633.

Båtstrand, S., 2015: More than Markets: a Comparative Study of Nine Conservative Parties on Climate Change. *Polit. Policy*, **43(4)**, 538–561, doi:10.1111/polp.12122.

Bättig, M.B. and T. Bernauer, 2009: National Institutions and Global Public Goods: Are Democracies More Cooperative in Climate Change Policy? *Int. Organ.*, **63(02)**, 281, doi:10.1017/S0020818309090092.

Bauer, N. et al., 2015: $CO_2$ emission mitigation and fossil fuel markets: Dynamic and international aspects of climate policies. *Technol. Forecast. Soc. Change*, **90(Part A)**, 243–256, doi:10.1016/j.techfore.2013.09.009.

Bauer, N. et al., 2016: Global fossil energy markets and climate change mitigation – an analysis with REMIND. *Clim. Change*, **136(1)**, 69–82, doi:10.1007/s10584-013-0901-6.

Baxter, J., R. Morzaria, and R. Hirsch, 2013: a case-control study of support/opposition to wind turbines: Perceptions of health risk, economic benefits, and community conflict. *Energy Policy*, **61**, 931–943, doi:10.1016/j.enpol.2013.06.050.

Bayer, P. and M. Aklin, 2020: The European Union Emissions Trading System reduced $CO_2$ emissions despite low prices. *Proc. Natl. Acad. Sci.*, **117(16)**, doi:10.1073/pnas.1918128117.

Baylis, K., D. Fullerton, and D.H. Karney, 2014: Negative Leakage. *J. Assoc. Environ. Resour. Econ.*, **1(1/2)**, 51–73, doi:10.1086/676449.

Bebbington, A. and J. Bury, eds., 2013: *Subterranean Struggles: New Dynamics of Mining, Oil, and Gas in Latin America*. University of Texas Press, Austin, Texas, USA, 361 pp.

Bechtel, M.M., F. Genovese, and K.F. Scheve, 2019: Interests, Norms and Support for the Provision of Global Public Goods: The Case of Climate Co-operation. *Br. J. Polit. Sci.*, **49(4)**, 1333–1355, doi:10.1017/S0007123417000205.

Becken, S., 2019: Decarbonising tourism: mission impossible? *Tour. Recreat. Res.*, **44(4)**, 419–433, doi:10.1080/02508281.2019.1598042.

Beermann, J., A. Damodaran, K. Jörgensen, and M.A. Schreurs, 2016: Climate action in Indian cities: an emerging new research area. *J. Integr. Environ. Sci.*, **13(1)**, 55–66, doi:10.1080/1943815X.2015.1130723.

Beg, N. et al., 2002: Linkages between climate change and sustainable development. *Clim. Policy*, **2(2–3)**, 129–144, doi:10.3763/cpol.2002.0216.

Bel, G. and J.J. Teixidó, 2020: The political economy of the Paris Agreement: Income inequality and climate policy. *J. Clean. Prod.*, **258**, 121002, doi:10.1016/j.jclepro.2020.121002.

Bennett, M., 2018: The role of National Framework Legislation in Implementing Australia's emission reduction commitments under the Paris Agreement. *Univ. West. Aust. Law Rev.*, **43(1)**, 240–263.

Bennett, W.L., 2013: *The Logic of Connective Action: Digital Media And The Personalization Of Contentious Politics*. Cambridge University Press, Cambridge, UK, 256 pp.

Bento, A., R. Kanbur, and B. Leard, 2016: On the importance of baseline setting in carbon offsets markets. *Clim. Change*, **137(3–4)**, 625–637, doi:10.1007/s10584-016-1685-2.

Bergquist, P., S. Ansolabehere, S. Carley, and D. Konisky, 2020: Backyard voices: How sense of place shapes views of large-scale energy transmission infrastructure. *Energy Res. Soc. Sci.*, **63**, 101396, doi:10.1016/j.erss.2019.101396.

Bernard, J.-T. and M. Kichian, 2019: The long and short run effects of British Columbia's carbon tax on diesel demand. *Energy Policy*, **131**(August 2019), 380–389, doi:10.1016/j.enpol.2019.04.021.

Bernard, J.-T. and M. Kichian, 2021: The Impact of a Revenue-Neutral Carbon Tax on GDP Dynamics: The Case of British Columbia. *Energy J.*, **42(3)**, doi:10.5547/01956574.42.3.jber.

Bernhagen, P., 2008: Business and International Environmental Agreements: Domestic Sources of Participation and Compliance by Advanced Industrialized Democracies. *Glob. Environ. Polit.*, **8(1)**, 78–110, doi:10.1162/glep.2008.8.1.78.

Bernstein, S. and M. Hoffmann, 2018: The politics of decarbonization and the catalytic impact of subnational climate experiments. *Policy Sci.*, **51(2)**, 189–211, doi:10.1007/s11077-018-9314-8.

Bernstein, S. and M. Hoffmann, 2019: Climate politics, metaphors and the fractal carbon trap. *Nat. Clim. Change*, **9(12)**, 919–925, doi:10.1038/s41558-019-0618-2.

Berry, P.M. et al., 2015: Cross-sectoral interactions of adaptation and mitigation measures. *Clim. Change*, **128(3–4)**, 381–393, doi:10.1007/s10584-014-1214-0.

Bertoldi, P., 2018: The Paris Agreement 1.5 C goal: what it does mean for energy efficiency? *Proceedings of the ACEEE Summer Study on Energy Efficiency in Buildings*. American Council for an Energy-Efficient Economy, Washington, DC, USA. www.aceee.org/files/proceedings/2018/#/paper/event-data/p268 (Accessed September 24, 2021).

Bertoldi, P. et al., 2020: *Covenant of Mayors: 2019 Assessment*. Publications Office of the European Union, Luxembourg, 63 pp. https://publications.jrc.ec.europa.eu/repository/handle/JRC118927.

Bertram, C. et al., 2015a: Carbon lock-in through capital stock inertia associated with weak near-term climate policies. *Technol. Forecast. Soc. Change*, **90**, 62–72, doi:10.1016/j.techfore.2013.10.001.

Bertram, C. et al., 2015b: Complementing carbon prices with technology policies to keep climate targets within reach. *Nat. Clim. Change*, **5(3)**, 235–239, doi:10.1038/nclimate2514.

Besanko, D., 1987: Performance versus design standards in the regulation of pollution. *J. Public Econ.*, **34(1)**, 19–44, doi:10.1016/0047-2727(87)90043-0.

Best, R., P.J. Burke, and F. Jotzo, 2020: Carbon Pricing Efficacy: Cross-Country Evidence. *Environ. Resour. Econ.*, **77(1)**, 69–94, doi:10.1007/s10640-020-00436-x.

Bevan, L.D., T. Colley, and M. Workman, 2020: Climate change strategic narratives in the United Kingdom: Emergency, Extinction, Effectiveness. *Energy Res. Soc. Sci.*, **69**, 101580, doi:10.1016/j.erss.2020.101580.

Bhagavathy, S.M. and M. McCulloch, 2020: *Electric Vehicle transition in the UK*. arXiv, Oxford, UK, 7 pp. https://arxiv.org/ftp/arxiv/papers/2007/2007.03745.pdf (Accessed November 28, 2020).

Bhamidipati, P.L., J. Haselip, and U. Elmer Hansen, 2019: How do energy policies accelerate sustainable transitions? Unpacking the policy transfer process in the case of GETFiT Uganda. *Energy Policy*, **132**, 1320–1332, doi:10.1016/j.enpol.2019.05.053.

Bhardwaj, A., M. Joshi, R. Khosla, and N.K. Dubash, 2019: More priorities, more problems? Decision-making with multiple energy, development and climate objectives. *Energy Res. Soc. Sci.*, **49**, 143–157, doi.org/10.1016/j.erss.2018.11.003.

Bhardwaj, C., J. Axsen, F. Kern, and D. McCollum, 2020: Why have multiple climate policies for light-duty vehicles? Policy mix rationales, interactions and research gaps. *Transp. Res. Part a Policy Pract.*, **135**, 309–326, doi:10.1016/j.tra.2020.03.011.

Biggs, E.M. et al., 2015: Sustainable development and the water-energy-food nexus: a perspective on livelihoods. *Environ. Sci. Policy*, **54**, 389–397, doi:10.1016/j.envsci.2015.08.002.

Black, R. et al., 2021: *Taking Stock: a global assessment of net zero targets*. The Real Press, London, UK, 30 pp. https://ca1-eci.edcdn.com/reports/ECIU-Oxford_Taking_Stock.pdf (Accessed October 20, 2021).

Blok, A., 2016: Assembling urban riskscapes. *City*, **20(4)**, 602–618, doi:10.1080/13604813.2016.1194000.

Blondeel, M. and T. Van de Graaf, 2018: Toward a global coal mining moratorium? a comparative analysis of coal mining policies in the USA, China, India and Australia. *Clim. Change*, **150(1–2)**, 89–101, doi:10.1007/s10584-017-2135-5.

Blondeel, M., J. Colgan, and T. Van de Graaf, 2019: What Drives Norm Success? Evidence from Anti-Fossil Fuel Campaigns. *Glob. Environ. Polit.*, **19(4)**, 63–84, doi:10.1162/glep_a_00528.

Blue, G. and J. Medlock, 2014: Public Engagement with Climate Change as Scientific Citizenship: a Case Study of World Wide Views on Global Warming. *Sci. Cult.*, **23(4)**, 560–579, doi:10.1080/09505431.2014.917620.

Boasson, E.L., 2015: *National Climate Policy: a Multi-field Approach*. 1st ed. Routledge, New York, NY, USA, 250 pp.

Boasson, E.L., 2019: Constitutionalization and Entrepreneurship: Explaining Increased EU Steering of Renewables Support Schemes. *Polit. Gov.*, **7(1)**, 70–80, doi:10.17645/pag.v7i1.1851.

Boasson, E.L. and J. Wettestad, 2013: *EU Climate Policy: Industry, Policy Interaction and External Environment*. 1st ed. Routledge, 236 pp.

Boasson, E.L., M.D. Leiren, and J. Wettestad, eds., 2021: *Comparative Renewables Policy: Political, Organizational and European Fields*. 1st ed. Routledge, London, UK, 278 pp.

Bocklet, J., M. Hintermayer, L. Schmidt, and T. Wildgrube, 2019: The reformed EU ETS – Intertemporal emission trading with restricted banking. *Energy Econ.*, **84**, doi:10.1016/j.eneco.2019.104486.

Bohari, A.A.M., M. Skitmore, B. Xia, and M. Teo, 2017: Green oriented procurement for building projects: Preliminary findings from Malaysia. *J. Clean. Prod.*, **148**, 690–700, doi:10.1016/j.jclepro.2017.01.141.

Böhmelt, T., M. Böker, and H. Ward, 2016: Democratic inclusiveness, climate policy outputs, and climate policy outcomes. *Democratization*, **23(7)**, 1272–1291, doi:10.1080/13510347.2015.1094059.

Böhringer, C., E.J. Balistreri, and T.F. Rutherford, 2012: The role of border carbon adjustment in unilateral climate policy: Overview of an Energy Modeling Forum study (EMF 29). *Energy Econ.*, **34(sup2)**, S97–S110, doi:10.1016/j.eneco.2012.10.003.

Böhringer, C., A. Keller, M. Bortolamedi, and A. Rahmeier Seyffarth, 2016: Good things do not always come in threes: On the excess cost of overlapping regulation in EU climate policy. *Energy Policy*, **94**, 502–508, doi:10.1016/j.enpol.2015.12.034.

Böhringer, C., B. Bye, T. Fæhn, and K.E. Rosendahl, 2017a: Targeted carbon tariffs: Export response, leakage and welfare. *Resour. Energy Econ.*, **50**, 51–73, doi:10.1016/j.reseneeco.2017.06.003.

Böhringer, C., A. Cuntz, D. Harhoff, and E. Asane-Otoo, 2017b: The impact of the German feed-in tariff scheme on innovation: Evidence based on patent filings in renewable energy technologies. *Energy Econ.*, **67**, 545–553, doi:10.1016/j.eneco.2017.09.001.

Böhringer, C., K.E. Rosendahl, and H.B. Storrøsten, 2017c: Robust policies to mitigate carbon leakage. *J. Public Econ.*, **149**, 35–46, doi:10.1016/j.jpubeco.2017.03.006.

Boli, J. and G.M. Thomas, eds., 1999: *Constructing World Culture: International Nongovernmental Organizations Since 1875*. Stanford University Press, Stanford, CA, USA, 380 pp.

Borenstein, S., J. Bushnell, F.A. Wolak, and M. Zaragoza-Watkins, 2019: Expecting the Unexpected: Emissions Uncertainty and Environmental Market Design. *Am. Econ. Rev.*, **109(11)**, 3953–3977, doi:10.1257/aer.20161218.

Borghesi, S., C. Franco, and G. Marin, 2020: Outward Foreign Direct Investment Patterns of Italian Firms in the European Union's Emission Trading Scheme. *Scand. J. Econ.*, **122(1)**, doi:10.1111/sjoe.12323.

Bosetti, V. and D.G. Victor, 2011: Politics and Economics of Second-Best Regulation of Greenhouse Gases: The Importance of Regulatory Credibility. *Energy J.*, **32(1)**, 1–24, doi:10.5547/ISSN0195-6574-EJ-Vol32-No1-1.

Bostrom, A., A.L. Hayes, and K.M. Crosman, 2018: Efficacy, Action, and Support for Reducing Climate Change Risks. *Risk Anal.*, **39(4)**, 805–828, doi:10.1111/risa.13210.

Boudet, H., D. Bugden, C. Zanocco, and E. Maibach, 2016: The effect of industry activities on public support for 'fracking.' *Env. Polit.*, **25(4)**, 593–612, doi:10.1080/09644016.2016.1153771.

Boulianne, S., M. Lalancette, and D. Ilkiw, 2020: 'School Strike 4 Climate': Social Media and the International Youth Protest on Climate Change. *Media Commun.*, **8(2)**, 208–218, doi:10.17645/mac.v8i2.2768.

Boussalis, C. and T.G. Coan, 2016: Text-mining the signals of climate change doubt. *Glob. Environ. Change*, **36**, 89–100, doi:10.1016/j.gloenvcha.2015.12.001.

Bouteligier, S., 2013: Inequality in new global governance arrangements: the North–South divide in transnational municipal networks. *Innov. Eur. J. Soc. Sci. Res.*, **26(3)**, 251–267, doi:10.1080/13511610.2013.771890.

Bouwer, K., 2018: The Unsexy Future of Climate Change Litigation. *J. Environ. Law*, **30(3)**, 483–506, doi:10.1093/jel/eqy017.

Bouzarovski, S., J. Frankowski, and S. Tirado Herrero, 2018: Low-Carbon Gentrification: When Climate Change Encounters Residential Displacement. *Int. J. Urban Reg. Res.*, **42(5)**, 845–863, doi:10.1111/1468-2427.12634.

Bowen, K.J., K. Ebi, and S. Friel, 2014: Climate change adaptation and mitigation: next steps for cross-sectoral action to protect global health. *Mitig. Adapt. Strateg. Glob. Change*, **19(7)**, 1033–1040, doi:10.1007/s11027-013-9458-y.

Bowman, M. and D. Wiseman, 2020: Finance actors and climate-related disclosure regulation: Logic, limits, and emerging accountability. In: *Criminology and Climate: Insurance, Finance and the Regulation of Harmscapes* [Holley, C., L. Phelan, and C. Shearing, (eds.)]. Routledge, London, UK, 26 pp.

Boykoff, M. and J. Farrell, 2019: Climate Change Countermovement Organizations and Media Attention in the United States. In: *Climate Change Denial and Public Relations. Strategic Communication and Interest Groups in Climate Inaction* [Almiron, N. and J. Xifra, (eds.)]. Routledge, London, pp. 121–139.

Boykoff, M. and B. Osnes, 2019: a Laughing matter? Confronting climate change through humor. *Polit. Geogr.*, **68**, 154–163, doi:10.1016/j.polgeo.2018.09.006.

Boykoff, M. et al., 2019: World Newspaper Coverage of Climate Change or Global Warming, 2004-2019. Media and Climate Change Observatory (MECCO) Data Sets. Cooperative Institute for Research in Environmental Sciences, University of Colorado, doi:10.25810/4c3b-b819.20.

Boykoff, M. et al., 2021: World Newspaper Coverage of Climate Change or Global Warming, 2004-2021. Media and Climate Change Observatory (MECCO) Data Sets. Cooperative Institute for Research in Environmental Sciences, University of Colorado, doi:10.25810/4c3b-b819.

Boykoff, M.T., 2011: *Who Speaks for the Climate? Making Sense of Media Reporting on Climate Change*. Cambridge University Press, Cambridge, UK, 240 pp.

Boykoff, M.T., 2013: Public Enemy No. 1? Understanding Media Representations of Outlier Views on Climate Change. *Am. Behav. Sci.*, **57(6)**, 796–817, doi:10.1177/0002764213476846.

Boykoff, M.T. and T. Yulsman, 2013: Political economy, media, and climate change: sinews of modern life. *Wiley Interdiscip. Rev. Clim. Change*, **4(5)**, 359–371, doi:10.1002/wcc.233.

Branger, F. and P. Quirion, 2014: Would border carbon adjustments prevent carbon leakage and heavy industry competitiveness losses? Insights from a meta-analysis of recent economic studies. *Ecol. Econ.*, **99**, 29–39, doi:10.1016/j.ecolecon.2013.12.010.

Branger, F., P. Quirion, and J. Chevallier, 2016: Carbon Leakage and Competitiveness of Cement and Steel Industries Under the EU ETS: Much Ado About Nothing. *Energy J.*, **37(3)**, 109–135, doi:10.5547/01956574.37.3.fbra.

Brannstrom, C. et al., 2017: Is Brazilian wind power development sustainable? Insights from a review of conflicts in Ceará state. *Renew. Sustain. Energy Rev.*, **67**, 62–71, doi:10.1016/j.rser.2016.08.047.

Brattland, C. and T. Mustonen, 2018: How Traditional Knowledge Comes to Matter in Atlantic Salmon Governance in Norway and Finland. *Arctic*, **71(4)**, 375–392, doi:10.14430/arctic4751.

Bratton, W. and J.A. McCahery, 2015: *Institutional Investor Activism: Hedge Funds and Private Equity, Economics and Regulation*. Penn Law: Legal Scholarship Repository, 49 pp. https://scholarship.law.upenn.edu/cgi/viewcontent.cgi?article=2646&context=faculty_scholarship.

Brewer, P.R. and J. McKnight, 2015: Climate as Comedy: The Effects of Satirical Television News on Climate Change Perceptions. *Sci. Commun.*, **37(5)**, 635–657, doi:10.1177/1075547015597911.

Broadbent, J. et al., 2016: Conflicting Climate Change Frames in a Global Field of Media Discourse. *Socius Sociol. Res. a Dyn. World*, **2**, 237802311667066, doi:10.1177/2378023116670660.

Bromley-Trujillo, R. and M.R. Holman, 2020: Climate Change Policymaking in the States: a View at 2020. *Publius J. Fed.*, **50(3)**, 446–472, doi:10.1093/publius/pjaa008.

Brown, D.M., 2012: Comparative Climate Change Policy and Federalism: An Overview. *Rev. Policy Res.*, **29(3)**, 322–333, doi:10.1111/j.1541-1338.2012.00562.x.

Brüggemann, M. and S. Engesser, 2017: Beyond false balance: How interpretive journalism shapes media coverage of climate change. *Glob. Environ. Change*, **42**, 58–67, doi:10.1016/j.gloenvcha.2016.11.004.

Brugnach, M., M. Craps, and A. Dewulf, 2017: Including indigenous peoples in climate change mitigation: addressing issues of scale, knowledge and power. *Clim. Change*, **140(1)**, 19–32, doi:10.1007/s10584-014-1280-3.

Brulle, R.J., 2000: *Agency, democracy, and nature: the U.S. environmental movement from a critical theory perspective*. 1st edltlo. MIT Press, Boston, MA, 360 pp.

Brulle, R.J., 2014: Institutionalizing delay: foundation funding and the creation of U.S. climate change counter-movement organizations. *Clim. Change*, **122(4)**, 681–694, doi:10.1007/s10584-013-1018-7.

Brulle, R.J., 2019: Networks of Opposition: a Structural Analysis of U.S. Climate Change Countermovement Coalitions 1989–2015. *Sociol. Inq.*, **91(3) 603–624**, doi:10.1111/soin.12333.

Brulle, R.J., J. Carmichael, and J.C. Jenkins, 2012: Shifting public opinion on climate change: An empirical assessment of factors influencing concern over climate change in the U.S., 2002-2010. *Clim. Change*, **114(2)**, 169–188, doi:10.1007/s10584-012-0403-y.

Bruninx, K., M. Ovaere, and E. Delarue, 2020: The long-term impact of the market stability reserve on the EU emission trading system. *Energy Econ.*, **89**, 104746, doi:10.1016/j.eneco.2020.104746.

Brunner, S., C. Flachsland, and R. Marschinski, 2012: Credible commitment in carbon policy. *Clim. Policy*, **12(2)**, 255–271, doi:10.1080/14693062.2011.582327.

Buchholz, W., L. Dippl, and M. Eichenseer, 2019: Subsidizing renewables as part of taking leadership in international climate policy: The German case. *Energy Policy*, **129**, 765–773, doi:10.1016/j.enpol.2019.02.044.

Büchs, M., C. Saunders, R. Wallbridge, G. Smith, and N. Bardsley, 2015: Identifying and explaining framing strategies of low-carbon lifestyle movement organisations. *Glob. Environ. Change*, **35**, 307–315, doi:10.1016/j.gloenvcha.2015.09.009.

Buckman, G., J. Sibley, and M. Ward, 2019: The large-scale feed-in tariff reverse auction scheme in the Australian Capital Territory 2012, to 2016. *Renew. Energy*, **132**, 176–185, doi:10.1016/j.renene.2018.08.011.

Buitenzorgy, M. and A.P.J. Mol, 2011: Does Democracy Lead to a Better Environment? Deforestation and the Democratic Transition Peak. *Environ. Resour. Econ.*, **48(1)**, 59–70, doi:10.1007/s10640-010-9397-y.

Bulfin, A., 2017: Popular culture and the 'new human condition': Catastrophe narratives and climate change. *Glob. Planet. Change*, **156**, 140–146, doi:10.1016/j.gloplacha.2017.03.002.

Bulkeley, H., 2000: Discourse Coalitions and the Australian Climate Change Policy Network. *Environ. Plan. C Gov. Policy*, **18(6)**, 727–748, doi:10.1068/c9905j.

Bulkeley, H., 2013: *Cities and Climate Change*. 1st ed. Routledge, London, UK, 280 pp.

Bulkeley, H. and V. Castán Broto, 2013: Government by experiment? Global cities and the governing of climate change. *Trans. Inst. Br. Geogr.*, **38(3)**, 361–375, doi:10.1111/j.1475-5661.2012.00535.x.

Bulkeley, H., J. Carmin, V. Castan Broto, G.A.S. Edwards, and S. Fuller, 2013: Climate justice and global cities: Mapping the emerging discourses. *Glob. Environ. Change*, **23(5)**, 914–925, doi:10.1016/j.gloenvcha.2013.05.010.

Bulkeley, H. et al., 2016: Urban living labs: governing urban sustainability transitions. *Curr. Opin. Environ. Sustain.*, **22**, 13–17, doi:10.1016/j.cosust.2017.02.003.

Bullard, R.D., P. Mohai, R. Saha, and B. Wright, 2008: Toxic Wastes and Race at Twenty: Why Race Still Matters After all of These Years. *Environ. Law*, **38(2)**, 371–411.

Burch, S., H. Schroeder, S. Rayner, and J. Wilson, 2013: Novel Multisector Networks and Entrepreneurship: The Role of Small Businesses in the Multilevel Governance of Climate Change. *Environ. Plan. C Gov. Policy*, **31(5)**, 822–840, doi:10.1068/c1206.

Burch, S. et al., 2014: Triggering transformative change: a development path approach to climate change response in communities. *Clim. Policy*, **14(4)**, 467–487, doi:10.1080/14693062.2014.876342.

Burch, S. et al., 2016: Governing and accelerating transformative entrepreneurship: exploring the potential for small business innovation on urban sustainability transitions. *Curr. Opin. Environ. Sustain.*, **22**, 26–32, doi:10.1016/j.cosust.2017.04.002.

Burck, J., U. Hagen, N. Höhne, L. Nasclmento, and C. Bals, 2019: *Climate Change Performance Index Results 2020*. Germanwatch, New Climate Institute and Climate Action Network International, 32 pp. https://newclimate.org/wp-content/uploads/2019/12/CCPI-2020-Results_Web_Version.pdf (Accessed October 18, 2021).

Burger, C., A. Froggatt, C. Mitchell, and J. Weimann (eds.), 2020a: *Decentralised Energy: a Global Game Changer*. Ubiquity Press, London, UK, 313 pp.

Burger, M., J. Wentz, and R. Horton, 2020b: The Law and Science of Climate Change Attribution. *Columbia J. Environ. Law*, **45(1)**, 57–240, doi:10.7916/cjel.v45i1.4730.

Burgers, L., 2020: Should Judges Make Climate Change Law? *Transnatl. Environ. Law*, **9(1)**, 55–75, doi:10.1017/S2047102519000360.

Burgers, L. and T. Staal, 2019: Climate Action as Positive Human Rights Obligation: The Appeals Judgment in Urgenda v the Netherlands. In: *Netherlands Yearbook of International Law 2018* [Nijman, J. and W. Werner (eds.)]. Asser Press, The Hague, Netherlands, pp. 223–244.

Burke, P.J., 2016: Undermined by Adverse Selection: Australia's Direct Action Abatement Subsidies. *Econ. Pap. a J. Appl. Econ. policy*, **35(3)**, 216–229, doi:10.1111/1759-3441.12138.

Burke, P.J. and S. Kurniawati, 2018: Electricity subsidy reform in Indonesia: Demand-side effects on electricity use. *Energy Policy*, **116**, 410–421, doi:10.1016/j.enpol.2018.02.018.

Burniaux, J.-M. and J. Chateau, 2014: Greenhouse gases mitigation potential and economic efficiency of phasing-out fossil fuel subsidies. *Int. Econ.*, **140**, 71–88, doi:10.1016/j.inteco.2014.05.002.

Busch, H., L. Bendlin, and P. Fenton, 2018: Shaping local response – The influence of transnational municipal climate networks on urban climate governance. *Urban Clim.*, **24**, 221–230, doi:10.1016/j.uclim.2018.03.004.

Busch, J., F. Godoy, W.R. Turner, and C.A. Harvey, 2011: Biodiversity co-benefits of reducing emissions from deforestation under alternative reference levels and levels of finance. *Conserv. Lett.*, **4(2)**, 101–115, doi:10.1111/j.1755-263X.2010.00150.x.

Bush, J. and A. Doyon, 2019: Building urban resilience with nature-based solutions: How can urban planning contribute? *Cities*, **95**(October), 102483, doi:10.1016/j.cities.2019.102483.

Bush, R.E., C.S.E. Bale, and P.G. Taylor, 2016: Realising local government visions for developing district heating: Experiences from a learning country. *Energy Policy*, **98**, 84–96, doi:10.1016/j.enpol.2016.08.013.

Bushley, B.R., 2014: REDD+ policy making in Nepal: toward state-centric, polycentric, or market-oriented governance? *Ecol. Soc.*, **19(3)**, 34, doi:10.5751/ES-06853-190334.

Bustamante, M.M.C. et al., 2019: Ecological restoration as a strategy for mitigating and adapting to climate change: lessons and challenges from Brazil. *Mitig. Adapt. Strateg. Glob. Change*, **24(7)**, 1249–1270, doi:10.1007/s11027-018-9837-5.

Butler-Sloss, S., K. Bond, and H. Benham, 2021: *Spiralling Disruption: The feedback loops of the energy transition*. Carbon Tracker, 23 pp. https://carbontracker.org/reports/spiralling-disruption/ (Accessed October 28, 2021).

C40 and ARUP, 2015: *Climate Action in Megacities 3.0: Networking works, there is no global solution without local action*. C40 and ARUP, London, 128 pp. https://www.arup.com/perspectives/publications/research/section/climate-action-in-megacities-cam-30 (Accessed March 3, 2021).

Caetano, T., H. Winker, and J. Depledge, 2020: Towards zero carbon and zero poverty: integrating national climate change mitigation and sustainable development goals. *Clim. Policy*, **20(7)**, 773–778, doi:10.1080/14693062.2020.1791404.

Calel, R., 2020: Adopt or Innovate: Understanding Technological Responses to Cap-and-Trade. *Am. Econ. J. Econ. Policy*, **12(3)**, doi:10.1257/pol.20180135.

Calel, R. and A. Dechezleprêtre, 2016: Environmental Policy and Directed Technological Change: Evidence from the European Carbon Market. *Rev. Econ. Stat.*, **98(1)**, 173–191, doi:10.1162/REST_a_00470.

Calzadilla, P.V., 2019: Climate Change Litigation: a Powerful Strategy for Enhancing Climate Change Communication. In: *Addressing the Challenges in Communicating Climate Change Across Various Audiences. Climate Change Management. Climate Change Management* [Leal Filho, W., B. Lackner and H. McGhie (eds.)]. Springer, Cham, Switzerland, pp. 231–246.

Camilleri, A.R., R.P. Larrick, S. Hossain, and D. Patino-Echeverri, 2019: Consumers underestimate the emissions associated with food but are aided by labels. *Nat. Clim. Change*, **9(1)**, 53–58, doi:10.1038/s41558-018-0354-z.

Campagnolo, L. and M. Davide, 2019: Can the Paris deal boost SDGs achievement? An assessment of climate mitigation co-benefits or side-effects on poverty and inequality. *World Dev.*, **122**, 96–109, doi:10.1016/j.worlddev.2019.05.015.

Candel, J.J.L. and R. Biesbroek, 2016: Toward a processual understanding of policy integration. *Policy Sci.*, **49(3)**, 211–231, doi:10.1007/s11077-016-9248-y.

Carattini, S., S. Kallbekken, and A. Orlov, 2019: How to win public support for a global carbon tax. *Nature*, **565**, 289–291, doi:10.1038/d41586-019-00124-x.

Carbone, J.C. and N. Rivers, 2017: The Impacts of Unilateral Climate Policy on Competitiveness: Evidence From Computable General Equilibrium Models. *Rev. Environ. Econ. Policy*, **11(1)**, 24–42, doi:10.1093/reep/rew025.

Carley, S. and D.M. Konisky, 2020: The justice and equity implications of the clean energy transition. *Nat. Energy*, **5(8)**, 569–577, doi:10.1038/s41560-020-0641-6.

Carley, S., L.L. Davies, D.B. Spence, and N. Zirogiannis, 2018: Empirical evaluation of the stringency and design of renewable portfolio standards. *Nat. Energy*, **3(9)**, 754–763, doi:10.1038/s41560-018-0202-4.

Carmichael, J.T. and R.J. Brulle, 2018: Media use and climate change concern. *Int. J. Media Cult. Polit.*, **14(2)**, 243–253, doi:10.1386/macp.14.2.243_7.

Carmichael, J.T., R.J. Brulle, and J.K. Huxster, 2017: The great divide: understanding the role of media and other drivers of the partisan divide in public concern over climate change in the USA, 2001–2014. *Clim. Change*, **141(4)**, 599–612, doi:10.1007/s10584-017-1908-1.

Carrington, D., 2019: School climate strikes: 1.4 million people took part, say campaigners. The Guardian, London, UK, March 19. https://www.theguardian.com/environment/2019/mar/19/school-climate-strikes-more-than-1-million-took-part-say-campaigners-greta-thunberg (Accessed March 30, 2021).

Carroll, J., C. Aravena, M. Boeri, and E. Denny, 2022: "Show Me the Energy Costs": Short and Long-term Energy Cost Disclosure Effects on Willingness-to-pay for Residential Energy Efficiency. *Energy J.*, **43(3)**, doi:10.5547/01956574.43.3.jcar.

Carter, N., 2006: Party Politicization Of The Environment In Britain. *Party Polit.*, **12(6)**, 747–767, doi:10.1177/1354068806068599.

Carter, N., 2013: Greening the mainstream: party politics and the environment. *Env. Polit.*, **22(1)**, 73–94, doi:10.1080/09644016.2013.755391.

Carter, N., 2014: The politics of climate change in the UK. *Wiley Interdiscip. Rev. Clim. Change*, **5(3)**, 423–433, doi:10.1002/wcc.274.

Carter, N. and M. Childs, 2018: Friends of the Earth as a policy entrepreneur: 'The Big Ask' campaign for a UK Climate Change Act. *Env. Polit.*, **27(6)**, 994–1013, doi:10.1080/09644016.2017.1368151.

Carter, S. et al., 2018: Climate-smart land use requires local solutions, transdisciplinary research, policy coherence and transparency. *Carbon Manag.*, **9(3)**, 291–301, doi:10.1080/17583004.2018.1457907.

Carter, A.V. and J. McKenzie, 2020: Amplifying "Keep It in the Ground" First-Movers: Toward a Comparative Framework. *Soc. Nat. Resour.*, **33(11)**, 1339–1358, doi:10.1080/08941920.2020.1772924.

Cashore, B., S. Bernstein, D. Humphreys, I. Visseren-Hamakers, and K. Rietig, 2019: Designing stakeholder learning dialogues for effective global governance. *Policy Soc.*, **38(1)**, 118–147, doi:10.1080/14494035.2019.1579505.

Castán Broto, V., 2017: Urban Governance and the Politics of Climate change. *World Dev.*, **93**, 1–15, doi:10.1016/j.worlddev.2016.12.031.

Castán Broto, V. and H. Bulkeley, 2013: a survey of urban climate change experiments in 100 cities. *Glob. Environ. Change*, **23(1)**, 92–102, doi:10.1016/j.gloenvcha.2012.07.005.

Castán Broto, V. and L. Westman, 2017: Just sustainabilities and local action: evidence from 400 flagship initiatives. *Local Environ.*, **22(5)**, 635–650, doi:10.1080/13549839.2016.1248379.

Castán Broto, V. and L. Westman, 2019: *Urban Sustainability and Justice: Just Sustainabilities and Environmental Planning*. Zed Books Ltd, Bloomsbury Publishing, London, UK, 208 pp.

Castrejon-Campos, O., L. Aye, and F.K.P. Hui, 2020: Making policy mixes more robust: An integrative and interdisciplinary approach for clean energy transitions. *Energy Res. Soc. Sci.*, **64**, 101425, doi:10.1016/j.erss.2020.101425.

Ćetković, S. and J.B. Skjærseth, 2019: Creative and disruptive elements in Norway's climate policy mix: the small-state perspective. *Env. Polit.*, **28(6)**, 1039–1060, doi:10.1080/09644016.2019.1625145.

Chailleux, S., 2020: Making the subsurface political: How enhanced oil recovery techniques reshaped the energy transition. *Environ. Plan. C Polit. Sp.*, **38(4)**, 733–750, doi:10.1177/2399654419884077.

Chaker, M., E. Berezowska-Azzag, and D. Perrotti, 2021: Exploring the performances of urban local symbiosis strategy in Algiers, between a potential of energy use optimization and $CO_2$ emissions mitigation. *J. Clean. Prod.*, **292**, 125850, doi:10.1016/j.jclepro.2021.125850.

Chamorel, P., 2019: Macron Versus the Yellow Vests. *J. Democr.*, **30(4)**, 48–62, doi:10.1353/jod.2019.0068.


13

Chan, G., A.P. Goldstein, A. Bin-Nun, L. Diaz Anadon, and V. Narayanamurti, 2017: Six principles for energy innovation. *Nature*, **552(7683)**, 25–27, doi:10.1038/d41586-017-07761-0.

Chan, N.W. and J.W. Morrow, 2019: Unintended consequences of cap-and-trade? Evidence from the Regional Greenhouse Gas Initiative. *Energy Econ.*, **80**(May 2019), 411–422, doi:10.1016/j.eneco.2019.01.007.

Chan, S. et al., 2015: Reinvigorating International Climate Policy: A Comprehensive Framework for Effective Nonstate Action. *Glob. Policy*, **6(4)**, 466–473, doi:10.1111/1758-5899.12294.

Chandrashekeran, S., B. Morgan, K. Coetzee, and P. Christoff, 2017: Rethinking the green state beyond the Global North: a South African climate change case study. *Wiley Interdiscip. Rev. Clim. Change*, **8(6)**, e473, doi:10.1002/wcc.473.

Charlery, L. and S.L.M. Trærup, 2019: The nexus between Nationally Determined Contributions and technology needs assessments: a global analysis. *Clim. Policy*, **19(2)**, 189–205, doi:10.1080/14693062.2018.1479957.

Chaton, C., A. Creti, and M.-E. Sanin, 2018: Assessing the implementation of the Market Stability Reserve. *Energy Policy*, **118**, 642–654, doi:10.1016/j.enpol.2018.03.027.

Chelminski, K., 2018: Fossil Fuel Subsidy Reform in Indonesia. In: *The Politics of Fossil Fuel Subsidies and their Reform* [Skovgaard, J. and H. Van Asselt, (eds.)]. Cambridge University Press, Cambridge, UK, pp. 193–211.

Chen, X., H. Huang, M. Khanna, and H. Önal, 2014: Alternative transportation fuel standards: Welfare effects and climate benefits. *J. Environ. Econ. Manage.*, **67(3)**, 241–257, doi:10.1016/j.jeem.2013.09.006.

Cherry, E., 2006: Veganism as a Cultural Movement: a Relational Approach. *Soc. Mov. Stud.*, **5(2)**, 155–170, doi:10.1080/14742830600807543.

Cherry, T.L., S. Kallbekken, and S. Kroll, 2012: The acceptability of efficiency-enhancing environmental taxes, subsidies and regulation: An experimental investigation. *Environ. Sci. Policy*, **16**, 90–96, doi:10.1016/j.envsci.2011.11.007.

Chèze, B., J. Chevallier, N. Berghmans, and E. Alberola, 2020: On the $CO_2$ Emissions Determinants During the EU ETS Phases I and II: a Plant-level Analysis Merging the EUTL and Platts Power Data. *Energy J.*, **41(4)**, doi:10.5547/01956574.41.4.bche.

Choi, C., P. Berry, and A. Smith, 2021: The climate benefits, co-benefits, and trade-offs of green infrastructure: a systematic literature review. *J. Environ. Manage.*, **291**(March), 112583, doi:10.1016/j.jenvman.2021.112583.

Claeys, P. and D. Delgado Pugley, 2017: Peasant and indigenous transnational social movements engaging with climate justice. *Can. J. Dev. Stud./Rev. Can. d'études du développement*, **38(3)**, 325–340, doi:10.1080/02255189.2016.1235018.

Climate Action Tracker, 2020: *Warming Projections Global Update, September 2020. Pandemic recovery: Positive intentions vs policy rollbacks, with just a hint of green*. NewClimate Institute, Cologne and Climate Analytics, Berlin, Germany, 26 pp. https://climateactiontracker.org/documents/790/CAT_2020-09-23_Briefing_GlobalUpdate_Sept2020.pdf (Accessed March 3, 2021).

Cludius, J., S. de Bruyn, K. Schumacher, and R. Vergeer, 2020: Ex-post investigation of cost pass-through in the EU ETS - an analysis for six industry sectors. *Energy Econ.*, **91**, 104883, doi:10.1016/j.eneco.2020.104883.

Clulow, Z., 2019: Democracy, electoral systems and emissions: explaining when and why democratization promotes mitigation. *Clim. Policy*, **19(2)**, 244–257, doi:10.1080/14693062.2018.1497938.

Coady, D., V. Flamini, and L. Sears, 2015: *The unequal benefits of fuel subsidies revisited: Evidence for developing countries*. Fiscal Affairs Department, IMF, 25 pp. https://www.imf.org/external/pubs/ft/wp/2015/wp15250.pdf (Accessed March 30, 2021).

Cobbinah, P.B., M.O. Asibey, M. Opoku-Gyamfi, and C. Peprah, 2019: Urban planning and climate change in Ghana. *J. Urban Manag.*, **8(2)**, 261–271, doi:10.1016/j.jum.2019.02.002.

Cock, J., 2019: Resistance to coal inequalities and the possibilities of a just transition in South Africa. *Dev. South. Afr.*, **36(6)**, 860–873, doi:10.1080/0376835X.2019.1660859.

Cohen, B. et al., 2019: Multi-criteria decision analysis in policymaking for climate mitigation and development. *Clim. Dev.*, **11(3)**, 212–222, doi:10.1080/17565529.2018.1445612.

Cointe, B., 2019: Mutualising sunshine: economic and territorial entanglements in a local photovoltaic project. *Local Environ.*, **24(11)**, 980–996, doi:10.1080/13549839.2018.1436044.

Cole, M.A., 2007: Corruption, income and the environment: An empirical analysis. *Ecol. Econ.*, **62(3–4)**, 637–647, doi:10.1016/j.ecolecon.2006.08.003.

Colgan, J.D., J.F. Green, and T.N. Hale, 2021: Asset Revaluation and the Existential Politics of Climate Change. *Int. Organ.*, **75(2)**, 586–610, doi:10.1017/S0020818320000296.

Colls, A., N. Ash, and N. Ikkala, 2009: *Ecosystem-based Adaptation: A natural response to climate change*. IUCN, Gland, Switzerland, 16 pp. https://portals.iucn.org/library/sites/library/files/documents/2009-049.pdf (Accessed March 3, 2021).

Colombo, E., 2021: From Bushfires to Misfires: Climate-related Financial Risk after McVeigh v. Retail Employees Superannuation Trust. *Transnatl. Environ. Law*, **11(1)**, 173–199, doi:10.1017/S204710252100025X.

Compston, H. and I. Bailey, 2016: Climate policy strength compared: China, the US, the EU, India, Russia, and Japan. *Clim. Policy*, **16(2)**, 145–164, doi:10.1080/14693062.2014.991908.

Conger, T. and S.E. Chang, 2019: Developing indicators to identify coastal green infrastructure potential: The case of the Salish Sea region. *Ocean Coast. Manag.*, **175**(April), 53–69, doi:10.1016/j.ocecoaman.2019.03.011.

Cook, M.J. and E.K. Chu, 2018: Between policies, programs, and projects: How local actors steer domestic urban climate adaptation finance in India. In: *Climate Change in Cities. The Urban Book Series* [Hughes, S., E.K. Chu, and S.G. Mason, (eds.)]. Springer, Cham, Switzerland, pp. 255–277.

Cooley, S.R. et al., 2019: Overlooked ocean strategies to address climate change. *Glob. Environ. Change*, **59** (November), 101968, doi:10.1016/j.gloenvcha.2019.101968.

Cornelis, E., 2019: History and prospect of voluntary agreements on industrial energy efficiency in Europe. *Energy Policy*, **132**, 567–582, doi:10.1016/j.enpol.2019.06.003.

Corner, A. and J. Clarke, 2017: *Talking Climate: From Research to Practice in Public Engagement*. Springer International Publishing, Cham, Switzerland, pp. 146.

Corry, O., 2012: Securitisation and 'Riskification': Second-order Security and the Politics of Climate Change. *Millenn. J. Int. Stud.*, **40(2)**, 235–258, doi:10.1177/0305829811419444.

Cory, J., M. Lerner, and I. Osgood, 2021: Supply Chain Linkages and the Extended Carbon Coalition. *Am. J. Pol. Sci.*, **65(1)**, 69–87, doi:10.1111/ajps.12525.

Coryat, D., 2015: Extractive Politics, Media Power, and New Waves of Resistance Against Oil Drilling in the Ecuadorian Amazon: The Case of Yasunidos. *Int. J. Commun.*, **9**, 3741–3760.

Cosbey, A., S. Droege, C. Fischer, and C. Munnings, 2019: Developing Guidance for Implementing Border Carbon Adjustments: Lessons, Cautions, and Research Needs from the Literature. *Rev. Environ. Econ. Policy*, **13(1)**, 3–22, doi:10.1093/reep/rey020.

Costantini, V., F. Crespi, and A. Palma, 2017: Characterizing the policy mix and its impact on eco-innovation: a patent analysis of energy-efficient technologies. *Res. Policy*, **46(4)**, 799–819, doi:10.1016/j.respol.2017.02.004.

Cox, R., 2016: a climate change litigation precedent: Urgenda Foundation v The State of the Netherlands. *J. Energy Nat. Resour. Law*, **34(2)**, doi:10.1080/02646811.2016.1147887.

Coxhead, I. and C. Grainger, 2018: Fossil Fuel Subsidy Reform in the Developing World: Who Wins, Who Loses, and Why? *Asian Dev. Rev.*, **35(2)**, 180–203, doi:10.1162/adev_a_00119.

Creasy, A., M. Lane, A. Owen, C. Howarth, and D. Van der Horst, 2021: Representing 'Place': City Climate Commissions and the Institutionalisation of Experimental Governance in Edinburgh. *Polit. Gov.*, **9(2)**, 64–75, doi:10.17645/pag.v9i2.3794.

Creutzig, F., 2016: Evolving Narratives of Low-Carbon Futures in Transportation. *Transp. Rev.*, **36(3)**, 341–360, doi:10.1080/01441647.2015.1079277.

Creutzig, F., G. Baiocchi, R. Bierkandt, P.-P. Pichler, and K.C. Seto, 2015: Global typology of urban energy use and potentials for an urbanisation mitigation wedge. *Proc. Natl. Acad. Sci.*, **112(20)**, 6283–6288, doi:10.1073/pnas.1315545112.

Creutzig, F. et al., 2018: Towards demand-side solutions for mitigating climate change. *Nat. Clim. Change*, **8(4)**, 268–271, doi:10.1038/s41558-018-0121-1.

Criqui, P., M. Jaccard, and T. Sterner, 2019: Carbon Taxation: a Tale of Three Countries. *Sustainability*, **11(22)**, 6280, doi:10.3390/su11226280.

Cronin, J.M., M.B. McCarthy, and A.M. Collins, 2014: Covert distinction: how hipsters practice food-based resistance strategies in the production of identity. *Consum. Mark. Cult.*, **17(1)**, 2–28, doi:10.1080/10253866.2012.678785.

Crowley, K., 2017: Up and down with climate politics 2013-2016: the repeal of carbon pricing in Australia. *Wiley Interdiscip. Rev. Clim. Change*, **8(3)**, e458, doi:10.1002/wcc.458.

Culwick, C. et al., 2019: CityLab reflections and evolutions: nurturing knowledge and learning for urban sustainability through co-production experimentation. *Curr. Opin. Environ. Sustain.*, **39**, 9–16, doi:10.1016/j.cosust.2019.05.008.

Dagnachew, A.G., P.L. Lucas, A.F. Hof, and D.P. van Vuuren, 2018: Trade-offs and synergies between universal electricity access and climate change mitigation in Sub-Saharan Africa. *Energy Policy*, **114**, 355–366, doi:10.1016/j.enpol.2017.12.023.

Dal Maso, M., K.H. Olsen, Y. Dong, M.B. Pedersen, and M.Z. Hauschild, 2020: Sustainable development impacts of nationally determined contributions: assessing the case of mini-grids in Kenya. *Clim. Policy*, **20(7)**, 815–831, doi:10.1080/14693062.2019.1644987.

Dale, A. et al., 2020: Meeting the climate change challenge: local government climate action in British Columbia, Canada. *Clim. Policy*, **20(7)**, 866–880, doi:10.1080/14693062.2019.1651244.

Damigos, D., A. Kontogianni, C. Tourkolias, and M. Skourtos, 2020: Behind the scenes: Why are energy efficient home appliances such a hard sell? *Resour. Conserv. Recycl.*, **158**, 104761, doi:10.1016/j.resconrec.2020.104761.

Damsø, T., T. Kjær, and T.B. Christensen, 2016: Local climate action plans in climate change mitigation – examining the case of Denmark. *Energy Policy*, **89**, 74–83, doi:10.1016/j.enpol.2015.11.013.

Dang, H.H., A. Michaelowa, and D.D. Tuan, 2003: Synergy of adaptation and mitigation strategies in the context of sustainable development: the case of Vietnam. *Clim. Policy*, **3**, S81–S96, doi:10.1016/j.clipol.2003.10.006.

David, D. and A. Venkatachalam, 2019: a Comparative Study on the Role of Public–Private Partnerships and Green Investment Banks in Boosting Low-Carbon Investments. In: *Handbook of Green Finance: Energy Security and Sustainable Development* [Sachs, J.D., W.T. Woo, N. Yoshino, and F. Taghizadeh-Hesary (eds.)]. Springer Singapore, Singapore, pp. 261–287.

David, M., 2017: Moving beyond the heuristic of creative destruction: Targeting exnovation with policy mixes for energy transitions. *Energy Res. Soc. Sci.*, **33**, 138–146, doi:10.1016/j.erss.2017.09.023.

Davidovic, D. and N. Harring, 2020: Exploring the cross-national variation in public support for climate policies in Europe: The role of quality of government and trust. *Energy Res. Soc. Sci.*, **70**, 101785, doi:10.1016/j.erss.2020.101785.

Davis, L.W. and G.E. Metcalf, 2016: Does better information lead to better choices? Evidence from energy-efficiency labels. *J. Assoc. Environ. Resour. Econ.*, **3(3)**, 589–625, doi:10.1086/686252.

Davis, L.W. and C.R. Knittel, 2019: Are Fuel Economy Standards Regressive? *J. Assoc. Environ. Resour. Econ.*, **6(S1)**, 7–36, doi:10.1086/701187.

Davis, S.J. et al., 2018: Net-zero emissions energy systems. *Science*, **360(6396)**, eaas9793, doi:10.1126/science.aas9793.

Dawson, N.L. and K. Segerson, 2008: Voluntary Agreements with Industries: Participation Incentives with Industry-Wide Targets. *Land Econ.*, **84(1)**, 97–114, doi:10.3368/le.84.1.97.

de Coninck, H., A. Revi, M. Babiker, P. Bertoldi, M. Buckeridge, A. Cartwright, W. Dong, J. Ford, S. Fuss, J.-C. Hourcade, D. Ley, R. Mechler, P. Newman, A. Revokatova, S. Schultz, L. Steg, and T. Sugiyama, 2018: Strengthening and Implementing the Global Response. In: *Global Warming of 1.5°C. An IPCC Special Report on the impacts of global warming of 1.5°C above pre-industrial levels and related global greenhouse gas emission pathways, in the context of strengthening the global response to the threat of climate change, sustainable development, and efforts to eradicate poverty* [Masson-Delmotte, V., P. Zhai, H.-O. Pörtner, D. Roberts, J. Skea, P.R. Shukla, A. Pirani, W. Moufouma-Okia, C. Péan, R. Pidcock, S. Connors, J.B.R. Matthews, Y. Chen, X. Zhou, M.I. Gomis, E. Lonnoy, T. Maycock, M. Tignor, and T. Waterfield (eds.)]. Cambridge University Press, Cambridge, UK and New York, NY, USA, pp. 313–444.

de Graaf, K.J. and J.H. Jans, 2015: The Urgenda Decision: Netherlands Liable for Role in Causing Dangerous Global Climate Change. *J. Environ. Law*, **27(3)**, 517–527, doi:10.1093/jel/eqv030.

De la Sota, C., V.J. Ruffato-Ferreira, L. Ruiz-García, and S. Alvarez, 2019: Urban green infrastructure as a strategy of climate change mitigation. a case study in northern Spain. *Urban For. Urban Green.*, **40**, 145–151, doi:10.1016/j.ufug.2018.09.004.

de Moor, J., M. De Vydt, K. Uba, and M. Wahlström, 2021: New kids on the block: taking stock of the recent cycle of climate activism. *Soc. Mov. Stud.*, **20(5)**, 619–625, doi:10.1080/14742837.2020.1836617.

De Oliveira Silva, R., L.G. Barioni, G. Queiroz Pellegrino, and D. Moran, 2018: The role of agricultural intensification in Brazil's Nationally Determined Contribution on emissions mitigation. *Agric. Syst.*, **161**(January), 102–112, doi:10.1016/j.agsy.2018.01.003.

DEA, 2015: *The National Climate Change Response Monitoring and Evaluation System Framework*. Department of Environmental Affairs (DEA) South Africa, and Climate Action Now, Pretoria, South Africa, 97 pp. https://cer.org.za/wp-content/uploads/2019/07/nationalclimatechange response_MESF.pdf (Accessed October 25, 2021).

DEA, 2019: *South Africa's 3rd Biennial Update Report to the United Nations Framework Convention on Climate Change*. Department of Environmental Affairs (DEA), Pretoria, South Africa, 219 pp. https://unfccc.int/sites/default/files/resource/Final 3rd BUR of South Africa 100.pdf.

Dean, B. et al., 2020: *Chilling prospects: Tracking Sustainable Cooling for All 2020*. Sustainable Energy for All (SEforALL), Vienna, Austria, 59 pp. https://www.seforall.org/system/files/2020-07/CP-2020-SEforALL.pdf.

Dejuán, Ó., F. Portella-Carbó, and M. Ortiz, 2020: Economic and environmental impacts of decarbonisation through a hybrid MRIO multiplier-accelerator model. *Econ. Syst. Res.*, **34 (1)**, 1–21, doi:10.1080/09535314.2020.1848808.

del Río, P., 2014: On evaluating success in complex policy mixes: the casex of renewable energy support schemes. *Policy Sci.*, **47(3)**, 267–287, doi:10.1007/s11077-013-9189-7.

del Río, P. and P. Linares, 2014: Back to the future? Rethinking auctions for renewable electricity support. *Renew. Sustain. Energy Rev.*, **35**, doi:10.1016/j.rser.2014.03.039.

del Río, P. and E. Cerdá, 2017: The missing link: The influence of instruments and design features on the interactions between climate and renewable electricity policies. *Energy Res. Soc. Sci.*, **33**, 49–58, doi:10.1016/j.erss.2017.09.010.

Delpiazzo, E., R. Parrado, and G. Standardi, 2015: *Phase-out of fossil fuel subsidies: implications for emissions, GDP and public budget.* Centro Euro-Mediterraneo sui Cambiamenti Climatici (CMCC), Lecce, Italy, 32 pp. https://www.cmcc.it/wp-content/uploads/2016/06/rp0275-ecip-12-2015.pdf.

Deng, H.-M., Q.-M. Liang, L.-J. Liu, and L.D. Anadon, 2017: Co-benefits of greenhouse gas mitigation: a review and classification by type, mitigation sector, and geography. *Environ. Res. Lett.*, **12(12)**, 123001, doi:10.1088/1748-9326/aa98d2.


13

Deng, Z., D. Li, T. Pang, and M. Duan, 2018: Effectiveness of pilot carbon emissions trading systems in China. *Clim. Policy*, **18(8)**, 992–1011, doi:10.1080/14693062.2018.1438245.

Dennis, A., 2016: Household welfare implications of fossil fuel subsidy reforms in developing countries. *Energy Policy*, **96**, 597–606, doi:10.1016/j.enpol.2016.06.039.

Denton, F., T.J.Wilbanks, A.C. Abeysinghe, I. Burton, Q. Gao, M.C. Lemos, T. Masui, K.L. O'Brien, and K.Warner, 2015: Climate-resilient Pathways: Adaptation, mitigation, and sustainable development. In: *Climate Change 2014: Impacts, Adaptation, and Vulnerability. Part A: Global and Sectoral Aspects. Contribution of Working Group II to the Fifth Assessment Report of the Intergovernmental Panel on Climate Change* [Field, C.B., V.R. Barros, D.J. Dokken, K.J. Mach, M.D. Mastrandrea, T.E. Bilir, M. Chatterjee, K.L. Ebi, Y.O. Estrada, R.C. Genova, B. Girma, E.S. Kissel, A.N. Levy, S. MacCracken, P.R. Mastrandrea, and L.L.White (eds.)]. Cambridge University Press, Cambridge, United Kingdom and New York, NY, USA, pp. 1101–1131.

Devaney, L., D. Torney, P. Brereton, and M. Coleman, 2020: Ireland's Citizens' Assembly on Climate Change: Lessons for Deliberative Public Engagement and Communication. *Environ. Commun.*, **14(2)**, 141–146, doi:10.1080/17524032.2019.1708429.

DFFE Republic of South Africa, 2021: *South Africa's 4th Biennial Update Report to the United Nations Framework Convention on Climate Change*. UNFCCC, 255 pp. https://unfccc.int/sites/default/files/resource/South%20Africa%20BUR4%20to%20the%20UNFCCC.pdf.

Di Gregorio, M. et al., 2016: *Integrating mitigation and adaptation in climate and land-use policies in Brazil: a policy document analysis*. University of Leeds and CIFOR, Leeds, UK and Bogor, Indonesia, 55 pp. http://eprints.whiterose.ac.uk/96279/ (Accessed July 12, 2019).

Di Gregorio, M. et al., 2017: Climate policy integration in the land-use sector: Mitigation, adaptation and sustainable development linkages. *Environ. Sci. Policy*, **67**, 35–43, doi:10.1016/j.envsci.2016.11.004.

Di Gregorio, M. et al., 2019: Multi-level governance and power in climate change policy networks. *Glob. Environ. Change*, **54**, 64–77, doi:10.1016/j.gloenvcha.2018.10.003.

Di Gregorio, M., K. Massarella, H. Schroeder, M. Brockhaus, and T.T. Pham, 2020: Building authority and legitimacy in transnational climate change governance: Evidence from the Governors' Climate and Forests Task Force. *Glob. Environ. Change*, **64**, (September), 102126, doi:10.1016/j.gloenvcha.2020.102126.

Diaz, G., F.D. Munoz, and R. Moreno, 2020: Equilibrium Analysis of a Tax on Carbon Emissions with Pass-through Restrictions and Side-payment Rules. *Energy J.*, **41(2)**, doi:10.5547/01956574.41.2.gdia.

Dickson, B. and V. Kapos, 2012: Biodiversity monitoring for REDD+. *Curr. Opin. Environ. Sustain.*, **4(6)**, 717–725, doi:10.1016/j.cosust.2012.09.017.

Diehl, T., B. Huber, H. Gil de Zúñiga, and J. Liu, 2021: Social Media and Beliefs about Climate Change: a Cross-National Analysis of News Use, Political Ideology, and Trust in Science. *Int. J. Public Opin. Res.*, **33(2)**, 197–213, doi:10.1093/ijpor/edz040.

Dietz, T., K.A. Frank, C.T. Whitley, J. Kelly, and R. Kelly, 2015: Political influences on greenhouse gas emissions from US states. *Proc. Natl. Acad. Sci.*, **112(27)**, 8254–8259, doi:10.1073/pnas.1417806112.

Dillon, L. et al., 2018: The Environmental Protection Agency in the Early Trump Administration: Prelude to Regulatory Capture. *Am. J. Public Health*, **108(S2)**, S89–S94, doi:10.2105/AJPH.2018.304360.

Dobrotkova, Z., K. Surana, and P. Audinet, 2018: The price of solar energy: Comparing competitive auctions for utility-scale solar PV in developing countries. *Energy Policy*, **118**, 133–148, doi:10.1016/j.enpol.2018.03.036.

Doll, C.N.H. and J.A.P. de Oliveira (eds.), 2017: *Urbanisation and climate co-benefits: Implementation of win-win interventions in cities*. Routledge, London, UK, 348 pp.

Domorenok, E., P. Graziano, and L. Polverari, 2021: Introduction: policy integration and institutional capacity: theoretical, conceptual and empirical challenges. *Policy Soc.*, **40(1)**, 1–18, doi:10.1080/14494035.2021.1902058.

Dorband, I.I., M. Jakob, M. Kalkuhl, and J.C. Steckel, 2019: Poverty and distributional effects of carbon pricing in low- and middle-income countries – a global comparative analysis. *World Dev.*, **115**, doi:10.1016/j.worlddev.2018.11.015.

Dorninger, C. et al., 2020: Leverage points for sustainability transformation: a review on interventions in food and energy systems. *Ecol. Econ.*, **171**, 106570, doi:10.1016/j.ecolecon.2019.106570.

Dorsch, M.J. and C. Flachsland, 2017: a Polycentric Approach to Global Climate Governance. *Glob. Environ. Polit.*, **17(2)**, 45–64, doi:10.1162/GLEP_a_00400.

Dorsch, M.J., C. Flachsland, and U. Kornek, 2020: Building and enhancing climate policy ambition with transfers: allowance allocation and revenue spending in the EU ETS. *Env. Polit.*, **29(5)**, 781–803, doi:10.1080/09644016.2019.1659576.

Doswald, N. and M. Osti, 2011: *Ecosystem-based approaches to adaptation and mitigation – good practice examples and lessons learned in Europe*. Bundesamt für Naturschutz (BfN), Bonn, Germany, 49 pp. https://bfn.bsz-bw.de/frontdoor/deliver/index/docId/491/file/Skript306.pdf (Accessed October 17, 2021).

Doswald, N. et al., 2014: Effectiveness of ecosystem-based approaches for adaptation: review of the evidence-base. *Clim. Dev.*, **6(2)**, 185–201, doi:10.1080/17565529.2013.867247.

Dotson, D.M., S.K. Jacobson, L.L. Kaid, and J.S. Carlton, 2012: Media Coverage of Climate Change in Chile: a Content Analysis of Conservative and Liberal Newspapers. *Environ. Commun.*, **6(1)**, 64–81, doi:10.1080/17524032.2011.642078.

Downie, C., 2018: Ad hoc coalitions in the U.S. energy sector: Case studies in the gas, oil, and coal industries. *Bus. Polit.*, **20(4)**, 643–668, doi:10.1017/bap.2018.18.

Doyle, J., 2016: *Mediating Climate Change*. Routledge, New York, USA 195 pp.

Drews, S. and J.C.J.M. van den Bergh, 2016: What explains public support for climate policies? a review of empirical and experimental studies. *Clim. Policy*, **16(7)**, 855–876, doi:10.1080/14693062.2015.1058240.

Duan, M., Z. Tian, Y. Zhao, and M. Li, 2017: Interactions and coordination between carbon emissions trading and other direct carbon mitigation policies in China. *Energy Res. Soc. Sci.*, **33**, 59–69, doi:10.1016/j.erss.2017.09.008.

Duarte, C.M., 2017: Reviews and syntheses: Hidden forests, the role of vegetated coastal habitats in the ocean carbon budget. *Biogeosciences*, **14(2)**, 301–310, doi:10.5194/bg-14-301-2017.

Duarte, C.M., I.J. Losada, I.E. Hendriks, I. Mazarrasa, and N. Marbà, 2013: The role of coastal plant communities for climate change mitigation and adaptation. *Nat. Clim. Change*, **3(11)**, 961–968, doi:10.1038/nclimate1970.

Dubash, N.K., 2021: Varieties of climate governance: the emergence and functioning of climate institutions. *Env. Polit.*, **30(sup1)**, 1–25, doi:10.1080/09644016.2021.1979775.

Dubash, N.K. and N.B. Joseph, 2016: Evolution of Institutions for Climate Policy in India. *Econ. Polit. Wkly.*, **51(3)**, 44–54.

Dubash, N.K., M. Hagemann, N. Höhne, and P. Upadhyaya, 2013: Developments in national climate change mitigation legislation and strategy. *Clim. Policy*, **13(6)**, 649–664, doi:10.1080/14693062.2013.845409.

Dubash, N.K., R. Khosla, U. Kelkar, and S. Lele, 2018: India and Climate Change: Evolving Ideas and Increasing Policy Engagement. *Annu. Rev. Environ. Resour.*, **43(1)**, 395–424, doi:10.1146/annurev-environ-102017-025809.

Dubash, N.K. et al., 2021: National climate institutions complement targets and policies: Institutions can affect coordination, consensus, and strategy. *Science* (in press).

Duguma, L.A., P.A. Minang, and M. Van Noordwijk, 2014a: Climate change mitigation and adaptation in the land use sector: From complementarity to synergy. *Environ. Manage.*, **5**, 420–432, doi:10.1007/s00267-014-0331-x.

Duguma, L.A., S.W. Wambugu, P.A. Minang, and M. van Noordwijk, 2014b: a systematic analysis of enabling conditions for synergy between climate change mitigation and adaptation measures in developing countries. *Environ. Sci. Policy*, **42**, 138–148, doi:10.1016/j.envsci.2014.06.003.

Duit, A. and V. Galaz, 2008: Governance and Complexity – Emerging Issues for Governance Theory. *Governance*, **21(3)**, 311–335, doi:10.1111/j.1468-0491.2008.00402.x.

Dundas, S.J. et al., 2020: Integrating oceans into climate policy: Any green new deal needs a splash of blue. *Conserv. Lett.*, **13(5)**, doi:10.1111/conl.12716.

Dunlap, R.E. and P.J. Jacques, 2013: Climate Change Denial Books and Conservative Think Tanks. *Am. Behav. Sci.*, **57(6)**, 699–731, doi:10.1177/0002764213477096.

Dunlap, R.E. and A.M. McCright, 2015: Challenging Climate Change. In: *Climate Change and Society* [Dunlap, R. and R. Brulle, (eds.)]. Oxford University Press, Oxford, UK, pp. 300–332.

Dupont, C., 2015: *Climate Policy Integration into EU Energy Policy: Progress and Prospects*. 1st ed. Routledge, London, UK, 208 pp.

Dupont, C., 2019: The EU's collective securitisation of climate change. *West Eur. Polit.*, **42(2)**, 369–390, doi:10.1080/01402382.2018.1510199.

Dupont, C. and S. Oberthür, 2012: Insufficient climate policy integration in EU energy policy: the importance of the long-term perspective. *J. Contemp. Eur. Res.*, **8(2)**, 228–247, doi:10.30950/jcer.v8i2.474.

Dupont, C., S. Oberthür, and I. von Homeyer, 2020: The Covid-19 crisis: a critical juncture for EU climate policy development? *J. Eur. Integr.*, **42(8)**, 1095–1110, doi:10.1080/07036337.2020.1853117.

Dzonzi-Undi, J. and S. Li, 2016: Policy influence on clean coal uptake in China, India, Australia, and USA. *Environ. Prog. Sustain. Energy*, **35(3)**, 906–913, doi:10.1002/ep.12288.

Eberhard, A. and T. Kåberger, 2016: Renewable energy auctions in South Africa outshine feed-in tariffs. *Energy Sci. Eng.*, **4(3)**, 190–193, doi:10.1002/ese3.118.

EC, 2019a: Communication from the commission to the European Parliament, The European Council, The Council, The European Economic and Social Committee and The Committee of the Regions: The European Green Deal. European Commission (EC), Brussels, Belgium, 11.12.2019, COM(2019) 640 final., 24 pp. https://eur-lex.europa.eu/legal-content/EN/TXT/?uri=CELEX%3A52019DC0640.

EC, 2019b: Indications of illegal HFC trade based on an analysis of data reported under the F-gas Regulation, Eurostat dataset and Chinese export data. European Commission (EC), Brussels, Belgium, 8 pp. https://ec.europa.eu/clima/system/files/2019-10/report_illegal_trade_hcf_en.pdf.

EC, 2020a: Communication from the commission to the European Parliament, The Council, The Economic and Social Committee and The Committee of the Regions: 'Stepping up Europe's 2030 climate ambition – Investing in a climate-neutral future for the benefit of our people. European Commission (EC), Brussels, Belgium, 26 pp. https://eur-lex.europa.eu/legal-content/EN/TXT/?uri=CELEX%3A52020DC0562.

EC, 2020b: Strategic EU Ecolabel Work Plan 2020–2024. European Commission (EC), Brussels, Belgium, 16 pp. https://ec.europa.eu/environment/ecolabel/documents/EU%20Ecolabel%20Work%20plan%202020-2024%20Dec%202020.pdf.

Eccles, R.G. and M.P. Krzus, 2018: Why companies should report financial risks from climate change. *MIT Sloan Management Review*, MIT, October 27.

ECIU, 2019: Countdown to Zero: Plotting progress towards delivering net zero emissions by 2050., London, UK, 16 pp. https://ca1eci.edcdn.com/reports/ECIU_Countdown_to_Net_Zero.pdf (Accessed October 21, 2021).

ECIU, 2021: Net Zero Tracker. https://eciu.net/netzerotracker (Accessed October 1, 2021).

Eckersley, R., 2013: Poles Apart?: The Social Construction of Responsibility for Climate Change in Australia and Norway. *Aust. J. Polit. Hist.*, **59(3)**, 382–396, doi:10.1111/ajph.12022.

Eckersley, R., 2016: National identities, international roles, and the legitimation of climate leadership: Germany and Norway compared. *Env. Polit.*, **25(1)**, 180–201, doi:10.1080/09644016.2015.1076278.

Edenhofer, O. et al., 2017: *Decarbonization and EU ETS Reform: Introducing a price floor to drive low-carbon investments*. MCC, Berlin, Germany, 20 pp. https://www.mcc-berlin.net/fileadmin/data/C18_MCC_Publications/Decarbonization_EU_ETS_Reform_Policy_Paper.pdf (Accessed March 3, 2021).

Edmondson, D.L., F. Kern, and K.S. Rogge, 2019: The co-evolution of policy mixes and socio-technical systems: Towards a conceptual framework of policy mix feedback in sustainability transitions. *Res. Policy*, **48(10)**, 103555, doi:10.1016/j.respol.2018.03.010.

Edmondson, D.L., K.S. Rogge, and F. Kern, 2020: Zero carbon homes in the UK? Analysing the co-evolution of policy mix and socio-technical system. *Environ. Innov. Soc. Transitions*, **35**, 135–161, doi:10.1016/j.eist.2020.02.005.

EEA, 2019: *Fluorinated greenhouse gases 2019: Data reported by companies on the production, import, export, destruction and feedstock use of fluorinated greenhouse gases in the European Union, 2007-2018*. European Environment Agency (EEA), Copenhagen, Denmark, 80 pp. https://www.eea.europa.eu/publications/fluorinated-greenhouse-gases-2019.

Egeberg, M., 1999: The Impact of Bureaucratic Structure on Policy Making. *Public Adm.*, **77(1)**, 155–170, doi:10.1111/1467-9299.00148.

Eilstrup-Sangiovanni, M. and T.N.P. Bondaroff, 2014: From Advocacy to Confrontation: Direct Enforcement by Environmental NGOs. *Int. Stud. Q.*, **58(2)**, 348–361, doi:10.1111/isqu.12132.

Eisenstat, F., 2011: American Electric Power Co. v. Connecticut: How One Less Legal Theory Available in the Effort to Curb Emissions is Actually One Step forward for the Cause. *Tulane Environ. Law J.*, **25(1)**, 221–230.

Ekwurzel, B. et al., 2017: The rise in global atmospheric $CO_2$, surface temperature, and sea level from emissions traced to major carbon producers. *Clim. Change*, **144(4)**, 579–590, doi:10.1007/s10584-017-1978-0.

Elkerbout, M. et al., 2020: *The European Green Deal after Corona: Implications for EU climate policy*. CEPS, Brussels, Belgium, 12 pp. https://www.ceps.eu/ceps-publications/the-european-green-deal-after-corona/ (Accessed December 13, 2020).

Elliott, T. and J. Earl, 2018: Organizing the Next Generation: Youth Engagement with Activism Inside and Outside of Organizations. *Soc. Media + Soc.*, **4(1)**, 205630511775072, doi:10.1177/2056305117750722.

Ellis, N.R. and P. Tschakert, 2019: Triple-wins as pathways to transformation? a critical review. *Geoforum*, **103**, 167–170, doi:10.1016/j.geoforum.2018.12.006.

Elsharkawy, H. and P. Rutherford, 2018: Energy-efficient retrofit of social housing in the UK: Lessons learned from a Community Energy Saving Programme (CESP) in Nottingham. *Energy Build.*, **172**, 295–306, doi:10.1016/j.enbuild.2018.04.067.

Engels, A., 2018: Understanding how China is championing climate change mitigation. *Palgrave Commun.*, **4(1)**, 101, doi:10.1057/s41599-018-0150-4.

Engels, A., O. Hüther, M. Schäfer, and H. Held, 2013: Public climate-change skepticism, energy preferences and political participation. *Glob. Environ. Change*, **23(5)**, 1018–1027, doi:10.1016/j.gloenvcha.2013.05.008.

EPA, 2017: Voluntary Energy and Climate Programs, Archive. https://19january2017snapshot.epa.gov/climatechange/voluntary-energy-and-climate-programs.html (Accessed December 11, 2019).

EPA, 2019: Global Non-CO2 Greenhouse Gas Emission Projections & Marginal Abatement Cost Analysis: Methodology Documentation. U.S. Environmental Protection Agency, Washington, DC, 287 pp. https://www.epa.gov/sites/production/files/2019-09/documents/nonco2_methodology_report.pdf.

Ergas, C., 2010: a Model of Sustainable Living: Collective Identity in an Urban Ecovillage. *Organ. Environ.*, **23(1)**, 32–54, doi:10.1177/1086026609360324.

Erickson, P., S. Kartha, M. Lazarus, and K. Tempest, 2015: Assessing carbon lock-in. *Environ. Res. Lett.*, **10(8)**, 084023, doi:10.1088/1748-9326/10/8/084023.

Eskander, S., S. Fankhauser, and J. Setzer, 2021: Global Lessons from Climate Change Legislation and Litigation. *Environ. Energy Policy Econ.*, **2**, 44–82, doi:10.1086/711306.

Eskander, S.M.S.U. and S. Fankhauser, 2020: Reduction in greenhouse gas emissions from national climate legislation. *Nat. Clim. Change*, **10(8)**, 750–756, doi:10.1038/s41558-020-0831-z.

Eskjær, M., 2013: The Regional Dimension: How Regional Media Systems Condition Global Climate-Change Communication. *J. Int. Intercult. Commun.*, **6(1)**, 61–81, doi:10.1080/17513057.2012.748933.

European Environment Agency, 2019: *Sustainability transitions: policy and practice*. European Environment Agency, Luxembourg, 184 pp. https://www.eea.europa.eu/publications/sustainability-transitions-policy-and-practice.

European Union, 2014: *Regulation (EU) No 517/2014 of the European Parliament and of the Council of 16 April 2014 on fluorinated greenhouse gases and repealing Regulation (EC) No 842/2006*. The European Parliament and the Council of the European Union, Brussels, Belgium, L150/195--230 pp. http://eur-lex.europa.eu/eli/reg/2014/517/oj.

Evans, J. and A. Karvonen, 2014: 'Give Me a Laboratory and I Will Lower Your Carbon Footprint!' - Urban Laboratories and the Governance of Low-Carbon Futures. *Int. J. Urban Reg. Res.*, **38(2)**, 413–430, doi:10.1111/1468-2427.12077.

Evensen, D., 2019: The rhetorical limitations of the #FridaysForFuture movement. *Nat. Clim. Change*, **9(6)**, 428–430, doi:10.1038/s41558-019-0481-1.

Eyre, N., S.J. Darby, P. Grünewald, E. McKenna, and R. Ford, 2018: Reaching a 1.5°C target: socio-technical challenges for a rapid transition to low-carbon electricity systems. *Philos. Trans. R. Soc. a Math. Phys. Eng. Sci.*, **376(2119)**, 20160462, doi:10.1098/rsta.2016.0462.

Fahy, D., 2018: Objectivity as Trained Judgment: How Environmental Reporters Pioneered Journalism for a 'Post-truth' Era. *Environ. Commun.*, **12(7)**, 855–861, doi:10.1080/17524032.2018.1495093.

Fairbrother, M., I. Johansson Sevä, and J. Kulin, 2019: Political trust and the relationship between climate change beliefs and support for fossil fuel taxes: Evidence from a survey of 23 European countries. *Glob. Environ. Change*, **59**, 102003, doi:10.1016/j.gloenvcha.2019.102003.

Falduto, C. and M. Rocha, 2020: *Aligning short-term climate action with long-term climate goals: Opportunities and options for enhancing alignment between NDCs and long-term strategies*. OECD, Boulogne-Billancourt, Paris, France, 50 pp. https://www.oecd.org/environment/cc/LEDS-NDC-linkages.pdf.

Falkner, R., 2008: *Business Power and Conflict in International Environmental Politics*. Palgrave Macmillan, London, UK, 242 pp.

Fankhauser, S. and T.K.J. McDermott, 2016: *The Economics of Climate-Resilient Development*. [Fankhauser, S. and T.K. McDermott, (eds.)]. Edward Elgar Publishing, Cheltenham, UK, 256 pp.

Fankhauser, S. et al., 2013: Who will win the green race? In search of environmental competitiveness and innovation. *Glob. Environ. Change*, **23(5)**, 902–913, doi:10.1016/j.gloenvcha.2013.05.007.

Fankhauser, S., C. Gennaioli, and M. Collins, 2015: The political economy of passing climate change legislation: Evidence from a survey. *Glob. Environ. Change*, **35**, 52–61, doi:10.1016/j.gloenvcha.2015.08.008.

Fankhauser, S., C. Gennaioli, and M. Collins, 2016: Do international factors influence the passage of climate change legislation? *Clim. Policy*, **16(3)**, 318–331, doi:10.1080/14693062.2014.1000814.

Fankhauser, S., A. Averchenkova, and J. Finnegan, 2018: *10 years of the UK Climate Change Act*. The Grantham Research Institute on Climate Change and the Environment and the Centre for Climate Change Economics and Policy, London, UK, 43 pp. https://www.lse.ac.uk/granthaminstitute/wp-content/uploads/2018/03/10-Years-of-the-UK-Climate-Change-Act_Fankhauser-et-al.pdf.

Farrell, J., 2016a: Network structure and influence of the climate change counter-movement. *Nat. Clim. Change*, **6(4)**, 370–374, doi:10.1038/nclimate2875.

Farrell, J., 2016b: Corporate funding and ideological polarization about climate change. *Proc. Natl. Acad. Sci.*, **113(1)**, 92–97, doi:10.1073/pnas.1509433112.

Farstad, F.M., 2018: What explains variation in parties' climate change salience? *Party Polit.*, **24(6)**, 698–707, doi:10.1177/1354068817693473.

Farstad, F.M., 2019: Does size matter? Comparing the party politics of climate change in Australia and Norway. *Env. Polit.*, **28(6)**, 997–1016, doi:10.1080/09644016.2019.1625146.

Fast, S. et al., 2016: Lessons learned from Ontario wind energy disputes. *Nat. Energy*, **1(2)**, 7, doi:10.1038/nenergy.2015.28.

Fekete, H. et al., 2021: a review of successful climate change mitigation policies in major emitting economies and the potential of global replication. *Renew. Sustain. Energy Rev.*, **137**, doi:10.1016/j.rser.2020.110602.

Feldman, L., P.S. Hart, and T. Milosevic, 2017: Polarizing news? Representations of threat and efficacy in leading US newspapers' coverage of climate change. *Public Underst. Sci.*, **26(4)**, 481–497, doi:10.1177/0963662515595348.

Fenwick, J., K.J. Miller, and D. McTavish, 2012: Co-governance or meta-bureaucracy? Perspectives of local governance 'partnership' in England and Scotland. *Policy Polit.*, **40(3)**, 405–422, doi:10.1332/147084411X581907.

Ferrante, L. and P.M. Fearnside, 2019: Brazil's new president and 'ruralists' threaten Amazonia's environment, traditional peoples and the global climate. *Environ. Conserv.*, **46(4)**, 261–263, doi:10.1017/S0376892919000213.

Ferreira, P.G., 2016: 'Common But Differentiated Responsibilities' in the National Courts: Lessons from Urgenda v. The Netherlands. *Transnatl. Environ. Law*, **5(2)**, 329–351, doi:10.1017/S2047102516000248.

Finnegan, J., 2018: *Changing prices in a changing climate: electoral competitiveness and fossil fuel taxation*. London School of Economics and Political Science, London, UK, 54 pp. http://www.lse.ac.uk/GranthamInstitute/wp-content/uploads/2018/10/working-paper-307-Finnegan.pdf (Accessed December 16, 2019).

Finnegan, J., 2019: *Institutions, climate change, and the foundations of longterm policymaking*. London School of Economics and Political Science, London, UK, 55 pp. http://www.lse.ac.uk/GranthamInstitute/wp-content/uploads/2019/04/working-paper-321-Finnegan-1.pdf (Accessed March 30, 2021).

Fischer, C., 2008: Comparing flexibility mechanisms for fuel economy standards. *Energy Policy*, **36(8)**, 3116–3124, doi:10.1016/j.enpol.2008.03.042.

Fischer, C. and R.G. Newell, 2008: Environmental and technology policies for climate mitigation. *J. Environ. Econ. Manage.*, **55(2)**, 142–162, doi:10.1016/j.jeem.2007.11.001.

Fischer, C. and A.K. Fox, 2012: Comparing policies to combat emissions leakage: Border carbon adjustments versus rebates. *J. Environ. Econ. Manage.*, **64(2)**, 199–216, doi:10.1016/j.jeem.2012.01.005.

Fisher, D., 2004: *National governance and the global climate change regime*. Rowman & Littlefield Publishers, Washington, DC, 206 pp.

Fisher, D., 2019: *American resistance: From the Women's March to the Blue Wave*. Columbia University Press, New York City, USA, 216 pp.

Fisher, D.R., 2010: COP-15 in Copenhagen: How the Merging of Movements Left Civil Society Out in the Cold. *Glob. Environ. Polit.*, **10(2)**, 11–17, doi:10.1162/glep.2010.10.2.11.

Fisher, D.R., 2019: The broader importance of #FridaysForFuture. *Nat. Clim. Change*, **9(6)**, 430–431, doi:10.1038/s41558-019-0484-y.

Fisher, D.R. and P. Leifeld, 2019: The polycentricity of climate policy blockage. *Clim. Change*, **155(4)**, 469–487, doi:10.1007/s10584-019-02481-y.

Fisher, D.R. and S. Nasrin, 2021a: Shifting Coalitions within the Youth Climate Movement in the US. *Polit. Gov.*, **9(2)**, 112–123, doi:10.17645/pag.v9i2.3801.

Fisher, D.R. and S. Nasrin, 2021b: Climate activism and its effects. *WIREs Clim. Change*, **12(1)**, e683, doi:10.1002/wcc.683.

Fisher, D.R., K. Stanley, D. Berman, and G. Neff, 2005: How Do Organizations Matter? Mobilization and Support for Participants at Five Globalization Protests. *Soc. Probl.*, **52(1)**, 102–121, doi:10.1525/sp.2005.52.1.102.

13

Fisher, E., E. Scotford, and E. Barritt, 2017: The Legally Disruptive Nature of Climate Change. *Mod. Law Rev.*, **80(2)**, 173–201, doi:10.1111/1468-2230.12251.

Fitch-Roy, O., 2016: An offshore wind union? Diversity and convergence in European offshore wind governance. *Clim. Policy*, **16(5)**, 586–605, doi:10.1080/14693062.2015.1117958.

Fitch-Roy, O., D. Benson, and C. Mitchell, 2019a: Wipeout? Entrepreneurship, policy interaction and the EU's 2030 renewable energy target. *J. Eur. Integr.*, **41(1)**, 87–103, doi:10.1080/07036337.2018.1487961.

Fitch-Roy, O.W., D. Benson, and B. Woodman, 2019b: Policy Instrument Supply and Demand: How the Renewable Electricity Auction Took over the World. *Polit. Gov.*, **7(1)**, 81–91, doi:10.17645/pag.v7i1.1581.

Flachsland, C. and S. Levi, 2021: Germany's Federal Climate Change Act. *Env. Polit.*, **30(sup1)**, 118–140, doi:10.1080/09644016.2021.1980288.

Fleig, A., N.M. Schmidt, and J. Tosun, 2017: Legislative Dynamics of Mitigation and Adaptation Framework Policies in the EU. *Eur. Policy Anal.*, **3(1)**, 101–124, doi:10.1002/epa2.1002.

Floater, G. et al., 2016: *Co-benefits of urban climate action: a framework for cities*. London School of Economics and Political Science, London, UK, 86 pp. http://eprints.lse.ac.uk/68876/ (Accessed March 30, 2021).

Flowers, M.E. et al., 2016: Climate impacts on the cost of solar energy. *Energy Policy*, **94**, 264–273, doi:10.1016/j.enpol.2016.04.018.

Foerster, A., 2019: Climate Justice and Corporations. *King's Law J.*, **30(2)**, 305–322, doi:10.1080/09615768.2019.1645447.

Foerster, A., J. Peel, H. Osofsky, and B. McDonnell, 2017: Keeping Good Company in the Transition to a Low Carbon Economy? An Evaluation of Climate Risk Disclosure Practices in Australia. *Co. Secur. Law J.*, **35(3)**, 154–183, doi: 10.1080/09615768.2019.1645447.

Ford, J.D. et al., 2016: Including indigenous knowledge and experience in IPCC assessment reports. *Nat. Clim. Change*, **6(4)**, 349–353, doi:10.1038/nclimate2954.

Fouquet, R., 2016: Path dependence in energy systems and economic development. *Nat. Energy*, **1(8)**, 16098, doi:10.1038/nenergy.2016.98.

Fowlie, M. and M. Reguant, 2018: Challenges in the Measurement of Leakage Risk. *AEA Pap. Proc.*, **108**, 124–129, doi:10.1257/pandp.20181087.

Fowlie, M., M. Greenstone, and C. Wolfram, 2018: Do Energy Efficiency Investments Deliver? Evidence from the Weatherization Assistance Program. *Q. J. Econ.*, **133(3)**, 1597–1644, doi:10.1093/qje/qjy005.

Fox, J., J. Axsen, and M. Jaccard, 2017: Picking Winners: Modelling the Costs of Technology-specific Climate Policy in the U.S. Passenger Vehicle Sector. *Ecol. Econ.*, **137**, 133–147, doi:10.1016/j.ecolecon.2017.03.002.

Fox, J.A. and L.D. Brown, eds., 1998: *The Struggle for Accountability: The World Bank, NGOs, and Grassroots Movements*. 1st ed. MIT Press, Cambridge, Massachusetts, USA, 548 pp.

Frank, D.J., A. Hironaka, and E. Schofer, 2000: The Nation-State and the Natural Environment over the Twentieth Century. *Am. Sociol. Rev.*, **65(1)**, 96–116, doi:10.2307/2657291.

Franta, B., 2017: Litigation in the Fossil Fuel Divestment Movement. *Law Policy*, **39(4)**, 393–411, doi:10.1111/lapo.12086.

Fraser, N., 1995: From Redistribution to Recognition? Dilemmas of Justice in a 'Post-Socialist' Age. *New Left Rev.*, **0(212)**, 68.

Fredriksson, P.G. and D.L. Millimet, 2004: Electoral rules and environmental policy. *Econ. Lett.*, **84(2)**, 237–244, doi:10.1016/j.econlet.2004.02.008.

Fredriksson, P.G. and E. Neumayer, 2016: Corruption and Climate Change Policies: Do the Bad Old Days Matter? *Environ. Resour. Econ.*, **63(2)**, 451–469, doi:10.1007/s10640-014-9869-6.

Freudenburg, W.R. and V. Muselli, 2010: Global warming estimates, media expectations, and the asymmetry of scientific challenge. *Glob. Environ. Change*, **20(3)**, 483–491, doi:10.1016/j.gloenvcha.2010.04.003.

Friedrichs, J. and O.R. Inderwildi, 2013: The carbon curse: Are fuel rich countries doomed to high $CO_2$ intensities? *Energy Policy*, **62**, 1356–1365, doi:10.1016/j.enpol.2013.07.076.

Friel, S. et al., 2011: Climate Change, Noncommunicable Diseases, and Development: The Relationships and Common Policy Opportunities. *Annu. Rev. Public Health*, **32(1)**, 133–147, doi:10.1146/annurev-publhealth-071910-140612.

Friman, M. and M. Hjerpe, 2015: Agreement, significance, and understandings of historical responsibility in climate change negotiations. *Clim. Policy*, **15(3)**, 302–320, doi:10.1080/14693062.2014.916598.

Froese, R. and J. Schilling, 2019: The Nexus of Climate Change, Land Use, and Conflicts. *Curr. Clim. Change Reports*, **5(1)**, 24–35, doi:10.1007/s40641-019-00122-1.

Frumhoff, P.C., R. Heede, and N. Oreskes, 2015: The climate responsibilities of industrial carbon producers. *Clim. Change*, **132(2)**, 157–171, doi:10.1007/s10584-015-1472-5.

FSR Climate, 2019: *a literature-based assessment of the EU ETS*. Florence School of Regulation, European University Institute, Florence, Italy, 224 pp.

Fudge, S., M. Peters, and B. Woodman, 2016: Local authorities as niche actors: the case of energy governance in the UK. *Environ. Innov. Soc. Transitions*, **18**, 1–17, doi:10.1016/j.eist.2015.06.004.

Fuhr, H., T. Hickmann, and K. Kern, 2018: The role of cities in multi-level climate governance: local climate policies and the 1.5°C target. *Curr. Opin. Environ. Sustain.*, **30**, 1–6, doi:10.1016/j.cosust.2017.10.006.

Fuller, S. and D. McCauley, 2016: Framing energy justice: perspectives from activism and advocacy. *Energy Res. Soc. Sci.*, **11**, 1–8, doi:10.1016/j.erss.2015.08.004.

Fuso Nerini, F. et al., 2019: Connecting climate action with other Sustainable Development Goals. *Nat. Sustain.*, **2(8)**, 674–680, doi:10.1038/s41893-019-0334-y.

Ganguly, G., J. Setzer, and V. Heyvaert, 2018: If at First You Don't Succeed: Suing Corporations for Climate Change. *Oxf. J. Leg. Stud.*, **38(4)**, 841–868, doi:10.1093/ojls/gqy029.

Gao, J., C. Guan, and B. Zhang, 2020: China's CH4 emissions from coal mining: a review of current bottom-up inventories. *Sci. Total Environ.*, **725**, 138295, doi:10.1016/j.scitotenv.2020.138295.

Garcetti, E., 2019: *L.A.'s Green New Deal: Sustainable City pLAn 2019*. City of Los Angeles, USA, 152 pp. https://plan.lamayor.org/sites/default/files/pLAn_2019_final.pdf.

Garcia-Lamarca, M. et al., 2021: Urban green boosterism and city affordability: For whom is the 'branded' green city? *Urban Stud.*, **58(1)**, 90–112, doi:10.1177/0042098019885330.

Garcia Hernandez, A.L. and S. Bolwig, 2020: Understanding climate policy integration in the global South through the multiple streams framework. *Clim. Dev.*, **13(1)**, 68–80, doi:10.1080/17565529.2020.1723471.

Gass, P. and D. Echeverria, 2017: *Fossil fuel subsidy reform and the just transition: Integrating approaches for complementary outcomes*. International Institute for Sustainable Development, Winnipeg, USA, 47 pp. https://www.iisd.org/system/files/publications/fossil-fuel-subsidy-reform-just-transition.pdf (Accessed March 30, 2021).

Gattuso, J. et al., 2018: Ocean Solutions to Address Climate Change and Its Effects on Marine Ecosystems. *Front. Mar. Sci.*, **5**(October), art337, doi:10.3389/fmars.2018.00337.

Gaulin, N. and P. Le Billon, 2020: Climate change and fossil fuel production cuts: assessing global supply-side constraints and policy implications. *Clim. Policy*, **20(8)**, 888–901, doi:10.1080/14693062.2020.1725409.

Gavard, C., S. Voigt, and A. Genty, 2018: *Using Emissions Trading Schemes to Reduce Heterogeneous Distortionary Taxes: the Case of Recycling Carbon Auction Revenues to Support Renewable Energy*. Centre for European Economic Research (ZEW), Mannheim, Germany, 35 pp. http://ftp.zew.de/pub/zew-docs/dp/dp18058.pdf.

Geall, S., W. Shen, and Gongbuzeren, 2018: Solar energy for poverty alleviation in China: State ambitions, bureaucratic interests, and local realities. *Energy Res. Soc. Sci.*, **41**, 238–248, doi:10.1016/j.erss.2018.04.035.

Gebara, M.F., L. Fatorelli, P. May, and S. Zhang, 2014: REDD+ policy networks in Brazil: constraints and opportunities for successful policy making. *Ecol. Soc.*, **19(3)**, 53, doi:10.5751/ES-06744-190353.

Geden, O., 2016: The Paris Agreement and the inherent inconsistency of climate policymaking. *Wiley Interdiscip. Rev. Clim. Change*, **7(6)**, 790–797, doi:10.1002/wcc.427.

Geden, O., G.P. Peters, and V. Scott, 2019: Targeting carbon dioxide removal in the European Union. *Clim. Policy*, **19(4)**, 487–494, doi:10.1080/14693062.2018.1536600.

Geels, F.W., 2014: Regime Resistance against Low-Carbon Transitions: Introducing Politics and Power into the Multi-Level Perspective. *Theory, Cult. Soc.*, **31(5)**, 21–40, doi:10.1177/0263276414531627.

Geels, F.W., B.K. Sovacool, T. Schwanen, and S. Sorrell, 2017a: Sociotechnical transitions for deep decarbonization. *Science*, **357(6357)**, 1242–1244, doi:10.1126/science.aao3760.

Geels, F.W., B.K. Sovacool, T. Schwanen, and S. Sorrell, 2017b: The Socio-Technical Dynamics of Low-Carbon Transitions. *Joule*, **1(3)**, 463–479, doi:10.1016/j.joule.2017.09.018.

Gençsü, I. et al., 2020: Phasing out public financial flows to fossil fuel production in Europe. *Clim. Policy*, **20(8)**, 1010–1023, doi:10.1080/14693062.2020.1736978.

Genus, A. and K. Theobald, 2016: Creating low-carbon neighbourhoods: a critical discourse analysis. *Eur. Urban Reg. Stud.*, **23(4)**, 782–797, doi:10.1177/0969776414546243.

Gerlagh, R. and O. Kuik, 2007: *Carbon Leakage with International Technology Spillovers*. https://www.econstor.eu/handle/10419/74024 (Accessed October 18, 2021).

Gerlagh, R., R.J.R.K. Heijmans, and K.E. Rosendahl, 2020: COVID-19 Tests the Market Stability Reserve. *Environ. Resour. Econ.*, **76(4)**, 855–865, doi:10.1007/s10640-020-00441-0.

German Federal Ministry of Finance, 2020: Emerging from the crisis with full strength. Federal Ministry of Finance, Germanyza, Berlin, Germany. https://www.bundesfinanzministerium.de/Content/EN/Standardartikel/Topics/Public-Finances/Articles/2020-06-04-fiscal-package.html (Accessed March 30, 2021).

Geroski, P.A., 1995: Markets for technology: knowledge, innovation and appropriability. In: *Handbook of the Economics of Innovation and Technological Change* [Stoneman, P., (ed.)]. Blackwell Publishers, Oxford, UK, pp. 90–131.

Gerrard, M.B., 2021: Climate Change Litigation in the United States: High Volume of Cases, Mostly About Statutes. In: *Climate Change Litigation: Global Perspectives* [Alogna, I., C. Bakker, and J.-P. Gauci, (eds.)]. Koninklijke Brill, Leiden, Netherlands, pp. 33–46.

Ghaleigh, N.S., 2021: Climate Constitutionalism of the UK Supreme Court. *J. Environ. Law*, **33(2)**, 441–447, doi:10.1093/jel/eqab013.

Giampieri, M.A. et al., 2019: Visions of resilience: lessons from applying a digital democracy tool in New York's Jamaica Bay watershed. *Urban Ecosyst.*, **22(1)**, 1–17, doi:10.1007/s11252-017-0701-2.

Gill, S., J. Handley, A. Ennos, and S. Pauleit, 2007: Adapting Cities for Climate Change: The Role of the Green Infrastructure. *Built Environ.*, **33(1)**, 115–133, doi:10.2148/benv.33.1.115.

Gillan, S.L. and L.T. Starks, 2007: The Evolution of Shareholder Activism in the United States. In: *Institutional Investor Activism: Hedge Funds and Private Equity, Economics and Regulation* [Bratton, W. and J. McCahery, (eds.)]. Vol. 19. Oxford University Press, Oxford, UK, pp. 39–72.

Gillard, R., A. Gouldson, J. Paavola, and J. Van Alstine, 2016: Transformational responses to climate change: beyond a systems perspective of social change in mitigation and adaptation. Wiley Interdisciplinary Reviews: Climate Change, 7(April), 251–265, doi:10.1002/wcc.384.

Giraudet, L.-G. and P. Quirion, 2008: Efficiency and distributional impacts of tradable white certificates compared to taxes, subsidies and regulations. *Rev. d'économie Polit.*, **118**, 885–914.

Givoni, M., J. Macmillen, D. Banister, and E. Feitelson, 2013: From Policy Measures to Policy Packages. *Transp. Rev.*, **33(1)**, 1–20, doi:10.1080/01441647.2012.744779.

Global Covenant of Mayors for Climate and Energy, 2018: *Implementing Climate Ambition: Global Covenant of Mayors 2018 Global Aggregation Report*. Global Covenant of Mayors for Climate and Energy, Brussels, Belgium, 6 pp. https://www.globalcovenantofmayors.org/wp-content/uploads/2018/09/2018_GCOM_report_web.pdf (Accessed December 12, 2019).

Glynn, P., T. Cadman, and T.N. Maraseni, 2017: *Business, Organized Labour and Climate Policy: Forging a Role at the Negotiating Table*. Edward Elgar, Cheltenham, UK, 256 pp.

Goeschl, T., 2019: Cold Case: The forensic economics of energy efficiency labels for domestic refrigeration appliances. *Energy Econ.*, **84**, 104468, doi:10.1016/j.eneco.2019.08.001.

Goldberg, M.H., J.R. Marlon, X. Wang, S. van der Linden, and A. Leiserowitz, 2020: Oil and gas companies invest in legislators that vote against the environment. *Proc. Natl. Acad. Sci.*, **117(10)**, 5111–5112, doi:10.1073/pnas.1922175117.

Goldthau, A. and L. Hughes, 2020: Protect global supply chains for low-carbon technologies. *Nature*, **585**, 3 doi: 10.1038/d41586-020-02499-8.

Gomez-Echeverri, L., 2018: Climate and development: enhancing impact through stronger linkages in the implementation of the Paris Agreement and the Sustainable Development Goals (SDGs). *Philos. Trans. R. Soc. A Math. Phys. Eng. Sci.*, **376(2119)**, doi:10.1098/rsta.2016.0444.

Gomez Echeverri, L., 2018: Investing for rapid decarbonization in cities. *Curr. Opin. Environ. Sustain.*, **30**, 42–51, doi:10.1016/j.cosust.2018.02.010.

Gomez, J., 2013: The limitations of climate change donor intervention as deus ex machina: evidence from Sorsogon, the Philippines. *Int. Dev. Plan. Rev.*, **35(4)**, 371–394, doi:10.3828/idpr.2013.26.

Gordon, D. and M. Acuto, 2015: If Cities Are the Solution, What Are the Problems? The Promise and Perils of Urban Climate Leadership. In: *The Urban Climate Challenge: Rethinking the Role of Cities in the Global Climate Regime* [Johnson, C., N. Toly, and H. Schroeder, (eds.)]. Routledge, London, UK, pp. 63–91.

Gordon, D.J., 2015: An Uneasy Equilibrium: The Coordination of Climate Governance in Federated Systems. *Glob. Environ. Polit.*, **15(2)**, 121–141, doi:10.1162/GLEP_a_00301.

Gössling, S. and R. Buckley, 2016: Carbon labels in tourism: persuasive communication? *J. Clean. Prod.*, **111**, 358–369, doi:10.1016/j.jclepro.2014.08.067.

Gössling, S., A. Humpe, and T. Bausch, 2020: Does 'flight shame' affect social norms? Changing perspectives on the desirability of air travel in Germany. *J. Clean. Prod.*, **266**, 122015, doi:10.1016/j.jclepro.2020.122015.

Goulder, L.H. and R.N. Stavins, 2011: Challenges from State-Federal Interactions in US Climate Change Policy. *Am. Econ. Rev.*, **101(3)**, 253–257, doi:10.1257/aer.101.3.253.

Goulder, L.H., M.A.C. Hafstead, G. Kim, and X. Long, 2019: Impacts of a carbon tax across US household income groups: What are the equity-efficiency trade-offs? *J. Public Econ.*, **175**, 44–64, doi:10.1016/j.jpubeco.2019.04.002.

Gouldson, A. et al., 2016: Cities and climate change mitigation: Economic opportunities and governance challenges in Asia. *Cities*, **54**, 11–19, doi:10.1016/j.cities.2015.10.010.

Grafakos, S. et al., 2018: Integrating mitigation and adaptation: Opportunities and challenges. In: *Climate Change and Cities: Second Assessment Report of the Urban Climate Change Research Network* [Rosenzweig, C., P. Romero-Lankao, S. Mehrotra, S. Dhakal, and S.A. Ibrahim, (eds.)]. Cambridge University Press, New York, USA, pp. 101–138.

Grafakos, S., K. Trigg, M. Landauer, L. Chelleri, and S. Dhakal, 2019: Analytical framework to evaluate the level of integration of climate adaptation and mitigation in cities. *Clim. Change*, **154(1–2)**, 87–106, doi:10.1007/s10584-019-02394-w.

Grafakos, S. et al., 2020: Integration of mitigation and adaptation in urban climate change action plans in Europe: a systematic assessment.

13

*Renew. Sustain. Energy Rev.*, **121**(January), 109623, doi:10.1016/j.rser.2019.109623.

Grainger, C.A. and C.D. Kolstad, 2010: Who Pays a Price on Carbon? *Environ. Resour. Econ.*, **46(3)**, doi:10.1007/s10640-010-9345-x.

Grandin, J., H. Haarstad, K. Kjærås, and S. Bouzarovski, 2018: The politics of rapid urban transformation. *Curr. Opin. Environ. Sustain.*, **31**, 16–22, doi:10.1016/j.cosust.2017.12.002.

Granoff, I., J.R. Hogarth, and A. Miller, 2016: Nested barriers to low-carbon infrastructure investment. *Nat. Clim. Change*, **6(12)**, 1065–1071, doi:10.1038/nclimate3142.

Grant, D. and I.B. Vasi, 2017: Civil Society in an Age of Environmental Accountability: How Local Environmental Nongovernmental Organizations Reduce U.S. Power Plants' Carbon Dioxide Emissions. *Sociol. Forum*, **32(1)**, 94–115, doi:10.1111/socf.12318.

Grant, D., A. Jorgenson, and W. Longhofer, 2018: Pathways to Carbon Pollution: The Interactive Effects of Global, Political, and Organizational Factors on Power Plants' $CO_2$ Emissions. *Sociol. Sci.*, **5**, 58–92, doi:10.15195/v5.a4.

Grashof, K., 2021: Who put the hammer in the toolbox? Explaining the emergence of renewable energy auctions as a globally dominant policy instrument. *Energy Res. Soc. Sci.*, **73**, 101917, doi:10.1016/j.erss.2021.101917.

Grashof, K., V. Berkhout, R. Cernusko, and M. Pfennig, 2020: Long on promises, short on delivery? Insights from the first two years of onshore wind auctions in Germany. *Energy Policy*, **140**, 111240, doi:10.1016/j.enpol.2020.111240.

Greater London Authority, 2018: *London Environment Strategy*. Greater London Authority, London, UK, 452 pp. https://www.london.gov.uk/sites/default/files/london_environment_strategy.pdf (Accessed October 17, 2021).

Green, F., 2018: Anti-fossil fuel norms. *Clim. Change*, **150(1–2)**, 103–116, doi:10.1007/s10584-017-2134-6.

Green, F. and A. Gambhir, 2020: Transitional assistance policies for just, equitable and smooth low-carbon transitions: who, what and how? *Clim. Policy*, **20(8)**, 902–921, doi:10.1080/14693062.2019.1657379.

Green, J.F., 2021: Does carbon pricing reduce emissions? a review of *ex-post* analyses. *Environ. Res. Lett.*, **16(4)**, doi:10.1088/1748-9326/abdae9.

Green, J.F. and T.N. Hale, 2017: Reversing the Marginalization of Global Environmental Politics in International Relations: An Opportunity for the Discipline. *PS Polit. Sci. Polit.*, **50(02)**, 473–479, doi:10.1017/S1049096516003024.

Greiner, S. and A. Michaelowa, 2003: Defining Investment Additionality for CDM projects—practical approaches. *Energy Policy*, **31(10)**, doi:10.1016/S0301-4215(02)00142-8.

Griffin, P.A., 2020: Energy finance must account for extreme weather risk. *Nat. Energy*, **5(2)**, 98–100, doi:10.1038/s41560-020-0548-2.

Grossman, E., 2019: France's Yellow Vests – Symptom of a Chronic Disease. *Polit. Insight*, **10(1)**, 30–34, doi:10.1177/2041905819838152.

Grubb, M. et al., 2015: a review of Chinese $CO_2$ emission projections to 2030: the role of economic structure and policy. *Clim. Policy*, **15(sup1)**, S7–S39, doi:10.1080/14693062.2015.1101307.

Grubb, M. et al., 2020: Consumption-oriented policy instruments for fostering greenhouse gas mitigation. *Clim. Policy*, **20(sup1)**, S58–S73, doi:10.1080/14693062.2020.1730151.

Guey, A. and A. Bilich, 2019: *Multi-level governance and coordination under Kenya's National Climate Change Act*. GIZ, in cooperation with UNDP, Bonn and Eschborn, Germany, 11 pp. https://transparency-partnership.net/system/files/document/200114_GPD_Kenya_RZ.pdf.

Gugler, K., A. Haxhimusa, and M. Liebensteiner, 2021: Effectiveness of climate policies: Carbon pricing vs. subsidizing renewables. *J. Environ. Econ. Manage.*, **106**, 1–22, doi:10.1016/j.jeem.2020.102405.

Gupta, D., F. Ghersi, S.S. Vishwanathan, and A. Garg, 2020: Macroeconomic assessment of India's development and mitigation pathways. *Clim. Policy*, **20(7)**, 779–799, doi:10.1080/14693062.2019.1648235.

Gurara, D., V. Klyuev, N. Mwase, and A. F. Presbitero, 2018: Trends and Challenges in Infrastructure Investment in Developing Countries. *Int. Dev. Policy Rev. Int. Polit. développement [Online]*, **10.1**, doi:10.4000/poldev.2802.

Guy, S., V. Henshaw, and O. Heidrich, 2015: Climate change, adaptation and Eco-Art in Singapore. *J. Environ. Plan. Manag.*, **58(1)**, 39–54, doi:10.1080/09640568.2013.839446.

Haarstad, H., 2016: Where are urban energy transitions governed? Conceptualizing the complex governance arrangements for low-carbon mobility in Europe. *Cities*, **54**, 4–10, doi:10.1016/j.cities.2015.10.013.

Haase, D. et al., 2017: Greening cities – To be socially inclusive? About the alleged paradox of society and ecology in cities. *Habitat Int.*, **64**, 41–48, doi:10.1016/j.habitatint.2017.04.005.

Haasnoot, M., J.H. Kwakkel, W.E. Walker, and J. ter Maat, 2013: Dynamic adaptive policy pathways: a method for crafting robust decisions for a deeply uncertain world. *Glob. Environ. Change*, **23(2)**, 485–498, doi:10.1016/j.gloenvcha.2012.12.006.

Habib, S., S. Abdelmonen, and M. Khaled, 2020: The Effect of Corruption on the Environmental Quality in African Countries: a Panel Quantile Regression Analysis. *J. Knowl. Econ.*, **11(2)**, 788–804, doi:10.1007/s13132-018-0571-8.

Hadden, J., 2014: Explaining Variation in Transnational Climate Change Activism: The Role of Inter-Movement Spillover. *Glob. Environ. Polit.*, **14(2)**, 7–25, doi:10.1162/GLEP_a_00225.

Hadden, J., 2015: *Networks in Contention: The Divisive Politics of Climate Change*. Cambridge University Press, Cambridge, UK, 224 pp.

Hadden, J. and L. Jasny, 2019: The Power of Peers: How Transnational Advocacy Networks Shape NGO Strategies on Climate Change. *Br. J. Polit. Sci.*, **49(2)**, 637–659, doi:10.1017/S0007123416000582.

Haenfler, R., B. Johnson, and E. Jones, 2012: Lifestyle Movements: Exploring the Intersection of Lifestyle and Social Movements. *Soc. Mov. Stud.*, **11(1)**, 1–20, doi:10.1080/14742837.2012.640535.

Hainsch, K. et al., 2020: *Make the European Green Deal real: Combining climate neutrality and economic recovery*. DIW Berlin: Politikberatung kompakt, Berlin, 77 pp. https://www.econstor.eu/bitstream/10419/222849/1/1701746166.pdf (Accessed March 3, 2021).

Haites, E., 2016: Experience with linking greenhouse gas emissions trading systems. *Wiley Interdiscip. Rev. Energy Environ.*, **5(3)**, 246–260, doi:10.1002/wene.191.

Haites, E., 2018: Carbon taxes and greenhouse gas emissions trading systems: what have we learned? *Clim. Policy*, **18(8)**, 955–966, doi:10.1080/14693062.2018.1492897.

Haites, E. et al., 2018: Experience with Carbon Taxes and Greenhouse Gas Emissions Trading Systems. *Duke Environ. Law Policy Forum*, **29**, 109–182, doi: 10.1002/wene.191.

Hakelberg, L., 2014: Governance by Diffusion: Transnational Municipal Networks and the Spread of Local Climate Strategies in Europe. *Glob. Environ. Polit.*, **14(1)**, 107–129, doi:10.1162/GLEP_a_00216.

Hall, P.A. and D. Soskice, eds., 2001: *Varieties of Capitalism: The Institutional Foundations of Comparative Advantage*. Oxford University Press, Oxford, UK, 560 pp.

Halsnæs, K. et al., 2014: Climate change mitigation policy paradigms—national objectives and alignments. *Mitig. Adapt. Strateg. Glob. Change*, **19(1)**, 45–71, doi:10.1007/s11027-012-9426-y.

Hamarat, C., J.H. Kwakkel, E. Pruyt, and E.T. Loonen, 2014: An exploratory approach for adaptive policymaking by using multi-objective robust optimization. *Simul. Model. Pract. Theory*, **46**, 25–39, doi:10.1016/j.simpat.2014.02.008.

Hamilton, J., R. Mayne, Y. Parag, and N. Bergman, 2014: Scaling up local carbon action: the role of partnerships, networks and policy. *Carbon Manag.*, **5(4)**, 463–476, doi:10.1080/17583004.2015.1035515.

Han, H., 2017: Singapore, a Garden City: Authoritarian Environmentalism in a Developmental State. *J. Environ. Dev.*, **26(1)**, 3–24, doi:10.1177/1070496516677365.

Han, H. and S.W. Ahn, 2020: Youth Mobilization to Stop Global Climate Change: Narratives and Impact. *Sustainability*, **12(10)**, 4127, doi:10.3390/su12104127.

Hanna, R., Y. Xu, and D.G. Victor, 2020: After COVID-19, green investment must deliver jobs to get political traction. *Nature*, **582**, 178–180, doi:10.1038/d41586-020-01682-1.

Hans, F., T. Day, F. Röser, J. Emmrich, and M. Hagemann, 2020: *Making Long-Term Low GHG Emissions Development Strategies a Reality*, GIZ, Bonn and NewClimate Institute, Cologne. https://newclimate.org/2020/05/28/making-long-term-low-ghg-emissions-development-strategies-a-reality/.

Hansen, A. and D. Machin, 2008: Visually branding the environment: climate change as a marketing opportunity. *Discourse Stud.*, **10(6)**, 777–794, doi:10.1177/1461445608098200.

Harbinson, R., 2006: *Whatever the weather Media attitudes to reporting climate change*. Panos Institute, London, UK, 16 pp. http://panoslondon.panosnetwork.org/wp-content/files/2011/03/whatever_weathermjwnSt.pdf (Accessed March 30, 2021).

Harring, N., 2013: Understanding the Effects of Corruption and Political Trust on Willingness to Make Economic Sacrifices for Environmental Protection in a Cross-National Perspective. *Soc. Sci. Q.*, **94(3)**, 660–671, doi:10.1111/j.1540-6237.2012.00904.x.

Harring, N., S.C. Jagers, and S. Matti, 2019: The significance of political culture, economic context and instrument type for climate policy support: a cross-national study. *Clim. Policy*, **19(5)**, 636–650, doi:10.1080/14693062.2018.1547181.

Harrison, K. and L.M. Sundstrom, 2010: Introduction: Global Commons, Domestic Decisions. In: *Global Commons, Domestic Decisions: The Comparative Politics of Climate Change* [Harrison, K. and L.M. Sundstrom, (eds.)]. MIT Press, Cambridge, MA, USA, pp. 328.

Harrison, T. and G. Kostka, 2014: Balancing Priorities, Aligning Interests: Developing Mitigation Capacity in China and India. *Comp. Polit. Stud.*, **47(3)**, 450–480, doi:10.1177/0010414013509577.

Hathaway, J.R., 2020: Climate Change, the Intersectional Imperative, and the Opportunity of the Green New Deal. *Environ. Commun.*, **14(1)**, 13–22, doi:10.1080/17524032.2019.1629977.

Hawkey, D. and J. Webb, 2014: District energy development in liberalised markets: situating UK heat network development in comparison with Dutch and Norwegian case studies. *Technol. Anal. Strateg. Manag.*, **26(10)**, 1228–1241, doi:10.1080/09537325.2014.971001.

Haynes, J., 1999: Power, politics and environmental movements in the Third World. *Env. Polit.*, **8(1)**, 222–242, doi:10.1080/09644019908414445.

Heede, R., 2014: Tracing anthropogenic carbon dioxide and methane emissions to fossil fuel and cement producers, 1854–2010. *Clim. Change*, **122(1–2)**, 229–241, doi:10.1007/s10584-013-0986-y.

Heidrich, O. et al., 2016: National climate policies across Europe and their impacts on cities strategies. *J. Environ. Manage.*, **168**(March 2016), 36–45, doi:10.1016/j.jenvman.2015.11.043.

Heikkinen, M., T. Ylä-Anttila, and S. Juhola, 2019: Incremental, reformistic or transformational: what kind of change do C40 cities advocate to deal with climate change? *J. Environ. Policy Plan.*, **21(1)**, 90–103, doi:10.1080/1523908X.2018.1473151.

Helm, D., C. Hepburn, and R. Mash, 2003: Credible Carbon Policy. *Oxford Rev. Econ. Policy*, **19(3)**, 438–450, doi:10.1093/oxrep/19.3.438.

Hennessey, R., J. Pittman, A. Morand, and A. Douglas, 2017: Co-benefits of integrating climate change adaptation and mitigation in the Canadian energy sector. *Energy Policy*, **111**(December), 214–221, doi:10.1016/j.enpol.2017.09.025.

Hepburn, C., K. Neuhoff, W. Acworth, D. Burtraw, and F. Jotzo, 2016: The economics of the EU ETS market stability reserve. *J. Environ. Econ. Manage.*, **80**(November), 1–5, doi:10.1016/j.jeem.2016.09.010.

Hepburn, C., B. O'Callaghan, N. Stern, J. Stiglitz, and D. Zenghelis, 2020: Will COVID-19 fiscal recovery packages accelerate or retard progress on climate change? *Oxford Rev. Econ. Policy*, **36(sup1)**, S359–S381, doi:10.1093/oxrep/graa015.

Hermwille, L., 2018: Making initiatives resonate: how can non-state initiatives advance national contributions under the UNFCCC? *Int. Environ. Agreements Polit. Law Econ.*, **18(3)**, 447–466, doi:10.1007/s10784-018-9398-9.

Hermwille, L., W. Obergassel, and C. Arens, 2015: The transformative potential of emissions trading. *Carbon Manag.*, **6(5–6)**, 261–272, doi:10.1080/17583004.2016.1151552.

Herr, D. and E. Landis, 2016: *Coastal blue carbon ecosystems: Opportunities for Nationally Determined Contributions*. IUCN, Gland, Switzerland, and TNC, Washington, DC, USA, 28 pp. https://portals.iucn.org/library/node/48422.

Hess, D.J., 2019: Cooler coalitions for a warmer planet: a review of political strategies for accelerating energy transitions. *Energy Res. Soc. Sci.*, **57**, 101246, doi:10.1016/j.erss.2019.101246.

Hess, D.J. and R.G. McKane, 2021: Making sustainability plans more equitable: an analysis of 50 U.S. Cities. *Local Environ.*, **26(4)**, 461–476, doi:10.1080/13549839.2021.1892047.

Hewitt, D. and S. Coakley, 2019: Transforming our buildings for a low-carbon era: Five key strategies. *Electr. J.*, **32(7)**, 106624, doi:10.1016/j.tej.2019.106624.

Hibbard, P.J., S.F. Tierney, P.G. Darling, and S. Cullinan, 2018: An expanding carbon cap-and-trade regime? a decade of experience with RGGI charts a path forward. *Electr. J.*, **31(5)**, 1–8, doi:10.1016/j.tej.2018.05.015.

High-Level Commission on Carbon Prices, 2017: *Report of the High-Level Commission on Carbon Prices*. Carbon Pricing Leadership Coalition, Washington, DC, 68 pp. https://static1.squarespace.com/static/54ff9c5ce4b0a53decccfb4c/t/59244eed17bffc0ac256cf16/1495551740633/CarbonPricing_Final_May29.pdf (Accessed March 30, 2021).

Hill, S.D., J. Knott, and J.K.S Hill, 2010: Too close for comfort: Social controversies surrounding wind farm noise setback policies in Ontario. *Renew. Energy Law Policy Rev.*, **2**, 153–168.

Hilson, C., 2019: Climate Populism, Courts, and Science. *J. Environ. Law*, **31(3)**, 395–398, doi:10.1093/jel/eqz021.

Hindery, D., 2013: *From Enron to Evo: Pipeline Politics, Global Environmentalism, and Indigenous Rights in Bolivia*. 1st edition. The University of Arizona Press, Tucson, USA, pp. 328.

Hirt, L.F., M. Sahakian, and E. Trutnevyte, 2021: What socio-technical regimes foster solar energy champions? Analysing uneven photovoltaic diffusion at a sub-national level in Switzerland. *Energy Res. Soc. Sci.*, **74**, 101976, doi:10.1016/j.erss.2021.101976.

Hmielowski, J.D., L. Feldman, T.A. Myers, A. Leiserowitz, and E. Maibach, 2014: An attack on science? Media use, trust in scientists, and perceptions of global warming. *Public Underst. Sci.*, **23(7)**, 866–883, doi:10.1177/0963662513480091.

Hochstetler, K., 2020: *Political Economies of Energy Transition: Wind and Solar Power in Brazil and South Africa*. 1st ed. Cambridge University Press, Cambridge, UK, 278 pp.

Hochstetler, K., 2021: Climate institutions in Brazil: three decades of building and dismantling climate capacity. *Env. Polit.*, **30(sup1)**, 49–70, doi:10.1080/09644016.2021.1957614.

Hochstetler, K. and G. Kostka, 2015: Wind and Solar Power in Brazil and China: Interests, State–Business Relations, and Policy Outcomes. *Glob. Environ. Polit.*, **15(3)**, 74–94, doi:10.1162/GLEP_a_00312.

Hodson, M. and S. Marvin, 2009: Cities mediating technological transitions: understanding visions, intermediation and consequences. *Technol. Anal. Strateg. Manag.*, **21(4)**, 515–534, doi:10.1080/09537320902819213.

Hodson, M., J. Evans, and G. Schliwa, 2018: Conditioning experimentation: The struggle for place-based discretion in shaping urban infrastructures. *Environ. Plan. C Polit. Sp.*, **36(8)**, 1480–1498, doi:10.1177/2399654418765480.

Hoewe, J. and L. Ahern, 2017: First-Person Effects of Emotional and Informational Messages in Strategic Environmental Communications


13

Campaigns. *Environ. Commun.*, **11(6)**, 810–820, doi:10.1080/17524032.2017.1371050.

Hoffmann, M.J., 2011: *Climate Governance at the Crossroads: Experimenting with a Global Response after Kyoto*. Oxford University Press, Oxford, UK, 237 pp.

Höhne, N. et al., 2021: Wave of net zero emission targets opens window to meeting the Paris Agreement. *Nat. Clim. Change*, **11(10)**, 820–822, doi:10.1038/s41558-021-01142-2.

Holden, M. and M.T. Larsen, 2015: Institutionalizing a policy by any other name: in the City of Vancouver's Greenest City Action Plan, does climate change policy or sustainability policy smell as sweet? *Urban Res. Pract.*, **8(3)**, 354–370, doi:10.1080/17535069.2015.1051382.

Holland, B., 2017: Procedural justice in local climate adaptation: political capabilities and transformational change. *Env. Polit.*, **26(3)**, 391–412, doi:10.1080/09644016.2017.1287625.

Holland, S.P., J.E. Hughes, C.R. Knittel, and N.C. Parker, 2015: Unintended consequences of carbon policies: Transportation fuels, land-use, emissions, and innovation. *Energy J.*, **36(3)**, 35–74, doi:10.5547/01956574.36.3.2.

Holmberg, A. and A. Alvinius, 2020: Children's protest in relation to the climate emergency: a qualitative study on a new form of resistance promoting political and social change. *Childhood*, **27(1)**, 78–92, doi:10.1177/0907568219879970.

Holmes, D. and C. Star, 2018: Climate Change Communication in Australia: The Politics, Mainstream Media and Fossil Fuel Industry Nexus. In: *Handbook of Climate Change Communication: Vol. 1. Climate Change Management* [Leal Filho, W., E. Manolas, A. Azul, U. Azeiteiro, and H. McGhie (eds.)]. Springer, Cham, Switzerland, pp. 151–170. doi.org/10.1007/978-3-319-69838-0_10.

Hölscher, K., J.M. Wittmayer, and D. Loorbach, 2018: Transition versus transformation: What's the difference? *Environ. Innov. Soc. Transitions*, **27**, 1–3, doi:10.1016/j.eist.2017.10.007.

Holtz, G. et al., 2018: Competences of local and regional urban governance actors to support low-carbon transitions: Development of a framework and its application to a case-study. *J. Clean. Prod.*, **177**, 846–856, doi:10.1016/j.jclepro.2017.12.137.

Hönle, S.E., C. Heidecke, and B. Osterburg, 2019: Climate change mitigation strategies for agriculture: an analysis of nationally determined contributions, biennial reports and biennial update reports. *Clim. Policy*, **19(6)**, 688–702, doi:10.1080/14693062.2018.1559793.

Hoornweg, D., L. Sugar, and C.L. Trejos Gómez, 2011: Cities and greenhouse gas emissions: moving forward. *Environ. Urban.*, **23(1)**, 207–227, doi:10.1177/0956247810392270.

Horne, M., 2020: *Climate Emergency Action Plan – RTS 13199*. City of Vancouver, 371 pp. https://council.vancouver.ca/20201103/documents/p1.pdf.

Horne, R. and S. Moloney, 2019: Urban low-carbon transitions: institution-building and prospects for interventions in social practice. *Eur. Plan. Stud.*, **27(2)**, 336–354, doi:10.1080/09654313.2018.1472745.

Hornsey, M.J., E.A. Harris, and K.S. Fielding, 2018: Relationships among conspiratorial beliefs, conservatism and climate scepticism across nations. *Nat. Clim. Change*, **8(7)**, 614–620, doi:10.1038/s41558-018-0157-2.

Houghton, A., 2011: Health Impact Assessments: a Tool for Designing Climate Change Resilience Into Green Building and Planning Projects. *J. Green Build.*, **6(2)**, 66–87, doi:10.3992/jgb.6.2.66.

Houle, D., E. Lachapelle, and M. Purdon, 2015: Comparative Politics of Sub-Federal Cap-and-Trade: Implementing the Western Climate Initiative. *Glob. Environ. Polit.*, **15(3)**, 49–73, doi:10.1162/GLEP_a_00311.

Howard, J. et al., 2017: Clarifying the role of coastal and marine systems in climate mitigation. *Front. Ecol. Environ.*, **15(1)**, 42–50, doi:10.1002/fee.1451.

Howlett, M. and P. del Rio, 2015: The parameters of policy portfolios: verticality and horizontality in design spaces and their consequences for policy mix formulation. *Environ. Plan. C Gov. Policy*, **33(5)**, 1233–1245, doi:10.1177/0263774X15610059.

Howlett, M.P. and S. Oliphant, 2010: Environmental Organizations and Climate Change Policy Capacity: An Assessment of the Canadian Case. *Can. Polit. Sci. Rev.*, **4(2–3)**, 18–35.

Hsu, A., A.J. Weinfurter, and K. Xu, 2017: Aligning sub-national climate actions for the new post-Paris climate regime. *Clim. Change*, **142(3–4)**, 419–432, doi:10.1007/s10584-017-1957-5.

Hsu, A. et al., 2018: *Bridging the Emissions Gap – The Role of Non- State and Sub-national Actors. In The Emissions Gap Report 2018. a UN Environment Synthesis Report*. UNEP, Nairobi, 27 pp. https://wedocs.unep.org/bitstream/handle/20.500.11822/26093/NonState_Emissions_Gap.pdf?isAllowed=y&sequence=1 (Accessed March 30, 2021).

Hsu, A. et al., 2019: a research roadmap for quantifying non-state and sub-national climate mitigation action. *Nat. Clim. Change*, **9(1)**, 11–17, doi:10.1038/s41558-018-0338-z.

Hsu, A. et al., 2020a: ClimActor, harmonized transnational data on climate network participation by city and regional governments. *Sci. Data*, **7(1)**, 374, doi:10.1038/s41597-020-00682-0.

Hsu, A., N. Höhne, T. Kuramochi, V. Vilariño, and B.K. Sovacool, 2020b: Beyond states: Harnessing sub-national actors for the deep decarbonisation of cities, regions, and businesses. *Energy Res. Soc. Sci.*, **70(3)**, 101738, doi:10.1016/j.erss.2020.101738.

Hsu, A. et al., 2020c: Performance determinants show European cities are delivering on climate mitigation. *Nat. Clim. Change*, **10(11)**, 1015–1022, doi:10.1038/s41558-020-0879-9.

Huang, P., V. Castán Broto, Y. Liu, and H. Ma, 2018: The governance of urban energy transitions: a comparative study of solar water heating systems in two Chinese cities. *J. Clean. Prod.*, **180**, 222–231, doi:10.1016/j.jclepro.2018.01.053.

Huber, R.A., L. Fesenfeld, and T. Bernauer, 2020: Political populism, responsiveness, and public support for climate mitigation. *Clim. Policy*, **20(3)**, 373–386, doi:10.1080/14693062.2020.1736490.

Huggel, C. et al., 2015: a framework for the science contribution in climate adaptation: Experiences from science-policy processes in the Andes. *Environ. Sci. Policy*, **47**, 80–94, doi:10.1016/j.envsci.2014.11.007.

Hughes, L., 2019a: The Rocky Hill decision: a watershed for climate change action? *J. Energy Nat. Resour. Law*, **37(3)**, 341–351, doi:10.1080/02646811.2019.1600272.

Hughes, S., 2013: Justice in urban climate change adaptation: Criteria and application to Delhi. *Ecol. Soc.*, **18(4)**, doi:10.5751/ES-05929-180448.

Hughes, S., 2019b: *Repowering Cities: Governing Climate Change Mitigation in New York City, Los Angeles and Toronto*. Cornell University Press, Ithaca; New York, USA, 224 pp.

Hughes, S. and P. Romero-Lankao, 2014: Science and institution building in urban climate-change policymaking. *Env. Polit.*, **23(6)**, 1023–1042, doi:10.1080/09644016.2014.921459.

Hughes, S. and M. Hoffmann, 2020: Just urban transitions: Toward a research agenda. *WIREs Clim. Change*, **11(3)**, 1–11, doi:10.1002/wcc.640.

Huitema, D. et al., 2011: The evaluation of climate policy: theory and emerging practice in Europe. *Policy Sci.*, **44(2)**, 179–198, doi:10.1007/s11077-011-9125-7.

Hultman, N.E. et al., 2020: Fusing subnational with national climate action is central to decarbonization: the case of the United States. *Nat. Commun.*, **11(1)**, 5255, doi:10.1038/s41467-020-18903-w.

Humby, T.-L., 2018: The Thabametsi Case: Case No 65662/16 Earthlife Africa Johannesburg v Minister of Environmental Affairs. *J. Environ. Law*, **30(1)**, 145–155, doi:10.1093/jel/eqy007.

Hurlstone, M.J., S. Lewandowsky, B.R. Newell, and B. Sewell, 2014: The Effect of Framing and Normative Messages in Building Support for Climate Policies. *PLoS One*, **9(12)**, e114335, doi:10.1371/journal.pone.0114335.

Huttunen, S., P. Kivimaa, and V. Virkamäki, 2014: The need for policy coherence to trigger a transition to biogas production. *Environ. Innov. Soc. Transitions*, **12**, 14–30, doi:10.1016/j.eist.2014.04.002.

Hvelplund, F. and S. Djørup, 2017: Multilevel policies for radical transition: Governance for a 100% renewable energy system. *Environ. Plan. C Polit. Sp.*, **35(7)**, 1218–1241, doi:10.1177/2399654417710024.
Iacobuta, G., N.K. Dubash, P. Upadhyaya, M. Deribe, and N. Höhne, 2018: National climate change mitigation legislation, strategy and targets: a global update. *Clim. Policy*, **18(9)**, 1114–1132, doi:10.1080/14693062.2018.1489772.
IEA, 2015: *Energy and climate change: World energy outlook special report*. International Energy Agency, (IEA) and Organisation for Economic Co-operation and Development (OECD), Paris, France, 200 pp. https://iea.blob.core.windows.net/assets/8d783513-fd22-463a-b57d-a0d8d608d86f/WEO2015SpecialReportonEnergyandClimateChange.pdf (Accessed March 20, 2021).
IEA, 2020: IEA Energy subsidies: Tracking the impact of fossil-fuel subsidies. International Energy Agency, (IEA). https://www.iea.org/topics/energy-subsidies (Accessed December 21, 2020).
IEA, 2021: IEA Policies Database. International Energy Agency, (IEA). https://www.iea.org/policies (Accessed October 28, 2021).
Ihobe – Environmental Management Agency, 2017: *Nature-based solutions for local climate adaptation in the Basque Country: Methodological guide for their identification and mapping*. Ministry of the Environment, Territorial Planning and Housing, Basque Government, Bilbao, 91 pp. https://www.ihobe.eus/Publicaciones/Ficha.aspx?IdMenu=97801056-cd1f-4503-bafa-f54fa80d9a44&Cod=adbf2e51-3d8c-4879-ab8d-9a7ab8d48e45&Idioma=en-GB.
IISD, 2019: *Raising Ambition Through Fossil Fuel Subsidy Reform: Greenhouse gas emissions modelling results from 26 countries*. International Institute for Sustainable Development (IISD), Winnipeg, Canada, and Geneva, Switzerland, 54 pp. https://www.iisd.org/system/files/publications/raising-ambition-through-fossil-fuel-subsidy-reform.pdf (Accessed March 30, 2021).
Ikeme, J., 2003: Equity, environmental justice and sustainability: incomplete approaches in climate change politics. *Glob. Environ. Change*, **13(3)**, 195–206, doi:10.1016/S0959-3780(03)00047-5.
Inderberg, T.H.J., 2019: Climate Change Acts as diffusing governance innovations: adoption and implications of the UK model in Finland and Norway. *Proceeds of 14th Nordic Environmental Social Science (NESS) Conference, Luleå University of Technology, Sweden*, Luleå, Sweden, 9–11.
Inglehart, R. and P. Norris, 2017: Trump and the Populist Authoritarian Parties: The Silent Revolution in Reverse. *Perspect. Polit.*, **15(2)**, 443–454, doi:10.1017/S1537592717000111.
Ingold, K. and M. Fischer, 2014: Drivers of collaboration to mitigate climate change: An illustration of Swiss climate policy over 15 years. *Glob. Environ. Change*, **24**, 88–98, doi:10.1016/j.gloenvcha.2013.11.021.
IPBES, 2019: Summary for policymakers of the global assessment report on biodiversity and ecosystem services of the Intergovernmental Science-Policy Platform on Biodiversity and Ecosystem Services. [Díaz, S., J. Settele, E. S. Brondízio, H.T. Ngo, M. Guèze et al., (eds.)]. Intergovernmental Science-Policy Platform on Biodiversity and Ecosystem Services (IPBES), Bonn, Germany, 56 pp.
IPCC, 2007: *Climate Change 2007: Impacts, Adaptation and Vulnerability. Contribution of Working Group II to the Fourth Assessment Report of the Intergovernmental Panel on Climate Change*. [Parry, M., O. Canziani, J. Palutikof, P. Van der Linden, and C. Hanson, (eds.)]. Cambridge University Press, Cambridge, UK and New York, USA, 976 pp.
IPCC, 2014a: Summary for Policymakers. In: *Climate Change 2014: Impacts, Adaptation and Vulnerability. Contribution of Working Group II to the Fifth Assessment Report of the Intergovernmental Panel on Climate Change*, [Field, C.B., V.R. Barros, D.J. Dokken, K.J. Mach, and M.D. Mastrandrea, (eds.)]. Cambridge University Press, Cambridge, UK and New York, NY, USA, 1–32 pp.
IPCC, 2014b: *Climate Change 2014: Synthesis Report. Contribution of Working Groups I, II and III to the Fifth Assessment Report of the Intergovernmental Panel on Climate Change* [Core Writing Team, R.K. Pachauri and L.A. Meyer (eds.)]. IPCC, Geneva, Switzerland, 151 pp.
IPCC, 2018: *Global Warming of 1.5°C. An IPCC Special Report on the impacts of global warming of 1.5°C above pre-industrial levels and related global greenhouse gas emission pathways, in the context of strengthening the global response to the threat of climate change, sustainable development, and efforts to eradicate poverty* [Masson-Delmotte, V., P. Zhai, H.-O. Pörtner, D. Roberts, J. Skea, P.R. Shukla, A. Pirani, W. Moufouma-Okia, C. Péan, R. Pidcock, S. Connors, J.B.R. Matthews, Y. Chen, X. Zhou, M.I. Gomis, E. Lonnoy, T. Maycock, M. Tignor, and T. Waterfield (eds.)]. Cambridge University Press, Cambridge, UK and New York, NY, USA.
IRENA, 2019: *Renewable energy auctions: Status and trends beyond price*. International Renewable Energy Agency (IRENA), Abu Dhabi, 104 pp. https://www.irena.org/publications/2019/Dec/Renewable-energy-auctions-Status-and-trends-beyond-price (Accessed October 27, 2021).
Irwin, A. and B. Wynne, eds., 1996: *Misunderstanding Science?The Public Reconstruction of Science and Technology*. Cambridge University Press, Cambridge, UK, 232 pp.
Islam, S. and M.Z.R. Khan, 2017: a Review of Energy Sector of Bangladesh. *Energy Procedia*, **110**, 611–618, doi:10.1016/j.egypro.2017.03.193.
Ismer, R., K. Neuhoff, and A. Pirlot, 2020: *Border Carbon Adjustments and Alternative Measures for the EU ETS: An Evaluation*. Deutsches Institut für Wirtschaftsforschung (DIW), Berlin, Germany, 23 pp. https://www.diw.de/documents/publikationen/73/diw_01.c.743698.de/dp1855.pdf (Accessed October 18, 2021).
Jaccard, M., R. Murphy, B. Zuehlke, and M. Braglewicz, 2019: Cities and greenhouse gas reduction: Policy makers or policy takers? *Energy Policy*, **134**, 110875, doi:10.1016/j.enpol.2019.07.011.
Jackson, A., 2020: Ireland's Climate Action and Low Carbon Development Act 2015: Symbolic Legislation, Systemic Litigation, Stepping Stone? In: *National Climate Change Acts: The Emergence, Form and Nature of National Framework Climate Legislation* [Muinzer, T.L., (ed.)]. Bloomsbury, London, UK, pp. 129–152.
Jacobs, B. and F. van der Ploeg, 2019: Redistribution and pollution taxes with non-linear Engel curves. *J. Environ. Econ. Manage.*, **95**, 198–226, doi:10.1016/j.jeem.2019.01.008.
Jacques, P.J., R.E. Dunlap, and M. Freeman, 2008: The organisation of denial: Conservative think tanks and environmental scepticism. *Env. Polit.*, **17(3)**, 349–385, doi:10.1080/09644010802055576.
Jaffe, A.B. and R.N. Stavins, 1995: Dynamic Incentives of Environmental Regulations: The Effects of Alternative Policy Instruments on Technology Diffusion. *J. Environ. Econ. Manage.*, **29(3)**, S43–S63, doi:10.1006/jeem.1995.1060.
Jaffe, A.B., R.G. Newell, and R.N. Stavins, 2002: Environmental policy and technological change. *Environ. Resour. Econ.*, **22**, 41–70, doi:10.1023/A:1015519401088.
Jahn, D., 2016: *The Politics of Environmental Performance: Institutions and Preferences in Industrialized Democracies*. Cambridge University Press, Cambridge, UK, 376 pp.
Jakob, M., C. Chen, S. Fuss, A. Marxen, and O. Edenhofer, 2015: Development incentives for fossil fuel subsidy reform. *Nat. Clim. Change*, **5(8)**, 709–712, doi:10.1038/nclimate2679.
Jang, M., J.-A. Kim, and S.-T. Sun, 2010: Development and Evaluation of Laws and Regulation for the Low-Carbon and Green Growth in Korea. *Int. J. Urban Sci.*, **14(2)**, 191–206, doi:10.1080/12265934.2010.9693676.
Jänicke, M. and R.K.W. Wurzel, 2019: Leadership and lesson-drawing in the European Union's multilevel climate governance system. *Env. Polit.*, **28(1)**, 22–42, doi:10.1080/09644016.2019.1522019.
Jasny, L. and D.R. Fisher, 2019: Echo chambers in climate science. *Environ. Res. Commun.*, **1(10)**, 101003, doi:10.1088/2515-7620/ab491c.
Jasny, L., J. Waggle, and D.R. Fisher, 2015: An empirical examination of echo chambers in US climate policy networks. *Nat. Clim. Change*, **5(8)**, 782–786, doi:10.1038/nclimate2666.
Jaumotte, F. et al., 2020: *Mitigating climate change – Growth- and distribution-friendly strategies*. International Monetary Fund, Washington, DC, USA,

13

85–114 pp. https://www.imf.org/en/Publications/WEO/Issues/2020/09/30/world-economic-outlook-october-2020.

Jaumotte, F., W. Liu, and W.J. McKibbin, 2021: *Mitigating Climate Change: Growth-Friendly Policies to Achieve Net Zero Emissions by 2050*. International Monetary Fund, Washington, DC, USA, https://www.imf.org/en/Publications/WP/Issues/2021/07/23/Mitigating-Climate-Change-Growth-Friendly-Policies-to-Achieve-Net-Zero-Emissions-by-2050-462136 (Accessed September 24, 2021).

Jenkins, K., D. McCauley, R. Heffron, H. Stephan, and R. Rehner, 2016: Energy Justice: a conceptual review. *Energy Res. Soc. Sci.*, **11**, 174–182, doi:10.1016/j.erss.2015.10.004.

Jensen, C.B. and J.-J. Spoon, 2011: Testing the 'Party Matters' Thesis: Explaining Progress Towards Kyoto Protocol Targets. *Polit. Stud.*, **59(1)**, 99–115, doi:10.1111/j.1467-9248.2010.00852.x.

Jewell, J. and A. Cherp, 2020: On the political feasibility of climate change mitigation pathways: Is it too late to keep warming below 1.5°C? *Wiley Interdiscip. Rev. Clim. Change*, **11(1)**, e621, doi:10.1002/wcc.621.

Jewell, J. et al., 2018: Limited emission reductions from fuel subsidy removal except in energy-exporting regions. *Nature*, **554(7691)**, 229–233, doi:10.1038/nature25467.

Jewell, J., V. Vinichenko, L. Nacke, and A. Cherp, 2019: Prospects for powering past coal. *Nat. Clim. Change*, **9(8)**, 592–597, doi:10.1038/s41558-019-0509-6.

Joas, F. and C. Flachsland, 2016: The (ir)relevance of transaction costs in climate policy instrument choice: an analysis of the EU and the US. *Clim. Policy*, **16(1)**, 26–49, doi:10.1080/14693062.2014.968762.

Jodoin, S., 2017: *Forest Preservation in a Changing Climate: REDD+ and Indigenous and Community Rights in Indonesia and Tanzania*. 1st ed. Cambridge University Press, Cambridge, UK, 252 pp.

Jodoin, S., S. Snow, and A. Corobow, 2020: Realizing the Right to Be Cold? Framing Processes and Outcomes Associated with the Inuit Petition on Human Rights and Global Warming. *Law Soc. Rev.*, **54(1)**, 168–200, doi:10.1111/lasr.12458.

Joffe, H. and N. Smith, 2016: City dweller aspirations for cities of the future: How do environmental and personal well-being feature? *Cities*, **59**, 102–112, doi:10.1016/j.cities.2016.06.006.

Jogesh, A. and N.K. Dubash, 2015: State-led experimentation or centrally-motivated replication? a study of state action plans on climate change in India. *J. Integr. Environ. Sci.*, **12(4)**, 247–266, doi:10.1080/1943815X.2015.1077869.

Johnsson, F., J. Kjärstad, and J. Rootzén, 2019: The threat to climate change mitigation posed by the abundance of fossil fuels. *Clim. Policy*, **19(2)**, 258–274, doi:10.1080/14693062.2018.1483885.

Johnstone, N., I. Haščič, J. Poirier, M. Hemar, and C. Michel, 2012: Environmental policy stringency and technological innovation: evidence from survey data and patent counts. *Appl. Econ.*, **44(17)**, 2157–2170, doi:10.1080/00036846.2011.560110.

Joltreau, E. and K. Sommerfeld, 2019: Why does emissions trading under the EU Emissions Trading System (ETS) not affect firms' competitiveness? Empirical findings from the literature. *Clim. Policy*, **19(4)**, 453–471, doi:10.1080/14693062.2018.1502145.

Jones, S., 2013: Climate Change Policies of City Governments in Federal Systems: An Analysis of Vancouver, Melbourne and New York City. *Reg. Stud.*, **47(6)**, 974–992, doi:10.1080/00343404.2011.585150.

Jones, S., 2014: Flirting with Climate Change: a Comparative Policy Analysis of Sub-national Governments in Canada and Australia. *J. Comp. Policy Anal. Res. Pract.*, **16(5)**, 424–440, doi:10.1080/13876988.2014.942570.

Jordaan, S.M., A. Davidson, J.A. Nazari, and I.M. Herremans, 2019: The dynamics of advancing climate policy in federal political systems. *Environ. Policy Gov.*, **29(3)**, 220–234, doi:10.1002/eet.1849.

Jordan, A., D. Huitema, H. Asselt and J. Forster, 2018a: *Governing Climate Change: Polycentricity in Action?* Cambridge University Press, Cambridge, UK.

Jordan, A., D. Huitema, H. van Asselt, and J. Forster, eds., 2018: *Governing Climate Change: Polycentricity in Action?* 1st ed. Cambridge University Press, Cambridge, UK, 408 pp.

Jordan, A.J. and B. Moore, 2020: *Durable by Design? Policy Feedback in a Changing Climate*. Cambridge University Press, Cambridge, UK, 276 pp.

Jordan, A.J. et al., 2015: Emergence of polycentric climate governance and its future prospects. *Nat. Clim. Change*, **5(11)**, 977–982, doi:10.1038/nclimate2725.

Jörgensen, K., A. Jogesh, and A. Mishra, 2015a: Multi-level climate governance and the role of the subnational level. *J. Integr. Environ. Sci.*, **12(4)**, 235–245, doi:10.1080/1943815X.2015.1096797.

Jörgensen, K., A. Mishra, and G.K. Sarangi, 2015b: Multi-level climate governance in India: the role of the states in climate action planning and renewable energies. *J. Integr. Environ. Sci.*, **12(4)**, 267–283, doi:10.1080/1943815X.2015.1093507.

Jorgenson, A.K., C. Dick, and J.M. Shandra, 2011: World Economy, World Society, and Environmental Harms in Less-Developed Countries*. *Sociol. Inq.*, **81(1)**, 53–87, doi:10.1111/j.1475-682X.2010.00354.x.

Jost, G.F. and K. Jacob, 2004: The climate change policy network in Germany. *Eur. Environ.*, **14(1)**, 1–15, doi:10.1002/eet.337.

Jotzo, F., T. Jordan, and N. Fabian, 2012: Policy Uncertainty about Australia's Carbon Price: Expert Survey Results and Implications for Investment. *Aust. Econ. Rev.*, **45(4)**, 395–409, doi:10.1111/j.1467-8462.2012.00709.x.

Jotzo, F. et al., 2018: China's emissions trading takes steps towards big ambitions. *Nat. Clim. Change*, **8**, 265–267, doi:10.1038/s41558-018-0130-0.

Jotzo, F., T. Longden, and Z. Anjum, 2020: *Fiscal stimulus for low-carbon compatible COVID-19 recovery: criteria for infrastructure investment*. Crawford School of Public Policy Centre for Climate & Energy Policy, ANU, Canberra, Australia, 43 pp. https://www.energy-transition-hub.org/files/resource/attachment/ccep2005_low-carbon_stimulus_-_jotzo_longden_anjum.pdf (Accessed March 30, 2021).

Jung, J., P. Petkanic, D. Nan, and J.H. Kim, 2020: When a Girl Awakened the World: a User and Social Message Analysis of Greta Thunberg. *Sustainability*, **12(7)**, 2707, doi:10.3390/su12072707.

Kadefors, A., S. Lingegård, S. Uppenberg, J. Alkan-Olsson, and D. Balian, 2020: Designing and implementing procurement requirements for carbon reduction in infrastructure construction – international overview and experiences. *J. Environ. Plan. Manag.*, **64(4)**, 1–24, doi:10.1080/09640568.2020.1778453.

Kahsar, R., 2021: The soft path revisited: Policies that drive decentralization of electric power generation in the contiguous U.S. *Energy Policy*, **156**, 112429, doi:10.1016/j.enpol.2021.112429.

Kalfagianni, A. and O. Kuik, 2017: Seeking optimality in climate change agri-food policies: stakeholder perspectives from Western Europe. *Clim. Policy*, **17**, S72–S92, doi:10.1080/14693062.2016.1244508.

Kamiya, G. and J. Teter, 2019: Shared, automated… and electric? – Analysis. International Energy Agency (IEA), Paris, France, https://www.iea.org/commentaries/shared-automated-and-electric (Accessed March 30, 2021).

Kammerer, M. and C. Namhata, 2018: What drives the adoption of climate change mitigation policy? a dynamic network approach to policy diffusion. *Policy Sci.*, **51(4)**, 477–513, doi:10.1007/s11077-018-9332-6.

Kammerer, M. et al., 2021: What Explains Collaboration in High and Low Conflict Contexts? Comparing Climate Change Policy Networks in Four Countries. *Policy Stud. J.*, **49(4)**, 1065–1086, doi:10.1111/psj.12422.

Kanger, L., B.K. Sovacool, and M. Noorkõiv, 2020: Six policy intervention points for sustainability transitions: a conceptual framework and a systematic literature review. *Res. Policy*, **49(7)**, 104072, doi:10.1016/j.respol.2020.104072.

Karani, P. and P. Failler, 2020: Comparative coastal and marine tourism, climate change, and the blue economy in African Large Marine Ecosystems. *Environ. Dev.*, **36**(January), 100572, doi:10.1016/j.envdev.2020.100572.

Karceski, S.M., N. Dolšak, A. Prakash, and T.N. Ridout, 2020: Did TV ads funded by fossil fuel industry defeat the Washington carbon tax? *Clim. Change*, **158(3–4)**, 301–307, doi:10.1007/s10584-019-02626-z.

Karlsson, M., E. Alfredsson, and N. Westling, 2020: Climate policy co-benefits: a review. *Clim. Policy*, **20(3)**, 292–316, doi:10.1080/14693062.2020.1724070.

Kashwan, P., 2015: Forest Policy, Institutions, and REDD+ in India, Tanzania, and Mexico. *Glob. Environ. Polit.*, **15(3)**, 95–117, doi:10.1162/GLEP_a_00313.

Keele, D.M., 2017: Climate Change Litigation and the National Environmental Policy Act. *J. Environ. Law*, **30(2)**, 285–309, doi:10.1093/jel/eqx030.

Keenan, R.J., L. Caripis, A. Foerster, L. Godden, and J. Peel, 2012: Science and the governance of Australia's climate regime. *Nat. Clim. Change*, **2(7)**, 477–478, doi:10.1038/nclimate1608.

Kennedy, C., S. Demoullin, and E. Mohareb, 2012: Cities reducing their greenhouse gas emissions. *Energy Policy*, **49**, 774–777, doi:10.1016/j.enpol.2012.07.030.

Kern, F. and K.S. Rogge, 2018: Harnessing theories of the policy process for analysing the politics of sustainability transitions: a critical survey. *Environ. Innov. Soc. Transitions*, **27**(November 2017), 102–117, doi:10.1016/j.eist.2017.11.001.

Kern, F., K.S. Rogge, and M. Howlett, 2019: Policy mixes for sustainability transitions: New approaches and insights through bridging innovation and policy studies. *Res. Policy*, **48(10)**, 103832, doi:10.1016/j.respol.2019.103832.

Kernecker, T. and M. Wagner, 2019: Niche parties in Latin America. *J. Elections, Public Opin. Parties*, **29(1)**, 102–124, doi:10.1080/17457289.2018.1464014.

Khan, F. and B.K. Sovacool, 2016: Testing the efficacy of voluntary urban greenhouse gas emissions inventories. *Clim. Change*, **139(2)**, 141–154, doi:10.1007/s10584-016-1793-z.

Khan, M., D. Mfitumukiza, and S. Huq, 2020: Capacity building for implementation of Nationally Determined Contributions under the Paris Agreement. *Clim. Policy*, **20(4)**, 499–510, doi:10.1080/14693062.2019.1675577.

Khreis, H., A.D. May, and M.J. Nieuwenhuijsen, 2017: Health impacts of urban transport policy measures: a guidance note for practice. *J. Transp. Heal.*, **6**, 209–227, doi:10.1016/j.jth.2017.06.003.

Kim, E., 2016: The politics of climate change policy design in Korea. *Env. Polit.*, **25(3)**, 454–474, doi:10.1080/09644016.2015.1104804.

Kim, S.E., J. Urpelainen, and J. Yang, 2016: Electric utilities and American climate policy: lobbying by expected winners and losers. *J. Public Policy*, **36(2)**, 251–275, doi:10.1017/S0143814X15000033.

Kinniburgh, C., 2019: Climate Politics after the Yellow Vests. *Dissent*, **66(2)**, 115–125, doi:10.1353/dss.2019.0037.

Kissinger, G., A. Gupta, I. Mulder, and N. Unterstell, 2019: Climate financing needs in the land sector under the Paris Agreement: An assessment of developing country perspectives. *Land-use policy*, **83**, 256–269, doi:10.1016/j.landusepol.2019.02.007.

Kivimaa, P. and F. Kern, 2016: Creative destruction or mere niche support? Innovation policy mixes for sustainability transitions. *Res. Policy*, **45(1)**, 205–217, doi:10.1016/j.respol.2015.09.008.

Kivimaa, P., D. Huitema, A. Jordan, and J. Newig, 2017a: Experiments in climate governance – a systematic review of research on energy and built environment transitions. *J. Clean. Prod.*, **169**, 17–29, doi:10.1016/J.JCLEPRO.2017.01.027.

Kivimaa, P., H.-L. Kangas, and D. Lazarevic, 2017b: Client-oriented evaluation of 'creative destruction' in policy mixes: Finnish policies on building energy efficiency transition. *Energy Res. Soc. Sci.*, **33(SI)**, 115–127, doi:10.1016/j.erss.2017.09.002.

Klausbruckner, C., H. Annegarn, L.R.F. Henneman, and P. Rafaj, 2016: A policy review of synergies and trade-offs in South African climate change mitigation and air pollution control strategies. *Environ. Sci. Policy*, **57**, 70–78, doi:10.1016/j.envsci.2015.12.001.

Klein, R.J.T., E.L.F. Schipper, and S. Dessai, 2005: Integrating mitigation and adaptation into climate and development policy: three research questions. *Environ. Sci. Policy*, **8(6)**, 579–588, doi:10.1016/j.envsci.2005.06.010.

Kleinschmidt, J., 2020: *HFC refrigerant prices fell again in the third quarter of 2019*. Öko_recherche, Frankfurt, 2 pp. https://www.oekorecherche.de/sites/default/files/publikationen/vdkf_jk_preismonitoring_3-2019.pdf (Accessed October 29, 2021).

Klenert, D. et al., 2018: Making carbon pricing work for citizens. *Nat. Clim. Change*, **8(8)**, 669–677, doi:10.1038/s41558-018-0201-2.

Klinsky, S., 2018: An initial scoping of transitional justice for global climate governance. *Clim. Policy*, **18(6)**, 752–765, doi:10.1080/14693062.2017.1377594.

Kluczek, A. and P. Olszewski, 2017: Energy audits in industrial processes. *J. Clean. Prod.*, **142(4)**, 3437–3453, doi:10.1016/j.jclepro.2016.10.123.

Kneese, A.V. and C.L. Schultze, 1975: *Pollution, prices, and public policy*. The Brookings Institution, Washington, DC, USA, 136 pp.

Knodt, M., M. Ringel, and R. Müller, 2020: 'Harder' soft governance in the European Energy Union. *J. Environ. Policy Plan.*, **22(6)**, 787–800, doi:10.1080/1523908X.2020.1781604.

Knox-Hayes, J., 2016: *The Cultures of Markets: The Political Economy of Climate Governance*. 1st ed. Oxford University Press, New York, USA, 352 pp.

Koch, N. and H. Basse Mama, 2019: Does the EU Emissions Trading System induce investment leakage? Evidence from German multinational firms. *Energy Econ.*, **81**, 479–492, doi:10.1016/j.eneco.2019.04.018.

Köhler, J. et al., 2019: An agenda for sustainability transitions research: State of the art and future directions. *Environ. Innov. Soc. Transitions*, **31**, 1–32, doi:10.1016/j.eist.2019.01.004.

Kollenberg, S. and L. Taschini, 2019: Dynamic supply adjustment and banking under uncertainty in an emission trading scheme: The market stability reserve. *Eur. Econ. Rev.*, **118**, 213–226, doi:10.1016/j.euroecorev.2019.05.013.

Kona, A. et al., 2016: *Covenant of Mayors: Greenhouse Gas Emissions Achievements and Projections*. Publications Office of the European Union, Luxembourg, 59 pp. https://op.europa.eu/en/publication-detail/-/publication/ae6b48ff-aa54-11e6-aab7-01aa75ed71a1/language-en.

Kong, N. and S. Hardman, 2019: *UC Davis Research Reports Title Electric Vehicle Incentives in 13 Leading Electric Vehicle Markets*. Plug-in Hybrid & Electric Vehicle Research Center, UC Davis, California, USA, 1–22 pp. https://escholarship.org/uc/item/0fm3x5bh (Accessed March 30, 2021).

Kongsager, R. and E. Corbera, 2015: Linking Mitigation and Adaptation in Carbon Forestry Projects: Evidence from Belize. *World Dev.*, **76**, 132–146, doi:10.1016/j.worlddev.2015.07.003.

Kongsager, R., B. Locatelli, and F. Chazarin, 2016: Addressing Climate Change Mitigation and Adaptation Together: a Global Assessment of Agriculture and Forestry Projects. *Environ. Manage.*, **57(2)**, 271–282, doi:10.1007/s00267-015-0605-y.

Kooij, H.-J. et al., 2018: Between grassroots and treetops: Community power and institutional dependence in the renewable energy sector in Denmark, Sweden and the Netherlands. *Energy Res. Soc. Sci.*, **37**, 52–64, doi:10.1016/j.erss.2017.09.019.

Koplow, D., 2018: Defining and Measuring Fossil Fuel Subsidies. In: *The Politics of Fossil Fuel Subsidies and their Reform* [Skovgaard, J. and H. Van Asselt, (eds.)]. Cambridge University Press, Cambridge, England, UK, pp. 23–46.

Kotilainen, K. et al., 2019: From path dependence to policy mixes for Nordic electric mobility: Lessons for accelerating future transport transitions. *Policy Sci.*, **52(4)**, 573–600, doi:10.1007/s11077-019-09361-3.

Kotze, L. and A. du Plessis, 2019: Putting Africa on the Stand: a Bird's Eye View of Climate Change Litigation on the Continent. *Univ. Oregon's J. Environ. Law Litig.*, **50(3)**, 615–663.

Krause, R.M., H. Yi, and R.C. Feiock, 2016: Applying Policy Termination Theory to the Abandonment of Climate Protection Initiatives by U.S. Local Governments. *Policy Stud. J.*, **44(2)**, 176–195, doi:10.1111/psj.12117.

Kriegler, E. et al., 2018: Short term policies to keep the door open for Paris climate goals. *Environ. Res. Lett.*, **13(7)**, 074022, doi:10.1088/1748-9326/aac4f1.

Krzywoszynska, A. et al., 2016: Co-producing energy futures: impacts of participatory modelling. *Build. Res. Inf.*, **44(7)**, 804–815, doi:10.1080/09613218.2016.1211838.

Kukkonen, A. et al., 2018: International organizations, advocacy coalitions, and domestication of global norms: Debates on climate change in Canada, the US, Brazil, and India. *Environ. Sci. Policy*, **81**, 54–62, doi:10.1016/J.ENVSCI.2017.12.008.

Kulovesi, K. and S. Oberthür, 2020: Assessing the EU's 2030 Climate and Energy Policy Framework: Incremental change toward radical transformation? *Rev. Eur. Comp. Int. Environ. Law*, **29(2)**, 151–166, doi:10.1111/reel.12358.

Kundzewicz, Z.W. et al., 2018: Uncertainty in climate change impacts on water resources. *Environ. Sci. Policy*, **79**(October 2017), 1–8, doi:10.1016/j.envsci.2017.10.008.

Kuramochi, T. et al., 2020: Beyond national climate action: the impact of region, city, and business commitments on global greenhouse gas emissions. *Clim. Policy*, **20(3)**, 275–291, doi:10.1080/14693062.2020.1740150.

Kuusi, T. et al., 2020: *Carbon Border Adjustment Mechanisms and Their Economic Impact on Finland and the EU*. ETLA Economic Research, Helsinki, Finland, https://www.etla.fi/en/publications/carbon-border-adjustment-mechanisms-and-their-economic-impact-on-finland-and-the-eu/ (Accessed October 18, 2021).

Kuzemko, C., M. Lockwood, C. Mitchell, and R. Hoggett, 2016: Governing for sustainable energy system change: Politics, contexts and contingency. *Energy Res. Soc. Sci.*, **12**, 96–105, doi:10.1016/j.erss.2015.12.022.

Kwakkel, J.H., W.E. Walker, and M. Haasnoot, 2016: Coping with the Wickedness of Public Policy Problems: Approaches for Decision-Making under Deep Uncertainty. *J. Water Resour. Plan. Manag.*, **142(3)**, 01816001, doi:10.1061/(ASCE)WR.1943-5452.0000626.

Kyle, J., 2018: Local Corruption and Popular Support for Fuel Subsidy Reform in Indonesia. *Comp. Polit. Stud.*, **51(11)**, 1472–1503, doi:10.1177/0010414018758755.

La Porta, R., F. Lopez-de-Silanes, A. Shleifer, and R.W. Vishny, 1998: Law and Finance. *J. Polit. Econ.*, **106(6)**, 1113–1155, doi:10.1086/250042.

Labeaga, J.M., X. Labandeira, and X. López-Otero, 2021: Energy taxation, subsidy removal and poverty in Mexico. *Environ. Dev. Econ.*, **26(3)**, 239–260, doi:10.1017/S1355770X20000364.

Lachapelle, E. and M. Paterson, 2013: Drivers of national climate policy. *Clim. Policy*, **13(5)**, 547–571, doi:10.1080/14693062.2013.811333.

Lachapelle, E., C.P. Borick, and B. Rabe, 2012: Public Attitudes toward Climate Science and Climate Policy in Federal Systems: Canada and the United States Compared. *Rev. Policy Res.*, **29(3)**, 334–357, doi:10.1111/j.1541-1338.2012.00563.x.

Ladrech, R. and C. Little, 2019: Drivers of political parties' climate policy preferences: lessons from Denmark and Ireland. *Env. Polit.*, **28(6)**, 1017–1038, doi:10.1080/09644016.2019.1625157.

Lægreid, O.M. and M. Povitkina, 2018: Do Political Institutions Moderate the GDP-$CO_2$ Relationship? *Ecol. Econ.*, **145**, 441–450, doi:10.1016/j.ecolecon.2017.11.014.

Lam, S.-P., 2015: Predicting support of climate policies by using a protection motivation model. *Clim. Policy*, **15(3)**, 321–338, doi:10.1080/14693062.2014.916599.

Lamb, W.F. and J.C. Minx, 2020: The political economy of national climate policy: Architectures of constraint and a typology of countries. *Energy Res. Soc. Sci.*, **64**, 101429, doi:10.1016/j.erss.2020.101429.

Landauer, M., S. Juhola, and M. Söderholm, 2015: Inter-relationships between adaptation and mitigation: a systematic literature review. *Clim. Change*, **131**, 505–517, doi:10.1007/s10584-015-1395-1.

Landauer, M., S. Juhola, and J. Klein, 2019: The role of scale in integrating climate change adaptation and mitigation in cities. *J. Environ. Plan. Manag.*, **62(5)**, 741–765, doi:10.1080/09640568.2018.1430022.

Langpap, C., 2015: Voluntary agreements and private enforcement of environmental regulation. *J. Regul. Econ.*, **47(1)**, 99–116, doi:10.1007/s11149-014-9265-8.

Laurent, B. and D. Pontille, 2018: Towards a study of city experiments. In: *Creating Smart Cities* [Coletta, C., L. Evans, L. Heaphy, and R. Kitchin, (eds.)]. Routledge, Oxfordshire, UK, pp. 90–103.

Le Treut, G., J. Lefèvre, F. Lallana, and G. Bravo, 2021: The multi-level economic impacts of deep decarbonization strategies for the energy system. *Energy Policy*, **156**, 1–16, doi:10.1016/j.enpol.2021.112423.

Leck, H. and D. Roberts, 2015: What lies beneath: Understanding the invisible aspects of municipal climate change governance. *Curr. Opin. Environ. Sustain.*, **13**, 61–67, doi:10.1016/j.cosust.2015.02.004.

Lecuyer, O. and P. Quirion, 2019: Interaction between $CO_2$ emissions trading and renewable energy subsidies under uncertainty: feed-in tariffs as a safety net against over-allocation. *Clim. Policy*, **19(8)**, doi:10.1080/14693062.2019.1625743.

Lee, T., 2013: Global Cities and Transnational Climate Change Networks. *Glob. Environ. Polit.*, **13(1)**, 108–127, doi:10.1162/GLEP_a_00156.

Lee, T., 2017: The effect of clean energy regulations and incentives on green jobs: panel analysis of the United States, 1998-2007. *Nat. Resour. Forum*, **41(3)**, 145–155, doi:10.1111/1477-8947.12125.

Lee, T., 2019: Network comparison of socialization, learning and collaboration in the C40 cities climate group. *J. Environ. Policy Plan.*, **21(1)**, 104–115, doi:10.1080/1523908X.2018.1433998.

Lee, T. and C. Koski, 2012: Building Green: Local Political Leadership Addressing Climate Change. *Rev. Policy Res.*, **29(5)**, 605–624, doi:10.1111/j.1541-1338.2012.00579.x.

Lee, T. and H.Y. Jung, 2018: Mapping city-to-city networks for climate change action: Geographic bases, link modalities, functions, and activity. *J. Clean. Prod.*, **182**, 96–104, doi:10.1016/j.jclepro.2018.02.034.

Lee, T.M., E.M. Markowitz, P.D. Howe, C.-Y. Ko, and A.A. Leiserowitz, 2015: Predictors of public climate change awareness and risk perception around the world. *Nat. Clim. Change*, **5(11)**, 1014–1020, doi:10.1038/nclimate2728.

Leipold, S., P.H. Feindt, G. Winkel, and R. Keller, 2019: Discourse analysis of environmental policy revisited: traditions, trends, perspectives. *J. Environ. Policy Plan.*, **21(5)**, 445–463, doi:10.1080/1523908X.2019.1660462.

Leiren, M.D. and I. Reimer, 2018: Historical institutionalist perspective on the shift from feed-in tariffs towards auctioning in German renewable energy policy. *Energy Res. Soc. Sci.*, **43**, 33–40, doi:10.1016/j.erss.2018.05.022.

Leiserowitz, A., 2006: Climate Change Risk Perception and Policy Preferences: The Role of Affect, Imagery, and Values. *Clim. Change*, **77(1–2)**, 45–72, doi:10.1007/s10584-006-9059-9.

Leiter, T. and P. Pringle, 2018: *Pitfalls and Potential of Measuring Climate Change Adaptation through Adaptation Metrics*. UNEP DTU Partnership, Copenhagen, Denmark, Denmark, 166 pp. https://resilientcities2018.iclei.org/wp-content/uploads/UDP_Perspectives-Adaptation-Metrics-WEB.pdf. (Accessed September 24, 2021).

Leiter, T. et al., 2019: *Adaptation metrics: current landscape and evolving practices*. UNEP DTU Partnership Global Commission on Adaptation, Rotterdam and Washington, DC, USA, 51 pp.

Lennon, M., 2020: Postcarbon Amnesia: Toward a Recognition of Racial Grief in Renewable Energy Futures. *Sci. Technol. Hum. Values*, **45(5)**, 934–962, doi:10.1177/0162243919900556.

Letnik, T., M. Marksel, G. Luppino, A. Bardi, and S. Božičnik, 2018: Review of policies and measures for sustainable and energy efficient urban transport. *Energy*, **163**, 245–257, doi:10.1016/j.energy.2018.08.096.

Levenda, A.M., J. Richter, T. Miller, and E. Fisher, 2019: Regional sociotechnical imaginaries and the governance of energy innovations. *Futures*, **109**, 181–191, doi:10.1016/j.futures.2018.03.001.

Levi, S., C. Flachsland, and M. Jakob, 2020: Political Economy Determinants of Carbon Pricing. *Glob. Environ. Polit.*, **20(2)**, 128–156, doi:10.1162/glep_a_00549.

Levin, K. et al., 2018: *Long-Term Low Greenhouse Gas Emission Development Strategies: Approaches and Methodologies for Their Design*. World Resources Institute (WRI), Washington DC and UNDP, New York, USA, 52 pp. https://files.wri.org/s3fs-public/long-term-low-greenhouse-gas-emission-development-strategies.pdf.

Levin, K., D. Rich, K. Ross, T. Fransen, and C. Elliott, 2020: *Designing and Communicating Net-Zero Targets*. World Resources Institute (WRI), Washington, DC, USA, 30 pp. https://www.wri.org/research/designing-and-communicating-net-zero-targets (Accessed October 20, 2021).

Levinson, A., 2019: Energy Efficiency Standards Are More Regressive Than Energy Taxes: Theory and Evidence. *J. Assoc. Environ. Resour. Econ.*, **6(S1)**, doi:10.1086/701186.

Lewis, J., D. Hernández, and A.T. Geronimus, 2020: Energy efficiency as energy justice: addressing racial inequities through investments in people and places. *Energy Effic.*, **13(3)**, 419–432, doi:10.1007/s12053-019-09820-z.

Lewison, R. et al., 2015: Dynamic Ocean Management: Identifying the Critical Ingredients of Dynamic Approaches to Ocean Resource Management. *Bioscience*, **65(5)**, 486–498, doi:10.1093/biosci/biv018.

Li, A., Y. Xu, and H. Shiroyama, 2019: Solar lobby and energy transition in Japan. *Energy Policy*, **134**, 110950, doi:10.1016/j.enpol.2019.110950.

Li, H., X. Zhao, Y. Yu, T. Wu, and Y. Qi, 2016: China's numerical management system for reducing national energy intensity. *Energy Policy*, **94**, 64–76, doi:10.1016/j.enpol.2016.03.037.

Li, H. et al., 2017: Reducing rebound effect through fossil subsidies reform: a comprehensive evaluation in China. *J. Clean. Prod.*, **141**, 305–314, doi:10.1016/j.jclepro.2016.09.108.

Li, M. et al., 2018: Air quality co-benefits of carbon pricing in China. *Nat. Clim. Change*, **8(5)**, 398, doi:10.1038/s41558-018-0139-4.

Liang Wong, I. and E. Krüger, 2017: Comparing energy efficiency labelling systems in the EU and Brazil: Implications, challenges, barriers and opportunities. *Energy Policy*, **109**, 310–323, doi:10.1016/j.enpol.2017.07.005.

Liefferink, D., B. Arts, J. Kamstra, and J. Ooijevaar, 2009: Leaders and laggards in environmental policy: a quantitative analysis of domestic policy outputs. *J. Eur. Public Policy*, **16(5)**, 677–700, doi:10.1080/13501760902983283.

Lieu, J. et al., 2018: Evaluating Consistency in Environmental Policy Mixes through Policy, Stakeholder, and Contextual Interactions. *Sustainability*, **10(6)**, 1896, doi:10.3390/su10061896.

Lilliestam, J., A. Patt, and G. Bersalli, 2021: The effect of carbon pricing on technological change for full energy decarbonization: a review of empirical ex-post evidence. *WIREs Clim. Change*, **12(1)**, e681, doi:10.1002/wcc.681.

Lin, J., 2012a: Climate Governance in China: Using the 'Iron Hand'. In: *Local Climate Change Law: Environmental Regulation in Cities and Other Localities* [Richardson, B.J., (ed.)]. Edward Elgar Publishing, Cheltenham, UK and Northampton, MA, USA, pp. 424.

Lin, J., 2012b: Climate change and the courts. *Leg. Stud.*, **32(1)**, 35–57, doi:10.1111/j.1748-121X.2011.00206.x.

Lin, J. and D.A. Kysar, eds., 2020: *Climate Change Litigation in the Asia Pacific*. Cambridge University Press, Cambridge, UK, 428 pp.

Liu, J.-Y. et al., 2019: Identifying trade-offs and co-benefits of climate policies in China to align policies with SDGs and achieve the 2°C goal. *Environ. Res. Lett.*, **14(12)**, 1–11, doi:10.1088/1748-9326/ab59c4.

Liu, T., Q. Wang, and B. Su, 2016: a review of carbon labeling: Standards, implementation, and impact. *Renew. Sustain. Energy Rev.*, **53**, 68–79, doi:10.1016/j.rser.2015.08.050.

Liu, W., W.J. McKibbin, A.C. Morris, and P.J. Wilcoxen, 2020: Global economic and environmental outcomes of the Paris Agreement. *Energy Econ.*, **90**, 104838, doi:10.1016/j.eneco.2020.104838.

Liu, X., E. Lindquist, and A. Vedlitz, 2011: Explaining Media and Congressional Attention to Global Climate Change, 1969-2005: An Empirical Test of Agenda-Setting Theory. *Polit. Res. Q.*, **64(2)**, 405–419, doi:10.1177/1065912909346744.

Lo, A.Y., 2013: Carbon trading in a socialist market economy: Can China make a difference? *Ecol. Econ.*, **87**, 72–74, doi:10.1016/j.ecolecon.2012.12.023.

Lo, A.Y. and R. Cong, 2017: After CDM: Domestic carbon offsetting in China. *J. Clean. Prod.*, **141**, 1391–1399, doi:10.1016/j.jclepro.2016.09.220.

Loaiza, T., M.O. Borja, U. Nehren, and G. Gerold, 2017: Analysis of land management and legal arrangements in the Ecuadorian Northeastern Amazon as preconditions for REDD+ implementation. *For. Policy Econ.*, **83**(June), 19–28, doi:10.1016/j.forpol.2017.05.005.

Locatelli, B., V. Evans, A. Wardell, A. Andrade, and R. Vignola, 2011: Forests and Climate Change in Latin America: Linking Adaptation and Mitigation. *Forests*, **2(1)**, 431–450, doi:10.3390/f2010431.

Locatelli, B. et al., 2017: Research on climate change policies and rural development in Latin America: Scope and gaps. *Sustain.*, **9(10)**, 1–17, doi:10.3390/su9101831.

Lockwood, M., 2013: The political sustainability of climate policy: The case of the UK Climate Change Act. *Glob. Environ. Change*, **23(5)**, 1339–1348, doi:10.1016/j.gloenvcha.2013.07.001.

Lockwood, M., 2021a: a hard Act to follow? The evolution and performance of UK climate governance. *Env. Polit.*, **30(sup1)**, 26–48, doi:10.1080/09644016.2021.1910434.

Lockwood, M., 2021b: Routes to credible climate commitment: the UK and Denmark compared. *Clim. Policy*, **21(9)**, 1–14, doi:10.1080/14693062.2020.1868391.

Lockwood, M., C. Kuzemko, C. Mitchell, and R. Hoggett, 2017: Historical institutionalism and the politics of sustainable energy transitions: a research agenda. *Environ. Plan. C Polit. Sp.*, **35(2)**, 312–333, doi:10.1177/0263774X16660561.

Long, J. and J.L. Rice, 2019: From sustainable urbanism to climate urbanism. *Urban Stud.*, **56(5)**, 992–1008, doi:10.1177/0042098018770846.

Longhofer, W. and A. Jorgenson, 2017: Decoupling reconsidered: Does world society integration influence the relationship between the environment and economic development? *Soc. Sci. Res.*, **65**, 17–29, doi:10.1016/j.ssresearch.2017.02.002.

Longhofer, W., E. Schofer, N. Miric, and D.J. Frank, 2016: NGOs, INGOs, and Environmental Policy Reform, 1970–2010. *Soc. Forces*, **94(4)**, 1743–1768, doi:10.1093/sf/sow031.

Lorenzoni, I. and D. Benson, 2014: Radical institutional change in environmental governance: Explaining the origins of the UK Climate Change Act 2008 through discursive and streams perspectives. *Glob. Environ. Change*, **29**, 10–21, doi:10.1016/j.gloenvcha.2014.07.011.

Loth, M.A., 2016: Climate Change Liability After All: a Dutch Landmark Case. *Tilbg. Law Rev.*, **21(1)**, doi:10.1163/22112596-02101001.

Lovelock, C.E. and R. Reef, 2020: Variable Impacts of Climate Change on Blue Carbon. *One Earth*, **3(2)**, 195–211, doi:10.1016/j.oneear.2020.07.010.

LSE Grantham Research Institute on Climate Change and the Environment, 2021: Climate Change Laws of the World. https://climate-laws.org/ (Accessed March 30, 2021).

Lück, J., H. Wessler, A. Wozniak, and D. Lycarião, 2018: Counterbalancing global media frames with nationally colored narratives: a comparative study of news narratives and news framing in the climate change coverage of five countries. *Journalism*, **19(12)**, 1635–1656, doi:10.1177/1464884916680372.

Lyon, T.P., 2016: Drivers and Impacts of Renewable Portfolio Standards. *Annu. Rev. Resour. Econ.*, **8(1)**, 141–155, doi:10.1146/annurev-resource-100815-095432.

Lyytimäki, J., 2011: Mainstreaming climate policy: the role of media coverage in Finland. *Mitig. Adapt. Strateg. Glob. Change*, **16(6)**, 649–661, doi:10.1007/s11027-011-9286-x.

Ma, S.-C., Y. Fan, and L. Feng, 2017: An evaluation of government incentives for new energy vehicles in China focusing on vehicle purchasing restrictions. *Energy Policy*, **110**(October), 609–618, doi:10.1016/j.enpol.2017.07.057.

MacArthur, J.L., C.E. Hoicka, H. Castleden, R. Das, and J. Lieu, 2020: Canada's Green New Deal: Forging the socio-political foundations of climate resilient infrastructure? *Energy Res. Soc. Sci.*, **65**, 101442, doi:10.1016/j.erss.2020.101442.

Macchi, C. and J. Zeben, 2021: Business and human rights implications of climate change litigation: Milieudefensie et al. vs Royal Dutch Shell. *Rev. Eur. Comp. Int. Environ. Law*, **30(3)**, 409–415, doi:10.1111/reel.12416.

MacNeil, R., 2021: Swimming against the current: Australian climate institutions and the politics of polarisation. *Env. Polit.*, **30(sup1)**, 162–183, doi:10.1080/09644016.2021.1905394.

Madden, N.J., 2014: Green means stop: veto players and their impact on climate-change policy outputs. *Env. Polit.*, **23(4)**, 570–589, doi:10.1080/09644016.2014.884301.

Maestre-Andrés, S., S. Drews, and J. van den Bergh, 2019: Perceived fairness and public acceptability of carbon pricing: a review of the literature. *Clim. policy*, **19(9)**, 1186–1204, doi.org/10.1080/14693062.2019.1639490.

Magro, E., M. Navarro, and J.M. Zabala-Iturriagagoitia, 2014: Coordination-Mix: The Hidden Face of STI Policy. *Rev. Policy Res.*, **31(5)**, 367–389, doi:10.1111/ropr.12090.

Mahdavi, P., C.B. Martinez-Alvarez, and M. Ross, 2020: *Why Do Governments Tax or Subsidize Fossil Fuels?* Center for Global Development, Washington, DC, USA, 72 pp. https://www.cgdev.org/sites/default/files/why-do-governments-tax-or-subsidize-fossil-fuels.pdf (Accessed October 19, 2021).

Maier, H.R. et al., 2016: An uncertain future, deep uncertainty, scenarios, robustness and adaptation: How do they fit together? *Environ. Model. Softw.*, **81**, 154–164, doi:10.1016/j.envsoft.2016.03.014.

Marjanac, S. and L. Patton, 2018: Extreme weather event attribution science and climate change litigation: an essential step in the causal chain? *J. Energy Nat. Resour. Law*, **36(3)**, 265–298, doi:10.1080/02646811.2018.1451020.

Marjanac, S., L. Patton, and J. Thornton, 2017: Acts of God, human influence and litigation. *Nat. Geosci.*, **10(9)**, 616–619, doi:10.1038/ngeo3019.

Markard, J., 2018: The next phase of the energy transition and its implications for research and policy. *Nat. Energy*, **3(8)**, 628–633, doi:10.1038/s41560-018-0171-7.

Markard, J., F.W. Geels, and R. Raven, 2020: Challenges in the acceleration of sustainability transitions. *Environ. Res. Lett.*, **15(8)**, 081001, doi:10.1088/1748-9326/ab9468.

Markell, D. and J.B. Ruhl, 2012: An Empirical Assessment of Climate Change in the Courts: a New Jurisprudence or Business as Usual. *Fla. Law Rev.*, **64(1)**, 15–86.

Markkanen, S. and A. Anger-Kraavi, 2019: Social impacts of climate change mitigation policies and their implications for inequality. *Clim. Policy*, **19(7)**, 827–844, doi:10.1080/14693062.2019.1596873.

Markowitz, E.M. and M.L. Guckian, 2018: Climate change communication: Challenges, insights, and opportunities. In: *Psychology and Climate Change: Human Perceptions, Impacts, and Responses* [Clayton S. and C. Manning (eds.)]. Academic Press, Elsevier, London, UK, pp. 35–63.

Marlon, J.R. et al., 2019: How Hope and Doubt Affect Climate Change Mobilization. *Front. Commun.*, **4**, 20 pp., doi:10.3389/fcomm.2019.00020.

Marlow, T., S. Miller, and J.T. Roberts, 2021: Bots and online climate discourses: Twitter discourse on President Trump's announcement of U.S. withdrawal from the Paris Agreement. *Clim. Policy*, **21(6)**, 765–777, doi:10.1080/14693062.2020.1870098.

Marquardt, J., 2014: a Struggle of Multi-level Governance: Promoting Renewable Energy in Indonesia. *Energy Procedia*, **58**, 87–94, doi:10.1016/j.egypro.2014.10.413.

Marris, E., 2019: Why young climate activists have captured the world's attention. *Nature*, **573(7775)**, 471–472, doi:10.1038/d41586-019-02696-0.

Martin, M. and M. Islar, 2021: The 'end of the world' vs. the 'end of the month': understanding social resistance to sustainability transition agendas, a lesson from the Yellow Vests in France. *Sustain. Sci.*, **16(2)**, 601–614, doi:10.1007/s11625-020-00877-9.

Martinez Romera, B. and R. Caranta, 2017: EU Public Procurement Law: Purchasing beyond Price in the Age of Climate Change. *Eur. Procure. Pub. Priv. Partnersh. L. Rev.*, **12**, 281, doi:10.21552/epppl/2017/3/10.

Martinez, S., S. Kralisch, O. Escolero, and M. Perevochtchikova, 2015: Vulnerability of Mexico City's water supply sources in the context of climate change. *J. Water Clim. Change*, **6(3)**, 518–533, doi:10.2166/wcc.2015.083.

Martiskainen, M., 2017: The role of community leadership in the development of grassroots innovations. *Environ. Innov. Soc. Transitions*, **22**, 78–89, doi:10.1016/j.eist.2016.05.002.

Martiskainen, M. et al., 2020: Contextualizing climate justice activism: Knowledge, emotions, motivations, and actions among climate strikers in six cities. *Glob. Environ. Change*, **65**, 102180, doi:10.1016/j.gloenvcha.2020.102180.

Marvin, S., H. Bulkeley, L. Mai, K. McCormick, and Y. Voytenko Palgan, eds., 2018: *Urban Living Labs: Experimenting with City Futures*. 1st ed. Routledge, Oxfordshire, England, UK, 279 pp.

Mason, C.F. and A.J. Plantinga, 2013: The additionality problem with offsets: Optimal contracts for carbon sequestration in forests. *J. Environ. Econ. Manage.*, **66(1)**, 1–14, doi:10.1016/j.jeem.2013.02.003.

Matsumoto, K., K. Morita, D. Mavrakis, and P. Konidari, 2017: Evaluating Japanese policy instruments for the promotion of renewable energy sources. *Int. J. Green Energy*, **14(8)**, 724–736, doi:10.1080/15435075.2017.1326050.

Matthes, F.C., 2017: Energy transition in Germany: a case study on a policy-driven structural change of the energy system. *Evol. Institutional Econ. Rev.*, **14(1)**, 141–169, doi:10.1007/s40844-016-0066-x.

Mauritsen, T. and R. Pincus, 2017: Committed warming inferred from observations. *Nat. Clim. Change*, **7(9)**, 652–655, doi:10.1038/nclimate3357.

Mayer, A., T.O. Shelley, T. Chiricos, and M. Gertz, 2017: Environmental Risk Exposure, Risk Perception, Political Ideology and Support for Climate Policy. *Sociol. Focus*, **50(4)**, 309–328, doi:10.1080/00380237.2017.1312855.

Mayer, B., 2019: The State of the Netherlands v. Urgenda Foundation: Ruling of the Court of Appeal of The Hague (9 October 2018). *Transnatl. Environ. Law*, **8(1)**, 167–192, doi:10.1017/S2047102519000049.

Mayer, J., 2016: *Dark Money: the Hidden History of the Billionaires Behind the Rise of the Radical Right*. Doubleday, New York, USA, 576 pp.

Mayrhofer, J.P. and J. Gupta, 2016: The science and politics of co-benefits in climate policy. *Environ. Sci. Policy*, **57**, 22–30, doi:10.1016/j.envsci.2015.11.005.

Mazmanian, D.A., J.L. Jurewitz, and H.T. Nelson, 2020: State Leadership in U.S. Climate Change and Energy Policy: The California Experience. *J. Environ. Dev.*, **29(1)**, 51–74, doi:10.1177/1070496519887484.

Mbeva, K., C. Ochieng, J. Atela, W. Khaemba, and C. Tonui, 2015: *Intended Nationally Determined Contributions as a Means to Strengthening Africa's Engagement in International Climate Negotiations*. African Centre for Technology Studies, Nairobi, Kenya 1–28 pp. https://media.africaportal.org/documents/INDC_PaperFin.pdf (Accessed March 30, 2021).

McAdam, D., 2017: Social Movement Theory and the Prospects for Climate Change Activism in the United States. *Annu. Rev. Polit. Sci.*, **20(1)**, 189–208, doi:10.1146/annurev-polisci-052615-025801.

McAdam, D. and H.S. Boudet, 2012: *Putting social movements in their place: Explaining opposition to energy projects in the United States, 2000–2005*. Cambridge University Press, Cambridge, UK, 1–266 pp.

McAllister, L. et al., 2021: Balance as bias, resolute on the retreat? Updates & analyses of newspaper coverage in the United States, United Kingdom, New Zealand, Australia and Canada over the past 15 years. *Environ. Res. Lett.*, **16(9)**, 094008, doi:10.1088/1748-9326/ac14eb.

McAllister, L.K., 2007: Revisiting a Promising Institution: Public Law Litigation in the Civil Law World. *Ga. State Univ. Law Rev.*, **24(3)**, 693–734.

McAllister, R.R.J., R. McCrea, and M.N. Lubell, 2014: Policy networks, stakeholder interactions and climate adaptation in the region of South East Queensland, Australia. *Reg. Environ. Change*, **14(2)**, 527–539, doi:10.1007/s10113-013-0489-4.

McBurney, M., 2021: Paying for Ecological Services in Ecuador: The Socio Bosque Program in Kichwa Communities of Chimborazo, Ecuador. The University of Guelph, Ontario, Canada, 280 pp. https://idl-bnc-idrc.dspacedirect.org/bitstream/handle/10625/59975/IDL – 59975.pdf?sequence=2&isAllowed=y.

McCauley, D. and R. Heffron, 2018: Just Transition: Integrating climate, energy and environmental justice. *Energy Policy*, **119**, 1–7, doi:10.1016/j.enpol.2018.04.014.

McCloskey, M., 1991: Twenty years of change in the environmental movement: An insider's view. *Soc. Nat. Resour.*, **4(3)**, 273–284, doi:10.1080/08941929109380760.

McCollum, D.L. et al., 2018: Connecting the sustainable development goals by their energy inter-linkages. *Environ. Res. Lett.*, **13(3)**, 033006, doi:10.1088/1748-9326/aaafe3.

McCord, G.C., D.R. Kanter, J. Sklarew, G. Wu, and M. Jacobson, 2020: *Chapter 5.5: Accelerating Sustainable Land Use Practices in the U.S.* Sustainable Development Solutions Network, New York, USA, 262–281 pp. https://www.unsdsn.org/Zero-Carbon-Action-Plan (Accessed October 20, 2021).

McCormick, S. et al., 2018: Strategies in and outcomes of climate change litigation in the United States. *Nat. Clim. Change*, **8(9)**, 829–833, doi:10.1038/s41558-018-0240-8.

McCright, A.M. and R.E. Dunlap, 2000: Challenging Global Warming as a Social Problem: An Analysis of the Conservative Movement's Counter-Claims. *Soc. Probl.*, **47(4)**, 499–522, doi:10.2307/3097132.

McCright, A.M. and R.E. Dunlap, 2003: Defeating Kyoto: The Conservative Movement's Impact on U.S. Climate Change Policy. *Soc. Probl.*, **50(3)**, 348–373, doi:10.1525/sp.2003.50.3.348.

McCright, A.M., R.E. Dunlap, and S.T. Marquart-Pyatt, 2016a: Political ideology and views about climate change in the European Union. *Env. Polit.*, **25(2)**, 338–358, doi:10.1080/09644016.2015.1090371.

McCright, A.M., S.T. Marquart-Pyatt, R.L. Shwom, S.R. Brechin, and S. Allen, 2016b: Ideology, capitalism, and climate: Explaining public views about climate change in the United States. *Energy Res. Soc. Sci.*, **21**, 180–189, doi:10.1016/j.erss.2016.08.003.

McGlade, C. and P. Ekins, 2015: The geographical distribution of fossil fuels unused when limiting global warming to 2°C. *Nature*, **517(7533)**, 187–190, doi:10.1038/nature14016.

McGregor, P.G., J. Kim Swales, and M.A. Winning, 2012: a review of the role and remit of the committee on climate change. *Energy Policy*, **41**, 466–473, doi:10.1016/j.enpol.2011.11.007.

Mcguirk, P., R. Dowling, C. Brennan, and H. Bulkeley, 2015: Urban Carbon Governance Experiments: The Role of Australian Local Governments. *Geogr. Res.*, **53(1)**, 39–52, doi:10.1111/1745-5871.12098.

McMeekin, A., F.W. Geels, and M. Hodson, 2019: Mapping the winds of whole system reconfiguration: Analysing low-carbon transformations across production, distribution and consumption in the UK electricity system (1990–2016). *Res. Policy*, **48(5)**, 1216–1231, doi:10.1016/j.respol.2018.12.007.

Meadowcroft, J., 2009: What about the politics? Sustainable development, transition management, and long term energy transitions. *Policy Sci.*, **42(4)**, 323–340, doi:10.1007/s11077-009-9097-z.

Meckling, J., 2011: *Carbon Coalitions: Business, Climate Politics, and the Rise of Emissions Trading*. MIT Press, Cambridge, MA, USA, 264 pp.

Meckling, J. and J. Nahm, 2018: The power of process: state capacity and climate policy. *Governance*, **31(4)**, 741–757, doi:10.1111/gove.12338.

Meckling, J., N. Kelsey, E. Biber, and J. Zysman, 2015: Winning coalitions for climate policy: Green industrial policy builds support for carbon regulations. *Science*, **349(6253)**, 1170–1171, doi:10.1126/science.aab1336.

Mehling, M.A., H. Van Asselt, K. Das, S. Droege, and C. Verkuijl, 2019: Designing Border Carbon Adjustments for Enhanced Climate Action. *Am. J. Int. Law*, **113(3)**, 433–481, doi:10.1017/ajil.2019.22.

Melamed, M.L., J. Schmale, and E. von Schneidemesser, 2016: Sustainable policy—key considerations for air quality and climate change. *Curr. Opin. Environ. Sustain.*, **23**, 85–91, doi:10.1016/j.cosust.2016.12.003.

Mercure, J.F. et al., 2019: System complexity and policy integration challenges: The Brazilian Energy- Water-Food Nexus. *Renew. Sustain. Energy Rev.*, **105**(February), 230–243, doi:10.1016/j.rser.2019.01.045.

Messori, G., F. Brocchieri, E. Morello, S. Ozgen, and S. Caserini, 2020: a climate mitigation action index at the local scale: Methodology and case study. *J. Environ. Manage.*, **260**, 110024, doi:10.1016/j.jenvman.2019.110024.

Metcalf, G.E. and J.H. Stock, 2017: Integrated Assessment Models and the Social Cost of Carbon: a Review and Assessment of U.S. Experience. *Rev. Environ. Econ. Policy*, **11(1)**, 80–99, doi:10.1093/reep/rew014.

Metcalf, G.E. and J.H. Stock, 2020: *The Macroeconomic Impact of Europe's Carbon Taxes*. 53 pp. https://www.nber.org/system/files/working_papers/w27488/w27488.pdf.

Michaelowa, A., M. Allen, and F. Sha, 2018: Policy instruments for limiting global temperature rise to 1.5°C – can humanity rise to the challenge? *Clim. Policy*, **18(3)**, 275–286, doi:10.1080/14693062.2018.1426977.

Michaelowa, A., L. Hermwille, W. Obergassel, and S. Butzengeiger, 2019a: Additionality revisited: guarding the integrity of market mechanisms under the Paris Agreement. *Clim. Policy*, **19(10)**, 1211–1224, doi:10.1080/14693062.2019.1628695.

Michaelowa, A., I. Shishlov, S. Hoch, P. Bofil, and A. Espelage, 2019b: *Overview and comparison of existing carbon crediting schemes*. Nordic Environment Finance Corporation (NEFCO), Helsinki, Finland, 54 pp. https://www.nefco.int/wp-content/uploads/2019/05/NICA-Crediting-Mechanisms-Final-February-2019.pdf (Accessed October 18, 2021).

Michaelowa, K. and A. Michaelowa, 2017: Transnational Climate Governance Initiatives: Designed for Effective Climate Change Mitigation? *Int. Interact.*, **43(1)**, 129–155, doi:10.1080/03050629.2017.1256110.

Middlemiss, L., 2011: The effects of community-based action for sustainability on participants' lifestyles. *Local Environ.*, **16(3)**, 265–280, doi:10.1080/13549839.2011.566850.

Mildenberger, M., 2020: *Carbon Captured: How Business and Labor Control Climate Politics*. MIT Press, Cambridge, Massachusetts, USA, 368 pp.

Mildenberger, M., 2021: The development of climate institutions in the United States. *Env. Polit.*, **30(sup1)**, 71–92, doi:10.1080/09644016.2021.1947445.

Mildenberger, M. and D. Tingley, 2019: Beliefs about Climate Beliefs: The Importance of Second-Order Opinions for Climate Politics. *Br. J. Polit. Sci.*, **49(4)**, 1279–1307, doi:10.1017/S0007123417000321.

Millard-Ball, A. and L. Ortolano, 2010: Constructing carbon offsets: The obstacles to quantifying emission reductions. *Energy Policy*, **38(1)**, 533–546, doi:10.1016/j.enpol.2009.10.005.

Miller, S.M. et al., 2019: China's coal mine methane regulations have not curbed growing emissions. *Nat. Commun.*, **10(1)**, 303, doi:10.1038/s41467-018-07891-7.

Mills-Novoa, M. and D.M. Liverman, 2019: Nationally Determined Contributions: Material climate commitments and discursive positioning in the NDCs. *WIREs Clim. Change*, **10(5)**, e589, doi:10.1002/wcc.589.

Minnerop, P. and F.E.L. Otto, 2020: Climate change and causation: joining law and climate science on the basis of formal logic. *Buffalo Environ. Law J.*, **27**, 49–86.

Minx et al., 2021: A comprehensive and synthetic dataset for global, regional, and national greenhouse gas emissions by sector 1970–2018 with an extension to 2019. *Earth Syst. Sci. Data*, **13**, 5213–5252, doi:10.5194/essd-13-5213-2021.

Mistry, R., 2014: *MAPS Provocateur Briefing Report Forum on Development and Mitigation-Hunger, Poverty, Inequality and Climate Change Mitigation*. Oxfam, South Africa, Johannesburg, 7 pp. https://www.africaportal.org/publications/maps-provocateur-briefing-report-forum-on-development-and-mitigation-hunger-poverty-inequality-and-climate-change-mitigation/.

13

Moallemi, E.A. and S. Malekpour, 2018: A participatory exploratory modelling approach for long-term planning in energy transitions. *Energy Res. Soc. Sci.*, **35**, 205–216, doi:10.1016/j.erss.2017.10.022.

Moe Singh, E., 2012: Structural Change, Vested Interests, and Scandinavian Energy Policy-Making: Why Wind Power Struggles in Norway and not in Denmark. *Open Renew. Energy J.*, **5(1)**, 19–31, doi:10.2174/1876387101205010019.

Mohtar, R.H. and B. Daher, 2014: *A Platform for Trade-off Analysis and Resource Allocation: The Water-Energy-Food Nexus Tool and its Application to Qatar's Food Security*. The Royal Institute of International Affairs, Chatham House, London, UK, 10 pp. https://agrilife.org/wefnexus/files/2015/01/20141216WaterEnergyFoodNexusQatarMohtarDaher.pdf.

Mol, A.P.J., D.A. Sonnenfeld, and G. Spaargaren, eds., 2009: *The Ecological Modernisation Reader: Environmental reform in theory and practice*. 1st ed. Routledge, Oxfordshire, England, UK, 560 pp.

Monasterolo, I. and M. Raberto, 2019: The impact of phasing out fossil fuel subsidies on the low-carbon transition. *Energy Policy*, **124**, 355–370, doi:10.1016/j.enpol.2018.08.051.

Montgomery, L., P.A. McLaughlin, T. Richards, and M. Febrizio, 2019: *Performance Standards vs. Design Standards: Facilitating a Shift toward Best Practices*. Mercatus Center, George Mason University Arlington, VA, USA, 37 pp.

Moreira, P.F. et al., 2019: South–south transnational advocacy: Mobilizing against Brazilian dams in the Peruvian amazon. *Glob. Environ. Polit.*, **19(1)**, 77–98, doi:10.1162/glep_a_00495.

Morris, C. and A. Jungjohann, 2016: *Energy Democracy: Germany's Energiewende to Renewables*. Palgrave Macmillan, London, UK, 437 pp.

Moss, T. and M. Bazilian, 2018: Signalling, governance, and goals: Reorienting the United States Power Africa initiative. *Energy Res. Soc. Sci.*, **39**, 74–77, doi:10.1016/j.erss.2017.11.001.

Mourao, P.R., 2019: The effectiveness of Green voices in parliaments: Do Green Parties matter in the control of pollution? *Environ. Dev. Sustain.*, **21(2)**, 985–1011, doi:10.1007/s10668-017-0070-2.

Mulugetta, Y. and V. Castán Broto, 2018: Harnessing deep mitigation opportunities of urbanisation patterns in LDCs. *Curr. Opin. Environ. Sustain.*, **30**, 82–88, doi:10.1016/j.cosust.2018.03.007.

Mundaca, L. and A. Markandya, 2016: Assessing regional progress towards a 'Green Energy Economy'. *Appl. Energy*, **179**, 1372–1394, doi:10.1016/j.apenergy.2015.10.098.

Muradian, R. and U. Pascual, 2020: Ecological economics in the age of fear. *Ecol. Econ.*, **169**, doi:10.1016/j.ecolecon.2019.106498.

Murray, B. and N. Rivers, 2015: British Columbia's revenue-neutral carbon tax: A review of the latest 'grand experiment' in environmental policy. *Energy Policy*, **86**, 674–683, doi:10.1016/j.enpol.2015.08.011.

Nachmany, M. and J. Setzer, 2018: *Global trends in climate change legislation and litigation: 2018 snapshot*. Grantham Research Institute on Climate Change, London, UK, 8 pp. https://www.lse.ac.uk/granthaminstitute/wp-content/uploads/2018/04/Global-trends-in-climate-change-legislation-and-litigation-2018-snapshot-3.pdf.

Nachmany, M. et al., 2015: *The 2015 Global Climate Legislation Study: a review of climate change legislation in 99 countries: summary for policy-makers*. Grantham Research Institute on Climate Change and the Environment, GLOBE and Inter-Parliamentary Union, London, UK, 45 pp.

Nago, M. and M. Krott, 2020: Systemic failures in north–south climate change knowledge transfer: a case study of the Congo Basin. *Clim. Policy*, **22 (5)** 1–14, doi:10.1080/14693062.2020.1820850.

Nagorny-Koring, N.C., 2019: Leading the way with examples and ideas? Governing climate change in German municipalities through best practices. *J. Environ. Policy Plan.*, **21(1)**, 46–60, doi:10.1080/1523908X.2018.1461083.

Nagorny-Koring, N.C. and T. Nochta, 2018: Managing urban transitions in theory and practice – The case of the Pioneer Cities and Transition Cities projects. *J. Clean. Prod.*, **175**, 60–69, doi:10.1016/j.jclepro.2017.11.072.

Nakano, R. et al., 2017: Low Carbon Governance in Indonesia and India: A Comparative Analysis with Recommendations. *Procedia Eng.*, **198**, 570–588, doi:10.1016/j.proeng.2017.07.112.

Nalau, J. and J. Handmer, 2015: When is transformation a viable policy alternative? *Environ. Sci. Policy*, **54**, 349–356, doi:10.1016/j.envsci.2015.07.022.

Narassimhan, E., K.S. Gallagher, S. Koester, and J.R. Alejo, 2018: Carbon pricing in practice: a review of existing emissions trading systems. *Clim. Policy*, **18(8)**, 967–991, doi:10.1080/14693062.2018.1467827.

Nascimento, L. et al., 2021: Twenty years of climate policy: G20 coverage increased, but important gaps remain (accepted for publication). *Clim. Policy*, **22(2)**, 158–174, doi.org/10.1080/14693062.2021.1993776.

Nash, S.L. and R. Steurer, 2019: Taking stock of Climate Change Acts in Europe: living policy processes or symbolic gestures? *Clim. Policy*, **19(8)**, 1–14, doi:10.1080/14693062.2019.1623164.

Nasiritousi, N., 2017: Fossil fuel emitters and climate change: unpacking the governance activities of large oil and gas companies. *Env. Polit.*, **26(4)**, 621–647, doi:10.1080/09644016.2017.1320832.

Nemet, G.F., T. Holloway, and P. Meier, 2010: Implications of incorporating air-quality co-benefits into climate change policymaking. *Environ. Res. Lett.*, **5(1)**, 014007, doi:10.1088/1748-9326/5/1/014007.

Nemet, G.F., M. Jakob, J.C. Steckel, and O. Edenhofer, 2017: Addressing policy credibility problems for low-carbon investment. *Glob. Environ. Change*, **42**, 47–57, doi:10.1016/j.gloenvcha.2016.12.004.

Neuhoff, K. and R.A. Ritz, 2019: *Carbon cost pass-through in industrial sectors*. 25 pp. http://www.jstor.org/stable/resrep30282 (Accessed March 30, 2021).

Neumayer, E., 2003: Are left-wing party strength and corporatism good for the environment? Evidence from panel analysis of air pollution in OECD countries. *Ecol. Econ.*, **45(2)**, 203–220, doi:10.1016/S0921-8009(03)00012-0.

Nevens, F., N. Frantzeskaki, L. Gorissen, and D. Loorbach, 2013: Urban Transition Labs: Co-creating transformative action for sustainable cities. *J. Clean. Prod.*, **50**, 111–122, doi:10.1016/j.jclepro.2012.12.001.

Newbery, D., 2018: Evaluating the case for supporting renewable electricity. *Energy Policy*, **120**, 684–696, doi:10.1016/j.enpol.2018.05.029.

NewClimate Institute, Data-Driven Lab, PBL Netherlands, German Development Institute, and Blavatnik School of Government, University of Oxford, 2019: *Global Climate Action from Cities, Regions and Businesses – 2019. Impact of Individual Actors and Cooperative Initiatives on Global an National Emissions.* 94 pp. https://newclimate.org/wp-content/uploads/2019/09/Report-Global-Climate-Action-from-Cities-Regions-and-Businesses_2019.pdf (Accessed March 30, 2021).

NewClimate Institute and Data Driven EnviroLab, 2020: *Navigating the nuances of net-zero targets.* [Day, S.M. et al. (eds.)]. NewClimate Institute, Cologne, Data-Driven Lab, Yale-NUS, Singapore80 pp. https://newclimate.org/2020/10/22/navigating-the-nuances-of-net-zero-targets/ (Accessed October 18, 2021).

Newell, P., 2021a: *Power Shift: The Global Political Economy of Energy Transitions*. 1st ed. Cambridge University Press, Cambridge, UK, 284 pp.

Newell, P., 2021b: The Business of Climate Transformation. *Curr. Hist.*, **120(829)**, 307–312, doi:10.1525/curh.2021.120.829.307.

Newell, P. and M. Paterson, 2010: *Climate Capitalism: Global Warming and the Transformation of the Global Economy*. Cambridge University Press, Cambridge, UK, 205 pp.

Newell, P. and A. Simms, 2020: How Did We Do That? Histories and Political Economies of Rapid and Just Transitions. *New Polit. Econ.*, **25**, 1–16, doi:10.1080/13563467.2020.1810216.

Newell, R.G., W.A. Pizer, and D. Raimi, 2013: Carbon markets 15 years after Kyoto: Lessons learned, new challenges. *J. Econ. Perspect.*, **27(1)**, 123–146, doi:10.1257/jep.27.1.123.

Ney, S. and M. Verweij, 2015: Messy institutions for wicked problems: How to generate clumsy solutions? *Environ. Plan. C Gov. Policy*, **33(6)**, 1679–1696, doi:10.1177/0263774X15614450.

Nogueira, L.P., F.D. Longa, and B. van der Zwaan, 2020: A cross-sectoral integrated assessment of alternatives for climate mitigation in Madagascar. *Clim. Policy*, **20(10)**, 1257–1273, doi:10.1080/14693062.2020.1791030.

Nordbeck, R. and R. Steurer, 2016: Multi-sectoral strategies as dead ends of policy integration: Lessons to be learned from sustainable development. *Environ. Plan. C Gov. Policy*, **34(4)**, 737–755, doi:10.1177/0263774X15614696.

Nordhaus, W., 2013: *The climate casino: Risk, uncertainty, and economics for a warming world*. Yale University Press, New Haven, USA, 392 pp.

Nordic Council of Ministers, 2017: *Mitigation & Adaptation Synergies in the NDCs*. 86 pp. http://norden.diva-portal.org/smash/get/diva2:1097909/FULLTEXT01.pdf (Accessed March 30, 2021).

Nosek, G., 2018: Climate change litigation and narrative: How to use litigation to tell compelling climate stories. *William Mary Environ. Law Policy Rev.*, **42(3)**, 733–804.

Nunan, F., A. Campbell, and E. Foster, 2012: Environmental mainstreaming: The organisational challenges of policy integration. *Public Adm. Dev.*, **32(3)**, 262–277, doi:10.1002/pad.1624.

Nyiwul, L.M., 2019: Climate Change Mitigation and Adaptation in Africa: Strategies, Synergies, and Constraints. In: *Climate Change and Global Development: Market, Global Players and Empirical Evidence* [Sequeira, T. and L. Reis, (eds.)]. Springer, Cham, Switzerland, pp. 219–241.

O'Brien, K., 2018: Is the 1.5°C target possible? Exploring the three spheres of transformation. *Curr. Opin. Environ. Sustain.*, **31**, 153–160, doi:10.1016/j.cosust.2018.04.010.

O'Brien, K., E. Selboe, and B.M. Hayward, 2018: Exploring youth activism on climate change: dutiful, disruptive, and dangerous dissent. *Ecol. Soc.*, **23(3)**, 42, doi:10.5751/ES-10287-230342.

O'Callaghan, B. and E. Murdock, 2021: *Are we building back better? Evidence from 2020 and Pathways to Inclusive Green Recovery Spending*. UNEP Nairobi, Kenya, 60 pp (Accessed October 27, 2021). https://www.unep.org/resources/publication/are-we-building-back-better-evidence-2020-and-pathways-inclusive-green.

O'Donnell, T., 2019: Contrasting land-use policies for climate change adaptation: A case study of political and geo-legal realities for Australian coastal locations. *Land use policy*, **88** (September), 104145, doi:10.1016/j.landusepol.2019.104145.

O'Dwyer, B. and J. Unerman, 2020: Shifting the focus of sustainability accounting from impacts to risks and dependencies: researching the transformative potential of TCFD reporting. *Accounting, Audit. Account. J.*, **33(5)**, 1113–1141, doi:10.1108/AAAJ-02-2020-4445.

O'Neill, S. et al., 2015: Dominant frames in legacy and social media coverage of the IPCC Fifth Assessment Report. *Nat. Clim. Change*, **5(4)**, 380–385, doi:10.1038/nclimate2535.

Obersteiner, M. et al., 2016: Assessing the land resource–food price nexus of the Sustainable Development Goals. *Sci. Adv.*, **2(9)**, e1501499, doi:10.1126/sciadv.1501499.

Oberthür, S., 2019: Hard or Soft Governance? The EU's Climate and Energy Policy Framework for 2030. *Polit. Gov.*, **7(1)**, 17–27, doi:10.17645/pag.v7i1.1796.

OECD, 2017: Insights from national adaptation monitoring and evaluation systems. **3**, OECD Publishing, Paris, France, doi:10.1787/da48ce17-en (Accessed March 30, 2021).

OECD, 2018: *Effective Carbon Rates 2018: Pricing Carbon Emissions Through Taxes and Emissions Trading*. OECD Publishing, Paris, France, 92 pp.

OECD, 2020: OECD Inventory of Support Measures for Fossil Fuels. https://www.oecd.org/fossil-fuels/ (Accessed December 21, 2020).

OECD, 2021a: *Effective Carbon Rates 2021: Pricing Carbon Emissions through Taxes and Emissions Trading*. OECD Publishing, Paris, France, 40 pp.

OECD, 2021b: *Scaling up Nature-based Solutions to Tackle Water-related Climate Risks: Insights from Mexico and the United Kingdom*. OECD Publishing, Paris, France, 84 pp.

OECD, 2021c: *Key findings from the update of the OECD Green Recovery Database*. OECD Publishing, Paris, France, 21 pp. https://www.oecd.org/coronavirus/policy-responses/key-findings-from-the-update-of-the-oecd-green-recovery-database-55b8abba/.

Oei, P.Y. et al., 2020: Coal phase-out in Germany – Implications and policies for affected regions. *Energy*, **196**, 117004, doi:10.1016/j.energy.2020.117004.

Oels, A., 2012: From 'Securitization' of Climate Change to 'Climatization' of the Security Field: Comparing Three Theoretical Perspectives. In: *Climate Change, Human Security and Violent Conflict. Hexagon Series on Human and Environmental Security and Peace, volume 8* [Scheffran, J., M. Brzoska, H.G. Brauch, P.M. Link, and J. Schilling, (eds.)]. Springer, Berlin, Germany, pp. 185–205.

Ohlendorf, N., M. Jakob, J.C. Minx, C. Schröder, and J.C. Steckel, 2021: Distributional Impacts of Carbon Pricing: A Meta-Analysis. *Environ. Resour. Econ.*, **78(1)**, 1–42, doi:10.1007/s10640-020-00521-1.

Okereke, C. et al., 2019: Governing green industrialisation in Africa: Assessing key parameters for a sustainable socio-technical transition in the context of Ethiopia. *World Dev.*, **115**, 279–290, doi:10.1016/j.worlddev.2018.11.019.

Olsen, K.H., 2007: The clean development mechanism's contribution to sustainable development: A review of the literature. *Clim. Change*, **84(1)**, 59–73, doi:10.1007/s10584-007-9267-y.

Olzak, S., S.A. Soule, M. Coddou, and J. Muñoz, 2016: Friends or Foes? How Social Movement Allies Affect the Passage of Legislation in the U.S. Congress. *Mobilization An Int. Q.*, **21(2)**, 213–230, doi:10.17813/1086-671X-21-2-213.

Oreskes, N. and E.M. Conway, 2012: *Merchants of Doubt: How a Handful of Scientists Obscured the Truth on Issues from Tobacco Smoke to Global Warming*. 1st ed. Bloomsbury Publishing, London, UK, 368 pp.

Ortega Díaz, A. and E.C. Gutiérrez, 2018: Competing actors in the climate change arena in Mexico: A network analysis. *J. Environ. Manage.*, **215**, 239–247, doi:10.1016/j.jenvman.2018.03.056.

Osofsky, H.M., 2007: The Intersection of Scale, Science, and Law in Massachusetts v. Epa. *Proc. ASIL Annu. Meet.*, **101**, 61–65, doi:10.1017/S0272503700025180.

Otto, I.M. et al., 2020: Social tipping dynamics for stabilizing Earth's climate by 2050. *Proc. Natl. Acad. Sci.*, **117(5)**, 2354–2365, doi:10.1073/pnas.1900577117.

Oulu, M., 2015: Climate Change Governance: Emerging Legal and Institutional Frameworks for Developing Countries. In: *Handbook of Climate Change Adaptation* [Leal Filho, W., (ed.)]. Springer Berlin Heidelberg, Germany, pp. 227–250.

Pahle, M., S. Pachauri, and K. Steinbacher, 2016: Can the Green Economy deliver it all? Experiences of renewable energy policies with socio-economic objectives. *Appl. Energy*, **179**(October 2016), 1331–1341, doi:10.1016/j.apenergy.2016.06.073.

Pahle, M. et al., 2018: Sequencing to ratchet up climate policy stringency. *Nat. Clim. Change*, **8(10)**, 861–867, doi:10.1038/s41558-018-0287-6.

Paiement, P., 2020: Urgent agenda: how climate litigation builds transnational narratives. *Transnatl. Leg. Theory*, **11(1–2)**, 121–143, doi:10.1080/20414005.2020.1772617.

Painter, J., 2019: Climate Change Journalism: Time to Adapt. *Environ. Commun.*, **13(3)**, 424–429, doi:10.1080/17524032.2019.1573561.

Painter, J. and N.T. Gavin, 2016: Climate Skepticism in British Newspapers, 2007–2011. *Environ. Commun.*, **10(4)**, 432–452, doi:10.1080/17524032.2014.995193.

Palermo, V., P. Bertoldi, M. Apostolou, A. Kona, and S. Rivas, 2020a: Data on mitigation policies at local level within the Covenant of Mayors' monitoring emission inventories. *Data Br.*, **32**, 106217, doi:10.1016/j.dib.2020.106217.

Palermo, V., P. Bertoldi, M. Apostolou, A. Kona, and S. Rivas, 2020b: Assessment of climate change mitigation policies in 315 cities in the Covenant of Mayors initiative. *Sustain. Cities Soc.*, **60**, 102258, doi:10.1016/j.scs.2020.102258.

Pals, B., 2021: Taxes v. Torts: Which Will Make Fossil Fuel Producers Share Climate Change Burdens? *N.Y.U Environ. Law J.*, **29**, 235.

Pan, X., M. den Elzen, N. Höhne, F. Teng, and L. Wang, 2017: Exploring fair and ambitious mitigation contributions under the Paris Agreement goals. *Environ. Sci. Policy*, **74**, 49–56, doi:10.1016/j.envsci.2017.04.020.

Papadis, E. and G. Tsatsaronis, 2020: Challenges in the decarbonization of the energy sector. *Energy*, **205**, 118025, doi:10.1016/j.energy.2020.118025.

Papadopoulos, S., B. Bonczak, and C.E. Kontokosta, 2018: Pattern recognition in building energy performance over time using energy benchmarking data. *Appl. Energy*, **221**, 576–586, doi:10.1016/j.apenergy.2018.03.079.

Parks, D., 2019: Energy efficiency left behind? Policy assemblages in Sweden's most climate-smart city. *Eur. Plan. Stud.*, **27(2)**, 318–335, doi:10.1080/09654313.2018.1455807.

Parry, I., S. Black, and N. Vernon, 2021: *Still Not Getting Energy Prices Right: A Global and Country Update of Fossil Fuel Subsidies*. IMF, Washington, DC, USA, 40 pp. https://www.imf.org/en/Publications/WP/Issues/2021/09/23/Still-Not-Getting-Energy-Prices-Right-A-Global-and-Country-Update-of-Fossil-Fuel-Subsidies-466004 (Accessed October 19, 2021).

Parry, I.A.N., C. Veung, and D. Heine, 2015: How Much Carbon Pricing is in the Countries' Own Interests? The Critical Role of Co-Benefits. *Clim. Change Econ.*, **6(4)**, 36 pp., doi:10.1142/S2010007815500190.

Pasimeni, M.R., D. Valente, G. Zurlini, and I. Petrosillo, 2019: The interplay between urban mitigation and adaptation strategies to face climate change in two European countries. *Environ. Sci. Policy*, **95**, 20–27, doi:10.1016/j.envsci.2019.02.002.

Patt, A., 2017: Beyond the tragedy of the commons: Reframing effective climate change governance. *Energy Res. Soc. Sci.*, **34**, 1–3, doi:10.1016/j.erss.2017.05.023.

Patt, A. and J. Lilliestam, 2018: The Case against Carbon Prices. *Joule*, **2(12)**, 2494–2498, doi:10.1016/j.joule.2018.11.018.

Pauw, P., K. Mbeva, and H. van Asselt, 2019: Subtle differentiation of countries' responsibilities under the Paris Agreement. *Palgrave Commun.*, **5(1)**, 1–7, doi:10.1057/s41599-019-0298-6.

Pauw, W.P. et al., 2018: Beyond headline mitigation numbers: we need more transparent and comparable NDCs to achieve the Paris Agreement on climate change. *Clim. Change*, **147(1–2)**, 23–29, doi:10.1007/s10584-017-2122-x.

Pauw, W.P., P. Castro, J. Pickering, and S. Bhasin, 2020: Conditional Nationally Determined Contributions in the Paris Agreement: foothold for equity or Achilles heel? *Clim. Policy*, **20(4)**, 468–484, doi:10.1080/14693062.2019.1635874.

Peel, J. and H.M. Osofsky, 2015: *Climate Change Litigation: Regulatory Pathways to Cleaner Energy*. 1st edition. Cambridge University Press, Cambridge, UK, 352 pp.

Peel, J. and H.M. Osofsky, 2018: A Rights Turn in Climate Change Litigation? *Transnatl. Environ. Law*, **7(1)**, 37–67, doi:10.1017/S2047102517000292.

Peel, J. and J.L. Lin, 2019: Transnational Climate Litigation: The Contribution of the Global South. *Am. J. Int. Law*, **113(4)**, 679–726, doi:10.1017/ajil.2019.48.

Peel, J. and H.M. Osofsky, 2020: Climate Change Litigation. *Annu. Rev. Law Soc. Sci.*, **16(1)**, 21–38, doi:10.1146/annurev-lawsocsci-022420-122936.

Peel, J. and R. Markey-Towler, 2021: A Duty to Care: The Case of Sharma v Minister for the Environment [2021] FCA 560. *J. Environ. Law*, **33(3)**, 727–736, doi:10.1093/jel/eqab022.

Peeters, M., 2016: Urgenda Foundation and 886 Individuals v. The State of the Netherlands: The Dilemma of More Ambitious Greenhouse Gas Reduction Action by EU Member States. *Rev. Eur. Comp. Int. Environ. Law*, **25(1)**, 123–129, doi:10.1111/reel.12146.

Penasco, C., D. Romero-Jordan, and P. del Rio, 2019: The Impact of Policy on the Efficiency of Solar Energy Plants in Spain: A Production-Frontier Analysis. *Econ. Energy Environ. Policy*, **8(2)**, pp. 99–116, -doi:10.5547/2160-5890.8.2.cpen.

Peñasco, C., L.D. Anadón, and E. Verdolini, 2021: Systematic review of the outcomes and trade-offs of ten types of decarbonization policy instruments. *Nat. Clim. Change*, **11(3)**, 257–265, doi:10.1038/s41558-020-00971-x.

Peng, W. et al., 2021: The surprisingly inexpensive cost of state-driven emission control strategies. *Nat. Clim. Change*, **11(9)**, 738–745, doi:10.1038/s41558-021-01128-0.

Peng, Y. and X. Bai, 2018: Experimenting towards a low-carbon city: Policy evolution and nested structure of innovation. *J. Clean. Prod.*, **174**, 201–212, doi:10.1016/j.jclepro.2017.10.116.

Pereira, L. et al., 2019: Transformative spaces in the making: key lessons from nine cases in the Global South. *Sustain. Sci.*, **15**, 161–178, doi:10.1007/s11625-019-00749-x.

Perino, G., 2018: New EU ETS Phase 4 rules temporarily puncture waterbed. *Nat. Clim. Change*, **8**, 262–264, doi:10.1038/s41558-018-0120-2.

Perrow, C. and S. Pulver, 2015: Organizations and Markets. In *Climate Change and Society: Sociological Perspectives*. 1st ed. [Dunlap, R.E. and R.J. Brulle, (eds.)]. Oxford University Press, Oxford, UK, 61–92 pp.

Perry, K.K., 2020: For politics, people, or the planet? The political economy of fossil fuel reform, energy dependence and climate policy in Haiti. *Energy Res. Soc. Sci.*, **63**, 101397, doi:10.1016/j.erss.2019.101397.

Petek, G., 2020: *Assessing California's Climate Policies—Electricity Generation*. 32 pp. https://autl.assembly.ca.gov/sites/autl.assembly.ca.gov/files/CalifLAO-ElectricityEmissions.pdf (Accessed March 3, 2021).

Petersen, J.P., 2016: Energy concepts for self-supplying communities based on local and renewable energy sources: A case study from northern Germany. *Sustain. Cities Soc.*, **26**, 1–8, doi:10.1016/j.scs.2016.04.014.

Pezzey, J.C.V., 2019: Why the social cost of carbon will always be disputed. *WIREs Clim. Change*, **10(1)**, e558, doi:10.1002/wcc.558.

Pham, T.T., M. Di Gregorio, R. Carmenta, M. Brockhaus, and D.N. Le, 2014: The REDD+ policy arena in Vietnam: Participation of policy actors. *Ecol. Soc.*, **19(2)**, pp. 22, doi:10.5751/ES-06389-190222.

Pierson, P., 2004: *Politics in Time: History, Institutions, and Social Analysis*. 1st ed. Princeton University Press, Princeton, NJ, USA, 208 pp.

Pillai, A.V. and N.K. Dubash, 2021: The limits of opportunism: the uneven emergence of climate institutions in India. *Env. Polit.*, **30(sup1)**, 93–117, doi:10.1080/09644016.2021.1933800.

Pizer, W.A., 2002: Combining price and quantity controls to mitigate global climate change. *J. Public Econ.*, **85(3)**, 409–434, doi:10.1016/S0047-2727(01)00118-9.

Poberezhskaya, M., 2015: Media coverage of climate change in Russia: Governmental bias and climate silence. *Public Underst. Sci.*, **24(1)**, 96–111, doi:10.1177/0963662513517848.

Postic, S. and M. Fetet, 2020: *Global Carbon Account in 2020*. Institute for Climate Economics (I4CE), Paris, France, 4 pp. https://www.i4ce.org/download/global-carbon-account-in-2020/ (Accessed March 3. 2021).

Prasad, N. et al., 2009: *Climate Resilient Cities: A Primer on Reducing Vulnerability to Disasters*. World Bank, Washington, DC, USA, 186 pp.

Prendeville, S., E. Cherim, and N. Bocken, 2018: Circular Cities: Mapping Six Cities in Transition. *Environ. Innov. Soc. Transitions*, **26**, 171–194, doi:10.1016/j.eist.2017.03.002.

Pretis, F., 2019: *Does a Carbon Tax Reduce CO2 Emissions? Evidence From British Columbia*. 52 pp. https://link.springer.com/article/10.1007/s10640-022-00679-w (Accessed October 18. 2021).

Prussi, M. et al., 2021: CORSIA: The first internationally adopted approach to calculate life-cycle GHG emissions for aviation fuels. *Renew. Sustain. Energy Rev.*, **150**, 111398, doi:10.1016/j.rser.2021.111398.

Pulver, S., 2007: Importing environmentalism: Explaining Petroleos Mexicanos' cooperative climate policy. *Stud. Comp. Int. Dev.*, **42(3–4)**, 233–255, doi:10.1007/s12116-007-9010-8.

Pulver, S. and T. Benney, 2013: Private-sector responses to climate change in the Global South. *WIREs Clim. Change*, **4(6)**, 479–496, doi:10.1002/wcc.240.

Puppim de Oliveira, J.A., 2013: Learning how to align climate, environmental and development objectives in cities: lessons from the implementation of climate co-benefits initiatives in urban Asia. *J. Clean. Prod.*, **58**, 7–14, doi:10.1016/j.jclepro.2013.08.009.

Purdon, M., 2015: Advancing comparative climate change politics: Theory and method. *Glob. Environ. Polit.*, **15(3)**, 1–26, doi:10.1162/GLEP_e_00309.

Pyrgou, A., A. Kylili, and P.A. Fokaides, 2016: The future of the Feed-in Tariff (FiT) scheme in Europe: The case of photovoltaics. *Energy Policy*, **95**, 94–102, doi:10.1016/j.enpol.2016.04.048.

Qi, Y. and T. Wu, 2013: The politics of climate change in China. *WIREs Clim. Change*, **4(4)**, 301–313, doi:10.1002/wcc.221.

Qian, H., L. Wu, and W. Tang, 2017: 'Lock-in' effect of emission standard and its impact on the choice of market based instruments. *Energy Econ.*, **63**, 41–50, doi:10.1016/j.eneco.2017.01.005.

Qian, H., Y. Zhou, and L. Wu, 2018: Evaluating various choices of sector coverage in China's national emissions trading system (ETS). *Clim. Policy*, **18(sup1)**, 7–26, doi:10.1080/14693062.2018.1464894.

Quak, H., N. Nesterova, and R. Kok, 2019: Public procurement as driver for more sustainable urban freight transport. *Transp. Res. Procedia*, **39**, 428–439, doi:10.1016/j.trpro.2019.06.045.

Rabe, B., 2011: Contested Federalism and American Climate Policy. *Publius J. Fed.*, **41(3)**, 494–521, doi:10.1093/publius/pjr017.

Rabe, B.G., 2007: Beyond Kyoto: Climate Change Policy in Multilevel Governance Systems. *Governance*, **20(3)**, 423–444, doi:10.1111/j.1468-0491.2007.00365.x.

Rabe, B.G., 2018: *Can we price carbon?* MIT Press, Cambridge, MA, USA, 376 pp.

Rafaty, R., 2018: Perceptions of Corruption, Political Distrust, and the Weakening of Climate Policy. *Glob. Environ. Polit.*, **18(3)**, 106–129, doi:10.1162/glep_a_00471.

Rafaty, R., G. Dolphin, and F. Pretis, 2020: *Carbon pricing and the elasticity of CO2 emissions*. Faculty of Economics, University of Cambridge, 76 pp, doi: 10.17863/CAM.62831 (Accessed October 18, 2021).

Raff, Z. and J.M. Walter, 2020: Regulatory Avoidance and Spillover: The Effects of Environmental Regulation on Emissions at Coal-Fired Power Plants. *Environ. Resour. Econ.*, **75(3)**, 387–420, doi:10.1007/s10640-019-00394-z.

Ramos-Castillo, A., E.J. Castellanos, and K. Galloway McLean, 2017: Indigenous peoples, local communities and climate change mitigation. *Clim. Change*, **140(1)**, doi:10.1007/s10584-016-1873-0.

Rashidi, K. and A. Patt, 2018: Subsistence over symbolism: the role of transnational municipal networks on cities' climate policy innovation and adoption. *Mitig. Adapt. Strateg. Glob. Change*, **23(4)**, 507–523, doi:10.1007/s11027-017-9747-y.

Rasul, G. and B. Sharma, 2016: The nexus approach to water–energy–food security: an option for adaptation to climate change. *Clim. Policy*, **16(6)**, 682–702, doi:10.1080/14693062.2015.1029865.

Räthzel, N., J. Cock, and D. Uzzell, 2018: Beyond the nature–labour divide: trade union responses to climate change in South Africa. *Globalizations*, **15(4)**, 504–519, doi:10.1080/14747731.2018.1454678.

Raymond, C.M. et al., 2017: A framework for assessing and implementing the co-benefits of nature-based solutions in urban areas. *Environ. Sci. Policy*, **77**(June), 15–24, doi:10.1016/j.envsci.2017.07.008.

Raymond, L., 2019: Policy perspective: Building political support for carbon pricing—Lessons from cap-and-trade policies. *Energy Policy*, **134(C)**, 110986, doi:10.1016/j.enpol.2019.110986.

Reckien, D. et al., 2017: Climate change, equity and the Sustainable Development Goals: an urban perspective. *Environ. Urban.*, **29(1)**, 159–182, doi:10.1177/0956247816677778.

Reckien, D. et al., 2018: How are cities planning to respond to climate change? Assessment of local climate plans from 885 cities in the EU-28. *J. Clean. Prod.*, **191**, 207–219, doi:10.1016/j.jclepro.2018.03.220.

Reed, P., 2011: REDD+ and the Indigenous Question: A Case Study from Ecuador. *Forests*, **2(2)**, 525–549, doi:10.3390/f2020525.

Rennkamp, B., 2019: Power, coalitions and institutional change in South African climate policy. *Clim. Policy*, **19(6)**, 756–770, doi:10.1080/14693062.2019.1591936.

Rentschler, J. and M. Bazilian, 2017: Reforming fossil fuel subsidies: drivers, barriers and the state of progress. *Clim. Policy*, **17(7)**, 891–914, doi:10.1080/14693062.2016.1169393.

Rey, L., A. Markandya, M. González-Eguino, and P. Drummond, 2013: *Assessing interaction between instruments and the 'optimality' of the current instrument mix*. CECILIA250 Project, Brussels, Belgium, 73 pp. https://cecilia2050.eu/publications/171.

Rezessy, S. and P. Bertoldi, 2011: Voluntary agreements in the field of energy efficiency and emission reduction: Review and analysis of experiences in the European Union. *Energy Policy*, **39(11)**, 7121–7129, doi:10.1016/j.enpol.2011.08.030.

RGGI, 2019: *The Investment of RGGI Proceeds in 2017*. Regional Greenhouse Gas Initiative, Inc., New York, USA, 46 pp. https://www.rggi.org/sites/default/files/Uploads/Proceeds/RGGI_Proceeds_Report_2017.pdf.

Rhodes, E., J. Axsen, and M. Jaccard, 2017: Exploring Citizen Support for Different Types of Climate Policy. *Ecol. Econ.*, **137**, 56–69, doi:10.1016/j.ecolecon.2017.02.027.

Rice, J.L., D.A. Cohen, J. Long, and J.R. Jurjevich, 2020: Contradictions of the Climate-Friendly City: New Perspectives on Eco-Gentrification and Housing Justice. *Int. J. Urban Reg. Res.*, **44(1)**, 145–165, doi:10.1111/1468-2427.12740.

Richerzhagen, C. and I. Scholz, 2008: China's Capacities for Mitigating Climate Change. *World Dev.*, **36(2)**, 308–324, doi:10.1016/j.worlddev.2007.06.010.

Rickards, L., J. Wiseman, and Y. Kashima, 2014: Barriers to effective climate change mitigation: The case of senior government and business decision makers. *Wiley Interdiscip. Rev. Clim. Change*, **5(6)**, 753–773, doi:10.1002/wcc.305.

Ridzuan, A.R. et al., 2019: The Impact of Corruption on Environmental Quality in the Developing Countries of ASEAN-3: The Application of the Bound Test. *Int. J. Energy Econ. Policy*, **9(6)**, 469–478, doi:10.32479/ijeep.8135.

Roberts, C. and F.W. Geels, 2019: Conditions for politically accelerated transitions: Historical institutionalism, the multi-level perspective, and two historical case studies in transport and agriculture. *Technol. Forecast. Soc. Change*, **140**, 221–240, doi:10.1016/j.techfore.2018.11.019.

Roberts, C. et al., 2018: The politics of accelerating low-carbon transitions: Towards a new research agenda. *Energy Res. Soc. Sci.*, **44**, 304–311, doi:10.1016/j.erss.2018.06.001.

Roberts, D., 2010: *Thinking Globally, Acting Locally: Institutionalising Climate Change within Durban's Local Government*. Cities Alliance Washington, DC, USA, 8 pp. https://www.citiesalliance.org/sites/default/files/CIVIS_3_June2010.pdf.

Roberts, D., 2020: Feed-in tariffs for renewable power and the role of auctions: the Chinese and global experience. *China Econ. J.*, **13(2)**, 152–168, doi:10.1080/17538963.2020.1752494.

Roberts, D. et al., 2016: Durban, South Africa. In: *Cities on a Finite Planet* [Bartlett, S. and D. Satterthwaite, (eds.)]. Routledge, Oxon, UK and New York, USA, pp. 96–115.

Rocha, M. and C. Falduto, 2019: *Key questions guiding the process of setting up long-term low-emissions development strategies*. OECD Publishing, Paris, France, 1–52 pp.

Rockström, J. et al., 2017: A roadmap for rapid decarbonization. *Science*, **355(6331)**, 1269–1271, doi:10.1126/science.aah3443.

Rodríguez-Garavito, C., 2020: Human Rights: The Global South's Route to Climate Litigation. *AJIL Unbound*, **114**, 40–44, doi:10.1017/aju.2020.4.

Roelfsema, M., 2017: *Assessment of US City Reduction Commitments, from a Country Perspective*. The Hague, The Netherlands, 26 pp, https://www.researchgate.net/profile/Mark-Roelfsema/publication/321920191_ASSESSMENT_OF_US_CITY_REDUCTION_COMMITMENTS_FROM_A_COUNTRY_PERSPECTIVE_Note_Mark_Roelfsema/links/5a3983cf458515889d2aca1c/ASSESSMENT-OF-US-CITY-REDUCTION-COMMITMENTS-FROM-A-COUNTRY-PERSPECTIVE-Note-Mark-Roelfsema.pdf (Accessed March 3, 2021).

Roelfsema, M., M. Harmsen, J.J.G. Olivier, A.F. Hof, and D.P. van Vuuren, 2018: Integrated assessment of international climate mitigation commitments outside the UNFCCC. *Glob. Environ. Change*, **48**, 67–75, doi:10.1016/j.gloenvcha.2017.11.001.

Rogelj, J. et al., 2015: Energy system transformations for limiting end-of-century warming to below 1.5°C. *Nat. Clim. Change*, **5(6)**, 519–527, doi:10.1038/nclimate2572.

Rogelj, J., O. Geden, A. Cowie, and A. Reisinger, 2021: Three ways to improve net-zero emissions targets. *Nature*, **591(7850)**, 365–368, doi:10.1038/d41586-021-00662-3.

Rogge, K.S., 2019: Policy mixes for sustainable innovation: conceptual considerations and empirical insights. In: *Handbook of Sustainable Innovation* [Boons, F. and A. McMeekin, (eds.)]. Edward Elgar Publishing, Cheltenham, UK and Northampton, MA, USA, pp. 165–185.

Rogge, K.S. and K. Reichardt, 2016: Policy mixes for sustainability transitions: An extended concept and framework for analysis. *Res. Policy*, **45(8)**, 1620–1635, doi:10.1016/j.respol.2016.04.004.

Rogge, K.S. and J. Schleich, 2018: Do policy mix characteristics matter for low-carbon innovation? A survey-based exploration of renewable power generation technologies in Germany. *Res. Policy*, **47(9)**, 1639-1654, doi:10.1016/j.respol.2018.05.011.

Rogge, K.S., F. Kern, and M. Howlett, 2017: Conceptual and empirical advances in analysing policy mixes for energy transitions. *Energy Res. Soc. Sci.*, **33**(September), 1–10, doi:10.1016/j.erss.2017.09.025.

Rogner, H.-H. and K.-K. Leung, 2018: The Effectiveness of Foreign Aid for Sustainable Energy and Climate Change Mitigation. In: *Aid Effectiveness for Environmental Sustainability* [Huang, Y., and U. Pascual (eds.)]. Springer Singapore, Singapore, pp. 81–124.

Rohlinger, D.A. and L.A. Bunnage, 2018: Collective identity in the digital age: thin and thick identities in moveon.org and the Tea Party movement. *Mobilization An Int. Q.*, **23(2)**, 135–157, doi:10.17813/1086-671X-23-2-135.

Romero-Lankao, P. and D. Gnatz, 2019: Risk Inequality and the Food-Energy-Water (FEW) Nexus: A Study of 43 City Adaptation Plans. *Front. Sociol.*, **4**, pp. 31, doi:10.3389/fsoc.2019.00031.

Romero-Lankao, P. and E. Nobler, 2021: *Energy Justice: Key Concepts and Metrics Relevant to EERE Transportation Projects*. 15 pp. https://afdc.energy.gov/files/pdfs/energy-justice-key-concepts.pdf (Accessed October 7, 2021).

Romero-Lankao, P., S. Hughes, A. Rosas-Huerta, R. Borquez, and D.M. Gnatz, 2013: Institutional Capacity for Climate Change Responses: An Examination of Construction and Pathways in Mexico City and Santiago. *Environ. Plan. C Gov. Policy*, **31(5)**, 785–805, doi:10.1068/c12173.

Romero-Lankao, P. et al., 2015: Multilevel Governance and Institutional Capacity for Climate Change Responses in Latin American Cities. In: *The Urban Climate Challenge Rethinking the Role of Cities in the Global Climate Regime* [Johnson, C., N. Toly, and H. Schroeder, (eds.)]. Routledge, New York, NY, USA, pp. 179–204.

Romero-Lankao, P., D. Gnatz, O. Wilhelmi, and M. Hayden, 2016: Urban Sustainability and Resilience: From Theory to Practice. *Sustainability*, **8(12)**, 1224, doi:10.3390/su8121224.

Romero-Lankao, P., T. McPhearson, and D. Davidson, 2017: The food-energy-water nexus and urban complexity. *Nat. Clim Change*, **7**, 233–235, doi:10.1038/nclimate3260.

Romero-Lankao, P. et al., 2018a: Governance and Policy. In: *Climate Change and Cities: Second Assessment Report of the Urban Climate Change Research Network* [Rosenzweig, C., W.D. Solecki, P. Romero-Lankao, S. Mehrotra, S. Dhakal, and S.A. Ibrahim (eds.)]. Cambridge University Press, Cambridge, UK, pp. 585–606.

Romero-Lankao, P. et al., 2018b: Urban transformative potential in a changing climate. *Nat. Clim. Change*, **8(9)**, 754–756, doi:10.1038/s41558-018-0264-0.

Romero-Lankao, P. et al., 2021: Of actors, cities and energy systems: advancing the transformative potential of urban electrification. *Prog. Energy*, **3(3)**, 032002, doi:10.1088/2516-1083/abfa25.

Romero Lankao, P. et al., 2019: Urban Electrification: Knowledge Pathway Toward an Integrated Research and Development Agenda. *SSRN Electron. J.*, doi:10.2139/ssrn.3440283.

Roppongi, H., A. Suwa, and J.A. Puppim De Oliveira, 2017: Innovating in sub-national climate policy: the mandatory emissions reduction scheme in Tokyo. *Clim. Policy*, **17(4)**, 516–532, doi:10.1080/14693062.2015.1124749.

Rosenbloom, D., 2018: Framing low-carbon pathways: A discursive analysis of contending storylines surrounding the phase-out of coal-fired power in Ontario. *Environ. Innov. Soc. Transitions*, **27**(November 2017), 129–145, doi:10.1016/j.eist.2017.11.003.

Rosenbloom, D. and A. Rinscheid, 2020: Deliberate decline: An emerging frontier for the study and practice of decarbonization. *WIREs Clim. Change*, **11(6)**, e669, doi:10.1002/wcc.669.

Rosenbloom, D., B. Haley, and J. Meadowcroft, 2018: Critical choices and the politics of decarbonization pathways: Exploring branching points surrounding low-carbon transitions in Canadian electricity systems. *Energy Res. Soc. Sci.*, **37**, 22–36, doi:10.1016/j.erss.2017.09.022.

Rosenbloom, D., J. Meadowcroft, and B. Cashore, 2019: Stability and climate policy? Harnessing insights on path dependence, policy feedback, and transition pathways. *Energy Res. Soc. Sci.*, **50**, 168–178, doi:10.1016/j.erss.2018.12.009.

Rosenbloom, D., J. Markard, F.W. Geels, and L. Fuenfschilling, 2020: Opinion: Why carbon pricing is not sufficient to mitigate climate change—and how 'sustainability transition policy' can help. *Proc. Natl. Acad. Sci.*, **117(16)**, 8664–8668, doi:10.1073/pnas.2004093117.

Rosendahl, K.E. and J. Strand, 2011: Carbon Leakage from the Clean Development Mechanism. *Energy J.*, **32(4)**, doi:10.5547/ISSN0195-6574-EJ-Vol32-No4-3.

Rosenow, J., T. Fawcett, N. Eyre, and V. Oikonomou, 2016: Energy efficiency and the policy mix. *Build. Res. Inf.*, **44(5–6)**, 562–574, doi:10.1080/09613218.2016.1138803.

Röser, F., O. Widerberg, N. Höhne, and T. Day, 2020: Ambition in the making: analysing the preparation and implementation process of the Nationally Determined Contributions under the Paris Agreement. *Clim. Policy*, **20(4)**, 415–429, doi:10.1080/14693062.2019.1708697.

Routledge, P., A. Cumbers, and K.D. Derickson, 2018: States of just transition: Realising climate justice through and against the state. *Geoforum*, **88**, 78–86, doi:10.1016/j.geoforum.2017.11.015.

Roy, J., P. Tschakert, H. Waisman, S. Abdul Halim, P. Antwi-Agyei, P. Dasgupta, B. Hayward, M. Kanninen, D. Liverman, C. Okereke, P.F. Pinho, K. Riahi, and A.G. Suarez Rodriguez, 2018: Sustainable Development, Poverty Eradication and Reducing Inequalities. In: *Global Warming of 1.5°C. An IPCC Special Report on the impacts of global warming of 1.5°C above pre-industrial levels and related global greenhouse gas emission pathways, in the context of strengthening the global response to the threat of climate change, sustainable development, and efforts to eradicate poverty* [Masson-Delmotte, V., P. Zhai, H.-O. Pörtner, D. Roberts, J. Skea, P.R. Shukla, A. Pirani, W. Moufouma-Okia, C. Péan, R. Pidcock, S. Connors, J.B.R. Matthews, Y. Chen, X. Zhou, M.I. Gomis, E. Lonnoy, T. Maycock, M. Tignor, and T. Waterfield (eds.)]. Cambridge University Press, Cambridge and New York, NY, USA, pp. 445–538.

Roy, S. and E. Woerdman, 2016: Situating Urgenda v the Netherlands within comparative climate change litigation. *J. Energy Nat. Resour. Law*, **34(2)**, 165–189, doi:10.1080/02646811.2016.1132825.

Rubio, F., 2017: Ecosystem-based adaptation to climate change: concept, scalability and a role for conservation science. *Perspect. Ecol. Conserv.*, **15(2)**, 65–73, doi:10.1016/j.pecon.2017.05.003.

Rucht, D., 1999: The impact of environmental movements in Western Societies. In: *How Social Movements Matter* [Guigni, M., D. McAdam, and C. Tilly, (eds.)]. University of Minnesota Press, Minneapolis, USA, pp. 204–224.

Rumble, O., 2019: Climate change legislative development on the African continent. In: *Law | Environment | Africa: Publication of the 5th Symposium | 4th Scientific Conference | 2018 of the Association of Environmental Law Lecturers from African Universities in cooperation with the Climate Policy and Energy Security Programme for Sub-Saharan Africa of the Konrad-Adenauer-Stiftung and UN Environment* [Kameri-Mbote, P., A. Paterson, O.C. Ruppel, B.B. Orubebe, and E.D. Kam Yogo, (eds.)]. Nomos Verlagsgesellschaft mbH & Co. KG, Baden-Baden, Germany, pp. 31–60.

Runhaar, H., B. Wilk, Å. Persson, C. Uittenbroek, and C. Wamsler, 2018: Mainstreaming climate adaptation: taking stock about 'what works' from empirical research worldwide. *Reg. Environ. Change*, **18(4)**, 201–1210, doi:10.1007/s10113-017-1259-5.

Ryan, D., 2015: From commitment to action: a literature review on climate policy implementation at city level. *Clim. Change*, **131(4)**, 519–529, doi:10.1007/s10584-015-1402-6.

Ryder, S.S., 2018: Developing an intersectionally-informed, multi-sited, critical policy ethnography to examine power and procedural justice in multiscalar energy and climate change decision-making processes. *Energy Res. Soc. Sci.*, **45**, 266–275, doi:10.1016/j.erss.2018.08.005.

Rydin, Y., 2013: *The Future of Planning Beyond Growth Dependence*. Policy Press, Bristol, UK, 208 pp.

Sachs, N., 2012: Can We Regulate Our Way to Energy Efficiency: Product Standards as Climate Policy. *Vanderbilt Law Rev.*, **65(6)**, 1631–1678.

Saelim, S., 2019: Carbon tax incidence on household demand: Effects on welfare, income inequality and poverty incidence in Thailand. *J. Clean. Prod.*, **234**, 521–533, doi:10.1016/J.JCLEPRO.2019.06.218.

Sahli, I. and J. Ben Rejeb, 2015: The Environmental Kuznets Curve and Corruption in the Mena Region. *Procedia – Soc. Behav. Sci.*, **195**, 1648–1657, doi:10.1016/j.sbspro.2015.06.231.

Saito, N., 2013: Mainstreaming climate change adaptation in least developed countries in South and Southeast Asia. *Mitig. Adapt. Strateg. Glob. Change*, **18(6)**, 825–849, doi:10.1007/s11027-012-9392-4.

Sallee, J.M., 2019: *Pigou creates losers: On the implausibility of achieving Pareto improvements from efficiency-enhancing policies*. National Bureau of Economic Research, Cambridge, MA, USA, 49 pp. https://haas.berkeley.edu/wp-content/uploads/WP302.pdf.

Sandover, R., A. Moseley, and P. Devine-Wright, 2021: Contrasting Views of Citizens' Assemblies: Stakeholder Perceptions of Public Deliberation on Climate Change. *Polit. Gov.*, **9(2)**, 76–86, doi:10.17645/pag.v9i2.4019.

Santamouris, M., 2014: Cooling the cities – a review of reflective and green roof mitigation technologies to fight heat island and improve comfort in urban environments. *Sol. Energy*, **103**, 682–703, doi:10.1016/j.solener.2012.07.003.

Santos-lacueva, R. and M.V. González, 2018: Policy coherence between tourism and climate policies: the case of Spain and the Autonomous Community of Catalonia. *J. Sustain. Tour.*, **26(10)**, 1708–1727, doi:10.1080/09669582.2018.1503672.

Sato, I. and J.-C. Altamirano, 2019: *Uncertainty, Scenario Analysis, and Long-Term Strategies: State of Play and a Way Forward*. 1–44 pp. https://files.wri.org/s3fs-public/uncertainty-scenario-analysis-long-term-strategies.pdf (Accessed March 30, 2021).

Saunders, C., M. Grasso, C. Olcese, E. Rainsford, and C. Rootes, 2012: Explaining Differential Protest Participation: Novices, Returners, Repeaters, and Stalwarts. *Mobilization An Int. Q.*, **17(3)**, 263–280, doi:10.17813/maiq.17.3.bqm553573058t478.

Saunders, C., M. Büchs, A. Papafragkou, R. Wallbridge, and G. Smith, 2014: Beyond the Activist Ghetto: A Deductive Blockmodelling Approach to Understanding the Relationship between Contact with Environmental Organisations and Public Attitudes and Behaviour. *Soc. Mov. Stud.*, **13(1)**, 158–177, doi:10.1080/14742837.2013.832623.

Savaresi, A. and J. Auz, 2019: Climate Change Litigation and Human Rights: Pushing the Boundaries. *Clim. Law*, **9(3)**, 244–262, doi:10.1163/18786561-00903006.

Savaresi, A. and J. Setzer, 2021: Rights-based litigation in the climate emergency: mapping the landscape and new knowledge frontiers. *J. Hum. Rights Environ.*, **3(1)**, 7–34.

Schäfer, M.S. and I. Schlichting, 2014: Media Representations of Climate Change: A Meta-Analysis of the Research Field. *Environ. Commun.*, **8(2)**, 142–160, doi:10.1080/17524032.2014.914050.

Schäfer, M.S. and J. Painter, 2021: Climate journalism in a changing media ecosystem: Assessing the production of climate change-related news around the world. *WIREs Clim. Change*, **12(1)**, doi:10.1002/wcc.675.

Schaffitzel, F., M. Jakob, R. Soria, A. Vogt-Schilb, and H. Ward, 2020: Can government transfers make energy subsidy reform socially acceptable? A case study on Ecuador. *Energy Policy*, **137**, 111120, doi:10.1016/j.enpol.2019.111120.

Schatzki, T. and R.N. Stavins, 2012: *Implications of policy interactions for California's climate policy*. Analysis Group, Boston, USA, 23 pp. https://www.analysisgroup.com/globalassets/content/insights/publishing/implications_policy_interactions_california_climate_policy.pdf.

Schifeling, T. and A.J. Hoffman, 2019: Bill McKibben's Influence on U.S. Climate Change Discourse: Shifting Field-Level Debates Through Radical Flank Effects. *Organ. Environ.*, **32(3)**, 213–233, doi:10.1177/1086026617744278.

Schipper, E.L.F., T. Tanner, O.P. Dube, K.M. Adams, and S. Huq, 2020: The debate: Is global development adapting to climate change? *World Dev. Perspect.*, **18**(July 2019), 100205, doi:10.1016/j.wdp.2020.100205.

Schleich, J., C. Schwirplies, and A. Ziegler, 2018: Do perceptions of international climate policy stimulate or discourage voluntary climate protection activities? A study of German and US households. *Clim. Policy*, **18(5)**, 568–580, doi:10.1080/14693062.2017.1409189.

Schlosberg, D., 2012: Climate Justice and Capabilities: A Framework for Adaptation Policy. *Ethics Int. Aff.*, **26(4)**, 445–461, doi:10.1017/S0892679412000615.

Schlozman, K.L., S. Verba, and H.E. Brady, 2012: *The Unheavenly Chorus: Unequal Political Voice and the Broken Promise of American Democracy*. Princeton University Press, Princeton, NJ, USA, 727 pp.

Schmidt, A., A. Ivanova, and M.S. Schäfer, 2013: Media attention for climate change around the world: A comparative analysis of newspaper coverage in 27 countries. *Glob. Environ. Change*, **23(5)**, 1233–1248, doi:10.1016/j.gloenvcha.2013.07.020.

Schmidt, N.M. and A. Fleig, 2018: Global patterns of national climate policies: Analyzing 171 country portfolios on climate policy integration. *Environ. Sci. Policy*, **84**, 177–185, doi:10.1016/j.envsci.2018.03.003.

Schmidt, T.S., T. Matsuo, and A. Michaelowa, 2017: Renewable energy policy as an enabler of fossil fuel subsidy reform? Applying a socio-technical perspective to the cases of South Africa and Tunisia. *Glob. Environ. Change*, **45**, 99–110, doi:10.1016/j.gloenvcha.2017.05.004.

Schmitz, A., M. Flemmen, and L. Rosenlund, 2018: Social class, symbolic domination, and Angst: The example of the Norwegian social space. *Sociol. Rev.*, **66(3)**, 623–644, doi:10.1177/0038026117738924.

Schmitz, H., 2017: Who drives climate-relevant policies in the rising powers? *New Polit. Econ.*, **22(5)**, 521–540, doi:10.1080/13563467.2017.1257597.

Schmitz, H., O. Johnson, and T. Altenburg, 2015: Rent Management – The Heart of Green Industrial Policy. *New Polit. Econ.*, **20(6)**, 812–831, doi:10.1080/13563467.2015.1079170.

Schofer, E. and A. Hironaka, 2005: The Effects of World Society on Environmental Protection Outcomes. *Soc. Forces*, **84(1)**, 25–47, doi:10.1353/sof.2005.0127.

Schoonees, T. et al., 2019: Hard Structures for Coastal Protection, Towards Greener Designs. *Estuaries and Coasts*, **42(7)**, 1709–1729, doi:10.1007/s12237-019-00551-z.

Schreurs, M.A., 1997: Japan's changing approach to environmental issues. *Env. Polit.*, **6(2)**, 150–156, doi:10.1080/09644019708414332.

Schroeder, H., S. Burch, and S. Rayner, 2013: Novel Multisector Networks and Entrepreneurship in Urban Climate Governance. *Environ. Plan. C Gov. Policy*, **31(5)**, 761–768, doi:10.1068/c3105ed.

Schwanen, T., 2021: Achieving just transitions to low-carbon urban mobility. *Nat. Energy*, **6(7)**, 685–687, doi:10.1038/s41560-021-00856-z.

Schwanitz, V.J., F. Piontek, C. Bertram, and G. Luderer, 2014: Long-term climate policy implications of phasing out fossil fuel subsidies. *Energy Policy*, **67**, 882–894, doi:10.1016/j.enpol.2013.12.015.

Schwartz, E., 2019: Autonomous Local Climate Change Policy: An Analysis of the Effect of Intergovernmental Relations Among Sub-national Governments. *Rev. Policy Res.*, **36(1)**, 50–74, doi:10.1111/ropr.12320.

Schwietzke, S. et al., 2016: Upward revision of global fossil fuel methane emissions based on isotope database. *Nature*, **538(7623)**, 88–91, doi:10.1038/nature19797.

Scobie, M., 2016: Policy coherence in climate governance in Caribbean Small Island Developing States. *Environ. Sci. Policy*, **58**, 16–28, doi:10.1016/j.envsci.2015.12.008.

Scordato, L., A. Klitkou, V.E. Tartiu, and L. Coenen, 2018: Policy mixes for the sustainability transition of the pulp and paper industry in Sweden. *J. Clean. Prod.*, **183**, 1216–1227, doi:10.1016/j.jclepro.2018.02.212.

Scotford, E. and S. Minas, 2019: Probing the hidden depths of climate law: Analysing national climate change legislation. *Rev. Eur. Comp. Int. Environ. Law*, **28(1)**, 67–81, doi:10.1111/reel.12259.

Scotford, E., S. Minas, and A. Macintosh, 2017: Climate change and national laws across Commonwealth countries. *Commonw. Law Bull.*, **43(3–4)**, 318–361, doi:10.1080/03050718.2017.1439361.

Scullion, J.J. et al., 2016: Designing conservation-development policies for the forest frontier. *Sustain. Sci.*, **11(2)**, 295–306, doi:10.1007/s11625-015-0315-7.

Secretariat of the Convention on Biological Diversity, 2009: *Connecting Biodiversity and Climate Change Mitigation: Report of the Second Ad Hoc Technical Expert Group on Biodiversity and Climate Change*. Secretariat of the Convention on Biological Diversity, Montreal, Canada, 127 pp. https://www.cbd.int/doc/publications/cbd-ts-41-en.pdf (Accessed October 17, 2021).

Secretariat of the Convention on Biological Diversity, 2019: *Voluntary guidelines for the design and effective implementation of ecosystem-based approaches to climate change adaptation and disaster risk reduction and supplementary information*. Montreal, Canada, 154 pp. https://www.cbd.int/doc/publications/cbd-ts-93-en.pdf (Accessed October 17, 2021).

Seddon, N. et al., 2020: Global recognition of the importance of nature-based solutions to the impacts of climate change. *Glob. Sustain.*, **3** (September), e15, doi:10.1017/sus.2020.8.

Sengers, F., A.J. Wieczorek, and R. Raven, 2019: Experimenting for sustainability transitions: A systematic literature review. *Technol. Forecast. Soc. Change*, **145**, 153–164, doi:10.1016/j.techfore.2016.08.031.

Seok, J.-E., J. Kim, and H.S. Park, 2021: Regulatory and social dynamics of voluntary agreement adoption: The case of voluntary energy efficiency and GHG reduction agreement in South Korea. *Energy Policy*, **148(Part B)**, 111903, doi:10.1016/j.enpol.2020.111903.

Setzer, J., 2015: Testing the boundaries of sub-national diplomacy: The international climate action of local and regional governments. *Transnatl. Environ. Law*, **4(2)**, 319–337, doi:10.1017/S2047102515000126.

Setzer, J., 2017: How Sub-national Governments are Rescaling Environmental Governance: The Case of the Brazilian State of São Paulo. *J. Environ. Policy Plan.*, **19(5)**, 503–519, doi:10.1080/1523908X.2014.984669.

Setzer, J. and L. Benjamin, 2019: Climate Litigation in the Global South: Constraints and Innovations. *Transnatl. Environ. Law*, **9(1)**, 1–25, doi:10.1017/S2047102519000268.

Setzer, J. and R. Byrnes, 2019: *Global trends in climate change litigation: 2019 snapshot*. Grantham Research Institute on Climate Change and the Environment and Centre for Climate Change Economics and Policy, London School of Economics and Political Science, London, UK, 14 pp. http://www.lse.ac.uk/GranthamInstitute/wp-content/uploads/2019/07/GRI_Global-trends-in-climate-change-litigation-2019-snapshot-2.pdf (Accessed October 25, 2021).

Setzer, J. and L.C. Vanhala, 2019: Climate change litigation: A review of research on courts and litigants in climate governance. *Wiley Interdiscip. Rev. Clim. Change*, **10(3)**, e580, doi:10.1002/wcc.580.

Setzer, J. and C. Higham, 2021: *Global trends in climate litigation: 2021 snapshot*. Grantham Research Institute on Climate Change and the Environment and Centre for Climate Change Economics and Policy, London School of Economics and Political Science London, UK, 45 pp. https://www.lse.ac.uk/granthaminstitute/publication/global-trends-in-climate-litigation-2021-snapshot/ (Accessed September 29, 2021).

Shanahan, M., 2009: Time to adapt? Media coverage of climate change in non-industrialised countries. In: *Climate Change and the Media*. Peter Lang Publishing, Bern, Switzerland, pp. 145–157.

Shapiro, J., 2020: *The Environmental Bias of Trade Policy*. National Bureau of Economic Research, Working paper 26845, doi: 10.3386/w26845.

Sharifi, A., 2020: Trade-offs and conflicts between urban climate change mitigation and adaptation measures: A literature review. *J. Clean. Prod.*, **276**, 122813, doi:10.1016/j.jclepro.2020.122813.

Sharifi, A. and Y. Yamagata, 2016: Principles and criteria for assessing urban energy resilience: A literature review. *Renew. Sustain. Energy Rev.*, **60**, 1654–1677, doi:10.1016/j.rser.2016.03.028.

Sharpe, S. and T.M. Lenton, 2021: Upward-scaling tipping cascades to meet climate goals: plausible grounds for hope. *Clim. Policy*, **21(4)**, 421–433, do i:10.1080/14693062.2020.1870097.

Shaw, A., S. Burch, F. Kristensen, J. Robinson, and A. Dale, 2014: Accelerating the sustainability transition: Exploring synergies between adaptation and mitigation in British Columbian communities. *Glob. Environ. Change*, **25(1)**, 41–51, doi:10.1016/j.gloenvcha.2014.01.002.

Sheikh, F.A. and S. Bhaduri, 2021: Policy space for informal sector grassroots innovations: towards a 'bottom-up' narrative. *Int. Dev. Plan. Rev.*, **43(1)**, 115–137, doi:10.3828/idpr.2019.34.

Shen, W., 2015: Chinese business at the dawn of its domestic emissions trading scheme: incentives and barriers to participation in carbon trading. *Clim. Policy*, **15(3)**, 339–354, doi:10.1080/14693062.2014.926263.

Shen, W., 2017: Who drives China's renewable energy policies? Understanding the role of industrial corporations. *Environ. Dev.*, **21**, 87–97, doi:10.1016/j.envdev.2016.10.006.

Shen, W. and L. Xie, 2018: The Political Economy for Low-carbon Energy Transition in China: Towards a New Policy Paradigm? *New Polit. Econ.*, **23(4)**, 407–421, doi:10.1080/13563467.2017.1371122.

Shirky, C., 2010: The Political Power of Social Media: Technology, the Public Sphere, and Political Change. *Foreign Affairs*, **90(1)**, 28–41. http://www.jstor.org/stable/25800379 (Accessed May 10, 2022).

Shishlov, I., R. Morel, and V. Bellassen, 2016: Compliance of the Parties to the Kyoto Protocol in the first commitment period. *Clim. Policy*, **16(6)**, 768–782, doi.org/10.1080/14693062.2016.1164658.

Shwom, R.L., 2011: A middle range theorization of energy politics: The struggle for energy efficient appliances. *Env. Polit.*, **20(5)**, 705–726, doi:10.1080/09644016.2011.608535.

Shwom, R.L. et al., 2015: Public Opinion on Climate Change. In: *Climate Change and Society: Sociological Perspectives* [Dunlap, R.E. and R.J. Brulle, (eds.)]. Oxford University Press, New York, USA, pp. 269–299.

Silvero, F., F. Rodrigues, S. Montelpare, E. Spacone, and H. Varum, 2019: The path towards buildings energy efficiency in South American countries. *Sustain. Cities Soc.*, **44**(April 2018), 646–665, doi:10.1016/j.scs.2018.10.028.

Simon Rosenthal, C., J.A. Rosenthal, J.D. Moore, and J. Smith, 2015: Beyond (and Within) City Limits: Climate Policy in an Intergovernmental System. *Rev. Policy Res.*, **32(5)**, 538–555, doi:10.1111/ropr.12136.

Sims Gallagher, K. and X. Xuan, 2019: *Titans of the Climate: Explaining Policy Process in the United States and China*. 1st ed. The MIT Press, Cambridge, MA, USA, 272 pp.

Sindico, F., M.M. Mbengue, and K. McKenzie, 2021: Climate Change Litigation and the Individual: An Overview. In: *Comparative Climate Change*

*Litigation: Beyond the Usual Suspects* [Sindico, F. and M.M. Mbengue, (eds.)]. Springer, Cham, Switzerland, pp. 1–33.

Singh, P.K. and H. Chudasama, 2021: Pathways for climate resilient development: Human well-being within a safe and just space in the 21st century. *Glob. Environ. Change*, **68**, 1–14, doi:10.1016/j.gloenvcha.2021.102277.

Skjærseth, J.B., 2017: The European Commission's Shifting Climate Leadership. *Glob. Environ. Polit.*, **17(2)**, 84–104, doi:10.1162/GLEP_a_00402.

Skjærseth, J.B. and T. Skodvin, 2010: *Climate change and the oil industry: Common problem, different strategies*. Manchester University Press, Manchester, UK, 256 pp.

Skovgaard, J., 2012: Learning about Climate Change: Finance Ministries in International Climate Change Politics. *Glob. Environ. Polit.*, **12(4)**, 1–8, doi:10.1162/GLEP_a_00136.

Skovgaard, J., 2018: Policy coherence and organizational cultures: Energy efficiency and greenhouse gas reduction targets. *Environ. Policy Gov.*, **28(5)**, 350–358, doi:10.1002/eet.1821.

Skovgaard, J. and H. van Asselt, eds., 2018: *The Politics of Fossil Fuel Subsidies and their Reform*. Cambridge University Press, Cambridge, UK, pp. 324.

Skovgaard, J. and H. van Asselt, 2019: The politics of fossil fuel subsidies and their reform: Implications for climate change mitigation. *Wiley Interdiscip. Rev. Clim. Change*, **10(4)**, e581, doi:10.1002/wcc.581.

Skurka, C., J. Niederdeppe, R. Romero-Canyas, and D. Acup, 2018: Pathways of Influence in Emotional Appeals: Benefits and Tradeoffs of UsingFear or Humor to Promote Climate Change-Related Intentions and Risk Perceptions. *J. Commun.*, **68(1)**, 169–193, doi:10.1093/joc/jqx008.

Slovic, A.D., M.A. de Oliveira, J. Biehl, and H. Ribeiro, 2016: How Can Urban Policies Improve Air Quality and Help Mitigate Global Climate Change: a Systematic Mapping Review. *J. Urban Heal.*, **93(1)**, doi:10.1007/s11524-015-0007-8.

Smeds, E. and M. Acuto, 2018: Networking Cities after Paris: Weighing the Ambition of Urban Climate Change Experimentation. *Glob. Policy*, **9(4)**, 549–559, doi:10.1111/1758-5899.12587.

Smith, J., 2000: *The daily globe: environmental change, the public and the media*. Earthscan Publications Ltd., London, UK, pp. 280.

Smith, J. et al., 2016: Balancing competing policy demands: the case of sustainable public sector food procurement. *J. Clean. Prod.*, **112**, 249–256, doi:10.1016/j.jclepro.2015.07.065.

Smith, J.M., 2019: Boom to bust, ashes to (coal) dust: the contested ethics of energy exchanges in a declining US coal market. *J.R. Anthropol. Inst.*, **25(S1)**, 91–107, doi:10.1111/1467-9655.13016.

Snell, D. and P. Fairbrother, 2010: Unions as environmental actors. *Transf. Eur. Rev. Labour Res.*, **16(3)**, 411–424, doi:10.1177/1024258910373874.

Sokona, Y., 2021: Building capacity for 'energy for development' in Africa: four decades and counting. *Clim. Policy*, 22 (5) 671–679, doi:10.1080/14693062.2020.1870915.

Solana, J., 2020: Climate Litigation in Financial Markets: A Typology. *Transnatl. Environ. Law*, **9(1)**, 103–135, doi:10.1017/S2047102519000244.

Solarin, S.A., 2020: An environmental impact assessment of fossil fuel subsidies in emerging and developing economies. *Environ. Impact Assess. Rev.*, **85**, 106443, doi:10.1016/j.eiar.2020.106443.

Solecki, W. et al., 2019: Extreme events and climate adaptation-mitigation linkages: Understanding low-carbon transitions in the era of global urbanisation. *Wiley Interdiscip. Rev. Clim. Change*, **10(6)**, 1–16, doi:10.1002/wcc.616.

Somerville, P., 2020: A critique of climate change mitigation policy. *Policy Polit.*, **48(2)**, 355–378, doi:10.1332/030557319X15661682426163.

Sovacool, B.K. and M.H. Dworkin, 2015: Energy justice: Conceptual insights and practical applications. *Appl. Energy*, **142**, 435–444, doi:10.1016/j.apenergy.2015.01.002.

Sovacool, B.K. et al., 2020: Imagining sustainable energy and mobility transitions: Valence, temporality, and radicalism in 38 visions of a low-carbon future. *Soc. Stud. Sci.*, **50(4)**, 642–679, doi:10.1177/0306312720915283.

Spencer, B. et al., 2017: Case studies in co-benefits approaches to climate change mitigation and adaptation. *J. Environ. Plan. Manag.*, **60(4)**, 647–667, doi:10.1080/09640568.2016.1168287.

Spencer, T. and N.K. Dubash, 2021: Scenarios for different 'Future Indias': sharpening energy and climate modelling tools. *Clim. Policy*, **22(1)**, 30–47, doi:10.1080/14693062.2021.1973361.

Spencer, T. et al., 2018: The 1.5°C target and coal sector transition: at the limits of societal feasibility. *Clim. Policy*, **18(3)**, 335–351, doi:10.1080/14693062.2017.1386540.

Stadelmann, M. and R. Schubert, 2018: How Do Different Designs of Energy Labels Influence Purchases of Household Appliances? A Field Study in Switzerland. *Ecol. Econ.*, **144**, 112–123, doi:10.1016/j.ecolecon.2017.07.031.

Stadelmann, M., A. Michaelowa, S. Butzengeiger-Geyer, and M. Köhler, 2014: Universal Metrics to Compare the Effectiveness of Climate Change Adaptation Projects. In: *Handbook of Climate Change Adaptation* [Leal Filho, W., (ed.)]. Springer Berlin Heidelberg, Berlin, Germany, pp. 1–15.

Stavins, R.N., 2019: *The future of U.S. carbon-pricing policy*. Harvard Environmental Economics Program, Cambridge, MA, USA, 56 pp.

Steffen, B., T.S. Schmidt, and P. Tautorat, 2019: Measuring whether municipal climate networks make a difference: the case of utility-scale solar PV investment in large global cities. *Clim. Policy*, **19(7)**, 908–922, doi:10.1080/14693062.2019.1599804.

Steiner, A., 2009: Global Green New Deal. *New Solut. A J. Environ. Occup. Heal. Policy*, **19(2)**, 185–193, doi:10.2190/NS.19.2.s.

Steinhardt, H.C. and F. Wu, 2016: In the Name of the Public: Environmental Protest and the Changing Landscape of Popular Contention in China. *China J.*, **75**, 61–82, doi:10.1086/684010.

Stevins, D., 2013: Green jobs? Good jobs? Just jobs? US labour unions confront climate change. In: *Trade Unions in the Green Economy: Working for the Environment* [Räthzel, N. and D. Uzzell, (eds.)]. Routledge, London, UK, pp. 179–185.

Stevis, D., 2018: US labour unions and green transitions: depth, breadth, and worker agency. *Globalizations*, **15(4)**, 454–469, doi:10.1080/14747731.2018.1454681.

Stiglitz, J.E., 2019: Addressing climate change through price and non-price interventions. *Eur. Econ. Rev.*, **119**, 594–612, doi:10.1016/j.euroecorev.2019.05.007.

Stoddard, I. et al., 2021: Three Decades of Climate Mitigation: Why Haven't We Bent the Global Emissions Curve? *Annu. Rev. Environ. Resour.*, **46(1)**, 653–689, doi:10.1146/annurev-environ-012220-011104.

Stoddart, M.C.J. and D.B. Tindall, 2015: Canadian news media and the cultural dynamics of multilevel climate governance. *Env. Polit.*, **24(3)**, 401–422, doi:10.1080/09644016.2015.1008249.

Stoerk, T., D.J. Dudek, and J. Yang, 2019: China's national carbon emissions trading scheme: lessons from the pilot emission trading schemes, academic literature, and known policy details. *Clim. Policy*, **19(4)**, 472–486, doi:10.1080/14693062.2019.1568959.

Stokes, L.C., 2016: Electoral Backlash against Climate Policy: A Natural Experiment on Retrospective Voting and Local Resistance to Public Policy. *Am. J. Pol. Sci.*, **60(4)**, 958–974, doi:10.1111/ajps.12220.

Stokes, L.C., 2020: *Short Circuiting Policy: Interest Groups and the Battle Over Clean Energy and Climate Policy in the American States*. Oxford University Press, NY, USA, 337 pp.

Stokes, L.C. and C. Warshaw, 2017: Renewable energy policy design and framing influence public support in the United States. *Nat. Energy*, **2(8)**, 17107, doi:10.1038/nenergy.2017.107.

Stokes, L.C. and H.L. Breetz, 2018: Politics in the U.S. energy transition: Case studies of solar, wind, biofuels and electric vehicles policy. *Energy Policy*, **113**, 76–86, doi:10.1016/j.enpol.2017.10.057.

Strauch, Y., 2020: Beyond the low-carbon niche: Global tipping points in the rise of wind, solar, and electric vehicles to regime scale systems. *Energy Res. Soc. Sci.*, **62**, doi:10.1016/j.erss.2019.101364.

13

Stuart-Smith, R.F. et al., 2021: Filling the evidentiary gap in climate litigation. *Nat. Clim. Change*, **11(8)**, 651–655, doi:10.1038/s41558-021-01086-7.

Stuart, A., E.F. Thomas, N. Donaghue, and A. Russell, 2013: 'We may be pirates, but we are not protesters': Identity in the Sea Shepherd Conservation Society. *Polit. Psychol.*, **34(5)**, 753–777, doi:10.1111/pops.12016.

Stuart, J., P. Collins, M. Alger, and G. Whitelaw, 2016: Embracing sustainability: the incorporation of sustainability principles in municipal planning and policy in four mid-sized municipalities in Ontario, Canada. *Local Environ.*, **21(2)**, 219–240, doi:10.1080/13549839.2014.936844.

Stucki, T., M. Woerter, S. Arvanitis, M. Peneder, and C. Rammer, 2018: How different policy instruments affect green product innovation: A differentiated perspective. *Energy Policy*, **114**, 245–261, doi:10.1016/j.enpol.2017.11.049.

Sugiyama, M. et al., 2021: EMF 35 JMIP study for Japan's long-term climate and energy policy: scenario designs and key findings. *Sustain. Sci.*, **16(2)**, 355–374, doi:10.1007/s11625-021-00913-2.

Sundström, A., 2016: Understanding illegality and corruption in forest governance. *J. Environ. Manage.*, **181**, 779–790, doi:10.1016/j.jenvman.2016.07.020.

Supran, G. and N. Oreskes, 2017: Assessing ExxonMobil's climate change communications (1977–2014). *Environ. Res. Lett.*, **12(8)**, 084019, doi:10.1088/1748-9326/aa815f.

Sustainable Purchasing Leadership Council, 2021: *Recommendations for climate friendly refrigerant management and procurement*. 42 pp. https://www.sustainablepurchasing.org/wp-content/uploads/2020/09/Climate-Friendly-Refrigerant-Management-and-Procurement-22-September-2021.pdf (Accessed October 18, 2021).

Svoboda, M., 2016: Cli-fi on the screen(s): patterns in the representations of climate change in fictional films. *Wiley Interdiscip. Rev. Clim. Change*, **7(1)**, 43–64, doi:10.1002/wcc.381.

Swim, J.K., N. Geiger, and M.L. Lengieza, 2019: Climate Change Marches as Motivators for Bystander Collective Action. *Front. Commun.*, pp. 4, doi:10.3389/fcomm.2019.00004.

Takahashi, B., 2011: Framing and sources: a study of mass media coverage of climate change in Peru during the V ALCUE. *Public Underst. Sci.*, **20(4)**, 543–557, doi:10.1177/0963662509356502.

Talaei, A., M.S. Ahadi, and S. Maghsoudy, 2014: Climate friendly technology transfer in the energy sector: A case study of Iran. *Energy Policy*, **64**, 349–363, doi:10.1016/j.enpol.2013.09.050.

Tan, H., E. Thurbon, S.-Y. Kim, and J.A. Mathews, 2021: Overcoming incumbent resistance to the clean energy shift: How local governments act as change agents in coal power station closures in China. *Energy Policy*, **149**, 112058, doi:10.1016/j.enpol.2020.112058.

Tan, J., 2018: Wo guo mei kuang wa si zong he li yong fa zhan xian zhuang ji jian yi (Development status and suggestion of comprehensive utilization of coal mine gas in China). *Coal Process. Compr. Util.*, **(8)**, 59-61, 66.

Tao, S., S. Chen, and Z. Pan, 2019: Current status, challenges, and policy suggestions for coalbed methane industry development in China: A review. *Energy Sci. Eng.*, **7(4)**, 1059–1074, doi:10.1002/ese3.358.

Tarrow, S., 2005: *The New Transnational Activism*. Cambridge University Press, New York, USA, 288 pp.

Taylor, M., J. Watts, and J. Bartlett, 2019: Climate crisis: 6 million people join latest wave of global protests. *The Guardian*, https://www.theguardian.com/environment/2019/sep/27/climate-crisis-6-million-people-join-latest-wave-of-worldwide-protests (Accessed March 30, 2021).

Tegegne, Y.T. et al., 2021: REDD+ and equity outcomes: Two cases from Cameroon. *Environ. Sci. Policy*, **124**(April 2020), 324–335, doi:10.1016/j.envsci.2021.07.003.

Temper, L., F. Demaria, A. Scheidel, D. Del Bene, and J. Martinez-Alier, 2018: The Global Environmental Justice Atlas (EJAtlas): ecological distribution conflicts as forces for sustainability. *Sustain. Sci.*, **13(3)**, 573–584, doi:10.1007/s11625-018-0563-4.

Temper, L. et al., 2020: Movements shaping climate futures: A systematic mapping of protests against fossil fuel and low-carbon energy projects. *Environ. Res. Lett.*, **15(12)**, 123004, doi:10.1088/1748-9326/abc197.

Teng, F. and P. Wang, 2021: The evolution of climate governance in China: drivers, features, and effectiveness. *Env. Polit.*, **30(sup1)**, 141–161, doi:10.1080/09644016.2021.1985221.

Teng, F., F. Jotzo, and X. Wang, 2017: Interactions between market reform and a carbon price in China's power sector. *Econ. Energy Environ. Policy*, **6(2)**, doi:10.5547/2160-5890.6.1.ften.

Teske, S., T. Pregger, S. Simon, and T. Naegler, 2018: High renewable energy penetration scenarios and their implications for urban energy and transport systems. *Curr. Opin. Environ. Sustain.*, **30**, 89–102, doi:10.1016/j.cosust.2018.04.007.

Testa, F., E. Annunziata, F. Iraldo, and M. Frey, 2016: Drawbacks and opportunities of green public procurement: an effective tool for sustainable production. *J. Clean. Prod.*, **112**, 1893–1900, doi:10.1016/j.jclepro.2014.09.092.

Thackeray, S.J. et al., 2020: Civil disobedience movements such as School Strike for the Climate are raising public awareness of the climate change emergency. *Glob. Change Biol.*, **26(3)**, 1042–1044, doi:10.1111/gcb.14978.

The Climate Group with CDP, 2020: States and regions: CDP. www.cdp.net/en/cities/states-and-regions (Accessed October 21, 2021).

The European Parliament and the Council of the European Union, 2018: Directive (EU) 2018/2001 of the European Parliament and of the Council of 11 December 2018 on the promotion of the use of energy from renewable sources. http://data.europa.eu/eli/dir/2018/2001/2018-12-21 (Accessed October 24, 2021).

Thøgersen, J. et al., 2021: Why do people continue driving conventional cars despite climate change? Social-psychological and institutional insights from a survey of Norwegian commuters. *Energy Res. Soc. Sci.*, **79**, 102168, doi:10.1016/j.erss.2021.102168.

Thompson, F.J., K.K. Wong, and B.G. Rabe, 2020: *Trump, the Administrative Presidency, and Federalism*. Brookings Institution Press, Washington, DC, USA, 256 pp.

Thonig, R. et al., 2021: Does ideology influence the ambition level of climate and renewable energy policy? Insights from four European countries. *Energy Sources, Part B Econ. Planning, Policy*, **16(1)**, 4–22, doi:10.1080/15567249.2020.1811806.

Thornton, T.F. and C. Comberti, 2017: Synergies and trade-offs between adaptation, mitigation and development. *Clim. Change*, **140(1)**, 5–18, doi:10.1007/s10584-013-0884-3.

Thunberg, G., 2019: Young people have led the climate strikes. Now we need adults to join us too. *The Guardian*, https://www.theguardian.com/commentisfree/2019/may/23/greta-thunberg-young-people-climate-strikes-20-september (Accessed May 23).

Timilsina, G.R., 2018: *Where is the carbon tax after thirty years of research?* The World Bank, Washington, DC, USA, 88 pp. https://openknowledge.worldbank.org/handle/10986/29946 (Accessed March 30, 2021).

Timperley, J., 2020: What can the world learn from New Zealand on climate? *Lancet Planet. Heal.*, **4(5)**, e176–e177, doi:10.1016/S2542-5196(20)30109-1.

Tindall, D.B., M.C.J. Stoddart, and C. Callison, 2018: The Relationships Between Climate Change News Coverage, Policy Debate, and Societal Decisions. In: *Oxford Research Encyclopedia of Climate Science*, Oxford University Press, Oxford, UK, doi: 10.1093/acrefore/9780190228620.013.370.

Tobin, P., 2017: Leaders and Laggards: Climate Policy Ambition in Developed States. *Glob. Environ. Polit.*, **17(4)**, 28–47, doi:10.1162/GLEP_a_00433.

Tobler, C., V.H.M. Visschers, and M. Siegrist, 2012: Addressing climate change: Determinants of consumers' willingness to act and to support policy measures. *J. Environ. Psychol.*, **32(3)**, 197–207, doi:10.1016/j.jenvp.2012.02.001.

Torabi Moghadam, S., C. Delmastro, S.P. Corgnati, and P. Lombardi, 2017: Urban energy planning procedure for sustainable development in the built environment: A review of available spatial approaches. *J. Clean. Prod.*, **165**, 811–827, doi:10.1016/j.jclepro.2017.07.142.

Torney, D., 2017: If at first you don't succeed: the development of climate change legislation in Ireland. *Irish Polit. Stud.*, **32(2)**, 247–267, doi:10.1080/07907184.2017.1299134.

Torney, D., 2019: Climate laws in small European states: symbolic legislation and limits of diffusion in Ireland and Finland. *Env. Polit.*, **28(6)**, 1124–1144, doi:10.1080/09644016.2019.1625159.

Torre-Schaub, M., 2021: Climate Change Litigation in France: New Perspectives and Trends. In: *Climate Change Litigation: Global Perspectives* [Alogna, I., C. Bakker, and J.P. Gauci (eds.)]. Brill, Nijhoff, Leiden, The Netherlands, pp. 124–142.

Torrens, J., J. Schot, R. Raven, and P. Johnstone, 2019: Seedbeds, harbours, and battlegrounds: On the origins of favourable environments for urban experimentation with sustainability. *Environ. Innov. Soc. Transitions*, **31**, 211–232, doi:10.1016/j.eist.2018.11.003.

Tosun, J., 2018: Investigating Ministry Names for Comparative Policy Analysis: Lessons from Energy Governance. *J. Comp. Policy Anal. Res. Pract.*, **20(3)**, 324–335, doi:10.1080/13876988.2018.1467430.

Tosun, J. and A. Lang, 2017: Policy integration: mapping the different concepts. *Policy Stud.*, **38(6)**, 553–570, doi:10.1080/01442872.2017.1339239.

Tosun, J. and J. Schnepf, 2020: Measuring change in comparative policy analysis: concepts and empirical approaches. In: *Handbook of Research Methods and Applications in Comparative Policy Analysis*, Edward Elgar Publishing, Cheltenham, UK, pp. 167–185.

Townshend, T. et al., 2013: How national legislation can help to solve climate change. *Nat. Clim. Change*, **3(5)**, 430–432, doi:10.1038/nclimate1894.

Tozer, L., 2020: Catalyzing political momentum for the effective implementation of decarbonization for urban buildings. *Energy Policy*, **136**, 111042, doi:10.1016/j.enpol.2019.111042.

Treisman, D., 2000: The causes of corruption: A cross-national study. *J. Public Econ.*, **76(3)**, 399–457, doi:10.1016/S0047-2727(99)00092-4.

Trencher, G. and J. van der Heijden, 2019: Instrument interactions and relationships in policy mixes: Achieving complementarity in building energy efficiency policies in New York, Sydney and Tokyo. *Energy Res. Soc. Sci.*, **54**, 34–45, doi:10.1016/j.erss.2019.02.023.

Trencher, G. et al., 2016: Innovative policy practices to advance building energy efficiency and retrofitting: Approaches, impacts and challenges in ten C40 cities. *Environ. Sci. Policy*, **66**, 353–365, doi:10.1016/j.envsci.2016.06.021.

Trihartono, A., N. Viartasiwi, and C. Nisya, 2020: The giant step of tiny toes: youth impact on the securitization of climate change. *IOP Conf. Ser. Earth Environ. Sci.*, **485**, 012007, doi:10.1088/1755-1315/485/1/012007.

Trope, Y., N. Liberman, and C. Wakslak, 2007: Construal levels and psychological distance: Effects on representation, prediction, evaluation, and behavior. *J. Consum. Psychol.*, **17(2)**, 83–95, doi:10.1016/S1057-7408(07)70013-X.

Tsai, W.-T., 2017: Green public procurement and green-mark products strategies for mitigating greenhouse gas emissions—experience from Taiwan. *Mitig. Adapt. Strateg. Glob. Change*, **22(5)**, 729–742, doi:10.1007/s11027-015-9695-3.

Tufekci, Z., 2017: *Twitter and Tear Gas: The Power and Fragility of Networked Protest*. Yale University Press, New Haven, CT, USA, 360 pp.

Turnheim, B., P. Kivimaa, and F. Berkhout, eds., 2018: *Innovating climate governance: moving beyond experiments*. Cambridge University Press, Cambridge, UK, 250 pp.

Tvinnereim, E. and M. Mehling, 2018: Carbon pricing and deep decarbonisation. *Energy Policy*, **121**, 185–189, doi:10.1016/j.enpol.2018.06.020.

Tyler, E. and K. Hochstetler, 2021: Institutionalising decarbonisation in South Africa: navigating climate mitigation and socio-economic transformation. *Env. Polit.*, **30(sup1)**, 184–205, doi:10.1080/09644016.2021.1947635.

Uittenbroek, C.J., 2016: From Policy Document to Implementation: Organizational Routines as Possible Barriers to Mainstreaming Climate Adaptation. *J. Environ. Policy Plan.*, **18(2)**, 161–176, doi:10.1080/1523908X.2015.1065717.

UN Environment, 2018: *Legislative and Policy Options to Control Hydrofluorocarbons*. OzonAction, United Nations Environment Economy Division, Paris, France, 104 pp. https://www.unenvironment.org/ozonaction/resources/publication/legislative-and-policy-options-control-hydrofluorocarbons (Accessed March 30, 2021).

UN Habitat, 2016: *World Cities Report Urbanization and Development: Emerging Futures*. UN-Habitat, Nairobi, Kenya, 262 pp. https://unhabitat.org/world-cities-report (Accessed May 7, 2021).

UNEP, 2018: Bridging the gap: The role of innovation policy and market creation. In: *Emissions Gap Report 2018*. United Nations Environment Programme, Nairobi, Kenya, pp. 52–59.

UNEP, 2019a: *Measuring Fossil Fuel Subsidies in the Context of the Sustainable Development Goals*. United Nations Environment Programme, Nairobi, Kenya, 94 pp. https://www.unep.org/resources/report/measuring-fossil-fuel-subsidies-context-sustainable-development-goals (Accessed October 19, 2021).

UNEP, 2019b: *Emissions Gap Report 2019*. United Nations Environment Programme, Nairobi, Kenya, 1–108. pp. https://wedocs.unep.org/bitstream/handle/20.500.11822/30797/EGR2019.pdf?sequence=1&isAllowed=y (Accessed March 30, 2021).

UNEP, 2020a: *Emissions Gap Report 2020*. United Nations Environment Programme, Nairobi, Kenya, 128 pp. https://www.unenvironment.org/emissions-gap-report-2020 (Accessed March 30, 2021).

UNEP, 2020b: Global Climate Litigation Report: 2020 Status Review. United Nations Environment Programme, Nairobi, Kenya, https://www.unep.org/resources/report/global-climate-litigation-report-2020-status-review, 52 pp.

UNEP DTU Partnership, 2020: CDM Pipeline (1 May 2020 version). http://www.cdmpipeline.org/ (Accessed March 30, 2021).

UNFCCC, 2012: *Annual report of the Executive Board of the clean development mechanism to the Conference of the Parties serving as the meeting of the Parties to the Kyoto Protocol*. United Nations Framework Convention on Climate Change, Bonn, Germany, 20 pp. http://unfccc.int/resource/docs/2012/cmp8/eng/03p01.pdf (Accessed March 30, 2021).

UNFCCC Secretariat, 2016: *Aggregate effect of the intended Nationally Determined Contributions: an update*. United Nations Framework Convention on Climate Change, Marrakech, Morocco, 75 pp. https://unfccc.int/sites/default/files/resource/docs/2016/cop22/eng/02.pdf (Accessed March 30, 2021).

UNFCCC Secretariat, 2021: *Nationally Determined Contributions under the Paris Agreement: Synthesis report by the secretariat*. 42 pp. https://unfccc.int/news/full-ndc-synthesis-report-some-progress-but-still-a-big-concern (Accessed October 7, 2021).

Upadhyaya, P., M.K. Shrivastava, G. Gorti, and S. Fakir, 2021: Capacity building for proportionate climate policy: Lessons from India and South Africa. *Int. Polit. Sci. Rev.*, **42(1)**, 130–145, doi:10.1177/0192512120963883.

Ürge-Vorsatz, D. and K.C. Seto, 2018: Editorial Overview: 1.5°C Climate change and urban areas. *Curr. Opin. Environ. Sustain.*, **30**, iv–vi, doi:10.1016/j.cosust.2018.07.004.

Ürge-Vorsatz, D., S.T. Herrero, N.K. Dubash, and F. Lecocq, 2014: Measuring the Co-Benefits of Climate Change Mitigation. *Annu. Rev. Environ. Resour.*, **39(1)**, 549–582, doi:10.1146/annurev-environ-031312-125456.

Ürge-Vorsatz, Di. et al., 2018: Locking in positive climate responses in cities. *Nat. Clim. Change*, **8(3)**, 174–177, doi:10.1038/s41558-018-0100-6.

Urpelainen, J., 2018: RISE to the occasion? A critique of the World Bank's Regulatory Indicators for Sustainable Energy. *Energy Res. Soc. Sci.*, **39**, 69–73, doi:10.1016/j.erss.2017.10.034.

Uyarra, E., P. Shapira, and A. Harding, 2016: Low carbon innovation and enterprise growth in the UK: Challenges of a place-blind policy mix. *Technol. Forecast. Soc. Change*, **103**, 264–272, doi:10.1016/j.techfore.2015.10.008.

Valentová, M., J. Karásek, and J. Knápek, 2019: *Ex post* evaluation of energy efficiency programs: Case study of Czech Green Investment Scheme. *WIREs Energy Environ.*, **8(2)**, doi:10.1002/wene.323.

Valenzuela, J.M., 2014: Climate Change Agenda at Sub-national Level in Mexico: Policy coordination or policy competition? *Environ. Policy Gov.*, **24(3)**, 188–203, doi:10.1002/eet.1638.

van Benthem, A. and S. Kerr, 2013: Scale and transfers in international emissions offset programs. *J. Public Econ.*, **107**, 31–46, doi:10.1016/j.jpubeco.2013.08.004.

Van Den Berg, L., E. Braun, and J. Van Der Meer, 2018: *National Urban Policies in the European Union*. 1st ed. Routledge, Oxon, UK.

van den Bergh, J. and I. Savin, 2021: Impact of Carbon Pricing on Low-Carbon Innovation and Deep Decarbonisation: Controversies and Path Forward. *Environ. Resour. Econ.*, **80**, 705–715, doi:10.1007/s10640-021-00594-6.

van den Bergh, J. et al., 2021: Designing an effective climate-policy mix: accounting for instrument synergy. *Clim. Policy*, **21(6)**, 745–764, doi:10.1080/14693062.2021.1907276.

van der Heijden, J., 2018: City and Sub-national Governance. In: *Governing Climate Change: Polycentricity in Action?* [Jordan, A., D. Huitema, H. Van Asselt, and J. Forster, (eds.)]. Cambridge University Press, Cambridge, UK, pp. 81–96.

van der Linden, S., E. Maibach, and A. Leiserowitz, 2015: Improving Public Engagement With Climate Change: Five 'Best Practice' Insights From Psychological Science. *Perspect. Psychol. Sci.*, **10(6)**, 758–763, doi:10.1177/1745691615598516.

van der Linden, S., A. Leiserowitz, S. Rosenthal, and E. Maibach, 2017: Inoculating the Public against Misinformation about Climate Change. *Glob. Challenges*, **1(2)**, 1600008, doi:10.1002/gch2.201600008.

Vanegas Cantarero, M.M., 2020: Of renewable energy, energy democracy, and sustainable development: A roadmap to accelerate the energy transition in developing countries. *Energy Res. Soc. Sci.*, **70**, 101716, doi:10.1016/j.erss.2020.101716.

Vedung, E., 2005: *Public Policy and Program Evaluation*. Routledge, Boca Raton, Florida, USA, 336 pp.

Venmans, F., J. Ellis, and D. Nachtigall, 2020: Carbon pricing and competitiveness: are they at odds? *Clim. Policy*, **20(9)**, 1070–1091, doi:10.1080/14693062.2020.1805291.

Verde, S.F., 2020: The impact of the EU emissions trading system on competitiveness and carbon leakage: The econometric evidence. *J. Econ. Surv.*, **34(2)**, 320–343, doi:10.1111/joes.12356.

Verschuuren, J., 2019: The State of the Netherlands v Urgenda Foundation: The Hague Court of Appeal upholds judgment requiring the Netherlands to further reduce its greenhouse gas emissions. *Rev. Eur. Comp. Int. Environ. Law*, **28(1)**, 94–98, doi:10.1111/reel.12280.

Vestergren, S., J. Drury, and E.H. Chiriac, 2018: How collective action produces psychological change and how that change endures over time: A case study of an environmental campaign. *Br. J. Soc. Psychol.*, **57(4)**, 855–877, doi:10.1111/bjso.12270.

Vestergren, S., J. Drury, and E. Hammar Chiriac, 2019: How participation in collective action changes relationships, behaviours, and beliefs: An interview study of the role of inter- and intragroup processes. *J. Soc. Polit. Psychol.*, **7(1)**, 76–99, doi:10.5964/jspp.v7i1.903.

Veum, K.C., 2018: *Long-Term Agreements on Energy Efficiency for the non-ETS sector*. 11 pp. https://epatee.eu/system/tdf/epatee_case_study_netherlands_mja3_voluntary_agreements_in_the_non-ets_sectors_ok_0_0.pdf?file=1&type=node&id=155 (Accessed March 3, 2021).

Viardot, E., 2013: The role of cooperatives in overcoming the barriers to adoption of renewable energy. *Energy Policy*, **63**, 756–764, doi:10.1016/j.enpol.2013.08.034.

Victor, D.G., F.W. Geels, and S. Sharpe, 2019: *Accelerating the Low Carbon Transition: The Case for Stronger, More Targeted and Coordinated International Action*. The Brookings Institution, San Diego, CA, USA, The UK Government Department for Business, Energy and Industrial Strategy, London, UK, and The Energy Transitions Commission (ETC), London, UK, 71 pp. https://www.brookings.edu/wp-content/uploads/2019/12/Coordinatedactionreport.pdf.

Viguié, V. and S. Hallegatte, 2012: Trade-offs and synergies in urban climate policies. *Nat. Clim. Change*, **2(5)**, 334–337, doi:10.1038/nclimate1434.

Vihma, A., 2011: India and the Global Climate Governance: Between Principles and Pragmatism. *J. Environ. Dev.*, **20(1)**, 69–94, doi:10.1177/1070496510394325.

Villarroel Walker, R., M.B. Beck, J.W. Hall, R.J. Dawson, and O. Heidrich, 2017: Identifying key technology and policy strategies for sustainable cities: A case study of London. *Environ. Dev.*, **21**(May 2016), 1–18, doi:10.1016/j.envdev.2016.11.006.

Vivid Economics, 2021: *Greenness of Stimulus Index. An assessment of COVID-19 stimulus by G20 countries and other major economies in relation to climate action and biodiversity goals. July 2021.* 93 pp. https://www.vivideconomics.com/wp-content/uploads/2021/07/Green-Stimulus-Index-6th-Edition_final-report.pdf (Accessed October 27, 2021).

Vogt-Schilb, A. and S. Hallegatte, 2017: Climate policies and Nationally Determined Contributions: reconciling the needed ambition with the political economy. *Wiley Interdiscip. Rev. Energy Environ.*, **6(6)**, e256, doi:10.1002/wene.256.

von Stechow, C. et al., 2015: Integrating Global Climate Change Mitigation Goals with Other Sustainability Objectives: A Synthesis. *Annu. Rev. Environ. Resour.*, **40(1)**, 363–394, doi:10.1146/annurev-environ-021113-095626.

von Stechow, C. et al., 2016: 2°C and SDGs: united they stand, divided they fall? *Environ. Res. Lett.*, **11(3)**, 034022, doi:10.1088/1748-9326/11/3/034022.

von Stein, J., 2020: Democracy, Autocracy, and Everything in Between: How Domestic Institutions Affect Environmental Protection. *Br. J. Polit. Sci.*, **52(1)**, 339–357, doi:10.1017/S000712342000054X.

Vonhedemann, N., Z. Wurtzebach, T.J. Timberlake, E. Sinkular, and C.A. Schultz, 2020: Forest policy and management approaches for carbon dioxide removal: Forest Policy and Management for CDR. *Interface Focus*, **10(5)**, doi:10.1098/rsfs.2020.0001.

Voß, J.-P., A. Smith, and J. Grin, 2009: Designing long-term policy: rethinking transition management. *Policy Sci.*, **42(4)**, 275–302, doi:10.1007/s11077-009-9103-5.

Voytenko, Y., K. McCormick, J. Evans, and G. Schliwa, 2016: Urban living labs for sustainability and low carbon cities in Europe: towards a research agenda. *J. Clean. Prod.*, **123**, 45–54, doi:10.1016/j.jclepro.2015.08.053.

Wagner, P. and T. Ylä-Anttila, 2018: Who got their way? Advocacy coalitions and the Irish climate change law. *Env. Polit.*, **27(5)**, 872–891, doi:10.1080/09644016.2018.1458406.

Wahlström, M., M. Wennerhag, and C. Rootes, 2013: Framing 'The Climate Issue': Patterns of Participation and Prognostic Frames among Climate Summit Protesters. *Glob. Environ. Polit.*, **13(4)**, 101–122, doi:10.1162/GLEP_a_00200.

Wahlström, M., P. Kocyba, M. De Vydt, de M. Joost (eds), 2019: *Protest for a future: Composition, mobilization and motives of the participants in Fridays For Future climate protests on 15 March, 2019 in 13 European cities*. 120 pp, https://eprints.keele.ac.uk/6571/7/20190709_Protest%20for%20a%20future_GCS%20Descriptive%20Report.pdf (Accessed March 3, 2021).

Waisman, H. et al., 2019: A pathway design framework for national low greenhouse gas emission development strategies. *Nat. Clim. Change*, **9(4)**, 261–268, doi:10.1038/s41558-019-0442-8.

Wakabayashi, M. and T.H. Arimura, 2016: Voluntary agreements to encourage proactive firm action against climate change: An empirical study of industry associations' voluntary action plans in Japan. *J. Clean. Prod.*, **112**, 2885–2895, doi:10.1016/j.jclepro.2015.10.071.

Walgrave, S., R. Wouters, J. Van Laer, J. Verhulst, and P. Ketelaars, 2012: Transnational Collective Identification: May Day and Climate Change Protesters' Identification with Similar Protest Events in Other Countries. *Mobilization An Int. Q.*, **17(3)**, 301–317, doi:10.17813/maiq.17.3.3nkh1p041013500q.

Walker, C., J. Baxter, and D. Ouellette, 2014: Beyond Rhetoric to Understanding Determinants of Wind Turbine Support and Conflict in Two Ontario,

Canada Communities. *Environ. Plan. A Econ. Sp.*, **43(3)**, 730–745, doi:10.1068/a130004p.

Walter, S., M. Brüggemann, and S. Engesser, 2018: Echo Chambers of Denial: Explaining User Comments on Climate Change. *Environ. Commun.*, **12(2)**, 204–217, doi:10.1080/17524032.2017.1394893.

Walwyn, D.R. and A.C. Brent, 2015: Renewable energy gathers steam in South Africa. *Renew. Sustain. Energy Rev.*, **41**, 390–401, doi:10.1016/j.rser.2014.08.049.

Wamsler, C. et al., 2020: Environmental and climate policy integration: Targeted strategies for overcoming barriers to nature-based solutions and climate change adaptation. *J. Clean. Prod.*, **247**, 119154, doi:10.1016/j.jclepro.2019.119154.

Wang, H. and W. Chen, 2019: Gaps between pre-2020 climate policies with NDC goals and long-term mitigation targets: analyses on major regions. *Energy Procedia*, **158**, 3664–3669, doi:10.1016/j.egypro.2019.01.894.

Wang, P., L. Liu, and T. Wu, 2018a: A review of China's climate governance: state, market and civil society. *Clim. Policy*, **18(5)**, 664–679, doi:10.1080/14693062.2017.1331903.

Wang, P., L. Liu, X. Tan, and Z. Liu, 2019: Key challenges for China's carbon emissions trading program. *Wiley Interdiscip. Rev. Clim. Change*, **10(5)**, e599, doi:10.1002/wcc.599.

Wang, Z., Danish, B. Zhang, and B. Wang, 2018b: The moderating role of corruption between economic growth and $CO_2$ emissions: Evidence from BRICS economies. *Energy*, **148**, 506–513, doi:10.1016/j.energy.2018.01.167.

Wanvik, T.I. and H. Haarstad, 2021: Populism, Instability, and Rupture in Sustainability Transformations. *Ann. Am. Assoc. Geogr.*, **111(7)**, 1–16, doi:10.1080/24694452.2020.1866486.

Warbroek, B. and T. Hoppe, 2017: Modes of Governing and Policy of Local and Regional Governments Supporting Local Low-Carbon Energy Initiatives; Exploring the Cases of the Dutch Regions of Overijssel and Fryslân. *Sustainability*, **9(1)**, 75, doi:10.3390/su9010075.

Ward, H., J.C. Steckel, and M. Jakob, 2019: How global climate policy could affect competitiveness. *Energy Econ.*, **84**, doi:10.1016/j.eneco.2019.104549.

Warren, B., P. Christoff, and D. Green, 2016: Australia's sustainable energy transition: The disjointed politics of decarbonisation. *Environ. Innov. Soc. Transitions*, **21**, 1–12, doi:10.1016/j.eist.2016.01.001.

Wasim, R., 2019: Corporate (Non)disclosure of Climate Change Information. *Columbia Law Rev.*, **119(5)**, 1311–1354.

Watkiss, P. and R. Klein, 2019: *Long-term Strategies in a Changing Climate*. Deutsche Gesellschaft für Internationale Zusammenarbeit (GIZ) GmbH, Bonn, Germany, 59 pp. https://www.adaptationcommunity.net/wp-content/uploads/2019/10/2019_GIZ_Long-term-Strategies-in-a-Changing-Climate.pdf.

We Are Still In coalition, 2020: We Are Still In. https://www.wearestillin.com/ (Accessed December 21, 2020).

Webb, J., M. Tingey, and D. Hawkey, 2017: *What We Know about Local Authority Engagement in UK Energy Systems: Ambitions, Activities, Business Structures & Ways Forward*. UK Energy Research Centre, London, England, Energy Technologies Institute, Loughborough, England, The University of Edinburgh, Edinburgh, 56 pp, http://www.ukerc.ac.uk/publications/what-we-know-about-local-authority-engagement-in-uk-energy-systems.html (Accessed March 3, 2021).

Wegener, L., 2020: Can the Paris Agreement Help Climate Change Litigation and Vice Versa? *Transnatl. Environ. Law*, **9(1)**, 17–36, doi:10.1017/S2047102519000396.

Weidner, H. and L. Mez, 2008: German Climate Change Policy. *J. Environ. Dev.*, **17(4)**, 356–378, doi:10.1177/1070496508325910.

Weitzel, M. et al., 2019: Model-based assessments for long-term climate strategies. *Nat. Clim. Change*, **9(5)**, 345–347.

Welsch, H., 2004: Corruption, growth, and the environment: A cross-country analysis. *Environ. Dev. Econ.*, **9(5)**, 663–693, doi:10.1017/S1355770X04001500.

Westman, L. and V. Castan Broto, 2018: Climate governance through partnerships: A study of 150 urban initiatives in China. *Glob. Environ. Change*, **50**, 212–221, doi:10.1016/j.gloenvcha.2018.04.008.

Westman, L.K., V. Castán Broto, and P. Huang, 2019: Revisiting multi-level governance theory: Politics and innovation in the urban climate transition in Rizhao, China. *Polit. Geogr.*, **70**(July 2018), 14–23, doi:10.1016/j.polgeo.2019.01.002.

Wettestad, J. and L.H. Gulbrandsen, 2018: *The Evolution of Carbon Markets: Design and diffusion*. [Wettestad, J. and L.H. Gulbrandsen, (eds.)]. Routledge, London, UK, 264 pp.

Wettestad, J. and T. Jevnaker, 2019: Smokescreen Politics? Ratcheting Up EU Emissions Trading in 2017. *Rev. Policy Res.*, **36(5)**, 635–659, doi:10.1111/ropr.12345.

Wewerinke-Singh, M. and D.H. Salili, 2020: Between negotiations and litigation: Vanuatu's perspective on loss and damage from climate change. *Clim. Policy*, **20(6)**, 681–692, doi:10.1080/14693062.2019.1623166.

White, L.V., L. Hughes, C. Lyons, and Y. Peng, 2021: Iterating localisation policies in support of energy transition: The case of the Australian Capital Territory. *Energy Policy*, **158**, 112568, doi:10.1016/j.enpol.2021.112568.

Whitmarsh, L. and A. Corner, 2017: Tools for a new climate conversation: A mixed-methods study of language for public engagement across the political spectrum. *Glob. Environ. Change*, **42**, 122–135, doi:10.1016/j.gloenvcha.2016.12.008.

Whyte, K., 2017: Indigenous Climate Change Studies: Indigenizing Futures, Decolonizing the Anthropocene. *Engl. Lang. Notes*, **55(1–2)**, doi:10.1215/00138282-55.1-2.153.

Whyte, K., 2020: Too late for indigenous climate justice: Ecological and relational tipping points. *WIREs Clim. Change*, **11(1)**, e603, doi:10.1002/wcc.603.

Wieczorek, A.J., R. Raven, and F. Berkhout, 2015: Transnational linkages in sustainability experiments: A typology and the case of solar photovoltaic energy in India. *Environ. Innov. Soc. Transitions*, **17**, 149–165, doi:10.1016/j.eist.2015.01.001.

Wiedenhofer, D., B. Smetschka, L. Akenji, M. Jalas, and H. Haberl, 2018: Household time use, carbon footprints, and urban form: a review of the potential contributions of everyday living to the 1.5°C climate target. *Curr. Opin. Environ. Sustain.*, **30**, 7–17, doi:10.1016/j.cosust.2018.02.007.

Wiese, C., A. Larsen, and L.-L. Pade, 2018: Interaction effects of energy efficiency policies: a review. *Energy Effic.*, **11(8)**, 2137–2156, doi:10.1007/s12053-018-9659-z.

Wilensky, M., 2015: Climate Change in the Courts: An Assessment of Non-U.S. Climate Litigation. *Duke Environ. Law Policy Forum*, **26(1)**, 131–179.

Williams III, R.C., H. Gordon, D. Burtraw, J.C. Carbone, and R.D. Morgenstern, 2015: The Initial Incidence of a Carbon Tax Across Income Groups. *Natl. Tax J.*, **68(1)**, 195–214, doi:10.17310/ntj.2015.1.09.

Williams, S. and A. Doyon, 2019: Justice in energy transitions. *Environ. Innov. Soc. Transitions*, **31**, 144–153, doi:10.1016/j.eist.2018.12.001.

Williamson, V., T. Skocpol, and J. Coggin, 2011: The tea party and the remaking of Republican conservatism. *Perspect. Polit.*, **9(1)**, 25–43, doi:10.1017/S153759271000407X.

Willis, R., 2017: Taming the Climate? Corpus analysis of politicians' speech on climate change. *Env. Polit.*, **26(2)**, 212–231, doi:10.1080/09644016.2016.1274504.

Willis, R., 2018: How Members of Parliament understand and respond to climate change. *Sociol. Rev.*, **66(3)**, 475–491, doi:10.1177/0038026117731658.

Winiwarter, W., L. Höglund-Isaksson, Z. Klimont, W. Schöpp, and M. Amann, 2018: Technical opportunities to reduce global anthropogenic emissions of nitrous oxide. *Environ. Res. Lett.*, **13(1)**, 014011, doi:10.1088/1748-9326/aa9ec9.

Winkler, H. and L. Rajamani, 2014: CBDR&RC in a regime applicable to all. *Clim. Policy*, **14(1)**, 102–121, doi:10.1080/14693062.2013.791184.

Winkler, H., A. Boyd, M. Torres Gunfaus, and S. Raubenheimer, 2015: Reconsidering development by reflecting on climate change. *Int.*

*Environ. Agreements Polit. Law Econ.*, **15(4)**, 369–385, doi:10.1007/s10784-015-9304-7.

Wittmayer, J.M., F. van Steenbergen, A. Rok, and C. Roorda, 2016: Governing sustainability: a dialogue between Local Agenda 21 and transition management. *Local Environ.*, **21(8)**, 939–955, doi:10.1080/13549839.2015.1050658.

Wolch, J.R., J. Byrne, and J.P. Newell, 2014: Landscape and Urban Planning Urban green space, public health, and environmental justice: The challenge of making cities 'just green enough'. *Landsc. Urban Plan.*, **125**, 234–244, doi:10.1016/j.landurbplan.2014.01.017.

Wolfram, M., 2016: Conceptualizing urban transformative capacity: A framework for research and policy. *Cities*, **51**, 121–130, doi:10.1016/j.cities.2015.11.011.

Wolfram, M. and N. Frantzeskaki, 2016: Cities and Systemic Change for Sustainability: Prevailing Epistemologies and an Emerging Research Agenda. *Sustainability*, **8(2)**, 144, doi:10.3390/su8020144.

Wong, A., 2018: *The Roots of Japan's International Environmental Policies.* Routledge, New York, USA, 358 pp.

Wonneberger, A. and R. Vliegenthart, 2021: Agenda-Setting Effects of Climate Change Litigation: Interrelations Across Issue Levels, Media, and Politics in the Case of Urgenda Against the Dutch Government. *Environ. Commun.*, **15(5)**, 699–714, doi:10.1080/17524032.2021.1889633.

Wood, P.B. and D.A. Rossiter, 2017: The politics of refusal: Aboriginal sovereignty and the Northern Gateway pipeline. *Can. Geogr.*, **61(2)**, 165–177, doi:10.1111/cag.12325.

World Bank, 2014: *State and Trends of Carbon Pricing 2014*. World Bank Group, Washington, DC, USA, 140 pp.

World Bank, 2019: *State and Trends of Carbon Pricing 2019*. World Bank Group, Washington, DC, USA, 97 pp.

World Bank, 2021a: *State and Trends of Carbon Pricing 2021*. World Bank Group, Washington, DC, USA, 87 pp.

World Bank, 2021b: *Report of the Task Force on Net Zero Goals & Carbon Pricing*. World Bank Group, Washington, DC, USA, 60 pp. https://static1.squarespace.com/static/54ff9c5ce4b0a53decccfb4c/t/614b3a242b48a65e02ccc978/1632320041214/CPLC+_NetZero_Report.pdf.

World Semiconductor Council, 2017: Best Practice Guidance for Semiconductor PFC Emission Reduction. (May 2017), pp. 8, http://www.semiconductorcouncil.org/wp-content/uploads/2017/07/Best-Practice-Guidance-of-PFC-Emission-Reduction.pdf.

WRI, 2020: *A Brief Guide For Reviewing Countries' Long-term Strategies*. World Resources Institute (WRI), Washington, DC, 12 pp. https://www.wri.org/climate/brief-guide-reviewing-countries-long-term-strategies (Accessed March 3, 2021).

Wright, R.A. and H.S. Boudet, 2012: To act or not to act: Context, capability, and community response to environmental risk. *Am. J. Sociol.*, **118(3)**, 728–777, doi:10.1086/667719.

Wurzel, R.K.W., A.R. Zito, and A.J. Jordan, 2013: *Environmental Governance in Europe: A Comparative Analysis of New Environmental Policy Instruments*. Edward Elgar Publishing Limited, Cheltenham, UK and Northampton, MA, USA, 304 pp.

Wynes, S. and K.A. Nicholas, 2017: The climate mitigation gap: Education and government recommendations miss the most effective individual actions. *Environ. Res. Lett.*, **12(7)**, 074024, doi:10.1088/1748-9326/aa7541.

Wynes, S., K.A. Nicholas, J. Zhao, and S.D. Donner, 2018: Measuring what works: Quantifying greenhouse gas emission reductions of behavioural interventions to reduce driving, meat consumption, and household energy use. *Environ. Res. Lett.*, **13(11)**, 113002, doi:10.1088/1748-9326/aae5d7.

Xiang, D. and C. Lawley, 2019: The impact of British Columbia's carbon tax on residential natural gas consumption. *Energy Econ.*, **80**, 206–218, doi:10.1016/j.eneco.2018.12.004.

Xie, L. and L. Xu, 2021: Environmental Public Interest Litigation in China: A Critical Examination. *Transnatl. Environ. Law*, **10(3)**, 441–465, doi:10.1017/S2047102520000448.

Xie, Y. et al., 2018: Co-benefits of climate mitigation on air quality and human health in Asian countries. *Environ. Int.*, **119**, 309–318, doi:10.1016/j.envint.2018.07.008.

Xu, L. et al., 2019: Identifying the trade-offs between climate change mitigation and adaptation in urban land-use planning: An empirical study in a coastal city. *Environ. Int.*, **133**(September), 105162, doi:10.1016/j.envint.2019.105162.

Yamazaki, A., 2017: Jobs and climate policy: Evidence from British Columbia's revenue-neutral carbon tax. *J. Environ. Econ. Manage.*, **83**, 197–216, doi:10.1016/j.jeem.2017.03.003.

Yeh, S., J. Witcover, G.E. Lade, and D. Sperling, 2016: A review of low carbon fuel policies: principles, program status and future directions. *Energy Policy*, **97**, 220–234, doi:10.1016/j.enpol.2016.07.029.

Yildiz, Ö. et al., 2015: Renewable energy cooperatives as gatekeepers or facilitators? Recent developments in Germany and a multidisciplinary research agenda. *Energy Res. Soc. Sci.*, **6**, 59–73, doi:10.1016/J.ERSS.2014.12.001.

Young, D. and J. Bistline, 2018: The costs and value of renewable portfolio standards in meeting decarbonization goals. *Energy Econ.*, **73**, 337–351, doi:10.1016/j.eneco.2018.04.017.

Yu, B., 2021: Bureaucratic Deliberation and China's Engagement with International Ideas: A Case Study on China's Adoption of Carbon Emissions Trading. *J. Contemp. China*, **31 (136)**, 558–573 doi:10.1080/10670564.2021.1985831.

Zabaloy, M.F., M.Y. Recalde, and C. Guzowski, 2019: Are energy efficiency policies for household context dependent? A comparative study of Brazil, Chile, Colombia and Uruguay. *Energy Res. Soc. Sci.*, **52**, 41–54, doi:10.1016/j.erss.2019.01.015.

Zachmann, G., G. Fredriksson, and G. Claeys, 2018: *Distributional Effects of Climate Policies*. Bruegel, Brussels, 110 pp. https://www.bruegel.org/wp-content/uploads/2018/11/Bruegel_Blueprint_28_final1.pdf (Accessed November 27, 2020).

Zangheri, Serrenho, and Bertoldi, 2019: Energy Savings from Feedback Systems: A Meta-Studies' Review. *Energies*, **12(19)**, 3788, doi:10.3390/en12193788.

Zárate-Toledo, E., R. Patiño, and J. Fraga, 2019: Justice, social exclusion and indigenous opposition: A case study of wind energy development on the Isthmus of Tehuantepec, Mexico. *Energy Res. Soc. Sci.*, **54**, 1–11, doi:10.1016/j.erss.2019.03.004.

Zen, I.S., A.Q. Al-Amin, and B. Doberstein, 2019: Mainstreaming climate adaptation and mitigation policy: Towards multi-level climate governance in Melaka, Malaysia. *Urban Clim.*, **30**, 100501, doi:10.1016/j.uclim.2019.100501.

Zeng, S. and Z. Chen, 2016: Impact of fossil fuel subsidy reform in China: estimations of household welfare effects based on 2007–2012 data. *Econ. Polit. Stud.*, **4(3)**, 299–318, doi:10.1080/20954816.2016.1218669.

Zhang, D., Y. Chen, and M. Tanaka, 2018: On the effectiveness of tradable performance-based standards. *Energy Econ.*, **74**, 456–469, doi:10.1016/j.eneco.2018.06.012.

Zhang, D. et al., 2019: Integrity of firms' emissions reporting in China's early carbon markets. *Nat. Clim. Change*, **9(2)**, 164–169, doi.org/10.1038/s41558-018-0394-4.

Zhang, L. and Q. Qin, 2018: China's new energy vehicle policies: Evolution, comparison and recommendation. *Transp. Res. Part A Policy Pract.*, **110**, 57–72, doi:10.1016/j.tra.2018.02.012.

Zhang, X. and X. Bai, 2017: Incentive policies from 2006 to 2016 and new energy vehicle adoption in 2010–2020 in China. *Renew. Sustain. Energy Rev.*, **70**, 24–43, doi:10.1016/j.rser.2016.11.211.

Zhang, Z., 2012: Competitiveness and Leakage Concerns and Border Carbon Adjustments. *Int. Rev. Environ. Resour. Econ.*, **6(3)**, 225–287, doi:10.1561/101.00000052.

Zhang, Z., 2015: Carbon emissions trading in China: the evolution from pilots to a nationwide scheme. *Clim. Policy*, **15(sup1)**, S104–S126, doi:10.1080/14693062.2015.1096231.

Zhang, Z. and Z. Zhang, 2017: Intermediate input linkage and carbon leakage. *Environ. Dev. Econ.*, **22(6)**, 725–746, doi:10.1017/S1355770X17000250.

Zhao, C. et al., 2018: Adaptation and mitigation for combating climate change – from single to joint. *Ecosyst. Heal. Sustain.*, **4(4)**, 85–94, doi:10.1080/20964129.2018.1466632.

Zhao, Y., S. Lyu, and Z. Wang, 2019: Prospects for Climate Change Litigation in China. *Transnatl. Environ. Law*, **8(2)**, 349–377, doi:10.1017/S2047102519000116.

Zhou, X. and H. Mori, 2011: National institutional response to climate change and stakeholder participation: a comparative study for Asia. *Int. Environ. Agreements Polit. Law Econ.*, **11(4)**, 297–319, doi:10.1007/s10784-010-9127-5.

Ziegler, A., 2017: Political orientation, environmental values, and climate change beliefs and attitudes: An empirical cross country analysis. *Energy Econ.*, **63**, 144–153, doi:10.1016/j.eneco.2017.01.022.

Zimmer, A., M. Jakob, and J.C. Steckel, 2015: What motivates Vietnam to strive for a low-carbon economy? — On the drivers of climate policy in a developing country. *Energy Sustain. Dev.*, **24**, 19–32, doi:10.1016/j.esd.2014.10.003.

Ziolo, M., M. Pawlaczyk, and P. Sawicki, 2019: Sustainable Development Versus Green Banking: Where Is the Link? In: *Financing Sustainable Development: Key Challenges and Prospects* [Ziolo, M. and B.S. Sergi, (eds.)]. Springer International Publishing, Cham, Switzerland, pp. 53–81.

Zürn, M. and B. Faude, 2013: Commentary: On Fragmentation, Differentiation, and Coordination. *Glob. Environ. Polit.*, **13(3)**, 119–130, doi:10.1162/GLEP_a_00186.

Zwierzchowska, I., K. Fagiewicz, L. Poniży, P. Lupa, and A. Mizgajski, 2019: Introducing nature-based solutions into urban policy – facts and gaps. Case study of Poznań. *Land use policy*, **85** (April), 161–175, doi:10.1016/j.landusepol.2019.03.025.

# 14 International Cooperation


**Coordinating Lead Authors:**
Anthony Patt (Switzerland), Lavanya Rajamani (India)

**Lead Authors:**
Preety Bhandari (India), Antonina Ivanova Boncheva (Mexico), Alejandro Caparrós (Spain), Kamal Djemouai (Algeria), Izumi Kubota (Japan), Jacqueline Peel (Australia), Agus Pratama Sari (Indonesia), Detlef F. Sprinz (Germany), Jørgen Wettestad (Norway)

**Contributing Authors:**
Mustafa Babiker (Sudan/Saudi Arabia), Govindasamy Bala (India), Christopher Bataille (Canada), Paolo Bertoldi (Italy), Felix Creutzig (Germany), Heleen de Coninck (the Netherlands), Navroz K. Dubash (India), Rachael Garrett (Switzerland/the United States of America), Oliver Geden (Germany), Veronika Ginzburg (the Russian Federation), Michael Grubb (United Kingdom), Erik Haites (Canada), Benjamin Hinder (United Kingdom), Janna Hoppe (Germany), Yong-Gun Kim (Republic of Korea), Katharine J. Mach (the United States of America), Gregory F. Nemet (the United States of America/Canada), Kilian Raiser (Germany), Yamina Saheb (France/Algeria), Sonia I. Seneviratne (Switzerland), Raphael Slade (United Kingdom), Masahiro Sugiyama (Japan), Christopher H. Trisos (South Africa), Maarten van Aalst (the Netherlands), Harro van Asselt (the Netherlands)

**Review Editors:**
Esther Badiola (Spain), Pasha Carruthers (Cook Islands)

**Chapter Scientist:**
Zyaad Boodoo (Mauritius)



**This chapter should be cited as:**
Patt, A., L. Rajamani, P. Bhandari, A. Ivanova Boncheva, A. Caparrós, K. Djemouai, I. Kubota, J. Peel, A.P. Sari, D.F. Sprinz, J. Wettestad, 2022: International cooperation. In IPCC, 2022: *Climate Change 2022: Mitigation of Climate Change. Contribution of Working Group III to the Sixth Assessment Report of the Intergovernmental Panel on Climate Change* [P.R. Shukla, J. Skea, R. Slade, A. Al Khourdajie, R. van Diemen, D. McCollum, M. Pathak, S. Some, P. Vyas, R. Fradera, M. Belkacemi, A. Hasija, G. Lisboa, S. Luz, J. Malley, (eds.)]. Cambridge University Press, Cambridge, UK and New York, NY, USA. doi: 10.1017/9781009157926.016

# Table of Contents

## Executive Summary

**International cooperation is having positive and measurable results (*high confidence*).** The Kyoto Protocol led to measurable and substantial avoided emissions, including in 20 countries with Kyoto first commitment period targets that have experienced a decade of declining absolute emissions. It also built national capacity for greenhouse gas (GHG) accounting, catalysed the creation of GHG markets, and increased investments in low-carbon technologies (*medium confidence*). Other international agreements and institutions have led to avoided carbon dioxide ($CO_2$) emissions from land use practices, as well as avoided emissions of some non-$CO_2$ greenhouse gases (*medium confidence*). {14.3, 14.5, 14.6}

**New forms of international cooperation have emerged since the Intergovernmental Panel on Climate Change's Fifth Assessment Report (IPCC AR5) in line with an evolving understanding of effective mitigation policies, processes, and institutions. Both new and pre-existing forms of cooperation are vital for achieving climate mitigation goals in the context of sustainable development (*high confidence*).** While previous IPCC assessments have noted important synergies between the outcomes of climate mitigation and achieving sustainable development objectives, there now appear to be synergies between the two processes themselves (*medium confidence*). Since AR5, international cooperation has shifted towards facilitating national-level mitigation action through numerous channels. These now include both processes established under the United Nations Framework Convention on Climate Change (UNFCCC) regime and regional and sectoral agreements and organisations. {14.2, 14.3, 14.5, 14.6}

**Participation in international agreements and transboundary networks is associated with the adoption of climate policies at the national and sub-national levels, as well as by non-state actors (*high confidence*).** International cooperation helps countries achieve long-term mitigation targets when it supports development and diffusion of low-carbon technologies, often at the level of individual sectors, which can simultaneously lead to significant benefits in the areas of sustainable development and equity (*medium confidence*). {14.2, 14.3, 14.5, 14.6}

**International cooperation under the United Nations (UN) climate regime has taken an important new direction with the entry into force of the 2015 Paris Agreement, which strengthened the objective of the UN climate regime, including its long-term temperature goal, while adopting a different architecture from that of the Kyoto Protocol to achieve it (*high confidence*).** The core national commitments under the Kyoto Protocol have been legally binding quantified emission targets for developed countries tied to well-defined mechanisms for monitoring and enforcement. By contrast, the commitments under the Paris Agreement are primarily procedural, extend to all Parties, and are designed to trigger domestic policies and measures, enhance transparency, stimulate climate investments, particularly in developing countries, and to lead iteratively to rising levels of ambition across all countries (*high confidence*). Issues of equity remain of central importance in the UN climate regime, notwithstanding shifts in the operationalisation of 'common but differentiated responsibilities and respective capabilities' from Kyoto to Paris (*high confidence*). {14.3}

**There are conflicting views on whether the Paris Agreement's commitments and mechanisms will lead to the attainment of its stated goals.** Arguments in support of the Paris Agreement are that the processes it initiates and supports will in multiple ways lead, and indeed have already led, to rising levels of ambition over time. The recent proliferation of national mid-century net zero GHG targets can be attributed in part to the Paris Agreement (*medium confidence*). Moreover, its processes and commitments will enhance countries' abilities to achieve their stated level of ambition, particularly among developing countries (*medium confidence*). Arguments against the Paris Agreement are that it lacks a mechanism to review the adequacy of individual Parties' Nationally Determined Contributions (NDCs), that collectively current NDCs are inconsistent in their level of ambition with achieving the Paris Agreement's temperature goal, that its processes will not lead to sufficiently rising levels of ambition in the NDCs, and that NDCs will not be achieved because the targets, policies and measures they contain are not legally binding at the international level (*medium confidence*). To some extent, arguments on both sides are aligned with different analytic frameworks, including assumptions about the main barriers to mitigation that international cooperation can help overcome (*medium confidence*). The extent to which countries increase the ambition of their NDCs and ensure they are effectively implemented will depend in part on the successful implementation of the support mechanisms in the Paris Agreement, and in turn will determine whether the goals of the Paris Agreement are met (*high confidence*). {14.2, 14.3, 14.4}

**International cooperation outside the UNFCCC processes and agreements provides critical support for mitigation in particular regions, sectors and industries, for particular types of emissions, and at the sub- and transnational levels (*high confidence*).** Agreements addressing ozone depletion, transboundary air pollution, and release of mercury are all leading to reductions in the emissions of specific greenhouse gases (*high confidence*). Cooperation is occurring at multiple governance levels including cities. Transnational partnerships and alliances involving non-state and sub-national actors are also playing a growing role in stimulating low-carbon technology diffusion and emissions reductions (*medium confidence*). Such transnational efforts include those focused on climate litigation; the impacts of these are unclear but promising. Climate change is being addressed in a growing number of international agreements operating at sectoral levels, as well as within the practices of many multilateral organisations and institutions (*high confidence*). Sub-global and regional cooperation, often described as climate clubs, can play an important role in accelerating mitigation, including the potential for reducing mitigation costs through linking national carbon markets, although actual examples of these remain limited (*high confidence*). {14.2, 14.4, 14.5, 14.6}

**International cooperation will need to be strengthened in several key respects in order to support mitigation action**

**consistent with limiting temperature rise to well below 2°C in the context of sustainable development and equity (*high confidence*).** Many developing countries' NDCs have components or additional actions that are conditional on receiving assistance with respect to finance, technology development and transfer, and capacity building, greater than what has been provided to date (*high confidence*). Sectoral and sub-global cooperation is providing critical support, and yet there is room for further progress. In some cases, notably with respect to aviation and shipping, sectoral agreements have adopted climate mitigation goals that fall far short of what would be required to achieve the temperature goal of the Paris Agreement (*high confidence*). Moreover, there are cases where international cooperation may be hindering mitigation efforts, namely evidence that trade and investment agreements, as well as agreements within the energy sector, impede national mitigation efforts (*medium confidence*). International cooperation is emerging but so far fails to fully address transboundary issues associated with Solar Radiation Modification and $CO_2$ removal (*high confidence*). {14.2, 14.3, 14.4, 14.5, 14.6}

## 14.1 Introduction

This chapter assesses the role and effectiveness of international cooperation in mitigating climate change. Such cooperation includes multilateral global cooperative agreements among nation states such as the 1992 United Nations Framework Convention on Climate Change (UNFCCC), and its related legal instruments, the 1997 Kyoto Protocol and the 2015 Paris Agreement, but also plurilateral agreements involving fewer states, as well as those focused on particular economic and policy sectors, such as components of the energy system. Moreover, this chapter assesses the role of transnational agreements and cooperative arrangements between non-state and sub-national actors, including municipal governments, private sector firms and industry consortia, and civil society organisations. This chapter does not assess international cooperation within the European Union, as this is covered in Chapter 13 of this report.

Past IPCC assessment reports have discussed the theoretical literature, providing insights into the rationale for international cooperation, as well as guidance as to its structure and implementation. This chapter limits such theoretical discussion primarily to the new developments since the Fifth Assessment Report (AR5). Important developments in this respect include attention to climate clubs (groups of countries and potentially non-state actors that can work together to achieve particular objectives), and the effects of framing the global climate change mitigation challenge as one of accelerating a socio-technical transition or transformation, shifting development pathways accordingly, in addition to (or rather than) solving a global commons problem. This chapter draws from theory to identify a set of criteria by which to assess the effectiveness of existing forms of international cooperation.

The rest of this chapter describes existing cooperative international agreements, institutions, and initiatives with a view to clarifying how they operate, what effects they have, and ultimately, whether they work. At the heart of this international institutional architecture lies the Paris Agreement, which sets the overall approach for international cooperation under the UNFCCC at the global level. In many ways, the Paris Agreement reshapes the structure of such cooperation, from one oriented primarily towards target setting, monitoring, and enforcement, to one that is oriented towards supporting and enabling nationally determined actions (including targets), monitoring as well as catalysing non-state and sub-national actions at multiple levels of governance. In addition to the Paris Agreement, many forms of cooperation have taken shape in parallel: those designed to address other environmental problems that have a significant impact on climate mitigation; those operating at the sub-global or sectoral level; and those where the main participants are non-state actors. The chapter ends with an overall assessment of the effectiveness of current international cooperation and identifies areas that would benefit from improved and enhanced action.

### 14.1.1 Key Findings From the Fifth Assessment Report

The AR5 found that two characteristics of climate change make international cooperation essential: that it is a global commons problem that needs to be addressed in a coordinated fashion at the global scale; and that given the global diversity with respect to opportunities for and cost of mitigation, there are economic efficiencies associated with cooperative solutions (Section 13.2.1.1). Consequently, AR5 found evidence to suggest that climate policies that are implemented across geographical regions would be more effective in terms of both their environmental consequences and their economic costs (Sections 13.6, 13.13 and 14.4). The AR5 also suggested that regional cooperation could offer opportunities beyond what countries may be able to achieve by themselves. These opportunities are due to geographic proximity, shared infrastructure and policy frameworks, trade, and cross-border investments, and examples included renewable energy pools across borders, networks of energy infrastructure and coordinated forestry policies (Sections 1.2, 6.6, 14.2 and 15.2). The AR5 also suggested that policy linkages exist across regional, national, and sub-national scales (Sections 13.3.1 and 13.5.1.3). For these reasons, AR5 suggested that although the UNFCCC remains the primary international forum for climate negotiations, many other institutions engaged at the global, regional, and local levels do and should play an active role (Sections 13.3.1, 13.4.1.4 and 13.5). AR5 also noted that the inclusion of climate change issues across a variety of forums often creates institutional linkages between mitigation and adaptation (Sections 13.3–13.5). In addition to centralised cooperation and governance, with a primary focus on the UNFCCC and its associated institutions, AR5 noted the emergence of new transnational climate-related institutions of decentralised authority such as public-private sector partnerships, private sector governance initiatives, transnational non-governmental organisation (NGO) programmes, and city-led initiatives (Sections 13.2, 13.3.1 and 13.12). It noted that these have resulted in a multiplicity of cooperative efforts in the form of multilateral agreements, harmonised national policies and decentralised but coordinated national and regional policies (Sections 13.3.2, 13.4.1 and 14.4). Finally, it suggested that international cooperation may also have a role in promoting active engagement of the private sector in technological innovation and cooperative efforts leading to technology transfer and the development of new technologies (Sections 13.3, 13.9 and 13.12).

### 14.1.2 Developments Since the Fifth Assessment Report

#### 14.1.2.1 Negotiation of the Paris Agreement

The key development since AR5 has been the negotiation and adoption of the Paris Agreement, which, building on the UNFCCC, introduces a new approach to global climate governance. This new approach, as discussed below (Section 14.3.1.1), is driven by the need to engage developing countries in emissions reductions beyond those they had taken on voluntarily under the Cancun Agreements, extend mitigation commitments to those developed countries that had rejected or withdrawn from the Kyoto Protocol, and to respond to the rapidly changing geopolitical context (Section 14.3.1.2).

#### 14.1.2.2 2030 Agenda for Sustainable Development and the Sustainable Development Goals

It has long been clear that a failure to mitigate climate change would exacerbate existing poverty, accentuate vulnerability and worsen inequality (Denton at al. 2014), but there is an emerging attempt to harmonise mitigation actions with those oriented towards social and economic development. A key development since AR5 is the adoption in 2015 of the 2030 Agenda for Sustainable Development, which contains 17 Sustainable Development Goals (SDGs). This Agenda offers an aspirational narrative, coherent framework and actionable agenda for addressing diverse issues of development through goals that balance the economic, social and environmental dimensions of sustainable development as well as issues of governance and institutions (ICSU ISSC 2015). Scholars have noted that these dimensions of sustainable development are inter-dependent (Nilsson et al. 2016), and, as such it is difficult if not impossible to achieve economic and social gains while neglecting environmental concerns, including climate change (Le Blanc 2015). The SDGs are closely linked to the Paris Agreement, adopted a few weeks later. There is a growing body of literature that examines the interlinkages between SDGs, including SDG 13 (taking urgent action to combat climate change) and others, concluding that without a proper response to climate change, success in many of the other SDGs would be difficult if not impossible (ICSU ISSC 2015; Le Blanc 2015; Nilsson et al. 2016; Weitz et al. 2018). Likewise, failure to achieve the SDGs will have a detrimental effect on the ability to limit climate change to manageable levels. Initiatives such as The World in 2050 (TWI2050 2018), a large research initiative by a global consortium of research and policy institutions, work on the premise that pursuing climate action and sustainable development in an integrated and coherent way, based on a sound understanding of development pathways and dynamics, is the strongest approach to enable countries to achieve their objectives in both agreements.

#### 14.1.2.3 IPCC Special Reports

Further key developments since AR5 include the release of three IPCC special reports. The first of these assessed the differential impacts of limiting climate change to 1.5°C global average warming compared to 2°C warming, indicated the emissions reductions and enabling conditions necessary to stay within this limit (IPCC 2018a). While the events that have unfolded since the report are not yet comprehensively documented in literature, arguably the report has led to a renewed perception of the urgency of climate mitigation (Wolf et al. 2019). In particular, the report appears to have crystalised media coverage in some parts of the world around a need to reduce emissions to net zero by 2050 (whether of GHGs or $CO_2$), rather than delaying such reductions until the latter half of the century, as had been previously understood and indicated in the Paris Agreement. Its release is hence one factor explaining the rise in transnational climate mobilisation efforts (Boykoff and Pearman 2019). It has also played a role, in addition to the Paris Agreement (Geden 2016a), in the numerous announcements, pledges and indications by governments, including by all G7 countries, of their adoption of net zero GHG targets for 2050. The other two special reports focused on ocean and the cryosphere (IPCC 2019a), and the potential of land-related responses to contribute to adaptation and mitigation (IPCC 2019b). There has been no literature directly tying the publication of these latter two reports to changes in international cooperation. However, the 25th UNFCCC Conference of Parties in Madrid in 2019 convened a dialogue on ocean and climate change to consider how to strengthen mitigation and adaptation action in this context (UNFCCC 2019a, para. 31).

## 14.2 Evaluating International Cooperation

This section describes recent insights from social science theory that can shed light on the need for and ideal structure of international cooperation. This section starts by describing developments in framing the underlying problem, moves towards a body of theory describing the benefits of multilateral sub-global action, and ends with a theory-based articulation of criteria to assess the effectiveness of international cooperation.

### 14.2.1 Framing Concepts for Assessment of the Paris Agreement

Previous IPCC reports have framed international climate cooperation, and indeed climate mitigation more generally, primarily as addressing a global commons problem (Stavins et al. 2014). In this report, by contrast, multiple framings are considered. Chapter 1 introduces four analytic frameworks: aggregated economic approaches such as cost-benefit analysis, which maps onto the global commons framing; ethical approaches; analysis of transitions and transformations; and psychology and politics of changing course. Here, we highlight some of the findings that are of relevance to international cooperation.

When applied to the international context, the public good (or global commons) framing stresses that the incentives for mitigation at the global level are greater than they are for any single country, since the latter does not enjoy the benefits of its own mitigation efforts that accrue outside its own borders (Stavins et al. 2014; Patt 2017). This framing does not preclude countries engaging in mitigation, even ambitious mitigation, but it suggests that these countries' level of ambition and speed of abatement would be greater if they were part of a cooperative agreement.

Theoretical economists have shown that reaching such a global agreement is difficult, due to countries' incentives to free-ride, namely benefit from other countries' abatement efforts while failing to abate themselves (Barrett 1994; Gollier and Tirole 2015). Numerical models that integrate game theoretic concepts, whether based on optimal control theory or on dynamic programming, consistently confirm this insight, at least in the absence of transfers (Germain et al. 2003; Lessmann et al. 2015; Chander 2017). Recent contributions suggest that regional or sectoral agreements, or agreements focused on a particular subset of GHGs, can be seen as building blocks towards a global approach (Asheim et al. 2006; Froyn and Hovi 2008; Sabel and Victor 2017; Stewart et al. 2017). In a dynamic context, this gradual approach through building blocks can alleviate the free-riding problem and ultimately lead to global cooperation (Caparrós and Péreau 2017). Much of this literature is subsumed under the concept of 'climate clubs' described in the next section. Other developments based on dynamic game theory suggest that the free-riding problem can be mitigated if the treaties do not prescribe countries' levels of green investment and the duration of the agreement, as countries can credibly threaten potential free-riders with a short-term agreement where green investments will be insufficient due to the hold-up problem (Battaglini and Harstad 2016). Finally, thresholds and potential climate catastrophes have also been shown, theoretically and numerically, to reduce free-riding incentives, especially for countries that may become pivotal in failing to avoid the threshold (Barrett 2013; Emmerling et al. 2020).

In addition to mitigation in the form of emissions abatement, innovation in green technologies also has public good features, leading for the same reasons to less innovation than would be globally ideal (Jaffe et al. 2005). Here as well, theory suggests that there are benefits from cooperation on technology development at the regional or sectoral levels, but also that cooperation on technology, especially for breakthrough technologies, may prove to be easier than for abatement (El-Sayed and Rubio 2014; Rubio 2017). In a dynamic context, the combination of infrastructure lock-in, network effects with high switching cost, and dynamic market failures suggests that deployment and adoption of clean technologies is path dependent (Acemoglu et al. 2012; Aghion et al. 2014), with a multiplicity of possible equilibria. This implies that no outcome is guaranteed, although the most likely pathway will depend on economic expectations and initial conditions of the innovation process (Krugman 1991). Therefore, the government has a role to play, either by shifting expectations (e.g., credibly committing to climate policy), or by changing initial conditions (e.g., investing in green infrastructure or subsidising clean energy research) (Acemoglu et al. 2012; Aghion et al. 2014). This result is exacerbated by the irreversibility of energy investments and the extremely long periods of operation of the typical energy investment (Caparrós et al. 2015; Baldwin et al. 2020).

While the public goods and global commons framing concentrates on free-riding incentives as the primary barrier to mitigation taking place at a pace that would be globally optimal, other factors arise across the four analytic frameworks. For example, within the political framework, Beiser-McGrath and Bernauer (2021) highlight that not just the incentive to free-ride, but also the knowledge that another major emitter is free-riding, could lessen a country's political incentive to mitigate. Aklin and Mildenberger (2020) present evidence to suggest that distributive conflict within countries, rather than free-riding across countries, is the primary barrier to ambitious national-level action. Another barrier could be a lack of understanding and experience with particular policy approaches; there is evidence that participation in cooperative agreements could facilitate information exchange across borders and lead to enhanced mitigation policy adoption (Rashidi and Patt 2018).

The analytic approach focusing on transitions and transformation focuses on path-dependent processes as an impediment to the shift to low-carbon technologies and systems. Cross-Chapter Box 12 on Transition Dynamics (Chapter 16) summarises the key points of this literature. This chapter describes how the two framings focus on different indicators of progress, and potentially different types of cooperative action within the international context. This chapter highlights in later sections conflicting views on whether the Paris Agreement is likely to prove effective (Section 14.3.3.2). To some extent, the dichotomy of views aligns with the two framings: analysis implicitly aligned with the global commons framing is negative about the Paris architecture, whereas that aligned with the transitions framing is more positive (Kern and Rogge 2016; Patt 2017; Roberts et al. 2018).

Within the global commons framing, the primary indicator of progress is the actual level of GHG emissions, and the effectiveness of policies can be measured in terms of whether such emissions rise or fall (Patt 2017; Hanna and Victor 2021). The fact that the sum of all countries' emissions has continued to grow (IPCC 2018a), even as there has been a global recognition that they should decline, is seen as being consistent with the absence of a strong global agreement. Within this framing, there is traditionally an emphasis on treaties' containing self-enforcing agreements (Olmstead and Stavins 2012), ideally through binding commitments, as a way of dealing with the overarching problem of free-ridership (Barrett 1994; Finus and Caparrós 2015; Tulkens 2019). However, as discussed above, the emphasis has now shifted to a gradual cooperation approach, either regional or sectoral, as an alternative way of dealing with free-riding incentives (Caparrós and Péreau 2017; Sabel and Victor 2017; Stewart et al. 2017). The gradual linkage of emissions trading systems (discussed in Section 14.4.4), goes in the same direction. There is also literature suggesting that the diversity of the countries involved may in fact be an asset to reduce the free-rider incentive (Pavlova and De Zeeuw 2013; Finus and McGinty 2019), which argues in favour of a system where all countries, irrespectively of their income levels, are fully involved in mitigation, unlike the Kyoto Protocol and in line with the Paris Agreement. Finally, recent efforts have discussed potential synergies between mitigation and adaptation efforts in a strategic context (Bayramoglu et al. 2018) (Section 14.5.1.2) In general, current efforts go beyond considering climate policy as a mitigation-only issue, much in line with the discussion about linkages between climate change and sustainable development policies described in detail in Chapters 1 and 4 of this report.

In the transitions framing, by contrast, global emissions levels are viewed as the end (and often greatly delayed) result of a large number of transformative processes. International cooperation may be effective at stimulating such processes, even if a change in global emissions is not yet evident, implying that short-term changes in emissions levels may be a misleading indicator of progress towards long-term goals (Patt 2017). Hanna and Victor (2021) suggest a particular focus on technical advances and deployment patterns in niche low-carbon technologies, such as wind and solar power, and electric vehicles. However, this is one among many suggestions: the literature does not identify a single clear indicator to use, and there are many metrics of technological progress and transformation, described in Section 16.3.3 of this report. These can include national-level emissions among countries participating in particular forms of cooperation, as well as leading indicators of such emissions such as changes in low-carbon technology deployment and cost.

Just as the transition framing highlights indicators of progress other than global emissions, it de-emphasises the importance of achieving cost-effectiveness with respect to global emissions. Hence, this strand of the literature does not generally support the use of international carbon markets, suggesting that these can delay transformative processes within countries that are key drivers of technological change (Cullenward and Victor 2020). For similar reasons, achieving cross-sectoral cost-effectiveness, a goal of many carbon markets, is not seen as a high priority. Instead, within the transitions framing, the emphasis with respect to treaty design is often on providing mechanisms to support Parties' voluntary actions, such as with financial and capacity-building support for new technologies and technology regimes (Victor et al. 2019). The transitions literature also highlights impediments to transformation as being sector specific, and hence the importance of international cooperation addressing sector-specific issues (Victor et al. 2019). While such attention often starts with promoting innovation and diffusion of low-carbon technologies that are critical to a sector's functioning, it often ends with policies aimed at phasing out the high-carbon technologies once they are no longer needed (Markard 2018). In line with this, many scholars have suggested value in supply-side international agreements, aimed at phasing out the production and use of fossil fuels (Collier and Venables 2014; Piggot et al. 2018; Asheim et al. 2019; Newell and Simms 2020).

Analytic approaches centred on equity and development figure prominently within this report, with many of the key concepts addressed in Chapter 4. Primarily the focus is on aligning climate policy at the international level with efforts to shift development pathways towards improved quality of life and greater sustainability (Cross-Chapter Box 5 in Chapter 4). There are also overlaps between the equity framework and the others. Within the global commons framing, the emphasis is on international carbon markets to reduce the costs from climate policies, and as way of generating financial flows to developing countries (Michaelowa et al. 2019a). The transitions framing, while focused empirically primarily on industrialised countries, nevertheless aligns with an understanding of climate mitigation taking place within a wider development agenda; in many cases it is a lack of development that creates a barrier to rapid system transformation, which international cooperation can address (Delina and Sovacool 2018) (Cross-Chapter Box 12 in Chapter 16).

### 14.2.2 Climate Clubs and Building Blocks

A recent development in the literature on international climate governance has been increased attention to the potential for climate clubs (Victor 2011). Hovi et al. (2016) define these as 'any international actor group that (1) starts with fewer members than the UNFCCC has and (2) aims to cooperate on one or more climate change-related activities, notably mitigation, adaptation, climate engineering or climate compensation'. While providing public goods (such as mitigation), they also offer member-only benefits (such as preferential tariff rates) to entice membership. In practice, climate clubs are sub-global arrangements, and formal agreement by interstate treaty is not a prerequisite. Actors do not have to be states, although in the literature on climate clubs states have hitherto dominated. The literature has an essentially static dimension that focuses on the incentives for actors to join such a club, and a dynamic one which focuses on the 'building blocks' for global cooperative agreements.

The literature focusing on the static aspects of clubs highlight that they represent 'coalitions of the willing' (Falkner 2016a; Gampfer 2016; Falkner et al. 2021), which offer a package of benefits, part of which are pure public goods (available also to non-club members), and others are club benefits that are only available to members (Hovi et al. 2016). The members-only or excludable part can be a system of transfers within the club to compensate the countries with higher costs. For example, the benefit from participating in the club can be to have access to a common emissions trading system, which in general is more attractive the larger the diversity of the countries involved, although this is not a general result, as discussed in detail in Doda and Taschini (2017). However, as costs and effort-sharing agreements are unsuccessful in a static context (Barrett 1994), mainly due to free-rider incentives, several studies have proposed using tariffs on trade or other forms of sanctions to reduce incentives for free-riding (Helm and Sprinz 2000; Eyland and Zaccour 2012; Anouliès 2015; Nordhaus 2015; Al Khourdajie and Finus 2020). For example, Nordhaus (2015) uses a coalition formation game model to show that a uniform percentage tariff on the imports from nonparticipants into the club region (at a relatively low tariff rate of about 2%) can induce high participation within a range of carbon price values. More recently, Al Khourdajie and Finus (2020) show that border carbon adjustments

**Table 14.1 | Key climate club static modelling results.**

| | Aakre et al. (2018) | Nordhaus (2015) | Hovi et al. (2017); Sprinz et al. (2018) | Sælen (2020); Sælen et al. (2020) |
|---|---|---|---|---|
| Scope | Transboundary black carbon and methane in the Arctic | Global emissions | Global emissions | Global emissions |
| Modelling method | TM5-FASST model ('reduced-form air quality and impact evaluation tool') | C-DICE (coalition formation game based on a static version of the multiregional DICE-RICE optimisation model) | Agent-based model | Agent-based model |
| Border tax adjustment | No | Yes | No | No |
| Key results | Black carbon can be more easily controlled than methane, based on self-interest; inclusion of non-Arctic Council major polluters desirable to control pollutants | For non-participants in mitigation efforts, modest tariffs on trade are advised to stabilise coalition formation for emissions reductions | Climate clubs can substantially reduce GHG emissions, provided club goods are present. The (potential) departure of a single major actor (e.g., USA) reduces emissions coverage, yet is rarely fatal to the existence of the club | The architecture of the Paris Agreement will achieve the 2°C goal only under a very fortunate constellation of parameters. Potential withdrawal (e.g., USA) further reduces these chances considerably |

and an open membership policy can lead to a large stable climate agreement, including full participation. Table 14.1 presents a number of key results related to climate clubs from a static context.

In a dynamic context, the literature on climate clubs highlights the co-called 'building blocks' approach (Stewart et al. 2013a,b, 2017). This is a bottom-up strategy designed to create an array of smaller-scale, specialised initiatives for transnational cooperation in particular sectors and/or geographic areas with a wide range of participants. As part of this literature, Potoski and Prakash (2013) provide a conceptual overview of voluntary environmental clubs, showing that many climate clubs do not require demanding obligations for membership and that a substantial segment thereof are mostly informational (Weischer et al. 2012; Andresen 2014). Also crafted onto the building blocks approach, Potoski (2017) demonstrates the theoretical potential for green certification and green technology clubs. Green (2017) further highlights the potential of 'pseudo-clubs' with fluid membership and limited member benefits to promote the diffusion and uptake of mitigation standards. Falkner et al. (2021) suggest a typology of normative, bargaining, and transformational clubs. Before the adoption of the Paris Agreement, some literature suggested that the emergence of climate clubs in parallel to the multilateral climate regime would lead to 'forum shopping', with states choosing the governance arrangement that best suits their interests (McGee and Taplin 2006; van Asselt 2007; Biermann et al. 2009; Oh and Matsuoka 2017). However, more recent literature suggests that climate clubs complement rather than challenge the international regime established by the UNFCCC (van Asselt and Zelli 2014; Falkner 2016a; Draguljić 2019).

In this dynamic context, one question is whether to negotiate a single global agreement or to start with smaller agreements in the hope that they will eventually evolve into a larger agreement. It has been debated extensively in the context of free trade whether a multilateral (global) negotiating approach is preferable to a regional approach, seen as a building block towards global free trade. Aghion et al. (2007) analysed this issue formally for trade, showing that a leader would always choose to move directly to a global agreement. In the case of climate change, it appears that even the mildest form of club discussed above (an efforts and costs sharing agreement, as in the case of the linkage of emissions trading systems) can yield global cooperation following a building blocks approach, and that the sequential path relying on building blocks may be the only way to reach global cooperation over time (Caparrós and Péreau 2017). While the existence of a nearly universal agreement such as the Paris Agreement may arguably have rendered this discussion less relevant, the Paris Agreement co-exists, and will likely continue to do so, with a multitude of sectoral and regional agreements, meaning that this discussion is still relevant for the evolution of these complementary regimes.

Results based on an agent-based model suggest that climate clubs result in major emissions reductions if there is a sufficiently high provision of the club good and if initial membership by several states with sufficient emissions weight materialises. Such configurations allow the club to grow over time to enable effective global action (Hovi et al. 2017). The departure of a major emitter (specifically the United States) triggered a scientific discussion on the stability of the Paris Agreement. Sprinz et al. (2018) explore whether climate clubs are stable against a leader willing to change its status, for example, from leader to follower, or even completely leaving the climate club, finding in most cases such stability to exist. Related studies on the macroeconomic incentives for climate clubs by Paroussos et al. (2019) show that climate clubs are reasonably stable, both internally and externally (i.e., no member willing to leave and no new member willing to join), and climate clubs that include obligations in line with the 2°C goal combined with financial incentives can facilitate technology diffusion. The authors also show that preferential trade arrangements for low-carbon goods can reduce the macroeconomic effects of mitigation policies. Aakre et al. (2018) show numerically that small groups of countries can limit black carbon in the Arctic, driven mainly for reasons of self-interest, yet reducing methane requires larger coalitions due to its larger geographical dispersal and requires stronger cooperation.

### 14.2.3 Assessment Criteria

This section identifies a set of criteria for assessing the effectiveness of international cooperation, which is applied later in the chapter. Lessons from the implementation of other multilateral environmental agreements (MEAs) can provide some guidance. There is considerable literature on this topic, most of which predates AR5, and which will therefore not be covered in detail. Issues include ways to enhance compliance, and the fact that a low level of compliance with an MEA does not necessarily mean that the MEA has no effect (Downs et al. 1996; Victor et al. 1998; Weiss and Jacobson 1998). Recent research examines effectiveness from the viewpoint of the extent to which an MEA influences domestic action, including the adoption of implementing legislation and policies (Brandi et al. 2019).

Many have pointed to the Montreal Protocol, addressing stratospheric ozone loss, as an example of a successful treaty because of its ultimate environmental effectiveness, and relevance for solving climate change. Scholarship emerging since AR5 emphasises that the Paris Agreement has a greater 'bottom-up' character than many other MEAs, including the Montreal or Kyoto Protocols, allowing for more decentralised 'polycentric' forms of governance that engage diverse actors at the regional, national and sub-national levels (Ostrom 2010; Jordan et al. 2015; Falkner 2016b; Victor 2016). Given the differences in architecture, lessons drawn from studies of MEA regimes need to be supplemented with assessments of the effectiveness of cooperative efforts at other governance levels and in other forums. Emerging research in this area proposes methodologies for this task (Hsu et al. 2019a). Findings highlight the persistence of similar imbalances between developed and developing countries as at the global level, as well as the need for more effective ways to incentivise private sector engagement in transnational climate governance (Chan et al. 2018).

While environmental outcomes and economic performance have been long-standing criteria for assessment of effectiveness, the other elements deserve some note. It is the case that the achievement of climate objectives, such as limiting global average warming to 1.5°C–2°C, will require the transition from high- to low-carbon technologies and the transformation of the sectors

Table 14.2 | Criteria for assessing effectiveness of international cooperation.

| Criterion | Description |
|---|---|
| Environmental outcomes | To what extent does international cooperation lead to identifiable environmental benefits, namely the reduction of economy-wide and sectoral emissions of greenhouse gases from pre-existing levels or 'business as usual' scenarios? |
| Transformative potential | To what extent does international cooperation contribute to the enabling conditions for transitioning to a zero-carbon economy and sustainable development pathways at the global, national, or sectoral levels? |
| Distributive outcomes | To what extent does international cooperation lead to greater equity with respect to the costs, benefits, and burdens of mitigation actions, taking into account current and historical contributions and circumstances? |
| Economic performance | To what extent does international cooperation promote the achievement of economically efficient and cost-effective mitigation activities? |
| Institutional strength | To what extent does international cooperation create the institutional framework needed for the achievement of internationally agreed-upon goals, and contribute to national, sub-national, and sectoral institutions needed for decentralised and bottom-up mitigation governance? |

and social environments within which those technologies operate. Such transformations are not linear processes, and hence many of the early steps taken – such as supporting early diffusion of new renewable energy technologies – will have little immediate effect on GHG emissions (Patt 2015; Geels et al. 2017). Hence, activities that contribute to transformative potential include technology transfer and financial support for low-carbon infrastructure, especially where the latter is not tied to immediate emissions reductions. Assessing the transformative potential of international cooperation takes these factors into account. Equity and distributive outcomes are of central importance to the climate change debate, and hence for evaluating the effects of policies. Equity encompasses the notion of distributive justice which refers to the distribution of goods, burdens, costs and benefits, as well as procedural-related issues (Kverndokk 2018).

Finally, the literature on the performance of other MEAs highlights the importance of institutional strength, which can include regulative quality, mechanisms to enhance transparency and accountability, and administrative capacity. Regulative quality includes guidance and signalling (Oberthür et al. 2017), as well as clear rules and standards to facilitate collective action (Oberthür and Bodle 2016). The literature is clear that legally-binding obligations (which require the formal expression of state consent) and non-binding recommendations can each be appropriate, depending on the particular circumstances (Skjærseth et al. 2006), and indeed it has been argued that for climate change non-binding recommendations may better fit the capacity of global governance organisations (Victor 2011). Mechanisms to enhance transparency and accountability are essential to collect, protect, and analyse relevant data about Parties' implementation of their obligations, and to identify and address challenges in implementation (Kramarz and Park 2016; Kinley et al. 2020). Administrative capacity refers to the strength of the formal bodies established to serve the Parties to the regime and help ensure compliance and goal attainment (Andler and Behrle 2009; Bauer et al. 2017).

In addition to building on the social science theory just described, we recognise that it is also important to strike a balance between applying the same standards developed and applied to international cooperation in AR5 and maintaining consistency with other chapters of this report (primarily Chapters 1, 4, 13 and 15). Table 14.2 presents a set of criteria that do this, and which are then applied later in the chapter.

## 14.3 The UNFCCC and the Paris Agreement

### 14.3.1 The UN Climate Change Regime

#### 14.3.1.1 Instruments and Milestones

The international climate change regime, in evolution for three decades, comprises the 1992 UNFCCC, the 1997 Kyoto Protocol, and the 2015 Paris Agreement. The UNFCCC is a 'framework' convention, capturing broad convergence among states on an objective, a set of principles, and general obligations relating to mitigation, adaptation, reporting and support. The UNFCCC categorises Parties into Annex I and Annex II. Annex I Parties, comprising developed country Parties, have a goal to return, individually or jointly, their GHG emissions to 1990 levels by 2000. Annex II Parties, comprising developed country Parties except for those with economies in transition, have additional obligations relating to the provision of financial and technology support. Parties including developing country Parties, characterised as non-Annex-I Parties, have reporting obligations, as well as obligations to take policies and measures on mitigation and adaptation. The UNFCCC also establishes the institutional building blocks for global climate governance. Both the 1997 Kyoto Protocol and the 2015 Paris Agreement are distinct but 'related legal instruments' in that only Parties to the UNFCCC can be Parties to these later instruments.

The Kyoto Protocol specifies GHG emissions reduction targets for the 2008–2012 commitment period for countries listed in its Annex B (which broadly corresponds to Annex I to the UNFCCC) (UNFCCC 1997, Art. 3 and Annex B). The Kyoto Protocol entered into force in 2005. Shortly thereafter, states began negotiating a second commitment period under the Protocol for Annex B Parties, as well as initiating a process under the UNFCCC to consider long-term cooperation among all Parties.

At the 13th Conference of the Parties to the UNFCCC (COP13) in Bali in 2007, Parties adopted the Bali Action Plan which launched negotiations aimed at an agreed outcome enhancing the UNFCCC's 'full, effective and sustained implementation'. The agreed outcome was to be adopted at COP15 in Copenhagen in 2009, but negotiations failed to deliver a consensus document. The result instead was the Copenhagen Accord, which was taken note of by the COP. While it was a political agreement with no formal legal status under the UNFCCC, it reflected significant progress on several fronts and set in place the building blocks for the Paris Agreement, namely: setting a goal of limiting global temperature increase to below 2°C; calling on all countries to put forward mitigation pledges; establishing broad new terms for the reporting and verification of countries' actions; setting a goal of mobilising USD100 billion a year by 2020 from a wide variety of sources, public and private, bilateral and multilateral, including alternative sources of finance; and, calling for the establishment of a new Green Climate Fund and Technology Mechanism (Rajamani 2010; Rogelj et al. 2010; UNFCCC 2010a). One hundred and forty states endorsed the Copenhagen Accord, with 85 countries entering pledges to reduce their emissions or constrain their growth by 2020 (Christensen and Olhoff 2019).

At COP16 in Cancun in 2010, Parties adopted a set of decisions termed the Cancun Agreements that effectively formalised the core elements of the Copenhagen Accord, and the pledges states made, under the UNFCCC. The Cancun Agreements were regarded as an interim arrangement through to 2020, and Parties left the door open to further negotiations, in line with negotiations launched in 2005, toward a legally-binding successor to the Kyoto Protocol (Freestone 2010; Liu 2011a). Collectively the G20 states are on track to meeting the mid level of their Cancun pledges, although there is uncertainty about some individual pledges. However, there is significant gap between annual

**Table 14.3 | Continuities in and differences between the UNFCCC, Paris Agreement and the Kyoto Protocol.**

| Feature | UNFCCC | Kyoto Protocol | Paris Agreement |
|---|---|---|---|
| Objective | To stabilise GHGs in the atmosphere at a level that would prevent dangerous anthropogenic interference with the climate system, in a timeframe to protect food security, enable natural ecosystem adaptability and permit economic development in a sustainable manner | Primarily mitigation-focused (although in pursuit of the UNFCCC objective) | Mitigation in line with a long-term temperature goal, adaptation and finance goals, as well as sustainable development and equity (also, in pursuit of the UNFCCC objective) |
| Architecture | 'Framework' agreement with agreement on principles such as 'common but differentiated responsibilities and respective capabilities'), division of countries into Annexes, with different groups of countries with differentiated commitments | Differentiated targets, based on national offers submitted to the multilateral negotiation process, and multilaterally negotiated common metrics | Nationally Determined Contributions subject to transparency, multilateral consideration of progress, common metrics in inventories and accounting |
| Coverage of mitigation-related commitments | Annex I Parties with a GHG stabilisation goal, all Parties to take policies and measures | UNFCCC Annex I/Kyoto Annex B Parties only | All Parties |
| Targets | GHG stabilisation goal for Annex I Parties ('quasi target') | Legally-binding, differentiated mitigation targets inscribed in treaty | Non-binding (in terms of results) contributions incorporated in Parties' NDCs, and provisions including those relating to highest possible ambition, progression and 'common but differentiated responsibilities and respective capabilities', in light of different national circumstances |
| Timetable | Aim to return to 1990 levels of GHGs by 2000 | Two commitment periods (2008–2012; 2013–2020) | Initial NDCs for timeframes from 2020 running through to 2025 or 2030 with new or updated NDCs every five years, and encouragement to submit long-term low-GHG emission development strategies |
| Adaptation | Parties to cooperate in preparing for adaptation to the impacts of climate change | Parties to formulate and implement national adaptation measures, share of proceeds from CDM to fund adaptation | Qualitative global goal on adaptation to enhance adaptative capacity and resilience, and reduce vulnerability, Parties to undertake national adaptation planning and implementation |
| Loss and Damage | Not covered | Not covered | Cooperation and facilitation to enhance understanding, action and support for loss and damage, including through the Warsaw International Mechanism on Loss and Damage under the UNFCCC |
| Transparency | National communications from Parties, with differing content and set to differing timeframes for different categories of Parties | Reporting and review – Annex B Parties only | Enhanced transparency framework and five-yearly global stocktake for a collective assessment of progress towards goals – all Parties |
| Support | Annex II commitments relating to provision of finance, development and transfer of technology to developing countries | Advances UNFCCC Annex II commitments relating to provision of finance, development and transfer of technology to developing countries | Enhances reporting in relation to support, expands the base of donors, and tailors support to the needs and capacities of developing countries |
| Implementation | National implementation, communication on implementation | Market mechanisms (International Emissions Trading, Joint Implementation, CDM) | Voluntary cooperation on mitigation (through market-based and non-market approaches); encouragement of REDD+ (guidance and rules under negotiation) |
| Compliance | Multilateral consultative process, never adopted | Compliance committee with facilitative and enforcement branches; sanctions for non-compliance | Committee to promote compliance and facilitate implementation; no sanctions |

emissions expected under full implementation of pledges and the level consistent with the 2°C goal (Christensen and Olhoff 2019).

At the 2011 Durban climate conference, Parties launched negotiations for 'a Protocol, another legal instrument or agreed outcome with legal force' with a scheduled end to the negotiations in 2015 (UNFCCC 2012, Dec. 1, para. 2). At the 2012 Doha climate conference, Parties adopted a second commitment period for the Kyoto Protocol, running from 2013–2020. The Doha Amendment entered into force on 31 December 2020. Given the subsequent adoption of the Paris Agreement, the Kyoto Protocol is unlikely to continue beyond 2020 (Bodansky et al. 2017a). At the end of the compliance assessment period under the Kyoto Protocol, Annex B Parties were in full compliance with their targets for the first commitment period; in some cases through the use of the Protocol's flexibility mechanisms (Shishlov et al. 2016).

Although both the Kyoto Protocol and Paris Agreement are under the UNFCCC, they are generally seen as representing fundamentally different approaches to international cooperation on climate change (Held and Roger 2018; Falkner 2016b). The Paris Agreement has been characterised as a 'decisive break' from the Kyoto Protocol (Keohane and Oppenheimer 2016). Some note that the mitigation efforts under the Kyoto Protocol take the form of targets that, albeit based on national self-selection, were part of the multilateral negotiation process, whereas under the Paris Agreement Parties make Nationally Determined Contributions. The different approaches have been characterised by some as a distinction between a 'top down' and 'bottom up' approach (Bodansky and Rajamani 2016; Bodansky et al. 2016; Chan et al. 2016; Doelle 2016) but others disagree with such a characterisation, pointing to continuities within the regime, for example, in terms of rules for reporting and review, and crossover and use of common institutional arrangements (Depledge 2017; Allan 2019). Some note, in any case, that the Kyoto Protocol's core obligations are substantive obligations of result, while many of the Paris Agreement's core obligations are procedural obligations, complemented by obligations of conduct (Rajamani 2016a; Mayer 2018a).

The differences between and continuities in the three treaties that comprise the UN climate regime are summarised in Table 14.3. The Kyoto targets apply only to Annex I Parties, but the procedural obligations relating to NDCs in the Paris Agreement apply to all Parties, with flexibilities in relation to some obligations for Least Developed Countries (LDCs), Small Island Developing States (SIDS), and developing countries that need them in light of their capacities. The Kyoto targets are housed in its Annex B, therefore requiring a formal process of amendment for revision, whereas the Paris NDCs are located in an online registry that is maintained by the Secretariat, but to which Parties can upload their own NDCs. The Kyoto Protocol allows Annex B Parties to use three market-based mechanisms – the Clean Development Mechanism (CDM), Joint Implementation and International Emissions Trading – to fulfil a part of their GHG targets. The Paris Agreement recognises that Parties may choose to cooperate voluntarily on markets, in the form of cooperative approaches under Article 6.2, and a mechanism with international oversight under Article 6.4, subject to guidance and rules that are yet to be adopted. These rules relate to integrity and accounting (La Hoz Theuer et al. 2019). Article 5 also provides explicit endorsement of reducing emissions from deforestation and forest degradation and fostering conservation (REDD+). The Kyoto Protocol contains an extensive reporting and review process, backed by a compliance mechanism. This mechanism includes an enforcement branch, to ensure compliance, and sanction non-compliance (through the withdrawal of benefits such as participation in market-based mechanisms), with its national system requirements, and GHG targets. By contrast, the Paris Agreement relies on informational requirements and flows to enhance the clarity of NDCs, and to track progress in the implementation and achievement of NDCs.

#### 14.3.1.2 Negotiating Context and Dynamics

The 2015 Paris Agreement was negotiated in a starkly different geopolitical context to that of the 1992 UNFCCC and the 1997 Kyoto Protocol (Streck and Terhalle 2013; Ciplet et al. 2015). The 'rupturing binary balance of superpowers' of the 1980s had given way to a multipolar world with several distinctive trends: emerging economies began challenging US dominance (Ciplet et al. 2015); industrialised countries' emissions peaked in the 2010s and started declining, while emissions from emerging economies began to grow (Falkner 2019); the EU stretched eastwards and became increasingly supra-national (Kinley et al. 2020); disparities within the group of developing countries increased (Ciplet et al. 2015); and the role of non-state actors in mitigation efforts has grown more salient (Bäckstrand et al. 2017; Kuyper et al. 2018b; Falkner 2019). The rise of emerging powers, many of whom now have 'veto power', however, some noted, did not detract from the unequal development and inequality at the heart of global environmental politics (Hurrell and Sengupta 2012).

In this altered context, unlike in the 1990s when the main cleavages were between the EU and the US (Hurrell and Sengupta 2012), US–China 'great power politics' came to be seen as determinative of outcomes in the climate change negotiations (Terhalle and Depledge 2013). The US–China joint announcement (Whitehouse 2014), for instance, before the 2014 Lima climate conference, brokered the deal on differentiation that came to be embodied in the Paris Agreement (Rajamani 2016a; Ciplet and Roberts 2017). Others have identified, on the basis of economic standing, political influence, and emissions levels, three influential groups – the first comprising the USA with Japan, Canada, and Russia, the second comprising the EU and the third comprising China, India and Brazil (Brenton 2013). The emergence of the Major Economies Fora, among other climate clubs (discussed in Section 14.2.2) reflects this development (Brenton 2013). It also represents a 'minilateral' forum, built on a recognition of power asymmetries, in which negotiating compromises are politically tested and fed into multilateral processes (Falkner 2016a).

Beyond these countries, in the decade leading up to the Paris climate negotiations, increasing differences within the group of developing countries divided the 134-strong developing country alliance of the G77/China into several interest-based coalitions (Vihma et al. 2011; Bodansky et al. 2017b). A division emerged between the vulnerable least developed and small island states on the one side and rapidly developing economies, the BASIC (Brazil, South Africa, India and China) on the other, as the latter are 'decidedly not developed but not wholly developing' (Hochstetler and Milkoreit 2013). This fissure in part led to the High Ambition Coalition in Paris between vulnerable countries and the more progressive industrialised countries (Ciplet and Roberts 2017). A division also emerged between the BASIC countries (Hurrell and Sengupta 2012), that each have distinctive identities and positions (Hochstetler and Milkoreit 2013). In the lead up to the Paris negotiations, China and India formed the Like-Minded Developing Countries with the Organization of the Petroleum Exporting Countries (OPEC) and the Bolivarian Alliance for the Peoples of our Americas (ALBA) countries, to resist the erosion of differentiation in the regime. Yet, the 'complex and competing' identities of India and China, with differing capacities, challenges and self-images, have also influenced the negotiations (Ciplet and Roberts 2017; Rajamani 2017). Other developing countries' coalitions also played an important role in striking the final deal in Paris. The Alliance of Small Island States, despite their lack of structural power, played a leading role, in particular in relation to the inclusion of the 1.5°C long-term temperature goal in the UN climate regime (Agueda Corneloup and Mol 2014; Ourbak and Magnan 2018). The Association of the Latin American and Caribbean Countries (AILAC) that emerged in 2012 also played a decisive role in fostering ambition (Edwards et al. 2017; Watts and Depledge 2018).

Leadership is essential to reaching international agreements and overcoming collective action problems (Parker et al. 2015). The Paris negotiations were faced, as a reflection of the multipolarity that had emerged, with a 'fragmented leadership landscape' with the USA, EU, and China being perceived as leaders at different points in time and to varying degrees (Karlsson et al. 2012; Parker et al. 2014). Small island states are also credited with demonstrating 'moral leadership' (Agueda Corneloup and Mol 2014), and non-state and sub-national actors are beginning to be recognised as pioneers and leaders (Wurzel et al. 2019). There is also a burgeoning literature on the emergence of diffused leadership and the salience of followers (Parker et al. 2014; Busby and Urpelainen 2020).

It is in the context of this complex, multipolar and highly differentiated world – with a heterogeneity of interests, constraints and capacities, increased contestations over shares of the carbon and development space, as well as diffused leadership – that the Paris Agreement was negotiated. This context fundamentally influenced the shape of the Paris Agreement, in particular on issues relating to its architecture, 'legalisation' (Karlas 2017) and differentiation (Bodansky et al. 2017b; Kinley et al. 2020), all of which are discussed below.

### 14.3.2 Elements of the Paris Agreement Relevant to Mitigation

The 2015 Paris Agreement to the UNFCCC, which entered into force on 4 November 2016, and has 193 Parties as of March 2022, is at the centre of international cooperative efforts for climate change mitigation and adaptation in the post-2020 period. Although its legal form was heavily disputed, especially in the initial part of its four-year

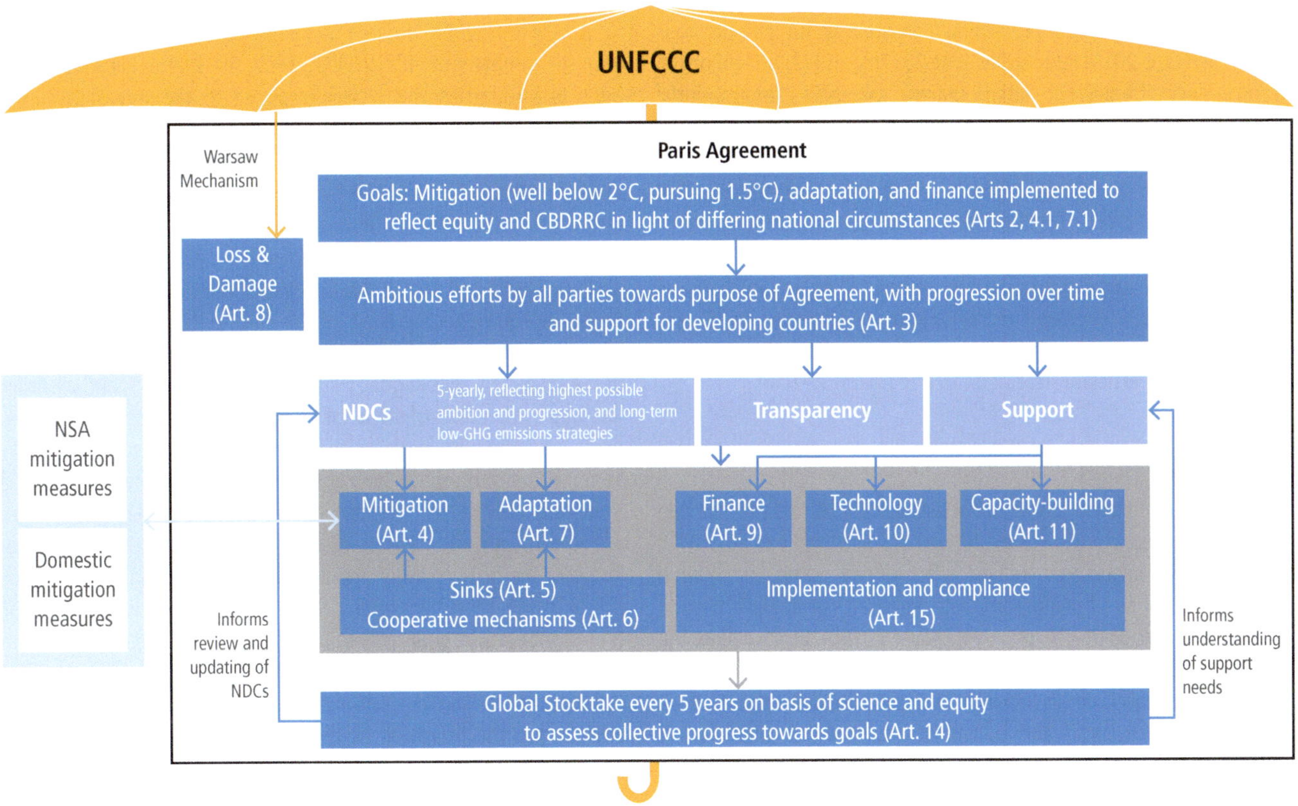


**Figure 14.1 | Key features of the Paris Agreement.** Arrows illustrate the interrelationship between the different features of the Paris Agreement, in particular between the Agreement's goals, required actions through NDCs, support (finance, technology and capacity building), transparency framework and global stocktake process. The figure also represents points of interconnection with domestic mitigation measures, whether taken by state Parties or by non-state actors (NSAs). This figure is illustrative rather than exhaustive of the features and interconnections.

negotiating process (Rajamani 2015; Maljean-Dubois and Wemaëre 2016; Bodansky et al. 2017b; Klein et al. 2017), the Paris Agreement is a treaty containing provisions of differing levels of 'bindingness' (Bodansky 2016; Oberthür and Bodle 2016; Rajamani 2016b). The legal character of provisions within a treaty, and the extent to which particular provisions lend themselves to assessments of compliance or non-compliance, depends on factors such as the normative content of the provision, the precision of its terms, the language used, and the oversight mechanisms in place (Werksman 2010; Bodansky 2015; Oberthür and Bodle 2016; Rajamani 2016b). Assessed on these criteria, the Paris Agreement contains the full spectrum of provisions, from hard to soft law (Rajamani 2016b; Pickering et al. 2019) and even 'non-law', provisions that do not have standard-setting or normative content but which play a narrative-building and context-setting role (Rajamani 2016b). The Paris Agreement, along with the UNFCCC and the Kyoto Protocol, can be interpreted in light of the customary international law principle of harm prevention according to which states must exercise due diligence in seeking to prevent activities within their jurisdiction from causing extraterritorial environmental harm (Mayer 2016a; Maljean-Dubois 2019). The key features of the Paris Agreement are set out in Box 14.1.

Figure 14.1 illustrates graphically the key features of the Paris Agreement. The Paris Agreement is based on a set of binding procedural obligations requiring Parties to 'prepare, communicate, and maintain' 'Nationally Determined Contributions' (NDCs) (UNFCCC 2015a, Art. 4.2) every five years (UNFCCC 2015a, Art. 4.9). These obligations are complemented by: (1) an 'ambition cycle' that expects Parties, informed by five-yearly global stocktakes (Art. 14), to submit successive NDCs representing a progression on their previous NDCs (UNFCCC 2015a; Bodansky et al. 2017b), and (2) an 'enhanced transparency framework' that places extensive informational demands on Parties, tailored to capacities, and establishes review processes to enable tracking of progress towards achievement of NDCs (Oberthür and Bodle 2016). In contrast to the Kyoto Protocol with its internationally inscribed targets and timetable for emissions reduction for developed countries, the Paris Agreement contains Nationally Determined Contributions embedded in an international system of transparency and accountability for all countries (Doelle 2016; Maljean-Dubois and Wemaëre 2016) accompanied by a shared global goal, in particular in relation to a temperature limit.

#### 14.3.2.1 Context and Purpose

The preamble of the Paris Agreement lists several factors that provide the interpretative context for the Agreement (Bodansky et al. 2017b; Carazo 2017), including a reference to human rights. The human rights implications of climate impacts garnered particular attention in the lead up to Paris (Duyck 2015; Mayer 2016b). In particular, the Human Rights Council, its special procedures mechanisms, and the Office of the High Commissioner for Human Rights, through a series of resolutions, reports, and activities, advocated a rights-based approach to climate impacts, and sought to integrate this approach in the climate change regime. The Paris Agreement's preambular recital on human rights recommends that Parties, 'when taking action to address human rights', take into account 'their respective obligations on human rights' (UNFCCC 2015a, preambular recital 14), a first for an environmental treaty (Knox 2016). The 'respective obligations' referred to in the Paris Agreement could potentially include those relating to the right to life (UNGA 1948, Art. 3, 1966, Art. 6), right to health (UNGA 1966b, Art. 12), right to development, right to an adequate standard of living, including the right to food (UNGA 1966b, Art. 11), which has been read to include the right to water and sanitation (CESCR 2002, 2010), the right to housing (CESCR 1991), and the right to self-determination, including as applied in the context of indigenous peoples (UNGA 1966a,b, Art. 1). In addition, climate impacts contribute to displacement and migration (Mayer and Crépeau 2016; McAdam 2016), and have disproportionate effects on women (Pearse 2017). There are differing views on the value and operational impact of the human rights recital in the Paris Agreement (Adelman 2018; Boyle 2018; Duyck et al. 2018; Rajamani 2018; Savaresi 2018; Knox 2019). Notwithstanding proposals from some Parties and stakeholders to mainstream and operationalise human rights in the climate regime post-Paris (Duyck et al. 2018), and references to human rights in COP decisions, the 2018 Paris Rulebook contains limited and guarded references to human rights (Duyck 2019; Rajamani 2019) (Section 14.5.1.2). In addition to the reference to human rights, the preamble also notes the importance of 'ensuring the integrity of all ecosystems, including oceans and the protection of biodiversity' which provides opportunities for integrating and mainstreaming other environmental protections.

The overall purpose of international cooperation through the Paris Agreement is to enhance the implementation of the UNFCCC, including its objective of stabilising atmospheric GHG concentrations 'at a level that would prevent dangerous anthropogenic interference with the climate system' (UNFCCC 1992, Art. 2). The Paris Agreement aims to strengthen the global response to the threat of climate change, in the context of sustainable development and efforts to eradicate poverty, by *inter alia* '[h]olding the increase in the global average temperature to well below 2°C above pre-industrial levels and pursuing efforts to limit the temperature increase to 1.5°C above pre-industrial levels' (UNFCCC 2015a, Art. 2(1)(a)). There is an ongoing structured expert dialogue under the UNFCCC in the context of the second periodic review of the long-term global goal (the first was held between 2013–2015) aimed at enhancing understanding of the long-term global goal, pathways to achieving it, and assessing the aggregate effect of steps taken by Parties to achieve the goal.

Some authors interpret the Paris Agreement's temperature goal as a single goal with two inseparable elements, the well below 2°C goal pressing towards 1.5°C (Rajamani and Werksman 2018), but others interpret the goal as a unitary one of 1.5°C with minimal overshoot (Mace 2016). Yet others interpret 1.5°C as the limit within the long-term temperature goal, and that it 'signals an increase in both the margin and likelihood by which warming is to be kept below 2°C' (Schleussner et al. 2016). Although having a long-term goal has clear advantages, the literature highlights the issue of credibility, given the lengthy timeframe involved (Urpelainen 2011), and stresses that future regulators may have incentives to relax current climate plans, which could have a significant effect on the achieved GHG stabilisation level (Gerlagh and Michielsen 2015).

As the risks of adverse climate impacts, even with a 'well below' 2°C increase, are substantial, the purpose of the Paris Agreement extends to increasing adaptive capacity and fostering climate resilience (UNFCCC 2015a, Art. 2(1)(b)), as well as redirecting investment and finance flows (UNFCCC 2015a, Art. (2)(1)(c); Thorgeirsson 2017). The finance and adaptation goals are not quantified in the Paris Agreement itself but the temperature goal and the pathways they generate may, some argue, enable a quantitative assessment of the resources necessary to reach these goals, and the nature of the impacts requiring adaptation (Rajamani and Werksman 2018). The decision accompanying the Paris Agreement resolves to set a new collective quantified finance goal prior to 2025 (not explicitly limited to developed countries), with USD100 billion $yr^{-1}$ as a floor (UNFCCC 2016a, para. 53; Bodansky et al. 2017b). Article 2 also references sustainable development and poverty eradication, and thus implicitly underscores the need to integrate the SDGs in the implementation of the Paris Agreement (Sindico 2016).

The Paris Agreement's purpose is accompanied by an expectation that the Agreement 'will be' implemented to 'reflect equity and the principle of common but differentiated responsibilities and respective capabilities (CBDRRC), in the light of different national circumstances' (UNFCCC 2015a, Art. 2.2). This provision generates an expectation that Parties will implement the agreement to reflect CBDRRC, and is not an obligation to do so (Rajamani 2016a). Further, the inclusion of the term 'in light of different national circumstances' introduces a dynamic element into the interpretation of the CBDRRC principle. As national circumstances evolve, the application of the principle will also evolve (Rajamani 2016a). This change in the articulation of the CBDRRC principle is reflected in the shifts in the nature and extent of differentiation in the climate change regime (Maljean-Dubois 2016; Rajamani 2016a; Voigt and Ferreira 2016a), including through a shift towards 'procedurally-oriented differentiation' for developing countries (Huggins and Karim 2016).

Although NDCs are developed by individual state Parties, the Paris Agreement requires that these are undertaken by Parties 'with a view' to achieving the Agreement's purpose and collectively 'represent a progression over time' (UNFCCC 2015a, Art. 3). The Paris Agreement also encourages Parties to align the ambition of their NDCs with the temperature goal through the Agreement's 'ambition cycle', thus imparting operational relevance to the temperature goal (Rajamani and Werksman 2018).

Article 4.1 contains a further non-binding requirement that Parties 'aim' to reach global peaking of GHG 'as soon as possible' and to undertake rapid reductions thereafter to achieve net zero GHG emissions 'in the second half of the century'. Some argue this implies a need to reach net zero GHG emissions in the third quarter of the 21st century (Rogelj et al. 2015; IPCC 2018b) (Chapter 2, Table 2.4 and Cross-Chapter Box 3 in Chapter 3). To reach net zero $CO_2$ around 2050, in the short-term global net human-caused $CO_2$ emissions would need to fall by about 45% to 60% from 2010 levels by 2030 (IPCC 2018b). Achieving the Paris Agreement's Article 4.1 aim potentially implies that global warming will peak and then follow a gradually declining path, potentially to below 1.5°C warming (Rogelj et al. 2021).

Albeit non-binding, Article 4.1 has acted as a catalyst for several national net-zero GHG targets, as well as net-zero $CO_2$ and GHG targets across local governments, sectors, businesses, and other actors (Day et al. 2020). There is a wide variation in the targets that have been adopted – in terms of their legal character (policy statement, executive order or national legislation), scope (GHGs or $CO_2$) and coverage (sectors or economy-wide). National net-zero targets could be reflected in the long-term strategies that states are urged to submit under Article 4.19, but only a few states have submitted such strategies thus far. The Paris Rulebook, agreed at the Agreement's first meeting of the Parties in 2018, further strengthens the operational relevance of the temperature goal by requiring Parties to provide information when submitting their NDCs on how these contribute towards achieving the objective identified in UNFCCC Article 2, and Paris Agreement Articles 2.1 (a) and 4.1 (UNFCCC 2019b, Annex I, para. 7). Parties could in this context include information on how their short-term actions align with their long-term net zero GHG or $CO_2$ targets, thereby enhancing the credibility of their long-term goals.

At last count 131 countries had adopted or had net zero targets (whether of carbon or GHG) in the pipeline, covering 72% of global emissions. If these targets are fully implemented some estimate that this could bring temperature increase down to 2°C–2.4°C by 2100 as compared to current policies which are estimated to lead to a temperature increase of 2.9–3.2°C, and NDCs submitted to the Paris Agreement which are estimated to lead to a temperature increase of 2.4°C–2.9°C (Höhne et al. 2021).

It is worth noting that Article 4.1 recognises that 'peaking will take longer for developing countries' and that the balance between emissions and removals needs to be on the 'basis of equity, and in the context of sustainable development and efforts to eradicate poverty'. This suggests that not all countries are expected to reach net zero GHG emissions at the same time, or in the same manner. If global cost-effective 1.5°C and 2°C scenarios from integrated assessment models are taken, without applying an equity principle, the results suggest that domestic net zero GHG and $CO_2$ emissions would be reached a decade earlier than the global average in Brazil and the USA and later in India and Indonesia (van Soest et al. 2021). By contrast, if equity principles are taken into account countries like Canada and the EU would be expected to phase out earlier than the cost-optimal scenarios indicate, and countries like China and Brazil could phase out emissions later, as well as other countries with lower per-capita emissions (van Soest et al. 2021). Some suggest that the application of such fairness considerations could bring forward the net zero GHG date for big emitting countries by up to 15 to 35 years as compared to the global least-cost scenarios (Lee et al. 2021b). In any case, reaching net zero GHG emissions requires to some extent the use of carbon dioxide removal (CDR) methods as there are important sources of non-$CO_2$ GHGs, such as methane and nitrous oxide, that cannot be fully eliminated (IPCC 2018b). However, there are divergent views on different CDR methods, policy choices determine the degree to which and the type of CDR methods that are considered and there is a patchwork of applicable regulatory instruments. There are also uncertainties and governance challenges associated with CDR methods which render tracking progress against net zero GHG emissions challenging (Mace et al. 2021). Researchers have noted that given the key role of CDR in net zero targets and 1.5°C compatible pathways, and the fact that it presents 'significant costs to current and future generations', it is important to consider what an equitable distribution of CDR might look like (UNFCCC 2019c; Day et al. 2020; Lee et al. 2021b).

#### 14.3.2.2 NDCs, Progression and Ambition

Each Party to the Paris Agreement has a procedural obligation to 'prepare, communicate and maintain' successive NDCs 'that it intends to achieve'. Parties have a further procedural obligation to 'pursue domestic mitigation measures' (UNFCCC 2015a, Art. 4.2). These procedural obligations are coupled with an obligation of conduct to make best efforts to achieve the objectives of NDCs (Rajamani 2016a; Mayer 2018b). Many states have adopted climate policies and laws, discussed in Chapter 13, and captured in databases (LSE 2020).

The framing and content of NDCs is thus largely left up to Parties, although certain normative expectations apply. These include developed country leadership through these Parties undertaking economy-wide absolute emissions reduction targets (UNFCCC 2015a, Art. 4.4), as well as 'progression' and 'highest possible ambition' reflecting 'common but differentiated responsibilities and respective capabilities in light of different national circumstances' (Art. 4.3). There is 'a firm expectation' that for every five-year cycle a Party puts forward a new or updated NDC that is 'more ambitious than their last' (Rajamani 2016a). While what represents a Party's highest possible ambition and progression is not prescribed by the Agreement or elaborated in the Paris Rulebook (Rajamani and Bodansky 2019), these obligations could be read to imply a due diligence standard (Voigt and Ferreira 2016b).

In communicating their NDCs every five years (UNFCCC 2015a, Art. 4.9), all Parties have an obligation to 'provide the information necessary for clarity, transparency and understanding' (UNFCCC 2015a, Art. 4.8). These requirements are further elaborated in the Paris Rulebook (Doelle 2019; UNFCCC 2019b). This includes requirements – for Parties' second and subsequent NDCs – to provide quantifiable information on the reference point, for example base year, reference indicators and target relative to the reference indicator (UNFCCC 2019b, Annex I, para. 1). It also requires Parties to provide information on how they consider their contribution 'fair and ambitious in light of different national circumstances', and how they address the normative expectations of developed country leadership, progression and highest possible ambition (UNFCCC 2019b, Annex I, para. 6). However, Parties are required to provide the enumerated information only 'as applicable' to their NDC (UNFCCC 2019b, Annex I, para. 7). This allows Parties to determine the informational requirements placed on them through their choice of NDC. In respect of Parties' first NDCs or NDCs updated by 2020, such quantifiable information 'may' be included, 'as appropriate', signalling a softer requirement, although Parties are 'strongly encouraged' to provide this information (UNFCCC 2019b, Annex I, para. 9).

Parties' first NDCs submitted to the provisional registry maintained by the UNFCCC Secretariat vary in terms of target type, reference year or points, timeframes, and scope and coverage of GHGs. A significant number of NDCs include adaptation, and several NDCs have conditional components, for instance, being conditional on the use of market mechanisms or on the availability of support (UNFCCC 2016b). There are wide variations across NDCs. Uncertainties are generated through interpretative ambiguities in the assumptions underlying NDCs (Rogelj et al. 2017). According to the assessment in this report, current policies lead to median global GHG emissions of 63 gigatonnes of $CO_2$ equivalent ($GtCO_2$-eq), with a full range of 57–70 by 2030 and unconditional and conditional NDCs to 59 (55–65) and 56 (52–61) $GtCO_2$-eq, respectively (Table 4.1). Many omit important mitigation sectors, provide little detail on financing implementation, and are not effective in meeting assessment and review needs (Pauw et al. 2018). Although, it is estimated that the land use sector could contribute as much as 20% of the full mitigation potential of all the intended NDC targets (Forsell et al. 2016), there are variations in how the land use component is included, and the related information provided, leading to large uncertainties on whether and how these will contribute to the achievement of the NDCs (Forsell et al. 2016; Grassi et al. 2017; Obergassel et al. 2017a; Benveniste et al. 2018; Fyson and Jeffery 2019). All these variations make it challenging to aggregate the efforts of countries and compare them to each other (Carraro 2016). Although Parties attempted to discipline the variation in NDCs, including whether they could be conditional, through elaborating the 'features' of NDCs in the Rulebook, no agreement was possible on this. Thus, Parties continue to enjoy considerable discretion in the formulation of NDCs (Rajamani and Bodansky 2019; Weikmans et al. 2020).

There are several approaches to evaluating NDCs, incorporating indicators such as $CO_2$ emissions, GDP, energy intensity of GDP, $CO_2$ per energy unit, $CO_2$ intensity of fossil fuels, and share of fossil fuels in total energy use (Peters et al. 2017). However, some favour approaches that use metrics beyond emissions such as infrastructure investment, energy demand, or installed power capacity (Iyer et al. 2017; Jeffery et al. 2018). One approach is to combine the comparison of aggregate NDC emissions using Integrated Assessment Model scenarios with modelling of NDC scenarios directly, and carbon budget analyses (Jeffery et al. 2018). Another approach is to engage in a comprehensive assessment of multiple indicators that reflect the different viewpoints of the Parties to the UNFCCC (Aldy et al. 2017; Höhne et al. 2018). These different approaches are described in greater depth in Section 4.2.2.

It is clear, however, that the NDCs communicated by Parties for the 2020–2030 period are insufficient to achieve the temperature goal (den Elzen et al. 2016; Rogelj et al. 2016; Schleussner et al. 2016; Robiou du Pont and Meinshausen 2018; UNEP 2018a; Alcaraz et al. 2019; UNEP 2019, 2020), and the emissions gap is larger than ever (Christensen and Olhoff 2019) (Chapter 4). The IPCC *Special Report on Global Warming of 1.5°C* (SR1.5) notes that pathways that limit global warming to 1.5°C with no or limited overshoot show up to 40–50% reduction of total GHG emissions from 2010 levels by 2030, and that current pathways reflected in the NDCs are consistent with cost-effective pathways that result in a global warming of about 3°C by 2100 (IPCC 2018b Summary for Policymakers D.1.1). Analysis by the UNFCCC Secretariat of the second round of those NDCs submitted by October 2021 suggests that 'total global GHG emission level, taking into account full implementation of all the latest NDCs

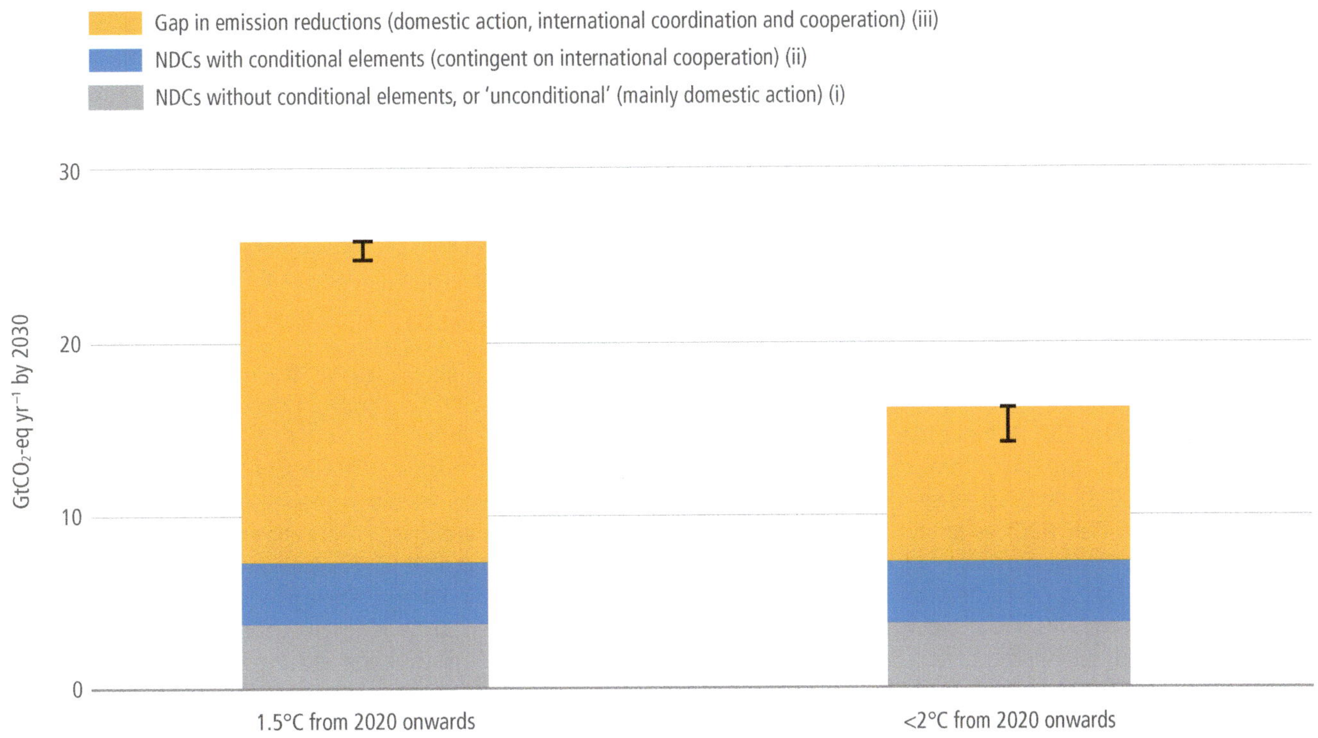


**Figure 14.2 | The role of international cooperation in the reductions in annual emissions by 2030 needed to follow a 1.5°C (respectively <2°C) cost-effective path from 2020 onwards.** The figure represents the additional contribution of pledges included in the NDCs over current policies at the global level, and the remaining gap in emissions reductions needed to move from current policies to pathways that limit warning to 1.5°C (>50%) with no or limited overshoot, and those to limit warming to 2°C (>67%). Median values are used, showing the confidence interval for the total effort. See Figure 1 in Cross-Chapter Box 4 in Chapter 4, and Tables 4.2 and 4.3 for details. (i) The grey share represents NDCs with abatement efforts pledged without any conditions (called 'unconditional' in the literature). They are based mainly on domestic abatement actions, although countries can use international cooperation to meet their targets. (ii) The blue share represents NDCs with conditional components. They require international cooperation, for example bilateral agreements under Article 6, financing or monetary and/or technological transfers. (iii) The remaining gap in emissions reductions – the yellow share – can potentially be achieved through national and international actions. International coordination of more ambitious efforts promotes global ambition and international cooperation provides the cost-saving basis for more ambitious NDCs.

(including their conditional elements), implies possibility of global emissions peaking before 2030'. However, such total global GHG emission level in 2030 is still expected to be 15.9% above the 2010 level. This 'implies an urgent need for either a significant increase in the level of ambition of NDCs between now and 2030 or a significant overachievement of the latest NDCs, or a combination of both.' (UNFCCC 2021a).

Many NDCs with conditional elements may not be feasible as the conditions are not clearly defined and existing promises of support are insufficient (Pauw et al. 2020). Moreover, 'leadership by conditional commitments' (when some states promise to take stronger commitments if others do so as well), and the system of pledge-and-review, may lead to decreasing rather than deeper contributions over time (Helland et al. 2017). Some note, however, that many of the NDCs are conservative and may be overachieved, that NDCs may be strengthened over time as expected under the Paris Agreement, and that there are significant non-state actions that have not been adequately captured in the NDCs (Höhne et al. 2017). Further, if all NDCs with and without conditional elements are implemented, net land use, land use change and forestry emissions will decrease in 2030 compared to 2010 levels, but large uncertainties remain on how Parties estimate, project and account for emissions and removals from this sector (Forsell et al. 2016; Fyson and Jeffery 2019). According to the estimates in Table 4.3, communicated unconditional commitments imply about a 7% reduction of world emissions by 2030, in terms of Kyoto GHGs, compared to a scenario where only current policies are in place. If conditional commitments are also included, the reduction in world emissions by 2030 would be about 12%.

In this context, it should be noted that many NDCs have been formulated with conditional elements, and such NDCs require international cooperation on finance, technology and capacity building (Kissinger et al. 2019), potentially including through Article 6 in the form of bilateral agreements and market mechanisms (UNFCCC 2016b). More broadly, some argue that there is a 'policy inconsistency' between the facilitative, 'bottom up' architecture of the Paris Agreement, and both the setting of the long-term temperature goal and expectations that it will be delivered (Geden 2016b). As Figure 14.2 shows, there is a large share of additional effort needed to reach a 1.5°C compatible path by 2030 (and even a 2°C compatible path). International coordination and cooperation are crucial in enhancing the ambition of current pledges, as countries will be more willing to increase their ambition if matched by other countries (coordination) and if cost-minimising agreements between developed and developing countries, through Article 6 and other means, are fully developed (cooperation) (Sælen 2020).

#### 14.3.2.3 NDCs, Fairness and Equity

The Paris Agreement encourages Parties, while submitting their NDCs, to explain how these are 'fair and ambitious' (UNFCCC 2015a, Art. 4.8 read with UNFCCC 2016a, para. 27). The Rulebook obliges Parties to provide information on 'fairness considerations, including reflecting on equity' as applicable to their NDC (Rajamani and Bodansky 2019; UNFCCC 2019b paras. 7a and 9, Annex, paras. 6(a) and (b)). Although equity within nations and between communities is also important, much of the literature on fairness and equity in the context of NDCs focuses on equity between nations.

In the first round of NDCs, most Parties declared their NDCs as fair (Robiou du Pont et al. 2017). Their claims, however, were largely unsubstantiated or drawn from analysis by in-country experts (Winkler et al. 2018). At least some of the indicators Parties have identified in their NDCs as justifying the 'fairness' of their contributions, such as a 'small share of global emissions', 'cost-effectiveness' and assumptions that privilege current emissions levels ('grandfathering') are not, according to one group of scholars, in accordance with principles of international environmental law (Rajamani et al. 2021). Moreover, the NDCs reveal long-standing institutional divisions and divergent climate priorities between Annex I and non-Annex I Parties, suggesting that equity and fairness concerns remain salient (Stephenson et al. 2019). Fairness concerns also affect the share of CDR responsibilities for major emitters if they delay near-term mitigation action (Fyson et al. 2020).

It is challenging, however, to determine 'fair shares', and address fairness and equity in a world of voluntary climate contributions (Chan 2016a), in particular, since these contributions are insufficient (Section 14.3.2.2.). Self-differentiation in contributions has also led to fairness and equity being discussed in terms of individual Nationally Determined Contributions rather than between categories of countries (Chan 2016a). In the climate change regime, one option is for Parties to provide more rigorous information under the Paris Agreement to assess fair shares (Winkler et al. 2018), and another is for Parties to articulate what equity principles they have adopted in determining their NDCs and how they have operationalised these principles, and to explain their mitigation targets in terms of the portion of the appropriated global carbon budget (Hales and Mackey 2018).

Equity is critical to addressing climate change, including through the Paris Agreement (Klinsky et al. 2017), however, since the political feasibility of developing equity principles within the climate change regime is low, the onus is on mechanisms and actors outside the regime to develop these (Lawrence and Reder 2019). Equity and fairness concerns are being raised in national and regional courts that are increasingly being asked to determine if the climate actions pledged by states are adequate in relation to their fair share (The Supreme Court of the Netherlands 2019; European Court of Human Rights 2020; German Constitutional Court 2021), as it is only in relation to such a 'fair share' that the adequacy of a state's contribution can be assessed in the context of a global collective action problem (Section 13.5.5). Some domestic courts have stressed that as climate change is a global problem of cumulative impact, all emissions contribute to the problem regardless of their relative size and there is a clear articulation under the UNFCCC and Paris Agreement for developed countries to 'take the lead' in addressing GHG emissions (Preston 2020). Given the limited avenues for multilateral determination of fairness, several researchers have argued that the onus is on the scientific community to generate methods to assess fairness (Herrala and Goel 2016; Lawrence and Reder 2019). Peer-to-peer comparisons also potentially create pressure for ambitious NDCs (Aldy et al. 2017).

There are a range of options to assess or introduce fairness. These include: adopting differentiation in financing rather than in mitigation (Gajevic Sayegh 2017); adopting a carbon budget approach (Hales and Mackey 2018; Alcaraz et al. 2019), which may occur through the transparency processes (Hales and Mackey 2018); quantifying national emissions allocations using different equity approaches, including those reconciling finance and emissions rights distributions (Robiou du Pont et al. 2017); combining equity concepts in a bottom-up manner using different sovereign approaches (Robiou du Pont and Meinshausen 2018), using data on adopted emissions targets to find an ethical framework consistent with the observed distribution (Sheriff 2019); adopting common metrics for policy assessment (Bretschger 2017); and developing a template for organising metrics on mitigation effort – emissions reductions, implicit prices, and costs – for both *ex-ante* and *ex-post* review (Aldy et al. 2017). The burden of agricultural mitigation can also be distributed using different approaches from effort sharing (responsibility, capability, need, equal cumulative per-capita emissions) (Richards et al. 2018). Further, there are temporal (inter-generational) and spatial (inter-regional) dimensions to the distribution of the mitigation burden, with additional emissions reductions in 2030 improving both inter-generational and inter-regional equity (Liu et al. 2016). Some of the equity approaches rely on 'grandfathering' as an allocation principle, which some argue has led to 'cascading biases' against developing countries (Kartha et al. 2018), and is morally 'perverse' (Caney 2011). While no country's NDC explicitly supports the grandfathering approach, many countries describe as 'fair and ambitious' NDCs that assume grandfathering as the starting point (Robiou du Pont et al. 2017). It is worth noting that the existence of multiple metrics associated with a range of equity approaches has implications for how the ambition and 'fair' share of each state is arrived at; some average out multiple approaches and indicators (Hof et al. 2012; Meinshausen et al. 2015; Robiou du Pont and Meinshausen 2018), others exclude indicators and approaches that do not, in their interpretation, accord with principles of international environmental law (Rajamani et al. 2021). One group of scholars has suggested that utilitarianism offers an 'ethically minimal and conceptually parsimonious' benchmark that promotes equity, climate and development (Budolfson et al. 2021).

#### 14.3.2.4 Transparency and Accountability

Although NDCs reflect a 'bottom-up', self-differentiated approach to climate mitigation actions, the Paris Agreement couples this to an international transparency framework designed, among other things, to track progress in implementing and achieving mitigation contributions (UNFCCC 2015a, Art. 13). This transparency framework builds on the processes that already exist under the UNFCCC. The transparency framework under the Paris Agreement is applicable to all Parties, although with flexibilities for developing country Parties that need it in light of their capacities (Mayer 2019). Each Party is required to submit a national inventory report, as well as 'the information necessary to track progress in implementing and achieving' its NDC (UNFCCC 2015a, Art. 13.7) biennially (UNFCCC 2016a, para. 90). The Paris Rulebook requires all Parties to submit their national inventory reports using the 2006 IPCC Guidelines (UNFCCC 2019b, Annex, para. 20).

In relation to the provision of information necessary to track progress towards implementation and achievement of NDCs, the Paris Rulebook allows each Party to choose its own qualitative or quantitative indicators (UNFCCC 2019k, Annex, para. 65), a significant concession to national sovereignty (Rajamani and Bodansky 2019). The Rulebook phases in common reporting requirements for developed and developing countries (except LDCs and SIDS) at the latest by 2024 (UNFCCC 2019k, para. 3), but offers flexibilities in 'scope, frequency, and level of detail of reporting, and in the scope of the review' for those developing countries that need it in light of their capacities (UNFCCC 2019k, Annex, para. 5). Some differentiation also remains for information on support provided to developing countries (Winkler et al. 2017), with developed country Parties required to report such information biennially, while others are only 'encouraged' to do so (UNFCCC 2015a, Art. 9.7).

The information provided by Parties in biennial transparency reports and GHG inventories will undergo technical expert review, which must include assistance in identifying capacity-building needs for developing country Parties that need it in light of their capacities. Each Party is also required to participate in a 'facilitative, multilateral consideration of progress' of implementation and achievement of its NDC. Although the aim of these processes is to expose each Party's actions on mitigation to international review, thus establishing a weak form of accountability for NDCs at the international level, the Rulebook circumscribes the reach of these processes (Rajamani and Bodansky 2019). The technical expert review teams are prohibited in mandatory terms from making 'political judgments' or reviewing the 'adequacy or appropriateness' of a Party's NDC, domestic actions, or support provided (UNFCCC 2019k, Annex, para. 149). This, among other such provisions, has led some to argue that the scope and practice of existing transparency arrangements reflect rather than mediate ongoing disputes around responsibility, differentiation and burden sharing, and thus there is limited answerability through transparency (Gupta and van Asselt 2019). There are also limits to the extent that the enhanced transparency framework will reduce ambiguities and associated uncertainties, for instance, in how land use, land-use change and forestry (LULUCF) are incorporated into the NDCs (Fyson and Jeffery 2019), and lead to increased ambition (Weikmans et al. 2020). More broadly, there has been 'weak' translation of transparency norms into accountability (Ciplet et al. 2018). Hence, the Paris Agreement's effectiveness in ensuring NDCs are achieved will depend on additional accountability pathways at the domestic level involving political processes and civil society engagement (Jacquet and Jamieson 2016; van Asselt 2016; Campbell-Duruflé 2018a; Karlsson-Vinkhuyzen et al. 2018).

#### 14.3.2.5 Global Stocktake

The Paris Agreement's transparency framework is complemented by the global stocktake, which will take place every five years (starting in 2023) and assess the collective progress towards achieving the Agreement's purpose and long-term goals (UNFCCC 2015a, Art. 14). The scope of the global stocktake is comprehensive – covering mitigation, adaptation and means of implementation and support – and the process is to be facilitative and consultative. The Paris Rulebook outlines the scope of the global stocktake to include social and economic consequences and impacts of response measures, and loss and damage associated with the adverse effects of climate change (UNFCCC 2019f, paras. 8–10).

The global stocktake is to occur 'in the light of equity and the best available science'. While the focus of the global stocktake is on collective and not individual progress towards the goals of the Agreement, the inclusion of equity in the global stocktake enables a discussion on equitable burden sharing (Rajamani 2016a; Winkler 2020), and for equity metrics to be factored in (Robiou du Pont and Meinshausen 2018). The Paris Rulebook includes consideration of the modalities and sources of inputs for the global stocktake (UNFCCC 2019f, paras. 1, 2, 13, 27, 31, 36h and 37g), which arguably will result in equity being factored into the outcome of the stocktake (Winkler 2020). The Rulebook does not, however, some argue, resolve the tension between the collective nature of the assessment that is authorised by the stocktake and the individual assessments required to determine relative 'fair share' (Rajamani and Bodansky 2019; Zahar 2019).

The global stocktake is seen as crucial to encouraging Parties to increase the ambition of their NDCs (Huang 2018; Hermwille et al. 2019; Milkoreit and Haapala 2019) as its outcome 'shall inform Parties in updating and enhancing, in a nationally determined manner, their actions and support' (Art. 14.3) (Rajamani 2016a; Friedrich 2017; Zahar 2019). The Rulebook provides for the stocktake to draw on a wide variety of inputs sourced from a full range of actors, including 'non-Party stakeholders' (UNFCCC 2019f, para. 37). However, the Rulebook specifies that the global stocktake will be 'a Party-driven process' (UNFCCC 2019f, para. 10), will not have an 'individual Party focus', and will include only 'non-policy prescriptive consideration of collective progress' (UNFCCC 2019f, para. 14).

#### 14.3.2.6 Conservation of Sinks and Reservoirs, Including Forests

Article 5 of the Paris Agreement calls for Parties to take action to conserve and enhance sinks and reservoirs of greenhouse gases, including biomass in terrestrial, coastal, and marine ecosystems, and encourages countries to take action to support the REDD+ framework under the Convention. The explicit inclusion of land use sector activities, including forest conservation, is potentially, while cautiously, a 'game changer' as it encourages countries to safeguard ecosystems for climate mitigation purposes (Grassi et al. 2017). Analyses of Parties' NDCs shows pledged mitigation from land use, and forests in particular, provides a quarter of the emissions reductions planned by Parties and, if fully implemented, would result in forests becoming a net sink of carbon by 2030 (Forsell et al. 2016; Grassi et al. 2017).

A key action endorsed by Article 5 is REDD+, which refers to initiatives established under the UNFCCC for reducing emissions from deforestation and forest degradation and the role of conservation, sustainable management of forests and enhancement of forest carbon stocks in developing countries. It remains an evolving concept and some identified weaknesses are being addressed, including the issues of scale (project-based vs sub-national jurisdictional approach), problems with leakage, reversal, and benefit sharing, as well as safeguards against potential impacts on local and indigenous communities. Nevertheless, REDD+ shows several innovations under the climate regime with regard to international cooperation. The legal system for REDD+ manages to reconcile flexibility (creating consensus) and legal security. It shows a high standard of effectiveness (Dellaux 2017).

Article 5.2 encourages Parties to implement and support the existing framework for REDD+, including through 'results-based payments', that is provision of financial payments for verified avoided or reduced forest carbon emissions (Turnhout et al. 2017). The existing REDD+ framework set up under decisions of the UNFCCC COP includes the Warsaw Framework for REDD+, which specifies modalities for measuring, reporting and verifying greenhouse gas emissions and removals. This provides an essential tool for linking REDD+ activities to results-based finance (Voigt and Ferreira 2015). Appropriate finance support for REDD+ is also considered critical to move from its inclusion in many countries' NDCs to implementation on the ground (Hein et al. 2018). Since public finance for REDD+ is limited, private sector participation is expected by some to leverage REDD+ (Streck and Parker 2012; Henderson et al. 2013; Pistorius and Kiff 2015; Seymour and Busch 2016; Ehara et al. 2019). Article 5.2 also encourages Parties' support for 'alternative policy approaches' to forest conservation and sustainable management such as 'joint mitigation and adaptation approaches'. It reaffirms the importance of incentivising, as appropriate, non-carbon benefits associated with such approaches (e.g., improvements in the livelihoods of forest-dependent communities, facilitating poverty reduction and sustainable development). This provision, along with the support for non-market mechanisms in Article 6 (discussed below), is seen as an avenue for cooperative joint mitigation–adaptation and non-market REDD+ activities with co-benefits for biodiversity conservation (Gupta and Dube 2018).

#### 14.3.2.7 Cooperative Approaches

Article 6 of the Paris Agreement provides for voluntary cooperative approaches. Its potential importance in terms of project-based cooperation should be viewed against the background of key lessons from the market-based mechanisms under the Kyoto Protocol, particularly the Clean Development Mechanism (CDM). The CDM has been used for implementing bilateral strategies and unilateral (non-market) actions for instance in India (Phillips and Newell 2013), hence arguably covering all the mechanisms now included in Article 6 of the Paris Agreement. As we describe in Section 14.3.3.1, below, *ex post* evaluation of the Kyoto market mechanisms, in particular the CDM, have been at best mixed. However, Article 6 goes beyond the project-based approach followed by the CDM, as hinted by the emerging landscape of activities based on Article 6 (Greiner et al. 2020), such as the bilateral treaty signed under the framework of Article 6 in October 2020 by Switzerland and Peru (Section 14.4.4).

This experience from the CDM is relevant to the implementation of Article 6 (4) of the Paris Agreement. It addresses a number of specific types of cooperative approaches, including those involving the use of internationally transferred mitigation outcomes (ITMOs) towards NDCs, a 'mechanism to contribute to mitigation and support sustainable development', and a framework for non-market approaches such as many aspects of REDD+.

Article 6.1 recognises the role that cooperative approaches can play, on a voluntary basis, in implementing Parties' NDCs 'in order to allow for higher ambition' in their mitigation actions and to promote sustainable development and environmental integrity. Article 6.2 indicates that ITMOs can originate from a variety of sources, and that Parties using ITMOs to achieve their NDCs shall promote sustainable development, ensure environmental integrity, ensure transparency, including in governance, and apply 'robust accounting' in accordance with Conference of the Parties serving as the Meeting of the Parties to the Paris Agreement (CMA) guidance to prevent double counting. While this provision, unlike Article 17 of the Kyoto Protocol, does not create an international carbon market, it enables Parties to pursue this option should they choose to do so, for example, through the linking of domestic or regional carbon markets (Marcu 2016; Müller and Michaelowa 2019). Article 6.2 could also be implemented in other ways, including direct transfers between governments, linkage of mitigation policies across two or more Parties, sectoral or activity crediting mechanisms, and other forms of cooperation involving public or private entities, or both (Howard 2017).

Assessments of the potential of Article 6.2 generally find that ITMOs are likely to result in cost reductions in achieving mitigation outcomes, with the potential for such reductions to enhance ambition and accelerate Parties' progression of mitigation pledges across NDC cycles (Fujimori et al. 2016; Gao et al. 2016; Mehling 2019). However, studies applying insights from the CDM highlight environmental integrity risks associated with using ITMOs under the Paris Agreement given the challenges that the diverse scope, metrics, types and timeframes of NDC targets pose for robust accounting (Schneider and La Hoz Theuer 2019) and the potential for transfers of 'hot air', as occurred under the Kyoto Protocol (La Hoz Theuer et al. 2019). These studies collectively affirm that robust governance on accounting for ITMOs, and for reporting and review, will be critical to ensuring the environmental integrity of NDCs making use of them (Mehling 2019; Müller and Michaelowa 2019).

Article 6.4 concerns the mitigation mechanism, with some similarities to the Kyoto Protocol's CDM. Unlike the CDM, there is no restriction on which Parties can host mitigation projects and which Parties can use the resulting emissions reductions towards their NDCs (Marcu 2016). This central mechanism will operate under the authority and guidance of the CMA, and is to be supervised by a body designated by the CMA (Marcu 2016).

The Article 6.4 central mechanism is intended to promote mitigation while fostering sustainable development. The decision adopting the Paris Agreement specifies experience with Kyoto market mechanisms as a basis for the new mitigation mechanism (UNFCCC 2016a, para. 37(f)). Compared with the CDM under the Kyoto Protocol, the central mechanism has a more balanced focus on both climate and development objectives, and a stronger political mandate to measure sustainable development impact and to verify that the impacts are 'real, measurable, and long-term' (Olsen et al. 2018). There are also opportunities to integrate human rights into the central mechanism (Obergassel et al. 2017b; Calzadilla 2018). It is further subject to the requirement that it must deliver 'an overall mitigation in global emissions', which is framed by the general objectives of Article 6 for cooperation to enhance ambition (Kreibich 2018).

Negotiations over rules to operationalise Article 6 have thus far proven intractable, failing to deliver both at COP24 in Katowice in 2018, where the rest of the Paris Rulebook was agreed, and in COP25 in Madrid in 2019. Ongoing points of negotiation have included: whether to permit the carryover and use of Kyoto CDM credits and assigned amount units into the Article 6.4 mechanism, whether to impose a mandatory share of proceeds on Article 6.2 mechanism to fund adaptation, like for Article 6.4; and whether and how credits generated under Article 6.4 should be subject to accounting rules under Article 6.2 (Michaelowa et al. 2020a).

#### 14.3.2.8 Finance Flows

Finance is the first of three means of support specified under the Paris Agreement to accomplish its objectives relating to mitigation (and adaptation) (UNFCCC 2015a, Art. 14.1). This sub-section discusses the provision made in the Paris Agreement for international cooperation on finance. Section 14.4.1 below considers broader cooperative efforts on public and private finance flows for climate mitigation, including by multilateral development banks and through instruments such as green bonds.

As highlighted above, the objective of the Paris Agreement includes the goal of '[m]aking finance flows consistent with a pathway towards low greenhouse gas emissions and climate-resilient development' (UNFCCC 2015a, Art. 2.1(c)). Alignment of financial flows, and in some cases provision of finance, will be critical to the achievement of many Parties' NDCs, particularly those that are framed in conditional terms (Zhang and Pan 2016; Kissinger et al. 2019) (Chapter 15).

International cooperation on climate finance represents 'a complex and fragmented landscape' with a range of different mechanisms and forums involved (Pickering et al. 2017; Roberts and Weikmans 2017). These include entities set up under the international climate change regime, such as the UNFCCC financial mechanism, with the Global Environment Facility (GEF) and Green Climate Fund (GCF) as operating entities; special funds, such as the Special Climate Change Fund, the Least Developed Countries Fund (both managed by the GEF), and the Adaptation Fund established under the Kyoto Protocol; the Standing Committee on Finance, a constituted body which assists the COP in exercising its functions with respect to the UNFCCC financial mechanism; and other bodies outside of the international climate change regime, such as the Climate Investment Funds administered through multilateral development banks (the role of these banks in climate finance is discussed further in Section 14.4.1 below).

Pursuant to decisions adopted at the Paris and Katowice conferences, Parties agreed that the operating entities of the financial mechanism – GEF and GCF – as well as the Special Climate Change Fund, the Least Developed Countries Fund, the Adaptation Fund and the Standing Committee on Finance, all serve the Paris Agreement (UNFCCC 2016a, paras. 58 and 63, 2019e,g). The GCF, which became operational in 2015, is the largest dedicated international climate change fund and plays a key role in channelling financial resources to developing countries (Antimiani et al. 2017; Brechin and Espinoza 2017).

Much of the current literature on climate finance and the Paris Agreement focuses on the obligations of developed countries to provide climate finance to assist the implementation of mitigation and adaptation actions by developing countries. The principal provision on finance in the Paris Agreement is the binding obligation on developed country Parties to provide financial resources to assist developing country Parties (UNFCCC 2015a, Art. 9.1). This provision applies to both mitigation and adaptation and is in continuation of existing developed country Parties' obligations under the UNFCCC. This signals that the Paris Agreement finance requirements must be interpreted in light of the UNFCCC (Yamineva 2016). The novelty introduced by the Paris Agreement is a further expansion in the potential pool of donor countries as Article 9.2 encourages 'other Parties' to provide or continue to provide such support on a voluntary basis. However, 'as part of the global effort, developed countries should continue to take the lead in mobilising climate finance', with a 'significant role' for public funds, and an expectation that such mobilisation of finance 'should represent a progression beyond previous efforts'. Beyond this, there are no new recognised promises (Ciplet et al. 2018). In the Paris Agreement, Parties formalised the continuation of the existing collective mobilisation goal to raise USD100 billion $yr^{-1}$ through to 2025 in the context of meaningful mitigation actions and transparency on implementation. The Paris Agreement decision also provided for the CMA by 2025 to set a new collective quantified goal from a floor of USD100 billion $yr^{-1}$, taking into account the needs and priorities of developing countries (UNFCCC 2016a, para. 53). This new collective goal on finance is not explicitly limited to developed countries and could therefore encompass finance flows from developing countries' donors (Bodansky et al. 2017b). Deliberations on setting a new collective quantified goal on finance is expected to be initiated at COP26 in 2021 (UNFCCC 2019g,e; Zhang 2019).

It is widely recognised that the USD100 billion $yr^{-1}$ figure is a fraction of the broader finance and investment needs of mitigation and adaptation embodied in the Paris Agreement (Peake and Ekins 2017). One estimate, based on a review of 160 Intended Nationally Determined Contributions ((I)NDCs), suggests the financial demand for both mitigation and adaptation needs of developing countries could reach USD474 billion $yr^{-1}$ by 2030 (Zhang and Pan 2016). The Organisation for Economic Co-operation and Development (OECD) reports that climate finance provided and mobilised by developed countries was USD79.6 billion in 2019. This finance included four components: bilateral public, multilateral public (attributed to developed countries), officially supported export credits and mobilised private finance (OECD 2021) (Section 15.3.2 and Box 15.4).

More broadly, there is recognition of the need for better accounting, transparency and reporting rules to allow evaluation of the fulfilment of finance pledges and the effectiveness of how funding is used (Xu et al. 2016; Roberts et al. 2017; Jachnik et al. 2019; Gupta and van Asselt 2019; Roberts et al. 2021). There is also a concern about climate finance being new and additional though the Paris Agreement does not make an explicit reference to it, nor is there a clear understanding of what constitutes new and additional (UNFCCC 2018; Carty et al. 2020; Mitchell et al. 2021). Some authors see the 'enhanced transparency framework' of the Paris Agreement (Section 14.3.2.4), and the specific requirements for developed countries to provide, biennially, indicative quantitative and qualitative information as well as report on financial support and mobilisation efforts (Articles 9.5 and 9.7), as promising marked improvements (Weikmans and Roberts 2019), including for the fairness of effort-sharing on climate finance provision (Pickering et al. 2015). Others offer a more circumspect view of the transformative capability of these transparency systems (Ciplet et al. 2018).

The more limited literature focusing on the specific finance needs of developing countries, particularly those expressed in NDCs conditional on international climate finance, suggests that once all countries have fully costed their NDCs, the demand for (public and private) finance to support NDC implementation is likely to be orders of magnitude larger than funds available from bilateral and multilateral sources. For some sectors, such as forestry and land use, this could leave 'NDC ambitions... in a precarious position, unless more diversified options are pursued to reach climate goals' (Kissinger et al. 2019). In addition, there is a need for fiscal policy reform in developing countries to ensure international climate finance flows are not undercut by public and private finance supporting unsustainable activities (Kissinger et al. 2019). During the 2018 Katowice conference, UNFCCC Parties requested the Standing Committee on Finance to prepare, every four years, a report on the determination of the needs of developing country Parties related to implementing the Convention and the Paris Agreement, for consideration by Parties at COP26 (UNFCCC 2019c).

#### 14.3.2.9 Technology Development and Transfer

Technology development and transfer is the second of three 'means of implementation and support' specified under the Paris Agreement to accomplish its objectives relating to mitigation (and adaptation) (UNFCCC 2015a, Art. 14.1). This sub-section discusses the provision made in the Paris Agreement for international cooperation on technology development and transfer. Section 14.4.2 below considers broader cooperative efforts on technology development and transfer under the UNFCCC. Both sections complement the discussion in Section 16.6 on the role of international cooperation in fostering transformative change.

The importance of technology as a means of implementation for climate mitigation obligations under the Paris Agreement is evident from Parties' NDCs. Of the 168 NDCs submitted as of June 2019, 109 were expressed as conditional upon support for technology development and transfer, with 70 Parties requesting technological support for both mitigation and adaptation, and 37 Parties for mitigation only (Pauw et al. 2020). Thirty-eight LDCs (79%) and 29 SIDS made their NDCs conditional on technology transfer, as did 50 middle-income countries (Pauw et al. 2020).

While technology is seen as a key means of implementation and support for Paris Agreement commitments, the issue of technology development and the transfer of environmentally sound technologies for climate mitigation was heavily contested between developed and developing countries in the Paris negotiations, and these differences are likely to persist as the Paris Agreement is implemented (Oh 2019). Contestations continued in negotiations for the Paris Rulebook, particularly regarding the meaning of technological innovation, which actors should be supported, and how support should be provided by the UNFCCC (Oh 2020a).

Article 10 of the Paris Agreement articulates a shared 'long-term vision on the importance of fully realising technology development and transfer in order to improve resilience to climate change and to reduce greenhouse gas emissions' (UNFCCC 2015, Art. 10.1). All Parties are required 'to strengthen cooperative action on technology development and transfer' (UNFCCC 2015, Art. 10.2). In addition, support, including financial support, 'shall be provided' to developing country Parties for the implementation of Article 10, 'including for strengthening cooperative action on technology development and transfer at different stages of the technology cycle, with a view to achieving a balance between support for mitigation and adaptation' (UNFCCC 2015, Art. 10.6). Available information on efforts related to support on technology development and transfer for developing country Parties is also one of the matters to be taken into account in the global stocktake (UNFCCC 2015, Art. 10.6) (Section 14.3.2.5).

The Paris Agreement emphasises that efforts to accelerate, encourage and enable innovation are 'critical for an effective long-term global response to climate change and promoting economic growth and sustainable development' and urges that they be supported, as appropriate, by the Technology Mechanism and Financial Mechanism of the UNFCCC (UNFCCC 2015, Art. 10.5). This support should be directed to developing country Parties 'for collaborative approaches to research and development, and facilitating access to technology, in particular for early stages of the technology cycle' (UNFCCC 2015, Art. 10.5). Inadequate support for research and development, particularly in developing countries, has been identified in previous studies of technology interventions by international institutions as a key technology innovation gap that might be addressed by the Technology Mechanism (de Coninck and Puig 2015).

To support Parties' cooperative action, the Technology Mechanism, established in 2010 under the UNFCCC (Section 14.4.2), will serve the Paris Agreement, subject to guidance of a new 'technology framework' (UNFCCC 2015, Art. 10.4). The latter was strongly advocated by the African group in the negotiations for the Paris Agreement (Oh 2020a), and was adopted in 2018 as part of the Paris Rulebook, with implementation entrusted to the component bodies of the Technology Mechanism. The guiding principles of the framework are coherence, inclusiveness, a results-oriented approach, a transformational approach and transparency. Its 'key themes' include innovation, implementation, enabling environment and capacity building, collaboration and stakeholder engagement, and support (UNFCCC 2019e, Annex). A number of 'actions and activities' are elaborated for each thematic area. These include: enhancing engagement and collaboration with relevant stakeholders, including local communities and authorities, national planners, the private sector and civil society organisations, in the planning and implementation of Technology Mechanism activities; facilitating Parties undertaking, updating and implementing technology needs assessments (TNAs) and aligning these with NDCs; and enhancing the collaboration of the Technology Mechanism with the Financial Mechanism for enhanced support for technology development and transfer. As regards TNAs, while some developing countries have already used the results of their TNA process in NDC development, other countries might benefit from following the TNA process, including its stakeholder involvement and multi-criteria decision analysis methodology, to strengthen their NDCs (Hofman and van der Gaast 2019).

#### 14.3.2.10 Capacity Building

Together with finance, and technology development and transfer, capacity building is the third of 'the means of implementation and support' specified under the Paris Agreement (UNFCCC 2015a, Art. 14.1). Capacity building has primarily been implemented through partnerships, collaboration and different cooperative activities, inside and outside the UNFCCC. This sub-section discusses the provision made in the Paris Agreement for international cooperation on capacity building. Section 14.4.3 below considers broader cooperative efforts on capacity building within the UNFCCC.

In its annual synthesis report for 2018, the UNFCCC secretariat stressed the importance of capacity building for the implementation of the Paris Agreement and NDCs, with a focus on measures already in place, regional and cooperative activities, and capacity-building needs for strengthening NDCs (UNFCCC 2019h). Of the 168 NDCs submitted as of June 2019, capacity building was the most frequently requested type of support (113 of 136 conditional NDCs) (Pauw et al. 2020). The focus of capacity-building activities is on enabling developing countries to take effective climate change action, given that many developing countries continue to face significant capacity challenges, undermining their ability to effectively or fully carry out the climate actions they intend to pursue (Dagnet et al. 2016). Content analysis of NDCs shows that capacity building for adaptation is prioritised over mitigation for developing countries, with the element of capacity building most indicated in NDCs being research and technology (Khan et al. 2020). In addition, developing countries' needs for education, training and awareness raising for climate change mitigation and adaptation feature prominently in NDCs, particularly those of LDCs (Khan et al. 2020). Differences are evident though between capacity-building needs expressed in the NDCs of LDCs (noting that Khan et al.s review was limited to NDCs in English) compared with those of upper-middle-income developing countries as categorised by the World Bank (World Bank 2021); the latter have more focus on mitigation with an emphasis on technology development and transfer (Khan et al. 2020).

The Paris Agreement urges all Parties to cooperate to enhance the capacity of developing countries to implement the Agreement (UNFCCC 2015a, Art. 11.3), with a particular focus on LDCs and SIDS (UNFCCC 2015a, Art. 11.1). Developed country Parties are specifically urged to enhance support for capacity-building actions in developing country Parties (UNFCCC 2015a, Art. 11.3). Article 12 of the Paris Agreement addresses cooperative measures to enhance climate change education, training, public awareness, public participation and public access to information, which can also be seen as elements of capacity building (Khan et al. 2020). Under the Paris Rulebook, efforts related to the implementation of Article 12 are referred to as 'Action for Climate Empowerment' and Parties are invited to develop and implement national strategies on this topic, taking into account their national circumstances (UNFCCC 2019i, para. 6). Actions to enhance climate change education, training, public awareness, public participation, public access to information, and regional and international cooperation may also be taken into account by Parties in the global stocktake process under Article 14 of the Paris Agreement (UNFCCC 2019i, para. 9).

Under the Paris Agreement, capacity-building can take a range of forms, including: facilitating technology development, dissemination and deployment; access to climate finance; education, training and public awareness; and the transparent, timely and accurate communication of information (UNFCCC 2015a, Art. 11.1) (Section 14.3.2.4). Principles guiding capacity-building support are that it should be: country-driven; based on and responsive to national needs; fostering country ownership of Parties at multiple levels; guided by lessons learned; and an effective, iterative process that is participatory, cross-cutting and gender-responsive (UNFCCC 2015a, Art. 11.2). Parties undertaking capacity building for developing country Parties must 'regularly communicate on these actions or measures'. Developing country Parties have a soft requirement ('should') to communicate progress made on implementing capacity-building plans, policies, actions or measures to implement the Paris Agreement (UNFCCC 2015a, Art. 11.4).

Article 11.5 provides that capacity-building activities 'shall be enhanced through appropriate institutional arrangements to support the implementation of this Agreement, including the appropriate institutional arrangements established under the Convention that serve this Agreement'. The COP decision accompanying the Paris Agreement established the Paris Committee on Capacity-building, with the aim to 'address gaps and needs, both current and emerging, in implementing capacity-building in developing country Parties and further enhancing capacity-building efforts, including with regard to coherence and coordination in capacity-building activities under the Convention' (UNFCCC 2016a, para. 71). The activities of the Committee are discussed further in Section 14.4.3 below. The relevant COP decision also established the Capacity-building Initiative for Transparency (UNFCCC 2016a, para. 84), which is managed by the GEF and designed to support developing country Parties in meeting the reporting and transparency requirements under Article 13 of the Paris Agreement (Robinson 2018).

Studies on past capacity-building support for climate mitigation offer some lessons for ensuring effectiveness of arrangements under the Paris Agreement. For example, Umemiya et al. (2020) suggest the need for a common monitoring system at the global level, and evaluation research at the project level, to achieve more effective capacity-building support. Khan et al. (2020) articulate 'four key pillars' of a sustainable capacity-building system for implementation of NDCs in developing countries: universities in developing countries as institutional hubs; strengthened civil society networks and partnerships; long-term programmatic finance support; and consideration of a capacity-building mechanism under the UNFCCC – paralleling the Technology Mechanism – to marshal, coordinate and monitor capacity-building activities and resources.

#### 14.3.2.11 Implementation and Compliance

The Paris Agreement establishes a mechanism to facilitate implementation and promote compliance under Article 15. This mechanism is to operate in a transparent, non-adversarial and non-punitive manner (Voigt 2016; Campbell-Duruflé 2018b; Oberthür and Northrop 2018) that distinguishes it from the more stringent compliance procedures of the Kyoto Protocol's Enforcement branch. The Paris Rulebook elaborated the modalities and procedures for the implementation and compliance mechanism, specifying the nature and composition of the compliance committee, the situations triggering its procedures, and the facilitative measures it can apply, which include a 'finding of fact' in limited situations, dialogue, assistance and recommendations (UNFCCC 2019e). The compliance committee is focused on ensuring compliance with a core set of binding procedural obligations (UNFCCC 2019j, Annex, Para. 22). This compliance committee, characterised as 'one of its kind' and an 'an important cornerstone' of the Agreement's legitimacy, effectiveness and longevity (Zihua et al. 2019), is designed to facilitate compliance rather than penalise non-compliance.

## Box 14.1 | Key Features of the Paris Agreement Relevant to Mitigation

The Paris Agreement's overall aim is to strengthen the global response to the threat of climate change, in the context of sustainable development and efforts to eradicate poverty. This aim is explicitly linked to enhancing implementation of the UNFCCC, including its objective in Article 2 of stabilising greenhouse gas concentrations at a level that would 'prevent dangerous anthropogenic interference with the climate system'. The Agreement sets three goals:

i. **Temperature:** holding the global average temperature increase to well below 2°C above pre-industrial levels and pursuing efforts to limit the temperature increase to 1.5°C above pre-industrial levels.
ii. **Adaptation and climate resilience:** increasing the ability to adapt to the adverse impacts of climate change and foster climate resilience and low greenhouse gas emissions development, in a manner that does not threaten food production.
iii. **Finance:** making finance flows consistent with a pathway towards low greenhouse gas emissions and climate-resilient development.

In order to achieve the long-term temperature goal, Parties aim to reach global peaking of emissions as soon as possible, recognising that peaking will take longer for developing countries, and then to undertake rapid reductions in accordance with the best available science. This is designed to reach global net zero GHG emissions in the second half of the century, with the emissions reductions effort to be determined on the basis of equity and in the context of sustainable development and efforts to eradicate poverty. In addition, implementation of the Agreement as a whole is expected to reflect equity and Parties' 'common but differentiated responsibilities and respective capabilities', in light of different national circumstances.

The core mitigation commitments of Parties under the Paris Agreement centre on preparing, communicating and maintaining successive 'Nationally Determined Contributions' (NDCs), the contents of which countries determine for themselves. All Parties must have NDCs and pursue domestic mitigation measures with the aim of achieving the objectives of their NDCs, but Parties' NDCs are neither subject to a review of adequacy (at an individual level) nor to legally binding obligations of result. The compliance mechanism is correspondingly facilitative.

The Paris Agreement establishes a global goal on adaptation, and recognises the importance of averting, minimising and addressing loss and damage associated with the adverse effects of climate change.

The efficacy of the Paris Agreement in achieving its goals is therefore dependent upon at least three additional elements:

i. **Ratcheting of NDCs:** Parties must submit a new or updated NDC every five years that is in line with the Paris Agreement's expectations of progression over time and the Party's highest possible ambition, reflecting common but differentiated responsibilities and respective capabilities in light of different national circumstances.

*Box 14.1 (continued)*

ii. **Enhanced transparency framework:** Parties' actions to implement their NDCs are subject to international transparency and review requirements, which will generate information that may also be used by domestic constituencies and peers to pressure governments to increase the ambition of their NDCs.
iii. **Collective global stocktake:** The global stocktake undertaken every five years, starting in 2023, will review the collective progress of countries in achieving the Paris Agreement's goals, in light of equity and best available science. The outcome of the global stocktake informs Parties in updating and enhancing their subsequent NDCs.

These international processes establish an iterative ambition cycle for the preparation, communication, implementation and review of NDCs.

For developing countries, the Paris Agreement recognises that increasing mitigation ambition and realising long-term low-emissions development pathways can be bolstered by the provision of financial resources, capacity building, and technology development and transfer. In continuation of existing obligations under the Convention, developed countries are obliged to provide financial assistance to developing countries with respect to mitigation and adaptation. The Paris Agreement also recognises that Parties may choose to voluntarily cooperate in the implementation of their NDCs to allow for higher ambition in their mitigation and adaptation actions and to promote sustainable development and environmental integrity.

## 14.3.3 Effectiveness of the Kyoto Protocol and the Paris Agreement

### 14.3.3.1 Ex-post Assessment of the Kyoto Protocol's Effects

Previous assessment reports have assessed the Kyoto Protocol with respect to each of the criteria identified in this chapter. However, at the time of AR5, it was premature to assess the impact of Kyoto on emissions, as these data had not been entirely compiled yet. Since AR5, a number of studies have done so. Chapter 2 of this report lists at least 18 countries that have sustained absolute emissions reductions for at least a decade, nearly all of which are countries that had Kyoto targets for the first commitment period. Most studies have concluded that Kyoto did cause emissions reductions. Such studies find a positive, statistically significant impact on emissions reductions in Annex I countries (Kim et al. 2020), Annex B countries (Grunewald and Martínez-Zarzoso 2012; Kumazawa and Callaghan 2012; Grunewald and Martínez-Zarzoso 2016; Maamoun 2019), or all countries respectively (Aichele and Felbermayr 2013; Iwata and Okada 2014). Overall, countries with emissions reduction obligations emit on average less $CO_2$ than similar countries without emissions reduction obligations – with estimates ranging from 3–50% (Grunewald and Martínez-Zarzoso 2012, 2016). Maamoun (2019) estimates that the Kyoto Protocol reduced GHG emissions of Annex B countries by 7% on average below a no-Kyoto scenario between 2005 and 2012. Aichele and Felbermayr (2013) conclude that Kyoto reduced $CO_2$ and GHG emissions by 10% compared to the counterfactual. By contrast, Almer and Winkler (2017) find no evidence for binding emission targets under Kyoto inducing significant and lasting emissions reductions for any of the Annex B or non-Annex B countries. The authors identify both negative and positive associations between Kyoto and emissions for several countries in several years, but no coherent picture emerges. Hartl (2019) calculates a Kyoto leakage share in global $CO_2$ trade of 4.3% for 2002–2009.

In terms of transformative potential, the Kyoto Protocol has been found to increase international patent applications for renewable energy technologies, especially in the case of solar energy technologies and especially in countries with more stringent emissions reduction targets, and has even led to an increase in patent applications in developing countries not obliged to reduçe emissions under Kyoto (Miyamoto and Takeuchi 2019). Kyoto also had a positive and statistically significant impact on the cost-effectiveness of renewable energy projects, as well as renewable energy capacity development, as it stimulated the introduction of domestic renewable energy policies (Liu et al. 2019).

The issue of institutional strength of Kyoto has been analysed by many authors, and much of this has been assessed in previous assessment reports. Since AR5, several papers question the environmental efficacy of the Kyoto Protocol based on its institutional design (Rosen 2015; Kuriyama and Abe 2018). Particular attention has focused on Kyoto's market mechanisms (Erickson et al. 2014; Kollmuss et al. 2015).

As described in previous IPCC reports and above, the 1997 Kyoto Protocol included three international market-based mechanisms. These operated among Annex I Parties (i.e., International Emissions Trading and Joint Implementation) and between Annex I Parties and non-Annex I countries (i.e., the CDM) (Grubb et al. 2014; World Bank 2018). Joint Implementation led to limited volumes of emissions credit transactions, mostly from economies in transition but also some Western European countries; International Emissions Trading also led only to limited transaction volumes (Shishlov et al. 2016).

Of the Kyoto Protocol's mechanisms, the CDM market has led to a greater amount of activity, with a 'gold rush' period between 2005 and 2012. The main buyers of CDM credits were private companies surrendering them within the European Union (EU) Emissions Trading System (ETS). Once the EU tightened its rules and restricted the use of CDM credits in 2011, there was a sharp drop in the price of CDM credits in 2012. This price never recovered, as the demand for CDM was very weak after 2012, in part because of the difficulties encountered in securing the entry into force of the Doha Amendment (Michaelowa et al. 2019b).

Assessing the effectiveness of Kyoto's market mechanisms is challenging, and the results have been mixed (Aichele and Felbermayr 2013; Iwata and Okada 2014; Kuriyama and Abe 2018). Kuriyama and Abe (2018) assessed emissions reduction quantities taking into account heightened criteria for additionality. They identified annual energy-related emissions reductions of 49 $MtCO_2$-eq $yr^{-1}$ flowing from the CDM, and non-energy related emissions reductions of 177 $MtCO_2$-eq $yr^{-1}$. Others have pointed to issues associated with non-energy related emissions reductions that suggest the latter estimate may be of questionable reliability, while also noting that regulatory tightening led later CDM projects to perform better with respect to the additionality criterion (Michaelowa et al. 2019b). The CDM's contribution to capacity building in some developing countries has been identified as possibly its most important achievement (Spalding-Fecher et al. 2012; Gandenberger et al. 2015; Murata et al. 2016; Xu et al. 2016; Dong and Holm Olsen 2017; Lindberg et al. 2018). There is evidence that the CDM lowered compliance costs for Annex 1 countries by at least USD3.6 billion (Spalding-Fecher et al. 2012). In host countries, the CDM led to the establishment of national approval bodies and the development of an ecosystem of consultants and auditors (Michaelowa et al. 2019b).

On the negative side, there are numerous findings that the CDM, especially at first, failed to lead to additional emissions cuts in host countries, meaning that the overall effect of CDM projects was to raise global emissions. Cames et al. (2016) concluded that over 70% of CDM projects led to emissions reductions that were likely less than projected, including the absence of additional reductions, while only 7% of projects led to actual additional emissions reductions that had a high likelihood of meeting or exceeding the *ex-ante* estimates. The primary reason the authors gave was associated with the low price for CDM credits; this meant that the contribution of the CDM to project finance was negligible, suggesting that most CDM projects would have been built anyway. A meta-analysis of *ex-post* studies of global carbon markets, which include the CDM, found net combined effects on emissions to be negligible (Green 2021). Across the board, CDM projects have been criticised for lack of 'additionality', problems of baseline determination, uneven geographic coverage (Michaelowa and Michaelowa 2011a; Cames et al. 2016; Michaelowa et al. 2019b), as well as failing to address human rights concerns (Schade and Obergassel 2014).

### 14.3.3.2 Effectiveness of the Paris Agreement

Given the comparatively recent conclusion of the Paris Agreement, evidence is still being gathered to assess its effectiveness in practice, in particular, since its long-term effectiveness hinges on states communicating more ambitious NDCs in successive cycles over time. Assessments of the Paris Agreement on paper are necessarily speculative and limited by the lack of credible counterfactuals. Despite these limitations, numerous assessments exist of the potential for international cooperation under the Paris Agreement to advance climate change mitigation.

These assessments are mixed and reflect uncertainty over the outcomes the Paris Agreement will achieve (Christoff 2016; Clémençon 2016; Keohane and Oppenheimer 2016; Young 2016; Dimitrov et al. 2019; Raiser et al. 2020). There is a divide between studies that do not expect a positive outcome from the Paris Agreement and those that do. The former base this assessment on factors such as: a lack of clarity in the expression of obligations and objectives; a lack of concrete plans collectively to achieve the temperature goal; extensive use of soft law (i.e., non-legally binding) provisions; limited incentives to avoid free-riding; and the Agreement's weak enforcement provisions (Allan 2019), as well as US non-cooperation under the Trump administration and the resulting gap in mitigation, finance and governance (Bang et al. 2016; Spash 2016; Tulkens 2016; Chai et al. 2017; Lawrence and Wong 2017; Thompson 2017; Barrett 2018; Kemp 2018). Studies expecting a positive outcome emphasise factors such as: the breadth of participation enabled by self-differentiated NDCs; the 'logic' of domestic climate policies driving greater national ambition; the multiplicity of actors engaged by the Paris Agreement's facilitative architecture; the falling cost of low-carbon technologies; provision for financial, technology and capacity-building support to developing country Parties; possibilities for voluntary cooperation on mitigation under Article 6; and the potential for progressive ratcheting up of Parties' pledges over time fostered by transparency of reporting and international scrutiny of national justifications of the 'fairness' of contributions (Caparrós 2016; Chan 2016a; Falkner 2016b; Victor 2016; Morgan and Northrop 2017; Urpelainen and Van de Graaf 2018; Hale 2020; Tørstad 2020). Turning to the assessment criteria articulated in this chapter, the following preliminary assessments of the Paris Agreement can be made.

In relation to the criterion of *environmental effectiveness*, the Paris Agreement exceeds the Kyoto Protocol in terms of coverage of GHGs and participation of states in mitigation actions. In terms of coverage of GHGs, the Kyoto Protocol limits its coverage to a defined basket of gases identified in its Annex A (carbon dioxide ($CO_2$), methane ($CH_4$), nitrous oxide ($N_2O$), hydrofluorocarbons (HFCs), perfluorocarbons (PFCs), sulphur hexafluoride ($SF_6$), as well as nitrogen trifluoride ($NF_3$)). The Paris Agreement does not specify the coverage of gases, thus Parties may cover the full spectrum of GHGs in their NDCs as encouraged by the accounting provisions in Annex II to Decision 18/CMA.1 (or conversely they may choose to exclude important mitigation sectors) and there is also the possibility to include other pollutants such as short-lived climate forcers like black carbon. Article 4.4 calls on developed countries to undertake economy-wide emissions reduction targets with the expectation that developing country Parties will also move to introduce these over time. Moreover, the Paris Agreement makes express reference to Parties taking action to conserve and enhance 'sinks and reservoirs of greenhouse gases' (Article 5). As under the UNFCCC and Kyoto Protocol, this allows for coverage of land use, land-use change and forestry and agriculture, forestry and other land use (AFOLU) emissions, both $CO_2$ and other Kyoto Annex A gases, as well as methane (Pekkarinen 2020). A few countries, particularly LDCs, include quantified non-$CO_2$ emissions reductions from the agricultural sector in their NDCs, and many others include agriculture in their economy-wide targets (Richards et al. 2018). Some studies find that agricultural development pathways with mitigation co-benefits can deliver 21–40% of needed mitigation for the 'well below 2°C' limit, thus necessitating 'transformative technical and policy options' (Wollenberg et al. 2016). Other studies indicate that broader 'natural climate solutions, including forests, can provide 37% of the cost-effective $CO_2$ mitigation needed through 2030 for a more than 66% chance of holding warming to below 2°C' (Griscom et al. 2017).

As Figure 14.2 illustrates graphically, communicated unconditional NDCs, if achieved, lead to a reduction of about 7% of world emissions by 2030 in relation to the Kyoto GHGs, and NDCs with conditional elements increase this reduction to about 12% (den Elzen et al. 2016). Although there are uncertainties in the extent to which countries will meet the conditional elements of their NDCs, the experience with the Cancun pledges has been positive, as countries will collectively meet their pledges by 2020, and even individual pledges will be met in most cases, although arguably helped by the COVID-19 pandemic (UNEP 2020). In any case, the main challenge that remains is to close the emissions gap, the difference between what has been pledged and what needs to be achieved by 2030 to reach a 1.5°C compatible path (respectively 2°C) (Roelfsema et al. 2020; UNEP 2020, see also Cross-Chapter Box 4 in Chapter 4). In terms of participation of states in mitigation actions, the Paris Agreement performs better than the Kyoto Protocol. The latter contains mitigation targets only for developed countries listed in its Annex B, while the Paris Agreement extends binding procedural obligations in relation to mitigation contributions to all states. It is noted, however, that the Paris Agreement represented a weakening of commitments for those industrialised countries that were Parties to the Kyoto Protocol, although a strengthening for those that were not, and for developing countries (Oberthür and Groen 2020). Finally, some analysts have suggested that the recent proliferation of national mid-century net-zero targets – currently 127 countries have considered or adopted such targets – can be attributed, at least in part, to participation in the Paris Agreement and having agreed to its Article 4 (Climate Action Tracker 2020a; Day et al. 2020).

In relation to the criterion of *transformative potential*, there is, as yet, limited empirical data or theoretical analysis on which to assess the Paris Agreement's transformative potential. The IPCC *Special Report on Global Warming of 1.5°C* concluded that pathways limiting global warming to 1.5°C would require systems transitions that are 'unprecedented in terms of scale' (IPCC 2018b). There is limited evidence to suggest that this is underway, although there are arguments made that Paris has the right structure to achieve this. The linking of the UNFCCC financial apparatus, including the GCF, to the Paris Agreement, and the provisions on technology support and capacity building, provide potential avenues for promoting increased investment flows into low-carbon technologies and development pathways, as Labordena et al. (2017) show in the case of solar energy development in Africa. Similarly, Kern and Rogge (2016) argue that the Paris Agreement's global commitment towards complete decarbonisation may play a critical role in accelerating underlying system transitions, by sending a strong signal as to the actions needed by national governments and other international support. Victor et al. (2019) argue that international cooperation that enhances transformative potential needs to operate at the sectoral level, as the barriers to transformation are highly specific to each sector; the Paris Agreement's broad consensus around a clear level of ambition sends a strong signal on what is needed in each sector, but on its own will do little unless bolstered with sector-specific action (Geels et al. 2019). On the less optimistic side, it is noted that the extent of the 'investment signal' sent by the Agreement to business is unclear (Kemp 2018), and it is also unclear to what extent the Paris Agreement is fostering investment in break-through technologies. United States non-cooperation from 2017 to 2020 posed a significant threat to adequate investment flows through the GCF (Chai et al. 2017; Urpelainen and Van de Graaf 2018).

In relation to the criterion of *distributive outcomes,* the Paris Agreement performs well in some respects but less well in others, and its performance relative to the Kyoto Protocol is arguably lower in respect of some indicators such as industrialised country leadership, and differentiation in favour of developing countries. While the Kyoto Protocol implemented a multilaterally agreed burden-sharing arrangement set out in the UNFCCC and reflected in Annex-based differentiation in mitigation obligations, the Paris Agreement relies on NDCs, accompanied by self-assessments of the fairness of these contributions; some of these do not accord with equity principles of international environmental law, although it is worth noting that the Kyoto Protocol was also not fully consistent with such principles. At present, mechanisms in the Paris Agreement for promoting equitable burden sharing and evaluating the fairness of Parties' contributions are undefined, although numerous proposals have been developed in the literature Herrala and Goel 2016; (Ritchie and Reay 2017; Robiou du Pont et al. 2017; Alcaraz et al. 2019; Sheriff 2019) (Section 14.3.2.3). Zimm and Nakicenovic (2020) analysed the first set of NDCs and concluded that they would result in a decrease in the inequality of per capita emissions across countries. In relation to other indicators, such as the provision of support, the distributive outcomes of the Paris Agreement are dependent on the availability of support through mechanisms such as the GCF to meet the mitigation and adaptation financing needs of developing countries (Antimiani et al. 2017; Chan et al. 2018). One study suggests that the implementation of the emissions reduction objectives stated in the NDCs implies trade-offs with poverty reduction efforts needed to achieve SDGs (Campagnolo and Davide 2019), while other studies offer evidence that the immediate economic, environmental, and social benefits of mitigation in line with developing countries' NDCs exceed those NDCs' costs, and ultimately align with the SDGs (Antwi-Agyei et al. 2018; Vandyck et al. 2018; Caetano et al. 2020) (Chapter 17). In relation to the promotion of co-benefits, the Paris Agreement has enhanced mechanisms for promoting co-benefits (e.g., in some cases for biodiversity conservation through the endorsement of REDD+ initiatives and activities) and linkages to sustainable development (e.g., through the Article 6.4 mechanism). Finally, in its preambular text the Paris Agreement endorses both a human rights perspective and the concept of just transitions, creating potential hooks for further elaboration and expansion of these principles in mitigation actions.

On the criterion of *economic performance*, the Paris Agreement's performance is potentially enhanced by the capacity for Parties to link mitigation policies, therefore improving aggregate cost-effectiveness. Voluntary cooperation under Article 6 of the Paris Agreement could facilitate such linkage of mitigation policies (Chan et al. 2018). A combination of common accounting rules and the absence of restrictive criteria and conditions on the use of ITMOs could accelerate linkage and increase the latitude of Parties to scale up the ambition of their NDCs. However, significant question marks remain over how the environmental integrity of traded emissions reductions can be ensured (Mehling 2019). The ability of Article 6 to contribute to the goal of the Paris Agreement will depend on the extent to which the rules ensure environmental integrity and avoid double counting, while utilising the full potential of cooperative efforts (Michaelowa et al. 2019a; Schneider et al. 2019).

In relation to the criterion of *institutional strength*, the Paris Agreement's signalling and guidance function is, however, arguably high. The Paris Agreement has the potential to interact with complementary approaches to climate governance emerging beyond it (Held and Roger 2018). It may also be used by public-sector organisations – organised and mobilised in many countries and transnationally – as a point of leverage in domestic politics to encourage countries to take costly mitigation actions (Keohane and Oppenheimer 2016). More broadly, the Paris Agreement's architecture provides flexibility for decentralised forms of governance (Jordan et al. 2015; Victor 2016) (Section 14.5). The Agreement has served a catalytic and facilitative role in enabling and facilitating climate action from non-state and sub-state actors (Chan et al. 2015; Chan et al. 2016; Hale 2016; Bäckstrand et al. 2017; Kuyper et al. 2018b). Such action could potentially 'bridge' the ambition gap created by insufficient NDCs from Parties (Hsu et al. 2019b). The 2018 UNEP Emissions Gap Report estimates that if 'cooperative initiatives are scaled up to their fullest potential', the impact of non-state and sub-national actors could be up to 1–23 $GtCO_2$-eq $yr^{-1}$ by 2030 compared to current policy, which could bridge the gap (Lui et al. 2021). However, at present such a contribution is limited (Michaelowa and Michaelowa 2017; UNEP 2018a). Non-state actors are also playing a role in enhancing the ambition of individual NDCs by challenging their adequacy in national courts (Chapter 13 and Section 14.5.3).

The Paris Agreement's institutional strength in terms of 'rules and standards to facilitate collective action' is disputed given the current lack of comparable information in NDCs (Peters et al. 2017; Pauw et al. 2018; Mayer 2019; Zihua et al. 2019), and the extent to which its language, as well as that of the Rulebook, strikes a balance in favour of discretion over prescriptiveness (Rajamani and Bodansky 2019). Similarly, in terms of 'mechanisms to enhance transparency and accountability', although detailed rules relating to transparency have been developed under the Paris Rulebook, these rules permit Parties considerable self-determination in the extent and manner of application (Rajamani and Bodansky 2019), and may not lead to further ambition (Weikmans et al. 2020). Further the Paris Agreement's compliance committee is facilitative and designed to ensure compliance with the procedural obligations in the Agreement rather than with the NDCs themselves, which are not subject to obligations of result. The Paris Agreement does, however, seek to support the building of transparency-related capacity of developing countries, potentially triggering institutional capacity-building at the national, sub-national and sectoral levels (Section 14.3.2.7).

Ultimately, the overall effectiveness of the Paris Agreement depends on its ability to lead to ratcheting up of collective climate action to meet the long-term global temperature goal (Bang et al. 2016; Christoff 2016; Young 2016; Dimitrov et al. 2019; Gupta and van Asselt 2019). As noted above, there is some evidence that this is already occurring. The design of the Paris Agreement, with 'nationally determined' contributions at its centre, countenances an initial shortfall in collective ambition in relation to the long-term global temperature goal on the understanding and expectation that Parties will enhance the ambition of their NDCs over time (Article 4). This is essential given the current shortfall in ambition. The pathways reflecting current NDCs, according to various estimates, imply global warming in the range of 3°C by 2100 (UNFCCC 2016b; UNEP 2018a) (Box 4.3). NDCs will need to be substantially scaled up if the temperature goal of the Paris Agreement is to be met (Rogelj et al. 2016; Rogelj et al. 2018; Höhne et al. 2017, 2018; UNEP 2020). The Paris Agreement's 'ambition cycle' is designed to trigger such enhanced ambition over time. Some studies find that like-minded climate mitigation clubs can deliver substantial emissions reductions (Hovi et al. 2017) and are reasonably stable despite the departure of a major emitter such as the United States (Sprinz et al. 2018); other studies find that conditional commitments in the context of a pledge and review mechanism are unlikely to substantially increase countries' contributions to emissions reductions (Helland et al. 2017), and hence need to be complemented by the adoption of instruments designed differently from the Paris Agreement (Barrett and Dannenberg 2016). In any case, high (but not perfect) levels of mean compliance rates with the Paris Agreement have to be assumed for reaching the 'well below 2°C' temperature goal (Sælen 2020; Sælen et al. 2020). This is by no means assured.

In conclusion, it remains to be seen whether the Paris Agreement will deliver the collective ambition necessary to meet the temperature goal. While the Paris Agreement does not contain strong and stringent obligations of result for major emitters, backed by a demanding compliance system, it establishes binding procedural obligations, lays out a range of normative expectations, and creates mechanisms for regular review, stock taking, and revision of NDCs. In combination with complementary approaches to climate governance, engagement of a wide range of non-state and sub-national actors, and domestic enforcement mechanisms, these have the potential to deliver the necessary collective ambition and implementation. Whether it will do so, remains to be seen.

## Cross-Chapter Box 10 | Policy Attribution – Methodologies for Estimating the Macro-level Impact of Mitigation Policies on Indices of Greenhouse Gas Mitigation

**Authors:** Mustafa Babiker (Sudan/Saudi Arabia), Paolo Bertoldi (Italy), Christopher Bataille (Canada), Felix Creutzig (Germany), Navroz K. Dubash (India), Michael Grubb (United Kingdom), Erik Haites (Canada), Ben Hinder (United Kingdom), Janna Hoppe (Switzerland), Yong-Gun Kim (Republic of Korea), Gregory F. Nemet (the United States of America/Canada), Anthony Patt (Switzerland), Yamina Saheb (France), Raphael Slade (United Kingdom)

This report notes both a growing prevalence of mitigation policies over the past quarter century (Chapter 13), and 'signs of progress' including various quantified indices of GHG mitigation (Table 2.4). Even though policies implemented and planned to date are clearly insufficient for meeting the Paris long-term temperature goals, a natural question is to what extent the observed macro-level changes (global, national, sectoral, technological) can be attributed to policy developments. This Assessment Report is the first to address that question. This box describes the methods for conducting such 'attribution analysis' as well as its key results, focusing on the extent to which polices have affected three main types of 'outcome indices':

- **GHG emissions:** emissions volumes and trends at various levels of governance including sub- and supra-national levels, and within and across sectors.
- **Proximate emission drivers:** trends in the factors that drive emissions, distinguished through decomposition analyses, notably: energy/GDP intensity and carbon/energy intensity (for energy-related emissions); indices of land use such as deforestation rates (for LULUCF/AFOLU); and more sector-specific component drivers such as the floor area per capita, or passenger kilometres per capita.
- **Technologies:** developments in key low-carbon technologies that are likely to have a strong influence on future emissions trends, notably levels of new investment and capacity expansions, as well as technology costs, with a focus on those highlighted in Figure 2.30.

*Policy attribution* examines the extent to which emission-relevant outcomes on these indices – charted for countries, sectors and technologies, particularly in Chapter 2 and the sectoral chapters – may be reasonably attributed to policies implemented prior to the observed changes. Such policies include regulatory instruments such as energy efficiency programmes or technical standards and codes, carbon pricing, financial support for low-carbon energy technologies and efficiency, voluntary agreements, and regulation of land-use practices. The sectoral chapters give more detail along with some accounts of policy, while trends in mitigation policy adoption are summarised in Chapter 13.

In reviewing hundreds of scientific studies cited in this report, the impacts of adopted policies on observed outcomes were assessed. The vast majority of these studies examine particular instruments in particular contexts, as covered in the sectoral chapters and Chapter 13; only a few have appraised global impacts of policies, directly or plausibly inferred (the most significant are cited in Figure 1 in this Cross-Chapter Box). Typically, studies consider 'mitigation policies' to be those adopted with either a primary objective of reducing GHG emissions or emissions reductions as one among multiple objectives.

Policies differ in design, scope, and stringency, may change over time as they require amendments or new laws, and often partially overlap with other instruments. Overall, the literature indicates that policy mixes are, theoretically and empirically, more effective in reducing emissions, stimulating innovation, and inducing behavioural change than stand-alone policy instruments (Sections 5.6 and 13.7) (Rosenow et al. 2017; Best and Burke 2018; Sethi et al. 2020). Nevertheless, these factors complicate analysis, because they give rise to the potential for double counting emissions reductions that have been observed, and which separate studies can attribute to different policy instruments.

Efforts to attribute observed outcomes to a policy or policy mix is also greatly complicated by the influence of many exogenous factors, including fossil fuel prices and socio-economic conditions. Likewise, technological progress can result from both exogenous causes, such as 'spillover' from other sectors, and policy pressure. Further, other policies, such as fossil fuel subsidies as well as trade-related policies, can partially counteract the effect of mitigation policies by increasing the demand for energy or carbon-intensive goods and services. In some cases, policies aimed at development, energy security, or air quality have climate co-benefits, while others increase emissions.

Studies have applied a number of methods to identify the actual effects of mitigation policies in the presence of such confounding factors. These include statistical attribution methodologies, including experimental and quasi-experimental design, instrumental variable approaches, and simple correlational methods. Typically, the relevant mitigation metric is the outcome variable, while measures of policies and other factors act as explanatory variables. Other methodologies include aggregations and extrapolations

*Cross-Chapter Box 10 (continued)*

from micro-level data evaluation, and inference from combining multiple lines of analysis, including expert opinion. Additionally, the literature contains reviews, many of them systematic in nature, that assess and aggregate multiple empirical studies.

With these considerations in mind, multiple lines of evidence, based upon the literature, support a set of high-level findings, as illustrated in Figure 1 in this Cross-Chapter Box, as follows.

**1. GHG Emissions.** There is robust evidence with a high level of agreement that mitigation policies have had a discernible impact on emissions. Several lines of evidence indicate that mitigation policies have led to avoided global emissions to date of several billion tonnes $CO_2$-eq annually. The figure in this box shows a selection of results giving rise to this estimate.

As a starting point, one methodologically sophisticated econometric study links global mitigation policies (defined as climate laws and executive orders) to emission outcomes; it estimates emission savings of 5.9 $GtCO_2\ yr^{-1}$ in 2016 compared to a no-policy world (Eskander and Fankhauser 2020) (Section 13.6.2).

A second line of evidence derives from analyses of the Kyoto Protocol. Countries which took on Kyoto Protocol targets accounted for about 24% of global emissions during the first commitment period (2008–12). The most recent robust econometric assessment (Maamoun 2019) estimates that these countries cut GHG emissions by about 7% on average over 2005–2012, rising over the period to around 12% (1.3 $GtCO_2$-eq $yr^{-1}$) *relative to a no-Kyoto scenario.* This is consistent with estimates of Grunewald and Martinez (2016) of about 800 $MtCO_2$-eq $yr^{-1}$ averaged to 2009. Developing countries' emissions reduction projects through the CDM (defined in Article 12 of the Kyoto Protocol) were certified as growing to over 240 $MtCO_2$-eq $yr^{-1}$ by 2012 (UNFCC 2021c). With debates about the full

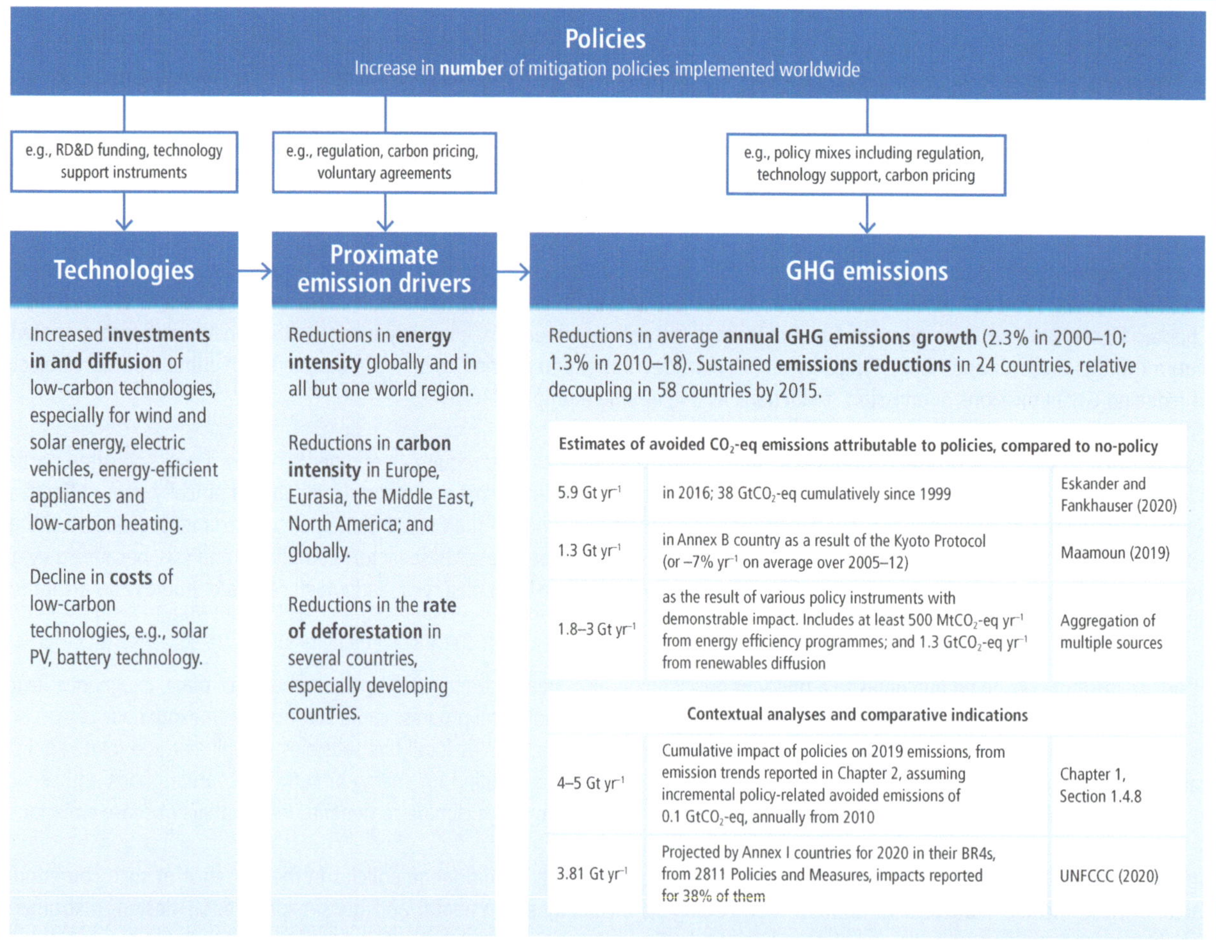


**Cross-Chapter Box 10, Figure 1 | Policy impacts on key outcome indices. The figure shows the impacts of policies on three indices: proximate emission drivers, technologies and GHG emissions, including several lines of evidence on GHG abatement attributable to policies.**

*Cross-Chapter Box 10 (continued)*

extent of 'additionality', academic assessments of savings from the CDM have been slightly lower, with particular concerns around some non-energy projects (Section 14.3.3.1).

A third line of evidence derives from studies that identify policy-related, absolute reductions from historical levels in particular countries and sectors through decomposition analyses (Le Quéré et al. 2019; Lamb et al. 2021), or evaluate the impact of particular policies, such as carbon pricing systems. From a wide range of estimates in the literature (Sections 2.8.2.2 and 13.6), many evaluations of the EU ETS suggest that it has reduced emissions by around 3% to 9% relative to unregulated firms and/or sectors (Schäfer 2019; Colmer et al. 2020), while other factors, both policy (energy efficiency and renewable support) and exogenous trends, played a larger role in the overall reductions seen (Haites 2018).

These findings derived from the peer-reviewed literature are also consistent with two additional sets of analysis. The first set concerns trends in emissions, drawing directly from Chapters 2, 6 and 11, showing that global annual emission growth has slowed, as evidenced by annual emission increments of 0.55 $GtCO_2$-eq $yr^{-1}$ between 2011 and 2019 compared to 1.014 $GtCO_2$-eq $yr^{-1}$ in 2000 and 2008. This suggests avoided emissions of 4–5 $GtCO_2$-eq $yr^{-1}$ (see also Figure 1.1d). The second set concerns emissions reductions projected by Annex I governments for 2020 in their fourth biennial reports to the UNFCCC. It is important to note that these are mostly projected annual savings from implemented policies (not *ex-post* evaluations), and there are considerable differences in countries' estimation methodologies. Nevertheless, combining estimates from 38% of the total of 2,811 reported policies and measures yields an overall estimate of 3.81 $GtCO_2$-eq $yr^{-1}$ emission savings (UNFCCC 2020d).

**2. Proximate emission drivers.** With less overt focus on emissions, studies of trends in energy efficiency, carbon intensity, or deforestation often point to associated policies. The literature includes an increasing number of studies on demonstrable progress in developing countries. For example, South and South-East Asia have seen energy intensity in buildings improving at about 5–6% $yr^{-1}$ since 2010 (Figure 2.22). In India alone, innovative programmes in efficient air conditioning, LED lighting, and industrial efficiency are reported as saving around 25 Mtoe in 2019–2020, thus leading to avoided emissions of over 150 $MtCO_2$ $yr^{-1}$ (Malhotra et al. 2021) (Box 16.3). Likewise, reductions in deforestation rates in several South and Central American and Asian countries are at least partly attributable to ecosystem payments, land-use regulation, and internal efforts (Section 7.6.2). Finally, the policy-driven displacement of fossil fuel combustion by renewables in energy has led to reductions in carbon intensity in several world regions (Chapters 2 and 6).

**3. Technologies.** The literature indicates unambiguously that the rapid expansion of low-carbon energy technologies is substantially attributable to policy (Sections 6.7.5 and 16.5). Technology-specific adoption incentives have led to a greater use of less carbon-intensive (e.g., renewable electricity) and less energy-intensive (especially in transport and buildings) technologies. As Chapters 2 and 6 of this report note that modern renewable energy sources currently satisfy over 9% of global electricity demand, and this is largely attributable to policy. There are no global-level studies estimating the avoided emissions due to renewable energy support policies, but there are methods that have been developed to link renewable energy penetration to avoided emissions, such as that of IRENA (2021). Using that method, and assuming that 70% of modern renewable energy expansion has been policy induced, yields an estimate of avoided emissions of 1.3 $GtCO_2$-eq $yr^{-1}$ in 2019. Furthermore, observed cost reductions are the result of policy-driven capacity expansion as well as publicly funded resarch and development, in individual countries and globally. These correspond with induced effects on number of patents, 'learning curve' correlations with deployed capacity, and cost component and related case study analyses (Kavlak et al. 2018; Nemet 2019; Popp 2019; Grubb et al. 2021).

## 14.4 Supplementary Means and Mechanisms of Implementation

As discussed above, the Paris Agreement sets in place a new framework for international climate policy albeit one that is embedded in the wider climate regime complex (Coen et al. 2020). Whereas international governance had earlier assumed centre stage, the Paris Agreement recognises the salience of domestic politics in the governance of climate change (Kinley et al. 2020). The new architecture also provides more flexibility for recognising the benefits of working in diverse forms and groups and allows for more decentralised 'polycentric' forms of governance (Jordan et al. 2015; Victor 2016). The next two sections address this complementarity between the Paris Agreement and other agreements and institutions.

The Paris Agreement identifies a number of pathways, or means of implementation, towards accomplishing rapid mitigation and the achieving of its temperature goal: finance; capacity building; technology and innovation; and cooperative approaches and markets (Sections 14.3.2.7–14.3.2.10 above). In this section, we examine each of these means and mechanisms of implementation, and the agreements and institutions lying outside of the Paris Agreement that contribute to each. In the following section, 14.5, we examine the agreements and institutions playing other governance roles:

| Type | Instrument/ organisation | Mitigation | Transparency | Sinks | Markets | Finance | Technology | Capacity building |
|---|---|---|---|---|---|---|---|---|
| Global treaties | Montreal Protocol | 14.5.1.1 | | | | 14.5.1.1 | | |
| | CBD | 14.5.1.1 | | 14.5.2.1 | | | | |
| | UNCCD | | | 14.5.2.1 | | | | 14.5.2.1 |
| | Minimata Mercury Convention | 14.5.1.1 | | | | | | |
| United Nations programmes and specialised agencies | UN REDD+ programme | 14.5.1.1 | | 14.5.2.1 | | 14.5.2.1 | | 14.4.3 |
| | UNEP | 14.5.1.1 | | | | | | 14.4.3 |
| | UNDP | | | | | | | 14.4.3 |
| | UNIDO | | | | | | | 14.4.1.2 |
| | UNOSSC | | | | | | | 14.4.1.2 |
| | FAO | | | 14.5.2.1 | | | | 14.4.1.2 |
| | ICAO | 14.5.2.3 | | | 14.5.2.3 | | 14.5.2.3 | |
| | IMO | 14.5.2.3 | 14.5.2.3 | | | | 14.5.2.3 | |
| Other global organisations | IEA | | | | | | 14.5.2.2 | |
| | IRENA | | | | | 14.5.2.2 | 14.5.2.2 | 14.5.2.2 |
| | MDBs | 14.4.1.2 | 14.4.1.2 | 14.5.4 | 14.4.4 | 14.4.1.2 | | 14.4.3 |
| Regional, multi- and bilateral agreements | LRTAP | 14.5.1.1 | | | | | | |
| | MIGA | | | | | 14.5.2.2 | | |
| | PPCA | 14.5.2.2 | | | | | | |
| | Regional trade agreements | 14.5.1.3 | | | 14.5.1.3 | | 14.5.1.3 | |
| | Bilateral development programmes | | | | 14.4.4 | 14.4.1.1 | 14.4.1.1 | 14.4.3 |
| | International science programmes | | | | | | 14.4.2 | |
| | South–South Cooperation | | | | | 14.5.1.4 | 14.5.1.4 | 14.4.3 |
| Non-state trans-national actors | Global city networks | 14.5.5 | | 14.5.5 | | 14.5.5 | 14.5.5 | 14.5.5 |
| | Environmental NGOs | 14.5.2.2 | 14.5.4 | | | 14.5.3 | | |
| | Social movements | 14.5.3 | | 14.5.3 | | | | |
| | Business partnerships | 14.5.4 | 14.5.4 | | | 14.5.4 | 14.5.4 | 14.5.4 |

**Figure 14.3 | Climate governance beyond the UNFCCC.** The figure shows those relationships, marked in blue, between international governance activities, described in the text, that relate to activities of the UNFCCC and Paris Agreement.

regulating activities in particular sectors; linking climate mitigation with other activities such as adaptation; and stimulating and coordinating the actions of non-state actors at a global scale.

Figure 14.3 maps out the interlinkages described in the text of Sections 14.4 and 14.5. It is an incomplete list, but illustrates clearly that across multiple types of governance, there are multiple instruments or organisations with activities connected to the different governance roles associated with the Paris Agreement and the UNFCCC more generally.

### 14.4.1 Finance

International cooperation on climate finance is underpinned by various articles of the UNFCCC including Articles 4.3, 4.4, 4.5, 4.7 and 11.5 (UNFCCC 1992). This was further amplified through the commitment by developed countries in the Copenhagen Accord and the Cancun Agreements to mobilise jointly through various sources USD100 billion $yr^{-1}$ by 2020 to meet the needs of the developing countries (UNFCCC 2010b). This commitment was made in the context of meaningful mitigation action and transparency of implementation. As mentioned in Section 14.3.2.8, in the Paris Agreement the binding obligation on developed country Parties to provide financial resources to assist developing country Parties applies to both mitigation and adaptation (UNFCCC 2015a, Art. 9.1). In 2019, climate finance provided and mobilised by developed countries was in the order of USD79.6 billion, coming from different channels including bilateral and multilateral channels, and also through mobilisation of the private sector attributable to these channels (OECD 2021). A majority (two-thirds) of these flows targeted mitigation action exclusively (Chapter 15). These estimates, however, have been criticised on various grounds, including that they are an overestimate and do not represent climate-specific net assistance only; that in grant equivalence terms the order of magnitude is lower; and the questionable extent of transparency of information on mobilised private finance, as well as the direction of these flows (Carty et al. 2020). On balance, such assessments need to be viewed in the context of the original commitment, the source of the data and the evolving guidance, and modalities and procedures from the UNFCCC processes. As mentioned in Chapter 15, the measurement of climate finance flows continues to face definitional, coverage and reliability issues, despite progress made by various data providers and collators (Section 15.3.2).

The multiplicity of actors providing financial support has resulted in a fragmented international climate finance architecture as indicated in Section 14.3.2.8. It is also seen as a system which allows for speed, flexibility and innovation (Pickering et al. 2017). However, the system is not yet delivering adequate flows given the needs of developing countries (Section 14.3.2.8). An early indication of these self-assessed needs is provided in the conditional NDCs. Of the 136 conditional NDCs submitted by June 2019, 110 have components or additional actions conditioned on financing support for mitigation and 79 have components or additional actions for support for adaptation (Pauw et al. 2020). While the Paris Agreement did not explicitly countenance conditionality for actions in developing countries, it is generally understood that the ambition and effectiveness of climate ambition in these countries is dependent on financial support (Voigt and Ferreira 2016b).

#### 14.4.1.1 Bilateral Finance

The Paris Agreement and the imperative for sustainable development reinforce the need to forge strong linkages between climate and development (Fay et al. 2015). This in turn has highlighted the urgent need for greater attention to the relationship between development assistance and finance, and climate change (Steele 2015).

The UNFCCC website cites some 20 bilateral development agencies providing support to climate change programmes in developing countries (UNFCCC 2020a). These agencies provide a mix of development cooperation, policy advice and support and financing for climate change projects. Since the year 2000, the OECD Development Assistance Committee has been tracking trends in climate-related development finance and assistance. The amount of bilateral development finance with climate relevance has increased substantially since 2000 (OECD 2019a). For 2019, it was reported to be USD28.8 billion in direct finance and USD2.6 billion through export credit agencies. Further, another USD34.1 billion of the climate finance provided through multilateral channels is attributable to the developed countries (OECD 2021). The OECD methodology has been critiqued as it uses Rio markers, the limitations of which could lead to erroneous reporting and assessment of finance provided as well as of the mitigation outcome (Michaelowa and Michaelowa 2011b; Weikmans and Roberts 2019). This issue is to be addressed through the modalities, procedures and guidance under the Enhanced Transparency Framework of the Paris Agreement (Section 14.3.2.4), through the mandate to the Subsidiary Body for Scientific and Technological Advice (SBSTA) to develop common tabular formats for the reporting of information on, *inter alia*, financial support provided, mobilised and received (UNFCCC 2019k). Until then, the Biennial Assessment Report prepared by the Standing Committee on Finance provides the best available information on financial support.

#### 14.4.1.2 Multilateral Finance

Multilateral development banks (MDBs) comprise six global development banks: the European Investment Bank, International Fund for Agricultural Development, International Investment Bank, New Development Bank, OPEC Fund for International Development, and the World Bank Group; six regional development banks: the African Development Bank, Asian Development Bank, Asian Infrastructure Investment Bank, European Bank for Reconstruction and Development, Inter-American Development Bank, and the Islamic Development Bank; and 13 sub-regional development banks: the Arab Bank for Economic Development in Africa, Arab Fund for Economic and Social Development, Black Sea Trade and Development Bank, Caribbean Development Bank, Central American Bank for Economic Integration, Development Bank of the Central African States, Development Bank of Latin America, East African Development Bank, Eastern and Southern African Trade and Development Bank, Economic Cooperation Organization Trade and Development Bank, Economic Community of West African States Bank for Investment and Development, Eurasian Development Bank, and the West African Development Bank. Together they play a key role in international cooperation at the global, regional and sub-regional levels because of their growing mandates and proximity to policymakers (Engen and Prizzon 2018). For many, climate change is a growing priority and for some, because of the needs of the regions or sub-regions in which they operate, climate change is embedded in many of their operations.

In 2015, 20 representative MDBs and members of the International Development Finance Club unveiled five voluntary principles to mainstream climate action in their investments: commitment to climate strategies, managing climate risks, promoting climate smart objectives, improving climate performance and accounting for their own actions (World Bank 2015a; Institute for Climate Economics 2017). The members subscribing to these principles had grown to 44 as of January 2020. Arguably, it is only through closer linkages between climate and development that significant inroads can be made in addressing climate change. MDBs can play a major role through the totality of their portfolios (Larsen et al. 2018).

The MDBs as a cohort have been collaborating and coordinating in reporting on climate financing following a commitment made in 2012 at the UN Conference on Sustainable Development in Rio de Janeiro (Inter-American Development Bank 2012). This has engendered other forms of collaboration among the MDBs, including, commitments to: collectively total at least USD65 billion annually by 2025 in climate finance, with USD50 billion for low- and middle-income economies; to mobilise a further USD40 billion annually by 2025 from private sector investors, including through the increased provision of technical assistance, use of guarantees, and other de-risking instruments; to help clients deliver on the goals of the Paris Agreement; to build a transparency framework on the impact of MDBs' activities; and to enable clients to move away from fossil fuels (Asian Development Bank 2019). While the share of MDBs in direct climate financing is small, their role in influencing national development banks and local financial institutions, and leveraging and crowding in private investments in financing sustainable infrastructure, is widely recognised (NCE 2016). However, with this recognition there is also an exhortation to do more to align with the goals of the Paris Agreement, including a comprehensive examination of their portfolios beyond investments that directly support climate action to also enabling the long-term net zero GHG emissions trajectory (Larsen et al. 2018; Cochran and Pauthier 2019). Further, a recent assessment has shown that MDBs perform relatively better in mobilising other public finance than private co-financing (Thwaites 2020). In addition, the banks have launched or are members of significant initiatives such as the Climate and Clean Air Coalition to reduce emissions of shortlived climate pollutants, the Carbon Pricing Leadership Coalition, the Coalition for Climate Resilient Investment and the Coalition of Finance Ministers for Climate Action. These help to spur action at different levels, from economic analysis to carbon financing, and convenors of finance and development ministers for climate action, with leadership of many of these initiatives led by the World Bank.

The multilateral climate funds also have a role in the international climate finance architecture. This includes, as mentioned in Section 14.3.2.8, those established under the UNFCCC's financial mechanism, its operating entities, the Global Environment Facility (GEF), which also manages two special funds, the Special Climate Change Fund and the Least Developed Countries Fund; and the Green Climate Fund (GCF), also an operating entity of the financial mechanism which in 2015, was given a special role in supporting the Paris Agreement. The GCF aims to provide funding at scale, balanced between mitigation and adaptation, using various financial instruments including grants, loans, equity, guarantees or others to activities that are aligned with the priorities of the countries compatible with the principle of country ownership (GCF 2011). The GCF faces many challenges. While some see the GCF as an opportunity to transform and rationalise what is now a complex and fragmented climate finance architecture with insufficient resources and overlapping remits (Nakhooda et al. 2014), others see it as an opportunity to address the frequent tensions which arise between mitigation-focused transformation and national priorities of countries. This tension is at the heart of the principle of country ownership and the need for transformational change (Winkler and Dubash 2016). Leveraging private funds and investments by the public sector and taking risks to unlock climate action are also expressed strategic aims of the GCF.

The UN system is also supporting climate action through much-needed technical assistance and capacity building, which is complementary to the financial flows insofar as it enables countries with relevant tools and methodologies to assess their needs, develop national climate finance roadmaps, establish relevant institutional mechanisms to receive support and track it, enhance readiness to access financing, and include climate action across relevant national financial planning and budgeting processes (UN 2017a). The United Nations Development Programme (UNDP) is the largest implementer of climate action among the UN Agencies, with others, such as the Food and Agriculture Organization (FAO), United Nations Environment Programme (UNEP), United Nations Industrial Development Organisation (UNIDO), and United Nations Office for South-South Cooperation (UNOSSC), providing relevant support.

The current architecture of climate finance is one that is primarily based on north-south, developed-developing country dichotomies. The Paris Agreement, however, has clearly recognised the role of climate finance flows across developing countries, thereby enhancing the scope of international cooperation (Voigt and Ferreira 2016b). Estimates of such flows, though, are not readily available. According to one estimate in 2020 the flows among non-OECD countries were of the order of USD29 billion (CPI 2021).

#### 14.4.1.3 Private Sector Financing

There is a growing recognition of the importance of mobilising private sector financing including for climate action (World Bank 2015b; Michaelowa et al. 2020b). An early example of the mobilisation of the private sector in a cooperative mode for mitigation outcomes is evidenced from the Clean Development Mechanism of the Kyoto Protocol and the linking with the European Union's Emissions Trading System, both triggered by relevant provisions in the Kyoto Protocol (Section 14.4.4) and lessons learned from this are relevant for development of market mechanisms in the post Paris Agreement period (Michaelowa et al. 2019b). In 2019 and 2020, on average for the two years, public and private climate financing was on the order of USD632 billion, of which USD310 billion originated from the private sector. However, as much as 76% of the (overall) finance stayed in the country of origin. This trends holds true also for private finance (CPI 2021). Figure 14.4 depicts the international climate finance flows totalling USD161 billion reported in 2020, about 19% of which were private flows. For (international) mitigation financing flows of USD116 billion, the share provided by private sources was 24%.

Foreign direct investments and their greening are seen as a channel for increasing cooperation. An assessment of the greenfield foreign direct investment in different sectors shows the growing share of renewable energy at USD92.2 billion (12% of the volume and 38% of the number of projects) (FDI Intelligence 2020). Coal, oil and gas sectors maintain the top spot for capital investments globally. Over the last decade there is growing issuance of green bonds with non-financial private sector issuance gaining ground (Almeida 2020). While it is questionable if green bonds have a significant impact on shifting capital from non-sustainable to sustainable investments, they do incentivise the issuing organisations to enhance their green ambition and have led to an appreciation within capital markets of

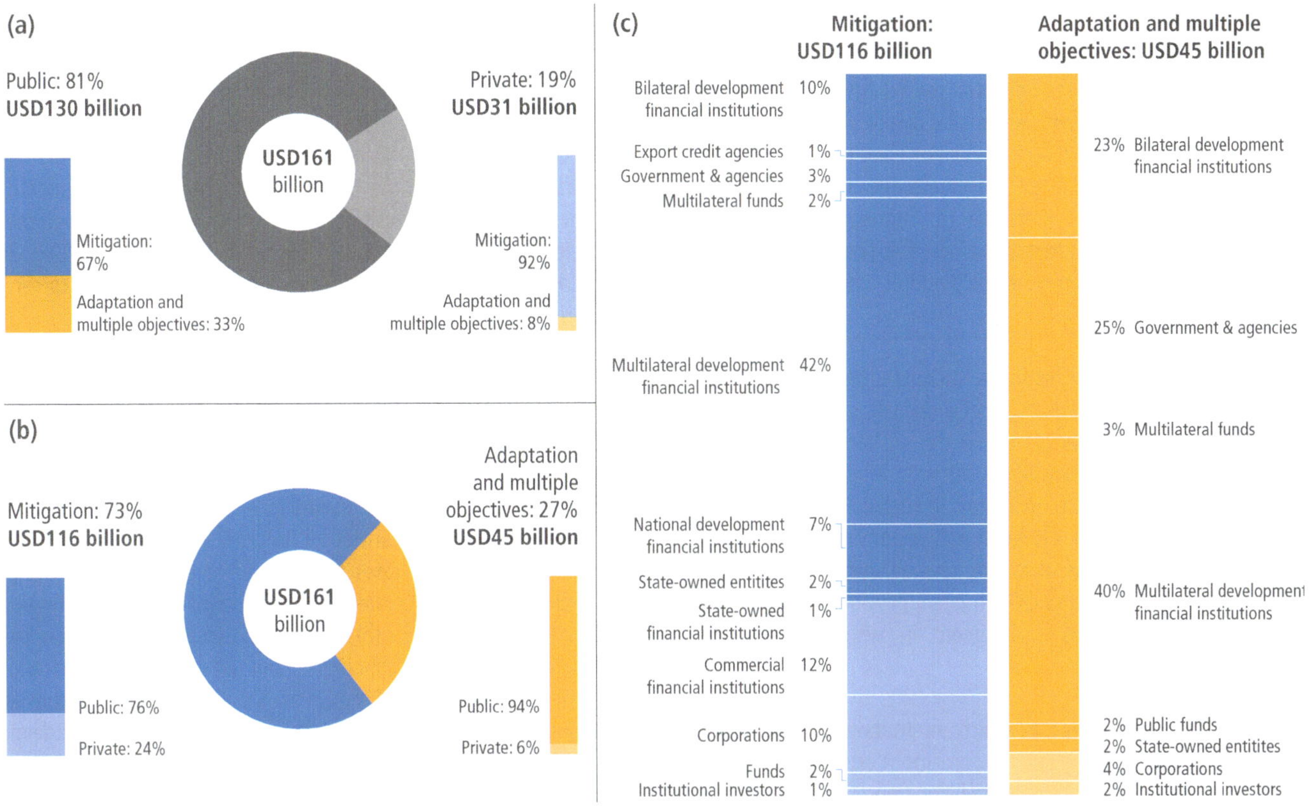


**Figure 14.4 | International finance flows.** Total international climate financial flows for 2020 were USD161 billion. By comparison, public sector bilateral and multilateral finance in 2017 for fossil fuel development, including gas pipelines, was roughly USD4 billion. Part (a) disaggregates total financial flows according to public and private sources, and indicates the breakdown between mitigation on the one hand, and adaptation and multiple objectives on the other, within each source. Part (b) disaggregates total financial flows according to intended purpose, namely mitigation or adaptation and multiple objectives, and disaggregates each type according to source. Part (c) provides additional detail on the relative contributions of different public and private sources. Sources: data from CPI 2021; OECD 2021.

green frameworks and guidelines and signalled new expectations (Maltais and Nykvist 2020). In parallel, institutional investors including pension funds are seeking investments that align with the Paris Agreement (IIGCC 2020). However, the readiness of institutional investors to make this transition is arguable (OECD 2019b; Ameli et al. 2020). This evidence suggests that international private financing could play an important role but this potential is yet to be realised (Chapter 15).

### 14.4.2 Science, Technology and Innovation

Science, technology and innovation are essential for the design of effective measures to address climate change and, more generally, for economic and social development (de Coninck and Sagar 2015a). The OECD finds that single countries alone often cannot provide effective solutions to today's global challenges, as these cross national borders and affect different actors (OECD 2012). Madani (2020) shows how conflict, including international sanctions, can reduce science and innovation capacity, which is not evenly distributed, particularly across the developed and the developing world. For this reason, many countries have introduced strategies and policies to enhance international cooperation in science and technology (Chen et al. 2019). Partnerships and international cooperation can play a role in establishing domestic innovation systems, which enable more effective science and technology innovation (de Coninck and Sagar 2015b,a).

International cooperation in science and technology occurs across different levels, with a growing number of international cooperation initiatives aimed at research and collaborative action in technology development. Weart (2012) finds that such global efforts are effective in advancing climate change science due to the international nature of the challenge. Global research programmes and institutions have also provided the scientific basis for major international environmental treaties. For example, the Long-Range Transboundary Air Pollution Convention and the Montreal Protocol were both informed by scientific assessments based on collaboration and cooperation of scientists across several geographies (Andresen et al. 2000). Furthermore, the Global Energy Assessment (GEA) provided the scientific basis and evidence for the 2030 Agenda for Sustainable Development, in particular SDG 7 to ensure access to affordable, reliable and sustainable modern energy for all (GEA 2012). The GEA drew on the expertise of scientists from over 60 countries and institutions. Several other platforms exist to provide scientists and policymakers an opportunity for joint research and knowledge sharing, such as The World in 2050, an initiative that brings together scientists from some 40 institutions from around the world to provide the science for SDG and Paris Agreement implementation (TWI2050 2018).

Non-state actors are also increasingly collaborating internationally. Such collaborations, referred to as international cooperative initiatives (ICIs), bring together multi-stakeholder groups across industry, communities, and regions, and operate both within and outside the UNFCCC process. Lui et al. (2021) find that such initiatives could make a major contribution to global emissions reduction, Bakhtiari (2018) finds that the impact on greenhouse gas reduction of these initiatives is hindered due to a lack of coordination between ICIs, overlap with other activities conducted by the UNFCCC and governments, and a lack of monitoring systems to measure impact. Increasing the exchange of information between ICIs, enhancing monitoring systems, and increasing collaborative research in science and technology would help address these issues (Boekholt et al. 2009; Bakhtiari 2018).

At the level of research institutes, there has been a major shift to a more structured and global type of cooperation in research; Wagner et al. (2017) found significant increases in both the proportion of papers written by author teams from multiple countries and in the number of countries participating in such collaboration, over the time period 1990–2013. Although only a portion of these scientific papers address the issue of climate change specifically, this growth of scientific collaboration across borders provides a comprehensive view of the conducive environment in which climate science collaboration has grown.

However, there are areas in which international cooperation can be strengthened. Both the Paris Agreement and the 2030 Agenda for Sustainable Development call for more creative forms of international cooperation in science that help bridge the science and policy interface, and provide learning processes and places to deliberate on possible policy pathways across disciplines on a more sustainable and long-lasting basis. Scientific assessments, such as the IPCC and Intergovernmental Science-Policy Platform on Biodiversity and Ecosystem Services (IPBES) offer this possibility, but processes need to be enriched for this to happen more effectively (Kowarsch et al. 2016).

A particular locus for international cooperation on technology development and innovation is found within institutions and mechanisms of the UN climate regime. The UNFCCC, in Article 4.1(c), calls on 'all Parties' to 'promote and cooperate in the development, application and diffusion, including transfer, of technologies, practices and processes that control, reduce or prevent anthropogenic emissions of greenhouse gases' and places responsibility on developed country Parties to 'take all practicable steps to promote, facilitate and finance, as appropriate, the transfer of, or access to environmentally sound technologies and know-how to other Parties, particularly developing country Parties, to enable them to implement the provisions of the Convention' (UNFCCC 1992, Art. 4.5). The issue of technology development and transfer has continued to receive much attention in the international climate policy domain since its initial inclusion in the UNFCCC in 1992 – albeit often overshadowed by dominant discourses around market-based mechanisms – and its role in reducing GHG emissions and adapting to the consequences of climate change 'is seen as becoming ever more critical' (de Coninck and Sagar 2015a). Milestones in the development of international cooperation on climate technologies under the UNFCCC have included: (i) the development of a technology transfer framework and establishment of the Expert Group on Technology Transfer (EGTT) under the SBSTA in 2001; (ii) recommendations for enhancing the technology transfer framework put forward at the Bali COP in 2007 and creation of the Poznan strategic programme on technology transfer under the GEF; and (iii) the establishment of the Technology Mechanism by the COP in 2010 as part of the Cancun Agreements (UNFCCC 2010b). The Technology Mechanism is presently the principal avenue within the UNFCCC for facilitating cooperation on the development and transfer of climate technologies to developing countries (UNFCCC 2015b). As discussed in Section 14.3.2.9 above, the Paris Agreement tasks the Technology Mechanism also to serve the Paris Agreement (UNFCCC 2015b, Art. 10.3).

The Technology Mechanism consists of the Technology Executive Committee (TEC) (replacing the EGTT), as its policy arm, and the Climate Technology Centre and Network (CTCN), as its implementation arm (UNFCCC 2015b). The TEC focuses on identifying and recommending policies that can support countries in enhancing and accelerating the development and transfer of climate technologies (UNFCCC 2020b). The CTCN facilitates the transfer of technologies through three core services: (i) providing technical assistance at the request of developing countries; (ii) creating access to information and knowledge on climate technologies; and (iii) fostering collaboration and capacity building (CTCN 2020a). The CTCN 'network' consists of a diverse set of climate technology stakeholders from academic, finance, non-government, private sector, public sector, and research entities, together with more than 150 National Designated Entities, which serve as CTCN national focal points. Through its network, the CTCN seeks to mobilise policy and technical expertise to deliver technology solutions, capacity-building and implementation advice to developing countries (CTCN 2020b). At the Katowice UNFCCC Conference of the Parties in 2018, the TEC and CTCN were requested to incorporate the technology framework developed pursuant to Article 10 of the Paris Agreement into their respective workplans and programmes of work (UNFCCC 2019f).

The Joint Annual Report of the TEC and CTCN for 2019 indicated that, as of July 2019, the CTCN had engaged with 93 developing country Parties regarding a total of 273 requests for technical assistance, including 11 multi-country requests. Nearly three-quarters (72.9%) of requests received by the CTCN had a mitigation component, with two-thirds of those mitigation requests related to either renewable energy or energy efficiency. Requests for decision-making or information tools are received most frequently (28% of requests), followed by requests for technology feasibility studies (20%) and technology identification and prioritisation (18%) (TEC and CTCN 2019).

The CTCN is presently funded from 'various sources, ranging from the [UNFCCC] Financial Mechanism to philanthropic and private sector sources, as well as by financial and in-kind contributions from the co-hosts of the CTCN and from participants in the Network' (TEC and CTCN 2019, para. 97). Oh (2020b) describes the institution as 'mainly financially dependent on bilateral donations from developed countries and multilateral support'. Nevertheless, inadequate funding of the CTCN poses a problem for its effectiveness and capacity to contribute to implementation of the Paris Agreement. A 2017 independent review of the CTCN identified 'limited availability of funding' as a key constraint on its ability to deliver services at the expected level and recommended that '[b]etter predictability and security over financial resources will ensure that the CTCN can continue to successfully respond to its COP mandate and the needs and expectations of developing countries' (Ernst & Young 2017, para. 84). The 2019 Joint Report of the TEC and CTCN indicates that resource mobilisation for the Network remains a challenge (TEC and CTCN 2019, pp. 23–24).

The importance of 'financial support' for strengthening cooperative action on technology development and transfer was recognised in Article 10.6 of the Paris Agreement. The technology framework established by the Paris Rulebook specifies actions and activities relating to the thematic area of 'support' as including: (i) enhancing the collaboration of the Technology Mechanism with the Financial Mechanism; (ii) identifying and promoting innovative finance and investment at different stages of the technology cycle; (iii) providing enhanced technical support to developing country Parties, in a country-driven manner, and facilitating their access to financing for innovation, enabling environments and capacity building, developing and implementing the results of TNAs, and engagement and collaboration with stakeholders, including organisational and institutional support; and (d) enhancing the mobilisation of various types of support, including pro bono and in-kind support, from various sources for the implementation of actions and activities under each key theme of the technology framework.

Notwithstanding the technology framework's directive for enhanced collaboration of the Technology and Financial Mechanisms of the UNFCCC, linkages between them, and particularly to the GCF, continue to engender political contestation between developing and developed countries (Oh 2020b). Developing countries sought to address concerns over the unsustainable funding status of the CTCN by advocating linkage through a funding arrangement or financial linkage, whereas developed countries favour the design of an institutional linkage maintaining the different and separate mandates of the CTCN and the GCF (Oh 2020a,b). With no resolution reached, the UNFCCC COP requested the Subsidiary Body for Implementation, at its fifty-third session, to take stock of progress in strengthening the linkages between the Technology Mechanism and the Financial Mechanism with a view to recommending a draft decision for consideration and adoption by the Glasgow COP, scheduled for 2021 (UNFCCC 2019l).

### 14.4.3 Capacity Building

International climate cooperation has long focused on supporting developing countries in building capacity to implement climate mitigation actions. While there is no universally agreed definition of capacity building and the UNFCCC does not define the term (Khan et al. 2020), elements of capacity building can be discerned from the Convention's provisions on education and training programmes (UNFCCC 1992, Art. 6), as well as the reference in Article 9(2)(d) to the SBSTA providing support for 'endogenous capacity-building in developing countries'.

Capacity building is generally conceived as taking place at three levels: individual (focused on knowledge, skills and training), organisational/institutional (focusing on organisational performance and institutional cooperation) and systemic (creating enabling environments through regulatory and economic policies (Khan et al. 2020; UNFCCC 2021b). In its annual synthesis report for 2018, the UNFCCC secretariat compiled information submitted by Parties on the implementation of capacity building in developing countries, highlighting cooperative and regional activities on NDCs, including projects to build capacity for implementation, workshops related to transparency under the Paris Agreement and collaboration to provide coaching and training (UNFCCC 2019h). A number of developing country Parties also highlighted their contributions to South–South cooperation (discussed further in Section 14.5.1.4), and identified capacity-building projects undertaken with others (e.g., capacity-building for risk management in Latin America and the Caribbean, improving capacity for measurement, reporting and verification through the Alliance of the Pacific and a climate action package launched by Singapore).

Beyond the UNFCCC, other climate cooperation and partnership activities on capacity building include climate-related bilateral cooperation and those organised by the OECD, IFDD (Francophonie Institute for Sustainable Development), UNDP National Communications Support Programme, UNEP and the World Bank.

Climate-related bilateral cooperation provides important human and institutional capacity building support for climate change actions and activities in developing countries, particularly through developed countries' bilateral cooperation structures, such as the French Development Agency (AFD), the German Development Agency (The Deutsche Gesellschaft für Internationale Zusammenarbeit – GIZ), the Japanese International Cooperation Agency (JICA) and others.

There are also a number of regional cooperative structures with capacity-building components, including ClimaSouth, Euroclima+, the UN-REDD Programme, the Caribbean Regional Strategic Programme for Resilience, the Caribbean Climate Online Risk and Adaptation Tool, a project on accelerating low carbon and resilient society realisation in the Southeast Asian region, the World Health Organisation's Global Salm-Surv network, the Red Iberoamericana de Oficinas de Cambio Climático network and the Africa Adaptation Initiative. Many climate-related capacity-building initiatives, including those coordinated or funded by international or regional institutions, are implemented at the national and sub-national levels, often with the involvement of universities, consultancy groups and civil society actors.

It is also noted that comprehensive support is provided by the GCF to developing countries (GCF, 2020). This support is made available and accessible for all developing countries through three different GCF tools: the Readiness Programme, the Project Preparation Facility, and the funding of transformative projects and programmes. The goal of the Readiness Programme is to strengthen institutional capacities, governance mechanisms, and planning and programming competencies in support of developing countries' transformational long-term climate policies (GCF, 2020). Despite a decades-long process of capacity-building efforts under many development and environmental regimes, including the UNFCCC, progress has been uneven and largely unsuccessful in establishing institution-based capacity in developing countries (Robinson 2018). In an effort to improve capacity-building efforts within the UNFCCC, in 2015, the Paris Committee on Capacity-building (PCCB) was established by the COP decision accompanying the Paris Agreement as the primary body for enhancing capacity-building efforts, including by improving coherence and coordination in capacity-building activities (UNFCCC 2016a, para. 71). The activities of the Committee include the provision of guidance and technical support on climate change training and capacity building, raising awareness and sharing climate information and knowledge. During 2020, the PCCB was able, despite the COVID-19 situation, to hold its fourth meeting, implement and assess its 2017–2020 work plan, and develop and agree on its future roadmap (2021–2024) (UNFCCC Subsidiary Body for Implementation 2020). Non-governmental organisations such as the Coalition on Paris Agreement Capacity-building provide expert input to the PCCB.

Quantifying the contribution of capacity-building efforts to climate mitigation is acknowledged to be 'difficult, if not impossible' (Hsu et al. 2019a). Nonetheless, such activities 'may play a valuable role in building a foundation for future reductions' by providing 'necessary catalytic linkages between actors' (Hsu et al. 2019a).

### 14.4.4 Cooperative Mechanisms and Markets

In theory, trading carbon assets can reduce the costs of global climate mitigation, by helping facilitate abatement of greenhouse gases at least-cost locations. This could help countries ratchet up their ambitions more than in a situation without such mechanisms (Mehling et al. 2018), particularly if mechanisms are scaled up from projects and programmes (Michaelowa et al. 2019b). Progress as to developing such mechanisms has however so far been moderate and uneven.

Of the three international market-based mechanisms under the 1997 Kyoto Protocol discussed in Section 14.3.2.7, and in previous IPCC reports, only the CDM or a similar mechanism may have a role to play under the Paris Agreement, although the precise terms are yet to be decided.

Article 6, also discussed in Section 14.3.2.7, is the main framework to foster enhanced cooperation within the Paris Agreement. Although there is an emerging global landscape of activities based on Article 6 (Greiner et al. 2020), such as the bilateral treaty signed under the framework of Article 6 in October 2020 by Switzerland and Peru, the possibilities of bilateral cooperation are yet to be fully exploited. As discussed above, adequate accounting rules are key to the success of Article 6. Sectoral agreements are also a promising cooperative mechanism, as discussed in Section 14.5.2. In fact, both bilateral and sectoral agreements have the potential to enhance the ambition of the Parties involved and can eventually serve as building blocks towards more comprehensive agreements (Section 14.2.2).

A relevant and promising new development is the international linkage of existing regional or national emissions trading systems (ETS). Several ETS are now operational in different jurisdictions, including the EU, Switzerland, China, South Korea, New Zealand, Kazakhstan and several US states and Canadian provinces (Wettestad and Gulbrandsen 2018). More systems are in the pipeline, including Mexico and Thailand (ICAP 2019). The link between the EU and Switzerland entered into force in January 2020 and other linkages are being negotiated. Scholars analyse the potential benefits of these multilateral linkages and demonstrate that these can be significant (Doda et al. 2019; Doda and Taschini 2017). Over time, the linkages of national emissions trading systems can be seen as building blocks to a strategic enlargement of international cooperation (Caparrós and Péreau 2017; Mehling 2019). The World Bank has emerged as an important lynchpin and facilitator of knowledge-building and sharing of lessons about the design and linking of carbon markets, through initiatives such as the Partnership for Market Readiness, Networked Carbon Markets and the Carbon Pricing Leadership Coalition (Wettestad et al. 2021).

However, it is important to distinguish between theory and practice. The practice of ETS linking so far demonstrates a few attempts that did not result in linkages due to shifts of governments and political preferences (for instance the process between the EU and Australia, and Ontario withdrawing from the Western Climate Initiative) (Bailey and Inderberg 2018). It is worth noting that the linking of carbon markets raises problems of distribution of costs and loss of political control and hence does not offer a politically easy alternative route to a truly international carbon market. Careful, piecemeal and incremental linking may be the most feasible approach forward (Green et al. 2014; Gulbrandsen et al. 2019). It is premature for any serious assessment of the practice of ETS linking to be conducted. Environmental effectiveness, transformative potential, economic performance, institutional strength and even distributional outcomes can potentially be significant and positive if linking is done carefully (Doda and Taschini 2017; Mehling et al. 2018; Doda et al. 2019), but are all marginal if one focuses on existing experiences (Spalding-Fecher et al. 2012; Haites 2016; Schneider et al. 2017; La Hoz Theuer et al. 2019; Schneider et al. 2019).

### 14.4.5 International Governance of SRM and CDR

While Solar Radiation Modification (SRM) and carbon dioxide removal (CDR) were often referred to as 'geoengineering' in earlier IPCC reports and in the literature, IPCC SR1.5 started to explore SRM and CDR more thoroughly and to highlight the differences between – but also within – both approaches more clearly. This section assesses international governance of both SRM and CDR, recognising that CDR, as a mitigation option, is covered elsewhere in this report, whereas SRM is not. Chapter 12 of this report covers the emerging national, sub-national and non-state governance of CDR, while Chapters 6, 7 and 12 also assess the mitigation potential, risks and co-benefits of some CDR options. Chapters 4 and 5 of AR6 WGI assess the physical climate system and biogeochemical responses to different SRM and CDR methods. Cross-Working Group Box 4 on SRM (AR6 WGII, Chapter 16; and Cross-Working Group Box 4 in this chapter) gives a brief overview of Solar Radiation Modification methods, risks, benefits, ethics and governance.

## Cross-Working Group Box 4 | Solar Radiation Modification

**Authors:** Govindasamy Bala (India), Heleen de Coninck (the Netherlands), Oliver Geden (Germany), Veronika Ginzburg (the Russian Federation), Katharine J. Mach (the United States of America), Anthony Patt (Switzerland), Sonia I. Seneviratne (Switzerland), Masahiro Sugiyama (Japan), Christopher H. Trisos (South Africa), Maarten van Aalst (the Netherlands)

### Proposed Solar Radiation Modification schemes

This cross-working group box assesses Solar Radiation Modification (SRM) proposals, their potential contribution to reducing or increasing climate risk, as well as other risks they may pose (categorised as risks from responses to climate change in the IPCC AR6 risk definition in 1.2.1.1), and related perception, ethics and governance questions.

SRM refers to proposals to increase the reflection of shortwave radiation (sunlight) back to space to counteract anthropogenic warming and some of its harmful impacts (de Coninck et al. 2018) (AR6 WGI Chapters 4 and 5). A number of SRM options have been proposed, including: stratospheric aerosol interventions (SAI), marine cloud brightening (MCB), ground-based albedo modifications (GBAM), and ocean albedo change (OAC). Although not strictly a form of SRM, cirrus cloud thinning (CCT) has been proposed to cool the planet by increasing the escape of longwave thermal radiation to space and is included here for consistency with previous assessments (de Coninck et al. 2018). SAI is the most-researched proposal. Modelling studies show SRM could reduce surface temperatures and potentially ameliorate some climate change risks (with more confidence for SAI than other options), but SRM could also introduce a range of new risks.

There is high agreement in the literature that for addressing climate change risks, SRM cannot be the main policy response to climate change and is, at best, a supplement to achieving sustained net zero or net negative $CO_2$ emission levels globally (de Coninck et al. 2018; MacMartin et al. 2018; Buck et al. 2020; National Academies of Sciences Engineering and Medecine 2021). SRM contrasts with climate change mitigation activities, such as emissions reductions and CDR, as it introduces a 'mask' to the climate change problem by altering the Earth's radiation budget, rather than attempting to address the root cause of the problem, which is the increase in GHGs in the atmosphere. In addition, the effects of proposed SRM options would only last as long as a deployment is maintained – for example, requiring a yearly injection of aerosols in the case of SAI as the lifetime of aerosols in the stratosphere is one to three years (Niemeier et al. 2011) or continuous spraying of sea salt in the case of MCB as the lifetime of sea salt aerosols in the atmosphere is only about 10 days – which contrasts with the long lifetime of $CO_2$ and its climate effects, with global warming resulting from $CO_2$ emissions likely remaining at a similar level for a hundred years or more (MacDougall et al. 2020) and long-term climate effects of emitted $CO_2$ remaining for several hundreds to thousands of years (Solomon et al. 2009).

### Which scenarios?

The choice of SRM deployment scenarios and reference scenarios is crucial in assessment of SRM risks and its effectiveness in attenuating climate change risks (Keith and MacMartin 2015; Honegger et al. 2021a). Most climate model simulations have used scenarios with highly stylised large SRM forcing to fully counteract large amounts of warming in order to enhance the signal-to-noise ratio of climate responses to SRM (Kravitz et al. 2015; Sugiyama et al. 2018a; Krishnamohan et al. 2019).

The effects of SRM fundamentally depend on a variety of choices about deployment (Sugiyama et al. 2018b), including: its position in the portfolio of human responses to climate change (e.g., the magnitude of SRM used against the background radiative forcing), governance of research and potential deployment strategies, and technical details (latitude, materials, and season, among others, see AR6 WGI Chapter 4.6.3.3). The plausibility of many SRM scenarios is highly contested and not all scenarios are equally plausible because of socio-political considerations (Talberg et al. 2018), as with, for example, CDR (Fuss et al. 2014, 2018). Development of scenarios and their selection in assessments should reflect a diverse set of societal values with public and stakeholder inputs (Sugiyama et al. 2018a; Low and Honegger 2020), as depending on the focus of a limited climate model simulation, SRM could look grossly risky or highly beneficial (Pereira et al. 2021).

In the context of reaching the long-term global temperature goal of the Paris Agreement, there are different hypothetical scenarios of SRM deployment: early, substantial mitigation with no SRM, more limited or delayed mitigation with moderate SRM, unchecked emissions with total reliance on SRM, and regionally heterogeneous SRM. Each scenario presents different levels and distributions of SRM benefits, side effects, and risks. The more intense the SRM deployment, the larger is the likelihood for the risks of side effects and environmental risks (e.g., Heutel et al., 2018). Regional disparities in climate hazards may result from both regionally-deployed SRM options such as GBAM, and more globally uniform SRM such as SAI (Jones et al. 2018; Seneviratne et al. 2018). There is an emerging literature on smaller forcings of SAI to reduce global average warming, for instance, to hold global warming to 1.5°C or 2°C alongside ambitious conventional mitigation (Jones et al. 2018; MacMartin et al. 2018), or bring down temperature after an overshoot

*Cross-Working Group Box 4 (continued)*

(Tilmes et al. 2020). If emissions reductions and CDR are deemed insufficient, SRM may be seen by some as the only option left to ensure the achievement of the Paris Agreement's temperature goal by 2100.

### SRM risks to human and natural systems and potential for risk reduction

Since AR5, hundreds of climate modelling studies have simulated effects of SRM on climate hazards (Kravitz et al. 2015; Tilmes et al. 2018). Modelling studies have shown SRM has the potential to offset some effects of increasing GHGs on the global and regional climate, including the increase in frequency and intensity of extremes of temperature and precipitation, melting of Arctic sea ice and mountain glaciers, weakening of Atlantic meridional overturning circulation, changes in frequency and intensity of tropical cyclones, and decrease in soil moisture (AR6 WGI, Chapter 4). However, while SRM may be effective in alleviating anthropogenic climate

**Cross-Working Group Box 4, Table 1 | SRM options and their potential climate and non-climate impacts. Description, potential climate impacts, potential impacts on human and natural systems, and termination effects of a number of SRM options: stratospheric aerosol interventions (SAI), marine cloud brightening (MCB), ocean albedo change (OAC), ground-based albedo modifications (GBAM), and cirrus cloud thinning (CCT).**

| SRM option | SAI | MCB | OAC | GBAM | CCT |
|---|---|---|---|---|---|
| Description | Injection of reflective aerosol particles directly into the stratosphere or a gas which then converts to aerosols that reflect sunlight | Spraying sea salt or other particles in marine clouds, making them more reflective | Increase surface albedo of the ocean (e.g., by creating microbubbles or placing reflective foam on the surface) | Whitening roofs, changes in land use management (e.g., no-till farming, bioengineering to make crop leaves more reflective), desert albedo enhancement, covering glaciers with reflective sheeting | Seeding to promote nucleation of cirrus clouds, reducing optical thickness and cloud lifetime to allow more outgoing longwave radiation to escape to space |
| Potential climate impacts *other than reduced warming* | Change precipitation and runoff pattern; reduced temperature and precipitation extremes; precipitation reduction in some monsoon regions; decrease in direct and increase in diffuse sunlight at surface; changes to stratospheric dynamics and chemistry; potential delay in ozone hole recovery; changes in surface ozone and UV radiation | Change in land–sea contrast in temperature and precipitation, regional precipitation and runoff changes | Change in land–sea contrast in temperature and precipitation, regional precipitation and runoff changes | Changes in regional precipitation pattern, regional extremes and regional circulation | Changes in temperature and precipitation pattern, altered regional water cycle, increase in sunlight reaching the surface |
| Potential impacts on human and natural systems | Changes in crop yields, changes in land and ocean ecosystem productivity, acid rain (if using sulphate), reduced risk of heat stress to corals | Changes in regional ocean productivity, changes in crop yields, reduced heat stress for corals, changes in ecosystem productivity on land, sea salt deposition over land | Unresearched | Altered photosynthesis and carbon uptake and side effects on biodiversity | Altered photosynthesis and carbon uptake |
| Termination effects | Sudden and sustained termination would result in rapid warming, and abrupt changes to water cycle. Magnitude of termination depends on the degree of warming offset. | Sudden and sustained termination would result in rapid warming, and abrupt changes to water cycle. Magnitude of termination depends on the degree of warming offset. | Sudden and sustained termination would result in rapid warming. Magnitude of termination depends on the degree of warming offset. | GBAM can be maintained over several years without major termination effects because of its regional scale of application. Magnitude of termination depends on the degree of warming offset. | Sudden and sustained termination would result in rapid warming. Magnitude of termination depends on the degree of warming offset. |
| References (also see main text of this box) | Visioni et al. (2017)<br>Tilmes et al. (2018)<br>Simpson et al. (2019) | Latham et al. (2012)<br>Ahlm et al. (2017)<br>Stjern et al. (2018) | Evans et al. (2010)<br>Crook et al. (2015) | Davin et al. (2014)<br>Crook et al. (2015)<br>Zhang et al. (2016)<br>Field et al. (2018)<br>Seneviratne et al. (2018) | Storelvmo and Herger (2014)<br>Crook et al. (2015)<br>Jackson et al. (2016)<br>Duan et al. (2020)<br>Gasparini et al. (2020) |

*Cross-Working Group Box 4 (continued)*

warming either locally or globally, it would not maintain the climate in a present-day state nor return the climate to a pre-industrial state (climate averaged over 1850–1900) (AR6 WGI, Box 1.2) in all regions and in all seasons even when used to fully offset the global mean warming (*high confidence*) (AR6 WGI Chapter 4). This is because the climate forcing and response to SRM options are different from the forcing and response to GHG increase. Because of these differences in climate forcing and response patterns, the regional and seasonal climates of a world with a global mean warming of 1.5°C or 2°C achieved via SRM would be different from a world with similar global mean warming but achieved through mitigation (MacMartin et al. 2018). At the regional scale and seasonal timescale there could be considerable residual climate change and/or overcompensating change (e.g., more cooling, wetting or drying than just what's needed to offset warming, drying or wetting due to anthropogenic greenhouse gas emissions), and there is *low confidence* in understanding of the climate response to SRM at the regional scale (AR6 WGI, Chapter 4).

SAI implemented to partially offset warming (e.g., offsetting half of global warming) may have potential to ameliorate hazards in multiple regions and reduce negative residual change, such as drying compared to present-day climate, that are associated with fully offsetting global mean warming (Irvine and Keith 2020), but may also increase flood and drought risk in Europe compared to unmitigated warming (Jones et al. 2021). Recent modelling studies suggest it is conceptually possible to meet multiple climate objectives through optimally designed SRM strategies (WGI, Chapter 4). Nevertheless, large uncertainties still exist for climate processes associated with SRM options (e.g., aerosol-cloud-radiation interaction) (AR6 WGI, Chapter 4) (Kravitz and MacMartin 2020).

Compared with climate hazards, many fewer studies have examined SRM risks – the potential adverse consequences to people and ecosystems from the combination of climate hazards, exposure and vulnerability – or the potential for SRM to reduce risk (Curry et al. 2014; Irvine et al. 2017). Risk analyses have often used inputs from climate models forced with stylised representations of SRM, such as dimming the sun. Fewer have used inputs from climate models that explicitly simulated injection of gases or aerosols into the atmosphere, which include more complex cloud-radiative feedbacks. Most studies have used scenarios where SAI is deployed to hold average global temperature constant despite high emissions.

There is *low confidence* and large uncertainty in projected impacts of SRM on crop yields due in part to a limited number of studies. Because SRM would result in only a slight reduction in $CO_2$ concentrations relative to the emissions scenario without SRM (AR6 WGI, Chapter 5), the $CO_2$ fertilisation effect on plant productivity is nearly the same in emissions scenarios with and without SRM. Nevertheless, changes in climate due to SRM are likely to have some impacts on crop yields. A single study indicates MCB may reduce crop failure rates compared to climate change from a doubling of $CO_2$ pre-industrial concentrations (Parkes et al. 2015). Models suggest SAI cooling would reduce crop productivity at higher latitudes compared to a scenario without SRM by reducing the growing season length, but benefit crop productivity in lower latitudes by reducing heat stress (Pongratz et al. 2012; Xia et al. 2014; Zhan et al. 2019). Crop productivity is also projected to be reduced where SAI reduces rainfall relative to the scenario without SRM, including a case where reduced Asian summer monsoon rainfall causes a reduction in groundnut yields (Xia et al. 2014; Yang et al. 2016). SAI will increase the fraction of diffuse sunlight, which is projected to increase photosynthesis in forested canopy, but will reduce the direct and total available sunlight, which tends to reduce photosynthesis. As total sunlight is reduced, there is a net reduction in crop photosynthesis with the result that any benefits to crops from avoided heat stress may be offset by reduced photosynthesis, as indicated by a single statistical modelling study (Proctor et al. 2018). SAI would reduce average surface ozone concentration (Xia et al. 2017) mainly as a result of aerosol-induced reduction in stratospheric ozone in polar regions, resulting in reduced downward transport of ozone to the troposphere (Pitari et al. 2014; Tilmes et al. 2018). The reduction in stratospheric ozone also allows more UV radiation to reach the surface. The reduction in surface ozone, together with an increase in surface UV radiation, would have important implications for crop yields but there is *low confidence* in our understanding of the net impact.

Few studies have assessed potential SRM impacts on human health and well-being. SAI using sulfate aerosols is projected to deplete the ozone layer, increasing mortality from skin cancer, and SAI could increase particulate matter due to offsetting warming, reduced precipitation and deposition of SAI aerosols, which would increase mortality, but SAI also reduces surface-level ozone exposure, which would reduce mortality from air pollution, with net changes in mortality uncertain and depending on aerosol type and deployment scenario (Effiong and Neitzel 2016; Eastham et al. 2018; Dai et al. 2020). However, these effects may be small compared to changes in risk from infectious disease (e.g., mosquito-borne illnesses) or food security due to SRM influences on climate (Carlson et al. 2022). Using volcanic eruptions as a natural analogue, a sudden implementation of SAI that forced the El Niño–Southern Oscillation (ENSO) system may increase risk of severe cholera outbreaks in Bengal (Trisos et al. 2018; Pinke et al. 2019). Considering only mean annual temperature and precipitation, SAI that stabilises global temperature at its present-day level is projected to reduce income inequality between countries compared to the highest warming pathway (RCP8.5) (Harding et al. 2020). Some integrated assessment model

*Cross-Working Group Box 4 (continued)*

scenarios have included SAI (Arino et al. 2016; Emmerling and Tavoni 2018; Heutel et al. 2018; Helwegen et al. 2019; Rickels et al. 2020) showing the indirect costs and benefits to welfare dominate, since the direct economic cost of SAI itself is expected to be relatively low (Moriyama et al. 2017; Smith and Wagner 2018). There is a general lack of research on the wide scope of potential risk or risk reduction to human health, well-being and sustainable development from SRM and on their distribution across countries and vulnerable groups (Honegger et al. 2021a; Carlson et al. 2022).

SRM may also introduce novel risks for international collaboration and peace. Conflicting temperature preferences between countries may lead to counter-geoengineering measures such as deliberate release of warming agents or destruction of deployment equipment (Parker et al. 2018). Game-theoretic models and laboratory experiments indicate a powerful actor or group with a higher preference for SRM may use SAI to cool the planet beyond what is socially optimal, imposing welfare losses on others although this cooling does not necessarily imply excluded countries would be worse off relative to a world of unmitigated warming (Ricke et al. 2013; Weitzman 2015; Abatayo et al. 2020). In this context, counter-geoengineering may promote international cooperation or lead to large welfare losses (Helwegen et al. 2019; Abatayo et al. 2020).

Cooling caused by SRM would increase the global land and ocean $CO_2$ sinks (*medium confidence*), but this would not stop $CO_2$ from increasing in the atmosphere or affect the resulting ocean acidification under continued anthropogenic emissions (*high confidence*) (AR6 WGI, Chapter 5).

Few studies have assessed potential SRM impacts on ecosystems. SAI and MCB may reduce risk of coral reef bleaching compared to global warming with no SAI (Latham et al. 2013; Kwiatkowski et al. 2015), but risks to marine life from ocean acidification would remain, because SRM proposals do not reduce elevated anthropogenic atmospheric $CO_2$ concentrations. MCB could cause changes in marine net primary productivity by reducing light availability in deployment regions, with important fishing regions off the west coast of South America showing both large increases and decreases in productivity (Partanen et al. 2016; Keller 2018).

There is large uncertainty in terrestrial ecosystem responses to SRM. By decoupling increases in atmospheric greenhouse gas concentrations and temperature, SAI could generate substantial impacts on large-scale biogeochemical cycles, with feedbacks to regional and global climate variability and change (Zarnetske et al. 2021). Compared to a high $CO_2$ world without SRM, global-scale SRM simulations indicate reducing heat stress in low latitudes would increase plant productivity, but cooling would also slow down the process of nitrogen mineralisation, which could decrease plant productivity (Glienke et al. 2015; Duan et al. 2020). In high latitude and polar regions SRM may limit vegetation growth compared to a high $CO_2$ world without SRM, but net primary productivity may still be higher than pre-industrial climate (Glienke et al. 2015). Tropical forests cycle more carbon and water than other terrestrial biomes but large areas of the tropics may tip between savanna and tropical forest depending on rainfall and fire (Beer et al. 2010; Staver et al. 2011). Thus, SAI-induced reductions in precipitation in Amazonia and central Africa are expected to change the biogeography of tropical ecosystems in ways different both from present-day climate and global warming without SAI (Simpson et al. 2019; Zarnetske et al. 2021). This would have potentially large consequences for ecosystem services (AR6 WGII, Chapters 2 and 9). When designing and evaluating SAI scenarios, biome-specific responses need to be considered if SAI approaches are to benefit rather than harm ecosystems. Regional precipitation change and sea salt deposition over land from MCB may increase or decrease primary productivity in tropical rainforests (Muri et al. 2015). SRM that fully offsets warming could reduce the dispersal velocity required for species to track shifting temperature niches whereas partially offsetting warming with SAI would not reduce this risk unless rates of warming were also reduced (Trisos et al. 2018; Dagon and Schrag 2019). SAI may reduce high fire-risk weather in Australia, Europe and parts of the Americas, compared to global warming without SAI (Burton et al. 2018). Yet SAI using sulphur injection could shift the spatial distribution of acid-induced aluminium soil toxicity into relatively undisturbed ecosystems in Europe and North America (Visioni et al. 2020). For the same amount of global mean cooling, SAI, MCB, and CCT would have different effects on gross and net primary productivity because of different spatial patterns of temperature, available sunlight, and hydrological cycle changes (Duan et al. 2020). Large-scale modification of land surfaces for GBAM may have strong trade-offs with biodiversity and other ecosystem services, including food security (Seneviratne et al. 2018). Although existing studies indicate SRM will have widespread impacts on ecosystems, risks and potential for risk reduction for marine and terrestrial ecosystems and biodiversity remain largely unknown.

*Cross-Working Group Box 4 (continued)*

A sudden and sustained termination of SRM in a high $CO_2$ emissions scenario would cause rapid climate change (*high confidence*) (AR6 WGI, Chapter 4). More scenario analysis is needed on the potential likelihood of sudden termination (Kosugi 2013; Irvine and Keith 2020). A gradual phase-out of SRM combined with emissions reduction and CDR could avoid these termination effects (*medium confidence*) (MacMartin et al. 2014; Keith and MacMartin 2015; Tilmes et al. 2016). Several studies find that large and extremely rapid warming and abrupt changes to the water cycle would occur within a decade if a sudden termination of SAI occurred (McCusker et al. 2014; Crook et al. 2015). The size of this 'termination shock' is proportional to the amount of radiative forcing being masked by SAI. A sudden termination of SAI could place many thousands of species at risk of extinction, because the resulting rapid warming would be too fast for species to track the changing climate (Trisos et al. 2018).

### Public perceptions of SRM

Studies on the public perception of SRM have used multiple methods: questionnaire surveys, workshops, and focus group interviews (Burns et al. 2016; Cummings et al. 2017). Most studies have been limited to Western societies with some exceptions. Studies have repeatedly found that respondents are largely unaware of SRM (Merk et al. 2015). In the context of this general lack of familiarity, the publics prefer carbon dioxide removal (CDR) to SRM (Pidgeon et al. 2012), are very cautious about SRM deployment because of potential environmental side effects and governance concerns, and mostly reject deployment for the foreseeable future. Studies also suggest conditional and reluctant support for research, including proposed field experiments, with conditions of proper governance (Sugiyama et al. 2020). Recent studies show that the perception varies with the intensity of deliberation (Merk et al. 2019), and that the public distinguishes different funding sources (Nelson et al. 2021). Limited studies for developing countries show a tendency for respondents to be more open to SRM (Visschers et al. 2017; Sugiyama et al. 2020), perhaps because they experience climate change more directly (Carr and Yung 2018). In some Anglophone countries, a small portion of the public believes in chemtrail conspiracy theories, which are easily found in social media (Tingley and Wagner 2017; Allgaier 2019). Since researchers rarely distinguish different SRM options in engagement studies, there remains uncertainty in public perception.

### Ethics

There is broad literature on ethical considerations around SRM, mainly stemming from philosophy or political theory, and mainly focused on SAI (Flegal et al. 2019). There is concern that publicly debating, researching and potentially deploying SAI could involve a 'moral hazard', with potential to obstruct ongoing and future mitigation efforts (Morrow 2014; Baatz 2016; McLaren 2016), while empirical evidence is limited and mostly at the individual, not societal, level (Burns et al. 2016; Merk et al. 2016; Merk et al. 2019). There is low agreement whether research and outdoors experimentation will create a 'slippery slope' toward eventual deployment, leading to a lock-in to long-term SRM, or whether it can be effectively regulated at a later stage to avoid undesirable outcomes (Hulme 2014; Parker 2014; Callies 2019; McKinnon 2019). Regarding potential deployment of SRM, procedural, distributive and recognitional conceptions of justice are being explored (Svoboda and Irvine 2014; Svoboda 2017; Preston and Carr 2018; Hourdequin 2019). With the SRM research community's increasing focus on distributional impacts of SAI, researchers have started more explicitly considering inequality in participation and inclusion of vulnerable countries and marginalised social groups (Flegal and Gupta 2018; Whyte 2018; Táíwò and Talati 2021), including considering stopping research (Stephens and Surprise 2020; National Academies of Sciences Engineering and Medecine 2021). There is recognition that SRM research has been conducted predominantly by a relatively small number of experts in the Global North, and that more can be done to enable participation from diverse peoples and geographies in setting research agendas and research governance priorities, and undertaking research, with initial efforts to this effect (Rahman et al. 2018), noting that unequal power relations in participation could influence SRM research governance and have potential implications for policy (Winickoff et al. 2015; Frumhoff and Stephens 2018; Whyte 2018; Biermann and Möller 2019; McLaren and Corry 2021; National Academies of Sciences Engineering and Medecine 2021; Táíwò and Talati 2021).

### Governance of research and of deployment

Currently, there is no dedicated, formal international SRM governance for research, development, demonstration, or deployment (AR6 WGIII, Chapter 14). Some multilateral agreements – such as the UN Convention on Biological Diversity or the Vienna Convention on the Protection of the Ozone Layer – indirectly and partially cover SRM, but none is comprehensive and the lack of robust and formal SRM governance poses risks (Ricke et al. 2013; Talberg et al. 2018; Reynolds 2019a). While governance objectives range broadly, from prohibition to enabling research and potentially deployment (Sugiyama et al. 2018b; Gupta et al. 2020), there is agreement that SRM governance should cover all interacting stages of research through to any potential, eventual deployment with rules, institutions, and norms (Reynolds 2019b). Accordingly, governance arrangements are co-evolving with respective SRM technologies across the interacting stages of research, development, demonstration, and – potentially – deployment (Rayner et al. 2013; Parker 2014; Parson 2014). Stakeholders are developing governance already in outdoors research; for example, for MCB and OAC experiments on the

*Cross-Working Group Box 4 (continued)*

Great Barrier Reef (McDonald et al. 2019). Co-evolution of governance and SRM research provides a chance for responsibly developing SRM technologies with broader public participation and political legitimacy, guarding against potential risks and harms relevant across a full range of scenarios, and ensuring that SRM is considered only as a part of a broader portfolio of responses to climate change (Stilgoe 2015; Nicholson et al. 2018). For SAI, large-scale outdoor experiments even with low radiative forcing could be transboundary and those with deployment-scale radiative forcing may not be distinguished from deployment, such that MacMartin and Kravitz (2019) argue for continued reliance on modelling until a decision on whether and how to deploy is made, with modelling helping governance development.

#### 14.4.5.1 Global Governance of Solar Radiation Modification and Associated Risks

Solar radiation modification, in the literature also referred to as 'solar geoengineering', refers to the intentional modification of the Earth's shortwave radiative budget, such as by increasing the reflection of sunlight back to space, with the aim of reducing warming. Several SRM options have been proposed, including stratospheric aerosol injection (SAI), marine cloud brightening (MCB), ground-based albedo modifications (GBAM), and ocean albedo change (OAC). SRM has been discussed as a potential response option within a broader climate risk management strategy, as a supplement to emissions reduction, carbon dioxide removal and adaptation (Crutzen 2006; Shepherd 2009; Caldeira and Bala 2017; Buck et al. 2020), for example as a temporary measure to slow the rate of warming (Keith and MacMartin 2015) or address temperature overshoot (MacMartin et al. 2018; Tilmes et al. 2020). SRM assessments of potential benefits and risks still primarily rely on modelling efforts and their underlying scenario assumptions (Sugiyama et al. 2018a), for example in the context of the Geoengineering Model Intercomparison Project GeoMIP6 (Kravitz et al. 2015). Recently, small-scale MCB and OAC experiments started to take place on the Great Barrier Reef (McDonald et al. 2019).

SAI – the most researched SRM method – poses significant international governance challenges since it could potentially be deployed uni- or minilaterally and alter the global mean temperature much faster than any other climate policy measure, at comparatively low direct costs (Parson 2014; Nicholson et al. 2018; Smith and Wagner 2018; Sugiyama et al. 2018b; Reynolds 2019a). While being dependent on the design of deployment systems, both geophysical benefits and adverse effects would potentially be unevenly distributed (AR6 WGI, Chapter 4). Perceived local harm could exacerbate geopolitical conflicts, not least depending on which countries are part of a deployment coalition (Maas and Scheffran 2012; Zürn and Schäfer 2013), but also because immediate attribution of climatic impacts to detected SAI deployment would not be possible. Uncoordinated or poorly researched deployment by a limited number of states, triggered by perceived climate emergencies, could create international tensions (Corry 2017; Lederer and Kreuter 2018). An additional risk is that of rapid temperature rise following an abrupt end of SAI activities (Parker and Irvine 2018; Rabitz 2019).

While there is room for national and even sub-national governance of SAI – for example on research (differentiating indoor from open-air) (Jinnah et al. 2018; Hubert 2020) and public engagement (Bellamy and Lezaun 2017; Flegal et al. 2019) – international governance of SAI faces the challenge that comprehensive institutional architectures designed too far in advance could prove either too restrictive or too permissive in light of subsequent political, institutional, geophysical and technological developments (Sugiyama et al. 2018a; Reynolds 2019a). Views on governance encompass a broad range, from aiming to restrict to wanting to enable research and potentially deployment; in between these poles, other authors stress the operationalisation of the precautionary approach: preventing deployment until specific criteria regarding scientific consensus, impact assessments and governance issues are met (Tedsen and Homann 2013; Wieding et al. 2020). Many scholars suggest that governance arrangements ought to co-evolve with respective SRM technologies (Parker 2014), including that it stay at least one step ahead of research, development, demonstration, and – potentially – deployment (Rayner et al. 2013; Parson 2014). With the modelling community's increasing focus on showing that, and in what ways, SAI could help to minimise climate change impacts in the Global South, the SRM governance literature has come to include considerations of how SAI could contribute to global equity (Horton and Keith 2016; Flegal and Gupta 2018; Hourdequin 2018).

Given that risks and potential benefits of SRM proposals differ substantially and their large-scale deployment is highly speculative, there is a wide array of concrete proposals for near-term anticipatory or adaptive governance. Numerous authors suggest a wide range of governance principles Nicholson et al. (2018) encapsulate most of these in suggesting a list of four: (i) Guard against potential risks and harm; (ii) Enable appropriate research and development of scientific knowledge; (iii) Legitimise any future research or policymaking through active and informed public and expert community engagement; (iv) Ensure that SRM is considered only as a part of a broader, mitigation-centred portfolio of responses to climate change. Regarding international institutionalisation, options range from formal integration into existing UN bodies like the UNFCCC (Nicholson et al. 2018) or the Convention on Biological Diversity (CBD) (Bodle et al. 2014) to the creation of specific, but less formalised global fora (Parson and Ernst 2013) to forms of club governance (Bodansky 2013; Lloyd and Oppenheimer 2014). Recent years have also seen the emergence of transnational non-state actors focusing on SRM governance, primarily expert networks and NGOs (Horton and Koremenos 2020).

Currently, there is no targeted international law relating to SRM, although some multilateral agreements – such as the Convention on Biological Diversity, the UN Convention on the Law of the Sea, the Environmental Modification Convention, and the Vienna Convention on the Protection of the Ozone Layer and its Montreal Protocol – contain provisions applicable to SRM (Bodansky 2013; Jinnah and Nicholson 2019; Reynolds 2019a).

#### 14.4.5.2 Carbon Dioxide Removal

Carbon dioxide removal (CDR) refers to a cluster of technologies, practices, and approaches that remove and sequester carbon dioxide from the ocean and atmosphere and durably store it in geological, terrestrial, or ocean reservoirs, or in products (Table 12.6). In contrast to SRM, CDR does not necessarily impose transboundary risks, except insofar as misleading accounting of its use and deployment could give a false picture of countries' overall mitigation efforts. CDR is clearly a form of climate change mitigation, and as described in Chapter 12 is needed to counterbalance residual GHG emissions that may prove hard to abate (e.g., from industry, aviation or agriculture) in the context of reaching net zero emissions both globally – in the context of Article 4 of the Paris Agreement – and nationally. CDR could also later be used for reducing atmospheric $CO_2$ concentrations by providing net negative emissions at the global level (Fuglestvedt et al. 2018; Bellamy and Geden 2019). Despite the common feature of removing carbon dioxide, technologies like afforestation/reforestation, soil carbon sequestration, bioenergy with carbon capture and storage, direct air capture with carbon storage, enhanced weathering, ocean alkalinity enhancement or ocean fertilisation are very different, as are the governance challenges. Chapter 12 highlights the sustainable development risks associated with land and water use that are connected to the biological approaches to CDR. As a public good which largely lacks incentives to be pursued as a business case, most types of CDR require a suite of dedicated policy instruments that address both near-term needs as well as long-term continuity at scale (Honegger et al. 2021b).

CDR methods other than afforestation/reforestation and soil carbon sequestration have only played a minor role in UNFCCC negotiations so far (Fridahl 2017; Rumpel et al. 2020). To accelerate, and indeed better manage CDR globally, stringent rules and practices regarding emissions accounting, measuring, reporting and verifying and project-based market mechanisms have been proposed (Honegger and Reiner 2018; Mace et al. 2018). Given their historic responsibility, it can be expected that developed countries would carry the main burden of researching, developing, demonstrating and deploying CDR, or finance such projects in other countries (Fyson et al. 2020; Pozo et al. 2020). McLaren et al. (2019) suggest that there is a rationale for separating the international commitments for net negative emissions from those for emissions reductions.

Specific regulations on CDR options have been limited to those posing transboundary risks, namely the use of ocean fertilisation. In a series of separate decisions from 2008 to 2013, Parties to the London Convention and Protocol limited ocean fertilisation activities to only those of a research character, and in 2012 the CBD made a non-legally-binding decision to do the same, further requiring such research activities to be limited scale, and carried out under controlled conditions, until more knowledge is gained to be able to assess the risks (GESAMP 2019; Burns and Corbett 2020). In doing so they have taken a precautionary approach (Sands and Peel, 2018). The London Convention and Protocol has also developed an Assessment Framework for Scientific Research Involving Ocean Fertilisation (London Convention/Protocol 2010) and in 2013 adopted amendments (which are not yet in force) to regulate marine carbon dioxide removal activities, including ocean fertilisation.

## 14.5 Multi-level, Multi-actor Governance

The Paris Agreement sets in place a new framework for international climate policy (Paroussos et al. 2019), which some cite as evidence of 'hybrid multilateralism' (Christoff 2016; Savaresi 2016; Bäckstrand et al. 2017). While a trend of widening involvement of non-state actors was evident prior to conclusion of the Paris Agreement, particularly at UNFCCC COPs, the 'new landscape of international climate cooperation' features an 'intensified interplay between state and non-state actors', including civil society and social movements, business actors, and sub-national or substate actors, such as local governments and cities (Bäckstrand et al. 2017, p. 562). This involvement of other actors beyond states in international climate cooperation is facilitated by the Paris Agreement's 'hybrid climate policy architecture' (Bodansky et al. 2016) (Section 14.3.1.1), which acknowledges the primacy of domestic politics in climate change and invites the mobilisation of international and domestic pressure to make the Agreement effective (Falkner 2016b). In this landscape, there is greater flexibility for more decentralised 'polycentric' forms of climate governance and recognition of the benefits of working in diverse forms and groups to realise global climate mitigation goals (Jordan et al. 2015; Oberthür 2016) (Section 1.9).

Increasing attention has focused on the role of multi-level, multi-actor cooperation among actors, groupings and agreements beyond the UNFCCC climate regime as potential 'building blocks' towards enhanced international action on climate mitigation (Falkner 2016a; Caparrós and Péreau 2017; Potoski 2017; Stewart et al. 2017). This can include agreements on emissions and technologies at the regional or sub-global level, what scholars often refer to as 'climate clubs' (Nordhaus 2015; Hovi et al. 2016; Green 2017; Sprinz et al. 2018). One forum through which such agreements are often discussed, in support of UNFCCC objectives, is high-level meetings of political leaders, such as the G7 and G20 states (Livingston 2016). It also includes cooperation on narrower sets of issues than are found within the Paris Agreement, for instance, other international environmental agreements dealing with a particular subset of GHGs; linkages with, or leveraging of, efforts or agreements in other spheres such as adaptation, human rights or trade; agreements within particular economic sectors; or transnational initiatives involving global cooperative efforts by different types of non-state actors. Cooperative efforts in each of these forums are reviewed in the following sections of the chapter. Section 14.5.1 discusses international cooperation at multiple governance levels (global, sub-global and regional); Section 14.5.2 discusses cooperation with international sectoral agreements and institutions such as in the forestry, energy and transportation sectors; and Sections 14.5.3–14.5.5 discuss transnational cooperation across civil society and social movements, business partnerships and investor coalitions, and between sub-national entities and cities, respectively.

A key idea underpinning this analysis is that decomposition of the larger challenge of climate mitigation into 'smaller units' may facilitate more effective cooperation (Sabel and Victor 2017) and complement cooperation in the UN climate regime (Stewart et al. 2017). However, it is recognised that significant uncertainty remains over the feasibility and costs of these efforts (Sabel and Victor 2017), as well as whether they ultimately strengthen progress on climate mitigation in the multilateral climate arena (Falkner 2016a).

### 14.5.1 International Cooperation at Multiple Governance Levels

#### 14.5.1.1 Role of Other Environmental Agreements

International cooperation on climate change mitigation takes place at multiple governance levels, including under a range of multilateral environmental agreements (MEAs) beyond those of the international climate regime.

The 1987 Montreal Protocol on Substances that Deplete the Ozone Layer (the Montreal Protocol) is the leading example of a non-climate MEA with significant implications for mitigating climate change (Barrett 2008). The Montreal Protocol regulates a number of substances that are both ozone-depleting substances (ODS) and GHGs with a significant global warming potential (GWP), including chlorofluorocarbons, halons and hydrochlorofluorocarbons (HCFCs). As a result, implementation of phase-out requirements for these substances under the Montreal Protocol has made a significant contribution to mitigating climate change (Molina et al. 2009) (Section 9.9.7.1). Velders et al. (2007) found that over the period from 1990 to 2010, the reduction in GWP100-weighted ODS emissions expected with compliance to the provisions of the Montreal Protocol was 8 $GtCO_2$-eq $yr^{-1}$, an amount substantially greater than the first commitment period Kyoto reduction target. Young et al. (2021) suggest that the Montreal Protocol may also be helping to mitigate climate change through avoided decreases in the land carbon sink.

The 2016 Kigali Amendment to the Montreal Protocol applies to the production and consumption of hydrofluorocarbons (HFCs). HFCs, which are widely used as refrigerants (Abas et al. 2018), have a high GWP100 of 14,600 for HFC-23, and are not ODS (Section 9.9.7.1). The Kigali Amendment addresses the risk that the phase-out of HCFCs under the Montreal Protocol and their replacement with HFCs could exacerbate global warming (Akanle 2010; Hurwitz et al. 2016), especially with the predicted growth in HFC usage for applications like air conditioners (Velders et al. 2015). In this way it creates a cooperative rather than a conflictual relationship between addressing ozone depletion and the climate protection goals of the UNFCCC regime (Hoch et al. 2019). The Kigali Amendment requires developed country Parties to phase down HFCs by 85% from 2011 to 2013 levels by 2036. Developing country Parties are permitted longer phase-down periods (out to 2045 and 2047), but must freeze production and consumption between 2024 and 2028 (Ripley and Verkuijl 2016; UN 2016). A ban on trade in HFCs with non-Parties will come into effect from 1 January 2033. For HFC-23, which is a by-product of HCFC production rather than an ODS, Parties are required to report production and consumption data, and to destroy all emissions of HFC-23 occurring as part of HCFCs or HFCs to the extent practicable from 2020 onwards using approved technologies (Ripley and Verkuijl 2016).

Full compliance with the Kigali Amendment is predicted to reduce HFC emissions by 61% of the global baseline by 2050 (Höglund-Isaksson et al. 2017), with avoided global warming in 2100 due to HFCs from a baseline of 0.3°C–0.5°C to less than 0.1°C (WMO 2018). Examining the interplay of the Kigali Amendment with the Paris Agreement, Hoch et al. (2019) show how the Article 6 mechanisms under the Paris Agreement could generate financial incentives for HFC mitigation and related energy efficiency improvements. Early action under Article 6 of the Paris Agreement could drive down baseline levels of HFCs for developing countries (calculated in light of future production and consumption in the early- and mid-2020s) thus generating long-term mitigation benefits under the Kigali Amendment (Hoch et al. 2019). However, achievement of the objectives of the Kigali Amendment is dependent on its ratification by key developed countries, such as the United States, and the provision of funds by developed countries through the Protocol's Multilateral Fund to meet developing countries' agreed incremental costs of implementation (Roberts 2017). The Kigali Amendment came into force on 1 January 2019 and has been ratified by 118 of the 198 Parties to the Montreal Protocol.

MEAs dealing with transboundary air pollution, such as the Convention on Long-Range Transboundary Air Pollution (CLRTAP) and its implementing protocols, which regulate non-GHGs like particulates, nitrogen oxides and ground-level ozone, can also have potential benefits for climate change mitigation (Erickson 2017). Studies have indicated that rigorous air quality controls targeting short-lived climate forcers, like methane, ozone and black carbon, could slow global mean temperature rise by about 0.5°C by mid-century (Schmale et al. 2014). Steps in this direction were taken with 2012 amendments to the CLRTAP Gothenburg Protocol (initially adopted in 1999) to include black carbon, which is an important driver of climate change in the Arctic region (Yamineva and Kulovesi 2018). The amended Protocol, which has 28 Parties including the US and EU, entered into force in October 2019. However, its limits on black carbon have been criticised as insufficiently ambitious in light of scientific assessments (Khan and Kulovesi 2018). There is still a non-negligible uncertainty in the assessment of radiative forcing of each short-lived climate forcer (SLCF), and the results of AR6 WGI have been updated since AR5. For example, the assessment of Emission-based Radiative Forcing from Black Carbon emissions was revised downward in AR6 (AR6 WGI Section 6.4.2). When discussing co-benefits with MEAs related to transboundary air pollution, attention should be paid to the uncertainty in radiative forcing of SLCFs and the update of relevant scientific knowledge.

Another MEA that may play a role in aiding climate change mitigation is the 2013 Minamata Convention on Mercury, which came into force on 16 August 2017. Coal burning for electricity generation represents the second largest source (behind artisanal and small-scale gold mining) of anthropogenic mercury emissions to air (UNEP 2013). Efforts to control and reduce atmospheric emissions of mercury from coal-fired power generation under the Minamata Convention may reduce GHG emissions from this source (Eriksen and Perrez 2014; Selin 2014). For instance, Giang et al. (2015) have modelled the implications of the Minamata Convention for mercury emissions from coal-fired power generation in India and China, concluding that reducing mercury emissions from present-day levels in these countries is likely to require 'avoiding coal consumption and transitioning toward less carbon-intensive energy sources' (Giang et al. 2015). Parties to the Minamata Convention include five of the six top global $CO_2$ emitters – China, the United States, the EU, India and Japan (Russia has not ratified the Convention). The Minamata Convention also establishes an Implementation and Compliance Committee to review compliance with its provisions on a 'facilitative' basis (Eriksen and Perrez 2014).

MEAs that require state Parties to conserve habitat (such as the Convention on Biological Diversity) or to protect certain ecosystems like wetlands (such as the Ramsar Convention on Wetlands of International Importance Especially as Waterfowl Habitat) may also have co-benefits for climate change mitigation through the adoption of well-planned conservation policies (Phelps et al. 2012; Gilroy et al. 2014). At a theoretical level, REDD+ activities have been identified as a particular opportunity for achieving climate mitigation objectives while also conserving tropical forest biodiversity and ecosystem services. Elements of REDD+ that promise greatest effectiveness for climate change mitigation (e.g., greater finance combined with reference levels which reduce leakage by promoting broad participation across countries with both high and low historical deforestation rates) also offer the greatest benefits for biodiversity conservation (Busch et al. 2011). However, actual biodiversity and ecosystem service co-benefits are dependent on the design and implementation of REDD+ programmes (Ehara et al. 2014; Panfil and Harvey 2016), with limited empirical evidence to date of emissions reductions from these programmes (Newton et al. 2016; Johnson et al. 2019), and concerns about whether they meet equity and justice considerations (Schroeder and McDermott 2014) (Section 7.6.1).

#### 14.5.1.2 Linkages with Sustainable Development, Adaptation, Loss and Damage, and Human Rights

As discussed in Chapter 1, the emerging framing for the issue of climate mitigation is that it is no longer to be considered in isolation but rather in the context of its linkages with other areas. Adaptation, loss and damage, human rights and sustainable development are all areas where there are clear or potential overlaps, synergies, and conflicts with the cooperation underway in relation to mitigation.

The IPCC defines adaptation as: 'in human systems, the process of adjustment to actual or expected climate and its effects, in order to moderate harm or exploit beneficial opportunities. In natural systems, the process of adjustment to actual climate and its effect; human intervention may facilitate adjustment to expected climate and its effects' (Annex I: Glossary).

Adaptation involves actions to lessen the harm associated with climate change, or take advantage of potential gains (Smit and Wandel 2006). It can seek to reduce present and future exposure to specific climate risks (Adger et al. 2003), mainstream climate information into existing planning efforts (Gupta et al. 2010; van der Voorn et al. 2012; van der Voorn et al. 2017), and reduce vulnerability (or increase resilience) of people or communities to the effects of climate change (Kasperson and Kasperson 2001). There is a body of literature highlighting potential synergies and conflicts between adaptation actions – in any of the three areas above – and mitigation actions – and potential strategies for resolving them (Locatelli et al. 2011; Casado-Asensio and Steurer 2014; Duguma et al. 2014; Suckall et al. 2015; Watkiss et al. 2015; van der Voorn et al. 2020). In a strategic context, this issue has been analysed in Bayramoglu et al. (2018), Eisenack and Kähler (2016) and Ingham et al. (2013), among others. Bayramoglu et al. (2018) analyse the strategic interaction between mitigation, as a public good, and adaptation, essentially a private good, showing that the fear that adaptation will reduce the incentives to mitigate carbon emissions may not be justified. On the contrary, adaptation can reduce free-rider incentives (lead to larger self-enforcing agreements), yielding higher global mitigation levels and welfare, if adaptation efforts cause mitigation levels between different countries to be complements instead of strategic substitutes (Ingham et al. 2013).

Distinct from project or programmatic level activities, however, international cooperation for adaptation operates to provide finance and technical assistance (Bouwer and Aerts 2006). In most cases it involves transboundary actions, such as in the case of transboundary watershed management (Wilder et al. 2010; Milman et al. 2013; van der Voorn et al. 2017). In others it involves the mainstreaming of climate change projections into existing treaties, such as for the protection of migratory species (Trouwborst et al. 2012).

International cooperation in mitigation and adaptation share many of the same challenges, including the need for effective institutions. The UNFCCC, for example, addresses international financial support for adaptation and for mitigation in the same general category, and subjects them to the same sets of institutional constraints (Peterson and Skovgaard 2019). Sovacool and Linnér (2016) argue that the history of the UNFCCC and its sub-agreements has been shaped by an implicit bargain that developing countries participate in global mitigation policy in return for receiving financial and technical assistance for adaptation and development from industrialised countries and international green funds. Khan and Roberts (2013) contend that this played out poorly under the Kyoto framework: the Protocol's basic architecture, oriented around legally binding commitments, was not amenable to merging the issues of adaptation and mitigation. Kuyper et al. Kuyper et al. (2018a) argue that the movement from the Kyoto Protocol to the Paris Agreement represents a shift in this regard; the Paris Agreement was designed not primarily as a mitigation policy instrument, but rather one encompassing mitigation, adaptation, and development concerns. While this argument suggests that the Paris architecture, involving voluntary mitigation actions and a greater attention to issues of financial support and transparency, functions better to leverage adaptation support into meaningful mitigation actions, there are only few papers that examine this issue. Stua (2017a,b) explores the relevance of the so-called 'share of proceeds' included in Article 6 of the Paris Agreement as a key tool for leveraging adaptation though mitigation actions.

There are recognised limits to adaptation (Dow et al. 2013), and exceeding these limits results in loss and damage, a topic that is gathering salience in the policy discourse. Roberts et al. (2014) focused on 'loss and damage', essentially those climate change impacts which cannot be avoided through adaptation. The Paris Agreement contains a free-standing article on loss and damage (UNFCCC 2015a), focused on cooperation and facilitation, under which Parties have established a clearing house on risk transfer, and a Task Force on Displacement (UNFCCC 2016a). The COP decision accompanying the Paris Agreement specifies that 'Article 8 does not involve or provide a basis for any liability or compensation' (UNFCCC 2016a). There is range of views on the treatment of loss and damage in the Paris Agreement, how responsibility for loss and damage should be allocated (Lees 2017; McNamara and Jackson 2019), and how it could be financed (Roberts et al. 2017; Gewirtzman et al. 2018). Some scholars argue that there are continuing options to pursue compensation and liability in the climate change regime (Mace and Verheyen 2016; Gsottbauer et al. 2018). There have also been efforts to establish accountability of companies – particularly 'carbon majors' – for climate damage in domestic courts (Ganguly et al. 2018; Benjamin 2021). For states that have suffered loss and damage there is also the option to pursue 'state responsibility' claims under customary international law and international human rights law (Wewerinke-Singh 2018; Wewerinke-Singh and Salili 2020).

One scholar argues that climate impacts are 'incremental violence structurally over-determined by international relations of power and control' that affect most those who have contributed the least to GHG emissions (Dehm 2020). Calls for compensation or reparation for loss and damage are therefore a demand for climate justice (Dehm 2020). Many small island states entered declarations on acceptance of the UNFCCC and Paris Agreement that they continue to have rights under international law regarding state responsibility for the adverse effects of climate change, and that no provision in these treaties can be interpreted as derogating from any claims or rights concerning compensation and liability due to the adverse effects of climate change.

The adoption in 2013 of the Warsaw International Mechanism on Loss and Damage as part of the UNFCCC occurred despite the historic opposition of the United States to this policy. Vanhala and Hestbaek (2016) examine the roles of 'frame contestation' (contestations over different framings of loss and damage, whether as 'liability and compensation' or 'risk management and insurance' or other) and ambiguity in accounting for the evolution and institutionalisation of the loss and damage norm within the UNFCCC. However, there is little international agreement on the scope of loss and damage programmes, and especially how they would be funded and by whom (Gewirtzman et al. 2018). Moreover, non-economic loss and damage (NELD) forms a distinct theme that refers to the climate-related losses of items both material and non-material that are not commonly traded in the market, but whose loss is still experienced as such by those affected. Examples of NELD include loss of cultural identity, sacred places, human health and lives (Serdeczny 2019). The Santiago Network is part of the Warsaw International Mechanism, to catalyse the technical assistance of relevant organisations, bodies, networks and experts, for the implementation of relevant approaches to avert, minimise and address loss and damage at the local, national and regional levels, in developing countries that are particularly vulnerable to the adverse effects of climate change (UNFCCC 2020c).

There are direct links between climate mitigation efforts, adaptation and loss and damage – the higher the collective mitigation ambition and the likelihood of achieving it, the lower the scale of adaptation ultimately needed and the lower the scale of loss and damage anticipated. The liability of states, either individually or collectively, for loss and damage is contested, and no litigation has yet been successfully launched to pursue such claims. The science of attribution, however, is developing (Otto et al. 2017; Skeie et al. 2017; Marjanac and Patton 2018; Patton 2021) and while it has the potential to address the thorny issue of causation, and thus compensation (Stuart-Smith et al. 2021), it could also be used to develop strategies for climate resilience (James et al. 2014).

There are also direct links between mitigation and sustainable development. The international agendas for mitigation and sustainable development have shaped each other, around concepts such as 'common but differentiated responsibilities and respective capabilities', as well as the distinction – in the UNFCCC and later the Kyoto Protocol – between Annex I and non-Annex I countries (Victor 2011; Patt 2015). The same implicit bargain that developing countries would support mitigation efforts in return for assistance with respect to adaptation also applies to support for development (Sovacool and Linnér 2016). That linkage between mitigation and sustainable development has become even more specific with the Paris Agreement and the 2030 Agenda for Sustainable Development, each of which explicitly pursues a set of goals that encompass both mitigation and development (Schmieg et al. 2017), reflecting the recognition that achieving sustainable development and climate mitigation goals are mutually dependent (Gomez-Echeverri 2018). It is well accepted that the long-term effects of climate mitigation will benefit sustainable development. A more contested finding is whether the mitigation actions themselves promote or hinder short-term poverty alleviation. One study, analysing the economic effects of developing countries' NDCs, finds that mitigation actions slow down poverty reduction efforts (Campagnolo and Davide 2019). Other studies suggest possible synergies between low-carbon development and economic development (Hanger et al. 2016; Labordena et al. 2017; Dzebo et al. 2019). These studies typically converge on the fact that financial assistance flowing from developed to developing countries enhances any possible synergies or lessens the conflicts. However, mitigation measures can also have negative impacts on gender equality, and peace and justice (Dzebo et al. 2019). The International Monetary Fund (IMF) has also taken on board the climate challenge and is examining the role of fiscal and macroeconomic policies to address the climate challenge for supporting its members with appropriate policy responses.

The literature also identifies institutional synergies at the international level, related to the importance of addressing climate change and development in an integrated, coordinated and comprehensive manner across constituencies, sectors and administrative and geographical boundaries (Le Blanc 2015). The literature also stresses the important role that robust institutions have in making this happen, including in international cooperation in key sectors for climate action as well for development (Waage et al. 2015). Since the publication of AR5, which emphasised the need for a type of development that combines both mitigation and adaptation as a way to strengthen resilience, much of the literature has focused on ways to address these linkages and the role institutions play in key sectors that are often the subject of international cooperation – for example, environmental and soil degradation, climate, energy, water resources, and forestry (Hogl et al. 2016). An assessment of thematic policy coherence between the voluntary domestic contributions regarding the Paris Agreement and the 2030 Agenda should be integrated in national policy cycles for sustainable and climate policymaking to identify overlaps, gaps, mutual benefits and trade-offs in national policies (Janetschek et al. 2020).

It is only since 2008 that the relationship between climate change and human rights has become a focus of international law and policymaking. It is not just climate impacts that threaten the enjoyment of human rights but also the mitigation responses to climate change that affect human rights (Shi et al. 2017). The issue of human rights–climate change linkages was first taken up by the UN Human Rights Council in 2008, but has since rapidly gained ground with UN human rights treaty bodies issuing comments (e.g., Human Rights Committee 2018), recommendations (e.g., Committee on the Elimination of Discrimination against Women 2018) and even a joint statement (Office of the High Commissioner for Human Rights 2019) on the impacts of climate change on the enjoyment of human rights. Climate change effects and related disasters have the potential to affect human rights broadly, for instance, by giving rise to deaths, disease or malnutrition (right to life, right to health), threatening food security or livelihoods (right to food), impacting upon water supplies and compromising access to safe drinking water (right to water), destroying coastal settlements through storm surge (right to adequate housing), and in some cases forcing relocation as traditional territories become uninhabitable (UNGA 2019). In addition, the right to a healthy environment, recognised in 2021 as an autonomous right at the international level by the Human Rights Council (UN Human Rights Council 2021), arguably extends to a right to a 'safe climate' shaped in part by the Paris Agreement (UNGA 2019).

As the intersections between climate impacts and human rights have become increasingly clear, litigants have begun to use human rights arguments, with a growing receptivity among courts towards such arguments in climate change cases (Peel and Osofsky 2018; Savaresi and Auz 2019; Macchi and van Zeben 2021). In the landmark Urgenda climate case in 2019, the Dutch Supreme Court interpreted the European Convention on Human Rights in light of customary international law and the UN climate change regime and ordered the state to reduce greenhouse gas emissions by 25% by 2020 compared to 1990 (The Supreme Court of the Netherlands 2019). In the Neubauer case in 2021, the German Federal Constitutional Court ordered the German legislature, in light of its obligations, including on rights protections, to set clear provisions for reduction targets from 2031 onward by the end of 2022 (German Constitutional Court 2021). There are cases in the Global South as well (Peel and Lin 2019; Setzer and Benjamin 2020), with the Supreme Court in Nepal in its 2018 decision in Shrestha ordering the government to amend its existing laws and introduce a new consolidated law to address climate mitigation and adaptation as this would protect the rights to life, food, and a clean environment, and give effect to the 2015 Paris Agreement (The Supreme Court of Nepal 2018). There are dozens of further cases in national and regional courts, increasingly based on human rights claims, and this trend is only likely to grow (Shi et al. 2017; Peel and Osofsky 2018; Beauregard et al. 2021). These cases face procedural hurdles, such as standing, as well as substantive difficulties, for instance, with regard to the primarily territorial scope of state obligations to protect human rights (Boyle 2018; Mayer 2021), however, there are increasing instances of successful outcomes across the world.

#### 14.5.1.3 Trade Agreements

As discussed in AR5, policies to open up trade can have a range of effects on GHG emissions, just as mitigation policies can influence trade flows among countries. Trade rules may impede mitigation action by limiting countries' discretion in adopting trade-related climate policies, but they also have the potential to stimulate the international adoption and diffusion of mitigation technologies and policies (Droege et al. 2017).

The mitigation impacts of trade agreements are difficult to ascertain, and the limited evidence is mixed. Examining the effects of three free trade agreements (FTAs) – Mercosur, the North American Free Trade Agreement (NAFTA) and the Australia–United States Free Trade Agreement – on GHG emissions, Nemati et al. (2019) find that these effects depend on the relative income levels of the countries involved, and that FTAs between developed and developing countries may increase emissions in the long run. However, studies also suggest that FTAs incorporating specific environmental or climate-related provisions can help reduce GHG emissions (Baghdadi et al. 2013; Sorgho and Tharakan 2020).

Investment agreements, which are often integrated in FTAs, seek to encourage the flow of foreign investment through investment protection. While international investment agreements hold potential to increase low-carbon investment in host countries (PAGE 2018), these agreements have tended to protect investor rights, constraining the latitude of host countries in adopting environmental policies (Miles 2019). Moreover, international investment agreements may lead to 'regulatory chill', which may lead to countries refraining from or delaying the adoption of mitigation policies, such as phasing out fossil fuels (Tienhaara 2018). More contemporary investment agreements seek to better balance the rights and obligations of investors and host countries, and in theory offer greater regulatory space to host countries (UNCTAD 2019), although it is unclear to what extent this will hold true in practice.

In their NDCs, Parties mention various trade-related mitigation measures, including import bans, standards and labelling schemes, border carbon adjustments (BCAs; see also Chapter 13), renewable energy support measures, fossil fuel subsidy reform, and the use of international market mechanisms (Brandi 2017). Some of these 'response measures' (Chan 2016b) may raise questions concerning their consistency with trade agreements of the World Trade Organization (WTO). Non-discrimination is one of the foundational rules of the WTO. This means, among others, that 'like' imported and domestic products are not treated differently ('national treatment') and that a WTO member should not discriminate between other members ('most-favoured-nation treatment'). These principles are elaborated in a set of agreements on the trade in goods and services, including the General Agreement on Tariffs and Trade (GATT), the General Agreement on Trade in Services(GATS), the Agreement on Technical Barriers to Trade (TBT), and the Agreement on Subsidies and Countervailing Measures (ASCM).

Several measures that can be adopted as part of carbon pricing instruments to address carbon leakage concerns have been examined in the light of WTO rules. For instance, depending on the specific design, the free allocation of emissions allowances under an ETS could be considered a subsidy inconsistent with the ASCM (Rubini and Jegou 2012; Ismer et al. 2021). The WTO compatibility of another measure to counter carbon leakage, BCAs, has also been widely discussed (Box 14.2). Alternatives to BCAs, such as consumption charges on carbon-intensive materials (Pollitt et al. 2020), can be consistent with WTO law, as they do not involve discrimination between domestic and foreign products based on their carbon intensity (Ismer and Neuhoff 2007; Tamiotti 2011; Pauwelyn 2013; Holzer 2014; Ismer and Haussner 2016; Cosbey et al. 2019; European Commission 2019; Mehling et al. 2019; Porterfield 2019; Ismer et al. 2020).

## Box 14.2 | Border Carbon Adjustments and International Climate and Trade Cooperation

Analyses of the WTO compatibility of BCAs (Ismer and Neuhoff 2007; Tamiotti 2011; Hillman 2013; Pauwelyn 2013; Holzer 2014; Trachtman 2017; Cosbey et al. 2019; Mehling et al. 2019; Porterfield 2019) gained new currency following the legislative proposal to introduce a 'carbon border adjustment mechanism' in the EU (European Commission 2021). BCAs can in principle be designed and implemented in accordance with international trade law, but the details matter (Tamiotti et al. 2009). To increase the likelihood that a BCA will be compatible with international trade law, studies suggest that it would need to: have a clear environmental rationale (i.e., reduce carbon leakage); apply to imports and exclude exports; consider the actual carbon intensity of foreign producers; account for the mitigation efforts by other countries; and provide for fairness and due process in its design and implementation (Pauwelyn 2013; Trachtman 2017; Cosbey et al. 2019; Mehling et al. 2019).

BCAs may also raise concerns regarding their consistency with international climate change agreements (Hertel 2011; Davidson Ladly 2012; Ravikumar 2020). To mitigate these concerns, BCAs could include special provisions (e.g., exemptions) for LDCs, or channel revenues from the BCA to developing countries to support low-carbon and climate-resilient development (Grubb 2011; Springmann 2013; Mehling et al. 2019). Moreover, international dialogue on principles and best practices guiding BCAs could help to ensure that such measures do not hinder international cooperation on climate change and trade (Bernasconi-Osterwalder and Cosbey 2021).

Other regulatory measures may also target the GHG emissions associated with the production of goods (Dobson 2018). These measures include bans on carbon-intensive materials, emissions standards for the production process of imported goods, and carbon footprint labels (Kloeckner 2012; Holzer and Lim 2020; Gerres et al. 2021). The compatibility of such measures with trade agreements remains subject to debate. While non-discriminatory measures targeting the emissions from a product itself (e.g., fuel efficiency standards for cars) are more likely to be allowed than measures targeting the production process of a good (Green 2005), some studies suggest that differentiation between products based on their production process may be compatible with WTO rules (Benoit 2011; McAusland and Najjar 2015). (Mayr et al. 2020) find that sustainability standards targeting the emissions from indirect land use change associated with the production of biofuels may be inconsistent with the TBT Agreement. Importantly, trade rules express a strong preference for the international harmonisation of standards over unilateral measures (Delimatsis 2016).

Renewable energy support measures may be at odds with the ASCM, the GATT, and the WTO Agreement on Trade-Related Investment Measures. In WTO disputes, measures adopted in Canada, India, and the United States to support clean energy generation were found to be inconsistent with WTO law due to the use of discriminatory local content requirements, such as the requirement to use domestically produced goods in the production of renewable energy (Cosbey and Mavroidis 2014; Kulovesi 2014; Lewis 2014; Wu and Salzman 2014; Charnovitz and Fischer 2015; Shadikhodjaev 2015; Espa and Marín Durán 2018).

Some measures may both lower trade barriers and potentially bring about GHG emissions reductions. An example is the liberalisation of trade in environmental goods (Hu et al. 2020). In 2012, the Asia-Pacific Economic Cooperation economies agreed to reduce tariffs for a list of 54 environmental goods (including, for example, solar cells; but excluding, for example, biofuels or batteries for electric vehicles). However, negotiations on an Environmental Goods Agreement under the WTO stalled in 2016 due in part to disagreement over which goods to include (de Melo and Solleder 2020). Another example is fossil fuel subsidy reform, which may reduce GHG emissions (Jewell et al. 2018; Chepeliev and van der Mensbrugghe 2020; Erickson et al. 2020) and lower trade distortions (Burniaux et al. 2011; Moerenhout and Irschlinger 2020). However, fossil fuel subsidies have largely remained unchallenged before the WTO due to legal and political hurdles (Asmelash 2015; De Bièvre et al. 2017; Meyer 2017; Steenblik et al. 2018; Verkuijl et al. 2019).

With limited progress in the multilateral trading system, some studies suggest that regional FTAs hold potential for strengthening climate governance. In some cases, climate-related provisions in such FTAs can go beyond provisions in the Kyoto Protocol and Paris Agreement, addressing for instance cooperation on carbon markets or electric vehicles (Gehring et al. 2013; van Asselt 2017; Morin and Jinnah 2018; Gehring and Morison 2020). However, Morin and Jinnah (2018) find that these provisions are at times vaguely formulated, not subject to third-party dispute settlement, and without sanctions or remedy in case of violations. Moreover, such provisions are not widely used in FTAs, and they are not adopted by the largest GHG emitters. For instance, the 2019 United States–Mexico–Canada Agreement, NAFTA's successor, does not include any specific provisions on climate change, although it could implement cooperative mitigation actions through its Commission for Environmental Cooperation (Laurens et al. 2019).

A trend in international economic governance has been the adoption of 'mega-regional' trade agreements involving nations responsible for a substantial share of world trade, such as the Comprehensive and Progressive Agreement for Trans-Pacific Partnership (CPTPP), the EU-Canada Comprehensive Economic and Trade Agreement (CETA), and the Regional Comprehensive Economic Partnership (RCEP) in East Asia. Given the size of the markets covered by these agreements, they hold potential to diffuse climate mitigation standards (Meltzer 2013; Holzer and Cottier 2015). While CETA includes climate-related provisions and Parties have made a broad commitment to implement the Paris Agreement (Laurens et al. 2019), and the CPTPP includes provisions promoting cooperation on clean energy and low-emissions technologies, the RCEP does not include specific provisions on climate change.

Studies have discussed various options to minimise conflicts, and strengthen the role of trade agreements in climate action, although the mitigation benefits and distributional effects of these options have yet to be assessed. Some options require multilateral action, including: (i) the amendment of WTO agreements to accommodate climate action; (ii) the adoption of a 'climate waiver' that temporarily relieves WTO members from their obligations; (iii) a 'peace clause' through which members commit to refraining from challenging each other's measures; (iv) an 'authoritative interpretation' by WTO members of ambiguous WTO provisions; (v) improved transparency of the climate impacts of trade measures; (vi) the inclusion of climate expertise in WTO disputes; and (vii) intensified institutional coordination between the WTO and UNFCCC (Hufbauer et al. 2009; Epps and Green 2010; Bacchus 2016; Droege et al. 2017; Das et al. 2019). In addition, issue-specific suggestions have been put forward, such as reinstating an exception for environmentally-motivated subsidies under the ASCM (Horlick and Clarke 2017).

Options can also be pursued at the plurilateral and regional levels. Several studies suggest that climate clubs (Section 14.2.2) could employ trade measures, such as lower tariffs for climate-related goods and services, or BCAs, to attract club members (Nordhaus 2015; Brewer et al. 2016; Keohane et al. 2017; Stua 2017a; Banks and Fitzgerald 2020). Another option is to negotiate a new agreement addressing both climate change and trade. Negotiations between six countries (Costa Rica, Fiji, Iceland, New Zealand, Norway, Switzerland) were launched in 2019 on a new Agreement on Climate Change, Trade and Sustainability (ACCTS), which, if successfully concluded, would liberalise trade in environmental goods and services, create new rules to remove fossil fuel subsidies, and develop guidelines for voluntary eco-labels (Steenblik and Droege 2019). At the regional level, countries could further opt for the inclusion of climate provisions in the (re)negotiation of FTAs (Morin and Jinnah 2018; Yamaguchi 2020). Moreover, the conduct of climate impact assessments of FTAs could help identify options to achieve both climate and trade objectives (Porterfield et al. 2017). In their assessment of the feasibility of various options for reform, Das et al. (2019) find that the near-term feasibility of options that require consensus at the multilateral level (notably amendments of WTO agreements) is low. By contrast, options involving a smaller number of Parties, as well as options that can be implemented by WTO members on a voluntary basis, face fewer constraints.

For international investment agreements, various other suggestions have been put forward to accommodate climate change concerns. These include incorporating climate change through ongoing reform processes, such as reform of investor-state dispute settlement under the UN Commission on International Trade Law; modernisation of the Energy Charter Treaty; the (re)negotiation of international investment agreements; and the adoption of a specific treaty to promote investment in climate action (Brauch et al. 2019; Tienhaara and Cotula 2020; Yamaguchi 2020; Cima 2021).

#### 14.5.1.4 South-South cooperation

South-South cooperation (SSC) and triangular cooperation (TrC) are bold, innovative, and rapidly developing means of strengthening cooperation for the achievement of the SDGs (FAO 2018). SSC is gaining momentum in achieving sustainable development and climate actions in developing countries (UN 2017b). Through SSC, countries are able to map their capacity needs and knowledge gaps and find sustainable, cost-effective, long-lasting and economically viable solutions (FAO 2019). In the UN Climate Change Engagement Strategy 2017 (UNOSSC 2017), South-South Cooperation Action Plan is identified as a substantive pillar to support.

In 2019, the role of South-South and triangular cooperation was further highlighted with the BAPA+40 Outcome document (UN 2019), noting outstanding contributions to alleviating global inequality, promoting sustainable development and climate actions, promoting gender equality and enriching multilateral mechanisms. Furthermore, the role of triangular cooperation was explicitly recognised in the document reflecting its increasingly relevant role in the implementation of the SDGs (UN 2019).

There has been a recent resurgence of South-South cooperation (Gray and Gills 2016). The South-South Cooperation Action Plan was adopted by the UN as a substantive pillar to support the implementation of the UN Climate Change Engagement Strategy 2017 (UNOSSC 2017). Liu et al. (2017a) explored prospects for South–South cooperation for large-scale ecological restoration, which is an important solution to mitigate climate change. Emphasis is given to experience and expertise sharing, co-financing, and co-development of new knowledge and know-how for more effective policy and practice worldwide, especially in developing and newly industrialised countries.

Janus et al. (2014) explore evolving development cooperation and its future governance architecture based on The Global Partnership for Effective Development Cooperation and The United Nations Development Cooperation Forum. Drawing on evidence from the hydropower, solar and wind energy industry in China, Urban (2018) introduces the concept of 'geographies of technology transfer and cooperation' and challenges the North–South technology transfer and cooperation paradigm for low-carbon innovation and climate change mitigation. While North–South technology transfer and cooperation (NSTT) for low-carbon energy technology has been implemented for decades, South–South technology transfer and cooperation (SSTT) and South–North technology transfer and cooperation (SNTT) have only recently emerged. Kirchherr and Urban (2018) provide a meta-synthesis of the scholarly writings on NSTT, SSTT and SNTT from the past 30 years. The discussion focuses on core drivers and inhibitors of technology transfer and cooperation, outcomes as well as outcome determinants. A case study of transfer of low-carbon energy innovation and its opportunities and barriers, based on the first large Chinese-funded and Chinese-built dam in Cambodia is presented by Hensengerth (2017).

Hensengerth (2017) explores the role that technology transfer/cooperation from Europe played in shaping firm-level wind energy technologies in China and India and discusses the recent technology cooperation between the Chinese, Indian, and European wind firms. The research finds that firm-level technology transfer/cooperation shaped the leading wind energy technologies in China and to a lesser extent in India. Thus, the technology cooperation between China, India, and Europe has become multi-faceted and increasingly Southern-led.

Rampa et al. (2012) focus on the manner in which African states understand and approach new opportunities for cooperation with emerging powers, especially China, India and Brazil, including the crucial issue of whether they seek joint development initiatives with both traditional partners and emerging powers. UN (2018) presents and analyses case studies of SSTT in Asia and Pacific and Latin America and Caribbean regions. Illustrative case studies on TrC can be consulted in Shimoda and Nakazawa (2012), and specific cases on biofuel SSC and TrC in UNCTAD (2012).

The central argument in the majority of these case studies is that South–South cooperation, which is value-neutral, is contributing to sustainable development and capacity building (Rampa et al. 2012; Shimoda and Nakazawa 2012; UN 2018). An important new development in SSC is that in relation to some technologies the cooperation is increasingly led by Southern countries (for instance, wind energy between Europe, India and China), challenging the classical North–South technology cooperation paradigm. More broadly, Parties should ensure the sustainability of cooperation, rather than focusing on short-term goals (Eyben 2013). The Belt and Road Initiative (BRI) is a classic example of a recent SSC initiative led by China. According to a joint study by Tsinghua University and Vivid Economics, the 126 countries in the BRI region, excluding China, currently account for about 28% of global GHG emissions, but this proportion may increase to around 66% by 2050 if the carbon intensity of these economies only decreases slowly (according to historical patterns shown by developing countries). In this context it is important to highlight that China has already outlined a vision for a green BRI, and recently increased its commitment through the Green Investment Principles initiative, announcing a new international coalition to improve sustainability and promote green infrastructure (Jun and Zadek 2019).

Information on triangular cooperation is more readily available than on South–South cooperation though some UN organisations such as UNDP and FAO have established platforms for the latter which also include climate projects. Further, although there are many South–South cooperation initiatives involving the development and transfer of climate technologies, the understanding of the motivations, approaches and designs is limited and not easily accessible. There is no dedicated platform for South–South and triangular cooperation on climate technologies. Hence, it is still too early to fully assess the achievements in the field of climate action (UNFCCC and UNOSSC 2018). In order to maximise its unique contribution to Agenda 2030, Southern providers recognise the benefits of measuring and monitoring South–South cooperation, and there is a clear demand for better information from partner countries. Di Ciommo (2017) argues that 'better data could support monitoring and evaluation, improve effectiveness, explore synergies with other resources, and ensure accountability' to a diverse set of stakeholders. Besharati et al. (2017) present a framework of 20 indicators, organised in five dimensions, that researchers and policymakers can use to access the quality and effectiveness of SSC and its contribution to sustainable development.

The global landscape of development cooperation has changed dramatically in recent years, with countries of the South engaging in collaborative learning models to share innovative, adaptable and cost-efficient solutions to their development and socio-economic-environmental challenges, ranging from poverty and education to climate change. The proliferation of new actors and cross-regional modalities has enriched the understanding and practice of development cooperation and generated important changes in the global development architecture towards a more inclusive, effective, and horizontal development agenda. South–South cooperation will grow in the future, while it is complimentary to North–South cooperation. However, there are knowledge gaps in relation to the precise volume, impact, effectiveness and quality of development cooperation from emerging development partners. This gap needs to be plugged, and evidence on such cooperation strengthened.

### 14.5.2 International Sectoral Agreements and Institutions

Sectors refer to distinct areas of economic activity, often subject to their own governance regimes; examples include energy production, mobility, and manufacturing. A sectoral agreement could include virtually any type of commitment with implications for mitigation. It could establish sectoral emission targets, on either an absolute or an indexed basis. It could also require states (or particular groups of states, if commitments are differentiated) to adopt uniform or harmonised policies and measures for a sector, such as technology-based standards, taxes, or best-practice standards, as well as providing for cooperation on technology research or deployment.

#### 14.5.2.1 Forestry, Land Use and REDD+

Since 2008, several, often overlapping, voluntary and non-binding international efforts and agreements have been adopted to reduce net emissions from the forestry sector. These initiatives have varying levels of private sector involvement and different objectives, targets, and timelines. Some efforts focus on reducing emissions from deforestation and degradation, while other focus on the enhancement of sinks through restoration of cleared or degraded landscapes. These initiatives do not elaborate specific policies, procedures, or implementation mechanisms. They set targets, frameworks, and milestones, aiming to catalyse further action, investment, and transparency in conservation and consolidate individual country efforts.

After the UN-sponsored Tropical Forestry Action Plan (Winterbottom 1990; Seymour and Busch 2016), among the longest standing programmes in the forestry sector are the World Bank-sponsored Forest Carbon Partnership Facility in 2007, which helps facilitate funding for REDD+ readiness and specific projects, in addition to preparing countries for results-based payments and future carbon markets while securing local communities' benefits managed sub-nationally, and the UN REDD+ Programme initiated in 2008, which aims to reduce forest emissions and enhance carbon stocks in forests while contributing to national sustainable development in developing countries, after the 2007 COP13 in Bali formally adopted REDD+ in the UNFCCC decisions and incorporated it in the Bali Plan of Action. As discussed above, Article 5 of the Paris Agreement encourages Parties to take action to implement and support REDD+. These efforts tend to focus on reducing emissions through the creation of protected areas, payments for ecosystem services, and/or land tenure reform (Pirard et al. 2019). The UNREDD+ programme supports national REDD+ efforts, inclusion of stakeholders in relevant dialogues, and capacity building toward REDD+ readiness in partner countries. To date the conservation and emissions impacts of REDD+ remain misunderstood (Pirard et al. 2019), but while existing evidence suggests that reductions in deforestation from sub-national REDD+ initiatives have been limited (Bos et al. 2017) it shows an increasing prominence (Maguire et al. 2021). Additionally, the Green Climate Fund has carried out results-based payments within REDD+. Eight countries have so far received significant funding (GCF 2021). The shift in the REDD+ focus from ecosystem service payment to domestic policy realignments and incentive structure has changed the way REDD+ was developed and implemented (Brockhaus et al. 2017). Large-scale market resources have not fully materialised as a global carbon market system that explicitly integrates REDD+ remains under development (Angelsen 2017). Public funding for REDD+ is also limited (Climate Focus 2017). Leading up to the adoption of the Paris Agreement, the governments of Germany, Norway, and the United Kingdom formed a partnership in 2014 called 'GNU' to support results-based financing for REDD+, with Norway emerging as one of, if not the single largest, major donor for REDD+ through its pledge in 2007 of approximately USD3 billion annually. Norway pledged USD1 billion for Brazil in 2008 and the same for Indonesia in 2010 (Schroeder et al. 2020). Meanwhile, REDD+ Early Movers was established with support from Germany, and the Central African Forest Initiative, a collaborative partnership between the European Union, Germany, Norway, France, and the United Kingdom was also set up. It supports six central African countries in fighting deforestation.

More recently, the Lowering Emissions by Accelerating Forest Finance (LEAF) Coalition was established, consisting of the governments of Norway, the UK, and the USA and initially nine companies, to accelerate REDD+ with a jurisdictional approach. LEAF uses the Architecture for REDD+ Transactions (ART)'s The REDD+ Environmental Excellence Standard (TREES), coordinated by Emergent, a non-profit intermediary between tropical countries and the private sector. Three jurisdictions in Brazil and two countries have already submitted concept notes to ART to receive results-based payments. REDD+ initiatives with a jurisdictional approach have also been adopted in various markets, such as the Carbon Offsetting and Reduction Scheme for International Aviation (CORSIA) (Maguire 2021). In addition to Brazil, Indonesia has attracted significant interest as a host country for REDD+. Indonesia ranks second, after Brazil, as the largest producer of deforestation-related GHG emissions (Zarin et al. 2016), but it has committed to a large reduction of deforestation in its NDC (Government of Indonesia 2016). Australia has collaborated on scientific research and emissions reduction monitoring (Tacconi 2017). It took a while, however, before emissions reductions were witnessed (Meehan et al. 2019). The expansion of commodity plantations, however, conflict with reduction ambitions (Anderson et al. 2016; Irawan et al. 2019) In addition to implementation at the site and jurisdictional levels, legal enforcement (Tacconi et al. 2019) as well as policy and regulatory reforms (Ekawati et al. 2019) appear to be needed.

Another relevant initiative is one under the 2015 United Nations Convention to Combat Desertification (UNCCD), which targets land degradation neutrality, that is, 'a *state whereby the amount and quality of land resources, necessary to support ecosystem functions and services and enhance food security, remains stable or increases within specified temporal and spatial scales and ecosystems*' (Orr et al. 2017). This overarching goal was recognised as also being critical to reaching the more specific avoided deforestation and degradation and restoration goals of the UNFCCC and UNCBD. The Land Degradation Neutrality (LDN) initiative from UNCCD includes target-setting programmes that assist countries by providing practical tools and guidance for the establishment of the voluntary targets and to formulate associated measures to achieve LDN and accelerate implementation of projects (Chasek et al. 2019).

Today, 124 countries have committed to their LDN national targets (UNCCD 2015). The LDN Fund is an investment vehicle launched in UNCCD COP 13 in 2017, which exists to provide long-term financing for private projects and programmes for countries to achieve their LDN targets. According to the UNCCD, most of the funds will be invested in developing countries.

Recent efforts towards the enhancement of sinks from the forestry sector have the overarching goal of reaching zero *gross* deforestation globally, that is, eliminating the clearing of all natural forests. The New York Declaration on Forests (NYDF) was the first international pledge to call for a halving of natural forest loss by 2020 and the complete elimination of natural forest loss by 2030 (Climate Focus 2016). It was endorsed at the United Nations Climate Summit in September 2014. By September 2019 the list of NYDF supporters included over 200 actors: national governments, sub-national governments, multi-national companies, groups representing indigenous communities, and non-government organisations. These endorsers committed to doing their part to achieve the NYDF's ten goals, which included ending deforestation for agricultural expansion by 2020, reducing deforestation from other sectors, restoring forests, and providing financing for forest action (Forest Declaration 2019). These goals are assessed and tracked through the NYDF Progress Assessment, which includes NYDF Assessment Partners that collect data, generate analysis, and release the finding based on the NYDF framework and goals.

The effectiveness of these agreements, which lack binding rules, can only be judged by the supplementary actions they have catalysed. The NYDF contributed to the development of several other zero-deforestation pledges, including the Amsterdam Declarations by seven European nations to achieve fully sustainable and deforestation-free agro-commodity supply chains in Europe by 2020 and over 150 individual company commitments to not source products associated with deforestation (Donofrio et al. 2017; Lambin et al. 2018). Recent studies indicate that these efforts currently lack the potential to achieve wide-scale reductions in clearing and associated emissions due to weak implementation (Garrett et al. 2019), although in some cases in Indonesia and elsewhere the commodity supply chain sustainability drive appears to contribute to lowering deforestation (Wijaya et al. 2019; Chain Reaction Research 2020; Schulte et al. 2020). The NYDF may have triggered small additional reductions in deforestation in some areas, particularly for soy, and to a lesser extent cattle, in the Brazilian Amazon (Lambin et al. 2018), but these effects were temporary, as efforts are being actively reversed and deforestation has increased again significantly. Deforestation rates have escalated in Brazil, with the rate in June 2019 (the first dry-season month in the new administration) up 88% over the 2018 rate in the same month (INPE 2019). Curtis et al. (2018) find global targets are clearly not being met. More recent increase in the deforestation rate remains to be assessed. NYDF confirms that the initiative did not reach its zero-deforestation goal (NYDF Assessment Partners 2020).

In 2010, the Parties to the CBD adopted the Strategic Plan for Biodiversity 2011–2020 which included 20 targets known as the Aichi Biodiversity targets (Marques et al. 2014). Of relevance to the forestry sector, Aichi Target 15 sets the goal of enhancing ecosystem resilience and the contribution of biodiversity to carbon stocks though conservation and restoration, including 'restoration of at least 15% of degraded ecosystems' (UNCBD 2010). The plan elaborates milestones, including the development of national plans for potential restoration levels and contributions to biodiversity protection, carbon sequestration, and climate adaptation to be integrated into other national strategies, including REDD+. In 2020, however, the CBD found that while progress was evident for the majority of the Aichi Biodiversity Targets, it was not sufficient for the achievement of the targets by 2020 (CBD 2020).

Recent efforts toward negative emissions through restoration include the Bonn Challenge, the African Forest Landscape Restoration Initiative (AFR100) and Initiative 20x20. The Bonn Challenge, initiated in 2011 by the Government of Germany and the International Union for Conservation of Nature, is intended to catalyse the existing international AFOLU commitments. It aimed to bring 150 million hectares (Mha) of the world's deforested and degraded land into restoration by 2020, and 350 Mha by 2030. AFR has the goal of restoring 100 Mha specifically in Africa (AUDA-NEPAD 2019), while 20x20 aims to restore 20 Mha in Latin America and the Caribbean (Anderson and Peimbert 2019). Increasing commitments for restoration have created momentum for restoration interventions (Chazdon et al. 2017; Mansourian et al. 2017; Djenontin et al. 2018). To date 97 Mha has been pledged in NDCs. Yet only a small part of this goal has been achieved. The Bonn Challenge Barometer – a progress-tracking framework and tool to support pledgers – indicates that 27 Mha (InfoFLR 2018) are currently being restored, equivalent to 1.379 $GtCO_2$-eq sequestered (Dave et al. 2019). A key challenge in scaling up restoration has been to mobilise sufficient financing (Liagre et al. 2015; Djenontin et al. 2018). This underscores the importance of building international financing for restoration (equivalent to the Forest Carbon Partnership Facility focused on avoided deforestation and degradation).

In sum, existing international agreements have had a small impact on reducing emissions from the AFOLU sector and some success in achieving the enhancement of sinks through restoration. However, these outcomes are nowhere near levels required to meet the Paris Agreement temperature goal – which would require turning land use and forests globally from a net anthropogenic source during 1990–2010 to a net sink of carbon by 2030, and providing a quarter of emissions reductions planned by countries (Grassi et al. 2017). The AFOLU sector has so far contributed only modestly to net mitigation (Chapter 7).

#### 14.5.2.2 Energy Sector

International cooperation on issues of energy supply and security has a long and complicated history. There exists a plethora of institutions, organisations, and agreements concerned with managing the sector. There have been efforts to map the relevant actors, with authors in one case identifying six primary organisations (Kérébel and Keppler 2009), in another 16 (Lesage et al. 2010), and in a third 50 (Sovacool and Florini 2012). At the same time, very little of that history has had climate mitigation as its core focus. Global energy governance has encompassed five broad goals – security of energy supply and demand, economic development, international security, environmental sustainability, and domestic good governance – and as only one of these provides an entry point for climate mitigation, effort in this direction has often been lost (van de Graaf and Colgan 2016). To take one example, during the 1980s and 1990s a combination of bilateral development support and lending practices from multilateral development banks pushed developing countries to adopt power market reforms consistent with the Washington Consensus: towards liberalised power markets and away from state-owned monopolies. The goals of these reforms did not include an environmental component, and among the results was new investment in fossil-fired thermal power generation (Foster and Rana 2020).

As Goldthau and Witte (2010) document, the majority of governance efforts, outside of oil and gas producing states, was oriented towards ensuring reliable and affordable access for oil and gas imports. For example, the original rationale for the creation of the International Energy Agency (IEA), during the oil crisis of 1973–74, was to manage a mechanism to ensure importing countries' access to oil (van de Graaf and Lesage 2009). On the other side of the aisle, oil exporting countries created the international institution OPEC to enable them to influence oil output, thereby stabilising prices and revenues for exporting countries (Fattouh and Mahadeva 2013). For years, energy governance was seen as a zero-sum game between these poles (Goldthau and Witte 2010). The only international governance agency focusing on low-carbon energy sources was the International Atomic Energy Agency, with a dual mission of promoting nuclear energy and nuclear weapons non-proliferation (Scheinman 1987).

More recently, however, new institutions have emerged, and existing institutions have realigned their missions, in order to promote capacity building and global investment in low-carbon energy technologies. Collectively, these developments may support the emergence of a nascent field of global sustainable energy governance, in which a broad range of global, regional, national, sub-national and non-state actors, in aggregate, shape, direct and implement the low carbon transition through climate change mitigation activities, which produce concomitant societal benefits (Bruce 2018). Beginning in the 1990s, for example, the IEA began to broaden its mission from one concerned primarily with security of oil supplies, which encompassed conservation of energy resources, to one also concerned with the sustainability of energy use, including work programmes on energy efficiency and clean energy technologies and scenarios (van de Graaf and Lesage 2009). Scholars have suggested that it was the widespread perception that the IEA was primarily interested in promoting the continued use of fossil fuels, and underplaying the potential role of renewable technologies, that led a number of IEA member states to successfully push for the creation of a parallel organisation, the International Renewable Energy Agency (IRENA), which was then established in 2009 (van de Graaf 2013). An assessment of IRENA's activities in 2015 suggested that the agency has a positive effect related to three core activities: offering advisory services to member states regarding renewable energy technologies and systems; serving as a focal point for data and analysis for renewable energy; and, mobilising other international institutions, such as multilateral development banks, promoting renewable energy (Urpelainen and Van de Graaf 2015). The United Nations, including its various agencies such as the Committee on Sustainable Energy within the United Nations Economic Commission for Europe, has also played a role in the realignment of global energy governance towards mitigation efforts. As a precursor to SDG 7, the United Nations initiated in 2011 the Sustainable Energy for All initiative, which in addition to aiming for universal access to modern energy services, included the goals of doubling the rate of improvement in energy efficiency, and doubling by 2030 the share of renewable energy in the global energy mix (Bruce 2018).

Sub-global agreements have also started to emerge, examples of issue-specific climate clubs. In 2015, 70 solar-rich countries signed a framework agreement dedicated towards promoting solar energy development (ISA 2015). In 2017 the Powering Past Coal Alliance was formed, uniting a set of states, businesses, and non-governmental organisations around the goal of eliminating coal-fired power generation by 2050 (Jewell et al. 2019; Blondeel et al. 2020). Scholars have argued that greater attention to supply-side agreements such as this – focusing on reducing and ultimately eliminating the supply of carbon-intensive energy sources – would strengthen the UNFCCC and Paris Agreement (Collier and Venables 2014; Piggot et al. 2018; Asheim et al. 2019; Newell and Simms 2020). Chapter 6 of this report, on energy systems, notes the importance of regional cooperation on electric grid development, seen as necessary to enable higher shares of solar and wind power penetration (RGI 2011). Finally, a number of transnational organisations and activities have emerged, such as REN21, a global community of renewable energy experts (REN21 2019), and RE100, an NGO-led initiative to enlist multilateral companies to shift towards 100% renewable energy in their value chains (RE100 2019).

Whether a result of the above activities or not, multilateral development banks' lending practices have shifted in the direction of renewable energy (Delina 2017), a point also raised in Chapter 15 of this report. Activities include new sources of project finance, concessional loans, as well as loan guarantees, the latter through the Multilateral Investment Guarantee Agency (Multilateral Investment Guarantee Agency 2019). This appears to matter. For example, Frisari and Stadelmann (2015) find concessional lending by multilateral development banks to solar energy projects in Morocco and India to have reduced overall project costs, due to more attractive financing conditions from additional lenders, as well as reducing the costs to local governments. Labordena et al. (2017) projected these results into the future, and found that with the drop in financing costs, renewable energy projects serving all major demand centres in sub-Saharan Africa could reach cost parity with fossil fuels by 2025, whereas without the drop in financing costs associated with concessional lending, this would not be the case. Similarly, Creutzig et al. (2017) suggest that greater international attention to finance could be instrumental in the full development of solar energy.

Despite improvements in the international governance of energy, it still appears that a great deal of this is still concerned with promoting further development of fossil fuels. One aspect of this is the development of international legal norms. A large number of bilateral and multilateral agreements, including the 1994 Energy Charter

Treaty, include provisions for using a system of investor–state dispute settlement (ISDS) designed to protect the interests of investors in energy projects from national policies that could lead their assets to be stranded. Numerous scholars have pointed to ISDS being able to be used by fossil-fuel companies to block national legislation aimed at phasing out the use of their assets (Tienhaara 2018; Bos and Gupta 2019). Another aspect is finance; Gallagher et al. (2018) examine the role of national development finance systems. While there has been a great deal of finance devoted to renewable energy, they find the majority of finance devoted to projects associated either with fossil fuel extraction or with fossil fuel-fired power generation.

Given the complexity of global energy governance, it is impossible to make a definitive statement about its overall contribution to mitigation efforts. Three statements, do however, appear to be robust. First, prior to the emergence of climate change on the global political agenda, international cooperation in the area of energy was primarily aimed at expanding and protecting the use of fossil energy, and these goals were entrenched in a number of multilateral organisations. Second, since the 1990s, international cooperation has gradually taken climate mitigation on board as one of its goals, seeing a realignment of many pre-existing organisations priorities, and the formation of a number of new international arrangements oriented towards the development of renewable energy resources. Third, the realignment is far from complete, and there are still examples of international cooperation having a chilling effect on climate mitigation, particularly through financing and investment practices, including legal norms designed to protect the interests of owners of fossil assets.

#### 14.5.2.3 Transportation

The transportation sector has been a particular focus of cooperative efforts on climate mitigation that extend beyond the sphere of the UNFCCC climate regime. A number of these cooperative efforts involve transnational public-private partnerships, such as the European-based Transport Decarbonisation Alliance, which brings together countries, regions, cities and companies working towards the goal of a 'net-zero emission mobility system before 2050' (TDA 2019). Other efforts are centred in specialised UN agencies, such as the International Civil Aviation Organization (ICAO) and the International Maritime Organization (IMO).

Measures introduced by the ICAO and IMO have addressed $CO_2$ emissions from international shipping and aviation. Emissions from these parts of the transportation sector are generally excluded from national emissions reduction policies and NDCs because the 'international' location of emissions release makes allocation to individual nations difficult (Bows-Larkin 2015; Lyle 2018; Hoch et al. 2019). The measures adopted by ICAO take the form of standards and recommended practices that are adopted in national legislation. IMO publishes 'regulations' but does not have a power of enforcement, with non-compliance a responsibility of flag states that issue a ship's 'MARPOL' certificate.

As discussed in Chapter 2 and Figure SPM.4, international aviation currently accounts for approximately 1% of global GHG emissions, with international shipping contributing 1.2% of global GHG emissions. These international transport emissions are projected to be between approximately 60% and 220% of global emissions of $CO_2$ in 2050, as represented by the four main illustrative model pathways in SR1.5 (Rogelj et al. 2018; UNEP 2020) Notably, however, the climate impact of aviation emissions is estimated to be two to four times higher due to non-$CO_2$ effects (Terrenoire et al. 2019; Lee et al. 2021a). Increases in trans-Arctic shipping and tourism activities with sea ice loss are also forecast to have strong regional effects due to ships' gas and particulate emissions (Stephenson et al. 2018).

The Kyoto Protocol required Annex I Parties to pursue emissions reductions from aviation and marine bunker fuels by working through IMO and ICAO (UNFCCC 1997, Art. 2.2). Limited progress was made by these organisations on emissions controls in the ensuing decades (Liu 2011b), but greater action was prompted by conclusion of the SDGs and Paris Agreement (Martinez Romera 2016), together with unilateral action, such as the EU's inclusion of aviation emissions in its Emissions Trading System (Dobson 2020).

The Paris Agreement neither explicitly addresses emissions from international aviation and shipping, nor repeats the Kyoto Protocol's provision requiring Parties to work through ICAO/IMO to address these emissions (Hoch et al. 2019). This leaves unclear the status of the Kyoto Protocol's Article 2.2 directive after 2020 (Martinez Romera 2016; Dobson 2020), potentially opening up scope for more attention to aviation and shipping emissions under the Paris Agreement (Doelle and Chircop 2019). Some commentators have suggested that emissions from international aviation and shipping should be part of the Paris Agreement (Gençsü and Hino 2015; Traut et al. 2018), and shipping and aviation industries themselves may prefer emissions to be treated under an international regime rather than a nationally-oriented one (Gilbert and Bows 2012). In the case of shipping emissions, there is nothing in the Paris Agreement to prevent a Party from including international shipping in some form in its NDC (Doelle and Chircop 2019) Under the Paris Rulebook, Parties 'should report international aviation and marine bunker fuel emissions as two separate entries and should not include such emissions in national totals but report them distinctly, if disaggregated data are available' (UNFCCC 2019d).

ICAO has an overarching climate goal to 'limit or reduce the impact of aviation greenhouse gas emissions on the global climate' with respect to international aviation. In order to achieve this, ICAO has two global aspirational goals for the international aviation sector, of 2% annual fuel efficiency improvement through 2050 and carbon neutral growth from 2020 onwards (ICAO 2016). In order to achieve these global aspirational goals, ICAO is pursuing a 'basket' of mitigation measures for the aviation sector consisting of technical and operational measures, such as a $CO_2$ emissions standard for new aircraft adopted in 2016, measures on sustainable alternative fuels and a market-based measure, known as the Carbon Offset and Reduction Scheme for International Aviation (CORSIA), which the triennial ICAO Assembly of 193 Member States resolved to establish in 2016 (ICAO 2016). In line with the 2016 ICAO Assembly Resolution that established CORSIA, in mid-2018, the ICAO's 36-member state governing Council adopted a series of Standards and Recommended

Practices (SARPs), now contained in Annex 16, Volume IV of the Chicago Convention (1944), as a common basis for CORSIA's implementation and enforcement by each state and its aeroplane operators. From 1 January 2019, the CORSIA SARPs require states and their operators to undertake an annual process of monitoring, verification, and reporting of emissions from all international flights, including to establish CORSIA's emissions baseline (ICAO 2019).

Based on this emissions data, CORSIA's carbon offsetting obligations commenced in 2021, with three-year compliance cycles, including a pilot phase in 2021–2023. States have the option to participate in the pilot phase and the subsequent voluntary three-year cycle in 2024–2026. CORSIA becomes mandatory from 2027 onwards for states whose share in the total international revenue tonnes per kilometre is above a certain threshold (Hoch et al. 2019). Under CORSIA, aviation $CO_2$ emissions are not capped, but rather emissions that exceed the CORSIA baseline are compensated through use of 'offset units' from emissions reduction projects in other industries (Erling 2018). However, it is unclear whether the goal of carbon neutral growth and further $CO_2$ emissions reduction in the sector will be sufficiently incentivised solely through the use of such offsets in combination with ICAO's manufacturing standards, programmes, and state action plans, without additional measures being taken, for example, constraints on demand (Lyle 2018). If countries such as China, Brazil, India and Russia do not participate in CORSIA's voluntary offsetting requirements this could significantly undermine its capacity to deliver fully on the sectoral goal by limiting coverage of the scheme to less than 50% of international aviation $CO_2$ emissions in the period 2021–2026 (Hoch et al. 2019; Climate Action Tracker 2020b). In addition, a wide range of offsets are approved as 'eligible emissions units' in CORSIA, including several certified under voluntary carbon offset schemes, which may go beyond those eventually agreed under the Paris Agreement Article 6 mechanism (Hoch et al. 2019). It is noted, however, that ICAO applies a set of 'Emissions Unit Eligibility Criteria', agreed in March 2019, which specify required design elements for eligible programmes. In June 2020, the ICAO Council decided to define 2019 emissions levels, rather than an average of 2019 and 2020 emissions, as the baseline year for at least the first three years of CORSIA, although there were significant reductions (45–60%) in aviation $CO_2$ emissions in 2020 compared with 2019 as a result of reductions in air travel associated with the COVID-19 pandemic (Climate Action Tracker 2020b).

Other measures adopted by ICAO include an aircraft $CO_2$ emissions standard that applies to new aircraft type designs from 2020, and to aircraft type designs already in production as of 2023 (Smith and Ahmad 2018). Overall, CORSIA and regional measures, such as the EU ETS, are estimated to reduce aviation carbon emissions by only 0.8% per year from 2017–2030 (noting, however, that 'if non-$CO_2$ emissions are included in the analysis, then emissions will increase') (Larsson et al. 2019). Accordingly, pathways consistent with the temperature goal of the Paris Agreement are likely to require more stringent international measures for the aviation sector (Larsson et al. 2019).

Similar to ICAO, the IMO has a stated vision of remaining committed to reducing greenhouse gas emissions from international shipping and, as a matter of urgency, aims to phase them out as soon as possible in this century. IMO has considered a range of measures to monitor and reduce shipping emissions. In 2016, the IMO's Marine Environment Protection Committee (MEPC) approved an amendment to the MARPOL Convention Annex VI for the introduction of a mandatory global data collection scheme for fuel oil consumption of ships (Dobson 2020). Other IMO measures have focused on energy efficiency (Martinez Romera 2016). The IMO's Energy Efficiency Design Index (EEDI), which is mandatory for new ships, is intended, over a ten-year period, to improve energy efficiency by up to 30% in several categories of ships propelled by diesel engines (Smith and Ahmad 2018). In May 2019, the MEPC approved draft amendments to the MARPOL Convention Annex VI, which if adopted, will bring forward the entry into force of the third phase of the EEDI requirements to 2022 instead of 2025 (IMO 2019; Joung et al. 2020).

However, it is unlikely that the EEDI and other IMO technical and operational measures will be sufficient to produce 'the necessary emissions reduction because of the future growth in international seaborne trade and world population' (Shi and Gullett 2018). Consequently, in 2018, the IMO adopted an initial strategy on reduction of GHG emissions from ships (IMO 2018). This includes a goal for declining carbon intensity of the sector by reducing $CO_2$ emissions per transport work, as an average across international shipping, by at least 40% by 2030, and pursuing efforts towards 70% by 2050, compared to 2008 levels (IMO 2018, Para. 3.1). The strategy also aims for peaking of total annual GHG emissions from international shipping as soon as possible and a reduction by at least 50% by 2050 compared to 2008 levels, while pursuing efforts towards phasing them out 'as soon as possible in this century' as a point 'on a pathway of $CO_2$ emissions reduction consistent with the Paris Agreement temperature goals' (IMO 2018, Para. 2, 3.1). The shipping industry is on track to overachieve the 2030 carbon intensity target but not its 2050 target (Climate Action Tracker 2020c). The initial IMO strategy is to be kept under review by the MEPC with a view to adoption of a revised strategy in 2023.

The IMO's initial strategy identifies a series of candidate short-term (2018–2023), medium-term (2023–2030) and long-term (beyond 2030) measures for achieving its emissions reduction goals, including possible market-based measures in the medium-to-long term (IMO 2018, paras. 4.7–4.9). Further progress on market-based measures faces difficulty in light of conflicts between the CBDRRC principle of the climate regime and the traditional non-discrimination approach and principle of no more favourable treatment enshrined in MARPOL and other IMO conventions (Zhang 2016). Both the CBDRRC and non-discrimination principles are designated as 'principles guiding the initial strategy' (IMO 2018, Para. 3.2). The challenges encountered in introducing global market-based measures for shipping emissions under the IMO have prompted regional initiatives such as the proposed extension of the EU ETS to emissions from maritime activities (Christodoulou et al. 2021), which was announced on 14 July 2021 by the EU Commission as part of its 'Fit for 55' legislative package (European Commission 2021).

While the IMO strategy is viewed as a reasonable first step that is ambitious for the shipping industry, achieving the 'vision' of alignment with the temperature goals of the Paris Agreement requires concrete implementation measures and strengthened targets in the next iteration in 2023 (Doelle and Chircop 2019; Climate Action Tracker 2020c). As a step towards this, in 2020, the IMO's MEPC put forward draft amendments to the MARPOL Convention that would require ships to combine a technical and an operational approach to reduce their carbon intensity. These amendments were formally adopted by the Committee at its session in June 2021.

### 14.5.3 Civil Society and Social Movements

Transnationally organised civil society actors have had long-standing involvement in international climate policy, with a particular focus on consulting or knowledge-sharing where they are present in transnational climate governance initiatives (Michaelowa and Michaelowa 2017). The term 'civil society' generally denotes 'the voluntary association of individuals in the public sphere beyond the realms of the state, the market and the family' (de Bakker et al. 2013, p. 575). Whereas civil society organisations are usually involved in lobbying or advocacy activities in a public arena, social movements focus on mobilisation and action for social change (Daniel and Neubert 2019). Examples of civil society groups involved in international climate policy include non-governmental organisations (NGOs) such as Greenpeace International, the World Wide Fund for Nature, the Environmental Defense Fund, the World Resources Institute, Friends of the Earth and Earthjustice among many others, as well as NGO networks such as the Climate Action Network, which has over 1300 NGO members in more than 130 countries, working to promote government and individual action to limit human-induced climate change to ecologically sustainable levels (Climate Action Network International 2020). The influence of civil society engagement in global climate governance is well acknowledged, with these organisations' globally dispersed constituencies and non-state status offering perspectives that differ in significant ways from those of many negotiating states (Derman 2014).

Historically, the issue of climate change did not give rise to intense, organised transnational protest characteristic of social movements (McAdam 2017). During the 1990s and early 2000s, the activities of the global climate movement were concentrated in developed countries and largely sought to exercise influence through participation in UNFCCC COPs and side events (Almeida 2019). The mid-2000s onwards, however, saw the beginnings of use of more non-institutionalised tactics, such as simultaneous demonstrations across several countries, focusing on a grassroots call for climate justice that grew out of previous environmental justice movements (Almeida 2019). Groups representing indigenous peoples, youth, women, and labour rights brought to the fore new tools of contention and new issues in the UNFCCC, such as questions of a just transition and gender equity (Allan 2020).

Climate justice has been variously defined, but centres on addressing the disproportionate impacts of climate change on the most vulnerable populations and calls for community sovereignty and functioning (Schlosberg and Collins 2014; Tramel 2016). Contemporary climate justice groups mobilise multiple strands of environmental justice movements from the Global North and South, as well as from distinct indigenous rights and peasant rights movements, and are organised as a decentralised network of semiautonomous, coordinated units (Claeys and Delgado Pugley 2017; Tormos-Aponte and García-López 2018). The climate justice movement held global days of protest in most of the world's countries in 2014 and 2015, and mobilised another large campaign in 2018 (Almeida 2019). The polycentric arrangement of the global climate movement allows simultaneous influence on multiple sites of climate governance, from the local to the global levels (Tormos-Aponte and García-López 2018).

Prominent examples of new climate social movements that operate transnationally are Extinction Rebellion and Fridays for Future, which collectively held hundreds of coordinated protests across the globe in 2019–2021, marking out 'the transnational climate justice movement as one of the most extensive social movements on the planet' (Almeida 2019). Fridays for Future is a children's and youth movement that began in August 2018, inspired by the actions of then 15-year old Greta Thunberg who pledged to strike in front of the Swedish parliament every Friday to protest against a lack of action on climate change in line with the Paris Agreement targets (Fridays for Future 2019). Fridays for Future events worldwide encompass more than 200 countries and millions of strikers. The movement is unusual for its focus on children and the rights of future generations, with children's resistance having received little previous attention in the literature. Fridays for Future is regarded as a progressive resistance movement that has quickly achieved global prominence (for example, Thunberg was invited to address governments at the UN Climate Summit in New York in September 2019) and is credited with helping to support the discourse about the responsibility of humanity as a whole for climate change (Holmberg and Alvinius 2019). Whereas Fridays for Future has focused on periodic protest action, Extinction Rebellion has pursued a campaign based on sustained non-violent direct citizen action that is focused on three key demands: declaration of a 'climate emergency', acting now to halt biodiversity loss and reduce greenhouse gas emissions to net zero by 2025, and creation of a citizen's assembly on climate and ecological justice (Booth 2019; Extinction Rebellion 2019). The movement first arose in the United Kingdom– where it claimed credit for adoption of a climate emergency declaration by the UK government – but now has a presence in 45 countries with some 650 groups having formed globally (Gunningham 2019).

The Paris Agreement's preamble explicitly recognises the importance of engaging 'various actors' in addressing climate change, and the decision adopting the Agreement created the Non-state Actor Zone for Climate Action platform to aid in scaling up these efforts. Specific initiatives have also been taken to facilitate participation of particular groups, such as the UNFCCC's Local Communities and Indigenous Peoples Platform, which commenced work in Katowice in 2019. Climate movements based in the Global South, as well as in indigenous territories, are playing an increasingly important role in transnational negotiations through networks such as the Indigenous Peoples Platform. These groups highlight the voices and perspectives of communities and peoples particularly affected by climate change. For instance, the Pacific Climate Warriors is a grassroots network of young people from various countries in the

Pacific Islands region whose activities focus on resisting narratives of future inevitability of their Pacific homelands disappearing, and re-envisioning islanders as warriors defending rights to homeland and culture (McNamara and Farbotko 2017). Youth global climate activism, particularly involving young indigenous climate activists, is another notable recent development. Although there remains little published literature on indigenous youth climate activism (MacKay et al. 2020), analysis of online sources indicates the emergence of several such groups, including the Pacific Climate Warriors and Te Ara Whatu from Aotearoa New Zealand (Ritchie 2021), as well as Seed Mob in Australia.

Transnational civil society organisations advocating for climate justice in global governance have articulated policy positions around rights protections, responsibility-based approaches to climate finance, and the need for transparency and accountability (Derman 2014). Another recent area of activity, which overlaps with that of emerging investor alliances (Section 14.5.4), is the sustainability of capital investment in fossil fuel assets. Efforts to shift away from fossil fuels led by civil society include the Beyond Coal Campaign (in the USA and Europe) and the organisation for a Fossil Fuel Non-proliferation Treaty. 350.org has supported mobilisation of youth and university students around a campaign of divestment that has grown into a global movement (Gunningham 2019). As Mormann (2020) notes, as of November 2020 'more than 1,200 institutional investors managing over USD14 trillion of assets around the world have committed to divest some or all of their fossil fuel holdings'. Studies suggest that the direct impacts of the divestment movement have so far been small, given a failure to differentiate between different types of fossil fuel companies, a lack of engagement with retail investors, and a lack of guidance for investors on clean energy re-investment (Osofsky et al. 2019; Mormann 2020). The movement has had a more significant impact on public discourse by raising the profile of climate change as a financial risk for investors (Bergman 2018). Blondeel et al. (2019) also find that broader appeal of the divestment norm was achieved when moral arguments were linked to financial ones, through the advocacy of economic actors, such as the Bank of England's governor.

Climate justice campaigns by transnational civil society organisations increasingly embrace action through the courts. Chapter 13 discusses the growth and policy impact of such 'climate litigation' brought by civil society actors in domestic courts, which is attracting increasing attention in the literature (Setzer and Vanhala 2019; Peel and Osofsky 2020). Transnational and international court actions focused on climate change, by contrast, have been relatively few in number (Peel and Lin 2019). This reflects – at least in part – the procedural hurdles to bringing such claims, as in many international courts and tribunals (outside of the area of human rights or investor–state arbitration) litigation can only be brought by states (Bruce 2017). However, there have been active discussions about seeking an advisory opinion from the International Court of Justice (ICJ) on states' international obligations regarding the reduction of greenhouse gas emissions (Sands 2016; Wewerinke-Singh and Salili 2020), or bringing a case to the International Tribunal for the Law of the Sea on marine pollution harms caused by climate change (Boyle 2019). In September 2021 the Government of Vanuatu announced a campaign to seek an advisory opinion from the ICJ. The aim of climate litigation more generally is to supplement other regulatory efforts by filling gaps and ensuring that interpretations of laws and policies are aligned with climate mitigation goals (Osofsky 2010).

The overall impact of transnationally-organised civil society action and social movements for international cooperation on climate change mitigation has not been comprehensively evaluated in the literature. This may reflect the polycentric organisation of the movement, which poses challenges for coordinating between groups operating in different contexts, acting with different strategies and around multiple issues, and lobbying multiple decision-making bodies at various levels of government in a sustainable way (Tormos-Aponte and García-López 2018). There is some literature emerging on environmental defenders and their need for protection against violence and repression, particularly in the case of indigenous environmental defenders who face significantly higher rates of violence (Scheidel et al. 2020). Scheidel et al. (2020) also find that combining strategies of preventive mobilisation, protest diversification and litigation can enhance rates of success for environmental defenders in halting environmentally destructive projects. In the area of climate litigation, commentators have noted the potential for activists and even researchers to suffer retaliation through the courts as a result of 'strategic lawsuits against public participation' and lawsuits against researchers brought by fossil fuel interests (Setzer and Byrnes 2019; Setzer and Benjamin 2020). Influence of social movements may be enhanced through taking advantage of 'movement spillover' (the involvement of activists in more than one movement) (Hadden 2014) and coordination of activities with a range of 'non-state governors', including cities, sub-national governments, and investor groups (Gunningham 2019). Studies of general societal change suggest that once 3.5% of the population are mobilised on an issue, far-reaching change becomes possible (Gladwell 2002; Chenoweth and Belgioioso 2019) – a tipping point that may be approaching in the case of climate change (Gunningham 2019). As noted in Chapter 5, in the particular case of low-carbon technologies, 'if 10–30% of the population were to demonstrate commitment to low-carbon technologies, behaviours, and lifestyles, new social norms would be established'.

### 14.5.4 Transnational Business and Public-Private Partnerships and Initiatives

Combined national climate commitments fall far short of the Paris Agreement's long-term temperature goals. Similar political ambition gaps persist across various areas of sustainable development. Many therefore argue that actions by non-state actors, such as businesses and investors, cities and regions, and NGOs, are crucial. However, non-state climate and sustainability actions may not be self-reinforcing but may heavily depend on supporting mechanisms. Governance risk-reduction strategies can be combined to maximize non-state potential in sustainable and climate-resilient transformations (Chan et al. 2019).

An important feature of the evolving international climate policy landscape of recent years is the entrepreneurship of UN agencies such as UNEP and UNDP, as well as international organisations such as the World Bank, in initiating public-private partnerships (PPPs). Andonova (2017) calls this 'governance entrepreneurship'. Such partnerships can be defined as 'voluntary agreements between public actors (international organisations, states, or sub-state public authorities) and non-state actors (non-governmental organisations (NGOs), companies, foundations, etc.) on a set of governance objectives and norms, rules, practices, and/or implementation procedures and their attainment across multiple jurisdictions and levels of governance' (Andonova 2017). Partnerships may carry out different main functions: first, *policy development*, establishing new agreements on norms, rules, or standards among a broader set of governmental and non-governmental actors; second, *enabling implementation and delivery of services*, by combining resources from governmental and non-governmental actors; and, third, *knowledge production and dissemination*, to for example, the evolution of relevant public policies.

An example of a prominent PPP in the area of climate mitigation is the Renewable Energy Network (REN21 2019), which is a global multi-stakeholder network focused on promoting renewable energy policies in support of the transition to renewable energy through knowledge, established in 2004. It includes members from industry, NGOs, intergovernmental organisations, and science and academia. Another example is the Green Economy Coalition founded in 2009 to bring to bear the perspectives of workers, business, poor people, the environment community, and academics in the transition to a greener and more sustainable economy. Another example is that in 2015, Peru, in collaboration with France and the UNFCCC Secretariat, launched the Non-state Actor Zone for Climate Action, an online platform to showcase commitments to climate action by companies, cities, regions and investors (Chan et al. 2016; Bertoldi et al. 2018). More recently, the UNFCCC Race to Zero initiative led by High-level Climate Champions Nigel Topping and Gonzalo Muñoz seeks to mobilise actors beyond national governments to join the Climate Ambition Alliance and pursue net zero $CO_2$ targets. Its membership includes 454 cities, 23 regions, 1391 businesses, 74 of the biggest investors, and 569 universities.

PPPs may also be developed to assist with implementation and support of states' climate mitigation commitments. For instance, UNEP has initiated a number of PPPs for climate change finance. These are designed to increase financing for the purposes of disseminating low-carbon technologies to tackle climate change and promote clean energy in many parts of developing countries (UNEP 2018b; Charlery and Traerup 2019).

In the same vein, in 2010 FAO delivered the Framework for Assessing and Monitoring Forest Governance. The Framework draws on several approaches currently in use or under development in major forest governance-related processes and initiatives, including the World Bank's Framework for Forest Governance Reform. The Framework builds on the understanding that governance is both the context and the product of the interaction of a range of actors and stakeholders with diverse interests (FAO 2010). For example, UNFCCC and the UN-REDD programme focus on REDD+ and UNEP focuses on The Economics of Ecosystems and Biodiversity (TEEB), institutional mechanisms that have been conceptualised as a 'win-win-win' for mitigating climate, protecting biodiversity and conserving indigenous culture by institutionalising payments on carbon sequestration and biodiversity conservation values of ecosystems services from global to local communities. These mechanisms include public-private partnership, and NGO participation. REDD+ and TEEB allocation policies will be interventions in a highly complex system, and will inevitably involve trade-offs; therefore, it is important to question the 'win-win-win' discourse (Zia and Kauffman 2018; Goulder et al. 2019). The initial investment and the longer periods of recovery of investment are sometimes barriers to private investment. In this sense, it is important to have government incentives and encourage public-private investment (Ivanova and Lopez 2013).

The World Bank has also established several partnerships since 2010, mainly in the field of carbon pricing. Prominent examples are the Networked Carbon Markets initiative (established 2013' spanning both governmental actors and experts' now entering a phase II) and the Carbon Pricing Leadership Coalition, established in 2015 and spanning a wide range of governmental and non-governmental actors, not least within business (World Bank 2018; World Bank 2019; Wettestad et al. 2021). These partnerships deal with knowledge production and dissemination and seek to enable implementation of carbon pricing policies. The leadership role of the international 'heavyweight' World Bank gives these partnerships additional comparative political weight, meaning also a potentially greater involvement of powerful finance ministries/ministers generally involved in Bank matters and meetings.

PPPs for cooperation on climate mitigation goals have emerged at multiple levels of governance beyond the realm of international organisations. For example, PPP funding for cities expanded rapidly in the 1990s and outpaced official external assistance almost tenfold. Most of the PPP infrastructure investment has been aimed at telecommunications, followed by energy. However, with the exception of the telecommunications sector, PPP investments have generally bypassed low-income countries (Ivanova 2017). It is therefore not surprising that PPPs have added relatively little to the financing of urban capital in developing countries over the past two decades (Bahl and Linn 2014). Liu and Waibel (2010) argue that the inherent risk of urban investment is the main obstacle to increasing the flow of private capital. Nevertheless, there have been cases where PPP investments have exceeded official external aid flows even for water and sanitation, and highly visible projects have been funded with PPPs in selected metropolitan areas of developing countries, including urban rail projects in Bangkok, Kuala Lumpur, and Manila (Liu and Waibel 2010).

Local governments are also creating cross-sector social partnerships (CSSPs) at the sub-national level, entities created for addressing social, economic, and/or environmental issues with partner organisations from the public, private and civil society sectors (Crane and Seitanidi 2014). In particular, with support from international networks such as ICLEI Local Governments for Sustainability, C40, Global Covenant of Mayors, and Global 100% Renewable Energy, local governments around the world are committing to aggressive carbon reduction targets for their cities (Ivanova et al. 2015; Clarke and Ordonez-Ponce 2017; Kona et al. 2018). Research on CSSPs implementing community sustainability plans shows that climate change is one of the four most common issues, after waste, energy and water (which are also highly relevant to climate mitigation) (MacDonald et al. 2017).

Community climate action plans consider all GHGs emitted within the local geographic boundaries, including from industry, home heating, burning fuel in vehicles, and so on. It is these community plans that require large multi-stakeholder partnerships to be successful. Partners in these partnerships generally include the local government departments, other government departments, utilities, large businesses, Chambers of Commerce, some small and medium-sized enterprises, universities, schools, and local civil society groups (Clarke and MacDonald 2016). Research shows that the partnership's structural features enable the achievement of plan outcomes, such as reducing GHG emissions, while also generating value for the partners (Austin and Seitanidi 2012; Clarke and MacDonald 2016; Clarke and Ordonez-Ponce 2017). Stua (2017b) explores the Mitigation Alliances (MAs) on the national level. The internal governance model of MAs consists of overarching authorities mandated to harmonise the overall organisational structure. These authorities guarantee an effective, equitable and transparent functioning of the MA's pillars (the demand, supply, and exchange of mitigation outcomes), in line with the principles and criteria of the Paris Agreement. This hybrid governance model relies upon its unique links with international climate institutions (Stua 2017a).

Transnational business partnerships are a growing feature of the landscape of multi-level, multi-actor governance of climate change. Many business leaders embraced the ethos of 'business cannot succeed in societies that fail'. Examples of this line of reasoning are: poverty limits consumer spending, political instability disrupts business activity, and climate change threatens the production and distribution of goods and services. Such situations endanger multinational enterprise investments, global asset management funds, and the core business of international insurance companies and pension funds (van Tulder et al. 2021).

A leading example is the World Business Council on Sustainable Development (WBCSD), a global, CEO-led organisation of over 200 leading businesses working together to accelerate the transition to a sustainable world. Member companies come from all business sectors and all major economies, representing a combined revenue of more than USD8.5 trillion and with 19 million employees. The WBCSD aims to enhance 'the business case for sustainability through tools, services, models and experiences'. It includes a Global Network of almost 70 national business councils across the globe. The overall vision is to create a world where more than 9 billion people are all living well and within the boundaries of our planet, by 2050. Vision 2050, released in 2010, explored what a sustainable world would look like in 2050, how such a world could be realised, and the role that business can play in making that vision a reality. A few years later, Action2020 took that Vision and translated it into a roadmap of necessary business actions and solutions (WBCSD 2019). WBCSD focuses on those areas where business operates and can make an impact. They identify six transformation systems that are critical in this regard: Circular Economy, Climate and Energy, Cities and Mobility, Food and Nature, People and Redefining Value. All have an impact on climate. An important initiative launched in September 2008 –Natural Climate Solutions – has the objective of leveraging business investment to capture carbon out of the atmosphere. This initiative has built strong cross-sectoral partnerships and is intended to tap into this immense emissions reduction solution potential through natural methods with the help of private investment.

The Global Methane Initiative (GMI) is a multilateral partnership launched in 2004 by the United States Environmental Protection Agency along with 36 other countries to generate a voluntary, non-binding agenda for global collaboration to decrease anthropogenic methane releases. The GMI builds on the Methane to Markets (M2M) Partnership, an international partnership launched in 2004. In addition to the GMI's own financial assistance, the initiative receives financial backing from the Global Methane Fund (GMF) for methane reduction projects. The GMF is a fund created by governments and private donors (Leonard 2014).

Another potentially influential type of transnational business partnership is investor coalitions or alliances formed for the purpose of pushing investee companies to adopt stronger measures for stranded asset management and climate change mitigation. MacLeod & Park (2011, p. 55) argue that these transnational groups 'attempt to re-orient and "regulate" the behaviour of business by holding corporations accountable via mechanisms of information sharing, monitoring of environmental impacts, and disclosure of activities related to the corporate climate footprint'. This favours a theory of active ownership (investor engagement with corporate boards) over capital divestment as the optimal pathway to shape the behaviour of corporate actors on climate risk (Kruitwagen et al. 2017; Krueger et al. 2020).

Transnational cooperative action by investors on climate mitigation has been facilitated by international standard-setting on issues of climate risk and disclosure. For example, in 2017 the Financial Stability Board's Taskforce on Climate-related Financial Disclosures (TCFD) adopted international recommendations for climate risk disclosure (TCFD 2017). These recommendations, which apply to all financial-sector organisations, including banks, insurance companies, asset managers, and asset owners, have received strong support from investor coalitions globally, including Climate Action 100+ (with 300 investors with more than USD33 trillion in assets under management), the Global Investor Coalition on Climate Change (a coalition of regional investor groups across Asia, Australia, Europe and North America) and the Institutional Investors Group on Climate Change (IIGCC). One of the key recommendations of the TCFD calls for stress-testing of investment portfolios taking into consideration different climate-related scenarios, including a 2°C or lower scenario. Broad adoption of the TCFD recommendations could provide a basis for decisions by investors to shift assets away from climate-risk exposed assets such as fossil fuel extraction projects (Osofsky et al. 2019). There is strong evidence showing the urgent need for scaling-up climate finance to mitigate greenhouse gases in line with pursuit of limiting the temperature increase to 1.5°C above pre-industrial levels, and to support adaptation to safeguard the international community from the consequences of a changing climate. While public actors have a responsibility to deploy climate finance, it is clear that the contribution from the private sector needs to be significant (Gardiner et al. 2016).

As most of these partnerships are of recent vintage an assessment of their effectiveness is premature. Instead, partnerships can be assessed on the basis of the three main functions introduced earlier. Starting with policy development, that is, establishing new agreements on norms, rules, or standards among a broader set of governmental and non-governmental actors, this is not the most prominent aspect of partnerships so far, although both the cities' networks and risk disclosure recommendations include some elements of this. The second element, enabling implementation and delivery of services, by combining resources from governmental and non-governmental actors, seems to be a more prominent part of the partnerships (Ivanova et al. 2020). Both UNEP financing, the WBCSD, the REDD+ and TEEB mechanisms, and PPP funding for cities are examples here. Finally, the third element, knowledge production and dissemination, for example, contributing to the evolution of relevant public policies, is the most prominent part of these partnerships, with the majority including such activities.

There is a relatively large volume of literature that assesses PPPs in general. Much of this applies to partnerships which, either by design or not, advance climate goals. This literature provides a good starting point for assessing these partnerships as they become operational. These can help assess whether such partnerships are worth the effort in terms of their performance and effectiveness (Liu et al. 2017b), their economic and social value added (Quélin et al. 2017), their efficiency (Estache and Saussier 2014) and the possible risks associated with them (Grimsey and Mervyn 2002).

What is less common, but gradually growing, is an important and more relevant literature on criteria to assess sustainability and impact on climate and development goals. Michaelowa and Michaelowa (2017) assess 109 transnational partnerships and alliances based on four design criteria: existence of mitigation targets; incentives for mitigation; definition of a baseline; and existence of a monitoring, reporting, and verification procedure . About half of the initiatives do not meet any of these criteria, and not even 15% satisfy three or more. A recent study using a systematic review of business and public administration literature on PPPs concludes that research in the past rarely incorporated sustainability concepts. The authors propose a research agenda and a series of success factors that, if appropriately managed, can contribute to sustainable development, and in so doing contribute to a more solid scientific evaluation of PPPs (Pinz et al. 2018). There is evidence that with the adoption of the Sustainable Development Goals, many of which are directly linked to climate goals, PPPs will become even more prominent as they will be called upon to provide resources, knowledge, expertise, and implementation support in a very ambitious agenda. PPPs in the developing world need to take into account different cultural and social decision-making processes, language differences, and unfamiliar bureaucracy (Gardiner et al. 2016). Having more evidence on what norms and standards in relation to sustainability are used and their governance is essential (Axel 2019). The issue of double counting should be revised. GHGs are accounted both at the national and sub-national level or company level (Schneider et al. 2014). Some recent studies aim to provide systems to assess the impact of PPPs beyond the much-used notion of value for money. One of these recent studies proposes a conceptual model that addresses six dimensions relevant to economic, social and environmental progress. These include resilience and environment, access of services to the population, scalability and replicability, economic impact, inclusiveness, and finally, degree of engagement of stakeholders (Berrone et al. 2019). These systems will most likely continue to evolve.

### 14.5.5 International Cooperation at the Sub-national and City Levels

Local and regional governments have an important role to play in global climate action, something recognised by the Paris Agreement, and also assessed in Sections 13.3.2 and 13.3.4 of this report. There are several ways they can be useful. First, sub-national governments can contribute insights and experience that provide valuable lessons to national governments, as well as offering needed implementation capacity (GIZ 2017; Leffel 2018). A great deal of policymaking has occurred at the level of city governments in particular. Cities have been responsible for more than 70% of global greenhouse gas emissions and generate over 80% of global income (World Bank 2010), and many of them have started to take their own initiative in enacting and developing mitigation policies (CDP 2015). Most of these activities aim at the reduction of GHG emissions in the sectors of energy, transportation, urban land use and waste (Bulkeley 2010; Xuemei 2007), and are motivated by concerns not only over climate, but also a consideration of local co-benefits (Rashidi et al. 2017, 2019). Second, sub-national governments can fill the void in policy leadership in cases where national governments are ineffectual, even to the point of claiming leadership and authority with respect to foreign affairs (Leffel 2018). International cooperation plays a role in such action. Several international networks, such as C40, ICLEI, Mayors for Climate Protection, and the Covenant of Mayors have played an important role in defining and developing climate-policy initiatives at the city level (Fünfgeld 2015). While the networks differ from each other, they generally are voluntary and non-hierarchical, intended to support the horizontal diffusion of innovative climate policies through information-sharing platforms linked to specific goals that member cities make (Kern and Bulkeley 2009). The literature has addressed the questions of why cities join the networks (Betsill and Bulkeley 2004; Pitt 2010), what recognition benefits cities can expect (Buis 2009; Kern and Bulkeley 2009), and how memberships can provide visibility to leverage international funding (Betsill and Bulkeley 2004; Heinrichs et al. 2013). Membership in the networks has been found to be a significant predictor of cities' adoption of mitigation policies, even when controlling for national-level policies that may be in place (Rashidi and Patt 2018). Kona et al. (2018) find that cities belonging to the Covenant of Mayors are engaging in emissions reductions at a rate consistent with achieving a 2°C global temperature target. Kona et al. (2021) document this trend continuing.

With respect to their role in formal international cooperation, however, it is unclear what authority, as a non-state actor, they actually have. Cities, for example, are members of transnational initiatives aimed at non-state actors, such as Global Climate Action, originally the Non-state Actor Zone for Climate Action, under the UNFCCC. While there is reason to believe that such membership can add value to mitigation efforts, one study suggests that the environmental effects have yet to be reliably quantified (Hsu et al. 2019a). By contrast, Kuramochi et al. (2020) provide evidence that non-state actors are leading to significant emissions reductions beyond what countries would otherwise be achieving. In terms of institutional strength, Michaelowa and Michaelowa (2017) suggest that few such networks fulfil governance criteria, and hence challenge their effectiveness. Several researchers suggest that their role is important in informal ways, given issues about the legitimacy of non-state actors (Chan et al. 2016; Nasiritousi et al. 2016). Bäckstrand et al. (2017) advance the concept of 'hybrid multilateralism' as a heuristic to capture this intensified interplay between state and non-state actors in the new landscape of international climate cooperation. The effectiveness of such non-state government actors should be measured not only by their contribution to mitigation, but also by their success to enhance the accountability, transparency and deliberative quality of the UNFCCC and the Paris Agreement (Chan et al. 2015; Busby 2016; Hale et al. 2016). In the post-Paris era, effectiveness also revolves around how to align non-state and intergovernmental action in a comprehensive framework that can help achieve low carbon futures (Chan et al. 2016). Stua (2017b) suggests that networks involving non-state actors can play an important role in enhancing transparency. Such effectiveness has to be complemented also by *normative questions*, applying a set of democratic values: participation, deliberation, accountability, and transparency (Bäckstrand and Kuyper 2017). Such concepts of polycentric governance offer new opportunities for climate action, but it has been argued that it is too early to judge their importance and effects (Jordan et al. 2015).

## 14.6 Synthesis

### 14.6.1 Changing Nature of International Cooperation

The main development since AR5 in terms of international climate cooperation has been the shift from the Kyoto Protocol to the Paris Agreement as the primary multilateral driver of climate mitigation policy worldwide (Section 14.3). Most *ex-post* assessments of the Kyoto Protocol suggest that it did lead to emissions reductions in countries with binding targets, in addition to changing investment patterns in low-carbon technologies. As noted earlier, the Paris Agreement is tailored to the evolving understanding of the climate mitigation challenge as well as shifting political imperatives and constraints. Whether the Paris Agreement will in fact be effective in supporting global action sufficient to achieve its objectives is contested, with competing arguments in the scientific literature supporting different views. To some extent these views align with the different analytic frameworks (Section 14.2.1): the Paris Agreement does not address the free-riding issue seen as important within the global commons framing, but may provide the necessary incentives and support mechanisms viewed as important under the political and transitions framings, respectively. The strongest critique of the Paris Agreement is that current NDCs themselves fail by a wide margin to add up to the level of aggregate emissions reductions necessary to achieve the objectives of holding global average warming well below 2°C, much less 1.5°C (Section 14.3.3 Figure 14.2), and that there is no legally binding obligation to achieve the NDCs. Arguments in support of Paris are that it puts in place the processes, and generates normative expectations, that nudge NDCs to become progressively more ambitious over time, including in developing countries. The growing number of countries with mid-century net-zero GHG or $CO_2$ targets, consistent with Article 4 of the Paris Agreement, lends support to this proposition, although there is as yet no empirical literature drawing an unambiguous connection. The collective quantified goal from a floor of USD100 billion a year in transfers to developing countries, the Green Climate Fund and other provisions on finance in the Paris Agreement have also been recognised as key to cooperation (Sections 14.3.2.8 and 14.4.1). But then these arguments are met with counter arguments, that even with Paris processes in place, given the logic of iterative, rising levels of ambition over time, this is unlikely to happen within the narrow window of opportunity that exists to avert dangerous levels of global warming (Section 14.3.3). The degree to which countries are willing to increase the ambition and secure the achievement of their NDCs over time will be an important indicator of the success of the Paris Agreement; evidence of this was expected by the end of 2020, but the COVID-19 pandemic has delayed the process of updating NDCs.

An increasing role is also played by other cooperative agreements, in particular (potentially) under Article 6 (Sections 14.3.2.10 and 14.4.4), transnational partnerships, and the institutions that support them. This fits both a transitions narrative that cooperation at the sub-global and sectoral levels is necessary to enable specific system transformations, and a recent emphasis in the public goods literature on club goods and a gradual approach to cooperation, also referred to as building blocks or incremental approach (Sections 14.2 and 14.5.1.4). There has been little analysis of whether these other agreements are of sufficient scale and scope to ensure that transformations happen quickly enough. This chapter, appraising them together, concludes that they are not. First, many agreements, such as those related to trade, may stand in the way of bottom-up mitigation efforts (Section 14.5.1.3). Second, many sectoral agreements aimed at decarbonisation – such as within the air travel sector – have not yet adopted targets comparable in scale, scope or legal character to those adopted under the Paris Agreement (Section 14.5.2.3). Third, there are many sectors for which there are no agreements in place. At the same time, there are some important bright spots, many in the area of transnational partnerships. A growing number of cities have committed themselves to adopting urban policies that will place them on a path to rapid decarbonisation, while learning from each other how to implement successful policies to realise climate goals (Section 14.5.5). An increasing number of large corporations have committed to decarbonising their industrial processes and supply chains (Section 14.5.4). And an ever-increasing number of non-state actors are adopting goals and initiating mitigation actions (Section 14.5.3). These goals and actions, some argue, could bridge the mitigation gap created by inadequate NDCs, although the empirical literature to date challenges this, suggesting that there is less transparency and limited accountability for such actions, and mitigation targets and incentives are also not clear (Sections 14.3.3 and 14.5).

### 14.6.2 Overall Assessment of International Cooperation

This section provides an overall assessment of international cooperation, taking into account the combined effects of cooperation within the UNFCCC process, other global agreements, as well as regional, sectoral, and transnational processes. Recent literature consistent with the transitions framing highlights that cooperation can be particularly effective when it addresses issues on a sector-by-sector basis (Geels et al. 2019). Table 14.4 below summarises the effects of international cooperation on mitigation efforts in each of the sectoral areas covered in Chapters 5 to 12 of this report. As it indicates, there are some strong areas of sector-specific cooperation, but also some important weaknesses. Formal agreements and programmes, both multilateral and bilateral, are advancing mitigation efforts in energy, AFOLU, and transportation, while transnational networks and partnerships are addressing issues in urban systems, industry, and buildings. Although many of the concerns relevant for buildings may be embedded in the energy sector with respect to their operation, and the industrial sector with respect to their materials, reinforcing the networks with more formal agreements could be vital to putting these sectors on a pathway to net zero GHG or $CO_2$ emissions. Several of the sectors have very little formal cooperation at the international level, and a common theme across many of them is a need for increased financial flows to achieve particular objectives.

Table 14.5 provides examples of mechanisms addressing each of the assessment criteria identified in Section 14.2.3. The effects of different forms of international cooperation are separated out, including not only UNFCCC and other multilateral processes, but also sub-global and sectoral agreements. Several points stand out. First, the Paris Agreement has the potential to significantly advance the UN climate regime's transformative potential. Second, the international market mechanisms under Article 6 – should an agreement on implementation deals be reached – allow a shift from projects and programmes to policy-based and sectoral generation of emissions credits. Moreover, the sectoral agreement CORSIA also makes use of such credits. Third, there is a lack of attention to both distributive outcomes and institutional support within sectoral agreements, representing a serious gap in efforts to harmonise mitigation with equity and sustainable development. Fourth, there are transnational partnerships and initiatives, representing the actions of non-state actors, addressing each of the assessment criteria, with the exception of economic effectiveness.

**Table 14.4 | Effects of international cooperation on sectoral mitigation efforts.**

| Sector | Key strengths | Key gaps and weaknesses |
|---|---|---|
| Demand, services, social aspects | Adoption of SDGs addressing social inequities and sustainable development in the context of mitigation | Little international attention to demand-side mitigation issues |
| Energy | Greater incorporation of climate goals into sectoral agreements and institutions; formation of new specialised agencies (e.g., IRENA, SE4All) devoted to climate-compatible energy | Need for enhanced financial support to place low-carbon energy sources on an equal footing with carbon-emitting energy in developing countries; investor–state dispute settlement mechanisms designed to protect the interests of companies engaged in high-carbon energy supply from national policies; ensuring just transition; and, addressing stranded assets |
| AFOLU | Bilateral support for REDD+ activities; transnational partnerships disincentivising use of products from degraded lands | Need for increased global finance for forest restoration projects and REDD+ activities; failure of national governments to meet internationally agreed upon targets with respect to deforestation and restoration; no cooperative mechanisms in place to address agricultural emissions |
| Urban systems | Transnational partnerships enhancing the capacity of municipal governments to design and implement effective policies | Need for increased financial support for climate-compatible urban infrastructure development |
| Buildings | Transnational initiative aimed at developing regional roadmaps | Need for formal international cooperation to enhance mitigation activities in buildings |
| Transport | Sectoral agreements in aviation and shipping begin to address climate concerns | Need to raise the level of ambition in sectoral agreements consistent with the Paris Agreement and complete decarbonisation, especially as emissions from international aviation and shipping continue to grow, unaccounted for in NDCs |
| Industry | Transnational partnerships and networks encouraging the adoption of zero-emission supply chain targets | No formal multilateral or bilateral cooperation to address issues of decarbonisation in industry |
| Cross-sectoral, including CDR and SRM | International agreements addressing risks of ocean-based CDR | Lack of cooperative mechanisms addressing risks and benefits of SRM; lack of cooperative mechanisms addressing financial and governance aspects of land- and technology-based CDR |

**Table 14.5 | Illustrative examples of multi-level governance addressing criteria of effectiveness.**

| | Environmental effectiveness | Transformative potential | Distributive outcomes | Economic effectiveness | Institutional strength |
|---|---|---|---|---|---|
| UNFCCC | Stabilisation goal, and quasi-targets for industrialised countries | Financial mechanism; technology mechanism, provisions for capacity building | Financial mechanism, transfers from developed to developing countries; leadership role for industrialised countries listed in Annex 1 | | Reporting requirements; capacity building for national climate change offices |
| Kyoto Protocol | Binding national targets for industrialised countries | | Adaptation Fund; targets restricted to industrialised countries | Market-based mechanisms | Emissions accounting and reporting requirements; institutional capacity building |
| Paris Agreement | NDCs and the global stocktake | Mechanisms for capacity building and technology development and transfer | Furthering financial commitments under the UNFCCC, including enhanced transparency on finance | Voluntary cooperation | Mechanism for enhanced transparency |
| Other multilateral agreements (Montreal Protocol, SDG 7, etc.) | Phase out of ozone-depleting substances with high global warming potential; significant effects on GHG mitigation | Ozone Fund; technology transfer; development and sharing of knowledge and expertise | SDGs embedding mitigation in sustainable development | | Processes for adjustment and amendment; reporting requirements |
| Multilateral and regional economic agreements and institutions | Harmonised lending practices of MDBs; mainstreaming climate change into IMF practices; liberalisation of trade in climate-friendly goods and services; negative effect from regulatory chill | | Concessional financing agreements | | Potentially negative results from dispute settlement processes |
| Sectoral agreements and institutions | Climate mitigation targets and actions in AFOLU, energy, and transport | Institutions devoted to developing and deploying zero-carbon energy technologies (e.g., IRENA) | | Use of carbon offsets to reduce growth in emissions from aviation | |
| Transnational networks and partnerships | Youth climate movement raising mitigation and fossil fuel divestment on political agendas and in financial sector | Non-state actor commitments to renewable energy-based supply chains | Climate justice legal initiatives | | City networks providing information exchange and technical support |

## 14.7 Knowledge Gaps

Any assessment of the effectiveness of international cooperation is limited by the methodological challenge of observing sufficient variance in cooperation in order to support inference on effects. There is little in the way of cross-sectional variance, given that most of the governance mechanisms assessed here are global in their geographical coverage. One exception is with respect to the effects of the Kyoto Protocol, which we have reported. Time series analysis is also challenging, given that other determinants of climate mitigation, including technology costs and the effects of national and sub-national level policies, are rapidly evolving. Thus, this chapter primarily reviews scholarship that compares observations with theory-based counterfactual scenarios.

Many of the international agreements and institutions discussed in this chapter, in particular the Paris Agreement, are new. The logic and architecture of the Paris Agreement, in particular, breaks new ground, and there is limited evaluation of prior experience in the form of analogous treaties to draw on. Such instruments have evolved in response to geopolitical and other drivers that are changing rapidly, and will continue to shape the nature of international cooperation under it and triggered by it. The Paris Agreement is also, in common with other multilateral agreements, a 'living instrument' evolving through interpretative and operationalising rules, and forms of implementation, that Parties continue to negotiate at conferences year on year. It is a constant 'work in progress' and thus challenging to assess at any given point in time. The Paris Agreement also engages a larger set of variables – given its privileging of national autonomy and politics, integration with the sustainable development agenda, and its engagement with actions and actors at multiple levels – than earlier international agreements, which further complicates the task of tracing causality between observed effects and international cooperation through the Paris Agreement.

Understanding of the effectiveness of international agreements and institutions is driven entirely by theory-driven prediction of how the world will evolve, both with these agreements in place and without them. The predictions in particular are problematic, because governance regimes are complex adaptive systems, making it impossible to predict how they will evolve over time, and hence what their effects will be. Time will cure this in part, as it will generate observations of the world with the new regime in place, which we can compare to the counterfactual situation of the new regime's being absent, which may be a simpler situation to model. But even here our modelling capacity is limited: it may simply never be possible to know with a high degree of confidence whether international cooperation, such as that embodied in the Paris Agreement, is having a significant effect, no matter how much data are accumulated.

Given the importance of theory for guiding assessments of the past and likely future impacts of policies, it is important to note that among the alternative theoretical frameworks for analysis, some have been much more extensively developed in the literature than others. This chapter has noted in particular the partial dichotomy between a global commons framing of climate change and a transitions framing, which include different indicators to be used to evaluate the effectiveness of policies. The latter framing is particularly under-developed. Greater development of theories resting in social science disciplines such as economic geography, sociology, and psychology could potentially provide a more complete picture of the nature and effectiveness of international cooperation.

Frequently Asked Questions (FAQs)

## FAQ 14.1 | Is international cooperation working?

Yes, to an extent. Countries' emissions were in line with their internationally agreed targets: the collective greenhouse gas (GHG) mitigation target for Annex I countries in the UNFCCC to return to their 1990 emissions levels by 2000, and their individual targets in the Kyoto Protocol for 2008–12. Numerous studies suggest that participation in the Kyoto Protocol led to substantial reductions in national GHG emissions, as well increased levels of innovation and investment in low-carbon technologies. In this latter respect, the Kyoto Protocol set in motion some of the transformational changes that will be required to meet the temperature goal of the Paris Agreement. It is too soon to tell whether the processes and commitments embodied in the Paris Agreement will be effective in achieving its stated goals with respect to limiting temperature rise, adaptation, and financial flows. There is, however, evidence that its entry into force has been a contributing factor to many countries' adopting mid-century targets of net-zero GHG or $CO_2$ emissions.

## FAQ 14.2 | What is the future role of international cooperation in the context of the Paris Agreement?

Continued international cooperation remains critically important both to stimulate countries' enhanced levels of mitigation ambition, and through various means of support to increase the likelihood that they achieve these objectives. The latter is particularly the case in developing countries, where mitigation efforts often rely on bilateral and multilateral cooperation on low-carbon finance, technology support, capacity building, and enhanced South-South cooperation. The Paris Agreement is structured around Nationally Determined Contributions that are subject to an international oversight system, and bolstered through international support. The international oversight system is designed to generate transparency and accountability for individual emissions reduction contributions, and regular moments for stock-taking of these efforts towards global goals. Such enhanced transparency may instil confidence and trust, and foster solidarity among nations, with theory-based arguments that this will lead to greater levels of ambition. Together with other cooperative agreements at the sub-global and sectoral levels, as well as a growing number of transnational networks and initiatives, the implementation of all of these mechanisms is likely to play an important role in making political, economic, and social conditions more favourable to ambitious mitigation efforts in the context of sustainable development and efforts to eradicate poverty.

## FAQ 14.3 | Are there any important gaps in international cooperation, which will need to be filled in order for countries to achieve the objectives of the Paris Agreement, such as holding temperature increase to well below 2°C and pursuing efforts towards 1.5°C above pre-industrial levels?

While international cooperation is contributing to global mitigation efforts, its effects are far from uniform. Cooperation has contributed to setting a global direction of travel, and to falling greenhouse gas emissions in many countries and avoided emissions in others. It remains to be seen whether it can achieve the kind of transformational changes needed to achieve the Paris Agreement's long-term global goals. There appears to be a large potential role for international cooperation to better address sector-specific technical and infrastructure challenges that are associated with such transformational changes. Finalising the rules to pursue voluntary cooperation, such as through international carbon market mechanisms and public climate finance in the implementation of NDCs, without compromising environmental integrity, may play an important role in accelerating mitigation efforts in developing countries. Finally, there is room for international cooperation to more explicitly address transboundary issues associated with carbon dioxide removal and solar radiation management.

## References


Aakre, S., S. Kallbekken, R. Van Dingenen, and D.G. Victor, 2018: Incentives for small clubs of Arctic countries to limit black carbon and methane emissions. *Nat. Clim. Change*, **8(1)**, 85–90, doi:10.1038/s41558-017-0030-8.

Abas, N. et al., 2018: Natural and synthetic refrigerants, global warming: A review. *Renew. Sustain. Energy Rev.*, **90**, 557–569, doi:10.1016/j.rser.2018.03.099.

Abatayo, A. et al., 2020: Solar geoengineering may lead to excessive cooling and high strategic uncertainty. *Proc. Natl. Acad. Sci.*, **117(24)**, 13393–13398, doi:10.1073/pnas.1916637117.

Acemoglu, D., P. Aghion, L. Bursztyn, and D. Hemous, 2012: The Environment and Directed Technical Change. *Am. Econ. Rev.*, **102(1)**, 131–166, doi:10.1257/aer.102.1.131.

Adelman, S., 2018: Human Rights in the Paris Agreement: Too Little, Too Late? *Transnatl. Environ. Law*, **7(1)**, 17–36, doi:10.1017/s2047102517000280.

Adger, W.N., S. Huq, K. Brown, and M. Hulme, 2003: Adaptation to climate change in the developing world. *Prog. Dev. Stud.*, **3(3)**, 179–195, doi:10.1191/1464993403ps060oa.

Aghion, P., P. Antràs, and E. Helpman, 2007: Negotiating free trade. *J. Int. Econ.*, **73(1)**, 1–30, doi:10.1016/j.jinteco.2006.12.003.

Aghion, P., C. Hepburn, A. Teytelboym, and D. Zenghelis, 2014: *Path dependence, innovation and the economics of climate change,* Working Paper, Centre for Climate Change Economics and Policy/Grantham Research Institute on Climate Change and the Environment Policy Paper & Contributing paper to New Climate Economy, 17 pp. www.lse.ac.uk/granthaminstitute/wp-content/uploads/2014/11/Aghion_et_al_policy_paper_Nov20141.pdf (Accessed October 31, 2021).

Ahlm, L. et al., 2017: Marine cloud brightening – As effective without clouds. *Atmos. Chem. Phys.*, **17(21)**, 13071–13087, doi:10.5194/acp-17-13071-2017.

Aichele, R. and G. Felbermayr, 2013: The Effect of the Kyoto Protocol on Carbon Emissions. *J. Policy Anal. Manag.*, **32(4)**, 731–757, doi:10.1002/pam.21720.

Akanle, T., 2010: Impact of Ozone Layer Protection on the Avoidance of Climate Change: Legal Issues and Proposals to Address the Problem. *Rev. Eur. Community Int. Environ. Law*, **19(2)**, 239–249, doi:10.1111/j.1467-9388.2010.00680.x.

Aklin, M. and M. Mildenberger, 2020: Prisoners of the Wrong Dilemma: Why Distributive Conflict, Not Collective Action, Characterizes the Politics of Climate Change. *Glob. Environ. Polit.*, **20(4)**, 4–27, doi:10.1162/glep_a_00578.

Al Khourdajie, A. and M. Finus, 2020: Measures to enhance the effectiveness of international climate agreements: The case of border carbon adjustments. *Eur. Econ. Rev.*, **124(103405)**, 1–18, doi:10.1016/j.euroecorev.2020.103405.

Alcaraz, O. et al., 2019: The global carbon budget and the Paris agreement. *Int. J. Clim. Change Strateg. Manag.*, **11(3)**, 310–325, doi:10.1108/ijccsm-06-2017-0127.

Aldy, J.E., W.A. Pizer, and K. Akimoto, 2017: Comparing emissions mitigation efforts across countries. *Clim. Policy*, **17(4)**, 501–515, doi:10.1080/14693062.2015.1119098.

Allan, J.I., 2019: Dangerous Incrementalism of the Paris Agreement. *Glob. Environ. Polit.*, **19(1)**, 4–11, doi:10.1162/glep_a_00488.

Allan, J.I., 2020: *The New Climate Activism: NGO Authority and Participation in Climate Change Governance*. University of Toronto Press, Toronto, Canada, 226 pp.

Allgaier, J., 2019: Science and Environmental Communication on YouTube: Strategically Distorted Communications in Online Videos on Climate Change and Climate Engineering. *Front. Commun.*, **4(36)**, 1–15, doi:10.3389/fcomm.2019.00036.

Almeida, M., 2020: *Global Green Bond State of the Market 2019. Climate Bonds Initiative, July 2020*. 16 pp. www.climatebonds.net/system/tdf/reports/cbi_sotm_2019_vol1_04d.pdf (Accessed December 9, 2020).

Almeida, P., 2019: Climate justice and sustained transnational mobilization. *Globalizations*, **16(7)**, 973–979, doi:10.1080/14747731.2019.1651518.

Almer, C. and R. Winkler, 2017: Analyzing the effectiveness of international environmental policies: The case of the Kyoto Protocol. *J. Environ. Econ. Manage.*, **82**(C), 125–151, doi:10.1016/j.jeem.2016.11.003.

Ameli, N., P. Drummond, A. Bisaro, M. Grubb, and H. Chenet, 2020: Climate finance and disclosure for institutional investors: why transparency is not enough. *Clim. Change*, **160(4)**, 565–589, doi:10.1007/s10584-019-02542-2.

Anderson, W. and N. Peimbert, 2019: Across Latin America, Restoring Land is Climate Action. *Initiat. 20X20*,. https://initiative20x20.org/es/node/386 (Accessed October 29, 2021).

Anderson, Z.R., K. Kusters, J. McCarthy, and K. Obidzinski, 2016: Green growth rhetoric versus reality: Insights from Indonesia. *Glob. Environ. Change*, **38**, 30–40, https://doi.org/10.1016/j.gloenvcha.2016.02.008.

Andler, L. and S. Behrle, 2009: *Managers of Global Change: The Influence of International Environmental Bureaucracies* [Biermann, F. and B. Siebenhüner, (eds.)]. MIT Press, Cambridge, MA, USA.

Andonova, L.B., 2017: *Governance Entrepreneurs: International Organizations and the Rise of Global Public-Private Partnerships*. Cambridge University Press, Cambridge, UK and New York, NY, USA, 275 pp.

Andresen, S., 2014: Exclusive Approaches to Climate Governance: More Effective than the UNFCCC? In: *Toward A New Climate Agreement: Conflict, Resolution and Governance* [Cherry, T., J. Hovi, and D.M. McEvoy, (eds.)]. Routledge, London and New York, pp. 167–181.

Andresen, S.,T. Skodvin, A. Underdal, and J. Wettestad, 2000: *Science and Politics in International Environmental Regimes*. [Andresen, S.,T. Skodvin, A. Underdal, and J. Wettestad, (eds.)]. Manchester University Press, Manchester, UK, 221 pp.

Angelsen, A., 2017: REDD+ as Result-based Aid: General Lessons and Bilateral Agreements of Norway. *Rev. Dev. Econ.*, **21(2)**, 237–264.

Anouliès, L., 2015: The Strategic and Effective Dimensions of the Border Tax Adjustment. *J. Public Econ. Theory*, **17(6)**, 824–847, doi:10.1111/jpet.12131.

Antimiani, A., V. Costantini, A. Markandya, E. Paglialunga, and G. Sforna, 2017: The Green Climate Fund as an effective compensatory mechanism in global climate negotiations. *Environ. Sci. Policy*, **77**, 49–68, doi:10.1016/j.envsci.2017.07.015.

Antwi-Agyei, P., A.J. Dougill, T.P. Agyekum, and L.C. Stringer, 2018: Alignment between nationally determined contributions and the sustainable development goals for West Africa. *Clim. Policy*, **18(10)**, 1296–1312, doi:10.1080/14693062.2018.1431199.

Arino, Y. et al., 2016: Estimating option values of solar radiation management assuming that climate sensitivity is uncertain. *Proc. Natl. Acad. Sci.*, **113(21)**, 5886–5891, doi:10.1073/pnas.1520795113.

Ascensão, F. et al., 2018: Environmental challenges for the Belt and Road Initiative. *Nat. Sustain.*, **1(5)**, 206–209, doi:10.1038/s41893-018-0059-3.

Asheim, G.B., C.B. Froyn, J. Hovi, and F.C. Menz, 2006: Regional versus global cooperation for climate control. *J. Environ. Econ. Manage.*, **51(1)**, 93–109, doi:10.1016/j.jeem.2005.04.004.

Asheim, G.B. et al., 2019: The case for a supply-side climate treaty. *Science*, **365(6451)**, 325–327, doi:10.1126/science.aax5011.

Asian Development Bank, 2019: *High Level MDB Statement – For Publication at the UNSG Climate Action Summit, 22 September 2019*. www.adb.org/sites/default/files/page/41117/climate-change-finance-joint-mdb-statement-2019-09-23.pdf (Accessed October 31, 2021).

Asmelash, H.B., 2015: Energy Subsidies and WTO Dispute Settlement: Why Only Renewable Energy Subsidies Are Challenged. *J. Int. Econ. Law*, **18(2)**, 261–285, doi:10.1093/jiel/jgv024.

AUDA-NEPAD, 2019: African Forest Landscape Restoration Initiative. *African Union Dev. Agency*.

Austin, J.E. and M.M. Seitanidi, 2012: Collaborative Value Creation. *Nonprofit Volunt. Sect. Q.*, **41(5)**, 726–758, doi:10.1177/0899764012450777.

Axel, M., 2019: Public-Private Partnerships for Sustainable Development: Exploring their design and Impact on Effectiveness. *Sustainability*, **11(1087)**, doi:10.3390/su11041087.

Baatz, C., 2016: Can we have it both ways? On potential trade-offs between mitigation and solar radiation management. *Environ. Values*, **25(1)**, 29–49, doi:10.3197/096327115x14497392134847.

Bacchus, J., 2016: *Global Rules for Mutually Supportive and Reinforcing Trade and Climate Regimes*. E15 Expert Group on Measures to Address Climate Change and the Trade System – Policy Options Paper. E15Initiative, International Centre for Trade and Sustainable Development, Geneva, Switzerland, 27 pp.

Bäckstrand, K., and J.W. Kuyper, 2017: The democratic legitimacy of orchestration: the UNFCCC, non-state actors, and transnational climate governance. *Env. Polit.*, **26(4)**, doi:10.1080/09644016.2017.1323579.

Bäckstrand, K., J.W. Kuyper, B.-O. Linnér, and E. Lövbrand, 2017: Non-state actors in global climate governance: from Copenhagen to Paris and beyond. *Env. Polit.*, **26(4)**, 561–579, doi:10.1080/09644016.2017.1327485.

Baghdadi, L., I. Martinez-Zarzoso, and H. Zitouna, 2013: Are RTA agreements with environmental provisions reducing emissions? *J. Int. Econ.*, **90(2)**, 378–390, doi.org/10.1016/j.jinteco.2013.04.001.

Bahl, R.W. and J.F. Linn, 2014: *Governing and Financing Cities in the Developing World*. 60 pp. Lincoln Institute of Land Policy", Cambridge, USA. It has 56 pages instead of 60. **ISBN-10 :** 1558442995, **ISBN-13 :** 978-1558442993.

Bailey, I. and T.H.J. Inderberg, 2018: Australia: domestic politics, diffusion and emissions trading design as a technical and political project. In: *The Evolution of Carbon Markets: Design and Diffusion* [Wettestad, J.; Gulbrandsen, L.H., (ed.)], Routledge, London, UK. pp. 124–144.

Bakhtiari, F., 2018: International cooperative initiatives and the United Nations Framework Convention on Climate Change. *Clim. Policy*, **18(5)**, 655–663, doi:10.1080/14693062.2017.1321522.

Baldwin, E., Y. Cai, and K. Kuralbayeva, 2020: To build or not to build? Capital stocks and climate policy. *J. Environ. Econ. Manage.*, **100**, 102235, doi:10.1016/j.jeem.2019.05.001.

Bang, G., J. Hovi, and T. Skodvin, 2016: The Paris Agreement: Short-Term and Long-Term Effectiveness. *Polit. Gov.*, **4(3)**, 209–218, doi:10.17645/pag.v4i3.640.

Banks, G.D. and T. Fitzgerald, 2020: A sectoral approach allows an artful merger of climate and trade policy. *Clim. Change*, **162(2)**, 165–173, doi:10.1007/s10584-020-02822-2.

Barrett, S., 1994: Self-Enforcing International Environmental Agreements. *Oxf. Econ. Pap.*, **46**, 878–894. www.jstor.org/stable/2663505

Barrett, S., 2008: Climate treaties and the imperative of enforcement. *Oxford Rev. Econ. Policy*, **24(2)**: 239–258, doi:10.1093/oxrep/grn015.

Barrett, S., 2013: Climate treaties and approaching catastrophes. *J. Environ. Econ. Manage.*, **66(2)**, 235–250, doi:10.1016/j.jeem.2012.12.004.

Barrett, S., 2018: Choices in the climate commons. *Science,* **362(6420)**, 1217 LP – 1217, doi:10.1126/science.aaw2116.

Barrett, S. and A. Dannenberg, 2016: An experimental investigation into 'pledge and review' in climate negotiations. *Clim. Change*, **138(1–2)**, 339–351, doi:10.1007/s10584-016-1711-4.

Battaglini, M. and B. Harstad, 2016: Participation and Duration of Environmental Agreements. *J. Polit. Econ.*, **124(1)**, 160–204, doi:10.1086/684478.

Bauer, M.W., C. Knill, and S. Eckhard, 2017: *International Bureaucracy: Challenges and Lessons for Public Administration Research*. Palgrave Macmillan, London, UK, 210 pp.

Bayramoglu, B., M. Finus, and J.-F. Jacques, 2018: Climate agreements in a mitigation-adaptation game. *J. Public Econ.*, **165**, 101–113, doi:10.1016/j.jpubeco.2018.07.005.

Beauregard, C., D. Carlson, S. Robinson, C. Cobb, and M. Patton, 2021: Climate justice and rights-based litigation in a post-Paris world. *Clim. Policy*, **21(5)**, 652–665, doi:10.1080/14693062.2020.1867047.

Beer, C. et al., 2010: Terrestrial Gross Carbon Dioxide Uptake: Global Distribution and Covariation with Climate. *Science,* **329(5993)**, 834–838, doi:10.1126/science.1184984.

Beiser-McGrath, L.F. and T. Bernauer, 2021: Domestic Provision of Global Public Goods: How Other Countries' Behavior Affects Public Support for Climate Policy. *Glob. Environ. Polit.*, **22(1)**, 117–138 doi:10.1162/glep_a_00612.

Bellamy, R. and J. Lezaun, 2017: Crafting a public for geoengineering. *Public Underst. Sci.*, **26(4)**, 402–417, doi:10.1177/0963662515600965.

Bellamy, R. and O. Geden, 2019: Govern $CO_2$ removal from the ground up. *Nat. Geosci.*, **12**, 874–876, doi:10.1038/s41561-019-0475-7.

Benjamin, L., 2021: *Companies and Climate Change: Theory and Law in the United Kingdom*. Cambridge University Press, Cambridge, UK and New York, NY, USA, 250 pp.

Benoit, C., 2011: Picking Tariff Winners: Non-Product Related PPMs and DSB Interpretations of Unconditionally within Article I:1. *Georg. J. Int. Law*, **42(2)**, 583–604.

Benveniste, H., O. Boucher, C. Guivarch, H. Le Treut, and P. Criqui, 2018: Impacts of nationally determined contributions on 2030 global greenhouse gas emissions: Uncertainty analysis and distribution of emissions. *Environ. Res. Lett.*, **13(1)**, doi:10.1088/1748-9326/aaa0b9.

Bergman, N., 2018: Impacts of the Fossil Fuel Divestment Movement: Effects on Finance, Policy and Public Discourse. *Sustainability*, **10(7)**, 2529, doi:10.3390/su10072529.

Bernasconi-Osterwalder, N. and A. Cosbey, 2021: Carbon and Controversy: Why we need global cooperation on border carbon adjustment. *Int. Inst. Sustain. Dev. Blog,*. www.iisd.org/articles/carbon-border-adjustment-global-cooperation (Accessed October 31, 2021).

Berrone, P. et al., 2019: EASIER: An evaluation Model for public private partnerships contributing to sustainable development. *Sustainability*, **11(8)**, 2339, doi:10.3390/su11082339.

Bertoldi, P., A. Kona, S. Rivas, and J.F. Dallemand, 2018: Towards a global comprehensive and transparent framework for cities and local governments enabling an effective contribution to the Paris climate agreement. *Curr. Opin. Environ. Sustain.*, **30**(C), 67–74, doi:10.1016/j.cosust.2018.03.009.

Besharati, N.A., C. Rawhani, and O.G. Rios, 2017: *A Monitoring And Evaluation Framework For South-South Cooperation*. NeST Africa, Johannesburg, 28 pp. www.jstor.org/stable/resrep25886 (Accessed October 31, 2021).

Best, R. and P.J. Burke, 2018: Adoption of solar and wind energy: The roles of carbon pricing and aggregate policy support. *Energy Policy*, **118**, 404–417, doi:10.1016/J.ENPOL.2018.03.050.

Betsill, M.M. and H. Bulkeley, 2004: Transnational Networks and Global Environmental Governance: The Cities for Climate Protection Program. *Int. Stud. Q.*, **48(2)**, 471–493, doi:10.1111/j.0020-8833.2004.00310.x.

Biermann, F. and I. Möller, 2019: Rich man's solution? Climate engineering discourses and the marginalization of the Global South. *Int. Environ. Agreements Polit. Law Econ.*, **19(2)**, 151–167, doi:10.1007/s10784-019-09431-0.

Biermann, F., P. Pattberg, H. van Asselt, and F. Zelli, 2009: The fragmentation of global governance architectures: A framework for analysis. *Glob. Environ. Polit.*, **9(4)**, doi:10.1162/glep.2009.9.4.14.

Blondeel, M., J. Colgan, and T. Van de Graaf, 2019: What drives norm success? Evidence from anti–fossil fuel campaigns. *Glob. Environ. Polit.*, **19(4)**, 63–84, doi:10.1162/glep_a_00528.

Blondeel, M., T. Van de Graaf, and T. Haesebrouck, 2020: Moving beyond coal: Exploring and explaining the Powering Past Coal Alliance. *Energy Res. Soc. Sci.*, **59**, 101304, doi:10.1016/j.erss.2019.101304.

Bodansky, D., 2013: The who, what, and wherefore of geoengineering governance. *Clim. Change*, **121(3)**, 539–551, doi:10.1007/s10584-013-0759-7.

Bodansky, D., 2015: Legally binding versus non-legally binding instruments. In: Towards a Workable and Effective Climate Regime [Barrett, S., C. Carraro and J. de Melo (eds.)]. CEPR Press,Centre for Economic Policy Research, London, pp. 155–165.

Bodansky, D., 2016: The legal character of the Paris agreement. *Rev. Eur. Comp. Int. Environ. Law*, **25(2)**, 142–150, doi:10.1111/reel.12154.

Bodansky, D. and L. Rajamani, 2016: The Evolution and Governance Architecture of the United Nations Climate Change Regime. In: *Global Climate Policy: Actors, Concepts, and Enduring Challenges* [Sprinz, D. and U. Luterbacher, (eds.)]. Cambridge University Press, Cambridge, UK and New York, NY, USA, pp. 13–66.

Bodansky, D., J. Brunnée, and L. Rajamani, 2017a: Introduction to International Climate Change Law. In: *International Climate Change Law*, Oxford University Press, pp. 1–34.

Bodansky, D., J. Brunnée, and L. Rajamani, 2017b: *International Climate Change Law*. First. Oxford University Press, Oxford, 374 pp.

Bodansky, D.M., S.A. Hoedl, G.E. Metcalf, and R.N. Stavins, 2016: Facilitating linkage of climate policies through the Paris outcome. *Clim. Policy*, **16**, 956–972, doi:10.1080/14693062.2015.1069175.

Bodle, R. et al., 2014: *Options and Proposals for the International Governance of Geoengineering*. Federal Environment Agency of Germany, Berlin, 215 pp.

Boekholt, P., J. Edler, P. Cunningham, and K. Flanagan, 2009: *Drivers of International Collaboration in Research. Final Report*. 56 pp. Publications Office of the European Union, Luxembourg.

Böhringer, C., A. Müller, and J. Schneider, 2015: Carbon tariffs revisited. *J. Assoc. Environ. Resour. Econ.*, **2(4)**, 629–672, doi:10.1086/683607.

Booth, E., 2019: Extinction Rebellion: social work, climate change and solidarity. *Crit. Radic. Soc. Work*, **7(2)**, 257–261, doi:10.1332/20498601 9x15623302985296.

Bos, A.B. et al., 2017: Comparing methods for assessing the effectiveness of subnational REDD+ initiatives. *Environ. Res. Lett.*, **12(7)**, 74007, doi:10.1088/1748-9326/aa7032.

Bos, K. and J. Gupta, 2019: Stranded assets and stranded resources: Implications for climate change mitigation and global sustainable development. *Energy Res. Soc. Sci.*, **56**, 101215, doi:10.1016/j.erss.2019.05.025.

Bouwer, L.M. and J.C.J.H. Aerts, 2006: Financing climate change adaptation. *Disasters*, **30(1)**, 49–63, doi:10.1111/j.1467-9523.2006.00306.x.

Bows-Larkin, A., 2015: All adrift: aviation, shipping, and climate change policy. *Clim. Policy*, **15(6)**, 681–702, doi:10.1080/14693062.2014.965125.

Boykoff, M. and O. Pearman, 2019: Now or Never: How Media Coverage of the IPCC Special Report on 1.5°C Shaped Climate-Action Deadlines. *One Earth*, **1(3)**, 285–288, doi:10.1016/j.oneear.2019.10.026.

Boyle, A., 2018: Climate Change, the Paris Agreement and Human Rights. *Int. Comp. Law Q.*, **67(4)**, doi:10.1017/S0020589318000222.

Boyle, A., 2019: Litigating climate change under Part XII of the LOSC. *Int. J. Mar. Coast. Law*, **34(3)**, 458–481, doi:10.1163/15718085-13431097.

Brandi, C., 2017: *Trade Elements in Countries' Climate Contributions under the Paris Agreement*. International Centre for Trade and Sustainable Development (ICTSD) Issue Paper, Geneva, Switzerland.

Brandi, C., D. Blümer, and J.-F. Morin, 2019: When Do International Treaties Matter for Domestic Environmental Legislation? *Glob. Environ. Polit.*, **19(4)**, 14–44, doi:10.1162/glep_a_00524.

Brauch, M.D. et al., 2019: Treaty on Sustainable Investment for Climate Change Mitigation and Adaptation: Aligning International Investment Law with the Urgent Need for Climate Change Action. *J. Int. Arbitr.*, **36(1)**, 7–35.

Brechin, S.R. and M.I. Espinoza, 2017: A case for further refinement of the Green Climate Fund's 50:50 ratio climate change mitigation and adaptation allocation framework: toward a more targeted approach. *Clim. Change*, **142(3–4)**, 311–320, doi:10.1007/s10584-017-1938-8.

Brenton, A., 2013: "Great Powers" in climate politics. *Clim. policy*, **13(5)**, 541–546, doi:10.1080/14693062.2013.774632.

Bretschger, L., 2017: Equity and the convergence of nationally determined climate policies. *Environ. Econ. Policy Stud.*, **19(1)**, 1–14, doi:10.1007/s10018-016-0161-6.

Brewer, T.L., H. Derwent, A. Błachowicz, and M. Grubb, 2016: *Carbon Market Clubs and the New Paris Regime*. World Bank, Washington, DC, USA. Brockhaus, M. et al., 2017: REDD+, transformational change and the promise of performance-based payments: a qualitative comparative analysis". *Clim. Policy*, **17(6)**, 708–730.

Bruce, S., 2017: The Project for an International Environmental Court. In: *Conciliation in International Law* [Tomuschat, C., R.P. Mazzeschi, and D. Thürer, (eds.)]. Brill Nijhoff, Leiden, The Netherlands, pp. 133–170.

Bruce, S., 2018: Global Energy Governance and International Institutions. *SSRN Electron. J.*, doi:10.2139/ssrn.3402057.

Buck, H.J. et al., 2020: Evaluating the efficacy and equity of environmental stopgap measures. *Nat. Sustain.*, **3**, 499–504, doi:10.1038/s41893-020-0497-6.

Budolfson, M.B. et al., 2021: Utilitarian benchmarks for emissions and pledges promote equity, climate and development. *Nat. Clim. Change*, **11(10)**, 827–833, doi:10.1038/s41558-021-01130-6.

Buis, H., 2009: The role of local government associations in increasing the effectiveness of city-to-city cooperation. *Habitat Int.*, **33(2)**, 190–194, doi:10.1016/j.habitatint.2008.10.017.

Bulkeley, H., 2010: Cities and the governing of climate change. *Annu. Rev. Environ. Resour.*, **35** (November 2010), 229–253. https://doi.org/10.1146/annurev-environ-072809-101747

Burniaux, J.-M., J. Château, and J. Sauvage, 2011: *The Trade Effects of Phasing Out Fossil-Fuel Consumption Subsidies*. OECD Publishing, Paris, 18 pp. https://doi.org/10.1787/5kg6lql8wk7b-en

Burns, E.T. et al., 2016: What do people think when they think about solar geoengineering? A review of empirical social science literature, and prospects for future research. *Earth's Future*, **4(11)**, 536–542, doi:10.1002/2016ef000461.

Burns, W. and C.R. Corbett, 2020: Antacids for the Sea? Artificial Ocean Alkalinization and Climate Change. *One Earth*, **3(2)**, 154–156, doi:10.1016/j.oneear.2020.07.016.

Burton, C., R.A. Betts, C.D. Jones, and K. Williams, 2018: Will Fire Danger Be Reduced by Using Solar Radiation Management to Limit Global Warming to 1.5°C Compared to 2.0°C? *Geophys. Res. Lett.*, **45(8)**, 3644–3652, doi:10.1002/2018gl077848.

Busby, J., 2016: After Paris: good enough climate governance. *Curr. Hist.*, **115(777)**, 3–9, doi:10.1525/curh.2016.115.777.3.

Busby, J. and J. Urpelainen, 2020: Following the Leaders? How to Restore Progress in Global Climate Governance. *Glob. Environ. Polit.*, **20(4)**, 99–121, doi:10.1162/glep_a_00562.

Busch, J., F. Godoy, W.R. Turner, and C.A. Harvey, 2011: Biodiversity co-benefits of reducing emissions from deforestation under alternative reference levels and levels of finance. *Conserv. Lett.*, **4(2)**, 101–116, doi:10.1111/j.1755-263x.2010.00150.x.

Caetano, T., H. Winker, and J. Depledge, 2020: Towards zero carbon and zero poverty: integrating national climate change mitigation and sustainable development goals. *Clim. Policy*, **20(7)**, 773–778, doi:10.1080/14693062.2020.1791404.

Caldeira, K. and G. Bala, 2017: Reflecting on 50 years of geoengineering research. *Earth's Future*, **5(1)**, 10–17, doi:10.1002/2016ef000454.

Callies, D.E., 2019: The Slippery Slope Argument against Geoengineering Research. *J. Appl. Philos.*, **36(4)**, 675–687, doi:10.1111/japp.12345.

Calzadilla, P.V., 2018: Human Rights and the New Sustainable Mechanism of the Paris Agreement: A New Opportunity to Promote Climate Justice. *Potchefstroom Electron. Law J.*, **21(1)**, 1–39, doi:10.17159/1727-3781/2018/v21i0a3189.

Cames, M. et al., 2016: *How additional is the Clean Development Mechanism? Analysis of the application of current tools and proposed alternatives*. Öko-Institut/INFRAS/SEI, Berlin, Germany, 173 pp.

Campagnolo, L. and M. Davide, 2019: Can the Paris deal boost SDGs achievement? An assessment of climate mitigation co-benefits or side-effects on poverty and inequality. *World Dev.*, **122**, 96–109, doi:10.1016/j.worlddev.2019.05.015.

Campbell-Duruflé, C., 2018a: Clouds or Sunshine in Katowice? Transparency in the Paris Agreement Rulebook. *Carbon Clim. Law Rev.*, **12(3)**, 209–217.

Campbell-Duruflé, C., 2018b: Accountability or Accounting? Elaboration of the Paris Agreement's Implementation and Compliance Committee at COP 23. *Clim. Law*, **8**, 1–38, doi:10.1163/18786561-00801001.

Caney, S., 2011: Climate change, energy rights, and equality. In: *The Ethics of Global Climate Change* [Arnold, D.G., (ed.)], Cambridge University Press, Cambridge, UK and New York, NY, USA, pp. 77–103.

Caparrós, A., 2016: The Paris Agreement as a step backward to gain momentum: Lessons from and for theory. *Rev. Econ. Polit.*, **126(3)**, 347, doi:10.3917/redp.263.0347.

Caparrós, A. and J.C. Péreau, 2017: Multilateral versus sequential negotiations over climate change. *Oxf. Econ. Pap.*, **69(2)**, 365–387, doi:10.1093/oep/gpw075.

Caparrós, A., R.E. Just, and D. Zilberman, 2015: Dynamic Relative Standards versus Emission Taxes in a Putty-Clay Model. *J. Assoc. Environ. Resour. Econ.*, **2(2)**, 277–308, doi:10.1086/681599.

Carazo, M.P., 2017: Contextual Provisions (Preamble and Article 1). In: *The Paris Agreement on Climate Change: Analysis and Commentary* [Daniel Klein, M.P. Carazo, M. Doelle, J. Bulmer, and A. Higham, (eds.)]. Oxford University Press, Oxford, UK, pp. 107–121.

Carlson, C.J. et al, 2022. Solar geoengineering could redistribute malaria risk in developing countries. Nat. Commun. **13**, 2150. https://doi.org/10.1038/s41467-022-29613-w.

Carr, W.A. and L. Yung, 2018: Perceptions of climate engineering in the South Pacific, Sub-Saharan Africa, and North American Arctic. *Clim. Change*, **147(1)**, 119–132, doi:10.1007/s10584-018-2138-x.

Carraro, C., 2016: A Bottom-Up, Non-Cooperative Approach to Climate Change Control: Assessment and Comparison of Nationally Determined Contributions (NDCs). *J. Sustain. Dev.*, **9(5)**, 175, doi:10.5539/jsd.v9n5p175.

Carty, T., J. Kowalzig, and B. Zagema, 2020: *Climate Finance Shadow Report 2020: Assessing progress towards the $100 billion commitment*. Oxfam International, Oxford, UK, 32 pp.

Casado-Asensio, J. and R. Steurer, 2014: Integrated strategies on sustainable development, climate change mitigation and adaptation in Western Europe: communication rather than coordination. *J. Public Policy*, **34(3)**, 437–473, doi:10.1017/s0143814x13000287.

CBD, 2020: *Global Biodiversity Outlook 5*. Secretariat of the Convention on Biological Diversity, Montreal, Canada, 208 pp. www.cbd.int/gbo/gbo5/publication/gbo-5-en.pdf (Accessed October 31, 2021).

CDP, 2015. Global Cities Report 2015. www.cdp.net/en/research/global-reports/global-cities-report-2015 (Accessed 31st October 2021).

CESCR, 1991: General comment No. 4: The right to adequate housing. *Committee on Economic, Social and Cultural Rights Sixth session*, E/1992/23.

CESCR, 2002: General Comment No. 15: The right to water. *Substantive Issues Arising in the Implementation of the International Covenant on Economic, Social and Cultural Rights*, E/C.12/2002/11.

CESCR, 2010: Statement on the Right to Sanitation. *Committee on Economic, Social and Cultural Rights 4th session*, E-C-12-2010-1.

Chai, Q., S. Fu, H. Xu, W. Li, and Y. Zhong, 2017: The gap report of global climate change mitigation, finance, and governance after the United States declared its withdrawal from the Paris Agreement. *Chinese J. Popul. Resour. Environ.*, **15(3)**, 196–208, doi:10.1080/10042857.2017.1365450.

Chain Reaction Research, 2020: The Chain: Detected Deforestation Within Oil Palm Concessions Has Decreased So Far in 2020. *Sustain. Risk Anal.* https://chainreactionresearch.com/the-chain-detected-deforestation-within-oil-palm-concessions-has-decreased-so-far-in-2020/ (Accessed August 26, 2021).

Chan, G., R. Stavins, and Z. Ji, 2018: International Climate Change Policy. *Annu. Rev. Resour. Econ.*, **10(1)**, 335–360, doi:10.1146/annurev-resource-100517-023321.

Chan, N., 2016a: Climate Contributions and the Paris Agreement: Fairness and Equity in a Bottom-Up Architecture. *Ethics Int. Aff.*, **30(3)**, 291–301, doi:10.1017/s0892679416000228.

Chan, N., 2016b: The 'New' Impacts of the Implementation of Climate Change Response Measures. *Rev. Eur. Comp. Int. Environ. Law*, **25(2)**, 228–237, doi.org/10.1111/reel.12161.

Chan, S. et al., 2015: Reinvigorating International Climate Policy: A Comprehensive Framework for Effective Nonstate Action. *Glob. Policy*, **6(4)**, 466–473, doi:10.1111/1758-5899.12294.

Chan, S., C. Brandi, and S. Bauer, 2016: Aligning Transnational Climate Action with International Climate Governance: The Road from Paris. *Rev. Eur. Comp. Int. Environ. Law*, **25(2)**, 238–247, doi:10.1111/reel.12168.

Chan, S. et al., 2019: Promises and risks of nonstate action in climate and sustainability governance. *Wiley Interdiscip. Rev. Clim. Change*, **10(3)**, e572, doi:10.1002/wcc.572.

Chander, P., 2017: Subgame-perfect cooperative agreements in a dynamic game of climate change. *J. Environ. Econ. Manage.*, **84**, 173–188, doi:10.1016/j.jeem.2017.03.001.

Charlery, L. and S.L.M. Traerup, 2019: The nexus between nationally determined contributions and technology needs assessments: a global analysis. *Clim. Policy*, **19**, 189–205, doi:10.1080/14693062.2018.1479957.

Charnovitz, S. and C. Fischer, 2015: Canada–Renewable Energy: Implications for WTO Law on Green and Not-So-Green Subsidies. *World Trade Rev.*, **14(2)**, 177–210, doi:10.1017/s1474745615000063.

Chasek, P. et al., 2019: Land degradation neutrality: The science-policy interface from the UNCCD to national implementation. *Environ. Sci. Policy*, **92**, 182-190, doi:10.1016/j.envsci.2018.11.017.

Chazdon, R.L. et al., 2017: A Policy-Driven Knowledge Agenda for Global Forest and Landscape Restoration: A policy-driven agenda for restoration. *Conserv. Lett.*, **10(1)**, 125–132, doi:10.1111/conl.12220.

Chen, K., Y. Zhang, and X. Fu, 2019: International research collaboration: An emerging domain of innovation studies? *Res. Policy*, **48(1)**, 149–168, doi:10.1016/j.respol.2018.08.005.

Chenoweth, E. and M. Belgioioso, 2019: The physics of dissent and the effects of movement momentum. *Nat. Hum. Behav.*, **3**, 1088–1095, doi:10.1038/s41562-019-0665-8.

Chepeliev, M. and D. van der Mensbrugghe, 2020: Global fossil-fuel subsidy reform and Paris Agreement. *Energy Econ.*, **85**, 104598, doi.org/10.1016/j.eneco.2019.104598.

Christensen, J. and A. Olhoff, 2019: *Lessons from a decade of emissions gap assessments*. United Nations Environment Programme, Nairobi, Kenya, 14 pp. https://wedocs.unep.org/bitstream/handle/20.500.11822/30022/EGR10.pdf. (Accessed October 31, 2021).

Christodoulou, A., D. Dalaklis, A.I. Ölçer, and P.G. Masodzadeh, 2021: Inclusion of Shipping in the EU-ETS: Assessing the Direct Costs for the Maritime Sector Using the MRV Data. *Energies*, **14(13)**, 3915, doi:10.3390/en14133915.

Christoff, P., 2016: The promissory note: COP 21 and the Paris Climate Agreement. *Env. Polit.*, **25(5)**, 765–787, doi:10.1080/09644016.2016.1191818.

Cima, E., 2021: Retooling the Energy Charter Treaty for climate change mitigation: lessons from investment law and arbitration. *J. World Energy Law Bus.*, **14(2)**, 75–87, doi:10.1093/jwelb/jwab007.

Ciplet, D. and J. Roberts, 2017: Splintering South: Ecologically Unequal Exchange Theory in a Fragmented Global Climate. *J. World – Syst. Res.*, **23(2)**, 372–398, doi:10.5195/jwsr.2017.669.

Ciplet, D., J.T. Roberts, and M.R. Khan, 2015: *Power in a warming world: The new global politics of climate change and the remaking of environmental inequality.* MIT Press, 342 pp.

Ciplet, D., K.M. Adams, R. Weikmans, and J.T. Roberts, 2018: The Transformative Capability of Transparency in Global Environmental Governance. *Glob. Environ. Polit.*, **18(3)**, 130–150, doi:10.1162/glep_a_00472.

Claeys, P. and D. Delgado Pugley, 2017: Peasant and indigenous transnational social movements engaging with climate justice. *Can. J. Dev. Stud.*, **38(3)**, 325–340, doi:10.1080/02255189.2016.1235018.

Clarke, A. and A. MacDonald, 2016: Outcomes to Partners in Multi-Stakeholder Cross-Sector Partnerships: A Resource-Based View. *Bus. Soc.*, **58(2)**, 298–332, doi:10.1177/0007650316660534.

Clarke, A. and E. Ordonez-Ponce, 2017: City scale: Cross-sector partnerships for implementing local climate mitigation plans. *Climate Change and Public Administration: A Blog Commentary Symposium*, [Dolšak, N. and A. Prakash, (eds.)]. Public Administration Review, **2**, 25–28 http://faculty.washington.edu/aseem/Public_Administration_Review_cliamte_change_symposium.pdf (Accessed October 31, 2021).

Clémençon, R., 2016: The Two Sides of the Paris Climate Agreement: Dismal Failure or Historic Breakthrough? *J. Environ. Dev.*, **25**(1), 3–24, doi:10.1177/1070496516631362.

Climate Action Network International, 2020: About CAN: Overview. *CAN Int.* https://climatenetwork.org/overview/ (Accessed October 27, 2021).

Climate Action Tracker, 2020a: Global update: Paris Agreement Turning Point. https://climateactiontracker.org/publications/global-update-paris-agreement-turning-point/ (Accessed August 25, 2021).

Climate Action Tracker, 2020b: International Aviation. https://climateactiontracker.org/sectors/aviation/ (Accessed October 27, 2021).

Climate Action Tracker, 2020c: International Shipping. https://climateactiontracker.org/sectors/shipping/ (Accessed October 27, 2021).

Climate Focus, 2016: *Progress on the New York Declaration on Forests: Eliminating Deforestation from the Production of Agricultural Commodities – Goal 2 Assessment Report*. 8 pp. www.climatefocus.com/sites/default/files/2016-NYDF-Goal-2-Assessment-Report-Executive-Summary.pdf (Accessed October 31, 2021).

Climate Focus, 2017: *Progress on the New York Declaration on Forests: Finance for Forests – Goals 8 and 9 Assessment Report*. 53 pp. https://www.climatefocus.com/sites/default/files/NYDF%20report%202017%20FINAL.pdf (Accessed October 31, 2021).

Cochran, I. and A. Pauthier, 2019: *A framework for Aligning with Paris Agreement. The Why, What and how for financial institutions. Discussion Paper.* Institute for Climate Economics, Paris, France, 51 pp. https://www.i4ce.org/wp-core/wp-content/uploads/2019/09/I4CE%E2%80%A2Framework_Alignment_Financial_Paris_Agreement_52p.pdf (Accessed October 31, 2021).

Coen, D., J. Kreienkamp, and T. Pegram, 2020: *Global Climate Governance*. Cambridge University Press, Cambridge, UK and New York, NY, USA. Collier, P. and A.J. Venables, 2014: Closing coal: economic and moral incentives. *Oxford Rev. Econ. Policy*, **30(3)**, 492–512, doi:10.1093/oxrep/gru024.

Colmer, J., Martin, R., Muûls, M. and Wagner, U. J., Does Pricing Carbon Mitigate Climate Change? Firm-Level Evidence from the European Union Emissions Trading Scheme (November 5, 2020). CRC TR 224, Discussion Paper No. 232, Published by Centre for Economic Performance, London School of Economics and Political Science, London, UK. Available at https://cep.lse.ac.uk/pubs/download/dp1728.pdf (Accessed October 31, 2021).

Committee on the Elimination of Discrimination against Women, 2018: *General Recommendation No. 37 on Gender-related dimensions of disaster risk reduction in the context of climate change.* UN Doc. CEDAW/C/GC/37. 22 pp. https://digitallibrary.un.org/record/1626306?ln=en (Accessed October 31, 2021).

Corry, O., 2017: The international politics of geoengineering: The feasibility of Plan B for tackling climate change. *Secur. Dialogue*, **48(4)**, 297–315, doi:10.1177/0967010617704142.

Cosbey, A. and P.C. Mavroidis, 2014: A Turquoise Mess: Green Subsidies, Blue Industrial Policy and Renewable Energy: The Case for Redrafting the Subsidies Agreement of the WTO. *J. Int. Econ. Law*, **17(1)**, 11–47, doi:10.1093/jiel/jgu003.

Cosbey, A., S. Droege, C. Fischer, and C. Munnings, 2019: Developing Guidance for Implementing Border Carbon Adjustments: Lessons, Cautions, and Research Needs from the Literature. *Rev. Environ. Econ. Policy*, **13(1)**, 3–22, doi:10.1093/reep/rey020.

CPI, 2021: *Preview: Global Landscape of Climate Finance 2021*. Climate Policy Initiative, 56 pp. www.climatepolicyinitiative.org/wp-content/uploads/2021/10/Global-Landscape-of-Climate-Finance-2021.pdf (Accessed October 31, 2021).

Crane, A. and M.M. Seitanidi, 2014: Social partnerships and responsible business: What, why and how? In: *Social Partnerships and Responsible Business: A Research Handbook* [Seitanidi, M.M. and Crane, A. (eds.)]. Routledge, London, UK, pp. 1–40.

Creutzig, F. et al., 2017: The underestimated potential of solar energy to mitigate climate change. *Nat. Energy*, **2(9)**, 17140, doi:10.1038/nenergy.2017.140.

Crook, J.A., L.S. Jackson, S.M. Osprey, and P.M. Forster, 2015: A comparison of temperature and precipitation responses to different Earth radiation management geoengineering schemes. *J. Geophys. Res. Atmos.*, **120(18)**, 9352–9373, doi:10.1002/2015jd023269.

Crutzen, P.J., 2006: Albedo Enhancement by Stratospheric Sulfur Injections: A Contribution to Resolve a Policy Dilemma? *Clim. Change*, **77(3–4)**, 211–220, doi:10.1007/s10584-006-9101-y.

CTCN, 2020a: About the Climate Technology Centre and Network (CTCN). *Connect. Ctries. to Clim. Technol. Solut.* www.ctc-n.org/about-ctcn (Accessed June 4, 2020).

CTCN, 2020b: Network. *Connecting countries to climate technology solutions*. www.ctc-n.org/network (Accessed June 4, 2020).

Cullenward, D. and D.G. Victor, 2020: *Making climate policy work*. John Wiley & Sons, 242 pp.

Cummings, C.L., S.H. Lin, and B.D. Trump, 2017: Public perceptions of climate geoengineering: a systematic review of the literature. *Clim. Res.*, **73(3)**, 247–264, doi:10.3354/cr01475.

Curry, C.L. et al., 2014: A multimodel examination of climate extremes in an idealized geoengineering experiment. *J. Geophys. Res. Atmos.*, **119(7)**, 3900–3923, doi:10.1002/2013jd020648.

Curtis, P.G. C.M. Slay, N.L. Harris, A. Tyukavina, and M.C. Hansen, 2018: Classifying drivers of global forest loss. *Science,* **361(6407)**, 1108–1111, doi:10.1126/science.aau3445.

Dagnet, Y. et al., 2016: *Staying on track from Paris: advancing the key elements of the Paris Agreement*. World Resources Institute, Washington DC, USA, 60 pp. www.wri.org/publication/staying-track-paris (Accessed October 31, 2021.

Dagon, K. and D.P. Schrag, 2019: Quantifying the effects of solar geoengineering on vegetation. *Clim. Change*, **153(1)**, 235–251, doi:10.1007/s10584-019-02387-9.

Dai, Z., D.K. Weisenstein, F.N. Keutsch, and D.W. Keith, 2020: Experimental reaction rates constrain estimates of ozone response to calcium carbonate geoengineering. *Commun. Earth Environ.*, **1(63)**, 1–9, doi:10.1038/s43247-020-00058-7.

Daniel, A. and D. Neubert, 2019: Civil society and social movements: conceptual insights and challenges in African contexts. *Crit. African Stud.*, **11(2)**, 176–192, doi:10.1080/21681392.2019.1613902.

Das, K., H. van Asselt, S. Droege, and M. Mehling, 2019: Making the International Trade System Work for the Paris Agreement: Assessing the Options. *Environ. Law Report.*, **49(6)**, 10553–10580.

Dave, R. et al., 2019: *Second Bonn Challenge progress report: application of the Barometer in 2018*. International Union for Conservation of Nature, Gland, Switzerland, 80 pp.

Davidson Ladly, S., 2012: Border carbon adjustments, WTO-law and the principle of common but differentiated responsibilities. *Int. Environ. Agreements Polit. Law Econ.*, **12(1)**, 63–84, doi:10.1007/s10784-011-9153-y.

Davin, E.L., S.I. Seneviratne, P. Ciais, A. Olioso, and T. Wang, 2014: Preferential cooling of hot extremes from cropland albedo management. *Proc. Natl. Acad. Sci.*, **111(27)**, 9757–9761, doi:10.1073/pnas.1317323111.

Day, T. et al., 2020: *Navigating the nuances of net-zero targets*. 74 pp. newclimate.org/wp-content/uploads/2020/10/NewClimate_NetZero Report_October2020.pdf (Accessed August 25, 2021).

de Agueda Corneloup, I. and A.P.J. Mol, 2014: Small island developing states and international climate change negotiations: the power of moral "leadership." *Int. Environ. Agreements Polit. Law Econ.*, **14(3)**, 281–297, doi:10.1007/s10784-013-9227-0.

de Bakker, F.G.A., F. den Hond, B. King, and K. Weber, 2013: Social Movements, Civil Society and Corporations: Taking Stock and Looking Ahead. *Organ. Stud.*, **34(5–6)**, 573–593, doi:10.1177/0170840613479222.

De Bièvre, D., I. Espa, and A. Poletti, 2017: No iceberg in sight: on the absence of WTO disputes challenging fossil fuel subsidies. *Int. Environ. Agreements Polit. Law Econ.*, **17(3)**, 411–425, doi:10.1007/s10784-017-9362-0.

de Coninck, H. and A. Sagar, 2015a: Making sense of policy for climate technology development and transfer. *Clim. Policy*, **15(1)**, 1–11, doi:10.10 80/14693062.2014.953909.

de Coninck, H. and A. Sagar, 2015b: *Technology in the 2015 Paris Climate Agreement and beyond*. Issue Paper No. 42, International Centre for Trade and Sustainable Development,, Geneva, Switzerland, 31 pp. www.ru.nl/publish/pages/749373/2015_-_technology_in_the_2015_paris_climate_agreement_and_beyond_-_ictsd_issue_paper_no_42.pdf (Accessed October 31, 2021).

de Coninck, H. and D. Puig, 2015: Assessing climate change mitigation technology interventions by international institutions. *Clim. Change*, **131(3)**, 417–433, doi:10.1007/s10584-015-1344-z.

de Coninck, H. A. Revi, M. Babiker, P. Bertoldi, M. Buckeridge, A. Cartwright, W. Dong, J. Ford, S. Fuss, J.-C. Hourcade, D. Ley, R. Mechler, P. Newman, A. Revokatova, S. Schultz, L. Steg, and T. Sugiyama, 2018: Strengthening and implementing the global response. In: *Global Warming of 1.5°C. An IPCC special report on the impacts of global warming of 1.5°C above pre-industrial levels and related global greenhouse gas emission pathways, in the context of strengthening the global response to the threat of climate change* [Masson-Delmotte, P. Zhai, H.-O. Pörtner, D. Roberts, J. Skea, P.R. Shukla, A. Pirani, W. Moufouma-Okia, C. Péan, R. Pidcock, S. Connors, J.B.R. Matthews, Y. Chen, X. Zhou, M.I. Gomis, E. Lonnoy, T. Maycock, M. Tignor, and T. Waterfield, (eds.)]. Cambridge University Press, Cambridge, UK and New York, NY, USA, pp. 313–443.

de Melo, J. and J.-M. Solleder, 2020: Barriers to trade in environmental goods: How important they are and what should developing countries expect from their removal. *World Dev.*, **130**, 104910, doi.org/10.1016/j.worlddev.2020.104910.

Dehm, J., 2020: Climate change, 'slow violence' and the indefinite deferral of responsibility for 'loss and damage.' *Griffith Law Rev.*, **29(2)**, 1–33, doi:10 .1080/10383441.2020.1790101.

Delimatsis, P., 2016: Sustainable Standard-setting, Climate Change and the TBT Agreement. In: *Research Handbook on Climate Change and Trade Law* [Delimatsis, P., (ed.)], Edward Elgar Publishing, Cheltenham, UK, pp. 148–180.

Delina, L., 2017: Multilateral development banking in a fragmented climate system: shifting priorities in energy finance at the Asian Development Bank. *Int. Environ. Agreements Polit. Law Econ.*, **17(1)**, 73–88, doi:10.1007/s10784-016-9344-7.

Delina, L.L. and B.K. Sovacool, 2018: Of temporality and plurality: an epistemic and governance agenda for accelerating just transitions for energy access and sustainable development. *Curr. Opin. Environ. Sustain.*, **34**, 1–6, doi:10.1016/j.cosust.2018.05.016.

Dellaux, J., 2017: Le mécanisme visant la conservation des forêts tropicales de la convention-cadre sur les changements climatiques (REDD+): illustration de l'adaptativité du droit international. PhD Thesis, Université Aix-Marseille, France, 737 pp.

den Elzen, M. et al., 2016: Contribution of the G20 economies to the global impact of the Paris agreement climate proposals. *Clim. Change*, **137(3)**, 655–665, doi:10.1007/s10584-016-1700-7.

Denton, F., T.J. Wilbanks, A.C. Abeysinghe, I. Burton, Q. Gao, M.C. Lemos, T. Masui, K.L. O'Brien, and K. Warner, 2014: *Climate-resilient pathways: adaptation, mitigation, and sustainable development. In: Climate Change 2014: Impacts, Adaptation, and Vulnerability. Part A: Global and Sectoral Aspects. Contribution of Working Group II to the Fifth Assessment Report of the Intergovernmental Panel on Climate Change* [Field, C.B., V.R. Barros, D.J. Dokken, K.J. Mach, M.D. Mastrandrea, T.E. Bilir, M. Chatterjee, K.L. Ebi, Y.O. Estrada, R.C. Genova, B. Girma, E.S. Kissel, A.N. Levy, S. MacCracken, P.R. Mastrandrea, and L.L. White (eds.)]. Cambridge University Press, Cambridge, United Kingdom and New York, NY, USA, pp. 1101–1131.

Depledge, J., 2017: The legal and policy framework of the United Nations Climate Change Regime, 27-42 pp. In: *The Paris Agreement on climate change: Analysis and commentary* [Klein, D., M.P. Carazo, M. Doelle, J. Bulmer, and A. Higham (eds.)], Oxford University Press, Oxford, UK. Derman, B.B., 2014: Climate governance, justice, and transnational civil society. *Clim. Policy*, **14(1)**, 23–41, doi:10.1080/14693062.2014.849492.

Di Ciommo, M., 2017: *Approaches to measuring and monitoring South-South cooperation*. Development Initiatives. Bristol, UK, 23 pp. http://devinit.org/wp-content/uploads/2017/02/Approaches-to-measuring-and-monitoring-South–South-cooperation.pdf (Accessed October 31, 2021).

Dimitrov, R., J. Hovi, D.F. Sprinz, H. Sælen, and A. Underdal, 2019: Institutional and Environmental Effectiveness: Will the Paris Agreement Work? *WIREs Clim. Change*, **10(4)**, e583, doi:10.1002/wcc.583.

Djenontin, I., S. Foli, and L. Zulu, 2018: Revisiting the Factors Shaping Outcomes for Forest and Landscape Restoration in Sub-Saharan Africa: A Way Forward for Policy, Practice and Research. *Sustainability*, **10(4)**, 906, doi:10.3390/su10040906.

Dobson, N.L., 2018: The EU's conditioning of the 'extraterritorial' carbon footprint: A call for an integrated approach in trade law discourse. *Rev. Eur. Comp. Int. Environ. Law*, **27(1)**, 75–89, doi.org/10.1111/reel.12226.

Dobson, N.L., 2020: Competing Climate Change Responses: Reflections on EU Unilateral Regulation of International Transport Emissions in Light of Multilateral Developments. *Netherlands Int. Law Rev.*, **67(2)**, 183–210, doi:10.1007/s40802-020-00167-2.

Doda, B. and L. Taschini, 2017: Carbon Dating: When Is It Beneficial to Link ETSs? *J. Assoc. Environ. Resour. Econ.*, **4(3)**, 701–730, https://doi.org/10.1086/691975

Doda, B., S. Quemin, and L. Taschini, 2019: Linking Permit Markets Multilaterally. *J. Environ. Econ. Manage.*, **98(1)**, 102259, doi:10.1016/j.jeem.2019.102259.

Doelle, M., 2016: The Paris Agreement: Historic Breakthrough or High Stakes Experiment? *Clim. Law*, **6(1–2)**, 1–20, doi:10.1163/18786561-00601001.

Doelle, M., 2019: The Heart of the Paris Rulebook: Communicating NDCs and Accounting for Their Implementation. *Clim. Law*, **9(1–2)**, 3–20, doi:10.1163/18786561-00901002.

Doelle, M. and A. Chircop, 2019: Decarbonizing international shipping: An appraisal of the IMO's Initial Strategy. *Rev. Eur. Comp. Int. Environ. Law*, **28(3)**, 268–277, doi:10.1111/reel.12302.

Dong, Y., and K. Holm Olsen, 2017: Stakeholder participation in CDM and new climate mitigation mechanisms: China CDM case study. *Clim. Policy*, **17(2)**, 171–188, doi:10.1080/14693062.2015.1070257.

Donofrio, S., P. Rothrock, and J. Leonard, 2017: *Supply Change: Tracking Corporate Commitments to Deforestation-free Supply Chain*. Forest Trends, Washington DC, 32 pp. www.forest-trends.org/wp-content/uploads/2017/03/2017SupplyChange_FINAL.pdf (Accessed October 31, 2021).
Dow, K. et al., 2013: Limits to adaptation. *Nat. Clim. Change*, **3(4)**, 305–307, doi:10.1038/nclimate1847.
Downs, G.W., D.M. Rocke, and P.N. Barsoom, 1996: Is the Good News About Compliance Good News About Cooperation? *Int. Organ.*, **50(3)**, 379–406, doi:10.1017/s0020818300033427.
Draguljić, G., 2019: The Climate Change Regime Complex: Path Dependence amidst Institutional Change. *Glob. Gov. A Rev. Multilater. Int. Organ.*, **25(3)**, 476–498, doi:10.1163/19426720-02503006.
Droege, S., H. van Asselt, K. Das, and M. Mehling, 2017: The Trade System and Climate Action: Ways Forward under the Paris Agreement. *South Carolina J. Int. Law Bus.*, **13(2)**, 195–276.
Duan, L., L. Cao, G. Bala, and K. Caldeira, 2020: A Model-Based Investigation of Terrestrial Plant Carbon Uptake Response to Four Radiation Modification Approaches. *J. Geophys. Res. Atmos.*, **125(9)**, e2019jd031883, doi:10.1029/2019jd031883.
Dubash, N.K., 2020: Climate laws help reduce emissions. *Nat. Clim. Change* **10(8)**, 709–710, doi:10.1038/s41558-020-0853-6.
Duguma, L.A., P.A. Minang, and M. van Noordwijk, 2014: Climate Change Mitigation and Adaptation in the Land Use Sector: From Complementarity to Synergy. *Environ. Manage.*, **54(3)**, 420–432, doi:10.1007/s00267-014-0331-x.
Duyck, S., 2015: The Paris Climate Agreement and the Protection of Human Rights in a Changing Climate. *Yearb. Int. Environ. Law*, **26**, 3–45, doi:10.1093/yiel/yvx011.
Duyck, S., 2019: Delivering on the Paris Promises? Review of the Paris Agreement's Implementing Guidelines from a Human Rights Perspective. *Clim. Law*, **9(3)**, 202–223, doi:10.1163/18786561-00903004.
Duyck, S., E. Lennon, W. Obergassel, and A. Savaresi, 2018: Human Rights and the Paris Agreement's Implementation Guidelines: Opportunities to Develop a Rights-based Approach. *Carbon Clim. Law Rev.*, **12(3)**, 191–202, doi:10.21552/cclr/2018/3/5.
Dzebo, A., H. Janetschek, C. Brandi, and G. Iacobuta, 2019: *Connections between the Paris Agreement and the 2030 Agenda The case for policy coherence*. Stockholm Environment Institute, Stockholm, Sweden, 38 pp. www.sei.org/wp-content/uploads/2019/08/connections-between-the-paris-agreement-and-the-2030-agenda.pdf (Accessed October 31, 2021).
Eastham, S.D., D.K. Weisenstein, D.W. Keith, and S.R.H. Barrett, 2018: Quantifying the impact of sulfate geoengineering on mortality from air quality and UV-B exposure. *Atmos. Environ.*, **187**, 424–434, doi:10.1016/j.atmosenv.2018.05.047.
Edwards, G., I. Cavelier Adarve, M.C. Bustos, and J.T. Roberts, 2017: Small group, big impact: how AILAC helped shape the Paris Agreement. *Clim. Policy*, **17(1)**, 71–85, doi:10.1080/14693062.2016.1240655.
Effiong, U. and R.L. Neitzel, 2016: Assessing the direct occupational and public health impacts of solar radiation management with stratospheric aerosols. *Environ. Heal.*, **15(1)**, 1–9, doi:10.1186/s12940-016-0089-0.
Ehara, M., K. Hyakumura, and Y. Yokota, 2014: REDD+ initiatives for safeguarding biodiversity and ecosystem services: harmonizing sets of standards for national application. *J. For. Res.*, **19**, 427–436, doi:10.1007/s10310-013-0429-7.
Ehara, M. et al., 2019: REDD+ engagement types preferred by Japanese private firms: The challenges and opportunities in relation to private sector participation. *For. Policy Econ.*, **106**, 101945, doi:10.1016/j.forpol.2019.06.002.
Eisenack, K. and L. Kähler, 2016: Adaptation to climate change can support unilateral emission reductions. *Oxf. Econ. Pap.*, **68(1)**, 258–278, doi:10.1093/oep/gpv057.
Ekawati, S., Subarudi, K. Budiningsih, G.K. Sari, and M.Z. Muttaqin, 2019: Policies affecting the implementation of REDD+ in Indonesia (cases in Papua, Riau and Central Kalimantan). *For. Policy Econ.*, **108**(C), 101939, doi:10.1016/j.forpol.2019.05.025.
El-Sayed, A. and S.J. Rubio, 2014: Sharing R and D investments in cleaner technologies to mitigate climate change. *Resour. Energy Econ.*, **38**, 168–180, doi:10.1016/j.reseneeco.2014.07.003.
Emmerling, J. and M. Tavoni, 2018: Exploration of the interactions between mitigation and solar radiation management in cooperative and non-cooperative international governance settings. *Glob. Environ. Change*, **53**, 244–251, doi:10.1016/j.gloenvcha.2018.10.006.
Emmerling, J., U. Kornek, V. Bosetti, and K. Lessmann, 2020: Climate thresholds and heterogeneous regions: Implications for coalition formation. *Rev. Int. Organ.*, **16(2)**, 293–316, doi:10.1007/s11558-019-09370-0.
Engen, L. and A. Prizzon, 2018: *A guide to multilateral development banks*. Overseas Development Institute, London, UK, 92 pp.
Epps, T. and A. Green, 2010: *Reconciling Trade and ClimateHow the WTO Can Help Address Climate Change*. Edward Elgar Publishing, Cheltenham, UK, 280 pp.
Erickson, L.E., 2017: Reducing greenhouse gas emissions and improving air quality: Two global challenges. *Environ. Prog. Sustain. Energy*, **36(4)**, 982–988, doi:10.1002/ep.12665.
Erickson, P., M. Lazarus, and R. Spalding-Fecher, 2014: Net climate change mitigation of the Clean Development Mechanism. *Energy Policy*, **72**(September), 146–154, doi:10.1016/j.enpol.2014.04.038.
Erickson, P. et al., 2020: Why fossil fuel producer subsidies matter. *Nature*, **578(7793)**, doi:10.1038/s41586-019-1920-x.
Eriksen, H.H. and F.X. Perrez, 2014: The Minamata Convention: A Comprehensive Response to a Global Problem. *Rev. Eur. Comp. Int. Environ. Law*, **23(2)**, 195–210, doi:10.1111/reel.12079.
Erling, U.M., 2018: How to Reconcile the European Union Emissions Trading System (EU ETS) for Aviation with the Carbon Offsetting and Reduction Scheme for International Aviation (CORSIA)? *Air Sp. Law*, **43(4)**, 371–386.
Ernst & Young, 2017: *Report on the independent review of the effective implementation of the Climate Technology Centre and Network presented to the Twenty-Third Conference of the Parties of the UNFCCC*. 72 pp. http://unfccc.int/resource/docs/2017/cop23/eng/03.pdf (Accessed October 31, 2021).
Eskander, S.M.S.U. and S. Fankhauser, 2020: Reduction in greenhouse gas emissions from national climate legislation. *Nat. Clim. Change*, **10**(8), 750–756, doi:10.1038/s41558-020-0831-z.
Espa, I. and G. Marín Durán, 2018: Renewable Energy Subsidies and WTO Law: Time to Rethink the Case for Reform Beyond Canada – Renewable Energy/Fit Program. *J. Int. Econ. Law*, **21**(3), 621–653, doi:10.1093/jiel/jgy031.
Estache, A. and S. Saussier, 2014: Public-Private Partnerships and Efficiency: A short assessment. *CESifo DICE Rep.*, **12**(3), 8–13.
European Commission, 2019: *Communication From The Commission To The European Parliament, The European Council, The Council, The European Economic And Social Committee And The Committee Of The Regions: The European Green Deal*. Brussels, Belgium, 24 pp.
European Commission, 2021: *Proposal for a directive of the European Parliament and of the Council amending Directive 2003/87/EC. No. COM/2021/551*. Brussels, Belgium, 65 pp. https://eur-lex.europa.eu/legal-content/EN/TXT/?uri=CELEX%3A52021PC0551.
European Court of Human Rights, 2020: *Duarte Agostinho and Others v. Portugal and Others, Application Number 39371/20*. http://climatecasechart.com/climate-change-litigation/wp-content/uploads/sites/16/non-us-case-documents/2020/20200902_3937120_complaint.pdf (Accessed October 31, 2021).
Evans, J.R.G., E.P.J. Stride, M.J. Edirisinghe, D.J. Andrews, and R.R. Simons, 2010: Can oceanic foams limit global warming? *Clim. Res.*, **42(2)**, 155–160, doi:10.3354/cr00885.
Extinction Rebellion, 2019: Our Demands. Extinction Rebellion webpage. https://rebellion.earth/ (Accessed December 17, 2019).
Eyben, R., 2013: *Building Relationships in Development Cooperation: Traditional Donors and the Rising Powers*, Institute of Development

Studies, Brighton, UK, 4 pp. https://assets.publishing.service.gov.uk/media/57a08a2fe5274a27b2000485/PB36_WEB.pdf (Accessed October 31, 2021).

Eyland, T. and G. Zaccour, 2012: Strategic effects of a border tax adjustment. *Int. Game Theory Rev.*, **14(3)**, 1250016, doi:10.1142/s0219198912500168.

Falkner, R., 2016a: A Minilateral Solution for Global Climate Change? On Bargaining Efficiency, Club Benefits, and International Legitimacy. *Perspect. Polit.*, **14(1)**, 87–101, doi:10.1017/s1537592715003242.

Falkner, R., 2016b: The Paris agreement and the new logic of international climate politics. *Int. Aff.*, **92(5)**, 1107–1125, doi:10.1111/1468-2346.12708.

Falkner, R., 2019: The unavoidability of justice – and order – in international climate politics: From Kyoto to Paris and beyond. *Br. J. Polit. Sci. Int. Relations*, **21**, 270–278, doi:10.1177/1369148118819069.

Falkner, R., N. Nasiritousi, and G. Reischl, 2021: Climate clubs: politically feasible and desirable? *Clim. Policy*, 1–8, Vol.2 Issue 4. doi:10.1080/14693062.2021.1967717.

FAO, 2010: *Managing forests for climate change*. FAO, Rome, Italy, 19 pp. www.fao.org/3/i1960e/i1960e00.pdf.

FAO, 2018: *FAO's south-south and triangular cooperation to achieve the sustainable development goals. Fostering partnership among the global South*. FAO, Rome, Italy, 16 pp. www.fao.org/3/ca1379en/CA1379EN.pdf.

FAO, 2019: *South-South and Triangular Cooperation in FAO – Strengthening partnerships to achieve the Sustainable Development Goals*. FAO, Rome, Italy, 84 pp. www.fao.org/3/ca3695en/CA3695EN.pdf.

Fattouh, B. and L. Mahadeva, 2013: OPEC: What Difference Has It Made? *Annu. Rev. Resour. Econ.*, **5(1)**, 427–443, doi:10.1146/annurev-resource-091912-151901.

Fay, M. et al., 2015: Getting the Finance Flowing. In: *Decarbonizing Development: Three Steps to a Zero-Carbon Future. Climate Change and Development.*, The World Bank, Washington DC, USA, pp. 119–136.

FDI Intelligence, 2020: *The fDi report 2020. Global Greenfield Investment Trends*. [McMillan, C. and G. Ewing, (eds.)]. The Financial Times Limited 2020, London, UK, pp.

Field, L. et al., 2018: Increasing Arctic Sea Ice Albedo Using Localized Reversible Geoengineering. *Earth's Future*, **6(6)**, 882–901, doi:10.1029/2018ef000820.

Finus, M. and A. Caparrós, 2015: *Game Theory and International Environmental Cooperation: Essential Readings*. Edward Elgar Publishing, Cheltenham, UK, 934 pp.

Finus, M., and M. McGinty, 2019: The anti-paradox of cooperation: Diversity may pay! *J. Econ. Behav. Organ.*, **157**, 541–559, doi:10.1016/j.jebo.2018.10.015.

Flegal, J.A. and A. Gupta, 2018: Evoking equity as a rationale for solar geoengineering research? Scrutinizing emerging expert visions of equity. *Int. Environ. Agreements Polit. Law Econ.*, **18**(1), 45–61, doi:10.1007/s10784-017-9377-6.

Flegal, J.A., A.M. Hubert, D.R. Morrow, and J.B. Moreno-Cruz, 2019: Solar Geoengineering: Social Science, Legal, Ethical, and Economic Frameworks. *Annu. Rev. Environ. Resour.*, **44**(1), 399–423, doi:10.1146/annurev-environ-102017-030032.

Forest Declaration, 2019: *Protecting and Restoring Forests: A Story of Large Commitments yet Limited Progres: Five Year Assessment Report. September 2019.* [Climate Focus, (ed.)]. 96 pp. www.climatefocus.com/sites/default/files/2019NYDFReport.pdf (Accessed October 31, 2021).

Forsell, N. et al., 2016: Assessing the INDCs' land use, land use change, and forest emission projections. *Carbon Balance Manag.*, **11**(1), 26, doi:10.1186/s13021-016-0068-3.

Foster, V. and A. Rana, 2020: *Rethinking Power Sector Reform in the Developing World*. World Bank, Washington, DC, USA, 359 pp.

Freestone, D., 2010: From Copenhagen to Cancun: Train Wreck or Paradigm Shift? *Environ. Law Rev.*, **12**(2), 87–93, doi:10.1350/enlr.2010.12.2.081.

Fridahl, M., 2017: Socio-political prioritization of bioenergy with carbon capture and storage. *Energy Policy*, **104** (May 2017), pp. 89-99, doi:10.1016/j.enpol.2017.01.050.

Fridays for Future, 2019: About Fridays for Future. www.fridaysforfuture.org/about (Accessed December 17, 2019).

Friedrich, J., 2017: Global Stocktake (Article 14). In: *The Paris Agreement on Climate Change – Analysis and Commentary* [Klein, D., M.P. Carazo, M. Doelle, J. Bulmer, and A. Higham, (eds.)]. Oxford University Press, Oxford, UK, pp. 319–337.

Frisari, G. and M. Stadelmann, 2015: De-risking concentrated solar power in emerging markets: The role of policies and international finance institutions. *Energy Policy*, **82**, 12–22, doi.org/10.1016/j.enpol.2015.02.011.

Froyn, C.B. and J. Hovi, 2008: A climate agreement with full participation. *Econ. Lett.*, **99(2)**, 317–319, doi:10.1016/j.econlet.2007.07.013.

Frumhoff, P.C. and J.C. Stephens, 2018: Towards legitimacy of the solar geoengineering research enterprise. *Philos. Trans. R. Soc. A Math. Phys. Eng. Sci.*, **376(2119)**, 1–12, doi:10.1098/rsta.2016.0459.

Fuglestvedt, J. et al., 2018: Implications of possible interpretations of 'greenhouse gas balance' in the Paris Agreement. *Philos. Trans. R. Soc. A Math. Phys. Eng. Sci.*, **376(2119)**, 20160445, doi:10.1098/rsta.2016.0445.

Fujimori, S. et al., 2016: Will international emissions trading help achieve the objectives of the Paris Agreement? *Environ. Res. Lett.*, **11(10)**, 104001, doi:10.1088/1748-9326/11/10/104001.

Fünfgeld, H., 2015: Facilitating local climate change adaptation through transnational municipal networks. *Curr. Opin. Environ. Sustain.*, **12**, 67–73, doi:10.1016/j.cosust.2014.10.011.

Fuss, S. et al., 2014: Betting on negative emissions. *Nat. Clim. Change*, **4(10)**, 850–853, doi:10.1038/nclimate2392.

Fuss, S. et al., 2018: Negative emissions—Part 2: Costs, potentials and side effects. *Environ. Res. Lett.*, **13(6)**, 63002, doi:10.1088/1748-9326/aabf9f.

Fyson, C.L. and M.L. Jeffery, 2019: Ambiguity in the Land Use Component of Mitigation Contributions Toward the Paris Agreement Goals. *Earth's Future*, **7(8)**, 873–891, doi:10.1029/2019ef001190.

Fyson, C.L., S. Baur, M. Gidden, and C.-F. Schleussner, 2020: Fair-share carbon dioxide removal increases major emitter responsibility. *Nat. Clim. Change*, **10(9)**, 836–841, doi:10.1038/s41558-020-0857-2.

Gajevic Sayegh, A., 2017: Climate justice after Paris: a normative framework. *J. Glob. Ethics*, **13(3)**, 344–365, doi:10.1080/17449626.2018.1425217.

Gallagher, K.P., R. Kamal, J. Jin, Y. Chen, and X. Ma, 2018: Energizing development finance? The benefits and risks of China's development finance in the global energy sector. *Energy Policy*, **122**, 313–321, doi:10.1016/j.enpol.2018.06.009.

Gampfer, R., 2016: Minilateralism or the UNFCCC? The Political Feasibility of Climate Clubs. *Glob. Environ. Polit.*, **16(3)**, 62–88, doi:10.1162/glep_a_00366.

Gandenberger, C., M. Bodenheimer, J. Schleich, R. Orzanna, and L. Macht, 2015: Factors driving international technology transfer: Empirical insights from a CDM project survey. *Clim. Policy*, **16(8)**, 1065–1084, doi:10.1080/14693062.2015.1069176.

Ganguly, G., J. Setzer, and V. Heyvaert, 2018: If at First You Don't Succeed: Suing Corporations for Climate Change. *Oxf. J. Leg. Stud.*, **38(4)**, 841–868, doi:10.1093/ojls/gqy029.

Gao, S., M. Smits, A.P.J. Mol, and C. Wang, 2016: New market mechanism and its implication for carbon reduction in China. *Energy Policy*, **98**, 221–231, doi:10.1016/j.enpol.2016.08.036.

Gardiner, A., M. Bardout, F. Grossi, and S. Dixson-Declève, 2016: *Public-Private Partnerships for Climate Finance*. TemaNord, Nordic Council of Ministers, Copenhagen, Denmark, 78 pp.

Garrett, R.D. et al., 2019: Criteria for effective zero-deforestation commitments. *Glob. Environ. Change*, **54**, 135–147, doi:10.1016/j.gloenvcha.2018.11.003.

Gasparini, B., Z. McGraw, T. Storelvmo, and U. Lohmann, 2020: To what extent can cirrus cloud seeding counteract global warming? *Environ. Res. Lett.*, **15(5)**, 54002, doi:10.1088/1748-9326/ab71a3.

GCF, 2011: *Governing Instrument for the Green Climate Fund*. Songdo, Republic of Korea, 17 pp. www.greenclimate.fund/sites/default/files/document/governing-instrument.pdf (Accessed December 11, 2020).

GCF, 2021: *Status of the GCF pipeline – Addendum IV Update on the REDD-plus Results-Based Payments*. Washington, DC, USA, 5 pp. www.greenclimate.fund/sites/default/files/document/gcf-b28-inf08-add04.pdf (Accessed October 31, 2021).

GEA, 2012: *Global Energy Assessment Toward a Sustainable Future*. Cambridge University Press, Cambridge and New York.

Geden, O., 2016a: An actionable climate target. *Nat. Geosci.*, **9(5)**, 340–342, doi:10.1038/ngeo2699.

Geden, O., 2016b: The Paris Agreement and the inherent inconsistency of climate policymaking. *WIREs Clim. Change*, **7(6)**, 790–797, doi.org/10.1002/wcc.427.

Geels, F., B.K. Sovacool, T. Schwanen, and S. Sorrell, 2017: Sociotechnical transitions for deep decarbonization. *Science,* **357(6357)**, 1242, doi:10.1126/science.aao3760.

Gehring, M.W. and E. Morison, 2020: *Climate and Energy Provisions in Trade Agreements with Relevance to the Commonwealth*. Commonwealth Secretariat, London, UK, 21 pp.

Gehring, M.W. et al., 2013: *Climate Change and Sustainable Energy Measures in Regional Trade Agreements (RTAs): An Overview. ICTSD Programme on Global Economic Policy and Institutions. Issue Paper No. 3.* International Centre for Trade and Sustainable Development, Geneva, Switzerland, 44 pp.

Gençsü, I. and M. Hino, 2015: *Raising Ambition to Reduce International Aviation and Maritime Emissions.* Contributing paper for *Seizing the Global Opportunity: Partnerships for Better Growth and a Better Climate*. [Davis, M. and S. Chatwin, (eds.)]. London, UK, and Washington, DC, USA, 24 pp.

Gerlagh, R. and T.O. Michielsen, 2015: Moving targets—cost-effective climate policy under scientific uncertainty. *Clim. Change*, **132(4)**, 519–529, doi:10.1007/s10584-015-1447-6.

Germain, M., P. Toint, H. Tulkens, and A. de Zeeuw, 2003: Transfers to sustain dynamic core-theoretic cooperation in international stock pollutant control. *J. Econ. Dyn. Control*, **28**, 79–99, doi:10.1016/S0165-1889(02)00107-0.

German Constitutional Court, 2021: *Neubauer et al. v. Germany. Order of the First Senate of 24 March 2021. paras. 1-270.* 78 pp. http://climatecasechart.com/wp-content/uploads/sites/16/non-us-case-documents/2021/20210324_11817_order-1.pdf (Accessed October 31, 2021).

Gerres, T., M. Haussner, K. Neuhoff, and A. Pirlot, 2021: To ban or not to ban carbon-intensive materials: A legal and administrative assessment of product carbon requirements. *Rev. Eur. Comp. Int. Environ. Law*, **30(2)**, 249–262, doi:10.1111/reel.12395.

GESAMP, 2019: *High level review of a wide range of proposed marine geoengineering techniques*. [Boyd, P.W. and C.M.G. Vivian, (eds.)]. International Maritime Organization, London, UK, 144 pp.

Gewirtzman, J. et al., 2018: Financing loss and damage: reviewing options under the Warsaw International Mechanism. *Clim. Policy*, **18(8)**, 1076–1086, doi:10.1080/14693062.2018.1450724.

Giang, A., L.C. Stokes, D.G. Streets, E.S. Corbitt, and N.E. Selin, 2015: Impacts of the Minamata Convention on Mercury Emissions and Global Deposition from Coal-Fired Power Generation in Asia. *Environ. Sci. Technol.*, **49(9)**, 5326–5335, doi:10.1021/acs.est.5b00074.

Gilbert, P. and A. Bows, 2012: Exploring the scope for complementary sub-global policy to mitigate $CO_2$ from shipping. *Energy Policy*, **50**, 613–622, doi:10.1016/j.enpol.2012.08.002.

Gilroy, J.J. et al., 2014: Cheap carbon and biodiversity co-benefits from forest regeneration in a hotspot of endemism. *Nat. Clim. Change*, **4**, 503–507, doi:10.1038/nclimate2200.

GIZ, 2017: *Enabling subnational climate action through multi-level governance*. GIZ, Bonn, Germany, 13 pp. http://e-lib.iclei.org/wp-content/uploads/2017/11/GIZ-ICLEI-UNHabitat_2017_EN_Enabling-subnational-climate-action.pdf (Accessed October 31, 2021).

Gladwell, M., 2002: *The Tipping Point: How little things can make a big difference.* Back Bay Books, Boston, New York and Lonond, 301 pp.

Glienke, S., P.J. Irvine, and M.G. Lawrence, 2015: The impact of geoengineering on vegetation in experiment G1 of the GeoMIP. *J. Geophys. Res. Atmos.*, **120(19)**, 10,110-196,213, doi:10.1002/2015jd024202.

Goldthau, A. and J. Witte, eds., 2010: *Global energy governance: The new rules of the game*. Brookings Institution Press, Washington, DC, USA, 372 pp.

Gollier, C. and J. Tirole, 2015: Negotiating effective institutions against climate change. *Econ. Energy Environ. Policy*, **4(2)**, 5–28, doi:10.5547/2160-5890.4.2.cgol.

Gomez-Echeverri, L., 2018: Climate and development: enhancing impact through stronger linkages in the implementation of the Paris Agreement and the Sustainable Development Goals (SDGs). *Philos. Trans. R. Soc. A Math. Eng. Sci.*, **376(2119)**, 20160444, doi:10.1098/rsta.2016.0444.

Goulder, L., M.A. Hafstead, G. Kim, and X. Long, 2019: Impacts of a Carbon Tax across US Household Income Groups: What Are the Equity-Efficiency Trade-Offs? *J. Public Econ.*, **175**, 44–64, doi:10.1016/j.jpubeco.2019.04.002.

Government of Indonesia, 2016: *First Nationally Determined Contribution. Republic Of Indonesia*. 18 pp. https://unfccc.int/sites/default/files/NDC/2022-06/First%20NDC%20Indonesia_submitted%20to%20UNFCCC%20Set_November%20%202016.pdf (Accessed January 9, 2021).

Grassi, G. et al., 2017: The key role of forests in meeting climate targets requires science for credible mitigation. *Nat. Clim. Change*, **7**, 220–226, doi:10.1038/nclimate3227.

Gray, K., and B.K. Gills, 2016: South–South cooperation and the rise of the Global South. *Third World Q.*, **37(4)**, 557–574, doi:10.1080/01436597.2015.1128817.

Green, A., 2005: Climate Change, Regulatory Policy and the WTO: How Constraining Are Trade Rules? *J. Int. Econ. Law*, **8(1)**, 143–189, doi:10.1093/jielaw/jgi008.

Green, J.F., 2017: The strength of weakness: pseudo-clubs in the climate regime. *Clim. Change*, **144(1)**, 41–52, doi:10.1007/s10584-015-1481-4.

Green, J.F., 2021: Does carbon pricing reduce emissions? A review of ex-post analyses. *Environ. Res. Lett.*, **16(4)**, 043004, doi:10.1088/1748-9326/abdae9.

Green, J.F., T. Sterner, and G. Wagner, 2014: A balance of "bottom–up" and "top–down" in linking climate policies. *Nat. Clim. Change*, **4(12)**, 1064–1067, doi:10.1038/nclimate2429.

Greiner, S. et al., 2020: *Article 6 Piloting: State of play and stakeholder experiences*. 135 pp. www.climatefinanceinnovators.com/wp-content/uploads/2020/12/Climate-Finance-Innovators_Article-6-piloting_State-of-play-and-stakeholder-experiences_December-2020.pdf (Accessed October 31, 2021).

Grimsey, D. and Mervyn, L., 2002: Evaluating the risks of public private partnerships for infrastructure project. *International J. Public Manag.*, **20(2)**, 107–118, doi:10.1016/s0263/d0263-7863(00)00040-5.

Griscom, B.W. et al., 2017: Natural climate solutions. *Proc. Natl. Acad. Sci.*, **114(44)**, 11645–11650, doi:10.1073/pnas.1710465114.

Grubb, M., 2011: International climate finance from border carbon cost levelling. *Clim. Policy*, **11(3)**, 1050–1057, doi:10.1080/14693062.2011.582285.

Grubb, M., J.-C. Hourcade, and K. Neuhoff, 2014: *Planetary economics: energy, climate change and the three domains of sustainable development.* Routledge, London, UK.

Grubb, M. et al., 2021: Induced innovation in energy technologies and systems: a review of evidence and potential implications for $CO_2$ mitigation. *Environ. Res. Lett.*, **16(4)**, 043007, doi:10.1088/1748-9326/abde07.

Grunewald, N. and I. Martínez-Zarzoso, 2012: How Well Did the Kyoto Protocol Work? A Dynamic-GMM Approach with External Instruments. *SSRN Electron. J.*, (February), 37, doi:10.2139/ssrn.2013086.

Grunewald, N. and I. Martínez-Zarzoso, 2016: Did the Kyoto Protocol fail? An evaluation of the effect of the Kyoto Protocol on CO2 emissions. *Environ. Dev. Econ.*, **21(1)**, 1–22, doi:10.1017/s1355770x15000091.

Gsottbauer, E., R. Gampfer, E. Bernold, and A.M. Delas, 2018: Broadening the scope of loss and damage to legal liability: an experiment. *Clim. Policy*, **18(5)**, 600–611, doi:10.1080/14693062.2017.1317628.

Gulbrandsen, L.H., J. Wettestad, D.G. Victor, and A. Underdal, 2019: The Political Roots of Diverging Carbon Market Design: implications for linking. *Clim. Policy*, **19(4)**, 427–438, doi:10.1080/14693062.2018.1551188.

Gunningham, N., 2019: Averting Climate Catastrophe: Environmental Activism, Extinction Rebellion and coalitions of Influence. *King's Law J.*, **30(2)**, 194–202, doi:10.1080/09615768.2019.1645424.

Gupta, A. and H. van Asselt, 2019: Transparency in multilateral climate politics: Furthering (or distracting from) accountability? *Regul. Gov.*, **13(1)**, 18–34, doi:10.1111/rego.12159.

Gupta, A. et al., 2020: Anticipatory governance of solar geoengineering: conflicting visions of the future and their links to governance proposals. *Curr. Opin. Environ. Sustain.*, **45**, 10–19, doi:10.1016/j.cosust.2020.06.004.

Gupta, H. and L.C. Dube, 2018: Addressing biodiversity in climate change discourse: Paris mechanisms hold more promise. *Int. For. Rev.*, **20(1)**, 104–114, doi:10.1505/146554818822824282.

Gupta, J. et al., 2010: Mainstreaming climate change in development cooperation policy: conditions for success. In: *Making climate change work for us* [Hulme, M. and H. Neufeldt, (eds.)], Cambridge University Press, Cambridge, UK and New York, NY, USA, pp. 319–339.

Hadden, J., 2014: Explaining Variation in Transnational Climate Change Activism: The Role of Inter-Movement Spillover. *Glob. Environ. Polit.*, **14(2)**, 7–25, doi:10.1162/glep_a_00225.

Haites, E., 2016: Experience with linking greenhouse gas emissions trading systems. *Wiley Interdiscip. Rev. Energy Environ.*, **5(3)**, 246–260, doi:10.1002/wene.191.

Haites, E., 2018: Carbon taxes and greenhouse gas emissions trading systems: what have we learned? *Climate Policy* **18(8)**, 955–966, doi:10.1080/14693062.2018.1492897.

Hale, T., 2016: "All Hands on Deck": The Paris Agreement and Nonstate Climate Action. *Glob. Environ. Polit.*, **16(3)**, 12–22, doi:10.1162/glep_a_00362.

Hale, T., 2020: Catalytic Cooperation. *Glob. Environ. Polit.*, **20(4)**, 73–98, doi:10.1162/glep_a_00561.

Hale, T. et al., 2016: Exploring links between national climate strategies and non-state and subnational climate action in nationally determined contributions (NDCs). *Clim. Policy*, **6(4)**, 1–15, doi:10.1080/14693062.2019.1624252.

Hales, R. and B. Mackey, 2018: Carbon budgeting post-COP21: The need for an equitable strategy for meeting CO2e targets. In: *Pathways to a Sustainable Economy: Bridging the Gap between Paris Climate Change Commitments and Net Zero Emissions* [Hossain, M., R. Hales, and T. Sarker, (eds.)], Springer International Publishing, Cham, Switzerland, pp. 209–220.

Hanger, S. et al., 2016: Community acceptance of large-scale solar energy installations in developing countries: Evidence from Morocco. *Energy Res. Soc. Sci.*, **14**, doi:10.1016/j.erss.2016.01.010.

Hanna, R. and D.G. Victor, 2021: Marking the decarbonization revolutions. *Nat. Energy*, **6(6)**, 568–571, doi:10.1038/s41560-021-00854-1.

Harding, A.R., K. Ricke, D. Heyen, D.G. MacMartin, and J. Moreno-Cruz, 2020: Climate econometric models indicate solar geoengineering would reduce inter-country income inequality. *Nat. Commun.*, **11(1)**, 1–9, doi:10.1038/s41467-019-13957-x.

Hartl, A., 2019: The effects of the Kyoto Protocol on the carbon trade balance. *Rev. World Econ.*, **155**, 539–574, doi:10.1007/s10290-019-00350-5.

Hein, J., A. Guarin, E. Frommé, and P. Pauw, 2018: Deforestation and the Paris climate agreement: An assessment of REDD + in the national climate action plans. *For. Policy Econ.*, **90**, 7–11, doi:10.1016/j.forpol.2018.01.005.

Heinrichs, D., K. Krellenberg, and M. Fragkias, 2013: Urban Responses to Climate Change: Theories and Governance Practice in Cities of the Global South. *Int. J. Urban Reg. Res.*, **37(6)**, 1865–1878, doi:10.1111/1468-2427.12031.

Held, D., and C. Roger, 2018: Three Models of Global Climate Governance: From Kyoto to Paris and Beyond. *Glob. Policy*, **9(4)**, 527–537, doi:10.1111/1758-5899.12617.

Helland, L., J. Hovi, and H. Sælen, 2017: Climate leadership by conditional commitments. *Oxf. Econ. Pap.*, **70(2)**, 417–442, doi:10.1093/oep/gpx045.

Helm, C. and D.F. Sprinz, 2000: Measuring the Effectiveness of International Environmental Regimes. *J. Conflict Resolut.*, **45(5)**, 630–652.

Helwegen, K.G., C.E. Wieners, J.E. Frank, and H.A. Dijkstra, 2019: Complementing CO2 emission reduction by solar radiation management might strongly enhance future welfare. *Earth Syst. Dyn.*, **10(3)**, 453–472, doi:10.5194/esd-10-453-2019.

Henderson, I., J. Coello, R. Fischer, I. Mulder, and T. Christophersen, 2013: *The Role of the Private Sector in REDD+: the Case for Engagement and Options for Intervention*. 12 pp. www.unredd.net/documents/redd-papers-and-publications-90/un-redd-publications-1191/policy-brief-series-3154/10509-private-sector-engagement-policy-brief-en-final-version-low-res-10509.html (Accessed October 31, 2021).

Hensengerth, O., 2017: Regionalism, Identity, and Hydropower Dams: The Chinese-Built Lower Sesan 2 Dam in Cambodia. *J. Curr. Chinese Aff.*, **46(3)**, 85–118, doi:10.1177/186810261704600304.

Hermwille, L., A. Siemons, H. Förster, and L. Jeffery, 2019: Catalyzing mitigation ambition under the Paris Agreement: elements for an effective Global Stocktake. *Clim. Policy*, **19(8)**, 988–1001, doi:10.1080/14693062.2019.1624494.

Herrala, R. and R.K. Goel, 2016: Sharing the emission reduction burden in an uneven world. *Energy Policy*, **94**, 29–39, doi:10.1016/j.enpol.2016.03.028.

Hertel, M., 2011: Climate-Change-Related Trade Measures and Article XX: Defining Discrimination in Light of the Principle of Common but Differentiated Responsibilities. *J. World Trade*, **45(3)**, 653–678.

Heutel, G., J. Moreno-Cruz, and S. Shayegh, 2018: Solar geoengineering, uncertainty, and the price of carbon. *J. Environ. Econ. Manage.*, **87**, 24–41, doi:10.1016/j.jeem.2017.11.002.

Hillman, J., 2013: *Changing Climate for Carbon Taxes: Who's Afraid of the WTO?* Climate & Energy Policy Paper Series, July 2013. The German Marshall Fund of the United States, Washington, DC, USA, 15 pp. https://scholarship.law.georgetown.edu/facpub/2030 (Accessed October 31, 2021).

Hoch, S., A. Michaelowa, A. Espelage, and A.K. Weber, 2019: Governing complexity: How can the interplay of multilateral environmental agreements be harnessed for effective international market-based climate policy instruments? *Int. Environ. Agreements Polit. Law Econ.*, **19**, 595–613, doi:10.1007/s10784-019-09455-6.

Hochstetler, K. and M. Milkoreit, 2013: Emerging Powers in the Climate Negotiations: Shifting Identity Conceptions. *Polit. Res. Q.*, **67(1)**, 224–235, doi:10.1177/1065912913510609.

Hof, A., C. Brink, A.M. Beltran, and M. den Elzen, 2012: *Greenhouse gas emission reduction targets for 2030: Conditions for an EU target of 40%*. PBL Publishers, The Hague, Netherlands, 52 pp.

Hofman, E, and W. van der Gaast, 2019: Enhancing ambition levels in nationally determined contributions-Learning from Technology Needs Assessments. *Wiley Interdiscip. Rev. – Energy Environ.*, **8**(e311), doi:10.1002/wene.311.

Hogl, K., D. Kleinschmit, and J. Rayner, 2016: Achieving policy integration across fragmented policy domains: Forests, agriculture, climate and energy. *Environ. Plan. C Gov. Policy*, **34(3)**, 399–414, doi:10.1177/0263774x16644815.

Höglund-Isaksson, L. et al., 2017: Cost estimates of the Kigali Amendment to phase-down hydrofluorocarbons. *Environ. Sci. Policy*, **75**, 138–147, doi:10.1016/j.envsci.2017.05.006.

Höhne, N. et al., 2017: The Paris Agreement: resolving the inconsistency between global goals and national contributions. *Clim. Policy*, **17**, doi:10.1080/14693062.2016.1218320.

Höhne, N., H. Fekete, M.G.J. den Elzen, A.F. Hof, and T. Kuramochi, 2018: Assessing the ambition of post-2020 climate targets: a comprehensive

framework. *Clim. Policy*, **18(4)**, 425–441, doi:10.1080/14693062.2017.1294046.
Höhne, N. et al., 2021: Wave of net zero emission targets opens window to meeting the Paris Agreement. *Nat. Clim. Change*, **11(10)**, 820–822, doi:10.1038/s41558-021-01142-2.
Holmberg, A. and A. Alvinius, 2019: Children's protest in relation to the climate emergency: A qualitative study on a new form of resistance promoting political and social change. *Childhood*, **27(1)**, 0907568219879970, doi:10.1177/0907568219879970.
Holzer, K., 2014: *Carbon-related Border Adjustment and WTO Law*. Edward Elgar Publishing, Cheltenham, UK, 352 pp.
Holzer, K. and T. Cottier, 2015: Addressing climate change under preferential trade agreements: Towards alignment of carbon standards under the Transatlantic Trade and Investment Partnership. *Glob. Environ. Change*, **35**, 514–522, doi:10.1016/j.gloenvcha.2015.06.006.
Holzer, K., and A.H. Lim, 2020: Trade and Carbon Standards: Why Greater Regulatory Cooperation is Needed. In: *Cool Heads in a Warming World: How Trade Policy Can Help Fight Climate Change* [Esty, D.C. and S. Biniaz, (eds.)], Yale Center for Environmental Law & Policy, New Haven, CT, USA, pp. 27.
Honegger, M. and D. Reiner, 2018: The political economy of negative emissions technologies: consequences for international policy design. *Clim. Policy*, **18(3)**, 306–321, doi:10.1080/14693062.2017.1413322.
Honegger, M., A. Michaelowa, and J. Pan, 2021a: Potential implications of solar radiation modification for achievement of the Sustainable Development Goals. *Mitig. Adapt. Strateg. Glob. Change*, **26(5)**, 1–20, doi:10.1007/s11027-021-09958-1.
Honegger, M., M. Poralla, A. Michaelowa, and H.-M. Ahonen, 2021b: Who Is Paying for Carbon Dioxide Removal? Designing Policy Instruments for Mobilizing Negative Emissions Technologies. *Front. Clim.*, **3** (June), 1–15, doi:10.3389/fclim.2021.672996.
Horlick, G. and P.A. Clarke, 2017: Rethinking Subsidy Disciplines for the Future: Policy Options for Reform. *J. Int. Econ. Law*, **20(3)**, 673–703, doi:10.1093/jiel/jgx022.
Horton, J. and D. Keith, 2016: Solar geoengineering and obligations to the global poor. In: *Climate justice and geoengineering: Ethics and policy in the atmospheric anthropocene* [Preston, C., (ed.)], Rowman Littlefield, Lanham, MD, USA, pp. 79–92.
Horton, J.B. and B. Koremenos, 2020: Steering and influence in transnational climate governance: nonstate engagement in solar geoengineering research. *Glob. Environ. Polit.*, **20(3)**, 93–111, doi:10.1162/glep_a_00572.
Hourdequin, M., 2018: Climate Change, Climate Engineering, and the 'Global Poor': What Does Justice Require? *Ethics, Policy Environ.*, **21(3)**, 270–288, doi:10.1080/21550085.2018.1562525.
Hourdequin, M., 2019: Geoengineering Justice: The Role of Recognition: *Sci. Technol. Hum. Values*, **44(3)**, 448–477, doi:10.1177/0162243918802893.
Hovi, J., D.F. Sprinz, H. Sælen, and A. Underdal, 2016: Climate change mitigation: A role for climate clubs? *Palgrave Commun.*, **2**(May), 1–9, doi:10.1057/palcomms.2016.20.
Hovi, J., D.F. Sprinz, H. Sælen, and A. Underdal, 2017: The Club Approach: A Gateway to Effective Climate Co-operation? *Br. J. Polit. Sci.*, **49(3)**, 1071–1096, doi:10.1017/s0007123416000788.
Howard, A., 2017: Voluntary Cooperation (Article 6). In: *The Paris Agreement on climate change: Analysis and commentary* [Klein, D., M. Pía Carazo, M. Doelle, J. Bulmer, and A. Higham, (eds.)], Oxford University Press, Oxford, UK, 178 pp.
Hsu, A. et al., 2019a: A research roadmap for quantifying non-state and subnational climate mitigation action. *Nat. Clim. Change*, **9(1)**, doi:10.1038/s41558-018-0338-z.
Hsu, A., J. Brandt, O. Widerberg, S. Chan, and A. Weinfurter, 2019b: Exploring links between national climate strategies and non-state and subnational climate action in nationally determined contributions (NDCs). *Clim. Policy*, **6(4)**, 1–15, doi:10.1080/14693062.2019.1624252.
Hu, X. et al., 2020: The impacts of the trade liberalization of environmental goods on power system and CO2 emissions. *Energy Policy*, **140**, 111173, doi:10.1016/j.enpol.2019.111173.
Huang, J., 2018: What Can the Paris Agreement's Global Stocktake Learn from the Sustainable Development Goals? *Carbon Clim. Law Rev.*, **12(3)**, 218–228, doi:10.21552/cclr/2018/3/8.
Hubert, A.M., 2020: A Code of Conduct for Responsible Geoengineering Research. *Glob. Policy*, **12**(S1), 82–96, doi:10.1111/1758-5899.12845.
Hufbauer, G.C., J. Kim, and S. Charnovitz, 2009: *Global Warming and the World Trading System*. Peterson Institute for International Economics, Washington, DC, USA, 166 pp.
Huggins, A. and M.S. Karim, 2016: Shifting Traction: Differential Treatment and Substantive and Procedural Regard in the International Climate Change Regime. *Transnatl. Environ. Law*, **5(2)**, 427–448, doi:10.1017/s2047102516000170.
Hulme, M., 2014: *Can science fix climate change?: A case against climate engineering*. Polity Press, Cambridge, UK, 144 pp.
Human Rights Committee, 2018: *General comment No. 36 (2018) on article 6 of the International Covenant on Civil and Political Rights, on the right to life (30 October 2018)*. UN Doc. CCPR/C/GC/36. 21 pp. https://tbinternet.ohchr.org/Treaties/CCPR/SharedDocuments/1_Global/CCPR_C_GC_36_8785_E.pdf (Accessed October 31, 2021).
Hurrell, A. and S. Sengupta, 2012: Emerging powers, North–South relations and global climate politics. *Int. Aff.*, **88(3)**, 463–484, doi:10.1111/j.1468-2346.2012.01084.x.
Hurwitz, M.M., E.L. Fleming, P.A. Newman, F. Li, and Q. Liang, 2016: Early action on HFCs mitigates future atmospheric change. *Environ. Res. Lett.*, **11(11)**, 114019, doi:10.1088/1748-9326/11/11/114019.
ICAO, 2016: *Resolution A39-3: Consolidated statement of continuing ICAO policies and practices related to environmental protection – Global Market-based Measure (MBM) scheme*. International Civil Aviation Organisation, pp. 25–32, www.icao.int/Meetings/a39/Documents/Resolutions/a39_res_prov_en.pdf (Accessed October 31, 2021).
ICAO, 2019: *SARPs – Annex 16 – Environmental Protection, Volume IV – Carbon Offsetting and Reduction Scheme for International Aviation (CORSIA)*. International Civil Aviation Organisation, 120 pp. www.icao.int/environmental-protection/CORSIA/Pages/SARPs-Annex-16-Volume-IV.aspx (Accessed October 31, 2021).
ICAP, 2019: *Emissions Trading Worldwide. International Carbon Action Partnership (ICAP) – Status Report 2019*. International Carbon Action Partnership, Berlin, 130 pp. https://icapcarbonaction.com/en/icap-status-report-2019 (Accessed October 31, 2021).
ICSU ISSC, 2015: *Review of Targets for the Sustainable Development Goals: The science perspective*. International Council for Science, Paris, 88 pp. https://council.science/wp-content/uploads/2017/05/SDG-Report.pdf (Accessed October 31, 2021).
IIGCC, 2020: *Paris Aligned Investment Initiative: Net Zero Investment Framework for Consultation*. The Institutional Investors Group on Climate Change, 40 pp. www.iigcc.org/download/net-zero-investment-framework-consultation/?wpdmdl=3602&masterkey=5f270ef146677 (Accessed December 9, 2020).
IMO, 2018: *Resolution MEPC.304(72): Initial IMO Strategy on Reduction of GHG Emissions from Ships. MEPC 72/17/Add.1 Annex 11*. International Maritime Organization (IMO), 11 pp. https://wwwcdn.imo.org/localresources/en/OurWork/Environment/Documents/ResolutionMEPC.304(72)_E.pdf (Accessed October 31, 2021).
IMO, 2019: UN agency pushes forward on shipping emissions reduction. International Maritime Organizations, www.imo.org/en/MediaCentre/PressBriefings/Pages/11-MEPC-74-GHG.aspx (Accessed October 27, 2021).
InfoFLR, 2018: The Bonn Challenge Barometer. Info Forest Landscape Restoration. International Union for Conservation of Nature, https://infoflr.org/bonn-challenge-barometer (Accessed October 27, 2021).

14

Ingham, A., J. Ma, and A.M. Ulph, 2013: Can adaptation and mitigation be complements? *Clim. Change*, **120**, 39–53, doi:10.1007/s10584-013-0815-3.

INPE, 2019: Monitoramento dos Focos Ativos por País, Programa Queimadas, Brazilian National Institute for Space Research (INPE). http://queimadas.dgi.inpe.br/queimadas/portal-static/estatisticas_paises/ (Accessed December 22, 2020).

Institute for Climate Economics, 2017: THE INITIATIVE – Climate Action in Financial Institutions. www.mainstreamingclimate.org/initiative/ (Accessed December 8, 2020).

Inter-American Development Bank, 2012: MDB Joint Statement for Rio +20. 2012. *Announcements*,. www.iadb.org/en/news/announcements/2012-06-19/mdb-joint-statement-for-rio20%2C10032.html (Accessed October 28, 2021).

IPCC, 2018a: *Global Warming of 1.5°C. An IPCC Special Report on the impacts of global warming of 1.5°C above pre-industrial levels and related global greenhouse gas emission pathways, in the context of strengthening the global response to the threat of climate change, sustainable development, and efforts to eradicate poverty* [Masson-Delmotte, V., P. Zhai, H.-O. Pörtner, D. Roberts, J. Skea, P.R. Shukla, A. Pirani, W. Moufouma-Okia, C. Péan, R. Pidcock, S. Connors, J.B.R. Matthews, Y. Chen, X. Zhou, M.I. Gomis, E. Lonnoy, T. Maycock, M. Tignor, and T. Waterfield (eds.)]. Cambridge University Press, Cambridge, UK and New York, NY, USA, 616 pp.

IPCC, 2018b: Summary for Policymakers. In: *Global Warming of 1.5°C. An IPCC Special Report on the impacts of global warming of 1.5°C above pre-industrial levels and related global greenhouse gas emission pathways, in the context of strengthening the global response to the threat of climate change, sustainable development, and efforts to eradicate poverty* [Masson-Delmotte, V., P. Zhai, H.-O. Pörtner, D. Roberts, J. Skea, P.R. Shukla, A. Pirani, W. Moufouma-Okia, C. Péan, R. Pidcock, S. Connors, J.B.R. Matthews, Y. Chen, X. Zhou, M.I. Gomis, E. Lonnoy, T. Maycock, M. Tignor, and T. Waterfield (eds.)]. Cambridge University Press, Cambridge, UK and New York, NY, USA.

IPCC, 2019a: Summary for Policymakers. In: *IPCC Special Report on the Ocean and Cryosphere in a Changing Climate* [Pörtner, H.-O., D.C. Roberts, V. Masson-Delmotte, P. Zhai, M. Tignor, E. Poloczanska, K. Mintenbeck, A. Alegría, M. Nicolai, A. Okem, J. Petzold, B. Rama, N.M. Weyer (eds.)]. Cambridge University Press, Cambridge, UK and New York, NY, USA.

IPCC, 2019b: *Climate Change and Land. An IPCC Special Report on climate change, desertification, land degradation, sustainable land management, food security, and greenhouse gas fluxes in terrestrial ecosystems* [Shukla, P.R., J. Skea, E. Calvo Buendia, V. Masson-Delmotte, H.-O. Pörtner, D. C. Roberts, P. Zhai, R. Slade, S. Connors, R. van Diemen, M. Ferrat, E. Haughey, S. Luz, S. Neogi, M. Pathak, J. Petzold, J. Portugal Pereira, P. Vyas, E. Huntley, K. Kissick, M. Belkacemi, J. Malley, (eds.)]. Cambridge University Press, Cambridge, UK and New York, NY, USA, 896 pp.

Irawan, S., T. Widiastomo, L. Tacconi, J.D. Watts, and B. Steni, 2019: Exploring the design of jurisdictional REDD+: The case of Central Kalimantan, Indonesia. *For. Policy Econ.*, **108**, 101853, doi:10.1016/j.forpol.2018.12.009.

IRENA, 2021: Avoided Emissions Calculator. International Renewable Energy Agency (IRENA). www.irena.org/climatechange/Avoided-Emissions-Calculator (Accessed November 3, 2021).

Irvine, P.J. and D.W. Keith, 2020: Halving warming with stratospheric aerosol geoengineering moderates policy-relevant climate hazards. *Environ. Res. Lett.*, **15(4)**, 44011, doi:10.1088/1748-9326/ab76de.

Irvine, P.J. et al., 2017: Towards a comprehensive climate impacts assessment of solar geoengineering. *Earth's Future*, **5(1)**, 93–106, doi:10.1002/2016ef000389.

ISA, 2015: Framework Agreement on the establishment of the International Solar Alliance. International Solar Alliance (ISA). https://isolaralliance.org/about/framework-agreement#book5/undefined (Accessed October 28, 2021).

Ismer, R. and K. Neuhoff, 2007: Border tax adjustment: a feasible way to support stringent emission trading. *Eur. J. Law Econ.*, **24(2)**, 137–164, doi:10.1007/s10657-007-9032-8.

Ismer, R. and M. Haussner, 2016: Inclusion of Consumption into the EU ETS: The Legal Basis under European Union Law. *Rev. Eur. Comp. Int. Environ. Law*, **25(1)**, 69–80, doi.org/10.1111/reel.12131.

Ismer, R., K. Neuhoff, and A. Pirlot, 2020: *Border Carbon Adjustments and Alternative Measures for the EU ETS: An Evaluation*. German Institute for Economic Research (DIW), Berlin, Germany, 21 pp. www.diw.de/documents/publikationen/73/diw_01.c.743698.de/dp1855.pdf (Accessed October 31, 2021).

Ismer, R. et al., 2021: *Climate Neutral Production, Free Allocation of Allowances under Emissions Trading Systems, and the WTO: How to Secure Compatibility with the ASCM*. German Institute for Economic Research (DIW), Berlin, Germany, 18 pp. www.diw.de/documents/publikationen/73/diw_01.c.818655.de/dp1948.pdf (Accessed October 31, 2021).

Ivanova, A., 2017: Green financing for cities: current options and future challenges. In: *Climate Change-Sensitive Cities: Building capacities for urban resilience, sustainability, and equity* [Delgado, G.C., (ed.)], Research Program on Climate Change of the National Autonomous University of Mexico (PINCC-UNAM), Mexico, pp. 283–306.

Ivanova, A. and C. Lopez, 2013: The energy crisis and the policies for implementation of renewable energies. In: *Strategies towards a sustainable development in front of the three crisis*, UAM-Iztapalapa & Miguel Angel Porrúa, pp. 267–282.

Ivanova, A., A. Bermudez, and A. Martinez, 2015: Climate action plan for the city of La Paz, Baja California Sur, Mexico: a tool for sustainability. In: *The Sustainable City X* [Brebbia, C.A. and W.F. Florez-Escobar, (eds.)], *WIT Transactions on Ecology and the Environment*, **194**, pp. 439–449.

Ivanova, A., A. Zia, P. Ahmad, and M. Bastos-Lima, 2020: Climate mitigation policies and actions: access and allocation issues. *Int. Environ. Agreements Polit. Law Econ.*, **20(2)**, 287–301, doi:10.1007/s10784-020-09483-7.

Iwata, H. and K. Okada, 2014: Greenhouse gas emissions and the role of the Kyoto Protocol. *Environ. Econ. Policy Stud.*, **16(4)**, 325–342, doi:10.1007/s10018-012-0047-1.

Iyer, G. et al., 2017: Measuring progress from nationally determined contributions to mid-century strategies. *Nat. Clim. Change*, **7(12)**, 871–874, doi:10.1038/s41558-017-0005-9.

Jachnik, R., M. Mirabile, and A. Dobrinevski, 2019: *Tracking finance flows towards assessing their consistency with climate objectives*. OECD Publishing, Paris, France, 41 pp.

Jackson, L.S., J.A. Crook, and P.M. Forster, 2016: An intensified hydrological cycle in the simulation of geoengineering by cirrus cloud thinning using ice crystal fall speed changes. *J. Geophys. Res. Atmos.*, **121(12)**, 6822–6840, doi:10.1002/2015jd024304.

Jacquet, J. and D. Jamieson, 2016: Soft but significant power in the Paris Agreement. *Nat. Clim. Change*, **6(7)**, 643–646, doi:10.1038/nclimate3006.

Jaffe, A.B., R.G. Newell, and R.N. Stavins, 2005: A tale of two market failures: Technology and environmental policy. *Ecol. Econ.*, **54(2–3)**, 164–174, doi:10.1016/j.ecolecon.2004.12.027.

James, R. et al., 2014: Characterizing loss and damage from climate change. *Nat. Clim. Change*, **4(11)**, 938–939, doi:10.1038/nclimate2411.

Janetschek, H., C. Brandi, A. Dzebo, and B. Hackmann, 2020: The 2030 Agenda and the Paris Agreement: voluntary contributions towards thematic policy coherence. *Clim. Policy*, **20(4)**, 430–442, doi:10.1080/14693062.2019.1677549.

Janus, H., S. Klingebiel, and T.C. Mahn, 2014: *How to Shape Development Cooperation? The Global Partnership and the Development Cooperation Forum (March 10, 2014). Briefing Paper, 3/2014*. German Development Institute, Bonn, Germany, 4 pp.

Jeffery, M.L., J. Gütschow, M.R. Rocha, and R. Gieseke, 2018: Measuring Success: Improving Assessments of Aggregate Greenhouse Gas Emissions Reduction Goals. *Earth's Future*, **6(9)**, 1260–1274, doi:10.1029/2018ef000865.

Jewell, J. et al., 2018: Limited emission reductions from fuel subsidy removal except in energy-exporting regions. *Nature*, **554(7691)**, 229–233, doi:10.1038/nature25467.

Jewell, J., V. Vinichenko, L. Nacke, and A. Cherp, 2019: Prospects for powering past coal. *Nat. Clim. Change*, **9(8)**, 592–597, doi:10.1038/s41558-019-0509-6.

Jinnah, S., and S. Nicholson, 2019: The hidden politics of climate engineering. *Nat. Geosci.*, **12(11)**, 876–879, doi:10.1038/s41561-019-0483-7.

Jinnah, S., S. Nicholson, and J. Flegal, 2018: Toward Legitimate Governance of Solar Geoengineering Research: A Role for Sub-State Actors. *Ethics, Policy Environ.*, **21(3)**, 362–381, doi:10.1080/21550085.2018.1562526.

Johnson, B.A., R. Dasgupta, A.D. Mader, and H. Scheyvens, 2019: Understanding national biodiversity targets in a REDD+ context. *Environ. Sci. Policy*, **92**, 27–33, doi:10.1016/j.envsci.2018.11.007.

Jones, A. et al., 2021: North Atlantic Oscillation response in GeoMIP experiments G6solar and G6sulfur: Why detailed modelling is needed for understanding regional implications of solar radiation management. *Atmos. Chem. Phys.*, **21(2)**, 1287–1304, doi:10.5194/acp-21-1287-2021.

Jones, A.C. et al., 2018: Regional Climate Impacts of Stabilizing Global Warming at 1.5 K Using Solar Geoengineering. *Earth's Future*, **6(2)**, 230–251, doi:10.1002/2017ef000720.

Jordan, A.J. et al., 2015: Emergence of polycentric climate governance and its future prospects. *Nat. Clim. Change*, **5(11)**, 977–982, doi:10.1038/nclimate2725.

Joung, T.-H., S.-G. Kang, J.-K. Lee, and J. Ahn, 2020: The IMO initial strategy for reducing Greenhouse Gas(GHG) emissions, and its follow-up actions towards 2050. *J. Int. Marit. Safety, Environ. Aff. Shipp.*, **4(1)**, 1–7, doi:10.1080/25725084.2019.1707938.

Jun, M. and S. Zadek, 2019: *Decarbonizing the Road and Belt. A Green Finance Roadmap.* Tsinghua University Center for Finance and Development, Vivid Economics and the ClimateWorks Foundation, 54 pp. https://www.climateworks.org/wp-content/uploads/2019/09/Decarbonizing-the-Belt-and-Road_report_final_lo-res.pdf (Accessed October 31, 2021).

Karlas, J., 2017: States, coalitions, and the legalization of the global climate regime: negotiations on the post-2020 architecture. *Env. Polit.*, **26(5)**, 1–22, doi:10.1080/09644016.2017.1324754.

Karlsson-Vinkhuyzen, S.I. et al., 2018: Entry into force and then? The Paris agreement and state accountability. *Clim. Policy*, **18(5)**, 593–599, doi:10.1080/14693062.2017.1331904.

Karlsson, C., M. Hjerpe, C. Parker, and B.-O. Linnér, 2012: The Legitimacy of Leadership in International Climate Change Negotiations. *Ambio*, **41(1)**, 46–55, doi:10.1007/s13280-011-0240-7.

Kartha, S. et al., 2018: Cascading biases against poorer countries. *Nat. Clim. Change*, **8(5)**, 348–349, doi:10.1038/s41558-018-0152-7.

Kasperson, R. and J.X. Kasperson, 2001: *Climate change, vulnerability, and social justice*. Stockholm Environment Institute, Stockholm, Sweden, 18 pp. http://stc.umsl.edu/essj/unit4/climate%20change%20risk.pdf (Accessed October 31, 2021).

Kavlak, G., J. McNerney, and J.E. Trancik, 2018: Evaluating the causes of cost reduction in photovoltaic modules. *Energy Policy* **123**(August), 700–710, doi:10.1016/j.enpol.2018.08.015.

Keith, D.W. and D.G. MacMartin, 2015: A temporary, moderate and responsive scenario for solar geoengineering. *Nat. Clim. Change*, **5**, 201–206, doi:10.1038/nclimate2493.

Keller, D.P., 2018: Marine Climate Engineering. In: *Handbook on Marine Environment Protection* [Salomon, M. and T. Markus, (eds.)], Springer International Publishing, Cham, Switzerland, pp. 261–276.

Kemp, L., 2018: A Systems Critique of the 2015 Paris Agreement on Climate. In: *Pathways to a Sustainable Economy* [Hossain, M., R. Hales and T. Sarker. (eds.)]. Springer International Publishing, Cham, Switzerland, pp. 25–41.

Keohane, N., A. Petsonk, and A. Hanafi, 2017: Toward a club of carbon markets. *Clim. Change*, **144(1)**, 81–95, doi:10.1007/s10584-015-1506-z.

Keohane, R.O. and M. Oppenheimer, 2016: Paris: Beyond the Climate Dead End through Pledge and Review? *Polit. Gov.*, **4(3)**, 142, doi:10.17645/pag.v4i3.634.

Kérébel, C. and J. Keppler, 2009: *La Gouvernance Mondiale De L'énergie. Gouvernance Européenne Et Géopolitique De L'énergie*. IFRI, Paris, France, 258 pp.

Kern, F. and K.S. Rogge, 2016: The pace of governed energy transitions: Agency, international dynamics and the global Paris agreement accelerating decarbonisation processes? *Energy Res. Soc. Sci.*, **22**, 13–17, doi:10.1016/j.erss.2016.08.016.

Kern, K. and H. Bulkeley, 2009: Cities, Europeanization and Multi-level Governance: Governing Climate Change through Transnational Municipal Networks. *JCMS J. Common Mark. Stud.*, **47(2)**, 309–332, doi:10.1111/j.1468-5965.2009.00806.x.

Khan, M., D. Mfitumukiza, and S. Huq, 2020: Capacity building for implementation of nationally determined contributions under the Paris Agreement. *Clim. Policy*, **20(4)**, 499–510, doi:10.1080/14693062.2019.1675577.

Khan, M.R. and J.T. Roberts, 2013: Adaptation and international climate policy. *Wiley Interdiscip. Rev. Clim. Change*, **4(3)**, 171–189, doi:10.1002/wcc.212.

Khan, S.A. and K. Kulovesi, 2018: Black carbon and the Arctic: Global problem-solving through the nexus of science, law and space. *Rev. Eur. Comp. Int. Environ. Law*, **27(1)**, 5–14, doi:10.1111/reel.12245.

Kim, Y., K. Tanaka, and S. Matsuoka, 2020: Environmental and economic effectiveness of the Kyoto Protocol. *PLoS One*, **15(7)**, e0236299, doi:10.1371/journal.pone.0236299.

Kinley, R., M.Z. Cutajar, Y. de Boer, and C. Figueres, 2020: Beyond good intentions, to urgent action: Former UNFCCC leaders take stock of thirty years of international climate change negotiations. *Clim. Policy*, **21(5)**, 593–603, doi:10.1080/14693062.2020.1860567.

Kirchherr, J. and F. Urban, 2018: Technology transfer and cooperation for low carbon energy technology: Analysing 30 years of scholarship and proposing a research agenda. *Energy Policy*, **119**, 600–609, doi:10.1016/j.enpol.2018.05.001.

Kissinger, G., A. Gupta, I. Mulder, and N. Unterstell, 2019: Climate financing needs in the land sector under the Paris Agreement: An assessment of developing country perspectives. *Land use policy*, **83**, 256–269, doi:10.1016/j.landusepol.2019.02.007.

Klein, D., M. Pia Carazo, M. Doelle, J. Bulmer, and A. Higham, (eds.), 2017: *The Paris agreement on climate change: analysis and commentary*. Oxford University Press, Oxford, UK, 328 pp.

Klinsky, S. et al., 2017: Why equity is fundamental in climate change policy research. *Glob. Environ. Change*, **44**, 170–173, doi:10.1016/j.gloenvcha.2016.08.002.

Kloeckner, J., 2012: The power of eco-labels: Communicating climate change using carbon footprint labels consistent with international trade regimes under the WTO. *Clim. Law*, **3(3–4)**, 209–230, doi:10.1163/cl-120064.

Knox, J.H., 2016: *Report of the Special Rapporteur on the issue of human rights obligations relating to the enjoyment of a safe, clean, healthy and sustainable environment*. UN Human Rights Council, 22 pp. https://digitallibrary.un.org/record/861173?ln=en (Accessed October 31, 2021).

Knox, J.H., 2019: The Paris Agreement as a Human Rights Treaty. In: *Human Rights and the 21st Century Challenges: Poverty, Conflict and the Environment*, Oxford University Press, Oxford, UK, pp. 323–347.

Kollmuss, A., L. Schneider, and V. Zhezherin, 2015: *Has Joint Implementation reduced GHG emissions? Lessons learned for the design of carbon market mechanisms*. Stockholm Environment Institute, 124 pp. www.sei.org/publications/has-joint-implementation-reduced-ghg-emissions-lessons-learned-for-the-design-of-carbon-market-mechanisms/ (Accessed October 31, 2021).

Kona, A., P. Bertoldi, F. Monforti-Ferrario, S. Rivas, and J.F. Dallemand, 2018: Covenant of mayors signatories leading the way towards 1.5 degree global warming pathway. *Sustain. Cities Soc.*, **41**, 568–575, doi:10.1016/j.scs.2018.05.017.

Kona, A. et al., 2021: Global Covenant of Mayors, a dataset of greenhouse gas emissions for 6200 cities in Europe and the Southern Mediterranean countries. *Earth Syst. Sci. Data*, **13(7)**, 3551–3564, doi:10.5194/essd-13-3551-2021.

Kosugi, T., 2013: Fail-safe solar radiation management geoengineering. *Mitig. Adapt. Strateg. Glob. Change*, **18(8)**, 1141–1166, doi:10.1007/s11027-012-9414-2.

Kowarsch, M. et al., 2016: Scientific assessments to facilitate deliberative policy learning. *Palgrave Commun.*, **2(1)**, 16092, doi:10.1057/palcomms.2016.92.

Kramarz, T. and S. Park, 2016: Accountability in Global Environmental Governance: A Meaningful Tool for Action? *Glob. Environ. Polit.*, **16(2)**, 1–21, doi:10.1162/glep_a_00349.

Kravitz, B. and D.G. MacMartin, 2020: Uncertainty and the basis for confidence in solar geoengineering research. *Nat. Rev. Earth Environ.*, **1(1)**, 64–75, doi:10.1038/s43017-019-0004-7.

Kravitz, B. et al., 2015: The Geoengineering Model Intercomparison Project Phase 6 (GeoMIP6): simulation design and preliminary results. *Geosci. Model Dev.*, **8(10)**, 3379–3392, doi:10.5194/gmd-8-3379-2015.

Kreibich, N., 2018: *Raising Ambition through Cooperation. Using Article 6 to bolster climate change mitigation.* JIKO Policy Paper. No. 02/2018. Wuppertal Institute, Wuppertal, Germany, 29 pp. https://epub.wupperinst.org/frontdoor/deliver/index/docId/7122/file/7122_Raising_Ambition.pdf (Accessed October 31, 2021).

Krishnamohan, K.P.S. P., G. Bala, L. Cao, L. Duan, and K. Caldeira, 2019: Climate system response to stratospheric sulfate aerosols: Sensitivity to altitude of aerosol layer. *Earth Syst. Dyn.*, **10(4)**, 885–900, doi:10.5194/esd-10-885-2019.

Krueger, P., Z. Sautner, and L.T. Starks, 2020: The Importance of Climate Risks for Institutional Investors. *Rev. Financ. Stud.*, **33(3)**, 1067–1111, doi:10.1093/rfs/hhz137.

Krugman, P., 1991: History Versus Expectations. *Q. J. Econ.*, **106(2)**, 651–667, doi:10.2307/2937950.

Kruitwagen, L., K. Madani, B. Caldecott, and M.H.W. Workman, 2017: Game theory and corporate governance: conditions for effective stewardship of companies exposed to climate change risks. *J. Sustain. Financ. Invest.*, **7(1)**, 14–36, doi:10.1080/20430795.2016.1188537.

Kulovesi, K., 2014: International Trade Disputes on Renewable Energy: Testing Ground for the Mutual Supportiveness of WTO Law and Climate Change Law. *Rev. Eur. Comp. Int. Environ. Law*, **23(3)**, 342–353, doi:10.1111/reel.12092.

Kumazawa, R, and M.S. Callaghan, 2012: The effect of the Kyoto Protocol on carbon dioxide emissions. *J. Econ. Financ.*, **36(1)**, 201–210, doi:10.1007/s12197-010-9164-5.

Kuramochi, T. et al., 2020: Beyond national climate action: the impact of region, city, and business commitments on global greenhouse gas emissions. *Clim. Policy*, **20(3)**, 275–291, doi:10.1080/14693062.2020.1740150.

Kuriyama, A. and N. Abe, 2018: Ex-post assessment of the Kyoto Protocol – quantification of CO2 mitigation impact in both Annex B and non-Annex B countries. *Appl. Energy*, **220**, 286–295, doi:10.1016/j.apenergy.2018.03.025.

Kuyper, J., H. Schroeder, and B.-O. Linnér, 2018a: The Evolution of the UNFCCC. *Annu. Rev. Environ. Resour.*, **43(1)**, 343–368, doi:10.1146/annurev-environ-102017-030119.

Kuyper, J.W., B.O. Linnér, and H. Schroeder, 2018b: Non-state actors in hybrid global climate governance: justice, legitimacy, and effectiveness in a post-Paris era. *Wiley Interdiscip. Rev. Clim. Change*, **9(1)**, 1–18, doi:10.1002/wcc.497.

Kverndokk, S., 2018: Climate Policies, Distributional Effects and Transfers Between Rich and Poor Countries. *Int. Rev. Environ. Resour. Econ.*, **12(2–3)**, 129–176, doi:10.1561/101.00000100.

Kwiatkowski, L., P. Cox, P.R. Halloran, P.J. Mumby, and A.J. Wiltshire, 2015: Coral bleaching under unconventional scenarios of climate warming and ocean acidification. *Nat. Clim. Change*, **5(8)**, 777–781, doi:10.1038/nclimate2655.

La Hoz Theuer, S., L. Schneider, and D. Broekhoff, 2019: When less is more: limits to international transfers under Article 6 of the Paris Agreement. *Clim. Policy*, **19(4)**, 401–413, doi:10.1080/14693062.2018.1540341.

Labordena, M., A. Patt, M. Bazilian, M. Howells, and J. Lilliestam, 2017: Impact of political and economic barriers for concentrating solar power in Sub-Saharan Africa. *Energy Policy*, **102**, 52–72, doi:10.1016/j.enpol.2016.12.008.

Lamb, W.F. et al., 2021: A review of trends and drivers of greenhouse gas emissions by sector from 1990 to 2018. *Environ. Res. Lett.*, **16(7)**, 073005, doi:10.1088/1748-9326/ABEE4E.

Lambin, E.F. et al., 2018: The role of supply-chain initiatives in reducing deforestation. *Nat. Clim. Change*, **8**, 109–116, doi:10.1038/s41558-017-0061-1.

Larsen, G. et al., 2018: *Towards Paris Alignment. How the Multilateral Development Banks Can Better Support the Paris Agreement*. World Resources Institute, Germanwatch, NewClimate Institute and Fundación Avina, 118 pp.

Larsson, J., A. Elofsson, T. Sterner, and J. Åkerman, 2019: International and national climate policies for aviation: a review. *Clim. Policy*, **19(6)**, 787–799, doi:10.1080/14693062.2018.1562871.

Latham, J. et al., 2012: Marine cloud brightening. *Philos. Trans. R. Soc. A Math. Phys. Eng. Sci.*, **370(1974)**, 4217–4262, doi:10.1098/rsta.2012.0086.

Latham, J., J. Kleypas, R. Hauser, B. Parkes, and A. Gadian, 2013: Can marine cloud brightening reduce coral bleaching? *Atmos. Sci. Lett.*, **14(4)**, 214–219, doi:10.1002/asl2.442.

Laurens, N., Z. Dove, J.-F. Morin, and S. Jinnah, 2019: NAFTA 2.0: The Greenest Trade Agreement Ever? *World Trade Rev.*, **18(4)**, 659–677, doi:10.1017/s1474745619000351.

Lawrence, P. and D. Wong, 2017: Soft law in the paris climate agreement: Strength or weakness? *Rev. Eur. Comp. Int. Environ. Law*, **26(3)**, 276–286, doi:10.1111/reel.12210.

Lawrence, P. and M. Reder, 2019: Equity and the Paris agreement: Legal and philosophical perspectives. *J. Environ. Law*, **31(3)**, 511–531, doi:10.1093/jel/eqz017.

Le Blanc, D., 2015: Towards Integration at Last? The Sustainable Development Goals as a Network of Targets. *Sustain. Dev.*, **23(3)**, 176–187, doi:10.1002/sd.1582.

Lederer, M. and J. Kreuter, 2018: Organising the unthinkable in times of crises: Will climate engineering become the weapon of last resort in the Anthropocene? *Organization*, **25(4)**, 472–490, doi:10.1177/1350508418759186.

Lee, D.S. et al., 2021a: The contribution of global aviation to anthropogenic climate forcing for 2000 to 2018. *Atmos. Environ.*, **244**, 117834, doi:10.1016/j.atmosenv.2020.117834.

Lee, K., C. Fyson, and C.-F. Schleussner, 2021b: Fair distributions of carbon dioxide removal obligations and implications for effective national net-zero targets. *Environ. Res. Lett.*, **16(9)**, 94001, doi:10.1088/1748-9326/ac1970.

Lees, E., 2017: Responsibility and liability for climate loss and damage after Paris. *Clim. Policy*, **17(1)**, 59–70, doi:10.1080/14693062.2016.1197095.

Leffel, B., 2018: *Subnational Diplomacy, Climate Governance & Californian Global Leadership*. USC Center on Public Diplomacy, 10 pp. www.uscpublicdiplomacy.org/sites/uscpublicdiplomacy.org/files/useruploads/u39301/Subnational_Diplomacy%2C_Climate_Governance_%26_Californian_Global_Leadership_-_Benjamin_Leffel_0.pdf (Accessed October 31, 2021).

Leonard, L., 2014: Tackling Climate Change in the Global South: An Analysis of the Global Methane Initiative Multilateral Partnership. *J. Soc. Dev. Sci.*, **5(4)**, 168–175, doi:10.22610/jsds.v5I4.817.

*Le Quéré C, Korsbakken J I, Wilson C, Tosun J, Andrew R, Andres R J, Canadell J G, Jordan A, Peters G P and van Vuuren D P, 2019: Drivers of declining $CO_2$ emissions in 18 developed economies Nat. Clim. Change 9 213–7.*

Lesage, D., T. Van de Graaf, and K. Westphal, 2010: *Global Energy Governance in a Multipolar World*. [Kirton, J.J. and M. Schreurs, (eds.)]. Ashgate Publishing Limited, Aldershot, UK, 221 pp.

Lessmann, K. et al., 2015: The Stability and Effectiveness of Climate Coalitions: A Comparative Analysis of Multiple Integrated Assessment Models. *Environ. Resour. Econ.*, **62(4)**, 811–836, doi:10.1007/s10640-015-9886-0.

Lewis, J.I., 2014: The Rise of Renewable Energy Protectionism: Emerging Trade Conflicts and Implications for Low Carbon Development. *Glob. Environ. Polit.*, **14(4)**, 10–35, doi:10.1162/glep_a_00255.

Liagre, L., P.L. Almuedo, C. Besacier, and M. Conigliaro, 2015: *Sustainable financing for forest and landscape restoration: Opportunities, challenges and the way forward*. FAO/UNCCD, Rome, Italy, 131 pp.

Lindberg, M.B., J. Markard, and A.D. Andersen, 2018: Policies, actors and sustainability transition pathways: A study of the EU's energy policy mix. *Res. Policy*, **48(10)**, 103668, doi:10.1016/j.respol.2018.09.003.

Liu, J.-Y., S. Fujimori, and T. Masui, 2016: Temporal and spatial distribution of global mitigation cost: INDCs and equity. *Environ. Res. Lett.*, **11(11)**, 114004, doi:10.1088/1748-9326/11/11/114004.

Liu, J., 2011a: The Cancun Agreements. *Environ. Law Rev.*, **13(1)**, 43–49, doi:10.1350/enlr.2011.13.1.112.

Liu, J., 2011b: The Role of ICAO in Regulating the Greenhouse Gas Emissions of Aircraft. *Carbon Clim. Law Rev.*, **5(4)**, 417–431.

Liu, J. et al., 2017a: South-south cooperation for large-scale ecological restoration. *Restor. Ecol.*, **25(1)**, 27–32, doi:10.1111/rec.12462.

Liu, J., P.E. Love, J. Smith, and M. Regan, 2017b: A new framework for evaluating public-private partnerships. *Second International Conference on Public-Private Partnerships. May 26–29, 2015*, [Zhang, Z., C. Queiroz, and C.M. Walton, (eds.)], American Society of Civil Engineers, Austin, TX, USA, pp. 290–301.

Liu, L. and M. Waibel, 2010: Managing Subnational Credit and Default Risks. In: *Sovereign Debt and the Financial Crisis* [Primo Braga, C.A. and G.A. Vincelette, (eds.)], The World Bank, Washington DC, USA, pp. 273–293.

Liu, W., X. Zhang, and S. Feng, 2019: Does renewable energy policy work? Evidence from a panel data analysis. *Renew. Energy*, **135**(May), 635–642, doi:10.1016/j.renene.2018.12.037.

Livingston, D., 2016: *The G7 Climate Mandate and the Tragedy of the Horizons*. Carnegie Endowment for International Peace, 34 pp. www.jstor.org/stable/resrep12840.

Lloyd, I.D. and M. Oppenheimer, 2014: On the Design of an International Governance Framework for Geoengineering. *Glob. Environ. Polit.*, **14(2)**, 45–63, doi:10.1162/glep_a_00228.

Locatelli, B., V. Evans, A. Wardell, A. Andrade, and R. Vignola, 2011: Forests and Climate Change in Latin America: Linking Adaptation and Mitigation. *Forests*, **2(1)**, 431–450, doi:10.3390/f2010431.

London Convention/Protocol, 2010: *Resolution LC-LP.2 on the assessment framework for scientific research involving ocean fertilization (14 October 2010)*. Document LC 32/15, Annex 5, 2 pp. https://wwwcdn.imo.org/localresources/en/KnowledgeCentre/IndexofIMOResolutions/LCLPDocuments/LC-LP.2(2010).pdf (Accessed October 31, 2021).

Low, S. and M. Honegger, 2020: A Precautionary Assessment of Systemic Projections and Promises From Sunlight Reflection and Carbon Removal Modeling. *Risk Anal.*, (Online), 15 pp, doi:10.1111/risa.13565.

LSE, 2020: Climate Change Laws of the World database, Grantham Research Institute on Climate Change and the Environment and Sabin Center for Climate Change Law. https://climate-laws.org/ (Accessed December 14, 2020).

Lui, S. et al., 2021: Correcting course: the emission reduction potential of international cooperative initiatives. *Clim. Policy*, **21(2)**, 232–250, doi:10.1080/14693062.2020.1806021.

Lyle, C., 2018: Beyond the ICAO's CORSIA: Towards a More Climatically Effective Strategy for Mitigation of Civil-Aviation Emissions. *Clim. Law*, **8(1–2)**, 104–127, doi:10.1163/18786561-00801004.

Maamoun, N., 2019: The Kyoto protocol: Empirical evidence of a hidden success. *J. Environ. Econ. Manage.*, **95**(May), 227–256, doi:10.1016/j.jeem.2019.04.001.

Maas, A. and J. Scheffran, 2012: Climate Conflicts 2.0? Climate Engineering as a Challenge for International Peace and Security. *Sicherheit und Frieden / Secur. Peace*, **30(4)**, 193–200, doi:10.2307/24233201.

Macchi, C. and J. van Zeben, 2021: Business and human rights implications of climate change litigation: Milieudefensie et al. v Royal Dutch Shell. *Rev. Eur. Comp. Int. Environ. Law*, **30(3)**, 1–7, doi:10.1111/reel.12416.

MacDonald, A., A. Clarke, L. Huang, M. Roseland, and M.M. Seitanidi, 2017: Multi-stakeholder Partnerships (SDG #17) as a Means of Achieving Sustainable Communities and Cities. In: *World Sustainability Series* [Leal Filho, W. (ed.)], Springer International Publishing, pp. 193–209.

MacDougall, A.H. et al., 2020: Is there warming in the pipeline? A multi-model analysis of the Zero Emissions Commitment from CO2. *Biogeosciences*, **17(11)**, 2987–3016, doi:10.5194/bg-17-2987-2020.

Mace, M.J., 2016: Mitigation Commitments under the Paris Agreement and the Way Forward. *Clim. Law*, **6(1–2)**, 21–39, doi:10.1163/18786561-00601002.

Mace, M.J., and R. Verheyen, 2016: Loss, damage and responsibility after COP21: All options open for the Paris agreement. *Rev. Eur. Comp. Int. Environ. Law*, **25(2)**, 197–214, doi:10.1111/reel.12172.

Mace, M.J., C. Fyson, M. Schaeffer, and B. Hare, 2018: *Governing large-scale carbon dioxide removal: are we ready?* Carnegie Climate Geoengineering Governance Initiative, New York, NY, USA, 46 pp. www.c2g2.net/wp-content/uploads/C2G2-2018-CDR-Governance-1.pdf (Accessed October 31, 2021).

Mace, M.J., C.L. Fyson, M. Schaeffer, and W.L. Hare, 2021: Large-Scale Carbon Dioxide Removal to Meet the 1.5°C Limit: Key Governance Gaps, Challenges and Priority Responses. *Glob. Policy*, **12**(S1), 67–81, doi:10.1111/1758-5899.12921.

MacKay, M., B. Parlee, and C. Karsgaard, 2020: Youth Engagement in Climate Change Action: Case Study on Indigenous Youth at COP24. *Sustainability*, **12(16)**, 6299, doi:10.3390/su12166299.

MacLeod, M. and J. Park, 2011: Financial Activism and Global Climate Change: The Rise of Investor-Driven Governance Networks. *Glob. Environ. Polit.*, **11(2)**, 54–74, doi:10.1162/glep_a_00055.

MacMartin, D.G. and B. Kravitz, 2019: Mission-driven research for stratospheric aerosol geoengineering. *Proc. Natl. Acad. Sci.*, **116(4)**, 1089–1094, doi:10.1073/pnas.1811022116.

MacMartin, D.G. K. Caldeira, and D.W. Keith, 2014: Solar geoengineering to limit the rate of temperature change. *Philos. Trans. R. Soc. A Math. Phys. Eng. Sci.*, **372(2031)**, doi:10.1098/rsta.2014.0134.

MacMartin, D.G., K.L. Ricke, and D.W. Keith, 2018: Solar geoengineering as part of an overall strategy for meeting the 1.5°C Paris target. *Philos. Trans. R. Soc. A Math. Phys. Eng. Sci.*, **376(2119)**, 20160454, doi:10.1098/rsta.2016.0454.

Madani, K., 2020: How International Economic Sanctions Harm the Environment. *Earth's Future*, **8(12)**, e2020ef001829, doi:10.1029/2020ef001829.

Maguire, P. et al., 2021: *A Green Growth Spurt: State of Forest Carbon Finance 2021*. Ecosystem Marketplace Insights Report, Forests Trends, Washington, DC, USA, 67 pp. www.ecosystemmarketplace.com/publications/state-of-forest-carbon-finance-2021/ (Accessed October 31, 2021).

Malhotra, A., A. Mathur, S. Diddi, and A.D. Sagar, 2021: Building institutional capacity for addressing climate and sustainable development goals: achieving energy efficiency in India. *Climate Policy*, Forthcoming Special Issue: Capacity building, pp. 1-19, doi:10.1080/14693062.2021.1984195.

Maljean-Dubois, S., 2016: The Paris Agreement: A New Step in the Gradual Evolution of Differential Treatment in the Climate Regime? *Rev. Eur. Community Int. Environ. Law*, **25(2)**, 151–160, doi:10.1111/reel.12162.

Maljean-Dubois, S., 2019: Climate change litigation. In: *Max Planck Encyclopedia of Procedural Law*, Oxford University Press, Oxford, UK, pp. 1–26.

Maljean-Dubois, S. and M. Wemaëre, 2016: The Paris Agreement: A Starting Point towards Achieving Climate Neutrality? *Carbon Clim. Law Rev.*, **10(1)**, 1–4.

Maltais, A. and B. Nykvist, 2020: Understanding the role of green bonds in advancing sustainability. *J. Sustain. Financ. Invest.*, (Online), 1–20, doi:10.1080/20430795.2020.1724864.

Mansourian, S., N. Dudley, and D. Vallauri, 2017: Forest Landscape Restoration: Progress in the Last Decade and Remaining Challenges. *Ecol. Restor.*, **35(4)**, 281–288, doi:10.3368/er.35.4.281.

Marcu, A., 2016: *Carbon Market Provisions in the Paris Agreement (Article 6)*. Centre for European Policy Studies, Brussels, Belgium, 26 pp. http://aei.pitt.edu/71012/1/SR_No_128_ACM_Post_COP21_Analysis_of_Article_6.pdf (Accessed October 31, 2021).

Marjanac, S. and L. Patton, 2018: Extreme weather event attribution science and climate change litigation: An essential step in the causal chain? *J. Energy Nat. Resour. Law*, **36(3)**, 265–298, doi:10.1080/02646811.2018.1451020.

Markard, J., 2018: The next phase of the energy transition and its implications for research and policy. *Nat. Energy*, **3(8)**, 628–633, doi:10.1038/s41560-018-0171-7.

Marques, A. et al., 2014: A framework to identify enabling and urgent actions for the 2020 Aichi Targets. *Basic Appl. Ecol.*, **15(8)**, 633–638, doi:10.1016/j.baae.2014.09.004.

Martinez Romera, B., 2016: The Paris Agreement and the Regulation of International Bunker Fuels. *Rev. Eur. Comp. Int. Environ. Law*, **25(2)**, 215–227, doi:10.1111/reel.12170.

Mayer, B., 2016a: The relevance of the no-harm principle to climate change law and politics. *Asia Pacific J. Environ. Law*, **19(1)**, 79–104, doi:10.4337/apjel.2016.01.04.

Mayer, B., 2016b: Human Rights in the Paris Agreement. *Clim. Law*, **6(1–2)**, doi:10.1163/18786561-00601007.

Mayer, B., 2018a: Obligations of conduct in the international law on climate change: A defence. *Rev. Eur. Comp. Int. Environ. Law*, **27(2)**, 130–140, doi:10.1111/reel.12237.

Mayer, B., 2018b: International Law Obligations Arising in relation to Nationally Determined Contributions. *Transnatl. Environ. Law*, **7(2)**, 251–275, doi:10.1017/s2047102518000110.

Mayer, B., 2019: Transparency Under the Paris Rulebook: Is the Transparency Framework Truly Enhanced? *Clim. Law*, **9(1–2)**, 40–64, doi:10.1163/18786561-00901004.

Mayer, B., 2021: Climate Change Mitigation as an Obligation Under Human Rights Treaties? *Am. J. Int. Law*, **115(3)**, 409–451, doi:10.1017/ajil.2021.9.

Mayer, B. and F. Crépeau, 2016: Introduction. In: *Research Handbook on Climate Change, Migration and the Law* [Mayer, B. and F. Crépeau, (eds.)], Elgar, Cheltenham, UK, pp. 1–26.

Mayr, S., B. Hollaus, and V. Madner, 2020: Palm Oil, the RED II and WTO Law: EU Sustainable Biofuel Policy Tangled up in Green? *Rev. Eur. Comp. Int. Environ. Law*, **30(2)**, 233–248, doi:10.1111/reel.12386.

McAdam, D., 2017: Social Movement Theory and the Prospects for Climate Change Activism in the United States. *Annu. Rev. Polit. Sci.*, **20**, 189–208, doi:10.1146/annurev-polisci-052615-025801.

McAdam, J., 2016: Climate Change-related Displacement of Persons. In: *The Oxford Handbook of International Climate Change Law* [Gray, K.R., R. Tarasofsky, and C. Carlarne, (eds.)]. Oxford University Press, Oxford, UK.

McAusland, C. and N. Najjar, 2015: The WTO Consistency of Carbon Footprint Taxes. *Georg. J. Int. Law*, **46(3)**, 765–801.

McCusker, K.E., K.C. Armour, C.M. Bitz, and D.S. Battisti, 2014: Rapid and extensive warming following cessation of solar radiation management. *Environ. Res. Lett.*, **9(2)**, 24005, doi:10.1088/1748-9326/9/2/024005.

McDonald, J., J. McGee, K. Brent, and W. Burns, 2019: Governing geoengineering research for the Great Barrier Reef. *Clim. Policy*, **19(7)**, 801–811, doi:10.1080/14693062.2019.1592742.

McGee, J. and R. Taplin, 2006: The Asia–Pacific partnership on clean development and climate: A complement or competitor to the Kyoto protocol? *Glob. Change, Peace Secur.*, **18(3)**, 173–192, doi:10.1080/14781150600960230.

McKinnon, C., 2019: Sleepwalking into lock-in? Avoiding wrongs to future people in the governance of solar radiation management research. *Env. Polit.*, **28(3)**, 441–459, doi:10.1080/09644016.2018.1450344.

McLaren, D., 2016: Mitigation deterrence and the "moral hazard" of solar radiation management. *Earth's Future*, **4(12)**, 596–602, doi:10.1002/2016ef000445.

McLaren, D. and O. Corry, 2021: The politics and governance of research into solar geoengineering. *Wiley Interdiscip. Rev. Clim. Change*, **12(3)**, e707, doi:10.1002/wcc.707.

McLaren, D.P., D.P. Tyfield, R. Willis, B. Szerszynski, and N.O. Markusson, 2019: Beyond "Net-Zero": A Case for Separate Targets for Emissions Reduction and Negative Emissions. *Front. Clim.*, **1**, 1–5, doi:10.3389/fclim.2019.00004.

McNamara, K.E. and C. Farbotko, 2017: Resisting a 'Doomed' Fate: an analysis of the Pacific Climate Warriors. *Aust. Geogr.*, **48(1)**, 17–26, doi:10.1080/00049182.2016.1266631.

McNamara, K.E. and G. Jackson, 2019: Loss and damage: A review of the literature and directions for future research. *Wiley Interdiscip. Rev. Clim. Change*, **10(2)**, e564, doi:10.1002/wcc.564.

Meehan, F., L. Tacconi, and K. Budiningsih, 2019: Are national commitments to reducing emissions from forests effective? Lessons from Indonesia. *For. Policy Econ.*, **108**(November), 101968, doi:10.1016/j.forpol.2019.101968.

Mehling, M.A., 2019: Governing Cooperative Approaches under the Paris Agreement. *Ecol. Law Q.*, **46(3)**, 765–827, doi:10.15779/z389g5gd97.

Mehling, M.A., G.E. Metcalf, and R.N. Stavins, 2018: Linking climate policies to advance global mitigation. *Science*, **359(6379)**, 997–998, doi:10.1126/science.aar5988.

Mehling, M.A., H. Van Asselt, K. Das, S. Droege, and C. Verkuijl, 2019: Designing Border Carbon Adjustments for Enhanced Climate Action. *Am. J. Int. Law*, **113(3)**, doi:10.1017/ajil.2019.22.

Meinshausen, M. et al., 2015: National post-2020 greenhouse gas targets and diversity-aware leadership. *Nat. Clim. Change*, **5(12)**, 1098–1106, doi:10.1038/nclimate2826.

Meltzer, J.P., 2013: The Trans-Pacific Partnership Agreement, the environment and climate change. In: *Trade Liberalisation and International Co-operation. A Legal Analysis of the Trans-Pacific Partnership Agreement* [Voon, T., (ed.)]. Edward Elgar Publishing, Cheltenham, UK, pp. 207–230.

Merk, C., G. Pönitzsch, C. Kniebes, K. Rehdanz, and U. Schmidt, 2015: Exploring public perceptions of stratospheric sulfate injection. *Clim. Change*, **130(2)**, 299–312, doi:10.1007/s10584-014-1317-7.

Merk, C., G. Pönitzsch, and K. Rehdanz, 2016: Knowledge about aerosol injection does not reduce individual mitigation efforts. *Environ. Res. Lett.*, **11(5)**, 54009, doi:10.1088/1748-9326/11/5/054009.

Merk, C., G. Pönitzsch, and K. Rehdanz, 2019: Do climate engineering experts display moral-hazard behaviour? *Clim. Policy*, **19(2)**, 231–243, doi:10.1080/14693062.2018.1494534.

Meyer, T., 2017: Explaining energy disputes at the World Trade Organization. *Int. Environ. Agreements Polit. Law Econ.*, **17(3)**, 391–410, doi:10.1007/s10784-017-9356-y.

Michaelowa, A. and K. Michaelowa, 2011a: Climate business for poverty reduction? The role of the World Bank. *Rev. Int. Organ.*, **6**, 259–286, doi:10.1007/s11558-011-9103-z.

Michaelowa, A. and K. Michaelowa, 2011b: Coding Error or Statistical Embellishment? The Political Economy of Reporting Climate Aid. *World Dev.*, **39(11)**, 2010–2020, doi:10.1016/j.worlddev.2011.07.020.

Michaelowa, A., L. Hermwille, W. Obergassel, and S. Butzengeiger, 2019a: Additionality revisited: guarding the integrity of market mechanisms under the Paris Agreement. *Clim. Policy*, **19(10)**, 1211–1224, doi:10.1080/14693062.2019.1628695.

Michaelowa, A., I. Shishlov, and D. Brescia, 2019b: Evolution of international carbon markets: lessons for the Paris Agreement. *Wiley Interdiscip. Rev. Clim. Change*, **10(6)**, doi:10.1002/wcc.613.

Michaelowa, A., A. Espelage, and B. Müller, 2020a: *2020 Update: Negotiating cooperation under Article 6 of the Paris Agreement*. [Sharma, A., (ed.)]. European Capacity Building Initiative, 30 pp. https://ecbi.org/sites/default/files/Article 6 2020_0.pdf (Accessed August 24, 2021).

Michaelowa, A., S. Hoch, A.K. Weber, R. Kassaye, and T. Hailu, 2020b: Mobilising private climate finance for sustainable energy access and climate change mitigation in Sub-Saharan Africa. *Clim. Policy*, **21(1)**, 47–62, doi:10.1080/14693062.2020.1796568.

Michaelowa, K. and A. Michaelowa, 2017: Transnational Climate Governance Initiatives: Designed for Effective Climate Change Mitigation? *Int. Interact.*, **43**, 129–155, doi:10.1080/03050629.2017.1256110.

Miles, K., 2019: *Research Handbook on Environment and Investment Law*. [Miles, K., (ed.)]. Edward Elgar Publishing, Cheltenham, UK, 576 pp.

Milkoreit, M. and K. Haapala, 2019: The global stocktake : design lessons for a new review and ambition mechanism in the international climate regime. *Int. Environ. Agreements Polit. Law Econ.*, **19(1)**, 89–106, doi:10.1007/s10784-018-9425-x.

Milman, A., L. Bunclark, D. Conway, and W.N. Adger, 2013: Assessment of institutional capacity to adapt to climate change in transboundary river basins. *Clim. Change*, **121(4)**, 755–770, doi:10.1007/s10584-013-0917-y.

Mitchell, I., E. Ritchie, and A. Tahmasebi, 2021: *Is Climate Finance Towards $100 Billion 'New and Additional'?* Washington, DC, USA, 14 pp. www.cgdev.org/sites/default/files/PP205-Mitchell-Ritchie-Tahmasebi-Climate-Finance.pdf (Accessed October 31, 2021).

Miyamoto, M. and K. Takeuchi, 2019: Climate agreement and technology diffusion: Impact of the Kyoto Protocol on international patent applications for renewable energy technologies. *Energy Policy*, **129**(June), 1331–1338, doi:10.1016/j.enpol.2019.02.053.

Moerenhout, T.S.H. and T. Irschlinger, 2020: *Exploring the Trade Impacts of Fossil Fuel Subsidies*. International Institute for Sustainable Development, Geneva, Switzerland, 50 pp. www.iisd.org/system/files/publications/trade-impacts-fossil-fuel-subsidies.pdf (Accessed October 31, 2021).

Molina, M. et al., 2009: Reducing abrupt climate change risk using the Montreal Protocol and other regulatory actions to complement cuts in $CO_2$ emissions. *Proc. Natl. Acad. Sci.*, **106(49)**, 20616–20621, doi:10.1073/pnas.0902568106.

Morgan, J. and E. Northrop, 2017: Will the Paris Agreement accelerate the pace of change? *Wiley Interdiscip. Rev. Clim. Change*, **8(5)**, e471, doi:10.1002/wcc.471.

Morin, J.-F. and S. Jinnah, 2018: The untapped potential of preferential trade agreements for climate governance. *Env. Polit.*, **27(3)**, 541–565, doi:10.1080/09644016.2017.1421399.

Moriyama, R. et al., 2017: The cost of stratospheric climate engineering revisited. *Mitig. Adapt. Strateg. Glob. Change*, **22(8)**, 1207–1228, doi:10.1007/s11027-016-9723-y.

Mormann, F., 2020: Why the divestment movement is missing the mark. *Nat. Clim. Change*, **10**, 1067–1068, doi:10.1038/s41558-020-00950-2.

Morrow, D.R., 2014: Ethical aspects of the mitigation obstruction argument against climate engineering research. *Philos. Trans. R. Soc. A Math. Phys. Eng. Sci.*, **372(2031)**, 1–14, doi:10.1098/rsta.2014.0062.

Müller, B. and A. Michaelowa, 2019: How to operationalize accounting under Article 6 market mechanisms of the Paris Agreement. *Clim. Policy*, **19(7)**, 812–819, doi:10.1080/14693062.2019.1599803.

Multilateral Investment Guarantee Agency, 2019: About MIGA. www.miga.org/about-us (Accessed October 28, 2021).

Murata, A. et al., 2016: Environmental co-benefits of the promotion of renewable power generation in China and India through clean development mechanisms. *Renew. Energy*, **87**(Part 1), 120–129, doi:10.1016/j.renene.2015.09.046.

Muri, H., U. Niemeier, and J.E. Kristjánsson, 2015: Tropical rainforest response to marine sky brightening climate engineering. *Geophys. Res. Lett.*, **42(8)**, 2951–2960, doi:10.1002/2015gl063363.

Nakhooda, S. et al., 2014: *Climate Finance – Is it making a difference? A review of the effectiveness of multilateral climate funds*. Overseas Development Institute, London, UK, 88 pp.

Nasiritousi, N., M. Hjerpe, and B.-O. Linnér, 2016: The roles of non-state actors in climate change governance: understanding agency through governance profiles. *Int. Environ. Agreements Polit. Law Econ.*, **16(1)**, 109–126, doi:10.1007/s10784-014-9243-8.

National Academies of Sciences Engineering and Medecine, 2021: *Reflecting Sunlight: Recommendations for Solar Geoengineering Research and Research Governance*. National Academies Press, Washington, DC, USA, 328 pp.

NCE, 2016: *The Sustainable Infrastructure Imperative, Financing for better growth and development. The 2016 New Climate Economy Report.* [Davis, M. (ed.)]. New Climate Economy, 150 pp. http://newclimateeconomy.report/2016/wp-content/uploads/sites/4/2014/08/NCE_2016Report.pdf (Accessed October 31, 2021).

Nelson, J.P., L. Kaplan, and D. Tomblin, 2021: Assessing solar geoengineering research funders: Insights from two US public deliberations: *Anthr. Rev.*, **8(1)**, 37–55, doi:10.1177/2053019620964845.

Nemati, M., W. Hu, and M. Reed, 2019: Are free trade agreements good for the environment? A panel data analysis. *Rev. Dev. Econ.*, **23(1)**, 435–453, doi:10.1111/rode.12554.

Nemet, G.F., 2019: *How solar energy became cheap a model for low-carbon innovation*. Routledge. London and New York, 260 pp.

Newell, P. and A. Simms, 2020: Towards a fossil fuel non-proliferation treaty. *Clim. Policy*, **20(8)**, 1043–1054, doi:10.1080/14693062.2019.1636759.

Newton, P., J. A Oldekop, G. Brodnig, B.K. Karna, and A. Agrawal, 2016: Carbon, biodiversity, and livelihoods in forest commons: synergies, trade-offs, and implications for REDD+. *Environ. Res. Lett.*, **11(4)**, 44017, doi:10.1088/1748-9326/11/4/044017.

Nicholson, S., S. Jinnah, and A. Gillespie, 2018: Solar radiation management: a proposal for immediate polycentric governance. *Clim. Policy*, **18(3)**, 322–334, doi:10.1080/14693062.2017.1400944.

Niemeier, U., H. Schmidt, and C. Timmreck, 2011: The dependency of geoengineered sulfate aerosol on the emission strategy. *Atmos. Sci. Lett.*, **12(2)**, 189–194, doi:10.1002/asl.304.

Nilsson, M., D. Griggs, and M. Visbeck, 2016: Policy: map the interactions between Sustainable Development Goals. *Nature*, **534(7607)**, 320–322, doi:10.1038/534320a.

Nordhaus, W., 2015: Climate Clubs: Overcoming Free-riding in International Climate Policy. *Am. Econ. Rev.*, **105(4)**, 1339–1370, doi:10.1257/aer.15000001.

NYDF Assessment Partners, 2020: *Progress on the New York Declaration on Forests (NYDF): Balancing forests and development – Addressing infrastructure and extractive industries, promoting sustainable livelihoods. Goals 3 & 4 Progress Report*. Washington, DC, USA, 109 pp. www.forestdeclaration.org/images/uploads/resource/2020NYDFReport.pdf (Accessed October 31, 2021).

Obergassel, W., F. Mersmann, and H. Wang-Helmreich, 2017a: Two for One: Integrating the Sustainable Development Agenda with International Climate Policy. *Gaia – Ecol. Perspect. Sci. Soc.*, **26**, 249–253, doi:10.14512/gaia.26.3.8.

Obergassel, W. et al., 2017b: Human rights and the clean development mechanism: lessons learned from three case studies. *J. Hum. Rights Environ.* **8(1)**, doi:10.4337/jhre.2017.01.03.

Oberthür, S., 2016: Reflections on Global Climate Politics Post Paris: Power, Interests and Polycentricity. *Int. Spect.*, **51**, 80–94, doi:10.1080/03932729.2016.1242256.

Oberthür, S. and R. Bodle, 2016: Legal Form and Nature of the Paris Outcome. *Clim. Law*, **6(1–2)**, 40–57, doi:10.1163/18786561-00601003.

Oberthür, S. and E. Northrop, 2018: Towards an Effective Mechanism to Facilitate Implementation and Promote Compliance under the Paris Agreement. *Clim. Law*, **8(1–2)**, 39–69, doi:10.1163/18786561-00801002.

Oberthür, S. and L. Groen, 2020: Hardening and softening of multilateral climate governance towards the Paris Agreement. *J. Environ. Policy Plan.*, **22(6)**, 801–813, doi:10.1080/1523908x.2020.1832882.

Oberthür, S. et al., 2017: *COP21: Results and Implications for Pathways and Policies for Low Emissions European Societies. Ref. Ares (2017)5316295 – 31/10/2017*. European Commission, Brussels, Belgium, 129 pp. https://cordis.europa.eu/project/id/730427/results (Accessed October 31, 2021).

OECD, 2012: *Meeting global challenges through better governance: international co-operation in science, technology and innovation*. OECD Publishing, Paris, France, 242 pp.

OECD, 2019a: Climate Change: OECD DAC External Development Finance Statistics. Organisation for Economic Co-operation and Development. OECD Publishing, Paris. www.oecd.org/dac/financing-sustainable-development/development-finance-topics/climate-change.htm (Accessed December 11, 2019).

OECD, 2019b: *Annual survey of large pension funds and public pension reserve funds*. Organisation for Economic Co-operation and Development, OECD Publishing, Paris, 45 pp. www.oecd.org/finance/survey-large-pension-funds.htm (Accessed December 9, 2020).

OECD, 2021: *Climate Finance Provided and Mobilised by Developed Countries: Aggregate Trends Updated with 2019 Data, Climate Finance and the USD 100 Billion Goal*. OECD Publishing, Paris, France, 21 pp.

Office of the High Commissioner for Human Rights, 2019: Five UN human rights treaty bodies issue a joint statement on human rights and climate change (16 September 2019). www.ohchr.org/EN/NewsEvents/Pages/DisplayNews.aspx?NewsID=24998 (Accessed October 25, 2021).

Oh, C., 2019: Political Economy of International Policy on the Transfer of Environmentally Sound Technologies in Global Climate Change Regime. *New Polit. Econ.*, **24(1)**, 22–36, doi:10.1080/13563467.2017.1417361.

Oh, C., 2020a: Discursive Contestation on Technological Innovation and the Institutional Design of the UNFCCC in the New Climate Change Regime. *New Polit. Econ.*, **25(4)**, 660–674, doi:10.1080/13563467.2019.1639147.

Oh, C., 2020b: Contestations over the financial linkages between the UNFCCC's Technology and Financial Mechanism: using the lens of institutional interaction. *Int. Environ. Agreements*, **20**, 559–575, doi:10.1007/s10784-020-09474-8.

Oh, C. and S. Matsuoka, 2017: The genesis and end of institutional fragmentation in global governance on climate change from a constructivist perspective. *Int. Environ. Agreements Polit. Law Econ.*, **17(2)**, 143–159, doi:10.1007/s10784-015-9309-2.

Olmstead, S.M., and R.N. Stavins, 2012: Three Key Elements of a Post-2012 International Climate Policy Architecture. *Rev. Environ. Econ. Policy*, **6(1)**, 65–85, doi:10.1093/reep/rer018.

Olsen, K.H., C. Arens, and F. Mersmann, 2018: Learning from CDM SD tool experience for Article 6.4 of the Paris Agreement. *Clim. Policy*, **18(4)**, 383–395, doi:10.1080/14693062.2016.1277686.

Orr, B.J. et al., 2017: *Scientific Conceptual Framework for Land Degradation Neutrality. A report of the science-policy interface*. United Nations Convention to Combat Desertification, Bonn, Germany, 136 pp.

Osofsky, H., J. Peel, B. McDonnell, and A. Foerster, 2019: Energy re-investment. *Indiana Law J.*, **94(2)**, 595–652.

Osofsky, H.M., 2010: The continuing importance of climate change litigation. *Clim. Law*, **1(1)**, 3–29, doi:10.1163/cl-2010-002.

Ostrom, E., 2010: Polycentric systems for coping with collective action and global environmental change. *Glob. Environ. Change*, **20(4)**, 550–557, doi:10.1016/j.gloenvcha.2010.07.004.

Otto, F.E.L., R.B. Skeie, J.S. Fuglestvedt, T. Berntsen, and M.R. Allen, 2017: Assigning historic responsibility for extreme weather events. *Nat. Clim. Change*, **7(11)**, 757–759, doi:10.1038/nclimate3419.

Ourbak, T. and A.K. Magnan, 2018: The Paris Agreement and climate change negotiations: Small Islands, big players. *Reg. Environ. Change*, **18(8)**, 2201–2207, doi:10.1007/s10113-017-1247-9.

PAGE, 2018: *International Investment Agreements and Sustainable Development: Safeguarding Policy Space and Mobilising Investment for a Green Economy*. United Nations Environment Programme, 35 pp. www.un-page.org/files/public/international_investment_agreements_sustainable_development_1.pdf (Accessed October 31, 2021).

Panfil, S.N. and C.A. Harvey, 2016: REDD+ and Biodiversity Conservation: A Review of the Biodiversity Goals, Monitoring Methods, and Impacts of 80 REDD+ Projects. *Conserv. Lett.*, **9(2)**, 143–150, doi:10.1111/conl.12188.

Parker, A., 2014: Governing solar geoengineering research as it leaves the laboratory. *Philos. Trans. R. Soc. A Math. Phys. Eng. Sci.*, **372(2031)**, 1–17, doi:10.1098/rsta.2014.0173.

Parker, A. and P.J. Irvine, 2018: The Risk of Termination Shock From Solar Geoengineering. *Earth's Future*, **6(3)**, 456–467, doi:10.1002/2017ef000735.

Parker, A., J.B. Horton, and D.W. Keith, 2018: Stopping Solar Geoengineering Through Technical Means: A Preliminary Assessment of Counter-Geoengineering. *Earth's Future*, **6(8)**, 1058–1065, doi:10.1029/2018ef000864.

Parker, C.F., C. Karlsson, and M. Hjerpe, 2014: Climate change leaders and followers: Leadership recognition and selection in the UNFCCC negotiations. *Int. Relations*, **29(4)**, 434–454, doi:10.1177/0047117814552143.

Parker, C.F., C. Karlsson, and M. Hjerpe, 2015: Climate change leaders and followers: Leadership recognition and selection in the UNFCCC negotiations. *Int. Relations*, **29(4)**, 434–454, doi:10.1177/0047117814552143.

Parkes, B., A. Challinor, and K. Nicklin, 2015: Crop failure rates in a geoengineered climate: impact of climate change and marine cloud brightening. *Environ. Res. Lett.*, **10(8)**, 84003, doi:10.1088/1748-9326/10/8/084003.

Paroussos, L. et al., 2019: Climate clubs and the macro-economic benefits of international cooperation on climate policy. *Nat. Clim. Change*, **9(7)**, 542–546, doi:10.1038/s41558-019-0501-1.

Parson, E.A., 2014: Climate Engineering in Global Climate Governance: Implications for Participation and Linkage. *Transnatl. Environ. Law*, **3(1)**, 89–110, doi:10.1017/s2047102513000496.

Parson, E.A. and L.N. Ernst, 2013: International Governance of Climate Engineering. *Theor. Inq. Law*, **14(1)**, 307–338, doi:10.1515/til-2013-015.

Partanen, A.-I., D.P. Keller, H. Korhonen, and H.D. Matthews, 2016: Impacts of sea spray geoengineering on ocean biogeochemistry. *Geophys. Res. Lett.*, **43(14)**, 7600–7608, doi:10.1002/2016gl070111.

Patt, A., 2015: *Transforming energy: solving climate change with technology policy*. Cambridge University Press, New York, USA.

Patt, A., 2017: Beyond the tragedy of the commons: Reframing effective climate change governance. *Energy Res. Soc. Sci.*, **34**, 1–3, doi:10.1016/j.erss.2017.05.023.

Patton, L.E., 2021: Litigation needs the latest science. *Nat. Clim. Change*, **11(8)**, 644–645, doi:10.1038/s41558-021-01113-7.

Pauw, W.P. et al., 2018: Beyond headline mitigation numbers: we need more transparent and comparable NDCs to achieve the Paris Agreement on climate change. *Clim. Change*, **147(1–2)**, 23–29, doi:10.1007/s10584-017-2122-x.

Pauw, W.P., P. Castro, J. Pickering, and S. Bhasin, 2020: Conditional nationally determined contributions in the Paris Agreement: foothold for equity or Achilles heel? *Clim. Policy*, **20(4)**, 468–484, doi:10.1080/14693062.2019.1635874.

Pauwelyn, J., 2013: Carbon leakage measures and border tax adjustments under WTO law. In: *Research handbook on Environment, Health and the WTO* [Van Calster, G. and D. Prévost, (eds.)], Edward Elgar Publishing, Cheltenham, UK, pp. 448–506.

Pavlova, Y. and A. De Zeeuw, 2013: Asymmetries in international environmental agreements. *Environ. Dev. Econ.*, **18(1)**, 51–68, doi:10.1017/s1355770x12000289.

Peake, S. and P. Ekins, 2017: Exploring the financial and investment implications of the Paris Agreement. *Clim. Policy*, **17(7)**, 832–852, doi:10.1080/14693062.2016.1258633.

Pearse, R., 2017: Gender and climate change. *Wiley Interdiscip. Rev. Clim. Change*, **8(2)**, e451, doi:10.1002/wcc.451.

Peel, J. and H.M. Osofsky, 2018: A Rights Turn in Climate Change Litigation? *Transnatl. Environ. Law*, **7(1)**, 37–67, doi:10.1017/S2047102517000292.

Peel, J. and J. Lin, 2019: Transnational Climate Litigation: The Contribution of the Global South. *Am. J. Int. Law*, **113(04)**, 679–726, doi:10.1017/ajil.2019.48.

Peel, J. and H.M. Osofsky, 2020: Climate Change Litigation. *Annu. Rev. Law Soc. Sci.*, **16(1)**, 21–38, doi:10.1146/annurev-lawsocsci-022420-122936.

Pekkarinen, V., 2020: Going beyond CO2: Strengthening action on global methane emissions under the UN climate regime. *Rev. Eur. Comp. Int. Environ. Law*, **29(3)**, 464–478, doi:10.1111/reel.12329.

Pereira, L. et al., 2021: From Fairplay to Climate Wars: Making climate change scenarios more dynamic, creative and integrative. *Ecol. Soc.*, (in press).

Peters, G.P. et al., 2017: Key indicators to track current progress and future ambition of the Paris Agreement. *Nat. Clim. Change*, **7(2)**, 118–122, doi:10.1038/nclimate3202.

Peterson, L. and J. Skovgaard, 2019: Bureaucratic politics and the allocation of climate finance. *World Dev.*, **117**, 72–97, doi:10.1016/j.worlddev.2018.12.011.

Phelps, J., E.L. Webb, and W.M. Adams, 2012: Biodiversity co-benefits of policies to reduce forest-carbon emissions. *Nat. Clim. Change*, **2**, 497–503, doi:10.1038/nclimate1462.

Phillips, J. and P. Newell, 2013: The governance of clean energy in India: The clean development mechanism (CDM) and domestic politics. *Energy Policy*, **59**, 654–662, doi:10.1016/j.enpol.2013.04.019.

Pickering, J., F. Jotzo, and P.J. Wood, 2015: Sharing the Global Climate Finance Effort Fairly with Limited Coordination. *Glob. Environ. Polit.*, **15(4)**, 39–62, doi:10.1162/GLEP_a_00325.

Pickering, J., C. Betzold, and J. Skovgaard, 2017: Special issue: managing fragmentation and complexity in the emerging system of international climate finance. *Int. Environ. Agreements Polit. Law Econ.*, **17**, 1–16, doi:10.1007/s10784-016-9349-2.

Pickering, J., J.S. McGee, S.I. Karlsson-Vinkhuyzen, and J. Wenta, 2019: Global climate governance between hard and soft law: Can the Paris agreement's "Crème Brûleé" approach enhance ecological reflexivity? *J. Environ. Law*, **31(1)**, 1–28, doi:10.1093/jel/eqy018.

Pidgeon, N. et al., 2012: Exploring early public responses to geoengineering. *Philos. Trans. R. Soc. A Math. Phys. Eng. Sci.*, **370(1974)**, 4176–4196, doi:10.1098/rsta.2012.0099.

Piggot, G., P. Erickson, H. van Asselt, and M. Lazarus, 2018: Swimming upstream: addressing fossil fuel supply under the UNFCCC. *Clim. Policy*, **18(9)**, 1189–1202, doi:10.1080/14693062.2018.1494535.

Pinke, Z., S. Pow, and Z. Kern, 2019: Volcanic mega-eruptions may trigger major cholera outbreaks. *Clim. Res.*, **79(2)**, 151–162, doi:10.3354/cr01587.

Pinz, A., N. Roudyani, and J. Thaler, 2018: Public–private partnerships as instruments to achieve sustainability-related objectives: the state of the art and a research agenda. *Public Manag. Rev.*, **20(1)**, 1–22, doi:10.1080/14719037.2017.1293143.

Pirard, R. et al., 2019: *Effectiveness of forest conservation interventions: An evidence gap map*. Green Climate Fund, Songdo, Republic of Korea, 57 pp. www.cbd.int/doc/presentations/tc-imrr/Learning-Paper-2.pdf (Accessed October 31, 2021).

Pistorius, T. and L. Kiff, 2015: The Politics of German Finance for REDD+. In: *CGD Working Paper 390, CGD Climate and Forest Paper Series #16*, Center for Global Development, Washington, DC, USA, 49 pp.

Pitari, G. et al., 2014: Stratospheric ozone response to sulfate geoengineering: Results from the Geoengineering Model Intercomparison Project (GeoMIP). *J. Geophys. Res. Atmos.*, **119(5)**, 2629–2653, doi:10.1002/2013jd020566.

Pitt, D., 2010: The impact of internal and external characteristics on the adoption of climate mitigation policies by US municipalities. *Environ. Plan. C Polit. Sp.*, **28(5)**, 851–871, doi:10.1068/c09175.

Pollitt, H., K. Neuhoff, and X. Lin, 2020: The impact of implementing a consumption charge on carbon-intensive materials in Europe. *Clim. Policy*, **20**(sup1), s74–s89, doi:10.1080/14693062.2019.1605969.

Pongratz, J., D.B. Lobell, L. Cao, and K. Caldeira, 2012: Crop yields in a geoengineered climate. *Nat. Clim. Change*, **2(2)**, 101–105, doi:10.1038/nclimate1373.

Popp, D., 2019: Environmental Policy and Innovation: A Decade of Research. *Int. Rev. Environ. Resour. Econ.*, **13(3–4)**, 265–337, doi:10.1561/101.00000111.

Porterfield, M., 2019: Border Adjustments for Carbon Taxes, PPMs, and the WTO. *Univ. Pennsylvania J. Int. Law*, **41(1)**, 1–42.

Porterfield, M., K.P. Gallagher, and J.C. Schachter, 2017: Assessing the Climate Impacts of U.S. Trade Agreements. *Michigan J. Environ. Adm. Law*, **7(1)**, 51–81.

Potoski, M., 2017: Green clubs in building block climate change regimes. *Clim. Change*, **144(1)**, 53–63, doi:10.1007/s10584-015-1517-9.

Potoski, M., and A. Prakash, 2013: Green Clubs: Collective Action and Voluntary Environmental Programs. *Annu. Rev. Polit. Sci.*, **16(1)**, 399–419, doi:10.1146/annurev-polisci-032211-211224.

Pozo, C., Á. Galán-Martín, D.M. Reiner, N. Mac Dowell, and G. Guillén-Gosálbez, 2020: Equity in allocating carbon dioxide removal quotas. *Nat. Clim. Change*, **10**, 640–646, doi:10.1038/s41558-020-0802-4.

Preston, B.J., 2020: The Influence of the Paris Agreement on Climate Litigation: Legal Obligations and Norms (Part I). *J. Environ. Law*, **33(1)**, 1–32, doi:10.1093/jel/eqaa020.

Preston, C. and W. Carr, 2018: Recognitional Justice, Climate Engineering, and the Care Approach. *Ethics, Policy Environ.*, **21(3)**, 308–323, doi:10.1080/21550085.2018.1562527.

Proctor, J., S. Hsiang, J. Burney, M. Burke, and W. Schlenker, 2018: Estimating global agricultural effects of geoengineering using volcanic eruptions. *Nature*, **560(7719)**, 480–483, doi:10.1038/s41586-018-0417-3.

Quélin, B.V., I. Kivleniece, and S. Lazzarini, 2017: Public-private collaboration, hybridity and social value: Towards new theoretical perspectives. *J. Manag. Stud.*, **54(6)**, 763–792, doi:10.1111/joms.12274.

Rabitz, F., 2019: Governing the termination problem in solar radiation management. *Env. Polit.*, **28(3)**, 502–522, doi:10.1080/09644016.2018.1519879.

Rahman, A.A., P. Artaxo, A. Asrat, and A. Parker, 2018: Developing countries must lead on solar geoengineering research. *Nature*, **556(7699)**, 22–24, doi:10.1038/d41586-018-03917-8.

Raiser, K., U. Kornek, C. Flachsland, and W.F. Lamb, 2020: Is the Paris Agreement effective? A systematic map of the evidence. *Environ. Res. Lett.*, **15(8)**, 83006, doi:10.1088/1748-9326/ab865c.

Rajamani, L., 2010: The making and unmaking of the Copenhagen accord. *Int. Comp. Law Q.*, **59(3)**, 824–843, doi:10.1017/s0020589310000400.

Rajamani, L., 2015: The Devilish Details: Key Legal Issues in the 2015 Climate Negotiations. *Mod. Law Rev.*, **78(5)**, 826–853, doi:10.1111/1468-2230.12145.

Rajamani, L., 2016a: Ambition and Differentiation in the Paris Agreement: Interpretative Possibilities and Underlying Politics. *Int. Comp. Law Q.*, **65(2)**, 493–514, doi:10.1017/s0020589316000130.

Rajamani, L., 2016b: The 2015 Paris Agreement: Interplay between hard, soft and non-obligations. *J. Environ. Law*, **28(2)**, 337–358, doi:10.1093/jel/eqw015.

Rajamani, L., 2017: India's approach to international law in the climate change regime. *Indian J. Int. Law*, **57(1)**, 1–23, doi:10.1007/s40901-018-0072-0.

Rajamani, L., 2018: Human Rights in the Climate Change Regime: From Rio to Paris and Beyond. In: *The Human Right to a Healthy Environment* [Knox, J.H. and R. Pejan, (eds.)], Cambridge University Press, Cambridge, UK and New York, NY, USA, 2018, pp. 236–251.

Rajamani, L., 2019: Integrating Human Rights in the Paris Climate Architecture: Contest, Context, and Consequence. *Clim. Law*, **9(3)**, 180–201, doi:10.1163/18786561-00903003.

Rajamani, L. and J. Werksman, 2018: The legal character and operational relevance of the Paris Agreement's temperature goal. *Philos. Trans. R. Soc. A Math. Phys. Eng. Sci.*, **376(2119)**, 20160458, doi:10.1098/rsta.2016.0458.

Rajamani, L. and D. Bodansky, 2019: The Paris Rulebook: Balancing Prescriptiveness with National Discretion. *Int. Comp. Law Q.*, **68(4)**, 1023–1040, doi:10.1017/s0020589319000320.

Rajamani, L. et al., 2021: National 'fair shares' in reducing greenhouse gas emissions within the principled framework of international environmental law. *Clim. Policy*, **21(8)**, 983–1004, doi:10.1080/14693062.2021.1970504.

Rampa, F., S. Bilal, and E. Sidiropoulos, 2012: Leveraging South–South cooperation for Africa's development. *South African J. Int. Aff.*, **19(2)**, 247–269, doi:10.1080/10220461.2012.709400.

Rashidi, K., and A. Patt, 2018: Subsistence over symbolism: the role of transnational municipal networks on cities' climate policy innovation and adoption. *Mitig. Adapt. Strateg. Glob. Change*, **23(4)**, 507–523, doi:10.1007/s11027-017-9747-y.

Rashidi, K., M. Stadelmann, and A. Patt, 2017: Valuing co-benefits to make low-carbon investments in cities bankable: The case of waste and transportation projects. *Sustain. Cities Soc.*, **34**, 69–78, doi:10.1016/j.scs.2017.06.003.

Rashidi, K., M. Stadelmann, and A. Patt, 2019: Creditworthiness and climate: Identifying a hidden financial co-benefit of municipal climate adaptation and mitigation policies. *Energy Res. Soc. Sci.*, **48**, 131–138, doi:10.1016/j.erss.2018.09.021.

Ravikumar, A., 2020: Opinion: Carbon border taxes are unjust. *MIT Technol. Rev.*,. www.technologyreview.com/2020/07/27/1005641/carbon-border-taxes-eu-climate-change-opinion/ (Accessed October 29, 2021).

Rayner, S. et al., 2013: The Oxford Principles. *Clim. Change*, **121(3)**, 499–512, doi:10.1007/s10584-012-0675-2.

RE100, 2019: RE100 About us. http://there100.org/re100 (Accessed December 18, 2019).

REN21, 2019: REN21: Renewables Now! www.ren21.net/about-us/who-we-are/ (Accessed December 18, 2019).

Reynolds, J.L., 2019a: *The governance of solar geoengineering: managing climate change in the Anthropocene*. Cambridge University Press, Cambridge, UK and New York, NY, USA, 268 pp.

Reynolds, J.L., 2019b: Solar geoengineering to reduce climate change: a review of governance proposals. *Proc. R. Soc. A*, **475(2229)**, 1–33, doi:10.1098/rspa.2019.0255.

RGI, 2011: *European Grid Declaration On Electricity Network Development and Nature Conservation in Europe.* 10 November 2011, Renewables Grid Initiative, 11 pp.

Richards, M.B., E. Wollenberg, and D. van Vuuren, 2018: National contributions to climate change mitigation from agriculture: allocating a global target. *Clim. Policy*, **18(10)**, 1271–1285, doi:10.1080/14693062.2018.1430018.

Ricke, K.L., J.B. Moreno-Cruz, and K. Caldeira, 2013: Strategic incentives for climate geoengineering coalitions to exclude broad participation. *Environ. Res. Lett.*, **8**, 014021, doi:10.1088/1748-9326/8/1/014021.

Rickels, W. et al., 2020: Who turns the global thermostat and by how much? *Energy Econ.*, **91**, 104852, doi:10.1016/j.eneco.2020.104852.

Ripley, K. and C. Verkuijl, 2016: "Ozone Family" Delivers Landmark Deal for the Climate. *Environ. Policy Law*, **46(6)**, 371–375.

Ritchie, H. and D.S. Reay, 2017: Delivering the two degree global climate change target using a flexible ratchet framework. *Clim. Policy*, **17(8)**, 1031–1045, doi:10.1080/14693062.2016.1222260.

Ritchie, J., 2021: Movement from the margins to global recognition: climate change activism by young people and in particular indigenous youth. *Int. Stud. Sociol. Educ.*, **30(1–2)**, 53–72, doi:10.1080/09620214.2020.1854830.

Roberts, C. et al., 2018: The politics of accelerating low-carbon transitions: Towards a new research agenda. *Energy Res. Soc. Sci.*, **44**, 304–311, doi:10.1016/j.erss.2018.06.001.

Roberts, E., K. van der Geest, K. Warner, and S. Andrei, 2014: Loss and Damage: When adaptation is not enough. *Environ. Dev.*, **11**, 219–227, doi:10.1016/j.envdev.2014.05.001.

Roberts, J.T. and R. Weikmans, 2017: Postface: fragmentation, failing trust and enduring tensions over what counts as climate finance. *Int. Environ. Agreements Polit. Law Econ.*, **17(1)**, 129–137, doi:10.1007/s10784-016-9347-4.

Roberts, J.T. et al., 2017: How Will We Pay for Loss and Damage? *Ethics Policy Environ.*, **20(2)**, 208–226, doi:10.1080/21550085.2017.1342963.

Roberts, J.T. et al., 2021: Rebooting a failed promise of climate finance. *Nat. Clim. Change*, **11(3)**, 180–182, doi:10.1038/s41558-021-00990-2.

Roberts, M.W., 2017: Finishing the job: The Montreal Protocol moves to phase down hydrofluorocarbons. *Rev. Eur. Comp. Int. Environ. Law*, **26(3)**, 220–230, doi:10.1111/reel.12225.

Robinson, S.-A., 2018: Capacity building and transparency under Paris. In: *The Paris Framework for Climate Change Capacity Building* [Khan, M.R., J.T. Roberts, S. Huq, and V. Hoffmeister, (eds.)]. Routledge, London, UK, pp. 203–222.

Robiou du Pont, Y., and M. Meinshausen, 2018: Warming assessment of the bottom-up Paris Agreement emissions pledges. *Nat. Commun.*, **9(1)**, 4810, doi:10.1038/s41467-018-07223-9.

Robiou du Pont, Y. et al., 2017: Equitable mitigation to achieve the Paris Agreement goals. *Nat. Clim. Change*, **7(1)**, 38–43, doi:10.1038/nclimate3186.

Roelfsema, M. et al., 2020: Taking stock of national climate policies to evaluate implementation of the Paris Agreement. *Nat. Commun.*, **11(1)**, 2096, doi:10.1038/s41467-020-15414-6.

Rogelj, J. et al., 2010: Analysis of the Copenhagen Accord pledges and its global climatic impacts—a snapshot of dissonant ambitions. *Environ. Res. Lett.*, **5(3)**, 034013, doi:10.1088/1748-9326/5/3/034013.

Rogelj, J. et al., 2015: Energy system transformations for limiting end-of-century warming to below 1.5°C. *Nat. Clim. Change*, **5(6)**, 519–527, doi:10.1038/nclimate2572.

Rogelj, J. et al., 2016: Paris Agreement climate proposals need a boost to keep warming well below 2°C. *Nature*, **534(7609)**, 631–639, doi:10.1038/nature18307.

Rogelj, J. et al., 2017: Understanding the origin of Paris Agreement emission uncertainties. *Nat. Commun.*, **8(15748)**, doi:10.1038/ncomms15748.

Rogelj, J. D. Shindell, K. Jiang, S. Fifita, P. Forster, V. Ginzburg, C. Handa, H. Kheshgi, S. Kobayashi, E. Kriegler, L. Mundaca, R. Séférian and M.V. Vilariño, 2018: Mitigation pathways compatible with 1.5°C in the context of sustainable development. In: *Global Warming of 1.5°C. An IPCC Special Report on the impacts of global warming of 1.5°C above pre-industrial levels and related global greenhouse gas emission pathways, in the context of strengthening the global response to the threat of climate change* [Masson-Delmotte, V. P. Zhai, H.-O. Pörtner, D. Roberts, J. Skea, P.R. Shukla, A. Pirani, W. Moufouma-Okia, C. Péan, R. Pidcock, S. Connors, J.B.R. Matthews, Y. Chen, X. Zhou, M.I. Gomis, E. Lonnoy, T. Maycock, M. Tignor, and T. Waterfield, (eds.)]. Cambridge University Press, Cambridge, UK and New York, NY, USA,pp. 93–174.

Rogelj, J., O. Geden, A. Cowie, and A. Reisinger, 2021: Net-zero emissions targets are vague: three ways to fix. *Nature*, **591(7850)**, 365–368, doi:10.1038/d41586-021-00662-3.

Rosen, A.M., 2015: The Wrong Solution at the Right Time: The Failure of the Kyoto Protocol on Climate Change. *Polit. Policy*, **43(1)**, 30–58, doi:10.1111/polp.12105.

Rosenow, J., F. Kern, and K. Rogge, 2017: The need for comprehensive and well targeted instrument mixes to stimulate energy transitions: The case of energy efficiency policy. *Energy Res. Soc. Sci.*, **33**, 95–104, doi:10.1016/J.ERSS.2017.09.013.

Rubini, L. and I. Jegou, 2012: Who'll Stop the Rain? Allocating Emissions Allowances for Free: Environmental Policy, Economics, and WTO Subsidy Law. *Transnatl. Environ. Law*, **1(2)**, 325–354, doi:10.1017/s2047102512000143.

Rubio, S.J., 2017: Sharing R & D investments in breakthrough technologies to control climate change. *Oxf. Econ. Pap.*, **69(2)**, 496–521, doi:10.1093/oep/gpw067.

Rumpel, C. et al., 2020: The 4p1000 initiative: Opportunities, limitations and challenges for implementing soil organic carbon sequestration as a sustainable development strategy. *Ambio*, **49(1)** pp. 350–360, doi:10.1007/s13280-019-01165-2.

Sabel, C.F. and D.G. Victor, 2017: Governing global problems under uncertainty: making bottom-up climate policy work. *Clim. Change*, **144(1)**, 15–27, doi:10.1007/s10584-015-1507-y.

Sælen, H., 2020: Under What Conditions Will the Paris Process Produce a Cycle of Increasing Ambition Sufficient to Reach the 2°C Goal? *Glob. Environ. Polit.*, **20(2)**, 83–104, doi:10.1162/glep_a_00548.

Sælen, H., J. Hovi, D.F. Sprinz, and A. Underdal, 2020: How US Withdrawal Might Influence Cooperation Under the Paris Climate Agreement. *Environ. Sci. Policy*, **108**, 121–132, doi:10.1016/j.envsci.2020.03.011.

Sands, P.Q.C., 2016: Climate change and the rule of law: Adjudicating the future in international law. *J. Environ. Law*, **28(1)**, 19–35, doi:10.1093/jel/eqw005.

Sands, P.Q.C. and J. Peel, 2018: *Principles of International Environmental Law*. 4th ed. Cambridge University Press, Cambridge, UK and New York, NY, USA, 1032 pp.

Savaresi, A., 2016: The Paris agreement: A new beginning? *J. Energy Nat. Resour. Law*, **34(1)**, 16–26, doi:10.1080/02646811.2016.1133983.

Savaresi, A., 2018: Climate change and human rights: Fragmentation, interplay, and institutional linkages. In: *Routledge Handbook of Human Rights and Climate Governance* [Duyck, S., S. Jodoin, and A. Johl, (eds.)], Routledge, London, UK, pp. 31–42.

Savaresi, A. and J. Auz, 2019: Climate change litigation and human rights: Pushing the boundaries. *Clim. Law*, **9(3)**, 244–262, doi:10.1163/18786561-00903006.

Schade, J., and W. Obergassel, 2014: Human rights and the Clean Development Mechanism. *Cambridge Rev. Int. Aff.*, **27(4)**, 717–735, doi:10.1080/09557571.2014.961407.

Schäfer, S., 2019: Decoupling the EU ETS from subsidized renewables and other demand side effects: lessons from the impact of the EU ETS on $CO_2$ emissions in the German electricity sector. *Energy Policy*, **133**, 110858, doi:10.1016/J.ENPOL.2019.06.066.

Scheidel, A. et al., 2020: Environmental conflicts and defenders: A global overview. *Glob. Environ. Change*, **63**, 102104, doi:10.1016/j.gloenvcha.2020.102104.

Scheinman, L., 1987: *The International Atomic Energy Agency and World Nuclear Order*. Resources for the Future, Washington, DC, USA, 320 pp.

Schleussner, C.F. et al., 2016: Science and policy characteristics of the Paris Agreement temperature goal. *Nat. Clim. Change*, **6(9)**, 827–835, doi:10.1038/nclimate3096.

Schlosberg, D. and L.B. Collins, 2014: From environmental to climate justice: Climate change and the discourse of environmental justice. *Wiley Interdiscip. Rev. Clim. Change*, **5(3)**, 359–374, doi:10.1002/wcc.275.

Schmale, J., D. Shindell, E. Von Schneidemesser, I. Chabay, and M. Lawrence, 2014: Air pollution: Clean up our skies. *Nature*, **515**, 335–337, doi:10.1038/515335a.

Schmieg, G. et al., 2017: Modeling normativity in sustainability: a comparison of the sustainable development goals, the Paris agreement, and the papal encyclical. *Sustain. Sci.*, **13**, 785–796, doi:10.1007/s11625-017-0504-7.

Schneider, L. and S. La Hoz Theuer, 2019: Environmental integrity of international carbon market mechanisms under the Paris Agreement. *Clim. Policy*, **19(3)**, 386–400, doi:10.1080/14693062.2018.1521332.

Schneider, L., A. Kollmuss, and M. Lazarus, 2014: *Addressing the risk of double counting emission reductions under the UNFCCC*. Stockholm Environment Institute, Stockholm, Sweden, 56 pp. https://mediamanager.sei.org/documents/Publications/Climate/SEI-WP-2014-02-Double-counting-risks-UNFCCC.pdf (Accessed October 31, 2021).

Schneider, L., M. Lazarus, C. Lee, and H. Van Asselt, 2017: Restricted linking of emissions trading systems: options, benefits and challenges. *Int. Environ. Agreements Polit. Law Econ.*, **17(6)**, 883–898, doi:10.1007/s10784-017-9370-0.

Schneider, L. et al., 2019: Double counting and the Paris Agreement rulebook. *Science*, **366(6462)**, 180–183, doi:10.1126/science.aay8750.

Schroeder, H. and C. McDermott, 2014: Beyond carbon: Enabling justice and equity in REDD+ across levels of governance. *Ecol. Soc.*, **19(1)**, 2–4, doi:10.5751/es-06537-190131.

Schroeder, H., M. Di Gregorio, M. Brockhaus, and T.T. Pham, 2020: Policy Learning in REDD+ Donor Countries: Norway, Germany, and the UK. *Glob. Environ. Change*, **63**, 102106, doi:10.1016/j.gloenvcha.2020.102106.

Schulte, I. et al., 2020: *Supporting Smallholder Farmers for a Sustainable Cocoa Sector: Exploring the Motivations and Role of Farmers in the Effective Implementation of Supply Chain Sustainability in Ghana and Cote d'Ivoire*. Meridian Institute, Washington, DC, USA, 59 pp. https://www.climatefocus.com/sites/default/files/Supporting%20Smallholder%20Farmers%20for%20a%20Sustainable%20Cocoa%20Sector%20June%202020.pdf (Accessed October 31, 2021).

Selin, H., 2014: Global Environmental Law and Treaty-Making on Hazardous Substances: The Minamata Convention and Mercury Abatement. *Glob. Environ. Polit.*, **14(1)**, 1–19, doi:10.1162/glep_a_00208.

Seneviratne, S.I. et al., 2018: Land radiative management as contributor to regional-scale climate adaptation and mitigation. *Nat. Geosci.*, **11(2)**, 88–96, doi:10.1038/s41561-017-0057-5.

Serdeczny, O., 2019: Non-economic Loss and Damage and the Warsaw International Mechanism. In: *Loss and Damage from Climate Change. Concepts, Methods and Policy Options. Climate Risk Management, Policy and Governance.* [Mechler, R., L. Bouwer, T. Schinko, S. Surminski, and J. Linnerooth-Bayer, (eds.)], Springer International Publishing, Cham, Switzerland, pp. 205–220.

Sethi, M., W. Lamb, J. Minx, and F. Creutzig, 2020: Climate change mitigation in cities: a systematic scoping of case studies. *Environ. Res. Lett.*, **15(9)**, 093008, doi:10.1088/1748-9326/AB99FF.

Setzer, J. and R. Byrnes, 2019: *Global trends in climate change litigation: 2019 snapshot*. London School of Economics and Political Science, London, UK, 14 pp. www.lse.ac.uk/GranthamInstitute/wp-content/uploads/2019/07/GRI_Global-trends-in-climate-change-litigation-2019-snapshot-2.pdf (Accessed 31 October 2021).

Setzer, J. and L.C. Vanhala, 2019: Climate change litigation: A review of research on courts and litigants in climate governance. *Wiley Interdiscip. Rev. Clim. Change*, **10(3)**, e580, doi:10.1002/wcc.580.

Setzer, J. and L. Benjamin, 2020: Climate litigation in the Global South: constraints and innovations. *Transnatl. Environ. Law*, **9(1)**, 77–101, doi:10.1017/s2047102519000268.

Seymour, F. and J. Busch, 2016: *Why forests? Why now?: The science, economics, and politics of tropical forests and climate change*. Brookings Institution Press, Washington, DC, USA, 429 pp.

Shadikhodjaev, S., 2015: Renewable Energy and Government Support: Time to 'Green' the SCM Agreement? *World Trade Rev.*, **14(3)**, 479–506, doi:10.1017/s1474745614000317.

Shepherd, J.G., 2009: *Geoengineering the climate: science, governance and uncertainty (RS Policy Document 10/09)*. The Royal Society, London, UK, 84 pp.

Sheriff, G., 2019: Burden Sharing under the Paris Climate Agreement. *J. Assoc. Environ. Resour. Econ.*, **6(2)**, 275–318, doi:10.1086/701469.

Shi, X., Y. Chen, and H. Liu, 2017: The Negative Impact of Climate Change Mitigation Measures on Human Rights and the Countermeasures. *J. Hunan Police Acad.*, **1**, 7.

Shi, Y. and W. Gullett, 2018: International Regulation on Low-Carbon Shipping for Climate Change Mitigation: Development, Challenges, and Prospects. *Ocean Dev. Int. Law*, **49(2)**, 134–156, doi:10.1080/00908320.2018.1442178.

Shimoda, Y. and S. Nakazawa, 2012: Flexible Cooperation for Indonesia's Multi-dimensional Challenges for South-South Cooperation Under A Shared Vision. In: *Scaling Up South-South and Triangular Cooperation* [Kato, H., (ed.)], Japan International Cooperation Agency Research Institute, Tokyo, Japan, pp. 149–172.

Shishlov, I., R. Morel, and V. Bellassen, 2016: Compliance of the Parties to the Kyoto Protocol in the first commitment period. *Clim. Policy*, **16(6)**, 768–782, doi:10.1080/14693062.2016.1164658.

Simpson, I.R. et al., 2019: The Regional Hydroclimate Response to Stratospheric Sulfate Geoengineering and the Role of Stratospheric Heating. *J. Geophys. Res. Atmos.*, **124(23)**, 12587–12616, doi:10.1029/2019jd031093.

Sindico, F., 2016: Paris, Climate Change, and Sustainable Development. *Clim. Law*, **6(1–2)**, 130–141, doi:10.1163/18786561-00601009.

Skeie, R.B. et al., 2017: Perspective has a strong effect on the calculation of historical contributions to global warming. *Environ. Res. Lett.*, **12(2)**, 024022, doi:10.1088/1748-9326/aa5b0a.

Skjærseth, J.B., O.S. Stokke, and J. Wettestad, 2006: Soft law, hard law, and effective implementation of international environmental norms. *Glob. Environ. Polit.*, **6(3)**, 104–120, doi:10.1162/glep.2006.6.3.104.

Smit, B. and J. Wandel, 2006: Adaptation, adaptive capacity and vulnerability. *Glob. Environ. Change*, **16(3)**, 282–292, doi:10.1016/j.gloenvcha.2006.03.008.

Smith, J.J. and M.T. Ahmad, 2018: Globalization's Vehicle: The Evolution and Future of Emission Regulation in the icao and imo in Comparative Assessment. *Clim. Law*, **8(1–2)**, 70–103, doi:10.1163/18786561-00801003.

Smith, W. and G. Wagner, 2018: Stratospheric aerosol injection tactics and costs in the first 15 years of deployment. *Environ. Res. Lett.*, **13(124001)**, doi:10.1088/1748-9326/aae98d.

Solomon, S., G.-K. Plattner, R. Knutti, and P. Friedlingstein, 2009: Irreversible climate change due to carbon dioxide emissions. *Proc. Natl. Acad. Sci.*, **106(6)**, 1704–1709, doi:10.1073/pnas.0812721106.

Sorgho, Z. and J. Tharakan, 2020: *Do PTAs with Environmental Provisions Reduce Emissions? Assessing the Effectiveness of Climate-related Provisions?* Fondation pour les études et recherches sur le développement international, Clermont-Ferrand, France, 47 pp. https://ferdi.fr/dl/df-nJ7WgYnTDp6qA2SGjnT25Y9F/ferdi-p274-do-ptas-with-environmental-provisions-reduce-emissions-assessing.pdf (Accessed October 31, 2021).

Sovacool, B.K. and A. Florini, 2012: Examining the Complications of Global Energy Governance. *J. Energy Nat. Resour. Law*, **30(3)**, 235–263, doi:10.1080/02646811.2012.11435295.

Sovacool, B.K. and B.-O. Linnér, 2016: The Perils of Climate Diplomacy: The Political Economy of the UNFCCC. In: *The Political Economy of Climate Change Adaptation*, Palgrave Macmillan, London, UK, pp. 110–135.

Spalding-Fecher, R. et al., 2012: *Assessing The Impact Of The Clean Development Mechanism. Final Report. Report Commissioned By The High-Level Panel On The CDM Policy Dialogue*. UNFCCC, Bonn, Germany, 172 pp.

Spash, C.L., 2016: This Changes Nothing: The Paris Agreement to Ignore Reality. *Globalizations*, **13(6)**, 928–933, doi:10.1080/14747731.2016.1161119.

Springmann, M., 2013: Carbon tariffs for financing clean development. *Clim. Policy*, **13(1)**, 20–42, doi:10.1080/14693062.2012.691223.

Sprinz, D.F., H. Sælen, A. Underdal, and J. Hovi, 2018: The Effectiveness of Climate Clubs under Donald Trump. *Clim. Policy*, **18(7)**, 828–838, doi:10.1080/14693062.2017.1410090.

Staver, A.C., S. Archibald, and S.A. Levin, 2011: The Global Extent and Determinants of Savanna and Forest as Alternative Biome States. *Science*, **334(6053)**, 230–232, doi:10.1126/science.1210465.

Stavins, R., J. Zou, T. Brewer, M. Conte Grand, M. den Elzen, M. Finus, J. Gupta, N. Höhne, M.-K. Lee, A. Michaelowa, M. Paterson, K. Ramakrishna, G. Wen, J. Wiener, and H. Winkler, 2014: International Cooperation: Agreements and Instruments. In: Climate Change 2014: Mitigation of Climate Change. Contribution of Working Group III to the Fifth Assessment of the Intergovernmental Panel on Climate Change [Edenhofer, O.R., Pichs-Madruga, Y. Sokona, E. Farahani, S. Kadner, K. Seyboth, A. Adler, I. Baum, S. Brunner, P. Eickemeier, B. Kriemann, J. Savolainen, S. Schlömer, C. von Stechow, T. Zwickel and J.C. Minx (eds.)], Cambridge University Press, Cambridge, UK, and New York, NY, USA, pp. 1001–1082.

Steele, P., 2015: *Development finance and climate finance: Achieving zero poverty and zero emissions*. IIED, London, UK, 31 pp.

Steenblik, R. and S. Droege, 2019: Time to ACCTS? Five countries announce new initiative on trade and climate change. *Int. Inst. Sustain. Dev. Blog*, www.iisd.org/articles/time-accts-five-countries-announce-new-initiative-trade-and-climate-change (Accessed October 29, 2021).

Steenblik, R., J. Sauvage, and C. Timiliotis, 2018: Fossil Fuel Subsidies and the Global Trade Regime. In: *The Politics of Fossil Fuel Subsidies and their Reform* [Van Asselt, H. and J. Skovgaard, (eds.)], Cambridge University Press, Cambridge, UK and New York, NY, USA, pp. 121–139.

Stephens, J.C. and K. Surprise, 2020: The hidden injustices of advancing solar geoengineering research. *Glob. Sustain.*, **3**(e2), 1–6, doi:10.1017/sus.2019.28.

Stephenson, S.R. et al., 2018: Climatic Responses to Future Trans-Arctic Shipping. *Geophys. Res. Lett.*, **45(18)**, 9898–9908, doi:10.1029/2018gl078969.

Stephenson, S.R., N. Oculi, A. Bauer, and S. Carhuayano, 2019: Convergence and Divergence of UNFCCC Nationally Determined Contributions. *Ann. Am. Assoc. Geogr.*, **109(4)**, 1240–1261, doi:10.1080/24694452.2018.1536533.

Stewart, R.B., M. Oppenheimer, and B. Rudyk, 2013a: A new strategy for global climate protection. *Clim. Change*, **120(1)**, 1–12, doi:10.1007/s10584-013-0790-8.

Stewart, R.B., M. Oppenheimer, and B. Rudyk, 2013b: Building Blocks for Global Climate Protection. *Stanford Environ. Law J.*, **32(2)**, 341–392, doi:10.2139/ssrn.2186541.

Stewart, R.B., M. Oppenheimer, and B. Rudyk, 2017: Building blocks: a strategy for near-term action within the new global climate framework. *Clim. Change*, **144(1)**, 1–13, doi:10.1007/s10584-017-1932-1.

Stilgoe, J., 2015: *Experiment earth: responsible innovation in geoengineering*. Taylor and Francis Ltd., London, UK, 222 pp.

Stjern, C.W. et al., 2018: Response to marine cloud brightening in a multi-model ensemble. *Atmos. Chem. Phys.*, **18(2)**, 621–634, doi:10.5194/acp-18-621-2018.

Storelvmo, T., and N. Herger, 2014: Cirrus cloud susceptibility to the injection of ice nuclei in the upper troposphere. *J. Geophys. Res. Atmos.*, **119(5)**, 2375–2389, doi:10.1002/2013jd020816.

Streck, C., and C. Parker, 2012: Financing REDD+. In: *Analysing REDD+: Challenges and choices* [Angelsen, A., M. Brockhaus, W.D. Sunderlin, and L. Verchot, (eds.)], CIFOR, Bogor, Indonesia, pp. 111–127.

Streck, C. and M. Terhalle, 2013: The changing geopolitics of climate change. *Clim. Policy*, **13(5)**, 533–537, doi:10.1080/14693062.2013.823809.

Stua, M., 2017a: A single mechanism for the certification of mitigation outcomes. In: *From the Paris Agreement to a Low-Carbon Bretton Woods: Rationale for the Establishment of a Mitigation Alliance*, Springer International Publishing, Cham, Switzerland, pp. 85–107.

Stua, M., 2017b: *From the Paris agreement to a low-carbon bretton woods: Rationale for the establishment of a mitigation alliance*. Springer International Publishing, Cham, Switzerland, 239 pp.

Stuart-Smith, R.F. et al., 2021: Filling the evidentiary gap in climate litigation. *Nat. Clim. Change*, **11(8)**, 651–655, doi:10.1038/s41558-021-01086-7.

Suckall, N., L.C. Stringer, and E.L. Tompkins, 2015: Presenting Triple-Wins? Assessing Projects That Deliver Adaptation, Mitigation and Development Co-benefits in Rural Sub-Saharan Africa. *Ambio*, **44(1)**, 34–41, doi:10.1007/s13280-014-0520-0.

Sugiyama, M., Y. Arino, T. Kosugi, A. Kurosawa, and S. Watanabe, 2018a: Next steps in geoengineering scenario research: limited deployment scenarios and beyond. *Clim. Policy*, **18(6)**, 681–689, doi:10.1080/14693062.2017.1323721.

Sugiyama, M., A. Ishii, S. Asayama, and T. Kosugi, 2018b: Solar Geoengineering Governance. *Oxford Res. Encycl. Clim. Sci.*, Oxford University Press, doi:10.1093/acrefore/9780190228620.013.647.

Sugiyama, M., S. Asayama, and T. Kosugi, 2020: The North–South Divide on Public Perceptions of Stratospheric Aerosol Geoengineering?: A Survey in Six Asia-Pacific Countries. *Environ. Commun.*, **14(5)**, 641–656, doi:10.1080/17524032.2019.1699137.

Svoboda, T., 2017: *The ethics of climate engineering : solar radiation management and non-ideal justice*. 1st ed. Routledge, New York, USA, 186 pp.

Svoboda, T. and P. Irvine, 2014: Ethical and Technical Challenges in Compensating for Harm Due to Solar Radiation Management Geoengineering. *Ethics, Policy Environ.*, **17(2)**, 157–174, doi:10.1080/21550085.2014.927962.

Tacconi, L., 2017: Strengthening policy research and development through foreign aid: the case of reducing deforestation and forest degradation in Indonesia. *Aust. For.*, **80(3)**, 188–194, doi:10.1080/00049158.2017.1335579.

Tacconi, L., R.J. Rodrigues, and A. Maryudi, 2019: Law enforcement and deforestation: Lessons for Indonesia from Brazil. *For. Policy Econ.*, **108**, 101943, doi:10.1016/j.forpol.2019.05.029.

Táíwò, O.O. and S. Talati, 2021: Who Are the Engineers? Solar Geoengineering Research and Justice. *Glob. Environ. Polit.*, **22(1)**, 1–7, doi:10.1162/glep_a_00620.

Talberg, A., S. Thomas, P. Christoff, and D. Karoly, 2018: How geoengineering scenarios frame assumptions and create expectations. *Sustain. Sci.*, **13(4)**, 1093–1104, doi:10.1007/s11625-018-0527-8.

Tamiotti, L., 2011: The legal interface between carbon border measures and trade rules. *Clim. Policy*, **11(5)**, 1202–1211, doi:10.1080/14693062.2011.592672.

Tamiotti, L. et al., 2009: *Trade and Climate Change, A report by the United Nations Environment Programme and the World Trade Organization*. 166 pp. www.wto.org/english/res_e/booksp_e/trade_climate_change_e.pdf.

TCFD, 2017: *Final Report: Recommendations of the Task Force on Climate-related Financial Disclosures*. 66 pp. https://assets.bbhub.io/company/sites/60/2020/10/FINAL-2017-TCFD-Report-11052018.pdf.

TDA, 2019: *Join us in raising ambition, action and advocacy towards the decarbonisation of the transport sector before 2050*. Transportation Decarbonisation Alliance, 4 pp. http://tda-mobility.org/wp-content/uploads/2018/08/TDA-Flyer.pdf.

TEC and CTCN, 2019: *Joint annual report of the Technology Executive Committee and the Climate Technology Centre and Network for 2019*. 26 pp. https://unfccc.int/sites/default/files/resource/sb2019_04E.pdf.

Tedsen, E. and G. Homann, 2013: Implementing the Precautionary Principle for Climate Engineering. *Carbon Clim. Law Rev.*, **7(2)**, 90–100, doi:10.21552/cclr/2013/2/250.

Terhalle, M. and J. Depledge, 2013: Great-power politics, order transition, and climate governance: insights from international relations theory. *Clim. policy*, **13(5)**, 572–588, doi:10.1080/14693062.2013.818849.

Terrenoire, E., D.A. Hauglustaine, T. Gasser, and O. Penanhoat, 2019: The contribution of carbon dioxide emissions from the aviation sector to future climate change. *Environ. Res. Lett.*, **14**, 084019, doi:10.1088/1748-9326/ab3086.

The Supreme Court of Nepal, 2018: *Advocate Padam Bahadur Shrestha Vs The office of the Prime Minister and Council of Ministers, Singhadurbar, Kathmandu and others.* Decision No. 10210, NKP, Part 61, Vol. 3 (25 December 2018). 15 pp. http://climatecasechart.com/wp-content/uploads/sites/16/non-us-case-documents/2018/20181225_074-WO-0283_judgment-2.pdf (Accessed October 31, 2021).

The Supreme Court of the Netherlands, 2019: *The State of the Netherlands v. Stichting Urgenda Foundation. Case 19/00135 (English translation)*.

Thompson, R., 2017: Whither climate change post-Paris? *Anthr. Rev.*, **4(1)**, 62–69, doi:10.1177/2053019616676607.

Thorgeirsson, H., 2017: Objective (Article 2.1). In: *The Paris Agreement on Climate Change: Analysis and commentary* [Klein, D., M. Pía Carazo, M. Doelle, J. Bulmer, and A. Higham, (eds.)]. Oxford University Press, Oxford, UK, pp. 123–130.

Thwaites, J., 2020: *The Good, the bad and the urgent in MDB Climate Finance 2019*. World Resources Institute, Washington DC, USA, www.wri.org/insights/good-bad-and-urgent-mdb-climate-finance-2019 (Accessed October 25, 2021).

Tienhaara, K., 2018: Regulatory Chill in a Warming World: The Threat to Climate Policy Posed by Investor-State Dispute Settlement. *Transnatl. Environ. Law*, **7(2)**, 229–250, doi:10.1017/s2047102517000309.

Tienhaara, K. and L. Cotula, 2020: *Raising the Cost of Climate Action? Investor-State Dispute Settlement and Compensation for Stranded Fossil Fuel Assets. IIED Land, Investment and Rights series.* International Institute for Environment and Development, London, UK, 49 pp. https://pubs.iied.org/pdfs/17660IIED.pdf (Accessed October 31, 2021).

Tilmes, S., B.M. Sanderson, and B.C. O'Neill, 2016: Climate impacts of geoengineering in a delayed mitigation scenario. *Geophys. Res. Lett.*, **43(15)**, 8222–8229, doi:10.1002/2016gl070122.

Tilmes, S. et al., 2018: CESM1(WACCM) Stratospheric Aerosol Geoengineering Large Ensemble Project. *Bull. Am. Meteorol. Soc.*, **99(11)**, 2361–2371, doi:10.1175/bams-d-17-0267.1.

Tilmes, S. et al., 2020: Reaching 1.5 and 2.0°C global surface temperature targets using stratospheric aerosol geoengineering. *Earth Syst. Dyn.*, **11(3)**, 579–601, doi:10.5194/esd-11-579-2020.

Tingley, D. and G. Wagner, 2017: Solar geoengineering and the chemtrails conspiracy on social media. *Palgrave Commun.*, **3(1)**, 1–7, doi:10.1057/s41599-017-0014-3.

Tormos-Aponte, F., and G.A. García-López, 2018: Polycentric struggles: The experience of the global climate justice movement. *Environ. Policy Gov.*, **28(4)**, 284–294, doi:10.1002/eet.1815.

Tørstad, V.H., 2020: Participation, ambition and compliance: can the Paris Agreement solve the effectiveness trilemma? *Env. Polit.*, **29(5)**, 761–780, doi:10.1080/09644016.2019.1710322.

Trachtman, J.P., 2017: WTO Law Constraints on Border Tax Adjustment and Tax Credit Mechanisms to Reduce the Competitive Effects of Carbon Taxes. *Natl. Tax J.*, **70(2)**, 469–493, doi:10.17310/ntj.2017.2.09.

Tramel, S., 2016: The Road Through Paris: Climate Change, Carbon, and the Political Dynamics of Convergence. *Globalizations*, **13(6)**, 960–969, doi:10.1080/14747731.2016.1173376.

Traut, M. et al., 2018: CO2 abatement goals for international shipping. *Clim. Policy*, **18(8)**, 1066–1075, doi:10.1080/14693062.2018.1461059.

Trisos, C.H. et al., 2018: Potentially dangerous consequences for biodiversity of solar geoengineering implementation and termination. *Nat. Ecol. Evol.*, **2(3)**, 475–482, doi:10.1038/s41559-017-0431-0.

Trouwborst, A., 2012: Transboundary Wildlife Conservation in A Changing Climate: Adaptation of the Bonn Convention on Migratory Species and Its Daughter Instruments to Climate Change. *Diversity*, **4(3)**, 258–300, doi:10.3390/d4030258.

Tulkens, H., 2016: COP 21 and Economic Theory: Taking Stock. *Rev. Econ. Polit.*, **126(4)**, 471–486, doi:10.3917/redp.264.0471.

Tulkens, H., 2019: *Economics, Game Theory and International Environmental Agreements. The Ca' Foscari Lectures.* World Scientific Publishing Co. Pte. Ltd., Singapore, 460 pp.

Turnhout, E. et al., 2017: Envisioning REDD+ in a post-Paris era: between evolving expectations and current practice. *Wiley Interdiscip. Rev. Clim. Change*, **8(1)**, e425, doi:10.1002/wcc.425.

TWI2050, 2018: *Transformations to achieve the Sustainable Development Goals.* Report prepared by the World in 2050 initiative. International Institute for Applied Systems Analysis, Laxenburg, Austria, 154 pp.

Umemiya, C., M. Ikeda, and M.K. White, 2020: Lessons learned for future transparency capacity building under the Paris Agreement: A review of greenhouse gas inventory capacity building projects in Viet Nam and Cambodia. *J. Clean. Prod.*, **245**, 118881, doi:10.1016/j.jclepro.2019.118881.

UN, 2016: *Montreal Protocol on Substances that Deplete the Ozone Layer, Kigali Amendment*. Kigali, Rwanda, 15 October 2016, 63 pp. https://treaties.un.org/doc/Treaties/2016/10/20161015%2003-23%20PM/Ch_XXVII-2.f.pdf (Accessed October 31, 2021).

UN, 2017a: *UN System Strategic Approach on Climate Change Action. Report of the High Level Committee on Programmes, Chief Executives Board for Coordination. CEB/2017/4/Add.1*. 15 pp. https://unsceb.org/sites/default/files/2021-01/CEB_2017_4_Add1.pdf (Accessed October 31, 2021).

UN, 2017b: *Catalyzing the Implementation of Nationally Determined Contributions in the Context of the 2030 Agenda through South-South Cooperation*. 68 pp. https://unfccc.int/files/resource_materials/application/pdf/ssc_ndc_report.pdf (Accessed October 31, 2021).

UN, 2018: *South-South and Triangular Cooperation on Climate Technologies*. 76 pp. https://www.iai.int/admin/site/sites/default/files/2018%20SouthSouth%20and%20Triangular%20Climate%20Tech.pdf (Accessed October 31, 2021).

UN, 2019: *Buenos Aires outcome document of the second High-level United Nations Conference on South-South Cooperation (A/RES/73/291)*. 11 pp. digitallibrary.un.org/record/3801900/files/A_RES_73_291-EN.pdf (Accessed October 31, 2021).

UN Human Rights Council, 2021: *Promotion and protection of all human rights, civil, political, economic, social and cultural rights, including the right to development*. 5 October 2021, UN Doc. A/HRC/48/L.23/Rev.1. 3 pp. https://undocs.org/a/hrc/48/l.23/rev.1 (Accessed October 31, 2021).

UNCBD, 2010: *Decision X/2. The Strategic Plan for Biodiversity 2011-2020 and the Aichi Biodiversity Targets*. UNEP/CBD/COP/DEC/X/2 (29 October 2010). United Nations Convention on Biological Diversity (UNCBD), Nagoya, Japan, 13 pp. www.cbd.int/doc/decisions/cop-10/cop-10-dec-02-en.pdf (Accessed October 31, 2021).

UNCCD, 2015: Land Degradation Neutrality: The Target Setting Programme. *Glob. Mech. United Nations Convention to Combat Desertification*, Bonn, Germany, www.unccd.int/actions/ldn-target-setting-programme (Accessed December 19, 2020).

UNCTAD, 2012: *State of South-South and Triangular Cooperation in the Production, Use and Trade of Sustainable Biofuels*. United Nations Conference on Trade and Development, New York and Geneva, 26 pp. https://unctad.org/en/PublicationsLibrary/ditcted2011d10_en.pdf (Accessed October 31, 2021).

UNCTAD, 2019: *Taking Stock of IIA Reform: Recent Developments. IIA Issue Note. Issue 3. June 2019.* United Nations Conference on Trade and Development, Geneva, Switzerland, 12 pp. https://unctad.org/system/files/official-document/diaepcbinf2019d5_en.pdf (Accessed October 31, 2021).

UNEP, 2013: *Global Mercury Assessment*. United Nations Environment Programme Chemicals Branch, Geneva, Switzerland, 32 pp. https://wedocs.unep.org/handle/20.500.11822/7984 (Accessed October 31, 2021).

UNEP, 2018a: *Emissions Gap Report 2018*. United Nations Environment Programme, Nairobi, Kenya, 85 pp. www.unenvironment.org/resources/emissions-gap-report-2018 (Accessed October 31, 2021).

UNEP, 2018b: *Report by the Secretariat on UN Environment Programme's Private Sector Engagement. 142nd meeting of the Committee of Permanent Representatives to the United Nations Environment Programme.* 14 pp.

UNEP, 2019: *Emissions Gap Report 2019*. United Nations Environment Programme, Nairobi, Kenya, 81 pp. www.unep.org/resources/emissions-gap-report-2019 (Accessed October 31, 2021).

UNEP, 2020: *Emissions Gap Report 2020*. United Nations Environment Programme, Nairobi, Kenya, 101 pp. www.unenvironment.org/emissions-gap-report-2020 (Accessed October 31, 2021).

UNFCCC, 1992: *United Nations Framework Convention on Climate Change*. 24 pp. https://unfccc.int/resource/docs/convkp/conveng.pdf (Accessed December 10, 2020).

UNFCCC, 1997: *Kyoto Protocol to the United Nations Framework Convention on Climate Change*. 20 pp. https://unfccc.int/resource/docs/convkp/kpeng.pdf. (Accessed October 31, 2021).

UNFCCC, 2010a: *Decision 2/CP.15. Copenhagen Accord*. 1pp. 4–9, https://unfccc.int/resource/docs/2009/cop15/eng/11a01.pdf#page=4 (Accessed October 31, 2021).

UNFCCC, 2010b: Decision 1/CP.16 The Cancun Agreements: Outcome of the work of the Ad Hoc Working Group on Long-term Cooperative Action under the Convention. *Report of the Conference of the Parties on its sixteenth session, held in Cancun from 29 November to 10 December 2010*, 31.

UNFCCC, 2012: *Report of the Conference of the Parties on its seventeenth session, held in Durban from 28 November to 11 December 2011. Addendum. Part Two: Action taken by the Conference of the Parties at its seventeenth session*. UNFCCC, Durban, 86 pp. https://unfccc.int/resource/docs/2011/cop17/eng/09a01.pdf (Accessed October 31, 2021).

UNFCCC, 2015a: *Paris Agreement*. 25 pp. https://unfccc.int/sites/default/files/english_paris_agreement.pdf (Accessed October 31, 2021).

UNFCCC, 2015b: *Technology Mechanism: Enhancing climate technology development and transfer*. United Nations Framework Convention on Climate Change, 8 pp.

UNFCCC, 2016a: *Decision 1/CP.21 Adoption of the Paris Agreement. Report of the Conference of the Parties on its twenty-first session, held in Paris from 30 November to 13 December 2015. FCCC/CP/2015/10/Add.1*. 36 pp. https://unfccc.int/resource/docs/2015/cop21/eng/10a01.pdf#page=2.

UNFCCC, 2016b: *Aggregate effect of the intended nationally determined contributions: an update. Synthesis report by the secretariat. FCCC/CP/2016/2*. 75 pp. https://unfccc.int/resource/docs/2016/cop22/eng/02.pdf (Accessed October 31, 2021).

UNFCCC, 2018: *UNFCCC Standing Committee on Finance: 2018 Biennial Assessment and Overview of Climate Finance Flows. Technical Report*. 170 pp. https://unfccc.int/sites/default/files/resource/2018%20BA%20Technical%20Report%20Final.pdf (Accessed October 31, 2021).

UNFCCC, 2019a: *Decision 1/CP.25. Report of the Conference of the Parties on its twenty-fifth session, Addendum, Part two*. 59 pp. https://unfccc.int/sites/default/files/resource/cp2019_13a01E.pdf (Accessed October 31, 2021).

UNFCCC, 2019b: Dec. 4/CMA.1 Further guidance in relation to the mitigation section of decision 1/CP.21. *Report of the Conference of the Parties serving as the meeting of the Parties to the Paris Agreement on the third part of its first session, held in Katowice from 2 to 15 December 2018. Addendum 1. Part two: Action taken by the Conference of the Parties s*, 4–13 https://unfccc.int/sites/default/files/resource/cma2018_03a01E.pdf (Accessed October 31, 2021).

UNFCCC, 2019c: Decision 4/CP.24 Report of the Standing Committee on Finance. FCCC/CP/2018/10/Add.1. *Report of the Conference of the Parties on its twenty-fourth session, held in Katowice from 2 to 15 December 2018. Addendum. Part two: Action taken by the Conference of the Parties at its twenty-fourth session*, UNFCCC 2019, 14–28 https://unfccc.int/sites/default/files/resource/10a1.pdf (Accessed October 31, 2021).

UNFCCC, 2019d: Decision 18/CMA.1 Modalities, procedures and guidelines for the transparency framework for action and support referred to in Article 13 of the Paris Agreement. *Report of the Conference of the Parties serving as the meeting of the Parties to the Paris Agreement on the third part of its first session, held in Katowice from 2 to 15 December 2018*, 18–52 https://unfccc.int/sites/default/files/resource/cma2018_3_add2_new_advance.pdf (Accessed October 31, 2021).

UNFCCC, 2019e: Dec. 19/CMA.1 Matters relating to Article 14 of the Paris Agreement and paragraphs 99–101 of decision 1/CP.21. *Report of the Conference of the Parties serving as the meeting of the Parties to the Paris Agreement on the third part of its first session, held in Katowice from 2 to*

*15 December 2018*, 53–58 https://unfccc.int/sites/default/files/resource/cma2018_3_add2_new_advance.pdf (Accessed October 31, 2021).

UNFCCC, 2019f: Decision 15/CMA.1 Technology framework under Article 10, paragraph 4, of the Paris Agreement. FCCC/PA/CMA/2018/3 Add.2. *Report of the Conference of the Parties serving as the meeting of the Parties to the Paris Agreement on the third part of its first session, held in Katowice from 2 to 15 December 2018 Addendum Part two: Action taken by the Conference of the Parties servi*, 4–10 https://unfccc.int/sites/default/files/resource/cma2018_3_add2_new_advance.pdf (Accessed October 31, 2021).

UNFCCC, 2019g: Dec. 14/CMA.1 Setting a new collective quantified goal on finance in accordance with decision 1/CP.21, paragraph 53. *Report of the Conference of the Parties serving as the meeting of the Parties to the Paris Agreement on the third part of its first session, held in Katowice from 2 to 15 December 2018*, 3 https://unfccc.int/sites/default/files/resource/cma2018_3_add2_new_advance.pdf (Accessed October 31, 2021).

UNFCCC, 2019h: *Implementation of the framework for capacity-building in developing countries. Synthesis report by the secretariat*. 22 pp. https://unfccc.int/sites/default/files/resource/03.pdf.

UNFCCC, 2019i: Decision 17/CMA.1 Ways of enhancing the implementation of education, training, public awareness, public participation and public access to information so as to enhance actions under the Paris Agreement. *Report of the Conference of the Parties serving as the meeting of the Parties to the Paris Agreement on the third part of its first session, held in Katowice from 2 to 15 December 2018*, 15–17 https://unfccc.int/sites/default/files/resource/cma2018_3_add2_new_advance.pdf. (Accessed October 31, 2021).

UNFCCC, 2019j: Decision 20/CMA.1. Modalities and procedures for the effective operation of the committee to facilitate implementation and promote compliance referred to in Article 15, paragraph 2, of the Paris Agreement. *Report of the Conference of the Parties serving as the meeting of the Parties to the Paris Agreement on the third part of its first session, held in Katowice from 2 to 15 December 2018 Addendum Part two: Action taken by the Conference of the Parties servi*, 59–64 https://unfccc.int/sites/default/files/resource/cma2018_3_add2_new_advance.pdf. (Accessed October 31, 2021).

UNFCCC, 2019k: *Common tabular formats for the electronic reporting of the information on financial, technology development and transfer and capacity-building support provided and mobilized, as well as support needed and received, under Articles 9–11 of the Paris Agreeme*. 19 pp. https://unfccc.int/sites/default/files/resource/SBSTA51.IN_.i11c.pdf (Accessed October 31, 2021).

UNFCCC, 2019l: Decision 14/CP.24 Linkages between the Technology Mechanism and the Financial Mechanism of the Convention. FCCC/CP/2018/10/Add.2. *Report of the Conference of the Parties on its twenty-fourth session, held in Katowice from 2 to 15 December 2018. Addendum. Part two: Action taken by the Conference of the Parties at its twenty-fourth session*, pg 5 https://unfccc.int/sites/default/files/resource/cp2018_10_add2_advance.pdf (Accessed October 31, 2021).

UNFCCC, 2020a: Bilateral and Multilateral Funding. *UNFCCC Top. Clim. Financ. Resour.*, https://unfccc.int/topics/climate-finance/resources/multilateral-and-bilateral-funding-sources (Accessed December 8, 2020).

UNFCCC, 2020b: Technology Executive Committee: Strengthening climate technology policies. *TT Clear*, https://unfccc.int/ttclear/tec (Accessed June 4, 2020).

UNFCCC, 2020c: Developing the Santiago Network for Loss and Damage. *Event.*, https://unfccc.int/event/developing-the-santiago-network-for-loss-and-damage (Accessed October 30, 2021).

UNFCCC, 2020d: Subsidiary Body for Implementation, Compilation and synthesis of fourth biennial reports of Parties included in Annex I to the Convention. https://unfccc.int/sites/default/files/resource/sbi2020_inf10.pdf (Accessed November 2, 2021).

UNFCCC, 2021a: *Nationally determined contributions under the Paris Agreement – Synthesis report by the secretariat (FCCC/PA/CMA/2021/8)*. 42 pp. https://unfccc.int/sites/default/files/resource/cma2021_08_adv_1.pdf (Accessed October 31, 2021).

UNFCCC, 2021b: Building capacity in the UNFCCC process. https://unfccc.int/topics/capacity-building/the-big-picture/capacity-in-the-unfccc-process (Accessed October 30, 2021).

UNFCCC and UNOSSC, 2018: *Potential of South South Cooperation and Triangular Cooperation on climate technologies for advancing the implementation of nationally determined contributions and national adaptation plans.* United Nations Framework Convention on Climate Change and United Nations Office for South-South Cooperation, 34 pp.

UNFCCC Subsidiary Body for Implementation, 2020: *Annual technical progress report of the Paris Committee on Capacity-building. FCCC/SBI/2020/13.* 34 pp.

UNGA, 1948: *The Universal Declaration of Human Rights. A/RES/217.*

UNGA, 1966a: *International Covenant on Civil and Political Rights. General Assembly Resolution 2200A (XXI) of 16 December 1966.*

UNGA, 1966b: *International Covenant on Economic, Social and Cultural Rights. General Assembly Resolution 2200A (XXI) of 16 December 1966.*

UNGA, 2019: *Report of the Special Rapporteur on the issue of human rights obligations relating to the enjoyment of a safe, clean, healthy and sustainable environment*. 15 July 2019, UN Doc. A/76/161. United Nations General Assembly (UNGA), 25 pp.

UNOSSC, 2017: United Nations Action Plan on South-South Climate Cooperation (2017-2021). www.unsouthsouth.org/south-south-cooperation-action-plan-for-climate-change-engagement-strategy-2017-2021/ (Accessed October 30, 2021).

Urban, F., 2018: China's rise: Challenging the North-South technology transfer paradigm for climate change mitigation and low carbon energy. *Energy Policy*, **113**, 320–330, doi:10.1016/j.enpol.2017.11.007.

Urpelainen, J., 2011: Can Unilateral Leadership Promote International Environmental Cooperation? *Int. Interact.*, **37(3)**, 320–339, doi:10.1080/03050629.2011.596018.

Urpelainen, J. and T. Van de Graaf, 2015: The International Renewable Energy Agency: a success story in institutional innovation? *Int. Environ. Agreements Polit. Law Econ.*, **15(2)**, 159–177, doi:10.1007/s10784-013-9226-1.

Urpelainen, J. and T. Van de Graaf, 2018: United States non-cooperation and the Paris agreement. *Clim. Policy*, **18(7)**, 839–851, doi:10.1080/14693062.2017.1406843.

van Asselt, H., 2007: From UN-ity to Diversity? The UNFCCC, the Asia-Pacific Partnership and the Future of International Law on Climate Change. *Carbon Clim. Law Rev.*, **1(1)**, 12, doi:10.21552/cclr/2007/1/9.

van Asselt, H., 2016: The Role of Non-State Actors in Reviewing Ambition, Implementation, and Compliance under the Paris Agreement. *Clim. Law*, **6(1–2)**, 91–108, doi:10.1163/18786561-00601006.

van Asselt, H., 2017: *Climate Change and Trade Policy Interaction: Implications of Regionalism.* OECD Trade and Environment Working Papers. 2017/03. OECD Publishing, Paris, France, 54 pp.

van Asselt, H. and F. Zelli, 2014: Connect the dots: managing the fragmentation of global climate governance. *Environ. Econ. Policy Stud.*, **16(2)**, 137–155, doi:10.1007/s10018-013-0060-z.

van de Graaf, T., 2013: Fragmentation in Global Energy Governance: Explaining the Creation of IRENA. *Glob. Environ. Polit.*, **13(3)**, 14–33, doi:10.1162/glep_a_00181.

van de Graaf, T, and D. Lesage, 2009: The International Energy Agency after 35 years: Reform needs and institutional adaptability. *Rev. Int. Organ.*, **4(3)**, 293–317, doi:10.1007/s11558-009-9063-8.

van de Graaf, T. and J. Colgan, 2016: Global energy governance: a review and research agenda. *Palgrave Commun.*, **2**, 15047, doi:10.1057/palcomms.2015.47.

van der Voorn, T., C. Pahl-Wostl, and J. Quist, 2012: Combining backcasting and adaptive management for climate adaptation in coastal regions: A methodology and a South African case study. *Futures*, **44(4)**, 346–364, doi:10.1016/j.futures.2011.11.003.

van der Voorn, T., J. Quist, C. Pahl-Wostl, and M. Haasnoot, 2017: Envisioning robust climate change adaptation futures for coastal regions: a comparative evaluation of cases in three continents. *Mitig. Adapt. Strateg. Glob. Change*, **22(3)**, 519–546, doi:10.1007/s11027-015-9686-4.

van der Voorn, T., Å. Svenfelt, K.E. Björnberg, E. Fauré, and R. Milestad, 2020: Envisioning carbon-free land use futures for Sweden: a scenario study on conflicts and synergies between environmental policy goals. *Reg. Environ. Change*, **20(2)**, 35, doi:10.1007/s10113-020-01618-5.

van Soest, H.L., M.G.J. den Elzen, and D.P. van Vuuren, 2021: Net-zero emission targets for major emitting countries consistent with the Paris Agreement. *Nat. Commun.*, **12(1)**, 2140, doi:10.1038/s41467-021-22294-x.

van Tulder, R., S.B. Rodrigues, H. Mirza, and K. Sexsmith, 2021: The UN's Sustainable Development Goals: Can multinational enterprises lead the Decade of Action? *J. Int. Bus. Policy*, **4(1)**, 1–21, doi:10.1057/s42214-020-00095-1.

Vandyck, T. et al., 2018: Air quality co-benefits for human health and agriculture counterbalance costs to meet Paris Agreement pledges. *Nat. Commun.*, **9(1)**, 4939, doi:10.1038/s41467-018-06885-9.

Vanhala, L. and C. Hestbaek, 2016: Framing Climate Change Loss and Damage in UNFCCC Negotiations. *Glob. Environ. Polit.*, **16(4)**, 111, doi:10.1162/glep_a_00379.

Velders, G.J.M., S.O. Andersen, J.S. Daniel, D.W. Fahey, and M. McFarland, 2007: The importance of the Montreal Protocol in protecting climate. *Proc. Natl. Acad. Sci.*, **104(12)**, 4814–4819, doi:10.1073/pnas.0610328104.

Velders, G.J.M., D.W. Fahey, J.S. Daniel, S.O. Andersen, and M. McFarland, 2015: Future atmospheric abundances and climate forcings from scenarios of global and regional hydrofluorocarbon (HFC) emissions. *Atmos. Environ.*, **123**, 200–209, doi:10.1016/j.atmosenv.2015.10.071.

Verkuijl, C., H. van Asselt, T. Moerenhout, L. Casier, and P. Wooders, 2019: Tackling Fossil Fuel Subsidies Through International Trade Agreements: Taking Stock, Looking Forward. *VA. J. Int. Law*, **58(2)**, 309–368.

Victor, D.G., 2011: *Global Warming Gridlock: Creating More Effective Strategies for Protecting the Planet*. Cambridge University Press, Cambridge, UK and New York, NY, USA, 392 pp.

Victor, D.G., 2016: What the Framework Convention on Climate Change Teaches Us About Cooperation on Climate Change. *Polit. Gov.*, **4(3)**, 133, doi:10.17645/pag.v4i3.657.

Victor, D.G., K. Raustiala, and E.B. Skolnikoff (eds.), 1998: *The Implementation and Effectiveness of International Environmental Commitments: Theory and Practice*. MIT Press, Cambridge, MA, USA, 737 pp.

Victor, D.G., Geels, F.W., and Sharpe, S. 2019: *Accelerating the Low Carbon Transition: The case for stronger, more targeted and coordinated international action*. Energy Transitions Commission, London and Manchester, UK, and San Diego, CA, USA, 138 pp.

Vihma, A., Y. Mulugetta, and S. Karlsson-Vinkhuyzen, 2011: Negotiating solidarity? The G77 through the prism of climate change negotiations. *Glob. Change, Peace Secur.*, **23(3)**, 315–334, doi:10.1080/14781158.2011.601853.

Visioni, D., G. Pitari, and V. Aquila, 2017: Sulfate geoengineering: a review of the factors controlling the needed injection of sulfur dioxide. *Atmos. Chem. Phys.*, **17(6)**, 3879–3889, doi:10.5194/acp-17-3879-2017.

Visioni, D. et al., 2020: What goes up must come down: impacts of deposition in a sulfate geoengineering scenario. *Environ. Res. Lett.*, **15(9)**, 94063, doi:10.1088/1748-9326/ab94eb.

Visschers, V.H.M., J. Shi, M. Siegrist, and J. Arvai, 2017: Beliefs and values explain international differences in perception of solar radiation management: insights from a cross-country survey. *Clim. Change*, **142(3)**, 531–544, doi:10.1007/s10584-017-1970-8.

Voigt, C., 2016: The compliance and implementation mechanism of the Paris agreement. *Rev. Eur. Comp. Int. Environ. Law*, **25(2)**, doi:10.1111/reel.12155.

Voigt, C. and F. Ferreira, 2015: The Warsaw Framework for REDD+: Implications for National Implementation and Access to Results-Based Finance. *Carbon Clim. Law Rev.*, **2**, 113.

Voigt, C. and F. Ferreira, 2016a: 'Dynamic Differentiation': The Principles of CBDR-RC, Progression and Highest Possible Ambition in the Paris Agreement. *Transnatl. Environ. Law*, **5(2)**, 285–303, doi:10.1017/s2047102516000212.

Voigt, C. and F. Ferreira, 2016b: Differentiation in the Paris Agreement. *Clim. Law*, **6(1–2)**, 58–74, doi:10.1163/18786561-00601004.

Waage, J. et al., 2015: Governing Sustainable Development Goals: interactions, infrastructures, and institutions. In: *Thinking Beyond Sectors for Sustainable Development*, Ubiquity Press, London, UK, pp. 79–88.

Wagner, C.S., T.A. Whetsell, and L. Leydesdorff, 2017: Growth of international collaboration in science: revisiting six specialties. *Scientometrics*, **110(3)**, 1633–1652, doi:10.1007/s11192-016-2230-9.

Watkiss, P., M. Benzie, and R.J.T. Klein, 2015: The complementarity and comparability of climate change adaptation and mitigation. *Wiley Interdiscip. Rev. Clim. Change*, **6(6)**, 541–557, doi:10.1002/wcc.368.

Watts, J. and J. Depledge, 2018: Latin America in the climate change negotiations: Exploring the AILAC and ALBA coalitions. *Wiley Interdiscip. Rev. Clim. Change*, **9(6)**, e533, doi:10.1002/wcc.533.

WBCSD, 2019: About us. World Business Council for Sustainable Development. www.wbcsd.org/Overview/About-us (Accessed December 12, 2019).

Weart, S.R., 2012: The evolution of international cooperation in climate science. *J. Int. Organ. Stud.*, **3(1)**, 41–59.

Weikmans, R., and J.T. Roberts, 2019: The international climate finance accounting muddle: is there hope on the horizon? *Clim. Dev.*, **11(2)**, 97–111, doi:10.1080/17565529.2017.1410087.

Weikmans, R., H. van Asselt, and J.T. Roberts, 2020: Transparency requirements under the Paris Agreement and their (un)likely impact on strengthening the ambition of nationally determined contributions (NDCs). *Clim. Policy*, **20(4)**, 511–526, doi:10.1080/14693062.2019.1695571.

Weischer, L., J. Morgan, and M. Patel, 2012: Climate Clubs: Can Small Groups of Countries make a Big Difference in Addressing Climate Change? *Rev. Eur. Community Int. Environ. Law*, **21(3)**, 177–192, doi:10.1111/reel.12007.

Weiss, E.B., and H.K. Jacobson (eds.), 1998: *Engaging Countries: Strengthening Compliance with Internatioal Environmental Accords*. MIT Press, Cambridge, MA, YSA 640 pp.

Weitz, N., H. Carlsen, M. Nilsson, and K. Skånberg, 2018: Towards systemic and contextual priority setting for implementing the 2030 Agenda. *Sustain. Sci.*, **13(2)**, 531–548, doi:10.1007/s11625-017-0470-0.

Weitzman, M.L., 2015: A Voting Architecture for the Governance of Free-Driver Externalities, with Application to Geoengineering. *Scand. J. Econ.*, **117(4)**, 1049–1068, doi:10.1111/sjoe.12120.

Werksman, J., 2010: Legal symmetry and legal differentiation under a future deal on climate. *Clim. Policy*, **10(6)**, 672–677, doi:10.3763/cpol.2010.0150.

Wettestad, J. and L.H. Gulbrandsen, 2018: *The Evolution of Carbon Markets: Design and Diffusion*. Routledge, London and New York, 264 pp.

Wettestad, J., L.H. Gulbrandsen, and S. Andresen, 2021: Calling in the heavyweights: Why the World Bank established the Carbon Pricing Leadership Coalition, and what it might achieve. *Int. Stud. Perspect.*, **22(2)**, 201–217, doi:10.1093/isp/ekaa013.

Wewerinke-Singh, M., 2018: *State Responsibility, Climate Change and Human Rights Under International Law*. Hart Publishing, Oxford, UK, 216 pp.

Wewerinke-Singh, M. and D.H. Salili, 2020: Between negotiations and litigation: Vanuatu's perspective on loss and damage from climate change. *Clim. Policy*, **20(6)**, 681–692, doi:10.1080/14693062.2019.1623166.

White House, 2014: US-China Joint Announcements on Climate Change. White House Press Briefings, Office of the US Press Secretary. https://obamawhitehouse.archives.gov/the-press-office/2014/11/11/us-china-joint-announcement-climate-change (Accessed January 11, 2021).

Whyte, K.P., 2018: Indigeneity in Geoengineering Discourses: Some Considerations. *Ethics, Policy Environ.*, **21(3)**, 289–307, doi:10.1080/21550085.2018.1562529.

Wieding, J., J. Stubenrauch, and F. Ekardt, 2020: Human Rights and Precautionary Principle: Limits to Geoengineering, SRM, and IPCC Scenarios. *Sustainability*, **12(21)**, 8858, doi:10.3390/su12218858.

Wijaya, A., T.N. Samadhi, and R. Juliane, 2019: Indonesia Is Reducing Deforestation, but Problem Areas Remain. World Resources Institute, Washington DC, USA, www.wri.org/insights/indonesia-reducing-deforestation-problem-areas-remain (Accessed August 26, 2021).

Wilder, M. et al., 2010: Adapting Across Boundaries: Climate Change, Social Learning, and Resilience in the U.S.–Mexico Border Region. *Ann. Assoc. Am. Geogr.*, **100(4)**, 917–928, doi:10.1080/00045608.2010.500235.

Winickoff, D.E., J.A. Flegal, and A. Asrat, 2015: Engaging the Global South on climate engineering research. *Nat. Clim. Change*, **5(7)**, 627–634, doi:10.1038/nclimate2632.

Winkler, H., 2020: Putting equity into practice in the global stocktake under the Paris Agreement. *Clim. Policy*, **1**, 124–132, doi:10.1080/14693062.2019.1680337.

Winkler, H. and N.K. Dubash, 2016: Who determines transformational change in development and climate finance? *Clim. Policy*, **16(6)**, 783–791, doi:10.1080/14693062.2015.1033674.

Winkler, H., B. Mantlana, and T. Letete, 2017: Transparency of action and support in the Paris Agreement. *Clim. Policy*, **17(7)**, 853–872, doi:10.1080/14693062.2017.1302918.

Winkler, H. et al., 2018: Countries start to explain how their climate contributions are fair: More rigour needed. *Int. Environ. Agreements Polit. Law Econ.*, **18**, 99–115, doi:10.1007/s10784-017-9381-x.

Winterbottom, R., 1990: *Taking Stock – The Tropical Forestry Action Plan After Five Years*. World Resources Institute, New York and Washington DC, pp.

WMO, 2018: *Scientific Assessment of Ozone Depletion: 2018*. World Meteorological Organization, Geneva, Switzerland, 588 pp. https://ozone.unep.org/sites/default/files/2019-05/SAP-2018-Assessment-report.pdf

Wolf, S., C. Jaeger, J. Mielke, F. Schuetze, and R. Rosen, 2019: Framing 1.5°C – Turning an Investment Challenge into a Green Growth Opportunity. *SSRN Electron. J.*, doi:10.2139/ssrn.3324509.

Wollenberg, E. et al., 2016: Reducing emissions from agriculture to meet the 2°C target. *Glob. Change Biol.*, **22(12)**, 3859–3864, doi:10.1111/gcb.13340.

World Bank, 2010: *Cities and Climate Change: An Urgent Agenda. Urban development series. Knowledge Papers no. 10.* World Bank, Washington, DC, USA, 81 pp. https://openknowledge.worldbank.org/handle/10986/17381 (Accessed October 31, 2021).

World Bank, 2015a: *Joint Statement by the Multilateral Development Banks at Paris, COP21 Delivering Climate Change Action at Scale: Our Commitment to Implementation*. World Bank, Washington, DC, USA, 3 pp. www.worldbank.org/content/dam/Worldbank/document/Climate/Joint MDB Statement Climate_NOV 28_final.pdf (Accessed December 8, 2020).

World Bank, 2015b: *From Billions to Trillions: Transforming Development Finance Post-2015 Financing for Development: Multilateral Development Finance. Development Committee Discussion Note*. World Bank, Washington, DC, USA, 23 pp. http://pubdocs.worldbank.org/en/622841485963735448/DC2015-0002-E-FinancingforDevelopment.pdf (Accessed December 9, 2020).

World Bank, 2018: *Carbon Markets for Greenhouse Gas Emission Reduction in a Warming World.* World Bank, Washington, DC, USA, 191 pp.

World Bank, 2019: *State and Trends of Carbon Pricing*. World Bank Group, Washington DC, USA, 94 pp. http://documents.worldbank.org/curated/en/191801559846379845/State-and-Trends-of-Carbon-Pricing-2019.

World Bank, 2021: World Bank Country and Lending Groups. *Data Blog*, World Bank, Washington, DC, USA https://datahelpdesk.worldbank.org/knowledgebase/articles/906519-world-bank-country-and-lending-groups (Accessed October 31, 2021).

Wu, M. and J. Salzman, 2014: The Next Generation of Trade and Environment Conflicts: The Rise of Green Industrial Policy. *Northwest. Univ. Law Rev.*, **108(2)**, 401–474.

Wurzel, R.K.W., D. Liefferink, and D. Torney, 2019: Pioneers, leaders and followers in multilevel and polycentric climate governance. *Env. Polit.*, **28(1)**, 1–21, doi:10.1080/09644016.2019.1522033.

Xia, L. et al., 2014: Solar radiation management impacts on agriculture in China: A case study in the Geoengineering Model Intercomparison Project (GeoMIP). *J. Geophys. Res. Atmos.*, **119(14)**, 8695–8711, doi:10.1002/2013jd020630.

Xia, L., J.P. Nowack, S. Tilmes, and A. Robock, 2017: Impacts of stratospheric sulfate geoengineering on tropospheric ozone. *Atmos. Chem. Phys.*, **17(19)**, 11913–11928, doi:10.5194/acp-17-11913-2017.

Xu, Y., Z. Dong, and Y. Wang, 2016: Establishing a measurement, reporting, and verification system for climate finance in post-Paris agreement period. *Chinese J. Popul. Resour. Environ.*, **14(4)**, 235–244, doi:10.1080/10042857.2016.1258802.

Xuemei, B., 2007: Integrating Global Environmental Concerns into Urban Management: The Scale and Readiness Arguments. *J. Ind. Ecol.*, **11(2)**, 15–29, doi:10.1162/jie.2007.1202.

Yamaguchi, S., 2020: *Greening Regional Trade Agreements: Subsidies Related to Energy and Environmental Goods*. OECD Publishing, Paris, France, 81 pp.

Yamineva, Y., 2016: Climate Finance in the Paris Outcome: Why Do Today What You Can Put Off Till Tomorrow? *Rev. Eur. Comp. Int. Environ. Law*, **25(2)**, 174–185, doi:10.1111/reel.12160.

Yamineva, Y. and K. Kulovesi, 2018: Keeping the Arctic White: The Legal and Governance Landscape for Reducing Short-Lived Climate Pollutants in the Arctic Region. *Transnatl. Environ. Law*, **7(2)**, 201–227, doi:10.1017/s2047102517000401.

Yang, H. et al., 2016: Potential negative consequences of geoengineering on crop production: A study of Indian groundnut. *Geophys. Res. Lett.*, **43(22)**, 11786–11795, doi:10.1002/2016gl071209.

Young, O.R., 2016: The Paris Agreement: Destined to Succeed or Doomed to Fail? *Polit. Gov.*, **4(3)**, 124–132, doi:10.17645/pag.v4i3.635.

Young, P.J. et al., 2021: The Montreal Protocol protects the terrestrial carbon sink. *Nature*, **596(7872)**, 384–388, doi:10.1038/s41586-021-03737-3.

Zahar, A., 2019: Collective Progress in the Light of Equity Under the Global Stocktake. *Clim. Law*, **9(1–2)**, 101–121, doi:10.1163/18786561-00901006.

Zarin, D.J. et al., 2016: Can carbon emissions from tropical deforestation drop by 50% in 5 years? *Glob. Change Biol.*, **22(4)**, 1336–1347, doi:10.1111/gcb.13153.

Zarnetske, P.L. et al., 2021: Potential ecological impacts of climate intervention by reflecting sunlight to cool Earth. *Proc. Natl. Acad. Sci.*, **118(15)**, 1921854118, doi:10.1073/pnas.1921854118.

Zhan, P., W. Zhu, T. Zhang, X. Cui, and N. Li, 2019: Impacts of Sulfate Geoengineering on Rice Yield in China: Results From a Multimodel Ensemble. *Earth's Future*, **7(4)**, 395–410, doi:10.1029/2018ef001094.

Zhang, H., 2016: Towards global green shipping: the development of international regulations on reduction of GHG emissions from ships. *Int. Environ. Agreements Polit. Law Econ.*, **16(4)**, 561–577, doi:10.1007/s10784-014-9270-5.

Zhang, H., 2019: Implementing Provisions on Climate Finance Under the Paris Agreement. *Clim. Law*, **9(1–2)**, 21–39, doi:10.1163/18786561-00901003.

Zhang, J., K. Zhang, J. Liu, and G. Ban-Weiss, 2016: Revisiting the climate impacts of cool roofs around the globe using an Earth system model. *Environ. Res. Lett.*, **11(8)**, 84014, doi:10.1088/1748-9326/11/8/084014.


14

Zhang, W. and X. Pan, 2016: Study on the demand of climate finance for developing countries based on submitted INDC. *Adv. Clim. Change Res.*, **7(1–2)**, 99–104, doi:10.1016/j.accre.2016.05.002.

Zia, A. and S. Kauffman, 2018: The Limits of Predictability in Predefining Phase Spaces of Dynamic Social–Ecological Systems: "Command and Control" Versus "Complex Systems"-Based Policy Design Approaches to Conserve Tropical Forests. *J. Policy Complex Syst.*, **4(2)**, doi:10.18278/jpcs.4.2.9.

Zihua, G., C. Voigt, and J. Werksman, 2019: Facilitating Implementation and Promoting Compliance With the Paris Agreement Under Article 15: Conceptual Challenges and Pragmatic Choices. *Clim. Law*, **9(1–2)**, 65–100, doi:10.1163/18786561-00901005.

Zimm, C. and N. Nakicenovic, 2020: What are the implications of the Paris Agreement for inequality? *Clim. Policy*, **20(4)**, 458–467, doi:10.1080/14693062.2019.1581048.

Zürn, M. and S. Schäfer, 2013: The Paradox of Climate Engineering. *Glob. Policy*, **4(3)**, 266–277, doi:10.1111/gpol.12004.

# 15

# Investment and Finance

**Coordinating Lead Authors:**
Silvia Kreibiehl (Germany), Tae Yong Jung (Republic of Korea)

**Lead Authors:**
Stefano Battiston (Switzerland/Italy), Pablo Esteban Carvajal Sarzosa (Ecuador), Christa Clapp (Norway/the United States of America), Dipak Dasgupta (India), Nokuthula Dube (Zimbabwe/United Kingdom), Raphaël Jachnik (France), Kanako Morita (Japan), Nahla Samargandi (Saudi Arabia), Mariama Williams (Jamaica/the United States of America)

**Contributing Authors:**
Myriam Bechtoldt (Germany), Christoph Bertram (Germany), Lilia Caiado Couto (Brazil), Jean-Charles Hourcade (France), Jean-François Mercure (United Kingdom), Sanusi Mohamed Ohiare (Nigeria), Mahesti Okitasari (Japan/Indonesia), Tamiksha Singh (India), Kazi Sohag (the Russian Federation), Mohamed Youba Sokona (Mali), Doreen Stabinsky (the United States of America)

**Review Editors:**
Amjad Abdulla (Maldives), María José López Blanco (Spain)

**Chapter Scientists:**
Michael König (Germany), Jongwoo Moon (Republic of Korea), Justice Issah Surugu Musah (Ghana)

**This chapter should be cited as:**
Kreibiehl, S., T. Yong Jung, S. Battiston, P. E. Carvajal, C. Clapp, D. Dasgupta, N. Dube, R. Jachnik, K. Morita, N. Samargandi, M. Williams, 2022: Investment and finance. In IPCC, 2022: *Climate Change 2022: Mitigation of Climate Change. Contribution of Working Group III to the Sixth Assessment Report of the Intergovernmental Panel on Climate Change* [P.R. Shukla, J. Skea, R. Slade, A. Al Khourdajie, R. van Diemen, D. McCollum, M. Pathak, S. Some, P. Vyas, R. Fradera, M. Belkacemi, A. Hasija, G. Lisboa, S. Luz, J. Malley, (eds.)]. Cambridge University Press, Cambridge, UK and New York, NY, USA. doi: 10.1017/9781009157926.017

15

# Table of Contents

## Executive Summary

**Finance to reduce net greenhouse gas (GHG) emissions and enhance resilience to climate impacts represents a critical enabling factor for the low carbon transition. Fundamental inequities in access to finance as well as its terms and conditions, and countries' exposure to physical impacts of climate change overall result in a worsening outlook for a global just transition (*high confidence*).** Decarbonising the economy requires global action to address fundamental economic inequities and overcome the climate investment trap that exists for many developing countries. For these countries the costs and risks of financing often represent a significant challenge for stakeholders at all levels. This challenge is exacerbated by these countries' general economic vulnerability and indebtedness. The rising public fiscal costs of mitigation, and of adapting to climate shocks, are affecting many countries and worsening public indebtedness and country credit ratings at a time when there were already significant stresses on public finances. The COVID-19 pandemic has made these stresses worse and tightened public finances still further. Other major challenges for commercial climate finance include: the mismatch between capital and investment needs,[1] home bias[2] considerations, differences in risk perceptions for regions, as well as limited institutional capacity to ensure safeguards represent. {15.2, 15.6.3}

**Investors, central banks, and financial regulators are driving increased awareness of climate risk. This increased awareness can support climate policy development and implementation (*high confidence*).** Climate-related financial risks arise from physical impacts of climate change (already relevant in the short term), and from a disorderly transition to a low-carbon economy. Awareness of these risks is increasing leading also to concerns about financial stability. Financial regulators and institutions have responded with multiple regulatory and voluntary initiatives by to assess and address these risks. Yet despite these initiatives, climate-related financial risks remain greatly underestimated by financial institutions and markets limiting the capital reallocation needed for the low-carbon transition. Moreover, risks relating to national and international inequity – which act as a barrier to the transformation – are not yet reflected in decisions by the financial community. Stronger steering by regulators and policy makers has the potential to close this gap. Despite the increasing attention of investors to climate change, there is limited evidence that this attention has directly impacted emission reductions. This leaves high uncertainty, both near-term (2021–30) and longer-term (2021–50), on the feasibility of an alignment of financial flows with the Paris Agreement (*high confidence*). {15.2, 15.6}

**Progress on the alignment of financial flows with low GHG emissions pathways remains slow. There is a climate financing gap which reflects a persistent misallocation of global capital (*high confidence*).** Persistently high levels of both public and private fossil-fuel related financing continue to be of major concern despite recent commitments. This reflects policy misalignment, the current perceived risk-return profile of fossil fuel-related investments, and political economy constraints (*high confidence*). {15.3}

Estimates of climate finance flows – which refers to local, national, or transnational financing from public, private, multilateral, bilateral and alternative sources, to support mitigation and adaptation actions addressing climate change – exhibit highly divergent patterns across regions and sectors and a slowing growth. {15.3}

When the perceived risks are too high the misallocation of abundant savings persists. Investors refrain from investing in infrastructure and industry in search of safer financial assets, even earning low or negative real returns. {15.2, 15.3}

Global climate finance is heavily focused on mitigation (more than 90% on average between 2017–2020). This is despite the significant economic effects of climate change's expected physical impacts, and the increasing awareness of these effects on financial stability. To meet the needs for rapid deployment of mitigation options, global mitigation investments are expected to need to increase by the factor of 3 to 6 (*medium confidence*). The gaps are wide for all sectors and represent a major challenge for developing countries,[3] especially Least-Developed Countries (LDCs), where flows have to increase by factor 4 to 7, for specific sectors like agriculture, forestry and other land use (AFOLU) in relative terms, and for specific groups with limited access to, and high costs of, climate finance (*high confidence*). {15.4, 15.5}

The actual size of sectoral and regional climate financing gaps is only one component driving the magnitude of the challenge, with financial and economic viability, access to capital markets, investment requirements for adaptation, reduction of losses and damages, climate-responsive social protection, appropriate regulatory frameworks and institutional capacity to attract and facilitate investments and ensure safeguards being decisive to scale-up financing. Financing needs for the creation and strengthening of regulatory environment and institutional capacity, upstream financing needs as well as R&D and venture capital for development of new technologies and business models are often overlooked despite their critical role to facilitate the deployment of scaled-up climate finance (*high confidence*). {15.4.1, 15.5.2}

**The relatively slow implementation of commitments by countries and stakeholders in the financial system to scale up climate finance reflects neither the urgent need for ambitious climate action, nor the economic rationale for ambitious climate action (*high confidence*).** Delayed climate investments and financing – and limited alignment of investment activity with the Paris Agreement – will result in significant carbon

---

1 The term Investment 'Needs' used in the chapter means equal to the term Investment Requirement used in SPM.

2 Most of climate finance stays within national borders, especially private climate flows (over 90%). Reasons are national policy support, differences in regulatory standards, exchange rate, political and governance risks, to information market failures.

3 In modelled pathways, regional investments are projected to occur when and where they are most cost-effective to limit global warming. The model quantifications help to identify high-priority areas for cost-effective investments, but do not provide any indication on who would finance the regional investments.

lock-ins, stranded assets, and other additional costs. This will particularly impact urban infrastructure and the energy and transport sectors (*high confidence*). A common understanding of debt sustainability and debt transparency, including negative implications of deferred climate investments on future GDP, and how stranded assets and resources may be compensated, has not yet been developed (*medium confidence*). {15.6}

The greater the urgency of action to remain on a 1.5°C pathway the greater need for parallel investment decisions in upstream and downstream parts of the value chain. Greater urgency also reduces the lead times to build trust in regulatory frameworks. Consequently, many investment decisions will need to be made based on the long-term global goals. This highlights the importance of trust in political leadership which, in turn, affects risk perception and ultimately financing costs (*high confidence*). {15.6.1, 15.6.2}

There is a mismatch between capital availability in the developed world and the future emissions expected in developing countries. This emphasises the need to recognise the explicit and positive social value of global cross-border mitigation financing. A significant push for international climate finance access for vulnerable and poor countries is particularly important given these countries' high costs of financing, debt stress and the impacts of ongoing climate change (*high confidence*). {15.2, 15.3.2.3, 15.5.2, 15.6.1, 15.6.7}

**Ambitious global climate policy coordination and stepped-up (public) climate financing over the next decade (2021–2030) can help address macroeconomic uncertainty and alleviate developing countries' debt burden post-COVID-19. It can also help redirect capital markets and overcome challenges relating to the need for parallel investments in mitigation and the up-front risks that deter economically sound low carbon projects. (*high confidence*).** Providing strong climate policy signals helps guide investment decisions. Credible and clear signalling by governments and the international community reduce uncertainty for financial decision-makers and help reduce transition risk. In addition to indirect and direct subsidies, the public sector's role in addressing market failures, barriers, provision of information, and risk sharing (equity, various forms of public guarantees) can encourage the efficient mobilisation of private sector finance (*high confidence*). {15.2, 15.6.1, 15.6.2}

The mutual benefits of coordinated support for climate mitigation and adaptation in the next decade for both developed and developing regions could potentially be very high in the post-COVID-19 era. Climate compatible stimulus packages could significantly reduce the macro-financial uncertainty generated by the pandemic and increase the sustainability of the world economic recovery. {15.2, 15.3.2.3, 15.5.2, 15.6.1, 15.6.7}

Political leadership and intervention remain central to addressing uncertainty as a fundamental barrier for a redirection of financial flows. Existing policy misalignments – for example in fossil fuel subsidies – undermine the credibility of public commitments, reduce perceived transition risks and limit financial sector action (*high confidence*). {15.2, 15.3.3, 15.6.1, 15.6.2, 15.6.3}

**Innovative financing approaches could help reduce the systemic underpricing of climate risk in markets and foster demand for Paris-aligned investment opportunities. Approaches include de-risking investments, robust 'green' labelling and disclosure schemes, in addition to a regulatory focus on transparency and reforming international monetary system financial sector regulations (*medium confidence*).** Markets for green bonds, ESG (environmental, social, and governance), and sustainable finance products have grown significantly since the Fifth Assessment Report of the Intergovernmental Panel on Climate Change (IPCC AR5) and the landscape continues to evolve. Underpinning this evolution is investors' preference for scalable and identifiable low-carbon investment opportunities. These relatively new labelled financial products will help by allowing a smooth integration into existing asset allocation models (*high confidence*). Markets for green bonds, ESG (environmental, social, and governance), and sustainable finance products have also increased significantly since AR5, but challenges nevertheless remain, in particular there are concerns about 'greenwashing' and the limited application of these markets to developing countries. New business models (e.g., pay-as-you-go) can facilitate the aggregation of small-scale financing needs and provide scalable investment opportunities with more attractive risk-return profiles. Support and guidance for enhancing transparency can promote capital markets' climate financing by providing quality information to price climate risks and opportunities. Examples include Sustainable Development Goals (SDG) and environmental, social and governance (ESG) disclosure, scenario analysis and climate risk assessments, including the Task Force on Climate-Related Financial Disclosures (TCFD). The outcome of these market-correcting approaches on capital flows cannot be taken for granted, however, without appropriate fiscal, monetary and financial policies. Mitigation policies will be required to enhance the risk-weighted return of low-emission and climate-resilient options, and – supported by progress in transparent and scientifically based projects' assessment methods – to accelerate the emergence and support for financial products based on real projects, such as green bonds, and phase out fossil fuel subsidies. Greater public-private cooperation can also encourage the private sector to increase and broaden investments, within a context of safeguards and standards, and this can be integrated into national climate change policies and plans. {15.1, 15.2.4, 15.3.1, 15.3.2, 15.3.3, 15.5.2, 15.6.1, 15.6.2, 15.6.6, 15.6.7, 15.6.8}.

**The following policy options can have important long-term catalytic benefits (*high confidence*).** (i) Stepped-up both the quantum and composition of financial, technical support and partnership in low-income and vulnerable countries alongside low-carbon energy access in low-income countries, such as in sub-Saharan Africa, which currently receives less than 5% of global climate financing flows; (ii) continued strong role of international and national financial institutions, including multilateral, especially location-based regional, and national development banks; (iii) de-risking cross-border investments in low-carbon infrastructure, development of local green bond markets, and the alignment of climate and non-climate policies, including direct and indirect supports on fossil fuels, consistent with the climate goals; (iv) lowering financing costs including transaction costs and addressing risks through funds and risk-sharing mechanisms for under-served groups; (v) accelerated finance for nature-based solutions, including mitigation in the forest sector (REDD+), and climate-responsive social protection; (vi) improved financing instruments for loss and damage events, including risk-pooling-transfer-sharing for climate risk insurance; (vii) economic instruments, such as phasing in carbon pricing and phasing out fossil fuel subsidies in a way that addresses equity and access; and (viii) gender-responsive and women-empowered programmes. {15.2.3, 15.2.4, 15.3.1, 15.3.2.2, 15.3.3, 15.4.1, 15.4.2, 15.4.3, 15.5.2, 15.6, 15.6.2, 15.6.4, 15.6.5, 15.6.6, 15.6.7, 15.6.8.2}

15

## 15.1 Climate Finance – Key Concepts and Scope

Finance for climate action (or climate finance), environmental finance (which also covers other environmental priorities such as water, air pollution and biodiversity), and sustainable finance (which encompasses issues relating to socio-economic impacts, poverty alleviation and empowerment) are interrelated rather than mutually exclusive concepts (UNEP Inquiry 2016a; ICMA 2020a). Their combination is needed to align mitigation investments with multiple SDGs, and at a minimum, minimise the conflicts between climate targets and SDGs not being targeted. From a climate policy perspective, climate finance refers to finance 'whose expected effect is to reduce net GHG emissions and/or enhance resilience to the impacts of climate variability and projected climate change' (UNFCCC 2018a). However, as pinpointed in the AR5, significant room for interpretation and context-specific considerations remains. Further, such definition needs to be put in perspective with the expectations of investors and financiers (see Box 15.2).

Specifying the scope of climate finance requires defining two terms: what qualifies as 'finance' and as 'climate' respectively. In terms of what type of finance to consider, options include considering investments or total costs (Box 15.1), stocks or flows, gross or net (the latter taking into account reflows and/or depreciation), and domestic or cross-border, public or private (Box 15.2). In terms of what may be considered as 'climate', a key difference relates to measuring climate-specific finance (only accounts for the portion of finance resulting in climate benefits) or climate-related finance (captures total project costs and aims to measure the mainstreaming of climate considerations). One should even consider the investments

### Box 15.1 | Core Terms

This box defines some core terms used in this chapter as well as in other chapters addressing finance issues: cost, investment, financing, public and private. The chapter makes broad use of the term *finance* to refer to all types of transactions involving monetary amounts. It avoids the use of the terms *funds* and *funding* to the extent possible, which should otherwise be understood as synonyms for *money* and *money provided*.

**Cost, investment and financing: different but intertwined concepts.** *Cost* encompasses capital expenditures (CAPEX or upfront investment value leveraged over the lifetime of a project) operating and maintenance expenditures (OPEX), as well as financing costs. Note that some projects e.g., related to technical assistance may only involve OPEX (e.g., staff costs) but no CAPEX, or may not incur direct financing costs (e.g., if fully financed via own funds and grants).

*Investment*, in an economic sense, is the purchase of (or CAPEX for) a physical asset (notably infrastructure or equipment) or intangible asset (e.g., patents, IT solutions) not consumed immediately but used over time. For financial investors, physical and intangible assets take the form of financial assets such as bonds or stocks which are expected to provide income or be sold at a higher price later. In practice, investment decisions are motivated by a calculation of risk-weighted expected returns that takes into account all expected costs, as well as the different types of risks, discussed in Section 15.6.1, that may impact the returns of the investment and even turn them into losses.

*Incremental cost* (or *investment*) accounts for the difference between the cost (or investment value) of a climate project compared to the cost (or investment value) of a counterfactual reference project (or investment). In cases where climate projects and investments are more cost effective than the counterfactual, the incremental cost will be negative.

Financing refers to the process of securing the money needed to cover an investment or project cost. Financing can rely on debt (e.g., through bond issuance or loan subscription), equity issuances (listed or unlisted shares), own funds (typically savings or auto-financing through retained earnings), as well as on grants and subsidies

**Public and private: statistical standard and grey zones.** International statistics classify economic actors as pertaining to the public or private sectors. Households always qualify as private and governmental bodies and agencies as public. Criteria are needed for other types of actors such as enterprises and financial institutions. Most statistics rely on the majority ownership and control principle. This is the case for the Balance of Payment, which records transactions between residents of a country and the rest of the world (IMF 2009).

Such a strict boundary between public and private sectors may not always be suitable for mapping and assessing investment and financing activities. On the one hand, some publicly owned entities may have a mandate to operate on a fully- or semi-commercial basis, for example state-owned enterprises, commercial banks, and pension funds, as well as sovereign wealth funds. On the other hand, some privately owned or controlled entities can pursue not-for-profit objectives, e.g., philanthropies and charities. The present chapter considers these nuances to the extent made possible by available data and information.

decided for reasons unrelated with climate objectives but which contribute to these objectives (hydroelectricity, rail transportation).

In many cases, the scope of what may be considered as 'climate finance' will also depend on the context of implementation such as priorities and activities listed in countries' Nationally Determined Contributions (NDCs) under the Paris Agreement (UNFCCC 2019a) as well as national development plans more broadly targeting the achievement of SDGs. Hence, rather than opposing the different options listed above, the choice of one or the other depends on the desired scope of measurement, which in turn depends on the policy objective being pursued. The increasingly diverse initiatives and body of grey literature address a range of different information needs. They provide analyses at the levels of domestic finance flows (e.g., UNDP 2015; Hainaut and Cochran 2018), international flows (e.g. OECD 2016; AfDB et al. 2018), global flows (UNFCCC 2018a; Buchner et al. 2019), the financial system (e.g., UNEP Inquiry 2016a) or specific financial instruments such as bonds (e.g., CBI 2018). Common frameworks, reporting transparency are, however, necessary in order to identify overlaps, commonalities and differences between these different measurements in terms of scope and underlying definitions. In that regard, the developments of national and international taxonomies, definitions and standards can help, as further discussed in Section 15.6, and Chapter 17 in AR6 WGII report.

Beyond the need to scale up levels of climate finance, the Paris Agreement provides a broad policy environment and momentum for a more systemic and transformational change in investment and financing strategies and patterns. Article 2.1c, which calls for 'making finance flows consistent with a pathway towards low greenhouse gas emissions and climate-resilient development', positions finance as one of the Agreement's three overarching goals (UNFCCC 2015). This formulation is a recognition that the mitigation and resilience goals cannot be achieved without finance, both in the real economy and in the financial system, being made consistent with these goals (Zamarioli et al. 2021). It has in turn contributed to the development of the concept of alignment (with the Paris Agreement) used in the financial sector (banks, institutional investors), businesses, and public institutions (development banks, public budgets). As a result, since AR5, in addition to measuring and analysing climate finance, an increasing focus has been placed on assessing the consistency or alignment, as well as respectively the inconsistency or misalignment, of finance with climate policy objectives, as for instance illustrated by the multilateral development banks' joint framework for aligning their activities with the goals of the Paris Agreements (MDBs 2018).

Assessing climate consistency or alignment implies looking at all investment and financing activities, whether they target, contribute to, undermine or have no particular impact on climate objectives. This all-encompassing scope notably includes remaining investments and financing for high-GHG emission activities that may be incompatible with remaining carbon budgets, but also activities that may play a transition role in climate mitigation pathways and scenarios (Section 15.3.2.3). As a result, any meaningful assessment of progress requires the use of different shades to assess activities based on their negative, neutral ('do no harm') or positive contributions, (e.g., CICERO 2015; Cochran and Pauthier 2019; Natixis 2019). Doing so in practice requires the development of robust definitions, assessment methods and metrics, an area of work and research that remained under development at the time of writing. A range

## Box 15.2 | International Climate Finance Architecture

International climate finance can flow through different bilateral, multilateral, and other channels, involving a range of different types of institutions both public (official) and private (commercial) with different mandates and focuses. In practice, the architecture of international public climate finance is rapidly evolving, with the creation by traditional donors of new public sources and channels over the years (Watson and Schalatek 2019), as well as emergence of new providers of development co-operation, both bilateral (Benn and Luijkx 2017) and multilateral (e.g., Asian Infrastructure Investment Bank), as well as of non-governmental actors such as philanthropies (OECD 2018a).

The operationalisation of the Green Climate Fund (GCF), which channels the majority of its funds via accredited entities, has notably attracted particular attention since AR5. Section 14.3.2 (in Chapter 14) provides a further assessment of progress and challenges of financial mechanisms under the United Nations Framework Convention on Climate Change (UNFCCC), such as the GCF, the Global Environment Facility (GEF) and the Adaptation Fund (AF).

The multiplication of sources and channels of international climate finance can help address growing climate-related needs, and partly results from increased decentralisation as well financial innovation, which in turn can increase the effectiveness of finance provided. There is, however, also evidence that increased complexity implies transaction costs (Brunner and Enting 2014), in part due to bureaucracy and intra-governmental factors (Peterson and Skovgaard 2019), which constitutes a barrier to low-carbon projects and are often not accounted for in assessments of international climate finance. On the ground, activities by international providers operating in the same countries may overlap, with sub-optimal coordination and hence duplication of efforts, both on the bilateral and multilateral sides (Ahluwalia et al. 2016; Gallagher et al. 2018; Humphrey and Michaelowa 2019), as well as risks of fragmentation of efforts (Watson and Schalatek 2020) which slows down coordination with international providers, national development banks and other domestic institutions.

of financial sector coalitions and civil society organisations as well as commercial services providers to the financial industry have developed frameworks, approaches and metrics, mainly focusing on investment portfolios (Institut Louis Bachelier et al.. 2020; IIGCC 2021; TCFD Portfolio Alignment Team 2021; UN-Convened Net-Zero Asset Owner Alliance 2021), and, to a lesser extent for real economy investments (Micale et al. 2020; Jachnik and Dobrinevski 2021).

**Key findings from AR5 and other IPCC publications.** For the first time the IPCC in AR5 (Clarke et al. 2014) elaborated on the role of finance in a dedicated chapter. In the following year, the Paris Agreement (UNFCCC 2015) recognised the transformative role of finance, as a means to achieving climate outcomes, and the need to align financial flows with the long-term global goals even as implementation issues were left unresolved (Bodle and Noens 2018). AR5 noted the absence of a clear definition and measurement of climate finance flows, a difficulty that continues (Weikmans and Roberts 2019) (Sections 15.2 and 15.3). The approach taken in AR5 was to report ranges of available information on climate finance flows from diverse sources, using a broad definition of climate finance, as in the Biennial Assessments in 2014 and again in 2018 (UNFCCC 2014a, 2018a) of the Standing Committee under the UNFCCC: Climate finance is taken to refer to local, national or transnational financing – drawn from public, private and alternative sources of financing – that seeks to support mitigation and adaptation actions that address climate change (UNFCCC 2014b). For this chapter, while the focus is primarily on mitigation, adaptation, resilience and loss and damage financing needs cannot be entirely separated because of structural relationships, synergies, trade-offs and policy coherence requirements between these sub-categories of climate finance (Box 15.1).

The AR5 concluded that published assessments of financial flows whose expected effect was to reduce net greenhouse gas (GHG) emissions and/or to enhance resilience to climate change aggregated USD343–385 billion[4] $yr^{-1}$ globally between 2010 and 2012 (*medium confidence*). Most (95% of total) went towards mitigation, which was nevertheless underfinanced and adaptation even more so. Measurement of progress towards the commitment by developed countries to provide USD100 billion $yr^{-1}$ by 2020 to developing countries, for both mitigation and adaptation (Bhattacharya et al. 2020) – a narrower goal than overall levels of climate finance – continued to be a challenge, given the lack of clear definition of such finance, although there remain divergent perspectives (Section 15.2.4). As against these flows, annual need for global aggregate mitigation finance between 2020 and 2030 was cited briefly in the AR5 to be about USD635 billion (mean annual), both public and private, implying that the reported 'gap' in mitigation financing of estimated flows during 2010 to 2012 was slightly under one-half of that required (IPCC 2014).

More recent published data from the Biennial Assessments (UNFCCC 2018a) and the Special Report on Global Warming of 1.5°C (IPCC 2018) have revised upwards the needs of financing between 2020 and 2030 to 2035 to contain global temperature rise to below

### Box 15.3 | Mitigation, Adaptation and Other Related Climate Finance Merit Joint Examination

Mitigation finance deals with investments that aim to reduce global carbon emissions, while adaptation finance deals with the consequences of climate change (Lindenberg and Pauw 2013). Mitigation affects the scale of adaptation needs and adaptation may have strong synergies and co-benefits as well as trade-offs with mitigation (Grafakos et al. 2019). If mitigation investments are inadequate to reducing global warming (as in the last decade) with asymmetric adverse impacts in lower latitudes and low-lying geographies, the scale of adaptation investments has to rise and the benefits of stronger adaptation responses may be high (Markandya and González-Eguino 2019). If adaptation investments build greater resilience, they might even moderate mitigation financing costs. Similar policy coherence considerations apply to disaster risk reduction financing, the scale of which depends on success with both adaptation and mitigation (Mysiak et al. 2018). The same financial actors, especially governments and the private sector, decide at any given time on their relative allocations of available financing for mitigation, adaptation and disaster-risk reduction from a constrained common pool of resources. The trade-offs and substitutability between closely-linked alternative uses of funds, therefore, make it essential for a simultaneous assessment of needs – as in parts of this chapter. Climate finance versus the financing of other Sustainable Development Goals (SDGs) faces a similar issue. A key agreement was that climate financing should be 'new and additional' and not at the cost of SDGs. Resources prioritising climate at the cost of non-climate development finance increase the vulnerability of a population for any given level of climate shocks, and additionality of climate financing is thus essential (Brown et al. 2010). Policy coherence is also the reason why mitigation finance cannot be separated from consideration of spending and subsidies on fossil fuels. Climate change may additionally cause the breaching of physical and social adaptation limits, resulting in climate-related residual risks (i.e., potential impacts after all feasible mitigation, adaptation, and disaster risk reduction measures have been implemented) (Mechler et al. 2020). Because these residual losses and damages from climate-related risks are related to overall mitigation and adaptation efforts, the magnitude of potential impacts is related to the overall quantum of mitigation, adaptation, and disaster risk reduction finance available (Frame et al. 2020). All categories of climate finance thus need to be considered together in discussions around climate finance.

---

4 In the chapter, USD units are used as reported in the original sources in general. Some monetary quantities have been adjusted selectively for achieving comparability by deflating the values to constant $USD_{2015}$. In such cases, the unit is explicitly expressed as $USD_{2015}$.

2°C and 1.5°C respectively by 2100: USD1.7 trillion $yr^{-1}$ (mean) in the Biennial Assessment 2018 for the former, and for the latter, USD2.4 trillion $yr^{-1}$ (mean) for the energy sector alone (and three times higher if transport and other sectors were to be included). The resulting estimated gaps in annual mitigation financing during 2014 to 2017, using reporting of climate financing from published sources, was about 67% for 2015, and 76% for the energy sector alone in 2017 (*medium confidence*), and greater if other sectors were to be included. While the annual reported flows of climate financing showed some moderate progress (Section 15.3), from earlier USD364 billion (mean 2010/2011) to about USD600 billion (mean 2017/2020), with a slowing in the most recent period 2014 to 2017, the gap in financing was reported to have widened considerably (Sections 15.4 and 15.5). In the context of policy coherence, it is also important to note that reported annual investments going into the fossil fuel sectors, oil and gas upstream and coal mining, during the same period were about the same size as global climate finance, although the absence of alternative financing and access to low-carbon energy is a complicating factor.

Adaptation financing needs, meanwhile, were rising rapidly. The Adaption Gap Report 2020 (UNEP 2021) reported that the current efforts are insufficient to narrow the adaptation finance gap, and additional adaptation finance is necessary, particularly in developing countries. The gap is expected to be aggravated by COVID-19 (*high confidence*). It reaffirmed earlier assessments that by 2030 (2050) the estimated costs of adaptation ranges between USD140 and 300 billion $yr^{-1}$ (USD280 and 500 billion $yr^{-1}$). Against this, the reported actual global public finance flows for adaptation in 2019/2020 were estimated at 46 billion (Naran et al. 2021). The costs of climate disasters meanwhile continued to rise, affecting low-income developing countries the most. Climate natural disasters – not all necessarily attributable to climate change – caused some USD300 billion $yr^{-1}$ economic losses and well-being losses of about USD520 billion $yr^{-1}$ (Hallegatte et al. 2017).

## 15.2 Background Considerations

The institutions under climate finance in this chapter refer to the set of financial actors, instruments and markets that are recognised to play a key role in financing decisions on climate mitigation and adaptation. For a definition of climate financial stock and flows see further Section 15.3 and the Glossary. The issue of climate finance is closely related to the conversation on international cooperation and the question of how cross-border investments can support climate mitigation and adaptation in developing countries. However, the issue is also related to more general questions of how financial institutions, both public and private, can assess climate risks and opportunities from all investments, and what roles states, policymakers, regulators and markets can play in making them more sustainable. In particular, the question of the respective roles of the public and private financial actors has become important in deliberations on climate finance in recent years. The broader macroeconomic context is an important starting point. Four major events and macro trends mark the developments in climate finance in the previous five years and likely developments in the near term.

- First, the 2015 Paris Agreement, with the engagement of the financial sector institutions in the climate agenda, has been followed by a series of related developments in financial regulation in relation to climate change and in particular to the disclosure of climate-related financial risk (*high confidence*) (Section 15.2.1).
- Second, the last five years have been characterised by a series of interconnected 'headwinds' (Section 15.2.2), including rising private and public debt and policy uncertainty which work against the objective of filling the climate investment gap (*high confidence*).
- Third, the 2020 COVID-19 pandemic crisis has put enormous additional strain on the global economy, debt and the availability of finance, which will be longer lasting (Section 15.2.3). At the same time, while it is still too early to draw positive conclusions, this crisis highlights opportunities in terms of political and policy feasibility and behavioural change in respect of realigning climate finance (*medium confidence*).
- Fourth, the sharp rise in global inequality and the effects of the pandemic have brought into renewed sharp focus the need for a Just Transition (Section 15.2.4) and a realignment of climate finance and policies that would be beneficial for a new social compact towards a more sustainable world that addresses energy equity and environmental justice (*high confidence*).

### 15.2.1 Paris Agreement and the Engagement of the Financial Sector in the Climate Agenda

This is the first IPCC Assessment Report chapter on investment and finance since the 2015 Paris Agreement, which represented a landmark event for climate finance because for the first time the key role of aligning financial flows to climate goals was spelled out. Since then, the financial sector has recognised the opportunity and has stepped up to centre-stage in the global policy conversation on climate change. While before the Paris Agreement, only few financial professionals and regulators were acquainted with climate change, today climate change is acknowledged as a strategic priority in most financial institutions. This is a major change in the policy landscape from AR5. However, this does not mean that finance necessarily plays an adequate enabling role for climate investments. On the contrary, the literature shows that without appropriate conditions, finance can represent a barrier to filling the climate investment gap (Hafner et al. 2020). Indeed, despite the enormous acceleration in policy initiatives (e.g., NGFS 2020) and coalitions of the willing in the private sectors, the effect in terms of closing the investment gap identified already in AR5 has been limited (Section 15.5.2).

Financial investors have started to account for climate risk in some contexts but they do so only to a limited extent (Monasterolo and de Angelis 2020; Alessi et al. 2021; Bolton and Kacperczyk 2021) and the reasons for these remain unclear. Two aspects are relevant here. The first is the endogenous nature of climate financial risk and opportunities (with the term 'risk' meaning here the potential for adverse financial impact, whether or not the distribution of losses is known). Academics and practitioners in finance are aware that financial risk can in certain contexts be endogenous, that is, the materialisation of losses is affected by the action of financial players themselves. However, the standard treatment of risk both in financial valuation models and in asset pricing assumes that risk is exogenous. In contrast, endogeneity is a key feature of climate risk because today's perception of climate risk affects climate investment, which in turn affects directly the future risk. This endogeneity leads to the fact that multiple and rather different mitigation scenarios are possible (Chapter 3). Moreover, the likelihood of occurrence of each alternative scenario is very hard to estimate. Further, the assessment of climate-related financial risk requires to combine information related to mitigation scenarios as well as climate impact scenarios, leaving open an important knowledge gap for the next years (Section 15.6.1).

The second aspect is that the multiplicity of equilibria results in a coordination problem whereby the majority of investors wait to move and reallocate their investments until they can follow a clear signal. Despite the initial momentum of the Paris Agreement, for many investors, both public and private, the policy signal seems not strong enough to induce them to align their investment portfolios to climate goals.

Analyses of the dynamics of the low-carbon transition suggest that it does not occur by itself and that it requires a policy signal credible enough in the perception of market players and investors (Battiston et al. 2021b). Credibility could require a policy commitment device (Brunner et al. 2012). The commitment would also need to be large enough (analogous to the 'whatever it takes' statement by the European Central Bank during the 2011–2012 European sovereign crisis (Kalinowski and Chenet 2020)). In principle, public investments in low-carbon infrastructures (or private-public partnerships) as well as regulation could provide credible signals if their magnitude and time horizon are appropriate (past experiences with feed-in-tariffs (FiTs) models across countries provide useful lessons).

### 15.2.2 Macroeconomic Context

Entering 2020, the world already faced large macroeconomic headwinds to meeting the climate finance gap in the near term – barring some globally coordinated action. While an understanding of the disaggregated country-by-country, sector-by-sector, project-by-project, and instrument-by-instrument approach to raising climate finances analysed in the later parts of this chapter remains important, macroeconomic drivers of finance remain crucial in the near term.

Near-term finance financial flows in aggregate often show strong empirically observed cycles over time, especially in terms of macroeconomic and financial cycles. By *near-term*, we mean here the likely cycle over the next five to ten years (2020–2025 and 2020–2030), as influenced by global macroeconomic real business cycles (output, investment and consumption), with periodic asymmetric downside impacts and crises (Gertler and Kiyotaki 2010; Borio 2014; Jordà et al. 2017; Borio et al. 2018). Financial cycles typically have strong co-movements (asset prices, credit growth, interest rates, leverage, risk factors, market fear, macro-prudential and central bank policies) (Coeurdacier and Rey 2013), they have large consequences for all types of financial flows such as equity, bond and banking credit markets, which in turn are likely to impact climate finance flows to all sub-sectors and geographies (with greater expected volatility in more risky and more leveraged regions). This is in contrast to *longer-term trend considerations* (2020–2050) that typically focus the attention on drivers of disaggregated flows of climate finance and policies. The upward trends of the cycles tend to favour speculative bubbles like real estates at the expense of investment in production and infrastructures whereas the asymmetric downsides raise uncertainty and risks for longer-term investments on newer climate technologies, and favour a flight to near-term safety (e.g., lowest risk non-climate short-term treasury investments, highest creditworthy countries, and away from cross-border investments (Section 15.5) – making the challenge of longer-term low-carbon transition more difficult. In this respect, the impact of financial regulation is unclear. On the one hand, it could be argued that the tighter bank regulations under Basel III, combined with an economic environment with higher uncertainty and flatter yield curve, can push banks to retrench from climate finance projects (Blended Finance Taskforce, 2018a), since banks tend to limit loan maturity to five or eight years, while infrastructure projects typically require the amortisation of debt over 15 to 20 years (Arezki et al. 2016). On the other hand, other studies report that stricter capital requirements are not a driving factor for moving away from sustainability projects (CISL and UNEP FI 2014).

Four key aspects of the global macroeconomy, each slightly different, pointed in a cascading fashion towards a deteriorating environment for stepped-up climate financing over the next crucial decade (2020–2030), even before COVID-19. The argument is often made that there is enough climate financing available if the right projects and enabling policy actions ('bankable projects') present themselves (Cuntz et al. 2017; Meltzer 2018). The attention to 'bankability' does not however address access and equity issues (Bayliss and Van Waeyenberge 2018). Some significant gains in climate financing at the sectoral and microeconomic levels were nevertheless happening in specific segments, such as solar energy financing and labelled green bonds (although how much of such labelled financing is incremental to unlabelled financing that might have happened anyway remains uncertain) (Tolliver et al. 2019). Issues of 'labelling' (Cornell 2020) apply even more to ESG (environmental, social and governance) investments, which started to grow rapidly after 2016 (Section 15.6.5). Overall, these increments for climate finance remained, however, small in aggregate relative to the size of the shifts in climate financing required in the coming decade. Annual energy investments in developing regions (other than China) which account for two-thirds of the world population, with least costs of mitigation per tonne of emissions (one-half that in developed regions), and for the bulk of future expected global GHG emissions, saw a 20% decline since 2016, and only a one-fifth share of global clean energy investment, reflecting persistent financing problems and costs of mobilising finance towards clean energy transition, even prior to the pandemic (IEA 2021a). In the words of a macroeconomic institution, 'tangible policy responses to reduce greenhouse gas emissions have been grossly insufficient to date' (IMF 2020a). The reason is in part global macroeconomic headwinds, which show a relative stagnation since 2016 and limited cross-border flows in particular (Yeo 2019).

**Slowing and more unstable GDP growth.** The first headwind was more unstable and slowing GDP growth at individual country levels and in aggregate because of worsening climate change impact events (Donadelli et al. 2019; Kahn et al. 2019). As each warmer year keeps producing more negative impacts – arising from greater and rising variability and intensity of rainfall, floods, droughts, forest fires and storms – the negative consequences have become more macroeconomically significant, and worst for the most climate-vulnerable developing countries (*high confidence*). Paradoxically, while these effects should have raised the social returns and incentives to invest more in future climate mitigation, a standard public policy argument, these macroeconomic shocks may work in the opposite direction for private decisions by raising the financing costs now (Cherif and Hasanov 2018). With some climate tipping points, potentially in the near-term reach (see AR6 WGI Chapter 4) the uncertainty with regard to the economic viability and growth prospects of selected macroeconomically critical sectors increases significantly (AR6 WGII Chapters 8 and 17). Taking account of other behavioural failures, this was creating a barrier for proactive and accelerated mitigation and adaptation action.

**Public finances.** The second headwind was rising public fiscal costs of mitigation and adapting to rising climate shocks affecting many countries, which were negatively impacting public indebtedness and country credit ratings (Cevik and Jalles 2020; Klusak et al. 2021) at a time of growing stresses on public finances and debt (Benali et al. 2018; Kling et al. 2018; Kose et al. 2020) (*high confidence*). Every climate shock and slowing growth puts greater pressures on public finances to offset these impacts. Crucially, the negative consequences were typically greater at the lower end of income distributions everywhere (Acevedo et al. 2018; Aggarwal 2019). As a result, the standard prescription of raising distributionally adverse carbon taxes and reducing fossil fuel subsidies to raise resources faced political pushback in several countries (Copland 2020; Green 2021), and low rates elsewhere. Reduced taxes on capital, by contrast, was viewed as a way to improve growth (Bhattarai et al. 2018; Font et al. 2018), and working against broader fiscal action. Progress with carbon pricing remained modest across 44 OECD and G20 countries, with 55–70% of all carbon emissions from energy use entirely unpriced as of 2018 (OECD 2021a). Climate-vulnerable countries meanwhile faced sharply rising cost of sovereign debt. Buhr et al. (2018) calculate the additional financing costs of Climate Vulnerable Forum countries of USD40 billion[5] on government debt over the past 10 years and USD62 billion for the next 10 years. Including private financing cost, the amount increases to USD146–168 billion over the next decade.

**Credit risks.** The third headwind is rising financial and insurance sector risks and stresses (distinct from real 'physical' climate risks above) arising from the impacts of climate change, and systematically affecting both national and international financial institutions and raising their credit risks (*high confidence*) (Dafermos et al. 2018; Rudebusch 2019; Battiston et al. 2021a). Central banks are beginning to take notice (Carney 2019; NGFS 2019). It is also the case that, even if at greater risk from stranded assets in the future, the large-scale financing of new fossil fuel projects by large global financial institutions rose significantly since 2016, because of perceived lower private risks and higher private returns in these investments and other factors than in alternative but perceived more risky low-carbon investments.

**Global growth.** The fourth headwind entering 2020 was the sharply slowing global macroeconomic growth, and prospects for near-term recession (which occurred in the pandemic). During global real and financial cycle downturns (Jordà et al. 2019), the perception of general financial risk rises, causing financial institutions and savers to reallocate their financing to risk-free global assets (*high confidence*). This 'flight to safety' was evident even before the recent pandemic, marked by an extraordinary tripling of financial assets to about USD16.5 trillion in negative-interest earning 'safer' assets in 2019 in world debt markets – enough to have nearly closed the total financing gap in climate finance over a decade.

### 15.2.3 Impact of COVID-19 Pandemic

The macroeconomic headwinds have worsened dramatically with the onset of COVID-19. Almost two years after the pandemic started, it is still too uncertain and early to conclude impacts of the pandemic until 2025–2030, especially as they affect climate finance. Multiple waves of the pandemic, new virus mutations, accumulating human toll, and growing vaccine coverage but vastly differing access across developed versus developing regions, are evident. They are causing divergent impacts across sectors and countries, which combined with the divergent ability of countries and regions to mount sufficient fiscal and monetary policy actions imply continued high uncertainty on the economic recovery paths from the crisis. The situation remains more precarious in middle- and low-income developing countries (IMF 2021a). While recovery is happening, the job losses have been large, poverty rates have climbed, public health systems are suffering long-term consequences, education gains have been set back, public debt levels are higher (5–10% of GDP higher), financial institutions have come under longer-term stress, a larger number of developing countries are facing debt distress, and many key high-contact sectors, such as tourism and trade, will take time to recover (Eichengreen et al. 2021). The implication is negative headwinds for climate finance with public attention focused on pandemic relief and recovery and limited (and divergent) fiscal headroom for a low-carbon transition, with considerable uncertainties ahead (Hepburn et al. 2020b; Maffettone and Oldani 2020; Steffen et al. 2020).

The larger and still open public policy choice question that COVID-19 now raises is whether there is room for public policy globally and in respect of their individual economies to integrate climate more centrally to their growth, jobs and sustainable development strategies worldwide for ecological and economic survival. The outcomes will depend on the robustness of recovery from the pandemic, and the still evolving public policy responses to the climate agenda in the recovery process. Private equity and asset markets have recovered

[5] In the chapter, USD units are used as reported in the original sources in general. Some monetary quantities have been adjusted selectively for achieving comparability by deflating the values to constant $USD_{2015}$. In such cases, the unit is explicitly expressed as $USD_{2015}$.

surprisingly rapidly during the pandemic (in response to the massive fiscal and central bank actions generating large excess savings with very low or negative yields boosting stock markets). On public spending, some early studies suggest that the immediate economic recovery packages were falling well short of being sufficiently climate sustainable (Gosens and Jotzo 2020; Kuzemko et al. 2020; O'Callaghan 2021) but several governments have also announced intentions to spend more on a green recovery, 'build back better' and Just Transition efforts (Section 15.2.4), although outcomes remain highly uncertain (Lehmann et al. 2021; Markandya et al. 2021).

An important immediate finding from the COVID-19 crisis was that the slowdown in economic activity is illustrating some of these choices: immediately after the onset, more costly and carbon-intensive coal use for energy use tumbled in major countries such as China and the USA, while the forced 'stay-at-home' policies adopted around the major economies of the world led to a –30–35% decline in individual country GDP, and was in turn associated with a decrease in daily global $CO_2$ emissions by –26% at their peak in individual countries, and –17% globally (–11 to –25% for $\pm 1\sigma$) by early April 2020 compared with the mean 2019 levels, with just under half coming from changes in surface transport, city congestion and country mobility (Le Quéré et al. 2020). Along with the carbon emissions drop was a dramatic improvement in other parameters such as clean air quality. Moreover, longer-term behavioural impacts are also possible: a dramatic acceleration of digital technologies in communications, travel, retail trade and transport. The question however is whether the world might revert to the earlier carbon-intensive path of recovery, or to a different future, and the choice of policies in shaping this future. Studies generally suggest that the gains from long-term impacts of the pandemic on future global warming will be limited and depend more on the nature of public policy actions and long-term commitments by countries to raise their ambitions, not just on climate but on sustainable development broadly (Barbier 2020; Barbier and Burgess 2020; Forster et al. 2020; Gillingham et al. 2020; Reilly et al. 2021). The positive lesson is clear: opportunities exist for accelerating structural change, and for a re-orientation of economic activity modes to a low-carbon use strategy in areas such as coal use in energy consumption and surface transport, city congestion and in-country mobility, for which lower-cost alternatives exist and offer potentially dramatic gains (Hepburn et al. 2020b).

A new consensus and compact towards such a structural change and economic stimulus instruments may therefore need to be redrawn worldwide, where an accelerated low-carbon transition is a priority; and accelerated climate finance to spur these investments may gain by becoming fully and rapidly integrated with near-term economic stimulus, growth and macroeconomic strategies for governments, central banks, and private financial systems alike. If that were to happen, COVID-19 may well be a turning point for sustainable climate policy and financing. Absent that, a return to 'business-as-usual' modes will mean a likely down-cycle in climate financing and investments in the near term.

Expectations that the recovery package stimulus will increase economic activity rely on the assumption that increased credit investment will have a positive effect on demand, the so-called demand-led policy (Mercure et al. 2019). The argument for a green recovery also draws on the experience from the post-global financial crisis in 2008–2009 recovery, in which large economies such as China, South Korea, the USA and the EU observed that green investments propelled the development of new industrial sectors. Noticeably, this had a positive net effect on job creation when compared to the investment in traditional infrastructure (UKERC 2014; Vona et al. 2018; Jaeger et al. 2020). For a more in-depth discussion on macroeconomic-finance possible response see Section 15.6.3. Here, we conclude with the options for reviving a better globally coordinated macroeconomic climate action. The options are some combinations of five possible elements:

1. Reaffirmation of a strong financial agenda in future UNFCCC Conference of Parties meetings, and a new collective finance target, which will need to be undertaken by 2025. Given that the shortfalls in financing are likely to be acute for developing regions and especially the more debt-stressed and vulnerable (Dibley et al. 2021; Elkhishin and Mohieldin 2021; Laskaridis 2021; Umar et al. 2021), developed countries may wish to step up their collective support (Resano and Gallego 2021). One possibility is to expedite the new Special Drawing Rights (SDR) issuance allocation rules for the USD650 billion recently (2021) approved, most of which will go to increase the reserves of G7 and other high-income countries unless voluntarily reallocated towards the needs of the most vulnerable low-income countries, raising resources potentially 'larger than the Marshall Plan in today's money' (IMF 2021b; Jensen 2021; Obstfeld and Truman 2021), with decisions to be taken. Ameli et al. (2021a) note the climate investment trap of the current high cost of finance that effectively lowers green electricity production possibilities in Africa for a cost optimal pathway. Other initiatives could also include G7 and G20 governments (especially with the lead taken by the developed members for cross-border support to avoid over-burdening public resources in developing countries) running coordinated fiscal deficits to accelerate the financing of low carbon investments ('green fiscal stimulus').
2. Introducing new actions, including regulatory, to take some of the risks off the table from institutional financial players investing in climate mitigation investment and insurance. This could include the provision of larger sovereign guarantees to such private finance, primarily from developed countries but jointly with developing countries to create a level playing field (Dafermos et al. 2021) backed by explicit and transparent recognition of the 'social value of mitigation actions' or SVMAs, as fiscally superior (because of bigger 'multipliers' of such fiscal action to catalyse private investment than direct public investment) and the bigger social value of such investments (Article 108, UNFCCC) (Hourcade et al. 2018; Krogstrup and Oman 2019).
3. Facilitating and incentivising much larger flows of cross-border climate financing which is especially crucial for such investments to happen in developing regions, where as much as two-thirds of collective investment may need to happen (IEA 2021a), and where the role of multilateral, regional and global institutions such as the International Monetary Fund (IMF) (including the expansion in availability of climate SDRs referred to earlier) could be important.

4. Global central banks acting in coordination to include climate finance as an intrinsic part of their monetary policy and stimulus (Carney 2019; Jordà et al. 2019; Hilmi et al. 2021; Schoenmaker 2021; Svartzman et al. 2021).
5. An acceleration of Just Transition initiatives, outlined further below (Section 15.2.4).

### 15.2.4 Climate Finance and Just Transition

Climate finance in support of a Just Transition is likely to be a key to a successful low-carbon transition globally (*high confidence*). Ambitious global climate agreements are likely to work far better by maximising cooperative arrangements (IPCC 2018; Gazzotti et al. 2021) with greater financing support from developed to developing regions in recognition of 'common but differentiated responsibilities and respective capabilities' and a greater ethical sense of climate justice (Khan et al. 2020; Sardo 2020; Warner 2020; Pearson et al. 2021). While Just Transition issues apply within developed countries as well (see later discussion), these are of relatively second-order significance to addressing climate justice issues between richer and poorer countries – given the scale of financing and existing social safety nets in the former and their absence in the latter. For example, over the past three decades drought in Africa has caused more climate-related mortality than all climate-related events combined from the rest of the world (Warner 2020). These issues can however serve both as a bridge and a barrier to greater cooperation on climate change. The key is to build greater mutual trust with clearer commitments and well-structured key decisions and instruments (Sardo 2020; Pearson et al. 2021).

The Just Transition discussion has picked up steam. It was explicitly recognised in the Paris Agreement and the 2018 Just Transition Declaration signed by 53 countries at COP24, which 'recognised the need to factor in the needs of workers and communities to build public support for a rapid shift to a zero-carbon economy.' Originally proposed by global trade unions in the 1980s, the recent discourse has become broader. It has coalesced into a more inclusive process to reduce inequality across all three areas of energy, environment and climate (McCauley and Heffron 2018; Bainton et al. 2021). It seeks accelerated public policy support to ensure environmental sustainability, decent work, social inclusion and poverty eradiation (Burrow 2017), widely shared benefits, and protection of indigenous rights, and livelihoods of communities and workers who stand to lose (including workers in fossil fuel sectors such as coal and oil and gas) (UNFCCC 2018b; EBRD 2020; Jenkins et al. 2020). Because the process involves 'climate justice' and equity within and across generations, it involves difficult political trade-offs (Newell and Mulvaney 2013). The implications for a Just Transition in climate finance are clear: expanding equitable and greater access to climate finance for vulnerable countries, communities and sectors, not just for the most profitable private investment opportunities, and a larger role for public finance in fulfilling existing finance commitments (Bracking and Leffel 2021; Kuhl 2021; Long 2021; Roberts et al. 2021).

Large shocks such as pandemics, and slow-growing ones such as climate, are typically known to worsen inequality (IMF and World Bank 2020). Evidence from 133 countries between 2001–2018 suggests that such shocks can cause social unrest, and migration pressures, especially when starting inequality is high and social transfers are low (Saadi Sedik and Xu 2020). Additionally, climate policies are more politically difficult to implement when the setting is one of high inequality but much less politically costly where incomes are more evenly distributed with stronger social safety nets (Furceri et al. 2021). A redrawn social compact incorporating climate (Beck et al. 2021) that would adopt redistributive taxes and lower carbon consumption, and strengthen state capacity to deliver safety nets, health and education with accelerated climate and environmental sustainability within and across countries, is increasingly recognised as important. Countries, regions and coordination bodies of the larger countries (G7, G20) have already begun such a shift to financing of a Just Transition, but primarily focused on the developed countries, although gaps remain (Krawchenko and Gordon 2021).

Such a redrawing of a social compact has happened significantly in the past, for example, after the 1860s 'gilded age of capital' with the enlargement of the franchise in democratisation waves in Europe and the Americas (Dasgupta and Ziblatt 2015, 2016). Not only was social conflict avoided but growth outcomes became more equitable and faster. Similarly, comprehensive modern social safety nets and progressive taxation, which started in the Great Depression and was extended in the post-war period, had both a positive pro-growth and lower inequality effects (Brida et al. 2020).

There are three levels at which policy attention on climate financing now may need to be focused. The first is the need to address the global equity issues in climate finance in a more carefully constructed globally cooperative public policy approach. The second is to address issues appropriately with enhanced support, at the national level. The third is to work it down further, to addressing needs at local community levels. Because private investors and financing mostly deal with allocation to climate finance at a global portfolio level, then to allocation by countries, and finally to individual projects, the challenge for them is to refocus attention to Just Transition issues at the country level, but also globally as well as locally (in other words, at all three levels).

Climate finance will likely face greater challenges in the post-pandemic context (Hanna et al. 2020; Henry et al. 2020). Evidence from the COVID-19 pandemic suggests that those in greatest vulnerability often had the least access to human, physical, and financial resources (Ruger and Horton 2020). It has also left in in its wake divergent prospects for economic recovery, with rising constraints on credit ratings and costly debt burden in many developing countries contrasted with the exceptionally low interest rate settings in developed economies driving the limited fiscal space in the former groups (Benmelech and Tzur-Ilan 2020). Similarly, monetary policies are likely to be much tighter in developing countries in part structurally because of the absence of 'exceptional privilege' of global reserve currencies in developed economies.

The result is a divergence in recovery prospects in the aftermath of the pandemic, with output losses (compared to potential) set to worsen in developing economies (excluding China) as compared to developed countries (IMF 2020b). In these circumstances, a coordinated and cooperative approach, instead of unilateralism, might work better (McKibbin and Vines 2020). In the case of climate, simulations clearly suggest the need and advantages of better coordinated climate action with stepped-up Paris Agreement envisaged transfers (IMF 2020b). Several options in international climate finance arrangements to support a Just Transition are both available and urgent.

As a first priority, measures might need to accelerate a mix of equitable financial grants, low-interest loans, guarantees and workable business models access across countries and borders, from developed countries to low-income countries. A big push on low-carbon energy access globally, especially in large low income regions such as Africa, with accelerated financial transfers, makes sense (Boamah 2020). For about one billion people globally at the base of the pyramid without access to modern low-carbon energy access, such an action, with enormous immediate leap-frogging potential, would be a key pathway to achieve the SDGs, ensure that high-carbon energy use is avoided, such as the burning of biomass and forests for charcoal, and improve air quality and public health, especially women's health (van der Zwaan et al. 2018; Nathwani and Kammen 2019; Dalla Longa and van der Zwaan 2021; Michaelowa et al. 2021; Osabuohien et al. 2021).

A second priority is to accelerate the implementation of the USD100 billion a year (and likely more, given growing financing gaps) in climate finance commitments expressed in the Copenhagen Agreement Accord (and reiterated since) from developed to developing countries, and to build greater confidence by agreeing rapidly on key definitions. Shifting to a grant equivalent net flows definition of climate finance, which is now universally accepted for all other aid flows by all parties since 2014 and which took effect since 2019 on every other public international good finance provision (under the SDGs), with the sole exception of climate finance, would resolve many uncertainties: the disbursement of climate finance flows on a grant equivalent basis that is comparable across institutions, instruments and countries, and measurement with greater accuracy about the effective transfer of resources. The journey to get to a clear and precise definition of net official overseas development assistance (ODA) took time. The original proposal was first initiated in the 1960s (Pincus 1963) but it was not till multilateral development banks (MDBs) and others laid out the compelling reasons why (Chang et al. 1998) that this was accomplished: especially to resolve decades of confusion and inconsistency between different types of financial flows and hence the perennial measurement problems and 'the compromise between political expediency and statistical reality' (Bulow and Rogoff 2005; Hynes and Scott 2013; Scott 2015, 2017).

A third related and increasingly crucial priority is to expedite the operational definition of blended finance and promote the use of public guarantee instruments. Private flows to accelerate the low-carbon transition in developing countries would benefit enormously, by gaining clearer access to public international funds and support defined on a grant equivalent basis, provided development and climate finance operational definitions and procedures were improved on an urgent basis (Blended Finance Taskforce 2018a; OECD-DAC 2021). When blended and supported by public finance and policy, the grant equivalency measure can easily and more accurately measure the value and benefit of blended public and private finance by comparing the effective interest cost (and volume) gain with such financing, against the benchmark costs without such blending. Here again, a pressing challenge is to improve the operational definitions of what counts as ODA within blended finance. Blended finance remains very poorly defined and accounted (Pereira 2017; Andersen et al. 2019; Attridge and Engen 2019; Basile and Dutra 2019). Guarantees are expressly not included in the definition of ODA (Garbacz et al. 2021). As a result, bilateral and multilateral agencies have no incentive or limited authority and basis to use such instruments, while multilateral development banks continue to approach guarantees with great caution because of the limits of their original charters (World Bank 2009) and require counter-indemnities by recipient countries, internal and historic agency inertia, perceived loss of control over the use of funds (compared to their preferred direct project-based lending) and employ restrictive accounting rules for capital provisioning of guarantees at 100% of their face value to maintain AAA ratings with credit rating agencies (Humphrey 2017; Pereira dos Santos and Kearney 2018; Bandura and Ramanujam 2019; Hourcade et al. 2021a). Largely because of such official uncertainty the actual flows of blended finance and guarantees continue to remain a very small share (typically, less than 5%) of official and multilateral finance flows to lower project risks and costs, and hence the potential for large-scale accelerated low-carbon private investments in developing countries. Public guarantees can offer a fifteen times multiplier effect on the scale of low-carbon investments generated with such support, compared to a 1:1 ratio in direct financing (Hourcade et al. 2021a).

It makes sense to expedite these operational procedures (Khan et al. 2020) which cannot be otherwise explained except in terms of avoiding responsibilities, even where the benefits would be high (Klöck et al. 2018). It also causes (unnecessary) fragmentation and complexity and often 'strategic' ambiguity by many actors (Pickering et al. 2017), which worsens the possibilities for international cooperation, a critical requirement to achieve the Paris goals (IPCC 2018). The world would gain collectively if these issues were to be decided soon. The absence of such a collective decision continues to be exceptionally costly for the implementation of the Paris Agreement because of the fractious and seemingly insoluble negotiating climate and a breakdown of trust that this has created (Roberts and Weikmans 2017).

A fourth priority is expanding jobs and dealing with job losses in the global low-carbon transition (Carley and Konisky 2020; Crowe and Li 2020; Pai et al. 2020; Cunningham and Schmillen 2021; Hanto et al. 2021), especially in coal and other sectors, as well as land and other effects for indigenous communities (Zografos and Robbins 2020). Many countries, especially low-income countries, remain dependent on fossil fuels for their energy and exports and jobs, and support for their transition to a low-carbon future will be essential. Global recovery from the pandemic will take longer than initially envisaged (IMF 2021c; OECD 2021b) and an accelerated climate action for a Build Back Better global infrastructure plan with better and more resilient jobs might play a key role as part of the Just Transitions. Already, there is substantial evidence (Sulich et al. 2020; Dell'Anna 2021; Dordmond et al. 2021) that a more sustainable climate path would generate many more net productive jobs (with much higher employment multipliers and mutual gains from given spending) than would any other large-scale alternative. But this would nevertheless require a carefully managed transition globally, including access to much larger volumes of climate financing in developing economies (Muttitt and Kartha 2020). The multilateral finance institutions have generally played a supportive role, expanding their financing to developing countries during the pandemic (even as bilateral aid flows have fallen sharply), but have been hampered by the constraints on their mandates and instruments (as noted earlier). Political leadership and direction will be again crucial to enhance their roles. The recent expansion of SDR quotas at the IMF similarly might help, but the current distributions of quota benefits flow primarily to the developed countries and do little to expand investment flows on a longer-term basis for a global expansion in growth and job opportunities in the low-carbon transition.

As a fifth priority, transformative climate financing options based on equity and global sustainability objectives may also need to consider a greater mix of public pricing and taxation options on the consumption side (Arrow et al. 2004; Folke et al. 2021). Two-thirds of global GHG emissions directly or indirectly are linked to household consumption, with average per capita carbon footprint of North America and Europe of 13.4 and 7.5 $tCO_2$-eq per capita, respectively, compared to 1.7 in Africa and Middle East (Gough 2020) and as high as 200 $tCO_2$-eq per capita among the top 1% in some high-income geographies versus 0.1 $tCO_2$-eq at the other end of the income distribution in some least-developed countries (Chancel and Piketty 2015). Globally, the highest-expenditure households account for eleven times the per capita emissions of lowest-expenditure households, with rising carbon income elasticities that suggest 'redistribution of carbon shares from global elites to global poor' as welfare efficient (Chancel and Piketty 2015; Hubacek et al. 2017). Within countries and regions, and within sectors, similar patterns hold. The top 10% of the population with the highest per capita footprints account for 27% of the EU carbon footprint, and the top 1% have a carbon footprint of 55 $tCO_2$-eq per capita, with air transport the most elastic, unequal and carbon-intensive consumption (Ivanova and Wood 2020). Similarly, within sectors, there are large differences in carbon-intensity in the building sector in North America (Goldstein et al. 2020) and across cities where consumption-based GHG emissions vary widely across the world (ranging from 1.8 to 25.9 $tCO_2$-eq per capita).

Numerous options exist (Broeks et al. 2020; Nyfors et al. 2020) for such carbon consumption reduction measures, while potentially improving societal well-being, for example: (i) inner-city zoning restrictions on private cars and promoting walking/bicycle use and improved shared low-carbon transport infrastructure; (ii) advertising regulation and carbon taxes and fees on high-carbon luxury status goods and services; (iii) subsidies and exemptions for low-carbon options, higher value-added taxes on specific high-carbon products and services, subsidies for public low-carbon options such as commuter transport, and other behavioural nudges (Reisch et al. 2021); and (iv) framing options (emphasising total cost of car over lifetimes), mandatory smart metering, collective goods and services (leasing, renting, sharing options) and others. Finally, reducing subsidies on fossil fuels, raising the progressivity of taxes and raising overall wealth taxes on the richest households, which have been sharply falling (Scheuer and Slemrod 2021) even as global income and wealth have risen, with regressive and falling overall taxes (Alvaredo et al. 2020; Saez and Zucman 2020), could effectively generate significant revenues (over 1% of GDP $yr^{-1}$), about the same size as the proposed global USD50 pertonne carbon price proposed and estimated by the IMF/OECD 2021 report to the G20 (IMF and OECD 2021) to cover expected net interest costs on overall decarbonisation initiatives and financing of green new deals (Schroeder 2021).

These five options identified above on near-term actions and priorities will however, require greater collective political leadership. A review of past crisis episodes suggests that collective actions to avoid large global or multi-country risks work well primarily when the problems are well defined, a small number of actors are involved, solutions are relatively well established scientifically, and public costs to address them are relatively small (Sandler 1998, 2015) (for example, dealing with early pandemic outbreaks such as Ebola, TB, and cholera; extending global vaccination programmes such as smallpox, measles and polio; early warning systems and actions for natural disasters such as tsunamis, hurricanes/cyclones and volcanic disasters; the Montreal Protocol for ozone-depleting refrigerants, and renewables wind and solar energy development). They do not appear to work as well for more complex global collective action problems which concern a number of economic actors, sectors, without inexpensive and mature technological options, and where political and institutional governance is fragmented. Greater political coordination is needed because the impacts are often not near term or imminent, but diffuse, slow moving and long term, and where preventive disaster avoidance is costly even when these costs are low compared to the longer-term damages – till tipping points are reached of the need for reduced 'stressors' and increasing 'facilitators' (Jagers et al. 2020). But by then, it may be too late.

Private institutional investors equally might equally wish to pay greater attention to the Just Transition finance issues. It would be useful for investors to identify ways to support to such initiatives, and more clearly identify the benefits of such transition measures envisaged by both countries and investment financing proposals, including incorporating Just Transition consideration in their support to broader ESG and green financing initiatives.

The second level of attention needed on Just Transitions has to do with inequities within a large country setting, developed or developing. The Just Transition issue exists within developed countries as well. As the ongoing pandemic illustrates, the first climate burden hit is often felt most acutely at the level of states and cities, with many smaller ones without enough fiscal capacity or ability to mount an adequate discretionary counter policy. Only national governments have the ability to borrow more in their fiscal accounts to address large collective problems, whether pandemics or climate change. Therefore, it is important that national policies and funds be available for programmes to address the Just Transition issues for larger subnational states, cities and regions. This would be helped by countries including Just Transition initiatives in their NDCs for financing (as South Africa has recently done), and attention by external financing agencies and MDBs to large-scale adverse impacts in their climate policies and investments. For example, the EU Green Deal plans (Nae and Panie 2021) include several initiatives (focusing on industries, regions and workers adversely affected, with explicit programmes to address them).

The third level of argument is for a shift in focus from an exclusive attention to financing of mitigation and low-carbon new investments projects to also better understanding and addressing the local adverse impacts of climate change on communities and people, who are vulnerable and increasingly dispossessed due to losses and damages from climate change or even those who are impacted by decarbonisation measures in the fossil fuel sectors and transportation, as well as those who are harmed by polluting sectors: indigenous men and women, minorities and generally the poor. It is evident that very few resources are available to countries, investors, civil society, and smaller development institutions seeking to achieve a just transition (Robins and Rydge 2020).

Finally, greater support is warranted for smaller towns and cities, local networks, small and medium-sized enterprises (SMEs), communities, local authorities and universities for projects, research ideas and proposals (Lubell and Morrison 2021; Moftakhari et al. 2021; Stehle 2021; Vedeld et al. 2021).

## 15.3 Assessment of Current Financial Flows

### 15.3.1 Financial Flows and Stocks: Orders of Magnitude

Assessments of finance for climate action need to be placed within the broader perspective of all investments and financing flows and stocks. This section provides aggregate level reference points of relevance to the remainder of this chapter, notably when assessing current levels of climate and fossil fuel-related investments and financing (Sections 15.3.2.3 and 15.3.2.4 respectively), as well as estimates of investment and financing needed to meet climate objectives (Section 15.4).

Measures of financial flows and stocks provide complementary and interrelated insights into trends over time: the accumulation of flows, measured per unit of time, results in stocks, observed at a given point in time (IMF 2009; UN and ECB 2015). On the flows side, GDP, a System of National Accounts (SNA) statistical standard that measures the monetary value of final goods and services produced in a country in a given period of time. In 2020, global GDP represented above $USD_{2015}$ 70 trillion[6] (down from around 80 trillion $USD_{2015}$ in 2019), out of which developed countries represented approximately 60% (Figure 15.1); a slowly decreasing share over the last years. The GDP metric is useful here as an indicator of the level of activity of an economy but gives no indication relating to human well-being or SDG achievements (Giannetti et al. 2015) as it counts positively activities that negatively impact the environment, without making deductions for the depletion and degradation of natural resources.

Gross-fixed capital formation (GFCF), another SNA standard that covers tangible assets (notably infrastructure and equipment) and intangible assets, is a good proxy for investment flows in the real economy. In 2019, global GFCF reached around 20 trillion $USD_{2015}$ compared to around 14 trillion $USD_{2015}$ in 2010, a more than 40% increase (Figure 15.2). Global GFCF represents about a quarter of global GDP, a relatively stable ratio since 2008. This share is, however, much higher for emerging economies, notably in Asia, which are building new infrastructure at scale. As analysed in Sections 15.4 and 15.5, infrastructure investment needs and gaps in developing countries are significant. How these are met over the next decade will critically influence the likelihood of reaching the Paris Agreement goals.

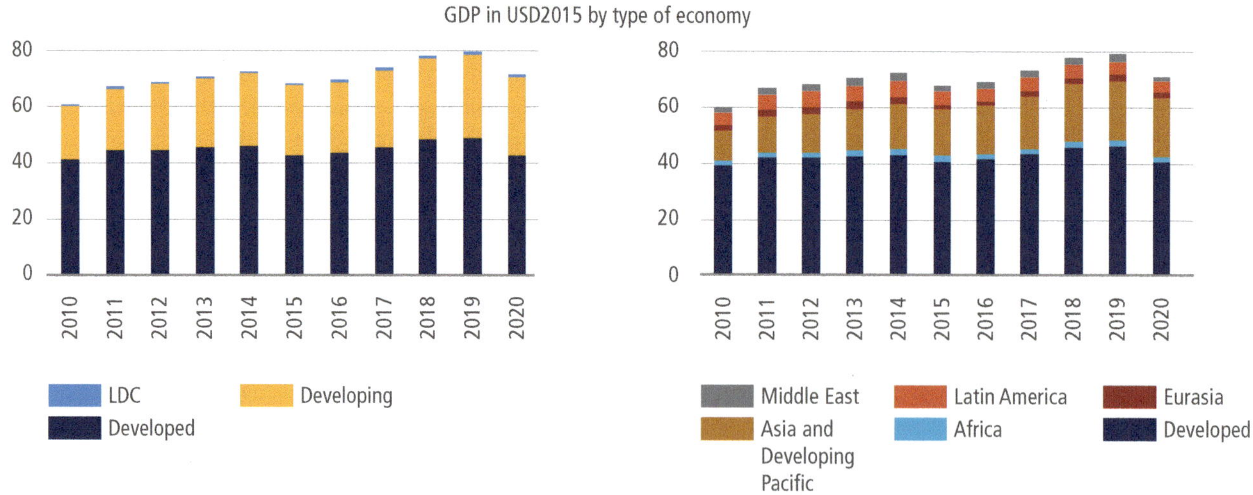


**Figure 15.1 | Financial flows – GDP (trillion $USD_{2015}$) by type of economy (left) and region (right).** Note: Regional breakdown based on official UN country classification. GDP in trillion $USD_{2015}$. Source: World Bank Data (2020a). Numbers represent aggregated country data. Last updated data on 15 September 2021. CC BY-4.0.

[6] In the chapter, USD units are used as reported in the original sources in general. Some monetary quantities have been adjusted selectively for achieving comparability by deflating the values to constant $USD_{2015}$. In such cases, the unit is explicitly expressed as $USD_{2015}$.

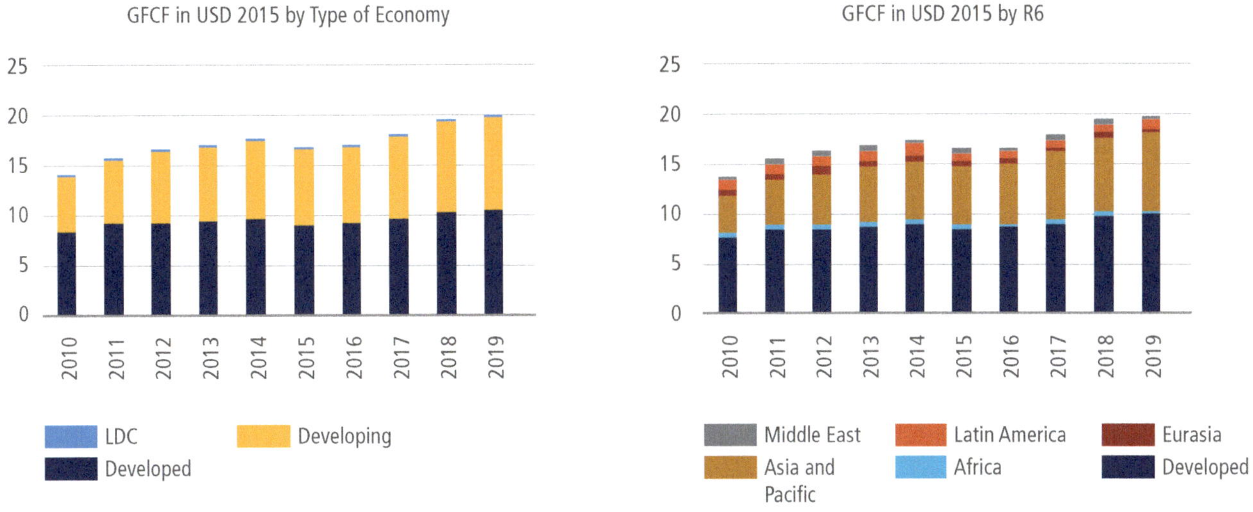


**Figure 15.2 | Financial flows – GFCF (trillion USD$_{2015}$) by type of economy (left) and region (right).** Note: Regional breakdown based on official UN country classification. GDP in trillion USD$_{2015}$. Gross fixed capital formation (GFCF) includes land improvements (fences, ditches, drains, and so on); plant, machinery, and equipment purchases; and the construction of roads, railways, and the like, including schools, offices, hospitals, private residential dwellings, and commercial and industrial buildings. Source: World Bank Data (2020b). Data for 2020 not available. Last updated data on 15 September 2021. CC BY-4.0.

On the stock side, an increasingly significant portion of the growing value of financial capital (stocks in particular) may be disconnected from the value of underlying productive capital in the real economy (Igan et al. 2020). This trend, however, remains uneven between developed countries, most of which have relatively deep capital markets, and developing countries at different stages of development (Section 15.6.7). Bonds, a form of debt financing, represent a significant share of total financial assets. As of August 2020, the overall size of the global bond markets (amount outstanding) was estimated at approximately USD128.3 trillion, out of which over two thirds was from 'supranational, sovereign, and agencies', and just under a third from corporations (ICMA 2020b). As discussed later in the chapter, since AR5, an increasing number and volume of bonds have been earmarked for climate action but these still only represent less than 1% of the total bond market. As of end-2020, climate-aligned bonds outstanding were estimated at USD0.9 trillion (Giorgi and Michetti 2021), though already raising concerns in terms of both underlying definitions (Section 15.6.6) and risks of increased climate-related indebtedness (Section 15.6.1, 15.6.3).

From the perspective of climate change action, these orders of magnitude make it possible to highlight the relatively small size of current climate finance flows and relatively larger size of remaining fossil fuel-related finance flows (discussed in the following two sub-sections), as well as, more generally, the significant overall scale of financial flows and stocks that have to be made consistent with climate goals. These orders of magnitude further make it possible to put in perspective climate-related investment needs (Section 15.4) and gaps (Section 15.5).

### 15.3.2 Estimates of Climate Finance Flows

The measurement of climate finance flows continues to face similar definitional, coverage and reliability issues as at the time of AR5 and the Special Report on Global Warming of 1.5°C, despite progress made (more sources, greater frequency, and some definitional improvements) by a range of data providers and collators. Based on available estimates (Table 15.1 and Figure 15.3), flows of annual

**Table 15.1 | Total climate finance flows between 2013 and 2020.**

| Source (type) | 2013 | 2014 | 2015 | 2016 | 2017 | 2018 | 2019 | 2020 |
|---|---|---|---|---|---|---|---|---|
| UNFCCC SCF (total high) | 687 | 584 | 680 | 681 | Published after lit. cut-off | | n/a | n/a |
| *Deflated to USD$_{2015}$* | *706* | *590* | *680* | *674* | | | | |
| UNFCCC SCF (total low/CPI) | 339 | 392 | 472 | 456 | /608 | /540 | /623 | /640 |
| *Deflated to USD$_{2015}$* | *349* | *396* | *472* | *451* | */590* | */513* | */581* | */590* |

Note: CPI: Climate Policy Initiative; SCF: Standing Committee on Finance. Numbers in current billion USD. Deflated to USD$_{2015}$ in *italic.* Given the variations in numbers reported by different entities, changes in data, definitions and methodologies over time, there is *low confidence* attached to the aggregate numbers presented here. The higher bound reported in the SCF's Biennial Assessment reports includes estimates from the International Energy Agency on energy efficiency investments, which are excludes from the lower bound and CPI's estimates. Sources: UNFCCC (2018a); Buchner et al. (2019); Naran et al. (2021).

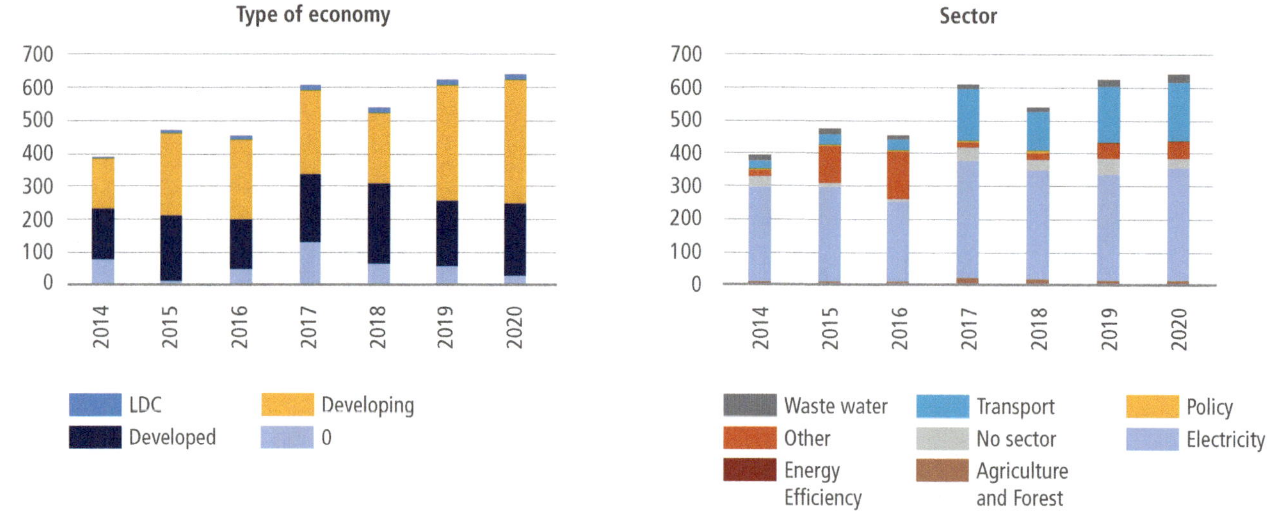


**Figure 15.3 | Available estimates of global climate finance between 2014 and 2020.** Note: Numbers in current billion USD. Deflated to $USD_{2015}$ see Table 15.1 in *italic*. Type of Economy figure **(left)**: Regional breakdown based on official UN country classification. '0' no regional mapping information available. Sectoral figure **(right)**: *Policy* includes Disaster Risk Management; Policy and national budget support and capacity building. *Transport* includes Sustainable/Low-carbon Transport. *Energy Efficiency* includes Industry, Extractive Industries, Manufacturing & Trade, Low-carbon Technologies, Information and Communications Technology, Buildings and Infrastructure. *Electricity* includes Renewable Energy Feneration, "Infrastructure, energy and other built environment", Transmission and Distribution Systems, and Energy Systems. *No sector* means no sector information available, or negligible flows. *Other* includes Non-energy GHG reductions, Coastal Protection. Source: own calculations, based on Naran et al. (2021).

global climate finance are on an upward trend since AR5, reaching a high-bound estimate of USD681 billion in 2016 (UNFCCC 2018a), representing USD674 billion 2015. Latest available estimates indicate a drop in 2018 (Buchner et al. 2019) and a rebound in 2019 and 2020 (*medium confidence*) (Naran et al. 2021). Although not directly comparable in terms of scope, current climate finance flows remain small (approx. 3%) compared to the GFCF reference point introduced in Section 15.3.1, and need to be put in perspective with remaining fossil fuel financing (*medium confidence*) (Section 15.3.2.3).

At an aggregate level, in both developed and developing countries, the vast majority of tracked climate finance is sourced from domestic or national markets rather than cross-border financing (Buchner et al. 2019). This reinforces the point that national policies and settings remain crucial (Section 15.6.2), along with the development of local capital markets (Section 15.6.7).

Climate finance in developing countries remains heavily concentrated in a few large economies (*high confidence*), with Brazil, India, China and South Africa accounting for around one-quarter to more than a third depending on the year, a share similar to that represented by developed countries. Least-developed countries (LDCs), on the other hand, continue to represent less than 5% year-on-year (*medium confidence*) (BNEF 2019; Buchner et al. 2019). Further, the relatively modest growth of climate finance in developed countries is a matter of concern given that economic circumstances are, in most cases, relatively more amenable to greater financing, savings and affordability than in developing countries.

At a global level, the majority of tracked climate finance is assessed as coming from private actors (Buchner et al. 2019), although, the boundaries between private and public finance include significant grey zones (Box 15.2), which implies that different definitions could lead to different conclusions (Yeo 2019; Weikmans and Roberts 2019). However, private investments in climate projects and activities often benefit from public support in the form of co-financing, guarantees or fiscal measures. In terms of financial instruments and mechanisms, debt as well as balance sheet financing (which can rely on both own resources and further debt) and project financing (combining a large debt portion and smaller equity portion) represent the lion's share. In this context, the rapid rise of climate-related bond issuances since AR5 (Giorgi and Michetti 2021) represents an opportunity for scaling up climate finance but also poses underlying issues of integrity (Nicol et al. 2018a; Shishlov et al. 2018) and additionality (Schneeweiss 2019), as further discussed in Section 15.6.5, and needs to be considered in the context of overall indebtedness and debt sustainability (Sections 15.6.1 and 15.6.3).

Mitigation continues to represent the lion's share of global climate finance (consistently above 90% between 2017 and 2020), and in particular renewable energy, followed by energy efficiency and transport (*high confidence*) (UNFCCC 2018a; Buchner et al. 2019). While capacity additions on the ground kept rising, falling technology costs in certain sectors (e.g., solar energy) has had a negative impact on the year-on-year trend that can be observed in terms of volumes of climate finance (BNEF 2019; IRENA 2019a). However, such cost reduction could free up investment and financing capacities for potential use in other climate-related activities.

Tracking adaptation finance continues to pose significant challenges in terms of data and methods. Notably, the mainstreaming of resilience into investments and business decisions makes it difficult to identify relevant activities within financial datasets (Agrawala et al. 2011; Brown et al. 2015; Averchenkova et al. 2016). Despite these limitations, evidence shows that finance for adaptation remains fragmented and significantly below rapidly rising needs (Section 15.4 and Cross-Chapter Box FINANCE: Finance for Adaptation and Resilience in Chapter 17 of AR6 WGII report).

## Box 15.4 | Measuring Progress Towards the USD100 Billion $yr^{-1}$ by 2020 Goal – Issues of Method

In 2009, at COP15, Parties to the UNFCCC agreed the following: 'In the context of meaningful mitigation actions and transparency on implementation, developed countries commit to a goal of mobilising jointly USD100 billion a year by 2020 to address the needs of developing countries. This funding will come from a wide variety of sources, public and private, bilateral and multilateral, including alternative sources of finance' (UNFCCC 2009).

This goal is further embedded as a target under SDG 13 Climate Action. While the parameters for what and how to count were not defined when the goal was set, progress in this area has been achieved under the UNFCCC (UNFCCC 2019b) and via a UN-driven independent expert review (Bhattacharya et al. 2020).

There remain well documented interpretations and debates on how to account for progress (Clapp et al. 2012; Stadelmann et al. 2013; Jachnik et al. 2015; Weikmans and Roberts 2019). Different interpretations relate mainly to the type and proportion of activities that may qualify as 'climate' on the one hand, and to how to account for different types of finance (and financial instruments) on the other hand. As an example, there are different points at which financing can be measured, for example, pledges, commitments, disbursements. There can be significant lags between these different points in time, for example disbursements may spread over time. Further, the choice of point of measurement can have an impact on both the volumes and on the characteristics (geographical origin, labelling as public or private) of the finance tracked. The enhanced transparency framework under the Paris Agreement may lead to improvements and more consensus in the way climate finance is accounted for and reported under the UNFCCC. Available analyses specifically aimed at assessing progress towards the USD100 billion goal remain rare, for example the UNFCCC SCF Biennial Assessments do not directly address this point (UNFCCC 2018a). Dedicated OECD reports provide figures based on accounting for gross flows of climate finance based on analysing activity-level data recorded by the UNFCCC (bilateral public climate finance) and the OECD (multilateral public climate finance, mobilised private climate finance and climate-related export credits) (OECD 2015a; OECD 2019a; OECD 2020b). For 2018, the OECD analysis resulted in a total of USD78.9 billion, out of which USD62.2 billion of public finance, USD2.1 billion of export credits and USD14.5 billion of private finance was mobilised. Mitigation represented 73% of the total, adaptation 19% and cross-cutting activities 8%.

Reports by Oxfam provide a complementary view on public climate finance, building on OECD figures and underlying data sources to translate gross flows of bilateral and multilateral public climate finance in grant equivalent terms, while also, for some activities, applying discounts to the proportion considered as climate finance (Carty et al. 2016; Carty and Le Comte 2018; Carty et al 2020). The resulting annual averages for 2015–2016 and 2017–2018 range between 32% (low bound) and 44% (high bound) of gross public climate finance. The difference with OECD figures stems from the high share represented by loans, both concessional and non-concessional, in public climate finance, that is, 74% in 2018 (OECD 2020b).

A point of method that attracts much attention relates to how to account for private finance mobilised. The OECD, through its Development Assistance Committee, established an international standard to measure private finance mobilised by official development finance, which consists in methods tailored to different financial mechanisms. These methods take into account the role of, risk taken, and/or amount provided by all official actors involved in a given project, including recipient country institutions, thereby also avoiding risks of double counting (OECD 2019b). MDBs apply a different method (World Bank 2018a) in their joint climate finance reporting (AfDB et al. 2020), which neither correspond to the geographical scope of the USD100 billion goal, nor address the issue of attribution to the extent required in that context.

Notwithstanding methodological discussions under the UNFCCC, there is still some distance from the USD100 billion a year commitment being achieved, including in terms of further prioritising adaptation. While the scope of the commitment corresponds to only a fraction of the larger sums needed (Section 15.4), its fulfilment can both contribute to climate action in developing countries as well as to trust building in international climate negotiations. Combined with further clarity on geographical and sectoral gaps, this can, in turn, facilitate the implementation of better coordinated and cooperative arrangements for mobilising funds (Peake and Ekins 2017).

Further, there is increasing awareness about the need to better understand and address the interlinkages between climate change adaptation and disaster risk reduction (DRR) towards achieving resilience (OECD 2020a). Watson et al. (2015) however, note that between 2003 and 2014, of the USD2 billion that flowed through dedicated climate change adaptation funds, only USD369 million explicitly went to DRR activities (Climate Funds Update 2014; Nakhooda et al. 2014a; Nakhooda et al. 2014b; Watson et al. 2015). For the private sector, insurance and reinsurance remain the dominant way to transfer risk as discussed in Section 15.6.4).

More generally, significant gaps remain to track climate finance comprehensively at a global level:

- Available estimates are based on a good coverage of investments in renewable energy and, where available, energy efficiency and transport, while other sectors remain more difficult to track, such as industry, agriculture and land use (*high confidence*) (UNFCCC 2018a; Buchner et al. 2019).
- In contrast to international public climate finance, domestic public finance data remain partial despite initiatives to track domestic climate finance (e.g., Hainaut and Cochran 2018) and public expenditures (*high confidence*) (for instance based on the UNDP's Climate Public Expenditure and Institutional Review approach).

Data on private and commercial finance remain very patchy, particularly for corporate financing (including debt financing provided by commercial banks), for which it is difficult to establish a link with activities and projects on the ground (*high confidence*). Further, as individual sources of aggregate reporting (UNFCCC 2018a; Buchner et al. 2019; FS-UNEP Centre and BNEF 2020) tend to rely on the same main data sources (notably the BNEF commercial database for renewable energy investments) as well as to cross-check numbers against similar other sources, there is a potential for 'group-think' and bias.

Such data gaps as well as varying definitions of what qualifies as 'climate' (or more broadly as 'green' and 'sustainable') not only pose a measurement challenge. They also result in a lack of clarity for investors and financiers seeking climate-related opportunities. Such uncertainty can lead both to reduced climate finance as well as to a lack of transparency in climate-related reporting (further discussed in Section 15.6.1), which in turn further hinders reliable measurement.

In terms of finance provided and mobilised by developed countries for climate action in developing countries, while accounting scope and methodologies continue to be debated (Box 15.4), progress has been achieved on these matters in the context of the UNFCCC (UNFCCC 2019b). A consensus, however, exists, on a need to further scale up public finance and improve its effectiveness in mobilising private finance (OECD 2020b), as well as to further prioritise adaptation financing, in particular towards the most vulnerable countries (Carty et al. 2020). The relatively low share of adaptation in international climate finance to date may in part be due to a low level of obligation and precision in global adaptation rules and commitments (Hall and Persson 2018). Further, providers of international climate finance may have more incentive to support mitigation over adaptation as mitigation benefits are global while the benefits of adaptation are local or regional (Abadie et al. 2013).

### 15.3.3 Fossil Fuel-related and Transition Finance

As called for by Article 2.1c of the Paris Agreement and introduced in Section 15.3.1, achieving the goal of the Paris Agreement of holding the increase in the global average temperature to well below 2°C above pre-industrial levels and pursuing efforts to limit the temperature increase to 1.5°C above pre-industrial levels requires making all finance consistent with this goal. Data on investments and financing to high GHG activities remain very partial and difficult to access, as relevant actors currently have little incentive or obligations to disclose such information compared to reporting on and communicating about their activities contributing to climate action. Further, the development of methodologies to assess finance for activities misaligned with climate mitigation goals, for hard- and costly-to-abate sectors such as heavy industries, as well as for activities that eventually need to be phased out but can play a transition role for a given period, remain work in progress. This results in limited empirical evidence to date.

In modelled pathways that limit warming to 1.5°C (>50%) with no or limited overshoot, however, make it clear that the share of fossil fuels in energy supply has to decrease (see Chapter 3). For instance, the International Energy Agency (IEA) Net Zero by 2050 scenario relies on halting sales of new internal combustion engine passenger cars by 2035, rapid and steady decrease of the production of coal (minus 90%), oil (minus 75%) and natural gas (minus 55%) by 2050, and phasing out all unabated coal and oil power plants by 2040 (IEA 2021b). To avoid locking GHG emissions incompatible with remaining carbon budgets, this implies a rapid scaling down of new fossil fuel-related investments, combined with a scaling up of financing to allow energy and infrastructure systems to transition (*high confidence*).

The IEA provides comprehensive analyses of global energy investments, estimated at about USD1.8 trillion a year over 2017–2019 (IEA 2019a, 2020a), and expected to reach that level again in 2021 after a drop to about 1.6 trillion in 2020 (IEA 2021c). Energy investments represent about 8% of global GFCF (Section 15.3.2.1). In the power sector, fossil fuel-related investments reached an estimated USD120 billion $yr^{-1}$ on average over 2019–2020, which remains well above the level that underpin the IEA's own Paris-compatible Sustainable Development Scenario (SDS) and Net Zero Emission (NZE) scenarios. The IEA observes a similar inconsistency for supply-side new investments: in 2019–2020 on average $yr^{-1}$, an estimated USD650 billion were invested in oil supply and close to USD100 billion in coal supply. These estimates also result in fossil fuel investments remaining larger in aggregate than the total tracked climate finance worldwide (Section 15.3.2.2). For oil and gas companies, which are amongst the world's largest corporations and sometimes government owned or backed, low-carbon solutions are estimated to represent less than 1% of capital expenditure (IEA 2020b). As discussed in the remainder of this chapter, shifting investments towards low-GHG solutions requires a combination of conducive public policies, attractive investment opportunities, as well as the availability of financing to finance such a transition.

In terms of financing provided to fossil fuel investments, available analyses point out a still significant role played by commercial banks and export credit agencies. Commercial banks provide both direct lending as well as underwriting services, the latter facilitating capital raising from investors in the form of bond or share issuance. Available estimates indicate that lending and underwriting extended over 2016– 2019 by 35 of the world's largest banks to 2100 companies active across the fossil fuel lifecycle reached USD687 billion $yr^{-1}$ on average (Rainforest Action Network et al. 2020). Official export credit agencies, which are owned or backed by their government, de-risk exports by providing guarantees and insurances or, less often, loans. In 2016–2018, available estimates indicate the provision of about USD31 billion $yr^{-1}$ worth of fossil fuel-related official export credits, out of which close to 80% was for oil and gas, and over 20% for coal (DeAngelis and Tucker 2020).

Finance for new fossil fuel-related assets lock in future GHG emissions that may be inconsistent with remaining carbon budgets and, as discussed above, with emission pathways to reach the Paris Agreement goals. This inconsistency exposes investors and asset owners to the risk of stranded assets, which results from potential sharp strengthening climate public policies, that is, transition risk. As a result, a growing number of investors and financiers are assessing climate-related risks with the aim to disclose information about their current level of exposure (to both transition and physical climate-related risks), as well as to inform their future decisions (TCFD 2017). Reporting to date is, however, inconsistent across geographies and jurisdictions (CDSB and CDP 2018; Perera et al. 2019), with also a wide variety of metrics, methodologies, and approaches developed by commercial providers that contribute to disparate outcomes (Kotsantonis and Serafeim 2019; Boffo and Patalano 2020). Further, as developed in Section 15.6.1, there is currently not enough evidence in order to conclude whether climate-related risk assessments result in increased climate action and alignment with the goals of the Paris Agreement (The 2° Investing Initiative and Wüest Partner 2020).

As developed in Section 15.6.3, the insufficient level of ambition and coherence of public policies at national and international levels remains the root cause of the still significant misalignment of investment and financing compared to pathways compatible with the Paris Agreement temperature goal (UNEP 2018). Such lack of coherence includes low pricing of carbon and of environmental externalities more generally, as well as misaligned policies in non-climate policy areas such as fiscal, trade, industrial and investment policy, and financial regulation (OECD 2015b), as further specified in the sectoral Chapters 6 to 12.

The most documented policy misalignment relates to the remaining very large scale of public direct and indirect financial support for fossil fuel-related production and consumption in many parts of the world (Bast et al. 2015; Coady et al. 2017; Climate Transparency 2020). Fossil fuel subsidies are embedded across economic sectors as well as policy areas, for example, from a trade policy perspective, in most countries, import tariffs and non-tariff barriers are substantially lower on relatively more $CO_2$ intensive industries (Shapiro 2020). Available inventories of fossil fuel subsidies (in the form of direct budgetary transfers, revenue forgone, risk transfers, or induced transfers), covering 76 economies, indicate a rise to USD340 billion in 2017, a 5% increase compared to 2016. Such trend is due to slowed down progress in reducing support among OECD and G20 economies in 2017 (OECD 2018b) and to a rise in fossil fuel subsidies for consumption in several developing economies (Matsumura and Adam 2019), which, in turn, reduces the efficiency of public instruments and incentives aimed at redirecting investments and financing towards low-GHG activities.

As a result, the demand for fossil fuels, especially in the energy production, transport and buildings sectors, remain high, and the risk-return profile of fossil fuel-related investments is still positive in many instance (Hanif et al. 2019). Political economy constraints of fossil fuel subsidy reform continue to be a major hurdle for climate action (Schwanitz et al. 2014; Röttgers and Anderson 2018), as further discussed in Section 15.5.2. and Chapter 13.

## 15.4 Financing Needs

### 15.4.1 Definitions of Financing Needs

Financing needs[7] are discussed in various contexts, only one being international climate politics and finance. Also, financing needs are used as an indicator for required system changes (when compared to current flows and asset bases) and an indicator for near- to long-term investment opportunities from the perspective of investors and corporates. Investment needs are widely used as an indicator focusing on initial investments required to realise new infrastructure. It compares relatively well with private sector flows dominated by return-generating investments but lacks comparability and explanatory power regarding the needs in the context of international climate cooperation, where considerations on economic costs play a more substantial role. Chapter 12 elaborates on global economic cost estimates for various technologies. This indicator includes both costs and benefits of options, of which investment-related costs make up only one component. Both analyses offer complementary insights. There are financing needs not directly related to the realisation of physical infrastructure and which are not covered in both investment and cost estimates. For instance, the needs for building institutional capacity to achieve social and economic goals and to strengthen knowledge, skills, national and international cooperation might not be significant, but an enabling environment for future investments would not be established without satisfying it. Moreover, comprehending financial needs for addressing economic losses due to climate change can hardly be measured in terms of the indicators introduced before.

Understanding the magnitude of the challenge to scale up finance in sectors and regions requires a more comprehensive (and qualitative)

7 The term Investment 'Needs' used in the chapter means equal to the term Investment Requirement used in SPM.

assessment of the needs. For finance to become an enabler of the transition, domestic and international public interventions can be needed to ensure enough supply of finance across sectors, regions and stakeholders. The location of financing needs and vicinity to capital matter given home bias (Fuchs et al. 2017; OECD 2017a; Ito and McCauley 2019) (prioritising own country or regions), transaction costs and risk considerations (Section 15.2). Most of the finance is mobilised domestically but the depth of capital markets is substantially greater in developed countries, increasing the challenges to mobilise substantial volumes of additional financing for many developing countries. The same applies to various stakeholders with limited connections into the financial sector. In addition, governments enabling financial market frameworks, guidelines and supportive infrastructure is crucial for inclusive finance for the bottom of the pyramid, especially disadvantaged and economically marginalised segments of society.

The attractiveness of a sector and region for capital markets depends on several factors. Some essential elements are the duration of loan and profile as long-term loans and heavily heterogeneous returns represent challenges in financing mitigation technologies and policies. After the financial crisis and restricted access to long-term debt, capital intensity of technologies and resulting long payback periods of investment opportunities for mitigation technologies have been a crucial challenge (Bertoldi et al. 2021). Also, implicit discount rates applied during the investment decision process vary depending on the payback profile, with research mainly covering the difference between the financing of assets generating revenues versus costs (Jaffe et al. 2004; Schleich et al. 2016). In addition, a low correlation between the climate projects and dominating asset classes might provide an opportunity in climate action by satisfying the appetite of institutional investors, which tend to manage portfolios with consideration of the Markowitz modern portfolio theory (optimising return and risk of a portfolio through diversification) (Marinoni et al. 2011). Transaction cost is a significant barrier to the diffusion and commercialisation of low-carbon technologies and business models and adaptation action. High transaction costs, attributed to various factors, such as complexity and limited standardisation of investments, limited pipelines, complex institutional and administrative procedures, create significant opportunity costs of green investments comparing with other standard investments (IRENA 2016; Nelson et al. 2016; Feldman et al. 2018). For example, transaction costs are commonly observed in small-scale, dispersed independent renewable energy systems, especially in rural areas, and energy efficiency projects (Hunecke et al. 2019). A more robust standardisation and alignment of Power Purchase Agreement (PPA) terms with best practices globally has led to a substantially increased interest in capital markets in developing countries (WBCSD 2016; Schmidt et al. 2019; World Bank 2021). Notably, PPA significantly increases the probability of more balanced investment and development outcomes and ultimately more sustainable independent power projects in developing countries. Therefore, lowering transaction costs would be essential for creating investor appetite. The role of intermediaries bundling demand for financing has been demonstrated to reduce transaction costs and to reach investors' critical size. In addition, new innovative approaches, such as fintech and blockchain (Section 15.6.8), have been discussed for providing new opportunities in the energy sector.

Economic viability of investments – ideally not relying on the pricing of positive externalities – has been a critical driver of momentum in the past. The falling technology costs and the competitiveness of renewable technologies, especially solar PV and wind, have accelerated the deployment of renewable technologies over the past years. Renewable energy technologies are now often competitive, and have even become the cheapest, in many countries, even without financial support (FS-UNEP Centre and BNEF 2015, 2016, 2017, 2018, 2019; IEA 2020c; IRENA 2020a) and without pricing of the avoided carbon emissions. In contrast, the dependency on regulatory interventions and public financial support to create financial viability has provided a source of volatile investor appetite. The annual volume of renewable investment by country is often volatile, reflecting ending and new regulations and policies (IEA 2019a).

For example, the recent Chinese policy direction towards tougher access to and a substantial cut in feed-in-tariffs in 2018 led to a significant drop in renewable investment and new capacity addition in China (FS-UNEP Centre and BNEF 2019; Hove 2020). However, the significant bouncing back of newly installed capacity (72 GW wind power and 47 GW solar power in 2020) shows the strong development of zero-carbon power generation driven by lower cost and policies to support them by energy revolution strategies in China. Investors had proven to be willing to work with transparent support mechanisms, such as with the Clean Development Mechanism (CDM), which stimulated emission reductions and allowed industrialised countries to implement emission-reduction projects in developing countries to meet their emission targets (Michaelowa et al. 2019). However, the collapse of carbon markets and prices, especially of the EU Emissions Trading System, led to the continuous decline of Certified Emission Reductions issuances from CDM in the past years (World Bank Group 2020). Also, the dependency on regulatory intervention to ensure fair market access only has proven to burden investor appetite.

A significant share of investment needs in heavily regulated sectors, such as electricity, public transport, and telecom, emphasises the importance of regulatory intervention, such as ownership and market access (OECD 2017b). For instance, energy-system developments require effective and credible commitments and action by policymakers to ensure an efficient capital allocation aligned with climate targets (Bertram et al. 2021).

There is a lot of discussion about the regulated ownership of the private sector (European Commission 2017) and the restructuring of electricity market contributed to low level of investment in baseline electricity capacity and in investment research and innovation. These changes create uncertainty of investment, and barriers to market entry and exit also potentially limit the competition in the market and restrict the entrance of new investment (Finon 2006; Joskow 2007; Grubb and Newbery 2018). This is also the case in developing countries (Foster and Rana 2020).

The positive development in the energy sector has benefitted from the evident stand-alone character of renewable energy generation projects. First movers realised these projects with investors and developers acting from conviction (Steffen et al. 2018). Such action is not possible to this extent in energy efficiency with related investment rather representing an add-on component and consequently requiring the support of decision-makers used to business-as-usual projects. Despite the benefits that improvement of energy efficiency has in contributing to curbing energy consumption, mitigating greenhouse gas emissions, and providing multiple co-benefits (IEA 2014a), investment in energy efficiency is a low priority for firms, and the financial environment is not favourable due to lack of awareness of energy efficiency by financial institutions, existing administrative barriers, lack of expertise to develop projects, asymmetric information, and split incentives (UNEP DTU 2017; Cattaneo 2019). While Energy Service Companies' (ESCO) business models are expected to facilitate the investment in energy efficiency by sharing a portion of financial risk and providing expertise, there has been limited progress made with ESCO business models, and only slightly over 20% of projects used financing through ESCOs (UNEP DTU 2017).

The investment needs and existing challenges differ by sector. Each sector has different characteristics along the arguments listed above making the supply of finance by commercial investors an enabling factor or barrier. In the transport sector, transformation towards green mobility would provide significant co-benefits for human health by reducing transport-related air pollution, so the transport sector cannot achieve such transformation in isolation from other sectors. However, a considerable involvement of the public sector in many transportation infrastructure projects is given, and the absence of a standard solution increases transaction costs (including bidding package, estimating, drawing up a contract, administering the contract, corruption, and so on). Financial constraints, including access to adequate finance, pose a significant challenge in the agriculture sector, especially for SMEs and smallholder farmers. The distortion created by government failure and a lack of effective policies create barriers to financing for agriculture. The inability to manage the impact of the agriculture-related risks, such as seasonality, increases uncertainty in financial management. Moreover, inadequate infrastructure, such as electricity and telecommunication, makes it difficult for financial institutions to reach agricultural SMEs and farmers and increases transaction costs (World Bank 2016). Low economies of scale, low bargaining power, poor connectivity to markets, and information asymmetry also lead to higher transaction costs (Pingali et al. 2019). In the industrial manufacturing and residential sector, gaining energy efficiency remains one of the critical challenges. Investment in achieving energy efficiency encounters some challenges when it may not necessarily generate direct or indirect benefits, such as increase in production capacity or productivity and improvement in product quality. Also, early-stage, high upfront cost and future, stable revenue stream structure suggest the needs for a better enabling environment, such as a robust financial market, awareness of financial institutions, and regulatory frameworks (e.g., stringent building codes, incentives for ESCOs) (IEA 2014a; Barnsley et al. 2015).

### 15.4.2 Quantitative Assessment of Financing Needs

Multiple stakeholders prepare and present quantitative financing needs assessments with methodologies applied to vary significantly representing a major challenge for aggregation of needs (e.g., Osama et al. (2021) for African countries), most of them with a focus on scenarios likely to limit warming to 2°C or lower. The differences relate to the scope of the assessments regarding sectors, regions and periods, top-down versus bottom-up approaches, and methodological issues around boundaries of climate-related investment needs, particularly full vs incremental costs and the exclusion or inclusion of consumer-level investments. Information on investment needs and financing options in NDCs mirrors this challenge and is heavily heterogeneous (Zhang and Pan 2016).

In particular, for global approaches, modelling assumptions are often heavily standardised, focusing on technology costs. Only limited global analysis is available on incremental costs and investments, reflecting the reality of developing countries, also considering the interplay with significant infrastructure finance gaps, and can hardly serve as a robust basis for negotiations about international public climate finance. The focus on investment irrespective of uncertainty as well as other qualitative aspects of needs does not allow for a straightforward analysis of the need for public finance to leverage private sector financing and of the country heterogeneity in terms of investment risks and access to capital (Clark et al. 2018).

One source of uncertainty about the investment estimates for the power sector is the evolution of the levelised cost of technical options in the future, for example the continuation of the observed declining costs trends of renewable energy (IRENA 2020b) which has been underestimated in many modelling exercises. The learning by doing processes and economies of scale might be at least partially outweighed, in all countries and more specifically in Small Island Developing States (SIDS) and other developing countries because of different risk factors, scales of installations, accessibility, and others (Lucas et al. 2015; van der Zwaan et al. 2018). These parameters, together with transaction costs/soft costs (Section 15.5), financing costs and the level of technical competences need to be better represented in the future to represent the 'climate investment trap' in many developing countries (García de Fonseca et al. 2019). This 'climate investment trap', as flagged by Ameli et al. (2021a), is created by existing and expected physical effects of climate change, higher financing costs and resulting lower investment levels in developing countries. Applying significantly standardised assumptions can consequently not provide robust insights for specific country groups. This will require progress in the spatiotemporal granularity of the models (Collins et al. 2016).

Another source of uncertainty about the financing needs is the interplays between (i) the baseline economic growth rates, (ii) the link between economic growth and energy demand, including rebound effects of energy efficiency gains, (iii) the evolution of microeconomic parameters such as fossil fuel prices, interest rates, currency exchange rates (iv) the level of integration between climate policies and sectoral policies and their efficacy, and (v) the impact of climate policies on growth and the capacity of fiscal and financial policies to offset their adverse effect (IPCC 2014; IPCC 2018). Integrated assessment models (IAMs) try to capture some of these interplays even though they typically do not capture the financial constraints and the structural causes of the infrastructure investment gap. Many of them rely on growth models with full exploitation of the means of production (labour and capital). They nevertheless provide useful indications of the orders of magnitude at play over the long run, and the determinants of their uncertainty. Global yearly average low-carbon investment needs until 2030 for electricity, transportation, AFOLU and energy efficiency measures including industry and buildings are estimated between 3% and 6% of the world's GDP according to the analysis in Section 15.5. The incremental costs of low-carbon options are less than that and their funding could be achieved without reducing global consumption by reallocating 1.4% to 3.9% of global savings. 2.4% on average (see Box 4.8 of SR1.5 (IPCC 2018)) currently flow towards real estate, land and liquid financial vehicles. For the short-term decisions, the major information they give is the uncertainty range because this is an indicator of the risks decision-makers need.

While the AR6 Scenarios Database provides good transparency with regard to technology costs for electricity generation, assumptions driving in particular investments in energy efficiency are rarely made available in both IAM-based assessments and also other studies. Taking into account the much broader range of tested and untested technologies the confidence levels, in particular for 2050 estimates, remain low but can provide an initial indication. Also, the ranges allow for a rough indication on possible 'green' investment volumes and respective asset allocation for financial sector stakeholders.

**Using global scenarios assessed in Chapter 3 for assessing investment requirements.** Tables 15.2 and 15.3 present the analysis of investment requirements in global modelled mitigation pathways assessed in Chapter 3 for key energy sub-sectors within modelled global pathways that limit warming to 2°C (>67%) or lower. These pathways explore the energy, land-use, and climate system interactions and thus help identify required energy sector transformations to reach specific long-term climate targets. However, reporting of investment needs outside the energy sector was scarce, reducing the explanatory power of the shown total investment need in the context of overall investment needs (Ekholm et al. 2013; IPCC 2018, Box 4.8; McCollum et al. 2018; Bertram et al. 2021). The modelling of these scenarios is done with a variation of scenario assumptions along different dimensions (*inter alia* policy, socio-economic development and technology availability), as well as with different modelling tools which represent different assumptions about the structural functioning of the energy-economy-land-use system (see Annex III: 'Scenarios and modelling methods' for details). Tables 15.2 and 15.3 focus on the near-term (2023–2032) investment requirements in the energy sector and how these differ depending on temperature category. Figures 3.36 and 3.37 present the data for the medium term (2023–2052). The results highlight both requirements for increased investments and a shift from fossil towards renewable technologies and efficiency for more ambitious temperature categories. The substantial ranges within each category reflect multiple pathways, differentiated by socio-economic assumptions, technology, and so on. It is necessary to open up these extra dimensions and contrast them with national and sub-regional analysis to understand how investment requirements depend on particular circumstances and assumptions within a country for a specific technology. Limiting peak temperature to levels of 1.5°C–2°C requires rapid decarbonisation of the global energy systems, with the fastest relative emission reductions occurring in the power generation sector (Hirth and Steckel 2016; Luderer et al. 2018).

This requires fast shifts of investment as infrastructures in the power sector generally have long lifetimes of a few decades. in global modelled pathways that limit warming to 1.5°C (>50%) with no or limited overshoot, investments into non-biomass renewables (especially solar and wind, but also including hydro, geothermal, and others not shown in Table 15.2) increase to over USD1 trillion $yr^{-1}$ in 2030, increasing by more than factor 3 over the values of around USD250–300 billion $yr^{-1}$ that have been relatively stable over the last decade (IEA 2019a). Overall, electricity generation investments increase considerably, reflecting the higher relevance of capital expenditures in decarbonised electricity systems. While decreasing technology costs have substantially reduced the challenge of high capital intensity, still remaining relative disadvantages in terms of capital intensity of low-carbon power technologies can especially create obstacles for fast decarbonisation in countries with high interest rates, which decrease the competitiveness of those technologies (Iyer et al. 2015; Hirth and Steckel 2016; Steckel and Jakob 2018; Schmidt et al. 2019). CCS as well as nuclear will not drive investment needs until 2030, given considerably longer lead-times for these technologies, and the lack of a significant project pipeline currently.

**Table 15.2 | Global average yearly investments from 2023–2032 for electricity supply in billion $USD_{2015}$.**

| Category | Fossil | Nuclear | Storage | Transmission and distribution | Non-Biomass Renewables | | |
|---|---|---|---|---|---|---|---|
| | | | | | All | Thereof | |
| | | | | | | Solar | Wind |
| C1 | 53 [50] | 127 [52] | 221 [39] | 549 [50] | 1190 [52] | 498 [52] | 390 [52] |
| (Range) | (34;115) | (85;165) | (88;295) | (422;787) | (688;1430) | (292;603) | (273;578) |
| C2 | 78 [100] | 116 [92] | 57 [66] | 489 [81] | 736 [96] | 312 [96] | 237 [96] |
| (Range) | (50;129) | (61;150) | (37;139) | (401;620) | (482;848) | (181;385) | (174;328) |
| C3 | 75 [221] | 96 [190] | 28 [129] | 389 [157] | 639 [207] | 220 [207] | 266 [207] |
| (Range) | (52;129) | (50;122) | (8;155) | (326;760) | (432;820) | (167;345) | (137;353) |

Note: Global average yearly investments from 2023–2032 (in $USD_{2015}$). Electricity subcomponents are not exhaustive. Hydro, geothermal, biomass and others are not shown, as these are shown to be of smaller magnitude (Chapter 3). Difference between non-biomass renewables and solar/wind represents hydro and in some scenarios geothermal, tidal, and ocean. Scenarios are grouped into common AR6 categories (vertical axis, C1–C3). The numbers represent medians across all scenarios within one category, and rounded brackets indicate inter-quartile ranges, while the numbers in squared brackets indicate number of scenarios. C6, C7, and C8 are not shown in Table 15.2. Reference C5 category for Transmission and Distribution (T&D) is 364bn (294bn to 445bn) [111] used for calculation of incremental needs in Figure 15.4. Data source: AR6 Scenarios Database.

**Table 15.3 | Regional average yearly investments from 2023–2032 for electricity supply in billion $USD_{2015}$.**

| | Africa | East Asia | Europe | South Asia | Latin America | Middle East | North America | Australia, Japan, and New Zealand | East. Eur. W.C. Asia | South East Asia |
|---|---|---|---|---|---|---|---|---|---|---|
| **Non-biomass renewables** | | | | | | | | | | |
| C1 | 41 [39] | 302 [41] | 130 [41] | 120 [41] | 69 [41] | 67 [41] | 177 [41] | 37 [41] | 48 [41] | 85 [41] |
| (Range) | (36;66) | (188;356) | (101;150) | (83;164) | (55;97) | (31;90) | (149;222) | (28;39) | (35;65) | (59;141) |
| C2 | 32 [77] | 179 [87] | 95 [87] | 69 [87] | 55 [87] | 28 [87] | 106 [87] | 19 [87] | 17 [87] | 63 [87] |
| (Range) | (27;42) | (124;255) | (64;104) | (35;84) | (27;73) | (19;43) | (73;134) | (12;29) | (10;37) | (35;78) |
| C3 | 17 [170] | 166 [185] | 91 [185] | 53 [182] | 53 [185] | 22 [182] | 119 [185] | 22 [179] | 15 [185] | 38 [182] |
| (Range) | (12;47) | (108;200) | (42;118) | (35;80) | (25;81) | (11;32) | (71;167) | (12;30) | (11;30) | (22;67) |
| **Thereof solar** | | | | | | | | | | |
| C1 | 16 [39] | 134 [41] | 43 [41] | 53 [41] | 22 [41] | 33 [41] | 81 [41] | 11 [41] | 20 [41] | 33 [41] |
| (Range) | (8;24) | (89;147) | (38;55) | (37;82) | (14;34) | (16;40) | (75;95) | (10;16) | (10;25) | (17;56) |
| C2 | 10 [77] | 83 [87] | 34 [87] | 37 [87] | 16 [87] | 15 [82] | 44 [87] | 7 [80] | 5 [81] | 20 [87] |
| (Range) | (6;14) | (54;125) | (19;47) | (17;41) | (8;21) | (10;23) | (18;69) | (4;10) | (1;12) | (9;33) |
| C3 | 7 [170] | 53 [185] | 28 [184] | 23 [182] | 12 [184] | 12 [164] | 32 [185] | 9 [157] | 8 [164] | 14 [182] |
| (Range) | (3;14) | (42;83) | (17;36) | (17;39) | (5;25) | (9;20) | (21;74) | (4;11) | (3;12) | (7;27) |
| **Thereof wind** | | | | | | | | | | |
| C1 | 10 [39] | 133 [41] | 59 [41] | 45 [41] | 19 [41] | 22 [41] | 58 [41] | 20 [41] | 17 [41] | 28 [41] |
| (Range) | (4;30) | (86;164) | (29;86) | (23;71) | (15;26) | (13;39) | (44;122) | (12;25) | (10;23) | (17;52) |
| C2 | 5 [77] | 63 [87] | 41 [83] | 23 [87] | 15 [87] | 8 [81] | 31 [87] | 8 [87] | 4 [81] | 19 [87] |
| (Range) | (4;14) | (44;102) | (9;59) | (14;30) | (7;18) | (3;16) | (19;75) | (5;12) | (2;12) | (6;23) |
| C3 | 3 [170] | 64 [185] | 59 [169] | 21 [182] | 12 [184] | 10 [160] | 52 [184] | 10 [179] | 4 [164] | 10 [182] |
| (Range) | (2;15) | (40;93) | (12;65) | (12;37) | (7;22) | (5;13) | (19;86) | (6;13) | (2;10) | (5;32) |
| **Storage** | | | | | | | | | | |
| C1 | 3 [27] | 68 [32] | 46 [32] | 27 [32] | 7 [29] | 13 [30] | 56 [30] | 4 [32] | 3 [24] | 15 [30] |
| (Range) | (0;8) | (30;80) | (9;54) | (24;45) | (2;11) | (3;19) | (30;62) | (2;6) | (0;4) | (1;30) |
| C2 | 2 [36] | 19 [60] | 18 [52] | 10 [57] | 3 [42] | 3 [31] | 13 [44] | 1 [43] | 0 [20] | 3 [41] |
| (Range) | (0;4) | (6;36) | (7;35) | (4;17) | (1;8) | (0;4) | (11;34) | (1;2) | (0;0) | (2;13) |
| C3 | 4 [78] | 20 [106] | 22 [92] | 9 [107] | 9 [85] | 4 [78] | 29 [81] | 1 [90] | 0 [78] | 9 [83] |
| (Range) | (0;6) | (1;33) | (3;41) | (1;21) | (0;13) | (0;9) | (2;42) | (0;2) | (0;1) | (0;16) |
| **Transmission and distribution** | | | | | | | | | | |
| C1 | 24 [39] | 147 [39] | 67 [39] | 51 [39] | 40 [39] | 27 [39] | 87 [39] | 16 [39] | 24 [39] | 64 [39] |
| (Range) | (13;39) | (96;250) | (61;105) | (46;97) | (29;62) | (22;40) | (70;120) | (13;19) | (18;35) | (26;94) |
| C2 | 24 [77] | 132 [77] | 60 [77] | 49 [77] | 36 [77] | 33 [77] | 70 [77] | 14 [77] | 26 [77] | 36 [77] |
| (Range) | (14;30) | (84;175) | (48;79) | (43;56) | (28;45) | (27;37) | (53;92) | (8;19) | (17;34) | (28;61) |
| C3 | 14 [150] | 93 [153] | 61 [153] | 46 [150] | 26 [153] | 25 [150] | 70 [153] | 14 [147] | 23 [153] | 26 [150] |
| (Range) | (10;37) | (74;190) | (52;86) | (38;86) | (21;62) | (17;40) | (52;90) | (11;16) | (17;27) | (17;87) |
| C5 | 13 [109] | 81 [110] | 55 [110] | 41 [109] | 25 [110] | 23 [109] | 58 [110] | 14 [109] | 23 [110] | 25 [109] |
| (Range) | (9;13) | (67;160) | (46;59) | (22;46) | (19;28) | (15;28) | (51;67) | (12;16) | (16;26) | (17;29) |

Note: Average yearly investments from 2023–2032 for electricity generation capacity, by aggregate regions (in billion $USD_{2015}$). Further notes see Table 15.2. Reference C5 category for Transmission and Distribution shown in Table 15.2 as it is used for calculation of incremental needs for Figure 15.4. Vertical axis, C4–C8 except Transmission and Distribution not shown. Data source: AR6 Scenarios Database.

What is apparent is that the bulk of investment requirements corresponds to medium- and low-income countries in Asia, Latin America, the Middle East and Africa, as these still have growing energy demand, and it is still considerably lower than the global average. This illustrates a vital opportunity to ensure the build-up of sustainable energy infrastructures in these regions and constitutes a risk of additional carbon lock-in if investments into fossil infrastructures, especially coal-fired power plants, and uncontrolled urban expansion, continue.

Investment needs in electrification derived from IAMs do not include systematically investments in end-use equipment and distribution (Box 4.8 in SR1.5 (IPCC 2018)). Model-based estimates of investment needs don't have the regional granularity to single out

15

LDCs, as model regions typically are defined based on geographic proximity and therefore aggregate LDCs and other countries. With the average electricity consumption per capita in Africa increasing to 0.68–0.87 (1.43–2.92) MWh in 2030 (2050) $yr^{-1}$ and remaining at the very low end of the global range [0.46 in Africa compared to the upper end of 12.02 in North America, MWh per capita and year in 2020], the targeted full electrification until 2030 appears unrealistic across all scenarios. SEforAll and IEA estimate assumed investment needs to decentralised end-user electrification to come in around USD40 billion on average until 2030 (SEforALL and CPI 2020; IEA 2021d).

**Quantitative analysis of investment needs in energy generation based on IRENA and IEA data and comparison to AR6 scenario database output.**

According to IRENA, the government plans in place today call for investing at least USD95 trillion in energy systems over the coming three decades (2016–2050) (IRENA 2020c). Redirecting and increasing investments to ensure a climate-safe future (Transforming Energy Scenario, TES) would require reaching on average around 1 trillion $USD_{2015}$ $yr^{-1}$ (average until 2030) for electricity generation as well as grids and storage, increasing to above 2 trillion $USD_{2015}$ $yr^{-1}$ (average until 2030) in the 1.5 scenario (IRENA 2021). IEA's respective SDS and NZE scenarios come in at average annual investments between USD1.0 trillion $yr^{-1}$ and USD1.6 trillion $yr^{-1}$ (average until 2030) (IEA 2021b). These additional data points for the C1 and C3 category underpin the range presented in the AR6 Scenarios Database for needs until 2032 despite the slightly varying periods.

In contrast to the IAMs, IRENA and IEA assessments do not allow for an analysis of mitigation-driven investment needs in transmission and distribution, which likely results in an overestimation of the mitigation-driven investment needs in their analysis.

It is worth highlighting that driven by technology cost assumptions, IRENA forecasts falling average annual investments needs for energy, but also energy efficiency, for the period 2030–2050 compared to 2020–2030. In the 1.5°C scenario (1.5-S) the total annual investment needs excluding fossils and nuclear decrease from 5.0 trillion $USD_{2015}$ until 2030 $yr^{-1}$ to 3.8 trillion $USD_{2015}$ $yr^{-1}$ for 2030–2050 (IRENA 2021). In IAM scenarios of Category C1, electricity supply investments (including generation, transmission and distribution, and storage) remain flat at 2.2 trillion $USD_{2015}$ $yr^{-1}$ through the coming three decades in absolute terms. Given rising GDP, the complementary methods and sources thus consistently point to a peak in electricity supply investments as a percentage of GDP in mitigation scenarios in the coming decade. This reflects the fact that the coming decade requires low-carbon power generation investments to both cover the demand increase and (partly premature) replacement of fossil generation capacities, both concentrated in emerging and developing countries. Relative investment numbers for electricity measured against GDP then decrease towards 2050, as they only need to cover natural replacement and increasing demands (which due to electrification will also pick up in developed countries), and due to further declining technology costs. Investments for low-carbon fuel supply like hydrogen and synthetic fuels, and for direct electrification equipment (heat pumps, electric vehicles (EV), etc.) scale up from much lower levels and will likely continue to grow as a share of GDP until mid-century, though uncertainties and accounting is still much more uncertain. (Bertram et al. 2021).

**Quantitative analysis of investment needs in other sectors.** As described above, investment needs in non-energy sectors tend to be ignored in many integrated assessment models with studies for individual countries or regions providing a more fragmented picture only. However, the quality of estimates is likely not to be less robust given the drawbacks of integrated assessment models.

Chapter 7 stresses the importance of opportunity costs for AFOLU mitigation options, in particular for afforestation and avoided deforestation projects, and derives net annual costs of around USD278 billion $yr^{-1}$ in the next several decades, mostly opportunity costs. Net costs of delivering 5-6 Gt $CO_2$ $yr^{-1}$ of forest related carbon sequestration and emission reduction around 2050 as assessed with sectoral models are estimated to reach to ~ USD400 billion $yr^{-1}$ by 2050, excluding externality costs (Chapter 7.4).

**Energy efficiency.** Estimates on energy investment needs vary significantly with a low level of transparency with regard to underlying technology cost assumptions burdening the confidence levels.

IRENA only selectively reports financing needs for energy efficiency in buildings and industry as separate categories. For the 1.5-S average $yr^{-1}$ needs until 2050 come in at 963 billion $USD_{2015}$ for buildings, 102 billion $USD_{2015}$ for heat pumps, and 354 billion $USD_{2015}$ for industry. Applying the relative share of these categories on higher total needs until 2030, around 1.8 trillion $USD_{2015}$ $yr^{-1}$ in buildings and industry are needed in the 1.5-S. For the TES cumulative energy efficiency investment needs until 2030 are stated at 29 trillion $USD_{2015}$ translating into an yearly average of around 1.7 trillion $USD_{2015}$ $yr^{-1}$, excluding transportation. IEA estimates come in at a much lower level at 0.6 and 0.8 billion $USD_{2015}$ $yr^{-1}$ on average between 2026–2030 for their SDS and NZE scenarios.

**Transportation.** For the transportation sector, OECD has presented the most comprehensive assessment of financing needs in the AR6 database based on IEA data with the annual average coming in at USD2.7 trillion between 2015 and 2035 i In modelled global pathways that limit warming to 2°C (>67%). The assessment comprises road, rail and airports/ports infrastructure, with only rail infrastructure being considered in this analysis.

On a regional level, Oxford Economics (2017) shows that annual infrastructure investments between 2016 and 2040 vary widely. For all available countries (n=50) estimates count close to 0.4 trillion $USD_{2015}$ $yr^{-1}$, including 0.217 trillion $USD_{2015}$ $yr^{-1}$ for China. Based on available data for nine African countries, investments in rail infrastructure range from USD0.1 billion in Senegal to USD1.6 billion in Nigeria. Osama et al. (2021) highlight a USD4.7 billion financing gap for African countries in the transport sector. In Latin America Oxford Economics (2017) identifies Brazil as frontrunner of required rail investments with USD8.3 billion, followed by Peru with USD2.3 billion.

In total, developed countries' financing needs mount up to almost USD120 billion $yr^{-1}$ (n=15, mean=7.97bn USD) for rail infrastructure. Financing needs in developing countries (excluding LDCs and excluding China) mount up to almost USD50 billion $yr^{-1}$ (n=27, mean=1.78bn USD, excluding China). Oxford Economics (2017) reports rail infrastructure financing needs for China of more than USD200 billion $yr^{-1}$ between 2016 and 2040.

Fisch-Romito and Guivarch (2019) show, by endogenising the impact of urban infrastructure policies on mobility needs and modal choices that transportation investment needs globally might be lower in low-carbon pathways compared with baselines, with lower investments in road and air infrastructure. This does mean that higher investments are not needed over the following two decades; this is confirmed by Rozenberg and Fay (2019) that strong policy integration between urban, transportation and energy policies reduce the total investment gap.

IRENA as well as IEA have presented estimates for energy efficiency investments in the transport sector. For the 1.5-S scenario, IRENA indicates average investment needs of $USD_{2015}$ 0.2 trillion $yr^{-1}$ for EV infrastructure, $USD_{2015}$ 0.2 trillion $yr^{-1}$ for transport energy efficiency and $USD_{2015}$ 0.3 trillion $yr^{-1}$ for EV batteries (average until 2030) (IRENA 2020d). IEA indicates a total of around 0.6 and 0.7 trillion $USD_{2015}$ $yr^{-1}$ for transport energy efficiency in the SDS and IEA scenarios for the 2026–2030 period (IEA 2021c). Many investment categories relating to mitigation options, in particular with regard to behavioural change and transport mode changes (Chapter 10, Figure SPM.8), are neglected in these analyses despite their significant mitigation potential.

**AFOLU.** The Food and Land Use Coalition estimates additional investment needs for ten critical transitions for the global food and land use systems to achieve the long-term global goal (LTGG) and SDGs. Additional annual investment needs until 2030 add up to USD300–350 billion. Considering the change in global diets as well as the land-based nature-based solutions only, annual investment needs would come in between USD110–135 billion. Chapter 7 stresses the importance of opportunity costs for AFOLU mitigation options, in particular for afforestation projects, and derives average yearly investment needs of around 278 billion $USD_{2015}$ $yr^{-1}$ until 2030 rising to 431 billion $USD_{2015}$ $yr^{-1}$ over the next several decades, including opportunity costs. The estimate is based on an assumption of emission reductions consistent with pathways C1–C4, leading to average abatement of 9.1 $GtCO_2$ $yr^{-1}$ (median range 6.7–12.3 $GtCO_2$ $yr^{-1}$) from 2020–2050 and marginal costs of USD100 per tonne $CO_2$, excluding investments in bioenergy with carbon capture and storage and changes in food consumption and food waste (Section 7.4). The largest investments are projected to occur in Latin America, South-East Asia, and Africa, constituting 61% of total expenditure. The implied change of land use might trigger negative effects on other SDGs which need to be addressed to offer robust safeguards and labelling for investors.

However, given the strong interlinkage of the presented transitions and accumulated effects, climate change related investments can hardly be separated (The Food and Land Use Coalition 2019). Shakhovskoy et al. (2019) present an overview of financing needs of small-scale farmers globally, however, without focusing on the required climate-related investments. According to their assessment, 270 million smallholder farmers in South and South-East Asia, sub-Saharan Africa and Latin America face approximately USD240 billion of financing needs, thereof USD100 billion short-term agricultural needs, USD88 billion long-term agricultural needs and USD50 billion non-agricultural needs (Shakhovskoy et al. 2019). These numbers can only provide 'an indication of the magnitude of the climate investments required in small-scale agriculture' (CPI 2020). Table 15.4 summarises the studies used as well as adjustments made to determine needs for the gap discussion in Section 15.5.2.

**Adaptation financing needs.** Financing needs for adaptation are even more difficult to define than those of mitigation because mobilising specific adaptation investments is only part of the challenge since ultimately improving societies' adaptive capacities depends on

**Table 15.4 | Sector studies to determine average financing needs.**

| Sector | Studies | Global ranges trillion USD $yr^{-1}$ – *Confidence Level* | | Regional breakdown | | Comment |
|---|---|---|---|---|---|---|
| Energy | IAM database, SEforAll (SEforALL and CPI 2020), IRENA 1.5-S and TES scenarios (IRENA 2021), IEA SDS and NZE scenarios (IEA 2021b) | 0.8–1.5 | *High confidence* | Detailed breakdown for R10 possible for IAM database and applied to the derived range | *Medium confidence* | Wide ranges primarily driven by varying assumptions with regard to grid investments relating to the increased renewable energy penetration. |
| Energy Efficiency | IRENA 1.5-S and TES scenarios, IEA SDS and NZE scenarios | 0.5–1.7 | *Medium confidence* | Adjustments required to regional categorisation by IEA and IRENA | *Low-medium confidence* | Medium confidence levels due to missing transparency with regard to underlying assumptions on technology costs. Low-to-medium confidence level on regional allocations due to required adjustments. |
| Transport | OECD/IEA (OECD 2017b) and Oxford Economics (2017) on rail investment data, IRENA 1.5-S and TES scenarios, IEA SDS and NZE scenarios for transport (energy efficiency) and electrification | 1.0–1.1 | *Medium confidence* | Adjustments required to regional categorisation by IEA and IRENA | *Low-medium confidence* | Needs including battery costs, not total costs, of electric vehicles, likely underestimation of needs due to missing data points on rail infrastructure. |
| AFOLU | Chapter 7 analysis, Section 7.4; The Food and Land Use Coalition (Land use Coalition (2019); (Shakhovskoy et al. 2019) | 0.1–0.3 | *High confidence* | Breakdown for R10 possible for Chapter 7 analysis | *Medium confidence* | Upper end of range includes opportunity costs as these likely increase costs of investment in land. |

Note: Total range USD2.3 trillion to USD4.5 trillion $yr^{-1}$.

the SDGs' fulfilment (Hallegatte et al. 2016). Bridging the investment gap on irrigation, water supply, health care, energy access, and quality buildings is an essential enabling condition for adapting to climate change. The scenario analysis conducted by Rozenberg and Fay (2019) show that fulfilling the SDGs to improve the adaptive capacity of low- and middle-income countries would require investments in water supply, sanitation, irrigation and flood protection that would account for about 0.5% of developing countries' GDP in a baseline scenario to 1.85% and 1% with a strong and anticipatory policy integration (USD664 billion and 351 billion on average by 2030).

Most studies choose to assess public sector projects, ignoring household-level investments as well as private sector adaptation (UNEP 2018; Buchner et al. 2019). UNEP's 2020 Adaptation Gap Report estimates adaptation costs amounting to 140–300 billion USD $yr^{-1}$ in 2030 and USD280–500 billion $yr^{-1}$ in 2050 (UNEP 2021). Over 100 countries included adaptation components in their intended NDCs (INDCs) and approximately 25% of these referenced national adaptation plans (NAPs) (GIZ 2017a) but estimates of the financing required for NAP processes is not available. These NAPs, as formally agreed under the UNFCCC in 2010, are iterative, continuous processes that have multiple stages with a developmental phase that requires country-specific financing of primarily which comprises grants, bond issuance or debt conversion (NDC Partnership 2020, NAP Global Network 2017). At the same time, multilateral climate funds such as the Green Climate Fund and the GEF/Least Developed Countries Fund offer 'readiness and preparatory support' and implementation for the NAPs and adaptation planning process (GCF 2020a; GEF 2021a,b). There has been no significant updating of adaptation cost estimates since UNEP's (UNEP 2016, 2018). The Global Commission on Adaptation makes the case that investing USD1.8 trillion in early warning system, climate-resilient infrastructure, global mangrove and resilient water resources would generate about USD1.7 trillion in benefits due to avoided cost and non-monetary and social resources (Verkooijen 2019; UNEP 2021).

There is increasing recognition of rising adaptation challenges and associated costs within and across developed countries. Undoubtedly many developed countries are spending more on a wide range of adaptation issues, both as preventive measures and building resilience (greening infrastructure, climate-proofing major projects and managing climate-related risks) against the impacts of climate change extreme weather events (US GCRP 2018a). Developed countries' climate change adaptation spending covers areas such as federal insurance programmes, federal, state and local property and infrastructure, supply chains, and water systems.

## 15.5 Considerations on Financing Gaps and Drivers

### 15.5.1 Definitions

The analysis of financing gaps in climate action, which is used to measure implementation action and mitigation impact (FS-UNEP Centre and BNEF 2019) cannot be carried out as a pure demand-side challenge, in isolation from the analysis of barriers to deploy funds (e.g., Ramlee and Berma 2013) and to take investment initiatives. These barriers are 'friction that prevents socially optimal investments from being commercially attractive' (Druce et al. 2016). They are at the root of the 'microeconomic paradox' of a deficit of infrastructure investments despite a real return between 4% and 8% (Bhattacharya et al. 2016), of the low share of carbon-saving potentials tapped by dedicated policies such as energy renovation programmes (Ürge-Vorsatz et al. 2018), and, more generally of a demand for climate finance lower than the volume of economically viable projects (de Gouvello and Zelenko 2010; Timilsina et al. 2010).

A few exercises tried assess the consequences of the perpetuation of these drivers on the magnitude of the financing gap. They suggest, comparing the evolution of the infrastructure investment trends (beyond energy) by comparison with what they should be in an optimal scenario, a cumulative deficit between 19% (Oxford Economics 2017) and 32% (Arezki et al. 2016). The volume of this gap is of the same order of magnitude as the incremental infrastructure investments (energy and beyond) for meeting a 1.5°C target (2.4% of the world GDP on average) (Box 4.8 of SR1.5 (IPCC 2018)) calculated by exercises assuming no pre-existing investment gap. This figure is consistent with the 1.5% to 1.8% assessed by the European Commission (2020) for Europe and the 2% of the IMF (2021d) for the G20, which do not encompass many developing countries for which economic take-off is today fossil fuels dependent. For low- and middle-income economies, Rozenberg and Fay's (2019) results suggest to increase the infrastructure investments by 2.5 to 6 percentage points of GDP to cover both the reduction of the structural investment gap and the specific additional costs for bridging it with low-carbon and climate-resilient options. These assessments indicate the challenge at stake but do not exist at very disaggregated sectoral and regional levels for sectors other than energy.

The below quantitative analysis does not differentiate between financing gaps driven by barriers within or outside the financial sector given that the IAM models as well as most other studies used do not incorporate actual risk ranges depending on policy strength and coherence and institutional capacity, low-carbon policy risks, lack of long-term capital, cross-border currency fluctuation, and pre-investment costs and barriers within the financial sector that discourage private sector financing. They comprise short-termism (UNEP Inquiry 2016b), high perceived risks for mitigation-relevant technologies and/or regions (information gap through incomplete/asymmetric information, (Kempa and Moslener 2017; Clark et al. 2018)), lack of carbon pricing effects (Best and Burke 2018), home bias (results in limited balancing for regional mismatches between current capital and needs distribution, (Boissinot et al. 2016)), and perceived high opportunity and transaction costs (results from limited visibility of future pipelines and policy interventions; SME financing tickets and the missing middle, (Grubler et al. 2016)). In addition, barriers outside the financial sector will have to be addressed to close future financing gaps. The mix and dominance of individual barriers might vary significantly across sectors and regions and is analysed below.

The interpretation of the quantitative analysis thus needs to be performed, taking into account the qualitative needs assessment in Section 15.4.1 and the evolution of parameters that determine the risk-weighted relative attractiveness of low-carbon and climate-resilient investments compared to other investment opportunities. With some institutions having announced climate finance commitments and/or targets (see also Box 15.4), the actual asset allocation of commercial financial sector players including sectoral and regional focus will respond to tangible and financially viable investment opportunities available in the short term. Robust long-term pathways to create such conditions for a significant private sector involvement rarely exist and expectations on private sector involvement in some critical sectors/regions might be too high (Clark et al. 2018).

### 15.5.2 Identified Financing Gaps for Sector and Regions

The following section compares recent climate finance flows as reported by CPI and IEA to needs derived in Section 15.4, ignoring the slight mismatch in time horizons. The analysis ignores interlinked gaps, in particular infrastructure investment gaps and other SDG-related investment gaps, which need to be addressed in parallel to reach the LTGG but also at least partially to facilitate green investments.



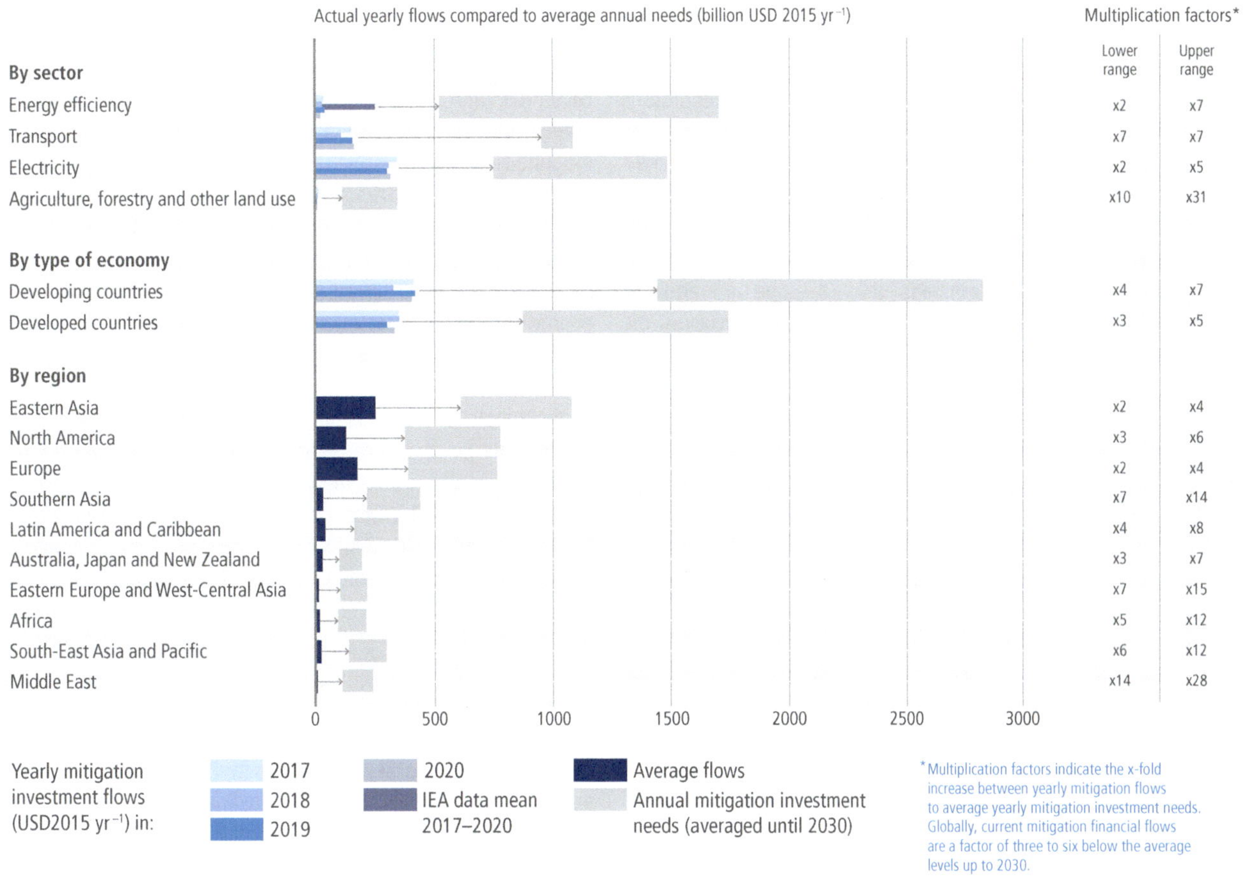


**Figure 15.4 | Breakdown of recent average (downstream) mitigation investments and model-based investment requirements for 2020–2030 (USD billion) in scenarios that likely limit warming to 2°C or lower.** Mitigation investment flows and model-based investment requirements by sector / segment (energy efficiency in buildings and industry, transport including efficiency, electricity generation, transmission and distribution including electrification, and agriculture, forestry and other land use), by type of economy, and by region (see Annex II Part I Section 1: By region is based on intermediate level (R10) classification scheme. By type of economy is based on intermediate level (R10) classification scheme, which considers 'North America', 'Europe', and 'Australia, Japan and New Zealand' as developed countries, and the other seven regions as developing countries). Breakdown by sector / segment may differ slightly from sectoral analysis in other contexts due to the availability of investment needs data. The granularity of the models assessed in Chapter 3, and other studies, do not allow for a robust assessment of the specific investment needs of LDCs or SIDSs. Investment requirements in developing countries might be underestimated due to missing data points as well as underestimated technology costs. In modelled pathways, regional investments are projected to occur when and where they are cost cost-effective to limit global warming. The model quantifications help to identify high-priority areas for cost-effective investments, but do not provide any indication on who would finance the regional investments. Investment requirements and flows covering downstream / mitigation technology deployment only. Data includes investments with a direct mitigation effect, and in the case of electricity, additional transmission and distribution investments. See section 15.4.2 Quantitative assessment of financing needs for detailed data on investment requirements. Data on mitigation investment flows are based on a single series of reports (Climate Policy Initiative, CPI) which assembles data from multiple sources. Investment flows for energy efficiency are adjusted based on data from the International Energy Agency (IEA). Data on mitigation investments do not include technical assistance (i.e., policy and national budget support or capacity building), other non-technology deployment financing. Adaptation only flows are also excluded. Data on mitigation investment requirements for electricity are based on emission pathways C1, C2 and C3 (Table SPM.1). For electricity investment requirements, the upper end refers to the mean of C1 pathways and the lower end to the mean of C3 pathways. Data points for energy efficiency, transport and AFOLU cannot always be linked to C1–C3 scenarios. Data do not include needs for adaptation or general infrastructure investment or investment related to meeting the SDGs other than mitigation, which may be at least partially required to facilitate mitigation. The multiplication factors show the ratio of average annual model-based mitigation investment requirements (2020–2030) and most recent annual mitigation investments (averaged for 2017–2020). The lower and upper multiplication factors refer to the lower and upper ends of the range of investment needs.

Given the multiple sources and lack of harmonised methodologies, the data can only be indicative of the size and pattern of investment gaps. The gap between most recent flows and required investments is only a single indicator. A more comprehensive (and qualitative) assessment is required in order to understand the magnitude of the challenge of scaling up investment in sectors and regions. The analysis also does not consider the effects of misaligned flows. {15.3, 15.4, 15.5, Table 15.2, Table 15.3, Table 15.4}

Total investments in mitigation need to increase by around three and six times with significant gaps existing across sectors and regions[8] (*high confidence*). The findings on still significant gaps and limited progress over the past few years to some extent seem to contradict the massive increase in commitments by financial institutions. As discussed in Section 15.6, the investment gap is not due to global scarcity of funds.

However, these investment gaps have little explanatory power in terms of the magnitude of the challenge to mobilise funding. In addition to measurement challenges from different definitions and data gaps, sectors and regions offer highly divergent financial risk-return profiles, in particular due to missing or weak regulatory environments consistent with ambitions levels, and economic costs as well as limited local capital markets, limited institutional capacity to ensure safeguard, standardisation, scalability and replicability of investment opportunities and financing models, and a pipeline ready for commercial investments. Moreover, soft costs and institutional capacity for enabling environment that can be prerequisite for addressing financing gaps are ignored when focusing on investment cost needs.

**Sectoral considerations.** The renewable energy sector attracted the highest level of financing in absolute and relative terms with business models in generation being proven and rapidly falling technology costs driving the competitiveness of solar photovoltaic and onshore wind, even without taking account of the mitigation component (FS-UNEP Centre and BNEF 2019; IRENA 2020a). This investment activity comes in line with the first generation of NDCs and their heavy focus on mitigation opportunities in the renewable energy sector (Pauw et al. 2016; Schletz et al. 2017). Still, the investment gap tends to remain stable with flows over the past years not showing an upward trend.

Comparing annual average total investments in global fuel supply and the power sector of approximately USD1.5 trillion[9] $yr^{-1}$ in 2019 (IEA 2020a) to the investment in the Stated Policies Scenario (approximately 1.7 trillion $USD_{2015}$ $yr^{-1}$) and the Sustainable Development Scenario (approximately 1.8 trillion $USD_{2015}$ $yr^{-1}$) in 2030 underlines the required shift of existing capital investment from fossil to renewables even more than the need to increase sector allocations (Granoff et al. 2016; McCollum et al. 2018).

Ensuring access to the heavily regulated electricity markets is a key driver for an accelerated private sector engagement (IFC 2016; FS-UNEP Centre and BNEF 2018; REN21 2019), with phasing out of support schemes and regulatory uncertainty being a major driver for reduced investment volumes in various regional markets in the past years (FS-UNEP Centre and BNEF 2015, 2016, 2017, 2018, 2020). Strategic investors and corporate investments by utilities dominate the investment activity in developed countries and countries in transition (BNEF 2019) based on the competitiveness of renewable energy sources. Reasonable auction results based on a substantial private-sector competition for investments have also been achieved in selected developing countries driven by rather standardised contract structures and the increased availability of risk mitigation instruments addressing political and regulatory risks and home bias constraints (FS-UNEP Centre and BNEF 2019; IRENA 2020a). Development finance institution (DFI) climate portfolios tend to be driven by concessional loans for renewable energy generation assets with equity often being provided by (semi-) commercial investors (Section 15.3) which will have to change to accelerate renewable energy investment activity.

Given the wide range of estimates on current investment flows into energy efficiency, substantial uncertainty exists with regard to the magnitude of the investment gaps. While CPI publishes investment levels of 41 billion $USD_{2015}$ in 2019 and 24 billion $USD_{2015}$ in 2020 for energy efficiency, counting majorly international flows, IEA results come in at a much higher level of around 250 billion $USD_{2015}$ annually between 2017 and 2020 (IEA 2021c) and IRENA (2020c) estimates energy efficiency investments in buildings between 2017–2019 at an average of USD139 billion $yr^{-1}$.

Public sector investments in the transport sector have increased significantly in the past years reflecting the increased interest of capital markets in renewable energy and the efficient and corresponding reallocation of public funding. Provision of funding by capital markets for public transport infrastructure among others heavily depends on suitable financing vehicles and increased funding for development of projects with a low level of standardisation (OECD 2015a).

Both IRENA and IEA include only incremental costs of EVs in their estimates on needs while CPI, when measuring actual flows, includes those at full costs. Total private flows for EVs included in CPI numbers amount to USD41 billion in 2018 (Buchner et al. 2019), representing more than 80% of private sector finance into the transport sector, around one third of total public and private funding to the transport sector in 2018. This likely results in an underestimation of the financing gap – in addition to the fact that estimates for investment needs for rail infrastructure are only available for selected countries.

Current financing of land-based mitigation options is less than USD1 billion $yr^{-1}$ representing only 2.5% of climate mitigation funding, significantly below the potential proportional contribution (Buchner et al. 2019). A stronger focus on deforestation-free value chain, including a stronger reflection in taxonomies and financial sector investment decision processes are necessary to *ensure* an alignment of financial flows with the LTGG. Taking into account the specifics of land-based mitigation (in particular long investment horizons, strong dependency on the monetisation of mitigation effects, strong public sector involvement) a significant scale-up of commercial financing to the sector can hardly be expected in the absence of strong climate

8 In modelled pathways, regional investments are projected to occur when and where they are most cost-effective to limit global warming. The model quantifications help to identify high-priority areas for cost-effective investments, but do not provide any indication on who would finance the regional investments.

9 In the chapter, USD units are used as reported in the original sources in general. Some monetary quantities have been adjusted selectively for achieving comparability by deflating the values to constant US Dollar 2015. In such cases, the unit is explicitly expressed as $USD_{2015}$.

policies (Clark et al. 2018). Agriculture is likely to develop more potential to mobilise private finance than the forest sector given its strong linkage to food security and hunger and shorter payback periods. The significant gap in land-based mitigation finance also indicates the crucial lack of finance to the bottom of the pyramid.

Agricultural support is an important source of distortions to agricultural incentives in both rich and poor countries (Mamun et al. 2019) ranging from the largest component of the support, market price supports, increased gross revenue to farmers as a result of higher prices due to market barriers created by government policies, to production payments and other support including input subsidy (e.g., fertiliser subsidy) (Searchinger et al. 2020). USD600 billion of annual governmental support for agriculture in the OECD database contributes only modestly to the related objectives of boosting crop yields and just transition (Searchinger et al. 2020). A review of NDCs of 40 developing countries which submitted a NDC to the UNFCCC Interim NDC Registry by April 2017, and include within their NDC efforts to REDD+ via support from the UN-REDD Programme and/or World Bank Forest Carbon Partnership Facility, indicates that none of the countries reviewed mention fiscal policy reform of existing finance flows to agricultural commodity production or other publicly supported programmes that affect the direct and underlying drivers of land use conversion (Kissinger et al. 2019).

**Analysis by region and type of economy.** The analysis of gaps by type of economy illustrates the challenge for developing countries. Estimated mitigation financing needs as a percentage of mean 2017–2020 GDP in $USD_{2015}$ comes in at around 2–4% for developed countries, and around 4-9% for developing countries (*high confidence*) (Figure 15.4). Climate finance flows have to increase by a factor of four to seven in developing countries and three to five in developed countries. This disparity is further exacerbated when considering adaptation, infrastructure and SDG-related investment needs (*high confidence*) (Hourcade et al. 2021a). However, differences across developing countries are significant. Flows to Eastern Asia, with its annual average flows (2017–2020) of 252 billion $USD_{2015}$ being dominated by China (more than 95% of total mitigation flows to Eastern Asia), would have to increase by a factor of two to four, a comparable level to developed countries. Section 15.6.2 elaborates on outlooks with regard to fiscal space and ability to tap capital markets, in particular for developing countries. In particular, attention must accelerate on low-income Africa. This large continent currently contributes very little to global emissions, but its rapidly rising energy demands and renewable energy potential versus its growing reliance on fossil fuels and 'cheap' biomass (especially fuelwood for cooking and charcoal, with impacts on deforestation) amid fast-rising urbanisation makes it imperative that institutional investors and policymakers recognise the very large 'leap-frog' potential for the renewable energy transition as well as risks of lock-in effects in infrastructure more generally in Africa that is critical to hold the global temperatures rise to well below 2°C in the longer term (2020–2050). Overlooking this transition opportunity, rivalling China, India, USA and Europe, would be costly. Policies centred around the accelerated development of local capital markets for energy transitions – with support from external grants, supra-national guarantees and recognition of carbon remediation assets – are crucial options here, as in other low-income countries and regional settings. Notably, climate finance flows to African countries might have even decreased for mitigation technology deployment (stagnated for adaptation between 2017 and 2020), widening the finance gap in African countries in the recent years (*high confidence*).

Over 80% of climate finance is reported to originate and stay within borders, and even higher for private climate flows (over 90%) (Boissinot et al. 2016). There are multiple reasons for such 'home bias' in finance – national policy support, differences in regulatory standards, exchange rate, political and governance risks, as well as information market failures. The extensive home bias means that even if national actions are announced and intended to be implemented unilaterally and voluntarily, the ability to implement them requires access to climate finance which is constrained by the relative ability of financial and capital markets at home to provide such financing, and access to global capital markets that requires supporting institutional policies in source countries. 'Enabling' public policies and actions locally (cities, states, countries and regions), to reduce investment risks and boost domestic climate capital markets financing, and to enlarge the pool of external climate financing sources with policy support from source capital countries thus matters at a general level. The biggest challenge in climate finance is likely to be in developing countries, even in the presence of enabling policies and quite apart from any other considerations such as equity and climate justice (Klinsky et al. 2017) or questions about the equitable allocations of future 'climate budgets' (Gignac and Matthews 2015). The differentiation between developed and developing countries matters most on financing. Most developed countries have already achieved very high levels of incomes, have the largest pool of capital stock and financial capital (which can be more easily redeployed within these countries given the home bias of financial markets), the most well-developed financial markets and the highest sovereign credit ratings, in addition to starting with very high levels of per capita carbon consumption – factors that should allow the fastest adjustment to low-carbon investments and transition in these countries from domestic policies alone. The financial and economic circumstances are more challenging in many developing countries, even within a heterogeneity of circumstances across countries. The dilemma, however, is that the fastest rates of the expected increase in future carbon emissions are in developing countries. The biggest challenge of climate finance globally is thus likely to be the constraints to climate financing because of the opportunity costs and relative under-development of capital markets and financing constraints (and costs) at home in developing countries, and the relative availability or absence of adequate financing policy support internationally from developed countries. The Paris Agreement and commitment by developed countries to support the climate financing needs of developing countries thus continue to matter a great deal.

**Soft costs/institutional capacity** (Osama et al. 2021). Most funding needs assessments focus on technology costs and ignore the cascade of financing needs as outlined above. International grant funding or national budget allocations for soft costs like the creation of a regulatory environment can be a prerequisite for the supply of commercial financing for the deployment of technologies. Such critical funding needs might represent a small share of overall investment needs but current (relatively small) gaps in funding of policy reforms can hinder or delay deployment of large volumes of funding in later years. The role, as well as the approximate volumes of such required timely international grant funding or national budget allocations, appear underestimated in research. The numbers available for the creation of an enabling environment for medium-sized renewable energy (RE) projects in Uganda (GET FiT Uganda 2018) are illustrative only and cannot be transferred as assumptions to other countries without taking into account potentially varying starting points in terms of institutional readiness, pipelines, as well as the general business environment. GET FiT Uganda supported 170 MWp of medium-scale RE capacity triggering investments of USD453 million (GET FiT Uganda 2018), international results-based incremental cost support amounted to USD92 million and project preparation, technical assistance, and implementation support, required USD8 million, excluding support from national agencies.

There is strong evidence of the correlation between institutional capacity of countries and international climate finance flows towards those economies (Adenle et al. 2017; Stender et al. 2019) and a strong need for robust institutional capacity to manage the transformation in a sustainable and human rights based way (Duyck et al. 2018). One example to consider unaddressed social concerns is the ongoing call for feedback by the European Commission and its platform on sustainable finance. It argues for a social taxonomy, that can support the identification of financing opportunities for economic activities contributing to social objectives (European Commission 2021b). SEforAll has highlighted the issue of investments not going to the countries with the greatest need, also partly driven by institutional capacity levels (SEforALL and CPI 2020). Also, most of the developing countries' NDCs are conditional upon international support for capacity building (Pauw et al. 2020). The Climate Technology Centre and Network (CTCN) was created as an operational arm of the UNFCCC Technology Mechanism with the mandate to respond to requests from developing countries. Initial evaluations of the mechanism underpin its importance and value for developing countries but stress long lead times and predictability of future international public finance to maintain operations as key challenges (UNFCCC 2017; DANIDA 2018). While limited pipelines, limited absorptive capacities as well as restricted institutional capacity of countries are often stated as challenges for an accelerated deployment of finance (Adenle et al. 2017), the question remains on the role of international public climate finance to address this gap and whether a concrete current financing gap exists for patient institutional capacity building. While current short-term, mostly project-related, capacity building often fails to meet needs but alternative, well-structured patient interventions and finance could play an important role (Saldanha 2006; Hope 2011) accepting other barriers than financing playing a role as well. One reason why international public climate finance is not sufficiently directed to such needs might be the complexity in measuring intangible, direct outcomes like improved institutional capacity (Clark et al. 2018).

**Early stage/venture capital financing/pilot project financing.** Early-stage companies in impact investment sectors with business solutions can contribute positively to climate impact. Figure SPM.8 highlights the need for new business models facilitating parts of the behavioural change. Also, SE4All has underpinned the need for an expansion of available business models to achieve universal access (SEforALL and CPI 2020). Further research and development needs range from resource efficiency of proven technologies and next generation technologies but also new technologies (Chapter 16). Access to early stage financing remains critical with performance in recent years being weak (Gaddy et al. 2016). This historically weak performance of clean tech start-ups burdens the interest of investors in the sector on the one hand and discourages experienced executive talent (Wang and Yee 2020). Besides that, the concentration of venture capital markets in the USA, Europe and India represents a major challenge (FS-UNEP Centre and BNEF 2019; Statistica 2021). With regard to commercial-scale demonstration projects, IEA estimates a need of USD90 billion of public sector finance before 2030 having around USD25 billion already planned by governments to 2030 (IEA 2021c).

**Need for parallel rather than sequential investment decisions.** The needs and gaps assessment does not include upstream investment needs required to facilitate the technology deployment as foreseen in the scenarios presented above. For example, for their transforming energy scenario IRENA estimates the number of EVs to increase from around 8 million units in 2019 to 269 million units in 2030 (IRENA 2020c). This would require investments in battery factories amounting to approximately USD207 billion with further investment requirements in the value chain (IRENA 2020d). This illustrates the extent of parallel investments based on goals rather than concrete regulatory interventions and/or demand and poses a problem of upfront investment risks for each industry in the chain in the absence of certainty of the presence of parallel decisions in the upstream and downstream links in the chain. This is a typical element of the 'valley of the death' of innovation (Scherer et al. 2000; Åhman et al. 2017). It discourages risk-taking and slows down the learning-by-doing processes, economies of scale and increasing returns to adoption needed for lowering the costs of systemic technical change (Kahouli-Brahmi 2009; Weiss et al. 2010). Implications for risk perception, financing costs as well as investment decision-making processes and ultimately for feasibility are rarely considered.

**Finance for adaptation and resilience.** As explained early, the reduction of the infrastructure gap to increase societies' resilience and the implementation of the NAPs will require more and higher levels of sustained financing. Activities mobilised for adaptation and resilience are often not marketable and their financing will continue coming from the public sector (Murphy and Parry 2020) and, at the international level, from grants-based technical assistance or through budgetary support or basket finance for large projects/programmes or sector-wide approaches or multilateral finance under (Non-) UNFCCC[10] that also anticipate supporting NAP implementation – particularly those involving incremental costs and co-benefits,

[10] Those under the UNFCCC, such as the GCF through its USD3 million per country readiness and preparatory support programme, the Least Developed Countries Fund (LDCF) and the Special Climate Change Fund (SCCF), the Pilot Program for Climate Resilience (PPCR) and the Adaptation for Smallholder Agriculture Programme (ASAP) are focused on supporting the preparatory process of the NAPs. But the Adaptation Fund will support the implementation of concrete projects up to USD10 million per country.

which will include sectoral approaches such as water, energy, infrastructures, and food production. According to the UNFCCC, 'in 2015–2016, 3% of international public adaptation finance flows was supplied by multilateral climate funds, while 84% came from development finance institutions and 13% from other government sources' (UNFCCC 2019c). Comprehensive reporting on adaptation finance by Murphy and Parry (2020) and Buchner et al. (2019) argues that flows of finance for adaptation action in developing countries in 2017 and 2018 were estimated to be approximately USD30 billion; this plus an additional estimated flow of USD12 billion for dual adaptation and mitigation actions totalled USD42 billion, accounting for 7.25% of the total estimated international public and private flows of climate finance (Buchner et al. 2019). They are far below the financing needs given in Section 15.4. To date, the private sector has limited involvement in NAPs and adaptation projects and planning but can be involved through public-private partnership (Section 15.6.2.1) and other incentives provided by governments (Schmidt-Traub and Sachs 2015; Druce et al. 2016; Koh et al. 2016; UNEP 2016; NAP Global Network 2017; Murphy and Parry 2020) and innovative private financing mechanisms such as green and blue bonds. However, adaptation financing is only about 2% of the share of green bond financing raised up to June 2019 (UNFCCC 2019c),[11] whereas it is about 10% of sovereign green bonds raised (UNFCCC 2019d). (Tuhkanen 2020), in a detailed review of green bond issuance in the Environmental Finance Data base 2019, found that between March 2010 to April 2019, '5% of all green bonds issued were categorised as adaptation and that 'the private sector accounts for a significant proportion of adaptation-related green bond issuances' (Tuhkanen 2020). However, GIZ (2017b), Nicol et al. (2017, 2018a), and Tuhkanen (2020) highlight that there is scepticism about this stream of finance for adaptation due to the factors that have thus far limited the private sector's involvement in adaptation: lack of resilience-related revenue streams, the small scale of some adaptation projects and the overall 'intangibility' of financing adaptation projects (Larsen et al. 2019).

Financing for resilience is limited, unpredictable, fragmented and focused on few projects or sectors and short term as opposed to programmatic and long term (10–15 years) finacing to build resilience (ISDR 2009, 2011; Kellett and Peters 2014; Watson et al. 2015). Market-based mechanisms are available but not equally accessible to all developing countries, particularly SIDS and LDCs, and such mechanisms can undermine debt sustainability (OECD and World Bank 2016). While resilience financing is mainly grant funding, concessional loans are increasing substantially and are key sources of financing for disaster and resilience, particularly for upper-middle-income countries (OECD and World Bank 2016). The combination of these trends can contribute to greater levels of indebtedness among many developing countries, many of which are already at or approaching debt distress.

Social protection systems can be linked with a number of the instruments already considered: reserve funds, insurance and catastrophe bonds, regional risk-sharing facilities, contingent credit, in addition to traditional international aid and disaster response. Hallegatte et al. (2017) recommend combining adaptive social protection with financial instruments in a consistent policy package, which includes financial instruments to deliver adequate liquidity and contingency plans for the disbursement of funds post disaster. Challenges related to financing residual climate-related losses and damages are particularly high for developing countries. Financing losses and damages from extreme events requires rapid pay-outs; the cost of financing for many developing countries is already quite high; and the expense of risk financing is expected to increase as disasters become more frequent, intense and more costly, not only due to climate change but also due to higher levels of exposure. Addressing both extreme and slow onset climate impacts requires designing adequate financial protection systems for reaching the most vulnerable. Moreover, some fraction of losses and damages, both material and non-material, are not commonly valued in monetary terms (non-economic loss) and hence financing requirements are hard to estimate. These non-market-based residual impacts include loss of cultural identity, sacred places, human health and lives (Ameli et al. 2021a; Paul 2019; Serdeczny 2019).

## 15.6 Approaches to Accelerate Alignment of Financial Flows with Long-term Global Goals

Near-term actions to shift the financial system over the next decade are critically important and possible with globally coordinated efforts. Taking into account the inertia of the financial system as well as the magnitude of the challenge to align financial flows with the long-term global goals, fast action is required to ensure the readiness of the financial sector as an enabler of the transition (*high confidence*). The following subsections elaborate on key areas which can have a catalytic effect in terms of addressing existing barriers – besides political leadership and interventions discussed in other Chapters of AR6.

Addressing knowledge gaps with regard to climate risk analysis and transparency will be one key driver for more appropriate climate risk assessment and efficient capital allocation (Section 15.6.1), efficient enabling environments to support the reduction of financing costs and reduce dependency on public financing (Section 15.6.2), a revised common understanding of debt sustainability, including that negative implications of deferred climate investments on future GDP, particularly stranded assets and resources to be compensated, can facilitate the stronger access to public climate finance, domestically and internationally (Section 15.6.3), climate risk pooling and insurance approaches are a key element of financing of a just transition (Section 15.6.4), the supply of finance to a widened focus on relevant actors can ensure transformational climate action at all levels (Section 15.6.5), new green asset classes and financial products can attract the attention of capital markets and support the scale up of financing by providing standardised investment opportunities which can be well integrated in existing investment processes (Section 15.6.6), a stronger focus on the development of local capital markets can help mobilise new investor groups and to some extent mitigate home bias effects (Section 15.6.7), new business models

[11] According to the climate bonds initiative, total green bond finance raised in 2018 was USD168.5 billion across 44 countries (UNFCCC 2019c).

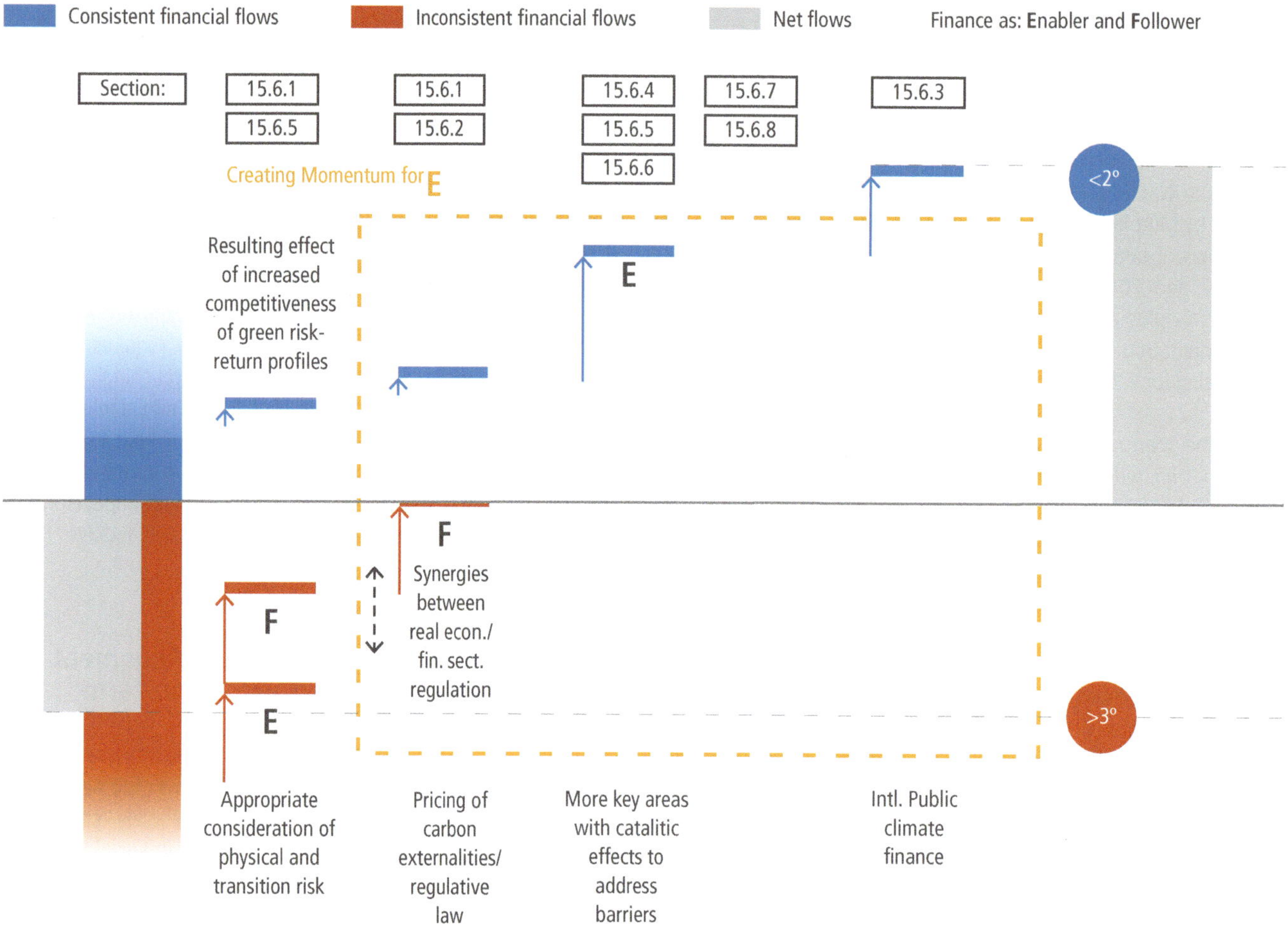


**Figure 15.5 | Visual abstract to address financing gaps in Section 15.6.**

and financing approaches can help to overcome barriers related to transactions costs by aggregating and/or transferring financing needs and establishing a supply of finance for needs of stakeholder groups lacking financial inclusion (Section 15.6.8).

### 15.6.1 Addressing Knowledge Gaps with Regard to Climate Risk Analysis and Transparency

Climate change as a source of financial risk.

Achieving climate mitigation and adaptation objectives requires ambitious climate finance flows in the near-term, that is, 5–10 years ahead. However, knowledge gaps in the assessment of climate-related financial risk are a key barrier to such climate finance flows. Therefore, this section discusses the main knowledge gaps that are currently being addressed in the literature and those that remain outstanding.

Climate-related financial risk is meant here as the potential adverse impact of climate change on the value of financial assets. A recent but remarkable development since AR5 is that climate change has been explicitly recognised by financial supervisors as a source of financial risk that matters both for financial institutions and citizens' savings (Bolton et al. 2020). Previously, climate change was mostly regarded in the finance community only as an ethical issue. The reasons why climate change implies financial risk are not new and are discussed more in detail below. What is new is that climate enters now as a factor in the assessment of financial institutions' risk (e.g., the European Central Bank or the European Banking Authority) and credit rating (Section 15.6.3), and, going forward, into stress-test exercises. This implies changes in incentives of the supervised financial actors, both public and private, and thus changes in the landscape of mitigation action by generating a new potential for climate finance flows. However, critical knowledge gaps remain. In particular, the underestimation of climate-related financial risk by public and private financial actors can explain that the current allocation of capital among financial institutions is often inconsistent with the mitigation objectives (Rempel et al. 2020). Moreover, even a correct assessment of risk, which could provide incentives for divesting from carbon-intensive activities, does not necessarily lead to investing in the technical options needed for deep decarbonisation. Therefore, understanding the dynamics of the low-carbon transition require to fill in at the same time gaps about risk and gaps about investments in enabling activities in a broader sense.

**Physical risk.** On the one hand, unmitigated climate change implies an increased potential for adverse socio-economic impacts especially in more exposed economic activities and areas (*high confidence*). Accordingly, *physical risk* refers to the component of financial risk associated with the adverse physical impact of hazards related to climate change (e.g., extreme weather events or sea level rise) on the financial value of assets such as industrial plants or real estate. In turn, these losses can translate into losses on the values of financial assets issued by exposed companies (e.g., equity/bonds) and or sovereign entities as well as losses for insurance companies. The assessment of climate financial physical risks poses challenges in terms of data, methods and scenarios. It requires cross-match scenarios of climate-related hazards at granular geographical scale, with the geolocation and financial value of physical assets. The relationship between the value of physical assets (such as plants or real estate) and the financial value of securities issued by the owners of those assets is not straightforward. Further, the repercussion of climate-related hazards on sovereign risk should also be accounted for.

**Transition risks and opportunities.** On the other hand, the mitigation of climate change, by means of a transition to a low-carbon economy, requires a transformation of the energy and production system at a pace and scale that implies adverse impacts on a range of economic activities, but also opportunities for some other activities (*high confidence*). If these impacts are factored in by financial markets, they are reflected in the value of financial assets. Thus, *transition risks and opportunities* refers to the component of financial risk (opportunities) associated with negative (positive) adjustments in assets' values resulting directly or indirectly from the low-carbon transition.

The concepts of *carbon stranded assets* (see e.g., Leaton and Sussams 2011), and *orderly* vs *disorderly transition* (Sussams et al. 2015) which emerged in the NGO community, have provided powerful metaphors to conceptualise transition risks and have evolved into concepts used also by financial supervisors (NGFS 2019)and academics. The term *carbon stranded assets* refers to fossil fuel-related assets (fuel or equipment) that become unproductive. An *orderly transition* is defined here as a situation in which market players are able to fully anticipate the price adjustments that could arise from the transition. In this case, there would still be losses associated with stranded assets, but it would be possible for market players to spread losses over time and plan ahead. In contrast, a *disorderly transition* is defined here as a situation in which a transition to a low-carbon economy on a 2°C path is achieved (i.e., by about 2040), but the impact of climate policies in terms of reallocation of capital into low-carbon activities and the corresponding adjustment in prices of financial assets (e.g., bonds and equity shares) is large, sudden and not fully anticipated by market players and investors. Note the impact could be unanticipated even if the date of the introduction is known in advance by the market players. There are several reasons why such adjustments could occur. One simple argument is that the political economy of the transition is characterised by forces pulling in different directions, including opposing interests within the industry, and mounting pressure from social awareness of unmitigated climate risks. Politics will have to find a synthesis and the outcome could remain uncertain until it suddenly unravels. Note also that, in order to be relevant for financial risk, the disorderly transition does not need to be a catastrophic scenario in terms of the fabric of markets. It also does not automatically entail systemic risk, as discussed below. Knowledge gaps in this area are related to emerging questions, including: What are, in detail, the transmission channels of physical and transition risk? How to assess the magnitude of the exposure to these risks for financial institutions and ultimately for people's savings? How do transition risk and opportunities depend on the future scenarios of climate change and climate policies? How to deal with the intrinsic uncertainty around the scenarios? To what extent could an underestimation of climate-related financial risk feed back on the alignment of climate finance flows and hamper the low-carbon transition? Should climate risk be explicitly accounted for in regulatory frameworks for financial institutions, such as Basel III for banks and national frameworks for insurance? What lessons from the 2008 financial crisis are relevant here, regarding moral hazard and the trustworthiness of credit risk ratings? The attention of both practitioners and the scientific community to these questions has grown since the Paris Agreement. In the following we review some of the findings from the literature, but the field is relatively young and many of the questions are still open.[12] Damages from climate change are expected to escalate dramatically in Europe (Forzieri et al. 2018) and in some EU countries there is already some evidence that banks, anticipating possible losses on the their loan books, lend proportionally less as a consequence.

**Assessment of physical risk.** There is a literature on estimates of economic losses on physical assets (see Cross-Working Group Box ECONOMIC in chapter 16 of AR6 WGII). Here we discuss some figures and mechanisms that are relevant for the financial system. Significant cost increases have been observed related to increases in frequency and magnitude of extreme events (*high confidence*) (Section 15.4.2). At the global level, the expected 'climate value at risk' (climate VaR) of financial assets has been estimated to be 1.8% along a business-as-usual emissions path (Dietz et al. 2016), with however, a concentration of risk in the tail (e.g., 99th VaR equals to 16.9%, or USD24.2[13] trillion, in 2016). Climate-related impacts are estimated to increase the frequency of banking crises (up over 200% across scenarios) while rescuing insolvent banks could increase the ratio of public debt to gross domestic product by a factor of two (Lamperti et al. 2019). Further assessments of physical risk for financial assets (Mandel 2020), accounting in particular for the propagation of losses through financial networks, estimate global yearly GDP losses at 7.1% (1.13%) in 2080, without adaptation (with adaptation), the former corresponding to a 10-fold increase with respect to the current yearly losses (0.76% of global GDP). Finally, climate physical risk can impact on the value of sovereign **bonds** (one of the top asset classes by size), in particular for vulnerable countries (Volz et al. 2020).

Insurance pay-outs for catastrophes have increased significantly over the last 10 years, with dramatic cost spikes in years with multiple

---

12 In context, while belonging to grey literature, reports from financial supervisors or non- academic stakeholders can be of interest for what they document in terms of changes in perception and incentives among the market players and hence of the dynamics of climate finance flows.

13 In the chapter, USD units are used as reported in the original sources in general. Some monetary quantities have been adjusted selectively for achieving comparability by deflating the values to constant $USD_{2015}$. In such cases, the unit is explicitly expressed as $USD_{2015}$.

major catastrophes (such as in 2018 with hurricanes Harvey, Irma, and Maria). This trend is expected to continue. The indirect costs of a climate-related flooding event can be up to 50% of the total costs, the majority of which is not covered by insurance (Alnes et al. 2018) (Section15.6.4). The gap between total damage losses and insurance pay-outs has increased over the past 10 years (Swiss Re Institute 2019). Indeed, the probability of 'extreme but plausible' scenarios will be progressively revised upwards in the 'value at risk'. As a result it becomes more difficult to find financial actors willing to provide insurance, as was observed for real estate in relation to flood and wildfires in California (Ouazad and Kahn 2019). This progressive adjustment would keep the financial system safe (Climate-Related Market Risk Subcommittee 2020; Keenan and Bradt 2020), but transfer to taxpayers the onus of damage compensation and the financing of adaptation investments (OECD 2021c) as well as build up latent liabilities.

**Assessment of transition risk. Carbon stranded assets.** Fossil fuel reserve and resource estimates exceed in equivalent quantity of $CO_2$ with virtual certainty the carbon budget available to reach the 1.5°C and 2°C targets (*high confidence*) (Meinshausen et al. 2009; McGlade and Ekins 2015; Millar et al. 2017). In relative terms, stranded assets of fossil fuel companies amount to 82% of global coal reserves, 49% of global gas reserves and 33% of global oil reserves (McGlade and Ekins 2015). This suggests that only less than the whole quantity of fossil fuels currently valued (either currently extracted, waiting for extraction as reserves or assets on company balance sheets) can yield economic return if the carbon budget is respected. The devaluation of fossil fuel assets implies financial losses for both the public sector (Section 15.6.8) and the private sector (Coffin and Grant 2019). Global estimates of potential stranded fossil fuel assets amount to at least 1 trillion, based on ongoing low-carbon technology trends and in the absence of climate policies (cumulated to 2035 with 10% discount rate applied; USD8 trillion without discounting (Mercure et al. 2018a)). With worldwide climate policies to achieve the 2°C target with 75% likelihood, this could increase to over USD4 trillion (until 2035, 10% discount rate; USD12 trillion without discounting). Other estimates indicate USD8–15 trillion (until 2050, 5% discount rate, (Bauer et al. 2015)) and USD185 trillion (cumulated to year 2115 using combined social and private discount rate (Linquiti and Cogswell 2016)). However the geographical distribution of potential stranded fossil fuel assets (also called 'unburnable carbon') is not even across the world due to differences in production costs (McGlade and Ekins 2015). In this context, a delayed deployment of climate finance and consequently limited alignment of investment activity with the Paris Agreement tend to strengthen carbon and thus to increase the magnitude of stranded assets.

**Assets directly and indirectly exposed to transition risk.** In terms of types of assets and economic activities, the focus of estimates of carbon stranded assets tends to be on physical reserves of fossil fuel (e.g., oil fields) and sometimes financial assets of fossil fuel companies (van der Ploeg and Rezai 2020). However, a precondition for a broader analysis of transition risks and opportunities is to go beyond the narrative of stranded assets and to consider a classification of sectors of all the economic activities that could be affected (Monasterolo 2020). This, in turn depends on their direct or indirect role in the GHG value chain, their level of substitutability with respect to fossil fuel and their role in the policy landscape. Moreover, such a classification needs to be replicable and comparable across portfolios and jurisdictions. One classification that meets these criteria is the Climate Policy Relevant Sectors (CPRS) (Battiston et al. 2017) which has been used in several studies by financial supervisors (EIOPA 2018; ECB 2019; EBA 2020; ESMA 2020). The CPRS classification builds on the international classification of economic activities (ISIC) to map the most granular level (4 digits) into a small set of categories characterised by differing types of risk: fossil fuel (i.e., all activities whose revenues depend mostly and directly on fossil fuel, including concession of reserves and operating industrial plants for extraction and refinement); electricity (affected in terms of input but that can in principle diversify their energy sources); energy intensive (e.g., steel or cement production plants, automotive manufacturing plants), which are affected in terms of energy cost but not in terms of the main input); and transport and buildings (affected in terms of both energy sources and specific policies). All financial assets (e.g., bonds, equity shares, loans) having as issuers or counterparties firms whose revenues depend significantly on the above activities are thus potentially exposed to transition risks and opportunities. Further, investors' portfolios have to be part of the analysis since changes in financial assets values affect the stability of financial institutions and can thus feed back into the transition dynamics itself (e.g., through cost of debt for firms and through costs for assisting the financial sector). One outstanding challenge for the analysis of investors' exposure to climate risks is the difficulty of gathering granular and standardised information on the breakdown of non-financial firms' revenues and CAPEX in terms of low-/high-carbon activities (*high confidence*).

Several financial supervisors have conducted assessments of transition risk for the financial system at the regional level. For instance, the European Central Bank (ECB) reported preliminary estimates of aggregate exposures of financial institutions to CPRS relative to their total debt securities holdings as ranging between 1% for banks to about 9% for investment funds (ECB 2019). The European Insurance and Occupational Pensions Authority (EIOPA) reported aggregate exposures to CPRS of EU insurance companies at about 13% of their total securities holdings (EIOPA 2018). Further analyses on the EU securities holdings indicate that among financial investments in bonds issued by non-financial corporations, EU institutions hold exposures to CPRS ranging between 36.8% for investment funds to 47.7% for insurance corporations; analogous figures for equity holdings range from 36.4% for banks to 43.1% for pension funds (Alessi et al. 2019). Another study indicates that losses on EU insurance portfolios of sovereign bonds could reach up to 1%, in conservative scenarios (Battiston et al. 2019).

Given the magnitude of the assets that are potentially exposed, reported in the previously cited studies, a delayed or uncoordinated transition risk can have implications for financial stability not only at the level of individual financial institutions, but also at the macro level. The possible systemic nature of climate financial risk has been highlighted on the basis of general equilibrium economic analysis (Stern and Stiglitz 2021).

Some financial authorities recognise that climate change represents a major source of systemic risk, particularly for banks with portfolios concentrated in certain economic sectors or geographical areas (de Guindos 2021). Specifically, the concern that central banks would have to act as 'climate rescuers of last resort' in a systemic financial crisis stemming from some combination of physical and transition risk has been raised in the financial supervisor community (Bolton et al. 2020). The systemic nature of climate risk is reinforced by the possible presence of moral hazard. Indeed, if a sufficient number of financial actors have an incentive to downplay climate-related financial risk, then systemic risk builds up in the financial system, eventually materialising for taxpayers (Climate-Related Market Risk Subcommittee 2020). While such type of risk may go undetected to standard market indicators for a while, it can materialise with a time delay, similarly to the developments observed in the run up to the 2008 financial crisis.

These considerations are part of an ongoing discussion on whether the current financial frameworks, including Basel III, should incorporate explicitly climate risk as a systemic risk. In particular, the challenges in quantifying the extent of climate risk, reviewed in this section, especially if risk is systemic, raise the question whether a combination of quantitative and qualitative restrictions on banks' portfolios could be put in place to limit the build-up of climate risks (Baranović et al. 2021).

**Endogeneity of risk and multiplicity of scenarios.** One fundamental challenge is that climate-related financial risk is endogenous (*high confidence*). This means that the perception of the risk changes the risk itself, unlike most contexts of financial risk. Indeed, transition risk depends on whether governments and firms continue on a business-as-usual pathway (i.e., misaligned with the Paris Agreement targets) or engage on a climate mitigation pathway. But the realisation of the transition pathway depends itself on how, collectively, society, including financial investors and supervisors, perceive the risk of taking or not taking the transition scenario. The circularity between perception of risk and realisation of the scenario implies that multiple scenarios are possible, and that which scenario is ultimately realised can depend on policy action. The coordination problem associated also with low-carbon investments opportunities increases the uncertainty. Further, not all low-carbon activities are directly functional to the transition (e.g., investments in pharmaceutical, IT companies, or financial intermediaries), thus not all reallocations of capital lead to the same path.

In this context, probabilities of occurrence of scenarios are difficult to assess and this is important because risks vary widely across the different scenarios. In this context a major challenge is the fat-tail nature of physical risk. One the one hand, forecasts of climate change and its impact on humans and ecosystems imply tail events (Weitzman 2014) and tipping points which cannot be overcome by model consensus (Knutti 2010). On the other hand, everything else the same, costs and benefits vary substantially with assumptions on agents' utility, productivity, and intertemporal discount rate, which ultimately depend on philosophical and ethical considerations (Nordhaus 2007; Stern 2008; Pindyck 2013). Thus, more knowledge is needed on the interaction of climate physical and transition risks, the possible reinforcing feedbacks and transmission channels to the economy and to finance. Moreover, models need to account for compound risk, that is, the interaction of climate physical and/or transition risk with other sources of risk such as pandemics, such as COVID-19.

**Challenges for climate transition scenarios.** The endogeneity of risk and its associated deep uncertainty implies that the standard approach to financial risk, consisting of computing expected values and risk based on historical values of market prices, is not adequate for climate risk (*high confidence*) (Bolton et al. 2020). To address this challenge, a recent stream of work has developed an approach to make use of climate policy scenarios to derive risk measures (e.g., expected shortfall) for financial assets and portfolios, conditioned to scenarios of disorderly transition (Battiston et al. 2017; Monasterolo and Battiston 2020; Roncoroni et al. 2020). In particular, climate policy shocks on the output of low-/high-carbon economic activities are calculated based on trajectories of energy technologies as provided by large-scale Integrated Assessment Models (Kriegler et al. 2015; McCollum et al. 2018) conditioned to the introduction of specific climate policies over time. This approach allows to conduct climate stress-tests both at the level of financial institutions and at the level of the financial system of a given jurisdiction.

In a similar spirit, recently, the community of financial supervisors in collaboration with the community of climate economics has identified a set of climate policy scenarios, based on large-scale IAM, as candidate scenarios for assessing transition risk (Monasterolo and Battiston 2020). These scenarios have been used, for instance, in an assessment of transition risk conducted at a national central bank (Allen et al. 2020). This development is key to mainstreaming the assessment of transition risk among financial institutions, but the following challenges emerge (*high confidence*). First, a consensus among financial supervisors and actors on scenarios of transition risk that are too mild could lead to a systematic underestimation of risk. The reason is that the default probability of leveraged financial institutions is sensitive to errors in the estimation of the loss distribution and hence sensitive on the choice of transition scenarios (Battiston and Monasterolo 2020). This in turn could lead to an allocation of capital across low-/high-carbon activities that is insufficient to cater for the investment needs of the low-carbon transition.

Second, IAM do not contain a description of the financial system in terms of actors and instruments and make assumptions on agents' expectations that could be inconsistent with the nature of a disorderly transition (Espagne 2018; Pollitt and Mercure 2018a; Battiston et al. 2020b). In particular, IAMs solve for least cost pathways to an emissions target in 2100 (AR4 WGIII SPM Box 3), while the financial sector's time horizon is much shorter and risk is an important factor in investment decisions.

Third, the current modelling frameworks used to develop climate mitigation scenarios, which are based on large-scale IAM, assume that the financial system acts always as an enabler and do not account for the fact that, under some condition (i.e., if there is underestimation of climate transition risk) can also act as a barrier to the transition

(Battiston et al. 2020a) because it invests disproportionately more in high-carbon activities.

**Macroeconomic implications of the technological transition.** Global macroeconomic changes that may affect asset prices are expected to take place as a result of a possible reduction in growth or contraction of fossil fuel demand, in scenarios in which climate targets are met according to carbon budgets, but also following ongoing energy efficiency changes (*high confidence*) (Clarke et al. 2014; Mercure et al. 2018a). A review of the economic mechanisms involved in the accumulation of systemic risk associated with declining industries, with focus on fossil fuels, is given by Semieniuk et al. (2021). An example is the transport sector, which uses around 50% of oil extracted (IEA 2018; Thomä 2018). A rapid diffusion of EV (and other alternative vehicle types) poses an important risk as it could lead to oil demand peaking far before mid-century (Mercure et al. 2018b; 2021). New technologies and fuel switching in aviation, heavy industry and shipping could further displace liquid fossil fuel demand (IEA 2017). A rapid diffusion of solar photovoltaic could displace electricity generation based predominantly on coal and gas (Sussams and Leaton 2017). A rapid diffusion of household and commercial indoor heating and cooling based on electricity could further reduce the demand for oil, coal and gas (Knobloch et al. 2019). Parallels can be made with earlier literature on great waves of innovation, eras of clustered technological innovation and diffusion between which periods of economic, financial and social instability have emerged (Freeman and Louca 2001; Perez 2009).

Due to the predominantly international nature of fossil fuel markets, assets may be at risk from regulatory and technological changes both domestically and in foreign countries (*medium confidence*). Fossil fuel exporting nations with lower competitiveness could lose substantial amounts of industrial activity and employment in scenarios of peaking or declining demand for fossil fuels. In scenarios of peaking oil demand, production is likely to concentrate towards the Middle East and OPEC countries (IEA 2017). Since state-owned fossil fuel companies tend to enjoy lower production costs, privately-owned fossil fuel companies are more at risk (Thomä 2018). Losses of employment may be directly linked to losses of fossil fuel-related industrial activity or indirectly linked through losses of large institutions, notably of government income from extraction royalties and export duties. A multiplier effect may take place making losses of employment spill out of fossil fuel extraction, transformation and transportation sectors into other supplying sectors (Mercure et al. 2018a).

**Main regulatory developments and voluntary responses to climate risk.** Framing climate risk as a financial risk (not just as an ethical issue) is key for it to become an actionable criterion for investment decision among mainstream investors (*high confidence*) (TCFD 2019). Since 2015 financial supervisors and central banks (e.g., the Financial Stability Board, the G20 Green Finance Study Group, and the Network for Greening the Financial System (NGFS)) have played a central role in raising awareness and increasing transparency of the potential material financial impacts of climate change within the financial sector (Bank of England 2015, 2018; TCFD 2019). The NGFS initiative has engaged, in particular, in the elaboration of climate financial risk scenarios.

Although disclosure has increased since the TCFD recommendations were published, the information is still insufficient for investors and more clarity is needed on potential financial impacts and how resilient corporate strategies are under different scenarios (TCFD 2019). Several efforts to provide guidance and tools for the application of the TCFD recommendations have been made (using Sustainability Accounting Standards Board (SASB) Standards and the Climate Disclosure Standards Board (CDSB) Framework to Enhance Climate-Related Financial Disclosures in Mainstream Reporting TCFD Implementation Guide (UNEP FI 2018; CDSB and SASB 2019). Results of voluntary reporting have been mixed, with one study pointing to unreliable and incomparable results reported by the US utilities sector to the CDP (Stanny 2018).

There have been also similar initiatives at the national level (DNB 2017; UK Government 2017; US GCRP 2018b). In particular, France was the first country to mandate climate risk disclosure from financial institutions (via Article 173 of the law on energy transition). However, disclosure responses have been so far mixed in scope and detail, with the majority of insurance companies not reporting on physical risk (Evain et al. 2018). In the UK, mandatory GHG emissions reporting for UK-listed companies has not led to substantial emissions reductions to date but could be laying the foundation for future mitigation (Tang and Demeritt 2018).

A key recent development is the EU Taxonomy for Sustainable Finance (TEG 2019), which provides a classification of economic activities that (among other dimensions) contribute to climate mitigation or can be enabling for the low-carbon transition. Indirectly, such classification provides useful information on investors' exposure to transition risk (Alessi et al. 2019; ESMA 2020). Finally, many consultancies have stepped forward offering services related to climate risk. However, the methods are typically proprietary, non-transparent, or based primarily on carbon footprinting, which is a necessary but insufficient measure of climate risk. Further, ESG (environmental, social and governance) metrics can be useful but are, alone, inadequate to assess climate risk.

### Illustrative mitigation pathways and financial risk for end-users of climate scenarios

Decision-makers in financial risk management make increasing use of climate policy scenarios, in line with the TCFD guidelines and the recommendations of the NGFS. In order to reduce the number of scenarios to consider, Illustrative Mitigation Pathways (IMPs, Chapter 3), have been elaborated to illustrate key features that characterise the possible climate (policy) futures. The following considerations can be useful for scenario end-users who carry out risk analyses on the basis of the scenarios described in Chapter 3. It is possible to associate climate policy scenarios with levels of physical and/or transition risk, but these are not provided with the scenario data themselves.

On the one hand, each scenario is associated with a warming path, which in turn, on the basis of the results from WGII, implies certain levels of physical risk (AR6 WGII Chapter 16). However, climate impacts are not accounted for in the scenarios. Moreover, levels of risk may vary with the reason for concern and with the speed of the implementation of adaptation. On the other hand, while mitigation can come with transition risk, in the case of lack of coordination among the actors, as discussed earlier in this section, this is not modelled explicitly in the trajectories, since the financial sector is not represented in underlying models. The scientific state of the art in climate-related financial risk offers an analysis that is not yet comprehensive of both the physical and transition risk dimensions in the same quantitative framework. However, decision-makers can follow a mixed approach where they can combine quantitative risk assessment for transition risk with more qualitative risk analysis related to physical risk.

Figure 15.6 represents sequences of events following along a scenario both in terms of physical risk (left) and transition risk (right). Four groups of IMPs (more are considered based on the warming level they lead to in 2100. Current Policies (CurPol) considers climate policies implemented in 2020 with only a gradual strengthening afterwards, leading to above 4°C warming (with respect to pre-industrial levels). Moderate Action (ModAct) explores the impact of implementing the NDCs (pledged mitigation targets) as formulated in 2020 and some further strengthening afterwards, thereby limiting warming to less than 4°C (>50%), but above 3°C (>50%). In these two scenarios, there is no stabilisation of temperature, meaning that further warming occurs after 2100 (and higher risk) even if stabilisation could be eventually achieved. They are referred to as pathways with higher emissions. The warming levels reached along these two scenarios imply physical risk levels that are 'Moderate' until 2050 and 'Very High' in 2050–2100 (with low levels of adaptation). Noting, that 'Moderate' physical risk can mean for some countries (i.e., SIDS) significant and even hardly absorbable consequences (i.e., reaching hard adaptation limits). Transition risk is not relevant for these scenarios, since a transition is not pursued.

Illustrative Mitigation Pathways include two groups of scenarios consistent with modelled global pathways that limit warming to 2°C (>67%) or lower, respectively. The two groups are representative for the IMPs defined in Chapter 3. In these scenarios, warming is stabilised before 2100. The warming levels along these paths imply 'Moderate' physical risk until 2050 and 'High' risk in 2050–2100 (with low levels of adaptation). Transition risk can arise along these trajectories from changes in expectations of economic actors about which of the scenarios is about to materialise. These changes imply, in turn, possible large variations in the financial valuation of securities and contracts, with losses on the portfolio of institutional investors and households. High policy credibility is key to avoiding transition risk, by making expectations consistent early on with the scenario. Low credibility can delay the adjustment of expectations by several years, leading either to a late and sudden adjustment. However, if the policy never becomes credible, this changes the scenario since the initial target is not met.

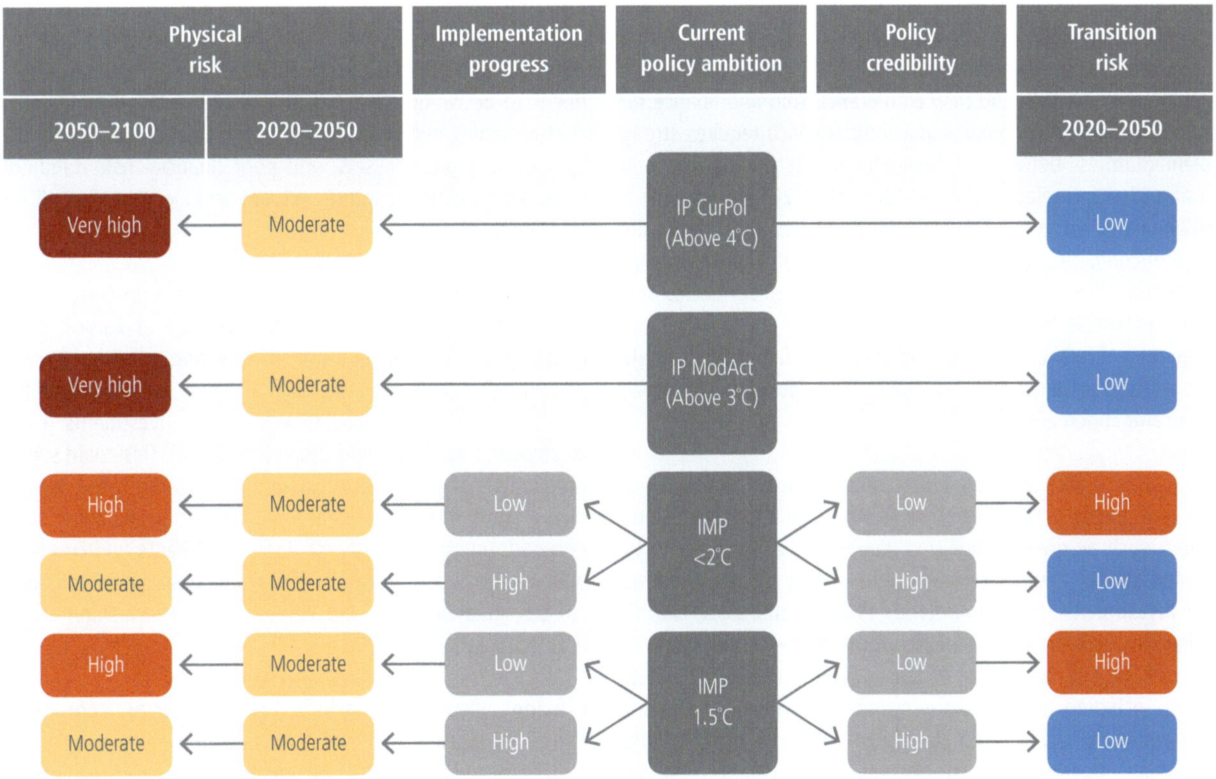


**Figure 15.6 | Schematic representation of climate scenarios in terms of both physical and transition risk.** While the figure does not cover all possible events, it maps out how the combination of stated targets can lead to different paths in terms of risk, depending on implementation progress and policy credibility. IMP 1.5°C and IMP <2°C are representative for IMP-GS (Sens. Neg; Ren), IMP-Neg, IMP-LD; IMP-Ren; IMP-SP. Note that the figure defines 'High' progress as higher, but it is important that the physical risk varies by region and country. This means, that 'Moderate' physical risk can be significant and even hardly absorbable for some countries.

### 15.6.2 Enabling Environments

The Paris Agreement recognised for the first time the key role of aligning financial flows to climate goals. It further emphasises the importance of making financial flows consistent with climate actions and SDGs (Zamarioli et al. 2021).This alignment has now to be operated in a specific environment where the scaling-up of climate policies is conditional upon their contribution to post-COVID-19 recovery packages (Sections 15.2.2 and 15.2.3 and Box 15.6). The enabling environments that are to be established account for the structural parameters of the underinvestment in long-term assets. The persistent gap between the 'propensity to save' and 'propensity to invest' (Summers 2016) obstructs the scaling up of climate investments, and it results from a short-term bias of economic and financial decision-making (Miles 1993; Bushee 2001; Black and Fraser 2002) that returns weighted on short-term risk dominate the investment horizon of financial actors. Overcoming this bias is the objective of an enabling environment apt to *launch of a self-reinforcing circle of trust* between project initiators, industry, institutional investors, the banking system, and governments.

The role of government is crucial for creating an enabling environment for climate (Clark 2018), and governments are critical in the launching and maintenance of this circle of trust by lowering the political, regulatory, macroeconomic and business risks (*high confidence*). The issue is not just to progressively enlarge the space of low-carbon investments but to replace one system (fossil fuels energy system) rapidly with another (low-carbon energy system). This is a wave of 'creative destruction' with the public support for developing new markets and new entrepreneurship and finance for green products and technologies in a context which requires strong complementarities between Schumpeterian (technological) and Keynesian (demand-related) policies (Dosi et al. 2017). However, it is challenging to overcome the constraint of public budget under the pressure of competing demands and of creditworthy constraints for countries that do not have an easy access to reserve currencies. It is needed to maximise, both at the national and international levels, the leverage ratio of public funds engaged in blended finance for climate change which is currently very low, especially in developing countries (Attridge and Engen 2019).

**Transparency:** Policy de-risking measures, such as robust policy design and better transparency, as well as financial de-risking measures, such as green bonds and guarantees, at both domestic and international levels, enhance the attractiveness of clean energy investments (*high confidence*) (Steckel and Jakob 2018). Organisations such as the Task Force on Climate-related Financial Disclosures (TCFD) can help increase capital markets' climate financing, including private sector, by providing financial markets with information to price climate-related risks and opportunities (TCFD 2020). However, risk disclosures alone would likely be insufficient as long as market failures that inhibit the emergence of low-carbon investment initiatives with positive risk-weighted returns (*high confidence*) (Christophers 2017; Ameli et al. 2020).

**Central banks and climate change.** Central banks in all economies will likely have to play a critical role in supporting the financing of fiscal operations, particularly in a post-COVID-19 world (*high confidence*). Instruments and institutional arrangements for better international monetary policy coordination will likely be necessary in the context of growing external debt stress and negative credit rating pressures facing both emerging and low-income countries. Central bankers have started examining the implications of disruptive risks of climate change, as part of their core mandate of managing the stability of the financial system (Chenet et al. 2021). Climate-related risk assessments and disclosure, including central banks' stress testing of climate change risks, can be considered as a first step (Rudebusch 2019), although such risk assessments and disclosure may not be enough by themselves to spur increased institutional low-carbon climate finance (Ameli et al. 2020).

Green quantitative easing (QE) is now being examined as a tool for enabling climate investments (Dafermos et al. 2018) in which central banks could explicitly conduct a programme of purchases of low-carbon assets (Aglietta et al. 2015). A green QE programme 'would have the benefit of providing large amounts of additional liquidity to companies interested' in green projects (*medium confidence*) (Campiglio et al. 2018). Green QE would have positive effects for stimulating a low-carbon transition, such as accelerating the development of green bond markets (Hilmi et al. 2021), encouraging investments and banking reserves, and reducing risks of stranded assets, while it might increase income inequality and financial instability (Monasterolo and Raberto 2017). While the short-term effectiveness would not be substantial, the central bank's purchase of green bonds could have a positive effect on green investment in the long run (Dafermos et al. 2018). However, the use of green QE needs to be cautious on potential issues, such as undermining the central bank's independence, affecting the central bank's portfolio by including green assets with poor financial risk standards, and potential regulatory capture and rent-seeking behaviours (Krogstrup and Oman 2019).

Additional monetary policies and macroprudential financial regulation may facilitate the expected role of carbon pricing on boosting low-carbon investments (*medium confidence*) (D'Orazio and Popoyan 2019). Commercial banks may not respond to the price signal and allocate credits to low-carbon investments due to the existence of market failure (Campiglio 2016). This could support the productivity of green capital goods and encourage green investments in the short term, but might cause financial instability by raising non-performing loans ratio of dirty investments and creating green bubbles (Dunz et al. 2021). Financial supervisors needs to implement stricter guidelines to overcome the greenwashing challenges (Caldecott 2020).

**Efficient financial markets and financial regulation.** An influential efficient financial markets hypothesis (Fama 1970, 1991, 1997) proceeds from the assumption that in well-developed financial markets, available information at any point of time is already well captured in capital markets with many participants. Despite increasing challenges to the theory (Sewell 2011), especially by repeated episodes of global financial crashes and crises, and other widely noted anomalies, a weaker form of the efficient markets hypothesis may still apply (*medium confidence*). It is arguable that accumulating scientific evidence of climate impacts is being accompanied by rising levels of climate finance. Banks and institutional investors are also progressively rebalancing their investment portfolios away from fossil fuels and towards low-carbon investments (IEA 2019b; Monasterolo and de Angelis 2020). In the meantime, the world runs the risk of sharp adjustments, crises and irreversible 'tipping points' (Lontzek et al. 2015) sufficiently destabilising climate outcomes. This leads to the policy prescription towards financial regulatory agencies requiring greater and swifter disclosure of information about rising climate risks faced by financial institutions in projects and portfolios and central bank attention to systemic climate risk problems as one possible route of policy action (Carney 2015; Dietz et al. 2016; Zenghelis and Stern 2016; Campiglio et al. 2018). However, disclosure requirements of risks and information in private settings remain mostly voluntary and difficult to implement (Battiston et al. 2017; Monasterolo et al. 2017).

Nevertheless, financial markets are innovating in search of solutions (Section 15.6.6). Recognising and dealing with stranded fossil fuel assets is also a key area of growing concern that financial institutions are beginning to grapple with. Larger institutions with more patient capital (pensions, insurance) are also increasingly beginning to enter the financing of projects and green bond markets. The case for efficient financial markets in developing countries is worse (Abbasi and Riaz 2016; Hong et al. 2019) because of weaker financial institutions (Hamid et al. 2017), heightened credit rationing behaviour (Bond et al. 2015), and high risk aversion as most markets are rated as junk, or below/barely investment grade (Hanusch et al. 2016). Other constraints such as limited long-term financial instruments and underdeveloped domestic capital markets, absence of significant domestic bond markets for investments other than sovereign borrowing, and inadequate term and tenor of financing, make the efficient markets thesis practically inapplicable for most developing countries.

**Markets, finance and creative destruction.** Branches of macro-innovation theory could be grouped into two principal classes (Mercure et al. 2016): 'equilibrium – optimisation' theories that treat innovators as rational perfectly informed agents and reaching equilibrium under market price signals; and 'non-equilibrium' theory where market choices are shaped by history and institutional forces and the role of public policy is to intervene in processes, given a historical context, to promote a better outcome or new economic trajectory. The latter suggests that new technologies might not find their way to the market without price or regulatory policies to reduce uncertainty on expected economic returns. A key issue is the perception of risk by investors and financial institutions. The financial system is part of complex policy packages involving multiple instruments (cutting subsidies to fossil fuels, supporting clean energy innovation and diffusion, levelling the institutional playing field and making risks transparent) (Polzin 2017) and the needed big systemic push (Kern and Rogge 2016) requires it takes on the role of 'institutional innovation intermediaries' (Polzin et al. 2016).

As far as climate finance is concerned, public R&D support had large cross-border knowledge spill-overs indicating that openness to trade was important, capacity expansion had positive effects on learning-by-doing on innovation over time, and that feed-in-tariffs (FiTs), in particular, had positive impacts on technology diffusion (Grafström and Lindman 2017) (Box 16.4). The FiTs programme has been associated with rapid increase in early renewables capacity expansion across the world by reducing market risks in financing and stability in project revenues (Menanteau et al. 2003; Jacobsson et al. 2009) (Section 9.9.5). Competitive auctions where the bidder with the lowest price or other criteria is selected for government's call for tender are increasingly being utilised as an alternative to FITs due to their strengths of flexibility, potential for real price discovery, ability to ensure greater certainty in price and quantity, and capability to guarantee commitments and transparency (IRENA and CEM 2015).

Outside of renewable energy, scattered but numerous examples are available on the role of innovative public policy to spur and create new markets and technologies (Arent et al. 2017): (i) proactive role of the state in energy transitions (e.g., the retirement of all coal-fired power plants in Ontario, Canada, between 2007 and 2014 (Kern and Rogge 2016; Sovacool 2016)); (ii) too early exit and design problems not considering the market acceptability and financing issues (e.g., energy-efficient retrofitting in housing in UK (Rosenow and Eyre 2016), low or negative returns in reality versus engineering estimates in weatherisation programmes in US (Fowlie et al. 2018)); and (iii) energy performance contracting for sharing the business risks and profits and improving energy efficiency (energy service companies (Bertoldi and Boza-Kiss 2017; Qin et al. 2017) and utility energy service contracts in the USA (Clark 2018)).

**Crowding out.** Literature has discussed the risks of low effectiveness of public interventions and of a crowding out effect of climate-targeted public support to other innovation sectors (Buchner et al. 2013). However, much academic literature suggests no strong evidence of crowding out. (Deleidi et al. 2020). Examining the effect of public investment on private investment into renewables in 17 countries over 2004–2014, showed that the concept of crowding out or in does not apply well to sectoral studies and found that public investments positively support private investments in general.

**Support climate action via carbon pricing, taxes, and emission trading systems.** Literature and evidence suggest that futures markets regarding climate are incomplete because they do not price in externalities (Scholtens 2017). As a result, low-carbon investments do not take place to socially and economically optimal levels, and the correct market signals would involve setting carbon prices high enough or equivalent trading in reduced carbon emissions by regulatory action to induce sufficient and faster shift towards low-carbon investments (*high confidence*) (Aghion et al. 2016). Nonetheless, durable carbon pricing in economic and political systems must be implemented and approached combining related elements to both price and quantity (Grubb 2014).

The introduction of fiscal measures, such as carbon taxes, or market-based pricing, such as emission trading schemes, to reflect carbon pricing have benefits and drawbacks that policymakers need to consider, taking account of both country-specific conditions and policy characteristics. Carbon tax can be a simpler and easier way to implement carbon pricing, especially in developing countries, because countries can utilise the existing fiscal tools and do not need concrete enabling conditions as market-based frameworks (*high confidence*). The reallocation of revenues from carbon taxes can be used for low-carbon investments, supporting poorer sections of society and fostering technological change (High-Level Commission on Carbon Prices 2017). In combination with other policies, such as subsidies and public R&D on resource-saving technologies, properly designed carbon taxes can facilitate the shift towards low-carbon, resource-efficient investments (Bovari et al. 2018; Naqvi and Stockhammer 2018; Dunz et al. 2021) (Section 9.9.3). The effectiveness of carbon pricing has been supported by various evidence. EU ETS has cut emissions by 42.8% in the main sectors covered (European Commission 2021a), and China had achieved emissions reductions and energy conservation through its pilot ETS between 2013 and 2015 (Zhang et al. 2019; Hu et al. 2020). Institutional learning, administrative prudence, appropriate carbon revenue management and stakeholder engagement are key ingredients for successful ETS regimes (Narassimhan et al. 2018).

The presence of carbon prices can promote low-carbon technologies and investments (Best and Burke 2018), and price signals, including carbon taxation, provide powerful and efficient incentives for households and firms to reduce $CO_2$ emissions (IMF 2019). The expansion of carbon prices is dependent on country-specific fiscal and social policies to hedge against regressive impacts on welfare, competitiveness, and employment (Michaelowa et al. 2018). Such impacts need to be offset using the proceeds of carbon taxes or auctioned emission allowances to reduce distortive taxation (Bovenberg and de Mooij 1994; Goulder 1995; de Mooij 2000; Chiroleu-Assouline and Fodha 2014) and fund compensating measures for the population sections that are most adversely impacted (Combet et al. 2010; Jaccard 2012; Klenert et al. 2018). This is more difficult for developing countries with a large share of energy-intensive activities, fossil fuel exporting countries and countries which have lower potential to mitigate impacts due to lower wages or existing taxes (Lefèvre et al. 2018).

Non-carbon price instruments, such as market-oriented regulation and public programmes involving low-carbon infrastructure, may be preferable in developing countries where market and regulatory failure and political economy constraints are more prevalent (Finon 2019). While carbon pricing was suggested by many economists and researchers (Nordhaus 2015; Pahle et al. 2018), overcoming the political and regulatory barriers would be necessary for the further implementation of an effective carbon pricing scheme nationally and internationally. Without strong political support, the effectiveness of carbon pricing would be limited to least-cost movements (Meckling et al. 2015).

**Role of domestic financing sources.** Efforts to address climate change can be scaled up through the mobilisation of domestic funds (Fonta et al. 2018). Publicly organised and supported low-carbon infrastructures through resurrected national development banks may be justified (Mazzucato and Penna 2016). It is important to efficiently allocate the public financing, and State Investment Banks (SIBs) can take up key roles (i) to provide capital to assist with overcoming financial barriers, (ii) to signal and direct investments towards green projects, and (iii) to attract private investors by taking up a de-risking role. Also, they can become a first mover by investing in new and innovative technologies or business models (Geddes et al. 2018). State-owned enterprises (SOEs) can also have an overall positive effect on renewables investments, outweighing any effect of crowding out private competitors (Prag et al. 2018). Green investment banks can assist in the green transition by developing valuable expertise in implementing effective public interventions to overcome investment barriers and mobilise private investment in infrastructure (OECD 2015c). De-risking measures may reduce investment risks, but lacking research and data availability hinders designing such measures (Dietz et al. 2016). Local governments' efforts to de-risk by securitisation might have negative effects by narrowing the scope for a green developmental state and encouraging privatisation of public services (Gabor 2019).

**The potential role of coordinated multilateral initiatives.** There is a growing awareness of the low leverage ratio of public to private capital in climate blended finance (Blended Finance Taskforce 2018b) and of a 'glass ceiling', caused by a mix of agencies' inertia and perceived loss of control over the use of funds, on the use of public guarantees by MDBs to increase it (*high confidence*) (Gropp et al. 2014; Schiff and Dithrich 2017; Lee et al. 2018). Many proposals have emerged for multilateral guarantee funds: Green Infrastructure Funds (de Gouvello and Zelenko 2010; Studart and Gallagher 2015), Multilateral Investment Guarantee Agency (Enhanced Green MIGA) (Déau and Touati 2018), guarantee funds to bridge the infrastructure investment gap (Arezki et al. 2016), and multi-sovereign guarantee mechanisms (Dasgupta et al. 2019). The obstacle of limited fiscal space for economic recovery and climate actions in low-income and some emerging economies can be overcome only in a multilateral setting. Several multilateral actions are being envisaged: G20's suspension of official bilateral debt payments, IMF's adoption of new SDRs allocation (IMF 2021b). However, any form of unconventional debt relief will generate development and climate benefits only if they credibly target bridging the countries' infrastructure gap with low-carbon climate-resilient options.

Of interest in multilateral settings is a credibility-enhancing effect provided by reciprocal gains for both the donor and the host country. Guarantor countries can compensate the public cost of their commitments with the fiscal revenues of induced exports. As to the host countries, they would benefit from new capital inflows and the grant equivalents of reduced debt service which might potentially go far beyond USD100 billion $yr^{-1}$ (Hourcade et al. 2021a). A second interest would be to support a learning process about agreed-upon assessment and monitoring methods using clear metrics. Developing standardised and science-based assessment methods at low transaction costs is essential to strengthen the credibility of green investments and the emergence of a pipeline of high-quality bankable projects which can be capitalised in the form of credible assets and supported with transparent and credible domestic spending. Multi-sovereign guarantees would provide a quality backing to developing

## Box 15.5 | The Role of Enabling Environments for Decreasing Economic Cost of Renewable Energy

A widely used indicator for the relative attractiveness of renewable energy but also development of price levels is the levelised cost of energy (LCOE). It is applied by a wide range of public and private stakeholders when tracking progress with regard to cost degression (Aldersey-Williams and Rubert 2019). LCOE calculation methodologies vary but in principle consider project-level costs only (NEA 1989). Besides other weaknesses, the LCOE concept usually does not consider societal costs resulting from de-risking instruments and/or other public interventions/support and therefore caution has to be applied when using the LCOE as the sole indicator of the success of enabling environments. The yearly IRENA mapping on renewable energy auction results demonstrates the extremely broad ranges of LCOEs (equal to the agreed tariffs) for renewable energy which can be observed (IRENA 2019a). For example, in 2018, solar PV LCOEs for utility-scale projects came in between USD0.04 $kWh^{-1}$ and USD0.35 $kWh^{-1}$ with a global weighted average of USD0.085 $kWh^{-1}$. However, comparative analysis taking into account societal costs is hardly available driven by challenges in the context of the quantification of public support.

The GET FiT concept argued that the mitigation of political and regulatory risk by sovereign and international guarantees is cost-efficient in developing countries, illustrating the estimated impact of such risk-mitigation instruments on equity and debt financing costs, and consequently required feed-in tariff levels (Deutsche Bank Climate Change Advisors 2011). The impact of financing costs on cost of renewable energy generation is well researched with significant differences across countries and technologies being observed, with major drivers being the regulatory framework as well as the availability and type of public support instruments (Geddes et al. 2018; Steffen 2019). With a focus on developing countries and based on a case study in Thailand Huenteler et al. (2016) demonstrate the significant effect of regulatory environments but also local learning and skilled workforce on cost of renewables. The effect of those exceeds the one of global technology learning curves.

Egli et al. (2018) identify macroeconomic conditions (general interest rate) and experience effects within the renewable energy finance industry as key drivers in developed countries with a stable regulatory environment, contributing 5% (PV) and 24% (wind) to the observed reductions in LCOEs in the German market with a relatively stable regulatory environment. They conclude that 'extant studies may overestimate technological learning and that increases in the general interest rate may increase renewable energies' LCOEs, casting doubt on the efficacy of plans to phase out policy support' (Egli et al. 2018). A rising general interest rate level could heavily impact LCOEs – for Germany, a rise of interest rates to pre-financial crisis levels in five years could increase LCOEs of solar and wind by 11–25% respectively (Schmidt et al. 2019).

countries and allow for expanding developing countries' access to capital markets at a lower cost and longer maturities, overcome the Basel III's liquidity impediment and the EU's Solvency II directive on liquidity (Blended Finance Taskforce 2018b), and accelerate the recognition of climate assets by investors seeking safe investment havens (Hourcade et al. 2021b). They would also strengthen the efficacy of climate disclosure through high grades climate assets and minimise the risks of 'greening' of the portfolios by investing in 'carbon neutral' activities and not in low-carbon infrastructures. Finally, they would free up grant capacities for SDGs and adaptation that mostly involve non-marketable activities by crowding in private investments for marketable mitigation activities.

#### 15.6.2.1 The Public-Private and Mobilisation Narrative and Current Initiatives

Financing by development finance institutions and development banks aims to address market failures and barriers related to limited access to capital as well as provide direct and indirect subsidisation by accepting higher risk, longer loan tenors and/or lower pricing. Many development and climate projects in developing and emerging countries have traditionally been supported with concessional loans by development finance institutions and/or international financial institutions (DFIs/IFIs). With an increasing number of sectors becoming viable and increasing complaints of private sector players with regard to crowding out (Bahal et al. 2018), a stronger separation and crowding in of commercial financing at the project/asset level is targeted. MDBs and IFIs were crucial for opening and growth in the early years of the green bonds, which represent a substantial share of issuances (CBI 2019a). Drivers of an efficient private sector involvement are stronger incentives to have projects delivered on time and in budget as well as market competition (Hodge et al. 2018). It remains key that the private sector mobilisation goes hand in hand with institutional capacity building as well as strong sectoral development in the host country, as a strong, knowledgeable public partner with the ability to manage the private sector is a dominating success factor for public-private cooperation (WEF 2013; Yescombe 2017; Hodge et al. 2018).

Limited research is available on the efficiency of mobilisation of the private sector at the various levels and/or the theory of change attached to the different approaches as applied in classical public-private partnerships. Also, transparency on current flows and private involvement at the various levels is limited with no differentiation being made in reporting (e.g., GCF co-financing reporting). Limited prioritisation and agreement on prioritisation of sectors and/or project categories being ready and/or preferred for direct private sector involvement might become a challenge in the coming years (*high confidence*) (Sudmant et al. 2017a; Sudmant et al. 2017b).

Public guarantees have been increasingly proposed to expand climate finance, especially from the private sector, with scarce public finance, by reducing the risk premium of the low-carbon investment opportunities (de Gouvello and Zelenko 2010; Emin et al. 2014; Studart and Gallagher 2015; Schiff and Dithrich 2017; Lee et al. 2018; Steckel and Jakob 2018). They have the advantage of a broad coverage including the 'macro' country risks and to tackle the up-front risks during the preparation, bidding and development phases of the project lifecycle that deter project initiators, especially for capital-intensive and immature options. Insurances are also powerful de-risking instruments (Déau and Touati 2018) but they entitle the issuer to review claims concerning events and cannot cope with up-front costs. Contractual arrangements like power purchase agreements are powerful instruments to reduce market risks through a guaranteed price but they weigh on public budgets. Risk-sharing that brings together public agencies, firms, local authorities, private corporates, professional cooperatives, and institutional financiers can reduce costs (UNEP 2011), and support the deployment of innovative business models (Déau and Touati 2018). Combined with emission taxes they can contribute to reducing credit rationing of immature and risky low-carbon technologies (Haas and Kempa 2020).

### 15.6.3 Considerations on Availability and Effectiveness of Public Sector Funding

The gap analysis as well as other considerations presented in this chapter illustrate the critical role of increased volumes and efficient allocation of public finance to reach the long-term global goals, both nationally and internationally.

**Higher public spending levels driven by the impacts of COVID-19 and related recovery packages.** Higher levels of public funding represent a massive chance but also a substantial risk. A missing alignment of public funding and investment activity with the Paris Agreement (and Sustainable Development Goals) would result in significant carbon lock-ins, stranded assets and thus increase transition risks and ultimately economic costs of the transition (*high confidence*). Using IMF data for stimulus packages, Andrijevic et al. (2020) estimated that COVID-19-related fiscal expenditure had surpassed USD12 trillion by October 2020 (80% in OECD countries), a third of which being spent in liquidity support and health care. Total stimulus pledged to date is ten times higher than low-Paris-consistent carbon investment needs from 2020–2024 (Andrijevic et al. 2020; Vivid Economics 2020). Overall, stimulus packages launched include USD3.5 trillion to sectors directly affecting future emissions, with overall fossil fuel investment flows outweighing low-carbon technology investment (Vivid Economics 2020).

Lessons from the global financial crises show that although deep economic crises create a sharp short-term emission drop, and green stimulus is argued to be the ideal response to tackle both the economic and the climate crises at once, disparities between regional strategies hinder the low-carbon transition (*high confidence*). Indeed, inconsistent policies within countries can also counterbalance emission reductions from green stimulus, as well as a lack of transparency and green spending pledged not materialising (Jaeger et al. 2020). Also, aggressive monetary policy as a response to the global financial crisis, including quantitative easing that did not target low-carbon sectors, has been heavily criticised (Jaeger et al. 2020). The COVID-19 crisis recovery, in contrast, benefits from developments which have taken place since, such as an emerging climate-risk awareness from the financial sector, reflected in the call from the Coalition of Finance Ministers for Climate Action (Coalition of Finance Ministers for Climate Action 2020), which unites 50 countries' finance ministers, for a climate-resilient recovery.

The steep decrease in renewable electricity costs since 2010 also represents a relevant driver for a low-carbon recovery (Jaeger et al. 2020). Many more sectors are starting to show similar opportunities for rapid growth with supportive public spending such as low-carbon transport and buildings (IEA 2020d). Expectations that the package will increase economic activity rely on the assumption that increased credit will have a positive effect on demand, the so-called demand-led policy (Mercure et al. 2019). Boosting investment should propel job creation, increasing household income and therefore demand across economic sectors (*high confidence*). A similar plan has also been proposed by the US administration and the European Union through the Next Generation EU (European Council 2020).

Nevertheless, three uncertainties remain. First, only those countries and regions with highest credit-ratings (AAA or AA) with access to deep financial markets and excess savings will be able to mount such counter-cyclical climate investment paths, typically high-income developed economies (*high confidence*). In more debt constrained developing countries lower access to global savings pools because of higher risk perceptions and lower credit ratings (BBB or less), exacerbated by COVID-19, are already leading to credit downgrades and defaults (Kose et al. 2020) and have long tended to be fiscally pro-cyclical (McManus and Ozkan 2015). These include the general class of virtually all major emerging and especially low-income developing countries, to which such demand-stimulating counter-cyclical climate-consistent borrowing path is likely. To access such funds, these countries would need globally coordinated fiscal policy and explicit supporting cross-border instruments, such as sovereign guarantees, strengthening local capital markets and boosting the USD100 billion annual climate finance commitment (Dasgupta et al. 2019).

Second, a strong assumption is that voters will be politically supportive of extended and increased fiscal deficit spending on climate on top of COVID-19-related emergency spending and governments will overcome treasury biases towards fiscal conservatism (to preserve credit ratings). However, evidence strongly suggests that voters (and credit rating agencies) tend to be fiscally conservative (Peltzman 1992; Lowry et al. 1998; Alesina et al. 2011; Borge and Hopland 2020), especially where expenditures involve higher taxes in the future and do not identifiably flow back to their local bases (the 'public good' problem) (*high confidence*). Such mistrust has been a reason for abortive return to fiscal austerity often in the past (most recently during global financial crisis) and may benefit for political support by consistently reframing the climate expenditures in terms of job creation benefits (Bougrine 2012), effectiveness of least-cost fiscal spending on climate for reviving private activity, and the avoidance of catastrophic losses (Huebscher et, al. 2020) from higher carbon emissions. A new understanding of debt sustainability including negative implications of deferred climate investments on future GDP has not yet been mainstreamed (see more on the debt sustainability discussion below (e.g., Buhr et al. 2018; Fresnillo 2020a). In addition, implications on the availability of international public finance flows are not yet clear since current additional funding prioritises urgent health care support rather than an increase in predictable mid-/long-term financial support. Heavy investment needs for recovery packages in developed countries on the one hand and their international climate finance commitments on the other might be perceived to compete for available 'perceived as appropriate' budgets.

## Box 15.6 | Macroeconomics and Finance of a Post-COVID-19 Green Stimulus Economic Recovery Path

Financial history suggests that capital markets may be willing to accommodate extended public borrowing for transient spending spikes (Barro 1987) when macroeconomic conditions suggest excess savings relative to private investment opportunities (Summers 2015) and when public spending is seen as timely, effective and productive, with governments able to repay when conditions improve as economic crisis conditions abate (*high confidence*). A surge in global climate mitigation spending in the post-pandemic recovery may be an important opportunity, which global capital markets are signalling (Global Investor Statement 2019). The standard 'neo-classical' macroeconomic model is often used in integrated energy-economy-climate assessments (Balint et al. 2016; Nordhaus 2018). This class of Computable General Equilibrium (CGE) models, however, has a limited treatment of the financial sector and assumes that all resources and factors of production are fully employed, there is no idle capacity and no inter-temporal financial intermediation (Pollitt and Mercure 2018b). Investment cannot assume larger values than the sum of previously determined savings, as a fixed proportion of income. Such constraint, as stressed by Mercure et al. (2019), implies that investment in low-carbon infrastructure, under the equilibrium assumptions, necessarily creates a (neo-Ricardian) crowding-out effect that contracts the remaining sectors. Box 15.6, Figure 1 shows the implications (in the red-shaded part of Figure 1).

Post-Keynesian demand-side macroeconomic models, with financial sectors and supply-side effects, in contrast, allow for the reality of non-equilibrium situations: persistent short- to medium-term underemployed economy-wide resources and excess savings over investment because of unexpected shocks, such as COVID-19. In these settings, economic stimulus packages allow a faster recovery

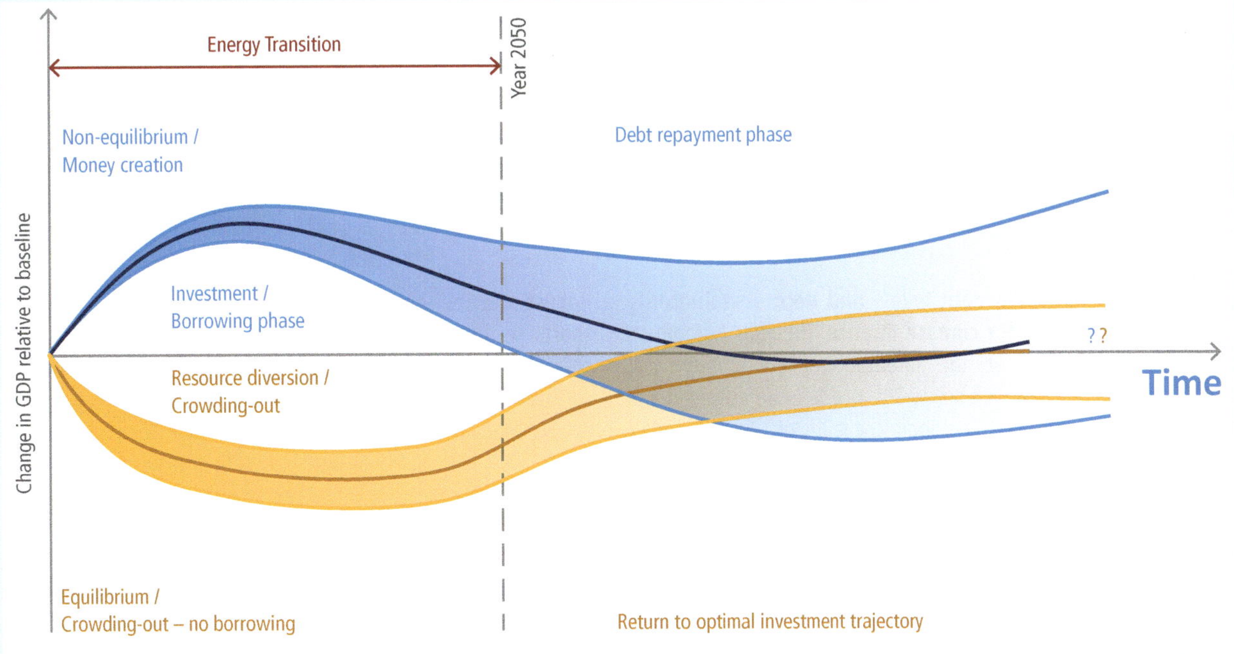


**Box 15.6, Figure 1 | Two worlds – energy transition outcomes under alternative model assumptions (Keynesian vs General Equilibrium).** Source: Mercure et al. (2019).

*Box 15.6 (continued)*

with demand-led effects: 'Economic multipliers are near zero when the economy operates near capacity. In contrast, during crises such as the GFC, economic multipliers can be high' (Blanchard and Leigh 2013; Hepburn et al. 2020b). The expected results are opposite to the standard supply-led equilibrium models as a response to investment stimulus (the green-shaded part of Box 15.6, Figure 1), as intended by 'green-stimulus' packages such as proposed by the EU (Balint et al. 2016; Mercure et al. 2019).

Even if demand-led models work better in depressions, the question nevertheless is whether the additional public borrowing for such 'green stimulus' can be undertaken by market borrowings given already high public debt levels and recovered in the future from taxes as the economy revives. The results of recent macroeconomic modelling work (Liu et al. 2021) represented by 10 major countries/regions suggests answers. It uses a non-standard macroeconomic framework, with Keynesian features such as financial and labour market rigidities and fiscal and monetary rules (McKibbin and Wilcoxen 2013). First, a global 'green stimulus' of about an average of 0.8% of GDP annually in additional fiscal spending between 2020–30 would be required to accelerate the emissions reduction path required for a 1.5°C transition. Second, such a stimulus would also accelerate the global recovery by boosting GDP growth rates by about 0.6% annually during the critical post-COVID period. Third, the optimal tax policy would be to backload the carbon taxes to later in the macroeconomic cycle, both because this would avoid dampening near-term growth while pre-announced carbon tax plans would incentivise long-term private energy transition investment decisions today and provide neutral borrowing. This macroeconomic modelling path thus replicates the 'green stimulus' impacts expected in theory (Box 15.6, Figure 1). There are also some other additional features of the modelled proposal: (i) fiscal stimulus – needed in the aftermath of the pandemic – can be an opportunity to boost green and resilient public infrastructure; (ii) green research and development 'subsidies' are feasible to boost technological innovations; and (iii) income transfers to lower income groups are necessary to offset negative impacts of rising carbon taxes.

Substantial effects of the COVID-19 pandemic, which is relatively unique in its public health impacts when combined with the consequences of deep economy-wide shocks (economic downturn, public finances, and debt), are expected to last for decades even in the absence of no significant future recurrence. A scenario where the pandemic recurs mildly every year for the foreseeable future further hinders GDP and investment recovery, where growth is unlikely to rebound to previous trajectories, even within OECD economies (McKibbin and Vines 2020) and with worse effects in developing regions. History is strongly supportive: studies on the longevity of pandemics' impacts indicate significant macroeconomic effects persisting for decades, with depressed real rates of return, increased precautionary savings (Jordà et al. 2020), unemployment (Rodríguez-Caballero and Vera-Valdés 2020) and social unrest (Barrett and Chen 2021). The direct effect on emissions is likely to be a small reduction from previous trajectories, but the longer-lasting impacts are more on the macroeconomic-finance side. Pandemic responses have increased sovereign debt across countries in all income bands (IMF 2021e). However, its sharp increase in most developing economies and regions has caused debt distress (Bulow et al. 2021), widening the gap in developing countries' access to capital (Hourcade et al. 2021b). While strong coordinated international recovery strategies with climate-compatible economic stimulus is justified (Barbier 2020; Barbier and Burgess 2020; IMF 2020c; Le Quéré et al. 2021; Pollitt et al. 2021), national recovery packages announced do not show substantial alignment with climate goals (D'Orazio 2021; Hourcade et al. 2021b; Rochedo et al. 2021; Shan et al. 2021). Contradictory post-COVID-19 investments in fossil fuel-based infrastructure may create new carbon lock-ins, which would either hinder climate targets or create stranded assets (Hepburn et al. 2020a; Le Quéré et al. 2021; Shan et al. 2021), whilst deepening global inequalities (Hourcade et al. 2021b).

**Considerations on global debt levels and debt sustainability as well as implications for climate finance.** The Paris Agreement marked the consensus of the international community that a temperature increase of well below 2°C needs to be achieved and the SR1.5 has demonstrated the economic viability of 1.5°C. However, in terms of increase of supply of, in particular, public finance, often the debate is still driven by the question on affordability, considerations around financial debt sustainability and budgetary constraints against the background of macroeconomic headwinds – even more in the (post-)COVID-19 world (*high confidence*). The level of climate alignment of debt is hardly considered in debt-related regulation and/or debt sustainability agreements like the Maastricht Treaty ceilings (3% of GDP government deficit and 60% of GDP (gross) government debt) not considering economic costs of deferred climate action as well as economic benefits of the transformation.

Robust studies on the economic costs and benefits in the short- to long-term of reaching the LTGG exist for only few countries and/or regions, primarily in the developed world (*high confidence*) (e.g. BCG 2018; McKinsey 2020a). With many studies underpinning the strong economic rationale for high investments in the short-term(e.g., McKinsey 2020a), regional differences are significant highlighting the need for extensive cooperation and solidarity initiatives.

For many developing countries, the focus of debt sustainability discussions is on the negative effect of climate change on the future GDP and the uncertainty with regard to short-term effects of climate change and their economic implications (*high confidence*). With long-term economic impacts of climate change being in the focus of the modelling community, the volatility of GDP in the short term driven by shocks is more difficult to analyse and requires country-specific deep-dives. IPCC scenario data is often not sufficient to perform such analysis with additional assumptions being needed (Acevedo 2016). For debt sustainability analysis, these more short-term impacts are, however, a crucial driver with transparency being limited to the significance of climate-related revision of estimates. The latter might result in a continued overestimation of future GDP as happened in the past, increasing the vulnerability of highly indebted countries (Guzman 2016; Mallucci 2020). While climate change considerations have already impacted country ratings and debt sustainability assessments (and financing costs), it is unclear whether current GDP forecasts are realistic. The review of the IMF debt sustainability framework leads to a stronger focus on vulnerability rather than only income thresholds when deciding upon eligibility for debt relief and/or concessional resources (Mitchell 2015), which could become a mitigation factor for the challenge described before.

Debt levels globally but particularly in developing and vulnerable countries have significantly increased over the past years with current and expected climate change impacts further burdening debt sustainability (*high confidence*). For low- and middle-income countries, 2018 marked a new peak of debt levels amounting to 51% of GDP; between 2010 and 2018, external debt payments as a percentage of government budget grew by 83% in low- and middle-income countries, from an average of 6.71% in 2010 to an average of 12.56% in 2018 (Fresnillo 2020b). COVID-19 has further reduced the fiscal space of many developing governments and/or increased the likelihood of debt stress. With many vulnerable countries already being burdened with higher financing costs, this limited fiscal space further shrinks their ability to actively steer the required transformation (Buhr et al. 2018). Limited progress in increasing debt transparency remains another burden (Section 15.6.7).

Considering the need for responses to both short-term liquidity issues and long-term fiscal space, current G20/IMF/World Bank debt service suspension initiatives are focused on the liquidity issue rather than underlying problems of more structural nature of many low-income countries (Fresnillo 2020a). In order to ensure fiscal space for climate action in the coming decade, a mix between debt relief, deferrals of liabilities, extended debt levels and sustainable lending practices including new solidarity structures need to be considered in addition to higher levels of bilateral and multilateral lending to reduce dependency on capital markets and to bridge the availability of sustainably structured loans for highly vulnerable and indebted countries. More standardised debt-for-climate swaps, a higher share of GDP-linked bonds or structures ensuring (partial) debt cancellation in case countries are hit by physical climate change impacts/shocks appear possible. The 'hurricane' clause introduced by Grenada, or wider natural disaster clauses provide issuers with an option to defer payments of interest and principal in the event of a qualifying natural disaster and can reduce short-term debt stress (UN Addis Ababa Action Agenda Art. 102) (UN 2015a). A mainstreaming of such clauses has been pushed by various international institutions. The collective action clause might be a good example of a loan/debt term which became market standard. Definition of triggers is likely the most complex challenge in this context.

The use of debt-for-nature and debt-for-climate-swaps is still very limited and not mainstreamed but offers significant potential if used correctly (*high confidence*).

An increasing number of debt-for-climate/nature swaps have been seen in recent years applied primarily in international climate cooperation and in bilateral contexts, however, not (yet) to an extent addressing severe and acute debt crises (Essers et al. 2021; Volz et al. 2021) offering significant potential if used correctly (Warland and Michaelowa 2015). Significant lead times, needs-based structuring, transparency with regard to the additionality of financed climate action, uncertainty with regard to own resource constraints and ODA accountability remain as barriers for a massive scale-up needed to make transactions relevant (Mitchell 2015; Fuller et al. 2018; Essers et al. 2021). At the same time, the limitation of the use of debt-based instruments as a response to climate-related disasters and counter-cyclical loans might be necessary (Griffith-Jones and Tyson 2010).

Ensuring efficient debt restructuring and debt relief in events of extreme shocks and imminent over-indebtedness and sovereign debt default are further crucial elements with a joint responsibility of debtors and creditors (UN 2015a). In this context, the Commonwealth Secretariat flagged that the diversification of the lender portfolio made debt restructuring more difficult with more and more heterogeneous stakeholders being involved (Mitchell 2015) and the UN AAAA raising concerns about non-cooperative creditors and disruption of timely completion of debt restructuring (UN 2015a). This is a side effect of a stronger use of capital markets, which needs to be carefully considered in the context of sovereign bond issuances (Section 15.6.7).

**Stranded assets.** The debate around stranded assets focuses strongly on the loss of value to financial assets for investors (Section 15.6.1), however, stranded assets and resources in the context of the transition towards a low-emission economy 'are expected to become a major economic burden for states and hence the tax payers' (*high confidence*) (EEAC 2016). Assets include not only financial assets but also infrastructure, equipment, contracts, know-how, jobs as well as stranded resources (Bos and Gupta 2019). Besides financial investors and fiscal budgets, consumers remain vulnerable to stranded investments. Against the background of the frequent simultaneousness of losses occurring for financial investors on the one hand and negative employment effects as well as regional development and fiscal effects on the other hand, negotiations about compensations and public support to compensate for negative effects of phasing out of polluting technologies often remain interlinked and compensation mechanisms and related redistribution effects untransparent.

Recent phase-out deals tend to aim for (partial or full) compensation rather than no relief for losses. In contrast to the line of argument in the tobacco industry, the backward-looking approach and a resulting obligation of compensation by investors in polluting assets can be observed rarely with the forward-looking approach of compensations by future winners for current losers dominating – despite the high level of awareness about carbon externalities and resulting climate change impacts among polluters for many years (van der Ploeg and

Rezai 2020). In particular, transactions in the energy sector show a high level of investor protection also against much needed climate action which is also well illustrated by the share of claims settled in favour of foreign investors under the Energy Charter Treaty and investor-state dispute settlement (Bos and Gupta 2019).

Late government action can delay action and consequently strengthen the magnitude of action needed at a later point in time with implications for employment and economic development in impacted regions requiring higher level of fiscal burden (*high confidence*). This has also been considered in the context of global climate cooperation with prolonged support for polluting infrastructure resulting in heavy lock-in effects and higher economic costs in the long run (Bos and Gupta 2019). Despite a significant share of fossil resources which need to become stranded in developing countries to reach the LTGG, REDD+ remains a singular example for international financial cooperation in the context of compensation for stranded resources.

### 15.6.4 Climate Risk Pooling and Insurance Approaches

Since 2000, the world has been experiencing significant increase in economic losses and damages from natural disasters and weather perils such as tropical cyclones, earthquakes, flooding and drought. Total global estimate of damage is about USD4210 billion, 2000–2018 (Aon Benfield UCL Hazard Research Centre 2019). The largest portion of this is attributed to tropical cyclones (USD1253 billion), followed by flooding (USD914 billion), earthquakes (USD757 billion) and drought (approximately USD372 billion, or about USD20 billion $yr^{-1}$ losses) (Aon Benfield UCL Hazard Research Centre 2019). In the period 2017–2018, natural catastrophe losses totalled approximately USD219 billion (Bevere 2019). According to the National Oceanic and Atmospheric Administration, 14 weather and climate disasters cost USD91 billion in 2018 (NOAA NCEI 2019). The European Environment Agency reports that 'disasters caused by weather and climate-related extremes accounted for some 83% of the monetary losses over the period 1980–2017' for EU Member States (EU-28) and that 'weather and climate-related losses amounted to EUR426 billion (at 2017 values)'. For the EEA member countries (EEA-33), the 'total reported economic losses caused by weather and climate-related extremes' over the same period amounted to approximately EUR453 billion (EEA 2019). Asia Pacific and Oceania has been particularly impacted by typhoon and flooding (China, India, the Philippines) resulting in economic losses of USD58 billion, 2000–2017, and a combination of flooding, typhoon and drought totalling USD89 billion in 2018 (inclusive of loss by private insurers and government sponsored programmes (Aon Benfield UCL Hazard Research Centre 2019). Based on past historical analysis, a region such as the Caribbean, which has experienced climate-related losses equal to 1% of GDP each year since 1960, is expected to have significant increases in such losses in the future leading to possibly upwards of 8% of projected GDP in 2080 (Commonwealth Secretariat 2016). Similarly, Latin American countries, such as Argentina, El Salvador and Guatemala, experienced severe losses in agriculture totalling about USD6 billion due to drought in 2018 (Aon Benfield UCL Hazard Research Centre 2019). In the African region, where climate is projected to get significantly warmer, continuing severe drought in parts of East Africa, Tropical Cyclone Idai, had devastating economic impacts for Mozambique, Zimbabwe and Malawi (WMO 2019). According to Munich Re, loss from about 100 significant events in 2018 for Africa are estimated at USD1.4 billion (Munich Re 2019).

**While there are questions about the sufficiency of insurance products to address the losses and damages of climate-related disasters, insurance can help to cover immediate needs directly, provide rapid response and transfer financial risk in times of extreme crisis (*high confidence*) (GIZ 2015; Lucas 2015; Schoenmaker and Zachmann 2015; Hermann et al. 2016; Wolfrom and Yokoi-Arai 2016; Kreft and Schäfer 2017; UNESCAP 2017; Matias et al. 2018; UNECA 2018; Broberg and Hovani-Bue 2019; EEA 2019; Martinez-Diaz et al. 2019). Commercial insurability is heavily driven by the predictability of losses and the resulting ability to calculate insurance premium levels properly. Climate change has become a major factor of increasing uncertainty. The previously strong reliance on historic data in calculation of premium levels may be but a starting point given the likely need for upward adjustment due to climate change and potential consequential economic damage. Different risk perceptions between policyholders and insurers will create contrary assessments on premium levels and consequently underinsurance. McKinsey (2020b) also stresses the systemic effect of climate change on insurers' business models and resulting availability of appropriate insurance products.**

The conventional approach to such protective or hedging position has been indemnity and other classical insurance micro-, meso- and macro-level schemes (Hermann et al. 2016). These include micro insurance schemes such as index insurance and weather derivative approaches that cover individuals' specific needs such as coverage for farm crops. Meso-level insurance schemes, which primarily benefit intermediary institutions, such as NGOs, credit unions, financial institutions and farmer credit entities, seek to reduce losses caused by credit default thereby 'enhancing investment potential', whereas macro-level insurance schemes 'allow both insured and uninsured individuals to be compensated for damages caused by extreme weather events' (Hermann et al. 2016). These macro-level insurance schemes include catastrophe bonds and weather derivatives and so on, that transfer risk to capital markets (Hermann et al. 2016). Over the last decades, there has been a trend towards weather-index insurance and other parametric insurance products based on predefined pay-out risk pooling instruments. It has gained favour with governments in developing regions such as Africa, the Caribbean and the Pacific because it provides certainty and predictability about funding – financial preparedness – for emergency actions and initial reconstruction and reduces moral hazard. This 'financial resilience' is also increasingly appealing to the business sector, particularly micro, small and medium enterprises (MSMEs), in developing countries (MEFIN Network and GI RFPI Asia 2016; Woods 2016; Schaer and Kuruppu 2018).

To date, sovereign parametric climate risk pooling as a way of managing climate risk does not seem to have much traction in developed countries and does not appear to be attractive to actors in the G20 countries. No G20 members are yet party to any climate risk pooling initiative (Kreft and Schäfer 2017). However, international bilateral donors such as the USAID and the UK Foreign, Commonwealth and Development Office (FCDO, formerly DFID), and the multilateral development banks are all, to different extent, supporters of the various climate risk pooling initiatives now operational in developing countries.

As noted also in IPCC AR5, risk sharing and risk transfer strategies provide 'pre-disaster financing arrangements that shift economic risk from one party to another' (IPCC 2012). Risk pooling among countries and regions is relatively advantageous when compared to conventional insurance because of the effective subsidising of 'affected regions' using revenues from unaffected regions which involve pooling among a large subset of countries (*high confidence*) (Lucas 2015). In general, the premiums are less costly than what an individual country or entity can achieve and disbursement is rapid and there are also fewer transaction costs (Lucas 2015; World Bank 2015). The World Bank argues that the experience with the Pacific Catastrophe Risk Insurance Pilot (PCRIP) and Africa Risk Capacity risk pooling (ARC) show savings of 50% in obtaining insurance cover for pooled risk compared with purchasing comparable coverage individually (Lucas 2015; World Bank 2015; ARC 2016). However, it requires, as noted by UNESCAP, 'extensive coordination across participating countries, and entities' (Lucas 2015).

At the same time, this approach has substantial basis risk (actual losses do not equal financial compensation) (*high confidence*) (Hermann et al. 2016). With parametric insurance, pay-outs are pre-defined and based on risk modelling rather than on-the-ground damage assessment so may be less than, equal to, or greater than the actual damage. It does not cover actual losses and damage and therefore, may be insufficient to meet the cost of rehabilitation and reconstruction. It may also be 'non-viable' or damaging to livelihoods in the long run (UNFCCC 2008; Hellmuth et al. 2009; Hermann et al. 2016). Additionally, if the required threshold is not met, there may be no pay-out, though a country may have experienced substantial damages from a climatic event. This occurred for the Solomon Islands in 2014 which discontinued its insurance with the Pacific Catastrophe Risk Insurance Pilot when neither its Santa Cruz earthquake nor the 2014 flash floods were eligible to receive a pay-out under the terms of the insurance (Lucas 2015).

Increasingly, climate risk insurance schemes are being blended into disaster risk management as part of a comprehensive risk management approach (*high confidence*). The best-known example is the Caribbean Catastrophe Risk Insurance Facility (CCRIF SPC 2018), which involves cooperation among Caribbean states, Japan, Canada, UK and France and international organisations such as the World Bank (UNESCAP 2017). But there are growing platforms of such an approach mainly under the umbrella of the G7's InsuResilience Initative (Deutsche Klimafinanzierung 2020), including, the Pacific Catastrophe Risk Assessment and Financing Initiative for the Pacific Islands (PCRAFI), the African Risk Capacity (ARC Agency and its financial affiliate), and the African Risk Capacity Limited (ARC Ltd/ the ARC Group) (ARC 2016) and in the Asian region, the South East Asian Disaster Risk Insurance Facility (SEADRIF) and the ASEAN Disaster Risk Financing and Insurance Program (ADRFI), (SEADRIF 2018; GIZ and World Bank 2019; Martinez-Diaz et al. 2019; Vyas et al. 2019; World Bank 2019a). The group of 20 vulnerable countries (V20) has also developed a Sustainable Insurance Facility (SIF), billed as a technical assistance facility for climate-smart[14] insurance for MSMEs in 48 developing countries as well as potentially to de-risk renewable energy in these countries and regions (ACT Alliance 2020; V20 2020; V20 2021).

However, as noted above, climate risk pooling is not a panacea. There are very obvious and significant challenges. According to Kreft and Schäfer (2017), limitations of insurance schemes include coordination challenges, limited scope, destabilisation due to exit of one or more members as premiums rise and inadequate attention to permanence (Schaeffer et al. 2014). There are also challenges with risk diversification, replication, and scalability (*high confidence*). For example, CCRIF is extending both its membership and diversifying its geographic dimensions into Central America in seeking to lower covariate risk (similar shocks among cohorts such as droughts or floods). Under the SPC portfolio, CCRIF is able to segregate risk across the regions. Risk insurance does not obviate from the need to engage in capacity building to scale-up as well as having process for addressing systemic risk. Currently, risk pools have limited sectoral reach and may cover agriculture but not other important sectors such as fisheries and public utilities. Only recently (July 2019) has CCRIF initiated coverage of fisheries with the development of its Caribbean Oceans and Aquaculture Sustainability Facility (COAST) instrument (CCRIF SPC 2019; ACT Alliance 2020). Historically, risk pool mechanisms, like CCRIF and ARC, only cover a small subset of perils, such as tropical cyclones, earthquakes and excess rainfall but do not include other perils such as drought. Since 2016, ARC has increased its scope to cover drought and in 2019 launched ARC Replica, which not only covers drought but offers premiums and coverage to NGOs and the World Food Programme through the START Network and a pastoral drought product for protecting small farmers and ensuring food security. In some regions and countries, there may also be limited access to reinsurance (Schaeffer et al. 2014; Lucas 2015). An important down-side of climate risk pooling is that it does not cover the actual cost of damage and losses. Though on the positive side, pay-out may exceed costs, but it may also be less than costs. Hence, the parametric approach is not a panacea and does not preclude having recourse to conventional indemnity insurance, which will cover full damage costs after a climate change event as it involves full on-the-ground assessment of factors such as the necessity and costs of repair versus, say, replacement value of damaged infrastructure. This may be important for governmental and publicly provided services such as schools, hospitals, roads, airports, communications equipment and water supply facilities. Given the growing popularity of parametric insurance and climate risk

[14] According to the V20, 'the term "climate-smart" captures the need for two types of climate-related insurance products for MSMEs in vulnerable economies: (1) Climate risk insurance (2) Insurance products which enable low carbon investments, and thereby contribute to increased efficiencies through cost-savings from cheaper low-carbon technologies' (V20 2021).

pooling, there are very ambitious attempts to expand this approach on several fronts (Scherer 2017). Schoenmaker and Zachmann (2015) have proposed a global climate risk pool to help the most vulnerable countries. The pathway to this includes capacity building in underdeveloped financing sectors of developing countries. They argue that as climate extremes become more normalised, they will wipe out significant parts of the infrastructure and productive capacity of developing countries. This will have knock-on impact on fiscal capacity due to lowered tax revenue and high rebuilding costs. 'Developing countries', Schoenmaker and Zachmann (2015) argue, 'cannot insure against such events on a market basis, nor would it be sensible to divert scarce fiscal resources away from infrastructure investment into accumulating a financial buffer for such situations'. In that context, Schoenmaker and Zachmann (2015) call for international risk pooling as 'the only sensible strategy', especially if it addresses the major gaps in climate risk insurance for poor and vulnerable communities by enhancing demand through 'smart support instrument' for premium support such as full or partial premium subsidies and investment in providing risk reduction (Schäfer et al. 2016; Le Quesne et al. 2017; MCII 2018; Vyas et al. 2019). This, it is argued, may help to smoothen out the limited uptake of regional institutions such as ARC and CCRIF SPC, which are only in three regions of the world (with missing mechanism in South America) (Kreft and Schäfer 2017). Existing regional mechanisms, while they may perform very well, only cover a portion of climatic hazards and tend to have limited subscribers. For example, across the key four sovereign risk pools (ARC, CRIFSPC, PCRAFI and SEADRIF), though there are 68 countries only one-third or 32% have purchased coverage in 2019 and 46% 'did not deploy disaster risk financing instruments' (ACT Alliance 2020).

Other gaps and challenges flagged by Kreft and Schäfer (2017) include limited coverage of the full spectrum of contingency risks experienced by countries, inadequate role of risk management as a standard for all regional pools, though there are some emerging best practices in terms of data provision on weather-related risks, and incentivisation of risk reduction (*high confidence*). Here, they recognise the work of Africa Risk Capacity for not only providing the infrastructure to trigger disbursement but for also promoting national risk analysis. Another important gap in the landscape of climate risk pooling is lack of attention to financial institutions' lending portfolios that are vulnerable to weather shocks. In this regard subsidies as part of innovative financing schemes facilitated by the donor community can encourage the uptake of meso-level climate risk insurance solutions (Kreft and Schäfer 2017).

In the literature, there are two attempts at systematic evaluation or comprehensive assessment of regional climate risk pools: a comprehensive study by Scherer (2017) and FCDO's ten-year evaluation (2015–2024). Overall, neither of these studies draw adverse conclusions about regional climate risk pooling initiatives/mechanisms. According to Scherer, 'it appears that insurances work in principle and there is certainly success' and 'initial experiences demonstrate regional climate risk insurances works'. The author cited the 28 pay-outs to 16 countries of USD106 million arguing that it provides cash-starved countries with much needed cash (Scherer 2017, p. 4). The FCDO study (Scott 2017) examines the uptake of ARC and its impact on reducing vulnerability to disasters. It notes that there is scarce literature on disaster risk insurance mechanisms in terms of impacts. In its current sample of 20 countries as of November 2017, four are projected to experience food security crisis (IPC Level 3) but are not signatories to the ARC, which may signal that ARC is not attractive to all food insecure countries and that there is no overwhelming appetite for ARC among poorer countries. Additionally, Panda and Surminski (2020) research the importance of indicators and frameworks for monitoring the performance and impact of Coalition for Disaster Resilient Infrastructure (CDRI) but make no final assessment of any of the regional climate risk pool. However, they propose mechanisms to improve the transparency and accountability of the system. Scherer (2017), Forest (2018) and Panda and Surminski (2020) seem to indicate that there is 'enthusiasm to support and scale-up regional climate risk insurance' (Scherer 2017, p. 4) Examples of this support include: the Germany Ministry for Economic Cooperation and Development (BMZ) has provided USD5.9 million for the World Food Programme (WFP) to protect 1.2 million vulnerable African farmers with climate risk insurance, through ARC Replica, and the G7 InsuResilience Vision 2025, which has committed to ensuring 400–500 million poor persons are covered against disaster shock by pre-arranged finance and insurance mechanism by 2025; some of this will be through ARC (WFP 2020). Of course, this does not mean that risk pools are without challenges or are not failing on specific sets of metrics. Forest (2018) flags three failing areas: policy holder and hazard coverage, the cost of premium and risk transfer parameters, and the use of pay-out, which in most cases are up to the government. Here, ARC is flagged among the three regional risk pools, as the only one with contingency plan requirements that can support effective use of pay-outs. Other research exploring climate risk pooling and its impacts flag lack of transparency around pay-out, premium or risk transfer parameters. Ultimately, climate risk pools are not full insurance; they offer only limited coverage. Entities such as the U4 Anti-Corruption Help Desk are exploring how to mitigate potential corruption with regard to climate risk insurance.

### 15.6.5 Widen the Focus of Relevant Actors: Role of Communities, Cities and Subnational Levels

There is an urgency and demand to meet the financial needs of the climate change actions not only at the national level but also at the subnational level, to achieve low-carbon and climate-resilient cities and communities (*high confidence*) (Barnard 2015; Moro et al. 2018). Scaling up subnational climate finance and investment is a necessary condition to achieve climate change mitigation and adaptation action (Ahmad et al. 2019).

**The importance of exploring effective subnational climate finance.** Stronger subnational climate action is indispensable to adapt cities to build more sustainable, climate-positive communities (Kuramochi et al. 2020). It has transformative potential as a key enabler of inclusive urban economic development through the building of resilient communities (*high confidence*) (Floater et al. 2017a; Colenbrander et al. 2018b; Ahmad et al. 2019). Yet the significant potential of subnational climate finance mechanisms remains unfulfilled. Policy frameworks, governance, and choices at higher levels underpin subnational climate investments (Colenbrander et al. 2018b; Hadfield and Cook 2019). To scale climate investment, a systematicunderstanding of the preconditions to mobilising high-potential financing instruments at the national and subnational levels is necessary.

**Subnational climate finance needs and flows.** Subnational climate finance covers financing mechanisms reaching or utilising subnational actors to develop climate positive investment in urban areas. The fragility of interconnected national and subnational finances affects subnational finance flows, including the impact of the social-economic crisis (Canuto and Liu 2010; Ahrend et al. 2013). The effect of deficit in investment for global infrastructure towards the growing subnational-level debt also creates pressure on subnational finances and constrains future access to financing (*high confidence*) (Smoke 2019).

The International Finance Corporation estimates a cumulative climate investment opportunity of USD29.4 trillion across six urban sectors (waste, renewable energy, public transportation, water, EVs, and green buildings) in emerging market cities, cities in developing countries with more than 500,000 population, to 2030 (IFC 2018). However, the State of Cities Climate Finance report estimated that an average of USD384 billion was invested in urban climate finance annually in 2017–2018 (Negreiros et al. 2021). The International Institute for Environment and Development estimates that out of the USD17.4 billion total investments in climate finance, less than 10% (USD1.5 billion) was approved for locally-focused climate change projects between 2003 and 2016 (Soanes et al. 2017).

**Subnational climate public and private finance.** Urban climate finance and investment are prominent in the subnational climate finance landscape (CCFLA 2015; Buchner et al. 2019). Finance mechanisms that can support climate investment for the urban sector include public-private partnerships (PPPs); international finance; national investment vehicles; pricing, regulation, standards; land value capture; debt finance; and fiscal decentralisation (Granoff et al. 2016; Floater et al. 2017b; Gorelick 2018; White and Wahba 2019). Among these mechanisms, PPPs, debt finance, and land value capture have the potential to mobilise private finance (Ahmad et al. 2019). Better standardisation in processes is needed, including those bearing on contracts and regulatory arrangement, to reflect local specificities (Bayliss and Van Waeyenberge 2018) (Section 15.6.1.1).

PPPs are particularly important in cities with mature financial systems as the effectiveness of PPPs depends on appropriate investment architecture at scale and government capacity (*high confidence*). Such cities can enable infrastructure such as renewable energy production and distribution, water networks, and building developments to generate consumer revenue streams that incentivise private investors to purchase equity as a long-term investment (Floater et al. 2017b).

National-level investment vehicles can provide leadership for subnational climate financing and crowd in private finance by providing early-stage market support to technologies or evidence related to asset performance and costs-benefits (*high confidence*). The use of carbon pricing is increasing at the subnational level along with regulation and standards on negative externalities, such as pollution, to steer investment towards climate financing (World Bank Group 2019).

Debt financing via subnational bonds and borrowing, including municipal bonds, is another potential tool for raising upfront capital, especially for rich cities (*high confidence*). The share of subnational, sub-sovereign, and sovereign bonds could grow over time, given efforts to expand the creditworthiness and ensure a sufficient supply of own-source revenue to reduce the default risk. As of now, subnational and sub-sovereign bonds are constrained by public finance limits and the fiscal capacities of governments. However, while green bonds have potential for growth at the subnational level and may result in a lower cost of capital in some cases, the market faces challenges related to scaling up and has been associated with limited measurable environmental impact to date (Section 15.6.8). Further, bonds with lower credit ratings drive higher issuance costs for climate risk cities, for example, costs related to disclosure and reporting (Painter 2020).

**Key challenges of subnational climate finance.** Across all types of cities, five key challenges constrain the flow of subnational climate finance (*high confidence*): (i) difficulties in mobilising and scaling-up private financing (Granoff et al. 2016); (ii) deficient existing architecture in providing investment on the scale and with the characteristics needed (Anguelovski and Carmin 2011; Brugmann 2012); (iii) political-economic uncertainties, primarily related to innovation and lock-in barriers that increase investment risks (Unruh 2002; Cook and Chu 2018; White and Wahba 2019); (iv) the deficit in investment for global infrastructure affects the growing subnational-level debt (Canuto and Liu 2010); and (v) insufficient positive value capture (Foxon et al. 2015).

**Different finance challenges between rich and poor cities.** Access to capital markets has been one of the major sources for subnational financing and is generally limited to rich cities, and much of this occurs through loans (*high confidence*). Different challenges to accessing capital markets associated with wealthy and poorer cities are compounded into three main issues: (i) scarcity and access of financial resources (Bahl and Linn 2014; Colenbrander et al. 2018b; Cook and Chu 2018; Gorelick 2018); (ii) the level of implication from the existing distributional uncertainties to the current financing of infrastructural decarbonisation across carbon markets (Silver 2015); and (iii) the policy and jurisdictional ambiguity in urban public finance institutions (Padigala and Kraleti 2014; Cook and Chu 2018). In poorer cities, these differing features continue to be inhibited by contextual characteristics of subnational finance, including gaps in domestic and foreign capital (Meltzer 2016), the mismatch between investment needs and available finance (Gorelick 2018), weak financial autonomy, insufficient financial maturity, investment-grade credit ratings in local debt markets (Bahl and Linn 2014), scarce diversified funding sources and stakeholders (Gorelick 2018; Zhan et al. 2018; Zhan and de Jong 2018) and weak enabling environments (Granoff et al. 2016).

The depth and character of the local capital market also affect cities differently in generating bonds (*high confidence*). Challenges facing cities in developing countries include insufficient appropriate institutional arrangements, the issues of minimum size, and high transaction costs associated with green bonds (Banga 2019). Green projects and project pipelines are generally smaller in scale feasible for a bond market transaction (Saha and D'Almeida 2017; DFID 2020). De-risking in the different phases of long-term project financing can be promoted to improve the appetite of capital markets (Section in 15.6.7).

**Climate investment and finance for communities.** There is insufficient evidence about which financing schemes contribute to climate change mitigation and adaptations at community level (*high confidence*). There is growing interest in the linkages between microfinance and adaptation in the agriculture sector (Agrawala and Carraro 2010; Fenton et al. 2015; Chirambo 2016; CIF 2018; Dowla 2018), the finance for community-based adaptation actions (Fenton et al. 2014; Sharma et al. 2014), and the relations between remittances and adaptation (Le De et al. 2013). However, there is less discussion on community finance aside from the benefits of community finance and village funds in contributing to close investment gaps and community-based mitigation in the renewable energy and forest sectors (Ebers Broughel and Hampl 2018; Bauwens 2019; Watts et al. 2019) The full potential and barriers of the community finance model are still unknown and research needs to expand understanding of favourable policy environments for community finance (Bauwens 2019; Watts et al. 2019).

**Implications for the transformation pathway.** Cities often have capacity constraints on planning and preparing capital investment plans. Integrated urban capital investment planning is an option to develop cross-sectoral solutions that reduce investment needs, boost coordination capacity, and increase climate-smart impacts (*high confidence*) (Negreiros et al. 2021). In countries with weak and poorly functioning intergovernmental systems, alliances and networks may influence their organisational ability to translate adaptive capacity for transformation into actions (Leck and Roberts 2015; Colenbrander et al. 2018a). Deepening understanding of country-specific enabling environment for mobilising urban climate finance among and within cities and communities, design of policy, institutional practices and intergovernmental systems are needed to reduce negative implications of transformation (Steele et al. 2015).

### 15.6.6 Innovative Financial Products

Innovative financial products with increased transparency on climate risk have attracted investor demand, and can facilitate investor identification of low-carbon investments (*high confidence*). Innovative products may not necessarily increase financial flows for climate solutions in the near term, however they can help build capacity on climate risk and opportunities within institutions and companies to pave the way for increased flows over time.

**Investor demand is driving developments in innovative financial products (*high confidence*).** Since AR5, innovative financial products such as sustainability and green-labelled financial products have proliferated (Section 15.3). These financial products are not necessarily 'new' in terms of financial design but are packaged or labelled in an innovative way to attract responsible and impact-oriented institutional investors.

The growth and diversity of the green bond market illustrates how innovative financial products can attract both public and private investors (*high confidence*). Demand for green financial products initially stemmed from public sector pension funds. Pension funds and insurance companies in OECD countries have traditionally favoured bonds as an asset class with lower risk (OECD and Bloomberg 2015).

Since AR5, labelled green bonds have grown significantly, exceeding USD290 billion issued in 2020 with a total of USD1.1 trillion in outstanding bonds (CBI 2021a) (Section 15.6.7). Corporates, financial institutions and government-backed entities (for example in real estate, retail, manufacturing, energy utilities) issued the largest volumes, with use of proceeds focused primarily on GHG mitigation in energy, buildings and transport projects (CBI 2021a). Given their focus on GHG mitigation, green bonds are also sometimes referred to as climate bonds, but the common market terminology is 'green'. Municipal green bond issuance has also been growing (Section 15.6.7). Beyond green bonds, additional products such as green loans, green commercial paper, green initial public offerings (IPOs), green commodities, and sustainability-linked bonds and loans have also been introduced in the market (CBI 2019a) (Section 15.6.7).

Investor demand for green bonds is evidenced by over-subscription of deals. Recent studies indicate an over-subscription for green-labelled bonds by an average of between three and five times, as compared to non-labelled bonds (Gore and Berrospi 2019; Nauman 2020). Results of a survey of global treasurers showed a higher demand for green bonds than non-labelled bonds for 70% of the respondents (CBI 2020a).

The financial crisis associated with COVID-19 has put increased pressure on debt issuers, and the extent to which the increase in indebtedness for sovereigns and corporates has been financed via climate-related-labelled debt products is not known. Further, at this time there is no identified literature assessing the degree to which international versus domestic investors are financing sovereign green debt in developing countries (Section 15.6.7) However, since the onset of the COVID-19 crisis, continued steady growth in issuance has been observed broadly across sustainable bonds (including green, social and sustainability bonds), with more significant growth in social bonds to support the COVID-19 recovery (Maltais and Nykvist 2020; CBI 2021a).

Index providers and exchanges can also play a supporting role in transparency for identification of benchmarks and innovative financial products for climate action. Low-carbon indices have proliferated in recent years, with varying approaches including reduced exposure to fossil, best-in-class performers within a sector, and fossil-free (UN PRI 2018) (see discussion on ESG index performance that follows in this section). Indices can provide transparency on low-carbon opportunities, making it simpler for funds and investors to identify green investment options. Exchanges can also play a supporting role to the uptake of green financial products through transparent listings and requirements to improve credibility of green labelling. The number of green or sustainability bond listing segments tripled from five in 2016 to 15 in 2018 (SSE 2018). Green security listings can also be used to enhance local capital markets (Section 15.6.7).

**Significant potential exists for continued growth in innovative financial products, though some challenges remain (*high confidence*).** Despite recent growth and diversification, green bonds face several challenges in scaling up. Issuance of green-labelled bonds constitutes approximately 1% of the global bond market issuance (ICMA 2020b; CBI 2021a) Potential exists to increase issuance amongst corporates, for instance, and across a broader regional scope (although subject to limitations of local capital markets). Yet there remain several challenges to growing the green bond market, including *inter alia* concerns about greenwashing and limitations in application to developing countries (Shishlov et al. 2018; Banga 2019).

There is no globally accepted definition of green bonds, and varied definitions of eligible green activities are evolving across regional bond markets. Beyond the most commonly used green label, other related labels such as blue, sustainable, transition, sustainable development goal (SDG), social and environmental, social and governance (ESG) have some overlapping applications (Schumacher 2020). The degree to which these labels represent climate-relevant investments depends on underlying criteria and how they are applied (Section 15.6.4).

There are several initiatives aimed at protecting the integrity of the green label. Guidance on use and management of proceeds established by the International Capital Markets Association's Green Bond Principles (GBP) is followed on a voluntary basis, which notes eligible use of proceeds as primarily climate mitigation and adaptation projects. The GBP also recommend independent external reviews at the time of issuance, with 89% of green bond issuers in 2020 having external reviews at the time of issuance (CBI 2021a). In addition to best practice based on voluntary principles, a further check on greenwashing, although insufficient on its own, is the fear of reputation risk on behalf of investors, issuers and intermediaries in the age of social media (Hoepner et al. 2017; Deschryver and de Mariz 2020). A report on post-issuance green bond impact reporting notes that despite concerns (Shishlov et al. 2018), greenwashing incidence is rare, with 77% of green bond issuers reporting on allocation and 59% reporting on impact, but with significant variance in quality and consistency of impact reporting (CBI 2021b).

Financial disclosure regulatory developments can help further align and specify definitions of green in the financial sector but are not a substitute for climate policy (*high confidence*). Developing a common basis for understanding a green label could further reduce uncertainty or concerns of greenwashing. Regulatory developments in some regions seek to further guard against greenwashing with more specific definitions. The EU sustainable finance package, including the EU Taxonomy and EU Green Bond Standard draft regulations, is the broadest reaching, but not the only, regional initiative focused on disclosure of climate risk (Section 15.6.3). Taxonomies across regions are not always aligned on what can constitute a green project, for example with respect to transition activities (Pfaff et al. 2021) (Section 15.6.7). While standardisation can help reduce uncertainty in markets with imperfect knowledge, the green bond market is currently developing and is expected to continue to reflect regional differences in economic governance approaches (Nedopil et al. 2021). Regulations may also have trade-offs in terms of transaction costs for green financial product issuers. Classification approaches can also face challenges, depending on how they are designed, in their ability to capture new technologies and social impacts (Section 15.4).

Green bonds have been primarily targeting climate mitigation projects, with far fewer projects identified as adaptation. Green bonds mainly finance projects in the energy, buildings and transportation sectors, which constituted 85% of the use of proceeds of green bonds in 2020 (CBI 2020b, 2021a). Agriculture and forestry projects, including adaptation projects, have been less suited to be financed in a bond structure, which could be in part due to the more dispersed and smaller nature of the projects and in part due to project 'bankability' or ability to contribute steady streams of financing to pay back the terms of a bond. However, adaptation projects may not be identified as such as resiliency becomes more mainstreamed into infrastructure planning (Section 15.3.2).

While green bonds have the potential to further support financial flows to developing countries, local capital markets can be at varying stages of development (Banga 2019) (Sections 15.6.2 and 15.6.7). While multilateral and bilateral development finance institutions have been active in the green bond market, global issuance in 2020 in the top 10 countries included only one developing country (CBI 2021a). Targeting international investors can be enhanced via de-risking activities (15.6.4).

**Identifying green financial products can increase uptake and may result in a lower cost of capital in certain parts of the market (*high confidence*).** Investors face a systematic underpricing of climate risk in financial markets (Krogstrup and Oman 2019; Kumar et al. 2019). Transparent identification of financial products can make it easier for investors to include low-carbon products in their portfolios. Investors with mandates that include or are focused on climate change are showing an interest in green-labelled financial products. Investors that identify themselves as green constitute approximately 53% of the investor base for green bonds in the first half of 2019 (CBI 2019b).

There is some evidence of a premium, or an acceptance of lower yields by the investor, for green bonds (*medium confidence*). A survey of recent literature finds some consensus of the existence of a green premium in 56% of the studies on the primary markets (with a wide variance of premium amount), and 70% of the studies on the secondary market (with an average premium of –1 to –9 basis points), particularly for government issued, investment grade and green bonds that follow defined governance and reporting practices (MacAskill et al. 2021). In the US municipal bond market, as credit quality for green-labelled bonds has increased in the past few years, some studies show a positive premium for green bonds is arising (Baker et al. 2018; Karpf and Mandel 2018), or appearing only in the secondary market (Partridge and Medda 2020), while others find no evidence of a premium (Hyun et al. 2019; Larcker and Watts 2020). Several studies also show a recent emergence of a premium and oversubscription for some green-labelled bonds denominated in EUR (CBI 2019b), in some cases for both USD or EUR green bonds (Ehlers and Packer 2017), with a wide variation in the range of the observed difference in basis points focusing on the secondary market (Gianfrate and Peri 2019; Nanayakkara and Colombage 2019; Zerbib 2019), with financial institution and corporate green bonds exhibiting a marginal premium compared with their non-green comparisons (Hachenberg and Schiereck 2018; Kempa et al. 2021).

Spillover effects of green bonds may also impact equity markets and other financing conditions. Stock prices have been shown to positively respond to green bond issuance (Tang and Zhang 2020). One study linked enhanced credit quality induced by issuing green-labelled bonds to a lower cost of capital for corporate issuers (Agliardi and Agliardi 2019). Issuers' reputation and use of third-party verification can also improve financing conditions for green bonds (Bachelet et al. 2019). Green bonds are strongly dependent on fixed income market movements and are impacted by significant price spillover from the corporate and treasury bond markets (Reboredo 2018). A simulation of future green sovereign bond issuances shows that this can promote green finance via firm's expectations and the credit market (Monasterolo and Raberto 2018).

**Financial flows via these instruments have limited measurable environmental impact to date, however they can support capacity building on climate risk and opportunities within institutions to realise future impacts (*high confidence*).** There is a lack of evidence to date that green and sustainable financial products have significant impacts in terms of climate change mitigation and adaptation Box 15.7). Further, new products must be coupled with tightened climate policy and a reduction in investments associated with GHG-emitting activities to make a difference on the climate (Section 15.3.3.2).

It is challenging to link specific emission reductions with specific instruments that mainly target climate activities such as green bonds. Data challenges point to an inability to link emission reductions, including Scope 3 GHG emissions, at the organisation or firm level with green bond use-of-proceeds issuance (Ehlers et al. 2020; Tuhkanen and Vulturius 2020). However one study found evidence of a signalling effect of issuing green bonds resulting in emission reductions at the corporate level following issuance (Flammer 2020), and another study characterised the lifecycle emissions of renewable energy financed by green bonds, indicating potentially substantive avoided emissions but with variance up to a factor of 12 across bonds depending on underlying assumptions (Gibon et al. 2020). There is also a lack of impact reporting requirements and consistency in the green bond market. Impact reporting is not typically required for green bond listings on specific exchanges, nor are there any requirements for independent reviews of impact reporting, however this could change in future if investors apply pressure.

Green-labelled products may not necessarily result in increased financial flows to climate projects, although there can be benefits from capacity building with issuing institutions. Green bonds can be used to finance new climate projects or refinance existing climate projects, and thus do not necessarily result in finance for new climate projects constituting additional GHG reductions (a framing used in the Clean Development Mechanism). The labelling process itself may not necessarily lead to additional financing (Dupre et al. 2018; Nicol et al. 2018b). However, the labelling process has merit in contributing to building capacity within issuing institutions on climate change (Schneeweiss 2019), which could support identification of new green projects in the pipeline.

Climate risk disclosure initiatives, some of which are voluntary in nature, may have a limited direct climate impact. Transparency on climate risk may not change investor decisions nor result in divestment, especially in the emerging economies, as support and clear direction from regulatory and policy mechanisms are required to drive institutional investors at large (Ameli et al. 2021b). On the other hand, there is evidence of reduced fossil fuel investments following mandatory climate risk disclosure requirements, indicating a broader signalling effect of transparency (Mésonnier and Nguyen 2021).

## Box 15.7 | Impact of ESG and Sustainable Finance Products and Strategies

While scaling up climate finance remains a challenge (Section 15.3.2), there is consensus that investments that are managed taking into account broader sustainability criteria have increased consistently and ESG integration into sustainable investment is increasingly being mainstreamed by the financial sector over recent years (Maiti 2021). The United Nations Principles for Responsible Investment (PRI) grew to over 3000 signatories in 2020, representing over USD100 trillion in assets under management (UN PRI 2020). And according to the 2018 biennial assessment by Global Sustainable Investment Alliance,[15] sustainable investments in five major developed economies grew by 34% in the two-year period following the 2016 assessment. The primary ESG approaches leveraged were exclusion criteria and ESG integration, which together amounted to over USD37 trillion, accounting for two-thirds of the assessed sustainable investments, with novel strategies such as best-in class screening and sustainability-themed investing showing significant growth, although together they accounted for around 6% of these investments (GSIA 2019). Shareholder activism or corporate engagement is the other key approach, which has been well established and continued to grow to nearly USD10 trillion (GSIA 2019).

[15] GSIA is an international collaboration of membership-based sustainable investment organisations.

*Box 15.7 (continued)*

However, research indicates that ESG strategies by themselves do not yield meaningful social or environmental outcomes (Kölbel et al. 2020). When it comes to the tangible impact of the financial sector on addressing climate change and sustainable development, there remains ambiguity. There is a growing need for more robust assessment of ESG scores, including establishing higher standardisation of scoring processes and a common understanding of the different ESG criteria and their tangible impact on addressing climate change. The issue was highlighted in an assessment of six of the leading ESG rating agencies' company ratings under the MIT Aggregate Confusion Project, which found the correlation among them to be 0.61, leading them to conclude that available ESG data was 'noisy and unreliable' (Berg et al. 2020). This need is reaffirmed by Drempetic et al. (2020), who claim that a thorough investigation of ESG scores remains a relatively neglected topic, with extraneous factors, such as firm size, influencing the score (Drempetic et al. 2020).

There continues to be a research gap in assessing the direct impact of ESG and sustainable investments on climate change indicators, with most existing studies assessing the co-relation between either the factors driving the sustainable finance trends and the impact on sustainable investments, or sustainable investments and the impact on corporate financial performance. Nevertheless, since the post-SDG adoption period, there has been a notable uptake on research linking sustainable business practices and financial performance (Muhmad and Muhamad 2020). This research shows that there is a growing business case for ESG investing, with evidence increasingly indicating a non-negative co-relation between ESG, SDG adoption and corporate financial performance (Friede et al. 2015; Muhmad and Muhamad 2020), and ESG performance having a positive relation with stock returns (Consolandi et al. 2020). Research focused on developed economies also indicates towards a positive relation between ESG criteria and disclosure, and economic sustainability of a firm (Giese et al. 2019; Alsayegh et al. 2020) and allays investor fears by showing that sustainable finance initiatives, such as divestment, do not adversely impact investment portfolio performance (Henriques and Sadorsky 2018; Trinks et al. 2018). It should be reiterated that this research assesses the co-relation between ESG criteria and corporate financial performance, with the researchers in some cases, such as Friede et al. (2015), including disclaimers of the results being inconclusive and highlighting the need for a deeper assessment for linking ESG criteria with impact on financial performance.

On the other hand, there is growing evidence for a sustainable investment lens having a broader positive impact on creating an enabling environment and strengthening the case for such investments. For instance, corporate social responsibility (CSR) activities and investments on the environment dimension, specifically in the areas of emission and resource reduction, were found to be profitable and a predictor of future abnormal returns in the longer term, from additional cash flow and additional demand (Dorfleitner et al. 2018). These factors could be contributing to the increasing trend of sustainable and green investments, and can be said to be further reiterated by the spate of investor-led collaborative initiatives and recent announcements by leading finance institutes in the developed economies, which is well recorded in a range of recent grey literature, including new climate-aligned investment strategies and ambition towards net zero targets.

Yet there is also a risk of companies announcing projected sustainability or net zero targets and claiming the associated positive reputational impact, while having no clear action plan in place to achieve these. The lack of mandatory reporting frameworks, which results in an over-reliance on self-reported carbon data by companies for ESG assessments, can be a primary contributor (In and Schumacher 2021).

While there is a lack of research on the impact of sustainable finance products, divestment impact has been assessed in more detail. Although the research here also points towards the ambiguous direct impact of divestment on reducing GHG emissions or on the financial performance of fossil fuel companies, its indirect impact on framing the narrative around sustainable finance decisions (Bergman 2018), and the inherent potential of the divestment movement for building awareness and mobilising broader public support for effective climate policies, have been better researched and could be considered to be the more relevant outcomes (Braungardt et al. 2019). Arguments against divestment point to its largely symbolic nature, but Braungardt et al. (2019) elaborate on the broader positive impacts of divestment, which include its ability to spur climate action as a moral imperative and stigmatise and reduce the power of the fossil fuel lobby, and the potential of the approach to mitigate systemic financial risks arising due to climate change and address the legal responsibilities of investors merging in this regard.

Challenges remain with regards to overlapping definitions of sustainable and ESG investment opportunities, which also vary depending on social norms and pathways. There is also a general need for more extensive ESG disclosure at a corporate level, against the background of emerging mandatory impact reporting for asset managers in some regions. A movement is building towards sustainable investment strategies and increased sustainable development awareness in the financial sector (Muhmad and Muhamad 2020; Maiti 2021), which points to the ability of civil society movements, such as divestment campaigns, to have some influence on investor behaviour, although there are other influences such as climate risk disclosure initiatives and regulations.

### 15.6.7 Development of Local Capital Markets

**International situational context.** Developing countries make up two-thirds of the world's population and carry carbon-intensive economies where 70% of investments (see Chapter 3) need to be conducted to limit warming to 2°C. The focus for climate investments has been on China, USA, Europe, India and the G20 (UNEP 2019) but studies highlight Paris and SDG attention should be devoted to Africa, LDCs and SIDS (African Union Commission 2015; Feindouno et al. 2020; GCA-AAI 2020; Warner 2020; AOSIS 2021). The 'special needs, circumstances and vulnerability' of African, LDC and SIDS nations are recognised under UNFCCC and UN agreements (UN 2009, 2015a,b,c; UNFCCC 2010, 2015; Pauw et al. 2019). These nations currently contribute very little to global emissions. Developing countries with their growing economies, including the vast African continent roughly the size of China, Europe, USA, and India combined (IEA 2014b, p. 20) with a 1 billion population expected to double by 2050, growing reliance on fossil fuels and 'cheap' biomass (charcoal use and deforestation) amid rising urbanisation and industrialisation ambitions – collectively these nations hold large leap-frog potential for the energy transition as well as risks of infrastructure lock-in. Accelerated international cooperation is a critical enabler (IPCC 2018) in recognising this potential. This could mobilise global savings, scale up development of local capital markets for accelerated low-carbon investment and adaptation in low- and lower-middle-income countries as well as tackle illicit finance including tax avoidance leakages that deprive developing countries of valuable resources (US DoJ 2009; Hearson 2014; Hanlon 2017; US DoJ 2019; IATFD 2021). Diversifying funding sources is important at a time hard-currency Eurobond issuances reach records (Panizza and Taddei 2020; Moody's Investors Service 2021). Otherwise, the structure of voluntary, nationally oriented, and financially fragmented arrangements under the Paris Agreement (Chapter 17) could lead to 'regional rivalry' (SSP 3) pathways (IPCC 2018; Gazzotti et al. 2021). The benefits are many times greater than apparent costs in terms of expected decline in global GHG emissions and attaining SDGs. These could even generate large 'win-win' opportunities back in capital source countries which will benefit from a flow back in import demand (Hourcade et al. 2021a).

**Lessons from literature on policy options in mobilising capital for Paris and SDGs in developing countries can be summarised as follows:**

1. development of national just transition strategies meet the USD100 billion commitment on a grant-equivalent basis to support NDCs that integrate policies on COVID-19 recovery, climate action, sustainable development and equity;
2. increase the leverage of public funds on diverse sources of private capital through de-risking investments and public-private partnerships involving location-based entities with AAA-rated players and institutional investors;
3. coordination of project preparation and development of project pipelines by infrastructure coordinator agencies, one-stop structuring and financing shops, project risk facilities provided by entities such as cities' development banks, green banks, a world climate bank, global guarantee mechanism, and global infrastructure investment platform;
4. development of local currency bond markets backed by cross-border guarantees, technical assistance, remediation assets, especially by regional and national players whose mandates include nurturing local capital markets to support bond yield curve development and exchange listing options;
5. adopting advances in science-based assessment methods to foster accountability;

   (a) for project assessment, measuring, reporting and verifying, and certification,
   (b) for disclosures in climate, fossil fuels, SDGs, debt transparency and debt sustainability, and
   (c) for progress on UN systems of national accounts particularly for public sector finance statistics.

**Whole-of-society approach to mobilising diverse capital.** There's no shortage of money globally: it is simply that it has yet to travel to where it's most needed. One challenge is unlocking unencumbered endowments to contribute to Paris and SDGs (*high confidence*). The aggregate global wealth figures exceed USD200 trillion (Davies et al. 2016; UBS 2017; Credit Suisse 2020; Heredia et al. 2020). Some developing countries have run pilots for investing in government bonds capitalising on fintech growth discussed (The Economist 2017; Akwagyiram and Ohuocha 2021) (Section 15.6.6). Others are developing green products to encourage uptake by middle class retail investors (Eurosif 2018; UK DMO 2021). Millennial-aged inheritors expected to receive intergenerational transfers mobilised by global citizen activism (Chapter 2) invest in green retail and tech products (Morgan Stanley 2017; UBS 2017; Capgemini 2021). Historic inequity and diaspora-related private and public resources pledged and debated during the COVID-19 pandemic might have potential to contribute towards Paris and SDGs (Olusoga 2015; Glueck and Friedman 2020; Hall 2020; Piketty 2020; Timsit 2020; Goldman Sachs 2021; Guthrie 2021; Mieu 2021; Wagner 2021). Philanthropic institutions use grants, debt, equity, guarantees and issue investment grade bonds in using unencumbered endowments (Manilla 2018; Covington 2020; Moody's Investors Service 2020) but only about 2% of their resources are dedicated to climate action (Williams T., 2015; Kramer 2017; Morena 2018; Delanoë et al. 2021). The pandemic exemplified the unprecedented collaboration and mobilisation of multilateral and scientific communities supported by the COVAX risk sharing mechanism for COVID-19 vaccines with pooling of financial and scientific resources (OECD 2021d). This momentum in international cooperation can be harnessed to galvanise resources, including for teaching of sciences in developing countries important in tackling society challenges, alleviating poverty (TWAS 2021) and inequity legacies compounded by climate impacts debated by many (Henochsberg 2016; Obregon 2018; Fernandez et al. 2021; The Economist 2021). Suggestions towards equitable models include 'global adaptation funding approaches' (Chancel and Piketty 2015), a 'world climate bank' to finance climate investments through long-term bonds (Foley 2009; Broome 2012; Broome and Foley 2016), a 'cities development bank' (Alexander et al. 2019), and 'public debt financing models' (Rendall 2021) for generations to share the burden which has precedence in history (Draper 2007; Fowler 2015).

**Local financial institutions with local markets knowledge could benefit from technical assistance and partnership to scale up their potential with institutional investors better mobilised** (*high confidence*). The Global South has some 260 public development banks/PDBs representing USD5 trillion in assets with a worldwide PDB capacity to provide more than USD400 billion $yr^{-1}$ of climate finance (IDFC and GCF 2020). Case studies discuss the potential for diaspora bond issuance being deployed for climate investments including securitisation of remittances as collateral for infrastructure bonds (Ketkar and Ratha 2010; Akkoyunlu and Stern 2012; Gelb et al. 2021). Such instruments could help harness diaspora remittances, whose flows rose from under USD 100 billion to USD530 billion during 1990–2018 (World Bank 2019c). PDBs could benefit from technical partnership with multilaterals and other local banks (Torres and Zeidan 2016). Their knowledge of local markets, can help build project pipelines (Figure 15.7) to channel local, domestic and international capital (Griffith-Jones et al. 2020). Institutional domestic and international investors have growing assets estimated to exceed USD100 trillion (*high confidence*) (Willis Towers Watson 2020; UN PRI 2020; Halland et al. 2021; Heredia et al. 2021; Inderst 2021) and could be better mobilised. Some 36% of total assets under management (AUM) by the 100 largest asset owners come from pensions and sovereign wealth funds in the Asia Pacific region, with the remainder split almost evenly across Europe, the Middle East, Africa and North America. The largest pension fund in South Africa held about USD130 billion AUM in 2019 and African institutional investors held USD1.8 trillion in 2020 (PwC 2015; GEPF 2019; Bagus et al. 2020; GCA 2021a). UK NGO War on Want's (2016) analysis of 101 fossil fuel and mining companies on the London Stock Exchange estimates these as holding USD1 trillion assets inside Africa. The Latin America and Caribbean region holds just about USD1 trillion AUM (Curtis 2016; Serebrisky et al. 2015; Cavallo and Powell 2019).

**Investors with accumulated private capital are reported as looking for climate investments to ensure Just Transition, alignment with Paris and SDGs. However, progress remains pilot, slow and piecemeal** (*high confidence*). Global investors have published statements on their possible contribution, with recommendations to governments on de-risking to accelerate private sector investment to support Paris-aligned NDCs in developing countries (IIGCC 2015; IIGCC 2017; Global Investor Statement 2018; IIGCC 2018; Global Investor Statement 2019; IIGCC 2020). In March 2020, the UN Principles for Responsible Investment (PRI), had 3038 members representing USD103 trillion (UN PRI 2020); another coalition of investors published COVID-19 recovery plans (Investor Agenda 2020) and the Net Zero Asset Managers initiative was launched in December 2020 (NZAM 2020). However, it is still unclear how these pronouncements will be transformed to adequate financial flows and volumes of investment pipelines (IEA 2021d) (Chapter 3). Rempel and Gupta (2020) posit that a proportion of institutional holding is in fossil fuels. Clean energy transition minerals raise ESG questions around inclusive development for indigenous populations and require changes to supply chains exploiting child labour (Herrington 2021; IEA 2021a; IEA 2021f).

Options to mobilise institutional investors currently remain small pilots, relative to Paris and SDG ambitions (*high confidence*). In terms of examples: in the *women of colour-led arena*, a Chicago pension fund invested in a developing country using a *private equity fund*; (Langhorne 2021, USAID 2021). Institutional BlackRock's blended finance vehicle with OECD MDB partners focuses on developing countries (BlackRock 2021). In regional AAA MDB partnerships, the African Development Bank (AfDB) collaborates with African nations through a *regional infrastructure fund* (Africa50 2019); the Asian Development Bank (ADB) collaborates with a Philippines state-owned pension fund and Dutch pension fund in using a *private equity fund* to catalyse private sector investment (ADB 2012). A UN entity with several pooled public-private investment platforms includes an SDG blended finance vehicle (UN CDF 2020a; 2020b). A multilateral International Finance Corporation (IFC) blended finance fund, supported by a sovereign guarantee from Sweden's SIDA, and separately a USD1 billion green bond fund by IFC and Europe's Amundi asset manager buy green securities issued by developing country banks financing local currency climate investments (IFC 2018, 2021; Amundi and IFC 2019). The key parameter is the *investment multiplier*, the *ratio of private investment mobilised by a given amount of public funds* which varies by product type. IFC's portfolio of blended finance investments point to a self-reported range of 3 to 15 times for project debt and even higher levels (10 to 30) for debt finance provided on concessional terms (IFC 2021a). Although an AAA-rated IFC blended finance fund was established in 2013, most investors joined in 2017 with insurers AXA and Swiss Re investing USD500 million each to bring the fund to USD7 billion raised from eight global investors (Attridge and Gouett 2021). Critics of blended finance mechanisms point to lack of data transparency hampering independent assessment on (i) value for public money and costs of blending versus other financial mechanisms, (ii) risks and benefits of de-risking private capital to collateralising climate-vulnerable Global South populations, (iii) lack of partnership with local players, and (iv) complex structures (Akyüz 2017; Mawdsley 2018; Convergence 2020; Attridge and Gouett 2021; Gabor 2021). Whilst blended finance transactions (BFTF 2018) are quite common in mature regulated markets with mandatory reporting requirements (Morse 2015; ICAEW 2021), the additional finance mobilised and their developmental impact remain unknown due to poor reporting that hammpers evidence-based policy making (Attridge and Gouett 2021). Projects that are aligned with blended finance principles in the UN Addis Agenda (UN 2015a), and take account of local contexts by partnering with local actors, are much more likely to have sustainable impacts.

**De-risking tools to lower capital costs and mobilise diverse investors.** Paris-aligned NDCs that integrate policies on COVID-19 pandemic recovery, climate action, sustainable development, just transition and equity can harness co-benefits including contribution to *Invisible UN SDG 7 energy poverty sectors* (*high confidence*). Developing countries require access to affordable finance for projects ranging from clean cooking solutions (Accenture 2018; World Bank et al. 2021); decentralised energy systems, intra-country power stations and regionally shared power pools with their associated energy distribution networks (IEA 2020d; IRENA 2020c). Close to 3 billion people in Africa and developing Asia have no access to clean cooking. For sub-Saharan Africa, the acute lack of electricity access lags behind all regions on SDG 7 indicators, impacting mostly women and children (IEA 2014b; IRENA 2020b,c; IEA et al. 2021; ESMAP 2020; Zhang 2021) (Box 6.1). These dire statistics remind of compounding tensions: historical inequities and the associated 'first comer' exploiting African resources for development elsewhere, the local climate change, 'latecomer' capacity development and technology transfer challenges, illicit mining finance and stranded assets (Curtis 2016; Bos and Gupta 2019; UNU-INRA 2019; Arezki 2021). The COVID-19 pandemic exacerbates this tension with more people pushed below the poverty line (Sumner et al. 2020) (section 15.6.4, Box 15.6 on post-COVID). Recent analysis points to the 60 largest banks providing USD3.8 trillion to fossil fuel companies since 2016, including inside Africa (Rainforest Action Network et al. 2021). IMF estimated fossil fuel subsidies totalling USD5.2 trillion or 6.5% of global GDP in 2017 (Coady et al. 2019) to be compared with the USD2.4 trillion yr$^{-1}$ energy investments over the next decade to limit global warming to 1.5°C (IPCC 2018). Analysts point to models in improvements to resources husbandry that include (i) developing strong minerals sector governance through sovereign wealth funds for domestic development (Wills et al. 2016) and (ii) compensation for Africa (Walsh et al. 2021) leaving fossil fuels underground (McGlade and Ekins 2015) in the *Just Transition* (Section 15.2.4) and *Right to Develop* debates as assets continue to be mined (IEA 2019c). In many developing regions, some of the world's best renewable energy sources remain out of reach due to high costs which can be up to seven times those in developed countries (IEA 2021d). Shifting some risks through financial de-risking approaches could be instrumental (Schmidt 2014; Sweerts et al. 2019; Drumheller et al. 2020; Matthäus and Mehling 2020).

**Combining approaches: (i) developed countries meeting UNFCCC USD100 billion commitment on a grant-equivalent basis, (ii) stepped up technical assistance, (iii) infrastructure coordination, (iv) knowledge sharing by project preparation entities, and (iv) harnessing project risk facilities such as guarantees could be instrumental for scaling climate finance for Paris-SDGs** (*high confidence*). Figure 15.7 illustrates the interplay between infrastructure project financing phases, bond refinancing and opportunities for developing bond yield curve benchmarks in nurturing local capital markets and mobilising diverse investors. These project financing phases have varying risk-return profiles and different benchmarks to track performance are often required by investors for different securities that might be created (Ketterer and Powell 2018).

An ODI (2018) survey of private and public project preparation facilities internationally showed high failure rates in *early project preparation phases* with recommendations on '*one-stop-shops*' and knowledge sharing on effective approaches. During the very high-risk *concept phase* (Figure 15.7) – grants and technical assistance de-risk with design concepts, project proposals and feasibility studies completed to 'kick-start' the right projects. The early-stage developmental phase is characterised by short-term debt in the two to five years phase to

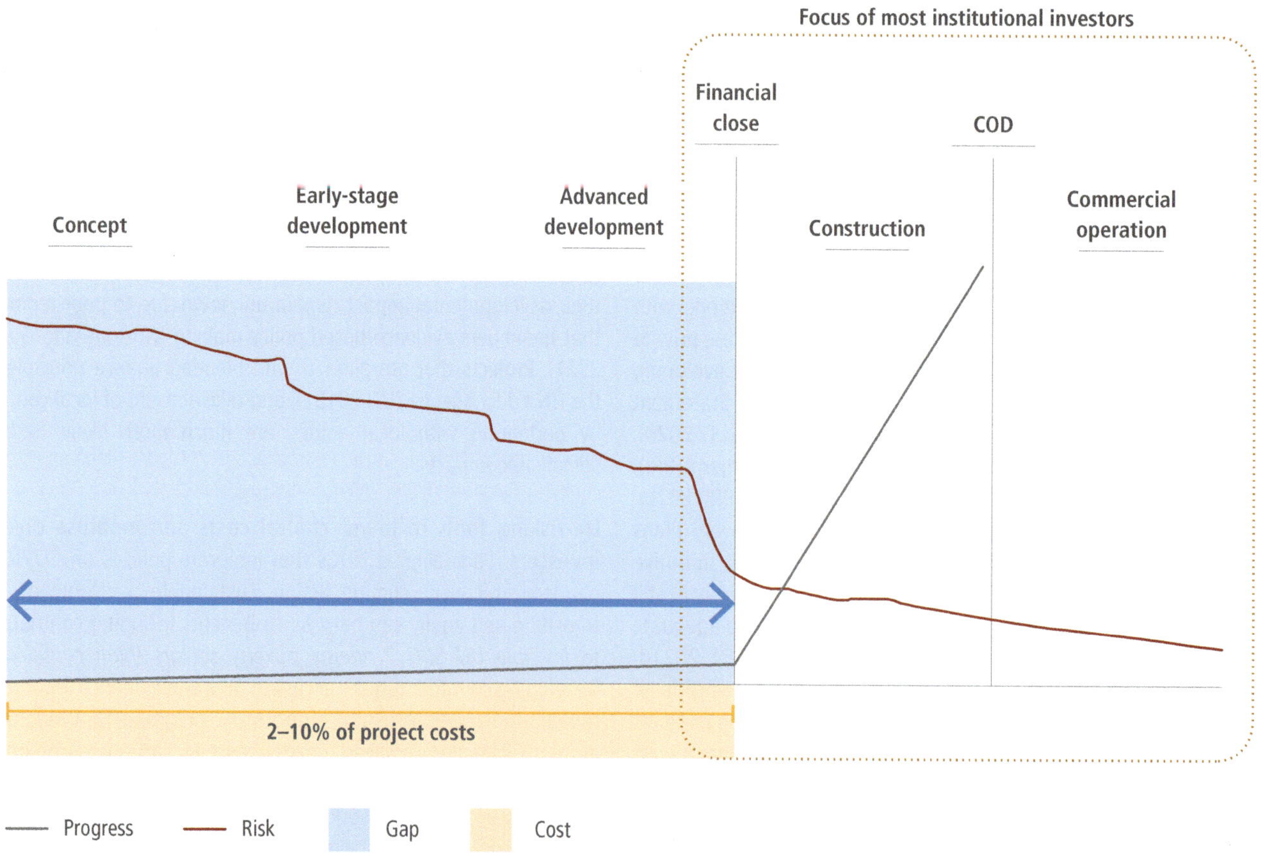


**Figure 15.7 | Bond refinancing mobilises institutional investors in mature project phase. De-risk early-stage infrastructure projects.** Source: adapted from PIDG (2019).

complete construction enabled by concession finance. Bank loans are paid back by issuing bonds once the construction phase is completed. Such bond refinancing over say, 15–25 years, in the low-risk *mature project phase* can provide a lower cost of capital. Market-making to develop a pipeline of investment opportunities uses a complimentary mix of high-risk capital options in the form of grants, guarantees, equity, and mezzanine financing that can help (Attridge and Gouett 2021): (i) reduce up-front risks in the early phases, (ii) allow banks to recycle loans to new projects, and (iii) galvanise multilateral technical assistance for building bond yield curve benchmarks and de-risking local currency bond issuance of long tenors such as green bonds/resilience bonds (Berensmann et al. 2015; CBI 2015; Mercer 2018; Dasgupta et al. 2019; PIDG 2019; Braga et al. 2021; CBI et al. 2021; Hourcade et al. 2021a,b). Convergence (2019) points to investment from commercial banks with commercial debt of 11–15 years maturity being covered by guarantees. To achieve scale, some have issued special purpose vehicle (SPV) green infrastructure project bonds combining tenors up to 15 years with credit ratings assigned to mobilise investors with community trusts for local participation (Kaminker and Stewart 2012; Mathews and Kidney 2012; Mbeng Mezui and Hundal 2013; Essers et al. 2016; Moody's Investors Service 2016; Ng and Tao 2016; Harber 2017). Bond refinancing could be facilitated through standardised national infrastructure style bonds, national infrastructure funds (Amonya 2009; Ketterer and Powell 2018) and country SPV infrastructure funds issuing bonds (Cavallo and Powell 2019) embedding MDBs.

**Existing project risk facilities including guarantees could benefit from coordination, scaling and better reporting frameworks** (*high confidence*). Individual and clubs of developed and developing countries currently provide public guarantees (ADB 2015, 2018; IIGCC 2015; Pereira Dos Santos 2018; GGGI 2019; Garbacz et al. 2021). However MDB business models impose limitations on use of guarantees and collaboration with other MDBs (Gropp et al. 2014; Schiff and Dithrich 2017; Lee et al. 2018; Pereira dos Santos and Kearney 2018). Loans continue to dominate as the financial instrument of choice by MDBs and DFIs, with guarantees mobilising the most private finance for OECD reported data, even if their use remains limited (IATFD 2020; OECD 2020c; Attridge and Gouett 2021). Ramping up the use of guarantees to mobilise private investment raises questions around understanding efficacy in the design as there is no one size that fits all and more research is required to better understand this aspect (Convergence 2019). Sample guarantee forms in literature: (i) single-country Sweden and USA DFI forms (SIDA 2016, DCA 2018), (ii) multilateral institution offerings (Pereira Dos Santos 2018; IRENA 2020e), (iii) multi-sovereign guarantees one-stop platforms such as those on the PIDG/GuarantCo (PIDG 2019) and Africa Guarantee Fund owned by DFIs, including the African Development Bank (AfDB), the French Development Agency (AFD), the Nordic Development Fund (NDF), and the KfW Development Bank (AGF 2020), (iv) MIGA, established to provide political risk guarantees (enhanced green MIGA) (Déau and Touati 2018), (v) multilateral partnerships with developing nations via infrastructure funds (Section 15.6.7.2) and green infrastructure options (de Gouvello and Zelenko 2010; Studart and Gallagher 2015), (vi) guarantees embedded in project risk facilities such as currency fund TCX established by 22 DFIs (TCX 2020), and (vii) ASEAN and African multi-sovereign regional local currency bond guarantee funds and a co-guarantee platform (GGGI 2019; Garbacz et al. 2021). Fossil fuels currently benefit from de-risking tools from export credit agencies (Lawrence and Archer 2021), with questions around sustainable development (Wright 2011); Gupta et al. (2020) argue that these could be deployed for renewable energy. Sample project facilities reflecting the diverse project types across developing country regions can include i) UNEP Seed Capital ii) C40 Cities Facility iii) Blue Natural Capital Facility (IUCN 2021); iv) Clean Cooking Fund (ESMAP 2021) v) opportunities for guarantees in LDCs (Garbacz et al. 2021) vi) World Bank's Renewables Risk Mitigation (GCF 2021) and World Bank's Global Infrastructure Facility (GGGI 2019).. Multilaterals offer credit enhancement to manage both actual and perceived risks: in India's corporate sector, renewable energy SPV project bonds have been guaranteed jointly by ADB and an infrastructure company raising the credit rating from sub-investment grade to investment grade to lower borrowing costs (ADB 2018; Agarwal and Singh 2018; Carrasco 2018).

Investment vehicles into green infrastructure come in various forms (*high confidence*) and can include indirect corporate investment such as bonds; semi-direct investment funds via pooled vehicles such as infrastructure funds and private equity funds and project investment (direct) in green projects through equity and debt including loans, project bonds and green bonds. For pension funds in Australia and Canada, direct investment in infrastructure is about 5% of total AUM (Inderst and Della Croce 2013) whilst less than 1% for OECD pension funds goes to green infrastructure (Kaminker et al. 2013). Some regional developing country institutional investors use a variety of investment vehicles that span SPVs, private equity, domestic and regional local currency bond markets with statutory level mandates to address historic inequities (GEPF 2019). Cross-border collaboration in regional power markets such as Europe's Nordpool; for developing countries could be led by repository of technical partnership from infrastructure funds and multilaterals (Oseni and Pollitt 2016; Juvonen et al. 2019; Chen et al. 2020; Nordpool 2021). Barriers to investments include non-standardised investment vehicles of scale and lack of national infrastructure road maps to give investor confidence in government commitment. Some have set up infrastructure coordinating entities embedding local science and engineering R&D (IPA 2021; National Infrastructure Commission 2021). Arezki et al. (2016) argue that coordination within existing platforms could create a global infrastructure investment platform for de-risking through guarantees and securitisation; Matthäus and (Mehling (2020) point to a global guarantee mechanism. Such AAA multilateral approaches create credibility-enhancing effects in developing capital markets. Hourcade et al. (2021a) suggest that the overall economic efficiency could be higher with guarantees calibrated per tonne on an agreed *'social, economic, and environmental value of mitigation actions [and] their co-benefits'* (Article 108, Paris Agreement) basis, which would operate as a notional carbon price (High-Level Commission on Carbon Prices 2017). The grant equivalent of guarantees and induced equity inflows could be far beyond the USD100 billion promise. Such cooperative solutions in adopting development of local capital markets would end the drawbacks of the current plethora of low-scale fragmented project-by-project and 'special-purpose' pilots and programmes.

**Harnessing existing bond markets and securities exchanges in nascent markets.** The G20 has an action plan to support strengthening local currency bond markets and development of local capital markets is also part of the option for financing UN SDGs in developing countries (UN 2015a, 2019, 2020; IATFD 2016, 2021). Primers are available on bond market development to support policy choices (World Bank and IMF 2001; Silva et al. 2020; World Bank 2020; Adrian et al 2021; IMF and World Bank 2021). Developing government bond yield curves with different maturities can be an important policy objective (*high confidence*). This can support pricing discovery, liquidity (Wooldridge 2001) and can be achieved through step by step tranches from shorter to longer maturities to boost confidence and encourage municipals and other quasi-sovereigns. Money market instruments (such as, green commercial paper) anchor the short end of the yield curve with bonds of varying maturity issued by sovereign/quasi-sovereign entities (national treasuries, SOEs, municipalities) to mobilise investors (Goodfriend 2011; LSEG 2018; Tolliver et al. 2019). A variety of bonds are being used for developing countries including green (Ketterer et al. 2019), blue-water (Roth et al. 2019), transition, SDG/social, biodiversity bonds (Aglionby 2019), green/resilience bonds (AAC 2021); gender bonds (Andrade and Prado 2020) diaspora (LSEG 2017) and infrastructure project bonds (CBK 2021). Local policymakers would gain from technical and financial assistance in building green yield curves, for example with support from multilaterals (EIB 2012; IATFD 2016; Shi 2017; EIB 2018; Impact Investing Institute 2021). Green bonds are one of the most readily accessible to help fund Paris goals (Tolliver et al. 2019; Tuhkanen and Vulturius 2020). Section 15.3.2 refers to the growth in labelled bond markets (CBI 2021a), low borrowing costs and yield curve building in Europe (Bahceli 2020; Serenelli 2021; Stubbington 2021; UK DMO 2021). For developing countries, labelled bonds have mostly been in hard currency (e.g. Smith 2021) despite local currency markets making up more than 80% total debt stock (IMF and World Bank 2016; Silva et al. 2020; Adrian et al 2021; Inderst 2021). The labelled bonds issuance by multilaterals do not currently mobilise the trillion levels needed. Research studies show that participating in green bond markets in part depends on a country having credible NDCs (Tolliver et al. 2020a; Tolliver et al. 2020b) and highlights diverse approaches working together to support local bond market development (Amacker and Donovan 2021; ICMA 2021; IMF and World Bank 2021).

**Technical assistance options would benefit from coordination. Labelled bond costs remain high. Developing countries are using fiscal incentives, grants, and guarantees to support nascent bond markets with most taxonomies under development** (*high confidence*). Technical assistance requirements to improve the investment climate and bond market development will vary across national capacities. These would benefit from the USD100 billion UNFCCC grant equivalent basis to develop (i) regulatory and policy frameworks; (ii) UN national statistical systems (Singh et al. 2016; MacFeely and Barnat 2017; Paris21 2018; Bleeker and Abdulkadri 2020); (iii) credible NDC and SDG investment plans; (iv) project assessment certification and taxonomies; (v) bond market guidelines; and (vi) public finance management (US DoJ 2009; US DoJ 2019). Other technical assistance channels include diaspora entities, universities and learned societies (ICEAW 2012; UNFCCC 2021). LDCs supported by humanitarian entities are least likely to have active capital markets (ICRC 2020; IDFC 2020; Cao et al. 2021b). Clubs of LDCs are partnering with AAA MDBs in aggregation approaches (AfDB 2020; GCF 2020b). Some UN entities provide technical assistance on municipal aggregation of projects (UN CDF 2021a), with Africa, LDC, SIDS nations and cities accessing green technical facilities and listings for labelled bonds (C40 Cities Climate Leadership Group 2016; Gorelick 2018; Jackson 2019; FSD Africa and CBI 2020; Gorelick and Walmsley 2020; MoE Fiji 2020; IFC 2021c). Elevated climate risks imperil developing country ability to repay debts (Schmidt 2014; Buhr et al. 2018; Volz et al. 2020; Dibley et al. 2021). To lower overall costs and achieve more, entities have accessed technical assistance, listed local currency labelled bonds, and used credit enhancing bond guarantees, regulatory treatments and philanthropy schemes (Europe 2020 Project Bond Initiative 2012; SBN 2018; Agliardi and Agliardi 2019; Banga 2019). In the regions, China issued guidelines for stock exchanges and regulatory support for green bonds (Cao and Ma 2021), India issued regulations for local issuance of green bonds (CBI 2019a), while in the Latin America and Caribbean region, both plain vanilla and labelled bonds use the same authority (Ketterer et al. 2019). African, LDC and SIDS nations are reviewing ways to harness local exchanges (SSE 2018; GCF 2019; World Bank et al. 2021b; UN CDF 2021b). For taxonomies, the differences reflect the multitude of local Just Transition pathways, some with a purely environmental focus and others incorporating livelihood improvements (ICMA 2021). The sustainable bond market has been expanding as transition bonds become listed in anticipation of future developments (Roos 2021).

**Progress towards transparency using scientific-based methods to build trust and accountability.** After 60 years of development finance, critics underline limits coming from i) multilaterals model, lack of transparency around aid and debt (Mkandawire 2010; Lee 2017, PWYF 2019; Bradlow 2021; Gianfagna et al. 2021) ii) illicit finance (Plank 1993; Sachs and Warner 2001; Hanlon 2016; US DoJ 2019) ) iii) lack of developed country commitment to pledges (Nhamo and Nhamo 2016) iv) unregulated players as financial intermediaries in blended finance (Pereira 2017; Donaldson and Hawkes 2018; Attridge and Engen 2019; Tan 2019) v) weak accountability reflected in soft SDG data and vi) burden of responsibility in mobilising Paris and SDG resources to countries with historically soft institutional capacity (Hickel 2015; Donald and Way 2016; Scheyvens et al. 2016; Liverman 2018). Literature around trust in blended finance pinpoints four progress areas in accountability. First, debt transparency through public debt registries, centralised UN legacy debt restructuring and science-centred UN national statistical systems (Donaldson and Hawkes 2018; Jubilee Debt Campaign 2019; Stiglitz and Rashid 2020). Second, international reporting bell-weathers could be called upon to produce harmonised mandatory reporting frameworks that capitalise on TCFD to capture climate, debt sustainability (Section 15.6.7.3), SDG and fossil fuels (GISD 2020). Third, standardisation of assessment by third parties of the quantity and values of carbon saved by green projects (Hourcade et al. 2012) and of their contribution to quantified performance biodiversity targets (Finance for Biodiversity Initiative 2021) to facilitate their bundling, securitisation and repackaging in standardised liquid products and bonds (Arezki et al. 2016; Blended Finance Taskforce 2018a).

### 15.6.8 Facilitating the Development of New Business Models and Financing Approaches

New and innovative business models and financing approaches have emerged to help overcome barriers related to transactions costs by aggregating and/or transferring financing needs and establishing supply of finance for stakeholder groups lacking financial inclusion (*high confidence*).

#### 15.6.8.1 Service-based Business Models in the Energy and Transport Sectors

**Energy as a service (EaaS)** is a business model whereby customers pay for an energy service without having to make any upfront capital investment (PwC 2014; Hamwi and Lizarralde 2017; Cleary and Palmer 2019). EaaS performance-based contracts can also be a form of 'creative financing; for capital improvement that makes it possible to fund energy upgrades from cost reductions and deployment of decentralised renewable energy (KPMG 2015; Moles-Grueso et al. 2021). Innovation in EaaS has started at the household level, where smart meters using real-time data are used to predict peak demand levels and optimise electricity dispatch (Chasin et al. 2020; Government of UK 2016; Smart Energy International 2018).

**Aggregators.** An aggregator is a grouping of agents in a power system to act as a single entity when engaging in power system markets (MIT 2016). Aggregators can use operation optimisation platforms to provide real-time operating reserve capacity and a range of balancing services to integrate higher shares of variable renewable energy (Zancanella et al. 2016; Ma et al. 2017; Enbala 2018; Research and Markets 2017; IRENA 2019b). This makes a business case for deferred investments in grid infrastructure (*medium confidence*). Aggregating and managing demand-response of heat systems (micro CHP and heat pumps) has shown reduction in peak demand (TNO 2016).

**Peer-to-peer (P2P) electricity trading.** Producers and consumers can directly trade electricity with other consumers in an online marketplace to avoid the relatively high tariffs and the relatively low buy-back rates of traditional utilities (Liu et al. 2019; IRENA 2020f). P2P models trading with distributed energy resources reduce transmission losses and congestion (Mengelkamp et al. 2018; SEDA 2020; Lumenaza 2020; Sonnen 2020; UNFCCC 2020).

**Community ownership models.** Community ownership models refer to the collective ownership and management of energy-related assets with lower levels of investment, usually distributed renewable energy resources but also recently in heating systems and energy services (e.g., storage and charging) (Gall 2018; IRENA 2018; Kelly and Hanna 2019; Singh et al. 2019; Bisello et al. 2021; Maclurcan and Hinton 2021). Community ownership projects may need significant upfront investments, and the ability of communities to raise the required financing might prove insufficient, which can be supported by microcredits in the initial stages of the projects (Aitken 2013; Federici 2014; REN21 2016; Rescoop 2020).

**Payment method: Pay-as-you-go (PayGo).** PayGo business models emerged to address the energy access challenge and provide chiefly solar energy at affordable prices, using mobile telecommunication to facilitate payment through instalments; Yadav et al. 2019). However, PayGo has the technology and product risk, requires a financially viable and large customer base, and the system supplier must provide a significant portion of the finance and requires substantial equity and working capital (C40 Cities Climate Leadership Group 2018).

**Transport sector business models.** Analog to EaaS, mobility as a service (MaaS) offers a business model whereby customers pay for a mobility service without making any upfront capital investment (e.g., buying a car). MaaS tends to deliver significant urban benefits (e.g., cleaner air) and brings in efficiency gains in the use of resources (*high confidence*). However, the switch to MaaS hardly improves the carbon footprint and further tempted on-demand mobility is likely to nurture carbon emissions (Suatmadi et al. 2019). Therefore, to support climate change mitigation, MaaS must be integrated with the deployment of smart charging of electric (autonomous) vehicles coupled to renewable energy sources (IRENA 2019d; Jones and Leibowicz 2019).

**Financial technology applications to climate change.** Financial technology, abbreviated as 'fintech', applies to data-driven technological solutions that aim to improve financial services (Dorfleitner et al. 2017; Lee and Shin 2018; Schueffel 2018). Fintech can enhance climate investment in innovative financial products and build trust through data, but also presents some challenges including potentially significant emissions from increased energy use with distributed transactions (Lei et al. 2021). Blockchain is a key fintech that secures individual transactions in a distributed system, which can have many applications with high impact potential but is also associated with uncertainty (OECD 2019c; World Energy Council 2019). Fintech applications with climate change mitigation potential have been growing recently, including tracking payment or asset history for credit scoring in AFOLU activities (Nassiry 2018; Davidovic et al. 2019), blockchain supported grid transactions (Livingston et al. 2018), carbon accounting throughout value chains (World Bank 2018b), or transparency and verification mechanisms for green financial instrument investors (Kyriakou et al. 2017; Stockholm Green Digital Finance 2017). Generally, blockchain and digital currency applications are not well covered by governance systems (Tapscott and Kirkland 2016; Nassiry 2018), which could lead to problems with security (Davidovic et al. 2019), and some licensing and prudential supervision frameworks are in flux.

#### 15.6.8.2 Nature-Based Solutions Including REDD+

Nature-based solutions are 'actions to protect, sustainably manage and restore natural or modified ecosystems that address societal challenges effectively and adaptively, simultaneously providing human well-being and biodiversity benefits' (Cohen-Shacham et al. 2016). Nature-based solutions consist of a wide range of measures including ecosystem-based mitigation and adaptation.

The studies on investment and finance for nature-based solutions is still limited. However, frameworks and schemes to incentivise the implementation of nature-based solutions, such as reducing emissions from deforestation and forest degradation and the role of conservation, sustainable management of forests and enhancement of forest carbon stocks in developing countries (REDD+), which contributes to climate change mitigation, has been actively discussed under the UNFCCC, with lessons from finance for REDD+ being available.

If effectively implemented, nature-based solutions can be cost-effective measures and able to provide multiple benefits, such as enhanced climate resilience, enhanced climate change mitigation, biodiversity habitat, water filtration, soil health, and amenity values (*high confidence*) (Griscom et al. 2017; Keesstra et al. 2018; OECD 2019d; Griscom et al. 2020; Dasgupta 2021).

Nature-based solutions have large potential to address climate change and other sustainable development issues (*high confidence*). Nature-based solutions are undercapitalised and the limited investment and finance, especially limited private capital, is widely recognised as one of the main barriers to the implementation and monitoring of the nature-based solutions (Seddon et al. 2020; Toxopeus and Polzin 2021; UNEP et al. 2021) Finance and investment models that generate their own revenues or consistently save costs are necessary to reduce dependency on grants (Schäfer et al. 2019; Wamsler et al. 2020).

**REDD+.** REDD+ can significantly contribute to climate change mitigation and also produce other co-benefits like climate change adaptation, biodiversity conservation, and poverty reduction, if well-implemented (*high confidence*) (Milbank et al. 2018; Morita and Matsumoto 2018). We use the term REDD+ broadly, not limited to REDD+ implemented under the UNFCCC decisions, including Warsaw Framework for REDD+ (Chapter 14), but include voluntary REDD+ projects, such as projects which utilise voluntary carbon markets. Finance is a core element that incentivises and implements REDD+ activities. Various financial sources are financing REDD+ activities, including bilateral and multilateral, public and private, and international and domestic sources, with linking with several finance approaches/mechanisms including results-based finance and voluntary carbon markets (FAO 2018). However, there is lack of sufficient finance for REDD+ (Lujan and Silva-Chávez 2018; Maguire et al. 2021). REDD+ under the UNFCCC is implemented in three phases: readiness, implementation, and results-based payment phases. The Ecosystem Marketplace identified that at least USD5.4 billion in REDD+ in three phases funding has been committed through multiple development finance institutions so far (Maguire et al. 2021), and public funds are main sources that are supporting three phases, and most of the REDD+ finance was spent on the readiness phase (Atmadja et al. 2018; Lujan and Silva-Chávez 2018; Watson and Schalatek 2021). There is a significant gap between the existing finance and finance needs of REDD+ in each phase (Lujan and Silva-Chávez 2018). Furthermore, private sector contributions to REDD+ are currently limited mostly to the project-scale payments for carbon offsets/units through voluntary carbon markets (McFarland 2015; Lujan and Silva-Chávez 2018).

Current main challenges of REDD+ finance include the uncertainty of compliance carbon markets (which allow regulated entities to obtain and surrender emissions allowances or offsets to meet regulatory emissions reduction targets) (Maguire et al. 2021), as well as limited engagement of the private sector in REDD+ finance (*high confidence*). With regard to the compliance carbon markets, at the international level, integrating climate cooperation through carbon markets into Article 6 of the Paris Agreement and including REDD+ has potential to enable emission reduction in more cost-effective ways, while the links between carbon markets and REDD+ under Article 6 is under discussion at the UNFCCC (Environmental Defense Fund 2019; Maguire et al. 2021) (Chapter 14). At the national and subnational levels, although compliance carbon markets such as in New Zealand, Australia and Colombia allow forest carbon units, how REDD+ will be dealt in the national and subnational government-led compliance carbon markets is uncertain (Streck 2020; Maguire et al. 2021). As for limited engagement of the private sector in REDD+ finance, there are various reasons why mobilising more private finance in REDD+ is difficult (Dixon and Challies 2015; Laing et al. 2016; Golub et al. 2018; Ehara et al. 2019; Streck 2020). The challenges include the needs of a clear understanding of carbon rights and transparent regulation on who can benefit from national REDD+ (Streck 2020); a clear regulatory framework and market certainty (Dixon and Challies 2015; Laing et al. 2016; Golub et al. 2018; Ehara et al. 2019); strong forest governance (Streck 2020), and implementation of REDD+ activities at national and subnational levels. Other challenges are associated with the nature of forest-based mitigation activities, the costs and complexity of monitoring, reporting and verification of REDD+ activities, because of the need to consider the risks of permanence, carbon leakage, and precisely determine and monitor the forest carbon sinks (van der Gaast et al. 2018; Yanai et al. 2020). Although REDD+ has many challenges to mobilise more private finance, there is discussion on exploring other finance opportunities for the forest sector, such as building new blended finance models combining different funding sources like public and private finance (Streck 2016; Rode et al. 2019), and developing enhanced bonds for forest-based mitigation activities (World Bank 2017).

**Private finance opportunities for nature-based solutions.** The development of nature-based solutions faces barriers that relate to the value proposition, value delivery and value capture of nature-based solutions business models and sustainable sources of public/private finance to tap into (*high confidence*) (Toxopeus and Polzin 2017; Mok et al. 2021). However, the demand for establishing new finance and business models to attract both public and private finance to nature-based solutions is increasing in a wide range of topics such as urban areas, forestry and agriculture sectors, and blue natural capital including mangroves and coral reefs (Toxopeus and Polzin 2017; EIB 2019; Cziesielski et al. 2021; Mok et al. 2021; Thiele et al. 2021; UNEP et al. 2021). Furthermore, the recognition of the needs of financial institutions to identify the physical, transition and reputational risks resulting from not only climate change but also loss of biodiversity is gradually increasing (De Nederlandsche Bank and PBL Netherlands Environmental Assessment Agency 2020; Dasgupta 2021; TNFD 2021). Development of finance and business models for nature-based solutions needs to be explored, for example through utilising a wide range of financial instruments (e.g., equity, loans, bonds, and insurance), and creating standard metrics, baselines and common characteristics for nature-based solutions to promote the creation of a new asset class (Thiele et al. 2021; UNEP et al. 2021).

#### 15.6.8.3 Exploring Gender-responsive Climate Finance

Global and national recognition of the lack of finance for women has led to increasing emphasis on financial inclusion for women (*high confidence*). Currently, it is estimated that 980 million women are excluded from formal financial system (Miles and Wiedmaier-Pfister 2018); and there is a 9% gender gap in financial access across developing countries (Demirguc-Kunt et al. 2018). This gender gap is the percentage difference between men and women with bank accounts as measured and reported in the Global Financial Inclusion (Global Findex) database. Policies and frameworks to expand and enhance financial inclusion also extend to the area of climate finance (*high confidence*). Since AR5, there remain many questions and not enough evidence on the gender, distribution and allocative effectiveness of climate finance in the context of gender equality and women's empowerment (Williams M., 2015; Chan et al. 2018; Wong et al. 2019). Nonetheless, the existing global policy framework (entry points, policy priorities, etc.) of climate funds is gradually improving in order to support women's financial inclusion in both the public and the private dimensions of climate finance/investment (Schalatek 2015; Chan et al. 2018; Schalatek 2020). At the level of public multilateral climate funds, there have been significant improvements in integrating gender equality and women's empowerment issues in the governance structures, policies, project approval and implementation processes of existing multilateral climate funds such as the UNFCCC's funds managed by the Global Environment Facility, the Green Climate Fund and the World Bank's CIFs (*high confidence*) (Schalatek 2015; Williams M., 2015; Sellers 2016; GCF 2017). But according to a recent evaluation report, the integration of gender into operational policies and programmes is fragmented and there is lack of an 'adequate, systematic and comprehensive gender equality approach for the allocation and distribution of funds for projects and programmes on the ground' (GEF Independent Evaluation Office 2017; Schalatek 2018). The review found that 'almost half of the analysed sample of 70 climate projects were judged to be largely gender-blind, and only 5% considered to have successfully mainstreamed gender, including in two Least Developed Countries Fund adaptation projects' (GEF Independent Evaluation Office 2017; Schalatek 2018). While the GCF requires funding proposals to consider gender impact as part of their investment framework,[16] the fund does not have its own funding stream targeted to women's project on the ground, nor is there as yet an evaluation as to how entities are actually implementing gender action plan in the projects. In the case of the CIFs, as noted by Schalatek (2018), 'gender is not included in the operational principles of the Pilot Program on Climate Resilience (PPCR), which funds programmatic adaptation portfolios in a few developing countries, although most pilot countries have included some gender dimensions'. And, 'gender is not integrated into the operations of the Clean Technology Fund (CTF), which finances large-scale mitigation in large economies and accounts for 70% of the CIFs' pledged funding portfolio of 8.2 billion USD' (Schalatek 2018). However, both the Forest Investment Program (FIP) and the Scaling-Up Renewable Energy in Low-Income Countries Program (SREP) have integrated gender equality as either a co-benefit or core criteria of these programmes (Schalatek 2018).

Overall, efforts to promote gender responsive/sensitive climate finance, at national and local levels, both in the public and private dimensions and more specifically in mitigation-oriented sectors such as clean and renewable energy, remain deficient (*high confidence*). Recent developments in the capital markets in the areas of social bond are focused around gender bonds – debt instruments targeted to activities and behaviours that are relevant to gender equality and women's empowerment. These bonds are aligned with Sustainability-linked Bonds as well as Social Bonds Principles of the International Capital Market Association. Issuances of gender-labelled bonds are increasing in the Asia Pacific region (the most comprehensive initiative is the Impact Investment Exchange's (IIX) multi-country USD150 million Women's Livelihood Bond[17]) and in Latin America, Colombia, Mexico and Panama each have gender bond issuances). Additionally, a few developing countries, such as Pakistan (May 2021) and Morocco (March 2021) have issued gender bond guidelines for financial market participants.

**Linkage to sectoral climate change issues and gender and climate finance.** Subsets of actions designed to enhance women's more formal integration into climate policies, programmes and actions by the global private sector include: investment in clean energy, redirecting funds to support women and vulnerable regions as a component of social and green bonds as well as insurance for climate risk management. In the latter context, insurance providers are arguing that 'given the fact that women are disproportionately affected by climate change, there could be new finance innovations to address this gap'.(Miles and Wiedmaier-Pfister 2018). AXA and IFC estimate that the global women's insurance market has the opportunity to grow to three times its current size, to UDS1.7 trillion by 2030 (AXA Group et al. 2015; GIZ et al. 2017). However, across the board, and in particular with regard to public funds, despite improvements in the substantive gender sensitisation and operational gender responsiveness of multilateral and bilateral climate finance funds operations, current flows of public and climate finance do not seem to be going to women and local communities in significant amounts (Chan et al. 2018; Schalatek 2020). At the same time, evaluations of the effectiveness of climate finance show that equitable flow of climate finance can play an important role in levelling the playing field and in enabling women and men to successfully respond to climate change and to enable the success and sustainability of local response in ensuring effective and sustainable

16 Notably, the GCF provides guidance to Accredited Entities submitting funding proposals on the inclusion of an initial gender and social assessment during the project planning, preparation and development stage and a gender and social inclusion action plan at the project preparation stage.

17 The Women's Livelihood Bond (WLB) series has been on the market since 2017 when WLB1 was launched. WLB2 issuance of USD12 million arrived January 2020. WLB3 was launched December 2020 to support 180,000 underserved women and women entrepreneurs in the Asia Pacific region to respond, to recover from, and to build resilience in the aftermath of the COVID-19 pandemic (Rockfeller Foundation and Shujog 2016; IIX 2020).

climate strategies that can contribute to the global goals of the Paris Agreement (Minniti and Naudé 2010; Bird et al. 2013; Barrett 2014; Eastin 2018). This is particularly, so in the case of female-owned MSMEs, who, the literature increasingly shows, are key to promoting resilience at micro and macro scale in many developing countries (Omolo et al. 2017; Atela et al. 2018; Crick, F. et al. 2018).

Frequently Asked Questions (FAQs)

## FAQ 15.1 | What's the role of climate finance and the finance sector for a transformation towards a sustainable future?

The Paris Agreement has widened the scope of all financial flows from climate finance only to the full alignment of finance flows with the long-term goals of the Paris Agreement. While climate finance relates historically to the financial support of developed countries to developing countries, the Paris Agreement and its Article 2.1(c) have developed a new narrative that goes much beyond traditional flows and relates to all sectors and actors. Finance flows are consistent when the effects are either neutral with or without positive climate co-benefits to climate objectives; or explicitly targeted to climate benefits in adaptation and/or mitigation result areas. Climate-related financial risk is still massively underestimated by financial institutions, financial decision-makers more generally and also among public sector stakeholders, limiting the sector's potential of being an enabler of the transition. The private sector has started to recognise climate-related risks and consequently redirect investment flows. Dynamics vary across sectors and regions with the financial sector being an enabler of transitions in only some selected (sub-)sectors and regions. Consistent, credible, timely and forward-looking political leadership remains central to strengthen the financial sector as enabler.

## FAQ 15.2 | What's the current status of global climate finance and the alignment of global financial flows with the Paris Agreement?

There is no agreed definition of climate finance. The term 'climate finance' is applied to the financial resources devoted to addressing climate change by all public and private actors from global to local scales, including international financial flows to developing countries to assist them in addressing climate change. Total climate finance includes all financial flows whose expected effect aims to reduce net greenhouse gas (GHG) emissions and/or to enhance resilience to the impacts of current and projected climate change. This includes private and public funds, domestic and international flows and expenditures. Tracking of climate finance flows faces limitations, in particular for national climate finance flows.

Progress on the alignment of financial flows with low GHG emissions pathways remains slow. Annual global climate finance flows are on an upward trend since the Fifth Assessment Report, according to the Climate Policy Initiative reaching more than USD630 billion in 2019/2020, however, growth has likely slowed down and flows remain significantly below needs. This is driven by barriers within and outside the financial sector. More than 90% of financing is allocated to mitigation activities despite the strong economic rationale of adaptation action. Adjusting for higher estimates on current flows for energy efficiency based on International Energy Agency data, the dominance of mitigation becomes even stronger. Persistently high levels of both public and private fossil-fuel related financing as well as other misaligned flows continue to be of major concern despite recent commitments. Significant progress has been made in the commercial finance sector with regard to the awareness of climate risks resulting from inadequate financial flows and climate action. However, a more consequent investment and policy decision-making that enables a rapid redirection of financial flows is needed. Regulatory support as a catalyser is an essential driver of such redirections. Dynamics across sectors and regions vary, with some being better positioned to close financing gaps and to benefit from an enabling role of finance in the short-term.

## FAQ 15.3 | What defines a financing gap, and where are the critically identified gaps?

A financing gap is defined as the difference between current flows and average needs to meet the long-term goals of the Paris Agreement. Gaps are driven by various barriers inside (short-termism, information gaps, home bias, limited visibility of future pipelines) and outside (e.g., missing pricing of externalities, missing regulatory frameworks) of the financial sector. Current mitigation financing flows come in significantly below average needs across all regions and sectors despite the availability of sufficient capital on a global basis. Globally, yearly climate finance flows have to increase by a factor between three and six to meet average annual needs between 2020 and 2030.

Gaps are in particular concerning for many developing countries, with COVID-19 exacerbating the macroeconomic outlook and fiscal space for governments. Also, limited institutional capacity represents a key barrier for many developing countries, burdening risk perceptions and access to appropriately priced financing as well as limiting their ability to actively manage the transformation. Existing fundamental inequities in access to finance, as well as its terms and conditions, and countries' exposure to physical impacts of climate change, overall result in a worsening outlook for a global just transition.

## References


Abadie, L.M., I. Galarraga, and D. Rübbelke, 2013: An analysis of the causes of the mitigation bias in international climate finance. *Mitig. Adapt. Strateg. Glob. Chang.*, **18(7)**, 943–955, doi:10.1007/s11027-012-9401-7.

Abbasi, F. and K. Riaz, 2016: CO2 emissions and financial development in an emerging economy: An augmented VAR approach. *Energy Policy*, **90**, 102–114, doi:10.1016/j.enpol.2015.12.017.

Accenture, 2018: *Financing Growth in the Clean Cookstoves and Fuels Market: An Analysis and Recommendations - Strengthening the Pipeline through Better Alignment of Financing with Enterprise Needs*. 42 pp. https://cleancooking.org/wp-content/uploads/2021/07/549-1.pdf (Accessed November 1, 2021).

Acevedo, S., 2016: *Gone with the Wind: Estimating Hurricane and Climate Change Costs in the Caribbean*. International Monetary Fund, Washington, DC, 40 pp. https://www.imf.org/en/Publications/WP/Issues/2016/12/31/Gone-with-the-Wind-Estimating-Hurricane-and-Climate-Change-Costs-in-the-Caribbean-44333 (Accessed November 1, 2021).

Acevedo, S., M. Mrkaic, N. Novta, E. Pugacheva, and P. Topalova, 2018: *The Effects of Weather Shocks on Economic Activity: What are the Channels of Impact?*, International Monetary Fund, Washington, DC, USA, 40 pp. https://www.imf.org/%7B~%7D/media/Files/Publications/WP/2018/wp18144.ashx (Accessed November 1, 2021).

ACT Alliance, 2020: *Climate Risk Insurance and Risk Financing in the Context of Climate Justice - A Manual for Development and Humanitarian Aid Practitioners*. ACT Alliance, Geneva, 72 pp. https://reliefweb.int/sites/reliefweb.int/files/resources/Climate-Risk-Insurance-Manual_English-1.pdf (Accessed November 1, 2021).

ADB, 2012: First-Ever Private Equity Fund for Philippine Infrastructure Investments. Asian Development Bank (ADB) News Release, 31 July.

ADB, 2015: *Local Currency Bonds and Infrastructure Finance in ASEA Countries*. Asian Development Bank (ADB), Manila, 152 pp. https://www.adb.org/sites/default/files/publication/167313/local-currency-bonds-and-infrastructure-finance-asean-3.pdf (Accessed August 18, 2021).

ADB, 2018: *India Infrastructure Finance Company Limited (IIFCL) guarantee facility credit enhancement of project bonds (India)*. Asian Development Bank (ADB), Manila, 2 pp.

Adenle, A.A. et al., 2017: Managing Climate Change Risks in Africa - A Global Perspective. *Ecol. Econ.*, **141**, 190–201, doi:10.1016/j.ecolecon.2017.06.004.

Adrian et al., 2021: Keeping it Local: A Secure and Stable Way to Access Financing. IMF Blog. https://blogs.imf.org/2021/03/12/keeping-it-local-a-secure-and-stable-way-to-access-financing/ (Accessed August 18, 2021).

AfDB, 2020: Sahel Group of Five: African Development Bank approves program to expand solar energy generation under Desert-to-Power scheme. African Development Bank Group (AfDB), press release, 18 December.

AfDB et al., 2018: *2017 Joint Report on Multilateral Development Banks' Climate Finance*.European Bank for Reconstruction and Development London, UK 44 pp.

AfDB et al., 2020: *2019 Joint Report on Multilateral Development Banks Climate Finance*. European Bank for Reconstruction and Development, London, UK 56 pp.

Africa50, 2020: *Annual Report 2019: Investing in infrastructure for Africa's growth*., Casablanca, 139 pp.

African Union Commission, 2015: *Agenda 2063: The Africa We Want. First Ten-Year Implementation Plan 2013–2023*. African Union Commission, Addis Ababa, Ethiopia, 20 pp.

Agarwal, S. and T. Singh, 2018: *Unlocking the Green Bond Potential in India*. The Energy and Resources Institute, New Delhi, 29 pp. https://www.teriin.org/projects/nfa/files/Green-Bond-Working-Paper.pdf (Accessed November 1, 2021).

AGF, 2020: *African Guarantee Fund: Annual Report 2019*. African Guarantee Fund (AGF), 109 pp. https://africanguaranteefund.com/wp-content/uploads/2021/04/AGF-Annual-Report-2019-Smaller-Version.pdf (Accessed August 18, 2021).

Aggarwal, R., 2019: The impact of climate shocks on consumption and the consumption distribution in India. PhD Thesis, Paris School of Economics, Paris, France, 60 pp.

Aghion, P., A. Dechezleprêtre, D. Hémous, R. Martin, and J. Van Reenen, 2016: Carbon Taxes, Path Dependency, and Directed Technical Change: Evidence from the Auto Industry. *J. Polit. Econ.*, **124(1)**, 1–51, doi:10.1086/684581.

Agliardi, E. and R. Agliardi, 2019: Financing environmentally-sustainable projects with green bonds. *Environ. Dev. Econ.*, **24(6)**, 608–623, doi:10.1017/S1355770X19000020.

Aglietta, M., É. Espagne, and F. Baptiste Perrissin, 2015: *A proposal to finance low carbon investment in Europe*. France Strategie, Paris, 1–7 pp. https://www.wmo.int/pages/publications/bulletin_fr/archives/61_1_fr/Fondsvertpourleclimat.html (Accessed November 1, 2021).

Aglionby, J., 2019: 'Rhino bond' breaks new ground in conservation finance. *Financial Times*, July 16.

Agrawala, S. and M. Carraro, 2010: Assessing the Role of Microfinance in Fostering Adaptation to Climate Change. *SSRN Electron. J.*, **(15)**, 20, doi:10.2139/ssrn.1646883.

Agrawala, S. et al., 2011: *Private Sector Engagement in Adaptation to Climate Change: Approaches to Managing Climate Risks*. OECD Publishing, Paris, France, 56 pp.

Ahluwalia, M.S., L. Summers, A. Velasco, N. Birdsall, and S. Morris, 2016: *Multilateral Development Banking for this Century's Development Challenges: Five Recommendations to Shareholders of the Old and New Multilateral Development Banks*. Center for Global Development, Washington, DC, USA, 56 pp. https://www.cgdev.org/publication/multilateral-development-banking-for-this-centurys-development-challenges (Accessed November 1, 2021).

Ahmad, E., D. Dowling, D. Chan, S. Colenbrander, and N. Godfrey, 2019: *Scaling up investment for sustainable urban infrastructure: A guide to national and subnational reform*. Coalition for Urban Transitions, London, UK and Washington, DC, USA 1–46 pp. Coalition for Urban Transitions: London and Washington. https://newclimateeconomy.report/workingpapers/wp-content/uploads/sites/5/2019/04/CUT2019_Scaling_up_investment_for_sustainable_urban_infrastructure.pdf.

Åhman, M., L.J. Nilsson, and B. Johansson, 2017: Global climate policy and deep decarbonization of energy-intensive industries. *Clim. Policy*, **17(5)**, 634–649, doi:10.1080/14693062.2016.1167009.

Ahrend, R., M. Curto-grau, and C. Vammalle, 2013: *Passing the Buck? Central and Sub-national Governments in Times of Fiscal Stress*. OECD, Paris, 33 pp.

Aitken, R., 2013: The Financialization of Micro-Credit. *Dev. Change*, **44(3)**, 473–499, doi:10.1111/dech.12027.

Akkoyunlu, S. and M. Stern, 2012: *An Empirical Analysis of Diaspora Bonds*. Programme for the Study of Global Migration, 42 pp.

Akwagyiram, A. and C. Ohuocha, 2021: African fintech firm Flutterwave eyes U.S. listing after raising $170 million. *Reuters*, March 10.

Akyüz, Y., 2017: *Playing with Fire: Deepened Financial Integration and Changing Vulnerabilities of the Global South*. Oxford University Press, Oxford, UK, 279 pp.

Aldersey-Williams, J. and T. Rubert, 2019: Levelised cost of energy – A theoretical justification and critical assessment. *Energy Policy*, **124**, 169–179, doi:10.1016/j.enpol.2018.10.004.

Alesina, A., D. Carloni, and G. Lecce, 2011: *The Electoral Consequences of Large Fiscal Consolidations*. National Bureau of Economic Research, Cambridge, MA, 19 pp. https://www.nber.org/system/files/working_papers/w17655/w17655.pdf (Accessed November 1, 2021).

Alessi, L., S. Battiston, A.S. Melo, and A. Roncoroni, 2019: The EU Sustainability Taxonomy: a financial impact assessment. Publications Office of the European Union, Luxembourg, 2019, doi:10.2760/347810 (Accessed November 1, 2021).

Alessi, L., E. Ossola, and R. Panzica, 2021: What greenium matters in the stock market? The role of greenhouse gas emissions and environmental disclosures. *J. Financ. Stab.*, **54**, 100869, doi:10.1016/j.jfs.2021.100869.

Alexander, J. et al., 2019: Financing the sustainable urban future Scoping a green cities development bank.ODI, London, UK, 44 pp.

Allen, T. et al., 2020: *Climate-Related Scenarios for Financial Stability Assessment: an Application to France*. Banque de France, Paris, France, 65 pp.

Alnes, K., A. Berg, C. Clapp, E. Lannoo, and K. Pillay, 2018: *Flood Risk for Investors. Are you prepared?* CICERO, Oslo, Norway, 12 pp. https://pub.cicero.oslo.no/cicero-xmlui/handle/11250/2497350 (Accessed November 1, 2021).

Alsayegh, M.F., R. Abdul Rahman, and S. Homayoun, 2020: Corporate Economic, Environmental, and Social Sustainability Performance Transformation through ESG Disclosure. *Sustainability*, **12(9)**, 3910, doi:10.3390/su12093910.

Alvaredo, F., L. Chancel, T. Piketty, E. Saez, and G. Zucman, 2020: Towards a System of Distributional National Accounts: Methods and Global Inequality Estimates from WID.world. *Econ. Stat./Econ. Stat.*, **(517-518–519)**, 41–59, doi:10.24187/ecostat.2020.517t.2018.

Amacker, J. and C. Donovan, 2021: *Marathon or Sprint?: The Race for Green Capital in Emerging Markets*. Centre for Climate Finance and Investment, London, UK, 41 pp. https://imperialcollegelondon.app.box.com/s/ge9thj52rcehma4bwxyt0fchi3q2niv5 (Accessed August 19, 2021).

Ameli, N., P. Drummond, A. Bisaro, M. Grubb, and H. Chenet, 2020: Climate finance and disclosure for institutional investors: why transparency is not enough. *Clim. Change*, **160(4)**, 565–589, doi:10.1007/s10584-019-02542-2.

Ameli, N. et al., 2021a: Higher cost of finance exacerbates a climate investment trap in developing economies. *Nat. Commun.*, **12(1)**, 4046, doi:10.1038/s41467-021-24305-3.

Ameli, N., S. Kothari, and M. Grubb, 2021b: Misplaced expectations from climate disclosure initiatives. *Nat. Clim. Chang.*, **11**, 917–924, doi:10.1038/s41558-021-01174-8.

Amonya, F., 2009: Infrastructure Sovereign Bonds in Sub-Saharan Africa Dissecting the Road Governance Challenge of the New Financing Instrument. *SSRN Electron. J.* doi:10.2139/ssrn.2653429 (Accessed November 1, 2021).

Amundi and IFC, 2019: Emerging Market Green Bonds Report 2019. Amundi Asset Management, Paris, France, 34 pp.

Andersen, O.W. et al., 2019: *Blended Finance Evaluation: Governance and Methodological Challenges*. OECD Development Co-operation Working Papers 51, OECD Publishing, Paris, France, 32 pp.

Andrade, G. and G. Prado, 2020: *Financial Innovation to Support Women-Led Businesses: Mexico's First Gender Bond and the Role of National Development Banks*. Inter-American Development Bank, Washington, DC, USA, 9 pp.

Andrijevic, M., C.-F. Schleussner, M.J. Gidden, D.L. McCollum, and J. Rogelj, 2020: COVID-19 recovery funds dwarf clean energy investment needs. *Science*, **370(6514)**, 298–300, doi:10.1126/science.abc9697.

Anguelovski, I. and J. Carmin, 2011: Something borrowed, everything new: innovation and institutionalization in urban climate governance. *Curr. Opin. Environ. Sustain.*, **3(3)**, 169–175, doi:10.1016/j.cosust.2010.12.017.

Aon Benfield UCL Hazard Research Centre, 2019: *Weather, Climate & Catastrophe Insight: 2018 Annual Report*. AON PLC, London, UK, 84 pp. http://thoughtleadership.aonbenfield.com/Documents/20190122-ab-if-annual-weather-climate-report-2018.pdf (Accessed November 1, 2021).

AOSIS, 2021: Alliance Of Small Island States (AOSIS) Leaders' Declaration. https://www.aosis.org/launch-of-the-alliance-of-small-island-states-leaders-declaration/ (Accessed October 27, 2021).

ARC, 2016: *ARC's Agenda for Action on Climate Resilience: $2 Billion of Insurance Coverage for Africa by 2020*.The African Risk Capacity Group, Johannesburg, South Africa, 21 pp.

Arent, D., C. Arndt, M. Miller, F. Tarp, and O. Zinaman (eds.), 2017: *The political economy of clean energy transitions*. 1st ed. Oxford University Press, Oxford, UKi, 632 pp.

Arezki, R., 2021: Climate finance for Africa requires overcoming bottlenecks in domestic capacity. *Nat. Clim. Chang.*, **11**, 888, doi:10.1038/s41558-021-01191-7.

Arezki, R., P. Bolton, S. Peters, F. Samama, and J. Stiglitz, 2016: *From Global Savings Glut to Financing Infrastructure: The Advent of Investment Platforms*. IMF, Washington DC, USA, 47 pp. https://www.imf.org/en/Publications/WP/Issues/2016/12/31/From-Global-Savings-Glut-to-Financing-Infrastructure-The-Advent-of-Investment-Platforms-43689 (Accessed November 1, 2021).

Arrow, K. et al., 2004: Are We Consuming Too Much? *J. Econ. Perspect.*, **18(3)**, 147–172, doi:10.1257/0895330042162377.

Atela, J., K.E. Gannon, and F. Crick, 2018: Climate Change Adaptation among Female-Led Micro, Small, and Medium Enterprises in Semiarid Areas: A Case Study from Kenya. In: Filho, W.L. (ed) *Handbook of Climate Change Resilience*, Springer International Publishing, Cham, Switzerland, pp. 1–18.

Atmadja, S.S., S. Arwida, C. Martius, and P.T. Thuy, 2018: Financing REDD+: A transaction among equals, or an uneven playing field? In: *Transforming REDD+: Lessons and new directions*, Center for International Forestry Research (CIFOR), Bogor, Indonesia, pp. 29–40.

Attridge, S. and L. Engen, 2019: *Blended finance in the poorest countries: The need for a better approach*. ODI, London, UK, 75 pp. https://cdn.odi.org/media/documents/12666.pdf (Accessed November 1, 2021).

Attridge, S. and M. Gouett, 2021: *Development finance institutions: the need for bold action to invest better*. ODI, London, UK, 74 pp. https://cdn.odi.org/media/documents/DPF_Blended_finance_report_tuMbRjW.pdf (Accessed October 30, 2021).

Averchenkova, A., F. Crick, A. Kocornik-Mina, H. Leck, and S. Surminski, 2016: Multinational and large national corporations and climate adaptation: are we asking the right questions? A review of current knowledge and a new research perspective. *Wiley Interdiscip. Rev. Clim. Chang.*, **7(4)**, 517–536, doi:10.1002/wcc.402.

AXA Group, Accenture, and IFC, 2015: *SheforShield: Insure Women to Better Protect All*. IFC, Washington, DC, USA, 178 pp. https://www.ifc.org/wps/wcm/connect/7a8950dd-db88-4d52-8b88-fbeb771cf89c/SheforShield_Final-Web2015.pdf?MOD=AJPERES&CVID=nZh46on.

Bachelet, M.J., L. Becchetti, and S. Manfredonia, 2019: The Green Bonds Premium Puzzle: The Role of Issuer Characteristics and Third-Party Verification. *Sustainability*, **11(4)**, 1098, doi:10.3390/su11041098.

Bagus, U., F.J. de Girancourt, R. Mahmood, and Q. Manji, 2020: *Africa's insurance market is set to take off*. McKinsey and Company, 10 pp, New York, USA. https://www.mckinsey.com/featured-insights/middle-east-and-africa/africas-insurance-market-is-set-for-takeoff (Accessed November 1, 2021).

Bahal, G., M. Raissi, and V. Tulin, 2018: Crowding-out or crowding-in? Public and private investment in India. *World Dev.*, **109**, 323–333, doi:10.1016/j.worlddev.2018.05.004.

Bahceli, Y., 2020: German 30-yr green bond bucks market selloff with record demand. *Reuters*. May 11.

Bahl, R.W. and J.F. Linn, 2014: *Governing and Financing Cities in the Developing World*. Lincoln Institute of Land Management, Cambridge, UK, 56 pp. https://www.lincolninst.edu/sites/default/files/pubfiles/governing-and-financing-cities-developing-world-full_0.pdf (Accessed November 1, 2021).

Bainton, N., D. Kemp, E. Lèbre, J.R. Owen, and G. Marston, 2021: The energy-extractives nexus and the just transition. *Sustain. Dev.*, **29(4)**, 624–634, doi:10.1002/sd.2163.

Baker, M.P., D.B. Bergstresser, G. Serafeim, and J.A. Wurgler, 2018: Financing the Response to Climate Change: The Pricing and Ownership of U.S. Green Bonds. *SSRN Electron. J.*, 3275327, 1–44, doi:10.2139/ssrn.3275327.

Balint, T. et al., 2016: *Complexity and the Economics of Climate Change: a Survey and a Look Forward*. https://halshs.archives-ouvertes.fr/halshs-01390694.

Balint, T, et al., Complexity and the Economics of Climate Change: A Survey and a Look Forward (July 08, 2016). LEM Working Paper Series, 2016/29, SSRN. https://ssrn.com/abstract=2807571 (Accessed November 1, 2021).

Bandura, R. and S.R. Ramanujam, 2019: *Innovations in guarantees for development*. Center for Strategic and International Studies, Washington, DC, USA, and Rowman & Littlefield, Lanham, MD, USA, 50 pp.

Banga, J., 2019: The green bond market: a potential source of climate finance for developing countries. *J. Sustain. Financ. Invest.*, **9(1)**, 17–32, doi:10.1080/20430795.2018.1498617.

Bank of England, 2015: *The impact of climate change on the UK insurance sector: A Climate Change Adaptation Report by the Prudential Regulation Authority*. Prudential Regulation Authority, London, UK, 85 pp. https://www.bankofengland.co.uk/-/media/boe/files/prudential-regulation/publication/impact-of-climate-change-on-the-uk-insurance-sector.pdf (Accessed November 1, 2021).

Bank of England, 2018: *Transition in thinking: The impact of climate change on the UK banking sector*. Prudential Regulation Authority, London, UK, 52 pp. https://www.bankofengland.co.uk/-/media/boe/files/prudential-regulation/report/transition-in-thinking-the-impact-of-climate-change-on-the-uk-banking-sector.pdf (Accessed November 1, 2021).

Baranović, I., I. Busies, W. Coussens, M. Grill, and H. Hempell, 2021: *The challenge of capturing climate risks in the banking regulatory framework: is there a need for a macroprudential response? Macroprudential Bulletin, 15, ECB,* Frankfurt am Main, Germany. https://www.ecb.europa.eu/pub/financial-stability/macroprudential-bulletin/html/ecb.mpbu202110_1~5323a5baa8.en.html (Accessed November 1, 2021).

Barbier, E.B., 2020: Greening the Post-pandemic Recovery in the G20. *Environ. Resour. Econ.*, **76(4)**, 685–703, doi:10.1007/s10640-020-00437-w.

Barbier, E.B. and J.C. Burgess, 2020: Sustainability and development after COVID-19. *World Dev.*, **135**, 105082, doi:10.1016/j.worlddev.2020.105082.

Barnard, S., 2015: *Climate finance for cities: How can international climate funds best support low-carbon and climate resilient urban development?* ODI, London, UK 27 pp. https://odi.org/en/publications/climate-finance-for-cities-how-can-climate-funds-best-support-low-carbon-and-climate-resilient-urban-development/ (Accessed November 1, 2021).

Barnsley, I., A. Blank, and A. Brown, 2015: *Enabling Renewable Energy and Energy Efficiency Technologies*. International Energy Agency, Paris, France, 115 pp. https://www.idaea.csic.es/medspring/sites/default/files/Enabling-Renewable-Energy-and-Energy-Efficiency-Technologies.pdf (Accessed November 1, 2021).

Barrett, P. and S. Chen, 2021: *Social Repercussions of Pandemics*. Working Paper No. 2021/021, IMF, Washington DC, USA, 24 pp. https://www.imf.org/-/media/Files/Publications/WP/2021/English/wpiea2021021-print-pdf.ashx (Accessed November 1, 2021).

Barrett, S., 2014: Subnational Climate Justice? Adaptation Finance Distribution and Climate Vulnerability. *World Dev.*, **58**, 130–142, doi:10.1016/j.worlddev.2014.01.014.

Barro, R.J., 1987: Government spending, interest rates, prices, and budget deficits in the United Kingdom, 1701–1918. *J. Monet. Econ.*, **20(2)**, 221–247, doi:10.1016/0304-3932(87)90015-8.

Basile, I. and J. Dutra, 2019: *Blended Finance Funds and Facilities: 2018 Survey Results*. OECD *Development Co-operation Working Papers*, No. 59, OECD Publishing, Paris, France, 69 pp. doi: 10.1787/806991a2-en.

Bast, E., A. Doukas, S. Pickard, L. Van Der Burg, and S. Whitley, 2015: *Empty promises: G20 subsidies to oil, gas and coal production*. ODI, London, UK, and Washington DC, USA, 103 pp. https://www.odi.org/publications/10058-empty-promises-g20-subsidies-oil-gas-and-coal-production (Accessed November 1, 2021).

Battiston, S. and I. Monasterolo, 2020: On the dependence of investor's probability of default on climate transition scenarios. *SSRN Electron. J.*, **3743647**, 25, doi: 10.2139/ssrn.3743647.

Battiston, S., A. Mandel, I. Monasterolo, F. Schütze, and G. Visentin, 2017: A climate stress-test of the financial system. *Nat. Clim. Change*, **7(4)**, 283–288, doi:10.1038/nclimate3255.

Battiston, S., P. Jakubik, I. Monasterolo, K. Riahi, and B. van Ruijven, 2019: Climate risk assessment of sovereign bonds' portfolio of European insurers. In: EIOPA Financial Stability Report, December 2019. European Insurance and Occupational pensions Authority (EIOPA), Publications Office of the European Union, Luxembourg (Accessed November 1, 2021).

Battiston, S., I. Monasterolo, J. Min, K. Riahi, and B. van Ruijven, 2020a: Enabling or hampering. Climate risk and the role of finance in the low-carbon transition. *SSRN Electron. J.*, **3748642**, http://dx.doi.org/10.2139/ssrn.3748642.

Battiston, S., I. Monasterolo, K. Riahi, and B. van Ruijven, 2020b: Climate mitigation pathways need to account for the ambivalent role of finance. *SSRN Electron. J.*, **3748041**, doi:10.2139/ssrn.3748041.

Battiston, S., Y. Dafermos, and I. Monasterolo, 2021a: Climate risks and financial stability. *J. Financ. Stab.*, **54**, 100867, doi:10.1016/j.jfs.2021.100867.

Battiston, S., I. Monasterolo, K. Riahi, and B.J. van Ruijven, 2021b: Accounting for finance is key for climate mitigation pathways. *Science*, **372(6545)**, 918–920, doi:10.1126/science.abf3877.

Bauer, N. et al., 2015: CO2 emission mitigation and fossil fuel markets: Dynamic and international aspects of climate policies. *Technol. Forecast. Soc. Change*, **90**, 243–256, doi:10.1016/j.techfore.2013.09.009.

Bauwens, T., 2019: Analyzing the determinants of the size of investments by community renewable energy members: Findings and policy implications from Flanders. *Energy Policy*, **129** (October 2018), 841–852, doi:10.1016/j.enpol.2019.02.067.

Bayliss, K. and E. Van Waeyenberge, 2018: Unpacking the Public Private Partnership Revival. *J. Dev. Stud.*, **54(4)**, 577–593, doi:10.1080/00220388.2017.1303671.

BCG, 2018: *Klimapfade Deutschland*. BCG, https://www.bcg.com/de-de/publications/2018/climate-paths-for-germany (Accessed November 1, 2021).

Beck, S., S. Jasanoff, A. Stirling, and C. Polzin, 2021: The governance of sociotechnical transformations to sustainability. *Curr. Opin. Environ. Sustain.*, **49**, 143–152, doi:10.1016/j.cosust.2021.04.010.

Benali, N., I. Abdelkafi, and R. Feki, 2018: Natural-disaster shocks and government's behavior: Evidence from middle-income countries. *Int. J. Disaster Risk Reduct.*, **27**, 1–6, doi:10.1016/j.ijdrr.2016.12.014.

Benmelech, E. and N. Tzur-Ilan, 2020: *The Determinants of Fiscal and Monetary Policies During the Covid-19* Crisis.,NBER working paper 27461, NBER, Cambridge, MA, USA, 42 pp. doi: 10.3386/w27461.

Benn, J. and W. Luijkx, 2017: *Emerging providers' international co-operation for development*., Paris, France, 26 pp.

Berensmann, K., F. Dafe, and U. Volz, 2015: Developing local currency bond markets for long-term development financing in Sub-Saharan Africa. *Oxford Rev. Econ. Policy*, **31(3–4)**, 350–378, doi:10.1093/oxrep/grv032.

Berg, F., J. Kölbel, and R. Rigobon, 2020: Aggregate Confusion: The Divergence of ESG Ratings. Forthcoming Review of Finance, *SSRN Electron. J.*, **3438533**, doi:10.2139/ssrn.3438533.Bergman, N., 2018: Impacts of the Fossil Fuel Divestment Movement: Effects on Finance, Policy and Public Discourse. *Sustainability*, **10(7)**, 2529, doi:10.3390/su10072529.

Bertoldi, P. and B. Boza-Kiss, 2017: Analysis of barriers and drivers for the development of the ESCO markets in Europe. *Energy Policy*, **107** (July 2016), 345–355, doi:10.1016/j.enpol.2017.04.023.

Bertoldi, P., M. Economidou, V. Palermo, B. Boza-Kiss, and V. Todeschi, 2021: How to finance energy renovation of residential buildings: Review of current and emerging financing instruments in the EU. *WIREs Energy Environ.*, **10(1)**, doi:10.1002/wene.384.

Bertram, C. et al., 2021: Energy system developments and investments in the decisive decade for the Paris Agreement goals. *Environ. Res. Lett.*, **16(7)**, 074020, doi:10.1088/1748-9326/ac09ae.

Best, R. and P.J. Burke, 2018: Adoption of solar and wind energy: The roles of carbon pricing and aggregate policy support. *Energy Policy*, **118**, 404–417, doi:10.1016/j.enpol.2018.03.050.

Bevere, L., 2019: sigma 2/2019: Secondary natural catastrophe risks on the front line. *sigma Res.*, Swiss Re Institute. https://www.swissre.com/institute/research/sigma-research/sigma-2019-02.html (Accessed November 1, 2021).

Bhattacharya, A. et al., 2020: Delivering on the $100 Billion Climate Finance Commitment and Transforming Climate Finance. Independent Expert Group on Climate Change. 70 pp. https://www.un.org/sites/un2.un.org/files/2020/12/100_billion_climate_finance_report.pdf (Accessed November 1, 2021).

Bhattarai, K. et al., 2018: Tax plan debates in the US presidential election: A dynamic CGE analysis of growth and redistribution trade-offs. *Econ. Model.*, **68**, 529–542, doi:10.1016/j.econmod.2017.08.031.

Bird, N., T. Beloe, S. Ockenden, J. Corfee-Morlot, and S. Zou, 2013: *Understanding Climate Change Finance Flows and Effectiveness – Mapping of Recent Initiatives*. ODI, London, 15 pp.

Bisello, A., D. Vettorato, D. Ludlow, and C. Baranzelli, (eds.), 2021: *Smart and Sustainable Planning for Cities and Regions*. Springer International Publishing, Cham, Switzerland, 307 pp.

Black, A. and P. Fraser, 2002: Stock market short-termism—an international perspective. *J. Multinatl. Financ. Manag.*, **12(2)**, 135–158, doi:10.1016/S1042-444X(01)00044-5.

BlackRock, 2021: Climate Finance Partnership – Institutional | BlackRock. https://www.blackrock.com/institutions/en-us/strategies/alternatives/real-assets/infrastructure/climate-finance-partnership (Accessed October 30, 2021).

Blanchard, O. and D. Leigh, 2013: *Growth Forecast Errors and Fiscal Multipliers*. IMF Working Paper 13/1, IMF, Washington DC, USA, 42 pp. https://www.imf.org/external/pubs/ft/wp/2013/wp1301.pdf (Accessed November 1, 2021).

Bleeker and Abdulkadri, 2020: *A review of Caribbean national statistical legislation in relation to the United Nations Fundamental Principles of Official Statistics*. Economic Commission for Latin America and the Caribbean (ECLAC). https://www.cepal.org/en/publications/45100-review-caribbean-national-statistical-legislation-relation-united-nations (Accessed October 27, 2021).

Blended Finance Taskforce, 2018a: Better finance better world. Consultation paper of the Blended Finance Taskforce. Business & Sustainable Development Commission - Blended Finance Taskforce, London, UK, 119 pp. http://s3.amazonaws.com/aws-bsdc/BFT_BetterFinance_final_01192018.pdf#asset:614:url (Accessed November 1, 2021).

Blended Finance Taskforce, 2018b: *Blended Finance Taskforce calls to scale up the issuance and use* of development guarantees. Business & Sustainable Development Commission - Blended Finance Taskforce, London, UK, 7 pp.

BNEF, 2019: *Clean Energy Investment Trend 2018*. Bloomberg NEF, New York, USA, 51 pp. https://data.bloomberglp.com/professional/sites/24/BNEF-Clean-Energy-Investment-Trends-2018.pdf (Accessed November 1, 2021).

Boamah, F., 2020: Desirable or debatable? Putting Africa's decentralised solar energy futures in context. *Energy Res. Soc. Sci.*, **62**, 101390, doi:10.1016/j.erss.2019.101390.

Bodle, R. and V. Noens, 2018: Climate Finance: Too Much on Detail, Too Little on the Big Picture? *Carbon & Clim. Law Rev.*, **12(3)**, 248–257, doi:10.21552/cclr/2018/3/11.

Boffo, R. and R. Patalano, 2020: *ESG Investing: Practices, Progress and Challenges*. OECD, Paris, France, 85 pp. https://www.oecd.org/finance/ESG-Investing-Practices-Progress-Challenges.pdf (Accessed November 1, 2021).

Boissinot, J., D. Huber, and G. Lame, 2016: Finance and climate. *OECD J. Financ. Mark. Trends*, **2015(1)**, 7–23, doi:10.1787/fmt-2015-5jrrz76d5td5.

Bolton, P. and M. Kacperczyk, 2021: Do investors care about carbon risk? *J. Financ. Econ.*, **142(2)**, 517–549, doi:10.1016/j.jfineco.2021.05.008.

Bolton, P., M. Despres, L.A. Pereira Da Silva, F. Samama, and R. Svartzman, 2020: *The green swan - Central banking and financial stability in the age of climate change*. Bank of International Settlement, Basel, Switzerland, 107 pp. https://www.bis.org/publ/othp31.pdf (Accessed November 1, 2021).

Bond, E.W., J. Tybout, and H. Utar, 2015: Credit Rationing, Risk Aversion, and Industrial Evoluation in Developing Countries. *Int. Econ. Rev.*, **56(3)**, 695–722, doi:10.1111/iere.12119.

Borge, L.-E. and A.O. Hopland, 2020: Less fiscal oversight, more adjustment. *Eur. J. Polit. Econ.*, **63** (June 2019), 101893, doi:10.1016/j.ejpoleco.2020.101893.

Borio, C., 2014: The financial cycle and macroeconomics: What have we learnt? *J. Bank. Financ.*, **45**, 182–198, doi:10.1016/j.jbankfin.2013.07.031.

Borio, C., M. Drehmann, and D. Xia, 2018: *The financial cycle and recession risk. BIS Quarterly Review December* 2018, BIS, Basel, Switzerland, 13 pp. https://www.bis.org/publ/qtrpdf/r_qt1812g.pdf (Accessed November 1, 2021).

Bos, K. and J. Gupta, 2019: Stranded assets and stranded resources: Implications for climate change mitigation and global sustainable development. *Energy Res. Soc. Sci.*, **56** (June), 101215, doi:10.1016/j.erss.2019.05.025.

Bougrine, H., 2012: Fiscal austerity, the Great Recession and the rise of new dictatorships. *Rev. Keynes. Econ.*, **0(1)**, 109–125, doi:10.4337/roke.2012.01.07.

Bovari, E., G. Giraud, and F. Mc Isaac, 2018: Coping With Collapse: A Stock-Flow Consistent Monetary Macrodynamics of Global Warming. *Ecol. Econ.*, **147**, 383–398, doi:10.1016/j.ecolecon.2018.01.034.

Bovenberg, A.L. and R.A. de Mooij, 1994: Environmental levies and distortionary taxation. *Am. Econ. Rev.*, **84(4)**, 1085–1089.

Bracking, S. and B. Leffel, 2021: Climate finance governance: Fit for purpose? *WIREs Clim. Change*, **12(4)**, doi:10.1002/wcc.709.

Braga, J.P., W. Semmler, and D. Grass, 2021: De-risking of green investments through a green bond market – Empirics and a dynamic model. *J. Econ. Dyn. Control*, **131**, 104201, doi:10.1016/J.JEDC.2021.104201.

Braungardt, S., J. van den Bergh, and T. Dunlop, 2019: Fossil fuel divestment and climate change: Reviewing contested arguments. *Energy Res. & Soc. Sci.*, **50**, 191–200, doi:10.1016/j.erss.2018.12.004.

Brida, J.G., E.J.S. Carrera, and V. Segarra, 2020: Clustering and regime dynamics for economic growth and income inequality. *Struct. Chang. Econ. Dyn.*, **52**, 99–108, doi:10.1016/j.strueco.2019.09.010.

Broberg, M. and E. Hovani-Bue, 2019: Disaster Risk Reduction through Risk Pooling: The Case of Hazard Risk Pooling Schemes. In: *The Cambridge Handbook of Disaster Risk Reduction and International Law* [Samuel, K.L.H., M. Aronsson-Storrier, and K.N. Bookmiller, (eds.)]. Cambridge University Press, Cambridge, UK, pp. 257–274.

Broeks, M.J. et al., 2020: A social cost-benefit analysis of meat taxation and a fruit and vegetables subsidy for a healthy and sustainable food consumption in the Netherlands. *BMC Public Health*, **20(1)**, 643, doi:10.1186/s12889-020-08590-z.

Broome, J., 2012: *Climate matters: Ethics in a warming world*. W.W. Norton & Company, New York, USA, 224 pp.

Broome, J. and D. Foley, 2016: A world climate bank. In: *Institutions for future generation* [González-Ricoy, I. and A. Gosseries, (eds.)]. Oxford University Press, Oxford, UK.

Brown, J., N. Bird, and L. Schalatek, 2010: *Climate Finance Additionality: Emerging Definitions and their* Implications. Climate Finance Policy Brief No.2, ODI and Heinrich Boll Foundation, London, UK, and Berlin, Germany., 11 pp.

Brown, J. et al., 2015: Estimating mobilized private finance for adaptation: exploring data and methods. Climate Policy Initiative and OECD. London, UK and Berlin, Germany. http://climatepolicyinitiative.org/wp-content/uploads/2015/11/Estimating-mobilized-private-finance-for-adaptation-Exploring-data-and-methods.pdf (2015) (Accessed November 1, 2021).

Brugmann, J., 2012: Financing the resilient city. *Environ. Urban.*, **24(1)**, 215–232, doi:10.1177/0956247812437130.

Brunner, S. and K. Enting, 2014: Climate finance: A transaction cost perspective on the structure of state-to-state transfers. *Glob. Environ. Chang.*, **27**, 138–143, doi:10.1016/j.gloenvcha.2014.05.005.

Brunner, S., C. Flachsland, and R. Marschinski, 2012: Credible commitment in carbon policy. *Clim. Policy*, **12(2)**, 255–271, doi:10.1080/14693062.2011.582327.

Buchner, B. et al., 2013: *The Global Landscape of Climate Finance 2013*. Climate Policy Initiative, London, UK, 54 pp. https://www.climatepolicyinitiative.org/wp-content/uploads/2013/10/The-Global-Landscape-of-Climate-Finance-2013.pdf (Accessed November 1, 2021).

Buchner, B. et al., 2019: *Global Landscape of Climate Finance 2019*. Climate Policy Initiative, London, UK, 36 pp. https://www.climatepolicyinitiative.org/wp-content/uploads/2019/11/2019-Global-Landscape-of-Climate-Finance.pdf (Accessed November 1, 2021).

Buhr, B. et al., 2018: *Climate Change and the Cost of Capital in Developing Countries*. Imperial College London, UK; SOAS University of London, UK; UN Environment, London, UK, and Geneva, Switzerland, 32 pp. https://eprints.soas.ac.uk/26038/ (Accessed November 1, 2021).

Bulow, J. and K. Rogoff, 2005: Grants versus Loans for Development Banks. *Am. Econ. Rev.*, **95(2)**, 393–397, doi:10.1257/000282805774669727.

Bulow, J., C.M. Reinhart, K. Rogoff, and C. Trebesch, 2020: The debt pandemic: new steps are needed to improve sovereign debt workouts. *IMF Finance & Development*. https://www.imf.org/external/pubs/ft/fandd/2020/09/debt-pandemic-reinhart-rogoff-bulow-trebesch.htm (Accessed November 1, 2021).

Burrow, S., 2017: Foreword: The Imperative of a Just Transition. In: *Just Transition: A Report for the OECD*, Just Transition Centre, ITUC, Brussels, Belgium, pp. 1–2.

Bushee, B.J., 2001: Do Institutional Investors Prefer Near-Term Earnings over Long-Run Value? *Contemp. Account. Res.*, **18(2)**, 207–246, doi:10.1506/J4GU-BHWH-8HME-LE0X.

C40 Cities Climate Leadership Group, 2016: C40 Cities Alliance Creditworthiness 2016. C40 Cities, London, UK, New York, NY, USA, and Rio de Janeiro, Brazil, 18 pp. http://c40-production-images.s3.amazonaws.com/good_practice_briefings/images/6_C40_GPG_Creditworthiness.original.pdf?1456788925 (Accessed November 1, 2021).

C40 Cities Climate Leadership Group, 2018: *Clean Energy Business Model Manual*. C40 Cities, London, UK, 66 pp (Accessed November 1, 2021).

Caldecott, B., 2020: Post Covid-19 stimulus and bailouts need to be compatible with the Paris Agreement. *J. Sustain. Financ. & Invest.*, **0(0)**, 1–8, doi:10.1080/20430795.2020.1809292.

Campiglio, E., 2016: Beyond carbon pricing: The role of banking and monetary policy in financing the transition to a low-carbon economy. *Ecol. Econ.*, **121**, 220–230, doi:10.1016/j.ecolecon.2015.03.020.

Campiglio, E. et al., 2018: Climate change challenges for central banks and financial regulators. *Nat. Clim. Change*, **8(6)**, 462–468, doi:10.1038/s41558-018-0175-0.

Canuto, O. and L. Liu, 2010: *Subnational Debt Finance and the Global Financial Crisis*. Economic Premise, World Bank, Washington, DC, USA, 7 pp. http://documents1.worldbank.org/curated/en/637331468159337803/pdf/545070BRI0EP130Box349420B01PUBLIC1.pdf (Accessed November 1, 2021).

Cao, X., C. Jin, and W. Ma, 2021: Motivation of Chinese commercial banks to issue green bonds: Financing costs or regulatory arbitrage? *China Econ. Rev.*, **66**, 101582, doi:10.1016/j.chieco.2020.101582.

Capgemini, 2021: *World Wealth Report 2021*. Capgemini, Paris, France, 51 pp. https://worldwealthreport.com/resources/world-wealth-report-2021/ (Accessed September 12, 2021).

Carley, S. and D.M. Konisky, 2020: The justice and equity implications of the clean energy transition. *Nat. Energy*, **5(8)**, 569–577, doi:10.1038/s41560-020-0641-6.

Carney, M., 2015: Breaking the Tragedy of the Horizon–climate change and financial stability. Speech at Lloyd's of London, UK, https://www.bankofengland.co.uk/.

Carney, M., 2019: A New Horizon. Speech at European Commission Conference, Brussels, Belgium: A global approach to sustainable finance. Bank of England, London, UK, https://www.bankofengland.co.uk/-/media/boe/files/speech/2019/a-new-horizon-speech-by-mark-carney (Accessed November 1, 2021).

Carrasco, B., 2018: ADB as a Responsible Development Partner: The India Infrastructure Finance Company Limited (IIFCL) Case Study. *The KDI-ADB Workshop on Corporate Governance and SOE Reform*, Seoul, Republic of Korea, https://events.development.asia/system/files/materials/2018/06/201806-adb-responsible-developments-partner-india-infrastructure-finance-company-limited-iifcl-case.pdf (Accessed November 1, 2021).

Carty, T. and A. Le Comte, 2018: *Climate Finance Shadow Report 2018: Assessing progress towards the $100 billion commitment*. Oxfam International, Oxford, 28 pp, https://www.oxfam.org/en/research/climate-finance-shadow-report-2016 (Accessed November 1, 2021).

Carty, T., J. Kowalzig, and A. Peterson, 2016: *Climate Finance Shadow Report: Lifting the Lid on Progress towards the $100 Billion Commitment*. Oxfam International, Oxford, UK, 24 pp. https://www.oxfam.org/sites/www.oxfam.org/files/file%7B%5C_%7Dattachments/bp-climate-finance-shadow-report-031116-en.pdf (Accessed November 1, 2021).

Carty, T., J. Kowalzig, and B. Zagema, 2020: *Climate Finance Shadow Report 2020: Assessing progress towards the $100 billion commitment*. Oxfam International, Oxford, 31 pp, https://www.oxfam.org/en/research/climate-finance-shadow-report-2020 (Accessed November 1, 2021).

Cattaneo, C., 2019: Internal and external barriers to energy efficiency: which role for policy interventions? *Energy Effic.*, **12(5)**, 1293–1311, doi:10.1007/s12053-019-09775-1.

Cavallo, E.A. and A. Powell, 2019: *2019 Latin American and Caribbean Macroeconomic Report: Building Opportunities to* Grow in a Challenging World. Inter-American Development Bank, Washington, DC, USA, doi: 10.18235/0001633.

CBI, 2015: *Scaling up green bond markets for sustainable development*. Climate Bonds Initiative (CBI), London, UK, 52 pp.

CBI, 2018: *Climate Bonds Taxonomy*. Climate Bonds Initiative (CBI), London, UK, 16 pp. https://www.climatebonds.net/files/files/CBI-Taxonomy-Sep18.pdf (Accessed November 1, 2021).

CBI, 2019a: *Green Bonds: The State of the Market 2018*. Climate Bonds Initiative (CBI), London, UK, 28 pp. https://www.climatebonds.net/resources/reports/green-bonds-state-market-2018 (Accessed November 1, 2021).

CBI, 2019b: *Green Bond Pricing in the Primary Market: January – June 2019*. Climate Bonds Initiative (CBI), London, UK, 24 pp. https://www.climatebonds.net/files/files/CBI_GB_Pricing_H1_2019_final.pdf (Accessed November 1, 2021).

CBI, 2020a: *Green Bond Treasurer Survey 2020*. Climate Bonds Initiative (CBI), London, UK, 24 pp. https://www.climatebonds.net/files/reports/climate-bonds-gb-treasurer-survey-2020-14042020final.pdf (Accessed November 1, 2021).

CBI, 2020b: *Green Bonds: The State of the Market 2019*. Climate Bonds Initiative (CBI), London, UK, 16 pp. https://www.climatebonds.net/files/reports/cbi%7B%5C_%7Dsotm%7B%5C_%7D2019%7B%5C_%7Dvol1%7B%5C_%7D04d.pdf (Accessed November 1, 2021).

CBI, 2021a: *Sustainable Debt: Global State of the Market 2020*. Climate Bonds Initiative (CBI), London, UK, 30 pp.

CBI, 2021b: *Post-issuance reporting in the green bond market*. Climate Bonds Initiative (CBI), London, UK, 58 pp. https://www.climatebonds.net/files/reports/cbi_post_issuance_2021_02f.pdf (Accessed November 1, 2021).

CBI, GCA, and EBRD, 2021: *Green Bonds for Climate Resilience. State of Play and Roadmap to Scale*. Climate Bonds Initiative (CBI), London, UK, Global Center on Adaptation (GCA), Rotterdam, The Netherlands, and the European Bank for Reconstruction and Development (EBRD), 10 pp.

CCFLA, 2015: *The State of City Climate Finance 2015*. Cities Climate Finance Leadership Alliance (CCFLA), New York, NY, USA, 68 pp. http://www.citiesclimatefinance.org/2015/12/the-state-of-city-climate-finance-2015-2/ (Accessed November 1, 2021).

CCRIF SPC, 2018: *Annual Report 2017-2018*. The Caribbean Catastrophe Risk Insurance Facility (CCRIF SPC), Grand Cayman, Cayman Islands, 106 pp.

CCRIF SPC, 2019: *The Caribbean Oceans and Aquaculture Sustainability Facility*. The Caribbean Catastrophe Risk Insurance Facility (CCRIF SPC), Grand Cayman, Cayman Islands, 13 pp. https://www.ccrif.org/sites/default/files/publications/CCRIFSPC_COAST_Brochure_July2019.pdf (Accessed November 1, 2021).

CDSB and CDP, 2018: *Ready or not: Are companies prepared for the TCFD recommendations? A geographical analysis of CDP 2017 responses*. Climate Disclosure Standards Board (CDSB) and Carbon Disclosure Project (CDP), 34 pp. https://www.cdsb.net/sites/default/files/tcfd%7B%5C_%7Dpreparedness%7B%5C_%7Dreport%7B%5C_%7Dfinal.pdf (Accessed November 1, 2021).

CDSB, and SASB, 2019: *Using SASB Standards and the CDSB Framework to Enhance Climate-Related Financial Disclosures in Mainstream Reporting TCFD Implementation Guide*. Climate Disclosure Standards Board (CDSB) and 61 pp. https://www.cdsb.net/tcfd-implementation-guidef (Accessed November 1, 2021).

Cevik, S. and J.T. Jalles, 2020: *Feeling the Heat: Climate Shocks and Credit Ratings*. IMF, Washington DC, USA, 22 pp. https://www.imf.org/en/Publications/WP/Issues/2020/12/18/Feeling-the-Heat-Climate-Shocks-and-Credit-Ratings-49945.

Chan, G., L. Forsberg, P. Garnaas-Halvorson, S. Holte, and D. Kim, 2018: *Issue Linkage in the Climate Regime: Gender policies in climate finance*. Center for Science, Technology, and Environmental Policy, Minneapolis, MN, USA, 49 pp. https://conservancy.umn.edu/bitstream/handle/11299/201625/Gender_in_Climate_Finance_Review_20180913_FULL.pdf?sequence=1&isAllowed=y (Accessed November 1, 2021).

Chancel, L. and T. Piketty, 2015: *Carbon inequality: from Kyoto to Paris - Trends in the global inequality of carbon emissions (1998-2013) & prospects for an equitable adaptation fund*. Paris School of Economics, Paris, France, 48 pp.

Chang, C.C., E. Fernandez-Arias, and L. Serven, 1998: Measuring Aid Flows: A New Approach. *SSRN Electron. J.*, **1817185**, doi:10.2139/ssrn.1817185.

Chasin, F., U. Paukstadt, T. Gollhardt, and J. Becker, 2020: Smart energy driven business model innovation: An analysis of existing business models and implications for business model change in the energy sector. *J. Clean. Prod.*, **269**, 122083, doi:10.1016/j.jclepro.2020.122083.

Chen, D. et al., 2020: *Bridging the gaps and mitigating the risks: Tackling the challengesofsustainablecross-borderenergyinfrastructurefinance*.G20Insights, 14 pp. https://www.g20-insights.org/policy%7B%5C_%7Dbriefs/bridging-the-gaps-and-mitigating-the-risks-tackling-the-challenges-of-sustainable-cross-border-energy-infrastructure-finance/ (Accessed November 1, 2021).

Chenet, H., J. Ryan-Collins, and F. van Lerven, 2021: Finance, climate-change and radical uncertainty: Towards a precautionary approach to financial policy. *Ecol. Econ.*, **183**, 106957, doi:10.1016/j.ecolecon.2021.106957.

Cherif, R. and F. Hasanov, 2018: The volatility trap: Precautionary saving, investment, and aggregate risk. *Int. J. Financ. & Econ.*, **23(2)**, 174–185, doi:10.1002/ijfe.1610.

Chirambo, D., 2016: Integrating Microfinance, Climate Finance and Climate Change Adaptation: A Sub-Saharan Africa Perspective. In: *Climate Change Adaptation, Resilience and Hazards* [Leal Filho, W., H. Musa, G. Cavan, P. O'Hare, and J. Seixas, (eds.)]. Climate Change Management series, Springer International Publishing, Cham, Switzerland, pp. 195–207.

Chiroleu-Assouline, M. and M. Fodha, 2014: From regressive pollution taxes to progressive environmental tax reforms. *Eur. Econ. Rev.*, **69**, 126–142, doi:10.1016/j.euroecorev.2013.12.006.

Christophers, B., 2017: Climate Change and Financial Instability: Risk Disclosure and the Problematics of Neoliberal Governance. *Ann. Am. Assoc. Geogr.*, **107(5)**, 1108–1127, doi:10.1080/24694452.2017.1293502.

CICERO, 2015: Shades of Green. https://www.cicero.oslo.no/en/posts/single/cicero-shades-of-green (Accessed August 8, 2019).

CIF, 2018: *Microfinance for Climate Adaptation: From Readiness to Resilience*. Climate Investment Funds (CIF), Washington DC, USA, 37 pp. https://www.climateinvestmentfunds.org/sites/cif_enc/files/knowledge-documents/micro-finance_research_brief.pdf.

CISL and UNEP FI, 2014: *Stability and sustainability in banking reform: Are environmental risks missing in Basel III?* Cambridge Institute for Sustainability Leadership (CISL), Cambridge, UK, and United Nations Environment Programme Finance Initiative (UNEP FI), Geneva, Switzerland, 39 pp. https://www.unepfi.org/fileadmin/documents/StabilitySustainability.pdf (Accessed November 1, 2021).

Clapp, C., J. Ellis, J. Benn, and J. Corfee-Morlot, 2012: *Tracking Climate Finance: What and How?*, OECD, Paris, 44 pp. https://www.oecd.org/env/cc/50293494.pdf (Accessed November 1, 2021).

Clark, C.E., 2018: *Energy Savings Performance Contracts (ESPCs) and Utility Energy Service Contracts (UESCs)*. LIBRARY OF CONGRESS WASHINGTON DC, Washington DC, USA, 16 pp. https://fas.org/sgp/crs/misc/R45411.pdf (Accessed November 1, 2021).Clark, R., J. Reed, and T. Sunderland, 2018: Bridging funding gaps for climate and sustainable development: Pitfalls, progress and potential of private finance. *Land use policy*, **71**, 335–346, doi:10.1016/j.landusepol.2017.12.013.

Clarke, L., K. Jiang, K. Akimoto, M. Babiker, G. Blanford, K. Fisher-Vanden, J.-C. Hourcade, V. Krey, E. Kriegler, A. Löschel, D. McCollum, S. Paltsev, S. Rose, P.R. Shukla, M. Tavoni, B.C.C. van der Zwaan, and D.P. van Vuuren, 2014: Assessing Transformation Pathways. In: *Climate Change 2014: Mitigation of Climate Change. Contribution of Working Group III to the Fifth Assessment Report of the Intergovernmental Panel on Climate Change* [Edenhofer, O., R. Pichs-Madruga, Y. Sokona, E. Farahani, S. Kadner, K. Seyboth, A. Adler, I. Baum, S. Brunner, P. Eickemeier, B. Kriemann, J. Savolainen, S. Schlömer, C. von Stechow, T. Zwickel and J.C. Minx (eds.)]. Cambridge University Press, Cambridge, UK, and New York, NY, USA, pp. 413–510.

Cleary, K. and K. Palmer, 2019: *Energy-as-a-Service: A Business Model for Expanding Deployment of Low-Carbon Technologies*. Resources for the Future, Washington DC, USA, 6 pp. https://media.rff.org/documents/IB_19-09_EaaS.pdf (Accessed November 1, 2021).

Climate-Related Market Risk Subcommittee, 2020: *Managing Climate Risk in the U.S. Financial System*. USCFTC, Washington DC, USA, 165 pp. https://www.cftc.gov/sites/default/files/2020-09/9-9-20 Report of the Subcommittee on Climate-Related Market Risk - Managing Climate Risk in the U.S. Financial System for posting.pdf (Accessed November 1, 2021).

Climate Funds Update, 2014: Climate Funds Update: Data Dashboard. https://climatefundsupdate.org/data-dashboard/ (Accessed October 10, 2019).

Climate Transparency, 2020: *The Climate Transparency Report 2020,* Climate Transparency, Berlin, Germany. 69 pp.

Coady, D., I. Parry, L. Sears, and B. Shang, 2017: How Large Are Global Fossil Fuel Subsidies? *World Dev.*, **91**, 11–27, doi:10.1016/j.worlddev.2016.10.004.

Coady, D., I. Parry, N.-P. Le, and B. Shang, 2019: *Global Fossil Fuel Subsidies Remain Large: An Update Based on Country-Level Estimates*. IMF, Washington DC, USA, 39 pp. https://www.imf.org/en/Publications/WP/Issues/2019/05/02/Global-Fossil-Fuel-Subsidies-Remain-Large-An-Update-Based-on-Country-Level-Estimates-46509 (Accessed August 18, 2021).

Coalition of Finance Ministers for Climate Action, 2020: *Better Recovery, Better World: Resetting Climate Action in the Aftermath of the COVID-19 Pandemic,* Coalition of Finance Ministers for Climate Action, Washington, DC, USA.

Cochran, I. and A. Pauthier, 2019: *A Framework for Alignment with the Paris Agreement: Why, What and How for Financial Insitutions? Institute of Climate Economics*, Paris, France 51 pp.

Coeurdacier, N. and H. Rey, 2013: Home Bias in Open Economy Financial Macroeconomics. *J. Econ. Lit.*, **51(1)**, 63–115, doi:10.1257/jel.51.1.63.

Coffin, M. and A. Grant, 2019: *Balancing the Budget: Why deflating the carbon bubble requires oil and gas companies to shrink*. Carbon Tracker Initiative, London, UK, 55 pp. https://www.carbontracker.org/reports/balancing-the-budget/ (Accessed November 1, 2021).

Cohen-Shacham, E., G. Walters, C. Janzen, and S. Maginnis (eds.), 2016: *Nature-based solutions to address global societal challenges*. *IUCN: Gland, Switzerland, 97*, pp.2016–2036.

Colenbrander, S., D. Dodman, and D. Mitlin, 2018a: Using climate finance to advance climate justice: the politics and practice of channelling resources to the local level. *Clim. Policy*, **18(7)**, 902–915, doi:10.1080/14693062.2017.1388212.

Colenbrander, S., M. Lindfield, J. Lufkin, and N. Quijano, 2018b: *Financing Low Carbon, Climate-Resilient Cities. Coalition for Urban Transitions. London and Washington, DC. Coalition for Urban Transitions C40 Climate Leadership Group, WRI Ross Center for Sustainable Cities*, London, UK, and Washington DC, USA, 44 pp. https://citiesipcc.org/wp-content/uploads/2018/03/IPCC-Background-Paper-Financing-Low-Carbon-Climate-Resilient-Cities.pdf (Accessed November 1, 2021).

Collins, S. et al., 2016: *Integrating short term variations of the power system into integrated energy system models: A methodological review. Renewable and Sustainable Energy Reviews*, *76*, pp.839–856. 35 pp.

Combet, E., F. Ghersi, J.-C. Hourcade, and C. Thubin, 2010: La fiscalité carbone au risque des enjeux d'équité. *Rev. française d'économie*, **25(2)**, 59–91, doi:10.3406/rfeco.2010.1805.

Commonwealth Secretariat, 2016: *Climate* Risk Management: Opportunities and Challenges for Risk Pooling. The Commonwealth Small States Digest, 28 pp. doi: 10.14217/5jln4p72226b-en.

Consolandi, C., R.G. Eccles, and G. Gabbi, 2020: How material is a material issue? Stock returns and the financial relevance and financial intensity of ESG materiality. *J. Sustain. Financ. Invest.*, 1–24, doi:10.1080/20430795.2020.1824889.

Convergence, 2019: The use of guarantees in blended finance. *Convergence*, Toronto, Canada, July 30.

Convergence, 2020: *The State of Blended Finance 2020*. Convergence, Toronto, Canada, 67 pp.

Cook, M.J. and E.K. Chu, 2018: Between Policies, Programs, and Projects: How Local Actors Steer Domestic Urban Climate Adaptation Finance in India. In: *Climate Change in Cities. The Urban Book Series* [Hughes, S., E. Chu, and S. Mason, (eds.)]. Springer, Cham, Switzerland, pp. 255–277.

Copland, S., 2020: Anti-politics and Global Climate Inaction: The Case of the Australian Carbon Tax. *Crit. Sociol.*, **46(4–5)**, 623–641, doi:10.1177/0896920519870230.

Cornell, B., 2020: ESG Investing: Conceptual Issues. *J. Wealth Manag.*, **23(3)**, 61–69, doi:10.3905/jwm.2020.1.117.

Covington, C., 2020: Loan Guarantees. Confluence Philanthropy. https://www.confluencephilanthropy.org/Loan-Guarantees (Accessed October 14, 2021).

CPI, 2020: *Examining the Climate Finance Gap for Small-Scale Agriculture*. Climate Policy Initiative (CPI), https://www.climatepolicyinitiative.org/wp-content/uploads/2020/11/Examining-the-Climate-Finance-Gap-in-Small-Scale-Agriculture.pdf (Accessed November 1, 2021).

Credit Suisse, 2020: *Global Wealth Report 2020*. Credit Suisse, Zurich, Switzerland, 55 pp.

Crick, F., S.M. Eskander, S. Fankhauser, and M. Diop, 2018: How do African SMEs respond to climate risks? Evidence from Kenya and Senegal. *World Dev.*, **108**, 157–168.

Crowe, J.A. and R. Li, 2020: Is the just transition socially accepted? Energy history, place, and support for coal and solar in Illinois, Texas, and Vermont. *Energy Res. Soc. Sci.*, **59**, 101309, doi:10.1016/j.erss.2019.101309.

Cunningham, W. and A. Schmillen, 2021: *The Coal Transition: Mitigating Social and Labor Impacts*. World Bank, Washington DC, USA, 42 pp. https://openknowledge.worldbank.org/bitstream/handle/10986/35617/The-Coal-Transition-Mitigating-Social-and-Labor-Impacts.pdf?sequence=1&isAllowed=y (Accessed November 1, 2021).

Cuntz, C., A. Afanador, N. Klein, F. Barrera, and R. Sharma, 2017: *Connecting multilateral climate finance to mitigation projects A guide to the multilateral climate finance landscape of NAMAs Mountain Ecosystem Services View project NBS and Sustainable Urbanization View project*. Ecofys, Utrecht, Netherlands, 48 pp. https://www.researchgate.net/publication/320197254 (Accessed November 1, 2021).

Curtis, M, 2016: *The New Colonialism – Britain's Scramble for Africa's energy and mineral resources*. War on Want, London, 37 pp. https://waronwant.org/resources/new-colonialism-britains-scramble-africas-energy-and-mineral-resources (Accessed November 1, 2021).

Cziesielski, M.J. et al., 2021: Investing in blue natural capital to secure a future for the Red Sea ecosystems. *Front. Mar. Sci.*, **7**(January), doi:10.3389/fmars.2020.603722.

D'Orazio, P., 2021: Towards a post-pandemic policy framework to manage climate-related financial risks and resilience. *Clim. Policy*, **0(0)**, 1–15, doi:10.1080/14693062.2021.1975623.

D'Orazio, P. and L. Popoyan, 2019: Fostering green investments and tackling climate-related financial risks: Which role for macroprudential policies? *Ecol. Econ.*, **160**(July 2018), 25–37, doi:10.1016/j.ecolecon.2019.01.029.

Dafermos, Y., M. Nikolaidi, and G. Galanis, 2018: Climate Change, Financial Stability and Monetary Policy. *Ecol. Econ.*, **152**, 219–234, doi:10.1016/j.ecolecon.2018.05.011.

Dafermos, Y., D. Gabor, and J. Michell, 2021: The Wall Street Consensus in pandemic times: what does it mean for climate-aligned development? *Can. J. Dev. Stud.*, **42(1–2)**, 238–251, doi:10.1080/02255189.2020.1865137.

Dalla Longa, F. and B. van der Zwaan, 2021: Heart of light: an assessment of enhanced electricity access in Africa. *Renew. Sustain. Energy Rev.*, **136**, 110399, doi:10.1016/j.rser.2020.110399.

DANIDA, 2018: *Review of the Climate Technology Centre and Network (CTCN)*. Danish International Development Agency (DANIDA), Copenhagen, Denmark, 46 pp.

Dasgupta, A. and D. Ziblatt, 2015: How Did Britain Democratize? Views from the Sovereign Bond Market. *J. Econ. Hist.*, **75(1)**, 1–29, doi:10.1017/S0022050715000017.

Dasgupta, A. and D.F. Ziblatt, 2016: Capital Meets Democracy: Representative Institutions and the Rise of Mass Suffrage in Sovereign Bond Markets. *SSRN Electron. J.*, **2768848**, doi:10.2139/ssrn.2768848.

Dasgupta, D., J.C. Hourcade, and S. Nafo, 2019: A Climate Finance Initiative To Achieve the Paris Agreement and Strenghten Sustainable Development. HAL, Lyon, France. https://hal.archives-ouvertes.fr/hal-02121231 (Accessed November 1, 2021).

Dasgupta, P., 2021: *The Economics of Biodiversity: The Dasgupta Review*. HM Treasury, London, UK, 602 pp. https://www.gov.uk/government/collections/the-economics-of-biodiversity-the-dasgupta-review (Accessed November 1, 2021).

Davidovic, S., M.E. Loukoianova, C. Sullivan, and H. Tourpe, 2019: *Strategy for Fintech Applications in the Pacific Island Countries*. International Monetary Fund, Washington, DC, USA, 66 pp. https://www.imf.org/%7B~%7D/media/Files/Publications/DP/2019/English/sfapicea.ashx.

Davies, J.B., R. Lluberas, and A.F. Shorrocks, 2016: *Estimating the level and distribution of global wealth, 2000–14*. United Nations University World Institute for Development Economics Research (UNU-WIDER), Helsinki, Finland, 23 pp.

de Gouvello, C. and I. Zelenko, 2010: *A Financing Facility for Low-Carbon Development in Developing Countries*. The World Bank, Washington DC, USA, 56 pp.

de Guindos, L., 2021: Shining a light on climate risks: the ECB's economy-wide climate stress test. European Central Bank Blog, March 18.

de Mooij, R.A., 2000: A survey of the double-dividend literature. In: *Environmental Taxation and the Double Dividend* [De Mooij, R.A., (ed.)]. Emerald Group Publishing Limited, Bingley, UK, pp. 11–28.

De Nederlandsche Bank, and PBL Netherlands Environmental Assessment Agency, 2020: Indebted to nature: Exploring biodiversity risks for the Dutch financial sector. Amsterdam, Netherlands and The Hague, Netherlands, https://www.dnb.nl/media/4c3fqawd/indebted-to-nature.pdf (Accessed November 1, 2021).

DeAngelis, K. and B. Tucker, 2020: Adding Fuel to the Fire: Export Credit Agencies and Fossil Fuel Finance. Oil Change International and Friends of the Earth — U.S. Washington, DC, USA. 15 pp. https://priceofoil.org/2020/01/30/g20-ecas-2020/ (Accessed November 1, 2021).

Déau, T. and J. Touati, 2018: Financing Sustainable Infrastructure. In: *Coping with the Climate Crisis* [Arezki, R., P. Bolton, K. El Aynaoui, and M. Obstfeld, (eds.)]. Columbia University Press, New York, NY, USA, pp. 167–178.

Delanoë, E., A. Gautier, and A.-C. Pache, 2021: What can philanthropy do for the climate? Strategic pathways for climate giving. *Alliance*, January 19.

Deleidi, M., M. Mazzucato, and G. Semieniuk, 2020: Neither crowding in nor out: Public direct investment mobilising private investment into renewable electricity projects. *Energy Policy*, **140** (February), 111195, doi:10.1016/j.enpol.2019.111195.

Dell'Anna, F., 2021: Green jobs and energy efficiency as strategies for economic growth and the reduction of environmental impacts. *Energy Policy*, **149**, 112031, doi:10.1016/j.enpol.2020.112031.

Demirguc-Kunt, A., L. Klapper, D. Singer, S. Ansar, and J. Hess, 2018: *The Global Findex Database 2017: Measuring Financial Inclusion and the Fintech revolution*. World Bank Publications, Washington, DC, USA, http://hdl.handle.net/10986/29510 (Accessed November 1, 2021).

Deschryver, P. and F. de Mariz, 2020: What Future for the Green Bond Market? How Can Policymakers, Companies, and Investors Unlock the Potential of the Green Bond Market? *J. Risk Financ. Manag.*, **13(3)**, 61, doi:10.3390/jrfm13030061.

Deutsche Bank Climate Change Advisors, 2011: *Get FiT Plus: De-Risking Clean Energy Business Models in a Developing Country Context*. Deutsche Bank Group, Frankfurt, Germany. https://www.osti.gov/etdeweb/servlets/purl/22090459 (Accessed November 1, 2021).

Deutsche Klimafinanzierung, 2020: InsuResilience Initiative and Global Partnership. Deutsche Klimafinanzierung. https://www.germanclimatefinance.de/overview-climate-finance/channels-german-climate-finance/insuresilience/ (Accessed November 1, 2021).

DFID, 2020: *International Development Infrastructure Commission: Recommendations Report*. Department for International Development (DFID), London, UK, https://www.gov.uk/government/publications/international-development-infrastructure-commission-report (Accessed November 1, 2021).

Dibley, A., T. Wetzer, and C. Hepburn, 2021: National COVID debts: climate change imperils countries' ability to repay. *Nature*, **592(7853)**, 184–187, doi:10.1038/d41586-021-00871-w.

Dietz, S., A. Bowen, C. Dixon, and P. Gradwell, 2016: 'Climate value at risk' of global financial assets. *Nat. Clim. Chang.*, **6(7)**, 676–679, doi:10.1038/nclimate2972.

Dixon, R. and E. Challies, 2015: Making REDD+ pay: Shifting rationales and tactics of private finance and the governance of avoided deforestation in Indonesia. *Asia Pac. Viewp.*, **56(1)**, 6–20, doi:10.1111/apv.12085.

DNB, 2017: *Waterproof?: An exploration of climate-related risks for the Dutch financial sector*. De Nederlandsche Bank (DNB), Amsterdam, The Netherlands, 64 pp. https://www.dnb.nl/en/binaries/Waterproof%7B%5C_%7Dtcm47-363851.pdf (Accessed November 1, 2021).

Donadelli, M., M. Jüppner, A. Paradiso, and C. Schlag, 2019: Temperature Volatility Risk. *SSRN Electron. J.*, **3333915**, doi:10.2139/ssrn.3333915.

Donald, K. and S.-A. Way, 2016: Accountability for the Sustainable Development Goals: A Lost Opportunity? *Ethics Int. Aff.*, **30(2)**, 201–213, doi:10.1017/S0892679416000083.

Donaldson, C. and S. Hawkes, 2018: *Open Books: How development finance institutions can be transparent in their financial intermediary lending, and why they should be*. Oxfam GB, Oxford, UK, 37 pp.

Dordmond, G., H.C. de Oliveira, I.R. Silva, and J. Swart, 2021: The complexity of green job creation: An analysis of green job development in Brazil. *Environ. Dev. Sustain.*, **23(1)**, 723–746, doi:10.1007/s10668-020-00605-4.

Dorfleitner, G., L. Hornuf, M. Schmitt, and M. Weber, 2017: Definition of FinTech and Description of the FinTech Industry. In: *FinTech in Germany* [Dorfleitner, G., L. Hornuf, M. Schmitt, and M. Weber (eds.)]. Springer International Publishing, Cham, Switzerland, pp. 5–10.

Dorfleitner, G., S. Utz, and M. Wimmer, 2018: Patience pays off – corporate social responsibility and long-term stock returns. *J. Sustain. Financ. Invest.*, **8(2)**, 132–157, doi:10.1080/20430795.2017.1403272.

Dosi, G., M. Napoletano, A. Roventini, and T. Treibich, 2017: Micro and macro policies in the Keynes+Schumpeter evolutionary models. *J. Evol. Econ.*, **27(1)**, 63–90, doi:10.1007/s00191-016-0466-4.

Dowla, A., 2018: Climate change and microfinance. *Bus. Strateg. & Dev.*, **1(2)**, 78–87, doi:10.1002/bsd2.13.

Draper, N., 2007: "Possessing Slaves": Ownership, Compensation and Metropolitan Society in Britain at the time of Emancipation 1834-40. *Hist. Work. J.*, **64(1)**, 74–102, doi:10.1093/hwj/dbm030.

Drempetic, S., C. Klein, and B. Zwergel, 2020: The Influence of Firm Size on the ESG Score: Corporate Sustainability Ratings Under Review. *J. Bus. Ethics*, **167(2)**, 333–360, doi:10.1007/s10551-019-04164-1.

Druce, L., U. Moslener, C. Gruening, P. Pauw, and R. Connell, 2016: *Demystifying adaptation finance for the private sector*. UNEPNairobi, 87 pp. https://www.unepfi.org/publications/climate-change-publications/demystifying-adaptation-finance-for-private-sector/ (Accessed. November 1, 2021).

Drumheller, C., S. Denison, R. Ebel, and M. Martin, 2020: De-risking Renewable Energy Financing. *First EAGE Workshop on Geothermal Energy and Hydro Power in Africa*, Vol. 2020. European Association of Geoscientists & Engineers, Houten, Netherlands, pp. 1–5.

Dunz, N., A. Naqvi, and I. Monasterolo, 2021: Climate sentiments, transition risk, and financial stability in a stock-flow consistent model. *J. Financ. Stab.*, **54**, 100872, doi:10.1016/j.jfs.2021.100872.

Dupre, S., T. Posey, T. Wang, and T. Jamison, 2018: *Shooting for the Moon in a Hot Air Ballon?* 2° Investing Initiative Discussion Paper May 2018, 2° Investing Initiative, Paris, France, 25 pp.

Duyck, S., S. Jodoin, and A. Johl, (eds.), 2018: *Routledge Handbook of Human Rights and Climate Governance*. Routledge, Abingdon, UK, 52 pp.

Eastin, J., 2018: Climate change and gender equality in developing states. *World Dev.*, **107**, 289–305, doi:10.1016/j.worlddev.2018.02.021.

EBA, 2020: *Risk assessment of the European banking system*. Publications Office of the European Union, Luxembourg, 94 pp. https://eba.europa.eu/sites/default/documents/files/document%7B%5C_%7Dlibrary/Risk, Accessed November 1, 2021).

Ebers Broughel, A. and N. Hampl, 2018: Community financing of renewable energy projects in Austria and Switzerland: Profiles of potential investors. *Energy Policy*, **123** (August), 722–736, doi:10.1016/j.enpol.2018.08.054.

EBRD, 2020: *The EBRD just transition initiative: Sharing the benefits of a green economy transition and protecting vulnerable countires, regions and people from falling behind*. EBRD, London, UK, 47 pp. https://www.ebrd.com/just-transition.

ECB, 2019: *Financial Stability Review, May 2019*. European Central Bank (ECB), Frankfurt, Germany, 156 pp. https://www.ecb.europa.eu/pub/financial-stability/fsr/html/ecb.fsr201905%7B~%7D266e856634.en.html (Accessed November 1, 2021).

EEA, 2019: *Economic losses from climate-related extremes in Europe*. European Environment Agency, Copenhagen, Denmark, 30 pp. https://www.eea.europa.eu/data-and-maps/indicators/direct-losses-from-weather-disasters-3/assessment-2#:, (Accessed November 1, 2021).

EEAC, 2016: International Scan 2016: Emerging Issues in an International Context. EEAC and Rli, The Hague, Netherlands, https://www.rli.nl/sites/default/files/international_scan_eeac-rli_webversie_def.pdf (Accessed November 1, 2021).

Egli, F., B. Steffen, and T.S. Schmidt, 2018: A dynamic analysis of financing conditions for renewable energy technologies. *Nat. Energy*, **3(12)**, 1084–1092, doi:10.1038/s41560-018-0277-y.

Ehara, M. et al., 2019: REDD+ engagement types preferred by Japanese private firms: The challenges and opportunities in relation to private sector participation. *For. Policy Econ.*, **106** (September 2018), 101945, doi:10.1016/j.forpol.2019.06.002.

Ehlers, T. and F. Packer, 2017: *Green bond finance and certification*, BIS Quarterly Review September 2017, BIS, Basel, Switzerland, 89–104 pp. https://www.bis.org/publ/qtrpdf/r_qt1709h.pdf (Accessed November 1, 2021).

Ehlers, T., B. Mojon, and F. Packer, 2020: *Green bonds and carbon emissions: exploring the case for a rating system at the firm-level*. BIS Quarterly Review September 2020, BIS, Basel, Switzerland. pp. 31–47. https://www.bis.org/publ/qtrpdf/r_qt2009c.pdf (Accessed November 1, 2021).

EIB, 2012: *An outline guide to Project Bonds Credit Enhancement and the Project Bond Initiative*. European Investment Bank (EIB), Luxembourg, 27pp.https://www.eib.org/attachments/documents/project_bonds_guide_en.pdf (Accessed November 1, 2021).

EIB, 2018: *European Investment Bank Financial Report 2017*. European Investment Bank (EIB), Luxembourg, 251 pp. https://www.eib.org/en/publications/financial-report-2017 (Accessed December 31, 2020).

EIB, 2019: *Investing in nature: Financing conservation and Nature-based Solutions*. European Investment Bank (EIB), Luxembourg.

Eichengreen, B., D. Park, and K. Shin, 2021: The shape of recovery: Implications of past experience for the duration of the COVID-19 recession. *J. Macroecon.*, **69**, 103330, doi:10.1016/j.jmacro.2021.103330.

EIOPA, 2018: European Insurance and Occupational Pensions Authority - Financial Stability Report December2018.EIOPA, Frankfurt, Germany.

Ekholm, T., H. Ghoddusi, V. Krey, and K. Riahi, 2013: The effect of financial constraints on energy-climate scenarios. *Energy Policy*, **59**, 562–572, doi:10.1016/j.enpol.2013.04.001.

Elkhishin, S. and M. Mohieldin, 2021: External debt vulnerability in emerging markets and developing economies during the COVID-19 shock. *Rev. Econ. Polit. Sci.*, **6(1)**, 24–47, doi:10.1108/REPS-10-2020-0155.

Emin, G., M. Lepetit, A. Grandjean, and O. Ortega, 2014: *Massive financing of the energy transition*. SFTE feasibility study: synthesis report. Energy renovation of public buildings. Association for the Financing of the Energy Transition, A.F.T.E.R, Paris, France, 37 pp.

Enbala, 2018: *VIRTUAL POWER PLANTS: Coming Soon to a Grid Near You*. Microgrid Knowledge, Newbury, NH, USA. https://microgridknowledge.com/virtual-power-plants-grid/ (Accessed November 1, 2021).

Environmental Defense Fund, 2019: *The power of markets to increase ambition*. Environmental Defense Fund, New York, USA. https://www.edf.org/sites/default/files/documents/Power_of_markets_to_increase_ambition.pdf (Accessed November 1, 2021).

ESMA, 2020: *Consultation Paper: Draft advice to European Commission under Article 8 of the Taxonomy Regulation*. European Securities and Markets Authority (ESMA), Paris, France, 102 pp. https://www.esma.europa.eu/press-news/consultations/consultation-paper-draft-advice-ec-under-article-8-taxonomy-regulation (Accessed November 1, 2021).

ESMAP, 2021: *The State of Access to Modern Energy Cooking Services*. Energy Sector Management Assistance Program (ESMAP), Washington DC, USA, 121 pp.

Espagne, E., 2018: Money, Finance and Climate: The Elusive Quest for a Truly Integrated Assessment Model. *Comp. Econ. Stud.*, **60(1)**, 131–143, doi:10.1057/s41294-018-0055-7.

Essers, D., H.J. Blommestein, D. Cassimon, and P.I. Flores, 2016: Local Currency Bond Market Development in Sub-Saharan Africa: A Stock-Taking Exercise and Analysis of Key Drivers. *Emerg. Mark. Financ. Trade*, **52(5)**, 1167–1194, doi:10.1080/1540496X.2015.1073987.

Essers, D., D. Cassimon, and M. Prowse, 2021: *Debt-for-climate swaps: Killing two birds with one stone? Global Environmental Change Volume 71, doi: 10.1016/j.gloenvcha.2021.102407*, Antwerp, Belgium, 4 pp.

Europe 2020 Project Bond Initiative, 2012: *Innovative Infrastructure financing: the Project Bond Initiative, EIB, London, UK*. 4 pp. https://www.eib.org/attachments/press-news-the-europe-2020-project-bond-initiative-07112012-en.pdf (Accessed November 1, 2021).

European Commission, 2020: *Impact Assessment accompanying the document Stepping up Europe's 2030 climate ambition. Investing in a climate-neutral future for the benefit of our people*. Commission Staff Working Document, European Commission, Brussels, Belgium, 141 pp.

European Commission, 2021a: Questions and Answers - Emissions Trading – Putting a Price on carbon, July 14 2021.. European Commission, Brussels, Belgium.

European Commission, 2021b: *Draft Report by Subgroup 4: Social Taxonomy*. 61 pp. Platform on Sustainable Finance, Brussels, Belgium, https://ec.europa.eu/info/sites/default/files/business_economy_euro/banking_and_finance/documents/sf-draft-report-social-taxonomy-july2021_en.pdf (Accessed November 1, 2021).

European Council, 2020: European Council conclusions, 17-21 July 2020 - Consilium. European Council, Brussels, Belgium, https://www.consilium.europa.eu/media/45109/210720-euco-final-conclusions-en.pdf (Accessed November 1, 2021).

Eurosif, 2018: *European SRI Study 2018*. The European Sustainable Investment Forum (Eurosif), Brussels, Belgium, 114 pp. https://www.eurosif.org/wp-content/uploads/2021/10/European-SRI-2018-Study.pdf (Accessed November 1, 2021).

Evain, J., M. Cardona, and M. Nicol, 2018: *Article 173: Overview of climate-related financial dislosure after two years of implementation*. Institute for Climate Economics, Paris, France, 4 pp. https://www.i4ce.org/en/publication/article-173-overview-of-climate-related-financial-dislosure-after-two-years-of-implementation/ (Accessed November 1, 2021).

Fama, E.F., 1970: Efficient Capital Markets: A Review of Theory and Empirical Work. *J. Finance*, **25(2)**, 383, doi:10.2307/2325486.

Fama, E.F., 1991: Efficient Capital Markets: II. *J. Finance*, **46(5)**, 1575, doi:10.2307/2328565.

Fama, E.F., 1997: Market Efficiency, Long-Term Returns, and Behavioral Finance. *SSRN Electron. J.*, doi:10.2139/ssrn.15108.

FAO, 2018: *REDD + finance and investments, FAO, Rome, Italy*.

Federici, S., 2014: From Commoning to Debt: Financialization, Microcredit, and the Changing Architecture of Capital Accumulation. *South Atl. Q.*, **113(2)**, 231–244, doi:10.1215/00382876-2643585.

Feindouno, S., P. Guillaumont, and C. Simonet, 2020: The Physical Vulnerability to Climate Change Index: An Index to Be Used for International Policy. *Ecol. Econ.*, **176**, 106752, doi:10.1016/J.ECOLECON.2020.106752.

Feldman, D., R. Jones-Albertus, and R. Margolis, 2018: *Impact of Research and Development, Analysis, and Standardization on PV Project Financing Costs*. National Renewable Energy Laboratory, Golden, CO, USA. https://www.nrel.gov/docs/fy18osti/70939.pdf (Accessed November 1, 2021).

Fenton, A., D. Gallagher, H. Wright, S. Huq, and C. Nyandiga, 2014: Up-scaling finance for community-based adaptation. *Clim. Dev.*, **6(4)**, 388–397, doi:10.1080/17565529.2014.953902.

Fenton, A., J. Paavola, and A. Tallontire, 2015: Microfinance and climate change adaptation: an overview of the current literature. *Enterp. Dev. Microfinance*, **26(3)**, 262–273, doi:10.3362/1755-1986.2015.023.

Fernandez, K., K. Casey, and N. Nikova, 2021: France's Overdue Debt to Haiti. *SSRN Electron. J.*, doi:10.2139/ssrn.3798841.

Finance for Biodiversity Initiative, 2021: *Greening Sovereign Debt: Building a Nature and Climate Sovereign Bond Facility*. Finance for Biodiversity Initiative, London, UK, 44 pp. https://a1be08a4-d8fb-4c22-9e4a-2b2f4cb7e41d.filesusr.com/ugd/643e85_021432a338a34c3e92237ffdd128404c.pdf (November 1, 2021).

FinansNorge, Forsikring & Pension, FFI, and Svensk Forsakring, 2013: *Weather related damage in the Nordic countries-from an insurance perspective*. FinansNorge, Forsikring & Pension, FFI, and Svensk Forsakring, Oslo, Norway, Hellerup, Denmark and Stockholm, Sweden. 42 pp. https://www.svenskforsakring.se/globalassets/rapporter/klimat/weather-related-damage-in-the-nordic-countries.pdf (Accessed November 1, 2021).

Finon, D., 2006: Incentives to invest in liberalised electricity industries in the North and South. Differences in the need for suitable institutional arrangements. *Energy Policy*, **34(5)**, 601–618, doi:10.1016/j.enpol.2005.11.012.

Finon, D., 2019: Carbon policy in developing countries: Giving priority to non-price instruments. *Energy Policy*, **132** (September 2018), 38–43, doi:10.1016/j.enpol.2019.04.046.

Fisch-Romito, V. and C. Guivarch, 2019: Transportation infrastructures in a low carbon world: An evaluation of investment needs and their determinants. *Transp. Res. Part D Transp. Environ.*, **72**, 203–219, doi:10.1016/j.trd.2019.04.014.

Flammer, C., 2020: Corporate Green Bonds. *SSRN Electron. J.*, doi:10.2139/ssrn.3125518.

Floater, G. et al., 2017a: *Financing The Urban Transition: Policymakers' Summary*. Coalition for Urban Transitions. London, UK and Washington, USA. https://newclimateeconomy.report/workingpapers/wp-content/uploads/sites/5/2017/10/NCE2017_PolicyMakers_02012018.pdf (Accessed November 1, 2021).

Floater, G. et al., 2017b: *Global Review of Finance For Sustainable Urban Infrastructure*. Coalition for Urban Transitions, London, UK, and Washington USA, 60 pp. https://newclimateeconomy.report/workingpapers/wp-content/uploads/sites/5/2018/01/NCE2017%7B%5C_%7DCUT%7B%5C_%7DGlobalReview%7B%5C_%7D02012018.pdf (Accessed November 1, 2021).

Foley, 2009: The economic fundamentals of global warming. SFI WORKING PAPER: 2007-12-044, Sante Fe Institute, Santa Fe, USA. https://citeseerx.ist.psu.edu/viewdoc/download?doi=10.1.1.79.1173&rep=rep1&type=pdf (Accessed November 1, 2021).

Folke, C. et al., 2021: Our future in the Anthropocene biosphere. *Ambio*, **50(4)**, 834–869, doi:10.1007/s13280-021-01544-8.

Font, B.C., P. Clerc, and M. Lemoine, 2018: Should euro area countries cut taxes on labour or capital in order to boost their growth? *Econ. Model.*, **71**, 279–288, doi:10.1016/j.econmod.2017.12.019.

Fonta, W.M., E.T. Ayuk, T. van Huysen, and T. van Huysen, 2018: Africa and the Green Climate Fund: current challenges and future opportunities. *Clim. Policy*, **18(9)**, 1210–1225, doi:10.1080/14693062.2018.1459447.

Forest, R., 2018: *Climate Risk Insurance: Transparency, Participation and Accountability. An overview Assessment of Regional Risk Pools.* Munich Climate Insurance Initiative, Bonn, Germany, 33 pp.

Forster, P.M. et al., 2020: Current and future global climate impacts resulting from COVID-19. *Nat. Clim. Change*, **10(10)**, 913–919, doi:10.1038/s41558-020-0883-0.

Forzieri, G. et al., 2018: Escalating impacts of climate extremes on critical infrastructures in Europe. *Glob. Environ. Change*, **48**, 97–107, doi:10.1016/j.gloenvcha.2017.11.007.

Foster, V. and A. Rana, 2020: *Rethinking Power Sector Reform in the Developing World*. World Bank, Washington DC, USA, 324 pp.

Fowler, N., 2015: Britain's Slave Owner Compensation Loan, reparations and tax havenry. *Tax Justice Network*, June 9.

Fowlie, M., M. Greenstone, and C. Wolfram, 2018: Do Energy Efficiency Investments Deliver? Evidence from the Weatherization Assistance Program*. *Q. J. Econ.*, **133(3)**, 1597–1644, doi:10.1093/qje/qjy005.

Foxon, T.J. et al., 2015: Low carbon infrastructure investment: extending business models for sustainability. *Infrastruct. Complex.*, **2(1)**, 4, doi:10.1186/s40551-015-0009-4.

Frame, D.J. et al., 2020: Climate change attribution and the economic costs of extreme weather events: a study on damages from extreme rainfall and drought. *Clim. Change*, **162(2)**, 781–797, doi:10.1007/s10584-020-02729-y.

Freeman, C. and F. Louçã, 2001: *As Time Goes by: From the Industrial Revolutions to the Information Revolution*. Oxford University Press, Oxford, UK.

Fresnillo, I., 2020a: *The G20 Debt Service Suspension Initiative: Draining out the Titanic with a bucket?* European Network on Debt and Development, Brussels, Belgium, https://dette-developpement.org/IMG/pdf/dssishadowreport.pdf (Accessed November 1, 2021).

Fresnillo, I., 2020b: *Out of service: How public services and human rights are being threatened by the growing debt crisis*. European Network on Debt and Development, Brussels, Belgium. https://www.eurodad.org/outofservice (Accessed November 1, 2021).

Friede, G., T. Busch, and A. Bassen, 2015: ESG and financial performance: aggregated evidence from more than 2000 empirical studies. *J. Sustain. Financ. Invest.*, **5(4)**, 210–233, doi:10.1080/20430795.2015.1118917.

FS-UNEP Centre and BNEF, 2015: *Global Trends in Renewable Energy Investment Report 2015*. Frankfurt School – United Nations Environment Programme (FS-UNEP) Frankfurt, Germany, and Bloomberg New Energy Finance (BNEF).

FS-UNEP Centre and BNEF, 2016: *Global Trends in Renewable Energy Investment Report 2016*. Frankfurt School – United Nations Environment Programme (FS-UNEP) Frankfurt, Germany, and Bloomberg New Energy Finance (BNEF).

FS-UNEP Centre and BNEF, 2017: *Global Trends in Renewable Energy Investment Report 2017*. Frankfurt School – United Nations Environment Programme (FS-UNEP) Frankfurt, Germany, and Bloomberg New Energy Finance (BNEF).

FS-UNEP Centre and BNEF, 2018: *Global Trends in Renewable Energy Investment Report 2018*. Frankfurt School – United Nations Environment Programme (FS-UNEP) Frankfurt, Germany, and Bloomberg New Energy Finance (BNEF), 76 pp. https://fs-unep-centre.org/fileadmin/gtr/Global%7B%5C_%7DTrends%7B%5C_%7DReport (Accessed November 1, 2021).

FS-UNEP Centre and BNEF, 2019: *Global Trends in Renewable Energy Investment Report 2019*. Frankfurt School – United Nations Environment Programme (FS-UNEP) Frankfurt, Germany, and Bloomberg New Energy Finance (BNEF). https://fs-unep-centre.org/fileadmin/user%7B%5C_%7Dupload/GTR%7B%5C_%7D2019.pdf (Accessed November 1, 2021).

FS-UNEP Centre and BNEF, 2020: *Global Trends in Renewable Energy Investment Report 2020*. Frankfurt School – United Nations Environment Programme (FS-UNEP) Frankfurt, Germany, and Bloomberg New Energy Finance (BNEF), 79 pp. https://www.fs-unep-centre.org/wp-content/uploads/2020/06/GTR_2020.pdf (Accessed November 1, 2021).

FSD Africa, and CBI, 2020: *Africa Green Bond Toolkit: A Practical Guide to Issuing Green Bonds for Africa*. FSD Africa, and CBI, Nairobi, Kenya and London, UK, 27 pp. https://www.climatebonds.net/standard/taxonomy (Accessed November 1, 2021).

Fuchs, A., K. Gehring, K.G.-J. of the E. Economic, and undefined 2017, 2017: The Home Bias in Sovereign Ratings. *J. Eur. Econ. Assoc.*, **15(6)**, 1386–1423, doi:10.1093/jeea/jvx009.

Fuller, F., L. Zamarioli, B. Kretschmer, A. Thomas, and L. De Marez, 2018: *Debt for Climate Swaps: Caribbean Outlook*. Climate Analytics, Berlin, Germany, 18 pp. https://climateanalytics.org/media/debt%7B%5C_%7Dfor%7B%5C_%7Dclimate%7B%5C_%7Dswap%7B%5C_%7Dimpact%7B%5C_%7Dbriefing.pdf (Accessed November 1, 2021).

Furceri, D., M. Ganslmeier, and J.D. Ostry, 2021: *Are climate change policies politically costly?* IMF, Washington DC, USA, 51 pp. https://www.imf.org/en/Publications/WP/Issues/2021/06/04/Are-Climate-Change-Policies-Politically-Costly-460565 (Accessed November 1, 2021).

Gabor, D., 2019: *Securitization for Sustainability: Does it help achieve the Sustainable Development Goals?* Heinrich Böll Stiftung, Washington, DC, USA, 30 pp.

Gabor, D., 2021: The Liquidity and Sustainability Facility for African sovereign bonds: who benefits? – Eurodad, Brussels, Belgium. https://www.eurodad.org/the_liquidity_and_sustainability_facility_for_african_sovereign_bonds_who_benefits (Accessed September 22, 2021).

Gaddy, B., V. Sivaram, and F. O'Sullivan, 2016: *Venture capital and cleantech: The wrong model for clean energy innovation*. MIT Energy Initiative, Cambridge, MA, USA, 22 pp. https://energy.mit.edu/wp-content/uploads/2016/07/MITEI-WP-2016-06.pdf (Accessed November 1, 2021).

Gall, J., 2018: *The benefits of community-owned renewable energy projects*. University of Saskatchewan, Saskatoon, Canada. https://renewableenergy.usask.ca/news-articles/the-benefits-of-community-owned-renewable-energy-projects.php (Accessed November 1, 2021).

Gallagher, K.P., R. Kamal, J. Jin, Y. Chen, and X. Ma, 2018: Energizing development finance? The benefits and risks of China's development finance in the global energy sector. *Energy Policy*, **122**, 313–321, doi:10.1016/j.enpol.2018.06.009.

Garbacz, W., D. Vilalta, and L. Moller, 2021: *The role of guarantees in blended finance*. OECD Development Co-operation Working Papers 97, OECD Publishing, Paris, France, 55 pp.

García de Fonseca, L., Parikh, M., Manghani, R., 2019: *Evolucion futura de costos de las energías renovables y almacenamiento en America* Latina. IDB, Washington, DC, USA. doi: 10.18235/0002101.

Gazzotti, P. et al., 2021: Persistent inequality in economically optimal climate policies. *Nat. Commun.*, **12(1)**, 3421, doi:10.1038/s41467-021-23613-y.

GCA-AAI, 2020: Integrated Responses To Building Climate And Pandemic Resilience In Africa. GCA, Rotterdam, Netherlands. https://gca.org/programs/africa-adaptation-acceleration-program/ (Accessed October 25, 2021).

GCA, 2019: Adapting to climate change could add $7 trillion to the global economy by 2030. GCA, Rotterdam, Netherlands. Global Center on Adaptation (GCA).

GCF, 2017: *Mainstreaming Gender in Green Climate Fund Projects*. Green Climate Fund (GCF), Incheon, Republic of Korea, 78 pp. https://www.greenclimate.fund/sites/default/files/document/guidelines-gcf-toolkit-mainstreaming-gender_0.pdf (Accessed November 1, 2021).

GCF, 2019: Facilitating an enabling environment for a Caribbean Green Bond Listingon the Jamaica Stock Exchange. Green Climate Fund (GCF), Incheon, Republic of Korea.

GCF, 2020a: *Independent Evaluation of the Green Climate Fund's Environmental and Social Safeguards (ESS) and the Environmental and Social Management System (ESMS)*. Green Climate Fund (GCF), Incheon, Republic of Korea, 204 pp. https://www.greenclimate.fund/sites/default/files/document/gcf-b27-13.pdf (Accessed November 1, 2021).

GCF, 2020b: *Desert to Power G5 Sahel Facility*. Green Climate Fund (GCF), Incheon, Republic of Korea, 19 pp. https://www.greenclimate.fund/sites/default/files/document/25180-afdb-desert-power.pdf (Accessed August 24, 2021).

GCF, 2021: *FP163: Sustainable Renewables Risk Mitigation Initiative (SRMI) Facility*. Green Climate Fund (GCF), Incheon, Republic of Korea, 72 pp.

Geddes, A., T.S. Schmidt, and B. Steffen, 2018: The multiple roles of state investment banks in low-carbon energy finance: An analysis of Australia, the UK and Germany. *Energy Policy*, **115** (December 2017), 158–170, doi:10.1016/j.enpol.2018.01.009.

GEF, 2021a: *Progress report on the Least Developed Countries Fund and the Special Climate Change Fund*. Global Environment Facility (GEF), Washington, DC, USA, 64 pp. https://www.thegef.org/sites/default/files/council-meeting-documents/EN_GEF.LDCF_.SCCF_.30_03_Progress Report on the LDCF and the SCCF.pdf (Accessed November 1, 2021).

GEF, 2021b: *Long-term vision on complementarity, coherence, and collaboration between the Green Climate Fund and the Global Environment Facility*. Global Environment Facility (GEF), Washington, DC, USA, 10 pp. https://www.thegef.org/sites/default/files/council-meeting-documents/EN_GEF_C.60_08_Long-Term Vision on Complementarity%2C Coherence and Collaboration between the Green Climate Fund and the Global Environment Facility.pdf.

GEF Independent Evaluation Office, 2017: *Evaluation on Gender Mainstreaming in the GEF*. Global Environment Facility (GEF), Washington, DC, USA, 82 pp. https://www.gefieo.org/evaluations/evaluation-gender-mainstreaming-gef-2017 (Accessed November 1, 2021).

Gelb, S., S. Kalantaryan, S. Mcmahon, and M. Perez Fernandez, 2021: *Diaspora finance for development: from remittances to investment*. Publications Office of the European Union, Luxembourg, 48 pp.

GEPF, 2019: *2019/2020 Annual Report*. GEPF, Pretoria, South Africa, 126 pp. https://www.gepf.co.za/wp-content/uploads/2020/11/GEPF-IR-2020.pdf (Accessed November 1, 2021).

Gertler, M. and N. Kiyotaki, 2010: Financial Intermediation and Credit Policy in Business Cycle Analysis. NYU and Princeton, New York, NY, USA and Princeton, USA. pp. 547–599.

GET FiT Uganda, 2018: *GET FiT Uganda - Annual Report 2018*. Get Fit Uganda, Kampala, Uganda. https://www.getfit-uganda.org/downloads/ (Accessed November 1, 2021).

GGGI, 2019: *Product Analysis of Diverse de-Risking Financial Instruments Available in Indonesia's Market*. 101 pp. Global Green Growth Institute, Seoul, South Korea, http://greengrowth.bappenas.go.id/wp-content/uploads/2019/12/Product-Analysis-of-Diverse-de-Risking-Financial_Report-1.pdf (Accessed August 18, 2021).

Gianfrate, G. and M. Peri, 2019: The green advantage: Exploring the convenience of issuing green bonds. *J. Clean. Prod.*, **219**, doi:10.1016/j.jclepro.2019.02.022.

Giannetti, B.F., F. Agostinho, C.M.V.B. Almeida, and D. Huisingh, 2015: A review of limitations of GDP and alternative indices to monitor human wellbeing and to manage eco-system functionality. *J. Clean. Prod.*, **87**, 11–25, doi:10.1016/j.jclepro.2014.10.051.

Gibon, T., I.Ş. Popescu, C. Hitaj, C. Petucco, and E. Benetto, 2020: Shades of green: Life cycle assessment of renewable energy projects financed through green bonds. *Environ. Res. Lett.*, **15(10)**, doi:10.1088/1748-9326/abaa0c.

Giese, G., L.-E. Lee, D. Melas, Z. Nagy, and L. Nishikawa, 2019: Foundations of ESG Investing: How ESG Affects Equity Valuation, Risk, and Performance. *J. Portf. Manag.*, **45(5)**, 69–83, doi:10.3905/jpm.2019.45.5.069.

Gignac, R. and H.D. Matthews, 2015: Allocating a 2 °C cumulative carbon budget to countries. *Environ. Res. Lett.*, **10(7)**, 75004, doi:10.1088/1748-9326/10/7/075004.

Gillingham, K.T., C.R. Knittel, J. Li, M. Ovaere, and M. Reguant, 2020: The Short-run and Long-run Effects of Covid-19 on Energy and the Environment. *Joule*, **4(7)**, 1337–1341, doi:10.1016/j.joule.2020.06.010.

Giorgi, A. and C. Michetti, 2021: *Climate Investment Opportunities: Climate-Aligned Bonds & Issuers 2020*. Climate Bonds Initiative, London, UK, 34 pp. https://www.climatebonds.net/files/reports/cbi_climate-aligned_bonds_issuers_2020.pdf (Accessed November 1, 2021).

GISD, 2020: *Renewed, Rechargedand Reinforced: Urgent actions to harmonize and scale sustainable finance*. Global Investors for Sustainable Development (GISD) Alliance, New York, NY, USA, 45 pp. https://www.un.org/development/desa/financing/sites/www.un.org.development.desa.financing/files/2020-08/Renewed%2C Recharged and Reinforced %28GISD 2020%29_vF.pdf (Accessed November 1, 2021).

GIZ, 2015: *Climate Risk Insurance: For strengthening climate resilience of poor people in vulnerable countries*. GIZ, Bonn, Germany, 20 pp.

GIZ, 2017a: *Financing strategies: A missing link to translate NDCs into action*. GIZ, Bonn, Germany.

GIZ, 2017b: *The Potential of Green Bonds. A Climate Finance Instrument for the Implementation of Nationally Determined Contributions?* GIZ, Bonn, Germany.

GIZ, and World Bank, 2019: *Central America & Caribbean Catastrophe Risk Insurance Program*. GIZ and World Bank Bonn, Germany, and Washington, DC, USA. 4 pp. https://www.preventionweb.net/files/66323_66323factsheet33wbcentralamericaweb.pdf (Accessed November 1, 2021).

GIZ, IFC, and Women's World Banking, 2017: *Mainstreaming Gender and Targeting Women in Inclusive Insurance: Perspectives and Emerging Lessons. A Compendium of Technical Notes and Case Studies*. GIZ, IFC, and Women's World Banking, Bonn and Eschborn, Germany, 96 pp. https://www.ifc.org/wps/wcm/connect/4dbd983e-2ecd-4cde-b63e-191ffb2d48e6/Full+Women+%7B%5C%25%7D26+Inclusive+Insurance+BMZ%7B%5C_%7DWeb.pdf?MOD=AJPERES%7B%5C&%7DCVID=lK1xhtq (Accessed November 1, 2021).

Global Investor Statement, 2019: *Global Investor Statement on Climate Change 2019*. IIGCC, London, UK.

Glueck, K. and L. Friedman, 2020: Biden announces $2 trillion climate plan. *The New York Times*, July 14.

15

Goldman Sachs, 2021: Goldman Sachs Commits $10 Billion in Investment Capital and $100 Million in Philanthropic Capital To Impact The Lives of One Million Black Women. Goldman Sachs Press Release, March 10, 2021.

Goldstein, B., D. Gounaridis, and J.P. Newell, 2020: The carbon footprint of household energy use in the United States. *Proc. Natl. Acad. Sci.*, **117(32)**, 19122–19130, doi:10.1073/pnas.1922205117.

Golub, A.A. et al., 2018: Escaping the climate policy uncertainty trap: options contracts for REDD+. *Clim. Policy*, **18(10)**, 1227–1234, doi:10.1080/14693062.2017.1422478.

Goodfriend, M., 2011: Money Markets. *Annu. Rev. Financ. Econ.*, **3(1)**, 119–137, doi:10.1146/annurev-financial-102710-144853.

Gore, G. and M. Berrospi, 2019: Rise of controversial transition bonds leads to call for industry standards. *Reuters*, September.

Gorelick, J., 2018: Supporting the future of municipal bonds in sub-Saharan Africa: the centrality of enabling environments and regulatory frameworks. *Environ. Urban.*, **30(1)**, 103–122, doi:10.1177/0956247817741853.

Gorelick, J. and N. Walmsley, 2020: The greening of municipal infrastructure investments: technical assistance, instruments, and city champions. *Green Financ.*, **2(2)**, 114–134, doi:10.3934/GF.2020007.

Gosens, J. and F. Jotzo, 2020: China's post-COVID-19 stimulus: No Green New Deal in sight. *Environ. Innov. Soc. Transitions*, **36**, 250–254, doi:10.1016/j.eist.2020.07.004.

Gough, I., 2020: Defining floors and ceilings: the contribution of human needs theory. *Sustain. Sci. Pract. Policy*, **16(1)**, 208–219, doi:10.1080/15487733.2020.1814033.

Goulder, L.H., 1995: Environmental taxation and the double dividend: A reader's guide. *Int. Tax Public Financ.*, **2(2)**, 157–183, doi:10.1007/BF00877495.

Government of UK, 2016: *Appendix 5.2: What is the evidence from the international experience of smart meters?* Government of the United Kingdom, London, https://assets.publishing.service.gov.uk/media/56ebdf6540f0b60385000002/Appendix_5.2_-_What_is_the_evidence_from_the_international_experience_of_smart_meters.pdf (Accessed November 1, 2021).

Grafakos, S., K. Trigg, M. Landauer, L. Chelleri, and S. Dhakal, 2019: Analytical framework to evaluate the level of integration of climate adaptation and mitigation in cities. *Clim. Change*, **154(1–2)**, 87–106, doi:10.1007/s10584-019-02394-w.

Grafström, J. and Å. Lindman, 2017: Invention, innovation and diffusion in the European wind power sector. *Technol. Forecast. Soc. Change*, **114(2017)**, 179–191, doi:10.1016/j.techfore.2016.08.008.

Granoff, I., J.R. Hogarth, and A. Miller, 2016: Nested barriers to low-carbon infrastructure investment. *Nat. Clim. Change*, **6(12)**, 1065–1071, doi:10.1038/nclimate3142.

Green, J.F., 2021: Beyond Carbon Pricing: Tax Reform is Climate Policy. *Glob. Policy*, **12(3)**, 372–379, doi:10.1111/1758-5899.12920.

Griffith-Jones, S. and J. Tyson, 2010: *Reform of the European Investment Bank: How to Upgrade the EIB's Role in Development*. European Parliament, Brussels, Belgium. 38 pp. http://www.europarl.europa.eu/RegData/etudes/etudes/join/2010/410213/EXPO-DEVE%7B%5C_%7DET(2010)410213%7B%5C_%7DEN.pdf (Accessed November 1, 2021).

Griffith-Jones, S., S. Attridge, and M. Gouett, 2020: Securing climate finance through national development banks.

Griscom, B.W. et al., 2017: Natural climate solutions. *Proceedings of the National Academy of Sciences, Washington, DC, USA*, **114(44)**, 11645–11650, doi:10.1073/pnas.1710465114.Griscom, B.W. et al., 2020: National mitigation potential from natural climate solutions in the tropics. *Philos. Trans. R. Soc. B Biol. Sci.*, **375(1794)**, 20190126, doi:10.1098/rstb.2019.0126.

Gropp, R., C. Gruendl, and A. Guettler, 2014: The Impact of Public Guarantees on Bank Risk-Taking: Evidence from a Natural Experiment*. *Rev. Financ.*, **18(2)**, 457–488, doi:10.1093/rof/rft014.

Grubb, M., 2014: *Planetary Economics: Energy, climate change and the three domains of sustainable development*. Routledge, Abingdon, UK. 548 pp.

Grubb, M. and D. Newbery, 2018: UK Electricity Market Reform and the Energy Transition: Emerging Lessons. *Energy J.*, **39(01)**, doi:10.5547/01956574.39.6.mgru.

Grubler, A., C. Wilson, and G. Nemet, 2016: Apples, oranges, and consistent comparisons of the temporal dynamics of energy transitions. *Energy Res. & Soc. Sci.*, **22**, 18–25, doi:10.1016/j.erss.2016.08.015.

GSIA, 2019: *2018 Global Sustainable Investment Review*. Global Sustainable Investment Alliance (GSIA), 29 pp. http://www.gsi-alliance.org/wp-content/uploads/2019/03/GSIR%7B%5C_%7DReview2018.3.28.pdf (Accessed November 1, 2021).

Gupta, J., A. Rempel, and H. Verrest, 2020: Access and allocation: the role of large shareholders and investors in leaving fossil fuels underground. *Int. Environ. Agreements Polit. Law Econ.*, **20(2)**, 303–322, doi:10.1007/s10784-020-09478-4.

Guthrie, J., 2021: Lex in depth: Examining the slave trade — 'Britain has a debt to repay.' *Financial Times*, June 28.

Guzman, M., 2016: *Definitional Issues in the IMF Debt Sustainablility Analysis Framework - A Proposal*. CIGI Policy Brief No 77, Waterloo, ON, Canada. 8 pp. https://www.cigionline.org/sites/default/files/pb%7B%5C_%7Dno.77%7B%5C_%7Dweb.pdf (Accessed November 1, 2021).

Haas, C. and K. Kempa, 2020: Low-Carbon Investment and Credit Rationing. *SSRN Electron. J.*, doi:10.2139/ssrn.3521332.

Hachenberg, B. and D. Schiereck, 2018: Are green bonds priced differently from conventional bonds? *J. Asset Manag.*, **19(6)**, doi:10.1057/s41260-018-0088-5.

Hadfield, P. and N. Cook, 2019: Financing the Low-Carbon City: Can Local Government Leverage Public Finance to Facilitate Equitable Decarbonisation? *Urban Policy Res.*, **37(1)**, 13–29, doi:10.1080/08111146.2017.1421532.

Hafner, S., A. Jones, A. Anger-Kraavi, and J. Pohl, 2020: Closing the green finance gap – A systems perspective. *Environ. Innov. Soc. Transitions*, **34**, 26–60, doi:10.1016/j.eist.2019.11.007.

Hainaut, H. and I. Cochran, 2018: The Landscape of domestic climate investment and finance flows: Methodological lessons from five years of application in France. *Int. Econ.*, **155**, 69–83, doi:10.1016/j.inteco.2018.06.002.

Hall, C., 2020: Opinion: The slavery business contributed to the building of modern Britain. Can we make amends? *UCL News*, June 24.

Hall, N. and Å. Persson, 2018: Global climate adaptation governance: Why is it not legally binding? *Eur. J. Int. Relations*, **24(3)**, 540–566, doi:10.1177/1354066117725157.

Halland, H., A. Dixon, S.Y. In, A. Monk, and R. Sharma, 2021: Mobilising institutional investor capital for climate-aligned development. OECD Development Policy Papers 35, OECD Publishing., Paris, 33 pp.

Hallegatte, S. et al., 2016: *Shock Waves: Managing the Impacts of Climate Change on Poverty*. World Bank, Washington DC, USA, 207 pp.

Hallegatte, S., A. Vogt-Schilb, M. Bangalore, and J. Rozenberg, 2017: *Unbreakable: Building the Resilience of the Poor in the Face of Natural Disasters*. The World Bank, Washington, DC, USA, 201 pp.

Hamid, K., M.T. Suleman, S.Z. Ali Shah, and R.S. Imdad Akash, 2017: Testing the Weak Form of Efficient Market Hypothesis: Empirical Evidence from Asia-Pacific Markets. *SSRN Electron. J.*, doi:10.2139/ssrn.2912908.

Hamwi, M. and I. Lizarralde, 2017: A Review of Business Models towards Service-Oriented Electricity Systems. *Procedia CIRP*, **64**, 109–114, doi:10.1016/j.procir.2017.03.032.

Hanif, I., S.M. Faraz Raza, P. Gago-de-Santos, and Q. Abbas, 2019: Fossil fuels, foreign direct investment, and economic growth have triggered CO2 emissions in emerging Asian economies: Some empirical evidence. *Energy*, **171**, 493–501, doi:10.1016/j.energy.2019.01.011.

Hanlon, J., 2017: Following the donor-designed path to Mozambique's US$2.2 billion secret debt deal. *Third World Q.*, **38(3)**, 753–770, doi:10.1080/01436597.2016.1241140.

Hanna, R., Y. Xu, and D.G. Victor, 2020: After COVID-19, green investment must deliver jobs to get political traction. *Nature*, **582(7811)**, 178–180, doi:10.1038/d41586-020-01682-1.

Hanto, J. et al., 2021: Effects of decarbonization on the energy system and related employment effects in South Africa. *Environ. Sci. Policy*, **124**, 73–84, doi:10.1016/j.envsci.2021.06.001.

Hanusch, M., S. Hassan, Y. Algu, L. Soobyah, and A. Kranz, 2016: *The Ghost of a Rating Downgrade*. World Bank, Washington, DC, USA.

Harber, M., 2017: The role of institutional investors in promoting long-term value creation: A South African perspective. *African Rev. Econ. Financ.*, **9(1)**, 272–291.

Hearson, M., 2014: Tax-motivated Illicit Financial Flows. A Guide for Development Practitioners. *U4 Issue*, **2**, 72.

Hellmuth, M.E., D.E. Osgood, U. Hess, A. Moorhead, and H. Bhojwani, 2009: *Index insurance and climate risk*. International Research Institute for Climate and Society, 122 pp.

Henochsberg, S., 2016: Public debt and slavery: the case of Haiti (1760-1915). Paris School of Economics, Paris, France.

Henriques, I. and P. Sadorsky, 2018: Investor implications of divesting from fossil fuels. *Glob. Financ. J.*, **38**, 30–44, doi:10.1016/j.gfj.2017.10.004.

Henry, M.S., M.D. Bazilian, and C. Markuson, 2020: Just transitions: Histories and futures in a post-COVID world. *Energy Res. Soc. Sci.*, **68**, 101668, doi:10.1016/j.erss.2020.101668.

Hepburn, C., B. O'Callaghan, N. Stern, J. Stiglitz, and D. Zenghelis, 2020a: Will COVID-19 fiscal recovery packages accelerate or retard progress on climate change? *Oxford Rev. Econ. Policy*, **36(20)**, 1–48, doi:10.1093/oxrep/graa015.

Hepburn, C., O'Callaghan, B., Stern, N., Stiglitz, J. and Zenghelis, D., 2020b: Will COVID-19 fiscal recovery packages accelerate or retard progress on climate change? Smith School Working Paper 20-02. https://www.inet.ox.ac.uk/publications/will-covid-19-fiscal-recovery-packages-accelerate-or-retard-progress-on-climate-change/ (Accessed November 1, 2021).

Heredia, L. et al., 2020: *Global Asset Management Report: Protect, Adapt and Innovate*. Boston Consulting Group, Boston, USA, 23 pp. https://image-src.bcg.com/Images/BCG-Global-Asset-Management-2020-May-2020-r_tcm9-247209.pdf (Accessed November 1, 2021).

Heredia, L. et al., 2021: *Global Asset Management 2021: The $100 trillion machine*. Boston Consulting Group, Boston, USA, 24 pp. https://web-assets.bcg.com/79/bf/d1d361854084a9624a0cbce3bf07/bcg-global-asset-management-2021-jul-2021.pdf (Accessed November 1, 2021).

Hermann, A., P. Köfer, and J.P. Mairhöfer, 2016: *Climate Risk Insurance: New Approaches and Schemes*. Allianze Working Paper September 2016, Allianze, Munich, Germany, 1–22 pp.

Herrington, R., 2021: Mining our green future. *Nat. Rev. Mater.*, **6(6)**, 456–458, doi:10.1038/s41578-021-00325-9.

Hickel, J., 2015: Five reasons to think twice about the UN's sustainable development goals. Africa at LSE, September 23.

High-Level Commission on Carbon Prices, 2017: *Report of the high-level commission on carbon prices* [Stern, N. and J.E. Stiglitz, (eds.)]. World Bank, Washington, DC, USA, Creative Commons Attribution CC BY 3.0 IGO, 69 pp.

Hilmi, N., S. Djoundourian, W. Shahin, and A. Safa, 2021: Does the ECB policy of quantitative easing impact environmental policy objectives? *J. Econ. Policy Reform*, **00(00)**, 1–13, doi:10.1080/17487870.2020.1855176.

Hirth, L. and J.C. Steckel, 2016: The role of capital costs in decarbonizing the electricity sector. *Environ. Res. Lett.*, **11(11)**, 114010, doi:10.1088/1748-9326/11/11/114010.

HM Treasury and UK DMO, 2021: *UK Government Green Financing Framework*. HM Treasury and UK DMO, London, UK, 30 pp.

Hodge, G., C. Greve, and M. Biygautane, 2018: Do PPP's work? What and how have we been learning so far? *Public Manag. Rev.*, **20(8)**, 1105–1121, doi:10.1080/14719037.2018.1428410.

Hoepner, A.G.F., S. Dimatteo, J. Schaul, P.-S. Yu, and M. Musolesi, 2017: *Tweeting about Sustainability: can Emotional Nowcasting discourage Greenwashing? Discussion Paper,* Henley Business School, University of Reading, Reading, UK. https://ssrn.com/abstract=2924088.

Hong, H., F.W. Li, and J. Xu, 2019: Climate risks and market efficiency. *J. Econom.*, **208(1)**, 265–281, doi:10.1016/j.jeconom.2018.09.015.

Hope, K.R., 2011: Investing in capacity development: towards an implementation framework. *Policy Stud.*, **32(1)**, 59–72, doi:10.1080/01442872.2010.529273.

Hourcade, J.-C., A. Pottier, and E. Espagne, 2018: Social value of mitigation activities and forms of carbon pricing. *Int. Econ.*, **155**, 8–18, doi:10.1016/j.inteco.2018.06.001.

Hourcade, J.-C., D. Dasgupta, and F. Ghersi, 2021a: Accelerating the speed and scale of climate finance in the post-pandemic context. *Clim. Policy*, **21(10)**, 1–15, doi:10.1080/14693062.2021.1977599.

Hourcade, J.-C. et al., 2021b: *Scaling up climate finance in the context of Covid-19*. Green Climate Fund, Incheon, Republic of Korea, 116 pp.

Hourcade, J.C., B. Perrissin Fabert, and J. Rozenberg, 2012: Venturing into uncharted financial waters: an essay on climate-friendly finance. *Int. Environ. Agreements Polit. Law Econ.*, **12(2)**, 165–186, doi:10.1007/s10784-012-9169-y.

Hove, A., 2020: *Current direction for renewable energy in China*. Oxford Institute for Energy Studies, Oxford, UK, 8 pp.

Hu, Y., S. Ren, Y. Wang, and X. Chen, 2020: Can carbon emission trading scheme achieve energy conservation and emission reduction? Evidence from the industrial sector in China. *Energy Econ.*, **85**, 104590, doi:10.1016/j.eneco.2019.104590.

Hubacek, K. et al., 2017: Global carbon inequality. *Energy, Ecol. Environ.*, **2(6)**, 361–369, doi:10.1007/s40974-017-0072-9.

Huenteler, J., C. Niebuhr, and T.S. Schmidt, 2016: The effect of local and global learning on the cost of renewable energy in developing countries. *J. Clean. Prod.*, **128**, 6–21, doi:10.1016/j.jclepro.2014.06.056.

Humphrey, C., 2017: He who pays the piper calls the tune: Credit rating agencies and multilateral development banks. *Rev. Int. Organ.*, **12(2)**, 281–306, doi:10.1007/s11558-017-9271-6.

Humphrey, C. and K. Michaelowa, 2019: China in Africa: Competition for traditional development finance institutions? *World Dev.*, **120**, 15–28, doi:10.1016/j.worlddev.2019.03.014.

Hunecke, K. et al., 2019: What role do transaction costs play in energy efciency improvements and how can they be reduced. *EECEE* Summer Study Proc., European Council for an Energy Efficient Economy, Stockholm, Sweden.

Hynes, W. and S. Scott, 2013: *The Evolution of Official Development Assistance: Achievements, Criticisms and a Way Forward*. OECD Development Co-operation Working Papers, No. 12, OECD Publishing. Paris, France. doi: 10.1787/5k3v1dv3f024-en.

Hyun, S., D. Park, and S. Tian, 2019: The price of going green: the role of greenness in green bond markets. *Account. Financ.*, Vol 60 Issue 1, doi:10.1111/acfi.12515.

IATFD, 2016: *Developing domestic capital markets*. Inter-agency Task Force on Financing for Development (IATFD). https://www.un.org/esa/ffd/wp-content/uploads/2016/01/Developing-domestic-capital-markets_IFC-World-Bank-Group_IATF-Issue-Brief.pdf (Accessed November 1, 2021).

IATFD, 2020: *Financing for Sustainable Development Report 2020*. Inter-agency Task Force on Financing for Development (IATFD). United Nations Publication, New York, 207 pp.

IATFD, 2021: *Financing for Sustainable Development Report 2021*. Inter-agency Task Force on Financing for Development (IATFD). United Nations Publication, P.50, New York, NY, USA, 189 pp. https://developmentfinance.un.org/sites/developmentfinance.un.org/files/FSDR_2021.pdf.

ICAEW, 2021: Accounting for financial guarantees in government accounts | ICAEW. Institute of Chartered Accountants in England and Wales (ICAEW), London, UK. https://www.icaew.com/insights/covid-19-global-recovery/fiscal-policy/accounting-for-financial-guarantees-in-government-accounts (Accessed October 30, 2021).

ICEAW, 2012: *Sustainable Public Finances: Global Views*. Institute of Chartered Accountants in England and Wales (ICAEW). London, UK, 19 pp.

ICMA, 2020a: *Sustainable Finance: High-level definitions*. International Capital Market Association, Zurich, Switzerland, 7 pp. https://www.icmagroup.org/assets/documents/Regulatory/Green-Bonds/Sustainable-Finance-High-Level-Definitions-May-2020-051020.pdf.

ICMA, 2020b: Bond Market Size. International Capital Market Association, Zurich, Switzerland.

ICMA, 2021: *Overview and Recommendations for Sustainable Finance Taxonomies*. International Capital Market Association, Zurich, London, Paris, Brussels, Hong Kong, 34 pp. www.icmagroup.org (Accessed August 24, 2021).

IDFC, and GCF, 2020: *The Green Climate Fund and the International Development Finance Club: A strategic alliance to realize the full potential of public development banks in financing the green and climate-resilient transition*. 28 pp. https://www.greenclimate.fund/document/gcf-idfc-working-paper (Accessed October 27, 2021).

IEA, 2014a: *Capturing the Multiple Benefits of Energy Efficiency*. OECD/IEA, Paris, France., 224 pp. https://iea.blob.core.windows.net/assets/28f84ed8-4101-4e95-ae51-9536b6436f14/Multiple_Benefits_of_Energy_Efficiency-148x199.pdf (Accessed November 1, 2021).

IEA, 2014b: *Africa Energy Outlook: A focus on energy prospects in sub-Saharan Africa*. OECD/IEA, Paris, France, 237 pp.

IEA, 2017: *World Energy Outlook 2017*. Flagship Report, November 2017, IEA, Paris, France.

IEA, 2018: World Energy Balances 2018. In: *International Energy Agency*, OECD/IEA, Paris, France.

IEA, 2019a: *World Energy Investment 2019*. Flagship Report, May 2019. OECD, Paris, France 176 pp.

IEA, 2019b: *Securing Investments in Low-Carbon Power Generation Sources*. OECD, Paris, 16 pp.

IEA, 2019c: *Africa Energy Outlook 2019*. 288 pp. https://www.iea.org/reports/africa-energy-outlook-2019 (Accessed November 1, 2021).

IEA, 2020a: *World Energy Investment 2020*. OECD Publishing, Paris, France, 176 pp.

IEA, 2020b: *The Oil and Gas Industry in Energy Transitions: Insights from IEA analysis*.. OECD Publishing, Paris, France.

IEA, 2020c: *Carbon pricing can help put clean energy at the heart of stimulus packages*. OECD Publishing, Paris, France, https://www.iea.org/commentaries/carbon-pricing-can-help-put-clean-energy-at-the-heart-of-stimulus-packages (Accessed November 1, 2021).

IEA, 2020d: *World Energy Outlook 2020*. OECD Publishing, Paris, France.

IEA, 2021a: *Financing Clean Energy Transitions in Emerging and Developing Economies*. OECD Publishing, Paris, France, 235 pp. https://iea.blob.core.windows.net/assets/6756ccd2-0772-4ffd-85e4-b73428ff9c72/FinancingCleanEnergyTransitionsinEMDEs_WorldEnergyInvestment2021SpecialReport.pdf (Accessed November 1, 2021).

IEA, 2021b: *Net Zero by 2050*. OECD Publishing, Paris, France, 224 pp.

IEA, 2021c: *World Energy Investment 2021*. OECD Publishing, Paris, France, 64 pp.

IEA, 2021d: *World energy model documentation*. OECD Publishing, Paris, France, 91 pp. https://iea.blob.core.windows.net/assets/bc4936dc-73f1-47c3-8064-0784ae6f85a3/WEM_Documentation_WEO2020.pdf.

IEA, 2021e: *World Energy Outlook 2021*. IEA Publications, Paris, 383 pp.

IEA, 2021f: *The role of critical minerals in clean energy transitions*. OECD Publishing, Paris, France, 283 pp. www.iea.org/t&c/ (Accessed September 19, 2021).

IEA, IRENA, UNSD, World Bank, and WHO, 2021: *Tracking SDG 7: The Energy Progress Report 2021*. IRENA, Washington DC, USA, 228 pp. https://www.irena.org/publications/2021/Jun/Tracking-SDG-7-2021 (Accessed August 18, 2021).

IFC, 2016: *Climate Investment Opportunities in Emerging Markets - An IFC Analysis*. International Finance Corporation, World Bank Group, Washington, DC, USA, 140 pp. https://www.ifc.org/wps/wcm/connect/59260145-ec2e-40de-97e6-3aa78b82b3c9/3503-IFC-Climate%7B%5C_%7DInvestment%7B%5C_%7DOpportunity-Report-Dec-FINAL.pdf?MOD=AJPERES%7B%5C&%7DCVID=lBLd6Xq (Accessed November 1, 2021).

IFC, 2018: *Climate Investment Opportunities in Cities: A IFC Analysis*. International Finance Corporation, 176 pp. https://www.ifc.org/wps/wcm/connect/875afb8f-de49-460e-a66a-dd2664452840/201811-CIOC-IFC-Analysis.pdf?MOD=AJPERES&CVID=mthPzYg (Accessed November 1, 2021).

IFC, 2021: *Green Bond Technical Assistance Program: Delivering Global Public Goods - Stimulating the Supply of Green Bonds in Emerging Markets*. International Finance Corporation, World Bank Group, Washington, DC, USA, 4 pp.

IFC 2018b, 2018: *MCPP Infrastructure: An innovative structure to mobilize institutional investment in emerging market infrastructure loans*. International Finance Corporation, World Bank Group, Washington, DC, USA, 1 pp. https://www.ifc.org/wps/wcm/connect/4c9e0868-1232-4212-b4f2-a5c39d177afa/MCPP+Infrastructure+Flyer+2018.pdf?MOD=AJPERES&CVID=mcoa4bt (Accessed November 1, 2021).

IFC 2021: *Using Blended Concessional Finance to Invest in Challenging Markets*. International Finance Corporation, World Bank Group, Washington, DC, USA, 59 pp. https://www.ifc.org/wps/wcm/connect/publications_ext_content/ifc_external_publication_site/publications_listing_page/using+blended+concessional+finance+to+invest+in+challenging+markets (Accessed August 17, 2021).

Igan, D., D. Kirti, and S.M. Peria, 2020: *The disconnect between financial markets and the real economy*. IMF, Washington D.C., 10 pp.

IIGCC, 2015: Climate finance for developing and emerging countries: Five recommendations to catalyse institutional investment Institutional Investors Group on Climate Change 2. Position paper, IIGCC, London, UK.

IIGCC, 2021: Global framework for investors to achieve net zero emissions alignment launched – $8 trillion investors put it into practice. Press Release, March 10.

IIX, 2020: IIX Accelerates Gender Bond Issuances with Pricing of Women's Livelihood Bond 3. Press Release, December 7.

IMF, 2009: *Balance of Payments and International Investment Position Manual Sixth Edition (BPM6)*. 6th ed. International Monetary Fund (IMF), Washington, DC, USA, 371 pp.

IMF, 2019: *Fiscal Monitor: How to Mitigate Climate Change*. International Monetary Fund (IMF), Washington, DC, USA, 96 pp. https://www.imf.org/en/Publications/FM/Issues/2019/10/16/Fiscal-Monitor-October-2019-How-to-Mitigate-Climate-Change-47027 (Accessed November 1, 2021).

IMF, 2020a: Mitigating Climate Change - Growth and Distribution Friendly Strategies. In: *World Economic Outlook: A Long and Difficult Ascent*, International Monetary Fund (IMF), Washington, DC, USA, pp. 85–113.

IMF, 2020b: *World Economic Outlook: A Long and Difficult Ascent*. International Monetary Fund (IMF), Washington, DC, USA.

IMF, 2020c: Greening the recovery. *Special Series on Fiscal Policies to Respond to COVID-19*, (October 2019), International Monetary Fund (IMF), Washington, DC, USA, 1–3.

IMF, 2021a: *World Economic Outlook: Managing Divergent Recoveries*. International Monetary Fund (IMF), Washington, DC, USA, 170 pp. https://www.imf.org/en/Publications/WEO/Issues/2021/03/23/world-economic-outlook-april-2021 (Accessed November 1, 2021).

IMF, 2021b: IMF Executive Directors Discuss a New SDR Allocation of US$650 billion to Boost Reserves, Help Global Recovery from COVID-19. Press Release March 23, 2021. International Monetary Fund (IMF), Washington, DC, USA.

IMF, 2021c: *World Economic Outlook Update (July 2021)*. International Monetary Fund (IMF), Washington, DC, USA, 21 pp. https://www.imf.org/en/Publications/WEO/Issues/2021/07/27/world-economic-outlook-update-july-2021 (Accessed November 1, 2021).

IMF, 2021d: *Reaching Net Zero Emissions*. International Monetary Fund (IMF), Washington, DC, USA, 32 pp. https://www.imf.org/external/np/g20/pdf/2021/062221.pdf (Accessed November 1, 2021).

IMF, 2021e: Gross debt position % of GDP. IMF Datamapper. https://www.imf.org/external/datamapper/G_XWDG_G01_GDP_PT@FM/ADVEC/FM_EMG/FM_LIDC.

IMF and World Bank, 2016: *Staff Note for the G20 IFAWG: Development of Local Currency Bond Markets Overview of Recent Developments and Key Themes*. IMF, Washington, DC, USA. 33 pp.

IMF and World Bank, 2020: *Enhancing Access to Opportunities*. IMF and World BankWashington DC, USA, 70 pp.

IMF and OECD, 2021: *Tax Policy and Climate Change: IMF/OECD Report for the G20 Finance Ministers and Central Bank Governors, April 2021, OECD, Italy*. 36 pp. https://www.oecd.org/tax/tax-policy/tax-policy-and-climate-change-imf-oecd-g20-report-april-2021.pdf (Accessed November 1, 2021).

IMF and World Bank, 2021: *Guidance Note for Developing Government Local Currency Bond Markets*. IMF, Washington DC, USA, 140 pp. https://www.imf.org/en/Publications/analytical-notes/Issues/2021/03/17/Guidance-Note-For-Developing-Government-Local-Currency-Bond-Markets-50256.

In, S.Y. and K. Schumacher, 2021: *Carbonwashing: A New Type of Carbon Data-related ESG Greenwashing*. Sustainable Finance Initiative, Precourt Institute for Energy, Stanford, CA, USA, 25 pp. https://energy.stanford.edu/sites/g/files/sbiybj9971/f/carbonwashing-_a_new_type_of_carbon_data-related_esg_greenwashing_working_paper.pdf (Accessed November 1, 2021).

Inderst, G., 2021: Financing Development: Private Capital Mobilization and Institutional Investors. ZBW - Leibniz Information Centre for Economics, Kiel, Hamburg, Germany, 41 pp.

Inderst, G. and R. Della Croce, 2013: *Pension Fund Investment in Infrastructure: A Comparison between Australia and Canada*. OECD, Paris, France, 54 pp.

Institut Louis Bachelier et al., 2020: *The Alignment Cookbook: A Technical Review of Methodologies Assessing a Portfolio's Alignment with Low-carbon Trajectories or Temperature Goal*. Institut Louis Bachelier, Paris, France, 173 pp. https://www.louisbachelier.org/wp-content/uploads/2020/10/cookbook.pdf (Accessed November 1, 2021).

Investor Agenda, 2020: *The Investor Agenda: A Sustainable Recovery from the COVID-19 Pandemic*. The Investor Agenga, https://www.iigcc.org/download/ia-a-sustainable-recovery-from-the-covid-19-pandemic/?wpdmdl=3289%7B%5C&%7Drefresh=5feff7e6a9f471609562086 (Accessed November 1, 2021).

IPA, 2021: *Annual Report on Major Projects 2020-21*. Infrastructure and Projects Authority, HM Government, Londin, UK, 58 pp.

IPCC, 2012: *Managing the Risks of Extreme Events and Disasters to Advance Climate Change Adaptation*. [Field, C.B., V. Barros, T.F. Stocker, D. Qin, D.J. Dokken, K.L. Ebi, M.D. Mastrandrea, K.J. Mach, G.-K. Plattner, S.K. Allen, M. Tignor, and P.M. Midgley (eds.)]. Cambridge University Press, Cambridge, UK, and New York, NY, USA, 582 pp.

IPCC, 2014: *Climate Change 2014: Impacts, Adaptation, and Vulnerability. Part A: Global and Sectoral Aspects. Contribution of Working Group II to the Fifth Assessment Report of the Intergovernmental Panel on Climate Change* [Field, C.B., V.R. Barros, D.J. Dokken, K.J. Mach, and M.D. Mastrandrea, (eds.)]. Cambridge University Press, Cambridge, UK, and New York, NY, USA, 833–868 pp.

IPCC, 2018: *Global Warming of 1.5°C: An IPCC special report on the impacts of global warming of 1.5 °C above pre-industrial levels and related global greenhouse gas emission pathways, in the context of strengthening the global response to the threat of climate change* [Masson-Delmotte, V., P. Zhai, H.-O. Pörtner, D. Roberts, J. Skea, P.R. Shukla, A. Pirani, W. Moufouma-Okia, C. Péan, R. Pidcock, S. Connors, J.B.R. Matthews, Y. Chen, X. Zhou, M.I. Gomis, E. Lonnoy, T. Maycock, M. Tignor, and T. Waterfield (eds.)]. Cambridge University Press, Cambridge, UK, and New York, NY, USA.

IRENA, 2016: *Unlocking Renewable Energy Investment: The Role of Risk Mitigation and Structured Finance*. International Renewable Energy Agency, Abu Dhabi, https://www.res4med.org/wp-content/uploads/2017/11/IRENA%7B%5C_%7DRisk%7B%5C_%7DMitigation%7B%5C_%7Dand%7B%5C_%7DStructured%7B%5C_%7DFinance%7B%5C_%7D2016.pdf (Accessed November 1, 2021).

IRENA, 2018: Community Energy. International Renewable Energy Agency, Abu Dhabi.

IRENA, 2019a: *Renewable Power Generation Costs in 2018*. International Renewable Energy Agency (IRENA), Abu Dhabi, 88 pp.

IRENA, 2019b: *Innovation landscape for a renewable-powered future*. International Renewable Energy Agency, Abu Dhabi.

IRENA, 2019c: *Business Models. Aggregators - Innovation Landscape Brief*. International Renewable Energy Agency, Abu Dhabi.

IRENA, 2019d: *Innovation landscape brief: Electric-vehicle smart charging*. International Renewable Energy Agency, Abu Dhabi.

IRENA, 2020a: *Renewable Power Generation Costs in 2019*. International Renewable Energy Agency (IRENA), Abu Dhabi, 143 pp. https://www.irena.org/publications/2020/Jun/Renewable-Power-Costs-in-2019 (Accessed November 1, 2021).

IRENA, 2020b: *Power system organisational structures for the renewable energy era*. International Renewable Energy Agency, Abu Dhabi.

IRENA, 2020c: *Global Renewables Outlook: Energy transformation 2050*. International Renewable Energy Agency (IRENA), Abu Dhabi.

IRENA, 2020d: *The post-COVID recovery: An agenda for resilience, development and equality*. International Renewable Energy Agency (IRENA), Abu Dhabi, 143 pp.

IRENA, 2020e: *Renewable energy finance: Sovereign guarantees*. IRENA, Abu Dhabi, 14 pp. www.irena.org/publications (Accessed October 29, 2021).

IRENA, 2020f: *Business Models: Innovation Landscape Briefs*. International Renewable Energy Agency, Abu Dhabi.

IRENA, 2021: *World Energy Transitions Outlook: 1.5C Pathway*. International Renewable Energy Agency (IRENA), Abu Dhabi, 311 pp.

IRENA and CEM, 2015: *Renewable energy auctions - A guide to design*. International Renewable Energy Agency (IRENA), Abu Dhabi, 200 pp. https://www.irena.org/publications/2015/Jun/Renewable-Energy-Auctions-A-Guide-to-Design (Accessed November 1, 2021).

Irving, J., 2020: *How the COVID-19 crisis is impacting African pension fund approaches to portfolio management*. International Finance Corporation, World Bank Group, Washington, D.C., USA. 15 pp. https://www.ifc.org/wps/wcm/connect/61e14b0d-b283-4f8f-8d17-4bce3f54d4a1/African+pension+funds%7B%5C_%7DFINAL-10-9-20.pdf?MOD=AJPERES%7B%5C&%7DCVID=nkeOGIJ (Accessed November 1, 2021).

ISDR, 2009: *Global Assessment Report on Disaster Risk Reduction 2009: Risk and Poverty in a Changing Climate*. International Strategy for Disaster Reduction (ISDR), Geneva, Switzerland.

ISDR, 2011: *Global Assessment Report on Disaster Risk Reduction 2011: Revealing Risk, Redefining Development*. International Strategy for Disaster Reduction (ISDR), Geneva, Switzerland.

Ito, H. and R.N. McCauley, 2019: *A disaster under-(re)insurance puzzle: Home bias in disaster risk-bearing*. BIS Working Paper No 808, Bank for International Settlements, Basel, Switzerland. 40 pp. https://www.bis.org/publ/work808.pdf (Accessed November 1, 2021).

IUCN, 2021: Blue Natural Capital Financing Facility. https://bluenaturalcapital.org/wp2018/wp-content/uploads/2018/10/BNCFF-brochure-FINAL-WEB-small.pdf.

Ivanova, D. and R. Wood, 2020: The unequal distribution of household carbon footprints in Europe and its link to sustainability. *Glob. Sustain.*, **3**, e18, doi:10.1017/sus.2020.12.

Iyer, G.C. et al., 2015: Improved representation of investment decisions in assessments of CO2 mitigation. *Nat. Clim. Change*, **5(5)**, 436–440, doi:10.1038/nclimate2553.

Jaccard, M., 2012: The political acceptability of carbon taxes: lessons from British Columbia. In: *Handbook of Research on Environmental Taxation* [Milne, J.E. and M.S. Andersen, (eds.)]. Edward Elgar Publishing, Cheltenham, UK, pp. 175–191.

Jachnik, R. and A. Dobrinevski, 2021: *Measuring the alignment of real economy investments with climate mitigation objectives*. OECD Environment Working Paper, OECD, Paris, France, 71 pp. https://doi.org/10.1787/8eccb72a-en.

Jachnik, R., R. Caruso, and A. Srivastava, 2015: *Estimating mobilised private climate finance: Methodological Approaches, Options and Trade-offs*. OECD Environment Working Paper, OECD, Paris, France. 66 pp.

Jackson, 2019: DEAL: Seychelles' sovereign blue bond. *International Financial Law Review*.World Bank, Washington, D.C., USA.

Jacobsson, S. et al., 2009: EU renewable energy support policy: Faith or facts? *Energy Policy*, **37(6)**, 2143–2146, doi:10.1016/j.enpol.2009.02.043.

Jaeger, J., M.I. Westphal, and C. Park, 2020: *Lessons Learned On Green Stimulus: Case studies from the global financial crisis*. World Resources Institute, Washington, D.C., USA. 32 pp. https://files.wri.org/s3fs-public/lessons-learned-on-green-stimulus-case-studies-from-the-global-financial-crisis.pdf?5gDebandVTpi8eM8XtXwLE7jjZem2DPI (Accessed November 1, 2021).

Jaffe, A.B., R.G. Newell, and R.N. Stavins, 2004: Economics of Energy Efficiency. In: *Encyclopedia of Energy* [Cleveland, C.J., (ed.)]. Vol. 2. Elsevier, Amsterdam, The Netherlands, pp. 79–90.

Jagers, S.C. et al., 2020: On the preconditions for large-scale collective action. *Ambio*, **49(7)**, 1282–1296, doi:10.1007/s13280-019-01284-w.

Jenkins, K.E.H., B.K. Sovacool, A. Błachowicz, and A. Lauer, 2020: Politicising the Just Transition: Linking global climate policy, Nationally Determined Contributions and targeted research agendas. *Geoforum*, **115**, 138–142, doi:10.1016/j.geoforum.2020.05.012.

Jensen, L., 2021: *An Unprecedented Opportunity to Boost Finance for Development: The Upcoming Special Drawing Rights Allocation*. UNDP, New York, NY, USA, 10 pp. https://www.undp.org/publications/unprecedented-opportunity-boost-finance-development#modal-publication-download.

Jones, E.C. and B.D. Leibowicz, 2019: Contributions of shared autonomous vehicles to climate change mitigation. *Transp. Res. Part D Transp. Environ.*, **72**, 279–298, doi:10.1016/j.trd.2019.05.005.

Jordà, Ò., M. Schularick, and A.M. Taylor, 2017: Macrofinancial History and the New Business Cycle Facts. *NBER Macroecon. Annu.*, **31(1)**, 213–263, doi:10.1086/690241.

Jordà, Ò., M. Schularick, A.M. Taylor, and F. Ward, 2019: Global Financial Cycles and Risk Premiums. *IMF Econ. Rev.*, **67(1)**, 109–150, doi:10.1057/s41308-019-00077-1.

Jordà, Ò., S.R. Singh, and A.M. Taylor, 2020: The Long Economic Hangover of Pandemics. *Financ. & Dev.*, **57(2)**.

Joskow, P.L., 2007: Market electricity markets and investment in new generating capacity. In: *The New Energy Paradigm* [Helm, D., (ed.)]. Oxford University Press, Oxford, UK.

Jubilee Debt Campaign, 2019: *Transparency of loans to governments: The public's right to know about their debts*. Jubilee Debt Campaign, London, UK, 4 pp. https://jubileedebt.org.uk/wp/wp-content/uploads/2019/04/Transparency-of-loans-to-governments_04.19.pdf (Accessed November 1, 2021).

Juvonen, K., A. Kumar, H. Ben Ayed, and A.O. Marin, 2019: *Unleashing the Potential of institutional investors in Africa*. ADBG Working Paper No 325, African Development Bank Group, Abidjan, Côte d'Ivoire, 42 pp. https://www.afdb.org/en/documents/publications/working-paper-series/.

Kahn, M. et al., 2019: Long-Term Macroeconomic Effects of Climate Change: A Cross-Country Analysis (July 1, 2019). USC-INET Research Paper No. 19-13SSRN Electronic Journal, Cambridge, MA, USA. doi: 10.2139/ssrn.3428610.

Kahouli-Brahmi, S., 2009: Testing for the presence of some features of increasing returns to adoption factors in energy system dynamics: An analysis via the learning curve approach. *Ecol. Econ.*, **68(4)**, 1195–1212, doi:10.1016/j.ecolecon.2008.08.013.

Kalinowski, W. and H. Chenet, 2020: *A "whatever it takes" climate strategy in central banking*. Institut Veblen, Paris, France. 28 pp. https://www.veblen-institute.org/IMG/pdf/whatever_it_takes_climatique_eng.pdf (Accessed November 1, 2021).

Kaminker, C. and F. Stewart, 2012: *The Role of Institutional Investors in Financing Clean Energy*. OECD Working Papers on Finance, Insurance and Private Pensions, No.23, OECD Publishing, Paris, France. 53 pp.

Kaminker, C., O. Kawanishi, F. Stewart, B. Caldecott, and N. Howarth, 2013: *Institutional Investors and Green Infrastructure Investments: Selected Case Studies*. OECD Working Papers on Finance, Insurance and Private Pensions, No. 35, OECD Publishing, Paris. 23 pp. doi: 10.1787/5k3xr8k6jb0n-en.

Karpf, A. and A. Mandel, 2018: The changing value of the 'green' label on the US municipal bond market. *Nat. Clim. Change*, **8(2)**, 161–165, doi:10.1038/s41558-017-0062-0.

Keenan, J.M. and J.T. Bradt, 2020: Underwaterwriting: from theory to empiricism in regional mortgage markets in the U.S. *Clim. Change*, **162(4)**, 2043–2067, doi:10.1007/s10584-020-02734-1.

Keesstra, S. et al., 2018: The superior effect of nature based solutions in land management for enhancing ecosystem services. *Sci. Total Environ.*, **610–611**, 997–1009, doi:10.1016/j.scitotenv.2017.08.077.

Kellett, J. and K. Peters, 2014: *Dare to prepare: taking risk seriously*. Overseas Development Institute, London, UK, 140 pp.

Kelly, M. and T.M. Hanna, 2019: Democratic ownership in the USA: a quiet revolution. *Int. J. Public Pol.*, **15(1/2)**, 92, doi:10.1504/IJPP.2019.099057.

Kempa, K., U. Moslener, and O. Schenker, 2021: The cost of debt of renewable and non-renewable energy firms. *Nat. Energy*, **6(2)**, 135–142, doi:10.1038/s41560-020-00745-x.

Kern, F. and K.S. Rogge, 2016: The pace of governed energy transitions: Agency, international dynamics and the global Paris agreement accelerating decarbonisation processes? *Energy Res. & Soc. Sci.*, **22**, 13–17, doi:10.1016/j.erss.2016.08.016.

Ketkar, S.L. and D. Ratha, 2010: Diaspora Bonds: Tapping the Diaspora during Difficult Times. *J. Int. Commer. Econ. Policy*, **01(02)**, 251–263, doi:10.1142/S1793993310000147.

Ketterer, J. and A. Powell, 2018: Financing Infrastructure: On the Quest for an Asset-Class. IDB, Washington, DC, USA, 25 pp. doi: 10.18235/0001323.

Ketterer, J.A., G. Andrade, M. Netto, and M.I. Haro, 2019: *Transforming Green Bond Markets: Using Financial Innovation and* Technology to Expand Green Bond Issuance in Latin America and the Caribbean. IDB, Washington D.C., 29 pp. doi: 10.18235/0001900.

Khan, M., S. Robinson, R. Weikmans, D. Ciplet, and J.T. Roberts, 2020: Twenty-five years of adaptation finance through a climate justice lens. *Clim. Change*, **161(2)**, 251–269, doi:10.1007/s10584-019-02563-x.

Kissinger, G., A. Gupta, I. Mulder, and N. Unterstell, 2019: Land Use Policy Climate financing needs in the land sector under the Paris Agreement: An assessment of developing country perspectives. *Land use policy*, **83** (October 2018), 256–269, doi:10.1016/j.landusepol.2019.02.007.

Klenert, D. et al., 2018: Making carbon pricing work for citizens. *Nat. Clim. Chang.*, **8(8)**, 669–677, doi:10.1038/s41558-018-0201-2.

Kling, G., Y. Lo, V. Murinde, and U. Volz, 2018: Climate Vulnerability and the Cost of Debt. *SSRN Electron. J.*, doi:10.2139/ssrn.3198093.

Klinsky, S. et al., 2017: Why equity is fundamental in climate change policy research. *Glob. Environ. Chang.*, **44**, 170–173, doi:10.1016/j.gloenvcha.2016.08.002.

Klöck, C., N. Molenaers, and F. Weiler, 2018: Responsibility, capacity, greenness or vulnerability? What explains the levels of climate aid provided by bilateral donors? *Env. Polit.*, **27(5)**, 892–916, doi:10.1080/09644016.2018.1480273.

Klusak, P., M. Agarwala, M. Burke, M. Kraemer, and K. Mohaddes, 2021: *Rising Temperatures, Falling* Ratings: The Effect of Climate Change on Sovereign Creditworthiness. Cambridge Working Papers in Economics 2127, Faculty of Economics, University of Cambridge, Cambridge, UK, 50 pp.

Knobloch, F., H. Pollitt, U. Chewpreecha, V. Daioglou, and J.-F. Mercure, 2019: Simulating the deep decarbonisation of residential heating for limiting global warming to 1.5 °C. *Energy Effic.*, **12(2)**, 521–550, doi:10.1007/s12053-018-9710-0.

Knutti, R., 2010: The end of model democracy? *Clim. Change*, **102(3)**, 395–404, doi:10.1007/s10584-010-9800-2.

Koh, J., E. Mazzacurati, and S. Swann, 2016: *Bridging the Adaptation Gap: Approaches to Measurement of Physical Climate Risk and Examples of Investment in Climate Adaptation and Resilience*. Global Adaptation & Resilience Investment Working Group, 65 pp. http://427mt.com/wp-content/uploads/2016/11/GARI-2016-Bridging-the-Adaptation-Gap.pdf (Accessed November 1, 2021).

Kölbel, J.F., F. Heeb, F. Paetzold, and T. Busch, 2020: Can Sustainable Investing Save the World? Reviewing the Mechanisms of Investor Impact. *Organ. Environ.*, **33(4)**, 554–574, doi:10.1177/1086026620919202.

Kose, M.A., F. Ohnsorge, P. Nagle, and N. Sugawara, 2020: *Caught by a Cresting Debt Wave*. IMF, Washington, DC, USA, 4 pp.

Kotsantonis, S. and G. Serafeim, 2019: Four Things No One Will Tell You About ESG Data. *J. Appl. Corp. Financ.*, **31(2)**, 50–58, doi:10.1111/jacf.12346.

KPMG, 2015: *The Utility as The Network Integrator*. KPMG, New York, 13 pp. https://assets.kpmg/content/dam/kpmg/pdf/2016/05/the-utility-as-the-network-integrator.pdf (Accessed November 1, 2021).

Kramer, L., 2017: Why philanthropy must do more on climate change. *William and Flora Hewlett Foundation*, March 14.

Krawchenko, T.A. and M. Gordon, 2021: How Do We Manage a Just Transition? A Comparative Review of National and Regional Just Transition Initiatives. *Sustainability*, **13(11)**, 6070, doi:10.3390/su13116070.

Kreft, S. and L. Schäfer, 2017: *The G20' s role on climate risk insurance & pooling: Weathering Climate Change through Climate Risk Transfer Solutions*.Munich Climate Insurance Initiative, Bonn, Germany, 1–8 pp. http://www.climate-insurance.org/fileadmin/user%7B%5C_%7Dupload/20170316%7B%5C_%7DMCII%7B%5C_%7DG20%7B%5C_%7DPosition%7B%5C_%7D2017.pdf (Accessed November 1, 2021).

Kriegler, E. et al., 2015: Making or breaking climate targets: The AMPERE study on staged accession scenarios for climate policy. *Technol. Forecast. Soc. Change*, **90**, 24–44, doi:10.1016/j.techfore.2013.09.021.

Krogstrup, S. and W. Oman, 2019: *Macroeconomic and Financial Policies for Climate Change Mitigation: A Review of the Literature*. IMF, Washington, DC, USA, 58 pp. https://www.imf.org/en/Publications/WP/Issues/2019/09/04/Macroeconomic-and-Financial-Policies-for-Climate-Change-Mitigation-A-Review-of-the-Literature-48612 (Accessed November 1, 2021).

Kuhl, L., 2021: Policy making under scarcity: reflections for designing socially just climate adaptation policy. *One Earth*, **4(2)**, 202–212, doi:10.1016/j.oneear.2021.01.008.

Kumar, A., W. Xin, and C. Zhang, 2019: Climate Sensitivity and Predictable Returns. *SSRN Electron. J.*, doi:10.2139/ssrn.3331872.

Kuramochi, T. et al., 2020: Beyond national climate action: the impact of region, city, and business commitments on global greenhouse gas emissions. *Clim. Policy*, **20(3)**, 275–291, doi:10.1080/14693062.2020.1740150.

Kuzemko, C. et al., 2020: Covid-19 and the politics of sustainable energy transitions. *Energy Res. Soc. Sci.*, **68**, 101685, doi:10.1016/j.erss.2020.101685.

Kyriakou, V., I. Garagounis, E. Vasileiou, A. Vourros, and M. Stoukides, 2017: Progress in the Electrochemical Synthesis of Ammonia. *Catal. Today*, **286**, 2–13, doi:10.1016/j.cattod.2016.06.014.

Laing, T., L. Taschini, and C. Palmer, 2016: Understanding the demand for REDD+ credits. *Environ. Conserv.*, **43(4)**, 389–396, doi:10.1017/S0376892916000187.

Lamperti, F., V. Bosetti, A. Roventini, and M. Tavoni, 2019: The public costs of climate-induced financial instability. *Nat. Clim. Change*, **9(11)**, 829–833, doi:10.1038/s41558-019-0607-5.

Langhorne, E., 2021: How one pension fund identified investment opportunities in Africa. *USAID Invest*, July 8. https://www.marketlinks.org/blogs/how-one-pension-fund-identified-investment-opportunities-africa (Accessed November 1, 2021).

Larcker, D.F. and E.M. Watts, 2020: Where's the greenium? *J. Account. Econ.*, **69(2–3)**, 101312, doi:10.1016/j.jacceco.2020.101312.

Larsen, G., G. Christianson, and N. Amerasinghe, 2019: So Far, Green Bonds Fail to Raise Much Money for Resilience. The Climate Resilience Principles Aim to Change That. *World Resources Institute Blog*, October 15.

Laskaridis, C., 2021: When push came to shove: COVID-19 and debt crises in low-income countries. *Can. J. Dev. Stud. / Rev. Can. d'études du développement*, **42(1–2)**, 200–220, doi:10.1080/02255189.2021.1894102.

Lawrence, A. and F. Archer, 2021: The project finance law review: Export credit agencies and insurers. *Law Review*, June 23.

Le De, L., J.C. Gaillard, and W. Friesen, 2013: Remittances and disaster: a review. *Int. J. Disaster Risk Reduct.*, **4**, 34–43, doi:10.1016/j.ijdrr.2013.03.007.

Le Quéré, C. et al., 2020: Temporary reduction in daily global CO2 emissions during the COVID-19 forced confinement. *Nat. Clim. Change*, **10(7)**, 647–653, doi:10.1038/s41558-020-0797-x.

Le Quéré, C. et al., 2021: Fossil CO2 emissions in the post-COVID-19 era. *Nat. Clim. Change*, **11(3)**, 197–199, doi:10.1038/s41558-021-01001-0.

Le Quesne, F. et al., 2017: *The role of insurance in integrated disaster & climate risk management: Evidence and lessons learned*. United Nations University, Institute for Environment and Human Security (UNU-EHS) UNU-EHS Publication Series Report 2017 No. 22. UNU, Bonn, Germany, 68 pp. http://collections.unu.edu/eserv/UNU:6312/MCII_ACRI_1710020_Online-meta.pdf (Accessed November 1, 2021).

Leaton, J. and L.L.B.-L. Sussams, 2011: *Unburnable carbon: Are the world's financial markets carrying a carbon bubble? 13th July 2011*. Carbon Tracker Initiative, London, UK. 33 pp. https://carbontracker.org/reports/carbon-bubble/ (Accessed November 1, 2021).

Leck, H. and D. Roberts, 2015: What lies beneath: understanding the invisible aspects of municipal climate change governance. *Curr. Opin. Environ. Sustain.*, **13**, 61–67, doi:10.1016/j.cosust.2015.02.004.

Lee, C., A. Betru, and P. Horrocks, 2018: *Guaranteeing the goals: Adapting public sector guarantees to unlock blended financing for the U.N. Sustainable Development Goals*. Milken Institute and OECD, Santa Monica, CA, USA, and Paris, France, 44 pp.

Lee, I. and Y.J. Shin, 2018: Fintech: Ecosystem, business models, investment decisions, and challenges. *Bus. Horiz.*, **61(1)**, 35–46, doi:10.1016/j.bushor.2017.09.003.

Lefèvre, J., W. Wills, and J.-C. Hourcade, 2018: Combining low-carbon economic development and oil exploration in Brazil? An energy–economy assessment. *Clim. Policy*, **18(10)**, 1286–1295, doi:10.1080/14693062.2018.1431198.

Lehmann, P. et al., 2021: Making the COVID-19 crisis a real opportunity for environmental sustainability. *Sustain. Sci.*, *16*(6), pp.2137–2145., doi:10.1007/s11625-021-01003-z.

Lei, N., E. Masaneta, and J. Koomey, 2021: Best practices for analyzing the direct energy use of blockchain technology systems: Review and policy recommendations. *Energy Policy*, Volume **156**. doi: 10.1016/j.enpol.2021.112422.

Lindenberg, N. and P. Pauw, 2013: *Don't lump together apples and oranges-Adaptation finance is different from mitigation finance*. German Institute of Development and Sustainability, Bonn, Germany, 3 pp. https://www.die-gdi.de/en/the-current-column/article/dont-lump-together-apples-and-oranges-adaptation-finance-is-different-from-mitigation-finance-1/ (Accessed November 1, 2021).

Linquiti, P. and N. Cogswell, 2016: The Carbon Ask: effects of climate policy on the value of fossil fuel resources and the implications for technological innovation. *J. Environ. Stud. Sci.*, **6(4)**, 662–676, doi:10.1007/s13412-016-0397-2.

Liu, W., W. McKibbin, and F. Jaumotte, 2021: Mitigating Climate Change: Growth-Friendly Policies to Achieve Net Zero Emissions by 2050. *IMF Working Paper*, **2021(195)**, 1, doi:10.5089/9781513592978.001.

Liu, Y., L. Wu, and J. Li, 2019: Peer-to-peer (P2P) electricity trading in distribution systems of the future. *Electr. J.*, **32(4)**, 2–6, doi:https://doi.org/10.1016/j.tej.2019.03.002.

Liverman, D.M., 2018: Geographic perspectives on development goals. *Dialogues Hum. Geogr.*, **8(2)**, 168–185, doi:10.1177/2043820618780787.

Livingston, D., V. Sivaram, M. Freeman, and M. Fiege, 2018: *Applying Blockchain Technology to Electric Power Systems*. Council on Foreign Relations, New York, NY, USA. 37 pp. https://cdn.cfr.org/sites/default/files/report%7B%5C_%7Dpdf/Discussion%7B%5C_%7DPaper%7B%5C_%7DLivingston%7B%5C_%7Det%7B%5C_%7Dal%7B%5C_%7DBlockchain%7B%5C_%7DOR%7B%5C_%7D0.pdf (Accessed November 1, 2021).

Long, J., 2021: Crisis Capitalism and Climate Finance: The Framing, Monetizing, and Orchestration of Resilience-Amidst-Crisis. *Polit. Gov.*, **9(2)**, 51–63, doi:10.17645/pag.v9i2.3739.

Lontzek, T.S., Y. Cai, K.L. Judd, and T.M. Lenton, 2015: Stochastic integrated assessment of climate tipping points indicates the need for strict climate policy. *Nat. Clim. Change*, **5(5)**, 441–444, doi:10.1038/nclimate2570.

Lowry, R.C., J.E. Alt, and K.E. Ferree, 1998: Fiscal Policy Outcomes and Electoral Accountability in American States. *Am. Polit. Sci. Rev.*, **92(4)**, 759–774, doi:10.2307/2586302.

LSEG, 2017: Nigeria's first Diaspora Bond starts trading on London Stock Exchange. London Stock Exchange Group (LSEG), LSEG Press Release, June 29.

LSEG, 2018: *London Stock Exchange Africa Advisory Group Report of Recommendations - Developing the Green Bond Market in Africa*. London Stock Exchange Group, London, UK, 16 pp. https://www.lseg.com/sites/default/files/content/documents/Africa%7B%5C_%7DGreenFinancing%7B%5C_%7DMWv10%7B%5C_%7D0.pdf (Accessed November 1, 2021).

Lubell, M. and T.H. Morrison, 2021: Institutional navigation for polycentric sustainability governance. *Nat.* Sustain volume 4, pp 664–671 (2021), doi:10.1038/s41893-021-00707-5.

Lucas, B., 2015: *Disaster risk financing and insurance in the Pacific (GSDRC Helpdesk Research Report 1314)*. GSDRC, Birmingham, 15 pp. http://www.gsdrc.org/wp-content/uploads/2016/01/HDQ1314.pdf (Accessed November 1, 2021).

Lucas, P.L. et al., 2015: Future energy system challenges for Africa: Insights from Integrated Assessment Models. *Energy Policy*, **86**, 705–717, doi:10.1016/j.enpol.2015.08.017.

Luderer, G. et al., 2018: Residual fossil CO2 emissions in 1.5–2 °C pathways. *Nat. Clim. Change*, **8(7)**, 626–633, doi:10.1038/s41558-018-0198-6.

Lujan, B. and G. Silva-Chávez, 2018: *Mapping Forest Finance: A Landscape of Available Sources of Finance for REDD+ and Climate Action in Forests*. Environmental Defense Fund and Forest Trends, New York, NY, USA and Washington, D.C., USA, 48 pp. https://www.forest-trends.org/wp-content/uploads/2018/04/doc_5734.pdf (Accessed November 1, 2021).

Lumenaza, 2020: Accelerating new energy. Together. https://www.lumenaza.de/en/home/ (Accessed November 1, 2021).

Ma, Z., J.D. Billanes, and B.N. Jørgensen, 2017: Aggregation Potentials for Buildings—Business Models of Demand Response and Virtual Power Plants. *Energies*, **10(10)**, 1646, doi:10.3390/en10101646.

MacAskill, S., E. Roca, B. Liu, R.A. Stewart, and O. Sahin, 2021: Is there a green premium in the green bond market? Systematic literature review revealing premium determinants. *J. Clean. Prod.*, **280**, 124491, doi:10.1016/j.jclepro.2020.124491.

MacFeely, S. and N. Barnat, 2017: Statistical capacity building for sustainable development: Developing the fundamental pillars necessary for modern national statistical systems. *Stat. J. IAOS*, **33(4)**, 895–909, doi:10.3233/SJI-160331.

Maclurcan, D. and J. Hinton, 2019: How on Earth: Flourishing in a Not-for-Profit World by 2050. Post Growth Publishing, 266 pp. doi.org/10.48550/arXiv.1902.01398.

Maffettone, P. and C. Oldani, 2020: COVID-19: A Make or Break Moment for Global Policy Making. *Glob. Policy*, **11(4)**, 501–507, doi:10.1111/1758-5899.12860.

Maguire, P. et al., 2021: *A green growth spurt: State of forest carbon finance 2021*. Ecosystem Marketplace Insights Report June 2021, Forest Trends, Washington D.C., USA.

Mahmud, S. and S. Huq, 2018: *Report on the Planning Meeting of the LUCCC Partners*. The Frati, Bellagio Study and Conference Center of the Rockefeller Foundation, Italy. ICCCAD. 17 pp. http://www.icccad.net/wp-content/uploads/2018/11/Final_LUCCC_Bellagio-Workshop-2018_-Report.pdf (Accessed August 24, 2021).

Maiti, M., 2021: Is ESG the succeeding risk factor? *J. Sustain. Financ. Invest.*, **11(3)**, 199–213, doi:10.1080/20430795.2020.1723380.

Mallucci, E., 2020: Natural Disasters, Climate Change, and Sovereign Risk. *Int. Financ. Discuss. Pap.*, **2020(1291r1)**, 1–34, doi:10.17016/IFDP.2020.1291r1.

Maltais, A. and B. Nykvist, 2020: Understanding the role of green bonds in advancing sustainability. *J. Sustain. Financ. Invest.*, doi:10.1080/20430795.2020.1724864.

Mamun, A., W. Martin, and S. Tokgoz, 2019: Reforming agricultural support for improved environmental outcomes. *IFPRI Discussion Paper*, December.

Mandel, A., 2020: Risks on Global Financial Stability Induced by Climate Change. *SSRN Electron. J.*, doi:10.2139/ssrn.3626936.

Manilla, R.J., 2018: Mission, Money & Markets: Making 'cents' of guarantees, the view from a CIO. *Kresge Found.*, Troy, MI, USA. https://kresge.org/news-views/mission-money-markets-making-cents-of-guarantees-the-view-from-a-cio/ (Accessed October 14, 2021).

Marinoni, O., P. Adkins, and S. Hajkowicz, 2011: Water planning in a changing climate: Joint application of cost utility analysis and modern portfolio theory. *Environ. Model. & Softw.*, **26(1)**, 18–29, doi:10.1016/j.envsoft.2010.03.001.

Markandya, A. and M. González-Eguino, 2019: Integrated Assessment for Identifying Climate Finance Needs for Loss and Damage: A Critical Review. In: Mechler, R., Bouwer, L., Schinko, T., Surminski, S., Linnerooth-Bayer, J. (eds) Loss *and Damage from Climate Change. Climate Risk Management,* Policy and Governance. Springer, Cham, Switzerland, pp. 343–362. doi: 10.1007/978-3-319-72026-5_14.

Markandya, A., J. Salcone, S. Hussain, A. Mueller, and S. Thambi, 2021: Covid, the Environment and Food Systems: Contain, Cope and Rebuild Better. *Front. Environ. Sci.*, **9**, doi:10.3389/fenvs.2021.674432.

Martinez-Diaz, L., L. Sidner, and J. Mcclamrock, 2019: *The Future of Disaster Risk Pooling for Developing Countries: Where Do We Go From Here?* Working Paper August 2019, WRI, Washington, DC, USA, 64 pp.

Mathews, J.A. and S. Kidney, 2012: Financing climate-friendly energy development through bonds. *Dev. South. Afr.*, **29(2)**, 337–349, doi:10.1080/0376835X.2012.675702.

Matias, D.M., R. Fernández, M.-L. Hutfils, and M. Winges, 2018: *Pro-Poor Climate Risk Insurance – The Role of Community-Based* Organisations (CBOs). Briefing Paper, German Development Institute, Bonn, Germany, 4 pp. doi: 10.23661/bp19.2018.

Matsumura, W. and Z. Adam, 2019: Fossil fuel consumption subsidies bounced back strongly in 2018. *Comment. - IEA*, Paris, France.

Matthäus, D. and M. Mehling, 2020: De-risking Renewable Energy Investments in Developing Countries: A Multilateral Guarantee Mechanism. *Joule*, **4(12)**, 2627–2645, doi:10.1016/j.joule.2020.10.011.

Mawdsley, E., 2018: 'From billions to trillions': Financing the SDGs in a world 'beyond aid.' *Dialogues Hum. Geogr.*, **8(2)**, 191–195, doi:10.1177/2043820618780789.

Mazzucato, M. and C.C.R. Penna, 2016: Beyond market failures: the market creating and shaping roles of state investment banks. *J. Econ. Policy Reform*, **19(4)**, 305–326, doi:10.1080/17487870.2016.1216416.

Mbeng Mezui, C.A. and B. Hundal, 2013: *Structured Finance: Conditions for Infrastructure Project Bonds in African Markets*. African Development Bank Group, Abidjan, Côte d'Ivoire, 420 pp.

McCauley, D. and R. Heffron, 2018: Just transition: Integrating climate, energy and environmental justice. *Energy Policy*, **119**, 1–7, doi:10.1016/j.enpol.2018.04.014.

McCollum, D.L. et al., 2018: Energy investment needs for fulfilling the Paris Agreement and achieving the Sustainable Development Goals. *Nat. Energy*, **3(7)**, 589–599, doi:10.1038/s41560-018-0179-z.

McFarland, B.J., 2015: International Finance for REDD+ Within the Context of Conservation Financing Instruments. *J. Sustain. For.*, **34(6–7)**, 534–546, doi:10.1080/10549811.2015.1017109.

McGlade, C. and P. Ekins, 2015: The geographical distribution of fossil fuels unused when limiting global warming to 2 °C. *Nature*, **517(7533)**, 187–190, doi:10.1038/nature14016.

MCII, 2018: *Submission on the type and nature of actions to address loss and damage for which finance may be required*. Munich Climate Insurance Initiative (MCII), Bonn, Germany, 12 pp. https://cop23.unfccc.int/sites/default/files/resource/MCII_Submission_to_the_Excom_Feb2018.pdf (Accessed November 1, 2021).

McKibbin, W. and D. Vines, 2020: Global Macroeconomic Cooperation in response to the COVID-19 pandemic: a roadmap for the G20 and IMFOxford Review of Economic Policy, Volume 36, Issue Supplement_1, 2020, Pages S297–S337, doi: 10.1093/oxrep/graa032.

McKibbin, W.J. and P.J. Wilcoxen, 2013: A Global Approach to Energy and the Environment. In: *Handbook of* Computable General Equilibrium Modeling, Handbook of Computable General Equilibrium Modeling. 2013; 1: 995–1068., pp. 995–1068. doi: 10.1016/B978-0-444-59568-3.00015-8McKinsey, 2020a: How the European Union could achieve net-zero emissions at net-zero cost. McKinsey Sustainability, London, UK, https://www.mckinsey.com/capabilities/sustainability/our-insights/how-the-european-union-could-achieve-net-zero-emissions-at-net-zero-cost (Accessed November 1, 2021).

McKinsey, 2020b: Climate change and P&C insurance: The threat and opportunity. McKinsey Sustainability, London, UK, https://www.mckinsey.com/industries/financial-services/our-insights/climate-change-and-p-and-c-insurance-the-threat-and-opportunity (Accessed November 1, 2021).

McManus, R. and F.G. Ozkan, 2015: On the consequences of pro-cyclical fiscal policy. *Fisc. Stud.*, **36(1)**, 29–50, doi:10.1111/j.1475-5890.2015.12044.x.

MDBs, 2018: Multilateral Development Banks (MDBs) Announced a Joint Framework for Aligning their Activities with the Goals of the Paris Agreement. World Bank Press Release, December 3 World Bank, Washington, D.C., USA.

Mechler, R. et al., 2020: Loss and Damage and limits to adaptation: recent IPCC insights and implications for climate science and policy. *Sustain. Sci.*, **15**, doi:10.1007/s11625-020-00807-9.

Meckling, J., N. Kelsey, E. Biber, and J. Zysman, 2015: Winning coalitions for climate policy. *Science*, **349(6253)**, 1170–1171, doi:10.1126/science.aab1336.

MEFIN Network and GI RFPI Asia, 2016: *Diagnostic toolkit for insurance against natural catastrophes for MSMEs in the agricultural and mining sectors*. GIZ, Bonn, Germany.

Meinshausen, M. et al., 2009: Greenhouse-gas emission targets for limiting global warming to 2 °C. *Nature*, **458(7242)**, 1158–1162, doi:10.1038/nature08017.

Meltzer, J.P., 2016: *Financing Low Carbon, Climate Resilient Infrastructure: The Role of Climate Finance and Green Financial Systems*. Global Economy & Development Working Paper 96 | September 2016. Brookings Institution, Washington, D.C., USA. doi:10.2139/ssrn.2841918.

Meltzer, J.P., 2018: *Blending Climate Funds to Finance Low-Carbon, Climate-Resilient Infrastructure*. Global Economy & Development Working Paper 120. Brookings Institution, Washington, D.C., USA doi:10.2139/ssrn.3205293.

Menanteau, P., D. Finon, and M.-L. Lamy, 2003: Prices versus quantities: choosing policies for promoting the development of renewable energy. *Energy Policy*, **31(8)**, 799–812, doi:10.1016/S0301-4215(02)00133-7.

Mengelkamp, E. et al., 2018: Designing microgrid energy markets: A case study: The Brooklyn Microgrid. *Appl. Energy*, **210**, 870–880, doi.org/10.1016/j.apenergy.2017.06.054.

Mercer, 2018: *Investment in African Infrastructure: Challenges and Opportunities*. Mercer, New York, NY, USA. 42 pp. https://www.mmc.com/content/dam/mmc-web/insights/publications/2018/november/Infrastructure-investing-in-Africa/gl-2018-wealth-investment-opportunities-in-african-infrastructure-summary-report-mercer-1.pdf (Accessed November 1, 2021).

Mercure, J.-F. et al., 2016: *Policy-induced energy technological innovation and finance for low-carbon economic growth*. European Commission, Brussels, Belgium. 78 pp. https://ec.europa.eu/energy/sites/ener/files/documents/ENER%20Macro-Energy_Innovation_D2%20Final%20%28Ares%20registered%29.pdf (Accessed November 1, 2021).

Mercure, J.-F. et al., 2018a: Macroeconomic impact of stranded fossil fuel assets. *Nat. Clim. Change*, **8(7)**, 588–593, doi:10.1038/s41558-018-0182-1.

Mercure, J.-F., A. Lam, S. Billington, and H. Pollitt, 2018b: Integrated assessment modelling as a positive science: private passenger road transport policies to meet a climate target well below 2 °C. *Clim. Change*, **151(2)**, 109–129, doi:10.1007/s10584-018-2262-7.

Mercure, J.-F. et al., 2019: Modelling innovation and the macroeconomics of low-carbon transitions: theory, perspectives and practical use. *Clim. Policy*, **19(8)**, 1019–1037, doi:10.1080/14693062.2019.1617665.

Mercure, J.-F. et al., 2021: Reframing incentives for climate policy action. *Nat. Energy*, **6**, 1133–1143, doi:10.1038/s41560-021-00934-2.

Mésonnier, J.-S. and B. Nguyen, 2021: *Showing off cleaner hands: mandatory climate-related disclosure by financial institutions and the financing of fossil energy*. Working Paper Series no. 800, Banque de France, Paris, France.

Micale, V., C. Wetherbee, R. Macquarie, and P. Rosane, 2020: *A Proposed Method for Measuring Paris Alignment of New Investment*. Climate Policy Initiative, San Francisco, CA, USA. 44 pp. https://www.climatepolicyinitiative.org/wp-content/uploads/2020/12/2.-A-Proposed-Method-for-Measuring-Paris-Alignment-of-New-Investment-3.pdf (Accessed November 1, 2021).

Michaelowa, A., M. Allen, and F. Sha, 2018: Policy instruments for limiting global temperature rise to 1.5°C – can humanity rise to the challenge? *Clim. Policy*, **18(3)**, 275–286, doi:10.1080/14693062.2018.1426977.

Michaelowa, A., I. Shishlov, and D. Brescia, 2019: Evolution of international carbon markets: lessons for the Paris Agreement. *Wiley Interdiscip. Rev. Clim. Chang.*, **10(6)**, doi:10.1002/wcc.613.

Michaelowa, A., S. Hoch, A.-K. Weber, R. Kassaye, and T. Hailu, 2021: Mobilising private climate finance for sustainable energy access and climate change mitigation in Sub-Saharan Africa. *Clim. Policy*, **21(1)**, 47–62, doi:10.1080/14693062.2020.1796568.

Mieu, B., 2021: France begins transfer of €5bn to BCEAO as part of CFA franc reform. *The Africa Report, May 5 2021*. https://www.theafricareport.com/85566/france-begins-transfer-of-e5bn-to-bceao-as-part-of-cfa-franc-reform/ (Accessed November 1, 2021).

Milbank, C., D. Coomes, and B. Vira, 2018: Assessing the progress of REDD+ projects towards the sustainable development goals. *Forests*, **9(10)**, 589, doi:10.3390/f9100589.

Miles, D., 1993: Testing for Short Termism in the UK Stock Market. *Econ. J.*, **103(421)**, 1379, doi:10.2307/2234472.

Miles, K. and M. Wiedmaier-Pfister, 2018: *Applying a gender lens to climate risk finance and insurance*. InsuResilience Global Partnership Secretariat, Bonn, Germany, 44 pp.

Millar, R.J. et al., 2017: Emission budgets and pathways consistent with limiting warming to 1.5 °C. *Nat. Geosci.*, **10(10)**, 741–747, doi:10.1038/ngeo3031.

Minniti, M. and W. Naudé, 2010: What Do We Know About The Patterns and Determinants of Female Entrepreneurship Across Countries? *Eur. J. Dev. Res.*, **22(3)**, 277–293, doi:10.1057/ejdr.2010.17.

MIT, 2016: *The value of Aggregators in Electricity Systems*. MITei working paper, MIT, Cambridge, MA, USA, 28 pp.

Mitchell, T., 2015: *Addressing the Financing and Debt Challenges of Commonwealth Small States*. Commonwealth Secretariat, London, UK, 12 pp. doi: 10.14217/5js65z34ffvl-en.

Mkandawire, T., 2010: Aid, Accountability, and Democracy in Africa. *Soc. Res.* **77(2)** 1149–1182.

MoE Fiji 2020, 2020: *Fiji Sovereign Green Bond Impact Report 2018*. MoE Fiji, Suva, Fiji, https://www.jstor.org/stable/23347123 (Accessed November 1, 2021).

Moftakhari, H. et al., 2021: Enabling incremental adaptation in disadvantaged communities: polycentric governance with a focus on non-financial capital. *Clim. Policy*, **21(3)**, 396–405, doi:10.1080/14693062.2020.1833824.

Mok, S., E. Mačiulytė, P.H. Bult, and T. Hawxwell, 2021: Valuing the Invaluable(?)—A Framework to Facilitate Stakeholder Engagement in the Planning of Nature-Based Solutions. *Sustainability*, **13(5)**, 2657, doi:10.3390/su13052657.

Moles-Grueso, S., P. Bertoldi, and B. Boza-Kiss, 2021: *Energy Performance Contracting in the Public Sector of the EU - 2020*. EUR 30614 EN, Publications Office of the European Union, Luxembourg, 133 pp. doi:10.2760/171970.

Monasterolo, I., 2020: Embedding finance in the macroeconomics of climate change: research challenges and opportunities ahead. *CESifo Working Paper*, **21(4)**, 25–32.

Monasterolo, I. and M. Raberto, 2017: Is There a Role for Central Banks in the Low-Carbon Transition? A Stock-Flow Consistent Modelling Approach. *SSRN Electron. J.*, doi:10.2139/ssrn.3075247.

Monasterolo, I. and M. Raberto, 2018: The EIRIN Flow-of-funds Behavioural Model of Green Fiscal Policies and Green Sovereign Bonds. *Ecol. Econ.*, **144** (July 2017), 228–243, doi:10.1016/j.ecolecon.2017.07.029.

Monasterolo, I. and S. Battiston, 2020: Assessing Forward-Looking Climate Risks in Financial Portfolios: A Science-Based Approach for Investors and Supervisors. In: *NGFS Occasional paper. Case Studies of Environmental Risk Analysis Methodologies*. Central Banks and Supervisors Network for Greening the Financial System (NGFS), pp. 52–72.

Monasterolo, I. and L. de Angelis, 2020: Blind to carbon risk? An analysis of stock market reaction to the Paris Agreement. *Ecol. Econ.*, **170** (October 2019), 106571, doi:10.1016/j.ecolecon.2019.106571.

Monasterolo, I., S. Battiston, A.C. Janetos, and Z. Zheng, 2017: Vulnerable yet relevant: the two dimensions of climate-related financial disclosure. *Clim. Change*, **145(3–4)**, 495–507, doi:10.1007/s10584-017-2095-9.

Moody's Investors Service, 2016: *CPV Power Plant No.1 Bond SPV (RF) Limited*. 9 pp. http://www.verdigris.co.za/wp-content/uploads/2017/12/20160526-Moodys-CPV-Credit-Opinion.pdf (Accessed November 1, 2021).

Moody's Investors Service, 2020: *The Rockefeller Foundation, NY: New Issuer*. 8 pp. www.moodys.com.

Moody's Investors Service, 2021: *Government of Ghana - B3 negative: Annual credit analysis*. 30 pp.

Morena, E., 2018: L'odeur de l'argent: les fondations philanthropiques dans le débat climatique international. *Rev. Int. Strat.*, **109(1)**, 115, doi:10.3917/ris.109.0115.

Morgan Stanley, 2017: *Sustainable Signals New Data from the Individual Investor*. Morgan Stanley Institute For Sustainable Investing, New York, NY, USA. 7 pp.

Morita, K. and K. Matsumoto, 2018: Synergies among climate change and biodiversity conservation measures and policies in the forest sector: A case study of Southeast Asian countries. *For. Policy Econ.*, **87** (September 2017), 59–69, doi:10.1016/j.forpol.2017.10.013.

Moro, A., O. Walker, and A. Hooks, 2018: *The Demand for Financing Climate Projects in Cities*. C40 Cities Finance Facility, London, UK, 17 pp. https://cff-prod.s3.amazonaws.com/storage/files/QMDd8kN4kaVWSbyCrR7zN4NveA1gh9NgOCOscnhG.pdf (Accessed November 1, 2021).

Morse, A., 2015: *UK Guarantees scheme for infrastructure*. National Audit Office, London, UK, 65 pp.

Moyo, D., 2009: Dead Aid:Why aid is not working. *Dead Aid Why aid is not Working and how there is a better Way for Africa* Farrar, Straus and Giroux, New York, NY, USA, pp. 29–47.

Muhmad, S.N. and R. Muhamad, 2020: Sustainable business practices and financial performance during pre- and post-SDG adoption periods: a systematic review. *J. Sustain. Financ. Invest.*, **11(4)**, 291–309, doi:10.1080/20430795.2020.1727724.

Munich RE, 2019: The natural disasters of 2018 in figures: Losses in 2018 dominated by wildfires and tropical storms. Munich RE, Munich, Germany.

Murphy, D. and J.-E. Parry, 2020: *Filling the Gap: A review of Multilateral Development Banks' efforts to scale up financing of climate adaptation*. IISD, Manitoba, Canada, 47 pp. https://www.iisd.org/system/files/2020-12/filling-gap-financing-climate-adaptation.pdf (Accessed November 1, 2021).

Muttitt, G. and S. Kartha, 2020: Equity, climate justice and fossil fuel extraction: principles for a managed phase out. *Clim. Policy*, **20(8)**, 1024–1042, doi:10.1080/14693062.2020.1763900.

Mysiak, J. et al., 2018: Brief communication: Strengthening coherence between climate change adaptation and disaster risk reduction. *Nat. Hazards Earth Syst. Sci.*, **18(11)**, 3137–3143, doi:10.5194/nhess-18-3137-2018.

Nae, T.-M. and N.-A. Panie, 2021: European Green Deal: The Recovery Strategy Addressing Inequalities. *J. East. Eur. Res. Bus. Econ.*, **2021**, 887980, doi:10.5171/2021.887980.

Nakhooda, S. et al., 2014a: *Climate Finance: Is it making a difference?* ODI, London, UK.

Nakhooda, S. et al., 2014b: *Ten things to know about Climate Finance in 2014*. ODI briefing/policy papers December 5 2014, ODI, London, UK.

Nanayakkara, M. and S. Colombage, 2019: Do investors in Green Bond market pay a premium? Global evidence. *Appl. Econ.*, **51(40)**, 4425–4437, doi:10.1080/00036846.2019.1591611.

NAP Global Network, 2017: *Financing National Adaptation Plan (NAP) Processes: Contributing to the achievement of nationally determined contribution (NDC) adaptation goals*. NAP Globa Netowrk, Winnipeg, Manitoba, Canada, 74 pp. http://napglobalnetwork.org/wp-content/uploads/2017/08/napgn-en-2017-financing-nap-processes-contributing-to-the-achievement-of-ndc-goals.pdf.

Naqvi, A. and E. Stockhammer, 2018: Directed Technological Change in a Post-Keynesian Ecological Macromodel. *Ecol. Econ.*, **154** (June), 168–188, doi:10.1016/j.ecolecon.2018.07.008.

Naran, B. et al., 2021: *Global landscape of climate finance 2021*. Climate Policy Initiative, London, UK, 55 pp. https://www.climatepolicyinitiative.org/wp-content/uploads/2021/10/Global-Landscape-of-Climate-Finance-2021.pdf (Accessed November 1, 2021).

Narassimhan, E., K.S. Gallagher, S. Koester, and J.R. Alejo, 2018: Carbon pricing in practice: a review of existing emissions trading systems. *Clim. Policy*, **18(8)**, 967–991, doi:10.1080/14693062.2018.1467827.

Nassiry, D., 2018: *The Role of Fintech in Unlocking Green Finance: Policy Insights for Developing Countries*. ADB Institute, Tokyo, Japan, 26 pp. https://www.adb.org/publications/role-fintech-unlocking-green-finance (Accessed November 1, 2021).

Nathwani, J. and D.M. Kammen, 2019: Affordable Energy for Humanity: A Global Movement to Support Universal Clean Energy Access. *Proc. IEEE*, **107(9)**, 1780–1789, doi:10.1109/JPROC.2019.2918758.

National Infrastructure Commission (NIC), 2021: UK National Infrastructure Assessment. https://nic.org.uk/studies-reports/national-infrastructure-assessment/ (Accessed August 18, 2021).

Natixis, 2019: Natixis rolls out its Green Weighting Factor and becomes the first bank to actively manage its balance sheet's climate impact. Groupe

BCPE, Paris, France. https://natixis.groupebpce.com/natixis/en/green-weighting-factor-and-climate-trajectory-rqaz5_111931.html (Accessed November 1, 2021).

Nauman, B., 2020: Green bonds set to keep flying off shelves in 2020. *Financial Times*, January 7.

NDC Partnership, 2020: *Adaptation and NDCs: From Analysis and Planning to Action and Ambition Raising (Extended Report)*. NDP Partnership, Washington DC, USA, 22 pp. https://ndcpartnership.org/sites/default/files/Extended_Report-Adaptation_and_NDCs_From_Analysis_and_Planning_to_Action_and_Ambition_Raising_October-2020.pdf (Accessed November 1, 2021).

NEA, 1989: *Projected costs of generating electricity from power stations for commissioning in the period 1995-2000*. OECD, Paris, France. https://inis.iaea.org/search/search.aspx?orig%7B%5C_%7Dq=RN:21072727 (Accessed November 1, 2021).

Nedopil, C., T. Dordi, and O. Weber, 2021: The nature of global green finance standards-evolution, differences, and three models. *Sustainability*, **13(7)**, doi:10.3390/su13073723.

Negreiros, P. et al., 2021: The State of Cities Climate Finance. June.

Nelson, D., M. Huxham, S. Muench, and B. O'Connell, 2016: *Policy and investment in German renewable energy*. Climate Policy Initiative, San Francisco, USA. 9 pp. https://climatepolicyinitiative.org/wp-content/uploads/2016/04/Policy-and-investment-in-German-renewable-energy-Summary.pdf (Accessed November 1, 2021).

Newell, P. and D. Mulvaney, 2013: The political economy of the 'just transition.' *Geogr. J.*, **179(2)**, 132–140, doi:10.1111/geoj.12008.

Ng, T.H. and J.Y. Tao, 2016: Bond financing for renewable energy in Asia. *Energy Policy*, **95**, 509–517, doi:10.1016/j.enpol.2016.03.015.

NGFS, 2019: *A call for action Climate change as a source of financial risk*. Central Banks and Supervisors Network for Greening the Financial System (NGFS), 42 pp. https://www.ngfs.net/sites/default/files/medias/documents/synthese_ngfs-2019_-_17042019_0.pdf (Accessed November 1, 2021).

NGFS, 2020: *Annual Report 2019*. Central Banks and Supervisors Network for Greening the Financial System (NGFS), Paris, France, 9 pp. https://www.ngfs.net/sites/default/files/medias/documents/ngfs_annual_report_2019.pdf (Accessed November 1, 2021).

Nhamo, G. and S. Nhamo, 2016: Paris (COP21) Agreement: Loss and damage, adaptation and climate finance issues. *Int. J. African Renaiss. Stud. - Multi-, Inter- Transdiscipl.*, **11(2)**, 118–138, doi:10.1080/18186874.2016.1212479.

Nicol, M., I. Shishlov, and I. Cochran, 2017: *Green Bonds: what contribution to the Paris Agreement and how to maximize it?* Institute for Climate Economics, Paris, France, 12 pp. https://www.i4ce.org/wp-core/wp-content/uploads/2017/12/2017-I4CE-Green-Bonds-Energy-Transition-Exec-Sum.pdf (Accessed November 1, 2021).

Nicol, M., I. Cochran, and I. Shishlov, 2018a: *Green Bonds: Improving their contribution to the low-carbon and climate resilient transition*. Institute for Climate Economics, Paris, France, https://www.i4ce.org/download/green-bonds-improving-their-contribution/.

Nicol, M., I. Cochran, and I. Shishlov, 2018b: *Green Bonds: Improving their contribution to the low-carbon and climate resilient transition*. Institute for Climate Economics, Paris, France, 57 pp. https://www.i4ce.org/download/green-bonds-improving-their-contribution/ (Accessed November 1, 2021).

NOAA NCEI, 2019: *Billion-Dollar Weather and Climate Disasters: Overview*. NOAA National Centers for Environmental Information (NCEI), Washington, D.C., USA. Nordhaus, W., 2015: Climate Clubs: Overcoming Free-riding in International Climate Policy. *Am. Econ. Rev.*, **105(4)**, 1339–1370, doi:10.1257/aer.15000001.

Nordhaus, W., 2018: Projections and uncertainties about climate change in an era of minimal climate policies. *Am. Econ. J. Econ. Policy*, **10(3)**, 333–360, doi:10.1257/pol.20170046.

Nordhaus, W.D., 2007: A review of the Stern Review on the economics of climate change. *J. Econ. Lit.*, **45(3)**, 686–702, doi:10.1257/jel.45.3.686.

Nordpool, 2021: *Annual Review 2020: Navigating a changed world*. Nordpool, Lysaker, Norway, 17 pp. https://www.nordpoolgroup.com/49eea7/globalassets/download-center/annual-report/annual-review-2020.pdf (Accessed November 1, 2021).

Nyfors, T. et al., 2020: Ecological Sufficiency in Climate Policy: Towards Policies for Recomposing Consumption. *Futura*, **(3)**, 1–23.

NZAM, 2020: Net Zero Asset Managers Initiative. https://www.netzeroassetmanagers.org/ (Accessed December 1, 2020).

O'Callaghan, B.J., 2021: *Are We Building Back Better?: Evidence from 2020 and Pathways for Inclusive Green Recovery Spending*. UNEP, Geneva, Switzerland, 57 pp. https://wedocs.unep.org/bitstream/handle/20.500.11822/35281/AWBBB.pdf (Accessed November 1, 2021).

Obregon, L., 2018: Empire, Racial Capitalism and International Law: The Case of Manumitted Haiti and the Recognition Debt. *Leiden J. Int. Law*, **31(3)**, 597–615, doi:10.1017/S0922156518000225.

Obstfeld, M. and E.M. Truman, 2021: The new SDR allocation will benefit all countries. *Realtime Economic Issues Watch*, March 25.

ODI, 2018: *Clean energy project preparation facilities: Mapping the global landscape*. ODI, London, UK, 39 pp. https://www.odi.org/publications/11229-clean-energy-project-preparation-facilities (Accessed November 1, 2021).

OECD-DAC, 2021: *OECD Blended Finance and Impact Week, Delivering on the 2030 Agenda in the COVID-19 era*. OECD Publishing, Paris, France, 18 pp.

OECD, 2015a: *Climate Finance in 2013-14 and the USD 100 billion Goal*. OECD Publishing, Paris, France, 68 pp.

OECD, 2015b: *Aligning Policies for a Low-carbon Economy*. OECD Publishing, Paris, France, 242 pp.

OECD, 2015c: *Green Investment Banks: Scaling up Private Investment in Low-carbon, Climate-resilient Infrastructure*. OECD Publishing, Paris, France, 19 pp. http://www.oecd.org/environment/cc/Green-Investment-Banks-POLICY-PERSPECTIVES-web.pdf (Accessed November 1, 2021).

OECD, 2016: *OECD DAC Rio Markers for Climate: Handbook*. OECD Publishing, Paris, France, 34 pp.

OECD, 2017a: *Pension markets in focus 2017*. OECD Publishing, Paris, France, 38 pp. https://www.oecd.org/daf/fin/private-pensions/Pension-Markets-in-Focus-2017.pdf (Accessed November 1, 2021).

OECD, 2017b: *Investing in Climate, Investing in Growth*. OECD Publishing, Paris, France, 314 pp.

OECD, 2018a: *Private Philanthropy for Development,* OECD Publishing, Paris, France.

OECD, 2018b: *OECD Companion to the Inventory of Support Measures for Fossil Fuels 2018*. OECD Publishing, Paris, France.

OECD, 2019a: *Climate Finance Provided and Mobilised by Developed Countries in 2013-2017*. OECD Publishing, Paris, 48 pp.

OECD, 2019b: DAC: Amounts Mobilised from the Private Sector for Development, OECD Publishing, Paris, France.

OECD, 2019c: *Blockchain Technologies as a Digital Enabler for Sustainable Infrastructure*. OECD Publishing, Paris, France, 4 pp. https://www.financialcapability.gov.au/files/blockchain-technologies-as-a-digital-enabler-for-sustainable-infrastructure.pdf (Accessed November 1, 2021).

OECD, 2019d: *Biodiversity: Finance and the Economic and Business Case for Action. A report prepared by the OECD for the French G7 Presidency and the G7 Environment Ministers' Meeting, 5-6 May 2019*. OECD Publishing, Paris, France. 96 pp. http://www.oecd.org/environment/resources/biodiversity/G7-report-Biodiversity-Finance-and-the-Economic-and-Business-Case-for-Action.pdf (Accessed November 1, 2021).

OECD, 2020a: *Climate Change Adaptation and Disaster Risk Reduction*. OECD Publishing, Paris, France.

OECD, 2020b: *Climate Finance Provided and Mobilised by Developed Countries in 2013-18*. OECD Publishing, Paris, France.

OECD, 2020c: Amounts mobilised from the private sector for development – OECD Publishing, Paris, France. https://www.oecd.org/dac/financing-sustainable-development/development-finance-standards/mobilisation.htm (Accessed October 21, 2021).

OECD, 2021a: *OECD Secretary-General Tax Report to G20 Finance Ministers and Central Bank Governors – April 2021*. OECD Publishing, Paris, France, 100 pp. https://www.oecd.org/tax/oecd-secretary-general-tax-report-g20-finance-ministers-april-2021.pdf (Accessed November 1, 2021).

OECD, 2021b: *OECD Economic Outlook, Volume 2021 Issue 1*. OECD Publishing, Paris, France, 218 pp.

OECD, 2021c: *Climate change and long term fiscal sustainability*. OECD Publishing, Paris, France. 46 pp. https://www.oecd.org/gov/budgeting/scoping-paper-on-fiscal-sustainability-and-climate-change.pdf (Accessed November 1, 2021).

OECD, 2021d: *OECD Science, Technology and Innovation Outlook 2020*. OECD, Paris, France, 207 pp.

OECD and Bloomberg, 2015: *Green bonds Mobilising the debt capital markets for a low-carbon transition*. OECD Publishing, Paris, France. 24 pp.

OECD and World Bank, 2016: *Climate and Disaster Resilience Financing in Small Island Developing States*. OECD Publishing, Paris, France, 70 pp.

Olusoga, D., 2015: The history of British slave ownership has been buried: now its scale can be revealed. *The Guardian*, July 12.

Omolo, N., P. Mafongoya, O. Ngesa, and K. Voi, 2017: Gender and Resilience to Climate Variability in Pastoralists Livelihoods System: Two Case Studies in Kenya. *Sustain. Dev.*, **10(2)**, 218–227.

Osabuohien, E.S., J.O. Ejemeyovwi, O.B. Ihayere, C.M.W. Gitau, and F.M. Oyebola, 2021: Post-Pandemic Renewable Energy Development in Sub-Saharan Africa. *Sustain. Clim. Change*, **14(3)**, 183–192, doi:10.1089/scc.2020.0077.

Osama, A., N. Eltouny, F. Adel, and M. Adel, 2021: *Needs of African countries related to implementing the UN Framework Convention on Climate Change and the Paris Agreement*. UNFCCC, Abidjan, Côte d'Ivoire, 179 pp. https://unfccc.int/sites/default/files/resource/Needs Report_African counties_AfDB_FINAL.pdf (Accessed November 1, 2021).

Oseni, M.O. and M.G. Pollitt, 2016: The promotion of regional integration of electricity markets: Lessons for developing countryies. *Energy Policy*, **88**, 628–638, doi:10.1016/j.enpol.2015.09.007.

Ouazad, A. and M. Kahn, 2019: *Mortgage Finance and Climate Change: Securitization Dynamics in the Aftermath of Natural Disasters*. National Bureau of Economic Research, Cambridge, MA, USA, 56 pp.

Oxford Economics, 2017: *Global infrastructure outlook: Infrastructure investment needs 50 countries, 7 sectors to 2040*. Oxford Economics, Sydney, Australia, 213 pp. https://cdn.gihub.org/outlook/live/methodology/Global+Infrastructure+Outlook+-+July+2017.pdf (Accessed November 1, 2021).

Padigala, B. and S. Kraleti, 2014: Financing low carbon urban development through clean development mechanism. *Int. J. Environ. Sci.*, **5(1)**, 98–116, doi:10.6088/ijes.2014050100009.

Pahle, M. et al., 2018: Sequencing to ratchet up climate policy stringency. *Nat. Clim. Change*, **8(10)**, 861–867, doi:10.1038/s41558-018-0287-6.

Pai, S., H. Zerriffi, J. Jewell, and J. Pathak, 2020: Solar has greater techno-economic resource suitability than wind for replacing coal mining jobs. *Environ. Res. Lett.*, **15(3)**, 034065, doi:10.1088/1748-9326/ab6c6d.

Painter, M., 2020: An inconvenient cost: The effects of climate change on municipal bonds. *J. Financ. Econ.*, **135(2)**, 468–482, doi:10.1016/j.jfineco.2019.06.006.

Panda, A. and S. Surminski, 2020: *Climate and disaster risk insurance in low income countries: Reflections on the importance of indicators and frameworks for monitoring the performance and impact of CDRI*.Centre for Climate Change Economics and Policy Working Paper 377/Grantham Research Institute on Climate Change and the Environment Working Paper 348. London School of Economics and Political Science, London, UK.

Panizza, U. and F. Taddei, 2020: *Local Currency Denominated Sovereign Loans A Portfolio Approach to Tackle Moral Hazard and Provide Insurance*. IHEID Working Papers 09-2020, Economics Section, The Graduate Institute of International Studies, Geneva, Switzerland.

Paris21, 2018: *Istanbul Programme of Action (2011-2020). Preliminary findings*. 19 pp. Paris21, Paris, France. https://www.un.org/ohrlls/sites/www.un.org.ohrlls/files/istanbul-programme-of-action-2011-2020-preliminary-findings-el-iza-mohamedou-paris-21-secretariat.pdf (Accessed November 1, 2021).

Partridge, C. and F.R. Medda, 2020: The evolution of pricing performance of green municipal bonds. *J. Sustain. Financ. & Invest.*, **10(1)**, 44–64, doi:10.1080/20430795.2019.1661187.

Paul, H., 2019: *Market solutions to help climate victims fail human rights test*. ActionAid International, Johannesburg, South Africa, 60 pp.

Pauw, P., K. Mbeva, and H. van Asselt, 2019: Subtle differentiation of countries' responsibilities under the Paris Agreement. *Palgrave Commun.*, **5(1)**, 86, doi:10.1057/s41599-019-0298-6.

Pauw, W.P. et al., 2016: NDC Explorer. UNFCCC, Bonn, Germany, doi:10.23661/ndc_explorer_2016_1.0.

Pauw, W.P., P. Castro, J. Pickering, and S. Bhasin, 2020: Conditional nationally determined contributions in the Paris Agreement: foothold for equity or Achilles heel? *Clim. Policy*, **20(4)**, 468–484, doi:10.1080/14693062.2019.1635874.

Peake, S. and P. Ekins, 2017: Exploring the financial and investment implications of the Paris Agreement. *Clim. Policy*, **17(7)**, 832–852, doi:10.1080/14693062.2016.1258633.

Pearson, A.R., C.G. Tsai, and S. Clayton, 2021: Ethics, morality, and the psychology of climate justice. *Curr. Opin. Psychol.*, **42**, 36–42, doi:10.1016/j.copsyc.2021.03.001.

Peltzman, S., 1992: Voters as Fiscal Conservatives. *Q. J. Econ.*, **107(2)**, 327–361, doi:10.2307/2118475.

Pereira dos Santos, P. and M. Kearney, 2018: *Multilateral Development Banks' Risk Mitigation Instruments for Infrastructure Investment*. IDB, Washington DC, USA, 40 pp. doi: 10.18235/0001008.

Pereira Dos Santos, P., 2018: *Introductory Guide to Infrastructure Guarantee Products from Multilateral Development Banks*. IDB, Washington, DC, USA, 48 pp. doi: 10.18235/0001517.

Pereira, J., 2017: *Blended Finance: What it is, how it works and how it is used*. Oxfam, Oxford, 51 pp. https://www.oxfam.org/en/research/blended-finance-what-it-how-it-works-and-how-it-used (Accessed September 25, 2021).

Perera, L., C. Jubb, and S. Gopalan, 2019: A comparison of voluntary and mandated climate change-related disclosure. *J. Contemp. Account. & Econ.*, **15(2)**, 243–266, doi:10.1016/j.jcae.2019.100157.

Perez, C., 2009: The double bubble at the turn of the century: technological roots and structural implications. *Cambridge J. Econ.*, **33(4)**, 779–805, doi:10.1093/cje/bep028.

Peterson, L. and J. Skovgaard, 2019: Bureaucratic politics and the allocation of climate finance. *World Dev.*, **117**, 72–97, doi:10.1016/j.worlddev.2018.12.011.

Pfaff, N., Altun, O., and Jia, Y., 2021: *Overview and Recommendations for Sustainable Finance Taxonomies*. International Capital Market Association, Zurich, Switzerland.

Pickering, J., C. Betzold, and J. Skovgaard, 2017: Special issue: managing fragmentation and complexity in the emerging system of international climate finance. *Int. Environ. Agreements Polit. Law Econ.*, **17(1)**, 1–16, doi:10.1007/s10784-016-9349-2.

Pickup, L and Mantero, C, 2017: *Transport Infrastructure: Expert group report*. Final Report On 'Transport Accessibility For The EU Outermost Regions' European Commission, Brussels, Belgium. https://ec.europa.eu/regional_policy/sources/policy/themes/outermost-regions/pdf/transport_report_en.pdf (Accessed November 1, 2021).

PIDG, 2019: *Five-Year Strategic Plan 2019-2023*. Private Infrastructure Development Group, London, UK, 56 pp. https://www.pidg.org/wp-content/uploads/2020/09/PIDG-Five-Year-Strategic-Plan%7B%5C_

%7D2019-2023%7B%5C_%7Dpublished%7B%5C_%7Dpages.pdf (Accessed November 1, 2021).
Piketty, P., 2020: Confronting racism, repairing history. *Le Monde Blog.*, June 16, 2021.
Pincus, J.A., 1963: The Cost of Foreign Aid. *Rev. Econ. Stat.*, **45(4)**, 360, doi:10.2307/1927920.
Pindyck, R.S., 2013: Climate change policy: What do the models tell us? *J. Econ. Lit.*, **51(3)**, 860–872, doi:10.1257/jel.51.3.860.
Pingali, P., A. Aiyar, M. Abraham, and A. Rahman, 2019: *Transforming Food Systems for a Rising India*. Springer International Publishing, Cham, Switzerland.
Plank, D.N., 1993: Aid, debt, and the end of sovereignty: Mozambique and its donors. *J. Mod. Afr. Stud.*, Volume 31, Issue 3, pp. 407 - 430, doi:10.1017/S0022278X00012015.
Pollitt, H. and J.-F. Mercure, 2018a: The role of money and the financial sector in energy-economy models used for assessing climate and energy policy. *Clim. Policy*, **18(2)**, 184–197, doi:10.1080/14693062.2016.1277685.
Pollitt, H. and J.-F. Mercure, 2018b: The role of money and the financial sector in energy-economy models used for assessing climate and energy policy. *Clim. Policy*, **18(2)**, 184–197, doi:10.1080/14693062.2016.1277685.
Pollitt, H., R. Lewney, B. Kiss-Dobronyi, and X. Lin, 2021: Modelling the economic effects of COVID-19 and possible green recovery plans: a post-Keynesian approach. *Clim. Policy*, **21(10)**, 1257–1271, doi:10.1080/14693062.2021.1965525.
Polzin, F., 2017: Mobilizing private finance for low-carbon innovation – A systematic review of barriers and solutions. *Renew. Sustain. Energy Rev.*, **77** (April), 525–535, doi:10.1016/j.rser.2017.04.007.
Polzin, F., P. von Flotow, and L. Klerkx, 2016: Addressing barriers to eco-innovation: Exploring the finance mobilisation functions of institutional innovation intermediaries. *Technol. Forecast. Soc. Change*, **103**, 34–46, doi:10.1016/j.techfore.2015.10.001.
Prag, A., D. Röttgers, and I. Scherrer, 2018: *State-Owned Enterprises and the Low-Carbon Transition*. OECD Environment Working Papers, No. 129, OECD Publishing, Paris, France, 57 pp. doi:10.1787/06ff826b-en.
PwC, 2014: *The road ahead: Gaining momentum from energy transformation.*, PwC global power and utilities, London, UK, https://www.pwc.com/gx/en/utilities/publications/assets/pwc-the-road-ahead.pdf (Accessed November 1, 2021).
PwC, 2015: *Africa Asset Management 2020*. 136 pp. PwC Asset Management, London, UK, http://www.amafrica2020.com/amafrica2020/docs/am-africa-2020.pdf (Accessed November 1, 2021).
PWYF, 2020: *Aid Transparency Index 2020*. Publish What You Find, London, UK, 29 pp.
Qin, Q., F. Liang, L. Li, and Y.M. Wei, 2017: Selection of energy performance contracting business models: A behavioral decision-making approach. *Renew. Sustain. Energy Rev.*, **72** (February 2016), 422–433, doi:10.1016/j.rser.2017.01.058.
Rainforest Action Network et al., 2020: *Banking on Climate Change - Fossil Fuel Finance Report 2020*. Rainforest Action Network, San Francisco, CA, USA.
Ramlee, S. and B. Berma, 2013: Financing gap in Malaysian small-medium enterprises: A supply-side perspective. *South African J. Econ. Manag. Sci.*, **16(5)**, 115–126.
Rainforest Action Network et al., 2021: *Banking on climate chaos*. Rainforest Action Network, San Francisco, CA, USA. 157 pp. https://www.ran.org/wp-content/uploads/2021/03/Banking-on-Climate-Chaos-2021.pdf (Accessed November 1, 2021).
Reboredo, J.C., 2018: Green bond and financial markets: Co-movement, diversification and price spillover effects. *Energy Econ.*, **74**, 38–50, doi:10.1016/j.eneco.2018.05.030.
Reilly, J.M., Y.-H.H. Chen, and H.D. Jacoby, 2021: The COVID-19 effect on the Paris agreement. *Humanit. Soc. Sci. Commun.*, **8(1)**, 16, doi:10.1057/s41599-020-00698-2.
Reisch, L.A. et al., 2021: Mitigating climate change via food consumption and food waste: A systematic map of behavioral interventions. *J. Clean. Prod.*, **279**, 123717, doi:10.1016/j.jclepro.2020.123717.
Rempel, A. and J. Gupta, 2020: Conflicting commitments? Examining pension funds, fossil fuel assets and climate policy in the organisation for economic co-operation and development (OECD). *Energy Res. Soc. Sci.*, **69**, 101736, doi:10.1016/J.ERSS.2020.101736.
REN21, 2016: *Renewables 2016 Global Status Report*. REN21 Secretariat., Paris, France.
REN21, 2019: *Renewables 2019 Global Status Report*. REN21 Secretariat., Paris, France. 336 pp.
Rendall, M., 2021: Public debt and intergenerational ethics: how to fund a clean technology 'Apollo program'? **21(7)**, 976–982, doi:10.1080/14693062.2021.1935679.
Resano, J.R.M. and S. Gallego, 2021: *G-20 Debt Relief Initiatives for Low-Income Countries During the Pandemic*. Economic Bulletin, Banco de España, issue 3/2021, Madrid, Spain, 12 pp.
REScoop, 2020: Mutual for Energy Communities. REScoop, Antwerp, Belgium.
Research and Markets, 2017: *Virtual Power Plant Market - Industry Forecast, 2017-2023*. Allied Market Research, Portland, OR, USA.
Roberts, J.T. and R. Weikmans, 2017: Postface: fragmentation, failing trust and enduring tensions over what counts as climate finance. *Int. Environ. Agreements Polit. Law Econ.*, **17(1)**, 129–137, doi:10.1007/s10784-016-9347-4.
Roberts, J.T. et al., 2021: Rebooting a failed promise of climate finance. *Nat. Clim. Change*, **11(3)**, 180–182, doi:10.1038/s41558-021-00990-2.
Robins, N. and J. Rydge, 2020: *Why a just transition is crucial for effective climate action. Vivid Economics, London, UK*. https://www.unpri.org/inevitable-policy-response/why-a-just-transition-is-crucial-for-effective-climate-action/4785.article (Accessed November 1, 2021).
Rochedo, P.R.R. et al., 2021: Is Green Recovery Enough? Analysing the Impacts of Post-COVID-19 Economic Packages. *Energies*, **14(17)**, 5567, doi:10.3390/en14175567.
Rockefeller Foundation and Shujog, 2016: *IIX women's livelihood bond. IIX sustainability bonds: Changing finance, financing change*. Rockefeller Foundation, Shujog Opportunity for Impact, Singapore, 57 pp.
Rode, J. et al., 2019: Why 'blended finance' could help transitions to sustainable landscapes: Lessons from the Unlocking Forest Finance project.' *Ecosyst. Serv.*, **37** (March), 100917, doi:10.1016/j.ecoser.2019.100917.
Rodríguez-Caballero, C.V. and J.E. Vera-Valdés, 2020: Long-lasting economic effects of pandemics: Evidence on growth and unemployment. *Econometrics*, **8(3)**, 1–16, doi:10.3390/econometrics8030037.
Roncoroni, A., S. Battiston, L.O.L. Escobar Farfan, and S. Martinez-Jaramillo, 2020: Climate risk and financial stability in the network of banks and investment funds.
Roncoroni, A., Battiston, S., Escobar-Farfán, L.O. and Martinez-Jaramillo, S., 2021. Climate risk and financial stability in the network of banks and investment funds. *Journal of Financial Stability*, **54**, 100870, doi: 10.1016/j.jfs.2021.100870.
Roos, M., 2021: Financing the climate transition: London Stock Exchange extends its Sustainable Bond Market. *London Stock Exchange News and Insights*, February 19.
Rosenow, J. and N. Eyre, 2016: A post mortem of the Green Deal: Austerity, energy efficiency, and failure in British energy policy. *Energy Res. & Soc. Sci.*, **21**, 141–144, doi:10.1016/j.erss.2016.07.005.
Roth, N., T. Thiele, and M. von Unger, 2019: *Blue Bonds: financing resilience of coastal ecosystems*. BNCFF, Gland, Switzerland. 68 pp. https://www.4climate.com/dev/wp-content/uploads/2019/04/Blue-Bonds_final.pdf (Accessed November 1, 2021).
Röttgers, D. and B. Anderson, 2018: *Power struggle: Decarbonising the electricity sector*. OECD Environment Working Paper No. 139, OECD, Paris, France. 51 pp.

15

Rozenberg, J. and M. Fay, 2019: *Beyond the Gap: How Countries Can Afford the Infrastructure They Need while Protecting the Planet*. World Bank, Washington DC, USA, 175 pp.

Rudebusch, G.D., 2019: *Climate Change and the Federal Reserve*. FRBSF Economic Letter. FRBSF, San Francisco, USA. https://www.frbsf.org/economic-research/files/el2019-09.pdf.

Ruger, J.P. and R. Horton, 2020: Justice and health: The Lancet–Health Equity and Policy Lab Commission. *Lancet*, **395(10238)**, 1680–1681, doi:10.1016/S0140-6736(20)30928-4.

Saadi Sedik, T. and R. Xu, 2020: *A Vicious Cycle: How Pandemics Lead to Economic Despair and Social Unrest*. Working Paper No. 2020/216, IMF, Washington, D.C., USA.

Sachs, J.D. and A.M. Warner, 2001: The curse of natural resources. *Eur. Econ. Rev.*, **45(4–6)**, 827–838, doi:10.1016/S0014-2921(01)00125-8.

Saez, E. and G. Zucman, 2020: The Rise of Income and Wealth Inequality in America: Evidence from Distributional Macroeconomic Accounts. *J. Econ. Perspect.*, **34(4)**, 3–26, doi:10.1257/jep.34.4.3.

Saha, D. and S. D'Almeida, 2017: Green Municipal Bonds. *In Finance for City Leaders Handbook: Improving Municipal Finance to Deliver Better Services* [Kamiya, Macro and L.-Y. Zhang, (eds.)]. UN-Habitat, Nairobi, Kenya. 98–118 pp.

Saldanha, C., 2006: Rethinking Capacity Development. *Int. Public Manag. Rev.*, **7(2)**.

Sandler, T., 1998: Global and Regional Public Goods: A Prognosis for Collective Action. *Fisc. Stud.*, **19(3)**, 221–247, doi:10.1111/j.1475-5890.1998.tb00286.x.

Sandler, T., 2015: Collective action: fifty years later. *Public Choice*, **164(3–4)**, 195–216, doi:10.1007/s11127-015-0252-0.

Sardo, M.C., 2020: Responsibility for climate justice: Political not moral. *Eur. J. Polit. Theory*, doi:10.1177/1474885120955148.

SBN, 2018: *Sustainable Banking Network Creating Green Bond Markets – Insights, Innovations, and Tools from Emerging Markets*. Sustainable Banking NetworkWashington DC, USA, 74 pp. https://www.ifc.org/wps/wcm/connect/topics%7B%5C_%7Dext%7B%5C_%7Dcontent/ifc%7B%5C_%7Dexternal%7B%5C_%7Dcorporate%7B%5C_%7Dsite/sustainability-at-ifc/publications/publications%7B%5C_%7Dreport%7B%5C_%7Dsbngreenbond2018 (Accessed November 1, 2021).

Schaeffer, M. et al., 2014: *Loss and Damage in Africa*. UNECA/ACPC, Addis Ababa, Ethiopia, 59 pp. https://climateanalytics.org/media/uneca__2014__loss_and_damage_in_africa.pdf (Accessed November 1, 2021).

Schaer, C. and N. Kuruppu, 2018: *Private-sector action in adaptation: Perspectives on the role of micro, small and medium size enterprises*. UNEP DTU Partnership, Copenhagen, Denmark, 198 pp. https://backend.orbit.dtu.dk/ws/files/162053774/MSME_Adaptation_updated_WEB.pdf (Accessed November 1, 2021).

Schäfer, L., E. Waters, S. Kreft, and M. Zissener, 2016: *Making climate risk insurance work for the most vulnerable: Seven guiding principles*. Unu-Ehs Publication Series Policy Report 2016 No.1, UNU-EHS, Bonn, Germany, 60 pp. https://collections.unu.edu/eserv/UNU:5830/MCII_ProPoor_161031_Online_meta.pdf (Accessed November 1, 2021).

Schäfer, L., K. Warner, and S. Kreft, 2019: Exploring and Managing Adaptation Frontiers with Climate Risk Insurance. In: *Loss and Damage from Climate Change*, Climate Risk Management, Policy and Governance. Springer, Cham, Luxemburg, pp. 317–341. doi: 10.1007/978-3-319-72026-5_13

Schalatek, L., 2015: *From Innovative Mandate to Meaningful Implementation: Ensuring Gender-Responsive Green Climate Fund (GCF) Projects and Programs*, Washington, DC, USA, 35 pp.

Schalatek, L., 2018: *Gender and Climate Finance*. 8 pp. Heinrich Böll Stiftung, Washington D.C., USA ttps://www.odi.org/sites/odi.org.uk/files/resource-documents/11046.pdf (Accessed November 1, 2021).

Schalatek, L., 2020: *Gender and Climate Finance*. Heinrich Böll Stiftung, Washington DC, USA, 8 pp. https://climatefundsupdate.org/wp-content/uploads/2021/03/CFF10-ENG-2020-Digital.pdf (Accessed November 1, 2021).

Scherer, F.M., D. Harhoff, and J. Kukies, 2000: Uncertainty and the size distribution of rewards from innovation. *J. Evol. Econ.*, **10(1–2)**, 175–200, doi:10.1007/s001910050011.

Scherer, N., 2017: *How to Advance Regional Climate Risk Insurances*. Policy Brief, Climate Diplomacy, Berlin, Germany.

Scheuer, F. and J. Slemrod, 2021: Taxing Our Wealth. *J. Econ. Perspect.*, **35(1)**, 207–230, doi:10.1257/jep.35.1.207.

Scheyvens, R., G. Banks, and E. Hughes, 2016: The Private Sector and the SDGs: The Need to Move Beyond 'Business as Usual.' *Sustain. Dev.*, **24(6)**, 371–382, doi:10.1002/sd.1623.

Schiff, H. and H. Dithrich, 2017: *Scaling the use of guarantees in U.S. community investing*. Global Impact Investing Network, New York, NY, USA, 54 pp.

Schleich, J., X. Gassmann, C. Faure, and T. Meissner, 2016: Making the implicit explicit: A look inside the implicit discount rate. *Energy Policy*, **97**, 321–331, doi:10.1016/j.enpol.2016.07.044.

Schletz, M.C., S. Konrad, F. Staun, and D.D.R. Desgain, 2017: Taking stock of the (I)NDCs of developing countries: regional (I)NDC coverage of mitigation sectors and measures. UNEP, Nairobi, Kenya, 42p.

Schmidt-Traub, G. and J.D. Sachs, 2015: *The Roles of Public and Private Development Finance*. Sustainable Development Solutions Network, New Work, NY, USA. https://irp-cdn.multiscreensite.com/be6d1d56/files/uploaded/The-Roles-of-Public-and-Private-Development-Finance.pdf (Accessed November 1, 2021).

Schmidt, T.S., 2014: Low-carbon investment risks and de-risking. *Nat. Clim. Change*, **4(4)**, 237–239, doi:10.1038/nclimate2112.

Schmidt, T.S. et al., 2019: Adverse effects of rising interest rates on sustainable energy transitions. *Nat. Sustain.*, **2(9)**, 879–885, doi:10.1038/s41893-019-0375-2.

Schneeweiss, A., 2019: *Great Expectations Credibility and Additionality of Green Bonds*. Südwind, Bonn, Germany. doi:10.13140/RG.2.2.15563.64808.

Schoenmaker, D., 2021: Greening monetary policy. *Clim. Policy*, **21(4)**, 581–592, doi:10.1080/14693062.2020.1868392.

Schoenmaker, D. and G. Zachmann, 2015: *Can a Global Climate Risk Pool Help the Most Vulnerable Countries?* Bruegel, Brussels, Belgium. 8 pp.

Scholtens, B., 2017: Why Finance Should Care about Ecology. *Trends Ecol. & Evol.*, **32(7)**, 500–505, doi:10.1016/j.tree.2017.03.013.

Schroeder, S., 2021: A Kaleckian wealth tax to support a Green New Deal. *Econ. Labour Relations Rev.*, **32(2)**, 190–208, doi:10.1177/10353046211017350.

Schueffel, P., 2018: Taming the Beast: A Scientific Definition of Fintech. *SSRN Electron. J.*, **4(4)**, 32–54, doi:10.2139/ssrn.3097312.

Schumacher, K., 2020: The Shape of Green Fixed Income Investing to Come. *J. Environ. Invest.*, **10(1)**, doi:10.2139/ssrn.3663308.

Schwanitz, V.J., F. Piontek, C. Bertram, and G. Luderer, 2014: Long-term climate policy implications of phasing out fossil fuel subsidies. *Energy Policy*, **67**, 882–894, doi:10.1016/j.enpol.2013.12.015.

Scott, S., 2015: *The accidental birth of "official development assistance." OECD, Paris, France.*

Scott, S., 2017: *The grant element method of measuring the concessionality of loans and debt relief*. Working Paper No. 339, OECD, Paris, France.

Scott, Z., C. Simon, J. McConnell, P.S. Villanueva, 2017: *Independent Evaluation of African Risk Capacity (ARC) Final Inception Report.* Commissioned by FCDO (ex DFID) and undertaken by Oxford Policy Management, Oxford, UK, 85 pp.

SEADRIF, 2018: Southeast Asia Disaster Risk Insurance Facility., Singapore, 17 pp. https://www.financialprotectionforum.org/file/1754/download?token=7Mcw2NT6.

Searchinger, T.D. et al., 2020: *Revising Public Agricultural Support to Mitigate Climate Change.* Working Paper, World Bank, Washington, D.C., USA *doi: 10.1596/33677.*

SEDA, 2020: *Malaysia's 1st Pilot Run of Peer-to-Peer (P2P) Energy Trading*. SEDA, Putrajaya, Malaysia.

Seddon, N. et al., 2020: Understanding the value and limits of nature-based solutions to climate change and other global challenges. *Philos. Trans. R. Soc. B Biol. Sci.*, **375(1794)**, 20190120, doi:10.1098/rstb.2019.0120.

Sellers, S., 2016: *Gender and Climate Change: A closer look at existence evidence*. Global Gender and Climate Alliance and Women's Environment and Development Organisation, New York, NY, USA. 27 pp.

SEforALL and CPI, 2020: *Energizing finance: Understanding the landscape 2020*. Sustainable Energy for All (SEforALL), Vienna, Austria, Washington, DC and New York, NY, USA, 109 pp. https://www.seforall.org/system/files/2020-11/EF-2020-UL-ES-SEforALL.pdf (Accessed November 1, 2021).

Semieniuk, G., E. Campiglio, J. Mercure, U. Volz, and N.R. Edwards, 2021: Low-carbon transition risks for finance. *WIREs Clim. Change*, **12(1)**, e678, doi:10.1002/wcc.678.

Serdeczny, O., 2019: Non-economic Loss and Damage and the Warsaw International Mechanism. pp. 205–220. In: Mechler, R., Bouwer, L., Schinko, T., Surminski, S., Linnerooth-Bayer, J. (eds) Loss and Damage from Climate Change. Climate Risk Management, Policy and Governance. Springer, Cham, Switzerland, doi: 10.1007/978-3-319-72026-5_8.

Serebrisky, T., A. Suárez-Alemán, D. Margot, and M.C. Ramirez, 2015: *Financing Infrastructure in Latin America and the Caribbean: How, How Much and by Whom?* IDB, Washington, D.C., 30 pp. https://publications.iadb.org/en/financing-infrastructure-latin-america-and-caribbean-how-how-much-and-whom (Accessed October 30, 2021).

Serenelli, L., 2021: Italy records high demand for first green bond offering. March 4, 2021. IPE. https://www.ipe.com/news/italy-records-high-demand-for-first-green-bond-offering/10051279.article (Accessed November 1, 2021).

Sewell, M., 2011: History of the Efficient Market Hypothesis. Research note **11(04)**. UCL Department of Computer Science, London, UK.

Shakhovskoy, M., C. Colina, and M.C. Höök, 2019: *Pathways to Prosperity: Rural and Agricultural Finance State of the Sector Report*. 60 pp. https://isfadvisors.org/wp-content/uploads/2019/11/2019_RAF-State-of-the-Sector-10.pdf (Accessed November 1, 2021).

Shan, Y. et al., 2021: Impacts of COVID-19 and fiscal stimuli on global emissions and the Paris Agreement. *Nat. Clim. Chang.*, **11(3)**, 200–206, doi:10.1038/s41558-020-00977-5.

Shapiro, J., 2020: *The Environmental Bias of Trade Policy*. NBER WORKING PAPER 26845, NBER, Cambridge, MA, USA. doi: 10.3386/w26845.

Sharma, V., V. Orindi, C. Hesse, J. Pattison, and S. Anderson, 2014: Supporting local climate adaptation planning and implementation through local governance and decentralised finance provision. *Dev. Pract.*, **24(4)**, 579–590, doi:10.1080/09614524.2014.907240.

Shi, L., 2017: *Masala bond programme: Nurturing a local currency bond market*. EMCompass, no. 30;. International Finance Corporation, Washington, DC, USA. https://openknowledge.worldbank.org/handle/10986/30356 (Accessed November 1, 2021). 3 pp. https://openknowledge.worldbank.org/bitstream/handle/10986/30356/112690-BRI-EMCompass-Note-30-Masala-Bond-Program-PUBLIC.pdf?sequence=1&isAllowed=y.

Shishlov, I., M. Nicol, and I. Cochran, 2018: *Environmental integrity of green bonds: stakes, status and next steps*. Institute for Climate Economics, Paris, France, 39 pp. https://www.i4ce.org/wp-core/wp-content/uploads/2018/03/I4CE-GreenBondsProgram-Environmental-Integrity-web.pdf (Accessed November 1, 2021).

Silva, A.C. et al., 2020: *Staff Note for the G20 International Financial Architecture Working Group: Recent Development on Local Currency Bond Markets in Emerging Economies*. World Bank, Washington DC, USA, 30 pp. http://documents.worldbank.org/curated/en/129961580334830825/Staff-Note-for-the-G20-International-Financial-Architecture-Working-Group-IFAWG-Recent-Developments-On-Local-Currency-Bond-Markets-In-Emerging-Economies (Accessed November 1, 2021).

Silver, J., 2015: The potentials of carbon markets for infrastructure investment in sub-Saharan urban Africa. *Curr. Opin. Environ. Sustain.*, **13**, 25–31, doi:10.1016/j.cosust.2014.12.004.

Singh, M.K., V. Kekatos, and C. Liu, 2019: Optimal Distribution System Restoration with Microgrids and Distributed Generators. *2019 IEEE Power Energy Society General Meeting (PESGM)*, 1–5.

Singh, N., J. Finnegan, and K. Levin, 2016: *MRV 101: Understanding Measurement, Reporting, and Verification of Climate Change Mitigation*. WRI, Washington, DC, USA, 27 pp. https://files.wri.org/d8/s3fs-public/MRV_101_0.pdf (Accessed November 1, 2021).

Smart Energy International, 2018: Global trends in smart metering. December 31, 2018. Smart Energy International. Maarssen, Netherlands.

Smith, G., 2021: *Where credit is due: how Africa's debt can be a benefit, not a burden*. C Hurst & Co Publishers Ltd, London, UK, 240 pp.

*Smoke, P., 2019: Improving Subnational Government Development Finance in Emerging and Developing Economies: Toward a Strategic Approach.* Working Paper No: 921. ADBI, Metro Manila, Philippines.

Soanes, M., N. Rai, P. Steele, C. Shakya, and J. Macgregor, 2017: *Delivering real change: Getting international climate finance to the local level*. International Institute for Environment and Development (IIED), 47 pp.

Sonnen, 2020: *What is the sonnen community?* Sonnen GmbH, Wildpoldsried, Germany. https://sonnengroup.com/sonnencommunity/ (Accessed November 1, 2021).

Sovacool, B.K., 2016: How long will it take? Conceptualizing the temporal dynamics of energy transitions. *Energy Res. Soc. Sci.*, **13**, 202–215, doi:10.1016/j.erss.2015.12.020.

SSE, 2018: *2018 Report On Progress: A paper prepared for the Sustainable Stock Exchanges 2018 Global Dialogue*. Sustainable Stock Exchanges Initiative, New York, NY, USA. 47 pp.

Stadelmann, M., A. Michaelowa, and J.T. Roberts, 2013: Difficulties in accounting for private finance in international climate policy. *Clim. Policy*, **13(6)**, 718–737, doi:10.1080/14693062.2013.791146.

Stanny, E., 2018: Reliability and Comparability of GHG Disclosures to the CDP by US Electric Utilities. *Soc. Environ. Account. J.*, **38(2)**, 111–130, doi:10.1080/0969160X.2018.1456949.

Statistica, 2021: Value of venture capital financing worldwide in 2020, by region (in billion U.S. dollars). Statistica, Hamburg, Germany. https://www.statista.com/statistics/1095957/global-venture-capita-funding-value-by-region/.

Steckel, J.C. and M. Jakob, 2018: The role of financing cost and de-risking strategies for clean energy investment. *Int. Econ.*, **155** (September 2017), 19–28, doi:10.1016/j.inteco.2018.02.003.

Steele, W., L. Mata, and H. Fünfgeld, 2015: Urban climate justice: creating sustainable pathways for humans and other species. *Curr. Opin. Environ. Sustain.*, **14**, 121–126, doi:10.1016/j.cosust.2015.05.004.

Steffen, B., 2019: *Estimating the Cost of Capital for Renewable Energy Projects*. Energy Economics 88 (2020): 104783 9 pp. doi: 10.1016/j.eneco.2020.104783.

Steffen, B., T. Matsuo, D. Steinemann, and T.S. Schmidt, 2018: Opening new markets for clean energy: The role of project developers in the global diffusion of renewable energy technologies. *Bus. Polit.*, **20(4)**, 553–587, doi:10.1017/bap.2018.17.

Steffen, B., F. Egli, M. Pahle, and T.S. Schmidt, 2020: Navigating the Clean Energy Transition in the COVID-19 Crisis. *Joule*, **4(6)**, 1137–1141, doi:10.1016/j.joule.2020.04.011.

Stehle, F., 2021: Governing Climate Change: Polycentricity in Action? *Glob. Environ. Polit.*, **21(1)**, 157–159, doi:10.1162/glep_r_00596.

Stender, F., U. Moslener, and W.P. Pauw, 2019: More than money: does climate finance support capacity building? *Appl. Econ. Lett.*, **27(15)**, 1247–1251, doi:10.1080/13504851.2019.1676384.

Stern, N., 2008: *The Economics of Climate Change*. American Economic Review, 98 (2): 1-37. 37 pp. doi: 10.1257/aer.98.2.1.

Stern, N. and J. Stiglitz, 2021: *The Social Cost of Carbon, Risk, Distribution, Market Failures: An Alternative Approach*. NBER WORKING PAPER 28472, Cambridge, MA, USA, 75 pp. DOI 10.3386/w28472.

15


Stiglitz, J. and H. Rashid, 2020: *Averting Catastrophic Debt Crises in Developing Countries: Extraordinary challenges call for extraordinary measures*. Policy Insight 104, CEPR, Washington, D.C., USA 29 pp. https://cepr.org/sites/default/files/policy_insights/PolicyInsight104.pdf (Accessed November 1, 2021).

Stockholm Green Digital Finance, 2017: *Unlocking the Potential of Green Fintech*. Stockholm Green Digital Finance Insight Brief #1 2017, Stockholm, Sweden.12 pp.

Streck, C., 2016: Mobilizing Finance for redd+ After Paris. *J. Eur. Environ. & Plan. Law*, **13(2)**, 146–166, doi:10.1163/18760104-01302003.

Streck, C., 2020: Who owns REDD+? carbon markets, carbon rights and entitlements to REDD+ finance. *Forests*, **11(9)**, 1–15, doi:10.3390/f11090959.

Stritzke, S. et al., 2021: Results-Based Financing (RBF) for Modern Energy Cooking Solutions: An Effective Driver for Innovation and Scale? *Energies*, **14(15)**, 4559, doi:10.3390/EN14154559.

Stubbington, T., 2021: UK's debut 'green gilt' sale draws blockbuster demand. *Financial Times*, September 21.

Studart, R. and K. Gallagher, 2015: Guaranteeing finance for sustainable infrastructure: A proposal. In: *Moving the trillions a debate on positive pricing of mitigation actions* [Sirkis, A. et al., (eds.)]. Brasil no Clima, pp. 91–113.

Suatmadi, A.Y., F. Creutzig, and I.M. Otto, 2019: On-demand motorcycle taxis improve mobility, not sustainability. *Case Stud. Transp. Policy*, **7(2)**, 218–229, doi:10.1016/j.cstp.2019.04.005.

Sudmant, A., S. Colenbrander, A. Gouldson, and N. Chilundika, 2017a: Private opportunities, public benefits? The scope for private finance to deliver low-carbon transport systems in Kigali, Rwanda. *Urban Clim.*, **20**, 59–74, doi:10.1016/j.uclim.2017.02.011.

Sudmant, A.H. et al., 2017b: Understanding the case for low-carbon investment through bottom-up assessments of city-scale opportunities. *Clim. Policy*, **17(3)**, 299–313, doi:10.1080/14693062.2015.1104498.

Sulich, A., M. Rutkowska, and Ł. Popławski, 2020: Green jobs, definitional issues, and the employment of young people: An analysis of three European Union countries. *J. Environ. Manage.*, **262**, 110314, doi:10.1016/j.jenvman.2020.110314.

Summers, L.H., 2015: Demand Side Secular Stagnation. *Am. Econ. Rev.*, **105(5)**, 60–65, doi:10.1257/aer.p20151103.

Summers, L.H., 2016: The age of secular stagnation: What it is and what to do about it. *Foreign Affairs, March/April 2016*. https://www.foreignaffairs.com/articles/united-states/2016-02-15/age-secular-stagnation (Accessed November 1, 2021).

Sumner, A., C. Hoy, and E. Ortiz-Juarez, 2020: *Estimates of the impact of COVID-19 on global poverty*. WIDER Working Paper 2020/43. UNU-WIDER, Helsinki, Finland, doi:10.35188/UNU-WIDER/2020/800-9.

Sussams, L. and J. Leaton, 2017: *Expect the Unexpected: The Disruptive Power of Low-carbon Technology*. Carbon Tracker, London, UK.

Sussams, L., A. Grant, and M. Gagliardi, 2015: *The US Coal Crash – Evidence for Structural Change*. Carbon Tracker Initiative, London, UK, 50 pp. http://www.carbontracker.org/report/the-us-coal-crash/ (Accesed November 1, 2021).

Svartzman, R., P. Bolton, M. Despres, L.A. Pereira Da Silva, and F. Samama, 2021: Central banks, financial stability and policy coordination in the age of climate uncertainty: a three-layered analytical and operational framework. *Clim. Policy*, **21(4)**, 563–580, doi:10.1080/14693062.2020.1862743.

Sweerts, B., F.D. Longa, and B. van der Zwaan, 2019: Financial de-risking to unlock Africa's renewable energy potential. *Renew. Sustain. Energy Rev.*, **102**, 75–82, doi:10.1016/j.rser.2018.11.039.

Swiss Re Institute, 2019: Natural catastrophes and man-made disasters in 2018: "secondary" perils on the frontline. No 2/2019. Swiss Re Institute, Zurich, Switzerland.

Tan, C., 2019: Creative cocktails or toxic brews? Blended finance and the regulatory framework for sustainable development. In: *Sustainable Trade, Investment and Finance*, [Gammage and Novitz (eds.)] Edward Elgar Publishing, Cheltenham, UK, pp. 300–330.

Tang, D.Y. and Y. Zhang, 2020: Do shareholders benefit from green bonds? *J. Corp. Financ.*, **61**, doi:10.1016/j.jcorpfin.2018.12.001.

Tang, S. and D. Demeritt, 2018: Climate Change and Mandatory Carbon Reporting: Impacts on Business Process and Performance. *Bus. Strateg. Environ.*, **27(4)**, 437–455, doi:10.1002/bse.1985.

Tapscott, D. and R. Kirkland, 2016: How Blockchains could Change the World. Interview, May 6, 2016. McKinsey and Company, London, UK.

TCFD, 2017: *Recommendations of the Task Force on Climate-related Financial Disclosures*. FSB, Basel, Switzerland, https://www.fsb-tcfd.org/wp-content/uploads/2017/06/FINAL-TCFD-Report-062817.pdf (Accessed November 1, 2021).

TCFD, 2019: *The Task Force on Climate-related Financial Disclosures Status Report*. FSB, Basel, Switzerland. 135 pp.

TCFD, 2020: *Task Force on Climate-related Financial Disclosures: 2020 Status Report*. FSB, Basel, Switzerland. 114 pp. https://assets.bbhub.io/company/sites/60/2020/09/2020-TCFD%7B%5C_%7DStatus-Report.pdf.

TCFD Portfolio Alignment Team, 2021: *Measuring portfolio alignment: Technical report*. FSB, Basel, Switzerland. 8 pp. https://www.tcfdhub.org/wp-content/uploads/2021/10/20211005-PAT-Summary-of-Responses.pdf (Accessed November 1, 2021).

TCX, 2020: *Annual Report. TCX protection against currency risk*. TCX Investment Management Company, Amsterdam, The Netherlands, 78 pp.

TEG, 2019: *Taxonomy Technical Report*.EU Technical Expert Group On Sutainable Finance, Brussels, Belgium.

The 2° Investing Initiative, 2017: *Out of the Fog: Quantifiying the alignment of Swiss pension funds and insurances with the Paris Agreement*. 2 Degrees Investing Initiative, Paris, France, 49 pp. https://2degrees-investing.org/out-of-the-fog-quantifiying-the-alignment-of-swiss-pension-funds-and-insurances-with-the-paris-agreement/ (Accessed November 1, 2021).

The 2° Investing Initiative and Wüest Partner, 2020: *Bridging the Gap: Measuring progress on the climate goal alignment and climate actions of Swiss Financial Institutions*. 2 Degrees Investing Initiative, Paris, France, 104 pp. https://2degrees-investing.org/wp-content/uploads/2020/11/Bridging-the-Gap.pdf (Accessed November 1, 2021).

The Economist, 2017: Kenya launches the world's first mobile-only sovereign bond. June 29, 2017. *The Economist*.

The Economist, 2021: Haiti's lack of preparedness makes bad disasters worse. August 18, 2021. *The Economist*.

The Food and Land Use Coalition, 2019: *Growing Better: Ten Critical Transitions to Transform Food and Land Use*. The Food and Land Use Coalition, London, 236 pp. https://www.foodandlandusecoalition.org/wp-content/uploads/2019/09/FOLU-GrowingBetter-GlobalReport.pdf.

Thiele, T., M. von Unger, and A. Mohan, 2021: *MDB Engagement: Mainstreaming blue nature-based solutions into infrastrucure finance*. BNCFF, Gland, Switzerland.

Thomä, J., 2018: The stranding of upstream fossil fuel assets in the context of the transition to a low-carbon economy. In: *Stranded Assets and the Environment [Caldecott, B (ed.)]*, Routledge, Abingdon, UK, and New York, NY, pp. 111–124.

Timilsina, G.R., C. de Gouvello, M. Thioye, and F.B. Dayo, 2010: Clean Development Mechanism Potential and Challenges in Sub-Saharan Africa. *Mitig. Adapt. Strateg. Glob. Change*, **15(1)**, 93–111, doi:10.1007/s11027-009-9206-5.

Timsit, A., 2020: The blueprint the US can follow to finally pay reparations. *Quartz*, October 13.

TNFD, 2021: *TNFD: Nature in scope*. Taskforce on Nature-related Financial Disasters (TNFD), 1–26 pp. https://tnfd.global/wp-content/uploads/2021/07/TNFD-Nature-in-Scope-2.pdf.

TNO, 2016: *PowerMatcher, Matching energy supply and demand to expand smart energy potential*., The Hague, The Netherlands.

Tolliver, C., A.R. Keeley, and S. Managi, 2019: Green bonds for the Paris agreement and sustainable development goals. *Environ. Res. Lett.*, **14(6)**, 64009, doi:10.1088/1748-9326/ab1118.

Tolliver, C., A.R. Keeley, and S. Managi, 2020a: Policy targets behind green bonds for renewable energy: Do climate commitments matter? *Technol. Forecast. Soc. Change*, **157**, 120051, doi:10.1016/j.techfore.2020.120051.

Tolliver, C., A.R. Keeley, and S. Managi, 2020b: Drivers of green bond market growth: The importance of Nationally Determined Contributions to the Paris Agreement and implications for sustainability. *J. Clean. Prod.*, **244**, 118643, doi:10.1016/J.JCLEPRO.2019.118643.

Torres, E. and R. Zeidan, 2016: The life-cycle of national development banks: The experience of Brazil's BNDES. *Q. Rev. Econ. Financ.*, **62**, 97–104, doi:10.1016/J.QREF.2016.07.006.

Toxopeus, H. and F. Polzin, 2017: *Characterizing nature-based solutions from a business model and financing perspective*. Naturvation, Berlin, Germany.

Toxopeus, H. and F. Polzin, 2021: Reviewing financing barriers and strategies for urban nature-based solutions. *J. Environ. Manage.*, **289** (August 2020), 112371, doi:10.1016/j.jenvman.2021.112371.

Trinks, A., B. Scholtens, M. Mulder, and L. Dam, 2018: Fossil Fuel Divestment and Portfolio Performance. *Ecol. Econ.*, **146**, 740–748, doi:10.1016/j.ecolecon.2017.11.036.

Tuhkanen, H., 2020: *Green bonds: a mechanism for bridging the adaptation gap?* Stockholm Environment Institute, Stockholm, Sweden, 23 pp. https://www.sei.org/wp-content/uploads/2020/02/sei-working-paper-green-bonds-tuhkanen.pdf.

Tuhkanen, H. and G. Vulturius, 2020: Are green bonds funding the transition? Investigating the link between companies' climate targets and green debt financing. *J. Sustain. Financ. Invest.*, Abingdon, UK, doi:10.1080/20430795.2020.1857634. pp. 1–23.

TWAS, 2021: *TWAS Annual reports*. The World Academy of Sciences, Trieste, Italy, 44 pp. https://twas.org/publications/annual-reports (Accessed October 24, 2021).

UBS, 2017: *Mobilizing private wealth for public good*. UBS, New York, NY, USA. 38 pp. https://www.longfinance.net/media/documents/ubs-wef-2017-whitepaper-mobilizing-private-wealth-for-public-good1.pdf (Accessed August 18, 2021).

UKERC, 2014: *Low carbon jobs: The evidence for net job creation from policy support for energy efficiency and renewable energy*.UK Energy Research Centre, London, UK. https://d2e1qxpsswcpgz.cloudfront.net/uploads/2020/03/low-carbon-jobs.pdf (Accessed November 1, 2021).

Umar, Z., Y. Manel, Y. Riaz, and M. Gubareva, 2021: Return and volatility transmission between emerging markets and US debt throughout the pandemic crisis. *Pacific-Basin Financ. J.*, **67**, 101563, doi:10.1016/j.pacfin.2021.101563.

UN, 2009: *Report of the Fourth United Nations Conference on the Least Developed Countries. Instanbul Programme of Action (IPoA)*: Istanbul, Turkey, 9-13 May 2011. United Nations, New York, NY, USA, 125 pp. https://undocs.org/en/A/CONF.219/7 (Accessed October 27, 2021).

UN, 2015a: *Addis Ababa Action Agenda of the Third International Conference on Financing for Development*, Addis Ababa, Ethiopia, 13-16th July 2015. United Nations, New York, NY, USA. 37 pp.

UN, 2015b: *Sendai Framework for Disaster Risk Reduction 2015-2030*. UN, Geneva, Switzerland, 35 pp.

UN, 2015c: *SIDS Accelerated Modalities of Action (SAMOA) Pathway*. UN-OHRLLS, Apia, Samoa, 30 pp. https://www.un.org/ga/search/view_doc.asp?symbol=A/RES/69/15&Lang=E.

UN, 2019: *Roadmap for Financing the 2030 Agenda for Sustainable Development 2019-2021*.United Nations, New York, NY, USA, 57 pp.

UN, 2020: *2020 Voluntary National Reviews Synthesis Report*. UN, New York, NY, USA, 102 pp. https://sustainabledevelopment.un.org/content/documents/27027VNR_Synthesis_Report_2020.pdf (Accessed August 19, 2021).

UN and ECB, 2015: *Financial Production, Flows and Stocks in the System of National Accounts - Series F No. 113 (Studies in Methods - Handbook of National Accopunting)*. Department of Economics and Social Affairs, New York, NY, USA.

UN, 2015: *Addis Ababa Action Agenda of the Third International Conference on Financing for Development*. UNCTAD, Addis Ababa, Ethiopia, 37 pp. https://unctad.org/meetings/en/SessionalDocuments/ares69d313%7B%5C_%7Den.pdf.

UN CDF, 2020a: *Making finance work for the poor: Supporting SDG achievement in the last mile (Annual Report 2019)*. UN, New York, NY, USA, 40 pp.

UN CDF, 2020b: The SDG500 Platform Investment Opportunity. UNCDF, New York, NY, USA. https://www.uncdf.org/article/5311/the-sdg500-platform-investment-opportunity (Accessed September 21, 2021).

UN CDF, 2021a: *Harnessing the true power of capital: Unlocking the growth potential of the last mile*. UNCDF, New York, NY, USA, 51 pp.

UN CDF, 2021b: *Cambodia Stock Exchange: Identifying Key Bottlenecks*. UNCDF, New York, NY, USA. 13 pp. https://www.uncdf.org/article/6866/cambodia-stock-exchange--identifying-key-bottlenecks (Accessed November 1, 2021).

UN ESCAP, 2015: *Tapping Capital Markets & Institutional Investors for Infrastructure Development*. UN ESCAP, Bangkok, Thailand.

*How to Invest in the Low-carbon Economy: An Institutional Investors' Guide*. UN PRI, London, UK. 31 pp.

UN PRI, 2020: Annual Report 2020. UN PRI, London, UK. https://www.unpri.org/pri/about-the-pri/annual-report (Accessed December 23, 2020).

UN-Convened Net-Zero Asset Owner Alliance, 2021: *Inaugural 2025 target setting protocol*. 75 pp. https://www.unepfi.org/wordpress/wp-content/uploads/2021/01/Alliance-Target-Setting-Protocol-2021.pdf.

UNDP, 2015: *A Methodological Guidebook: Climate Public Expenditure and Institutional Review (CPEIR)*. UNDP, Bangkok, Thailand, 72 pp.

UNECA, 2018: *Africa Sustainable Development Report 2018* [African Union Commission, UN Economic Commission of Africa, African Development Bank, and UN Development Programme, (eds.)]. UN ECA, Addis Ababa, Ethiopia, 154 pp.

UNEP, 2011: *Innovative Climate Finance Products - Examples from the UNEP Bilateral Finance Institutions Climate Change Working Group*. United Nations Environment Programme (UNEP), 28 pp.

UNEP, 2016: *Adaptation Finance Gap Report 2016*. United Nations Environment Programme (UNEP), Nairobi, Kenya, 50 pp.

UNEP, 2018: *Adaptation Gap Report 2018*. United Nations Environment Programme (UNEP), Nairobi, Kenya, 104 pp. https://www.unenvironment.org/resources/adaptation-gap-report (Accessed November 1, 2021).

UNEP, 2019: *Emissions Gap Report 2019*. United Nations Environment Programme (UNEP), Nairobi, Kenya, 81 pp.

UNEP, 2021: *Adaptation Gap Report 2020*. United Nations Environment Programme (UNEP), Nairobi, Kenya, 99 pp.

UNEP, WEF, ELD, and Vivid Economics, 2021: *State of Finance for Nature*. United Nations Environment Programme (UNEP), Nairobi, Kenya, 60 pp. https://www.unep.org/resources/state-finance-nature (Accessed November 1, 2021).

UNEP DTU, 2017: *Overcoming Barriers to Investing in Energy Efficiency*. UNEP DTU, Copenhagen, Denmark. 42 pp.

UNEP FI, 2018: *Extending our Horizons: Assessing Credit Risk and Opportunity in a Changing Climate*. UNEP FI, Geneva, Switzerland. 76 pp. https://www.unepfi.org/publications/banking-publications/extending-our-horizons/ (Accessed November 1, 2021).

UNEP Inquiry, 2016a: *Inquiry into the Design of a Sustainable Financial System: Definitions and Concepts – Background Note*. UNEP, Nairobi, Kenya. 19 pp. https://unepinquiry.org/publication/definitions-and-concepts-background-note/ (Accessed November 1, 2021).

UNEP Inquiry, 2016b: *Greening the banking system; taking stock of G20 green banking market practice*. UNEP, Geneva, Switzerland, 30 pp.

UNESCAP, 2017: *Economic and Social Survey of Asia and the Pacific 2017, UNESCAP, Bangkok, Thailand*.

UNFCCC, 2008: *Mechanisms to manage financial risks from direct impacts of climate change in developing countries*. United Nations Framework Convention on Climate Change (UNFCCC), Bonn, Germany. 114 pp. https://unfccc.int/resource/docs/2008/tp/09.pdf.

UNFCCC, 2009: *Report of the Conference of the Parties on its fifteenth session, held in Copenhagen from 7 to 19 December 2009. Part Two: Action taken by the Conference of the Parties at its fifteenth session*. United Nations Framework Convention on Climate Change (UNFCCC). https://unfccc.int/sites/default/files/resource/docs/2009/cop15/eng/11a01.pdf.

UNFCCC, 2010: *Part Two: Action taken by the Conference of the Parties at its fifteenth session*. United Nations Framework Convention on Climate Change (UNFCCC), Copenhagen, Denmark, 43 pp. https://unfccc.int/resource/docs/2009/cop15/eng/11a01.pdf (Accessed November 1, 2019).

UNFCCC, 2014a: *2014 Biennial Assessment and Overview of Climate Finance Flows Technical Report*. United Nations Framework Convention on Climate Change (UNFCCC), Bonn, Germany. https://unfccc.int/files/cooperation_and_support/financial_mechanism/standing_committee/application/pdf/2014_biennial_assessment_and_overview_of_climate_finance_flows_report_web.pdf (Accessed November 1, 2021).

UNFCCC, 2014b: Introduction to Climate Finance. United Nations Framework Convention on Climate Change (UNFCCC), Bonn, Germany. https://unfccc.int/topics/introduction-to-climate-finance (Accessed November 1, 2021).

UNFCCC, 2015: *Adoption of the Paris Agreement*. United Nations Framework Convention on Climate Change. United Nations Framework Convention on Climate Change (UNFCCC), Paris, France, 32 pp. https://unfccc.int/resource/docs/2015/cop21/eng/l09r01.pdf (Accessed November 1, 2021).

UNFCCC, 2017: *Report on the independent review of the effective implementation of the Climate Technology Centre and Network*. United Nations Framework Convention on Climate Change (UNFCCC), Bonn, Gernamy.

UNFCCC, 2018a: *2018 Biennial Assessment and Overview of Climate Finance Flows Technical Report*. UN Standing Committee on Finance, United Nations Framework Convention on Climate Change (UNFCCC), Bonn, Germany. https://unfccc.int/sites/default/files/resource/2018%20BA%20Technical%20Report%20Final.pdf (Accessed November 1, 2019).

UNFCCC, 2018b: *Solidarity and Just Transition Silesia Declaration*. United Nations Framework Convention on Climate Change (UNFCCC), Katowice, Poland, 3 pp. https://cop24.gov.pl/fileadmin/user_upload/Solidarity_and_Just_Transition_Silesia_Declaration_2_.pdf (Accessed November 1, 2021).

UNFCCC, 2019a: Interim NDC registry. United Nations Framework Convention on Climate Change (UNFCCC), Bonn, Germany. https://unfccc.int/news/ndc-interim-registry (Accessed November 1, 2021).

UNFCCC, 2019b: *Report of the Conference of the Parties on its twenty-fourth session, held in Katowice from 2 to 15 December 2018*. United Nations Framework Convention on Climate Change (UNFCCC), Katowice, Poland, 46 pp. https://unfccc.int/sites/default/files/resource/cma2018_3_add2_new_advance.pdf#page=18 (Accessed November 1, 2021).

UNFCCC, 2019c: *Opportunities and options for adaptation finance, including in relation to the private sector*. United Nations Framework Convention on Climate Change (UNFCCC) Bonn, Germany. https://unfccc.int/sites/default/files/resource/tp2019%7B%5C_%7D03E.pdf (Accessed November 1, 2021).

UNFCCC, 2019d: *25 Years of Adaptation Under the UNFCCC*. United Nations Framework Convention on Climate Change (UNFCCC), Bonn, Germany, 34 pp. https://unfccc.int/sites/default/files/resource/AC_25 Years of Adaptation Under the UNFCCC_2019.pdf (Accessed November 1, 2021).

UNFCCC, 2020: ME SOLshare: Peer-to-Peer Smart Village Grids | Bangladesh. United Nations Framework Convention on Climate Change (UNFCCC).

UNFCCC, 2021: UNFCC Universities Climate Change Partnership. United Nations Framework Convention on Climate Change (UNFCCC), Bonn, Germany. https://www4.unfccc.int/sites/NWPStaging/Pages/university-partnerships.aspx (Accessed November 1, 2021).

Unruh, G.C., 2002: Escaping carbon lock-in. *Energy Policy*, **30(4)**, 317–325, doi:10.1016/S0301-4215(01)00098-2.

UNU-INRA, 2019: Africa's Development In The Age Of Stranded Assets. UNU-INRA, Ghana. https://i.unu.edu/media/inra.unu.edu/publication/5247/DIscussion-paper-Africas-Development-in-the-age-of-stranded-Assets_INRAReport2019.pdf (Accessed November 1, 2021).

Ürge-Vorsatz, D. et al., 2018: Locking in positive climate responses in cities. *Nat. Clim. Chang.*, **8(3)**, 174–177, doi:10.1038/s41558-018-0100-6.

US DoJ, 2009: Charting the future course of international technical assistance at the Federal Trade Commission and U.S. Department of Justice.

US DoJ, 2019: Mozambique's Former Finance Minister Indicted Alongside Other Former Mozambican Officials, Business Executives, and Investment Bankers in Alleged $2 Billion Fraud and Money Laundering Scheme that Victimized U.S. Investors | OPA | Department of Justice. https://www.justice.gov/opa/pr/mozambique-s-former-finance-minister-indicted-alongside-other-former-mozambican-officials (Accessed August 21, 2021).

US GCRP, 2018a: *Impacts, Risks, and Adaptation in the United States: The Fourth National Climate Assessment, Volume II*. [Reidmiller, D.R., C.W. Avery, D.R. Easterling, K.E. Kunkel, K.L.M. Lewis, T.K. Maycock, and B.C. Stewart, (eds.)]., Washington, DC,.

US GCRP, 2018b: *Fourth National Climate Assessment Volume II Impacts, Risks, and Adaptation in the United States Report-in-Brief*., Washington, D.C., 1515 pp. https://nca2018.globalchange.gov/downloads/NCA4%7B%5C_%7D2018%7B%5C_%7DFullReport.pdf.

V20, 2020: *Sustainable Insurance Facility (SIF): Solutions to build resilient micro, small and medium enterprises*. Vulnerable Twenty Group, Geneva, Switzerland. 18 pp. https://climate-insurance.org/wp-content/uploads/2020/04/SIF_Kick-Off_Summary_Sept_2019.pdf (Accessed November 1, 2021).

V20, 2021: *The V20-led Sustainable Insurance Facility at a Glance*. Vulnerable Twenty Group, Geneva, Switzerland. 30 pp. https://climate-insurance.org/wp-content/uploads/2021/01/The-V20-led-SIF-at-a-Glance_January-2021.pdf (Accessed November 1, 2021).

van der Gaast, W., R. Sikkema, and M. Vohrer, 2018: The contribution of forest carbon credit projects to addressing the climate change challenge. *Clim. Policy*, **18(1)**, 42–48, doi:10.1080/14693062.2016.1242056.

van der Ploeg, F. and A. Rezai, 2020: Stranded Assets in the Transition to a Carbon-Free Economy. *Annu. Rev. Resour. Econ.*, **12(1)**, 281–298, doi:10.1146/annurev-resource-110519-040938.

van der Zwaan, B., T. Kober, F.D. Longa, A. van der Laan, and G. Jan Kramer, 2018: An integrated assessment of pathways for low-carbon development in Africa. *Energy Policy*, **117**, 387–395, doi:10.1016/j.enpol.2018.03.017.

Vedeld, T., H. Hofstad, H. Solli, and G.S. Hanssen, 2021: Polycentric urban climate governance: Creating synergies between integrative and interactive governance in Oslo. *Environ. Policy Gov.*, **31(4)**, 347–360, doi:10.1002/eet.1935.

Verkooijen, P., 2019: Adapting to climate change could add $7 trillion to the global economy by 2030. *GCA Blog*, September 10.

Vivid Economics, 2020: *Greenness of Stimulus Index - Vivid Economics*. https://www.vivideconomics.com/casestudy/greenness-for-stimulus-index/ (Accessed November 1, 2021).

Volz, U. et al., 2020: *Climate Change and Sovereign Risk*., London, Tokyo, Singapore, Berkeley, 133 pp

Volz, U. et al., 2021: *Debt relief for a green and inclusive recovery: Securing private-sector participation and creating policy space for sustainable development*. Boston University Global Development Policy Center, the Heinrich Böll Foundation and the Centre for Sustainable Finance at SOAS, University of London, Boston, MA, USA, Berlin, Germany, and London, UK, 42 pp. https://www.boell.de/sites/default/files/2021-07/E-PAPER - Debt Relief for a Green and Inclusive Recovery.pdf?dimension1=division_ip (Accessed November 1, 2021).

Vona, F., G. Marin, and D. Consoli, 2018: Measures, drivers and effects of green employment: evidence from US local labor markets, 2006–2014. *J. Econ. Geogr.*, **19(5)**, 1021–1048, doi:10.1093/jeg/lby038.

Vyas, S., V. Seifert, L. Schaefer, and S. Kreft, 2019: *Climate Risk Insurance Solutions: Understanding the Drivers of Cost-effectivess*. Munich Climate Insurance Initiative, Bonn, Germany, 57 pp. https://www.insuresilience.org/wp-content/uploads/2019/03/MCII-Discussion-Paper-3_Understanding-cost-effectiveness.pdf (Accessed November 1, 2021).

Wagner, K., 2021: Apologies for historical atrocities fall short of a proper reckoning. *Financial Times*, June 4.

Walsh, G., I. Ahmed, J. Said, and M.F.E. Maya, 2021: *A Just Transition for Africa: Championing a Fair and Prosperous Pathway to Net Zero*. Tony Blair Institute for Climate Change, London, UK. 28 pp. https://institute.global/advisory/just-transition-africa- (Accessed October 15, 2021).

Wamsler, C. et al., 2020: Environmental and climate policy integration: Targeted strategies for overcoming barriers to nature-based solutions and climate change adaptation. *J. Clean. Prod.*, **247**, 119154, doi:10.1016/j.jclepro.2019.119154.

Wang, H. and C. Yee, 2020: Climate Tech's Four Valleys of Death and Why We Must Build a Bridge. *RMI*, June 17.

Warland, L. and A. Michaelowa, 2015: *Can debt for climate swaps be a promising climate finance instrument? Lessons from the past and recommendations for the future*. Perspectives, Zurich, doi: 10.5167/uzh-159661.

Warner, K., 2020: Climate justice: Who bears the burden and pays the price? *Soc. Altern.*, **39(2)**, 19–25.

Watson, C. and L. Schalatek, 2019: The Global Climate Finance Architecture.. Overseas Development Institute, London, UK, and Heinrich Böll Stiftung North America, Washington DC, USA. 6.

Watson, C. and L. Schalatek, 2020: *The global climate finance architecture*. Overseas Development Institute, London, UK, and Heinrich Böll Stiftung North America, Washington DC, USA, 6 pp. https://climatefundsupdate.org/publications/the-global-climate-finance-architecture-2/ (Accessed November 1, 2021).

Watson, C. and L. Schalatek, 2021: *Climate Finance Thematic Briefing: REDD+ Finance*. Overseas Development Institute, London, UK, and Heinrich Böll Stiftung North America, Washington DC, USA. 1–4 pp. https://climatefundsupdate.org/wp-content/uploads/2021/03/CFF5-ENG-2020-Digital.pdf (Accessed November 1, 2021).

Watson, C., A. Caravani, T. Mitchell, J. Kellett, and K. Peters, 2015: *Finance for reducing disaster risk: 10 things to know*. Overseas Development Institute, London, UK, 18 pp. https://cdn.odi.org/media/documents/9480.pdf (Accessed November 1, 2021).

Watts, J.D., L. Tacconi, S. Irawan, and A.H. Wijaya, 2019: Village transfers for the environment: Lessons from community-based development programs and the village fund. *For. Policy Econ.*, **108** (January), 101863, doi:10.1016/j.forpol.2019.01.008.

WBCSD, 2016: Corporate renewable power purchase agreements (PPAs). World Business Council for Sustainable Development (WBCSD). https://www.wbcsd.org/Programs/Climate-and-Energy/Energy/REscale/Corporate-renewable-power-purchase-agreements-PPAs (Accessed November 1, 2021).

WEF, 2013: *The Green Investment Report: The ways and means to unlock private finance for green growth*. World Economic Forum, Geneva, Switzerland, 40 pp.

Weikmans, R. and J.T. Roberts, 2019: The international climate finance accounting muddle: is there hope on the horizon? *Clim. Dev.*, **11(2)**, 97–111, doi:10.1080/17565529.2017.1410087.

Weiss, M., M. Junginger, M.K. Patel, and K. Blok, 2010: A review of experience curve analyses for energy demand technologies. *Technol. Forecast. Soc. Change*, **77(3)**, 411–428, doi:10.1016/j.techfore.2009.10.009.

Weitzman, M.L., 2014: Fat tails and the social cost of carbon. *American Economic Review*, **104**, 544–546.

WFP, 2020: *Amid COVID-19 crises, Germany donates to protect African countries with climate insurance*. World Food Programme, Berlin, Germany. https://www.wfp.org/news/amid-covid-19-crises-germany-donates-protect-african-countries-climate-insurance (Accessed November 1, 2021).

White, R. and S. Wahba, 2019: Addressing constraints to private financing of urban (climate) infrastructure in developing countries. *Int. J. Urban Sustain. Dev.*, **11(3)**, 245–256, doi:10.1080/19463138.2018.1559970.

Williams, M., 2015: *Gender and Climate Change Financing: Coming out of the margin*. 1st ed. Routledge, London, UK, 554 pp.

Williams, T., 2015: Where the Hell Is All the Climate Funding? Inside Philanthropy, April 22, 2015. Santa Monica, CA, USA. https://www.insidephilanthropy.com/home/2015/4/22/where-the-hell-is-all-the-climate-funding.html (Accessed November 1, 2021).

Willis Towers Watson, 2020: *The world's largest asset managers - 2020*. Thinking Ahead Institute, London, UK. 56 pp. https://www.thinkingaheadinstitute.org/research-papers/the-worlds-largest-asset-managers-2020/ (Accessed August 16, 2021).

Wills, S.E., L.W. Senbet, and W. Simbanegavi, 2016: Sovereign Wealth Funds and Natural Resource Management in Africa: Table A1: *J. Afr. Econ.*, **25** (suppl 2), ii3–ii19, doi:10.1093/jae/ejw018.

WMO, 2019: Tropical Cyclone Idai hits Mozambique. World Meteorological Organization (WMO), Geneva, Switzerland.

Wolfrom, L. and M. Yokoi-Arai, 2016: Financial instruments for managing disaster risks related to climate change. *OECD J. Financ. Mark. Trends*, **2015(1)**, 25–47, doi:10.1787/fmt-2015-5jrqdkpxk5d5.

Wong, G.Y. et al., 2019: Narratives in REDD+ benefit sharing: examining evidence within and beyond the forest sector. *Clim. Policy*, **19(8)**, 1038–1051, doi:10.1080/14693062.2019.1618786.

Woods, K., 2016: How rising insurance costs are squeezing small Irish businesses dry. *Fora*, July 24, 2016. Ireland.

Wooldridge, P.D., 2001: The emergence of new benchmark yield curves. *BIS Quarterly Review. December 2001.* Bank for International Settlements (BIS), Basel, Switzerland.

World Bank, 2009: *The World Bank Group guarantee instruments 1990-2007: An independent evaluation*. World Bank, Washington DC, USA, 115 pp.

World Bank, 2015: Pacific Catastrope Risk Insurance Pilot: Lessons Learned and Next Steps. World Bank, Washington, D.C., USA.

World Bank, 2016: *Making Climate Finance Work in Agriculture*. World Bank, Washington, D.C., USA.

World Bank, 2017: *The Potential Role of Enhanced Bond Structures in Forest Climate Finance*. World Bank, Washington, DC, USA, 95 pp.

World Bank, 2018a: *MDB Methodology for Private Investment Mobilization: Reference Guide*. 17 pp.

World Bank, 2018b: *Blockchain and emerging digital technologies for enhancing post-2020 climate markets*. World Bank, Washington, DC, USA, 32 pp. http://documents.worldbank.org/curated/en/942981521464296927/Blockchain-and-emerging-digital-technologies-for-enhancing-post-2020-climate-markets (Accessed November 1, 2021).

World Bank, 2019a: Southeast Asia Disaster Risk Insurance Facility (SEADRIF) Technical Briefing for Japanese Insurance Industry. Feature Story, January 17, 2019.

World Bank, 2019b: *Leveraging Economic Migration for Development: A Briefing for the World Bank Board*. World Bank, Washington, DC, USA, 83 pp.

World Bank, 2020: *Capital Markets Development: A Primer for Policymakers*. World Bank Group, Washington DC, USA, 23 pp. https://openknowledge.worldbank.org/handle/10986/36058 (Accessed August 18, 2021).

World Bank, 2021: Power Purchase Agreements (PPAs) and Energy Purchase Agreements (EPAs). World Bank, Washington, D.C., USA. https://ppp.worldbank.org/public-private-partnership/sector/energy/energy-power-agreements/power-purchase-agreements (Accessed November 1, 2021).

World Bank and IMF, 2001: *Developing Government Bond Markets: A Handbook*. World Bank, Washington DC, USA, 440 pp.

World Bank, Clean Cooking Alliance, and ESMAP, 2021: *Understanding market-based solutions and access to finance options for clean-cooking technologies in Bangladesh*. World Bank Group, Washington DC, USA, and Dhaka, Bangladesh, 74 pp. https://documents1.worldbank.org/curated/

en/311561624871430872/pdf/Understanding-Market-Based-Solutions-and-Access-to-Finance-Options-for-Clean-Cooking-Technologies-in-Bangladesh.pdf (Accessed September 21, 2021).

World Bank Data, 2020a: Gross Domestic Product (current US$). https://data.worldbank.org/indicator/NY.GDP.MKTP.CD.

World Bank Data, 2020b: Gross Capital Formation (current US$). https://data.worldbank.org/indicator/NY.GDP.MKTP.CD.

World Bank Group, 2019: *State and Trends of Carbon Pricing 2019*. World Bank Group, Washington, DC, USA. https://openknowledge.worldbank.org/handle/10986/31755 (Accessed November 1, 2021).

World Bank Group, 2020: *State and Trends of Carbon Pricing 2020*. World Bank Group, Washington, DC, USA, 105 pp. https://openknowledge.worldbank.org/handle/10986/33809 (Accessed November 1, 2021).

World Energy Council, 2019: *World Energy Scenarios 2019: Exploring Innovation Pathways to 2040*. World Energy Council, London, 148 pp. https://www.worldenergy.org/assets/downloads/Scenarios_Report_FINAL_for_website.pdf (Accessed November 1, 2021).

Wright, C., 2011: Export Credit Agencies and Global Energy: Promoting National Exports in a Changing World. *Glob. Policy*, **2(sup1)**, 133–143, doi:10.1111/J.1758-5899.2011.00132.X.

Yadav, P., A.P. Heynen, and D. Palit, 2019: Energy for Sustainable Development Pay-As-You-Go financing: A model for viable and widespread deployment of solar home systems in rural India. *Energy Sustain. Dev.*, **48**, 139–153, doi:10.1016/j.esd.2018.12.005.

Yanai, R.D. et al., 2020: Improving uncertainty in forest carbon accounting for REDD+ mitigation efforts. *Environ. Res. Lett.*, **15**, 124002.

Yeo, S., 2019: Where climate cash is flowing and why it's not enough. *Nature*, **573(7774)**, 328–331, doi:10.1038/d41586-019-02712-3.

Yescombe, E.R., 2017: *Public-Private Partnerships in Sub-Saharan Africa - Case Studies for Policymakers 2017*. UONGOZI Institute, Dar es Salaam, Tanzania. https://www.africaportal.org/publications/public-private-partnerships-in-sub-saharan-africa-case-studies-for-policymakers-2017/ (Accessed November 1, 2021).

Zamarioli, L.H., P. Pauw, M. König, and H. Chenet, 2021: The climate consistency goal and the transformation of global finance. *Nat. Clim. Change*, **11(7)**, 578–583, doi:10.1038/s41558-021-01083-w.

Zancanella, P., P. Bertoldi, and B. Boza-Kiss, 2016: *Demand response status in EU Member States*. European Commission Joint Research Centre, Luxembourg, 153 pp.

Zenghelis, D. and N. Stern, 2016: *The importance of looking forward to manage risks: submission to the Task Force on Climate-Related Financial Disclosures*. The London School of Economics and Political Science, and the Grantham Research Institute on Climate Change and the Environment, London, UK.

Zerbib, O.D., 2019: The effect of pro-environmental preferences on bond prices: Evidence from green bonds. *J. Bank. Financ.*, **98**, doi:10.1016/j.jbankfin.2018.10.012.

Zhan, C. and M. de Jong, 2018: Financing eco cities and low carbon cities: The case of Shenzhen International Low Carbon City. *J. Clean. Prod.*, **180**, 116–125, doi:10.1016/j.jclepro.2018.01.097.

Zhan, C., M. de Jong, and H. de Bruijn, 2018: Funding Sustainable Cities: A Comparative Study of Sino-Singapore Tianjin Eco-City and Shenzhen International Low-Carbon City. *Sustainability*, **10(11)**, 4256, doi:10.3390/su10114256.

Zhang, H., M. Duan, and Z. Deng, 2019: Have China's pilot emissions trading schemes promoted carbon emission reductions?– the evidence from industrial sub-sectors at the provincial level. *J. Clean. Prod.*, **234**, 912–924, doi:10.1016/j.jclepro.2019.06.247.

Zhang, W. and X. Pan, 2016: Study on the demand of climate finance for developing countries based on submitted INDC. *Adv. Clim. Chang. Res.*, **7(1–2)**, 99–104, doi:10.1016/j.accre.2016.05.002.

Zhang, Y., 2021: Accelerating Access to Clean Cooking Will Require a Heart-Head-and-Hands Approach. *Development*, **65**, 59–62, doi:10.1057/s41301-021-00297-x.

Zografos, C. and P. Robbins, 2020: Green Sacrifice Zones, or Why a Green New Deal Cannot Ignore the Cost Shifts of Just Transitions. *One Earth*, **3(5)**, 543–546, doi:10.1016/j.oneear.2020.10.012.

16

# Innovation, Technology Development and Transfer


**Coordinating Lead Authors:**
Gabriel Blanco (Argentina), Heleen de Coninck (the Netherlands)

**Lead Authors:**
Lawrence Agbemabiese (Ghana/the United States of America), El Hadji Mbaye Diagne (Senegal), Laura Diaz Anadon (Spain/United Kingdom), Yun Seng Lim (Malaysia), Walter Alberto Pengue (Argentina), Ambuj D. Sagar (India), Taishi Sugiyama (Japan), Kenji Tanaka (Japan), Elena Verdolini (Italy), Jan Witajewski-Baltvilks (Poland)

**Contributing Authors:**
Maarten van Aalst (the Netherlands), Lara Aleluia Reis (Portugal), Mustafa Babiker (Sudan/Saudi Arabia), Xuemei Bai (Australia), Rudi Bekkers (the Netherlands), Paolo Bertoldi (Italy), Sara Burch (Canada), Luisa F. Cabeza (Spain), Clara Caiafa (Brazil/the Netherlands), Brett Cohen (South Africa), Felix Creutzig (Germany), Renée van Diemen (the Netherlands/United Kingdom), María Josefina Figueroa Meza (Venezuela/Denmark), Clara Galeazzi (Argentina), Frank Geels (United Kingdom/the Netherlands), Michael Grubb (United Kingdom), Kirsten Halsnæs (Denmark), Joni Jupesta (Indonesia/Japan), Şiir Kilkiş (Turkey), Michael König (Germany), Jonathan Köhler (Germany), Abhishek Malhotra (India), Eric Masanet (the United States of America), William McDowall (United Kingdom), Nikola Milojevic-Dupont (France), Catherine Mitchell (United Kingdom), Gregory F. Nemet (the United States of America/Canada), Lars J. Nilsson (Sweden), Anthony Patt (Switzerland), Patricia Perkins (Canada), Joyashree Roy (India/Thailand), Karolina Safarzynska (Poland), Yamina Saheb (France/Algeria), Ayyoob Sharifi (Iran/Japan), Kavita Surana (India), Harald Winkler (South Africa)

**Review Editors:**
Nagmeldin Mahmoud (Sudan), Emi Mizuno (Japan)

**Chapter Scientists:**
Muneki Adachi (Japan), Clara Caiafa (Brazil/the Netherlands), Daniela Keesler (Argentina), Eriko Kiriyama (Japan)



**This chapter should be cited as:**
Blanco, G., H. de Coninck, L. Agbemabiese, E. H. Mbaye Diagne, L. Diaz Anadon, Y. S. Lim, W.A. Pengue, A.D. Sagar, T. Sugiyama, K. Tanaka, E. Verdolini, J. Witajewski-Baltvilks, 2022: Innovation, technology development and transfer. In IPCC, 2022: *Climate Change 2022: Mitigation of Climate Change. Contribution of Working Group III to the Sixth Assessment Report of the Intergovernmental Panel on Climate Change* [P.R. Shukla, J. Skea, R. Slade, A. Al Khourdajie, R. van Diemen, D. McCollum, M. Pathak, S. Some, P. Vyas, R. Fradera, M. Belkacemi, A. Hasija, G. Lisboa, S. Luz, J. Malley, (eds.)]. Cambridge University Press, Cambridge, UK and New York, NY, USA. doi: 10.1017/9781009157926.018

# Table of Contents

## Executive Summary

**Innovation in climate mitigation technologies has seen enormous activity and significant progress in recent years. Innovation has also led to, and exacerbated, trade-offs in relation to sustainable development (*high confidence*).** Innovation can leverage action to mitigate climate change by reinforcing other interventions. In conjunction with other enabling conditions, innovation can support system transitions to limit warming and help shift development pathways. The currently widespread implementation of solar photovoltaic (solar PV) and light-emitting diodes (LEDs), for instance, could not have happened without technological innovation (*high confidence*). Technological innovation can also bring about new and improved ways of delivering services that are essential to human well-being. At the same time as delivering benefits, innovation can result in trade-offs that undermine both progress on mitigation and progress towards other Sustainable Development Goals (SDGs). Trade-offs include negative externalities – for instance, greater environmental pollution and social inequalities – rebound effects leading to lower net emission reductions or even increases in emissions, and increased dependency on foreign knowledge and providers (*high confidence*). Effective governance and policy has the potential to avoid and minimise such misalignments (*medium evidence, high agreement*). {16.1, 16.2, 16.3, 16.4, 16.5.1, 16.6}

**A systemic view of innovation to direct and organise the processes has grown over the last decade. This systemic view of innovation takes into account the role of actors, institutions and their interactions, and can inform how innovation systems that vary across technologies, sectors and countries, can be strengthened (*high confidence*).** Where a systemic view of innovation has been taken, it has enabled the development and implementation of indicators that are better able to provide insights into innovation processes. This, in turn, has enabled the analysis and strengthening of innovation systems. Traditional quantitative innovation indicators mainly include research and development (R&D) investments and patents. Systemic indicators of innovation, however, go well beyond these approaches. They include structural innovation system elements including actors and networks, as well as indicators for how innovation systems function, such as access to finance, employment in relevant sectors, and lobbying activities. For example, in Latin America, monitoring systemic innovation indicators for the effectiveness of agroecological mitigation approaches has provided insights on the appropriateness and social alignment of new technologies and practices. Climate-energy-economy models, including integrated assessment models, generally employ a stylised and necessarily incomplete view of innovation, and have yet to incorporate a systemic representation of innovation systems. {16.2, 16.2.4, 16.3, 16.3.4, 16.5, Table 16.7, Box 16.1, Box 16.3, Box 16.10}

**A systemic perspective on technological change can provide insights to policymakers supporting their selection of effective innovation policy instruments (*high confidence*).** A combination of scaled-up innovation investments with demand-pull interventions can achieve faster technology unit cost reductions and more rapid scale-up than either approach in isolation (*high confidence*). These innovation policy instruments would nonetheless have to be tailored to local development priorities, to the specific context of different countries, and to the technology being supported. The timing of interventions and any trade-offs with sustainable development also need to be addressed. Public R&D funding and support, as well as innovation procurement, have proven valuable for fostering innovation in small to medium cleantech firms. Innovation outcomes of policy instruments not necessarily aimed at innovation, such as feed-in tariffs, auctions, emissions trading schemes, taxes and renewable portfolio standards, vary from negligible to positive for climate change mitigation. Some specific designs of environmental taxation can also result in negative distributional outcomes. Most of the available literature and evidence on innovation systems come from industrialised countries and larger developing countries. However, there is a growing body of evidence from developing countries and Small Island Developing States (SIDS). {16.4, 16.4.4.3, 16.4.4.4, 16.5, 16.7}

**Experience and analyses show that technological change is inhibited if technological innovation system functions are not adequately fulfilled. This inhibition occurs more often in developing countries (*high confidence*).** Examples of such functions are knowledge development, resource mobilisation, and activities that shape the needs, requirements and expectations of actors within the innovation system (guidance of the search). Capabilities play a key role in these functions, the build-up of which can be enhanced by domestic measures, but also by international cooperation (*high confidence*). For instance, innovation cooperation on wind energy has contributed to the accelerated global spread of this technology. As another example, the policy guidance by the Indian government, which also promoted development of data, testing capabilities and knowledge within the private sector, has been a key determinant of the success of an energy-efficiency programme for air conditioners and refrigerators in India. {16.3, 16.5, 16.6, Cross-Chapter Box 12 in this chapter, Box 16.2}

**Consistent with innovation system approaches, the sharing of knowledge and experiences between developed and developing countries can contribute to addressing global climate and SDGs. The effectiveness of such international cooperation arrangements, however, depends on the way they are developed and implemented (*high confidence*).** The effectiveness and sustainable development benefits of technology sharing under market conditions appear to be determined primarily by the complexity of technologies, local capabilities and the policy regime. This suggests that the development of planning and innovation capabilities remains necessary, especially in least-developed countries and SIDS. International diffusion of low-emission technologies is also facilitated by knowledge spillovers from regions engaged in clean R&D (*medium confidence*). {16.6}

**The evidence on the role of intellectual property rights (IPR) in innovation is mixed. Some literature suggests that it is a barrier, while other sources suggest that it is an enabler to the diffusion of climate-related technologies (*medium confidence*).** There is agreement that countries with well-developed institutional capacity may benefit from a strengthened IPR regime, but that countries with limited capabilities might face greater barriers to innovation as a consequence. This enhances the continued need for capacity building. Ideas to improve the alignment of the global IPR regime and address climate change include specific arrangements for least-developed countries, case-by-case decision-making and patent-pooling institutions. {16.2.3.3, 16.5, Box 16.9}

**Although some initiatives have mobilised investments in developing countries, gaps in innovation cooperation remain, including in the Paris Agreement instruments. These gaps could be filled by enhancing financial support for international technology cooperation, by strengthening cooperative approaches, and by helping build suitable capacity in developing countries across all technological innovation system functions (*high confidence*).** The implementation of current arrangements of international cooperation for technology development and transfer, as well as capacity building, are insufficient to meet climate objectives and contribute to sustainable development. For example, despite building a large market for mitigation technologies in developing countries, the lack of a systemic perspective in the implementation of the Clean Development Mechanism, operational since the mid-2000s, has only led to some technology transfer, especially to larger developing countries, but limited capacity building and minimal technology development (*medium confidence*). In the current climate regime, a more systemic approach to innovation cooperation could be introduced by linking technology institutions, such as the Technology Mechanism, and financial actors, such as the financial mechanism. {16.5.3}

**Countries are exposed to sustainable development challenges in parallel with the challenges that relate to climate change. Addressing both sets of challenges simultaneously presents multiple and recurrent obstacles that systemic approaches to technological change could help resolve, provided they are well managed (*high confidence*).** Obstacles include both entrenched power relations dominated by vested interests that control and benefit from existing technologies, and governance structures that continue to reproduce unsustainable patterns of production and consumption (*medium confidence*). Studies also highlight the potential for cultural factors to strongly influence the pace and direction of technological change. Sustainable solutions require adoption and mainstreaming of locally novel technologies that can meet local needs, and simultaneously address the SDGs. Acknowledging the systemic nature of technological innovation, which involves many levels of actors, stages of innovation and scales, can lead to new opportunities to shift development pathways towards sustainability. {16.4, 16.5, 16.6}

**An area where sustainable development, climate change mitigation and technological change interact is digitalisation. Digital technologies can promote large increases in energy efficiency through coordination and an economic shift to services, but they can also greatly increase energy demand because of the energy used in digital devices. System-level rebound effects may also occur (*high confidence*).** Digital devices, including servers, increase pressure on the environment due to the demand for rare metals and end-of-life disposal. The absence of adequate governance in many countries can lead to harsh working conditions and unregulated disposal of electronic waste. Digitalisation also affects firms' competitiveness, the demand for skills, and the distribution of, and access to, resources. The existing digital divide, especially in developing countries, and the lack of appropriate governance of the digital revolution can hamper the role that digitalisation could play in supporting the achievement of stringent mitigation targets. At present, the understanding of both the direct and indirect impacts of digitalisation on energy use, carbon emissions and potential mitigation, is limited (*medium confidence*). {Cross-Chapter Box 11 in this chapter, 16.2}

**Strategies for climate change mitigation can be most effective in accelerating transformative change when actions taken to strengthen one set of enabling conditions also reinforce and strengthen the effectiveness of other enabling conditions (*medium confidence*).** Applying transition or system dynamics to decisions can help policymakers take advantage of such high-leverage intervention points, address the specific characteristics of technological stages, and respond to societal dynamics. Inspiration can be drawn from the global unit cost reductions of solar PV, which were accelerated by a combination of factors interacting in a mutually reinforcing way across a limited group of countries (*high confidence*). {Box 16.4, Cross-Chapter Box 12 in this chapter}

**Better and more comprehensive data on innovation indicators can provide timely insights for policymakers and policy design locally, nationally and internationally, especially for developing countries, where such insights are missing more often.** Data needed include those that can show the strength of technological, sectoral and national innovation systems. It is also necessary to validate current results and generate insights from theoretical frameworks and empirical studies for developing countries contexts. Innovation studies on adaptation and mitigation other than energy and ex-post assessments of the effectiveness of various innovation-related policies and interventions, including R&D, would also provide benefits. Furthermore, methodological developments to improve the ability of integrated assessment models (IAMs) to capture energy innovation system dynamics, and the relevant institutions and policies (including design and implementation), would allow for more realistic assessment. {16.2, 16.3, 16.7}

## 16.1 Introduction

Technological change and innovation are considered key drivers of economic growth and social progress (Brandão Santana et al. 2015; Heeks and Stanforth 2015). Increased production and consumption of goods and services creates economic benefits through higher demands for improved technologies (Gossart 2015). Since the Industrial Revolution, however, and notwithstanding the benefits, this production and consumption trend and the technological changes associated with it have also come at the cost of long-term damage to the life support systems of our planet (Alarcón and Vos 2015; Steffen et al. 2015). The significance of such impacts depends on the technology, but also on the intrinsic characteristics of the country or region analysed (Brandão Santana et al. 2015).



Other chapters in this volume have discussed technological change in various ways, including as a framing issue (Chapter 1), in the context of specific sectors (Chapters 6–11), for specific purposes (Chapter 12) and as a matter of policy, international cooperation and finance (Chapters 13–15). Chapter 2 discusses past trends in technological change and chapters 3 and 4 discuss it in the context of future modelling. In general, implicitly or explicitly, technological change is assigned an important role in climate change mitigation and achieving sustainable development (Thacker et al. 2019), as also discussed in past IPCC reports (IPCC 2014, 2018a). Chapter 16 describes how a well-established innovation system at a national level, guided by well-designed policies, can contribute to achieving mitigation and adaptation targets along with broader Sustainable Development Goals (SDGs), while avoiding undesired consequences of technological change.

The environmental impacts of social and economic activities, including emissions of greenhouse gases (GHGs), are greatly influenced by the rate and direction of technological changes (Jaffe et al. 2000). Technological changes usually designed and used to increase productivity and reduce the use of natural resources can lead to increased production and consumption of goods and services through different rebound effects that diminish the potential benefits of reducing the pressure on the environment (Kemp and Soete 1990; Grübler 1998; Sorrell 2007; Barker et al. 2009; Gossart 2015).

Those environmental impacts depend not only on which technologies are used, but also on how they are used (Grübler et al. 1999a). Technological change is not exogenous to social and economic systems; technologies are not conceived, selected, and applied autonomously (Grubler et al. 2018). Underlying driving forces of the problem, such as more resource-intensive lifestyles and larger populations (Hertwich and Peters 2009; UNEP 2014), remain largely unchallenged. Comprehensive knowledge of the direct and indirect effects of technological changes on physical and social systems could improve decision-making, including in those cases where technological change mitigates environmental impacts.

A sustainable global future for people and nature requires rapid and transformative societal change by integrating technical, governance (including participation), financial and societal aspects of the solutions to be implemented (Sachs et al. 2019; Pörtner et al. 2021). A growing body of interdisciplinary research from around the world can inform implementation of adaptive solutions that address the benefits and drawbacks of linkages in social-ecological complexity, including externalities and rebound effects from innovation and technological transformation (Balvanera et al. 2017; Pörtner et al. 2021).

Technological change and transitional knowledge can reinforce each other. The value of traditional wisdom and its technological practices provide examples of sustainable and adaptive systems that could potentially adapt to and mitigate climate change (Kuoljok 2019; Singh et al. 2020). Peasants and traditional farmers have been able to respond well to climate changes through their wisdom and traditional practices (Nicholls and Alteri 2013). The integration of the traditional wisdom with new technologies can offer new and effective solutions (Galloway McLean 2010).

Achieving climate change mitigation and other SDGs thus also requires rapid diffusion of knowledge and technological innovations. However, these are hampered by various barriers, some of which are illustrated in Table 16.1 (Markard et al. 2020).

The literature has been growing rapidly over the past decades on how, in a systemic way, the barriers to sustainability transition can be overcome in various circumstances. A central element is that national systems of innovation can help achieve both climate change goals and SDGs, by integrating new ideas, devices, resources, new and traditional knowledge, and technological changes for more effective and adaptive solutions (Lundvall 1992). At the organisational level, innovation is seen

**Table 16.1 | Overview of challenges to accelerated diffusion of technological innovations.** Source: based on Markard et al. (2020).

| Challenges | Description | Examples |
|---|---|---|
| Innovations in whole systems | Since entire systems are changing, changes in system architecture are also needed, which may not keep pace. | Decentralisation of electricity supply and integration of variable sources. |
| Interaction between multiple systems and subsystems | Simultaneous, accelerating changes multiple systems or sectors, vying for the same resources and showing other interactions. | Electrification of transport, heating and industry all using the same renewable electricity source. |
| Industry decline and incumbent resistance | Decline of existing industries and businesses can lead to incumbents slowing down change, and resistance, e.g., from unions or workers. | Traditional car industry leading to facture closures, demise of coal mining and coal-fired power generation leading to local job loss. |
| Consumers and social practices | Consumers need to change practices and demand patterns. | Reduced car ownership in a sharing economy, trip planning for public and non-motorised transport, fuelling practices in electric driving. |
| Coordination in governance and policy | Increasing complexity of governance requires coordination between multiple levels of government and a multitude of actors relevant to the transition, e.g., communities, financial institutions, private sector. | Multilevel governance between European Commission and member states in Energy Union package. |

as a process that can bring value by means of creating more effective products, services, processes, technologies, policies and business models that are applicable to commercial, business, financial and even societal or political organisations (Brooks 1980; Arthur 2009).

The literature refers to the terms 'technology push', 'market pull', 'regulatory push-pull', and 'firm specific factors' as drivers for innovation, mostly to inform policymakers (Zubeltzu-Jaka et al. 2018). There has also been growing interest in social drivers, motivated by the recognition of social issues, such as unemployment and public health, linked to the deployment of innovative low-carbon technologies (Altantsetseg et al. 2020). Policy and social factors and the diverse trajectories of innovation are influenced by regional and national conditions (Tariq et al. 2017), and such local needs and purposes need to be considered in crafting international policies aimed at fostering the global transition towards increased sustainability (Caravella and Crespi 2020). From this standpoint, a multidimensional, multi-actor, systemic innovation approach would be needed to enhance global innovation diffusion (de Jesus and Mendonça 2018), especially if this is to lead to overall sustainability improvements rather than result in new sustainability challenges.

Policies to mitigate climate change do not always take into account the effects of mitigation technologies on other environmental and social challenges (Arvesen et al. 2011). Policies also often disregard the strong linkages between technological innovation and social innovation; the latter is understood to be the use of soft technologies that brings about transformation through establishing new institutions, new practices, and new models to create a positive societal impact, characterised by collaboration that crosses traditional roles and boundaries, between citizens, civil society, the state, and the private sector (Reynolds et al. 2017). Market forces do not provide sufficient incentives for investment in development or diffusion of technologies, leaving a role for public policy to create the conditions to assure a systemic innovation approach (Popp 2010; Popp and Newell 2012). Moreover, public action is more than just addressing market failure, it is an unalienable element of an innovation system (Mazzucato 2013).

Coupling technological innovation with sustainable development and the SDGs would need to address overall social, environmental, and economic consequences, given that public policy is intertwined with innovation, technological changes and other factors in a complex manner. Chapter 16 is organised in the following manner to provide an overview of innovation and technology development and transfer for climate change and sustainable development.

Section 16.2 discusses drivers of innovation process, including macro factors that can redirect technological change towards low-carbon options. Representations of these drivers in mathematical and statistical models allow for explaining the past and constructing projections of future technological change. They also integrate the analysis of drivers and consequences of technological change within economic-energy-economy (or integrated assessment) models (Chapter 3). The section also describes the different phases of innovation and metrics, such as the widely used but also criticised technology readiness levels (TRLs).

Section 16.3 discusses innovation as a systemic process based on recent literature. While the innovation process is often stylised as a linear process, innovation is now predominantly seen as a systemic process in that it is a result of actions by, and interactions among, a large set of actors, whose activities are shaped by, and shape, the context in which they operate and the user group with which they are engaging.

Section 16.4 presents innovation and technology policy, including technology push (e.g., publicly funded R&D) and demand-pull (e.g., governmental procurement programmes) instruments that address potential market failures related to innovation and technology diffusion. The section also assesses the cost-effectiveness of innovation policies as well as other policy assessment criteria introduced in Chapter 13.

Section 16.5 assesses the role of international cooperation in technology development and transfer, in particular the mechanisms established under the UN Framework Convention on Climate Change (UNFCCC), but also other international initiatives for technology cooperation. The discussion on international cooperation includes information exchange, research, development and demonstration cooperation, access to financial instruments, intellectual property rights, as well as promotion of domestic capacities and capacity building.

Section 16.6 describes the role of technology in sustainable development, including unintended effects of technological changes, and synthesises the chapter.

Finally, Section 16.7 discusses gaps in knowledge emerging from this chapter.

## 16.2 Elements, Drivers and Modelling of Technology Innovation

Models of the innovation process, its drivers and incentives provide a tool for technology assessment, constructing projections of technological change and identifying which macro conditions facilitate development of low-carbon technologies. The distinction between stages of the innovation process allows for assessment of technology readiness (Section 16.2.1). Qualitative and quantitative analysis of the main elements underpinning innovation – research and development (R&D), learning by doing, and spillovers – allows for an explanation of past and projected future technological changes (Section 16.2.2). In addition, general purpose technologies can play a role in climate change mitigation.

In the context of mitigation pathways, the feasibility of any emission reduction targets depends on the ability to promote innovation in low- and zero-carbon technologies, as opposed to any other technology. For this reason, Section 16.2.3 reviews the literature of the levers influencing the *direction* of technological change in favour of low- and zero-carbon technologies. Moreover, representation of drivers in mathematical and statistical models from Section 16.2.2 allows integration of its analysis with economic and climate effects within integrated assessment models (IAMs), hence permitting more precise modelling of decarbonisation pathways (Section 16.2.4).

In addition to technological innovation, other innovation approaches are relevant in the context of climate mitigation and more broadly sustainable development (Section 16.6). Frugal innovations, that is, 'good enough' innovations that fulfil the needs of non-affluent consumers mostly in developing countries (Hossain 2018), are characterised by low costs, concentration on core functionalities, and optimised performance level (Weyrauch and Herstatt 2016) and are hence often associated with (ecological and social) sustainability (Albert 2019). Grassroots innovations are products, services and processes developed to address specific local challenges and opportunities, and which can generate novel, bottom-up solutions responding to local situations, interests and values. (Pellicer-Sifres et al. 2018; Dana et al. 2021).



### 16.2.1 Stages of the Innovation Process

The innovation cycle is commonly thought of as having three distinct innovation phases on the path between basic research and commercial application: Research and development (R&D); demonstration; and deployment and diffusion (IPCC 2007). Each of these phases differs with respect to the kind of activity carried out, the type of actors involved and their roles, financing needs, and the associated risks and uncertainties. All phases involve a process of trial and error, and failure is common; the share of innovation that successfully reaches the deployment phase is small. The path occurring between basic research and commercialisation is not linear (Section 16.3); it often requires a long time and is characterised by significant bottlenecks and roadblocks. Furthermore, technologies may regress in the innovation cycle, rather than move forward (Skea et al. 2019). Successfully passing from each stage to the next one in the innovation cycle requires overcoming 'valleys of deaths' (Auerswald and Branscomb 2003; UNFCCC 2017), most notably the demonstration phase (Frank et al. 1996; Weyant 2011; Nemet et al. 2018). Over time, new and improved technologies are discovered; this often makes the dominant technology obsolete, but this is not discussed in this report.

Table 16.2 summarises the different innovation stages and main funding actors, and maps phases into the technology readiness levels (TRLs) discussed in Section 16.2.1.4.

#### 16.2.1.1 Research and Development

This phase of the innovation process focuses on generating knowledge or solving particular problems by creating a combination of artefacts to perform a particular function, or to achieve a specific goal. R&D activities comprise basic research, applied research and technology development. Basic research is experimental or theoretical work undertaken primarily to acquire new knowledge of the underlying foundations of phenomena and observable facts, without any particular application or use in view. Applied research is original investigation undertaken in order to acquire new knowledge, primarily directed towards a specific, practical aim or objective (OECD 2015a). Importantly, R&D activities can be incremental – that is, focused on addressing a specific need by marginally improving an existing technology – or radical, representing a paradigm shift, promoted by new opportunities arising with the accumulation of new knowledge (Mendonça et al. 2018). Technology development, often leading to prototyping, consists of generating a working model of the technology that is usable in the real world, proving the usability and customer desirability of the technology, and giving an idea of its design, features and function (OECD 2015a). These early stages

**Table 16.2 | Stages of the innovation process (Section 16.2.1) mapped onto technology readiness levels (Section 16.2.1.4).** Source: adapted from Auerswald and Branscomb (2003), TEC (2017), IEA (2020a).

<table>
<tr><th>Stage</th><th>Main funding actors</th><th>Phases</th><th>Related technology readiness levels (TRLs)</th></tr>
<tr><td rowspan="5">Research and development</td><td rowspan="5">Governments<br>Firms</td><td>Basic research</td><td>1 – Initial idea (basic principles defined)</td></tr>
<tr><td rowspan="4">Applied research and technology development</td><td>2 – Application formulated (technology concept and application of solution formulated)</td></tr>
<tr><td>3 – Concept needs validation (solutions need to be prototyped and applied)</td></tr>
<tr><td>4 – Early prototype (prototype proven in test conditions)</td></tr>
<tr><td>5 – Full prototype at scale (components proven in conditions to be deployed)</td></tr>
<tr><td rowspan="4">Demonstration</td><td rowspan="4">Governments<br>Firms<br>Venture Capital<br>Angel investors</td><td rowspan="4">Experimental pilot project or full-scale testing</td><td>6 – Full prototype at scale (prototype proven at scale in conditions to be deployed)</td></tr>
<tr><td>7 – Pre-commercial demonstration (solutions working in expected conditions)</td></tr>
<tr><td>8 – First-of-a-kind commercial (commercial demonstration, full-scale deployment in final form)</td></tr>
<tr><td rowspan="3">9 – Commercial operation in early environment (solution is commercial available, needs evolutionary improvement to stay competitive)<br>10 – Integration needed at scale (solution is commercial and competitive but needs further integration efforts)<br>11 – Proof of stability reached (predictable growth)</td></tr>
<tr><td rowspan="2">Deployment and diffusion</td><td>Firms<br>Private equity<br>Commercial banks<br>Mutual funds</td><td>Commercialisation and scale-up<br>(business)</td></tr>
<tr><td>International organisations and financial institutions<br>Non-governmental organisations (NGOs)</td><td>Transfer</td></tr>
</table>

of technological innovation are referred to as the 'formative phase', during which the conditions are shaped for a technology to emerge and become established in the market (Wilson and Grubler 2013) and the constitutive elements of the innovation system emerging around a particular technology are set up (Bento and Wilson 2016; Bento et al. 2018) (Section 16.3).

The outcomes of R&D are uncertain: the amount of knowledge that will result from any given research project or investment is unknown *ex ante* (Rosenberg 1998). This risk to funders (Goldstein and Kearney 2020) translates into underinvestment in R&D due to low appropriability (Weyant 2011; Sagar and Majumdar 2014). In the case of climate mitigation technologies, low innovation incentives for the private sector also result from a negative environmental externality (Jaffe et al. 2005). Furthermore, in the absence of stringent climate policies and targets, incumbent fossil-based energy technologies are characterised by lower financing risk, are heavily subsidised (Davis 2014; Kotchen 2021), and depreciate slowly (Arrow 1962a; Nanda et al. 2016; Semieniuk et al. 2021) (Section 16.2.3). In this context, public research funding plays a key role in supporting high-risk R&D, both in developed and developing economies: it can provide patient and steady funding not tied to short-term investment returns (Kammen and Nemet 2007; Anadon et al. 2014; Mazzucato 2015a; Chan and Diaz Anadon 2016; Anadón et al. 2017; Howell 2017; Zhang et al. 2019) (Section 16.4). Public policies also play a role in increasing private incentives in energy research and development funding (Nemet 2013). R&D statistics are an important indicator of innovation and are collected following the rules of the *Frascati Manual* (OECD 2015a) (Section 16.3.3, Box 16.3 and Table 16.7).

#### 16.2.1.2 Demonstration

Demonstration is carried out through pilot projects or large-scale testing in the real world. Successfully demonstrating a technology shows its utility and that it is able to achieve its intended purpose and, consequently, that the risk of failure is reduced (i.e., that it has market potential) (Hellsmark et al. 2016). Demonstration projects are an important step to promote the deployment of low-carbon energy and industrial technologies in the context of the transition. Government funding often plays a large role in energy technology demonstration projects because scaling up hardware energy technologies is expensive and risky (Brown and Hendry 2009; Hellsmark et al. 2016). Governments' engagement in low-carbon technology demonstration also signals support for businesses willing to take the investment risk (Mazzucato 2016). Venture capital, traditionally not tailored for energy investment, can also play an increasingly important role, thanks to the incentives (e.g., through de-risking) provided by public funding and policies (Gaddy et al. 2017; IEA 2017a).

#### 16.2.1.3 Deployment and Diffusion

Deployment entails producing a technology at large scale and scaling up its adoption and use across individual firms or households in a given market, and across different markets (Jaffe 2015). In the context of climate change mitigation and adaptation technologies, the purposeful diffusion to developing countries, is referred to as 'technology transfer'. Most recently, the term 'innovation cooperation' has been proposed to indicate that technologies needs to be co-developed and adapted to local contexts (Pandey et al. 2021). Innovation cooperation is an important component of stringent mitigation strategies as well as international agreements (Section 16.5).

Diffusion is often sluggish due to lock-in of dominant technologies (Liebowitz and Margolis 1995; Unruh 2000; Ivanova et al. 2018), as well as the time needed to diffuse information about the technologies, heterogeneity among adopters, the incentive to wait until costs fall even further, the presence of behavioural and institutional barriers, and the uncertainty surrounding mitigation policies and long-term commitments to climate targets (Gillingham and Sweeney 2012; Corey 2014; Jaffe 2015; Haelg et al. 2018). In addition, novel technology has been hindered by the actions of powerful incumbents who accrue economic and political advantages over time, as in the case of renewable energy generation (Unruh 2002; Supran and Oreskes 2017; Hoppmann et al. 2019).

Technologies have been shown to penetrate the market with a gradual non-linear process in a characteristic logistic (S-shaped) curve (Grübler 1996; Rogers 2003). The time needed to reach widespread adoption varies greatly across technologies relevant for adaptation and mitigation (Gross et al. 2018); in the case of energy technologies, the time needed for technologies to get from a 10–90% market share of saturation ranges between 5 to over 70 years (Wilson 2012). Investment in commercialisation of low-emission technology is largely provided by private financiers; however, governments play a key role in ensuring incentives through supportive policies, including R&D expenditures providing signals to private investors (Haelg et al. 2018), pricing carbon dioxide emissions, public procurement, technology standards, information diffusion and the regulation for end-lifecycle treatment of products (Cross and Murray 2018) (Section 16.4).

#### 16.2.1.4 Technology Readiness Levels

Technology readiness levels (TRLs) are a categorisation that enables consistent, uniform discussions of technical maturity across different types of technology. They were developed by the National Aeronautics and Space Administration (NASA) in the 1970s (Mankins 1995, 2009) and originally used to describe the readiness of components forming part of a technological system. Over time, more classifications of TRLs have been introduced, notably the one used by the European Union (EU). Most recently, the International Energy Agency (IEA) extended previous classifications to include the later stages of the innovation process (IEA 2020b) and applied it to compare the market readiness of clean energy technologies and their components (OECD 2015a; IEA 2020b). TRLs are currently widely used by engineers, business people, research funders and investors, often to assess the readiness of whole technologies rather than single components. To determine a TRL for a given technology, a technology readiness assessment (TRA) is carried out to examine programme concepts, technology requirements, and demonstrated technology capabilities. In the most recent version of the IEA (IEA 2020b), TRLs range from 1 to 11, with 11 indicating the most mature (Table 16.2).

The purpose of TRLs is to support decision-making. They are applied to avoid the premature application of technologies, which would lead to increased costs and project schedule extensions (US Department of Energy 2011). They are used for risk management, and can also be used to make decisions regarding technology funding, and to support the management of the R&D process within a given organisation or country (De Rose et al. 2017).

In practice, the usefulness of TRLs is limited by several factors. These include limited applicability in complex technologies or systems, the fact that they do not define obsolescence, nor account for manufacturability, commercialisation or the readiness of organisations to implement innovations (European Association of Research Technology Organisations 2014) and do not consider any type of technology-system mismatch or the relevance of the products' operation environment to the system under consideration (Mankins 2009). Many of these limitations can be eased by using TRLs in combination with other indicators such as system readiness levels and other economic indicators on, for example, investments and returns (IEA 2020b).

### 16.2.2 Sources of Technological Change

The speed of technological change could be explained with the key drivers of innovations process: R&D effort; learning by doing; and spillover effects. In addition, new innovations are sometimes enabled by the development of general purpose technologies, such as digitalisation.

#### 16.2.2.1 Learning by Doing and Research and Development

Learning by doing and R&D efforts are two factors commonly used by the literature to explain past and projected future speed of technological change (Klaassen et al. 2005; Mayer et al. 2012; Bettencourt et al. 2013). Learning by doing is the interaction of workers with new machines or processes that allows more efficient use (Arrow 1962b). R&D effort is dedicated to looking for new solutions (e.g., blueprints) that could increase the efficiency of existing production methods or result in entirely new methods, products or services (Section 16.2.1.1).

Learning by doing and R&D are interdependent. Young (1993) postulates that learning by doing cannot continue forever without R&D because it is bounded by an upper physical productivity limit of an existing technology. R&D can shift this limit because it allows for replacing the existing technology with a new one. On the other hand, incentives to invest in R&D depend on the future cost of manufacturing, which in turn depends on the scale of learning by doing. The empirical evidence for virtuous circle between cost reduction, market growth and R&D were found in the case of the photovoltaic (PV) market (Watanabe et al. 2000) (Box 16.4), but could also lead to path dependency and lock-in (Erickson et al. 2015). Sections 16.4.4 and 13.7.3.1 discuss how simultaneous use of technology push and pull policies could amplify the effects of research and learning.

The benefits of R&D and learning by doing are larger at the economy level than at the firm level (Arrow 1962b; Romer 1990;). As a result, when left to its own, the market tends to generate less investment than socially optimal. For instance, if the cost of a technology is too high before a large amount of learning by doing has occurred, there is a risk that it will not be adopted by the market, even if it is economically advantageous for the society. Indeed, initially new technologies are often expensive and cannot compete with the incumbent technologies (Cowan 1990). Large numbers of adopters could lower this cost via learning by doing to a level sufficient to beat the incumbent technology (Gruebler et al. 2012). However, firms could hesitate to be the first adopter and bear the high cost (Isoard and Soria 2001). If this disadvantage overwhelms the advantages of being a first mover[1] and if adopters are not able to coordinate, it will lead to situation of a lock-in (Gruebler et al. 2012).

The failure of markets to deliver the size of R&D investment and learning by doing that would be socially optimal is one of the justifications of government intervention. Policies to address these market failures can be categorised as technology-push and demand-pull policies. The role of these policies is explained in Table 16.3.

Section 16.4 discusses individual policy instruments in greater detail.

**Table 16.3 | Categories of policies and interventions accelerating technological changes, the factors promoting them and slowing them down, illustrated with examples.**

| | What it refers to | What promotes technological change | What slows down technological change | Examples |
|---|---|---|---|---|
| Technology push | Support the creation of new knowledge to make it easier to invest in innovation | Research and development (R&D), funding and performance of early demonstrations (Brown and Hendry 2009; Hellsmark et al. 2016) | Inadequate supply of trained scientists and engineers (Popp and Newell 2012); gap with demand pull (Grübler et al. 1999b) | Japan's Project Sunshine, the US Project Independence in the 1970s. Breakthrough Energy Coalition and Mission Innovation, respectively private- and public-sector international collaborations to respectively focus energy innovation and double energy R&D, both initiated concurrently with the Paris Agreement in 2015 (Sanchez and Sivaram 2017) |
| Demand pull | Instruments creating market opportunities | Enlarging potential markets, increasing adoption of new fuels and mitigation technology<br>Digital innovations<br>Social innovation and awareness | Willingness of consumers to accept new technology<br>Policy and political volatility can deter investment | Subsidies for wind power California, the German feed-in tariff for photovoltaic, quotas for electric vehicles in China (F. Wang et al. 2017) and Norway (Pereirinha et al. 2018)<br>Biofuels (Brazil)<br>Social innovation with wind energy (Denmark, Germany) |

1 For example, see Spence (1981) and Bhattacharya (1984) for a discussion of first-mover advantages.

The size of the learning-by-doing effect is quantified in literature using learning rates, that is estimates of negative correlation between costs and size of deployment of technologies. The results from this literature include estimates for energy technologies (McDonald and Schrattenholzer 2001), electricity generation technologies (Rubin et al. 2015; Samadi 2018), for storage (Schmidt 2017), for end-of-pipe control (Kang et al. 2020) and for energy demand and energy supply technologies (Weiss et al. 2010). Meta-analyses find that learning rates vary across technologies, within technologies, and over time (Nemet 2009a; Rubin et al. 2015; Wei et al. 2017). Moreover, different components of one technology have different learning rates (Elshurafa et al. 2018). Central tendencies are around 20% cost reduction for each doubling of deployment (McDonald and Schrattenholzer 2001).

Studies of correlation between cumulative deployment of technologies and costs are not sufficiently precise to disentangle the causal effect of increase in deployment from the causal effects of R&D and other factors (Nemet 2006). Numerous subsequent studies attempted to, among others issues, separate the effect of learning by doing and R&D (Klaassen et al. 2005; Mayer et al. 2012; Bettencourt et al. 2013), economies of scale (Arce 2014), and knowledge spillovers (Nemet 2012). Once those other factors are accounted for, some empirical studies find that the role of learning by doing in driving down the costs becomes minor (Nemet 2006; Kavlak et al. 2018). In addition, the relation could reflect reverse causality: increase in deployment could be an effect (and not a cause) of a drop in price (Nordhaus 2014; Witajewski-Baltvilks et al. 2015). Nevertheless, in some applications, learning curves can be a useful proxy and heuristic (Nagy et al. 2013).

The negative relation between costs and experience is a reason to invest in a narrow set of technologies; the uncertainty regarding the parameters of this relation is the reason to invest in wider ranges of technologies (Fleming and Sorenson 2001; Way et al. 2019). Concentrating investment in narrow sets of technologies (specialisation) enables fast accumulation of experience for these technologies and large cost reductions. However, when the potency of technology is uncertain, one does not know which technology is truly optimal in the long run. The narrower the set, the higher the risk that the optimal technology will not be supported, and hence will not benefit from learning by doing. Widening the set of supported technologies would reduce this risk (Way et al. 2019). Uncertainty is present because noise in historical data hides the true value of learning rates, and due to unanticipated future shocks to technology costs (Lafond et al. 2018). Ignoring uncertainty in integrated assessment models implies that these model results are biased towards supporting a narrow set of technologies, neglecting the benefits of decreasing risk through diversification (Sawulski and Witajewski-Baltvilks 2020).

#### 16.2.2.2 Knowledge Spillovers

Knowledge spillovers drive continuous technological change (Romer 1990; Rivera-Batiz and Romer 1991) and are for that reason relevant to climate technologies as well as incumbent, carbon-intensive technologies. Knowledge embedded in innovations by one innovator gives an opportunity for others to create new innovations and increase the knowledge stock even further. The constant growth of knowledge stock through spillovers translates into constant growth of productivity and cost reduction.

By allowing for experimenting with existing knowledge and combining different technologies, knowledge spillovers can result in the emergence of novel technological solutions, which has been referred to as 'recombinant innovation' (Weitzman 1998; Fleming and Sorenson 2001; Olsson and Frey 2002; Tsur and Zemel 2007; Arthur 2009). Recombinant innovations speed up technological change by combining different technological solutions, and make things happen that would be impossible with only incremental innovations (van den Bergh 2008; Safarzyńska and van den Bergh 2010; Frenken et al. 2012). It has been shown that 77% of all patents granted between 1790 and 2010 in the USA are coded by a combination of at least two technology codes (Youn et al. 2015). Spillovers related to energy and low-carbon technologies have been documented by a number of empirical studies (*high confidence*) (Popp 2002; Verdolini and Galeotti 2011; Aghion et al. 2016; Witajewski-Baltvilks et al. 2017; Conti et al. 2018). The presence of spillovers can have both positive and negative impacts on climate change mitigation (*high confidence*).

The spillover effect associated with innovation in carbon-intensive technologies may lead to lock-in of fossil-fuel technologies. Continuous technological change of carbon-intensive industry raises the bar for clean technologies: a larger drop in clean technologies' cost is necessary to become competitive (Acemoglu et al. 2012; Aghion et al. 2016). The implication is that delaying climate policy increases the cost of that policy (Aghion 2019).

On the other hand, the spillover effect associated with innovation in low-emission technologies increases the potency of climate policy (Aghion 2019). For instance, a policy that encourages clean innovation leads to accumulation of knowledge in clean industry which, through spillover effects, encourages further innovation in clean industries. Once the stock of knowledge is sufficiently large, the value of clean industries will be so high that technology firms will invest there, even without policy incentives. Once this point is reached, the policy intervention can be discontinued (Acemoglu et al. 2012).

In addition, the presence of spillovers implies that a unilateral effort to reduce emissions in one region could reduce emissions in other regions (*medium confidence*) (Golombek and Hoel 2004; Gerlagh and Kuik 2014). For instance, in the presence of spillovers, a carbon tax that incentivises clean technological change increases the competitiveness of clean technologies not only locally, but also abroad. The size of this effect depends on the size of the spillovers. If they are sufficiently strong, the reduction of emissions abroad due to clean technological change could be larger than the increase of emissions due to carbon leakage (Gerlagh and Kuik 2014). Different types of carbon leakage are discussed in Chapter 13, Section 13.7.1, and other consequences of spillovers for the design of policy are discussed in Chapter 13, Section 13.7.3.

16

#### 16.2.2.3 General-purpose Technologies and Digitalisation

General-purpose technologies (GPTs) provide solutions that could be applied across sectors and industries (Goldfarb 2011) by creating technological platforms for a growing number of interrelated innovations. Examples of GPTs relevant to climate change mitigation are hydrogen and fuel cell technology, which may find applications in transport, industry and distributed generation (Hanley et al. 2018), and nanotechnology which played a significant role in advancement of all the different types of renewable energy options (Hussein 2015). Assessing the environmental, social and economic implications of such technologies, including increased emissions through energy use, is challenging (Section 5.3.4.1 and Cross-Chapter Box 11 in this chapter).

Several GPTs relevant for climate mitigation and adaptation emerged as a result of digitalisation, namely the adoption or increase in the use of information and communication technologies (ICTs) by citizens, organisations, industries or countries, and the associated restructuring of several domains of social life and of the economy around digital technologies and infrastructures (Brennen and Kreiss 2016; IEA 2017b). The digital revolution is underpinned by innovation in key technologies, for example, ubiquitous connected consumer devices such as mobile phones (Grubler et al. 2018), rapid expansions of global internet infrastructure and access (World Bank 2014), and steep cost reductions and performance improvements in computing devices, sensors, and digital communication technologies (Verma et al. 2020). The increasing pace at which the physical and digital worlds are converging increases the relevance of disruptive digitalisation in the context of climate mitigation and sustainability challenges (European Commission 2020) (Cross-Chapter Box 11 in this chapter and Chapter 4, Section 4.4.1).

Digital technologies require energy, but increase efficiency, potentially offering technology-specific greenhouse gas (GHG) emission savings; they also have larger system-wide impacts (Kaack et al. 2021). In industrial sectors, robotisation, smart manufacturing (SM), internet of things (IoT), artificial intelligence (AI), and additive manufacturing (AM or 3D printing) have the potential to reduce material demand and promote energy management (Section 11.3.4.2). Smart mobility is changing transport demand and efficiency (Section 10.2.3). Smart devices in buildings, the deployment of smart grids and the provision of renewable energy increase the role of demand-side management (Serrenho and Bertoldi 2019) (Sections 9.4 and 9.5), and support the shift away from asset redundancy (Section 6.4.3). Digital solutions are equally important on the supply side, for example, by accelerating innovation with simulations and deep learning (Rolnick et al. 2021) or realising flexible and decentralised opportunities through energy-as-a-service concepts and particularly with pay-as-you-go (Section 15.6.8).

Yet, increased digitalisation could increase energy demand, thus wiping away potential efficiency benefits, unless appropriately governed (IPCC 2018a). Moreover, digital technologies could negatively impact labour demand and increase inequality (Cross-Chapter Box 11 in this chapter).

### Cross-Chapter Box 11 | Digitalisation: Efficiency Potentials and Governance Considerations

**Authors:** Felix Creutzig (Germany), Elena Verdolini (Italy), Paolo Bertoldi (Italy), Luisa F. Cabeza (Spain), María Josefina Figueroa Meza (Venezuela/Denmark), Kirsten Halsnæs (Denmark), Joni Jupesta (Indonesia/Japan), Şiir Kilkiş (Turkey), Michael König (Germany), Eric Masanet (the United States of America), Nikola Milojevic-Dupont (France), Joyashree Roy (India/Thailand), Ayyoob Sharifi (Iran/Japan)

**Digital technologies impact positively and negatively on GHG emissions through: their own carbon footprint; technology application for mitigation; and induced larger social change. Digital technologies also raise broader sustainability concerns due to their use of rare materials and associated waste, and their potential negative impact on inequalities and labour demand.**

**Direct impacts emerge because digital technologies consume large amounts of energy, but also have the potential to steeply increase energy efficiency in all end-use sectors through material input savings and increased coordination (*medium evidence, medium agreement*)** (Horner et al. 2016; Huang et al. 2016; IEA 2017b; Jones 2018). Global energy demand from digital appliances reached 7.14 EJ in 2018 (Chapter 9, Box 9.5), implying higher related carbon emissions. However, a small smartphone offers services previously requiring many different devices (Grubler et al. 2018). Demand for data services is increasing rapidly; quantitative estimates of the growth of associated energy demand range from slow and marginal to rapid and sizeable, depending the efficiency trends of digital technologies (Avgerinou et al. 2017; Vranken 2017; Stoll et al. 2019; Masanet et al. 2020) (Section 5.3.4.1). Renewable energy can serve as a low-carbon energy provider for the operation of a data centre, which in turn can provide waste heat for other purposes. Digital technologies can markedly increase the energy efficiency of mobility and residential and public buildings, especially in the context of systems integration (IEA 2020a). Reduction in energy demand and associated GHG emissions from buildings and industry, while maintaining service levels is estimated at 5 to 10%, with larger savings possible. Approaches include building energy management systems (BEMS), home energy management system (HEMS), demand response

*Cross-Chapter Box 11 (continued)*

and smart charging (Cross-Chapter Box 11, Table 1). Data centres can also play a role in energy system management, for example, by increasing renewable energy generation through predictive control (Dabbagh et al. 2019), and by helping to drive the market for battery storage and fuel cells (Riekstin et al. 2014). Temporal and spatial scheduling of electricity demand can provide about 10 GW in demand response in the European electricity system in 2030 (Wahlroos et al. 2017, 2018; Koronen et al. 2020; Laine et al. 2020).

**However, system-wide effects may endanger energy and GHG emission savings (*high evidence, high agreement*).** Economic growth resulting from higher energy and labour productivities can increase energy demand (Lange et al. 2020) and associated GHG emissions. Importantly, digitalisation can also benefit carbon-intensive technologies (Victor 2018). Impacts on GHG emissions are varied in smart and shared mobility systems, as ride hailing increases GHG emissions due to deadheading, whereas shared pooled mobility and shared cycling reduce GHG emissions, as occupancy levels and/or weight per person km transported improve (Section 5.3). Energy and GHG emission impacts from the ubiquitous deployment of smart sensors and service optimisation applications in smart cities are insufficiently assessed in the literature (Milojevic-Dupont and Creutzig 2021). Systemic effects have wider boundaries of analysis, including broader environmental impacts (e.g., demand for rare materials, disposal of digital devices). These need to be



**Cross-Chapter Box 11, Table 1 | Selected sector approaches for reducing GHG emissions that are supported by new digital technologies.** Contributions of digitalisation include a) supporting role (+), b) necessary role in mix of tools (++), c) necessary unique contribution (+++), but digitalisation may also increase emissions (–). (Chapters 5, 8, 9 and 11).

| Sector | Approach | Quantitative evidence | Contribution of digitalisation | Systems perspective and broader societal impacts | References |
|---|---|---|---|---|---|
| **Residential energy use** | Nudges (feedback, information, etc.) | 2–4% reduction in global household energy use possible | + In combination with monetary incentives, non-digital information | New appliances increase consumption | Zangheri et al. (2019); Buckley (2020); Nawaz et al. (2020); Khanna et al. (2021) |
| **Smart mobility** | Shared mobility and digital feedback (ecodriving) | Reduction for shared cycling and shared pooled mobility; increase for ride hailing/ ride sourcing; reduction for ecodriving | – or ++ Apps together with big data and machine learning algorithm key precondition for new shared mobility | Ride hailing increases GHG emissions, especially due to deadheading | Zeng et al. (2017); OECD and ITF (2020) |
| **Smart cities** | Using digital devices and big data to make urban transport and building use more efficient | Precise data about roadway use can reduce material intensity and associated GHG emissions by 90% | ++ Big data analysis necessary for optimisation | Efficiency gains are often compensated by more driving and other rebound effects; privacy concerns linked with digital devices in homes | Milojevic-Dupont and Creutzig (2021) (Chapter 10, Box 10.1) |
| **Agriculture** | Precision agriculture through sensors and satellites providing information on soil moisture, temperature, crop growth and livestock feed levels | Very high potential for variable-rate nitrogen application, moderate potential for variable-rate irrigation | + ICTs provide information and technologies which enables farmers to increase yields, optimise crop management, reduce fertilisers and pesticides, feed and water; increases efficiency of labour-intensive tasks | The digital divide is growing fast, especially between modern and subsistence farming; Privacy and data may erode trust in technologies | Deichmann et al. (2016); Chlingaryan et al. (2018); Soto Embodas et al. (2019); Townsend et al. (2019) |
| **Industry** | Industrial internet of things (IIoT) | Process, activity and functional optimisation increases energy and carbon efficiency | ++ Increased efficiency<br>++ 1.3 $GtCO_2$-eq estimated abatement potential in manufacturing<br>+ Promote sustainable business models | Optimisation in value chains can reduce wasted resources | GeSI (2012); Wang et al. (2016); Parida et al. (2019); Rolnick et al. (2021) |
| **Load management and battery storage optimisation** | Big data analysis for optimising demand management and using flexible load of appliances with batteries | Reduces capacity intended for peak demand, shifts demand to align with intermittent renewable energy availability | + Accelerated experimentation in material science with artificial intelligence<br>++ / +++ Forecast and control algorithms for storage and dispatch management | Facilitate integration of renewable energy sources Improve utilisation of generation assets System-wide rebound effects possible | Akorede et al. (2010); Aghaei and Alizadeh (2013); de Sisternes et al. (2016); Voyant et al. (2017); Gür (2018); Hirsch et al. (2018); Sivaram (2018a); Vázquez-Canteli and Nagy (2019) (Chapter 6, Section 6.4) |

*Cross-Chapter Box 11 (continued)*

integrated holistically within policy design (Kunkel and Matthess 2020), but they are difficult to quantify and investigate (Bieser and Hilty 2018). Policies and adequate infrastructures and choice architectures can help manage and contain the negative repercussions of systemic effects (Sections 5.4, 5.6 and 9.9).

**Broader societal impacts of digitalisation can also influence climate mitigation because of induced demand for consumption goods, impacts on firms' competitiveness, changes the demand for skills and labour, worsening of inequality – including reduced access to services due to the digital divide – and governance aspects (*low evidence, medium agreement*)** (Sections 4.4, 5.3 and 5.6). Digital technologies expand production possibilities in sectors other than ICTs through robotics, smart manufacturing, and 3D printing, and have major implications on consumption patterns (Matthess and Kunkel 2020). Initial evidence suggests that robots displace routine jobs and certain skills, change the demand for high-skilled and low-skilled workers, and suppress wages (Acemoglu and Restrepo 2019). Digitalisation can thus reduce consumers' liquidity and consumption (Mian et al. 2020) and contribute to global inequality, including across the gender dimension, raising fairness concerns (Kerras et al. 2020; Vassilakopoulou and Hustad 2021). Digital technologies can lead to additional concentration in economic power (e.g., Rikap 2020) and lower competition; however, open source digital technologies can counter this tendency (e.g., Rotz et al. 2019). Digital technologies play a role in mobilising citizens for climate and sustainability actions (Segerberg 2017; Westerhoff et al. 2018).

**Whether the digital revolution will be an enabler or a barrier for decarbonisation will ultimately depend on the governance of both digital decarbonisation pathways and digitalisation in general (*medium evidence, high agreement*).** The understanding of the disruptive potential of the wide range of digital technologies is limited due to their ground-breaking nature, which makes it hard to extrapolate from previous history/experience. Municipal and national entities can make use of digital technologies to manage and govern energy use and GHG emissions in their jurisdiction (Bibri 2019a,b) and break down solution strategies to specific infrastructures, building, and places, relying on remote sensing and mapping data, and contextual machine learning about their use (Milojevic-Dupont and Creutzig 2021). Mobility apps can provide mobility-as-a-service access to cities, ensuring due preference to active and healthy modes (Section 9.9 for the example of the Finnish city of Lahti). Trusted data governance can promote the implementation of local climate solutions, supported by available big data on infrastructures and environmental quality (Hansen and Porter 2017; Hughes et al. 2020). Governance decisions, such as taxing data, prohibiting surveillance technologies, or releasing data that enable accountability, can change digitalisation pathways, and thus underlying GHG emission (Hughes et al. 2020).

**Closing the digital gap in developing countries and rural communities enables an opportunity for leapfrogging (*medium evidence, medium agreement*).** Communication technologies (such as mobile phones) enable the participation of rural communities, especially in developing countries, and promote technological leapfrogging, for example, decentralised renewable energies and smart farming (Ugur and Mitra 2017; Foster and Azmeh 2020; Arfanuzzaman 2021). Digital technologies have sector-specific potentials and barriers, and may benefit certain regions/areas/socio-economic groups more than others. For example, integrated mobility services benefit cities more than rural and peripheral areas (OECD 2017).

**Appropriate mechanisms also need to be designed to govern digitalisation as a megatrend (*medium evidence, high agreement*).** Digitalisation is expected to be a fast process, but this transformation takes place against entrenched individual behaviours, existing infrastructure, the legacy of time frames, vested interest and slow institutional processes, and requires trust from consumers, producers and institutions. A core question relates to who controls and manages data created by everyday operations (calls, shopping, weather data, service use, and so on). Regulations that limit or ban the expropriation and exploitation of behavioural data, sourced via smartphones, represent crucial aspects in digitalisation pathways, alongside the possibility to create climate movements and political pressure from the civil society. Governance mechanisms need to be developed to ensure that digital technologies such as AI take over ethical choices (Craglia et al. 2018; Rahwan et al. 2019). Appropriate governance is necessary for digitalisation to effectively work in tandem with established mitigation technologies and choice architectures. Consideration of system-wide effects and overall management is essential to avoid runaway effects. Overall governance of digitalisation remains a challenge, and will have large-scale repercussions on energy demand and GHG emissions.

#### 16.2.2.4 Explaining Past and Projecting Future Technology Cost Changes

Researchers and policymakers alike are interested in using observed empirical patterns of learning to project future reductions in costs of technologies. Studies cutting across a wide range of industrial sectors (not just energy) have tried to relate cost reductions to different functional forms, including cost reductions as a function of time (Moore's law) and cost reductions as a function of production or deployment (Wright's law, also known as Henderson's law), finding that those two forms perform better than alternatives combining different factors, with costs as a function of production (Wright's law) performing marginally better (Nagy et al. 2013). A comparison of expert elicitation and model-based forecasts of the future cost of technologies for the energy transition indicates that model-based forecast medians were closer to the average realised values in 2019 (Meng et al. 2021).

Recent studies attempt to separate the influence of learning by doing (which is a basis of Wright's law) versus other factors in explaining cost reductions, specifically in energy technologies. Some studies explain cost reductions with two factors: cumulative deployment (as proxy for experience); and R&D investment – see the 'two factor' learning curve (Klaassen et al. 2005). However, reliable information on public energy R&D investments for developing countries is not systematically collected. Available data for OECD countries cannot be precisely assigned to specific industrial sectors or sub-technologies (Verdolini et al. 2018). Some learning-curve studies take into account that historical variation in technology costs could be explained by variation in key materials and fuel costs – for example, steel costs for wind turbines (Qiu and Anadon 2012), silicon costs (Nemet 2006; Kavlak et al. 2018) as well as coal and coal plant construction costs (McNerney et al. 2011). Economies of scale played a significant role in the PV cost reductions since the early 2000s (Yu et al. 2011) (Box 16.4), which can also become the case in organic PV technologies (Gambhir et al. 2016; Kavlak et al. 2018).

### 16.2.3 Directing Technological Change

Technological change is characterised not only by its speed, but also its direction. The early works that considered the role of technology in economic and productivity growth (Solow 1957; Nelson and Phelps 1966) assumed that technology can move forward along only one dimension – every improvement led to an increase in efficiency and increased demand for all factors of production. This view, however, ignores the potency of technological change to alter the otherwise fixed relation between economic growth and the use of resources.

Technological change that saves fossil fuels could decouple economic growth and $CO_2$ emissions (Acemoglu et al. 2012, 2014; Hémous 2016; Greaker et al. 2018). Saving of fossils could be obtained with increasing efficiency of producing alternatives to fossils (Acemoglu et al. 2012, 2014). This is the case of oil consumption by combustion engine cars which could be substituted with electric cars (Aghion et al. 2016). If there is no close substitute for a 'dirty resource', then its intensity in production could still be reduced by increasing the efficiency of the dirty resource relative to the efficiency of other inputs (Hassler et al. 2012; André and Smulders 2014; Witajewski-Baltvilks et al. 2017). For instance, energy efficiency improvement leads to a drop in relative demand for energy (Hassler et al. 2012; Witajewski-Baltvilks et al. 2017).

#### 16.2.3.1 Determinants of Technological Change Direction: Prices, Market Size and Government

Firms change their choice of technology upon change in prices: when one input (e.g., energy) becomes relatively expensive, firms pick technologies that allow them to economise on that input, according to price-induced technological change theory (Reder and Hicks 1965; Samuelson 1965; Sue Wing 2006). For example, an increase in oil price will lead to a choice of fuel-saving technologies. Such a response of technological change was evident during the oil-price shocks in the 1970s (Hassler et al. 2012). Technological change that is induced by an increase in price of a resource can never lead to an increase in use of that resource. In other words, rebound effects associated with induced technological change can never offset the saving effect of that technological change (Antosiewicz and Witajewski-Baltvilks 2021).

The impact of energy prices on the size of low-carbon technological change is supported by large number of empirical studies (Popp 2019; Grubb and Wieners 2020). Studies document that higher energy prices are associated with a higher number of low-carbon energy or energy efficiency patents (Newell et al. 1999; Popp 2002; Verdolini and Galeotti 2011; Noailly and Smeets 2015; Ley et al. 2016; Witajewski-Baltvilks et al. 2017; Lin and Chen 2019). Sue Wing (2008) finds that innovation induced by energy prices had a minor impact on the decline in US energy intensity in the last decades of the 20th century, and that autonomous technological change played a more important role. Several studies explore the impact of a carbon tax on green innovation (Section 16.4). However, disentangling the effect of policy tools is complex because the presence of some policies could distort the functioning of other policies (Böhringer and Rosendahl 2010; Fischer et al. 2017) and because the impact of policies could be lagged in time (Antosiewicz and Witajewski-Baltvilks 2021).

The direction of technological change depends also on the market size for dirty technologies relative to the size of other markets (Acemoglu et al. 2014). Due to this dependence, climate and trade policy choices in a single region can alter the direction of technological change at the global level (Section 16.2.3.3).

The value of the market for clean technologies is determined not only by current profit, but also by a firm's expectation of future profits (Alkemade and Suurs 2012; Greaker et al. 2018; Aghion 2019). One implication is that bolstering the credibility and durability of policies related to low-carbon technology is crucial to accelerating technological change and inducing the private sector investment required (Helm et al. 2003), especially in the rapidly growing economies of Asia and Africa which are on the brink of making major decisions about the type of infrastructure they build as they grow, develop, and industrialise (Nemet et al. 2017).

If governments commit to climate policies, firms expect that the future size of markets for clean technologies will be large and they are eager to redirect research effort towards development of these technologies today. The commitment would also incentivise acquiring skills that could further reduce the costs of those technologies (Aghion 2019). However, historical evidence shows that policies related to energy and climate over the long term have tended to change (Taylor 2012; Nemet et al. 2013; Koch et al. 2016). Still, where enhancing policy durability has proven infeasible, multiple uncorrelated potentially overlapping policies can provide sufficient incentives (Nemet 2010).

#### 16.2.3.2 Determinants of Direction of Technological Change: Financial Markets

The challenges of investing in innovation in energy when compared to other important areas, such as ICT and medicine are also reflected in the trends in venture capital funding. Research found that early-stage investments in cleantech companies were more likely to fail and returned less capital than comparable investments in software and medical technology (Gaddy et al. 2017). This led to investors retreating from hardware technologies required for renewable energy generation and storage, and moving to software-based technologies and demand-side solutions (Bumpus and Comello 2017).

The preference for particular types of investments in renewable energy technologies depends on investors attitude to risk (Mazzucato and Semieniuk 2018). Some investors invest in only one technology, others may spread their investments, or invest predominantly in high-risk technologies. The distribution of different types of investors will affect whether finance goes to support deployment of new high-risk technologies, or diffusion of more mature, less-risky technologies characterised by incremental innovations. The role of finance in directing investment is further discussed in Chapter 15, Section 15.6.2.

#### 16.2.3.3 Internationalisation of Green Technological Change

A unilateral effort to reduce emissions (via a combination of climate, industrial and trade policies) in a coalition of regions that are technology leaders will reduce the cost of clean technologies, and induce emissions reduction in the countries outside the coalition (Golombek and Hoel 2004; Di Maria and Smulders 2005; Di Maria and van der Werf 2008; Hémous 2016; van den Bijgaart 2017). The literature suggests various mechanisms leading to this result. Di Maria and van der Werf (2008) argue that the effort to reduce emissions in one region reduces global demand for 'dirty goods'. This will redirect global innovation towards clean technologies, leading to a drop in the cost of clean production in every region.

The model in Hemous (2016) predicts that such a coalition could induce acceleration of clean technological change through a mix of carbon taxation, clean R&D subsidies and trade policies in that region leading to reduction of cost of clean production inside the coalition. Export of goods produced with clean technologies to a region outside the coalition reduces demand for dirty goods in that region. In the model by van den Bijgaart (2017) local advancements of clean technologies by a coalition with strong R&D potential are imitated outside the coalition. Furthermore, advancements of clean technologies will incentivise future clean R&D outside the coalition due to intertemporal knowledge spillovers. In Golombek and Hoel (2004) an increase in environmental concern in one region increases abatement R&D in that region. Part of this knowledge spills over to other regions, increasing their incentive to increase abatement too, provided that the latter regions did not invest in abatement before.

However, this chain breaks if the regions that are behind the technological frontier (i.e., technological followers) are not able to absorb the solutions developed by regions at the frontier. New technologies might fail due to deficiencies of political, commercial, industrial, and financial institutions, which we list in Table 16.4. For instance, countries might not benefit fully from international knowledge spillovers due to insufficient domestic R&D investment, since local knowledge is needed to determine the appropriateness of technologies for the local market, adapting them, installing and using effectively (Gruebler et al. 2012). From the policy perspective, this implies that simple transfer of technologies could be insufficient to guarantee adoption of new technologies (Gruebler et al. 2012).

Research relying on patent citations has indicated that Foreign Direct Investment (FDI) is a mechanism for firms to contribute to the recipient country's innovation output as well as benefit from the recipient country in industrialised countries (Branstetter 2006) and in developing countries (Newman et al. 2015). However, insights specific for energy or climate change mitigation areas are not available, nor is there much information about how other innovation metrics may react to FDI.

Finally, technologies could be not efficient in developing countries, even if they are efficient in countries at the technological frontier. For instance, technologies that are highly capital intensive and labour saving will be efficient only in countries where costs of capital are low and costs of labour are high. Similarly, technologies which require a large number of skilled labour will be more competitive in a country where skilled labour is abundant (and hence cheap) than where it is scarce (Basu and Weil 1998; Caselli and Coleman 2006).

**Table 16.4 | Examples of institutional deficiencies preventing deployment of new technologies in countries behind the technological frontier.**

| Institutions | Examples of deficiencies | Literature reference |
|---|---|---|
| Industrial | Inability to benefit fully from international knowledge spillover due to insufficient domestic R&D investment | Mancusi (2008); Unel (2008); Gruebler et al. (2012) |
| Commercial | Insufficient experience with the organisation and management of large-scale enterprise | Abramovitz (1986); Aghion et al. (2005) |
| Political | Vested interests and customary relations among firms and between employers and employees | Olson (1982); Abramovitz (1986) |
| Financial | Financial markets incapable of mobilising capital for individual firms at large scale | Abramovitz (1986); Aghion et al. (2005) |

#### 16.2.3.4 Market Failures in Directing Technological Change

Market forces alone cannot deliver Pareto optimal (i.e., social) efficiency due to at least two types of externalities: GHG emissions that cause climate damage; and knowledge spillovers that benefit firms other than the inventor. Nordhaus (2011) argues that these two problems would have to be tackled separately: once the favourable intellectual property right regimes (i.e., the laws or rules or regulation on protection and enforcement) are in place, a price on carbon that corrects the emission externality is sufficient to induce optimal level of green technological change. Acemoglu et al. (2012) demonstrates that subsidising clean technologies (and not dirty ones) is also necessary to break the lock-in of dirty technological change. Recommendations for technical changes are often based on climate considerations only and neglect secondary externalities and environmental costs of technology choices (such as loss of biodiversity due to inappropriate scale-up of bioenergy use). The scale of adverse side effects and co-benefits varies considerably between low-carbon technologies in the energy sector (Luderer et al. 2019).

### 16.2.4 Representation of the Innovation Process in Modelled Decarbonisation Pathways

A variety of models are used to generate climate mitigation pathways, compatible with 2°C and well below 2°C targets. These include integrated assessment models (IAMs), energy system models, computable general equilibrium models, and agent based models. They range from global (Chapter 3) to national models and include both top-down and bottom-up approaches (Chapter 4). Innovation in energy technologies, which comprises the development and diffusion of low-, zero- and negative-carbon energy options, but also investments to increase energy efficiency, is a key driver of emissions reductions in model-based scenarios.

#### 16.2.4.1 Technology Cost Development

Assumptions on energy technology cost developments is one of the factors that determine the speed and magnitude of the deployment in climate-energy-economy models. The modelling is informed by the empirical literature that estimates the rates of cost reduction for energy technologies. A first strand of literature relies on the extrapolation of historical data, assuming that costs decrease either as a power law of cumulative production, exponentially with time (Nagy et al. 2013) or as a function of technical performance metrics (Koh and Magee 2008). Another approach relies on expert estimates of how future costs will evolve, including expert elicitations (Verdolini et al. 2018).

In these models, technology costs may evolve exogenously or endogenously (Mercure et al. 2016; Krey et al. 2019). In the first case, technology costs are assumed to vary over time at some predefined rate, generally extrapolated from past observed patterns or based on expert estimates. This formulation of cost dynamics generally underestimates future costs (Meng et al. 2021) as, among other things, it does not capture any policy-induced carbon-saving technological change or any spillover arising from the accumulation of national and international knowledge (Sections 16.2.2 and 16.2.3) or positive macroeconomic effects of a transition (Karkatsoulis et al. 2016). The influence of cost and diffusion assumptions may be evaluated through sensitivity analysis. In the second case, costs are a function of a choice variable within the model. For instance, technology costs decrease as a function of either cumulative installed capacity (learning by doing) (Seebregts et al. 1998; Kypreos and Bahn 2003) or R&D investments or spillovers from other sectors and countries.

One factor in this 'learning by researching' is applied to a wide range of energy technologies but also to model improvements in the efficiency of energy use (Goulder and Schneider 1999; Popp 2004). More complex formulations include two-factor learning processes (Criqui et al. 2015; Emmerling et al. 2016; Paroussos et al. 2020) (Section 16.2.2.1), multifactor learning curves (Kahouli 2011; Yu et al. 2011), or other drivers of cost reduction such as economies of scale and markets (Elia et al. 2021). The application of two-factor learning curves to model energy technology costs is often constrained by the lack of information on public and/or private energy R&D investments in many fast-developing and developing countries (Verdolini et al. 2018). The approach used to model energy technology cost reductions varies across technologies, even within the same model, depending on the availability of data and/or the level of maturity. Less mature technologies generally depend highly on learning by research, whereas learning by doing dominates in more mature technologies (Jamasb 2007).

In addition to learning, knowledge spillover effects are also integrated in climate-energy-economy models to reflect the fact that innovation in a given country depends also on knowledge generated elsewhere (Emmerling et al. 2016; Fragkiadakis et al. 2020). Models with a more detailed representation of sectors (Paroussos et al. 2020) can use spillover matrices to include bilateral spillovers and compute learning rates that depend on the human capital stock and the regional and/or sectoral absorption rates (Fragkiadakis et al. 2020). Accounting for knowledge spillovers in the EU for PV, wind turbines, electric vehicles, biofuels, industry materials, batteries and advanced heating and cooking appliances can lead to the following results in a decarbonisation scenario over the period 2020–2050 as compared to the reference scenario: an increase of 1.0–1.4% in GDP, 2.1–2.3% in investment, and 0.2–0.4% in employment by clean energy technologies (Paroussos et al. 2017). When comparing two possible EU transition strategies – being a first-mover with strong unilateral emission reduction strategy until 2030 versus postponing action for the period after 2030 – endogenous technical progress in the green technologies sector can alleviate most of the negative effects of pioneering low-carbon transformation associated with loss of competitiveness and carbon leakage (Karkatsoulis et al. 2016).

#### 16.2.4.2 Technology Deployment and Diffusion

To simulate possible paths of energy technology diffusion for different decarbonisation targets, models rely on assumptions about the cost of a given technology relative to the costs of other technologies, and its ability to supply the energy demand under the relevant energy system and physical constraints. These assumptions include, for example, considerations regarding renewable intermittency, inertia on technology lifetime (for instance, under less stringent temperature scenarios, early retirement of fossil plants does not take place), distribution, capacity and market growth constraints, as well as the presence of policies. These factors change the relative price of technologies. Furthermore, technological diffusion in one country is also influenced by technology advancements in other regions (Kriegler et al. 2015).

Technology diffusion may also be strongly influenced, either positively or negatively, by a number of non-cost, non-technological barriers or enablers regarding behaviours, society and institutions (Knobloch and Mercure 2016). These include network or infrastructure externalities, the co-evolution of technology clusters over time ('path dependence'), the risk-aversion of users, personal preferences and perceptions and lack of adequate institutional framework which may negatively influence the speed of (low-carbon) technological innovation and diffusion, heterogeneous agents with different preferences or expectations, multi-objectives and/or competitiveness advantages and uncertainty around the presence and the level of environmental policies and institutional and administrative barriers (Marangoni and Tavoni 2014; Baker et al. 2015; Iyer et al. 2015; Napp et al. 2017; Biresselioglu et al. 2020; van Sluisveld et al. 2020). These types of barriers to technology diffusion are currently not explicitly detailed in most of the climate-energy-economy models. Rather, they are accounted for in models through scenario narratives, such as the ones in the *Shared Socioeconomic Pathways* (Riahi et al. 2017), in which assumptions about technology adoption are spanned over a plausible range of values. Complementary methods are increasingly used to explore their importance in future scenarios (Turnheim et al. 2015; Geels et al. 2016; Doukas et al. 2018; Gambhir et al. 2019; Trutnevyte et al. 2019). It takes a very complex modelling framework to include all aspects affecting technology cost reductions and technology diffusion, such as heterogeneous agents (Lamperti et al. 2020), regional labour costs (Skelton et al. 2020), materials cost and trade and perfect foresight multi-objective optimisation (Aleluia Reis et al. 2021). So far, no model can account for all these interactions simultaneously.

Another key aspect of decarbonisation regards issues of acceptability and social inclusion in decision-making. Participatory processes involving stakeholders can be implemented using several methods to incorporate qualitative elements in model-based scenarios on future change (van Vliet et al. 2010; Nikas et al. 2017, 2018; Doukas and Nikas 2020; van der Voorn et al. 2020).

#### 16.2.4.3 Implications for the Modelling of Technical Change in Decarbonisation Pathways

Although the debate is still ongoing, preliminary conclusions indicate that integrated assessment models tend to underestimate innovation on energy supply but overestimate the contributions by energy efficiency (IPCC 2018b). Scenarios emerging from cost-optimal climate-energy-economy models are too pessimistic, especially in the case of rapidly changing technologies such as wind and batteries in the past decade. Conversely, they tend to be too optimistic regarding the timing of action, or the availability of a given technology and its speed of diffusion (Shiraki and Sugiyama 2020). Furthermore, some technological and economic transformations may emerge as technically feasible from IAMs, but are not realistic if taking into account political economy, international politics, human behaviours, and cultural factors (Bosetti 2021).

There is a range of projected energy technology supply costs included in the IPCC's Sixth Assessment Report (AR6) Scenario Database (Box 16.1). Variations of costs over time and across scenarios are within ranges comparable to those observed in recent years. Conversely, model results show that limiting warming to 2°C or 1.5°C will require faster diffusion of installed capacity of low-carbon energy options and a rapid phase-out of fossil-based options. This points to the importance of focusing on overcoming real-life barriers to technology deployment.

### Box 16.1 | Comparing Observed Energy Technology Costs and Deployment Rates with Projections from AR6 Global Modelled Pathways

Currently observed costs and deployment for electricity supply technologies from a variety of sources are compared with projections from two different sets of scenarios contained in the AR6 Scenario database: (i) scenarios that limit warming to 3°C (>50%) and scenarios that limit warming to 4°C (>50%), and (ii) scenarios that limit warming to 2°C (>67%) or lower (AR6 Scenarios Database). Global aggregate costs are shown for the following technologies: coal with carbon dioxide capture and storage (CCS), gas with CCS, nuclear, solar PV, onshore and offshore wind.

The decrease in forecasted capital costs is not large compared to current capital costs for most technologies, and does not differ much between the two sets of scenarios (Box 16.1, Figure 1a). For offshore wind some of the models are more optimistic than the current reality (Timilsina 2020). Several sources of current solar PV costs report values that are at the low end of the AR6 Scenario Database. By 2050, the median technology cost forecasts decrease by between 5% for nuclear and 45–52% for solar (Box 16.1, Figure 1c).

Median values of renewables installed capacity increase with respect to 2020 capacity in scenarios that limit warming to 3°C (>50%) and in scenarios that limit warming to 4°C (>50%) (Box 16.1 Figure 1b), where energy and climate policies are implemented in line with NDCs announced prior to COP26. More stringent targets (2°C) are achieved through a higher deployment of renewable technologies: by 2050 solar (wind) capacity is estimated to increase by a factor of 15 (10) (Box 16.1, Figure 1c). This is accompanied by an almost complete phase-out of coal (–87%). The percentage of median changes in installed capacity in scenarios that limit warming to 3°C (>50%) and in scenarios that limit warming to 4°C (>50%)is within comparable ranges of that observed in the last decade. In the case of scenarios that limit warming to 2°C (>67%) or lower, capacity installed is higher for renewable technologies and nuclear, and lower for fossil-based technologies (Box 16.1, Figure 1c).

*Box 16.1 (continued)*

The higher deployment in scenarios that limit warming to 2°C (>67%) or lower cannot be explained solely as a result of technology cost dynamics. In IAMs, technology deployment is also governed by system constraints that characterise both 3°C (>50%) and 4°C (>50%) scenarios, for example, the flexibility of the energy system, the availability of storage technologies. From a modelling point of view, implementing more stringent climate policies to meet the 2°C limit forces models to find solutions, even if costly, to meet those intermittency and flexibility constraints and temperature target constraints.

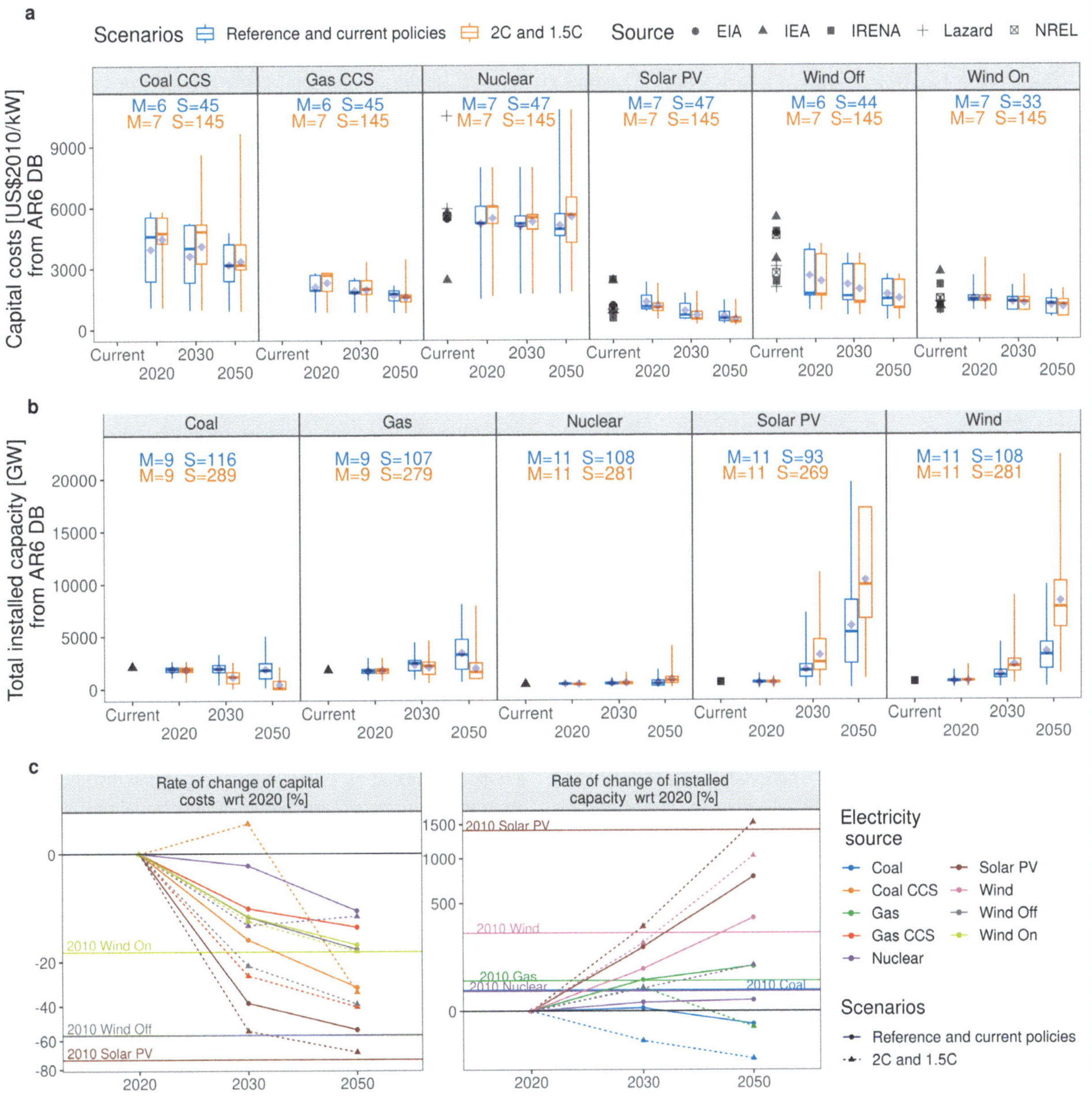


**Box 16.1, Figure 1 | Global technology cost and deployment in two groups of AR6 scenarios: (i) scenarios that limit warming to 3°C (>50%) and scenarios that limit warming to 4°C (>50%) ("Reference and current policies"), and (ii) scenarios that limit warming to 2°C (>67%) or lower ("2°C and 1.5°C").** Panel **(a)** Current capital costs are sourced from Table 1 (Timilsina 2020); distribution of capital costs in 2030 and 2050 (AR6 Scenarios Database). Blue symbols represent the mean. 'Current' capital costs for coal and gas plants with CCS are not available; Panel **(b)** Total installed capacity in 2019 (IEA 2020c; IRENA 2020a, b); distribution of total installed capacity in 2030 and 2050 (AR6 Scenario Database). Blue symbols represent the mean; Panel **(c)** Percentage of change in capital costs and installed capacity between (2010–2020) and percentage of median change (2020–2030 and 2020–2050) ($Median_{year}$–$Median_{2020}$)/$Median_{2020}$*100. 'M' indicates the number of models, 'S' the number of scenarios for which this data is available. 'Reference and current policies' are scenarios that limit warming to 3°C (>50%) and scenarios that limit warming to 4°C (>50%) (C6 and C7 AR6 scenario categories). '2C and 1.5C' are scenarios that limit warming to 2°C (>67%) or lower (C1, C2 and C3 AR6 scenario categories). Each model may have submitted data for more than one model version.

## 16.3 A Systemic View of Technological Innovation Processes

The innovation process, which consists of a set of sequential phases (Section 16.2.1), is often simplified to a linear process. Yet, it is now well understood that it is also characterised by numerous kinds of interactions and feedbacks between the domains of knowledge generation, knowledge translation and application, and knowledge use (Kline and Rosenberg 1986). Furthermore, it is not just invention that leads to technological change; the cumulative contribution of incremental innovations over time can be very significant (Kline and Rosenberg 1986). Innovations can come, not just from formal research and development (R&D) but also sources such as production engineers and the shop floor (Kline and Rosenberg 1986; Freeman 1995).

This section reviews the literature focusing on innovation as a systemic process. This now predominant view enriches the understanding of innovation as presented in Section 16.2; it conceptualises innovation as the result of actions by, and interactions among, a large set of actors, whose activities are shaped by, and shape, the context in which they operate and the user group with which they are engaging. This section aligns with the discussion of socio-technical transitions (Section 1.7.3, Chapter 5 Supplementary Material, and Cross-Chapter Box 12 in this chapter).

### 16.3.1 Frameworks for Analysing Technological Innovation Processes

The resulting overarching framework that is commonly used in the innovation scholarship and in policy analyses is termed an 'innovation system', where the key constituents of the systems are actors, their interactions, and the institutional landscape, including formal rules, such as laws, and informal restraints, such as culture and codes of conduct, that govern the behaviour of the actors (North 1991).

One application of this framework, *national innovation systems (NIS)*, highlight the importance of national and regional relationships for determining the technological and industrial capabilities and development of a country (Lundvall 1992; Nelson 1993; Freeman 1995). Nelson (1993) and Freeman (1995) highlight the role of institutions that determine the innovative performance of national firms as a way to understand differences across countries, while Lundvall (1992) focuses on the 'elements and relationships which interact in the production, diffusion and use of new, and economically useful, knowledge' – that is, notions of interactive learning, in which user-producer relationships are particularly important (Lundvall 1988). Building on this, various other applications of the 'innovation system' framework have emerged in the literature.

*Technological innovation systems (TIS)*, with a technology or a set of technologies (more narrowly or broadly defined in different cases) as the unit of analysis, focus on explaining what accelerates or hinders their development and diffusion. Carlsson and Stankiewicz (1991) define a technological system as 'a dynamic network of agents interacting in a specific economic/industrial area under a particular institutional infrastructure and involved in the generation, diffusion, and utilisation of technology'. More recent work takes a 'functional approach' to TIS (Hekkert et al. 2007; Bergek et al. 2008), which was later expanded with explanations of how some of the sectoral, geographical and political dimensions intersect with technology innovation systems (Bergek et al. 2015; Quitzow 2015).

*Sectoral innovation systems (SIS)* are based on the understanding that the constellation of relevant actors and institutions will vary across industrial sectors, with each sector operating under a different technological regime and under different competitive or market conditions. A sectoral innovation, thus, can be defined as 'that system (group) of firms active in developing and making a sector's products and in generating and utilising a sector's technologies' (Breschi and Malerba 1997).

*Regional innovation systems (RIS) and global innovation systems (GIS)*, recognise that the many innovation processes have a spatial dimension, where the development of system resources such as knowledge, market access, financial investment, and technology legitimacy may well draw on actors, networks, and institutions within a region (Cooke et al. 1997). In other cases, the distribution of many innovation processes are highly internationalised and therefore outside specific territorial boundaries (Binz and Truffer 2017). Importantly, Binz and Truffer (2017) note that the GIS framework 'differentiates between an industry's dominant innovation mode...

**Table 16.5 | Functions that the literature identified as key for well-performing technological innovation systems.** Source: based on Hekkert et al. (2007) and Bergek et al. (2008).

| Functions | Description |
|---|---|
| Entrepreneurial activities and experimentation | Entrepreneurial activities and experimentation for translating new knowledge and/or market opportunities into real-world application |
| Knowledge development | Knowledge development includes both learning by searching and learning by doing |
| Knowledge diffusion | Knowledge diffusion through networks, both among members of a community (e.g., scientific researchers) and across communities (e.g., universities, business, policy, and users) |
| Guidance of search | Guidance of search directs the investments in innovation in consonance with signals from the market, firms or government |
| Market formation | Market formation through customers or government policy is necessary to allow new technologies to compete with incumbent technologies |
| Resource mobilisation | Resource mobilisation pertains to the basic inputs – human and financial capital – to the innovation process |
| Creation of legitimacy/counteract resistance to change | Creation of legitimacy or counteracting resistance to change, through activities that allow a new technology to become accepted by users, often despite opposition by incumbent interests |
| Development of external economies | Development of external economies, or the degree to which other interests benefit from the new technology |

and the economic system of valuation in which markets for the innovation are constructed'.

The relevance of *mission-oriented innovation systems (MIS),* comes into focus with the move towards mission-oriented programmes as part of the increasing innovation policy efforts to address societal challenges. Accordingly, an MIS is seen as consisting of 'networks of agents and sets of institutions that contribute to the development and diffusion of innovative solutions with the aim to define, pursue and complete a societal mission' (Hekkert et al. 2020).

Notably the innovation systems approach has been used in a number of climate-relevant areas such as agriculture (Echeverría 1998; Horton and Mackay 2003; Brooks and Loevinsohn 2011; Klerkx et al. 2012), energy (Sagar and Holdren 2002; OECD 2006; Gallagher et al. 2012; Wieczorek et al. 2013; Darmani et al. 2014; Mignon and Bergek 2016), industry (Koasidis et al. 2020b) and transport (Koasidis et al. 2020a), and sustainable development (Anadon et al. 2016b; Clark et al. 2016; Bryden and Gezelius 2017; Nikas et al. 2020).

A number of functions can be used to understand and characterise the performance of technological innovation systems (Hekkert et al. 2007; Bergek et al. 2008). The most common functions are listed in Table 16.5.

Evidence from empirical case studies indicates that all the above functions are important and that they interact with one another (Hekkert and Negro 2009). The approach therefore serves as both a rationale for and a guide to innovation policy (Bergek et al. 2010).

A much-used, complementary systemic framework is the Multi-Level Perspective (MLP) (Geels 2002), which focuses mainly on the diffusion of technologies in relation to incumbent technologies in their sector and the overall economy. A key point of MLP is that new technologies need to establish themselves in a stable 'socio-technical regime' and are therefore generally at a disadvantage, not just because of their low technological maturity, but also because of an unwelcoming system. The MLP highlights that the uptake of technologies in society is an evolutionary process, which can be best understood as a combination of 'variation, selection and retention' as well as 'unfolding and reconfiguration' (Geels 2002). Thus, new technologies in their early stages need to be selected and supported at the micro-level by niche markets, possibly through a directed process that has been termed 'strategic niche management' (Kemp et al. 1998). As, at the landscape level, pressures on incumbent regimes mount, and those regimes destabilise, the niche technologies get a chance to get established in a new socio-technical regime. This allows these technologies to grow and stabilise, shaping a changed or sometimes radically renewed socio-technical regime. The MLP takes a systematic and comprehensive view about how to nurture and shape technological transitions by understanding them as evolutionary, multidirectional and cumulative socio-technical processes playing out at multiple levels over time, with a concomitant expansion in the scale and scope of the transition (Elzen et al. 2004; Geels 2005). There have been numerous studies that draw on the MLP to understand different aspects of climate technology innovation and diffusion (van Bree et al. 2010; Geels 2012; Geels et al. 2017).

Systemic analyses of innovation have predominantly focused on industrialised countries There have been some efforts to use the innovation systems lens for the developing country context (Jacobsson and Bergek 2006; Altenburg 2009; Lundvall et al. 2009; Tigabu et al. 2015; Tigabu 2018; Choi and Zo 2019) and specific suggestions on ways for developing countries to strengthening their innovation systems (e.g., by universities taking on a 'developmental' role (Arocena et al. 2015), or industry associations acting as intermediaries to build institutional capacities (Watkins et al. 2015; Khan et al. 2020), including specifically for addressing climate challenges (Sagar et al. 2009; Ockwell and Byrne 2016). But the conditions in developing countries are quite different, leading to suggestions that different theoretical conceptualisations of the innovation systems approach may be needed for these countries (Arocena and Sutz 2020), although a system perspective would still be appropriate (Boodoo et al. 2018).



### 16.3.2 Identifying Systemic Failures to Innovation in Climate-related Technologies

Traditional perspectives on innovation policy were mostly science-driven, and focused on strengthening invention and its translation into application in a narrow sense. Also, a second main traditional perspective on innovation policy was focused on correcting for 'market failures' (Weber and Truffer 2017) (Section 16.2). The more recent understanding of, and shift of focus to, the systemic nature on the innovation and diffusion of technologies has implications for innovation policy, since innovation outcomes depend not just on inputs such as R&D, but much more on the functioning of the overall innovation system (see Sections 16.3.1 and 16.4). Policies can therefore be directed at innovation systems components and processes that need the greatest attention or support. This may include, for example, strengthening the capabilities of weak actors and improving interactions between actors (Jacobsson et al. 2017; Weber and Truffer 2017). At the same time, a systemic perspective also brings into sharp relief the notion of 'system failures' (Weber and Truffer 2017).

Systemic failures include: infrastructural failures; hard (e.g., laws, regulation) and soft (e.g., culture, social norms) institutional failures; interaction failures (strong and weak network failures); capability failures relating to firms and other actors; lock-in; and directional, reflexivity, and coordination failures (Klein Woolthuis et al. 2005; Chaminade and Esquist 2010; Negro et al. 2012; Weber and Rohracher 2012; Wieczorek and Hekkert 2012). Most of the literature that unpacks such failures and explores ways to overcome them is on energy-related innovation policy. For example, Table 16.6 summarises a meta-study (Negro et al. 2012) that examined cases of renewable energy technologies trying to disrupt incumbents across a range of countries to understand the roles, and relative importance, of the 'systemic problems' highlighted in Section 16.3.1.

Depending on the sector, specific technology characteristics, and national and regional context, the relevance of these systemic problems varies (Trianni et al. 2013; Bauer et al. 2017; Wesseling and Van der Vooren 2017; Koasidis et al. 2020a, b), suggesting that the innovation policy mix has to be tailor-made to respond to the diversity of systemic failures (Rogge et al. 2017). An illustration of how such systemic failures have been addressed is given in Box 16.2, which shows how the Indian government designed its standards and labelling programme for energy-efficient air conditioners and refrigerators. The success of this programme resulted from the careful attention to bring on board and coordinate the relevant actors and resources, the design of the standards, and ensuring effective administration and enforcement of the standards (Malhotra et al. 2021).

**Table 16.6 | Examination of systemic problems preventing renewable energy technologies from reaching their potential, including number of case studies in which the particular 'systemic problem' was identified.** Source: Negro et al. (2012).

| Systemic problems | Empirical sub-categories | No. of cases |
|---|---|---|
| Hard institutions | – 'Stop and go policy': lack of continuity and long-term regulations; inconsistent policy and existing laws and regulations<br>– 'Attention shift': policymakers only support technologies if they contribute to the solving of a current problem<br>– 'Misalignment' between policies on sector level such as agriculture, waste, and on governmental levels, i.e., EU, national, regional level, etc.<br>– 'Valley of Death': lack of subsidies, feed-in tariffs, tax exemption, laws, emission regulations, venture capital to move technology from experimental phase towards commercialisation phase | 51 |
| Market structures | – Large-scale criteria<br>– Incremental/near-to-market innovation<br>– Incumbent's dominance | 30 |
| Soft institutions | – Lack of legitimacy<br>– Different actors opposing change | 28 |
| Capabilities/ capacities | – Lack of technological knowledge of policymakers and engineers<br>– Lack of ability of entrepreneurs to pack together, to formulate clear message, to lobby to the government<br>– Lack of users to formulate demand<br>– Lack of skilled staff | 19 |
| Knowledge infrastructure | – Wrong focus or not specific courses at universities knowledge institutes<br>– Gap/misalignment between knowledge produced at universities and what is needed in practice | 16 |
| Too weak interactions | – Individualistic entrepreneurs<br>– No networks, no platforms<br>– Lack of knowledge diffusion between actors<br>– Lack of attention for learning by doing | 13 |
| Too strong interactions | – Strong dependence on government action or dominant partners (incumbents)<br>– Networks allows no access to new entrants | 8 |
| Physical infrastructure | – No access to existing electricity or gas grid for renewable energy technologies<br>– No decentralised, small-scale grid<br>– No refill infrastructure for biofuels, hydrogen, biogas | 2 |

## Box 16.2 | Standards and Labelling for Energy Efficient Refrigerators and Air Conditioners in India[2]

Energy efficiency is often characterised as a 'low-hanging fruit' for reducing energy use. However, systemic failures such as lack of access to capital, hidden costs of implementation, and imperfect information can result in low investments into adoption and innovation in energy efficiency measures (Sorrell et al. 2004). To address such barriers, India's governmental Bureau of Energy Efficiency (BEE) introduced the Standards and Labelling (S&L) programme to promote innovation in energy efficient appliances in 2006 (Sundaramoorthy and Walia 2017). While context-dependent, the programme's design, policies and scale-up contain lessons for addressing systemic failures elsewhere too.

### Programme design and addressing of early systemic barriers

To design the S&L programme, BEE drew on the international experiences and technical expertise of the Collaborative Labelling and Appliance Standards Program (CLASP) – a non-profit organisation that provides technical and policy support to governments in implementing S&L programmes. For example, since there was no data on the efficiency of appliances in the Indian market, CLASP assisted with early data collection efforts, resulting in a focus on refrigerators and air conditioners (McNeil et al. 2008).

2 This section draws on *The role of capacity-building in policies for climate change mitigation and sustainable development: The case of energy efficiency in India*, (Malhotra et al. 2021).

*Box 16.2 (continued)*

Besides drawing from international knowledge, the involvement of manufacturers, testing laboratories, and customers was crucial for the functioning of the innovation system.

To involve manufacturers, BEE employed three strategies to set the standards at an ambitious yet acceptable level. First, BEE enlisted the Indian Institute of Technology (IIT) Delhi (a public technical university) to engage with manufacturers and to demonstrate cost-effective designs of energy-efficient appliances. Second, BEE agreed to make the standards voluntary from 2006 to 2010. In return, the manufacturers agreed to mandatory and progressively more stringent standards starting in 2010. Third, BEE established a multistakeholder committee with representation from BEE, the Bureau of Indian Standards, appliance manufacturers, test laboratories, independent experts, and consumer groups (Jairaj et al. 2016) to ensure that adequately stringent standards are negotiated every two years.

At this time, India had virtually no capacity for independent testing of appliances. Here, too, BEE used multiple approaches towards creating the actors and resources needed for the innovation system to function. First, BEE funded the Central Power Research Institute (CPRI) – a national laboratory for applied research, testing and certification of electrical equipment – to set up refrigerator and AC testing facilities. Second, they invited bids from private laboratories, thus creating a demand for testing facilities. Third, BEE developed testing protocols in partnership with universities. Australian standards for testing frost-free refrigerators were adopted until local standards were developed. Thus, once the testing laboratories, protocols and benchmark prices for testing were in place, the appliance manufacturers could employ their services.

Finally, a customer outreach programme was conducted from 2006 to 2008 to inform customers about energy-efficient appliances, to enable them to interpret the labels correctly, and to understand their purchase decisions and information sources (Jain et al. 2018; Joshi et al. 2019). BEE initiated a capacity-building programme for retailers to be an information source for customers. A comprehensive document with details of different models and labels was provided to retailers, together with a condensed booklet to be shared with customers.

### Adapting policies to technologies and local context

While many of India's standards and testing protocols were based on international standards, they needed to be adapted to the Indian context. For example, because of higher temperatures in India, the reference outside temperature of 32°C for refrigerators was changed to 36°C.

AC testing protocols also had to be adapted because of the emergence of inverter-based ACs. Existing testing done only at a single temperature did not value inverter-based ACs' better average performance as compared to fixed-speed ACs over a range of temperatures. Thus, the Indian Seasonal Energy Efficiency Ratio (ISEER) was developed for Indian temperature conditions in 2015 by studying International Organization for Standardization (ISO) standards and through consultations with manufacturers (Mukherjee et al. 2020).

These measures had multiple effects on technological change. As a result of stringent standards, India has some of the most efficient refrigerators globally. In the case of ACs, the ISEER accelerated technological change by favouring inverter-based ACs over fixed-speed ACs, driving down their costs and increasing their market shares (BEE 2020).

### Scaling up policies for market transformation

As the S&L programme was expanded, BEE took measures to standardise, codify and automate it. For example, to process a high volume of applications for labels efficiently, an online application portal with objective and transparent certification criteria was created. This gave certainty to the manufacturers, enabling diversity and faster diffusion of energy-efficient appliances. Thus by 2019, the programme expanded to cover thousands of products across 23 appliance types (BEE 2020).

Besides issuing labels, the enforcement of standards also needed to be scaled up efficiently. BEE developed protocols for randomly sampling appliances for testing. Manufacturers were given a fixed period to rectify products that did not meet the standards, failing which they would be penalised and the test results would be made public.

### 16.3.3 Indicators for Technological Innovation

Assessing the state of technological innovation helps in understanding the progress of current efforts and policies in meeting stated objectives, and how we might design policies to do better.

Traditionally, input measures such as research, development and demonstration (RD&D) investments, and output measures such as scientific publication and patents were used to characterise innovation activities (Freeman and Soete 2009). This is partly because of the successes of specialised R&D efforts (Freeman 1995), the predominant linear model of innovation, and because such measures can (relatively) easily be obtained and compared. In the realm of energy-related innovation, RD&D investments remain the single most-used indicator to measure inputs into the innovation process (Box 16.3). Patent counts are a widely used indicator of the outputs of the innovation process, especially because they are detailed enough to provide information on specific adaptation and mitigation technologies. Mitigation and adaptation technologies have their own classification (Y02) with the European Patent Office (EPO) (Veefkind et al. 2012; Angelucci et al. 2018), which can be complemented with keyword search and manual inspection (Persoon et al. 2020; Surana et al. 2020b). However, using energy-related patents as an indicator of innovative activities is complicated by several issues (de Rassenfosse

16

## Box 16.3 | Investments in Public Energy Research and Development

Public energy R&D investments are a crucial driver of energy technology innovation (Sections 16.2.1.1 and 16.4.1). Box 16.3, Figure 1 shows the time profile of energy-related RD&D budgets in OECD countries as well as some key events which coincided with developments of spending (IEA 2019). Such data on other countries, in particular developing countries, are not available, although recent evidence suggests that expenditures are increasing there (IEA 2020c). The IEA collected partial data from China and India in the context of Mission Innovation, but this is only available starting from 2014 and thus not included in Figure 1.

The figure illustrates two points. First, energy-related RD&D has risen slowly in the last 20 years, and is now reaching levels comparable with the peak of energy RD&D investments following the two oil crises. Second, over time there has been a reorientation of the portfolio of funded energy technologies away from nuclear energy. In 2019, around 80% of all public energy RD&D spending was on low-emission technologies – energy efficiency, carbon dioxide capture, use and storage, renewables, nuclear, hydrogen, energy storage and cross-cutting issues such as smart grids. A more detailed discussion of the time profile of RD&D spending in IEA countries, including as a share of GDP, is available in IEA (2020b).

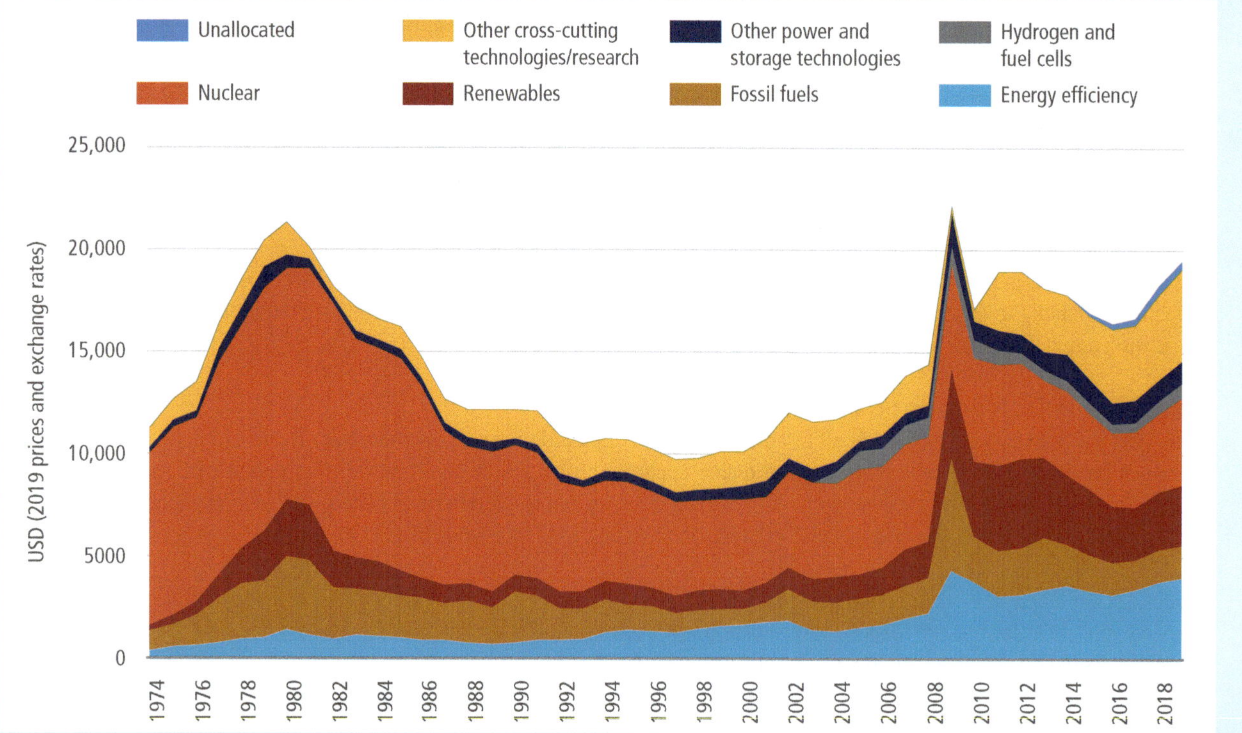


**Box 16.3, Figure 1 | Fraction of public energy RD&D spending by technology over time for IEA (largely OECD) countries between 1974 and 2018.** Sources: RD&D Database (2019), IEA (2019) (extracted on November 11, 2020).

et al. 2013; Haščič and Migotto 2015; Jaffe and de Rassenfosse 2017), including the fact that the scope of what are considered climate mitigation inventions is not always clear or straightforward.

Conversely, private energy R&D investments and investments by financing firms cannot be precisely assessed for a number of reasons, including limited reporting and the difficulty of singling out energy-related investments. This inability to precisely quantify private investments in energy R&D leads to a patchy understanding of the energy innovation system, and how private energy R&D investments responds to public energy R&D investments. Overall, evidence shows that some of the industrial sectors that are important for meeting climate goals (electricity, agriculture and forestry, mining, oil and gas, and other energy-intensive industrial sectors) are investing relatively small fractions of sales on R&D (*medium evidence*, *high agreement*) (Jasmab and Pollitt 2005; Jamasb and Pollitt 2008; Sanyal and Cohen 2009; European Commission 2015; American Energy Innovation Council 2017; Gaddy et al. 2017; National Science Board 2018).

Financing firms also play an important role in the energy innovation process, but data availability is limited. The venture capital (VC) financing model, used to overcome the 'valley of death' in the biotech and IT space (Frank et al. 1996), has not been as suitable for hardware start-ups in the energy space: for example, the percentage of exit outcomes in cleantech start-ups was almost half of that in medical start-ups, and less than a third of software investments (Gaddy et al. 2017). The current VC model and other private finance do not sufficiently cover the need to demonstrate energy technologies at scale (Anadón 2012; Mazzucato 2013; Nemet et al. 2018). This greater difficulty in reaching the market compared to other sectors may have contributed to a reduction in private equity and venture capital finance for renewable energy technologies after the boom of the late 2000s (Frankfurt School-UNEP Centre/BNEF 2019).

Quantitative indicators such as energy-related RD&D spending are insufficient for the assessment of innovation systems (David and Foray 1995): they only provide a partial view into innovation activities, and one that is potentially misleading (Freeman and Soete 2009). Qualitative indicators measuring the more intangible aspects of the innovation process and system are crucial to fully understand the innovation dynamics in a climate or energy technologies or sectors (Gallagher et al. 2006), including in relation to adopting an adaptive learning strategy and supporting learning through demonstration projects (Chan et al. 2017).

In Table 16.7, both quantitative and qualitative indicators for systemic innovation are outlined, using clean energy innovation as an illustrative example, and drawing on a broad literature base, taking into account both the input-output-outcome classification and its variations (Freeman and Soete 1997; Sagar and Holdren 2002; Hu et al. 2018), combined with the functions of technological innovation systems (Miremadi et al. 2018), while also being cognisant of the specific role of key actors and institutions (Gallagher et al. 2012). A specific assessment of innovation may focus on part of such a list of indicators, depending on what aspect of innovation is being studied, whether the analysis takes a more or less systemic perspective, and the specific technology and geography considered. Similarly, innovation policies may be designed to specifically boost only some of these aspects, depending on whether a given country/region is committed to strengthen a given technology or phase.

**Table 16.7 | Commonly used quantitative innovation metrics, organised by inputs, outputs and outcomes.** Sources: based on Sagar and Holdren (2002); Gallagher et al. (2006, 2011, 2012); Hekkert et al. (2007); Gruebler et al. (2012); Hu et al. (2018); Miremadi et al. (2018); Avelino et al. (2019).

| Function | Input indicators | Output indicators | Outcome indicators | Actors | Policies | Structural and systemic indicators |
|---|---|---|---|---|---|---|
| **Knowledge development** | Higher education investments<br>Research and development (R&D) investments<br>Number of researchers<br>R&D projects over time | Scientific publications<br>Highly-cited publications<br>Patents<br>New product configurations | Number of technologies developed (proof-of-concept/prototypes)<br>Increase in number of researchers<br>Learning rates | Governments<br>Private corporations<br>Universities | Research programmes and strategies<br>Intellectual Property Rights (IPR) policies<br>International technical norms (e.g., standards)<br>Higher education policies | Well-defined processes to define research priorities<br>Stakeholder involvement in priority-setting |
| **Knowledge diffusion** | R&D networks<br>Number of research agreements<br>Number of research exchange programmes<br>Number of scientific conferences | Citations to literature or patents<br>Public-private co-publications<br>Co-patenting<br>Number of co-developed products<br>International scientific co-publications<br>Number of workshops and conferences | Number of licensed patents<br>Number of technologies transferred<br>Knowledge-intensive services exports<br>Number of patent applications by foreigners<br>Number of researchers working internationally | Governments<br>Private corporations<br>Scientific societies<br>Universities | Development of communication centres<br>Facilitation of the development of networks<br>Open-access publication policies<br>IPR policies<br>International policy: e.g., treaties, clean development mechanism | Accessibility to exchange programmes<br>Strength of linkage among key stakeholders<br>Participation to framework agreements<br>ICT access |

| Function | Input indicators | Output indicators | Outcome indicators | Actors | Policies | Structural and systemic indicators |
|---|---|---|---|---|---|---|
| Guidance of search | Policy action plans and long-term targets<br>Shared strategies and roadmaps<br>Articulation of interest from lead customers<br>Expectations of markets/profits | Level of media coverage<br>Scenarios and foresight projects | Budget allocations<br>Mission-oriented innovation programmes | Governments<br>Interest groups<br>Media | Targets set by government for industry<br>Innovation policies<br>Credible political support | Media strength |
| Resource mobilisation | Access to finance<br>Graduate in Science, Technology, Engineering, and Mathematics (STEM)<br>Gross expenditure on R&D/total expenditure<br>Domestic credit to private sector<br>Number of researchers in R&D per capita<br>Public energy R&D expenditure/total expenditure<br>Expenditure on education<br>Investment in complementary assets and/or infrastructure (e.g., charging infrastructure for electric vehicles, smart grids)<br>Venture capital on deals | Number of green projects/technologies funded<br>Share of domestic credit granted to low-carbon technology projects<br>Share of domestic credit granted to projects developing complementary assets/infrastructure | Employment in knowledge-intensive activities<br>Employment in relevant industries<br>Scale of innovative activities<br>Rate of growth of dedicated investment<br>Availability of complementary assets and infrastructure | Governments<br>Private firms<br>Private investors (angel, venture capital, private equity)<br>Banks | Financial resources support<br>Development of innovative financing<br>International agreements (e.g., technology agreements)<br>Infrastructure support<br>Project/programme evaluation<br>Innovation policies<br>Higher education policies | |
| Entrepreneurial activities | Number of new entrants<br>Percentage of clean energy start-ups/incumbents<br>Access to finance for cleantech start-ups | Small and medium-sized enterprises (SMEs) introducing product or process innovation<br>Market introduction of new technological products<br>Number of new businesses<br>Experimental application projects<br>Creative goods exports | | Private firms<br>Government<br>Risk-capital providers<br>Philanthropists | Ease of starting a business<br>Risk-capital policies<br>Start-up support programmes<br>Incubator programmes | Start-up support services |
| Market formation | Public market support<br>High-tech imports | Market penetration of new technologies<br>Increase in installed capacity<br>Number of niche markets<br>Number of technologies commercialised | Environmental performance<br>Level of environmental impact on society<br>Renewable energy jobs<br>Renewable energy production<br>Trade of energy technology and equipment<br>High-tech exports | Private firms<br>Governments<br>institutions regulating trade, finance, investment, environment, development, security, and health issues | Environmental and energy regulation<br>Fiscal and financial incentives<br>Cleantech-friendly policy processes<br>Transparency<br>Specific tax regimes | Resource endowments<br>Attractiveness of renewable energy infrastructure<br>Coordination across relevant actors (e.g., renewable energy producers, grid operators, and distribution companies) |
| Creation of legitimacy | Youth and public demonstration<br>Lobbying activities<br>Regulatory acceptance and integration<br>Technology support | Level of discussion/debate among key stakeholders (public, firms, policymakers, etc.)<br>Greater recognition of benefits | Public opinion<br>Policymaker opinion<br>Executive opinion on regulation<br>Environmental standards and certification | Governments<br>Stakeholders<br>Citizens<br>Philanthropists | Regulatory quality<br>Regulatory instruments<br>Political consistency | Participatory processes |

The systemic approach to innovation and transition dynamics (Cross-Chapter Box 12 in this chapter) has advanced our understanding of the complexity of the innovation process, pointing to the importance of assessing the efficiency and effectiveness in producing, diffusing and exploiting knowledge (Lundvall 1992), including how the existing stock of knowledge may be recombined and used for new applications (David and Foray 1995). There remains a crucial need for more relevant and comprehensive approaches of assessing innovation (Freeman and Soete 2009; Dziallas and Blind 2019). In the context of climate mitigation, innovation is a means to an end; therefore, there is the need to consider the processes by which the output of innovation (e.g., patents) are translated into real-world outcomes (e.g., deployment of low-carbon technologies) (Freeman and Soete 1997; Sagar and Holdren 2002). Currently, there is no available set of quantitative metrics that, collectively, can help get a picture of innovation in a particular energy technology or set of energy technologies. Also we are still lacking an understanding of how to systematically use qualitative indicators to characterise the more intangible aspects of the energy innovation system and to improve front-end innovation decisions (Dziallas and Blind 2019).

### 16.3.4 Emerging Policy Perspectives on Systemic Transformations

Because of the multiple market, government, system, and other failures that are associated with the energy system, a range of policy interventions are usually required to enable the development and introduction of new technologies in the market (Jaffe et al. 2005; Bürer and Wüstenhagen 2009; Negro et al. 2012; Twomey 2012; Veugelers 2012; Weber and Rohracher 2012) used in what is termed as 'policy mixes' (Rogge and Reichardt 2016; Edmondson et al. 2019, 2020; Rogge et al. 2020). Empirical research shows that, in the energy and environment space, when new technologies were developed and introduced in the market, it was usually at least partly as a result of a range of policies that shaped the socio-technical system (*robust evidence*, *high agreement*) (Bunn et al. 2014; Bergek et al. 2015; Rogge and Reichardt 2016; Nemet 2019). An example of this systemic and dynamic nature of policies is the 70-year innovation journey of solar photovoltaic (PV), covering multiple countries, which is reviewed in Box 16.4.

16

## Box 16.4 | Sources of Cost Reductions in Solar Photovoltaics

**No single country persisted in developing solar photovoltaic (PV): five countries each made a distinct contribution, with each leader relinquishing its lead. The free flow of ideas, people, machines, finance, and products across countries explains the success of solar PVs. Barriers to knowledge flow delay innovation.**

Solar PV has attracted interest for decades, and until recently was seen as an intriguing novelty, serving a niche, but widely dismissed as a serious answer to climate change and other social problems associated with energy use. Since the IPCC's Fifth Assessment Report (AR5), PV has become a substantial global industry – a truly disruptive technology that has generated trade disputes among superpowers, threatened the solvency of large energy companies, and prompted reconsideration of electric utility regulation rooted in the 1930s. More favourably, its continually falling costs and rapid adoption are improving air quality and facilitating climate change mitigation. PV is now so inexpensive that it is important in an expanding set of countries. In 2020, 41 countries, in six continents, had each installed at least 1GW of solar (IRENA 2020a).

The cost of generating electricity from solar PV is now lower in sunny locations than running existing fossil fuel power plants (IEA 2020c) (Chapter 6). Prices in 2020 were below where even the most optimistic experts expected they would be in 2030.

The costs of solar PV modules have fallen by more than a factor of 10,000 since they were first commercialised in 1957. This four orders of magnitude cost reduction from the first commercial application in 1958 until 2018 can be summarised as the result of distinct contributions by the USA, Japan, Germany, Australia, and China – in that sequence (Green 2019; Nemet 2019). As shown in Box 16.4, Figure 1, PV improved as the result of:

i. scientific contributions in the 1800s and early 1900s, in Europe and the USA, that provided a fundamental understanding of the ways that light interacts with molecular structures, leading to the development of the p-n junction to separate electrons and holes (Einstein 1905; Ohl 1941);
ii. a breakthrough at a corporate laboratory in the USA in 1954 that made a commercially available PV device available and led to the first substantial orders, by the US Navy in 1957 (Ohl 1946; Gertner 2013);
iii. a government R&D and public procurement effort in the 1970s in the USA, that enlisted skilled scientists and engineers into the effort and stimulated the first commercial production lines (Christensen 1985; Blieden 1999; Laird 2001);
iv. Japanese electronic conglomerates, with experience in semiconductors, serving niche markets in the 1980s and in 1994 launching the world's first major rooftop subsidy programme, with a declining rebate schedule, and demonstrating there was substantial consumer demand for PV (Kimura and Suzuki 2006);
v. Germany passing a feed-in tariff in 2000 that quadrupled the market for PV, catalysing development of PV-specific production equipment that automated and scaled PV manufacturing (RESA 2001; Lauber and Jacobsson 2016);

*Box 16.4 (continued)*

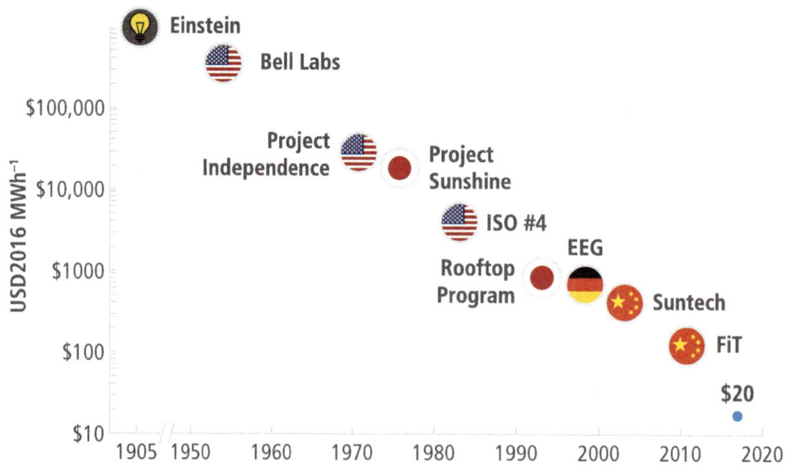


**Box 16.4, Figure 1 | Milestones in the development of low-cost solar photovoltaics.** Source: Nemet (2019).

vi. Chinese entrepreneurs, almost all trained in Australia and using Australian-invented passivated emitter rear cell technology, building supply chains and factories of gigawatt scale in the 2000s. China became the world's leading installer of PVs from 2013 onward (Quitzow 2015; Helveston and Nahm 2019); and

vii. a cohort of adopters with high willingness to pay, accessing information from neighbours, and installer firms that learnt from their installation experience as well as that of their competitors, to lower soft costs (Ardani and Margolis 2015; Gillingham et al. 2016).

As this evolution makes clear, no individual country persisted in leading the technology, and every world-leading firm lost its lead within a few years (Green 2019). Solar followed an overlapping but sequential process of technology creation, market creation and cost reduction (comparable to emergence, early adoption, diffusion and stabilisation in Cross-Chapter Box 12 in this chapter). In the technology creation phase, examples of central processes include flows of knowledge from one person to another, between firms, and between countries as well as US and Japanese R&D funding in the 1970s and early 1980s. During market creation, PVs modular scale allowed it to serve a variety of niche markets from satellites in the 1950s to toys in the 1980s, when Germany transformed the industry from niche to mass market with its subsidy programme that began in 2000 and became important for PV in 2004. The dramatic increase in size combined with its 20-year guaranteed contracts reduced risk for investors and created confidence in PV's long-term growth. Supportive policies also emerged outside Germany, in Spain, Italy, California, and China, which spread the risk, even as national policy support was more volatile. Rapid and deep cost reductions were made possible by: learning by doing in the process of operating, optimising, and combining production equipment; investing and improving each manufacturing line to gradually scale up to massive sizes; and incremental improvements in the PV devices themselves.

Central to PV development has been its modularity, which provided two distinct advantages: access to niche markets, and iterative improvement. Solar has been deployed as a commercial technology across nine orders of magnitude: from a 1W cell in a calculator to a 1GW plant in the Egyptian desert, and almost every scale in between. This modular scale enabled PV to serve a sequence of policy-independent niche markets (such as satellites and telecoms applications), which generally increased in size and decreased in willingness to pay, in line with the technology cost reductions. This modular scale also enabled a large number of iterations, such that in 2020 over three billion solar panels had been produced. Compared to, for instance, approximately 1000 nuclear reactors that were ever constructed, a million times more opportunities for learning by doing were available to solar PV: to make incremental improvements, to introduce new manufacturing equipment, to optimise that equipment, and to learn from failures. More generally, recent work has pointed to the benefits of modularity in the speed of adoption (Wilson et al. 2020) and learning rates (Sweerts et al. 2020).

While many technologies do not fit into the solar model, some – including micro nuclear reactors and direct air capture – also have modular characteristics that make them suitable for following solar's path and benefit from solar's drivers. However, it took solar PV 60 years to become cheap, which is too slow for addressing climate change if a technology is now still at the lab scale. A challenge in learning from the solar model is therefore how to use public policy to speed up innovation over much shorter time frames, for example, 15 or fewer years.

There are many definitions of policy mixes from various disciplines (Rogge et al. 2017), including environmental economics (Lehmann 2012), policy studies (Kern and Howlett 2009) and innovation studies. Generally speaking, a policy mix can be characterised by a combination of building blocks, namely elements, processes and characteristics, which can be specified using different dimensions (Rogge and Reichardt 2016). Elements include: (i) the policy strategy with its objectives and principal plans; (ii) the mix of policy instruments; and (iii) instrument design. The content of these elements is the result of policy processes. Both elements and processes can be described by their characteristics in terms of the consistency of the elements, the coherence of the processes, and the credibility and comprehensiveness of the policy mix in different policy, governance, geography and temporal context (Rogge and Reichardt 2016). Other aspects in the evaluation of policy mixes include framework conditions, the type of policy instrument and the lower level of policy granularity, namely design elements or design features (del Río 2014; del Río and Cerdá 2017). In addition, many have argued for the need to craft policies that affect different actors in the transition, some supporting and some 'destabilising' (Geels 2002; Kivimaa and Kern 2016).

Learning from the innovation systems literature, some of the recent policy focus is not only directed on innovation policies that can optimise the innovation system to improve economic competitiveness and growth, but also policies that can induce strategic directionality and guide processes of transformative changes towards desired societal objectives (Mitcham 2003; Steneck 2006). Therefore, the aim is to connect innovation policy with societal challenges and transformative changes through engagement with a variety of actors and ideas and incorporating equity, nowadays often referred to as a 'just transition' (Newell and Mulvaney 2013; Swilling et al. 2016; Heffron and McCauley 2018; Jasanoff 2018) (Chapters 1 and 17). This new policy paradigm is opening up a new discursive space, shaping policy outcomes, and giving rise to the emerging idea of transformative innovation policy (Fagerberg 2018; Diercks et al. 2019).

Transformative innovation policy has a broader coverage of the innovation process with a much wider participation of actors, activities and modes of innovation. It is often expressed as socio-technical transitions (Elzen et al. 2004; Turnheim and Sovacool 2020) or societal transformations (Scoones 2015; Roberts et al. 2018). Transformative innovation policy encompasses different ideas and concepts that aim to address the societal challenges involving a variety of discussions, including social innovation (Mulgan 2012), complex adaptive systems (Gunderson and Holling 2002), eco-innovation (Kemp 2011) and a framework for responsible innovation (Stilgoe et al. 2013), value-sensitive design (Friedman and Hendry 2019) and social-technical integration (Fisher et al. 2006).

## 16.4 Innovation Policies and Institutions

Building on the frameworks for identifying market failures (Section 16.2) and systemic failures (Section 16.3) in the innovation system for climate-related technologies, Section 16.4 proceeds as follows. First, it considers some of the policy instruments introduced in Chapter 13 that are particularly relevant for the pace and direction of innovation in technologies for climate change mitigation and adaptation. Second, it explains why governments put in place policies to promote innovation in climate-related technologies. Third, it takes stock of the overall empirical and theoretical evidence regarding the relationship between policy instruments with a direct and an indirect impact on innovation outcomes (including intellectual property regimes) and also other outcomes (competitiveness and distributional outcomes). Fourth, it assesses the evidence on the impact of trade-related policies and of sub-national policies aiming to develop cleantech industrial clusters.

This section focuses on innovation policies and institutions which are implemented at the national level. Whenever relevant, this section highlights examples of policies or initiatives that delve more deeply into the main high-level sectors: power, transport, industry, buildings, and agriculture, forestry and other land-use (AFOLU). Whenever possible, this section also discusses issues in policy selection, design, and implementation that have been identified as more relevant in developing countries and emerging economies.

Overall, this section shows that national and subnational policies and institutions are one of the main factors determining the redirection and acceleration of technological innovation and low-emission technological change (Anadon et al. 2016b; Rogge and Reichardt 2016; Åhman et al. 2017; Anadón et al. 2017; Roberts et al. 2018) (*robust evidence, high agreement*). Both technology push (e.g., scientific training, research and development (R&D)) and demand pull (e.g., economic and fiscal support and regulatory policy instruments), as well as instruments promoting knowledge flows and especially research-firm technology transfer, can be part of the mix (*robust evidence, medium agreement*) (Sections 16.2 and 16.3).

Public R&D investments in energy and climate-related technologies have a positive impact on innovation outcomes (*medium evidence, high agreement*). The evidence on procurement is generally positive, but limited. The economic policy instruments that can be classified as market pull instruments when it comes to the competitiveness outcome (at least in the short term) is more mixed. The review of the literature in this section shows that market pull policy instruments had positive but also some negative impacts on outcomes in some instances on some aspects of competitiveness and distributional outcomes (*medium evidence, medium agreement*) (Peñasco et al. 2021). For several of them – such as carbon taxes or feed-in tariffs – the evidence of a positive impact on innovation is more consistent than the others. Evidence suggests that complementary policies or improved policy design can mitigate such short-term negative distributional impacts.

### 16.4.1 Overview of Policy Instruments for Climate Technology Innovation

Government policies can influence changes in technologies, as well as changes to the systems they support (Somanathan et al. 2014) (Chapter 13 and Sections 16.2 and 16.3).

Technology-push policy instruments stimulate innovation by increasing the supply of new knowledge through funding and performing research; increasing the supply of trained scientists and engineers which contribute to knowledge-generation and provide technological opportunities, which private firms can decide to commercialise (Mowery and Rosenberg 1979; Anadon and Holdren 2009; Nemet 2009b; Mazzucato 2013).

Governments can also stimulate technological change through demand-pull (or market-pull) instruments which support market creation or expansion and technology transfer, and thus promote learning by doing, economies of scale, and automation (Section 16.2). Demand-pull policy instruments include regulation, carbon prices, subsidies that reduce the cost of adoption, public procurement, and intellectual property regulation. Typically, technology push is especially important for early-stage technologies, characterised by higher uncertainty and lower appropriability (Section 16.2); demand-pull instruments become more relevant in the later stages of the innovation process (Mowery and Rosenberg 1979; Anadon and Holdren 2009; Nemet 2009b) (Section 16.2).

The second column of Table 16.8 summarises the set of policies shaping broader climate outcomes over the past few decades in many countries outlined in Chapter 13, Section 13.6, which groups them into economic and financial, regulatory, and soft instruments. Other policies, such as monetary, banking and trade policies, for instance, can also shape innovation, but most government action to shape energy has not focused on them. As Table 16.8 shows, this section discusses the set of policy instruments on innovation outcomes, or a subset of the 'Transformative Potential' criterion presented in Chapter 13, and thus complements the more general discussion presented there. Table 16.8 specifically prioritises the impact of the subset of policy instruments on innovation outcomes for which evidence is available. This focus is complemented by a discussion of the impact of the same policy instruments on competitiveness (a subcomponent of the economic effectiveness evaluation criterion) and on distributional outcomes. Many of the policy instrument types listed in Table 16.8 have been implemented or proposed to address the different types of market or systemic failures or bottlenecks described in Sections 16.2 and 16.3 (OECD 2011a).

Section 16.3 characterised technological innovation as a systemic, non-linear and dynamic process. Figure 16.1 below presents a stylised (and necessarily incomplete) view connecting the innovation process stages presented in Section 16.2, some of the key mechanisms in technology innovation systems, and some of the decarbonisation policy instruments that have been assessed in terms of their impact on technological innovation outcomes in Section 16.4.4. As noted in the caption and discussed in Section 16.4.4, regulatory policy instruments also shape the early stages of technology development.

**Table 16.8 | Overview of policy instrument types covered in Chapter 13 and their correspondence to the subset of policy instrument types reviewed in Chapter 16 with a focus on innovation outcomes.**

| High-level categorisation | Lower-level policy instrument type in Chapter 13 | Policy instrument types reviewed in Section 16.4 (for definitions see Peñasco et al. 2021) |
|---|---|---|
| **Economic or financial policy instrument types** | Research and development (R&D) investments | R&D investments (including demonstration) (Box 16.3) |
| | Subsidies for mitigation | Feed-in tariffs or premia (set administratively) |
| | | Energy auctions |
| | | Other public financing options (public investment banks, loans, loan guarantees) |
| | Emissions trading schemes | Emissions trading scheme |
| | Carbon taxes | Taxes/tax relief (including carbon taxes, energy taxes and congestion taxes) |
| | Government provision | Government provision (focus on innovation procurement) |
| | Removing fossil fuel subsidies | *Not covered* |
| | Border carbon adjustments | *Not covered* |
| | Offsets | *Not covered* |
| **Regulatory policy instrument types** | Performance standards (including with tradeable credits) | Renewable obligations with tradeable green certificates |
| | | Efficiency obligations with tradeable white certificates |
| | | Clean energy or renewable portfolio standards (electricity) |
| | | Building codes (building efficiency codes) |
| | | Fuel efficiency standards |
| | | Appliance efficiency standards |
| | Technology standards | *Not covered* |
| **Soft policy instruments** | Divestment and disclosure | *Not covered* |
| | Voluntary agreements (public voluntary programmes and negotiated agreements) | Voluntary agreements |
| | | Energy labels |

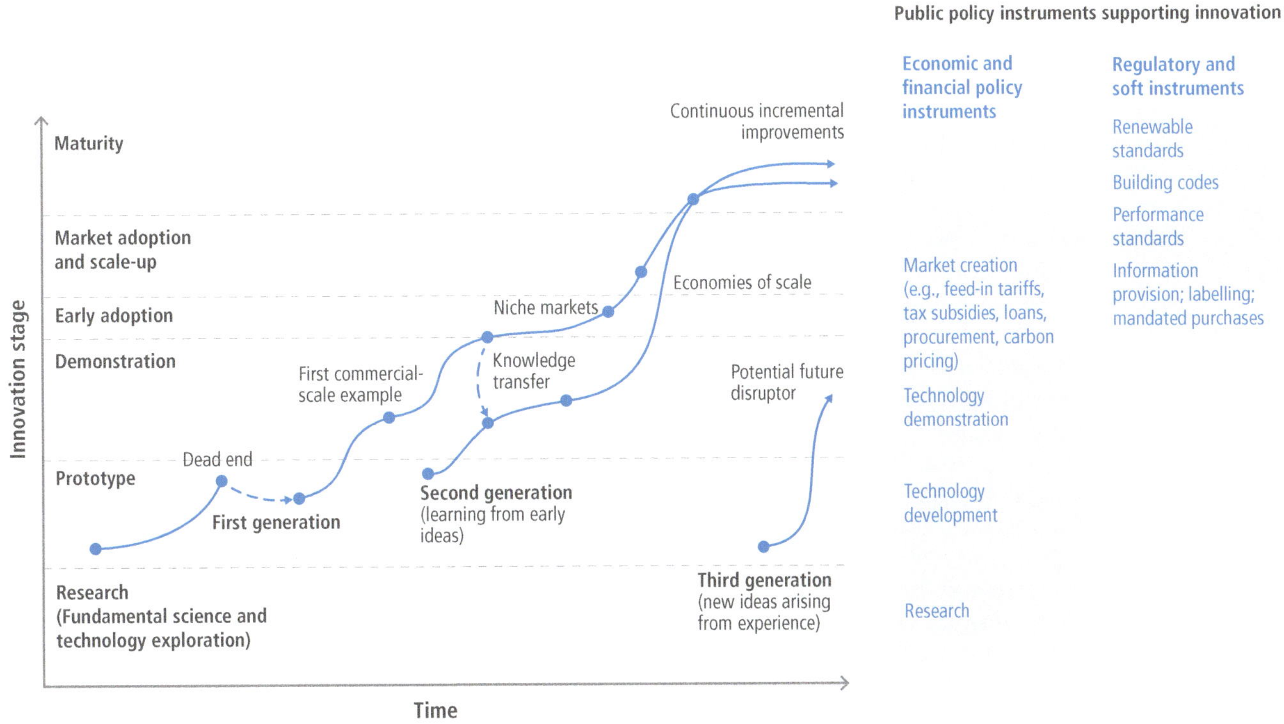


**Figure 16.1 | Technology innovation process and the (illustrative) and role of different public policy instruments (on the right-hand side).** Source: adapted from IEA (2020a). Note that, as shown in Section 16.4.4, demand-pull instruments in the regulatory instrument category, for instance, can also shape the early stages of the innovation process. Their position on the latter stages is highlighted in this figure because typically these instruments have been introduced in latter stages of the development of the technology.

### 16.4.2 The Drivers and Politics of National Policies for Climate Change Mitigation and Adaptation

Governments around the world implement innovation policies in the energy and climate space with the aim of simultaneously advancing environmental, industrial policy (or competitiveness), and security goals (Anadón 2012; Surana and Anadon 2015; Meckling et al. 2017; Matsuo and Schmidt 2019; Peñasco et al. 2021) (*medium evidence*, *medium agreement*). Co-benefits of policies shaping technological innovation in climate-related technologies, including competitiveness, health, and improved distributional impacts can be drivers of climate mitigation policy in the innovation sphere (Stokes and Warshaw 2017; Deng et al. 2018; Probst et al. 2020). For instance, this was the case for climate and air pollution policies with local content requirements for different types of renewable energy projects in places including China (Qiu and Anadon 2012; Lewis 2014), India (Behuria 2020), South Africa (Kuntze and Moerenhout 2012), and Canada (Genest 2014) (*robust evidence*, *medium agreement*).

The emergence of industries and support groups can lead to more sustained support for innovation policies (Meckling et al. 2015; Schmidt and Sewerin 2017 Stokes and Breetz 2018; Meckling 2019; Meckling and Nahm 2019; Schmid et al. 2020). Conversely, policies shaping technology innovation contribute to the creation and evolution of different stakeholder groups (*robust evidence*, *high agreement*). Most of the literature on the role of the politics and interest groups has focused on renewable energy technologies, although there is some work on heating in buildings (Wesche et al. 2019).

As novel technologies become cost-competitive, opposition of incumbents usually grows, as well as the dangers of lock-in that can be posed by the new winner. Addressing this involves adapting policy (*robust evidence*, *high agreement*).

Three phases of politics in the development of policies to meet climate and industrial objectives can be identified, at the top, the middle and the bottom of the experience curve (Breetz et al. 2018) (see also Figure 16.1, and Geels 2002). In the first phase of 'niche market diffusion', the politics of more sustained support for a technology or set of technologies become possible after a group of economic winners and 'clean energy constituencies' are created (Meckling et al. 2015). When technologies grow out of the niche (second phase), they pose a more serious competition to incumbents who may become more vocal opponents of additional support for innovation in the competing technologies (Geels 2014; Stokes 2016). In a third phase, path-dependence in policymaking and lock-in in institutions need to change to accommodate new infrastructure, the integration of technologies, the emergence of complementary technologies and of new regulatory regimes (Levin et al. 2012; Aklin and Urpelainen 2013).

16

### 16.4.3 Indicators to Assess the Innovation, Competitiveness and Distributional Outcomes of Policy Instruments

If policy instruments are created to (at least partly) shape innovation for systemic transitions to a zero-carbon future, they also need to be evaluated on their impact on the whole socio-technical system (Neij and Åstrand 2006) and a wide range of goals, including distributional impacts and competitiveness and jobs (Stern 2007; Peñasco et al. 2021). Given this and the current policy focus on green recovery and green industrial policy, we assess impacts on competitiveness and equity, although we primarily focus on innovation outcomes. Table 16.9 lists the selected set of indicators used to assess the impact of the policy instrument types covered in the right-hand column in Table 16.8. The table does not include technology diffusion or deployment because these are covered in the technological effectiveness evaluation criterion in Chapter 13. As noted in section 16.2, it is very difficult to measure or fully understand innovation with one or even several indicators. In addition, all indicators have strengths and weaknesses, and may be more relevant in some countries and sectors than in others. The literature assessing the impact of different policy instruments on innovation often covers just one of the various indicators listed in the second column of Table 16.9.

### 16.4.4 Assessment of Innovation and Other Impacts of Innovation Policy Instruments

While it is very difficult to attribute a causal relationship between a particular policy instrument implementation and different innovation indicators, given the complexity of the innovation system (Section 16.3), there is a large volume of quantitative and qualitative literature aiming to identify such an impact.

#### 16.4.4.1 Assessment of the Impact on Innovation of Technology Push Policy Instruments: Public RD&D Investments, Other R&D Incentives and Public Procurement

Economic and direct investment policy instrument types are typically associated with a direct focus on technological innovation: research and development (R&D) grants, R&D tax credits, prizes, national laboratories, technology incubators (including support for business development, plans), novel direct funding instruments (e.g., Advanced Research Projects Agency–Energy (ARPA-E)), and innovation procurement.

Public research, development and demonstration (RD&D) investments have been found to have a positive impact on different innovation in energy- and climate-related technologies (*robust evidence, high agreement*), but the assessment relies almost entirely on evidence from industrialised countries. Out of 17 publications focusing on this assessment, only three found no relationship between R&D funding and innovation metrics (Doblinger et al. 2019; Goldstein et al. 2020; Peñasco et al. 2021). Sixteen of them used *ex post* quantitative methods, and one relied on theoretical *ex ante* assessment; only two of them included some non-industrialised countries, with one being the theoretical analysis. The evidence available does not point to public R&D funding for climate-related technologies crowding out private R&D (an important driver of innovation) but instead crowding it in. Box 16.6 summarises the evidence available of the impact of ARPA-E (a public institution created in the USA in 2009 to allocate public R&D funding in energy) on innovation and competitiveness outcomes. Another institution supporting energy R&D that is the subject of much interest is the institutions of the Fraunhofer Society.

No evidence has been found regarding the specific impact of R&D tax credits on climate mitigation or adaptation technologies, but it is worth noting that, generally speaking, R&D tax credits are found to incentivise innovation in firms, with a greater impact on small and medium firms (OECD 2020). This is consistent with the fact that most of the evidence on the positive impact of public R&D support schemes covers small and medium firms (Howell 2017; Doblinger et al. 2019; Goldstein et al. 2020). Although there is a high level of agreement in the literature regarding the impact of R&D investments on innovation outcomes in climate-related technologies, it is important to note that this evidence comes from industrialised countries. This does not mean that public R&D investments in energy have been found to have no impact on developing countries innovation or competitiveness outcomes, but rather that we were not able to find such studies focussing on developing countries.

Overall, public procurement has high potential to incentivise innovation in climate technologies, but the evidence is mixed, particularly in developing countries (*limited evidence, medium agreement*). Public procurement accounted for 13% of gross domestic product (GDP) in OECD in 2013 and much more in some

**Table 16.9 | Outcomes (first row) and indicators (second row) to evaluate the impact of policies shaping innovation to foster carbon neutral economies.** Sources: innovation outcomes indicators are sourced from Del Rio and Cerdá (2014), Grubb et al. (2021) and Peñasco et al. (2021); the indicators under the competitiveness and distributional effects criteria are sourced from Peñasco et al. (2021).

| Policy instrument Outcomes | Innovation (Part of Chapter 13 'Transformative potential' evaluation criterion) | Competitiveness (Part of Chapter 13 'Economic effectiveness' evaluation criterion) | Distributional impacts (Defined in the same way as in Chapter13) |
|---|---|---|---|
| Examples of indicators used for each outcome in the literature | R&D investments, cost improvements, learning rates, patents, publications, reductions in abatement costs, energy efficiency improvements, other performance characteristics, firms reporting carbon saving innovation | Industry creation, net job creation, export of renewable energy technology equipment, economic growth (GNP, GDP), productivity, other investments | Level and incidence of support costs, change in spending on electricity as a percentage of total household spending, participation of different stakeholders, international equity (e.g., $tCO_2$-eq per capita), unequal access between large vs. small producers or firms |

## Box 16.5 | Green Public Procurement in The Netherlands

In 2005, the Dutch national government acknowledged a move in the House of Representatives to utilise their annual spending power to promote the market for sustainable goods and services, as well as to act as a role model. Hence, a policy for environmentally-friendly procurement was developed and implemented across the national, local and provincial governments. Subsequently, sustainable public procurement has expanded into a multidimensional policy in The Netherlands, accommodating policies on green public procurement, bio-based public procurement, international social criteria, social return on investment, innovation-oriented public procurement and circular economy.

The Green Public Procurement (GPP) policy is targeted at minimising the negative impacts of production and consumption on the nature environment (Melissen and Reinders 2012; Cerutti et al. 2016). It includes a wide range of environmental criteria for different product groups that public organisations frequently procure, such as office equipment, uniforms, road works and catering. There are 45 product groups (Melissen and Reinders, 2012) and six product clusters as part of the government's purchasing in terms of sustainability (PIANOo Expertisecentrum 2020). The six product clusters are: i) automation and telecommunications; ii) energy; iii) ground, road and hydraulic engineering; iv) office facilities and services; v) office buildings; and vi) transport (PIANOo Expertisecentrum 2020).The GPP 2020 Tender Implementation Plan spells out the terms and conditions for green public procurement. Some of these are confidential documents and are not shared online. Others are available for download. The tender implementation plan for The Netherlands is available on https://gpp2020.eu/low-carbon-tenders/open-tenders/. One of the important scenarios is that the public procurers need the details of Life Cycle Analysis (LCA) carried out in a tool called DuboCalc, which calculates the environmental impacts of the materials and methods of an infrastructural projects. GPP 2020 has reported that three million tonnes of $CO_2$ would be saved in The Netherlands alone if all Dutch public authorities applied the national Sustainable Public Procurement Criteria.

Research has been carried out to determine the prime mover for implementing Green Public Procurement. An online survey was administered among public procurement officers who subscribed to the newsletters of two Dutch associations that provide advice and training to public procurers. This yielded a sample size of more than 200 (Grandia and Voncken 2019). The first association is called Nevi which is the only organisation in The Netherlands that offers certified procurement training programmes. The second association is called PIANOo which is a public procurement expertise centre paid by the Dutch national government to bring together relevant information regarding public procurement and provide public procurers with useful tools through their websites, workshops, meetings and annual conferences. The data from the survey was then analysed using structural equations modelling (SEM) and the results show that ability, motivation and opportunities affect the implementation of GPP. Particularly, opportunity was found to affect GPP, innovation-oriented public procurement and the circular economy, but not the other types of public procurement.

emerging and developing economies (Baron 2016). Its main goal is to acquire products or services to improve public services, infrastructures and facilities and, in some cases, to also incentivise innovation. It is important to implement several steps in the public procurement procedure to improve transparency, minimise waste, fraud and corruption of public fund. These steps range from the assessment of a need, issuance of a tender, to the monitoring of delivery of the good or service. Box 16.5 outlines a public procurement programme that was implemented in The Netherlands in 2005 with a focus on green technologies. In spite of the fact that green procurement policies have been implemented, the literature assessing the innovation impact of public procurement programmes is relatively limited, and suggests either a positive impact or no impact (Alvarez and Rubio 2015; Baron 2016; Fernández-Sastre and Montalvo-Quizhpi 2019; Peñasco et al. 2021). The majority of cases where the impact is positive are analyses of industrialised countries, while no impact emerges in the case of a developing country (Ecuador). More empirical research in both developing and developed countries is needed to understand the impact of public procurement, which has the potential to support the achievement of other societal challenges (Edler and Georghiou 2007; Henderson and Newell 2011; Baron 2016; ICLEI 2018).

#### 16.4.4.2 Assessment of the Impact on Competitiveness of Technology Push Policy Instruments: Public RD&D Investments, Other R&D Incentives and Public Procurement

Public R&D investments in the energy, renewables, and environment space are generally associated with positive impacts on industrial development or 'competitiveness outcome' (*robust evidence*, *medium agreement*). In a number of cases, negligible or negative impacts emerge (Doblinger et al. 2019; Goldstein et al. 2020; Peñasco et al. 2021). The majority of the 15 analyses rely on *ex post* quantitative methods, while only four use *ex ante* modelling approaches. Also, in this case, the vast majority of the evidence is from industrialised countries.

There is limited and mixed evidence regarding the (positive or negative) impact of public procurement for low-carbon or climate technologies in developed countries (*limited evidence*, *low agreement*), and none from developing countries. All of the four evaluations identified in the Peñasco et al. (2021) review relied on qualitative methods. One found a positive impact, another a negative impact and two others found no impact. All of the studies covered European country experiences.

R&D and procurement policies have a positive impact on distributional outcomes (*limited evidence*, *high agreement*). Peñasco et al. (2021) identify three evaluations of the impact of RD&D funding on distributional outcomes (two using quantitative methods and one *ex ante* theoretical methods) and one of procurement on distributional outcomes (relying on qualitative analysis).

#### 16.4.4.3 Emerging Insights on Different Public R&D and Demonstration Funding Schemes

The ability of a given R&D policy instrument to impact innovation and competitiveness depends to some extent on policy design features (*limited evidence*, *high agreement*). As discussed in Section 16.4.4.4, this is not unique to R&D funding. Most of these assessments use a limited number of indicators (e.g., patents and publications and follow-on private financing, firm growth and survival, respectively), and are focused on the energy sector, and on the USA and other industrialised countries. Extrapolating to emerging economies and low-income countries is difficult. There is no evidence on the impact of different ways of allocating public energy R&D investments in the context of developing countries.

Block funding, which tends to be more flexible, can lead to research that is more productive or novel, but there are other factors that can affect the extent to which block funding can lead to more or less novel outcomes (*limited evidence*, *medium agreement*). Research on national research laboratories, which conduct at least 30% of all research in 68 countries around the world (Anadon et al. 2016a), are a widespread mechanism to carry out public R&D and allocate funds, but assessments of their performance is limited to developed countries. R&D priorities are also guided by institutions, and research focused on general technology innovation policy finds that institutions often do not embody the goals of the poor or marginalised (Anadon et al. 2016b).

In the case of the US Department of Energy, block funding that can be quickly allocated to novel projects (such as that allocated to National Labs as part of the Laboratory Directed Research and Development funding) has been found to be associated with improved innovation indicators (Anadon et al. 2016a). Research in Japan on R&D funding in general (not for climate-related technologies) however, indicates that R&D funds allocated competitively result in higher novelty for 'high status' (the term used in the paper to refer to senior male researchers), while block funding was associated with research of higher novelty for 'lower status' researchers (e.g., junior female researchers) (Wang et al. 2018).

### Box 16.6 | ARPA-E – A Novel R&D Funding Allocation Mechanism Focused on an Energy Mission

One approach for allocating public R&D funds in energy involves relying on active programme managers and having clear technology development missions that focus on high-risk high-reward areas and projects. This approach can be exemplified by a relatively new energy R&D funding agency in the USA, the Advanced Research Projects Agency for Energy (ARPA-E). This agency was created in 2009 and it was modelled on the experience of Defense Advanced Research Projects Agency (DARPA) – a US government agency funding high-risk, high-reward research in defence-related areas (Bonvillian and Van Atta 2011; US National Academies of Sciences Engineering and Medicine 2017; Bonvillian 2018). DARPA programme managers had a lot of discretion for making decisions about funding projects, but since energy R&D funding is usually more politically vulnerable than defence R&D funding, the ARPA-E model involved programme managers requesting external review as an informational input (Azoulay et al. 2019).

As for DARPA, ARPA-E programme managers use an active management approach that involves empowering programme manages to make decisions about funding allocation, milestones and goals. ARPA-E managers also differ from other R&D allocation mechanisms in that ARPA-E staff retain some control on the funded projects after the allocation of funds. As argued by Azoulay et al. (2019), even though this relative control over the project can result in a reduction in the flexibility of funded researchers, some 'exploration' happens at the programme manager level.

Research on ARPA-E also sheds light on the process of project selection, or how programme managers decide what projects to fund. Programme managers do not just follow the rankings of peer reviewers (sometimes projects with very disparate rankings were funded) and in many cases programme managers reported using information from review comments instead of the rankings (Goldstein and Kearney 2020). Azoulay et al. (2019) suggest that, if expert disagreement is a useful proxy for uncertainty in research, then the use of individual discretion in ARPA-E would result in a portfolio of projects with a higher level of uncertainty, as defined by disagreement among reviewers. Moreover, under the premise that uncertainty is a corollary to novelty, individual discretion is an antidote to novelty bias in peer review.

While innovation is notoriously hard to track and, particularly for emerging technologies, it can take a lot time to assess, early analysis has shown that this mission-orientation and more 'actively managed' R&D funding programme may yield greater innovation patenting outcomes than other US energy R&D funding programmes, and a greater or similar rate of academic publications when compared to other public funding agencies in energy in the USA, ranging from the Office of Science, the more applied Office of Energy Efficiency and Renewable Energy, or the small grants office (US National Academies of Sciences Engineering and Medicine 2017; Goldstein and

*Box 16.6 (continued)*

Narayanamurti 2018). In addition, research analysing the first cohort of cleantech start-ups has found that start-ups supported by ARPA-E had more innovative outcomes when compared to those that had applied but not received funding, with others that had not received any government support, and with others that had received other types of government R&D support (Goldstein et al. 2020). Overall, the mission-oriented ARPA-E approach has been successful in the USA when it comes to innovation outcomes. The extent to which it can yield the same outcomes in other geographies with different innovation and financing environments remains unknown. (*limited evidence*, *high agreement*).



Public financing for R&D and research collaboration in the energy sector is important for small firms, at least in industrialised countries, and it does not seem to crowd out private investment in R&D (*medium evidence*, *high agreement*). Small US and UK firms accrue more patents and financing when provided with cash incentives for R&D in the form of grants (Howell 2017; Pless 2019). US cleantech start-ups which partner with government partners for joint technology development or licensing partnerships accrue more patents and follow-on financing (Doblinger et al. 2019).

Overall, the body of literature on public R&D funding design in energy- and climate-related technologies provides some high-level guidance on how to make the most of these direct RD&D investments in energy technologies in the climate change mitigation space, including: giving researchers and technical experts autonomy and influence over funding decisions; incorporating technology transfer in research organisations; focusing demonstration projects on learning; incentivising international collaboration in energy research; adopting an adaptive learning strategy; and making funding stable and predictable (Narayanamurti et al. 2009; Narayanamurti and Odumosu 2016; Chan et al. 2017) (*medium evidence*, *high agreement*).

Without carefully designed public funding for demonstration efforts, often in a cost-shared manner with industry, the experimentation at larger scales needed for more novel technologies needed for climate change mitigation may not take place. (*medium evidence*, *high agreement*). Government funding, specifically for technology demonstration projects, for RD&D in energy technologies, plays a crucial supporting role (Section 16.2.1). Governments can facilitate knowledge spillovers between firms, between countries, and between technologies (Cohen et al. 2002; Baudry and Bonnet 2019) (Section 16.2).

### 16.4.4.4 Assessment of the Impact on Innovation and on Competitiveness and Distributional Outcomes of Market Pull Policy Instruments

Demand-pull policies such as tradeable green certificates, taxes, or auctions, are essential to support scaling-up efforts (Remer and Mattos 2003; Wilson 2012; Nahm and Steinfeld 2014). Just as for R&D investments, research has indicated that effective demand pull needs to be credible, durable, and aligned with other policies (Nemet et al. 2017) and that the effectiveness of different demand-pull instruments depends on policy design (del Río and Kiefer 2021).

Historical analyses of the relative importance of demand pull and technology push are clear: both are needed to provide robust incentives for investment in innovation. Interactions between them are central as their combination enables innovators to connect a technical opportunity with a market opportunity (Freeman 1995; Jacobsson et al. 2004; Grubler and Wilson 2013). It is important to note that these market pull policies are often put in place primarily to meet security and/or environmental goals, although innovation and competitiveness are sometimes also pursued explicitly.

*Emissions trading schemes*

Overall evidence suggests that the emissions trading schemes, as currently designed, have not significantly contributed to innovation outcomes (*medium evidence*, *medium/high agreement*).

Penasco et al. (2021) review 20 evaluations: eight identified a positive impact (although in at least two cases, the paper indicated that the impact was small or negligible); 11 no impact; and one was associated with a negative impact on innovation indicators. The studies that found no impact and the studies that found some impact covered all three methods (quantitative *ex post*, qualitative and theoretical and *ex ante* analysis). Another review focused only on empirical studies (mainly quantitative but also qualitative), covered a slightly longer period and identified 19 studies (15 using quantitative methods) (Lilliestam et al. 2021). With a narrower set of indicators of innovation, they concluded that there was very little empirical evidence linking innovation with the emissions trading schemes studied to date (Lilliestam et al. 2021). This review focused mainly on papers evaluating the earlier stages of the European Emissions Trading Scheme, which featured relatively low carbon dioxide prices, and covered a small set of firms, showing that carbon pricing policy design is an important determinant of innovation outcomes. Combining both reviews, there are a total of 27 individual studies, some of them providing mixed evidence of impact, and 23 of them suggest there was no impact or that (in a couple of cases) it was small. It is important to note that some researchers note that, for particular subsectors and actors, emissions trading schemes have had an impact on patenting trends (Calel and Dechezleprêtre 2016). Overall the expectation is that higher prices and coverage would result in higher impacts and that, over time, the impact on innovation would grow.

16

*Carbon and environmental taxes*

The impact of carbon taxes on innovation outcomes is more positive than that for emissions trading schemes, but the evidence is more limited (*limited evidence, medium agreement*). Assessments of their impact on innovation metrics have been very limited, with only four studies (three quantitative and one *ex ante*). Three of the studies found a positive impact of carbon taxes on innovation outcomes, and one found no impact (Peñasco et al. 2021).

Depending on the design (including the value and coverage of the tax), carbon taxes can either have positive, negative or null impact on competitiveness and distributional outcomes (*medium evidence, medium agreement*). The evidence on the impact of carbon taxes on competitiveness is significant (a total of 27 evaluations) and mixed, with six of them reporting some positive impacts, 10 reporting no impact, and 11 reporting negative impacts (so 59% were not associated with negative impacts). Most of the evaluations reporting negative impacts were theoretical assessments, and only three *ex post* quantitative analysis (Peñasco et al. 2021). Twenty-four evaluations covered distributional impacts of carbon taxes and other environmental taxes, the majority (15) found the existence of some negative distributional impacts, six found positive impacts, and three found no distributional impacts. Differences in the assessment results stem from the design of the taxes (Peñasco et al. 2021). It is important to note that, once again, the evidence comes from industrialised countries and emerging economies.

*Feed-in-tariffs*

Many factors affect the impacts of feed-in tariffs (FITs) on outcomes other than innovation (*robust evidence, high agreement*). While FITs have been generally associated with positive innovation outcomes, some of the differences found in the literature may arise from differences in the evaluation method (Peñasco et al. 2021) or differences in policy design (e.g., the level and the rate of decrease of the tariff) (Hoppmann et al. 2014), the policy mixes (Rogge et al. 2017), the technologies targeted and their stage of development (Huenteler et al. 2016b), and the geographical and temporal context of where the policy was put in place (Section 16.3). Research has also found that, particularly for less mature technologies, a higher technology specificity in the design of FITs is associated with more innovation (Del Río 2012). FITs yield better results if they account for the specificities of the country; or else, the technology and the policy could result in negative distributional and (to a lesser extent) competitiveness impacts. Meckling et al. (2017) indicate that an 'enduring challenge' of technology-specific industrial policy such as some FITs is to avoid locking in suboptimal clean technologies – a challenge which, among other options, could be overcome with targeted niche procurement for next-generation technologies. Other authors have cautioned that the move from renewable FITs to auctions may favour existing PVs (e.g., polysilicon) over more novel solar power technologies (Sivaram 2018b) such as thin-film PV, amorphous PV, and perovskites.

Policy design, policy mixes, and domestic capacity and infrastructure are important factors determining the extent to which economic policy instruments in industrialised countries and emerging economies can also lead to positive (or at least not negative) competitiveness outcomes and distributional outcomes (*medium evidence, medium agreement*) (Section 16.3). Prioritising low-cost energy generation in the design of FIT schemes can result in a lower focus of innovation efforts on more novel technologies and greater barriers to incumbents in less mature technologies (Hoppmann et al. 2013). Similarly, case study research from Mexico and South Africa indicates that focusing on low-cost renewable energy generation can only result in a greater reliance on existing foreign value chains and capital, and thus in lower or negative impacts on domestic competitiveness. In other words, some approaches can hinder the development of the local capabilities that could result in greater long-term benefits domestically (Matsuo and Schmidt 2019). Evidence for developing countries indicates that local and absorptive capacity also play an important role, in particular, on the ability of policies to contribute to competitiveness or industrial policy goals (Binz and Anadon 2018). Research comparing China's and India's policies and outcomes on wind energy also suggest that policy durability and systemic approaches can affect industrial outcomes (Surana and Anadon 2015).

*Energy auctions*

The evidence of the impact of renewable energy auctions on innovation outcomes is very small and provides mixed results (*limited evidence, low agreement*). Out of six evaluations, three identify positive impacts, two no impacts, and one negative impacts. All of the evaluations but one were qualitative or theoretical, and the quantitative assessment indicated no impact (Peñasco et al. 2021). There is more evidence covering emerging economies analysing the impacts of auctions when compared to other policy instrument types. For example, there is work comparing the approaches to renewable energy auctions in South Africa and Denmark (Toke 2015) finding a positive impact on the latter stages of innovation (mainly deployment), and broader work on auctions covering OECD countries as well as Brazil, South Africa and China not finding a significant impact on innovation (Wigand et al. 2016). Work comparing renewable energy auctions in different countries in South America generally finds a positive impact on innovation outcomes (Mastropietro et al. 2014). The body of evidence on the impact of auctions on competitiveness is also limited (six evaluations) and indicates negative outcomes of renewable auctions of competitiveness (*limited evidence, low agreement*). As with other policies, the design of the auctions can affect innovation outcomes (del Río and Kiefer 2021). Only two studies investigated distributional outcomes, and both were negative.

*Other financial instruments*

There is no explicit literature on the ability of green public banks, and targeted loans, and loan guarantees to lead to upstream innovation investments and activities, although there is evidence on their role in deployment (Geddes et al. 2018). This notwithstanding, the key role of these institutions is in the innovation system (OECD 2015b; Geddes et al. 2018) (Sections 16.2.1 and 16.3) and the belief that they can de-risk scale-up and the testing of business models (Geddes et al. 2018; Probst et al. 2021) (Chapter 17).

*Renewable obligations with tradeable green certificates*

There is mixed evidence of the impact of tradeable green certificates (TGCs) on innovation (*limited evidence, low agreement*) and competitiveness (*limited evidence, low agreement*). Out of the 11 evaluations in Peñasco et al. (2021), six found no impact, two a positive impact, and three a negative impact. All of them used a qualitative research approach. Of the six studies focusing on competitiveness outcomes, three conclude that TGCs have had no impact on competitiveness, while two indicate a negative impact and one a positive impact. Only one of the studies was quantitative, and did not identify an impact on competitiveness.

TGCs are associated with the existence of negative distributional impacts in most applications (*medium evidence, high agreement*). Ten out of 12 studies identify the existence of some negative impacts. All but one of these studies (which focused on India) are based on analysis of policies implemented in industrialised countries.

*Clean energy and renewable portfolio standards*

The impact of renewable portfolio standards without tradeable credits on innovation outcomes is negligible or very small (*medium evidence, medium agreement*). Out of the nine studies, seven reported no impact on innovation outcomes and two a positive impact (Peñasco et al. 2021). Most of these papers focused on patenting and private R&D innovation indicators and not cost reductions. Impact on competitiveness is found to be negligible or positive (*limited evidence, medium agreement*). Out of eight evaluations, five report a positive impact and three a negligible impact; only two are quantitative studies (Peñasco et al. 2021). Negative distributional impacts from renewable portfolio standards can emerge in some cases (*limited evidence, low agreement*). Out of eight evaluations, four identified positive impacts, and four negative impacts; all of the studies identifying a positive impact were theoretical. There are efforts focused on clean energy portfolio standards which include technologies beyond renewables.

*Efficiency obligations with tradeable credits*

The impact of tradeable white certificates in innovation is largely positive, but the evidence is limited (*limited evidence, medium/high agreement*). Out of four evaluations, only one of which was quantitative, three report a positive impact and one reports no impact (Peñasco et al. 2021). The impact of white certificates on competitiveness is positive (*limited evidence, high agreement*) while the impact on distributional outcomes is very mixed (*limited evidence, low agreement*). Two theoretical studies report positive competitiveness impacts. Out of 11 evaluations of distributional outcomes, eight rely on theoretical *ex ante* approaches. Of the 11 evaluations: seven reported positive impacts (four of them using theoretical methods); three indicated negative impacts (using theoretical methods); and one reported no impact.

*Building codes*

There is evidence of the impact of building codes on innovation outcomes (Peñasco et al. 2021). Only two studies assessed competitiveness impacts (one identified positive impacts and one negligible ones) and three studies identified distributional impacts, all positive.

Overall, the evidence on the impact of the market pull policy instruments covered in Section 16.4.4.4 when it comes to the competitiveness outcome (at least in the short term) is more mixed. For some of them, the evidence of a positive impact on innovation is more consistent than the others (for carbon taxes or FITs, for example). Peñasco et al. (2021) found that the disagreements in the evidence regarding the positive, negative or no impact of a policy on competitiveness or distributional outcomes can often be explained by differences in policy design, differences in geographical or temporal context (since the review included evidence from countries from all over the world), or on how policy mixes may have affected the ability of the research design of the underlying papers to separate the impact of the policy under consideration from the others.

#### 16.4.4.5 Assessment of the Impact on Innovation, Competitiveness and Distributional Outcomes of Regulatory Policy Instruments Targeting Efficiency Improvements

There is medium evidence that the introduction of flexible, performance-based environmental regulation on energy efficiency in general (e.g., efficiency standards) can stimulate innovative responses in firms (Ambec et al. 2013; Popp 2019) (*medium evidence, high agreement*). Evidence comes from both observational studies that examine patenting, R&D or technological responses to regulatory interventions, and from surveys and qualitative case studies in which firms report regulatory compliance as a driving force for the introduction of environmentally-beneficial innovations (Grubb et al. 2021). While the literature examining the impact of environmental regulation on innovation is large, there have been fewer studies on the innovation effects of minimum energy or emissions performance regulations specifically relating to climate mitigation. We discuss in turn two types of efficiency regulations: on vehicles, and on appliances.

*Relationship between automotive efficiency regulations and innovation*

The announcement, introduction and tightening of vehicle fleet efficiency or greenhouse gas (GHG) emission standards either at the national or sub-national level positively impacts innovation as measured by patents (Barbieri 2015) or vehicle characteristics (Knittel 2011; Kiso 2019) as summarised in a review by Grubb et al. (2021). Detailed studies on the innovation effects of national pollutant (rather than energy) regulations on automotive innovation also indicate that introducing or tightening performance standards has driven technological change (Lee et al. 2010). Some studies in the USA that examine periods in which little regulatory change took place have found that the effects of performance standards on fuel economy have been small (Knittel 2011) or not significant relative to the innovation effects of prices (Crabb and Johnson 2010). This is at least in part because ongoing efficiency improvements during this period were offset by increases in other product attributes. For example, a study by Knittel (2011) observed that size and power increased without a corresponding increase in fuel consumption. It has also been observed that regulatory design may introduce distortions that affect automotive innovation choices: in particular, fuel economy standards based on weight classes have been observed to distort light-weighting strategies for fuel efficiency in both China (Hao et al. 2016) and Japan (Ito and Sallee 2018).

A number of studies have focused on the impacts of a sub-national technology-forcing policy: the California Zero Emission Vehicle (ZEV) mandate. When it was introduced in 1990, this policy required automotive firms to ensure that 2% of the vehicles they sold in 1998 would be zero-emission. In the years immediately after introduction of the policy, automotive firms reported that it was a significant stimulus to their R&D activity in electric vehicles (Brown et al. 1995). Quantitative evidence examining patents and prototypes has indicated that the stringency of the policy was a significant factor in stimulating innovation, though this was, in part, dependent on firm strategy (Sierzchula and Nemet 2015). As for the previous instruments, most of the evidence comes from industrialised countries, and additional research on other countries would be beneficial.

*Relationship between appliance efficiency standards and innovation*

Regulation-driven deployment of existing technologies can generate innovation in those technologies through learning by- doing, induced R&D and other mechanisms, although not in all cases (*medium evidence, medium agreement*) (Grubb et al. 2021). The introduction or tightening of minimum energy performance standards for appliances (and for buildings, in Noailly (2012)) have driven innovation responses, using direct measures of product attributes (Newell et al. 1999) and patents (Noailly 2012; Kim and Brown 2019), though not all studies have found a significant relationship (Girod et al. 2017). There is also evidence of a correlation between regulation-driven deployment of energy-efficient products with accelerated learning in those technologies (Van Buskirk et al. 2014; Wei et al. 2017).

In addition to observational studies, evidence on the relationship between innovation and regulation comes from surveys in which respondents are asked whether they have engaged in innovation leading to energy saving or reduced GHG emissions, and what the motivations were for such innovation. Survey evidence has found that expected or current regulation can drive both R&D investment and decisions to adopt or introduce innovations that reduce energy consumption or $CO_2$ emissions (Horbach et al. 2012; Grubb et al. 2021). Survey-based studies, however, tend not to specify the type of regulation.

*Competitiveness and distributional impacts associated with vehicles and appliance performance standards*

Minimum energy performance standards and appliance standards have been known to result in negative distributional impacts (*limited evidence, medium/high agreement*). Several studies focused on the USA have highlighted that minimum energy performance standards for vehicles tend to be regressive, with poorer households disproportionately affected (Jacobsen 2013; Levinson 2019), particularly when second-hand vehicles are taken into account (Davis and Knittel 2019). Similar arguments, though with less evidence, have been made for appliance standards (Sutherland 2006).

Overall, the extent to which regulations in energy efficiency result in positive or negative competitiveness impacts in firms is mixed (*limited evidence, high disagreement*). A meta-analysis of 107 studies, of which 13 focused on regulations relating to energy consumption or GHG emissions, found that around half showed that regulations resulted in competitiveness impacts, while half did not (Cohen and Tubb 2018). Cohen and Tubb (2018) also found that studies examining performance-based regulations were less likely to find positive competitiveness impacts than those that examined market-based instruments.

*Insights into causal mechanisms and co-evolutionary dynamics from case studies on efficiency regulations*

While most of the literature addresses the extent to which regulation can induce innovation, a number of case studies highlight that innovation can also influence regulation, as the costs of imposing regulation are reduced and political interests emerge that seek to exploit competitive advantages conferred by successfully developing energy-efficient or low-carbon technologies (*medium evidence, high agreement*). Case studies map the causal mechanisms relating regulations and innovation responses in specific firms or industries (Gann et al. 1998; Kemp 2005; Ruby 2015; Wesseling et al. 2015).

#### 16.4.4.6 Assessment of the Impact on Innovation and on Competitiveness and Distributional Outcomes of Soft Instruments

*Energy labels and innovation*

The literature specifically focusing on the impacts of labels is very limited and indicates positive outcomes (*limited evidence, high agreement*). Energy labels may accompany a minimum energy performance standard, and the outcomes of these policies are often combined in literature (IEA 2015). But again, given the limited evidence, more research is needed. Although there are many studies on energy efficiency more broadly and for both standards and labels, only eight studies specifically focus on labels. Furthermore, seven of them report positive outcomes and one negative outcomes. Six of the studies used qualitative methods mentioning the impacts of labelling on the development of new products (Wiel et al. 2006). Research specifically comparing voluntary labels with other mechanisms found a significant and positive relationship between labels and the number of energy-efficient inventions (Girod et al. 2017). More research is needed, especially in developing countries, that have extensive labelling programmes in place, and also with quantitative methods, to develop evidence on the impacts of labelling on innovation. Box 16.7 discusses an example of a combination of policy instruments in China including labelling, sale bans and financial support.

## Box 16.7 | China Energy Labelling Policies, Combined with Sale Bans and Financial Subsidies

From 1970 to 2001, China was able to significantly limit energy demand growth through energy-efficiency programmes. Energy use per unit of gross domestic product (GDP) declined by approximately 5% $yr^{-1}$ during this period. However, between 2002 and 2005, energy demand per unit of GDP increased on average by 3.8% $yr^{-1}$. To curb this energy growth, in 2005, the Chinese government announced a mandatory goal of 20% reduction of energy intensity between 2006 and 2010 (Zhou et al. 2010; Lo 2014).

An energy labelling system was passed in 2004. It requires manufacturers to provide information about the efficiency of their electrical appliances to consumers. From 2004 to 2010, 23 electrical appliances (including refrigerators, air conditioners and flat-screen TVs) being labelled as energy efficient with five different grades – grade 1 being the most energy efficient and grade 5 the least efficient. Any appliances with an efficiency grade higher than 5 cannot be sold in the market.

In addition to providing information to consumers, the National Development and Reform Commission, (which was in charge of designing the policies), and the Ministry of Finance launched in 2009 the 'energy-saving products and civilian-benefiting project' (Zhan et al. 2011). It covered air conditioners, refrigerators, flat panel televisions, washing machines, electrical efficient lighting, energy saving and new energy vehicles with the energy grades at 1 or 2. The project also included financial subsidies for the enterprises producing these products. The standard design of these financial subsidies involved the government paying for the price difference of energy-efficient products and general products. The manufacturers that produce the energy-efficient products receive financial subsidies directly from the government (Z. Wang et al. 2017).

Before 2008, the market share of grade 1 and grade 2 air conditioners was about 5%, and about 70% of all air conditioners were grade 5 (the most inefficient). Driven by the financial subsidies, the selling price of the highly efficient air conditioners became competitive with that of the general air conditioners. Hence, the sales of energy-efficient air conditioners increased substantially, making the market share of grade 1 and 2 air conditioners about 80% in 2010 (Z. Wang et al. 2017). According to the information from China's National Institute of Standardization, the energy label system saved more than 1.5 hundred billion kWh power between 2005 and March 2010, equivalent to more than 60 million tonnes of standard coal, 1.4 billion tonnes of carbon dioxide emissions, and 60 tonnes of sulphur dioxide emissions (Zhan et al. 2011), which significantly contributed to energy saving goals of China's 11th Five-Year Plan.

*Voluntary approaches and innovation*

Voluntary approaches have a largely positive impact on innovation for those that choose to participate (*robust evidence*, *medium agreement*). Research on voluntary approaches focuses on firms adopting voluntary environmental management systems that can be certified based on standards of the widely adopted International Organization for Standardization (ISO 14001 – standard for environmental management) or the European Union's Eco-Management and Auditing Scheme (EMAS), which is partly mandatory. Out of 16 analyses: 70% report positive innovation outcomes in terms of patents, products or processes; 17% report negligible impacts; and 13% report negative impacts. Positive innovation outcomes have been linked to firms' internal resource management practices and were found to be strengthened in firms with mature environmental management systems and in the presence of other environmental regulations (Inoue et al. 2013; He and Shen 2019; Li et al. 2019a). Overall, studies are concentrated in a few countries that do not fully capture where environmental management systems have been actually adopted (Boiral et al. 2018). There is a need for research in analyses of such instruments in emerging economies, including China and India, and methodologically in qualitative and longitudinal analyses (Boiral et al. 2018).

*Competitiveness and distributional outcomes of soft instruments*

The outcomes for performance or endorsement labels have been associated with positive competitiveness outcomes (*medium evidence*, *medium agreement*). Out of 19 studies, 89% report positive impact and 11% negligible impact. Although there are several studies analysing competitiveness-related metrics, evidence on most individual metrics is sporadic, except for housing premiums. A large number of studies quantitatively assessing competitiveness find that green labels in buildings are associated with housing price premiums in multiple countries and regions (Fuerst and McAllister 2011; Kahn and Kok 2014; Zhang et al. 2017). Of those studies, 32% were qualitative, associating appliance labelling programmes with employment and industry development (European Commission 2018). There is a research gap in analyses of developing countries, and also in quantitatively assessing outcomes beyond housing price premiums.

A few studies on the distributional outcomes of voluntary labelling programmes point to positive impacts (*limited evidence*, *high agreement*). All four studies that focus on benefits for consumers and tenants report positive impacts (Devine and Kok 2015). Although there are benefits for utility companies and other stakeholders, more research is needed to specifically attribute these benefits to voluntary labels rather than energy efficiency programmes in general.

Voluntary agreements are associated with positive competitiveness outcomes (*medium evidence*, *medium agreement*): 14 out of 19 evaluations identified were associated with positive outcomes, while three were associated with negligible outcomes, and two with negative outcomes. Research found an increase in perceived firm financial performance (de Jong et al. 2014; Moon et al. 2014). Studies also show an association with higher exports as more environmentally-conscious trade partners increasingly value environmental certifications (Bellesi et al. 2005). More research is needed to develop evidence on metrics of competitiveness besides firms' financial performance, and especially in developing countries.

Voluntary agreements are associated with a positive impact on distributional outcomes (*limited evidence*, *high agreement*). Five studies, mainly using qualitative approaches, report a positive association between a firm adopting an environmental management system and impacts on its supply chains. There is a need for more studies with quantitative assessments and geographical diversity.

#### 16.4.4.7 Summary of the Size and Direction of the Evidence of All Policy Instrument Types on Innovation Outcomes

Positive impacts have been identified more frequently in some policies than in others. There is also a lot of variation in the density of the literature. Developing countries are severely underrepresented in the decarbonisation policy instrument evaluation literature aiming to understand the impact on innovation. (*high evidence, high agreement).*

Figure 16.2 below indicates the extent to which some decarbonisation policy instruments have been more or less investigated in terms of their impact on innovation outcomes (as described in Table 16.9). For example, it indicates the extent to which there has been a greater focus of evaluations of the impact of R&D investments, emissions trading schemes and voluntary approaches on innovation. It also shows a limited amount of evidence on procurement, efficiency obligations with tradeable green certificates (TGCs), building codes and auctions.

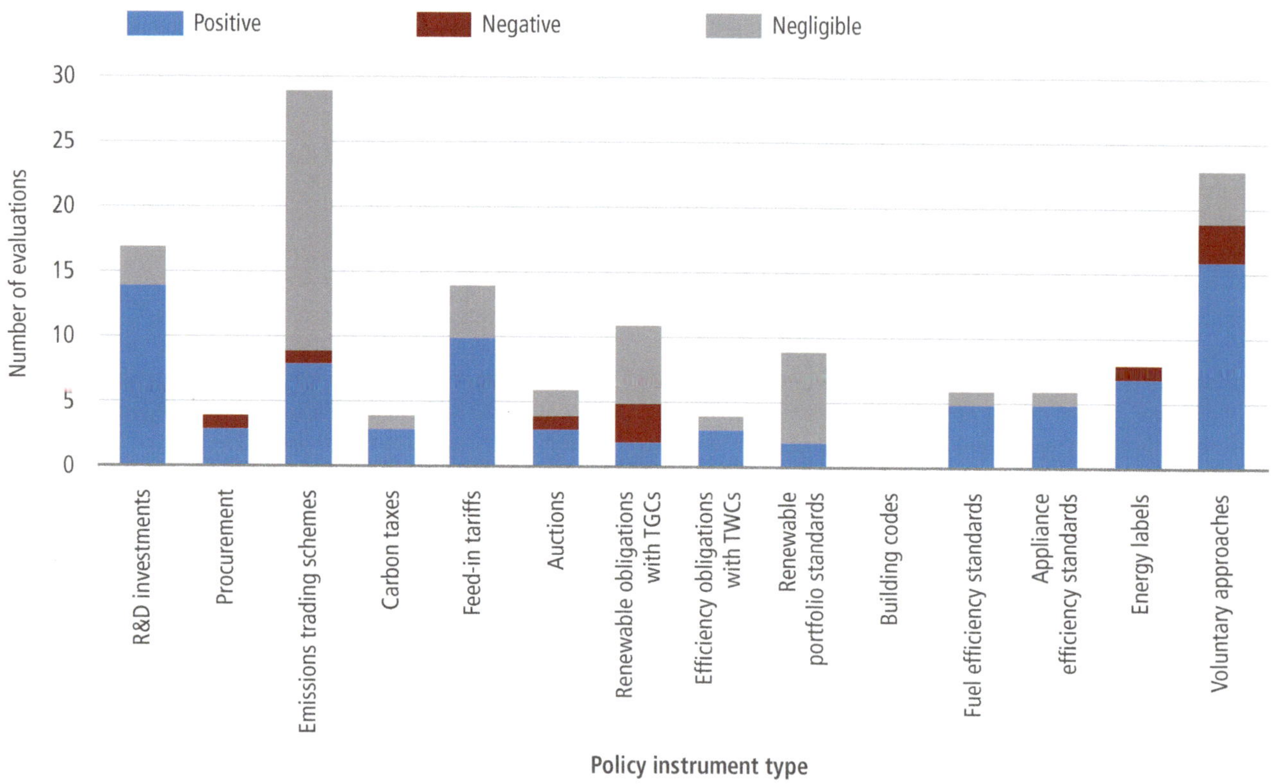


**Figure 16.2 | Number of evaluations available for each policy instrument type covered regarding their impact on innovation and direction of the assessment.** The vertical axis displays the number of evaluations claiming to isolate the impact of each policy instrument type on innovation outcomes as listed in Table 16.9. The colour indicates whether each evaluation identified a positive impact on the innovation outcome (blue), the existence of a negative impact (in red), and no impact (in grey). It builds on Grubb et al. (2021), Lilliestam et al. (2021) and Peñasco et al. (2021), and additional studies identified as part of these reviews. TGC stands for tradeable green certificates. TWC stands for tradeable white certificates.

### 16.4.5 Trade Instruments and their Impact on Innovation

There has been long-standing interest on the impact of Foreign Direct Investment (FDI) on domestic capacity, innovation and environmental outcomes. While this section looks at the impact of trade instruments on innovation, it does not cover the much larger body of evidence on the relationship between FDI and economic development and growth.

Overall, research indicates that trade can facilitate the entrance of new technologies, but the impact on innovation is less clear (*limited evidence*, *low agreement*). A recent study indicates that, for countries with high environmental performance, FDI has a negligible impact on environmental performance, while countries with a lower environmental performance may benefit from FDI in terms of their environmental performance (Li et al. 2019b). One analysis on China links FDI with improved environmental performance and energy efficiency and also innovation outcomes in general (Gao and Zhang 2013). Other work links FDI with increased productivity across firms (not just those engaged in climate-related technologies) through spillovers (Newman et al. 2015). In addition, Brandão and Ehrl (2019) indicate that productivity of the electric power industry is more influenced by the transfer of embodied technology from other industries than by investments of the power industry. Also, they find that countries with high R&D stocks are the main sources of international technology spillovers and the source countries may also benefit from the spillover.

Other emerging work investigates the role of local content requirements on innovation outcomes and suggests that it can lead to increased power costs (negative distributional impacts). The benefits to the domestic innovation system, measured by patents or exports, are unclear if the policies are not part of a holistic and longer-lasting policy framework (Probst et al. 2020).

### 16.4.6 Intellectual Property Rights, Legal Framework and the Impact on Innovation

Virtually all countries around the world have instituted systems for the protection of creations and inventions, known as intellectual property rights (IPR) systems (WIPO 2021). While several types of intellectual property exist – patents, copyright, design rights, trademarks, and more – this section will focus on patents, as the most relevant property right for technological innovations (WIPO 2008), and hence the most relevant for policy instruments in this context.

Patent systems aim to promote innovation and economic growth, by stimulating both the creation of new knowledge and diffusion of that knowledge (*high evidence*, *high agreement*). National patent systems, as institutions, play a central role in theories on national innovation systems (*high evidence*, *strong agreement*). Patent systems are usually instituted to promote innovation and economic growth (Machlup and Penrose 1950; Nelson and Mazzoleni 1996; Encaoua et al. 2006). Some countries explicitly refer to this purpose in their law or legislation – for instance, the US Constitution states the purpose of the US IP rights system to 'promote the progress of science and useful arts'. Patent systems aim to reach their goals by trying to strike a balance between the creation of new knowledge and diffusion of that knowledge (Scotchmer and Green 1990; Devlin 2010; Anadon et al. 2016b). They promote the creation of new knowledge (e.g., technological inventions) by providing a temporary, exclusive right to the holder of the patent, thus providing incentives to develop such new knowledge and helping parties to justify investments in R&D. They promote the diffusion of this new knowledge via the detailed disclosure of the invention in the patent publication, and by enabling a 'market for knowledge' via trading patents and issuing licences (Arora et al. 2004). Although IP protections provide incentives to invest in innovation, they can also restrict the use of new knowledge by raising prices or blocking follow-on innovation (Wallerstein et al. 1993; Stiglitz 2008). As institutions, national patent systems feature prominently in models and theories of national innovation systems (Edquist 1997; Klein Woolthuis et al. 2005).



The degree to which patent systems actually promote innovation is subject to debate. Patent protection has been found to have a positive impact on R&D activities in patent-intensive industries, but this effect was found to be conditional on access to finance (Maskus et al. 2019). Patents are believed to be especially important to facilitate innovation in selected areas such as pharmaceuticals, where investments in developments and clinical trials are high, imitation costs are low, and there is often a one-to-one relationship between a patent and a product, referred to as a 'discrete' product industry (Cohen et al. 2000). At the same time, an increasing body of theoretical and empirical literature suggests that the proliferation of patents also discourages innovation (*medium evidence*, *low agreement*). Theoretical contributions note that a appropriability regime that is too stringent may greatly limit the diffusion of advanced technological knowledge and eventually block the development of differentiated technological capabilities within an industry, in what is called an 'appropriability trap' (Edquist 1997; Klein Woolthuis et al. 2005). There has been a long-standing debate on the impact of patents and other IP rights on innovation and economic development (Machlup 1958; Hall and Helmers 2019). Jaffe and Lerner (2004) and Bessen and Meurer (2009) highlight how IP rights also hamper innovation in a variety of ways. Other contributions in the literature focus on more specific factors. For example, Shapiro (2001) discusses 'patent thickets', where overlapping sets of patent rights mean that those seeking to commercialise new technology need to obtain licences from multiple patentees. Heller and Eisenberg (1998) argue that a 'tragedy of the anticommons' is likely to emerge when too many parties obtain the right to exclude others from using fragmented and overlapping pieces of knowledge – ultimately leading to no one having the privilege of using the results of biomedical research. Reitzig et al. (2007) describe the damaging effects of extreme business strategies employing patents, such as 'patent trolling'.

In general, IP protection and enforcement may have different impacts on economic growth in different types of countries (*limited evidence*, *high agreement*). There has been a significant degree of harmonisation and cooperation between national IP systems over time. The most recent milestone is the World Trade Organization (WTO) 1994 Trade-Related Aspects of Intellectual Property Rights (TRIPS) Agreement, entered into by all WTO members, which sets down minimum standards for the regulation by national governments of many forms of IP as applied to nationals of other WTO member nations (WTO 1994). Developing countries successfully managed to include some flexibilities into TRIPS, both in terms of timing of legislative reform, and the content of the reforms. In an attempt to understand the effects of the introduction of TRIPS, Falvey et al. (2006) find that the effect of IP protection on growth is positively and significantly related to growth for low- and high-income countries, but not for middle-income countries. They argue that low-income countries benefit from increased technology flows, but middle-income countries may have offsetting losses from the reduced scope for imitation. Note that Falvey et al. (2006) do not break down their results in different technological areas, and they do not focus on innovation, but instead on growth. It has been argued that the increasingly globalised IP regime through initiatives such as the TRIPS agreement will diminish prospects for technology transfer and competition in developing countries, particularly for several important technology areas related to meeting sustainable development needs (Maskus and Reichman 2017).

In principle, patent holders are not required to take their protected invention into use, and neither have the obligation to allow (i.e., license) others to use the inventions in question (*high evidence, high agreement*). Studies have shown that the way patent holders use their patent differs considerably across industrial sectors: in pharmaceutics, patents are typically used to enable exclusive production of a certain good (and obtain monopoly rents), while in industries such as computers, semiconductors, and communications, patents are often used to strengthen positions in cross-licensing negotiations and to generate licensing income (Cohen et al. 2000; Foray 2004). There are also companies that predominantly obtain patents for defensive reasons: they seek freedom to design and manufacture, and by owning a patent portfolio themselves, they hope to prevent becoming the target of litigation by other patent holders (Hall and Ziedonis 2001). Patents are often used strategically to impede the development and diffusion of competing, alternative products, processes or services, by employing strategies known as 'blanketing' and 'fencing' (Grandstrand 2000), although the research is not specific to the climate space.

There are notable but specific exceptions to the general principle that patent holders are not obliged to license their patent to others. These exceptions include the compulsory licence, fair, reasonable and non-discriminatory (FRAND) policies, and statement on licences of right (*high evidence, high agreement*). While patent holders are, in principle, free to choose not to license their innovation, there are three important exceptions to this. First, most national patent laws have provisions for compulsory licensing, meaning that a government allows someone else to produce a patented product or process without the consent of the patent holder, or plans to use the patent-protected invention itself (WTO 2020). Compulsory licences may be issued in cases of public interest or events of abuse of the patent (WIPO 2008; Biadgleng 2009). Compulsory licensing is explicitly allowed in the WTO TRIPS agreement, and its use in context of medicine (for instance, to control diseases of public health importance, including HIV, tuberculosis and malaria) is further clarified in the 'DOHA Declaration' from 2001 (Reichman 2009; WHO 2020). Second, standard-setting organisations have policies to include patented inventions in their standards only if the patent holder is willing to commit FRAND licensing conditions for those patents (Contreras 2015). While a patent holder can choose not to make such a commitment, by doing so, its patent is no longer a candidate for inclusion in the standard. In the (many) fields where standards are of key importance, it is very unusual for patent holders not to be willing to enter into FRAND commitments (Bekkers 2017). Third, when a patent holder files at the patent office and opts for the 'licence of right' regime, in return for reduced patent fees, they enter into a contractual agreement that obliges them to license the patent to those who request it. While not all national patent systems feature this regime, it is a feature present in the new European Community patent (EPO 2017), and may therefore increase in importance.

For a discussion on the impact of intellectual property rights (IPR) on international technology diffusion, see Box 16.9 in Section 16.5.

### 16.4.7 Sub-national Innovation Policies and Industrial Clusters

Research examining the impacts of sub-national policies on innovation and competitiveness is sporadic – regional variations have been quantitatively assessed in the USA or China, or with case studies in these and other countries. Research on wind energy in the USA, distributed PV balance of systems in China, and renewable energy technologies in Italy have found that policies that incentivised local demand were associated with inducing innovation, measured with patents (Corsatea 2016; Fu et al. 2018; Gao and Rai 2019). Different policies may have different impacts – for example, in the USA, state-level tax incentives and subsidies induced innovation within the state; but for renewable portfolio standards, policies in other states were associated with innovation because of impact on demand, but own state policies were not (Fu et al. 2018). Research has also noted that the outcomes of policy and regulation on innovation are spatially heterogenous, because of differences in local planning authorities and capabilities (Corsatea 2016; Song et al. 2019).

Sub-national deployment policies have been associated with different impacts on competitiveness metrics (*limited evidence, medium agreement*). Research on green jobs shows positive association between sub-national policies and green jobs or green firms at the metropolitan level as well as the state of provincial level, in both China and the USA (Yi 2013; Yi and Liu 2015; Lee 2017), while others find no impact of renewable portfolio standards on green job growth in the state (Bowen et al. 2013). Other examples of competitiveness are in the impact of regional green industrial policy in Brazil's Rio Grande do Sul region in attracting auctioned contracts for wind energy (Adami et al. 2017) or in the changes in net positive state revenues associated with removing tax incentives for wind producers in Idaho in the USA (Black et al. 2014).

Sub-national policies also directly support innovation and competitiveness through green incubators and direct grants or R&D funding for local companies working on clean energy, intending to promote local economic development (*limited evidence, medium agreement*). The literature on the impacts of such policies on innovation and competitiveness is sparse. Some case studies and programme evaluation reports, primarily in the USA, have identified the impacts of sub-national policies on competitiveness — for example, job creation from direct R&D funding in North Carolina (Hall and Link 2015), perceptions for local industry development and support for follow-on financing for companies receiving state-funded grants in Colorado (Surana et al. 2020b), and return on investments for the state in research and innovation spending from the New York state's energy agency (NYSERDA 2020). There is a general paucity of metrics on innovation and competitiveness for systematic assessments of such programmes in developed countries, and even more so in India and other developing countries where such programmes have been increasing (Gonsalves and Rogerson 2019; Surana et al. 2020a).

Although states and local governments increasingly support clean energy deployment as well as directly support innovation, given its link with economic development goals, there is a lack of systematic research on the impacts of these policies at the subnational level. More research – qualitative and quantitative, and in developed and developing countries – is needed to systematically develop evidence on these impacts and to understand the reasons behind regional differences in terms of the type of policy as well as the capabilities in the region.

### 16.4.8 System-oriented Policies and Instruments

Although previous sections summarised the research disentangling the role of individual policies in advancing or hindering innovation (as well as impacts on other objectives), other research has tried to characterise the impact of a policy mix on a particular outcome. Although the outcome studied was not innovation, but diffusion (technology effectiveness is in the set of criteria outlined in Chapter 13), it seems relevant to discuss overall findings. Research reviewing renewable energy policies in nine OECD countries concludes that, over time, a broad set of policies characterised by a 'balance' metric has been put in place. This research also identifies a significant negative association between the balance of policies in renewable energy and the diffusion of total renewable energy capacity, but no significant effect of the overall intensity (coded as the 46 weighted average of six indicators) on renewable capacity (Schmidt and Sewerin 2019). This indicates that a neutral conception of balance across all possible policies may not be desirable, and that policy mix intensity by itself does not explain technology diffusion.

A growing body of research aims to understand how different policies interact and how to characterise policy mixes (del Río 2010; Howlett and del Rio 2015; Rogge and Reichardt 2016; del Río and Cerdá 2017). The empirical impact on the innovation outcomes is not yet discussed. A more detailed discussion of this literature is located in Chapter 13.

An emerging stream of research in complex systems suggests that relatively small changes in policy near a possible tipping point in climate impacts in areas, including changing strategies related to investments in innovation, could trigger large positive societal feedbacks in the long term (Farmer et al. 2019; Otto et al. 2020).

## 16.5 International Technology Transfer and Cooperation for Transformative Change

This section covers international transfer and cooperation in relation to climate-related technologies, 'the flows of know-how, experience and equipment for mitigating and adapting to climate change amongst different stakeholders' (IPCC 2000) as well as innovation to support transformative change compared to AR5 (IPCC 2014) and the IPCC Special Report on Global Warming of 1.5°C (SR1.5) (IPCC 2018a). This complements the discussion on international cooperation on science and technology in Chapter 14.

This section first outlines the needs and opportunities for international transfer and cooperation on low-emission technologies. It then describes the main objectives and roles of these activities, and then reviews recent institutional approaches within and outside the UN Framework Convention on Climate Change (UNFCCC) to support international technology transfer and cooperation. Finally, it discusses emerging ideas for international technology transfer and cooperation, and possible modifications to support the achievement of climate change and Sustainable Development Goals (SDGs), building up to Section 16.6.



### 16.5.1 International Cooperation on Technology Development and Transfer: Needs and Opportunities

With the submission of their Nationally Determined Contributions (NDCs) as part of the Paris Agreement, most developing countries are now engaged in climate mitigation and adaptation. While technology is seen as one of the 'means of implementation' of climate action, developing countries often have relatively limited technology innovation capabilities, which requires them to access technologies developed in higher-income countries with stronger innovation systems (Popp 2011; Binz et al. 2012; Urban 2018). In many cases, these technologies require adaptation for the local context and needs (Sagar 2009; Anadon et al. 2016b), and innovation capabilities are required to suitably adapt these technologies for local use and also to create new markets and business models that are required for successful deployment (Sagar 2009; Ockwell et al. 2015; Ockwell and Byrne 2016). This can lead to dependencies on foreign knowledge and providers (Ockwell and Byrne 2016), negative impacts in terms of higher costs (Huenteler et al. 2016a), balance of payments constraints, and vulnerability to external shocks (Ebeling 2020).

The climate technology transition can also yield other development benefits, for instance better health, increased energy access, poverty alleviation and economic competitiveness (Deng et al. 2018), including industrial development, job creation and economic growth (Porter and Van der Linde 1995; Altenburg and Rodrik 2017; Lema et al. 2020; Pegels and Altenburg 2020) (Section 16.6). The growing complexity of technologies and global competition have made technology development a globalised process involving the flow of knowledge and products across borders (Lehoux et al. 2014; Koengkan et al. 2020). For instance, in electronics production, Asian economies have captured co-location synergies and dominate production and assembly of product components, whereas American firms have adopted 'design-only' strategies (Tassey 2014). In the context of renewable energy technologies, 'green global division of labour' has been observed, with countries specialising in investments in research and development (R&D), manufacturing or deployment of renewables (Lachapelle et al. 2017). In the case of solar photovoltaic (PV), for example, while many technical innovations emerged from the USA, Japan and China emphasised the manufacture of physical modules (Deutch and Steinfeld 2013) (Box 16.4).

Such globalisation of production and supply chains opens up economic development opportunities for developing countries (Lema et al. 2020). At the same time, not all countries benefit from the globalisation of innovation – barriers remain related to finance, environmental performance, human capabilities and cost (Weiss and Bonvillian 2013; Egli et al. 2018), with developing countries being particularly disadvantaged at leveraging these opportunities. The gap in low-carbon technology innovation between countries appears to have reduced only among OECD countries (Yan et al. 2017; Du and Li 2019; Du et al. 2019) and the lower-income countries are not able to benefit as much from low-carbon technologies. For instance, in the case of agriculture, Fuglie (2018) notes that international R&D spillovers seem to have benefitted developed countries more than developing countries. Gross et al. (2018) also argue that the development timescales for new energy technologies can extend up to 70 years, even within one country. They recommend that innovation efforts be balanced between early-stage R&D spending, and commercialising already low-emission technologies in the demonstration phase and diffusing them globally.

Thus international cooperation on technology development and transfer can enable developing countries to achieve their climate goals more effectively, while also addressing other SDGs – taking advantage, where possible, of the globalisation of innovation and production (Lema et al. 2020). Earlier assessments in AR5 and SR1.5 have made it clear that international technology transfer and cooperation could play a role in climate policy at both the international and the domestic policy level (Somanathan et al. 2014; Stavins et al. 2014; IPCC 2018b) and for low-carbon development at the regional level (Agrawala et al. 2014). The Paris Agreement also reflects this view by noting that countries shall strengthen cooperative action on technology development and transfer regarding two main aspects: (i) promoting collaborative approaches to R&D; and (ii) facilitating access to technology to developing country Parties (UNFCCC 2015). Furthermore, both in literature and in UNFCCC deliberations, South-South technology transfer is highlighted (Khosla et al. 2017) as a complement to the transfer of technology and know-how from the North to the South.

This is consistent with literature that suggests that greenhouse gas (GHG) mitigation in developing countries can be enhanced by: (i) technology development and transfer collaboration and a 'needs-driven' approach; (ii) development of the specific types of capacity required across the entire innovation chain; and (iii) strengthening of the coordination and agendas across and between governance levels (including domestic and international levels) (Khosla et al. 2017; Zhou 2019; Upadhyaya et al. 2020).

### 16.5.2 Objectives and Roles of International Technology Transfer and Cooperation Efforts

International efforts involving technology transfer can have different objectives and roles. These include access to knowledge and financial resources as well as promotion of new industries in both the developed and recipient country (Huh and Kim 2018). Based on an econometric analysis of international technology transfer factors and characteristics of Clean Development Mechanism (CDM) projects, Gandenberger et al. (2016) find that complexity and novelty of technologies explain whether a CDM project includes hardware technology transfer, and that factors like project size and absorptive capacity of the host country do not seem to be drivers. Halleck Vega and Mandel (2018) argue that 'long-term economic relations', for instance being part of a customs union, affect technological diffusion between countries in the case of wind energy, and indicate that this has resulted in low-income countries being largely overlooked.

There is some literature studying whether technology cooperation could complement or replace international cooperation based on emission reductions, such as in the Kyoto Protocol, and whether that would have positive impacts on climate change mitigation and compliance. A handful of papers conducted game-theoretic analysis on technology cooperation, sometimes as an alternative for cooperation on emission reductions, and found partially positive effects (Bosetti et al. 2017; Narita and Wagner 2017; Rubio 2017; Verdolini and Bosetti 2017). However, Sarr and Swanson (2017) model that, due to the rebound effect, technology development and transfer of resource-saving technologies may not lead to envisioned emission reductions.

While technology cooperation can be aimed at emission reduction through mitigation projects, as indicated above, not all cooperative actions directly result in mitigation outcomes. Overall, technology transfer broadly has focused on: (i) enhanced climate technology absorption and deployment in developing countries; and (ii) enhanced research, development and demonstration (RD&D) through cooperation and knowledge spillovers.

#### 16.5.2.1 Enhancing Low-emission Technology Uptake in Developing Countries

Real-world outcomes in terms of low-emission technology deployment in developing countries may vary significantly, depending on the nature of the international engagement and the domestic context. While there has been some success in the enhancement of technology deployment through technology transfer in some developing countries (de la Tour et al. 2011; Zhang and Gallagher 2016), many others, and particularly least-developed countries, are lagging behind (Glachant and Dechezleprêtre 2017). Glachant and Dechezleprêtre (2017) indicate that this is due to the lack of participation in economic globalisation and that climate negotiations could facilitate technology transfer to those countries through the creation of global demand for low-emission technologies through stronger mitigation targets that will result in lowering of costs and therefore enhanced technology diffusion. A broader perspective presents a host of other factors that govern technology diffusion and commercialisation in developing countries, including: investment;

social, cultural and behavioural, marketing and market building; macroeconomics; and support policy (Bakhtiar et al. 2020). Ramos Mejía et al. (2018) indicate that the governance of low-emission technology transfer and deployment in developing countries is frequently negatively affected by a mixture of well- and ill-functioning institutions – for instance, in a context of market imperfection, clientelist and social exclusive communities and patrimonial and/or marketised states. Furthermore, existing interests, such as fossil fuel production, may also impede the deployment of low-emission technologies, as highlighted in case studies of Vietnam and Indonesia (Dorband et al. 2020; Ordonez et al. 2021). It is for such reasons that both domestic efforts and international engagement are seen as necessary to facilitate technology transfer as well as deployment in developing countries (Boyd 2012). The same has been seen as true in the case of agriculture, where the very successful international research efforts of the CGIAR – with remarkably favourable benefit-cost ratios (Alston et al. 2021) – were complemented by the national agricultural research systems for effective uptake of high-yielding varieties of crops (Evenson and Gollin 2003).

One key area for underpinning effective technology uptake in developing countries relates to capabilities for managing technological change. This includes the capabilities to innovate, implement, and undertake integrated planning. There is much research to indicate that the ability of a country's firms to adopt new technologies is determined by its absorptive capacity, which includes its own R&D activities, human capacity (e.g., technical personnel), government involvement (including institutional capacity), the infrastructure in the country (Kumar et al. 1999), and knowledge and capacity as part of its 'intangible assets' or the 'software' (Ockwell et al. 2015; da Silva et al. 2019; Corsi et al. 2020). For sustainable development, the capacity to plan in an integrated way and implement the SDGs (Khalili et al. 2015; Elder et al. 2016), including using participatory approaches (Disterheft et al. 2015), is a conditional means of implementation. It also is argued that, if human capital were the focus of international climate negotiations as well as national climate policy, it could change the political economy in favour of climate mitigation, which is needed for developing such capabilities in advance to keep up with the required speed of transformation (Ockwell et al. 2015; Hsu 2017; IPCC 2018b; Upadhyaya et al. 2020). In a global analysis of wind energy using econometric analysis, Halleck-Vega et al. (2018) lend quantitative credibility to the claim that a technology skill base is a key determinant of technological diffusion. Activities to enhance capabilities include informational contacts, research activities, consulting, education and training, and activities related to technical facilities (Huh and Kim 2018; Khan et al. 2020).

There are multiple studies drawing on empirical work that also support this conclusion. For South-South technology transfer between India and Kenya, not just technical characteristics, but also mutual learning on how to address common problems of electricity access and poverty, was suggested as an important condition for success (Ulsrud et al. 2018). Olawuyi (2018) discusses the specific capability gap in Africa, despite decades of technology transfer efforts under various mechanisms and programmes of the UNFCCC. The study suggests that barriers need to be resolved by African countries themselves, in particular: inadequate access to information about imported climate technologies; lack of domestic capacities to deploy and maintain imported technologies; the weak regulatory environment to stimulate clean technology entrepreneurship; the absence or inadequacy of climate change laws; and weak legal protection for imported technologies. Moreover, Ziervogel et al. (2021) indicate that, for transformative adaptation, transdisciplinary approaches and capacity-building shifting, 'the co-creation of contextual understandings' instead of top-down transfer of existing knowledge would deliver better results. Despite the understanding of the importance of the capacity issue, significant gaps still remain on this front (TEC 2019) (Section 16.5.4).

#### 16.5.2.2 Enhancing RD&D and Knowledge Spillovers

As mentioned earlier, RD&D can aid the development of new technologies as well as their adoption for new use contexts. Therefore, it is not surprising that international cooperation on RD&D is identified as a mechanism to promote low-carbon innovation (Suzuki 2015; Mission Innovation 2019; TEC 2021). This has resulted in a variety of international initiatives to cooperate on technology in order to create knowledge spillovers and develop capacity. For example, the UNFCCC Technology Mechanism, among other things, aims to facilitate finance for RD&D of climate technologies by helping with readiness activities for developing country actors. In particular preparing early-stage technologies for a smoother transition to deployment and commercialisation has been emphasised in the context of the Technology Executive Committee (TEC) (TEC 2017). There are numerous multilateral, bilateral and private programmes that have facilitated RD&D, biased mostly towards mitigation (as opposed to adaptation) activities. Many programmes that seemed to be about RD&D were in reality dialogues about research coordination (Ockwell et al. 2015). There are also a variety of possible bilateral and multilateral models and approaches for engaging in joint R&D (Mission Innovation 2019). An update by the TEC (2021) reviewing good practices in international cooperation of technology confirmed the conclusions of Ockwell et al. (2015), and moreover highlighted that most initiatives are led by the public sector, and that the private sector tended to get involved only in incubation, commercialisation and diffusion phases. It also concluded that, although participation of larger, higher-income developing countries seems to have increased, participation of least-developed countries is still very low.

### 16.5.3 International Technology Transfer and Cooperation: Recent Institutional Approaches

The sections below discuss the literature on various categories of international technology cooperation and transfer.

#### 16.5.3.1 UNFCCC Technology and Capacity-building Institutions

Technology development and transfer have been a part of UNFCCC discussions and developments in the context of the international climate negotiations ever since its agreement in 1992, as assessed in AR5 (Stavins et al. 2014). Support on 'Technology Needs Assessment' to developing countries was the first major action undertaken by the UNFCCC, and this has undergone different cycles of learning

## Box 16.8 | Capacity Building and Innovation for Early Warning Systems in Small Island Developing States

One of the areas of international cooperation on capacity building is adaptation, which has been highlighted by both the Technology Executive Committee (TEC) (Ockwell et al. 2015; TEC 2015) and the Paris Committee on Capacity-building (UNFCCC 2020b) as an area where capacity gaps remain, especially in Small Island Developing States (SIDS).

While adaptation was initially conceived primarily in terms of infrastructural adjustments to long-term changes in average conditions (e.g., rising sea levels), a key innovation in recent years has been to couple such long-term risk management to existing efforts to manage disaster risk, specifically including early warning systems, enabling early action in the face of climate- and weather-risk at much shorter timescales (IPCC 2012), with potentially significant rates of return (Rogers and Tsirkunov 2010; Hallegatte 2012; Global Commission on Adaptation 2019).

In recent years, deliberate international climate finance investments have focused on ensuring that developing countries (and especially SIDS and least-developed countries) have access to improvements in hydrometeorological observations, modelling, and prediction capacity, sometimes with a particular focus on the people intended to benefit from the information produced (CREWS 2016). For instance, on the Eastern Caribbean SIDS of Dominica, researchers took a community-based approach to identify the mediating factors affecting the challenges to coastal fishing communities in the aftermath of two extreme weather events (in particular hurricane Maria in 2017) (Turner et al. 2020). Adopting an adaptive capacity framework (Cinner et al. 2018), they identified 'intangible resources' that people relied on in their post-disaster response as important for starting up fishery, but also went beyond that framework to conclude that the response ability on the part of governmental organisations as well as other actors (e.g., fish vendors) in the supply chain is also a requirement for rebuilding and restarting income-generating activity (Turner et al. 2020). Numerous other studies have highlighted capacity-building as adaptation priorities (Basel et al. 2020; Kuhl et al. 2020; Sarker et al. 2020; Vogel et al. 2020; Williams et al. 2020).

One of several helpful innovations in these efforts is impact-based forecasting (Harrowsmith et al. 2020), which provides forecasts targeted at the impact of the hazard rather than simply the meteorological variable. This enables a much easier coupling to early action in response to the information, and a more appropriate response afterwards. Automatic responses to warnings have also been adopted in the humanitarian field for anticipatory action ahead of (rather than simply in response to) disasters triggered by natural hazards (Coughlan de Perez et al. 2015). This has resulted in a rapid scale-up of such anticipatory financing mechanisms to tens of countries over the past few years, and emerging evidence of its effectiveness. Still, the response is lacking in coherence and comprehensiveness, resulting in calls for a more systematic evidence agenda for anticipatory action (Weingärtner et al. 2020).

(Nygaard and Hansen 2015; Hofman and van der Gaast 2019). Since 2009, the UNFCCC discussions on technology development and transfer have focused on the Technology Mechanism under the Cancun Agreements of 2010, which can be seen as the global climate governance answer to redistributive claims by developing countries (McGee and Wenta 2014). The Technology Mechanism consists of the TEC and the Climate Technology Centre & Network (CTCN). An independent review of CTCN, evaluated it on five dimensions – relevance, effectiveness, efficiency, impacts and sustainability – and indicated that the organisation is achieving its mandate in all these dimensions, although there are some possible areas of improvement. The review also specifically noted that 'the lack of predictability and security over financial resources significantly affected the CTCN's ability to deliver services at the expected level, as did the CTCN's lack of human and organizational resources and the capacity of NDEs [National Designated Entities].' (TEC 2017). The CTCN has overcome some of the limitations imposed by resource constraints by acting as a matchmaker from an open-innovation perspective (Lee and Mwebaza 2020). The CTCN's lack of financial sustainability has been a recurring issue, which may potentially be resolved by deepening the linkage between the CTCN and Green Climate Fund (Oh 2020). In the meanwhile, the Green Climate Fund is planning to establish the Climate Innovation Facility to support and accelerate early-stage innovations and climate technologies through the establishment of regional innovation hubs and climate accelerators as well as a climate growth fund (Green Climate Fund 2020).

The 'technology' discussion has been further strengthened by the Paris Agreement, in which Article 10 is fully devoted to technology development and transfer (UNFCCC 2015). However, the political discussions around technology continue to be characterised by viewing technology mostly as hardware (Haselip et al. 2015), and relatively limited in scope (de Coninck and Sagar 2017). The workplans of the TEC and the CTCN do, however, indicate a broadening of the perspective on technology (CTCN 2019; TEC 2019).

Since the Kyoto Protocol's CDM has been operational, studies have assessed its hypothesised contribution to technology transfer, including transfer of knowledge. Though not an explicit objective of the CDM, numerous papers have investigated whether CDM projects contribute to technology transfer (Michaelowa et al. 2019). The literature varies in its assessment. Some find extensive use of domestic technology and hence lower levels of international technology transfer (Doranova et al. 2010), while others indicate that around 40% of projects feature hardware or other types of international transfer of technology (Seres et al. 2009; Murphy et al. 2015), depending on the nature of technology, the host country and region (Cui et al. 2020) and the project type (Karakosta et al. 2012). The CDM was generally positively evaluated on its contribution to technology transfer. However, it was also regarded critically as the market-responsiveness and following of export implies a bias to larger, more advanced economies rather than those countries most in need of technology transfer (Gandenberger et al. 2016), although some countries have managed to correct that by directing the projects, sub-nationally, to provinces with the greatest need (Bayer et al. 2016). Also, the focus on hardware in evaluations of technology transfer under the CDM has been criticised (Haselip et al. 2015; Michaelowa et al. 2019). Indeed, although many studies do go beyond hardware in their evaluations (e.g., Murphy et al. 2015), the degree to which the project leads to a change in the national system of innovation or institutional capacity development is not commonly assessed, or has been assessed as limited (de Coninck and Puig 2015).

There is significantly less literature on capacity building under the UNFCCC, especially as it relates to managing the technology transition. In a legal analysis, D'Auvergne and Nummelin (2017) indicate the nature, scope and principles of Article 11 on capacity building of the Paris Agreement as being demand- and country-driven, following a needs approach, fostering national, subnational and local ownership, and being iterative, incorporating the lessons learnt, as well as participatory, cross-cutting and gender-response. They also highlight that it is novel that least-developed countries and Small Island Developing States (SIDS) are called out as the most vulnerable and most in need of capacity building, and that it raises a 'legal expectation' that all parties 'should' cooperate to enhance the capacity in developing countries to implement the Paris Agreement. These aspects are reflected in the terms of reference of the Paris Committee on Capacity-building (PCCB) that was established in 2015 at the 21st Conference of the Parties (UNFCCC 2016; D'Auvergne and Nummelin 2017), and was extended by five years at the 25th Conference of the Parties in 2019 (UNFCCC 2020a, b). In its work plan for 2020–2024, its aims include 'identifying capacity gaps and needs, both current and emerging, and recommending ways to address them'.

An example of how innovative technologies combined with capacity development, and how institutional innovation is combined in the context of adaptation to extreme weather in SIDS can be found in Box 16.8.

From the broader assessment above, despite limitations of available information, it is clear that the number of initiatives and activities on international cooperation and technology transfer and capacity building seem to have been enhanced since the Cancun Agreements and the Paris Agreement (TEC 2021). However, much more can be done, given the complexity and magnitude of the requirements in terms of coverage of activities, the amount of committed funding, and its effectiveness. Some assessments of UNFCCC instruments specifically for technology transfer to developing countries have indicated that functions such as knowledge development, market formation and legitimacy in developing countries' low-emission technological innovation systems would need much more support to fulfil the Paris Agreement goals (de Coninck and Puig 2015; Ockwell et al. 2015); such areas would benefit from continued attention, given their role in the overall climate technology transition.

#### 16.5.3.2 International RD&D Cooperation and Capacity-building Initiatives

Besides the UNFCCC mechanisms, there are numerous other initiatives that promote international cooperation on RD&D as well as capacity building. Some of them are based on the notion of 'mission-oriented innovation policy' (Mazzucato and Semieniuk 2017; Mazzucato 2018), which shapes markets rather than merely corrects market failures.

For instance, Mission Innovation is a global initiative consisting of 23 member countries and the European Commission working together to reinvigorate and accelerate global clean energy innovation with the objective to make clean energy widely affordable with improved reliability and secured supply of energy. The goal is to accelerate clean energy innovation in order to limit the rise in the global temperature to well below 2°C. The members seek to foster international collaboration among its members and increase public investments in clean energy R&D with the engagement of the private sector. A recent assessment shows that, although

### Box 16.9 | Intellectual Property Rights (IPR) Regimes and Technology Transfer

In the global context of climate mitigation technologies, it has been noted that technologies have been developed primarily in industrialised countries but are urgently required in fast-growing emerging economies (Dechezleprêtre et al. 2011). International technology transfers can take place via three primary channels: (i) trade in goods, where technology is embedded in products; (ii) Foreign Direct Investment (FDI), where enterprises transfer firm-specific technology to foreign affiliates; and (iii) patent licences, where third parties obtain the right to use technologies. IPRs are relevant for all these three channels.

Not surprisingly, the role of IPRs in international transfer of climate mitigation technologies has been much discussed but also described as particularly controversial (Abdel-Latif 2015). The relationships between IPR, innovation, international technology transfer and local mitigation and adaptation are complex (Maskus 2010; Abdel-Latif 2015; Li et al. 2020) and there is no clear consensus on what kind of an IPR regime will be most beneficial for promoting technology transfer.

*Box 16.9 (continued)*

Several studies argue that, particularly in developing nations, the global IPR regime has resulted in delayed access, reduced competition and higher prices (Littleton 2008; Zhuang 2017) and that climate-change-related technology transfer is insufficiently stimulated under the current IPR regime. Compulsory licensing (as already used in medicine) is one of the routes proposed to repair this (Littleton 2008; Abdel-Latif 2015).

There is little systematic evidence that patents and other IPRs restrict access to environmentally-sound technologies, since these technologies are mostly in sectors based on mature technologies where numerous substitutes among global competitors are available (Maskus 2010). This might, however, change in the future – for instance, with new technologies based on plants, via biotechnologies and synthetic fuels (Maskus 2010), for which Correa et al. (2020) already find some evidence.

There is also literature suggesting that weak IPR regimes have a 'strong and negative impact on the international diffusion of patented knowledge' (Dechezleprêtre et al. 2013; Glachant and Dechezleprêtre 2017). Also, patents may support market transactions in technology, including international technology transfer, especially to middle-income countries and larger developing countries (Maskus 2010; Hall and Helmers 2019) but least-developed countries may be better served by building capacity to absorb and implement technology (Hall and Helmers 2010; Maskus 2010; Sanni et al. 2016; Glachant and Dechezleprêtre 2017). It is also argued that it is not even clear that the patent system as it exists today is the most appropriate vehicle for encouraging international access (Hall and Helmers 2010; Maskus 2010; Sanni et al. 2016; Glachant and Dechezleprêtre 2017). Given the large variation in perspectives on the role of IPRs in technology transfer, there is a need for more evidence and analysis to better understand if, and under what conditions, IPR may hinder or promote technology transfer (TEC 2012).

In terms of ways forward to meet the challenge of climate change, different suggestions are made in the context of IPR that can help to further improve international technology transfer of climate mitigation technologies, including through the Trade-Related Aspects of Intellectual Property Rights (TRIPS) Agreement, by making decisions on IPR to developing countries on a case-by-case basis, by developing countries experimenting more with policies on IPR protection, or through brokering or patent-pooling institutions (Littleton 2009; Maskus and Reichman 2017; Dussaux et al. 2018). Others also suggest that distinctions among country groups be made on the basis of levels of technological and economic development, with least-developed countries getting particular attention (Zhuang 2017; Abbott 2018).

expenditures are rising, the aims were not met by 2020 (Myslikova and Gallagher 2020). Gross et al. (2018) caution against too much focus on R&D efforts for energy technologies to address climate change, including for Mission Innovation. They argue that, given the timescales of commercialisation, developing new technologies now would mean they would be commercially too late for addressing climate change. Huh and Kim (2018) discuss two 'knowledge and technology transfer' projects that were eventually not pursued beyond the feasibility study phase due to cooperation and commitment problems between national and local governments, and they highlight the need for ownership and engagement of local residents and recipient governments.

Intellectual property rights (IPR) regimes (Box 16.9) can be an enabler or a barrier to energy transition. For more background on IPR and impact on innovation, see Section 16.4.6.

### 16.5.4 Emerging Ideas for International Technology Transfer and Cooperation

As with the broader innovation literature (Section 16.3), and drawing on such literature, there has been an emergence of a greater understanding of, and emphasis on, the role of innovation systems (at national, sectoral, and technological levels) as a way to help developing countries with the climate technology transition (TEC 2015; Ockwell and Byrne 2016). This has given rise to several proposals, discussed here and summarised in Figure 16.3.

Enhancing deployment and diffusion of climate technologies in developing countries would require a variety of actors with sufficient capabilities (*robust evidence*, *medium agreement*) (Kumar et al. 1999; Sagar et al. 2009; Ockwell et al. 2018). This may include strengthening existing actors (Malhotra et al. 2021), supporting science, technology, and innovation-based start-ups to meet social goals (Surana et al. 2020b), and developing entities and programmes that are intended to address specific gaps relating to technology development and deployment (Sagar et al. 2009; Ockwell et al. 2018).

There is also an increasing emphasis on the relevance of participative social innovation, local grounding and policy learning as a replacement of the expert-led technological change (Chaudhary et al. 2012; Disterheft et al. 2015; Kowarsch et al. 2016). Others have suggested a shift to international innovation cooperation rather than technology transfer, which implies a donor-recipient relationship. The notion of innovation cooperation also makes more explicit the focus on innovation processes and systems (Pandey et al. 2021). A broad transformative agenda therefore proposes that contemporary societal challenges are complex and multivariegated in scope and will require the actions of a diverse set of actors to formulate and address the

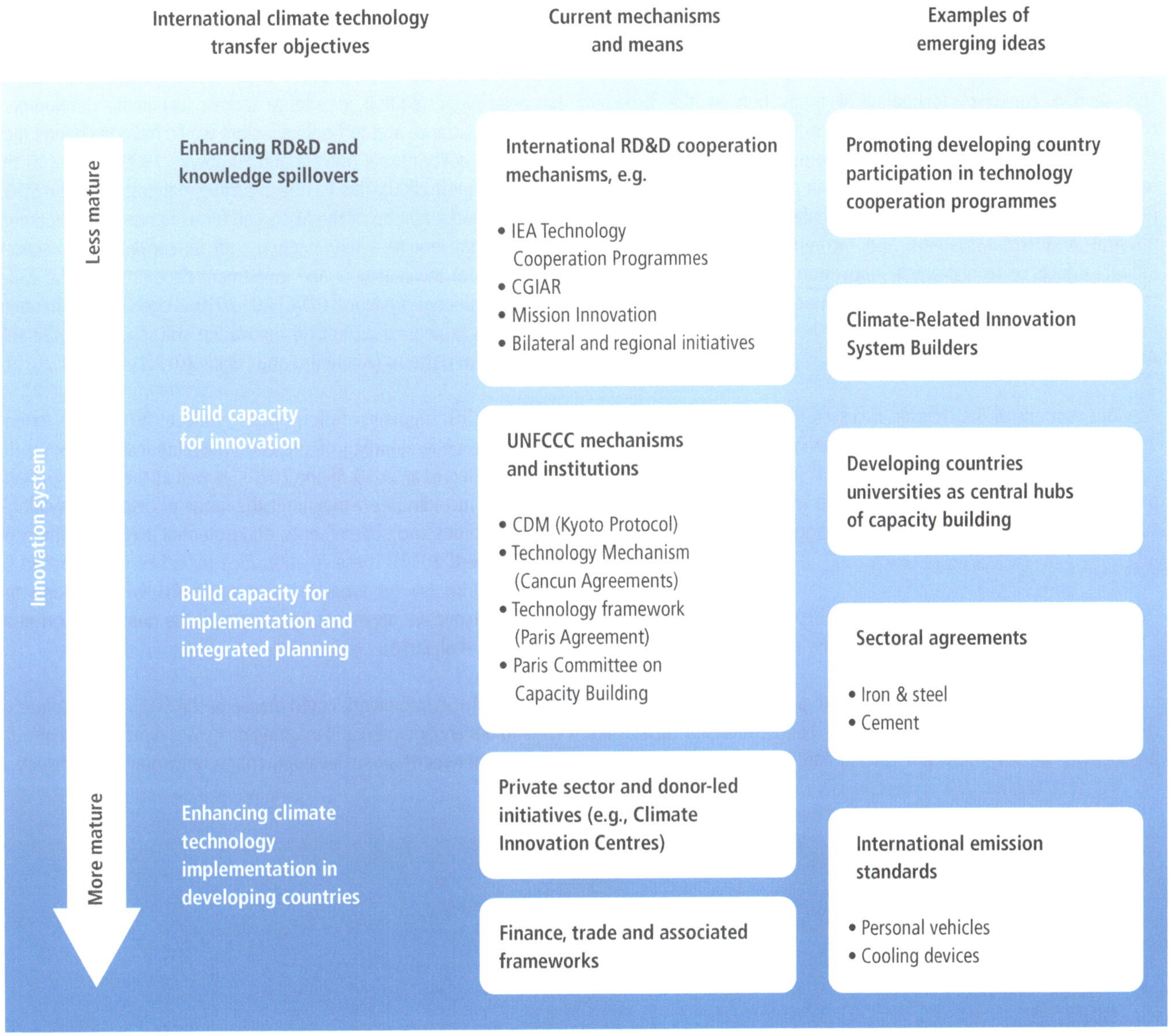


**Figure 16.3 | Examples of recent mechanisms and emerging ideas (right column) in relation to level of maturity of the national or technological innovation system, objectives of international climate technology transfer efforts and current mechanisms and means.** Sources: Sagar (2009); Ockwell and Byrne (2016); Khan et al. (2020); Oberthür et al. (2021).

policy, implying that social, institutional and behavioural changes next to technological innovations are the possible solutions (Geels 2004) (see also Cross-Chapter Box 12 in this chapter).

Several authors have proposed new mechanisms for international cooperation on technology. Ockwell and Byrne (2016) argue that a role for the UNFCCC Technology Mechanism could be to support Climate Relevant Innovation-system Builders (CRIBs) in developing countries, institutions locally that develop capabilities that 'form the bedrock of transformative, climate-compatible, technological change and development'. Khan et al. (2020) propose a specific variant with universities in developing countries serving as 'central hubs' for capacity building to implement the NDCs as well as other climate policy and planning instruments; they also suggest that developing countries outline their capacity-building needs more clearly in their NDCs.

Building on an earlier discussion of technology-oriented and sectoral agreements (Meckling and Chung 2009) and the potential for international cooperation in energy-intensive industry (Åhman et al. 2017), where deep emission reduction measures require transformative changes (Chapter 11), Oberthür et al. (2021) propose that that a way forward for the global governance for energy-intensive industry could be through sub-sector 'clubs' that include governmental, private and societal actors (Oberthür et al. 2021).

Figure 16.3 summarises examples of emerging ideas for international cooperation on climate technology, their relation to the objectives and existing efforts, and the level of development of the innovation system around a technology (Hekkert et al. 2007; Bergek et al. 2008) or in nations (Lundvall et al. 2009).

## 16.6 Technological Change and Sustainable Development

This section considers technological innovation in the broader context of sustainable development, recognising that technological change happens within social and economic systems, and therefore technologies are conceived and applied in relation to those systems (Grübler 1998). Simplifications of complex interactions between physical and social systems and incomplete knowledge of the indirect effects of technological innovation may systematically lead to underestimation of environmental impacts and overestimation of our ability to mitigate climate change (Hertwich and Peters 2009; Arvesen et al. 2011).

Previous sections of the chapter discussed how a systemic approach, appropriate public policies and international cooperation on innovation can enhance technological innovation. This section provides more details on how innovation and technological change, sustainable development and climate change mitigation intertwine.

### 16.6.1 Linking Sustainable Development and Technological Change

Sustainable development and technological change are deeply related (UNCTAD 2019). Technology has been critical for increasing productivity as the dominant driving force for economic growth. Also, the concentration of technology in few hands has boosted consumption of goods and services which are not necessarily aligned with the Sustainable Development Goals (SDGs) (Walsh et al. 2020). It has been suggested that, in order to address sustainable development challenges, science and technology actors would have to change their relation to policymakers (Ravetz and Funtowicz 1999) as well as the public (Jasanoff 2003). This has been further elaborated for the SDGs. The scale and ambition of the SDGs call for a change in development patterns that require a fundamental shift in: current best practices; guidelines for technological and investment decisions; and the wider socio-institutional systems (UNCTAD 2019; Pegels and Altenburg 2020). This is needed as not all innovation will lead to sustainable development patterns (Altenburg and Pegels 2012; Lema et al. 2015).

Current SDG implementation gaps reflect, to some extent, inadequate understanding of the complex relationships among the goals (Waiswa et al. 2019; Skene 2020), as well as their synergies and trade-offs, including how they limit the range of responses available to communities and governments, and potential injustices (Thornton and Comberti 2017). These relationships have been approached by focusing primarily on synergies and trade-offs while lacking the holistic perspective necessary to achieve all the goals (Nilsson et al. 2016; Roy et al. 2018).

A more holistic framework could envisage the SDGs as outcomes of stakeholder engagement and learning processes directed at achieving a balance between human development and environmental protection

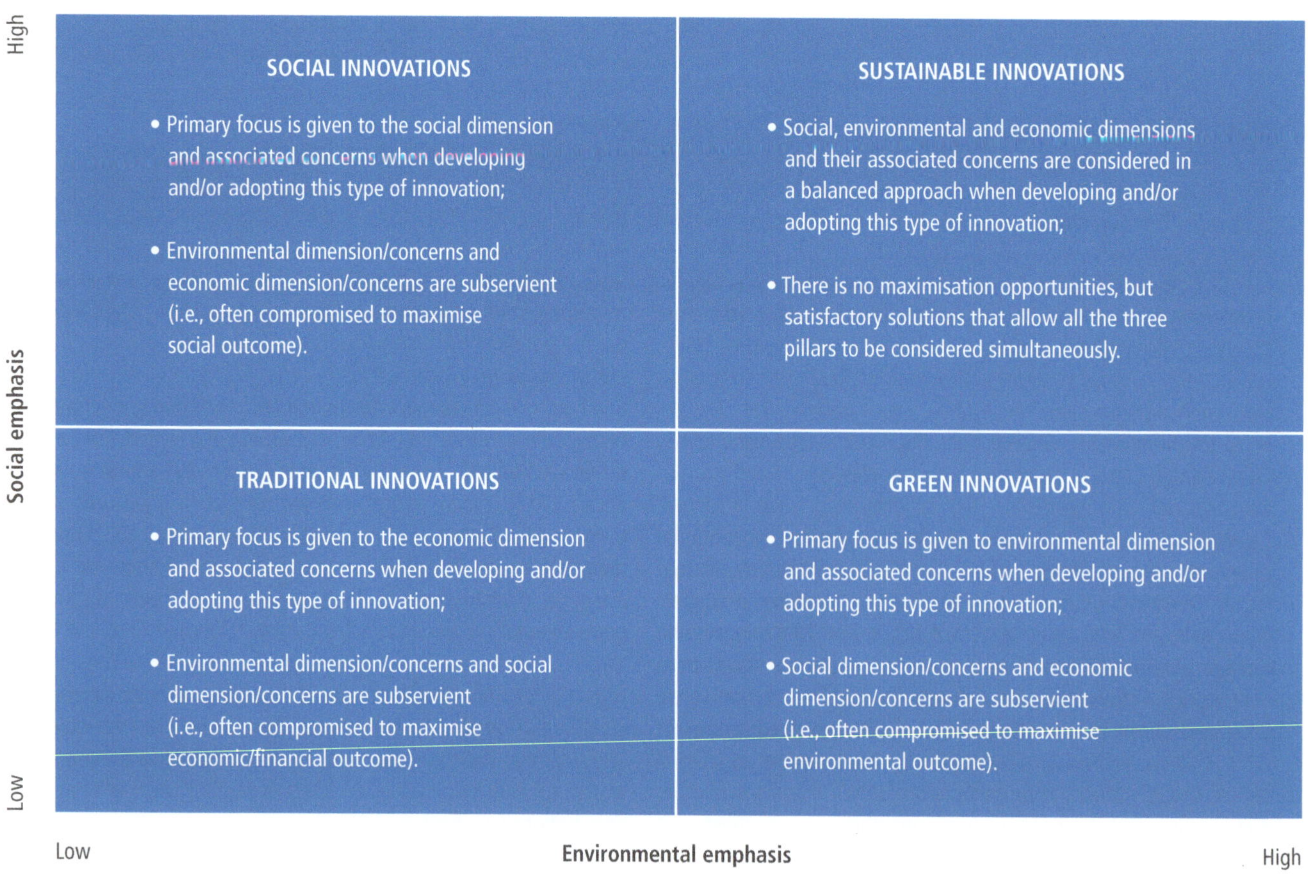


**Figure 16.4 | Considerations and typology of innovations for sustainable development.** Source: Silvestre and Ţîrcă (2019).

(Gibbons 1999; Jasanoff 2003), to the extent that the two can be separated. From a science, technology and innovation perspective, Fu et al. (2019) distinguish three categories of SDGs. The first category comprises those SDGs representing essential human needs for which inputs that put pressure on sustainable development would need to be minimised. These include Zero hunger (SDG 2), Clear water and sanitation (SDG 6) and Affordable and clean energy (SDG 7) resources, which continue to rely on production technologies and practices that are eroding ecosystem services, potentially hampering the realisation of SDGs 15 (Life on land) and 14 (Life below water) (Díaz et al. 2019). The second category includes those related to governance and which compete with each other for scarce resources, such as Industry, innovation and infrastructure (SDG 9) and Climate action (SDG 13), which require an interdisciplinary perspective. The third category are those that require maximum realisation, include No poverty (SDG 1), Quality education (SDG 4) and Gender equality (SDG 5) (Fu et al. 2019).

Resolving tensions between the SDGs requires adoption and mainstreaming of novel technologies that can meet needs while reducing resource waste and improving resource-use efficiency, and acknowledging the systemic nature of technological innovation, which involves many levels of actors, stages of innovation and scales (Anadon et al. 2016b). Changes in production technology have been found effective to overcome trade-offs between food and water goals (Gao and Bryan 2017). Innovative technologies at the food, water and energy nexus are transforming production processes in industrialised and developing countries, such as developments in agrivoltaics, which is co-development of land for agriculture and solar with water conservation benefits (Barron-Gafford et al. 2019; Lytle et al. 2020; Schindele et al. 2020), and other renewably powered low- to zero-carbon food, water and energy systems (He et al. 2019). Silvestre and Țîrcă (2019) indicate that maximising both social and environmental aims is not possible, but that sustainable innovations include satisfactory solutions for social, environmental and economic pillars (Figure 16.4).

There is evidence that technological changes can catalyse implementation of the reforms needed to the manner in which goods and services are distributed among people (Fu et al. 2019). A recently developed theoretical framework based on a capability approach (CA) has been used to evaluate the quality of human life and the process of development (Haenssgen and Ariana 2018). Variations of the CA have been applied to exploratory studies of the link between technological change, human development, and economic growth (Mayer 2001; Mormina 2019). This suggests that the transformative potential of technology as an enabling condition is not intrinsic, but is assigned to it by people within a given technological context. A failure to recognise and account for this property of technology is a root cause of many failed attempts at techno-fixing sustainable development projects (Stilgoe et al. 2013; Fazey et al. 2020).

The basic rationale for governance of technological change is the creation and maintenance of an enabling environment for climate and SDG-oriented technological change (Avelino et al. 2019). Such an environment poses high demands on governance and policy to coordinate with actors and provide a direction for innovation and technological change. Cross-Chapter Box 12 illustrates how the dynamics of socio-technical transitions and shifting development pathways towards sustainable development offer options for policymakers and other actors to accelerate the system transitions needed for both climate change mitigation and sustainable development. Governance interventions to implement the SDGs will need to be operationalised at sub-national, national and global levels and support integration of resource concerns in policy, planning and implementation (UNEP 2015; Williams et al. 2020).



## Cross-Chapter Box 12 | Transition Dynamics

**Authors:** Anthony Patt (Switzerland), Heleen de Coninck (the Netherlands), Xuemei Bai (Australia), Paolo Bertoldi (Italy), Sarah Burch (Canada), Clara Caiafa (Brazil/the Netherlands), Felix Creutzig (Germany), Renée van Diemen (the Netherlands/United Kingdom), Frank Geels (United Kingdom/the Netherlands), Michael Grubb (United Kingdom), María Josefina Figueroa Meza (Venezuela/Denmark), Şiir Kilkiş (Turkey), Jonathan Köhler (Germany), Catherine Mitchell (United Kingdom), Lars J. Nilsson (Sweden), Patricia Perkins (Canada), Yamina Saheb (France/Algeria), Harald Winkler (South Africa)

### Introduction

Numerous studies suggest that transformational changes would be required in many areas of society if climate change is to be limited to 2°C warming or less. Many of these involve shifts to low-carbon technologies, such as renewable energy, which typically involve changes in associated regulatory and social systems; others more explicitly concern behavioural shifts, such as towards plant-based diets or cleaner cooking fuels, or, at the broadest level, a shift in development pathways. Chapter 1 establishes an analytic framework focusing on transitions, which chapters 5, 13, 14, 15 and 16 further develop. In this Cross-Chapter Box, we provide a complementary overview of the dynamics of different kinds of transformational changes for climate mitigation and sustainable development. We first focus on insights from socio-technical transitions approaches, and then expand to broader system transitions.

### Dynamics of socio-technical transitions

A large volume of literature documents the processes associated with transformational changes in technology and the social systems associated with their production and use (Geels 2019; Köhler et al. 2019). Transformational technological change typically goes hand in hand with shifts in knowledge, behaviour, institutions, and markets (Geels and Schot 2010; Markard et al. 2012); stickiness in these

*Cross-Chapter Box 12 (continued)*

factors often keeps society 'locked in' to those technologies already in widespread use, rather than allowing a shift to new ones – even those that offer benefits (David 1985; Arthur 1994). Exceptions often follow consistent patterns (Geels 2002; Unruh 2002); since AR5 a growing number of scholars have suggested using these insights to design more effective climate policies and actions (Geels et al. 2017). Chapter 1 (Section 1.7 and Figure 1.6) represents technology diffusion and a corresponding shift in policy emphasis as a continuous process; it is also useful to identify a sequence of distinct stages that typically occur, associating each stage with a distinct set of processes, challenges, and effective policies (Patt and Lilliestam 2018; Victor et al. 2019). Consistent with elsewhere in this report (Section 5.5.2 and Supplementary Material 5.5.3 in Chapter 5, and Section 16.3 in Chapter 16), Cross-Chapter Box 12 Figure 1 elaborates on four distinct stages: it portrays these as occurring in a cycle, recognising that even transformative technologies will eventually be replaced with newer ones.

The *emergence* stage is marked by experimentation, innovation in the laboratory, and demonstration in the field, to produce technologies and system architectures (Geels 2005). By its very nature, experimentation includes both successes and failures, and implies high risks. Because of these risks, especially in the case of fundamentally new technologies, government funding for research, development and demonstration (RD&D) projects is crucial to sustaining development (Mazzucato 2015b).

The second stage is *early adoption*, during which successful technologies jump from the laboratory to limited commercial application (Pearson and Foxon 2012). Reaching this stage is often described as crossing the 'Valley of Death', because the cost/performance ratio

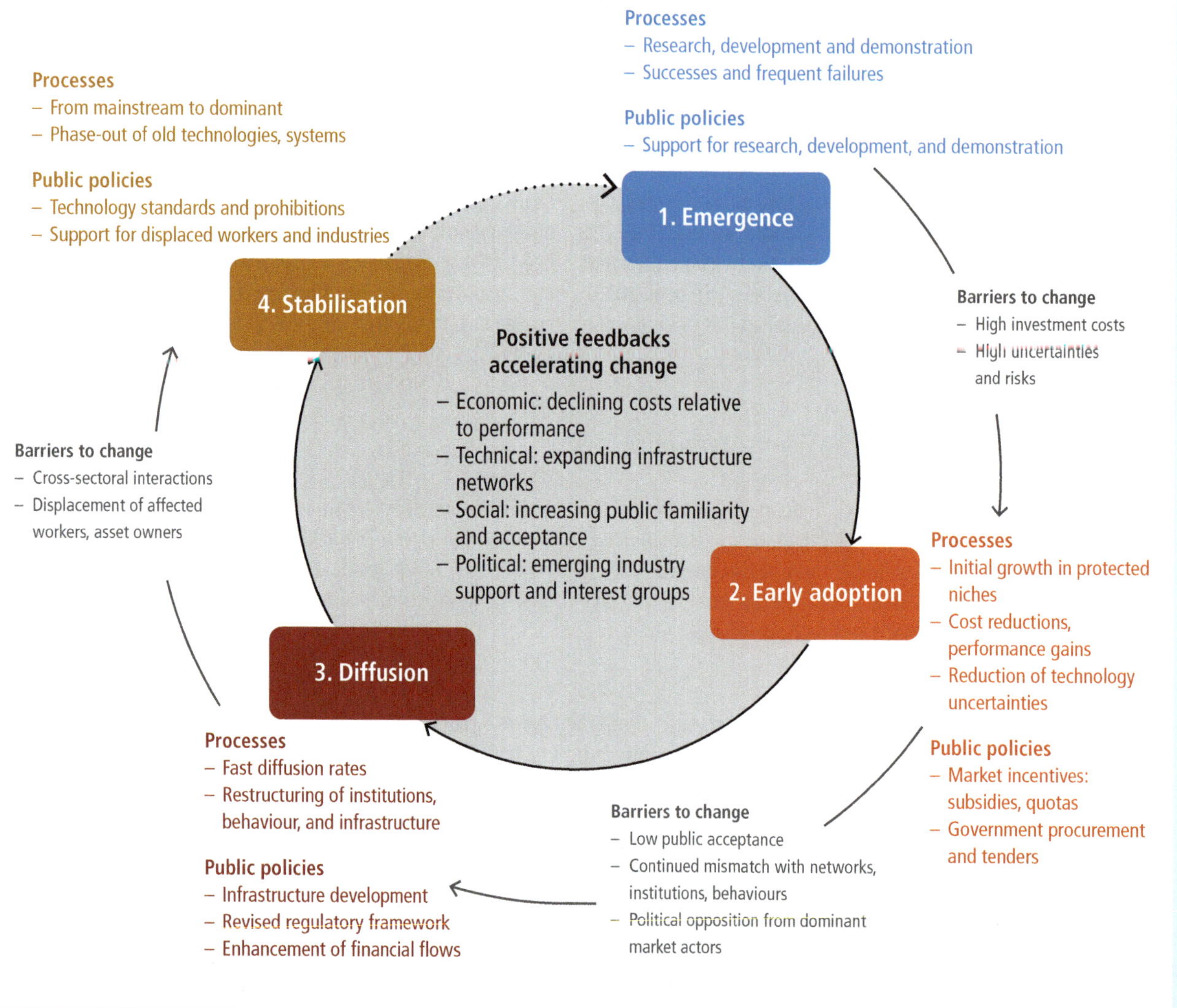


**Cross-Chapter Box 12, Figure 1 | Stages of socio-technical transition processes.**

*Cross-Chapter Box 12 (continued)*

for these new market entrants is too low for them to appear viable to investors (Murphy and Edwards 2003). A key process in the early adoption phase is induced innovation, a result of incremental improvements in both design and production processes, and of mass-production of a growing share of key components (Nemet 2006; Grubb et al. 2021). There is diversity across classes of technologies, and learning tends to occur faster for technologies that are modular (Wilson et al. 2020) – such as photovoltaics – and slower for those that require site- or context-specific engineering, such as in the shift to low-carbon materials production (Malhotra and Schmidt 2020). Public policies that create a secure return on investment for project developers can lead to learning associated with industry expansion (Chapter 16, Figure 16.1); typically these are economically and politically viable when they promote growth within a market niche, causing little disruption to the mainstream market (Roberts et al. 2018). Direct support mechanisms are effective, including cross-subsidies (such as feed-in tariffs) and market quotas (such as renewable portfolio standards) (Geels et al. 2017b; Patt and Lilliestam 2018; and Chapter 9 for assessment of early adoption policies in the building sector). The value of these policies is less in their immediate emissions reductions, but more in generating the conditions for self-sustaining transformational change to take place as technologies later move from niche to mainstream (Hanna and Victor 2021).



The third stage, *diffusion*, is where niche technologies become mainstream, with accelerating diffusion rates (Sections 1.7 and 16.4), and is marked by changes to the socio-technical 'regime', including infrastructure networks, value chains, user practices, and institutions. This stage is often the most visible and turbulent, because more widespread adoption of a new technology gives rise to structural changes in institutions and actors' behaviour (e.g., increased adoption of smartphones to new payment systems and social media), and because when incumbent market actors become threatened, they often contest policies promoting the new technologies (Köhler et al. 2019). In the diffusion stage, policy emphasis is shifted from financial support during the early adoption stage, towards supporting regime-level factors needed to sustain, or cope with, rapid and widespread diffusion (Markard 2018). These factors and policies are context specific. For example, Patt et al. (2019) document that the policies needed to expand residential charging networks for electric vehicles depend on the local structure of the housing market.

The fourth stage is *stabilisation*, in which the new technologies, systems, and behaviours are both standardised and insulated from rebound effects and backsliding (Andersen and Gulbrandsen 2020). Sectoral bans on further investment in high-carbon technologies may become politically feasible at this point (Breetz et al. 2018; Economidou et al. 2020). The decline of previously dominant products or industries can lead to calls for policymakers to help those negatively affected, enabling a just transition (McCauley and Heffron 2018; Newell and Simms 2020). Political opposition to the system reconfiguration that comes with integration and stabilisation can also be overcome by offering incumbent actors an attractive exit strategy (de Gooyert et al. 2016).

Because different sectors are at different stages of low-carbon transitions, and because the barriers that policies need to address are stage- and often context-specific, effective policies stimulating socio-technical transitions operate primarily at the sectoral level (Victor et al. 2019). This is particularly the case during early adoption, where economic barriers predominate; during diffusion, policies that address regime-level factors often need to deal with cross-sectoral linkages and coupling, such as those between power generation, transportation, and heating (Patt 2015; Bloess 2019; Fridgen et al. 2020). The entire cycle can take multiple decades. However, later stages can go faster by building on the earlier stages that have taken place elsewhere. For example, early RD&D into wind energy took place primarily in Denmark, was followed by early adoption in Denmark, Germany, and Spain, before other countries, including the USA, India, and China, leapfrogged directly to the diffusion stage (Chaudhary et al. 2015; Dai and Xue 2015; Lacal-Arántegui 2019). A similar pattern played out for solar power (Nemet 2019). International cooperation, geared towards technology transfer, capacity and institution-building, and finance, can help ensure that developing countries leapfrog to low-carbon technologies that have undergone commercialisation elsewhere (Adenle et al. 2015; Fankhauser and Jotzo 2018) (see also Chapter 5, Box 5.9, Chapter 15, Section 15.5, and Section 16.5 in this chapter).

This report contains numerous examples of the positive feedbacks in the centre of Cross-Chapter Box 12, Figure 1, predominantly arising during the early adoption and diffusion stages, and leading to rapid or unexpected acceleration of change. For example, public acceptance of meat alternatives leads to firms improving the products, increasing political and economic feedbacks (Section 5.4 and Box 5.5). Declining costs in solar and wind cause new investment in the power-generation sector being dominated by those technologies, leading to increased political support and further cost reductions (Chapter 6). In buildings (Chapter 9) and personal mobility (Chapter 10), low-carbon heating systems and electric vehicles are gaining public acceptance, leading to improved infrastructure and human resources, more employment in those sectors, and behavioural contagion. Some have argued that technologies cross societal tipping points on account of these feedbacks (Obama 2017; Sharpe and Lenton 2021).

**Dynamics between enabling conditions for system transitions**

Abson et al. (2017) argue that it is possible to make use of 'leverage points' inherent in system dynamics in order to accelerate sustainability transitions. Otto et al. (2020) argue that interventions geared towards the social factors driving change can 'activate

*Cross-Chapter Box 12 (continued)*

contagious processes' leading to the transformative changes required for climate mitigation. These self-reinforcing dynamics involve the interaction of enabling conditions, including public policy and governance, institutional and technological innovation capacity, behaviour change, and finance. For example, Mercure et al. (2018) simulated financial flows into fossil-fuel extraction, and showed how investors taking into account transition risk in combination with technological innovation would lead to the enhancement of investments in low-carbon assets and further enhanced innovation. As another example, behaviour, lifestyle, and policy can also initiate demand-side transitions (Tziva et al. 2020) (Chapter 5), such as with food systems (Rust et al. 2020) (Section 7.4.5), and can contribute to both resilience and carbon storage (Sendzimir et al. 2011) (Box 16.5).



In the urban context, the concept of sustainability experiments has been used to examine innovative policies and practices adopted by cities that have significant impact on transition towards low-carbon and sustainable futures (Bai et al. 2010; Castán Broto and Bulkeley 2013). Individual innovative practices can potentially be upscaled to achieve low-carbon transition in cities (Peng and Bai 2018), leading to a process of broadening and scaling innovative practices in other cities (Peng et al. 2019). Such sustainability experiments give rise to new actor networks, which in some cases may accelerate change, and in others may lead to conflict (Bulkeley et al. 2014). As in the diffusion phase in Cross-Chapter Box 12, Figure 1, contextual factors play a strong role. Examining historical transitions to cycling across European cities, Oldenziel et al. (2016) found that contextual factors, including specific configurations of actors, can lead to very different outcomes. Kraus and Koch (2021) found a short-term social shock – such as the COVID-19 crisis – to lead to differential increases in cycling behaviour, contingent on other enabling conditions.

**Linking system dynamics to development pathways and broader societal goals**

Transition dynamics insights can be broadened to shifting development pathways. Development paths are characterised by particular sets of interlinking regime rules and behaviours, including inertia and cascading effects over time, and are reinforced at multiple levels, with varied capacities and constraints on local agency occurring at each level (Burch et al. 2014) (Cross-Chapter Box 5 in Chapter 4). This is also observed by Schot and Kanger (2018), who identify a needed change in a 'meta-regime', crossing sectoral lines in linking value chains or infrastructure and overall development objectives. In the context of the UN climate change regime, international cooperation can bring together such best practices and lessons learnt (Adenle et al. 2015; Pandey et al. 2021). This is especially relevant for developing countries, which often depend on technologies and financial resources from abroad, witnessing their pace and direction influenced by transnational actors (Marquardt et al. 2016; Bhamidipati et al. 2019), and benefitting little in terms of participating in high value-added activities (Whittaker et al. 2020).

System transitions differ according to context, such as across industrialised and developing countries (Ramos-Mejía et al. 2018), and within countries. Lower levels of social capital and trust negatively impact niche commercialisation (Lepoutre and Oguntoye 2018). In contexts of poverty and inequality, stakeholders' – including users' – capabilities for meaningful participation are limited, and transition outcomes can end up marginalising or further excluding social groups (Osongo and Schot 2017; Hansen et al. 2018). Many studies of transitions in developing countries make note of the importance of innovation in the informal sector (Charmes 2016) (Box 5.10 in Chapter 5). Facilitating informal sector access to renewable energy sources, safe and sustainable buildings, and finance can advance low-carbon transitions (McCauley et al. 2019; Masuku and Nzewi 2021). On the contrary, disregarding its importance can result in misleading or ineffective innovation and climate strategies (Maharajh and Kraemer-Mbula 2010; Mazhar and Ummad 2014; de Beer et al. 2016; Masuku and Nzewi 2021).

Policies shifting innovation in climate-compatible directions can also reinforce other development benefits, for instance better health, increased energy access, poverty alleviation and economic competitiveness (Deng et al. 2018; IPCC 2018a; Karlsson et al. 2020). Development benefits, in turn, can create feedback effects that sustain public support for subsequent policies, and hence help to secure effective long-term climate mitigation (Geels 2014; Meckling et al. 2015; Schmidt and Sewerin 2017; Breetz et al. 2018), increasing legitimacy of environmental sustainability actions (Hansen et al. 2018; Herslund et al. 2018; van Welie and Romijn 2018) and addressing negative socio-economic impacts (Deng et al. 2018; McCauley and Heffron 2018; Eisenberg 2019; Henry et al. 2020).

**Summary and gaps in knowledge**

Strategies to accelerate climate mitigation can be most effective at accelerating and achieving transformative change when they are synchronised with transition processes in systems. They address technological stage characteristics, take advantage of high-leverage intervention points, and respond to societal dynamics (Abson et al. 2017; Geels et al. 2017; Köhler et al. 2019). Gaps in knowledge remain on how to tailor policy mixes, the interaction of enabling conditions, the generalisability of socio-technical transition insights to other types of systems, and how to harness these insights to better shift development pathways.

### 16.6.2 Sustainable Development and Technological Innovation: Synergies, Trade-offs and Governance

#### 16.6.2.1 Synergies and Trade-offs

Policies that shift innovation in climate compatible directions can promote other development benefits, for instance, better health, increased energy access, poverty alleviation and economic competitiveness (Deng et al. 2018) (Cross-Chapter Box 12). Economic competitiveness co-benefits can emerge as climate mitigation policies trigger innovation that can be leveraged for promoting industrial development, job creation and economic growth, both in terms of localising low-emission energy technologies value chains as well as increased energy efficiency and avoided carbon lock-ins (Section 16.4). However, without adequate capabilities, co-benefits at the local level would be minimal, and they would probably materialise far from where activities take place (Ockwell and Byrne 2016; Vasconcellos and Caiado Couto 2021). Innovation and technological change can also empower citizens. Grass-roots innovation promotes the participation of grass-roots actors, such as social movements and networks of academics, activists and practitioners, and facilitate experimenting with alternative forms of knowledge creation (Seyfang and Smith 2007; UNCTAD 2019). Examples of ordinary people and entrepreneurs adopting and adapting technologies to local needs to address locally defined needs have been documented in the development literature (van Welie and Romijn 2018) (Box 16.10). Digital technologies can empower citizens and communities in decentralised energy systems, contributing not only to a more sustainable but also to a more democratic and fairer energy system (Van Summeren et al. 2021) (Section 5.4 in Chapter 5, and Cross-Chapter Box 11 in this chapter).

Therefore, even though science, technology and innovation is an explicit focus of SDG 9, it is an enabler of most SDGs (UNCTAD 2019). Striving for synergies between innovation and technological change for climate change mitigation with other SDGs can help to secure effective long-term climate mitigation, as development benefits can create feedback effects that sustain public and political support for subsequent climate mitigation policies (Geels 2014; Meckling et al. 2015; Cross-Chapter Box 12 in this chapter). However, innovation is not always geared to sustainable development – for instance, firms tend to know how to innovate when value chains are left intact (Hall and Martin 2005), which is usually not the case in systemic transitions.

A comprehensive study of these effects distinguishes among '... anticipated-intended, anticipated-unintended, and unanticipated-unintended consequences' (Tonn and Stiefel 2019). Theoretical and empirical studies have demonstrated that unintended consequences are typical of complex adaptive systems, and while a few are predictable, a much larger number are not (Sadras 2020). Even when unintended consequences are unanticipated, they can be prevented through actor responses, for instance, rebound effects following the introduction of energy-efficient technologies. Other examples of unintended consequences include worse-than-expected physical damage to infrastructure and resistance from communities in the rapidly growing ocean renewable energy sector (Quirapas and Taeihagh 2020), and gaps between expected and actual performance of building-integrated photovoltaic (BIPV) technology (Boyd and Schweber 2018; Gram-Hanssen and Georg 2018). In the agricultural sector, new technologies and associated practices that target the fitness of crop pests have been found to favour resistant variants. Unintended consequences of digitalisation are reported as well (Lynch et al. 2019) (Cross-Chapter Box 11 in this chapter).

Innovation and climate mitigation policies can also have negative socio-economic impacts, and not all countries, actors and regions around the world benefit equally from rapid technological change (Deng et al. 2018; McCauley and Heffron 2018; Eisenberg 2019; UNCTAD 2019; Henry et al. 2020). In fact, socio-technical transitions often create winners and losers (Roberts et al. 2018). Technological change can reinforce existing divides between women and men, rural and urban populations, and rich and poor communities: older workers displaced by technological change will not qualify for jobs if they were unable to acquire new skills; weak educational systems may not prepare young people for emerging employment opportunities; and disadvantaged social groups, including women in many countries, often have fewer opportunities for formal education (McCauley and Heffron 2018; UNCTAD 2019). That is a risk regarding technological change for climate change mitigation, as emerging evidence suggests that the energy transition can create jobs and productivity opportunities in the renewable energy sector, but will also lead to job losses in fossil fuel and exposed sectors (Le Treut et al. 2021). At the same time, these new jobs may use more intensively high-level cognitive and interpersonal skills compared to regular, traditional jobs, requiring higher levels of human capital dimensions such as formal education, work experience and on-the-job training (Consoli et al. 2016). Despite the empowerment potentials of decentralised energy systems, not all societal groups are equally positioned to benefit from energy community policies, with issues of energy justice taking place within initiatives, between initiatives and related actors, as well as beyond initiatives (Calzadilla and Mauger 2018; van Bommel and Höffken 2021).

The opportunities and challenges of technological change can also differ within country regions and between countries (Garcia-Casals et al. 2019). Within countries, Vasconcellos and Caiado Couto (2021) show that, in the absence of policies and capacity-building activities which promote local recruiting, a significant part of total benefits of wind projects, especially high-income jobs and high value-added activities, is captured by already higher-income regions. Between countries, developing countries usually have lower innovation capabilities, which means they need to import low-emission technology from abroad and are also less able to adapt these technologies to local conditions and create new markets and business models. This can lead to external dependencies and limit opportunities to leverage economic benefits from technology transfer (Section 16.5.1).

This means that, in countries below the technological frontier, the contribution of technological change to climate change mitigation can happen primarily through the adoption and less through the development of new technologies, which can reduce potential economic and welfare benefits from rapid technological change (UNCTAD 2019). The adoption of consumer information and communication technology (ICT) (Baller et al. 2016) or renewable energy technology (Lema et al. 2021) cannot bring least-developed economies close to the technological frontier without appropriate technological capabilities in other sectors, and an enabling innovation system (Ockwell and Mallett 2012; Sagar and Majumdar 2014; Ockwell et al. 2018; UNCTAD 2019; Malhotra et al. 2021; Vasconcellos and Caiado Couto 2021). It has been argued widely that both hard and soft infrastructure, as well as appropriate policy frameworks and capability building, would facilitate developing countries' engagement in long-term technological innovation and sustainable industrial development, and eventually in achieving the SDGs (Ockwell and Byrne 2016; Altenburg and Rodrik 2017; UNCTAD 2019).

#### 16.6.2.2 Challenges to Governing Innovation for Sustainable Development

Dominant economic systems and centralised governance structures continue to reproduce unsustainable patterns of production and consumption, reinforcing many economic and governance structures from local through national and global scales (Johnstone and Newell 2018). Technological change, as an inherently complex process (Funtowicz 2020), poses governance challenges (Bukkens et al. 2020) requiring social innovation (Repo and Matschoss 2019) (Section 5.6 and Chapter 13).

Prospects for effectively governing SDG-oriented technological transformations require, at a minimum, balanced views and new tools for securing the scientific legitimacy and credibility to connect public policy and technological change in society (Jasanoff 2018; Sadras 2020). Many frameworks of governance have been proposed, such as reflexive governance (Voss et al. 2006), polycentric governance (Ostrom 2010), collaborative governance (Bodin 2017), adaptive governance (Munene et al. 2018) and transformative governance (Rijke et al. 2013; Westley et al. 2013) (Chapters 13 and 14).

A particular class of barriers to the development and adoption of new technologies comprises entrenched power relations dominated by vested interests that control and benefit from existing technologies (Chaffin et al. 2016; Dorband et al. 2020). Such interests can generate balancing feedbacks within multilevel social-technological regimes that are related to technological lock-in, including allocations of investment between fossil and renewable energy technologies (Unruh 2002; Sagar et al. 2009; Seto et al. 2016).

Weaker coordination and implementation capacity in some developing countries can undermine the ability to avoid trade-offs with other development objectives – such as reinforced inequalities or excessive indebtedness and increased external dependency – and can limit the potential of leveraging economic benefits from technologies transferred from abroad (Section 16.5 and Cross-Chapter Box 12 in this chapter). Van Welie and Romijn (2018) show that, in a low-income setting, the exclusion of some local stakeholders from the decision-making process may undermine sustainability transitions efforts. Countries with high levels of inequality can be more prone to elite capture, non-transparent political decision-making processes, relations based on clientelism and patronage, and no independent judiciary (Jasanoff 2018), although in particular contexts, non-elites manage to exert influence (Moldalieva and Heathershaw 2020). The dominance of incumbents, however, implies that sustainable technological transitions could be achieved without yielding any social and democratic benefits (Hansen et al. 2018). In the cultural domain, a recurrent policy challenge that has been observed in most countries is the limited public support for development and deployment of low-carbon technologies (Bernauer and McGrath 2016). The conventional approach to mobilising such support has been to portray technological change as a means of minimising climate change. Empirical studies show that simply reframing climate policy is highly unlikely to build and sustain public support (Bernauer and McGrath 2016).

Finally, there is a link between social and technological innovation; any innovation is grounded in complex socio-economic arrangements, to which governance arrangements would need to respond (Sections 5.5 and 5.6, Chapter 13, and Cross-Chapter Box 12 in this chapter). Social innovation can contribute to maximising synergies and minimising trade-offs in relation to technological and other innovative practices, but for this to materialise, national, regional and local circumstances need to be taken into account and, if needed, changed. Even in circumstances of high capabilities, the extent that social innovation might help to promote synergies and avoid trade-offs is not easy to evaluate (Grimm et al. 2013).

### 16.6.3 Actions that Maximise Synergies and Minimise Trade-offs Between Innovation and Sustainable Development

Technological innovation may bring significant synergy in pursuing SDGs, but it may also create challenges to the economy, human well-being, and the environment (Schillo and Robinson 2017; Thacker et al. 2019; Walsh et al. 2020). The degree of potential synergies and trade-offs among SDGs differs from country to country and over time (Section 16.6.1.1). These potentials will depend on available resources, geographical conditions, development stage and policy measures. Even though synergies and trade-offs related to technological innovation have received the least attention from researchers (Deng et al. 2018), literature show that higher synergy was found where countries' policies take into account the linkages between sectors (Mainali et al. 2018). For technology innovation to be effective in enhancing synergies and reducing trade-offs, its role and nature in production and consumption patterns, as well as in value chains and in the wider economy, requires clarification. Technology ownership and control together with its current orientation and focus towards productivity, needs to be revised if a meaningful contribution to the implementation of the SDGs is to be achieved in a transformative way (Walsh et al. 2020). Responsible innovation, combining anticipation, reflexivity, inclusion and responsiveness, has been suggested as a framework for conducting innovation (Stilgoe et al. 2013). Also inclusive innovation (Hoffecker 2021) could make sure that unheard voices and interests are included in decision-making, and that methods for this have been implemented in practice (Douthwaite and Hoffecker 2017).

There are several examples of how to maximise synergies and avoid or minimise trade-offs when bringing technological innovation to the ground. When implementing off-grid solar energy in Rwanda, synergies were found between 80 of the 169 SDG targets, demonstrating how mainstreaming off-grid policies and prioritising investment in the off-grid sector can realise human development and well-being, build physical and social infrastructures, and achieve sustainable management of environmental resources (Bisaga et al. 2021). Another example is related to wind power in Northeast of Brazil where the creation of direct and indirect jobs has been demonstrated in areas where capabilities are high, as well as associated improvements in wholesale and retail trade and real estate activities, though this also emphasises the need for capacity development along with international collaboration projects (Vasconcellos and Caiado Couto 2021). Other examples include studies raising awareness on solar energy and women's empowerment (Winther et al. 2018) and recycling and waste (Cross and Murray 2018).

Other actions with the potential to maximise synergies are those related to community or grassroots technological innovation. The importance of the link between technological innovation and community action and its contribution to sustainable development is usually underestimated. Further research is needed on this and, most importantly, its inclusion in the political agenda on sustainable development (Seyfang and Smith 2007). On the other hand, when technological innovation occurs far from where is implemented and participation in the production, and hence training activities of local actors is minimal, co-benefits and synergies among SDGs are limited and usually far below expectations (Bhamidipati and Hansen 2021; Vasconcellos and Caiado Couto 2021). Actions by policymakers that safeguard environmental and social aspects can boost synergies and maximise those co-benefits (Lema et al. 2021). Given that technological change impacts countries, regions and social groups differently, transition policies can be designed to ensure that all regions and communities are able to take advantage of the energy and other transitions (McCauley and Heffron 2018; Henry et al. 2020).



Box 16.10 provides insights on how a systemic approach to technological innovation can contribute to reconcile synergies and trade-offs to achieve sustainable development and mitigation goals.

## Box 16.10 | Agroecological Approaches: The Role of Local and Indigenous Knowledge and Innovation

Major improvements in agricultural productivity have been recorded over recent decades (FAO 2018a). However, progress has also come with social and environmental costs, high levels of greenhouse gas (GHG) emissions, and rising demand for natural resources (UNEP 2013; UNEP 2017; FAO 2018a; Bringezu 2019; Díaz et al. 2019).

Trend analysis indicates that a large share of the global demand for land is projected to be supplied by South America, in particular the Amazon (Lambin and Meyfroidt 2011; TEEB 2018) and Gran Chaco forests (Grau et al. 2015). In developing countries, land use change for satisfying international meat demand is leading to deforestation. In Brazil, the amount of GHGs emitted by the beef cattle sector alone represents 65% of the agricultural sector's emissions and 15% of the country's overall emissions (May 2019).

Agricultural and food systems are complex and diverse; they include traditional food systems, mixed food systems and modern food systems (Pengue et al. 2018). Multiple forms of visible and invisible flows of natural resources exist in global food systems (Pascual et al. 2017; TEEB 2018; IPBES 2019).

Technological practices, management and changes in the food chain could help adapt to climate change, reduce emissions and absorb carbon in soil, thus contributing to carbon dioxide removal (IPCC, 2018, 2019). A range of technologies can be implemented – from highly technological options, such as transgenic crops resistant to drought (González et al. 2019), salt or pesticides (OECD 2011b; Kim and Kwak 2020) or smart and 4.0 agriculture (Klerkx et al. 2019), to more frugal, low-cost technologies such as agroecological approaches adapted to local circumstances (Francis et al. 2003; FAO 2018b). These agroecological approaches are the subject of this box.

For developing countries, agroecological approaches could tackle climate change challenges and food security (WGII-report, Chapter 5, Box 5.10). Small Island Developing States (SIDS) support livelihoods to develop local food value chains that can promote sustainable management of natural resources, preserve biodiversity and help build resilience to climate change impacts and natural disasters (FAO 2019). Other advantages of agroecological practices include their adaptation to different social, economic and ecological environments (Altieri and Nicholls 2017), the fact that they are physical and financial capital-extensive, and are well-integrated with the social and cultural capital of rural territories and local resources (knowledge, natural resources, etc.), without leading to technological dependencies (Côte et al. 2019).

Agroecology is a dynamic concept that has gained prominence in scientific, agricultural and political discourses in recent years (Wezel et al. 2020; Anderson et al. 2021) (Chapter 7, Chapter 5, WGII Box 5.10). Three of the different agroecological approaches are briefly discussed here: agroecological intensification; agroforestry; and biochar use in rice paddy fields.

*Box 16.10 (continued)*

Agricultural intensification provides ways to use land, water and energy resources to ensure adequate food supply while also addressing concerns about climate change and biodiversity (Cassman and Grassini 2020). The term ecological intensification (Tittonell 2014) focuses on biological and ecological processes and functions in agroecosystems. In line with the development of the concept of agroecology, agroecological intensification integrates social and cultural perspectives (Wezel et al. 2015). Agroecological intensification (Mockshell and Villarino 2019) for sub-Saharan Africa aims to address employment and food security challenges (Pretty et al. 2011; Altieri et al. 2015).

Another example of an agroecological approach is agroforestry. Agroforestry provides examples of positive agroecological feedbacks, such as 'the regreening of the Sahel' in Niger. The practice is based on the assisted natural regeneration of trees in cultivated fields, an old method that was slowly dying out, but which innovative public policies (the transfer of property rights over trees from the state to farmers) helped restore (Sendzimir et al. 2011).

Rice paddy fields are a major source of methane. Climate change impacts and adaptation strategies can affect rice production and rice farmers' net income. Biochar use in rice paddy fields has been advocated as a potential strategy to reduce GHG emissions from soils, enhance soil carbon stocks and nitrogen retention, and improve soil function and crop productivity (Mohammadi et al. 2020).

The contributions of indigenous people (Díaz et al. 2019), heritage agriculture (Koohafkan and Altieri 2010) and peasants' agroecological knowledge (Holt-Giménez 2002) to technological innovation offer a wide array of options for management of land, soils, biodiversity and enhanced food security without depending on modern, foreign agricultural technologies (Denevan 1995). In farming agriculture and food systems, innovation and technology based on nature could help to reduce climate change impacts (Griscom et al. 2017). Evidence suggests that there are benefits to integrating tradition with new technologies in order to design new approaches to farming, and that these are greatest when they are tailored to local circumstances (Nicholls and Altieri 2018).

### 16.6.4 Climate Change, Sustainable Development and Innovation

This section gives a synthesis of this chapter on innovation and technology development and transfer, connecting it to sustainable development.

In conjunction with other enabling conditions, technological innovation can support system transitions to limit warming, help shift development pathways, and bring about new and improved ways of delivering goods and services that are essential to human well-being (*high confidence*). At the same time, however, innovation can result in trade-offs that undermine progress on mitigation and towards other SDGs. Trade-offs include negative externalities, such as environmental impacts and social inequalities, rebound effects leading to lower net emission reductions or even increases in emissions, and increased dependency on foreign knowledge and providers (*high confidence*). Digitalisation, for example, holds both opportunity for emission reduction and emission-saving behaviour change, but at the same time causes significant environmental, social and greenhouse gas (GHG) impacts (*high confidence*).

A systemic view of innovation that takes into account the roles of actors, institutions, and their interactions, can contribute to enhanced understanding of processes and outcomes of technological innovation, and to interventions and arrangements that can help innovation. It can also play a role in clarifying the synergies and trade-offs between technological innovation and the SDGs. Effective governance and policy, implemented in an inclusive, responsible and holistic way, could make innovation policy more effective, and avoid and minimise misalignments between climate change mitigation, technological innovation, and other societal goals (*medium evidence, high agreement*).

A special feature is the dynamics of transitions. Like other enabling conditions, technological innovation plays a balancing role – by inhibiting change as innovation strengthens incumbent technologies and practices – and a reinforcing role, by allowing new technologies and practices to disrupt the existing socio-technical regimes (*high confidence*). Appropriate innovation policies can help to better organise innovation systems, while other policies (technology push and demand pull) can provide suitable resources and incentives to support and guide these innovation systems towards societally-desirable outcomes, ensure the innovations are deployed at scale, and direct these dynamics towards system transitions for climate change mitigation, and also towards addressing other SDGs. This means taking into account the full lifecycle or value chain as well as analysis of synergies and trade-offs.

Against this backdrop, international cooperation on technological innovation is one of the enablers of climate action in developing countries on both mitigation and adaptation (*high confidence*). Experiences with international cooperation on technology development and deployment suggest that such activities are most effective when they: are approached as 'innovation cooperation' that engenders a holistic, systemic view of innovation requirements; are an equitable partnership between donors and recipients; and develop local innovation capabilities (*medium evidence, high agreement*).

Chapter 17, in particular Section 17.4, connects technological innovation with other enabling conditions, such as behaviour, institutional capacity and multilevel governance, to clarify the actions that could be taken, holistically and in conjunction, to strengthen and accelerate the system transitions required to limit warming to be in line with the Paris Agreement and to place countries in sustainable development pathways.

## 16.7 Knowledge Gaps

Filling gaps in literature availability, data collection, modelling, application of frameworks and further analysis in several sectors will improve knowledge on innovation and technology development and transfer, including research and development (R&D) to support policymaking in climate change mitigation as well as adaptation. These policies and related interventions need to benefit from data and methodologies for the *ex post* evaluation of their effectiveness.

This section addresses identified knowledge gaps related to: what extent developing countries are represented in studies on innovation and technology development and transfer; national contexts and local innovation capacity; potential and actual contributions of businesses; literature emphasis on mitigation; indicators to assess innovation systems; non-technical barriers for the feasibility of decarbonisation pathways; the role of domestic intellectual property rights (IPR) policy; digitalisation in low-emissions pathways; and Paris Agreement compliance regarding technology and capacity building.

### Representation of developing countries

One of knowledge gaps identified when assessing the literature is on the representation of developing countries in studies on innovation and technology development and transfer. This includes the conceptual core disciplines of the economics of innovation, innovation systems and sustainability transitions. This is true for studies on developing countries, and for authors originating from, or active in, developing country contexts. The evidence of the impact of decarbonisation policy instruments applied to developing countries or Small Island Developing States (SIDS) is limited. Expanding the knowledge base with studies that focus on developing countries would not only allow for testing whether the theories (developed by predominantly by developed-country researchers for industrialised countries) hold in developing country contexts, but also yield policy insights that could help both domestic and international policymakers working on climate-related technology cooperation.

### National contexts and local innovation capacity

While a growing body of literature has shown how technology characteristics and complexity, national context and innovation capacity can influence the capacity of a country's innovation ecosystem as a result of incentive and attraction policies, more research is needed to help prioritise and design policies in different national contexts. Important knowledge gaps need to be filled regarding the impact of 'green' public procurement, lending, 'green' public banking, and building code policies on innovation outcomes.

There is also a superficial understanding of the potential and actual contributions of businesses, educational institutions and socially responsible programmes, particularly in developing countries, as sources of innovation and early adopters of new technologies, and a notable lack of knowledge about indigenous practices.

### Emphasis on mitigation

Current literature has a strong bias to studies originating from and based on developed countries. Also, innovation and technology literature is skewed to mitigation and, specifically, energy. Literature on technology innovation for adaptation is largely missing.

In the area of innovation studies, data are limited on the different indicators used to assess the strength of the innovation system, (even for energy), including global figures on R&D and demonstration spending, also for developing countries, and their effectiveness. There is also a lack of a comprehensive framework and detailed data to assess the strengths of low-emission innovation systems, including interactions among actors, innovation policy implementation, and strength of institutions.

### Indicators to assess innovation systems

Another gap in knowledge remains between the results from energy-climate-economy models and those emerging from systems and sustainability transition approaches, empirical case studies, and the innovation system literature. If this gap is filled, understanding could be improved of the feasibility of decarbonisation pathways in light of the many non-technical barriers to technology deployment and diffusion.

### Non-technical barriers for the feasibility of decarbonisation pathways

In the field of policy instruments, existing evaluations provide insufficient evidence to assess the impact of decarbonisation policy instruments on innovation, as these evaluations mainly focus on environmental or technological effects.

### Domestic IPR policy

The potential positive or negative role of domestic IPR policy in technology transfer to least-developed countries remains unclear as the literature does not show agreement. Moreover, gaps remain in impact evaluations of sub-national green industrial policies, which are of growing importance. The interaction between subnational and national decarbonisation policies to advance innovation would also benefit from further research, particularly in developing countries.

### Digitalisation in low-emissions pathways and digitalisation

The understanding of the role of digitalisation in decarbonisation pathways is lacking and needs to be studied from several angles. Existing studies do not sufficiently take into account knowledge on the energy impact of digital technologies, in particular the increase in energy demand by digital devices, and the increase in energy efficiency. Studies would benefit from being technology/sector/country-specific.

Further exploration is needed into the way digitalisation influences the framework conditions that cause decarbonisation, the socio-economic and behavioural barriers influencing the diffusion of technologies in the long-term scenarios, and the relationship with society and its effects.

Given the implications of the digital revolution for sustainability, a better characterisation of governance aspects would increase understanding of the implications for policymakers of digitalisation and the possibilities for it and other general-purpose technologies.

Research (theoretical and empirical) on the impacts of imitation, or adaptation of new technological solutions invented in one region and used in other regions, could fill knowledge gaps and accelerate diffusion of climate-related technologies, while taking care not to reduce the incentive for inventors to search for new solutions.

### Paris Agreement compliance

An independent assessment is underway to look at the compliance of the Paris Agreement with regard to technology and capacity building as means of implementation. The Enhanced Transparency Framework for action and support is developing a methodology for monitoring, reporting and verification. There is a lack of analysis of the full landscape of international cooperation, of the effectiveness of the UN Framework Convention on Climate Change (UNFCCC) and the Paris Agreement, and what is needed to meet their objectives.

Frequently Asked Questions (FAQs)

## FAQ 16.1 | Will innovation and technological changes be enough to meet the Paris Agreement objectives?

The Paris Agreement stressed the importance of development and transfer of technologies to improve resilience to climate change and to reduce greenhouse gas emissions. However, innovation and even fast technological change will not be enough to achieve Paris Agreement mitigation objectives. Other changes are necessary across the production and consumption system and the society in general, including behavioural changes.

Technological changes never happen in a vacuum; they are always accompanied by, for instance, people changing habits, companies changing value chains, or banks changing risk profiles. Therefore, technological changes driven by holistic approaches can contribute to accelerate and spread those changes towards the achievement of climate and sustainable development goals.

In innovation studies, such systemic approaches are said to strengthen the functions of technological or national innovation systems, so that climate-friendly technologies can flourish. Innovation policies can help respond to local priorities and prevent unintended and undesirable consequences of technological change, such as unequal access to new technologies across countries and between income groups, environmental degradation and negative effects on employment.



## FAQ 16.2 | What can be done to promote innovation for climate change and the widespread diffusion of low-emission and climate-resilient technology?

The speed and success of innovation processes could be enhanced with the involvement of a wider range of actors from the industry, research and financial communities working in partnerships at national, regional and international levels. Public policies play a critical role to bring together these different actors and create the necessary enabling conditions, including financial support, through different instruments as well as institutional and human capacities.

The increasing complexity of technologies requires cooperation if their widespread diffusion is to be achieved. Cooperation includes the necessary knowledge flow within and between countries and regions. This knowledge flow can take the form of exchanging experiences, ideas, skills, and practices, among others.

## FAQ 16.3 | What is the role of international technology cooperation in addressing climate change?

Technologies that are currently known but not yet widely used need to be spread around the world, and adapted to local preferences and conditions. Innovation capabilities are required not only to adapt new technologies for local use, but also to create new markets and business models. International technology cooperation can serve that purpose.

In fact, evidence shows that international cooperation on technology development and transfer can help developing countries to achieve their climate goals more effectively and, if this is done properly, can also help to addressing other sustainable development goals. Many initiatives exist both regionally and globally to help countries in achieving technology development and transfer through partnerships and research collaboration that include developed and developing countries, with a key role for technological institutions and universities. Enhancing current activities would help an effective, long-term global response to climate change, while promoting sustainable development.

Globalisation of production and supply of goods and services, including innovation and new technologies, may open up opportunities for developing countries to advance technology diffusion; however, so far not all countries have benefitted from the globalisation of innovation due to different barriers, such as access to finance and technical capabilities. These asymmetries between countries in the globalisation process can also lead to dependencies on foreign knowledge and providers.

Not all technology cooperation directly results in mitigation outcomes. Overall, technology transfer broadly has focused on enhancing climate technology absorption and deployment in developing countries as well as research, development and demonstration, and knowledge spillovers.

The Paris Agreement also reflects this view by noting that countries shall strengthen cooperative action on technology development and transfer regarding two main aspects: (i) promoting collaborative approaches to research and development; and (ii) facilitating access to technology to developing country Parties.

## References


Abbott, F.M., 2018: Intellectual Property Rights and Climate Change: Interpreting the TRIPS Agreement for Environmentally Sound Technologies. By Wei Zhuang. *J. Int. Econ. Law*, **21(1)**, 233–236, doi:10.1093/jiel/jgy011.

Abdel-Latif, A., 2015: Intellectual property rights and the transfer of climate change technologies: Issues, challenges, and way forward. *Clim. Policy*, **15(1)**, 103–126, doi:10.1080/14693062.2014.951919.

Abramovitz, M., 1986: Catching Up, Forging Ahead, and Falling Behind. *J. Econ. Hist.*, **46(2)**, 385–406, doi:10.1017/S0022050700046209.

Abson, D.J. et al., 2017: Leverage points for sustainability transformation. *Ambio*, **46(1)**, 30–39, doi:10.1007/s13280-016-0800-y.

Acemoglu, D. and P. Restrepo, 2019: Automation and New Tasks: How Technology Displaces and Reinstates Labor. *J. Econ. Perspect.*, **33(2)**, 3–30, doi:10.1257/jep.33.2.3.

Acemoglu, D., P. Aghion, L. Bursztyn, and D. Hemous, 2012: The Environment and Directed Technical Change. *Am. Econ. Rev.*, **102(1)**, 131–166, doi:10.1257/aer.102.1.131.

Acemoglu, D., U. Akcigit, D. Hanley, and W.R. Kerr, 2014: Transition to Clean Technology. *SSRN Electron. J.*, PIER Working Paper No. 14-044, doi:10.2139/ssrn.2534407.

Adami, V.S., J.A.V. Antunes Júnior, and M.A. Sellitto, 2017: Regional industrial policy in the wind energy sector: The case of the State of Rio Grande do Sul, Brazil. *Energy Policy*, **111**, 18–27, doi:10.1016/j.enpol.2017.08.050.

Adenle, A.A., H. Azadi, and J. Arbiol, 2015: Global assessment of technological innovation for climate change adaptation and mitigation in developing world. *J. Environ. Manage.*, **161**, 261–275, doi:10.1016/j.jenvman.2015.05.040.

Aghaei, J. and M.-I. Alizadeh, 2013: Demand response in smart electricity grids equipped with renewable energy sources: A review. *Renew. Sustain. Energy Rev.*, **18**, 64–72, doi:10.1016/j.rser.2012.09.019.

Aghion, P., 2019: Path dependence, innovation and the economics of climate change. In: *Handbook on Green Growth*,[Fouquet, R. (ed.)]. Edward Elgar Publishing, Cheltenham, UK, pp. 67–83.

Aghion, P., P. Howitt, and D. Mayer-Foulkes, 2005: The Effect of Financial Development on Convergence: Theory and Evidence. *Q. J. Econ.*, **120(1)**, 173–222, doi:10.1162/0033553053327515.

Aghion, P., A. Dechezleprêtre, D. Hemous, R. Martin, and J.M. Van Reenen, 2016: Carbon Taxes, Path Dependency and Directed Technical Change: Evidence from the Auto Industry. Journal of Political Economy, Vol 124:1, pp. 1-51. doi: 10.1086/684581

Agrawala S., S. Klasen, R. Acosta Moreno, L. Barreto, T. Cottier, D. Guan, E.E. Gutierrez-Espeleta, A.E. Gámez Vázquez, L. Jiang, Y.G. Kim, J. Lewis, M. Messouli, M. Rauscher, N. Uddin, and A. Venables, 2014: Regional Development and Cooperation. In: *Climate Change 2014: Mitigation of Climate Change. Contribution of Working Group III to the Fifth Assessment Report of the Intergovernmental Panel on Climate Change* [Edenhofer, O., R. Pichs-Madruga, Y. Sokona, E. Farahani, S. Kadner, K. Seyboth, A. Adler, I. Baum, S. Brunner, P. Eickemeier, B. Kriemann, J. Savolainen, S. Schlömer, C. von Stechow, T. Zwickel and J.C. Minx (eds.)]. Cambridge University Press, Cambridge, United Kingdom and New York, NY, USA, 1083–1140.

Åhman, M., L.J. Nilsson, and B. Johansson, 2017: Global climate policy and deep decarbonization of energy-intensive industries. *Clim. Policy*, **17(5)**, 634–649, doi:10.1080/14693062.2016.1167009.

Aklin, M. and J. Urpelainen, 2013: Political Competition, Path Dependence, and the Strategy of Sustainable Energy Transitions. *Am. J. Pol. Sci.*, **57(3)**, 643–658, doi:10.1111/ajps.12002.

Akorede, M.F., H. Hizam, and E. Pouresmaeil, 2010: Distributed energy resources and benefits to the environment. *Renew. Sustain. Energy Rev.*, **14(2)**, 724–734, doi:10.1016/j.rser.2009.10.025.

Alarcón, D. and R. Vos, 2015: Introduction: The Imperative of Sustainable Development. In: *Technology and Innovation for Sustainable Development*,[Vos, R. and Alarcón, D. (eds.)]. Bloomsbury Collections, London, UK, pp. 1–14.

Albert, M., 2019: Sustainable frugal innovation – The connection between frugal innovation and sustainability. *J. Clean. Prod.*, **237**, 117747, doi:10.1016/j.jclepro.2019.117747.

Aleluia Reis, L., L. Drouet, and M. Tavoni, 2022: Internalising health-economic impacts of air pollution into climate policy: a global modelling study. *Lancet Planetary Health*, **6(1)**, 40-48, https://doi.org/10.1016/S2542-5196(21)00259-X.

Alkemade, F. and R.A.A. Suurs, 2012: Patterns of expectations for emerging sustainable technologies. *Technol. Forecast. Soc. Change*, **79(3)**, 448–456, doi:10.1016/j.techfore.2011.08.014.

Alston, J.M., P.G. Pardey, and X. Rao, 2021: Payoffs to a half century of CGIAR research. *Am. J. Agric. Econ.*,**104 (2)**, doi:https://doi.org/10.1111/ajae.12255.

Altantsetseg, P., A. Dadvari, T. Munkhdelger, G.-O. Lkhagvasuren, and M. Moslehpour, 2020: Sustainable Development of Entrepreneurial Orientation through Social Drivers. *Sustainability*, **12(21)**, 8816, doi:10.3390/SU12218816.

Altenburg, T., 2009: Building inclusive innovation systems in developing countries: Challenges for IS research. In: *Handbook of Innovation Systems and Developing Countries: Building Domestic Capabilities in a Global Setting* [Lundvall, B.-Å., K.J. Joseph, C. Chaminade, and J. Vang, (eds.)]. Edward Elgar Publishing, London, UK, pp. 33–56.

Altenburg, T. and A. Pegels, 2012: Sustainability-oriented innovation systems – managing the green transformation. *Innov. Dev.*, **2(1)**, 5–22, doi:10.1080/2157930X.2012.664037.

Altenburg, T. and D. Rodrik, 2017: Green Industrial Policy: Accelerating Structural Change Towards Wealthy Green Economies. In: *Green Industrial Policy. Concept, Policies, Country Experiences* [Altenburg, T. and C. Assman, (eds.)]. UN Environment; German Development Institute / Deutsches Institut für Entwicklungspolitik (DIE)., Geneva, Bonn, pp. 1–20.

Altieri, M.A. and C.I. Nicholls, 2017: The adaptation and mitigation potential of traditional agriculture in a changing climate. *Clim. Change*, **140(1)**, 33–45, doi:10.1007/s10584-013-0909-y.

Altieri, M.A., C.I. Nicholls, A. Henao, and M.A. Lana, 2015: Agroecology and the design of climate change-resilient farming systems. *Agron. Sustain. Dev.*, **35(3)**, 869–890, doi:10.1007/s13593-015-0285-2.

Alvarez, S. and A. Rubio, 2015: Carbon footprint in Green Public Procurement: a case study in the services sector. *J. Clean. Prod.*, **93**, 159–166, doi:10.1016/j.jclepro.2015.01.048.

Ambec, S., M.A. Cohen, S. Elgie, and P. Lanoie, 2013: The Porter Hypothesis at 20: Can Environmental Regulation Enhance Innovation and Competitiveness? *Rev. Environ. Econ. Policy*, **7(1)**, 2–22, doi:10.1093/reep/res016.

American Energy Innovation Council, 2017: *The power of innovation: Inventing the Future*Bipartisan Policy Center, USA, 52 pp.

Anadón, L.D., 2012: Missions-oriented RD&D institutions in energy between 2000 and 2010: A comparative analysis of China, the United Kingdom, and the United States. *Res. Policy*, **41(10)**, 1742–1756, doi:10.1016/j.respol.2012.02.015.

Anadon, L.D. and J.P. Holdren, 2009: Policy for Energy Technology Innovation. In: *Acting in Time on Energy Policy* [Gallagher, K.S. (ed.)]. Brookings Institution Press, Washington D.C., USA.

Anadon, L.D., M. Bunn, and V. Narayanamurti, 2014: *Transforming U.S. Energy Innovation*. Cambridge University Press, New York, USA.

Anadon, L.D. et al., 2016a: The pressing energy innovation challenge of the US National Laboratories. *Nat. Energy*, **1(10)**, 16117, doi:10.1038/nenergy.2016.117.

Anadon, L.D. et al., 2016b: Making technological innovation work for sustainable development. *Proc. Natl. Acad. Sci.*, **113(35)**, 9682–9690, doi:10.1073/pnas.1525004113.

Anadón, L.D., E. Baker, and V. Bosetti, 2017: Integrating uncertainty into public energy research and development decisions. *Nat. Energy*, **2(5)**, 17071, doi:10.1038/nenergy.2017.71.

Andersen, A.D. and M. Gulbrandsen, 2020: The innovation and industry dynamics of technology phase-out in sustainability transitions: Insights from diversifying petroleum technology suppliers in Norway. *Energy Res. Soc. Sci.*, **64**, 101447, doi:10.1016/J.ERSS.2020.101447.

Anderson, K., J. Bruil, J. Chappell, C. Kiss, and M.P. Pimbert, 2021: *Agroecology Now! Transformations Towards More Just and Sustainable Food Systems*. Palgrave Macmillan-Springer Nature Switzerland, Cham, Switzerland, 204 pp. https://doi.org/10.1007/978-3-030-61315-0.

André, F.J. and S. Smulders, 2014: Fueling growth when oil peaks: Directed technological change and the limits to efficiency. *Eur. Econ. Rev.*, **69**, 18–39, doi:10.1016/j.euroecorev.2013.10.007.

Angelucci, S., F.J. Hurtado-Albir, and A. Volpe, 2018: Supporting global initiatives on climate change: The EPO's "Y02-Y04S" tagging scheme. *World Pat. Inf.*, **54**, S85–S92, doi:10.1016/j.wpi.2017.04.006.

Antosiewicz, M. and J. Witajewski-Baltvilks, 2021: Short- and long-run dynamics of energy demand. *Energy Econ.*, **103**, 105525, doi:10.1016/j.eneco.2021.105525.

Arce, D., 2014: Experience, learning and returns to scale. *South. Econ. J.*, **80(4)**, 938–947.doi:10.4284/0038-4038-2013-088.

Ardani, K. and R. Margolis, 2015: *Decreasing Soft Costs for Solar Photovoltaics by Improving the Interconnection Process: A Case Study of Pacific Gas and Electric*. National Renewable Energy Laboratory (NREL), Golden, CO, USA, 19 pp. https://www.nrel.gov/docs/fy15osti/65066.pdf.

Arfanuzzaman, M., 2021: Big Data for Smart Cities and Inclusive Growth. In: *The Palgrave Encyclopedia of Urban and Regional Futures* [Brears, R. (ed.)]. Palgrave Macmilan, Cham, Switzerland. doi:10.1007/978-3-030-51812-7_92-1.

Arocena, R. and J. Sutz, 2020: The need for new theoretical conceptualizations on National Systems of Innovation, based on the experience of Latin America. *Econ. Innov. New Technol.*, **29(7)**, 814–829, doi:10.1080/10438599.2020.1719640.

Arocena, R., B. Göransson, and J. Sutz, 2015: Knowledge policies and universities in developing countries: Inclusive development and the "developmental university." *Technol. Soc.*, **41**, 10–20, doi:10.1016/j.techsoc.2014.10.004.

Arora, A., A. Fosfuri, and A. Gambardella, 2004: *Markets for Technology: The Economics of Innovation and Corporate Strategy*. [Arora, A., A. Fosfuri, and A. Gambardella (eds.)]. MIT Press, Cambridge, MA, USA, 352 pp.

Arrow, K.J., 1962a: Economic welfare and the allocation of resources for invention. In: *The Rate and Direction of Inventive Activity: Economic and Social Factors* [National Bureau of Economic Research, (ed.)]. Princeton University Press, Princeton, NJ, USA, pp. 609–626.

Arrow, K.J., 1962b: The economic implications of learning by doing. *Rev. Econ. Stud.*, **29(3)**, 155, doi:10.2307/2295952.

Arthur, W., 2009: *The Nature of Technology*. Free Press, New York, NY, USA, 246 pp.

Arthur, W.B., 1994: *Increasing Returns and Path Dependence in the Economy*. University of Michigan Press, Anne Arbor, MI, USA, 201 pp.

Arvesen, A., R.M. Bright, and E.G. Hertwich, 2011: Considering only first-order effects? How simplifications lead to unrealistic technology optimism in climate change mitigation. *Energy Policy*, **39(11)**, 7448–7454, doi:10.1016/j.enpol.2011.09.013.

Auerswald, P.E. and L.M. Branscomb, 2003: Valleys of Death and Darwinian Seas: Financing the Invention to Innovation Transition in the United States. *J. Technol. Transf.*, **28**, 227–239, doi:https://doi.org/10.1023/A:1024980525678.

Avelino, F. et al., 2019: Transformative social innovation and (dis)empowerment. *Technol. Forecast. Soc. Change*, **145**, 195–206, doi:10.1016/j.techfore.2017.05.002.

Avgerinou, M., P. Bertoldi, and L. Castellazzi, 2017: Trends in Data Centre Energy consumption under the European Code of Conduct for Data Centre Energy Efficiency. *Energies*, **10(10)**, 1470, doi:10.3390/en10101470.

Azoulay, P., E. Fuchs, A.P. Goldstein, and M. Kearney, 2019: Funding Breakthrough Research: Promises and Challenges of the "ARPA Model." *Innov. Policy Econ.*, **19**, 69–96, doi:10.1086/699933.

Bai, X., B. Roberts, and J. Chen, 2010: Urban sustainability experiments in Asia: Patterns and pathways. *Environ. Sci. Policy*, **13(4)**, 312–325, doi:10.1016/J.ENVSCI.2010.03.011.

Baker, E., O. Olaleye, and L. Aleluia Reis, 2015: Decision frameworks and the investment in R&D. *Energy Policy*, **80**, 275–285, doi:10.1016/j.enpol.2015.01.027.

Bakhtiar, A., A. Aslani, and S.M. Hosseini, 2020: Challenges of diffusion and commercialization of bioenergy in developing countries. *Renew. Energy*, **145**, 1780–1798, doi:10.1016/j.renene.2019.06.126.

Baller, S., S. Dutta, and B. Lanvin, 2016: *The Global Information Technology Report 2016 Innovating in the Digital Economy*. World Economic Forum, Geneva, Switzerland, 307 pp.

Balvanera, P. et al., 2017: Interconnected place-based social–ecological research can inform global sustainability. *Curr. Opin. Environ. Sustain.*, **29**, 1–7, doi:10.1016/J.COSUST.2017.09.005.

Barbieri, N., 2015: Investigating the impacts of technological position and European environmental regulation on green automotive patent activity. *Ecol. Econ.*, **117**, 140–152, doi:10.1016/j.ecolecon.2015.06.017.

Barker, T., A. Dagoumas, and J. Rubin, 2009: The macroeconomic rebound effect and the world economy. *Energy Effic.*, **2(4)**, 411–427, doi:10.1007/s12053-009-9053-y.

Baron, R., 2016: *The Role of Public Procurement in Low-carbon Innovation*. OECD, Paris, France, 32 pp. https://www.oecd.org/sd-roundtable/papersandpublications/The Role of Public Procurement in Low-carbon Innovation.pdf.

Barron-Gafford, G.A. et al., 2019: Agrivoltaics provide mutual benefits across the food–energy–water nexus in drylands. *Nat. Sustain.*, **2(9)**, 848–855, doi:10.1038/s41893-019-0364-5.

Basel, B., G. Goby, and J. Johnson, 2020: Community-based adaptation to climate change in villages of Western Province, Solomon Islands. *Mar. Pollut. Bull.*, **156**, 111266, doi:10.1016/j.marpolbul.2020.111266.

Basu, S. and D.N. Weil, 1998: Appropriate Technology and Growth. *Q. J. Econ.*, **113(4)**, 1025–1054, doi:10.1162/003355398555829.

Baudry, M. and C. Bonnet, 2019: Demand-Pull Instruments and the Development of Wind Power in Europe: A Counterfactual Analysis. *Environ. Resour. Econ.*, **73(2)**, 385–429, doi:10.1007/s10640-018-0267-3.

Bauer, F., L. Coenen, T. Hansen, K. McCormick, and Y.V. Palgan, 2017: Technological innovation systems for biorefineries: A review of the literature. *Biofuels, Bioprod. Biorefining*, **11(3)**, 534–548, doi:10.1002/bbb.1767.

Bayer, P., J. Urpelainen, and A. Xu, 2016: Explaining differences in sub-national patterns of clean technology transfer to China and India. *Int. Environ. Agreements Polit. Law Econ.*, **16(2)**, 261–283, doi:10.1007/s10784-014-9257-2.

BEE, 2020: *Impact of Energy Efficiency Measures for the Year 2018–19*. Bureau of Energy Efficiency (BEE), New Dehli, India, 148 pp. https://beeindia.gov.in/sites/default/files/BEE Final Report_1.pdf.

Behuria, P., 2020: The politics of late late development in renewable energy sectors: Dependency and contradictory tensions in India's National Solar Mission. *World Dev.*, **126**, 104726, doi:10.1016/j.worlddev.2019.104726.

Bekkers, R., 2017: Where patents and standards come together. In: *Handbook of Standards and Innovation* [Hawkins, R., K. Blind, and R. Page (eds.)]. Edward Elgar Publishing, London, UK, and New York, NY, USA, pp. 227–251.

Bellesi, F., D. Lehrer, and A. Tal, 2005: Comparative advantage: The impact of ISO 14001 environmental certification on exports. *Environ. Sci. Technol.*, **39(7)**, 1943–1953, doi:10.1021/es0497983.

Bento, N. and C. Wilson, 2016: Measuring the duration of formative phases for energy technologies. *Environ. Innov. Soc. Transitions*, **21**, 95–112, doi:10.1016/j.eist.2016.04.004.

Bento, N., C. Wilson, and L.D. Anadon, 2018: Time to get ready: Conceptualizing the temporal and spatial dynamics of formative phases for energy technologies. *Energy Policy*, **119**, 282–293, doi:10.1016/j.enpol.2018.04.015.

Bergek, A., S. Jacobsson, B. Carlsson, S. Lindmark, and A. Rickne, 2008: Analyzing the functional dynamics of technological innovation systems: A scheme of analysis. *Res. Policy*, **37(3)**, 407–429, doi:10.1016/j.respol.2007.12.003.

Bergek, A., S. Jacobsson, M. Hekkert, and K. Smith, 2010: Functionality of innovation systems as a rationale for and guide to innovation policy. In: *The Theory and Practice of Innovation Policy: An International Research Handbook*.[Smits, R., Kulhman, S., and Shapira, P. (eds.)]. Edward Elgar Publishing, Cheltenham, UK, 496pp. ISBN 978-1 84542-848-8.

Bergek, A. et al., 2015: Technological innovation systems in contexts: Conceptualizing contextual structures and interaction dynamics. *Environ. Innov. Soc. Transitions*, **16**, 51–64, doi:10.1016/j.eist.2015.07.003.

Bernauer, T. and L.F. McGrath, 2016: Simple reframing unlikely to boost public support for climate policy. *Nat. Clim. Chang.*, **6(7)**, 680–683, doi:10.1038/nclimate2948.

Bessen, J. and M.J. Meurer, 2009: Patent failure: How judges, bureaucrats, and lawyers put innovators at risk. *Choice Rev. Online*, **46(06)**, 46-3523-46–3523, doi:10.5860/CHOICE.46-3523.

Bettencourt, L.M.A., J.E. Trancik, and J. Kaur, 2013: Determinants of the Pace of Global Innovation in Energy Technologies. *PLoS One*, **8(10)**, e67864, doi:10.1371/journal.pone.0067864.

Bhamidipati, P.L. and U.E. Hansen, 2021: Unpacking local agency in China–Africa relations: Frictional encounters and development outcomes of solar power in Kenya. *Geoforum*, **119**, 206–217, doi:10.1016/j.geoforum.2020.12.010.

Bhamidipati, P.L., U. Elmer Hansen, and J. Haselip, 2019: Agency in transition: The role of transnational actors in the development of the off-grid solar PV regime in Uganda. *Environ. Innov. Soc. Transitions*, **33**, 30–44, doi:10.1016/j.eist.2019.02.001.

Bhattacharya, G., 1984: Learning and the Behavior of Potential Entrants. *RAND J. Econ.*, **15(2)**, 281, doi:10.2307/2555681.

Biadgleng, E.T., 2009: *Compulsory Licenses & Government Use*. UNCTAD, Paris, France, pp 24.

Bibri, S.E., 2019a: Data-driven smart sustainable urbanism: The intertwined societal factors underlying its materialization, success, expansion, and evolution. *Geoj. 2019861*, **86(1)**, 43–68, doi:10.1007/S10708-019-10061-X.

Bibri, S.E., 2019b: The anatomy of the data-driven smart sustainable city: Instrumentation, datafication, computerization and related applications. *J. Big Data 2019 61*, **6(1)**, 1–43, doi:10.1186/S40537-019-0221-4.

Bieser, J. and L. Hilty, 2018: Assessing Indirect Environmental Effects of Information and Communication Technology (ICT): A Systematic Literature Review. *Sustainability*, **10(8)**, 2662, doi:10.3390/su10082662.

Binz, C. and B. Truffer, 2017: Global Innovation Systems – A conceptual framework for innovation dynamics in transnational contexts. *Res. Policy*, **46(7)**, 1284–1298, doi:10.1016/j.respol.2017.05.012.

Binz, C. and L.D. Anadon, 2018: Unrelated diversification in latecomer contexts: Emergence of the Chinese solar photovoltaics industry. *Environ. Innov. Soc. Transitions*, **28**, 14–34, doi:10.1016/j.eist.2018.03.005.

Binz, C., B. Truffer, L. Li, Y. Shi, and Y. Lu, 2012: Conceptualizing leapfrogging with spatially coupled innovation systems: The case of onsite wastewater treatment in China. *Technol. Forecast. Soc. Change*, **79(1)**, 155–171, doi:10.1016/j.techfore.2011.08.016.

Biresselioglu, M.E., M.H. Demir, M. Demirbag Kaplan, and B. Solak, 2020: Individuals, collectives, and energy transition: Analysing the motivators and barriers of European decarbonisation. *Energy Res. Soc. Sci.*, **66**, 101493, doi:10.1016/J.ERSS.2020.101493.

Bisaga, I., P. Parikh, J. Tomei, and L.S. To, 2021: Mapping synergies and trade-offs between energy and the sustainable development goals: A case study of off-grid solar energy in Rwanda. *Energy Policy*, **149**, 112028, doi:10.1016/J.ENPOL.2020.112028.

Black, G., D. Holley, D. Solan, and M. Bergloff, 2014: Fiscal and economic impacts of state incentives for wind energy development in the Western United States. *Renew. Sustain. Energy Rev.*, **34**, 136–144, doi:10.1016/j.rser.2014.03.015.

Blieden, R., 1999: *Cherry Hill Revisited –A retrospective on the creation of a national plan for photovoltaic conversion of solar energy for terrestrial applications.* AIP Conference Proceedings,**462 (1)**, pp.796–799. https://doi.org/10.1063/1.57945.

Bloess, A., 2019: Impacts of heat sector transformation on Germany's power system through increased use of power-to-heat. *Appl. Energy*, **239**, 560–580, doi:https://doi.org/10.1016/j.apenergy.2019.01.101.

Bodin, Ö., 2017: Collaborative environmental governance: Achieving collective action in social-ecological systems. *Science*, **357(6352)**, eaan1114, doi:10.1126/science.aan1114.

Böhringer, C. and K.E. Rosendahl, 2010: Green promotes the dirtiest: On the interaction between black and green quotas in energy markets. *J. Regul. Econ.*, **37(3)**, 316–325, doi:10.1007/s11149-010-9116-1.

Boiral, O., L. Guillaumie, I. Heras-Saizarbitoria, and C.V. Tayo Tene, 2018: Adoption and outcomes of ISO 14001: A systematic review. *Int. J. Manag. Rev.*, **20(2)**, 411–432, doi:10.1111/ijmr.12139.

Bonvillian, W.B., 2018: DARPA and its ARPA-E and IARPA clones: A unique innovation organization model. *Ind. Corp. Chang.*, **27(5)**, 897–914, doi:10.1093/icc/dty026.

Bonvillian, W.B. and R. Van Atta, 2011: ARPA-E and DARPA: Applying the DARPA model to energy innovation. *J. Technol. Transf.*, **36(5)**, 469–513, doi:10.1007/s10961-011-9223-x.

Boodoo, Z., F. Mersmann, and K.H. Olsen, 2018: The implications of how climate funds conceptualize transformational change in developing countries. *Clim. Dev.*, **10(8)**, 673–686, doi:10.1080/17565529.2018.1442788.

Bosetti, V., 2021: Integrated Assessment Models for Climate Change. In: *Oxford Research Encyclopedia of Economics and Finance*, Oxford University Press, Oxford, UK.

Bosetti, V., M. Heugues, and A. Tavoni, 2017: Luring others into climate action: Coalition formation games with threshold and spillover effects. *Oxf. Econ. Pap.*, **69(2)**, 410–431, doi:10.1093/oep/gpx017.

Bowen, W.M., S. Park, and J.A. Elvery, 2013: Empirical Estimates of the Influence of Renewable Energy Portfolio Standards on the Green Economies of States. *Econ. Dev. Q.*,**27(4)**, doi:10.1177/0891242413491316.

Boyd, A., 2012: Informing international UNFCCC technology mechanisms from the ground up: Using biogas technology in South Africa as a case study to evaluate the usefulness of potential elements of an international technology agreement in the UNFCCC negotiations process. *Energy Policy*, **51**, 301–311, doi:10.1016/j.enpol.2012.08.020.

Boyd, P. and L. Schweber, 2018: Unintended consequences: Institutional artefacts, closure mechanisms and the performance gap. *Build. Res. Inf.*, **46(1)**, 10–22, doi:10.1080/09613218.2017.1331096.

Brandão, L.G.L. and P. Ehrl, 2019: International R&D spillovers to the electric power industries. *Energy*, **182**, 424–432, doi:10.1016/j.energy.2019.06.046.

Brandão Santana, N., D.A.D.N. Rebelatto, A.E. Périco, H.F. Moralles, and W. Leal Filho, 2015: Technological innovation for sustainable development: An analysis of different types of impacts for countries in the BRICS and G7 groups. *Int. J. Sustain. Dev. World Ecol.*, **22(5)**, 1–12, doi:10.1080/13504509.2015.1069766.

Branstetter, L., 2006: Is foreign direct investment a channel of knowledge spillovers? Evidence from Japan's FDI in the United States. *J. Int. Econ.*, **68(2)**, 325–344, doi:10.1016/j.jinteco.2005.06.006.

Breetz, H., M. Mildenberger, and L. Stokes, 2018: The political logics of clean energy transitions. *Bus. Polit.*, **20(4)**, 492–522, doi:10.1017/bap.2018.14.

Brennen, J.S. and D. Kreiss, 2016: Digitalization. In: *The International Encyclopedia of Communication Theory and Philosophy* [Jensen, K.B. and R.. Craig, (eds.)], Wiley Online Library, pp. 1–11. https://doi.org/10.1002/9781118766804.wbiect111.

Breschi, S. and F. Malerba, 1997: Sectoral Innovation Systems: Technological Regimes, Schumpeterian Dynamics, and Spatial Boundaries. In: *Systems of Innovation Technologies, Institutions and Organisations* [Edquist, C., (ed.)], Pinter Publishers, London, pp. 41–60.

Bringezu, S., 2019: Toward Science-Based and Knowledge-Based Targets for Global Sustainable Resource Use. *Resources*, **8(3)**, 140, doi:10.3390/resources8030140.

Brooks, H., 1980: Technology, Evolution and Purpose. *Daedalus*, **109(1)**, 65–81.

Brooks, S. and M. Loevinsohn, 2011: Shaping agricultural innovation systems responsive to food insecurity and climate change. *Nat. Resour. Forum*, **35(3)**, 185–200, doi:10.1111/j.1477-8947.2011.01396.x.

Brown, J. and C. Hendry, 2009: Public demonstration projects and field trials: Accelerating commercialisation of sustainable technology in solar photovoltaics. *Energy Policy*, **37(7)**, 2560–2573, doi:10.1016/j.enpol.2009.01.040.

Brown, M.B., W. Canzler, F. Fischer, and A. Knie, 1995: Technological Innovation through Environmental Policy: California's Zero-Emission Vehicle Regulation. *Public Product. Manag. Rev.*, **19(1)**, 77, doi:10.2307/3380822.

Bryden, J. and S.S. Gezelius, 2017: Innovation as if people mattered: The ethics of innovation for sustainable development. *Innov. Dev.*, **7(1)**, 101–118, doi:10.1080/2157930X.2017.1281208.

Buckley, P., 2020: Prices, information and nudges for residential electricity conservation: A meta-analysis. *Ecol. Econ.*, **172**, 106635, doi:10.1016/j.ecolecon.2020.106635.

Bukkens, S. et al., 2020: *The Nexus Times*. [Bukkens, S., T. Dunlop, L.J. Di Felice, Z. Kovacic, I.W. Nilsen, R. Strand, T. Völker, and L. Zamarioli, (eds.)]. Megaloceros Press, Bergen, Norway, pp.189. ISBN-13: 978-82-91851-05-1.

Bulkeley, H., V.C. Broto, and A. Maassen, 2014: Low-carbon transitions and the reconfiguration of urban infrastructure. *Urban Stud.*, **51(7)**, 1471–1486, doi:10.1177/0042098013500089.

Bumpus, A. and S. Comello, 2017: Emerging clean energy technology investment trends. *Nat. Clim. Chang.*, **7(6)**, 382–385, doi:10.1038/nclimate3306.

Bunn, M., L.D. Anadon, and V. Narayanamurti, 2014: The need to transform U.S. energy innovation. In: *Transforming U.S. Energy Innovation* [Anadon, L.D., M. Bunn, and V. Narayanamurti, (eds.)], Cambridge University Press, Cambridge, UK, pp. 14–16.

Burch, S., A. Shaw, A. Dale, and J. Robinson, 2014: Triggering transformative change: A development path approach to climate change response in communities. *Climate Policy,* **14(4)**, 467–487, doi:10.1080/14693062.2014.876342.

Bürer, M.J. and R. Wüstenhagen, 2009: Which renewable energy policy is a venture capitalist's best friend? Empirical evidence from a survey of international cleantech investors. *Energy Policy*, **37(12)**, 4997–5006, doi:10.1016/j.enpol.2009.06.071.

Calel, R. and A. Dechezleprêtre, 2016: Environmental Policy and Directed Technological Change: Evidence from the European Carbon Market. *Rev. Econ. Stat.*, **98(1)**, 173–191, doi:10.1162/REST_A_00470.

Calzadilla, P.V. and R. Mauger, 2018: The UN's new sustainable development agenda and renewable energy: The challenge to reach SDG7 while achieving energy justice. *J. Energy \& Nat. Resour. Law*, **36(2)**, 233–254, doi:10.1080/02646811.2017.1377951.

Caravella, S. and F. Crespi, 2020: Unfolding heterogeneity: The different policy drivers of different eco-innovation modes. *Environ. Sci. Policy*, **114**, 182–193, doi:10.1016/j.envsci.2020.08.003.

Carlsson, B. and R. Stankiewicz, 1991: On the nature, function and composition of technological systems. *J. Evol. Econ.*, **1(2)**, 93–118, doi:10.1007/BF01224915.

Caselli, F. and W.J. Coleman, 2006: The World Technology Frontier. *Am. Econ. Rev.*, **96(3)**, 499–522, doi:10.1257/aer.96.3.499.

Cassman, K.G. and P. Grassini, 2020: A global perspective on sustainable intensification research. *Nat. Sustain.*, **3(4)**, 262–268, doi:10.1038/s41893-020-0507-8.

Castán Broto, V. and H. Bulkeley, 2013: A survey of urban climate change experiments in 100 cities. *Glob. Environ. Chang.*, **23(1)**, 92–102, doi:10.1016/J.GLOENVCHA.2012.07.005.

Cerutti, A.K., S. Contu, F. Ardente, D. Donno, and G.L. Beccaro, 2016: Carbon footprint in green public procurement: Policy evaluation from a case study in the food sector. *Food Policy*, **58**, 82–93, doi:10.1016/j.foodpol.2015.12.001.

Chaffin, B.C. et al., 2016: Transformative Environmental Governance. *Annu. Rev. Environ. Resour.*, **41(1)**, 399–423, doi:10.1146/annurev-environ-110615-085817.

Chaminade, C. and C. Esquist, 2010: Rationales for Public Policy Intervention in the Innovation Process: Systems of Innovation Approach. In: *The Theory and Practice of Innovation Policy* [Smits, R.E. and P. Shapira, (eds.)], Edward Elgar Publishing, London, UK and New York, NY, USA, 20 pp.

Chan, G. and L. Diaz Anadon, 2016: *Improving Decision Making for Public R&D Investment in Energy: Utilizing Expert Elicitation in Parametric Models*. Cambridge Working Papers in Economics, Cambridge University, Cambridge, UK, 55 pp.

Chan, G., A.P. Goldstein, A. Bin-Nun, L. Diaz Anadon, and V. Narayanamurti, 2017: Six principles for energy innovation. *Nature*, **552(7683)**, 25–27, doi:10.1038/d41586-017-07761-0.

Charmes, J., 2016: The Informal Economy: Definitions, Size, Contribution and Main Characteristics. In: *The Informal Economy in Developing Nations* [Kraemer-Mbula, E. and S. Wunsch-Vincent, (eds.)], Cambridge University Press, Cambridge, UK, pp. 13–52.

Chaudhary, A., A.D. Sagar, and A. Mathur, 2012: Innovating for energy efficiency: A perspective from India. *Innov. Dev.*, **2(1)**, 45–66, doi:10.1080/2157930X.2012.667212.

Chaudhary, A., C. Krishna, and A. Sagar, 2015: Policy making for renewable energy in India: Lessons from wind and solar power sectors. *Clim. Policy*, **15(1, SI)**, 58–87, doi:10.1080/14693062.2014.941318.

Chlingaryan, A., S. Sukkarieh, and B. Whelan, 2018: Machine learning approaches for crop yield prediction and nitrogen status estimation in precision agriculture: A review. *Comput. Electron. Agric.*, **151**, 61–69, doi:10.1016/j.compag.2018.05.012.

Choi, H. and H. Zo, 2019: Assessing the efficiency of national innovation systems in developing countries. *Sci. Public Policy*, **46(4)**, 530–540, doi:10.1093/scipol/scz005.

Christensen, E., 1985: *Electricity from photovoltaic solar cells. Flat-Plate Solar Array Project of the US Department of Energy's National Photovoltaics Program: 10 years of progress*. NASA Technical Reports Server, US, 102 pp. https://ntrs.nasa.gov/citations/19880002806.

Cinner, J.E. et al., 2018: Building adaptive capacity to climate change in tropical coastal communities. *Nat. Clim. Chang.*, **8(2)**, 117–123, doi:10.1038/s41558-017-0065-x.

Clark, W.C., L. van Kerkhoff, L. Lebel, and G.C. Gallopin, 2016: Crafting usable knowledge for sustainable development. *Proc. Natl. Acad. Sci.*, **113(17)**, 4570–4578, doi:10.1073/pnas.1601266113.

Cohen, M.A. and A. Tubb, 2018: The Impact of Environmental Regulation on Firm and Country Competitiveness: A Meta-analysis of the Porter Hypothesis. *J. Assoc. Environ. Resour. Econ.*, **5(2)**, 371–399, doi:10.1086/695613.

Cohen, W., R. Nelson, and J. Walsh, 2000: *Protecting Their Intellectual Assets: Appropriability Conditions and Why U.S. Manufacturing Firms Patent (or Not)*. National Bureau of Economic Research, Cambridge, MA, USA, 50 pp.

Cohen, W.M., A. Goto, A. Nagata, R.R. Nelson, and J.P. Walsh, 2002: R&D spillovers, patents and the incentives to innovate in Japan and the United States. *Res. Policy*,, doi:10.1016/S0048-7333(**02**)00068-9.

Consoli, D., G. Marin, A. Marzucchi, and F. Vona, 2016: Do green jobs differ from non-green jobs in terms of skills and human capital? *Res. Policy*, **45(5)**, 1046–1060, doi:10.1016/j.respol.2016.02.007.

Conti, C., M.L. Mancusi, F. Sanna-Randaccio, R. Sestini, and E. Verdolini, 2018: Transition towards a green economy in Europe: Innovation and knowledge integration in the renewable energy sector. *Res. Policy*, **47(10)**, 1996–2009, doi:10.1016/j.respol.2018.07.007.

Contreras, J.L., 2015: A brief history of frand: Analyzing current debates in standard setting and antitrust through a historical lens. *Antitrust Law J.*, **80(39)**, 82, doi:10.2139/ssrn.2374983.

Cooke, P., M. Gomez Uranga, and G. Etxebarria, 1997: Regional innovation systems: Institutional and organisational dimensions. *Res. Policy*, **26(4–5)**, 475–491, doi:10.1016/S0048-7333(**97**)00025-5.

Corey, A., 2014: Diffusion of Green Technology: A Survey. *Int. Rev. Environ. Resour. Econ.*, **7(1)**, 1–33, doi:10.1561/101.00000055.

Correa, C.M., J.I. Correa, and B. De Jonge, 2020: The status of patenting plants in the Global South. *J. World Intellect. Prop.*, **23(1–2)**, 121–146, doi:10.1111/jwip.12143.

Corsatea, T.D., 2016: Localised knowledge, local policies and regional innovation activity for renewable energy technologies: Evidence from Italy. *Pap. Reg. Sci.*, **95(3)**, 443–466, doi:10.1111/pirs.12136.

Corsi, A., R.N. Pagani, J.L. Kovaleski, and V. Luiz da Silva, 2020: Technology transfer for sustainable development: Social impacts depicted and some other answers to a few questions. *J. Clean. Prod.*, **245**, 118522, doi:10.1016/j.jclepro.2019.118522.

Côte, F.-X. et al., 2019: *The agroecological transition of agricultural systems in the Global South*. [Côte, F.-X., E. Poirier-Magona, S. Perret, P. Roudier, B. Rapidel, and M.-C. Thirion, (eds.)]. éditions Quae, Versailles, France, 360pp.

Coughlan de Perez, E. et al., 2015: Forecast-based financing: An approach for catalyzing humanitarian action based on extreme weather and climate forecasts. *Nat. Hazards Earth Syst. Sci.*, **15(4)**, 895–904, doi:10.5194/nhess-15-895-2015.

Cowan, R., 1990: Nuclear Power Reactors: A Study in Technological Lock-in. *J. Econ. Hist.*, **50(3)**, 541–567, doi:10.1017/S0022050700037153.

Crabb, J.M. and D.K.N. Johnson, 2010: Fueling Innovation: The Impact of Oil Prices and CAFE Standards on Energy-Efficient Automotive Technology. *Energy J.*, **31(1)**, doi:10.5547/ISSN0195-6574-EJ-Vol31-No1-9.

Craglia, M. et al., 2018: *Artificial Intelligence: A European Perspective*. Publications Office of the European Union, Brussels, pp 137.

CREWS, 2016: *CREWS Operational Procedures Note No.1: Programming and Project Development*.Climate Risk & Early Warning Systems (CREWS), Geneva, 11 pp. https://ane4bf-datap1.s3-eu-west-1.amazonaws.com/wmocrews/s3fs-public/ckeditor/files/Revised_Operational_Procedures_Note_No1_Programming_and_Project_Development.pdf?OIO3Yfu.Sp1.aliGNEBPbphkoloVjhnN.

Criqui, P., S. Mima, P. Menanteau, and A. Kitous, 2015: Mitigation strategies and energy technology learning: An assessment with the POLES model. *Technol. Forecast. Soc. Change*, **90**, 119–136, doi:10.1016/j.techfore.2014.05.005.

Cross, J. and D. Murray, 2018: The afterlives of solar power: Waste and repair off the grid in Kenya. *Energy Res. Soc. Sci.*, **44**, 100–109, doi:10.1016/j.erss.2018.04.034.

CTCN, 2019: *Programme of Work 2019-2022: Climate Technology Centre and Network*. UN Environment, UNFCCC, UNIDO. Published Online. 22 pp. https://www.ctc-n.org/sites/www.ctc-n.org/files/ctcn_programme_of_work_2019-2022.pdf (Accessed August 25, 2021).

Cui, J., X. Liu, Y. Sun, and H. Yu, 2020: Can CDM projects trigger host countries' innovation in renewable energy? Evidence of firm-level dataset from China. *Energy Policy*, **139**, 111349, doi:10.1016/j.enpol.2020.111349.

D'Auvergne, C. and M. Nummelin, 2017: Capacity-Building (Article 11). In: *The Paris Agreement on Climate Change: Analysis and Commentary* [Klein, D., M. Pia Carazo, M. Doelle, J. Bulmer, and A. Higham, (eds.)], Oxford University Press, Oxford, UK, pp 277–291.

da Silva, V.L., J.L. Kovaleski, and R.N. Pagani, 2019: Technology transfer in the supply chain oriented to industry 4.0: A literature review. *Technol. Anal. Strateg. Manag.*, **31(5)**, 546–562, doi:10.1080/09537325.2018.1524135.

Dabbagh, M., B. Hamdaoui, A. Rayes, and M. Guizani, 2019: Shaving Data Center Power Demand Peaks Through Energy Storage and Workload Shifting Control. *IEEE Trans. Cloud Comput.*, **7**, 1095–1108.

Dai, Y. and L. Xue, 2015: China's policy initiatives for the development of wind energy technology. *Clim. Policy*, **15(1)**, 30–57, doi:10.1080/14693062.2014.863549.

Dana, L.P., C. Gurău, F. Hoy, V. Ramadani, and T. Alexander, 2021: Success factors and challenges of grassroots innovations: Learning from failure. *Technol. Forecast. Soc. Change*, **164**, 119600, doi:10.1016/J.TECHFORE.2019.03.009.

Darmani, A., N. Arvidsson, A. Hidalgo, and J. Albors, 2014: What drives the development of renewable energy technologies? Toward a typology for the systemic drivers. *Renew. Sustain. Energy Rev.*, **38**, 834–847, doi:10.1016/j.rser.2014.07.023.

David, P. and D. Foray, 1995: *Accessing and expanding the science and technology knowledge base*. OECD, Paris, France, 13–68 pp. https://www.oecd-ilibrary.org/docserver/sti_rev-v1995-1-en.pdf?expires=1636825312&id=id&accname=guest&checksum=8EF2C3D4C4CFDA3532CC73D10337415E.

David, P.A., 1985: Clio and the Economics of QWERTY. *Am. Econ. Rev.*, **75(2)**, 332–337, doi:http://dx.doi.org/10.2307/1805621.

Davis, L.W., 2014: The Economic Cost of Global Fuel Subsidies. *Am. Econ. Rev.*, **104(5)**, 581–585, doi:10.1257/aer.104.5.581.

Davis, L.W. and C.R. Knittel, 2019: Are Fuel Economy Standards Regressive? *J. Assoc. Environ. Resour. Econ.*, **6**(S1), S37–S63, doi:10.1086/701187.

de Beer, J., K. Fu, and S. Wunsch-Vincent, 2016: Innovation in the informal economy. In: *The informal economy in developing nations: Hidden engine of innovation?* [Wunsch-Vincent, S. and E. Kraemer-Mbula, (eds.)], Cambridge University Press, Cambridge, UK, 439 pp.

de Coninck, H. and D. Puig, 2015: Assessing climate change mitigation technology interventions by international institutions. *Clim. Change*, **131(3)**, 417–433, doi:10.1007/s10584-015-1344-z.

de Coninck, H. and A. Sagar, 2017: Technology Development and Transfer (Article 10). In: *The Paris Agreement on Climate Change: Analysis and Commentary* [Klein, D., M. Pía Carazo, M. Doelle, J. Bulmer, and A. Higham, (eds.)], Oxford University Press, Oxford, UK, pp. 258–276.

de Gooyert, V., E. Rouwette, H. van Kranenburg, E. Freeman, and H. van Breen, 2016: Sustainability transition dynamics: Towards overcoming policy resistance. *Technol. Forecast. Soc. Change*, **111**, 135–145, doi:10.1016/J.TECHFORE.2016.06.019.

de Jesus, A. and S. Mendonça, 2018: Lost in Transition? Drivers and Barriers in the Eco-innovation Road to the Circular Economy. *Ecol. Econ.*, **145**, 75–89, doi:10.1016/J.ECOLECON.2017.08.001.

de Jong, P., A. Paulraj, and C. Blome, 2014: The Financial Impact of ISO 14001 Certification: Top-Line, Bottom-Line, or Both? *J. Bus. Ethics*, **119(1)**, 131–149, doi:10.1007/s10551-012-1604-z.

de la Tour, A., M. Glachant, and Y. Ménière, 2011: Innovation and international technology transfer: The case of the Chinese photovoltaic industry. *Energy Policy*, **39(2)**, 761–770, doi:10.1016/J.ENPOL.2010.10.050.

de Rassenfosse, G., H. Dernis, D. Guellec, L. Picci, and B. van Pottelsberghe de la Potterie, 2013: The worldwide count of priority patents: A new indicator of inventive activity. *Res. Policy*, **42(3)**, 720–737, doi:10.1016/j.respol.2012.11.002.

De Rose, A. et al., 2017: *Technology Readiness Level: Guidance Principles for Renewable Energy Technologies*. European Commission, Brussels, 48 pp.

de Sisternes, F.J., J.D. Jenkins, and A. Botterud, 2016: The value of energy storage in decarbonizing the electricity sector. *Appl. Energy*, **175**, 368–379, doi:https://doi.org/10.1016/j.apenergy.2016.05.014.

Dechezleprêtre, A., M. Glachant, I. Haščič, N. Johnstone, and Y. Ménière, 2011: Invention and Transfer of Climate Change–Mitigation Technologies: A Global Analysis. *Rev. Environ. Econ. Policy*, **5(1)**, 109–130, doi:10.1093/reep/req023.

Dechezleprêtre, A., M. Glachant, and Y. Ménière, 2013: What Drives the International Transfer of Climate Change Mitigation Technologies? Empirical Evidence from Patent Data. *Environ. Resour. Econ.*, **54(2)**, 161–178, doi:10.1007/s10640-012-9592-0.

Deichmann, U., A. Goyal, and D. Mishra, 2016: *Will Digital Technologies Transform Agriculture in Developing Countries?* World Bank, Washington, DC, USA, 30 pp.

del Río, P., 2010: Analysing the interactions between renewable energy promotion and energy efficiency support schemes: The impact of different instruments and design elements. *Energy Policy*, **38(9)**, 4978–4989, doi:10.1016/j.enpol.2010.04.003.

del Río, P., 2014: On evaluating success in complex policy mixes: The case of renewable energy support schemes. *Policy Sci.*, **47(3)**, 267–287, doi:10.1007/s11077-013-9189-7.

del Río, P. and E. Cerdá, 2014: The policy implications of the different interpretations of the cost-effectiveness of renewable electricity support. *Energy Policy*, **64**, 364–372, doi:10.1016/j.enpol.2013.08.096.

del Río, P. and E. Cerdá, 2017: The missing link: The influence of instruments and design features on the interactions between climate and renewable electricity policies. *Energy Res. Soc. Sci.*, **33**, 49–58, doi:10.1016/j.erss.2017.09.010.

del Río, P. and C. Kiefer, 2021: *Analysing the effects of auctions on technological innovation: Report D4.3 of AURES II project*. Aures II. Published Online. 99 pp. http://aures2project.eu/wp-content/uploads/2021/04/AURES_II_D4_3_technological_innovation.pdf (Accessed July 22, 2021).

del Río, P., 2012: The dynamic efficiency of feed-in tariffs: The impact of different design elements. *Energy Policy*, **41**, 139–151, doi:10.1016/J.ENPOL.2011.08.029.

Denevan, W.M., 1995: 2 Prehistoric agricultural methods as models for sustainability. *Advances in Plant Pathology*. **11**, pp. 21–43. https://doi.org/10.1016/S0736-4539(06)80004-8.

Deng, H.M., Q.M. Liang, L.J. Liu, and L.D. Anadon, 2018: Co-benefits of greenhouse gas mitigation: A review and classification by type, mitigation sector, and geography. *Environ. Res. Lett.*, **12(12)**, 123001, doi:10.1088/1748-9326/aa98d2.

Deutch, J. and E. Steinfeld, 2013: *A Duel in the Sun: The Solar Photovoltaics Technology Conflict between China and the United States*. Energy Initiative Massachusetts Institute of Technology. Cambridge, US, 32 pp. http://mitei.mit.edu/futureofsolar (Accessed August 30, 2021).

Devine, A. and N. Kok, 2015: Green Certification and Building Performance: Implications for Tangibles and Intangibles. *J. Portf. Manag.*, **41(6)**, 151–163, doi:10.3905/jpm.2015.41.6.151.

Devlin, A., 2010: The Misunderstood Function of Disclosure in Patent Law. *Harvard Journa Technol. Law*, **23**, 401.

Di Maria, C. and S.A. Smulders, 2005: Trade Pessimists vs Technology Optimists: Induced Technical Change and Pollution Havens. *Adv. Econ. Anal. Policy*, **3(2)**, doi:10.2202/1538-0637.1344.

Di Maria, C. and E. van der Werf, 2008: Carbon leakage revisited: Unilateral climate policy with directed technical change. *Environ. Resour. Econ.*, **39(2)**, 55–74, doi:10.1007/s10640-007-9091-x.

Díaz, S. et al., 2019: Pervasive human-driven decline of life on Earth points to the need for transformative change. *Science*, **366(6471)**, 1-10, doi:10.1126/science.aax3100.

Diercks, G., H. Larsen, and F. Steward, 2019: Transformative innovation policy: Addressing variety in an emerging policy paradigm. *Res. Policy*, **48(4)**, 880–894, doi:10.1016/j.respol.2018.10.028.

Disterheft, A., S. Caeiro, U.M. Azeiteiro, and W.L. Filho, 2015: Sustainable universities – a study of critical success factors for participatory approaches. *J. Clean. Prod.*, **106**, 11–21, doi:10.1016/j.jclepro.2014.01.030.

Doblinger, C., K. Surana, and L.D. Anadon, 2019: Governments as partners: The role of alliances in U.S. cleantech startup innovation. *Res. Policy*, **48(6)**, 1458–1475, doi:10.1016/j.respol.2019.02.006.

Doranova, A., I. Costa, and G. Duysters, 2010: Knowledge base determinants of technology sourcing in clean development mechanism projects. *Energy Policy*, **38(10)**, 5550–5559, doi:10.1016/j.enpol.2010.04.055.

Dorband, I.I., M. Jakob, and J.C. Steckel, 2020: Unraveling the political economy of coal: Insights from Vietnam. *Energy Policy*, **147**, 111860, doi:10.1016/j.enpol.2020.111860.

Doukas, H. and A. Nikas, 2020: Decision support models in climate policy. *Eur. J. Oper. Res.*, **280(1)**, 1–24, doi:10.1016/j.ejor.2019.01.017.

Doukas, H., A. Nikas, M. González-Eguino, I. Arto, and A. Anger-Kraavi, 2018: From Integrated to Integrative: Delivering on the Paris Agreement. *Sustainability*, **10(7)**, 2299, doi:10.3390/su10072299.

Douthwaite, B. and E. Hoffecker, 2017: Towards a complexity-aware theory of change for participatory research programs working within agricultural innovation systems. *Agric. Syst.*, **155**, 88–102, doi:https://doi.org/10.1016/j.agsy.2017.04.002.

Du, K. and J. Li, 2019: Towards a green world: How do green technology innovations affect total-factor carbon productivity. *Energy Policy*, **131**, 240–250, doi:10.1016/j.enpol.2019.04.033.

Du, K., P. Li, and Z. Yan, 2019: Do green technology innovations contribute to carbon dioxide emission reduction? Empirical evidence from patent data. *Technol. Forecast. Soc. Change*, **146**, 297–303, doi:10.1016/j.techfore.2019.06.010.

Dussaux, D., A. Dechezleprêtre, and M. Glachant, 2018: *Intellectual property rights protection and the international transfer of low-carbon technologies*. Centre for Climate Change Economics and Policy Working Paper No. 323, Grantham Research Insitute on Climate Change and the Environment Working Paper No.288, Published Online, 63 pp. https://personal.lse.ac.uk/dechezle/Working-Paper-288-Dussaux-et-al.pdf.ISSN 2515-5709.

Dziallas, M. and K. Blind, 2019: Innovation indicators throughout the innovation process: An extensive literature analysis. *Technovation*, **80–81**, 3–29, doi:10.1016/j.technovation.2018.05.005.

Ebeling, F., 2020: Assessing the macroeconomic effects of sustainable energy transitions in Costa Rica and Chile: A multisectoral balance-of-payments-constrained growth approach. In: *The Regulation and Policy of Latin American Energy Transitions* [Guimaraes, L.N., (ed.)], Elsevier Science, Amsterdam, Netherlands, pp. 39–58.

Echeverría, R.G., 1998: Agricultural research policy issues in Latin America: An overview. *World Dev.*, **26(6)**, 1103–1111, doi:10.1016/S0305-750X**(98)**00036-9.

Economidou, M. et al., 2020: Review of 50 years of EU energy efficiency policies for buildings. *Energy Build.*, **225**, 110322, doi:https://doi.org/10.1016/j.enbuild.2020.110322.

Edler, J. and L. Georghiou, 2007: Public procurement and innovation – Resurrecting the demand side. *Res. Policy*, **36(7)**, 949–963, doi:10.1016/j.respol.2007.03.003.

Edmondson, D.L., F. Kern, and K.S. Rogge, 2019: The co-evolution of policy mixes and socio-technical systems: Towards a conceptual framework of policy mix feedback in sustainability transitions. *Res. Policy*, **48(10)**, 103555, doi:10.1016/j.respol.2018.03.010.

Edmondson, D.L., K.S. Rogge, and F. Kern, 2020: Zero carbon homes in the UK? Analysing the co-evolution of policy mix and socio-technical system. *Environ. Innov. Soc. Transitions*, **35**, 135–161, doi:10.1016/j.eist.2020.02.005.

Edquist, C., 1997: *Systems of Innovation: Technologies, Institutions and Organizations*. [Edquist, C., (ed.)]. Pinter Publishers/Cassell Academic, London, UK, 432 pp.

Egli, F., B. Steffen, and T.S. Schmidt, 2018: A dynamic analysis of financing conditions for renewable energy technologies. *Nat. Energy*, **3(12)**, 1084–1092, doi:10.1038/s41560-018-0277-y.

Einstein, A., 1905: Über einen die Erzeugung und Verwandlung des Lichtes betreffenden heuristischen Gesichtspunkt. *Ann. Phys.*, **322(6)**, 132–148, doi:10.1002/andp.19053220607.

Eisenberg, A.M., 2019: Just transitions. *South. Calif. Law Rev.*, **92(2)**, 273–330.

Elder, M., M. Bengtsson, and L. Akenji, 2016: An Optimistic Analysis of the Means of Implementation for Sustainable Development Goals: Thinking about Goals as Means. *Sustainability*, **8(9)**, 962, doi:10.3390/su8090962.

Elia, A., M. Kamidelivand, F. Rogan, and B. Ó Gallachóir, 2021: Impacts of innovation on renewable energy technology cost reductions. *Renew. Sustain. Energy Rev.*, **138**, 110488, doi:10.1016/j.rser.2020.110488.

Elshurafa, A.M., S.R. Albardi, S. Bigerna, and C.A. Bollino, 2018: Estimating the learning curve of solar PV balance-of-system for over 20 countries: Implications and policy recommendations. *J. Clean. Prod.*, **196**, 122–134, doi:10.1016/J.JCLEPRO.2018.06.016.

Elzen, B., F. Geels, and K. Green, 2004: *System Innovation and the Transition to Sustainability: Theory, evidence and policy*. 1st ed. [Elzen, B., F. Geels, and K. Green, (eds.)]. Edward Elgar Publishing, London, UK and New York, NY, USA, 336 pp.

Emmerling, J. et al., 2016: *The WITCH 2016 Model – Documentation and Implementation of the Shared Socioeconomic Pathways. FEEM Working Paper No. 42.2016*. Fondazione Eni Enrico Mattei, Milan, Italy, 63 pp.

Encaoua, D., D. Guellec, and C. Martínez, 2006: Patent systems for encouraging innovation: Lessons from economic analysis. *Res. Policy*, **35(9)**, 1423–1440, doi:10.1016/j.respol.2006.07.004.

EPO, 2017: *Unitary Patent Guide*. European Patent Office, Munich, https://www.epo.org/applying/european/guide-up/html/e/index.html

Erickson, P., S. Kartha, M. Lazarus, and K. Tempest, 2015: Assessing carbon lock-in. *Environ. Res. Lett.*, **10(8)**, 084023, doi:10.1088/1748-9326/10/8/084023.

European Association of Research Technology Organisations, 2014: *The TRL Scale as a Research & Innovation Policy Tool*. EARTO, Published Online, 17pp. http://www.earto.eu/fileadmin/content/03_Publications/ The_TRL_Scale_as_a_R_I_Policy_Tool_-_EARTO_Recommendations_-_Final.pdf.

European Commission, 2015: *EU R&D Survey: The 2015 EU Survey on Industrial R&D Investment Trends*. European Commission Joint Research Centre, Seville, Spain, 54 pp.

European Commission, 2018: *Ecodesign Impacts Accounting, Status October 2018*. European Union, Published Online, 382pp https://ec.europa.eu/energy/sites/ener/files/documents/eia_status_report_2017_-_v20171222.pdf.

European Commission, 2020: *An SME Strategy for a sustainable and digital Europe*. COM/2020/103. European Union, Brussels, 19pp.

Evenson, R.E. and D. Gollin, 2003: Assessing the Impact of the Green Revolution, 1960 to 2000. *Science*, **300(5620)**, 758–762, doi:10.1126/science.1078710.

Fagerberg, J., 2018: Mobilizing innovation for sustainability transitions: A comment on transformative innovation policy. *Res. Policy*, **47(9)**, 1568–1576, doi:10.1016/j.respol.2018.08.012.

Falvey, R., N. Foster, and D. Greenaway, 2006: Intellectual Property Rights and Economic Growth. *Rev. Dev. Econ.*, **10(4)**, 700–719, doi:10.1111/j.1467-9361.2006.00343.x.

Fankhauser, S. and F. Jotzo, 2018: Economic growth and development with low-carbon energy. *WIREs Clim. Chang.*, **9(1)**, e495, doi:10.1002/wcc.495.

FAO, 2018a: *Transforming Food and Agriculture to Achieve the SDGs: 20 interconnected actions to guide decision-makers*. FAO, Rome, Italy, 132 pp.

FAO, 2018b: *Los 10 Elementos De La Agroecología Guía para la Transición Hacia Sistemas Alimentarios y Agrícolas Sostenibles*. FAO, Rome, Italy, 15 pp.

FAO, 2019: *FAO activities in SIDS*. FAO, Published Online, 16 pp. http://www.fao.org/sids/activities-in-sids/en/.

Farmer, J.D. et al., 2019: Sensitive intervention points in the post-carbon transition. *Science*, **364(6436)**, 132–134, doi:10.1126/science.aaw7287.

Fazey, I. et al., 2020: Transforming knowledge systems for life on Earth: Visions of future systems and how to get there. *Energy Res. Soc. Sci.*, **70**, 101724, doi:https://doi.org/10.1016/j.erss.2020.101724.

Fernández-Sastre, J. and F. Montalvo-Quizhpi, 2019: The effect of developing countries' innovation policies on firms' decisions to invest in R&D. *Technol. Forecast. Soc. Change*, **143**(February), 214–223, doi:10.1016/j.techfore.2019.02.006.

Fischer, C., L. Preonas, and R.G. Newell, 2017: Environmental and Technology Policy Options in the Electricity Sector: Are We Deploying Too Many? *J. Assoc. Environ. Resour. Econ.*, **4(4)**, 959–984, doi:10.1086/692507.

Fisher, E., R.L. Mahajan, and C. Mitcham, 2006: Midstream Modulation of Technology: Governance From Within. *Bull. Sci. Technol. Soc.*, **26(6)**, 485–496, doi:10.1177/0270467606295402.

Fleming, L. and O. Sorenson, 2001: Technology as a complex adaptive system: Evidence from patent data. *Res. Policy*, **30(7)**, 1019–1039, doi:10.1016/S0048-7333**(00)**00135-9.

Foray, D., 2004: The patent system and the dynamics of innovation in Europe. *Sci. Public Policy*, **31(6)**, 449–456, doi:10.3152/147154304781779732.

Foster, C. and S. Azmeh, 2020: Latecomer Economies and National Digital Policy: An Industrial Policy Perspective. *J. Dev. Stud.*, **56(7)**, 1247–1262, doi:10.1080/00220388.2019.1677886.

Fragkiadakis, K., P. Fragkos, and L. Paroussos, 2020: Low-Carbon R&D Can Boost EU Growth and Competitiveness. *Energies*, **13(19)**, 5236, doi:10.3390/en13195236.

Francis, C. et al., 2003: Agroecology: The Ecology of Food Systems. *J. Sustain. Agric.*, **22(3)**, 99–118, doi:10.1300/J064v22n03_10.

Frank, C., C. Sink, L. Mynatt, R. Rogers, and A. Rappazzo, 1996: Surviving the "valley of death": A comparative analysis. *J. Technol. Transf.*, **21(1–2)**, 61–69, doi:10.1007/BF02220308.

Frankfurt School-UNEP Centre/BNEF, 2019: *Global Trends in Renewable Energy Investment 2020*. Frankfurt School-UNEP Centre/BNEF, Frankfurt, Germany, 76 pp. https://wedocs.unep.org/bitstream/handle/20.500.11822/29752/GTR2019.pdf.

Freeman, C., 1995: The 'National System of Innovation' in historical perspective. *Cambridge J. Econ.*, **19(1)**, 5–14, doi:10.1093/oxfordjournals.cje.a035309.

Freeman, C. and L. Soete, 1997: *The Economics of Industrial Innovation*. 3rd ed. [Freeman, C. and L. Soete, (eds.)]. Routledge, Taylor & Francis Group, London, UK, 480 pp.

Freeman, C. and L. Soete, 2009: Developing science, technology and innovation indicators: What we can learn from the past. *Res. Policy*, **38(4)**, 583–589, doi:10.1016/j.respol.2009.01.018.

Frenken, K., L.R. Izquierdo, and P. Zeppini, 2012: Branching innovation, recombinant innovation, and endogenous technological transitions. *Environ. Innov. Soc. Transitions*, **4**, 25–35, doi:10.1016/j.eist.2012.06.001.

Fridgen, G., R. Keller, M.-F. Körner, and M. Schöpf, 2020: A holistic view on sector coupling. *Energy Policy*, **147**, 111913, doi:https://doi.org/10.1016/j.enpol.2020.111913.

Friedman, B. and D.G. Hendry, 2019: *Value Sensitive Design*. The MIT Press, Cambridge, MA, USA, 17–23 pp.

Fu, B., S. Wang, J. Zhang, Z. Hou, and J. Li, 2019: Unravelling the complexity in achieving the 17 sustainable-development goals. *Natl. Sci. Rev.*, **6(3)**, 386–388, doi:10.1093/nsr/nwz038.

Fu, W., C. Li, J. Ondrich, and D. Popp, 2018: *Technological Spillover Effects of State Renewable Energy Policy: Evidence from Patent Counts*. National Bureau of Economic Research, Cambridge, MA, USA, 41 pp.

Fuerst, F. and P. McAllister, 2011: Eco-labeling in commercial office markets: Do LEED and Energy Star offices obtain multiple premiums? *Ecol. Econ.*, **70(6)**, 1220–1230, doi:10.1016/j.ecolecon.2011.01.026.

Fuglie, K., 2018: R&D Capital, R&D spillovers, and productivity growth in world agriculture. *Appl. Econ. Perspect. Policy*, **40(3)**, 421–444, doi:10.1093/AEPP/PPX045.

Funtowicz, S., 2020: From risk calculations to narratives of danger. *Clim. Risk Manag.*, **27**, 100212, doi:10.1016/j.crm.2020.100212.

Gaddy, B.E., V. Sivaram, T.B. Jones, and L. Wayman, 2017: Venture capital and cleantech: The wrong model for energy innovation. *Energy Policy*, **102**(July), 385–395, doi:10.1016/j.enpol.2016.12.035.

Gallagher, K.S., J.P. Holdren, and A.D. Sagar, 2006: Energy-technology innovation. *Annu. Rev. Environ. Resour.*, **31(1)**, 193–237, doi:10.1146/annurev.energy.30.050504.144321.

Gallagher, K.S., L.D. Anadon, R. Kempener, and C. Wilson, 2011: Trends in investments in global energy research, development, and demonstration. *Wiley Interdiscip. Rev. Clim. Chang.*, **2(3)**, 373–396, doi:10.1002/wcc.112.

Gallagher, K.S., A. Grübler, L. Kuhl, G. Nemet, and C. Wilson, 2012: The energy technology innovation system. *Annu. Rev. Environ. Resour.*, **37(1)**, 137–162, doi:10.1146/annurev-environ-060311-133915.

Galloway McLean, K., 2010: *Advance Guard: Climate Change Impacts, Adaptation, Mitigation and Indigenous Peoples – A Compendium of Case Studies*. United Nations University, Traditional Knowledge Initiative, Darwin, Australia, 128 pp. http://www.unutki.org/ (Accessed July 22, 2021).

Gambhir, A., P. Sandwell, and J. Nelson, 2016: The future costs of OPV – A bottom-up model of material and manufacturing costs with uncertainty analysis. *Sol. Energy Mater. Sol. Cells*, **156**, 49–58, doi:10.1016/j.solmat.2016.05.056.

Gambhir, A., I. Butnar, P.-H. Li, P. Smith, and N. Strachan, 2019: A Review of Criticisms of Integrated Assessment Models and Proposed Approaches to Address These, through the Lens of BECCS. *Energies*, **12(9)**, 1747, doi:10.3390/en12091747.

Gandenberger, C., M. Bodenheimer, J. Schleich, R. Orzanna, and L. Macht, 2016: Factors driving international technology transfer: Empirical insights from a CDM project survey. *Clim. Policy*, **16(8)**, 1065–1084, doi:10.1080/14693062.2015.1069176.

Gann, D.M., Y. Wang, and R. Hawkins, 1998: Do regulations encourage innovation? The case of energy efficiency in housing. *Build. Res. Inf.*, **26(5)**, 280–296, doi:10.1080/096132198369760.

Gao, L. and B.A. Bryan, 2017: Finding pathways to national-scale land-sector sustainability. *Nature*, **544(7649)**, 217–222, doi:10.1038/nature21694.

Gao, X. and W. Zhang, 2013: Foreign investment, innovation capacity and environmental efficiency in China. *Math. Comput. Model.*, **58(5–6)**, 1040–1046, doi:10.1016/j.mcm.2012.08.012.

Gao, X. and V. Rai, 2019: Local demand-pull policy and energy innovation: Evidence from the solar photovoltaic market in China. *Energy Policy*, **128**(February 2018), 364–376, doi:10.1016/j.enpol.2018.12.056.

Garcia-Casals, X., R. Ferroukhi, and B. Parajuli, 2019: Measuring the socio-economic footprint of the energy transition. *Energy Transitions*, **3(1–2)**, 105–118, doi:10.1007/s41825-019-00018-6.

Geddes, A., T.S. Schmidt, and B. Steffen, 2018: The multiple roles of state investment banks in low-carbon energy finance: An analysis of Australia, the UK and Germany. *Energy Policy*, **115**, 158–170, doi:10.1016/j.enpol.2018.01.009.

Geels, F.W. and J. Schot, 2010: *The Dynamics of Transitions: A Socio-Technical* Perspective. In: Transitions to Sustainable Development: New Directions in the Study of Long Term Transformative Change [Grin, J. Rotmans, J. and Schot, J. (eds.)]Routledge, New York US and Oxon UK, pp. 11–104.

Geels, F.W., 2002: Technological transitions as evolutionary reconfiguration processes: A multi-level perspective and a case-study. *Res. Policy*, **31(8–9)**, 1257–1274, doi:10.1016/S0048-7333**(02)**00062-8.

Geels, F.W., 2004: From sectoral systems of innovation to socio-technical systems. *Res. Policy*, **33(6–7)**, 897–920, doi:10.1016/j.respol.2004.01.015.

Geels, F.W., 2005: *Technological transition and system innovations: A co-evolutionary and socio-technical analysis*. Edward Elgar Publishing, Cheltenham, UK, 328 pp.

Geels, F.W., 2012: A socio-technical analysis of low-carbon transitions: Introducing the multi-level perspective into transport studies. *J. Transp. Geogr.*, **24**, 471–482, doi:10.1016/j.jtrangeo.2012.01.021.

Geels, F.W., 2014: Regime Resistance against Low-Carbon Transitions: Introducing Politics and Power into the Multi-Level Perspective. *Theory, Cult. Soc.*, **31(5)**, 21–40, doi:10.1177/0263276414531627.

Geels, F.W., 2019: Socio-technical transitions to sustainability: A review of criticisms and elaborations of the Multi-Level Perspective. *Curr. Opin. Environ. Sustain.*, **39**, 187–201, doi:10.1016/J.COSUST.2019.06.009.

Geels, F.W., F. Berkhout, and D.P. van Vuuren, 2016: Bridging analytical approaches for low-carbon transitions. *Nat. Clim. Chang.*, **6(6)**, 576–583, doi:10.1038/nclimate2980.

Geels, F.W., B.K. Sovacool, T. Schwanen, and S. Sorrell, 2017: The Socio-Technical Dynamics of Low-Carbon Transitions. *Joule*, **1(3)**, 463–479, doi:10.1016/j.joule.2017.09.018.

Genest, A., 2014: The Canada-FIT case and the WTO subsidies agreement: Failed fact-finding, needless complexity, and missed judicial economy. *McGill J. Sustain. Dev. Law Policy*, **10(2)**, 239–257.

Gerlagh, R. and O. Kuik, 2014: Spill or leak? Carbon leakage with international technology spillovers: A CGE analysis. *Energy Econ.*, **45**, 381–388, doi:10.1016/j.eneco.2014.07.017.

Gertner, J., 2013: *The Idea Factory: Bell Labs and the great age of American innovation*. Penguin Books, London, UK, 422 pp.

GeSI, 2012: *GeSI SMARTer 2020: The Role of ICT in Driving a Sustainable Future*. GeSI, Published Online. https://gesi.org/research/gesi-smarter2020-the-role-of-ict-in-driving-a-sustainable-future.

Gibbons, M., 1999: Science's new social contract with society. *Nature*, **402(6761)**, C81–C84, doi:10.1038/35011576.

Gillingham, K. et al., 2016: Deconstructing solar photovoltaic pricing: The role of market structure, technology, and policy. *Energy J.*,**37(3)**,pp 231–250, doi:10.5547/01956574.37.3.kgil.

Gillingham, K.T. and J. Sweeney, 2012: Barries to implementing low-carbon technologies. *Clim. Chang. Econ.*, **03(04)**, 1250019, doi:10.1142/S2010007812500194.

Girod, B., T. Stucki, and M. Woerter, 2017: How do policies for efficient energy use in the household sector induce energy-efficiency innovation? An evaluation of European countries. *Energy Policy*, **103**, 223–237, doi:10.1016/j.enpol.2016.12.054.

Glachant, M. and A. Dechezleprêtre, 2017: What role for climate negotiations on technology transfer? *Clim. Policy*, **17(8)**, 962–981, doi:10.1080/14693062.2016.1222257.

Global Commission on Adaptation, 2019: *Adapt Now. A global call for leadership on climate resilience*. Global Commission on Adaptation, Rotterdam and Washington, 81 pp. https://cdn.gca.org/assets/2019-09/GlobalCommission_Report_FINAL.pdf.

Goldfarb, B., 2011: Economic Transformations: General Purpose Technologies and Long-Term Economic Growth. By Richard G. Lipsey, Kenneth I. Carlaw, and Clifford T. Bekar. Oxford: Oxford University Press, UK,. *J. Econ. Hist.*, **71(3)**, 820–823, doi:10.1017/S0022050711002099.

Goldstein, A., C. Doblinger, E. Baker, and L.D. Anadón, 2020: Patenting and business outcomes for cleantech startups funded by the Advanced Research Projects Agency-Energy. *Nat. Energy*, **5(10)**, 803–810, doi:10.1038/s41560-020-00683-8.

Goldstein, A.P. and V. Narayanamurti, 2018: Simultaneous pursuit of discovery and invention in the US Department of Energy. *Res. Policy*, **47(8)**, 1505–1512, doi:10.1016/j.respol.2018.05.005.

Goldstein, A.P. and M. Kearney, 2020: Know when to fold 'em: An empirical description of risk management in public research funding. *Res. Policy*, **49(1)**, 103873, doi:10.1016/j.respol.2019.103873.

Golombek, R. and M. Hoel, 2004: Unilateral Emission Reductions and Cross-Country Technology Spillovers. *Adv. Econ. Anal. Policy*, **3(2)**, 57–83, doi:10.2202/1538-0637.1318.

Gonsalves, M. and J.M. Rogerson, 2019: Business incubators and green technology: The Gauteng Climate Innovation Centre, South Africa. Urbani izziv, **30**, 212–224, doi:10.5379/urbani-izziv-en-2019-30-supplement-014.

González, F.G. et al., 2019: Field-grown transgenic wheat expressing the sunflower gene HaHB4 significantly outyields the wild type. *J. Exp. Bot.*, **70(5)**, 1669–1681, doi:10.1093/jxb/erz037.


16

Gossart, C., 2015: *ICT Innovations for Sustainability*. [Hilty, L.M. and B. Aebischer, (eds.)]. Springer International Publishing, Cham, Switzerland, 435–448 pp.

Goulder, L.H. and S.H. Schneider, 1999: Induced technological change and the attractiveness of $CO_2$ abatement policies. *Resour. Energy Econ.*, **21(3–4)**, 211–253, doi:10.1016/S0928-7655**(99)**00004-4.

Gram-Hanssen, K. and S. Georg, 2018: Energy performance gaps: Promises, people, practices. *Build. Res. Inf.*, **46(1)**, 1–9, doi:10.1080/09613 218.2017.1356127.

Grandia, J. and D. Voncken, 2019: Sustainable Public Procurement: The Impact of Ability, Motivation, and Opportunity on the Implementation of Different Types of Sustainable Public Procurement. *Sustainability*, **11(19)**, 5215, doi:10.3390/su11195215.

Grandstrand, O., 2000: *The Economics and Management of Intellectual Property: Towards Intellectual Capitalism*. [Granstrand, O., (ed.)]. Edward Elgar Publishing, Cheltenham, UK, 480 pp.

Grau, H.R. et al., 2015: Natural grasslands in the Chaco. A neglected ecosystem under threat by agriculture expansion and forest-oriented conservation policies. *J. Arid Environ.*, **123**, 40–46, doi:10.1016/j.jaridenv.2014.12.006.

Greaker, M., T. Heggedal, and K.E. Rosendahl, 2018: Environmental policy and the direction of technical change. *Scand. J. Econ.*, **120(4)**, 1100–1138, doi:10.1111/sjoe.12254.

Green, M.A., 2019: How Did Solar Cells Get So Cheap? *Joule*, **3(3)**, 631–633, doi:10.1016/j.joule.2019.02.010.

Green Climate Fund, 2020: *GCF Support to Climate Technologies: 16th Meeting of the Advisory Board of the CTCN*. 19 pp. https://www.ctc-n.org/sites/www.ctc-n.org/files/Agenda_4.3_Green_Climate_Fund.pdf (Accessed August 25, 2021).

Grimm, R., C. Fox, S. Baines, and K. Albertson, 2013: Social innovation, an answer to contemporary societal challenges? Locating the concept in theory and practice.

*Innovation: The European Journal of Social Science Research* **26(4)**, 436–455, doi:10.1080/13511610.2013.848163.

Griscom, B.W. et al., 2017: Natural climate solutions. *Proc. Natl. Acad. Sci.*, **114(44)**, 11645–11650, doi:10.1073/pnas.1710465114.

Gross, R., R. Hanna, A. Gambhir, P. Heptonstall, and J. Speirs, 2018: How long does innovation and commercialisation in the energy sectors take? Historical case studies of the timescale from invention to widespread commercialisation in energy supply and end use technology. *Energy Policy*, **123**, 682–699, doi:10.1016/j.enpol.2018.08.061.

Grubb, M. and C. Wieners, 2020: Modeling Myths: On the Need for Dynamic Realism in DICE and other Equilibrium Models of Global Climate Mitigation. *Inst. New Econ. Think. Work. Pap. Ser.*,112, 1–29, doi:10.36687/inetwp112.

Grubb, M. et al., 2021: Induced innovation in energy technologies and systems: A review of evidence and potential implications for $CO_2$ mitigation. *Environ. Res. Lett.*, **16(4)**, 043007, doi:10.1088/1748-9326/abde07.

Grubler, A. and C. Wilson, 2013: *Energy Technology Innovation: Learning from historical successes and failures*. Cambridge University Press, Cambridge, UK, 1–387 pp.

Grubler, A. et al., 2018: A low energy demand scenario for meeting the 1.5°C target and sustainable development goals without negative emission technologies. *Nat. Energy*, **3(6)**, 515–527, doi:10.1038/s41560-018-0172-6.

Grübler, A., 1996: Time for a change: On the patterns of diffusion of innovation. *Daedalus*, **125(3)**, 19–42.

Grübler, A., 1998: *Technology and Global Change*. Cambridge University Press, Cambridge, UK, 452 pp.

Grübler, A., N. Nakićenović, and D.G. Victor, 1999a: Modeling technological change: Implications for the global environment. *Annu. Rev. Energy Environ.*, **24(1)**, 545–569, doi:10.1146/annurev.energy.24.1.545.

Grübler, A., N. Nakićenović, and D.G. Victor, 1999b: Dynamics of energy technologies and global change. *Energy Policy*, **27(5)**, 247–280, doi:10.1016/S0301-4215**(98)**00067-6.

Gruebler, A. et al., 2012: Energy technology innovation systems. In: *The Global Energy Assessment* [Nakicenovic, N. et al., (ed.)], Cambridge University Press, Cambridge, UK, pp 11-29 doi:10.1017/CBO9781139150880.004.

Gunderson, L.H. and C.S. Holling, 2002: *Understanding Transformations in Human and Natural Systems*. Island Press, Washington DC, USA, 509 pp.

Gür, T.M., 2018: Review of electrical energy storage technologies, materials and systems: Challenges and prospects for large-scale grid storage. *Energy Environ. Sci.*, **11(10)**, 2696–2767, doi:10.1039/C8EE01419A.

Haelg, L., M. Waelchli, and T.S. Schmidt, 2018: Supporting energy technology deployment while avoiding unintended technological lock-in: A policy design perspective. *Environ. Res. Lett.*, **13(10)**, 104011, doi:10.1088/1748-9326/aae161.

Haenssgen, M.J. and P. Ariana, 2018: The place of technology in the capability approach. *Oxford Dev. Stud.*, **46(1)**, 98–112, doi:10.1080/136 00818.2017.1325456.

Hall, B. and C. Helmers, 2010: *The role of patent protection in (clean/green) technology transfer*. National Bureau of Economic Research,16323, 39 pp. http://www.nber.org/papers/w16323.pdf.

Hall, B.H. and R.H. Ziedonis, 2001: The Patent Paradox Revisited: An Empirical Study of Patenting in the U.S. Semiconductor Industry, 1979–1995. *RAND J. Econ.*.**32(1)**,101-128, doi:10.2307/2696400.

Hall, B.H. and C. Helmers, 2019: The impact of international patent systems: Evidence from accession to the European Patent Convention. *Res. Policy*, **48(9)**, 103810, doi:10.1016/j.respol.2019.103810.

Hall, J.K.K. and M.J.C. Martin, 2005: Disruptive technologies, stakeholders and the innovation value-added chain: A framework for evaluating radical technology development. *R&D Manag.*, **35(3)**, 273–284, doi:https://doi.org/10.1111/j.1467-9310.2005.00389.x.

Hall, M.J. and A.N. Link, 2015: Technology-based state growth policies: The case of North Carolina's Green Business Fund. *Ann. Reg. Sci.*, **54(2)**, 437–449, doi:10.1007/s00168-015-0661-5.

Halleck Vega, S., A. Mandel, and K. Millock, 2018: Accelerating diffusion of climate-friendly technologies: A network perspective. *Ecol. Econ.*, **152**(June), 235–245, doi:10.1016/j.ecolecon.2018.05.007.

Halleck Vega, S. and A. Mandel, 2018: Technology Diffusion and Climate Policy: A Network Approach and its Application to Wind Energy. *Ecol. Econ.*, **145**(December 2017), 461–471, doi:10.1016/j.ecolecon.2017.11.023.

Hallegatte, S., 2012: *A cost effective solution to reduce disaster losses in developing countries: Hydro-meterological services, early warning and evacuation*. Policy Research Working Paper; No. 6058. World Bank, Washington, DC, USA, 22 pp. http://hdl.handle.net/10986/9359.

Hanley, E.S., J. Deane, and B.Ó. Gallachóir, 2018: The role of hydrogen in low carbon energy futures – A review of existing perspectives. *Renew. Sustain. Energy Rev.*, **82**, 3027–3045, doi:10.1016/j.rser.2017.10.034.

Hanna, R. and D.G. Victor, 2021: Marking the decarbonization revolutions. *Nat. Energy*, **6(6)**, 568–571, doi:10.1038/s41560-021-00854-1.

Hansen, H.K. and T. Porter, 2017: What Do Big Data Do in Global Governance? *Glob. Gov. A Rev. Multilater. Int. Organ.*, **23(1)**, 31–42, doi:https://doi.org/10.1163/19426720-02301004.

Hansen, U.E. et al., 2018: Sustainability transitions in developing countries: Stocktaking, new contributions and a research agenda. *Environ. Sci. Policy*, **84**, 198–203, doi:10.1016/j.envsci.2017.11.009.

Hao, H., S. Wang, Z. Liu, and F. Zhao, 2016: The impact of stepped fuel economy targets on automaker's light-weighting strategy: The China case. *Energy*, **94**, 755–765, doi:10.1016/j.energy.2015.11.051.

Harrowsmith, M. et al., 2020: *The future of forecasts: Impact-based forecasting for early action*.ARRCC, Met Office, Climate Centre, UK Aid, Anticipation Hub, REAP, Geneva, 81 pp. https://www.forecast-based-financing.org/wp-content/uploads/2020/09/Impact-based-forecasting-guide-2020.pdf.

Haščič, I. and M. Migotto, 2015: Measuring environmental innovation using patent data. *OECD Environ. Work. Pap.*, **89**, 58, doi:https://doi.org/10.1787/19970900.

Haselip, J. et al., 2015: Governance, enabling frameworks and policies for the transfer and diffusion of low carbon and climate adaptation technologies in developing countries. *Clim. Change*, **131(3)**, 363–370, doi:10.1007/s10584-015-1440-0.

Hassler, J., P. Krusell, and C. Olovsson, 2012: Energy-Saving Technical Change. *National Bureau of Economic Research (NBER) Working Paper Series*, **18456**, 41, doi:10.3386/w18456.

He, W. and R. Shen, 2019: ISO 14001 Certification and Corporate Technological Innovation: Evidence from Chinese Firms. *J. Bus. Ethics*, **158(1)**, 97–117, doi:10.1007/s10551-017-3712-2.

He, X. et al., 2019: Solar and wind energy enhances drought resilience and groundwater sustainability. *Nat. Commun.*, **10(1)**, 4893, doi:10.1038/s41467-019-12810-5.

Heeks, R. and C. Stanforth, 2015: Technological change in developing countries: Opening the black box of process using actor–network theory. *Dev. Stud. Res.*, **2(1)**, 33–50, doi:10.1080/21665095.2015.1026610.

Heffron, R.J. and D. McCauley, 2018: What is the 'Just Transition'? *Geoforum*, **88**, 74–77, doi:10.1016/J.GEOFORUM.2017.11.016.

Hekkert, M., R. Suurs, S. Negro, S. Kuhlmann, and R. Smits, 2007: Functions of innovation systems: A new approach for analysing technological change. *Technol. Forecast. Soc. Change*, **74(4)**, 413–432.

Hekkert, M.P. and S.O. Negro, 2009: Functions of innovation systems as a framework to understand sustainable technological change: Empirical evidence for earlier claims. *Technol. Forecast. Soc. Change*, **76(4)**, 584–594, doi:10.1016/j.techfore.2008.04.013.

Hekkert, M.P., M.J. Janssen, J.H. Wesseling, and S.O. Negro, 2020: Mission-oriented innovation systems. *Environ. Innov. Soc. Transitions*, **34**, 76–79, doi:10.1016/j.eist.2019.11.011.

Heller, M.A. and R.S. Eisenberg, 1998: Can Patents Deter Innovation? The Anticommons in Biomedical Research. *Science*, **280(5364)**, 698–701, doi:10.1126/science.280.5364.698.

Hellsmark, H., J. Frishammar, P. Söderholm, and H. Ylinenpää, 2016: The role of pilot and demonstration plants in technology development and innovation policy. *Res. Policy*, **45(9)**, 1743–1761, doi:10.1016/j.respol.2016.05.005.

Helm, D., C. Hepburn, and R. Mash, 2003: Credible carbon policy. *Oxford Rev. Econ. Policy*, **19(3)**, 438–450. https://doi.org/10.1093/oxrep/19.3.438.

Helveston, J. and J. Nahm, 2019: China's key role in scaling low-carbon energy technologies. *Science*, **366(6467)**, 794–796, doi:10.1126/science.aaz1014.

Hémous, D., 2016: The dynamic impact of unilateral environmental policies. *J. Int. Econ.*, **103**(C), 80–95, doi:10.1016/j.jinteco.2016.09.001.

Henderson, R. and R. Newell, 2011: Introduction. In: *Accelerating Energy Innovation* [Henderson, R. and R. Newell, (eds.)], Chicago University Press, Chicago, USA, pp. 288.

Henry, M.S., M.D. Bazilian, and C. Markuson, 2020: Just transitions: Histories and futures in a post-COVID world. *Energy Res. Soc. Sci.*, **68**, 101668, doi:10.1016/j.erss.2020.101668.

Herslund, L. et al., 2018: Conditions and opportunities for green infrastructure – Aiming for green, water-resilient cities in Addis Ababa and Dar es Salaam. *Landsc. Urban Plan.*, **180**, 319–327, doi:10.1016/j.landurbplan.2016.10.008.

Hertwich, E.G. and G.P. Peters, 2009: Carbon Footprint of Nations: A Global, Trade-Linked Analysis. *Environ. Sci. Technol.*, **43(16)**, 6414–6420, doi:10.1021/es803496a.

Hirsch, A., Y. Parag, and J. Guerrero, 2018: Microgrids: A review of technologies, key drivers, and outstanding issues. *Renew. Sustain. Energy Rev.*, **90**, 402–411, doi:https://doi.org/10.1016/j.rser.2018.03.040.

Hoffecker, E., 2021: Understanding inclusive innovation processes in agricultural systems: A middle-range conceptual model. *World Dev.*, **140**, 105382, doi:https://doi.org/10.1016/j.worlddev.2020.105382.

Hofman, E. and W. van der Gaast, 2019: Enhancing ambition levels in nationally determined contributions – Learning from Technology Needs Assessments. *Wiley Interdiscip. Rev. Energy Environ.*, **8(1)**, e311, doi:10.1002/wene.311.

Holt-Giménez, E., 2002: Measuring farmers' agroecological resistance after Hurricane Mitch in Nicaragua: A case study in participatory, sustainable land management impact monitoring. *Agric. Ecosyst. Environ.*, **93(1-3)**, 87–105 doi:10.1016/S0167-8809**(02)**00006-3.

Hoppmann, J., M. Peters, M. Schneider, and V.H. Hoffmann, 2013: The two faces of market support – How deployment policies affect technological exploration and exploitation in the solar photovoltaic industry. *Res. Policy*, **42(4)**, 989–1003, doi:10.1016/j.respol.2013.01.002.

Hoppmann, J., J. Huenteler, and B. Girod, 2014: Compulsive policy-making – The evolution of the German feed-in tariff system for solar photovoltaic power. *Res. Policy*, **43(8)**, 1422–1441, doi:10.1016/j.respol.2014.01.014.

Hoppmann, J., F. Naegele, and B. Girod, 2019: Boards as a Source of Inertia: Examining the Internal Challenges and Dynamics of Boards of Directors in Times of Environmental Discontinuities. *Acad. Manag. J.*, **62(2)**, 437–468, doi:10.5465/amj.2016.1091.

Horbach, J., C. Rammer, and K. Rennings, 2012: Determinants of eco-innovations by type of environmental impact – The role of regulatory push/pull, technology push and market pull. *Ecol. Econ.*, **78**, 112–122, doi:10.1016/j.ecolecon.2012.04.005.

Horner, N.C., A. Shehabi, and I.L. Azevedo, 2016: Known unknowns: Indirect energy effects of information and communication technology. *Environ. Res. Lett.*, **11(10)**, 103001, doi:10.1088/1748-9326/11/10/103001.

Horton, D. and R. Mackay, 2003: Using evaluation to enhance institutional learning and change: Recent experiences with agricultural research and development. *Agric. Syst.*, **78(2)**, 127–142, doi:10.1016/S0308-521X**(03)**00123-9.

Hossain, M., 2018: Frugal innovation: A review and research agenda. *J. Clean. Prod.*, **182**, 926–936, doi:10.1016/J.JCLEPRO.2018.02.091.

Howell, S.T., 2017: Financing innovation: Evidence from R&D Grants. *Am. Econ. Rev.*, **107(4)**, 1136–1164, doi:10.1257/aer.20150808.

Howlett, M. and P. del Rio, 2015: The parameters of policy portfolios: Verticality and horizontality in design spaces and their consequences for policy mix formulation. *Environ. Plan. C Gov. Policy*, **33(5)**, 1233–1245, doi:10.1177/0263774X15610059.

Hsu, S.-L., 2017: Capital Transitioning: An International Human Capital Strategy for Climate Innovation. *Transnatl. Environ. Law*, **6(1)**, 153–176, doi:10.1017/S2047102516000169.

Hu, R., J. Skea, and M.J. Hannon, 2018: Measuring the energy innovation process: An indicator framework and a case study of wind energy in China. *Technol. Forecast. Soc. Change*, **127**, 227–244, doi:10.1016/j.techfore.2017.09.025.

Huang, R. et al., 2016: Energy and emissions saving potential of additive manufacturing: The case of lightweight aircraft components. *J. Clean. Prod.*, **135**, 1559–1570, doi:10.1016/j.jclepro.2015.04.109.

Huenteler, J., C. Niebuhr, and T.S. Schmidt, 2016a: The effect of local and global learning on the cost of renewable energy in developing countries. *J. Clean. Prod.*, **128**, 6–21, doi:10.1016/j.jclepro.2014.06.056.

Huenteler, J., T.S. Schmidt, J. Ossenbrink, and V.H. Hoffmann, 2016b: Technology life-cycles in the energy sector – Technological characteristics and the role of deployment for innovation. *Technol. Forecast. Soc. Change*, **104**, 102–121, doi:10.1016/j.techfore.2015.09.022.

Hughes, S., S. Giest, and L. Tozer, 2020: Accountability and data-driven urban climate governance. *Nat. Clim. Chang.*, **10(12)**, 1085–1090, doi:10.1038/s41558-020-00953-z.

Huh, T. and H.-J. Kim, 2018: Korean Experimentation of Knowledge and Technology Transfer to Address Climate Change in Developing Countries. *Sustainability*, **10(4)**, 1263, doi:10.3390/su10041263.

Hussein, A.K., 2015: Applications of nanotechnology in renewable energies – A comprehensive overview and understanding. *Renew. Sustain. Energy Rev.*, **42**, 460–476, doi:10.1016/j.rser.2014.10.027.

ICLEI, 2018: *Energy Innovation Procurement: A guide for city authorities*. ICLEI, Freiburg, Germany, 48 pp.

IEA, 2015: *4E: Achievements of Appliance Energy Efficiency Standards and Labelling Programs*. IEA, Paris, France, pp 25. https://www.iea.org/reports/achievements-of-appliance-energy-efficiency-standards-and-labelling-programs.

IEA, 2017a: *Early-Stage Venture Capital for Energy Innovation*. IEA, Paris, France, 25 pp. https://www.iea.org/reports/early-stage-venture-capital-for-energy-innovation (Accessed December 18, 2020).

IEA, 2017b: *Digitalisation and Energy*. IEA, Paris, France, 181 pp.

IEA, 2019: *Energy technology RD&D – Data services*. IEA, Paris, France, https://www.iea.org/subscribe-to-data-services/energy-technology-rdd (Accessed December 12, 2020).

IEA, 2020a: *Energy Technology Perspectives 2020*. IEA, Paris, France, 400 pp. https://www.iea.org/reports/energy-technology-perspectives-2020 (Accessed July 21, 2021).

IEA, 2020b: *Energy Technology Perspectives: Special report on Clean Energy Innovation*. IEA, Paris, France, 185 pp. https://iea.blob.core.windows.net/assets/04dc5d08-4e45-447d-a0c1-d76b5ac43987/Energy_Technology_Perspectives_2020_-_Special_Report_on_Clean_Energy_Innovation.pdf.

IEA, 2020c: *World Enery Outlook*. IEA, Paris, France, 464 pp.

Inoue, E., T.H. Arimura, and M. Nakano, 2013: A new insight into environmental innovation: Does the maturity of environmental management systems matter? *Ecol. Econ.*, **94**, 156–163, doi:10.1016/j.ecolecon.2013.07.014.

IPBES, 2019: *Summary for policymakers of the global assessment report on biodiversity and ecosystem services of the Intergovernmental Science-Policy Platform on Biodiversity and Ecosystem Services*. IPBES, Bonn, Germany, 56 pp.

IPCC, 2000: *Methodological and Technological Issues in Technology Transfer*. [Metz, B., D. Ogunlade, J.-W. Martens, S. Van Rooijen, and L. Van Wie Mcgrory, (eds.)]. Cambridge University Press, Cambridge, UK, 432 pp.

IPCC, 2007: *Climate Change 2007: Mitigation of Climate Change*. Contribution of Working Group III to the Fourth Assessment Report of the Intergovernmental Panel on Climate Change, 2007. [Metz, B., O.R. Davidson, P.R. Bosch, R. Dave, and L.A. Meyer, (eds.)]. Cambridge University Press, Cambridge, UK and New York, NY, USA, 863 pp.

IPCC, 2012: *Managing the Risks of Extreme Events and Disasters to Advance Climate Change Adaptation. A Special Report of Working Groups I and II of the Intergovernmental Panel on Climate Change.* [Field, C.B., V. Barros, T.F. Stocker, D. Qin, D.J. Dokken, K.L. Ebi, M.D. Mastrandrea, K.J. Mach, G.-K. Plattner, S.K. Allen, M. Tignor, and P.M. Midgley (eds.)]. Cambridge University Press, Cambridge, UK and New York, NY, USA, 582 pp.

IPCC, 2014: *Climate Change 2014: Mitigation of Climate Change. Contribution of Working Group III to the Fifth Assessment Report of the Intergovernmental Panel on Climate Change* [Edenhofer, O., R. Pichs-Madruga, Y. Sokona, E. Farahani, S. Kadner, K. Seyboth, A. Adler, I. Baum, S. Brunner, P. Eickemeier, B. Kriemann, J. Savolainen, S. Schlömer, C. von Stechow, T. Zwickel and J.C. Minx (eds.)]. Cambridge University Press, Cambridge, UK, and New York, NY, USA, 1454 pp.

IPCC, 2018a: *Global Warming of 1.5°C. An IPCC Special Report on the impacts of global warming of 1.5°C above pre-industrial levels and related global greenhouse gas emission pathways, in the context of strengthening the global response to the threat of climate change, sustainable development, and efforts to eradicate poverty* [Masson-Delmotte, V., P. Zhai, H.-O. Pörtner, D. Roberts, J. Skea, P.R. Shukla, A. Pirani, W. Moufouma-Okia, C. Péan, R. Pidcock, S. Connors, J.B.R. Matthews, Y. Chen, X. Zhou, M.I. Gomis, E. Lonnoy, T. Maycock, M. Tignor, and T. Waterfield (eds.)]. Cambridge University Press, Cambridge, UK and New York, NY, USA.

IPCC, 2018b: Summary for Policymakers. In: *Global Warming of 1.5°C. An IPCC Special Report on the impacts of global warming of 1.5°C above pre-industrial levels and related global greenhouse gas emission pathways, in the context of strengthening the global response to the threat of climate change, sustainable development, and efforts to eradicate poverty* [Masson-Delmotte, V., P. Zhai, H.-O. Pörtner, D. Roberts, J. Skea, P.R. Shukla, A. Pirani, W. Moufouma-Okia, C. Péan, R. Pidcock, S. Connors, J.B.R. Matthews, Y. Chen, X. Zhou, M.I. Gomis, E. Lonnoy, T. Maycock, M. Tignor, and T. Waterfield (eds.)]. Cambridge University Press, Cambridge, UK and New York, NY, USA.

IRENA, 2020a: *Renewable energy statistics*. IRENA, Abu Dhabi, 408 pp. https://www.irena.org/publications/2020/Jul/Renewable-energy-statistics-2020.

IRENA, 2020b: *Renewable Power Generation Costs in 2019*. IRENA, Abu Dhabi, 144 pp. https://www.irena.org/publications/2020/Jun/Renewable-Power-Costs-in-2019 (Accessed October 30, 2021).

Isoard, S. and A. Soria, 2001: Technical change dynamics: Evidence from the emerging renewable energy technologies. *Energy Econ.*, **23(6)**, 619–636, doi:10.1016/S0140-9883**(01)**00072-X.

Ito, K. and J.M. Sallee, 2018: The Economics of Attribute-Based Regulation: Theory and Evidence from Fuel Economy Standards. *Rev. Econ. Stat.*, **100(2)**, 319–336, doi:10.1162/REST_a_00704.

Ivanova, D. et al., 2018: Carbon mitigation in domains of high consumer lock-in. *Glob. Environ. Chang.*, **52**, 117–130, doi:10.1016/j.gloenvcha.2018.06.006.

Iyer, G. et al., 2015: Diffusion of low-carbon technologies and the feasibility of long-term climate targets. *Technol. Forecast. Soc. Change*, **90**(PA), 103–118, doi:10.1016/j.techfore.2013.08.025.

Jacobsen, M.R., 2013: Evaluating US Fuel Economy Standards in a Model with Producer and Household Heterogeneity. *Am. Econ. J. Econ. Policy*, **5(2)**, 148–187, doi:10.1257/pol.5.2.148.

Jacobsson, S. and A. Bergek, 2006: A Framework for Guiding Policy-makers Intervening in Emerging Innovation Systems in 'Catching-Up' Countries. *Eur. J. Dev. Res.*, **18(4)**, 687–707, doi:10.1080/09578810601094902.

Jacobsson, S., B. Sandén, and L. Bångens, 2004: Transforming the Energy System — the Evolution of the German Technological System for Solar Cells. *Technol. Anal. Strateg. Manag.*, **16(1)**, 3–30, doi:10.1080/0953732032000199061.

Jacobsson, S., A. Bergek, and B. Sandén, 2017: Improving the European Commission's analytical base for designing instrument mixes in the energy sector: Market failures versus system weaknesses. *Energy Res. Soc. Sci.*, **33**, 11–20, doi:10.1016/j.erss.2017.09.009.

Jaffe, A. and J. Lerner, 2004: *Innovation and its discontents: How our broken patent system is endangering innovation and progress, and what to do about it*. Princeton University Press, Princeton, New Jersey, USA, 256 pp.

Jaffe, A.B., 2015: Technology diffusion. In: *Emerging Trends in the Social and Behavioral Sciences* [Scott, R. and Kosslyn, (eds.)], John Wiley and Sons Inc., Hoboken, New Jersey, USA. doi: 10.1002/9781118900772.

Jaffe, A.B. and G. de Rassenfosse, 2017: Patent citation data in social science research: Overview and best practices. *J. Assoc. Inf. Sci. Technol.*, **68(6)**, 1360–1374, doi:10.1002/asi.23731.

Jaffe, A.B., R.G. Newell, and R.N. Stavins, 2000: Technological Change and the Environment. *SSRN Electron. J.*, **1**, doi:10.2139/ssrn.252927.

Jaffe, A.B., R.G. Newell, and R.N. Stavins, 2005: A tale of two market failures: Technology and environmental policy. *Ecol. Econ.*, **54(2–3)**, 164–174, doi:10.1016/j.ecolecon.2004.12.027.

Jain, M., A.B. Rao, and A. Patwardhan, 2018: Appliance labeling and consumer heterogeneity: A discrete choice experiment in India. *Appl. Energy*, **226**, 213–224, doi:10.1016/j.apenergy.2018.05.089.

Jairaj, B., A. Agarwal, T. Parthasarathy, and S. Martin, 2016: *Strengthening Governance of India's Appliance Efficiency Standards and Labeling Program*. World Resources Institute, New Delhi, India, 28 pp.

Jamasb, T. and M. Pollitt, 2005: *Deregulation and R&D in Network Industries: The Case of the Electricity Industry*. Cambridge Working Papers in Economics (CWPE), 0533, Cambridge, UK, 45 pp.

Jamasb, T., 2007: Technical Change Theory and Learning Curves: Patterns of Progress in Energy Technologies. *Energy J.*, **28(3)**, 51–72, doi:10.2307/41323109.

Jamasb, T. and M. Pollitt, 2008: Liberalisation and R&D in network industries: The case of the electricity industry. *Res. Policy*, **37(6–7)**, 995–1008, doi:10.1016/j.respol.2008.04.010.

Jasanoff, S., 2003: Technologies of Humility: Citizen Participation in Governing Science. *Minerva*, **41(3)**, 223–244, doi:10.1023/A:1025557512320.

Jasanoff, S., 2018: Just transitions: A humble approach to global energy futures. *Energy Res. Soc. Sci.*, **35**, 11–14, doi:10.1016/j.erss.2017.11.025.

Johnstone, P. and P. Newell, 2018: Sustainability transitions and the state. *Environ. Innov. Soc. Transitions*, **27**, 72–82, doi:10.1016/j.eist.2017.10.006.

Jones, N., 2018: How to stop data centres from gobbling up the world's electricity. *Nature*, **561(7722)**, 163–166, doi:10.1038/d41586-018-06610-y.

Joshi, G.Y., P.A. Sheorey, and A.V. Gandhi, 2019: Analyzing the barriers to purchase intentions of energy efficient appliances from consumer perspective. *Benchmarking An Int. J.*, **26(5)**, 1565–1580, doi:10.1108/BIJ-03-2018-0082.

Kaack, L. et al., 2022: *Aligning artificial intelligence with climate change mitigation*. Nature Climate Change, 12, pp.518–527.

Kahn, M.E. and N. Kok, 2014: The capitalization of green labels in the California housing market. *Reg. Sci. Urban Econ.*, **47**, 25–34, doi:10.1016/j.regsciurbeco.2013.07.001.

Kahouli, S., 2011: Effects of technological learning and uranium price on nuclear cost: Preliminary insights from a multiple factors learning curve and uranium market modeling. *Energy Econ.*, **33(5)**, 840–852, doi:10.1016/J.ENECO.2011.02.016.

Kammen, D.M. and G.F. Nemet, 2007: Energy Myth Eleven – Energy R&D Investment Takes Decades to Reach the Market. In: *Energy and American Society – Thirteen Myths* [Sovacool, B.K. and M.A. Brown, (eds.)], Springer Netherlands, Dordrecht, Netherlands, pp. 289–309.

Kang, J.-N. et al., 2020: The Prospects of Carbon Capture and Storage in China's Power Sector under the 2°C Target: A Component-based Learning Curve Approach. *Int. J. Greenh. Gas Control*, **101**, 103149, doi:10.1016/j.ijggc.2020.103149.

Karakosta, C., H. Doukas, and J. Psarras, 2012: Carbon market and technology transfer: Statistical analysis for exploring implications. *Int. J. Sustain. Dev. World Ecol.*, **19(4)**, 311–320, doi:10.1080/13504509.2011.644638.

Karkatsoulis, P., P. Capros, P. Fragkos, L. Paroussos, and S. Tsani, 2016: First-mover advantages of the European Union's climate change mitigation strategy. *Int. J. Energy Res.*, **40(6)**, 814–830, doi:10.1002/er.3487.

Karlsson, M., E. Alfredsson, and N. Westling, 2020: Climate policy co-benefits: A review. *Clim. Policy*, **20(3)**, 292–316, doi:10.1080/14693062.2020.1724070.

Kavlak, G., J. McNerney, and J.E. Trancik, 2018: Evaluating the causes of cost reduction in photovoltaic modules. *Energy Policy*, **123**, 700–710, doi:10.1016/j.enpol.2018.08.015.

Kemp, R., 2005: Zero emission vehicle mandate in California: Misguided policy or example of enlightened leadership. In: *Time Strategies, Innovation and Environmental Policy* [Sartorius, C. and S. Zundel, (eds.)], Edward Elgar Publishing, Cheltenham, UK and Northhampton, MA, USA, pp. 169–191.

Kemp, R., 2011: Ten themes for eco-innovation policies in Europe. *Sapiens*, **4(2)**, 1–20.

Kemp, R. and L. Soete, 1990: Inside the "green box": On the economics of technological change and the environment. In: New Explor. Econ. Technol. Chang.[Freeman, C. and Soete, L. (eds.)], Pinter, London, UK, pp 244–257.

Kemp, R., J. Schot, and R. Hoogma, 1998: Regime shifts to sustainability through processes of niche formation: The approach of strategic niche management. *Technol. Anal. Strateg. Manag.*, **10(2)**, 175–198, doi:10.1080/09537329808524310.

Kern, F. and M. Howlett, 2009: Implementing transition management as policy reforms: A case study of the Dutch energy sector. *Policy Sci.*, **42(4)**, 391–408, doi:10.1007/s11077-009-9099-x.

Kerras, H., J.L. Sánchez-Navarro, E.I. López-Becerra, and M.D. de-M. Gómez, 2020: The Impact of the Gender Digital Divide on Sustainable Development: Comparative Analysis between the European Union and the Maghreb. *Sustain.*, **12(8)**, 3347, doi:10.3390/SU12083347.

Khalili, N.R., S. Duecker, W. Ashton, and F. Chavez, 2015: From cleaner production to sustainable development: The role of academia. *J. Clean. Prod.*, **96**, 30–43, doi:10.1016/j.jclepro.2014.01.099.

Khan, M., D. Mfitumukiza, and S. Huq, 2020: Capacity building for implementation of nationally determined contributions under the Paris Agreement. *Clim. Policy*, **20(4)**, 499–510, doi:10.1080/14693062.2019.1675577.

Khanna, T.M. et al., 2021: A multi-country meta-analysis on the role of behavioural change in reducing energy consumption and $CO_2$ emissions in residential buildings. *Nat. Energy*, **6(9)**, 925–932, doi:10.1038/s41560-021-00866-x.

Khosla, R., A. Sagar, and A. Mathur, 2017: Deploying low-carbon technologies in developing countries: A view from India's buildings sector. *Environ. Policy Gov.*, **27(2)**, 149–162, doi:10.1002/eet.1750.

Kim, H.S. and S.-S. Kwak, 2020: Crop biotechnology for sustainable agriculture in the face of climate crisis. *Plant Biotechnol. Rep.*, **14(2)**, 139–141, doi:10.1007/s11816-020-00619-4.

Kim, Y.J. and M. Brown, 2019: Impact of domestic energy-efficiency policies on foreign innovation: The case of lighting technologies. *Energy Policy*, **128**(January), 539–552, doi:10.1016/j.enpol.2019.01.032.

Kimura, O. and T. Suzuki, 2006: 30 years of solar energy development in Japan: Co-evolution process of technology, policies, and the market. *Berlin Conference on the Human Dimensions of Global Environmental Change: "Resource Policies: Effectiveness, Efficiency and Equity"*, 17–18 November 2006. Berlin, Germany.

Kiso, T., 2019: Environmental Policy and Induced Technological Change: Evidence from Automobile Fuel Economy Regulations. *Environ. Resour. Econ.*, **74(2)**, 785–810, doi:10.1007/s10640-019-00347-6.

Kivimaa, P. and F. Kern, 2016: Creative destruction or mere niche support? Innovation policy mixes for sustainability transitions. *Res. Policy*, **45(1)**, 205–217, doi:10.1016/j.respol.2015.09.008.

Klaassen, G., A. Miketa, K. Larsen, and T. Sundqvist, 2005: The impact of R&D on innovation for wind energy in Denmark, Germany and the United Kingdom. *Ecol. Econ.*, **54(2–3)**, 227–240, doi:10.1016/j.ecolecon.2005.01.008.

Klein Woolthuis, R., M. Lankhuizen, and V. Gilsing, 2005: A system failure framework for innovation policy design. *Technovation*, **25(6)**, 609–619, doi:10.1016/j.technovation.2003.11.002.

Klerkx, L., B. van Mierlo, and C. Leeuwis, 2012: Evolution of systems approaches to agricultural innovation: Concepts, analysis and interventions. In: *Farming Systems Research into the 21st Century: The New Dynamic* [Darnhofer, I., D. Gibbon, and B. Dedieu, (eds.)], Springer Netherlands, Dordrecht, Netherlands, pp. 457–483.

Klerkx, L., E. Jakku, and P. Labarthe, 2019: A review of social science on digital agriculture, smart farming and agriculture 4.0: New contributions and a future research agenda. *NJAS - Wageningen J. Life Sci.*, **90–91**, 100315, doi:10.1016/j.njas.2019.100315.

Kline, S. and N. Rosenberg, 1986: An overview of innovation. In: *The Positive Sum Strategy: Harnessing Technology for Economic Growth* [Landau, R. and N. Rosenberg, (eds.)], The National Academies Press, Washington D.C., USA, pp. 275–306.

Knittel, C.R., 2011: Automobiles on Steroids: Product Attribute Trade-Offs and Technological Progress in the Automobile Sector. *Am. Econ. Rev.*, **101(7)**, 3368–3399, doi:10.1257/aer.101.7.3368.

Knobloch, F. and J.-F. Mercure, 2016: The behavioural aspect of green technology investments: A general positive model in the context of heterogeneous agents. *Environ. Innov. Soc. Transitions*, **21**, 39–55, doi:10.1016/j.eist.2016.03.002.

Koasidis, K. et al., 2020a: Many Miles to Paris: A Sectoral Innovation System Analysis of the Transport Sector in Norway and Canada in Light of the Paris Agreement. *Sustainability*, **12(14)**, 5832, doi:10.3390/su12145832.

Koasidis, K. et al., 2020b: The UK and German Low-Carbon Industry Transitions from a Sectoral Innovation and System Failures Perspective. *Energies*, **13(19)**, 4994, doi:10.3390/en13194994.

Koch, N., G. Grosjean, S. Fuss, and O. Edenhofer, 2016: Politics matters: Regulatory events as catalysts for price formation under cap-and-trade. *J. Environ. Econ. Manage.*, **78**, 121–139, doi:10.1016/j.jeem.2016.03.004.

Koengkan, M., Y.E. Poveda, and J.A. Fuinhas, 2020: Globalisation as a motor of renewable energy development in Latin America countries. *GeoJournal*, **85(6)**, 1591–1602, doi:10.1007/s10708-019-10042-0.

Koh, H. and C.L. Magee, 2008: A functional approach for studying technological progress: Extension to energy technology. *Technol. Forecast. Soc. Change*, **75(6)**, 735–758, doi:10.1016/j.techfore.2007.05.007.

Köhler, J. et al., 2019: An agenda for sustainability transitions research: State of the art and future directions. *Environ. Innov. Soc. Transitions*, **31**, 1–32, doi:10.1016/j.eist.2019.01.004.

Koohafkan, P. and M. Altieri, 2010: *Globally important agricultural heritage systems: A legacy for the future*. FAO, Rome, 41pp.

Koronen, C., M. Åhman, and L.J. Nilsson, 2020: Data centres in future European energy systems – energy efficiency, integration and policy. *Energy Effic.*, **13(1)**, 129–144, doi:10.1007/s12053-019-09833-8.

Kotchen, M.J., 2021: The producer benefits of implicit fossil fuel subsidies in the United States. *Proc. Natl. Acad. Sci.*, **118(14)**, e2011969118, doi:10.1073/pnas.2011969118.

Kowarsch, M. et al., 2016: Scientific assessments to facilitate deliberative policy learning. *Palgrave Commun.*, **2(1)**, 16092, doi:10.1057/palcomms.2016.92.

Kraus, S. and N. Koch, 2021: Provisional COVID-19 infrastructure induces large, rapid increases in cycling. *Proc. Natl. Acad. Sci.*, **118(15)**, doi:10.1073/pnas.2024399118.

Krey, V. et al., 2019: Looking under the hood: A comparison of techno-economic assumptions across national and global integrated assessment models. *Energy*, **172**, 1254–1267, doi:10.1016/j.energy.2018.12.131.

Kriegler, E. et al., 2015: Making or breaking climate targets: The AMPERE study on staged accession scenarios for climate policy. *Technol. Forecast. Soc. Change*, **90**, 24–44, doi:10.1016/j.techfore.2013.09.021.

Kuhl, L., K. Van Maanen, and S. Scyphers, 2020: An analysis of UNFCCC-financed coastal adaptation projects: Assessing patterns of project design and contributions to adaptive capacity. *World Dev.*, **127**, 104748, doi:10.1016/j.worlddev.2019.104748.

Kumar, V., U. Kumar, and A. Persaud, 1999: Building Technological Capability Through Importing Technology: The Case of Indonesian Manufacturing Industry. *J. Technol. Transf.*, **24(1)**, 81–96, doi:10.1023/A:1007728921126.

Kunkel, S. and M. Matthess, 2020: Digital transformation and environmental sustainability in industry: Putting expectations in Asian and African policies into perspective. *Environ. Sci. Policy*, **112**, 318–329, doi:https://doi.org/10.1016/j.envsci.2020.06.022.

Kuntze, J.-C. and T. Moerenhout, 2012: Local Content Requirements and the Renewable Energy Industry - A Good Match? *SSRN Electron. J.*, September, 19, doi:10.2139/ssrn.2188607.

Kuoljok, K., 2019: Without land we are lost: t=Traditional knowledge, digital technology and power relations. *Altern. An Int. J. Indig. Peoples*, **15(4)**, 349–358, doi:10.1177/1177180119890134.

Kypreos, S. and O. Bahn, 2003: A MERGE Model with Endogenous Technological Progress. *Environ. Model. Assess.*, **8**, 249–259, doi:https://doi.org/10.1023/A:1025551408939.

Lacal-Arántegui, R., 2019: Globalization in the wind energy industry: Contribution and economic impact of European companies. *Renew. Energy*, **134**, 612–628, doi:10.1016/J.RENENE.2018.10.087.

Lachapelle, E., R. MacNeil, and M. Paterson, 2017: The political economy of decarbonisation: From green energy 'race' to green 'division of labour.' *New Polit. Econ.*, **22(3)**, 311–327, doi:10.1080/13563467.2017.1240669.

Lafond, F. et al., 2018: How well do experience curves predict technological progress? A method for making distributional forecasts. *Technol. Forecast. Soc. Change*, **128**, 104–117, doi:10.1016/j.techfore.2017.11.001.

Laine, J., K. Kontu, J. Heinonen, and S. Junnila, 2020: Uncertain greenhouse gas implication in waste heat utilization – A case study with a data center. *J. Sustain. Dev. Energy, Water Environ. Syst.*, **8(2)**, 360–372, doi:10.13044/j.sdewes.d7.0301.

Laird, F.N., 2001: *Solar Energy, Technology Policy, and Institutional Values*. Cambridge University Press, Cambridge, UK, 248 pp.

Lambin, E.F. and P. Meyfroidt, 2011: Global land use change, economic globalization, and the looming land scarcity. *Proc. Natl. Acad. Sci.*, **108(9)**, 3465–3472, doi:10.1073/pnas.1100480108.

Lamperti, F., G. Dosi, M. Napoletano, A. Roventini, and A. Sapio, 2020: Climate change and green transitions in an agent-based integrated assessment model. *Technol. Forecast. Soc. Change*, **153**, 119806, doi:10.1016/j.techfore.2019.119806.

Lange, S., J. Pohl, and T. Santarius, 2020: Digitalization and energy consumption. Does ICT reduce energy demand? *Ecol. Econ.*, **176**, 106760, doi:10.1016/j.ecolecon.2020.106760.

Lauber, V. and S. Jacobsson, 2016: The politics and economics of constructing, contesting and restricting socio-political space for renewables – The German Renewable Energy Act. *Environ. Innov. Soc. Transitions*, **18**, 147–163, doi:10.1016/j.eist.2015.06.005.

Le Treut, G., J. Lefèvre, F. Lallana, and G. Bravo, 2021: The multi-level economic impacts of deep decarbonization strategies for the energy system. *Energy Policy*, **156**, 112423, doi:10.1016/j.enpol.2021.112423.

Lee, J., F.M. Veloso, D.A. Hounshell, and E.S. Rubin, 2010: Forcing technological change: A case of automobile emissions control technology development in the US. *Technovation*, **30(4)**, 249–264, doi:10.1016/j.technovation.2009.12.003.

Lee, T., 2017: The effect of clean energy regulations and incentives on green jobs: Panel analysis of the United States, 1998–2007. *Nat. Resour. Forum*, **41(3)**, 145–155, doi:10.1111/1477-8947.12125.

Lee, W.J. and R. Mwebaza, 2020: The role of the Climate Technology Centre and Network as a Climate Technology and Innovation Matchmaker for Developing Countries. *Sustainability*, **12(19)**, 7956, doi:10.3390/su12197956.

Lehmann, P., 2012: Justifying a policy mix for pollution control: A review of economic literature. *J. Econ. Surv.*, **26(1)**, 71–97, doi:10.1111/j.1467-6419.2010.00628.x.

Lehoux, N., S. D'Amours, and A. Langevin, 2014: Inter-firm collaborations and supply chain coordination: Review of key elements and case study. *Prod. Plan. Control*, **25(10)**, 858–872, doi:10.1080/09537287.2013.771413.

Lema, R., M. Iizuka, and R. Walz, 2015: Introduction to low-carbon innovation and development: Insights and future challenges for research. *Innov. Dev.*, **5(2)**, 173–187, doi:10.1080/2157930X.2015.1065096.

Lema, R., X. Fu, and R. Rabellotti, 2020: Green windows of opportunity: Latecomer development in the age of transformation toward sustainability. *Ind. Corp. Chang.*, **29(5)**, 1193–1209, doi:10.1093/icc/dtaa044.

Lema, R., P.L. Bhamidipati, C. Gregersen, U.E. Hansen, and J. Kirchherr, 2021: China's investments in renewable energy in Africa: Creating co-benefits or just cashing-in? *World Dev.*, **141**, 105365, doi:10.1016/j.worlddev.2020.105365.

Lepoutre, J. and A. Oguntoye, 2018: The (non-)emergence of mobile money systems in Sub-Saharan Africa: A comparative multilevel perspective of Kenya and Nigeria. *Technol. Forecast. Soc. Change*, **131**, 262–275, doi:10.1016/j.techfore.2017.11.010.

Levin, K., B. Cashore, S. Berstein, and G. Auld, 2012: Overcoming the tragedy of super wicked problems. *Policy Sci.*, **45(2)**, 123–152.

Levinson, A., 2019: Energy Efficiency Standards Are More Regressive Than Energy Taxes: Theory and Evidence. *J. Assoc. Environ. Resour. Econ.*, **6**(S1), S7–S36, doi:10.1086/701186.

Lewis, J.I., 2014: Industrial policy, politics and competition: Assessing the post-crisis wind power industry. *Bus. Polit.*, **16(4)**, 511–547, doi:10.1515/bap-2014-0012.

Ley, M., T. Stucki, and M. Woerter, 2016: The Impact of Energy Prices on Green Innovation. *Energy J.*, **37(1)**, doi:10.5547/01956574.37.1.mley.

Li, D., F. Tang, and J. Jiang, 2019a: Does environmental management system foster corporate green innovation? The moderating effect of environmental regulation. *Technol. Anal. Strateg. Manag.*, **31(10)**, 1242–1256, doi:10.1080/09537325.2019.1602259.

Li, J. et al., 2020: Does intellectual property rights protection constitute a barrier to renewable energy? An econometric analysis. *Natl. Inst. Econ. Rev.*, **251**, R37–R46, doi:10.1017/nie.2020.5.

Li, Z., H. Dong, Z. Huang, and P. Failler, 2019b: Impact of Foreign Direct Investment on Environmental Performance. *Sustainability*, **11(13)**, 3538, doi:10.3390/su11133538.

Liebowitz, S.J. and S.E. Margolis, 1995: Path Dependence, Lock-In, and history. *SSRN Electron. J.*, **11(1)**, 205–226, doi:10.2139/ssrn.1706450.

Lilliestam, J., A. Patt, and G. Bersalli, 2021: The effect of carbon pricing on technological change for full energy decarbonization: A review of empirical ex-post evidence. *WIREs Clim. Chang.*, **12(1)**, doi:10.1002/wcc.681.

Lin, B. and Y. Chen, 2019: Does electricity price matter for innovation in renewable energy technologies in China? *Energy Econ.*, **78**, 259–266, doi:10.1016/j.eneco.2018.11.014.

Littleton, M., 2008: *The TRIPS Agreement and Transfer of Climate- Change-Related Technologies to Developing Countries*. UNDESA, New York, NY, USA, 48 pp.

Littleton, M., 2009: The TRIPS Agreement and transfer of climate-change-related technologies to developing countries. *Nat. Resour. Forum*, **33(3)**, 233–244, doi:10.1111/j.1477-8947.2009.01228.x.

Lo, K., 2014: A critical review of China's rapidly developing renewable energy and energy efficiency policies. *Renew. Sustain. Energy Rev.*, **29**, 508–516, doi:10.1016/j.rser.2013.09.006.

Luderer, G. et al., 2019: Environmental co-benefits and adverse side-effects of alternative power sector decarbonization strategies. *Nat. Commun.*, **10(1)**, 5229, doi:10.1038/s41467-019-13067-8.

Lundvall, B.-Å., 1988: Innovation as an interactive process: From user-producer interaction to the national system of innovation. In: *Technical Change and Economic Theory* [Dosi, G., C. Freeman, R. Nelson, G. Silverberg, and L. Soete, (eds.)], Pinter Publishers, London UK and New York, NY, USA, pp 349–369.

Lundvall, B.-Å., 1992: *National Systems of Innovation: Toward a Theory of Innovation and Interactive Learning*. [Lundvall, B.-A., (ed.)]. Pinter Publishers, London, UK, 342 pp.

Lundvall, B.-Å., K. Joseph, C. Chaminade, and J. Vang, 2009: *Handbook of Innovation Systems and Developing Countries*. [Lundvall, B.-Å., K. Joseph, C. Chaminade, and J. Vang, (eds.)]. Edward Elgar Publishing, London, UK and New York, NY, USA, 416 pp.

Lynch, J.K., J. Glasby, and S. Robinson, 2019: If telecare is the answer, what was the question? Storylines, tensions and the unintended consequences of technology-supported care. *Crit. Soc. Policy*, **39(1)**, 44–65, doi:10.1177/0261018318762737.

Lytle, W. et al., 2020: Conceptual Design and Rationale for a New Agrivoltaics Concept: Pasture-Raised Rabbits and Solar Farming. *J. Clean. Prod.*, 124476, doi:10.1016/j.jclepro.2020.124476.

Machlup, F., 1958: *An economic review of the patent system, Subcommittee on Patents, Trademarks & Copyrights of the Senate Committee on the Judiciary, 85th Cong., 2d Sess, study number 15*. United States Government Printing Office, Washington D.C, USA, 94 pp.

Machlup, F. and E. Penrose, 1950: The Patent Controversy in the Nineteenth Century. *J. Econ. Hist.*, **10(1)**, 1–29, doi:10.1017/S0022050700055893.

Maharajh, R. and E. Kraemer-Mbula, 2010: Innovation strategies in developing countries. In: *Innovation and the Development Agenda*, OECD Publishing, Paris, France, pp. 133–151.

Mainali, B., J. Luukkanen, S. Silveira, and J. Kaivo-oja, 2018: Evaluating Synergies and Trade-Offs among Sustainable Development Goals (SDGs): Explorative Analyses of Development Paths in South Asia and Sub-Saharan Africa. *Sustain.* **10(3)**, 815, doi:10.3390/SU10030815.

Malhotra, A. and T.S. Schmidt, 2020: Accelerating Low-Carbon Innovation. *Joule*, **4(11)**, 2259–2267, doi:https://doi.org/10.1016/j.joule.2020.09.004.

Malhotra, A., A. Mathur, S. Diddi, and A.D. Sagar, 2021: Building institutional capacity for addressing climate and sustainable development goals: Achieving energy efficiency in India. *Clim. Policy*, 22(5), 1–19, doi:10.1080/14693062.2021.1984195.

Mancusi, M.L., 2008: International spillovers and absorptive capacity: A cross-country cross-sector analysis based on patents and citations. *J. Int. Econ.*, **76(2)**, 155–165, doi:10.1016/j.jinteco.2008.06.007.

Mankins, J., 1995: *Technology Readiness Level – A White Paper*. Advanced Concepts Office, Office of Space Access and Technology, National Aeronautics and Space Administration (NASA), Washington, USA, 6 pp.

Mankins, J.C., 2009: Technology readiness assessments: A retrospective. *Acta Astronaut.*, **65(9–10)**, 1216–1223, doi:10.1016/j.actaastro.2009.03.058.

Marangoni, G. and M. Tavoni, 2014: The clean energy strategy for 2°C. *Clim. Chang. Econ.*, **05(01)**, 1440003, doi:10.1142/S201000781440003X.

Markard, J., 2018: The next phase of the energy transition and its implications for research and policy. *Nat. Energy*, **3(8)**, 628–633, doi:10.1038/s41560-018-0171-7.

Markard, J., R. Raven, and B. Truffer, 2012: Sustainability transitions: An emerging field of research and its prospects. *Res. Policy*, **41(6)**, 955–967, doi:10.1016/j.respol.2012.02.013.

Markard, J., F.W. Geels, and R. Raven, 2020: Challenges in the acceleration of sustainability transitions. *Environ. Res. Lett.*, **15(8)**, 081001, doi:10.1088/1748-9326/AB9468.

Marquardt, J., K. Steinbacher, and M. Schreurs, 2016: Driving force or forced transition? *J. Clean. Prod.*, **128**, 22–33, doi:10.1016/j.jclepro.2015.06.080.

Masanet, E., A. Shehabi, N. Lei, S. Smith, and J. Koomey, 2020: Recalibrating global data center energy-use estimates. *Science*, **367(6481)**, 984–986, doi:10.1126/science.aba3758.

Maskus, K., 2010: Differentiated Intellectual Property Regimes for Environmental and Climate Technologies. *OECD Environment Working Papers*, No. 17, OECD Publishing, Paris, France. doi:10.1787/5kmfwjvc83vk-en.

Maskus, K.E. and J.H. Reichman, 2017: The globalization of private knowledge goods and the privatization of global public goods. In: *Globalization and Intellectual Property*, Vol. 7, Routledge, London, UK, pp. 335–377.

Maskus, K.E., S. Milani, and R. Neumann, 2019: The impact of patent protection and financial development on industrial R&D. *Res. Policy*, **48(1)**, 355–370, doi:10.1016/j.respol.2018.09.005.

Mastropietro, P., C. Batlle, L.A. Barroso, and P. Rodilla, 2014: Electricity auctions in South America: Towards convergence of system adequacy and RES-E support. *Renew. Sustain. Energy Rev.*, **40**, 375–385, doi:10.1016/j.rser.2014.07.074.

Masuku, B. and O. Nzewi, 2021: The South African informal sector's socio-economic exclusion from basic service provisions: A critique of Buffalo City Metropolitan Municipality's approach to the informal sector. *J. Energy South. Africa*, **32(2)**, 59–71, doi:10.17159/2413-3051/2021/v32i2a5856.

Matsuo, T. and T.S. Schmidt, 2019: Managing tradeoffs in green industrial policies: The role of renewable energy policy design. *World Dev.*, **122**, 11–26, doi:10.1016/j.worlddev.2019.05.005.

Matthess, M. and S. Kunkel, 2020: Structural change and digitalization in developing countries: Conceptually linking the two transformations. *Technol. Soc.*, **63**, 101428, doi:https://doi.org/10.1016/j.techsoc.2020.101428.

May, P., 2019: *Valuing* externalities of cattle and soy-maize systems in the Brazilian Amazon: A Test of the TEEBAgriFood *Framework*, TEEB for Agriculture and Food, UN Environment Programme (UNEP), Nairobi, Kenya, 128 pp.

Mayer, J., 2001: *Technology Diffusion, Human Capital and Economic Growth in Developing Countries*. UNCTAD, Published Online. 47 pp. https://unctad.org/system/files/official-document/dp_154.en.pdf.

Mayer, T., D. Kreyenberg, J. Wind, and F. Braun, 2012: Feasibility study of 2020 target costs for PEM fuel cells and lithium-ion batteries: A two-factor experience curve approach. *Int. J. Hydrogen Energy*, **37(19)**, 14463–14474, doi:10.1016/j.ijhydene.2012.07.022.

Mazhar, U., 2014: Environmental Regulation and the Informal Sector: Empirical Evidence – Legislazione ambientale ed economia sommersa: evidenze empiriche. *Econ. Internazionale / Int. Econ.*, **67(4)**, 471–491.

Mazzucato, M., 2013: *The Entrepreneurial State*. Anthem Press, London, UK, 288 pp.

Mazzucato, M., 2015a: 6. Innovation, the State and Patient Capital. *Polit. Q.*, **86**, 98–118, doi:10.1111/1467-923X.12235.

Mazzucato, M., 2015b: *The Entrepreneurial State*. Anthem Press, London, UK,266 pp.

Mazzucato, M., 2016: From market fixing to market-creating: A new framework for innovation policy. *Ind. Innov.*, **23(2)**, 140–156, doi:10.1080/13662716.2016.1146124.

Mazzucato, M., 2018: *Mission-Oriented Research & Innovation in the European Union: A problem-solving approach to fuel innovation-led growth*. European Commission, Brussels, 36 pp.

Mazzucato, M. and G. Semieniuk, 2017: Public financing of innovation: New questions. *Oxford Rev. Econ. Policy*, **33(1)**, 24–48, doi:10.1093/oxrep/grw036.

Mazzucato, M. and G. Semieniuk, 2018: Financing renewable energy: Who is financing what and why it matters. *Technol. Forecast. Soc. Change*, **127**, 8–22, doi:10.1016/j.techfore.2017.05.021.

McCauley, D. and R. Heffron, 2018: Just transition: Integrating climate, energy and environmental justice. *Energy Policy*, **119**, 1–7, doi:10.1016/j.enpol.2018.04.014.

McCauley, D. et al., 2019: Energy justice in the transition to low carbon energy systems: Exploring key themes in interdisciplinary research. *Appl. Energy*, **233–234**, 916–921, doi:10.1016/j.apenergy.2018.10.005.

McDonald, A. and L. Schrattenholzer, 2001: Learning rates for energy technologies. *Energy Policy*, **29(4)**, 255–261, doi:10.1016/S0301-4215**(00)**00122-1.

McGee, J. and J. Wenta, 2014: Technology Transfer Institutions in Global Climate Governance: The Tension between Equity Principles and Market Allocation. *Rev. Eur. Comp. Int. Environ. Law*, **23(3)**, 367–381, doi:10.1111/reel.12075.

McNeil, M.A., M. Iyer, S. Meyers, V.E. Letschert, and J.E. McMahon, 2008: Potential benefits from improved energy efficiency of key electrical products: The case of India. *Energy Policy*, **36(9)**, 3467–3476, doi:10.1016/j.enpol.2008.05.020.

McNerney, J., J. Doyne Farmer, and J.E. Trancik, 2011: Historical costs of coal-fired electricity and implications for the future. *Energy Policy*, **39(6)**, 3042–3054, doi:10.1016/j.enpol.2011.01.037.

Meckling, J., 2019: A New Path for U.S. Climate Politics: Choosing Policies That Mobilize Business for Decarbonization. *Ann. Am. Acad. Pol. Soc. Sci.*, **685(1)**, 82–95, doi:10.1177/0002716219862515.

Meckling, J. and J. Nahm, 2019: The politics of technology bans: Industrial policy competition and green goals for the auto industry. *Energy Policy*, **126**, 470–479, doi:10.1016/j.enpol.2018.11.031.

Meckling, J., N. Kelsey, E. Biber, and J. Zysman, 2015: Winning coalitions for climate policy: Green industrial policy builds support for carbon regulation. *Science*, **349(6253)**, 1170–1171, doi:10.1126/science.aab1336.

Meckling, J., T. Sterner, and G. Wagner, 2017: Policy sequencing toward decarbonization. *Nat. Energy*, **2(12)**, 918–922, doi:10.1038/s41560-017-0025-8.

Meckling, J.O. and G.Y. Chung, 2009: Sectoral approaches for a post-2012 climate regime: A taxonomy. *Clim. Policy*, **9(6)**, 652–668, doi:10.3763/cpol.2009.0629.

Melissen, F. and H. Reinders, 2012: A reflection on the Dutch Sustainable Public Procurement Programme. *J. Integr. Environ. Sci.*, **9(1)**, 27–36, doi:10.1080/1943815X.2012.658815.

Mendonça, H.L., T.D.L. van Aduard de Macedo-Soares, and M.V. de A. Fonseca, 2018: Working towards a framework based on mission-oriented practices for assessing renewable energy innovation policies. *J. Clean. Prod.*, **193**, 709–719, doi:10.1016/j.jclepro.2018.05.064.

Meng, J., R. Way, E. Verdolini, L. Diaz Anadon, and L. Anadon, 2021: Comparing expert elicitation and model-based probabilistic technology cost forecasts for the energy transition. *Proc. Natl. Acad. Sci.*, **118(27)**, e1917165118, doi:10.1073/pnas.1917165118.

Mercure, J.-F. et al., 2018: Macroeconomic impact of stranded fossil fuel assets. *Nat. Clim. Chang.*, **8(7)**, 588–593, doi:10.1038/s41558-018-0182-1.

Mercure, J.F. et al., 2016: *Policy-induced energy technological innovation and finance for low-carbon economic growth: Deliverable D2 study on the macroeconomics of energy and climate policies*. European Union, Brussels, Belgium, 78 pp.

Mian, A., L. Straub, and A. Sufi, 2020: *Indebted Demand*. National Bureau of Economic Research, Cambridge, MA, USA, 70 pp.

Michaelowa, A., I. Shishlov, and D. Brescia, 2019: Evolution of international carbon markets: Lessons for the Paris Agreement. *Wiley Interdiscip. Rev. Clim. Chang.*, **10(6)**, doi:10.1002/wcc.613.

Mignon, I. and A. Bergek, 2016: System-and actor-level challenges for diffusion of renewable electricity technologies: An international comparison. *J. Clean. Prod.*, **128**, 105–115. https://doi.org/10.1016/j.jclepro.2015.09.048.

Milojevic-Dupont, N. and F. Creutzig, 2021: Machine learning for geographically differentiated climate change mitigation in urban areas. *Sustain. Cities Soc.*, **64**, 102526, doi:10.1016/j.scs.2020.102526.

Miremadi, I., Y. Saboohi, and S. Jacobsson, 2018: Assessing the performance of energy innovation systems: Towards an established set of indicators. *Energy Res. Soc. Sci.*, **40**, 159–176, doi:10.1016/j.erss.2018.01.002.

Mission Innovation, 2019: *Collaborative Models for International Co-operation in Clean-energy Research and Innovation*. Mission Innovation, Published Online, 23 pp. http://mission-innovation.net/wp-content/uploads/2019/09/AJR-Paper-on-Multilateral-Collaboration-models_FINAL.pdf.

Mitcham, C., 2003: Co-responsibility for research integrity. *Sci. Eng. Ethics*, **9(2)**, 273–290, doi:10.1007/s11948-003-0014-0.

Mockshell, J. and M.E.J. Villarino, 2019: Agroecological Intensification: Potential and Limitations to Achieving Food Security and Sustainability. In: *Encyclopedia of Food Security and Sustainability*,[Ferranti, P., Berry, E. and Anderson, J. (eds)]. Elsevier Inc, Amsterdam, Netherlands, pp. 64–70.

Mohammadi, A., B. Khoshnevisan, G. Venkatesh, and S. Eskandari, 2020: A critical review on advancement and challenges of biochar application in paddy fields: Environmental and life cycle cost analysis. *Processes*, **8(10)**, doi:10.3390/pr8101275.

Moldalieva, J. and J. Heathershaw, 2020: Playing the "Game" of Transparency and Accountability: Non-elite Politics in Kyrgyzstan's Natural Resource Governance. *Post-Soviet Aff.*, **36(2)**, 171–187, doi:10.1080/1060586X.2020.1721213.

Moon, S.G., S. Bae, and M.-G. Jeong, 2014: Corporate Sustainability and Economic Performance: an Empirical Analysis of a Voluntary Environmental Program in the USA. *Bus. Strateg. Environ.*, **23(8)**, 534–546, doi:10.1002/bse.1800.

Mormina, M., 2019: Science, Technology and Innovation as Social Goods for Development: Rethinking Research Capacity Building from Sen's Capabilities Approach. *Sci. Eng. Ethics*, **25(3)**, 671–692, doi:10.1007/s11948-018-0037-1.

Mowery, D. and N. Rosenberg, 1979: The influence of market demand upon innovation: A critical review of some recent empirical studies. *Res. Policy*, **8(2)**, 102–153, doi:10.1016/0048-7333**(79)**90019-2.

Mukherjee, P.K., E. Gibbs, A. Walia, and C. Taylor, 2020: Staying cool: The development of India's pioneering energy efficiency policy for chillers. *WIREs Energy Environ.*, **9(4)**, doi:10.1002/wene.372.

Mulgan, G., 2012: The Theoretical Foundations of Social Innovation. In: *Social Innovation* [Nicholls, A. and A. Murdock, (eds.)], Palgrave Macmillan, London, UK, pp. 33–65.

Munene, M.B., Å.G. Swartling, and F. Thomalla, 2018: Adaptive governance as a catalyst for transforming the relationship between development and disaster risk through the Sendai Framework? *Int. J. Disaster Risk Reduct.*, **28**, 653–663, doi:10.1016/j.ijdrr.2018.01.021.

Murphy, K., G.A. Kirkman, S. Seres, and E. Haites, 2015: Technology transfer in the CDM: an updated analysis. *Clim. Policy*, **15(1)**, 127–145, doi:10.1080/14693062.2013.812719.

Murphy, L.M. and P.L. Edwards, 2003: *Bridging the Valley of Death: Transitioning from Public to Private Sector Financing*. National Renewable Energy Laboratory (NREL), Golden, CO, USA 58 pp.

Myslikova, Z. and K.S. Gallagher, 2020: Mission Innovation is mission critical. *Nat. Energy*, **5(10)**, 732–734, doi:10.1038/s41560-020-00694-5.

Nagy, B., J.D. Farmer, Q.M. Bui, and J.E. Trancik, 2013: Statistical Basis for Predicting Technological Progress. *PLoS One*, **8(2)**, e52669, doi:10.1371/journal.pone.0052669.

Nahm, J. and E.S. Steinfeld, 2014: Scale-up Nation: China's Specialization in Innovative Manufacturing. *World Dev.*,54, doi:10.1016/j.worlddev.2013.09.003.

Nanda, R., K. Younge, and L. Fleming, 2016: Economic Value Creation in Mobile Applications. In: *The Changing Frontier* [Jaffe, A.B. and B.F. Jones, (eds.)], University of Chicago Press, Chicago, USA, pp. 233–286.

Napp, T. et al., 2017: Exploring the Feasibility of Low-Carbon Scenarios Using Historical Energy Transitions Analysis. *Energies*, **10(1)**, 116, doi:10.3390/en10010116.

Narayanamurti, V. and T. Odumosu, 2016: *Cycles of Invention and Discovery: Rethinking the endless frontier*. [Narayanamurti, V. and T. Odumosu, (eds.)]. Harvard University Press, Cambridge, MA, USA, 176 pp.

Narayanamurti, V., L.D. Anadon, and A.D. Sagar, 2009: Transforming Energy Innovation. *Issues Sci. Technol.*, **XXIV(1)**, doi:https://issues.org/narayanamurti/.

Narita, D. and U.J. Wagner, 2017: Strategic uncertainty, indeterminacy, and the formation of international environmental agreements. *Oxf. Econ. Pap.*, **69(2)**, 432–452, doi:10.1093/oep/gpx001.

National Science Board, 2018: *Research and Development – Research Trends and Comparisons*. National Science Board, USA, 108 pp. Published Online: https://www.nsf.gov/statistics/2018/nsb20181/assets/1038/research-and-development-u-s-trends-and-international-comparisons.pdf.

Nawaz, A. et al., 2020: An Intelligent Integrated Approach for Efficient Demand Side Management With Forecaster and Advanced Metering Infrastructure Frameworks in Smart Grid. *IEEE Access*, **8**, 132551–132581, doi:10.1109/ACCESS.2020.3007095.

Negro, S.O., F. Alkemade, and M.P. Hekkert, 2012: Why does renewable energy diffuse so slowly? A review of innovation system problems. *Renew. Sustain. Energy Rev.*, **16(6)**, 3836–3846, doi:10.1016/j.rser.2012.03.043.

Neij, L. and K. Åstrand, 2006: Outcome indicators for the evaluation of energy policy instruments and technical change. *Energy Policy*, **34(17)**, 2662–2676, doi:10.1016/j.enpol.2005.03.012.

Nelson, R., 1993: *National Innovation Systems: A Comparative Analysis*. Oxford University Press, Oxford, UK, 552 pp.

Nelson, R. and E. Phelps, 1966: Investment in Humans, Technology Diffusion and Economic Growth. *Am. Econ. Rev.*, **56(1/2)**, 69–75.

Nelson, R. and R. Mazzoleni, 1996: Economic Theories About the Costs and Benefits of Patents. In: *Intellectual Property Rights and the Dissemination of Research Tools in Molecular Biology: Summary of a Workshop Held at the National Academy of Sciences, February 15–16, 1996.* National Research Council, The National Academies Press, Washington D.C, USA, pp. 17–27.

Nemet, G.F., 2006: Beyond the learning curve: Factors influencing cost reductions in photovoltaics. *Energy Policy*, **34(17)**, 3218–3232, doi:10.1016/j.enpol.2005.06.020.

Nemet, G.F., 2009a: Interim monitoring of cost dynamics for publicly supported energy technologies. *Energy Policy*, **37(3)**, 825–835, doi:10.1016/j.enpol.2008.10.031.

Nemet, G.F., 2009b: Demand-pull, technology-push, and government-led incentives for non-incremental technical change. *Res. Policy*, **38(5)**, 700–709, doi:10.1016/j.respol.2009.01.004.

Nemet, G.F., 2010: Robust incentives and the design of a climate change governance regime. *Energy Policy*, **38(11)**, 7216–7225, doi:10.1016/j.enpol.2010.07.052.

Nemet, G.F., 2012: Inter-technology knowledge spillovers for energy technologies. *Energy Econ.*, **34(5)**, doi:10.1016/j.eneco.2012.06.002.

Nemet, G.F., 2013: Technological change and climate change policy. In: *Encyclopedia of Energy, Natural Resource and Environmental Economics* [Shogren, J., (ed.)], Elsevier, Amsterdam, the Netherlands, pp. 107–116.

Nemet, G.F., 2019: *How Solar Became Cheap*. Routledge, London, UK and New York, NY, USA, 14–18 pp.

Nemet, G.F., P. Braden, and E. Cubero, 2013: *Credibility, Ambition, and Discretion in Long-term U.S. Energy Policy Targets from 1973 to 2011*. University of Wisconsin-Madison La Follette School of Public Affairs, Working Paper 2013-007, Madison, WI, USA, 29 pp.

Nemet, G.F., M. Jakob, J.C. Steckel, and O. Edenhofer, 2017: Addressing policy credibility problems for low-carbon investment. *Glob. Environ. Chang.*, **42**, 47–57, doi:10.1016/j.gloenvcha.2016.12.004.

Nemet, G.F., V. Zipperer, and M. Kraus, 2018: The valley of death, the technology pork barrel, and public support for large demonstration projects. *Energy Policy*, **119**, 154–167, doi:10.1016/j.enpol.2018.04.008.

Newell, P. and D. Mulvaney, 2013: The political economy of the 'just transition.' *Geogr. J.*, **179(2)**, 132–140, doi:10.1111/geoj.12008.

Newell, P. and A. Simms, 2020: How Did We Do That? Histories and Political Economies of Rapid and Just Transitions. *New Polit. Econ.*, **26(6)**, 1–16, doi:10.1080/13563467.2020.1810216.

Newell, R.G., A.B. Jaffe, and R.N. Stavins, 1999: The Induced Innovation Hypothesis and Energy-Saving Technological Change. *Q. J. Econ.*, **114(3)**, 941–975, doi:10.1162/003355399556188.

Newman, C., J. Rand, T. Talbot, and F. Tarp, 2015: Technology transfers, foreign investment and productivity spillovers. *Eur. Econ. Rev.*, **76**, 168–187, doi:10.1016/j.euroecorev.2015.02.005.

Nicholls, C., and M. Alteri, 2013: Agroecología y Cambio Climático Metodologías para evaluar la resiliencia socio-ecológica en comunidades rurales. FAO, Lima, Peru, 99 pp.

Nicholls, C., and M. Altieri, 2018: Pathways for the amplification of agroecology. *Agroecol. Sustain. Food Syst.*, **42(10)**, 1170–1193, doi:10.1080/21683565.2018.1499578.

Nikas, A. et al., 2017: Managing stakeholder knowledge for the evaluation of innovation systems in the face of climate change. *J. Knowl. Manag.*, **21(5)**, 1013–1034, doi:10.1108/JKM-01-2017-0006.

Nikas, A., H. Doukas, and L. Martínez López, 2018: A group decision making tool for assessing climate policy risks against multiple criteria. *Heliyon*, **4(3)**, e00588, doi:10.1016/j.heliyon.2018.e00588.

Nikas, A., H. Neofytou, A. Karamaneas, K. Koasidis, and J. Psarras, 2020: Sustainable and socially just transition to a post-lignite era in Greece: A multi-level perspective. *Energy Sources, Part B: Economics, Planning, and Policy*, **15(10–12)**, 513–544, doi:10.1080/15567249.2020.1769773.

Nilsson, M., D. Griggs, and M. Visbeck, 2016: Policy: Map the interactions between Sustainable Development Goals. *Nature*, **534(7607)**, 320–322, doi:10.1038/534320a.

Noailly, J., 2012: Improving the energy efficiency of buildings: The impact of environmental policy on technological innovation. *Energy Econ.*, **34(3)**, 795–806, doi:10.1016/j.eneco.2011.07.015.

Noailly, J. and R. Smeets, 2015: Directing technical change from fossil-fuel to renewable energy innovation: An application using firm-level patent data. *J. Environ. Econ. Manage.*, **72**, 15–37, doi:10.1016/j.jeem.2015.03.004.

Nordhaus, W., 2011: Designing a friendly space for technological change to slow global warming. *Energy Econ.*, **33(4)**, 665–673, doi:10.1016/j.eneco.2010.08.005.

Nordhaus, W.D., 2014: The Perils of the Learning Model for Modeling Endogenous Technological Change. *Energy J.*, **35(1)**, doi:10.5547/01956574.35.1.1.

North, D.C., 1991: Institutions. *J. Econ. Perspect.*, **5(1)**, 97–112. doi: 10.1257/jep.5.1.97.

Nygaard, I. and U.E. Hansen, 2015: The conceptual and practical challenges to technology categorisation in the preparation of technology needs assessments. *Clim. Change*, **131(3)**, 371–385, doi:10.1007/s10584-015-1367-5.

NYSERDA, 2020: *Innovation & Research Demonstration Project Impact Evaluation*. NYSERDA, Albany, NY, USA, 41 pp. https://www.nyserda.ny.gov/-/media/Files/Publications/PPSER/Program-Evaluation/2020-Innovation-Research-Impact-Evaluation-Final-Report.pdf.

Obama, B., 2017: The irreversible momentum of clean energy. *Science*, **355(6321)**, 126–129, doi:10.1126/science.aam6284.

Oberthür, S., G. Khandekar, and T. Wyns, 2021: Global governance for the decarbonization of energy-intensive industries: Great potential underexploited. *Earth Syst. Gov.*, **8**, 100072, doi:10.1016/j.esg.2020.100072.

Ockwell, D. and R. Byrne, 2016: Improving technology transfer through national systems of innovation: Climate relevant innovation-system builders (CRIBs). *Clim. Policy*, **16(7)**, 836–854, doi:10.1080/14693062.2015.1052958.

Ockwell, D., A. Sagar, and H. de Coninck, 2015: Collaborative research and development (R&D) for climate technology transfer and uptake in developing countries: Towards a needs driven approach. *Clim. Change*, **131(3)**, 401–415, doi:10.1007/s10584-014-1123-2.

Ockwell, D., R. Byrne, U.E. Hansen, J. Haselip, and I. Nygaard, 2018: The uptake and diffusion of solar power in Africa: Socio-cultural and political insights on a rapidly emerging socio-technical transition. *Energy Res. Soc. Sci.*, **44**, 122–129, doi:10.1016/j.erss.2018.04.033.

Ockwell, D.G. and A. Mallett, 2012: *Low-carbon Technology Transfer*. [Ockwell, D.G. and A. Mallett, (eds.)]. Routledge, London, UK and New York, NY, USA, 403 pp.

OECD, 2006: *Innovation in energy technology: Comparing national innovation systems at sectoral level*. OECD, Paris, France, 318pp. https://doi.org/10.1787/9789264014084-en.

OECD, 2011a: *Tools for delivering green growth*. OECD, Paris, France, 29pp. https://www.oecd.org/greengrowth/48012326.pdf.

OECD, 2011b: *Industrial Biotechnology and Climate Change: Opportunities and Challenges*. OECD, Paris, France, 41pp. http://www.oecd.org/science/emerging-tech/49024032.pdf.

OECD, 2015a: *Frascati Manual 2015: Guidelines for Collecting and Reporting Data on Research and Experimental Development*. OECD, Paris, France, 400 pp.

OECD, 2015b: *Green Investment Banks*. OECD, Paris, France, 1–20 pp.

OECD, 2017: *OECD Digital Economy Outlook 2017*. OECD, Paris, France, 324pp.

OECD, 2020: *The effects of R&D tax incentives and their role in the innovation policy mix*. OECD, Paris, France, 96 pp.

OECD, and ITF, 2020: *Good to go? Assessing the environmental performance of new mobility*. OECD, Paris, France, 87 pp. https://www.itf-oecd.org/sites/default/files/docs/environmental-performance-new-mobility.pdf.

Oh, C., 2020: Contestations over the financial linkages between the UNFCCC's Technology and Financial Mechanism: Using the lens of institutional interaction. *Int. Environ. Agreements Polit. Law Econ.*, **20(3)**, 559–575, doi:10.1007/s10784-020-09474-8.

Ohl, R.S., 1941: Light-sensitive electric device. *United States Pat. Off. 2402662*,. United States Patent Office, USA, 14pp.

Ohl, R.S., 1946: Light-sensitive electric device. United States Patent Office 2402662, 14pp.

Olawuyi, D.S., 2018: From technology transfer to technology absorption: Addressing climate technology gaps in Africa. *J. Energy Nat. Resour. Law*, **36(1)**, 61–84, doi:10.1080/02646811.2017.1379667.

Oldenziel, R., M. Emanuel, A.A. Albert de la Bruheze, and F. Veraart, 2016: *Cycling Cities: The European Experience: Hundred Years of Policy and Practice*. Foundation for the History of Technology, Eindhoven, Netherlands, 256 pp.

Olson, M., 1982: *The Rise and Decline of Nations: Economic Growth, Stagflation and Social Rigidities*. Yale University Press, New Haven, CT, USA and London, UK, 276 pp.

Olsson, O. and B.S. Frey, 2002: Entrepreneurship as recombinant growth. *Small Bus. Econ.*, **19**, 69–80, doi:10.1023/A:1016261420372.

Ordonez, J.A., M. Jakob, J.C. Steckel, and A. Fünfgeld, 2021: Coal, power and coal-powered politics in Indonesia. *Environ. Sci. Policy*, **123**, 44–57, doi:10.1016/j.envsci.2021.05.007.

Osongo, E. and J. Schot, 2017: Inclusive Innovation and Rapid Sociotechnical Transitions: The Case of Mobile Money in Kenya. *SPRU Work. Pap. Ser.*, **March 23, 2017**. SWPS 2017-07, doi:10.2139/ssrn.2940184.

Ostrom, E., 2010: Beyond Markets and States: Polycentric Governance of Complex Economic Systems. *Am. Econ. Rev.*, **100(3)**, 641–672, doi:10.1257/aer.100.3.641.

Otto, I.M. et al., 2020: Social tipping dynamics for stabilizing Earth's climate by 2050. *Proc. Natl. Acad. Sci.*, **117(5)**, 2354–2365, doi:10.1073/pnas.1900577117.

Pandey, N., H. de Coninck, and A.D. Sagar, 2021: Beyond technology transfer: Innovation cooperation to advance sustainable development in developing countries. *WIREs Energy Environ.*, **11(2)**, 1-25. doi:10.1002/wene.422.

Parida, V., D. Sjödin, and W. Reim, 2019: Reviewing Literature on Digitalization, Business Model Innovation, and Sustainable Industry: Past Achievements and Future Promises. *Sustainability*, **11(2)**, 391, doi:10.3390/su11020391.

Paroussos, L., F. Panagiotis, V. Zoi, and F. Kostas, 2017: *A technical case study on R&D and technology spillovers of clean energy technologies*. European Commission, Brussels, 87 pp. https://ec.europa.eu/energy/sites/ener/files/documents/case_study_3_technical_analysis_spillovers.pdf.

Paroussos, L., K. Fragkiadakis, and P. Fragkos, 2020: Macro-economic analysis of green growth policies: The role of finance and technical progress in Italian green growth. *Clim. Change*, **160(4)**, 591–608, doi:10.1007/s10584-019-02543-1.

Pascual, U. et al., 2017: Valuing nature's contributions to people: The IPBES approach. *Curr. Opin. Environ. Sustain.*, **26–27**, 7–16, doi:10.1016/j.cosust.2016.12.006.

Patt, A., 2015: *Transforming energy: Solving climate change with technology policy*. Cambridge University Press, New York, NY, USA, 360 pp.

Patt, A. and J. Lilliestam, 2018: The Case against Carbon Prices. *Joule*, **2(12)**, doi:10.1016/j.joule.2018.11.018.

Patt, A., D. Aplyn, P. Weyrich, and O. van Vliet, 2019: Availability of private charging infrastructure influences readiness to buy electric cars. *Transp. Res. Part A Policy Pract.*, **125**, 1–7, doi:10.1016/J.TRA.2019.05.004.

Pearson, P.J.G. and T.J. Foxon, 2012: A low carbon industrial revolution? Insights and challenges from past technological and economic transformations. *Energy Policy*, **50**, 117–127, doi:10.1016/J.ENPOL.2012.07.061.

Pegels, A. and T. Altenburg, 2020: Latecomer development in a "greening" world: Introduction to the Special Issue. *World Dev.*, **135**, 105084, doi:10.1016/j.worlddev.2020.105084.

Pellicer-Sifres, V., S. Belda-Miquel, I. Cuesta-Fernandez, and A. Boni, 2018: Learning, transformative action, and grassroots innovation: Insights from the Spanish energy cooperative Som Energia. *Energy Res. Soc. Sci.*, **42**, 100–111, doi:10.1016/J.ERSS.2018.03.001.

Peñasco, C., L.D. Anadón, and E. Verdolini, 2021: Systematic review of the outcomes and trade-offs of ten types of decarbonization policy instruments. *Nat. Clim. Chang.*, **11(3)**, 274–274, doi:10.1038/s41558-021-00992-0.

Peng, Y. and X. Bai, 2018: Experimenting towards a low-carbon city: Policy evolution and nested structure of innovation. *J. Clean. Prod.*, **174**, 201–212, doi:10.1016/J.JCLEPRO.2017.10.116.

Peng, Y., Y. Wei, and X. Bai, 2019: Scaling urban sustainability experiments: Contextualization as an innovation. *J. Clean. Prod.*, **227**, 302–312, doi:10.1016/J.JCLEPRO.2019.04.061.

Pengue, W. et al., 2018: 'Eco-agri-food systems': Today's realities and tomorrow's challenges. In: *TEEB for Agriculture & Food: Scientific and Economic Foundations Report*, UN Environment, Geneva, pp. 57–109.

Pereirinha, P.G. et al., 2018: Main trends and challenges in road transportation electrification. *Transp. Res. Procedia*, **33**, 235–242, doi:10.1016/j.trpro.2018.10.096.

Persoon, P.G.J., R.N.A. Bekkers, and F. Alkemade, 2020: The science base of renewables. *Technol. Forecast. Soc. Change*, **158**, 120121, doi:10.1016/j.techfore.2020.120121.

PIANOo Expertisecentrum, 2020: Biobased Inkopen. *Dutch Minist. Econ. Aff. Clim. Policy*,. https://www.pianoo.nl/en.

Pless, J., 2019: Are "Complementary Policies" Substitutes? Evidence from R&D Subsidies in the UK. *SSRN Electron. J.*, 57, doi:10.2139/ssrn.3379256.

Popp, D., 2002: Induced Innovation and Energy Prices. *Am. Econ. Rev.*, **92(1)**, 160–180, doi:10.1257/000282802760015658.

Popp, D., 2004: ENTICE: Endogenous technological change in the DICE model of global warming. *J. Environ. Econ. Manage.*, **48(1)**, 742–768, doi:10.1016/j.jeem.2003.09.002.

Popp, D., 2010: Innovation and climate policy. *Annu. Rev. Resour. Econ.*, **2**,275-298 doi:10.1146/annurev.resource.012809.103929.

Popp, D., 2011: International Technology Transfer, Climate Change, and the Clean Development Mechanism. *Rev. Environ. Econ. Policy*, **5(1)**, 131–152, doi:10.1093/reep/req018.

Popp, D., 2019: Environmental Policy and Innovation: A Decade of Research. *Int. Rev. Environ. Resour. Econ.*, **13(3–4)**, 265–337, doi:10.1561/101.00000111.

Popp, D. and R. Newell, 2012: Where does energy R&D come from? Examining crowding out from energy R&D. *Energy Econ.*, **34(4)**, 980–991, doi:10.1016/j.eneco.2011.07.001.

Porter, M. and C. Van der Linde, 1995: Green and competitive: Ending the stalemate. *Harvard Business Review, September 1995, 1-15.*Pörtner, H.O. et al., 2021: *Scientific outcome of the IPBES-IPCC co-sponsored workshop on biodiversity and climate change*. IPBES secretariat, Bonn, Germany, 256pp. doi:10.5281/zenodo.5101125.

Pretty, J., C. Toulmin, and S. Williams, 2011: Sustainable intensification in African agriculture. *Int. J. Agric. Sustain.*, **9(1)**, doi:10.3763/ijas.2010.0583.

Probst, B., V. Anatolitis, A. Kontoleon, and L.D. Anadón, 2020: The short-term costs of local content requirements in the Indian solar auctions. *Nat. Energy*, **5(11)**, 842–850, doi:10.1038/s41560-020-0677-7.

Probst, B., L. Westermann, L.D. Anadón, and A. Kontoleon, 2021: Leveraging private investment to expand renewable power generation: Evidence on financial additionality and productivity gains from Uganda. *World Dev.*, **140**, 105347, doi:10.1016/j.worlddev.2020.105347.

Qiu, Y. and L.D. Anadon, 2012: The price of wind power in China during its expansion: Technology adoption, learning-by-doing, economies of scale, and manufacturing localization. *Energy Econ.*, **34(3)**, 772–785, doi:10.1016/j.eneco.2011.06.008.

Quirapas, M.A.J.R. and A. Taeihagh, 2020: Ocean renewable energy development in Southeast Asia: Opportunities, risks and unintended consequences. *Renew. Sustain. Energy Rev.*, **137**(C), 110403, doi:10.1016/j.rser.2020.110403.

Quitzow, R., 2015: Dynamics of a policy-driven market: The co-evolution of technological innovation systems for solar photovoltaics in China and Germany. *Environ. Innov. Soc. Transitions*, **17**, 126–148, doi:10.1016/j.eist.2014.12.002.

Rahwan, I. et al., 2019: Machine behaviour. *Nature*, **568(7753)**, 477–486, doi:10.1038/s41586-019-1138-y.

Ramos-Mejía, M., M.-L. Franco-Garcia, and J.M. Jauregui-Becker, 2018: Sustainability transitions in the developing world: Challenges of socio-technical transformations unfolding in contexts of poverty. *Environ. Sci. Policy*, **84**, 217–223, doi:10.1016/j.envsci.2017.03.010.

Ravetz, J. and S. Funtowicz, 1999: Editorial. *Futures*, **31(7)**, 641–646, doi:https://doi.org/10.1016/S0016-3287(99)00023-3.

Reder, M.W. and J.R. Hicks, 1965: The Theory of Wages. *Economica*, **32(125)**, 88, doi:10.2307/2552450.

Reichman, J., 2009: Compulsory Licensing of Patented Pharmaceutical Inventions: Evaluating the Options. *J. Law, Med. Ethics*, **37(2)**, 247–263. 10.1111/j.1748-720X.2009.00369.x

Reitzig, M., J. Henkel, and C. Heath, 2007: On sharks, trolls, and their patent prey – Unrealistic damage awards and firms' strategies of "being infringed." *Res. Policy*, **36(1)**, 134–154, doi:10.1016/j.respol.2006.10.003.

Remer, D.S. and F.B. Mattos, 2003: Cost and scale-up factors, international inflation indexes and location factors. *Int. J. Prod. Econ.*, **84(1)**, 1–16, doi:10.1016/S0925-5273**(02)**00374-2.

Repo, P. and K. Matschoss, 2019: Social Innovation for Sustainability Challenges. *Sustainability*, **12(1)**, 319, doi:10.3390/su12010319.

RESA, 2001: Act on granting priority to renewable energy sources (Renewable Energy Sources Act, Germany, 2000). *Sol. Energy*, **70(6)**, 489–504, doi:10.1016/S0038-092X**(00)**00144-4.

Reynolds, S., M. Gabriel, and C. Heales, 2017: *Social innovation policy in Europe: Where next?* D5.3: Annual State of the Union Report – Part 1, Social Innovation Community, European Commission, Brussels, Belgium, 37 pp. https://www.siceurope.eu/sites/default/files/field/attachment/social_innovation_policy_in_europe_-_where_next.pdf (Accessed July 22, 2021).

Riahi, K. et al., 2017: The Shared Socioeconomic Pathways and their energy, land use, and greenhouse gas emissions implications: An overview. *Glob. Environ. Chang.*, **42**, 153–168, doi:10.1016/j.gloenvcha.2016.05.009.

Riekstin, A.C., S. James, A. Kansal, J. Liu, and E. Peterson, 2014: No More Electrical Infrastructure: Towards Fuel Cell Powered Data Centers. *SIGOPS Oper. Syst. Rev.*, **48(1)**, 39–43, doi:10.1145/2626401.2626410.

Rijke, J., M. Farrelly, R. Brown, and C. Zevenbergen, 2013: Configuring transformative governance to enhance resilient urban water systems. *Environ. Sci. Policy*, **25**, 62–72, doi:10.1016/j.envsci.2012.09.012.

Rikap, C., 2020: Amazon: A story of accumulation through intellectual rentiership and predation. *Compet. Change*, **26(3-4)**, 436-466 doi:10.1177/1024529420932418.

Rivera-Batiz, L.A. and P.M. Romer, 1991: Economic Integration and Endogenous Growth. *Q. J. Econ.*, **106(2)**, 531, doi:10.2307/2937946.

Roberts, C. et al., 2018: The politics of accelerating low-carbon transitions: Towards a new research agenda. *Energy Res. Soc. Sci.*, **44**(February), 304–311, doi:10.1016/j.erss.2018.06.001.

Rogers, D. and V. Tsirkunov, 2010: Costs and benefits of Early Warning Systems. In: *Global Assessment Report on Disaster Risk Reduction*, International Stragey for Disaster Reduction (ISDR) – The World Bank, Geneva, pp. 1–17.

Rogers, E.M., 2003: *Diffusion of Innovations*. 5th ed. Free Press, New York, NY, USA, 576 pp.

Rogge, K.S. and K. Reichardt, 2016: Policy mixes for sustainability transitions: An extended concept and framework for analysis. *Res. Policy*, **45(8)**, 1620–1635, doi:10.1016/j.respol.2016.04.004.

Rogge, K.S., F. Kern, and M. Howlett, 2017: Conceptual and empirical advances in analysing policy mixes for energy transitions. *Energy Res. Soc. Sci.*, **33**, 1–10, doi:10.1016/j.erss.2017.09.025.

Rogge, K.S., B. Pfluger, and F.W. Geels, 2020: Transformative policy mixes in socio-technical scenarios: The case of the low-carbon transition of the German electricity system **(2010–2050)**. *Technol. Forecast. Soc. Change*, **151**, 119259, doi:10.1016/J.TECHFORE.2018.04.002.

Rolnick, D. et al., 2021: Tackling Climate Change with Machine Learning. *ACM Cumput. Surv.*, **1(1)**, 95. doi: 10.1145/3485128.

Romer, P.M., 1990: Endogenous Technological Change. *J. Polit. Econ.*, **98(5, Part 2)**, S71–S102, doi:10.1086/261725.

Rosenberg, N., 1998: Uncertainty and Technological Change. In: *The Economic Impact of Knowledge* [Neef, D., G.. Siesfeld, and J. Cefola, (eds.)], Butterworth-Heinemann, Woburn, MA, USA, pp. 17–34.

Rotz, S. et al., 2019: The Politics of Digital Agricultural Technologies: A Preliminary review. *Sociol. Ruralis*, **59(2)**, 203–229, doi:10.1111/SORU.12233.

Roy, J., P. Tschakert, H. Waisman, S. Abdul Halim, P. Antwi-Agyei, P. Dasgupta, B. Hayward, M. Kanninen, D. Liverman, C. Okereke, P.F. Pinho, K. Riahi, and A.G. Suarez Rodriguez, 2018: Sustainable Development, Poverty Eradication and Reducing Inequalities. In: *Global Warming of 1.5°C an IPCC special report on the impacts of global warming of 1.5°C above pre-industrial levels and related global greenhouse gas emission pathways, in the context of strengthening the global response to the threat of climate change*, [Masson-Delmotte, V., P. Zhai, H.-O. Pörtner, D. Roberts, J. Skea, P.R. Shukla, A. Pirani, W. Moufouma-Okia, C. Péan, R. Pidcock, S. Connors, J.B.R. Matthews, Y. Chen, X. Zhou, M.I. Gomis, E. Lonnoy, T. Maycock, M. Tignor, and T. Waterfield (eds.)]. Cambridge University Press, Cambridge, UK and New York, NY, USA, pp. 445–538.

Rubin, E.S., I.M.L. Azevedo, P. Jaramillo, and S. Yeh, 2015: A review of learning rates for electricity supply technologies. *Energy Policy*, **86**, 198–218, doi:10.1016/j.enpol.2015.06.011.

Rubio, S.J., 2017: Sharing R&D investments in breakthrough technologies to control climate change. *Oxford Econ. Pap. Ser.*, **69(2, SI)**, 496–521, doi:10.1093/oep/gpw067.

Ruby, T.M., 2015: Innovation-enabling policy and regime transformation towards increased energy efficiency: The case of the circulator pump industry in Europe. *J. Clean. Prod.*, **103**, 574–585, doi:10.1016/j.jclepro.2015.02.017.

Rust, N.A. et al., 2020: How to transition to reduced-meat diets that benefit people and the planet. *Sci. Total Environ.*, **718**, 137208, doi:10.1016/j.scitotenv.2020.137208.

Sachs, J.D. et al., 2019: Six transformations to achieve the Sustainable Development Goals. *Nat. Sustain. 2019 29*, **2(9)**, 805–814, doi:10.1038/s41893-019-0352-9.

Sadras, V.O., 2020: Agricultural technology is unavoidable, directional, combinatory, disruptive, unpredictable and has unintended consequences. *Outlook Agric.*, **49(4)**, 293–297, doi:10.1177/0030727020960493.

Safarzyńska, K. and J.C.J.M. van den Bergh, 2010: Evolutionary models in economics: A survey of methods and building blocks. *J. Evol. Econ.*, **20(3)**, 329–373, doi:10.1007/s00191-009-0153-9.

Sagar, A., 2009: *Technology development and transfer to meet climate and developmental challenges*. Background note UNDESA Background Paper. Delhi High Level Conference on Climate Change, UN Division for Sustainable Development (UN DESA), New Delhi, India, 22–23 Oct. 2009 32 pp.

Sagar, A. and A. Majumdar, 2014: *Facilitating a sustainability transition in developing countries: Proposal for a global Advanced Research Project Agency for Sustainable Development*. Rio+20 Working Paper No.3, UN Division for Sustainable Development (UN DESA), Published Online, 30 pp. https://sustainabledevelopment.un.org/content/documents/1498SagarMajumdar.pdf.

Sagar, A.D. and J.P. Holdren, 2002: Assessing the global energy innovation system: Some key issues. *Energy Policy*, **30(6)**, 465–469, doi:10.1016/S0301-4215**(01)**00117-3.

Sagar, A.D., C. Bremner, and M. Grubb, 2009: Climate Innovation Centres: A partnership approach to meeting energy and climate challenges. *Nat. Resour. Forum*, **33(4)**, 274–284, doi:10.1111/j.1477-8947.2009.001252.x.

Samadi, S., 2018: The experience curve theory and its application in the field of electricity generation technologies – A literature review. *Renew. Sustain. Energy Rev.*, **82**, 2346–2364, doi:10.1016/j.rser.2017.08.077.

Samuelson, P.A., 1965: A Theory of Induced Innovation along Kennedy-Weisacker Lines. *Rev. Econ. Stat.*, **47(4)**, 343, doi:10.2307/1927763.

Sanchez, D.L. and V. Sivaram, 2017: Saving innovative climate and energy research: Four recommendations for Mission Innovation. *Energy Res. Soc. Sci.*, **29**, 123–126, doi:10.1016/j.erss.2017.05.022.

Sanni, M. et al., 2016: Climate change and intellectual property rights in Africa: Environmental necessity-economic opportunity. *African J. Sci. Technol. Innov. Dev.*, **8(5–6)**, 377–385, doi:10.1080/20421338.2016.1219482.

Sanyal, P. and L.R. Cohen, 2009: Powering Progress: Restructuring, Competition, and R&D in the U.S. Electric Utility Industry. *Energy J.*, **30(2)**, 41–79.

Sarker, M.N.I., M. Wu, G.M. Alam, and R.C. Shouse, 2020: Livelihood resilience of riverine island dwellers in the face of natural disasters: Empirical evidence from Bangladesh. *Land use policy*, **95**, 104599, doi:10.1016/j.landusepol.2020.104599.

Sarr, M. and T. Swanson, 2017: Will Technological Change Save the World? The Rebound Effect in International Transfers of Technology. *Environ. Resour. Econ.*, **66(3)**, 577–604, doi:10.1007/s10640-016-0093-4.

Sawulski, J. and J. Witajewski-Baltvilks, 2020: Optimal Diversity in Auctions for Renewable Energy Sources under Technological Uncertainty. *Int. Rev. Environ. Resour. Econ.*, **14(2–3)**, 299–347, doi:10.1561/101.00000118.

Schillo, R.S.S. and R.M.M. Robinson, 2017: Inclusive Innovation in Developed Countries: The Who, What, Why, and How. *Technol. Innov. Manag. Rev.*, **7**, 34–46, doi:http://doi.org/10.22215/timreview/1089.

Schindele, S. et al., 2020: Implementation of agrophotovoltaics: Techno-economic analysis of the price-performance ratio and its policy implications. *Appl. Energy*, **265**, 114737, doi:10.1016/j.apenergy.2020.114737.

Schmid, N., S. Sewerin, and T.S. Schmidt, 2020: Explaining Advocacy Coalition Change with Policy Feedback. *Policy Stud. J.*, **48(4)**, 1109–1134, doi:10.1111/psj.12365.

Schmidt, O. et al., 2017: Future cost and performance of water electrolysis: An expert elicitation study. *Int. J. Hydrog. Energ*, **42**, 30470–30492. https://doi.org/10.1016/j.ijhydene.2017.10.045

Schmidt, T.S. and S. Sewerin, 2017: Technology as a driver of climate and energy politics. *Nat. Energy*, **2(6)**, 17084, doi:10.1038/nenergy.2017.84.

Schmidt, T.S. and S. Sewerin, 2019: Measuring the temporal dynamics of policy mixes – An empirical analysis of renewable energy policy mixes' balance and design features in nine countries. *Res. Policy*, **48(10)**, 103557, doi:10.1016/j.respol.2018.03.012.

Schot, J. and L. Kanger, 2018: Deep transitions: Emergence, acceleration, stabilization and directionality. *Res. Policy*, **47(6)**, 1045–1059, doi:10.1016/j.respol.2018.03.009.

Scoones, I., 2015: *The Politics of Green Transformations*. 1st Edition. Routledge, London, UK, 238 pp.

Scotchmer, S. and J. Green, 1990: Novelty and Disclosure in Patent Law. *RAND J. Econ.*, **21(1)**, 131, doi:10.2307/2555499.

Seebregts, A. et al., 1998: *Endogenous Technological Change in Energy Systems Models: Synthesis of Experience with ERIS, MARKAL, and MESSAGE*. Paul Scherrer Institute, International Institute for Applied Systems Analysis (IIASA), Villigen, Switzerland and Laxenburg, Austria, 29pp., doi:ECN-C-99-025.

Segerberg, A., 2017: Online and Social Media Campaigns For Climate Change Engagement. *Oxford Res. Encycl. Clim. Sci.*, doi:10.1093/acrefore/9780190228620.013.398.

Semieniuk, G., E. Campiglio, J.-F. Mercure, U. Volz, and N.R. Edwards, 2021: Low-carbon transition risks for finance. *WIREs Clim. Chang.*, **12(1)**, doi:10.1002/wcc.678.

Sendzimir, J., C.P. Reij, and P. Magnuszewski, 2011: Rebuilding Resilience in the Sahel: Regreening in the Maradi and Zinder regions of Niger. *Ecol. Soc.*,**16(3)**, doi:10.5751/ES-04198-160301.

Seres, S., E. Haites, and K. Murphy, 2009: Analysis of technology transfer in CDM projects: An update. *Energy Policy*, **37(11)**, 4919–4926, doi:10.1016/j.enpol.2009.06.052.

Serrenho, T. and P. Bertoldi, 2019: *Smart home and appliances: State of the art*. 1st ed. [European Commission Joint Research Center, (ed.)]. Publications Office of the European Union, Luxembourg, 1–59 pp.

Seto, K.C. et al., 2016: Carbon Lock-In: Types, Causes, and Policy Implications. *Annu. Rev. Environ. Resour.*, **41(1)**, 425–452, doi:10.1146/annurev-environ-110615-085934.

Seyfang, D.G. and D.A. Smith, 2007: Grassroots innovations for sustainable development: Towards a new research and policy agenda. *Environmental Politics*, **16(4)**, 584–603, doi:10.1080/09644010701419121.

Shapiro, C., 2001: Navigating the Patent Thicket: Cross Licences, Patent Pools and Standard Setting. *Innov. Policy Econ.*, **1**, 119–150. https://doi.org/10.1086/ipe.1.25056143.

Sharpe, S. and T.M. Lenton, 2021: Upward-scaling tipping cascades to meet climate goals: Plausible grounds for hope. *Clim. Policy*, **21(4)**, 421–433, doi:10.1080/14693062.2020.1870097.

16

Shiraki, H. and M. Sugiyama, 2020: Back to the basic: Toward improvement of technoeconomic representation in integrated assessment models. *Clim. Change*, **162(1)**, 13–24, doi:10.1007/s10584-020-02731-4.

Sierzchula, W. and G. Nemet, 2015: Using patents and prototypes for preliminary evaluation of technology-forcing policies: Lessons from California's Zero Emission Vehicle regulations. *Technol. Forecast. Soc. Change*, **100**, 213–224, doi:10.1016/j.techfore.2015.07.003.

Silvestre, B.S.S. and D.M.M. Ţîrcă, 2019: Innovations for sustainable development: Moving toward a sustainable future. *J. Clean. Prod.*, **208**, 325–332, doi:https://doi.org/10.1016/j.jclepro.2018.09.244.

Singh, N.P. et al., 2020: Dynamics of socio-economic factors affecting climate vulnerability and technology adoption: Evidence from Jodhpur district of Rajasthan. *Indian J. Tradit. Knowl.*, **19(1)**, 192–196.

Sivaram, V., 2018a: *Digital Decarbonization. Promoting Digital Innovations to Advance Clean Energy Systems*. Council on Foreign Relations, New York, USA, 138 pp. https://www.cfr.org/report/digital-decarbonization.

Sivaram, V., 2018b: *Taming the Sun: Innovations to Harness Solar Energy and Power the Planet*. MIT Press, Cambridge, MA, USA, 50–52 pp.

Skea, J., R. van Diemen, M. Hannon, E. Gazis, and A. Rhodes, 2019: *Energy Innovation for the Twenty-First century: Accelerating the Energy Revolution*. 1st ed., Edward Elgar Publishing, Cheltenham, UK, 464 pp.

Skelton, A.C.H., L. Paroussos, and J.M. Allwood, 2020: Comparing energy and material efficiency rebound effects: An exploration of scenarios in the GEM-E3 macroeconomic model. *Ecol. Econ.*, **173**, 106544, doi:10.1016/j.ecolecon.2019.106544.

Skene, K.R., 2020: No goal is an island: The implications of systems theory for the Sustainable Development Goals. *Environ. Dev. Sustain.*, **23**, 9993–10012 doi:10.1007/s10668-020-01043-y.

Solow, R.M., 1957: Technical change and the aggregate production function. *Rev. Econ. Stat.*, **39(3)**, 312–320. doi: https://doi.org/10.2307/1926047.

Somanathan, E. T. Sterner, T. Sugiyama, D. Chimanikire, N.K. Dubash, J. Essandoh-Yeddu, S. Fifita, L. Goulder, A. Jaffe, X. Labandeira, S. Managi, C. Mitchell, J.P. Montero, F. Teng, and T. Zylicz, 2014: National and Sub-national Policies and Institutions. In: *Climate Change 2014: Mitigation of Climate Change. Contribution of Working Group III to the Fifth Assessment Report of the Intergovernmental Panel on Climate Change* [Edenhofer, O., R. Pichs-Madruga, Y. Sokona, E. Farahani, S. Kadner, K. Seyboth, A. Adler, I. Baum, S. Brunner, P. Eickemeier, B. Kriemann, J. Savolainen, S. Schlömer, C. von Stechow, T. Zwickel and J.C. Minx, (eds.)]. Cambridge University Press, Cambridge, UK and New York, USA, pp. 1141–1205.

Song, Y., T. Yang, and M. Zhang, 2019: Research on the impact of environmental regulation on enterprise technology innovation – an empirical analysis based on Chinese provincial panel data. *Environ. Sci. Pollut. Res.*, **26(21)**, 21835–21848, doi:10.1007/s11356-019-05532-0.

Sorrell, S., 2007: *The Rebound Effect: An assessment of the evidence for economy-wide energy savings from improved energy efficiency*. Energy Research Centre. London, UK,169 pp.

Sorrell, S. et al., 2004: *The Economics of Energy Efficiency*. Edward Elgar Publishing, Cheltenham, UK, 360 pp.

Soto Embodas, I. et al., 2019: *The contribution of precision agriculture technologies to farm productivity and the mitigation of greenhouse gas emissions in the EU*. Joint Research Centre of the European Commission, Luxembourg, 447 pp.

Spence, A.M., 1981: The Learning Curve and Competition. *Bell J. Econ.*, **12(1)**, 49, doi:10.2307/3003508.

Stavins, R., J. Zou, T. Brewer, M. Conte Grand, M. den Elzen, M. Finus, J. Gupta, N. Höhne, M.-K. Lee, A. Michaelowa, M. Paterson, K. Ramakrishna, G. Wen, J. Wiener, and H. Winkler, 2014: International Cooperation: Agreements and Instruments. In: *Climate Change 2014: Mitigation of Climate Change. Contribution of Working Group III to the Fifth Assessment Report of the Intergovernmental Panel on Climate Change* [Edenhofer, O., R. Pichs-Madruga, Y. Sokona, E. Farahani, S. Kadner, K. Seyboth, A. Adler, I. Baum, S. Brunner, P. Eickemeier, B. Kriemann, J. Savolainen, S. Schlömer, C. von Stechow, T. Zwickel and J.C. Minx (eds.)]. Cambridge University Press, Cambridge, UK, and New York, NY, USA, pp. 1001–1082.

Steffen, W. et al., 2015: Planetary boundaries: Guiding human development on a changing planet. *Science*, **347(6223)**, doi:10.1126/SCIENCE.1259855.

Steneck, N.H., 2006: Fostering integrity in research: Definitions, current knowledge, and future directions. *Science and Engineering Ethics*, **12**, 53–74.doi: doi:10.1007/PL00022268.

Stern, N., 2007: *The Economics of Climate Change*. Cambridge University Press, Cambridge, UK, 692 pp.

Stiglitz, J.E., 2008: Economic foundations of intellectual property rights. *Duke Law J.*, **57(6)**, 1693–1724.

Stilgoe, J., R. Owen, and P. Macnaghten, 2013: Developing a framework for responsible innovation. *Res. Policy*, **42(9)**, 1568–1580, doi:10.1016/j.respol.2013.05.008.

Stokes, L.C., 2016: Electoral Backlash against Climate Policy: A Natural Experiment on Retrospective Voting and Local Resistance to Public Policy. *Am. J. Pol. Sci.*, **60(4)**, 958–974, doi:10.1111/ajps.12220.

Stokes, L.C. and C. Warshaw, 2017: Renewable energy policy design and framing influence public support in the United States. *Nat. Energy*, **2(8)**, 17107, doi:10.1038/nenergy.2017.107.

Stokes, L.C. and H.L. Breetz, 2018: Politics in the U.S. energy transition: Case studies of solar, wind, biofuels and electric vehicles policy. *Energy Policy*, **113**, 76-86 doi:10.1016/j.enpol.2017.10.057.

Stoll, C., L. Klaaßen, and U. Gallersdörfer, 2019: The Carbon Footprint of Bitcoin. *Joule*, **3(7)**, 1647–1661, doi:10.1016/j.joule.2019.05.012.

Sue Wing, I., 2006: Representing induced technological change in models for climate policy analysis. *Energy Econ.*, **28(5–6)**, 539–562, doi:10.1016/j.eneco.2006.05.009.

Sue Wing, I., 2008: Explaining the declining energy intensity of the U.S. economy. *Resour. Energy Econ.*, **30(1)**, 21–49, doi:10.1016/j.reseneeco.2007.03.001.

Sundaramoorthy, S. and A. Walia, 2017: India's experience in implementing strategic schemes to enhance appliance energy efficiency & futuristic integrated policy approaches to adopt most efficient technologies. Panel: 2. Policy: governance, design, implementation and evaluation challenges, *ECEEE Summer Study Proceedings*. European Council for an Energy Efficient Economy, Stockholm, Sweden, pp. 297–284.

Supran, G. and N. Oreskes, 2017: Assessing ExxonMobil's climate change communications **(1977–2014)**. *Environ. Res. Lett.*, **12(8)**, 084019, doi:10.1088/1748-9326/aa815f.

Surana, K. and L.D. Anadon, 2015: Public policy and financial resource mobilization for wind energy in developing countries: A comparison of approaches and outcomes in China and India. *Glob. Environ. Chang.*, **35**, 340–359, doi:10.1016/j.gloenvcha.2015.10.001.

Surana, K., A. Singh, and A.D. Sagar, 2020a: Strengthening science, technology, and innovation-based incubators to help achieve Sustainable Development Goals: Lessons from India. *Technol. Forecast. Soc. Change*, **157**, 120057, doi:10.1016/j.techfore.2020.120057.

Surana, K. et al., 2020b: *Regional Clean Energy Innovation*. Energy Futures Initiative, Washington DC and University of Maryland Global Sustainability Initiative, Maryland, USA, 122 pp.

Sutherland, R.J., 2006: The Distributional Effects of Direct Regulation: A Case Study of Energy Efficiency Appliance Standards. In: *The Distributional Effects of Environmental Policy* [Serret; and P. Johnstone, (eds.)], Edward Elgar Publishing, Cheltenham, UK, pp 171–198.

Suzuki, M., 2015: Identifying roles of international institutions in clean energy technology innovation and diffusion in the developing countries: Matching barriers with roles of the institutions. *J. Clean. Prod.*, **98**, 229–240, doi:10.1016/j.jclepro.2014.08.070.

Sweerts, B., R.J. Detz, and B. van der Zwaan, 2020: Evaluating the Role of Unit Size in Learning-by-Doing of Energy Technologies. *Joule*, **4(5)**, 967–970, doi:10.1016/j.joule.2020.03.010.

Swilling, M., J. Musango, and J. Wakeford, 2016: Developmental States and Sustainability Transitions: Prospects of a Just Transition in South Africa. *J. Environ. Policy Plan.*, **18(5)**, 650–672, doi:10.1080/1523908X.2015.1107716.

Tariq, A., Y.F. Badir, W. Tariq, and U.S. Bhutta, 2017: Drivers and consequences of green product and process innovation: A systematic review, conceptual framework, and future outlook. *Technol. Soc.*, **51**, 8–23, doi:10.1016/J.TECHSOC.2017.06.002.

Tassey, G., 2014: Competing in Advanced Manufacturing: The Need for Improved Growth Models and Policies. *J. Econ. Perspect.*, **28(1)**, 27–48, doi:10.1257/jep.28.1.27.

Taylor, M.R., 2012: Innovation under cap-and-trade programs. *Proc. Natl. Acad. Sci.*, **109(13)**, 4804–4809, doi:10.1073/pnas.1113462109.

TEC, 2012: *Report on activities and performance of the Technology Executive Committee for 2012*. UNFCCC, Bonn, Germany, 13 pp. https://unfccc.int/resource/docs/2012/sb/eng/02.pdf.

TEC, 2015: *TEC Brief #7: Strengthening National Systems of Innovation to Enhance Action on Climate Change*. UNFCCC, Bonn, Germany, 12 pp. https://unfccc.int/ttclear/misc_/StaticFiles/gnwoerk_static/TEC_documents/5be1bf880cc34d52a4315206d54a711b/60d1580f741a4bc783da5a00cf64a879.pdf.

TEC, 2017: *Enhancing financing for the research, development and demonstration of climate technologies*. UNFCCC, Bonn, Germany, Germany, 26 pp.

TEC, 2019: *Rolling Workplan of the Technology Executive Committee for 2019–2022*. UNFCCC, Bonn, Germany, 13 pp. https://unfccc.int/ttclear/misc_/StaticFiles/gnwoerk_static/TEC_Members_doc/9ca11435f3d94446a917a01a5d52a64a/20deb28a9b454a978b17299aa0bbf64c.pdf (Accessed August 25, 2021).

TEC, 2021: *Compilation of good practices and lessons learned on international collaborative research, development and demonstration initiatives of climate technology*. UNFCCC, Bonn, Germany, 68 pp.

TEEB, 2018: *TEEB for Agriculture & Food: Scientific and Economic Foundations*. UN Environment, Geneva, 414 pp.

Thacker, S. et al., 2019: Infrastructure for sustainable development. *Nat. Sustain. 2019 24*, **2(4)**, 324–331, doi:10.1038/s41893-019-0256-8.

Thornton, T.F. and C. Comberti, 2017: Synergies and trade-offs between adaptation, mitigation and development. *Clim. Change*, **140(1)**, 5–18, doi:10.1007/s10584-013-0884-3.

Tigabu, A.D., 2018: Analysing the diffusion and adoption of renewable energy technologies in Africa: The functions of innovation systems perspective. *African J. Sci. Technol. Innov. Dev.*, **10(5)**, 615–624, doi:10.1080/20421338.2017.1366130.

Tigabu, A.D., F. Berkhout, and P. van Beukering, 2015: Functional evolution and accumulation of technological innovation systems: The case of renewable energy in East Africa. *Sci. Public Policy*, **42(5)**, 614–631, doi:10.1093/scipol/scu073.

Timilsina, R., 2020: *Demystifying the Costs of Electricity Generation Technologies*. World Bank, Washington, USA, 39 pp.

Tittonell, P., 2014: Ecological intensification of agriculture – sustainable by nature. *Curr. Opin. Environ. Sustain.*, **8**, 53–61, doi:10.1016/j.cosust.2014.08.006.

Toke, D., 2015: Renewable Energy Auctions and Tenders: How good are they? *Int. J. Sustain. Energy Plan. Manag.*, **8**, 43–56, doi:https://doi.org/10.5278/ijsepm.2015.8.5.

Tonn, B.E. and D. Stiefel, 2019: Anticipating the Unanticipated-Unintended Consequences of Scientific and Technological Purposive Actions. *World Futur. Rev.*, **11(1)**, 19–50, doi:10.1177/1946756718789413.

Townsend, R. et al., 2019: *Future of Food: Harnessing Digital Technologies to Improve Food System Outcomes*. International Bank for Reconstruction and Development / The World Bank, Washington D.C, USA, 44 pp.

Trianni, A., E. Cagno, and E. Worrell, 2013: Innovation and adoption of energy efficient technologies: An exploratory analysis of Italian primary metal manufacturing SMEs. *Energy Policy*, **61**, 430–440, doi:10.1016/j.enpol.2013.06.034.

Trutnevyte, E. et al., 2019: Societal Transformations in Models for Energy and Climate Policy: The Ambitious Next Step. *One Earth*, **1(4)**, 423–433, doi:10.1016/j.oneear.2019.12.002.

Tsur, Y. and A. Zemel, 2007: Towards endogenous recombinant growth. *J. Econ. Dyn. Control*, **31(11)**, 3459–3477, doi:10.1016/j.jedc.2006.12.002.

Turner, R., P. McConney, and I. Monnereau, 2020: Climate Change Adaptation and Extreme Weather in the Small-Scale Fisheries of Dominica. *Coast. Manag.*, **48(5)**, 436–455, doi:10.1080/08920753.2020.1795970.

Turnheim, B. and B.K. Sovacool, 2020: Exploring the role of failure in socio-technical transitions research. *Environ. Innov. Soc. Transitions*, **37**, 267–289, doi:10.1016/J.EIST.2020.09.005.

Turnheim, B. et al., 2015: Evaluating sustainability transitions pathways: Bridging analytical approaches to address governance challenges. *Glob. Environ. Chang.*, **35**, 239–253, doi:10.1016/j.gloenvcha.2015.08.010.

Twomey, P., 2012: Rationales for Additional Climate Policy Instruments under a Carbon Price. *Econ. Labour Relations Rev.*, **23(1)**, 7–31, doi:10.1177/103530461202300102.

Tziva, M., S.O. Negro, A. Kalfagianni, and M.P. Hekkert, 2020: Understanding the protein transition: The rise of plant-based meat substitutes. *Environ. Innov. Soc. Transitions*, **35**, 217–231, doi:10.1016/J.EIST.2019.09.004.

Ugur, M. and A. Mitra, 2017: Technology Adoption and Employment in Less Developed Countries: A Mixed-Method Systematic Review. *World Dev.*, **96**, 1–18, doi:10.1016/j.worlddev.2017.03.015.

Ulsrud, K., H. Rohracher, and C. Muchunku, 2018: Spatial transfer of innovations: South-South learning on village-scale solar power supply between India and Kenya. *Energy Policy*, **114**, 89–97, doi:10.1016/j.enpol.2017.11.064.

UNCTAD, 2019: *The Impact of Rapid Technological Change on Sustainable Development*. United Nations Publications, New York, NY, USA, 46 pp.

Unel, B., 2008: R&D spillovers through trade in a panel of OECD industries. *J. Int. Trade Econ. Dev.*, **17(1)**, 105–133, doi:10.1080/09638190701728024.

UNEP, 2013: *Assessing Global Land Use: Balancing Consumption with Sustainable Supply. A Report of the Working Group on Land and Soils of the International Resource Panel*. [Bringezu, S. et al., (eds.)]. UN Environment Programme, Nairobi, Kenya, 132 pp. https://www.resourcepanel.org/reports/assessing-global-land-use.

UNEP, 2014: *Decoupling 2: Technologies, opportunities and policy options. A Report of the Working Group on Decoupling to the International Resource Panel.* [Von Weizsäcker, E.U., J. De Larderel, K. Hargroves, C. Hudson, M. Smith, and M. Rodrigues, (eds.)]. UNEP, Nairobi, Kenya, 158 pp.

UNEP, 2015: *Policy Coherence of the Sustainable Development Goals A Natural Resource Perspective*. UNEP, Paris, France, 29 pp.

UNEP, 2017: *Assessing global resource use: A systems approach to resource efficiency and pollution reduction*. [Bringezu, S. et al., (eds.)]. United Nations Environment Programme, Nairobi, Kenya, 99 pp.

UNFCCC, 2015: *Paris Agreement*. United Nations, New York, NY, USA, 16 pp.

UNFCCC, 2016: *Report of the Conference of the Parties on its twenty-first session, held in Paris from 30 November to 13 December 2015. FCCC/CP/2015/10/Add.1*. United UNFCCC, Bonn, Germany, 42 pp.

UNFCCC, 2017: *Enhancing financing for the research, development and demonstration of climate technologies*. UNFCCC, Bonn, Germany, 26 pp.

UNFCCC, 2020a: *Paris Committee on Capacity-building: Review report on the status and progress of work under the Strategic Plan for Stakeholder Engagement, Communications and Resource Mobilization (June 2019–June 2020)*. UNFCCC, Bonn, Germany, 13 pp.

UNFCCC, 2020b: *Workplan of the Paris Committee on Capacity-building for 2021–2024*. UNFCCC, Bonn, Germany,13 pp.

Unruh, G.C., 2000: Understanding carbon lock-in. *Energy Policy*, **28(12)**, 817–830, doi:10.1016/S0301-4215**(00)**00070-7.

Unruh, G.C., 2002: Escaping carbon lock-in. *Energy Policy*, **30(4)**, 317–325, doi:10.1016/S0301-4215**(01)**00098-2.

Upadhyaya, P., M.K. Shrivastava, G. Gorti, and S. Fakir, 2020: Capacity building for proportionate climate policy: Lessons from India and South Africa. *Int. Polit. Sci. Rev.*, **0(0)**, 019251212096388, doi:10.1177/0192512120963883.

Urban, F., 2018: China's rise: Challenging the North-South technology transfer paradigm for climate change mitigation and low carbon energy. *Energy Policy*, **113**(November 2017), 320–330, doi:10.1016/j.enpol.2017.11.007.

US Department of Energy, 2011: *DOE G 413.3-4A, Technology Readiness Assessment Guide*. DOE, Washington D.C, USA, 73 pp. https://www.directives.doe.gov/directives-documents/400-series/0413.3-EGuide-04a.

US National Academies of Sciences Engineering and Medicine, 2017: *An Assessment of ARPA-E*. Washington, DC: The National Academies Press. https://doi.org/10.17226/24778.

van Bommel, N. and J.I. Höffken, 2021: Energy justice within, between and beyond European community energy initiatives: A review. *Energy Res. Soc. Sci.*, **79**, 102157, doi:10.1016/j.erss.2021.102157.

van Bree, B., G.P.J. Verbong, and G.J. Kramer, 2010: A multi-level perspective on the introduction of hydrogen and battery-electric vehicles. *Technol. Forecast. Soc. Change*, **77(4)**, 529–540, doi:10.1016/j.techfore.2009.12.005.

Van Buskirk, R.D., C.L.S. Kantner, B.F. Gerke, and S. Chu, 2014: A retrospective investigation of energy efficiency standards: Policies may have accelerated long term declines in appliance costs. *Environ. Res. Lett.*, **9(11)**, 114010, doi:10.1088/1748-9326/9/11/114010.

van den Bergh, J.C.J.M., 2008: Optimal diversity: Increasing returns versus recombinant innovation. *J. Econ. Behav. Organ.*, **68(3–4)**, 565–580, doi:10.1016/j.jebo.2008.09.003.

van den Bijgaart, I., 2017: The unilateral implementation of a sustainable growth path with directed technical change. *Eur. Econ. Rev.*, **91**, 305–327, doi:10.1016/j.euroecorev.2016.10.005.

van der Voorn, T., Å. Svenfelt, K.E. Björnberg, E. Fauré, and R. Milestad, 2020: Envisioning carbon-free land use futures for Sweden: A scenario study on conflicts and synergies between environmental policy goals. *Reg. Environ. Chang.*, **20(2)**, 35, doi:10.1007/s10113-020-01618-5.

van Sluisveld, M.A.E. et al., 2020: Aligning integrated assessment modelling with socio-technical transition insights: An application to low-carbon energy scenario analysis in Europe. *Technol. Forecast. Soc. Change*, **151**(October 2017), 119177, doi:10.1016/j.techfore.2017.10.024.

Van Summeren, L.F.M., A.J. Wieczorek, and G.P.J. Verbong, 2021: The merits of becoming smart: How Flemish and Dutch energy communities mobilise digital technology to enhance their agency in the energy transition. *Energy Res. Soc. Sci.*, **79**, 102160, doi:10.1016/j.erss.2021.102160.

van Vliet, M., K. Kok, and T. Veldkamp, 2010: Linking stakeholders and modellers in scenario studies: The use of Fuzzy Cognitive Maps as a communication and learning tool. *Futures*, **42(1)**, 1–14, doi:10.1016/j.futures.2009.08.005.

van Welie, M.J. and H.A. Romijn, 2018: NGOs fostering transitions towards sustainable urban sanitation in low-income countries: Insights from Transition Management and Development Studies. *Environ. Sci. Policy*, **84**, 250–260, doi:10.1016/j.envsci.2017.08.011.

Vasconcellos, H.A.S. and L. Caiado Couto, 2021: Estimation of socioeconomic impacts of wind power projects in Brazil's Northeast region using Interregional Input-Output Analysis. *Renew. Sustain. Energy Rev.*, **149**, 111376, doi:10.1016/j.rser.2021.111376.

Vassilakopoulou, P. and E. Hustad, 2021: Bridging Digital Divides: a Literature Review and Research Agenda for Information Systems Research. *Inf. Syst. Front.*,, 1–15, doi:10.1007/s10796-020-10096-3.

Vázquez-Canteli, J.R. and Z. Nagy, 2019: Reinforcement learning for demand response: A review of algorithms and modeling techniques. *Appl. Energy*, **235**, 1072–1089, doi:10.1016/j.apenergy.2018.11.002.

Veefkind, V., J. Hurtado-Albir, S. Angelucci, K. Karachalios, and N. Thumm, 2012: A new EPO classification scheme for climate change mitigation technologies. *World Pat. Inf.*, **34(2)**, 106–111, doi:10.1016/j.wpi.2011.12.004.

Verdolini, E. and M. Galeotti, 2011: At home and abroad: An empirical analysis of innovation and diffusion in energy technologies. *J. Environ. Econ. Manage.*, **61(2)**, 119–134, doi:10.1016/j.jeem.2010.08.004.

Verdolini, E. and V. Bosetti, 2017: Environmental Policy and the International Diffusion of Cleaner Energy Technologies. *Environ. Resour. Econ.*, **66(3)**, 497–536, doi:10.1007/s10640-016-0090-7.

Verdolini, E., L.D. Anadón, E. Baker, V. Bosetti, and L. Aleluia Reis, 2018: The Future Prospects of Energy Technologies: Insights from Expert Elicitations. *Rev. Environ. Econ. Policy*, **12(1)**, 133–153, doi:10.1093/reep/rex028.

Verma, P. et al., 2020: *Digitalization: Enabling the new phase of energy efficiency*. Group of Experts on Energy Efficiency Seventh session Geneva, 22 and 25 September 2020 Item 5 of the Annotated provisional agenda – Regulatory and policy dialogue addressing barriers to improve energy efficiency (GEEE-7/2020/INF.3), https://unece.org/sites/default/files/2020-12/GEEE-7.2020.INF_.3.pdf.

Veugelers, R., 2012: Which policy instruments to induce clean innovating? *Res. Policy*, **41(10)**, 1770–1778, doi:10.1016/j.respol.2012.06.012.

Victor, D.G., 2018: Digitalization: An Equal Opportunity Wave of Energy Innovation. In: *Digital Decarbonization Promoting Digital Innovations to Advance Clean Energy Systems* [Sivaram, V., (ed.)], Council on Foreign Relations Press, New York, USA, pp. 25–32.

Victor, D.G., F.W. Geels, and S. Sharpe, 2019: *Accelerating the Low Carbon Transition: The Case for Stronger, More Targeted and Coordinated International Action.* Commissioned by the UK Department for Business, Energy & Industrial Strategy; Supported by the Energy Transitions Commission.Brookings Institute, London, UK and San Diego, USA, 138 pp.

Vogel, B., D. Henstra, and G. McBean, 2020: Sub-national government efforts to activate and motivate local climate change adaptation: Nova Scotia, Canada. *Environ. Dev. Sustain.*, **22(2)**, 1633–1653, doi:10.1007/s10668-018-0242-8.

Voss, J.-P., D. Bauknecht, and R. Kemp, 2006: *Reflexive governance for sustainable development*. Edward Elgar Publishing, Cheltenham, UK, 457 pp.

Voyant, C. et al., 2017: Machine learning methods for solar radiation forecasting: A review. *Renew. Energy*, **105**, 569–582, doi:10.1016/j.renene.2016.12.095.

Vranken, H., 2017: Sustainability of bitcoin and blockchains. *Curr. Opin. Environ. Sustain.*, **28**, 1–9, doi:10.1016/j.cosust.2017.04.011.

Wahlroos, M., M. Pärssinen, J. Manner, and S. Syri, 2017: Utilizing data center waste heat in district heating – Impacts on energy efficiency and prospects for low-temperature district heating networks. *Energy*, **140**, 1228–1238, doi:10.1016/j.energy.2017.08.078.

Wahlroos, M., M. Pärssinen, S. Rinne, S. Syri, and J. Manner, 2018: Future views on waste heat utilization – Case of data centers in Northern Europe. *Renew. Sustain. Energy Rev.*, **82**, 1749–1764, doi:10.1016/j.rser.2017.10.058.

Waiswa, P. et al., 2019: Using research priority-setting to guide bridging the implementation gap in countries – a case study of the Uganda newborn research priorities in the SDG era. *Heal. Res. Policy Syst.*, **17(1)**, 54, doi:10.1186/s12961-019-0459-5.

Wallerstein, M., M. Mogee, R. Schoen, and P. David, 1993: Intellectual property institutions and the panda's thumb: Patents, copyrights, and trade secrets in economic theory and history. In: *Global Dimensions of Intellectual Property Rights in Science and Technology* [Wallerstein, M., M. Mogee, and R. Schoen, (eds.)], National Academies Press, Washington DC, USA, pp. 19–62.

Walsh, P.P.P., E. Murphy, and D. Horan, 2020: The role of science, technology and innovation in the UN 2030 agenda. *Technol. Forecast. Soc. Change*, **154**, 119957, doi:10.1016/J.TECHFORE.2020.119957.

Wang, F., J. Yu, P. Yang, L. Miao, and B. Ye, 2017: Analysis of the Barriers to Widespread Adoption of Electric Vehicles in Shenzhen China. *Sustainability*, **9(4)**, 522, doi:10.3390/su9040522.

Wang, J., Y.-N. Lee, and J.P. Walsh, 2018: Funding model and creativity in science: Competitive versus block funding and status contingency effects. *Res. Policy*, **47(6)**, 1070–1083, doi:10.1016/j.respol.2018.03.014.

Wang, X., H. Cai, and H.K. Florig, 2016: Energy-saving implications from supply chain improvement: An exploratory study on China's consumer goods retail system. *Energy Policy*, **95**, 411–420, doi:https://doi.org/10.1016/j.enpol.2016.04.044.

Wang, Z., X. Wang, and D. Guo, 2017: Policy implications of the purchasing intentions towards energy-efficient appliances among China's urban residents: Do subsidies work? *Energy Policy*, **102**, 430–439, doi:10.1016/j.enpol.2016.12.049.

Watanabe, C., K. Wakabayashi, and T. Miyazawa, 2000: Industrial dynamism and the creation of a "virtuous cycle" between R&D, market growth and price reduction. *Technovation*, **20(6)**, 299–312, doi:10.1016/S0166-4972**(99)**00146-7.

Watkins, A., T. Papaioannou, J. Mugwagwa, and D. Kale, 2015: National innovation systems and the intermediary role of industry associations in building institutional capacities for innovation in developing countries: A critical review of the literature. *Res. Policy*, **44(8)**, 1407–1418, doi:10.1016/j.respol.2015.05.004.

Way, R., F. Lafond, F. Lillo, V. Panchenko, and J.D. Farmer, 2019: Wright meets Markowitz: How standard portfolio theory changes when assets are technologies following experience curves. *J. Econ. Dyn. Control*, **101**, 211–238, doi:10.1016/j.jedc.2018.10.006.

Weber, K.M. and H. Rohracher, 2012: Legitimizing research, technology and innovation policies for transformative change. *Res. Policy*, **41(6)**, 1037–1047, doi:10.1016/j.respol.2011.10.015.

Weber, K.M. and B. Truffer, 2017: Moving innovation systems research to the next level: Towards an integrative agenda. *Oxford Rev. Econ. Policy*, **33(1)**, 101–121, doi:10.1093/oxrep/grx002.

Wei, M., S.J. Smith, and M.D. Sohn, 2017: Non-constant learning rates in retrospective experience curve analyses and their correlation to deployment programs. *Energy Policy*, **107**, 356–369, doi:10.1016/j.enpol.2017.04.035.

Weingärtner, L., T. Pforr, and E. Wilkinson, 2020: *The Evidence Base on Anticipatory Action*. World Food Programme, Rome, 48 pp. https://www.wfp.org/publications/evidence-base-anticipatory-action.

Weiss, C. and W.B. Bonvillian, 2013: Legacy sectors: Barriers to global innovation in agriculture and energy. *Technol. Anal. Strateg. Manag.*, **25(10)**, 1189–1208, doi:10.1080/09537325.2013.843658.

Weiss, M., M. Junginger, M.K. Patel, and K. Blok, 2010: A review of experience curve analyses for energy demand technologies. *Technol. Forecast. Soc. Change*, **77(3)**, 411–428, doi:10.1016/j.techfore.2009.10.009.

Weitzman, M.L., 1998: Recombinant Growth. *Q. J. Econ.*, **113(2)**, 331–360, doi:10.1162/003355398555595.

Wesche, J.P., S.O. Negro, E. Dütschke, R.P.J.M. Raven, and M.P. Hekkert, 2019: Configurational innovation systems – Explaining the slow German heat transition. *Energy Res. Soc. Sci.*, **52**, 99–113, doi:10.1016/j.erss.2018.12.015.

Wesseling, J.H. and A. Van der Vooren, 2017: Lock-in of mature innovation systems: The transformation toward clean concrete in the Netherlands. *J. Clean. Prod.*, **155**, 114–124, doi:https://doi.org/10.1016/j.jclepro.2016.08.115.

Wesseling, J.H., J.C.M. Farla, and M.P. Hekkert, 2015: Exploring car manufacturers' responses to technology-forcing regulation: The case of California's ZEV mandate. *Environ. Innov. Soc. Transitions*, **16**, 87–105, doi:10.1016/j.eist.2015.03.001.

Westerhoff, L., S.R.J. Sheppard, D. Mathew Iype, S. Cote, and J. Salter, 2018: Social mobilization on climate change and energy: An evaluation of research projects in British Columbia, Canada. *Energy Res. Soc. Sci.*, **46**, 368–380, doi:10.1016/J.ERSS.2018.07.022.

Westley, F.R. et al., 2013: A Theory of Transformative Agency in Linked Social-Ecological Systems. *Ecol. Soc.*, **18(3)**, art27, doi:10.5751/ES-05072-180327.

Weyant, J.P., 2011: Accelerating the development and diffusion of new energy technologies: Beyond the "valley of death." *Energy Econ.*, **33(4)**, 674–682, doi:10.1016/j.eneco.2010.08.008.

Weyrauch, T. and C. Herstatt, 2016: What is frugal innovation? Three defining criteria. *J. Frugal Innov. 2016 21*, **2(1)**, 1–17, doi:10.1186/S40669-016-0005-Y.

Wezel, A., G. Soboksa, S. McClelland, F. Delespesse, and A. Boissau, 2015: The blurred boundaries of ecological, sustainable, and agroecological intensification: A review. *Agron. Sustain. Dev.*, **35(4)**, 1283–1295, doi:10.1007/s13593-015-0333-y.

Wezel, A. et al., 2020: Agroecological principles and elements and their implications for transitioning to sustainable food systems. A review. *Agron. Sustain. Dev.*, **40(6)**, 40, doi:10.1007/s13593-020-00646-z.

Whittaker, D.H., T. Sturgeon, T. Okita, and T. Zhu, 2020: *Compressed Development*. Oxford University Press, Oxford, UK, 304pp.

WHO, 2002: *The implications of DOHA Declaration on the TRIPS Agreement and Public Health*. Document Number WHO/EDM/PAR/2002.3. World Health Organization, Geneva, Switzerland, 56pp. https://apps.who.int/iris/handle/10665/67345.

Wieczorek, A.J. and M.P. Hekkert, 2012: Systemic instruments for systemic innovation problems: A framework for policy makers and innovation scholars. *Sci. Public Policy*, **39(1)**, 74–87, doi:10.1093/scipol/scr008.

Wieczorek, A.J. et al., 2013: A review of the European offshore wind innovation system. *Renew. Sustain. Energy Rev.*, **26**, 294–306, doi:10.1016/j.rser.2013.05.045.

Wiel, S., C. Egan, and M. delta Cava, 2006: Energy efficiency standards and labels provide a solid foundation for economic growth, climate change mitigation, and regional trade. *Energy Sustain. Dev.*, **10(3)**, 54–63, doi:10.1016/S0973-0826**(08)**60544-X.

Wigand, F. et al., 2016: *Auctions for Renewable Energy Support: Effective use and efficient implementation options.* Aures Research Project Report D4.2 1–46 pp. https://ec.europa.eu/research/participants/documents/downloadPublic?documentIds=080166e5aa83001d&appId=PPGMS.

Williams, D.S., S. Rosendo, O. Sadasing, and L. Celliers, 2020: Identifying local governance capacity needs for implementing climate change adaptation in Mauritius. *Clim. Policy*, **20(5)**, 548–562, doi:10.1080/14693062.2020.1745743.

Wilson, C., 2012: Up-scaling, formative phases, and learning in the historical diffusion of energy technologies. *Energy Policy*, **50**, 81–94, doi:10.1016/j.enpol.2012.04.077.

Wilson, C. and A. Grubler, 2013: Energy Technology Innovation. In: *Energy Technology Innovation: Learning from Historical Successes and Failures* [Grubler, A. and C. Wilson, (eds.)], Cambridge University Press, Cambridge, UK, pp. 3–10.

Wilson, C. et al., 2020: Granular technologies to accelerate decarbonization. *Science*, **368(6486)**, 36–39, doi:10.1126/science.aaz8060.

Winther, T., K. Ulsrud, and A. Saini, 2018: Solar powered electricity access: Implications for women's empowerment in rural Kenya. *Energy Res. Soc. Sci.*, **44**, 61–74, doi:https://doi.org/10.1016/j.erss.2018.04.017.

WIPO, 2008: *WIPO Intellectual Property Handbook: Policy, Law and Use*. 2nd ed. World Intellectual Property Organization (WIPO), New York, NY, USA, 488 pp.

WIPO, 2021: WIPO Member States. https://www.wipo.int/members/en/ (Accessed November 1, 2021).

Witajewski-Baltvilks, J., E. Verdolini, and M. Tavoni, 2015: Bending the learning curve. *Energy Econ.*, **52**, S86–S99, doi:10.1016/j.eneco.2015.09.007.

Witajewski-Baltvilks, J., E. Verdolini, and M. Tavoni, 2017: Induced technological change and energy efficiency improvements. *Energy Econ.*, **68**, 17–32, doi:10.1016/j.eneco.2017.10.032.

World Bank, 2014: *World Development Indicators 2014*. The World Bank, Washington, D.C., USA115pp.

WTO, 1994: *Trade-Related Aspects of Intellectual Property Rights, Annex 1C of the Marrakesh Agreement Establishing the World Trade Organization*. World Trade Organization (WTO), Marrakesh, Morocco, pp 319-351.

WTO, 2020: *Compulsory licensing of pharmaceuticals and TRIPS*. https://www.wto.org/english/tratop_e/trips_e/public_health_faq_e.htm.

Yan, Z., K. Du, Z. Yang, and M. Deng, 2017: Convergence or divergence? Understanding the global development trend of low-carbon technologies. *Energy Policy*, **109**, 499–509, doi:10.1016/j.enpol.2017.07.024.

Yi, H., 2013: Clean energy policies and green jobs: An evaluation of green jobs in U.S. metropolitan areas. *Energy Policy*, **56**, 644–652, doi:10.1016/j.enpol.2013.01.034.

Yi, H. and Y. Liu, 2015: Green economy in China: Regional variations and policy drivers. *Glob. Environ. Chang.*, **31**, 11–19, doi:10.1016/j.gloenvcha.2014.12.001.

Youn, H., D. Strumsky, L.M.A. Bettencourt, and J. Lobo, 2015: Invention as a combinatorial process: Evidence from US patents. *J. R. Soc. Interface*, **12(106)**, 20150272, doi:10.1098/rsif.2015.0272.

Young, A., 1993: Invention and Bounded Learning by Doing. *J. Polit. Econ.*, **101(3)**, 443–472, doi:10.1086/261882.

Yu, C.F., W. van Sark, and E.A. Alsema, 2011: Unraveling the photovoltaic technology learning curve by incorporation of input price changes and scale effects. *Renew. Sustain. Energy Rev.*, **15(1)**, 324–337, doi:10.1016/j.rser.2010.09.001.

Zangheri, P., T. Serrenho, and P. Bertoldi, 2019: Energy Savings from Feedback Systems: A Meta-Studies' Review. *Energies*, **12(19)**, 3788, doi:10.3390/en12193788.

Zeng, W., T. Miwa, and T. Morikawa, 2017: Application of the support vector machine and heuristic k-shortest path algorithm to determine the most eco-friendly path with a travel time constraint. *Transp. Res. Part D Transp. Environ.*, **57**, 458–473, doi:10.1016/j.trd.2017.10.001.

Zhan, L., M. Ju, and J. Liu, 2011: Improvement of China Energy Label System to Promote Sustainable Energy Consumption. *Energy Procedia*, **5**, 2308–2315, doi:10.1016/j.egypro.2011.03.397.

Zhang, F. and K.S. Gallagher, 2016: Innovation and technology transfer through global value chains: Evidence from China's PV industry. *Energy Policy*, **94**, 191–203, doi:10.1016/J.ENPOL.2016.04.014.

Zhang, L., H. Liu, and J. Wu, 2017: The price premium for green-labelled housing: Evidence from China. *Urban Stud.*, **54(15)**, 3524–3541, doi:10.1177/0042098016668288.

Zhang, N., Y. Choi, and W. Wang, 2019: Does energy research funding work? Evidence from the Natural Science Foundation of China using TEI@I method. *Technol. Forecast. Soc. Change*, **144**, 369–380, doi:10.1016/j.techfore.2018.02.001.

Zhou, C., 2019: Enabling Law and Policy Environment for Climate Technology Transfer: From Perspectives of Host Countries. *J. East Asia Int. Law*, **12(1)**, 45–70, doi:10.14330/jeail.2019.12.1.03.

Zhou, N., M.D. Levine, and L. Price, 2010: Overview of current energy-efficiency policies in China. *Energy Policy*, **38(11)**, 6439–6452, doi:10.1016/j.enpol.2009.08.015.

Zhuang, W., 2017: *Intellectual Property Rights and Climate Change*. Cambridge University Press, Cambridge, UK, 428 pp.

Ziervogel, G., J. Enqvist, L. Metelerkamp, and J. van Breda, 2021: Supporting transformative climate adaptation: Community-level capacity building and knowledge co-creation in South Africa. *Clim. Policy*, **22(5)**, 1–16, doi:10.1080/14693062.2020.1863180.

Zubeltzu-Jaka, E., A. Erauskin-Tolosa, and I. Heras-Saizarbitoria, 2018: Shedding light on the determinants of eco-innovation: A meta-analytic study. *Bus. Strateg. Environ.*, **27(7)**, 1093–1103, doi:10.1002/BSE.2054.

16

17

# Accelerating the Transition in the Context of Sustainable Development


**Coordinating Lead Authors:**
Fatima Denton (the Gambia), Kirsten Halsnæs (Denmark)

**Lead Authors:**
Keigo Akimoto (Japan), Sarah Burch (Canada), Cristobal Diaz Morejon (Cuba), Fernando Farias (Chile), Joni Jupesta (Indonesia/Japan), Ali Shareef (Maldives), Petra Schweizer-Ries (Germany), Fei Teng (China), Eric Zusman (the United States of America)

**Contributing Authors:**
Fátima Antonethe Castaneda Mena (Guatemala), Morten Andreas Dahl Larsen (Denmark), Shreya Some (India)

**Review Editors:**
Diriba Korecha Dadi (Ethiopia), Hermann Held (Germany)

**Chapter Scientist:**
Fátima Antonethe Castaneda Mena (Guatemala)



**This chapter should be cited as:**
Denton, F., K. Halsnæs, K. Akimoto, S. Burch, C. Diaz Morejon, F. Farias, J. Jupesta, A. Shareef, P. Schweizer-Ries, F. Teng, E. Zusman, 2022: Accelerating the transition in the context of sustainable development. In IPCC, 2022: *Climate Change 2022: Mitigation of Climate Change. Contribution of Working Group III to the Sixth Assessment Report of the Intergovernmental Panel on Climate Change* [P.R. Shukla, J. Skea, R. Slade, A. Al Khourdajie, R. van Diemen, D. McCollum, M. Pathak, S. Some, P. Vyas, R. Fradera, M. Belkacemi, A. Hasija, G. Lisboa, S. Luz, J. Malley, (eds.)]. Cambridge University Press, Cambridge, UK and New York, NY, USA. doi: 10.1017/9781009157926.019

# Table of Contents

## Executive Summary

**Accelerating climate actions and progress towards a just transition is essential to reducing climate risks and addressing sustainable development priorities, including water, food and human security (*robust evidence, high agreement*).** Accelerating action in the context of sustainable development involves not only expediting the pace of change (speed) but also addressing the underlying drivers of vulnerability and high emissions (quality and depth of change) and enabling diverse communities, sectors, stakeholders, regions and cultures (scale and breadth of change) to participate in just, equitable and inclusive processes that improve the health and well-being of people and the planet. Looking at climate change from a justice perspective means placing the emphasis on (i) the protection of vulnerable populations and low-income countries from the impacts of climate change, (ii) mitigating the effects of the transformations, and (iii) ensuring an equitable decarbonised world. {17.1.1}

**While transition pathways will vary across countries, they are likely to be challenging in many contexts (*robust evidence, high agreement*).** Climate change is the result of decades of unsustainable production and consumption patterns (for example, energy production and land use), as well as governance arrangements and political economic institutions that lock in resource-intensive development patterns (*robust evidence, high agreement*). Reframing development objectives and shifting development pathways towards sustainability can help transform these patterns and practices, allowing space for transitions to transform unsustainable systems (*medium evidence, high agreement*). {17.1.1.2}

**Sustainable development can enhance sectoral integration and social inclusion (*robust evidence, high agreement*).** Inclusion merits attention because equity within and across countries is critical to transitions that are not simply rapid but also sustainable and just. Resource shortages, social divisions, inequitable distributions of wealth, poor infrastructure and limited access to advanced technologies can constrain the options and capacities for developing countries to achieve sustainable and just transitions (*medium evidence, high agreement*). {17.1.1.2}

**Concrete actions aligning sustainable development and climate mitigation and partnerships can support transitions. Strengthening different stakeholders' 'response capacities' to mitigate and adapt to a changing climate will be critical for a sustainable transition (*robust evidence, high agreement*).** Response capacities can be increased by means of alignment across multiple stakeholders at different levels of decision-making. This alignment will also help achieve synergies and manage trade-offs between climate and sectoral policies by breaking down sectoral silos and overcoming the multiple barriers that prevent transitions from gaining traction and gathering momentum (*medium evidence, high agreement*). {17.1.1.1}

**Economics, psychology, governance, and systems research have pointed to a range of factors that influence the speed, scale and quality of transitions (*robust evidence, high agreement*).** Views nonetheless differ on how much market-correcting policies; shift preferences (economics); shifts in individual and collective mindsets (psychology); and multi-level governance arrangements and inclusive political institutions (governance) contribute to system transitions (*medium evidence, high agreement*). {17.2}

**While economics, psychology, governance and systems thinking emphasise different enablers of transitions, they often share a view that strengthening synergies and avoiding trade-offs between climate and sustainable development priorities can overcome barriers to transitions (*medium evidence, high agreement*).** A growing body of research and evidence can show which factors in the views from economics, psychology, governance and systems affect how interrelationships are managed between climate, mitigation policies and sustainable development. Greater integration between studies based on different methodological approaches can show how to construct an enabling environment that increases the feasibility and sustainability of transitions. {17.2, 17.3, 17.4}

**Short- and long-term studies of transformations using macroeconomic models and integrated assessment models (IAMs) have identified synergies and trade-offs of mitigation options in the context of development pathways that align sustainable development and climate change (*robust evidence, high agreement*).** IAMs often look at climate change mitigation and Sustainable Development Goals (SDGs) in an aggregate manner: supplementing this aggregate view with detail-rich studies involving SDGs can build support for transitions within and across countries (*medium evidence, medium agreement*). {17.3.2}

**The impacts of climate change mitigation and adaptation responses, are highly context-specific and scale-dependent. There are synergies and trade-offs between adaptation and mitigation as well as synergies and trade-offs with sustainable development (*robust evidence, high agreement*).** A strong link exists between sustainable development, vulnerability and climate risks, as limited economic, social and institutional resources often result in low adaptive capacities and high vulnerability, especially in developing countries. Resource limitations in these countries can similarly weaken the capacity for climate mitigation and adaptation. The move towards climate-resilient societies requires transformational or deep systemic change. This has important implications for countries' sustainable development pathways (*medium evidence, high agreement*). {17.3.3.6}

**Sectoral mitigation options present synergies with the SDGs, but there are also trade-offs, which can become barriers to implementation. Such trade-offs are particularly identified in relation to the use of land for bioenergy crops, water and food access, and competition for land between forest or food production (*robust evidence, high agreement*).** Many industrial mitigation options, such as efficiency improvements, waste management and the circular economy, have synergies with the SDGs relating to access to food, water and energy (*robust evidence, high agreement*). The promotion of renewable energy in some industrial sectors can imply stranded energy supply investments, which need to be taken into consideration (*medium evidence, medium agreement*). The agriculture, forestry, and other land uses (AFOLU) sector offers many low-cost mitigation options, but actions aimed at producing bioenergy, extending food access and protecting biodiversity can also create trade-offs between different land uses (*robust evidence, high agreement*). Some options can help to minimise these trade-offs, for example, integrated land management, cross-sectoral policies and efficiency improvements. Lifestyle changes, including dietary changes and reduced food waste, have several synergies with climate mitigation and the SDGs (*medium evidence, medium agreement*). Cross-sectoral policies are important in avoiding trade-offs, to ensure that synergies between mitigation and SDGs are captured, and to ensure local people are involved in the development of new products, as well as production and consumption practices. There can be many synergies in urban areas between mitigation policies and the SDGs, but capturing these depends on the overall planning of urban structures and on local integrated policies, where, for example, affordable housing and spatial planning as a climate mitigation measure are combined with walkable urban areas, green electrification and clean renewable energy. Such integrated options can also reduce the pressures on agricultural land by reducing urban growth, thus improving food security. Access to green electricity can also support quality education (*medium evidence, medium agreement*). {17.3.3, 17.3.3.1, 17.3.3.3}

**Digitalisation could facilitate a fast transition to sustainable development and low-emission pathways by contributing to efficiency improvements, cross-sectoral coordination and a circular economy with new IT services and decreasing resource use (*low evidence, medium agreement*).** Several synergies with SDGs could emerge in terms of energy, food and water access, health and education, as well as trade-offs, for example, in relation to reduced employment, increasing energy demand and increasing demand for services, all implying increased GHG emissions. However, developing countries with limited internet access and poor infrastructure could be excluded from the benefits of digitalisation (*medium evidence, medium agreement*). {17.3.3}

**Actions aligning sustainable development and climate mitigation and partnerships can support transitions. Strengthening different stakeholders' 'response capacities' to mitigate and adapt to a changing climate will be critical for a sustainable transition (*robust evidence, high agreement*).** Response capacities can be increased by means of alignment across multiple stakeholders at different levels of decision-making. This alignment will also help achieve synergies and manage trade-offs between climate and sectoral policies by breaking down sectoral silos and overcoming the multiple barriers that prevent transitions from gaining traction and gathering momentum (*medium evidence, high agreement*). {17.1.1.1}

**The landscape of transitions to sustainable development is changing rapidly, with multiple transitions already underway. This creates the room to manage these transitions in ways that prioritise the needs for workers in vulnerable sectors (land, energy) to secure their jobs and maintain secure and healthy lifestyles, especially as the risks multiply for those exposed to heavy industrial jobs and associated outcomes (*medium evidence, high agreement*).** A just transition incorporates key principles, such as respect and dignity for vulnerable groups, the creation of decent jobs, social protection, employment rights, fairness in energy access and use, and social dialogue and democratic consultation with the relevant stakeholders, while coping with the effects of asset-stranding and the transition to green and clean economies (*medium evidence, medium agreement*). The economic implications of the transition will be felt especially strongly by developing countries, with high dependence on hydrocarbon products for revenue streams, as they will be exposed to reduced fiscal incomes given a low demand for oil and consequent fall in oil prices (*limited evidence, medium agreement*). {17.3.2, 17.3.2.3}

**Countries with assets that are at risk of becoming stranded may lack the relevant resources, knowledge, autonomy or agency to reorientate, or to decide on the speed, scale and quality of the transition (*limited evidence, medium agreement*).** The urgency of mitigation might overshadow some of the other priorities related to the transition, like climate change adaptation and its inherent vulnerabilities. Consequently, the transition imperative could reduce the scope and autonomy for local priority-setting and could ignore the additional risks in countries with a low capacity to adapt. A just transition will depend on local contexts, regional priorities, the starting points of different countries in the transition and the speed at which they want to travel. Both mitigation and adaptation warrant urgent and prompt action given current and continuing greenhouse gas (GHG) emissions and associated negative impacts on humanity and ecosystems (*limited evidence, medium agreement*). {17.3.2}

**A wide range of factors have been found to enable sustainability transitions, ranging from technological innovations to shifts in markets, and from policies and governance arrangements to shifts in belief systems and market forces (*robust evidence, high agreement*).** Many of these factors come together in a co-evolutionary process that has unfolded globally, internationally and locally over several decades (*low evidence, high agreement*). Those same conditions that may serve to impede the transition (i.e., organisational structure, behaviour, technological lock-in) can also 'flip' to enable both it and the framing of sustainable development policies to create a stronger basis and policy support (*robust evidence, high agreement*). It is important to note that strong shocks to these systems, including accelerating climate change impacts, economic crises and political changes, may provide crucial openings for accelerated transitions to sustainable systems. For example, rebuilding more sustainably after an extreme event, or renewed public debate about the drivers of social and economic vulnerability to multiple stressors (*medium evidence, medium agreement*). {17.4}

**Sustainable development and deep decarbonisation will involve people and communities being connected through various means, including globally via the internet and digital technologies, in ways that prompt shifts in thinking and behaviour consistent with climate change goals (*medium evidence, medium agreement*).** Individuals and organisations like institutional entrepreneurs can function to build transformative capacity through collective action (*robust evidence*, *high agreement*), but private-sector entrepreneurs can also play an important role in fostering and accelerating the transitions to sustainable development (*robust evidence*, *medium agreement*). Ultimately, the adoption of coordinated, multi-sectoral policies targeting new and rapid innovation can help national economies take advantage of widespread decarbonisation. Green industrial policies that focus on building domestic supply chains and capacities can help states prepare for the influx of renewable CDR-methods, or mechanisms for carbon capture and storage (CCS) (*medium evidence*, *medium agreement*). {17.4.2}

**Accelerating the transition to sustainability will be enabled by explicit consideration being given to the principles of justice, equality and fairness. Interventions to promote sustainability transitions that account for local context (including unequal access to resources, capacity and technology) in the development process are necessary but not sufficient in creating a just transition (*low evidence*, *high agreement*).** {17.4.6}

## 17.1 Introduction

This chapter focuses on the opportunities and challenges for 'accelerating the transition in the context of sustainable development'. The chapter suggests that accelerating transitions in the context of sustainable development requires more than concentrating on speed. Rather, it involves expediting the pace of change (speed) while also removing the underlying drivers of vulnerability and high emissions (quality and depth), and aligning the interests of different communities, regions, sectors, stakeholders and cultures (scale and breadth). One key to enabling deep and broad transitions is integrating the views of different government agencies, businesses and non-governmental organizations (NGOs) in transition processes. Another critical driver of deep and broad transitions is engaging and empowering workers, youth, women, the poor, minorities and marginalised stakeholders in just, equitable and inclusive processes. The result of such processes will be the transformation of large-scale socio-economic systems to restore the health and well-being of the planet and the people on it.

Section 17.1 begins by reviewing how climate and sustainability issues have been discussed in the Intergovernmental Panel on Climate Change (IPCC), as well as international climate change and sustainable development processes at different levels. It further introduces key themes addressed in the chapter's remaining subsections. Section 17.2 provides an overview of how key theories understand transitions and transformation, and notes a shared concern over leveraging synergies and managing trade-offs between climate change and sustainable development across different disciplines. Section 17.3 provides an assessment of the mitigation options that can help achieve these synergies and avoid trade-offs. Section 17.4 pulls together the theoretical and empirical aspects by detailing the essential elements of an enabling environment that helps drive forward transitions that are quick, deep, broad and, ultimately, sustainable.

### 17.1.1 Integrating Climate Change and Sustainable Development in International Assessments

Climate change not only poses a profound challenge to sustainable development, it is inexorably linked to it. From the early stages of the IPCC assessment process, this challenge and the inherent link between climate change and sustainable development have been well recognised. For example, the First Assessment Report (FAR) highlighted the relevance of sustainable development for climate policy. The Second Assessment Report (SAR) went further to include equity issues in its presentation of sustainable development. The Third Assessment Report (TAR) (Banuri et al. 2001) made the link even stronger, noting that 'parties have a right to and should promote sustainable development' (as stated in the text of the UNFCCC 2015 (Article 3.4)), and offering an early review of studies integrating sustainable development and climate change. The Fourth Assessment Report (AR4) (Sathaye et al. 2007) added an additional perspective to these interconnections, acknowledging the existence of a two-way relationship between sustainable development and climate change.

The Fifth Assessment Report (AR5) (Denton et al. 2014; Fleurbaey et al. 2014) and the Special Report on Global Warming of 1.5°C (SR1.5) (IPCC 2018; Roy et al. 2018a) have arguably made the strongest links between climate and sustainable development to date. One of the key messages of AR5 was that the implementation of climate mitigation and adaptation actions could help promote sustainable development, and it emphasised the need for transformational changes in this regard. The AR5 also concluded that the link between climate change and sustainable development is cross-cutting and complex, and that thus the impacts of climate change are threatening the efforts being made to achieve sustainable development. The SR1.5 helped systematise these links by mapping the synergies and trade-offs between selected SDG indicators and climate mitigation (IPCC 2018; Roy et al. 2018b) (Section 17.3).

Despite the clear links between sustainable development and climate change being recognised from the early stages of the IPCC, climate change has often been portrayed as an environmental problem to be addressed chiefly by environmental ministries (Brown et al. 2007; Munasinghe 2007; Swart and Raes 2007). However, this perception has evolved over time. It is now increasingly common to see governments and other actors understand the wider ramifications of a changing climate for sustainable development. In a growing number of studies, work on climate policies and just transitions towards sustainable development are framed as going hand in hand (Fuso Nerini et al. 2019; Dugarova and Gülasan 2017; Sanchez Rodriguez et al. 2018; Schramade 2017; Zhenmin and Espinosa 2019).

### 17.1.2 Integrating Climate Change and Sustainable Development in International Policymaking Processes

Among the reasons for the growing realisation of these interdependencies are milestones in international climate and sustainable development processes. As outlined in Chapter 14, the year 2015 was a turning point due to two agreements: (i) the Paris Agreement; and (ii) the 2030 Agenda for Sustainable Development and its 17 Sustainable Development Goals (SDGs) (Farzaneh et al. 2021).

Following a long history of references to sustainable development in the UNFCCC and related agreements, the Paris Agreement helped to strengthen the links between climate and sustainable development by emphasising that sustainability is related to its objectives (Sindico 2016; UNFCCC 2016). One of the ways that it helped tighten this link is by institutionalising bottom-up pledges and the review architecture. Toward this end, the Paris Agreement instituted Nationally Determined Contributions (NDCs) as vehicles through which countries make pledges and demonstrate their commitment to climate action. Although there was no clear guidance on what should be included in the NDCs, some of the requirements were elaborated in the Paris Rulebook . Some of the submitted NDCs included only mitigation efforts, but others set out mitigation and adaptation goals aligning NDC commitments to national planning processes, while yet others mentioned links with the SDGs.

Another way that the Paris Agreement and the NDCs could strengthen their links to sustainable development is to update country-specific climate pledges. Countries are free to choose their targets and the means and instruments with which to implement them. A core feature of the NDCs was that countries submit NDCs every five years, giving them an opportunity to assess themselves relative to other countries, raise their ambitions and learn from their peers. Moreover, it was emphasised that countries should not 'backslide' in subsequent NDCs, thus ensuring that countries should always be forward-looking in respect of increasing their ambitions to deliver the Paris Goals. (Höhne et al. 2017) found that, in developing countries especially, the NDC preparation process has improved national climate policymaking.

Despite some favourable reviews, several assessments of specific countries' NDCs (Andries et al. 2017; Rogelj et al. 2016; Vandyck et al. 2016) have assessed that those submitted for 2020–2030 are insufficient for delivering on the Paris goals. Updated and/or new NDCs were therefore submitted by the end of 2020. However, an assessment of those NDCs revealed that the level of ambition was significantly lower than the goals of the Paris Agreement (UNFCCO 2020) (see also this chapter). One of the urgent calls in Paris was to assess the impacts and efforts that need to be undertaken to keep global warming well below 2°C in relation to pre-industrial levels and evaluate related global GHG emission pathways (UNFCCC 2015). Although the initial NDCs fell short of these goals, the idea was that NDCs would be living documents that could ratchet up climate action and ambition.

Countries have also started to take actions on the SDGs themselves (Antwi-Agyei et al. 2018a; UNDESA 2016, 2017, 2018). The SDGs were perceived as a novel approach to development and as establishing a universal agenda for the transformation of development patterns and socio-economic systems. At their core, the SDGs hold that building an integrated framework for action necessitates addressing the economic, social and environmental dimensions of sustainable development in an integrated manner (Biermann et al. 2017; Kanie and Biermann 2017). The SDGs take multiple elements of development into account in aiming to offer coherent, well-integrated, overarching approaches to a range of sustainability challenges, including climate change.

One way a link is made between climate and the SDGs is through Voluntary National Reviews (VNRs). Paralleling the bottom-up orientation of the Paris Agreement and the NDCs, every year approximately forty countries voluntarily share their VNRs with the international community at the High-Level Political Forum (HLPF). Even more flexible than the NDCs, the VNRs can include content such as a summary of key policies and measures that are intended to achieve the SDGs, a list of the means of implementation that support the SDGs, and related challenges and needs. The VNRs also often cover SDG 13 (climate action) as well as many other issues connected with climate change. Even with these links, implementation of the SDGs should be mentioned as part of national development processes reflecting different countries' different priorities, visions and plans (Hanson and Korbla P. Puplampu 2018; Marcotullio et al. 2018; OECD 2016; P. Puplampu et al. 2017; Srikanth 2018).

Yet another way that the 2030 Agenda for Sustainable Development underlines the importance of capturing synergies is its calls for policy coherence (SDGs 14 and 17). Policy coherence and integration between sectors are two of the most critical factors in breaking down the silo mode of working of different sectors. Working across climate and other sustainability agendas is essential to coherence.

A final way that the sustainability and climate agendas have been linked is through vertical integration. Following a similar trend that appeared with Agenda 21, for which many cities adopted local plans, a growing number of cities have introduced Voluntary Local Reviews (VLRs). The VLRs resemble the VNRs, but place the emphasis on local actions and needs regarding the SDGs (and some links to climate change) (Ortíz-Moya et al. 2021). The 2019 SDG Report shows that 150 countries have developed national urban plans, almost half of them also being in the implementation phase (United Nations General Assembly 2019).

### 17.1.3 Integrating Climate Change and Sustainable Development in Other Policymaking Processes

Other non-UN-led initiatives involving international organisations or clusters of countries have also helped to raise the issue of sustainable development as a framework for mitigation. The OECD, for instance, assesses different types of investments and economic activities with reference to their significance for environmental sustainability (OECD 2020), while G20 countries have drawn up action agendas with sustainable development (UToronto 2016). Meanwhile, the Petersberg Climate Dialogue, a political movement convened by major country-group representatives and launched in 2010 by the German government, has also called for sustainability to be an intrinsic part of the transition (UNFCCO 2020) (BMU 2018).

Due in part to the shifting orientation of these international processes, there is growing evidence of action on climate change and sustainable development at other levels of decision-making. National policies often aim to implement climate change policies in the context of sustainable development (Chimhowu et al. 2019; Chirambo 2018; ECLAC 2017; Fuseini and Kemp 2015; Galli et al. 2018; Haywood et al. 2019; Ministry of Environment of Jordan 2016; McKenzie and Abdulkadri 2018; UNDESA 2016, 2017, 2018; UN Women 2017). Some countries are adjusting their existing policies to build on themes familiar to sustainable development (Lucas et al. 2016), including renewable energy and energy efficiency (Fastenrath and Braun 2018; Kousksou et al. 2015), urban planning (Gorissen et al. 2018; Loorbach et al. 2016; Mendizabal et al. 2018), health systems (Pencheon 2018; Roschnik et al. 2017) and agricultural systems (Lipper and Zilberman 2018; Shaw and Roberts 2017). Cross-cutting and integrated approaches, such as the circular economy, have also been gaining traction in some European countries (EESC 2015) and G20 countries (Noura et al. 2020). Many of these efforts have also extended up to the regional and down to the local level (Gorissen et al. 2018; Hess 2014; Shaw and Roberts 2017).

There has also been a shift to actors outside government aligning climate with sustainable development. An assessment by (Hoyer 2020) found that collective action against climate change by businesses, governments and civil society, reinforced through partnerships and coalitions across departments, industries and supply chains, can deliver significant development impacts. In order for this diverse collection of stakeholders to take action, a fundamental paradigm shift is needed from a linear model of knowledge-generation to an interdisciplinary model that co-produces knowledge (Liu et al. 2019). In fact, some have argued that accelerating just transitions for purposes of sustainable development requires the involvement of several actors, institutions and disciplines (Delina and Sovacool 2018). Not only do these roles need to be discussed more thoroughly (Kern and Rogge 2016); (den Elzen et al. 2019), but it is also important to survey different views on transitions and transformations. A variety of theories that are useful for explaining the causes and constraints regarding transitions are examined in Section 17.2.

## 17.2 Accelerating Transitions in the Context of Sustainable Development: Definitions and Theories

This section focuses on how different theoretical frameworks can help us understand and explain what is meant by accelerating transitions in the context of sustainable development. As suggested in Section 17.1, the reference to 'in the context of sustainable development' suggests that sustainable transitions require more than speed, also necessitating removing the underlying drivers of vulnerability and high emissions (quality and depth of transitions), while also aligning the interests of different individuals, communities, sectors, stakeholders and cultures (scale and breadth of transitions).

The outcome of sustainable transitions is a sustainable transformation. While transitions involve 'processes that shift development pathways and reorient energy, transport, urban and other subsystems' (Loorbach et al. 2017) (Chapter 16), transformation is the resulting 'fundamental reorganisation of large-scale socio-economic systems' (Hölscher et al. 2018). Such a fundamental reorganisation often requires dynamic multi-stage transition processes that change everything from public policies and prevailing technologies to individual lifestyles, and social norms to governance arrangements and institutions of political economy. This set of factors can lock-in development pathways and prevent transitions from gathering the momentum needed for transformations. Chapter 16 provides an overview of the multi-stage transition dynamics involved in moving from experimentation to commercialisation to integration to stabilisation. That overview describes how transitions can break through lock-ins and result in a transformation.

While there may be a relatively consistent set of transition dynamics for all countries, pathways are likely to vary across and even within countries. This variation is due to different development levels, starting points, capacities, agencies, geographies, power dynamics, political economies, ecosystems and other contextual factors. Given the diversity of contributing factors, a sustainable transition is likely to be a complex and multi-faceted process which cannot be reduced to a single dimension (Köhler et al. 2019). Even with this multi-dimensionality, transition processes are likely to gain speed and become more sustainable as decision-makers adopt targeted policies and other interventions. Many disciplines have reflected on the roles of and relative influence on the policies and interventions that can drive transitions. The following discussion describes this diversity of views with a survey of how prominent lines of economic, psychological, institutional and systems thinking explain transitions. Though these disciplines differ greatly, they often stress that leveraging synergies and managing trade-offs between climate change and sustainable development can help advance a transition.

### 17.2.1 Economics

This section concentrates on economic explanations for transitions. At the core of many of these explanations is the assumption that economic development can deliver multiple economic, social and environmental benefits. Many modern economic systems may nonetheless struggle to deliver these benefits due to major disruptions and shocks such as climate change (Heal 2020). One way to limit disruptions to free markets are targeted interventions in free markets such as taxes or regulation. These targeted interventions motivate firms and other entities to internalise GHGs and other pollutants, potentially paving the way for a sustainable transition (Arrow et al. 2004; Chichilnisky and Heal 1998).

A related line of thought common to economic explanations involves the principles of 'weak sustainability'. These principles suggest that the substitution of exhaustible resources is, to some extent, feasible (Arrow et al. 2004). One way to capitalise on this substitution is to target investments at technological change, green growth, and research and development. Targeted investments in the form of subsidies can encourage the substitution of exhaustible by non-exhaustible resources. To illustrate with a concrete example, investments in renewable energy can not only mitigate climate change but also offset the use of exhaustible fossil fuels and boost energy security (Heal 2020). It is nonetheless important to note that the principle of 'weak sustainability' contrasts with 'strong sustainability' or 'integrated sustainability' principles. These stronger principles suggest that constraints on resources restrict such substitutions (Rockström et al. 2009). These constraints merit attention because some scarce non-substitutable forms of natural capital can be exhausted (Bateman and Mace 2020). There is hence a need to capitalise on possible synergies such as those with other development priorities and trade-offs, for example, the exhaustion of non-substitutable resources. Capturing these synergies and managing these trade-offs is consistent with sustainable development, a state where the needs of the present generation do not compromise the ability of future generations to meet their own needs (Bruntland, WCED 1987).

As suggested above, aligning climate investments with other sustainable development objectives is critical to a transition. In order to support better investments in sustainable development, financing schemes, including environmental, social and governance (ESG) disclosure schemes and the Task Force on Climate-related Financial Disclosures (TCFD), can play important roles (Executive Summary in Chapter 15 of this report). After COVID-19, economic recovery packages have increased government-led investments (Section 1.3.3), which could potentially be aligned with sustainable development. Technological change and innovation are considered key drivers of economic growth and of many aspects of social progress (Section 16.1), but if technological innovation policies are coordinated with the shift to sustainable development pathways, then the economic benefits of technological change could come at the cost of increasing climate risks (Gossart 2015) Alarcón and Vos 2015). The environmental impacts of social and economic activities, including emissions of GHGs, are greatly influenced by the rate and direction of technological changes. Innovation and technological transformations present trade-offs that create externalities and rebound effects. This suggests that a sustainable future for people and nature requires rapid, radical and transformative societal change by integrating the technical, governance, financial and societal aspects (Pörtner et al. 2021) (Section 16.1).

One area that is pertinent to transitions and has received considerable attention in economic modelling involving climate change is innovation. In particular, some studies have shown how low-cost innovations and improvements in end-use technologies have significant potential for emissions reductions as well as sustainable development (Wilson et al. 2019). Currently information technologies are improving rapidly, and the internet of things (IoT), AI and Big Data can all contribute to other development needs. This is often the case in end-use sectors, as the benefits accrue directly to the individuals who use the new innovations. The achievement and widespread deployment of fully autonomous cars, for example, will bring about broader car- and ride-sharing with negative or low additional costs compared to more conventional approaches to car ownership, with their typically very low load factors. (Grubler et al. 2018) estimate that the Low Energy Demand (LED) scenario which assumes information technology innovations and induced social changes, including a sharing economy, have considerable potential for harmonising the multiple achievements of SDGs with low marginal abatement costs compared with other scenarios (IPCC 2018).

It is nonetheless important to highlight a caveat to the above logic on innovation. Whether a technological innovation is wholly sustainable or not becomes less clear when considering its effects on the wider economy. To illustrate, some models predict that $CO_2$ marginal abatement costs in the power sector will be USD240 and USD565 $tCO_2$ for the 2°C and below 2°C goals, respectively (IEA 2017).

In theory, if marginal abatement costs meet marginal climate damage, mitigation measures are economically optimal in the long run. Yet marginal damage from climate change is notoriously uncertain, and economic theories do not always reflect climate-related damage. On the other hand, marginal abatement mitigation costs impose additional costs in the short term. These added costs can cause productivity in capital to decline through increases in the prices of energy and products in which the energies are embodied. These increased costs can restrict the ability to invest in and achieve the sustainable development priorities. However, precisely the opposite can occur when innovation reduces additional costs or achieves negative costs. If technological innovation leads to the accumulation of capital and productivity increases due to the substitution of energy, material and labour, these are likely to deliver sustainable development and climate mitigation benefits.

### 17.2.2 Institutions, Governance and Political Economy

This subsection focuses on institutions, governance and the political economy. Institutional and governance arrangements can influence which actors possess authority, as well as how motivated they are to cooperate in transition processes that are directed at finding solutions to climate change and other sustainability challenges. Often cooperation is enabled when policy frameworks or institutions align climate change with the political and economic interests of national governments, cities or businesses, and when institutional and governance arguments that support that alignment expand the scale of the transitions. However, there may also be political and economic interests and structures that can lock-in unsustainable development patterns, frustrate this alignment and slow down transitions (Haas 2021; Mattioli et al. 2020; Newell and Mulvaney 2013; Power 2016).

An extensive literature has examined how the international climate agreements and architecture influence collaboration across countries regarding climate and sustainable development to support a transition (Bradley 2005). For example, international institutions offer opportunities for governments and other actors to share new perspectives on integrated solutions (Cole 2015). For some observers, however, decades of difficulties in crafting a comprehensive climate change agreement and the resulting fragmented climate policy landscape have been inimical to the collaboration needed for a transition (van Asselt 2014; Nasiritousi and Bäckstrand 2019) (Chapters 1 and 13). Yet others see the potential for more incremental cooperation across countries, even without a single, integrated form of climate governance (Keohane and Victor 2016).

A related argument suggests that fragmentation at the global level provides opportunities for cooperation at the national level (Kanie and Biermann 2017). For example, in contrast to the relatively top-down Kyoto Protocol, the bottom-up pledge and review architecture of the Paris Agreement has prompted national governments to integrate climate change with other sustainable development priorities (Nachmany and SetzerJoana 2018; Townshend et al. 2013). Concrete examples included incorporating the SDGs into the NDCs as an international response to climate change (The Energy and Resources Institute 2017) or bringing climate into sustainable development strategies and so-called Voluntary National Reviews (VNRs) as part of the SDGs and the 2030 Agenda process (Elder and King 2018; Elder and Bartalini 2019).

Another branch of institutional research is concerned with the interactions between multiple levels of governance. In this multi-level governance perspective, cities and other sub-national governments often lead transitions by devising innovative solutions to contribute to climate and local energy, transport, the environment, resilience and other forms of sustainability (Bellinson and Chu 2019; Doll and Puppim De Oliveira 2017; Geels 2011; Koehn 2008; Rabe 2007; van der Heijden et al. 2019). A complementary perspective suggests that national governments can help scale up transitions by allocating resources and can provide the technical support that can spread innovative solutions (Bowman et al. 2017; Corfee-Morlot et al. 2009; Gordon 2015). Such support has become increasingly important during the pandemic, as national governments transfer funds for investments in climate-friendly infrastructure, transport systems and energy systems. This line of thinking is supported by calls to strengthen vertical and horizontal integration within and across government agencies and stakeholders in ways that can enhance policy coherence (Amanuma et al. 2018; OECD 2018, 2019). The incoherence or misalignment between national and local fiscal institutions and policies can restrict the ability of local governments to secure resources for climate-friendly investments. Such investments are particularly likely to flow, as more local governments have adopted net-zero targets, climate emergency declarations and action plans that can stimulate innovations (Davidson et al. 2020). Others have seen greater potential for collaboration and innovation, with more multi-centred or polycentric forms of governance that lead to the formulation and dissemination of transformative solutions to climate and other environmental challenges (Ostrom 2008). Though much of the above governance research has focused on western countries, there are some applications in other regions and countries such as China (Gu et al. 2020).



Yet another set of channels facilitating integration between climate and other concerns are networks of like-minded actors working across administrative borders and physical boundaries. For instance, city networks such as the Global Covenant of Mayors for Climate & Energy (Covenant of Mayors 2019), the World Mayors Council on Climate Change (ICLEI 2019; C40 Cities 2019) and the United Nations Office for Disaster Risk Reduction (UNDRR 2019) have agreed to share decision-making tools and good practices, and to sponsor ambition-raising campaigns that help align climate and sustainable development concerns within and across cities (Betsill and Bulkeley 2006) (Chapter 8 and Section 17.3.3.5). This can be particularly important for less capable 'following' and 'laggard' cities needing greater financing and other forms of support to move a transition forward (Fuhr et al. 2018).

Furthermore, sub-national governments may often work together with civil-society groups to create new networked forms of governance (Biermann et al. 2012). Other forms of multi-stakeholder partnerships focusing on issues with strong climate synergies, such as forms of air pollution known as short-lived climate pollutants (Climate and Clean Air Coalition (CCAC)) or transport (Sustainable Low Carbon Transport Partnership (SLoCaT)), take their cue from global scientific communities or civic-minded advocacy groups that transmit knowledge across boundaries (Keck and Sikkink 1999). There is also scope for suggesting that the international climate regime serves a Global Framework for Climate Action (GFCA) in helping orchestrate the multilateral climate regime and non-state and sub-national initiatives (Chan and Pauw 2014), though questions remain about its actual impacts on mitigation (Michaelowa and Michaelowa 2017).

Policymaking institutions and networks are themselves policies. A significant literature has looked at integrated policy frameworks and efforts across sectors, including climate adaptation and mitigation, as drivers of transitions (Landauer et al. 2015; Favretto et al. 2018; Obersteiner et al. 2016; Steen and Weaver 2017; Thornton and Comberti 2017). Policy coherence between climate and other development objectives is often considered essential to sustainable development (Sovacool 2018). A similar discussion about synergies and conflicts has been raised on the relationship between resilience and sustainability (Marchese et al. 2018). To help achieve coherence, there have been some efforts to develop suitable tools and decision-making frameworks (Scobie 2016).

A related line of reasoning has suggested that sustainable development often requires not one but a mix of policy instruments to bring about the multiple policy effects needed for social and technological change (Edmondson et al. 2019; Rogge and Johnstone 2017). Following these calls, some governments have aimed to address climate change and sustainability jointly with coherent and integrated approaches to achieving these agendas (Chimhowu et al. 2019), although for some countries Small Island Developing States (SIDS) this has proven more challenging (Scobie 2016).

Though the above work tends to downplay politics and business, others suggest that political economy should feature prominently in transitions. Some branches of political-economy research underline how resource-intensive and fossil-fuel industries leverage their resources and positions to undermine transitions (Jones, C.A. and Levy 2009; Newell and Paterson 2010; Zhao et al. 2013; Geels 2014; Moe 2014) (Chapter 1). These vested interests can lock-in status quo policies in countries where political systems offer interest groups more opportunities to veto or overturn climate- or eco-friendly proposals (Madden 2014). Companies with a strong interest in earning profits and building competitiveness from conventional fossil fuel-based energy systems have particularly strong incentives to capture politicians and agencies (Meckling and Nahm 2018). Such strategies can be particularly powerful when combined with concerns over job losses and dislocation, preventing transitions from gaining traction (Haas 2021; Mattioli et al. 2020; Newell and Mulvaney 2013; Power 2016).

This suggests that politics can be an impediment to change: other studies argue instead that politics can be harnessed to drive transitions forward. For example, some observers contend that building coalitions around green industrial policies and sequencing reforms to reward industries in such coalitions can align otherwise divergent interests and inject momentum into transitions (Meckling et al. 2015). Others see the effects of political economy varying over time depending upon external market conditions. To illustrate, renewable feed-in tariffs in Europe persisted for over two decades and were crucial in wind and solar power technologies making the breakthrough. But once competition from China led to the demise of European technology providers, and once European populations started to oppose surcharges on their electricity bills, feed-in tariffs were abolished by politicians in the purely national interest (Michaelowa and Michaelowa 2017).

### 17.2.3 Psychology, Individual Beliefs and Social Change

This subsection draws on value- and action-oriented research that employs inter- or transdisciplinary methods such as transactional psychology, transformative science and similarly focused disciplines (Wamsler et al. 2021). These approaches frequently encourage researchers to participate in transitions that induce changes in the researcher's own beliefs while triggering wider shifts in social norms (including human stewardship for the natural environment) (Adger et al. 2013; Hulme 2009; Ives et al. 2019; O'Brien 2018). This research also emphasises how changes in individual beliefs could lead to climate actions that contribute to more sustainable, equitable and just societies (e.g., 'the mind- & paradigm shifts') (Göpel et al. 2016). They further suggest the potential for virtuous cycles of individual-level and wider social changes that ultimately benefit the climate (Banks 2007; Day et al. 2014; Lockhart 2011; Montuori and Donnelly 2018; Power 2016).

The starting point for this virtuous circle are inner transitions. Inner transitions occur within individuals, organisations and even larger jurisdictions that alter beliefs and actions involving climate change (Woiwode et al. 2021). An inner transition within an individual (see e.g., Parodi and Tamm 2018) typically involves a person gaining a deepening sense of peace and a willingness to help others, as well as protecting the climate and the planet (see e.g., Banks 2007; Power 2016). Inner transition can imply that individuals become sympathetic to concerns that include climate issues and values connected to nature. For instance, they may include a desire to become a steward of nature (Buijs et al. 2018); 'live according to the principles of integrated sustainability' (Schweizer-Ries 2018); 'achieve the good life' (Asara et al. 2015; Escobar 2015; Kallis 2017; Latouche 2018) (Chapter 5 and Section 1.6.2); or protect the well-being of other living creatures (Chapter 5 and Section 1.6.3.1).

Examples have also been seen in relation to a similar set of inner transitions to individuals, organisations and societies, which involve embracing post-development, degrowth, or non-material values that challenge carbon-intensive lifestyles and development models (D'Alisa 2014; Kothari 2019; Neuteleers and Engelen 2015; Paech 2017). These shifts in values can occur when humans reconnect with nature, deepen their consciousness and take responsibility for protecting the planet and its climate (Cross et al. 2019; Martinez-Juarez et al. 2015; Speldewinde et al. 2015). Changes in both values and beliefs may also emerge through consciousness-raising processes where people cooperate in ways that would protect the climate ((Banks 2007; Hedlund-de Witt et al. 2014; Woiwode and Woiwode 2019) (Section 1.6.4).

Many of the above-mentioned beliefs and values that support climate actions have spread through expanding interests in conservationist world views, indigenous cultures (see, for example, Lockhart 2011) and branches of neuroscience and psychology that suggest different notions of the self (Hüther 2018; Lewis 2016; Seligman and Csikszentmihalyi 2014). These beliefs and values can also be spread through meditation, yoga or other social practices that encourage lower-carbon lifestyles (Woiwode and Woiwode 2019). Another channel for spreading climate concerns is sustainability culture, which is premised on connecting people and communities, and has also benefited from the internet and digital technologies that support these connections (see e.g., Bradbury 2015; Scharmer 2018). The spread of this culture, in turn, has led to the creation of social fields that allow changes to happen (see e.g., Gillard et al. 2016) or has promoted low-carbon thinking and related behavioural changes (O'Brien 2018; Veciana and Ottmar 2018). Studies of social contagions may also offer insights into the mechanisms that lead to the adoption of new values and related climate actions (see e.g., Iacopini et al. 2019). It is nonetheless worth highlighting that communication networks and other mechanisms promoting the spread of interpersonal communication that can spread pro-climate views may also lead to the proliferation of climate scepticism and denial (Leombruni 2015). At the same time, some studies suggest that such scepticism can be countered by the generation of more credible information on climate change (Samantray and Pin 2019).

One of the more direct channels through which transitions spread are climate change education and action-oriented research (Fazey et al. 2018; Ives et al. 2019; Scharmer 2018; Schäpke et al. 2018; Schneidewind et al. 2016). For instance, research using 'social experiments' or 'real-world labs' has helped give rise to shifts in mindsets on energy, food, transport and other systems that can benefit the climate (Bernstein and Hoffmann 2018; Berkhout et al. 2010; Bulkeley et al. 2015; Hoffmann 2010). In much the same way, the acquisition of transformational knowledge and transformative learning (Lange 2018; O'Neil and Boyce 2018; Pomeroy and Oliver 2018; Walsh et al. 2020; Williams 2013) contributes to thinking and acting that open climate-friendly development pathways (Berkhout et al. 2010; Lo and Castán Broto 2019; Roberts et al. 2018; Turnheim and Nykvist 2019)) (Section 1.7.2). First-person and action research can also facilitate similar changes that bring about climate actions (see e.g., Dick 2007; Streck 2007; Hutchison and Walton 2015; Bradbury et al. 2019).



### 17.2.4 System-level Explanations

Systems explanations help explain the dynamics of transitions toward sustainable development while explicitly uncovering links between the human and natural worlds, the socio-cultural embeddedness of technology, and the inertia behind high-carbon development pathways. This line of thinking often envisages transitions emerging from complex systems in which many different elements interact at small scales and spontaneously self-organise to produce behaviour that is unexpected, unmanaged and fundamentally different from the sum of the system's constituent parts.

Social-ecological systems theory describes the processes of exchange and interaction between human and ecological systems, investigating in particular non-linear feedback occurring across different scales (Folke 2006; Holling 2001). This approach has informed subsequent theoretical and empirical developments, including the 'planetary boundaries' approach (Rockström et al. 2009), conceptualisations of vulnerability and adaptive capacity (Hinkel 2011; Pelling 2010) and more recent explorations of urban resilience (Romero-Lankao et al. 2016) and regenerative sustainability (Clayton and Radcliffe 2018; Robinson and Cole 2015). Employing a systems lens to address the 'root causes' of unsustainable development pathways (such as dysfunctional social or economic arrangements) rather than the 'symptoms' (dwelling quality, vehicle efficiency, etc.) can trigger the non-linear change needed for a transformation to take place (Pelling et al. 2015). Exploring synergies between climate change adaptation, mitigation and other sustainability priorities (such as biodiversity and social equity, for instance) (Beg 2002; Burch et al. 2014; Shaw et al. 2014) may help to yield these transformative outcomes, though data regarding the specific nature of these synergies is still emerging.

Socio-technical transition theory, on the other hand, explores the ways in which technologies such as low-carbon vehicles or regenerative buildings are bound up in a web of social practices, physical infrastructure, market rules, regulations, norms and habits (see, e.g, (Loorbach et al. 2017). Radical social and technical innovations can emerge that ultimately challenge destabilised or increasingly ineffective and undesirable incumbents, but path dependencies often stymie these transition processes, suggesting an important role for governance actors (Burch 2017; Frantzeskaki et al. 2012; Holscher et al. 2019).

This also reveals the large-scale macroeconomic, political and cultural trends (or contexts) that may reinforce or call into question the usefulness of current systems of production and consumption. One branch of this theory, transition management (Kern and Smith 2008; Loorbach 2010), explores ways of guiding a socio-technical system from one path to another. In particular, it highlights interactions between actors, technologies and institutions, and the complex governance mechanisms that facilitate them (Smith et al. 2005). The challenge, in part, becomes linking radical short-term innovations with longer-term visions of sustainability (Loorbach and Rotmans 2010) and creating opportunities for collaborative course-correction in light of new information or unexpected outcomes (Burch 2017).

### 17.2.5 Conclusions

This section has surveyed several explanations for interventions that can give rise to transitions. The review suggests that there are several differences between these various perspectives. Whether individuals, organisations, markets or socio-technical systems drive or undermine transitions is a key distinction. These differences have implications for the evidence these claims draw on in support of their arguments. For instance, some of the explanations tend to employ qualitative evidence to explain changes in attitudes at the individual or community levels as paving the way for broader changes to cultures and belief systems. Others assess how institutional arrangements can be reformed in order to align climate with the sustainable development agenda to enable a transition.

While there are indeed significant differences between explanations, there are also important parallels. Such parallels begin with a shared emphasis on synergies and trade-offs between climate and sustainable development. Most explanations tend to underline the importance of synergies in aligning the climate with broader sustainability agendas. Most importantly, many of the explanations are complementary with the systems-level discussion in that they offer a broad framework, while economic, psychological and governance theories offer more specific insights. Moving a transition forward will often require drawing upon insights from multiple schools of thought. Though it is unlikely that a one-size-fits-all set of factors will drive a transition, there is a growing body of empirical evidence shedding light on the factors that can strengthen synergies between climate and the broader sustainable development agenda.

## 17.3 Assessment of the Results of Studies Where Decarbonisation Transitions are Framed Within the Context of Sustainable Development

### 17.3.1 Introduction

This section assesses studies based on the links between sustainable development and climate change mitigation in order to facilitate robust conclusions on synergies and trade-offs between different policy objectives across methodologies, scenarios and sectors. Conclusions are drawn based on national and sub-national, sectoral and cross-sectoral, short- and long-term transition studies presented in this and other sections of the report as a basis for establishing an overall picture of how sustainable development and climate change policies can be linked as a basis for accelerated transitions.

This section focuses initially on issues related to short- and long-term transitions to meet climate change and sustainable development goals in the context of the UNFCCC and the UN 2030 Agenda for Sustainable Development. Global-modelling results and economy-wide studies are then assessed, followed by a discussion of specific challenges in relation to renewable-energy penetration and phasing out fossil fuels, stranded assets and just transitions. Key synergies and trade-offs between meeting the UN 2030 Sustainable Development Goals (SDGs) and mitigation are then illustrated by means of cross-sectoral examples. Finally, this section presents an overview of the assessment of SDG synergies and trade-offs based on all sectoral chapters in this report for a range of key mitigation options.

### 17.3.2 Short-term and Long-term Transitions

It is increasingly being recognised that sustainable development policy goals and meeting short- and long-term climate policy goals are closely linked (IPCC 2018). It is also being realised that, under the Paris Agreement, climate change policies should be integrated into sustainable development agendas, while the UN 2030 Agenda as well includes SDG 13 on climate actions. In this way, both UN agreements provide joint opportunities for systematic transitions in support of both climate change and sustainable development. Achievement of the Paris Agreement's goals will require a rapid and deep worldwide transition in all GHG emissions sectors, including land use, energy, industry, buildings, transport and cities, as well as in consumption and behaviour (UNEP 2019). Meeting the goals of such a transformation requires that the long-term targets and pathways to fulfil the stabilisation scenarios play an important role in guiding the direction and pathways of short-term transitions. There is therefore a need for long- and short-term policies and investment decisions to be closely coordinated.

In the context of the Paris Agreement, countries have submitted their initial plans for the decarbonisation of their economies to the UNFCCC in the form of their so-called National Determined Contributions (NDCs). The ambitions of the NDCs are closely related to the ongoing UNFCCC negotiations over the financial measures and forms of compensation. Although the Paris Agreement emphasises the links between climate policies and sustainable development, the UN's 2030 Agenda for Sustainable Development and the SDGs are not very well represented at present in the NDCs, according to Fuso Nerini et al. (2019). Very few of the NDCs include any reference to the SDGs, which (Fuso Nerini et al. 2019) highlight as a barrier to the successful implementation of the Paris Agreement, which induces them to call for a more holistic policy approach. Campagnolo and Davide (2019) have assessed the impacts of the submitted NDCs on poverty eradication and inequalities of income based on empirical research and a global Computable General Equilibrium (CGE) model. One conclusion is that the NDCs of less developed countries would tend to reduce poverty alleviation, but this can be offset if international financial support is provided for the mitigation actions.

The alignment of climate-policy targets in the NDCs with sustainable development has been assessed by means of integrated assessment models (IAMs), macroeconomic and sectoral modelling. (Iyer et al. 2018) based on IAM-based studies, the implications of framing NDCs being placed more narrowly on mitigation targets rather than on a framing in which the impacts on sustainable development were explicitly taken into consideration. It was thus concluded that some SDGs would be directly supported as a side benefit of the climate policy targets included in the NDCs, while other SDGs needed a special policy design going beyond narrow climate policy objectives. (Iyer et al. 2018) also assessed the regional distribution of efforts in terms of domestic mitigation costs and SDG impacts and concluded that the geographical distribution of mitigation costs and SDG benefits were not similar, so a special effort would be needed to match climate policies and policies to meet the SDGs. Accordingly, a national decision-making perspective suggests that SDGs should be integrated into national climate policies.

The NDCs submitted to the Paris Agreement have demonstrated a lack of progress in meeting the long-term temperature goals. In the context of the UN's 2030 Agenda for Sustainable Development, the UN Sustainable Development Report 2019 (Sachs et al. 2019) also concluded that there is a particular lack of progress in achieving SDG 13 (climate action), SDG 14 (life below water) and SDG 15 (life on land). Given the close link between the SDGs and climate change policies, the current obstacles in meeting the former could also be a barrier to realising transitions to low-carbon societies. Conversely, opportunities to leverage the SDGs could in many cases involve climate actions, since policies enabling climate adaptation and mitigation could also support food and energy security and water conservation if they were well designed (see the detailed discussion in the section on synergies and trade-offs between climate policies and meeting the SDGs in Section 17.3.3.7, Chapter 3, and IPCC 2018). These findings point to a specific need to align economic and social development perspectives, climate change and natural systems. While all countries share the totality of the SDGs, development priorities differ across countries and over time. These priorities are strongly linked to local contexts and depend on which dimension of the improvement in the well-being of people is considered to be the most urgent. Eradicating poverty and reducing inequality are key development priorities for many low- and middle-income countries (Section 4.3.2.1).

A key barrier to the development of national plans and policies to meet the UN 2030 SDGs is the lack of finance. (Sachs et al. 2019) conclude that meeting the SDGs to achieve social transformations worldwide would require 2–3% of global GDP and that it would be a huge challenge to ensure that finance is targeted to the world's poorest countries and people. The UN Secretary-General has called for the allocation of finance to meet the UN's 2030 Agenda with a strong emphasis on the private sector, but to date no governance frameworks or associated financial modalities have been established in the UN or the UNFCCC context for the formal alignment of sustainable development and transitions to take place in accordance with the low global temperature-stabilisation targets in the Paris Agreement. Accelerating investments, particularly in low-income countries, will be required to meet both the Paris goals and the SDGs (Section 15.6.7). The mismatch between capital and investment needs, home-bias considerations and differences in risk perceptions between rich and poor represent major challenges for private finance. Green bond markets and markets for sustainable financal products have increased significantly, and the landscape has continued to evolve since AR5 (Executive Summary in Chapter 15). Special efforts and activities are particularly required for raising finance in developing countries.

Based on the Paris Agreement, the UNFCCC has invited countries to communicate their mid-century and long-term low-GHG emission-development strategies by 2020 (UNFCCC 2019). National long-term low-emission development strategies and their global stocktake in the UNFCCC context provide a platform for informing the long-term strategic thinking on transitions towards low-carbon societies. One specific value of these plans is that they reflect how specific transition pathways, policies and measures can work in different parts of the world in a very context-specific way, that is, by taking context-specific issues and stakeholder perspectives into consideration. Many nations have submitted national long-term strategies to the UNFCCC, including sustainable development perspectives (see Section 4.2.4 for a review of the plans and scientific assessments).

#### 17.3.2.1 Model Assessments on the Sustainable Development Pathways for Decarbonisation

This section assesses the model evaluations of the sustainable development pathways for decarbonisation, including the co-benefits and trade-offs involving explorations of alternative future development pathways as a basis for clarifying societal objectives and understanding the restrictions. Shifting development pathways to increased sustainability involves a number of complex issues, which are difficult to integrate into models. For a more detailed discussion about this, see Section 4.4.1 and Cross-Chapter Box 5 in Chapter 4.

Development pathways that focus narrowly on climate mitigation or economic growth will not lead to the SDGs and long-term climate-stabilisation objectives being achieved. The best chances of doing this lie in development pathways that can maximise the synergies between climate mitigation and sustainable development more broadly (Section 1.3.2). Areas of focal modelling include green investments, technological change, employment generation and the performance of policy instruments, such as green taxes, subsidies, emission permits, investments and finance. Short- and long-term macroeconomic models have been used to assess the impacts of such policy instruments. Jaumotte et al. (2021) analyse the economic impacts on net zero emissions by 2050 with a focus on short-term economic policies and the integration of climate policies such as $CO_2$ taxes with green reform policies. This may imply the co-creation of benefits between climate policy objectives, and macroeconomic policy goals such as employment creation.

There is an emerging modelling literature focusing on the synergies and trade-offs between low-carbon development pathways and various aspects of sustainable development. The early literature, including that on IAMs, and macroeconomic and sectoral models, mainly focused on the co-benefits of mitigation policies in terms of reduced air pollution, energy security and to some extent employment generation security (IPCC 2014, 2018c) (Chapter 6). Some models have been developed further with assessments of a broader range of the joint benefits of mitigation, health, water, land use and food security (Clarke et al. 2014; IPCC 2014, 2018; Kolstad et al. 2014). According to Chapter 1, there is a need to incorporate issues and enablers further, including a wide range of non-climate risks, varying forms of innovation, possibilities for behavioural and social change, feasible policies and equity issues (Executive Summary in Chapter 1).

IAMs and macroeconomic models typically calculate mitigation costs based on the assumption that markets internalise externalities like GHG emissions through carbon prices (Barker et al. 2016; IEA 2017, 2019). Yet, there are legitimate questions to be asked about whether carbon pricing will be efficient if markets are inefficient (World Bank 2019). However, market inefficiencies are difficult to integrate into the models. How GHG emissions taxes would actually work is thus quite uncertain based on the modelling studies (Barker et al. 2016; Fontana and Sawyer 2016; Meyer et al. 2018). Despite these limitations, the use of GHG emission taxes as an effective instrument based on modelling results in practice has implications for public policies and private-sector investments.

Despite the shortcomings of conventional economic thought and models already pointed out, improved models have demonstrated new perspectives on how mitigation costs can be assessed in macroeconomic models. For instance, while a conventional perspective might suggest that climate change mitigation costs can limit investments in sustainability because they reduce the productivity of capital by increasing energy prices and the products in which energies are embodied, another perspective is that innovation can imply increases in efficiency and that the substitution of energy, material and labour can lead to the accumulation of capital and productivity gains. This appears to occur with innovations in end-use energy applications generating emissions reductions and delivering on other sustainable development benefits (Wilson et al. 2019). Similarly, IAM models have been applied to model the potential for Low Energy Demand (LED) scenarios associated with demand-side innovations in the service sector. (Grubler et al. 2018) have developed a climate-friendly LED scenario which assumes information technology innovations such as the internet of things (IoT) and induced social changes such as the sharing economy. Nonetheless there are still very important limits on the degree to which highly aggregated IAM models and macroeconomic models can integrate ethics, equity and several other key policy-relevant aspects of sustainable development (Easterlin et al. 2010; Koch 2020). A key limitation in this context is that, while all countries share the totality of the SDGs, development priorities differ across countries and over time. Moreover, these priorities are strongly linked to local contexts, and this can only be reflected directly in national models (Section 4.3.2).

An example of a project that assesses the economy-wide impacts of linking sustainable development with deep decarbonisation is the Deep Decarbonisation Pathways Project (DDPP) (Bataille et al. 2016), which is undertaking a comparative assessment of studies of 16 countries representing more than 74% of global energy-related emissions for the pathway to 2°C stabilisation scenarios. The DDPP's methodology is to combine scenario analysis in different national contexts using macroeconomic models and sectoral models and to facilitate a consistent cross-country analysis using a set of common assumptions.

The key conclusions of the DDPP team on the economy-wide impacts are that country-based studies such as South Africa's demonstrate that it is possible to improve income distribution, alleviate poverty and reduce unemployment while simultaneously transitioning to a low-carbon economy (Altieri et al. 2016). The DDPP in Japan explores whether energy security can be enhanced through increases in renewable energy (Oshiro et al. 2016). The reduction of uncontrolled fossil fuel emissions has significant public-health benefits according to the Chinese and Indian DDPPs, as fossil fuel combustion is the major source of air pollution.

For example, in the Chinese DDPP, deep decarboniation scenarios have resulted in reductions of 42–79% in primary air pollutants (e.g., $SO_2$, $NO_x$, particulate matter (PM2.5), volatile organic compounds (VOCs), and $NH_3$), thus meeting air-quality standards in major cities. The deep decarbonisation scenarios include the large and fast energy-efficient improvements required to improve energy access and affordability. The DDPP studies are thus an example of an approach in which national deep-carbonisation scenarios are linked to the development goals of income generation, energy access and affordability, employment, health and environmental policy.

Sustainable development scenarios have also been developed by the Low-Carbon Society's (LCS) assessments (Kainuma et al. 2012), in which multiple sustainable development and climate change mitigation goals were assessed jointly. The scenario analysis was conducted for Asian countries such as South Korea, Japan, India, China and Nepal with a soft linked IAM using economy-wide and sectoral models and linked to very active stakeholder engagement in order to reflect national policy perspectives and priorities. Some of the models are economy-wide global IAMs, while others are national partial equilibrium models.

The LCS scenarios also include a specific attempt to include ongoing dialogues with policymakers and stakeholders in order to reflect governance and enabling factors, and to enable the modelling processes to reflect political realism as far as possible. Diverse stakeholders who acted as validators of the scientific process were included, stakeholder preferences were revealed, and recipients and users of the LCS outputs were included in ongoing dialogues on outputs and in interpreting the results. The aim of the stakeholder interactions was thus to fill the gap between typical laboratory-style IAMs and down-scaled but unaligned practical assessments performed at disaggregated geographical and sector-specific scales.

Energy scenarios for sustainable development were included in The World Energy Outlook of the IEA (IEA 2019, 2020) in terms of a Sustainable Development Scenario (SDS), which assessed not only SDG 13 (climate action) but also SDG 7 (affordable and clean energy) and SDG 3.9 (air pollution). This scenario takes as its starting point the policy goal of meeting these SDGs and then assesses the costs of meeting an emissions reduction target of 70% of $CO_2$ from the energy system by 2030. The scenario concludes that retrofitting coal-fired power plants with pollution controls is the cheapest option for dealing with local pollution in the short term, but that this is not consistent with meeting the long-term emissions goals of the Paris Agreement. The SDS scenario combines the goal of reducing the amount of $CO_2$ in the energy system by 70%, with large decreases in energy-related emissions of $NO_X$, $SO_2$ and $PM_{2.5}$, leading to a fall of 40–60% by 2030, and to 2.5 million fewer premature deaths from air pollution in 2030 than in the Stated Policies Scenario (STEPS), which represent a continuation of current trends in the energy system (IEA 2020).

The costs of energy-system transitions have been assessed by several energy-system studies. The economic costs of meeting the different goals depend on the stringency of the mitigation target, as well as economic (fuel prices, etc.) and technological developments (technology availability, capital costs, etc.). In addition, changes in infrastructure and behavioural patterns and lifestyles matter. Model-based assessments vary, depending on these assumptions and differences in modelling approaches (Krey et al. 2019) (Section 6.7.7). Country characteristics determine the social, economic and technical priorities for low-emission pathways. Domestic policy circumstances impact on pathways and costs, for example, when affordability and energy-security concerns are emphasised (Oshiro et al. 2016).

Mitigation policies can have important distributive effects between and within countries, and may affect impact on the poorest through their effects on energy and food prices (Hasegawa et al. 2018; Fujimori et al. 2019) (Section 3.6.4), while higher levels of warming are projected to generate higher inequality between countries as well as within them (Chapter 16). Mitigation thus can reduce economic inequalities and poverty by avoiding such impacts (Section 3.6.4).

Improved air quality and the associated health effects are the co-benefit category dominating model-based assessments of co-benefits, but a few studies have also covered other aspects, such as the health effects of dietary change and biodiversity impacts (Sections 3.6.3 and 17.3). Mitigation has implications for global economic inequalities through different channels and can compound or lessen inequalities, avoid impacts and create co-benefits that reduce inequalities (Section 3.6.4). There are, however, several challenges involved in balancing the dilemmas associated with meeting the SDGs, such as, for example, energy access, equity and sustainability. Fossil fuel-dependent developing countries cannot transition to low-carbon economics without considering the wider impacts on development by doing so (Section 3.7.3).

Climate change has negative impacts on agricultural productivity in general, including unequal geographical distribution (Chapter 3). On top of that, there is also a risk that climate change mitigation aimed at achieving stringent climate goals could negatively affect food access and food security (Akimoto et al. 2012; Fujimori et al. 2019; Hasegawa et al. 2018). If not managed properly, the risk of hunger due to climate policies such as large-scale bioenergy production increases remarkably if the 2°C and 1.5°C targets are implemented (Section 3.7.1). Taking the highest median values from different IAMs for given classes of scenarios, up to 14.9 $GtCO_2\ yr^{-1}$ carbon dioxide removal (CDR) from BECCS is required in 2100, and 2.4 $GtCO_2\ yr^{-1}$ for afforestation. Across the different scenarios, median changes in global forest area throughout the 21st century reach the required 7.2 $Mkm^2$ increases between 2010 and 2100, and agricultural land used for second-generation bioenergy crop production may require up to 6.6 $Mkm^2$ in 2100, increasing the competition for land and potentially affecting sustainable development (AR6 scenarios database).

Reducing climate change can reduce the share of the global population exposed to increased stress from reductions in water resources (Arnell and Lloyd-Hughes 2014) and therefore to water scarcity as defined by a cumulative abstraction-to-demand ratio (Hanasaki et al. 2013). (Byers et al. 2018), show that 8–14% of the population will be exposed to severe reductions in water supply if average temperatures increase between 1.5°C and 2°C (Section 3.7.2). (Hayashi et al. 2018) assess the water availability for different emission pathways, including the 2°C and 1.5°C targets, in light of the various factors governing availability. There are very different impacts among nations. In Afghanistan, Pakistan and South Africa, water stress is estimated to increase by 2050 mainly due to increases in irrigation water associated with the rising demand for food, and climate change will already increase water stress within the next decades. Other factors, such as changes in the demand for municipal water, water for electricity generation, other industrial water, and water for livestock due to climate change mitigation, are of limited importance.

(Vandyck et al. 2018) estimate that the 2°C pathway would reduce air pollution and avoid 0.7–1.5 million premature deaths in 2050 compared to current levels. It is generally agreed that in both developed and developing countries there are additional benefits of climate change mitigation in terms of improved air quality (Section 3.7.4). (Markandya et al. 2018) assessed the health co-benefits of air pollution reductions and the mitigation costs of the Paris Agreement using global scenarios for up to 2050. They concluded that the health co-benefits substantially outweighed the policy costs of achieving the NDC targets and either 2°C or 1.5°C stabilisation. The ratio of health co-benefits to the mitigation costs ranged from 1.4 to 2.45, depending on the scenario. The extra effort of trying to pursue the 1.5°C target instead of the 2°C target would generate a substantial net benefit in some areas. In India, the co-health benefits were valued at USD3.28–8.4 trillion and those in China at USD0.27–2.31 trillion. (Gi et al. 2019) also show that developing countries such as India have a huge potential to produce co-benefits. In addition, this implies that while the cost advantages of simultaneously achieving reductions of $CO_2$ emissions and of $PM_{2.5}$ are clear, the advantages for integrated measures could be limited, as the costs greatly depend on the $CO_2$ emissions reduction target.

(Grubler et al. 2018) models a pathway leading to global temperature change of less than 1.5°C without carbon capture and storage (CCS), taking end-use changes into account, including innovations in information technologies and changes to consumer behaviour apart from passive consumption. The pathway estimates global final-energy demand of 245 EJ $yr^{-1}$ in 2050, which is much lower than in existing studies (Section 5.3.3). It also shows the possibilities of creating synergies between multiple SDGs, including hunger, health, energy access and land use. Integrated technological and social innovations will increase the opportunity to achieve sustainable development. (Millward-Hopkins et al. 2020) estimate global final energy at 149 EJ $yr^{-1}$ in 2050 as required to provide decent material living standards, which is much lower than the 1.5°C scenario ranges (330–480 EJ $yr^{-1}$ in 2050) of IAMs (IPCC 2018) and the 390 EJ $yr^{-1}$ in the IEA SDS (IEA 2019), and also lower than (Grubler et al. 2018). The conclusion is that, although providing material living standards does not guarantee that every person will live a good life, there are large potentials in achieving low energy demand with sustainable development.

An overview of the co-benefits and trade offs of several SDGs based on modelling results is provided in Figure 3.39 (Section 3.7). Selected mitigation co-benefits and trade-offs are provided in relation to meeting the 1.5°C temperature goal based on a subset of models and scenarios, despite many IAMs so far not having comprehensive coverage of the Sustainable Development Goals (Rao et al. 2017; van Soest et al. 2019). There are several co-benefits of mitigation policies, including increased forest cover (SDG 15) and reduced mortality from ambient $PM_{2.5}$ pollution (SDG 3) compared to reference scenarios. However, mitigation policies can also cause higher food prices and thus increase the share of the global population at risk from hunger (SDG 2), while also relying on solid fuels (SDGs 7 and 3) as side effects. It is then concluded in Section 3.7 that these trade-offs can be balanced through targeted support measures and/or additional SD policies (Bertram et al. 2018; Cameron et al. 2016; Fujimori et al. 2019).

The World in 2050 Initiative (TWI2050) includes a comprehensive assessment of technologies, economies and societies embodied in the SDGs (IIASA 2018). The assessment addresses social dynamics, governance and sustainable development pathways within the areas of human capacity and demography, consumption and production, decarbonisation and energy, food, the biosphere and water, smart cities and digitalisation. The report concludes that the 17 SDGs are integrated and complementary and need to be addressed in unison. Studies using global IAMs that were presented in the GEO6 report (United Nations Environment Programme 2019, Chapter 22) concluded that transitions to low-carbon pathways will require a broad portfolio of measures, including a mixture of technological improvements, lifestyle changes and localised solutions. The many different challenges require dedicated measures to improve access to, for example, food, water and energy, while at the same time reducing the pressure on environmental resources and ecosystems. A key contribution may be a redistribution of access to resources, where both physical access and affordability play a role. The IAMs cover large countries and regions, and localised solutions are not properly addressed in the modelling results. This implies that, for example, trade-offs between energy access and affordability are not fully represented in aggregate modelling results.

There are also several country-level studies for deep emissions reductions (see Chapter 4 for an overview of the results). The studies find significant impacts of mitigation policies at the sectoral level, reflecting the fact that the sectoral scope does not allow for as much flexibility in mitigation measures despite macroeconomic impacts being assessed to be small (Executive Summary in Chapter 4). Another key lesson is that the detailed design of mitigation policies is critical for the distributional impacts (Executive Summary in Chapter 4). The potential mitigation measures, the potential economic growth, the political priorities and so forth are different among nations, and there may be several emissions-reduction transition pathways to long-term goals among nations (Figure 4.2).

#### 17.3.2.2 Renewable Energy Penetration and Fossil Fuel Phase-out

As pointed out in Chapter 6, the achievement of long-term temperature goals in line with the Paris Agreement requires the rapid penetration of renewable energy and a timely phasing out of fossil fuels, especially coal, from the global energy system. Limiting warming to 1.5°C (>50%) with no or limited overshoot means that global $CO_2$ emissions must reach 'net zero' in 2050/2060 (IPCC 2018). Net zero emissions imply that fossil fuel use is minimised and replaced by renewables and other low-carbon primary forms of energy, or that the residual emissions from fossil fuels are offset by carbon dioxide removal (CDR). The 1.5°C scenario requires a 2–3% annual improvement rate in carbon intensities till 2050. The historical record only shows a slight improvement in the carbon intensity rate of global energy supplies, far from what is required to limit global warming to 2°C (>67%), or limit warming to 1.5°C (>50%) with no or limited overshoot.

The role of coal in the global energy system is changing fast. Given the global temperature goals of the Paris Agreement, the global coal sector needs a transition to near zero by 2050 – earlier in some regions (Bauer et al. 2018; IEA 2017; IPCC 2018). Other global trends, including air quality, water shortages, the improved cost efficiencies of renewables, the technical availability of energy storage and the economic rebalancing of emerging countries, are also driving global coal consumption to a plateau followed by a reverse (Sator 2018; Spencer et al. 2018). The world should be prepared for a managed transition away from coal and should identify appropriate transition options for the future of coal, which can include both the penetration of renewable energy and improvements in energy efficiency (Shah et al. 2015).

Phasing out fossil fuels from energy systems is technically possible and is estimated to be relatively low in cost (Chapter 6). The cost of low-carbon alternatives, including onshore and offshore wind, solar photovoltaic (PV) and electric vehicles, has been reduced substantially in recent years and has become competitive with fossil fuels (Shen et al. 2020). However, studies show that replacing fossil fuels with renewables can have major synergies and trade-offs with a broader agenda of sustainable development (Swain and Karimu 2020), including land use and food security (McCollum et al. 2018), decent jobs and economic growth (Swain and Karimu 2020). Clarke et al. (AR5 WG III Table 6.7) provides detailed mapping of the sectoral co-benefits and adverse side-impacts of and links to transformation pathways. In Section 17.3.3.7, this is supplemented with a mapping of the synergies and trade-offs between the deployment of renewable energy and the SDGs.

The general conclusion is that the potential co-benefits of renewable-energy end-use measures outweigh the adverse impacts in most sectors and in relation to the SDGs, though this is not the case for the AFOLU (agriculture, forestry and other land use) sectors. Some locally negative economic impacts can result in increased energy costs and competition over land areas and water resources. Some sectors may also experience increasing unemployment as a consequence of the transition process. Although the deployment of renewable energy will generate a new industry and associated jobs and benefits in some areas and economies, these impacts will often not directly replace or offset activities in areas that have been heavily dependent on the fossil fuel industry.

The transition to low-emission pathways will require policy efforts that also address the emissions that are locked-in to existing infrastructure such as power plants, factories, cargo ships and other infrastructure already in use: for example, today coal-fired power plants account for 30% of all energy-related emissions (IEA 2019). Over the past twenty years, Asia has accounted for 90% of all coal-fired capacity built worldwide, and these plants have potentially long operational lifetimes ahead of them. In developing economies in Asia, existing coal-fired plants are just twelve years old on average. There are three options for bringing down emissions from the existing stock of plants: to retrofit them with carbon capture and storage (CCS) or biomass co-firing equipment; to repurpose them to focus on providing system adequacy and flexibility while reducing operations; and to retire them early. In the IEA Sustainable Development Scenario, most of the 2080 GW of existing coal-fired capacity would be affected by one of these three options.

Even though the transition away from fossil fuels is desirable and technically feasible, it is still largely constrained by existing fossil fuel-based infrastructure and stranded investments. The 'committed' emissions from existing fossil fuel infrastructure may consume all the remaining carbon budget in the 1.5°C scenario, or two thirds of the carbon budget in the 2°C scenario (Tong et al. 2019). (Kefford et al. 2018) assess the early retirement of fossil fuel power plants in the US, EU, China and India based on the IEA 2°C scenario and



## Box 17.1 | Case Study: Coal Transitions

The coal transition will pose challenges not only to the power sector, but even more importantly to coal mining. A less diversified local economy, low labour mobility and heavy dependence on coal revenues will make closing down coal production particularly challenging from a political economy perspective. Policy is needed to support and invest in impacted areas to smooth the transition, absorb the impact and incentivise new opportunities. A supportive policy for the transition could include both short-term support and long-term investment. Short-term compensation could be helpful for local workers, communities, companies and governments to manage the consequences of coal closures. Earlier involvement with local stakeholders using a structured approach is crucial and will make the transition policy more targeted and better administered. The long-term policy should target support to the local economy and workers to move beyond coal, including a strategic plan to transform the impacted area, investment in local infrastructure and education, and preference policies to incentivise emerging businesses. Most importantly, *ex ante* policy implementation is far better than *ex post* compensation. Even without the climate imperative, historical evidence shows that coal closures can happen surprisingly fast.

Presently, coal-fired power plants play a key role in the German energy system, providing almost 46% of the electricity consumed in Germany. These coal power plants play a crucial role in balancing fluctuations in producing electricity form renewables (Parra et al. 2019). Political and economic considerations, at least regionally, are also of great importance in the coal sector due to the approximately 35,000 people employed within it (including coal mining and the power stations themselves). For a long time, coal-fired power plants were able to protect their position in Germany, but against the background of decreasing public acceptance, economic problems resulting from the growing use of renewables and ambitious GHG reduction targets, the sector cannot resist the political pressure against it any longer. The governing parties have agreed to establish a commission called 'Growth, structural change and employment' to develop a strategy for phasing out coal-fired power plants (E3G Annual Review 2018). This Commission consists of experts and stakeholders from industry, associations, unions, the scientific community, pressure groups and politicians. Its establishment shows that the phasing-out process deserves close attention and that management policies must be implemented to ensure a soft landing for the electricity sector.

conclude that a massive early retirement of coal-fired power plants is needed, and that two to three standard 500 MW generators will need to come offline every week for fifteen years. This high rate is the result of a very large deployment of coal-fired power plants from 2004 to 2012. The early phasing out of this infrastructure will result in a significant share of stranded assets (Ansari and Holz 2020) with an impact on workers, local communities, companies and governments (van der Ploeg and Rezai 2020). The challenge is thus to manage a transition which delivers the rapid phasing out of existing fossil fuel-based infrastructure while also developing a new energy system based on low-carbon alternatives within a very short window of opportunity.

Chapter 6 similarly concludes that the transition towards a high penetration of renewable systems faces various challenges in the technical, environmental and socio-economic fields. The integration of renewables into the grid requires not only sufficient flexibility in power grids and intensive coordination with other sources of generation, but also a fundamental change in long-term planning and grid operation (see Chapter 6 for more detail on these issues).

Examples from various countries show that, compared with top-down decision-making, bottom-up policymaking involving local stakeholders could enable regions to benefit and reduce their resistance to transitions. (Kainuma et al. 2012) conclude that social dialogue is a critical condition for engaging local workers and communities in managing the transitions with the necessary support from transition assistance. They also point out that macro-level policies, training programmes, participatory processes and specific programmes to support employment creation for workers in fossil fuel-dependent industries are needed.

Examples of challenges in transitions away from using coal are given in Box 17.1

The transition towards a high-penetration renewable system also raises concerns over the availability of rare metals for batteries like lithium and cobalt. While metal reserves are unlikely to limit the growth rate or total amount of solar and wind energy, used battery technologies and the known reserves currently being exploited are not compatible with the transition scenario due to insufficient cobalt and lithium reserves (Månberger and Stenqvist 2018). Global lithium production rose by roughly 13% from 2016 to 2017, to 43,000 Mt in 2018 (Golberg 2021). Africa has rich reserves of lithium and is expected to produce 15% of the world's supply soon (Rosenberg et al. 2019). Such reserves are found in Zimbabwe, Botswana, Mozambique, Namibia, South Africa (Steenkamp 2017) and the Democratic Republic of Congo (Roker 2018).

The demand for these resources as ingredients in rechargeable batteries is growing rapidly, with global demand for cobalt set to quadruple to over 190,000 tons by 2026. The DRC is a mineral-rich country (Smith et al. 2019a) with rich reserves of fossil fuels (coal and oil) (Buzananakova 2015). The extraction of lithium and cobalt can be environmentally and socially damaging, though its use as a principal component in most rechargeable batteries for electric vehicles and electronic smart grids affords it high sustainability value. Chapter 10 includes a more detailed assessment of the issues with mining these rare metals, as well as the associated social problems, including exploitative working conditions and child labour, the latter a major issue that needs to be taken into consideration in transitions. Recycling batteries is also highlighted as a major supplementary policy if negative environmental side impacts are to be avoided (Rosendahl and Rubiano 2019). In the future, more attention should be paid to reducing vulnerability through subsidising R&D in rare-metals recycling, establishing systems to incentivise the collection of rare-metal waste and promoting technological progress using abundant metals as a replacement for rare metals (Rosendahl and Rubiano 2019).

#### 17.3.2.3 Stranded Assets, Inequality and Just Transitions

As the momentum towards achieving carbon neutrality grows, the risk of certain assets becoming stranded is on the increase. International policies and the push for low-carbon technologies in the context of climate change are reducing the demand for and value of fossil fuel products. Stranded assets become devalued before the end of their economic life or can no longer be monetised due to changes in policies and regulatory frameworks, technological change, security, or environmental disruption. In short, stranded assets are 'assets that have suffered from unanticipated or premature write-down, devaluations or conversions to liabilities' (Caldecott et al. 2013).

Stranded assets are likely to 'lose economic value ahead of their anticipated useful life' (Bos and Gupta 2019). They are often described as creative when they become stranded because of innovation, competition or economic growth (Gupta et al. 2020). Divestment refers to 'the action or process of selling off subsidiary business interests or investments'. This often occurs due to changing social norms and perceptions of climate change.

Indeed, pressure is mounting on fossil fuel industries to remove their capital from heavy carbon industries. As the former Governor of the Bank of England, Mark Carney, remarked, a wholesale reassessment of prospects, especially if it were to occur suddenly, could potentially destabilise markets, sparking a pro-cyclical crystallisation of losses and a persistent tightening of financial conditions. In other words, an abrupt resolution to the tragedy of horizons itself poses a risk to financial stability (OECD 2015). The divestment narrative is also based on the view that a shift away from intensive carbon resources will be significant, as the 'less value will be destroyed, [...] the more can be re-invested in low carbon infrastructure' (OECD 2015). Social movements are critical to triggering rapid transformational change and moving away from dangerous levels of climate change (Mckibben 2012). Although divestment is hailed as a necessary action to decouple fossil fuel from growth and force carbon-intensive industries to go out of business, there is the sense that there is no shortage of investors who are willing to buy shares, so that such resources are not stranded, but simply relocated. Criticism has been levelled at the divestment movement for not having a significant impact on funding fossil fuels and not being sufficiently in tune with other wide-ranging complexities that go beyond the moral dimensions (Bergman 2018). Despite being labelled a 'moral entrepreneur', the divestment movement has the potential to disrupt current practices in the fossil fuel industry, shape a 'disruptive innovation' and contribute to a strategy for decarbonising economies globally (Bergman 2018). Divestment is contributing to the political situation that is 'weakening the political and economic stronghold of the fossil fuel industry' (Grady-Benson and Sarathy 2016).

The risks attached to the stranding of fossil fuel assets have increased with the recent and sustained plunge in oil prices because of the global health pandemic (COVID-19) and the concomitant economic downturn, forcing demand to plummet to unprecedentedly low levels. (Oil prices have recently increased.) Many economies in transition and countries dependent on fossil fuels are going through turbulent times where asset and transition management will be critical (UNEP/SEI 2020). However, COVID-19 provides a foretaste of what a low-carbon transition could look like, especially if assets become stranded in an effort to respond to the call for action in 'building back better' and putting clean energy jobs and the just transition at the heart of the post-COVID-19 recovery (IEA 2020; United Nations General Assembly 2021). COVID-19 provides a useful proxy for issuing two alerts. First, it is a reminder of the urgency of addressing climate change, given that delaying the move away from stranded assets will further worsen climate change. Second, failure to recognise the threat from stranded assets will result in new assets becoming stranded (Rempel and Gupta 2021). Hence, the momentum towards a transformational push is resting on a new opportunity ushered in by COVID-19 to emphasise the urgency for a new departure towards rapid emissions reductions (Cronin et al. 2021).

The stranded assets narrative has focused overwhelmingly on consumption by companies: not much emphasis has been placed on the commercialisation- and investment-related aspects. In addition, other carbon-intensive activities can also run the risk of being stranded, such as cement, petrochemicals, steel and aviation (Baron and David 2015). This is why stranded assets are often referred to as having a cascading impact on several other sectors.

Transitions are broad-based and complex, involving governance structures, institutions and climate vulnerabilities, and there is a need to include historical responsibility, resource intensity and capacity differentials, thus relegating the debate across simplistic binary lines of developed versus developing countries (Carney 2016). Hence, transition processes will have to respond to several preconditions and structural inequalities related to climate finance, energy poverty, vulnerabilities and the broader macroeconomic implications associated with managing the debt burden, fiscal deficits and uneven terms of development in developing countries. In addition to structural inequalities, the COVID-19 pandemic has severely disrupted energy and food systems, and reduced the speed at which developing countries can procure new low-carbon technologies and decouple economic growth from fossil fuels (Winkler 2020). For instance, global supply-chain transition costs might be lower when compared to in-country supply chains, as became evident when COVID-19 created further disruption to renewable-energy projects (Cronin et al. 2021). Moreover, developing countries can experience difficulties in phasing out old technologies, especially if the latter has a cost disadvantage, has not benefitted from an established track record and its performance is uncertain (Bos and Gupta 2019). There is the risk of lock-in effects related to grandfathering when emitters comply with less stringent standards.

Despite their efforts in deploying renewable energies, many developing countries are still contending with problems related to the immaturity of the current technologies and the challenges of battery storage. In short, the transition to low-carbon development must consider the challenges of renewable-energy penetration and existing energy-related vulnerabilities and inequalities. There are power asymmetries between first-comers and latecomers, especially in cases where mature technologies can be located in countries with less stringent laws and standards. Carbon leakage has implications for just transitions, as carbon-intensive industries can move their dirty industries to developing countries as a way of outsourcing the production of carbon (Bos and Gupta 2019; UNU-INRA 2020). When the challenge of climate mitigation is transferred to developing countries in the form of carbon leakage, the risks of carbon lock-in for developing countries are heightened (Bos and Gupta 2019).

Overcoming the carbon lock-in is not simply a matter of the right policies or switching to low-carbon technologies. Indeed, it would mean a radical change in the existing power relations between fossil fuel industries and their governments and social structural behaviour (Seto et al. 2016). Some actions to fix the climate change problem can themselves create injustices, thereby challenging sustainable development (Cronin et al. 2021). Not paying sufficient attention to perceptions of injustice related to the rights to development, energy and resource sovereignty can further create resistance to climate action (Cronin et al. 2021).

The shrinking carbon budget has raised questions over whether to meet our commitment to 2°C if fossil fuel resources were to be mined or left stranded, as McGlade and Ekins argue: '… [a] large portion of the reserve base and an even more significant proportion of the resource base should not be produced if the temperature rise is to remain below 2 degrees C' (McGlade and Ekins 2015). This logic means that developing countries that rely on fossil fuel extraction will need to replace their hydrocarbon revenues with other income-generating activities. Stranded assets remind most oil-producing governments that fossil fuel assets do not have a durable value and are vulnerable to politico-economic forces and fluctuations. The goal of staying within the 1.5°C temperature goal, in line with the Paris Agreement, is already part of the policy vision and planning of large fossil fuel-consuming economies. For early fossil-fuel producers, however, the reality that their resources may not yield the desired returns is often perceived as bad news, particularly in the context of the increasing depreciation of fossil fuel products.

Stranded assets raise fundamental questions related to issues of equity and just transitions:

- Who decides which resources should be stranded?
- Who shoulders the burden of the transition and losses incurred from moving away from heavy industries with associated compensation?
- How should the advantages of short-term fossil fuel exploitation be shared based on the principle of distributive justice?

The transition to a low-carbon development is wired in issues of justice and equity: how do you align carbon reductions to meet the needs of humanity? Distributive justice calls for a fairer sharing of the benefits and burdens of the transition process, while procedural justice is essentially about ensuring that the demands of vulnerable groups are not ignored in the pull to the transition. The impacts of climate change and the mitigation burdens are experienced differently by different social actors, with indigenous communities facing multiple threats and being subjected to unequal power dynamics (Sovacool 2021).

Nonetheless, the production of fossil fuels is central to many economies with numerous development implications related to rents associated with export revenues, energy security and poverty alleviation (Lazarus and van Asselt 2018). The central question is: who decides which types of carbon should be burnable or non-burnable? Hence, social equality is at the heart of the transition process, but it falls short of a response on how to chart a new road map towards carbon neutrality, especially given that fossil fuel producers and investors tend to belong to large, powerful companies and wield a great deal of influence and power, especially when their entrenched interests are at stake (Lazarus and van Asselt 2018). The question of whether developing countries should be compensated for foregoing their resources in light of their current development needs has not yielded many results and had only limited success in mobilising international finance, as demonstrated by the case of Yasuni-ITT in Ecuador (Sovacool and Scarpaci 2016). According to (Sovacool et al. 2021), affected communities and their views may be discounted and excluded from planning, which can neglect important matters such as rights, recognition and representation (Sovacool 2021).

Fossil fuel-dependent countries are doubly exposed to the vulnerability related to climate change impacts and are being targeted in the global effort to address the problem (Peszko et al. 2020). Countries that are heavily reliant on oil, coal and gas are also those most at risk from a low-carbon transition that may curtail the activities of their fossil fuel industries and render the value chains and economies associated with the exploitation of fossil fuels unviable (Peszko et al. 2020).

Developing countries in Latin America and Africa that are reliant on revenue streams from fossil fuels may not see these returns converted into much-needed infrastructure and other social and economic amenities that can reduce poverty. However, given the falling prices of renewables, developing countries do not have to face the burden of retrofitting their infrastructure to align with new low-carbon industries, since they can leapfrog technologies and shape a sustainable trajectory that is more resilient and fit for the future.

However, the transition towards a carbon-neutral world is complex and non-linear, and it will likely result in some disruptions, with manifest equality implications, given the scale of the transformation envisaged. There are parallel movements that can be observed. On the one hand, divestment initiatives are underway to move away from carbon-intensive investments. On the other hand, hydrocarbon-rich countries in some parts of the developing world are identifying new opportunities to reduce the fiscal loss associated with the loss of fossil fuel revenues. Indeed, with global investment in energy expected to shrink by 20% in 2021, this has created fiscal challenges for countries that are heavily reliant on fossil fuel products as their main source of revenue.

Other disruptions are linked to redundant contracts and postponed or cancelled explorations, as many oil companies are diversifying their production in the wake of the pandemic and are cutting back on planned hydrocarbon investments (Denton et al. 2021). These failed concessions and disruptions have implications for the just transition, especially in developing countries without the financial ability to pull out of fossil fuels and to diversify with the same urgency as the industrialised nations (Peszko et al. 2020). For instance, in South Africa, which is seeking to divest away from coal and decarbonise its energy sector, if the transition is not properly managed, this could lead to a loss in revenue of R1.8 trillion (USD125 billion), thus compromising the government's ability to support social spending (Huxham et al. 2019). Emerging oil producers like Uganda are having to postpone the start of production. Eni and Total, two of the largest international oil and gas majors in Africa, have already signalled they are making 25% cuts to their investment in exploration and production projects in 2020, representing a EUR4 billion reduction in foreign direct investment for Total and a USD2 billion reduction for Eni (Le Bec 2020).

A poorly managed transition will reproduce inequalities, thus contradicting the very essence of a just, sustainable, inclusive transition. Revenues from oil and gas have been ploughed into social safety nets and are supporting free senior high-school education in countries such as Ghana, thus enabling the realisation of SDG 4 (quality education) (UNU-INRA 2020). The move from fossil fuels towards a low-carbon economy has economic implications for lower-income countries that are dependent on hydrocarbon resources, are endowed with significant untapped oil and gas reserves, and may not have the transitional tools to move towards low-carbon technologies or economies (Peszko et al. 2020).

The energy transition landscape is changing rapidly, and we are witnessing multiple transitions. This creates room to manage the transition in ways that will prioritise the need for workers in vulnerable sectors (land, energy) to secure their jobs and to maintain a secure and healthy lifestyle, especially as the risks multiply for those who are exposed to heavy industrial jobs and all the associated outcomes. The shift to carbon neutrality is being driven by convergent factors related to energy security and the benefits of climate mitigation, including the health impacts of air pollution and consumer demand (Svobodova et al. 2020).

Climate change is high on the global agenda, as is energy's role in decarbonising the economy, giving rise to a number of equality issues. (Oswald et al. 2020) have shown that economic inequality translates into inequality in energy consumption, as well as emissions. This is largely because people with different levels of purchasing power make use of different goods and services, which are sustained by different energy quantities and carriers (Oswald et al. 2020; Poblete-Cazenave et al. 2021).

A study by (Bai et al. 2020) shows that an increase in income inequality in China hinders the carbon abatement effect of innovations in renewable-energy technologies, possibly even leading to an increase in carbon emissions, while a decrease in inequality of incomes is conducive to giving play to the role of this carbon abatement effect, thereby indicating that there is an important correlation between the goals of 'sustainable social development' and 'sustainable ecological development'.

India is home to one sixth of world's population but accounts for only 6.8% of global energy use and consumes only 5.25% of electricity produced globally. During the period 1990–1991 to 2014–2015, overall energy intensity in India declined from 0.007 Mtoe per billion INR of GDP to 0.004 Mtoe per billion INR of GDP, an annual average decline of 2%. The industrial sector is making the highest contribution $CO_2$ mitigation by reducing its energy intensity (Roy et al. 2021).

Household carbon emissions are mainly affected by incomes and other key demographic factors. Understanding the contribution of these factors can inform climate responsibilities and potential demand-side climate-mitigation strategies. A study by (Feng et al. 2021) on inequalities in household carbon the in USA shows that the per-capita carbon footprint (CF) of the highest income group (>USD200,000 $yr^{-1}$) with 32.3 tonnes is about 2.6 times the per-capita CF of the lowest income group (<USD15,000 $yr^{-1}$) with 12.3 tonnes. Most contributors of high carbon footprints across income groups in the US are heating, cooling and private transport, which reflects US settlement structures and lifestyles, heavily reliant as they are on cars and living in large houses.

Studies by (Jaccard et al. 2021) on energy in Europe shown a top-to-bottom decile ratio (90:10) of 7.2 for expenditure, 3.1 for net energy and 2.6 for carbon. Given such inequalities, these two targets can only be met through the use of carbon capture and storage (CCS), large efficiency improvements and an extremely low minimum final energy use of 28 GJ per adult equivalent. Assuming a more realistic minimum energy use of about 55 GJ per adult equivalent and no CCS deployment, the 1.5°C target can only be achieved at near full equality. The authors conclude that achieving both stated goals is an immense and widely underestimated challenge, the successful management of which requires far greater room for manoeuvre in monetary and fiscal terms than is reflected in the current European political discourse.

The 'Just Transition' concept has evolved over the years (Sweeney and Treat 2018) and is still undergoing further evolution. It emphasises the key principles of respect and dignity for vulnerable groups, the creation of decent jobs, social protection, employment rights, fairness in energy access and use, and social dialogue and democratic consultation with relevant stakeholders, whilst coping with the effects of asset-stranding or the transition to green and clean economies. The concept has come under increased scrutiny, with its protagonists emphasising the need to focus on the equality of the transition, not simply on its speed (Forsyth 2014). The emphasis on justice is also gaining in momentum, with a growing recognition that the sustainability transition is about justice in the transition and not simply about economics (Newell and Mulvaney 2013; Swilling, M. Annecke 2010; Williams and Doyon 2020). Scholars are increasingly of the view that a transition involving low-carbon development should not replace old forms of injustice with new ones (Setyowati 2021).

The economic implications of the transition will be felt by developing countries with high degrees of dependence on hydrocarbon products as a revenue stream, as they are exposed to reduced fiscal incomes, given the low demand for oil and low oil prices, and the associated economic fallout of the pandemic. This link with stranded assets is important, but it may be overlooked, as countries whose assets are becoming stranded may not have the relevant resources, knowledge, autonomy or agency to design a fresh orientation or decide on the transition. In addition, some developing countries are dependent not only on fossil fuel revenues, but also on foreign exchange earnings from exports. This dependence comes into sharp focus when one considers that 30% of the Malaysian government's revenues are linked to petroleum products, and that Mozambique, by exploiting its newly discovered natural-gas reserves, can earn seven times the country's current GDP over a period of 25 years (Cronin et al. 2021). Thus, any attempt to accelerate the transition to low-carbon development must take into account foreign exchange, domestic revenue and employment generation, which are precisely what ensure the attractiveness of fossil fuel industries (Addison and Roe 2018).

Energy use and its deployment are sovereign matters. State responsibilities over the control and use of natural resources concern both current and future generations (Carney 2016). Climate change impacts will disable the food, water and energy systems of the most vulnerable. Therefore, the resources required to enable a just transition are predicated on good leadership and governance institutions that will support quality and justice-based transitions. Beyond energy systems, changes to land systems can benefit from sustainable land management in ways that will reduce the pressure on land for food and at the same time support carbon storage. With land coming under increased pressure, land and forest management are critical for carbon sequestration, as well as other ecosystem benefits. Extractive processes have impacts on land, and often there are few if any redistributive benefits for communities in regions where extraction takes place. In addition, extraction of strategic minerals such as cobalt, copper and lithium have been linked to violence, human rights abuses and conflict (Cronin et al. 2021).

However, in the race to achieve carbon neutrality by 2050, some of the other priorities of the transition, like climate change adaptation and its inherent vulnerabilities, might become muted, given the urgency to mitigate at all costs. Consequently, the transition imperative reduces the scope for local priority-setting and ignores the additional risks faced by countries with the least capacity to adapt. Equally, the 'just transition' is often seen through the prism of job losses and the attendant retooling and reskilling imperatives necessary to re-dynamise local businesses, especially those that may fail as a result of mine closures. It is equally important to consider current disparities in knowledge and capacity which could maintain the existing inequalities in the global regional distribution of costs and benefits. One striking example is the manufacturing of PV in India when compared to manufacturing PV in China. In China, manufacturing costs are lower than in India, as are import tariffs (Behuria 2020). Similarly, a solar industry might have greater development prospects in one region than another given existing regional disparities in human capital, infrastructure, finance and technological development (Cronin et al. 2021).

Low-carbon transitions and equality implications will depend on local contexts, regional priorities, the points of departure of different countries in the transition and the speed at which they will want to travel. Hence, timing and scope are important elements that are associated more with a quality transition than a race to the bottom. To date, the debate has had some obvious blind spots, not least considerations of power, politics and political economy (Denton et al. 2021). Certainly, the transition will create winners and losers, as well as stakeholders that can frame their economic interests so as to determine the orientation, pace, timing and scope of the transition.

The determination of a just transition is complex and not simply dependent on the allocation of perceived risks or solutions, but rather on how risks and solutions are defined (Forsyth 2014). Acting urgently to achieve environmental solutions or meet transition imperatives has certain risks given the need to go beyond commonplace definitions of the just transition by emphasising the distributive or procedural aspects. The framing of policies to align with fast and low-cost mitigation without paying sufficient attention to social and economic resilience creates its own potential risks and can enhance social vulnerability rather than address it. The need to distribute climate change solutions must not delegitimise appropriate economic growth strategies, nor indeed create the additional risks of policy imposition. Perceptions of justice with regard to environmental problems and solutions matter equally. Hence, the types of transition pathway that are chosen may have equality implications. Mitigation at all costs, if done 'cheaply and crudely', can create additional problems for social justice and inclusive development (Forsyth 2014).



The assumption that the benefits of mitigation are enough to offset trade-offs with other policy objectives can be questioned. If one accepts the argument that not all adaptation addresses vulnerability concerns (Kjellén 2006), and that some adaptation strategies can heighten vulnerabilities if there are flaws in their design and implementation, then the same logic applies, namely that not all mitigation is necessarily beneficial. Hence the emphasis on the transition resulting from mitigation should be placed not only on speed or cost-effectiveness, but also on the legitimacy of the actions, and whether the transition is well designed or not. In short, justice is not always a shorthand for acting ethically, but rather a point of reasoning on what is considered legitimate. Planning for the transition often discounts human rights and social inclusivity that can occur as the result of a rapid transition. The emphasis should be placed on the management of the transition rather than the speed – for instance, if in the rush to build new hydropower energy sources implies that populations are displaced, then this constitutes a human rights violation (Castro et al. 2016; Piggot et al. 2019).

Ambitious climate goals can increase the urgency of mitigation and accelerate the speed at which carbon neutrality is achieved. However, if the transition is done with speed, then this will leave diversification efforts stymied, particularly in developing countries that are highly dependent on fossil fuel revenue streams (UNEP/SEI 2020). Transition decisions and policies may also have far-reaching gendered implications, as the closure of mines is often linked to several ancillary business impacts where men are laid off and women may have to take on multiple jobs to compensate for the reduction in the household's income (Piggot et al. 2019; UNU-INRA 2020).

A just transition holds out the prospects for alternative high-quality jobs, public-health improvements and an opportunity to focus on well-being and prosperity, with spillover benefits to urban areas and economic systems. Nonetheless, countries that transition from fossil fuels experience different challenges, different levels of dependence and have different capacities to transition. There will be countries with lower capacity and higher dependence, and vice versa (UNEP/SEI 2020).

Deciding on matters of justice is essential to the transition, and there are several inherent questions to consider when thinking through the allocation of costs and benefits, as is the case with distributive justice. How matters are defined and who defines matters such as the timing of phasing out, prioritising which energy sources need to be phased out and who might be affected are all political economy questions (Piggot et al. 2019).

Similarly, when considering issues of procedural justice, there are matters related to interests, participation and power dynamics that are essential to the process, but that might also subvert the process, depending on whose rights, whose participation and whose power are being put in jeopardy (Forsyth 2014; Piggot et al. 2019). Hence, both distribution and procedure matter, as do inter-generational and intra-generational equity in planning transitions. Six critical variables can shape or inhibit the transition process. These are dependence, timing, capacity, agency, scope and inclusion (Denton et al. 2021).

**Dependence,** or the extent to which a country may depend on revenue streams from fossil fuels, will determine its ability to manage the transition from fossil fuels. Countries who rely on the proceeds from hydrocarbon resources as economic rents to support fiscal income and spending on public service-related needs such as education, health and infrastructure, export earnings and foreign exchange reserves will have greater difficulties in foregoing their fossil fuel resources.

**Timing:** the transition pathway has to be aligned with a timetable which is anchored in national development priorities. For example, South Africa's Integrated Resource Planning indicates that the transition away from coal, if not aligned with national development priorities, will reproduce new forms of inequality. In addition, if the transition is imposed and its timing is not organic, then this might also produce social inequalities.

**Capacity:** transitions need to reflect spaces and planning. If knowledge about the transition pathway is not adequately mastered or in place, this can disable the process or steer it in the wrong direction. Capacity also relates to several attributes, including technical, governance, institutional, technologies, and economic resources to manage the transition. Poorer countries will have difficulties in managing all these resources, as well as absorbing the costs associated with the transition (UNEP/SEI 2020).

**Agency:** transitions are inherently about the sovereign right to determine one's orientation towards low-carbon development. However, given the urgency to stick to the Paris Agreement and the new conditionalities related to post-COVID stimulus packages, the absence of agency to deal with the transition might jeopardise its flow, orientation and pace (Newell and Mulvaney 2013).

**Scope:** the extent to which the transition is rolled out and its potential impacts. If transition policies are ambitious in making commensurate diversification investments, this may enable job creation, but it may also affect employees who are insufficiently prepared to undertake new jobs and skills.

**Inclusion:** who is considered in the transition process and how their interests and risks are assessed are important aspects of transition pathways. Stakeholders with strong vested interests may resist the transition, especially as it moves towards diversification activities and policies.

### 17.3.3 Cross-sectoral Transitions

Transitions will involve multiple sectoral- and cross-sectoral policies. Section 17.3.3 presents a range of studies and conclusions on the relationship between climate change mitigation goals and meeting the SDGs in order to identify major synergies and trade-offs. The interactions are manifold and complex (Nilsson et al. 2016; Pradhan et al. 2017) (Section 4.3.1.2). Here we draw on conclusions from sectoral chapters and add additional studies as a basis for drawing more general conclusions about agriculture, food and land use, the water-energy-food nexus, industry, cities, infrastructure and transportation, cross-sectoral digitalisation, and mitigation and adaptation relations.

#### 17.3.3.1 Agriculture, Forestry and Other Land Uses (AFOLU)

Sustainable development and mitigation policies are closely linked in the agriculture, food and land-use sectors. We assess synergies and trade-offs between meeting the SDGs and reducing GHG emissions within the sectors based on modelling studies and case studies illustrating how trade-offs between SDG 2 (zero hunger, biomass for energy) and SDG 15 (life on land) can be addressed by cross-sectoral mitigation options.

Chapter 7 emphasises the high expectations on land to deliver mitigation, yet the pressures on land have grown with population, dietary changes, the impacts of climate change and the conversion of uncultivated land to agriculture and other land uses. Agriculture, forestry and other land uses (AFOLU) are expected to play a vital role in the portfolio of mitigation options across all sectors. The AFOLU sector is also the only one in which it is currently feasible to achieve carbon dioxide removal (CDR) from the atmosphere, including afforestarion/reforestation (A/R), improved forest management and soil carbon sequestration (SCR) (Chapters 7 and 12). The AFOLU sector has a significant mitigation potential, with many scenarios showing a shift to net-negative $CO_2$ emissions during the 21st century. Total cumulative AFOLU $CO_2$ sequestration varies widely across scenarios, with as much as 415 $GtCO_2$ being sequestered between 2010 and 2100 in the most stringent mitigation scenarios. The largest share of net-GHG emissions reductions from AFOLU in both the 1.5°C and 2°C scenarios is from forestry-related measures, such as afforestation, reforestation and reduced deforestation. Afforestation, reforestation and forest management result in substantial CDR in many scenarios. $CO_2$ and $CH_4$ show larger and more rapid declines than $N_2O$, an indication of the difficulties of reducing $N_2O$ emissions in agriculture (Chapter 3).

The Global Assessment on Biodiversity and Ecosystem Services Report (IPBES 2019, Chapter 5) assessed the relationship between meeting the goals of the Paris Agreement and SDGs 2 (zero hunger), 7 (affordable and clean energy) and 15 (life on land). It concluded that a large expansion of the amount of land used for bioenergy production would not be compatible with these SDGs. However, combining bioenergy options with other mitigation options, like more efficient land management and the restoration of nature, could contribute to welfare improvements and to accessing food and water. Demand-side climate-mitigation measures, like energy-efficiency improvements, reduced meat consumption and reduced food waste, were considered to be the most economically attractive and efficient options in order to support low GHG emissions, food security and biodiversity objectives. Implementing such options, however, can involve challenges in terms of lifestyle changes (IPBES 2019).

The potential joint contribution of food and land-use systems to sustainable development and climate change has also been addressed in policy programmes by the UN, local governments and the private sector. These programmes address options for pursuing sustainable development and climate change jointly, such as agroforestry, agricultural intensification, better agriculture practices and avoided deforestation. (Griggs and Stafford-Smith 2013) assess production- and consumption-based methods of achieving joint sustainability and climate-change mitigation in food systems, concluding that efficiency improvements in agricultural production systems can provide large benefits. Given the expectations of high levels of population growth and the strong increase in the demand for meat and dairy products, there is also a need for the careful management of dietary changes, as well for those areas which could be used most effectively for livestock and plant production.

Loss of biodiversity has been highlighted in several studies as a major trade-off of the low stabilisation scenarios (Prudhomme et al. 2020). A wide range of mitigation and adaptation responses – for example, preserving natural ecosystems such as peatland, coastal lands and forests, reducing the competition for land, fire management, soil management and most risk-management options – have the potential to make positive contributions to sustainable development, ecosystems services and other social goals (McElwee et al. 2020). (Smith et al. 2019a) also stressed that agricultural practices (e.g., improving yields, agroforestry), forest conservation (e.g., afforestation, reforestation), soil carbon sequestration (e.g., biochar addition to soils) and the removal of carbon dioxide (e.g., BECCS) could contribute to climate change mitigation (Smith et al. 2019a). However, there are also options that could improve biodiversity if they were implemented jointly with climate change mitigation in AFOLOU. In their study, (Leclère et al. 2020) show that increasing conservation management, restoring degraded land and generalised landscape-level conservation planning could be positive for biodiversity. In general, the ambitious conservation efforts and transformations of food systems are central to an effective post-2020 biodiversity strategy.

The IPCC Special Report on Climate Change and Land (IPCC 2019) emphasises the need for governance in order to avoid conflict between sustainable development and land-use management. It states: 'Measuring progress towards goals is important in decision-making and adaptive governance to create common understanding and advance policy effectiveness'. The report concludes that measurable indicators are very useful in linking land-use policies, the NDCs and the SDGs.

One example of an area where special governance efforts have been called for is the protection of forestry, ecosystem services and local livelihoods in a context of the large-scale deployment of high-value crops like palm oil, short-term, high income-generating activities and sustainable development. Serious challenges are already being seen within these areas according to (IPBES 2019).

Palm oil is one example of a product with potentially major trade-offs between meeting the SDGs and climate change mitigation in the agriculture, forest and other land uses (AFOLU) sector. Currently the area under oil palms is showing a tremendous increase, mostly in forest conversions to oil-palm plantations (Austin et al. 2019; Gaveau et al. 2016; Schoneveld et al. 2019). The conversion of peat swamp forest and mineral forest to oil palms will yield different amounts of $CO_2$. A study by (Novita et al. 2020) shows that the carbon stock of primary peat-swamp forest was 1770 MgC $ha^{-1}$ compared to a carbon stock of oil palm of 759 MgC $ha^{-1}$. The study conducted by Guillaume et al. shows that the carbon stock in mineral soils was 284 MgC $ha^{-1}$ compared to that in rainforest, which was 110.76 Mg C $ha^{-1}$ (Guillaume et al. 2018).

Restoring peatlands is one of the most promising strategies for achieving nature-based CDR (Girardin et al. 2021; Seddon et al. 2021). A study by (Novita et al. 2021) shows that significantly different $CO_2$ emissions for different land-use categories are influenced more by the water-table depth and latitude position for those locations relative to other observed parameters, such as bulk density, air temperature and rainfall.

Given that the frequent peatland fires in Indonesia were caused by land clearances in the replanting season, multi-stakeholder collaboration between oil-palm plantations, local communities and local governments over practices such as zero burning when clearing land might be one of the most effective ways to reduce the deforestation impact of oil palm (Jupesta et al. 2020). Behavioural changes as a mitigation option have been suggested as a major factor in aligning sustainable development, climate change and land management. In the absence of the policy intervention, the expansion of oil-palm plantations has provided limited benefits to indigenous and Afro-descended communities. Even when oil-palm expansion improves rural livelihoods, the benefits are unevenly distributed across the rural population (Andrianto et al. 2019; Castellanos-Navarrete et al. 2021). In any case, while oil-palm production can improve smallholders' livelihoods in certain circumstances, this sector offers limited opportunities for agricultural labourers, especially women (Castellanos-Navarrete et al. 2019).

Economy-wide mitigation costs can be effectively limited by lifestyle, technology and policy choices, as well as benefitting from synergies with the SDGs. Synergies come from the consumption side *by* managing demand. For example, reducing food waste leads to resources being saved because water, land use, energy consumption and greenhouse gas emissions are all reduced (Chapter 3).

Chapter 12 emphasised that diets high in plant protein and low in meat, in particular red meat, are associated with lower GHG emissions. Emerging food-chain technologies such as microbial, plant, or insect-based protein promise substantial reductions in direct GHG emissions from food production. The full mitigation potential of such technologies can only be realised in low-GHG energy systems.

(Springmann et al. 2018) conclude that reductions in food waste could be a very important option for reducing agricultural GHG emissions, the demand for agricultural land and water, and nitrogen and phosphorous applications. In addition to the possibility to reduce food waste, their study analysed several other options for reducing the environmental effects of the food system, including dietary changes in the direction of healthier, more plant-based diets and improvements in technologies and management. It was concluded that, relative to a baseline scenario for 2050, dietary changes in the direction of healthier diets could reduce GHG emissions by 29% and 5–9% respectively in a dietary-guideline scenario, and by 56% and 6–22% respectively in a more plant-based diet scenario. Demand-side, service-oriented solutions vary between and within countries and regions, according to living conditions and context. Avoiding food waste reduces GHG emissions substantially. Dietary shifts to plant-based nutrition lead to healthier lives and reduce GHG emissions (Section 5.3).

A similar study also found a positive impact form zero food waste. The 'no food waste' scenario could decrease global average food calorie availability by 120 kcal $person^{-1}$ $d^{-1}$ and protein availability by 4.6 g protein $person^{-1}$ $d^{-1}$ relative to their baseline levels, thus reducing required crop and livestock production by 490 and 190 Mt respectively. This lower level of production reduces agricultural land use by 57 Mha and thus mitigates the associated side effects on the environment. The lower levels of production also reduce the requirements for fertilisers and water by 10 Mt and 110 $km^3$ respectively, and GHG emissions are reduced by 410 $MtCO_2$-eq $yr^{-1}$ relative to the 2030 baseline. Reducing food waste can contribute to lessening the demand for food, feed and other resources such as water and nitrogen, reducing the pressure on land and the environment while ending hunger (Hasegawa et al. 2019).

In 2007, Britain launched a nationwide initiative to reduce household food waste, which achieved a 21% reduction within five years

(FAO 2019). The basis of this initiative was the 'Love Food, Hate Waste' radio, TV, print and online media campaign run by a non-profit organisation, the Waste and Resources Action Programme (WRAP). The campaign raised awareness among consumers about how much food they waste, how it affects their household budgets and what they can do about it. This initiative collaborated with food manufacturers and retailers to stimulate innovation, such as resealable packaging, shared meal-planning and food-storage tips. The total implementation costs during the five-year period were estimated at GBP26 million, from which it was households that derived the most benefit, estimated to be worth GBP6.5 billion. Local authorities also realised a substantial GBP86 million worth of savings in food-waste disposal costs. As for the private sector, the benefits took the form of increased product shelf lives and reduced product loss. While households started to consume more efficiently and companies may have experienced a decline in food sales, the latter also stated that the non-financial benefits, such as strengthened consumer relationships, had offset the costs.

The Asia Pacific Economic Cooperation (APEC) group of countries has also created several types of public-private partnership to tackle food waste and reduce losses. Most of these partnerships are focused on food-waste recycling in both developed and developing countries (Rogelj et al. 2018). APEC members stated that knowledge-sharing and improved policy and project management were the most important advantages of public-private partnerships.

The inextricably intertwined factors in decision-making are influenced by the characteristics of the person, in interaction with the characteristics of more sustainable practices and products, which interact with a particular context that includes the immediate environment (e.g., household, farm), the indirect environment (e.g., community) and macro-environmental factors (e.g., the political, financial and economic contexts) (Hoek et al. 2021). Hence, to influence people to make decisions in favour of sustainable food production or consumption, a wider perspective is needed on decision-making processes and behavioural change, in which individuals are not targeted in isolation, but in interaction with this wider systemic environment.

In conclusion, the AFOLU sector offers many low-cost mitigation options, which, however, can also create trade-offs between land use for food, energy, forest and biodiversity. Some options can help to mitigate such trade-offs, like agricultural practices (e.g., improved yields, agroforestry), forest conservation (e.g., afforestation, reforestation), soil carbon sequestration (e.g., biochar addition to soils) and the removal of carbon dioxide (e.g., BECCS), which could contribute to climate change mitigation. Lifestyle changes, including dietary changes and reduced food waste, are tightly embedded in modes of behaviour that are influenced by the immediate environment (e.g., household, farm), the indirect environment (e.g., community) and macro-environmental factors (e.g., political, financial and economic contexts). Achieving zero food waste could reduce the demands for land (SDG 15), water use (SDG 6) and chemical fertilisers (SDG 9), leading to GHG emissions reductions (SDG 13) by encouraging sustainable consumption and production practices (SDG 12).

#### 17.3.3.2 Water-Energy-Food Nexus

This section addresses the links between water, energy and food in the context of sustainable development and the associated synergies and trade-offs, with links to related chapters. The focus outline includes scoping and the relationship with the SDGs, general climate change impacts on global water resources, energy-system impacts and the relationship to renewables, enabling strategies, trade-offs and cross-sectoral implications (see also Chapter 12), nexus-management tools and strategies, and a box with examples from India and South Africa.

The continually increasing pressures on natural resources, such as land and water, due to the rising demands from increases in population and living standards, which also require more energy, emphasises the need to integrate sustainable planning and exploitation (Bleischwitz et al. 2018).

The water-energy-food nexus (WEFN) is at the epicentre of these challenges, which are of global relevance and are the focus of policies and planning at all levels and sectors of global society. The nexus between water, energy and food (Zhang et al. 2018) is tight and complex, and needs careful attention and deciphering across spatio-temporal scales, sectors and interests to balance proper management and trade-offs and to pursue sustainable development (Biggs et al. 2015; Dai et al. 2018; Hamiche et al. 2016). The WEFN touches upon the majority of the UN's SDGs, such as SDG 2, SDG 6, SDG 7 and SDGs 11–15 (Bleischwitz et al. 2018), and deals with basic commodities, thus guaranteeing the basic livelihoods of the global population.

The task of gaining an improved understanding of WEFN processes across disciplines such as the natural sciences, economics, the social sciences and politics has been further exacerbated by climate change, population growth and resource depletion. In light of the system of interlinkages involved, the WEFN concept essentially also covers land (Ringler et al. 2013) and climate (Brouwer et al. 2018; Sušnik et al. 2018), and can be further assessed in light of the relevant economic, ecological, social and SDG aspects (Fan et al. 2019a). The nexus approach was introduced in the early 2010s, when it was argued that advantages could be gained by adopting a nexus approach with regard to cross-sectoral and human–nature dependencies and by taking externalities into account (Hoffmann 2011). Hence, within the nexus, obvious trade-offs exist with competing interests, such as water availability versus food production.

Climate change is projected to impact on the distribution, magnitude and variability of global water resources. A yearly increase in precipitation of 7% globally is expected by 2100 in a high-emissions scenario (RCP8.5), although with significant inter-model, inter-regional and inter-temporal differences (Giorgi et al. 2019). Similarly, extreme events related to the water balance, such as droughts and extreme precipitation, are projected to shift in the future (RCP4.5) towards 2100: for example, the number of consecutive dry days is projected to increase in the Mediterranean region, southern Africa, Australia and the Amazon (Chen et al. 2014). In impact terms, an increase of 20–30% in global water use is expected by 2050 due to the industrial and domestic demand for water. Already 4 billion people experience severe water scarcity for at least one month per year (WWAP-UNESCO 2019).

Globally, climate change has been shown to cause increases of 4%, 8% and 10% in the share of population being exposed to water scarcities under the 1.5°C, 2°C and 3°C scenarios for global warming respectively (RCP8.5) (Koutroulis et al. 2019). At the same time, climate change is projected to cause a general increase in extreme events and climate variability, placing a substantial burden on society and the economy (Hall et al. 2014). Other than the human influence on the global hydro-climate, human activities have been shown to surpass even the impact of climate change in low to moderate emission scenarios of the water balance (Haddeland et al. 2014). Similar conclusions have been found by (Destouni et al. 2013; Koutroulis et al. 2019).

An obvious consequence of the impact of climate change on future hydro-climatic patterns is the fact that the energy system is projected to experience vast impacts through climate change (Fricko et al. 2016; Van Vliet et al. 2016a; van Vliet et al. 2016b) (Chapter 6). In the short run, where fossil fuel sources make up a significant share of the global energy grid, climate impacts related to water availability and water temperatures will affect thermoelectric power generation, which relies mainly on water cooling (Larsen and Drews 2019; Pan et al. 2018); water is also used for pollution and dust control, cleaning, and so on (Larsen et al. 2019). Currently, 98% of electricity generation relies on thermoelectric power (81%) and hydropower (17%) (van Vliet et al. 2016a).



Of these thermoelectric sources, the vast majority employ substantial amounts of water for cooling purposes, although there is a trend currently towards implementing more hybrid or drier forms of cooling (Larsen et al. 2019).

The renewable-energy conversion technologies that are currently dominant globally and are projected to remain so are less vulnerable to water deficiencies than fossil-based technologies, since no cooling is used. These renewable-energy conversion sources include, for example, wind, solar PV and wave energy. The implementation of such sources will, in the longer run, have the potential to reduce water usage by the energy sector substantially (Lohrmann et al. 2019). Also, an increasing share of renewables within desalination, as well as improved irrigation efficiencies, have been shown to potentially improve the inter-sectorial WEFN water balance (Lohrmann et al. 2019; Caldera and Breyer 2020). Some less dominant renewable-energy technologies do use water for cooling, such as geothermal energy and concentrating solar power (CSP), if wet cooling is employed. Despite the general detachment from water resources, wind and solar PV, for example, are highly dependent on climate change patterns, including variability depending on future energy-storage capacities and on-/off-grid solutions (Schlott et al. 2018). Furthermore, regardless of whether or not they are based on renewables, climate change will affect energy usage across sectors, such as heating and cooling in the building stock. The energy systems in question need to be able to handle variations and extremes in demand (Larsen et al. 2020).

For the 2080s compared to 1971–2000, an increase of 2.4% to 6.3% in the global gross hydropower potential, from the hydrological side alone, is seen across all scenarios (van Vliet et al. 2016a) (Chapter 6). Alongside the global increase in hydropower potential, the global mean water-discharge cooling capacity, which also relates to water temperatures, experiences a decrease of 4.5% to 15% across the scenarios. In very general and global terms, when combined, these changes support the shift towards sources of renewable energy, including hydropower, in the energy mix. When it comes to ensuring stability in the management of the electricity grid, hydro-climatological extremes have the potential to pose vast difficulties in certain regions and/or seasons depending on the nature of the energy mix (Van Vliet et al. 2016c). Van Vliet et al. (2016b) showed significant reductions in both thermoelectric and hydropower electricity capacities, exemplified by the 2003 European drought, which resulted in reductions of 4.7% and 6.6%, respectively.

The energy sector is vulnerable to production losses caused mainly by heatwaves and droughts, whereas coastal and fluvial floods are also responsible for a large relative share of the energy sector's vulnerability, as assessed by (Forzieri et al. 2018) for Europe in 2100. In total, heatwaves and droughts will be responsible for 94% of the damage costs to the European energy system compared to 40% today. Similarly, (Craig et al. 2018) show that, despite potentially minor spatio-temporally aggregated differences for various energy-system components, such as demand, thermoelectric power, wind, and so on, the aggregated impact of climate change across these components will cause a significant impact on the energy system, as currently exemplified by the USA. In terms of investments and management, it is important to unravel these cross-component relations in light of the projected nature of the future climate.

In the ongoing transition towards renewable sources of energy (see also Chapters 3, 4 and 6), the impact of the hydro-climate on energy production continues to be highly relevant (Jones and Warner 2016). As the shares of thermoelectric energy production in the energy grid go down along with the introduction of thermoelectric cooling technologies using smaller amounts of water, new energy sources and technologies are being introduced, and existing sources scaled up. Of these, hydropower, wind and solar energy are the key energy sources currently and will be in the near future, making up 2.5% and 1.8% of the total global primary energy supply in 2017 respectively (IEA 2019). Wind and solar energy are directly independent of water in themselves, but are dependent on atmospheric conditions related to processes that also drive the water balance and circulation. Hydropower, on the other hand, is directly influenced by and dependent on the supply of water, while at the same time being an essential counter-component to seasonality and climatological variation, as well as to current and future demand curves and diurnal variations, as against wind and solar energy (De Barbosa et al. 2017).

Furthermore, policy instruments in power-system management, here exemplified by hydropower in a climate-change scenario, have been shown to enhance energy production during droughts (Gjorgiev and Sansavini 2018). The significant influence of variation in the planning of renewable energy for the 21st century has also been highlighted by (Bloomfield et al. 2016). At the same time, the integration of renewables must account for lower thermoelectric efficiencies and capacities due to increases in temperature (van Vliet et al. 2016a), power-plant closures during extreme weather events due to a lack of cooling capacity (Forzieri et al. 2018), and further efficiency reductions and penalties following the implementation of CCS technologies in the effort to reach the GHG mitigation targets (Byers et al. 2015). However, more recent studies find more promising amounts of water being used for energy conversion (IEAGHG 2020; Magneschi et al. 2017).

The extraction, distribution and wastewater processes of anthropogenic water-management systems similarly use vast amounts of energy, making the proper management of water essential to reduce energy usage and GHG emissions (Nair et al. 2014)Chapter 11). One study reports that the water sector accounts for 5% of total US GHG emissions (Rothausen and Conway 2011). Within the WEFN, there is an obvious trade-off between water availability and food production, competing demands that pose a risk to the supply of the basic commodities of food, energy and water in line with the SDGs (Bleischwitz et al. 2018; Gao et al. 2019), all of which have the potential for inter-sectorial or inter-regional conflicts (Froese and Schilling 2019). Currently, 24% of the global population live in regions with constant water-scarce food production, and 19% experience occasional water scarcities (Kummu et al. 2014). To counterbalance the demand for food and comestibles in regions that experience constant or intermittent supplies, transportation is needed, which in itself requires suitable infrastructure, energy supplies, a well-functioning trading environment and support policies. Of the 2.6 billion people who experience constant or occasional water scarcities in food production, 55% rely on international trade, 21% on domestic trade and the remainder on water stocks (Kummu et al. 2014).

The relations between the influence of hydro-climatic variability, socio-economic conditions and patterns of water scarcity have been addressed by (Veldkamp et al. 2015). A key finding of this study was the ability of the hydro-climate and the socio-economy to interact, enforcing or attenuating each other, though with the former acting as the key immediate driver, and the influence of the latter emerging after six to ten years.

The trade-offs between competing demands have been investigated on a continental scale in the US Great Plains, highlighting the influence of irrigation in mitigating reductions in crop yields (Zhang et al. 2018). Despite crop-yield reductions of 50% in dry years compared to wet years, a key conclusion was that the irrigation should be counterbalanced against general water and energy savings within the context of trade-offs. In East Asia, the WEFN has been quantified, highlighting obvious trade-offs between economic growth, environmental issues and food security (White et al. 2018). This same study also highlights the concept of a virtual WEFN that includes water embodied within products that are traded and shipped. (Liu et al. 2019) find an urgent need for proper assessment methods, including of trade within the WEFN, due to the significant resource allocations.

Within the WEFN, the implementation of policies to achieve low stabilisation targets is strongly linked to sustainable development within the water sector with regard to water management and water conservation, indicating that additional coherence in policies affecting the water, energy and food sectors (among others) will be critical in achieving the SDGs (Chapter 7). Subsidised fertilisers, energy and crops can drive unsustainable levels of water usage and pollution in agriculture. More than half the world's population, roughly 4.3 billion people in 2016, live in areas where the demand for water resources outstrips sustainable supplies for at least part of the year. Irrigated agriculture is already using around 70% of the available freshwater, and the large seasonal variations in water supply and the needs of different crops can create conflicts between water needs across sectors at different time scales (Wada et al. 2016). However, as there is little potential for increasing irrigation or expanding cropland (Steffen et al. 2015), gaps in food production gaps must be closed by increasing productivity and cropping densities on currently harvested land by increasing either rain-fed yields or water-use efficiency (Alexandratos and Bruinsma 2012).

It has been argued that applying an integrated approach to water-energy-climate-food resource management and policymaking is highly beneficial in properly addressing the co-benefits and trade-offs (Brouwer et al. 2018; Howells et al. 2013), accommodating the SDGs (Rasul 2016) and, in general, assessing enabling strategies to improve resource efficiency (Dai et al. 2018). For an integrated approach to analysing the WEFN, a number of modelling approaches, tools and frameworks have been proposed (Brouwer et al. 2018; de Strasser et al. 2016; Gao et al. 2019; Larsen et al. 2019; Smajgl et al. 2016), often involving multi-objective calibration. Such tools enable decision-makers to evaluate the optimal water-allocation and energy-saving solutions for the specific geography in question. As an example, (Scott 2011) found the higher transportability of electricity, compared to water, pivotal in water-energy adaptation solutions in the USA, while arguing for the additional coordination of water and energy policies as a key instrument in balancing the trade-offs.

Common to all these integrated efforts is the challenge involved in making comparisons across studies due to the combined complexities of assumptions, model codes, regions, variables, forcings, and so on. To accommodate these challenges, (Larsen et al. 2019) suggest employing shared criteria and forcing data to enable cross-model comparisons and uncertainty estimates, as also highlighted by (Brouwer et al. 2018). Other limitations in current WEFN research are partial system descriptions, the failure to address uncertainties, system boundaries, and evaluation methods and metrics (Zhang et al. 2018). The lack of proper access to WEFN data and data quality has been highlighted by (D'Odorico et al. 2018; Larsen et al. 2019). Furthermore, gaps have been identified between theory and end-user applications in the lack of any focus on food nutritional values as opposed to calories alone, in the understanding of water availability in relation to management practices, in integrating new energy technologies and in the resulting environmental issues (D'Odorico et al. 2018).

Therefore, looking ahead, future fields of WEFN research should provide greater insights into all these aspects. Holistic frameworks have been put forward to facilitate methods of WEFN management by focusing on, for example, the geographical complexities with regard to transboundary challenges within hydrological catchments (de Strasser et al. 2016), aligning policy incentives (Rasul 2016) and making synergies and trade-offs in relation to WEFN SDG targets (Fader et al. 2018), and so on. The roles of all levels of government in optimal WEFN management are also highlighted in (Kurian 2017), especially with regard to shaping the behaviour of individuals. Furthermore, (Kurian 2017) highlights the challenges involved in science and policy communicating with one another and in the provision of optimal instruments and guidelines. Engaging non-experts and end-users in scientific processes is seen as essential to capturing previous failures and successes, and to ensure that understanding the challenges is updated to help shape the research questions.

Coordination of water use across different sectors and deltas is an important factor in sustainable water management. Examples of instruments and policies that support this from India and Sub-Saharan Africa in relation to the groundwater crisis are given below. India is the world's largest user of groundwater for irrigation, which covers more than half of the country's total irrigated agricultural area, is responsible for 70% of food production and supports more than 50% of the population (700 million people) (Chapter 7). However, excessive extraction of groundwater is depleting aquifers across the country, and falls in the water table have become pervasive. Improved water-use efficiency in irrigated agriculture is being considered, both globally and in India, as a way of meeting future food requirements with increasingly scarce water resources (Fishman et al. 2015).

The entirety of Sub-Saharan Africa has an undeveloped potential for groundwater exploitation, despite the general perception of a global groundwater crisis, this being due to the absence of services to support groundwater development (Cobbing 2020). It is estimated that most Sub-Saharan countries in Africa utilise less than 5% of their national sustainable yields (Cobbing and Hiller 2019). The initial tool for driving sustainable groundwater exploitation is a change in the narrative of a lack of resources in order to stimulate increased agricultural production and increased fulfilment of the SDGs (Cobbing 2020). Quantitative measures of actual groundwater vulnerability based on multiple indicators have been calculated by, for example, (van Rooyen et al. 2020), showing that 20.4% of South Africa's current water resources are highly vulnerable and are projected to worsen fifty years into the future.

Despite the positive perspectives regarding Sub-Saharan groundwater resources, the 2015–2017 water crisis in South Africa, including in Cape Town, clearly predicts vulnerability to climate variability (Carvalho Resende et al. 2019), which is predicted to increase. Serving as inspiration for the future mitigation of water depletion, (Olivier and Xu 2019) suggest certain governance tools to improve the diversification of water sources and the management of existing supplies.

#### 17.3.3.3 Industry

Industrial transformation is a core component in achieving sustainable development. Across all industrial sectors, the development and deployment of innovative technologies, business models and policy approaches at scale will be essential in accelerating progress both with meeting the economic and social development goals and with achieving low emissions. In this section, we assess the synergies and trade-offs between mitigation options and the SDGs, with a specific focus on asking whether economic growth and employment creation can work jointly with climate actions and other SDGs in least developed and developing countries. Examples of synergies and trade-offs are provided based on the conclusions of Chapter 9 on the building sector and Chapter 11 on industry. The potential for greening industry is discussed in relation to eco-industrial parks, with examples from Ethiopia, China, South Africa and Ghana.

Chapter 11 concludes that achieving net zero emissions from the industrial sector are possible. This will require the provision of electricity free from greenhouse gas (GHG) emissions, including from other energy carriers, increased electrification, low-carbon feedstocks, and a combination of energy efficiency, reduced demand for materials, a more circular economy, electrification and carbon capture, use and storage (CCUS).

The potential co-benefits of mitigation options in industry has been mapped out in Chapter 11 in relation to five categories of mitigation options: material efficiency and reductions in the demand for materials, the circular economy and industrial waste, carbon capture and storage, energy efficiency, and electrification and fuel switching (Figure 11.15). In particular, the first two categories of options are assessed as having several co-benefits for the SDGs, including SDGs 3, 5, 7, 8, 9 11, 12, and 15. Some studies also point out the potential trade-offs in respect of employment and the costs of cleaner production. The other options primarily impact on climate actions, decent work and employment, and industry as such.

(Okereke et al. 2019) offer important generic conclusions on green industrialisation and the transition based on a study of socio-technical transition in Ethiopia. The importance of drivers for changes in terms of clear policy goals and government support for green growth and climate policies, as well as support from a strong culture of innovation, is emphasised. The study also identifies key barriers in relation to stakeholder interactions, the availability of resources and the ongoing tensions between ambitions for high economic growth and climate change. Green innovation in industry critically depends on regulations. (Gramkow and Anger-Kraavi 2018) have assessed the role of fiscal policies in greening Brazilian industry based on an econometric analysis of 24 manufacturing sectors. They conclude that instruments like low-cost finance for innovation and support to sustainable practices effectively promote green innovation.

(Luken 2019) have assessed the drivers, barriers and enablers for green industry in Sub-Saharan Africa, concluding that major barriers exist related to material and input costs, as well as product requirements in foreign markets, and that as a result there are trade-offs between economic and environmental performance. Studies of ten countries are reviewed, and although they suffer from limited information, they conclude similarly that further progress is being hindered by poor access to finance and weak government regulation. (Greenberg and Rogerson 2014). They similarly conclude that the greening of industry in South Africa is lagging behind due to economic barriers and weak governance, despite its high priority in government planning and among international partners.

Ghana has launched a 'One District One Factory' (1D1F) initiative, aimed at establishing at least one factory or enterprise in each of Ghana's 216 districts as a means of creating economic growth poles to accelerate the development of these areas and create jobs for the country's increasingly youthful population. The policy aims to transform the structure of the economy from one dependent on the production and export of raw materials to a value-added industrialised economy driven primarily by the private sector (Yaw 2018). As has been pointed out by (Mensah et al. 2021), in its initial design the programme did not take environmental quality into consideration. Although it was successful in creating economic growth, exports and employment, the environmental impacts have been negative. It has therefore been recommended that environmental regulations be imposed on foreign investments. Similar conclusions have been drawn by (Solarin et al. 2017).

Chapter 11 concludes that eco-industrial parks, in which businesses cooperate with each other in order to avoid environmental pressure and support sustainable development, have delivered several benefits in relation to overall reductions in both virgin materials and final wastes, implying significant reductions in industrial GHG emissions. Due to these advantages, eco-industrial parks have been actively promoted, especially in East Asian countries such as China, Japan and in the Republic of Korea (South Korea), where national indicators and governance exist (Geng et al. 2019; Geng and Hengxin 2009).

(Zeng et al. 2020) have assessed the role of eco-industrial parks in China's green transformation for 33 development zones in relation to contributions to GDP, industrial value added, exports, water and energy consumption, $CO_2$ levels and sulphur emissions. They concluded that industrial parks have played a very important role in China's industrialisation, and that this structure has supported the decoupling of economic growth and energy and water consumption from the environmental impacts. However, improved environmental performance would require better access to finance and a higher priority by management.

Eco-industrial parks have been promoted in Ethiopia by the government and UNIDO, based on the expectation that they could help to boost the economy (UNIDO 2018). One of the success stories is an industrial park in Hawassa, a nation-level textile and garment industrial park with a 'zero emissions commitment' based on renewable energy and energy-efficient technologies. However, the concept of the industrial park, including feasible policies and institutional arrangements, is new to Ethiopia's regulatory processes, and this has created problems for management, knowledge and governance, hindering their fast implementation.

A number of business associations have developed strategies for sustainable development and climate change, including corporate social responsibility (CSR). International initiatives have included the promotion of CSR initiatives by international investors in low-income countries to support a broad range of development priorities, including social working conditions, eliminating child labour and climate change (Lamb et al. 2017). (Leventon et al. 2015) evaluated the role of mining industries in Zambia in supporting climate-compatible development and concluded that, although the industry has played a positive role in avoiding migration and pressure on forest resources, there is a lack of coordination between government and industry initiatives.

It can be concluded that most of the mitigation options in industry considered in this section could have synergies with the SDGs, but also that some of the renewable-energy options could indicate some trade-offs in relation to land use, with implications for food- and water security and costs. Carbon capture and storage (CCS) could play an enabling role in the provision of reliable, sustainable and modern energy and could support decarbonisation, but it can also be costly (IEAGHG 2020; Mikunda et al. 2021). The provision of water for CCS can include both synergies and trade-offs with the SDGs due to recent progress in water-management technologies (Giannaris et al. 2020; IEAGHG 2020; Mikunda et al. 2021).

#### 17.3.3.4 Cities, Infrastructure and Transportation

With 80% of the global population expected to be urban by 2050, cities will shape development paths for the foreseeable future (United Nations 2018). The challenge for many policymakers is to construct development paths that make cities clean, prosperous and liveable while mitigating climate change and building resilience to heatwaves, flooding and other climate risks. The IPCC SR1.5 report sees achieving these objectives as feasible: cities could potentially realise significant climate and sustainable-development benefits from shifting development paths (Wiktorowicz et al. 2018). This section assesses the synergies and trade-offs between meeting the SDGs and climate change mitigation, as well as providing a general overview of mitigation options in cities and of enabling factors, including city networks and plans for jointly addressing the SDGs and climate change mitigation.

Chapter 8 concludes that urban areas potentially offer several joint benefits between mitigation and the SDGs, and that since AR5, evidence of the co-benefits of urban mitigation continues to grow. In developing countries, a co-benefits approach that frames climate objectives alongside other development benefits arise increasingly being seen as an important concept justifying and driving climate change actions in developing countries (Sethi and Puppum De Oliveria 2018; Seto et al. 2016).

Evidence of the co-benefits of urban mitigation measures on human health has increased significantly since the IPCC AR5, especially through the use of health-impact assessments in cities like Geneva, where energy savings and cleaner energy-supply structures based on measures for urban planning, heating and transport have reduced $CO_2$, $NO_x$ and $PM_{10}$ emissions and increased the opportunities for physical activity for the prevention of cardiovascular diseases (Diallo et al. 2016).

There is increasing evidence that climate-mitigation measures can lower health risks that are related to energy poverty, especially in vulnerable groups, such as the elderly (Monforti-Ferrario et al. 2019). Moreover, the use of urban forestry and green infrastructure as both a climate mitigation and an adaptation measure can reduce heat stress (Kim and Coseo 2019; Privitera and La Rosa 2017) while removing air pollutants to improve air quality (Scholz et al. 2018; De la Sota et al. 2019) and enhancing well-being, including contributions to local development and possible reductions of inequalities (Lwasa et al. 2015). Other studies evidence the potential to reduce premature mortality by up to 7000 in 53 towns and cities, to create 93,000 net new jobs and lower global climate costs, as well as reduce personal energy costs based on road maps for renewable-energy transformations (Jacobson et al. 2018).

The co-benefits of energy-saving measures described by 146 signatories to a city climate network due to improved air quality have been quantified as 6596 avoided premature deaths (with a 95% confidence interval of 4356 to 8572 avoided premature deaths) and 68,476 years of life saved (with a 95% confidence interval of 45,403 and 89,358 years of life saved) (Monforti-Ferrario et al. 2019). Better air quality further reinforces the health co-benefits of climate-mitigation measures based on walking and cycling, since the evidence suggests that increased physical activity in urban outdoor settings with low levels of black carbon improves lung function (Laeremans et al. 2018). Chapter 9 shows that mitigation actions in buildings have multiple co-benefits resulting in substantial social and economic value beyond their direct impacts on reducing energy consumption and GHG emissions, thus contributing to the achievement of almost all the UN's SDGs. Most studies agree that the value of these multiple benefits is greater than the value of the energy savings, while their quantification and inclusion in decision-making processes will strengthen the adoption of ambitious reduction targets and improve coordination across policy areas.



There are several examples of cities that have developed plans for meeting both the SDGs and mitigation, which demonstrates the feasibility of meeting these objectives jointly. Quito, Ecuador, a city with large carbon footprints (Global Opportunity Explorer 2019) and climate vulnerabilities, has adopted low carbon plans that aim to achieve the climate goals while introducing net-zero energy buildings and reducing water stress (Ordoñez et al. 2019; Marcotullio et al. 2018). Several cities in China, Indonesia and Japan have invested in green-city initiatives by means of green infrastructural investments, which is claimed to be a form of smart investment. Through this type of investment, economic growth and greenhouse gas (GHG) emissions reductions can be achieved in cities (Jupesta et al. 2016). Multi-level governance arrangements, public-private cooperation and robust urban-data platforms are among the factors enabling the pursuit of these objectives within countries (Corfee-Morlot et al. 2009; Gordon 2015; Creutzig et al. 2019; Yarime 2017).

In addition to the mostly domestic enablers listed previously, some cities have also benefited from working with international networks. The Global Covenant of Mayors for Climate & Energy (Covenant of Mayors 2019), the World Mayors Council on Climate Change, ECLEI, C40, and UNDRR (C40 Cities 2019; ECLEI 2019; UNDRR 2019) have provided targeted support, disseminated information and tools, and sponsored campaigns (Race to Zero) to motivate cities to embrace climate and sustainability objectives. Despite this support, it should be stressed that most cities are in the early stages of climate planning (Eisenack and Reckien 2013; Reckien et al. 2018; Climate-ADAPT 2019). Furthermore, in some cases city policymakers may fail to highlight the synergies and trade-offs between climate and sustainable development or rebrand GHG-intensive practices as 'sustainable' in relevant plans (Tozer 2018).

With regard to city networks, Section 8.5 concludes that the importance of urban-scale policies for sustainability has increasingly been recognised by international organisations and national and regional governments. For example, in 2015, more than 150 national leaders adopted the UN's 2030 Sustainable Development Agenda, including stand-alone SDG 11 (sustainable cities and communities) (UN 2015 p. 14). The following year, 170 countries agreed to the UN New Urban Agenda (NUA), a central part of which is recognising the importance of national urban policies (NUPs) as a key to achieving national economic, social and environmental goals (United Nations 2015a 2017). Similarly, the Sendai Framework for Disaster Risk Reduction identifies the need to focus on unplanned and rapid urbanisation to reduce exposure and vulnerability to the risks of disasters (United Nations 2015b).

For many cities, a key to reorienting development paths will be investing in sustainable, low-carbon infrastructure. Because infrastructure has a long lifetime and influences everything from lifestyle choices to consumption patterns, decisions over an estimated USD90 trillion of infrastructure investment (from now to 2030) will be critical in order to avoid becoming locked-in to unsustainable paths (WRI 2016). This is particularly true in developing countries, where demands for new buildings, roads, energy and waste-management systems are already surging. To some extent, policies that accelerate building renovation rates, including voluntary programmes (Van der Heijden 2018), can support transitions down more sustainable paths (Kuramochi et al. 2018). Factoring climate and sustainable development considerations into policy tools that facilitate the quantitative emission performance standard (EPS) and the inclusion of climate and sustainable development benefits and risks in infrastructure assessments or risk-adjusted returns on investments in development banks could also prove useful (Rydge et al. 2015). Strong policy signals from the UNFCCC and from national climate policies and strategies (including NDCs) could facilitate uptake of the relevant policies and the use of these tools.

Infrastructural investments will also have wide-ranging implications for sustainable, low-carbon urban development, namely transport and mobility. To some extent, decision-making frameworks such as Avoid-Shift-Improve (ASI) could help make these patterns low carbon and sustainable (Dalkmann and Brannigan 2007; Wittneben et al. 2009). Mixed land-use planning and compact cities can not only help avoid emissions or shift travellers into cleaner modes (Cervero 2009), they can also improve air quality, reduce commuting times, enhance energy security and improve connectivity (Zusman et al. 2011; Pathak and Shukla 2016).

#### 17.3.3.5 Mitigation-adaptation Relations

The section will consider the links between mitigation and adaptation options in the context of sustainable development and the associated synergies and trade-offs. Cross-cutting conclusions will be drawn based on Chapter 3 and the sectoral chapters of AR6 WGIII and

Chapter 18 of AR6 WGII. The focus will be on the following sectors: agriculture, food and land use; water-energy-food; industry and the circular economy; and urban areas.

IPCC AR6 WGII, concludes that coherent and integrated policy planning is needed in order to support integrated climate change adaptation and mitigation policies, and that this is a key component of climate-resilient development pathways. Section 4.5.2 assesses development pathways and the specific links between mitigation and adaptation, concluding that there can be co-benefits, and trade-offs, where mitigation implies maladaptation. However, adaptation can also be a prerequisite for mitigation. It is therefore concluded that making development pathways more sustainable can build the capacity for both mitigation and adaptation.

Climate actions, including climate change mitigation and adaptation, are highly scale-dependent, and solutions are very context-specific. Especially in developing countries, a strong link exists between sustainable development, vulnerability and climate risks, as limited economic, social and institutional resources often result in low adaptive capacities and high vulnerability. Similarly, the limitations in resources also constitute key elements weakening the capacity for climate change mitigation (Jakob et al. 2014). The change to climate-resilient societies requires transformational or systemic changes, which also have important implications for the suite of available sustainable-development pathways (Kates et al. 2012; Lemos et al. 2013). Thornton and Comberti (2017) point to the need for social-ecological transformations to take place if synergies between mitigation and adaptation are to be captured, based on the argument that incremental adaptation will not be sufficient when climate change impacts can be extreme or rapid and when deep decarbonisation simultaneously involves social change (Chapter 18 in AR6 WGII).

As discussed in AR6 WGII, Section 18.4, there are synergies and trade-offs between adaptation and sustainable development, as well as between mitigation and sustainable development, which is supported by comprehensive assessments such as that by Dovie (2019) and Sharifi (2020). Links between mitigation and adaptation options are identified in Chapter 18 in AR6 WGII, such as expected changes in energy demand due to climate change interacting with energy-system development and mitigation options, changes to agricultural production practices to manage the risks of potential changes in weather patterns affecting land-based emissions and mitigation strategies, or mitigation strategies that place additional demands on resources and markets. This increases the pressures on and costs of adaptation or ecosystem restoration linked to carbon sequestration and the benefits in terms of the resilience of natural and managed ecosystems, but it also could restrict mitigation options and increase costs. Chapter 3 of AR6 WGIII similarly concludes that the connectedness and coherence of actions to mitigate climate change could support the conservation and adaptation of ecosystems and meet the Sustainable Development Goals more widely.

Options to reduce agricultural demand (e.g., dietary change, reducing food waste) can have co-benefits for adaptation through reductions in the demand for land and water (Smith et al. 2019b). For example, Grubler et al. (2018) show that stringent climate-mitigation pathways without reliance on BECCS can be achieved through efficiency improvements and reduced energy service and consumption levels in high-income countries.

Agriculture, food and land use is the sector where most climate policy options can simultaneously generate impacts on mitigation, adaptation and the SDGs (Locatelli et al. 2015; Kongsager et al. 2016). Bryan et al. (2013) identified a range of synergies and trade-offs across adaptation, mitigation and the SDGs in Kenya, given the diversity of its climatic and ecological conditions. Improved management of soil fertility and improved livestock-feeding practices could provide benefits to both climate change mitigation and adaptation, as well as increase income generation from farming. However, other improvements to agricultural management in Kenya, for example, soil water conservation, could only provide benefits across all three domains in some specific sub-regions.

Conservation agriculture can yield mitigation co-benefits through improved fertiliser use or the efficient use of machinery and fossil fuels (Harvey et al. 2014; Pradhan et al. 2018; Cui et al. 2019). Climate-smart agriculture (CSA) ties mitigation to adaptation through its three pillars of increased productivity, mitigation and adaptation (Lipper et al. 2014), although managing trade-offs among the three pillars requires care (Kongsager et al. 2016; Thornton and Comberti 2017; Soussana et al. 2019). Sustainable intensification also complements CSA (Campbell et al. 2014). Enhanced sustainable adaption can lead to effective emission-reduction benefits, such as climate-smart agricultural technologies (Nefzaoui et al. 2012; Poudel 2014) and ecosystem-based adaptation. (Berry, P et al. 2015; Geneletti and Zardo 2016; Warmenbol and Smith 2018) have shown how increases in livelihoods can contribute to climate change mitigation in Europe.

Agroforestry can sustain or increase food production in some systems and increase farmers' resilience to climate change (Jones et al. 2013). Some sustainable agricultural practices have trade-offs, and their implementation can have negative effects on adaptation or other ecosystem services. Agricultural practices can aid both mitigation and adaptation on the ground, but yields may be lower, so there may be a trade-off between resilience to climate change and efficiency. Interconnections within the global agricultural system may also lead to deforestation elsewhere (Erb et al. 2016). Implementation of sustainable agriculture can increase or decrease yields, depending on context (Pretty et al. 2006) (Chapter 4).

Land-based mitigation and adaptation will not only help reduce greenhouse gas (GHG) emissions in the AFOLU sector, but also help augment the sector's role as a carbon sink by increasing forest and tree cover through afforestation and agroforestry activities, and other eco-system-based approaches. Some of these options, however, can also have negative impacts on GHG emissions in the form of indirect impacts on land use (Córdova 2019) (for a more detailed discussion, see Chapter 7). If managed and regulated appropriately, the land use, land-use change and forestry (LULUCF) sector could play a key role in mitigation and be a key sector for emissions reductions beyond 2025 instead of contributing substantially to emissions reductions beyond 2025 (Córdova et al. 2019; Keramidas et al. 2018). However, the large-scale deployment of intensive bioenergy plantations, including monocultures, replacing natural forests and subsistence farmlands are likely to have negative impacts on biodiversity and can threaten food and water security, as well as local livelihoods, partly by intensifying social conflicts, partly by reducing resilience (Díaz et al. 2019). Expansion on to abandoned or unused croplands and pastures nonetheless presents significant global potential, and will avoid the sustainability risks of expanding agriculture into natural vegetation (Næss et al. 2021).

Based on a literature review, (Berry, P et al. 2015) identified water-saving and irrigation techniques in agriculture as attractive adaptation options that have positive synergies with mitigation in increasing soil carbon, reducing energy consumption and reducing $CH_4$ emissions from intermittent rice-paddy irrigation. These measures could, however, reduce water flows in rivers and adversely affect wetlands and biodiversity. The study also concluded that afforestation could reduce peak water flows and increase carbon sequestration, but trade-offs could emerge in relation to the increased demand for water.



Fast-growing tree monocultures or biofuel crops may enhance carbon stocks but reduce downstream water availability and the availability of agricultural land (Harvey et al. 2014). Similarly, in some dry environments, agroforestry can increase competition with crops and pastureland, decreasing productivity and reducing the yields of catchment water (Schrobback et al. 2011) (Chapter 7).

Hydropower dams are among the low-cost mitigation options, provided the cost of constructing the plant is taken into account, but they could have serious trade-offs in relation to key sustainable-development aspects, since in respect of water and land availability dams can have negative effects on ecosystems and livelihoods, thereby implying increased vulnerabilities. Section 17.3.3.2 on the water-energy-food nexus includes examples of trade-offs between the benefits of producing electricity from hydropower dams and the trade-offs with ecosystem services and using land for agriculture and livelihoods.

There are several potentially strong links between climate change adaptation in industry and climate change mitigation. Various supply chains can be affected by climate change, energy supply and water supply, and other resources can be disrupted by climate events. Adaptation measures can influence GHG emissions in their turn and thus mitigation because of the demand for basic materials, for example, as well as by influencing outdoor environments and labour productivity (Section 11.17.1.4).

Implementing adaptation options in industry can also imply increasing the demand for packaging materials such as plastics and for access to refrigeration. These are among the adaptation options that are dependent on temperature and storage possibilities, as well as being major sources of GHG emissions.

An increasing number of cities are becoming involved in voluntary actions and networks aimed at drawing up integrated plans for sustainable development and climate change mitigation and adaptation, including cities in both high- and low-income countries around the world. (Grafakos et al. 2019; Sanchez Rodriguez et al. 2018) concluded that cities are an obvious place for the development of plans that can capture several synergies between sustainable development and climate-resilient pathways. (Kim and Grafakos 2019; Landauer et al. 2019) similarly concluded that cities are an obvious platform for the development of integrated planning efforts because of the scale of policies and actions, which could potentially match the different policy domains. (Kim and Grafakos 2019) assessed the level of integration of mitigation and adaptation in urban climate change plans across 44 major Latin American cities, concluding that the integration of climate change mitigation and adaption plans was very weak in about half the cities and that limited donor finance was a main barrier. The authors also mention barriers in relation to governance and the weakness or lack of legal frameworks. The integration of SDGs with adaptation could help increase the willingness of politicians to implement climate actions, as well as provide stronger arguments for investing the required resources (Sanchez Rodriguez et al. 2018).

The local integration of planning and policy implementation practices was also examined by (Newell et al. 2018) in a study of 11 Canadian communities. It was concluded that, in order to put plans into practice, a deeper understanding needs to be established of the potential synergies and trade-offs between sustainable development and climate change mitigation and adaptation. A model was applied to the evaluation of key impacts, including energy innovation, transportation, the greening of cities and city life. The impact assessment came to the conclusion that multiple benefits, costs and conflicting areas could be involved, and that bringing a broad range of stakeholders into policy implementation was therefore to be recommended.

There are several links between mitigation and adaptation options in the building sector, as pointed out in Chapter 9. Adaptation can increase energy consumption and associated GHG emissions (Kalvelage et al. 2013; Campagnolo and Davide 2019), for example, in relation to the demand for energy to meet indoor thermal comfort requirements in a future warmer climate (de Wilde and Coley 2012; Li and Yao 2012; Clarke et al. 2018). Mitigation alternatives using passive approaches may increase resilience to the impacts of climate change on thermal comfort and could reduce cooling needs (Wan et al. 2012; Andrić et al. 2019). However, climate change may reduce their effectiveness (Ürge-Vorsatz et al. 2014).

Mitigation and the co-benefits of adaptation in urban areas in relation to air quality, health, green jobs and equality issues are dealt with in Section 8.2, where it is concluded that most mitigation options will have positive impacts on adaptation, with the exception of compact cities, with trade-offs between mitigation and adaptation. This is because decreasing urban sprawl can increase the risks of flooding and heat stress. Detailed mapping between mitigation and adaptation in urban areas shows that there are many, very close interactions between the two policy domains and that coordinated governance across sectors is therefore called for.

Rebuilding and refurbishment after climate hazards can increase energy consumption and GHG emissions in the construction and building materials sectors, as it could make the existing building stock more climate-resilient (Hallegatte 2009; de Wilde and Coley 2012; Pyke et al. 2012) and thus also support implementation of the Sendai Framework on Disaster Risk Reduction (United Nations 2015b). Climate change in the form of extremely high temperatures, intense rainfall leading to flooding, more intense winds and/or storms and sea level rises (SLRs) can seriously impact transport infrastructure, including the operations and mobility of road, rail, shipping and aviation; Chapter 10 assesses the impacts on subsectors within transportation. At the same time, these sectors are major targets for GHG mitigation options, and many countries are currently examining what to do in terms of combined mitigation-adaptation efforts, using the need to mitigate climate change through transport-related GHG emissions reductions and pollutants as the basis for adaptation action (Thornbush et al. 2013; Wang and Chen 2019). For example, urban sprawl indirectly affects climate processes, increasing emissions and vulnerability, which worsens the ability to adapt (Congedo and Munafò 2014). Hence greater use of rail by passengers and freight will reduce the pressures on the roads, while having less urban sprawl will reduce the impacts on new infrastructure, often in more vulnerable areas (IPCC 2019; Newman et al. 2017).

Despite many links between mitigation and adaptation options, including synergies and trade-offs, Chapter 13 concludes that there are few frameworks for integrated policy implementation. One review of climate legislation in Europe found a lack of coordination between mitigation and adaptation, their implementation varying according to different national circumstances (Nachmany et al. 2015).

In developing and least-developed countries (LDCs), there are many examples of climate policies in the NDCs that have been drawn up in the context of sustainable development and that cover both mitigation and adaptation (Beg 2002; Duguma et al. 2014)) (Chapter 13). However, there are many barriers to joint policy implementation. Despite the emphasis on both mitigation and adaptation policies, there is very limited literature on how to design and implement integrated policies (Di Gregorio et al. 2017; Shaw et al. 2014). For example, the links within the water-energy-food nexus require coordination among sectoral institutions and capacity-building in innovative frameworks linking science, practice and policy at multiple levels (Cook and Chu 2018; Nakano 2017; Shaw et al. 2014).

Another challenge is the shortage of financial, technical and human resources for implementing joint adaptation and mitigation policies (Antwi-Agyei et al. 2018b; Chu 2018; David and Venkatachalam 2019; Kedia 2016; Satterthwaite 2017). Several studies have stressed that the lack of finance for integrating policy implementation between sustainable development and climate change mitigation and adaptation may constitute barriers to the implementation of adaptation projects to protect least-developed countries (LDCs) with many vulnerabilities.

(Locatelli et al. 2016) come to similar conclusions regarding finance based on interviews with multilateral development banks, green funds and government organisations in respect of the agricultural and forestry sectors. International climate finance has been totally dominated by mitigation projects. Those who were interviewed were asked about their willingness to change this balance and to commit more resources to projects that address both climate change mitigation and adaptation. More than two thirds of those interviewed, however, raised concerns that integrated projects could be too complicated and that a greater alignment of financial models across different policy domains could entail greater financial risks. Another barrier mentioned in respect of finance was that mitigation projects were primarily aimed at GHG emissions reductions, while adaptation projects had more national benefits and were also more suitable for community development and promoting equality and fairness. In an assessment of 201 projects in the forestry and agricultural sectors in the tropics, (Kongsager et al. 2016), found that a majority of the projects contributed to both adaptation and mitigation or at least had the potential to do so, despite the separation between these two objectives by international and national institutions.

#### 17.3.3.6 Cross-sectoral Digitalisation

In this section, the potential role of digitalisation as a facilitator of a fast transition to sustainable development and low-emission pathways is assessed based on sectoral examples. The contributions of digital technology could contribute to efficiency improvements, cross-sectoral coordination, including new IT services, and decreasing resource use, implying several synergies with the SDGs, as well as trade-offs, for example, in relation to reduced employment, increasing energy demand and the increasing demand for services, possibly increasing GHG emissions.

The COVID-19 pandemic caused radical temporary breaks with past energy-use trends. How post-pandemic recovery will impact on the longer-term energy transition is unclear. Recovering from the pandemic with energy-efficient practices embedded in new patterns of travel, work, consumption and production reduces climate mitigation challenges (Kikstra et al. 2021). The potential of digital contact tracing to slow the spread of a virus had been quietly explored for over a decade before the COVID-19 pandemic thrust the technology into the spotlight (Cebrian 2021). The COVID-19 crisis is among the most disruptive events in recent decades and has had consequences for consumer behaviour. During the lockdowns in most countries, consumers have turned to online shopping for food products, personal hygiene and disinfection (Cruz-Cárdenas et al. 2021), making society more digitally literate.

The cost of new services provided by digitalisation can be high, and this could imply barriers for low-income countries in joining new global information-sharing systems and markets. Altogether this implies that any assessment of the contribution of digitalisation to support the SDGs and low-carbon pathways will only be able to provide very context-specific results. Digital technologies could potentially disrupt production processes in nearly every sector of the economy. However, as an emerging area experiencing the rapid penetration of many sectors, there could be a window of opportunity for integrating sustainable development and low-emission pathways. (IIASA 2020) concludes that the digital revolution is characterised by many innovative technologies, which can create both synergies and trade-offs with the SDGs (IIASA 2020).

Digital technologies could potentially disrupt production processes in nearly every sector of the economy. However, as an emerging area experiencing the rapid penetration of many sectors, there could be a window of opportunity for integrating sustainable development and low-emission pathways. TWI2050 (2020) concludes that the digital revolution is characterised by many innovative technologies, which can create both synergies and trade-offs with the SDGs (IIASA 2020).

WBSD (2019) has assessed the potential of communication technologies (ICT) to contribute to the transition to a global low-carbon economy in the energy, transportation, building, industry, and other sectors. The potential is estimated to be around 15% $CO_2$-eq emissions reductions in 2020 compared with a business-as-usual scenario. A range of ICT solutions have been highlighted, including smart motors and industrial process-management in industry, traffic-flow management, efficient engines for transport, smart logistics and smart-energy systems.



The TWI2050 2019 report (IIASA 2019) assessed both the positive and negative impacts of digitalisation in the context of sustainable development. It found that efficiency improvements, reduced resource consumption and new services can support the SDGs, but also that there were challenges, including in relation to equality, facing the least-developed and developing countries because of their low level of access to technologies. The necessary preconditions for successful digital transformation include prosperity, social inclusion, environmental sustainability, protection of jobs and good governance of sustainability transitions. One negative impact of digitalisation could be the rebound effects, where easier access to services could increase demand and with it GHG emissions. Digitalisation in the manufacturing sector could also provide a comparative advantage to developed countries due to the falling importance of labour costs, while the barriers to emerging economies seeking to enter global markets could accordingly be increased.

In respect of governance, (Krishnan et al. 2020) point out that the creation of synergies between sustainable development and low-emission urbanisation based on digitalisation could face barriers in the form of inadequate knowledge of structures and value creation through ecosystems that would need to be addressed by means of smart digitalising, requiring organisational measures to support transformation processes.

Urban areas are one of the main arenas for new digital solutions due to rapid urbanisation rates and high concentrations of settlements, businesses and supply systems, which offer great potential for large-scale digital systems. The emergence of smart cities has supported the uptake of smart integrated energy, transportation, water and waste-management systems, while synergies have been created in terms of more flexible and efficient systems. In its 2018 Policy and Action document, the Japanese Business Federation (Keidanren) launched Society 5.0, which includes plans for smart-city development (Carraz and Yuko 2019; Narvaez Rojas et al. 2021). To achieve smart cities, Society 5.0 aimed to facilitate diverse lifestyles and business success, while the quality of life offered by these options will be enhanced. It also aims to offer high-standard medical and educational services. Autonomous vehicles will be available and integrated with smart-grid systems in order to facilitate mobility and flexibility in energy supply with a high share of renewable energy.

Chapter 6 of this report on 'Energy Systems' points out that there are many smart-energy options with the potential to support sustainable development by facilitating the integration of high shares of fluctuating renewable energy in electricity systems, potentially storing energy in electric vehicle (EV) batteries or fuel cells, and applying load shifting by varying prices over time. It is concluded that very large efficiency gains are expected to emerge from digitalisation in the energy sector (Figure 6.18).

Section 9.9.2 in Chapter 9 concludes that the improved energy efficiency and falling costs in the building sector that could result from digitalisation could have rebound effects in increasing both energy consumption and comfort levels. Increasing GHG emissions could be the result, but if low-income consumers are given faster access to affordable energy, this could agree with the SDGs, making it desirable to integrate policies targeting mitigation.

Section 10.1.2 in Chapter 10 discusses how the sharing economy, which, for example, could be facilitated by ICT platforms, could influence both mitigation and the SDGs. On the one hand, sharing has the potential to save transport emissions, especially if EVs are supplied with decarbonised grid electricity. However, an increase in transport emissions could result from this if increasing demand and higher comfort levels are facilitated, for example, by making access to EVs relatively easy compared with mass transit. Another possible trade-off is that the supply of public transport services would be limited to the elderly and other user groups.

Green innovation in agriculture is another emerging area in which digitalisation is making huge progress. From the perspective of water provision, weather data can be used to predict rain amounts so that farmers can better manage the application of farm chemicals to minimise polluting aquifers and surface-water systems used for drinking water. Meanwhile, smart meters, on-site and remote sensors and satellite data connected to mobile devices allow real-time monitoring of crop-water and optimal irrigation requirements. On the supply side, remote tele-control systems and efficient irrigation technologies enable farmers to control and optimise the quantity and timing of water applications, while minimising the energy-consumption trade-offs of pressurised irrigation in both rural and urban agricultural contexts (Germer et al. 2011; Ruiz-Garcia et al. 2009).

Technology-driven precision agriculture, which combines geomorphology, satellite imagery, global positioning and smart sensors, enables enormous increases in efficiency and productivity. Taken together, these technologies provide farmers with a decision-support system in real time for the whole farm. Arguably, the world could feed the projected rise in population without radical changes to current agricultural practices if food waste can be minimised or eliminated. Digital technologies will contribute to minimising these losses through increased efficiencies in supply chains, better shipping and transit systems, and improved refrigeration.

In conclusion, in most cases digitalisation options may have both positive synergistic impacts on mitigation and the SDGs and some negative trade-offs. Energy-sector options are assessed primarily as having synergies, while some digitalisation options in transport could increase the demand for emission-intensive modes of transport. Digital platforms for the sharing economy could have both positive and negative impacts depending on the goods and services that are actually exchanged (Cross-Chapter Box 6 in Chapter 7). Options related to agriculture and the water-energy-food nexus (WEFN) could help manage resources more efficiently across sectors, which could create synergies. Digitalisation can also raise a number of ethical challenges according to (Clark et al. 2019). Wider public discussion of internet-based activities was accordingly recommended, including topics such as the negotiation of online consent and the use of data for which consent has not been obtained.

#### 17.3.3.7 Cross-sectoral Overview of Synergies and Trade-offs Between Climate Change Mitigation and the SDGs

Based on a qualitative assessment in the sectoral Chapters 6, 7, 8, 9, 10, and 11, Figure 17.1 below provides an overview of the most likely links between sectoral mitigation options and SDGs in terms of synergies and trade-offs. The general overview provided in the figure is supplemented by specific sector-by-sector comments on how the synergies and trade-offs mapped depend on the scale of implementation and the overall development context of places where the mitigation options are implemented. For some mitigation options these scaling and context-specific issues imply that there can be both synergies and trade-offs in relation to specific SDGs. In addition to the information provided in Figure 17.1, Supplementary Material Table 17.SM.1 includes the detailed background material provided by the sectoral chapters in terms of qualitative information for each of the synergies and trade-offs mapped.

The assessment of synergies and trade-offs presented in Figure 17.1 depends on the underlying literature assessed by the sectoral chapters. In cases where no information about the links between specific mitigation options and SDGs are indicated, this does not imply that there are no links, but rather that the links have not been assessed by the literature.

Most of the energy-sector options are assessed as having synergies with several SDGs, but there could be mixed synergies and trade-offs between SDG 2 (zero hunger) for wind and solar energy, and for hydropower due to land-use conflicts and fishery damage. Offshore wind could also have both synergies and trade-offs with SDG 14 (life below water) dependent on scale and implementation site, and it is emphasised that land-use should be coordinated with biodiversity concerns. Both wind and solar energy are assessed as having trade-offs with SDG 12 (responsible production and consumption) due to significant material consumption and disposal needs.

Geothermal energy is assessed as having synergies with SDG 1 (no poverty) due to energy access, and mixed synergies and trade-offs in relation to SDG 3 (good health and well-being) due to reduced air pollution, but with some risks in relation to water pollution, and in relation to SDG 6 (clean water and sanitation), if it is not well managed. Nuclear power is assessed as having synergies with SDG 3 (good health and well-being) due to reduced air pollution, but potential trade-offs in relation to SDG 6 (clean water and sanitation) due to high water consumption, and water consumption issues are also possible in relation to many of the other mitigation options in the energy sector. Synergies are identified in relation to SDG 12 (responsible production and consumption) for nuclear power due to low material consumption. CCUS has been assessed as having trade-offs in relation to SDG 1 (no poverty) due to high costs and SDG 6 (clean water and sanitation) due to high water consumption. Synergies are related to SDG 3 (good health and well-being), and to SDG 9 (industry, innovation and infrastructure) due to the facilitation of decarbonisation of industrial processes. Both synergies and trade-offs could arrive in relation to SDG 12 (responsible production and consumption), since some rare chemicals and other inputs could in some cases be used with large-scale applications.

Bioenergy use as a fuel is assessed as one of the energy-sector mitigation options with most synergies and trade-offs with the SDGs. There could be synergies with SDG 1 (no poverty), with SDG 8 (decent work and economic growth) and SDG 9 (industry, innovation and infrastructure). This option, however, if combined with CCS, can be expensive and can compromise SDG 1 (no poverty) due to the high costs involved.

Agriculture, forestry and other land use (AFOLU) mitigation options are very closely linked to the SDGs and offer both synergies and trade-offs, which in many cases are highly dependent on the scale of implementation. All the mitigation options included in Figure 17.1 are assessed as potentially having synergies with SDG 1 (no poverty), but trade-offs could also happen if large areas are used for biocrops and taken away from other activities, thus causing poverty, as well as in relation to food costs if healthier diets are made more expensive. In relation to SDG 2 (zero hunger), most of the mitigation options are assessed as being associated with both synergies and trade-offs. Trade-offs are particularly a risk with large-scale applications of afforestation projects, bioenergy crops and other land-hungry activities, which can crowd out food production.

SDG 3 (good health and well-being) can be supported by many mitigation options in the agriculture, forestry and food sectors, primarily due to the reduced environmental impacts, and the same is the case with SDG 14 (life below water) due to decreased nutrient loads, and SDG 15 (life on land) due to increased biodiversity, with the caveat however, that SDGs 14 and 15 could have both synergies and trade-offs dependent on land use. It is considered that there could be both synergies and trade-offs in relation to SDG 8 (decent work and economic growth) due to competition over land use related to the mitigation options reducing deforestation and reforestation and restoration, and the same is the case in relation to SDG 7 (affordable and clean energy) depending on the economic outcome of the mitigation options. Similarly, the mitigation option of reduced $CH_4$ and $N_2O$ emissions from agriculture are assessed as having mixed impacts on SDG 8 (decent work and economic growth), and SDG 9 (industry, innovation and infrastructure) depending on innovative food production. The mitigation options of reforestation and forest management are assessed as having mixed impacts on

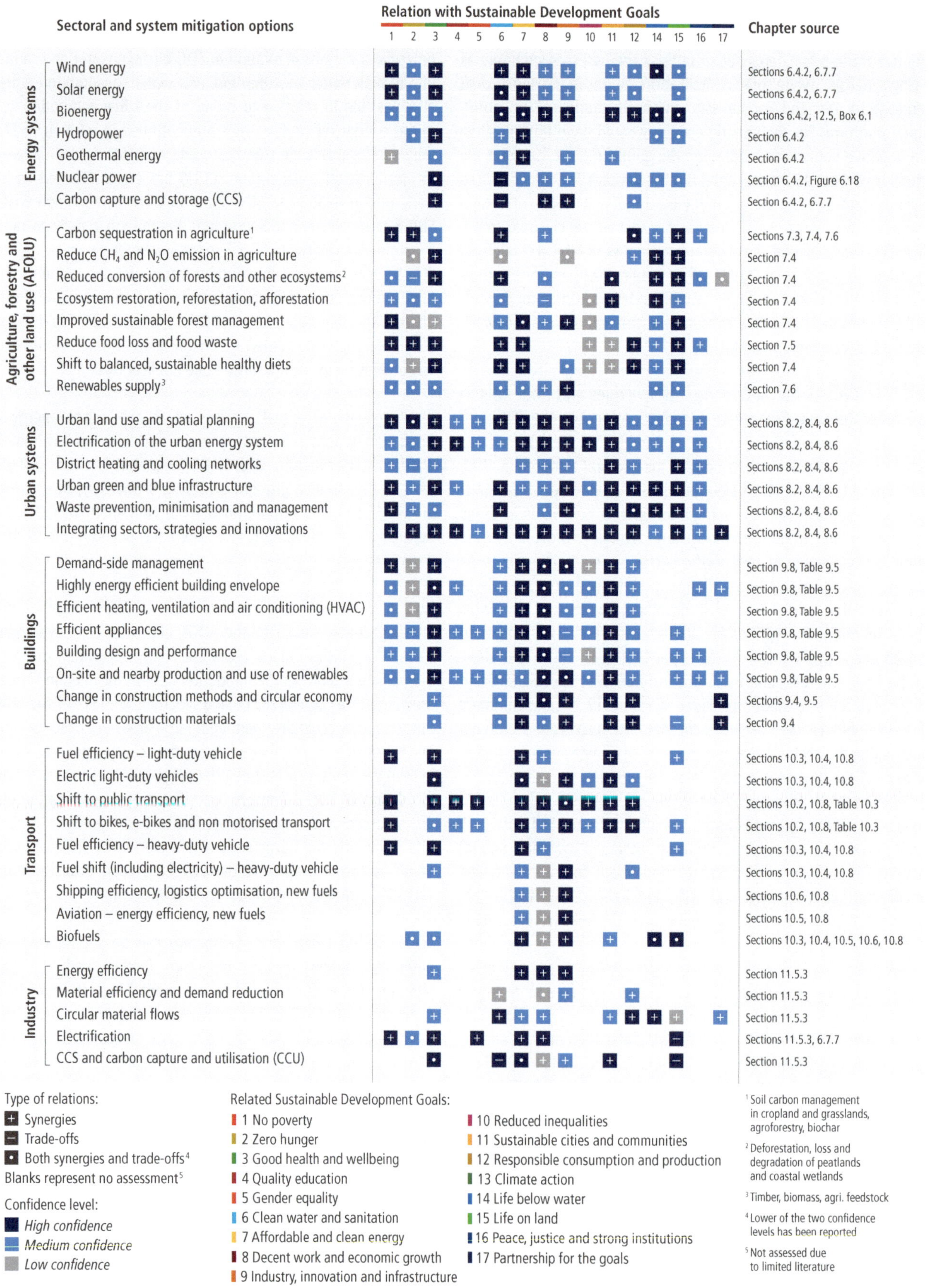


**Figure 17.1 | Trade-offs and synergies between sectoral mitigation options and the Sustainable Development Goals (SDGs).**

SDG 10 (reduced inequalities) depending on the involvement of local communities in projects. The assessment emphasises that the synergies and trade-offs of the mitigation options with the SDGs in this sector are very context- and scale-dependent, depending on how measures are carried out, for example, in relation to the enhanced production of renewables needed to replace fossil fuel-based products. If done on a massive scale and not adapted to local circumstances, there are adverse implications for food security, livelihoods and biodiversity.

All the urban mitigation options that have been assessed are considered to have synergies with the SDGs, and in a few cases both synergies and trade-offs are identified. In general, many links between mitigation options in the urban area and the SDGs have been identified in the literature. Urban land use and spatial planning, for example, can support SDG 1 (no poverty), and can also reduce vulnerability to climate change if integrated planning is undertaken, while access to food (SDG 2: zero hunger), and water (SDG 6: clean water and sanitation) can also be achieved if supported by integrated planning. Electrification, district heating, and green-and-blue infrastructure in urban areas are expected to have synergies with all the SDGs addressed by the reviewed studies.

Mitigation options like waste-prevention minimisation and management are also assessed as having many synergies with the SDGs, but trade-offs could depend on the application of air-pollution control technologies, and on the character of informal waste-recycling activities. The impacts of the possible synergies and/or trade-offs with the SDGs will change according to the specific urban context. Synergies and/or trade-offs may be more significant in certain contexts than others. Regarding the SDGs, urban mitigation can support shifting pathways of urbanisation towards sustainability. The feasibility of urban mitigation options is also malleable and can increase with more enablers. Strengthened institutional capacity that also supports the scale and coordination of the mitigation options can increase the synergies between urban mitigation options and the SDGs.

As for the urban mitigation options, the reviewed building-sector studies reveal a lot of links between mitigation and the SDGs. Highly efficient building envelopes are expected to have synergies with the SDGs in all cases except those with potential trade-offs in relation to SDG 10 (reduced inequalities). Many SDG synergies are also identified for the building design and performance, heating, ventilation and air conditioning, and efficient appliances mitigation options. However, some trade-offs could appear in relation to SDG 8 (decent work and economic growth) due to macroeconomic impacts of reduced energy consumption, decreasing prices and stranded investments. Similar issues related to the economic impacts of reduced energy demand are also highlighted for all the other mitigation options, including for the building sector. In relation to construction materials and the circular economy, some trade-offs have been identified in relation to SDG 6 (clean water and sanitation) and SDG 15 (life on land) related to the use of bio-based materials.

Consideration of the building sector highlights important context-specific issues related to synergies and trade-offs between mitigation options and SDGs such as the economic impacts (synergies and trade-offs) associated with reduced energy demand, resulting in lower energy prices, energy-efficiency investments, the fostering of innovation and improvements in labour productivity. Furthermore, the distributional costs of some mitigation policies may hinder the implementation of these measures. In this case, appropriate access policies should be designed to shield poor households efficiently from the burden of carbon taxation. Under real-world conditions, improved cookstoves have shown smaller, and in many cases limited, long-term health and environmental impacts than expected, as the households use these stoves irregularly and inappropriately, and fail to maintain them, so that their usage declines over time. Specific distributional issues are highlighted in relation to various cookstove programmes.

The mitigation options in the transportation sector are assessed as having synergies with SDG 1 (no poverty) and SDG 3 (good health and well-being) due to reduced environmental pollution, with exceptions in relation to pollution from biofuels and the risks of traffic accidents. Trade-offs are also mentioned in relation SDG 2 (zero hunger) where the production of biofuels takes land away from food production. Synergies are assessed in relation to SDG 7 (affordable and clean energy), SDG 8 (decent work and economic growth) and SDG 9 (industry, innovation and infrastructure). It is emphasised that some mitigation options, like the increased penetration of electric vehicles, require innovative business models, and that digitalisation and automatic vehicles will support the socio-economic structures that impede adoption of EVs and the urban structures that enable reduced car dependence. In conclusion, there is a need for investments in infrastructure that can support alternative fuels for light-duty vehicles (LDVs). The large-scale electrification of LDVs requires the expansion of low-carbon power systems, while charging or battery-swapping infrastructure is needed for some segments.

The mitigation options in the industrial sector have been assessed primarily as having synergies with meeting the SDGs. Several options, including energy efficiency, material recycling and electrification, are assessed has being able to create increased employment and business opportunities related to SDG 8 (decent work and economic growth), but material-efficiency improvements could reduce tax revenues. Electrification is assessed as having many synergies with SDGs, such as supporting SDG 1 (no poverty), SDG 2 (zero hunger), and SDG 3 (good health and well-being). CCS applied in industry is assessed as having synergies in terms of the control of non-$CO_2$ pollutants (such as sulphur dioxide), but increases in non-$CO_2$ pollutants (such as particulate matter, nitrogen oxide and ammonia). The conclusion is that 15–25% additional energy will be required by CCS technologies compared with conventional plants, implying that production costs could increase significantly. For the industrial sector in general, it is concluded that the balance between synergies and trade-offs between mitigation options and SDGs in industry depends on technology and the scale of the sharing of co-benefits across regions, as well as on the sharing of benefits in business models over whole value chains.

Thus, a number of cross-sectoral conclusions on synergies and trade-offs between mitigation options and the SDGs appear from the overview provided in Figure 17.1. There are many synergies in all sectors between mitigation options and the SDGs, and in a few cases there are also significant trade-offs that it is very important to address, since they can compromise major SDGs including SDG 1 (no poverty), SDG 2 (zero hunger), and in some cases SDG 14 (life below water) and SDG 15 (life on land). In particular, mitigation options in relation to land use, such as afforestation and reforestation and bioenergy crops, can in some cases imply trade-offs with access to food and local sharing of benefits, but synergies can also exist if proper land management and cross-sectoral policies take sustainable land use into account. The impacts and trade-offs for this sector are highly scale- and context-dependent, so the final outcome of mitigation policies should be considered in detail.

The urban systems and transportation could potentially achieve many synergies between mitigation policies and the SDGs, but integrated planning and infrastructure management are critical to avoiding trade-offs. Similarly, the buildings sector and industry have identified many potential synergies between mitigation options and the SDGs, but that raises issues related to the costs of new technologies, and in relation to households and buildings, important equity issues are emerging in relation to the ability of low-income groups to afford the introduction of new technologies. Altogether these cross-sectoral conclusions call for a need to support policies that aid coordination between different sectoral domains and that include context-specific assessments of the sharing of benefits and costs related to the implementation of mitigation options.



## 17.4 Key Barriers and Enablers of the Transition: Synthesising Results

This section provides a deep and broad synthesis of theory (Section 17.2) and evidence (Section 17.3) in order to identify the conditions that either enable or inhibit transitions to sustainable low-carbon futures. Following the literature on sustainability transitions (Cross-Chapter Box 12 in Chapter 16), the section finds that there is rarely any one single factor promoting or preventing such transitions. Rather, marked departures from business as usual typically involve several factors, including technological innovations, shifts in markets, concerted efforts by scientists and civil-society organisations to raise awareness of the costs of continued emissions, social movements, policies and governance arrangements, and changes in belief systems and values.

All of this comes together in a co-evolutionary process that has unfolded globally, internationally and locally over several decades (Hansen and Nygaard 2014; Rogge et al. 2017; Sorman et al. 2020), and that may be guided or facilitated by interventions that target leverage points in the underlying development path (Burch and Di Bella 2021; Leventon et al. 2021). While transitions necessarily follow context-specific trajectories, more general lessons can be drawn by comparing the empirical details with both system-level and narrower explanations of change.

Sections 17.2 and 17.3 show that transitions often face multiple barriers, including infrastructure lock-in, behavioural, cultural and institutional inertia (Markard et al. 2020), trade-offs between transitions and other social or political priorities (Chu 2016), cost and a reliable (and growing) supply of renewable-energy technologies and constituent materials (García-Olivares et al. 2018). Transitions away from fossil fuels and toward renewable energy-based systems, for instance, will require significant land-use decisions to avoid negative trade-offs with biodiversity and food security (Capellán-Pérez et al. 2017). Previous sections underline a related need to move beyond focusing on 'rational' assessments of the costs and benefits of policies and technologies to involve people at all levels in order to overcome these multiple barriers. A transition to a lower carbon system is unlikely to happen even if models find it technically feasible and cost-effective. Rather, achieving a transition requires breaking locked-in high-carbon technological trajectories, path dependencies and resistance to change from the industries and actors that are benefiting from the current system (Rogge et al. 2017). Lock-in effects may be weaker in sectors and policy areas where fewer technologies exist, potentially opening the door to innovations that embed the climate in broader sustainability objectives (e.g., technologies and innovations that support the integration of food, water and energy goals). Such effects may still happen when there are significant information asymmetries and high-cost barriers to action, as can occur when working across multiple climate and development-related sectors (Kemp and Never 2017).

However, the same conditions that may serve to impede a transition (i.e., organisational structure, behaviour, technological lock-in) can also be 'flipped' to enable it (Burch 2010; Lee et al. 2017), while the framing of policies that are relevant to the sustainable development agenda can also create a stronger basis and stronger policy support. The technological developments and broader cultural changes that may generate new social demands on infrastructure to contribute to sustainable development will involve a process of social learning and awareness building (Naber et al. 2017; Sengers et al. 2019). However, it is also important to note that strong shocks to these systems, including accelerated climate change impacts, economic crises and political changes, may provide crucial openings for accelerated transitions to sustainable systems through fundamental institutional changes (Broto et al. 2014). The global COVID-19 pandemic is one such shock that has sparked widespread conversations about recovery that is fundamentally more sustainable, equitable and resilient (McNeely and Munasinghe 2021). Key enabling conditions appear to be individual and collective actions, including leadership and education; financial, material, social and technical drivers that foster innovation; robust national and regional innovation systems that enhance technological diffusion (Wieczorek 2018); supportive policy and governance dynamics at multiple levels that permit both agility and coherence (Göpel et al. 2016); measures to recognise and address the challenges to equality inherent in the transition; and long-range, holistic planning that explicitly seeks synergies between climate change and sustainable development while avoiding trade-offs. The sections that follow seek to assess and integrate these key categories of the barriers to and enablers of an accelerated transition to sustainable development pathways.

### 17.4.1 Behavioural and Lifestyle Changes

Transitions toward more sustainable development pathways are both an individual and a collective challenge, requiring an examination of the role of values, attitudes, beliefs and structures that shape behaviour, and of the dynamics of social movements and education at the local community, regional and global levels. Labelling the carbon included in products, for example, could help the decision-making process and increase awareness and knowledge. Individual action suggests aggregated but uncoordinated actions taken by individuals, whereas collective sustainability actions involve coordination, a process of participation and governance that may ensure more efficient, equitable and effective outcomes. There is evidence that the behaviour of individuals and households are part of a more encompassing collective action (Section 5.4.1).

Indeed, individual actions are necessary but insufficient to deliver transformative mitigation, and it is suggested that this be coupled with collective actions to accelerate the transition to sustainable development (Dugast et al. 2019). Actors with conflicting interests will compete to frame mitigation technologies that either 'build or erode' the legitimacy of the technology, contested framing sites that can occur between incumbent and emerging actors or between actors in new but competing spaces (Rosenbloom et al. 2016). How narratives are built around desired development pathways and specific emerging technologies, as well as how local values are integrated into visions of the future, have relevance for how these experiments are managed and enabled to expand (Horcea-Milcu et al. 2020; Lam et al. 2020).

#### 17.4.1.1 Social Movements and Education

Sustainable development and deep decarbonisation will involve people and communities being connected locally through various means – including globally via the internet and digital technologies (Bradbury 2015; Scharmer 2018; Scharmer, C, Kaufer 2015) – in ways that form social fields that allow sustainability to unfold (Gillard et al. 2016), and that prompt other shifts in thinking and behaviour that are consistent with the 1.5°C goal (O'Brien 2018; Veciana and Ottmar 2018). Indeed, social movements serve to develop collective identities, foster collective learning and accelerate collective action ranging from energy justice (Campos and Marín-González 2020) (Section 17.4.5) to restricting fossil fuel extraction and supply (Piggot 2018). This does not apply only to adults: as seen in the 'Fridays for Future' marches, the young are also involving themselves politically (Peterson et al. 2019). Many initiatives have started with these marches, including 'science for future' and new forms of sustainability science (Shrivastava et al. 2020).

It was Theory-U (Scharmer 2018), building on the work of scholars such as Schein, Lewin and Senge) that inspired a so-called 'massive open online course' (MOOC) jointly initiated by the Bhutan Happiness Institute and German Technical Assistance (GIZ) in 2015, since when it has been developed further and adapted to transform business, society and self as one example of how social movements can go together with science and education. It brings together people from different professions, cultures and continents in shared discussions and practices of sustainability. It also included marginalised communities and is shifting towards more sustainable lifestyles in all sectors (Nikas et al. 2020), including climate action.

Moreover, approaches like the 'Art of Hosting' (Sandfort and Quick 2015) and qualitative research methods such as storytelling and first-person research, as well as second-person inquiries, for example (Scharmer, C, Kaufer 2015; Trullen and Torbert 2004; Varela 1999), have been employed to bridge differences in cultures and sciences, as well as to forge connections between those working on climate change and sustainable development. Likewise, experiential tools, simulations and role-playing games have been shown to increase knowledge of the causes and consequences of climate change, the sense of urgency around action and the desire to pursue further learning (Ahamer 2013; Eisenack and Reckien 2013; Hallinger et al. 2020; Rooney-Varga et al. 2020).

The results from these research communities reveal how experiential learning takes place and how it encourages bonding between people, society and nature. This can be achieved by going jointly and consciously into nature (Gioacchino 2019), by creating spaces for intensive-dialogue sessions with colleagues (Goldman-Schuyler et al. 2017) and forming, for example, a very practical u.lab hub, which involves following the MIT-u.lab course with a local community and is accompanied scientifically (Pomeroy and Oliver 2018). Others have pointed to social networks such as the 'transition initiative' (Hopkins 2010), eco-village networks (Barani et al. 2018), civil-society movements (Seyfang and Smith 2007) and intentional communities (Grinde et al. 2018; Veciana and Ottmar 2018) as ways of generating the shared understandings that are central to inner and outer transitions, as well as the broader development of social movements. In some cases, these networks build on principles like permaculture to encourage people to 'observe and interact', 'produce no waste' and 'design from patterns to details', not only in agriculture and gardening, but also in sustainable businesses and technologies to reduce $CO_2$ emissions (Ferguson and Lovell 2014; Lessem 2018).

A related line of inquiry involves education for sustainable development (ESD). This builds on the UNESCO programme, 'ESD for 2030', and involves core values like peace culture, valuing cultural diversity and living global citizenship. One of the core insights from research on ESC is lifelong education continuing outside the classroom, a lifelong learning process that involves sustained actions by all ages and social segments (Hume and Barry 2015) and achieving collaboration (Munger and Riemer 2012). Some authors have pointed to good levels of communication either directly or through the internet as the key to facilitating this learning (Sandfort and Quick 2015). Others have noted that transformative learning – that is, deepening the learning process – is critical because it helps to induce both shared awareness and collective actions (Brundiers et al. 2010; Singleton 2015; Wamsler and Brink 2018).

A final area of work points to the importance of moving toward the knowledge production that underpins awareness-raising (Pelling et al. 2015). The accumulation of applied knowledge is leading increasingly to the co-design of participatory research with local stakeholders who are investigating and transforming their own situations in line with climate action and sustainable development (Wiek et al. 2012; Abson et al. 2017; Fazey et al. 2018).

#### 17.4.1.2 Habits, Values and Awareness

Many of the cases that explore transitions to sustainable development point to ingrained habits, values and awareness levels as the most persistent yet least visible barriers to a transition. For example, in the transport sector, individuals can quickly become accustomed to personal vehicles, making it difficult for them to transition to sustainable, low-carbon modes of public transport. Demand for high-carbon transportation may also be locked-in, and habits reinforced, if low-cost housing (for instance) is not sufficiently served by more sustainable (i.e., mass transit, safe cycling and walking infrastructure) transportation options (Mattioli et al. 2020).

This is made all the more challenging because car-manufacturing 'incumbents' utilise information campaigns directed at the public, pursue lobbying and consulting with policymakers, and set technical standards that privilege the status quo and prevent the entry of more sustainable innovations (Smink et al. 2015; Turnheim and Nykvist 2019). Tools such as congestion pricing, however, have been shown to be effective in motivating the switch from single-occupancy vehicle use to public transit, thus improving air quality and reducing traffic delays in dense city centres (Baghestani et al. 2020).

Complicating the problem further is that even well-intentioned top-down programmes initiated by an external actor may in some cases ultimately hinder transformative change (Breukers et al. 2017). For instance, in Delhi, India, attempts to introduce ostensibly more sustainable bus rapid transit (BRT) systems failed in part due to an arguably top-down approach that had limited public support. It may nonetheless be difficult to win public support (Bachus and Vanswijgenhoven 2018), and even grassroots initiatives may themselves be contested and dynamic, making it difficult to generate the collective push to drive a bottom-up transition forward (Hakansson 2018).

However, dominant, top-down approaches and local, grassroots 'alternative' approaches and values do overlap and interact. For example, in Manchester, UK, dominant and alternative discourses interact with each other to create sustainable transformations through re-scaling (decentralising) energy generation, creating local engagement with sustainability, supporting green infrastructure to reduce costs, reclaiming local land, transforming industrial infrastructure and creating examples of sustainable living (Hodson et al. 2017).

Embedding local values in higher-level policy frameworks is also significant for forest communities in Nepal and Uganda. Even so, policy intermediaries are not confident that these values will be advanced due largely to an emphasis on carbon accounting and the distribution of benefits (Reckien et al. 2018). In this case, however, norm entrepreneurs were able to promote the importance of local values through the formation of grassroots associations, media campaigns and international support networks (Reckien et al. 2018).

### 17.4.2 Technological and Social Innovation

Individuals and organisations, like institutional entrepreneurs, can function to build transformative capacity through collective action (Brodnik and Brown 2018). The transition from a traditional water-management system to the Water Sensitive Urban Design (WSUD) model in Melbourne offers an illustration of how whole systems can be changed in an urban system.

Private-sector entrepreneurs also play an important role in fostering and accelerating transitions to sustainable development (Burch et al. 2016; Ehnert et al. 2018a; Dale et al. 2017). Sustainable entrepreneurs (SEs), for instance, are described as those who participate in the development of an innovation while simultaneously being rooted in the incumbent energy-intensive system. SE actors who have developed longer-term relationships, both formal and informal, with the public authorities can have considerable influence on developing novel renewable-energy technologies (Gasbarro et al. 2017). Institutions and policies that nurture the activities of sustainable entrepreneurs, in particular small- and medium-sized enterprises (Burch et al. 2016), can facilitate and strengthen transitions toward more sustainable development pathways, as can more fundamental adjustments to underlying business models, rather than relying only on incremental adjustments in the efficiency with which resources are used (Burch and Di Bella 2021).

The creation and growth of sustainable energy and clean-tech clusters enable economic development and transformation on regional scales. Such clusters can put pressure on incumbent technologies and rules to accelerate energy transitions. Successful clusters are nurtured by multi-institutional and multi-stakeholder actors building institutional support networks, facilitating collaboration between sectors and actors, and promoting learning and social change. Notably, regional economic clusters generate a buzz, which can have a strong influence on public acceptance, support and enthusiasm for socio-technical transitions (McCauley and Stephens 2012).

In Norway, many incumbent energy firms have already expanded their operations into the alternative-energy sector as both producers and suppliers (who often follow the lead of producers). Producers are responding to perceptions of larger-scale changes in the energy landscape (e.g., the green shift), along with uncertainties in their own sectors, and innovation can spill across actors in multiple sectors (Koasidis et al. 2020). While these firms are expanding out of self-interest, the expansion provides more legitimacy to new forms of technology and enables transfers of knowledge and resources to be introduced within this developing niche (Steen and Weaver 2017). Many large, well-established firms are pursuing sustainability agendas and opting for transparency with regard to their greenhouse gas emissions (Kolk et al. 2008; Guenther et al. 2016), supply-chain management (Formentini and Taticchi 2016) and sustainable technology or service development (Dangelico et al. 2016).

Experiments with the transition open up pathways that can lead to energy transitions on broader scales. Experiments can build capacity by developing networks and building bridges between diverse actors, leveraging capital from government funds, de-risking private- and public-sector investment, and acting as hubs for public education and engagement (Rosenbloom et al. 2018).

Material barriers and spatial dynamics (Coenen et al. 2012; Hansen and Coenen 2015) are other critical obstacles to innovation: often, infrastructure and built environments change more slowly than policies and institutions due to the inherently long lifespans of fixed assets (Turnheim and Nykvist 2019). The example of transport infrastructure in Ontario, Canada, illustrates the need to integrate climate change into these infrastructural decisions in the very short term to combat the risk of being left with unsustainable planning features long into the future, especially combustion engines, significant road networks and suburbanisation (Birch 2016).

### 17.4.3 Financial Systems and Economic Instruments

Market-oriented policies, such as carbon taxes and green finance, can promote low-carbon technology and encourage both private and public investment in enabling transitions. Policies that are currently being tested include loan guarantees for renewable-energy investments in Mali, policy insurance to reduce credit defaults within the feed-in tariff regime in Germany, or pledged funding to fully finance or partner private firms in order to advance renewable-energy projects (Roy et al. 2018a). However, there may be some limitations in using carbon pricing alone (rather than in combination with flexible regulations and incentives) where market failures hinder low-carbon investments (Campiglio 2016; World Bank 2019) and high political costs are incurred (Van Der Ploeg 2011).

Many forms of transformational change to energy systems are not possible when financial systems still privilege investing in unsustainable, carbon-intensive sectors. One of the root causes of the failure of traditional financial systems is the undervaluation of natural capital and unsettled property-right issues that are associated with it. The exclusion of proper rents for scarcities or for global and local externalities, including climate change, can undermine larger-scale changes to energy systems (Clark et al. 2018). But even smaller-scale low-carbon energy and infrastructure projects can fail to get off the ground if uncertainty and investment risk discourage project planning and bank-lending programmes (Bolton et al. 2016). The EU's previous actions regarding the 'shareholder maximisation norm' and non-binding measures have created path dependencies, limiting its flexibility in creating sustainable financial legislation. However, the Sustainable Finance Initiative and the Single Market may prove to be 'policy hotspots' in encouraging sustainable finance (Ahlström 2019). Taking advantage of these hotspots may be crucial in overcoming path dependencies and setting new ones in motion.

One possible positive turn in this regard is the acceleration in investing in the environment (impact and ESG) globally: for instance, there is evidence that some institutional investors are divesting from coal, potentially auguring well for the future (Richardson 2017). The encouragement of governance and policy reforms that could facilitate similar expansions of investment in sustainable firms and sectors (Clark et al. 2018; Owen et al. 2018) could contribute to the dynamic feedback that gives a transition lift and injects momentum into it. Also, the degrowth movement, with its focus on sustainability over profitability, has the potential to speed up transformations using alternative practices such as fostering the exchange of non-monetary goods and services if large numbers of stakeholders want to invest in these areas (Chiengkul 2017).

### 17.4.4 Institutional Capacities and Multi-level Governance

Capable institutions and multi-level governance often support the inter-agency coordination and stakeholder coalitions that drive sustainable transitions. Such institutions and governance arrangements are frequently required to formulate and implement the multi-sectoral policies that spur the adoption and scaling of innovative solutions to climate change and other sustainable development challenges. For example, such institutional and governance conditions have helped support the industrial policies that will be needed to spread renewables through the creation of domestic supply chains (Zenghelis 2020) or to pilot CDR methods (Quarton and Samsatli 2020).

However, government agencies with climate and other remits do not always work well together: the absence of coordination and consensus-building mechanisms can further deepen inter-agency conflicts that stall a transition. These challenges appear not only within but also between levels of decision-making. Studies of developing megacities, for instance, have found the lack of mechanisms promoting vertical cross-level integration to be a sizeable constraint on decarbonisation (Canitez 2019). Differences in perspectives across non-state actors can similarly frustrate transitions in areas such as green buildings (Song et al. 2020).

Here coordination complicates matters: coalition-building may require mutually reinforcing changes to institutions and policies. For example, decentralised renewable energy has made progress in Argentina, but consumer electricity subsidies give agencies and firms supporting conventional energy an advantage over those promoting renewable energy. Similarly, the lack of concrete guidance in green finance policies can deprive government agencies and other stakeholders of the information needed to balance ecological and financial goals (Wang and Zhi 2016). Many of these challenges can be particularly formidable in developing countries, where agencies lack sufficient financial and other capacities. A lack of government funds to cover ongoing maintenance costs along with resource shortages in rural locations can pose constraints on sustainable energy (Schaube et al. 2018).

Building inter-agency or multiple stakeholders is frequently challenging because of the mutually reinforcing interactions between institutions and ideas. The imperceptible embedding of long-standing development paradigms (such as 'grow now, clean up later') in agency rules and standard operating procedures can make changes to governance arrangements challenging. This is partly because these rules and procedures can also shape the interests of key decision-makers (e.g., the head of an environmental agency). For some, this suggests a need to look not just at changing prevailing ideas and interests, but also at broader institutional and governance arrangements (Kern 2011).

However, institutional and governance reforms can be more than a technical exercise. Political, economic and other power relations can lock-in dominant institutional and economic structures, making the integration of climate and sustainable development agendas exceedingly difficult. For example, though there have been recent reforms, the initial lack of early progress in Australia's energy transition is partly attributable to institutions of political economy being oriented to providing steady supplies of affordable fossil fuels (Warren et al. 2016).

This suggests that it is important to look closely at the pre-existing political economic system as well as the institutional context and capacities in assessing the prospects for transitions to sustainability. Furthermore, this is how existing institutions interact with ideas that often strengthen lock-ins. To illustrate, studies have shown that the status-quo orientations of leaders (including decision-makers' disciplinary backgrounds, world views and perceptions of risk) (Willis 2018), as well as the organisational culture and management paradigms within which they operate, affect the speed and ambitions of climate policies (Rickards et al. 2014).



Some studies have focused on factors that can break institutional and ideational lock-ins (Arranz 2017), while others have found that intentional higher-level (or, in the language of socio-technical transitions, 'landscape') pressures can be the destabilising force needed to move transitions forward (Falcone and Sica 2015). Often the state or national government (as the sovereign that determines how resources are used and allocated) can play a key role in destabilising incumbent energy regimes, a role that is significantly strengthened by public support (Arranz 2017; Avelino et al. 2016). However, this role is not limited to government insiders. In some contexts, regime outsiders have also played a pivotal role in destabilising regimes by combining persuasive narratives that gain market influence (Arranz 2017). Carbon-intensive luxury goods and services for wealthy consumers, for instance, especially if applied at the 'acceleration' phase of a transition, can help transform long-term social practices and behaviour and dissolve the 'structural imperative for growth' (Wiedmann et al. 2020). In a similar fashion, environmental taxes can remove 'locked-in' technology and place pressure on dominant regimes to become more sustainable (Bachus and Vanswijgenhoven 2018).

In many contexts, it is not multiple institutional and policy variables that come together to break unsustainable inertias. In South Korea, where the state was an initiator and enabler of change, the clean-energy transition took much longer than anticipated due to private-sector resistance. However, when policymakers focused on incorporating adaptive learning and flexibility into their decision-making, public- and private-sector interests gradually converged and joined with top-down policymaking to drive the transition forward (Lee et al. 2019). Thus, a political strategy can help align the interests and institutions needed to break lock-ins.

This becomes clear in studies that show that political coalitions can affect the speed of transitions (Hess 2014). These same studies show that incumbent industry coalitions are now competing with 'green' coalitions in terms of campaign spending over environmentally friendly ballot proposals (Hess 2014). Another way of shifting political-economic incentives is by offering a realistic exit strategy for incumbents, like interventions that provide long-term incentives for renewable-energy firms (de Gooyert et al. 2016; Hamman 2019).

Overall, the previous subsection suggests that complementary policies and institutions that simultaneously integrate across multiple sectors and scales and also alter political economic structures that lock in a carbon-intensive energy system are more likely to move a sustainable transition forward (Burch 2010). Yet, despite a trend in climate governance towards greater integration and inclusivity and certain other novel governance approaches, traditional approaches to governance and a tendency to incrementalism remain dominant (Holscher et al. 2019). Building the governance arrangements and capacities that prioritise climate change across all sectors and scales while destabilising entrenched interests and putting pressure on existing norms, rules and practices is still needed in many contexts (Holscher et al. 2019).

At least three themes require further research in the scholarship on the governance of transitions: (i) the role of coalitions in supporting and hindering acceleration; (ii) the role of feedback, through which policies may shape actor preferences, which in turn create stronger policies; and (iii) the role of broader contexts (political economies, institutions, cultural norms, and technical systems) in creating conditions for acceleration (Roberts et al. 2018). Importantly, these themes may serve as both barriers to and opportunities for transitions (ibid.).

### 17.4.5 Equity in a Just Transition

Energy justice, although increasingly being emphasised (Pellegrini-Masini et al. 2020), has been under-represented in the literature on sustainability and in debates on energy transitions, and it remains a contested term with multiple meanings (Green and Gambhir 2020). Energy justice includes affordability, sustainability, equality (accessibility for current and future households) and respect (ensuring that innovations do not impose further burdens on particular groups) (Fuso Nerini et al. 2019). Furthermore, it suggests that a just transition is a shared responsibility among countries that are making more rapid progress towards net-negative emissions and those economies that are focused on pressing development priorities related to improved health, well-being and prosperity (van den Berg et al. 2020).

Looking at climate change from a justice perspective means placing the emphasis on (i) the protection of vulnerable populations from the impacts of climate change; (ii) mitigating the effects of the transformations themselves, including easing the transition for those whose livelihoods currently rely on fossil fuel-based sectors;

and (iii) envisaging an equitable decarbonised world. Neglecting issues of justice risks a backlash against climate action generally, particularly from those who stand to lose from such actions (Patterson et al. 2018), and it will also have implications for the pace, scale and quality of the transition. Explicit interventions to promote sustainability transitions that integrate local spaces into the whole development process are necessary but not sufficient in creating a just transition (Breukers et al. 2017; Ehnert et al. 2018b).

Renewable energy transitions in rural, impoverished locations can simultaneously reinforce and disrupt local power structures and inequalities. Policy interventions to help the most impoverished individuals in a community gain access to the new energy infrastructure are critical in ensuring that existing inequalities are not reinforced. Individuals who are empowered by energy development projects can influence the onward extension of sustainable energy to other communities (Ahlborg 2017). In Denmark in the 1970s, for example, grassroots windmill cooperatives opened a pathway to the creation of one of the world's largest wind-energy markets. The unique dynamics of grassroots-led changes mean that new technologies and low-carbon initiatives develop strong foundations by being designed, tested and improved in the early stages with reference to the socio-political contexts in which they will grow later (Ornetzeder and Rohracher 2013).

Intersectional theory can shine a light on the hidden costs of resource extraction, as well as renewable-energy development (see, for instance, (Chatalova and Balmann 2017), which go beyond environmental or health risks to include the socio-cultural impacts on both communities adjacent to these sites and those who work in them (Daum 2018). Indeed, development decisions often do not properly integrate the burdens and risks placed on marginalised groups, such as indigenous peoples, while risk assessments tend to reinforce existing power imbalances by failing to differentiate between how benefits and risks might impact on certain groups (Healy et al. 2019; Kojola 2019). In some cases, such as the deployment of small-scale solar power in Tanzania by a non-profit organisation, an explicit gender lens on the impacts of energy poverty revealed the significant socio-economic benefits of improving access to renewable energy (Gray et al. 2019).

### 17.4.6 Holistic Planning and the Nexus Approach

Poor sectoral coordination and institutional fragmentation have triggered a wide range of unsustainable uses of resources and threatened the long-term sustainability of food, water and energy security (Rasul 2016). Greater policy coherence among the three sectors is critical to moving to a sustainable and efficient use of resources (United Nations 2019), given that political ambition, values, the energy mix, infrastructure and innovation capacities collectively shape transition outcomes (Neofytou et al. 2020). Capacity- and coalition-building, particularly among sub-national and non-state actors (e.g., non-governmental organisations) is a particularly important enabler of greater coherence (Bernstein and Hoffmann 2018). The nexus approach, a systems-based methodology that focuses attention on the many ways in which natural resources are deeply interwoven and mutually interdependent, can strengthen coordination and help to avoid maladaptive pathways (Cremades et al. 2016).

A major shift is required in the decision-making process in the direction of taking a holistic view, developing institutional mechanisms to coordinate the actions of diverse actors and strengthening complementarities and synergies (Nikas et al. 2020; Rasul 2016). Currently, nexus approaches have moved from purely conceptual arguments to application and implementation. (Liu et al. 2018) suggest the need for a systematic procedure and provide perspectives on future directions. These include expanding nexus frameworks that take into account interaction linkages with the SDGs, incorporating overlooked drivers and regions, diversifying nexus toolboxes and making these strategies central to policymaking and governance in integrating and implementing the SDGs.

In respect of processes, (Seyfang and Haxeltine 2012) found a lack of realistic and achievable expectations among both members (internally) and the wider public (externally), which hampers the acceleration of transitions. This movement could concentrate strategically on developing and promoting short-term steps towards shared long-term visions, including clearly identifiable goals and end-points. Sustainability science must link research on problem structures with a solutions-oriented approach that seeks to understand, conceptualise and foster experiments in how socio-technical innovations for sustainability develop, are diffused and are scaled up (Miller et al. 2014).

Various strategies and processes have been explored that might facilitate the translation of barriers into enablers, thus accelerating transitions to sustainable development. Common themes include frequent monitoring and system evaluation to reveal the barriers in the first place, the collaborative co-creation and envisioning of pathways toward sustainable development, ambitious goal-setting, the strategic tackling of sources of path dependence or inertia, iterative evaluations of progress and risk management, adaptive management and building in opportunities for agile course-correction at multiple levels of governance (Burch et al. 2014; Halbe et al. 2015). Given the political infeasibility of stable, long-term climate policies, the better choice may be to embrace uncertainty in specific policies but entrench the low-carbon transition as the overarching goal. Framing climate policy too narrowly, rather than taking a more holistic, sustainable development-oriented approach, may tie success to single policies, rather than allowing for system-wide change.

Decarbonisation may be encouraged by embedding the transition in a broader socio-economic agenda, focusing on constructing social legitimacy to justify the transformation, encouraging municipalities with a material interest in the transition and reforming institutions to support the long-term transition goals (Rosenbloom et al. 2019). In jurisdictions where climate and energy policy have been integrated and harmonised, such as the UK, progress has been made in transitioning to sustainable energy (Warren et al. 2016).

Developing countries that are rich in fossil fuels now have an opportunity to reset their development trajectories by focusing on those opportunities that will offer resilient development in land-use change, low-carbon energy generation and not least more efficient resource-planning (UNDRR 2019). Resource-rich developing countries can choose an alternative pathway by deciding to monetise carbon capital and diversifying away from the high-carbon aspects of risk. Countries rich in hydrocarbons can diversify their energy mix and maximise their renewable-energy potential. For instance, Namibia, a net importer of electricity, is seeking to reduce its current dependence on hydrocarbons by promoting solar energy. The government has issued permits allowing independent power producers (IPPs) to sell directly to consumers, thus ending the monopoly hitherto enjoyed by the state utility company NamPower (Kruger et al. 2019).

Cities are important spaces where the momentum to achieve low-carbon transitions can be built (Burch 2010; Holscher et al. 2019; Shaw et al. 2014), especially where centralised energy structures and national governance and politics are posing deep-rooted challenges to change (Dowling et al. 2018; Meadowcroft 2011). Cities can enter networks and partnerships with other cities and multi-level actors, spaces that are important for capacity-building and accelerating change (Dale et al. 2020; Heikkinen et al. 2019; Westman et al. 2021).

Addressing the uncertainties and complexities associated with locally, regionally and nationally sustainable development pathways requires creative methods and participatory processes. These may include powerful visualisations that make the implications of climate change (and decarbonisation) clear locally (Shaw et al. 2014; Sheppard et al. 2011), other visual aids or 'progress wheels' that effectively communicate the relevant contexts (Glaas et al. 2019), storytelling and mapping, and both analogue and digital games (Mangnus et al. 2019).

## 17.5 Conclusions

This chapter has been concerned to assess the opportunities and challenges for acceleration *in the context of sustainable development*. As such, many of the claims reviewed involve not only increasing the speed of the transition but also ensuring that it is just, equitable and delivers a wider range of environmental and social benefits. A sustainability transition requires removing the underlying drivers of vulnerability and high emissions (quality and depth) while aligning the interests of different communities, regions, sectors, stakeholders and cultures (scale and breadth).

Interest in a sustainability transition has grown steadily over the history of the IPCC and of climate and related policy processes. That interest hit a high point in 2015 with the Paris Agreement and the UN 2030 Agenda for Sustainable Development and its 17 SDGs. It has continued to remain high as countries have issued NDCs on climate change, VNRs on the SDGs and, in some instances, integrated climate and SDG plans (or similarly themed integrated actions, e.g., circular economy plans). Interest has also gained momentum as local governments, businesses and other stakeholders have followed suit with climate change- or SDG-related plans.

Implementing many of the recent pledges, however, has proved challenging. Part of the challenge is a need to address everything from public policies and prevailing technologies to individual lifestyles and social norms, to governance arrangements and institutions with associated political economy implications. These factors can lock-in development pathways and prevent transitions from gathering the momentum needed for large-scale transformations of socio-economic systems. Another consideration is that transition pathways are likely to vary across and within countries due to different development levels, starting points, differential vulnerabilities, capacities, agencies, geographies, power dynamics, political economies, ecosystems and other contextual factors.

Even with this diversity, prominent lines of economic, institutional, psychological and systems thinking have reflected on interventions that can enable transitions. Because these disciplines often focus on different levels of analysis and draw upon diverse analytical methods and empirical evidence, the recommended interventions also tend to vary. For instance, economic arguments often point to the need for targeted regulation or investments, institutional claims centre on multi-level governance reforms, and psychology encourages participation to change mindsets and social norms. Systems-level perspectives offer a useful frame for bringing together these views, but may not capture the richness and details of them treated separately. Greater inter- and transdisciplinary research is needed to integrate the more focused interventions and show how they work together in a system. Such research will be particularly important for working on the concern running through these studies: strengthening synergies between climate and the broader sustainable development agenda.

National and sub-national, sectoral and cross-sectoral, short- and long-term transition studies have assessed the links between sustainable development and mitigation policies and synergies and the trade-offs between the different policy domains. Some general conclusions can be drawn on synergies and trade-offs, despite the actual impacts of policy implementation depending on scale, context and the development starting point.

From a cross-sectoral perspective, it can be concluded that the AFOLU sector offers many low-cost mitigation options with synergetic SDG impacts, which, however, can also create trade-offs between land use for food, energy, forest and biodiversity. Some options can help to mitigate such trade-offs, like agricultural practices, forest conservation and soil carbon sequestration. Lifestyle changes, including dietary changes and reduced food waste, could jointly support the SDGs and mitigation. Industry also offers several mitigation options with SDG synergies, for example, related to energy efficiency and the circular economy. Some of the renewable-energy options in industry could indicate some trade-offs in relation to land use, with implications for food- and water security and costs. Cities provide a promising basis for implementing mitigation with SDG synergies, particularly if urban planning, transportation, infrastructure and settlements are coordinated jointly. Similarly, studies of the building sector have identified many synergies between the SDGs and mitigation, but there are issues related to the costs of new technologies. Also, in relation to households and buildings, important equity issues emerge due to the ability of low-income groups to afford the introduction of new technologies. Altogether these cross-sectoral conclusions create a need for policies to address both synergies and trade-offs, as well as for coordination between different sectoral domains. Context-specific assessments of synergies and trade-offs are here important, as is sharing the benefits and costs associated with mitigation policies.

Several opportunities for creating SDG synergies and avoiding trade-offs have also been identified in relation to integrated adaptation and mitigation policies. The AFOLU sector has a large potential for integrating adaptation and mitigation policies related to agriculture, bioenergy crops, forestry and water use. As was concluded for mitigation options, integrated adaptation and mitigation policies also entail the risks of creating trade-offs in relation to food, water, energy access and biodiversity. There are several potentially strong links between climate change adaptation in industry and climate change adaptation more generally. Various supply chains can be affected by climate change, and mitigation options related to energy and water supply can be disrupted by climate events, implying that great benefits may come from integrating adaptation in industrial planning efforts. Adaptation options in industry can imply increasing the demand for packaging materials such as plastics and for access to refrigeration, which are also major sources of GHG emissions, which then would require further mitigation options. Mitigation and the co-benefits of adaptation in urban areas in relation to air quality, health, green jobs and equality issues can in most cases be synergetic and can also support the SDGs. One exception are compact cities, with their trade-offs between mitigation and adaptation because decreasing urban sprawl can increase the risks of flooding and heat stress. Detailed mapping of mitigation and adaptation in urban areas shows that there are many, very close interactions between the two policy domains and that coordinated governance across sectors is therefore called for.

Meeting the ambitions of the Paris Agreement will require phasing out fossil fuels from energy systems, which is technically possible and is estimated to be relatively low in cost. However, studies also show that replacing fossil fuels with renewables can have major synergies and trade-offs with a broader agenda of sustainable development if a balance is established in relation to land use, food security and job creation (McCollum et al. 2018). Furthermore, the transition to low-emission pathways will require policy efforts that also address the emissions locked-in to existing infrastructure, like power plants, factories, cargo ships and other infrastructure already in use: for example, today coal-fired power plants account for 30% of all energy-related emissions. Thus, even though the transition away from fossil fuels is desirable and technically feasible, it is still largely constrained by existing fossil fuel-based infrastructure and the existence of stranded investments. The 'committed' emissions from existing fossil fuel infrastructure may consume all the remaining carbon budget in the 1.5°C scenario or two thirds of the carbon budget in the 2°C scenario.

Stranded hydrocarbon assets, including hydrocarbon resources and the infrastructure from which they are produced, and investments made in exploration and production activities, are likely to become unusable, lose value or may end up as liabilities before the end of the anticipated economic lifetime. This phenomenon is rapidly becoming a global reality as social norms change and the pressure to reduce emissions mounts. Energy and other forms of structural inequities are likely to make the transition planning more challenging, especially given stranded assets.

Countries dependent on fossil fuel income will need to forego these revenues to keep well within the Paris Agreement requirements and align with the rapidly growing divestment movement. Climate injustice, energy poverty and COVID-19 have reduced the space and manoeuverability for developing countries to innovate and use surplus funds to procure new and clean technologies. A rising debt burden already hamstrings many. Decisions on how to spend the remaining carbon budget and who has the right to decide on what to do with existing fossil fuels reflect the complexity of the transition and its non-linear character. Given the asymmetrical dimension of energy production, distribution and use, it is likely that stranded assets will have implications for oil-producing countries, especially for early producers who perceive that new-found oil and gas will open doors to new forms of prosperity.

While the transitional drivers are not in place in some developing countries, that is, technology, infrastructure, knowledge, and finance, among others, investing in new forms of renewable energy for the land, energy, or water sectors will see the emergence of a more diversified economy and one less vulnerable to carbon and other exogenous risks. The transition away from fossil fuels will come with hard choices. Still, these choices can enable a sustainable development world and reduce the many asymmetries and injustices inherent in the current system, not least the gaping energy disparities that divide the developed and the developing world.

Equality and justice are central dimensions of transitions in the context of sustainable development. Viewing climate change through the lens of justice requires a focus on the protection of vulnerable populations from the impacts of climate change, addressing the unequal distribution of the costs and consequences of the transitions themselves, including for those whose livelihoods are rooted in fossil fuel-based sectors, and developing more creative and participatory processes for envisioning an equitable decarbonised world. Neglecting issues of justice will have implications for the pace, scale and quality of the transition.

Ultimately, the evidence demonstrates that there is rarely any one single factor promoting or preventing transitions. A constellation of elements come into play, including technological innovations, shifts in markets, social and behavioural dynamics, and governance arrangements. Indeed, transitions require an examination of the role of values, attitudes, beliefs and the structures that shape behaviour, as well as the dynamics of social movements and education at multiple levels. Likewise, technological and social innovation both play an important role in enabling transitions, highlighting the importance of multi-institutional and multi-stakeholder actors building institutional support networks, facilitating collaboration between sectors and actors, and promoting learning and social change. Financial tools and economic instruments are crucial enablers, since many forms of transformational change to energy systems are not possible when financial systems still privilege investing in unsustainable, carbon-intensive sectors. These instruments are deployed within the context of the multi-level governance of climate change, which suggests the importance of complementary policies and institutions that simultaneously integrate across multiple sectors and scales to address the multiple sources of lock-in that are shaping the current carbon-intensive energy system. Systems-oriented approaches, which holistically address the intersections among climate, water and energy (for instance), have significant potential to reveal and help avoid trade-offs, foster experimentation, and deliver a range of co-benefits on the path towards sustainable development.

Frequently Asked Questions (FAQs)

### FAQ 17.1 | Will decarbonisation efforts slow or accelerate sustainable development transitions?

Sustainable development offers a comprehensive pathway to achieving ambitious climate change mitigation goals. Sustainable development requires the pursuit of synergies and the avoidance of trade-offs between the economic, social and environmental dimensions of development. It can thus provide pathways that accelerate progress towards ambitious climate change mitigation goals. Factoring in equality and distributional effects will be particularly important in the pursuit of sustainable policies and partnerships, and in accelerating the transition to sustainable development. Using climate change as a key conduit can only work if synergies across sectors are exploited and if policy implementation is supported by national and international partnerships.

The speed, quality, depth and scale of the transition will depend on the developmental starting point, that is, on explicit goals as well as the enabling environment consisting of individual behaviour, mindsets, beliefs and actions, social cohesion, governance, policies, institutions, social and technological innovations, and so on. The integration of both climate change mitigation and adaptation policies in sustainable development is also essential in the establishment of fair and robust transformation pathways.

### FAQ 17.2 | What role do considerations of justice and inclusivity play in the transition towards sustainable development?

Negative economic and social impacts in some regions could emerge as a consequence of ambitious climate change mitigation policies if these are not aligned with key sustainable development aspirations such as those represented by the Sustainable Development Goals (SDGs) on 'no poverty, energy-, water- and food access', and so on, which could in turn slow down the transition process. Nonetheless, many climate change mitigation policies could generate incomes, new jobs and other benefits. Capturing these benefits could require specific policies and investments to be targeted directly towards including all parts of society in the new activities and industries created by the climate change mitigation policies, and that activities that are reduced in the context of transitions to a low-carbon future, including industries and geographical areas, are seeing new opportunities. Poor understanding of how governance at multiple levels can meet these challenges to the transition may fail to make significant progress in relation to national policies and a global climate agreement. It may therefore either support or weaken the climate architecture, thus constituting a limiting factor.

### FAQ 17.3 | How critical are the roles of institutions in accelerating the transition and what can governance enable?

Institutions are critical in accelerating the transition towards sustainable development: they can help to shape climate change response strategies in terms of both adaptation and mitigation. Local institutions are the custodians of critical adaptation services, ranging from the mobilisation of resources, skills development and capacity-building to the dissemination of critical strategies. Transitions towards sustainable development are mediated by actors within particular institutions, the governance mechanisms they use as implementing tools and the political coalitions they form to enable action. Patterns of production and consumption have implications for a low-carbon development, and many of these patterns can act as barriers or opportunities towards sustainable development. Trade policies, international economic issues and international financial flows can positively support the speed and scale of the transition; alternatively, they can have negative impacts on policies that may inhibit the process. Nonetheless, contextual factors are a fundamental part of the change process, and institutions and their governance systems provide pathways that can influence contextual realities on the ground. For instance, politically vested interests may lead powerful lobby groups or coalition networks to influence the direction of the transition, or they could put pressure on a given political elite through the imposition of regulatory standards, taxation, incentives and policies that may speed or delay the transition process. Civil-society institutions, such as NGOs or research centres, can act as effective governance 'watchdogs' in the transition process, particularly when they exercise a challenge function and question government actions in respect of transitions related to sustainable development.

## References


Abson, D.J. et al., 2017: Leverage points for sustainability transformation. *Ambio*, **46(1)**, 30–39, doi:10.1007/s13280-016-0800-y.

Addison, T. and A.R. Roe, 2018: *Extractive Industries*. 1st ed. Oxford University Press, Oxford, UK, 766 pp.

Adger, W.N., J. Barnett, K. Brown, N. Marshall, and K. O'Brien, 2013: Cultural dimensions of climate change impacts and adaptation. *Nat. Clim. Change*, **3(2)**, doi:10.1038/nclimate1666.

Ahamer, G., 2013: Game, Not Fight: Change Climate Change! *Simul. Gaming*, **44(2–3)**, 272–301, doi:10.1177/1046878112470541.

Ahlborg, H., 2017: Towards a conceptualization of power in energy transitions. *Environ. Innov. Soc. TRANSITIONS*, **25**, 122–141, doi:10.1016/j.eist.2017.01.004.

Ahlström, H., 2019: Policy hotspots for sustainability: Changes in the EU regulation of sustainable business and finance. *Sustainability*, **11(2)**, 499, doi:10.3390/su11020499.

Akimoto, K. et al., 2012: Consistent assessments of pathways toward sustainable development and climate stabilization. *Nat. Resour. Forum*, **36(4)**, doi:10.1111/j.1477-8947.2012.01460.x.

Alarcón, D. and R. Vos, 2015: Introduction: The Imperative of Sustainable Development. In: *Technology and Innovation for Sustainable Development* [Vos, R. and D. Alarcón, (eds.)]. *The United Nations Series on Development*, Bloomsbury Academic, London, UK, pp. 1–14.

Alexandratos, N. and J. Bruinsma, 2012: *World Agriculture towards 2030/2050*.,FAO, Rome, Italy, 154 pp.

Altieri, K.E. et al., 2016: Achieving development and mitigation objectives through a decarbonization development pathway in South Africa. *Clim. Policy*, **16(1)**, 78–91, doi:10.1080/14693062.2016.1150250.

Amanuma, N. et al., 2018: *Governance for Integrated Solutions to Sustainable Development and Climate Change: From Linking Issues to Aligning Interests*. [Zusman, E. and N. Amanuma, (eds.)]. IGES, Hayama, Japan, 140 pp.

Andrianto, A., H. Komarudin, and P. Pacheco, 2019: Expansion of oil palm plantations in Indonesia's frontier: Problems of Externalities and the Future of Local and Indigenous Communities. *Land*, **8(4)**, 16, doi:10.3390/land8040056.

Andrić, I., M. Koc, and S.G. Al-Ghamdi, 2019: A review of climate change implications for built environment: Impacts, mitigation measures and associated challenges in developed and developing countries. *J. Clean. Prod.*, **211**, 83–102, doi:10.1016/j.jclepro.2018.11.128.

Andries, H.F. et al., 2017: Global and regional abatement costs of Nationally Determined Contributions (NDCs) and of enhanced action to levels well below 2°C and 1.5°C. *Environ. Sci. Policy*, **71**, 30–40, doi.org/10.1016/j.envsci.2017.02.008.

Ansari, D. and F. Holz, 2020: Between stranded assets and green transformation: Fossil-fuel-producing developing countries towards 2055. *World Dev.*, **130**, 18, doi:10.1016/j.worlddev.2020.104947.

Antwi-Agyei, P., A.J. Dougill, T.P. Agyekum, and L.C. Stringer, 2018a: Alignment between nationally determined contributions and the sustainable development goals for West Africa. *Clim. Policy*, **18(10)**, 1296–1312, doi:10.1080/14693062.2018.1431199.

Antwi-Agyei, P., A.J. Dougill, L.C. Stringer, and S.N.A. Codjoe, 2018b: Adaptation opportunities and maladaptive outcomes in climate vulnerability hotspots of northern Ghana. *Clim. Risk Manag.*, **19**, 83–93, doi.org/10.1016/j.crm.2017.11.003.

Arnell, N.W. and B. Lloyd-Hughes, 2014: The global-scale impacts of climate change on water resources and flooding under new climate and socio-economic scenarios. *Clim. Change*, **122(1–2)**, 127–140, doi:10.1007/s10584-013-0948-4.

Arranz, A.M., 2017: Lessons from the past for sustainability transitions? A meta-analysis of socio-technical studies. *Glob. Environ. Change Policy Dimens.*, **44**, 125–143, doi:10.1016/j.gloenvcha.2017.03.007.

Arrow, K. et al., 2004: Are we consuming too much? *J. Econ. Perspect.*, **18(3)**, 147–172, doi:10.1257/0895330042162377.

Asara, V., I. Otero, F. Demaria, and E. Corbera, 2015: Socially sustainable degrowth as a social–ecological transformation: repoliticizing sustainability. *Sustain. Sci.*, **10(3)**, 375–384, doi:10.1007/s11625-015-0321-9.

Austin, K.G., A. Schwantes, Y. Gu, and P.S. Kasibhatla, 2019: What causes deforestation in Indonesia? *Environ. Res. Lett.*, **14**, doi:10.1088/1748-9326/aaf6db.

Avelino, F., J. Grin, B. Pel, and S. Jhagroe, 2016: The politics of sustainability transitions. *J. Environ. Policy Plan.*, **18(5)**, 557–567, doi:10.1080/1523908X.2016.1216782.

Bachus, K. and F. Vanswijgenhoven, 2018: The use of regulatory taxation as a policy instrument for sustainability transitions: old wine in new bottles or unexplored potential? *J. Environ. Plan. Manag.*, **61(9)**, 1469–1486, doi:10.1080/09640568.2017.1358155.

Baghestani, A., M. Tayarani, M. Allahviranloo, and H.O. Gao, 2020: Evaluating the Traffic and Emissions Impacts of Congestion Pricing in New York City. *Sustainability*, **12(9)**, 3655, doi:10.3390/su12093655.

Bai, C. et al., 2020: Will income inequality influence the abatement effect of renewable energy technological innovation on carbon dioxide emissions? *J. Environ. Manage.*, **264(15)**, 110482, doi:10.1016/J.JENVMAN.2020.110482.

Banks, S., 2007: The Heart and Soul of Transition—Creating a Low Carbon Future with Psychological and Spiritual Awareness. *Self Soc.*, **35(2)**, 5–14, doi:10.1080/03060497.2007.11083972.

Banuri, T., J. Weyant, G. Akumu, A. Najam, L.P. Rosa, S. Rayner, W. Sachs, R. Sharma, and G. Yohe, 2001: Setting the Stage: Climate Change and Sustainable Development. In: *Climate Change 2001: Mitigation. Contribution of Working Group III to the Third Assessment Report of the Intergovernmental Panel on Climate Change* [Banuri, T., T. Barker, I. Bashmakov, K. Blok, D. Bouille, R. Christ, O. Davidson, J. Edmonds, K. Gregory, M. Grubb, K. Halsnaes, T. Heller, J.-C. Hourcade, C. Jepma, P. Kauppi, A. Markandya, B. Metz, W. Moomaw, J.R. Moreira, T. Morita, N. Nakicenovic, L. Price, R. Richels, J. Robinson, H.H. Rogner, J. Sathaye, R. Sedjo, P. Shukla, L. Srivastava, R. Swart, R. Toth, and J. Weyant (eds.)] Cambridge University Press, Cambridge, UK, pp. 47–114.

Barani, S., A.H. Alibeygi, and A. Papzan, 2018: A framework to identify and develop potential ecovillages: Meta-analysis from the studies of world's ecovillages. *Sustain. Cities Soc.*, **43**, 275–289, doi:10.1016/j.scs.2018.08.036.

Barker, T., E. Alexandri, J.F. Mercure, Y. Ogawa, and H. Pollitt, 2016: GDP and employment effects of policies to close the 2020 emissions gap. *Clim. Policy*, **16(4)**, 393–414, doi:10.1080/14693062.2014.1003774.

Baron, R. and F. David, 2015: *Divestment and Stranded Assets in the Low-carbon Transition*. OECD, Paris, France, 26 pp. https://www.oecd.org/sd-roundtable/papersandpublications/Divestment and Stranded Assets in the Low-carbon Economy 32nd OECD RTSD.pdf (Accessed July 10, 2021).

Bataille, C. et al., 2016: The need for national deep decarbonization pathways for effective climate policy. *Clim. Policy*, **16(No. S1)**, S7–S26, doi:10.1080/14693062.2016.1173005.

Bateman, I.J. and G.M. Mace, 2020: The natural capital framework for sustainably efficient and equitable decision making. *Nat. Sustain.*, **3(10)**, 776–783, doi:10.1038/s41893-020-0552-3.

Bauer, N., C. McGlade, J. Hilaire, and P. Ekins, 2018: Divestment prevails over the green paradox when anticipating strong future climate policies. *Nat. Clim. Change*, **8(2)**, 130–134, doi:10.1038/s41558-017-0053-1.

Beg, N., 2002: Linkages between climate change and sustainable development. *Clim. Policy*, **2(2–3)**, 129–144, doi:10.1016/s1469-3062(02)00028-1.

Behuria, P., 2020: The politics of late late development in renewable energy sectors: Dependency and contradictory tensions in India's


17

National Solar Mission. *World Dev.*, **126**, 104726, doi.org/10.1016/j.worlddev.2019.104726.

Bellinson, R. and E. Chu, 2019: Learning pathways and the governance of innovations in urban climate change resilience and adaptation. *Environ. Policy Plan*, **21(1)**, 76–89, doi.org/10.1080/1523908X.2018.1493916.

Bergman, N., 2018: Impacts of the Fossil Fuel Divestment Movement: Effects on Finance, Policy and Public Discourse. *Sustainability*, **10(7)**, 2529, doi:10.3390/su10072529.

Berkhout, F. et al., 2010: Sustainability experiments in Asia: Innovations shaping alternative development pathways? *Environ. Sci. Policy*, **13(4)**, 261–271, doi:10.1016/j.envsci.2010.03.010.

Bernstein, S. and M. Hoffmann, 2018: The politics of decarbonization and the catalytic impact of subnational climate experiments. *Policy Sci.*, **51(2)**, 189–211, doi:10.1007/s11077-018-9314-8.

Berry, P et al., 2015: Cross-sectoral interactions of adaptation and mitigation measures. *Clim. Change*, **128**, 381–393, doi.org/10.1007/s10584-014-1214-0.

Bertram, C. et al., 2018: Targeted policies can compensate most of the increased sustainability risks in 1.5°C mitigation scenarios. **13(6)**, 64038, doi:10.1088/1748-9326/aac3ec.

Betsill, M.M. and H. Bulkeley, 2006: Cities and the multilevel governance of global climate change. *Glob. Gov.*, **12(2)**, 141–159, doi:10.1163/19426720-01202004.

Biermann, F. et al., 2012: Transforming governance and institutions for global sustainability: Key insights from the Earth System Governance Project. *Curr. Opin. Environ. Sustain.*, **4(1)**, 51–60, doi:10.1016/j.cosust.2012.01.014.

Biermann, F., N. Kanie, and R.E. Kim, 2017: Global governance by goal-setting: the novel approach of the UN Sustainable Development Goals. *Curr. Opin. Environ. Sustain.*, **26–27**, 26–31, doi:10.1016/j.cosust.2017.01.010.

Biggs, E.M. et al., 2015: Sustainable development and the water-energy-food nexus: A perspective on livelihoods. *Environ. Sci. Policy*, **54**, 389–397, doi:10.1016/j.envsci.2015.08.002.

Birch, K., 2016: Materiality and sustainability transitions: integrating climate change in transport infrastructure in Ontario, Canada. *Prometheus*, **34(3–4)**, 191–206, doi:10.1080/08109028.2017.1331612.

Bleischwitz, R. et al., 2018: Resource nexus perspectives towards the United Nations Sustainable Development Goals. *Nat. Sustain.*, **1**, 737–743, doi:10.1038/s41893-018-0173-2.

Bloomfield, H.C., D.J. Brayshaw, L.C. Shaffrey, P.J. Coker, and H.E. Thornton, 2016: Quantifying the increasing sensitivity of power systems to climate variability. *Environ. Res. Lett.*, **11(124025)**, 1–12, doi:10.1088/1748-9326/11/12/124025.

BMU, 2018: *Conclusions of the Petersberg Climate Dialogue IX – BMU.* Federal Ministry for the Environment, Nature Conservation, Berlin, Germany, 5 pp. https://www.bmu.de/en/topics-1/climate-energy/climate/international-climate-policy/12th-petersberg-climate-dialogue-pcd-xii/petersberg-climate-dialogue-ix (Accessed May 14, 2020).

Bos, K. and J. Gupta, 2019: Stranded assets and stranded resources: Implications for climate change mitigation and global sustainable development. *Energy Res. Soc. Sci.*, **56**, 101215, doi.org/10.1016/j.erss.2019.05.025.

Bolton, R., T.J. Foxon, and S. Hall, 2016: Energy transitions and uncertainty: Creating low carbon investment opportunities in the UK electricity sector. *Environ. Plan. C-Government Policy*, **34(8)**, 1387–1403, doi:10.1177/0263774X15619628.

Bowman, A., O. Portney, and J. Berry, 2017: Multilevel Governance and City Sustainability Policies: Does It Exist? Does it Matter? Paper delivered at the Local Governance and Sustainability Conference, New Orleans, USA. Southern Political Science Association, 118 pp. https://localgov.fsu.edu/sites/g/files/upcbnu1196/files/Bowman Portney and Berry SPSA 2017.final.pdf (Accessed October 15, 2021).

Bradbury, D., 2015: Internet of things: How can privacy survive in the era of the internet of things? *Guardian*, https://www.theguardian.com/technology/2015/apr/07/how-can-privacy-survive-the-internet-of-things (Accessed August 10, 2021).

Bradbury, H. et al., 2019: A call to Action Research for Transformations: The times demand it. *Action Res.*, **17(1)**, 8, doi:10.1177/1476750319829633.

Bradley, H., 2005: Class, Self and Culture. *Br. J. Sociol.*, **56(4)**, 677–678, doi.org/10.1111/j.1468-4446.2005.00088_13.x.

Breukers, S., R.M. Mourik, L.F.M. van Summeren, and G.P.J. Verbong, 2017: Institutional 'lock-out' towards local self-governance? Environmental justice and sustainable transformations in Dutch social housing neighbourhoods. *Energy Res. Soc. Sci.*, **23**, 148–158, doi:10.1016/j.erss.2016.10.007.

Brodnik, C. and R. Brown, 2018: Strategies for developing transformative capacity in urban water management sectors: The case of Melbourne, Australia. *Technol. Forecast. Soc. Change*, **137**, 147–159, doi:10.1016/j.techfore.2018.07.037.

Broto, V.C., S. Glendinning, E. Dewberry, C. Walsh, and M. Powell, 2014: What can we learn about transitions for sustainability from infrastructure shocks? *Technol. Forecast. Soc. Change*, **84**, 186–196, doi:10.1016/j.techfore.2013.08.002.

Brouwer, F. et al., 2018: Energy modelling and the Nexus concept. *Energy Strateg. Rev.*, **19**, 1–6, doi:10.1016/j.esr.2017.10.005.

Brown, O., A. Hammill, and R. McLeman, 2007: Climate change as the 'new' security threat: Implications for Africa. *Int. Aff.*, **83(6)**, 1141–1154, doi:10.1111/j.1468-2346.2007.00678.x.

Brundiers, K., A. Wiek, and C.L. Redman, 2010: Real-world learning opportunities in sustainability: from classroom into the real world. *Int. J. Sustain. High. Educ.*, **11(4)**, 308–324, doi:10.1108/14676371011077540.

Bryan, E. et al., 2013: Can agriculture support climate change adaptation, greenhouse gas mitigation and rural livelihoods? insights from Kenya. *Clim. Change*, **118**, 151–165, doi.org/10.1007/s10584-012-0640-0.

Buijs, A., A. Fischcher, and A. Muhar, 2018: From urban gardening to planetary stewardship: human–nature relationships and their implications for environmental management. *J. Environ. Plan. Manag.*, **61(5–6)**, 10, doi.org/10.1080/09640568.2018.1429255.

Bulkeley, H., V. Castán Broto, and G.A.S. Edwards, 2015: *An Urban Politics of Climate Change: Experimentation and the Governing of Socio-Technical Transitions*. Routledge, London, UK and New York, USA, 282 pp.

Burch, S., 2010: Transforming barriers into enablers of action on climate change: Insights from three municipal case studies in British Columbia, Canada. *Glob. Environ. Change*, **20(2)**, 287–297, doi:10.1016/j.gloenvcha.2009.11.009.

Burch, S., 2017: The governance of transformative change: Tracing the pathway of the sustainability transition in Vancouver, Canada. *Urban Sustain. Transitions*, 50–64, doi:10.4324/9781315228389.

Burch, S. and J. Di Bella, 2021: Business models for the Anthropocene: accelerating sustainability transformations in the private sector. *Sustain. Sci.*, **16(6)**, 1963–1976, doi:10.1007/s11625-021-01037-3.

Burch, S., A. Shaw, A. Dale, and J. Robinson, 2014: Triggering transformative change: A development path approach to climate change response in communities. *Clim. Policy*, **14(4)**, 467–487, doi:10.1080/14693062.2014.876342.

Burch, S. et al., 2016: Governing and accelerating transformative entrepreneurship: exploring the potential for small business innovation on urban sustainability transitions. *Curr. Opin. Environ. Sustain.*, **22**, 26–32, doi:10.1016/j.cosust.2017.04.002.

Buzananakova, A., 2015: *Democratic Republic of the Congo Coal Reserves*. Government of Democratic Republic of the Congo, Democratic Republic of the Congo, 3 pp. https://drcongo.opendataforafrica.org/gkocbhe/democratic-republic-of-the-congo-coal-reserves?lang=en (Accessed October 18, 2021).

17

Byers, E. et al., 2018: Global exposure and vulnerability to multi-sector development and climate change hotspots. *Environ. Res. Lett.*, **13(5)**, 55012, doi:10.1088/1748-9326/aabf45.

Byers, E.A. et al., 2015: Cooling water for Britain's future electricity supply. *Proc. Inst. Civ. Eng. Energy*, **168(3)**, 188–204, doi:10.1680/ener.14.00028.

C40 Cities, 2019: Around the world, C40 cities are taking bold climate action, leading the way towards a healthier and more sustainable future. C40 Cities, London, UK. https://www.c40.org/ (Accessed August 16, 2021).

Caldecott, B., N. Howarth, and P. McSharry, 2013: *Stranded Assets in Agriculture: Protecting Value from Environment-Related Risks*. University of Oxford, Oxford, UK, 109 pp. https://www.smithschool.ox.ac.uk/publications/reports/stranded-assets-agriculture-report-final.pdf (Accessed June 3, 2021).

Caldera, U. and C. Breyer, 2020: Strengthening the global water supply through a decarbonised global desalination sector and improved irrigation systems. *Energy*, **200**, 117507, doi.org/10.1016/j.energy.2020.117507.

Cameron, C. et al., 2016: Policy trade-offs between climate mitigation and clean cook-stove access in South Asia. *Nat. Energy*, **1(1)**, 15010, doi:10.1038/nenergy.2015.10.

Campagnolo, L. and M. Davide, 2019: Can the Paris deal boost SDGs achievement? An assessment of climate mitigation co-benefits or side-effects on poverty and inequality. *World Dev.*, **122**, 96–109, doi:10.1016/j.worlddev.2019.05.015.

Campiglio, E., 2016: Beyond carbon pricing: The role of banking and monetary policy in financing the transition to a low-carbon economy. *Ecol. Econ.*, **121**, 220–230, doi:10.1016/j.ecolecon.2015.03.020.

Campos, I. and E. Marín-González, 2020: People in transitions: Energy citizenship, prosumerism and social movements in Europe. *Energy Res. Soc. Sci.*, **69**, 101718, doi.org/10.1016/j.erss.2020.101718.

Campbell, B.M., P. Thornton, R. Zougmoré, P. van Asten, and L. Lipper, 2014: Sustainable intensification: What is its role in climate smart agriculture? *Curr. Opin. Environ. Sustain.*, **8**, 5, doi:10.1016/j.cosust.2014.07.002.

Canitez, F., 2019: Pathways to sustainable urban mobility in developing megacities: A socio-technical transition perspective. *Technol. Forecast. Soc. Change*, **141**, 319–329, doi:10.1016/j.techfore.2019.01.008.

Capellán-Pérez, I., C. de Castro, and I. Arto, 2017: Assessing vulnerabilities and limits in the transition to renewable energies: Land requirements under 100% solar energy scenarios. *Renew. Sustain. Energy Rev.*, **77**, 760–782, doi:10.1016/j.rser.2017.03.137.

Carney, M., 2016: Resolving the climate paradox, 13pp. BIS Central Bankers speeches, Berlin, Germany. https://www.bis.org/review/r160926h.pdf (Accessed August 9, 2021).

Carraz, R. and H. Yuko, 2019: Japan's Innovation Systems at the Crossroads: Society 5.0. *Panor. Insights into Asian Eur. Aff.*, **3**, 33–45.

Carvalho Resende, T. et al., 2019: Assessment of the impacts of climate variability on total water storage across Africa: implications for groundwater resources management. *Hydrogeol. J.*, **27(2)**, 493–512, doi:10.1007/s10040-018-1864-5.

Castellanos-Navarrete, A., W.V. Tobar-Tomás, and C.E. López-Monzón, 2019: Development without change: Oil palm labour regimes, development narratives, and disputed moral economies in Mesoamerica. *J. Rural Stud.*, **71**, 169–180, doi:10.1016/J.JRURSTUD.2018.08.011.

Castellanos-Navarrete, A., F. de Castro, and P. Pacheco, 2021: The impact of oil palm on rural livelihoods and tropical forest landscapes in Latin America. *J. Rural Stud.*, **81**, 294–304, doi:10.1016/J.JRURSTUD.2020.10.047.

Castro, M.C. et al., 2016: Examples of coupled human and environmental systems from the extractive industry and hydropower sector interfaces. *Proc. Natl. Acad. Sci.*, **113(51)**, 14528–14535, doi:10.1073/pnas.1605678113.

Cebrian, M., 2021: The past, present and future of digital contact tracing. *Nat. Electron.*, **4(1)**, 2–4, doi:10.1038/s41928-020-00535-z.

Cervero, R., 2009: Public transport and sustainable urbanism: Global lessons. *Transit Oriented Dev. Mak. it Happen*, pp. 43–56.

Chan, S. and P. Pauw, 2014: A global framework for climate action: orchestrating non-state and subnational initiatives for more effective global climate governance. Discussion Paper. German Development Institute, Discussion Paper 34/2014, **(34)**, 54.

Chatalova, L. and A. Balmann, 2017: The hidden costs of renewables promotion: The case of crop-based biogas. *J. Clean. Prod.*, **168**, 893–903, doi:10.1016/j.jclepro.2017.09.031.

Chen, H., J. Sun, and X. Chen, 2014: Projection and uncertainty analysis of global precipitation-related extremes using CMIP5 models. *Int. J. Climatol.*, **34**, 2730–2748, doi:10.1002/joc.3871.

Chichilnisky, G. and G. Heal, 1998: Economic returns from the biosphere. *Nature*, **391**, 629–630, doi:10.1038/35481.

Chiengkul, P., 2017: *The Political Economy of the Agri-food system in Thailand: Hegemony, Counter-hegemony, and Co-optation of oppositions*. 8th ed. Routledge, London, UK, 228 pp.

Chimhowu, A.O., D. Hulme, and L.T. Munro, 2019: The 'New' national development planning and global development goals: Processes and partnerships. *World Dev.*, **120**, 76–89, doi.org/10.1016/j.worlddev.2019.03.013.

Chirambo, D., 2018: Towards the achievement of SDG 7 in Sub-Saharan Africa: Creating synergies between Power Africa, Sustainable Energy for All and climate finance in-order to achieve universal energy access before 2030. *Renew. Sustain. Energy Rev.*, **94**, 600–608, doi.org/10.1016/j.rser.2018.06.025.

Chu, E., 2016: The political economy of urban climate adaptation and development planning in Surat, India. *Environ. Plan. C Gov. Policy*, **34(2)**, 281–298, doi:10.1177/0263774X15614174.

Chu, E.K., 2018: Urban climate adaptation and the reshaping of state–society relations: The politics of community knowledge and mobilisation in Indore, India. *Urban Stud.*, **55(8)**, 1766–1782, doi:10.1177/0042098016686509.

Clark, K. et al., 2019: Advancing the ethical use of digital data in human research: challenges and strategies to promote ethical practice. *Ethics Inf. Technol.*, **21(1)**, 59–73, doi:10.1007/s10676-018-9490-4.

Clark, R., J. Reed, and T. Sunderland, 2018: Bridging funding gaps for climate and sustainable development: Pitfalls, progress and potential of private finance. *Land Use Policy*, **71**, 335–346, doi:10.1016/j.landusepol.2017.12.013.

Clarke, L., K. Jiang, K. Akimoto, M. Babiker, G. Blanford, K. Fisher-Vanden, J.-C. Hourcade, V. Krey, E. Kriegler, A. Löschel, D. McCollum, S. Paltsev, S. Rose, P.R. Shukla, M. Tavoni, B.C.C. van der Zwaan, and D.P. van Vuuren, 2014: Assessing Transformation Pathways. In: *Climate Change 2014: Mitigation of Climate Change. Contribution of Working Group III to the Fifth Assessment Report of the Intergovernmental Panel on Climate Change* [Edenhofer, O., R. Pichs-Madruga, Y. Sokona, E. Farahani, S. Kadner, K. Seyboth, A. Adler, I. Baum, S. Brunner, P. Eickemeier, B. Kriemann, J. Savolainen, S. Schlömer, C. von Stechow, T. Zwickel and J.C. Minx (eds.)]. Cambridge University Press, Cambridge, UK and New York, NY, USA, pp. 413–510.

Clarke, L. et al., 2018: Effects of long-term climate change on global building energy expenditures. *Energy Econ.*, **72**, 667–677, doi.org/10.1016/j.eneco.2018.01.003.

Clayton, T. and N. Radcliffe, 2018: *Sustainability: A Systems Approach*. 1st ed. Routledge, Edinburgh, UK, 253 pp.

Climate-ADAPT, 2019: Cities and towns. The European Environment Agency, Copenhagen, Denmark, https://climate-adapt.eea.europa.eu/countries-regions/cities (Accessed August 2, 2021).

Cobbing, J., 2020: Groundwater and the discourse of shortage in Sub-Saharan Africa. *Hydrogeol. J.*, **28(4)**, 1143–1154, doi:10.1007/s10040-020-02147-5.

Cobbing, J. and B. Hiller, 2019: Waking a sleeping giant: Realizing the potential of groundwater in Sub-Saharan Africa. *World Dev.*, **122**, 597–613, doi:10.1016/j.worlddev.2019.06.024.

Coenen, L., P. Benneworth, and B. Truffer, 2012: Toward a spatial perspective on sustainability transitions. *Res. Policy*, **41(6)**, 968–979, doi.org/10.1016/j.respol.2012.02.014.

Congedo, L. and M. Munafò, 2014: Urban Sprawl as a Factor of Vulnerability to Climate Change: Monitoring Land Cover Change in Dar es Salaam. In:

*Climate Change Vulnerability in Southern African Cities* [Macchi, S. and M. Tiepolo, (eds.)]. Springer, Cham, Switzerland, pp. 73–88.

Cook, M. and E. Chu, 2018: Between Policies, Programs, and Projects: How Local Actors Steer Domestic Urban Climate Adaptation Finance in India. In: *Climate Change in Cities. The Urban Book Series* [Hughes, S. and M. Chu, (eds.)]. Springer, Cham, Switzerland, pp. 255–277.

Córdova, R., N.J. Hogarth, and M. Kanninen, 2019: Mountain Farming Systems' Exposure and Sensitivity to Climate Change and Variability: Agroforestry and Conventional Agriculture Systems Compared in Ecuador's Indigenous Territory of Kayambi People. *Sustainability*, **11(9)**, doi:10.3390/su11092623.

Corfee-Morlot, J. et al., 2009: *Cities, Climate Change and Multilevel Governance*. OECD, Paris, France, 125 pp. http://www.oecd.org/dataoecd/10/1/44242293.pdf (Accessed May 4, 2020).

Covenant of Mayors, 2019: Covenant of Mayors for Climate Energy. Bonn, Germany. https://iclei.org/en/GCoM.html?gclid=CjwKCAjwx8iIBhBwEiwA2quaqy1lUqrcbjvH4tTdiFiUM7W-bOsP77Tbn2KyicroS-uOaaeMLzJHRBoCmqwQAvD_BwE (Accessed August 15, 2021).

Craig, M.T. et al., 2018: A review of the potential impacts of climate change on bulk power system planning and operations in the United States. *Renew. Sustain. Energy Rev.*, **98**, 255–267, doi:10.1016/j.rser.2018.09.022.

Cremades, R. et al., 2016: Co-benefits and trade-offs in the water–energy nexus of irrigation modernization in China. *Enviromental Res. Lett.*, **11(2)**, 054007, doi.org/10.1088/1748-9326/11/5/054007.

Creutzig, F. et al., 2019: Upscaling urban data science for global climate solutions. *Glob. Sustain.*, **2(2)**, 1–25, doi:10.1017/sus.2018.16.

Cronin, J. et al., 2021: Embedding justice in the 1.5°C transition: A transdisciplinary research agenda. *Renew. Sustain. Energy Transit.*, **1**, 100001, doi.org/10.1016/j.rset.2021.100001.

Cross, A.T., P.G. Nevill, K.W. Dixon, and J. Aronson, 2019: Time for a paradigm shift toward a restorative culture. *Restor. Ecol.*, **27(5)**, 924–928, doi:10.1111/rec.12984.

Cruz-Cárdenas, J., E. Zabelina, J. Guadalupe-Lanas, A. Palacio-Fierro, and C. Ramos-Galarza, 2021: COVID-19, consumer behavior, technology, and society: A literature review and bibliometric analysis. *Technol. Forecast. Soc. Change*, **173**, 121179, doi.org/10.1016/j.techfore.2021.121179.

Cui, R.Y. et al., 2019: Quantifying operational lifetimes for coal power plants under the Paris goals. *Nat. Commun.*, **10(1)**, 9, doi:10.1038/s41467-019-12618-3.

D'Alisa, G., Demaria, F. and Kallis. G., 2014: Degrowth: A Vocabulary for a New Era 2014 Routledge 248 pp.

D'Odorico, P. et al., 2018: The Global Food-Energy-Water Nexus. *Rev. Geophys.*, **56**, 456–531, doi:10.1029/2017RG000591.

Dai, J. et al., 2018: Water-energy nexus: A review of methods and tools for macro-assessment. *Appl. Energy*, **210**, 393–408, doi:10.1016/j.apenergy.2017.08.243.

Dale, A., S. Burch, and J. Robinson, 2017: Multilevel Governance of Sustainability Transitions in Canada: Policy Alignment, Innovation, and Evaluation. In: *Climate Change and Cities* [Chu, E., S. Manson, and S. Higues, (eds.)]. Springer, Cham, Switzerland, pp. 343–358.

Dale, A. et al., 2020: Meeting the climate change challenge: local government climate action in British Columbia, Canada. *Clim. Policy*, **20(7)**, 866–880, doi:10.1080/14693062.2019.1651244.

Dalkmann, H. and C. Brannigan, 2007: *Urban Transport and Climate Change: Module 5e Sustainable Transport*. First Edition. GIZ, Eschborn, Germany, 87 pp.

Dangelico, R.M., D. Pujari, and P. Pontrandolfo, 2016: Green product innovation in manufacturing firms: A sustainability-oriented dynamic capability perspective. *Bus. Strateg. Environ.*, **26(4)**, 490–506.

Daum, T., 2018: We need more than apps. 1. The International Journal of Rural Development, Bonn, Germany. https://www.rural21.com/english/home.html?no_cache=1 (Accessed October 5, 2020).

David, D. and A. Venkatachalam, 2019: A Comparative Study on the Role of Public-Private Partnerships and Green Investment Banks in Boosting Low-Carbon Investments. In: *Handbook of Green Finance: Energy Security and Sustainable Development* [Sachs, J.D., W.T. Woo, N. Yoshino, and F. Taghizadeh-Hesary, (eds.)]. Springer Singapore, Singapore, pp. 261–287.

Davidson, K. et al., 2020: The making of a climate emergency response: Examining the attributes of climate emergency plans. *Urban Clim.*, **33**, 100666, doi.org/10.1016/j.uclim.2020.100666.

Day, D.V., J.W. Fleenor, L.E. Atwater, R.E. Sturm, and R.A. McKee, 2014: Advances in leader and leadership development: A review of 25 years of research and theory. *Leadersh. Q.*, **25**, 1–182, doi:10.1016/j.leaqua.2013.11.004.

De Barbosa, L.S.N.S., D. Bogdanov, P. Vainikka, and C. Breyer, 2017: Hydro, wind and solar power as a base for a 100% renewable energy supply for South and Central America. *PLoS One*, **12(3)**, 1–28, doi:10.1371/journal.pone.0173820.

de Gooyert, V., E. Rouwette, H. van Kranenburg, E. Freeman, and H. van Breen, 2016: Sustainability transition dynamics: Towards overcoming policy resistance. *Technol. Forecast. Soc. Change*, **111**, 135–145, doi:10.1016/j.techfore.2016.06.019.

De la Sota, C., V.J. Ruffato-Ferreira, L. Ruiz-García, and S. Alvarez, 2019: Urban green infrastructure as a strategy of climate change mitigation. A case study in northern Spain. *Urban For. Urban Green.*, **40**, 145–151, doi:10.1016/j.ufug.2018.09.004.

de Strasser, L., A. Lipponen, M. Howells, S. Stec, and C. Bréthaut, 2016: A Methodology to Assess the Water Energy Food Ecosystems Nexus in Transboundary River Basins. *Water*, **8(2)**, 59, doi.org/10.3390/w8020059.

de Wilde, P. and D. Coley, 2012: The implications of a changing climate for buildings. *Build. Environ.*, **55**, 1–7, doi:10.1016/j.buildenv.2012.03.014.

Delina, L.L. and B.K. Sovacool, 2018: Of temporality and plurality: an epistemic and governance agenda for accelerating just transitions for energy access and sustainable development. *Curr. Opin. Environ. Sustain.*, **34**, 1–6, doi:10.1016/j.cosust.2018.05.016.

den Elzen, M. et al., 2019: Are the G20 economies making enough progress to meet their NDC targets? *Energy Policy*, **126** (November 2018), 238–250, doi:10.1016/j.enpol.2018.11.027.

Denton, F., T.J.Wilbanks, A.C. Abeysinghe, I. Burton, Q. Gao, M.C. Lemos, T. Masui, K.L. O'Brien, and K.Warner, 2014: Climate-resilient pathways: adaptation, mitigation, and sustainable development. In: *Climate Change 2014: Impacts, Adaptation, and Vulnerability. Part A: Global and Sectoral Aspects. Contribution of Working Group II to the Fifth Assessment Report of the Intergovernmental Panel on Climate Change*, Cambridge University Press, [Field, C.B., V.R. Barros, D.J. Dokken, K.J. Mach, M.D. Mastrandrea, T.E. Bilir, M. Chatterjee, K.L. Ebi, Y.O. Estrada, R.C. Genova, B. Girma, E.S. Kissel, A.N. Levy, S. MacCracken, P.R. Mastrandrea, and L.L. White (eds.)]. Cambridge, UK and New York, NY, USA, pp. 1101–1131.

Denton, F, T. et al.: 2021: *Blind Alleys and Bright Prospects: Africa Navigating Stranded Assets and Just Transitions*, 89 pp., UNU-INRA Publications, Africa. https://i.unu.edu/media/inra.unu.edu/attachment/5662/SA-Report-Final_2021web.pdf.

Destouni, G., F. Jaramillo, and C. Prieto, 2013: Hydroclimatic shifts driven by human water use for food and energy production. *Nat. Clim. Change*, **(3)**, 213–217, doi:10.1038/nclimate1719.

Di Gregorio, M. et al., 2017: Climate policy integration in the land use sector: Mitigation, adaptation and sustainable development linkages. *Environ. Sci. Policy*, **67**, 35–43, doi:10.1016/j.envsci.2016.11.004.

Diallo, I. et al., 2016: Projected changes of summer monsoon extremes and hydroclimatic regimes over West Africa for the twenty-first century. *Clim. Dyn.*, **47(12)**, 3931–3954, doi:10.1007/s00382-016-3052-4.

Díaz, S. et al., 2019: Pervasive human-driven decline of life on Earth points to the need for transformative change. *Science*, **366(6471)**, eaax3100, doi:10.1126/science.aax3100.

Dick, B., 2007: Action Research as an Enhancement of Natural Problem Solving. *Int. J. Action Res.*, **3(1/2)**, 149–167, doi:10.1177/1541344613490997.

Doll, C.N.H. and J.A. Puppim De Oliveira, 2017: *Urbanization and climate co-benefits: Implementation of win-win interventions in cities*. Taylor and Francis, London, UK, 324 pp.

Dovie, D.B.K., 2019: Case for equity between Paris Climate agreement's co-benefits and adaptation. *Sci. Total Environ.*, **656**, 732–739, doi:10.1016/j.scitotenv.2018.11.333.

Dowling, R., P. McGuirk, and S. Maalsen, 2018: Multiscalar governance of urban energy transitions in Australia: The cases of Sydney and Melbourne. *Energy Res. Soc. Sci.*, **44**, 260–267, doi:10.1016/j.erss.2018.05.027.

Dugarova, E. and N. Gülasan, 2017: *Global Trends: Challenges and Opportunities in the Implementation of the Sustainable Development Goals.*, United Nations Development Programme and United Nations Research Institute for Social Development, New York, USA, 101 pp. https://www.undp.org/publications/global-trends-challenges-and-opportunities-implementation-sdgs#modal-publication-download.

Dugast, J., S. Uslu, and P.-O. Weill, 2019: *A Theory of Participation in OTC and Centralized Markets*. National Bureau of Economic Research, Massachusetts, USA, 45 pp.

Duguma, L.A., P.A. Minang, and M. van Noordwijk, 2014: Climate Change Mitigation and Adaptation in the Land Use Sector: From Complementarity to Synergy. *Environ. Manage.*, **54(3)**, 420–432, doi:10.1007/s00267-014-0331-x.

E3G Annual Review, 2018: E3G Annual Review 2018. https://www.e3g.org/library_asset/e3g-annual-review-2018/ (Accessed August 26, 2021).

Easterlin, R.A., L.A. McVey, M. Switek, O. Sawangfa, and J.S. Zweig, 2010: The happiness - Income paradox revisited. *Proc. Natl. Acad. Sci.*, **107(52)**, 1–6, doi:10.1073/pnas.1015962107.

ECLAC, 2017: *Annual report on regional progress and challenges in relation to the 2030 Agenda for Sustainable Development in Latin America and the Caribbean*, CEPAL 114 pp. https://repositorio.cepal.org/handle/11362/41189.

ECLEI, 2019: Cities and towns gathered around one shared ambition. ECLEI, Bonn, Germany. https://iclei.org/en/GCoM.html?gclid=CjwKCAjwx8iIBhBwEiwA2quaqy1lUqrcbjvH4tTdiFiUM7W-bOsP77Tbn2KyicroS-uOaaeMLzJHRBoCmqwQAvD_BwE (Accessed August 15 2021).

Edmondson, D.L., F. Kern, and K.S. Rogge, 2019: The co-evolution of policy mixes and socio-technical systems: Towards a conceptual framework of policy mix feedback in sustainability transitions. *Res. Policy*, **48(10)**, 14, doi:10.1016/j.respol.2018.03.010.

EESC, 2015: *Building the Europe We Want: Models for civil society involvement in the implementation of the Post-2015 agenda at the EU level*. 54 pp. European Economical Social Committee. https://www.eesc.europa.eu/ru/our-work/publications-other-work/publications/building-europe-we-want-models-civil-society-involvement-implementation-post-2015-agenda-eu-level.

Ehnert, F. et al., 2018a: The Acceleration of Urban Sustainability Transitions: A Comparison of Brighton, Budapest, Dresden, Genk, and Stockholm. *Sustainability*, **10(3)**, 612, doi:10.3390/su10030612.

Ehnert, F. et al., 2018b: Urban sustainability transitions in a context of multi-level governance: A comparison of four European states. *Environ. Innov. Soc. Transitions*, **26**, 101–116, doi:10.1016/j.eist.2017.05.002.

Eisenack, K. and D. Reckien, 2013: Climate Change and Simulation/Gaming. *Simul. Gaming*, **44(2–3)**, 245–252, doi:10.1177/1046878113490568.

Elder, M. and P. King, 2018: *Realising the Transformative Potential of the SDGs*. IGES, Hayama, Japan, 188 pp. https://www.iges.or.jp/jp/pub/realising-transformative-potential-sdgs/en.

Elder, M. and A. Bartalini, 2019: *Assessment of the G20 Countries' Concrete SDG Implementation Efforts: Policies and Budgets Reported in Their 2016-2018 Voluntary National Reviews*. IGES, Hayama, Japan, 92 pp.

Erb, K.-H. et al., 2016: Exploring the biophysical option space for feeding the world without deforestation. *Nat. Commun.*, **7(1)**, 11382, doi:10.1038/ncomms11382.

Escobar, A., 2015: Degrowth, postdevelopment, and transitions: a preliminary conversation. *Sustain. Sci.*, **10(3)**, 451–462, doi:10.1007/s11625-015-0297-5.

Fader, M., C. Cranmer, R. Lawford, and J. Engel-Cox, 2018: Toward an understanding of synergies and trade-offs between water, energy, and food SDG targets. *Front. Environ. Sci.*, **6**, 112, doi:10.3389/fenvs.2018.00112.

Falcone, P.M. and E. Sica, 2015: How much does economic crisis affect sustainability transitions? A social network analysis of the italian biofuel sector. *Econ. Reg.*, **1(1)**, 10, doi:10.17059/2015-1-23.

Fan, J.L., L.S. Kong, X. Zhang, and J.-D. Wang, 2019a: Energy-water nexus embodied in the supply chain of China: Direct and indirect perspectives. *Energy Convers. Manag.*, **183**, 126–136, doi.org/10.1016/j.enconman.2018.12.095.

Fan, J.L., L.S. Kong, H. Wang, and X. Zhang, 2019b: A water-energy nexus review from the perspective of urban metabolism. *Ecol. Modell.*, **392**, 128–136, doi:10.1016/j.ecolmodel.2018.11.019.

FAO, 2019: *The State of Food and Agriculture: Moving forward on food loss and waste reduction.* Food and Agriculture Organization of the United Nations, Rome, Italy, 182 pp.

Farzaneh, H., E. Zusman, and Y. Chae, 2021: *Aligning Climate Change and Sustainable Development Policies in Asia*. 1st ed. [Farzaneh, H., E. Zusman, and Y. Chae, (eds.)]. Springer, Singapore, 185 pp.

Fastenrath, S. and B. Braun, 2018: Sustainability transition pathways in the building sector: Energy-efficient building in Freiburg (Germany). *Appl. Geogr.*, **90**, 339–349, doi:10.1016/j.apgeog.2016.09.004.

Favretto, N., A.J. Dougill, L.C. Stringer, S. Afionis, and C.H. Quinn, 2018: Links between climate change mitigation, adaptation and development in land policy and ecosystem restoration projects: lessons from South Africa. *Sustainability*, **10(3)**, 19. doi:10.3390/su10030779.

Fazey, I. et al., 2018: Community resilience for a 1.5°C world. *Curr. Opin. Environ. Sustain.*, **31**, 30–40, doi.org/10.1016/j.cosust.2017.12.006.

Feng, K., K. Hubacek, and K. Song, 2021: Household carbon inequality in the U.S. *J. Clean. Prod.*, **278**, 123994, doi:10.1016/j.jclepro.2020.123994.

Ferguson, R.S. and S. T. Lovell, 2014: Permaculture for agroecology: Design, movement, practice, and worldview. A review. *Agron. Sustain. Dev.*, **34**, 251–274, doi:10.1007/s13593-013-0181-6.

Fishman, R., N. Devineni, and S. Raman, 2015: Can improved agricultural water use efficiency save India's groundwater? *Environ. Res. Lett.*, **10(080422)**, doi:10.1088/1748-9326/10/8/084022.

Fleurbaey, M., S. Kartha, S. Bolwig, Y.L. Chee, Y. Chen, E. Corbera, F. Lecocq, W. Lutz, M.S. Muylaert, R.B. Norgaard, C. Okereke, and A.D. Sagar, 2014: Sustainable Development and Equity. In: *Climate Change 2014: Mitigation of Climate Change. Contribution of Working Group III to the Fifth Assessment Report of the Intergovernmental Panel on Climate Change* [Edenhofer, O., R. Pichs-Madruga, Y. Sokona, E. Farahani, S. Kadner, K. Seyboth, A. Adler, I. Baum, S. Brunner, P. Eickemeier, B. Kriemann, J. Savolainen, S. Schlömer, C. von Stechow, T. Zwickel and J.C. Minx (eds.)]. Cambridge University Press, Cambridge, UK and New York, NY, USA, pp. 283–350.

Folke, C., 2006: Resilience: The emergence of a perspective for social-ecological systems analyses. *Glob. Environ. Chang.*, **16(3)**, 253–267, doi:10.1016/j.gloenvcha.2006.04.002.

Fontana, G. and M. Sawyer, 2016: Towards post-Keynesian ecological macroeconomics. *Ecol. Econ.*, **121**, 186–195, doi:10.1016/j.ecolecon.2015.03.017.

Formentini, M. and P. Taticchi, 2016: Corporate sustainability approaches and governance mechanisms in sustainable supply chain management. *J. Clean. Prod.*, **112**, 1920–1933.

Forsyth, T., 2014: Climate justice is not justice. *Geoforum*, **54**, 230–232, doi:10.1016/j.geoforum.2012.12.008.

Forzieri, G. et al., 2018: Escalating impacts of climate extremes on critical infrastructures in Europe. *Glob. Environ. Change*, **48**, 97–107, doi:10.1016/j.gloenvcha.2017.11.007.

Frantzeskaki, N., D. Loorbach, and J. Meadowcroft, 2012: Governing societal transitions to sustainability. *Int. J. Sustain. Dev.*, **15(1)**, doi:10.1504/IJSD.2012.044032.

Fricko, O. et al., 2016: Energy sector water use implications of a 2°C climate policy. *Environ. Res. Lett.*, **11(034011)**, 1–11, doi:10.1088/1748-9326/11/3/034011.

Froese, R. and J. Schilling, 2019: The Nexus of Climate Change, Land Use, and Conflicts. *Curr. Clim. Change Reports*, **5(1)**, 24–35, doi:10.1007/s40641-019-00122-1.

Fuhr, H., T. Hickmann, and K. Kern, 2018: The role of cities in multi-level climate governance: local climate policies and the 1.5°C target. *Curr. Opin. Environ. Sustain.*, **30**, p.1-6, doi:10.1016/j.cosust.2017.10.006.

Fujimori, S. et al., 2019: A multi-model assessment of food security implications of climate change mitigation. *Nat. Sustain.*, **2(5)**, 386–396, doi:10.1038/s41893-019-0286-2.

Fuseini, I. and J. Kemp, 2015: A review of spatial planning in Ghana's socio-economic development trajectory: A sustainable development perspective. *Land Use Policy*, **47**, 309–320, doi.org/10.1016/j.landusepol.2015.04.020.

Fuso Nerini, F. et al., 2019: Connecting climate action with other Sustainable Development Goals. *Nat. Sustain.*, **2(8)**, 674–680, doi:10.1038/s41893-019-0334-y.

Galli, A., G. Đurović, L. Hanscom, and J. Knežević, 2018: Think globally, act locally: Implementing the sustainable development goals in Montenegro. *Environ. Sci. Policy*, **84**, 159–169, doi.org/10.1016/j.envsci.2018.03.012.

Gao, J. et al., 2019: Insights into water-energy cobenefits and trade-offs in water resource management. *J. Clean. Prod.*, **213**, 1–22, doi:10.1016/j.jclepro.2018.12.126.

García-Olivares, A., J. Solé, and O. Osychenko, 2018: Transportation in a 100% renewable energy system. *Energy Convers. Manag.*, **158**, 266–285, doi.org/10.1016/j.enconman.2017.12.053.

Gasbarro, F., F. Iraldo, and T. Daddi, 2017: The drivers of multinational enterprises' climate change strategies: A quantitative study on climate-related risks and opportunities. *J. Clean. Prod.*, **160**, 8–26, doi.org/10.1016/j.jclepro.2017.03.018.

Gaveau, D.L.A. et al., 2016: Rapid conversions and avoided deforestation: Examining four decades of industrial plantation expansion in Borneo. *Sci. Rep.*, **6**, 32017, doi:10.1038/srep32017.

Geels, F.W., 2011: The multi-level perspective on sustainability transitions: Responses to seven criticisms. *Environ. Innov. Soc. Transitions*, **1(1)**, 24–40, doi.org/10.1016/j.eist.2011.02.002.

Geels, F.W., 2014: Regime Resistance against Low-Carbon Transitions: Introducing Politics and Power into the Multi-Level Perspective. *Theory, Cult. Soc.*, **31(5)**, doi:10.1177/0263276414531627.

Geneletti, D. and L. Zardo, 2016: Ecosystem-based adaptation in cities: An analysis of European urban climate adaptation plans. *Land Use Policy*, **50**, 38–47, doi.org/10.1016/j.landusepol.2015.09.003.

Geng, Y. and Z. Hengxin, 2009: Industrial park management in the Chinese environment. *J. Clean. Prod.*, **17(14)**, 1289–1294, doi:10.1016/j.jclepro.2009.03.009.

Geng, Y., J. Sarkis, and R. Bleischwitz, 2019: How to globalize the circular economy. *Nature*, **565(7738)**, 153–155, doi:10.1038/d41586-019-00017-z.

Germer, J. et al., 2011: Skyfarming an ecological innovation to enhance global food security. *J. fur Verbraucherschutz und Leb.*, **6(2)**, doi:10.1007/s00003-011-0691-6.

Gi, K., F. Sano, A. Hayashi, and K. Akimoto, 2019: A model-based analysis on energy systems transition for climate change mitigation and ambient particulate matter 2.5 concentration reduction. *Mitig. Adapt. Strateg. Glob. Change*, **24(2)**, 181–204, doi:10.1007/s11027-018-9806-z.

Giannaris, S., C. Bruce, B. Jacobs, W. Srisang, and Janowczyk, 2020: Implementing a second generation CCS facility on a coal fired power station – results of a feasibility study to retrofit SaskPower's Shand power station with CCS. *Green Gases Sci. and Technol.*, **10**, 506–518, doi.org/10.1002/ghg.1989.

Gillard, R., A. Gouldson, J. Paavola, and J. Van Alstine, 2016: Transformational responses to climate change: Beyond a systems perspective of social change in mitigation and adaptation. *Wiley Interdiscip. Rev. Clim. Change*, **7(2)**, 251–265, doi:10.1002/wcc.384.

Gioacchino, G., 2019: You defend what you feel: 'Presencing' nature as 'experiential knowing.' *Action Res.*, **1(1)**, 108–129, doi:10.1177/1476750319829208.

Giorgi, F., F. Raffaele, and E. Coppola, 2019: The response of precipitation characteristics to global warming from climate projections. *Earth Syst. Dyn.*, **10(1)**, 73–89, doi:10.5194/esd-10-73-2019.

Girardin, C.A.J. et al., 2021: Nature-based solutions can help cool the planet – if we act now. *Nature*, **593(7858)**, 191–194, doi:10.1038/d41586-021-01241-2.

Gjorgiev, B. and G. Sansavini, 2018: Electrical power generation under policy constrained water-energy nexus. *Appl. Energy*, **210**, 568–579, doi:10.1016/j.apenergy.2017.09.011.

Glaas, E. et al., 2019: Developing transformative capacity through systematic assessments and visualization of urban climate transitions. *Ambio*, **48(5)**, 515–528, doi:10.1007/s13280-018-1109-9.

Global Opportunity Explorer, 2019: Quito: Planning For Smaller $CO_2$ And Water Footprints. *Global Opportunity Explorer*. https://goexplorer.org/quito-planning-for-smaller-co2-and-water-footprints/ (Accessed August 20, 2021).

Golberg, S., 2021: Lithium Mining: Dirty Investment or Sustainable Business? *Investopedia*. https://www.investopedia.com/investing/lithium-mining-dirty-investment-or-sustainable-business/ (Accessed June 8, 2021).

Goldman-Schuyler, K., S. Skjei, J. Sanzgiri, and V. Koskela, 2017: 'Moments of Waking Up: A Doorway to Mindfulness and Presence. *J. Manag. Inq.*, **26(1)**, 86–100, doi:10.1177/1056492616665171.

Göpel, M., V. Hermelingmeier, K. Kehl, D. Vallentin, and T. Wehnert, 2016: *System Innovation Lab*. Wuppertal Institut and the Centre for Social Investment, Berlin, Germany, pp. 1–41. https://epub.wupperinst.org/frontdoor/deliver/index/docId/6538/file/6538_System_Innovation_Lab.pdf (Accessed June 20, 2021).

Gordon, D.J., 2015: An uneasy equilibrium: The coordination of climate governance in federated systems. *Glob. Environ. Polit.*, **15(2)**, 121–141, doi:10.1162/GLEP_a_00301.

Gorissen, L., F. Spira, E. Meynaerts, P. Valkering, and N. Frantzeskaki, 2018: Moving towards systemic change? Investigating acceleration dynamics of urban sustainability transitions in the Belgian City of Genk. *J. Clean. Prod.*, **173**, 171–185, doi.org/10.1016/j.jclepro.2016.12.052.

Gossart, C., 2015: ICT Innovations for Sustainability. In: *ICT Innovations for Sustainability* [Hilty, L. and B. Aebischer, (eds.)]. Springer Link, Zurich, Switzerland, pp. 435–448.

Grady-Benson, J. and B. Sarathy, 2016: Fossil fuel divestment in US higher education: student-led organising for climate justice. *Local Environ.*, **21(6)**, 661–681, doi:10.1080/13549839.2015.1009825.

Grafakos, S., K. Trigg, M. Landauer, L. Chelleri, and S. Dhakal, 2019: Analytical framework to evaluate the level of integration of climate adaptation and mitigation in cities. *Clim. Change*, **154(1–2)**, 87–106, doi:10.1007/s10584-019-02394-w.

Gramkow, C. and A. Anger-Kraavi, 2018: Could fiscal policies induce green innovation in developing countries? The case of Brazilian manufacturing sectors. *Clim. Policy*, **18(2)**, 246–257, doi:10.1080/14693062.2016.1277683.

Gray, L., A. Boyle, E. Francks, and V. Yu, 2019: The power of small-scale solar: gender, energy poverty, and entrepreneurship in Tanzania. *Dev. Pract.*, **29**(1), 26–39, doi:10.1080/09614524.2018.1526257.

Green, F. and A. Gambhir, 2020: Transitional assistance policies for just, equitable and smooth low-carbon transitions: who, what and how? *Clim. Policy*, **20(8)**, 902–921, doi:10.1080/14693062.2019.1657379.

Greenberg, D.A. and J.M. Rogerson, 2014: The greening of industrial property developments in South Africa. *Urbani Izziv*, **25** (special issue), 122–133, doi:10.5379/urbani-izziv-en-2014-25-supplement-009.

Griggs, D. and Stafford-Smith, 2013: Sustainable development goals for people and planet. *Nature*, **495** (21 March), 305–307, doi:10.1038/495305a.

Grinde, B., R.B. Nes, I.F. MacDonald, and D.S. Wilson, 2018: Quality of Life in Intentional Communities. *Soc. Indic. Res.*, **137**, 625–640, doi:10.1007/s11205-017-1615-3.

Grubler, A. et al., 2018: A low energy demand scenario for meeting the 1.5°C target and sustainable development goals without negative emission technologies. *Nat. Energy*, **3**, 515–527, doi:10.1038/s41560-018-0172-6.

Gu, Y., Y. Wu, J. Liu, M. Xu, and T. Zuo, 2020: Ecological civilization and government administrative system reform in China. *Resour. Conserv. Recycl.*, **155**, 104654, doi.org/10.1016/j.resconrec.2019.104654.

Guillaume, T., M. Kotowska, and D. Hertel, 2018: Carbon costs and benefits of Indonesian rainforest conversion to plantations. *Nat Commun*, **9**, 2388, doi.org/10.1038/s41467-018-04755-y.

Gupta, J., A. Rempel, and H. Verrest, 2020: Access and allocation: the role of large shareholders and investors in leaving fossil fuels underground. *Int. Environ. Agreements Polit. Law Econ.*, **20(2)**, 303–322, doi:10.1007/s10784-020-09478-4.

Guenther, E., T. Guenther, F. Schiemann, and G. Weber, 2016: Stakeholder Relevance for Reporting: Explanatory Factors of Carbon Disclosure. *Bus. Soc.*, **5**(3), 361–397, doi:10.1177/0007650315575119.

Haas, T., 2021: From Green Energy to the Green Car State? The Political Economy of Ecological Modernisation in Germany. *New Polit. Econ.*, **26(4)**, 660–673, doi:10.1080/13563467.2020.1816949.

Haddeland, I. et al., 2014: Global water resources affected by human interventions and climate change. *Proc. Natl. Acad. Sci.*, **111(9)**, 3251–3256, doi:10.1073/pnas.1222475110.

Hakansson, I., 2018: The socio-spatial politics of urban sustainability transitions: Grassroots initiatives in gentrifying Peckham. *Environ. Innov. Soc. Transitions*, **29**, 34–46, doi:10.1016/j.eist.2017.10.003.

Halbe, J., C. Pahl-Wostl, M.A. Lange, and C. Velonis, 2015: Governance of transitions towards sustainable development – the water-energy-food nexus in Cyprus. *Water Int.*, **40(5–6)**, doi:10.1080/02508060.2015.1070328.

Hall, J.W. et al., 2014: Coping with the curse of freshwater variability: Institutions, infrastructure, and information for adaptation. *Science*, **346(6208)**, 429–430, doi:10.1126/science.1257890.

Hallegatte, S., 2009: Strategies to adapt to an uncertain climate change. *Glob. Environ. Change*, **19(2)**, 240–247, doi.org/10.1016/j.gloenvcha.2008.12.003.

Hallinger, P., R. Wang, C. Chatpinyakoop, V.T. Nguyen, and U.P. Nguyen, 2020: A bibliometric review of research on simulations and serious games used in educating for sustainability, 1997–2019. *J. Clean. Prod.*, **256**, doi:10.1016/j.jclepro.2020.120358.

Hamiche, A.M., A.B. Stambouli, and S. Flazi, 2016: A review of the water-energy nexus. *Renew. Sustain. Energy Rev.*, **65**, 319–331, doi:10.1016/j.rser.2016.07.020.

Hamman, P., 2019: Local governance of energy transition: sustainability, transactions and social ties. A case study in Northeast France. *Int. J. Sustain. Dev. World Ecol.*, **26(1)**, 1–10, doi:10.1080/13504509.2018.1471012.

Hanasaki, N. et al., 2013: A global water scarcity assessment under Shared Socio-economic Pathways – Part 2: Water availability and scarcity. *Hydrol. Earth Syst. Sci.*, **17(7)**, 2393–2413, doi:10.5194/hess-17-2393-2013.

Hansen, T. and L. Coenen, 2015: The geography of sustainability transitions: Review, synthesis and reflections on an emergent research field. *Environ. Innov. Soc. Transitions*, **17**, 92–109, doi.org/10.1016/j.eist.2014.11.001.

Hansen, U.E. and I. Nygaard, 2014: Sustainable energy transitions in emerging economies: The formation of a palm oil biomass waste-to-energy niche in Malaysia 1990–2011. *Energy Policy*, **66**, 666–676, doi:10.1016/j.enpol.2013.11.028.

Hanson, K.T. and T.M.S. Korbla P. Puplampu, 2018: *From Millennium Development Goals to Sustainable Development Goals Rethinking African Development*. Routledge, Oxford, UK, 180 pp.

Harvey, C.A. et al., 2014: Climate-Smart Landscapes: Opportunities and Challenges for Integrating Adaptation and Mitigation in Tropical Agriculture. *Conserv. Lett.*, **7(2)**, 14, doi:10.1111/conl.12066.

Hasegawa, T. et al., 2018: Risk of increased food insecurity under stringent global climate change mitigation policy. *Nat. Clim. Change*, **8(8)**, 699–703, doi:10.1038/s41558-018-0230-x.

Hasegawa, T., P. Havlík, S. Frank, A. Palazzo, and H. Valin, 2019: Tackling food consumption inequality to fight hunger without pressuring the environment. *Nat. Sustain.*, **2(12)**, 826–833, doi:10.1038/s41893-019-0439-3.

Hayashi, A., F. Sano, Y. Nakagami, and K. Akimoto, 2018: Changes in terrestrial water stress and contributions of major factors under temperature rise constraint scenarios. *Mitig. Adapt. Strateg. Glob. Change*, **23(8)**, 1179–1205, doi:10.1007/s11027-018-9780-5.

Haywood, L.K., N. Funke, M. Audouin, C. Musvoto, and A. Nahman, 2019: The Sustainable Development Goals in South Africa: Investigating the need for multi-stakeholder partnerships. *Dev. South. Afr.*, **36(5)**, 555–569, doi:10.1080/0376835X.2018.1461611.

Heal, G., 2020: *Economic aspects of the energy transitions*, National Bureau of Economic Research, New York, NY, USA, 22 pp. https://www.nber.org/papers/w27766.

Healy, N., J.C. Stephens, and S.A. Malin, 2019: Embodied energy injustices: Unveiling and politicizing the transboundary harms of fossil fuel extractivism and fossil fuel supply chains. *Energy Res. Soc. Sci.*, **48**, 219–234, doi:10.1016/j.erss.2018.09.016.

Hedlund-de Witt, A., J. de Boer, and J.J. Boersema, 2014: Exploring inner and outer worlds: A quantitative study of worldviews, environmental attitudes, and sustainable lifestyles. *J. Environ. Psychol.*, **37**, 40–54, doi:10.1016/j.jenvp.2013.11.005.

Heikkinen, M., T. Ylä-Anttila, and S. Juhola, 2019: Incremental, reformistic or transformational: what kind of change do C40 cities advocate to deal with climate change? *J. Environ. Policy & Plan.*, **21(1)**, 90–103, doi:10.1080/1523908X.2018.1473151.

Hess, D.J., 2014: Sustainability transitions: A political coalition perspective. *Res. Policy*, **43(2)**, 278–283, doi:10.1016/j.respol.2013.10.008.

Hinkel, J., 2011: Indicators of vulnerability and adaptive capacity: towards a clarification of the science-policy interface. *Glob. Environ. Change*, **21(1)**, 198–208, doi.org/10.1016/j.gloenvcha.2010.08.002.

Hodson, G., C. MacInnis, and M. Busseri, 2017: Bowing and kicking: Rediscovering the fundamental link between generalized authoritarianism and generalized prejudice. *Personality and Individual Differences*, **104**, 104–243, doi.org/10.1016/j.paid.2016.08.018.

Hoek, A., M. Shirin, Rob Raven, Eli Court, and Emily Byrne, 2021: Towards environmentally sustainable food systems: decision-making factors in sustainable food production and consumption. *Sci. Direct*, 610–626.

Hoffmann, I., 2010: Climate Change in Context: Implications for Livestock Production and Diversity. In: *Sustainable improvement of animal production and health* [Odongo, N.E., M. Garcia, and G.J. Viljoen, (eds.)]. FAO-IAEA, Rome, Italy, pp. 33–44.

Hoffmann, M.J., 2011: *Climate Governance at the Crossroads: Experimenting with a Global Response after Kyoto*. Oxford University Press, New York, USA, 240 pp.

Höhne, N. et al., 2017: The Paris Agreement: resolving the inconsistency between global goals and national contributions. *Clim. Policy*, **17(1)**, 16–32, doi:10.1080/14693062.2016.1218320.

Holling, C.S., 2001: Understanding the complexity of economic, ecological and social systems. *Ecosystems*, **4(5)**, 390–405, doi.org/10.1007/s10021-001-0101-5.

Holscher, K., N. Frantzeskaki, T. McPhearson, and D. Loorbach, 2019: Tales of transforming cities: Transformative climate governance capacities in New

York City, US and Rotterdam, Netherlands. *J. Environ. Manage.*, **231(1)**, 843–857, doi:10.1016/j.jenvman.2018.10.043.

Hölscher, K., J.M. Wittmayer, and D. Loorbach, 2018: Transition versus transformation: What's the difference? *Environ. Innov. Soc. Transitions*, **27**, 1–3, doi:10.1016/j.eist.2017.10.007.

Hopkins, R., 2010: What can communities do? In: *The Post Carbon Reader: Managing the 21st Century's Sustainability Crisis*, Richard Heinberg and Daniel Lerch Healdsburg, California, USA, pp. 442–454.

Horcea-Milcu, A.I., B. Martín-López, D.P.M. Lam, and D.J. Lang, 2020: Research pathways to foster transformation: Linking sustainability science and social-ecological systems research. *Ecol. Soc.*, **25(1)**, 13, doi:10.5751/ES-11332-250113.

Howells, M. et al., 2013: Integrated analysis of climate change, land-use, energy and water strategies. *Nat. Clim. Change*, **3**, 621–626, doi:10.1038/NCLIMATE1789.

Hoyer, C., 2020: Accelerating Climate Action Beyond Company Gates. In: *Handbook of Climate Change Management: Research, Leadership, Transformation* [Leal Filho, W., J. Luetz, and D. Ayal, (eds.)]. Springer International Publishing, Cham, Switzerland, pp. 1–22.

Hulme, M., 2009: *Why We Disagree About Climate Change: Understanding Controversy, Inaction and Opportunity*. Cambridge University Press, Cambridge, UK, 436 pp.

Hume, T. and J. Barry, 2015: Environmental Education and Education for Sustainable Development. In: *International Encyclopedia of the Social & Behavioral Sciences*, Elsevier, Sligo, pp. 733–739.

Hutchison, S. and J. Walton, 2015: An inquiry into the significance of first-person research for the creation of knowledge to reduce suffering and enhance human flourishing. In: *Action Learning and Action Research Association (ALARA) World Congress*. University of Cumbria, Pretoria, p.1–9.

Hüther, G., 2018: *Co-creativity and Community*. 1st ed. Vandenhoeck & Ruprecht, 184 pp.

Huxham, M., M. Anwar, and D. Nelson, 2019: *Understanding the impact of a low carbon transition on South Africa*. Climate Policy Initiative, South Africa, 114 pp.

Iacopini, I., G. Petri, A. Barrat, and V. Latora, 2019: Simplicial models of social contagion. *Nat. Commun.*, **10(1)**, 2485, doi:10.1038/s41467-019-10431-6.

ICLEI, 2019: Local Governments for Sustainability. https://www.iclei.org/ (Accessed September 20, 2021).

IEA, 2017: *World Energy Outlook 2017*. IEA, Paris, France, 782 pp. https://www.iea.org/reports/world-energy-outlook-2017 (Accessed July 13, 2021).

IEA, 2019: *World Energy Outlook 2020*. IEA, Paris, France, 810 pp. https://www.iea.org/reports/world-energy-outlook-2019 (Accessed July 13, 2021).

IEA, 2020: *World Energy Outlook 2020, Executive Summary*. IEA, Paris, France, 12 pp. https://iea.blob.core.windows.net/assets/80d64d90-dc17-4a52-b41f-b14c9be1b995/WEO2020_ES.PDF.

IEAGHG, 2020: CCS and the Sustainable Development Goals. *Technical Reports*, no.14. https://ieaghg.org/ (Accessed November 10, 2021).

IIASA, 2018: *Transformations to Achieve the Sustainable Development Goals*. Report prepared by The World in 2050 initiative, Laxenburg, Austria, 157 pp.

IIASA, 2019: *The Digital Revolution and Sustainable Development: Opportunities and Challenges*. Report prepared by the World in 2050 initiative, Laxenburg, Austria, 100 pp.

IIASA, 2020: *Innovations for Sustainability: Pathways to an Efficient and Sufficient Post-pandemic Future*. IIASA, Laxenburg, Austria, 118 pp. http://www.twi2050.org/.

IPBES, 2019: *Global assessment report on biodiversity and ecosystem services of the Intergovernmental Science-Policy Platform on Biodiversity and Ecosystem Services*. IPBES, Bonn, Germany,1148 pp.

IPCC, 2014: *Climate Change 2014: Synthesis Report. Contribution of Working Groups I, II and III to the Fifth Assessment Report of the Intergovernmental Panel on Climate Change*. [Core Writing Team, R.K. Pachauri, and L.A. Meyers, (eds.)]. IPCC, Geneva, Switzerland, 151 pp.

IPCC, 2018: *Global Warming of 1.5°C. An IPCC Special Report on the impacts of global warming of 1.5°C above pre-industrial levels and related global greenhouse gas emission pathways, in the context of strengthening the global response to the threat of climate change*. [Masson-Delmotte, V., P. Zhai, H.-O. Pörtner, D. Roberts, J. Skea, P.R. Shukla, A. Pirani, W. Moufouma-Okia, C. Péan, R. Pidcock, S. Connors, J.B.R. Matthews, Y. Chen, X. Zhou, M.I. Gomis, E. Lonnoy, T. Maycock, M. Tignor, and T. Waterfield (eds.)]. Cambridge University Press, Cambridge, UK, and New York, NY, USA, 630 pp.

IPCC, 2019: *Climate Change and Land: An IPCC special report on climate change, desertification, land degradation, sustainable land management, food security, and greenhouse gas fluxes in terrestrial ecosystems* [P.R. Shukla, J. Skea, E. Calvo Buendia, V. Masson-Delmotte, H.-O. Pörtner, D.C. Roberts, P. Zhai, R. Slade, S. Connors, R. van Diemen, M. Ferrat, E. Haughey, S. Luz, S. Neogi, M. Pathak, J. Petzold, J. Portugal Pereira, P. Vyas, E. Huntley, K. Kissick, M. Belkacemi, J. Malley (eds.)]. Cambridge University Press, Cambridge, UK, and New York, NY, USA, 874 pp.

Ives, C.D., R. Freeth, and J. Fischer, 2019: Inside-out sustainability: The neglect of inner worlds. *Ambio*, **(49)**, 208–2017, doi:10.1007/s13280-019-01187-w.

Iyer, G. et al., 2018: Implications of sustainable development considerations for comparability across nationally determined contributions. *Nat. Clim. Change*, **8**, 124–129, doi:10.1038/s41558-017-0039-z.

Jaccard, I.S., P.P. Pichler, J. Többen, and H. Weisz, 2021: The energy and carbon inequality corridor for a 1.5 C compatible and just Europe. *Environ. Res. Lett.*, **16(6)**, 064082, doi:10.1088/1748-9326/abfb2f.

Jacobson, M.Z. et al., 2018: 100% clean and renewable Wind, Water, and Sunlight (WWS) all-sector energy roadmaps for 53 towns and cities in North America. *Sustain. Cities Soc.*, **42**, 22–37, doi.org/10.1016/j.scs.2018.06.031.

Jakob, M. et al., 2014: Feasible mitigation actions in developing countries. *Nat. Clim. Change*, **4**, 961–698, doi.org/10.1038/nclimate2370.

Jaumotte, F., W. Liu, and W. McKibbin, 2021: *Mitigating Climate Change: Growth-Friendly Policies to Achieve Net Zero Emissions by 2050*. The Centre for Applied Macroeconomic Analysis (CAMA), Australia, 46 pp. https://cama.crawford.anu.edu.au/sites/default/files/publication/cama_crawford_anu_edu_au/2021-09/75_2021_jaumotte_liu_mckibbin00.pdf (Accessed July 17, 2021).

Jones, C.A. and Levy, D.L., 2009: Business strategies and climate change. Changing climates in North American politics. In: *Business strategies and climate change. Changing climates in North American politics* [Pulver, S., B.G. Rabe, and P.J. Stoett, (eds.)]. University of Massachusetts, Boston, USA, pp. 219–240.

Jones, G.A. and K.J. Warner, 2016: The 21st century population-energy-climate nexus. *Energy Policy*, **93**, 206–212, doi:10.1016/j.enpol.2016.02.044.

Jones, L. et al., 2013: *The political economy of local adaptation planning: Exploring barriers to flexible and forward-looking decision making in three districtis in Ethiopia, Uganda and Mozambique*. 1st ed. Overseas Development Institute, Africa Climate Change Resilience Alliance, Accra, Ghana, 29 pp.

Jupesta, J., T. Wakiyama, and A. Abdullah, 2016: Introduction. In: *Low Carbon Urban Infrastructure Investment in Asian Cities*. Palgrave Macmillan, London, UK, pp. 1–6.

Jupesta, J. et al., 2020: Establishing Multi-Partnerships Environmental Governance in Indonesia: Case of Desa Makmur Perduli Api (Fire Free Village) Program. In: *Food security and land use change under conditions of climatic variability: A multidimensional perspective* [Squire, V. and M.K. Gaur, (eds.)]. Springer Nature Publisher, pp. 181–196.

Kainuma, M., P.R. Shukla, and K. Jiang, 2012: Framing and modeling of a low carbon society: An overview. *Energy Econ.*, **34**, 316–324, doi:10.1016/j.eneco.2012.07.015.

Kallis, G., 2017: Socialism Without Growth. *Capital. Nat. Social.*, **30(2)**, 189–206, doi:10.1080/10455752.2017.1386695.

17

Kalvelage, T. et al., 2013: Nitrogen cycling driven by organic matter export in the South Pacific oxygen minimum zone. *Nat. Geosci.*, **6(3)**, 228–234, doi:10.1038/ngeo1739.

Kanie, N. and F. Biermann, 2017: *Governing through Goals: Sustainable Development Goals as Governance Innovation*. MIT Press, Cambridge, MA, USA, 352 pp.

Kates, R., W. Travis, and T. Wilbanks, 2012: Transformational adaptation when incremental adaptations to climate change are insufficient. *Proc. Natl. Acad. Sci.*, **109**, 7156–7161, doi.org/10.1073/pnas.1115521109.

Keck, M.E. and K. Sikkink, 1999: Transnational advocacy networks in international and regional politics. *Int. Soc. Sci. J.*, **51(159)**, 89–101, doi:10.1111/1468-2451.00179.

Kedia, S., 2016: Approaches to low carbon development in China and India. *Adv. Clim. Chang. Res.*, **4**, 213–221, doi:10.1016/j.accre.2016.11.001.

Kefford, B.M., B. Ballinger, D.R. Schmeda-Lopez, C. Greig, and S. Smart, 2018: The early retirement challenge for fossil fuel power plants in deep decarbonisation scenarios. *Energy Policy*, **119**, 294–306, doi:10.1016/j.enpol.2018.04.018.

Kemp, R. and B. Never, 2017: Green transition, industrial policy, and economic development. *Oxford Rev. Econ. Policy*, **33(1)**, 66–84, doi:10.1093/oxrep/grw037.

Keohane, R.O. and D.G. Victor, 2016: Cooperation and discord in global climate policy. *Nat. Clim. Change*, **6(6)**, doi:10.1038/nclimate2937.

Keramidas, K. et al., 2018: *Global Energy and Climate Outlook 2018: Sectoral mitigation options towards a low-emissions economy – Global context to the EU strategy for long-term greenhouse gas emissions reduction*. Publications Office of the European Union, Luxembourg, 200 pp.

Kern, F., 2011: Ideas, institutions, and interests: Explaining policy divergence in fostering 'system innovations' towards sustainability. *Environ. Plan. C Gov. Policy*, **29(6)**, 1116–1134, doi:10.1068/c1142.

Kern, F. and A. Smith, 2008: Restructuring energy systems for sustainability? Energy transition policy in the Netherlands. *Energy Policy*, **36(11)**, doi:10.1016/j.enpol.2008.06.018.

Kern, F. and K.S. Rogge, 2016: The pace of governed energy transitions: Agency, international dynamics and the global Paris Agreement accelerating decarbonisation processes? *Energy Res. Soc. Sci.*, **22**, 13–17, doi:10.1016/j.erss.2016.08.016.

Kikstra, J.S. et al., 2021: Climate mitigation scenarios with persistent COVID-19-related energy demand changes. *Nat. Energy*, **6**, 1–10, doi:10.1038/s41560-021-00904-8.

Kim, G., and P. Coseo, 2019: Urban Park Systems to Support Sustainability: The Role of Urban Park Systems in Hot Arid Urban Climates. In: *Growth and Ecosystem Services of Urban Trees* [Rotzer, T., (ed.)]. MDPI, Basel, Switzerland, pp. 142–163.

Kim, H. and S. Grafakos, 2019: Which are the factors influencing the integration of mitigation and adaptation in climate change plans in Latin American cities? *Environ. Res. Lett.*, **14(105008)**, doi.org/10.1088/1748-9326/ab2f4c.

Kjellén, M., 2006: From Public Pipes to Private Hands: Water Access and Distribution in Dar es Salaam, Tanzania. https://finna.fi/Record/fikka.2007220 (Accessed February 22, 2021).

Koasidis, K. et al., 2020: Many Miles to Paris: A Sectoral Innovation System Analysis of the Transport Sector in Norway and Canada in Light of the Paris Agreement. *Sustainability*, **12(14)**, 5832, doi:10.3390/su12145832.

Koch, M., 2020: The state in the transformation to a sustainable postgrowth economy. *Env. Polit.*, **29(1)**, 115–133, doi:10.1080/09644016.2019.1684738.

Koehn, P.H., 2008: Underneath Kyoto: Emerging subnational government initiatives and incipient issue-bundling opportunities in China and the United States. *Glob. Environ. Polit.*, **8(1)**, 53–77, doi:10.1162/glep.2008.8.1.53.

Köhler, J. et al., 2019: An agenda for sustainability transitions research: State of the art and future directions. *Environ. Innov. Soc. Transitions*, **31**, 1–32, doi.org/10.1016/j.eist.2019.01.004.

Kojola, E., 2019: Indigeneity, gender and class in decision-making about risks from resource extraction. *Environ. Sociol.*, **5(2)**, 130–148, doi:10.1080/23251042.2018.1426090.

Kolstad, C., K. Urama, J. Broome, A. Bruvoll, M. Cariño Olvera, D. Fullerton, C. Gollier, W.M. Hanemann, R. Hassan, F. Jotzo, M.R. Khan, L. Meyer, and L. Mundaca, 2014: Social, Economic, and Ethical Concepts and Methods. In: *Cimate Change 2014: Mitigation of Climate Change. Contribution of Working Group III to the Fifth Assessment Report of the Intergovernmental Panel on Climate Change* [Edenhofer, O., R. Pichs-Madruga, Y. Sokona, E. Farahani, S. Kadner, K. Seyboth, A. Adler, I. Baum, S. Brunner, P. Eickemeier, B. Kriemann, J. Savolainen, S. Schlömer, C. von Stechow, T. Zwickel and J.C. Minx (eds.)]. Cambridge University Press, Cambridge UK, New York, NY, USA, pp. 207–282.

Kolk, A., D. Levy, and J. Pinkse, 2008: Corporate responses in an emerging climate regime: the institutionalization and commensuration of carbon disclosure. *Eur. Account. Rev.*, **17(4)**, 719–745. doi:10.1080/09638180802489121.

Kongsager, R., B. Locatelli, and F. Chazarin, 2016: Addressing Climate Change Mitigation and Adaptation Together: A Global Assessment of Agriculture and Forestry Projects. *Environ. Manage.*, **57(2)**, 271–282, doi:10.1007/s00267-015-0605-y.

Kothari, C., 2019: *Research Methodology: Methods and Techniques*. 4th ed. New Age International, New Delhi, India, 240 pp.

Kousksou, T. et al., 2015: Renewable energy potential and national policy directions for sustainable development in Morocco. *Renew. Sustain. Energy Rev.*, **47**, 46–57, doi.org/10.1016/j.rser.2015.02.056.

Koutroulis, A.G. et al., 2019: Global water availability under high-end climate change: A vulnerability based assessment. *Glob. Planet. Change*, **175**, 52–63, doi:10.1016/j.gloplacha.2019.01.013.

Krey, V. et al., 2019: Looking under the hood: A comparison of techno-economic assumptions across national and global integrated assessment models. *Elsevier*, **172**, 1254–1267, doi:10.1016/j.energy.2018.12.131.

Krishnan, B., S. Arumugam, and K. Maddulety, 2020: Critical success factors for the digitalization of smart cities. *Int. J. Technol. Manag. Sustain. Dev.*, **19**, 69–86.

Kruger, W., S. Stritzke, and P.A. Trotter, 2019: De-risking solar auctions in sub-Saharan Africa – A comparison of site selection strategies in South Africa and Zambia. *Renew. Sustain. Energy Rev.*, **104**, 429–438, doi.org/10.1016/j.rser.2019.01.041.

Kummu, M., D. Gerten, J. Heinke, M. Konzmann, and O. Varis, 2014: Climate-driven interannual variability of water scarcity in food production potential: A global analysis. *Hydrol. Earth Syst. Sci.*, **18**, 447–461, doi:10.5194/hess-18-447-2014.

Kuramochi, T. et al., 2018: Ten key short-term sectoral benchmarks to limit warming to 1.5°C. *Clim. Policy*, **18(3)**, 287–305, doi:10.1080/14693062.2017.1397495.

Kurian, M., 2017: The water-energy-food nexus: Trade-offs, thresholds and transdisciplinary approaches to sustainable development. *Environ. Sci. Policy*, **68**, 97–106, doi:10.1016/j.envsci.2016.11.006.

Laeremans, M. et al., 2018: Short-term effects of physical activity, air pollution and their interaction on the cardiovascular and respiratory system. *Environ. Int.*, **117**, 82–90, doi:10.1016/j.envint.2018.04.040.

Lam, D. et al., 2020: Scaling the impact of sustainability initiatives: a typology of amplification processes. *Urban Transform.*, **2(3)**, doi:10.1186/s42854-020-00007-9.

Lamb, S., J. Jennings, and P. Calain, 2017: The evolving role of CSR in international development: Evidence from Canadian extractive companies' involvement in community health initiatives in low-income countries. *Extr. Ind. Soc.*, **4(3)**, 614–621, doi:10.1016/j.exis.2017.05.011.

Landauer, M., S. Juhola, and M. Söderholm, 2015: Inter-relationships between adaptation and mitigation: a systematic literature review. *Clim. Change*, **131(4)**, 505–517, doi:10.1007/s10584-015-1395-1.

Landauer, M., S. Juhola, and J. Klein, 2019: The role of scale in integrating climate change adaptation and mitigation in cities. *Cities. J. Environ. Plan. Manag.*, **5**, 741–765. doi: 10.1080/09640568.2018.1430022.

Lange, E.A., 2018: Transforming Transformative Education Through Ontologies of Relationality. *J. Transform. Educ.*, **16(4)**, 280–301, doi:10.1177/1541344618786452.

Larsen, M., S. Petrović, A.M. Radoszynski, R. Mckenna, and O. Balyk, 2020: Climate change impacts on trends and extremes in future heating and cooling demands over Europe. *Energy Build.*, **226**, 110397, doi:10.1016/j.enbuild.2020.110397.

Larsen, M.A.D. and M. Drews, 2019: Water use in electricity generation for water-energy nexus analyses: The European case. *Sci. Total Environ.*, **651**, 2044–2058, doi:10.1016/j.scitotenv.2018.10.045.

Larsen, M.A.D. et al., 2019: Challenges of data availability: Analysing the water-energy nexus in electricity generation. *Energy Strateg. Rev.*, **26**, 100426, doi:10.1016/j.esr.2019.100426.

Latouche, S., 2018: The Path to Degrowth for a Sustainable Society. In: *Actor X. Eco-Efficiency in Industry and Science* [Lehmann, H., (ed.)]. Springer, Cham, Switzerland, pp. 277–284.

Lazarus, M. and H. van Asselt, 2018: Fossil fuel supply and climate policy: exploring the road less taken. *Clim. Change*, **150(1)**, 1–13, doi:10.1007/s10584-018-2266-3.

Le Bec, C., 2020: Pandemic puts strain 30 major African oil and gas projects. The Africa Report, https://www.theafricareport.com/27566/coronavirus-pandemic-puts-strain-30-major-african-oil-and-gas-projects/ (Accessed September 22, 2021).

Leclère, D. et al., 2020: Bending the curve of terrestrial biodiversity needs an integrated strategy. *Nature*, **585(7826)**, 551–556, doi:10.1038/s41586-020-2705-y.

Lee, H., E.Y. Jung, and J.D. Lee, 2019: Public–private co-evolution and niche development by technology transfer: A case study of state-led electricity system transition in South Korea. *Energy Res. Soc. Sci.*, **49**, 103–113, doi:10.1016/j.erss.2018.11.001.

Lee, S.-Y., E. Zusman, and S. Lee, 2017: Tracing Sustainability Transitions in Seoul Governance: Enabling and Scaling Grassroots Innovations. *Urban Transitions Conference*, [Seto, K., D. Robinson, H. and Virji, Z. Kovacs, J. Zhai, N. Sami, C. Pettit, and K. Sridhar, (eds.)]. Vol. 198 of *Procedia Engineering*, pp. 293–304. doi: 10.1016/j.proeng.2017.07.162.

Lemos, M. et al., 2013: Building Adaptive Capacity to Climate Change in Less Developed Countries. In: *Climate Science for Serving Society* [Asrar, G. and J. Hurrell, (eds.)]. Springer Netherlands, Dordrecht, Netherlands, pp. 437–457.

Leombruni, L. V, 2015: How you talk about climate change matters: A communication network perspective on epistemic skepticism and belief strength. *Glob. Environ. Change*, **35**, 148–161, doi.org/10.1016/j.gloenvcha.2015.08.006.

Lessem, R., 2018: *Community Activation for Integral Development*. 1st ed. Routledge, 156 pp.

Leventon, J., J.C. Dyer, and J.D. Van Alstine, 2015: The private sector in climate governance: Opportunities for climate compatible development through multilevel industry-government engagement. *J. Clean. Prod.*, **102**, 316–323, doi:10.1016/j.jclepro.2015.04.125.

Leventon, J., D.J. Abson, and D.J. Lang, 2021: Leverage points for sustainability transformations: nine guiding questions for sustainability science and practice. *Sustain. Sci.*, **16(3)**, 721–726, doi:10.1007/s11625-021-00961-8.

Lewis, S., 2016: *Positive Psychology and Change: How Leadership, Collaboration and Appreciative Inquiry Drive Transformational Results*, Wiley-Blackwell, 272 pp.

Li, B. and R. Yao, 2012: Building energy efficiency for sustainable development in China: challenges and opportunities. *Build. Res. & Inf.*, **40(4)**, 417–431, doi:10.1080/09613218.2012.682419.

Lipper, L. and D. Zilberman, 2018: A Short History of the Evolution of the Climate Smart Agriculture Approach and Its Links to Climate Change and Sustainable Agriculture Debates. In: *Climate Smart Agriculture: Building Resilience to Climate Change* [Lipper, L., N. McCarthy, D. Zilberman, S. Asfaw, and G. Branca, (eds.)]. Springer International Publishing, Cham, Switzerland, pp. 13–30.

Lipper, L. et al., 2014: Climate-smart agriculture for food security. *Nat. Clim. Change*, **4(12)**, 1068–1072, doi:10.1038/nclimate2437.

Liu, J. et al., 2018: Nexus approaches to global sustainable development. *Nat. Sustain.*, **1**, 466–476, doi.org/10.1038/s41893-018-0135-8.

Liu, J. et al., 2019: On knowledge generation and use for sustainability. *Nat. Sustain.*, **2(2)**, 80–82, doi:10.1038/s41893-019-0229-y.

Lo, K. and V. Castán Broto, 2019: Co-benefits, contradictions, and multi-level governance of low-carbon experimentation: Leveraging solar energy for sustainable development in China. *Glob. Environ. Change*, **59**, 18, doi:10.1016/j.gloenvcha.2019.101993.

Locatelli, B., C. Pavageau, E. Pramova, and M. Di Gregorio, 2015: Integrating climate change mitigation and adaptation in agriculture and forestry: Opportunities and trade-offs. *Wiley Interdiscip. Rev. Clim. Change*, **6(6)**, 585–598, doi:10.1002/wcc.357.

Locatelli, B., G. Fedele, V. Fayolle, and A. Baglee, 2016: Synergies between adaptation and mitigation in climate change finance. *Clim. Change Strateg. Manag*, **8**, 112–128, doi.org/10.1108/IJCCSM-07-2014-0088.

Lockhart, H., 2011: *Spirituality & Nature in the Transformation to a More Sustainable World: Perspectives of South African Change Agents.* Stellenbosch University, South Africa, 502 pp. https://scholar.sun.ac.za/handle/10019.1/18075.

Lohrmann, A., J. Farfan, U. Caldera, C. Lohrmann, and C. Breyer, 2019: Global scenarios for significant water use reduction in thermal power plants based on cooling water demand estimation using satellite imagery. *Nat. Energy*, **4(12)**, 1040–1048, doi:10.1038/s41560-019-0501-4.

Loorbach, D., 2010: Transition management for sustainable development: A prescriptive, complexity-based governance framework. *Governance*, **23(1)**, 161–183, doi:10.1111/j.1468-0491.2009.01471.x.

Loorbach, D. and J. Rotmans, 2010: The practice of transition management: Examples and lessons from four distinct cases. *Futures*, **42(3)**, 237–246, doi:10.1016/j.futures.2009.11.009.

Loorbach, D., H. Shiroyama, J.M. Wittmayer, J. Fujino, and S. Mizuguchi, 2016: *Governance of Urban Sustainability Transitions – European and Asian Experiences*. 202 pp. Springer Nature.

Loorbach, D., N. Frantzeskaki, and F. Avelino, 2017: Sustainability Transitions Research: Transforming Science and Practice for Societal Change. In: *Annual Review Of Environment And Resources, Vol 42* [Gadgil, A and T. Tomich (eds.)]. Dutch Research, Institute for Transitions, Erasmus University, Rotterdam. pp. 599–626.

Lucas, P., K. Ludwig, M. Kok, and S. Kruitwagen, 2016: *Sustainable Development Goals in the Netherlands: Building blocks for environmental policy for 2030*. PBL Netherlands Environmental Assessment Agency, Hague, The Netherlands, 56 pp.

Luken, T., 2019: Easy does it: an innovative view on developing career identity and self-direction. *Career Dev. Int.*, **25(2)**, 120–145, doi:10.1108/CDI-05-2019-0110.

Lwasa, S. et al., 2015: A meta-analysis of urban and peri-urban agriculture and forestry in mediating climate change. *Curr. Opin. Environ. Sustain.*, **13**, 68–73, doi.org/10.1016/j.cosust.2015.02.003.

Madden, N.J., 2014: Green means stop: Veto players and their impact on climate-change policy outputs. *Env. Polit.*, **23(4)**, doi:10.1080/09644016.2014.884301.

Magneschi, G., T. Zhang, and R. Munson, 2017: The Impact of $CO_2$ Capture on Water Requirements of Power Plants. *Energy Procedia*, **114**, 6337–6347, doi.org/10.1016/j.egypro.2017.03.1770.

Månberger, A. and B. Stenqvist, 2018: Global metal flows in the renewable energy transition: Exploring the effects of substitutes, technological mix and development. *Energy Policy*, **119**, 226–241, doi:10.1016/j.enpol.2018.04.056.

Mangnus, A.C. et al., 2019: New pathways for governing food system transformations: A pluralistic practice-based futures approach using visioning, back-casting, and serious gaming. *Ecol. Soc.*, **24(4)**, 2, doi:10.5751/ES-11014-240402.

Marchese, D. et al., 2018: Resilience and sustainability: Similarities and differences in environmental management applications. *Sci. Total Environ.*, **613–614**, 1275–1283, doi:10.1016/J.SCITOTENV.2017.09.086.

Marcotullio, P. et al., 2018: Energy transformation in cities. In: *Climate Change and Cities: Second Assessment Report of the Urban Climate Change Research Network*. [Rosenzweig, C., W. Solecki, P. Romero-Lankao, S. Mehrotra, S. Dhakal, and A. Ibrahim, (eds.)]. Cambridge University Press, New York, USA, pp. 443–490.

Markandya, A. et al., 2018: Health co-benefits from air pollution and mitigation costs of the Paris Agreement: a modelling study. *Lancet Planet. Heal.*, **2(3)**, 126–133, doi:10.1016/S2542-5196(18)30029-9.

Markard, J., F.W. Geels, and R. Raven, 2020: Challenges in the acceleration of sustainability transitions. **15**(8), 81001, doi:10.1088/1748-9326/ab9468.

Martinez-Juarez, P., A. Chiabai, T. Taylor, and S. Quiroga Gómez, 2015: The impact of ecosystems on human health and well-being: A critical review. *J. Outdoor Recreat. Tour.*, **10**, 63–69, doi.org/10.1016/j.jort.2015.06.008.

Mattioli, G., C. Roberts, J.K. Steinberger, and A. Brown, 2020: The political economy of car dependence: A systems of provision approach. *Energy Res. Soc. Sci.*, **66**, 101486, doi:10.1016/j.erss.2020.101486.

McCauley, S.M. and J.C. Stephens, 2012: Green energy clusters and socio-technical transitions: Analysis of a sustainable energy cluster for regional economic development in Central Massachusetts, USA. *Sustain. Sci.*, **7(2)**, 213–225, doi:10.1007/s11625-012-0164-6.

McCollum, D.L. et al., 2018: Energy investment needs for fulfilling the Paris Agreement and achieving the Sustainable Development Goals. *Nat. Energy*, **3(7)**, 589–599, doi:10.1038/s41560-018-0179-z.

McElwee, P. et al., 2020: The impact of interventions in the global land and agri-food sectors on Nature's Contributions to People and the UN Sustainable Development Goals. *Glob. Chang. Biol.*, **26(9)**, 4691–4721, doi:10.1111/gcb.15219.

McGlade, C. and P. Ekins, 2015: The geographical distribution of fossil fuels unused when limiting global warming to 2°C. *Nature*, **517**, 187–190, doi:10.1038/nature14016.

McKenzie, S. and A. Abdulkadri, 2018: *Mechanisms to accelerate the implementation of the Sustainable Development Goals in the Caribbean*. ECLAC, Santiago, Chile, 32 pp. https://repositorio.cepal.org/bitstream/handle/11362/43362/1/S1701302_en.pdf (Accessed July 1, 2021).

Mckibben, B., 2012: Global Warming's Terrifying New Math. Rolling Stone. https://www.rollingstone.com/politics/politics-news/global-warmings-terrifying-new-math-188550/ (Accessed August 15, 2021).

McNeely, J.A. and M. Munasinghe, 2021: Early lessons from COVID-19: An overview. *Ambio*, **50(4)**, 764–766, doi:10.1007/s13280-021-01541-x.

Meadowcroft, J., 2011: Engaging with the politics of sustainability transitions. *Environ. Innov. Soc. Transitions*, **1(1)**, 70–75, doi:10.1016/j.eist.2011.02.003.

Meadows, D., 2008: *Thinking in Systems*. [Wright, D., (ed.)]. Chelsea Green Publishing, USA, 240 pp.

Meckling, J. and J. Nahm, 2018: The power of process: State capacity and climate policy. *Governance*, **31(4)**, 741–757, doi:10.1111/gove.12338.

Meckling, J., N. Kelsey, E. Biber, and J. Zysman, 2015: Winning coalitions for climate policy. *Science*, doi:10.1126/science.aab1336.

Mendizabal, M., O. Heidrich, E. Feliu, G. García-Blanco, and A. Mendizabal, 2018: Stimulating urban transition and transformation to achieve sustainable and resilient cities. *Renew. Sustain. Energy Rev.*, **94**, 410–418, doi.org/10.1016/j.rser.2018.06.003.

Mensah, C.N., L. Dauda, K.B. Boamah, and M. Salman, 2021: One district one factory policy of Ghana, a transition to a low-carbon habitable economy? *Environ. Dev. Sustain.*, **23**, 703–721, doi:10.1007/s10668-020-00604-5.

Meyer, M., M. Hirschnitz-Garbers, and M. Distelkamp, 2018: Contemporary resource policy and decoupling trends-lessons learnt from integrated model-based assessments. *Sustainability*, **10(6)**, doi:10.3390/su10061858.

Michaelowa, K. and A. Michaelowa, 2017: Transnational Climate Governance Initiatives: Designed for Effective Climate Change Mitigation? *Int. Interact.*, **43(1)**, 129–155, doi:10.1080/03050629.2017.1256110.

Mikunda, T. et al., 2021: Carbon capture and storage and the sustainable development goals. *Int. J. Greenhouse Gas Control*, **108**, 103318, doi.org/10.1016/j.ijggc.2021.103318.

Miller, T. et al., 2014: The future of sustainability science: a solutions-oriented research agenda. *Sustain. Sci.*, **9(2)**, 239–246, doi.org/10.1007/s11625-013-0224-6.

Millward-Hopkins, J., J.K. Steinberger, N.D. Rao, and Y. Oswald, 2020: *Providing decent living with minimum energy: A global scenario*. IIASA, Laxemburg, 10 pp.

Ministry of Environment of Jordan, 2016: *National Strategy and Action Plan for Sustainable Consumption and Production in Jordan (2016–2025)*. Ministry of Environment of Jordan, Jordan, 86 pp. https://www.greengrowthknowledge.org/sites/default/files/downloads/resource/National_Strategy_and_Action_Plan_for_Sustainable_Consumption_and_Production_in_Jordan_%282016-2025%29.pdf (Accessed June 17, 2021)

Moe, E., 2014: Vested Interests, Energy Policy and Renewables in Japan, China, Norway and Denmark. In: *The Political Economy of Renewable Energy and Energy Security*, [Moe, E. and P. Midford (eds.)], Palgrave Macmillan, London, UK, pp. 276–317.

Monforti-Ferrario, F., C. Astorga-Llorens, and E. Pisoni, 2019: *Policy Pressures on Air – Anticipating Unforeseen Effects of EU Policies on Air Quality*. European Commission, Ispra, Italy, 50 pp. https://op.europa.eu/es/publication-detail/-/publication/de7643dd-0f34-11ea-8c1f-01aa75ed71a1/language-en/format-PDF (Accessed September 5, 2021)

Montuori, A. and G. Donnelly, 2018: Transformative leadership. In: *Handbook of Personal and Organizational Transformation* [Neal, J., (ed.)]. Springer International Publishing, 319–350 pp.

Munasinghe, M., 2007: The importance of social capital: Comparing the impacts of the 2004 Asian Tsunami on Sri Lanka, and Hurricane Katrina 2005 on New Orleans. *Ecol. Econ.*, **64(1)**, 9–11, doi:10.1016/j.ecolecon.2007.05.003.

Munger, F. and M. Riemer, 2012: A process model for research collaborations in environmental and sustainability fields. *Umweltpsychologie*, **1**, 112–142.

Naber, R., R. Raven, M. Kouw, and T. Dassen, 2017: Scaling up sustainable energy innovations. *Energy Policy*, **110**, 342–354, doi.org/10.1016/j.enpol.2017.07.056.

Nachmany, M. and J. Setzer, 2018: Global trends in climate change legislation and litigation: 2018 snapshot. Grantham Research Institute on Climate Change and the Environment, London, UK, 8 pp.

Nachmany, M. et al., 2015: *The 2015 Global Climate Legislation Study: a review of climate change legislation in 99 countries: summary for policy-makers*. Globe International, London, 1–45 pp. http://eprints.lse.ac.uk/65347/ (Accessed July 11, 2021).

Næss, J.S., O. Cavalett, and F. Cherubini, 2021: The land–energy–water nexus of global bioenergy potentials from abandoned cropland. *Nat. Sustain.*, **4(6)**, 525–536, doi:10.1038/s41893-020-00680-5.

Nair, S., B. George, H.M. Malano, M. Arora, and B. Nawarathna, 2014: Water-energy-greenhouse gas nexus of urban water systems: Review of concepts, state-of-art and methods. *Resour. Conserv. Recycl.*, **89**, 1–10, doi:10.1016/j.resconrec.2014.05.007.

Nakano, K., 2017: Screening of climatic impacts on a country's international supply chains: Japan as a case study. *Mitig. Adapt. Strateg. Glob. Change*, **22(4)**, 651–667, doi:10.1007/s11027-015-9692-6.

Narvaez Rojas, C., G.A. Alomia Peñafiel, D.F. Loaiza Buitrago, and C.A. Tavera Romero, 2021: Society 5.0: A Japanese Concept for a Superintelligent Society. *Sustainability*, **13(12)**, 6566, doi:10.3390/su13126567.

Nasiritousi, N. and K. Bäckstrand, 2019: *International Climate Politics in the Post-Paris Era*. Stockholm University, Stockholm, pp. 1–19. https://nordregio.org/wp-content/uploads/2018/10/International-Climate-Politics-in-the-post-Paris-era_Nasiritousi.pdf (Accessed May 14, 2021).

Nefzaoui, A., H. Ketata, and M. El Mourid, 2012: Agricultural Technological and Institutional Innovations for Enhanced Adaptation to Environmental Change in North Africa. In: *International Perspectives on Global Environmental Change* [Young, S. and S. Silvern (eds.)]. IntechOpen. doi:10.5772/27175.

Neofytou, H., A. Nikas, and H. Doukas, 2020: Sustainable energy transition readiness: A multicriteria assessment index. *Renew. Sustain. Energy Rev.*, **131**, 109988, doi.org/10.1016/j.rser.2020.109988.

Neuteleers, S. and B. Engelen, 2015: Talking money: How market-based valuation can undermine environmental protection. *Ecol. Econ.*, **117**, 253–260, doi:10.1016/j.ecolecon.2014.06.022.

Newell, P. and M. Paterson, 2010: *Climate capitalism: Global warming and the transformation of the global economy*. 1st ed. Cambridge University Press, London, UK, 223 pp.

Newell, P. and D. Mulvaney, 2013: The political economy of the 'just transition.' *Geogr. J.*, **179(2)**, 132–140, doi:10.1111/geoj.12008.

Newell, R., A. Dale, and M. Roseland, 2018: Climate Action Co-benefits and Integrated Community Planning: Uncovering the Synergies and Trade-Offs. *Clim. Change Impacts Responses*, **10**, 1–23, doi.org/10.18848/1835-7156/CGP/v10i04/1-23.

Newman, P., T. Beatley, and H. Boyer, 2017: Create Sustainable Mobility Systems. In: *Resilient Cities*, Island Press, Washington, DC, USA, pp. 53–87.

Nikas, A. et al., 2020: The desirability of transitions in demand: Incorporating behavioural and societal transformations into energy modelling. *Energy Res. Soc. Sci.*, **70**, 101780, doi:10.1016/J.ERSS.2020.101780.

Nilsson, M., D. Griggs, and M. Visbeck, 2016: Policy: Map the interactions between Sustainable Development Goals. *Nature*, **534(7607)**, 320–322, doi:10.1038/534320a.

Noura, M. et al., 2020: *A carbon management system of innovation: towards a circular carbon economy*. Saudi Arabia Think, Saudi Arabia, 26 pp. https://t20saudiarabia.org.sa/en/briefs/Documents/T20_TF2_PB4.pdf (Accessed July 30, 2021).

Novita, N. et al., 2020: Carbon Stocks from Peat Swamp Forest and Oil Palm Plantation in Central Kalimantan, Indonesia. In: *Climate Change Research, Policy and Actions in Indonesia: Science, Adaptation and Mitigation* [Djalante, R., J. Jupesta, and E. Aldrian, (eds.)]. Springer Nature Publisher, 203–227 pp.

Novita, N. et al., 2021: Geographic Setting and Groundwater Table Control Carbon Emission from Indonesian Peatland: A Meta-Analysis. *Forests*, **12(7)**, 832, doi:10.3390/f12070832.

O'Brien, K., 2018: Is the 1.5°C target possible? Exploring the three spheres of transformation. *Curr. Opin. Environ. Sustain.*, **31**, 153–160, doi:10.1016/J.COSUST.2018.04.010.

O'Neil, K. and B.A. Boyce, 2018: Improving Teacher Effectiveness in Physical Education Teacher Education Through Field-Based Supervision. *Phys. Educ.*, **75(5)**, 835–849, doi:10.18666/tpe-2018-v75-i5-7739.

Obersteiner, M. et al., 2016: Assessing the land resource food price nexus of the Sustainable Development Goals. *Sci. Adv.*, **2(9)**, doi:10.1126/sciadv.1501499.

OECD, 2015: Pathway to a low-carbon economy: Remarks at OECD-University of Cambridge COP21 side event. https://www.oecd.org/environment/pathway-to-a-low-carbon-economy-remarks-at-oecd-university-of-cambridge-cop21-side-event.htm (Accessed September 17, 2021).

OECD, 2016: *Better Policies for Sustainable Development 2016: A New Framework for Policy Coherence*. OECD Publishing, Paris, France, 295 pp.

OECD, 2018: *Development Co-operation Report 2018*. 472 pp. https://www.oecd-ilibrary.org/content/publication/dcr-2018-en.

OECD, 2019: *Development Co-operation Report 2019*. 156 pp. https://www.oecd-ilibrary.org/content/publication/9a58c83f-en.

OECD, 2020: Editorial: Turning hope into reality. *OECD*, **2(2)**, 3, doi.org/10.1787/fb94a4fb-en.

Okereke, C. et al., 2019: Governing green industrialisation in Africa: Assessing key parameters for a sustainable socio-technical transition in the context of Ethiopia. *World Dev.*, **115**, 279–290, doi:10.1016/j.worlddev.2018.11.019.

Olivier, D.W. and Y. Xu, 2019: Making effective use of groundwater to avoid another water supply crisis in Cape Town, South Africa. *Hydrogeol. J.*, **27(3)**, 823–826, doi:10.1007/s10040-018-1893-0.

Ordoñez, E., D. Mora, and K. Gaudry, 2019: Roadmap Toward NZEBs in Quito. *IOP Conf. Ser. Mater. Sci. Eng.*, **609**, 072040, doi:10.1088/1757-899X/609/7/072040.

Ornetzeder, M. and H. Rohracher, 2013: Of solar collectors, wind power, and car sharing: Comparing and understanding successful cases of grassroots innovations. *Glob. Environ. Change*, **23**(5), 856–867, doi:10.1016/j.gloenvcha.2012.12.007.

Oqubay, A. and J. Yifu Lin, 2020: *The Oxford Handbook of Industrial Hubs and Economic Development*. 1st ed. Oxford University Press, Oxford, UK, 800 pp.

Ortíz-Moya, F., E. Saraff, Y. Kataoka, and J. Fujino, 2021: *State of the Voluntary Local Reviews 2021: From Reporting to Action*. 1st ed. IGES, Kamiyamaguchi, 66 pp.

Oshiro, K., M. Kainuma, and T. Masui, 2016: Assessing decarbonization pathways and their implications for energy security policies in Japan. *Clim. Policy*, **16(1)**, 63–77, doi:10.1080/14693062.2016.1155042.

Ostrom, E., 2008:. *Clim. Change and Sustainable Development: New Challenges Poverty Reduction*, 22pp, doi:10.2139/ssrn.1304697.

Oswald, Y., A. Owen, and J.K. Steinberger, 2020: Publisher Correction: Large inequality in international and intranational energy footprints between income groups and across consumption categories. *Nat. Energy*, **5**, 231–239, doi:10.1038/s41560-020-0606-9.

Owen, R., G. Brennan, and F. Lyon, 2018: Enabling investment for the transition to a low carbon economy: government policy to finance early stage green innovation. *Curr. Opin. Environ. Sustain.*, **31**, 137–145, doi:10.1016/j.cosust.2018.03.004.

P. Puplampu, K., K. Hanson, and T. Shaw, 2017: 'From MDGs to SDGs: African Development Challenges and Prospects.' Kobena Hanson, Korbla Puplampu, Timothy Shaw (eds.), Routledge, pp. 1–10, United Nations University.

Paech, N., 2017: Woher kommt der Wachstumszwang? *GAIA – Ecol. Perspect. Sci. Soc.*, **16(4)**, doi:10.14512/gaia.16.4.13.

Pan, S.-Y., S.W. Snyder, A.I. Packman, Y. J. Lin, and P.-C. Chiang, 2018: Cooling water use in thermoelectric power generation and its associated challenges for addressing water-energy nexus. *Water-Energy Nexus*, **1(1)**, 26–41, doi:10.1016/j.wen.2018.04.002.

Parodi, O. and K. Tamm, 2018: *Personal Sustainability Exploring the Far Side of Sustainable Development*. 1st ed. [Parodi, O. and K. Tamm, (eds.)]. Routledge, London, UK, 224 pp.

Parra, D., L. Valverde, F.J. Pino, and M.K. Patel, 2019: A review on the role, cost and value of hydrogen energy systems for deep decarbonisation. *Renew. Sustain. Energy Rev.*, **101**, 279–294, doi:10.1016/j.rser.2018.11.010.

Pathak, M. and P.R. Shukla, 2016: Co-benefits of low carbon passenger transport actions in Indian cities: Case study of Ahmedabad. *Transp. Res. Part D Transp. Environ.*, **44**, 303–316, doi:10.1016/J.TRD.2015.07.013.

Patterson, J.J. et al., 2018: Political feasibility of 1.5°C societal transformations: the role of social justice. *Curr. Opin. Environ. Sustain.*, **31**, 1–9, doi:10.1016/j.cosust.2017.11.002.

Pellegrini-Masini, G., A. Pirni, and S. Maran, 2020: Energy justice revisited: A critical review on the philosophical and political origins of equality. *Energy Res. Soc. Sci.*, **59**, 101310, doi:10.1016/j.erss.2019.101310.

Pelling, B. et al., 2015: From research design to meta analysis guidelines. Université Libre de Bruxelles, Brussels, Belgium, 39 pp.

Pelling, M., 2010: *Adaptation to climate change: From resilience to transformation*. 1st ed. Taylor & Francis e-Library, London, UK, 274 pp. http://www.transitsocialinnovation.eu/content/original/Book covers/Local PDFs/163 TRANSIT D5.1 From research design to meta analysis guidelines.pdf.


17

Pencheon, D., 2018: Developing a sustainable health care system: the United Kingdom experience. *Med. J. Aust.*, **208(7)**, 284–285, doi:10.5694/mja17.01134.

Peszko, G., D. Mensbrugghe van der, and A. Golub, 2020: *Diversification and Cooperation Strategies in a Decarbonizing World*. World Bank, Washington, DC, pp. 1–36. https://openknowledge.worldbank.org/bitstream/handle/10986/34056/Diversification-and-Cooperation-Strategies-in-a-Decarbonizing-World.pdf?sequence=4&isAllowed=y (Accessed 30 October 2021).

Peterson, M.N., K.T. Stevenson, and D.F. Lawson, 2019: Reviewing how intergenerational learning can help conservation biology face its greatest challenge. *Biol. Conserv.*, **235** (2019), 290–294, doi:10.1016/j.biocon.2019.05.013.

Piggot, G., 2018: The influence of social movements on policies that constrain fossil fuel supply. *Clim. Policy*, **18(7)**, 942–954, doi:10.1080/14693062.2017.1394255.

Piggot, G., M. Boyland, A. Down, and A.R. Torre, 2019: Realizing a just and equitable transition away from fossil fuels. *Stock. Environ. Inst.*, (January), pp. 1–12. https://cdn.sei.org/wp-content/uploads/2019/01/realizing-a-just-and-equitable-transition-away-from-fossil-fuels.pdf (Accessed September 22, 2021).

Poblete-Cazenave, M., S. Pachauri, E. Byers, A. Mastrucci, and B. van Ruijven, 2021: Global scenarios of household access to modern energy services under climate mitigation policy. *Nat. Energy*, **6(8)**, 824–833, doi:10.1038/s41560-021-00871-0.

Pomeroy, E. and K. Oliver, 2018: Pushing the boundaries of self-directed learning: research findings from a study of u.lab participants in Scotland. *Int. J. Lifelong Educ.*, **37(6)**, 719–733, doi:10.1080/02601370.2018.1550447.

Pörtner, H. et al., 2021: *Launch of IPBES-IPCC Co-Sponsored Workshop Report on Biodiversity and Climate Change*. IPBES-IPCC, Bonn, Bremen-Geneva, 256 pp.

Poudel, S., 2014: Climate Smart Agriculture: A Development Alternative. https://sanjokpoudel.wordpress.com/2014/12/21/climate-smart-agriculture-a-development-alternative/ (Accessed September 2, 2021).

Power, C., 2016: The Integrity of Process: Is Inner Transition Sufficient? *J. Soc. Polit. Psychol.*, **4(1)**, 347–363, doi:10.5964/jspp.v4i1.538.

Pradhan, B.B., R.M. Shrestha, A. Pandey, and B. Limmeechokchai, 2018: Strategies to Achieve Net Zero Emissions in Nepal. *Carbon Manag.*, **9(5)**, 533–548, doi:10.1080/17583004.2018.1536168.

Pradhan, P., L. Costa, D. Rybski, W. Lucht, and J.P. Kropp, 2017: A Systematic Study of Sustainable Development Goal (SDG) Interactions. *Earth's Future*, **5(11)**, 1169–1179, doi.org/10.1002/2017EF000632.

Pretty, J.N. et al., 2006: Resource-conserving agriculture increases yields in developing countries. *Environ. Sci. Technol.*, **40(4)**, 1114–1119, doi:10.1021/es051670d.

Privitera, R. and D. La Rosa, 2017: Enhancing carbon sequestration potential of urban green spaces through transfer of development rights strategy. *Acta Geobalcanica*, **4(1)**, 17–23, doi:10.18509/agb.2018.02.

Prudhomme, R. et al., 2020: Combining mitigation strategies to increase co-benefits for biodiversity and food security. *Environ. Res. Lett.*, **15(11)**, 114005, doi:10.1088/1748-9326/abb10a.

Pyke, C.R., S. McMahon, L. Larsen, N.B. Rajkovich, and A. Rohloff, 2012: Development and analysis of Climate Sensitivity and Climate Adaptation opportunities indices for buildings. *Build. Environ.*, **55**, 141–149, doi.org/10.1016/j.buildenv.2012.02.020.

Quarton, C.J. and S. Samsatli, 2020: The value of hydrogen and carbon capture, storage and utilisation in decarbonising energy: Insights from integrated value chain optimisation. *Appl. Energy*, **257**, 113936, doi:10.1016/j.apenergy.2019.113936.

Rabe, B.G., 2007: Beyond Kyoto: Climate change policy in multilevel governance systems. *Governance*, **20(3)**, 423–444, doi:10.1111/j.1468-0491.2007.00365.x.

Rao, S. et al., 2017: Future air pollution in the Shared Socio-economic Pathways. *Glob. Environ. Change*, **42**, 346–358, doi.org/10.1016/j.gloenvcha.2016.05.012.

Rasul, G., 2016: Managing the food, water, and energy nexus for achieving the Sustainable Development Goals in South Asia. *Environ. Dev.*, **18**, 14–25, doi:10.1016/j.envdev.2015.12.001.

Reckien, D. et al., 2018: How are cities planning to respond to climate change? Assessment of local climate plans from 885 cities in the EU-28. *J. Clean. Prod.*, **191**, 207–219, doi:10.1016/j.jclepro.2018.03.220.

Rickards, L., J. Wiseman, and Y. Kashima, 2014: Barriers to effective climate change mitigation: The case of senior government and business decision makers. *Wiley Interdiscip. Rev. Clim. Change*, **5(6)**, 753–773, doi:10.1002/wcc.305.

Rempel, A. and J. Gupta, 2021: Fossil fuels, stranded assets and COVID-19: Imagining an inclusive and transformative recovery. *World Dev.*, **146**, 105608, doi:10.1016/J.WORLDDEV.2021.105608.

Richardson, B., 2017: Divesting from climate change: the road to influence. *Law Policy*, **39(4)**, 325–348, doi.org/10.1111/lapo.12081.

Ringler, C., A. Bhaduri, and R. Lawford, 2013: The nexus across water, energy, land and food (WELF): Potential for improved resource use efficiency? *Curr. Opin. Environ. Sustain.*, **6**, 617–624, doi:10.1016/j.cosust.2013.11.002.

Roberts, C. et al., 2018: The politics of accelerating low-carbon transitions: Towards a new research agenda. *Energy Res. Soc. Sci.*, **44**, 304–311, doi.org/10.1016/j.erss.2018.06.001.

Robinson, J. and R.J. Cole, 2015: Theoretical underpinnings of regenerative sustainability. *Build. Res. Inf.*, **43(2)**, 133–143, doi:10.1080/09613218.2014.979082.

Rockström, J. et al., 2009: A safe operating space for humanity. *Nature*, **461(7263)**, 472, doi:10.1038/461472a.

Rogelj, J. et al., 2016: Paris Agreement climate proposals need a boost to keep warming well below 2°C. *Nature*, **534(7609)**, 631–639, doi:10.1038/nature18307.

Rogelj, J., D. Shindell, K. Jiang, S. Fifita, P. Forster, V. Ginzburg, C. Handa, H. Kheshgi, S. Kobayashi, E. Kriegler, L. Mundaca, R. Séférian, and M.V.Vilariño, 2018: Mitigation Pathways Compatible with 1.5°C in the Context of Sustainable Development. In: *Global Warming of 1.5°C. An IPCC Special Report on the impacts of global warming of 1.5°C above pre-industrial levels and related global greenhouse gas emission pathways, in the context of strengthening the global response to the threat of climate change, sustainable development, and efforts to eradicate poverty* [Masson-Delmotte, V., P. Zhai, H.-O. Pörtner, D. Roberts, J. Skea, P.R. Shukla, A. Pirani, W. Moufouma-Okia, C. Péan, R. Pidcock, S. Connors, J.B.R. Matthews, Y. Chen, X. Zhou, M.I. Gomis, E. Lonnoy, T. Maycock, M. Tignor, and T. Waterfield (eds.)]. Cambridge University Press, Cambridge, UK and New York, NY, USA, pp. 93–174.

Rogge, K.S. and P. Johnstone, 2017: Exploring the role of phase-out policies for low-carbon energy transitions: The case of the German Energiewende. *Energy Res. Soc. Sci.*, **33**, 128–137, doi:10.1016/j.erss.2017.10.004.

Rogge, K.S., F. Kern, and M. Howlett, 2017: Conceptual and empirical advances in analysing policy mixes for energy transitions. *Energy Res. Soc. Sci.*, **33**, 1–10, doi:10.1016/j.erss.2017.09.025.

Roker, S., 2018: Democratic Republic of Congo home to largest hard rock lithium resource. Global Mining Review, https://www.globalminingreview.com/exploration-development/02082018/democratic-republic-of-congo-home-to-largest-hard-rock-lithium-resource/.

Romero-Lankao, P., D.M. Gnatz, O. Wilhelmi, and M. Hayden, 2016: Urban sustainability and resilience: From theory to practice. *Sustainability*, **8(12)**, 1224, doi:10.3390/su8121224.

Rooney-Varga, J.N. et al., 2020: The Climate Action Simulation. *Simul. Gaming*, **51(2)**, doi:10.1177/1046878119890643.

Roschnik, S. et al., 2017: Transitioning to environmentally sustainable health systems: the example of the NHS in England. *Public Heal. Panor.*, **3(02)**, 229–236. https://apps.who.int/iris/handle/10665/325309.

Rosenberg, L., J.D. Ouelette, and D.J. Dowgiallo, 2019: Analysis of S-band Passive Bistatic Sea Clutter. *2019 IEEE Radar Conf.*, 1–6, doi:10.1109/RADAR.2019.8835730.

Rosenbloom, D., H. Berton, and J. Meadowcroft, 2016: Framing the sun: A discursive approach to understanding multi-dimensional interactions within socio-technical transitions through the case of solar electricity in Ontario, Canada. *Res. Policy*, **45(6)**, 1275–1290, doi:10.1016/j.respol.2016.03.012.

Rosenbloom, D., J. Meadowcroft, S. Sheppard, S. Burch, and S. Williams, 2018: Transition experiments: Opening up low-carbon transition pathways for Canada through innovation and learning. *Can. Public Policy*, **44(4)**, 368–383, doi:10.3138/cpp.2018-020.

Rosenbloom, D., J. Meadowcroft, and B. Cashore, 2019: Stability and climate policy? Harnessing insights on path dependence, policy feedback, and transition pathways. *Energy Res. Soc. Sci.*, **50**, 168–178, doi:10.1016/j.erss.2018.12.009.

Rosendahl, K.E. and D.R. Rubiano, 2019: How Effective is Lithium Recycling as a Remedy for Resource Scarcity? *Environ. Resour. Econ.*, **74(3)**, doi:10.1007/s10640-019-00356-5.

Rothausen, S.G.S.A. and D. Conway, 2011: Greenhouse-gas emissions from energy use in the water sector. *Nat. Clim. Change*, **1**, 210–219, doi:10.1038/nclimate1147.

Roy, J., P. Tschakert, H. Waisman, S. Abdul Halim, P. Antwi-Agyei, P. Dasgupta, B. Hayward, M. Kanninen, D. Liverman, C. Okereke, P.F. Pinho, K. Riahi, and A.G. Suarez Rodriguez, 2018: Sustainable Development, Poverty Eradication and Reducing Inequalities. In: *Global Warming of 1.5°C: An IPCC Special Report on the impacts of global warming of 1.5°C above pre-industrial levels and related global greenhouse gas emission pathways, in the context of strengthening the global response to the threat of climate change, sustainable development and efforts to eradicate poverty* [MassonDelmotte, P. Zhai, H.-O. Pörtner, D. Roberts, J. Skea, P.R. Shukla, A. Pirani, W. Moufouma-Okia, C. Péan, R. Pidcock, S. Connors, J.B.R. Matthews, Y. Chen, X. Zhou, M.I. Gomis, E. Lonnoy, T. Maycock, M. Tignor, and T. Waterfield, (eds.)]. Cambridge University Press, Cambridge, UK, and New York, NY, USA, pp. 445–538.

Roy, J. et al., 2018b: Where is the hope? Blending modern urban lifestyle with cultural practices in India. *Curr. Opin. Environ. Sustain.*, **31**, 96–103, doi:10.1016/J.COSUST.2018.01.010.

Roy, J., N. Das, S. Some, and H. Mahmud, 2021: Fast-Growing Developing Countries: Dilemma and Way Forward in a Carbon-Constrained World. In: *Sustainable Development Insights from India*, Springer, Kolkata, India, pp. 23–41.

Ruiz-Garcia, L., L. Lunadei, P. Barreiro, and I. Robla, 2009: A Review of Wireless Sensor Technologies and Applications in Agriculture and Food Industry: State of the Art and Current Trends. *Sensors*, **9(6)**, 4728–4750, doi:10.3390/s90604728.

Rydge, J., M. Jacobs, and I. Granoff, 2015: *Ensuring New Infrastructure is Climate-Smart. Contributing paper for Seizing the Global Opportunity: Partnerships for Better Growth and a Better Climate*. The New Climate Economy, London, UK and Washington, USA, 1–20 pp. https://newclimateeconomy.report/2015/wp-content/uploads/sites/3/2014/08/Ensuring-infrastructure-is-climate-smart.pdf.

Sachs, J., G. Schmidt-Traub, C. Kroll, G. Lafortune, and G. Fuller, 2019: *Sustainable Development Report 2019*. Bertelsmann Stiftung and Sustainable Development Solutions Network, New York, USA. 478 pp.

Samantray, A. and P. Pin, 2019: Credibility of climate change denial in social media. *Palgrave Commun.*, **5(1)**, 127, doi:10.1057/s41599-019-0344-4.

Sanchez Rodriguez, R., D. Ürge-Vorsatz, and A. Barau, 2018: Sustainable Development Goals and climate change adaptation in cities. *Nat. Clim. Change*, **8**, 181–183, doi.org/10.1038/s41558-018-0098-9.

Sandfort, J. and K. Quick, 2015: Building deliberative capacity to create public value: the practices and artifacts of the Art of Hosting. In: *Public Administration and Public Value* [Bryson, J.M., L. Bloomberg, and B.C. Crosby (eds.)]. Georgetown University Press, Washington, DC, USA, pp. 39–52.

Sathaye, J., A. Najam, C. Cocklin, T. Heller, F. Lecocq, J. Llanes-Regueiro, J. Pan, G. Petschel-Held, S. Rayner, J. Robinson,R. Schaeffer, Y. Sokona, R. Swart, H. Winkler, 2007: Sustainable Development and Mitigation. In: *Climate Change 2007: Mitigation. Contribution of Working Group III to the Fourth Assessment Report of the Intergovernmental Panel on Climate Change* [Metz, B., O.R. Davidson, P.R. Bosch, R. Dave, and L.A. Meyer, (eds.)]. Cambridge University Press, Cambridge UK and New York, NY, USA, pp. 1–54.

Sator, O., 2018: *Insights from case studies of major coal-consuming economies*. IDDRI and Climate Strategies, France, 42 pp. https://coaltransitions.files.wordpress.com/2018/09/coal_synthesis_final.pdf (Accessed October 31, 2021).

Satterthwaite, D., 2017: Successful, safe and sustainable cities: towards a New Urban Agenda. *Commonwealth Journal of Local Governance*, 3–18, doi:10.5130/cjlg.v0i19.5446.

Schäpke, N., M. Bergmann, F. Stelzer, and D. Lang, 2018: Labs in the Real World: Advancing Transdisciplinary Research and Sustainability Transformation: Mapping the Field and Emerging Lines of Inquiry. *GAIA – Ecological Perspect. Sci. Soc.*, **27(1)**, 8–11, doi.org/10.14512/gaia.27.S1.4.

Scharmer, C. and Kaufer, K., 2015: Awareness-Based Action Research: Catching Social Reality Creation in Flight. In: *The SAGE Handbook of Action Research* [Bradbury, H., (ed.)]. SAGE Publications Ltd, London, UK, pp. 199–210.

Scharmer, O., 2018: *The Essentials of Theory U: Core Principles and Applications*. 2nd ed. Berret-Koehler, Oakland, USA, 248 pp.

Schaube, P., W. Ortiz, and M. Recalde, 2018: Status and future dynamics of decentralised renewable energy niche building processes in Argentina. *Energy Res. Soc. Sci.*, **35**, 57–67, doi:10.1016/j.erss.2017.10.037.

Schlott, M., A. Kies, T. Brown, S. Schramm, and M. Greiner, 2018: The impact of climate change on a cost-optimal highly renewable European electricity network. *Appl. Energy*, **230**, 1645–1649, doi:10.1016/j.apenergy.2018.09.084.

Schneidewind, U., M. Singer-Brodowski, K. Augenstein, and F. Stelzer, 2016: *Pledge for a transformative science: A conceptual framework*. Wuppertal Institut für Klima, Umwelt, Energie, Wuppertal, 30 pp. http://hdl.handle.net/10419/144815 (Accessed October 24, 2021).

Scholz, L., A. Ortiz Perez, B. Bierer, J. Wöllenstein, and S. Palzer, 2018: Gas sensors for climate research. *J. Sensors Sens. Syst.*, **7(2)**, 535–541, doi:10.5194/jsss-7-535-2018.

Schoneveld, G.C., D. Ekowati, A. Andrianto, and S. Van Der Haar, 2019: Modeling peat- and forestland conversion by oil palm smallholders in Indonesian Borneo. *Environ. Res. Lett.*, **14(1)**, 15, doi:10.1088/1748-9326/aaf044.

Schramade, W., 2017: Investing in the UN Sustainable Development Goals: Opportunities for Companies and Investors. *J. Appl. Corp. Financ.*, **29(2)**, 87–99, doi:10.1111/jacf.12236.

Schrobback, P., D. Adamson, and J. Quiggin, 2011: Turning Water into Carbon: Carbon Sequestration and Water Flow in the Murray-Darling Basin. *Environ. Resour. Econ.*, **49(1)**, 23–45, doi:10.1007/s10640-010-9422-1.

Schweizer-Ries, P., 2018: Sustainability Science and its contribution to IAPS: seeking for integrated sustainability. *IAPS Bull.*, **40**, 9–12. https://cupdf.com/document/iaps-bulletin-no40.html (Accessed August 13, 2021).

Scobie, M., 2016: Policy coherence in climate governance in Caribbean Small Island Developing States. *Environ. Sci. Policy*, **58**, 16–28, doi.org/10.1016/j.envsci.2015.12.008.

Scott, J., 2011: Social network analysis: developments, advances, and prospects. *Soc. Netw. Anal. Min.*, **1(1)**, 21–26, doi:10.1007/s13278-010-0012-6.

Seddon, N. et al., 2021: Getting the message right on nature-based solutions to climate change. *Glob. Chang. Biol.*, **27(8)**, 1518–1546, doi:10.1111/gcb.15513.

Seligman, M. and M. Csikszentmihalyi, 2014: Positive Psychology: An Introduction. In: *Flow and the Foundations of Positive Psychology* [Csikszentmihalyi, M., (ed.)]. Springer, Dordrecht, Netherlands, pp. 279–298.

Sengers, F., A.J. Wieczorek, and R. Raven, 2019: Experimenting for sustainability transitions: A systematic literature review. *Technol. Forecast. Soc. Change*, **145**, 153–164, doi.org/10.1016/j.techfore.2016.08.031.

Sethi, M. and J. Puppum De Oliveria, 2018: *Mainstreaming Climate Co-Benefits in Indian Cities*. 1st ed. [Sethi, M. and J. Puppum De Oliveria, (eds.)]. Springer, Singapore, 373 pp.

Seto, K.C. et al., 2016: Carbon Lock-In: Types, Causes, and Policy Implications. *Annu. Rev. Environ. Resour.*, **41**, 425–452, doi:10.1146/annurev-environ-110615-085934.

Setyowati, A.B., 2021: Mitigating inequality with emissions? Exploring energy justice and financing transitions to low carbon energy in Indonesia. *Energy Res. Soc. Sci.*, **71**, 101817, doi.org/10.1016/j.erss.2020.101817.

Seyfang, G. and A. Smith, 2007: Grassroots innovations for sustainable development: Towards a new research and policy agenda. *Env. Polit.*, **16(4)**, 584–603, doi:10.1080/09644010701419121.

Seyfang, G. and A. Haxeltine, 2012: Growing grassroots innovations: Exploring the role of community-based initiatives in governing sustainable energy transitions. *Environ. Plan. C Gov. Policy*, **30(3)**, 381–400, doi:10.1068/c10222.

Sgouridis, S., M. Carbajales-Dale, D. Csala, M. Chiesa, and U. Bardi, 2019: Comparative net energy analysis of renewable electricity and carbon capture and storage. *Nat. Energy*, **4(6)**, 456–465, doi:10.1038/s41560-019-0365-7.

Shah, N., M. Wei, V.E. Letschert, and A.A. Phadke, 2015: *Benefits of Leapfrogging to Superefficiency and Low Global Warming Potential Refrigerants in Room Air Conditioning*. Ernest Orlando Lawrence Berkeley National Laboratory, California, USA, 57 pp. https://www.osti.gov/servlets/purl/1235571 (Accessed 24 January 2020).

Sharifi, A., 2020: Co-benefits and synergies between urban climate change mitigation and adaptation measures: A literature review. *Sci. Total Environ.*, **750**, 141642, doi.org/10.1016/j.scitotenv.2020.141642.

Shaw, A., S. Burch, F. Kristensen, J. Robinson, and A. Dale, 2014: Accelerating the sustainability transition: Exploring synergies between adaptation and mitigation in British Columbian communities. *Glob. Environ. Change*, **25(1)**, 41–51, doi:10.1016/j.gloenvcha.2014.01.002.

Shaw, D. and P. Roberts, 2017: *Regional Planning and Development in Europe*. 1st ed. Routledge, London, UK, 314 pp.

Shen, W. et al., 2020: A comprehensive review of variable renewable energy levelized cost of electricity. *Renew. Sustain. Energy Rev.*, **133**, 110301, doi:10.1016/j.rser.2020.110301.

Sheppard, S.R.J. et al., 2011: Future visioning of local climate change: A framework for community engagement and planning with scenarios and visualisation. *Futures*, **43(4)**, 400–412, doi:10.1016/j.futures.2011.01.009.

Shrivastava, P., M. Stafford Smith, K. O'Brien, and L. Zsolnai, 2020: Transforming Sustainability Science to Generate Positive Social and Environmental Change Globally. *One Earth*, **2(4)**, 329–340, doi:10.1016/J.ONEEAR.2020.04.010.

Sindico, F., 2016: Paris, Climate Change, and Sustainable Developmen, Climate Law. *Clim. Law*, **6(1–2)**, 130–141, doi.org/10.1163/18786561-00601009.

Singleton, J., 2015: Head, heart and hands model for transformative learning: place as context for changing sustainability values Transforming eco-paradigms for sustainable values. *J. Sustain. Educ.*, **9** (March), 1–16.

Smajgl, A., J. Ward, and L. Pluschke, 2016: The water-food-energy Nexus – Realising a new paradigm. *J. Hydrol.*, **533**, 530–540, doi:10.1016/j.jhydrol.2015.12.033.

Smink, M.M., M.P. Hekkert, and S.O. Negro, 2015: Keeping sustainable innovation on a leash? Exploring incumbents' institutional strategies. *Bus. Strateg. Environ.*, **24(2)**, 86–101, doi:10.1002/bse.1808.

Smith, A., A. Stirling, and F. Berkhout, 2005: The governance of sustainable socio-technical transitions. *Res. Policy*, **34(10)**, 1491–1510, doi:10.1016/j.respol.2005.07.005.

Smith, P. et al., 2019a: Land-Management Options for Greenhouse Gas Removal and Their Impacts on Ecosystem Services and the Sustainable Development Goals. *Annu. Rev. Environ. Resour.*, **44(1)**, 255–286, doi:10.1146/annurev-environ-101718-033129.

Smith, P., J. Nkem, K. Calvin, D. Campbell, F. Cherubini, G. Grassi, V. Korotkov, A.L. Hoang, S. Lwasa, P. McElwee, E. Nkonya, N. Saigusa, J.-F. Soussana, M.A. Taboada, 2019: Interlinkages Between Desertification, Land Degradation, Food Security and Greenhouse Gas Fluxes: Synergies, Trade-offs and Integrated Response Options. In: *Climate Change and Land: an IPCC special report on climate change, desertification, land degradation, sustainable land management, food security, and greenhouse gas fluxes in terrestrial ecosystems* [P.R. Shukla, J. Skea, E. Calvo Buendia, V. Masson-Delmotte, H.- O. Portner, D. C. Roberts, P. Zhai, R. Slade, S. Connors, R. van Diemen, M. Ferrat, E. Haughey, S. Luz, S. Neogi, M. Pathak, J. Petzold, J. Portugal Pereira, P. Vyas, E. Huntley, K. Kissick, M. Belkacemi, J. Malley, (eds.)]. Cambridge University Press, Cambridge, UK, and New York, NY, USA, pp. 551–672.

Solarin, S.A., U. Al-Mulali, I. Musah, and I. Ozturk, 2017: Investigating the pollution haven hypothesis in Ghana: An empirical investigation. *Energy*, **124**, 706–719, doi:10.1016/j.energy.2017.02.089.

Song, L. et al., 2020: Contested energy futures, conflicted rewards? Examining low-carbon transition risks and governance dynamics in China's built environment. *Energy Res. Soc. Sci.*, **59**, 101306, doi:10.1016/J.ERSS.2019.101306.

Sorman, A.H., E. Turhan, and M. Rosas-Casals, 2020: Democratizing Energy, Energizing Democracy: Central Dimensions Surfacing in the Debate. *Front. Energy Res.*, **8**, 279, doi:10.3389/fenrg.2020.499888.

Soussana, J.F. et al., 2019: Matching policy and science: Rationale for the '4 per 1000 – soils for food security and climate' initiative. *Soil Tillage Res.*, **188**, 1–59, doi:10.1016/j.still.2017.12.002.

Sovacool, B.K., 2018: Bamboo Beating Bandits: Conflict, Inequality, and Vulnerability in the Political Ecology of Climate Change Adaptation in Bangladesh. *World Dev.*, **102**, 183–194, doi.org/10.1016/j.worlddev.2017.10.014.

Sovacool, B.K., 2021: Who are the victims of low-carbon transitions? Towards a political ecology of climate change mitigation. *Energy Res. Soc. Sci.*, **73**, 101916, doi.org/10.1016/j.erss.2021.101916.

Sovacool, B.K. and J. Scarpaci, 2016: Energy justice and the contested petroleum politics of stranded assets: Policy insights from the Yasuní-ITT Initiative in Ecuador. *Energy Policy*, **95**, 158–171, doi.org/10.1016/j.enpol.2016.04.045.

Sovacool, B.K., D.J. Hess, and R. Cantoni, 2021: Energy transitions from the cradle to the grave: A meta-theoretical framework integrating responsible innovation, social practices, and energy justice. *Energy Res. Soc. Sci.*, **75**, 102027, doi.org/10.1016/j.erss.2021.102027.

Speldewinde, P.C., D. Slaney, and P. Weinstein, 2015: Is restoring an ecosystem good for your health? *Sci. Total Environ.*, **502**, 276–279, doi:10.1016/j.scitotenv.2014.09.028.

Spencer, T. et al., 2018: The 1.5°C target and coal sector transition: at the limits of societal feasibility. *Clim. Policy*, **18(3)**, 335–351, doi:10.1080/14693062.2017.1386540.

Springmann, M. et al., 2018: Options for keeping the food system within environmental limits. *Nature*, **562**, 519–525, doi:10.1038/s41586-018-0594-0.

Srikanth, R., 2018: India's sustainable development goals – Glide path for India's power sector. *Energy Policy*, **123**, 325–336, doi:10.1016/j.enpol.2018.08.050.

Steen, M. and T. Weaver, 2017: Incumbents' diversification and cross-sectorial energy industry dynamics. *Res. Policy*, **46(6)**, 1071–1086, doi:10.1016/j.respol.2017.04.001.

Steenkamp, L.A., 2017: A review of policy options for clean electricity supply in South Africa. 2017. 6th International Conference on Clean Electrical Power: Renewable Energy Resources Impact, ICCEP 2, Santa Marguerita, Italy, pp. 94–102.


17

Steffen, W. et al., 2015: Planetary boundaries: Guiding human development on a changing planet. *Science*, **347(6223)**, 736–745, doi:10.1126/science.1259855.

Streck, D., 2007: Research and Social Transformation -Notes about Method and Methodology in Participatory Research. *Int. J. Action Res.*, **3(1+2)**, 112–130, doi:10.1688/1861-9916_ijar_2007_01_streck.

Sušnik, J. et al., 2018: Multi-stakeholder development of a serious game to explore the water-energy-food-land-climate nexus: The SIM4NEXUS approach. *Water (Switzerland)*, **2(10)**, 139, doi:10.3390/w10020139.

Svobodova, K., J.R. Owen, J. Harris, and S. Worden, 2020: Complexities and contradictions in the global energy transition: A re-evaluation of country-level factors and dependencies. *Appl. Energy*, **265**, 114778, doi:10.1016/j.apenergy.2020.114778.

Swain, R.B. and A. Karimu, 2020: Renewable electricity and sustainable development goals in the EU. *World Dev.*, **125**, 1046693, doi:10.1016/j.worlddev.2019.104693.

Swart, R. and F. Raes, 2007: Making integration of adaptation and mitigation work: Mainstreaming into sustainable development policies? *Clim. Policy*, **7(4)**, 288–303, doi:10.1080/14693062.2007.9685657.

Sweeney, S. and J. Treat, 2018: *Trade unions and just transitions. The Search for a Transformative Politics*. Trade Unions for Energy Democracy, New York, USA, 52 pp. http://unionsforenergydemocracy.org/wp-content/uploads/2018/04/TUED-Working-Paper-11.pdf (Accessed September 15, 2021).

Swilling, M. Annecke, E., 2010: *Just Transitions: Explorations of Sustainability in an Unfair World*. United Nations University Press, Tokyo, Japan, 448 pp.

Teng, F., X. Su, and X. Wang, 2019: Can China Peak Its Non-$CO_2$ GHG Emissions before 2030 by Implementing Its Nationally Determined Contribution? *Environ. Sci. Technol.*, **53(21)**, 12168–12176, doi:10.1021/acs.est.9b04162.

The Energy and Resources Institute, 2017: *Energy & Environment Data Diary and Yearbook (TEDDY)*. TERI, New Delhi, India, 350 pp.

The National BIM Report 2018, 2018: *The National BIM Report 2018*. https://www.thenbs.com/knowledge/the-national-bim-report-2018 (Accessed September 21, 2021).

The National People's Congress of the People's Republic of China, 2016: *The 13th Five-Year Plan for Economic and Social Development of the People's Republic of China*. The National People's Congress of the People's Republic of China, Beijing, China, 2019 pp.

Thornbush, M., O. Golubchikov, and S. Bouzarovski, 2013: Sustainable cities targeted by combined mitigation–adaptation efforts for future-proofing. *Sustain. Cities Soc.*, **9**, 1–9, doi.org/10.1016/j.scs.2013.01.003.

Thornton, T.F. and C. Comberti, 2017: Synergies and trade-offs between adaptation, mitigation and development. *Clim. Change*, **140(1)**, 5–18, doi:10.1007/s10584-013-0884-3.

Tong, D. et al., 2019: Committed emissions from existing energy infrastructure jeopardize 1.5°C climate target. *Nature*, **572(7769)**, 373–377, doi:10.1038/s41586-019-1364-3.

Townshend, T. et al., 2013: How national legislation can help to solve climate change. *Nat. Clim. Change*, **3(5)**, 430–432, doi:10.1038/nclimate1894.

Tozer, L., 2018: Urban climate change and sustainability planning: an analysis of sustainability and climate change discourses in local government plans in Canada. *J. Environ. Plan. Manag.*, **61(1)**, 176–194, doi:10.1080/09640568.2017.1297699.

Trullen, J. and B. Torbert, 2004: First, Second and Third person research in practice. *First, Second Third Pers. Res. Pract.*, The Systems Thinker, **15(1)**, 1–2.

Turnheim, B. and B. Nykvist, 2019: Opening up the feasibility of sustainability transitions pathways (STPs): Representations, potentials, and conditions. *Res. Policy*, **48(3)**, 775–788, doi:10.1016/j.respol.2018.12.002.

UN Women, 2017: *Approaches by African Countries in the Implementation and Localization of SDGs*. UN Women, Nairobi, Kenya, 60 pp. https://www2.unwomen.org/-/media/field office africa/attachments/publications/2017/11/approaches by african countries in the implementation and localization of sdgs- web.pdf?la=en&vs=5752 (Accessed April 3, 2021).

UNDRR, 2019: *UN Office for Disaster Risk Reduction, Sendai Framework*. 425 pp. https://www.undrr.org/publication/global-assessment-report-disaster-risk-reduction-2019 (Accessed August 14, 2021).

UNDESA, 2016: *Voluntary National Reviews – Synthesis Report 2016*. 71 pp. New York, USA, https://sustainabledevelopment.un.org/content/documents/127761701030E_2016_VNR_Synthesis_Report_ver3.pdf (Accessed October 25, 2021).

UNDESA, 2017: *Voluntary National Reviews – Synthesis Report 2017*. 78 pp. New York, USA, https://sustainabledevelopment.un.org/content/documents/17109Synthesis_Report_VNRs_2017.pdf (Accessed October 25, 2021).

UNDESA, 2018: *Voluntary National Reviews – Synthesis Report 2018*. 80 pp. New York, USA, https://sustainabledevelopment.un.org/content/documents/210732018_VNRs_Synthesis_compilation_11118_FS_BB_Format_FINAL_cover.pdf (Accessed October 25, 2021).

UNEP/SEI, 2020: *Production Gap Report 2020:The discrepancy between countries' planned fossil fuel production and global production levels consistent with limiting warming to 1.5°C or 2°C*. 76 pp. New York, USA.

UNFCCC, 2015: *Synthesis Report on the Aggregate Effect of the Intended Nationally Determined Contributions. Report No. FCCC/CP/2015/7*. The Secretariat of the United Nations Framework Convention on Climate Change (UNFCCC), New York, USA. http://unfccc.int/resource/docs/2015/cop21/eng/07.pdf (Accessed November 2, 2021).

UNFCCC, 2016: *Report of the Conference of the Parties on its twenty-first session, held in Paris from 30 November to 13 December 2015*. The Secretariat of the United Nations Framework Convention on Climate Change (UNFCCC), Paris, France, https://unfccc.int/resource/docs/2015/cop21/eng/10.pdf (Accessed June 22, 2020).

UNFCCC, 2019: *Communication of long-term strategies*. The Secretariat of the United Nations Framework Convention on Climate Change (UNFCCC), New York, USA, https://unfccc.int/process/the-paris-agreement/long-term-strategies (Accessed July 8, 2021).

UNFCCO, 2020: *Petersberg Climate Dialogue Seeks Climate-Resilient Economic Recovery*. The Secretariat of the United Nations Framework Convention on Climate Change (UNFCCC), New York, USA, https://unfccc.int/news/petersberg-climate-dialogue-seeks-climate-resilient-economic-recovery (Accessed August 11, 2021).

UNIDO, 2018: *Industrial park development in Ethiopia case of study report*. United Nations Industrial Development Organization (UNIDO), Vienna, Austria, pp. 2–85. https://www.unido.org/api/opentext/documents/download/10694802/unido-file-10694802.

United Nations, 2015a: Summit Charts New Era of Sustainable Development. *United Nations Press release*, New York, USA, 20 pp. https://sustainabledevelopment.un.org/post2015/summit (Accessed September 5, 2021).

United Nations, 2015b: *Sendai Framework for Disaster Risk Reduction 2015–2030*. United Nations, Geneva, Switzerland, 37 pp. https://www.preventionweb.net/files/43291_sendaiframeworkfordrren.pdf (Accessed June 22, 2020).

United Nations, 2017: *The Sustainable Development Goals Report*. United Nations, New York, USA, 1–64, https://www.undp.org/publications/un-sustainable-development-goals-report-2017?utm_source=EN&utm_medium=GSR&utm_content=US_UNDP_PaidSearch_Brand_English&utm_campaign=CENTRAL&c_src=CENTRAL&c_src2=GSR&gclid=Cj0KCQjw5oiMBhDtARIsAJi0qk210sGqvjhlX1DwK1NNf4WGlk (Accessed October 13, 2021).

United Nations, 2018: *Sustainable Development Goal 6: Synthesis Report on Water and Sanitation*. New York, USA, https://www.unwater.org/publications/sdg-6-synthesis-report-2018-on-water-and-sanitation/ (Accessed June 22, 2020).

United Nations, 2019: *Global Sustainable Development Report 2019: The Future is Now – Science for Achieving Sustainable Development*. United Nations,

17

New York, USA 252 pp. https://sustainabledevelopment.un.org/content/documents/24797GSDR_report_2019.pdf (Accessed 14 January 2020).

United Nations Environment Programme, 2019: *Emissions Gap Report 2019*. United Nations Environment Programme (UNEP), Nairobi, Kenya, 108 pp. https://www.unenvironment.org/resources/emissions-gap-report-2019.

United Nations General Assembly, 2019: Resolution adopted by the General Assembly on 15 October. New York, USA, 2pp, https://undocs.org/pdf?symbol=en/A/RES/74/5 (Accessed September 5, 2021).

United Nations General Assembly, 2021: High-Level Meetings of the 76th Session. New York, USA, https://www.un.org/en/ga/76/meetings/ (Accessed September 28, 2021).

UNU-INRA, 2020: Accra, Ghana. UNU Institute for Natural Resources in Africa (UNU-INRA). https://unu.edu/about/unu-system/inra (Accessed August 25, 2020).

Ürge-Vorsatz, D., S.T. Herrero, N.K. Dubash, and F. Lecocq, 2014: Measuring the Co-Benefits of Climate Change Mitigation. *Annu. Rev. Environ. Resour.*, **39(1)**, 549–582, doi:10.1146/annurev-environ-031312-125456.

UToronto, 2016: *G20 Action Plan on the 2030 Agenda for Sustainable Development*. Toronto, Canada, pp. 3–48. https://www.b20germany.org/fileadmin/user_upload/G20_Action_Plan_on_the_2030_Agenda_for_Sustainable_Development.pdf (Accessed November 28, 2021).

van Asselt, H., 2014: *The fragmentation of global climate governance: Consequences and management of regime interactions*. Edward Elgar Publishing Ltd., Cheltenham, UK, 335 pp.

van den Berg, N.J. et al., 2020: Implications of various effort-sharing approaches for national carbon budgets and emission pathways. *Clim. Change*, **162(4)**, 1805–1822, doi:10.1007/s10584-019-02368-y.

van der Heijden, J., J. Patterson, S. Juhola, and M. Wolfram, 2019: Special section: advancing the role of cities in climate governance–promise, limits, politics. *J. Environ. Plan. Manag.*, **62(3)**, 365–373, doi:10.1080/09640568.2018.1513832.

van der Heijden, J., 2018: The limits of voluntary programs for low-carbon buildings for staying under 1.5°C. *Curr. Opin. Environ. Sustain.*, **30**, 59–66, doi:10.1016/j.cosust.2018.03.006.

van der Ploeg, F., 2011: Natural resources: Curse or blessing? *J. Econ. Lit.*, **49(2)**, 366–420, doi:10.1257/jel.49.2.366.

van der Ploeg, F. and A. Rezai, 2020: Stranded assets in the transition to a carbon-free economy. *Annu. Rev. Resour. Econ.*, **12**, 281–298, doi:10.1146/annurev-resource-110519-040938.

van Rooyen, J.D., A.P. Watson, and J.A. Miller, 2020: Combining quantity and quality controls to determine groundwater vulnerability to depletion and deterioration throughout South Africa. *Environ. Earth Sci.*, **79(11)**, 255, doi:10.1007/s12665-020-08998-1.

van Soest, H.L. et al., 2019: Analysing interactions among Sustainable Development Goals with Integrated Assessment Models. *Glob. Transitions*, **1**, 210–225, doi.org/10.1016/j.glt.2019.10.004.

van Vliet, M.T.H. et al., 2016a: Multi-model assessment of global hydropower and cooling water discharge potential under climate change. *Glob. Environ. Change*, **40**, 156–170, doi:10.1016/j.gloenvcha.2016.07.007.

van Vliet, M.T.H., J. Sheffield, D. Wiberg, and E.F. Wood, 2016b: Impacts of recent drought and warm years on water resources and electricity supply worldwide. *Environ. Res. Lett.*, 1–11, doi:10.1088/1748-9326/11/12/124021.

van Vliet, M.T.H., D. Wiberg, S. Leduc, and K. Riahi, 2016c: Power-generation system vulnerability and adaptation to changes in climate and water resources. *Nat. Clim. Change*, , **6**, 375–380 pp. doi:10.1038/nclimate2903.

Vandyck, T., K. Keramidas, B. Saveyn, A. Kitous, and Z. Vrontisi, 2016: A global stocktake of the Paris pledges: Implications for energy systems and economy. *Glob. Environ. Change*, **41**, 46–63, doi.org/10.1016/j.gloenvcha.2016.08.006.

Vandyck, T. et al., 2018: Air quality co-benefits for human health and agriculture counterbalance costs to meet Paris Agreement pledges. *Nat. Commun.*, **9(1)**, 4939, doi:10.1038/s41467-018-06885-9.

Varela, F.J., 1999 and Jonathan Shear: *First-person Methodologies: What, Why, How ?* Semantic Scholar, https://www.semanticscholar.org/paper/and-Jonathan-Shear-First-person-Methodologies-%3A-%2C-%2C-Varela/3852a7981815f05f0a23e0710bbc7d6c52086ca3?p2df (Accessed September 1, 2021).

Veciana, S. and K. Ottmar, 2018: Inner conflict resolution and self-empowerment as contribution for personal sustainability on the case of intentional community practices. In: *Personal Sustainability: Exploring the Far Side of Sustainable Development*, Taylor & Francis Group, Routledge, 20 pp.

Veldkamp, T.I.E. et al., 2015: Changing mechanism of global water scarcity events: Impacts of socioeconomic changes and inter-annual hydro-climatic variability. *Glob. Environ. Change*, **32**, 18–29, doi:10.1016/j.gloenvcha.2015.02.011.

Wada, Y. et al., 2016: Modeling global water use for the 21st century: The Water Futures and Solutions (WFaS) initiative and its approaches. *Geosci. Model Dev.*, **9**, 175–222, doi:10.5194/gmd-9-175-2016.

Walsh, Z., J. Böhme, and C. Wamsler, 2020: Towards a relational paradigm in sustainability research, practice, and education. *Ambio*, **50**, doi:10.1007/s13280-020-01322-y.

Wamsler, C. and E. Brink, 2018: Mindsets for Sustainability: Exploring the Link Between Mindfulness and Sustainable Climate Adaptation. *Ecol. Econ.*, **151**, 55–61, doi.org/10.1016/j.ecolecon.2018.04.029.

Wamsler, C., G. Osberg, W. Osika, H. Herndersson, and L. Mundaca, 2021: Linking internal and external transformation for sustainability and climate action: Towards a new research and policy agenda. *Glob. Environ. Change*, **71**, 102373, doi.org/10.1016/j.gloenvcha.2021.102373.

Wan, K.K.W., D.H.W. Li, W. Pan, and J.C. Lam, 2012: Impact of climate change on building energy use in different climate zones and mitigation and adaptation implications. *Appl. Energy*, **97**, 274–282, doi.org/10.1016/j.apenergy.2011.11.048.

Wang, H. and W. Chen, 2019: Gaps between pre-2020 climate policies with NDC goals and long-term mitigation targets: Analyses on major regions. *Energy Procedia*, **158**, 22–25, doi:10.1016/j.egypro.2019.01.894.

Wang, Y. and Q. Zhi, 2016: The Role of Green Finance in Environmental Protection: Two Aspects of Market Mechanism and Policies. *Energy Procedia*, **104**, 311–316, doi.org/10.1016/j.egypro.2016.12.053.

Warmenbol, C. and M. Smith, 2018: International union for conservation of nature (IUCN). In: *The Wetland Book: I: Structure and Function, Management, and Methods*, [C. Max Finlayson, Mark Everard, Kenneth Irvine, Robert J. McInnes, Beth A. Middleton, Anne A. van Dam, Nick C. Davidson (eds.)], Springer Netherlands, pp. 665–669.

Warren, B., P. Christoff, and D. Green, 2016: Australia's sustainable energy transition: The disjointed politics of decarbonisation. *Environ. Innov. Soc. Transitions*, **21**, 1–12, doi:10.1016/j.eist.2016.01.001.

WCED, 1987: *World Commission on Environment and Development. 1987. Our Common Future.* WCED, Geneva, Switzerland, 190 pp. https://idl-bnc-idrc.dspacedirect.org/bitstream/handle/10625/8942/WCED_79365.pdf?sequence=1&isAllowed=y (Accessed May 31, 2020).

Westman, L., E. Moores, and S.L. Burch, 2021: Bridging the governance divide: The role of SMEs in urban sustainability interventions. *Cities*, **108**, 102944, doi.org/10.1016/j.cities.2020.102944.

White, David J. & Hubacek, Klaus & Feng, Kuishuang & Sun, Laixiang & Meng, Bo, 2018. "The Water-Energy-Food Nexus in East Asia: A tele-connected value chain analysis using inter-regional input-output analysis," Applied Energy, Elsevier, vol. 210(C), pages 550–567.

Wieczorek, A.J., 2018: Sustainability transitions in developing countries: Major insights and their implications for research and policy. *Environ. Sci. Policy*, **84**, 204–216, doi:10.1016/J.ENVSCI.2017.08.008.

Wiedmann, T., M. Lenzen, L.T. Keyßer, and J.K. Steinberger, 2020: Scientists' warning on affluence. *Nat. Commun.*, **11(1)**, 3107, doi:10.1038/s41467-020-16941-y.

Wiek, A., F. Farioli, K. Fukushi, and M. Yarime, 2012: Sustainability science: Bridging the gap between science and society. *Sustain. Sci.*, **7(1)**, 1–4, doi:10.1007/s11625-011-0154-0.

Wiktorowicz, J. et al., 2018: WGV: An Australian urban precinct case study to demonstrate the 1.5°C agenda including multiple SDGs. *Urban Plan.*, **3(2)**, 64–81, doi:10.17645/up.v3i2.1245.

Williams, L., 2013: Deepening Ecological Relationality Through Critical Onto-Epistemological Inquiry: Where Transformative Learning Meets Sustainable Science. *J. Transform. Educ.*, **11(2)**, 95–113, doi:10.1177/1541344613490997.

Williams, S. and A. Doyon, 2020: The Energy Futures Lab: A case study of justice in energy transitions. *Environ. Innov. Soc. Transitions*, **37**, 29–301, doi:10.1016/j.eist.2020.10.001.

Willis, R., 2018: How Members of Parliament understand and respond to climate changett. *Sociol. Rev.*, **66(3)**, doi:10.1177/0038026117731658.

Wilson, C., H. Pettifor, E. Cassar, L. Kerr, and M. Wilson, 2019: The potential contribution of disruptive low-carbon innovations to 1.5°C climate mitigation. *Energy Effic.*, **12(2)**, 423–440, doi:10.1007/s12053-018-9679-8.

Winkler, H., 2020: Putting equity into practice in the global stocktake under the Paris Agreement. *Clim. Policy*, **20(1)**, 124–132, doi:10.1080/14693062.2019.1680337.

Wittneben, B., D. Bongardt, H. Dalkmann, W. Sterk, and C. Baatz, 2009: Integrating Sustainable Transport Measures into the Clean Development Mechanism. *Transp. Rev.*, **29(1)**, 91–113, doi:10.1080/01441640802133494.

Woiwode, C. et al., 2021: Inner transformation to sustainability as a deep leverage point: fostering new avenues for change through dialogue and reflection Leverage Points for Sustainability Transformations. *Sustain. Sci.*, **16**, 841–858, doi:10.1007/s11625-020-00882-y.

Woiwode, D. and N. Woiwode, 2019: Off the beaten tracks: The neglected significance of interiority for sustainable urban development. In: *Practical Spirituality and the Contemporary City: Awakening the Transformative Power for Sustainable Living* [Giri, A., (ed.)]. Palgrave Mcmillan, Simgapore, pp. 129–151.

World Bank, 2019: *Competitiveness, Report of the higth-level commission on carbon princing and competitiveness*. World Bank, Washington D.C., 53 pp. https://openknowledge.worldbank.org/bitstream/handle/10986/32419/141917.pdf?sequence=4&isAllowed=y (Accessed March 5, 2020).

World Business Council for Sustainable Developmet, 2010: *Vision 2050*. 80 pp. https://www.wbcsd.org/Overview/About-us/Vision_2050/Resources/Vision-2050-The-new-agenda-for-business (Accessed 20 January 2020).

WRI, 2016: *The sustainable infrastructure imperative*. London, UK, 152 pp. http://newclimateeconomy.report/2016/wp-content/uploads/sites/4/2014/08/NCE_2016Report.pdf (Accessed September 1, 2020).

WWAP-UNESCO, 2019: *World Water Development Report. Leaving No One Behind*. WWAP-UNESCO, Paris, France, 201 pp. https://unesdoc.unesco.org/ark:/48223/pf0000367306 (Accessed November 1, 2020).

Yang, X. and F. Teng, 2018: The air quality co-benefit of coal control strategy in China. *Resour. Conserv. Recycl.*, **129**, 373–382, doi.org/10.1016/j.resconrec.2016.08.011.

Yarime, M., 2017: Facilitating data-intensive approaches to innovation for sustainability: opportunities and challenges in building smart cities. *Sustain. Sci.*, **12**, 881–885, doi:10.1007/s11625-017-0498-1.

Yaw, N., 2018: One district, one factory. *smartlinewestafrica*, https://1d1f.gov.gh/ (Accessed August 20, 2020).

Zeng, D.Z., L. Cheng, L. Shi, and W. Luetkenhorst, 2020: China's green transformation through eco-industrial parks. *World Dev.*, **140**, 105249, doi:10.1016/j.worlddev.2020.105249.

Zenghelis, D., 2020: Can we be green and grow? https://www.lombardodier.com/contents/corporate-news/responsible-capital/2019/november/can-we-be-green-and-grow.html (Accessed October 5, 2021).

Zhang, C., X. Chen, Y. Li, W. Ding, and G. Fu, 2018a: Water-energy-food nexus: Concepts, questions and methodologies. *J. Clean. Prod.*, doi:10.1016/j.jclepro.2018.05.194.

Zhang, J. et al., 2018b: The water-food-energy nexus optimization approach to combat agricultural drought: a case study in the United States. *Appl. Energy*, **227**, 449–464, doi:10.1016/j.apenergy.2017.07.036.

Zhang, R. et al., 2019: A Review of the Role of the Atlantic Meridional Overturning Circulation in Atlantic Multidecadal Variability and Associated Climate Impacts. *Rev. Geophys.*, **57(2)**, 316–375, doi:10.1029/2019RG000644.

Zhao, X., S. Zhang, Y. Zou, and J. Yao, 2013: To what extent does wind power deployment affect vested interests? A case study of the Northeast China Grid. *Energy Policy*, **63**, doi:10.1016/j.enpol.2013.08.092.

Zhenmin, L. and P. Espinosa, 2019: Tackling climate change to accelerate sustainable development. *Nat. Clim. Change*, **9(7)**, 494–496, doi:10.1038/s41558-019-0519-4.

Zusman, E., A. Srinivasan, and S. Dhakal, 2011: *Low carbon transport in Asia: Strategies for optimizing co-benefits*. 1st ed. [Zusman, E., A. Srinivasan, and S. Shobhakar, (eds.)]. Institute for Environmental Studies, Tsukuba, Japan, 296 pp.

# Annexes

# AI

# Annex I: Glossary

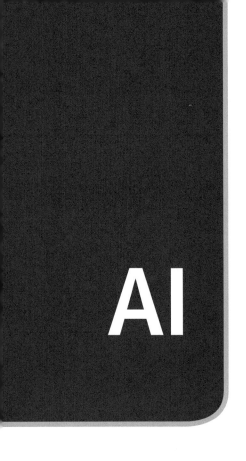

**Coordinating Editors:**
Renée van Diemen (the Netherlands/United Kingdom), J.B. Robin Matthews (France/United Kingdom), Vincent Möller (Germany), Jan S. Fuglestvedt (Norway), Valérie Masson-Delmotte (France), Carlos Méndez (Venezuela), Andy Reisinger (New Zealand), Sergey Semenov (the Russian Federation)

**Editorial Team:**
Elin Lerum Boasson (Norway), Antonethe Castaneda (Guatemala), Lazarus Chapungu (Zimbabwe), Lilia Caiado Couto (Brazil), Annette Cowie (Australia), Nandini Das (India), El Hdaji Mbaye Madien Diagne (Senegal), Oliver Geden (Germany), James S. Gerber (the United States of America), Siyue Guo (China), Shan Hu (China), Kiane de Kleijne (the Netherlands), Michael König (Germany), David Simon Lee (United Kingdom), Andreas Löschel (Germany), André F.P. Lucena (Brazil), Eduardo Miranda Casseres (Brazil), Jacqueline Peel (Australia), Patricia E. Perkins (Canada), Priyadarshi Shukla (India), Jim Skea (United Kingdom), Brent Sohngen (the United States of America), Enock Ssekuubwa (Uganda), Linda Steg (the Netherlands)

**Note:**
This glossary defines some specific terms as the Lead Authors intend them to be interpreted in the context of this report. Italicised words in definitions indicate that the italicised term is defined in the Glossary.

Subterms appear in *italics* beneath main terms.

**This chapter should be cited as:**
IPCC, 2022: Annex I: Glossary [van Diemen, R., J.B.R. Matthews, V. Möller, J.S. Fuglestvedt, V. Masson-Delmotte, C. Méndez, A. Reisinger, S. Semenov (eds)]. In IPCC, 2022: *Climate Change 2022: Mitigation of Climate Change. Contribution of Working Group III to the Sixth Assessment Report of the Intergovernmental Panel on Climate Change* [P.R. Shukla, J. Skea, R. Slade, A. Al Khourdajie, R. van Diemen, D. McCollum, M. Pathak, S. Some, P. Vyas, R. Fradera, M. Belkacemi, A. Hasija, G. Lisboa, S. Luz, J. Malley, (eds.)]. Cambridge University Press, Cambridge, UK and New York, NY, USA. doi: 10.1017/9781009157926.020

AI

**1.5°C pathway** See *Pathways*.

**Acceptability of policy or system change** The extent to which a policy or system change is evaluated unfavourably or favourably, or rejected or supported, by members of the general public (public acceptability) or politicians or governments (political acceptability). Acceptability may vary from totally unacceptable/fully rejected to totally acceptable/fully supported; individuals may differ in how acceptable policies or system changes are believed to be.

**Access to modern energy services** Access to clean, reliable and affordable energy services for cooking, heating, lighting, communications, and productive uses.

**Adaptation** In *human systems*, the process of adjustment to actual or expected *climate* and its effects, in order to moderate harm or exploit beneficial opportunities. In natural systems, the process of adjustment to actual *climate* and its effects; human intervention may facilitate adjustment to expected climate and its effects. See also *Adaptation options*, *Adaptive capacity*, and *Maladaptive actions (Maladaptation)*.

*Adaptation limits*
The change in climate where adaptation is unable to prevent damaging impacts and further risk. Soft limits occur when additional adaptation may be possible if constraints are able to be overcome. Hard limits occur when no additional adaptation is possible.

*Incremental adaptation*
Adaptation that maintains the essence and integrity of a system or process at a given scale (Park et al. 2012). In some cases, incremental adaptation can accrue to result in transformational adaptation (Tàbara et al. 2019; Termeer et al. 2017). Incremental adaptations to change in climate are understood as extensions of actions and behaviours that already reduce the losses or enhance the benefits of natural variations in extreme weather/climate events.

*Transformational adaptation*
Adaptation that changes the fundamental attributes of a social-ecological system in anticipation of climate change and its impacts.

**Adaptation options** The array of strategies and measures that are available and appropriate for addressing *adaptation*. They include a wide range of actions that can be categorised as structural, *institutional*, ecological or behavioural.

**Adaptation pathways** See *Pathways*.

**Adaptive capacity** The ability of systems, *institutions*, humans and other organisms to adjust to potential damage, to take advantage of opportunities, or to respond to consequences (MA 2005).

**Adaptive governance** See *Governance*.

**Additionality** The property of being additional. Mitigation is additional if the *greenhouse gas* emission reductions or removals would not have occurred in the absence of the associated policy intervention or activity.

[Note: Additionality is one of several key criteria used to ensure the environmental integrity of *Offsets (in climate change mitigation)*].

See also *Greenhouse gas emission metric*.

**Adverse side-effect** A negative effect that a policy or measure aimed at one objective has on another objective, thereby potentially reducing the net benefit to society or the environment. See also *Co-benefits*, *Risk*, and *Trade-off*.

**Aerosol** A suspension of airborne solid or liquid particles, with typical particle size in the range of a few nanometres to several tens of micrometres and atmospheric lifetimes of up to several days in the troposphere and up to years in the stratosphere. The term aerosol, which includes both the particles and the suspending gas, is often used in this report in its plural form to mean 'aerosol particles'. Aerosols may be of either natural or anthropogenic origin in the troposphere; stratospheric aerosols mostly stem from volcanic eruptions. Aerosols can cause an effective radiative forcing directly through scattering and absorbing radiation (aerosol–radiation interaction), and indirectly by acting as cloud condensation nuclei or ice nucleating particles that affect the properties of clouds (aerosol–cloud interaction), and upon deposition on snow- or ice-covered surfaces. Atmospheric aerosols may be either emitted as primary particulate matter or formed within the atmosphere from gaseous precursors (secondary production). Aerosols may be composed of sea salt, organic carbon, black carbon (BC), mineral species (mainly desert dust), sulphate, nitrate and ammonium or their mixtures. See also *Short-lived climate forcers (SLCFs)*.

**Afforestation** Conversion to *forest* of land that historically has not contained forests.

[Note: For a discussion of the term *forest* and related terms such as *afforestation*, *reforestation* and *deforestation*, see the 2006 IPCC Guidelines for National Greenhouse Gas Inventories and their 2019 Refinement, and information provided by the *United Nations Framework Convention on Climate Change* (IPCC 2006, 2019; UNFCCC 2021a,b).]

See also *Deforestation*, *Reducing Emissions from Deforestation and Forest Degradation (REDD+)*, *Reforestation*, *Anthropogenic Removals*, and *Carbon dioxide removal (CDR)*.

**Agreement** In this report, the degree of agreement within the scientific body of knowledge on a particular finding is assessed based on multiple lines of *evidence* (e.g., mechanistic understanding, theory, data, models, expert judgement) and expressed qualitatively (Mastrandrea et al. 2010). See also *Confidence*, *Likelihood*, and *Uncertainty*.

**Agriculture, Forestry and Other Land Use (AFOLU)** In the context of national *greenhouse gas (GHG)* inventories under the *United Nations Framework Convention on Climate Change (UNFCCC)*, AFOLU is the sum of the GHG inventory sectors Agriculture and Land Use, Land-Use Change and Forestry (LULUCF); see the 2006 IPCC Guidelines for National GHG Inventories for details. Given the difference in estimating the 'anthropogenic' *carbon dioxide ($CO_2$)* removals between countries and the global modelling community, the land-related net GHG emissions from global models included in this report are not necessarily directly comparable with LULUCF estimates in national GHG Inventories. See also *Land use*, *Land-use change and forestry (LULUCF)* and *Land-use change (LUC)*.

**Agroecology** The science and practice of applying ecological concepts, principles and knowledge (i.e., the interactions of, and explanations for, the diversity, abundance and activities of organisms) to the study, design and management of sustainable agroecosystems.

It includes the roles of human beings as a central organism in agroecology by way of social and economic processes in farming systems. Agroecology examines the roles and interactions among all relevant biophysical, technical and socio-economic components of farming systems and their surrounding landscapes (IPBES 2019).

**Air pollution** Degradation of air quality with negative effects on human health or the natural or built environment due to the introduction, by natural processes or human activity, into the *atmosphere* of substances (gases, *aerosols*) which have a direct (primary pollutants) or indirect (secondary pollutants) harmful effect. See also *Short-lived climate forcers (SLCFs)*.

**Albedo** The proportion of sunlight (solar radiation) reflected by a surface or object, often expressed as a percentage. Clouds, snow and ice usually have high albedo; soil surfaces cover the albedo range from high to low; vegetation in the dry season and/or in arid zones can have high albedo, whereas photosynthetically active vegetation and the ocean have low albedo. The Earth's planetary albedo changes mainly through changes in cloudiness and of snow, ice, leaf area and *land cover*.

**Anomaly** The deviation of a variable from its value averaged over a *reference period*.

**Anthropogenic** Resulting from or produced by human activities.

**Anthropogenic emissions** Emissions of *greenhouse gases (GHGs)*, *precursors* of GHGs and *aerosols* caused by human activities. These activities include the burning of *fossil fuels*, *deforestation*, *land use and land-use changes (LULUC)*, livestock production, fertilisation, waste management, and industrial processes. See also *Anthropogenic* and *Anthropogenic removals*.

**Anthropogenic removals** The withdrawal of *greenhouse gases* (GHGs) from the *atmosphere* as a result of deliberate human activities. These include enhancing biological sinks of $CO_2$ and using chemical engineering to achieve long-term removal and storage. *Carbon capture and storage (CCS)*, which alone does not remove $CO_2$ from the atmosphere, can help reduce atmospheric $CO_2$ from industrial and energy-related sources if it is combined with bioenergy production (*BECCS*), or if $CO_2$ is captured from the air directly and stored (*DACCS*).

[Note: In the 2006 IPCC Guidelines for National GHG Inventories (IPCC 2006), which are used in reporting of emissions to the UNFCCC, 'anthropogenic' land-related GHG fluxes are defined as all those occurring on 'managed land', i.e., 'where human interventions and practices have been applied to perform production, ecological or social functions'. However, some removals (e.g., removals associated with $CO_2$ fertilisation and N deposition) are not considered as 'anthropogenic', or are referred to as 'indirect' anthropogenic effects, in some of the scientific literature assessed in this report. As a consequence, the land-related net GHG emission estimates from global models included in this report are not necessarily directly comparable with LULUCF estimates in national GHG Inventories.]

See also *Carbon dioxide removal (CDR)*, *Afforestation*, *Biochar*, *Enhanced weathering*, *Ocean alkalinisation/Ocean alkalinity enhancement*, *Reforestation*, and *Soil carbon sequestration (SCS)*.

**Atmosphere** The gaseous envelope surrounding the Earth, divided into five layers – the troposphere which contains half of the Earth's atmosphere, the stratosphere, the mesosphere, the thermosphere, and the exosphere, which is the outer limit of the atmosphere. The dry atmosphere consists almost entirely of nitrogen (78.1% volume mixing ratio) and oxygen (20.9% volume mixing ratio), together with a number of trace gases, such as argon (0.93 % volume mixing ratio), helium and radiatively active *greenhouse gases (GHGs)* such as *carbon dioxide ($CO_2$)* (0.04% volume mixing ratio), *methane ($CH_4$)*, *nitrous oxide ($N_2O$)* and *ozone ($O_3$)*. In addition, the atmosphere contains the GHG water vapour ($H_2O$), whose concentrations are highly variable (0–5% volume mixing ratio) as the sources (*evapotranspiration*) and sinks (precipitation) of water vapour show large spatio-temporal variations, and atmospheric temperature exerts a strong constraint on the amount of water vapour an air parcel can hold. The atmosphere also contains clouds and *aerosols*.

**Avoid, Shift, Improve (ASI)** Reducing *greenhouse gas* emissions by avoiding the use of an emissions-producing service entirely, shifting to the lowest-emission mode of providing the service, and/or improving the technologies and systems for providing the service in ways that reduce emissions.

**Baseline/reference** See *Reference period* and *Reference scenario*.

**Baseline period** See *Reference period*.

**Biochar** Relatively stable, carbon-rich material produced by heating *biomass* in an oxygen-limited environment. Biochar is distinguished from charcoal by its application: biochar is used as a soil amendment with the intention to improve soil functions and to reduce *greenhouse gas* emissions from *biomass* that would otherwise decompose rapidly (IBI 2018). See also *Anthropogenic removals* and *Carbon dioxide removal (CDR)*.

**Biodiversity** Biodiversity or biological diversity means the variability among living organisms from all sources including, among other things, terrestrial, marine and other aquatic *ecosystems*, and the ecological complexes of which they are part; this includes diversity within species, between species, and of ecosystems (UN 1992). See also *Bioenergy* and *Biomass*.

**Bioenergy** Energy derived from any form of *biomass* or its metabolic by-products. See also *Biofuel*.

**Bioenergy with carbon dioxide capture and storage (BECCS)** *Carbon dioxide capture and storage (CCS)* technology applied to a *bioenergy* facility. Note that, depending on the total emissions of the BECCS supply chain, *carbon dioxide ($CO_2$)* can be removed from the *atmosphere*. See also *Anthropogenic removals* and *Carbon dioxide removal*.

**Biofuel** A fuel, generally in liquid form, produced from *biomass*. Biofuels include bioethanol from sugarcane, sugar beet or maize, and biodiesel from canola or soybeans. See also *Bioenergy*.

**Biogenic carbon emissions** Carbon released as *carbon dioxide* or *methane* from combustion or decomposition of *biomass* or biobased products.

**Biomass** Organic material excluding the material that is fossilised or embedded in geological formations. Biomass may refer to the mass of organic matter in a specific area (ISO 2014). See also *Bioenergy* and *Biofuel*.

*Traditional biomass*
The combustion of wood, charcoal, agricultural residues and/or animal dung for cooking or heating in open fires or in inefficient stoves as is common in low-income countries.

**Black carbon (BC)** A relatively pure form of carbon, also known as soot, arising from the incomplete combustion of *fossil fuels*, *biofuel*, and *biomass*. It only stays in the *atmosphere* for days or weeks. BC is a climate forcing agent with strong warming effect, both in the atmosphere and when deposited on snow or ice. See also *Aerosol*.

**Blue carbon** Biologically-driven carbon fluxes and storage in marine systems that are amenable to management. Coastal blue carbon focuses on rooted vegetation in the coastal zone, such as tidal marshes, mangroves and seagrasses. These *ecosystems* have high carbon burial rates on a per unit area basis and accumulate carbon in their soils and sediments. They provide many non-climatic benefits and can contribute to *ecosystem-based adaptation*. If degraded or lost, coastal blue carbon ecosystems are likely to release most of their carbon back to the *atmosphere*. There is current debate regarding the application of the blue carbon concept to other coastal and non-coastal processes and ecosystems, including the open ocean. See also *Sequestration*.

**Blue infrastructure** See *Infrastructure*.

**Business as usual (BAU)** The term *business as usual* scenario has been used to describe a scenario that assumes no additional policies beyond those currently in place and that patterns of socio-economic development are consistent with recent trends. The term is now used less frequently than in the past. See also *Reference scenario* (under *Scenario*).

**Carbon budget** Refers to two concepts in the literature: (i) an assessment of carbon cycle *sources* and *sinks* on a global level, through the synthesis of evidence for *fossil fuel* and cement emissions, emissions and removals associated with *land use* and *land-use change*, ocean and natural land sources and sinks of *carbon dioxide* ($CO_2$), and the resulting change in atmospheric $CO_2$ concentration. This is referred to as the global carbon budget; (ii) the maximum amount of cumulative net global *anthropogenic* $CO_2$ emissions that would result in limiting *global warming* to a given level with a given probability, taking into account the effect of other anthropogenic climate *forcers*. This is referred to as the Total Carbon Budget when expressed starting from the *pre-industrial* period, and as the Remaining Carbon Budget when expressed from a recent specified date.

[Note 1: Net anthropogenic $CO_2$ emissions are anthropogenic $CO_2$ emissions minus anthropogenic $CO_2$ removals. See also *Carbon Dioxide Removal (CDR)*.

Note 2: The maximum amount of cumulative net global anthropogenic $CO_2$ emissions is reached at the time that annual net anthropogenic $CO_2$ emissions reach zero.

Note 3: The degree to which anthropogenic climate forcers other than $CO_2$ affect the Total Carbon Budget and Remaining Carbon Budget depends on human choices about the extent to which these forcers are mitigated and their resulting *climate* effects.

Note 4: The notions of a Total Carbon Budget and Remaining Carbon Budget are also being applied in parts of the scientific literature and by some entities at regional, national, or sub-national level. The distribution of global budgets across individual different entities and emitters depends strongly on considerations of equity and other value judgements.]

**Carbon cycle** The flow of carbon (in various forms, e.g., as *carbon dioxide* ($CO_2$), carbon in *biomass*, and carbon dissolved in the ocean as carbonate and bicarbonate) through the atmosphere, hydrosphere, terrestrial and marine biosphere and lithosphere. In this report, the reference unit for the global carbon cycle is $GtCO_2$ or GtC (one Gigatonne = 1 Gt = $10^{15}$ grams; 1GtC corresponds to 3.664 $GtCO_2$).

**Carbon dioxide ($CO_2$)** A naturally occurring gas, $CO_2$ is also a by-product of burning *fossil fuels* (such as oil, gas and coal), of burning *biomass*, of *land-use changes* (LUCs) and of industrial processes (e.g., cement production). It is the principal *anthropogenic* greenhouse gas (GHG) that affects the Earth's radiative balance. It is the reference gas against which other GHGs are measured and therefore has a *global warming potential* (GWP) of 1.

**Carbon dioxide capture and storage (CCS)** A process in which a relatively pure stream of *carbon dioxide* ($CO_2$) from industrial and energy-related sources is separated (captured), conditioned, compressed and transported to a storage location for long-term isolation from the *atmosphere*. Sometimes referred to as Carbon Capture and Storage. See also *Anthropogenic removals*, *Bioenergy with carbon dioxide capture and storage (BECCS)*, *Carbon dioxide capture and utilisation (CCU)*, *Carbon dioxide removal (CDR)*, and *Sequestration*.

**Carbon dioxide capture and utilisation (CCU)** A process in which *carbon dioxide* ($CO_2$) is captured and the carbon then used in a product. The *climate* effect of CCU depends on the product lifetime, the product it displaces, and the $CO_2$ source (fossil, *biomass* or *atmosphere*). CCU is sometimes referred to as Carbon Dioxide Capture and Use, or Carbon Capture and Utilisation. See also *Anthropogenic removals*, *Carbon dioxide capture and storage (CCS)*, and *Carbon dioxide removal (CDR)*.

**Carbon dioxide removal (CDR)** *Anthropogenic* activities removing *carbon dioxide* ($CO_2$) from the *atmosphere* and durably storing it in geological, terrestrial, or ocean reservoirs, or in products. It includes existing and potential anthropogenic enhancement of biological or geochemical $CO_2$ *sinks* and *direct air carbon dioxide capture and storage (DACCS)*, but excludes natural $CO_2$ *uptake* not directly caused by human activities. See also *Anthropogenic removals*, *Afforestation*, *Biochar*, *Bioenergy with carbon dioxide capture and storage (BECCS)*, *Carbon dioxide capture and storage (CCS)*, *Enhanced weathering*, *Ocean alkalinisation/Ocean alkalinity enhancement*, *Reforestation*, and *Soil carbon sequestration (SCS)*.

**Carbon footprint** Measure of the exclusive total amount of emissions of *carbon dioxide* ($CO_2$) that is directly and indirectly caused by an activity or is accumulated over the lifecycle stages of a product (Wiedmann and Minx 2008).

*Household carbon footprint*
The carbon footprint of an individual household, inclusive of the direct and indirect *carbon dioxide* ($CO_2$) emissions associated with home energy use, transportation, food provision, and consumption of other goods and services associated with household expenditures.

**Carbon intensity** The amount of emissions of *carbon dioxide ($CO_2$)* released per unit of another variable such as *gross domestic product (GDP)*, output energy use or transport.

**Carbon leakage** See *Leakage*.

**Carbon neutrality** Condition in which *anthropogenic carbon dioxide ($CO_2$)* emissions associated with a subject are balanced by anthropogenic $CO_2$ removals. The subject can be an entity such as a country, an organisation, a district or a commodity, or an activity such as a service and an event. Carbon neutrality is often assessed over the lifecycle including indirect ('scope 3') emissions, but can also be limited to the emissions and removals, over a specified period, for which the subject has direct control, as determined by the relevant scheme.

[Note 1: Carbon neutrality and *net-zero $CO_2$ emissions* are overlapping concepts. The concepts can be applied at global or sub-global scales (e.g., regional, national and sub-national). At a global scale, the terms carbon neutrality and *net-zero $CO_2$ emissions* are equivalent. At sub-global scales, *net-zero $CO_2$ emissions* is generally applied to emissions and removals under direct control or territorial responsibility of the reporting entity, while carbon neutrality generally includes emissions and removals within and beyond the direct control or territorial responsibility of the reporting entity. Accounting rules specified by *greenhouse gas (GHG)* programmes or schemes can have a significant influence on the quantification of relevant $CO_2$ emissions and removals.

Note 2: In some cases, achieving carbon neutrality may rely on the supplementary use of *offsets* to balance emissions that remain after actions by the reporting entity are taken into account.]

See also *Greenhouse gas neutrality*, *Land use, land-use change and forestry (LULUCF)* and *Net-zero $CO_2$ emissions*.

**Carbon price** The price for avoided or released *carbon dioxide ($CO_2$)* or $CO_2$-equivalent emissions. This may refer to the rate of a carbon tax, or the price of emission permits. In many models that are used to assess the economic costs of *mitigation*, carbon prices are used as a proxy to represent the level of effort in mitigation policies.

**Carbon sink** See *Sink*.

**Carbon stock** The quantity of carbon in a carbon *pool*.

**Choice architecture** The presentation of choices to consumers, and the impact that presentation has on consumer decision-making.

**Circular economy** A system with minimal input and operational losses of materials and energy through extensive reduce, reuse, recycling, and recovery activities. Ten strategies for circularity include: Refuse, Rethink, Reduce, Reuse, Repair, Refurbish, Remanufacture, Repurpose, Recycle, Recover.

**Cities** Cities are open systems, continually exchanging resources, products and services, waste, people, ideas, and finances with the hinterlands and broader world. Cities are complex, self-organising, adaptive, and constantly evolving. Cities also encompass multiple actors with varying responsibilities, capabilities and priorities, as well as processes that transcend the institutional sector-based approach to city administration. Cities are embedded in broader ecological, economic, technical, institutional, legal, and governance structures that enable or often constrain their systemic function, which cannot be separated from wider power relations. Urban processes of physical, social, and economic nature are causally interlinked, with interactions and feedbacks that result in both intended and unintended impacts on emissions. See also *City region*, *Peri-urban areas* and *Urban*.

**Citizen science** A voluntary participation of the public in the collection and/or processing of data as part of a scientific study (Silvertown 2009).

**City region** The areal extent of an individual city's material associations and economic or political influence. The city region concept accepts that rural livelihoods and land uses can be incorporated within the functional activities of a city. This will include dormitory settlements, sources for critical inputs of water, some food, and waste disposal. See also *Cities*, *Region* and *Urban systems*.

**Climate** Climate in a narrow sense is usually defined as the average weather, or more rigorously, as the statistical description in terms of the mean and variability of relevant quantities over a period of time ranging from months to thousands or millions of years. The classical period for averaging these variables is 30 years, as defined by the World Meteorological Organization (WMO). The relevant quantities are most often surface variables such as temperature, precipitation and wind. Climate in a wider sense is the state, including a statistical description, of the *climate system*.

**Climate change** A change in the state of the *climate* that can be identified (e.g., by using statistical tests) by changes in the mean and/or the variability of its properties and that persists for an extended period, typically decades or longer. Climate change may be due to natural internal processes or external forcings such as modulations of the solar cycles, volcanic eruptions and persistent *anthropogenic* changes in the composition of the *atmosphere* or in *land use*. Note that the *United Nations Framework Convention on Climate Change (UNFCCC)*, in its Article 1, defines climate change as: 'a change of climate which is attributed directly or indirectly to human activity that alters the composition of the global atmosphere and which is in addition to natural climate variability observed over comparable time periods'. The UNFCCC thus makes a distinction between climate change attributable to human activities altering the atmospheric composition and *climate variability* attributable to natural causes.

**Climate change commitment** The unavoidable future *climate change* resulting from inertia in the geophysical and socio-economic systems. Different types of climate change commitment are discussed in the literature. Climate change commitment is usually quantified in terms of the further change in temperature, but it includes other future changes, for example in the hydrological cycle, in *extreme weather events*, in extreme climate events, and in sea level.

*Zero emissions commitment*

The zero emissions commitment is an estimate of the subsequent *global warming* that would result after *anthropogenic emissions* are set to zero. It is determined by both inertia in physical *climate system* components (ocean, cryosphere, land surface) and *carbon cycle* inertia. In its widest sense it refers to emissions of each climate *forcer* including *greenhouses gases*, *aerosols* and their precursors. The climate response to this can be complex due to the different climate response time scale of each climate forcer. A specific sub-category of zero emissions commitment is the Zero $CO_2$ Emissions Commitment which refers to the climate system response to $CO_2$ emissions after setting these to net zero. The $CO_2$-only definition is of specific use in estimating *remaining carbon budgets*.

**Climate extreme (extreme weather or climate event)** The occurrence of a value of a weather or *climate* variable above (or below) a threshold value near the upper (or lower) ends of the range of observed values of the variable. By definition, the characteristics of what is called *extreme weather* may vary from place to place in an absolute sense. When a pattern of extreme weather persists for some time, such as a season, it may be classified as an extreme climate event, especially if it yields an average or total that is itself extreme (e.g., high temperature, *drought*, or heavy rainfall over a season). For simplicity, both extreme weather events and extreme climate events are referred to collectively as 'climate extremes'.

**Climate finance** There is no agreed definition of climate finance. The term 'climate finance' is applied to the financial resources devoted to addressing climate change by all public and private actors from global to local scales, including international financial flows to developing countries to assist them in addressing climate change. Climate finance aims to reduce net *greenhouse gas* emissions and/or to enhance adaptation and increase resilience to the impacts of current and projected climate change. Finance can come from private and public sources, channelled by various intermediaries, and is delivered by a range of instruments, including grants, concessional and non-concessional debt, and internal budget reallocations.

**Climate governance** See *Governance*.

**Climate justice** See *Justice*.

**Climate model** A qualitative or quantitative representation of the *climate* system based on the physical, chemical and biological properties of its components, their interactions and feedback processes and accounting for some of its known properties. The *climate system* can be represented by models of varying complexity; that is, for any one component or combination of components a spectrum or hierarchy of models can be identified, differing in such aspects as the number of spatial dimensions, the extent to which physical, chemical or biological processes are explicitly represented, or the level at which empirical parametrisations are involved. There is an evolution towards more complex models with interactive chemistry and biology. Climate models are applied as a research tool to study and simulate the climate and for operational purposes, including monthly, seasonal and interannual climate predictions. See also *Simple climate model (SCM)* and *Emulators*.

**Climate projection** Simulated response of the *climate system* to a *scenario* of future emissions or concentrations of *greenhouse gases (GHGs)* and *aerosols* and changes in *land use*, generally derived using *climate models*. Climate projections are distinguished from climate predictions by their dependence on the emission/concentration/radiative forcing *scenario* used, which is in turn based on assumptions concerning, for example, future socio-economic and technological developments that may or may not be realised.

**Climate sensitivity** The change in the surface temperature in response to a change in the *atmospheric carbon dioxide ($CO_2$)* concentration or other *radiative forcing*.

*Transient climate response (TCR)*
The surface temperature response for the hypothetical scenario in which atmospheric *carbon dioxide ($CO_2$)* increases at 1% $yr^{-1}$ from *pre-industrial* to the time of a doubling of atmospheric $CO_2$ concentration (year 70).

*Transient climate response to cumulative CO2 emissions (TCRE)*
The transient surface temperature change per unit cumulative *carbon dioxide ($CO_2$)* emissions, usually 1000 GtC. TCRE combines both information on the airborne fraction of cumulative $CO_2$ emissions (the fraction of the total $CO_2$ emitted that remains in the *atmosphere*, which is determined by *carbon cycle* processes) and on the *transient climate response (TCR)*.

**Climate services** Climate services involve the provision of climate information in such a way as to assist decision-making. The service includes appropriate engagement from users and providers, is based on scientifically credible information and expertise, has an effective access mechanism, and responds to user needs (Hewitt et al. 2012).

**Climate system** The global system consisting of five major components: the *atmosphere*, the hydrosphere, the cryosphere, the lithosphere and the biosphere and the interactions between them. The climate system changes in time under the influence of its own internal dynamics and because of external forcings such as volcanic eruptions, solar variations, orbital forcing, and *anthropogenic* forcings such as the changing composition of the atmosphere and *land-use change*.

**Climate variability** Deviations of climate variables from a given mean state (including the occurrence of extremes, etc.) at all spatial and temporal scales beyond that of individual weather events. Variability may be intrinsic, due to fluctuations of processes internal to the *climate system* (*internal variability*), or extrinsic, due to variations in natural or anthropogenic external forcing (forced variability). See also *Climate change*.

**Co-benefits** A positive effect that a policy or measure aimed at one objective has on another objective, thereby increasing the total benefit to society or the environment. Co-benefits are also referred to as ancillary benefits. See also *Adverse side-effect* and *Trade-off*.

**$CO_2$ equivalent ($CO_2$-eq) emission** The amount of *carbon dioxide ($CO_2$)* emission that would have an equivalent effect on a specified key measure of *climate change*, over a specified time horizon, as an emitted amount of another *greenhouse gas (GHG)* or a mixture of other GHGs. For a mix of GHGs, it is obtained by summing the $CO_2$-equivalent emissions of each gas. There are various ways and time horizons to compute such equivalent emissions (see *greenhouse gas emission metric*). $CO_2$-equivalent emissions are commonly used to compare emissions of different GHGs, but should not be taken to imply that these emissions have an equivalent effect across all key measures of climate change.

[Note: Under the Paris Rulebook [Decision 18/CMA.1, annex, paragraph 37], parties have agreed to use GWP100 values from the IPCC AR5 or GWP100 values from a subsequent IPCC Assessment Report to report aggregate emissions and removals of GHGs. In addition, parties may use other metrics to report supplemental information on aggregate emissions and removals of GHGs.]

**Concentrations scenario** See *Scenario*.

**Conference of the Parties (COP)** The supreme body of UN conventions, such as the *United Nations Framework Convention on*

*Climate Change (UNFCCC)*, comprising parties with a right to vote that have ratified or acceded to the convention.

**Confidence** The robustness of a finding based on the type, amount, quality and consistency of *evidence* (e.g., mechanistic understanding, theory, data, models, expert judgement) and on the degree of *agreement* across multiple lines of evidence. In this report, confidence is expressed qualitatively (Mastrandrea et al. 2010).

**Conservation agriculture** A farming system that promotes minimum soil disturbance (e.g., by using no till practices), maintenance of a permanent soil cover, and diversification of plant species. It aims to prevent *land degradation* and regenerate degraded lands by enhancing *biodiversity* and natural biological processes above and below the ground surface, that contribute to increased water and nutrient use efficiency and improved and sustained crop production (FAO 2016).

**Consumption-based emissions** Emissions released to the atmosphere in order to generate the goods and services consumed by a certain entity (e.g., a person, firm, country, or region). See also *Production-based emissions*.

**Coping capacity** The ability of people, *institutions*, organisations, and systems, using available skills, values, beliefs, resources, and opportunities, to address, manage, and overcome adverse conditions in the short to medium term (UNISDR 2009; IPCC 2012). See also *Resilience*.

**Cost-benefit analysis** A type of economic evaluation that compares all monetised negative and positive impacts associated with a given action. Cost-benefit analysis enables comparison of different interventions, investments or strategies, and reveals how a given investment or policy effort pays off for a particular person, company or country, or at a global scale. Cost-benefit analyses representing society's point of view are important for *climate change* decision-making, but there are difficulties in aggregating costs and benefits across different actors and across time scales. See also *Discounting*.

**Cost-effectiveness analysis (CEA)** A type of economic evaluation that compares the costs of different courses of action reaching the same outcome. In this report, CEA focuses on comparing the costs of *mitigation* strategies designed to meet a prespecified *climate change* mitigation goal (e.g., an emission-reduction target or a temperature stabilisation target).

**Cumulative emissions** The total amount of emissions released over a specified period of time. See also *Carbon budget* and *Transient climate response to cumulative $CO_2$ emissions (TCRE)*.

**Decarbonisation** Human actions to reduce *carbon dioxide* emissions from human activities.

**Decent Living Standard** A set of minimal material requirements essential for achieving basic human *well-being* including nutrition, shelter, basic living conditions, clothing, healthcare, education, and mobility (Rao and Baer 2012; Rao and Min 2018; O'Neill et al. 2018).

**Decoupling** Decoupling (in relation to climate change) is where economic growth is no longer strongly associated with another relevant indicator such as *greenhouse gas* emissions. Relative decoupling is where both these indicators grow but the other indicators grow more slowly than the economy. Absolute decoupling is where there is economic growth but there is a decline in the other indicator.

**Deforestation** Conversion of *forest* to non-forest.

[Note: For a discussion of the term *forest* and related terms such as *afforestation*, *reforestation* and *deforestation*, see the 2006 IPCC Guidelines for National Greenhouse Gas Inventories and their 2019 Refinement, and information provided by the United Nations Framework Convention on Climate Change (IPCC 2006, 2019; UNFCCC 2021a,b).]

See also *Reducing Emissions from Deforestation and Forest Degradation (REDD+)*.

**Deliberative governance** See *Governance*.

**Demand** Disciplinary approaches use the term in different ways. In economics, demand by a consumer is willingness and ability to purchase in a marketplace. However, the motivation for purchase may vary and can include economic utility, welfare, *Decent standard of living (DSL)*, or for the good/services.

**Demand- and supply-side measures**

*Demand-side measures*
Policies and programmes for influencing the *demand* for goods and/or services. In the energy sector, demand-side mitigation measures aim at reducing the amount of *greenhouse gas* emissions emitted per unit of energy service used.

*Supply-side measures*
Policies and programmes for influencing how a certain *demand* for goods and/or services is met. In the energy sector, supply-side mitigation measures aim at reducing the amount of *greenhouse gas* emissions emitted per unit of energy service produced.

**Demand-side management** See *Demand-side measures*.

**Desertification** *Land degradation* in arid, semi-arid, and dry sub-humid areas resulting from many factors, including climatic variations and human activities (UNCCD 1994).

**Developed/developing countries (Industrialised/developed/developing countries)** There is a diversity of approaches for categorising countries on the basis of their level of development, and for defining terms such as 'industrialised', 'developed', or 'developing'. Several categorisations are used in this report: (i) In the United Nations (UN) system, there is no established convention for the designation of developed and developing countries or areas. (ii) The UN Statistics Division specifies developed and developing regions based on common practice. In addition, specific countries are designated as least developed countries, landlocked developing countries, *Small Island Developing States (SIDS)*, and transition economies. Many countries appear in more than one of these categories. (iii) The World Bank uses income as the main criterion for classifying countries as low, lower middle, upper middle, and high income. (iv) The UN Development Programme (UNDP) aggregates indicators for life expectancy, educational attainment, and income into a single composite Human Development Index (HDI) to classify countries as low, medium, high, or very high human development.

**Development pathway** See *Pathways*.

**Diet** The kinds of food that follow a particular pattern that a person or community eats (FAO and Alliance of Bioversity International and CIAT, 2021).

**Direct air capture (DAC)** Chemical process by which a pure *carbon dioxide* ($CO_2$) stream is produced by capturing $CO_2$ from the ambient air. See also *Anthropogenic removals*, *Carbon dioxide removal (CDR)* and *Direct air carbon dioxide capture and storage (DACCS)*.

**Direct air carbon dioxide capture and storage (DACCS)** Chemical process by which *carbon dioxide* ($CO_2$) is captured directly from the ambient air, with subsequent storage. Also known as direct air capture and storage (DACS). See also *Anthropogenic removals*, *Carbon dioxide removal (CDR)* and *Direct air capture (DAC)*.

**Direct and indirect services** Direct Services: Services (e.g., passenger mobility) required by end-users (consumers). Indirect services: Services required (e.g., goods transport, manufacturing) for provisioning systems of direct services.

**Direct emissions** Emissions that physically arise from activities within well-defined boundaries of, for instance, a *region*, an economic sector, a company, or a process. See also *Indirect emissions*.

**Disaster** A 'serious disruption of the functioning of a community or a society at any scale due to hazardous events interacting with conditions of exposure, vulnerability and capacity, leading to one or more of the following: human, material, economic and environmental losses and impacts' (UNGA 2016). See also *Exposure*, *Hazard*, *Risk* and *Vulnerability*.

**Disaster risk management (DRM)** Processes for designing, implementing, and evaluating strategies, policies, and measures to improve the understanding of current and future disaster *risk*, foster *disaster* risk reduction and transfer, and promote continuous improvement in disaster preparedness, prevention and protection, response, and recovery practices, with the explicit purpose of increasing human security, *well-being*, quality of life, and *sustainable development (SD)*.

**Discount rate** See *Discounting*.

**Discounting** A mathematical operation that aims to make monetary (or other) amounts received or expended at different times (years) comparable across time. If the discount rate is positive, future values are given less weight than those today. The choice of discount rate(s) is debated as it is a judgement based on hidden and/or explicit values.

**Disruptive innovation** Demand-led technological change that leads to significant system change and is characterised by strong exponential growth.

**Distributive equity** See *Equity*.

**Drought** An exceptional period of water shortage for existing *ecosystems* and the human population (due to low rainfall, high temperature, and/or wind).

**Ecosystem** A functional unit consisting of living organisms, their non-living environment and the interactions within and between them. The components included in a given ecosystem and its spatial boundaries depend on the purpose for which the ecosystem is defined: in some cases they are relatively sharp, while in others they are diffuse. Ecosystem boundaries can change over time. Ecosystems are nested within other ecosystems and their scale can range from very small to the entire biosphere. In the current era, most ecosystems either contain people as key organisms, or are influenced by the effects of human activities in their environment. See also *Ecosystem services*.

**Ecosystem services** Ecological processes or functions having monetary or non-monetary value to individuals or society at large. These are frequently classified as: (i) supporting services such as productivity or biodiversity maintenance; (ii) provisioning services such as food or fibre; (iii) regulating services such as climate regulation or carbon *sequestration*; and (iv) cultural services such as tourism or spiritual and aesthetic appreciation. See also *Ecosystem* and *Nature's Contribution to People*.

**Ecosystem-based adaptation (EBA)** The use of *ecosystem* management activities to increase the *resilience* and reduce the *vulnerability* of people and *ecosystems* to *climate change* (Campbell et al. 2009).

**Embodied (embedded) [emissions, water, land]** The total emissions [water use, *land use*] generated [used] in the production of goods and services regardless of the location and timing of those emissions [water use, land use] in the production process. This includes emissions [water use, land use] within the country used to produce goods or services for the country's own use, but also includes the emissions [water use, land use] related to the production of such goods or services in other countries that are then consumed in another country through imports. Such emissions [water, land] are termed 'embodied' or 'embedded' emissions, or, in some cases, (particularly with water) as 'virtual water use' (Davis and Caldeira 2010; Allan 2005; MacDonald et al. 2015).

**Emission and Socioe-conomic Scenario Ensemble** A set of modelled emission and socio-economic *scenarios* collected in a database. The scenarios can come from a single multi-model study with systematic variation of harmonised scenario designs (structured ensemble) or from multiple studies in the literature (unstructured ensemble). Depending on the scope of the ensemble, variation of the results across the scenarios in the ensemble give an indication of the spread of results in the literature (unstructured ensemble), or an estimate of uncertainties due to different modelling structures and methodologies (structured ensemble).

**Emission factor/Emissions intensity** A coefficient that quantifies the emissions or removals of a gas per unit activity. Emission factors are often based on a sample of measurement data, averaged to develop a representative rate of emission for a given activity level under a given set of operating conditions.

**Emission pathways** See *Pathways*.

**Emission trajectories** A projected development in time of the emission of a *greenhouse gas (GHG)* or group of GHGs, *aerosols*, and GHG *precursors*. See also *Pathways*.

**Emissions** See *Anthropogenic emissions*, *Direct emissions*, *Cumulative emissions*, *Indirect emissions*, *Consumption-based emissions*, *Production-based emissions* and *Embodied (embedded) [emissions, water, land]*.

**Emissions scenario** See *Scenario*.

**Emulation** Reproducing the behaviour of complex, process-based models – namely, Earth System Models (ESMs) – via simpler approaches, using either *emulators* or *simple climate models (SCMs)*. The computational efficiency of emulating approaches opens new analytical possibilities, given that ESMs take a lot of computational resources for each simulation.

**Emulators** A broad class of heavily parametrised models ('simple climate models'), statistical methods like neural networks, genetic algorithms or other artificial intelligence approaches, designed to reproduce the responses of more complex, process-based Earth System Models (ESMs). The main application of emulators is to extrapolate insights from ESMs and observational constraints to a larger set of emission scenarios. See also *Emulation* and *Simple climate models (SCMs)*.

**Enabling conditions (for adaptation and mitigation options)** Conditions that enhance the *feasibility* of *adaptation* and *mitigation* options. Enabling conditions include finance, technological innovation, strengthening policy instruments, *institutional capacity*, *multi-level governance*, and changes in *human behaviour* and lifestyles.

**Energy access** Access to clean, reliable and affordable energy services for cooking and heating, lighting, communications, and productive uses (with special reference to *Sustainable Development Goal* 7) (AGECC 2010). See also *Traditional biomass*.

**Energy efficiency** The ratio of output or useful energy or energy services or other useful physical outputs obtained from a system, conversion process, transmission or storage activity to the input of energy (measured as kWh $kWh^{-1}$, tonnes $kWh^{-1}$ or any other physical measure of useful output like tonne-km transported). Energy efficiency is often described by energy intensity.

**Energy poverty** The absence of sufficient choice in accessing adequate, affordable, reliable, high quality, safe and environmentally benign energy services to support economic and human development (Reddy 2000). See also *Fuel poverty*.

**Energy security** The goal of a given country, or the global community as a whole, to maintain an adequate, stable and predictable energy supply. Measures encompass safeguarding the sufficiency of energy resources to meet national energy demand at competitive and stable prices and the resilience of the energy supply; enabling development and deployment of technologies; building sufficient infrastructure to generate, store and transmit energy supplies and ensuring enforceable contracts of delivery.

**Energy services** A benefit or amenity (e.g., mobility, communication, thermal comfort) received as a result of energy or other resources use.

**Enhanced weathering** A proposed method to increase the natural rate of removal of *carbon dioxide ($CO_2$)* from the *atmosphere* using silicate and carbonate rocks. The active surface area of these minerals is increased by grinding, before they are actively added to soil, beaches or the open ocean. See also *Anthropogenic removals* and *Carbon dioxide removal (CDR)*.

**Ensemble** A collection of comparable datasets that reflect variations within the bounds of one or more sources of *uncertainty*, and that when averaged can provide a more robust estimate of underlying behaviour. Ensemble techniques are used by the observational, reanalysis and modelling communities. See also *Emission and Socio-economic Scenario Ensemble* and *Integrated Assessment Scenario Ensemble*.

**Enteric fermentation** A natural part of the digestion process in ruminant animal species (domesticated and wild), such as cattle, buffalo, sheep, goats, antelope, etc. Microorganisms (bacteria, archaea, fungi, protozoa and viruses) present in the fore-stomach (reticulorumen or rumen) breakdown plant *biomass* to produce substrates that can be used by the animal for energy and growth with methane produced as a by-product. Fermentation end-products such as hydrogen, *carbon dioxide*, formate and methyl-containing compounds are important substrates for the production of methane by the rumen's methane-forming archaea (known as methanogens).

**Equality** A principle that ascribes equal worth to all human beings, including equal opportunities, rights, and obligations, irrespective of origins.

*Inequality*
Uneven opportunities and social positions, and processes of discrimination within a group or society, based on gender, class, ethnicity, age, and (dis)ability, often produced by uneven development. Income inequality refers to gaps between highest and lowest income earners within a country and between countries.

See also *Equity* and *Fairness*.

**Equity** The principle of being fair and impartial, and a basis for understanding how the *impacts* and responses to *climate change*, including costs and benefits, are distributed in and by society in more or less equal ways. Often aligned with ideas of *equality*, *fairness* and *justice* and applied with respect to equity in the responsibility for, and distribution of, *climate* impacts and policies across society, generations, and gender, and in the sense of who participates and controls the processes of decision-making.

*Distributive equity*
Equity in the consequences, outcomes, costs and benefits of actions or policies. In the case of *climate change* or climate policies for different people, places and countries, including equity aspects of sharing burdens and benefits for mitigation and adaptation.

*Gender equity*
Equity between women and men with regard to their rights, resources and opportunities. In the case of climate change, gender equity recognises that women are often more vulnerable to the impacts of climate change and may be disadvantaged in the process and outcomes of climate policy.

*Inter-generational equity*
Equity between generations. In the context of climate change, inter-generational equity acknowledges that the effects of past and present emissions, vulnerabilities and policies impose costs and benefits for people in the future and of different age groups.

**Evidence** Data and information used in the scientific process to establish findings. In this report, the degree of evidence reflects the amount, quality and consistency of scientific/technical information on which the Lead Authors are basing their findings. See also *Agreement*, *Confidence*, *Likelihood*, and *Uncertainty*.

**Exergy** Capacity of energy flows to perform useful work. Exergy is a quality (versatility) indicator of energy flows which ranges from low (e.g., low-temperature heat, biomass) to high (e.g., electricity). Exergy efficiency describes how much useful work can be performed by a particular energy flow in relation to the thermodynamic maximum possible. It can be determined for all energy flows and energy conversion steps, also including alternative service delivery systems (Grubler et al. 2012).

**Exposure** The presence of people; *livelihoods*; species or *ecosystems*; environmental functions, services, and resources; infrastructure; or economic, social, or cultural assets in places and settings that could be adversely affected.

**Extreme weather event** An event that is rare at a particular place and time of year. Definitions of 'rare' vary, but an extreme weather event would normally be as rare as, or rarer than, the 10th or 90th percentile of a probability density function estimated from observations. By definition, the characteristics of what is called extreme weather may vary from place to place in an absolute sense. See also *Climate extreme (extreme weather or climate event)*.

**Fairness** Impartial and just treatment without favouritism or discrimination in which each person is considered of equal worth with equal opportunity. See also *Equality* and *Equity*.

**Feasibility** In this report, feasibility refers to the potential for a mitigation or adaptation option to be implemented. Factors influencing feasibility are context-dependent, temporally dynamic, and may vary between different groups and actors. Feasibility depends on geophysical, environmental-ecological, technological, economic, socio-cultural and institutional factors that enable or constrain the implementation of an option. The feasibility of options may change when different options are combined, and increase when enabling conditions are strengthened. See also *Enabling conditions (for adaptation and mitigation options)*.

**Final energy** The energy delivered to final users (firms, individuals, *institutions*), where it becomes usable energy in supplying energy services (e.g., light, heat, mobility). See also *Primary energy*.

**Flexibility (demand and supply)** Adjustment of energy load characteristics by technical and/or non-technical change to balance energy demand and supply.

**Flexible governance** See *Governance*.

**Flood** The overflowing of the normal confines of a stream or other water body, or the accumulation of water over areas that are not normally submerged. Floods can be caused by unusually heavy rain, for example, during storms and cyclones. Floods include river (fluvial) floods, flash floods, *urban* floods, rain (pluvial) floods, sewer floods, coastal floods, and glacial lake outburst floods (GLOFs).

**Food loss and waste** 'The decrease in quantity or quality of food'. Food waste is part of food loss and refers to discarding or alternative (non-food) use of food that is safe and nutritious for human consumption along the entire food supply chain, from primary production to end household consumer level. Food waste is recognised as a distinct part of food loss because the drivers that generate it and the solutions to it are different from those of food losses (FAO 2015).

**Food security** A situation that exists when all people, at all times, have physical, social and economic access to sufficient, safe and nutritious food that meets their dietary needs and food preferences for an active and healthy life. The four pillars of food security are: availability; access; utilisation; and stability. The nutritional dimension is integral to the concept of food security (FAO 2009, 2018).

*Access*
Economic and/or physical access to food. Economic access is determined by disposable income, food prices and the provision of and access to social support. Physical access is determined by the availability and quality of land and other infrastructure, property rights or the functioning of markets.

*Availability*
Physical availability of food. Food availability addresses the supply side of food security and is determined by the levels of food production, stocks and net trade.

*Stability*
The stability of the other three dimensions over time. Even if individuals' food intake is adequate today, they are still considered food-insecure if periodically they have inadequate access to food, risking deterioration of their nutrition status. Adverse weather conditions, political instability or economic factors (unemployment, rising food prices) may have an impact on individuals' food security status.

*Utilisation*
The way in which the body uses the various nutrients in food. Individuals achieve sufficient energy and nutrient intake through good care and feeding practices, food preparation, diet diversity and intra-household distribution of food. Combined with biological utilisation of the food consumed, energy and nutrient intake determine the nutrition status of individuals.

**Food system** All the elements (environment, people, inputs, processes, *infrastructures*, *institutions*, etc.) and activities that relate to the production, processing, distribution, preparation and consumption of food, and the output of these activities, including socio-economic and environmental outcomes (HLPE 2017).

[Note: Whilst there is a global food system (encompassing the totality of global production and consumption), each location's food system is unique, being defined by that place's mix of food produced locally, nationally, regionally or globally.]

**Forest** A vegetation type dominated by trees. Many definitions of the term forest are in use throughout the world, reflecting wide differences in biogeophysical conditions, social structure and economics.

[Note: For a discussion of the term forest in the context of National GHG inventories, see the 2006 IPCC Guidelines for National GHG Inventories and their 2019 Refinement, and information provided by the United Nations Framework Convention on Climate Change (IPCC 2006, 2019; UNFCCC 2021a,b).]

**Fossil fuels** Carbon-based fuels from fossil hydrocarbon deposits, including coal, oil, and natural gas.

**Fuel poverty** A condition in which a household is unable to guarantee a certain level of consumption of domestic energy services (especially heating) or suffers disproportionate expenditure burdens to meet these needs. See also *Energy poverty*.

**Fugitive emissions (oil and natural gas systems)** The release of *greenhouse gases* that occur during the exploration, processing and delivery of *fossil fuels* to the point of final use. This excludes *greenhouse gas emissions* from fuel combustion for the production of useful heat or power. It encompasses venting, flaring, and leaks.

**Gender equity** See *Equity*.

**Geothermal energy** Accessible thermal energy stored in the Earth's interior, in both rock and trapped steam or liquid water (hydrothermal resources), which may be used to generate electric energy in a thermal power plant, or to supply heat to any process requiring it. The main sources of geothermal energy are the residual energy available from planet formation and the energy continuously generated from radionuclide decay. See also *Renewable energy*.

**Gini coefficient** A statistical measure of dispersion in a distribution and degree of mathematical measure of *inequality*. For example, it can be used for measuring inequality in income, wealth, carbon emissions, and access to well-being defining services. The dimensionless GINI coefficient ranges between 0 (absolute *equality*) and 1 (absolute inequality).

**Global carbon budget** See *Carbon budget*.

**Global mean surface air temperature (GSAT)** Global average of near-surface air temperatures over land, oceans and sea ice. Changes in GSAT are often used as a measure of global temperature change in *climate models*. See also *Global mean surface temperature (GMST)*.

**Global mean surface temperature (GMST)** Estimated global average of near-surface air temperatures over land and sea ice, and *sea surface temperature (SST)* over ice-free ocean regions, with changes normally expressed as departures from a value over a specified *reference period*. See also *Global mean surface air temperature (GSAT)*.

**Global warming** Global warming refers to the increase in *global surface temperature* relative to a baseline *reference period*, averaging over a period sufficient to remove interannual variations (e.g., 20 or 30 years). A common choice for the baseline is 1850–1900 (the earliest period of reliable observations with sufficient geographic coverage), with more modern baselines used depending upon the application. See also *Climate change* and *Climate variability*.

**Global warming potential (GWP)** An index measuring the *radiative forcing* following an emission of a unit mass of a given substance, accumulated over a chosen time horizon, relative to that of the reference substance, *carbon dioxide* ($CO_2$). The GWP thus represents the combined effect of the differing times these substances remain in the atmosphere, and their effectiveness in causing radiative forcing. See also *Greenhouse gas emission metric*.

**Governance** The structures, processes, and actions through which private and public actors interact to address societal goals. This includes formal and informal *institutions* and the associated norms, rules, laws and procedures for deciding, managing, implementing and monitoring policies and measures at any geographic or political scale, from global to local.

*Adaptive governance*
Adjusting to changing conditions, such as climate change, through governance interactions that seek to maintain a desired state in a social-ecological system.

*Climate governance*
The structures, processes, and actions through which private and public actors seek to mitigate and adapt to climate change.

*Deliberative governance*
Deliberative governance involves decision-making through inclusive public conversation which allows opportunity for developing policy options through public discussion rather than collating individual preferences through voting or referenda (although the latter governance mechanisms can also be preceded and legitimated by public deliberation processes).

*Flexible governance*
Strategies of governance at various levels, which prioritise the use of social learning and rapid feedback mechanisms in planning and policymaking, often through incremental, experimental and iterative management processes.

*Multi-level governance*
The dispersion of governance across multiple levels of jurisdiction and decision-making, including, global, regional, national and local, as well as trans-regional and trans-national levels.

*Participatory governance*
A governance system that enables direct public engagement in decision-making using a variety of techniques, for example, referenda, community deliberation, citizen juries or participatory budgeting. The approach can be applied in formal and informal institutional contexts from national to local, but is usually associated with devolved decision-making (Fung and Wright 2003; Sarmiento and Tilly 2018).

**Governance capacity** The ability of governance *institutions*, leaders, and non-state and civil society to plan, coordinate, fund, implement, evaluate and adjust policies and measures over the short, medium and long term, adjusting for uncertainty, rapid change and wide-ranging impacts and multiple actors and demands. See also *Governance*.

**Grazing land** The sum of rangelands and pastures not considered as cropland, and subject to livestock grazing or hay production. It includes a wide range of *ecosystems*, for example, systems with vegetation that fall below the threshold used in the *forest* land category, silvo-pastoral systems, as well as natural, managed grasslands and semi-deserts.

**Green Climate Fund (GCF)** The GCF was established by the 16th Session of the *Conference of the Parties (COP)* in 2010 as an operating entity of the financial mechanism of the *United Nations Framework Convention on Climate Change (UNFCCC)*, in accordance with Article 11 of the Convention, to support projects, programmes and policies and other activities in developing country Parties. The Fund is governed by a Board and will receive guidance of the COP. See also *Climate finance*.

**Green infrastructure** See *Infrastructure*.

**Greenhouse gas emission metric** A simplified relationship used to quantify the effect of emitting a unit mass of a given *greenhouse gas (GHG)* on a specified key measure of *climate change*. A relative GHG emission metric expresses the effect from one gas relative to the effect of emitting a unit mass of a reference GHG on the same measure of climate change. There are multiple emission metrics, and the most appropriate metric depends on the application. GHG emission metrics may differ with respect to: (i) the key measure of climate change they consider; (ii) whether they consider climate outcomes for a specified point in time or integrated over a specified time horizon; (iii) the time horizon over which the metric is applied; (iv) whether they apply to a single emission pulse, emissions sustained over a period of time, or a combination of both; and (v) whether they consider the climate effect from an emission compared to the absence of that emission or compared to a reference emissions level or climate state.

[Note: Most relative GHG emission metrics (such as the *global warming potential (GWP)*, global temperature change potential (GTP), global damage potential, and GWP*), use carbon dioxide *($CO_2$)* as the reference gas. Emissions of non-$CO_2$ gases, when expressed using such metrics, are often referred to as 'carbon dioxide equivalent' emissions. A metric that establishes equivalence regarding one key measure of the *climate system* response to emissions does not imply equivalence regarding other key measures. The choice of a metric, including its time horizon, should reflect the policy objectives for which the metric is applied.]

**Greenhouse gas neutrality** Condition in which metric-weighted anthropogenic *greenhouse gas (GHG)* emissions associated with a subject are balanced by metric-weighted *anthropogenic* GHG removals. The subject can be an entity such as a country, an organisation, a district or a commodity, or an activity such as a service and an event. GHG neutrality is often assessed over the lifecycle, including indirect ('scope 3') emissions, but can also be limited to the emissions and removals, over a specified period, for which the subject has direct control, as determined by the relevant scheme. The quantification of GHG emissions and removals depends on the GHG emission metric chosen to compare emissions and removals of different gases, as well as the time horizon chosen for that metric.

[Note 1: Greenhouse gas neutrality and net-zero greenhouse gas emissions are overlapping concepts. The concepts can be applied at global or sub-global scales (e.g., regional, national and sub-national). At a global scale, the terms greenhouse gas neutrality and net-zero greenhouse gas emissions are equivalent. At sub-global scales, net-zero GHG emissions is generally applied to emissions and removals under direct control or territorial responsibility of the reporting entity, while GHG neutrality generally includes emissions and removals within and beyond the direct control or territorial responsibility of the reporting entity. Accounting rules specified by GHG programmes or schemes can have a significant influence on the quantification of relevant emissions and removals.

Note 2: Under the Paris Rulebook (Decision 18/CMA.1, annex, paragraph 37), parties have agreed to use GWP100 values from the IPCC AR5 or GWP100 values from a subsequent IPCC Assessment Report to report aggregate emissions and removals of GHGs. In addition, parties may use other metrics to report supplemental information on aggregate emissions and removals of GHGs.

Note 3: In some cases, achieving greenhouse gas neutrality may rely on the supplementary use of *offsets* to balance emissions that remain after actions by the reporting entity are taken into account.]

See also *Carbon neutrality*, *Greenhouse gas emission metric*, *Land use*, *Land-use change and forestry (LULUCF)* and *Net-zero greenhouse gas emissions*.

**Greenhouse gases (GHGs)** Gaseous constituents of the *atmosphere*, both natural and *anthropogenic*, that absorb and emit radiation at specific wavelengths within the spectrum of radiation emitted by the Earth's surface, by the atmosphere itself, and by clouds. This property causes the *greenhouse effect*. Water vapour ($H_2O$), *carbon dioxide ($CO_2$)*, *nitrous oxide ($N_2O$)*, *methane ($CH_4$)* and *ozone ($O_3$)* are the primary GHGs in the Earth's atmosphere. Human-made GHGs include sulphur hexafluoride ($SF_6$), hydrofluorocarbons (HFCs), chlorofluorocarbons (CFCs) and perfluorocarbons (PFCs); several of these are also $O_3$-depleting (and are regulated under the Montreal Protocol).

**Grey infrastructure** See *Infrastructure*.

**Gross domestic product (GDP)** The sum of gross value added, at purchasers' prices, by all resident and non-resident producers in the economy, plus any taxes and minus any subsidies not included in the value of the products in a country or a geographic region for a given period, normally one year. GDP is calculated without deducting for depreciation of fabricated assets or depletion and degradation of natural resources.

**Halocarbons** A collective term for the group of partially halogenated organic species, which includes the chlorofluorocarbons (CFCs), hydrochlorofluorocarbons (HCFCs), hydrofluorocarbons (HFCs), halons, methyl chloride and methyl bromide. Many of the halocarbons have large *global warming potentials*. The chlorine and bromine-containing halocarbons are also involved in the depletion of the ozone layer.

**Human behaviour** The responses of persons or groups to a particular situation, here likely to relate to *climate change*. Human behaviour covers the range of actions by individuals, communities, *organisations*, governments and at the international level.

**Human rights** Rights that are inherent to all human beings, universal, inalienable, and indivisible, typically expressed and guaranteed by law. They include the right to life, economic, social, and cultural rights, and the right to development and self-determination (OHCHR 2018).

**Human security** A condition that is met when the vital core of human lives is protected, and when people have the freedom and capacity to live with dignity. In the context of *climate change*, the vital core of human lives includes the universal and culturally specific, material and non-material elements necessary for people to act on behalf of their interests and to live with dignity.

**Human system** Any system in which human organisations and *institutions* play a major role. Often, but not always, the term is synonymous with society or social system. Systems such as agricultural systems, urban systems, political systems, technological systems and economic systems are all human systems in the sense applied in this report.

**Hydropower** Power harnessed from the flow of water. See also *Renewable energy*.

**Impacts** The consequences of realised *risks* on natural and *human systems*, where risks result from the interactions of climate-related *hazards* (including *extreme weather/climate events*), *exposure*, and *vulnerability*. Impacts generally refer to effects on lives, *livelihoods*, health and *well-being*, *ecosystems* and species, economic, social and cultural assets, services (including *ecosystem services*), and *infrastructure*. Impacts may be referred to as consequences or outcomes, and can be adverse or beneficial. See also *Adaptation*, *Loss and Damage, and losses and damages*.

**Indigenous knowledge** The understandings, skills and philosophies developed by societies with long histories of interaction with their natural surroundings. For many indigenous peoples, indigenous knowledge informs decision-making about fundamental aspects of life, from day-to-day activities to longer-term actions. This knowledge is integral to cultural complexes, which also encompass language, systems of classification, resource use practices, social interactions, values, ritual and spirituality. These distinctive ways of knowing are important facets of the world's cultural diversity (UNESCO 2018). See also *Local knowledge*.

**Indirect emissions** Emissions that are a consequence of the activities within well-defined boundaries of, for instance, a *region*, an economic sector, a company or process, but which occur outside the specified boundaries. For example, emissions are described as indirect if they relate to the use of heat but physically arise outside the boundaries of the heat user, or to electricity production but physically arise outside of the boundaries of the power supply sector. See also *Direct emissions*.

**Indirect land-use change (iLUC)** See *Land-use change (LUC)*.

**Industrial revolution** A period of rapid industrial growth with far-reaching social and economic consequences, beginning in Britain during the second half of the 18th century and spreading to Europe and later to other countries including the United States. The invention of the steam engine was an important trigger of this development. The industrial revolution marks the beginning of a strong increase in the use of *fossil fuels*, initially coal, and hence emission of *carbon dioxide ($CO_2$)*.

**Inequality** See *Equality*.

**Infrastructure** The designed and built set of physical systems and corresponding *institutional* arrangements that mediate between people, their communities, and the broader environment to provide services that support economic growth, health, quality of life, and safety (Chester 2019; Dawson et al. 2018).

*Blue infrastructure*
Blue infrastructure includes bodies of water, watercourses, ponds, lakes and storm drainage, that provide ecological and hydrological functions including evaporation, transpiration, drainage, infiltration, and temporary storage of runoff and discharge.

*Green infrastructure*
The strategically planned interconnected set of natural and constructed ecological systems, green spaces and other landscape features that can provide functions and services including air and water purification, temperature management, floodwater management and coastal defence often with co-benefits for people and biodiversity. Green infrastructure includes planted and remnant native vegetation, soils, wetlands, parks and green open spaces, as well as building and street-level design interventions that incorporate vegetation (Bobbins and Culwick 2016).

*Grey infrastructure*
Engineered physical components and networks of pipes, wires, roads, tracks that underpin energy, transport, communications (including digital), built form, water and sanitation and solid waste management systems.

*Social infrastructure*
The social, cultural, and financial activities and institutions as well as associated property, buildings and artefacts and policy domains such as social protection, health and education that support well-being and public life (Latham and Layton 2019; Frolova et al. 2016).

**Institutional capacity** Building and strengthening individual organisations and providing technical and management training to support integrated planning and decision-making processes between organisations and people, as well as empowerment, social capital, and an enabling environment, including the culture, values and power relations (Willems and Baumert 2003). See also *Governance* and *Institutions*.

**Institutions** Rules, norms and conventions that guide, constrain or enable human behaviours and practices. Institutions can be formally established, for instance through laws and regulations, or informally established, for instance by traditions or customs. Institutions may spur, hinder, strengthen, weaken or distort the emergence, adoption and implementation of climate action and climate governance.

[Note: Institutions can also refer to a large organisation.]

See also *Institutional capacity*.

**Integrated assessment** A method of analysis that combines results and models from the physical, biological, economic and social sciences and the interactions among these components in a consistent framework to evaluate the status and the consequences of environmental change and the policy responses to it. See also *Integrated assessment model (IAM)*.

**Integrated assessment model (IAM)** Models that integrate knowledge from two or more domains into a single framework. They are one of the main tools for undertaking integrated assessments. One class of IAM used with respect to climate change *mitigation* may include representations of: multiple sectors of the economy, such as energy, *land use* and *land-use change*; interactions between sectors; the economy as a whole; associated *greenhouse gas (GHG)* emissions and *sinks*; and reduced representations of the *climate system*. This class of model is used to assess linkages between economic, social and technological development and the evolution of the climate system. Another class of IAM additionally includes representations of the costs associated with climate change *impacts*, but includes less detailed representations of economic systems. These can be used to assess impacts and mitigation in a cost–benefit framework and have been used to estimate the *social cost of carbon*. See also *Integrated Assessment Scenario Ensemble*.

**Integrated Assessment Scenario Ensemble** A set of modelled scenarios from an intercomparison of *integrated assessment models (IAMs)* based on a systematic variation of harmonised scenario designs.

**Inter-generational equity** See *Equity*.

**Internet of Things (IoT)** The network of computing devices embedded in everyday objects such as cars, phones and computers, connected via the internet, enabling them to send and receive data.

**Irreversibility** A perturbed state of a dynamical system is defined as irreversible on a given time scale if the recovery from this state due to natural processes takes substantially longer than the time scale of interest. See also *Tipping point*.

**Just transitions** A set of principles, processes and practices that aim to ensure that no people, workers, places, sectors, countries or regions are left behind in the transition from a high-carbon to a low-carbon economy. It stresses the need for targeted and proactive measures from governments, agencies, and authorities to ensure that any negative social, environmental or economic impacts of economy-wide transitions are minimised, whilst benefits are maximised for those disproportionally affected. Key principles of just transitions include: respect and dignity for vulnerable groups; fairness in energy access and use, social dialogue and democratic consultation with relevant stakeholders; the creation of decent jobs; social protection; and rights at work. Just transitions could include fairness in energy, land use and climate planning and decision-making processes; economic diversification based on low-carbon investments; realistic training/retraining programmes that lead to decent work; gender-specific policies that promote equitable outcomes; the fostering of international cooperation and coordinated multilateral actions; and the eradication of poverty. Lastly, just transitions may embody the redressing of past harms and perceived injustices (ILO 2015; UNFCCC 2016).

**Justice** Justice is concerned with ensuring that people get what is due to them, setting out the moral or legal principles of *fairness* and *equity* in the way people are treated, often based on the ethics and values of society.

*Climate justice*
Justice that links development and human rights to achieve a human-centred approach to addressing *climate change*, safeguarding the rights of the most vulnerable people and sharing the burdens and benefits of climate change and its impacts equitably and fairly (MRFCJ 2018).

**Kaya identity** In this identity, global emissions are equal to the population size, multiplied by per capita output (gross world product), multiplied by the energy intensity of production, multiplied by the carbon intensity of energy.

**Land** The terrestrial portion of the biosphere that comprises the natural resources (soil, near-surface air, vegetation and other biota, and water), the ecological processes, topography, and *human settlements* and infrastructure that operate within that system (FAO 2007; UNCCD 1994).

**Land cover** The biophysical coverage of *land* (e.g., bare soil, rocks, forests, buildings and roads or lakes). Land cover is often categorised in broad land-cover classes (e.g., deciduous forest, coniferous forest, mixed forest, grassland, bare ground).

[Note: In some literature, land cover and land use are used interchangeably, but the two represent distinct classification systems. For example, the land cover class woodland can be under various land uses such as livestock grazing, recreation, conservation, or wood harvest.]

**Land cover change** Change from one *land cover* class to another, due to change in *land use* or change in natural conditions (Pongratz et al. 2018).

**Land degradation** A negative trend in land condition, caused by direct or indirect human-induced processes including *anthropogenic* climate change, expressed as long-term reduction or loss of at least one of the following: biological productivity, ecological integrity or value to humans.

[Note: This definition applies to *forest* and non-forest land. Changes in land condition resulting solely from natural processes (such as volcanic eruptions) are not considered to be land degradation. Reduction of biological productivity or ecological integrity or value to humans can constitute degradation, but any one of these changes need not necessarily be considered degradation.]

See also *Desertification*.

**Land degradation neutrality** A state whereby the amount and quality of land resources necessary to support *ecosystem* functions and services and enhance *food security* remain stable or increase within specified temporal and spatial scales and ecosystems (UNCCD 2020).

**Land management** Sum of land-use practices (e.g., sowing, fertilising, weeding, harvesting, thinning, clear-cutting) that take place within broader *land-use* categories (Pongratz et al. 2018).

*Land management change*
A change in land management that occurs within a *land-use* category.

**Land potential** The inherent, long-term potential of the *land* to sustainably generate *ecosystem services*, which reflects the capacity and *resilience* of the land-based natural capital, in the face of ongoing environmental change (UNEP 2016).

**Land rehabilitation** Direct or indirect actions undertaken with the aim of reinstating a level of *ecosystem* functionality, where the goal is provision of goods and services rather than ecological restoration (McDonald et al. 2016).

**Land restoration** The process of assisting the recovery of *land* from a degraded state (IPBES 2018; McDonald et al. 2016).

**Land use** The total of arrangements, activities and inputs applied to a parcel of *land*. The term land use is also used in the sense of the social and economic purposes for which land is managed (e.g., grazing, timber extraction, conservation and city dwelling). In national *greenhouse gas (GHG)* inventories, land use is classified according to the IPCC land-use categories of forest land, cropland, grassland, wetlands, settlements, other lands (see the 2006 IPCC Guidelines for National GHG Inventories and their 2019 Refinement for details (IPCC 2006, 2019)).

**Land use, land-use change and forestry (LULUCF)** In the context of national greenhouse gas (GHG) inventories under the *United Nations Framework Convention on Climate Change* (UNFCCC 2019), LULUCF is a GHG inventory sector that covers *anthropogenic* emissions and removals of GHG in managed lands, excluding non-$CO_2$ agricultural emissions. Following the 2006 IPCC Guidelines for National GHG Inventories and their 2019 Refinement, 'anthropogenic' land-related GHG fluxes are defined as all those occurring on 'managed land', that is, 'where human interventions and practices have been applied to perform production, ecological or social functions'. Since managed land may include *carbon dioxide ($CO_2$)*, removals not considered as 'anthropogenic' in some of the scientific literature assessed in this report (e.g., removals associated with $CO_2$ fertilisation and N deposition), the land-related net GHG emission estimates from global models included in this report are not necessarily directly comparable with LULUCF estimates in National GHG Inventories (IPCC 2006, 2019).

**Land-use change (LUC)** The change from one *land use* category to another. Note that, in some scientific literature, land-use change encompasses changes in land-use categories as well as changes in land management. See also *Afforestation*, *Agriculture*, *Forestry and Other Land Use (AFOLU)*, *Deforestation*, *Land use*, *land-use change and forestry (LULUCF)*, and *Reforestation*.

*Indirect land-use change (iLUC)*
Land-use change outside the area of focus that occurs as a consequence of change in use or management of land within the area of focus, such as through market or policy drivers. For example, if agricultural land is diverted to *biofuel* production, forest clearance may occur elsewhere to replace the former agricultural production.

**Latent heat flux** The turbulent *flux* of heat from the Earth's surface to the *atmosphere* that is associated with evaporation or condensation of water vapour at the surface; a component of the surface energy budget.

**Leakage** The effects of policies that result in a displacement of the environmental impact, thereby counteracting the intended effects of the initial policies.

**Leapfrogging** The ability of developing countries to bypass intermediate technologies and jump straight to advanced clean technologies.

**Lifecycle assessment (LCA)** Compilation and evaluation of the inputs, outputs and the potential environmental impacts of a product or service throughout its lifecycle (ISO 2018).

**Likelihood** The chance of a specific outcome occurring, where this might be estimated probabilistically. Likelihood is expressed in this report using a standard terminology (Mastrandrea et al. 2010). See also *Agreement*, *Confidence*, *Evidence*, and *Uncertainty*.

**Livelihood** The resources used and the activities undertaken in order for people to live. Livelihoods are usually determined by the entitlements and assets to which people have access. Such assets can be categorised as human, social, natural, physical or financial.

**Local knowledge (LK)** The understandings and skills developed by individuals and populations, specific to the places where they live. Local knowledge informs decision-making about fundamental aspects of life, from day-to-day activities to longer-term actions. This knowledge is a key element of the social and cultural systems which influence observations of and responses to climate change; it also informs *governance* decisions (UNESCO 2018). See also *Indigenous knowledge*.

**Lock-in** A situation in which the future development of a system, including *infrastructure*, technologies, investments, *institutions*, and behavioural norms, is determined or constrained ('locked in') by historic developments. See also *Path dependence*.

**Long-lived greenhouse gases (LLGHGs)** A set of well-mixed *greenhouse gases* with long atmospheric lifetimes. This set of compounds includes *carbon dioxide ($CO_2$)* and *nitrous oxide ($N_2O$)*, together with some fluorinated gases. They have a warming effect on *climate*. These compounds accumulate in the *atmosphere* at decadal to centennial time scales, and their effect on *climate* hence persists for decades to centuries after their emission. On time scales of decades to a century already emitted emissions of long-lived climate forcers can only be abated by greenhouse gas removal (GGR).

**Loss and Damage, and losses and damages** Research has taken Loss and Damage (capitalised letters) to refer to political debate under the *United Nations Framework Convention on Climate Change (UNFCCC)* following the establishment of the Warsaw International Mechanism for Loss and Damage in 2013, which is to 'address loss and damage associated with impacts of climate change, including extreme events and slow onset events, in developing countries that are particularly vulnerable to the adverse effects of climate change.' Lowercase letters (losses and damages) have been taken to refer broadly to harm from (observed) impacts and (projected) risks, and can be economic or non-economic (Mechler et al. 2018).

**Maladaptive actions (Maladaptation)** Actions that may lead to increased risk of adverse climate-related outcomes, including via increased *greenhouse gas (GHG)* emissions, increased *vulnerability* to climate change, or diminished welfare, now or in the future. Maladaptation is usually an unintended consequence.

**Malnutrition** Deficiencies, excesses, or imbalances in a person's intake of energy and/or nutrients. The term malnutrition addresses three broad groups of conditions: undernutrition, which includes wasting (low weight-for-height), stunting (low height-for-age) and underweight (low weight-for-age); micronutrient-related malnutrition, which includes micronutrient deficiencies (a lack of important vitamins and minerals) or micronutrient excess; and overweight, obesity and diet-related noncommunicable diseases (such as heart disease, stroke, diabetes and some cancers) (WHO 2018). Micronutrient deficiencies are sometimes termed 'hidden hunger' to emphasise that people can be malnourished in the sense of deficient without being deficient in calories. Hidden hunger can apply even where people are obese.

**Managed forest** *Forests* subject to human interventions (notably silvicultural management such as planting, pruning, thinning), timber and fuelwood harvest, protection (fire suppression, insect suppression) and management for amenity values or conservation, with defined geographical boundaries (Ogle et al. 2018). See also *Managed land*.

[Note: For a discussion of the term 'forest' in the context of National GHG inventories, see the 2006 IPCC Guidelines for National GHG Inventories (IPCC 2006).]

**Managed grassland** Grasslands on which human interventions are carried out, such as grazing domestic livestock or hay removal.

**Managed land** In the context of national *greenhouse gas (GHG)* inventories under the *United Nations Framework Convention on Climate Change (UNFCCC)*, the 2006 IPCC Guidelines for National GHG Inventories (IPCC 2006) defines managed land 'where human interventions and practices have been applied to perform production, ecological or social functions'. IPCC (2006) defines *anthropogenic* GHG emissions and removals in the *LULUCF* sector as all those occurring on 'managed land'. The key rationale for this approach is that the preponderance of anthropogenic effects occurs on managed lands.

[Note: More details can be found in IPCC 2006 Guidelines for National GHG Inventories, Volume 4, Chapter 1.]

**Market failure** When private decisions are based on market prices that do not reflect the real scarcity of goods and services but rather reflect market distortions, they do not generate an efficient allocation of resources but cause welfare losses. A market distortion is any event in which a market reaches a market clearing price that is substantially different from the price that a market would achieve while operating under conditions of perfect competition and state enforcement of legal contracts and the ownership of private property. Examples of factors causing market prices to deviate from real economic scarcity are environmental externalities, public goods, monopoly power, information asymmetry, transaction costs, and non-rational behaviour.

**Material substitution** Replacement of one material (including an energy carrier used as a feedstock) by another, due to scarcity, price, technological change, or because of lower environmental impacts or *greenhouse gas emissions*.

**Measurement, Reporting and Verification (MRV)**

*Measurement*
'Processes of data collection over time, providing basic datasets, including associated accuracy and precision, for the range of relevant variables. Possible data sources are field measurements, field observations, detection through remote sensing and interviews' (UN-REDD 2009).

*Reporting*
'The process of formal reporting of assessment results to the UNFCCC, according to predetermined formats and established standards, especially the Intergovernmental Panel on Climate Change (IPCC) Guidelines and GPG (Good Practice Guidance)' (UN-REDD 2009).

*Verification*
'The process of formal verification of reports, for example, the established approach to verify national communications and national inventory reports to the UNFCCC' (UN-REDD 2009).

**Megacity** Urban agglomerations with 10 million inhabitants or more. See also *City*.

**Methane ($CH_4$)** The *greenhouse gas (GHG)* methane is the major component of natural gas and associated with all hydrocarbon fuels. Significant *anthropogenic* emissions also occur as a result of animal husbandry and paddy rice production. Methane is also produced naturally where organic matter decays under anaerobic conditions, such as in wetlands. Under future *global warming*, there is potential for increased methane emissions from thawing permafrost, wetlands and sub-sea gas hydrates. See also *Short-lived climate forcers (SLCFs)*.

**Migrant** Any person who is moving or has moved across an international border or within a State away from his/her habitual place of residence, regardless of: (1) the person's legal status; (2) whether the movement is voluntary or involuntary; (3) what the causes for the movement are; or (4) what the length of the stay is (IOM 2018).

**Migration (of humans)** Movement of a person or a group of persons, either across an international border, or within a State. It is a population movement, encompassing any kind of movement of people, whatever its length, composition and causes; it includes migration of refugees, displaced persons, economic migrants, and persons moving for other purposes, including family reunification (IOM 2018).

**Mitigation (of climate change)** A human intervention to reduce emissions or enhance the *sinks* of *greenhouse gases*.

**Mitigation measures** In climate policy, mitigation measures are technologies, processes or practices that contribute to *mitigation*, for example, *renewable energy* technologies, waste minimisation processes, and public transport commuting practices.

**Mitigation option** A technology or practice that reduces *greenhouse gas* emissions or enhances *sinks*.

**Mitigation pathways** See *Pathways*.

**Mitigation potential** The quantity of net *greenhouse gas* emission reductions that can be achieved by a given *mitigation option* relative to specified emission baselines.

[Note: Net greenhouse gas emissions reduction is the sum of reduced emissions and/or enhanced *sinks*.]

See also *Sequestration potential*.

*Biogeophysical potential*
The mitigation potential constrained by biological, geophysical and geochemical limits and thermodynamics, without taking into account technical, social, economic and/or environmental considerations.

*Economic potential*
The portion of the technical potential for which the social benefits exceed the social costs, taking into account a social discount rate and the value of externalities.

*Technical potential*
The mitigation potential constrained by biogeophysical limits as well as availability of technologies and practices. Quantification of technical potentials takes into account primarily technical considerations, but social, economic and/or environmental considerations are occasionally also included, if these represent strong barriers for the deployment of an option.

**Mitigation scenario** See *Scenario*.

**Multi-level governance** See *Governance*.

**Narrative** See *Storyline*.

**Nature's contributions to people (NCP)** All the contributions, both positive and negative, of living nature (i.e., diversity of organisms, *ecosystems*, and their associated ecological and evolutionary processes) to the quality of life for people. Beneficial contributions from nature include such things as food provision, water purification, flood control, and artistic inspiration, whereas detrimental contributions include disease transmission and predation that damages people or their assets. Many NCP may be perceived as benefits or detriments depending on the cultural, temporal or spatial context (Díaz et al. 2018). See also *Ecosystem services*.

**Nature-based solutions** Actions to protect, sustainably manage and restore natural or modified ecosystems that address societal challenges effectively and adaptively, simultaneously providing human well-being and biodiversity benefits (IUCN 2016). See also *Biodiversity* and *Ecosystem*.

**Net negative greenhouse gas emissions** A situation of net negative greenhouse gas emissions is achieved when metric-weighted *anthropogenic greenhouse gas (GHG)* removals exceed metric-weighted anthropogenic GHG emissions. Where multiple GHG are involved, the quantification of net emissions depends on the metric chosen to compare emissions of different gases (such as *global warming potential*, global temperature change potential, and others, as well as the chosen time horizon). See also *Carbon dioxide removal (CDR)*, *Greenhouse gas emission metric*, *Net-zero $CO_2$ emissions; Net-zero greenhouse gas emissions*, and *Negative greenhouse gas emissions*.

**Net-zero $CO_2$ emissions** Condition in which *anthropogenic carbon dioxide ($CO_2$)* emissions are balanced by anthropogenic $CO_2$ removals over a specified period.

[Note: *Carbon neutrality* and net-zero $CO_2$ emissions are overlapping concepts. The concepts can be applied at global or sub-global scales (e.g., regional, national and sub-national). At a global scale, the terms *carbon neutrality* and net-zero $CO_2$ emissions are equivalent. At sub-global scales, net-zero $CO_2$ emissions is generally applied to emissions and removals under direct control or territorial responsibility of the reporting entity, while *carbon neutrality* generally includes emissions and removals within and beyond the direct control or territorial responsibility of the reporting entity. Accounting rules specified by GHG programmes or schemes can have a significant influence on the quantification of relevant $CO_2$ emissions and removals.]

See also *Carbon neutrality*, *Land use, land-use change and forestry (LULUCF)* and *Net-zero greenhouse gas emissions*.

**Net-zero greenhouse gas emissions** Condition in which metric-weighted *anthropogenic greenhouse gas (GHG)* emissions are balanced by metric-weighted anthropogenic GHG removals over a specified period. The quantification of net-zero GHG emissions depends on the *GHG emission metric* chosen to compare emissions and removals of different gases, as well as the time horizon chosen for that metric.

[Note 1: Greenhouse gas neutrality and net-zero GHG emissions are overlapping concepts. The concept of net-zero GHG emissions can be applied at global or sub-global scales (e.g., regional, national and sub-national). At a global scale, the terms GHG neutrality and net-zero GHG emissions are equivalent. At sub-global scales, net-zero GHG emissions is generally applied to emissions and removals under direct control or territorial responsibility of the reporting entity, while GHG neutrality generally includes anthropogenic emissions and anthropogenic removals within and beyond the direct control or territorial responsibility of the reporting entity. Accounting rules specified by GHG programmes or schemes can have a significant influence on the quantification of relevant emissions and removals.

Note 2: Under the Paris Rulebook (Decision 18/CMA.1, annex, paragraph 37), parties have agreed to use GWP100 values from the IPCC AR5 or GWP100 values from a subsequent IPCC Assessment Report to report aggregate emissions and removals of GHGs. In addition, parties may use other metrics to report supplemental information on aggregate emissions and removals of GHGs.]

See also *Greenhouse gas neutrality*, *Net-zero $CO_2$ emissions*, and *Land use, land-use change and forestry (LULUCF)*.

**Nitrous oxide ($N_2O$)** The main *anthropogenic* source of $N_2O$, a *greenhouse gas (GHG)*, is agriculture (soil and animal manure management), but important contributions also come from sewage treatment, *fossil fuel* combustion, and chemical industrial processes. $N_2O$ is also produced naturally from a wide variety of biological sources in soil and water, particularly microbial action in wet tropical *forests*.

**Non-overshoot pathways** See *Pathways*.

**Ocean alkalinisation/Ocean alkalinity enhancement** A proposed *carbon dioxide removal (CDR)* method that involves deposition of alkaline minerals or their dissociation products at the ocean surface. This increases surface total alkalinity, and may thus increase ocean *carbon dioxide ($CO_2$)* uptake and ameliorate surface ocean acidification. See also *Anthropogenic removals*.

**Ocean fertilisation** A proposed *carbon dioxide removal (CDR)* method that relies on the deliberate increase of nutrient supply to the near-surface *ocean* with the aim of *sequestering* additional $CO_2$ from the *atmosphere* through biological production. Methods include direct addition of micro-nutrients or macro-nutrients. To be successful, the additional carbon needs to reach the deep ocean where it has the potential to be sequestered on climatically relevant time scales. See also *Anthropogenic removals*.

**Offset (in climate policy)** The reduction, avoidance or removal of a unit of *greenhouse gas (GHG)* emissions by one entity, purchased by another entity to counterbalance a unit of GHG emissions by that other entity. Offsets are commonly subject to rules and environmental integrity criteria intended to ensure that offsets achieve their stated mitigation outcome. Relevant criteria include, but are not limited to, the avoidance of double counting and *leakage*, use of appropriate baselines, *additionality*, and permanence or measures to address impermanence. See also *Greenhouse gas emission metric* and *Carbon neutrality*.

**Organic farming** An agricultural production system that aims to utilise natural processes and cycles to limit off-farm and notably synthetic inputs, while also aiming to enhance agroecosystems and society. Organic farming is often legally defined and governed by standards, typically guided by principles outlined by the International Federation of Organic Agriculture Movements (IFOAM – Organics International) (IFOAM – Organics International 2014).

**Overshoot pathways** See *Pathways*.

**Ozone ($O_3$)** The triatomic form of oxygen, and a gaseous *atmospheric* constituent. In the troposphere, $O_3$ is created both naturally and by photochemical reactions involving gases resulting from human activities (e.g., smog). Tropospheric $O_3$ acts as a *greenhouse gas (GHG)*. In the stratosphere, $O_3$ is created by the interaction between solar ultraviolet radiation and molecular oxygen

($O_2$). Stratospheric $O_3$ plays a dominant role in the stratospheric radiative balance. Its concentration is highest in the ozone layer.

**Pareto optimum** A state in which no one's welfare can be increased without reducing someone else's welfare.

**Participatory governance** See *Governance*.

**Particulate matter (PM)** Atmospheric aerosols involved in air pollution issues. Of greatest concern for health are particles of aerodynamic diameter less than or equal to 10 micrometers, usually designated as PM10 and particles of diameter less than or equal to 2.5 micrometers, usually designated as PM2.5.

**Path dependence** The generic situation where decisions, events, or outcomes at one point in time constrain *adaptation*, *mitigation*, or other actions or options at a later point in time. See also *Lock-in*.

**Pathways** The temporal evolution of natural and/or human systems towards a future state. Pathway concepts range from sets of quantitative and qualitative *scenarios* or narratives of potential futures to solution-oriented decision-making processes to achieve desirable societal goals. Pathway approaches typically focus on biophysical, techno-economic, and/or socio-behavioural trajectories and involve various dynamics, goals, and actors across different scales. See also *Scenario* and *Storyline*.

*1.5°C pathway*
A pathway of emissions of *greenhouse gases* and other climate *forcers* that provides an approximately one-in-two to two-in-three chance, given current knowledge of the climate response, of global warming either remaining below 1.5°C or returning to 1.5°C by around 2100 following an overshoot.

*Adaptation pathways*
A series of *adaptation* choices involving trade-offs between short-term and long-term goals and values. These are processes of deliberation to identify solutions that are meaningful to people in the context of their daily lives and to avoid potential *maladaptation*.

*Climate-resilient pathways*
Iterative processes for managing change within complex systems in order to reduce disruptions and enhance opportunities associated with climate change.

*Development pathways*
Development pathways evolve as the result of the countless decisions being made and actions being taken at all levels of societal structure, as well due to the emergent dynamics within and between institutions, cultural norms, technological systems and other drivers of behavioural change.

See also *Shifting development pathways (SDPs)* and *Shifting development pathways to sustainability (SDPS)*.

*Emission pathways*
Modelled trajectories of global *anthropogenic emissions* over the 21st century.

*Mitigation pathways*
A temporal evolution of a set of *mitigation scenario* features, such as *greenhouse gas* emissions and socio-economic development.

*Non-overshoot pathways*
Pathways that stay below a specified concentration, *forcing*, or global warming level during a specified period of time (e.g., until 2100).

*Overshoot pathways*
*Pathways* that first exceed a specified concentration, *forcing*, or global warming level, and then return to or below that level again before the end of a specified period of time (e.g., before 2100). Sometimes the magnitude and likelihood of the overshoot is also characterised. The overshoot duration can vary from one pathway to the next, but in most overshoot pathways in the literature and referred to as overshoot pathways in the AR6, the overshoot occurs over a period of at least one decade and up to several decades.

*Representative Concentration Pathways (RCPs)*
*Scenarios* that include time series of *emissions* and concentrations of the full suite of *greenhouse gases (GHGs)* and *aerosols* and chemically active gases, as well as *land use/land cover* (Moss et al. 2010). The word representative signifies that each RCP provides only one of many possible scenarios that would lead to the specific *radiative forcing* characteristics. The term pathway emphasises that not only the long-term concentration levels are of interest, but also the trajectory taken over time to reach that outcome (Moss et al. 2010).

RCPs usually refer to the portion of the concentration pathway extending up to 2100, for which *integrated assessment models* produced corresponding emission scenarios. Extended concentration pathways describe extensions of the RCPs from 2100 to 2300 that were calculated using simple rules generated by stakeholder consultations, and do not represent fully consistent scenarios. Four RCPs produced from integrated assessment models were selected from the published literature and used in the Fifth IPCC Assessment, and are also used in this Assessment for comparison, spanning the range from approximately below 2°C warming to high (>4°C) warming best-estimates by the end of the 21st century: RCP2.6, RCP4.5 and RCP6.0 and RCP8.5.

- RCP2.6: One pathway where radiative forcing peaks at approximately 3 W $m^{-2}$ and then declines to be limited at 2.6 W $m^{-2}$ in 2100 (the corresponding Extended Concentration Pathway, or ECP, has constant emissions after 2100).
- RCP4.5 and RCP6.0: Two intermediate stabilisation pathways in which radiative forcing is limited at approximately 4.5 W $m^{-2}$ and 6.0 W $m^{-2}$ in 2100 (the corresponding ECPs have constant concentrations after 2150).
- RCP8.5: One high pathway which leads to >8.5 W $m^{-2}$ in 2100 (the corresponding ECP has constant emissions after 2100 until 2150 and constant concentrations after 2250).

See also *Shared socio-economic pathways (SSPs)* (under *Pathways*).

*Shared Socio-economic Pathways (SSPs)*
SSPs have been developed to complement the *Representative Concentration Pathways (RCPs)*. By design, the RCP emission and concentration pathways were stripped of their association with a certain socio-economic development. Different levels of *emissions* and *climate change* along the dimension of the RCPs can hence be explored against the backdrop of different socio-economic development pathways (SSPs) on the other dimension in a matrix. This integrative SSP-RCP framework is now widely used in the climate *impact* and policy analysis literature, where *climate projections*

obtained under the RCP scenarios are analysed against the backdrop of various SSPs. As several emission updates were due, a new set of emission scenarios was developed in conjunction with the SSPs. Hence, the abbreviation SSP is now used for two things: On the one hand SSP1, SSP2, …, SSP5 are used to denote the five socio-economic scenario families. On the other hand, the abbreviations SSP1-1.9, SSP1-2.6, …, SSP5-8.5 are used to denote the newly developed emission scenarios that are the result of an SSP implementation within an integrated assessment model. Those SSP scenarios are bare of climate policy assumption, but in combination with so-called shared policy assumptions (SPAs), various approximate *radiative forcing* levels of 1.9, 2.6, …, or 8.5 W $m^{-2}$ are reached by the end of the century, respectively.

*Transformation pathways*
Trajectories describing consistent sets of possible futures of *greenhouse gas (GHG)* emissions, atmospheric concentrations, or *global mean surface temperatures* implied from mitigation and adaptation actions associated with a set of broad and irreversible economic, technological, societal, and behavioural changes. This can encompass changes in the way energy and infrastructure are used and produced, natural resources are managed and institutions are set up, and in the pace and direction of technological change.

**Peri-urban areas** Dynamic transition zones that have intense interaction between rural and *urban* economies, activities, households, and lifestyles. Neither fully rural or urban (Seto et al. 2010).

**Policies (for climate change mitigation and adaptation)** Strategies that enable actions to be undertaken to accelerate *adaptation* and *mitigation*. Policies include those developed by national and subnational public agencies, and with the private sector. Policies for adaptation and mitigation often take the form of economic incentives, regulatory instruments, and decision-making and engagement processes.

**Political economy** The set of interlinked relationships between people, the state, society and markets as defined by law, politics, economics, customs and power that determine the outcome of trade and transactions and the distribution of wealth in a country or economy.

**Pool, carbon and nitrogen** A reservoir in the Earth System where elements, such as carbon and nitrogen, reside in various chemical forms for a period of time. See also *Sequestration*, *Sink*, *Source* and *Uptake*.

**Poverty** A complex concept with several definitions stemming from different schools of thought. It can refer to material circumstances (such as need, pattern of deprivation or limited resources), economic conditions (such as standard of living, *inequality* or economic position) and/or social relationships (such as social class, dependency, exclusion, lack of basic security or lack of entitlement).

**Poverty eradication** A set of measures to end poverty in all its forms everywhere. See also *Sustainable Development Goals (SDGs)*.

**Precursors** Atmospheric compounds that are not *greenhouse gases (GHGs)* or *aerosols*, but that have an effect on GHG or aerosol concentrations by taking part in physical or chemical processes regulating their production or destruction rates.

**Pre-industrial (period)** The multi-century period prior to the onset of large-scale industrial activity around 1750. The *reference period* 1850–1900 is used to approximate pre-industrial *global mean surface temperature (GMST)*. See also *Industrial revolution*.

**Primary energy** The energy that is embodied in resources as they exist in nature (e.g., coal, biomass uranium, solar radiation, wind, ocean currents) (Grubler et al. 2012).

[Note: Primary energy is defined in several alternative ways. The method used in this report is the direct equivalent method, which counts one unit of secondary energy provided from non-combustible sources as one unit of primary energy. For more details on the methodology, see Section 7 in Working Group III Annex II.]

See also *Final energy*.

**Primary production** The synthesis of organic compounds by plants and microbes, on land or in the ocean, primarily by photosynthesis using light and *carbon dioxide ($CO_2$)* as sources of energy and carbon respectively. It can also occur through chemosynthesis, using chemical energy, for example, in deep sea vents.

**Private costs** Costs carried by individuals, companies or other private entities that undertake an action, whereas *social costs* include additionally the external costs on the environment and on society as a whole. Quantitative estimates of both private and social costs may be incomplete, because of difficulties in measuring all relevant effects.

**Production-based emissions** Emissions released to the *atmosphere* for the production of goods and services by a certain entity (e.g., a person, firm, country, or region). See also *Consumption-based emissions*.

**Projection** A potential future evolution of a quantity or set of quantities, often computed with the aid of a model. Unlike predictions, projections are conditional on assumptions concerning, for example, future socio-economic and technological developments that may or may not be realised. See also *Climate projection*, *Pathways* and *Scenario*.

**Prosumers** A consumer that also produces energy and inputs energy to the system, for which it is an active agent in the energy system and market.

**Radiative forcing** The change in the net, downward minus upward, radiative flux (expressed in W $m^{-2}$) due to a change in an external driver of *climate change*, such as a change in the concentration of *carbon dioxide ($CO_2$)*, the concentration of volcanic *aerosols* or in the output of the Sun. The stratospherically adjusted radiative forcing is computed with all tropospheric properties held fixed at their unperturbed values, and after allowing for stratospheric temperatures, if perturbed, to readjust to radiative-dynamical equilibrium. Radiative forcing is called instantaneous if no change in stratospheric temperature is accounted for. The radiative forcing once both stratospheric and tropospheric adjustments are accounted for is termed the 'effective radiative forcing'.

**Rebound effect** Phenomena whereby the reduction in energy consumption or emissions (relative to a baseline) associated with the implementation of *mitigation measures* in a jurisdiction is offset to some degree through induced changes in consumption, production, and prices within the same jurisdiction. The rebound effect is most typically ascribed to technological energy efficiency improvements.

**Reducing Emissions from Deforestation and Forest Degradation (REDD+)** REDD+ refers to reducing emissions from *deforestation*; reducing emissions from *forest* degradation; conservation of forest carbon stocks; sustainable management of forests; and enhancement of forest carbon stocks (see UNFCCC decision 1/CP.16, para. 70).

**Reference period** A time period of interest, or a period over which some relevant statistics are calculated. A reference period can be used as a *baseline period* or as a comparison to a baseline period.

*Baseline period*
A time period against which differences are calculated (e.g., expressed as anomalies relative to a baseline).

**Reference scenario** See *Scenario*.

**Reforestation** Conversion to *forest* of land that has previously contained forests but that has been converted to some other use.

[Note: For a discussion of the term forest and related terms such as *afforestation*, reforestation and *deforestation*, see the 2006 IPCC Guidelines for National Greenhouse Gas Inventories and their 2019 Refinement, and information provided by the United Nations Framework Convention on Climate Change (IPCC 2006, 2019; UNFCCC 2021a,b).]

See also *Anthropogenic removals*, *Carbon dioxide removal (CDR)* and *Reducing Emissions from Deforestation and Forest Degradation (REDD+)*.

**Regenerative agriculture** A universally agreed definition of this relatively new farming approach has yet to be established, but regenerative agriculture broadly refers to the implementation of varying combinations of agricultural management practices, to ensure the continued restoration and enhancement of soil health, *biodiversity* and *ecosystem* functioning, in conjunction with profitable agricultural production.

**Region** *Land* and/or *ocean* area characterised by specific geographical and/or climatological features. The *climate* of a region emerges from a multi-scale combination of its own features, remote influences from other regions, and global climate conditions.

**Remaining carbon budget** See *Carbon budget*.

**Renewable energy (RE)** Any form of energy that is replenished by natural processes at a rate that equals or exceeds its rate of use.

*Variable renewable energy (VRE)*
Renewable energy sources such as *wind* and *solar energy* whose output is determined by weather, in contrast to 'dispatchable' generators that adjust their output as a reaction to economic incentives. Variable renewables have also been termed intermittent, fluctuating, or non-dispatchable (Hirth 2013).

**Representative Concentration Pathways (RCPs)** See *Pathways*.

**Resilience** The capacity of interconnected social, economic and ecological systems to cope with a hazardous event, trend or disturbance, responding or reorganising in ways that maintain their essential function, identity and structure. Resilience is a positive attribute when it maintains capacity for *adaptation*, learning and/or transformation (Arctic Council 2016). See also *Hazard*, *Risk* and *Vulnerability*.

**Resource cascade** Tracking resource use (materials, energy, water, etc.), efficiency and losses through all conversion steps from primary resource extraction to various conversion steps, all the way to final service delivery.

**Risk** The potential for adverse consequences for human or ecological systems, recognising the diversity of values and objectives associated with such systems. In the context of *climate change*, risks can arise from potential *impacts* of climate change as well as human responses to climate change. Relevant adverse consequences include those on lives, *livelihoods*, health and *well-being*, economic, social and cultural assets and investments, *infrastructure*, services (including *ecosystem services*), *ecosystems* and species.

In the context of climate change impacts, risks result from dynamic interactions between climate-related *hazards* with the *exposure* and *vulnerability* of the affected human or ecological system to the hazards. Hazards, exposure and vulnerability may each be subject to uncertainty in terms of magnitude and *likelihood* of occurrence, and each may change over time and space due to socio-economic changes and human decision-making (see also *risk management*, *adaptation* and *mitigation*).

In the context of climate change responses, risks result from the potential for such responses not achieving the intended objective(s), or from potential trade-offs with, or negative side-effects on, other societal objectives, such as the *Sustainable Development Goals (SDGs)* (see also *risk trade-off*). Risks can arise, for example, from uncertainty in implementation, effectiveness or outcomes of *climate policy*, climate-related investments, technology development or adoption, and system transitions.

See also *Hazard* and *Impacts*.

**Risk assessment** The qualitative and/or quantitative scientific estimation of *risks*. See also *Risk management* and *Risk perception*.

**Risk management** Plans, actions, strategies or policies to reduce the *likelihood* and/or magnitude of adverse potential consequences, based on assessed or perceived *risks*. See also *Risk assessment*, and *Risk perception*.

**Risk perception** The subjective judgement that people make about the characteristics and severity of a *risk*. See also *Risk assessment*, and *Risk management*.

**Risk trade-off** The change in the portfolio of *risks* that occurs when a countervailing risk is generated (knowingly or inadvertently) by an intervention to reduce the target risk (Wiener and Graham 2009).

**Sea surface temperature (SST)** The subsurface bulk temperature in the top few metres of the ocean, measured by ships, buoys and drifters. From ships, measurements of water samples in buckets were mostly switched in the 1940s to samples from engine intake water. Satellite measurements of skin temperature (uppermost layer; a fraction of a millimetre thick) in the infrared or the top centimetre or so in the microwave are also used, but must be adjusted to be compatible with the bulk temperature.

**Scenario** A plausible description of how the future may develop based on a coherent and internally consistent set of assumptions about key driving forces (e.g., rate of technological change, prices) and relationships. Note that scenarios are neither predictions nor forecasts, but are used to provide a view of the implications of developments and actions.

*Baseline scenario*
See *Reference Scenario* (under *Scenario*).

*Concentrations scenario*
A plausible representation of the future development of atmospheric concentrations of substances that are radiatively active (e.g., *greenhouse gases*, *aerosols*, tropospheric *ozone*), plus human-induced *land cover changes* that can be radiatively active via *albedo* changes, and often used as input to a *climate model* to compute *climate projections*.

*Emissions scenario*
A plausible representation of the future development of emissions of substances that are radiatively active (e.g., *greenhouse gases* or *aerosols*), plus human-induced land-cover changes that can be radiatively active via *albedo* changes, based on a coherent and internally consistent set of assumptions about driving forces (such as demographic and socio-economic development, technological change, energy and *land use*) and their key relationships. *Concentration scenarios*, derived from emission scenarios, are often used as input to a *climate model* to compute *climate projections*.

*Mitigation scenario*
A plausible description of the future that describes how the (studied) system responds to the implementation of *mitigation* policies and measures.

*Reference scenario*
Scenario used as starting or reference point for a comparison between two or more scenarios.

[Note 1: In many types of *climate change* research, reference scenarios reflect specific assumptions about patterns of socio-economic development and may represent futures that assume no climate policies or specified climate policies, for example, those in place or planned at the time a study is carried out. Reference scenarios may also represent futures with limited or no climate impacts or adaptation, to serve as a point of comparison for futures with impacts and adaptation. These are also referred to as 'baseline scenarios' in the literature.

Note 2: Reference scenarios can also be climate policy or impact scenarios, which in that case are taken as a point of comparison to explore the implications of other features, for example, of delay, technological options, policy design and strategy or to explore the effects of additional impacts and adaptation beyond those represented in the reference scenario.

Note 3: The term *business as usual* scenario has been used to describe a scenario that assumes no additional policies beyond those currently in place, and where patterns of socio-economic development are consistent with recent trends. The term is now used less frequently than in the past.

Note 4: In climate change attribution or impact attribution research reference scenarios may refer to counterfactual historical scenarios assuming no anthropogenic *greenhouse gas (GHG)* emissions (climate change attribution) or no climate change (impact attribution).]

*Socio-economic scenario*
A scenario that describes a plausible future in terms of population, *gross domestic product (GDP)*, and other socio-economic factors relevant to understanding the implications of *climate change*.

**Scenario storyline** See *Storyline*.

**Sequestration** The process of storing carbon in a carbon pool. See also *Pool, carbon and nitrogen*, *Sequestration potential*, *Sink*, *Soil carbon sequestration (SCS)*, *Source*, and *Uptake*.

**Sequestration potential** The quantity of *greenhouse gases* that can be removed from the *atmosphere* by anthropogenic enhancement of *sinks* and stored in a pool. See *Mitigation potential* for different subcategories of sequestration potential. See also *Pool, carbon and nitrogen*, *Sequestration*, *Sink*, *Source*, and *Uptake*.

**Service provisioning** Various services (such as illumination and mobility) can be provided by 'systems' through the use of energy, materials, and other resources comprising: (i) Resource flows (e.g., energy); (ii) Technologies for resource use and energy conversion (e.g., vehicles and their engines); and (iii) Social/organisational forms of service delivery (e.g., publicly owned companies, or privately owned companies, e-commerce).

**Services** Activities that help satisfy human wants or needs. While they usually involve relationships between producers and consumers, services are less tangible and less storable than goods since they represent flows not stocks, and when their regeneration conditions are protected they may be reused over time.

**Settlements** Places of concentrated human habitation. Settlements can range from isolated rural villages to *urban regions* with significant global influence. They can include formally planned and informal or illegal habitation and related infrastructure. See also *Cities* and *Urban*.

**Shared policy assumptions (SPAs)** See *Shared Socio-economic Pathways (SSPs)*.

**Shared Socio-economic Pathways (SSPs)** See *Pathways*.

**Sharing economy** A system which allows people to share goods and services by enabling collaborative use, access or ownership.

**Shifting development pathways (SDPs)** In this report, shifting development pathways describes transitions aimed at re-directing existing developmental trends. Societies may put in place *enabling conditions* to influence their future development pathways, when they endeavour to achieve certain outcomes. Some outcomes may be common, while others may be context-specific, given different starting points. See also *Development pathways* (under *Pathways*), and *Shifting development pathways to sustainability*.

**Shifting development pathways to sustainability** Shifting development pathways to sustainability involves transitions aligned with a shared aspiration in the *Sustainable Development Goals (SDGs)* agreed globally, though sustainability may be interpreted differently in various contexts as societies pursue a variety of sustainable development objectives. See also *Development pathways* (under *Pathways*), and *Shifting development pathways (SDPs)*.

**Short-lived climate forcers (SLCFs)** A set of chemically reactive compounds with short (relative to *carbon dioxide*) atmospheric lifetimes (from hours to about two decades) but characterised by different physiochemical properties and environmental effects. Their emission or formation has a significant effect on radiative forcing over a period determined by their respective atmospheric lifetimes. Changes in their *emissions* can also induce long-term *climate* effects via, in particular, their interactions with some biogeochemical cycles. SLCFs are classified as direct or indirect, with direct SLCFs exerting climate effects through their *radiative forcing* and indirect SLCFs being the *precursors* of other direct climate forcers. Direct SLCFs include *methane ($CH_4$)*, *ozone ($O_3$)*, primary *aerosols* and some halogenated species. Indirect SLCFs are *precursors* of ozone or secondary aerosols. SLCFs can be cooling or warming through interactions with radiation and clouds. They are also referred to as near-term climate forcers. Many SLCFs are also air pollutants. A subset of exclusively warming SLCFs is also referred to as short-lived climate pollutants (SLCPs), including methane, ozone, and *black carbon (BC)*.

**Short-lived climate pollutants (SLCP)** See *Short-lived climate forcers (SLCFs)*.

**Simple climate model (SCM)** A broad class of lower-dimensional models of the energy balance, radiative transfer, *carbon cycle*, or a combination of such physical components. SCMs are also suitable for performing emulations of climate-mean variables of Earth System Models (ESMs), given that their structural flexibility can capture both the parametric and structural uncertainties across process-oriented ESM responses. They can also be used to test consistency across multiple lines of evidence with regard to *climate sensitivity* ranges, *transient climate responses (TCRs)*, *transient climate response to cumulative $CO_2$ emissions (TCREs)* and *carbon cycle* feedbacks. See also *Emulators*.

**Sink** Any process, activity or mechanism which removes a *greenhouse gas*, an *aerosol* or a *precursor* of a *greenhouse gas* from the *atmosphere* (UNFCCC Article 1.8 (UNFCCC 1992)). See also *Pool, carbon and nitrogen*, *Sequestration*, *Source* and *Uptake*.

**Small Island Developing States (SIDS)** SIDS, as recognised by the United Nations OHRLLS (UN Office of the High Representative for the Least Developed Countries, Landlocked Developing Countries and Small Island Developing States), are a distinct group of developing countries facing specific social, economic and environmental vulnerabilities (UN-OHRLLS 2011). They were recognised as a special case, both for their environment and development, at the Rio Earth Summit in Brazil in 1992. Fifty-eight countries and territories are presently classified as SIDS by the UN OHRLLS, with 38 being UN member states and 20 being non-UN members or associate members of the Regional Commissions (UN-OHRLLS 2018).

**Smart grids** A smart grid uses information and communications technology to gather data on the behaviours of suppliers and consumers in the production, distribution, and use of electricity. Through automated responses or the provision of price signals, this information can then be used to improve the efficiency, reliability, economics, and sustainability of the electricity network.

**Social cost of carbon (SCC)** The net present value of aggregate climate damages (with overall harmful damages expressed as a number with positive sign) from one more tonne of carbon in the form of *carbon dioxide ($CO_2$)*, conditional on a global emissions trajectory over time.

**Social costs** The full costs of an action in terms of social welfare losses, including external costs associated with the impacts of this action on the environment, the economy (*GDP*, employment) and on the society as a whole.

**Social group** A collective of people who share similar characteristics and collectively may have a sense of unity (Forsyth 2010).

**Social identity** The portion of an individual's self-concept derived from perceived membership in a relevant *social group* (Tajfel and Turner 1986).

**Social inclusion** A process of improving the terms of participation in society, particularly for people who are disadvantaged, through enhancing opportunities, access to resources, and respect for rights (UNDESA 2018).

**Social infrastructure** See *Infrastructure*.

**Social learning** A process of social interaction through which people learn new behaviours, capacities, values, and attitudes.

**Social-ecological system** An integrated system that includes human societies and *ecosystems*, in which humans are part of nature. The functions of such a system arise from the interactions and interdependence of the social and ecological subsystems. The system's structure is characterised by reciprocal feedbacks, emphasising that humans must be seen as a part of, not apart from, nature (Berkes and Folke 1998; Arctic Council 2016).

**Socio-economic scenario** See *Scenario*.

**Socio-technical transitions** Where technological change is associated with social systems and the two are inextricably linked.

**Soil carbon sequestration (SCS)** *Land management* changes which increase the *soil organic carbon* content, resulting in a net removal of *carbon dioxide ($CO_2$)* from the *atmosphere*. See also *Anthropogenic removals* and *Carbon dioxide removal (CDR)*.

**Soil organic carbon** Carbon contained in *soil organic matter*.

**Soil organic matter** The organic component of soil, comprising plant and animal residue at various stages of decomposition, and soil organisms.

**Solar energy** Energy from the Sun. Often the phrase is used to mean energy that is captured from solar radiation either as heat, as light that is converted into chemical energy by natural or artificial photosynthesis, or by photovoltaic panels and converted directly into electricity. See also *Renewable energy*.

**Solar radiation modification (SRM)** Refers to a range of radiation modification measures not related to *greenhouse gas (GHG)* mitigation that seek to limit *global warming*. Most methods involve reducing the amount of incoming solar radiation reaching the surface, but others also act on the longwave radiation budget by reducing optical thickness and cloud lifetime.

**Source** Any process or activity which releases a *greenhouse gas (GHG)*, an *aerosol* or a *precursor* of a GHG into the *atmosphere*

(UNFCCC Article 1.9 (UNFCCC 1992)). See also *Sink*, *Pool, carbon and nitrogen*, *Sequestration*, *Sequestration Potential* and *Uptake*.

**Spill-over effect** The effects of domestic or sector mitigation measures on other countries or sectors. Spill-over effects can be positive or negative and include effects on trade, (carbon) *leakage*, transfer of innovations, and diffusion of environmentally sound technology and other issues.

**Storyline** A way of making sense of a situation or a series of events through the construction of a set of explanatory elements. Usually, it is built on logical or causal reasoning. In *climate* research, the term storyline is used both in connection to *scenarios* as related to a future trajectory of the climate and human systems or to a weather or climate event. In this context, storylines can be used to describe plural, conditional possible futures or explanations of a current situation, in contrast to single, definitive futures or explanations.

*Scenario storyline*
A narrative description of a *scenario* (or family of scenarios), highlighting the main scenario characteristics, relationships between key driving forces and the dynamics of their evolution.

**Stranded assets** Assets exposed to devaluations or conversion to 'liabilities' because of unanticipated changes in their initially expected revenues due to innovations and/or evolutions of the business context, including changes in public regulations at the domestic and international levels.

**Subnational actors** State/provincial, regional, metropolitan and local/municipal governments as well as non-party stakeholders, such as civil society, the private sector, *cities* and other subnational authorities, local communities and indigenous peoples.

**Sufficiency** A set of measures and daily practices that avoid demand for energy, materials, land and water while delivering human well-being for all within planetary boundaries.

**Sustainability** A dynamic process that guarantees the persistence of natural and human systems in an equitable manner.

**Sustainable development (SD)** Development that meets the needs of the present without compromising the ability of future generations to meet their own needs (WCED 1987) and balances social, economic and environmental concerns. See also *Development pathways* and *Sustainable Development Goals (SDGs)*.

**Sustainable Development Goals (SDGs)** The 17 global goals for development for all countries established by the United Nations through a participatory process and elaborated in the 2030 Agenda for Sustainable Development, including ending poverty and hunger; ensuring health and well-being, education, gender equality, clean water and energy, and decent work; building and ensuring resilient and sustainable infrastructure, cities and consumption; reducing inequalities; protecting land and water ecosystems; promoting peace, justice and partnerships; and taking urgent action on climate change. See also *Sustainable development*.

**Sustainable forest management** The stewardship and use of *forests* and forest lands in a way, and at a rate, that maintains their *biodiversity*, productivity, regeneration capacity, vitality and their potential to fulfil, now and in the future, relevant ecological, economic and social functions, at local, national, and global levels, and that does not cause damage to other *ecosystems* (Forest Europe 1993).

**Sustainable intensification (of agriculture)** Increasing yields from the same area of land while decreasing negative environmental impacts of agricultural production and increasing the provision of environmental services (CGIAR 2019).

[Note: This definition is based on the concept of meeting demand from a finite land area, but it is scale-dependent. Sustainable intensification at a given scale (e.g., global or national) may require a decrease in production intensity at smaller scales and, in particular, places (often associated with previous, unsustainable, intensification) to achieve *sustainability* (Garnett et al. 2013).]

**Sustainable land management** The stewardship and use of *land* resources, including soils, water, animals and plants, to meet changing human needs, while simultaneously ensuring the long-term productive potential of these resources and the maintenance of their environmental functions (WOCAT, no date).

**Systems of Innovation (SI)** The set of public and private sector organisations (i.e., formally organised entities such as firms and universities; 'actors') and *institutions*, whose activities and interactions generate, modify and deploy new technologies. The SI approach has been used to understand and analyse innovation at the national, regional, and technological levels, and in transnational contexts (Lundvall 1992, 1988).

**Technology deployment** The act of bringing technology into effective application, involving a set of actors and activities to initiate, facilitate and/or support its implementation. See also *Technology diffusion*.

**Technology diffusion** The spread of a technology across different groups users/markets over time. See also *Technology deployment* and *Technology transfer*.

**Technology transfer** The exchange of knowledge, hardware and associated software, money and goods among stakeholders, which leads to the spread of technology for *adaptation* or *mitigation*. The term encompasses both diffusion of technologies and technological cooperation across and within countries. See also *Technology diffusion*.

**Teleconnection** Association between *climate* variables at widely separated, geographically fixed locations related to each other through physical processes and oceanic and/or atmospheric dynamical pathways. Teleconnections can be caused by several climate phenomena, such as Rossby wave-trains, mid-latitude jet and storm track displacements, fluctuations of the Atlantic Meridional Overturning Circulation (AMOC), fluctuations of the Walker circulation, etc. They can be initiated by modes of climate variability, thus providing the development of remote climate anomalies at various temporal lags.

**Temperature overshoot** Exceedance of a specified global warming level, followed by a decline to or below that level during a specified period of time (e.g., before 2100). Sometimes the magnitude and likelihood of the overshoot is also characterised. The overshoot duration can vary from one *pathway* to the next, but in most *overshoot pathways* in the literature and as referred to as overshoot pathways in the AR6, the overshoot occurs over a period of at least one decade and up to several decades.

**Tipping point** A critical threshold beyond which a system reorganises, often abruptly and/or irreversibly. See also *Irreversibility*.

**Total carbon budget** See *Carbon budget*.

**Trade-off** A competition between different objectives within a decision situation, where pursuing one objective will diminish achievement of other objective(s). A trade-off exists when a policy or measure aimed at one objective (e.g., reducing *greenhouse gas* emissions) reduces outcomes for other objective(s) (e.g., *biodiversity* conservation, *energy security*) due to *adverse side effects*, thereby potentially reducing the net benefit to society or the environment. See also *Co-benefit*.

**Transformation** A change in the fundamental attributes of natural and human systems.

**Transformation pathways** See *Pathways*.

**Transient climate response (TCR)** See *Climate sensitivity*.

**Transient climate response to cumulative $CO_2$ emissions (TCRE)** See *Climate sensitivity*.

**Transition** The process of changing from one state or condition to another in a given period of time. Transition can occur in individuals, firms, cities, regions and nations, and can be based on incremental or transformative change.

**Uncertainty** A state of incomplete knowledge that can result from a lack of information or from disagreement about what is known or even knowable. It may have many types of sources, from imprecision in the data to ambiguously defined concepts or terminology, incomplete understanding of critical processes, or uncertain projections of *human behaviour*. Uncertainty can therefore be represented by quantitative measures (e.g., a probability density function) or by qualitative statements (e.g., reflecting the judgement of a team of experts) (Moss and Schneider 2000; Mastrandrea et al. 2010). See also *Confidence* and *Likelihood*.

**United Nations Convention to Combat Desertification (UNCCD)** A legally binding international agreement linking environment and development to sustainable land management, established in 1994. The Convention's objective is 'to combat desertification and mitigate the effects of drought in countries experiencing drought and/or desertification'. The Convention specifically addresses the arid, semi-arid and dry sub-humid areas, known as the drylands, and has a particular focus on Africa. As of September 2020, the UNCCD had 197 Parties. See also *Desertification*, *Drought* and *Land degradation*.

**United Nations Framework Convention on Climate Change (UNFCCC)** The UNFCCC was adopted in May 1992 and opened for signature at the 1992 Earth Summit in Rio de Janeiro. It entered into force in March 1994 and, as of September 2020, had 197 Parties (196 States and the European Union). The Convention's ultimate objective is the 'stabilisation of greenhouse gas concentrations in the atmosphere at a level that would prevent dangerous anthropogenic interference with the climate system' (UNFCCC 1992). The provisions of the Convention are pursued and implemented by two further treaties: the Kyoto Protocol and the Paris Agreement.

**Uptake** The transfer of substances (such as carbon) or energy (e.g., heat) from one compartment of a system to another; for example, in the Earth System from the atmosphere to the ocean or to the land. See also *Pool, carbon and nitrogen*, *Sequestration*, *Sequestration potential*, *Sink* and *Source*.

**Urban** The categorisation of areas as 'urban' by government statistical departments is generally based either on population size, population density, economic base, provision of services, or some combination of the above. *Urban systems* are networks and nodes of intensive interaction and exchange including capital, culture, and material objects. Urban areas exist on a continuum with rural areas and tend to exhibit higher levels of complexity, higher populations and population density, intensity of capital investment, and a preponderance of secondary (processing) and tertiary (service) sector industries. The extent and intensity of these features varies significantly within and between urban areas. Urban places and systems are open with much movement and exchange between more rural areas as well as other urban regions. Urban areas can be globally interconnected facilitating rapid flows between them – of capital investment, of ideas and culture, human migration, and disease. See also *Cities*, *Peri-urban areas*, and *Urbanisation*.

**Urban heat island** The relative warmth of a *city* compared with surrounding rural areas, associated with heat trapping due to land use, the configuration and design of the built environment, including street layout and building size, the heat-absorbing properties of urban building materials, reduced ventilation, reduced greenery and water features, and domestic and industrial heat emissions generated directly from human activities. See also *City region*, *Urban*, and *Urban System*.

**Urban Systems** Urban systems refer to two interconnected systems: first, the comprehensive collections of city elements with multiple dimensions and characteristics: a) encompass physical, built, socio-economic-technical, political, and ecological subsystems; b) integrate social agent/constituency/processes with physical structure and processes; and c) exist within broader spatial and temporal scales and governance and institutional contexts; and second, the global system of cities and towns. See also *City region*, and *Urban*.

**Urbanisation** Urbanisation is a multi-dimensional process that involves at least three simultaneous changes: (i) land-use change: transformation of formerly rural settlements or natural land into urban settlements; (ii) demographic change: a shift in the spatial distribution of a population from rural to urban areas; and (iii) infrastructure change: an increase in provision of infrastructure services including electricity, sanitation, etc. Urbanisation often includes changes in lifestyle, culture, and behaviour, and thus alters the demographic, economic, and social structure of both urban and rural areas (Stokes and Seto 2019; Seto et al. 2014; UNDESA 2018). See also *Urban*, and *Urban Systems*.

**Variable renewable energy (VRE)** See *Renewable energy*.

**Vulnerability** The propensity or predisposition to be adversely affected. Vulnerability encompasses a variety of concepts and elements including sensitivity or susceptibility to harm and lack of capacity to cope and adapt. See also *Exposure*, *Hazard* and *Risk*.

**Well-being** A state of existence that fulfils various human needs, including material living conditions, meaningful social and community relationships and quality of life, as well as the ability to pursue one's goals, to thrive, and feel satisfied with one's life. Ecosystem well-being refers to the ability of *ecosystems* to maintain their diversity and quality.

*Eudaimonic*
Relational well-being concept based on the premise that experiencing life purpose, challenges and growth leads to flourishing, self-realisation, personal expression, and full functioning (Niemiec 2014; Lamb and Steinberger 2017).

*Hedonic*
Subjective well-being concept based on the idea that attaining pleasure and avoiding pain leads to happiness (Ryan and Deci 2001).

**Wind energy** Kinetic energy from airflow arising from the uneven heating of the Earth's surface. The wind's kinetic energy is converted to mechanical shaft energy and electricity by a wind turbine, a rotating machine. A wind farm, wind project, wind park, or wind power plant is a group of wind turbines interconnected to a common utility system through a system of transformers, distribution lines, and (usually) one substation. See also *Renewable energy*.

**Zero emissions commitment** See *Climate change commitment*.

## References


AGECC, 2010: *Energy for a Sustainable Future. The secretary-general's Advisory group on Energy and climate Change*, New York, NY, USA.

Allan, J.A., 2005: Virtual water: A strategic resource global solutions to regional deficits. *Groundwater*, **36(4)**, 545–546, doi:10.1111/j.1745- 6584. 1998.tb02825.x.

Arctic Council, 2016: *Arctic Resilience Report 2016*. [Carson, M. and G. Peterson (eds.)]. Stockholm Environment Institute and Stockholm Resilience Centre, Stockholm, Sweden, 218 pp. https://oaarchive.arctic-council.org/handle/11374/1838 (Accessed October 2021).

Berkes, F. and C. Folke, 1998: *Linking Social and Ecological Systems: Management Practices and Social Mechanisms for Building Resilience*. [Berkes, F. and C. Folke (eds.)]. Cambridge University Press, Cambridge, UK and New York, NY, USA, 459 pp.

Bobbins, K. and C. Culwick, 2016: *A Framework for a Green Infrastructure Planning Approach in the Gauteng City-Region*. Gauteng City-Region Observatory (GRCO), Johannesburg, South Africa, 127 pp. http://hdl.handle.net/10539/23595.

Campbell, A. et al., 2009: *Review of the Literature on the Links between Biodiversity and Climate Change: Impacts, Adaptation and Mitigation*. Secretariat of the Convention on Biological Diversity, Montreal, QC, Canada, 124 pp. https://www.cbd.int/doc/publications/cbd-ts-42-en.pdf (Accessed October 11, 2021).

CGIAR, 2019: Sustainable intensification of agriculture: oxymoron or real deal? https://wle.cgiar.org/thrive/big-questions/sustainable-intensification-agriculture-oxymoron-or-real-deal/sustainable-1 (Accessed October 11, 2021).

Chester, M.V, 2019: Sustainability and infrastructure challenges. *Nat. Sustain.*, **2**(4), 265–266, doi:10.1038/s41893-019-0272-8.

Davis, S.J. and K. Caldeira, 2010: Consumption-based accounting of $CO_2$ emissions. *Proc. Natl. Acad. Sci.*, **107(12)**, 5687–5692, doi:10.1073/pnas.0906974107.

Dawson, R.J. et al., 2018: A systems framework for national assessment of climate risks to infrastructure. *Philos. Trans. R. Soc. A Math. Phys. Eng. Sci.*, **376(2121)**, 20170298, doi:10.1098/rsta.2017.0298.

Díaz, S. et al., 2018: Assessing nature's contributions to people. *Science*, **359(6373)**, 270–272, doi:10.1126/science.aap8826.

FAO, 2007: *Land evaluation: Towards a revised framework. Land and water discussion paper*. Food and Agriculture Organization of the United Nations (FAO), Rome, Italy.

FAO, 2009: *Declaration of the World Summit on Food Security. WSFS 2009/2*. Food and Agriculture Organization of the United Nations (FAO), Rome, Italy.

FAO, 2015: Food waste. Food and Agriculture Organization of the United Nations (FAO). http://www.fao.org/platform-food-loss-waste/food-waste/definition.

FAO, 2016: Conservation agriculture. https://www.fao.org/conservation-agriculture/overview/what-is-conservation-agriculture/en/ (Accessed October 11, 2021).

FAO, 2018: *The State of Food Security and Nutrition in the World: Building Climate Resilience for Food Security and Nutrition*. Food and Agriculture Organization of the United Nations (FAO), Rome, Italy, 181 pp.

FAO and Alliance of Bioversity International and CIAT, 2021: *Indigenous Peoples' food systems: Insights on sustainability and resilience from the front line of climate change*, Rome, doi: 10.4060/cb5131en.

Forest Europe, 1993: *Resolution H1 General Guidelines for the Sustainable Management of Forests in Europe*. Second Ministerial Conference on the Protection of Forests in Europe, 16–17 June 1993, Helsinki, https://www.foresteurope.org/docs/MC/MC_helsinki_resolutionH1.pdf.

Forsyth, D.R., 2010: *Group Dynamics*. Fifth Edition. Wadswoth Cengage Learning, Inc., Belmont, CA, USA.

Frolova, E.V, M.K. Vinichenko, A.V Kirillov, O.V Rogach, and E.E. Kabanova, 2016: Development of social infrastructure in the management practices of local authorities: trends and factors. *Int. J. Environ. Sci. Educ.*, **11(15)**, 7421–7430.

Fung, A. and E.O. Wright, 2003: *Deepening Democracy: Institutional Innovations in Empowered Participatory Governance*. [Fung, A. and E.O. Wright (eds.)]. Verso, London, UK, 312 pp.

Garnett, T. et al., 2013: Sustainable intensification in agriculture: Premises and policies. *Science.*, **341(6161)**, 33, doi:10.1126/science.1234485.

Grubler, A. et al., 2012: Energy Primer. In: *Global Energy Assessment*, pp. 99–150. Cambridge University Press, Cambridge, UK, doi:10.1017/CBO9780511793677.007.

Hewitt, C., S. Mason, and D. Walland, 2012: The Global Framework for Climate Services. *Nat. Clim. Change*, **2(12)**, 831–832, doi:10.1038/nclimate1745.

Hirth, L., 2013: The market value of variable renewables: The effect of solar wind power variability on their relative price. *Energy Econ.*, **38**, 218–236, doi:10.1016/j.eneco.2013.02.004.

HLPE, 2017: *Nutrition and food systems*. High Level Panel of Experts on Food Security and Nutrition (HLPE), Rome, Italy, 152 pp.

IBI, 2018: Frequently Asked Questions About Biochar: What is biochar? International Biochar Initiative (IBI). https://biochar-international.org/faqs/ (Accessed October 11, 2021).

IFOAM-Organics International, 2014: *The IFOAM Norms for Organic Production and Processing*. IFOAM-Organics International, Germany. https://www.ifoam.bio/sites/default/files/2020-09/IFOAM Norms July 2014 Edits 2019.pdf (Accessed October 11, 2021).

ILO, 2015: *Guidelines for a just transition towards environmentally sustainable economies and societies for all*. International Labour Organization (ILO), Switzerland.

IOM, 2018: Key Migration Terms. Retrieved from: https://www.iom.int/key-migration-terms (Accessed May 15, 2018).

IPBES, 2018: *The IPBES assessment report on land degradation and restoration*. [Montanarella, L., R. Scholes, and A. Brainich (eds.)]. Secretariat of the Intergovernmental Science-Policy Platform on Biodiversity and Ecosystem Services (IPBES), Bonn, Germany, 744 pp.

IPBES, 2019: Glossary. *Global assessment report on biodiversity and ecosystem services of the Intergovernmental Science-Policy Platform on Biodiversity and Ecosystem Services.* Secretariat of the Intergovernmental Science-Policy Platform on Biodiversity and Ecosystem Services (IPBES), Bonn, Germany.

IPCC, 2006: *2006 IPCC Guidelines for National Greenhouse Gas Inventories, Prepared by the National Greenhouse Gas Inventories Programme*. [Eggleston, H.S., L. Buendia, K. Miwa, T. Ngara, and K. Tanabe (eds.)]. Institute for Global Environmental Strategies (IGES), Hayama, Japan, https://www.ipcc-nggip.iges.or.jp/public/2006gl/index.html.

IPCC, 2012: *Managing the Risks of Extreme Events and Disasters to Advance Climate Change Adaptation. A Special Report of Working Groups I and II of the Intergovernmental Panel on Climate Change*. [Field, C.B., V. Barros, T.F. Stocker, D. Qin, D.J. Dokken, K.L. Ebi, M.D. Mastrandrea, K.J. Mach, G.-K. Plattner, S.K. Allen, M. Tignor, and P.M. Midgley (eds.)]. Cambridge University Press, Cambridge, UK and New York, NY, USA, 582 pp.

IPCC, 2019: *Refinement to the 2006 IPCC Guidelines for National Greenhouse gas Inventories*. [Buendia, E.C., S. Guendehou, B. Limmeechokchai, R. Pipatti, Y. Rojas, R. Sturgiss, K. Tanabe, T. Wirth, D. Romano, J. Witi, A. Garg, M.M. Weitz, B. Cai, D.A. Ottinger, H. Dong, J.D. MacDonald, S.M. Ogle, M.T. Rocha, M. José Sanz Sanchez, D.M. Bartram, S. Towprayoon (eds.)]. Intergovernmental Panel on Climate Change (IPCC), Geneva, Switzerland. https://www.ipcc-nggip.iges.or.jp/public/2019rf/index.html (Accessed October 11, 2021).

ISO, 2014: ISO 16559:2014(en). Solid biofuels – Terminology, definitions and descriptions. http://www.iso.org/obp/ui/#iso:std:iso:16559:ed-1:v1:en (Accessed October 11, 2021).

ISO, 2018: ISO 14044:2006. Environmental management – Life cycle assessment – Requirements and guidelines. https://www.iso.org/standard/38498.html (Accessed May 31, 2018).

IUCN, 2016: Defining Nature-based Solutions. WCC-2016-Res-069-EN, World Conservation Congress.

Lamb, W.F. and J.K. Steinberger, 2017: Human well-being and climate change mitigation. *Wiley Interdiscip. Rev. Clim. Change.*, **8(6)**, doi:10.1002/wcc.485.

Latham, A. and J. Layton, 2019: Social infrastruture and the public life of cities: Studying urban sociality and public spaces. *Geogr. Compass*, **13(7)**, 12444.

Lundvall, B.-Å., 1988: Innovation as an interactive process: from user-producer interaction to the national system of innovation. In: *Technical Change and Economic Theory* [Dosi, G., C. Freeman, R. Nelson, G. Silverberg, and L. Soete (eds.)]. Pinter Publishers, London, UK and New York, NY, USA.

Lundvall, B.-Å., 1992: *National Systems of Innovation: Toward a Theory of Innovation and Interactive Learning*. [Lundvall, B.-A., (ed.)]. Pinter Publishers, London, UK, 342 pp.

MA, 2005: Appendix D: Glossary. In: *Ecosystems and Human Well-being: Current States and Trends. Findings of the Condition and Trends Working Group* [Hassan, R., R. Scholes, and N. Ash (eds.)], Millennium Ecosystem Assessment (MA). Island Press, Washington, DC, USA, pp. 893–900.

MacDonald, G.K. et al., 2015: Rethinking agricultural trade relationships in an era of globalization. *Bioscience*, **65(3)**, 275–289, doi:10.1093/biosci/biu225.

Mastrandrea, M.D. et al., 2010: *Guidance Note for Lead Authors of the IPCC Fifth Assessment Report on Consistent Treatment of Uncertainties*. Intergovernmental Panel on Climate Change (IPCC), Geneva, Switzerland, 6 pp.

McDonald, T., J. Jonson, and K.W. Dixon, 2016: National standards for the practice of ecological restoration in Australia. *Restor. Ecol.*, **24**(S1), S4–S32, doi:10.1111/rec.12359.

Mechler, R., L. Bouwer, T. Schinko, S. Surminski, and J. Linnerooth-Bayer, 2018: *Loss and Damage from Climate Change: Concepts, Methods and Policy Options*. [Mechler, R., L.M. Bouwer, T. Schinko, S. Surminski, and J. Linnerooth-Bayer (eds.)]. Springer International Publishing, Cham, Switzerland, 561 pp.

Moss, R.H. and S.H. Schneider, 2000: Uncertainties in the IPCC TAR: Recommendations to Lead Authors for More Consistent Assessment and Reporting. In: *Guidance Papers on the Cross Cutting Issues of the Third Assessment Report of the IPCC* [Pachauri, R., T. Taniguchi, and K. Tanaka (eds.)], Intergovernmental Panel on Climate Change (IPCC), Geneva, Switzerland, pp. 33–51.

Moss, R.H. et al., 2010: The next generation of scenarios for climate change research and assessment. *Nature*, **463(7282)**, 747–756, doi:10.1038/nature08823.

MRFCJ, 2018: Principles of Climate Justice. Mary Robinson Foundation – Climate Justice (MRFCJ). http://www.mrfcj.org/principles-of-climate-justice (Accessed May 15, 2018).

Niemiec, C.P., 2014: Eudaimonic Well-Being. In: *Encyclopedia of Quality of Life and Well-Being Research* [Michalos, A.C. (ed.)]. Springer, Dordrecht, The Netherlands, pp. 2004–2005, doi:10.1007/978-94-007-0753-5_929.

O'Neill, D.W., A.L. Fanning, W.F. Lamb, and J.K. Steinberger, 2018: A good life for all within planetary boundaries. *Nat. Sustain.*, **1(2)**, 88–95, doi:10.1038/s41893-018-0021-4.

Ogle, S.M. et al., 2018: Delineating managed land for reporting national greenhouse gas emissions and removals to the United Nations framework convention on climate change. *Carbon Balance Manag.*, **13(1)**, 9, doi:10.1186/s13021-018-0095-3.

OHCHR, 2018: What are Human rights? Office of the High Commissioner for Human Rights (OHCHR). https://www.ohchr.org/en/what-are-human-rights (Accessed October 11, 2021).

Park, S.E. et al., 2012: Informing adaptation responses to climate change through theories of transformation. *Glob. Environ. Change.*, **22(1)**, 115–126, doi:10.1016/j.gloenvcha.2011.10.003.

Pongratz, J. et al., 2018: Models meet data: Challenges and opportunities in implementing land management in Earth system models. *Glob. Change. Biol.*, **24(4)**, 1470–1487, doi:10.1111/gcb.13988.

Rao, N.D. and P. Baer, 2012: "Decent Living" emissions: A conceptual framework. *Sustainability*, **4(4)**, 656–681, doi:10.3390/su4040656.

Rao, N.D. and J. Min, 2018: Decent Living Standards: Material Prerequisites for Human Wellbeing. *Soc. Indic. Res.*, **138(1)**, 225–244, doi:10.1007/s11205-017-1650-0.

Reddy, A.K.N., 2000: *Energy and social issues*. World Energy Council, New York, NY, USA.

Ryan, R.M. and E.L. Deci, 2001: On happiness and human potentials: a review of research on hedonic and eudaimonic well-being. *Annu. Rev. Psychol.*, **52**, 141–166, doi:10.1146/annurev.psych.52.1.141.

Sarmiento, H. and C. Tilly, 2018: Governance Lessons from Urban Informality. *Polit. Gov.*, **6(1)**, 199–202, doi:10.17645/pag.v6i1.1169.

Seto, K.C., R. Sánchez-Rodríguez, and M. Fragkias, 2010: The New Geography of Contemporary Urbanization and the Environment. *Annu. Rev. Environ. Resour.*, **35(1)**, 167–194, doi:10.1146/annurev-environ-100809-125336.

Seto, K.C., S. Dhakal, A. Bigio, H. Blanco, G.C. Delgado, D. Dewar, L. Huang, A. Inaba, A. Kansal, S. Lwasa, J.E. McMahon, D.B. Müller, J. Murakami, H. Nagendra, and A. Ramaswami, 2014: Human Settlements, Infrastructure and Social Planning. In: *Climate Change 2014: Mitigation of Climate Change. Contribution of Working Group III to the Fifth Assessment Report of the Intergovernmental Panel on Climate Change* [Edenhofer, O., R. Pichs-Madruga, Y. Sokona, E. Farahani, S. Kadner, K. Seyboth, A. Adler, I. Baum, S. Brunner, P. Eickemeier, B. Kriemann, J. Savolainen, S. Schlömer, C. von Stechow, T. Zwickel and J.C. Minx (eds.)]. Cambridge University Press, Cambridge, UK and New York, NY, USA, pp. 923–1000.

Silvertown, J., 2009: A new dawn for citizen science. In: *Trophic Cascades: Predators, Prey, and the Changing Dynamics of Nature* [Terborgh, J. and J.A. Estes (eds.)]. Island Press, Washington DC, USA, pp. 337–352.

Stokes, E.C. and K.C. Seto, 2019: Characterizing and measuring urban landscapes for sustainability. *Environ. Res. Lett.*, **14(4)**, 045002, doi:10.1088/1748-9326/aafab8.

Tàbara, D.J., J. Jäger, D. Mangalagiu, and M. Grasso, 2019: Defining transformative climate science to address high-end climate change. *Reg. Environ. Change.*, **19(3)**, 807–818, doi:10.1007/s10113-018-1288-8.

Tajfel, H. and J.C. Turner, 1986: The social identity theory of intergroup behaviour. In: *Psychology of Intergroup Relations* [Worschel, S. and W.G. Austin (eds.)], IL: Nelson-Hall, Chicago, USA, pp. 7–24.

Termeer, C.J.A.M., A. Dewulf, and G.R. Biesbroek, 2017: Transformational change: governance interventions for climate change adaptation from a continuous change perspective. *J. Environ. Plan. Manag.*, **60(4)**, 558–576, doi:10.1080/09640568.2016.1168288.

UN, 1992: Article 2: Use of Terms. In: *Convention on Biological Diversity*, United Nations (UN), pp. 3–4.

UN-OHRLLS, 2011: *Small Island Developing States: Small Islands Big(ger) Stakes*. Office for the High Representative for the Least Developed Countries, Landlocked Developing Countries and Small Island Developing States (UN-OHRLLS), New York, NY, USA, 32 pp.

UN-OHRLLS, 2018: Small Island Developing States: Country profiles. https://www.un.org/ohrlls/content/list-sids.

UN-REDD, 2009: *Measurement, Assessment, Reporting and Verification (MARV): Issues and Options for REDD*. United Nations Collaborative Programme on Reducing Emissions from Deforestation and Forest Degradation in Developing Countries (UN-REDD), Geneva, Switzerland, 12 pp. https://unredd.net/index.php?option=com_docman&task=doc_download&gid=148&Itemid=53 (Accessed October 11, 2021).

UNCCD, 1994: *Elaboration of an international convention to combat desertification in countries experiencing serious drought and/or desertification, particularly in Africa*. United Nations Convention to Combat Desertification (UNCCD), Paris, France, 1–58 pp.

UNCCD, 2020: Achieving Land Degradation Neutrality. United Nations Convention to Combat Desertification (UNCCD). https://www.unccd.int/.

UNDESA, 2018: *World Urbanization Prospects, the 2018 Revision, Technical Report*. United Nations Department of Economic and Social Affairs (UNDESA), New York, NY, USA.

UNEP, 2016: *Unlocking the sustainable potential of land resources: evaluating systems, strategies and tools*. United Nations Environment Programme (UNEP). https://www.unep.org/resources/report/unlocking-sustainable-potential-land-resources-evaluating-systems-strategies-and.

UNESCO, 2018: Local and Indigenous Knowledge Systems. United Nations Educational, Scientific and Cultural Organization (UNESCO). https://en.unesco.org/links.

UNFCCC, 1992: *United Nations Framework Convention on Climate Change*. United Nations Framework Convention on Climate Change (UNFCCC), 24 pp. https://unfccc.int/resource/docs/convkp/conveng.pdf.

UNFCCC, 2016: *Just transition of the workforce, and the creation of decent work and quality jobs*. United Nations Framework Convention on Climate Change (UNFCCC).

UNFCCC, 2019: Land Use, Land-Use Change and Forestry (LULUCF). United Nations Framework Convention on Climate Change (UNFCCC). https://unfccc.int/topics/land-use/workstreams/land-use--land-use-change-and-forestry-lulucf.

UNFCCC, 2021a: Reporting and Review under the Paris Agreement. United Nations Framework Convention on Climate Change (UNFCCC). https://unfccc.int/process-and-meetings/transparency-and-reporting/reporting-and-review-under-the-paris-agreement.

UNFCCC, 2021b: Reporting and accounting of LULUCF activities under the Kyoto Protocol. United Nations Framework Convention on Climate Change (UNFCCC). https://unfccc.int/topics/land-use/workstreams/land-use-land-use-change-and-forestry-lulucf/reporting-and-accounting-of-lulucf-activities-under-the-kyoto-protocol.

UNGA, 2016: *Report of the open-ended intergovernmental expert working group on indicators and terminology relating to disaster risk reduction*. United Nations General Assembly (UNGA), 41 pp. https://www.preventionweb.net/files/50683_oiewgreportenglish.pdf.

UNISDR, 2009: *2009 UNISDR Terminology on Disaster Risk Reduction*. United Nations International Strategy for Disaster Reduction (UNISDR), Geneva, Switzerland, 30 pp. https://www.unisdr.org/we/inform/publications/7817.

WCED, 1987: *Our Common Future*. World Commission on Environment and Development (WCED) Oxford University Press, Oxford, UK, 400 pp.

WHO, 2018: Malnutrition. World Health Organization (WHO). https://www.who.int/health-topics/malnutrition#tab=tab_1 (Accessed October 11, 2021).

Wiedmann, T. and J. Minx, 2008: A Definition of "Carbon Footprint." In: *Ecological Economics Research Trends* [C.C. Pertsova (ed.)], Nova Science Publishers, New York, NY, USA, pp. 1–11.

Wiener, J.B. and J.D. Graham, 2009: *Risk vs Risk: Tradeoffs in Protecting Health and the Environment*. [Wiener, J.B. and J.D. Graham (eds.)]. Harvard University Press, Cambridge, MA, USA, 352 pp.

Willems, S. and K. Baumert, 2003: *Institutional Capacity and Climate Actions*. Organisation for Economic Co-operation and Development (OECD) International Energy Agency (IEA), Paris, France, 50 pp. Retrieved from: http://www.oecd.org/env/cc/21018790.pdf.

WOCAT, [undated]: Glossary. World Overview of Conservation Approaches and Technologies (WOCAT). https://www.wocat.net/en/glossary.

AII

# Annex II: Definitions, Units and Conventions

**Coordinating Lead Author:**
Alaa Al Khourdajie (United Kingdom/Syria)

**Lead Authors:**
Renée van Diemen (the Netherlands/United Kingdom), William F. Lamb (Germany/United Kingdom), Minal Pathak (India), Andy Reisinger (New Zealand), Stéphane de la Rue du Can (the United States of America), Jim Skea (United Kingdom), Raphael Slade (United Kingdom), Shreya Some (India), Linda Steg (the Netherlands)

**This chapter should be cited as:**
IPCC, 2022: Annex II: Definitions, Units and Conventions [Al Khourdajie, A., R. van Diemen, W.F. Lamb, M. Pathak, A. Reisinger, S. de la Rue du Can, J. Skea, R. Slade, S. Some, L. Steg (eds)]. In IPCC, 2022: *Climate Change 2022: Mitigation of Climate Change. Contribution of Working Group III to the Sixth Assessment Report of the Intergovernmental Panel on Climate Change* [P.R. Shukla, J. Skea, R. Slade, A. Al Khourdajie, R. van Diemen, D. McCollum, M. Pathak, S. Some, P. Vyas, R. Fradera, M. Belkacemi, A. Hasija, G. Lisboa, S. Luz, J. Malley, (eds.)]. Cambridge University Press, Cambridge, UK and New York, NY, USA. doi: 10.1017/9781009157926.021

# Table of Contents

This annex on *Definitions, Units and Conventions* provides background information on material used in the Working Group III contribution to the Intergovernmental Panel on Climate Change (IPCC) Sixth Assessment Report (AR6 WGIII). The material presented in this annex documents metrics and common datasets that are typically used across multiple chapters of the report. In a few instances there are no updates to what was adopted by WGIII during the production of the Fifth Assessment Report (AR5), in which case this annex refers to Annex II of AR5 (Krey et al. 2014).

The annex comprises four parts: Part I introduces standards, metrics and common definitions adopted in the report; Part II presents methods to derive or calculate certain quantities and identities used in the report; Part III provides more detailed background information about common data sources; and Part IV presents integrative methodologies used in the assessment. While this structure may help readers to navigate through the annex, it is not possible in all cases to unambiguously assign a certain topic to one of these parts, naturally leading to some overlap between the parts.

## Part I: Definitions and Units

### A.II.1 Classification Schemes for Countries and Areas

In this report, two different levels of classification are used as a standard to present the results of analysis. The basis for the classification is the UN Statistics Division *Standard Country or Area Codes for Statistical Use*, also known as the M49 Standard (UNSD 1999). This covers geographical regions and, at the time of the literature cut-off date, identified developed regions, developing regions and least developed countries.

The high-level classification has six categories (Table 1): one covering North America, Europe, and Australia, Japan and New Zealand, labelled 'developed countries', and five covering other countries, all classified as developing using the M49 standard at the cut-off date. The high-level classification is an expansion of the RC5 (Regional Categorisation 5) adopted in AR5 WGIII, with Africa and the Middle East now identified separately. The low-level classification (ten categories) divides developed countries into three geographical regions, and Asia and Pacific into three sub-regions.

The high- and low-level classification schemes reflect schemes used in many global models and statistical sources. Where the report synthesises data, only these standard classification schemes have been used. On occasions, the underlying literature may deviate from the standard classification scheme and direct citations may unavoidably refer to alternative classifications. This is dealt with on a case-by-case basis and does not imply any endorsement of the scheme used in the underlying literature by the IPCC or the authors of this report.

The detailed allocation of countries and areas to the low-level classification is shown in Section 1.1. Following AR5, the classification scheme deviates from the UN regional classification with the result that Annex I, Annex II and non-Annex I countries as defined under the UN Framework Convention on Climate Change (UNFCCC) are distinguished. Some Annex I countries in Western Asia and countries in Eastern Europe which are not members of the European Union are allocated to Eastern Europe and West-Central Asia (EEA). In AR5, these formed part of the Economies in Transition group. The remainder of Western Asia (non-Annex I) is allocated to the Middle East.

Following the practice of the UN Statistics Division, we note that the designations employed and the presentation of material in this report do not imply the expression of any opinion by the United Nations, the IPCC or the authors of this report concerning the legal status of any country, territory, city or area or of its authorities, or concerning the delimitation of its frontiers or boundaries. The term 'country' as used in this material also refers, as appropriate, to territories or areas.

#### A.II.1.1 Low Level of Regional Groupings

**Africa:** Algeria, Angola, Benin, Botswana, Burkina Faso, Burundi, Côte d'Ivoire, Cabo Verde, Cameroon, the Central African Republic, Chad, the Comoros, the Congo, the Democratic Republic of the Congo, Djibouti, Egypt, Equatorial Guinea, Eritrea, Eswatini, Ethiopia, Gabon, the Gambia, Ghana, Guinea, Guinea-Bissau, Kenya, Lesotho, Liberia, Libya, Madagascar, Malawi, Mali, Mauritania, Mauritius, Morocco, Mozambique, Namibia, the Niger, Nigeria, Rwanda, São Tomé and Príncipe, Senegal, Seychelles, Sierra Leone, Somalia, South Africa, the South Sudan, the Sudan, Togo, Tunisia, Uganda, the United Republic of Tanzania, Zambia, Zimbabwe.

**Middle East:** Bahrain, Islamic Republic of Iran, Iraq, Israel, Jordan, Kuwait, Lebanon, Oman, Qatar, Saudi Arabia, the State of Palestine, the Syrian Arab Republic, the United Arab Emirates, Yemen.

**Latin America and Caribbean:** Antigua and Barbuda, Argentina, the Bahamas, Barbados, Belize, Plurinational State of Bolivia, Brazil, Chile, Colombia, Costa Rica, Cuba, Dominica, the Dominican Republic, Ecuador, El Salvador, Grenada, Guatemala, Guyana, Haiti, Honduras, Jamaica, Mexico, Nicaragua, Panama, Paraguay, Peru, Saint Kitts and Nevis, Saint Lucia, Saint Vincent and the Grenadines, Suriname, Trinidad and Tobago, Uruguay, Bolivarian Republic of Venezuela.

**North America:** Canada, the United States of America.

**Eastern Asia:** China, the Republic of Korea, the Democratic People's Republic of Korea, Mongolia.

**Southern Asia:** Afghanistan, Bangladesh, Bhutan, India, Maldives, Nepal, Pakistan, Sri Lanka.

**South-East Asia and Pacific:** Brunei Darussalam, Cambodia, Cook Islands, Fiji, Indonesia, Kiribati, the Lao People's Democratic Republic, Malaysia, the Marshall Islands, Federated States of Micronesia, Myanmar, Nauru, Niue, Palau, Papua New Guinea, the Philippines, Samoa, Singapore, Solomon Islands, Thailand, Timor-Leste, Tonga, Tuvalu, Vanuatu, Viet Nam.

**Europe:** Albania, Andorra, Austria, Belgium, Bosnia and Herzegovina, Bulgaria, Croatia, Cyprus, Czechia, Denmark, Estonia, Finland, France, Germany, Greece, Hungary, Iceland, Ireland, Italy, Latvia, Liechtenstein, Lithuania, Luxembourg, Malta, Monaco, Montenegro, the Netherlands,

North Macedonia, Norway, Poland, Portugal, Romania, San Marino, Serbia, Slovakia, Slovenia, Spain, Sweden, Switzerland, Turkey, the United Kingdom of Great Britain and Northern Ireland.

**Australia, Japan, and New Zealand**

**Eastern Europe and West-Central Asia:** Armenia, Azerbaijan, Belarus, Georgia, Kazakhstan, Kyrgyzstan, the Republic of Moldova, the Russian Federation, Tajikistan, Turkmenistan, Ukraine, Uzbekistan.

**International Shipping and Aviation**

### A.II.1.2 High, Low Levels of Regional Groupings

**Table 1 | Classification schemes for countries and areas.**

| WGIII AR6 | |
|---|---|
| **High Level (6)** | **Low Level (10)** |
| Developed Countries (DEV) | North America |
| | Europe |
| | Australia, Japan and New Zealand |
| Eastern Europe and West-Central Asia (EEA) | Eastern Europe and West-Central Asia |
| Latin America and Caribbean (LAM) | Latin America and Caribbean |
| Africa (AFR) | Africa |
| Middle East (ME) | Middle East |
| Asia and Pacific (APC) | Eastern Asia |
| | Southern Asia |
| | South-East Asia and Pacific |
| International Shipping and Aviation | |

## A.II.2 Standard Units and Unit Conversions

The following sections introduce standard units and unit conversions used throughout this report.

### A.II.2.1 Standard Units

Standard units of measurements include Système International (SI) units, SI-derived units, and other non-SI units as well the standard prefixes for basic physical units.

**Table 2 | Système International (SI) units.**

| Physical quantity | Unit | Symbol |
|---|---|---|
| Length | metre | m |
| Mass | kilogram | kg |
| Time | second | s |
| Thermodynamic temperature | kelvin | K |
| Amount of substance | mole | mol |

**Table 3 | Special names and symbols for certain SI-derived units.**

| Physical quantity | Unit | Symbol | Definition |
|---|---|---|---|
| Force | Newton | N | $kg\ m\ s^2$ |
| Pressure | Pascal | Pa | $kg\ m^{-1}\ s^{-2}$ ($= N\ m^{-2}$) |
| Energy | Joule | J | $kg\ m^2\ s^{-2}$ |
| Power | Watt | W | $kg\ m^2\ s^{-3}$ ($= J\ s^{-1}$) |
| Frequency | Hertz | Hz | $s^{-1}$ (cycles per second) |
| Ionizing radiation dose | sievert | Sv | $J\ kg^{-1}$ |

**Table 4 | Non-SI standard units.**

| Monetary units | Unit | Symbol |
|---|---|---|
| Currency (market exchange rate, MER) | Constant US Dollar 2015 | USD2015 |
| Currency (purchasing power parity, PPP) | Constant International Dollar 2015 | Int$2015 |
| **Emission- and climate-related units** | **Unit** | **Symbol** |
| Emissions | Metric tonnes | t |
| $CO_2$ emissions | Metric tonnes $CO_2$ | $tCO_2$ |
| $CO_2$-equivalent emissions[1] | Metric tonnes $CO_2$-equivalent | $tCO_2$-eq |
| Abatement costs and emissions prices/taxes | Constant US dollar 2015 per metric tonne | USD2015 $t^{-1}$ |
| $CO_2$ concentration or mixing ratio ($\mu mol\ mol^{-1}$) | Parts per million ($10^6$) | Ppm |
| $CH_4$ concentration or mixing ratio ($nmol\ mol^{-1}$) | Parts per billion ($10^9$) | ppb |
| $N_2O$ concentration or mixing ratio ($nmol\ mol^{-1}$) | Parts per billion ($10^9$) | ppb |
| Radiative forcing | Watts per square meter | $W/m^2$ |
| **Energy-related units** | **Unit** | **Symbol** |
| Energy | Joule | J |
| Electricity and heat generation | Watt hours | Wh |
| Power (peak capacity) | Watt (Watt thermal, Watt electric) | W (Wth, We) |
| Capacity factor | Percent | % |
| Technical and economic lifetime | Years | yr |
| Specific energy investment costs | US dollar 2015 per kW (peak capacity) | USD2015/kW |
| Energy costs (e.g., LCOE) and prices | Constant US dollar 2015 per GJ or US cents 2015 per kWh | USD2015/GJ and USct2015/kWh |
| Passenger-distance | Passenger-kilometre | pkm |
| Payload-distance[2] | Tonne-kilometre | tkm |
| **Land-related units** | **Unit** | **Symbol** |
| Area | Hectare | ha |

Note that all monetary and monetary-related units are expressed in constant US Dollar 2015 (*USD*2015) or constant International Dollar 2015 (*Int*$2015).

1 A measure of aggregate greenhouse gas (GHG) emissions. This report uses the GHG metric Global Warming Potential with a time horizon of 100 years (GWP100); for details see Section 8.

2 The is a unit of measure of freight transport which represents the transport of one tonne of goods (including packaging and tare weights of intermodal transport units) by a given transport mode (road, rail, air, sea, inland waterways, pipeline etc.) over a distance of one kilometre. The tonne measure here is not the same unit of measure as metric tonnes earlier in the third row of Table 4.

Table 5 | Prefixes for basic physical units.

| Multiple | Prefix | Symbol | Fraction | Prefix | Symbol |
|---|---|---|---|---|---|
| 1E+21 | zeta | Z | 1E-01 | deci | d |
| 1E+18 | exa | E | 1E-02 | centi | c |
| 1E+15 | peta | P | 1E-03 | milli | m |
| 1E+12 | tera | T | 1E-06 | micro | μ |
| 1E+09 | giga | G | 1E-09 | nano | n |
| 1E+06 | mega | M | 1E-12 | pico | p |
| 1E+03 | kilo | k | 1E-15 | femto | f |
| 1E+02 | hecto | h | 1E-18 | atto | a |
| 1E+01 | deca | da | 1E-21 | zepto | z |

## A.II.2.2 Physical Units Conversion

Table 6 | Conversion table for common mass units (IPCC 2001).

| *To:* | | kg | t | lt | st | lb |
|---|---|---|---|---|---|---|
| *From:* | Multiply by: | | | | | |
| Kilogram | kg | 1 | 1.00E-03 | 9.84E-04 | 1.10E-03 | 2.20E+00 |
| Tonne | t | 1.00E+03 | 1 | 9.84E-01 | 1.10E+00 | 2.20E+03 |
| Long ton | lt | 1.02E+03 | 1.02E+00 | 1 | 1.12E+00 | 2.24E+03 |
| Short Ton | st | 9.07E+02 | 9.07E-01 | 8.93E-01 | 1 | 2.00E+03 |
| Pound | lb | 4.54E-01 | 4.54E-04 | 4.46E-04 | 5.00E-04 | 1 |

Table 7 | Conversion table for common volumetric units (IPCC 2001).

| *To:* | | gal US | gal UK | bbl | $ft^3$ | l | $m^3$ |
|---|---|---|---|---|---|---|---|
| *From:* | Multiply by: | | | | | | |
| US gallon | gal US | 1 | 8.33E-01 | 2.38E-02 | 1.34E-01 | 3.79E+00 | 3.80E-03 |
| UK/imperial gallon | gal UK | 1.20E+00 | 1 | 2.86E-02 | 1.61E-01 | 4.55E+00 | 4.50E-03 |
| Barrel | bbl | 4.20E+01 | 3.50E+01 | 1 | 5.62E+00 | 1.59E+02 | 1.59E-01 |
| Cubic foot | $Ft^3$ | 7.48E+00 | 6.23E+00 | 1.78E-01 | 1 | 2.83E+01 | 2.83E-02 |
| Litre | L | 2.64E-01 | 2.20E-01 | 6.30E-03 | 3.53E-02 | 1 | 1.00E-03 |
| Cubic metre | $M^3$ | 2.64E+02 | 2.20E+02 | 6.29E+00 | 3.53E+01 | 1.00E+03 | 1 |

Table 8 | Conversion table for common energy units (NAS 2007; IEA 2019).

| *To:* | | TJ | Gcal | Mtoe | Mtce | MBtu | GWh |
|---|---|---|---|---|---|---|---|
| *From:* | Multiply by: | | | | | | |
| Tera joule | TJ | 1 | 2.39E+02 | 2.39E-05 | 3.41E-05 | 9.48E+02 | 2.78E-01 |
| Giga calorie | Gcal | 4.19E-03 | 1 | 1.0E-06 | 1.43E-07 | 3.97E+00 | 1.16E-03 |
| Mega tonne oil equivalent | Mtoe | 4.19E+04 | 1.0E+08 | 1 | 1.43E+00 | 3.97E+07 | 1.16E+04 |
| Mega tonne coal equivalent | Mtce | 2.93E+04 | 7.0E+06 | 7.00E-01 | 1 | 2.78E+07 | 8.14E+03 |
| Million british thermal units | MBtu | 1.06E-03 | 2.52E-01 | 2.52E-08 | 3.60E-08 | 1 | 2.93E-04 |
| Giga watt hours | GWh | 3.60E+00 | 8.60E+02 | 8.60E-05 | 1.23E-4 | 3.41E+03 | 1 |

In addition to the above physical units, datasets often report carbon emissions in either units of carbon (C) or carbon dioxide ($CO_2$). In this report we report carbon dioxide ($CO_2$) emissions where possible, using the conversion factor (44/12) to convert from units of C into $CO_2$. Finally, we note that the conversion from GJ to kWh is as follows: 1 GJ = ~277.78 kWh.

Where aggregate greenhouse gas emissions are reported, this report uses the Global Warming Potential with a time horizon of 100 years (GWP100); for details see Section 8.

### A.II.2.3 Monetary Unit Conversion

To achieve comparability across cost und price information from different regions, where possible monetary quantities reported in the AR6 WGIII have been expressed in constant US Dollar 2015 (*USD*2015) or constant International Dollar 2015 (*Int*$2015), as suitable.

To facilitate a consistent monetary unit conversion process, a simple and transparent procedure to convert different monetary units from the literature to USD2015 is established and described below.

In order to convert from year X local currency unit ($LCU_X$) to 2015 US Dollars (*USD*2015) two steps are needed:

1. Inflating or deflating from year X to 2015, and
2. Converting from *LCU* to *USD*.

In practice, the order of applying these two steps will lead to different results. In this report, the conversion route adopted is $LCU_X \rightarrow LCU2015 \rightarrow USD2015$, i.e., national or regional deflators are used to measure country- or region-specific inflation between year X and 2015 in local currency, then current (2015) exchange rates are used to convert to *USD*2015. The reason for adopting this route is when the economy's GDP deflator is used to convert to a common base year, that is, 2015, it captures the changes in prices of all goods and services that the economy produces. To convert from *LCU*2015 to *USD*2015, the official 2015 exchange rates are used. Note that exchange rates often fluctuate significantly in the short term.

In order to be consistent with the choice of the World Bank databases as the primary source for gross domestic product (GDP) and other financial data throughout the report, deflators and exchange rates from the World Bank Development Indicators are used.[3]

To summarise, the following procedure has been adopted to convert monetary quantities reported in $LCU_X$ to *USD*2015:

1. Use the country-/region-specific deflator and multiply with the deflator value to convert from $LCU_X$ to *LCU*2015. In case national/regional data are reported in non-LCU units (e.g., $USD_X$ or $Euro_X$), which is often the case in multi-national or global studies, apply the corresponding currency deflator to convert to 2015 currency (i.e., the US deflator and the Eurozone deflator in the examples above).

   Example of converting GDP from *LCU*2010 prices to *LCU*2015 prices:

   $$GDP2015 \text{ (in } LCU2015 \text{ prices)} = GDP2010 \text{ (in } LCU2010 \text{ prices)} * \frac{LCU_{2010}\, GDP\, deflator}{LCU_{2015}\, GDP\, deflator}$$

2. Use the appropriate 2015 exchange rate to convert from *LCU*2015 to *USD*2015.

## Part II: Conventions

### A.II.3 Levelised Cost Metrics

Across this report, a number of different metrics to characterise cost of climate change mitigation are employed. To facilitate a meaningful economic comparison across diverse options at the technology level, the metric of 'levelised costs' is used throughout several chapters of this report in various forms. The most used metrics are the levelised cost of energy (LCOE), the levelised cost of conserved energy (LCCE), and the levelised cost of conserved carbon (LCCC). These metrics are used throughout the AR6 WGIII to provide a benchmark for comparing different technologies or practices of achieving the respective output. Each comes with a set of context-specific caveats that need to be taken into account for correct interpretation. Various literature sources caution against drawing too strong conclusions from these metrics. Annex II in AR5, namely Section A.II.3.1, includes a detailed discussion on interpretations and caveats. Below is an introduction to each of these metrics and how they are derived.

#### A.II.3.1 Levelised Cost of Energy

The levelised cost of energy (LCOE) can be defined as the unique break-even cost-price where discounted revenues (price x quantities) are equal to the discounted net expenses (Moomaw et al. 2011), which is expressed as follows:

$$\sum_{t=0}^{n} \frac{E_t * LCOE}{(1+i)^t} = \sum_{t=0}^{n} \frac{Expenses_t}{(1+i)^t} \qquad (1)$$

where $E_t$ is the energy delivered in year *t* (might vary from year to year), expenses cover all (net) expenses in the year *t*, *i* is the discount rate and *n* the lifetime of the project.

solving for *LCOE*:

$$LCOE = \frac{\sum_{t=0}^{n} \frac{Expenses_t}{(1+i)^t}}{\sum_{t=0}^{n} \frac{E_t}{(1+i)^t}} \qquad (2)$$

The lifetime expenses comprise investment costs *I*, operation and maintenance cost *O&M* (including waste management costs), fuel costs *F*, carbon costs *C*, and decommissioning costs *D*. In this case, levelised cost can be determined by (IEA 2010):

$$LCOE = \frac{\sum_{t=0}^{n} \frac{I_t + O\&M_t + F_t + C_t + D_t}{(1+i)^t}}{\sum_{t=0}^{n} \frac{E_t *}{(1+i)^t}} \qquad (3)$$

[3] For instance, the data for GDP deflators for all countries can be downloaded following this link: https://data.worldbank.org/indicator/NY.GDP.DEFL.ZS?locations=US.

Assuming energy $E$ provided annually is constant during the lifetime of the project, one can rewrite (3) as follows:

$$LCOE = \frac{CRF \cdot NPV\,(Lifetime\ Expenses)}{E} = \frac{Annuity\,(Lifetime\ Expenses)}{E} \quad (4)$$

where $CRF = \frac{i}{1-(1-i)^{-n}}$ is the capital recovery factor and *NPV* the net present value of all lifetime expenditures (Suerkemper et al. 2012).

For the simplified case, where the annual costs are also assumed constant over time, this can be further simplified to (*O&M* costs and fuel costs *F* constants):

$$LCOE = \frac{CRF \cdot I + O\&M + F}{E} \quad (5)$$

Where *I* is the upfront investment, *O&M* are the annual operation and maintenance costs, *F* are the annual fuel costs, and *E* is the annual energy provision. The investment *I* should be interpreted as the sum of all capital expenditures needed to make the investment fully operational discounted to $t = 0$. These might include discounted retrofit payments during the project lifetime and discounted decommissioning costs at the end of the lifetime. Where applicable, annual *O&M* costs have to take into account revenues for by-products and existing carbon costs must be added or treated as part of the annual fuel costs.

### A.II.3.2 Levelised Cost of Conserved Energy

The levelised cost of conserved energy (LCCE) annualises the investment and operation and maintenance cost differences between a baseline technology and the energy-efficient alternative and divides this quantity by the annual energy savings.

The conceptual formula for *LCCE* is essentially the same as Equation (4) above, with $\Delta E$ measuring in this context the amount of energy saved annually (Suerkemper et al. 2012):

$$LCCE = \frac{CRF.NPV(\Delta Lifetime\ Expenses)}{\Delta E} = \frac{Annuity(\Delta Lifetime\ Expenses)}{\Delta E} \quad (6)$$

In the case of assumed annually constant *O&M* costs over the project lifetime, one can rewrite (6) as follows:

$$LCCE = \frac{CRF \cdot \Delta I + \Delta O\&M}{\Delta E} \quad (7)$$

where $\Delta I$ is the difference in investment costs of an energy saving measure (e.g., in USD) as compared to a baseline investment; $\Delta O\&M$ is the difference in annual operation and maintenance costs of an energy saving measure (e.g., in USD) as compared to the baseline in which the energy-saving measure is not implemented; $\Delta E$ is the annual energy conserved by the measure (e.g., in *kWh*) as compared to the usage of the baseline technology; and *CRF* is the capital recovery factor depending on the discount rate and the lifetime of the measure in years as defined above. It should be stressed once more that this equation is only valid if $\Delta O\&M$ and $\Delta E$ are constant over the project lifetime. As *LCCE* are designed to be compared with complementary levelised cost of energy supply, they do not include the annual fuel cost difference. Any additional monetary benefits that are associated with the energy-saving measure must be taken into account as part of the *O&M* difference.

### A.II.3.3 Levelised Cost of Conserved Carbon

The levelised cost of conserved carbon can be used for comparing mitigation costs per unit of avoided carbon emissions and comparing these specific emission reduction costs for different options. This concept can be applied to other pollutants.

The conceptual formula for *LCCC* is similar to Equation (6) above, with $\Delta C$ is the annual reduction in carbon emissions, which can be expressed as follows:

$$LCCC = \frac{CRF.NPV(\Delta Lifetime\ Expenses)}{\Delta C} = \frac{Annuity(\Delta Lifetime\ Expenses)}{\Delta C} \quad (8)$$

In the case of assumed annually constant *O&M* costs over the lifetime, one can rewrite (8) as follows:

$$LCCC = \frac{CRF \cdot \Delta I + \Delta O\&M - \Delta B}{\Delta C} \quad (9)$$

where $\Delta I$ is the difference in investment costs of a mitigation measure (e.g., in USD) as compared to a baseline investment; $\Delta O\&M$ is the difference in annual operation and maintenance costs (e.g., in USD) and $\Delta B$ denotes the annual benefits, all compared to a baseline for which the option is not implemented. Note that annual benefits include reduced expenditures for fuels, if the investment project reduces emissions via a reduction in fuel use. As such *LCCC* depend on energy prices. An important characteristic of this equation is that *LCCC* can become negative if $\Delta B$ is bigger than the sum of the other two terms in the numerator.

## A.II.4 Growth Rates

### A.II.4.1 Emissions Growth Rates

In order to ensure consistency throughout the reported growth rates for emissions in AR6 WGIII, this section establishes the convention for calculating these rates.

The annual growth rate of emissions in percent per year for adjacent years is given by:

$$r = \frac{(E_{FF}(t_0 - 1) - E_{FF}(t_0))}{E_{FF}(t_0)} * 100 \quad (10)$$

where $E_{FF}$ stands for fossil fuel $CO_2$ emissions, but can also be applied to other pollutants.

When relevant a leap-year adjustment is required in order to ensure valid interpretation of annual growth rates in the case of adjacent years. A leap-year affects adjacent years growth rate by approximately 0.3% $yr^{-1}$ $\left(\frac{1}{365}\right)$ which causes growth rates to go up approximately 0.3% if the first year is a leap year, and down 0.3% if the second year is a leap year (Friedlingstein et al. 2019).

The relative growth rate of $E_{FF}$ over time periods of greater than one year is derived as follows.

Starting from:

$$E_{FF}(t+n) = E_{FF}(t) * (1+r)^n \tag{11}$$

solving for $r$:

$$r = \left(\frac{E_{FF}(t+n)}{E_{FF}(t)}\right)^{1/n} - 1 \tag{12}$$

### A.II.4.2 Economic Growth Rates

A number of different methods exist for calculating economic growth rates (e.g., GDP), all of which lead to slightly different numerical results. If not stated otherwise, the annual growth rates shown in the report are derived using the *Log Difference Regression* technique *or Geometric Average* techniques which can be shown to be equivalent.

The Log Difference Regression growth rate $r_{LD}$ is calculated as follows:

$$r_{LD} = e^{\beta} - 1 \text{ with } \beta = \frac{1}{T-1}\sum_{t=2}^{T} \Delta ln X_t \tag{13}$$

The Geometric Average growth rate $r_{GEO}$ is calculated as shown below:

$$r_{GEO} = \left(\frac{X_T}{X_1}\right)^{\frac{1}{T-1}} - 1 \tag{14}$$

Other methods that are used to calculate annual growth rates include the Ordinary Least Square technique and the Average Annual Growth Rate technique.

## A.II.5 Trends Calculations Between Years and Over Decades

In order to compare or contrast trends between two different years, for instance comparing 2000 and 2010 cumulative $CO_2$ emissions, the year 2000 runs from 1st of January to 31st of December and similarly the year 2010 runs from 1st of January to 31st of December.

In order to undertake a timeseries calculation over a decade, the 10-year period should be defined as follows: from 1st of January 2001 to 31st of December 2010, that is 2001–2010.

## A.II.6 Primary Energy Accounting

Primary energy accounting methods are used to report primary energy from non-combustible energy sources, in other words, nuclear energy and all renewable energy sources except biomass. Annex II of AR5, namely Section A.II.4, includes a detailed discussion of the three main methods dominant in the literature. The method adopted in AR6 is the *direct equivalent method* which counts one unit of secondary energy provided from non-combustible sources as one unit of primary energy, that is, 1 kWh of electricity or heat is accounted for as 1 kWh = 3.6 MJ of primary energy. This method is mostly used in the long-term scenarios literature, including multiple IPCC reports (IPCC 1995, Morita et al. 2001, Fisher et al. 2007, Fischedick et al. 2011), because it deals with fundamental transitions of energy systems that rely to a large extent on low-carbon, non-combustible energy sources.

## A.II.7 The Concept of Risk

The concept of risk is a key aspect of how the IPCC assesses and communicates to decision-makers the potential adverse impacts of, and response options to, climate change. For the AR6 cycle, the definition of risk was revised (see below). Authors and IPCC Bureau members from all three Working Groups produced a Guidance (Reisinger et al. 2020) for authors on the concept of risk in order to ensure a consistent and transparent application across Working Groups.

This section summarises this Guidance briefly with a focus on issues related to WGIII, in other words, with focus on mitigation.

### A.II.7.1 The Definition of Risk

Definition (see Annex I: Glossary):

Risk is the potential for adverse consequences for human or ecological systems, recognising the diversity of values and objectives associated with such systems. In the context of climate change, risks can arise from potential impacts of climate change as well as *human responses to climate change*. Relevant adverse consequences include those on lives, livelihoods, health and well-being, economic, social and cultural assets and investments, infrastructure, services (including ecosystem services), ecosystems and species.

- In the context of climate change impacts, risks result from dynamic interactions between climate-related hazards with the exposure and vulnerability of the affected human or ecological system to the hazards. Hazards, exposure and vulnerability may each be subject to uncertainty in terms of magnitude and likelihood of occurrence, and each may change over time and space due to socio-economic changes and human decision-making (see also risk management, adaptation, mitigation).

- In the context of climate change responses, risks result from the potential for such responses not achieving the intended objective(s), or from potential trade-offs with, or negative side-effects on, other societal objectives, such as the Sustainable Development Goals. Risks can arise for example from uncertainty in implementation, effectiveness or outcomes of climate policy, climate-related investments, technology development or adoption, and system transitions.

### A.II.7.2 The Definition of Risk Management

Plans, actions, strategies or policies to reduce the likelihood and/or magnitude of adverse potential consequences, based on assessed or perceived risks (see also risk assessment, risk perception, risk transfer).

### A.II.7.3 The Uses of the Term Risk and Risk Management

In this report, with the aim of improving the ability of decision-makers to understand and manage risk, the term is used when considering the potential for adverse outcomes and the uncertainty relating to these outcomes.

The term risk is not used as a simple substitute for probability or chance, to describe physical hazards, or as generic term for 'anything bad that may happen in future'. While the probability of an adverse outcome does not necessarily have to be quantified, it needs to be characterised in some way to allow a risk assessment to inform responses via risk management.

In the AR6, risk refers to the potential for *adverse* consequences only. The term hazard is used where climatic events or trends has an identified potential for having adverse consequences to specific elements of an affected system. The contribution of Working Group I to the AR6 uses the more general term 'climatic impact driver' where a specific change in climate could have positive or negative consequences, and where a given climatic change may therefore act as a driver of risk or of an opportunity.

### A.II.7.4 Examples of Application in the Context of Mitigation

*Food Security*

Climate-related risk to food security arises from multiple drivers that include both climate change impacts, responses to climate change and other stressors.

In the context of responses to climate change, drivers of risk include the demand for land from climate change responses (both adaptation and mitigation), the role of markets (e.g., price spikes related to biofuel demand in other countries), governance (how are conflicts about access to land and water resolved) and human behaviour more generally (e.g., trade barriers, dietary preferences).

Given the multitude of drivers, the risk to food security depends on assumptions about what drivers of risk are changing and which are assumed to remain constant. Such assumptions are important for analytical robustness and are stated where relevant.

*Risk in the Investment and Finance Literature*

The investment and finance literature and practitioner community broadly distinguish between 'physical risk' and 'transition risk'. The term 'physical risk' generally refers to risks arising from climate change impacts and climate-related hazards, while the term 'transition risk' typically refers to risks associated with the transition to a low carbon economy. These two types of risk may interact and create cascading or compounding risks.

*Physical Risk*

In much of the business and financial literature, the term 'physical risk' relates to those derived from the hazard × exposure × vulnerability framework. Physical risks arise from the potential for climate change impacts on the financial value of assets such as industrial plants or real estate, risks to facilities and infrastructure, impact on operations, water and raw material availability and supply chain disruptions. Physical risks have direct financial consequences for organisations where those risks are realised, as well as up-front insurance and investment related costs and downstream effects for users of relevant goods and services.

*Transition Risk*

Transition risks typically refer to risks associated with transition to a low carbon economy, which can entail extensive policy, legal, technology, and market changes to address mitigation and adaptation requirements related to climate change. Depending on the nature, speed, and focus of these changes, transition risks may pose varying levels of financial and reputational risk to organisations. Transition risks, if realised, can result in stranded assets, loss of markets, reduced returns on investment, and financial penalties, as well as adverse outcomes for governance and reputation.

A key issue is the stranding of assets that may not provide the expected financial returns and may end up as large financial liabilities.

Examples of types of transition risk relating to business, finance and investments:

- Risk related to an asset losing its value: the potential for loss of investment in infrastructure.
- Risk related to losing some or all of the principal of an investment (or invested capital).
- Solvency risk: the risk from reduction in credit ratings due to potential adverse consequences of climate change or climate policy. This includes liquidity risk or the risk of not being able to access funds. Another example is suffering a downgraded credit rating.
- Risk of lower-than-expected return on investment.
- Liability risk: lack of response to climate change creates risk of liability for failure to accurately assess risk of climate change to infrastructure and people.

AII

- Technology risk: reliance on a particular technology to achieve an outcome creates the potential for adverse consequences if the technology fails to be developed or deployed.
- Policy risk: changes in policy or regulations in response to climate change could result in the loss of value of some assets.
- Market risk: changes in relative prices from increased prices of $CO_2$ for instance, could reduce financial returns and hence increase risks to investors.
- Residual risk: in parts of the financial literature, this concept refers to adverse consequences that cannot be quantified in probabilistic terms. Note that this is different from how the term 'residual risk' is generally used in IPCC, especially Working Group II, where it means the risk remaining after adaptation and risk reduction efforts.



## A.II.8 GHG Emission Metrics

Comprehensive mitigation policy relies on consideration of all anthropogenic forcing agents, which differ widely in their atmospheric lifetimes and impacts on the climate system. GHG emission metrics[4] provide simplified information about the effect that emissions of different GHGs have on global temperature or other aspects of climate, usually expressed relative to the effect of emitting $CO_2$. An assessment of different GHG emission metrics from a mitigation perspective is provided in Cross-Chapter Box 2 and Chapter 2 Supplementary Material, building on the assessment of GHG emission metrics from a physical science perspective in AR6 WGI (Forster et al., 2021, Section 7.6).

The WGIII contribution to the AR6 reports aggregate emissions and removals using updated values for the Global Warming Potential with a time horizon of 100 years (GWP100) from AR6 WGI unless stated otherwise. These updated GWP100 values reflect updated scientific understanding of the response of the climate system to emissions of different gases, and include a methodological update to incorporate climate-carbon cycle feedbacks associated with the emission of non-$CO_2$ gases (Forster et al. 2021). For the second-most important anthropogenic greenhouse gas, methane, the updated GWP100 value of 27 is similar but slightly lower than the value of 28 reported in the AR5 without climate-carbon cycle feedbacks. A full set of GWP100 values used in this report, based on the assessment of WGI (Forster et al. 2021, Section 7.6 and Table 7.SM.7), is provided in Table 9.

GWP100 was chosen in the WGIII contribution to the AR6 as the default GHG emissions metric for both procedural and scientific reasons.

Procedural reasons are to provide continuity with the use of GWP100 in past IPCC reports and the dominant use of GWP100 in the literature assessed by WGIII, and to match decisions made by Governments as part of the Paris Agreement Rulebook. Parties to the Paris Agreement decided to report aggregated emissions and removals (expressed as $CO_2$-eq) based on the Global Warming Potential with a time horizon of 100 years (GWP100), using values from IPCC AR5 or from a subsequent IPCC report as agreed upon by the CMA,[5] and to account for future nationally determined contributions (NDCs) in accordance with this approach. Parties may also report supplemental information on aggregate emissions and removals, expressed as $CO_2$-eq, using other GHG emission metrics assessed by the IPCC (4/CMA.1 and 18/CMA.1: UNFCCC 2019).

Scientific reasons for the use of GWP100 as default GHG emission metric in WGIII are that GWP100 approximates the relative damages caused by the two most important anthropogenic GHGs $CO_2$ and $CH_4$ for social discount rates around 3%. In addition, for pathways that limit warming to 2°C (>67%) or lower, using GWP100 to inform cost-effective abatement choices between gases would achieve these long-term temperature goals at close to least global cost within a few percent (*high confidence*) (see Cross-Chapter Box 2 in Chapter 2).

However, all emission metrics have limitations and uncertainties, given that they simplify the complexity of the physical climate system and its response to past and future GHG emissions. The most suitable metric for any given climate policy application, depends on judgements about the specific context, policy objectives and the way in which a metric would be used.

Wherever emissions, removals and mitigation potentials are expressed as $CO_2$-eq in this report, efforts have been made to recalculate those values consistently in terms of GWP100 values from AR6 WGI. However, in some cases it was not possible or feasible to disentangle conclusions from the existing literature into individual gases and then re-aggregate those emissions using updated GWP100 values. The existing literature assessed by WGIII uses a range of GWP100 values from previous IPCC reports; for $CH_4$, these values vary between 21 (based on the *IPCC Second Assessment Report*) to 28 or even 34 (based on the *IPCC Fifth Assessment Report* and depending on whether the study included or excluded climate-carbon cycle feedbacks). Consistent application of any metric is challenging as individual GHG emission species are not always provided in the literature assessed by WGIII. Where a full recalculation of $CO_2$-eq emissions or mitigation potentials into GWP100 AR6 values was not possible or feasible, and especially if non-$CO_2$ emissions constitute only a minor fraction of total emissions or abatement, individual chapters note this inconsistency and provide an indication of the potential magnitude of inconsistency.

To further reduce ambiguity regarding actual climate outcomes over time from any given set of emissions, the WGIII contribution to the AR6 reports emissions and mitigation options for individual gases where possible based on the available literature, and reports $CO_2$-eq emissions where this is judged to be policy relevant by author teams in addition to, not instead of individual gases.

---

[4] Emission metrics also exist for aerosols, but these are not commonly used in climate policy. This assessment focuses on GHG emission metrics only.

[5] The CMA is the Conference of the Parties serving as the meeting of the Parties to the Paris Agreement.

Table 9 | GWP100 values and atmospheric lifetimes for a range of GHGs, based on AR6 WGI (Forster et al. 2021).

| Gas | AR6 – GWP100 | Lifetime |
|---|---|---|
| $CO_2$ | 1 | N/A |
| $CH_4$ (biogenic) | 27.0 | 11.8 |
| $CH_4$ (fossil – combustion)[6] | 27.0 | 11.8 |
| $CH_4$ (fossil – fugitive and process) | 29.8 | 11.8 |
| $N_2O$ | 273 | 109 |
| HFC-32 | 770 | 5.4 |
| HFC-143a | 5807 | 51 |
| CF4 | 7379 | 50,000 |
| C2F6 | 12,410 | 10,000 |
| C3F8 | 9289 | 2600 |
| C4F10 | 10,022 | 2600 |
| C5F12 | 9218 | 4100 |
| C6F14 | 8617 | 3100 |
| C7F16 | 8409 | 3000 |
| c-C4F8 | 13,902 | 3000 |
| HFC-125 | 3744 | 30 |
| HFC-134a | 1526 | 14 |
| HFC-152a | 164 | 1.6 |
| HFC-227ea | 3602 | 36 |
| HFC-23 | 14,590 | 228 |
| HFC-236fa | 8689 | 213 |
| HFC-245fa | 962 | 7.9 |
| HFC-365mfc | 913 | 8.9 |
| HFC-43-10-mee | 1599 | 17 |
| SF6 | 25,184 | 3200 |
| NF3 | 17,423 | 569 |

## Part III: Emissions Datasets

In this section we report on the historical emissions data used in the report (Section 9), the sectoral mapping on emissions sources (Section 9.1), the methane emissions sources (Section 9.2), and indirect emissions (Section 10).

### A.II.9 Historical Data

Historic emissions data for countries, regions and sectors are presented throughout the report, but especially in Chapters 2, 6–7, 9–11, the Technical Summary and Summary for Policymakers. To ensure consistency and transparency we use the same emissions data across these chapters, with a single methodology, division of emissions sources, and following the classification scheme of countries and areas in Section 1 above.

Our primary data source is the Emissions Database for Global Atmospheric Research (EDGAR) (Crippa et al. 2021, Minx et al. 2021). This dataset provides annual $CO_2$, $CH_4$, $N_2O$ and F-gas emissions on a country and emissions source level for the time span 1970 to 2019. The fossil fuel combustion component of EDGAR is closely linked to and sourced from International Energy Agency (IEA 2021) energy and emissions estimates. Section 2.2.1 in Chapter 2 of this report describes the differences between and coverage of different global emissions datasets.

In addition to EDGAR, land-use $CO_2$ emissions are sourced as the mean of three bookkeeping models, in a convention established by the Global Carbon Project (Friedlingstein et al. 2020) and consistent with the Working Group I approach. The bookkeeping models are BLUE (Bookkeeping of Land Use Emissions), Hansis et al. (2015), Houghton and Nassikas (2017) and OSCAR (Gasser et al. 2020).

Global total greenhouse gas emissions reported throughout AR6 are the sum of EDGAR and land-use $CO_2$ emissions. Significant uncertainties are associated with each gas and emissions source. These uncertainties are comprehensively treated in Section 2.2.1 of Chapter 2.

#### A.II.9.1 Mapping of Emission Sources to Sectors

The list below shows how emission sources in EDGAR are mapped to sectors throughout the AR6 WGIII. This defines unambiguous system boundaries for the sectors as represented in Chapters 6, 7 and 9–11 in the report and enables a discussion and representation of emission sources without double-counting.

Emission sources follows the definitions by the IPCC Task Force on National Greenhouse Gas Inventories (TFI) (IPCC 2019). EDGARv6 identifies each source as either 'Fossil' or 'Bio'. The 'Bio' label indicates the biomass component of fuel combustion, while 'Fossil' is the default label for all other emissions sources (including, for example, agricultural GHG emissions).

6 The biogenic $CH_4$ GWP100 value applies here, given Tier 1 IPCC $CO_2$ emissions factors which are based on total carbon content. The associated emissions are estimated on the bases of complete (100%) oxidation to $CO_2$ of carbon contained in combusted mass.

**Table 10 | Mapping emission sources to sectors.**

| Chapter title | Subsector title | EDGAR code | IPCC 2019 | Gases |
|---|---|---|---|---|
| AFOLU | Biomass burning ($CO_2$, $CH_4$) | 4F1 (bio), 4F2 (bio), 4F3 (bio), 4F4 (bio), 4F5 (bio) | 3.C.1.b (bio) | $CH_4$, $N_2O$ |
| AFOLU | Enteric fermentation ($CH_4$) | 4A1-d (fossil), 4A1-n (fossil), 4A2 (fossil), 4A3 (fossil), 4A4 (fossil), 4A5 (fossil), 4A6 (fossil), 4A7 (fossil), 4A8 (fossil) | 3.A.1.a.i (fossil), 3.A.1.a.ii (fossil), 3.A.1.b (fossil), 3.A.1.c (fossil), 3.A.1.d (fossil), 3.A.1.e (fossil), 3.A.1.f (fossil), 3.A.1.g (fossil), 3.A.1.h (fossil) | $CH_4$ |
| AFOLU | Managed soils and pasture ($CO_2$, $N_2O$) | 4D12 (fossil), 4D13 (fossil), 4D14 (fossil), 4D15 (fossil), 4D2 (fossil), 4D3a (fossil), 4D3b (fossil), 4D4a (fossil), 4D4b (fossil) | 3.C.4 (fossil), 3.C.5 (fossil), 3.C.6 (fossil), 3.C.3 (fossil), 3.C.2 (fossil) | $N_2O$, $CO_2$ |
| AFOLU | Manure management ($N_2O$, $CH_4$) | 4B1-d (fossil), 4B1-n (fossil), 4B2 (fossil), 4B3 (fossil), 4B4 (fossil), 4B5 (fossil), 4B6 (fossil), 4B7 (fossil), 4B8 (fossil), 4B9 (fossil) | 3.A.2.a.i (fossil), 3.A.2.a.ii (fossil), 3.A.2.b (fossil), 3.A.2.c (fossil), 3.A.2.i (fossil), 3.A.2.d (fossil), 3.A.2.e (fossil), 3.A.2.f (fossil), 3.A.2.g (fossil), 3.A.2.h (fossil) | $CH_4$, $N_2O$ |
| AFOLU | Rice cultivation ($CH_4$) | 4C (fossil) | 3.C.7 (fossil) | $CH_4$ |
| AFOLU | Synthetic fertiliser application ($N_2O$) | 4D11 (fossil) | 3.C.4 (fossil) | $N_2O$ |
| Buildings | Non-$CO_2$ (all buildings) | 2F3 (fossil), 2F4 (fossil), 2F9a (fossil), 2F9c (fossil) | 2.F.3 (fossil), 2.F.4 (fossil), 2.G.2.c (fossil) | c-C4F8, C4F10, CF4, HFC-125, HFC-227ea, HFC-23, HFC-236fa, HFC-134a, HFC-152a |
| Buildings | Non-residential | 1A4a (bio), 1A4a (fossil) | 1.A.4.a (bio), 1.A.4.a (fossil) | $CH_4$, $N_2O$, $CO_2$ |
| Buildings | Residential | 1A4b (bio), 1A4b (fossil) | 1.A.4.b (bio), 1.A.4.b (fossil) | $CH_4$, $N_2O$, $CO_2$ |
| Energy systems | Coal mining fugitive emissions | 1B1a1 (fossil), 1B1a1r (fossil), 1B1a2 (fossil), 1B1a3 (fossil), 1B1b2 (fossil), 1B1b4 (fossil) | 1.B.1.a (fossil), 1.B.1.c (fossil) | $CO_2$, $CH_4$ |
| Energy systems | Electricity and heat | 1A1a1 (bio), 1A1a1 (fossil), 1A1a2 (bio), 1A1a2 (fossil), 1A1a3 (bio), 1A1a3 (fossil), 1A1a4 (bio), 1A1a4 (fossil), 1A1a5 (bio), 1A1a5 (fossil), 1A1a6 (bio), 1A1a6 (fossil), 1A1a7 (bio), 1A1a7 (fossil) | 1.A.1.a.i (bio), 1.A.1.a.i (fossil), 1.A.1.a.ii (bio), 1.A.1.a.ii (fossil), 1.A.1.a.iii (bio), 1.A.1.a.iii (fossil) | $CO_2$, $CH_4$, $N_2O$ |
| Energy systems | Oil and gas fugitive emissions | 1B2a1 (bio), 1B2a1 (fossil), 1B2a2 (fossil), 1B2a3-l (fossil), 1B2a4-l (fossil), 1B2a4-t (fossil), 1B2a5(e) (fossil), 1B2b1 (fossil), 1B2b3 (fossil), 1B2b4 (fossil), 1B2b5 (fossil), 1B2c (fossil) | 1.B.2.a.iii.2 (bio), 1.B.2.a.iii.2 (fossil), 1.B.2.a.iii.3 (fossil), 1.B.2.a.iii.4 (fossil), 1.B.2.b.iii.2 (fossil), 1.B.2.b.iii.4 (fossil), 1.B.2.b.iii.5 (fossil), 1.B.2.b.iii.3 (fossil), 1.B.2.b.ii (fossil), 1.B.2.a.ii (fossil) | $CO_2$, $CH_4$, $N_2O$ |
| Energy systems | Other (energy systems) | 1A1c3 (bio), 1A1c3 (fossil), 1A1c4 (bio), 1A1c5 (bio), 1A1c5 (fossil), 1A4c1 (bio), 1A4c1 (fossil), 1A4d (bio), 1A4d (fossil), 1B1b3 (bio), 2F8b (fossil), 7A1 (fossil), 7A2 (fossil), 7B1 (fossil), 7C1 (fossil) | 1.A.1.c.ii (bio), 1.A.1.c.ii (fossil), 1.A.1.c.i (bio), 1.A.1.c.i (fossil), 1.A.4.c.i (bio), 1.A.4.c.i (fossil), 1.A.5.a (bio), 1.A.5.a (fossil), 1.B.1.c (bio), 2.G.1.b (fossil), 5.B (fossil), 5.A (fossil) | $CO_2$, $CH_4$, $N_2O$, SF6 |
| Energy systems | Petroleum refining | 1A1b (bio), 1A1b (fossil) | 1.A.1.b (bio), 1.A.1.b (fossil) | $CO_2$, $CH_4$, $N_2O$ |
| Industry | Cement | 2A1 (fossil) | 2.A.1 (fossil) | $CO_2$ |
| Industry | Chemicals | 1A2c (bio), 1A2c (fossil), 2A2 (fossil), 2A3 (fossil), 2A4a (fossil), 2A4b (fossil), 2A7a (fossil), 2B1g (fossil), 2B1s (fossil), 2B2 (fossil), 2B3 (fossil), 2B4a (fossil), 2B4b (fossil), 2B5a (fossil), 2B5b (fossil), 2B5d (fossil), 2B5e (fossil), 2B5f (fossil), 2B5g (fossil), 2B5g2 (fossil), 2B5h1 (fossil), 2E (fossil), 2E1 (fossil), 3A (fossil), 3B (fossil), 3C (fossil), 3D (fossil), 3D1 (fossil), 3D3 (fossil) | 1.A.2.c (bio), 1.A.2.c (fossil), 2.A.2 (fossil), 2.A.4.d (fossil), 2.A.4.b (fossil), 2.A.3 (fossil), 2.B.1 (fossil), 2.B.2 (fossil), 2.B.3 (fossil), 2.B.5 (fossil), 2.B.8.f (fossil), 2.B.8.b (fossil), 2.B.8.c (fossil), 2.B.8.a (fossil), 2.B.4 (fossil), 2.B.6 (fossil), 2.B.9.b (fossil), 2.D.3 (fossil), 2.G.3.a (fossil), 2.G.3.b (fossil) | $CH_4$, $N_2O$, $CO_2$, c-C4F8, C2F6, C3F8, C4F10, C5F12, C6F14, CF4, HFC-125, HFC-134a, HFC-143a, HFC-152a, HFC-227ea, HFC-32, HFC-365mfc, NF3, SF6, HFC-23 |
| Industry | Metals | 1A1c1 (fossil), 1A1c2 (fossil), 1A2a (bio), 1A2a (fossil), 1A2b (bio), 1A2b (fossil), 1B1b1 (fossil), 2C1a (fossil), 2C1b (fossil), 2C1d (fossil), 2C2 (fossil), 2C3a (fossil), 2C3b (fossil), 2C4a (fossil), 2C4b (fossil), 2C5lp (fossil), 2C5mp (fossil), 2C5zp (fossil) | 1.A.1.c.i (fossil), 1.A.1.c.ii (fossil), 1.A.2.a (bio), 1.A.2.a (fossil), 1.A.2.b (bio), 1.A.2.b (fossil), 1.B.1.c (fossil), 2.C.1 (fossil), 2.C.2 (fossil), 2.C.3 (fossil), 2.C.4 (fossil), 2.C.5 (fossil), 2.C.6 (fossil) | $CO_2$, $CH_4$, $N_2O$, C2F6, CF4, SF6 |
| Industry | Other (industry) | 1A2d (bio), 1A2d (fossil), 1A2e (bio), 1A2e (fossil), 1A2f (bio), 1A2f (fossil), 1A2f1 (fossil), 1A2f2 (fossil), 1A5b1 (fossil), 2F1a (fossil), 2F1b (fossil), 2F1c (fossil), 2F1d (fossil), 2F1e (fossil), 2F1f (fossil), 2F2a (fossil), 2F2b (fossil), 2F5 (fossil), 2F6 (fossil), 2F7a (fossil), 2F7b (fossil), 2F7c (fossil), 2F8a (fossil), 2F9 (fossil), 2F9d (fossil), 2F9e (fossil), 2F9f (fossil), 2G1 (fossil), 7B2 (fossil), 7C2 (fossil) | 1.A.2.d (bio), 1.A.2.d (fossil), 1.A.2.e (bio), 1.A.2.e (fossil), 1.A.2.f (bio), 1.A.2.f (fossil), 1.A.2.k (fossil), 1.A.2.i (fossil), 1.A.5.b.iii (fossil), 2.F.1.a (fossil), NA (fossil), 2.F.5 (fossil), 2.E.1 (fossil), 2.E.2 (fossil), 2.E.3 (fossil), 2.G.1.a (fossil), 2.G.2.c (fossil), 2.G.2.b (fossil), 2.G.2.a (fossil), 2.D.1 (fossil), 5.A (fossil) | $CH_4$, $N_2O$, $CO_2$, HFC-125, HFC-134a, HFC-143a, HFC-152a, HFC-227ea, HFC-236fa, HFC-245fa, HFC-32, HFC-365mfc, C3F8, C6F14, CF4, HFC-43-10-mee, HFC-134, HFC-143, HFC-23, HFC-41, c-C4F8, C2F6, NF3, SF6, HCFC-141b, HCFC-142b, C4F10 |
| Industry | Waste | 6A1 (fossil), 6B1 (fossil), 6B2 (fossil), 6C (fossil), 6Ca (bio), 6Cb1 (fossil), 6Cb2 (fossil), 6D (fossil) | 4.A.1 (fossil), 4.D.2 (fossil), 4.D.1 (fossil), 4.C.1 (fossil), 4.C.2 (bio), 4.C.2 (fossil), 4.B (fossil) | $CH_4$, $N_2O$, $CO_2$ |

| Chapter title | Subsector title | EDGAR code | IPCC 2019 | Gases |
|---|---|---|---|---|
| Transport | Domestic Aviation | 1A3a (fossil) | 1.A.3.a.ii (fossil) | $CO_2$, $CH_4$, $N_2O$ |
| Transport | Inland Shipping | 1A3d (bio), 1A3d (fossil) | 1.A.3.d.ii (bio), 1.A.3.d.ii (fossil) | $CH_4$, $N_2O$, $CO_2$ |
| Transport | International Aviation | 1C1 (fossil) | 1.A.3.a.i (fossil) | $CO_2$, $CH_4$, $N_2O$ |
| Transport | International Shipping | 1C2 (bio), 1C2 (fossil) | 1.A.3.d.i (bio), 1.A.3.d.i (fossil) | $CH_4$, $N_2O$, $CO_2$ |
| Transport | Other (transport) | 1A3e (bio), 1A3e (fossil), 1A4c2 (fossil), 1A4c3 (bio), 1A4c3 (fossil) | 1.A.3.e.i (bio), 1.A.3.e.i (fossil), 1.A.4.c.ii (fossil), 1.A.4.c.iii (bio), 1.A.4.c.iii (fossil) | $CH_4$, $N_2O$, $CO_2$ |
| Transport | Rail | 1A3c (bio), 1A3c (fossil) | 1.A.3.c (bio), 1.A.3.c (fossil) | $CH_4$, $N_2O$, $CO_2$ |
| Transport | Road | 1A3b (bio), 1A3b (fossil) | 1.A.3.b_RES (bio), 1.A.3.b_RES (fossil) | $CH_4$, $N_2O$, $CO_2$ |

### A.II.9.2 Methane Emissions Sources

In order to identify emission trends and mitigation opportunities by sector WGIII allocates each emission source to a sector and subsequently a subsector (check Section 9 above). These trends and mitigation opportunities are, in most cases and whenever possible, reported in the native unit of gases as well as in $CO_2$-eq using IPCC AR6 GWP100 values (Section 8). In the case of methane ($CH_4$), it has two different GWP100 values according to its source. The relevant sources of methane are: biogenic methane, fossil methane (source: combustion) and fossil methane (source: fugitive and process).

The majority of biogenic methane emissions result from the AFOLU sector due to livestock and other agricultural practices, but also from the energy systems, building, transport and industry (waste) sectors. Meanwhile, fossil methane (combustion) emissions result from electricity and heat generation in the energy systems sector as well as various combustion activities in all other sectors. Finally, fossil methane (fugitive and process) is emitted from the extraction and transportation of fossil fuels (fugitive methane), in addition to some activities in the industry sector (fugitive and process methane). See Table 12 below for a comprehensive list.

There are two GWP100 values assigned to methane depending on its source: a GWP100 value of 27 for biogenic methane and fossil methane (combustion), and a higher GWP100 value of 29.8 for fossil methane (fugitive and process), see Table 11 below. The difference between these two GWP100 values arises from treatment of the effect of methane conversion into $CO_2$ during its chemical decay in the atmosphere. The higher GWP100 value takes account of the warming caused by $CO_2$ that methane decays into, which adds to the warming caused by methane itself, while the lower GWP100 value does not.

In the case of biogenic methane, the correct GWP100 value is always the low value irrespective of the specific source. This is because all $CO_2$ originated from biomass is either already estimated and reported as $CO_2$ emissions from AFOLU sector, or in the case of short-rotation biomass, the original removal of $CO_2$ from the atmosphere is not reported and hence neither does the release of $CO_2$ back into the atmosphere need to be reported.

For fossil methane, the correct GWP100 value depends on the source, in other words, combustion source vs fugitive and process sources. Fossil methane (fugitive and process) should use the higher GWP100 value because $CO_2$ converted from methane in the atmosphere is not estimated anywhere else.

For fossil methane (combustion), despite it being fossil, the correct GWP100 value is always the low one, for the dataset reported here. This is due to the fact that the emissions data provider EDGAR (Section 9) considers a complete oxidation to $CO_2$ of all the carbon contained in the fossil fuel upon combustion, which is then reflected in the $CO_2$ emissions factors for the different sources based on the carbon content of fuels. In other words, IPCC (IPCC 2019) methods and defaults (Tier 1 IPCC $CO_2$ emissions factors) have been used where the associated $CO_2$ emissions are estimated on the basis of complete (100%) oxidation to $CO_2$ of carbon contained in combusted mass, which includes not only $CO_2$ directly released to the atmosphere but also $CO_2$ generated in the atmosphere from the carbon released as methane and converted to $CO_2$ only subsequently.

There are two exceptions applied to the above categorisation, both belong to the industry sector, sector codes 6Cb1 (Waste incineration – uncontrolled municipal solid waste (MSW) burning) and 6D (other waste). Uncontrolled MSW burning (6Cb1) includes both biogenic and fossil material, with incomplete oxidation for this source even when the IPCC Tier 1 default emission/oxidation factor is used. The GWP100 value adopted for this source is the low one, given that the fossil-origin methane component is unlikely to be very large. The 'other waste' (6D) source may also include both biogenic and fossil methane. However, it is unclear what type of waste handling is included here. Furthermore, the associated $CO_2$ emissions are not estimated. Therefore, the high GWP100 value is used.

In total, the estimation of EDGAR methane emissions in 2019 using a GWP100 value of 27 across all related sources results in 10.2 $GtCO_2$-eq, compared to 10.6 $GtCO_2$-eq using the higher GWP100 value as described. This is primarily driven by the readjustment of methane emissions from hard coal mining, gas production, and venting and flaring (sectors 1B1a1, 1B2b1 and 1B2c).

**Table 11 | Summary of methane GWP100 values in AR6 depending on type and source.**

| $CH_4$ | GWP100 value |
|---|---|
| $CH_4$ (biogenic) | 27 |
| $CH_4$ (fossil – combustion) | 27 |
| $CH_4$ (fossil – fugitive and process) | 29.8 |

Table 12 | Methane sources and types.

| Sector code | Description | Sector | Subsector | $CH_4$ type |
|---|---|---|---|---|
| 1A1a1 | Public Electricity Generation (biomass) | Energy systems | Electricity and heat | $CH_4$ Biogenic |
| 1A1a1 | Public Electricity Generation | Energy systems | Electricity and heat | $CH_4$ Fossil (Combustion) |
| 1A1a2 | Public Combined Heat and Power gen. (biom.) | Energy systems | Electricity and heat | $CH_4$ Biogenic |
| 1A1a2 | Public Combined Heat and Power gen. | Energy systems | Electricity and heat | $CH_4$ Fossil (Combustion) |
| 1A1a3 | Public Heat Plants (biomass) | Energy systems | Electricity and heat | $CH_4$ Biogenic |
| 1A1a3 | Public Heat Plants | Energy systems | Electricity and heat | $CH_4$ Fossil (Combustion) |
| 1A1a4 | Public Electricity Gen. (own use) (biom.) | Energy systems | Electricity and heat | $CH_4$ Biogenic |
| 1A1a4 | Public Electricity Generation (own use) | Energy systems | Electricity and heat | $CH_4$ Fossil (Combustion) |
| 1A1a5 | Electricity Generation (autoproducers) (biom.) | Energy systems | Electricity and heat | $CH_4$ Biogenic |
| 1A1a5 | Electricity Generation (autoproducers) | Energy systems | Electricity and heat | $CH_4$ Fossil (Combustion) |
| 1A1a6 | Combined Heat and Power gen. (autopr.) (biom.) | Energy systems | Electricity and heat | $CH_4$ Biogenic |
| 1A1a6 | Combined Heat and Power gen. (autoprod.) | Energy systems | Electricity and heat | $CH_4$ Fossil (Combustion) |
| 1A1a7 | Heat Plants (autoproducers) (biomass) | Energy systems | Electricity and heat | $CH_4$ Biogenic |
| 1A1a7 | Heat Plants (autoproducers) | Energy systems | Electricity and heat | $CH_4$ Fossil (Combustion) |
| 1A1b | Refineries (biomass) | Energy systems | Petroleum refining | $CH_4$ Biogenic |
| 1A1b | Refineries | Energy systems | Petroleum refining | $CH_4$ Fossil (Combustion) |
| 1A1c1 | Fuel combustion coke ovens | Industry | Metals | $CH_4$ Fossil (Combustion) |
| 1A1c2 | Blast furnaces (pig iron prod.) | Industry | Metals | $CH_4$ Fossil (Combustion) |
| 1A1c3 | Gas works (biom.) | Energy systems | Other (energy systems) | $CH_4$ Biogenic |
| 1A1c3 | Gas works | Energy systems | Other (energy systems) | $CH_4$ Fossil (Combustion) |
| 1A1c4 | Fuel comb. charcoal production (biom.) | Energy systems | Other (energy systems) | $CH_4$ Biogenic |
| 1A1c5 | Other transf. sector (BKB, etc.) (biom.) | Energy systems | Other (energy systems) | $CH_4$ Biogenic |
| 1A1c5 | Other transformation sector (BKB, etc.) | Energy systems | Other (energy systems) | $CH_4$ Fossil (Combustion) |
| 1A2a | Iron and steel (biomass) | Industry | Metals | $CH_4$ Biogenic |
| 1A2a | Iron and steel | Industry | Metals | $CH_4$ Fossil (Combustion) |
| 1A2b | Non-ferrous metals (biomass) | Industry | Metals | $CH_4$ Biogenic |
| 1A2b | Non-ferrous metals | Industry | Metals | $CH_4$ Fossil (Combustion) |
| 1A2c | Chemicals (biomass) | Industry | Chemicals | $CH_4$ Biogenic |
| 1A2c | Chemicals | Industry | Chemicals | $CH_4$ Fossil (Combustion) |
| 1A2d | Pulp and paper (biomass) | Industry | Other (industry) | $CH_4$ Biogenic |
| 1A2d | Pulp and paper | Industry | Other (industry) | $CH_4$ Fossil (Combustion) |
| 1A2e | Food and tobacco (biomass) | Industry | Other (industry) | $CH_4$ Biogenic |
| 1A2e | Food and tobacco | Industry | Other (industry) | $CH_4$ Fossil (Combustion) |
| 1A2f | Other industries (stationary) (biom.) | Industry | Other (industry) | $CH_4$ Biogenic |
| 1A2f | Other industries (stationary) (fos.) | Industry | Other (industry) | $CH_4$ Fossil (Combustion) |
| 1A2f1 | Off-road machinery: construction (diesel) | Industry | Other (industry) | $CH_4$ Fossil (Combustion) |
| 1A2f2 | Off-road machinery: mining (diesel) | Industry | Other (industry) | $CH_4$ Fossil (Combustion) |
| 1A3a | Domestic air transport | Transport | Domestic Aviation | $CH_4$ Fossil (Combustion) |
| 1A3b | Road transport (incl. evap.) (biom.) | Transport | Road | $CH_4$ Biogenic |
| 1A3b | Road transport (incl. evap.) (foss.) | Transport | Road | $CH_4$ Fossil (Combustion) |
| 1A3c | Non-road transport (rail, etc.) (biom.) | Transport | Rail | $CH_4$ Biogenic |
| 1A3c | Non-road transport (rail, etc.) (fos.) | Transport | Rail | $CH_4$ Fossil (Combustion) |
| 1A3d | Inland shipping (biom.) | Transport | Inland Shipping | $CH_4$ Biogenic |
| 1A3d | Inland shipping (fos.) | Transport | Inland Shipping | $CH_4$ Fossil (Combustion) |
| 1A3e | Non-road transport (biom.) | Transport | Other (transport) | $CH_4$ Biogenic |
| 1A3e | Non-road transport (fos.) | Transport | Other (transport) | $CH_4$ Fossil (Combustion) |
| 1A4a | Commercial and public services (biom.) | Buildings | Non-residential | $CH_4$ Biogenic |
| 1A4a | Commercial and public services (fos.) | Buildings | Non-residential | $CH_4$ Fossil (Combustion) |
| 1A4b | Residential (biom.) | Buildings | Residential | $CH_4$ Biogenic |
| 1A4b | Residential (fos.) | Buildings | Residential | $CH_4$ Fossil (Combustion) |

| Sector code | Description | Sector | Subsector | $CH_4$ type |
|---|---|---|---|---|
| 1A4c1 | Agriculture and forestry (biom.) | Energy systems | Other (energy systems) | $CH_4$ Biogenic |
| 1A4c1 | Agriculture and forestry (fos.) | Energy systems | Other (energy systems) | $CH_4$ Fossil (Combustion) |
| 1A4c2 | Off-road machinery: agric./for. (diesel) | Transport | Other (transport) | $CH_4$ Fossil (Combustion) |
| 1A4c3 | Fishing (biom.) | Transport | Other (transport) | $CH_4$ Biogenic |
| 1A4c3 | Fishing (fos.) | Transport | Other (transport) | $CH_4$ Fossil (Combustion) |
| 1A4d | Non-specified other (biom.) | Energy systems | Other (energy systems) | $CH_4$ Biogenic |
| 1A4d | Non-specified other (fos.) | Energy systems | Other (energy systems) | $CH_4$ Fossil (Combustion) |
| 1A5b1 | Off-road machinery: mining (diesel) | Industry | Other (industry) | $CH_4$ Fossil (Combustion) |
| 1B1a1 | Hard coal mining (gross) | Energy systems | Coal mining fugitive emissions | $CH_4$ Fossil (Fugitive) |
| 1B1a1r | Methane recovery from coal mining | Energy systems | Coal mining fugitive emissions | $CH_4$ Fossil (Fugitive) |
| 1B1a2 | Abandoned mines | Energy systems | Coal mining fugitive emissions | $CH_4$ Fossil (Fugitive) |
| 1B1a3 | Brown coal mining | Energy systems | Coal mining fugitive emissions | $CH_4$ Fossil (Fugitive) |
| 1B1b1 | Fuel transformation coke ovens | Industry | Metals | $CH_4$ Fossil (Fugitive) |
| 1B1b3 | Fuel transformation charcoal production | Energy systems | Other (energy systems) | $CH_4$ Biogenic |
| 1B2a1 | Oil production (biom.) | Energy systems | Oil and gas fugitive emissions | $CH_4$ Biogenic |
| 1B2a1 | Oil production | Energy systems | Oil and gas fugitive emissions | $CH_4$ Fossil (Fugitive) |
| 1B2a2 | Oil transmission | Energy systems | Oil and gas fugitive emissions | $CH_4$ Fossil (Fugitive) |
| 1B2a3-l | Tanker loading | Energy systems | Oil and gas fugitive emissions | $CH_4$ Fossil (Fugitive) |
| 1B2a4-l | Tanker oil transport (crude and NGL) | Energy systems | Oil and gas fugitive emissions | $CH_4$ Fossil (Fugitive) |
| 1B2a4-t | Transport by oil trucks | Energy systems | Oil and gas fugitive emissions | $CH_4$ Fossil (Fugitive) |
| 1B2a5(e) | Oil refineries (evaporation) | Energy systems | Oil and gas fugitive emissions | $CH_4$ Fossil (Fugitive) |
| 1B2b1 | Gas production | Energy systems | Oil and gas fugitive emissions | $CH_4$ Fossil (Fugitive) |
| 1B2b3 | Gas transmission | Energy systems | Oil and gas fugitive emissions | $CH_4$ Fossil (Fugitive) |
| 1B2b4 | Gas distribution | Energy systems | Oil and gas fugitive emissions | $CH_4$ Fossil (Fugitive) |
| 1B2c | Venting and flaring during oil and gas production | Energy systems | Oil and gas fugitive emissions | $CH_4$ Fossil (Fugitive) |
| 1C1 | International air transport | Transport | International Aviation | $CH_4$ Fossil (Combustion) |
| 1C2 | International marine transport (biom.) | Transport | International Shipping | $CH_4$ Biogenic |
| 1C2 | International marine transport (bunkers) | Transport | International Shipping | $CH_4$ Fossil (Combustion) |
| 2B4a | Silicon carbide production | Industry | Chemicals | $CH_4$ Fossil (Process) |
| 2B5a | Carbon black production | Industry | Chemicals | $CH_4$ Fossil (Process) |
| 2B5b | Ethylene production | Industry | Chemicals | $CH_4$ Fossil (Process) |
| 2B5d | Styrene production | Industry | Chemicals | $CH_4$ Fossil (Process) |
| 2B5e | Methanol production | Industry | Chemicals | $CH_4$ Fossil (Process) |
| 2B5g | Other bulk chemicals production | Industry | Chemicals | $CH_4$ Fossil (Process) |
| 2C1d | Sinter production | Industry | Metals | $CH_4$ Fossil (Process) |
| 2C2 | Ferroy Alloy production | Industry | Metals | $CH_4$ Fossil (Process) |
| 4A1-d | Dairy cattle | AFOLU | Enteric Fermentation ($CH_4$) | $CH_4$ Biogenic |
| 4A1-n | Non-dairy cattle | AFOLU | Enteric Fermentation ($CH_4$) | $CH_4$ Biogenic |
| 4A2 | Buffalo | AFOLU | Enteric Fermentation ($CH_4$) | $CH_4$ Biogenic |
| 4A3 | Sheep | AFOLU | Enteric Fermentation ($CH_4$) | $CH_4$ Biogenic |
| 4A4 | Goats | AFOLU | Enteric Fermentation ($CH_4$) | $CH_4$ Biogenic |
| 4A5 | Camels and Lamas | AFOLU | Enteric Fermentation ($CH_4$) | $CH_4$ Biogenic |
| 4A6 | Horses | AFOLU | Enteric Fermentation ($CH_4$) | $CH_4$ Biogenic |
| 4A7 | Mules and asses | AFOLU | Enteric Fermentation ($CH_4$) | $CH_4$ Biogenic |
| 4A8 | Swine | AFOLU | Enteric Fermentation ($CH_4$) | $CH_4$ Biogenic |
| 4B1-d | Manure Man.: Dairy Cattle (confined) | AFOLU | Manure management ($N_2O$, $CH_4$) | $CH_4$ Biogenic |
| 4B1-n | Manure Man.: Non-Dairy Cattle (confined) | AFOLU | Manure management ($N_2O$, $CH_4$) | $CH_4$ Biogenic |
| 4B2 | Manure Man.: Buffalo (confined) | AFOLU | Manure management ($N_2O$, $CH_4$) | $CH_4$ Biogenic |
| 4B3 | Manure Man.: Sheep (confined) | AFOLU | Manure management ($N_2O$, $CH_4$) | $CH_4$ Biogenic |
| 4B4 | Manure Man.: Goats (confined) | AFOLU | Manure management ($N_2O$, $CH_4$) | $CH_4$ Biogenic |
| 4B5 | Manure Man.: Camels and llamas (confined) | AFOLU | Manure management ($N_2O$, $CH_4$) | $CH_4$ Biogenic |

| Sector code | Description | Sector | Subsector | $CH_4$ type |
|---|---|---|---|---|
| 4B6 | Manure Man.: Horses (confined) | AFOLU | Manure management ($N_2O$, $CH_4$) | $CH_4$ Biogenic |
| 4B7 | Manure Man.: Mules and asses (confined) | AFOLU | Manure management ($N_2O$, $CH_4$) | $CH_4$ Biogenic |
| 4B8 | Manure Man.: Swine (confined) | AFOLU | Manure management ($N_2O$, $CH_4$) | $CH_4$ Biogenic |
| 4B9 | Manure Man.: Poultry (confined) | AFOLU | Manure management ($N_2O$, $CH_4$) | $CH_4$ Biogenic |
| 4C | Rice cultivation (CH4) | AFOLU | Rice cultivation ($CH_4$) | $CH_4$ Biogenic |
| 4F1 | Field burning of agric. res.: cereals | AFOLU | Biomass burning ($CH_4$, $N_2O$) | $CH_4$ Biogenic |
| 4F2 | Field burning of agric. res.: pulses | AFOLU | Biomass burning ($CH_4$, $N_2O$) | $CH_4$ Biogenic |
| 4F3 | Field burning of agric. res.: tuber and roots | AFOLU | Biomass burning ($CH_4$, $N_2O$) | $CH_4$ Biogenic |
| 4F4 | Field burning of agric. res.: sugar cane | AFOLU | Biomass burning ($CH_4$, $N_2O$) | $CH_4$ Biogenic |
| 4F5 | Field burning of agric. res.: other | AFOLU | Biomass burning ($CH_4$, $N_2O$) | $CH_4$ Biogenic |
| 6A1 | Managed waste disposal on land | Industry | Waste | $CH_4$ Biogenic |
| 6B1 | Industrial wastewater | Industry | Waste | $CH_4$ Biogenic |
| 6B2 | Domestic and commercial wastewater | Industry | Waste | $CH_4$ Biogenic |
| 6C | Waste incineration – hazardous | Industry | Waste | $CH_4$ Fossil (Combustion) |
| 6Ca | Waste incineration – biogenic | Industry | Waste | $CH_4$ Biogenic |
| 6Cb1 | Waste incineration – uncontrolled MSW burning | Industry | Waste | $CH_4$ Fossil (Combustion) |
| 6Cb2 | Waste incineration – other non-biogenic | Industry | Waste | $CH_4$ Fossil (Combustion) |
| 6D | Other waste | Industry | Waste | $CH_4$ Fossil (Process) |
| 7A1 | Coal fires (underground) | Energy systems | Other (energy systems) | $CH_4$ Fossil (Combustion) |
| 7A2 | Oil fires (Kuwait) | Energy systems | Other (energy systems) | $CH_4$ Fossil (Combustion) |



## A.II.10 Indirect Emissions

Carbon dioxide emissions resulting from fuel combusted to produce electricity and heat are traditionally reported in the energy sector. An indirect emissions accounting principle allocates these emissions to the end-use sectors (industry, buildings, transport, and agriculture) where the electricity and heat are ultimately consumed. Attributing indirect emissions to consuming sectors makes it possible to assess the full potential impact of demand-side mitigation actions that reduce electricity and heat consumption (de la Rue du Can et al. 2015).

In order to estimate the indirect emissions of sectors and subsectors, the $CO_2$ Emissions from Fuel Combustion dataset of the International Energy Agency (IEA 2020a) is used. This database reports direct and indirect $CO_2$ emissions for IEA sectors, which are related to the IPCC (IPCC 2019) classification of emissions sources. The IEA adopted a new methodology in 2020 that is in line with the methodology used in Annex II of the WGIII contribution to AR5 (Krey et al. 2014), namely Section A.II.4. The IEA now estimates individual electricity and heat specific emission factors and allocates indirect emissions related to electricity and heat in the sectors where these forms of energy are used respectively (IEA 2020b). In order to estimate the share of energy input that results in the production of heat from the share that results in the production of electricity in Combined heat and Power plants, the IEA fixes the efficiency for heat production equal to 90%, which is the typical efficiency of a heat boiler and then allocates the remaining inputs to electricity production (IEA 2020b).

The base data for total global, regional and sectoral emissions in this report is the EDGAR database (see Section 9). Since there are some discrepancies between the electricity and heat emissions totals in EDGAR and IEA, we make some adjustments in order to estimate indirect emissions in EDGAR using the IEA data. First, we match the sectors in EDGAR and IEA. Second, for each country and emissions source available in the IEA database, we take the IEA indirect emissions value and divide it by the total IEA value for electricity and heat. Third, we multiply these values through by the EDGAR value for electricity and heat. This procedure ensures that indirect emissions, in principle, sum to the correct total (EDGAR) value of electricity and heat that we use elsewhere in the reporting. However, total indirect emissions still do not sum to the total electricity and heat sector. This is due to an incomplete allocation of electricity and heat emissions in the IEA dataset, equal to 0.008 $GtCO_2$ in 2018, or about 0.06% of the total electricity and heat generation.

Additionally, a couple of adjustments were made to allocate emissions from IEA sector categories to IPCC categories from IPCC Task force definition as described in IPCC (2019) Guidelines (see Section 9). These include:

- Other non-specified sector: the IEA energy statistics report final energy and electricity use for three end-use sectors: industry, transport, and other. The 'other' category is further subdivided into agriculture, fishing, commercial and public services, residential, and non-specified other. The 'non-specified other' category includes energy used for agriculture, fishing, commercial and public services, and residential sectors that has not been allocated to these end-use sectors by the submitting countries. In most cases, there is no entry in the non-specified other category, indicating that all end-use energy consumption has been allocated to other end-use sectors. However, for some countries the energy reported in the non-specified other category needed to be allocated to the appropriate end-use sectors. To perform this allocation, the energy use in the non-specified

Table 13 | Feasibility dimensions and indicators to assess the barriers and enablers of implementing mitigation options.

| Metric | Indicators |
|---|---|
| Geophysical feasibility | – Physical potential: physical constraints to implementation.<br>– Geophysical resource availability (including geological storage capacity): availability of resources needed to implementation.<br>– Land use: claims on land when option would be implemented. |
| Environmental-ecological feasibility | – Air pollution: increase or decrease in air pollutants, such as $NH_4$, $CH_4$ and fine dust.<br>– Toxic waste, mining, ecotoxicity and eutrophication.<br>– Water quantity and quality: changes in amount of water available for other uses, including groundwater.<br>– Biodiversity: changes in conserved primary forest or grassland that affect biodiversity, and management to conserve and maintain land carbon stocks. |
| Technological feasibility | – Simplicity: is the option technically simple to operate, maintain and integrate.<br>– Technology scalability: can the option be scaled up, quickly.<br>– Maturity and technology readiness: R&D and time needed to implement to option. |
| Economic feasibility | – Costs now, in 2030 and in the long term, including investment costs, costs in USD $tCO_2\text{-eq}^{-1}$, and hidden costs.<br>– Employment effects and economic growth. |
| Socio-cultural feasibility | – Public acceptance: extent to which the public supports the option and changes behavior accordingly.<br>– Effects on health and well-being.<br>– Distributional effects: equity and justice across groups, regions, and generations, including security of energy, water, food and poverty. |
| Institutional feasibility | – Political acceptance: extent to which politicians and governments support the option.<br>– Institutional capacity and governance, cross-sectoral coordination: capability of institutions to implement and handle the option, and to coordinate it with other sectors, stakeholder and civil society.<br>– Legal and administrative capacity: extent to which supportive legal and administrative changes can be achieved. |

other category was allocated to the other end-use sectors based on the share of energy allocated to each of these sub-sectors for each region.

- Other energy industry own use: emissions from this category in the IEA statistics corresponds to the IPCC Source/Sink categories 1A1b and 1A1c (see Section 9) and contains emissions from fuel combusted in energy transformation industries that are not producing heat and/or power and therefore include oil refineries, coal mining, oil and gas extraction and other energy-producing industries. These emissions were not reallocated to the end use sectors where final products are ultimately consumed due to the lack of data.

Finally, it is also worth noting that indirect emissions only cover $CO_2$ emissions and that a small portion of non-$CO_2$ are not included in the IEA dataset and therefore have not been allocated to the end use sectors. Non-$CO_2$ emissions from total electricity and heat generation represents 0.55% of all GHG emissions from that sector.

# Part IV: Assessment Methods

In this section we report on assessment methods adopted in the report. Section 11 describes the methodology adopted for assessing the feasibility of mitigation response options. Section 12 describes the methodology adopted for assessing synergies and trade-offs between mitigation options and the SDGs.

## A.II.11 Methodology Adopted for Assessing the Feasibility of Mitigation Response Options

The feasibility assessment aims to identify barriers and enablers of the deployment of mitigation options and pathways. The assessment organises evidence to support decision making on actions and policies that would improve the feasibility of mitigation options and pathways, by removing relevant barriers and strengthening enablers of change.

### A.II.11.1 Feasibility of mitigation response options

The sectoral chapters in AR6 WGIII assess six dimensions of feasibility, with each dimension comprising a key set of indicators that can be evaluated by combining various strands of literature (see Table 13). The feasibility of systems-level changes is addressed in Chapter 3 of this report.

The sectoral chapters in this report assess to what extent the indicators in Table 13 would be enablers or barriers to implementation using the following scores (Nilsson et al. 2016):

| | |
|---|---|
| – | The indicator has a negative impact on the feasibility of the option, for example, it is associated with prohibitively high costs, levels of pollution or land use, or low public or political acceptance. |
| ± | Mixed evidence: the indicator has mixed positive and negative impacts on the feasibility of the option (e.g., more land use in some regions, while lower in other regions). |
| + | The indicator has a positive impact on the feasibility of the option, for example, it is associated with low costs, pollution, land use, or high public or political acceptance. |
| 0/NA | The indicator does not affect the feasibility of the option/criterion is not applicable for the option. |
| NE | No evidence available to assess the impact on the feasibility of the option. |
| LE | Limited evidence available to assess the impact on the feasibility the option. |

### A.II.11.2 Assessment

Each sectoral chapter assesses to what extent the indicators listed above would be an enabler or barrier to the implementation of selected mitigation options, by using the above scores. Then the total number of minus and plus points were computed, relative to the maximum possible number of points, per feasibility dimensions, for each option; a + counts as two plus points, a – as two minus points, and a ± as one plus and one minus point. The resulting scores reveal the extent to which each feasibility dimension enables or inhibits the deployment of the relevant option, and indicates which type of additional effort would be needed to reduce or remove barriers as to improve the feasibility of relevant options.



The assessment is based on the literature, which is reflected in a line of sight. When appropriate, it is indicated whether the feasibility of an option varies across context (e.g., region), scale (e.g., small, medium, full scale), time (e.g., implementation in 2030 versus 2050) and warming level (e.g., 1.5°C versus 2°C).

Synergies and trade-offs may occur between the feasibility dimensions, and between specific mitigation options. Therefore, Chapters 3 and 4 employ a systems perspective and discuss the feasibility of mitigation scenarios and pathways in the long term and near to mid-term, respectively, on the basis of the feasibility assessments in the sectoral chapters taking into account such synergies and trade-offs. Chapter 5 (demand, services and social aspects of mitigation), Chapter 13 (national and sub-national policies and institutions), Chapter 14 (international cooperation), Chapter 15 (investment and finance) and Chapter 16 (innovation, technology development and transfer) address technological, economic, socio-cultural and institutional enabling conditions that can enhance the feasibility of options and remove relevant barriers.

## A.II.12 Methodology Adopted for Assessing Synergies and Trade-offs Between Mitigation Options and the SDGs

Adopting climate mitigation options can generate multiple positive (synergies) and negative (trade-offs) interactions with sustainable development. Understanding these are crucial for selecting mitigation options and policy choices that maximise the synergies, minimise trade-offs, and potentially offset trade-offs (Roy et al. 2018). Chapter 5 in the IPCC's Special Report on Global Warming of 1.5°C examines the synergies and trade-offs of adaptation and mitigation measures with sustainable development and UN's Sustainable Development Goals (SDGs). Building on this, the sectoral chapters in the WGIII contribution to the AR6 include a qualitative assessment of the synergies and trade-offs between mitigation options in different sectors and the SDGs based on existing literature. All these assessments are collated and presented in Chapter 17 with a supplementary table including the details of the synergies and trade-offs with a line of sight (Section 17.3.3.7, Figure 17.1 and Supplementary Material Table 17.1). The assessment also recognises that interactions of mitigation options with the SDGs are context-specific and therefore provides a detailed explanation in the supplementary table of Chapter 17.

For the assessment, the mitigation options were shortlisted from each of the sectoral chapters. The sectoral chapters assessed the literature in terms of the impacts of each of these mitigation options on the 17 SDGs. The assessment uses three signs:

+ to denote positive interaction only (synergies),

– to denote negative interaction only (trade-offs) and

± to denote mixed interactions.

In some cases, where there is gap in literature, these are left blank denoting that these impacts have not been assessed in the literature included in the sectoral chapters. To support these signs, brief statements are provided followed by uncertainty qualifiers in the supplementary table of Chapter 17. These uncertainty qualifiers denote the confidence levels (low, medium and high).

## References


Crippa, M. et al., 2021: EDGAR v6.0 Greenhouse Gas Emissions. European Commission, Joint Research Centre (JRC) [Dataset] PID: http://data.europa.eu/89h/97a67d67-c62e-4826-b873-9d972c4f670b.

de la Rue du Can, S., L. Price, and T. Zwickel, 2015: Understanding the full climate change impact of energy consumption and mitigation at the end-use level: A proposed methodology for allocating indirect carbon dioxide emissions. *Appl. Energy*, **159**, 548–559, doi:10.1016/j.apenergy.2015.08.055.

Fischedick, M., R. Schaeffer, A. Adedoyin, M. Akai, T. Bruckner, L. Clarke, V. Krey, I. Savolainen, S. Teske, D. Ürge-Vorsatz, R. Wright, 2011: *Mitigation Potential and Costs*. In: *IPCC Special Report on Renewable Energy Sources and Climate Change Mitigation* [Edenhofer, O., R. Pichs-Madruga, Y. Sokona, K. Seyboth, P. Matschoss, S. Kadner, T. Zwickel, P. Eickemeier, G. Hansen, S. Schlömer, C. von Stechow, (eds.)]. Cambridge University Press, Cambridge, UK and New York, NY, USA.

Fisher, B.S., N. Nakicenovic, K. Alfsen, J. Corfee Morlot, F. de la Chesnaye, J.-Ch. Hourcade, K. Jiang, M. Kainuma, E. La Rovere, A. Matysek, A. Rana, K. Riahi, R. Richels, S. Rose, D. van Vuuren, R. Warren, 2007: Issues related to mitigation in the long-term context. In *Climate Change 2007: Mitigation of Climate Change. Contribution of Working Group III to the Fourth Assessment Report of the Intergovernmental Panel of Climate Change* [Metz, B., O.R. Davidson, P.R. Bosch, R. Dave, and L.A. Meyer, (eds.)]. Cambridge University Press, Cambridge, UK and New York, NY, USA, 169–250.

Forster, P., T. Storelvmo, K. Armour, W. Collins, J.-L. Dufresne, D. Frame, D.J. Lunt, T. Mauritsen, M.D. Palmer, M. Watanabe, M. Wild, and H. Zhang, 2021: The Earth's Energy Budget, Climate Feedbacks, and Climate Sensitivity. In: *Climate Change 2021: The Physical Science Basis. Contribution of Working Group I to the Sixth Assessment Report of the Intergovernmental Panel on Climate Change* [Masson-Delmotte, V., P. Zhai, A. Pirani, S.L. Connors, C. Péan, S. Berger, N. Caud, Y. Chen, L. Goldfarb, M.I. Gomis, M. Huang, K. Leitzell, E. Lonnoy, J.B.R. Matthews, T.K. Maycock, T. Waterfield, O. Yelekçi, R. Yu, and B. Zhou (eds.)]. Cambridge University Press, Cambridge, United Kingdom and New York, NY, USA, pp. 923–1054, doi:10.1017/9781009157896.009.

Friedlingstein, P. et al., 2019: Global Carbon Budget 2019. *Earth Syst. Sci. Data*, **11(4)**, 1783–1838, doi:10.5194/essd-11-1783-2019.

Friedlingstein, P. et al., 2020: Global Carbon Budget 2020. *Earth Syst. Sci. Data*, **12(4)**, 3269–3340, doi:10.5194/essd-12-3269-2020.

Gasser, T. et al., 2020: Historical $CO_2$ emissions from land use and land cover change and their uncertainty. *Biogeosciences*, **17(15)**, 4075–4101, doi:10.5194/BG-17-4075-2020.

Hansis, E., S.J. Davis, and J. Pongratz, 2015: Relevance of methodological choices for accounting of land use change carbon fluxes. *Global Biogeochem. Cycles*, **8(9)**, doi:10.1002/2014GB004997.

Houghton, R.A. and A.A. Nassikas, 2017: Global and regional fluxes of carbon from land use and land cover change 1850–2015. *Global Biogeochem. Cycles*, **31(3)**, 456–472, doi:10.1002/2016GB005546.

IEA, 2010: *Projected Costs of Generating Electricity – 2010 Edition*. International Energy Agency, Paris, France.

IEA, 2019: *World Energy Outlook 2019*. International Energy Agency (IEA) and Organisation for Economic Co-operation and Development (OECD), Paris, France, 810 pp.

IEA, 2020a: *$CO_2$ Emissions from Fuel Combustion: Database Documentation (2020 edition)*. International Energy Agency, Paris, France.

IEA, 2020b: *Emissions Factors: Database Documentation (2020 edition)*. International Energy Agency (IEA), Paris, France.

IEA, 2021: *World Energy Balances*. 1st ed. International Energy Agency (IEA), Paris, France.

IPCC, 1995: *Climate Change 1995: Impacts, Adaptations and Mitigation of Climate Change: Scientific-Technical Analyses. Contribution of Working Group II to the Second Assessment Report of the Intergovernmental Panel on Climate Change* [Watson, R., M.C. Zinyowera, and R. Moss, (eds.)]. Cambridge University Press, Cambridge, UK, 861 pp.

IPCC, 2001: Appendix – IV Units, Conversion Factors, and GDP Deflators. In: *Climate Change 2001: Mitigation. Contribution of Working Group III to the Third Assessment Report of the Intergovernmental Panel on Climate Change* [Metz, B., O. Davidson, R. Swart, and J. Pan, (eds.)]. Cambridge University Press, Cambridge, UK, pp. 727–732.

IPCC, 2019: *Refinement to the 2006 IPCC Guidelines for National Greenhouse gas Inventories*. [Buendia, E.C., S. Guendehou, B. Limmeechokchai, R. Pipatti, Y. Rojas, R. Sturgiss, K. Tanabe, and T. Wirth (eds.)]. Intergovernmental Panel on Climate Change (IPCC), Geneva, Switzerland.

Krey V., O. Masera, G. Blanford, T. Bruckner, R. Cooke, K. Fisher-Vanden, H. Haberl, E. Hertwich, E. Kriegler, D. Mueller, S. Paltsev, L. Price, S. Schlömer, D. Ürge-Vorsatz, D. van Vuuren, and T. Zwickel, 2014: Annex II: Metrics & Methodology. In: *Climate Change 2014: Mitigation of Climate Change. Contribution of Working Group III to the Fifth Assessment Report of the Intergovernmental Panel on Climate Change* [Edenhofer, O., R. Pichs-Madruga, Y. Sokona, E. Farahani, S. Kadner, K. Seyboth, A. Adler, I. Baum, S. Brunner, P. Eickemeier, B. Kriemann, J. Savolainen, S. Schlömer, C. von Stechow, T. Zwickel and J.C. Minx (eds.)]. Cambridge University Press, Cambridge, United Kingdom and New York, NY, USA.

Minx, J.C. et al., 2021: A comprehensive dataset for global, regional and national greenhouse gas emissions by sector 1970–2019. *Earth Syst. Sci. Data*, **13(11)**, doi:10.5194/essd-2021-228.

Moomaw, W., P. Burgherr, G. Heath, M. Lenzen, J. Nyboer, A. Verbruggen, et al., 2011: Annex II: Methodology. In: *IPCC Special Report on Renewable Energy Sources and Climate Change Mitigation* [Edenhofer, O., R. Pichs-Madruga, Y. Sokona, K. Seyboth, P. Matschoss, S. Kadner, T. Zwickel, P. Eickemeier, G. Hansen, S. Schlömer, C. von Stechow (eds.)]. Cambridge University Press, Cambridge, United Kingdom and New York, NY, USA, pp. 973–1000.

Morita, T., J.B. Robertson, A. Adegbulugbe, J. Alcamo, 2001: Greenhouse gas emission mitigation scenarios and implications. In: *Climate Change 2001: Mitigation. Contribution of Working Group III to the Third Assessment Report of the Intergovernmental Panel on Climate Change* [Metz, B., O. Davidson, R. Swart, and J. Pan, (eds.)]. Cambridge University Press, Cambridge, UK, 115–166.

NAS, 2007: *Coal: Research and Development to Support National Energy Policy*. [Committee on Coal Research and Technology and National Research Council, (ed.)]. The National Academies Press, Washington, DC, USA.

Nilsson, M., D. Griggs, and M. Visbeck, 2016: Policy: Map the interactions between Sustainable Development Goals. *Nature*, **534(7607)**, 320–322, doi:10.1038/534320a.

Reisinger, A., M. Howden, C. Vera, et al., 2020: The Concept of Risk in the IPCC Sixth Assessment Report: A Summary of Cross-Working Group Discussions. Intergovernmental Panel on Climate Change, Geneva, Switzerland, 15 pp. https://www.ipcc.ch/site/assets/uploads/2021/02/Risk-guidance-FINAL_15Feb2021.pdf.

Roy, J., P. Tschakert, H. Waisman, S. Abdul Halim, P. Antwi-Agyei, P. Dasgupta, B. Hayward, M. Kanninen, D. Liverman, C. Okereke, P.F. Pinho, K. Riahi, and A.G. Suarez Rodriguez, 2018: Sustainable Development, Poverty Eradication and Reducing Inequalities. In: *Global Warming of 1.5°C. An IPCC Special Report on the impacts of global warming of 1.5°C above pre-industrial levels and related global greenhouse gas emission pathways, in the context of strengthening the global response to the threat of climate change, sustainable development, and efforts to eradicate poverty* [Masson-Delmotte, V., P. Zhai, H.-O. Pörtner, D. Roberts, J. Skea, P.R. Shukla, A. Pirani, W. Moufouma-Okia, C. Péan, R. Pidcock, S. Connors, J.B.R. Matthews, Y. Chen, X. Zhou, M.I. Gomis, E. Lonnoy, T. Maycock, M. Tignor, and T. Waterfield (eds.)]. Cambridge University

Press, Cambridge, UK and New York, NY, USA, pp. 445–538. https://doi.org/10.1017/9781009157940.007.

Suerkemper, F., S. Thomas, D. Osso, and P. Baudry, 2012: Cost-effectiveness of energy efficiency programmes—evaluating the impacts of a regional programme in France. *Energy Effic.*, **5(1)**, 121–135, doi:10.1007/s12053-011-9112-z.

UNFCCC, 2019: *Report of the Conference of the Parties serving as the meeting of the Parties to the Paris Agreement on the third part of its first session, held in Katowice from 2 to 15 December 2018. Addendum, Part 2: Action taken by the Conference of the Parties servi*. Katowice, Poland.

UNSD, 1999: *Standard Country or Area Codes for Statistical Use, Revision 4*. https://unstats.un.org/unsd/methodology/m49/.

AIII

# Annex III: Scenarios and Modelling Methods


**Coordinating Lead Authors:**
Céline Guivarch (France), Elmar Kriegler (Germany), Joana Portugal-Pereira (Brazil).

**Lead Authors:**
Valentina Bosetti (Italy), James Edmonds (the United States of America), Manfred Fischedick (Germany), Petr Havlík (Austria/Czech Republic), Paulina Jaramillo (the United States of America), Volker Krey (Germany/Austria), Franck Lecocq (France), André F.P. Lucena (Brazil), Malte Meinshausen (Australia/Germany), Sebastian Mirasgedis (Greece), Brian O'Neill (the United States of America), Glen P. Peters (Norway/Australia), Joeri Rogelj (Belgium/United Kingdom), Steven Rose (the United States of America), Yamina Saheb (France/Algeria), Goran Strbac (Serbia/United Kingdom), Anders Hammer Strømman (Norway), Detlef P. van Vuuren (the Netherlands), Nan Zhou (the United States of America).

**Contributing Authors:**
Alaa Al Khourdajie (United Kingdom/Syria), Hossein Ameli (Germany), Cornelia Auer (Germany), Nico Bauer (Germany), Edward Byers (Austria/Ireland), Michael Craig (the United States of America), Bruno Cunha (Brazil), Stefan Frank (Austria), Jan S. Fuglestvedt (Norway), Mathijs Harmsen (the Netherlands), Alan Jenn (the United States of America), Jarmo Kikstra (Austria/the Netherlands), Paul Kishimoto (Canada), Robin Lamboll (United Kingdom/the United States of America), Julien Lefèvre (France), Eric Masanet (the United States of America), David McCollum (the United States of America), Zebedee Nicholls (Australia), Aleksandra Novikova (Germany/the Russian Federation), Simon Parkinson (Canada), Pedro Rochedo (Brazil), Sasha Samadi (Germany), David Vérez (Cuba/Spain), Sonia Yeh (Sweden).



**This chapter should be cited as:**
IPCC, 2022: Annex III: Scenarios and modelling methods [Guivarch, C., E. Kriegler, J. Portugal-Pereira, V. Bosetti, J. Edmonds, M. Fischedick, P. Havlík, P. Jaramillo, V. Krey, F. Lecocq, A. Lucena, M. Meinshausen, S. Mirasgedis, B. O'Neill, G.P. Peters, J. Rogelj, S. Rose, Y. Saheb, G. Strbac, A. Hammer Strømman, D.P. van Vuuren, N. Zhou (eds)]. In IPCC, 2022: *Climate Change 2022: Mitigation of Climate Change. Contribution of Working Group III to the Sixth Assessment Report of the Intergovernmental Panel on Climate Change* [P.R. Shukla, J. Skea, R. Slade, A. Al Khourdajie, R. van Diemen, D. McCollum, M. Pathak, S. Some, P. Vyas, R. Fradera, M. Belkacemi, A. Hasija, G. Lisboa, S. Luz, J. Malley, (eds.)]. Cambridge University Press, Cambridge, UK and New York, NY, USA. doi: 10.1017/9781009157926.022

# Table of Contents

## Preamble

The use of scenarios and modelling methods are pillars in IPCC Working Group III (WGIII) Assessment Reports. Past WGIII assessment report cycles identified knowledge gaps about the integration of modelling across scales and disciplines, mainly between global integrated assessment modelling methods and bottom-up modelling insights of mitigation responses. The need to improve the transparency of model assumptions and enhance the communication of scenario results was also recognised.

This annex on *Scenarios and Modelling Methods* aims to address some of these gaps by detailing the modelling frameworks applied in the WGIII Sixth Assessment Report (AR6) chapters and disclose scenario assumptions and its key parameters. It has been explicitly included in the Scoping Meeting Report of the WGIII contribution to the AR6 and approved by the IPCC Panel at the 46th Session of the Panel.

The annex includes two parts: Part I on *Modelling Methods* summarises methods and tools available to evaluate sectoral, technological and behavioural mitigation responses as well as integrated assessment models (IAMs) for the analysis of 'whole system' transformation pathways; Part II on *Scenarios* sets out the portfolio of climate change scenarios and mitigation pathways assessed in the AR6 WGIII chapters, its underlying principles and interactions with scenario assessments by WGI and WGII.

# Part I: Modelling Methods

## A.III.I.1 Overview of Modelling Tools

Modelling frameworks vary vastly among themselves, and several key characteristics can be used as basis for model classification (Scrieciu et al. 2013; Dodds et al. 2015; Hardt and O'Neill 2017; Capellán-Pérez et al. 2020). Broadly, literature characterises models along three dimensions: (i) level of detail and heterogeneity, (ii) mathematical algorithm concepts, and (iii) temporal and spatial system boundaries (Krey 2014).

Commonly climate mitigation models are referred to as bottom-up and top-down depending upon their degree of detail (van Vuuren et al. 2009). Generally, bottom-up approaches present more systematic individual technological details about a reduced number of mitigation strategies of a specific sector or sub-sector. These models tend to disregard relations between specific sectors/technologies and miss evaluating interactions with the whole system. On the other hand, top-down approaches present a more aggregated and global analysis, in detriment of less detailed technological heterogeneity. They tend to focus on interactions within the whole system, such as market and policy instrument interactions within the global economy systems. Studies using top-down models are more capable of representing economic structural change than adopting technology-explicit decarbonisation strategies (van Vuuren et al. 2009; Kriegler et al. 2015a). Integrated assessment models (IAMs) typically use a top-down approach to model sectoral mitigation strategies.

Although this dichotomic classification has been mentioned in the literature, since the IPCC's Fifth Assessment Report (AR5), climate mitigation models have evolved towards a more hybrid approach incorporating attributes of both bottom-up and top-down approaches. This is partly due to different modelling communities having different understandings of these two approaches' principles, which can be misleading.

One of the most basic aspects of a modelling tool is how it approaches the system modelled from a solution perspective. A broad interpretation of mathematical algorithm concepts classifies models as simulation and optimisation models. **Simulation models** are based on the evaluation of the dynamic behaviour of a system (Lund et al. 2017). They can be used to determine the performance of a system under alternative options of key parameters in a plausible manner. Most often, simulation models require comprehensive knowledge of each parameter, in order to choose a specific path under several alternatives. On the other hand, **optimisation models** seek to maximise or minimise a mathematical objective function under a set of constraints (Baños et al. 2011; Iqbal et al. 2014). Most often, the objective function represents the total cost or revenue of a given system or the total welfare of a given society. One major aspect of optimisation models is that the solution is achieved by simultaneously binding a set of constraints, which can be used to represent real-life limitations on the system, such as: constraints on flows, resource and technology availability, labour and financial limitations, environmental aspects, and many other characteristics that the model may require (Fazlollahi et al. 2012; Pfenninger et al. 2014; Cedillos Alvarado et al. 2016). Specifically, when modelling climate mitigation responses, limiting carbon budgets is often used to represent future temperature level pathways (Rogelj et al. 2016; Millar et al. 2017; Peters 2018; Gidden et al. 2019).

Another major distinction among modelling tools is related to the solution methodology from a temporal perspective. They can have a perfect foresight intertemporal assumption or a recursive-dynamic assumption. Intertemporal optimisation with **perfect foresight** is an optimisation method for achieving an overall optimal solution over time. It is based on perfect information on all future states of a system and assumptions (such as technology availability and prices) and, as such, today's and future decisions are made simultaneously, resulting in a single path of optimal actions that lead to the overall optimal solution (Keppo and Strubegger 2010; Gerbaulet et al. 2019). Such a modelling approach can present an optimal trajectory of the set of actions and policies that would lead to the overall first-best solution. However, real-life decisions are not always based on optimal solutions (Ellenbeck and Lilliestam 2019) and, therefore, solutions from perfect foresight models can be challenging to be implemented by policymakers (Pindyck 2013, 2017). For instance, perfect foresight implies perfect knowledge of the future states of the system, such as future demand for goods and products and availability of production factors and technology.

**Recursive-dynamic** models, also known as myopic or limited foresight models, make decisions over sequential periods of time. For each time step, the solution is achieved without information on future

AIII

time steps. Therefore, the solution path is a series of solutions in short trajectories that, ultimately, is very unlikely to achieve the overall optimal solution over the whole time period considered (Fuso Nerini et al. 2017). Nonetheless, the solution represents a set of possible and plausible policies and behavioural choices of the agents that could be taken in short-term cycles, without perfect information (Heuberger et al. 2018; Hanna and Gross 2020). In between, some models consider **imperfect or adaptive expectations**, where economic decisions are based on past, current and imperfectly anticipated future information (Keppo and Strubegger 2010; Kriegler et al. 2015a; Löffler et al. 2019). Modelling tools can also be differentiated by their level of representation of economic agents and sectors: they can have a full representation of all agents of the economy and their interactions with each other (**general equilibrium**) or focus on a more detailed representation of a subset of economic sectors and agents (**partial equilibrium**) (Cheng et al. 2015; Babatunde et al. 2017; Hanes and Carpenter 2017; Sanchez et al. 2018; Guedes et al. 2019; Pastor et al. 2019) (Annex III.I.2).

The most basic aspect to differentiate models is their main objective function, which includes the detail at which they represent key sectors, systems and agents. This affects the decision on methodology and other coverage aspects. Several models have been developed for different sectoral representation, such as the energy (Annex III.I.3), buildings (Annex III.I.4), transport (Annex III.I.5), industry (Annex III.I.6) and land-use (Annex III.I.7) models.

Modelling exercises vary considerably in terms of key characteristics, including geographical scales, time coverage, environmental variables, technologies portfolios, and socio-economic assumptions. A detailed comparison of key characteristics of global and national models used in this report is presented in Annex III.I.9. Geographical coverage ranges from sub-national (Cheng et al. 2015; Feijoo et al. 2018; Rajão et al. 2020), national (Li et al. 2019; Sugiyama et al. 2019; Vishwanathan et al. 2019; Schaeffer et al. 2020), regional (Vrontisi et al. 2016; Hanaoka and Masui 2020) and global (Gidden et al. 2018; Kriegler et al. 2018a; McCollum et al. 2018; Rogelj et al. 2019b; Drouet et al. 2021) models. Even models with the same geographical coverage can still be significantly different from each other, for instance, due to the number of regions within the model. Models can also have spatially implicit and explicit formulations, which in turn can have different spatial resolution. This distinction is especially important for land-use models, which account for changes in land use and agricultural practices (Annex III.I.7: Land-use modelling). The time horizon, time steps and time resolution are major aspects that differ across models. Model horizon can range from short- to long-term, typically reaching from a few years to up until the end of the century (Fujimori et al. 2019b; Gidden et al. 2019; Rogelj et al. 2019a; Ringkjøb et al. 2020). Time resolution is particularly relevant for specific applications, such as power sector models, which have detailed representation of power technologies dispatch and operation (Soria et al. 2016; Abujarad et al. 2017; Guan et al. 2020).

Life Cycle Assessment (LCA) is an integrated technique to evaluate the sustainability of a product throughout its life cycle. It quantifies the environmental burdens associated with all stages from the extraction of raw materials, through the production of the product itself, its utilisation, and end-life, either via reuse, recycling or final disposal (Rebitzer et al. 2004; Finnveden et al. 2009; Guinée et al. 2011; Curran 2013; Hellweg and Milà i Canals 2014). The environmental impacts covered include all types of loads on the environment through the extraction of natural resources and emission of hazardous substances. For this reason, LCA has the flexibility to evaluate an entire product system, hence avoiding sub-optimisation in a single process and identifying the products and processes that result in the least environmental impact. Thus, it allows for the quantification of possible trade-offs between different environmental impacts (e.g., eliminating air emissions by increasing non-renewable energy resources) (Hawkins et al. 2013; Nordelöf et al. 2014; Gibon et al. 2017) and/or from one stage to other (e.g., reuse or recycling a product to bring it back in at the raw material acquisition phase) (Hertwich and Hammitt 2001a,b). It gives a holistic view of complex systems and reduces the number of parameters for which decisions have to be taken, while not glossing over technical and economical details. In recent years, LCA has been widely used in both retrospective and prospective analysis of product chains in various climate mitigation fields, namely comparing existing energy technologies with planned alternatives (Cetinkaya et al. 2012; Portugal-Pereira et al. 2015), product innovation and development (Wender et al. 2014; Portugal-Pereira et al. 2015; Sharp and Miller 2016), certification schemes (Prussi et al. 2021), or supply chain management (Hagelaar 2001; Blass and Corbett 2018).

Two different types of LCA approaches can be distinguished: Attributional Life Cycle Assessment (ALCA) and Consequential Life Cycle Assessment (CLCA). ALCA aims at describing the direct environmental impacts of a product. It typically uses average and historical data to quantify the environmental burden during a product's life cycle, and it tends to exclude market effects or other indirect effects of the production and consumption of products (Baitz 2017). CLCA, on the other hand, focuses on the effects of changes due to product life cycle, including both consequences inside and outside the product life cycle (Earles and Halog 2011). Thus, the system boundaries are generally expanded to represent direct and indirect effects of products' outputs. CLCA tends to describe more complex systems than ALCA, which are highly sensitive to data assumptions (Plevin et al. 2014; Weidema et al. 2018; Bamber et al. 2020).

**Integrated assessment models** (IAMs) are simplified representations of complex physical and social systems, focusing on the interaction between economy, society and the environment (Annex III.I.9). They represent the coupled energy-economy-land-climate system to varying degrees. In a way, IAMs differ among themselves on all the topics discussed in this section: significant variation in geographical, sectoral, spatial and time resolution; they rely greatly on socio-economic assumptions; different technological representation; partial or general equilibrium assumptions; differentiated between perfect foresight or recursive-dynamic methodology. The difficulty in fully representing the extent of climate damages in monetary terms may be the most important and challenging limitation of IAMs and it is mostly directed to cost-benefit IAMs. However, all categories of IAMs present important limitations (Annex III.I.9).

Following this brief synopsis of modelling taxonomies, Section I.2 details key aspects of economic frameworks and principles used to model climate mitigation responses and estimate their costs. Sections I.3, I.4, I.5, I.6, and I.7 present key aspects of sectoral modelling approaches in energy systems, buildings, transport, industry, and land use, respectively. Interactions between WGI climate emulators and WGIII mitigation models are described in Section I.8. A review of integrated assessment model approaches, their components and limitations, is presented in Section I.9. Sections I.10 and I.11 present comparative tables of key characteristics and measures of national and global models that contributed to the AR6 WGIII scenario database.

### A.III.I.2 Economic Frameworks and Concepts Used in Sectoral Models and Integrated Assessment Models

Several types of 'full-economy' frameworks are used in integrated assessment models. The **general equilibrium** framework – often referred to as Computable General Equilibrium (CGE) – represents the economic interdependencies between multiple sectors and agents, and the interaction between supply and demand on multiple markets (Robinson et al. 1999). It captures the full circularity of economic flows through income and demand relationships and feedbacks including the overall balance of payments. Most CGE approaches used are neoclassical supply-led models with market clearing based on price adjustment. Representative agents usually minimise production costs or maximise utility under given production and utility function, although optimal behaviours are not a precondition *per se*. Most CGE models also include assumptions of perfect markets with full employment of factors although market imperfections and underemployment of factors (e.g., unemployment) can be assumed (Babiker and Eckaus 2007; Guivarch et al. 2011). CGE frameworks can either be static or dynamic and represent pathways as a sequence of equilibria in the second case.

**Macro-econometric** frameworks represent similar sectoral interdependence with balance of payments as general equilibrium, and are sometimes considered a subset of the general equilibrium framework. They differ from standard neoclassical CGE models in the main aspect that economic behaviours are not micro-founded optimising behaviours but are represented by macroeconomic and sectoral functions estimated through econometric techniques (Barker and Scrieciu 2010). In addition, they usually adopt a demand-led post-Keynesian approach where final demand and investment determine supply and not the other way around. Prices also do not instantaneously clear markets and adjust with lag.

**Macro-economic growth** frameworks are also full-economy approaches derived from aggregated growth models. They are based on a single macroeconomic production function combining capital, labour and sometimes energy to produce a generic good for consumption and investment. They are used as the macroeconomic component of cost-benefit IAMs (Nordhaus 1993) and some detailed-process IAMs.

The **disaggregation of economic actors and sectors and the representation of their interaction** differ across full-economy frameworks. A main distinction is between models based on full Social Accounting Matrix (SAM) and aggregated growth approaches. On the one hand, SAM-based frameworks – CGE and macro-econometric – follow a multi-sectoral approach distinguishing from several to a hundred different economic sectors or production goods and represent sector-specific value-added, final consumption and interindustry intermediary consumption (Robinson 1989). They also represent economic agents (firms, households, public administration, etc.) with specific behaviours and budget constraints. On the other hand, macro-economic growth frameworks are reduced to a single macro-economic agent producing, consuming and investing a single macroeconomic good without considering interindustry relationships. In some detailed process IAMs, the aggregated growth approach is combined with a detailed representation of energy supply and demand systems that surmises different economic actors and subsectors. However, the energy system is driven by an aggregated growth engine (Bauer et al. 2008).

**Partial equilibrium** frameworks do not cover the full economy but only represent a subset of economic sectors and markets disconnected from the rest of the economy. They basically represent market balance and adjustments for a subset of sectors under *ceteris paribus* assumptions about other markets (labour, capital, etc.), income, and so on, ignoring possible feedbacks. Partial equilibrium frameworks are used in sectoral models, as well as to model several sectors and markets at the same time – for example, energy and agriculture markets – in energy system models and some detailed process IAMs but still without covering the full economy.

In most models the treatment of **economic growth** follows Solow or Ramsey growth approach based on the evolution through time of production factors, endowment and productivity. Classically, labour endowment and demography are exogenous, and capital accumulates through investment. Partial equilibrium frameworks do not model economic growth but use exogenous growth assumptions derived from growth models. Factors' productivity evolution is assumed exogenous in most cases that is, general technical progress is assumed to be an autonomous process. A few models feature endogenous growth aspects where factor productivity increases with cumulated macroeconomic investment. Models also differ in the content of technical progress and alternatively consider unbiased total factor productivity improvement or labour-specific factor-augmenting productivity. In multi-sectoral macroeconomic models, economic growth comes with endogenous changes of the sectoral composition of GDP known as structural change. **Structural change** results from the interplay between differentiated changes of productivity between sectors and of the structure of final demand as income grows (Herrendorf et al. 2014). If general technical progress is mostly assumed exogenous and autonomous at an aggregated level, **innovation in relation to energy demand and technical systems** follow more detailed specifications in models. Energy efficiency can be assumed an autonomous process at different levels – macroeconomic, sector or technology – or energy technical change can be endogenous and induced as a learning by doing process or as a result of R&D investments (learning-by-searching) (Löschel 2002).

Multi-regional models consider interactions between regions through **trade** in energy goods, non-energy goods and services – depending on model scope – and emission permits in the context of climate policy. For each type of goods, trade is usually represented as a common pool where regions interact with the pool through supply (exports) or demand (imports). A few models consider bilateral trade flows between regions. Traded goods can be assumed as perfectly substitutable between regions of origin (Heckscher-Ohlin assumption), such as is often the case for energy commodities, or as imperfectly substitutable (e.g., Armington goods) for non-energy goods. The representation of trade and capital imbalances at the regional level and their evolution through time vary across models and imbalances are either not considered (regional current accounts are balanced at each point in time), or a constraint for intertemporal balance is included (an export surplus today will be balanced by an import surplus in the future), or else trade imbalances follow other rules such as a convergence towards zero in the long run (Foure et al. 2020).

**Strategic interaction** can also occur between regions, especially in the presence of externalities such as climate change, energy prices or technology spillovers. Intertemporal models can include several types of strategic interaction: (i) a cooperative Pareto optimal solution where all externalities are internalised and based on the maximisation of a global discounted welfare with weighted regional welfare (Negishi weights), (ii) a non-cooperative solution that is strategically optimal for each region (Nash equilibrium) (Leimbach et al. 2017b), and (iii) partially cooperative solutions (Eyckmans and Tulkens 2003; Yang 2008; Bréchet et al. 2011; Tulkens 2019), akin to climate clubs (Nordhaus 2015).

Models cover different **investment** flows depending on the economic framework used. Partial equilibrium models compute energy system and/or sectoral (transport, building, industry, etc.) technology-specific investment flows associated with productive capacities and equipment. Full-economy models compute both energy system and macroeconomic investment, the second being used to increase macroeconomic capital stock. Full-economy multi-sectoral models compute sector-specific (energy and non-energy sectors) investment and capital flows with some details about the investments goods involved.

Full-economy models differ in the representation of **macro-finance**. In most CGE and macro-economic growth frameworks financial mechanisms are only implicit and total financial capacity and investment are constrained by savings. Consequently, investment in a given sector (e.g., low-carbon energy) fully crowds out investment in other sectors. In macro-econometric frameworks, macro-finance is sometimes explicit, and investments can be financed by credit on top of savings, which implies more limited crowding out of investments (Mercure et al. 2019). Macro-financial constraints are usually not accounted for in partial equilibrium models.

Models compare economic flows over time through **discounting**. Table 5 summarises key characteristics of different models assessed in AR6, including the uses of discounting. In cost-benefit analysis (CBA), discounting enables the comparison of mitigation costs and climate change damage. In the context of mitigation and in cost-effectiveness analysis (CEA), discounting allows the comparison of mitigation costs over time.

In optimisation models a social discount rate is used to compare costs and benefits over time. In the case of partial equilibrium optimisation models, the objective is typically to minimise total discounted system cost. The social discount rate is then an exogenous parameter, which can be assumed constant or changing (generally decreasing) over time (e.g., Gambhir et al. (2017), where a 5% discount rate is used). In the case of intertemporal welfare optimisation models, a Ramsey intertemporal optimisation framework is generally used, considering a representative agent who decides how to allocate her consumption, and hence saving, over time, subject to a resource constraint. Ramsey (1928) shows that the solution must always satisfy the Ramsey Equation, which provides the determinants of the social discount rate. The Ramsey Equation is given as follows:

$$\rho = \delta + \eta g_t$$

where $\rho$ is the consumption discount rate (also known as the social discount rate), $\delta$ is the utility discount rate (also known as the pure time discount rate, or time preferences rate) which is a value judgement that determines the present value of a change in the utility experienced in the future and hence it is an ethical parameter, $g_t$ is the growth rate of consumption per capita over time, and $\eta$ is the elasticity of marginal utility of consumption, which is also a value judgement and hence an ethical parameter. The parameter $\eta$ is also a measure of risk aversion and of society's aversion to inequality within and across generations. The pure time preference rate is an exogenous parameter, but the social discount rate is endogenously computed by the model itself and depends on the growth rate of consumption per capita over time. Note that more complex frameworks disentangle inequality aversion from risk aversion, and introduce uncertainty, leading to extensions of the social discount rate equation (see, for instance, Gollier 2013).

Discounting is also used for *ex post* comparison of mitigation cost pathways across models and scenarios. Values typically used for such *ex post* comparison are 2–5% (e.g., Admiraal et al. 2016). Across this report, whenever discounting is used for *ex post* comparisons, the discount rate applied is stated explicitly.

The choice of the appropriate social discount rate (and the appropriate rate of pure time preference when applicable) is highly debated (e.g., Arrow et al. 2013; Gollier and Hammitt 2014; Polasky and Dampha 2021) and two general approaches are commonly used. Based on ethical principles, the prescriptive approach states that the discount rate should reflect how costs and benefits supported by different generations should be weighted. The descriptive approach identifies the social discount rate to the risk-free rate of return to capital as observed in the real economy, which generally yields higher values.

In CBA the choice of discount rate is crucial for the balance of mitigation costs and avoided climate damages in the long run and a lower discount rate yields more abatement effort and lower global temperature increases (Stern 2006; Hänsel et al. 2020). In CEA, the choice of social discount rate influences the timing of emission reductions to limit warming to a given temperature level. A lower discount rate increases short-term emissions reductions, lowers temperature overshoot, favours currently available mitigation options (energy efficiency, renewable energy, etc.) over future deployment of net negative emission options and distributes mitigation effort more evenly between generations (Emmerling et al. 2019; Strefler et al. 2021b).

Outside social discounting for intertemporal optimisation, discounting is used in simulation models to compute the life cycle costs of investment decisions (e.g., energy efficiency choices, choices between different types of technologies based on their levelised costs). In this case, the discount rate can be interpreted as the cost of capital faced by investors. The cost of capital influences the merit order of technologies and lower capital cost favours capital-intensive technologies over technologies with higher variable costs. Models can reflect regional, sectoral or technology-specific cost of capital – through heterogeneous discount rates for life cycle cost estimates in simulation models (Iyer et al. 2015) or as hurdle rates in energy optimisation models (Ameli et al. 2021). In some cases, simulation models may also produce mitigation pathways following the Hotelling principle and assuming that the carbon price rises at the social discount rate (e.g., Global Change Assessment Model (GCAM) scenarios in the Shared Socio-economic Pathways (SSP) study with carbon prices increasing at 5% yearly (Guivarch and Rogelj 2017)).

## A.III.I.3 Energy System Modelling

In the literature, the energy system models are categorised based on different criteria, such as (i) energy sectors covered, (ii) geographical coverage, (iii) time resolution, (iv) methodology, and (v) programming techniques. In the following sections, examples on different types of energy system models applied in Chapter 6 are presented.

### A.III.I.3.1 Bottom-up Models

#### *A.III.I.3.1.1 Modelling Electricity System Operation and Planning with Large-scale Penetration of Renewables*

A number of advanced grid modelling approaches have been developed (Sani Hassan et al. 2018), such as robust optimisation (Jiang et al. 2012), interval optimisation (Dvorkin et al. 2015), or stochastic optimisation (Meibom et al. 2011; Monforti et al. 2014) to optimally schedule the operation of the future low-carbon systems with high penetration of variable renewable energies. Advanced stochastic models demonstrated that this would not only lead to significantly higher cost of system management but may eventually limit the ability of the system to accommodate renewable generation (Bistline and Young 2019; Hansen et al. 2019; Perez et al. 2019; Badesa et al. 2020). Modelling tools such as *European Model for Power System Investment with Renewable Energy (EMPIRE)* (Skar et al. 2016), *Renewable Energy Mix for Sustainable Electricity Supply (REMix)* (Scholz et al. 2017), *European Unit Commitment And Dispatch model (EUCAD)* (Després 2015), *SWITCH* (Fripp 2012), *GenX* (TNO 2021), and *Python for Power System Analysis (PyPSA)* (Brown et al. 2018) investigated these issues. SWITCH is a stochastic model, in which investments in renewable and conventional power plants are optimised over a multi-year period (Fripp 2012). In GenX the operational flexibility as well as capacity planning is optimised from a system-wide perspective (TNO 2021). PyPSA is an optimisation model for modern electricity systems, including unit commitment of generation plants, renewable sources, storage, and interaction with other energy vectors (Brown et al. 2018).

Furthermore, advanced modelling tools have been developed for the purpose of providing estimations of system-wide inertial frequency response that would assist system operators in maintaining adequate system inertia (Sharma et al. 2011; Teng and Strbac 2017). These innovative models also provide fundamental evidence regarding the role and value of advanced technologies and control systems in supporting cost-effective operation of future electricity systems with very high penetration of renewable generation. In particular, the importance of enhancing the control capabilities of renewable generation and applying flexible technologies, such as energy storage (Hall and Bain 2008; Obi et al. 2017; Arbabzadeh et al. 2019), demand-side response, interconnection (Aghajani et al. 2017) and transmission grid extensions (Schaber et al. 2012) to provide system stability control, is demonstrated through novel system integration models (Lund et al. 2015; Sinsel et al. 2020).



A novel modelling framework is proposed to deliver inertia and support primary frequency control through variable-speed wind turbines (Morren et al. 2006) and PVs (Waffenschmidt and Hui 2016; Liu et al. 2017), including quantification of the value of this technology in future renewable generation-dominated power grids (Chu et al. 2020). Advanced models for controlling distributed energy storage systems to provide an effective virtual inertia have been developed, demonstrating the provision of virtual synchronous machine capabilities for storage devices with power electronic converters, which can support system frequency management following disturbances (Hammad et al. 2019; Markovic et al. 2019). Regarding the application of interconnection for exchange of balancing services between neighbouring power grids, alternative control schemes for high-voltage direct current (HVDC) converters have been proposed, in order to demonstrate that this would reduce the cost of balancing (Tosatto et al. 2020).

#### *A.III.I.3.1.2 Modelling the Interaction between Different Energy Sectors*

Several integrated models have been developed in order to study the interaction between different energy vectors and whole-system approaches, such as *Integrated Energy System Simulation model (IESM)* (NREL 2020), *Integrated Whole-Energy System (IWES)* (Strbac et al. 2018), *UK TIMES* (Daly and Fais 2014), and *Calliope* (Pfenninger and Pickering 2018).

IESM is an approach in which the multi-system energy challenge is investigated holistically rather than looking at each of the systems in isolation. IESM capabilities include co-optimisation across multiple energy systems, including electricity, natural gas, hydrogen, and water systems. These provide the opportunity to perform hydro, thermal, and gas infrastructure investment and resource use coordination for time horizons ranging from sub-hourly (markets and operations) to multi-year (planning) (NREL 2020).

The IWES model incorporates detailed modelling of electricity, gas, transport, hydrogen, and heat systems and captures the complex interactions across those energy vectors. The IWES model also considers the short-term operation and long-term investment timescales (from seconds to years) simultaneously, while coordinating operation of and investment in local district and national/international level energy infrastructures (Strbac et al. 2018).

The UK TIMES Model ('The Integrated MARKAL-EFOM System') uses linear programming to produce a least-cost energy system, optimised according to a number of user constraints, over medium- to long-term time horizons. It portrays the UK energy system, from fuel extraction and trading to fuel processing and transport, electricity generation and all final energy demands (Taylor et al. 2014; Daly and Fais 2014). The model generates scenarios for the evolution of the energy system based on different assumptions around the evolution of demand and future technology costs, measuring energy system costs and all greenhouse gases (GHGs) associated with the scenario. UKTM is built using the TIMES model generator: as a partial equilibrium energy system and technologically detailed model, it is well suited to investigate the economic, social, and technological trade-offs between long-term divergent energy scenarios.

Calliope is an open source Python-based toolchain for developing energy system models, focusing on flexibility, and high temporal and spatial granularities. This model has the ability to execute many runs on the same base model, with clear separation of model (data) and framework (code) (Pfenninger and Pickering 2018).

#### A.III.I.3.2 Modelling of Energy Systems in the Context of the Economy

To study the impact of low-carbon energy systems on the economy, numerous integrated assessment modelling tools (top-down models) are applied, such as: *General Equilibrium Model for Economy-Energy-Environment (GEM-E3)* (Capros et al. 2013), *ENV-Linkages* (Burniaux and Chateau 2010), and *Emissions Prediction and Policy Analysis (EPPA)* (Chen et al. 2016).

GEM-E3 is a recursive dynamic computable general equilibrium model that covers the interactions between the economy, the energy system and the environment. It is specially designed to evaluate energy, climate, and environmental policies. GEM-E3 can evaluate consistently the distributional and macro-economic effects of policies for the various economic sectors and agents across the countries/regions (Capros et al. 2013).

The modelling work based on ENV-Linkages (as a successor to the OECD GREEN model) provides insights to policymakers in identifying least-cost policies by taking into account environmental issues, such as phasing out fossil fuel subsidies, and climate change mitigation (Burniaux and Chateau 2010).

In the EPPA model, different processes (e.g., economic and technological) which have impacts on the environment from regional to global at multiple scales are simulated. The outputs of this modelling (e.g., greenhouse gas emissions, air and water pollutants) are provided to the MIT Earth System (MESM), which investigates the interaction between sub-models of physical, dynamical and chemical processes in different systems (Chen et al. 2016).

#### A.III.I.3.3 Hybrid Models

Hybrid models are a combination of macro-economic models (i.e., top-down) with at least one energy sector model (i.e., bottom-up) that could benefit from the advantages of both mentioned approaches. In this regard, linking these two models can be carried out either manually through transferring the data from one model to the other (soft linking), or automatically (hard linking) (Prina et al. 2020). In this section, some of these models are presented including *World Energy Model (WEM)* (IEA 2020a) and the *National Energy Modelling System (NEMS)* (Fattahi et al. 2020).

The WEM is a simulation model covering energy supply, energy transformation and energy demand. The majority of the end-use sectors use stock models to characterise the energy infrastructure. In addition, energy-related $CO_2$ emissions and investments related to energy developments are specified. The model is focused on determining the share of alternative technologies in satisfying energy service demand. This includes investment costs, operating and maintenance costs, fuel costs and in some cases costs for emitting $CO_2$ (IEA 2020a).

The NEMS is an energy-economy modelling system applied for the USA through 2030. NEMS projects consider the production, import, conversion, consumption, and prices of energy, subject to assumptions on macroeconomic and financial factors, world energy markets, resource availability and costs, behavioural and technological choice criteria, cost and performance characteristics of energy technologies, and demographics. NEMS was designed and implemented by the Energy Information Administration (EIA) of the US Department of Energy. NEMS is used by EIA to project the energy, economic, environmental, and security impacts on the United States considering alternative energy policies and assumptions related to energy markets (Fattahi et al. 2020).

### A.III.I.4 Building Sector Models

#### A.III.I.4.1 Models: Purpose, Scope and Types

GHG emissions and mitigation potentials in the building sector are modelled using either a top-down, a bottom-up or a hybrid approach (Figure 1).

The top-down models are used for assessing economy-wide responses of building policies. These models are either economic or technological and have low granularity.

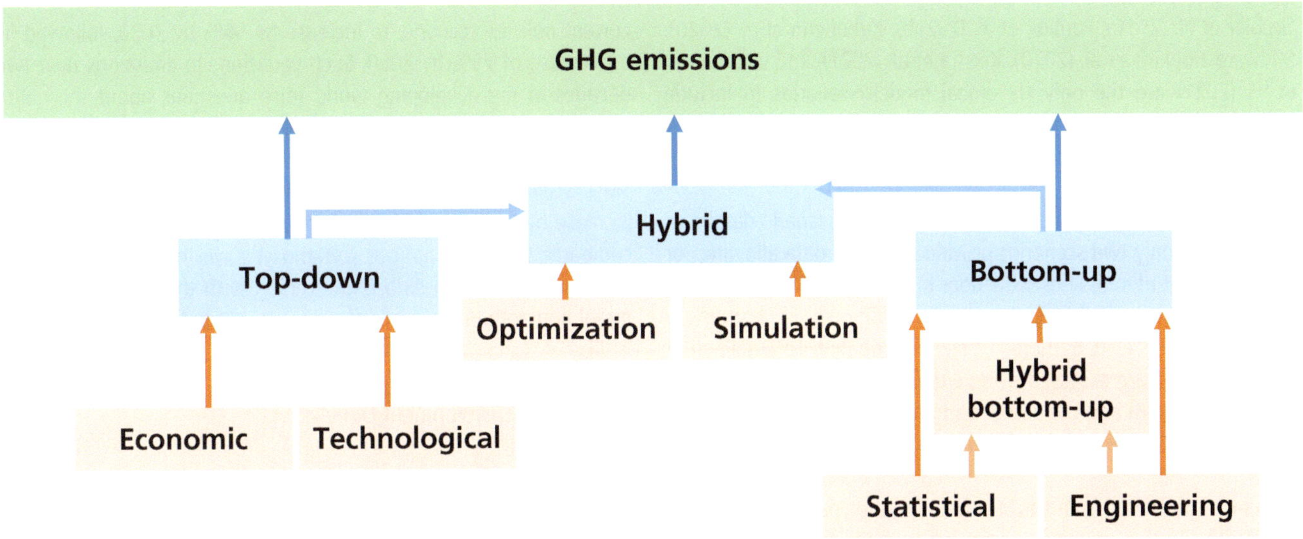


**Figure 1 | Modelling approaches for GHG emissions used in the building sector.**

The bottom-up models are data intensive and based on microscopic data of individual end uses and the characteristics of each component of buildings. Bottom-up models can be either physics-based, also known as engineering models; data-driven, also known as statistical models; or a combination of both, also known as hybrid bottom-up models. Bottom-up models are useful to assess the technico-economic potentials of the overall building stock by extrapolating the estimated energy consumption of a representative set of invidual buildings (Duerinck et al. 2008; Hall and Buckley 2016; Bourdeau et al. 2019).

Hybrid models used for buildings can be either optimisation or simulation models (Duerinck et al. 2008; Hall and Buckley 2016; Bourdeau et al. 2019) (Figure 1). The latter can also be agent-based models and could be combined with building performance models to allow for an assessment of occupants' behaviour (Papadopoulos and Azar 2016; Sachs et al. 2019a; Niamir et al. 2020). Hybrid models are used for exploring the impacts of resource constraints and for investigating the role of specific technological choices as well as for analysing the impact of specific building policies.

The use of geographical information systems (GIS) layers (Reinhart and Cerezo Davila 2016) combined with machine learning techniques (Bourdeau et al. 2019) allows the creation of detailed datasets of building characteristics while optimising the computing time, thus, leading to a better representation of energy demand of buildings and a more accurate assessment of GHG mitigation potential.

### A.III.I.4.2 Representation of Energy Demand and GHG Emissions

Comprehensive models represent energy demand per energy carrier and end use for both residential and non-residential buildings, for different countries or sets of countries, further disaggregated across urban/rural and income groups. Drivers of energy demand considered include population, the floor area per capita, appliances ownership and to some extent occupants' behaviour in residential buildings. The former are included in top-down, hybrid and bottom-up models while the latter is usually included in bottom-up and agent-based models (Niamir et al. 2020; IEA 2021). In non-residential buildings, value added is considered among the drivers.

GHG emissions from buildings are usually modelled on the basis of the estimated energy demand per energy carrier and appropriate emissions factors. The purpose of most building models is to assess the impact of mitigation measures on energy demand in the use phase of buildings and for a given assumption on the per-capita floor area and technological improvement (IEA 2021; Pauliuk et al. 2021b). After decades of ignoring material cycles and embodied emissions (Pauliuk et al. 2017), a few IAMs are now including material stocks and flows (Deetman et al. 2020; IEA 2021; Zhong et al. 2021). However, the top-down nature of these models and the modelling methodology of embodied emissions, which are added onto the emissions estimated in the use phase, questions the policy relevance of these estimates. As of today, the resource efficiency and climate change (RECC) scenario (IRP 2020; Fishman et al. 2021; Pauliuk and Heeren 2021; Pauliuk et al. 2021b; ) is the only global scenario identified which includes measures to limit, in the first place, embodied emissions from buildings. The scenario is modelled using the bottom-up ODYM-RECC model.

### A.III.I.4.3 Representation of Mitigation Options

The assessment conducted in Chapter 9 was based on the SER (Sufficiency, Efficiency, Renewable), framework with sufficiency being all the measures and daily practices which avoid, in the first place, the demand for energy, materials, water, land and other natural resources over the life cycle of buildings and appliances/equipment, while providing decent living standards for all within the planetary boundaries. By contrast to efficiency, sufficiency measures do not consume energy in the use phase. Efficiency improvement of the building envelope and appliances/equipment are the main mitigation options considered in the existing models and scenarios. They are, usually, combined with market-based and information instruments and to some extent with behaviour change. As of today,

Grubler et al. (2018), Pauliuk et al. (2021b), Kuhnhenn et al. (2020), Millward-Hopkins et al. (2020), Kikstra et al. (2021), and van Vuuren et al. (2021) are the only six global models/scenarios to include sufficiency measures, out of which detailed data were available only for two scenarios (Pauliuk et al. 2021b; van Vuuren et al. 2021).

In total, 931 scenarios were submitted to the AR6 scenario database, out of which only two scenarios provided detailed data allowing for an assessment of emissions reductions based on the SER framework considered in the building chapter. An additional 78 bottom-up models/scenarios were gathered (Table 1). Mitigation potentials from these scenarios are assessed using either a decomposition analysis (Section 9.3) or an aggregation of bottom-up potential estimates for different countries into regional and then global figures (Section 9.6).

Scenarios considered in the illustrative mitigation pathways included in Chapter 3 were assessed, compared to current policy scenarios. The assessment was possible for only the combined direct $CO_2$ emissions for both residential and non-residential buildings due to lack of data on other gases as well as on indirect and embodied emissions. The assessment shows mitigation potentials, compared to current policies scenarios, at a global level ranging from 9% to 13% by 2030 and from 58% to 89% in 2050 (Figure 2-b).

There are great discrepancies in the projected potentials by the IAMs across regions and scenarios. In the deep electrification and high renewable scenario, emissions in Africa are projected to increase by 88% by 2030, followed by a decrease of 97% by 2050, compared to current policies scenario. Similarly, in the sustainable development scenario, emissions in developing Asia are projected, compared to current policies scenario, to increase by 56% by 2030, followed by a decrease of 75% by 2050. Such variations in emissions over two decades in the developing world raise questions about the policy relevance of these scenarios. In developed countries, emissions are projected to go down in all regions across all scenarios, except in SSP2 scenario in Asia-Pacific, where emissions are projected to increase by 18% by 2030 followed by a decrease of 25% by 2050, compared to current policies scenario. It is worth noting that, across all scenarios, Eastern Asia is the region with the lowest estimated mitigation potential compared to the current policies (Figure 2-b).

#### A.III.I.4.4 Representation of Sustainable Development Dimensions

The link to the Sustainable Development Goals (SDGs) is not always explicit in buildings models/scenarios. However, some models include requirements to ensure access to decent living standards for all (Grubler et al. 2018); Millward-Hopkins et al. 2020; Kikstra et al. 2021) or to specifically meet the 2030 SDG 7 goal (IEA 2020a, 2021).

#### A.III.I.4.5 Models Underlying the Assessment in Chapter 9

The AR6 scenario database received 101 models with a building component, out of which 96 were IAM models and five building specific models. This is equivalent to 931 scenarios. After an initial screening, quality control and further vetting to assess if they sufficiently represented historical trends and climate goals, 43 models (42 IAMs and 1 building-specific model) were kept for the assessment, thus reducing the number of scenarios to assess to 554. The unvetted scenarios are still available in the database.

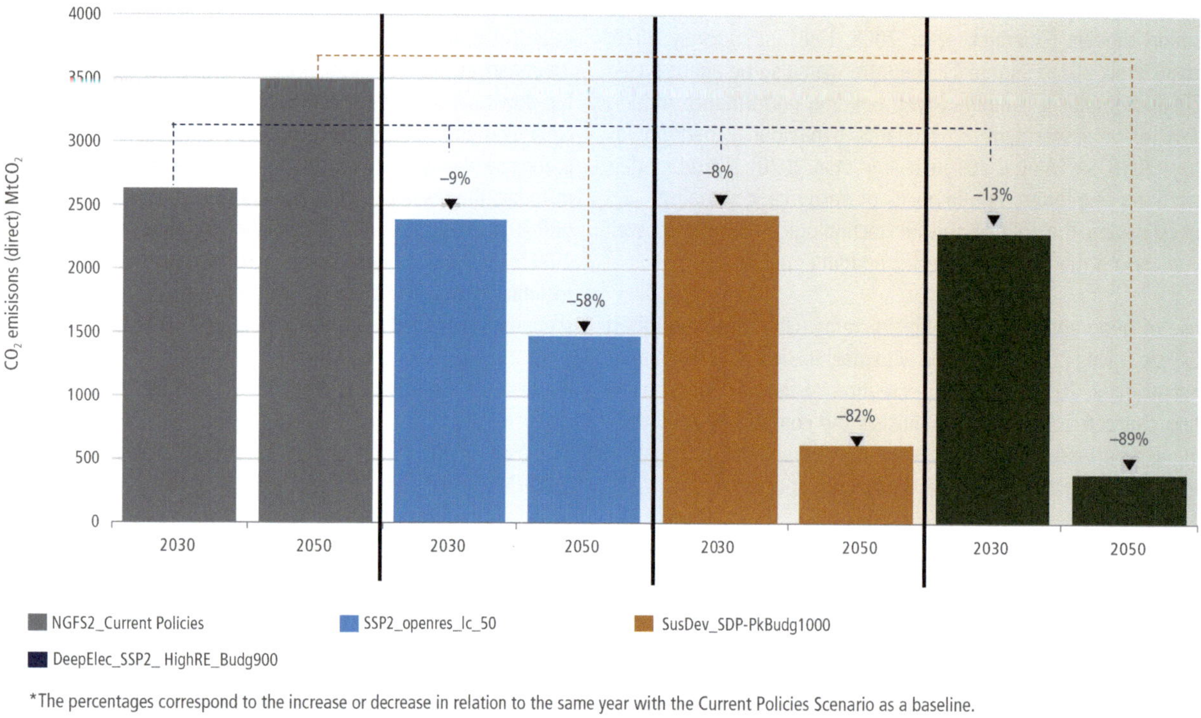


a) Global

**Figure 2 | GHG emissions reductions in the building sector (direct emissions) in scenarios considered as illustrative mitigation pathways in Chapter 3.**

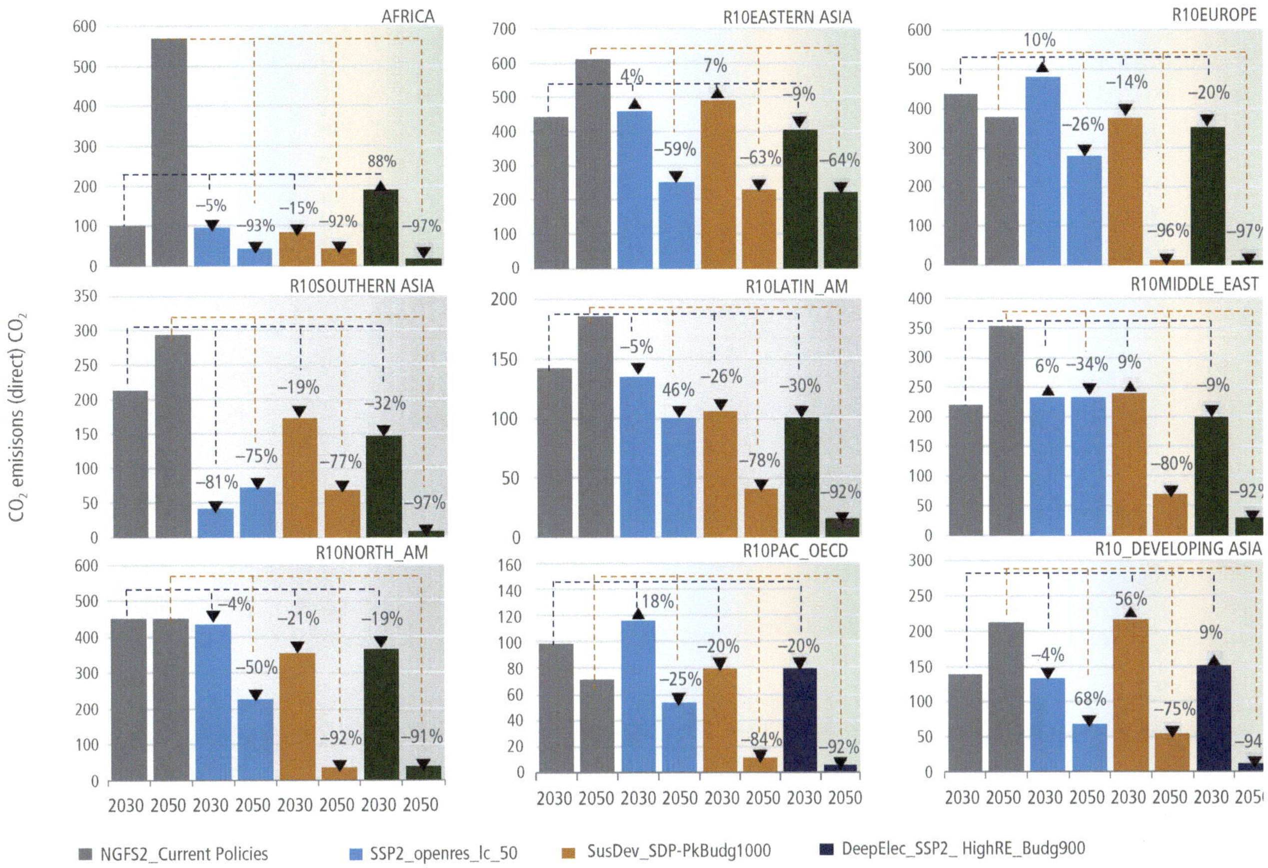


* The percentages correspond to the increase or decrease in relation to the same year with the Current Policies Scenario as a baseline.

**b) Regional**

**Figure 2 (continued): GHG mitigation potentials of scenarios considered in the illustrative mitigation pathways considered in Chapter 3.**

After a final screening based on the SER (Sufficiency, Efficiency, Renewable) framework, only two IAMs were kept. Given the top-down nature of IAMs and their weaknesses in assessing mitigation measures, especially sufficiency measures, 78 bottom-up models with technological representation have been included in the assessment (Table 1). These additional bottom-up models were not submitted to the AR6 scenario database. However, scenario owners supplied Chapter 9 with the underlying assumptions and data.

## A.III.I.5 Transport Models

### A.III.I.5.1 Purpose and Scope of Models

GHG emissions from transport are largely a function of **travel demand, transport mode,** and **transport technology and fuel**. The purpose of transportation system models is to describe how future **demand** for transport can be fulfilled through different **modes** and **technologies** under different climate change mitigation targets or policies. Within a given transport mode, technologies differ by efficiency and fuel use.

Common components of transportation energy systems models mirror these main drivers of GHG emissions. Most models will also quantify how much movement occurs, or the **travel demand** associated with each mode. Models commonly quantify demand through **transportation mode** (e.g., active transit, passenger vehicles, trucks, boats, planes, etc.) or how movement occurs (e.g., passenger travel distance pkm and freight distance tkm). Higher fidelity models provide more nuanced breakdowns of demand by trips of various lengths such as short-, medium-, and long-distance trips or by region (e.g., kilometres or tkm per region). The scope of the model often determines how much information it provides on where and when movement occurs. While larger scale models typically provide aggregate travel demand, higher resolution travel demand models can be integrated into transportation system models and provide much more information on origin and destination of trips, when and where trips occur, and the route of travel taken. This level of detail is not often characterised in the output of system models but can be employed as a 'base' model to determine how travel occurs before aggregation (Edelenbosch et al. 2017a; Yeh et al. 2017).

A key distinguishing feature between different model types is how they control the above components. Our review of the transport energy system models can be broadly divided into three main categories: (i) optimisation models, (ii) simulation models, and (iii) accounting and exploratory models.

Table 1 | Models underlying the assessment in Chapter 9.

| Model name/ institution using the model | Model description | Geographic scope | Building type included | Energy demand | Example of publications |
|---|---|---|---|---|---|
| World Energy Model (WEM)/International Energy Agency (IEA) | A simulation model with detailed bottom-up building stock model | Global | Residential and non-residential | The building module includes a stock model with detailed technologies, end uses and energy carriers. Activity variables such as floor area and appliance ownership are projected by end use. A cost-based approach, influenced by policy and other constraints, is used to allocate between almost 100 technologies. Energy demand projections are based on country-level historical data for both residential and non-residential buildings. The buildings module is integrated within the wider World Energy Model. | IEA (2020a); IEA (2021) |
| IMAGE 3.2 model/ Netherlands Environmental Assessment Agency | A modular integrated assessment model using a simulation model for energy demand | Global | Residential and non-residential buildings | Energy demand is calculated as a function of household expenditure and population growth, disaggregating across urban/rural and income groups. The model includes a building stock model (residential) with a detailed description of end uses, energy carrier use and building technologies for both residential and non-residential buildings. A scenario analysis assessing assumptions on lifestyle changes has also been conducted. | van Vuuren et al. (2021) |
| Resource Efficiency and Climate Change (RECC) model. Research Institutions: Norwegian University of Science & Technology and University of Freiburg. Funding Institutions: UNEP and International Resource Panel | Bottom-up building stock-flow model estimating material and energy flows associated with housing stock growth, driven by input parameters of population and floor area per capita | Global | Residential buildings | Energy demand is calculated the model BuildME, a physical model using the EnergyPlus simulation engine, incorporating country/region-specific projections of envelope and equipment efficiency. | IRP (2020); Fishman et al. (2021); Pauliuk and Heeren (2021); Pauliuk et al. (2021a); Pauliuk et al. (2021b) |
| A total of 77 bottom-up models out of which 67 were technology-rich and 10 sufficiency-focused | Bottom-up technology-rich models with detailed building and other technology stock models | Three global (all sufficiency models), six regional (regions here refer to regions including several countries), two subnational, and the rest national | Residential and/or non-residential buildings | In most cases, energy demand was modelled by multiplying units of energy consumption of technologies/product/buildings by stocks of corresponding technologies/products and/or buildings at national level. The projected stocks of buildings and/or technologies/products are modelled based on past levels. The potential is demonstrated by replacing the business-as-usual technologies and practices with demonstrated best available or commercially feasible technologies and practices. The studies rely on some or all of the following mitigation options: the construction of new high-performance buildings using building design, forms, and passive construction methods; the thermal efficiency improvement of building envelopes of the existing stock; the installation of advanced heating, ventilation air conditioning systems, equipment and appliances; the exchange of lights, appliances, and office equipment, including ICT, water heating, and cooking; active and passive demand-side management measures; as well as on-site production and use of renewable energy. Many bottom-up studies considered the measures as an integrated package due to their technological complementarity and interdependence, rather than the penetration of individual technologies applied in an incremental manner in or to these buildings. | Department of Environmental Affairs (2014); Alaidroos and Krarti (2015); de Melo and de Martino Jannuzzi (2015); Kusumadewi and Limmeechokchai (2015); Markewitz et al. (2015); Prada-Hernández et al. (2015); Csoknyai et al. (2016); Energetics (2016); Gagnon et al. (2016); Horváth et al. (2016); Nadel (2016); Oluleye et al. (2016); Timilsina et al. (2016); Trottier (2016); Virage-Energie Nord-Pas-de-Calais. (2016); Yeh et al. (2016); ADB (2017); Bashmakov (2017); Chaichaloempreecha et al. (2017); Iten et al. (2017); Khan et al. (2017); Krarti et al. (2017); Kusumadewi and Limmeechokchai (2017); Momonoki et al. (2017); négaWatt (2017); Ploss et al. (2017); Radpour et al. (2017); Streicher et al. (2017); Subramanyam et al. (2017a,b); Wakiyama and Kuramochi (2017); Wilson et al. (2017); de la Rue du Can et al. (2018); Grubler et al. (2018); Novikova et al. (2018a,b); Oluleye et al. (2018); Ostermeyer et al.(2018a,b,c); Tan et al. (2018); Toleikyte et al. (2018); Yu et al. (2018); Zhou et al. (2018); Bierwirth and Thomas (2019); Bürger et al. (2019); Cabrera Serrenho et al. (2019); Colenbrander et al. (2019); de la Rue du Can et al. (2019); Dioha et al. (2019); Duscha et al. (2019); González-Mahecha et al. (2019); Kamal et al. (2019); Krarti (2019); Kwag et al. (2019); Levesque et al. (2019); Minami et al. (2019); Onyenokporo and Ochedi (2019); Ostermeyer et al. (2019b); Butler et al. (2020); Filippi Oberegger et al. (2020); Grande-Acosta and Islas-Samperio (2020); Merini et al. (2020); Millward-Hopkins et al. (2020); Roca-Puigròs et al. (2020); Rosas-Flores and Rosas-Flores (2020); Roscini et al. (2020); Sugiyama et al. (2020b); Brugger et al. (2021); Calise et al. (2021); Sandberg et al. (2021); Xing et al. (2021); Zhang et al. (2020) |

i. **Optimisation models:** Identify least cost pathways to meet policy targets (such as $CO_2$ emission targets of transport modes or economy-wide) given constraints (such as rate of adoption of vehicle technologies or vehicle efficiency standards). For example MessageIX-TransportV5 (Krey et al. 2016) and TIMES (Daly et al. 2014).
ii. **Simulation models:** Simulate behaviour of consumers and producers given prices, policies, and other factors by using parameters calibrated to historically observed behaviours such as demand price elasticity and consumer preferences. For example models by Barter et al. (2015), Brooker et al. (2015) and Schäfer (2017).
iii. **Accounting and exploratory models:** Track the outcomes (such as resources use and emissions) of key decisions (such as the adoption of advanced fuels or vehicle technologies) that are based on 'what-if' scenarios. The major difference between accounting models versus optimisation and simulation models is that key decision variables such as new technologies adoptions typically follow modellers' assumptions as opposed to being determined by mathematical formulations as in optimisation and simulation models. See models in Fulton et al. (2009), IEA (2020a), Gota et al. (2019) and Khalili et al. (2019).

Due to the model types' relative strengths and weaknesses, they are commonly applied to certain problem types (Table 2). Models can do **forecasting,** which makes projections of how futures may evolve, or **backcasting**, which makes projections of a future that meets a predefined goal such as a policy target of 80% reduction in GHG emissions from a historical level by a certain year. Models are often also used to explore 'what-if' questions, to confirm the **feasibility** of certain assumptions/outcomes, and to quantify the **impacts** of a change such as a policy under different conditions. Enhancing fuel efficiency standards, banning internal combustion engines, setting fuel quality standards, and the impacts of new technologies are typical examples of problem types analysed in energy system models.

While these four model types drive the component dynamics in different ways, they commonly include modules that include: learning and diffusion (via exogenous, e.g., autonomous learning, or endogenous learning regarding costs and efficiency: i.e., cost decreases and/or efficiency increases as a function of adoption, and increased diffusion due to lower costs) (Jochem et al. 2018), stock turnover (the performance and characteristics of vehicle fleets including survival ages, mileages, fuel economies and loads/occupancy rates are tracked for each new sales/vehicle stocks), consumer choice (theories of how people invest in new technology and utilise different modes of transport based on their individual preferences given the characteristics of mode or technology) (Daly et al. 2014; Schäfer 2017), or other feedback loops (Linton et al. 2015).

IAMs (Krey et al. 2016; Edelenbosch et al. 2017a) are typically global in scope and seek to solve for feasible pathways meeting a global temperature target (Annex III.I.9). This implies finding least-cost mitigation options within and across sectors. In contrast, global and national transport energy system models (GTEMs/NTEMs) typically only assess feasible pathways within the transport sector (Yeh et al. 2017). The range of feasible pathways can be determined through optimisation, simulation, accounting and exploratory methods, as outlined in Table 2. Some GTEMs are linked to an IAMs model (Krey et al. 2016; Edelenbosch et al. 2017a; Roelfsema et al. 2020). The key difference between IAMs and GTEMs or NTEMs is whether the transportation system is integrated with the rest of the energy systems, specifically regarding energy and fuel production and use, fuel prices, economic drivers such as GDP, and mitigation options given a policy goal. IAMs can endogenously determine these factors because the transport sector is just one of many sectors captured by the IAM. While this gives IAMs certain advantages, IAMs sacrifice resolution and complexity for this broader scope. For example, most IAMs lack a sophisticated travel demand model that reflects the heterogeneity of demands and consumer preferences, whereas GTEM/NTEMs can incorporate greater levels of details regarding travel demands, consumer choices, and the details of transport policies. Consequently, what GTEMs/NTEMs lack in integration with other sectors they make up through more detailed analyses of travel patterns, policies, and impacts (Yeh et al. 2017).

Several noteworthy recent active research areas in long-term transportation energy systems modelling involves the consideration of infrastructure investment and consumer acceptance for non-fossil fuel vehicles including charging for electric vehicles (Jochem et al. 2019; Statharas et al. 2021) and refuelling stations for hydrogen vehicles (Rose and Neumann 2020); and the greater integration of the electric, transport, residential, and industrial sectors in fuel production, storage, and utilisation (Bellocchi et al. 2020; Lester et al. 2020; Olovsson et al. 2021; Rottoli et al. 2021). While national and regional transport energy models have the advantage of exploring these relationships in greater spatial, temporal, and policy details for specific countries or regions (Jochem et al. 2019; Bellocchi et al. 2020; Lester et al. 2020; Rottoli et al. 2021; Statharas et al. 2021), the IAMs have the advantage of examining these interactions across the entire economy at the global level (Brear et al. 2020; Rottoli et al. 2021).

### A.III.I.5.2 Inventory of Transportation Models included in AR6

AIII

**Table 2 | Taxonomy of transport models by method (modelling type) and application (problem type).**

| Problem Type | Optimisation model | Simulation model | Accounting model | Heuristic model |
|---|---|---|---|---|
| Backcasting | ● | | | ● |
| Forecasting | ● | ● | ● | |
| Exploring feasibility space | | ● | ● | ● |
| Impact analysis | ● | ● | ● | |

**Table 3 | GTEM/NTEMs models evaluated in Chapter 10.**

| Model name | Organisation | Scope | Resolution | Period | Economy-wide | Method |
|---|---|---|---|---|---|---|
| Mobility model (MoMo) | International Energy Agency (IEA) | Global | Country groups | 2050 | Soft link | Accounting model |
| Global Transportation Roadmap | International Council on Clean Transportation (ICCT) | Global | Country groups | 2050 | No | Accounting model |
| MESSAGE-Transport V.5 | International Institute for Applied Systems Analysis (IIASA) | Global | Country groups | 2100 | Yes | Optimisation model |
| Global Change Assessment Model (GCAM) | Pacific Northwest National Laboratory (PNNL) | Global | Country groups | 2100 | Yes | Partial equilibrium model |

The global/national transport energy system models included in the transportation chapter (Chapter 10) are listed below in Table 3.

## A.III.I.6 Industry Sector Models

### A.III.I.6.1 Types of Industry Sector Models

Industry sector modelling approaches can vary considerably from one another. As with other types of models, a key characteristic of industry sector models is related to their geographical scope. While IAMs are often global in scope, many bottom-up sector models are limited to individual countries or regions. The models' system boundaries also differ, with some models fully considering the use of energy for feedstock purposes and other models focusing only on the use of energy for energetic purposes. Differences between models also exist in regard to differentiation between the industry sector and the energy transformation sector, concerning, for example, refineries and industrial power plants.

### A.III.I.6.2 Representation of Demand for Industrial Products

Industry sector models vary in regard to their representation of demand for industrial goods or products. A more detailed representation of demand in a model allows for a more explicit discussion of different types of drivers of industrial demand and therefore a more detailed representation of demand-side strategies such as material recycling, longer use of products or sharing of products.

Particularly, demand for industrial products is often considered in more detail in bottom-up models of the industry sector than in top-down models by taking more drivers into account. These drivers can be, inter alia, population, gross value added, construction activity, transport activity, but also changes in material efficiency, recycling rates and scrap rates as well as product use efficiency (e.g., through longer use of products or sharing of products) (Fleiter et al. 2018; Material Economics 2019; IEA 2020b).

### A.III.I.6.3 Representation of Mitigation Options

In most top-down IAMs, some energy-intensive sectors, such as iron and steel or cement, are included separately, at least in a generalised manner, but typically few if any sector-specific technologies are explicitly represented. Instead, energy efficiency improvements in the industry sector and its subsectors are often either determined by exogenous assumptions or are a function of energy prices. Likewise, fuel switching occurs primarily as a result of changes in relative fuel prices, which in turn are influenced by $CO_2$ price developments. In IAMs that include specific technologies, fuel switching can be constrained based on the characteristics of those technologies, while in IAMs with no technological detail, more generic

**Table 4 | Models underlying specific assessments in Chapter 11.**

| Model name and institution using the model | Model description | Geo-graphic scope | Industrial sectors included/ distinguished | Demand for industrial products | Examples of publications |
|---|---|---|---|---|---|
| Industry sector model of the Energy Technology Perspectives model (IEA) | The bottom-up industry sector model is one of four soft-linked models making up the ETP model. The four models are an energy supply optimisation model and three end-use sector models (transport, industry, buildings). Technologies and fuels in the industry sector model are chosen based on cost optimisation. | Global | Aluminium, iron and steel, chemical and petrochemical, cement, pulp and paper and other industry sectors | Demand for industrial products is derived based on country-level historical data on per capita consumption. This per capita consumption is projected forward by using population projections and industry value-added projections. Demand for materials is derived by also taking the build-up of material stocks into account. | IEA (2020b, 2021) |
| World Energy Model (IEA) | Simulation model consisting inter alia of technologically detailed bottom-up representations of several industry sectors | Global | See ETP model | See ETP model | IEA (2020a, 2021) |
| Material Economics modelling framework | Modelling tool consisting of several separate bottom-up models | European Union | Steel, chemicals (plastics and ammonia), cement | Demand for industrial products is derived based on scenarios of future activity levels in key segments such as construction, mobility and food production. Separate models additionally explore opportunities for improving materials efficiency and increasing materials circulation. | Material Economics (2019) |

constraints on fuel switching in the industry sector are embedded (Edelenbosch et al. 2017b).

In bottom-up models, individual technological mitigation options are represented in detail, especially for energy-intensive sectors such as iron and steel, cement and chemicals. Typically, for each considered technology, not only specific energy demand but also investment and operating costs are included in these models. Investment costs can change over time, either based on an exogenous assumption or on an endogenised process such as a learning rate. While bottom-up models often consider technology-specific learning, IAMs cover technological progress in a more general way associated to industry branches. The uptake of new technologies is typically restricted in bottom-up models, for example by assuming a minimum lifetime for existing stock or by assuming S-shaped diffusion curves (Fleiter et al. 2018). The industrial sector models included in the industry chapter (Chapter 11) are listed in Table 4.

#### A.III.I.6.4 Limitations and Critical Analysis

Aggregated, top-down models of the industry sector, as used in most IAMs, are typically calibrated based on long-term historical data, for example on the diffusion of new technologies or on new fuels. These models are therefore able to implicitly consider real-life restrictions of the whole sector that bottom-up models (with their focus on individual technologies) may not fully take into account. These restrictions may arise, inter alia, from delays in the construction of infrastructure or market actors possessing incomplete information about new technologies. Furthermore, as IAMs also model the climate system, these models can principally take into account potential repercussions of climate change impacts on the growth rate and structure of economies.

However, a downside of top-down models is that they are typically limited in their representation of individual technologies and processes in the industry sector and particularly of technology-driven structural change. This lack of technological detail limits the usefulness of these models to analyse technology-specific and sector-specific mitigation measures and related policies. Top-down models also tend to have a relatively aggregated representation of industrial energy demand, meaning demand-side mitigations strategies such as recycling, product-service efficiency and demand reduction options are difficult to assess with these models (Pauliuk et al. 2017).

In contrast, technology-rich bottom-up models allow detailed analysis of the potential of new technologies, processes and fuels in individual industrial sectors to reduce GHG emissions. Their often-detailed analysis of the demand side allows demand-side mitigation strategies to be evaluated. Furthermore, radical future changes in technology, climate policy or social norms can more easily be reflected in bottom-up models than in top-down models which are calibrated on past observations. Both types of models are typically not able to account for product substitution (e.g., steel vs plastics) arising from changing production cost differentials or changing product quality due to new production processes. In principle, technology-rich input-output models could fill this gap.

### A.III.I.7 Land-use Modelling

Land use related IAM modelling results as presented in Chapter 7 are based on comprehensive land-use models (LUMs) that are either integrated directly, or through emulators, into the integrated assessment framework. Given the increasing awareness of the importance of the land use sector to achieve ambitious climate mitigation targets, LUMs and their integration into IAM systems was one of the key innovations to the integrated assessment over the past decade to allow for an economy-wide quantification of climate stabilisation pathways.

LUMs allow developments in the land-use sector to be projected over time and the impacts of mitigation policies on different economic (markets, trade, prices, demand, supply, etc.) and environmental (land use, emissions, fertiliser, irrigation water use, etc.) indicators to be assessed. The following models submitted scenarios to the AR6 database: AIM (Fujimori et al. 2014, 2017; Hasegawa et al. 2017), EPPA (Chen et al. 2016), GCAM (Calvin et al. 2019), IMAGE (Stehfest et al. 2014), MERGE, MESSAGE-GLOBIOM (Havlík et al. 2014; Fricko et al. 2017; Huppmann et al. 2019), POLES (Keramidas et al. 2017), REMIND-MAgPIE (Kriegler et al. 2017; Dietrich et al. 2019), WITCH (Emmerling et al. 2016).



#### A.III.I.7.1 Modelling of Land Use and Land-use Change

LUMs represent different land use activities for managed land (agriculture including cropland and pastures, managed forests, and dedicated energy crops) while natural lands (primary forests, natural grasslands, shrubland, savannahs, etc.) act as land reserves that can be converted to management depending on other constraints (Popp et al. 2014a; Schmitz et al. 2014). Typically, the agricultural sector has the greatest level of detail across land use sectors. LUMs include different crop and livestock production activities, some even at the spatially explicit level and differentiated by production system (Havlík et al. 2014; Weindl et al. 2015). Forestry is covered with varying degrees of complexity across LUMs. While some models represent only afforestation/deforestation activities dynamically, others have detailed representation of forest management activities and/or forest industries (Lauri et al. 2017). The models endogenously determine the land allocation of different land use activities as well as land-use changes according to different economic principles (land rent, substitution elasticities, etc.) and/or considering biophysical characteristics such as land suitability (Schmitz et al. 2014; Weindl et al. 2017).

#### A.III.I.7.2 Demand for Food, Feed, Fibre and Agricultural Trade

LUMs project demand for food, feed, other industrial or energy uses for different agriculture and forestry commodities over time. While partial equilibrium models typically use reduced-form demand functions with greater level of detail at the commodity level, however limited agriculture and forestry, Computable General Equilibrium (CGE) models represent demand starting from utility functions from which it is possible to derive demand functions, and functional forms for income and price elasticities, however for a more limited set of agricultural and forestry commodities but with full coverage of all economic sectors (Valin et al. 2014; von Lampe et al. 2014). Over time, demand for food, feed, and other industrial uses is projected conditional on population and income growth while bioenergy demand is typically informed in partial equilibrium (PE) models by linking with IAMs/energy systems models, and is usually endogenous in CGE/IAMs (Hasegawa et al. 2020). Depending on the model, demand projections are sensitive to price changes (Valin et al. 2014). International trade is often represented in LUMs using either Armington or spatial equilibrium approaches (von Lampe et al. 2014).

### A.III.I.7.3 Treatment of Land-based Mitigation Options

Two broad categories of land-based mitigation options are represented in LUMs: (i) reduction of GHG ($CO_2$, $CH_4$ and $N_2O$) emissions from land use, and (ii) carbon sink enhancement options including biomass supply for bioenergy. Each of these categories is underpinned by a portfolio of mitigation options with varying degrees of complexity and parameterisation across LUMs. The representation of mitigation measures is influenced on the one hand by the availability of data for its techno-economic characteristics and future prospects as well as the computational challenge, for example in terms of spatial and process detail, to represent the measure, and on the other hand, by structural differences and general focus of the different LUMs, and prioritisation of different mitigation options by the modelling teams. While GHG emission reduction and $CO_2$ sequestration options such as afforestation are typically covered directly in LUMs (Hasegawa et al. 2021), carbon sequestration from biomass supplied for bioenergy with carbon capture and storage (BECCS) is usually not accounted for in LUMs but in the energy sector and hence is taken care of directly in the IAMs. Yet, LUMs provide estimates of available biomass for energy production and the impacts of its production.

AIII

#### *A.III.I.7.3.1 Treatment of GHG Emissions Reduction*

Agricultural non-$CO_2$ emissions covered in LUMs include $CH_4$ from enteric fermentation, manure management and cultivation of rice paddies, and $N_2O$ emissions from soils (fertiliser and manure application, crop residues) and manure management and are based on IPCC accounting guidelines (IPCC 2019a). For each of those sources, LUMs typically represent a (sub)set of technical, structural and demand-side mitigation options. Technical options refer to technologies such as anaerobic digesters, feed supplements or nitrogen inhibitors that are either explicitly represented (Frank et al. 2018) or implicitly via the use of marginal abatement cost curves (MACC) (Lucas et al. 2007; Beach et al. 2015; Harmsen et al. 2019). Emission savings from structural changes refer to more fundamental changes in the agricultural sector, for example through international trade, production system changes or reallocation and substitution effects (Havlík et al. 2014). Demand-side options include dietary changes and reduction of food waste (Springmann et al. 2016; Creutzig et al. 2018; Ritchie et al. 2018; Frank et al. 2019; Mbow et al. 2019; Clark et al. 2020; Ivanova et al. 2020; Popp et al. 2010; Rosenzweig et al. 2020). For the forest sector, emission reduction options are mainly targeting $CO_2$ from deforestation (Overmars et al. 2014; Hasegawa et al. 2017; Rochedo et al. 2018; Bos et al. 2020; Doelman et al. 2020; Eriksson 2020). Mitigation/restoration options for wetlands to reduce emissions from drained organic soils are typically not represented in LUMs (Humpenöder et al. 2020).

There are significant differences between UNFCCC nationally reported GHG inventories and analytical global land use models. According to Grassi et al. (2017), this discrepancy results in a 3Gt$CO_2$-eq difference in estimates between country reports and global models. The difference relies on different methods to classify and assess managed forests and forest management fluxes (Houghton et al. 2012; Pongratz et al. 2014; Smith et al. 2014; Tubiello et al. 2015; Grassi et al. 2017, 2021). While global models account for GHG emissions from indirect human-induced effects and natural effects in unmanaged land, countries only consider fluxes of land use and land-use change in managed land. In order to produce policy-relevant land-use model exercises, reconciling these differences is needed by harmonising definitions and approaches of anthropogenic land and the treatment of indirect environmental change (Grassi et al. 2017).

#### *A.III.I.7.3.2 Treatment of Terrestrial Carbon Dioxide Removal Options including Biomass Supply for Bioenergy*

Terrestrial carbon dioxide removal options are only partially included in LUMs and mostly rely on afforestation and bioenergy with carbon capture and storage (BECCS) (Fuss et al. 2014, 2018; Minx et al. 2018; Smith et al. 2019; Butnar et al. 2020). Especially some nature-based solutions (Griscom et al. 2017) such as soil carbon management (Paustian et al. 2016), which have the potential to alter the contribution of land-based mitigation in terms of timing, potential and sustainability consequences, are only recently being implemented in LUMs (Frank et al. 2017; Humpenöder et al. 2020). The representation of bioenergy feedstocks varies across models but typically LUMs have comprehensive representation of a series of crops (starch, sugar, oil, wood/lignocellulosic feedstocks) or residues/byproducts that can be used for liquid and solid bioenergy production (Hanssen et al. 2019).

### A.III.I.7.4 Treatment of Environmental and Socio-economic Impacts of Land Use

Aside reporting the implications on agriculture, forestry and other land use (AFOLU) GHG emissions, LUMs can provide a set of environmental and socio-economic impact indicators to assess the quantified climate stabilisation pathways in a broader sustainable development agenda (van Vuuren et al. 2015; Obersteiner et al. 2016; van Vuuren et al. 2019; Frank et al. 2021; Soergel et al. 2021). These indicators typically span from land use area developments (Popp et al. 2017; Stehfest et al. 2019), fertiliser use, irrigation water use and environmental flows (Bonsch et al. 2015; Pastor et al. 2019; Chang et al. 2021; de Vos et al. 2021), and on biodiversity (Leclère et al. 2020; Marquardt et al. 2021), to market impacts on commodity prices and food consumption, or impact on undernourishment (Hasegawa et al. 2018; Doelman et al. 2019; Fujimori et al. 2019a; Hasegawa et al. 2020; Soergel et al. 2021).

## A.III.I.8 Reduced Complexity Climate Modelling

Climate model emulators (often referred to as reduced complexity or simple climate models) are used to integrate the WGI knowledge of physical climate science into the WGIII assessment. Hence, emulators are used to assess the climate implications of the GHG and other emissions trajectories that IAMs produce (van Vuuren et al. 2008; Rogelj et al. 2011; Clarke et al. 2014; Schaeffer et al. 2015; Rogelj et al. 2018a). The IAM literature typically uses one of two approaches: comprehensive emulators such as MAGICC (Meinshausen et al. 2011) or Hector (Hartin et al. 2015) or minimal complexity representations such as the representation used in DICE (Nordhaus 2018), PAGE (Yumashev et al. 2019; Kikstra et al. 2021c) and Fund (Waldhoff et al. 2014). In physical science research, a wider range of different emulators are used (Nicholls et al. 2020b, 2021a).

A key application of emulators within IPCC WGIII is the classification of emission scenarios with respect to their global mean temperature outcomes (Clarke et al. 2014; Rogelj et al. 2018a). WGIII relies on emulators to assess the full range of carbon-cycle and climate response uncertainty of thousands of scenarios, as assessed by AR6 WGI. An exercise of such amplitude is currently infeasible with more computationally demanding state-of-the-art Earth system models. Cross-Chapter Box 7.1 in AR6 WGI documents how emulators used in WGIII are consistent with the physical science assessment of WGI (Forster et al. 2021).

Previous IPCC Assessment Reports relied either on the climate output from each individual IAM (IPCC 2000) or a more streamlined approach, where one consistent emulator set-up was used to assess all scenarios. For instance, in AR5 and the Special Report on Global Warming of 1.5°C (SR1.5), MAGICC was used for scenario classification (Clarke et al. 2014; Rogelj et al. 2018a). In recent years, numerous other emulators have been developed and increased confidence and understanding can thus be gained by combining insights from more than one emulator. For example, SR1.5 used MAGICC for its scenario classification, with additional insights provided by the FaIR model (Smith et al. 2018). The SR1.5 experience highlighted that the veracity of emulators 'is a substantial knowledge gap in the overall assessment of pathways and their temperature thresholds' (Rogelj et al. 2018a). Since SR1.5, international research efforts have demonstrated tractable ways to compare emulator performance (Nicholls et al. 2020b), as well as their ability to accurately represent a set of uncertainty ranges in physical parameters (Nicholls et al. 2021b), such as those reported by AR6 WGI (Forster et al. 2021).

Finally, the recently developed OpenSCM-Runner package (Nicholls et al. 2020a) provides users with the ability to run multiple emulators from a single interface. OpenSCM-Runner has been built in collaboration with the WGIII research community and forms part of the WGIII assessment (Annex III.II.2.5.1).

### A.III.I.9 Integrated Assessment Modelling

Process-based integrated assessment models (IAMs) describe the coupled energy-land-economy-climate system (Weyant 2009; Krey 2014; Weyant 2017). They typically capture all greenhouse gas (GHG) emissions induced by human activities and, in many cases, emissions of other climate forcers like sulphate aerosols. Process-based IAMs represent most GHG and climate pollutant emissions by modelling the underlying processes in energy and land use. Those models are able to endogenously describe the change in emissions due to changes in energy and land use activities, particularly in response to climate action. But IAMs differ in the extent to which all emissions and the corresponding sources, processes and activities are represented endogenously and, thus, can be subjected to policy analysis.[1] IAMs also differ regarding the scope of representing carbon removal options and their interlinkage with other vital systems such as the energy and land-use sectors.

Typically, IAMs consider multi-level systems of global, regional, national and local constraints and balance equations for different categories such as emissions, material and energy flows, financial flows, and land availability that are solved simultaneously. Intertemporal IAMs can fully incorporate not only flow constraints that are satisfied in each period, but also stock constraints that are aggregated over time and require to balance activities over time. Changes of activities, for example induced by policies to reduce emissions, are connected to a variety of balance equations and constraints and therefore such policies lead to system-wide changes that can be analysed with IAMs. Many IAMs also contain gridded components to capture, for example, land-use and climate change processes where the spatial distribution matters greatly for the dynamics of the system. Processes that operate on smaller spatial and temporal scales than resolved by IAMs, such as temporal variability of renewables, are included by parameterisation and statistical modelling approaches that capture the impact of these subscale processes on the system dynamics at the macro level (Pietzcker et al. 2017).

Global IAMs are used to analyse global emissions scenarios extrapolating current trends under a variety of assumptions and climate change action pathways under a variety of global goals. In recent years, a class of national and regional IAMs have emerged that describe the coupled energy-land-economy system in a given geography. They typically have higher sectoral, policy and technology resolution than global models and make assumptions about boundary conditions set by global markets and international policy regimes. These IAMs are used to study trends and transformation pathways for a given region (Shukla and Chaturvedi 2011; Capros et al. 2014; Lucena et al. 2016).

[1] See the common IAM documentation at www.iamcdocumentation.eu.

### A.III.I.9.1 Types of Integrated Assessment Models

IAMs include a variety of model types that can be distinguished into two broad classes (Weyant 2017). The first class comprises *cost-benefit IAMs* that fully integrate a stylised socio-economic model with a reduced-form climate model to simultaneously account for the costs of mitigation and the damages of global warming using highly aggregate cost functions derived from more detailed models. In the model context, these functions do not explicitly represent the underlying processes, but map mitigation efforts and temperature to costs. This closed-loop approach between climate and socio-economic systems enables cost-benefit analysis by balancing the cost of mitigation and the benefits of avoided climate damages. This can be done in a globally cooperative setting to derive the globally optimal climate policy where no region can further improve its welfare without reducing the welfare of another region (Pareto optimum). Alternatively, it can be assumed that nations do not engage in emission mitigation at all or mitigate in a non-cooperative way, only considering the marginal benefit of their own action (Nash equilibrium). Also, differing degrees of partial cooperation are possible.



The second class of IAMs, called *process-based IAMs*, focuses on the analysis of transformation processes depending on a broad set of activities that induce emissions as side effects. They describe the interlinkages between economic activity, energy use, land use, and emissions with emission reductions and removals as well as broader sustainable development targets. GHGs and other climate pollutants are caused by a broad range of activities that are driven by socio-economic developments (Riahi et al. 2017) and also induce broader environmental consequences such as land-use change (Popp et al. 2017) and air pollution (Rao et al. 2017b). With few exceptions, these models typically do not close the loop with climate change and damages that affect the economy, but focus on emission scenarios and climate change mitigation pathways. Due to the process-based representations of emission sources and alternatives, it is not only possible to investigate the implications of policies on GHG emissions, but also the trade-offs and synergies with social and environmental sustainability criteria (von Stechow et al. 2015) (Annex III.I.9.3). The analysis of different cross-sectoral synergies and trade-offs is frequently termed a nexus analysis, such as the energy-water-land nexus. The analysis can also address socio-economic sustainability criteria such as energy access and human health. Process-based IAMs are also used to explore the synergies and trade-offs of 'common, but differentiated responsibilities' by analysing issues of burden sharing, equity, international cooperation, policy differentiation and transfer measures (Tavoni et al. 2015; Fujimori et al. 2016; Leimbach and Giannousakis 2019; Bauer et al. 2020b).

There exists a broad range of process-based IAMs that differ regarding the economic modelling approaches (Annex III.I.2) as well as the methodology and detail of sector representation (Annex III.I.3–7) and how they are interlinked with each other.

This leads to differences in model results regarding global aggregates as well as sectoral and regional outputs. Several approaches have been used to evaluate the performance of IAMs and understand differences in IAM behaviour (Schwanitz 2013; Wilson et al. 2021), including sensitivity analysis (McJeon et al. 2011; Luderer et al. 2013; Rogelj et al. 2013a; Bosetti et al. 2015; Marangoni et al. 2017; Giannousakis et al. 2021), model comparisons (Clarke et al. 2009; Kriegler et al. 2014a, 2015a; Riahi et al. 2015; Tavoni et al. 2015; Kriegler et al. 2016; Riahi et al. 2017; Luderer et al. 2018; Roelfsema et al. 2020; Riahi et al. 2021; van Soest et al. 2021), model diagnostics (Kriegler et al. 2015a; Wilkerson et al. 2015; Harmsen et al. 2021), and comparison with historical patterns (Wilson et al. 2013; van Sluisveld et al. 2015; Napp et al. 2017).

### A.III.I.9.2 Components of Integrated Assessment Models

#### *A.III.I.9.2.1 Energy-economy Component*

Typically, IAMs comprise a model of energy flows, emissions and the associated costs (Krey 2014). The demand for exploring the Paris Agreement climate goals led to model developments to make the challenges and opportunities of the associated transformation pathways more transparent. Since AR5 much progress has been achieved to improve the representation of mitigation options in the energy supply sector (e.g., renewable energy integration (Pietzcker et al. 2017), energy trade (Bauer et al. 2016; McCollum et al. 2016; Bauer et al. 2017; Jewell et al. 2018), capacity inertia, carbon removals, and decarbonisation bottlenecks (Luderer et al. 2018)) and technological and behavioural change measures in energy demand sectors such as transport (van Sluisveld et al. 2016; Edelenbosch et al. 2017a; McCollum et al. 2017). An energy sector model can be run as a partial equilibrium model using exogenous demand drivers for final energy and energy services. These models derive mitigation policy costs in terms of additional energy sector costs and area under the marginal abatement costs curve.

Energy models can be also embedded into a broader, long-term macroeconomic context in a general equilibrium model (Bauer et al. 2008; Messner and Schrattenholzer 2000). The demands for final energy and energy services are endogenously driven by an economic growth model that also endogenises the economic allocation problem of macroeconomic resources for the energy sector that crowd out with alternatives. This allows impact analysis of climate policies on economic growth and structural change, investment financing and crowding out as well as income distribution and tax revenue recycling (Guivarch et al. 2011). Moreover, general equilibrium models also derive mitigation costs in terms of GDP losses and consumption losses, which comprise the full macroeconomic impacts rather than only the narrow energy-related costs (Paltsev and Capros 2013).

#### *A.III.I.9.2.2 Land System Component*

In recent years, substantial efforts have been devoted to improve and integrate land-use sector models in IAMs (Popp et al. 2014b, 2017). This acknowledges the importance of land-use GHG emissions of the agricultural and forestry sectors as well as the role of bioenergy, afforestation and other land-based mitigation measures. The integration is particularly important in light of the long-term climate goals of the Paris Agreement, for four reasons (IPCC 2019b). First, the GHG emissions from land-use change account for more than 10% of global GHG emissions (Kuramochi et al. 2020) and some sources of $CH_4$ and $N_2O$ constitute serious mitigation bottlenecks. Second, bioenergy is identified as a crucial primary energy source for low-emission energy supply and carbon removal (Bauer et al. 2020a; Butnar et al. 2020; Calvin et al. 2021). Third, land use-based mitigation measures such as afforestation and reduced deforestation have substantial mitigation potentials. Finally, land-cover changes alter the Earth surface albedo, which has implications for regional and global climate. Pursuing the Paris Agreement climate goals requires the inclusion of a broad set of options regarding GHG emissions and removals, which will intensify the interaction between the energy sector, the economy and the land use sector. Consequently, intersectoral policy coordination becomes more important and the land-related synergies and trade-offs with sustainable development targets will intensify (Calvin et al. 2014b; Kreidenweis et al. 2016; Frank et al. 2017; van Vuuren et al. 2017a; Humpenöder et al. 2018; Bauer et al. 2020d). IAMs used by the IPCC in the AR6 have continuously improved the integration of land-use models with energy models to explore climate mitigation scenarios under varying policy and technology conditions (Rogelj et al. 2018a; Smith et al. 2019). However, feedbacks from changes in climate variables are not included in the land-use sector models, or are only included to a limited degree.

#### *A.III.I.9.2.3 Climate System Component*

Reduced complexity climate models (often called simple climate models or emulators) are used for communicating WGI physical climate science knowledge to the research communities associated with other IPCC working groups (Annex III.I.8). They are used by IAMs to model the climate outcome of the multi-gas emissions trajectories that IAMs produce (van Vuuren et al. 2011a). A main application of such models is related to scenario classifications in WGIII (Clarke et al. 2014; Rogelj et al. 2018a). Since WGIII assesses a large number of scenarios, it must rely on the use of these simple climate models; more computationally demanding models (as used by WGI) will not be feasible to apply. For consistency across the AR6 reports, it is important that these reduced-complexity models are up to date with the latest assessments from WGI. This relies on calibrating these models so that they match, as closely as possible, the assessments made by WGI (Annex III.II.2.5.1). The calibrated models can then be used by WGIII in various parts of its assessment.

### A.III.I.9.3 Representation of Nexus Issues and Sustainable Development Impacts in Integrated Assessment Models

An energy-water-land nexus approach integrates the analysis of linked resources and infrastructure systems to provide a consistent platform for multi-sector decision-making (Howells et al. 2013). Many of the IAMs that contributed to the assessment incorporate a nexus approach that considers simultaneous constraints on land, water and energy, as well as important mutual dependencies (Fricko et al. 2017; Fujimori et al. 2017; Calvin et al. 2019; Dietrich et al. 2019; van Vuuren et al. 2019). Recently IAMs have also been integrated with life cycle assessment tools in assessing climate mitigation policies to better understand the relevance of life cycle GHG emissions in cost-optimal mitigation scenarios (Portugal-Pereira et al. 2016; Pehl et al. 2017; Arvesen et al. 2018; Tokimatsu et al. 2020). This holistic perspective ensures mitigation pathways do not exacerbate challenges for other sectors or environmental indicators. At the same time, pathways are leveraging potential synergies along the way towards achieving multiple goals.

IAMs rely on biophysical models with a relatively high degree of spatial and temporal resolution to inform coarser-scale economic models of the potentials and costs for land, water and energy systems (Johnson et al. 2019). IAMs leverage population, GDP and urbanisation projections to generate consistent water, energy and crop demand projections across multiple sectors (e.g., agriculture, livestock, domestic, manufacturing and electricity generation) (Mouratiadou et al. 2016). The highly distributed nature of decisions and impacts across sectors, particularly for land and water, has been addressed using multi-scale frameworks that embed regional and sub-regional models within global IAMs (Mosnier et al. 2014; Hejazi et al. 2015; Bijl et al. 2018; Portugal-Pereira et al. 2018). These analyses have demonstrated how local constraints and policies interact with national and international strategies aimed at reducing emissions.

Sustainable development impacts extending beyond climate outcomes have been assessed by the IAMs that contributed to the assessment, particularly in the context of the targets and indicators consistent with the Sustainable Development Goals (SDGs). The representation of individual SDGs is diverse (Figure 3), and recent model development has focused mainly on improving capabilities to assess climate change mitigation policy combined with indicators for economic growth, resource access, air pollution and land use (van Soest et al. 2019). Synergies and trade-offs across sustainable development objectives can be quantified by analysing multi-sector impacts across ensembles of IAM scenarios generated from single or multiple models (McCollum et al. 2013; Mouratiadou et al. 2016). Modules have also been developed for IAMs with the specific purpose of incorporating policies that address non-climatic sustainability outcomes (Cameron et al. 2016; Fujimori et al. 2018; Parkinson et al. 2019). Similar features have been utilised to incorporate explicit adaptation measures and targeted policies that balance mitigation goals with other sustainability criteria (Bertram et al. 2018; McCollum et al. 2018).

### A.III.I.9.4 Policy Analysis with IAMs

A key purpose of IAMs is to provide orientation knowledge for the deliberation of future climate action strategies by policymakers, civil society and the private sector. This is done by presenting different courses of actions (climate change and climate action pathways) towards a variety of long-term climate outcomes under a broad range of assumptions about future socio-economic, institutional and technological developments. The resulting climate change and climate action pathways can be analysed in terms of their outcomes towards a set of societal goals (such as the SDGs) and the resulting trade-offs between different pathways. Key trade-offs that have been investigated in the IAM literature are between (i) no, moderate, and ambitious mitigation pathways (Riahi et al. 2017), (ii) early vs delayed mitigation action (Riahi et al. 2015; Luderer et al. 2018), (iii) global action with a focus on economic efficiency equalising

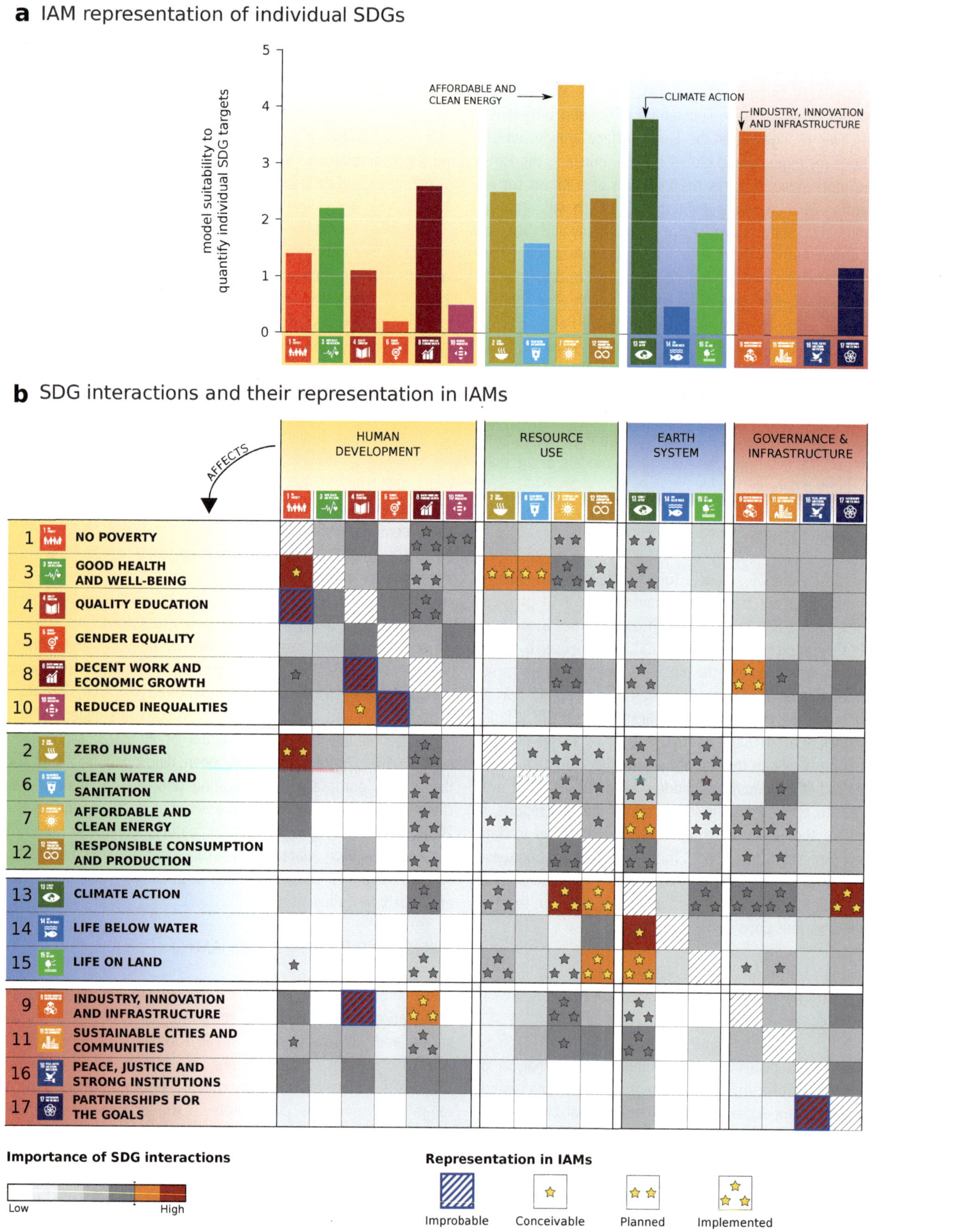


**Figure 3 | The representation of Sustainable Development Goals by Integrated Assessment Models. (a)** Individual target coverage from a multi-model survey; and **(b)** SDG interactions and coverage by IAM models according to a combination of expert and model surveys. The strength dimension of SDG interactions is indicated by grey shading: darker shades represent strong interactions while white represents no interactions. Orange cells indicate where there is the highest agreement between the importance of interactions and model representation, while blue coloured cells show the most important interactions without model representation. Source: van Soest et al. (2019).

marginal abatement costs across countries and sectors vs regionally and sectorally fragmented action (Blanford et al. 2014a; Bertram et al. 2015; Kriegler et al. 2015b, 2018b; Bauer et al. 2020b; Roelfsema et al. 2020), (iv) pathways with different emphasis on supply-side vs demand-side mitigation measures (Grubler et al. 2018; van Vuuren et al. 2018) or more broadly different sustainable development strategies (Riahi et al. 2012; van Vuuren et al. 2015; Soergel et al. 2021), and (v) pathways with different preferences about technology deployment, in particular with regard to carbon capture and storage (CCS) and carbon dioxide removal (CDR) (Krey 2014; Kriegler et al. 2014a; Riahi et al. 2015; Strefler et al. 2018; Rose et al. 2020; Luderer et al. 2021; Strefler et al. 2021b). Key uncertainties that were explored in the IAM literature are between (i) different socio-economic futures as, for example, represented by the Shared Socio-economic Pathways (SSPs) (Bauer et al. 2017; Popp et al. 2017; Riahi et al. 2017), (ii) different technological developments (Bosetti et al. 2015) and (iii) different resource potentials (Kriegler et al. 2016).

Policy analysis with IAMs follows the approach that a baseline scenario is augmented by some kind of policy intervention. To address the uncertainties in baseline projections, the scientific community has developed the Shared Socio-economic Pathways (SSPs) that provide a set of vastly different future developments as reference cases (Annex III.II.1.3,2). Most scenarios used in AR6 are based on the middle-of-the-road reference system (SSP2). Depending on the research interest, the baseline can be defined as a no-policy baseline or it can include policies that either address GHG emissions like the nationally determined contributions (NDCs) or other pre-existing policies such as energy subsidies and taxes. There is no standard definition for baseline scenarios regarding the inclusion of policies. The baseline scenario is augmented by additional policies like a carbon tax aiming towards a long-term climate goal. Hence, the IAM-based policy analysis assumes a reference system like SSP2 within which policy scenarios are compared with a baseline scenario.

Most policy analysis with process-based IAMs apply a mix of short-term policy evaluation and long-term policy optimisation. Policy evaluation applies an exogenous set of policies such as the stated NDCs and evaluates the emission outcomes. Policy optimisation is mostly implemented as a cost-effectiveness analysis: a long-term climate stabilisation target is set to derive the optimal mitigation strategy that equalises marginal abatement cost across sectors, GHGs and countries. This optimal mitigation strategy can be implemented by a broad set of well-coordinated sector-specific policies or by comprehensive carbon-pricing policies.

Most commonly, the baseline scenario is either a no-policy baseline or based on the NDCs applying an extrapolation beyond 2030 (Grant et al. 2020; Roelfsema et al. 2020). The climate policy regimes most commonly applied include a long-term target to be reached. The optimal climate strategy can be phased in gradually or applied immediately after 2020. It can focus on a global carbon price equalising marginal abatement costs across countries or policy intensities can vary across countries and sectors in the near to medium term. The climate policy regime can include or exclude effort-sharing mechanisms and transfers between regions. Also, it can be extended to include additional sector policies such as improved forest protection or fossil fuel subsidy removal. If certain technologies or activities are related to spill-overs such as technology learning, carbon pricing might be complemented by technology support (Schultes et al. 2018). If carbon-pricing policies are fragmented or delayed, additional and early sector policies can help reduce distortions and carbon leakage effects (Bauer et al. 2020b). All these variations to the policy regime can lead to very different transformation pathways and policy costs, which is a core result of the IAM analysis.

By applying sensitivity analysis, IAMs can be used to assess the importance of strategically developing new technologies and options for mitigation and identifying sticking points in climate policy frameworks. The sensitivity analysis evaluates differences in outcomes subject to changes in assumptions. For instance, the assumption about the timing and costs of CCS and CDR availability can be varied (Bauer et al. 2020a). The differences in mitigation costs and the transformation pathways support the assessment of policy prioritisation by identifying and quantifying crucial levers for achieving long-term climate mitigation targets such as R&D efforts and timing of policies.

#### A.III.I.9.5 Limitations of IAMs

The application of IAMs and their results for providing knowledge on climate change response strategies have been criticised based on four arguments (Gambhir et al. 2019; Keppo et al. 2021). First, there are concerns that IAMs are missing important dynamics, for example with regard to climate damages and economic co-benefits of mitigation (Stern 2016), demand-side responses (Wilson et al. 2012), bioenergy, land degradation and management (Creutzig et al. 2015; IPCC 2019b), carbon dioxide removal (Smith et al. 2016), rapid technological progress in the renewable energy sector (Creutzig et al. 2017), actor heterogeneity, and distributional impacts of climate change and climate policy. This has given rise to criticism that IAMs lack credibility in a set of crucial assumptions, among which stands out the critique on the availability of carbon dioxide removal technologies (Anderson and Peters 2016; Bednar et al. 2019).

These concerns spur continuous model development and improvements in scenario design (Keppo et al. 2021), particularly with regard to improved representations of energy demand, renewable energy, carbon dioxide removal technologies, and land management. IAMs are aiming to keep pace with the development of sector-specific models, including latest advances in estimating and modelling climate damages (Piontek et al. 2019). In places, where dynamic modelling approaches are lacking, scenarios are being used to explore relevant futures (Grubler et al. 2018). Moreover, sector-specific model comparison studies have brought together domain experts and modellers to improve model representations in these areas (Edelenbosch et al. 2017a; Pietzcker et al. 2017; Bauer et al. 2020a; Harmsen et al. 2020; Rose et al. 2020). Although most models are still relying on the concept of a single representative household representing entire regions, efforts are under way to better represent agent heterogeneity and distributional impacts of climate change and climate mitigation policies (Rao et al. 2017a; Peng et al. 2021).

AIII

Second, concerns have been raised that IAMs are non-transparent and thus make it difficult to grasp the context and meaning of their results (Skea et al. 2021). These concerns have facilitated a substantial increase in model documentation (see the common IAM documentation at www.iamcdocumentation.eu as an entry point) and open-source models. Nonetheless, more communication tools and co-production of knowledge formats will be needed to contextualise IAM results for users (Auer et al. 2021). When projecting over a century, uncertainties are large and cannot be ignored. Efforts have been undertaken (Marangoni et al. 2017; Gillingham et al. 2018; Harmsen et al. 2021; Wilson et al. 2021) to diagnose key similarities and differences between models and better gauge robust findings from these models and how much they depend on key assumptions (such as, for example, long-term growth of the economy, the monetary implication of climate damages, or the diffusion and cost of key mitigation technologies).

Third, there are concerns that IAMs are describing transformative change on the level of energy and land use, but are largely silent about the underlying socio-cultural transitions that could imply restructuring of society and institutions. Weyant (2017) notes the inability of IAMs to mimic extreme and discontinuous outcomes related to these underlying drivers as one of their major limitations. This is relevant when modelling extreme climate damages as well as when modelling disruptive changes. Dialogues and collaborative work between IAM researchers and social scientists have explored ways to bridge insights from the various communities to provide a more complete picture of high-impact climate change scenarios and, on the other end, deep transformation pathways (Turnheim et al. 2015; Geels et al. 2016; Trutnevyte et al. 2019). The extension of IAM research to sustainable development pathways is giving rise to further inter-disciplinary research on underlying transformations towards the Paris climate goals and other sustainable development goals (Kriegler et al. 2018c; Sachs et al. 2019b).



Finally, there are concerns that IAM analysis could focus on only a subset of relevant futures and thus push society in certain directions without sufficient scrutiny (Beck and Mahony 2017). IAMs aim to explore a wide range of socio-economic, technology and policy assumptions (Riahi et al. 2017), but it remains a constant challenge to capture all relevant perspectives (O'Neill et al. 2020). These concerns can be addressed by adopting an iterative approach between researchers and societal actors in shaping research questions and IAM applications (Edenhofer and Kowarsch 2015). IAM research is constantly taking up concerns about research gaps and fills it with new pathway research, as, for example, occurred for low energy demand and limited bioenergy with CCS scenarios (Grubler et al. 2018; van Vuuren et al. 2018).

## A.III.I.10 Key Characteristics of Models that Contributed Mitigation Scenarios to the Assessment[2]

**Table 5 | Comparison of modelling characteristics as stated by contributing modelling teams to the AR6 database.** Attributes include regional scope, sectoral coverage, type of baseline or benchmark setup as a basis for mitigation policies comparison, technology diffusion, capital vintaging and 'sunsetting' of technologies and variety of discount rates approaches.

| | | Global integrated and energy models | | | | | | | | | | | | | | | | | | | |
|---|---|---|---|---|---|---|---|---|---|---|---|---|---|---|---|---|---|---|---|---|---|
| | | AIM | C3IAM 2.0 | COFFEE 1.1 | EPPA 6 | IMAGE 3.0 & 3.2 | IMACLIM | GCAM | GENeSYS-MOD | GMM (Global MARKAL Model) | McKinsey 1.0 | MERGE-ETL | MESSAGEix-GLOBIOM 1.1 | MUSE 1.0 | POLES | PROMETHEUS | TIAM-ECN 1.1 | REmap GRO2020 | REMIND 2.1 - MAgPIE 4.2 | WEM (World Energy Model) | WITCH |
| Regional scope | Global | ● | ● | ● | ● | ● | ● | ● | ● | ● | ● | ● | ● | ● | ● | ● | ● | ● | ● | ● | ● |
| | National | ● | | | | | | | | | | | | | | | | | | | |
| | Non-global multi-region | | | | | | | | | | | | | | | | | | | | |
| Sectoral coverage | Full system (covering all GHGs from all sectors) | ● | ● | ● | ● | ● | ● | ● | | | ● | ● | ● | | ● | | ● | | ● | | ● |
| | Energy | | | | | | | | ● | ● | | | | ● | | ● | | ● | | ● | |
| | Buildings | | | | | | | | ● | ● | | | | | | ● | | ● | | ● | |
| | Transport | | | | | | | | ● | ● | | | | | | ● | | ● | | ● | |
| | Industry | | | | | | | | ● | ● | | | | | | ● | | ● | | ● | |
| Characteristics of baseline/ benchmark setup | Well-functioning markets in equilibrium | ● | ● | ● | ● | ● | | ● | | ● | ● | ● | ● | ● | ● | | ● | ● | ● | ● | ● |
| | Regulatory and/or pricing policies | ● | ● | ● | ● | ● | ● | | | ● | ● | ● | ● | | ● | ● | ● | ● | ● | ● | ● |
| | Socioeconomic costs & benefits of climate change impacts | | ● | | | | | | | | | ● | | | | | | ● | ● | | |
| | Physical impacts of climate change on key processes | ● | ● | | | | | | | | | | | | | | | ● | | | |

| | | National integrated models | | | | | | | | | | | |
|---|---|---|---|---|---|---|---|---|---|---|---|---|---|
| | | 7see6-20_GB | AIM/Enduse-Japan | BLUES 2.0 | China DREAM | CONTO-RUS 1.0 | E4SMA-EU-TIMES 1.0 | STEM (Swiss TIMES Energy Systems Model) | JRC-EU-TIMES | TIMES-China 2.0 | TIMES-France | TIMES_PT | TIMES-Sweden 2.0 |
| Regional scope | Global | | | | | | | | | | | | |
| | National | ● | ● | ● | ● | ● | | ● | | ● | ● | ● | ● |
| | Non-global multi-region | | | | | | ● | | ● | | | | |
| Sectoral coverage | Full system (covering all GHGs from all sectors) | | | ● | ● | ● | | | ● | ● | | | ● |
| | Energy | ● | ● | | | | ● | ● | | | ● | ● | |
| | Buildings | ● | ● | | | | ● | ● | | | | | |
| | Transport | ● | ● | | | | ● | ● | | | | | |
| | Industry | ● | ● | | | | ● | ● | | | | | |
| Characteristics of baseline/ benchmark setup | Well-functioning markets in equilibrium | ● | | ● | ● | ● | ● | ● | ● | ● | | ● | ● |
| | Regulatory and/or pricing policies | | ● | ● | ● | ● | ● | ● | ● | ● | | ● | ● |
| | Socioeconomic costs & benefits of climate change impacts | | | | ● | | | | | | | | ● |
| | Physical impacts of climate change on key processes | | | | | | | ● | | | | | |

2 The tables are limited to the integrated models that provided the information in response to a survey circulated in 2021, and therefore do not have a comprehensive coverage of all models that submitted scenarios to the AR6 scenario database.

**Global integrated and energy models**

| | | AIM | C3IAM 2.0 | COFFEE 1.1 | EPPA 6 | IMAGE 3.0 & 3.2 | IMACLIM | GCAM | GENeSYS-MOD | GMM (Global MARKAL Model) | McKinsey 1.0 | MERGE-ETL | MESSAGEix-GLOBIOM 1.1 | MUSE 1.0 | POLES | PROMETHEUS | TIAM-ECN 1.1 | REmap GRO2020 | REMIND 2.1 - MAgPIE 4.2 | WEM (World Energy Model) | WITCH |
|---|---|---|---|---|---|---|---|---|---|---|---|---|---|---|---|---|---|---|---|---|---|
| Technology diffusion | Logit substitution | ● | | | | ● | ● | ● | | | | | | | ● | ● | | | | | |
| | Constant elasticity of substitution | | ● | | | | | | | | | ● | | | | | | | | | ● |
| | Lowest marginal cost w/ expansion constraints | | | ● | ● | | | | ● | ● | ● | ● | ● | | | | ● | ● | ● | ● | ● |
| | Technology choice depends on agents' preferences | | | | | | | | | | | | | ● | | | | | | | |
| | Technologies w/o constraints or marginal cost w/ expansion constraints | | | | | | | | | | | | | | | | | | | | |
| Capital vintaging and "sunsetting" of technologies | Single capital stock with fixed lifetime and load factor, early retirement via reduction in load factor possible | | | | | | | | | | | | | | | | | | | | ● |
| | Capital vintaging with fixed lifetime and load factors, early retirement of vintages or reduction in load factors possible | | ● | | ● | | | ● | | ● | | ● | ● | | | ● | ● | ● | ● | ● | |
| | Single capital stock with fixed lifetime and load factor, without early retirement | | | | | | | | | | | | | | ● | | | | | | |
| | Mix of the above for different technologies | ● | | ● | | ● | ● | | ● | | ● | | | | | | | | | | |
| Discount rates | As a property of an intertemporal welfare function (social discount rate) | | | | | | | | | | | ● | | | | | | | ● | | ● |
| | In an objective function of an intertemporal optimization, to sum values at different times | | ● | ● | | ● | | | ● | ● | | | ● | | | | ● | | | ● | |
| | To compute lifecycle costs of investment decisions or return on investments, in functions representing agents investment choices | | | | ● | ● | ● | ● | ● | ● | | ● | | ● | ● | ● | | ● | | | |

**National integrated models**

| | | 7see6-20_GB | AIM/Enduse-Japan | BLUES 2.0 | China DREAM | CONTO-RUS 1.0 | E4SMA-EU-TIMES 1.0 | STEM (Swiss TIMES Energy Systems Model) | JRC-EU-TIMES | TIMES-China 2.0 | TIMES-France | TIMES_PT | TIMES-Sweden 2.0 |
|---|---|---|---|---|---|---|---|---|---|---|---|---|---|
| Technology diffusion | Logit substitution | ● | | | | | | | | | | | |
| | Constant elasticity of substitution | | | | | | | | | | | | |
| | Lowest marginal cost w/ expansion constraints | | ● | ● | ● | | ● | ● | ● | ● | ● | ● | |
| | Technology choice depends on agents' preferences | | | | | | | | | | | | |
| | Technologies w/o constraints or marginal cost w/ expansion constraints | | | | | | | | | | | | ● |
| Capital vintaging and "sunsetting" of technologies | Single capital stock with fixed lifetime and load factor, early retirement via reduction in load factor possible | | | | ● | | | | | | | | |
| | Capital vintaging with fixed lifetime and load factors, early retirement of vintages or reduction in load factors possible | | ● | | | | ● | ● | | | | | ● |
| | Single capital stock with fixed lifetime and load factor, without early retirement | | | | | ● | | | | | ● | | |
| | Mix of the above for different technologies | ● | | ● | | | | | ● | ● | | ● | |
| Discount rates | As a property of an intertemporal welfare function (social discount rate) | | | | | | | | ● | | | | |
| | In an objective function of an intertemporal optimization, to sum values at different times | | | ● | | | ● | ● | | ● | ● | | ● |
| | To compute lifecycle costs of investment decisions or return on investments, in functions representing agents investment choices | | ● | | | | | ● | | | | ● | |

**Table 6 | Overview of evaluated GHG emissions as stated by contributing modelling teams to the AR6 database: carbon dioxide ($CO_2$) from energy, industrial processes and land-use change, methane ($CH_4$) from fossil fuel combustion, from fugitive and process activities, and agricultural biogenic fluxes, nitrous oxide ($N_2O$), hydrofluorocarbons (HFCs), perfluorocarbons (PFCs), sulphur hexafluoride ($SF_6$), sulphur dioxide ($SO_2$), black and organic carbon, and non-methane volatile organic compounds (NMVOC).** Levels of emission factor (EF) evaluation were classified in four categories: linked to explicit technology but for average fuel, linked to the evolution of other emissions, dependent on average technology classes, and based on an average activity sector.

| Type of GHG emissions evaluation | Global integrated and energy models | | | | | | | | | | | | | | | | | | | |
|---|---|---|---|---|---|---|---|---|---|---|---|---|---|---|---|---|---|---|---|---|
| | AIM | C3IAM 2.0 | COFFEE 1.1 | EPPA 6 | IMAGE 3.0 & 3.2 | IMACLIM | GCAM | GENeSYS-MOD | GMM (Global MARKAL Model) | McKinsey 1.0 | MERGE-ETL | MESSAGEix-GLOBIOM 1.1 | MUSE 1.0 | POLES | PROMETHEUS | TIAM-ECN 1.1 | REmap GRO2020 | REMIND 2.1 - MAgPIE 4.2 | WEM (World Energy Model) | WITCH |
| $CO_2$ energy | a | a | a | a | a | a | a | a | a | a | a | a | a | a | a | a | a | a | a | a |
| $CO_2$ industrial processes | a | d | a | a | a | b | a | e | d | a | a | a | a | a | c | a | a | d | a | b |
| $CO_2$ land-use change | a | d | a | a | a | b | a | e | e | c | d | a | e | d | e | a | d | a | e | c |
| $CH_4$ fossil (combustion) | a | a | a | a | a | b | a | e | e | c | a | a | a | a | c | a | e | a | a | a |
| $CH_4$ fossil (fugitive and process) | a | d | a | a | a | b | a | e | e | a | a | a | e | a | c | a | e | c | e | d |
| $CH_4$ biogenic | a | e | a | a | a | b | a | e | e | a | d | a | e | d | b | a | e | d | e | c |
| $N_2O$ | a | d | a | a | a | b | a | e | e | a | d | a | a | d | c | a | e | a | e | c |
| HFCs | d | e | e | a | a | e | a | e | e | e | d | d | e | c | d | e | e | e | e | d |
| PFCs | d | e | e | a | a | e | a | e | e | e | d | e | e | c | d | e | e | e | e | d |
| $SF_6$ | d | e | e | a | a | e | a | e | e | e | d | d | e | c | d | e | e | e | e | d |
| $SO_2$ | a | a | e | d | a | e | a | e | e | e | d | a | e | a | e | e | a | a | e | a |
| Black carbon | a | d | e | d | a | e | a | e | e | e | e | a | e | a | e | e | e | a | e | a |
| Organic carbon | a | d | e | d | a | e | a | e | e | e | e | a | e | a | e | e | e | a | e | a |
| Non-methane volatile organic compounds (NMVOC) | a | a | e | d | a | e | a | e | e | e | d | a | e | a | e | e | a | a | e | a |

| Type of GHG emissions evaluation | National integrated models | | | | | | | | | | | |
|---|---|---|---|---|---|---|---|---|---|---|---|---|
| | 7see6-20_GB | AIM/Enduse-Japan | BLUES 2.0 | China DREAM | CONTO-RUS 1.0 | E4SMA-EU-TIMES 1.0 | STEM (Swiss TIMES Energy Systems Model) | JRC-EU-TIMES | TIMES-China 2.0 | TIMES-France | TIMES_PT | TIMES-Sweden 2.0 |
| $CO_2$ energy | a | a | a | a | a | a | a | a | a | a | a | a |
| $CO_2$ industrial processes | a | a | a | a | a | a | c | a | e | e | a | a |
| $CO_2$ land-use change | e | e | a | e | d | e | e | e | e | e | e | e |
| $CH_4$ fossil (combustion) | e | a | a | a | a | e | e | e | e | e | a | e |
| $CH_4$ fossil (fugitive and process) | e | d | a | a | c | e | e | e | e | e | a | e |
| $CH_4$ biogenic | e | a | a | e | d | e | e | e | e | e | a | e |
| $N_2O$ | e | a | a | a | a | e | e | e | e | e | a | e |
| HFCs | e | c | e | a | b | e | e | e | e | e | e | e |
| PFCs | e | c | e | a | b | e | e | e | e | e | e | e |
| $SF_6$ | e | c | e | a | b | e | e | e | e | e | e | e |
| $SO_2$ | e | e | a | e | b | e | e | e | e | e | e | a |
| Black carbon | e | e | e | a | e | e | e | e | e | e | e | a |
| Organic carbon | e | e | e | e | e | e | e | e | e | e | e | a |
| Non-methane volatile organic compounds (NMVOC) | e | e | a | e | e | e | e | e | e | e | e | a |

EF linked to explicit technology w/ or w/o fuel representation: a
Not represented: e
EF linked to evolution of other emissions: b
Average EF for technology class: c
EF for sector: d

AIII

## A.III.I.11 Comparison of Mitigation and Removal Measures Represented by Models that Contributed Mitigation Scenarios to the Assessment[3]

**Table 7 | Overview of demand- and supply-side mitigation and removal measures in the energy, transport, building, industry and AFOLU sectors, as stated by contributing modelling teams to the AR6 database.** Levels of inclusion were classified in two dimensions of explicit versus implicit and endogenous or exogenous. An explicit level suggests that the measure is directly represented in the model, while an implicit level refers to measures that are estimated indirectly by a proxy. An endogenous level reflects measures that are included in the dynamics of the model framework, whereas an exogenous level refers to measures that are not part of the model dynamics.

**Global integrated and energy models**

| Level of inclusion | AIM | C3IAM 2.0 | COFFEE 1.1 | EPPA 6 | IMAGE 3.0 & 3.2 | IMACLIM | GCAM | GENeSYS-MOD | GMM (Global MARKAL Model) | McKinsey 1.0 | MERGE-ETL | MESSAGEix-GLOBIOM 1.1 | MUSE 1.0 | POLES | PROMETHEUS | TIAM-ECN 1.1 | REmap GRO2020 | REMIND 2.1 - MAgPIE 4.2 | WEM (World Energy Model) | WITCH |
|---|---|---|---|---|---|---|---|---|---|---|---|---|---|---|---|---|---|---|---|---|
| **Demand-side measures** | | | | | | | | | | | | | | | | | | | | |
| Energy efficiency improvements in energy end uses | A | B | A | C | A | A | C | B | A | A | C | C | A | C | A | A | A | A | A | C |
| Electrification of transport demand | A | C | A | A | A | A | A | A | A | A | A | A | A | A | A | A | A | A | A | A |
| Electrification of energy demand for buildings | A | C | A | C | A | C | A | A | A | A | A | A | A | A | A | A | A | A | A | C |
| Electrification of industrial energy demand | A | C | A | C | A | C | C | A | A | A | A | A | A | C | A | A | A | A | A | C |
| CCS in industrial process applications | A | A | A | A | A | E | A | E | A | A | A | A | A | A | A | A | A | A | A | C |
| Higher share of useful energy in final energy | C | B | A | C | A | D | A | D | A | A | C | C | A | A | C | A | C | C | A | C |
| Reduced energy and service demand in industry | C | C | A | C | A | C | C | D | C | B | C | C | D | C | C | C | B | C | B | C |
| Reduced energy and service demand in buildings | C | C | D | C | A | D | C | D | C | B | C | C | D | A | A | C | B | C | B | C |
| Reduced energy and service demand in transport | C | C | A | C | A | A | A | A | D | B | D | C | D | A | C | C | B | C | B | D |
| Reduced energy and service demand in international transport | C | E | A | C | C | C | C | D | D | B | D | C | D | A | C | C | B | C | B | D |
| Reduced material demand | C | B | B | C | C | D | E | E | E | A | E | E | E | E | E | E | B | E | B | E |
| Urban form | E | E | B | E | C | D | E | D | E | E | E | E | E | E | E | E | E | E | C | E |
| Switch from traditional biomass and modern fuels | B | A | A | B | A | E | A | C | E | B | A | D | A | A | A | A | A | B | A | D |
| Dietary changes (e.g., reducing meat consumption) | B | E | B | A | B | B | A | E | E | A | E | A | E | E | E | E | E | B | E | E |
| Food processing | A | E | A | C | B | B | E | E | E | E | A | E | E | E | E | E | E | E | E | E |
| Reduction of food waste | B | E | E | E | B | E | C | E | E | B | E | B | E | E | E | E | E | B | E | E |
| Substitution of livestock-based products with plant-based products | A | E | B | A | B | D | E | E | E | B | E | E | E | E | E | E | E | B | E | E |

**National integrated models**

| Level of inclusion | 7see6-20_GB | AIM/Enduse-Japan | BLUES 2.0 | China DREAM | CONTO-RUS 1.0 | E4SMA-EU-TIMES 1.0 | STEM (Swiss TIMES Energy Systems Model) | JRC-EU-TIMES | TIMES-China 2.0 | TIMES-France | TIMES_PT | TIMES-Sweden 2.0 |
|---|---|---|---|---|---|---|---|---|---|---|---|---|
| **Demand-side measures** | | | | | | | | | | | | |
| Energy efficiency improvements in energy end uses | B | A | A | B | A | A | A | A | A | A | A | A |
| Electrification of transport demand | B | A | A | B | B | A | A | A | A | A | A | A |
| Electrification of energy demand for buildings | B | A | A | B | B | A | A | A | A | A | A | A |
| Electrification of industrial energy demand | B | A | A | B | B | A | A | A | A | A | A | A |
| CCS in industrial process applications | B | A | A | E | B | A | A | A | A | A | A | A |
| Higher share of useful energy in final energy | B | A | A | B | B | A | A | A | A | A | A | A |
| Reduced energy and service demand in industry | B | B | A | A | C | C | A | C | C | C | B | A |
| Reduced energy and service demand in buildings | B | B | A | B | D | C | A | C | C | C | B | B |
| Reduced energy and service demand in transport | B | B | A | B | D | C | A | C | C | C | B | A |
| Reduced energy and service demand in international transport | B | E | A | B | E | C | E | C | C | C | E | B |
| Reduced material demand | D | B | B | B | D | E | E | C | C | B | B | B |
| Urban form | D | B | B | B | E | E | A | E | E | E | E | D |
| Switch from traditional biomass and modern fuels | B | E | A | B | B | A | A | A | A | E | A | E |
| Dietary changes (e.g., reducing meat consumption) | E | E | B | E | E | E | E | E | E | E | E | E |
| Food processing | D | E | A | E | E | E | E | E | E | E | E | E |
| Reduction of food waste | E | E | D | E | E | E | E | E | E | E | E | E |
| Substitution of livestock-based products with plant-based products | E | E | B | E | E | E | E | E | E | E | E | E |

| | Explicit | Implicit |
|---|---|---|
| Endogenous | A | C |
| Exogenous | B | D |
| Not represented | E | |

3 The tables are limited to the integrated models that provided the information in responses to a survey circulated in 2021, and therefore do not have a comprehensive coverage of all models that submitted scenarios to the AR6 scenario database.

| Level of inclusion | Global integrated and energy models | | | | | | | | | | | | | | | | | | | |
|---|---|---|---|---|---|---|---|---|---|---|---|---|---|---|---|---|---|---|---|---|
| | AIM | C3IAM 2.0 | COFFEE 1.1 | EPPA 6 | IMAGE 3.0 & 3.2 | IMACLIM | GCAM | GENeSYS-MOD | GMM (Global MARKAL Model) | McKinsey 1.0 | MERGE-ETL | MESSAGEix-GLOBIOM 1.1 | MUSE 1.0 | POLES | PROMETHEUS | TIAM-ECN 1.1 | REmap GRO2020 | REMIND 2.1 - MAgPIE 4.2 | WEM (World Energy Model) | WITCH |
| **Supply-side measures** | | | | | | | | | | | | | | | | | | | | |
| *Decarbonisation of electricity* | | | | | | | | | | | | | | | | | | | | |
| Solar PV | A | A | A | A | A | A | A | A | A | A | A | A | A | A | A | A | A | A | A | A |
| Solar CSP | E | E | A | E | A | A | A | A | A | A | A | A | A | A | A | A | A | A | A | A |
| Hydropower | A | A | A | A | A | A | B | A | A | A | A | A | A | A | A | A | A | A | A | D |
| Nuclear energy | A | A | A | A | A | A | A | A | A | A | A | A | A | A | A | A | A | A | A | A |
| Advanced, small modular nuclear reactor designs (SMR) | E | E | E | C | C | E | E | E | A | E | A | E | E | E | C | A | D | E | C | E |
| Fuel cells (hydrogen) | E | A | A | A | A | E | A | A | A | A | A | A | A | A | A | A | B | A | A | A |
| CCS at coal and gas-fired power plants | A | A | A | A | A | A | A | A | A | A | A | A | A | A | A | A | A | A | A | A |
| Ocean energy (incl. tidal and current energy) | E | E | E | E | C | E | E | A | A | D | A | E | E | A | E | A | A | E | A | E |
| High-temperature geothermal heat | A | E | A | E | C | E | A | A | A | D | A | A | A | A | C | A | A | A | A | E |
| Wind (on-shore and off-shore lumped together) | A | A | E | A | A | A | E | A | E | E | A | A | E | E | A | A | A | A | A | A |
| Wind (on-shore and off-shore represented individually) | E | E | A | E | A | A | A | A | A | A | A | A | A | A | A | A | A | E | A | A |
| Bio-electricity, including biomass co-firing, without CCS | A | A | A | A | A | A | A | A | A | A | A | A | A | A | A | A | A | A | A | A |
| Bio-electricity, including biomass co-firing, with CCS | A | A | A | A | A | A | A | A | A | A | A | A | A | A | A | A | A | A | A | A |
| *Decarbonisation of non-electric fuels* | | | | | | | | | | | | | | | | | | | | |
| 1st generation biofuels | A | E | A | A | A | A | A | E | A | C | A | B | A | A | A | A | A | B | A | A |
| 2nd generation biofuels (grassy/woody biomass to liquids) without CCS | A | E | A | A | A | A | A | A | A | C | A | A | A | A | C | A | A | A | A | A |
| 2nd generation biofuels (grassy/woody biomass to liquids) with CCS | A | E | A | A | A | A | A | A | A | C | A | A | A | A | C | A | E | A | A | A |
| Solar and geothermal heating | A | E | A | E | C | E | E | A | A | C | A | A | E | A | C | A | A | A | A | E |
| Nuclear process heat | E | E | E | E | C | E | E | E | A | E | A | E | E | E | C | E | E | E | A | E |
| Hydrogen from fossil fuels with CCS | E | E | A | A | A | E | C | A | A | A | A | A | A | A | A | A | A | A | A | A |
| Hydrogen from electrolysis | E | E | A | A | A | E | A | A | A | A | A | A | A | A | A | A | A | A | A | A |
| Hydrogen from biomass without CCS | E | E | A | A | A | A | A | E | A | D | A | A | A | A | A | A | A | A | A | E |

| Level of inclusion | National integrated models | | | | | | | | | | | |
|---|---|---|---|---|---|---|---|---|---|---|---|---|
| | 7see6-20_GB | AIM/Enduse-Japan | BLUES 2.0 | China DREAM | CONTO-RUS 1.0 | E4SMA-EU-TIMES 1.0 | STEM (Swiss TIMES Energy Systems Model) | JRC-EU-TIMES | TIMES-China 2.0 | TIMES-France | TIMES_PT | TIMES-Sweden 2.0 |
| **Supply-side measures** | | | | | | | | | | | | |
| *Decarbonisation of electricity* | | | | | | | | | | | | |
| Solar PV | B | A | A | A | A | A | A | A | A | A | A | A |
| Solar CSP | B | E | A | A | E | A | A | A | A | A | A | E |
| Hydropower | B | A | A | A | A | A | A | A | A | A | A | A |
| Nuclear energy | B | A | A | A | A | A | A | A | A | A | A | A |
| Advanced, small modular nuclear reactor designs (SMR) | B | E | E | E | E | A | A | E | E | E | E | E |
| Fuel cells (hydrogen) | B | A | A | B | E | A | A | A | A | A | A | A |
| CCS at coal and gas-fired power plants | B | A | A | E | B | A | A | A | A | A | A | A |
| Ocean energy (incl. tidal and current energy) | B | A | E | E | E | A | A | A | A | A | A | E |
| High-temperature geothermal heat | B | A | E | E | B | A | A | A | A | A | A | E |
| Wind (on-shore and off-shore lumped together) | B | E | E | A | A | A | A | E | E | E | E | E |
| Wind (on-shore and off-shore represented individually) | B | A | A | E | A | A | E | A | A | A | A | A |
| Bio-electricity, including biomass co-firing, without CCS | B | A | A | A | B | A | A | A | A | A | A | A |
| Bio-electricity, including biomass co-firing, with CCS | B | A | A | E | E | A | A | A | A | A | A | A |
| *Decarbonisation of non-electric fuels* | | | | | | | | | | | | |
| 1st generation biofuels | B | A | A | B | E | A | A | A | A | A | A | A |
| 2nd generation biofuels (grassy/woody biomass to liquids) without CCS | B | A | A | E | E | A | A | A | A | A | A | A |
| 2nd generation biofuels (grassy/woody biomass to liquids) with CCS | B | A | A | E | E | A | A | A | A | A | A | A |
| Solar and geothermal heating | B | A | A | B | B | A | A | A | A | A | A | A |
| Nuclear process heat | B | E | E | E | B | A | A | A | E | E | E | E |
| Hydrogen from fossil fuels with CCS | B | A | A | E | E | A | A | A | A | A | A | A |
| Hydrogen from electrolysis | B | A | A | B | E | A | A | A | A | A | A | A |
| Hydrogen from biomass without CCS | B | A | A | E | E | A | A | A | E | A | A | A |

| | Explicit | Implicit |
|---|---|---|
| Endogenous | A | C |
| Exogenous | B | D |
| Not represented | E | |

AIII

| Level of inclusion | AIM | C3IAM 2.0 | COFFEE 1.1 | EPPA 6 | IMAGE 3.0 & 3.2 | IMACLIM | GCAM | GENeSYS-MOD | GMM (Global MARKAL Model) | McKinsey 1.0 | MERGE-ETL | MESSAGEix-GLOBIOM 1.1 | MUSE 1.0 | POLES | PROMETHEUS | TIAM-ECN 1.1 | REmap GRO2020 | REMIND 2.1 - MAgPIE 4.2 | WEM (World Energy Model) | WITCH |
|---|---|---|---|---|---|---|---|---|---|---|---|---|---|---|---|---|---|---|---|---|
| | **Global integrated and energy models** | | | | | | | | | | | | | | | | | | | |
| Hydrogen from biomass with CCS | E | E | A | A | A | E | A | E | A | D | A | A | A | A | A | A | A | A | A | E |
| Algae biofuels without CCS | E | E | E | E | E | E | E | E | E | E | E | E | A | E | E | E | E | E | C | E |
| Algae biofuels with CCS | E | E | E | E | E | E | E | E | E | E | E | E | E | E | E | E | E | E | C | E |
| Power-to-gas, methanisation, synthetic fuels, fed with fossil $CO_2$ | E | E | A | A | C | E | E | A | A | E | A | E | E | A | A | A | A | E | A | E |
| Power-to-gas, methanisation, syn-fuels, fed with biogenic or atmospheric $CO_2$ | E | E | A | E | C | E | E | A | A | E | A | E | E | A | A | A | A | E | A | E |
| Fuel switching and replacing fossil fuels by electricity in end-use sectors | C | A | A | A | A | A | A | A | A | A | A | A | A | A | A | A | A | A | A | A |
| *Other processes* | | | | | | | | | | | | | | | | | | | | |
| Substitution of halocarbons for refrigerants and insulation | E | E | E | E | A | E | C | E | E | E | E | D | E | C | C | E | E | E | E | C |
| Reduced gas flaring and leakage in extractive industries | C | E | A | B | C | E | C | E | E | A | E | D | E | A | E | B | E | C | A | C |
| Electrical transmission efficiency improvements, including smart grids | E | E | A | C | C | E | E | D | C | E | D | E | E | E | C | B | A | E | A | C |
| Grid integration of intermittent renewables | C | E | A | C | A | C | A | C | C | E | C | A | A | A | C | A | A | A | A | A |
| Electricity storage | C | D | A | A | A | E | A | A | C | A | C | A | A | A | A | A | A | A | A | A |
| **AFOLU measures** | | | | | | | | | | | | | | | | | | | | |
| Reduced deforestation, forest protection, avoided forest conversion | A | D | A | A | A | B | A | E | E | A | E | A | E | C | E | B | D | A | E | C |
| Methane reductions in rice paddies | A | E | A | C | A | C | C | E | E | A | E | A | E | C | E | B | E | C | E | C |
| Livestock and grazing management | A | E | A | C | A | A | C | E | E | A | E | A | E | C | E | B | E | C | E | C |
| Increasing agricultural productivity | A | C | A | C | A | A | A | E | E | A | E | A | A | C | E | D | D | C | E | E |
| Nitrogen pollution reductions | A | E | B | C | A | A | A | E | E | A | E | A | E | C | E | D | E | C | E | E |
| Changing agricultural practices enhancing soil carbon | E | E | E | C | A | E | E | E | E | A | E | E | A | C | E | B | E | E | E | E |
| Agroforestry and silviculture | E | C | A | E | D | E | E | E | E | B | E | E | E | E | E | B | E | E | E | E |
| Land-use planning | E | D | A | E | B | E | E | E | E | E | E | E | E | E | E | E | E | E | E | E |
| Urban and peri-urban agriculture and forestry | E | E | E | E | D | E | E | E | E | E | E | E | E | E | E | E | E | E | E | E |

| Level of inclusion | 7see6-20_GB | AIM/Enduse-Japan | BLUES 2.0 | China DREAM | CONTO-RUS 1.0 | E4SMA-EU-TIMES 1.0 | STEM (Swiss TIMES Energy Systems Model) | JRC-EU-TIMES | TIMES-China 2.0 | TIMES-France | TIMES_PT | TIMES-Sweden 2.0 |
|---|---|---|---|---|---|---|---|---|---|---|---|---|
| | **National integrated models** | | | | | | | | | | | |
| Hydrogen from biomass with CCS | B | A | A | E | E | A | A | A | E | A | A | A |
| Algae biofuels without CCS | B | E | E | E | E | E | E | E | E | E | E | E |
| Algae biofuels with CCS | B | E | E | E | E | E | E | E | E | E | E | E |
| Power-to-gas, methanisation, synthetic fuels, fed with fossil $CO_2$ | B | A | A | E | E | A | A | A | E | A | A | A |
| Power-to-gas, methanisation, syn-fuels, fed with biogenic or atmospheric $CO_2$ | B | A | A | E | E | A | A | A | E | A | A | A |
| Fuel switching and replacing fossil fuels by electricity in end-use sectors | B | A | A | B | B | A | A | A | A | A | A | A |
| *Other processes* | | | | | | | | | | | | |
| Substitution of halocarbons for refrigerants and insulation | E | A | E | B | E | E | E | E | E | E | E | E |
| Reduced gas flaring and leakage in extractive industries | E | E | A | B | B | E | E | E | E | E | E | E |
| Electrical transmission efficiency improvements, including smart grids | D | E | A | B | B | B | C | E | A | E | B | E |
| Grid integration of intermittent renewables | D | A | A | A | E | A | A | C | E | E | A | C |
| Electricity storage | B | A | A | A | D | A | A | C | A | A | A | A |
| **AFOLU measures** | | | | | | | | | | | | |
| Reduced deforestation, forest protection, avoided forest conversion | E | E | A | E | B | E | E | E | E | E | E | E |
| Methane reductions in rice paddies | E | A | A | B | E | E | E | E | E | E | E | E |
| Livestock and grazing management | E | A | A | B | D | E | E | E | E | E | E | E |
| Increasing agricultural productivity | E | E | A | E | D | E | E | E | E | E | E | E |
| Nitrogen pollution reductions | E | A | B | B | B | E | E | E | E | E | E | E |
| Changing agricultural practices enhancing soil carbon | E | E | A | E | D | E | E | E | E | E | E | E |
| Agroforestry and silviculture | E | E | A | E | D | E | E | E | E | E | E | E |
| Land-use planning | E | E | A | E | B | E | E | E | E | E | E | E |
| Urban and peri-urban agriculture and forestry | E | E | E | E | E | E | E | E | E | E | E | E |

| | Explicit | Implicit |
|---|---|---|
| Endogenous | A | C |
| Exogenous | B | D |
| Not represented | E | |

| Level of inclusion | Global integrated and energy models | | | | | | | | | | | | | | | | | | | |
|---|---|---|---|---|---|---|---|---|---|---|---|---|---|---|---|---|---|---|---|---|
| | AIM | C3IAM 2.0 | COFFEE 1.1 | EPPA 6 | IMAGE 3.0 & 3.2 | IMACLIM | GCAM | GENeSYS-MOD | GMM (Global MARKAL Model) | McKinsey 1.0 | MERGE-ETL | MESSAGEix-GLOBIOM 1.1 | MUSE 1.0 | POLES | PROMETHEUS | TIAM-ECN 1.1 | REmap GRO2020 | REMIND 2.1 - MAgPIE 4.2 | WEM (World Energy Model) | WITCH |
| Fire management and (ecological) pest control | C | E | E | E | D | E | D | E | E | E | E | E | E | E | E | E | E | E | E | E |
| Conservation agriculture | E | E | A | E | D | E | E | E | E | E | E | A | E | E | E | D | E | E | E | E |
| Influence on land albedo of land-use change | E | E | E | E | A | E | E | E | E | E | E | E | E | E | E | E | E | D | E | E |
| Manure management | A | E | E | E | A | C | C | E | E | A | E | A | E | E | E | B | E | C | E | C |
| Reduce food post-harvest losses | B | D | E | E | D | E | D | E | E | E | E | B | E | E | E | E | E | E | E | E |
| Recovery of forestry and agricultural residues | E | E | A | E | A | B | A | E | E | E | E | A | E | C | E | E | D | E | E | E |
| Forest management – increasing forest productivity | C | E | E | C | C | B | D | E | E | E | E | A | E | C | E | E | D | E | E | C |
| Forest management – increasing timber/biomass extraction | C | E | E | E | C | B | D | E | E | E | E | A | E | C | E | E | D | E | E | C |
| Forest management – remediating natural disturbances | E | E | E | E | B | B | E | E | E | E | E | E | E | E | E | E | E | E | E | C |
| Forest management – conservation for carbon sequestration | E | D | E | E | B | B | D | E | E | A | E | A | E | E | E | E | D | E | E | C |
| **Carbon dioxide removal** | | | | | | | | | | | | | | | | | | | | |
| Bioenergy production with carbon capture and sequestration (BECCS) | A | A | A | A | A | A | A | A | A | A | A | A | A | A | A | A | A | A | A | A |
| Direct air capture and storage (DACS) | E | E | A | A | A | E | E | A | A | A | A | E | E | A | A | A | A | A | A | A |
| Mineralisation of atmospheric $CO_2$ through enhanced weathering of rocks | E | E | E | E | E | E | E | E | E | C | E | E | E | E | E | E | E | A | E | E |
| Afforestation/Reforestation | A | A | A | A | A | B | A | E | E | C | E | A | E | C | C | B | C | A | E | A |
| Restoration of wetlands | E | E | E | E | C | E | E | E | E | E | E | E | E | E | E | E | E | E | E | E |
| Biochar | E | E | E | E | D | E | E | E | E | E | A | E | E | E | E | E | E | E | E | E |
| Soil carbon enhancement, enhancing carbon sequestration in biota and soils | E | E | A | C | D | D | E | E | E | E | A | A | E | C | E | E | E | C | E | E |
| Material substitution of fossil $CO_2$ with bio-$CO_2$ in industrial application | E | E | A | C | A | E | E | E | E | E | A | E | E | E | E | D | E | E | A | E |
| Ocean iron fertilisation | E | E | E | E | E | E | E | E | E | E | E | E | E | E | E | E | E | E | E | E |
| Ocean alkalinisation | E | E | E | E | E | E | E | E | E | E | E | E | E | E | E | E | E | E | E | E |
| **Carbon capture and usage (CCU)** | | | | | | | | | | | | | | | | | | | | |
| Bioplastics, carbon fibre and other construction materials | E | E | A | A | E | E | C | E | A | D | A | E | E | E | A | E | A | E | A | E |

| Level of inclusion | National integrated models | | | | | | | | | | | |
|---|---|---|---|---|---|---|---|---|---|---|---|---|
| | 7see6-20_GB | AIM/Enduse-Japan | BLUES 2.0 | China DREAM | CONTO-RUS 1.0 | E4SMA-EU-TIMES 1.0 | STEM (Swiss TIMES Energy Systems Model) | JRC-EU-TIMES | TIMES-China 2.0 | TIMES-France | TIMES_PT | TIMES-Sweden 2.0 |
| Fire management and (ecological) pest control | E | E | E | E | D | E | E | E | E | E | E | E |
| Conservation agriculture | E | E | A | E | E | E | E | E | E | E | E | E |
| Influence on land albedo of land-use change | E | E | E | E | E | E | E | E | E | E | E | E |
| Manure management | E | A | A | B | E | E | E | E | E | E | E | B |
| Reduce food post-harvest losses | E | E | E | E | E | E | E | E | E | E | E | E |
| Recovery of forestry and agricultural residues | E | E | A | E | B | E | E | E | E | E | E | E |
| Forest management – increasing forest productivity | E | E | E | E | D | E | E | E | E | E | E | A |
| Forest management – increasing timber/biomass extraction | E | E | E | E | D | E | E | E | E | E | E | A |
| Forest management – remediating natural disturbances | E | E | E | E | D | E | E | E | E | E | E | E |
| Forest management – conservation for carbon sequestration | E | E | E | E | E | E | E | E | E | E | E | E |
| **Carbon dioxide removal** | | | | | | | | | | | | |
| Bioenergy production with carbon capture and sequestration (BECCS) | B | A | A | E | E | A | A | A | A | A | A | A |
| Direct air capture and storage (DACS) | B | A | A | E | E | E | A | A | A | E | A | E |
| Mineralisation of atmospheric $CO_2$ through enhanced weathering of rocks | E | E | E | E | E | E | E | E | E | E | E | E |
| Afforestation/Reforestation | E | E | A | E | B | E | E | E | E | E | E | E |
| Restoration of wetlands | E | E | E | E | E | E | E | E | E | E | E | E |
| Biochar | E | E | E | E | E | E | A | E | E | E | E | E |
| Soil carbon enhancement, enhancing carbon sequestration in biota and soils | E | E | A | E | E | E | E | E | E | E | E | E |
| Material substitution of fossil $CO_2$ with bio-$CO_2$ in industrial application | D | E | A | E | E | A | E | E | E | E | E | E |
| Ocean iron fertilisation | E | E | E | E | E | E | E | E | E | E | E | E |
| Ocean alkalinisation | E | E | E | E | E | E | E | E | E | E | E | E |
| **Carbon capture and usage (CCU)** | | | | | | | | | | | | |
| Bioplastics, carbon fibre and other construction materials | E | E | A | E | B | A | E | A | E | E | A | E |

| | Explicit | Implicit |
|---|---|---|
| Endogenous | A | C |
| Exogenous | B | D |
| Not represented | E | |

AIII

# Part II: Scenarios

## A.III.II.1 Overview on Climate Change Scenarios

Scenarios are descriptions of alternative future developments. They are used to explore the potential implications of possible future developments and how they might depend on alternative courses of action. They are particularly useful in the context of deep uncertainty. Scenarios are conditional on the realisation of external assumptions and can be used to explore possible outcomes under a variety of assumptions.

Future climate change is a prime example for the application of scenarios. It is driven by human activities across the world and thus can be altered by human agency. It affects all regions over many centuries to come. Humankind's response to climate change touches not only on the way we use energy and land, but also on socio-economic and institutional layers of societal development. Climate change scenarios provide a central tool to analyse this wicked problem.



### A.III.II.1.1 Purposes of Climate Change Scenarios

Climate change scenarios are developed for a number of purposes (O'Neill et al. 2020). First, they are constructed to explore possible climate change futures covering the causal chain from (i) socio-economic developments to (ii) energy and land use to (iii) greenhouse gas emissions to (iv) changes in the atmospheric composition of greenhouse gases and short-lived climate forcers and the associated radiative forcing to (v) changes in temperature and precipitation patterns to (vi) bio-physical impacts of climate change and finally to (vii) impacts on socio-economic developments, thus closing the loop. Quantitative scenarios exploring possible climate change futures are often called climate change projections and climate change impact projections.

Second, climate change scenarios are developed to explore pathways towards long-term climate goals. Goal-oriented scenarios often carry the word 'pathway' in their name, such as climate change mitigation pathway, climate change adaptation pathway, or more generally climate change transition or transformation pathway. They are sometimes called 'backcasting'[4] scenarios, or 'short backcasts', in the literature, particularly when contrasted with forecasts (Robinson 1982). Goal-oriented/backcasting scenarios are inherently normative and intricately linked to human intervention. They can be used to compare and contrast different courses of actions. For example, they are applied in climate change mitigation analysis by comparing reference scenarios without or with only moderate climate policy intervention, sometimes called baseline scenarios, with mitigation pathways that achieve certain climate goals (Grant et al. 2020). Transformation pathways to climate goals are examples of backcasting scenarios. Among other things, they can be used to learn about the multi-dimensional trade-offs between raising or lowering ambition (Clarke et al. 2014; Schleussner et al. 2016). In addition, different transformation pathways to the same goal are often used to analyse trade-offs between different routes towards this goal (Rogelj et al. 2018a). These scenarios need to be looked at as a set to understand attainable outcomes and the trade-offs between them. With scenarios, context matters.

Third, climate change scenarios are used to integrate knowledge and analysis between the three different climate change research communities working on the climate system and its response to human interference (linked to WG I of the IPCC), climate change impacts, adaptation and vulnerability (linked to WGII) and climate change mitigation (linked to WGIII) (O'Neill et al. 2016; IPCC 2000; van Vuuren et al. 2011b) (Annex III.II.1.3). This involves the adoption of common scenario frameworks that allow the consistent use of, for example, shared emissions scenarios, socio-economic development scenarios and climate change projections (Moss et al. 2010; Kriegler et al. 2012; van Vuuren et al. 2012; O'Neill et al. 2014; van Vuuren et al. 2014). The integrative power of scenarios extends beyond the climate change research community into neighbouring fields such as the social sciences and ecology (Pereira et al. 2020; Rosa et al. 2020). To foster such integration, underlying scenario narratives have proven extremely useful as they allow researchers to develop and link quantitative scenario expressions in very different domains of knowledge (O'Neill et al. 2020).

Fourth, climate change scenarios and their assessment aim to inform society (Kowarsch et al. 2017; Weber et al. 2018; Auer et al. 2021). To achieve this, it is important to connect climate change scenarios to broader societal development goals (Riahi et al. 2012; van Vuuren et al. 2015; Kriegler et al. 2018c; Soergel et al. 2021) and relate them to social, sectoral and regional contexts (Absar and Preston 2015; Frame et al. 2018; Kok et al. 2019; Aguiar et al. 2020). To this end, scenarios can be seen as tools for societal discourse and decision-making to coordinate perceptions about possible and desirable futures between societal actors (Edenhofer and Kowarsch 2015; Beck and Mahony 2017).

### A.III.II.1.2 Types of Climate Change Mitigation Scenarios

Different types of climate change scenarios are linked to different purposes and knowledge domains and different models are used to construct them (Annex III.I). Global reference and mitigation scenarios and their associated emissions projections, which are often called emission scenarios, and national, sector and service transition scenarios are key types of scenarios assessed in the Working Group III report. They are briefly summarised below.[5]

A brief description of the common climate change scenario framework with relevance for all three IPCC Working Groups is provided in Annex III.II.1.3, and a discussion how the WGI and WGII assessments relate to the WGIII scenario assessment is given in Annex III.II.2.5.

4 Backcasting is different from hindcasting. Hindcasting refers to testing the ability of a mathematical model to reproduce past events. In contrast, backcasting begins with a desired future outcome and calculates a pathway from the present to that outcome consistent with constraints.

5 The terms mitigation/transition/transformation scenarios and mitigation/transition/transformation pathways are used interchangeably, as they refer to goal-oriented scenarios.

*A.III.II.1.2.1 Global mitigation scenarios*

Global mitigation scenarios are mostly derived from global integrated assessment models (Annex III.I.9) and have been developed in single model studies as well as multi-model comparison studies. The research questions of these studies have evolved together with the climate policy debate and the knowledge about climate change, drivers, and response measures. The assessment of global mitigation pathways in the Fifth Assessment Report (AR5) (Clarke et al. 2014) was informed, inter alia, by a number of large-scale multi-model studies comparing overshoot and not-to-exceed scenarios for a range of concentration stabilisation targets (Energy Modelling Forum (EMF) study 22: EMF22) (Clarke et al. 2009), exploring the economics of different decarbonisation strategies and robust characteristics of the energy transition in global mitigation pathways (EMF27, RECIPE) (Luderer et al. 2012; Krey and Riahi 2013; Kriegler et al. 2014a), and analysing co-benefits and trade-offs of mitigation strategies with energy security, energy access, and air quality objectives (Global Energy Assessment: GEA) (McCollum et al. 2011; Riahi et al. 2012; McCollum et al. 2013; Rao et al. 2013; Rogelj et al. 2013b). They also investigated the importance of international cooperation for reaching ambitious climate goals (EMF22, EMF27, AMPERE) (Clarke et al. 2009; Blanford et al. 2014b; Kriegler et al. 2015b), the implications of collective action towards the 2°C goal from 2020 onwards vs delayed mitigation action (AMPERE, LIMITS) (Kriegler et al. 2014b; Riahi et al. 2015), and the distribution of mitigation costs and burden-sharing schemes in global mitigation pathways (LIMITS) (Tavoni et al. 2014, 2015). Scenarios from these and other studies were collected in a scenario database supporting the AR5 assessment (Krey et al. 2014). With a shelf life of 8 to 14 years, they are now outdated and no longer part of this assessment.

Since AR5, many new studies published global mitigation pathways and associated emissions projections. After the adoption of the Paris Agreement, several large-scale multi-model studies newly investigated pathway limiting warming to 1.5°C (ADVANCE: Luderer et al. (2018); CD-LINKS: McCollum et al. (2018a); ENGAGE: Riahi et al. (2021); SSPs: Rogelj et al. (2018b)), allowing this report to conduct a robust assessment of 1.5°C pathways. Most scenario studies took the hybrid climate policy architecture of the Paris Agreement with global goals, nationally determined contributions (NDCs) and an increasing number of implemented national climate policies as a starting point, including hybrid studies with participation of global and national modelling teams to inform the global stocktake (ENGAGE: Fujimori et al. (2021); COMMIT: van Soest et al. (2021); CD-LINKS: Schaeffer et al. (2020), Roelfsema et al. (2020)). Multi-model studies covered a range of scenarios from extrapolating current policy trends and the implementation of NDCs, respectively, to limiting warming to 1.5°C–2°C with immediate global action and after passing through the NDCs in 2030, respectively. These scenarios are used to investigate, among others, the end-of-century warming implications of extrapolating current policy trends and NDCs (Perdana et al. 2020); the ability of the NDCs to keep limiting warming to 1.5°C–2°C in reach (Luderer et al. 2018; Vrontisi et al. 2018; Roelfsema et al. 2020), the scope for global accelerated action to go beyond the NDCs in 2030 (van Soest et al. 2021), and the benefits of early action vs the risk of overshoot and the use of net negative $CO_2$ emissions in the long-term (Bertram et al. 2021; Hasegawa et al. 2021; Riahi et al. 2021). Other large-scale multi-model studies looked into specific topics: the international economic implications of the NDCs in 2030 (EMF36) (Böhringer et al. 2021), the impact of mitigating short-lived climate forcers on warming and health co-benefits in mitigation pathways (EMF30) (Harmsen et al. 2020; Smith et al. 2020b) and the role and implications of large-scale bioenergy deployment in global mitigation pathways (EMF33) (Bauer et al. 2020a; Rose et al. 2020).

A large variety of recent modelling studies, mostly based on individual models, deepened research on a diverse set of questions (Annex III.II.3.2). Selected examples are the impact of peak vs end-of-century targets on the timing of action in mitigation pathways (Rogelj et al. 2019a; Strefler et al. 2021a); demand-side driven deep mitigation pathways with sustainable development co-benefits (Bertram et al. 2018; Grubler et al. 2018; van Vuuren et al. 2018); synergies and trade-offs between mitigation and sustainable development goals (Fujimori et al. 2020; Soergel et al. 2021); and the integration of climate impacts into mitigation pathways (Schultes et al. 2021). There have also been a number of recent sectoral studies with global integrated assessment models and other global models across all sectors, for example the energy sector (IRENA 2020; Kober et al. 2020; IEA 2021) and transport sector (Edelenbosch et al. 2017a; Mercure et al. 2018; Zhang et al. 2018; Fisch-Romito and Guivarch 2019; Rottoli et al. 2021; Lam and Mercure 2021; Paltsev et al. 2022). Very recent work investigated the impact of COVID-19 on mitigation pathways (Kikstra et al. 2021a) and co-designed global scenarios for users in the financial sector (NGFS 2021). In addition to these policy-, technology- and sector-oriented studies, a few diagnostic studies developed mitigation scenarios to diagnose model behaviour (Harmsen et al. 2021) and explore model harmonisation (Giarola et al. 2021).

The scenarios from most of these and many other studies were collected in the AR6 scenario database (Annex III.II.3.2) and are primarily assessed in Chapter 3 of the report. However sectoral chapters have also used the scenarios, including their climate mitigation categorisations, to ensure consistent cross-chapter treatment. Only a small fraction of these scenarios were already available to the assessment of global mitigation pathways in the Special Report on Global Warming of 1.5°C (SR1.5) (Rogelj et al. 2018a) and were included in the supporting SR1.5 database (Huppmann et al. 2018).

*A.III.II.1.2.2 National Transition Scenarios*

A large number of transition scenarios is developed on a national/regional level by national integrated assessment, energy-economy or computable general equilibrium models, among others. These aim to analyse the implications of current climate plans of countries and regions, as well as long-term strategies until 2050 investigating different degrees of low-carbon development. National/regional transition scenarios are assessed in Chapter 4 of the report.

Recent research has focused on several different types of national transition scenarios that focus on accelerated climate mitigation pathways in the near term to 2050. These include scenarios considered by the authors as tied to meeting specific global climate goals[6] and scenarios tied to specific policy targets (e.g., carbon neutrality or 80–95% reduction from a certain baseline year). A majority of the accelerated national transition modelling studies up to 2050 evaluate pathways that the authors consider compatible with a 2°C global warming limit, with fewer scenarios defined as compatible with 1.5°C global pathways. Regionally, national transition scenarios have centred on countries in Asia (particularly in China, India, Japan), in the European Union, and in North America, with fewer and more narrowly focused scenario studies in Latin America and Africa (Lepault and Lecocq 2021).

*A.III.II.1.2.3 Sector Transition Scenarios*

There are also a range of sector transition scenarios, both on the global and the country level. These include scenarios for the transition of the electricity, buildings, industry, transport and AFOLU sectors until 2050. Due to the accelerated electrification in mitigation pathways, sector coupling plays an increasingly important role to overcome decarbonisation bottlenecks, complicating a separate sector-by-sector scenario assessment. Likewise, the energy-water-land nexus limits the scope of a separate assessment of the energy and agricultural sectors. Nevertheless, sector transition scenarios play an important role for this assessment as they can usually offer much more technology, policy and behaviour detail than integrated assessment models. They are primarily assessed in the sector chapters of the report. Their projections of emissions reductions in the sectors in the near to medium term is used to check the sector dynamics of global models in Chapter 3 of the report.

Recent transition scenarios considered overarching accelerated climate mitigation strategies across multiple sectors, including demand reduction, energy efficiency improvement, electrification and switching to low-carbon fuels. The sectoral strategies considered are often specific to national resource availability, political, economic, climate, and technological conditions. Many sectoral transition strategies have focused on the energy supply sectors, particularly the power sector, and the role for renewable and bio-based fuels in decarbonising energy supply and carbon capture and sequestration (CCS). Some studies present comprehensive scenarios for both supply-side and demand-side sectors, including sector-specific technologies, strategies, and policies. Nearly all demand sector scenarios have emphasised the need for energy efficiency, conservation and reduction through technological changes, with a limited number of models also exploring possible behavioural changes enabled by new technological and societal innovations.

*A.III.II.1.2.4 Service Transition Scenarios*

A central feature of service transition pathways is a focus on the provision of adequate energy services to provide decent standards of living for all as the main scenario objective. Energy services are proxies for well-being, with common examples being provision of shelter (expressed as $m^2$ per capita), mobility (expressed as passenger-kilometres), nutrition (expressed as kCal per capita), and thermal comfort (expressed as degree-days) (Creutzig et al. 2018). Service transition pathways seek to meet adequate levels of such services with minimal carbon emissions, using combinations of demand- and supply-side options. Ideally this is done by improving the efficiency of service provision systems to minimise overall final energy and resource demand, thereby reducing pressure on supply-side and carbon dioxide removal technologies (Grubler et al. 2018). Specifically, this includes providing convenient access to end-use services (health care, education, communication, etc.), while minimising both primary and end-use energy required. Service transition pathways provide a compelling scenario narrative focused on well-being, resulting in technology and policy pathways that give explicit priority to decent living standards. Furthermore, more efficient service provision often involves combinations of behavioural, infrastructural and technological change, expanding the options available to policymakers for achieving mitigation goals (van Sluisveld et al. 2016, 2018). These dimensions are synergistic, in particular in that behavioural and lifestyle changes often require infrastructures adequately matching lifestyles. Service transition scenarios are primarily assessed in Chapter 5 of the report.

### A.III.II.1.3 Scenario Framework for Climate Change Research

*A.III.II.1.3.1 History of Scenario Frameworks used by the IPCC*

For the first three assessment reports, the IPCC directly commissioned emission scenarios with social, economic, energy and partially policy aspects as drivers of projected GHG emissions. The first set of scenarios, the 'SA90' of the IPCC First Assessment Report (IPCC 1990), had four distinct scenarios, 'business-as-usual' and three policy scenarios of increasing ambition. The set of 'IS92' scenarios used in the Second Assessment Report investigated variations of business-as-usual scenarios with respect to uncertainties about the key drivers of economic growth, technology and population (Leggett et al. 1992). The SRES scenarios from the IPCC Special Report on Emission Scenarios (SRES) (IPCC 2000) were produced by multiple modelling organisations and were used in the Third and Fourth Assessment reports. Four distinct scenario families were characterised by narratives and projections of key drivers like population development and economic growth (but no policy measures) to examine their influence on a range of GHG and air pollutant emissions. Until the Fourth Assessment Report, the IPCC organised the scenario development process centrally. Since then, scenarios are developed by the research community and the IPCC limited its role to catalysing and assessing scenarios. To shorten development times, a parallel approach was chosen (Moss et al. 2010) and representative concentration pathways (RCPs) were developed (van Vuuren et al. 2011b) to inform the next generation of climate modelling for the Fifth Assessment Report. RCPs explored four different emissions and atmospheric composition pathways structured to result in different levels of radiative forcing in 2100: 2.6, 4.5, 6.0 and 8.5 W $m^{-2}$. They were used as an input to the Climate Model Intercomparison Project Phase 5 (CMIP5) (Taylor et al. 2011) and its results were assessed in AR5 (Collins et al. 2013).

[6] National emission pathways in the near or mid-term cannot be linked to long-term mitigation goals without making additional assumptions about emissions by other countries up to the mid-term, and assumptions by all countries up to 2100 (see Chapter 4, Box 4.1).

*A.III.II.1.3.2 Current Scenario Framework and SSP-based Emission Scenarios*

The current scenario framework for climate change research (Kriegler et al. 2014c; O'Neill et al. 2014; van Vuuren et al. 2014) is based on the concept of Shared Socio-economic Pathways (SSPs) (Kriegler et al. 2012; O'Neill et al. 2014). Unlike their predecessor scenarios from the SRES (IPCC 2000), their underlying narratives are motivated by the purpose of using the framework for mitigation and adaptation policy analysis. Hence the narratives are structured to cover the space of socio-economic challenges to both adaptation and mitigation. They tell five stories of sustainability (SSP1), middle of the road development (SSP2), regional rivalry (SSP3), inequality (SSP4) and fossil-fuelled development (SSP5) (O'Neill et al. 2017). SSP1, SSP2, and SSP3 were structured to explore futures with socio-economic challenges to adaptation and mitigation increasing from low to high with increasing number of SSP. SSP4 was structured to explore a world with high socio-economic challenges to adaptation but low socio-economic challenges to mitigation, while SSP5 explored a world with low challenges to adaptation but high challenges to mitigation. The five narratives have been translated into population and education (Kc and Lutz 2017), economic growth (Crespo Cuaresma 2017; Dellink et al. 2017; Leimbach et al. 2017a), and urbanisation projections (Jiang and O'Neill 2017) for each of the SSPs.

The SSP narratives and associated projections of socio-economic drivers provide the core components for building SSP-based scenario families. These basic SSPs are not scenarios or goal-oriented pathways themselves (despite carrying 'pathway' in the name), but building blocks from which to develop full-fledged scenarios. In particular, their basic elements do not make quantitative assumptions about energy and land use, emissions, climate change, climate impacts and climate policy. Even though including these aspects in the scenario-building process may alter some of the basic elements, such as projections of economic growth, the resulting scenario remains associated with its underlying SSP. To improve the ability of SSPs to capture socio-economic environments, basic SSPs have been extended in various ways, including the addition of quantitative projections on further key socio-economic dimensions like inequality (Rao et al. 2019), governance (Andrijevic et al. 2020b), and gender equality (Andrijevic et al. 2020a). Extensions also included spatially downscaled projections of, for example, population developments (Jones and O'Neill 2016). By now, the SSPs have been widely used in climate change research ranging from projections of future climate change to mitigation, impact, adaptation and vulnerability analysis (O'Neill et al. 2020).

The integrated assessment modelling community has used the SSPs to provide a set of global integrated energy-land use-emissions scenarios (Bauer et al. 2017; Calvin et al. 2017; Fricko et al. 2017; Fujimori et al. 2017; Kriegler et al. 2017; Popp et al. 2017; Rao et al. 2017b; Riahi et al. 2017; van Vuuren et al. 2017b; Rogelj et al. 2018b) in line with the matrix architecture of the scenario framework (van Vuuren et al. 2014) (Figure 4). It is structured along two dimensions: socio-economic assumptions varied along the SSPs, and climate (forcing) outcomes varied along the Representative Concentration Pathways (RCPs) (van Vuuren et al. 2011b). To distinguish resulting emission scenarios from the original four RCPs (RCP2.6, RCP4.5, RCP6.0, and RCP8.5), they are typically named SSPx–y with x = {1,…,5} the SSP label and y = {1.9, **2.6**, 3.4, **4.5**, **6.0**, 7.0, **8.5**} W $m^{-2}$ the nominal forcing level in 2100. The four forcing levels that were already covered by the original RCPs are bolded here.

The new SSP-based emissions and concentrations pathways provided the input for CMIP6 (Eyring et al. 2015; O'Neill et al. 2016) and its climate change projections are assessed in AR6 (WGI Cross-chapter Box 1.2, WGI Chapter 4). From the original set of more than 100 SSP-based energy-land use-emissions scenarios produced by six IAMs (Figure 4), five Tier 1 scenarios (SSP1-1.9, SSP1-2.6, SSP2-4.5, SSP3-7.0, SSP5-8.5), and four Tier 2 scenarios (SSP4-3.4, SSP4-6.0, variants of SSP7-3.0, SSP5-3.4) were selected[7] (O'Neill et al. 2016), further processed and harmonised with historic emissions and land-use change estimates (Gidden et al. 2019; Hurtt et al. 2020), and then taken up by CMIP6 models. WGI focuses its assessment of CMIP6 climate change projections on the five Tier 1 scenarios (WGI Chapter 4), but also uses the Tier 2 scenarios where they allow assessment of specific aspects like air pollution. All SSP-based IAM scenarios from the original studies are included in the AR6 emissions scenario database and are part of the assessment of global mitigation pathways in Chapter 3.

IAMs could not identify SSP-based emissions scenarios for all combinations of SSPs and RCPs (Riahi et al. 2017; Rogelj et al. 2018b) (Figure 4). The highest emission scenarios leading to forcing levels similar to RCP8.5 could only be obtained in a baseline without climate policy in SSP5 (SSP5-8.5). Since by now climate policies are implemented in many countries around the world, the likelihood of future emission levels as high as in SSP5-8.5 has become small (Ho et al. 2019). Baselines without climate policies for SSP1 and SSP4 reach up to 6.0–7.0 W $m^{-2}$, with baselines for SSP2 and SSP3 coming in higher at around 7.0 W $m^{-2}$. On the lower end, no 1.5°C (RCP1.9) and likely 2°C scenarios (RCP2.6) could be identified for SSP3 due to the lack of cooperative action in this world of regional rivalry. 1.5°C scenarios (RCP1.9) could only be reached by all models under SSP1 assumptions. Models struggled to limit warming to 1.5°C under SSP4 assumptions due to limited ability to sustainably manage land, and under SSP5 assumptions due to their high dependence on ample fossil fuel resources in the baseline (Rogelj et al. 2018b).

A.III.II.1.4 Key Design Choices and Assumptions in Mitigation Scenarios

The development of a scenario involves design choices, in addition to the selection of the model. This section will focus on key choices related to scenario design, and the respective socio-economic, technical, and

7 Each SSPx-y combination was calculated by multiple IAMs. The specific scenarios developed by the marker models for the associated SSPs (SSP1: IMAGE; SSP2: MESSAGE-GLOBIOM; SSP3: AIM; SSP4: GCAM; SSP5: REMIND-MAgPIE) were selected as Tier 1/Tier 2 scenarios for use in CMIP6. Tier 2 variants include SSP7-3.0 with high emissions of short–lived climate forcers and SSP5-3.4 with high overshoot from following SSP5-8.5 until 2040.

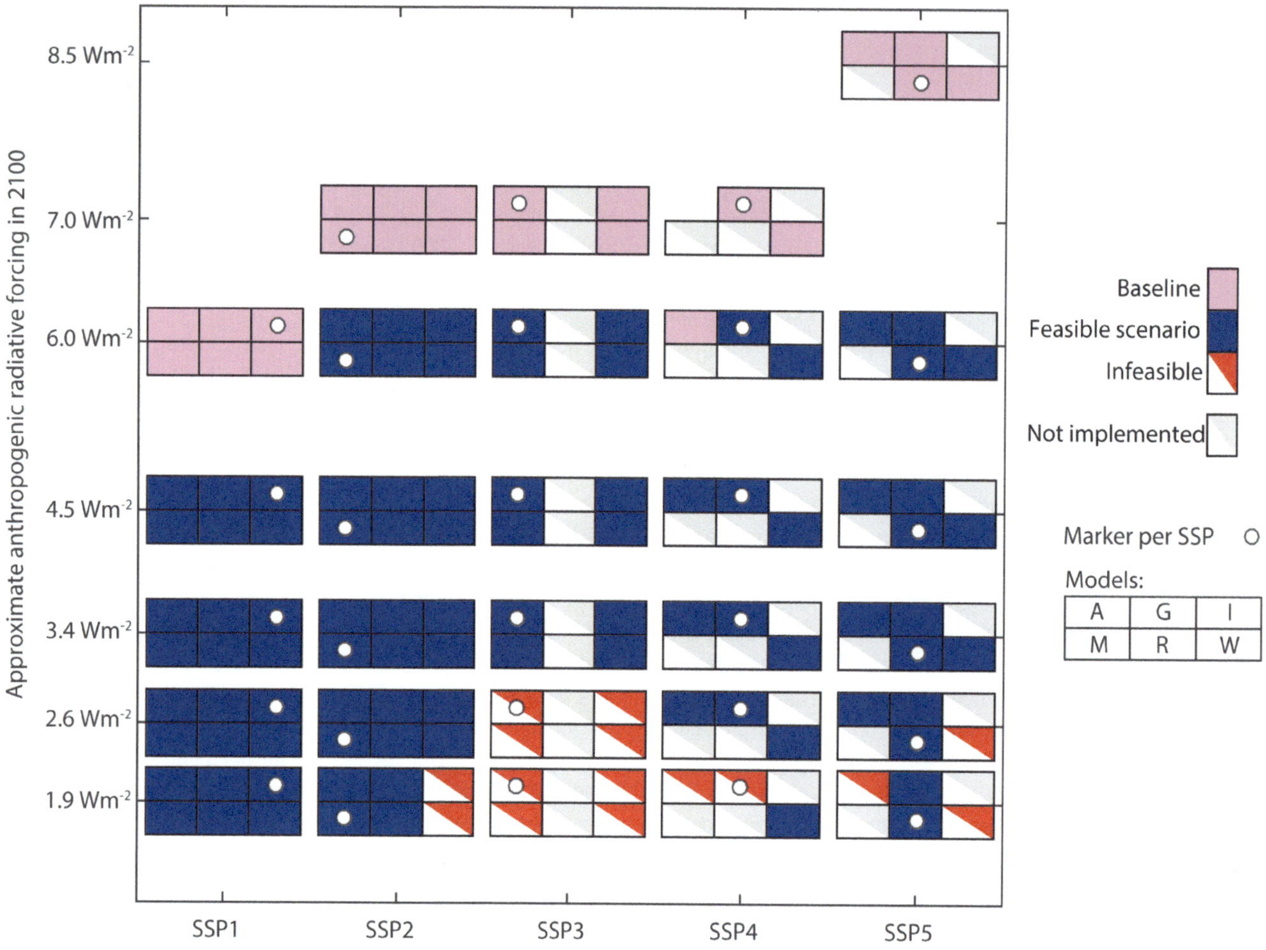


**Figure 4 | The SSP/RCP matrix showing the SSPs on the horizontal axis and the forcing levels on the vertical axis.** A = AIM, G = GCAM; I = IMAGE, M = MESSAGE-GLOBIOM, R = REMIND-MAgPIE, W = WITCH]. Not all SSP/RCP combinations are feasible (red triangles), and not all combinations were tried (grey triangles). Source: adapted with permission from Figure 5 of Rogelj et al. (2018b). Corresponding scenarios were published in Riahi et al. (2017) and Rogelj et al. (2018b) and included in the AR6 scenario database.

policy assumptions. Model selection cannot be separated from these choices. The various advantages and disadvantages of models are described in Annex III, Part I (Modelling Methods).

**Target setting:** Goal-oriented scenarios in the climate scenario literature initially focused on concentration stabilisation but have now shifted towards temperature limits and associated carbon budgets. In early model intercomparisons, climate targets were often specified as a $CO_2$-equivalent concentration level, for example, 450ppm $CO_2$-eq or 550ppm $CO_2$-eq (Clarke et al. 2009). These targets were either applied as not-to-exceed or overshoot targets. In the latter case, concentration levels could be returned to the target level by 2100. Overshoot was particularly allowed for low concentration and temperature targets as many models could not find a solution otherwise (Clarke et al. 2009; Blanford et al. 2014b; Kriegler et al. 2014a; Rogelj et al. 2018b). Bioenergy with carbon capture and storage (BECCS) was an important technology that facilitated aggressive targets to be met in 2100. Due to its ability to remove $CO_2$ from the atmosphere and produce net negative $CO_2$ emissions, it enabled overshoot of the target, leading to a distinctive peak-and-decline behaviour in concentration, radiative forcing, and temperature (Clarke et al. 2014; Fuss et al. 2014). The mitigation scenarios based on the SSP-RCP framework also applied radiative forcing levels in 2100 (Riahi et al. 2017). Temperature targets were often implemented by imposing end-of-century carbon budgets, that is, cumulative emissions up until 2100. In the case of 2°C pathways, those budgets were usually chosen such that the 2°C limit was not overshoot with some pre-defined probability (Luderer et al. 2018). Arguably, the availability of net negative $CO_2$ emissions has led to high levels of carbon dioxide removal (CDR) in the second half of the century, although CDR deployment is often already substantial to compensate residual emissions (Rogelj et al. 2018a).

Recent literature has increasingly focused on alternative approaches such as peak warming or peak $CO_2$ budget constraints to implement targets (Rogelj et al. 2019b; Johansson et al. 2020; Riahi et al. 2021). Nevertheless, due to the availability of net negative $CO_2$ emissions and the assumption of standard (exponentially increasing) emissions-pricing profiles from economic theory, peak and decline temperature profiles still occurred in a large number of mitigation pathways in the literature even in the presence of peak warming and carbon budget targets (Strefler et al. 2021b). This has led to proposals to combine peak targets with additional assumptions affecting the timing of emissions reductions like a constraint on net negative $CO_2$

emissions (Obersteiner et al. 2018; Rogelj et al. 2019a; Riahi et al. 2021) and different carbon pricing profiles (Strefler et al. 2021b). These proposals are aiming at a stabilisation rather than a peak and decline of warming under a given warming limit. However, arguments in support of peak and decline warming profiles also exist: the goal of hedging against positive feedback loops in the Earth system (Lenton et al. 2019) and the aim of increasing the likelihood of staying below a temperature limit towards the end of the century (Schleussner et al. 2016). It is also noteworthy that peak and decline temperature pathways are connected to achieving net-zero GHG emissions (with $CO_2$-eq emissions calculated using GWP100) in the second half of the century (Rogelj et al. 2021).

**Efficiency considerations:** Process-based IAMs typically calculate cost-effective mitigation pathways towards a given target as a benchmark case (Clarke et al. 2014). In these pathways, global mitigation costs are minimised by exploiting the abatement options with the least marginal costs across all sectors and regions at any time, implicitly assuming a globally integrated and harmonised mitigation regime. This idealised benchmark is typically compared across different climate targets or with reference scenarios extrapolating current emissions trends (UNEP 2019). It naturally evolves over time as the onset of cost-effective action is being set to the immediate future of respective studies. This onset was pushed back from 2010–2015 in studies assessed by AR5 (Clarke et al. 2014) to the first modelling time step after 2020 in studies assessed by AR6.

The notion of cost-effectiveness is sensitive to economic assumptions in the underlying models, particularly concerning the assumptions on pre-existing market distortions (Guivarch et al. 2011; Clarke et al. 2014; Krey et al. 2014) and the discount rate on future values. Those assumptions are often not clearly expressed. Most models have a discount rate of 3–5%, though the range of alternatives is larger. Cost-benefit IAMs have had a tradition of exploring the importance of discount rates, but process-based IAMs have generally not. A lower discount rate brings mitigation forward in time and uses less net negative $CO_2$ emissions in cases where target overshoot is allowed (Emmerling et al. 2019; Realmonte et al. 2019). While most models report discount rates in documentation, there is arguably too little sensitivity analysis of how the discount rate affects modelled outcomes.

Cost-effective pathways typically do not account for climate impacts below the temperature limit, although recent updates to climate damage estimates suggest a strengthening of near-term action in cost-effective mitigation pathways (Schultes et al. 2021). Recently, the research community has begun to combine mitigation pathway analysis with *ex post* analysis of associated climate impacts and the benefits of mitigation (Drouet et al. 2021). Cost-effective pathways that tap into least cost abatement options globally without considering compensation schemes to equalise the mitigation burden between countries are not compatible with equity considerations. There is a large body of literature exploring international burden-sharing regimes to accompany globally cost-effective mitigation pathways (Tavoni et al. 2015; Pan et al. 2017; van den Berg et al. 2020).

**Policy assumptions:** Cost-effective mitigation scenarios assume that climate policies are globally uniform. There is a substantial literature contrasting these benchmark cases with pathways derived under the assumption of regionally fragmented and heterogeneous mitigation policy regimes (Blanford et al. 2014b; Kriegler et al. 2015b, 2018b; Roelfsema et al. 2020; van Soest et al. 2021; Bauer et al. 2020b). For example, the Shared Policy Assumptions (Kriegler et al. 2014c) used in the SSP-RCP framework allow for some fragmentation of policy implementation, and many scenarios follow current policies or emission pledges until 2030 before implementing stringent policies (Riahi et al. 2015; Vrontisi et al. 2018; Roelfsema et al. 2020). Other studies assume a gradual strengthening of emissions pledges and regulatory measures converging to a globally harmonised mitigation regime slowly over time (Kriegler et al. 2018b; van Soest et al. 2021). With increasing announcements of mid-century strategies and the rise of net-zero $CO_2$ or GHG targets, global mitigation scenario analysis has begun to build in nationally-specific policy targets until mid-century (NGFS 2021).

Scenarios limiting warming to below 2°C phase in climate policies in all regions and sectors. Almost all converge to a harmonised global mitigation regime before the end of century (with the exception of Bauer et al. (2020b)). In practice, policies are often a mix of regulations, standards, or subsidies. Implementing these real-world policies can give different outcomes to optimal uniform carbon pricing (Mercure et al. 2019). Modelled carbon prices will generally be lower when other policies are implemented (Calvin et al. 2014a; Bertram et al. 2015). As countries implement more and a diverse set of policies, the need to further develop the policy assumptions in models is becoming apparent (Grant et al. 2020; O'Neill et al. 2020; Keppo et al. 2021).



**Socio-economic drivers:** Key socio-economic drivers of emission scenarios are assumptions on population and economic activity. There are other socio-economic assumptions, often included in underlying narratives (O'Neill et al. 2017), that strongly affect energy demand per capita or unit of GDP and dietary choices (Bauer et al. 2017; Popp et al. 2017; Grubler et al. 2018; van Vuuren et al. 2018). The SSPs are often used to help harmonise socio-economic assumptions, and further explore the scenario space. Many studies focus on the middle-of-the-road SSP2 as their default assumption, and many use SSP variations to explore the sensitivity of their results to socio-economic drivers (Marangoni et al. 2017; Riahi et al. 2017; Rogelj et al. 2017). While the SSPs help harmonisation, they are not unique and do not fully explore the scenario space (O'Neill et al. 2020). A wider range of narratives describing alternative worlds is also conceivable. The sustainability world (SSP1), for example, is a world with strong economic growth, but sustainability worlds with low growth or even elements of degrowth in developed countries could also be explored. Thus, standardisation of scenario narratives and drivers has advantages, but can also risk narrowing the scenario space that is explored by the literature. Consequently, many studies in the literature have adopted other socio-economic assumptions, for example with regard to population and GDP (Kriegler et al. 2016; Gillingham et al. 2018) and sustainable development trends (Soergel et al. 2021).

**Technology availability and costs:** Technology assumptions are a key component of IAMs, with some models representing hundreds or thousands of technologies. Despite the importance of technology costs (Creutzig et al. 2017), there has been limited comparison of technology assumptions across models (Kriegler et al. 2015b; Krey et al. 2019). There is, however, a substantial literature on the sensitivity of mitigation scenarios to technology assumptions, including model comparisons (Kriegler et al. 2014a; Riahi et al. 2015), single-model sensitivity studies (McJeon et al. 2011; Krey and Riahi 2013; Giannousakis et al. 2021) and multi-model sensitivity studies (Bosetti et al. 2015). Not only are the initial technology costs important, but also how these costs evolve over time either exogenously or endogenously. Since IAMs have so many interacting technologies, assumptions on one technology can affect the deployment of another. For example, limits on solar energy expansion rates, or integration, may lead to higher levels of deployment for alternative technologies. Because of these interactions, it can be difficult to determine what factors affect deployment across a range of models.

Within these key scenario design choices, model choice cannot be ignored. Not all models can implement aspects of a scenario in the same way. Alternative target implementations are difficult for some model frameworks, and implementation issues also arise around technological change and policy implementation. Certain scenario designs may lock out certain modelling frameworks. These issues indicate the need for a diversity of scenario designs (Johansson et al. 2020) to ensure that model diversity can be fully exploited.



It is possible for many assumptions to be harmonised, depending on the research question. The SSPs were one project aimed at increasing harmonisation and comparability. It is also possible to harmonise emission data, technology assumptions, and policies (Giarola et al. 2021). While harmonisation facilitates greater comparability between studies, it also limits scenario and model diversity. The advantages and disadvantages of harmonisation need to be discussed for each model study.

## A.III.II.2 Use of Scenarios in the Assessment

### A.III.II.2.1 Use of Scenario Literature and Database

The WGIII assessment draws on the full literature on mitigation scenarios. To support the assessment, as many mitigation scenarios in the literature as possible were collected in a scenario database with harmonised output reporting (Annex III.II.3). The collection of mitigation pathways in a common database is motivated by a number of reasons: First, to establish comparability of quantitative scenario information in the literature which is often only sporadically available from tables and figures in peer-reviewed publications, reports and electronic supplementary information. Moreover, this information is often reported using different output variables and definitions requiring harmonisation. Second, to increase latitude of the assessment by establishing direct access to quantitative information underlying the scenario literature. Third, to improve transparency and reproducibility of the assessment by making the quantitative information underlying the scenario figures and tables shown in the report available to the readers of AR6. The use of such scenario databases in AR5 WGIII (Krey et al. 2014) and SR1.5 (Huppmann et al. 2018) proved its value for the assessment as well as for broad use of the scenario information by researchers and stakeholders. This is now being continued for AR6.

### A.III.II.2.2 Treatment of Scenario Uncertainty

The calls for scenarios issued in preparation of this assessment report allowed the collection of a large ensemble of scenarios, coming from many modelling teams using various modelling frameworks in many different studies. Although a large ensemble of scenarios was gathered, it should be acknowledged that only a portion of the full uncertainty space is investigated, and that the distribution of the scenarios within the ensemble reflects the context of the studies the scenarios were developed in. This introduces 'biases' in the ensemble, for example, (i) the topics of the scenario studies collected in the database determine coverage of the scenario space, with large model-comparison studies putting large weight on selected topics over less explored topics explored by individual models, (ii) some models are more represented than others, (iii) only 'optimistic' models (i.e., models finding lower mitigation costs) reach the lowest mitigation targets (Tavoni and Tol 2010). Where appropriate, sampling bias was recognised in the assessment, but formal methods to reduce bias were not employed due to conceptual limitations.

Furthermore, although attempts have been made to elicit scenario likelihoods from expert knowledge (Christensen et al. 2018), scenarios are difficult to associate with probabilities as they typically describe a situation of deep uncertainty (Grübler and Nakicenovic 2001). This and the non-statistical nature of the scenario ensemble collected in the database do not allow a probabilistic interpretation of the distribution of output variables in the scenario database. Throughout the report, descriptive statistics are used to describe the spread of scenario outcomes across the scenarios ensemble. The ranges of results and the position of scenarios outcomes relative to some thresholds of interest are analysed. In some figures, the median of the distribution of results is plotted together with the interquartile range and possibly other percentiles (5th-10th-90th-95th) to facilitate the assessment of results, but these should not be interpreted in terms of likelihood of outcomes.

### A.III.II.2.3 Feasibility of Mitigation Scenarios

In order to develop feasibility metrics of mitigation scenarios (Chapter 3, Section 3.8), the assessment relied on the multidimensional feasibility framework developed in Brutschin et al. (2021), considering five feasibility dimensions: (i) geophysical, (ii) technological, (iii) economic, (iv) institutional and (v) socio-cultural. For each dimension, a set of indicators was developed, capturing not only the scale but also the timing and the disruptiveness of transformative change (Kriegler et al. 2018b). All AR6 scenarios (C1–C3 climate categories) were categorised through this framework to quantify feasibility challenges by climate category, time, policy architecture and by feasibility dimension, summarised in Figure 3.43 (Chapter 3).

Scenarios were categorised into three levels of concerns: (i) low levels of concern where transformation is similar to the past or identified in the literature as feasible/plausible, (ii) medium levels of

Table 8 | Feasibility dimensions, associated indicators and thresholds for the onset of medium and high concerns about feasibility (Chapter 3.8).

| | | Indicators | Computation | Medium | High | Source |
|---|---|---|---|---|---|---|
| Geophysical | | Biomass potential | Total primary energy generation from biomass in a given year | 100 EJ $yr^{-1}$ | 245 EJ $yr^{-1}$ | Frank et al. (2021); Creutzig et al. (2014) |
| Geophysical | | Wind potential | Total secondary energy generation from wind in a given year | 830 EJ $yr^{-1}$ | 2000 EJ $yr^{-1}$ | Deng et al. (2015); Eurek et al. (2017) |
| Geophysical | | Solar potential | Total primary energy generation from solar in a given year | 1600 EJ $yr^{-1}$ | 50 000 EJ $yr^{-1}$ | Rogner et al. (2012); Moomaw et al. (2011) |
| Economic | | GDP loss | Decadal percentage difference in GDP in mitigation vs baseline scenario | 5% | 10% | Analogy to current COVID-19 spending Andrijevic et al. (2020c) |
| Economic | | Carbon price | Carbon price levels (NPV) and decadal increases | USD60 | USD120 and 5× | Brutschin et al. (2021); OECD (2021) |
| Economic | | Energy investments | Ratio between investments in mitigation vs baseline in a given decade | 1.2 | 1.5 | McCollum et al. (2018) |
| Economic | | Stranded coal assets | Share of prematurely retired coal power generation in a given decade | 20% | 50% | Brutschin et al. (202)1; Global Energy Monitor (2021) |
| Technological | Established | Wind/solar scale-up | Decadal percentage point increase in the wind/solar share in electricity generation | 10 pp | 20 pp | Brutschin et al. (2021); Wilson et al. (2020) |
| Technological | Established | Nuclear scale-up | Decadal percentage point increase in the nuclear share in electricity generation | 5 pp | 10 pp | Brutschin et al. (2021); Markard et al. (2020); Wilson et al. (2020) |
| Technological | New Technologies | BECCS scale-up | Amount of $CO_2$ captured in a given year | 3 $GtCO_2$ $yr^{-1}$ | 7 $GtCO_2$ $yr^{-1}$ | Warszawski et al. (2021) |
| Technological | New Technologies | Fossil CCS scale-up | Amount of $CO_2$ captured in a given year | 3.8 $GtCO_2$ $yr^{-1}$ | 8.8 $GtCO_2$ $yr^{-1}$ | Budinis et al. (2018) |
| Technological | New Technologies | Biofuels in transport scale-up | Decadal percentage point increase in the share of biofuels in the final energy demand of the transport sector | 5 pp | 10 pp | Nogueira et al. (2020) |
| Technological | New Technologies | Electricity in transport scale-up | Decadal percentage point increase in the share of electricity in the final energy demand of the transport sector | 10 pp | 15 pp | Muratori et al. (2021) |
| Socio-cultural | | Total/transport/industry/residential energy demand decline | Decadal percentage decrease in demand | 10 % | 20 % | Grubler et al. (2018) |
| Socio-cultural | | Decline of livestock share in food demand | Decadal percentage decrease in the livestock share in total food demand | 0.5 pp | 1 pp | Grubler et al. (2018); Bajželj et al. (2014) |
| Socio-cultural | | Forest cover increase | Decadal percentage increase in forest cover | 2 % | 5 % | Brutschin et al. (2021) |
| Socio-cultural | | Pasture cover decrease | Decadal percentage decrease in pasture cover | 5 % | 10 % | Brutschin et al. (2021) |
| Institutional | | Governance level and decarbonisation rate | Governance levels and per capita $CO_2$ emission reductions over a decade | >0.6 and <20% | <0.6 and >20% | Brutschin et al. (2021); Andrijevic et al. (2020b) |

concern that might be challenging but within reach, given certain enablers, (iii) high levels of concern representing unprecedented levels of transformation, attainable only under consistent enabling conditions. Indicator thresholds defining these three levels of concern were obtained from the available literature and developed with additional empirical literature. Table 8 summarises the main indicators used and the associated thresholds for medium and high levels of concern. Finally, we aggregated feasibility concerns for each dimension and each decade, employing the geometric mean, a non-compensatory method which limits the degree of substitutability between indicators, and used for example by the United Nations for the Human Development Index (HDI). Alternative aggregation scores such as the counting of scenarios exceeding the thresholds were also implemented.

### A.III.II.2.4 Illustrative Mitigation Pathways

In the IPCC Special Report on Global Warming of 1.5°C (SR1.5), illustrative pathways (IPs) were used in addition to descriptions of the key characteristics of the full set of scenarios in the database to assess and communicate the results from the scenario literature. While the latter express the spread in scenario outcomes highlighting uncertain vs robust outcomes, IPs can be used to contrast different stories of mitigating climate change (Rogelj et al. 2018a).

Following the example of the SR1.5, IPs have also been selected for the AR6 of WGIII. In contrast to SR1.5, the selection needed to cover a larger range of climate outcomes while keeping the number of IPs limited. The selection focused on a range of critical themes that emerged from the AR6 assessment: (i) the level of ambition of climate policy, (ii) the different mitigation strategies, (iii) timing of mitigation actions, and (iv) the combination of climate policy with sustainable development policies. The IPs consist of narratives (Table 9) as well as possible quantifications. The IPs are illustrative and denote implications of different societal choices for the development of future emissions and associated transformations of main GHG-emitting sectors. For Chapter 3, for each of the IPs a quantitative scenario was selected from the AR6 scenario database to have particular characteristics and from diverse modelling frameworks (Table 10).

In total, two reference pathways with warming above 2°C and five Illustrative Mitigation Pathways (IMPs) limiting warming in the 1.5–2°C range were selected. The first reference pathway follows current policies as formulated around 2018 (Current Policies, CurPol) through to 2030 and then continues to follow a similar mitigation effort to 2100. The associated quantitative scenario (NGFS 2021) selected by Chapter 3 leads to about 3°C–4°C warming at the end of the century. The second reference pathway follows emission pledges to 2030 (NDCs) and then continues with moderate climate action over time (Moderate Action, ModAct).

The five IMPs are deep mitigation pathways with warming in the 1.5°C–2°C range. The first IMP pursues gradual strengthening beyond NDC ambition levels until 2030 and then acts to likely limit warming to 2°C (Climate Category C3) (IMP-GS) (van Soest et al. 2021) (Chapter 3.5.3). Three others follow different mitigation strategies focusing on low energy demand (IMP-LD) (Grubler et al. 2018),

**Table 9 | Storylines for the two reference pathways and five Illustrative Mitigation Pathways (IMPs) limiting warming to 1.5°C–2°C considered in the report.**

| | | General char. | Policy | Innovation | Energy | Land use, food biodiversity | Lifestyle |
|---|---|---|---|---|---|---|---|
| **CurPol** | | Continuation of current policies and trends | Implementation of current climate policies and neglect of stated goals and objectives; grey COVID-19 recovery | Business as usual; slow progress in low-carbon technologies | Fossil fuels remain important; lock-in | Further expansion of western diets; further slow expansion of agriculture area | Demand will continue to grow; no significant changes in current habits |
| **ModAct** | | NDCs in 2030 as announced in 2020, fragmented policy landscape; post-2030 action consistent with modest action until 2030 | Strengthening of policies to implement NDCs; some further post-2030 strengthening and mixed COVID-19 recovery | Modest change compared to Cur-Pol | Mostly moving away from coal; growth of renewables; some lock-in in fossil investments | Afforestation/ reforestation policies as in NDCs | Modest change compared to Cur-Pol |
| **IMP** | **Neg** | Mitigation in all sectors includes a heavy reliance on carbon dioxide removal that results in net negative global GHG emissions | Successful international climate policy regime with a focus on a long-term temperature goal | Further development of CDR options | Heavy reliance on CDR in power sector and industry; CDR used to compensate fossil fuel emissions | Afforestation/ reforestation, BECCS, increased competition for land | Not critical – some induced via price increases |
| | **Ren** | Greater emphasis on renewables: rapid deployment and technology development of renewables; electrification | Successful international climate policy regime; policies and financial incentives favouring renewable energy | Rapid further development of innovative electricity technologies and policy regimes | Renewable energy; electrification; sector coupling; storage or power-to-X technologies; better interconnections | | Service provisioning and demand changes to better adapt to high renewable energy supply |
| | **LD** | Efficient resource use as well as shifts in consumption patterns globally, leading to low demand for resources, while ensuring a high level of services and satisfying basic needs | | Social innovation; efficiency; across all sectors | low demand for energy, while ensuring a high level of energy services and meeting energy needs; modal shifts in transport; rapid diffusion of best available technology in buildings and industry | Lower food and agricultural waste; less meat-intensive lifestyles | Service provisioning and demand changes; behavioural changes |
| | **GS** | less rapid introduction of mitigation measures followed by a subsequent gradual strengthening | Until 2030, primarily current NDCs are implemented and gradually strengthened moving gradually towards a strong, universal climate policy regime post-2030 | | Similar to IMP-Neg, but with some delay | Similar to IMP-Neg, but with some delay | |
| | **SP** | Shifting the global pathway towards sustainable development. including reduced inequality and deep GHG emissions reduction | SDG policies in addition to climate policy (poverty reduction; environmental protection) | | low demand for energy, while ensuring a high level of energy services and meeting energy needs; renewable energy | Lower food and agricultural waste; less meat-intensive lifestyles; afforestation | Service provisioning and demand changes |

**Table 10 | Quantitative scenario selection to represent the two reference pathways and five Illustrative Mitigation Pathways warming to 1.5°C–2°C for the assessment in Chapter 3.** These quantitative representations of the IMPs have also been taken up by a few other chapters where suitable. The warming profile of IMP-Neg peaks around 2060 and declines to below 1.5°C (50% likelihood) shortly after 2100. While technically classified as a C3, it exhibits the characteristics of C2 high overshoot pathways.

| Acronym | Climate Category (II.3.2) | Model | Scenario name in the AR6 scenario database (III.II.3) | Reference |
|---|---|---|---|---|
| **CurPol** | C7 | GCAM 5.3 | NGFS2_Current Policies | NGFS (2021) |
| **ModAct** | C6 | IMAGE 3.0 | EN_INDCi2030_3000f | Riahi et al. (2021) |
| | | **Illustrative Mitigation Pathways (IMPs)** | | |
| **Neg** | C2* | COFFEE 1.1 | EN_NPi2020_400f_lowBECCS | Riahi et al. (2021) |
| **Ren** | C1 | REMIND-MAgPIE 2.1-4.3 | DeepElec_SSP2_ HighRE_Budg900 | Luderer et al. (2021) |
| **LD** | C1 | MESSAGEix-GLOBIOM 1.0 | LowEnergyDemand_1.3_IPCC | Grubler et al. (2018) |
| **GS** | C3 | WITCH 5.0 | CO_Bridge | van Soest et al. (2021) |
| **SP** | C1 | REMIND-MAgPIE 2.1-4.2 | SusDev_SDP-PkBudg1000 | Soergel et al. (2021) |
| | | **Sensitivity cases** | | |
| **Neg-2.0** | C3 | AIM/CGE 2.2 | EN_NPi2020_900f | Riahi et al. 2(021) |
| **Ren-2.0** | C3 | MESSAGEix-GLOBIOM_GEI 1.0 | SSP2_openres_lc_50 | Guo et al. (2021) |

renewable electricity (IMP-Ren) (Luderer et al. 2021) and large-scale deployment of carbon dioxide removal measures resulting in net negative $CO_2$ emissions in the second half of the century (IMP-Neg). The fifth IMP explicitly pursues a broad sustainable development agenda and follows SSP1 socio-economic assumptions (IMP-SP) (Soergel et al. 2021). IMP-LD, IMP-Ren and IMP-SP limit warming to 1.5°C (>50%) with no or limited overshoot (C1), while IMP-Neg has a higher overshoot and only returns to nearly 1.5°C (50% chance) by 2100 (close to C2). In addition, two sensitivity cases for IMP-Ren and IMP-Neg are considered that limit warming to 2°C (>67%) (C3) rather than pursuing limiting warming to 1.5°C.

The IMPs are used in different parts of the report. We just mention some examples here. In Chapter 3, they are used to illustrate key differences between the mitigation strategies, for instance in terms of timing and sectoral action. In Chapter 6, Box 6.9 discusses the consequences for energy systems. Chapter 7 discusses some of the land-use consequences. In Chapter 8, the implications of the IMPs are further explored for urban systems where the elements of energy, innovation, policy, land use and lifestyle interact (Chapter 8, Sections 8.3 and 8.4). In Chapter 10, the consequences of different mitigation strategies for mobility are highlighted in different figures. The IMPs are discussed further in Chapter 1, Section 1.3; Chapter 3, Section 3.2; and the respective sector chapters.

## A.III.II.2.5 Scenario Approaches to Connect WGIII with the WGI and WGII assessments

### *A.III.II.2.5.1 Assessment of WGIII Scenarios Building on WGI Physical Climate Knowledge*

A transparent assessment pipeline has been set up across WGI and WGIII to ensure integration of the WGI assessment in the climate assessment of emission scenarios in WGIII. This pipeline consists of a step where emissions scenarios are harmonised with historical emissions (harmonisation), a step in which species not reported by an IAM are filled in (infilling), and a step in which the emission evolutions are assessed with three climate model emulators (Annex III.I.8) calibrated to the WGI assessment. These three steps ensure a consistent and comparable assessment of the climate response across emission scenarios from the literature.

**Harmonisation:** IAMs may use different historical datasets, and emission scenarios submitted to the AR6 WGIII scenario database (Annex III.II.3) are therefore harmonised against a common source of historical emissions. To be consistent with WGI, we use the same historical emissions that were used for CMIP6 and RCMIP (Gidden et al. 2018; Nicholls et al. 2020b). This dataset comprises many different emission harmonisation sources (Velders et al. 2015; Gütschow et al. 2016; Le Quéré et al. 2016; van Marle et al. 2017; Meinshausen et al. 2017; Hoesly et al. 2018), including estimates of $CO_2$ emissions from agriculture, forestry, and land-use change (mainly CEDS, (Hoesly et al. 2018)) which are on the lower end of historical observation uncertainty as assessed in Chapter 2. The harmonisation is performed so that different climate futures resulting from two different scenarios are a result of different future emission evolutions within the scenarios, not due to different historical definitions and starting points. Sectoral $CO_2$ emissions from energy and industrial processes and $CO_2$ from agriculture, forestry, and land-use change were harmonised separately. All other emissions species are harmonised based on the total reported emissions per species. For $CO_2$ from energy and industrial processes we use a ratio-based method with convergence in 2080, in line with CMIP6 (Gidden et al. 2018, 2019). For $CO_2$ from agriculture, forestry, and land-use change and other emissions species with high historical interannual variability, we use an offset method with convergence target 2150, to avoid strong harmonisation effects resulting from uncertainties in historical observations. For all remaining fluorinated gases (F-gases), constant ratio harmonisation is used. For all other emissions species, we use the default settings of Gidden et al. (2018, 2019a).

**Infilling missing species:** Infilling ensures that scenarios include all relevant anthropogenic emissions. This reduces the risk of a biased climate assessment and is important because not all IAMs report all climatically active emission species. Infilling was only performed for scenarios where models provided native reporting of energy and industrial process $CO_2$, land use $CO_2$, $CH_4$, and $N_2O$ emissions to avoid infilling gases that have large individual radiative forcing contributions and cannot be infilled with high confidence. Models that did not meet this minimum reporting requirement were not included in the climate assessment. Infilling is performed following the methods and guidelines in Lamboll et al. (2020). Missing species are infilled based on the relationship with $CO_2$ from energy and industrial processes as found in the harmonised set of all scenarios reported to the WGIII scenario database that pass the vetting requirements. To ensure high stability to small changes, we apply a Quantile Rolling Window method (Lamboll et al. 2020) for aerosol precursor emissions, volatile organic compounds and greenhouse gases other than F-gases, based on the quantile of the reported $CO_2$ from energy and industrial processes in the database at each time point. F-fases and other gases with small radiative forcing are infilled based on a pathway with lowest root mean squared difference, allowing for consistency in spite of limited independently modelled pathways in the database.

**WGI-calibrated emulators:** Using expert judgement, emulators that reproduce the best estimates and uncertainties of the majority of AR6 WGI assessed metrics are recommended for scenario classification use by WGIII (see WGI Cross-Chapter Box 7.1). MAGICC (v7) was used for the main scenario classification, with FaIR (v1.6.2) being used to provide additional uncertainty ranges on reported statistics to capture climate model uncertainty. The WGI emulators' probabilistic parameter ensembles are derived such that they match a range of key climate metrics assessed by WGI and the extent to which agreement is achieved is evaluated (WGI Cross-Chapter Box 7.1). Of particular importance to this evaluation is the verification against the WGI temperature assessment of the five scenarios assessed in Chapter 4 of WGI (SSP1-1.9, SSP1-2.6, SSP2-4.5, SSP3-7.0, and SSP5-8.5). The inclusion of the temperature assessment as a benchmark for the emulators provides the strongest verification that WGIII's scenario classification reflects the WGI assessment. The comprehensive nature of the evaluation is a clear improvement on previous reports and ensures that multiple components of the emulators, from their climate response to effective radiative forcing through to their carbon cycles, have been examined before they are deemed fit for use by WGIII.

**Scenario climate assessment:** For the WGIII scenario climate assessment, emulators are run hundreds to thousands of times per scenario, sampling from an emulator-specific probabilistic parameter set, which incorporates carbon cycle and climate system uncertainty in line with the WGI assessment (WGI Cross-Chapter Box 7.1). Percentiles for different output variables provide information about the spread in individual variables for a given scenario, but the set of variables for a given percentile do not form an internally consistent climate change projection. Instead, joint distributions of these parameter sets are employed by the calibrated emulators. Consistent climate change projections are represented by individual ensemble member runs and the whole ensemble of these individual member runs. To facilitate analysis, multiple percentiles of these large (hundred to thousand member) ensemble distributions of projected climate variables are provided in the AR6 scenario database. The emulators provide an assessment of global surface air temperature (GSAT) response to emission scenarios and its key characteristics like peak warming and year of peak warming, ocean heat uptake, atmospheric $CO_2$, $CH_4$ and $N_2O$ concentrations and effective radiative forcing from a range of species including $CO_2$, $CH_4$, $N_2O$ and aerosols for each emissions scenario, as well as an estimate of $CO_2$ and non-$CO_2$ contributions to the temperature increase. The climate emulator's GSAT projections are normalised to match the WGI Chapter 2 assessed total warming between 1850–1900 and 1995–2014 of 0.85°C.

The GSAT projections from the emulator runs are used for classifying those emissions scenarios in the AR6 database that passed the initial vetting and allowed a robust climate assessment. MAGICC (v7) was selected as emulator for the climate classification of scenarios, as it happens to be slightly warmer than the other considered climate emulator, particularly for the higher and long-term warming scenarios – reflecting long-term warming in line with Earth system models (ESMs) (WGI Cross-Chapter Box 7.1). This means that scenarios identified to stay below a given warming limit with a given probability by MAGICC will in general be identified to have this property by the other emulator as well. There is the possibility that the other emulator would classify a scenario in a lower warming class based on their slightly cooler emulation of the temperature response. Unlike during the assessment of the SR1.5 database in the IPCC SR1.5 report, the updated versions of FaIR and MAGICC are however very close, providing robustness to the climate assessment. MAgiCC and FaIR were both used to assess the overall uncertainty in the warming response for a single scenario or a set of scenarios, including both parametric and model uncertainty. Specifically, the 5th to 95th percentile range across the two emulators is calculated, characterising the joint climate uncertainty range of the two models.

**Carbon budgets in WGI and WGIII:** The remaining carbon budget corresponding to a certain level of future warming depends on non-$CO_2$ emissions of modelled pathways. Box 3.4 in Chapter 3 highlighted this key uncertainty in estimating carbon budgets. In this section (Figure 5), we put this into the context of the dependence of carbon budgets on two aspects of the non-$CO_2$ warming contribution: (i) assumptions on historical non-$CO_2$ emissions and how they can impact future non-$CO_2$ warming estimates relative to a recent reference period (2010–2019) (Panel a) and (ii) the scenario set underlying estimates of non-$CO_2$ warming at the time of reaching net zero $CO_2$ (Panel b). Both aspects affect the estimated remaining carbon budget by changing the non-$CO_2$ warming contribution from the base year to the time of reaching net zero $CO_2$. MAGICC7 is used in WGI in conjunction with different input files for the historical warming. For the reported remaining carbon budget estimates (WGI CB) WGI used the non-$CO_2$ warming contributions from MAGICC7 in line with Meinshausen et al. (2020) and in line with the CMIP6 GHG concentration projections, while the WGI emulator setup in line with WGI Cross-Chapter Box 7.1 was used for the WGIII climate assessment. The WGIII assessment uses MAGICC7 in line with Nicholls et al. (2021) in line with the emission harmonisation process employed in WGIII (see above). The difference in historical assumptions changes the estimated non-$CO_2$ contribution by up to about 0.05°C for the lower temperature levels, or slightly more than 10% of the warming until 1.5°C relative to 2010–2019. For peak warming around 2°C relative to pre-industrial levels (about 0.97°C warming relative to 2010–2019 in Figure 5 plots), the difference is

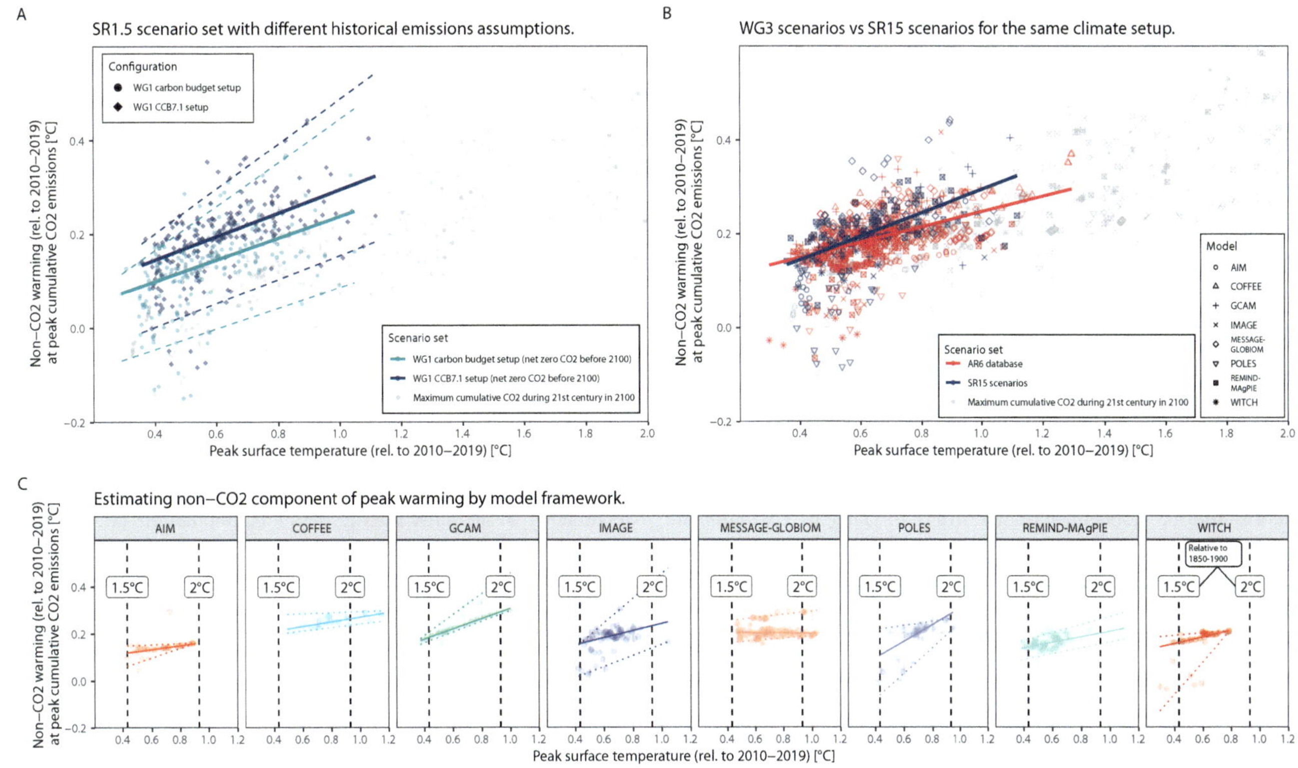


**Figure 5 | Comparison of non-$CO_2$ warming relevant for the derivation of cumulative carbon budgets – and its sensitivity to (a) assumptions on historical emissions and (b) the set of investigated scenarios (right). Panel (c)** shows how the relationship across scenarios between peak surface temperature and non-$CO_2$ warming and peak cumulative $CO_2$ is different for modelling frameworks. All dashed regression lines are at the 5th and 95th percentiles, solid lines are a regression at the median.

All panels depict non-$CO_2$ warming in relation to 2010–2019 at the time of peak cumulative $CO_2$, using MAGICC7. Scenarios that reach net-zero $CO_2$ this century are coloured, with dots in grey indicating scenarios that do not reach net-zero $CO_2$ but still remain below 2°C median peak warming relative to 2010–2019 levels in this century. The scenario set 'AR6 database' in **(b)** includes only scenarios of those model frameworks that are shown in panel (c)) which have a detailed land-use model and enough scenarios to imply a relationship.

**Panel (a)** The WGI remaining carbon budget takes into account the non-$CO_2$ warming in dependence of peak surface temperatures via a regression line approach (lighter blue-coloured solid line). For the same scenario set, with historical emissions assumptions as used in Cross-Chapter Box 7.1 (darker blue-coloured solid line), a relationship is found with a difference of approximately 0.05°C.

**Panel (b)** The WGIII database of scenarios tends to imply very similar non-$CO_2$ warming at peak cumulative $CO_2$ to the SR15 scenario database, especially around 1.5°C above pre-industrial (0.43°C above 2010–2019 levels), though with slightly lower non-$CO_2$ warming for higher peak temperatures.

**Panel (c)** Regressions at the 5th, 50th, and 95th percentiles indicate a model framework footprint affecting the relationship between peak warming and non-$CO_2$ warming at peak cumulative $CO_2$.

offset by the difference arising from using either the SR1.5 or AR6 scenario databases (see panel (b) in Figure 5).

Estimates of the remaining carbon budget that take into account non-$CO_2$ uncertainty are not only dependent on historical assumptions, but also on future non-$CO_2$ scenario characteristics, which are different across the various scenarios in the AR6 database. In panel (b) of Figure 5, we show how the SR15 database of scenarios, which was used to inform the WGI remaining carbon budget, differs from the larger set considered in the WGIII report (both using MAGICC7 using input files in line with Nicholls et al. (2021)). Overall, there is limited difference in the covered range of non-$CO_2$ warming at different peak surface temperature levels, leading to no clear change in estimated carbon budgets compared to SR1.5 based on the full scenario database. However, as discussed in Cross-Working Group Box 1 in Chapter 3, and shown in panel (c) of Figure 5, mitigation strategies expressed by both the IAM footprint and scenario design (e.g., dietary change scenarios) can have strong effects on estimated carbon budgets for staying below 1.5°C.

### *A.III.II.2.5.2 Relating the WGII and WGIII Assessments by use of Warming Levels*

WGII sets out common climate dimensions to help contextualise and facilitate consistent communication of impacts and synthesis across WGII, as well as to facilitate WGI and WGII integration, with the dimensions adopted when helpful and possible across WGII (AR6 WGII Cross-Chapter Box 1.1). 'Common climate dimensions' are defined as common global warming levels (GWLs), time periods, and levels of other variables as needed by WGII authors (see below for a list of variables associated with these dimensions). Projected ranges for associated climate variables were derived from the AR6 WGI report and supporting resources and help contextualise and inform the projection of potential future climate impacts and key risks. The information enables the mapping of climate variable levels to climate projections by WGI (AR6 WGI Table SPM.1) and vice versa, with ranges of results provided to characterise the physical uncertainties relevant to assessing climate impacts risk. Common socio-economic dimensions are not adopted in WGII due to a desire to draw on the full literature, inform the broad ranges of relevant possibilities (climate, development, adaptation, mitigation), and be flexible. The impacts literature is wide-ranging and diverse, with a fraction based on global socio-economic scenarios. WGII's approach allows chapters and cross-chapter boxes to assess how impacts and ranges depend on socio-economic factors affecting exposure, vulnerability, and adaptation independently, as appropriate for their literature. For example, WGII Chapter 16 assesses how Representative Key Risks vary under low vs high exposure/vulnerability conditions by drawing on impact literature based on Shared Socio-economic Pathways (SSPs). In general, WGII chapters, when possible and conducive with their literature, used GWLs or climate projections based on Representative Concentration Pathways or SSPs to communicate information and facilitate integration and synthesis, with impacts results characterised according to other drivers when possible and relevant, such as socio-economic conditions.



In the context of common climate dimensions, WGII considers common projected GWL ranges by time period, the timing for when GWLs might be reached, and projected continental-level result ranges for select temperature and precipitation variables by GWL (average and extremes), as well as sea surface temperature changes by GWL and ocean biome. Where available, WGII considers the assessed WGI ranges as well as the raw CMIP5 and CMIP6 climate change projections (ranges and individual projections) from Earth system models (Hauser et al. 2019). With WGII's climate impacts literature based primarily on climate projections available at the time of AR5 (CMIP5) and earlier, or assumed temperature levels, it was important to be able to map climate variable levels to climate projections of different vintages and vice versa. WGII's common GWLs are based on AR6 WGI's proposed 'Tier 1' dimensions of integration range – 1.5, 2.0, 3.0, and 4.0°C (relative to the 1850 to 1900 period), which are simply proposed common GWLs to facilitate integration across and within WGs (WGI Chapter 1). Within WGII, GWLs facilitate comparison of climate states across climate change projections, assessment of the full impacts literature, and cross-chapter comparison. Across AR6, GWLs facilitate integration across Working Groups of climate change projections, climate change risks, adaptation opportunities, and mitigation.

For facilitating integration with WGIII, GWLs need to be related to WGIII's classification of mitigation efforts by temperature outcome. WGIII's Chapter 3 groups full century emissions projections resulting from a large set of assessed mitigation scenarios into temperature classes (Chapter 3, Sections 3.2 and 3.3, Annex III.II.2.5.1, and Annex III.II.3.2.1). Scenarios are classified by median peak global mean temperature increase since 1850–1900 in the bands <2°C, 2°C–2.5°C, 2.5°C–3°C, 3°C–4°C, and >4°C, with the range below 2°C broken out in greater detail using estimates of warming levels at peak and in 2100 for which the warming response is projected to be likely higher (33th percentile), as likely higher as lower (median), and likely lower (67th percentile) (Chapter 3, Section 3.2 and Annex III.II.3.2.1). WGII's common GWLs and WGIII's global warming scenario classes are relatable but differ in several important ways. While GWLs represent temperature change that occurs at some point in time, emissions scenarios in a temperature class result in an evolving warming response over time. The emissions scenario warming also has a likelihood attached to the warming level at any point in time, that is, actual warming outcomes can be lower or higher than median warming projections within the range of the estimated uncertainty. Thus, multiple WGII results across GWLs will be relevant to any particular WGIII emissions pathway, including at the peak temperature level.

However, socio-economic conditions are an important factor defining both impacts exposure, vulnerability, and adaptation, as well as mitigation opportunity and costs, that needs special considerations. The WGIII scenario assessment is using additional classifications relating to, inter alia, near-term policy developments, technology availability, energy demand, population and economic growth (Chapter 3, Section 3.3 and Annex III.II.3.2.2), and a set of illustrative mitigation pathways with varying socio-techno-economic assumptions (Annex III.II.2.4, Chapter 3, Section 3.2). Synthesising WGII assessments of climate change impacts and WGIII assessments of climate change mitigation efforts for similar GWLs/global warming scenario classes would have to address how socio-techno-economic conditions affect impacts, adaptation, and mitigation outcomes. Furthermore, a synthesis of mitigation costs and mitigation benefits in terms of avoided climate change impacts would require a framework that ensures consistency in socio-economic development assumptions and emissions and adaptation dynamics and allows for consideration of benefits and costs along the entire pathway (O'Neill et al. 2020) (Cross-Working Group Box 1 in Chapter 3).

### A.III.II.3 WGIII AR6 Scenario Database

*[Note: The scenario numbers documented in this section refer to all scenarios that were submitted and not retracted by the literature acceptance deadline of 11 October 2021, and that fulfilled the requirement of being supported by an eligible literature source. Not all those scenarios were used in the assessment, for example some did not pass the vetting process as documented in II.3.1.]*

As for previous IPCC reports of Working Group III, including the Special Report on Global Warming of 1.5°C (SR1.5) (Huppmann et al. 2018; Rogelj et al. 2018a) and the Fifth Assessment Report (AR5) (Clarke et al. 2014; Krey et al. 2014), quantitative information on mitigation pathways is collected in a dedicated AR6 scenario database[8] to underpin the assessment.

By the time of the AR6 literature acceptance deadline of IPCC WGIII (11 October 2021) the AR6 scenario database comprised 191 unique modelling frameworks (including different versions and country setups) from 95+ model families – of which 98 were globally comprehensive, 71 national or multi-regional, and 20 sectoral models – with in total 3,131 scenarios, summarised in Tables 11–17

[8] https://data.ece.iiasa.ac.at/ar6/.

(global mitigation pathways), Table 18 (national and regional mitigation pathways) and Table 19 (sector transition pathways).

### A.III.II.3.1 Process of Scenario Collection and Vetting

To facilitate the AR6 assessment, modelling teams were invited to submit their available emissions scenarios to a web-based database hosted by the International Institute for Applied Systems Analysis (IIASA).[9] The co-chairs of Working Group III as well as a range of scientific institutions, including the Integrated Assessment Modelling Consortium (IAMC), University of Cape Town and the Centre International de Recherche sur l'Environnement et le Développement (CIRED), supported the open call for scenarios which was subdivided into four dedicated calls:

i. a call for global long-term scenarios to underpin the assessment in Chapter 3 as well as facilitating integration with sectoral Chapters 6, 7, 8, 9, 10 and 11,
ii. a call for short- to medium-term scenarios at the national and regional scales underpinning the assessment in Chapter 4, and
iii. a call for building-focused scenarios to inform the assessment in Chapter 9, and
iv. a call for transport-focused scenarios to inform the assessment in Chapter 10.

A common data reporting template with a defined variable structure was used and all teams were required to register and submit detailed model and scenario metadata. Scenarios were required to come from a formal quantitative model and the scenarios must be published in accordance with IPCC literature requirements. The calls for scenarios were open for a period of 22 months (September 2019 to July 2021), with updates possible until October 2021 in line with the literature acceptance deadline. The data submission process included various quality control procedures to increase accuracy and consistency in reporting. Additional categorisation and processing of metadata over the full database provided a wide range of indicators and categories that were made centrally available to authors of the report to enhance consistency of the assessment, such as: climate, policy and technology categories; characteristics about emissions, energy, socio-economics and carbon sequestration; metadata such as literature references, model documentation and related projects.

For all scenarios reporting global data, a vetting process was undertaken to ensure that key indicators were within reasonable ranges for the baseline period – primarily for indicators relating to emissions and the energy sector (Table 11). As part of the submission process, model teams were contacted individually with information on the vetting outcome with regard to their submitted scenarios, giving them the opportunity to verify the reporting of their data. Checks on technology-specific variables for nuclear, solar and wind energy, and CCS, screen not only for accuracy with respect to recent developments, but also indicate reporting errors relating to different primary energy accounting methods. While the criteria ranges appear to be large, the focus of these scenarios is the medium to long term and there is also uncertainty in the historical values. For vetting of the Illustrative Mitigation Pathways, the same criteria were used, albeit with narrower ranges (Table 11). Selected future values were also vetted and the result of the vetting reported to authors, but not used as exclusion criterion. Where possible the latest values available were used, generally 2019, and if necessary extrapolated to 2020 as most

**Table 11 | Summary of the vetting criteria and ranges applied to the global scenarios for the climate assessment and preliminary screening for Illustrative Mitigation Pathways.**

| | Reference value | Range (IP range) | Pass | Fail | Not reported |
|---|---|---|---|---|---|
| Historical emissions (sources: EDGAR v6 IPCC and CEDS, 2019 values) | | | | | |
| $CO_2$ total (EIP + AFOLU) emissions | 44,251 $MtCO_2\ yr^{-1}$ | ±40% (±20%) | 1848 | 23 | 395 |
| $CO_2$ EIP emissions | 37,646 $MtCO_2\ yr^{-1}$ | ±20% (±10%) | 2162 | 55 | 49 |
| $CH_4$ emissions | 379 $MtCH_4\ yr^{-1}$ | ±20% (±20%) | 1651 | 139 | 476 |
| $CO_2$ emissions EIP 2010–2020 % change | – | +0 to +50% | 1742 | 74 | 450 |
| CCS from energy 2020 | – | 0-250 (100) $MtCO_2\ yr^{-1}$ | 1624 | 77 | 565 |
| Historical energy production (sources: IEA 2019; IRENA; BP; EMBERS; trends extrapolated to 2020) | | | | | |
| Primary energy (2020, IEA) | 578 EJ | ±20% (±10%) | 1813 | 73 | 380 |
| Electricity: nuclear (2020, IEA) | 9.77 EJ | ±30% (±20%) | 1603 | 266 | 397 |
| Electricity: solar and wind (2020. IEA, IRENA, BP, EMBERS). | 8.51 EJ | ±50% (±25%) | 1459 | 377 | 430 |
| Overall | | | **1686** | **580** | – |
| Future criteria (not used for exclusion in climate assessment but flagged to authors as potentially problematic) | | | | | |
| No net negative $CO_2$ emissions before 2030 | $CO_2$ total in 2030 >0 | | 1867 | 4 | 395 |
| CCS from energy in 2030 | <2000 $MtCO_2\ yr^{-1}$ | | 1518 | 183 | 565 |
| Electricity from nuclear in 2030 | <20 $EJ\ yr^{-1}$ | | 1595 | 274 | 397 |
| $CH_4$ emissions in 2040 | 100–1000 $MtCH_4\ yr^{-1}$ | | 1775 | 15 | 476 |

Rows do not sum to the same total of scenarios as not all scenarios reported all variables. EIP stands for energy and industrial process emissions.

9 https://data.ene.iiasa.ac.at/ar6/#/about.

AIII

models report only at five- to 10-year intervals. 2020 as reported in most scenarios collected in the database does not include the impact of the COVID-19 pandemic.

Almost three-quarters of submitted global scenarios passed the vetting. The remaining quarter comprised a fraction of scenarios that were rolled over from the SR1.5 database, and were no longer up to date with recent developments (excluding the COVID-19 shock). This included scenarios that started stringent mitigation action already in 2015. Other scenarios were expected to deviate from historical trends due to their diagnostic design. All historical criteria for reported variables needed to be met in order to pass the vetting.

2266 global scenarios were submitted to the scenario database that fulfilled a minimum requirement of reporting at least one global emission or energy variable covering multiple sectors. 1686 global scenarios passed the vetting criteria described in Table 11. These scenarios were subsequently flagged as meeting minimum quality standards for use in long-term scenarios assessment. Additional criteria for inclusion in the Chapter 3 climate assessment are described in Annex III.II.3.2.1.

### A.III.II.3.2 Global Pathways

Scenarios were submitted by both individual studies and model inter-comparisons. The main model inter-comparisons submitting scenarios are shown in Table 12. Model inter-comparisons have a shared experimental design and assess research questions across different modelling platforms to enable more structured and systematic assessments. The model comparison projects thus help to understand the robustness of the insights.

The number of submitted scenarios varies considerably by study, for example from 10 to almost 600 scenarios for the model inter-comparison studies (Table 12). The number of scenarios also varies substantially by model (Table 15), highlighting the fact that the global scenario set collected in the AR6 scenario database is not a statistical sample (Section II.2.2).



**Table 12 | Model inter-comparison studies that submitted global scenarios to the AR6 scenario database and for which at least one scenario passed the vetting.** Scenario counts refer to all scenarios submitted by a study (in brackets), those that passed vetting (centre) and those that passed the vetting and received a climate assessment (left).

| Project | Description | Publication year | Key references | Website | Number of scenarios |
|---|---|---|---|---|---|
| SSP model-comparison | The SSPs are part of a new framework that the climate change research community has adopted to facilitate the integrated analysis of future climate impacts, vulnerabilities, adaptation, and mitigation (II.1.3) | 2017 / 2018 | Riahi et al. (2017); Rogelj et al. (2018b) | https://tntcat.iiasa.ac.at/SspDb | 70 / 77 (126) |
| ADVANCE | Developed a new generation of advanced IAMs and applied the improved models to explore different climate mitigation policy options in the post-Paris framework | 2018 | Luderer et al. (2018); Vrontisi et al. (2018) | http://www.fp7-advance.eu/ | 37 / 40 (72) |
| | Industry sector study | 2017 | Edelenbosch et al. (2017b) | http://www.fp7-advance.eu/ | 0 / 6 (6) |
| CD-LINKS | Exploring the complex interplay between climate action and development, while simultaneously taking both global and national perspectives and thereby informing the design of complementary climate-development policies | 2018 | McCollum et al. (2018); Roelfsema et al. (2020) | https://www.cd-links.org/ | 41 / 52 (77) |
| COMMIT | Exploring new climate policy scenarios at the global level and in different parts of the world | 2021 | van Soest et al. (2021) | https://themasites.pbl.nl/commit/ | 41 / 59 (68) |
| ENGAGE | Exploring new climate policy scenarios at the global level and in different parts of the world | 2021 | Riahi et al. (2021) | http://www.engage-climate.org/ | 591 / 591 (603) |
| EMF30 | Energy Modelling Forum study into the role of non-$CO_2$ climate forcers | 2020 | Smith et al. (2020a); Harmsen et al. (2020) | https://emf.stanford.edu/projects/emf-30-short-lived-climate-forcers-air-quality | 61 / 69 (149) |
| EMF33 | Energy Modelling Forum study into the role of bioenergy | 2020 | Rose et al. (2020); Bauer et al. (2020a) | https://emf.stanford.edu/projects/emf-33-bio-energy-and-land-use | 67 / 68 (173) |
| EMF36 | Energy Modelling Forum study into the role of carbon pricing and economic implications of NDCs | 2021 | Böhringer et al. (2021) | https://emf.stanford.edu/projects/emf-36-carbon-pricing-after-paris-carpri | 0 / 305 (320) |

| Project | Description | Publication year | Key references | Website | Number of scenarios |
|---|---|---|---|---|---|
| NGFS | Study for scenario-based financial risk assessment with details on impacts, and sectoral and regional granularity | 2021 | NGFS (2020, 2021) | https://www.ngfs.net/ngfs-scenarios-portal | 24 / 24 (24)<br>2 / 2 (2)[10] |
| PARIS REINFORCE | Study on the long-term implications of current policies and NDCs | 2020 | Perdana et al. (2020) | https://paris-reinforce.eu | 3 / 25 (39) |
| PARIS REINFORCE | Study with a focus on harmonising socio-economics and techno-economics in baselines | 2021 | Giarola et al. (2021) | https://paris-reinforce.eu | 0 / 8 (16) |
| CLIMACAP-LAMP | Study on the role of climate change mitigation in Latin America | 2016 | van der Zwaan et al. (2016) | n.a. | 0 / 10 (22) |
| | | | | **Total** | **937 / 1336 (1697)** |

**Table 13 | Single-model studies that submitted global scenarios to the AR6 scenario database and for which at least one scenario passed the vetting.** Scenario counts refer to all scenarios submitted by a study (in brackets), those that passed vetting (centre) and those that passed the vetting and received a climate assessment (left).

| Title of study | Literature reference[11] | Number of scenarios |
|---|---|---|
| Quantification of an efficiency–sovereignty trade-off in climate policy | Bauer et al. (2020b) | 4 / 4 (4) |
| Transformation and innovation dynamics of the energy-economic system within climate and sustainability limits | Baumstark et al. (2021) | 18 / 18 (18) |
| Tracing international migration in projections of income and inequality across the Shared Socio-economic Pathways | Benveniste et al. (2021) | 0 / 10 (10) |
| Targeted policies can compensate most of the increased sustainability risks in 1.5°C mitigation scenarios | Bertram et al. (2018) | 3 / 3 (12) |
| Long term, cross country effects of buildings insulation policies | Edelenbosch et al. (2021) | 0 / 8 (8) |
| The role of the discount rate for emission pathways and negative emissions | Emmerling et al. (2019) | 4 / 4 (28) |
| Studies with the EPPA model on the costs of low-carbon power generation, the cost and deployment of CCS, the economics of BECCS, the global electrification of light duty vehicles, the 2018 food, water, energy and climate outlook and the 2021 global change outlook | Reilly et al. (2018); Morris et al. (2019, 2021); Smith et al. (2021); Fajardy et al. (202)1; Paltsev et al. (2021, 2022) | 7 / 7 (10) |
| Transportation infrastructures in a low carbon world: An evaluation of investment needs and their determinants | Fisch-Romito and Guivarch (2019) | 0 / 24 (32) |
| Measuring the sustainable development implications of climate change mitigation | Fujimori et al. (2020) | 5 / 5 (5) |
| How uncertainty in technology costs and carbon dioxide removal availability affect climate mitigation pathways | Giannousakis et al. (2021) | 9 / 9 (9) |
| A low energy demand scenario for meeting the 1.5°C target and sustainable development goals without negative emission technologies | Grubler et al. (2018) | 1 / 1 (1) |
| Global Energy Interconnection: A scenario analysis based on the MESSAGEix-GLOBIOM Model | Guo et al. (2021) | 20 / 20 (20) |
| Climate–carbon cycle uncertainties and the Paris Agreement | Holden et al. (2018) | 0 / 5 (5) |
| Ratcheting ambition to limit warming to 1.5 °C – trade-offs between emission reductions and carbon dioxide removal | Holz et al. (2018) | 6 / 6 (6) |
| Peatland protection and restoration are key for climate change mitigation | Humpenöder et al. (2020) | 0 / 3 (3) |
| Energy Technology Perspectives 2020 | IEA (2020b) | 0 / 1 (1) |
| World Energy Outlook 2020 – Analysis – IEA | IEA (2020a) | 0 / 1 (1) |
| Net Zero by 2050 – A Roadmap for the Global Energy Sector | IEA (2021) | 0 / 1 (1) |
| Global Renewables Outlook: Energy transformation 2050 | IRENA (2020) | 0 / 2 (2) |
| Climate mitigation scenarios with persistent COVID-19-related energy demand changes | Kikstra et al. (2021a) | 19 / 19 (19) |
| Global anthropogenic emissions of particulate matter including black carbon | Klimont et al. (2017) | 0 / 2 (2) |
| Global energy perspectives to 2060 – WEC's World Energy Scenarios 2019 | Kober et al. (2020) | 0 / 4 (4) |
| Prospects for fuel efficiency, electrification and fleet decarbonisation | Kodjak and Meszler (2019) | 0 / 4 (4) |
| Short term policies to keep the door open for Paris climate goals | Kriegler et al. (2018b) | 18 / 18 (18) |
| Deep decarbonisation of buildings energy services through demand and supply transformations in a 1.5°C scenario | Levesque et al. (2021) | 4 / 4 (4) |
| Designing a model for the global energy system – GENeSYS-MOD: An application of the Open-Source Energy Modelling System (OSeMOSYS) | Löffler et al. (2017) | 0 / 1 (1) |
| Impact of declining renewable energy costs on electrification in low emission scenarios | Luderer et al. (2021) | 8 / 8 (8) |

10 The first NGFS scenario publication in 2020 comprised 15 scenarios from the literature and 2 newly developed scenarios. The 15 scenarios are also contained in the database under their original study name.

11 Publication date of scenarios coincides with year of publication.

| Title of study | Literature reference[11] | Number of scenarios |
|---|---|---|
| The road to achieving the long-term Paris targets: energy transition and the role of direct air capture | Marcucci et al. (2017) | 1 / 1 (3) |
| The transition in energy demand sectors to limit global warming to 1.5°C | Méjean et al. (2019) | 0 / 3 (27) |
| Deep mitigation of $CO_2$ and non-$CO_2$ greenhouse gases toward 1.5°C and 2°C futures | Ou et al. (2021) | 34 / 35 (36) |
| Alternative electrification pathways for light-duty vehicles in the European transport sector | Rottoli et al. (2021) | 8 / 8 (8) |
| Economic damages from on-going climate change imply deeper near-term emission cuts | Schultes et al. (2021) | 24 / 24 (24) |
| A sustainable development pathway for climate action within the UN 2030 Agenda | Soergel et al. (2021) | 8 / 8 (8) |
| Delayed mitigation narrows the passage between large-scale CDR and high costs | Strefler et al. (2018) | 7 / 7 (7) |
| Alternative carbon price trajectories can avoid excessive carbon removal | Strefler et al. (2021b) | 9 / 9 (9) |
| Carbon dioxide removal technologies are not born equal | Strefler et al. (2021a) | 8 / 8 (8) |
| The Impact of U.S. Re-engagement in Climate on the Paris Targets | van de Ven et al. (2021) | 0 / 10 (10) |
| The 2021 SSP scenarios of the IMAGE 3.2 model | Müller-Casseres et al. (2021); van Vuuren et al. (2014, 2021) | 40 / 40 (40) |
| Pathway comparison of limiting global warming to 2°C | Wei et al. (2021) | 0 / 5 (5) |
| | **Total** | **265 / 350 (421)** |



### *A.III.II.3.2.1 Climate Classification of Global Pathways*

The global scenarios underpinning the assessment in Chapter 3 have been classified, to the degree possible, by their warming outcome. The definition of the climate categories and the distribution of scenarios in the database across these categories is shown in Table 14 (Chapter 3, Section 3.2). The first four of these categories correspond to the ones used in the IPCC SR1.5 (Rogelj et al. 2018a) while the latter four have been added as part of the AR6 to capture a broader range of warming outcomes.

For inclusion in the climate assessment, in addition to passing the vetting (Section II.3.1), scenarios needed to run until the end of century and report as a minimum $CO_2$ (total and for energy and industrial processes (EIP)), $CH_4$ and $N_2O$ emissions to 2100. Where $CO_2$ for AFOLU was not reported, the difference between total and EIP in 2020 must be greater than 500 $MtCO_2$. Of the total 2266 global scenarios submitted, 1574 could be assessed in terms of their associated climate response, and 1202 of those passed the vetting process.

**Table 14 | Classification of global pathways into warming levels using MAGICC (Chapter 3, Section 3.2).**

| Description | Definition | Scenarios | |
|---|---|---|---|
| | | Passed vetting | All |
| **C1: Limit warming to 1.5°C (>50%) with no or limited overshoot** | Reach or exceed 1.5°C during the 21st century with a likelihood of ≤67%, and limit warming to 1.5°C in 2100 with a likelihood >50%. Limited overshoot refers to exceeding 1.5°C by up to about 0.1°C and for up to several decades. | 97 | 160 |
| **C2: Return warming to 1.5°C (>50%) after a high overshoot** | Exceed warming of 1.5°C during the 21st century with a likelihood of >67%, and limit warming to 1.5°C in 2100 with a likelihood of >50%. High overshoot refers to temporarily exceeding 1.5°C global warming by 0.1°C–0.3°C for up to several decades. | 133 | 170 |
| **C3: Limit warming to 2°C (>67%)** | Limit peak warming to 2°C throughout the 21st century with a likelihood of >67%. | 311 | 374 |
| **C4: Limit warming to 2°C (>50%)** | Limit peak warming to 2°C throughout the 21st century with a likelihood of >50%. | 159 | 213 |
| **C5: Limit warming to 2.5°C (>50%)** | Limit peak warming to 2.5°C throughout the 21st century with a likelihood of >50%. | 212 | 258 |
| **C6: Limit warming to3°C (>50%)** | Limit peak warming to 3°C throughout the 21st century with a likelihood of >50%. | 97 | 129 |
| **C7: Limit warming to 4°C (>50%)** | Limit peak warming to 4°C throughout the 21st century with a likelihood of >50%. | 164 | 230 |
| **C8: Exceed warming of 4°C (≥50%)** | Exceed warming of 4°C during the 21st century with a likelihood of ≥50%. | 29 | 40 |
| **No climate assessment** | Scenario time horizon <2100; insufficient emissions species reported. | 484 | 692 |
| | Total: | 1686 | 2266 |

**Table 15 | Global scenarios by modelling framework and climate category.** Table includes number of scenarios that passed all vetting checks and number of all scenarios that received a climate categorisation (in brackets, including those not passing vetting). Unique model versions have been grouped into modelling frameworks for presentation in this table.[12] For a full list of unique model versions, please see the AR6 scenario database.

| Model group | C1: Limit to 1.5°C (>50%) with no or limited OS | C2: Return to 1.5°C (>50%) after high OS | C3: Limit to 2°C (>67%) | C4: Limit to 2°C (>50%) | C5: Limit to 2.5°C (>50%) | C6: Limit to 3.0°C (>50%) | C7: Limit to 4.0°C (>50%) | C8: Exceed 4.0°C (≥50%) | No climate assessment | Total with climate categorisation |
|---|---|---|---|---|---|---|---|---|---|---|
| AIM/ CGE+Hub | 4 (18) | 3 (7) | 17 (37) | 8 (23) | 13 (23) | 4 (7) | 6 (32) | – (8) | 7 (7) | 55 (155) |
| C-ROADS | 3 (3) | 2 (2) | | | | | | 1 (1) | | 6 (6) |
| COFFEE | 1 (1) | 4 (7) | 14 (16) | 15 (22) | 21 (24) | 9 (11) | 1 (3) | | | 65 (84) |
| DNE21+ | – (4) | | – (7) | – (10) | – (3) | – (4) | – (8) | | 9 (10) | – (36) |
| EPPA | | | 1 (3) | 3 (4) | | 1 (1) | 2 (2) | | | 7 (10) |
| En-ROADS | – (2) | | | | | | | – (1) | | – (3) |
| GCAM | 6 (10) | 6 (9) | 13 (17) | 9 (16) | 6 (13) | – (1) | 4 (6) | 1 (1) | 18 (63) | 45 (73) |
| GCAM-PR | | | | | – (1) | 1 (3) | 2 (3) | | 13 (14) | 3 (7) |
| GEM-E3 | 2 (2) | 10 (10) | 12 (12) | 6 (6) | 5 (5) | 3 (3) | 3 (3) | | 4 (11) | 41 (41) |
| GRAPE-15 | | | | – (1) | – (7) | – (8) | – (2) | | | – (18) |
| IMAGE | 7 (16) | 9 (9) | 34 (34) | 18 (18) | 22 (22) | 16 (16) | 34 (34) | 2 (2) | 2 (2) | 142 (151) |
| MERGE-ETL | – (1) | | | 1 (1) | | | | – (1) | | 1 (3) |
| MESSAGE | | – (1) | – (4) | – (3) | | | – (1) | | – (1) | – (9) |
| MESSAGE-GLOBIOM | 20 (20) | 43 (48) | 59 (61) | 39 (40) | 57 (59) | 20 (22) | 28 (33) | – (1) | | 266 (284) |
| POLES | 4 (14) | 10 (15) | 26 (26) | 24 (26) | 20 (21) | 11 (12) | 19 (23) | | 1 (1) | 114 (137) |
| REMIND | 13 (15) | 12 (19) | 34 (39) | 1 (1) | 7 (8) | 6 (6) | 22 (24) | 9 (9) | | 104 (121) |
| REMIND-MAgPIE | 28 (36) | 32 (33) | 50 (50) | 15 (15) | 27 (27) | 13 (13) | 26 (26) | 2 (2) | | 193 (202) |
| TIAM-ECN | | | 20 (20) | 6 (6) | 10 (10) | 4 (4) | 5 (5) | | – (13) | 45 (45) |
| TIAM-UCL | | | – (4) | – (1) | | | – (2) | | | – (7) |
| TIAM-WORLD | | | | | – (3) | – (2) | – (4) | | – (2) | – (9) |
| WITCH | 5 (13) | 1 (9) | 29 (35) | 14 (16) | 24 (24) | 9 (9) | 4 (4) | 4 (4) | | 90 (114) |
| WITCH-GLOBIOM | 4 (5) | 1 (1) | 2 (9) | – (4) | – (8) | – (7) | 8 (15) | 10 (10) | | 25 (59) |
| Total | 97 (160) | 133 (170) | 311 (374) | 159 (213) | 212 (258) | 97 (129) | 164 (230) | 29 (40) | 54 (124) | 1202 (1574) |



12 Scenario numbers by modelling framework combine submissions from different model versions of the same model (indicated by version number or project name in the AR6 scenario database). For the AIM, MESSAGE and REMIND modelling frameworks, the grouping covers the following distinct models (including different versions):
**AIM/CGE+Hub:** AIM/CGE, AIM/Hub
**MESSAGE:** MESSAGE, MESSAGE-Transport
**MESSAGE-GLOBIOM:** MESSAGE-GLOBIOM, MESSAGEix-GLOBIOM.
**REMIND:** REMIND, REMIND-H13, REMIND-Buildings, REMIND-Transport, REMIND_EU

**Table 16 | Global scenarios by modelling framework that were not included in the climate assessment due to a time horizon shorter than 2100 or a limited reporting of emissions species that did not include $CO_2$ (total emissions or emissions from energy and industry), $CH_4$ or $N_2O$.** Unique model versions have been grouped into modelling frameworks for presentation in this table.[13] For a full list of unique model versions, please see the AR6 scenario database.

| Model framework | Time horizon | Passed vetting | Total |
|---|---|---|---|
| BET | 2100 | 0 | 16 |
| C-GEM | 2030 | 32 | 32 |
| C3IAM | 2100 | 5 | 14 |
| CGE-MOD | 2030 | 32 | 32 |
| DART | 2030 | 17 | 32 |
| E3ME | 2050 | 10 | 10 |
| EC-MSMR | 2030 | 32 | 32 |
| EDF-GEPA | 2030 | 32 | 32 |
| EDGE-Buildings | 2100 | 8 | 8 |
| ENV-Linkages | 2060 | 7 | 15 |
| ENVISAGE | 2030 | 32 | 32 |
| FARM | 2100 | 0 | 13 |
| GAINS | 2050 | 2 | 2 |
| GEMINI-E3 | 2050 | 6 | 6 |
| GENeSYS-MOD | 2050 | 1 | 1 |
| Global TIMES | 2050 | 0 | 14 |
| GMM | 2060 | 4 | 4 |
| Global Transportation Roadmap | 2050 | 4 | 4 |
| ICES | 2030/2050 | 32 | 43 |
| IEA ETP | 2070 | 1 | 1 |
| IEA WEM | 2050 | 2 | 2 |
| IRENA REmap GRO2020 | 2050 | 2 | 2 |
| IMACLIM | 2050/2080 | 30 | 68 |
| IMACLIM-NLU | 2100 | 1 | 3 |
| LUT-ESTM | 2050 | 0 | 1 |
| MAgPIE | 2100 | 3 | 3 |
| MIGRATION | 2100 | 10 | 10 |
| MUSE | 2100 | 5 | 11 |
| McKinsey | 2050 | 0 | 3 |
| PROMETHEUS | 2050 | 7 | 7 |
| SNOW | 2030 | 32 | 32 |
| TEA | 2030 | 32 | 32 |
| TIAM-Grantham | 2100 | 17 | 19 |
| WEGDYN | 2030 | 32 | 32 |
| Total | | 430 | 568 |

[13] Scenario numbers by modelling framework combine submissions from different model versions of the same model (indicated by version number or project name in the AR6 scenario database).

**Changes in climate classification of scenarios since SR1.5:** Since the definition of warming classes was unchanged from SR1.5 for the lower range of scenarios limiting warming to 2°C or lower, changes in overall emissions characteristics of scenarios in these classes from SR1.5 to AR6 would need to come from the substantially larger ensemble of deep mitigation scenarios collected in the AR6 database compared to the SR1.5 database and from updates in the methodology of the climate assessment. Updates since SR1.5 include the methodology for infilling and harmonisation and the use of an updated climate emulator (MAGICC v7) to provide consistency with AR6 WGI assessment (Annex III.II.2.5.1). Out of the full set of SR1.5, 57% of the 411 scenarios that were represented with global temperature assessments in SR1.5 also have been assessed in AR6. Some SR1.5 scenarios could not be taken on board since they are outdated (too early emissions reductions) and failed the vetting or do not provide sufficient information/data to be included in AR6.

Comparison between SR1.5 and AR6 scenarios and associated climate responses are shown in Figure 6, bottom panel. We show that changes in the climate assessment pipeline are minor compared to climate model uncertainty ranges in WGI (in the order of 0.1°C), but show considerable variation due to different scenario characteristics. The updated harmonisation and infilling together have a small cooling effect compared to raw modelled emissions for the subset of 95 scenarios in C1, C2, and C3 that also were assessed in SR15 (SR1.5 Chapter 2, Table 2.4). This is due to both applying more advanced harmonisation methods consistent with the CMIP6 harmonisation used for WGI, and changing the historical harmonisation year from 2010 to 2015. Together with the update in the climate emulator, we find that the total AR6 assessment is remarkably consistent with SR1.5, albeit slightly cooler (in the order of 0.05°C at peak temperature, 0.1°C in 2100).

The lowest temperature category (C1, limiting warming to 1.5°C with no or low overshoot) used for classifying the most ambitious climate mitigation pathways in the literature, indicates that emissions are on average higher in AR6 in the near term (e.g., 2030) and the time of net-zero $CO_2$ is later by about five years compared to SR1.5 (Figure 6, middle panel). These differences can in part be ascribed to the fact that historical emissions in scenarios, especially among those that passed the vetting, have risen since SR1.5 in line with inventories. This increase has moved the attainable near-term emissions reductions upwards. As a result, the scenarios in the lowest category have also a lower probability of staying below 1.5°C peak warming. Using the WGI emulators, we find that the median probability of staying below 1.5°C in the lowest category (C1) has dropped from about 46% in the SR1.5 scenarios to 38% among the AR6 scenarios. Note that the likelihood of the SR1.5 scenarios limiting warming to 1.5°C with no or limited overshoot has changed from 41% in SR1.5 to 46% in AR6 due to the updated climate assessment using the WGI AR6 climate emulator. Within C1, the vast majority of scenarios that were submitted to AR6 but were not assessed in SR1.5 have median peak temperatures close to 1.6°C. The AR6 scenarios in the lowest category show higher emissions and have a lower chance of keeping warming below 1.5°C, as indicated by the panels showing the distribution of peak warming and exceedance probability in AR6 vs SR1.5, with for instance C1 median peak temperature warming going from 1.55°C in SR1.5 (1.52°C if reassessed with AR6 assessment pipeline) to 1.58°C in AR6.

*A.III.II.3.2.2 Policy Classification of Global Scenarios*

Global scenarios were also classified based on their assumptions regarding climate policy. This information can be deduced from study protocols or the description of scenario designs in the published literature. It has also been elicited as meta-information for scenarios that were submitted to the AR6 database. There are multiple purposes for a policy classification, including controlling for the level of near-term action (Chapter 3, Section 3.5) and estimating costs and other differences between two policy classes (Chapter 3, Section 3.6). Policy classes can be combined with climate classes, for example to identify scenarios that follow the NDC until 2030 and limit warming to 2°C (>67%).

Table 17 presents the policy classification that was chosen for this assessment and the distribution of scenarios across the policy classes. There is a top-level distinction between diagnostic scenarios, scenarios from cost-benefit analyses, scenarios without globally coordinated action, scenarios with immediate such action, and hybrid scenarios that move to globally coordinated action after a period of diverse and uncoordinated national action. On the second hierarchy level, scenarios are classified along distinctive features of scenarios in each class. Scenarios without globally coordinated action are often used as reference scenarios and come as baselines without climate policy efforts, as an extrapolation of current policy trends or as implementation and extrapolation of NDCs (Grant et al. 2020). Scenarios that act immediately to limit warming to some level can be distinguished by whether or not they include transfers to reflect equity considerations (Tavoni et al. 2015; Bauer et al. 2020c; van den Berg et al. 2020) or by whether or not they assume additional policies augmenting a global carbon price (Soergel et al. 2021). Scenarios that delay globally coordinated action until 2030 can differ in their assumptions about the level of near-term action (Roelfsema et al. 2020; van Soest et al. 2021).

To identify the policy classification of each global scenario in the AR6 database, classes are first assigned via text pattern matching on all the metadata collected when submitting the scenarios to the database. The algorithm first looks for keywords and text patterns to establish whether a scenario represents a global, fragmented, diagnostic or CBA policy setup. Then it looks for evidence on the presence of specific regional policies, delayed actions and transfers of permits. Eventually the different pieces of evidence are harmonised into a single policy categorisation decision. The process has been calibrated on the best-known scenarios belonging to the larger model intercomparison projects, and fine-tuned on the other scenarios via further validation against the related literature, consistency checks on reported emission and carbon price trajectories, exchanges with modellers and supervision by the involved IPCC authors. If the information available is enough to identify a policy category number but not sufficient for a subcategory, then only the number is retained (e.g., P2 instead of P2a/b/c). A suffix added after P0 further qualifies a diagnostic scenario as one of the other policy categories.

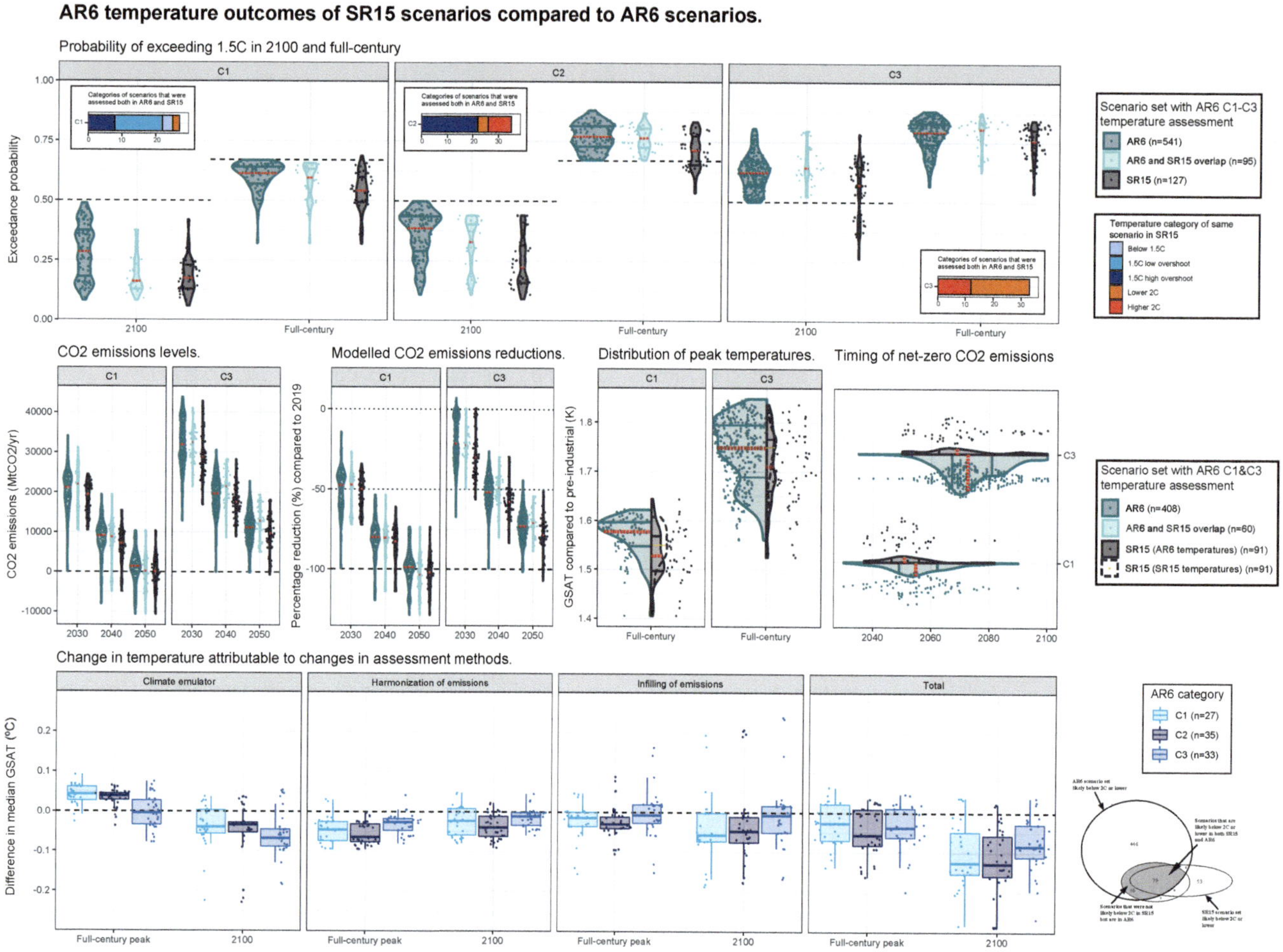


**Figure 6 | Comparing multiple characteristics of scenarios underlying SR1.5 Table 2.4 to the AR6 assessment.**

**Top row: The probability of exceeding 1.5C for scenarios using the AR6 climate assessment pipeline for C1, C2, and C3.** All scenarios in AR6 that pass vetting requirements and get climate classification C1, C2, or C3, are labelled as 'AR6' (n=541). The scenarios that are both in the AR6 database (passing the vetting) and were used for SR1.5 Table 2.4, and are classified as C1, C2 and C3 using the AR6 assessment, are labelled as 'AR6 and SR1.5 overlap (n=95). 'SR1.5 (n=127)' shows all SR1.5 scenarios (except five that were not resubmitted for the AR6 report), including those that failed AR6 vetting, that are classified C1, C2, C3 with the updated AR6 temperature assessment. Dashed lines indicate cut-off temperature exceedance probabilities that align with AR6 category definitions. The violin area is proportional to the number of scenarios. Coloured lines indicate the 25th and 75th percentiles, while the dashed black line indicates the median. The insets in each figure show how the temperature category classifications have changed from SR1.5 to AR6 for those scenarios that are in both databases.

**Middle row: Characteristics of $CO_2$ emissions pathways and the distribution of median peak temperature assessments for C1 and C3.** From left to right: (i) Change in $CO_2$ emissions levels and reductions in 2030, 2040 and 2050 between the AR6 (n=408), AR6 and SR1.5 overlap (n=60) and SR15 sets (n=91). (ii) distribution of scenarios with different median peak temperature scenario outcomes for C1 and C3 for AR6 and SR1.5 (both with AR6 temperature assessment as a solid line and with SR1.5 temperature assessment as a dashed line, with median in yellow). (iii) Year of net-zero $CO_2$ for C1 and C3 for AR6 and SR1.5. Within C3, 27 AR6 scenarios and 2 SR1.5 scenarios with no net-zero year before 2100 have not been visualised. The violin area is proportional to the number of scenarios. Coloured lines indicate the 25th and 75th percentiles, while the dashed black line indicates the median.

**Bottom row: Change in median global mean surface air temperature (GSAT) between the AR6 and SR1.5 climate assessments for both 2100 values and peak temperature values during the 21st century.** Positive values indicate that the temperature assessment is higher for the same scenario than the SR1.5 climate assessment. From left to right: (i) the effect of using MAGICCv7 calibrated to the WGI assessment compared with MAGICC6 as used in SR1.5; (ii) the effect of more advanced emissions harmonisation methods; (iii) the effect of more advanced emissions infilling methods; (iv) the total effect which is the sum of the three components. Boxplots show the median and interquartile range, with the whiskers indicating the 95% range.

**Table 17 | Policy classification of global scenarios.** If the total for a class exceeds the sum of the subclasses, there are scenarios in the class that could not be assigned to a subclass.

| Class | Definition | Number of scenarios | | |
|---|---|---|---|---|
| | | Passed vetting, with climate assessment | Passed vetting | All |
| P0 | Diagnostic scenario | 73 | 99 | 138 |
| **P1** | **No globally coordinated climate policy and either** | **207** | **500** | 632 |
| P1a | – no climate mitigation efforts | 72 | 124 | 179 |
| P1b | – current national mitigation efforts | 51 | 59 | 72 |
| P1c | – NDCs | 56 | 160 | 184 |
| P1d | – other policy assumptions | 24 | 153 | 189 |
| **P2** | **Globally coordinated climate policies with immediate (i.e., before 2030) action and** | **579** | **634** | 992 |
| P2a | – without any transfer of emission permits | 403 | 435 | 610 |
| P2b | – with transfers | 70 | 70 | 143 |
| P2c | – with additional policy assumptions | 45 | 55 | 83 |
| **P3** | **Globally coordinated climate policies with delayed (i.e., from 2030 onwards or after 2030) action, preceded by** | **341** | **451** | 502 |
| P3a | – no mitigation commitment or current national policies | 3 | 7 | 9 |
| P3b | – NDCs | 322 | 426 | 464 |
| P3c | – NDCs and additional policies | 16 | 18 | 29 |
| **P4** | **Cost-benefit analysis** | 2 | 2 | 2 |
| | Total | **1202** | **1686** | **2266** |

AIII

### A.III.II.3.3 National and Regional Pathways

National and regional pathways have been collected in the AR6 scenario database to support the Chapter 4 assessment. In total more than 500 pathways for 24 countries/regions have been submitted to the AR6 scenario database by integrated assessment, energy-economic and computable general equilibrium modelling research teams. This represents a limited sample of the overall literature on mitigation pathways at the national level. The majority of these pathways originate from a set of larger model intercomparison projects, JMIP/EMF35 (Sugiyama et al. 2020a) focusing on Japan, CD-LINKS (Roelfsema et al. 2020; Schaeffer et al. 2020), COMMIT (van Soest et al. 2021), ENGAGE (Fujimori et al. 2021), and Paris Reinforce (Perdana et al. 2020; Nikas et al. 2021) each covering several countries/regions from the following: Australia, Brazil, China, EU, India, Indonesia, Japan, Korea, Russia, Thailand, USA, Vietnam. The remaining pathways stem from individual modelling studies that submitted scenarios to the database (Table 18).

**Table 18 | National and regional mitigation pathways by modelling framework, region and scenario type.**

| Country/ region[a] | Model | CP | NDC | Other | Total |
|---|---|---|---|---|---|
| ARG | IMACLIM-ARG | | 1 | 2 | 3 |
| AUS | TIMES-Australia | 1 | | 7 | 8 |
| BRA | BLUES-Brazil | 2 | 2 | 15 | 19 |
| BRA | COPPE_MSB-Brazil | | | 8 | 8 |
| BRA | IMACLIM-BRA | | | 5 | 5 |
| CHE | STEM-Switzerland | 1 | | 11 | 12 |
| CHN | AIM/Hub-China | 1 | 1 | 7 | 9 |
| CHN | C3IAM | | 3 | 11 | 14 |
| CHN | DREAM-China | | | 1 | 1 |
| CHN | GENeSYS-MOD-CHN | | | 3 | 3 |
| CHN | IPAC-AIM/technology-China | 1 | 1 | 11 | 13 |
| CHN | PECE-China | | | 2 | 2 |
| CHN | TIMES-Australia | | 1 | | 1 |
| CHN | TIMES-China | 1 | 2 | 8 | 11 |
| ECU | ELENA-Ecuador | | | 2 | 2 |
| ETH | TIAM-ECN ETH | 1 | | 1 | 2 |
| EU | E4SMA-EU-TIMES | 1 | | | 1 |
| EU | eTIMES-EU | | | 23 | 23 |
| EU | JRC-EU-TIMES | | | 8 | 8 |
| EU | PRIMES | 2 | 2 | 9 | 13 |
| EU | REMIND_EU | | | 9 | 9 |
| FRA | TIMES-France | | | 8 | 8 |
| GBR | 7see | | | 11 | 11 |
| IDN | AIM/Hub-Indonesia | | | 2 | 2 |
| IDN | DDPP Energy | | | 4 | 4 |
| IND | AIM/Enduse India | 1 | 1 | 5 | 7 |
| IND | AIM/Hub-India | 1 | 1 | 7 | 9 |
| IND | MARKAL-INDIA | 2 | 3 | 13 | 18 |
| JPN | AIM/CGE-Enduse-Japan | | | 6 | 6 |
| JPN | AIM/Enduse-Japan | 3 | 3 | 69 | 75 |

| Country/ region[a] | Model | CP | NDC | Other | Total |
|---|---|---|---|---|---|
| JPN | AIM/Hub-Japan | 1 | 2 | 42 | 45 |
| JPN | DNE21-Japan | | 1 | 30 | 31 |
| JPN | DNE21+ V.14 (national) | 1 | 1 | 4 | 6 |
| JPN | IEEJ-Japan | | 1 | 34 | 35 |
| KEN | TIAM-ECN KEN | 1 | 1 | 2 | 4 |
| KOR | AIM/CGE-Korea | 1 | 1 | 6 | 8 |
| KOR | AIM/Hub-Korea | 1 | 1 | 7 | 9 |
| MDG | TIAM-ECN MDG | 1 | 2 | | 3 |
| MEX | GENeSYS-MOD-MEX | | | 4 | 4 |
| PRT | TIMES-Portugal | | 1 | 3 | 4 |
| RUS | RU-TIMES | 1 | 1 | 4 | 6 |
| SWE | TIMES-Sweden | | | 4 | 4 |
| THA | AIM/Hub-Thailand | 1 | 2 | 19 | 22 |
| USA | GCAM-USA | 2 | 2 | 9 | 13 |
| USA | RIO-USA | | | 12 | 12 |
| VNM | AIM/Hub-Vietnam | 1 | 2 | 14 | 17 |
| ZAF | TIAM-ECN AFR | | | 4 | 4 |
| | Total | 29 | 39 | 466 | 534 |

[a] Countries are abbreviated by their ISO 3166-1 alpha-3 letter codes. EU denotes the European Union.

Notes: CP = current policies, NDC = implementation of Nationally Determined Contributions by 2025/30, Other = all other scenarios.

### A.III.II.3.4 Sector Transition Pathways

Sectoral transition pathways based on the AR6 scenario database are addressed in a number of Chapters, primarily Chapter 6 (Energy systems), Chapter 7 (AFOLU), Chapter 9 (Buildings), Chapter 10 (Transport) and Chapter 11 (Industry). These analyses cover both contributions from global IAMs and from sector-specific models with regional or global coverage. The assessments cover a variety of perspectives, including long-term global and macro-region trends for the sectors, sectoral analysis of the Illustrative Pathways, and comparison of the scenarios between full-economy IAMs and sector-specific models on shorter time horizons. These perspectives have a bi-directional utility – to understand how well IAMs are representing sectoral trends from more granular models, and to position sectoral models in the context of full-economy transitions to verify consistency with different climate outcomes.

**Table 19 | Overview of how models and scenarios were used in sectoral chapters.** All scenario and model counts listed in the table are contained in the AR6 scenario database, with the exception of Chapter 9 (Buildings), which supplemented its dataset with a large number of scenarios separately pulled from the sectoral literature. Scenario counts represents unique model-scenario combinations in the database.

| Sector | Number of models | Number of scenarios | Key sections | Key perspectives |
|---|---|---|---|---|
| Energy systems (Chapter 6) | 12<br>18<br>13 | 476<br>536<br>776 | 6.6<br>6.7<br>6.7.1 | Regional and global energy system characteristics along mitigation pathways and at net-zero emissions specifically: $CO_2$ and GHG emissions; energy resource shares; electricity and hydrogen shares of final energy; energy intensity; per-capita energy use; peak emissions; energy investments |
| AFOLU (Chapter 7) | 11<br>14<br>13<br>3 | 384<br>572<br>559<br>4 | 7.5.1<br>7.5.2<br>7.5.4<br>7.5.5 | Regional and global GHG emissions and land use dynamics; economic mitigation potential for different GHGs; integrated mitigation pathways |
| Buildings (Chapter 9) | 80<br>*(of which 2 are in AR6 scenario database)* | 82<br>*(of which 4 are in AR6 scenario database)* | 9.3, 9.6 | A mixture of top-down and bottom-up models. The former were either national, regional or global while the latter were global only with a breakdown per end use, building type, technologies and energy carrier |
| Transport (Chapter 10) | 24 | 1210 | 10.7 | Global and regional transport demand, activity, modes, vehicles, fuels, and mitigation options |
| Industry (Chapter 11) | 14 | 508 | 11.4.2 | Global final energy use, $CO_2$ emissions, carbon sequestration, fuel shares |

Note 1: The number of models and scenarios reported in the table cannot be summed across chapters, as there is considerable overlap in selected model-scenario combinations across chapters, depending on the filtering processes used for relevant analyses. Moreover, the numbers in the table – and certainly not their sum – are not intended to match those reported for the global pathways assessed by Chapter 3 in Section II.3.2.

Note 2: Numbers shown in the model-count column are arrived at through the authors' best judgement. This has to do with the overlapping nature of unique model versions (within a given model family) as models evolve over time. In this case, model versions with substantial overlap were considered the same model, whereas model versions that differ significantly were counted as unique. For example, MESSAGEix-GLOBIOM 1.0 and MESSAGEix-GLOBIOM_1.1 are counted as the same model, while MESSAGEix-GLOBIOM 1.0 and MESSAGE are counted as different. If instead counting all model versions uniquely, then the following counts would apply to each chapter: Energy systems (30/38/29), AFOLU (18/27/25/4), Buildings (80), Transport (50), Industry (32).

Note 3: The Transport chapter figures in Chapter 10, Section 10.7, are produced from the final AR6 scenario database by the code accompanying this report. The set of model and scenario names appearing in each plot or figure of Section 10.7 varies, depending on whether particular submissions to the database included the specific variables appearing in that plot. Authors advise inspecting the data files accompanying each figure for the set of models/scenarios specific to that figure, or running the code against the final database snapshot to reproduce the figures in question.

## References


Absar, S.M. and B.L. Preston, 2015: Extending the Shared Socioeconomic Pathways for sub-national impacts, adaptation, and vulnerability studies. *Glob. Environ. Change*, **33**, 83–96, doi:10.1016/j.gloenvcha.2015.04.004.

Abujarad, S.Y., M.W. Mustafa, and J.J. Jamian, 2017: Recent approaches of unit commitment in the presence of intermittent renewable energy resources: A review. *Renew. Sustain. Energy Rev.*, **70**, 215–223, doi:10.1016/j.rser.2016.11.246.

ADB, 2017: *Pathways to Low-Carbon Development for Viet Nam*. Asian Development Bank, Phillippines, 138 pp.

Admiraal, A.K., A.F. Hof, M.G.J. den Elzen, and D.P. van Vuuren, 2016: Costs and benefits of differences in the timing of greenhouse gas emission reductions. *Mitig. Adapt. Strateg. Glob. Change*, **21(8)**, doi:10.1007/s11027-015-9641-4.

Aghajani, G.R., H.A. Shayanfar, and H. Shayeghi, 2017: Demand side management in a smart micro-grid in the presence of renewable generation and demand response. *Energy*, **126**, 622–637, doi:10.1016/j.energy.2017.03.051.

Aguiar, A.P.D. et al., 2020: Co-designing global target-seeking scenarios: A cross-scale participatory process for capturing multiple perspectives on pathways to sustainability. *Glob. Environ. Change*, **65**, 102198, doi:10.1016/j.gloenvcha.2020.102198.

Alaidroos, A. and M. Krarti, 2015: Optimal design of residential building envelope systems in the Kingdom of Saudi Arabia. *Energy Build.*, **86**, 104–117, doi:10.1016/j.enbuild.2014.09.083.

Ameli, N. et al., 2021: Higher cost of finance exacerbates a climate investment trap in developing economies. *Nat. Commun.*, **12(1)**, doi:10.1038/s41467-021-24305-3.

Anderson, K. and G. Peters, 2016: The trouble with negative emissions. *Science*, **354(6309)**, 182–183, doi:10.1126/science.aah4567.

Andrijevic, M., J. Crespo Cuaresma, T. Lissner, A. Thomas, and C.F. Schleussner, 2020a: Overcoming gender inequality for climate resilient development. *Nat. Commun.*, **11(1)**, 1–8, doi:10.1038/s41467-020-19856-w.

Andrijevic, M., J. Crespo Cuaresma, R. Muttarak, and C.-F. Schleussner, 2020b: Governance in socioeconomic pathways and its role for future adaptive capacity. *Nat. Sustain.*, **3(1)**, 35–41, doi:10.1038/s41893-019-0405-0.

Andrijevic, M., C.-F. Schleussner, M.J. Gidden, D.L. McCollum, and J. Rogelj, 2020c: COVID-19 recovery funds dwarf clean energy investment needs. *Science*, **370(6514)**, 298–300, doi:10.1126/science.abc9697.

Arbabzadeh, M., R. Sioshansi, J.X. Johnson, and G.A. Keoleian, 2019: The role of energy storage in deep decarbonization of electricity production. *Nat. Commun.*, **10**, doi:10.1038/s41467-019-11161-5.

Arrow, K. et al., 2013: Determining benefits and costs for future generations. *Science*, **341(6144)**, 349–350, doi:10.1126/science.1235665.

Arvesen, A., G. Luderer, M. Pehl, B.L. Bodirsky, and E.G. Hertwich, 2018: Deriving life cycle assessment coefficients for application in integrated assessment modelling. *Environ. Model. Softw.*, **99**, 111–125, doi:10.1016/j.envsoft.2017.09.010.

Auer, C. et al., 2021: Climate change scenario services: From science to facilitating action. *One Earth*, **4(8)**, 1074–1082, doi:10.1016/j.oneear.2021.07.015.

Babatunde, K.A., R.A. Begum, and F.F. Said, 2017: Application of computable general equilibrium (CGE) to climate change mitigation policy: A systematic review. *Renew. Sustain. Energy Rev.*, **78**, 61–71, doi:10.1016/j.rser.2017.04.064.

Babiker, M.H. and R.S. Eckaus, 2007: Unemployment effects of climate policy. *Environ. Sci. Policy*, **10(7–8)**, 600–609, doi:10.1016/j.envsci.2007.05.002.

Badesa, L., F. Teng, and G. Strbac, 2020: Optimal Portfolio of Distinct Frequency Response Services in Low-Inertia Systems. *IEEE Trans. Power Syst.*, **35(6)**, 4459–4469, doi:10.1109/TPWRS.2020.2997194.

Baitz, M., 2017: Attributional Life Cycle Assessment. In: *Goal and Scope Definition in Life Cycle Assessment*, Springer, Dordrecht, The Netherlands, pp. 123–143.

Bajželj, B. et al., 2014: Importance of food-demand management for climate mitigation. *Nat. Clim. Change*, **4(10)**, 924–929, doi:10.1038/nclimate2353.

Bamber, N. et al., 2020: Comparing sources and analysis of uncertainty in consequential and attributional life cycle assessment: review of current practice and recommendations. *Int. J. Life Cycle Assess.*, **25(1)**, doi:10.1007/s11367-019-01663-1.

Baños, R. et al., 2011: Optimization methods applied to renewable and sustainable energy: A review. *Renew. Sustain. Energy Rev.*, **15(4)**, 1753–1766, doi:10.1016/j.rser.2010.12.008.

Barker, T. and S. Scrieciu, 2010: Modeling low climate stabilization with E3MG: Towards a "new economics" approach to simulating energy-environment-economy system dynamics. *Energy J.*, **31** (special issue), 137–164, doi:10.5547/issn0195-6574-ej-vol31-nosi-6.

Barter, G.E., M.A. Tamor, D.K. Manley, and T.H. West, 2015: Implications of modeling range and infrastructure barriers to adoption of battery electric vehicles. *Transp. Res. Rec.*, **2502(1)**, 80–88, doi:10.3141/2502-10.

Bashmakov, I., 2017: Improving the Energy Efficiency of Russian Buildings. *Probl. Econ. Transit.*, **58(11–12)**, 1096–1128, doi:10.1080/10611991.2016.1316099.

Bauer, N., O. Edenhofer, and S. Kypreos, 2008: Linking energy system and macroeconomic growth models. *Comput. Manag. Sci.*, **5(1–2)**, 95–117, doi:10.1007/s10287-007-0042-3.

Bauer, N. et al., 2016: Global fossil energy markets and climate change mitigation – an analysis with REMIND. *Clim. Change*, **136(1)**, 69–82, doi:10.1007/s10584-013-0901-6.

Bauer, N. et al., 2017: Shared Socio-Economic Pathways of the Energy Sector – Quantifying the Narratives. *Glob. Environ. Change*, **42**, 316–330, doi:10.1016/j.gloenvcha.2016.07.006.

Bauer, N. et al., 2020a: Global energy sector emission reductions and bioenergy use: overview of the bioenergy demand phase of the EMF-33 model comparison. *Clim. Change*, **163(3)**, 1553–1568, doi:10.1007/s10584-018-2226-y.

Bauer, N. et al., 2020b: Quantification of an efficiency–sovereignty trade-off in climate policy. *Nature*, **588(7837)**, 261–266, doi:10.1038/s41586-020-2982-5.

Bauer, N. et al., 2020c: Bio-energy and $CO_2$ emission reductions: an integrated land-use and energy sector perspective. *Clim. Change*, **163(3)**, doi:10.1007/s10584-020-02895-z.

Baumstark, L. et al., 2021: REMIND2.1: transformation and innovation dynamics of the energy-economic system within climate and sustainability limits. *Geosci. Model Dev.*, **14(10)**, 6571–6603, doi:10.5194/gmd-14-6571-2021.

Beach, R.H. et al., 2015: Global mitigation potential and costs of reducing agricultural non-$CO_2$ greenhouse gas emissions through 2030. *J. Integr. Environ. Sci.*, **12**, 87–105, doi:10.1080/1943815X.2015.1110183.

Beck, S. and M. Mahony, 2017: The IPCC and the politics of anticipation. *Nat. Clim. Change*, **7(5)**, 311–313, doi:10.1038/nclimate3264.

Bednar, J., M. Obersteiner, and F. Wagner, 2019: On the financial viability of negative emissions. *Nat. Commun.*, **10(1)**, 1783, doi:10.1038/s41467-019-09782-x.

Bellocchi, S., M. Manno, M. Noussan, M.G. Prina, and M. Vellini, 2020: Electrification of transport and residential heating sectors in support of renewable penetration: Scenarios for the Italian energy system. *Energy*, **196**, 117062, doi:10.1016/j.energy.2020.117062.

Benveniste, H., J.C. Cuaresma, M. Gidden, and R. Muttarak, 2021: Tracing international migration in projections of income and inequality across the Shared Socioeconomic Pathways. *Clim. Change*, **166(39)**, 1–22, doi:10.1007/S10584-021-03133-W.

Bertram, C. et al., 2015: Complementing carbon prices with technology policies to keep climate targets within reach. *Nat. Clim. Change*, **5(3)**, 235–239, doi:10.1038/nclimate2514.

Bertram, C. et al., 2018: Targeted policies can compensate most of the increased sustainability risks in 1.5 °C mitigation scenarios. *Environ. Res. Lett.*, **13(6)**, doi:10.1088/1748-9326/aac3ec.

Bertram, C. et al., 2021: Energy system developments and investments in the decisive decade for the Paris Agreement goals. *Environ. Res. Lett.*, **16(7)**, 74020, doi:10.1088/1748-9326/ac09ae.

Bierwirth, A. and S. Thomas, 2019: *Estimating the sufficiency potential in buildings: the space between under-dimensioned and oversized.* ECEEE Summer Study Proceedings.

Bijl, D.L., P.W. Bogaart, S.C. Dekker, and D.P. van Vuuren, 2018: Unpacking the nexus: Different spatial scales for water, food and energy. *Glob. Environ. Change..*, **48**, 22–31, doi:10.1016/J.GLOENVCHA.2017.11.005.

Bistline, J.E.T. and D.T. Young, 2019: Economic drivers of wind and solar penetration in the US. *Environ. Res. Lett.*, **14(12)**, 1–12, doi:10.1088/1748-9326/ab4e2d.

Blanford, G., J. Merrick, R. Richels, and S. Rose, 2014a: Trade-offs between mitigation costs and temperature change. *Clim. Change*, **123(3–4)**, 527–541, doi:10.1007/s10584-013-0869-2.

Blanford, G.J., E. Kriegler, and M. Tavoni, 2014b: Harmonization vs. fragmentation: Overview of climate policy scenarios in EMF27. *Clim. Change*, **123(3–4)**, 383–396, doi:10.1007/s10584-013-0951-9.

Blass, V. and C.J. Corbett, 2018: Same Supply Chain, Different Models: Integrating Perspectives from Life Cycle Assessment and Supply Chain Management. *J. Ind. Ecol.*, **22(1)**, doi:10.1111/jiec.12550.

Böhringer, C., S. Peterson, T.F. Rutherford, J. Schneider, and M. Winkler, 2021: Climate policies after Paris: Pledge, Trade and Recycle: Insights from the 36th Energy Modeling Forum Study (EMF36). *Energy Econ.*, **103**, doi:10.1016/j.eneco.2021.105471.

Bonsch, M. et al., 2015: Environmental flow provision: Implications for agricultural water and land-use at the global scale. *Glob. Environ. Change*, **30**, 113–132, doi:10.1016/j.gloenvcha.2014.10.015.

Bos, A.B. et al., 2020: Integrated assessment of deforestation drivers and their alignment with subnational climate change mitigation efforts. *Environ. Sci. Policy*, **114**, 352–365, doi:10.1016/j.envsci.2020.08.002.

Bosetti, V. et al., 2015: Sensitivity to energy technology costs: A multi-model comparison analysis. *Energy Policy*, **80**, 244–263, doi:10.1016/j.enpol.2014.12.012.

Bourdeau, M., X. qiang Zhai, E. Nefzaoui, X. Guo, and P. Chatellier, 2019: Modeling and forecasting building energy consumption: A review of data-driven techniques. *Sustain. Cities Soc.*, **48**, doi:10.1016/j.scs.2019.101533.

Brear, M.J., R. Baldick, I. Cronshaw, and M. Olofsson, 2020: Sector coupling: Supporting decarbonisation of the global energy system. *Electr. J.*, **33(9)**, doi:10.1016/j.tej.2020.106832.

Bréchet, T., F. Gerard, and H. Tulkens, 2011: Efficiency vs. Stability in Climate Coalitions: A Conceptual and Computational Appraisal. *Energy J.*, **32(1)**, doi:10.5547/ISSN0195-6574-EJ-Vol32-No1-3.

Brooker, A., J. Gonder, S. Lopp, and J. Ward, 2015: ADOPT: A Historically Validated Light Duty Vehicle Consumer Choice Model. *SAE Technical Papers*, Vol. 2015 (April).

Brown, T., J. Hörsch, and D. Schlachtberger, 2018: PyPSA: Python for Power System Analysis. *J. Open Res. Softw.*, **6**, doi:10.5334/jors.188.

Brugger, H., W. Eichhammer, N. Mikova, and E. Dönitz, 2021: Energy Efficiency Vision 2050: How will new societal trends influence future energy demand in the European countries? *Energy Policy*, **152**, doi:10.1016/j.enpol.2021.112216.

Brutschin, E. et al., 2021: A multidimensional feasibility evaluation of low-carbon scenarios. *Environ. Res. Lett.*, **16(6)**, doi:10.1088/1748-9326/abf0ce.

Budinis, S., S. Krevor, N. Mac Dowell, N. Brandon, and A. Hawkes, 2018: An assessment of CCS costs, barriers and potential. *Energy Strateg. Rev.*, **22**, 61–81, doi:10.1016/j.esr.2018.08.003.

Bürger, V., T. Hesse, B. Köhler, A. Palzer, and P. Engelmann, 2019: German Energiewende—different visions for a (nearly) climate neutral building sector in 2050. *Energy Effic.*, **12(1)**, doi:10.1007/s12053-018-9660-6.

Burniaux, J.-M. and J. Chateau, 2010: *An Overview of the OECD ENV-Linkages model*. OECD, Paris, France, 17 pp.

Butler, C. et al., 2020: *Decarbonisation futures: Solutions, actions and benchmarks for a net zero emissions Australia*. ClimateWorks Australia, Melbourne, Australia, 138 pp.

Butnar, I. et al., 2020: A deep dive into the modelling assumptions for biomass with carbon capture and storage (BECCS): A transparency exercise. *Environ. Res. Lett.*, **15(8)**, 84008, doi:10.1088/1748-9326/ab5c3e.

Calise, F. et al., 2021: Energy and economic assessment of energy efficiency options for energy districts: Case studies in Italy and Egypt. *Energies*, **14(4)**, 1–24, doi:10.3390/en14041012.

Calvin, K. et al., 2014a: EU 20-20-20 energy policy as a model for global climate mitigation. *Clim. Policy*, **14(5)**, 581–598, doi:10.1080/14693062.2013.879794.

Calvin, K. et al., 2014b: Trade-offs of different land and bioenergy policies on the path to achieving climate targets. *Clim. Change*, **123(3–4)**, 691–704, doi:10.1007/s10584-013-0897-y.

Calvin, K. et al., 2017: The SSP4: A world of deepening inequality. *Glob. Environ. Change*, **42**, 284–296, doi:10.1016/j.gloenvcha.2016.06.010.

Calvin, K. et al., 2019: GCAM v5.1: representing the linkages between energy, water, land, climate, and economic systems. *Geosci. Model Dev.*, **12(2)**, 677–698, doi:10.5194/gmd-12-677-2019.

Calvin, K. et al., 2021: Bioenergy for climate change mitigation: Scale and sustainability. *GCB Bioenergy*, **13(9)**, 1346–1371, doi:10.1111/gcbb.12863.

Cameron, C. et al., 2016: Policy trade-offs between climate mitigation and clean cook-stove access in South Asia. *Nat. Energy*, **1(1)**, 1–5, doi:10.1038/nenergy.2015.10.

Capellán-Pérez, I. et al., 2020: MEDEAS: A new modeling framework integrating global biophysical and socioeconomic constraints. *Energy Environ. Sci.*, **13(3)**, 986–1017, doi:10.1039/c9ee02627d.

Capros, P. et al., 2013: *GEM-E3 Model Documentation*. European Commission, Seville, Spain, 154 pp.

Capros, P. et al., 2014: European decarbonisation pathways under alternative technological and policy choices: A multi-model analysis. *Energy Strateg. Rev.*, **2(3–4)**, 231–245, doi:10.1016/j.esr.2013.12.007.

Cedillos Alvarado, D., S. Acha, N. Shah, and C.N. Markides, 2016: A Technology Selection and Operation (TSO) optimisation model for distributed energy systems: Mathematical formulation and case study. *Appl. Energy*, **180**, 491–503, doi:10.1016/j.apenergy.2016.08.013.

Cetinkaya, E., I. Dincer, and G.F. Naterer, 2012: Life cycle assessment of various hydrogen production methods. *Int. J. Hydrogen Energy*, **37(3)**, doi:10.1016/j.ijhydene.2011.10.064.

Chaichaloempreecha, A., P. Winyuchakrit, and B. Limmeechokchai, 2017: Long-term energy savings and GHG mitigations in Thailand's building sector: Impacts of energy efficiency plan. *Energy Procedia*, **138**, 847–852, doi:10.1016/j.egypro.2017.10.110.

Chang, J. et al., 2021: Reconciling regional nitrogen boundaries with global food security. *Nat. Food*, **2(9)**, doi:10.1038/s43016-021-00366-x.

Chen, Y.-H.H., S. Paltsev, J.M. Reilly, J.F. Morris, and M.H. Babiker, 2016: Long-term economic modeling for climate change assessment. *Econ. Model.*, **52**, doi:10.1016/j.econmod.2015.10.023.

Cheng, R. et al., 2015: A multi-region optimization planning model for China's power sector. *Appl. Energy*, **137**, 413–426, doi:10.1016/j.apenergy.2014.10.023.

Christensen, P., K. Gillingham, and W. Nordhaus, 2018: Uncertainty in forecasts of long-run economic growth. *Proc. Natl. Acad.*, **115(21)**, 5409–5414, doi:10.1073/pnas.1713628115.

Chu, Z., U. Markovic, G. Hug, and F. Teng, 2020: Towards optimal system scheduling with synthetic inertia provision from wind turbines. *IEEE Trans. Power Syst.*, **35(5)**, 4056–4066, doi:10.1109/TPWRS.2020.2985843.

Clark, M.A. et al., 2020: Global food system emissions could preclude achieving the 1.5° and 2°C climate change targets. *Science*, **370(6517)**, 705–708, doi:10.1126/science.aba7357.

Clarke, L. et al., 2009: International climate policy architectures: Overview of the EMF 22 International Scenarios. *Energy Econ.*, **31**, S64–S81, doi:10.1016/j.eneco.2009.10.013.

Clarke L., K. Jiang, K. Akimoto, M. Babiker, G. Blanford, K. Fisher-Vanden, J.-C. Hourcade, V. Krey, E. Kriegler, A. Löschel, D. McCollum, S. Paltsev, S. Rose, P.R. Shukla, M. Tavoni, B.C.C. van der Zwaan, and D.P. van Vuuren, 2014: Assessing Transformation Pathways. In: *Climate Change 2014: Mitigation of Climate Change. Contribution of Working Group III to the Fifth Assessment Report of the Intergovernmental Panel on Climate Change* [Edenhofer, O., R. Pichs-Madruga, Y. Sokona, E. Farahani, S. Kadner, K. Seyboth, A. Adler, I. Baum, S. Brunner, P. Eickemeier, B. Kriemann, J. Savolainen, S. Schlömer, C. von Stechow, T. Zwickel and J.C. Minx (eds.)]. Cambridge University Press, Cambridge, UK and New York, NY, USA.

Colenbrander, S., A. Sudmant, N. Chilundika, and A. Gouldson, 2019: The scope for low-carbon development in Kigali, Rwanda: An economic appraisal. *Sustain. Dev.*, **27(3)**, 349–365, doi:10.1002/sd.1906.

Collins, M., R. Knutti, J. Arblaster, J.-L. Dufresne, T. Fichefet, P. Friedlingstein, X. Gao, W.J. Gutowski, T. Johns, G. Krinner, M. Shongwe, C. Tebaldi, A.J. Weaver and M. Wehner, 2013: Long-term Climate Change: Projections, Commitments and Irreversibility. In: *Climate Change 2013: The Physical Science Basis. Contribution of Working Group I to the Fifth Assessment Report of the Intergovernmental Panel on Climate Change* [Stocker, T.F., D. Qin, G.-K. Plattner, M. Tignor, S.K. Allen, J. Boschung, A. Nauels, Y. Xia, V. Bex and P.M. Midgley (eds.)]. Cambridge University Press, Cambridge, UK and New York, NY, USA.

Crespo Cuaresma, J., 2017: Income projections for climate change research: A framework based on human capital dynamics. *Glob. Environ. Change*, **42**, 226–236, doi:10.1016/j.gloenvcha.2015.02.012.

Creutzig, F. et al., 2015: Bioenergy and climate change mitigation: an assessment. *GCB Bioenergy*, **7(5)**, 916–944, doi:10.1111/gcbb.12205.

Creutzig, F. et al., 2017: The underestimated potential of solar energy to mitigate climate change. *Nat. Energy*, **2(9)**, doi:10.1038/nenergy.2017.140.

Creutzig, F. et al., 2018: Towards demand-side solutions for mitigating climate change. *Nat. Clim. Change*, **8(4)**, 260–263, doi:10.1038/s41558-018-0121-1.

Csoknyai, T. et al., 2016: Building stock characteristics and energy performance of residential buildings in Eastern-European countries. *Energy Build.*, **132**, 39–52, doi:10.1016/j.enbuild.2016.06.062.

Curran, M.A., 2013: Life Cycle Assessment: a review of the methodology and its application to sustainability. *Curr. Opin. Chem. Eng.*, **2(3)**, doi:10.1016/j.coche.2013.02.002.

Daly, H.E. and B. Fais, 2014: UK TIMES Model Overview. UCL Energy Institute. https://www.ucl.ac.uk/drupal/site_energy-models/sites/energy-models/files/uk-times-overview.pdf.

Daly, H.E. et al., 2014: Incorporating travel behaviour and travel time into TIMES energy system models. *Appl. Energy*, **135**, 429–439, doi:10.1016/j.apenergy.2014.08.051.

de la Rue du Can, S., D. Pudleiner, and K. Pielli, 2018: Energy efficiency as a means to expand energy access: A Uganda roadmap. *Energy Policy*, **120**, 354–364, doi:10.1016/j.enpol.2018.05.045.

de la Rue du Can, S. et al., 2019: Modeling India's energy future using a bottom-up approach. *Appl. Energy*, **238**(December 2018), 1108–1125, doi:10.1016/j.apenergy.2019.01.065.

de Melo, C.A. and G. de Martino Jannuzzi, 2015: Cost-effectiveness of $CO_2$ emissions reduction through energy efficiency in Brazilian building sector. *Energy Effic.*, **8(4)**, 815–826, doi:10.1007/s12053-014-9322-2.

de Vos, L., H. Biemans, J.C. Doelman, E. Stehfest, and D.P. van Vuuren, 2021: Trade-offs between water needs for food, utilities, and the environment – a nexus quantification at different scales. *Environ. Res. Lett.*, **16(11)**, doi:10.1088/1748-9326/ac2b5e.

Deetman, S. et al., 2020: Modelling global material stocks and flows for residential and service sector buildings towards 2050. *J. Clean. Prod.*, **245(118658)**, doi:10.1016/j.jclepro.2019.118658.

Dellink, R., J. Chateau, E. Lanzi, and B. Magné, 2017: Long-term economic growth projections in the Shared Socioeconomic Pathways. *Glob. Environ. Change*, **42**, 200–214, doi:10.1016/j.gloenvcha.2015.06.004.

Deng, Y.Y. et al., 2015: Quantifying a realistic, worldwide wind and solar electricity supply. *Glob. Environ. Change*, **31**, 239–252, doi:10.1016/j.gloenvcha.2015.01.005.

Department of Environmental Affairs, 2014: *South Africa's Greenhouse Gas (GHG) Mitigation Potential Analysis*. Department of Environmental Affairs, Pretoria, South Africa, 152 pp.

Després, J., 2015: Modelling the long-term deployment of electricity storage in the global energy system. PhD Thesis, Université Grenoble Alpes, Grenoble, France, https://tel.archives-ouvertes.fr/tel-01231455v1/document.

Dietrich, J.P. et al., 2019: MAgPIE 4 – a modular open-source framework for modeling global land systems. *Geosci. Model Dev.*, **12(4)**, 1299–1317, doi:10.5194/gmd-12-1299-2019.

Dioha, M.O., N.V. Emodi, and E.C. Dioha, 2019: Pathways for low carbon Nigeria in 2050 by using NECAL2050. *Renew. Energy Focus*, **29**(June), 63–77, doi:10.1016/j.ref.2019.02.004.

Dodds, P.E., I. Keppo, and N. Strachan, 2015: Characterising the Evolution of Energy System Models Using Model Archaeology. *Environ. Model. Assess.*, **20(2)**, 83–102, doi:10.1007/s10666-014-9417-3.

Doelman, J.C., E. Stehfest, A. Tabeau, and H. van Meijl, 2019: Making the Paris agreement climate targets consistent with food security objectives. *Glob. Food Sec.*, **23**, 93–103, doi:10.1016/j.gfs.2019.04.003.

Doelman, J.C. et al., 2020: Afforestation for climate change mitigation: Potentials, risks and trade-offs. *Glob. Change.,. Biol.*, **26(3)**, 1576–1591, doi:10.1111/GCB.14887.

Drouet, L. et al., 2021: Net zero-emission pathways reduce the physical and economic risks of climate change. *Nat. Clim. Change*, **11(12)**, 1070–1076, doi:10.1038/s41558-021-01218-z.

Duerinck, J. et al., 2008: *Assessment and improvement of methodologies used for Greenhouse Gas projections*. Vlaamse Instelling voor Technologisch Onderzoek Mol, Belgium, Öko-Institut e.V., Berlin, Germany, and Institute for European Environmental Policy, Brussels, Belgium, https://ec.europa.eu/clima/document/download/1cf69fe3-f5c6-40a0-b5f1-efb9e2b2d8df_en?filename=assessing_methodologies_for_ghg_projections_en.pdf.

Duscha, V., J. Wachsmuth, J. Eckstein, and B. Pfluger, 2019: *GHG-neutral EU2050 – a scenario of an EU with net-zero greenhouse gas emissions and its implications*. Umweltbundesamt, Dessau-Roßlau, Germany, 81 pp.

Dvorkin, Y., H. Pandžić, M.A. Ortega-Vazquez, and D.S. Kirschen, 2015: A hybrid stochastic/interval approach to transmission-constrained unit commitment. *IEEE Trans. Power Syst.*, **30(2)**, 621–631, doi:10.1109/TPWRS.2014.2331279.

Earles, J.M. and A. Halog, 2011: Consequential life cycle assessment: a review. *Int. J. Life Cycle Assess.*, **16(5)**, doi:10.1007/s11367-011-0275-9.

Edelenbosch, O., D. Rovelli, A. Levesque, G. Marangoni, and M. Tavoni, 2021: Long term, cross-country effects of buildings insulation policies. *Technol. Forecast. Soc. Change*, **170**, 120887, doi:10.1016/j.techfore.2021.120887.

Edelenbosch, O.Y. et al., 2017a: Decomposing passenger transport futures: Comparing results of global integrated assessment models. *Transp. Res. Part D Transp. Environ.*, **55**, 281–293, doi:10.1016/j.trd.2016.07.003.

Edelenbosch, O.Y. et al., 2017b: Comparing projections of industrial energy demand and greenhouse gas emissions in long-term energy models. *Energy*, **122**, 701–710, doi:10.1016/j.energy.2017.01.017.


AIII

Edenhofer, O. and M. Kowarsch, 2015: Cartography of pathways: A new model for environmental policy assessments. *Environ. Sci. Policy*, **51**, 56–64, doi:10.1016/j.envsci.2015.03.017.

Ellenbeck, S. and J. Lilliestam, 2019: How modelers construct energy costs: Discursive elements in Energy System and Integrated Assessment Models. *Energy Res. Soc. Sci.*, **47**, 69–77, doi:10.1016/j.erss.2018.08.021.

Emmerling, J. et al., 2016: The WITCH 2016 Model – Documentation and Implementation of the Shared Socioeconomic Pathways. FEEM Working Paper No. 42.2016, Fondazio Eni Enrico Mattei, Milan, Italy, doi:10.2139/ssrn.2800970.

Emmerling, J. et al., 2019: The role of the discount rate for emission pathways and negative emissions. *Environ. Res. Lett.*, **14(10)**, 104008, doi:10.1088/1748-9326/ab3cc9.

Energetics, 2016: *Modelling and analysis of Australia's abatement opportunities*. Report to the Department of Environment. Energetics, 60 p. pp. Australia.

Eriksson, M., 2020: Afforestation and avoided deforestation in a multi-regional integrated assessment model. *Ecol. Econ.*, **169**, doi:10.1016/j.ecolecon.2019.106452.

Eurek, K. et al., 2017: An improved global wind resource estimate for integrated assessment models. *Energy Econ.*, **64**, 552–567, doi:10.1016/j.eneco.2016.11.015.

Eyckmans, J. and H. Tulkens, 2003: Simulating coalitionally stable burden sharing agreements for the climate change problem. *Resour. Energy Econ.*, **25(4)**, doi:10.1016/S0928-7655(03)00041-1.

Eyring, V. et al., 2015: Overview of the Coupled Model Intercomparison Project Phase 6 (CMIP6) experimental design and organisation. *Geosci. Model Dev. Discuss.*, **8(12)**, 10539–10583, doi:10.5194/gmdd-8-10539-2015.

Fajardy, M. et al., 2021: The economics of bioenergy with carbon capture and storage (BECCS) deployment in a 1.5°C or 2°C world. *Glob. Environ. Change*, **68**, 102262, doi:10.1016/j.gloenvcha.2021.102262.

Fattahi, A., J. Sijm, and A. Faaij, 2020: A systemic approach to analyze integrated energy system modeling tools: A review of national models. *Renew. Sustain. Energy Rev.*, **133**, doi:10.1016/j.rser.2020.110195.

Fazlollahi, S., P. Mandel, G. Becker, and F. Maréchal, 2012: Methods for multi-objective investment and operating optimization of complex energy systems. *Energy*, **45(1)**, 12–22, doi:10.1016/j.energy.2012.02.046.

Feijoo, F. et al., 2018: The future of natural gas infrastructure development in the United States. *Appl. Energy*, **228**, 149–166, doi:10.1016/j.apenergy.2018.06.037.

Filippi Oberegger, U., R. Pernetti, and R. Lollini, 2020: Bottom-up building stock retrofit based on levelized cost of saved energy. *Energy Build.*, **210**, 109757, doi:10.1016/j.enbuild.2020.109757.

Finnveden, G. et al., 2009: Recent developments in Life Cycle Assessment. *J. Environ. Manage.*, **91(1)**, 1–21, doi:10.1016/j.jenvman.2009.06.018.

Fisch-Romito, V. and C. Guivarch, 2019: Transportation infrastructures in a low carbon world: An evaluation of investment needs and their determinants. *Transp. Res. Part D Transp. Environ.*, **72**(May), 203–219, doi:10.1016/j.trd.2019.04.014.

Fishman, T. et al., 2021: A comprehensive set of global scenarios of housing, mobility, and material efficiency for material cycles and energy systems modeling. *J. Ind. Ecol.*, **25(2)**, 305–320, doi:10.1111/jiec.13122.

Fleiter, T. et al., 2018: A methodology for bottom-up modelling of energy transitions in the industry sector: The FORECAST model. *Energy Strateg. Rev.*, **22**, doi:10.1016/j.esr.2018.09.005.

Forster, P., T. Storelvmo, K. Armour, W. Collins, J.-L. Dufresne, D. Frame, D.J. Lunt, T. Mauritsen, M.D. Palmer, M. Watanabe, M. Wild, and H. Zhang: 2021, The Earth's Energy Budget, Climate Feedbacks, and Climate Sensitivity. In *Climate Change 2021: The Physical Science Basis. Contribution of Working Group I to the Sixth Assessment Report of the Intergovernmental Panel on Climate Change* [Masson-Delmotte, V., P. Zhai, A. Pirani, S.L. Connors, C. Péan, S. Berger, N. Caud, Y. Chen, L. Goldfarb, M.I. Gomis, M. Huang, K. Leitzell, E. Lonnoy, J.B.R. Matthews, T.K. Maycock, T. Waterfield, O. Yelekçi, R. Yu, and B. Zhou (eds.)]. Cambridge University Press, Cambridge, UK and New York, NY, USA.

Foure, J. et al., 2020: Macroeconomic drivers of baseline scenarios in dynamic CGE models: review and guidelines proposal. *J. Glob. Econ. Anal.*, **5(1)**, 28–62, doi:10.21642/jgea.050102af.

Frame, B., J. Lawrence, A.-G. Ausseil, A. Reisinger, and A. Daigneault, 2018: Adapting global shared socio-economic pathways for national and local scenarios. *Clim. Risk Manag.*, **21**, 39–51, doi:10.1016/j.crm.2018.05.001.

Frank, S. et al., 2017: Reducing greenhouse gas emissions in agriculture without compromising food security? *Environ. Res. Lett.*, **12(10)**, 105004, doi:10.1088/1748-9326/aa8c83.

Frank, S. et al., 2018: Structural change as a key component for agricultural non-$CO_2$ mitigation efforts. *Nat. Commun.*, **9(1)**, doi:10.1038/s41467-018-03489-1.

Frank, S. et al., 2019: Agricultural non-$CO_2$ emission reduction potential in the context of the 1.5°C target. *Nat. Clim. Change*, **9(1)**, 66–72, doi:10.1038/s41558-018-0358-8.

Frank, S. et al., 2021: Land-based climate change mitigation potentials within the agenda for sustainable development. *Environ. Res. Lett.*, **16(2)**, 24006, doi:10.1088/1748-9326/ABC58A.

Fricko, O. et al., 2017: The marker quantification of the Shared Socioeconomic Pathway 2: A middle-of-the-road scenario for the 21st century. *Glob. Environ. Change*, **42**, 251–267, doi:10.1016/j.gloenvcha.2016.06.004.

Fripp, M., 2012: Switch: A Planning Tool for Power Systems with Large Shares of Intermittent Renewable Energy. *Environ. Sci. Technol.*, **46(11)**, doi:10.1021/es204645c.

Fujimori, S., T. Masui, and Y. Matsuoka, 2014: Development of a global computable general equilibrium model coupled with detailed energy end-use technology. *Appl. Energy*, **128**, doi:10.1016/j.apenergy.2014.04.074.

Fujimori, S. et al., 2016: Will international emissions trading help achieve the objectives of the Paris Agreement? *Environ. Res. Lett.*, **11(10)**, 104001, doi:10.1088/1748-9326/11/10/104001.

Fujimori, S. et al., 2017: SSP3: AIM implementation of Shared Socioeconomic Pathways. *Glob. Environ. Change*, **42**, 268–283, doi:10.1016/j.gloenvcha.2016.06.009.

Fujimori, S. et al., 2018: Inclusive climate change mitigation and food security policy under 1.5°C climate goal. *Environ. Res. Lett.*, **13(7)**, doi:10.1088/1748-9326/aad0f7.

Fujimori, S. et al., 2019a: A multi-model assessment of food security implications of climate change mitigation. *Nat. Sustain.*, **2(5)**, 386–396, doi:10.1038/s41893-019-0286-2.

Fujimori, S., J. Rogelj, V. Krey, and K. Riahi, 2019b: A new generation of emissions scenarios should cover blind spots in the carbon budget space. *Nat. Clim. Change*, **9(11)**, 798–800, doi:10.1038/s41558-019-0611-9.

Fujimori, S. et al., 2020: Measuring the sustainable development implications of climate change mitigation. *Environ. Res. Lett.*, **15(8)**, doi:10.1088/1748-9326/ab9966.

Fujimori, S. et al., 2021: A framework for national scenarios with varying emission reductions. *Nat. Clim. Change*, **11(6)**, 472–480, doi:10.1038/s41558-021-01048-z.

Fulton, L., P. Cazzola, and F. Cuenot, 2009: IEA Mobility Model (MoMo) and its use in the ETP 2008. *Energy Policy*, **37(10)**, 3758–3768, doi:10.1016/j.enpol.2009.07.065.

Fuso Nerini, F., I. Keppo, and N. Strachan, 2017: Myopic decision making in energy system decarbonisation pathways. A UK case study. *Energy Strateg. Rev.*, **17**, 19–26, doi:10.1016/j.esr.2017.06.001.

Fuss, S. et al., 2014: Betting on negative emissions. *Nat. Clim. Change*, **4(10)**, 850–853, doi:10.1038/nclimate2392.

Fuss, S. et al., 2018: Negative emissions – Part 2: Costs, potentials and side effects. *Environ. Res. Lett.*, **13(6)**, 063002, doi:10.1088/1748-9326/aabf9f.

Gagnon, P., R. Margolis, J. Melius, C. Phillips, and R. Elmore, 2016: *Photovoltaic Technical Potential in the United States: A Detailed Assessment*. Technical

Report NREL/TP-6A20-65298. National Renewable Energy Laboratory. U.S. Department of Energy. 82 pp.

Gambhir, A. et al., 2017: The Contribution of Non-$CO_2$ Greenhouse Gas Mitigation to Achieving Long-Term Temperature Goals. *Energies*, **10(5)**, 602, doi:10.3390/en10050602.

Gambhir, A., I. Butnar, P.-H. Li, P. Smith, and N. Strachan, 2019: A Review of Criticisms of Integrated Assessment Models and Proposed Approaches to Address These, through the Lens of BECCS. *Energies*, **12(9)**, 1747, doi:10.3390/en12091747.

Geels, F.W., F. Berkhout, and D.P. van Vuuren, 2016: Bridging analytical approaches for low-carbon transitions. *Nat. Clim. Change*, **6(6)**, 576–583, doi:10.1038/nclimate2980.

Gerbaulet, C., C. von Hirschhausen, C. Kemfert, C. Lorenz, and P.Y. Oei, 2019: European electricity sector decarbonization under different levels of foresight. *Renew. Energy*, **141**, 973–987, doi:10.1016/j.renene.2019.02.099.

Giannousakis, A. et al., 2021: How uncertainty in technology costs and carbon dioxide removal availability affect climate mitigation pathways. *Energy*, **216**, 119253, doi:10.1016/j.energy.2020.119253.

Giarola, S. et al., 2021: Challenges in the harmonisation of global integrated assessment models: A comprehensive methodology to reduce model response heterogeneity. *Sci. Total Environ.*, **783**, 146861, doi:10.1016/j.scitotenv.2021.146861.

Gibon, T., A. Arvesen, and E.G. Hertwich, 2017: Life cycle assessment demonstrates environmental co-benefits and trade-offs of low-carbon electricity supply options. *Renew. Sustain. Energy Rev.*, **76**, 1283–1290, doi:10.1016/j.rser.2017.03.078.

Gidden, M.J. et al., 2018: A methodology and implementation of automated emissions harmonization for use in Integrated Assessment Models. *Environ. Model. Softw.*, **105**, 187–200, doi:10.1016/j.envsoft.2018.04.002.

Gidden, M.J. et al., 2019: Global emissions pathways under different socioeconomic scenarios for use in CMIP6: a dataset of harmonized emissions trajectories through the end of the century. *Geosci. Model Dev.*, **12(4)**, 1443–1475, doi:10.5194/gmd-12-1443-2019.

Gillingham, K. et al., 2018: Modeling uncertainty in integrated assessment of climate change: A multimodel comparison. *J. Assoc. Environ. Resour. Econ.*, **5(4)**, 791–826, doi:10.1086/698910.

Global Energy Monitor, 2021: Global Energy Monitor. https://globalenergymonitor.org/. Acessed November 1 2021.

Gollier, C., 2013: *Pricing the Planet's Future: The Economics of Discounting in an Uncertain World*. Princeton University Press, Princeton, NJ, USA.

Gollier, C. and J.K. Hammitt, 2014: The Long-Run Discount Rate Controversy. *Annu. Rev. Resour. Econ.*, **6(1)**, doi:10.1146/annurev-resource-100913-012516.

González-Mahecha, R.E. et al., 2019: Greenhouse gas mitigation potential and abatement costs in the Brazilian residential sector. *Energy Build.*, **184**, 19–33, doi:10.1016/j.enbuild.2018.11.039.

Gota, S., C. Huizenga, K. Peet, N. Medimorec, and S. Bakker, 2019: Decarbonising transport to achieve Paris Agreement targets. *Energy Effic.*, **12(2)**, 363–386, doi:10.1007/s12053-018-9671-3.

Grande-Acosta, G.K. and J.M. Islas-Samperio, 2020: Boosting energy efficiency and solar energy inside the residential, commercial, and public services sectors in Mexico. *Energies*, **13(21)**, doi:10.3390/en13215601.

Grant, N., A. Hawkes, T. Napp, and A. Gambhir, 2020: The appropriate use of reference scenarios in mitigation analysis. *Nat. Clim. Change*, **10(7)**, 605–610, doi:10.1038/s41558-020-0826-9.

Grassi, G. et al., 2017: The key role of forests in meeting climate targets requires science for credible mitigation. *Nat. Clim. Change*, **7**, 220–226, doi:10.1038/nclimate3227.

Grassi, G. et al., 2021: Critical adjustment of land mitigation pathways for assessing countries' climate progress. *Nat. Clim. Change*, **11(5)**, 425–434, doi:10.1038/s41558-021-01033-6.

Griscom, B.W. et al., 2017: Natural climate solutions. *Proc. Natl. Acad. Sci.* **114(44)**, 11645–11650, doi:10.1073/pnas.1710465114.

Grubler, A. et al., 2018: A low energy demand scenario for meeting the 1.5°C target and sustainable development goals without negative emission technologies. *Nat. Energy*, **3(6)**, 515–527, doi:10.1038/s41560-018-0172-6.

Grübler, A. and N. Nakicenovic, 2001: Identifying dangers in an uncertain climate. *Nature*, **412(6842)**, 15, doi:10.1038/35083752.

Guan, J., H. Tang, K. Wang, J. Yao, and S. Yang, 2020: A parallel multi-scenario learning method for near-real-time power dispatch optimization. *Energy*, **202**, doi:10.1016/j.energy.2020.117708.

Guedes, F. et al., 2019: Climate-energy-water nexus in Brazilian oil refineries. *Int. J. Greenh. Gas Control*, **90**, doi:10.1016/j.ijggc.2019.102815.

Guinée, J.B. et al., 2011: Life Cycle Assessment: Past, Present, and Future. *Environ. Sci. Technol.*, **45(1)**, doi:10.1021/es101316v.

Guivarch, C. and J. Rogelj, 2017: Carbon price variations in 2°C scenarios explored. Carbon Pricing Leadership Coalition, 15 pp, http://pure.iiasa.ac.at/14685 (Accessed October 28, 2021).

Guivarch, C., R. Crassous, O. Sassi, and S. Hallegatte, 2011: The costs of climate policies in a second-best world with labour market imperfections. *Clim. Policy*, **11(1)**, 768–788, doi:10.3763/cpol.2009.0012.

Guo, F., B. van Ruijven, B. Zakeri, V. Krey, and K. Riahi, 2021: *Global Energy Interconnection: A scenario analysis based on the MESSAGEix-GLOBIOM Model.* IIASA Report, IIASA, Laxenburg, Austria, 20 pp. http://pure.iiasa.ac.at/id/eprint/17487/.

Gütschow, J. et al., 2016: The PRIMAP-hist national historical emissions time series. *Earth Syst. Sci. Data*, **8(2)**, 571–603, doi:10.5194/essd-8-571-2016.

Hagelaar, G., 2001: Environmental supply chain management: using life cycle assessment to structure supply chains. *Int. Food Agribus. Manag. Rev.*, **4(4)**, doi:10.1016/S1096-7508(02)00068-X.

Hall, L.M.H. and A.R. Buckley, 2016: A review of energy systems models in the UK: Prevalent usage and categorisation. *Appl. Energy*, **169**, 607–628, doi:10.1016/j.apenergy.2016.02.044.

Hall, P.J. and E.J. Bain, 2008: Energy-storage technologies and electricity generation. *Energy Policy*, **36(12)**, 4352–4355, doi:10.1016/j.enpol.2008.09.037.

Hammad, E., A. Farraj, and D. Kundur, 2019: On Effective Virtual Inertia of Storage-Based Distributed Control for Transient Stability. *IEEE Trans. Smart Grid*, **10(1)**, 327–336, doi:10.1109/TSG.2017.2738633.

Hanaoka, T. and T. Masui, 2020: Exploring effective short-lived climate pollutant mitigation scenarios by considering synergies and trade-offs of combinations of air pollutant measures and low carbon measures towards the level of the 2°C target in Asia. *Environ. Pollut.*, **261**, doi:10.1016/j.envpol.2019.113650.

Hanes, R.J. and A. Carpenter, 2017: Evaluating opportunities to improve material and energy impacts in commodity supply chains. *Environ. Syst. Decis.*, **37(1)**, 6–12, doi:10.1007/s10669-016-9622-5.

Hanna, R. and R. Gross, 2020: How do energy systems model and scenario studies explicitly represent socio-economic, political and technological disruption and discontinuity? Implications for policy and practitioners. *Energy Policy*, **149**, doi:10.1016/j.enpol.2020.111984.

Hänsel, M.C. et al., 2020: Climate economics support for the UN climate targets. *Nat. Clim. Change*, **10(8)**, 781–789, doi:10.1038/s41558-020-0833-x.

Hansen, K., C. Breyer, and H. Lund, 2019: Status and perspectives on 100% renewable energy systems. *Energy*, **175**, 471–480, doi:10.1016/j.energy.2019.03.092.

Hanssen, S.V. et al., 2020: Biomass residues as twenty-first century bioenergy feedstock – a comparison of eight integrated assessment models. *Clim. Change*, **163(3)**, 1569–1586, doi:10.1007/s10584-019-02539-x.

Hardt, L. and D.W. O'Neill, 2017: Ecological Macroeconomic Models: Assessing Current Developments. *Ecol. Econ.*, **134**, 198–211, doi:10.1016/j.ecolecon.2016.12.027.

Harmsen, J.H.M. et al., 2019: Long-term marginal abatement cost curves of non-$CO_2$ greenhouse gases. *Environ. Sci. Policy*, **99**(March), 136–149, doi:10.1016/j.envsci.2019.05.013.

Harmsen, M. et al., 2020: The role of methane in future climate strategies: mitigation potentials and climate impacts. *Clim. Change*, **163(3)**, 1409–1425, doi:10.1007/s10584-019-02437-2.

Harmsen, M. et al., 2021: Integrated assessment model diagnostics: key indicators and model evolution. *Environ. Res. Lett.*, **16(5)**, 54046, doi:10.1088/1748-9326/abf964.

Hartin, C.A., P. Patel, A. Schwarber, R.P. Link, and B.P. Bond-Lamberty, 2015: A simple object-oriented and open-source model for scientific and policy analyses of the global climate system – Hector v1.0. *Geosci. Model Dev.*, **8(4)**, 939–955, doi:10.5194/gmd-8-939-2015.

Hasegawa, T., S. Fujimori, A. Ito, K. Takahashi, and T. Masui, 2017: Global land-use allocation model linked to an integrated assessment model. *Sci. Total Environ.*, **580**, 787–796, doi:10.1016/j.scitotenv.2016.12.025.

Hasegawa, T. et al., 2018: Risk of increased food insecurity under stringent global climate change mitigation policy. *Nat. Clim. Change*, **8(8)**, 699–703, doi:10.1038/s41558-018-0230-x.

Hasegawa, T. et al., 2020: Food security under high bioenergy demand toward long-term climate goals. *Clim. Change*, **163(3)**, doi:10.1007/s10584-020-02838-8.

Hasegawa, T. et al., 2021: Land-based implications of early climate actions without global net-negative emissions. *Nat. Sustain.*, **4(12)**, 1052–1059, doi:10.1038/s41893-021-00772-w.

Havlik, P. et al., 2014: Climate change mitigation through livestock system transitions. *Proc. Natl. Acad. Sci.*, **111(10)**, 3709–3714, doi:10.1073/pnas.1308044111.

Hawkins, T.R., B. Singh, G. Majeau-Bettez, and A.H. Strømman, 2013: Comparative Environmental Life Cycle Assessment of Conventional and Electric Vehicles. *J. Ind. Ecol.*, **17(1)**, doi:10.1111/j.1530-9290.2012.00532.x.

Hejazi, M. et al., 2014: Long-term global water projections using six socioeconomic scenarios in an integrated assessment modeling framework. *Technol. Forecast. Soc. Change*, **81(1)**, 205–226, doi:10.1016/j.techfore.2013.05.006.

Hejazi, M.I. et al., 2015: 21st century United States emissions mitigation could increase water stress more than the climate change it is mitigating. *Proc. Natl. Acad. Sci.*, **112(34)**, 10635–10640, doi:10.1073/pnas.1421675112.

Hellweg, S. and L. Milà i Canals, 2014: Emerging approaches, challenges and opportunities in life cycle assessment. *Science*, **344(6188)**, 1109–1113, doi:10.1126/science.1248361.

Herrendorf, B., R. Rogerson, and Á. Valentinyi, 2014: Growth and Structural Transformation. In: *Handbook of Economic Growth*, Vol. 2, pp. 855–941.

Hertwich, E.G., and J.K. Hammitt, 2001a: A decision-analytic framework for impact assessment part I: LCA and decision analysis. *Int. J. Life Cycle Assess.*, **6(1)**, doi:10.1007/BF02977588.

Hertwich, E.G. and J.K. Hammitt, 2001b: A decision-analytic framework for impact assessment part 2 - Midpoints, endpoints, and criteria for method development. *Int. J. Life Cycle Assess.*, **6(5)**, doi:10.1007/BF02978787l.

Heuberger, C.F., I. Staffell, N. Shah, and N. Mac Dowell, 2018: Impact of myopic decision-making and disruptive events in power systems planning. *Nat. Energy*, **3(8)**, 634–640, doi:10.1038/s41560-018-0159-3.

Ho, E., D. V Budescu, V. Bosetti, D.P. van Vuuren, and K. Keller, 2019: Not all carbon dioxide emission scenarios are equally likely: a subjective expert assessment. *Clim. Change*, **155(4)**, 545–561, doi:10.1007/s10584-019-02500-y.

Hoesly, R.M. et al., 2018: Historical (1750–2014) anthropogenic emissions of reactive gases and aerosols from the Community Emissions Data System (CEDS). *Geosci. Model Dev.*, **11(1)**, 369–408, doi:10.5194/gmd-11-369-2018.

Holden, P.B. et al., 2018: Climate–carbon cycle uncertainties and the Paris Agreement. *Nat. Clim. Change, 2018 87*, **8(7)**, 609–613, doi:10.1038/s41558-018-0197-7.

Holz, C., L.S. Siegel, E. Johnston, A.P. Jones, and J. Sterman, 2018: Ratcheting ambition to limit warming to 1.5°C-trade-offs between emission reductions and carbon dioxide removal. *Environ. Res. Lett.*, **13(6)**, doi:10.1088/1748-9326/aac0c1.

Horváth, M., D. Kassai-Szoó, and T. Csoknyai, 2016: Solar energy potential of roofs on urban level based on building typology. *Energy Build.*, **111**, 278–289, doi:10.1016/j.enbuild.2015.11.031.

Houghton, R.A. et al., 2012: Carbon emissions from land use and land-cover change. *Biogeosciences*, **9(12)**, 5125–5142, doi:10.5194/bg-9-5125-2012.

Howells, M. et al., 2013: Integrated analysis of climate change, land-use, energy and water strategies. *Nat. Clim. Change*, **3(7)**, 621–626, doi:10.1038/nclimate1789.

Humpenöder, F. et al., 2018: Large-scale bioenergy production: How to resolve sustainability trade-offs? *Environ. Res. Lett.*, **13(2)**, 24011, doi:10.1088/1748-9326/aa9e3b.

Humpenöder, F. et al., 2020: Peatland protection and restoration are key for climate change mitigation. *Environ. Res. Lett.*, **15(10)**, doi:10.1088/1748-9326/abae2a.

Huppmann, D., J. Rogelj, E. Kriegler, V. Krey, and K. Riahi, 2018: A new scenario resource for integrated 1.5 °C research. *Nat. Clim. Change*, **8(12)**, 1027–1030, doi:10.1038/s41558-018-0317-4.

Huppmann, D. et al., 2019: The MESSAGEix Integrated Assessment Model and the ix modeling platform (ixmp): An open framework for integrated and cross-cutting analysis of energy, climate, the environment, and sustainable development. *Environ. Model. Softw.*, **112**, 143–156, doi:10.1016/J.ENVSOFT.2018.11.012.

Hurtt, G.C. et al., 2020: Harmonization of global land use change and management for the period 850–2100 (LUH2) for CMIP6. *Geosci. Model Dev.*, **13(11)**, 5425–5464, doi:10.5194/gmd-13-5425-2020.

IEA, 2020a: *World Energy Model*. International Energy Agency, Paris, France, 112 pp.

IEA, 2020b: *Energy Technology Perspectives 2020*. International Energy Agency, Paris, France, 400 pp.

IEA, 2021: *Net Zero by 2050: A Roadmap for the Global Energy Sector*. International Energy Agency, Paris, France, 224 pp.

IPCC, 1990: *Climate change: the IPCC scientific assessment* [J.T. Houghton, G.J. Jenkins and J.J. Ephraums (eds.)]. Cambridge University Press, Cambridge, UK and New York, NY, USA.

IPCC, 2000: *Special report on emissions scenarios (SRES): A special report of Working Group III of the Intergovernmental Panel on Climate Change* [N. Nakicenovic, J. Alcamo, G. Davis, B. de Vries, J. Fenhann, S. Gaffin, K. Gregory, A. Griibler, T.Y. Jung, T. Kram, E.L. La Rovere, L. Michaelis, S. Mori, T. Morita, W. Pepper, H. Pitcher, L. Price, K. Riahi, A. Roehrl, H.H. Rogner, A. Sankovski, M. Schlesinger, P. Shukla, S. Smith, R. Swart, S. van Rooijen, N. Victor, Z. Dadi (eds.)]. Cambridge University Press, Cambridge, UK and New York, NY, USA.

IPCC, 2019a: *2019 Refinement to the 2006 IPCC Guidelines for National Greenhouse Gas Inventories* [Calvo Buendia, E., Tanabe, K., Kranjc, A., Baasansuren, J., Fukuda, M., Ngarize, S., Osako, A., Pyrozhenko, Y., Shermanau, P. and Federici, S. (eds). IPCC, Switzerland.

IPCC, 2019b: *Climate Change and Land: an IPCC special report on climate change, desertification, land degradation, sustainable land management, food security, and greenhouse gas fluxes in terrestrial ecosystems* [P.R. Shukla, J. Skea, E. Calvo Buendia, V. Masson-Delmotte, H.-O. Pörtner, D.C. Roberts, P. Zhai, R. Slade, S. Connors, R. van Diemen, M. Ferrat, E. Haughey, S. Luz, S. Neogi, M. Pathak, J. Petzold, J. Portugal Pereira, P. Vyas, E. Huntley, K. Kissick, M. Belkacemi, J. Malley, (eds.)]. Cambridge University Press, Cambridge, UK, and New York, NY, USA.

Iqbal, M., M. Azam, M. Naeem, A.S. Khwaja, and A. Anpalagan, 2014: Optimization classification, algorithms and tools for renewable energy: A review. *Renew. Sustain. Energy Rev.*, **39**, 640–654, doi:10.1016/j.rser.2014.07.120.

AIII

IRENA, 2020: *IRENA 2020 – Global Renewables Outlook: Energy transformation 2050*. International Renewable Energy Agency, Abu Dhabi, 291 pp.

IRP, 2020: *Resource Efficiency and Climate Change: Material Efficiency Strategies for a Low-Carbon Future*. Hertwich, E., Lifset, R., Pauliuk, S., Heeren, N. A report of the International Resource Panel. United Nations Environment Programme, Nairobi, Kenya.

Iten R. et al., 2017: *Auswirkungen eines subsidiären Verbots fossiler Heizungen*. Infras, TEP Energy, Zürich/Bern, Switzerland.

Ivanova, D. et al., 2020: Quantifying the potential for climate change mitigation of consumption options. *Environ. Res. Lett.*, **15(9)**, 093001, doi:10.1088/1748-9326/ab8589.

Iyer, G.C. et al., 2015: Improved representation of investment decisions in assessments of $CO_2$ mitigation. *Nat. Clim. Change*, **5(5)**, 436–440, doi:10.1038/nclimate2553.

Jewell, J. et al., 2018: Limited emission reductions from fuel subsidy removal except in energy-exporting regions. *Nature*, **554(7691)**, 229–233, doi:10.1038/nature25467.

Jiang, L. and B.C. O'Neill, 2017: Global urbanization projections for the Shared Socioeconomic Pathways. *Glob. Environ. Change*, **42**, 193–199, doi:10.1016/j.gloenvcha.2015.03.008.

Jiang, R., J. Wang, and Y. Guan, 2012: Robust unit commitment with wind power and pumped storage hydro. *IEEE Trans. Power Syst.*, **27(2)**, 800–810, doi:10.1109/TPWRS.2011.2169817.

Jochem, P., J.J. Gómez Vilchez, A. Ensslen, J. Schäuble, and W. Fichtner, 2018: Methods for forecasting the market penetration of electric drivetrains in the passenger car market. *Transp. Rev.*, **38(3)**, 322–348, doi:10.1080/01441647.2017.1326538.

Jochem, P., E. Szimba, and M. Reuter-Oppermann, 2019: How many fast-charging stations do we need along European highways? *Transp. Res. Part D Transp. Environ.*, **73**, doi:10.1016/j.trd.2019.06.005.

Johansson, D.J.A., C. Azar, M. Lehtveer, and G.P. Peters, 2020: The role of negative carbon emissions in reaching the Paris climate targets: The impact of target formulation in integrated assessment models. *Environ. Res. Lett.*, **15(12)**, 124024, doi:10.1088/1748-9326/abc3f0.

Johnson, N. et al., 2019: Integrated Solutions for the Water-Energy-Land Nexus: Are Global Models Rising to the Challenge? *Water*, **11(11)**, 2223, doi:10.3390/w11112223.

Jones, B. and B.C. O'Neill, 2016: Spatially explicit global population scenarios consistent with the Shared Socioeconomic Pathways. *Environ. Res. Lett.*, **11(8)**, 084003, doi:10.1088/1748-9326/11/8/084003.

Kamal, A., S.G. Al-Ghamdi, and M. Koç, 2019: Role of energy efficiency policies on energy consumption and $CO_2$ emissions for building stock in Qatar. *J. Clean. Prod.*, **235**, 1409–1424, doi:10.1016/j.jclepro.2019.06.296.

KC, S. and W. Lutz, 2017: The human core of the shared socioeconomic pathways: Population scenarios by age, sex and level of education for all countries to 2100. *Glob. Environ. Change*, **42**, 181–192, doi:10.1016/j.gloenvcha.2014.06.004.

Keppo, I. and M. Strubegger, 2010: Short term decisions for long term problems – The effect of foresight on model based energy systems analysis. *Energy*, **35(5)**, 2033–2042, doi:10.1016/j.energy.2010.01.019.

Keppo, I. et al., 2021: Exploring the possibility space: taking stock of the diverse capabilities and gaps in integrated assessment models. *Environ. Res. Lett.*, **16(5)**, 053006, doi:10.1088/1748-9326/abe5d8.

Keramidas, K., A. Kitous, J. Després, and A. Schmitz, 2017: *POLES-JRC Model Documentation*. Publications Office of the European Union, Luxembourg, 78 pp.

Khalili, S., E. Rantanen, D. Bogdanov, and C. Breyer, 2019: Global Transportation Demand Development with Impacts on the Energy Demand and Greenhouse Gas Emissions in a Climate-Constrained World. *Energies*, **12(20)**, 3870, doi:10.3390/en12203870.

Khan, M.M.A., M. Asif, and E. Stach, 2017: Rooftop PV potential in the residential sector of the kingdom of Saudi Arabia. *Buildings*, **7(2)**, doi:10.3390/buildings7020046.

Kikstra, J.S. et al., 2021a: Climate mitigation scenarios with persistent COVID-19-related energy demand changes. *Nat. Energy*, **6(12)**, 1114–1123, doi:10.1038/s41560-021-00904-8.

Kikstra, J.S., A. Mastrucci, J. Min, K. Riahi, and N.D. Rao, 2021b: Decent living gaps and energy needs around the world. *Environ. Res. Lett.*, **16(9)**, doi:10.1088/1748-9326/ac1c27.

Kikstra, J.S. et al., 2021c: The social cost of carbon dioxide under climate-economy feedbacks and temperature variability. *Environ. Res. Lett.*, **16(9)**, doi:10.1088/1748-9326/ac1d0b.

Klimont, Z. et al., 2017: Global anthropogenic emissions of particulate matter including black carbon. *Atmos. Chem. Phys.*, **17(14)**, 8681–8723, doi:10.5194/ACP-17-8681-2017.

Kober, T., H.W. Schiffer, M. Densing, and E. Panos, 2020: Global energy perspectives to 2060 – WEC's World Energy Scenarios 2019. *Energy Strateg. Rev.*, **31**, 100523, doi:10.1016/J.ESR.2020.100523.

Kodjak, D. and D. Meszler, 2019: *Prospects for fuel efficiency, electrification and fleet decarbonisation*. Global Fuel Economy Initiative, London, UK, 31 pp.

Kok, K., S. Pedde, M. Gramberger, P.A. Harrison, and I.P. Holman, 2019: New European socio-economic scenarios for climate change research: operationalising concepts to extend the shared socio-economic pathways. *Reg. Environ. Change*, **19(3)**, 643–654, doi:10.1007/s10113-018-1400-0.

Kowarsch, M. et al., 2017: A road map for global environmental assessments. *Nat. Clim. Change*, **7(6)**, 379–382, doi:10.1038/nclimate3307.

Krarti, M., 2019: Evaluation of Energy Efficiency Potential for the Building Sector in the Arab Region. *Energies*, **12(22)**, 4279, doi:10.3390/en12224279.

Krarti, M., F. Ali, A. Alaidroos, and M. Houchati, 2017: Macro-economic benefit analysis of large scale building energy efficiency programs in Qatar. *Int. J. Sustain. Built Environ.*, **6(2)**, 597–609, doi:10.1016/j.ijsbe.2017.12.006.

Kreidenweis, U. et al., 2016: Afforestation to mitigate climate change: Impacts on food prices under consideration of albedo effects. *Environ. Res. Lett.*, **11(8)**, doi:10.1088/1748-9326/11/8/085001.

Krey, V., 2014: Global energy-climate scenarios and models: a review. *Wiley Interdiscip. Rev. Energy Environ.*, **3(4)**, 363–383, doi:10.1002/wene.98.

Krey, V. and K. Riahi, 2013: Risk hedging strategies under energy system and climate policy uncertainties. *Int. Ser. Oper. Res. Manag. Sci.*, **199**(August), 435–474, doi:10.1007/978-1-4614-9035-7_17.

Krey V., O. Masera, G. Blanford, T. Bruckner, R. Cooke, K. Fisher-Vanden, H. Haberl, E. Hertwich, E. Kriegler, D. Mueller, S. Paltsev, L. Price, S. Schlömer, D. Ürge-Vorsatz, D. van Vuuren, and T. Zwickel, 2014: Annex II: Metrics & Methodology. In: *Climate Change 2014: Mitigation of Climate Change. Contribution of Working Group III to the Fifth Assessment Report of the Intergovernmental Panel on Climate Change* [Edenhofer, O., R. Pichs-Madruga, Y. Sokona, E. Farahani, S. Kadner, K. Seyboth, A. Adler, I. Baum, S. Brunner, P. Eickemeier, B. Kriemann, J. Savolainen, S. Schlömer, C. von Stechow, T. Zwickel and J.C. Minx (eds.)]. Cambridge University Press, Cambridge, UK and New York, NY, USA.

Krey, V. et al., 2016: *MESSAGE-GLOBIOM 1.0 Documentation*. International Institute for Applied Systems Analysis, Laxenburg, Austria.

Krey, V. et al., 2019: Looking under the hood: A comparison of techno-economic assumptions across national and global integrated assessment models. *Energy*, **172**, 1254–1267, doi:10.1016/j.energy.2018.12.131.

Kriegler, E. et al., 2012: The need for and use of socio-economic scenarios for climate change analysis: A new approach based on shared socio-economic pathways. *Glob. Environ. Change*, **22(4)**, 807–822, doi:10.1016/j.gloenvcha.2012.05.005.

Kriegler, E. et al., 2014a: The role of technology for achieving climate policy objectives: Overview of the EMF 27 study on global technology and climate policy strategies. *Clim. Change*, **123(3–4)**, 353–367, doi:10.1007/s10584-013-0953-7.

Kriegler, E. et al., 2014b: What does the 2°C target imply for a global climate agreement in 2020? The limits study on Durban Platform scenarios. *Clim. Change, Econ.*, **4(4)**, doi:10.1142/S2010007813400083.

Kriegler, E. et al., 2014c: A new scenario framework for climate change research: The concept of shared climate policy assumptions. *Clim. Change*, **122(3)**, 401–414, doi:10.1007/s10584-013-0971-5.

Kriegler, E. et al., 2015a: Diagnostic indicators for integrated assessment models of climate policy. *Technol. Forecast. Soc. Change*, **90**(PA), 45–61, doi:10.1016/j.techfore.2013.09.020.

Kriegler, E. et al., 2015b: Making or breaking climate targets: The AMPERE study on staged accession scenarios for climate policy. *Technol. Forecast. Soc. Change*, **90**, 24–44, doi:10.1016/j.techfore.2013.09.021.

Kriegler, E. et al., 2016: Will economic growth and fossil fuel scarcity help or hinder climate stabilization?: Overview of the RoSE multi-model study. *Clim. Change*, **136(1)**, 7–22, doi:10.1007/s10584-016-1668-3.

Kriegler, E. et al., 2017: Fossil-fueled development (SSP5): An energy and resource intensive scenario for the 21st century. *Glob. Environ. Change*, **42**(Supplement C), 297–315, doi:10.1016/j.gloenvcha.2016.05.015.

Kriegler, E. et al., 2018a: Pathways limiting warming to 1.5°C: a tale of turning around in no time? *Philos. Trans. R. Soc. A Math. Phys. Eng. Sci.*, **376(2119)**, 20160457, doi:10.1098/rsta.2016.0457.

Kriegler, E. et al., 2018b: Short term policies to keep the door open for Paris climate goals. *Environ. Res. Lett.*, **13(7)**, 074022, doi:10.1088/1748-9326/aac4f1.

Kriegler, E. et al., 2018c: *Transformations to Achieve the Sustainable Development Goals – Report prepared by The World in 2050 Initiative*. International Institute for Applied Systems Analysis, Laxenburg, Austria, 157 pp.

Kuhnhenn, K., L. Costa, E. Mahnke, L. Schneider, and S. Lange, 2020: *A Societal Transformation Scenario for Staying Below 1.5°C*. Heinrich Böll Foundation, Berlin, Germany, 100 pp.

Kuramochi, T. et al., 2020: Global Emissions Trends and G20 Status and Outlook. *In Emissions Gap Report 2020*. UNEP, Nairobi, Kenya.

Kusumadewi, T.V. and B. Limmeechokchai, 2015: Energy Efficiency Improvement and $CO_2$ Mitigation in Residential Sector: Comparison between Indonesia and Thailand. *Energy Procedia*, **79**, 994–1000, doi:10.1016/j.egypro.2015.11.599.

Kusumadewi, T.V. and B. Limmeechokchai, 2017: $CO_2$ Mitigation in Residential Sector in Indonesia and Thailand: Potential of Renewable Energy and Energy Efficiency. *Energy Procedia*, **138**, 955–960, doi:10.1016/j.egypro.2017.10.006.

Kwag, B.C., B.M. Adamu, and M. Krarti, 2019: Analysis of high-energy performance residences in Nigeria. *Energy Effic.*, **12(3)**, 681–695, doi:10.1007/s12053-018-9675-z.

Lam, A. and J.-F. Mercure, 2021: Which policy mixes are best for decarbonising passenger cars? Simulating interactions among taxes, subsidies and regulations for the United Kingdom, the United States, Japan, China, and India. *Energy Res. Soc. Sci.*, **75**, 101951, doi:10.1016/j.erss.2021.101951.

Lamboll, R.D., Z.R.J. Nicholls, J.S. Kikstra, M. Meinshausen, and J. Rogelj, 2020: Silicone v1.0.0: an open-source Python package for inferring missing emissions data for climate change research. *Geosci. Model Dev.*, **13(11)**, 5259–5275, doi:10.5194/gmd-13-5259-2020.

Lauri, P. et al., 2017: Impact of the 2°C target on global woody biomass use. *For. Policy Econ.*, **83**, doi:10.1016/j.forpol.2017.07.005.

Le Quéré, C. et al., 2016: Global Carbon Budget 2016. *Earth Syst. Sci. Data*, **8(2)**, 605–649, doi:10.5194/essd-8-605-2016.

Leclère, D. et al., 2020: Bending the curve of terrestrial biodiversity needs an integrated strategy. *Nature*, **585(7826)**, 551–556, doi:10.1038/s41586-020-2705-y.

Leggett, J., W.J. Pepper and R.J. Swart, 1992: Emissions scenarios for the IPCC: an update. In: *Climate Change 1992 Supplementary Report to the IPCC Scientific Assessessment* [J.T. Houghton, B.A. Callander and S.K. Varney (eds.)], Cambridge University Press, Cambridge, UK and New York, USA, pp. 69–95.

Leimbach, M. and A. Giannousakis, 2019: Burden sharing of climate change mitigation: global and regional challenges under shared socio-economic pathways. *Clim. Change*, **155(2)**, 273–291, doi:10.1007/s10584-019-02469-8.

Leimbach, M., E. Kriegler, N. Roming, and J. Schwanitz, 2017a: Future growth patterns of world regions – A GDP scenario approach. *Glob. Environ. Change*, **42**, 215–225, doi:10.1016/j.gloenvcha.2015.02.005.

Leimbach, M., A. Schultes, L. Baumstark, A. Giannousakis, and G. Luderer, 2017b: Solution algorithms for regional interactions in large-scale integrated assessment models of climate change. *Ann. Oper. Res.*, **255(1–2)**, 29–45, doi:10.1007/s10479-016-2340-z.

Lenton, T.M. et al., 2019: Climate tipping points — too risky to bet against. *Nature*, **575(7784)**, 592–595, doi:10.1038/d41586-019-03595-0.

Lepault, C. and F. Lecocq, 2021: Mapping forward-looking mitigation studies at country level. *Environ. Res. Lett.*, **16(8)**, 83001, doi:10.1088/1748-9326/ac0ac8.

Lester, M.S., R. Bramstoft, and M. Münster, 2020: Analysis on Electrofuels in Future Energy Systems: A 2050 Case Study. *Energy*, **199**, doi:10.1016/j.energy.2020.117408.

Levesque, A., R.C. Pietzcker, and G. Luderer, 2019: Halving energy demand from buildings: The impact of low consumption practices. *Technol. Forecast. Soc. Change*, **146**, 253–266, doi:10.1016/j.techfore.2019.04.025.

Levesque, A., R.C. Pietzcker, L. Baumstark, and G. Luderer, 2021: Deep decarbonisation of buildings energy services through demand and supply transformations in a 1.5°C scenario. *Environ. Res. Lett.*, **16(5)**, 54071, doi:10.1088/1748-9326/ABDF07.

Li, N. et al., 2019: Air Quality Improvement Co-benefits of Low-Carbon Pathways toward Well Below the 2 °C Climate Target in China. *Environ. Sci. Technol.*, **53(10)**, 5576–5584, doi:10.1021/acs.est.8b06948.

Linton, C., S. Grant-Muller, and W.F. Gale, 2015: Approaches and Techniques for Modelling $CO_2$ Emissions from Road Transport. *Transp. Rev.*, **35(4)**, 533–553, doi:10.1080/01441647.2015.1030004.

Liu, J. et al., 2017: PV-based virtual synchronous generator with variable inertia to enhance power system transient stability utilizing the energy storage system. *Prot. Control Mod. Power Syst.*, **2(1)**, doi:10.1186/s41601-017-0070-0.

Löffler, K. et al., 2017: Designing a model for the global energy system-GENeSYS-MOD: An application of the Open-Source Energy Modeling System (OSeMOSYS). *Energies*, **10(10)**, 1468, doi:10.3390/en10101468.

Löffler, K., T. Burandt, K. Hainsch, and P.Y. Oei, 2019: Modeling the low-carbon transition of the European energy system – A quantitative assessment of the stranded assets problem. *Energy Strateg. Rev.*, **26**, doi:10.1016/j.esr.2019.100422.

Löschel, A., 2002: Technological change in economic models of environmental policy: A survey. *Ecol. Econ.*, **43(2–3)**, 105–126, doi:10.1016/S0921-8009(02)00209-4.

Lucas, P.L., D.P. van Vuuren, J.G.J. Olivier, and M.G.J. den Elzen, 2007: Long-term reduction potential of non-$CO_2$ greenhouse gases. *Environ. Sci. Policy*, **10(2)**, 85–103, doi:10.1016/j.envsci.2006.10.007.

Lucena, A.F.P. et al., 2016: Climate policy scenarios in Brazil: A multi-model comparison for energy. *Energy Econ.*, **56**, 564–574, doi:10.1016/j.eneco.2015.02.005.

Luderer, G. et al., 2012: The economics of decarbonizing the energy system – results and insights from the RECIPE model intercomparison. *Clim. Change*, **114(1)**, 9–37, doi:10.1007/s10584-011-0105-x.

Luderer, G. et al., 2013: Economic mitigation challenges: how further delay closes the door for achieving climate targets. *Environ. Res. Lett.*, **8(3)**, 34033, doi:10.1088/1748-9326/8/3/034033.

Luderer, G. et al., 2018: Residual fossil $CO_2$ emissions in 1.5–2°C pathways. *Nat. Clim. Change*, **8(7)**, 626–633, doi:10.1038/s41558-018-0198-6.

Luderer, G. et al., 2021: Impact of declining renewable energy costs on electrification in low emission scenarios. *Nat. Energy*, accepted, doi:10.1038/s41560-021-00937-z.

Lund, H. et al., 2017: Simulation versus optimisation: Theoretical positions in energy system modelling. *Energies*, **10(7)**, doi:10.3390/en10070840.

Lund, P.D., J. Lindgren, J. Mikkola, and J. Salpakari, 2015: Review of energy system flexibility measures to enable high levels of variable renewable electricity. *Renew. Sustain. Energy Rev.*, **45**, 785–807, doi:10.1016/j.rser.2015.01.057.

Marangoni, G. et al., 2017: Sensitivity of projected long-term $CO_2$ emissions across the Shared Socioeconomic Pathways. *Nat. Clim. Change*, **7(2)**, 113–117, doi:10.1038/nclimate3199.

Marcucci, A., S. Kypreos, and E. Panos, 2017: The road to achieving the long-term Paris targets: energy transition and the role of direct air capture. *Clim. Change*, **144(2)**, 181–193, doi:10.1007/s10584-017-2051-8.

Markard, J., N. Bento, N. Kittner, and A. Nuñez-Jimenez, 2020: Destined for decline? Examining nuclear energy from a technological innovation systems perspective. *Energy Res. Soc. Sci.*, **67**, 101512, doi:10.1016/j.erss.2020.101512.

Markewitz, P., P. Hansen, W. Kuckshinrichs, and J.F. Hake, 2015: Strategies for a low carbon building stock in Germany. 8th International Scientific Conference on Energy and Climate Change, Athens, Institute of Energy and Climate Research, Systems Analysis and Technology Evaluation (IEK-STE), Jülich, Germany.

Markovic, U., Z. Chu, P. Aristidou, and G. Hug, 2019: LQR-Based Adaptive Virtual Synchronous Machine for Power Systems With High Inverter Penetration. *IEEE Trans. Sustain. Energy*, **10(3)**, 1501–1512, doi:10.1109/TSTE.2018.2887147.

Marquardt, S.G. et al., 2021: Identifying regional drivers of future land-based biodiversity footprints. *Glob. Environ. Change*, **69**, doi:10.1016/j.gloenvcha.2021.102304.

Material Economics, 2019: *Industrial Transformation 2050 - Pathways to Net-Zero Emissions from EU Heavy Industry*. University of Cambridge Institute for Sustainability Leadership, Cambridge, UK, 208 pp.

Mbow, C., C. Rosenzweig, L.G. Barioni, T.G. Benton, M. Herrero, M. Krishnapillai, E. Liwenga, P. Pradhan, M.G. Rivera-Ferre, T. Sapkota, F.N. Tubiello, Y. Xu, 2019: Food Security. In: *Climate Change and Land: an IPCC special report on climate change, desertification, land degradation, sustainable land management, food security, and greenhouse gas fluxes in terrestrial ecosystems*. [P.R. Shukla, J. Skea, E. Calvo Buendia, V. Masson-Delmotte, H.-O. Pörtner, D.C. Roberts, P. Zhai, R. Slade, S. Connors, R. van Diemen, M. Ferrat, E. Haughey, S. Luz, S. Neogi, M. Pathak, J. Petzold, J. Portugal Pereira, P. Vyas, E. Huntley, K. Kissick, M. Belkacemi, J. Malley, (eds.)]. Cambridge University Press, Cambridge, UK and New York, NY, USA.

McCollum, D.L., V. Krey, and K. Riahi, 2011: An integrated approach to energy sustainability. *Nat. Clim. Change, 2011 19*, **1(9)**, 428–429, doi:10.1038/nclimate1297.

McCollum, D.L. et al., 2013: Climate policies can help resolve energy security and air pollution challenges. *Clim. Change*, **119(2)**, 479–494, doi:10.1007/s10584-013-0710-y.

McCollum, D.L. et al., 2016: Quantifying uncertainties influencing the long-term impacts of oil prices on energy markets and carbon emissions. *Nat. Energy*, **1(7)**, 16077, doi:10.1038/nenergy.2016.77.

McCollum, D.L. et al., 2017: Improving the behavioral realism of global integrated assessment models: An application to consumers' vehicle choices. *Transp. Res. Part D Transp. Environ.*, **55**, 322–342, doi:10.1016/j.trd.2016.04.003.

McCollum, D.L. et al., 2018: Energy investment needs for fulfilling the Paris Agreement and achieving the Sustainable Development Goals. *Nat. Energy*, **3(7)**, 589–599, doi:10.1038/s41560-018-0179-z.

McJeon, H.C. et al., 2011: Technology interactions among low-carbon energy technologies: What can we learn from a large number of scenarios? *Energy Econ.*, **33(4)**, 619–631, doi:10.1016/j.eneco.2010.10.007.

Meibom, P. et al., 2011: Stochastic optimization model to study the operational impacts of high wind penetrations in Ireland. *IEEE Trans. Power Syst.*, **26(3)**, 1367–1379, doi:10.1109/TPWRS.2010.2070848.

Meinshausen, M., S.C.B. Raper, and T.M.L. Wigley, 2011: Emulating coupled atmosphere-ocean and carbon cycle models with a simpler model, MAGICC6 – Part 1: Model description and calibration. *Atmos. Chem. Phys.*, **11(4)**, 1417–1456, doi:10.5194/acp-11-1417-2011.

Meinshausen, M. et al., 2017: Historical greenhouse gas concentrations for climate modelling (CMIP6). *Geosci. Model Dev.*, **10(5)**, doi:10.5194/gmd-10-2057-2017.

Meinshausen, M. et al., 2020: The shared socio-economic pathway (SSP) greenhouse gas concentrations and their extensions to 2500. *Geosci. Model Dev.*, **13(8)**, 3571–3605, doi:10.5194/gmd-13-3571-2020.

Méjean, A., C. Guivarch, J. Lefèvre, and M. Hamdi-Cherif, 2019: The transition in energy demand sectors to limit global warming to 1.5 °C. *Energy Effic.*, **12(2)**, 441–462, doi:10.1007/s12053-018-9682-0.

Mercure, J.-F., A. Lam, S. Billington, and H. Pollitt, 2018: Integrated assessment modelling as a positive science: private passenger road transport policies to meet a climate target well below 2 °C. *Clim. Change*, **151(2)**, 109–129, doi:10.1007/s10584-018-2262-7.

Mercure, J.-F. et al., 2019: Modelling innovation and the macroeconomics of low-carbon transitions: theory, perspectives and practical use. *Clim. Policy*, **19(8)**, 1019–1037, doi:10.1080/14693062.2019.1617665.

Merini, I., A. Molina-García, M. Socorro García-Cascales, M. Mahdaoui, and M. Ahachad, 2020: Analysis and comparison of energy efficiency code requirements for buildings: A Morocco-Spain case study. *Energies*, **13(22)**, doi:10.3390/en13225979.

Messner, S. and L. Schrattenholzer, 2000: MESSAGE-MACRO: Linking an energy supply model with a macroeconomic module and solving it iteratively. *Energy*, **25(3)**, 267–282, doi:10.1016/S0360-5442(99)00063-8.

Millar, R.J. et al., 2017: Emission budgets and pathways consistent with limiting warming to 1.5°C. *Nat. Geosci.*, **10(10)**, 741–747, doi:10.1038/ngeo3031.

Millward-Hopkins, J., J.K. Steinberger, N.D. Rao, and Y. Oswald, 2020: Providing decent living with minimum energy: A global scenario. *Glob. Environ. Change*, **65**, 102168, doi:10.1016/j.gloenvcha.2020.102168.

Minx, J.C. et al., 2018: Negative emissions – Part 1: Research landscape and synthesis. *Environ. Res. Lett.*, **13(6)**, 063001, doi:10.1088/1748-9326/aabf9b.

Momonoki, T., A. Taniguchi-Matsuoka, Y. Yamaguchi, and Y. Shimoda, 2017: Evaluation of the greenhouse gas reduction effect in the Japanese residential sector considering the characteristics of regions and households. *Build. Simul. Conf. Proc.*, **1**, 494–501, doi:10.26868/25222708.2017.718.

Monforti, F. et al., 2014: Assessing complementarity of wind and solar resources for energy production in Italy. A Monte Carlo approach. *Renew. Energy*, **63**, 576–586, doi:10.1016/j.renene.2013.10.028.

Moomaw, W., F. Yamba, M. Kamimoto, L. Maurice, J. Nyboer, K. Urama, T. Weir, 2011: Introduction. In *Special Report on Renewable Energy Sources and Climate Change Mitigation* [O. Edenhofer, R. Pichs-Madruga, Y. Sokona, K. Seyboth, P. Matschoss, S. Kadner, T. Zwickel, P. Eickemeier, G. Hansen, S. Schlömer, C.von Stechow (eds.)], pp. 161–208, Cambridge University Press, Cambridge, UK and New York, NY, USA.

Morren, J., S.W.H. de Haan, W.L. Kling, and J.A. Ferreira, 2006: Wind turbines emulating inertia and supporting primary frequency control. *IEEE Trans. Power Syst.*, **21(1)**, 433–434, doi:10.1109/TPWRS.2005.861956.

Morris, J. et al., 2019: Representing the costs of low-carbon power generation in multi-region multi-sector energy-economic models. *Int. J. Greenh. Gas Control*, **87**, 170–187, doi:10.1016/j.ijggc.2019.05.016.

Morris, J., H. Kheshgi, S. Paltsev, and H. Herzog, 2021: Scenarios for the deployment of carbon capture and storage in the power sector in a portfolio of mitigation options. *Clim. Change Econ.*, **12(01)**, 2150001, doi:10.1142/S2010007821500019.

Mosnier, A. et al., 2014: Modeling Impact of Development Trajectories and a Global Agreement on Reducing Emissions from Deforestation on Congo Basin Forests by 2030. *Environ. Resour. Econ.*, **57(4)**, 505–525, doi:10.1007/s10640-012-9618-7.

Moss, R.H. et al., 2010: The next generation of scenarios for climate change research and assessment. *Nature*, **463(7282)**, 747–756, doi:10.1038/nature08823.

Mouratiadou, I. et al., 2016: The impact of climate change mitigation on water demand for energy and food: An integrated analysis based on the Shared Socioeconomic Pathways. *Environ. Sci. Policy*, **64**, 48–58, doi:10.1016/j.envsci.2016.06.007.

Müller-Casseres, E., O.Y. Edelenbosch, A. Szklo, R. Schaeffer, and D.P. van Vuuren, 2021: Global futures of trade impacting the challenge to decarbonize the international shipping sector. *Energy*, **237**, 121547, doi:10.1016/J.ENERGY.2021.121547.

Muratori, M. et al., 2021: The rise of electric vehicles – 2020 status and future expectations. *Prog. Energy*, **3(2)**, 22002, doi:10.1088/2516-1083/abe0ad.

Nadel, S., 2016: *Pathway to Cutting Energy Use and Carbon Emissions in Half*. American Council for an Energy-Efficient Economy, Washington, D.C, USA, 43 pp.

Napp, T. et al., 2017: Exploring the Feasibility of Low-Carbon Scenarios Using Historical Energy Transitions Analysis. *Energies*, **10(1)**, 116, doi:10.3390/en10010116.

négaWatt, 2017: *Scénario négaWatt : Un scénario de transition énergétique*. négaWatt, 4 pp.

NGFS, 2020: *NGFS climate scenarios for central banks and supervisors*. Network for Greening the Financial System, https://www.ngfs.net/sites/default/files/medias/documents/820184_ngfs_scenarios_final_version_v6.pdf (Accessed August 22, 2022).

NGFS, 2021: *NGFS Climate Scenarios for central banks and supervisors*. Network for Greening the Financial System, https://www.ngfs.net/en/ngfs-climate-scenarios-central-banks-and-supervisors-june-2021 (Accessed October 29, 2021).

Niamir, L., O. Ivanova, T. Filatova, A. Voinov, and H. Bressers, 2020: Demand-side solutions for climate mitigation: Bottom-up drivers of household energy behavior change in the Netherlands and Spain. *Energy Res. Soc. Sci.*, **62(101356)**, 101356, doi:10.1016/j.erss.2019.101356.

Nicholls, Z. et al., 2020a: OpenSCM-Runner: Thin wrapper to run simple climate models (emissions driven runs only). *GitHub Repos.*, https://github.com/openscm/openscm-runner (Accessed August 22, 2022).

Nicholls, Z. et al., 2021: Reduced Complexity Model Intercomparison Project Phase 2: Synthesizing Earth System Knowledge for Probabilistic Climate Projections. *Earth's Future*, **9(6)**, 29, doi:10.1029/2020EF001900.

Nicholls, Z. et al., 2020b: Reduced Complexity Model Intercomparison Project Phase 1: introduction and evaluation of global-mean temperature response. *Geosci. Model Dev.*, **13(11)**, 5175–5190, doi:10.5194/gmd-13-5175-2020.

Nikas, A. et al., 2021: Where is the EU headed given its current climate policy? A stakeholder-driven model inter-comparison. *Sci. Total Environ.*, **793**, 148549, doi:10.1016/J.SCITOTENV.2021.148549.

Nogueira, L.A.H., G.M. Souza, L.A.B. Cortez, and C.H. de Brito Cruz, 2020: Biofuels for Transport. In: *Future Energy* [Letcher, T.M., (ed.)], Elsevier, Amsterdam, The Netherlands, pp. 173–197.

Nordelöf, A., M., Messagie, A.-M. Tillman, M. Ljunggren Söderman, and J. Van Mierlo, 2014: Environmental impacts of hybrid, plug-in hybrid, and battery electric vehicles – what can we learn from life cycle assessment? *Int. J. Life Cycle Assess.*, **19(11)**, doi:10.1007/s11367-014-0788-0.

Nordhaus, W., 1993: Optimal greenhouse-gas reductions and tax policy in the DICE model. *Am. Econ. Rev. (United States)*, **83:2(2)**, 313–317.

Nordhaus, W., 2015: Climate Clubs: Overcoming Free-Riding in International Climate Policy. *Am. Econ. Rev.*, **105(4)**, 1339–1370, doi:10.1257/aer.15000001.

Nordhaus, W., 2018: Evolution of modeling of the economics of global warming: changes in the DICE model, 1992–2017. *Clim. Change*, **148(4)**, 623–640, doi:10.1007/s10584-018-2218-y.

Novikova, A., T. Csoknyai, M. Jovanovic-Popovic, B. Stankovic, and Z. Szalay, 2018a: Assessment of decarbonisation scenarios for the residential buildings of Serbia. *Therm. Sci.*, **22**(Suppl. 4), 1231–1247, doi:10.2298/TSCI171221229N.

Novikova, A., T. Csoknyai, and Z. Szalay, 2018b: Low carbon scenarios for higher thermal comfort in the residential building sector of South Eastern Europe. *Energy Effic.*, **11(4)**, 845–875, doi:10.1007/s12053-017-9604-6.

NREL, 2020: Integrated Energy System Simulation. The National Renewable Energy Laboratory, US Department of Energy, https://www.nrel.gov/grid/integrated-energy-system-simulation.html (Accessed December 19, 2020).

O'Neill, B.C. et al., 2014: A new scenario framework for climate change research: the concept of shared socioeconomic pathways. *Clim. Change*, **122(3)**, 387–400, doi:10.1007/s10584-013-0905-2.

O'Neill, B.C. et al., 2016: The Scenario Model Intercomparison Project (ScenarioMIP) for CMIP6. *Geosci. Model Dev.*, **9(9)**, 3461–3482, doi:10.5194/gmd-9-3461-2016.

O'Neill, B.C. et al., 2017: The roads ahead: Narratives for shared socioeconomic pathways describing world futures in the 21st century. *Glob. Environ. Change*, **42**, 169–180, doi:10.1016/J.GLOENVCHA.2015.01.004.

O'Neill, B.C. et al., 2020: Achievements and needs for the climate change scenario framework. *Nat. Clim. Change*, **10(12)**, 1074–1084, doi:10.1038/s41558-020-00952-0.

Obersteiner, M. et al., 2016: Assessing the land resource–food price nexus of the Sustainable Development Goals. *Sci. Adv.*, **2(9)**, e1501499, doi:10.1126/sciadv.1501499.

Obersteiner, M. et al., 2018: How to spend a dwindling greenhouse gas budget. *Nat. Clim. Change*, **8(1)**, 7–10, doi:10.1038/s41558-017-0045-1.

Obi, M., S.M. Jensen, J.B. Ferris, and R.B. Bass, 2017: Calculation of levelized costs of electricity for various electrical energy storage systems. *Renew. Sustain. Energy Rev.*, **67**, 908–920, doi:10.1016/j.rser.2016.09.043.

OECD, 2021: *Effective Carbon Rates 2021*. OECD Publishing, Paris, France. https://www.oecd-ilibrary.org/taxation/effective-carbon-rates-2021_0e8e24f5-en.

Olovsson, J., M. Taljegard, M. Von Bonin, N. Gerhardt, and F. Johnsson, 2021: Impacts of Electric Road Systems on the German and Swedish Electricity Systems – An Energy System Model Comparison. *Front. Energy Res.*, **9**, doi:10.3389/fenrg.2021.631200.

Oluleye, G., L. Vasquez, R. Smith, and M. Jobson, 2016: A multi-period Mixed Integer Linear Program for design of residential distributed energy centres with thermal demand data discretisation. *Sustain. Prod. Consum.*, **5**, 16–28, doi:10.1016/j.spc.2015.11.003.

Oluleye, G., J. Allison, N. Kelly, and A.D. Hawkes, 2018: An optimisation study on integrating and incentivising Thermal Energy Storage (TES) in a dwelling energy system. *Energies*, **11(5)**, 1–17, doi:10.3390/en11051095.

Onyenokporo, N.C. and E.T. Ochedi, 2019: Low-cost retrofit packages for residential buildings in hot-humid Lagos, Nigeria. *Int. J. Build. Pathol. Adapt.*, **37(3)**, 250–272, doi:10.1108/IJBPA-01-2018-0010.

Ostermeyer, Y. et al., 2018a: *Building Market Brief. United Kingdom*. CUES Foundation, Delft, The Netherlands, 70 pp.

Ostermeyer, Y. et al., 2018b: *Building Market Brief. France*. CUES Foundation, Delft, The Netherlands, 70 pp.

Ostermeyer, Y. et al., 2018c: *Building Market Brief. The Netherlands*. CUES Foundation, Delft, The Netherlands, 70 pp.

Ostermeyer, Y. et al., 2019a: *Building Market Brief. Poland*. CUES Foundation, Delft, The Netherlands, 64pp.

Ostermeyer, Y. et al., 2019b: *Building Market Brief. Germany*. CUES Foundation, Delft, The Netherlands, 70 pp.

Ou, Y. et al., 2021: Deep mitigation of $CO_2$ and non-$CO_2$ greenhouse gases toward 1.5°C and 2°C futures. *Nat. Commun.*, **12(1)**, 6245, doi:10.1038/s41467-021-26509-z.

Overmars, K.P. et al., 2014: Estimating the opportunity costs of reducing carbon dioxide emissions via avoided deforestation, using integrated assessment modelling. *Land use policy*, **41**, 45–60, doi:10.1016/j.landusepol.2014.04.015.

Paltsev, S. and P. Capros, 2013: Cost Concepts for Climate Change Mitigation. *Clim. Change Econ.*, **04**(supp01), 1340003, doi:10.1142/S2010007813400034.

AIII

Paltsev, S. et al., 2021: *2021 Global Change Outlook: Charting the Earth's Future Energy, Managed Resources, Climate, and Policy Prospects*. MIT, Cambridge, MA, USA, 52 pp.

Paltsev, S., A. Ghandi, J. Morris, and H. Chen, 2022: Global Electrification of Light-duty Vehicles: Impacts of Economics and Climate Policy. *Econ. Energy Environ. Policy*, **11(1)**, (In press), doi:10.5547/2160-5890.11.1.spal.

Pan, X., M. den Elzen, N. Höhne, F. Teng, and L. Wang, 2017: Exploring fair and ambitious mitigation contributions under the Paris Agreement goals. *Environ. Sci. Policy*, **74**, 49–56, doi:10.1016/j.envsci.2017.04.020.

Papadopoulos, S. and E. Azar, 2016: Integrating building performance simulation in agent-based modeling using regression surrogate models: A novel human-in-the-loop energy modeling approach. *Energy Build.*, **128**, 214–223, doi:10.1016/j.enbuild.2016.06.079.

Parkinson, S. et al., 2019: Balancing clean water-climate change mitigation trade-offs. *Environ. Res. Lett.*, **14(1)**, 014009, doi:10.1088/1748-9326/aaf2a3.

Pastor, A.V. et al., 2019: The global nexus of food–trade–water sustaining environmental flows by 2050. *Nat. Sustain.*, **2(6)**, 499–507, doi:10.1038/s41893-019-0287-1.

Pauliuk, S. and N. Heeren, 2021: Material efficiency and its contribution to climate change mitigation in Germany: A deep decarbonization scenario analysis until 2060. *J. Ind. Ecol.*, **25(2)**, 479–493, doi:10.1111/jiec.13091.

Pauliuk, S., A. Arvesen, K. Stadler, and E.G. Hertwich, 2017: Industrial ecology in integrated assessment models. *Nat. Clim. Change*, **7(1)**, 13–20, doi:10.1038/nclimate3148.

Pauliuk, S. et al., 2021a: Linking service provision to material cycles: A new framework for studying the resource efficiency–climate change (RECC) nexus. *J. Ind. Ecol.*, **25(2)**, 260–273, doi:10.1111/jiec.13023.

Pauliuk, S. et al., 2021b: Global scenarios of resource and emission savings from material efficiency in residential buildings and cars. *Nat. Commun.*, **12(1)**, doi:10.1038/s41467-021-25300-4.

Paustian, K. et al., 2016: Climate-smart soils. *Nature*, **532(7597)**, 49–57, doi:10.1038/nature17174.

Pehl, M. et al., 2017: Understanding future emissions from low-carbon power systems by integration of life-cycle assessment and integrated energy modelling. *Nat. Energy*, **2(12)**, 939–945, doi:10.1038/s41560-017-0032-9.

Peng, W. et al., 2021: Climate policy models need to get real about people – here's how. *Nature*, **594**, 174–176, doi:10.1038/d41586-021-01500-2.

Perdana, S.P. et al., 2020: A multi-model analysis of long-term emissions and warming implications of current mitigation efforts. *Nat. Clim. Change*,

Pereira, L.M. et al., 2020: Developing multiscale and integrative nature–people scenarios using the Nature Futures Framework. *People Nat.*, **2(4)**, 1172–1195, doi:10.1002/pan3.10146.

Perez, M., R. Perez, K.R. Rábago, and M. Putnam, 2019: Overbuilding & curtailment: The cost-effective enablers of firm PV generation. *Sol. Energy*, **180**, 412–422, doi:10.1016/j.solener.2018.12.074.

Peters, G.P., 2018: Beyond carbon budgets. *Nat. Geosci.*, **11(6)**, 378–380, doi:10.1038/s41561-018-0142-4.

Pfenninger, S. and B. Pickering, 2018: Calliope: a multi-scale energy systems modelling framework. *J. Open Source Softw.*, **3(29)**, 825, doi:10.21105/joss.00825.

Pfenninger, S., A. Hawkes, and J. Keirstead, 2014: Energy systems modeling for twenty-first century energy challenges. *Renew. Sustain. Energy Rev.*, **33**, 74–86, doi:10.1016/j.rser.2014.02.003.

Pietzcker, R.C. et al., 2017: System integration of wind and solar power in integrated assessment models: A cross-model evaluation of new approaches. *Energy Econ.*, **64**, 583–599, doi:10.1016/j.eneco.2016.11.018.

Pindyck, R.S., 2013: Climate change policy: What do the models tell us? *J. Econ. Lit.*, **51(3)**, 860–872, doi:10.1257/jel.51.3.860.

Pindyck, R.S., 2017: The use and misuse of models for climate policy. *Rev. Environ. Econ. Policy*, **11(1)**, 100–114, doi:10.1093/reep/rew012.

Piontek, F. et al., 2019: Economic Growth Effects of Alternative Climate Change Impact Channels in Economic Modeling. *Environ. Resour. Econ.*, **73(4)**, 1357–1385, doi:10.1007/s10640-018-00306-7.

Plevin, R.J., M.A. Delucchi, and F. Creutzig, 2014: Using Attributional Life Cycle Assessment to Estimate Climate-Change Mitigation Benefits Misleads Policy Makers. *J. Ind. Ecol.*, **18(1)**, doi:10.1111/jiec.12074.

Ploss, M., T. Hatt, C. Schneider, T. Rosskopf, and M. Braun, 2017: *Modellvorhaben "KliNaWo": Klimagerechter Nachhaltiger Wohnbau*. Dornbirn, Austria.

Polasky, S. and N.K. Dampha, 2021: Discounting and Global Environmental Change. *Annu. Rev. Environ. Resour.*, **46(1)**, doi:10.1146/annurev-environ-020420-042100.

Pongratz, J., C.H. Reick, R.A. Houghton, and J.I. House, 2014: Terminology as a key uncertainty in net land use and land cover change carbon flux estimates. *Earth Syst. Dyn.*, **5(1)**, 177–195, doi:10.5194/esd-5-177-2014.

Popp, A., H. Lotze-Campen, and B. Bodirsky, 2010: Food consumption, diet shifts and associated non-$CO_2$ greenhouse gases from agricultural production. *Glob. Environ. Change*, **20(3)**, 451–462, doi:10.1016/j.gloenvcha.2010.02.001.

Popp, A. et al., 2014a: Land-use transition for bioenergy and climate stabilization: Model comparison of drivers, impacts and interactions with other land use based mitigation options. *Clim. Change*, **123(3–4)**, 495–509, doi:10.1007/s10584-013-0926-x.

Popp, A. et al., 2014b: Land-use protection for climate change mitigation. *Nat. Clim. Change*, **4(12)**, 1095–1098, doi:10.1038/nclimate2444.

Popp, A. et al., 2017: Land-use futures in the shared socio-economic pathways. *Glob. Environ. Change*, **42**, 331–345, doi:10.1016/j.gloenvcha.2016.10.002.

Portugal-Pereira, J., J. Nakatani, K.H. Kurisu, and K. Hanaki, 2015: Comparative energy and environmental analysis of Jatropha bioelectricity versus biodiesel production in remote areas. *Energy*, **83**, doi:10.1016/j.energy.2015.02.022.

Portugal-Pereira, J. et al., 2016: Overlooked impacts of electricity expansion optimisation modelling: The life cycle side of the story. *Energy*, **115**, 1424–1435, doi:10.1016/j.energy.2016.03.062.

Portugal-Pereira, J. et al., 2018: Interactions between global climate change strategies and local air pollution: lessons learnt from the expansion of the power sector in Brazil. *Clim. Change*, **148(1–2)**, 293–309, doi:10.1007/s10584-018-2193-3.

Prada-Hernández, A., H. Vargas, A. Ozuna, and J.L. Ponz-Tienda, 2015: Marginal Abatement Costs Curve ( MACC ) for Carbon Emissions Reduction from Buildings: An Implementation for Office Buildings in Colombia. *Int. J. Civ. Struct. Eng.*, **2(1)**, 175–183.

Prina, M.G., G. Manzolini, D. Moser, B. Nastasi, and W. Sparber, 2020: Classification and challenges of bottom-up energy system models – A review. *Renew. Sustain. Energy Rev.*, **129**, doi:10.1016/j.rser.2020.109917.

Prussi, M. et al., 2021: CORSIA: The first internationally adopted approach to calculate life-cycle GHG emissions for aviation fuels. *Renew. Sustain. Energy Rev.*, **150**, doi:10.1016/j.rser.2021.111398.

Radpour, S., M.A. Hossain Mondal, and A. Kumar, 2017: Market penetration modeling of high energy efficiency appliances in the residential sector. *Energy*, **134**, 951–961, doi:10.1016/j.energy.2017.06.039.

Rajão, R. et al., 2020: The rotten apples of Brazil's agribusiness. *Science*, **369(6501)**, 246–248, doi:10.1126/science.aba6646.

Ramsey, F.P., 1928: A Mathematical Theory of Saving. *Econ. J.*, **38(152)**, doi:10.2307/2224098.

Rao, N.D., B.J. Van Ruijven, K. Riahi, and V. Bosetti, 2017a: Improving poverty and inequality modelling in climate research. *Nat. Clim. Change*, **7(12)**, 857–862, doi:10.1038/s41558-017-0004-x.

Rao, N.D., P. Sauer, M. Gidden, and K. Riahi, 2019: Income inequality projections for the Shared Socioeconomic Pathways (SSPs). *Futures*, **105**, 27–39, doi:10.1016/j.futures.2018.07.001.

Rao, S. et al., 2013: Better air for better health: Forging synergies in policies for energy access, climate change and air pollution. *Glob. Environ. Change*, **23(5)**, 1122–1130, doi:10.1016/J.GLOENVCHA.2013.05.003.

Rao, S. et al., 2017b: Future air pollution in the Shared Socio-economic Pathways. *Glob. Environ. Change*, **42**, 346–358, doi:10.1016/j.gloenvcha.2016.05.012.

Realmonte, G. et al., 2019: An inter-model assessment of the role of direct air capture in deep mitigation pathways. *Nat. Commun.*, **10(1)**, 1–12, doi:10.1038/s41467-019-10842-5.

Rebitzer, G. et al., 2004: Life cycle assessment. *Environ. Int.*, **30(5)**, doi:10.1016/j.envint.2003.11.005.

Reilly, J. et al., 2018: *Food, Water, Energy, Climate Outlook: Perspectives from 2018*. MIT, Cambridge, MA, USA, 48 pp.

Reinhart, C.F. and C. Cerezo Davila, 2016: Urban building energy modeling – A review of a nascent field. *Build. Environ.*, **97**, 196–202, doi:10.1016/j.buildenv.2015.12.001.

Riahi, K. et al., 2012: Chapter 17: Energy Pathways for Sustainable Development. In: *Global Energy Assessment – Toward a Sustainable Future*, Cambridge University Press, Cambridge, UK and New York, NY, USA and the International Institute for Applied Systems Analysis, Laxenburg, Austria, pp. 1203–1306.

Riahi, K. et al., 2015: Locked into Copenhagen pledges – Implications of short-term emission targets for the cost and feasibility of long-term climate goals. *Technol. Forecast. Soc. Change*, **90**, 8–23, doi:10.1016/j.techfore.2013.09.016.

Riahi, K. et al., 2017: The Shared Socioeconomic Pathways and their energy, land use, and greenhouse gas emissions implications: An overview. *Glob. Environ. Change*, **42**, 153–168, doi:10.1016/J.GLOENVCHA.2016.05.009.

Riahi, K. et al., 2021: Long-term economic benefits of stabilizing warming without overshoot – the ENGAGE model intercomparison. *Rev.,*. doi:10.21203/rs.3.rs-127847/v1.

Ringkjøb, H.K. et al., 2020: Short-term solar and wind variability in long-term energy system models – A European case study. *Energy*, **209**, doi:10.1016/j.energy.2020.118377.

Ritchie, H., D.S. Reay, and P. Higgins, 2018: The impact of global dietary guidelines on climate change. *Glob. Environ. Change*, **49** (June 2017), 46–55, doi:10.1016/j.gloenvcha.2018.02.005.

Robinson, J.B., 1982: Energy backcasting A proposed method of policy analysis. *Energy Policy*, **10(4)**, 337–344, doi:10.1016/0301-4215(82)90048-9.

Robinson, S., 1989: Chapter 18: Multisectoral models. In: *Handbook of Development Economics*, Vol. 2, Chenery, H. and T.N. Srinivasan (eds.), Elsevier, Amsterdam, The Netherlands, pp. 885–947.

Robinson, S., A. Yúnez-Naude, R. Hinojosa-Ojeda, J.D. Lewis, and S. Devarajan, 1999: From stylized to applied models: Building multisector CGE models for policy analysis. *North Am. J. Econ. Financ.*, **10(1)**, 5–38, doi:10.1016/S1062-9408(99)00014-5.

Roca-Puigròs, M., R.G. Billy, A. Gerber, P. Wäger, and D.B. Müller, 2020: Pathways toward a carbon-neutral Swiss residential building stock. *Build. Cities*, **1(1)**, 579–593, doi:10.5334/bc.61.

Rochedo, P.R.R. et al., 2018: The threat of political bargaining to climate mitigation in Brazil. *Nat. Clim. Change*, **8(8)**, 695–698, doi:10.1038/s41558-018-0213-y.

Roelfsema, M. et al., 2020: Taking stock of national climate policies to evaluate implementation of the Paris Agreement. *Nat. Commun.*, **11(1)**, 2096, doi:10.1038/s41467-020-15414-6.

Rogelj, J. et al., 2011: Emission pathways consistent with a 2°C global temperature limit. *Nat. Clim. Change*, **1(8)**, doi:10.1038/nclimate1258.

Rogelj, J., D.L. McCollum, A. Reisinger, M. Meinshausen, and K. Riahi, 2013a: Probabilistic cost estimates for climate change mitigation. *Nature*, **493(7430)**, 79–83, doi:10.1038/nature11787.

Rogelj, J., D.L. McCollum, and K. Riahi, 2013b: The UN's "Sustainable Energy for All" initiative is compatible with a warming limit of 2 °C. *Nat. Clim. Change, 2013 36*, **3(6)**, 545–551, doi:10.1038/nclimate1806.

Rogelj, J. et al., 2016: Differences between carbon budget estimates unravelled. *Nat. Clim. Change*, **6(3)**, 245–252, doi:10.1038/nclimate2868.

Rogelj, J. et al., 2017: Understanding the origin of Paris Agreement emission uncertainties. *Nat. Commun.*, **8**, 15748, doi:10.1038/ncomms15748.

Rogelj, J., D. Shindell, K. Jiang, S. Fifita, P. Forster, V. Ginzburg, C. Handa, H. Kheshgi, S. Kobayashi, E. Kriegler, L. Mundaca, R. Séférian, and M.V.Vilariño, 2018a: Mitigation Pathways Compatible with 1.5°C in the Context of Sustainable Development. In: *Global Warming of 1.5°C. An IPCC Special Report on the impacts of global warming of 1.5°C above pre-industrial levels and related global greenhouse gas emission pathways, in the context of strengthening the global response to the threat of climate change, sustainable development, and efforts to eradicate poverty* [Masson-Delmotte, V., P. Zhai, H.-O. Pörtner, D. Roberts, J. Skea, P.R. Shukla, A. Pirani, W. Moufouma-Okia, C. Péan, R. Pidcock, S. Connors, J.B.R. Matthews, Y. Chen, X. Zhou, M.I. Gomis, E. Lonnoy, T. Maycock, M. Tignor, and T. Waterfield (eds.)]. Cambridge University Press, Cambridge, UK and New York, NY, USA.

Rogelj, J. et al., 2018b: Scenarios towards limiting global mean temperature increase below 1.5°C. *Nat. Clim. Change*, **8(4)**, 325–332, doi:10.1038/s41558-018-0091-3.

Rogelj, J., P.M. Forster, E. Kriegler, C.J. Smith, and R. Séférian, 2019a: Estimating and tracking the remaining carbon budget for stringent climate targets. *Nature*, **571(7765)**, 335–342, doi:10.1038/s41586-019-1368-z.

Rogelj, J. et al., 2019b: A new scenario logic for the Paris Agreement long-term temperature goal. *Nature*, **573(7774)**, 357–363, doi:10.1038/s41586-019-1541-4.

Rogelj, J., O. Geden, A. Cowie, and A. Reisinger, 2021: Net-zero emissions targets are vague: three ways to fix. *Nature*, **591(7850)**, 365–368, doi:10.1038/d41586-021-00662-3.

Rogner, H.-H. et al., 2012: Chapter 7: Energy Resources and Potentials. In: *Global Energy Assessment – Toward a Sustainable Future*, Cambridge University Press, Cambridge, UK and New York, NY, USA and the International Institute for Applied Systems Analysis, Laxenburg, Austria, pp. 423–512.

Rosa, I.M.D. et al., 2020: Challenges in producing policy-relevant global scenarios of biodiversity and ecosystem services. *Glob. Ecol. Conserv.*, **22**, e00886, doi:10.1016/j.gecco.2019.e00886.

Rosas-Flores, J.A. and D. Rosas-Flores, 2020: Potential energy savings and mitigation of emissions by insulation for residential buildings in Mexico. *Energy Build.*, **209**, doi:10.1016/j.enbuild.2019.109698.

Roscini, A.V., O. Rapf, and J. Kockat, 2020: *On the way to a climate-neutral Europe. Contribution from the building sector to a strengthened 2030 climate target*. Buildings Performance Institute Europe, Brussels, Belgium, 24 pp.

Rose, P.K. and F. Neumann, 2020: Hydrogen refueling station networks for heavy-duty vehicles in future power systems. *Transp. Res. Part D Transp. Environ.*, **83**, doi:10.1016/j.trd.2020.102358.

Rose, S.K. et al., 2020: An overview of the Energy Modeling Forum 33rd study: assessing large-scale global bioenergy deployment for managing climate change. *Clim. Change*, **163(3)**, 1539–1551, doi:10.1007/s10584-020-02945-6.

Rosenzweig, C. et al., 2020: Climate change responses benefit from a global food system approach. *Nat. Food*, **1(2)**, 94–97, doi:10.1038/s43016-020-0031-z.

Rottoli, M., A. Dirnaichner, R. Pietzcker, F. Schreyer, and G. Luderer, 2021: Alternative electrification pathways for light-duty vehicles in the European transport sector. *Transp. Res. Part D Transp. Environ.*, **99**, 103005, doi:10.1016/J.TRD.2021.103005.

Sachs, J., Y. Meng, S. Giarola, and A. Hawkes, 2019a: An agent-based model for energy investment decisions in the residential sector. *Energy*, **172**, 752–768, doi:10.1016/j.energy.2019.01.161.

Sachs, J.D. et al., 2019b: Six Transformations to achieve the Sustainable Development Goals. *Nat. Sustain.*, **2(9)**, 805–814, doi:10.1038/s41893-019-0352-9.

Sanchez, D.L., N. Johnson, S.T. McCoy, P.A. Turner, and K.J. Mach, 2018: Near-term deployment of carbon capture and sequestration from biorefineries in the United States. *Proc. Natl. Acad. Sci.*, **115(19)**, 4875–4880, doi:10.1073/pnas.1719695115.

Sandberg, N.H., J.S. Næss, H. Brattebø, I. Andresen, and A. Gustavsen, 2021: Large potentials for energy saving and greenhouse gas emission reductions from large-scale deployment of zero emission building technologies in a national building stock. *Energy Policy*, **152**, doi:10.1016/j.enpol.2020.112114.

Sani Hassan, A., L. Cipcigan, and N. Jenkins, 2018: Impact of optimised distributed energy resources on local grid constraints. *Energy*, **142**, 878–895, doi:10.1016/j.energy.2017.10.074.

Schaber, K., F. Steinke, and T. Hamacher, 2012: Transmission grid extensions for the integration of variable renewable energies in Europe: Who benefits where? *Energy Policy*, **43**, 123–135, doi:10.1016/j.enpol.2011.12.040.

Schaeffer, M. et al., 2015: Mid- and long-term climate projections for fragmented and delayed-action scenarios. *Technol. Forecast. Soc. Change*, **90**(PA), 257–268, doi:10.1016/j.techfore.2013.09.013.

Schaeffer, R. et al., 2020: Comparing transformation pathways across major economies. *Clim. Change*, **162(4)**, 1787–1803, doi:10.1007/s10584-020-02837-9.

Schäfer, A.W., 2017: Long-term trends in domestic US passenger travel: the past 110 years and the next 90. *Transportation (Amst).*, **44(2)**, 293–310, doi:10.1007/s11116-015-9638-6.

Schleussner, C.-F. et al., 2016: Differential climate impacts for policy-relevant limits to global warming: the case of 1.5 °C and 2 °C. *Earth Syst. Dyn.*, **7(2)**, 327–351, doi:10.5194/esd-7-327-2016.

Schmitz, C. et al., 2014: Land-use change trajectories up to 2050: Insights from a global agro-economic model comparison. *Agric. Econ. (United Kingdom)*, **45(1)**, 69–84, doi:10.1111/agec.12090.

Scholz, Y., H.C. Gils, and R.C. Pietzcker, 2017: Application of a high-detail energy system model to derive power sector characteristics at high wind and solar shares. *Energy Econ.*, **64**, 568–582, doi:10.1016/j.eneco.2016.06.021.

Schultes, A. et al., 2018: Optimal international technology cooperation for the low-carbon transformation. *Clim. Policy*, **18(9)**, 1165–1176, doi:10.1080/14693062.2017.1409190.

Schultes, A. et al., 2021: Economic damages from on-going climate change imply deeper near-term emission cuts. *Environ. Res. Lett.*, **16(10)**, 104053, doi:10.1088/1748-9326/AC27CE.

Schwanitz, V.J., 2013: Evaluating integrated assessment models of global climate change. *Environ. Model. Softw.*, **50**, 120–131, doi:10.1016/j.envsoft.2013.09.005.

Scrieciu, S., A. Rezai, and R. Mechler, 2013: On the economic foundations of green growth discourses: The case of climate change mitigation and macroeconomic dynamics in economic modeling. *Wiley Interdiscip. Rev. Energy Environ.*, **2(3)**, 251–268, doi:10.1002/wene.57.

Serrenho, A.C., M. Drewniok, C. Dunant, and J.M. Allwood, 2019: Testing the greenhouse gas emissions reduction potential of alternative strategies for the English housing stock. *Resour. Conserv. Recycl.*, **144**, 267–275, doi:10.1016/j.resconrec.2019.02.001.

Sharma, S., S.H. Huang, and N.D.R. Sarma, 2011: System inertial frequency response estimation and impact of renewable resources in ERCOT interconnection. IEEE Power and Energy Society General Meeting, IEEE, doi:10.1109/PES.2011.6038993.

Sharp, B.E. and S.A. Miller, 2016: Potential for Integrating Diffusion of Innovation Principles into Life Cycle Assessment of Emerging Technologies. *Environ. Sci. Technol.*, **50(6)**, doi:10.1021/acs.est.5b03239.

Shukla, P.R. and V. Chaturvedi, 2011: Sustainable energy transformations in India under climate policy. *Sustain. Dev.*, **21(1)**, 48–59, doi:10.1002/sd.516.

Sinsel, S.R., R.L. Riemke, and V.H. Hoffmann, 2020: Challenges and solution technologies for the integration of variable renewable energy sources – a review. *Renew. Energy*, **145**, 2271–2285, doi:10.1016/j.renene.2019.06.147.

Skar, C., G. Doorman, G.A. Pérez-Valdés, and A. Tomasgard, 2016: *A multi-horizon stochastic programming model for the European power system*. CenSES working paper 2/2016, Centre for Sustainable Energy Studies, Norwegian University of Science and Technology, Trondheim, Norway, 30 pp. https://www.ntnu.no/documents/7414984/202064323/1_Skar_ferdig.pdf/855f0c3c-81db-440d-9f76-cfd91af0d6f0.

Skea, J., P. Shukla, A. Al Khourdajie, and D. McCollum, 2021: Intergovernmental Panel on Climate Change: Transparency and integrated assessment modeling. *WIREs Clim. Change*, **12(5)**, e727–e727, doi.org/10.1002/wcc.727.

Smith, C.J. et al., 2018: FAIR v1.3: A simple emissions-based impulse response and carbon cycle model. *Geosci. Model Dev.*, **11(6)**, 2273–2297, doi:10.5194/gmd-11-2273-2018.

Smith, E. et al., 2021: The cost of $CO_2$ transport and storage in global integrated assessment modeling. *Int. J. Greenh. Gas Control*, **109**, 103367, doi:10.1016/j.ijggc.2021.103367.

Smith P., M. Bustamante, H. Ahammad, H. Clark, H. Dong, E.A. Elsiddig, H. Haberl, R. Harper, J. House, M. Jafari, O. Masera, C. Mbow, N.H. Ravindranath, C.W. Rice, C. Robledo Abad, A. Romanovskaya, F. Sperling, and F. Tubiello, 2014: Agriculture, Forestry and Other Land Use (AFOLU). In: *Climate Change 2014: Mitigation of Climate Change. Contribution of Working Group III to the Fifth Assessment Report of the Intergovernmental Panel on Climate Change* [Edenhofer, O., R. Pichs-Madruga, Y. Sokona, E. Farahani, S. Kadner, K. Seyboth, A. Adler, I. Baum, S. Brunner, P. Eickemeier, B. Kriemann, J. Savolainen, S. Schlömer, C. von Stechow, T. Zwickel and J.C. Minx (eds.)]. Cambridge University Press, Cambridge, UK and New York, NY, USA.

Smith, P. et al., 2016: Biophysical and economic limits to negative $CO_2$ emissions. *Nat. Clim. Change*, **6(1)**, 42–50, doi:10.1038/nclimate2870.

Smith, P., J. Nkem, K. Calvin, D. Campbell, F. Cherubini, G. Grassi, V. Korotkov, A.L. Hoang, S. Lwasa, P. McElwee, E. Nkonya, N. Saigusa, J.-F. Soussana, M.A. Taboada, 2019: Interlinkages Between Desertification, Land Degradation, Food Security and Greenhouse Gas Fluxes: Synergies, Trade-offs and Integrated Response Options. In: *Climate Change and Land: an IPCC special report on climate change, desertification, land degradation, sustainable land management, food security, and greenhouse gas fluxes in terrestrial ecosystems* [P.R. Shukla, J. Skea, E. Calvo Buendia, V. Masson-Delmotte, H.-O. Portner, D.C. Roberts, P. Zhai, R. Slade, S. Connors, R. van Diemen, M. Ferrat, E. Haughey, S. Luz, S. Neogi, M. Pathak, J. Petzold, J. Portugal Pereira, P. Vyas, E. Huntley, K. Kissick, M. Belkacemi, J. Malley, (eds.)]. Cambridge University Press, Cambridge, UK and New York, NY, USA.

Smith, S.J. et al., 2020a: Impact of methane and black carbon mitigation on forcing and temperature: a multi-model scenario analysis. *Clim. Change*, **163(3)**, 1427–1442, doi:10.1007/s10584-020-02794-3.

Smith, S.J. et al., 2020b: The Energy Modeling Forum (EMF) –30 study on short-lived climate forcers: introduction and overview. *Clim. Change*, **163(3)**, 1399–1408, doi:10.1007/s10584-020-02938-5.

Soergel, B. et al., 2021: A sustainable development pathway for climate action within the UN 2030 Agenda. *Nat. Clim. Change*, **11(8)**, 656–664, doi:10.1038/s41558-021-01098-3.

Soria, R. et al., 2016: Modelling concentrated solar power (CSP) in the Brazilian energy system: A soft-linked model coupling approach. *Energy*, **116**, 265–280, doi:10.1016/j.energy.2016.09.080.

Springmann, M., H.C.J. Godfray, M. Rayner, and P. Scarborough, 2016: Analysis and valuation of the health and climate change cobenefits of dietary change. *Proc. Natl. Acad. Sci.*, **113(15)**, 4146–4151, doi:10.1073/pnas.1523119113.

Statharas, S., Y. Moysoglou, P. Siskos, and P. Capros, 2021: Simulating the Evolution of Business Models for Electricity Recharging Infrastructure Development by 2030: A Case Study for Greece. *Energies*, **14(9)**, doi:10.3390/en14092345.

Stehfest, E. et al., 2014: *Integrated Assessment of Global Environmental Change with IMAGE 3.0 – Model description and policy applications*. PBL Netherlands Environmental Assessment Agency, The Hague, The Netherlands, 370 pp.

AIII

Stehfest, E. et al., 2019: Key determinants of global land-use projections. *Nat. Commun.*, **10(1)**, 1–10, doi:10.1038/s41467-019-09945-w.

Stern, N. 2006: The Stern Review on the Economic Effects of Climate Change. *Popul. Dev. Rev.*, **32(4)**, doi:10.1111/j.1728-4457.2006.00153.x.

Stern, N., 2016: Economics: Current climate models are grossly misleading. *Nature*, **530(7591)**, 407–409, doi:10.1038/530407a.

Strbac, G. et al., 2018: *Value of Flexibility in a Decarbonised Grid and System Externalities of Low-Carbon Generation Technologies: For the Committee on Climate Change*. Imperial College and NERA, London, UK.

Strefler, J. et al., 2018: Between Scylla and Charybdis: Delayed mitigation narrows the passage between large-scale CDR and high costs. *Environ. Res. Lett.*, **13(4)**, 044015, doi:10.1088/1748-9326/aab2ba.

Strefler, J. et al., 2021a: Carbon dioxide removal technologies are not born equal. *Environ. Res. Lett.*, **16(7)**, 74021, doi:10.1088/1748-9326/ac0a11.

Strefler, J. et al., 2021b: Alternative carbon price trajectories can avoid excessive carbon removal. *Nat. Commun.*, **12(1)**, 2264, doi:10.1038/s41467-021-22211-2.

Streicher, K.N., D. Parra, M.C. Buerer, and M.K. Patel, 2017: Techno-economic potential of large-scale energy retrofit in the Swiss residential building stock. *Energy Procedia*, **122**, 121–126, doi:10.1016/j.egypro.2017.07.314.

Subramanyam, V., M. Ahiduzzaman, and A. Kumar, 2017a: Greenhouse gas emissions mitigation potential in the commercial and institutional sector. *Energy Build.*, **140**, 295–304, doi:10.1016/j.enbuild.2017.02.007.

Subramanyam, V., A. Kumar, A. Talaei, and M.A.H. Mondal, 2017b: Energy efficiency improvement opportunities and associated greenhouse gas abatement costs for the residential sector. *Energy*, **118**, 795–807, doi:10.1016/j.energy.2016.10.115.

Sugiyama, M. et al., 2019a: Japan's long-term climate mitigation policy: Multi-model assessment and sectoral challenges. *Energy*, **167**, 1120–1131, doi:10.1016/j.energy.2018.10.091.

Sugiyama, M., A. Taniguchi-Matsuoka, Y. Yamaguchi, and Y. Shimoda, 2019b: Required Specification of Residential End-use Energy Demand Model for Application to National GHG Mitigation Policy Making – Case Study for the Japanese Plan for Global Warming Countermeasures. *Proceedings of Building Simulation 2019: 16th Conference of IBPSA*, Vol. 6, pp. 3706–3713.

Sugiyama, M. et al., 2020: EMF 35 JMIP study for Japan's long-term climate and energy policy: scenario designs and key findings. *Sustain. Sci.*, **Rev**.

Tan, X., H. Lai, B. Gu, Y. Zeng, and H. Li, 2018: Carbon emission and abatement potential outlook in China's building sector through 2050. *Energy Policy*, **118**(March), 429–439, doi:10.1016/j.enpol.2018.03.072.

Tavoni, M. and R.S.J. Tol, 2010: Counting only the hits? The risk of underestimating the costs of stringent climate policy: A letter. *Clim. Change*, **100(3)**, 769–778, doi:10.1007/s10584-010-9867-9.

Tavoni, M. et al., 2014: The distribution of the major economies' effort in the Durban Platform scenarios. *Clim. Change Econ.*, **4(4)**, doi:10.1142/S2010007813400095.

Tavoni, M. et al., 2015: Post-2020 climate agreements in the major economies assessed in the light of global models. *Nat. Clim. Change*, **5(2)**, 119–126, doi:10.1038/nclimate2475.

Taylor, K.E., R.J. Stouffer, and G.A. Meehl, 2011: An Overview of CMIP5 and the Experiment Design. *Bull. Am. Meteorol. Soc.*, **93(4)**, 485–498, doi:10.1175/BAMS-D-11-00094.1.

Taylor, P.G., P. Upham, W. McDowall, and D. Christopherson, 2014: Energy model, boundary object and societal lens: 35 years of the MARKAL model in the UK. *Energy Res. Soc. Sci.*, **4**(C), 32–41, doi:10.1016/j.erss.2014.08.007.

Teng, F. and G. Strbac, 2017: Full Stochastic Scheduling for Low-Carbon Electricity Systems. *IEEE Trans. Autom. Sci. Eng.*, **14(2)**, 461–470, doi:10.1109/TASE.2016.2629479.

Timilsina, G., A. Sikharulidze, E. Karapoghosyan, and S. Shatvoryan, 2016: *How Do We Prioritize the GHG Mitigation Options? Development of a Marginal Abatement Cost Curve for the Building Sector in Armenia and Georgia*. World Bank, Washington, DC, USA, 37 pp.

TNO, 2021: *GenX: Configurable Capacity Expansion Model*. MIT. https://tlo.mit.edu/technologies/genx-configurable-capacity-expansion-model.

Tokimatsu, K. et al., 2020: Toward more comprehensive environmental impact assessments: interlinked global models of LCIA and IAM applicable to this century. *Int. J. Life Cycle Assess.*, **25(9)**, 1710–1736, doi:10.1007/s11367-020-01750-8.

Toleikyte, A., L. Kranzl, and A. Müller, 2018: Cost curves of energy efficiency investments in buildings – Methodologies and a case study of Lithuania. *Energy Policy*, **115**, 148–157, doi:10.1016/j.enpol.2017.12.043.

Tosatto, A., G. Misyris, A. Junyent-Ferré, F. Teng, and S. Chatzivasileiadis, 2020: Towards Optimal Coordination between Regional Groups: HVDC Supplementary Power Control. *IEEE Trans. Power Syst.*, **37**(1), 1–8.

Trottier, 2016: *Canada's challenge & opportunity. Transformations for major reductions in GHG emissions*. David Suzuki Foundation and partners, Ottawa, Canada, 321 pp.

Trutnevyte, E. et al., 2019: Societal Transformations in Models for Energy and Climate Policy: The Ambitious Next Step. *One Earth*, **1(4)**, 423–433, doi:10.1016/j.oneear.2019.12.002.

Tubiello, F.N. et al., 2015: The Contribution of Agriculture, Forestry and other Land Use activities to Global Warming, 1990–2012. *Glob. Change Biol.*, **21(7)**, 2655–2660, doi:10.1111/gcb.12865.

Tulkens, H., 2019: *Economics, Game Theory and International Environmental Agreements*. World Scientific Publishing Co. Pte. Ltd. ISBN: 9789813141223.

Turnheim, B. et al., 2015: Evaluating sustainability transitions pathways: Bridging analytical approaches to address governance challenges. *Glob. Environ. Change*, **35**, 239–253, doi:10.1016/j.gloenvcha.2015.08.010.

UNEP, 2019: Bridging the Gap: Enhancing Mitigation Ambition and Action at G20 Level and Globally in UNEP *Emissions Gap Report 2019*. UNEP, Nairobi, Kenya.

Valin, H. et al., 2014: The future of food demand: Understanding differences in global economic models. *Agric. Econ. (United Kingdom)*, **45(1)**, 51–67, doi:10.1111/agec.12089.

van de Ven, D.-J. et al., 2021: The Impact of U.S. Re-engagement in Climate on the Paris Targets. *Earth's Future*, **9(9)**, e2021EF002077-e2021EF002077, doi:10.1029/2021EF002077.

van den Berg, N.J. et al., 2020: Implications of various effort-sharing approaches for national carbon budgets and emission pathways. *Clim. Change*, **162(4)**, 1805–1822, doi:10.1007/s10584-019-02368-y.

van der Zwaan, B.C.C., K. V Calvin, and L.E. Clarke, 2016: Climate Mitigation in Latin America: Implications for Energy and Land Use: Preface to the Special Section on the findings of the CLIMACAP-LAMP project. *Energy Econ.*, **56**, 495–498, doi:10.1016/J.ENECO.2016.05.005.

van Marle, M.J.E. et al., 2017: Historic global biomass burning emissions for CMIP6 (BB4CMIP) based on merging satellite observations with proxies and fire models (1750–2015). *Geosci. Model Dev.*, **10(9)**, 3329–3357, doi:10.5194/gmd-10-3329-2017.

van Sluisveld, M.A.E. et al., 2015: Comparing future patterns of energy system change in 2°C scenarios with historically observed rates of change. *Glob. Environ. Change*, **35**, 436–449, doi:10.1016/j.gloenvcha.2015.09.019.

van Sluisveld, M.A.E., S.H. Martínez, V. Daioglou, and D.P. van Vuuren, 2016: Exploring the implications of lifestyle change in 2°C mitigation scenarios using the IMAGE integrated assessment model. *Technol. Forecast. Soc. Change*, **102**, 309–319, doi:10.1016/j.techfore.2015.08.013.

van Sluisveld, M.A.E. et al., 2018: Comparing future patterns of energy system change in 2°C scenarios to expert projections. *Glob. Environ. Change*, **50**(March), 201–211, doi:10.1016/j.gloenvcha.2018.03.009.

van Soest, H.L. et al., 2019: Analysing interactions among Sustainable Development Goals with Integrated Assessment Models. *Glob. Transitions*, **1**, 210–225, doi:10.1016/j.glt.2019.10.004.

van Soest, H.L., Aleluia Reis, L., Baptista, L.B. *et al.* Global roll-out of comprehensive policy measures may aid in bridging emissions gap. *Nat Commun* **12**, 6419 (2021). https://doi.org/10.1038/s41467-021-26595-z

van Vuuren, D. et al., 2021: *The 2021 SSP scenarios of the IMAGE 3.2*

*model*. PBL Netherlands Environmental Assessment Agency, The Hague, The Netherlands.
van Vuuren, D.P. et al., 2008: Temperature increase of 21st century mitigation scenarios. *Proc. Natl. Acad. Sci.*, **105(40)**, doi:10.1073/pnas.0711129105.
van Vuuren, D.P. et al., 2009: Comparison of top-down and bottom-up estimates of sectoral and regional greenhouse gas emission reduction potentials. *Energy Policy*, **37(12)**, 5125–5139, doi:10.1016/j.enpol.2009.07.024.
van Vuuren, D.P. et al., 2011a: How well do integrated assessment models simulate climate change? *Clim. Change*, **104(2)**, 255–285, doi:10.1007/s10584-009-9764-2.
van Vuuren, D.P. et al., 2011b: The representative concentration pathways: an overview. *Clim. Change*, **109(1–2)**, 5–31, doi:10.1007/s10584-011-0148-z.
van Vuuren, D.P. et al., 2012: A proposal for a new scenario framework to support research and assessment in different climate research communities. *Glob. Environ. Change*, **22(1)**, 21–35, doi:10.1016/j.gloenvcha.2011.08.002.
van Vuuren, D.P. et al., 2014: A new scenario framework for Climate Change Research: Scenario matrix architecture. *Clim. Change*, **122(3)**, 373–386, doi:10.1007/s10584-013-0906-1.
van Vuuren, D.P. et al., 2015: Pathways to achieve a set of ambitious global sustainability objectives by 2050: Explorations using the IMAGE integrated assessment model. *Technol. Forecast. Soc. Change*, **98**, 303–323, doi:10.1016/j.techfore.2015.03.005.
van Vuuren, D.P. et al., 2017a: Energy, land-use and greenhouse gas emissions trajectories under a green growth paradigm. *Glob. Environ. Change*, **42**, 237–250, doi:10.1016/j.gloenvcha.2016.05.008.
van Vuuren, D.P. et al., 2017b: The Shared Socio-economic Pathways: Trajectories for human development and global environmental change. *Glob. Environ. Change*, **42**, 148–152, doi:10.1016/j.gloenvcha.2016.10.009.
van Vuuren, D.P. et al., 2018: Alternative pathways to the 1.5°C target reduce the need for negative emission technologies. *Nat. Clim. Change*, **8(5)**, 391–397, doi:10.1038/s41558-018-0119-8.
Van Vuuren, D.P. et al., 2019: Integrated scenarios to support analysis of the food–energy–water nexus. *Nat. Sustain.*, **2(12)**, 1132–1141, doi:10.1038/s41893-019-0418-8.
Velders, G.J.M., D.W. Fahey, J.S. Daniel, S.O. Andersen, and M. McFarland, 2015: Future atmospheric abundances and climate forcings from scenarios of global and regional hydrofluorocarbon (HFC) emissions. *Atmos. Environ.*, **123**, 200–209, doi:10.1016/J.ATMOSENV.2015.10.071.
Virage-Energie Nord-Pas-de-Calais., 2016: *Mieux Vivre en Région Nord-Pas-de-Calais – Pour un virage énergétique et des transformations sociétales*. 28 pp.
Vishwanathan, S.S., P. Fragkos, K. Fragkiadakis, L. Paroussos, and A. Garg, 2019: Energy system transitions and macroeconomic assessment of the Indian building sector. *Build. Res. Inf.*, **47(1)**, 38–55, doi:10.1080/09613218.2018.1516059.
von Lampe, M. et al., 2014: Why do global long-term scenarios for agriculture differ? An overview of the AgMIP Global Economic Model Intercomparison. *Agric. Econ.*, **45(1)**, doi:10.1111/agec.12086.
Von Stechow, C. et al., 2015: Integrating Global Climate Change Mitigation Goals with Other Sustainability Objectives: A Synthesis. *Annu. Rev. Environ. Resour.*, **40(1)**, 363–394, doi:10.1146/annurev-environ-021113-095626.
Vrontisi, Z., J. Abrell, F. Neuwahl, B. Saveyn, and F. Wagner, 2016: Economic impacts of EU clean air policies assessed in a CGE framework. *Environ. Sci. Policy*, **55**, 54–64, doi:10.1016/j.envsci.2015.07.004.
Vrontisi, Z. et al., 2018: Enhancing global climate policy ambition towards a 1.5 °C stabilization: A short-term multi-model assessment. *Environ. Res. Lett.*, **13(4)**, 44039, doi:10.1088/1748-9326/aab53e.
Waffenschmidt, E. and R.S.Y. Hui, 2016: Virtual inertia with PV inverters using DC-link capacitors. 2016 18th European Conference on Power Electronics and Applications, EPE 2016 ECCE Europe, IEEE.
Wakiyama, T. and T. Kuramochi, 2017: Scenario analysis of energy saving and $CO_2$ emissions reduction potentials to ratchet up Japanese mitigation target in 2030 in the residential sector. *Energy Policy*, **103**, 1–15, doi:10.1016/j.enpol.2016.12.059.
Waldhoff, S., D. Anthoff, S. Rose, and R.S.J. Tol, 2014: The Marginal Damage Costs of Different Greenhouse Gases: An Application of FUND. *Economics*, **8(1)**, 1–33, doi:10.5018/economics-ejournal.ja.2014-31.
Warszawski, L. et al., 2021: All options, not silver bullets, needed to limit global warming to 1.5 °C: a scenario appraisal. *Environ. Res. Lett.*, **16(6)**, 64037, doi:10.1088/1748-9326/abfeec.
Weber, C. et al., 2018: Mitigation scenarios must cater to new users. *Nat. Clim. Change*, **8(10)**, 845–848, doi:10.1038/s41558-018-0293-8.
Wei, Y.-M. et al., 2021: Pathway comparison of limiting global warming to 2°C. *Energy Clim. Chane.*, **2**, 100063, doi:10.1016/J.EGYCC.2021.100063.
Weidema, B.P., M. Pizzol, J. Schmidt, and G. Thoma, 2018: Attributional or consequential Life Cycle Assessment: A matter of social responsibility. *J. Clean. Prod.*, **174**, doi:10.1016/j.jclepro.2017.10.340.
Weindl, I. et al., 2015: Livestock in a changing climate: production system transitions as an adaptation strategy for agriculture. *Environ. Res. Lett.*, **10(9)**, 094021, doi:10.1088/1748-9326/10/9/094021.
Weindl, I. et al., 2017: Livestock and human use of land: Productivity trends and dietary choices as drivers of future land and carbon dynamics. *Glob. Planet. Change*, **159**, 1–10, doi:10.1016/J.GLOPLACHA.2017.10.002.
Wender, B.A. et al., 2014: Anticipatory life-cycle assessment for responsible research and innovation. *J. Responsible Innov.*, **1(2)**, doi:10.1080/23299460.2014.920121.
Weyant, J., 2017: Some Contributions of Integrated Assessment Models of Global Climate Change. *Rev. Environ. Econ. Policy*, **11(1)**, 115–137, doi:10.1093/reep/rew018.
Weyant, J.P., 2009: A perspective on integrated assessment. *Clim. Change*, **95(3–4)**, 317–323, doi:10.1007/s10584-009-9612-4.
Wilkerson, J.T., B.D. Leibowicz, D.D. Turner, and J.P. Weyant, 2015: Comparison of integrated assessment models: Carbon price impacts on U.S. energy. *Energy Policy*, **76**, 18–31, doi:10.1016/j.enpol.2014.10.011.
Wilson, C., A. Grubler, K.S. Gallagher, and G.F. Nemet, 2012: Marginalization of end-use technologies in energy innovation for climate protection. *Nat. Clim. Change*, **2(11)**, 780–788, doi:10.1038/nclimate1576.
Wilson, C., A. Grubler, N. Bauer, V. Krey, and K. Riahi, 2013: Future capacity growth of energy technologies: Are scenarios consistent with historical evidence? *Clim. Change*, **118(2)**, 381–395, doi:10.1007/s10584-012-0618-y.
Wilson, C. et al., 2020: Granular technologies to accelerate decarbonization. *Science*, **368(6486)**, 36–39, doi:10.1126/science.aaz8060.
Wilson, C. et al., 2021: Evaluating process-based integrated assessment models of climate change mitigation. *Clim. Change*, **166(1–2)**, 3, doi:10.1007/s10584-021-03099-9.
Wilson, E. et al., 2017: *Energy Efficiency Potential in the U. S. Single-Family Housing Stock*. National Renewable Energy Laboratory, Golden, CO, USA, 157 pp.
Xing, R., T. Hanaoka, and T. Masui, 2021: Deep decarbonization pathways in the building sector: China's NDC and the Paris agreement. *Environ. Res. Lett.*, **16(4)**, doi:10.1088/1748-9326/abe008.
Yang, Z., 2008: *Strategic Bargaining and Cooperation in Greenhouse Gas Mitigations – An Integrated Assessment Modeling Approach*. MIT Press, London, UK.
Yeh, S. et al., 2016: A modeling comparison of deep greenhouse gas emissions reduction scenarios by 2030 in California. *Energy Strateg. Rev.*, **13–14**(August 2016), 169–180, doi:10.1016/j.esr.2016.10.001.
Yeh, S. et al., 2017: Detailed assessment of global transport-energy models' structures and projections. *Transp. Res. Part D Transp. Environ.*, **55**, 294–309, doi:10.1016/j.trd.2016.11.001.
Yu, S. et al., 2018: Implementing nationally determined contributions: building energy policies in India's mitigation strategy. *Environ. Res. Lett.*, **13(3)**, 034034, doi:10.1088/1748-9326/aaad84.

Yumashev D. et al. 2019: Climate policy implications of nonlinear decline of Arctic land permafrost and other cryosphere elements. *Nat. Commun.*, **10(1)**, 1900, doi:10.1038/s41467-019-09863-x.

Zhang, R., S. Fujimori, and T. Hanaoka, 2018: The contribution of transport policies to the mitigation potential and cost of 2°C and 1.5°C goals. *Environ. Res. Lett.*, **13(5)**, 054008, doi:10.1088/1748-9326/aabb0d.

Zhang, S. et al., 2020: Scenarios of energy reduction potential of zero energy building promotion in the Asia-Pacific region to year 2050. *Energy*, **213**, doi:10.1016/j.energy.2020.118792.

Zhong, X. et al., 2021: Global greenhouse gas emissions from residential and commercial building materials and mitigation strategies to 2060. *Nat. Commun.*, **12(1)**, 6126, doi:10.1038/s41467-021-26212-z.

Zhou, N., N. Khanna, W. Feng, J. Ke, and M. Levine, 2018: Scenarios of energy efficiency and $CO_2$ emissions reduction potential in the buildings sector in China to year 2050. *Nat. Energy*, **3(11)**, 978–984, doi:10.1038/s41560-018-0253-6.

AIV

# Annex IV: Contributors to the IPCC WGIII Sixth Assessment Report

**This chapter should be cited as:**

IPCC, 2022: Annex IV: Contributors to the IPCC Working Group III Sixth Assessment Report. In *Climate Change 2022: Mitigation of Climate Change. Contribution of Working Group III to the Sixth Assessment Report of the Intergovernmental Panel on Climate Change* [P.R. Shukla, J. Skea, R. Slade, A. Al Khourdajie, R. van Diemen, D. McCollum, M. Pathak, S. Some, P. Vyas, R. Fradera, M. Belkacemi, A. Hasija, G. Lisboa, S. Luz, J. Malley, (eds.)]. Cambridge University Press, Cambridge, UK and New York, NY, USA. doi: 10.1017/9781009157926.023

**AASEN, Marianne**
Centre for International Climate and Environmental Research (CICERO)
Norway

**ABDEL-AZIZ, Amr**
Integral Consult
Egypt

**ABDULLA, Amjad**
IPCC Working Group III Vice-Chair
Maldives

**ABU HATAB, Assem**
Department of Economics, The Swedish University of Agricultural Sciences
Egypt/Sweden

**ACQUAYE, Adolf**
Rochester Institute of Technology
Ghana/United Kingdom

**ADACHI, Muneki**
Japanese Ministry of the Environment
Japan

**AGBEMABIESE, Lawrence**
University of Delaware
Ghana/the United States of America

**AKIMOTO, Keigo**
Research Institute of Innovative Technology for the Earth (RITE)
Japan

**AL KHOURDAJIE, Alaa**
IPCC WG III TSU, Imperial College London
United Kingdom/Syria

**ALELUIA REIS, Lara**
European Institute on Economics and the Environment
Portugal

**AMELI, Hossein**
Imperial College London
Germany

**AMON, Barbara**
Leibniz Institute for Agricultural Engineering and Bioeconomy (ATB)
Germany

**ANDREW, Robbie M.**
Centre for International Climate and Environmental Research (CICERO)
Norway

**ANGERS, Denis**
Agriculture and Agri-Food Canada
Canada

**ANGLIVIEL DE LA BEAUMELLE, Nils**
Stanford University
France/the United States of America

**AOKI, Lillian**
University of Oregon
the United States of America

**ARANGO, Jacobo**
International Center for Tropical Agriculture (CIAT)/Tropical Forages Program
Colombia

**ARIMA, Jun**
Graduate School of Public Policy, University of Tokyo
Japan

**ARNETH, Almut**
Karlsruhe Institute of Technology
Germany

**AUER, Cornelia**
Potsdam Institute for Climate Impact
Germany

**AYALA-NIÑO, Fernando**
Laboratorio de Edafología Aplicada y Servicios Ambientales, Facultad de Estudios Superiores Iztacala, Universidad Nacional Autónoma de México
Mexico

**BABIKER, Mustafa**
Saudi Aramco
the Sudan/Saudi Arabia

**BADIOLA, Esther**
European Investment Bank (EIB)
Spain

**BAI, Quan**
Energy Research Institute of National Development and Reform Commission
China

**BAI, Xuemei**
Australian National University
Australia

**BAIOCCHI, Giovani**
The University of Maryland
the United States of America

**BALA, Govindasamy**
Indian Institute of Science
India

**BASHMAKOV, Igor Alexeyevich**
Center for Energy Efficiency – XXI (CENEF-XXI)
the Russian Federation

**BATAILLE, Christopher**
Institute for Sustainable Development and International Relations (IDDRI)/ Simon Fraser University
Canada

**BATTISTON, Stefano**
University of Zurich/University Ca' Foscari, Venice
Switzerland/Italy

**BAUER, Fredric**
Lund University
Sweden

**BAUER, Nico**
Potsdam Institute for Climate Impact
Germany

**BECHTOLDT, Myriam**
EBS Universität
Germany

**BEERLING, David**
School of Biosciences, The University of Sheffield
United Kingdom

**BEKKERS, Rudi**
Eindhoven University of Technology
the Netherlands

**BERNDES, Göran**
Chalmers University of Technology
Sweden

**BERRILL, Peter**
Mercator Research Institute on Global Commons and Climate Change (MCC) gGmbH
Germany/Ireland

**BERTOLDI, Paolo**
European Commission
Italy

AIV

**BERTRAM, Christoph**
Potsdam Institute for Climate
Impact Research (PIK)
Germany

**BEZNER KERR, Rachel**
Cornell University
the United States of America/Canada

**BHANDARI, Preety**
Asian Development Bank (ADB)
India

**BHATIA, Parth**
Centre for Policy Research, Initiative
on Climate, Energy and Environment (ICEE)
India

**BISTLINE, John**
Electric Power Research Institute (EPRI)
the United States of America

**BIZEUL, Alexandre**
International Energy Agency (IEA)
France

**BLANCO, Gabriel**
Universidad Nacional del Centro (UNICEN)
Argentina

**BLANCO, Hilda**
University of Southern California
the United States of America

**BLOK, Kornelis**
Delft University of Technology
the Netherlands

**BOASSON, Elin Lerum**
Centre for International Climate
and Environmental Research (CICERO)/
Department of Political Science,
University of Oslo
Norway

**BOODOO, Zyaad**
Government of Mauritius
Mauritius

**BORBOR-CÓRDOVA, Mercy J.**
Escuela Superior Politecnica del Litoral
Ecuador

**BOSETTI, Valentina**
Bocconi University
Italy

**BOYKOFF, Maxwell**
Cooperative Institute for Research
in Environmental Sciences,
University of Colorado Boulder
the United States of America

**BRADLEY, Jessie**
Delft University of Technology
the Netherlands

**BRITTON, Jessica**
University of Exeter
United Kingdom

**BRUTSCHIN, Elina**
International Institute for Applied
Systems Analysis (IIASA)
Austria

**BUCK, Holly Jean**
State University of New York at Buffalo
the United States of America

**BURCH, Sara**
University of Waterloo
Canada

**BURNS, Charlotte**
University of Sheffield
United Kingdom

**BUSTAMANTE, Mercedes**
Department of Ecology IB,
Universidade de Brasilia
Brazil

**BYERS, Edward**
International Institute for Applied
Systems Analysis (IIASA)
Austria/Ireland

**CABEZA, Luisa F.**
University of Lleida
Spain

**CAIADO COUTO, Lilia**
Chatham House
Brazil

**CAIAFA, Clara**
Eindhoven University of Technology
Brazil/the Netherlands

**CALDAS, Lucas R.**
Universidade Federal do Rio de Janeiro
Brazil

**CALVIN, Katherine**
Pacific Northwest National Laboratory
the United States of America

**CAMPBELL, Donovan**
University of the West Indies
Jamaica

**CAMPBELL, Nick**
International Chamber of Commerce
France

**CAPARRÓS, Alejandro**
University of Durham and Spanish
Research Council
Spain

**CARLETON, Tamma**
Bren School of Environmental
Science and Management, UCSB
the United States of America

**CARNICER COLS, Jofre**
Universitat de Barcelona
Spain

**CARRARO, Carlo**
IPCC Working Group III Vice-Chair
Italy

**CARRUTHERS, Pasha**
Cook Islands Red Cross Society
Cook Islands

**CARVAJAL SARZOSA, Pablo Esteban**
International Renewable Energy Agency
Ecuador

**CASTÁN BROTO, Vanesa**
University of Sheffield
United Kingdom/Spain

**CASTANEDA, Antonethe**
UNESCO CONT E ECT
Guatemala

**CAVALETT, Otavio**
Norwegian University of Science
and Technology (NTNU)
Brazil

**CHÀFER, Marta**
Universitat de Lleida
Spain

**CHAPUNGU, Lazarus**
Great Zimbabwe University
Zimbabwe

**CHATURVEDI, Vaibhav**
Council on Energy, Environment and Water
India

**CHAWLA, Kiran**
Stanford University
India

**CHEAH, Lynette**
Singapore University of Technology and Design
Singapore

**CHEN, Wenying**
Tsinghua University
China

**CHEN, Ying**
Institute for Urban and Environmental Studies (IUE), Chinese Academy of Social Sciences (CASS)
China

**CHOW, Winston**
Singapore Management University
Singapore

**CHURKINA, Galina**
Berlin University of Technology
the Russian Federation/Germany

**CIESZEWSKA, Basia**
University of Exeter
Poland/United Kingdom

**CLAPP, Christa**
CICERO Shades of Green
Norway/the United States of America

**CLARK, Harry**
New Zealand Agricultural Greenhouse Gas Research Centre
New Zealand

**CLARKE, Leon**
Bezos Earth Fund
the United States of America

**COHEN, Brett**
University of Cape Town
South Africa

**COLLINS, William**
University of Reading
United Kingdom

**COMPOSTO, Jordana**
Princeton University
the United States of America

**COWIE, Annette**
NSW Department of Primary Industries; Science-Policy Interface UNCCD; Scientific and Technical Advisory Panel Global Environment Facility
Australia

**CRAIG, Michael**
School of Environment and Sustainability, University of Michigan
the United States of America

**CREUTZIG, Felix**
Mercator Research Institute on Global Commons and Climate Change (MCC) gGmbH
Germany

**CRIPPA, Monica**
European Commission
Italy

**CUI, Yiyun (Ryna)**
Centre for Global Sustainability, University of Maryland
China

**CULLEN, Jonathan M.**
Department of Engineering, University of Cambridge
New Zealand/United Kingdom

**CUMPSTY, Nick**
Imperial College London
United Kingdom

**CUNHA, Bruno**
Universidade Federal do Rio de Janeiro
Brazil

**DADI, Diriba Korecha**
IPCC Working Group III Vice-Chair
Ethiopia

**DAGNACHEW, Anteneh Getnet**
Netherlands Environmental Assessment Agency (PBL)/Utrecht University
the Netherlands/Ethiopia

**DAHL LARSEN, Morten Andreas**
The Danish Technical University (DTU)
Denmark

**DAI, Hancheng**
Peking University
China

**DAIOGLOU, Vassilis**
Copernicus Institute of Sustainable Development
Greece

**DAKA, Julius Partson**
Zambia Environmental Management Authority (ZEMA)
Zambia

**DAS, Nandini**
Climate Analytics/Jadavpur University
India

**DASGUPTA, Dipak**
The Energy and Resources Institute (TERI)
India

**DAVIS, Steven J.**
University of California Irvine
the United States of America

**DE CONINCK, Heleen**
Eindhoven University of Technology
the Netherlands

**DE KLEIJNE, Kiane**
Radboud University
the Netherlands

**DE LA RUE DU CAN, Stéphane**
Lawrence Berkeley National Laboratory
the United States of America

**DE LA VEGA NAVARRO, Angel**
National Autonomous University of Mexico (UNAM)
Mexico

**DEANGELO, Julianne**
Department of Earth System Science, University of California, Irvine
the United States of America

**DEN ELZEN, Michel**
Netherlands Environmental Assessment Agency (PBL)
the Netherlands

**DENTON, Fatima**
United Nations University
the Gambia

**DEPPERMANN, Andre**
International Institute for Applied Systems Analysis (IIASA)
Germany

**DEVINE-WRIGHT, Patrick**
University of Exeter
United Kingdom/Ireland

**DHAKAL, Shobhakar**
Asian Institute of Technology
Nepal/Thailand

**DHAR, Subash**
UNEP DTU Partnership
India/Denmark

**DIAGNE, El Hadji Mbaye Madien**
National Climate Change Committee/Afrique Energy Environnement
Senegal

**DIAZ, Delavane**
Electric Power Research Institute (EPRI)
the United States of America

**DIAZ ANADON, Laura**
Department of Land Economy, University of Cambridge; Centre for Environment, Energy and Natural Resource Governance (C-EENRG)
Spain/United Kingdom

**DIAZ MOREJON, Cristobal**
Ministry of Science, Technology and the Environment
Cuba

**DIAZ-JOSÉ, Julio**
Universidad Veracruzana
Mexico

**DIEMUODEKE, Ogheneruona E.**
University of Port Harcourt
Nigeria

**DJEMOUAI, Kamal**
Algeria

**DLAMINI, Reuben**
University of Cape Town
eSwatini

**DÖBBELING, Niklas**
Mercator Research Institute on Global Commons and Climate Change (MCC) gGmbH
Germany

**DODMAN, David**
International Institute for Environment and Development
Jamaica/United Kingdom

**DROUET, Laurent**
RFF-CMCC European Institute on Economics and the Environment
Italy/France

**DUBASH, Navroz K.**
Centre for Policy Research
India

**DUBE, Nokuthula**
Zimbabwe/United Kingdom

**DUBEUX, Carolina Burle Schmidt**
Universidade Federal do Rio de Janeiro
Brazil

**EDGE, Jacqueline Sophie**
Imperial College London
South Africa/United Kingdom

**EDMONDS, James**
Pacific Northwest National Laboratory
the United States of America

**ELGIZOULI IDRIS, Ismail**
the Sudan

**EMMET-BOOTH, Jeremy P.**
UCD School of Biosystems & Food Engineering
Ireland/New Zealand

**ESSANDOH-YEDDU, Joseph Kow**
Energy Commission
Ghana

**EYRE, Nicholas**
University of Oxford
United Kingdom

**FARIAS, Fernando**
Ministry of Environment
Chile

**FENG, Kuishuang**
University of Maryland
the United States of America

**FIFITA, Solomone**
Pacfic Centre for Renewable Energy and Energy Efficiency
Tonga

**FIGUEROA MEZA, Maria Josefina**
Copenhagen Business School
Venezuela/Denmark

**FISCHEDICK, Manfred**
Wuppertal Institute for Climate, Environment, Energy
Germany

**FISH-ROMITO, Vivien**
Centre International de Recherche sur l'Environnement et le Développement (CIRED)
France

**FISHER, Dana R.**
University of Maryland
the United States of America

**FORSTER, Piers M.**
School of Earth and Environment, University of Leeds
United Kingdom

**FRANK, Robert H.**
Climate One/Cornell SC Johnson College of Business
the United States of America

**STEFAN Frank**
International Institute for Applied Systems Analysis (IIASA)
Austria

**FU, Sha**
National Center for Climate Change Strategy and International Cooperation
China

**FUGLESTVEDT, Jan S.**
Centre for International Climate and Environmental Research (CICERO)
Norway

**FUJIMORI, Shinichiro**
Kyoto University
Japan

**GALEAZZI, Clara**
University of Cambridge
Argentina

**GALEOTTI, Marzio Domenico**
Department of Environmental Science and Policy, University of Milan
Italy

**GAO, Yuan**
School of Public Administration, Zhengzhou University
China

**GARCÍA TAPIA, Victor**
International Energy Agency (IEA)
Spain

**GARRETT, Rachael**
ETH Zürich
Switzerland/the United States of America

**GARG, Amit**
Indian Institute of Management Ahmedabad
India

**GEDEN, Oliver**
German Institute for International and Security Affairs
Germany

**GEELS, Frank W.**
Manchester Institute of Innovation Research, University of Manchester
United Kingdom/the Netherlands

**GENG, Yong**
School of Environmental Science and Engineering, Shanghai JiaoTong University
China

**GERBER, James S.**
Institute on the Environment, University of Minnesota
the United States of America

**GERMESHAUSEN, Robert**
Leibniz Centre for European Economic Research (ZEW)
Germany

**GHOSH, Bipashyee**
Science Policy Research Unit, University of Sussex
India/United Kingdom

**GINZBURG, Veronika**
Institute of Global Climate and Ecology Roshydromet and RAS
the Russian Federation

**GODOY, Alex**
Sustainability Research Centre & Strategic Resource Management, School of Engineering, Universidad del Desarrollo
Chile

**GRASSI, Giacomo**
Joint Research Centre, European Commission
Italy/European Union

**GRUBB, Michael**
Institute of Sustainable Resources, University College London
United Kingdom

**GRÜBLER, Arnulf**
International Institute for Applied Systems Analysis (IIASA)
Austria

**GU, Baihe**
Institutes of Science and Development, Chinese Academy of Sciences
China

**GUIVARCH, Céline**
International Centre for Development and Environment (CIRED)
France

**GUIZZARDI, Diego**
European Commission
Italy

**GÜNERALP, Burak**
Texas A&M University
Turkey/the United States of America

**GUO, Siyue**
Institute for Energy, Environment & Economy, Tsinghua University
China

**GUPTA, Shreekant**
Delhi School of Economics, University of Delhi
India

**GURNEY, Kevin R.**
Arizona State University
the United States of America

**HAHMANN, Andrea N.**
Department of Wind Energy, Technical University of Denmark
Chile/Denmark

**HAITES, Erik**
Margaree Consultants Inc.
Canada

**HALSNÆS, Kirsten**
The Danish Technical University (DTU)
Denmark

**HAMDI, Rafiq**
Royal Meteorological Institute of Belgium
Belgium

**HAN, Rong**
Beijing Institute of Technology
China

**HARMSEN, Mathijs**
Netherlands Environmental Assessment Agency (PBL)
the Netherlands

**HASANBEIGI, Ali**
Global Efficiency Intelligence
the United States of America

**HASEGAWA, Tomoko**
Ritsumeikan University
Japan

**HAVLIK, Petr**
International Institute for Applied Systems Analysis (IIASA)
Czech Republic

**HAYWARD, Bronwyn**
University of Canterbury
New Zealand

**HEALY, Noel**
Salem State University
the United States of America

**HEEREN, Niko**
Norwegian University of Science and Technology (NTNU)
Switzerland/Norway

**HEINREICH, Viola**
University of Bristol
United Kingdom/Germany

**HEJAZI, Mohamad**
King Abdullah Petroleum Studies
and Research Center (KAPSARC)
Syria/the United States of America

**HELD, Hermann**
University of Hamburg
Germany

**HELD, Maximilian**
ETH Zürich
Switzerland

**HINDER, Benjamin**
Heriot-Watt University
United Kingdom

**HÖGLUND-ISAKSSON, Lena**
International Institute for Applied
Systems Analysis (IIASA)
Sweden

**HÖHNE, Niklas**
New Climate Institute
Germany

**HOPPE, Janna**
ETH Zürich
Germany

**HOURCADE, Jean-Charles**
Centre International de Recherche
sur l'Environnement et le Développement
(CIRED)
France

**HOUSE, Joanna I.**
University of Bristol
United Kingdom

**HSU, Angel**
Yale University (Singapore)
the United States of America/Singapore

**HU, Shan**
Building Energy Research Centre,
Tsinghua University
China

**HUBACEK, Klaus**
University of Maryland
the Netherlands/the United States of America

**HUMPENÖDER, Florian**
Potsdam Institute for Climate
Impact Research (PIK)
Germany

**HUNG, Christine**
Norwegian University of Science
and Technology (NTNU)
Norway

**HUTYRA, Lucy**
Institute for Sustainable Energy,
Boston University
the United States of America

**IACOBUTA, Gabriela**
German Institute of Development
and Sustainability (IDOS)
Germany

**IVANOVA BONCHEVA, Antonina**
Universidad Autónoma de Baja California Sur
Mexico

**IYER, Gokul**
Pacific Northwest National Laboratory
India

**JACCARD, Mark**
Simon Fraser University
Canada

**JACHNIK, Raphaël**
Organisation for Economic Cooperation
and Development (OECD)
France

**JÄGER-WALDAU, Arnulf**
Joint Research Centre, European Commission
Italy/Germany

**JANNUZZI, Gilberto**
University of Campinas
Brazil

**JARAMILLO, Paulina**
Carnegie Mellon University
the United States of America

**JEFFERY, Louise**
New Climate Institute
United Kingdom

**JENN, Alan**
University of California, Davis
the United States of America

**JIANG, Kejun**
Energy Research Institute, National
Development and Reform Commission
China

**JODOIN, Sébastien**
McGill University
Canada

**JONCKHEERE, Inge G.C.**
Food and Agriculture Organisation
of the United Nations (FAO)
Italy/Belgium

**JOTZO, Frank**
Crawford School of Public Policy,
Australian National University
Australia

**JUNG, Tae Yong**
Graduate School of International
Studies, Yonsei University
Republic of Korea

**JUPESTA, Joni**
Research Institute of Innovative
Technology for the Earth (RITE)
Indonesia/Japan

**KABISCH, Nadja**
Humboldt Universität zu Berlin
Germany

**KAHN RIBEIRO, Suzana**
Universidade Federal do Rio de Janeiro
Brazil

**KAJINO, Tsutomu**
Toyota Central R&D Labs Inc.
Japan

**KANHYE, Vimla**
Ministry of Environment, Solid Waste
Management and Climate Change
Mauritius

**KARTHA, Sivan**
Stockholm Environment Institute
the United States of America

**KAUFFMAN, J. Boone**
Oregon State University
the United States of America

**KEENAN, Jesse**
Tulane University
the United States of America/Austria

**KEESLER, Daniela**
Universidad Nacional de Salta
Argentina

**KELLER, David**
GEOMAR Helmholtz Centre
for Ocean Research
Germany/the United States of America

**KELLER, Meredith**
Yale University
the United States of America

**KHANNA, Nina**
Lawrence Berkeley National Laboratory
the United States of America

**KHENNAS, Smail**
Energy and Climate Change Consultant
Algeria

**KHOSLA, Radhika**
University of Oxford
United Kingdom/India

**KIHILA, Jacob M.**
Ardhi University
the United Republic of Tanzania

**KIKSTRA, Jarmo**
International Institute for Applied
Systems Analysis (IIASA)
Austria/the Netherlands

**KILKIŞ, Şiir**
The Scientific and Technological
Research Council of Turkey (TÜBİTAK)
Turkey

**KIM, Yong-Gun**
Korea Environment Institute
Republic of Korea

**KIRIYAMA, Eriko**
Tokyo Institute of Technology
Japan

**KISHIMOTO, Paul**
International Institute for Applied
Systems Analysis (IIASA)
Canada

**KLIMONT, Zbigniew**
International Institute for Applied
Systems Analysis (IIASA)
Austria/Poland

**KÖBERLE, Alexandre**
Universidade Federal do Rio de Janeiro/
Imperial College London
Brazil/United Kingdom

**KÖHLER, Jonathan**
Fraunhofer Institute for Systems
and Innovation Research ISI
Germany

**KOIVISTO, Matti Juhani**
Department of Wind Energy,
Technical University of Denmark
Finland/Denmark

**KÖNIG, Michael**
Frankfurt School of Finance & Management
GmbH, UNEP Collaborating Centre for
Climate and Sustainable Energy Finance
Germany

**KREIBIEHL, Silvia**
Frankfurt School – UNEP Collaborating
Centre for Climate and Sustainable
Energy Finance
Germany

**KREY, Volker**
International Institute for Applied
Systems Analysis (IIASA)
Germany/Austria

**KRIEGLER, Elmar**
Potsdam Institute for Climate
Impact Research (PIK)
Germany

**KUBOTA, Izumi**
National Institute for Environmental Studies
Japan

**KUGELBERG, Susanna**
Copenhagen Business School
Sweden

**KUZUYA, Takashi**
Toyota Central R&D Labs Inc.
Japan

**KVERNDOKK, Snorre**
Frisch Centre
Norway

**LABANDEIRA, Xavier**
University of Vigo
Spain

**LAMB, William F.**
Mercator Research Institute
on Global Commons and Climate
Change (MCC) gGmbH
Germany/United Kingdom

**LAMBOLL, Robin**
Imperial College London
United Kingdom/the United States of America

**LAURANCE, William**
Centre for Tropical Environmental
and Sustainability Science,
James Cook University
Australia

**LEAHY, Sinead**
New Zealand Agricultural Greenhouse
Gas Research Centre
New Zealand

**LECOCQ, Franck**
Centre International de Recherche
sur l'Environnement et le Développement
(CIRED)
France

**LEE, Arthur**
Chevron Corporation
the United States of America

**LEE, David S.**
Manchester Metropolitan University
United Kingdom

**LEFÈVRE, Julien**
Centre International de Recherche
sur l'Environnement et le Développement
(CIRED)
France

**LEIP, Adrian**
European Commission –
Joint Research Centre (JRC)
Italy/Germany

**LEMPERT, Robert**
Frederick S. Pardee Center for Longer
Range Global Policy and the Future
Human Condition
the United States of America

**LEPAULT, Claire**
Paris School of Economics
France

**LEVI, Peter**
International Energy Agency (IEA)
United Kingdom

**LEWIS, Jared**
Climate Resource
Australia/New Zealand

**LIM, Yun Seng**
Universiti Tunku Abdul Rahman
Malaysia

**LIMA DE AZEVEDO, Inês Margarida**
School of Earth, Energy & Environmental Sciences, Standord University
Portugal/the United States of America

**LOCKWOOD, Matthew**
University of Sussex
United Kingdom

**LOFGREN, Hans**
Independent Researcher
Sweden/the United States of America

**LÓPEZ BLANCO, María José**
Gauss International S.L.
Spain

**LÖSCHEL, Andreas**
University of Münster
Germany

**LOWE, Jason**
Met Office Hadley Centre
United Kingdom

**LUCENA, André F.P.**
Universidade Federal do Rio de Janeiro
Brazil

**LUCON, Oswaldo**
S. Paulo State Environment Secretariat
Brazil

**LUDERER, Gunnar**
Potsdam Institute for Climate Impact Research (PIK)
Germany

**LUYSSAERT, Sebastiaan**
VU University Amsterdam
Belgium

**LWASA, Shuaib**
Makerere University
Uganda

**MACH, Katharine J.**
University of Miami
the United States of America

**MAHMOUD, Nagmeldin**
IPCC Working Group III Vice-Chair
the Sudan

**MAÏZI, Nadia**
MINES Paris – PSL, Université PSL (Paris Sciences et Lettres)
France/Algeria

**MAKAROV, Igor**
HSE University
the Russian Federation

**MALHOTRA, Abhishek**
Indian Institute of Technology Delhi
India

**MALIK, Arunima**
The University of Sydney
Australia

**MARANGONI, Giacomo**
Polytechnic University of Milano
Italy

**MARTÍNEZ-BARÓN, Deissy**
International Center for Tropical Agriculture (CIAT)
Colombia

**MASANET, Eric**
University of California, Santa Barbara/ Northwestern University
the United States of America

**MASSON-DELMOTTE, Valérie**
IPCC WG II Co-Chair
France

**MASUI, Toshihiko**
National Institute for Environmental Studies
Japan

**MATA, Érika**
IVL Swedish Environmental Research Institute
Spain/Sweden

**MATHUR, Ritu**
The Energy and Resources Institute (TERI)
India

**MATTAUCH, Linus**
University of Oxford
Germany/United Kingdom

**MATTHEWS, J. B. Robin**
IPCC WG I TSU, Université Paris Saclay
France

**MATTHEWS, Robert**
Forest Research
United Kingdom

**MATTIOLI, Giulio**
TU Dortmund University
Germany/Italy

**MATTION, Francesco**
International Energy Agency (IEA)
Italy

**MBEVA, Kennedy**
Blavatnik School of Government, University of Oxford
Kenya

**MBOW, Cheikh**
University of Pretoria
Senegal

**MCCOLLUM, David**
IPCC Working Group III Technical Support Unit – Oak Ridge National Laboratory
the United States of America

**MCDOWALL, William**
University College London
United Kingdom

**MCPHEARSON, Timon**
Stockholm Resilience Centre
the United States of America

**MEINSHAUSEN, Malte**
The University of Melbourne
Australia/Germany

**MÉJEAN, Aurélie**
Centre International de Recherche sur l'Environnement et le Développement (CIRED)
France

**MÉNDEZ, Carlos**
IPCC WG II Vice-Chair
Venezuela

**MERCURE, Jean-François**
World Bank
United Kingdom

**MEZA, Francisco**
Pontifica Universdad Catolica de Chile
Chile

**MILLWARD-HOPKINS, Joel**
School of Earth and Environment,
University of Leeds
United Kingdom

**MILOJEVIC-DUPONT, Nikola**
Mercator Research Institute
on Global Commons and Climate
Change (MCC) gGmbH
France

**MINX, Jan Christoph**
Mercator Research Institute on Global
Commons and Climate Change/Priestley
International Centre for Climate,
University of Leeds
Germany

**MIR, Kaleem Anwar**
Global Change Impact Studies Centre,
Ministry of Climate Change, Government
of Pakistan
Pakistan

**MIRASGEDIS, Sebastian**
National Observatory of Athens
Greece

**MITCHELL, Catherine**
University of Exeter
United Kingdom

**MITTAL, Shivika**
Ahdmedabad University
India

**MIZUNO, Emi**
Sustainable Energy for All
Japan

**MÖLLER, Vincent**
IPCC WGII TSU/Alfred Wegener Institute
Germany

**MOON, Jongwoo**
Yonsei University
Republic of Korea

**MOORE, Brendan**
University of East Anglia
the United States of America/United Kingdom

**MORECROFT, Michael**
Natural England
United Kingdom

**MORITA, Kanako**
Forestry and Forest Products
Research Institute
Japan

**MRABET, Rachid**
Institut National de la Recherche
Agronomique (INRA)
Morocco

**MÜLLER-CASSERES, Eduardo**
Universidade Federal do Rio de Janeiro
Brazil

**MULUGETTA, Yacob**
University College London
Ethiopia/United Kingdom

**MURAKAMI, Jin**
Singapore University
of Technology and Design
Japan

**MURATORI, Matteo**
National Renewable Energy
Laboratory (NREL)
Italy

**MURI, Helene**
Norwegian University of Science
and Technology (NTNU)
Norway

**MUSAH, Justice Issah Surugu**
Institute for Environment and Human
Security (UNU-EHS), United Nations
University (UNU)
Ghana

**MYLAN, Josephine**
University of Manchester
United Kingdom

**MYSHAK, Anna**
Centre for Energy Efficiency –
XXI (CENEF-XXI)
the Russian Federation

**NABUURS, Gert-Jan**
Wageningen University and Research
the Netherlands

**NAIDOO, Sasha**
Council for Scientific and
Industrial Research (CSIR)
South Africa

**NDLOVU, Vinnet**
Curtin University
Zimbabwe/Australia

**NEMET, Gregory F.**
University of Wisconsin-Madison
the United States of America/Canada

**NEOGI, Suvadip**
Global Centre for Environment
and Energy, Ahmedabad University
India

**NEPAL, Mani**
South Asian Network for Development
and Environmental Economics (SANDEE),
International Center for Integrated
Mountain Development (ICIMOD)
Nepal

**NEUFELDT, Henry**
UNEP DTU Partnership
Denmark/Germany

**NEWMAN, Peter**
Curtin University
Australia

**NIAMIR, Leila**
Mercator Research Institute on Global
Commons and Climate Change (MCC)
gGmbH/International Institute for
Applied Systems Analysis (IIASA)
Iran/Germany

**NICHOLLS, Zebedee**
Climate Energy College,
University of Melbourne
Australia

**NILSSON, Lars J.**
Lund University
Sweden

**NINAN, Karachepone N.**
Centre for Economics,
Environment and Society
India

**NOVIKOVA, Aleksandra**
Institute for Climate Protection,
Energy, and Mobility (IKEM)
Germany/the Russian Federation

**NOWAK, David**
Northern Research Station
the United States of America

**NUGROHO, Sudarmanto Budi**
Institute for Global Environmental Strategies
Indonesia

**OHIARE, Sanusi Mohamed**
Rural Electrification Agency
Nigeria

**OKEREKE, Chukwumerije**
Alex Ekwueme Federal
University Ndufu-Alike
Nigeria/United Kingdom

**OKITASARI, Mahesti**
United Nations University, Institute for
the Advanced Study of Sustainability
Japan/Indonesia

**OLIVIER, Jos**
Netherlands Environmental
Assessment Agency (PBL)
the Netherlands

**O'NEILL, Brian**
Pacific Northwest National Laboratory
the United States of America

**ONYIGE, Chioma Daisy**
Department of Sociology,
University of Port Harcourt
Nigeria

**ORGANSCHI, Alan**
Gray Organschi Architecture
the United States of America

**OSCHLIES, Andreas**
GEOMAR Helmholtz Centre
for Ocean Research
Germany

**O'SULLIVAN, Michael**
University of Exeter
United Kingdom

**OU, Xunmin**
Tsinghua University
China

**OWEN, Anne**
University of Leeds
United Kingdom

**PACHAURI, Shonali**
International Institute for Applied
Systems Analysis (IIASA)
India

**PACHECO-ROJAS, Daniel Alejandro**
Universidad Nacional Autónoma de México
Mexico

**PAN, Jiahua**
Institute for Urban & Environmental Studies,
Chinese Academy of Social Sciences
China

**PARKINSON, Simon**
Institute for Integrated Energy Systems,
University of Victoria
Canada

**PARMESAN, Camille**
CNRS Ecology Station (SETE)
United Kingdom/the United States of America

**PATANGE, Omkar**
International Institute for Applied
Systems Analysis (IIASA)
India

**PATERSON, Matthew**
University of Manchester
United Kingdom

**PATHAK, Minal**
IPCC WG III TSU, Ahmedabad University
India

**PATT, Anthony**
ETH Zürich
Switzerland

**PEDACE, Roque**
Climate Action Network – Latin America
(CAN-LA)/Buenos Aires University
Argentina

**PEEL, Jacqueline**
Melbourne Law School,
University of Melbourne
Australia

**PELLING, Mark**
King's College London
United Kingdom

**PENGUE, Walter Alberto**
Universidad de Buenos Aires, GEPAMA/
Universidade de General Sarmiento, ICO
Argentina

**PERCZYK, Daniel**
Fundación Torcuato di Tella
Argentina

**PERKINS, Patricia E.**
Faculty of Urban and Environmental
Studies, York University
Canada

**PESZKO, Grzegorz**
World Bank
Poland/the United States of America

**PETERS, Glen P.**
Centre for International Climate
and Environmental Research (CICERO)
Norway/Australia

**PHILIBERT, Cedric**
Institut Français des Relations Internationales
France

**PICHS-MADRUGA, Ramón**
IPCC Working Group III Vice-Chair
Cuba

**PIDGEON, Nick**
University of Cardiff
United Kingdom

**PIONTEK, Franziska**
Potsdam Institute for Climate
Impact Research (PIK)
Germany

**PONGRATZ, Julia**
Max Planck Institute for Meteorology
Germany

**POORE, Joseph**
Department of Biology, University of Oxford
United Kingdom

**POPP, Alexander**
Potsdam Institute for Climate
Impact Research (PIK)
Germany

**PORTUGAL-PEREIRA, Joana**
Universidade Federal do Rio de Janeiro
Brazil

**POSEN, Daniel**
Department of Civil & Mineral
Engineering, University of Toronto
Canada

**POSTIGO, Julio C.**
Indiana University
the United States of America/Peru

**POVITKINA, Marina**
Department of Political Science, University of Gothenburg
Sweden

**PRADHAN, Prajal**
Potsdam Institute for Climate Impact Research (PIK)
Germany/Nepal

**PREGITZER, Clara**
Natural Areas Conservancy
the United States of America

**PU, Wang**
Chinese Academy of Sciences
China

**PULVER, Simone**
UC Santa Barbara
the United States of America

**QUADRELLI, Roberta**
International Energy Agency (IEA)
Italy

**RAISER, Kilian**
Centre for Sustainability, Hertie School
Germany

**RAJAMANI, Lavanya**
Faculty of Law, University of Oxford
India

**RAMASWAMI, Anu**
Civil and Environmental Engineering, Princeton University
the United States of America

**RAO, Anand**
Indian Institute of Technology Bombay
India

**RAO, Narasimha D.**
Yale School of the Environment
the United States of America

**RAVINDRANATH, Nijavalli Hanumantharao**
Indian Institure of Science
India

**REBACK, Mia**
Rocky Mountain Institute (RMI)
the United States of America

**RECKIEN, Diana**
University of Twente, Faculty of Geo-Information Science and Earth Observation (ITC), Department of Urban and Regional Planning and Geo-Information Management
Germany

**REISCH, Lucia A.**
Cambridge Judge Business School, University of Cambridge
United Kingdom

**REISINGER, Andy**
IPCC Working Group III Vice-Chair
New Zealand

**RENFORTH, Phil**
Heriot Watt University
United Kindgom

**RIAHI, Keywan**
International Institute for Applied Systems Analysis (IIASA)
Austria

**RICKE, Katherine**
Scripps Institution of Oceanography, UC San Diego
the United States of America

**RIGBY, Matthew**
School of Chemistry, University of Bristol
United Kingdom

**RIVERA-FERRE, Marta G.**
INGENIO (CSIC-UPVV)
Spain

**ROBERTSON, Simon**
The University of New South Wales
Australia

**ROCHEDO, Pedro**
Universidade Federal do Rio de Janeiro|
Brazil

**ROE, Stephanie**
University of Virginia/Climate Focus
the Philippines/the United States of America

**ROGELJ, Joeri**
Imperial College London
Belgium/United Kingdom

**ROGGE, Karoline**
University of Sussex
Germany/United Kingdom

**ROMERO-LANKAO, Patricia**
National Renewable Energy Laboratory/ University of Chicago
Mexico/the United States of America

**ROSE, Steven**
Electric Power Research Institute (EPRI)
the United States of America

**ROSENSTOCK, Todd**
World Agroforestry Centre (ICRAF)
the United States of America

**ROY, Joyashree**
Asian Institute of Technology/ Jadavpur University
India/Thailand

**RUNGE, Carlisle Ford**
University of Minnesota
the United States of America

**SAFARZYNSKA, Karolina**
Faculty of Economic Sciences, University of Warsaw
Poland

**SAGAR, Ambuj D.**
Indian Institute of Technology Delhi
India

**SAHEB, Yamina**
OpenExp, Ecole des Mines de Paris
France/Algeria

**SAMADI, Sascha**
Wuppertal Institute for Climate, Environment, Energy
Germany

**SAMARGANDI, Nahla**
King Abdulaziz University
Saudi Arabia

**SANCHES PEREIRA, Alessandro**
University of São Paulo/Institute of Energy and Environment (IEE)
Brazil

**SANDSTAD, Marit**
Centre for International Climate and Environmental Research (CICERO)
Norway

**SANTILLÁN VERA, Mónica**
Facultad de Economía, Universidad Nacional Autónoma de México
Mexico

**SANTOS, Andrea**
Universidade Federal do Rio de Janeiro
Brazil

**SAPKOTA, Tek**
Climate Change, Agriculture and Food Security (CGIAR)
Nepal/Canada

**SARI, Agus Pratama**
Landscape Indonesia
Indonesia

**SAUNOIS, Marielle**
Laboratoire des Sciences du Climat et de l'Environnement
France

**SCHAEFFER, Roberto**
Universidade Federal do Rio de Janeiro
Brazil

**SCHWEIZER-RIES, Petra**
University of Applied Sciences Bochum
Germany

**SEBBIT, Adam Mohammed**
Makerere University
Uganda

**SEBOTHOMA, Dimakatso**
University of Cape Town
South Africa

**SEMENOV, Sergey**
IPCC WG II Vice-Chair
the Russian Federation

**SENEVIRATNE, Sonia I.**
ETH Zürich
Switzerland

**SERRANO DINÁ, María Isabel**
Pontificia Universidad Catolica Madre y Maestra
the Dominican Republic

**SETO, Karen C.**
Yale University
the United States of America

**SETZER, Joana**
London School of Economics
Brazil/United Kingdom

**SHAN, Yuli**
University of Groningen China
China

**SHAREEF, Ali**
Ministry of Environment and Energy
Maldives

**SHARIFI, Ayyoob**
Hiroshima University
Iran/Japan

**SHARMA, Rohit**
Centre for Science and Policy, University of Cambridge
India

**SHEN, Wei**
Institute of Development Studies
China/United Kingdom

**SHIN, Jen**
Yale School of Architecture
the United States of America

**SHLAPAK, Mykola**
Consultant
Ukraine

**SHUKLA, Priyadarshi**
IPCC Working Group III Co-Chair
India

**SIMS, Ralph E.H.**
Centre for Energy Research, School of Engineering and Advanced Technology, Massey University
New Zealand

**SINGH, Ajay Kumar**
Independent Researcher
India

**SINGH, Pramod K.**
Institute of Rural Management Anand
India

**SINGH, Tamiksha**
Energy and Resources Institute (TERI)
India

**SINGH, Udayan**
Argonne National Laboratory
India

**SKEA, Jim**
IPCC Working Group III Co-Chair
United Kingdom

**SLADE, Raphael**
IPCC WG III TSU, Imperial College London
United Kingdom

**SMITH, Chris**
School of Earth and Environment, University of Leeds
Austria/United Kingdom

**SMITH, Pete**
University of Aberdeen
United Kingdom

**SMITH, Stephen M.**
Smith School of Enterprise and the Environment, University of Oxford
United Kingdom

**SMITH, Steven J.**
Pacific Northwest National Laboratory
the United States of America

**SOERGEL, Björn**
Potsdam Institute for Climate Impact Research (PIK)
Germany

**SOHAG, Kazi**
Ural Federal University
the Russian Federation

**SOHNGEN, Brent**
Ohio State University
the United States of America

**SOKONA, Mohamed Youba**
African Development Bank
Mali

**SOLAZZO, Efisio**
European Commission – Joint Research Centre (JRC)
Italy

**SOME, Shreya**
IPCC WG III TSU, Asian Institute of Technology
India

**SOVACOOL, Benjamin Kenneth**
Aarhus University/University of Sussex
Denmark/United Kingdom

**SPRINZ, Detlef F.**
Potsdam Institute for Climate Impact Research (PIK)
Germany

**SSEKUUBWA, Enock**
University of Makerere
Uganda

**STABINSKY, Doreen**
College of the Atlantic
the United States of America

**STANKEVICIUTE, Loreta**
International Atomic Energy Agency (IAEA)
Lithuania

**STECKEL, Jan**
Mercator Research Institute on Global Commons and Climate Change (MCC) gGmbH
Germany

**STEG, Linda**
Instituto 17/Curtin University Sustainability Policy Institute
the Netherlands

**STEINBERGER, Julia**
Institute of Geography and Sustainability, University of Lausanne
Switzerland/United Kingdom

**STEINFELD, Jonas**
Wageningen University and Research
Germany/Brazil

**STERN, David**
Australian National University
Australia

**STRBAC, Goran**
Imperial College London
Serbia/United Kingdom

**STRØMMAN, Anders Hammer**
Industrial Ecology Programme, Norwegian University of Science and Technology
Norway

**SUGIYAMA, Masahiro**
Institute for Future Initiatives, The University of Tokyo
Japan

**SUGIYAMA, Taishi**
The Canon Institute for Global Studies
Japan

**SULISTIAWATI, Linda Yanti**
Universitas Gadjah Mada
Indonesia

**SUN, Laixiang**
Department of Geographical Sciences, University of Maryland
the United States of America

**SUNSTEIN, Cass R.**
Harvard University and Harvard Law School
the United States of America

**SURANA, Kavita**
Centre for Global Sustainability, School of Public Policy, University of Maryland
India

**TAHAVI-MOHARAMLI, Pouya**
International Energy Agency (IEA)
Canada

**TAN, Xianchun C.**
Institute of Science and Development, Chinese Academy of Sciences
China

**TANAKA, Kanako**
National Institute of Advanced Industrial Science and Technology (AIST)
Japan

**TANAKA, Kenji**
Department of Technology Management for Innovation, Graduate School of Engineering, University of Tokyo
Japan

**TAVONI, Massimo**
Politechnic Institute of Milan
Italy

**TENG, Fei**
Tsinghua University
China

**TIAN, Hanqin**
School of Forestry and Wildlife Sciences, Auburn University
the United States of America

**TIRADO VON DER PAHLEN, Maria Cristina**
University of California Los Angeles
the United States of America/Spain

**TIRANA, Florian**
France

**TORIBIO RAMIREZ, Daniela**
University of Amsterdam
Mexico

**TORRES MARTÍNEZ, Julio**
Center for World Economic Research (CIEM)
Cuba

**TOSUN, Jale**
University of Heidelberg
Germany

**TOTH, Ferenc L.**
International Institute for Applied Systems Analysis (IIASA)
Austria/Hungary

**TOWPRAYOON, Sirintornthep**
King Mongkut's University of Technology Thonburi
Thailand

**TRISOS, Christopher H.**
University of Maryland
South Africa

**TUBIELLO, Francesco N.**
Food and Agriculture Organisation of the United Nations (FAO)
Italy

**UECKERDT, Falko**
Potsdam Institute for Climate Impact Research (PIK)
Germany

**ÜRGE-VORSATZ, Diana**
IPCC Working Group III Vice-Chair
Hungary

**USAI, Lorenzo**
Norwegian University of Science and Technology (NTNU)
Italy

**UVO, Cintia B.**
Lund University
Brazil/Sweden/Italy

**VAN AALST, Maarten**
Red Cross Red Crescent Climate Centre/
IRI/Columbia University & STEaPP/UCL/
University of Twente
the Netherlands

**VAN ASSELT, Harro**
University of Eastern Finland
the Netherlands

**VAN DER WIJST, Kaj-Ivar**
Netherlands Environmental
Assessment Agency (PBL)
the Netherlands

**VAN DIEMEN, Renée**
IPCC WG III TSU, Imperial College London
the Netherlands/United Kingdom

**VAN RUIJVEN, Bastiaan**
International Institute for Applied
Systems Analysis (IIASA)
the Netherlands

**VAN SOEST, Heleen**
Netherlands Environmental
Assessment Agency (PBL)
the Netherlands

**VAN VUUREN, Detlef P.**
Netherlands Environmental
Assessment Agency (PBL)
the Netherlands

**VÁSQUEZ-ARROYO, Eveline María**
Universidade Federal do Rio de Janeiro
Peru/Brazil

**VELDSTRA, Janet**
University of Groningen
the Netherlands

**VENKATESH, Aranya**
Carnegie Mellon University
the United States of America/India

**VERDOLINI, Elena**
Fondazione Eni Enrico Mattei (FEEM)/
CMCC Foundation – Euro-Mediterranean
Center on Climate Change
Italy

**VÉREZ, David**
University of Lleida
Cuba/Spain

**VERKERK, Pieter Johannes**
European Forest Institute
the Netherlands

**VILARIÑO, Maria Virginia**
Argentinean Business Council
for Sustainable Development
Argentina

**VYAS, Purvi**
IPCC Working Group III Technical
Support Unit – Ahmedabad University
India

**WADA, Kenichi**
Research Institute of Innovative
Technology for the Earth (RITE)
Japan

**WANEMARK, Joel**
Swedish Environmental Research Institute
Sweden

**WANG, Can**
Tsinghua University
China

**WEBB, Jeremy**
University College London
New Zealand

**WEBER, Elke Ursula**
Andlinger Center for Energy and
the Environment, Princeton University
the United States of America

**WEI, Yi-Ming**
Center for Energy and Environmental Policy
Research, Beijing Institute of Technology
China

**WESTPHAL, Michael**
Pacific Northwest National Laboratory
the United States of America

**WETTESTAD, Jørgen**
The Fridtjof Nansen Institute
Norway

**WHITE, Lee**
Australian National University
Australia

**WHITEHEAD, Jake**
The University of Queensland
Australia

**WIEDENHOFER, Dominik**
University of Natural Resources
and Life Sciences
Austria

**WIEDMANN, Thomas**
University of New South Wales
(UNSW Sydney)
Australia

**WILLIAMS, Mariama**
Integrated Policy Research Institute (IPRI)/
Development Alternatives With Women
for a New Era (DAWN)
Jamaica/the United States of America

**WILSON, Charlie**
Environmental Change Institute,
University of Oxford/International Institute
for Applied Systems Analysis (IIASA)
United Kingdom

**WINKLER, Harald**
Energy Research Centre,
University of Cape Town
South Africa

**WITAJEWSKI-BALTVILKS, Jan**
Faculty of Economic Sciences,
University of Warsaw
Poland

**WU, Libo**
School of Economics, Fudan University
China

**YAMAGATA, Yoshiki**
Graduate School of System Design
and Management, Keio University
Japan

**YAMBA, Francis**
Centre for Energy, Environment
and Engineering of Zambia
Zambia

**YEH, Sonia**
Chalmers University of Technology
Sweden

**YU, Biying**
Beijing Institute of Technology
China

**ZHOU, Nan**
Lawrence Berkeley National Laboratory
the United States of America

**ZIMM, Caroline**
International Institute for Applied
Systems Analysis (IIASA)
Austria

**ZUSMAN, Eric**
Institute for Global Environmental Strategies
the United States of America

AV

# Annex V: Expert Reviewers of the IPCC WGIII Sixth Assessment Report

**This chapter should be cited as:**

IPCC, 2022: Annex V: Expert Reviewers of the IPCC Working Group III Sixth Assessment Report. In *Climate Change 2022: Mitigation of Climate Change. Contribution of Working Group III to the Sixth Assessment Report of the Intergovernmental Panel on Climate Change* [P.R. Shukla, J. Skea, R. Slade, A. Al Khourdajie, R. van Diemen, D. McCollum, M. Pathak, S. Some, P. Vyas, R. Fradera, M. Belkacemi, A. Hasija, G. Lisboa, S. Luz, J. Malley, (eds.)]. Cambridge University Press, Cambridge, UK and New York, NY, USA. doi: 10.1017/9781009157926.024

**ABDELREHIM, Ahmed**
Centre for Environment & Development for the Arab Region and Europe – CEDARE
Egypt

**ABECASSIS, Adrien**
Harvard Kennedy School of Government
The United States of America

**ABIKOYE, Semilore**
Department of Chemical Engineering, University of Cape Town
South Africa

**ABRAHAM, Julie**
U.S. Department of Transportation
The United States of America

**ABUHIJLEH, Bassam**
The British University in Dubai
United Arab Emirates

**ABUJE, Sunday**
Kenya

**ABU-SAMHA, Mahmoud**
College of Engineering and Technology, American University of the Middle East
Kuwait

**ADACHI, Muneki**
Japan

**ADEGOKE, Abiodun**
Samsung electronics West Africa
Nigeria

**ADEN, Nathaniel**
World Resources Institute (WRI)
The United States of America

**ADOJOH, Onema**
Missouri University of Science and Technology, Rolla
The United States of America

**ADRIANO, Sciacovelli**
University of Birmingham
United Kingdom (of Great Britain and Northern Ireland)

**AGUILAR-AMUCHASTEGUI, Naikoa**
WWF-US
The United States of America

**AGUNG SUGARDIMAN, Ruandha**
Ministry of Environment and Forestry
Indonesia

**AGUS, Cahyono**
Universitas Gadjah Mada
Yogyakarta Indonesia
Indonesia

**AHAMER, Gilbert**
Environment Agency Austria
Austria

**AHENKORAH, Alfred**
Ofosu Ahenkorah and Partners Energy and Engineering Services Ltd
Ghana

**AHMAD FARID, Mohammed**
Ministry for the Environment and Water, Climate Change Division
Malaysia

**AHMADI NAMIN, Mahnaz**
Meteorology Organization of Iran
Iran

**AHMED, Atiq Kainan**
Asian Disaster Preparedness Center (ADPC)
Thailand

**AHN, Young-Hwan Sookmyung**
Women's University
The Republic of Korea

**AHNERT, Carolina**
Universidad Politécnica de Madrid (UPM)
Spain

**AIBA, Takao**
Japan Automobile Manufacturers Association/Toyota
Japan

**AIKING, Harry**
Institute for Environmental Studies, Vrije Universiteit
The Netherlands

**AKHTAR, Farhan**
U.S. Department of State
The United States of America

**AKIYAMA, Hiroko**
National Agriculture and Food Research Organization
Japan

**AKWANGO, Damalie**
National Agricultural Research Organisation
Uganda

**AL KHOURDAJIE, Alaa**
Imperial College London
United Kingdom (of Great Britain and Northern Ireland)

**ALAM, Edris**
Rabdan Acadmey
The United Arab Emirates

**ALATAŞ, Sedat**
Aydın Adnan Menderes University
Turkey

**ALBEROLA, Emilie**
Ecoact
France

**ALBUQUERQUE DE ANDRADE FLEURY, Marina**
Instituto Universitário de Lisboa
Portugal

**ALCALDE, Juan**
Geosciences Barcelona, CSIC
Spain

**ALDERMAN, Delton**
USFS
The United States of America

**ALDRIAN, Edvin**
BPPT Indonesia
Indonesia

**ALESSANDRO, Chiodi**
E4SMA
Italy

**ALEXANDROS, Nikas**
National Technical University of Athens
Greece

**ALEXEEVA, Victoria**
International Atomic Energy Agency (IAEA)
Austria

**ALGO, John Leo YOUNGO**
Cities Working Group
(Living Laudato Si' Philippines)
The Philippines

**ALLAM, Zaheer**
Deakin University
Mauritius

**ALLAN, Jennifer**
Cardiff University
United Kingdom (of Great Britain and Northern Ireland)

**ALLEN, Grant**
University of Manchester
United Kingdom (of Great Britain and Northern Ireland)

**ALLEN, Myles**
University of Oxford
United Kingdom (of Great Britain and Northern Ireland)

**ALPERT, Alice**
U.S. Department of State
The United States of America

**ALTERMATT, Pietro**
R&D Center of Trinasolar
Germany

**AMANULLAH, Amanullah**
Department of Agronomy, The University of Agriculture Peshawar
Pakistan

**AMELI, Nadia**
University College London
United Kingdom (of Great Britain and Northern Ireland)

**AMONYA, Fred**
Lyciar
United Kingdom (of Great Britain and Northern Ireland)

**AN, Qi**
Energy Research Institute, National Development and Reform Commission of China
China

**ANDERSEN, Rayner**
Department of Fisheries and Oceans
Canada

**ANDREI, Belyi**
University of Eastern Finland, Centre for Climate Change, Energy and Environmental Law
Estonia

**ANJOS, Lucia Helena**
UFRRJ/ITPS/FAO
Brazil

**ANORUO, Chukwuma**
University of Nigeria, Nsukka
Nigeria

**ANTAL, Miklós**
Eötvös Loránd University
Hungary

**ANTONICH, Beate**
Center for Climate Change, Energy and Environmental Law (CCEEL) School of Law, University of Eastern Finland
The United States of America

**AOKI, Naoki**
Japan Cement Association
Japan

**AOYAGI, Midori**
National Instituute for Environmental Studies
Japan

**APOSTOLAKI, Penny**
Independent Expert
United Kingdom (of Great Britain and Northern Ireland)

**ARA BEGUM, Rawshan**
Universiti kebangsaan Malaysia (UKM)
Malaysia

**ARAUJO, Kathleen**
CAES Eenrgy Policy Institute/ Boise State University
The United States of America

**AREGBESHOLA, Rafiu Adewale**
University of South Africa
South Africa

**ARENT, Doug**
U.S. Department of Energy
The United States of America

**ARSHAD, Adnan**
China Agricultural University
China

**ARTETXE, Ainara**
NEIKER-Basque Institute of Agricultural Research
Spain

**ARTHUR, Ebenezer**
Wangara Green Ventures
Ghana

**ASARPOTA, Karishma**
YOUNGO Cities Working Group (ICLEI)
Germany/India

**ASFAW, Solomon**
LUT University
Finland

**ATAII, Edward**
KPMG
United Kingdom (of Great Britain and Northern Ireland)

**ATHAR, Ghulam Rasul**
Pakistan Atomic Energy Commission
Pakistan

**AUBINET, Marc**
University of Liège
Belgium

**AUKEMA, Juliann**
U.S. Agency for International Development
The United States of America

**AULIYANI, Diah**
Watershed Management Technology Centre
Indonesia

**AVEN, Terje**
University of Stavanger
Norway

**AZZARA, Alyson**
U.S. Department of Transportation
The United States of America

**BABIKER, Mustafa**
Saudi Aramco
Saudi Arabia

**BADIOLA, Esther**
European Investment Bank
The United States of America

**BAGHERI, Mahsa**
Fraunhofer Institute for Systems and Innovation Research ISI
Germany

**BAILEY, Ian**
University of Plymouth
United Kingdom (of Great Britain and Northern Ireland)

**BAIMAN, Ron**
Benedictine University
The United States of America

**BAIRD, Kervelle**
YOUNGO Cities Working Group
Trinidad and Tobago

**BAK, Celine**
IISD
Canada

**BAKAR, Khalil**
Ministry of Sustainability and the Environment (MSE)
Singapore

**BAKER, Keith**
Built Environment Asset Management (BEAM) Centre, Glasgow Caledonian University
United Kingdom (of Great Britain and Northern Ireland)

**BAKKER, Stefan**
KiM Netherlands Institute for Transport Policy Assessment
The Netherlands

**BALA, Govindasamy**
Indian Institute of Science
India

**BALBUS, John M.**
National Institute of Environmental Health Sciences
The United States of America

**BALDWIN, Sam**
U.S. Department of Energy
The United States of America

**BALTRUSZEWICZ, Marta**
University of Leeds
United Kingdom (of Great Britain and Northern Ireland)

**BANDIVADEKAR, Srushti**
Manipal School of Architecture and Planning
India

**BANNON-GODFREY, Rachel**
Stantec
The United States of America

**BARBIERI, Alisson**
Universidade Federal de Minas Gerais (UFMG)
Brazil

**BARBOZA, Maria**
Columbia University
The United States of America

**BARDHAN, Suchandra**
Jadavpur University
India

**BARGER, Nichole**
UNCCD-SPI
The United States of America

**BARK, Glenn**
Luleå University of Technology
Sweden

**BARKER, Timothy**
Keele University
United Kingdom (of Great Britain and Northern Ireland)

**BARKHOUSE, Aaron**
SunPower Corporation
The United States of America

**BARRANTES, Olivia**
Universidad de Zaragoza
Spain

**BARRETO, Leonardo**
Austrian Energy Agency
Austria

**BARRETT, Brendan**
Osaka University
Japan

**BARTOCCI, Pietro**
University of Perugia
Italy

**BATAILLE, Christopher**
IDDRI/Simon Fraser University
Canada

**BATES, Albert**
Global Village Institute
The United States of America

**BATISHA, Ayman**
International Sustainability Institute (ISI)
Egypt

**BATTISTON, Stefano**
University of Zurich
Switzerland

**BAUER, Jan**
Copenhagen Business School
Denmark

**BAUER, Nico**
Potsdam Institute for Climate Impact Research (PIK)
Germany

**BAUER, Susanne**
National Aeronautics and Space Administration
The United States of America

**BAUGHCUM, Steven**
Boeing Company
The United States of America

**BAYLISS, Kate**
SOAS, University of London
United Kingdom (of Great Britain and Northern Ireland)

**BAZILIAN, Morgan**
Colorado School of Mines
The United States of America

**BEACH, Robert**
RTI International
The United States of America

**BEAUCHEMIN, Karen A.**
Agriculture and Agri-Food Canada
Canada

**BECCARI BARRETO, Beatriz**
Politecnico di Milano
Brazil

**BECHTA, Sevostian**
KTH Royal Institute of Technology
Sweden

**BEEMER, Emil**
Dutch Research Institute For Transitions, Erasmus University Rotterdam
The Netherlands

**BEJARANO CASTILLO, Marylin**
National Water Comission of Mexico
Mexico

**BELLAMY, Rob**
University of Manchester
United Kingdom (of Great Britain and Northern Ireland)

**BELLASSEN, Valentin**
INRAE
France

**BELYI, Andrei**
University of Eastern Finland
Finland

**BENKEBLIA, Noureddine**
The University of the West Indies
Jamaica

**BENKO, Bernadett**
Ministry of Innovation and Technology, Climate Policy Department
Hungary

**BENNETT, Helen**
Department of Industry, Science, Energy and Resources
Australia

**BENTSEN, Niclas Scott**
University of Copenhagen, Department of Geosciences and Natural Resource Management
Denmark

**BERGER, Mitchell**
U.S. Department of Health and Human Services
The United States of America

**BERK, Marcel**
Ministry of Economic Affairs and Climate Policy
The Netherlands

**BERNIER, Pierre**
Natural Resources Canada
Canada

**BERRILL, Peter**
Yale University
The United States of America

**BERTOLDI, Paolo**
European Commission
Italy

**BETTS, Richard**
Met Office Hadley Centre
United Kingdom (of Great Britain and Northern Ireland)

**BEUTTLER, Christoph**
Climeworks AG, Risk Dialogue Foundation
Switzerland

**BEZNER KERR, Rachel**
Cornell University
The United States of America

**BHADURI, Budhu**
U.S. Department of Energy
The United States of America

**BHARAT, Alka**
Maulana Azad National Institute of Technology, Bhopal
India

**BHARDWAJ, Amar**
Stanford University
The United States of America

**BHATIA, Parth**
Centre for Policy Research
India

**BHATT, Jayavardhan Ramanlal**
Ministry of Environment, Forests and Climate Change
India

**BIBERIAN, Jean-Paul**
Aix-Marseille University
France

**BIEH, Erin**
Johns Hopkins Center for a Livable Future
The United States of America

**BIESBROEK, Robbert**
Wageningen University
The Netherlands

**BIGANO, Andrea**
Fondazione Centro Euro-Mediterraneo sui Cambiamenti Climatici
Italy

**BIGIO, Anthony Gad**
George Washington University
The United States of America

**BILBAO Y LEON, Sama**
World Nuclear Association
United Kingdom (of Great Britain and Northern Ireland)

**BISHOP, Justin**
Arup
United Kingdom (of Great Britain and Northern Ireland)

**BISHOP, Kevin**
The Royal Swedish Academy of Agriculture and Forestry
Sweden

**BIZEUL, Alexandre**
International Energy Agency (IEA)
France

**BJURLIN, Curt**
Stantec Consulting
The United States of America

**BLACK, Iain**
University of Stirling
United Kingdom (of Great Britain and Northern Ireland)

**BLAHUT, Nina**
Joint Global Change Research Institute
The United States of America

**BLAINE, Tegan**
Blue Cairn Climate Consulting
The United States of America

**BLAKERS, Andrew**
Australian National University
Australia

**BLANCO, Herib**
International Renewable Energy Agency (IRENA)
Germany

**BLAZOWSKI, Adam**
FOTA4Climate.org
Poland

**BLENMAN, Carol-Anne**
Violeta Consulting Services
Barbados

**BLOK, Kornelis**
Delft University of Technology
The Netherlands

**BLONDEL, Ana**
Environment and Climate Change Canada
Canada

**BLUM, Megan**
U.S. Department of Transportation
The United States of America

**BLUWSTEIN, Jevgeniy**
University of Fribourg
Switzerland

**BOBIN, Jean Louis**
Sorbonne universités Paris
France

**BOCCARELLA, Ermenegilda**
Caelex
Belgium

**BOCHICCHIO, Juliet**
U.S. Department of Transportation
The United States of America

**BOER, Dieter**
Universitat Rovira i Virgili
Spain

**BOHAN, Richard**
Portland Cement Association
The United States of America

**BOIX-FAYOS, Carolina**
CEBAS-CSIC
Spain

**BOJARIU, Roxana**
National Meteorological Administration
Romania

**BONDOUELLE, Antoine**
Climate Action Network
France

**BOODOO, Zyaad**
Government of Mauritius
Mauritius

**BORE, Nelly**
University of Nairobi
Kenya

**BORENER, Sherry**
U.S. Department of Transportation
The United States of America

**BORGHESI, Simone**
European University Institute – Florence School of Regulation Climate, and University of Siena
Italy

**BORMANN, Jeanne**
Ministry of Agriculture
Luxembourg

**BOSE STYCZYNSKI, Annika**
O.P. Jindal Global University
India

**BOSETTI, Valentina**
BOCCONI -eiee
Italy

**BRADSHAW, Michael**
University of Warwick
United Kingdom (of Great Britain and Northern Ireland)

**BRAND CORREA, Lina**
University of Leeds
United Kingdom (of Great Britain and Northern Ireland)

**BRECHA, Robert**
Climate Analytics
Germany

**BREON, Francois-Marie**
CEA
France

**BREYER, Christian**
LUT University
Finland

**BRIAND, Yann**
IDDRI
France

**BROCKWAY, Paul**
University of Leeds
United Kingdom (of Great Britain and Northern Ireland)

**BROWN, Reed**
U.S. Department of State
The United States of America

**BROWN, Tom**
Karlsruhe Institute of Technology
Germany

**BRUCE, Stuart**
KPMG/International Chamber of Commerce Energy and Environment Committee Co-Chair/IUCN Energy Transition Project
United Kingdom (of Great Britain and Northern Ireland)

**BRÜGGER, Adrian**
University of Bern, Dep. of Consumer Behavior
Switzerland

**BRUHWILER, Lori**
National Aeronautics and Space Administration
The United States of America

**BRULLE, Robert**
Brown University
The United States of America

**BRUN, Eric**
Ministère de la Transition écologique et solidaire
France

**BRUNNER, Beat**
Lightning MultiCom SA
Switzerland

**BU, Maoliang**
Nanjing University
China

**BUCHANAN, Kent**
Department of Environment, Forestry and Fisheries
South Africa

**BUCHS, Milena**
University of Leeds
United Kingdom (of Great Britain and Northern Ireland)

**BUDINIS, Sara**
International Energy Agency (IEA)
France

**BUHR, Robert**
Green Planet Consulting Ltd.
United Kingdom (of Great Britain and Northern Ireland)

**BUONGIORNO, Jacopo**
Massachusetts Institute of Technology
The United States of America

**BUONOCORE, Jonathan**
Harvard University
The United States of America

**BURAK, Adam**
University of Michigan
The United States of America

**BURGERS, Laura**
University of Amsterdam
The Netherlands

**BURLOT, Alan**
Commissariat à l'Energie Atomique et aux Energies Alternatives
France

**BURNETT, Wesley**
U.S. Department of Agriculture
The United States of America

**BURNS, Wil**
Institute for Carbon Removal Law & Policy, American University
The United States of America

**BUSCH, Jonah**
Earth Innovation Institute
The United States of America

**BUSCHIAZZO, Daniel**
Instituto de Ciencias del la Tierra y Ambientales de La Pampa (INCITAP-CONICET)
Argentina

**BUSH, Elizabeth**
Environment and Climate Change Canada
Canada

**BUSTOS-SALVAGNO, Javier**
Universidad del Desarrollo
Chile

**BUTLER, James H.**
National Oceanic and Atmospheric Administration
The United States of America

**CABRAL, Marco Tulio S.**
Ministry of Foreign Affairs of Brazil
Brazil

**CABRERIZO, Marco J.**
Department of Ecology and Animal Biology, University of Vigo
Spain

**CAI, Ouchen**
China Waterborne Transport Research Institute
China

**CAIADO COELHO BELTRAO COUTO, Lilia**
University College London
United Kingdom (of Great Britain and Northern Ireland)

**CAIAN, Mihaela**
National Meteorological Administration Romania
Romania

**CAIN, Michelle**
Cranfield University
United Kingdom (of Great Britain and Northern Ireland)

**CALDEIRA, Ken**
Carnegie Institution for Science
The United States of America

**CALDEIRA, Sandra**
European Commission
Italy

**CALLIHOO, Christine**
Canadian Institute of Planners
Canada

**CAMES, Martin**
Oeko-Institut
Germany

**CAMOBRECO, Vincent**
U.S. Environmental Protection Agency
The United States of America

**CAMPBELL, Kristin**
Institute for Governance & Sustainable Development
The United States of America

**CAMPUS, Lorenzo**
Ca' Foscari University of Venice
The United States of America

**CANEILL, Jean-Yves**
ERCST
France

**CAPELLÁN-PÉREZ, Iñigo**
University of Valladolid
Spain

**CAPOOR, Karan**
World Bank
The United States of America

**CAPPELLI, Federica**
Roma Tre University
Italy

**CAPSTICK, Stuart**
Cardiff University
United Kingdom (of Great Britain and Northern Ireland)

**CARDENAS, Laura**
Rothamsted Research
United Kingdom (of Great Britain and Northern Ireland)

**CARDINAEL, Rémi**
CIRAD
France

**CARLSEN, Henrik**
Stockholm Environment Institute
Sweden

**CARPINTERO, Óscar**
University of Valladolid
Spain

**CARTER, Michael**
U.S. Department of Transportation
The United States of America

**CASARES GUILLÉN, Blanca**
EfecTo TP
Spain

**CASERINI, Stefano**
Politecnico di Milano
Italy

**CASERMEIRO, Miguel Ángel**
Universidad Complutense de Madrid
Spain

**CASSEN, Christophe**
CNRS – CIRED
France

**CASSISA, Cyril**
International Energy Agency (IEA)
France

**CASTAN BROTO, Vanesa**
University of Sheffield
United Kingdom (of Great Britain and Northern Ireland)

**CASTANEDA, Antonethe**
UNESCO CONT E ECT
Guatemala

**CAZZOLA, Pierpaolo**
International Transport Forum
France

**CERUTTI, Furio**
University of Florence
Italy

**CHADWICK, Cristian**
University of Chile
Chile

**CHAN, Hoy Yen**
ASEAN Centre for Energy
Malaysia

**CHANG, Ladislaus**
Tanzania Meteorological Authority (TMA)
The United Republic of Tanzania

**CHANG, Shiyan**
Tsinghua University
China

**CHAPUNGU, Lazarus**
Great Zimbabwe University
Zimbabwe

**CHARLENE, Watson**
ODI
Switzerland

**CHAURASIA, Reetesh**
Department of Atomic Energy, Government of India
India

**CHEAH, Singfoong**
Independent consultant
The United States of America

**CHEN, Thomas**
U.S. Technology Policy Committee
The United States of America

**CHENET, Hugues**
University College London, Institute for Sustainable Resources
France

**CHEPELIEV, Maksym**
Purdue University
The United States of America

**CHERYL, Jeffers**
Department of Environment, Ministry of Agriculture, Marine Resources, Cooperatives, Environment and Human Settlements
Saint Kitts and Nevis

**CHESTNOY, Sergey**
UC RUSAL
The Russian Federation

**CHHABRA, ABHA**
Space Applications Centre, Indian Space Research Organisation
India

**CHOI, Eun Jung**
National institute of agricultural sciences
The Republic of Korea

**CHOI, Tony**
U.S. Department of Transportation
The United States of America

**CHOI, Young-jin**
Phineo gAG
Germany

**CHORLEY, Hanna**
Ministry for the Environment
New Zealand

**CHOU, Cleo**
U.S. Agency for International Development
The United States of America

**CHRISTENSEN, Tina**
Danish Meteorological Institute
Denmark

**CHRISTINE, Merk**
Kiel Institute for the World Economy
Germany

**CHRISTOPHERSEN, Oyvind**
Norwegian Environment Agency
Norway

**CIGLIANO, Bernardo**
SIX srl
Italy

**CLAPP, Christa**
CICERO Center for International Climate Research
Norway

**CLARK, Emily**
Goldsmiths, University of London
United Kingdom (of Great Britain and Northern Ireland)

**CLARK, Michael**
University of Oxford
United Kingdom (of Great Britain and Northern Ireland)

**COBB, Jonathan**
World Nuclear Association
United Kingdom (of Great Britain and Northern Ireland)

**COHEN, Brett**
The Green House consultants
South Africa

**COINTE, Béatrice**
Centre de Sociologie de l'Innovation i3, CNRS UMR9217
France

**COLLINS, William**
University of Reading
United Kingdom (of Great Britain and Northern Ireland)

**COMBES, Jean-Louis**
University of Clermont Auvergne, Center for Studies and Resarch on Development Economics (CERDI)
France

**COMBES MOTEL, Pascale**
University of Clermont Auvergne
France

**CONNORS, Sarah**
IPCC WGI TSU
France

**CONSTANTIN, Marin**
RATEN ICN
Romania

**CONTEJEAN, Arthur**
International Energy Agency (IEA)
France

**COOK, Jolene**
Department for Business,
Energy & Industrial Strategy
United Kingdom (of Great Britain
and Northern Ireland)

**COOK, Lindsey**
Quaker United Nations Office/Friends World
Committee for Consultation
Germany

**CORBETT, Charles**
University of California,
Los Angeles School of Law
The United States of America

**CORNELIUS, Stephen**
WWF
United Kingdom (of Great Britain
and Northern Ireland)

**CORTES, Pedro**
Universidade de São Paulo
Brazil

**COSTA LEITE, João**
Universidade Fernando Pessoa
Portugal

**COUGHLIN, Jennifer**
U.S. Department of Defense
The United States of America

**COULSTON, John**
U.S. Forest Service
The United States of America

**COWIE, Annette**
New South Wales Department
of Primary Industries/University
of New England
Australia

**CRAGLIA, Matteo**
International Transport Forum
France

**CREASON, Jared**
U.S. Environmental Protection Agency
The United States of America

**CREMADES, Roger**
GERICS
Germany

**CRETNEY, Raven**
University of Waikto
New Zealand

**CREUTZIG, Felix**
MCC Berlin/TU Berlin
Germany

**CRIMMINS, Allison**
U.S. Environmental Protection Agency
The United States of America

**CRISPIN, Euan**
YOUNGO Cities Working Group
United Kingdom (of Great Britain
and Northern Ireland)

**CRITCHFIELD, James**
U.S. Environmental Protection Agency
The United States of America

**CROCI, Edoardo**
Bocconi University
Italy

**CROW, Daniel**
International Energy Agency (IEA)
France

**CULP, Michael**
U.S. Department of Transportation
The United States of America

**CZEPKIEWICZ, Michał**
University of Iceland
Poland

**CZERNIAK, Michael**
Atlas Copco – Edwards
United Kingdom (of Great Britain
and Northern Ireland)

**D'IORIO, Marc**
Environment and Climate Change Canada
Canada

**DAIOGLOU, Vassilis**
Copernicus Institute
of Sustainable Development
The Netherlands

**DAKA, Julius**
Zambia Institute of Environmental
Management
Zambia

**DALAEI, Hamideh**
Islamic Republic of Iran
Meteorological Organisation
Iran

**D'ALESSANDRO, Simone**
University of Pisa
Italy

**DARRAS, Marc**
Association 4D
France

**DAS, Kasturi**
Institute of Management Technology,
Ghaziabad/Climate Strategies
India

**DAS, Satyaprakash**
Manipal School of Architecture and Planning
India

**DASH, Shanta Pragyan**
Manipal School of Architecture and Planning
India

**DAVE, Lakshmi**
Massey University
New Zealand

**DAVIDSSON KURLAND, Simon**
Chalmers University of Technology
Sweden

**DE FELICE, Matteo**
European Commission
The Netherlands

**DE GERLACHE, Jacques**
GreenFacts
Belgium

**DE JONG, Johan**
Wageningen University and Research
The Netherlands

**DE KONING, Dirk Jan**
The Royal Swedish Academy
of Agriculture and Forestry
Sweden

**DE LA VEGA NAVARRO, Angel**
UNAM – National Autonomous
University of Mexico
Mexico

**DE MIGLIO, Rocco**
Energy analyst and modeller
Italy

**DE SMEDT, Guillaume**
Hydrogen Council
France

**DEANGELO, Benjamin**
National Oceanic and Atmospheric
Administration
The United States of America

**DEASON, Jeff**
U.S. Department of Energy
The United States of America

**DEB BURMAN, Pramit Kumar**
Indian Institute of Tropical
Meteorology Pune
India

**DEBSARKAR, Anupam**
University
India

**DEĞER, Saygın**
SHURA Energy Transition Center
Turkey

**DEISSENBERG, Christophe**
Institute for non-linear dynamic inference
Luxembourg

**DEKKER, Sabrina**
Dublin City University
Ireland

**DEL RIO GONZÁLEZ, Pablo**
Consejo Superior de Investigaciones
Científicas (CSIC)
Spain

**DELL, Rebecca**
ClimateWorks Foundation
The United States of America

**DEMARIA, Federico**
Environmental Science and Technology
Institute, Autonomous University
of Barcelona
Spain

**DEMENOIS, Julien**
CIRAD
France

**DEN ELZEN, Michel**
PBL Netherlands Environmental
Assessment Agency
The Netherlands

**DENG, Xiangzheng**
UNCCD-SPI
China

**DEPLEDGE, Joanna**
Centre for Environment, Energy
and Natural Resource Governance
(CEENRG), University of Cambridge
United Kingdom (of Great Britain
and Northern Ireland)

**DESPORT, Lucas**
MINES ParisTech, Total
France

**DEVAEY, John**
Trinity College Dublin
Ireland

**DEVANE, Eoin**
United Kingdom Climate
Change Committee
United Kingdom (of Great Britain
and Northern Ireland)

**DEVEZEAUX DE LAVERGNE, Jean-Guy**
Université Paris-Dauphine & Société
Française d'Energie Nucléaire
France

**DEVINE-WRIGHT, Patrick**
University of Exeter
United Kingdom (of Great Britain
and Northern Ireland)

**DEWI, Ova Candra**
Universitas Indonesia
Indonesia

**DHAR, Subash**
UNEP DTU Partnership, DTU
Denmark

**DHAUNDIYAL, Alok**
Szent Istvan University
Hungary

**DI VITTORIO, Alan**
Lawrence Berkeley National Laboratory
The United States of America

**DIAZ, Juan**
Association
The United States of America

**DIAZ MOREJON, Cristobal Felix**
Environmental Directorate/Ministry of
Science, Technology and the Environment
Cuba

**DIAZ-BONE, Harald**
First Climate AG
Switzerland

**DIBENEDETTO, Angela**
University of Bari
Italy

**DICKSON, Neil**
ICAO
Canada

**DIOP, Salif**
National Sciences & Techniques
Academy of Senegal
Senegal

**DIXON, Tim**
IEAGHG
United Kingdom (of Great Britain
and Northern Ireland)

**DOD, David**
U.S. Agency for International Development
The United States of America

**DOHERTY, Eric**
Ecopath Planning
Canada

**DOHERTY, Jane**
U.S. Department of Transportation
The United States of America

**DOKKEN, David Jon**
U.S. Global Change Research Program
The United States of America

**DOLININA, Yulia**
UC RUSAL
The Russian Federation

**DOMEC, Jean-Christophe**
Bordeaux Sciences Agro
France

**DOMKE, Grant**
U.S. Department of Agriculture
The United States of America

**DONG, Leo**
University of California – Los Angeles
The United States of America

**DORIZAS, Paraskevi**
BPIE
Belgium

**DOUKAS, Haris**
School of Electrical and Computer Engineering, National Technical University of Athens
Greece

**DOUNMBIA, Mamadou Lamine**
University of Quebec
Canada

**DOYLE, Bridget**
Tsleil-Waututh Nation
Canada

**DREYFUS, Gabrielle**
Institute for Governance & Sustainable Development
The United States of America

**DROEGE, Susanne**
German Institute for International and Security Affairs
Germany

**DRUCKMAN, Angela**
University of Surrey
United Kingdom (of Great Britain and Northern Ireland)

**DRUPP, Moritz**
University of Hamburg
Germany

**DUAN, Hongxia**
Institute for Environment and Sustainable Development
China

**DUBASH, Navroz**
Centre for Policy Research
India

**DUBE, Lokesh Chandra**
TERI School of Advanced Studies
India

**DUBE, Smile**
California State University, Sacramento
The United States of America

**DUIC, Neven**
University of Zagreb
Croatia

**DUMBLE, Paul**
Paul's Environment Ltd
United Kingdom (of Great Britain and Northern Ireland)

**DUNHAM MACIEL, André**
Ministry of Foreign Affairs
Brazil

**DUNN, Seth**
ServiceMax/GE
The United States of America

**DUPAR, Mairi**
Overseas Development Institute
United Kingdom (of Great Britain and Northern Ireland)

**DUPONT, Claire**
Ghent University
Belgium

**DURGA, Siddarth**
PNNL
The United States of America

**DUSPIVA, Jiri**
Czech Nuclear Society
The Czech Republic

**DUTROW, Elizabeth**
U.S. Environmental Protection Agency
The United States of America

**EBHUOMA, Eromose**
University of South Africa
South Africa

**EBI, Kristie**
University of Washington
The United States of America

**EDLING, Peter**
The Royal Swedish Academy of Agriculture and Forestry
Sweden

**EDMONDSON, Beth**
Federation University
Australia

**EDWARDS, Larry**
Larry Edwards Environmental Consulting
The United States of America

**EHARA, Makoto**
Forestry and Forest Products Research Institute
Japan

**EHSAN, Taghavinejad**
NIOC
Iran

**EK, Lena**
The Royal Swedish Academy of Agriculture and Forestry
Sweden

**EKARDT, Felix**
Research Unit Sustainability and Climate Policy
Germany

**EKDAHL, Asa**
World Steel Association
Belgium

**EKHOLM, Tommi**
Finnish Meteorological Institute (FMI)
Finland

**EL BILALI, Hamid**
International Centre for Advanced Mediterranean Agronomic Studies (CIHEAM-Bari)
Italy

**EL-FEIAZ, Amira**
Technische Universiteit Eindhoven
The Netherlands

**ELLIOTT, Lorraine**
The Australian National University
Australia

**ELLIOTT III, R. Neal**
American Council for an
Energy-Efficient Economy
The United States of America

**ELOFSSON, Katarina**
Aarhus University
Denmark

**ENGELEN, Peter-Jan**
University of Antwerp
Belgium

**ENGLISCH, Michael**
Austrian Research Centre for Forests
Austria

**EORY, Vera**
Scotland's Rural College
United Kingdom (of Great Britain
and Northern Ireland)

**ERB, Karlheinz**
Institute of Social Ecology, University of
Natural Resources and Life Sciences, Vienna
Austria

**ERELL, Evyatar**
Ben-Gurion University of the Negev
Israel

**ERICKSON, Peter**
Stockholm Environment Institute
The United States of America

**ERIKSEN, Siri**
Norwegian University of Life Sciences
Norway

**ERLICH, David**
Orange
France

**ERLWEIN, Alfredo**
Soil Science Institute, Universidad
Austral de Chile
Chile

**ESCOTT, Susan**
Escott Hunt Limited
United Kingdom (of Great Britain
and Northern Ireland)

**ESCRIBANO, Gonzalo**
Universidad Nacional Educación a Distancia
(UNED) and The Elcano Royal Institute
Spain

**ESCUDERO, Luis**
INIAP
Ecuador

**ESSANDOH-YEDDU, Joseph**
Energy Commission
Ghana

**EUGENE, maguy**
French National Institute on Agriculture
Food and Environment (INRAE)
France

**EVANS, Meredydd**
U.S. Department of Energy
The United States of America

**EYCKMANS, Johan**
KU Leuven
Belgium

**FABIAN, Levihn**
KTH Royal Institute of Technology
Sweden

**FÆHN, Taran**
Statistics Norway, Research Department
Norway

**FAGEL, Nathalie**
AGEs, Departement of Geology,
University of Liege
Belgium

**FALCONER, Ryan**
Auckland Council, New Zealand
Australia

**FALK, Donald**
University of Arizona
The United States of America

**FANG, Andrew**
U.S. Agency for International Development
The United States of America

**FARAJZADEH ASL, Manuchehr**
Tarbiat Modares University
Iran

**FARIA, Sergio Henrique**
Basque Centre for Climate Change (BC3)
Spain

**FARRELL, Aidan**
The University of the West Indies
Trinidad and Tobago

**FARSTAD, Fay**
CICERO Center for International
Climate Research
Norway

**FARZAD, Hosseini Hossein Abadi**
Sharif University of Technology
Iran

**FAUDON, valerie**
SFEN
France

**FAWCETT, Allen**
U.S. Environmental Protection Agency
The United States of America

**FELGENHAUER, Tyler**
Duke University
The United States of America

**FELICIANO, Diana**
University of Aberdeen
United Kingdom (of Great Britain
and Northern Ireland)

**FELLMANN, Thomas**
European Commission, Joint
Research Centre
Spain

**FENNELL, Paul**
Imperial College London
United Kingdom (of Great Britain
and Northern Ireland)

**FERNANDES, Alexandre**
Belgian Science Policy Office (BELSPO)
Belgium

**FERNANDEZ, Ana Ines**
University of Barcelona
Spain

**FERREIRA, Luana**
Federal University of Recôncavo
of Bahia
Brazil

**FETZER, Ingo**
Stockholm Resilience Centre
Sweden

**FILIPPELLI, Gabriel**
Indiana University
The United States of America

**FINNVEDEN, Göran**
KTH Royal Institute of Technology
Sweden

**FINUS, Michael**
University of Graz
Austria

**FISCHLIN, Andreas**
IPCC Vice-chair WGII/ETH Zurich
Switzerland

**FITCH-ROY, Oscar**
University of Exeter
United Kingdom (of Great Britain and Northern Ireland)

**FLANNERY, Aimee**
U.S. Department of Transportation
The United States of America

**FLAVELL, Joanna**
University of Manchester
United Kingdom (of Great Britain and Northern Ireland)

**FLEMING, Gregg**
U.S. Department of Transportation
The United States of America

**FLEMING, Sean**
Oregon State University, University of British Columbia, US Department of Agriculture
The United States of America

**FLORIAN, Pichler**
Energie-Control Austria für die Regulierung der Elektrizitäts- und Erdgaswirtschaft
Austria

**FLOWER, Jean-Marie**
Fleur de Carbone SARL
France

**FODA, Rabiz**
Hydro One Networks Inc.
Canada

**FOLTESCU, Valentin**
Climate and Clean Air Coalition Secretariat/UNEP
India

**FONBEYIN, Henry Abanda**
Oxford Institute for Sustainable development, Oxford Brookes University
United Kingdom (of Great Britain and Northern Ireland)

**FORGIONI, Fernando**
Universidad Nacional de Villa María
Argentina

**FOUQUET, Roger**
London School of Economics and Political Science
United Kingdom (of Great Britain and Northern Ireland)

**FRA PALEO, Urbano**
University of Extremadura
Spain

**FRACASSI, Eduardo Pedro**
ITBA Instituto Tecnologico de Buenos Aires
Argentina

**FRAME, Dave**
University of Wellington
New Zealand

**FRANCESCO, Pomponi**
Edinburgh Napier University
United Kingdom (of Great Britain and Northern Ireland)

**FRANCKE-CAMPAÑA, Samuel**
Ministry of Agriculture, Chilean Forestry Service
Chile

**FRANSEN, Taryn**
World Resources Institute
The United States of America

**FREEMAN, Rachel**
University College London, Energy Institute
United Kingdom (of Great Britain and Northern Ireland)

**FRICK, Natalie**
U.S. Department of Energy
The United States of America

**FRIDAHL, Mathias**
Linköping University
Sweden

**FRISCHKNECHT, Rolf**
treeze Ltd.
Switzerland

**FRITSCHE, Uwe**
IINAS
Germany

**FRYER, Emma**
techUK
United Kingdom (of Great Britain and Northern Ireland)

**FUESSLER, Juerg**
INFRAS
Switzerland

**FUGLESTVEDT, Jan**
CICERO Center for International Climate Research
Norway

**FUJII, Mai**
Sasakawa Peace Foundation, Ocean Policy Research Institute
Japan

**FUJIMORI, Shinichiro**
Kyoto University
Japan

**FUKUI, Hiroyuki**
Company
Japan

**FULTON, Lewis**
University of California, Davis
The United States of America

**FUSS, Sabine**
MCC Berlin
Germany

**FUZZI, Sandro**
ISAC CNR
Italy

**GAINO, Bruna**
UCLouvain
Belgium

**GALE, David**
Gale & Snowden Architedcts Ltd
United Kingdom (of Great Britain and Northern Ireland)

**GALEAZZI, Clara**
University of Cambridge, Center for Energy, Environment and Natural Resource Governance
United Kingdom (of Great Britain and Northern Ireland)

**GALLAGHER, Kelly**
The Fletcher School, Tufts University
The United States of America

**GAMBHIR, Ajay**
Imperial College London
United Kingdom (of Great Britain and Northern Ireland)

**GANESH, Sahana**
Manipal School of Architecture and Planning
India

**GANGOLI RAO, Arvind**
Delft University of Technology
The Netherlands

**GAO, Yuan**
Zhengzhou University
China

**GARBOLINO, Emmanuel**
Climpact Data Science
France

**GARCIA, Maria Luisa**
Green Architecture Advocacy Philippines; United Nations Development Program; Climate change Commission of the Philippines
The Philippines

**GARCIA, Thiago**
Potsdam Institute for Climate Impact Research (PIK)
Germany

**GARCIA MATEO, Maria Carmen**
MCG Research & Innovation Sustainability Architecture/Urban Planning
Spain

**GARCIA TAPIA, Victor**
International Energy Agency (IEA)
France

**GARCIA-MARTINEZ, Antonio**
Universidad de Sevilla
Spain

**GASIA, Jaume**
Jose Antonio Romero Polo SA
Spain

**GAVARD, Claire**
ZEW Mannheim
Germany

**GAVAZZI, Bruno**
UMR7516 Institut de Physique du Globe de Strasbourg (CNRS)/Université de Strasbourg
France

**GEDEN, Oliver**
German Institute for International and Security Affairs
Germany

**GEHL, Georges**
Ministère du développement durable et des infrastructures, Département de l'Environnement
Luxembourg

**GEMEDA, Adugna**
Ethiopian Public Health Institute
Ethiopia

**GERHARD, Reese**
University of Koblenz-Landau
Germany

**GERRARD, Emily**
Comhar Group Pty Limited (law firm)
Australia

**GIBEK, Jakub**
Ministry of Environment, Department of Air Protection and Climate
Poland

**GIBON, Thomas**
Luxembourg Institute of Science and Technology (LIST)
France

**GIDDEN, Matthew**
Climate Analytics
Germany

**GIESE, Monique**
KPMG AG
Germany

**GIESEKE, Robert**
Independent
Germany

**GIL-CRUZ, José Manuel**
World Resources Institute (WRI)
Mexico

**GILLER, Ken**
Wageningen University
The Netherlands

**GILLETT, Nathan**
Environment and Climate Change Canada
Canada

**GIRARDIN, Cecile**
University of Oxford
United Kingdom (of Great Britain and Northern Ireland)

**GIRAUDET, Louis-Gaëtan**
CIRED, Ecole des Ponts ParisTech
France

**GIROD, Bastien**
ETH Zürich
Switzerland

**GOEDEKING, Nicholas**
University of California, Berkeley
The United States of America

**GOETZKE, Frank**
University of Louisville
The United States of America

**GOGGINS, Gary**
National Univeristy of Ireland Galway
Ireland

**GOHEER, Muhammad Arif**
GCISC
Pakistan

**GOLDSWORTHY, Michael**
Drax
United Kingdom (of Great Britain and Northern Ireland)

**GOMEZ VILCHEZ, Jonatan J.**
European Commission, Joint Research Centre
Italy

**GONELLA, Francesco**
Ca' Foscari University of Venice
Italy

**GOODSON, Timothy**
International Energy Agency (IEA)
France

**GOPALAKRISHNAN, Ranjith**
University of Eastern Finland
Finland

**GORNER, Marine**
International Energy Agency (IEA), former
France

**GOSWAMI, Prashant**
Institute of Frontier Science
and Application
India

**GOTA, Sudhir**
Independent Consultant/Researcher
India

**GOUDA, Jambavati**
Manipal School of Architecture and Planning
India

**GOUGH, Ian**
London School of Economics
and Political Science
United Kingdom (of Great Britain
and Northern Ireland)

**GRAÇA, João**
Institute of Social Sciences,
University of Lisbon
Portugal

**GRAGNANI, Patrizia**
YOUNGO Cities Working Group
(Sant'Anna School of Advanced Studies)
Italy

**GRANDERSON, Jessica**
U.S. Department of Energy
The United States of America

**GRASSI, Giacomo**
Joint Research Centre, European
Commission
Italy

**GRASSO, Giacomo**
ENEA
Italy

**GREAVES, Deborah**
University of Plymouth
United Kingdom (of Great Britain
and Northern Ireland)

**GREEN, Fergus**
Utrecht University
The Netherlands

**GREEN, John**
Royal Aeronautical Society
United Kingdom (of Great Britain
and Northern Ireland)

**GREEN, Martin**
UNSW Sydney
Australia

**GREEN, Tom**
Far Away Projects
The United States of America

**GREWE, Volker**
DLR-Oberpfaffenhofen
Germany

**GRIFFIN, Emer**
Department of Communications, Climate
Action and Environment, Climate Mitigation
and Awareness Division
Ireland

**GRIGERA NAÓN, Juan Jose**
Sociedad Rural Argentina (member
of ICC Argentine branch)
Argentina

**GROEN, Lisanne**
Open University of the Netherlands
Belgium

**GROOS, Ulf**
Fraunhofer ISE
Germany

**GROSS, Robert**
Imperial College London/UKERC
United Kingdom (of Great Britain
and Northern Ireland)

**GROVER, Ravi B**
Homi Bhabha National Institute
India

**GROVE-SMITH, Jessica**
Passive House Institute
Germany

**GRUBB, Michael**
University College London (UCL),
Institute of Sustainable Resources
United Kingdom (of Great Britain
and Northern Ireland)

**GRUDDE, Vivian**
Climate Change & Transformation Advisory
Germany

**GUARATO, Pietro**
Université de Lausanne
Switzerland

**GUENET, Bertrand**
CNRS
France

**GUENTHER, Genevieve**
Tishman Enviornment and Design Center,
The New School
The United States of America

**GUERRA, Flávia**
REN21
Germany

**GUINEY, Itchell**
Department of Environment,
Forestry and Fisheries
South Africa

**GUIVARCH, Celine**
CIRED
France

**GÜNGÖR, Görkem**
Middle East Technical University
Turkey

**GUO, Jie**
China Academy of Transportation Sciences
China

**GUO, Ru**
Tongji University
China

**GUPTA, Himangana**
Institute for the Advanced Study
of Sustainability, United Nations
University, Tokyo
Japan

**GUPTA, Rajat**
Oxford Brookes University
United Kingdom (of Great Britain
and Northern Ireland)

**GURWICK, Noel**
U.S. Agency for International Development
The United States of America

**GUTIÉRREZ VILLALPANDO, Verónica**
Consejo Nacional de Ciencia
y Tecnología comisionada en el
Colegio de Postgraduados
Mexico

**GWAMBENE, Brown**
Marian University College
The United Republic of Tanzania

**HABERL, Helmut**
Institute of Social Ecology, University
of Natural Resources and Life
Sciences, Vienna
Austria

**HABERT, Guillaume**
ETH Zürich
Switzerland

**HAGEN, Achim**
Humboldt-Universität zu Berlin
Germany

**HAHNEL, Ulf**
University of Geneva
Switzerland

**HAINES, Andrew**
London School of Hygiene
and Tropical Medicine
United Kingdom (of Great Britain
and Northern Ireland)

**HAITES, Erik**
Margaree Consultants Inc.
Canada

**HAJDU, Flora**
Swedish Univeristy of Agricultural Sciences
Sweden

**HAJIMOLANA, Yashar**
University of Twente
The Netherlands

**HALCOMB, Jacob Stuart**
Consultant to Cities and Adaptation
Units of the UN Environment Programme
France

**HALE, Galina**
UC Santa Cruz
The United States of America

**HALENKA, Tomáš**
Charles University
The Czech Republic

**HALVERSON, Mark**
U.S. Department of Energy
The United States of America

**HAMEIRI, Ziv**
The University of New South Wales
Australia

**HAMILTON, Ian**
University College London
United Kingdom (of Great Britain
and Northern Ireland)

**HAMILTON, Kirsty**
Chatham House (Associate Fellow, unpaid)
United Kingdom (of Great Britain
and Northern Ireland)

**HAMLAT, Abdelkader**
University of Laghouat
Algeria

**HAMPSHIRE, Robert**
U.S. Department of Transportation
The United States of America

**HAMRICK, Kelley**
The Nature Conservancy
The United States of America

**HAN, Sang-Min**
Hallym University
The Republic of Korea

**HANA, Kim**
KAIST
The Republic of Korea

**HANCOCK, Linda**
Centre of Excellence on Electromaterials
Science, Deakin University
Australia

**HANCOCK, Susana**
University of Oxford
The United States of America

**HANG, Yun**
Emory University
The United States of America

**HANNON, Matthew**
University of Strathclyde
United Kingdom (of Great Britain
and Northern Ireland)

**HANSEN, Gerrit**
Robert Bosch Stiftung
Germany

**HAPPEL, Robin**
Yale Center for Environmental Law & Policy
The United States of America

**HAQ, Zia**
U.S. Department of Energy
The United States of America

**HARBY, Atle**
SINTEF Energy Research
Norway

**HARGROVES, Karlson**
Curtin University, Sustainability Policy
Institute
Australia

**HARJO, Bekki**
NOAA/National Weather Service
The United States of America

**HARNISCH, Jochen**
KfW
Germany

**HARRELL, Greg**
Milligan Engineering
The United States of America

**HART, David**
George Mason University
The United States of America

**HARTLEY, Chloe**
Tsleil-Waututh Nation
Canada

**HARTLEY, Tilman**
Institute of Environmental Science
and Technology, Autonomous University
of Barcelona (ICTA-UAB)
Spain

**HARTMANN, Jens**
Universität Hamburg
Germany

**HARTMANN, Johann**
SustainableFamily
Switzerland

**HASANBEIGI, Ali**
Global Efficiency Intelligence
The United States of America

**HASANEIN, Amin**
Islamic Relief Deutschland
Germany

**HASEGAWA, Tomoko**
Ritsumeikan University
Japan

**HASHIMOTO, Shoji**
Forestry and Forest Products Research Institute/The University of Tokyo
Japan

**HAUGHEY, Eamon**
Department of Natural Resources & the Environment, Atlantic Technological University
Ireland

**HAUSKER, Karl**
World Resources Institute (WRI)
The United States of America

**HAXTAUSEN, Eric**
Independent Consultant
The United States of America

**HAYAT, Jameel**
AECOM
United Kingdom (of Great Britain and Northern Ireland)

**HAYWARD, Bronwyn**
University of Canterbury
New Zealand

**HEFFRON, Raphael**
Centre for Energy, Petroleum, Mineral Law & Policy
United Kingdom (of Great Britain and Northern Ireland)

**HEGDE, Gajanana**
UNFCCC (Climate Change Secretariat)
Germany

**HEGERL, Gabriele**
University of Edinburgh
United Kingdom (of Great Britain and Northern Ireland)

**HELIOUI, Khalil**
EDF Lab Paris Saclay
France

**HELMAN, Daniel**
College of Micronesia
Federated States of Micronesia

**HÉNAULT, Catherine**
INRAE
France

**HENDRY, David**
Nuffiled College, Oxford University
United Kingdom (of Great Britain and Northern Ireland)

**HENSON, Stephanie**
National Oceanography Centre
United Kingdom (of Great Britain and Northern Ireland)

**HEREDIA-FRAGOSO, Marco**
National Institute of Ecology and Climate Change
Mexico

**HERNANDEZ, Michelle D.**
YOUNGO Cities Working Group (UNFCCC Official Youth Constituency)
The United States of America

**HEROLD, Anke**
Oeko-Institut e.V.
Germany

**HEROLD, Martin**
Wageningen University
The Netherlands

**HERTWICH, Edgar**
Norwegian University of Science and Technology
Norway

**HESPEL, Bertrand**
University of Namur
Belgium

**HESS, David**
World Nuclear Association
United Kingdom (of Great Britain and Northern Ireland)

**HETTIARACHCHI, Suresh**
UNSW Sydney
Australia

**HEUTTE, Fred**
Sierra Club
The United States of America

**HEYD, Thomas**
University of Victoria
Canada

**HEYMANN, Fabian**
INESC TEC
Switzerland

**HICKEL, Jason**
Goldsmiths, University of London
United Kingdom (of Great Britain and Northern Ireland)

**HILMI, Nathalie**
Centre Scientifique de Monaco
France

**HILSON, Chris**
University of Reading
United Kingdom (of Great Britain and Northern Ireland)

**HINNELLS, Mark**
Ricardo Energy and Environment
United Kingdom (of Great Britain and Northern Ireland)

**HIRVIJOKI, Eero**
Aalto University
Finland

**HITAJ, Claudia**
Luxembourg Institute of Science and Technology (LIST)
Luxembourg

**HOBART, Stacey**
New Buildings Institute
The United States of America

**HOEKSTRA, Auke**
Eindhoven University of Technology
The Netherlands

**HOFFERBERTH, Elena**
University of Leeds
United Kingdom (of Great Britain and Northern Ireland)

**HÖHNE, Niklas**
NewClimate Institute
Germany

**HOLLENDER, Lina**
Germany

**HOMBU, Kazuhiko**
Graduate School of Public Policy, The University of Tokyo
Japan

**HOMMA, Takashi**
Research Institute of Innovative Technology for the Earth (RITE)
Japan

**HONEGGER, Matthias**
Utrecht University, Perspectives climate research, IASS-Potsdam
Germany

**HONG, Tianzhen**
U.S. Department of Energy
The United States of America

**HONGO, Takashi**
Mitsui & Co. Global Strategic Studies Institute
Japan

**HOPKINS, Asa**
Synapse Energy Economics
The United States of America

**HOPWOOD, Jerry**
University Network of Excellence in Nuclear Engineering (UNENE)
Canada

**HORIE, Junichi**
Advantage Partnership Lawyers
Japan

**HORTON, Joshua**
Harvard University
The United States of America

**HOULE, David**
Environment and Climate Change Canada
Canada

**HOULIHAN WIBERG, Aoife**
The Belfast School of Architecture and the Built Environment, Ulster University
United Kingdom (of Great Britain and Northern Ireland)

**HOWES, Tom**
International Energy Agency (IEA)
France

**HOYOS-SANTILLAN, Jorge**
University of Magallanes
Chile

**HU, Guoquan**
National Climate Center, China Meteorological Administration
China

**HUANG, Hsin**
International Meat Secretariat
France

**HUANG, Lei**
Chinese Meteorological Administration
China

**HUBER, Pia Paola**
Austria

**HUGHES, Helen**
University of Edinburgh
United Kingdom (of Great Britain and Northern Ireland)

**HUGHES, Llewelyn**
Australian National University
Australia

**HUIZENGA, Cornie**
CESG
Germany

**HUNTER, Cutting**
Climate Nexus
The United States of America

**HURLBERT, Margot**
University of Regina
Canada

**HURMEKOSKI, Elias**
University of Helsinki
Finland

**HURTADO ALBIR, Francisco Javier**
European Patent Office
Germany

**HUYNH, Thi Lan Huong**
Viet Nam Institute of Meteorology, Hydrology and Climate change
Vietnam

**HYMAN, Eric**
U.S. Agency for International Development
The United States of America

**IAKOVOGLOU, Valasia**
International Hellenic University
Greece

**IBRAHIM, Zelina**
Universiti Putra Malaysia
Malaysia

**IDOHOU, Alix Frank Rodrigue**
National University of Agriculture
Benin

**IGARASHI, Keiichi**
Mitsubishi UFJ Research and Consulting Co., Ltd.
Japan

**ILACQUA, Vito**
U.S. Environmental Protection Agency
The United States of America

**ILLESCAS, Martin**
Ministry of Environment and Sustainable Development of Argentina
Argentina

**ILLESCAS, Nelson**
Fundación INAI – Bolsa de Cereales de Buenos Aires
Argentina

**INDRAWAN, Mochamad**
Research Center for Climate Change – Universitas Indonesia (RCCC-UI)
Indonesia

**IOVANNA, Rich**
U.S. Department of Agriculture
The United States of America

**ISLAM, Md. Sirajul**
Department of Civil and Environmental Engineering, North South University
Bangladesh

**ITURREGUI, Patricia**
Universidad Científica del Sur
Peru

**IVANOVA, Diana**
University of Leeds
United Kingdom (of Great Britain and Northern Ireland)

**IVES, Matthew**
University of Oxford
United Kingdom (of Great Britain and Northern Ireland)

**JACOBSON, Mark**
Stanford University
The United States of America

**JADRIJEVIC GIRARDI, Maritza**
Ministry of Environment
Chile

**JAFARI, Mostafa**
Head of TPS for LFCCs
Iran

**JAFARZADEH, Sanaz**
Thermal Power Plants Holding Company
Iran

**JÄGER-WALDAU, Arnulf**
European Commission, Joint Research Centre
Italy

**JAIN, Hetal**
U.S. Department of Transportation
The United States of America

**JAIN, Niveta**
ICAR-Indian Agricultural Research Institute
India

**JAIN, Sanjeev**
IIT DELHI
India

**JAISWAL, Komal**
Manipal School of Architecture and Planning
India

**JAKOB, Michael**
MCC Berlin
Germany

**JAMEA, El Mostafa**
MENA Renewables and Sustainability – MENARES
Morocco

**JAMIESON, Craig**
Straw Innovations Ltd
The Philippines

**JANSSENS, Charlotte**
KU Leuven
Belgium

**JAUMOTTE, Florence**
International Monetary Fund
The United States of America

**JAXA-ROZEN, Marc**
University of Geneva
France

**JELLEY, Nick**
Physics Department University of Oxford
United Kingdom (of Great Britain and Northern Ireland)

**JENKINS, Kirsten**
University of Edinburgh
United Kingdom (of Great Britain and Northern Ireland)

**JEON, Eui-Chan**
Sejong University
The Republic of Korea

**JEONG, Seokhwan**
Kongju National University Graduate School
The Republic of Korea

**JEONG, Young Sun**
Korea Institute of Civil Engineering and Building Technology (KICT)
The Republic of Korea

**JESÚS ALBERTO, Pulido Arcas**
The University of Tokyo
Japan

**JOCHEM, Patrick**
German Aerospace Center (DLR)
Germany

**JOHANSSON, Daniel**
Chalmers University of Technology
Sweden

**JOHNSEN, Michael**
U.S. Department of Transportation
The United States of America

**JOHNSON, Ann Jessica**
FORATOM (European Atomic Forum)
Belgium

**JONES, Billy**
Lund University
Sweden

**JONES, Ceris**
National farmers union/world farmers organisation
United Kingdom (of Great Britain and Northern Ireland)

**JONES, Sarah**
JBA Risk Management
United Kingdom (of Great Britain and Northern Ireland)

**JONSSON, Anders**
The Royal Swedish Academy of Agriculture and Forestry
Sweden

**JORDAN, Andrew**
Tyndall Centre for Climate Change Research
United Kingdom (of Great Britain and Northern Ireland)

**JOSEPH, Benise**
Caribbean Cooperative MRV Hub
Saint Lucia

**JOSHI, Kirti**
Tribhuvan University
Nepal

**JOSHI, Rishikesh**
TU Delft
The Netherlands

**JOTZO, Frank**
ANU
Australia

**JOYCE, Adrian**
Catholic University of Louvain-le-Neuve
Belgium

**JUANCHICH, Marie**
University of Essex
United Kingdom (of Great Britain and Northern Ireland)

**JUNG, Jione**
Korea Institute for International Economic Policy (KIEP)
The Republic of Korea

**JUNO, Edith**
National Wildlife Federation
The United States of America

**KABIRI-MARIAL, Stella**
National Agricultural Research Organisation
Uganda

**KADIR, Sabaruddin**
Soil Science Department, Fac. of Agriculture, Universitas Sriwijaya, Inderalaya, South Sumatra
Indonesia

**KADITI, Eleni**
Organization of the Petroleum Exporting Countries (OPEC)
Austria

**KAINUMA, Mikiko**
Institute for Global Environmental Strategies
Japan

**KALLBEKKEN, Steffen**
CICERO Center for International Climate Research
Norway

**KALVANE, Gunta**
University of Latvia
Latvia

**KAMEYAMA, Yasuko**
National Institute for Environmental Studies
Japan

**KANAYA, Yugo**
Japan Agency for Marine-Earth Science and Technology (JAMSTEC)
Japan

**KANG, Suil**
Gwangju Institute of Science and Technology
The Republic of Korea

**KANNAN, Umasankari**
Bhabha Atomic Research Centre
India

**KANNINEN, Markku**
University of Helsinki
Finland

**KAPLAN MINTZ, Keren**
Shamir Research Institute, University of Haifa
Israel

**KAPOVIC SOLOMUN, Marijana**
UNCCD-SPI
Bosnia and Herzegovina

**KAPUR, Rituka**
Manipal School of Architecture and Planning
India

**KARAKAYA, Etem**
Independent researcher
Turkey

**KARIM, Taufiq Ramdani**
University of Mataram
Indonesia

**KARLSSON, Mikael**
KTH Royal Institute of Technology
Sweden

**KASHONGWE, Olivier**
Leibniz Institute for Agricultural Engineering and Bioeconomy
Germany

**KASPAR, Frank**
Deutscher Wetterdienst
Germany

**KATHRADA, Idriss**
Novasirhe
France

**KATO, Etsushi**
Institute of Applied Energy
Japan

**KATSUMASA, Tanaka**
Laboratoire des Sciences du Climat et de l'Environnement (LSCE)/ National Institute for Environmental Studies (NIES)
France

**KAUROLA, Jussi**
Finnish Meteorological Institute (FMI)
Finland

**KAWAGISHI, Shunsuke**
Mitsubishi Research Institute
Japan

**KEATING-BITONI, Caitlin**
U.S. Department of State
The United States of America

**KEKANA, Maesela**
International Climate Change Cooperation Department of Environmental Affairs
South Africa

**KELLER, Meredith**
Yale University
The United States of America

**KEMPA, Karol**
Frankfurt School of Finance and Management
Germany

**KEMPER, Jasmin**
IEA Greenhouse Gas R&D Programme (IEAGHG)
United Kingdom (of Great Britain and Northern Ireland)

**KENDALL, Gary**
Nedbank
South Africa

**KENNISH, Michael**
Rutgers University
The United States of America

**KENT DE GREY, Robert G.**
University of Utah
The United States of America

**KERRY, Constabile**
Oxford University School of Geography
The United States of America

**KESTERNICH, Martin**
ZEW, Leibniz Centre for European Economic Research
Germany

**KHAJEHPOUR, Hossein**
Energy Engineering Department, Sharif University of Technology
Iran

**KHALIL, Mohammad Ibrahim**
University College Dublin
Ireland

**KHANNA, Manav**
YOUNGO Cities Working Group (Sant'Anna School of Advanced Studies)
Italy/India

**KHENNAS, SMAIL**
Energy and Climate Change Consultant
United Kingdom (of Great Britain and Northern Ireland)

**KHESHGI, Haroon**
ExxonMobil Research and Engineering Company
The United States of America

**KICHEV, Emil**
Technical University of Sofia
Bulgaria

**KII, Masanobu**
Kagawa university
Japan

**KIKSTRA, Jarmo**
International Institute for Applied Systems Analysis (IIASA)
Austria

**KILIAN, Raiser**
Hertie School
Germany

**KILKIS, Siir**
The Scientific and Technological Research Council of Turkey
Turkey

**KIM, Raehyun**
National Institute of Forest Science
The Republic of Korea

**KIM, Schumacher**
Tokyo Institute of Technology
Japan

**KIM, Suyi**
Hongik University
The Republic of Korea

**KIM, Yeong Jae**
RFF-CMCC European Institute on Economics and the Environment
Italy

**KIM, Youngsun**
Korea Institute of Civil Engineering and Building Technology (KICT)
The Republic of Korea

**KIMENGSI, Jude Ndzifon**
Department of Geography and Environmental Studies, Catholic University of Cameroon (CATUC)
Cameroon

**KING-OKUMU, Caroline**
UNCCD-SPI
United Kingdom (of Great Britain and Northern Ireland)

**KIRIYAMA, Eriko**
Tokyo Institute of Technology
Japan

**KIRSCHNER, Yoel**
U.S. Agency for International Development
The United States of America

**KIRSTEN, Gram-Hanssen**
Aalborg University
Denmark

**KLINE, Keith**
U.S. Department of Energy
The United States of America

**KLØVERPRIS, Jesper**
Novozymes
Denmark

**KNOOPE, Marlinde**
KiM Netherlands Institute for Transport Policy Analysis
The Netherlands

**KNUCKEY, Deborah**
Published energy journalist plus consultant for renewable energy development and finance companies through Kiterocket
The United States of America

**KO, Malcolm**
Distinguished Research Associate, NASA Langley Research Center
The United States of America

**KOBAYASHI, Masanori**
New Energy and Industrial Technology Development Organization (NEDO)
Japan

**KOBAYASHI, Shigeki**
Transport Institute of Central Japan
Japan

**KOBERLE, Alexandre**
COPPE/UFRJ
Brazil

**KÖHLER, Jonathan**
Fraunhofer Institute for Systems and Innovation Research ISI
Germany

**KOLB, Laura**
U.S. Environmental Protection Agency
The United States of America

**KOLPAKOV, Andrey**
Institute of Economic Forecasting of the Russian Academy of Sciences
The Russian Federation

**KONDO, Hiroaki**
National Institute of Advanced Industrial Sciences
Japan

**KONGSAGER, Rico**
University College Copenhagen
Denmark

**KÖNIG, Michael**
Frankfurt School of Finance and Management
Germany

**KÖNIG, Sebastian**
Federal Office for the Environment FOEN
Switzerland

**KOOMEY, Jonathan**
Koomey Analytics
Canada

**KOPP, Robert**
Rutgers University
The United States of America

**KORHONEN, Janne M.**
Lappeenranta University of Technology
Finland

**KORNELSEN, Kurt**
Ontario Power Generation
Canada

**KORTEKANGAS, Otso**
KTH Royal Institute of Technology
Sweden

**KOSMAL, Ann**
U.S. General Services Administration
The United States of America

**KOSONEN, Kaisa**
Greenpeace
Finland

**KOTHAWADE, Gajanan**
Washington State University
The United States of America

**KOVALEVSKY, Dmitry**
Climate Service Center Germany (GERICS), Helmholtz-Zentrum Geesthacht
Germany

**KOWALZIG, Jan**
Oxfam Germany
Germany

**KRAEVOY, Aleksandr**
UC RUSAL
The Russian Federation

**KRAM, Tom**
PBL (Fellow)
The Netherlands

**KRELL, Matthew**
University of the West Indies
Barbados

**KRIEGLER, Elmar**
Potsdam Institute for Climate Impact Research (PIK)
Germany

**KROOK-RIEKKOLA, Anna**
Luleå University of Technology
Sweden

**KUCINOV, Vladimir**
National Research Nuclear University, Moscow Engineering Physics Institute (MEPHI)
The Russian Federation

**KUHNHENN, Kai**
Konzeptwerk Neue Ökonomie
Germany

**KULIONIS, Viktoras**
ETH Zürich, Ecological Systems Design
Switzerland

**KUMAR, Arvind**
India Water Foundation
India

**KUNKEL, Stefanie**
Institute for Advanced Sustainability Studies (IASS), Potsdam
Germany

**KURAMOCHI, Takeshi**
NewClimate Institute
Germany

**KURTZ, Sarah**
University of California Merced
The United States of America

**KUSCH-BRANDT, Sigrid**
University of Padua
Germany

**KUTTAN, Sanjay**
Singapore Maritime Institute
Singapore

**KVALEVÅG, Maria Malene**
Norwegian Environment Agency
Norway

**KVERNDOKK, Snorre**
Frisch Centre
Norway

**KYEREMATENG, Kate**
University of Ghana
Ghana

**LABERTEAUX, Kenneth**
Toyota Motor North America-R&D
The United States of America

**LAFOND, François**
University of Oxford
United Kingdom (of Great Britain and Northern Ireland)

**LAFRANCE, Marc**
U.S. Department of Energy
The United States of America

**LAGER, Pierre-yves**
LGM
France

**LAMB, William**
Mercator Research Institute on Global Commons and Climate Change (MCC)
Germany

**LAMBOLL, Robin**
Imperial College London
United Kingdom (of Great Britain and Northern Ireland)

**LAMERS, Patrick**
National Renewable Energy Laboratory
The United States of America

**LAMERS, Vanessa**
Public Health Foundation
The United States of America

**LAMY, Damien**
Mines Saint-Etienne
France

**LANDRY, Jean-Sébastien**
Environment and Climate Change Canada
Canada

**LANGEVIN, Jared**
U.S. Department of Energy
The United States of America

**LANTZ, Mattias**
Uppsala university
Sweden

**LARSON, Ronal**
Larson Consulting
The United States of America

**LATKA, Catharina**
University of Bonn
Germany

**LAWRENCE, Deborah**
University of Virginia
The United States of America

**LAY, Jann**
German Institute for Global and Area Studies (GIGA)
Germany

**LAYEGHI, Behzad**
IRIMO
Iran

**LE GALLIC, Thomas**
CNRS – CIRED
France

**LE STRAT, Florent**
ELECTRICTE DE FRANCE
France

**LEARY, David**
University of Technology Sydney
Australia

**LEBLANC, Florian**
Centre International de Recherche sur l'Environnement et le Développement
France

**LEBLOIS, Antoine**
INRA
France

**LECOCQ, Noé**
Inter-Environnement Wallonie
Belgium

**LEE, Arthur**
Chevron Corporation
The United States of America

**LEE, Chungkook**
Korea Research Institute on Climate Change
The Republic of Korea

**LEE, Jae Yoon**
Korea Institute for Industrial Economics & Trade (KIET)
The Republic of Korea

**LEE, Junhee**
Korea Meteorological Administration (KMA)
The Republic of Korea

**LEE, King**
World Nuclear Association
United Kingdom (of Great Britain and Northern Ireland)

**LEE, Sai Ming**
Hong Kong Observatory
China

**LEE, Taedong**
Yonsei University
The Republic of Korea

**LEGGETT, Jane**
Congressional Research Service
The United States of America

**LEHMANN, Tahina**
Alterna SA
Switzerland

**LEHOCZKY, Annamária**
Fauna and Flora International
United Kingdom (of Great Britain and Northern Ireland)

**LEIBOWICZ, Benjamin**
The University of Texas at Austin
The United States of America

**LEILA, Zamani**
Department of Environment of Iran
Iran

**LEITE DE FARIA COELHO DA SILVA, Mafalda**
International Energy Agency (IEA)
France

**LELONG, Christian**
Kayrros
United Kingdom (of Great Britain and Northern Ireland)

**LENGEFELD, Michael**
Goucher College, Maryland
The United States of America

**LENNOX, Kym**
Climate Change Equity
Australia

**LENOX, Carol**
U.S. Environmental Protection Agency
The United States of America

**LEVENTON, Julia**
Global Change Institute of the Czech Academy of Sciences CzechGlobe
The Czech Republic

**LEVESQUE, Antoine**
Potsdam Institute for Climate Impact Research (PIK)
Germany

**LEVI, Peter**
International Energy Agency (IEA)
France

**LEWANDROWSKI, Jan**
U.S. Department of Agriculture
The United States of America

**LEWIS, Matt**
Bangor University
United Kingdom (of Great Britain and Northern Ireland)

**LI, Jia**
U.S. Environmental Protection Agency
The United States of America

**LIDDLE, Brantley**
Energy Studies Institute, NUS
Singapore

**LIFKA, Roland**
Synto GmbH
Austria

**LILJELUND, Lars-Erik**
The Royal Swedish Academy of Agriculture and Forestry
Sweden

**LILLIESTAM, Johan**
Institute for Advanced Sustainability Studies & University of Potsdam
Germany

**LIM, Jinsun**
International Energy Agency (IEA)
France

**LIMAYE, Vijay**
Natural Resources Defense Council
The United States of America

**LIMBOURG, Sabine**
HEC-Uliege
Belgium

**LIN, Jungmin**
Korea Energy Economics Institute
The Republic of Korea

**LINDKVIST, Mathias**
KTH Royal Institute of Technology, Department of Sustainable Development, Environmental Science and Engineering (SEED)
Sweden

**LIPKA, Maciej**
National Centre for Nuclear Research
Poland

**LITMAN, Todd**
Victoria Transport Policy Institute
Canada

**LITSKAS, Vassilis**
Cyprus University of Technology/ Open University of Cyprus
Cyprus

**LIU, Beibei**
School of the Environment,
Nanjing University
China

**LIU, Changyi**
GEIDCO
China

**LIU, Junguo**
Southern University of Science
and Technology
China

**LIU, Lingna**
China University of Geosciences
China

**LIU, Wenling**
Beijing Institute of Technology
China

**LIU, Yaming**
China Meteorological Administration
China

**LIVET, Frédéric**
CNRS
France

**LOCKLEY, Andrew**
Andrew Lockley
United Kingdom (of Great Britain
and Northern Ireland)

**LOESCHER, Anne**
University of Leeds
United Kingdom (of Great Britain
and Northern Ireland)

**LOMBARD, Eric**
Stay Grounded
France

**LONGDEN, Thomas**
Australian National University
Australia

**LOOMANS, Naud**
Eindhoven University of Technology
The Netherlands

**LOREA, Claude**
Global Cement and Concrete Association
Belgium

**LOUREIRO, Luis**
INUMET
Uruguay

**LOVERING, Jessica**
Carnegie Mellon University
The United States of America

**LOVINS, Amory B.**
Rocky Mountain Institute/Environmental
& Civil Engineering, Stanford University
The United States of America

**LU, Hongyou**
U.S. Department of Energy
The United States of America

**LU, Rong**
Stanford University
The United States of America

**LUCASH, Melissa**
Portland State University
The United States of America

**LUDERER, Gunnar**
Potsdam Institute for Climate
Impact Research (PIK)
Germany

**LUGOVOY, Oleg**
Environmental Defense Fund
The United States of America

**LUIS FRANCISCO, Miranda**
University of Magdalena
Colombia

**LUND, Marianne Tronstad**
CICERO Center for International
Climate Research
Norway

**LUO, Fei**
VU Amsterdam
The Netherlands

**LÜTZKENDORF, Thomas**
Karlsruhe Institute of Technology
(KIT) – Research University of
Helmholtz Association
Germany

**LYND, Lee**
Dartmouth College
The United States of America

**LYNN, Jonathan**
IPCC
Switzerland

**LYU (FORMALLY LU), Zheng**
Shanghai Advanced Research Institute,
Chinese Academy of Sciences
China

**MA, Leiming**
Shanghai Central Meteorological
Observatory
China

**MAARFIELD, Cornelia**
Climate Action Network (CAN) Europe
Germany

**MACEY, Adrian**
Victoria University of Wellington
New Zealand

**MACHADO, Ana**
Faculdade de Ciências e Tecnologia,
Universidade Nova de Lisboa
Portugal

**MACLURCAN, Donald**
Post Growth Institute; Economics
department, Southern Oregon University;
Institute for Sustainable Futures, University
of Technology Sydney; Shumacher Institute
The United States of America

**MACOMBE, Catherine**
INRAE
France

**MACRI, Daniel**
U.S. Environmental Protection Agency
The United States of America

**MADKOUR, Khaled Mohamed**
Ain Shams University, Cairo
Egypt

**MAFOLE, Tshepiso**
University of Cape Town
South Africa

**MAGNE, Bertrand**
International Atomic Energy Agency
Austria

**MAHARAJ, Shobha**
Independent Consultant
Germany

**MAIDOWSKI, Paul**
Fletcher School, Tufts/
independent researcher
Germany

**MAIN, Anna**
Ministry of Foreign Affairs and Trade
Samoa

**MAINALY, Jony**
Vidhigya Legal Services and Research Center
Nepal

**MAJER, Stefan**
German Biomass Research Centre – DBFZ
Germany

**MAJOR, Mark**
Partnership on Sustainable Low Carbon
Transport
Spain

**MAJUMDAR, Suvra**
United Nations Development Programme
India

**MAKI, Alexander**
AAAS Science Policy Fellow
The United States of America

**MALAHAYATI, Marissa**
National Institute for Environmental Studies
Japan

**MALASHOCK, Danny**
U.S. Environmental Protection Agency
The United States of America

**MALATINSKÁ, Lenka**
State Advisor, Climate Change Policy
Department Ministry of the Environment
Slovakia

**MALIK, Jamaludin**
FORDA, Ministry of Environment
and Forestry
Indonesia

**MALJEAN-DUBOIS, Sandrine**
CNRS/Aix-Marseille University
France

**MALLABURN, Peter**
University College London
United Kingdom (of Great Britain
and Northern Ireland)

**MANNING, David**
Newcastle University
United Kingdom (of Great Britain
and Northern Ireland)

**MANTILLA-MELUK, Hugo**
Universidad del Quindio
Colombia

**MANYCH, Niccolò**
MCC Berlin
Germany

**MARBAIX, Philippe**
Université catholique de Louvain
Belgium

**MARCHESE, April**
U.S. Department of Transportation
The United States of America

**MARKUSSON, Nils**
Lancaster University
United Kingdom (of Great Britain
and Northern Ireland)

**MARQUES, Fabio**
Brazilian Forestry Industry/
Plantar Carbon Consulting
Brazil

**MARTANOVIČ, Tomáš**
Ministry of Industry and Trade
The Czech Republic

**MARTINERIE, Patricia**
Centre National de la Recherche Scientifique
France

**MARTINO, Yomayra**
GreEnergy Dominicana
The Dominican Republic

**MASANET, Eric**
Northwestern University
The United States of America

**MASSON-DELMOTTE, Valérie**
CEA, IPSL/LSCE
France

**MASUYAMA, Toshimasa**
International Renewable Energy
Agency (IRENA)
Germany

**MATARA ACHCHIGE, Melinda Yasaranji**
Qatar Green Leaders
Qatar

**MATHEW, Paul**
U.S. Department of Energy
The United States of America

**MATSUMURA, Yukari**
Institution
Japan

**MATSUTARO, Xavier**
Government of Palau
Palau

**MATTAUCH, LInus**
University of Oxford
United Kingdom (of Great Britain
and Northern Ireland)

**MATTIOLI, Giulio**
TU Dortmund University
Germany

**MATZNER, Eric**
Project Vesta
The United States of America

**MAURIZIO, Cellura**
University of Palermo
Italy

**MAWER, Danielle**
Bates Wells Braithwaite
United Kingdom (of Great Britain
and Northern Ireland)

**MAXWELL, Keely**
U.S. Environmental Protection Agency
The United States of America

**MAYER, Benoit**
The Chinese University of Hong Kong
China

**MAZAURIC, Vincent**
Schneider Electric/International Chamber
of Commerce (ICC)
France

**MCCREADY, Heather**
Environment and Climate Change Canada
Canada

**MCADAM, Jane**
University of New South Wales
Australia

**MCCARL, Bruce**
Texas A & M University
The United States of America

**MCDOWALL, Will**
University College London (UCL)
United Kingdom (of Great Britain and Northern Ireland)

**MCKAY, Jennifer**
University
Australia

**MCKINNON, Alan**
Kuehne Logistics University
United Kingdom (of Great Britain and Northern Ireland)

**MCLEOD, Christie**
Miller Thomson LLP
Canada

**MEBRATU, Desta**
Centre for Complex Systems in Transition, Stellenbosch University
Ethiopia

**MEDIMOREC, Nikola**
SLOCAT Partnership on Sustainable, Low Carbon Transport
The Republic of Korea

**MEHTA, Prakhar**
YOUNGO Cities Working Group (Friedrich-Alexander University of Erlangen-Nuremberg)
Germany/India

**MEJIA, Alvin**
Wuppertal Institute
Germany

**MENARD, Frederic**
Agir pour le climat
France

**MENG, Jing**
University College London
United Kingdom (of Great Britain and Northern Ireland)

**MENON, Induja P**
Manipal School of Architecture and Planning
India

**MERABET, Hamza**
Ministère de l'Enseignement supérieur et de la Recherche Scientific
Algeria

**MERK, Christine**
Kiel Institute for the World Economy
Germany

**MERRIFIELD, Jeffrey**
Pillsbury Law Firm
The United States of America

**METZ, Bert**
European Climate Foundation
The Netherlands

**MEYA, Jasper**
German Centre for Integrative Biodiversity Research
Germany

**MEYER, Leo**
Retired
The Netherlands

**MICHAELOWA, Axel**
University of Zurich
Switzerland

**MICHALEC, Ola Aleksandra**
The University of Bristol
United Kingdom (of Great Britain and Northern Ireland)

**MIDDLEBROOK, Craig**
U.S. Department of Transportation
The United States of America

**MIDDLEMISS, Lucie**
Sustainability Research Institute, University of Leeds
United Kingdom (of Great Britain and Northern Ireland)

**MIDGLEY, Pauline**
Independent Consultant
Germany

**MIGNONE, Bryan**
ExxonMobil Research and Engineering Company
The United States of America

**MIGUEL, Heleno**
Lawrence Berkeley National Laboratory
The United States of America

**MIGUEL GONZÁLEZ, Luis Javier**
University of Valladolid
Spain

**MIJEON, Charlotte**
Réseau "Sortir du nucléaire" – member of the French Réseau Action Climat
France

**MILLER, Andy**
U.S. Environmental Protection Agency
The United States of America

**MILLI, Manuela**
Government official
Italy

**MILLIEZ, Théo**
Alterna
Switzerland

**MILLSTONE, Carina**
Feedback Global
United Kingdom (of Great Britain and Northern Ireland)

**MINCHIN, Stuart**
The Pacific Community
Australia

**MINER, Reid**
Private Consultant
The United States of America

**MINGUEZ, Emilio**
Universidad Politécnica de Madrid (UPM)
Spain

**MINNEROP, Petra**
University of Dundee
United Kingdom (of Great Britain and Northern Ireland)

**MIRANDA ALFARO, Victor Arturo**
Gobierno
Peru

**MIRSHADIEV, Mirzokhid**
Wageningen University and Research
The Netherlands

**MITCHELL, Catherine**
University of Exeter
United Kingdom (of Great Britain and Northern Ireland)

**MITCHELL, Ronald**
University of Oregon
The United States of America

**MITONDO, louis lubango**
United Nations
Ethiopia

**MITTAL, Shivika**
Ahdmedabad University
India

**MOCCIA, Luigi**
Consiglio Nazionale delle Ricerche
Italy

**MOGELGAARD, Kathleen**
KAM Consulting
The United States of America

**MOGI, Kazuhisa**
Company
Japan

**MOHAMED, Abuleif Khalid**
Ministry of Petroleum and Mineral Resources
Saudi Arabia

**MOHANTY, Debadutta**
CSIR – Central Institute of Mining and Fuel Research, Dhanbad
India

**MOLINA, Tomas**
Televisió de Catalunya and Universitat de Barcelona
Spain

**MOLINA RODRÍGUEZ, Cintya Berenice**
El Colegio de México A.C
Mexico

**MONASTEROLO, Irene**
Vienna University of Economics and Business (WU)
Austria

**MONDEJAR, Maria E.**
Technical University of Denmark
Sweden

**MONTOUTE, Anita**
Department of Sustainable Development – Ministry of Education, Innovation, Gender Relations and Sustainable Developement
Saint Lucia

**MONTSERRAT, Francesc**
University of Amsterdam
The Netherlands

**MONTZKA, Stephen**
National Oceanic and Atmospheric Administration
The United States of America

**MOORE, Brendan**
University of East Anglia
United Kingdom (of Great Britain and Northern Ireland)

**MOORMAN, Saeda**
KiM Netherlands Institute for Transport Policy Analysis
The Netherlands

**MORA PERIS, Pedro**
University
Spain

**MORDENTE, Adriana**
United Nations Convention to Combat Desertification
Germany

**MORECROFT, Michael**
Natural England
United Kingdom (of Great Britain and Northern Ireland)

**MOREIRA, Marcelo**
UNICAMP – Agroicone
Brazil

**MORENO, Pablo**
International Monetary Fund
The United States of America

**MORFELDT, Johannes**
Chalmers University of Technology
Sweden

**MORGAVI, Diego**
INRAE
France

**MORIMOTO, Soichi**
The Institute of Energy Economics
Japan

**MORITA, Kanako**
Forestry and Forest Products Research Institute
Japan

**MORRESI, Maria del Valle**
University
Argentina

**MORROW, David**
American University
The United States of America

**MOSTAFAVI DARANI, Sayed Masoud**
Iran Meteorological Organization
Iran

**MOSTEFAOUI, Mounia**
LMD – ENS – Sorbonne
France

**MOTA-BABILONI, Adrián**
University Jaume I of Castellon
Spain

**MSOKA, Werner**
University of Dar es Salaam
The United Republic of Tanzania

**MSONGALELI, Barnabas**
University of Dodoma
The United Republic of Tanzania

**MUELBERT, Mônica M.C.**
UNIFESP
Brazil

**MUKHOPADHYAY, Chandrima**
Summer Winter School, CEPT University
India

**MULCHAN, Neil M.**
Adventure Physics, LLC
The United States of America

**MUNKÁCSY, Béla**
ELTE University
Hungary

**MÜNSTER, Marie**
Technical University of Denmark
Denmark

**MURATORI, Matteo**
NREL
The United States of America

**MURDOCK, Hannah E.**
REN21
France

**MUSYOKA, Nicholas**
Council for Scientific and Industrial Research (CSIR)
South Africa

**NADAR, Deshni**
Environmental Consultant
Qatar

**NAGABHUSHANA, Vinay**
U.S. Department of Transportation
The United States of America

**NAIK, Vaishali**
NOAA GFDL
The United States of America

**NAIK-DHUNGEL, Neeharika**
U.S. Environmental Protection Agency
The United States of America

**NAIRESIAE, Everlyne**
UNCCD-SPI
Kenya

**NAJIBZADEH, Mohammadreza**
ANRC/Tehran University
Iran

**NAKATA, Masami**
Shiga University
Japan

**NAKAYAMA, Sumie**
Tokyo Institute of Technology
Japan

**NASCIMENTO, Leonardo**
NewClimate Institute/Wageningen University and Research (WUR)
Germany

**NASER, Mostafa Mahmud**
Edith Cowan University
Australia

**NAYAK, Arun kumar**
Bhabha Atomic Research Centre
Trombay Mumbai
India

**NAZIR, Shareq Mohd**
KTH Royal Institute of Technology
Sweden

**NDIRITU, George Gatere**
University
Kenya

**NEEF, Mara**
Volkswagen AG
Germany

**NEETZOW, Paul**
Humboldt-Universität zu Berlin
Germany

**NEMA, Ashok**
Nuclear Power Corporation of India Limited
India

**NEOGI, Suvadip**
Global Centre for Environment and Energy, Ahmedabad University
India

**NEPAL, Mani**
International Center for Integrated Mountain Development
Nepal

**NESJE, Frikk**
Department of Economics, University of Copenhagen
Denmark

**NESJE, Frikk**
Heidelberg University
Germany

**NESLER, Clay**
Johnson Controls
The United States of America

**NEUFELDT, Henry**
UNEP DTU Partnership
Denmark

**NEWMAN, Peter**
Curtin University
Australia

**NG, Cheng Yee**
Universiti Teknologi PETRONAS
Malaysia

**NGO, Saik Peng Casey**
R-SYNC TECHNICAL RESOURCES SDN. BHD.
Malaysia

**NIAMIR, Leila**
MCC Berlin
Germany

**NICOLAI, Maike**
Helmholtz Centre Geesthacht
Germany

**NIFENECKER, Herve**
SLC/GISOC/Université Interage du Dauphine
France

**NILES, Meredith**
University of Vermont
The United States of America

**NILSSON, David**
KTH Royal Institute of Technology
Sweden

**NIMBALKAR, Sachin**
U.S. Department of Energy
The United States of America

**NISHIOKA, Shuzo**
Institute for Global Environmental Strategies
Japan

**NIZOU, Sylvain**
CEA
France

**NKUBA, Michael**
University of Botswana
Botswana

**NNKO, Happiness**
The University of Dodoma
The United Republic of Tanzania

**NOBOA, Sharl**
INOCAR
Ecuador

**NOH, Dong-Woon**
Korea Energy Economics Institute
The Republic of Korea

**NORTH, Michelle**
University of KwaZulu-Natal (UKZN)
South Africa

**NORTH, Peter**
Calorem Ltd/Imperial College London
United Kingdom (of Great Britain and Northern Ireland)

**NOUSSAN, Michel**
Fondazione Eni Enrico Mattei
Italy

**NOVAK, David**
DIPLOMA Fachhochschule Nordhessen
Germany

**NOVIKOVA, Victoria**
Initiative for Climate Action Transparency
Belarus

**NUGROHO, Bayu Dwi Apri**
Universitas Gadjah Mada
Indonesia

**NUNES, Ana Raquel**
University of Warwick
United Kingdom (of Great Britain and Northern Ireland)

**NYINGURO, Patricia**
Kenya Meteorological Service
Kenya

**OAKES, Robert**
United Nations University Institute for Environment and Human Security
United Kingdom (of Great Britain and Northern Ireland)

**OBERDABERNIG, Doris A.**
Leopold-Franzens Universität Innsbruck
Switzerland

**OBERHEITMANN, Andreas**
FOM University of Applied Sciences
Germany

**OBREKHT, Aglaia**
Environment and Climate Change Canada
Canada

**O'BRIEN, Jim**
Ireland

**OCALLAGHAN, Donal**
Teagasc (retired member)
Ireland

**OETTLÉ, Noel**
UNCCD-SPI
South Africa

**OGAWA, Junko**
The Institute of Energy Economics
Japan

**OGRIS, Manfred**
Federal Ministry of Agriculture, Forestry
Austria

**OH, Chaewoon**
Green Technology Center
The Republic of Korea

**OH, Yae Won**
Korea Meteorological Administration (KMA)
The Republic of Korea

**O'HANLON, Shane**
Stantec Consulting Ltd.
Canada

**OHREL, Sara**
U.S. Environmental Protection Agency
The United States of America

**OKABE, Masaaki**
AGC Inc.
Japan

**OKPALA, Denise**
Nigeria

**OLHOFF, Anne**
UNEP DTU Partnership, Technical University of Denmark
Denmark

**OLIVEIRA ALMEIDA MACHADO, Pedro Luiz**
Embrapa – Brazilian Agricultural Research Corporation
Brazil

**OLSON, Carolyn**
U.S. Geological Survey (retired)
The United States of America

**ÖNENLI, Özge**
Engie
Turkey

**ONO, Eiichi**
Toyota Technological Institute
Japan

**ÖRLANDER, Göran**
The Royal Swedish Academy of Agriculture and Forestry
Sweden

**ORTIZ GONZALO, Daniel**
University of Copenhagen
Denmark

**ÖRWALL LOVÉN, Ebba**
Betongindustri AB
Sweden

**OSBORNE-SAPONJA, Alex**
Sustainalytics
Canada

**OSHIRO, Ken**
Kyoto University
Japan

**OTAKA, Junichiro**
Climate Change Division, Ministry of Foreign Affairs
Japan

**OWENS, Allison**
U.S. Department of Agriculture
The United States of America

**OWINO, Fredrick**
Forest Resources International
Kenya

**ÖZDEMIR, Eray**
General directorate of Forestry
Turkey

**PACAÑOT, Vince Davidson**
University of the Philippines Diliman
The Philippines

**PACHECO ROJAS, Daniel Alejandro**
National Autonomous University of Mexico (UNAM)
Mexico

**PAESSLER, Dirk**
Paessler AG/Carbon Drawdown Initiative GmbH
Germany

**PAGLIANO, Lorenzo**
Politecnico di Milano, Energy Deparment, end-use Efficiency Research Group
Italy

**PAILLERE, Henri**
International Atomic Energy Agency (IAEA)
Austria

**PALAU, MFA**
Ministry of State National Authority
Bureau of Foreign Affairs
Palau

**PALERMO, Valentina**
JRC
Italy

**PÁLVÖLGYI, Tamás**
Budapest University of Technology and Economics, Department of Environmental Economics
Hungary

**PAN, Dongyang**
University College London
United Kingdom (of Great Britain and Northern Ireland)

**PAN, Jiahua**
Beijing University of Technology
China

**PAN, Jiahua**
Chinese Academy of Social Science
China

**PAN, Xunzhang**
School of Economics and Management, China University of Petroleum
China

**PANDEY, Neeraj**
National Institute of Industrial Engineering (NITIE), Mumbai
India

**PANT, Deepak**
Flemish Institute for Technological Research (VITO)
Belgium

**PARADELO, Remigio**
Universidade de Santiago de Compostela
Spain

**PARIKH, Jaimin**
United Nations SDSN/European Institute of Policy Research and Human Rights/University of Delhi
India

**PARK, Hwanil**
Science and Technology Policy Institution
The Republic of Korea

**PARK, Jungyu**
International Energy Agency (IEA)
France

**PARK, Nyun-bae**
Korea Institute of Energy Research
The Republic of Korea

**PARKER, Robert**
Nuclear for Climate Australia
Australia

**PARRA DE LA TORRE, Antonio**
University of Castilla-La Mancha
Spain

**PARTANEN, Rauli**
Think Atom
Finland

**PARTHASARATHY, Girija**
Thermo King
The United States of America

**PARTOWAZAM, Kevin**
U.S. Department of Transportation
The United States of America

**PASSER, Alexander**
Graz University of Technology
Austria

**PATERSON-JONES, Gregor**
Gregor Paterson-Jones Consulting
United Kingdom (of Great Britain and Northern Ireland)

**PATHAK, Minal**
Ahmedabad University
India

**PATRA, Sudhanwa**
Utkal University
India

**PATRICK, Jason**
New Zealand Green Investment Finance
New Zealand

**PATT, Anthony**
ETH Zürich
Switzerland

**PAUDEL, Jayash**
Boise State University
The United States of America

**PAUL, Mark**
New College of Florida
The United States of America

**PAULIUK, Stefan**
University
Germany

**PAWLOFF, Adam**
Greenpeace
Austria

**PEACOCK, Jacob**
The Humane League Labs
The United States of America

**PEDACE, Roque**
UBA.Buenos Aires University
Argentina

**PEDERSEN, Jiesper**
University of Lisbon
Portugal

**PEDRO, Linares**
Universidad Pontificia Comillas
Spain

**PEDRO, Mora Peris**
Universidad de la ETSI Minas y Energía de la Universidad Politécnica de Madrid
Spain

**PEEL, Jacqueline**
University of Melbourne
Australia

**PEETERS, Paul**
Breda University of applied sciences
The Netherlands

**PEGRAM, Tom**
University College London
United Kingdom (of Great Britain and Northern Ireland)

**PEGRUM-HARAM, Albertine**
European Climate Foundation
United Kingdom (of Great Britain and Northern Ireland)

**PELEJERO, Carles**
Institut de Ciències del Mar, CSIC
Spain

**PENG, Yuan**
The Australian National University
Australia

**PERIMENIS, Anastasios**
CO2 Value Europe
Belgium

**PERKINS, Patricia**
York University
Canada

**PERL, Kelly**
U.S. Energy Information Administration
The United States of America

**PERSSON, Jonas**
Malmö Stad
Sweden

**PETERS, Glen**
CICERO Center for International Climate Research
Norway

**PETERS, Ian Marius**
Forschungszentrum Jülcih
Germany

**PETERSON, Eric**
University of Leeds
United Kingdom (of Great Britain and Northern Ireland)

**PETROPOULOS, Apostolos**
International Energy Agency (IEA)
France

**PETTERSSON, Eva**
The Royal Swedish Academy of Agriculture and Forestry
Sweden

**PEUPORTIER, Bruno**
MINES ParisTech
France

**PFENNIG, Andreas**
University of Liège
Belgium

**PFLIEGER, Géraldine**
University of Geneva
Switzerland

**PHELPS, Peter**
University of Leeds
United Kingdom (of Great Britain and Northern Ireland)

**PHILIBERT, Cédric**
Institut Français des Relations Internationales
France

**PICHAT, Sylvain**
University of Lyon, Ecole normale supérieure de Lyon, Laboratoire de Géologie (LGL-TPE)
Germany

**PIETTE, Mary Anne**
U.S. Department of Energy
The United States of America

**PINA, Jorge**
ENEL
Spain

**PINO MAESO, Don Alfonso**
Area de Estrategias de Adaptacion – Oficina de Cambio Climatico – Ministerio de la Transicion Ecologica
Spain

**PIRGMAIER, Elke**
University of Lausanne
Switzerland

**PISELLO, Anna Laura**
Department of Engineering, University of Perugia
Italy

**PLINKE, Charlotte**
Climate Analytics
Germany

**PLOMTEUX, Adrien**
University College London (UCL)
United Kingdom (of Great Britain and Northern Ireland)

**POCH, Rosa M.**
ITPS and UdL
Spain

**POE, Carson**
U.S. Department of Transportation
The United States of America

**POERTNER, Hans**
Alfred-Wegener-Institut
Germany

**POLLITT, Hector**
Cambridge Econometrics
United Kingdom (of Great Britain and Northern Ireland)

**POLOCZANSKA, Elvira**
Alfred-Wegener-Institut
Germany

**POON, Otto**
Hong Kong Academy of Engineering Sciences
China

**PORSE, Sean**
U.S. Department of Energy
The United States of America

**POULTER, Benjamin**
National Aeronautics and Space Administration
The United States of America

**POWLSON, David**
Rothamsted Research
United Kingdom (of Great Britain and Northern Ireland)

**PREGNOLATO, Maria**
University of Bristol
United Kingdom (of Great Britain and Northern Ireland)

**PRESTELE, Reinhard**
Institute of Meteorology and Climate Research, Atmospheric Environmental Research, Karlsruhe Institute of Technology
Germany

**PRESTEMON, Jeff**
U.S. Forest Service
The United States of America

**PRESTON, Kurt**
U.S. Department of Defense
The United States of America

**PRESTON ARAGONES, Mark**
Bellona Europa
Belgium

**PRETENTERIS, Constantine**
Engaged Tracking
United Kingdom (of Great Britain and Northern Ireland)

**PRICE, Lynn**
Lawrence Berkeley National Laboratory
The United States of America

**PROUST, Eric**
European Nuclear Society (ENS)
France

**PRUDHOMME, Rémi**
CIRED
France

**PSOMOPOULOS, Constantinos**
University of West Attica, Department of Electrical and Electronics Engineering
Greece

**PUGNALINI, Chiara**
Altran Italy (Energy, Industry, Life Science division), European Commission
Italy

**PULWARTY, Roger**
National Oceanic and Atmospheric Administration
The United States of America

**PUPPIM DE OLIVEIRA, Jose Antonio**
FGV
Brazil

**PUTT, Margaret**
Enviromental Paper Network International
Australia

**PYROZHENKO, Yurii**
IPCC TFI TSU
Japan

**QUANTE, Lennart**
Potsdam Institute for Climate Impact Research (PIK)
Germany

**QUIRION, Philippe**
CNRS
France

**RABANI, Eli**
NanoCybernetics Corp.
The United States of America

**RADUNSKY, Klaus**
Austrian Standard Institute
Austria

**RAHAMAN, Hasibur**
Indian National Centre for Ocean Information Services (INCOIS), Ministry of Earth Sciences,Govt. of India
India

**RAHMAWATY, Rahmawaty**
universitas sumatera utara
Indonesia

**RAKONCZAY, Zoltán**
European Commission, Directorate General for Research
Belgium

**RAMADAN, Mohammad Fahmy**
Civil & Architecture Branch, MTC, Cairo
Egypt

**RAMCHANDRAN, Neeraj**
Independent Researcher
India

**RAMIG, Christopher**
U.S. Environmental Protection Agency
The United States of America

**RAMIREZ, Carlos**
AFA-ANDALUCIA
Spain

**RAMÍREZ, Betty**
INAMEH VENEZUELA
Venezuela

**RAMIREZ SANCHEZ-MAROTO, Carlos**
AFA-ANDALUCIA
Spain

**RAMOS QUISPE, Fany**
Environmental Engineers Associaton of La Paz
Bolivia

**RANAIVOSON, Andriamihaja**
Directorate General of Meteorology
Madagascar

**RANALDER, Lea**
REN21
France

**RANGEL MORENO, Karla Fabiola**
National Institute of Public Health
Mexico

**RAU, Greg**
IMS/Univ. Calif. Santa Cruz
The United States of America

**RAPF, Oliver**
BPIE – Buildings Performance Institute Europe
Belgium

**RASHEDI, A M Mabrur Ahmad**
Charles Darwin University
Australia

**RASHIDIAN, Leila**
Meteorological
Iran

**RASTOGI, Parag**
arbnco Ltd.
United Kingdom (of Great Britain and Northern Ireland)

**RAU, Alex**
Climate Wedge LLC
The United States of America

**RAUZIER, Emmanuel**
NGO Association negaWatt
France

**RAVINDRANATH, Nijavalli**
UNCCD-SPI
India

**REAY, Dave**
University of Edinburgh
United Kingdom (of Great Britain and Northern Ireland)

**REED, Tennant**
Australian Industry Group
Australia

**REISINGER, Andy**
Ministry for the Environment
New Zealand

**RETELSKA, Dorota**
FIBL Biological Agriculture Research Laboratory
Switzerland

**REYES, Julian**
U.S. Department of State
The United States of America

AV

**REYNOLDS, Jesse**
University of California, Los Angeles
The Netherlands

**RHOSANNA, Jenkins**
University of East Anglia
United Kingdom (of Great Britain and Northern Ireland)

**RIBEIRO, Silvia**
ETC Group
Mexico

**RICE, Jennie S.**
U.S. Department of Energy
The United States of America

**RICKELS, Wilfried**
Kiel Institute for the World Economy
Germany

**RIEDE, Moritz**
University of Oxford
United Kingdom (of Great Britain and Northern Ireland)

**RIGHI, Mattia**
Deutsches Zentrum für Luft- und Raumfahrt (DLR), Institut für Physik der Atmosphäre, Oberpfaffenhofen
Germany

**RIISE, Jan**
Chalmers University of Technology/ Gothenburg Centre for Sustainable Development
Sweden

**RISK, Mounir Wahba Labib**
National Academy of Science
Egypt

**RISSMAN, Jeffrey**
Energy Innovation: Policy and Technology LLC
The United States of America

**RITCHIE, Justin**
University of British Columbia, Institute for Resources, Environment and Sustainability
Canada

**RIVERA, Yvette**
U.S. Department of Transportation
The United States of America

**RIYAZ, Mahmood**
Maldivian Coral Reef Society
Maldives

**RIZEA, Lavinia**
SN Nuclearelectrica SA
Romania

**ROBERTS, Debra**
eThekwini Municipality
South Africa

**ROBERTSON, Simon**
University of New South Wales
Australia

**ROBIOU DU PONT, Yann**
Climate Analytics
France

**ROBOCK, Alan**
Rutgers University
The United States of America

**ROCK, Joachim**
Thuenen-Institute of Forest Ecosystems
Germany

**RÖCK, Martin**
KU Leuven
Austria

**RODHE, Lena**
The Royal Swedish Academy of Agriculture and Forestry
Sweden

**RODRIGUES, Mónica**
University of Coimbra
Portugal

**ROELANDT, Caroline**
Federal Agency for Nuclear Control
Belgium

**ROELKE, Luisa**
Federal Ministry for the Environment, Nature Conservation and Nuclear Safety International Climate Policy
Germany

**ROGELJ, Joeri**
Imperial College London
United Kingdom (of Great Britain and Northern Ireland)

**ROGGE, Karoline**
SPRU, University of Sussex
Germany

**ROGSTADIUS, Jakob**
RISE – Research Institutes of Sweden
Sweden

**ROJAS, Maisa**
University of Chile
Chile

**ROMANAK, Katherine**
The Univeristy of Texas at Austin
The United States of America

**ROMANOVSKAYA, Anna**
Institute of Global Climate and Ecology
The Russian Federation

**ROMERI, Mario Valentino**
Independent Consultant
Italy

**ROMERO, José**
Swiss Federal Office for the Environment, Federal Department of the Environment, Transport, Energy and Communications (DETEC)
Switzerland

**ROMERO-LANKAO, Paty**
U.S. Department of Energy
The United States of America

**ROSEN, Richard**
Tellus Institute (retired)
The United States of America

**ROSERO ABAD, Sofia**
University
The Netherlands

**ROSS, Kelsey**
The Center for Global Development
The United States of America

**ROTH, Amir**
U.S. Department of Energy
The United States of America

**ROUSE, Paul**
Carnegie Climate Governance Initiative
United Kingdom (of Great Britain and Northern Ireland)

**ROY, Joyashree**
Asian Institute of Technology/Jadavpur University
Thailand

**ROZITE, Vida**
International Energy Agency (IEA)
France

**RUANE, Alex**
NASA Goddard Institute for Space Studies
The United States of America

**RUDDIGKEIT, Dana**
German Environment Agency
Germany

**RUDGE RAMOS RIBEIRO, Rodrigo**
Getulio Vargas Foundation
Botswana

**RUDLOFF, Bettina**
German Institute for foreign and security affairs (SWP)
Germany

**RUIZ, Andrea Cristina**
Abdul Latif Jameel Poverty Action Lab/ Committee on Extreme Weather and Climate Change Adaptation Transportation Review Board – National Academy of Science
The United States of America

**RUIZ GARVIA, Carlos**
UNFCCC
Panama

**RUIZ MARTÍNEZ, Pamela**
UNID
Mexico

**RUMMUKAINEN, Markku**
Lund University
Sweden

**RUNDQVIST, Jan**
The Royal Swedish Academy of Agriculture and Forestry
Sweden

**RUSTAGI, Neha**
U.S. Department of Energy
The United States of America

**RUTH, Urs**
Robert Bosch GmbH
Germany

**RYAN, Lisa**
University College Dublin
Ireland

**RYDBERG, Ingrid**
The Royal Swedish Academy of Agriculture and Forestry
Sweden

**RYDSTEDT, Therese**
Environmental Department at the City of Stockholm
Sweden

**SAADE, Marcella**
Graz University of Technology
Austria

**SABATÉ, Santiago (Santi)**
University of Barcelona and CREAF
Spain

**SADEGH, Zeyaeyan**
Islamic Republic of Iran Meteorological Organization (IRIMO)
Iran

**SAGAR, Aarsi**
Global Green Growth Institute
The Republic of Korea

**SAGER, Lutz**
Georgetown University
The United States of America

**SAHAKIAN, Marlyne**
University of Geneva
Switzerland

**SAHEB, Yamina**
OpenExp, Ecole des Mines de Paris
France

**SAHLBERG, Per-Åke**
The Royal Swedish Academy of Agriculture and Forestry
Sweden

**SAKURAI, Keiichiro**
National Institute of Advanced Industrial Science and Technology
Japan

**SALANAVE, Jean-Luc**
Ecole Centrale-Supelec
France

**SALAS, Renee**
Harvard Global Health Institute
The United States of America

**SALAS REYES, Raul**
University of Toronto
Canada

**SALMI, Frederico**
UFRGS
Brazil

**SALVATORE, Philip**
Independent Consultant
The United States of America

**SÁNCHEZ, Gonzalo**
European Environmental Bureau
Belgium

**SANDS, Ron**
U.S. Department of Agriculture
The United States of America

**SANJUÁN, Miguel Angel**
Technical University of Madrid
Spain

**SANTALLA, Estela**
Facultad de Ingeniería UNICEN
www.fio.unicen.edu.ar
Argentina

**SANTO, Raychel**
Johns Hopkins Center for a Livable Future, Bloomberg School of Public Health
The United States of America

**SANWAL, Mukul**
Expert Group Strategic knowledging on Climate Change, Ministry of Science and Technology, Government of India
India

**SANZ SANCHEZ, Maria Jose**
Basque Centre for Climate Change (BC3)
Spain

**SANZ-COBENA, Alberto**
Universidad Politécnica de Madrid (UPM)
Spain

**SAPART, Célia**
Université Libre de Bruxelles/ CO2 Value Europe
Belgium

**SARDESHPANDE, Mallika**
Rhodes University
South Africa

**SARKAR, Pinaki**
CSIR – Central Institute of Mining and Fuel Research, Dhanbad
India

**SARLOV HERLIN, Ingrid**
The Royal Swedish Academy of Agriculture and Forestry
Sweden

**SASAKI, Midori**
industrial organization
Japan

**SATO, Atsushi**
Mitsubishi UFJ Research and Consulting Co.,Ltd.
Japan

**SAUJOT, Mathieu**
IDDRI
France

**SAUNDERS, Harry**
Carnegie Institution for Science
The United States of America

**SAVARESI, Annalisa**
University of Stirling
United Kingdom (of Great Britain and Northern Ireland)

**SAVCHUK, Olga**
Instituto Superior Tecnico
Portugal

**SAVILLE, Geoffrey**
Willis Towers Watson
United Kingdom (of Great Britain and Northern Ireland)

**SAVOLAINEN, Ilkka**
VTT Technical Research Centre of Finland
Finland

**SAWYER, Karma**
U.S. Department of Energy
The United States of America

**SCHADE, Wolfgang**
M-Five GmbH Mobility, Futures, Innovation, Economics
Germany

**SCHAEFFER, Michiel**
Climate Analytics
The Netherlands

**SCHÄFER, Laura**
Environmental NGO, Germanwatch
Germany

**SCHECHTMAN, Drew**
Voya Investment Management
The United States of America

**SCHEIDEL, Arnim**
Institute of Environmental Science and Technology (ICTA), Autonomous University of Barcelona (UAB)
Spain

**SCHENKER, Oliver**
Frankfurt School of Finance and Management
Germany

**SCHENUIT, Felix**
University Hamburg
Germany

**SCHERTZER, Daniel**
Hydrology Meteorology and Complexity, Ecole des Ponts ParisTech
France

**SCHINKO, Thomas**
International Institute for Applied Systems Analysis (IIASA)
Austria

**SCHLEUSNER, Carl**
Climate Analytics
Germany

**SCHNEIDER, Linda**
Heinrich Boell Foundation
Germany

**SCHOEMAN, David**
University of the Sunshine Coast
Australia

**SCHOEMAN, Laurie**
Enterprise Community Partners
The United States of America

**SCHOENEFELD, Jonas**
Institute of Political Science, Technical University of Darmstadt
Germany

**SCHÖNHART, Martin**
University of Natural Resources and Life Sciences, Vienna
Austria

**SCHROEDER, Andreas**
International Energy Agency (IEA)
France

**SCHUMACHER, Kim**
School of Environment and Society, Tokyo Institute of Technology
Japan

**SCHWANDNER FERREIRA, Luciana**
São Paulo City – Secretary of the Environment
Brazil

**SCHWELA, Dietrich**
Stockholm Environment Institute at University of York
Germany

**SCOTT, Spencer**
OneGeneration, Pachamama Institute
The United States of America

**SEALE, Jeffrey**
Bayer Crop Science
The United States of America

**SEGUINEAUD, Cécile**
Independent Consultant
France

**SEKEDOUA, Kouadio**
Université Gustave Eiffel
France

**SEKI, Shigetaka**
Consumer Product Safery Association
Japan

**SELLERS, Samuel**
U.S. Agency for International Development
The United States of America

**SELOSSE, Sandrine**
PSL Research University, MINES ParisTech, Centre for Applied Mathematics
France

**SEMENOV, Sergey**
Institute of Global Climate and Ecology
The Russian Federation

**SEMIENIUK, Gregor**
University of Massachusetts Amherst
The United States of America

**SEMPREVIVA, Anna Maria**
Wind Energy Department, Technical University of Denmark, Roskilde
Denmark

**SENTHILVADIVU, Hemamalini**
Manipal School of Architecture and Planning
India

**SERDECZNY, Oliwia**
Climate Analytics
Germany

**SEROA DA MOTTA, Ronaldo**
State University of Rio de Janeiro (UERJ)
Brazil

**SESSA, Emilio**
Carbon Credits Consulting
Italy

**SETO, Karen**
Yale University
The United States of America

**SHANG, li**
CAS shanghai advanced research institute
China

**SHARMA, Anjali**
Research, Projects and Collaborative initiatives, Delhi.
India

**SHELLENBERGER, Michael**
Environmental Progress
The United States of America

**SHEN, Bo**
U.S. Department of Energy
The United States of America

**SHETTY, Deepika**
Manipal School of Architecture and Planning
India

**SHIMIZU, Noriko**
Research Institution
Japan

**SHIMODA, Yoshiyuki**
Division of Sustainable Energy and Environmental Engineering, Osaka University
Japan

**SHINE, Keith**
University of Reading
United Kingdom (of Great Britain and Northern Ireland)

**SHIYAN, Chang**
Tsinghua University
China

**SHLAPAK, Mykola**
Ukraine

**SHOAI-TEHRANI, Bianka**
RTE Réseau de Transport d'Electricité, CentraleSupelec Paris Saclay University
France

**SHOLL, David**
Georgia Institute of Technology
The United States of America

**SHRESTHA, Rudra**
Tyndall Centre for Climate Change Research
United Kingdom (of Great Britain and Northern Ireland)

**SHRIVASTAVA, Manish Kumar**
TERI School of Advanced Studies
India

**SHUKLA, Anoop Kumar**
Manipal School of Architecture and Planning
India

**SHUKLA, Priyadarshi**
Ahmedabad University
India

**SIEBENHÜNER, Bernd**
Carl von Ossietzky University of Oldenburg
Germany

**SILVA, Adriana**
Venezuelan Institute for Scientific Research (IVIC)
Venezuela

**SILVA HERRAN, Diego**
National Institute for Environmental Studies
Japan

**SIMMONS, Aaron**
NSW Department of Primary Industries
Australia

**SIMON, Michel**
Vice Président SFENRAL
France

**SIMS, Ralph**
Massey University
New Zealand

**SINGH, Harjeet**
ActionAid International
India

**SINGH, Pramod K.**
Institute of Rural Management Anand
India

**SINGH, Reshma**
U.S. Department of Energy
The United States of America

**SINHA, Bhaskar**
Indian Institute of Forest Management
India

**SINNIGER, Kathryn**
U.S. Department of Transportation
The United States of America

**SKEA, Jim**
Imperial College London
United Kingdom (of Great Britain and Northern Ireland)

**SKLAREW, Jennifer**
George Mason University
The United States of America

**SLADE, Raphael**
Imperial College London
United Kingdom (of Great Britain and Northern Ireland)

**SLAMERŠAK, Aljoša**
The Institute of Environmental Science and Technology (ICTA-UAB)
Spain

**SLOAN, Andy**
Guernsey Finance
United Kingdom (of Great Britain and Northern Ireland)

**SMITH, Aaron**
Norwegian Institute of Bioeconomy Research
Norway

**SMITH, Chris**
University of Leeds
United Kingdom (of Great Britain and Northern Ireland)

**SMITH, Donald**
McGill University
Canada

**SMITH, Matthew**
Harvard University
The United States of America

**SMITH, Robert**
U.S. Department of Transportation
The United States of America

**SMITH, Sharon**
Geological Survey of Canada, Natural Resources Canada
Canada

**SMITH, Steven**
PNNL/JGCRI
The United States of America

**SMITH, Steven R**
CES, University of Surrey
United Kingdom (of Great Britain and Northern Ireland)

**SMITH, Thomas**
Masaryk University
The Czech Republic

**SOARES, Fabio Rubens**
Universidade de São Paulo
Brazil

**SODERBERG, Lena**
The Royal Swedish Academy of Agriculture and Forestry
Sweden

**SOGNNAES, Ida Andrea Braathen**
CICERO Center for International Climate Research
Norway

**SOHM, Stefanie**
Plateforme Mobilité Durable Maroc
Morocco

**SOICHI, Morimoto**
The Institute of Energy Economics
Japan

**SOIMAKALLIO, Sampo**
Finnish Environment Institute
Finland

**SOKONA, Youba**
South Centre
Switzerland

**SOLAR OBETO, Johannes**
University of Dar es Salaam
The United Republic of Tanzania

**SOLÉ, Jordi**
Universitat Rovira i Virgili
Spain

**SOLÉ-OLLÉ, Jordi**
Spanish National Research Council (CSIC)
Spain

**SOLIS, Karla**
UNFCC
Germany

**SOME, Shreya**
Ahmedabad University
India

**SONG, Su**
Young Crane Consulting
China

**SONG, Yongze**
Curtin University
Australia

**SONWA, Denis Jean**
CIFOR (Center for International Forestry Reseacrh)
Cameroon

**SOORA, Naresh Kumar**
Indian Agricultural Research Institute
India

**SOYSA, Ramesh**
Biomass Group & World Bank
Sri Lanka

**SPAISER, Viktoria**
University of Leeds
United Kingdom (of Great Britain and Northern Ireland)

**SPAJIC, Luke**
University of Adelaide (graduate student researcher), University of Oxford (visiting student researcher)
Australia

**SPALLINA, Vincenzo**
University of Manchester
United Kingdom (of Great Britain and Northern Ireland)

**SPENCE, Jacqueline**
Meteorological Service Division
Jamaica

**SPRINZ, Detlef**
Potsdam Institute for Climate Impact Research (PIK)
Germany

**SREENIVAS, Ashok**
Prayas (Energy Group)
India

**SRIVASTAVA, Siddharth**
YOUNGO Cities Working Group
India

**STANITSKI, Diane**
National Oceanic and Atmospheric Administration
The United States of America

**STECKEL, Jan**
MCC Berlin
Germany

**STEFANO, Carattini**
Georgia State University
The United States of America

**STEFFEN, Bjarne**
ETH Zürich
Switzerland

**STEINBERGER, Julia**
University of Lausanne
Switzerland

**STEPANIK, Humberto Edgardo**
Catedra libre ingeniería. UBA volunteer work for villa 31
Argentina

**STERN, David**
Australian National University
Australia

**STERNER, Thomas**
University of Gothenburg
Sweden

**STRANKS, Samuel**
University of Cambridge
United Kingdom (of Great Britain and Northern Ireland)

**STRAUTMANN, Maya**
RWTH University, Chair of Electrochemical Energy Conversion and Storage Systems
Germany

**STRECK, Charlotte**
University of Potsdam
Germany

**STRICKLAND, Francella**
Ministry of Foreign Affairs and Trade
Samoa

**STUA, Michele**
APE-FVG
United Kingdom (of Great Britain and Northern Ireland)

**SUAREZ, Avelino G.**
Research Centre for the World Economy
Cuba

**SUDDERTH, Erika**
U.S. Department of Transportation
The United States of America

**SUGIYAMA, Masahiro**
University of Tokyo
Japan

**SUHENDY, Christy Cecilia Veronica**
Pattimura University
Indonesia

**SULISTIAWATI, Linda Yanti**
Universitas Gadjah Mada
Indonesia

**SULPIS, Olivier**
Universiteit Utrecht
The Netherlands

**SUNDBERG, Cecilia**
Swedish University of Agricultural Sciences
Sweden

**SUPRABA, Intan**
Universitas Gadjah Mada
Indonesia

**SURAWSKI, Nicholas**
University of Technology Sydney
Australia

**SURDU, Dmytro**
Kray Technologies
Ukraine

**SUSCA, Tiziana**
Italian National Agency for New Technologies, Energy and Sustainable Economic Development
Italy

**SUSSMAN, Reuven**
American Council for an Energy-Efficient Economy
The United States of America

**SUTTON-GRIER, Ariana**
University of Maryland
The United States of America

**SVARSTAD, Hanne**
OsloMet – Oslo Metropolitan University
Norway

**SVENSSON, Harald**
The Royal Swedish Academy of Agriculture and Forestry
Sweden

**SVOBODA, Radek**
Czech Nuclear Society
The Czech Republic

**SYED, Md Abu**
Bangladesh Centre for Advanced Studies (BCAS)
Bangladesh

**SZOPA, Sophie**
Commissariat à l'Energie Atomique et aux Energies Alternatives
France

**SZUM, Carolyn**
U.S. Department of Energy
The United States of America

**TAGAMI, Takahiko**
Institute of Energy Economics, Japan
Japan

**TAHA, Mohyeldeen**
University of Bahri
Sudan

**TAIMAR, Ala**
Estonian Meteorological & Hydrological Institute
Estonia

**TAJBAKHSH MOSALMAN, Sahar**
Islamic Republic of Iran Meteorological Organization (IRIMO)
Iran

**TAKAHASHI, Kiyoshi**
National Institute for Environmental Studies
Japan

**TAKARINA, Noverita**
Universitas Indonesia
Indonesia

**TALLEY, Trigg**
U.S. Department of State
The United States of America

**TAMBKE, Jens**
Umweltbundesamt
Germany

**TAMM, James**
U.S. Department of Transportation
The United States of America

**TAMME, Eve**
Global Carbon Capture and Storage Institute
Belgium

**TANAKA, Kenji**
the University of Tokyo
Japan

**TANGRI, Neil**
GAIA
The United States of America

**TANTOH, Henry**
University of South Africa
South Africa

**TASICO, Reyneir**
UpScale PH/Psychological Association of the Philippines (Member)
The Philippines

**TAYARA, Danny**
Seattle University/Ulysses VR
The United States of America

**TAYLOR, Graeme**
BEST Futures
Australia

**TETSUYA, Deguchi**
Research Institute of Innovative Technology for the Earth (RITE)
Japan

**TEZUKA, Hiroyuki**
JFE Steel Corporation
Japan

**THALER, Thomas**
University of Natural Resources and Life Sciences
Italy

**THISTLETHWAITE, David Glen**
Ricardo
United Kingdom (of Great Britain and Northern Ireland)

**THONIG, Richard**
IASS Potsdam
Germany

**THORN, Dominic**
Ministry for the Environment
New Zealand

**THRELFALL, Richard**
KPMG/Institution of Civil Engineers
United Kingdom (of Great Britain and Northern Ireland)

**TIBIG, Lourdes**
Climate Change Commission
The Philippines

**TILLOU, Michael**
U.S. Department of Energy
The United States of America

**TIMMERMAN, Jos**
Waterframes
The Netherlands

**TIRADO, Reyes**
Greenpeace/University of Exeter
Spain

**TISCHLER, Stephan**
University of Innsbruck
Austria

**TISELJ, Iztok**
Jozef Stefan Institute & University of Ljubljana
Slovenia

**TOE, Doris**
Universiti Teknologi Malaysia
Malaysia

**TOLASZ, Radim**
Czech Hydrometeorological Institute
The Czech Republic

**TOLLIVER, Clarence**
University of Michigan Law School
The United States of America

**TOMOHIRO, Kuwae**
Port and Airport Research Institute
Japan

**TORI IN, Daniele**
The Open University Business School
United Kingdom (of Great Britain and Northern Ireland)

**TORNEY, Diarmuid**
Dublin City University
Ireland

**TOSHI, Arimura**
Waseda University
Japan

**TOSHIHIRO, Hasegawa**
National Agricultural and Food Research Organization
Japan

**TOURAY, Lamin Mai**
Department of Water Resources
Gambia

**TOWNSEND, Dina**
University of Witwatersrand
Austria

**TOWNSEND-SMALL, Amy**
University of Cincinnati
The United States of America

**TRAINER, Ted**
University of NSW (retired)
Australia

**TRAN, Martino**
University of British Columbia
Canada

**TREBER, Manfred**
Germanwatch
Germany

**TREGUIER, Anne Marie**
CNRS
France

**TROCHET, Jean-Michel**
EDF group (French Utility)
France

**TRONTL, Krešimir**
University of Zagreb, Faculty of Electrical Engineering and Computing
Croatia

**TROSCHKE, Manuela**
Scientists for Future Germany
Germany

**TSHEOLE, Thapelo**
Botswana Stock Exchange
Botswana

**TSUTSUI, Junichi**
Central Research Institute of Electric Power Industry
Japan

**TUBIELLO, Francesco**
Food and Agriculture Organization of the United Nations (FAO)
Italy

**TULIP, Robert**
Australian National University
Australia

**TULKENS, Henry**
CORE, Université catholique de Louvain
Belgium

**TULKENS, Philippe**
European Union (EU) – DG Research & Innovation
Belgium

**TUNINETTI, Luis**
Universidad Nacional de Villa María
Argentina

**TUNNESSEN, Walt**
U.S. Environmental Protection Agency
The United States of America

**TUOMISTO, Hanna**
University of Helsinki
Finland

**TYLECOTE, Andrew**
University of Sheffield
United Kingdom (of Great Britain and Northern Ireland)

**UEDA, Tatsuki**
National Agriculture and Food Research Organization
Japan

**UGOM, Michael**
University of Nigeria, Nsukka
Nigeria

**UMEMIYA, Chisa**
Institute for Global Environmental Strategies
Japan

**URBAN, Frauke**
KTH Royal Institute of Technology
Sweden

**UTTURKAR, Mayuri**
University of Delaware
The United States of America

**UVAROVA, Galyna**
Green Climate Fund, Independent Evaluation Unit
The Republic of Korea

**UZZAMAN, Md Arfan**
Food and Agriculture Organization of the United Nations (FAO)
Bangladesh

**VAKKILAINEN, Esa**
LUT University, Lappeenranta
Finland

**VALDOVINOS, Claudio**
University of Concepcion
Chile

**VAN ASSELT, Harro**
University of Eastern Finland
The Netherlands

**VAN BERKEL, Dennis**
Urgenda
The Netherlands

**VAN DE GRAAF, Thijs**
Ghent University
Belgium

**VAN DEN BROEK, Machteld**
Utrecht University
The Netherlands

**VAN DER VOORN, Tom**
Institute for Environmental Systems Research
The Netherlands

**VAN DIEMEN, Renee**
WG III TSU
United Kingdom (of Great Britain and Northern Ireland)

**VAN DRUTEN, Emiel**
Technical University of Eindhoven
The Netherlands

**VAN GOETHEM, Georges**
Royal Academy of Overseas Sciences of Belgium (ARSOM – KAOW)
Belgium

**VAN HERPEN, Maarten**
Acacia Impact Innovation BV
The Netherlands

**VAN SLUISVELD, Mariesse**
PBL Netherlands Environmental Assessment Agency
The Netherlands

**VAN 'T WOUT, Tamara**
Food and Agriculture Organization of the United Nations (FAO)
Qatar

**VAN WIJK, Ad**
Technical University Delft
The Netherlands

**VAN YPERSELE, Jean-Pascal**
Université catholique de Louvain
Belgium

**VANDERSTRAETEN, Martine**
Belgian Science Policy Office (BELSPO)
Belgium

**VANDYCK, Toon**
European Commission, Joint Research Centre
Spain

**VASQUEZ ARROYO, Eveline Maria**
COPPE/UFRJ
Brazil

**VÁVRA, Jan**
University of South Bohemia
The Czech Republic

**VELAZQUEZ, Abad**
Transport Research Laboratory
United Kingdom (of Great Britain and Northern Ireland)

**VENMANS, Frank**
UMons
Belgium

**VERA, Leonor**
Instituto Oceanográfico de la Armada
Ecuador

**VERBRUGGEN, Aviel**
University of Antwerp
Belgium

**VERCHOT, Louis**
International Center for Tropical Agriculture
Colombia

**VERDOLINI, Elena**
University of Brescia and Euro-Mediterranean Centre on Climate Change
Italy

**VETHMAN, Paul**
PBL Netherlands Environmental Assessment Agency
The Netherlands

**VEYSEY, Jason**
Stockholm Environment Institute
The United States of America

**VICCA, Sara**
University of Antwerp
Belgium

**VICENTE-SERRANO, Sergio**
UNCCD-SPI
Spain

**VICTORIA, Marta**
Aarhus University
Denmark

**VIGUIE, Vincent**
CIRED, Ecole des Ponts ParisTech
France

**VIJAYA KUMAR, Sowbarnika**
Manipal School of Architecture and Planning
India

**VILELLA, Mariel**
Zero Waste Europe/University of Manchester
United Kingdom (of Great Britain and Northern Ireland)

**VILLAVICENCIO-CALZADILLA, Paola**
Universitat Rovira i Virgili
The Netherlands

**VINER, David**
Green Investment Group
United Kingdom (of Great Britain and Northern Ireland)

**VITA, Gibran**
Open University of the Netherlands
The Netherlands

**VITALE, Domenico**
Centro-euro Mediterraneo sui Cambiamenti Climatici (CMCC)
Italy

**VIVIAN, Christopher**
Cefas (retired)
United Kingdom (of Great Britain and Northern Ireland)

**VLADU, Florin**
UNFCCC Secretariat
Germany

**VO, Hung**
U.S. Agency for International Development
The United States of America

**VOGEL, Jefim**
University of Leeds
United Kingdom (of Great Britain and Northern Ireland)

**VOGLER, Ingrid**
GdW Bundesverband deutscher Wohnungs- und Immobilienunternehmen e.V.
Germany

**VOÏTA, Thibaud**
IFRI
Germany

**VON HERZEN, Brian**
Climate Foundation
The United States of America

**VON MALTITZ, Graham**
University of Stellenbosch/UNCCD Science Policy Interface
South Africa

**WAGAI, Rota**
National Agriculture and Food Research Organization, Institute for Agro-Environmental Sciences, Division of Climate Change
Japan

**WAGNER, Lucrecia**
CONICET
Argentina

**WAGNER, Nicholas**
International Renewable Energy Agency (IRENA)
Germany

**WALDHOFF, Stephanie**
U.S. Department of Energy
The United States of America

**WALDTEUFEL, Philippe**
CNRS/IPSL/LATMOS
France

**WALIMBE, Sonali**
Manipal School of Architecture and Planning
India

**WALKER, Chad**
University of Exeter
United Kingdom (of Great Britain and Northern Ireland)

**WALKER, Iain**
Lawrence Berkeley National Laboratory
The United States of America

**WALLOE HANSEN, Aksel Walløe**
Niels Bohr Intsitute, University of Copenhagen
Denmark

**WANG, Binbin**
Institute of Climate Change and Sustainable Development at Tsinghua University
China

**WANG, Changke**
National Climate Center, China Meteorological Administration
China

**WANG, Lining**
Economics and Technology Research Institute, CNPC
China

**WANG, Xinfang**
University of Birmingham
United Kingdom (of Great Britain and Northern Ireland)

**WANG, Yang**
Beijing Climate Center
China

**WANSER, Kelly**
SilverLining
The United States of America

**WARD, Robert**
London School of Economics and Political Science
United Kingdom (of Great Britain and Northern Ireland)

**WARNER, Koko**
UNFCCC
Germany

**WAROUX, Véronique**
Planète-A
Belgium

**WARREN, Adam**
U.S. Department of Energy
The United States of America

**WASKO, Conrad**
University of Melbourne
Australia

**WEBER, Adam**
Lawrence Berkeley National Laboratory
The United States of America

**WEI, Max**
Lawrence Berkeley National Laboratory
The United States of America

**WEI, Taoyuan**
CICERO Center for International
Climate Research
Norway

**WEICHSELGARTNER, Juergen**
HWR Berlin
Germany

**WEIJIE, Zhang**
Ministry of Environment and
Natural Resources
Singapore

**WEIKMANS, Romain**
Université Libre de Bruxelles/
Free University of Brussels
Belgium

**WEIMER, Kerstin**
Mid Sweden University, Department
of Psychology and Social Work
Sweden

**WEISZ, Ulli**
University of Natural Resources
and Life Sciences Vienna, Institute
of Social Ecology
Austria

**WEITZEL, Matthias**
European Commission, Joint
Research Centre
Spain

**WELSH, Kevin**
U.S. Department of Transportation
The United States of America

**WEN, Philippe**
Ministère de l'Économie, des Finances
et de la Relance
France

**WERNECKE, Bianca**
South African Medical Research Council
South Africa

**WESTLÉN, Daniel**
Liberal party Swedish parliament
Sweden

**WETTER, Michael**
U.S. Department of Energy
The United States of America

**WEWERINKE-SINGH, Margaretha**
Leiden University/University
of the South Pacific
The Netherlands

**WHITE, Dave**
Climate Change Truth Inc.
The United States of America

**WHITE, Lee**
Australian National University
Australia

**WHITE, Robin**
Environment & Climate Change Canada,
Government of Canada
Canada

**WIEDMANN, Thomas**
UNSW Sydney
Australia

**WIKANDER, Johanna**
Company
Sweden

**WIKSTROM, Lennart**
The Royal Swedish Academy
of Agriculture and Forestry
Sweden

**WILKE, Nicole**
Federal Ministry for the Environment,
Nature Conservation and Nuclear Safety
International Climate Policy
Germany

**WILKINSON, Stephen**
University of Wollongong in Dubai
United Arab Emirates

**WILSON, Charlie**
Tyndall Centre for Climate Change Research
United Kingdom (of Great Britain
and Northern Ireland)

**WILSON, Grant**
University of Birmingham
United Kingdom (of Great Britain
and Northern Ireland)

**WIN, San**
Environmental Conservation Department,
Ministry of Natural Resources and
Environmental Conservation
Myanmar

**WINICK, Jeff**
U.S. Department of Energy
The United States of America

**WINKLER, Armin**
University of Applied Sciences Upper Austria
Austria

**WINTHROP, Robert**
University of Maryland
The United States of America

**WISSENBURG, Marcel**
Radboud University Nijmegen
The Netherlands

**WITAJEWSKI-BALTVILKS, Jan**
University of Warsaw
Poland

**WITHAM, Fred**
Rolls-Royce
United Kingdom (of Great Britain
and Northern Ireland)

**WITTENBRINK, Heinz**
FH Joanneum University of Applied
Sciences, Graz, Austria
Austria

**WOHLAND, Jan**
ETH Zürich
Switzerland

**WOLFRAM, Paul**
Yale University
The United States of America

**WONG, Liwah**
EIT Climate KIC, EIT RawMaterials
Germany

**WU, Jianguo**
Chinese Research Academy
of Environmental Sciences
China

**WU, Wenchao**
National Institute for Environmental Studies
Japan

**XINZHU, Zheng**
China University of Petroleum
China

**XU, Yinlong**
Institute of Environment and Sustainable Development in Agriculture, Chinese Academy of Agricultural Sciences
China

**YAMAGUCHI, Mitsutsune**
Research Institute of Innovative Technology for the Earth (RITE)
Japan

**YANG, Qing**
Harvard University
China

**YANG, Xinyan**
China Academy of Building Research
China

**YANKWA DJOBO, Jean Noël**
Ministry of Scientific Research and Innovation/Local Materials Promotion Authority (MIPROMALO)
Cameroon

**YAO, Yuan**
Yale University
The United States of America

**YARIME, Masaru**
Hong Kong University of Science and Technology
China

**YAZDANI, Alireza**
Shiraz University
Iran

**YIN, Feijia**
Delft University of Technology
The Netherlands

**YOKOHATA, Tokuta**
National Institute for Environmental Studies
Japan

**YONG, Kum Weng**
KW Yong Architect (Professional architect practice)
Malaysia

**YONG YOOK, Kim**
KIT Valley Inc.
The Republic of Korea

**YOON, Soonuk**
Green Technology Center
The Republic of Korea

**YOSHINO, Madoka**
United Nations University Institute for the Advanced Study on Sustainability
Japan

**YOUNG, Virginia**
Australian Rainforest Conservation Society/ Griffith University/CAN Ecosystems
Australia

**YU, Sha**
U.S. Department of Energy
The United States of America

**YU, Xiang**
Institute for Urban and Environmental Studies,Chinese Academy of Social Sciences
China

**YULIZAR, Yulizar**
Universitas Pertamina
Indonesia

**YUN SENG, Lim**
Universiti Tunku Abdul Rahman
Malaysia

**ZAELKE, Durwood**
Institute for Governance & Sustainable Development
The United States of America

**ŽAGAR, Tomaž**
Faculty for Energy Technology, University of Maribor
Slovenia

**ZAHARIA, Raymond**
Le Club des Argonautes
France

**ZAHLANDER, Frida**
DanChurchAid
Denmark

**ZAMBRANA, Jordan**
U.S. Environmental Protection Agency
The United States of America

**ZAMUDA, Craig**
U.S. Department of Energy
The United States of America

**ZARAGOZA, Aniceto**
Oficemen
Spain

**ZECCHINI CANTINHO, Roberta**
UNDP/UnB
Brazil

**ZEYAEYAN, Sadegh**
National Center for Forecasting and Weather Hazards Management, Islamic Republic of Iran Meteorological Organization (IRIMO)
Iran

**ZHANG, Jingyong**
Institute of Atmospheric Physics, Chinese Academy of Sciences
China

**ZHANG, Shining**
GEIDCO
China

**ZHANG, Wen**
Institute of Atmospheric Physics, Chinese Academy of Sciences
China

**ZHANG, Yongxiang**
National Climate Center, China Meteorological Administration
China

**ZHAO, Xiusheng**
Tsinghua University
China

**ZHIFU, Mi**
University College London
United Kingdom (of Great Britain and Northern Ireland)

**ZHOU, Minghua**
Institute of Mountain Hazards and Environment, Chinese Academy of Sciences
China

**ZHOU, Nan**
U.S. Department of Energy
The United States of America

**ZHOU, Yuyu**
Iowa State University
The United States of America

**ZHUANG, Guotai**
China Meteorological Administration
China

**ZIMM, Caroline**
International Institute for Applied Systems Analysis (IIASA)
Austria

**ZOBAA, Ahmed**
Brunel University London
United Kingdom (of Great Britain and Northern Ireland)

**ZOMMERS, Zinta**
United Nations Office for Disaster Risk Reduction
Germany

**ZUHAIB, Sheikh**
Buildings Performance Institute Europe asbl (BPIE)
Germany

## Note of thanks

The Co-Chairs of Working Group III would like to acknowledge the following individuals who assisted the Working Group III Technical Support Unit in reviewing the consistency and accuracy of scientific references:

**BOWLES, Alex**

**BRANDL, Patrick**

**CLUBE, Rebecca**

**CORADA, Karina**

**CROKER, Abigail**

**FOLKARD-TAPP, Hollie**

**GRIFFITHS, Tom**

**HARRIS, Catrin**

**HINSON, Caitlin**

**KIRKPATRICK, Liam**

**MOYA, Diego**

**ORTEGA ARRIAGA, Paloma**

**O'SHEA, Ryan**

**PIROLI, Erika**

**SUKPANIT, Paisan**

**XIE, Judy**

**YASIN, Liam**

**ZEA-REYES, Leonardo**

AVI

# Annex VI: Acronyms

**This chapter should be cited as:**
IPCC, 2022: Annex VI: Acronyms. In *Climate Change 2022: Mitigation of Climate Change. Contribution of Working Group III to the Sixth Assessment Report of the Intergovernmental Panel on Climate Change* [P.R. Shukla, J. Skea, R. Slade, A. Al Khourdajie, R. van Diemen, D. McCollum, M. Pathak, S. Some, P. Vyas, R. Fradera, M. Belkacemi, A. Hasija, G. Lisboa, S. Luz, J. Malley, (eds.)]. Cambridge University Press, Cambridge, UK and New York, NY, USA. doi: 10.1017/9781009157926.025

| | |
|---|---|
| **A/R** | afforestation/reforestation |
| **AB** | Assembly Bill |
| **AC** | alternating current |
| **ACCTS** | Agreement on Climate Change, Trade and Sustainability |
| **ACF** | areal carbon footprint |
| **ADEME** | Agence de l'Environnement et de la Maîtrise de l'Energie (French Environment and Energy Management Agency) |
| **AF** | Adaptation Fund |
| **AFR** | Africa |
| **AFD** | French Development Agency |
| **AFOLU** | agriculture, forestry and other land use |
| **AGAGE** | Advanced Global Atmospheric Gases Experiment |
| **AI** | artificial intelligence |
| **AILAC** | Association of the Latin American and Caribbean Countries |
| **ALBA** | Alianza Bolivariana para los Pueblos de Nuestra América (Bolivarian Alliance for the Peoples of our Americas) |
| **ALCA** | Attributional Life Cycle Assessment |
| **AM** | additive manufacturing |
| **APEC** | Asia-Pacific Economic Cooperation |
| **AR4** | Fourth Assessment Report of the Intergovernmental Panel on Climate Change |
| **AR5** | Fifth Assessment Report of the Intergovernmental Panel on Climate Change |
| **AR6** | Sixth Assessment Report of the Intergovernmental Panel on Climate Change |
| **AR7** | Seventh Assessment Cycle of the Intergovernmental Panel on Climate Change |
| **ARC** | African Risk Capacity |
| **ARPA-E** | Advanced Research Projects Agency-Energy |
| **ART** | Architecture for REDD+ Transactions |
| **Art.** | Article (e.g., of the UNFCCC) |
| **ASAP** | Adaptation for Smallholder Agriculture Programme |
| **ASCM** | Agreement on Subsidies and Countervailing Measures |
| **ASI** | Avoid-Shift-Improve |
| **ASK** | available seat kilometres |
| **AUM** | assets under management |
| **BASIC** | Brazil, South Africa, India and China |
| **BAT** | best available technology |
| **BAU** | business as usual |
| **BC** | black carbon |
| **BCA** | border carbon adjustment |
| **BECCS** | bioenergy with carbon capture and storage |
| **BEES** | building energy efficiency standards |
| **BEMS** | building energy management systems |
| **BEV** | battery electric vehicle |
| **BF-BOF** | blast furnace-basic oxygen furnace |
| **BIM** | Building Information Modelling |
| **BIPV** | building-integrated photovoltaic |
| **BLUE** | Bookkeeping of land-use emissions |
| **BR** | biennial report |
| **BRI** | Belt and Road Initiative |
| **BRT** | bus rapid transport |
| **BTR** | biennial transparency report |
| **BTU** | British thermal units |
| **BUR** | biennial update report |
| **CA** | capability approach |
| **CAT** | Climate Action Tracker |
| **CAGR** | compound annual growth rate |
| **CAIT** | Climate Analysis Indicators Tool |
| **CAPEX** | capital expenditure |
| **CAR** | Climate Action Reserve |
| **CBA** | cost-benefit analysis |
| **CBAM** | carbon border adjustment mechanism |
| **CBCF** | consumption-based carbon footprint (accounting) |
| **CBD** | Convention on Biological Diversity |
| **CBDRRC** | common but differentiated responsibilities and respective capabilities |
| **CBEs** | consumption-based emissions |
| **CCAC** | Climate and Clean Air Coalition |
| **CCD** | climate-compatible development |
| **CCPI** | Climate Change Performance Index |
| **CCRIF** | Caribbean Catastrophe Risk Insurance Facility |
| **CCS** | carbon capture and storage |
| **CCT** | cirrus cloud thinning |



This annex, which was not part of the approved outline of the Working Group III contribution to the IPCC Sixth Assessment Report, provides a list of acronyms used in the report.

| | |
|---|---|
| **CCU** | carbon capture and utilisation |
| **CCUS** | carbon capture, use and storage |
| **CCX** | Chicago Climate Exchange |
| **CD** | charge depleting |
| **CDD** | cooling degree-days |
| **CDIAC** | Carbon Dioxide Information Analysis Center |
| **CDM** | Clean Development Mechanism |
| **CDR** | carbon dioxide removal |
| **CE** | circular economy |
| **CEA** | cost-effectiveness analysis |
| **CEDS** | Community Emissions Data System |
| **CETA** | EU-Canada Comprehensive Economic and Trade Agreement |
| **CFCs** | chlorofluorocarbons |
| **CfD** | contract for difference |
| **CFL** | compact fluorescent lamp [/lighting] |
| **CGE** | Computable General Equilibrium |
| **CGTP** | combined global temperature change potential |
| **$CH_4$** | methane |
| **CHP** | combined heat and power |
| **CII** | Carbon Intensity Indicator |
| **CLASP** | Collaborative Labelling and Appliance Standards Program |
| **CLC** | constant land cover |
| **CLCA** | Consequential Life Cycle Assessment |
| **CLIMI** | Climate Laws, Institutions and Measures Index |
| **CLRTAP** | Convention on Long-Range Transboundary Air Pollution |
| **CLT** | cross-laminated timber |
| **CMA** | Conference of the Parties serving as the meeting of the Parties to the Paris Agreement |
| **CMIP6** | Coupled Model Intercomparison Project Phase 6 |
| **CNG** | compressed natural gas |
| **CO** | carbon monoxide |
| **$CO_2$** | carbon dioxide |
| **$CO_2$-eq** | carbon dioxide equivalent |
| **COMMIT** | Climate policy assessment and Mitigation Modelling to Integrate national and global Transition pathways |
| **CoP** | coefficient of performance |
| **COP** | Conference of the Parties to the UNFCCC |
| **CORSIA** | Carbon Offsetting and Reduction Scheme for International Aviation |
| **CPRS** | Climate Policy Relevant Sectors |
| **CPTPP** | Comprehensive and Progressive Agreement for Trans-Pacific Partnership |
| **CRD** | climate-resilient development |
| **CREMAs** | Community Resource Management Area Mechanisms (Ghana) |
| **CRF** | common reporting format |
| **CRIBs** | Climate Relevant Innovation-system Builders |
| **CS** | charge sustaining |
| **CSC** | climate-smart cocoa |
| **CSI** | Cement Sustainability Initiative |
| **CSP** | concentrating solar power |
| **CSR** | corporate social responsibility |
| **CSSP** | cross-sector social partnership |
| **CTCN** | Climate Technology Centre and Network |
| **CurPol** | Current Policies scenario |
| **DAC** | direct air capture |
| **DACCS** | direct air carbon capture and storage |
| **DACCU** | direct air capture carbon and utilisation |
| **DALY** | disability adjusted life year |
| **DBH** | diameter at breast height |
| **DC** | direct current |
| **DGVM** | dynamic global vegetation model |
| **DLS** | decent living standards |
| **DRI** | direct reduced iron |
| **DSM** | demand-side management |
| **DWM** | down woody material |
| **EaaS** | energy as a service |
| **EAF** | electric arc furnace |
| **EBEs** | extraction-based emissions |
| **EDGAR** | Emissions Database for Global Atmospheric Research |
| **EDLC** | electrochemical double layer capacitor |
| **EEA** | Eastern Europe and West Central Asia |
| **EED** | Energy Efficiency Directive |
| **EEDI** | Energy Efficiency Design Index |
| **EEE** | emissions embodied in exports |
| **EEI** | emissions embodied in imports |

| | |
|---|---|
| **EEM** | Energy Efficient Mortgage |
| **EES** | electrical energy storage |
| **EET** | emissions embodied in trade |
| **EEXI** | Energy Efficiency Existing Ship Index |
| **EF** | emission factor |
| **EGR** | exhaust gas recirculation |
| **EGTT** | Expert Group on Technology Transfer |
| **EIMs** | Energy Improvement Mortgages |
| **EIP** | energy and industrial processes |
| **EJ** | exajoule |
| **$E_{LUC}$** | land-use change emissions |
| **EMAS** | Eco-Management and Auditing Scheme |
| **EPD** | Environmental Product Declaration |
| **EPBD** | Energy Performance Buildings Directive |
| **EPCs** | Energy Performance Certificates |
| **EPS** | Emissions Performance Standard |
| **EPR** | extended producer responsibility |
| **ERF** | effective radiative forcing |
| **ERIA** | Economic Research Institute for ASEAN and East Asia |
| **ES-FiT** | Energy Savings Feed-in Tariff |
| **ESCO** | Energy Service Company |
| **ESA** | energy services agreement |
| **ESD** | education for sustainable development |
| **ESG** | environmental, social and governance |
| **ESM** | energy systems model |
| **ETP** | *Energy Technology Perspectives* (IEA report) |
| **ETS** | Emissions Trading System |
| **EU** | European Union |
| **EU-27** | European Union member states [excluding UK] |
| **EU-28** | European Union member states [including UK] |
| **EU ETS** | European Union Emissions Trading Scheme |
| **EU-RED** | EU Renewable Energy Directive |
| **EV** | electric vehicle |
| **EW** | enhanced weathering |
| **FaIR** | Finite Amplitude Impulse Response |
| **FAQ** | frequently asked question |
| **FAO** | Food and Agriculture Organization of the United Nations |
| **FBDG** | food-based dietary guidelines |
| **FCDO** | UK Foreign, Commonwealth and Development Office |
| **FCV** | fuel cell vehicle |
| **FDI** | Foreign Direct Investment |
| **FFI** | fossil fuel and industry |
| **F-gas** | fluorinated gas |
| **FIC** | Faster Innovation Case |
| **FiT** | feed-in tariff |
| **FiTP** | feed-in premium |
| **FLEGT** | Forest Law Enforcement, Governance and Trade |
| **FLW** | food loss and waste |
| **FRAND** | fair, reasonable and non-discriminatory |
| **FSC** | Forest Sustainability Council |
| **FT** | Fischer-Tropsch |
| **FTA** | free trade agreement |
| **FWM** | fine woody material |
| **GATS** | General Agreement on Trade in Services |
| **GATT** | General Agreement on Tariffs and Trade |
| **GBAM** | ground-based albedo modifications |
| **GCAM** | Global Change Assessment Model |
| **GCCA** | Global Cement and Concrete Association |
| **GCF** | Green Climate Fund |
| **GCoM** | Global Covenant of Mayors |
| **GCP** | Global Carbon Project |
| **GDP** | gross domestic product |
| **GEA** | Global Energy Assessment |
| **GEF** | Global Environment Facility |
| **GFBI** | Global Forest Biodiversity Initiative |
| **GFCA** | Global Framework for Climate Action |
| **GFCF** | Gross-fixed capital formation |
| **GFED** | Global Fire Emissions Database |
| **GHG** | greenhouse gas |
| **GIS** | geographic information system |
| **GIS** | global innovation system |
| **GIZ** | the German Development Agency (*Deutsche Gesellschaft für Internationale Zusammenarbeit*) |
| **GJ** | gigajoule |
| **GMF** | Global Methane Fund |

AVI

| | |
|---|---|
| **GMI** | Global Methane Initiative |
| **GMRIO** | global multi-region input-output |
| **GNI** | gross national income |
| **GPP** | Green Public Procurement |
| **GPT** | general-purpose technologies |
| **GSAT** | Global Surface Air Temperature |
| **Gt** | gigatonne |
| **$GtCO_2$-eq** | gigatonnes of $CO_2$ equivalent |
| **GTEM** | global transport energy sectoral models |
| **GTP** | global temperature change potential |
| **GWP** | global warming potential |
| **GWP100** | 100-year global warming potential |
| **HAP** | household air pollution |
| **HCE** | historical cumulative emission |
| **HCFCs** | hydrochlorofluorocarbons |
| **HCS** | High Carbon Stock |
| **HDI** | Human Development Index |
| **H-DRI** | Hydrogen-based direct reduced iron |
| **HDV** | Heavy-duty vehicles |
| **HEMS** | home energy management system |
| **HES** | Hybrid energy storage |
| **HEV** | hybrid electric vehicle |
| **HFC** | hydrofluorocarbon |
| **HFCV** | hydrogen fuel cell vehicle |
| **HIHD** | Historical Index of Human Development |
| **HLPF** | High-Level Political Forum |
| **HN** | Houghton and Nassikas |
| **HSR** | high-speed rail |
| **HVAC** | heating, ventilation and air conditioning |
| **HVO** | hydrotreated vegetable oil |
| **HYDE** | History database of the Global Environment |
| **IAGA** | International Air Transport Association |
| **IAM** | integrated assessment model |
| **IBE** | income-based emission accounting |
| **ICAO** | International Civil Aviation Organization |
| **ICCT** | International Council on Clean Transportation |
| **ICE** | internal combustion engine |
| **ICEV** | internal combustion engine vehicles |
| **ICI** | international cooperative initiative |
| **ICJ** | International Court of Justice |
| **ICT** | information and communication technology |
| **IDDRI** | Institute for Sustainable Development and International Relations |
| **IEA** | International Energy Agency |
| **IFC** | International Finance Corporation |
| **IFDD** | Institut de la Francophonie pour le Développement Durable (Francophonie Institute for Sustainable Development) |
| **IFI** | international financial institution |
| **IGCC** | International Green Construction Code |
| **IIASA** | International Institute for Applied Systems Analysis |
| **IIGCC** | Institutional Investors Group on Climate Change |
| **IIASA** | International Institute for Applied Systems Analysis |
| **IIoT** | industrial internet of things |
| **ILB** | incandescent light bulb |
| **ILM** | intrusive load monitoring |
| **IMF** | International Monetary Fund |
| **IMO** | International Maritime Organization |
| **IMP** | Illustrative Mitigation Pathway |
| **IMP-GS** | Illustrative Mitigation Pathway – Gradual Strengthening |
| **IMP-LD** | Illustrative Mitigation Pathway – Low Demand |
| **IMP-Neg** | Illustrative Mitigation Pathway – Net Negative Emissions |
| **IMP-Ren** | Illustrative Mitigation Pathway – Renewable Electricity |
| **IMP-SP** | Illustrative Mitigation Pathway – Shifting Pathways |
| **INDC** | Intended Nationally Determined Contributions |
| **IoT** | internet of things |
| **IPBES** | Intergovernmental Science-Policy Platform on Biodiversity and Ecosystem Services |
| **IPCC** | Intergovernmental Panel on Climate Change |
| **IP** | Illustrative Pathway |
| **IPP** | independent power producers |
| **IPPU** | Industrial processes and product use |
| **IPR** | intellectual property rights |
| **IRENA** | International Renewable Energy Agency |
| **IRP** | UN International Resource Panel |

| | |
|---|---|
| **ISDS** | investor–state dispute settlement |
| **ITF** | International Transport Forum |
| **ITMO** | internationally transferred mitigation outcome |
| **ITUC** | International Trade Union Confederation |
| **JICA** | Japanese International Cooperation Agency |
| **JRC** | GECO Joint Research Centre – Global Energy and Climate Outlook |
| **KR** | key risks |
| **kWh** | kilowatt hour |
| **LAM** | Latin America and the Caribbean |
| **LCA** | life cycle assessment *or* life cycle analysis |
| **LCC** | lifecycle costs |
| **LCCE** | levelised cost of conserved energy |
| **LCCC** | levelised cost of conserved carbon |
| **LCOE** | levelised cost of electricity *or* levelised cost of energy |
| **LCS** | low-carbon society |
| **LDCs** | Least-Developed Countries |
| **LDCF** | Least Developed Countries Fund |
| **LDN** | Land Degradation Neutrality |
| **LDV** | light-duty vehicle |
| **LEAF** | Lowering Emissions by Accelerating Forest Finance |
| **LED** | light-emitting diode |
| **LED scenario** | Low Energy Demand scenario |
| **LEDS** | low-emissions development strategies |
| **LIB** | lithium-ion battery |
| **LiRE** | IMAGE-Lifestyle-Renewable (IEA scenario) |
| **LNG** | liquefied natural gas |
| **LPG** | liquefied petroleum gas |
| **LTGG** | long-term global goal (to hold the increase in the global average temperature to well below 2°C above pre-industrial levels and to pursue efforts to limit the temperature increase to 1.5°C above pre-industrial levels) |
| **LTO** | long-term operation |
| **LULUCF** | land use, land-use change and forestry |
| **LUM** | land-use model |
| **MA** | Mitigation Alliance |
| **MaaS** | Mobility as a Service |
| **MAC** | marginal abatement cost |
| **MACC** | marginal abatement cost curve |
| **MAPS** | Mitigation Action Plans and Scenarios |
| **mbpd** | million barrels per day |
| **MCB** | marine cloud brightening |
| **MDB** | multilateral development bank |
| **ME** | material efficiency |
| **MES** | material efficiency scenario |
| **MEA** | multilateral environmental agreement |
| **MEPC** | Marine Environment Protection Committee |
| **MEPSs** | Minimum Energy Performance Standards |
| **Mha** | million hectares |
| **MIGA** | Multilateral Investment Guarantee Agency |
| **MIS** | mission-oriented innovation systems |
| **MJ** | megajoule |
| **$Mkm^2$** | million square kilometres |
| **MLP** | multi-level perspective |
| **ModAct** | Moderate Action scenario |
| **MOE** | molten oxide electrolysis |
| **MOOC** | massive open online course |
| **MPa** | megapascal |
| **MRV** | measuring, reporting and verifying *or* measuring, reporting and verification |
| **MS** | member state |
| **MSME** | micro, small and medium enterprises |
| **Mt** | megatonne |
| **MTA** | methanol-to-aromatics |
| **MTO** | methanol-to-olefins |
| **MWh** | megawatt hour |
| **$N_2O$** | nitrous oxide |
| **NAFTA** | North American Free Trade Agreement |
| **NAMA** | Nationally Appropriate Mitigation Actions |
| **NAP** | national adaptation plan |
| **NAZCA** | Non-State Actor Zone for Climate Action |
| **NBS** | nature-based solutions |
| **NDC** | Nationally Determined Contribution |
| **NEDO** | New Energy and Industrial Technology Development Organisation, Japan |
| **NELD** | non-economic loss and damage |
| **$NF_3$** | nitrogen trifluoride |
| **NGFS** | Network for Greening the Financial System |
| **NGHGI** | national greenhouse gas inventories |

| | |
|---|---|
| **NGO** | non-governmental organisation |
| **NiCD** | nickel–cadmium |
| **NILM** | non-intrusive load monitoring |
| **Nimby** | Not in my back yard |
| **NiMH** | nickel-metal hydride |
| **NIS** | national innovation system |
| **NMVOC** | non-methane volatile organic compounds |
| **NOAA** | National Oceanic and Atmospheric Administration |
| **NRG** | natural regrowth |
| **NR** | Non-Residential |
| **NSA** | non-state actor |
| **NTEM** | national transport -energy models |
| **NT** | Non-technological |
| **NZE** | net zero emissions |
| **NZE scenario** | Net-Zero Emissions by 2050 (IEA scenario) |
| **NZEB** | net zero energy building |
| **nZEB** | nearly zero energy building |
| **NSTT** | North–South technology transfer and cooperation |
| **NUA** | New Urban Agenda |
| **NYDF** | New York Declaration on Forests |
| **OA** | ocean alkalinity |
| **OAC** | ocean albedo change |
| **OAE** | ocean alkalinity enhancement |
| **ODA** | overseas development assistance |
| **ODS** | ozone-depleting substance |
| **OECD** | Organisation for Economic Co-operation and Development |
| **OF** | ocean fertilisation |
| **OPEC** | Organization of the Petroleum Exporting Countries |
| **OPEX** | operating and maintenance expenditures |
| **OS** | overshoot |
| **OSS** | one-stop shop |
| **P2P** | peer-to-peer |
| **PA** | The Paris Agreement |
| **PACE** | Property Assessed Clean Energy |
| **PBEs** | production-based emissions |
| **PCCB** | Paris Committee on Capacity-building |
| **PCF** | personal carbon footprint |
| **PCRAFI** | Pacific Catastrophe Risk Assessment and Financing Initiative |
| **PDB** | public development bank |
| **PEFC** | Programme for the Endorsement of Forest Certification |
| **PEMFC** | proton-exchange membrane fuel cells |
| **PES** | payment for ecosystem services |
| **PFCs** | perfluorocarbons |
| **PHEV** | plug-in hybrid electric vehicle |
| **pkm** | passenger-kilometres |
| **PM** | particulate matter |
| **PPA** | Power Purchase Agreement |
| **PPCA** | Powering Past Coal Alliance |
| **PPCR** | Pilot Program for Climate Resilience |
| **PPI** | pulp and paper industry |
| **PPP** | public-private partnership |
| **PPP** | purchasing power parity |
| **PRI** | Principles for Responsible Investment |
| **PV** | photovoltaic |
| **QE** | quantitative easing |
| **R&D** | research and development |
| **RCEP** | Regional Comprehensive Economic Partnership |
| **RCM** | reduced complexity model |
| **RCP** | Representative Concentration Pathway |
| **RD&D** | research, development and demonstration |
| **RDI** | Research, Development and Innovation |
| **RECC** | Resource Efficiency and Climate Change |
| **RECC-LED** | Resource Efficiency and Climate Change-Low Energy Demand (IEA scenario) |
| **REDD+** | reducing emissions from deforestation and forest degradation and the role of conservation, sustainable management of forests and enhancement of forest carbon stocks |
| **REEs** | rare earth elements |
| **ReSOLVE** | Regenerate, Share, Optimise, Loop, Virtualise, Exchange framework |
| **RGGI** | Regional Greenhouse Gas Initiative |
| **RIMAP** | Real-time Integrated Model for probabilistic Assessment of emissions Paths |
| **RIS** | regional innovation systems |
| **RKR** | Representative Key Risks |

| | |
|---|---|
| **RPK** | revenue passenger-kilometres |
| **RSD** | relative standard deviation |
| **RSPO** | Roundtable on Sustainable Palm Oil |
| **RTS** | Reference Technology Scenario |
| **S&L** | standards and labelling |
| **SAF** | sustainable aviation fuel |
| **SAI** | stratospheric aerosol interventions |
| **SAM** | Social Accounting Matrix |
| **SAR** | Second Assessment Report |
| **SARPs** | Standards and Recommended Practices |
| **SASB** | Sustainability Accounting Standards Board |
| **SBSTA** | Subsidiary Body for Scientific and Technological Advice |
| **SBT** | science-based target |
| **SCC** | social cost of carbon |
| **SCCF** | Special Climate Change Fund |
| **SCS** | soil carbon sequestration |
| **SDG** | Sustainable Development Goal |
| **SDPS** | shifting development pathways to increased sustainability |
| **SDR** | Special Drawing Rights |
| **SDS** | Sustainable Development Scenario (IEA scenario) |
| **SDSN** | Sustainable Development Solutions Network |
| **SE** | sustainable entrepreneur |
| **SEA** | strategic environmental assessment |
| **SEADRIF** | South East Asian Disaster Risk Insurance Facility |
| **SEC** | specific energy consumption |
| **SECA** | sulphur emission control area |
| **SEEMP** | Ship Energy Efficiency Management Plan |
| **SEM** | structural equations modelling |
| **SER** | Sufficiency, Efficiency, Renewal |
| **SETAC** | Society of Environmental Toxicology and Chemistry (UNEP-SETAC) |
| **$SF_6$** | sulphur hexafluoride |
| **SI** | sustainable intensification |
| **SIDS** | Small Island Developing States |
| **SIS** | sectoral innovation system |
| **SLCF** | short-lived climate forcer |
| **SLCP** | short-lived climate pollutant |
| **SLoCaT** | Sustainable Low Carbon Transport Partnership |
| **SLM** | sustainable land management |
| **SLR** | sea level rise |
| **SM** | smart manufacturing |
| **SMEs** | small and medium-sized enterprises |
| **SNA** | System of National Accounts |
| **SNTT** | South–North technology transfer and cooperation |
| **$SO_2$** | sulphur dioxide |
| **SOE** | state-owned enterprise |
| **SOFC** | solid oxide fuel cell |
| **SPM** | Summary for Policymakers |
| **SPV** | special purpose vehicle |
| **SR1.5** | IPCC Special Report on Global Warming of 1.5°C |
| **SRCCL** | IPCC Special Report on Climate Change and Land |
| **SRI** | Sustainable and Responsible Investment |
| **SRM** | solar radiation modification |
| **SROCC** | IPCC Special Report on the Ocean and Cryosphere in a Changing Climate |
| **SSC** | South-South cooperation |
| **SSP** | Shared Socio-economic Pathway |
| **SSTT** | South–South technology transfer and cooperation |
| **STEM** | science, technology, engineering and mathematics |
| **STEPS** | Stated Policies Scenario |
| **SUV** | sport utility vehicle |
| **TA** | territorial accounting |
| **TABS** | thermally activated building systems |
| **TBT Agreement** | WTO Agreement on Technical Barriers to Trade |
| **TCBA** | technology-adjusted consumption-based emission accounting |
| **TCFD** | Task Force on Climate-related Financial Disclosures |
| **TCRE** | transient climate response to cumulative emissions of carbon dioxide |
| **TDR** | travel demand reduction |
| **TEEB** | The Economics of Ecosystems and Biodiversity |
| **TEC** | Technology Executive Committee |
| **TES** | Transforming Energy Scenario |

| | |
|---|---|
| **TIS** | technological innovation system |
| **TFC** | total final energy consumption |
| **TGC** | tradeable green certificate |
| **tkm** | tonne-kilometre |
| **TNA** | technology needs assessment |
| **TOD** | transit-oriented development |
| **TPES** | total primary energy supply |
| **TRA** | technology readiness assessment |
| **TrC** | triangular cooperation |
| **TGCs** | Tradable Green Certificates |
| **TRIPS Agreement** | Trade-Related Aspects of Intellectual Property Rights Agreement |
| **TRL** | technology readiness level |
| **TW** | terawatt |
| **UF** | utility factor |
| **UHI** | urban heat island |
| **UKCCC** | United Kingdom Climate Change Committee |
| **ULCS** | ultra-low carbon steel |
| **UNCCD** | United Nations Convention to Combat Desertification |
| **UNCRD** | United Nations Centre for Regional Development |
| **UNDP** | United Nations Development Programme |
| **UNEP** | United Nations Environment Programme |
| **UNFCCC** | United Nations Framework Convention on Climate Change |
| **UNOSSC** | United Nations Office for South-South Cooperation |
| **USD** | US dollar |
| **US DOE** | United States Department of Energy |
| **US EPA** | United States Environmental Protection Agency |
| **UV** | ultraviolet |
| **V1G** | controlled charging (of an electric vehicle) |
| **V2G** | vehicle-to-grid |
| **VC** | venture capital |
| **VCS** | Verified Carbon Standard of the Verra programme |
| **vkm** | vehicle-kilometre |
| **VKT** | vehicle kilometres travelled |
| **VLR** | Voluntary Local Review |
| **VMT** | vehicle miles travelled |
| **VNR** | Voluntary National Review |
| **WBCSD** | World Business Council on Sustainable Development |
| **WEFN** | water-energy-food nexus |
| **WEO** | World Energy Outlook |
| **WFP** | World Food Programme |
| **WG** | Working Group |
| **WHO** | World Health Organization |
| **WHP** | waste heat to power |
| **WMO** | World Meteorological Organisation |
| **WRAP** | Waste and Resources Action Programme |
| **WSUD** | Water Sensitive Urban Design |
| **WTO** | World Trade Organization |
| **WTP** | willingness to pay |
| **ZEV** | zero emission vehicle |

# Index

**This chapter should be cited as:**

IPCC, 2022: Index. In IPCC, 2022: *Climate Change 2022: Mitigation of Climate Change. Contribution of Working Group III to the Sixth Assessment Report of the Intergovernmental Panel on Climate Change* [P.R. Shukla, J. Skea, R. Slade, A. Al Khourdajie, R. van Diemen, D. McCollum, M. Pathak, S. Some, P. Vyas, R. Fradera, M. Belkacemi, A. Hasija, G. Lisboa, S. Luz, J. Malley, (eds.)]. Cambridge University Press, Cambridge, UK and New York, NY, USA. doi: 10.1017/9781009157926.026

*Note: * indicates the term also appears in the Glossary and n indicates a footnote. Italicised page numbers denote tables, figures, associated captions and boxed material.* **Bold** *page numbers indicate entire chapter spans..*



## A

## B

Index

## D

# E

Index

Index

## G



Index

## H



## I

## J

## K



## L

## M

## N

## O

## S

Index

## T

Index

## U

## V

## W



## Y



## Z